WHY YOU SHOULD CONSIDER

Algebra and Trigonometry, Structure and Method, Book 2

■ **This sound, carefully developed course** meets the needs of a wide range of students. (See Contents, pages v–xiii.)

■ **Helpful worked-out examples** supported by clear explanations help students understand concepts and skills. (See pages 118–120, 307–309.)

■ An abundance of **carefully developed exercise sets** provides practice with skills, applications, and theory at A, B, and C levels of difficulty. (See pages 116–117, 280–281, 410–411.) The worked-out examples provide clear models for A and early B level exercises. *Challenge* exercises provide extra motivation for very capable students. (See page 492.)

■ **Real-life applications** throughout the text and problem sets show the usefulness of mathematics. (See pages 52–54, 131–132, 343–344.) Special *Application* sections connect algebra with the real world. (See pages 159–161, 493–494, 752–753.)

■ **Frequent reviews** boost students' retention. Short *Mixed Reviews* follow every other lesson. (See pages 263, 273.) *Chapter Reviews*, *Cumulative Reviews*, and *Mixed Problem Solving* sets provide both mixed and sequential review at the ends of chapters. (See pages 498, 545–546, 547.)

■ **Calculators and computers** help students explore new concepts and solve realistic problems. Suggestions for computer graphing techniques and calculator use are included throughout. (See pages 114, 327, 386–387.) *Calculator Key-Ins*, *Computer Key-Ins*, and *Computer Exercises* support and extend the lessons. (See pages 263, 321, 389.)

■ **Explorations** provide students with activities for exploring concepts using manipulatives, scientific and graphing calculators, and software. (See pages 832–847.)

■ **Porfolio Projects** offer students nonroutine problem-solving experiences and help students develop their ability to communicate mathematical ideas. (See pages 848–855.)

Supplementary materials (See pages T4–T5 for a description.)

The **Teacher's Edition** includes six interleaved pages of teaching material for each chapter, *Strategies for Teaching*, *Lesson Commentary*, and annotated student book pages with side-column notes. (See Contents, page T1, and pages T8–T9.)

The **Study Guide for Reteaching and Practice** accomodates students who need reteaching or who miss a class lesson. Colorful **Overhead Visuals** help you present abstract concepts in a more concrete way. The disk **Algebra Plotter Plus** and the booklet **Using Algebra Plotter Plus** help you and your students graph with ease and explore concepts on a computer.

Other ancillaries include **Teacher's Resource Files, Resource Book, Practice Masters, Tests, Computer Activities, Test Generator** software with **Test Bank,** and **Solution Key.**

Algebra and Trigonometry

Structure and Method

Book 2

Teacher's Edition

Richard G. Brown
Mary P. Dolciani
Robert H. Sorgenfrey
Robert B. Kane

Contributing Authors
Sandra K. Dawson
Barbara Nunn

Teacher Consultants
Keith Bangs
Doris Holland
Ouidasue Nash
Al Varness

McDougal Littell
A HOUGHTON MIFFLIN COMPANY
Evanston, Illinois • Boston • Dallas

Authors

Richard G. Brown, Mathematics Teacher, Phillips Exeter Academy, Exeter, New Hampshire

Mary P. Dolciani, formerly Professor of Mathematical Sciences, Hunter College of the City University of New York

Robert H. Sorgenfrey, Professor Emeritus of Mathematics, University of California, Los Angeles

Robert B. Kane, Professor of Mathematics Education, Purdue University, Lafayette, Indiana

Contributing Authors

Sandra K. Dawson, Mathematics Teacher, Glenbrook South High School, Glenview, Illinois

Barbara Nunn, Mathematics Department Chairperson, Marjory Stoneman Douglas High School, Parkland, Florida

Teacher Consultants

Keith Bangs, Mathematics Teacher, Owatonna High School, Owatonna, Minnesota

Ouidasue Nash, Mathematics Teacher, Columbia High School, Columbia, South Carolina

Doris Holland, Mathematics Teacher, O.D. Wyatt High School, International Baccalaureate Magnet Program, Fort Worth, Texas

Al Varness, former Mathematics Teacher, Lake Washington High School, Kirkland, Washington

The authors wish to thank Richard Galbraith, Mathematics Department Chair, Bryant High School, Bryant, Arkansas; James M. Sconyers, Math/Science Resource Teacher, Garrett County, Maryland; and Dorothy Peterson, Mathematics Department Chair, Jefferson High School, Cedar Rapids, Iowa, for their valuable contributions to this Teacher's Edition. The authors also wish to thank David L. Myers, Computer Coordinator and Mathematics Teacher, Winsor School, Boston, Massachusetts, for writing the Portfolio Projects.

Printed in U.S.A.

ISBN: 0-395-77119-6

123456789-DC-00 99 98 97 96

Contents

Introduction to the Program T2–T9d

Algebra for the 90's T2
A description of the philosophy of this edition.

Teaching the Course T4
Featuring the Teacher's Edition, Overhead Visuals, Study Guide for Reteaching and Practice, Algebra Plotter Plus, and Solution Key.

Additional Support T5
Featuring the Teacher's Resource Files, Resource Book, Practice Masters, Tests, Computer Activities, Test Generator, and Texas Instruments Calculators.

Organization of the Textbook T6
Featuring the student textbook and its organization.

Using the Teacher's Edition T8
Helpful information about using this Teacher's Edition.

Key Topics and Approaches T9a
Student textbook references to Problem Solving, Reasoning, Communication, Connections, and Technology.

Tests and Reviews T10–T46

Diagnostic Test T10
A test, keyed to the textbook, on the Algebra 1 topics reviewed in Chapters 1–4.

Chapter Tests T13–T28
A set of alternate chapter tests, one for each chapter, with permission to reproduce. Complete answers start on page T41.

Cumulative Reviews T29
These five cumulative reviews, which may be reproduced, cover groups of several chapters. Complete answers start on page T44.

Topical Reviews T35
Three cumulative topical reviews with permission to reproduce. Complete answers start on page T45.

Strategies for Teaching T47–T57

Varying the Mode of Instruction T47
Using manipulatives, cooperative learning, and technology.

Communicating in Mathematics T49

Learning Strategies T51

Problem Solving Strategies T53

Nonroutine Problems T53a

Making Connections T53b

Thinking Skills T54

Error Analysis T55

Reteaching T57

Assignment Guide T58
Average and extended courses, with or without trigonometry.

Supplementary Materials Guide T73
A guide for using the supplementary Tests, Practice Masters, Resource Book, and Study Guide.

Software Guide T77
Using Algebra Plotter Plus and the Test Generator.

Lesson Commentary T79
Teaching suggestions for extensions, using manipulatives, group activities, and reading algebra.

Interleaved Teacher Pages a
Six pages before each chapter, with objectives, guides to scheduling and available resources, strategies for teaching, and facsimiles of key resources.

Student Text with Annotations 1

Challenges of the Future

Experts predict that in the 21st century many jobs will require employees to have greater mathematical knowledge and better problem solving skills than they do today. In addition, citizens will need strong quantitative-reasoning skills to make effective decisions in their daily lives. Preparing students to meet the challenges of the future is our mutual goal. No textbook can guide or inspire students in the same way as a teacher. But we can help make your teaching job easier by providing a comprehensive textbook program that helps students build their thinking and problem solving skills as well as their understanding of algebra.

Contemporary curriculum

A Course for Today's Students

In planning this new edition, we have spoken with many algebra teachers throughout the country and have been guided by their suggestions. We have also been guided by the recommendations of professional organizations, such as the *Standards* of the National Council of Teachers of Mathematics. The result is a contemporary course that works in the classroom.

In this book, **problem solving** is introduced early and is integrated throughout. **Applications** of algebra are presented in chapter-opening pages, lesson developments, special *Application* sections, and interesting and varied word problems. **Reasoning skills** such as analyzing information, making conjectures, and giving convincing arguments are developed throughout the course. Interesting **nonroutine problems** are included to challenge capable students. Suggestions for appropriate use of **technology** are included in the text and the exercises. **Explorations** activities help students develop understanding of concepts.

Accessible to a wide range of students

Puts Algebra Within Reach

This book has been designed to make algebra accessible to a wide range of students — without sacrificing complete content and challenge for capable students. The following features help bring algebra within reach: **Readable text,** with clear, concise, language; **visual highlighting** of important results; **numerous worked-out examples**; **exercise sets** that start with straightforward skill development and build gradually in difficulty; **frequent mixed reviews**; special sections on **reading algebra**.

Classroom-tested for effectiveness

A Proven Program Improved

The *Algebra, Structure and Method*, series has been used successfully in many algebra classrooms. This new edition builds on the strengths of earlier editions that teachers have valued, such as **complete topic coverage**, **clear development of mathematical concepts**, and **carefully sequenced exercise sets.** At the same time, the books have been enhanced to provide a more contemporary and more accessible course — a course that will help prepare a wide range of students for the challenges of the future.

Algebra for the 90's

Algebra
and Trigonometry
Structure and Method
Book 2
2.71
Houghton Mifflin

Teaching the Course

Teacher's Edition

- Designed to help you teach algebra.
- Special articles discuss a variety of teaching strategies.
- Pages interleaved between chapters provide alternative strategies for teaching lessons.
- Easily accessible assignment and supplementary materials guides interleaved between chapters.
- Contains reduced facsimiles of key supplementary materials.
- Includes extra tests and reviews.

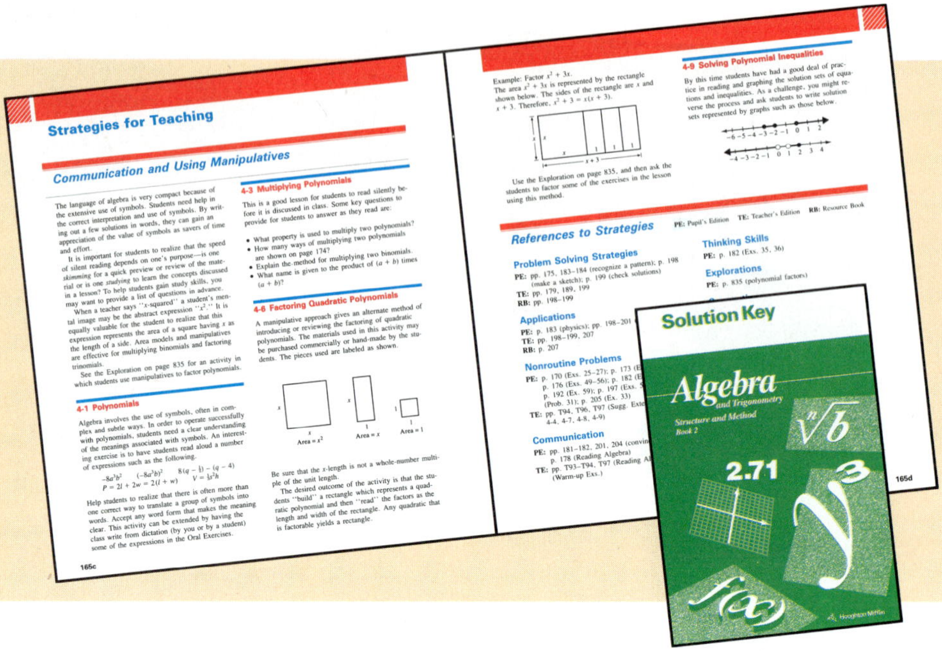

Overhead Visuals

- Full-color overhead transparencies, some with moving parts, provide concrete modeling to enhance presentations.
- Printed folders include objectives, textbook references, and questions and answers for using the visuals.

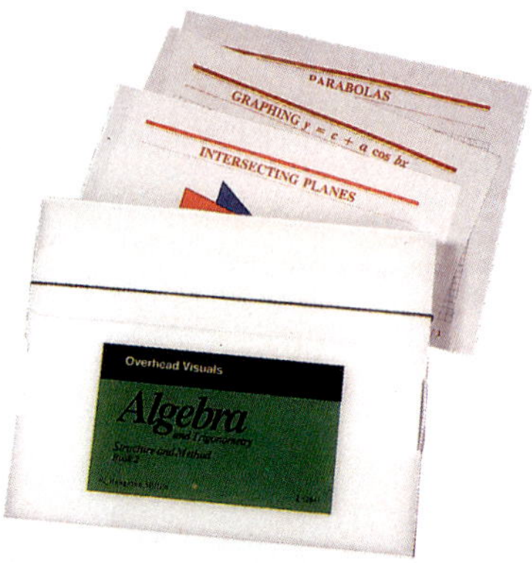

Study Guide for Reteaching and Practice

- Contains alternative two-page lessons for each textbook lesson.
- Provides vocabulary review.
- Has worked-out examples, exercises mainly at the "A" level, and mixed reviews.
- Accompanied by a separate *Answer Key*.

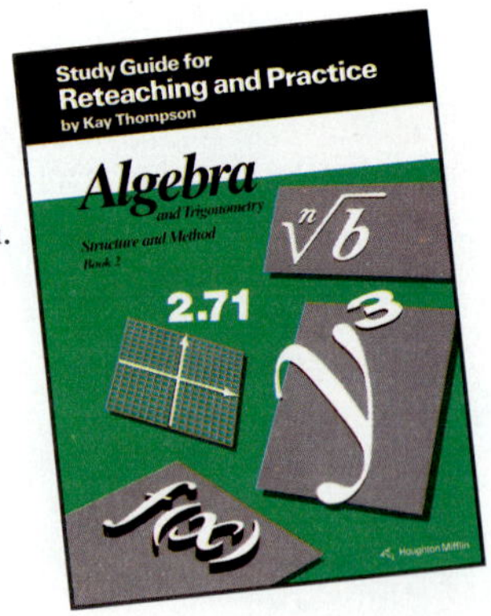

Algebra Plotter Plus

- *Disk* can be used independently by students or by the teacher for classroom presentations.
    - Plots lines, conics, functions, inequalities, and absolute value.
    - Utilities includes matrix reducer, statistics spreadsheet, and a sampling experiment simulation.
    - Apple II and IBM versions.
    - Available for 5 1/4" or 3 1/2" disk drives.
- *Using Algebra Plotter Plus* booklet includes worked examples, scripted demonstrations, enrichment topics, activity sheets, and user's manual.

Additional Support

Teacher's Resource Files

- *Teachers Resource Book* tests, practice, mixed review, problem solving, applications, enrichment, technology, thinking skills, warm-up exercises, diagram masters
- *Study Guide for Reteaching and Practice*
- *Algebra Plotter Plus* Demo disk and booklet
- *Overhead Visuals* sample
- *Tests* (blackline)
- *Practice Masters* (blackline)
- Topic and Chapter folders

Practice Masters

offer concentrated practice in worksheet format including periodic cumulative reviews on blackline masters and duplicating masters.

Tests

contain quizzes, chapter tests (in two parallel forms), and cumulative tests on blackline masters and duplicating masters.

Computer Activities

provide worksheets that extend and reinforce students' understanding of algebraic concepts through BASIC programming.

Test Generator

software for generating your own tests either manually or by random selection, with options for customizing the format and level of difficulty.

Texas Instruments Calculators

further enhance the technology strand.

- TI-34 a scientific calculator with fraction capability.
- TI-81 a graphing calculator with powerful mathematical capabilities.

Organization of the Textbook

Lesson Development

- **Chapters** divided into several groups of related lessons.
- **Objectives** stated at the start of lessons.
- **Lesson Text** focuses on the underlying structure of algebraic content, as well as methods and applications.
- **Worked-Out Examples** illustrate and reinforce the content and skills taught.
- **Vocabulary** words are boldfaced in the text and appear in the glossary.
- **Important Information** is boxed for easy reference and review.

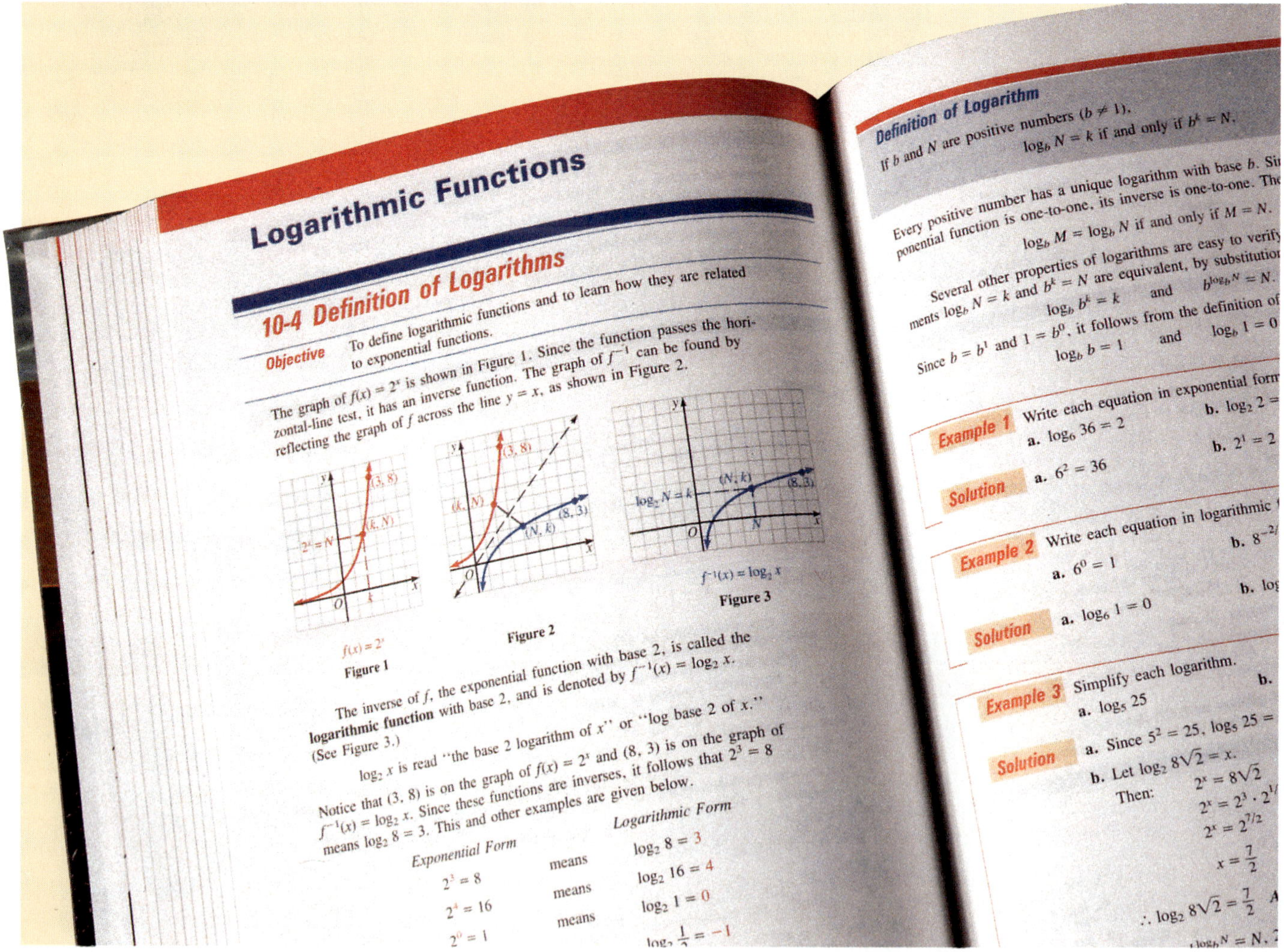

Exercises

Oral Exercises provide students and teachers with immediate feedback of lesson comprehension.

Written Exercises are graded for varying levels of student ability.

Problems put skills practice in a problem-solving context.

Exercises include suggestions for using *calculators* and *computers*.

Computer Exercises provide students opportunities to use programs for lesson enrichment.

Mixed Review exercises at the end of lessons maintain previously taught skills.

Applications

For example, p. 159

Applications relate algebra to other disciplines and everyday life, enriching the course for all students.

Study Skills

For example, p. 178

Reading Algebra features help students read and use their textbooks.

Challenges and Extras

For examples, pp. 140 and 273

Challenges are high interest problems where students utilize their problem solving and thinking skills.

Extras are for those students who are able to extend lesson content.

Enrichment

For examples, pp. 337, 573, and 618

Career Notes relate algebra to everyday life and the work place.

Historical and **Biographical Notes** provide historical background and human interest for all students.

Technology

For examples, pp. 12 and 193

Calculator Key-In features indicate ways of using calculators to enhance algebra.

Computer Key-In features show students how computers can also be used to enhance content.

Testing and Review

Self-Test lists important vocabulary and tests key concepts every 3-4 lessons.

Chapter Summary capsulizes important information highlighted within chapters.

Chapter Reviews and **Chapter Tests** provide checks of lesson objective comprehension.

Cumulative Reviews review skills and concepts from previous chapters.

Maintaining Skills review arithmetic and basic algebra skills at the end of each chapter.

Preparing for College Entrance Exams help students develop test-taking skills.

Using the Teacher's Edition

This Teacher's Edition provides pupil pages with answers and side columns. As illustrated below, the side columns include time-saving material for your lesson presentation, from Warm-up Exercises to lesson summary. Other side-column entries include **Computer** and **Calculator Key-In Commentaries, Additional Answers,** and **Application** descriptions and activities.

For each lesson, page references are given for the corresponding Lesson Commentary, which includes teaching suggestions and ideas for extensions, group activities, reading algebra, and using manipulatives.

Warm-Up Exercises practice skills taught in previous lessons.

Motivating the Lesson connects new concepts to prior ones or to real-world applications.

Chalkboard Examples provide additional examples to use in presenting the lesson.

Thinking Skills, Common Errors, and **Problem Solving Strategies** are identified.

Check for Understanding suggests a use of the Oral Exercises to assess how well students have understood one lesson concept before you move on to the next.

Teaching Suggestions p. T115

Suggested Extensions p. T115

Warm-Up Exercises

Identify each of the following as an equation of a circle, parabola, ellipse, or hyperbola.

1. $y = (x + 5)^2$ parabola

Motivating the Lesson

Remind the students that they have already dealt with circles having centers *not* at the origin. In today's lesson the centers of ellipses and hyperbolas will be shifted, or translated, off the origin.

Chalkboard Examples

1. Find an equation of the ellipse having foci at (2, 1) and (2, −3) and sum of focal radii 6.
   Center is at (2, −1); distance from center to each focus is 2. Thus $a = 3$, $c = 2$, and $b^2 = a^2 - c^2 = 9 - 4 = 5$. The equation is $\frac{(x-2)^2}{5} + \frac{(y+1)^2}{9} = 1$.

Thinking Skills

Conics can be identified directly from their equations without first completing the square in x and y. The *analysis* of the general equation $Ax^2 + By^2 + Cx + Dy + E = 0$ is outlined in Written Exercise 21 on page 435.

Check for Understanding

Here is a suggested use of the Oral Exercises to check students' understanding as you teach the lesson.
Oral Exs. 1–8: use before Example 1.

9-6 More on Central Conics

Objective To find an equation of a conic section with center not at the origin and to identify a conic as a circle, ellipse, or hyperbola.

SIMULATED PAGE

Circles, ellipses, and hyperbolas are called **central conics** because they have centers. Now you'll study central conics with centers not at the origin.

As you learned in Lesson 9-2, you can translate a graph centered at the origin by sliding every point of the graph h units horizontally and k units vertically. Replacing x by $x - h$ and y by $y - k$ in the equation of a central conic with center at the origin gives the equation of the same conic with center now at (h, k). Using this fact, the formulas of Lessons 9-4 and 9-5 can be rewritten in the following more general forms. The constants a and b have the same meaning here as in those lessons.

Ellipses with Center (h, k)	Hyperbolas with Center (h, k)
Horizontal major axis: $\frac{(x-h)^2}{a^2} + \frac{(y-k)^2}{b^2} = 1$	*Horizontal major axis:* $\frac{(x-h)^2}{a^2} - \frac{(y-k)^2}{b^2} = 1$
Foci at $(h - c, k)$ and $(h + c, k)$, where $c^2 = a^2 - b^2$.	Foci at $(h - c, k)$ and $(h + c, k)$, where $c^2 = a^2 + b^2$.
Vertical major axis: $\frac{(x-h)^2}{b^2} + \frac{(y-k)^2}{a^2} = 1$	*Vertical major axis:* $\frac{(y-k)^2}{a^2} - \frac{(x-h)^2}{b^2} = 1$
Foci at $(h, k - c)$ and $(h, k + c)$, where $c^2 = a^2 - b^2$.	Foci at $(h, k - c)$ and $(h, k + c)$, where $c^2 = a^2 + b^2$.

Example 1 Find an equation of the ellipse having foci (−3, 4) and (9, 4) and sum of focal radii 14.

Solution The sum of the focal radii, $2a$, is 14. So $a = 7$. The center is halfway between the foci, at (3, 4). The distance from the center to each focus is 6, so $c = 6$. Substituting 7 for a and 6 for c in the equation $b^2 = a^2 - c^2$ (from Lesson 9-4) gives $b^2 = 49 - 36 = 13$.

$\therefore$ an equation of the ellipse is $\frac{(x-3)^2}{49} + \frac{(y-4)^2}{13} = 1$. ***Answer***

Example 2 Find an equation of the hyperbola having foci (−3, −2) and (−3, 8) and difference of focal radii 8.

462 (Exs. 37–39), 467 (Exs. 24–26), 472 (Ex. 43), 482 (Exs. 41–42), 491 (Exs. 50–51), 505 (Exs. 31–35), 514 (Ex. 40), 515 (Ex. 14), 528 (Exs. 33–38), 534 (Exs. 27–30), 539 (Ex. 24), 543 (Exs. 28–29), 560 (Exs. 30–34), 577–578 (Exs. 20–24), 583 (Exs. 18–19), 617 (Exs. 28–32), 623 (Exs. 29–30), 629 (Exs. 39–48), 633 (Ex. 29), 644 (Exs. 34–37), 649 (Ex. 43), 653 (Exs. 33–35), 670 (Exs. 31–32), 684 (Exs. 35–37), 688 (Ex. 23), 696 (Ex. 29), 712 (Exs. 17–18), 718 (Ex. 20), 723 (Exs. 11–12), 728–729 (Exs. 9–13), 732–733 (Exs. 17–20), 741 (Exs. 27–30), 773 (Exs. 21–28), 778 (Exs. 27–30), 783–784 (Exs. 17–18), 789 (Exs. 19–20), 797 (Ex. 19), 800 (Exs. 13–15), 805 (Exs. 17–19)

Problems, 71–72 (Probs. 11–16), 106 (Probs. 9–14), 134 (Probs. 18–20), 201 (Prob. 31), 224–225 (Probs. 9–15), 252 (Probs. 21–23), 357 (Prob. 20), 362 (Probs. 13–14), 405–406 (Probs. 9–14), 443 (Probs. 13–14), 488 (Probs. 21–22), 530 (Probs. 13–15), 534–536 (Probs. 7–14), 579 (Probs. 15–17), 584 (Prob. 12), 590 (Prob. 13), 596 (Probs. 8–14), 612 (Probs. 11–13)

Challenges, 140, 158, 254, 325, 363, 411, 425, 492, 494, 516, 601, 623, 635, 647, 732–733, 753

Reasoning

Thinking Skills

Proofs, 16 (Exs. 18–19) 81–91, 182 (Ex. 35–36), 214 (Ex. 35), 220 (Exs. 53–58), 237 (Ex. 39), 269 (Exs. 67–70), 276 (Ex. 47), 296 (Exs. 53–56), 314 (Ex. 42), 355 (Exs. 13–22), 385 (Ex. 28), 403, 405–406, 560 (Exs. 30–31), 640 (Exs. 49–58), 644 (Exs. 15–20, 25–30, 35), 649 (Exs. 33–40), 653 (Exs. 25–28, 32, 34–35), 670 (Exs. 31–32), 684 (Ex. 36), 692 (Exs. 24–25), 696 (Ex. 29), 741 (Ex. 28), 773 (Exs. 21–28), 778 (Exs. 27–30), 805 (Exs. 17–19)

Logic, 20

Formula derivation, 482 (Ex. 42)

Induction, 523–524

Explorations

Logic, 20

Challenges, 48, 140, 158, 325, 363, 411, 492, 601, 623, 733

Boolean Algebra, 95–97

Linear Programming, 159–160

Graphing Rational Functions, 230–231

Electrical Circuits, 253–254

Irrationality of $\sqrt{2}$, 273

Conjugates and Absolute Value, 298–300

Growth of Functions, 462

Graphing Sequences, 517

Induction, 523–524

Spirals, 679

Random Numbers, 753

Exploring Irrational Numbers, 832

Exploring Inequalities, 833

Exploring Functions, 834

Exploring Polynomial Factors, 835

Exploring Continued Fractions, 836

Exploring Radicals, 837

Exploring Quadratic Equations, 838

Exploring Direct Variation, 839

Exploring Circles and Ellipses, 840

Exploring Powers and Roots, 841

Exploring Pascal's Triangle, 842

Exploring Trigonometric Ratios, 843

Exploring Sine Curves, 844

Exploring Polar Coordinate Equations, 845

Exploring Probability with Experiments, 846

Exploring Matrices in Geometry, 847

Communication

Reading

Understanding definition format, 6, 7
Reading Algebra/Independent Study, 26
Reading Algebra/Symbols, 178
Reading Algebra/Problem Solving, 371
Reading Algebra/Making a Sketch, 585
Reading Algebra/Probability, 751

Discussion

Discussion, 62 (Ex. 17), 63 (Ex. 33)

Writing

Translating, 2–3, 6–7, 9, 43, 70, 327–328
Convincing Argument, 62 (Ex. 17), 87 (Exs. 24–25), 117 (Exs. 49–54), 157 (Ex. 37), 181–182, 201, 204, 287 (Exs. 34, 38–42), 321 (Ex. 43c), 332 (Exs. 35–36), 367 (Ex. 34), 379 (Exs. 7–9), 389 (Ex. 28), 416 (Exs. 35, 37, 38), 482 (Ex. 41), 530 (Exs. 14–15), 534 (Ex. 27), 600, 617 (Ex. 27), 684, 699, 723, 729
Indirect Proof, 273

Connections

Mathematics

Analytic geometry, 401–453
Arithmetic, 21–23, 292–296, 368
Data analysis, xx, 58, 134, 221
Discrete math, 838–839
Geometry, xviii–xix, 39–41, 42, 44, 47, 71–72, 124, 198–201, 273, 276, 281, 312–315, 330–331, 343–345, 357, 362, 436–438, 500, 506, 534–536, 563, 586, 591–600, 608, 618, 647, 651, 680–684, 832, 846, 847
Mathematical modeling, 147–148, 150–152, 159–161, 183, 198–199, 340–341, 483–485, 779–782
Statistics, 337, 391–395, 709–729, 734–741, 743, 745, 752–753, 755–757, 846
Trigonometry, 555–572, 578–579, 583–584, 589–590, 595–596, 599, 611–617, 619, 624–633, 636–654, 659–705, 843, 844, 845

Other Disciplines

Architecture, 100
Art, 423, 766
Astronomy, 224–225, 350, 424–425, 436, 578
Biology, 454, 484–488, 500, 522
Business, 340–341
Chemistry, 244, 493–494
Communication, 779–782
Electricity, 253–254, 258, 718
Engineering, 64, 567
Geography, 41, 45, 611
Geology, 493–494
History, 42, 68, 152, 177, 182, 220, 297, 316, 385, 431, 472, 544, 573, 601, 618, 684, 793
Light, 166, 210
Linear programming, 159–161
Money, 102–103, 105–106, 448–449, 483–484
Music, 606, 634–635
Navigation, 426, 658, 661–663
Physics, 39, 40, 44, 131, 151, 199, 201, 246, 248–251, 253, 344, 351–354, 356–357, 359–362, 443, 472, 579, 609, 611–612, 634, 662, 672–674, 718, 844

Technology

Calculator

Exercises, 24, 30, 35, 38, 222, 224, 240, 267–268, 285, 312–314, 316, 319–320, 324, 419, 427, 437, 459, 479–482, 485–489, 514, 528–530, 534–536, 550, 568–572, 575, 577–583, 588–590, 594–596, 599–600, 608, 614–615, 683, 698, 712
Explorations, 833, 836, 837, 838, 839, 841, 843
Key-Ins, 12, 225–226, 269, 492, 543, 554

Computer

Exercises, 5, 32, 64, 75, 111, 114, 119, 122, 138, 161, 187, 231, 321, 327, 331, 336, 386–388, 395, 406, 410, 416, 421, 430, 435–436, 441, 460, 462, 464, 466, 471, 584, 606, 624–625, 629, 631, 633, 645, 670, 710, 742, 750, 766, 786
Explorations, 838, 840, 841, 845, 846
Key-Ins, 80, 193, 215, 389–390, 701–702, 762

Algebra Plotter Plus Disk

Line Plotter, 107, 110, 112, 114, 118, 119, 124, 125
Line Quiz, 112, 119
Parabola Plotter, 326, 327, 336, 412
Parabola Quiz, 326, 327, 412
Absolute Value Plotter, 153
Function Plotter, 141, 194, 231, 259, 333, 386, 410, 416, 421, 430, 441, 459, 460, 463, 465, 468, 471, 624, 625, 630, 650, 697
Inequality Plotter, 135, 160
Conics Plotter, 410, 416, 421, 430, 432, 435, 436, 441
Conics Quiz, 432, 435
Circular Function Quiz, 629, 633
Matrix Reducer, 786, 790
Statistics Spreadsheet, 351, 468, 711, 713, 716, 724, 727
Sampling Experiment, 745, 752, 753

Using Algebra Plotter Plus

Scripted Demonstrations

Linear Inequalities in Two Variables, 135
Quadratic Functions, 333

Enrichment Topics

Diophantine Equations, 107
Determining the Domain and Range of Functions by Graphing, 141
Solving Absolute Value Inequalities, 153
Solving Polynomial Equations by Graphing, 194
Graphing to Find *n*th Roots of Real Numbers, 259
Curve Fitting, 351
Approximations with Greater Accuracy, 386
Exploring Parabolas, 412
Functions with Inverses on a Restricted Domain, 463
Graphing Logarithmic Equations, 468
Curve Fitting Using Exponential Functions, 468
Graphing the Taylor Series for Sine, 624
Verifying Trigonometric Identities by Graphing, 650
Solving Trigonometric Equations by Graphing, 697
Finding the Inverse of a Matrix, 790

Activity Sheets

Exploring Slopes, 112
Parallel and Perpendicular Lines, 118
Estimating Solutions of Linear Systems, 124
Parabolas, 326
Approximating Roots of Polynomial Equations, 386
Identifying Central Conics, 432
Estimating Solutions of Quadratic Systems, 436
Exponential Growth, 459
Functions and Their Inverses, 463
Sine and Cosine Curves, 624
Tangent, Cotangent, Secant, and Cosecant, 630
Histograms, Means, and Standard Deviation, 713
Correlation and Line of Best Fit, 724
Sampling, 745

Diagnostic Test on Algebra 1

Chapters 1–4 review materials from first-year algebra. The following test can help to determine what review materials your students are able to omit. Each item of this test is keyed to a lesson of the student textbook.

1-1 **1.** Graph $-\frac{5}{2}$ and -3 on a number line. Write an inequality statement comparing these numbers.

2. List the numbers from least to greatest: -5.7, $|-6.2|$, -4.7, -6, -5.65.

1-2 **3.** Simplify $[2^3 + 4(7 - 3)] \div 8$.

4. Evaluate $\dfrac{(a^2 + b^2)(a^2 - b^2)}{(a + b)^2(2a^2 - 3ab + b^2)}$ if $a = 4$ and $b = 3$.

1-3 **5.** Give reasons for the steps in the following proof.
If $3x + (-1) = 0$, then $x = \frac{1}{3}$.

1.	$3x + (-1) = 0$	1. Given
2.	$[3x + (-1)] + 1 = 0 + 1$	2.
3.	$3x + [(-1) + 1] = 0 + 1$	3.
4.	$3x + 0 = 0 + 1$	4.
5.	$3x = 1$	5.
6.	$\frac{1}{3}(3x) = \frac{1}{3}(1)$	6.
7.	$(\frac{1}{3} \cdot 3)x = \frac{1}{3}(1)$	7.
8.	$1 \cdot x = \frac{1}{3} \cdot 1$	8.
9.	$x = \frac{1}{3}$	9.

1-4 **6.** Simplify $-12 - |7 - 11| - [3 - (-9)]$.

7. Susan opened a savings account and deposited \$32.45 and \$15.67. Later, she withdrew \$20.65 from it. Jerry opened an account and deposited \$61.50. He then withdrew \$13.71 and \$16.43 from it. Which account has the greater balance and what is that balance?

1-5 **8.** Multiply $(-\frac{1}{5})(15r^3 - 10r^2 + 20)$.

9. Evaluate $3y^3 + 2y^2 - y + 7$ if $y = -4$.

1-6 **10.** Simplify $\dfrac{-(7 - 9)(7 + 9)}{(2 - 6) \cdot 4^2}$.

11. Evaluate $\dfrac{s + 3}{s - 2}$ if $s = -8$.

1-7 **12.** Solve and check $5(3z - 7) = 4(2z + 7)$.

13. Solve $s = -\frac{1}{2}gt^2 + vt$ for g.

1-8 **14.** A car traveled for $\frac{1}{2}$ h at r mi/h, then increased the speed by 10 mi/h and traveled for $1\frac{1}{2}$ h more. How far did the car go? (Express the answer in terms of r.)

1-9 **15.** The measure of an angle is 8° less than three times the measure of the angle's supplement. Find the measure of the angle.

16. Two cars leave City A at the same time to travel to City B. One car travels at 72 km/h and the other at 78 km/h. If the slower car arrives 20 min after the faster car, how far apart are City A and City B?

2-1 17. Solve $5x - 4(6 + 2x) < 2(1 - x)$ and graph the solution set.

18. Tell whether each statement is true or false.
a. If $a < b$, then $a^3 < b^3$. **b.** If $a > b$ and $c > d$, then $c - a < d - b$.

2-2 19. Solve and graph the solution set.
a. $5y + 3 \leq -2$ or $y - 7 \geq 0$ **b.** $x - 9 < 3x - 4 < x + 8$

2-3 20. Laura has an equal number of dimes, nickels, and quarters. If their total value is less than \$9, what is the maximum number of each that she could have?

2-4 21. Solve and graph the solution set.
a. $|4q - 18| \geq 0$ **b.** $12 - |7 - 2m| = 9$ **c.** $|r + 5| + 3 \leq 3$

2-5 22. Translate the following statement into an open sentence involving absolute value: the numbers whose distance from 3 is less than 2.

2-6 23. Prove: For all real numbers a, b, and c, if $a + c = b + c$, then $a = b$.

2-7 24. Supply the reason for each step of the proof:
If a and b are negative numbers and $a < b$, then $a^2 > b^2$.
1. a and b are negative numbers; $a < b$.
2. $a \cdot a > b \cdot a$
3. $a \cdot a > a \cdot b$
4. $a \cdot b > b \cdot b$
5. $a \cdot a > b \cdot b$
6. $\therefore a^2 > b^2$

3-1 25. Determine the constant k so that $(-2, 3)$ will satisfy the equation $kx - 4y = -k$.

26. An isosceles triangle has perimeter 13 cm. Find all integral possibilities for the lengths of its sides.

3-2 27. Graph $2x - y = 6$ and $x + 3y = 3$ on the same coordinate plane. What is the point of intersection?

3-3 28. Find the slope of the line:
a. through $(5, -2)$ and $(-1, -9)$
b. with equation $4x - 5y = 10$

29. Determine k so that the line through $(k + 2, k)$ and $(3, 2)$ has slope 3.

3-4 30. Find an equation in standard form of the line:
a. through $(2, -2)$ and $(4, 4)$
b. perpendicular to $y = x + 5$ and through $(0, -8)$

3-5 31. Solve the system: $3c - d = 1$
$3d + c = 1$

3-6 **32.** Three loaves of bread and two jars of peanut butter cost \$5.49. Four loaves of bread and three jars of peanut butter cost \$7.75. Find the price of a loaf of bread and a jar of peanut butter.

3-7 **33.** Graph the system of inequalities: $y \geq x$
$x < 3$

3-8 **34.** Graph $f: x \rightarrow |x| + 2$ with domain $D = \{-2, -1, 0, 1, 2\}$.

35. If $f(x) = 1 - 3x$ and $g(x) = \dfrac{1}{x^2 + 1}$, find $f(g(0))$ and $g(f(1))$.

3-9 **36.** If f is a linear function and $f(0) = 10$ and $f(10) = 14$, find $f(5)$ and $f(-5)$.

3-10 **37.** If x and y are integers, find the domain of $\{(x, y): |y| = x \text{ and } x \leq 4\}$. Graph the relation. Is it a function?

4-1 **38.** Simplify $2(8x^2y - 7xy + 3xy^2) - (3x^2y - 2xy - 9xy^2)$.

4-2 **39.** Simplify, assuming that variable exponents denote positive integers.
a. $(3cd^2)^2(-2c^3d)^3$ **b.** $h^{m-n}(h^{m+n} + h^m)$

4-3 **40.** Multiply.
a. $(2r - 3s)(5r + 7s)$ **b.** $(h^2 + 6h - 7)(2h^2 + 7h + 10)$

4-4 **41.** For $56x^2y^3z$, $42x^2y^2z^2$, and $70x^3y^2$, find the **(a)** GCF and **(b)** LCM.

4-5 **42.** Factor completely. If prime, say so.
a. $40x^3 - 64x^2y$ **b.** $125r^3 - 8s^3$
4-6 **c.** $13 - 14y + y^2$ **d.** $5c^2 + 2c - 3$
e. $9x^2 - 30x + 25$ **f.** $16u^4 - v^4$

4-7 **43.** Solve. Identify all double roots.
a. $8y^2 + 1 = 6y$ **b.** $x^4 - 25x^2 = 0$ **c.** $(x - 3)^3 - (x - 3)^2 = 0$

4-8 **44.** Find two consecutive odd integers whose product is 195.

45. A projectile is fired from the ground with an upward speed of 245 m/s. After how many seconds will it return to the ground? Use the formula $h = vt - 4.9t^2$.

4-9 **46.** Find and graph the solution set.
a. $x^3 - 16x > 0$ **b.** $x^2 - 5x \leq -6$

Chapter Tests

Chapter 1 Test

1. On a number line, point M has coordinate -6.4 and point N is 5.3 units to the left of M. Find the coordinate of N.
2. List 5, $-1\frac{1}{5}$, $3\frac{4}{5}$, $-\frac{3}{5}$, and $2\frac{2}{5}$ in order from least to greatest.
3. Simplify $5 \cdot 6 - (8 + 2)$.
4. Simplify $3^3 \div 4^2 \times 2^5 - 4 \times 3$.
5. Evaluate $\frac{a^2 + b^2}{a - b} + \frac{bc}{a}$ if $a = 5$, $b = 3$, and $c = 15$.
6. Name the property used in each step of the simplification.

$$\begin{aligned} ab + ac + (-ca) &= ab + (ac + (-ca)) && \textbf{a. } \underline{\ ?\ } \\ &= ab + (ac + (-ac)) && \textbf{b. } \underline{\ ?\ } \\ &= ab + 0 && \textbf{c. } \underline{\ ?\ } \\ &= ab && \textbf{d. } \underline{\ ?\ } \end{aligned}$$

Simplify.

7. $-9(-6 + 2) - 7[3 + 2(8 + 6)]$
8. $|-26 + 10| - |13 - 6 + 3|$
9. $3(x + 4) - x$
10. $(-2)^3(5^2)\left(-\frac{1}{10}\right)$
11. $\frac{(5 - 8)^2}{-24 \div 3(2^3)}$
12. $\frac{16a^3 - 4a^2 + 64a + 36}{4}$

Solve.

13. $\frac{3}{5}a + 3 = 2a - 11$
14. $2(x - 1) = 5 - (3 - 2x)$

Express your answer in simplest form in terms of the given variable.

15. If you fill half of a container whose capacity is t liters and then take 2 liters out, how full is the container?
16. The perimeter of a rectangle with length l cm is 80 cm. What is the width?

Solve.

17. A theater has 600 tickets to sell for a show. Of these tickets 225 sell for \$2 apiece more than the others. If all tickets are sold and \$2250 is taken for the show, what price is each type of ticket?
18. At 1:30 P.M. two planes leave Chicago, one flying east at 540 km/h and the other flying west at 620 km/h. At what time will they be 1450 km apart?

Chapter 2 Test

Solve.

1. $5x - 4 \geq 3x$

2. $3(5t + 4) < 13t - 10$

3. $-3 < 3 + 2w < 6$

4. $2 > \frac{t}{5}$ or $2t + 4 \leq -6$

5. Find the largest possible values for a set of three consecutive even integers whose sum is less than 105.

Solve.

6. $|4x - 10| = 12$

7. $4|x - 7| \leq 12$

8. $|2 - 3x| > 1$

Graph the solution set of each open sentence.

9. $|2t + 5| \leq 7$

10. $|w - 5| > 3$

11. Write the converse of the statement: "If an integer is odd, then it is not an even integer."

12. Give reasons for the steps shown in the proof of the statement: "For all real numbers a, b, and c, if $c + a = c + b$, then $a = b$."

1. a, b, and c are real numbers; $c + a = c + b$
2. $-c$ is a real number.
3. $-c + (c + a) = -c + (c + b)$
4. $(-c + c) + a = (-c + c) + b$
5. $0 + a = 0 + b$
6. $\therefore \quad a = b$

Tell whether the statement is true for all real numbers. If it is not, give a counterexample.

13. If $a \geq b$, then $a^2 \geq b^2$.

14. If $a < b$, then $\frac{1}{a} < \frac{1}{b}$.

15. If $a < b$ and $c > 0$, then $\frac{a}{c} < \frac{b}{c}$.

16. If $|a| > |b|$, then $a > b$.

Chapter 3 Test

1. Determine the constant k so that $(-2, 5)$ will be a solution of $2x + ky = 3(k - 1)$.
2. Find the solution set of $3x + 2y = 14$ if each variable represents a whole number.
3. Graph each equation.

   a. $y - 3 = 0$ b. $3x - y = -5$ c. $\frac{x}{3} + \frac{y}{2} = 0$
4. Find the slope of each line described.
   a. Passes through $(-2, -7)$ and $(0, 9)$ b. Has equation $3x + 4y = 9$
5. Determine the constant k so that the graph of $3x + (k - 1)y = k + 1$ will have slope 5.
6. Find equations in standard form of the lines described.
   a. Passes through $(2, 5)$ and has slope -2
   b. Passes through $(-3, 0)$ and $(0, 6)$.
7. Find equations in standard form of the lines through $(4, -3)$ that are **(a)** parallel to and **(b)** perpendicular to the line $3x - y = 5$.
8. Solve the system: $2x + 3y = 4$
   $5x + 4y = 3$
9. It takes 6 h for a plane to travel 720 km with a tail wind and 8 h to make the return trip with a head wind. Find the air speed of the plane and the speed of the wind current.
10. Graph the system: $x - 3y > 6$
    $x + y \leq 2$
11. The domain of the function $g: t \rightarrow |t + 3| - 5$ is $D = \{0, 1, 2, 3, 4, 5\}$. Find the range of g.
12. If $f(x) = x^2 + 2$ and $g(x) = 2x - 1$, find $f(g(1))$ and $g(f(1))$.
13. Find $g(3)$, given that g is a linear function with $g(1) = -2$ and $g(4) = 8$.
14. A load of 5 kg stretches a coil spring to a length of 24 cm, and a load of 8 kg stretches it to a length of 30 cm. Find the length of the spring when there is no load.
15. Graph the relation $\{(-3, 1), (-2, 0), (0, 0), (1, 4), (-2, 3)\}$. Is the relation a function?

Chapter 4 Test

Simplify.

1. $4(x^2 + 3) + 5(2 - 3x^2)$
2. $(4p^2q + 5p^3q^2 - 6pq) - (-2p^3q^2 + 5pq - 8p^2q)$
3. $(-6x^2r^2y)(5xr^3y^2)$
4. $3w^2v(w^4 - v^4)$
5. $(2r - s)(3r + 2s)$
6. $b^2x^2(2b + x^2)^2$

Find (a) the GCF and (b) the LCM of the following.

7. 420, 504
8. $15r^2s^3$, $25rs^2$, $45r^2s$

Factor each polynomial completely.

9. $64r^2 + 16r + 1$
10. $6d^2 + 3ed + 10d + 5e$
11. $s^2 + sb - 6b^2$
12. $9x^2 - 15x + 6$
13. $x^4 + 6x^2 + 9$
14. $a^2x^3 - 12a^2x^2 + 20a^2x$

Solve.

15. $(x + 5)(2x - 1)(3x + 4) = 0$
16. $6t^2 - t - 2 = 0$
17. The height of a triangle is 3 cm less than the length of its base, and its area is 20 cm^2. Find the height.
18. Find and graph the solution set of $x^2 + 12 \geq 13x$.

Chapter 5 Test

Simplify. Use only positive exponents.

1. $\left(\frac{4b^2u}{3u^2}\right)\left(\frac{2u^3}{b^2}\right)^2$

2. $\left(\frac{x^3y^{-2}}{x^{-3}y^2}\right)^{-1}$

3. Simplify, assuming that the factors are approximations. Give the answer in scientific notation with two significant digits.

$$\frac{(3.8 \times 10^4)(7.8 \times 10^{-3})}{(2.7 \times 10^{-5})(3.5 \times 10^4)}$$

4. Simplify $\frac{5x^2 - 20}{3x^2 + 5x - 2}$.

5. For the function $f(x) = (x^2 - 4)(x + 3)^{-2}$, find **(a)** the domain and **(b)** its zeros, if any.

Simplify.

6. $\frac{5t^3}{4s} \div \frac{16s^2}{25t^3} \cdot \frac{16s}{25t^3}$

7. $\frac{a^2 + 10a + 25}{a^2 + 5a + 6} \div \frac{a^2 - 10a + 25}{a^2 - 2a - 15} \cdot \frac{a^2 - 3a - 10}{a^2 - 25}$

8. $\frac{x}{x + y} + \frac{x}{x - y}$

9. $\frac{3x + 1}{3x - 15} - \frac{x^2 - 9}{x^2 - 25}$

10. Simplify $\frac{2 - a^{-1} - 6a^{-2}}{2 - 7a^{-1} - 15a^{-2}}$.

11. Solve $\frac{3x - 4}{4} - 3 \leq \frac{4x - 5}{3}$.

12. A library crew can shelve books in six hours. Another crew can do the job in four hours. If the first crew begins and is joined by the second crew one hour later, how many hours will it take for the job to be completed?

13. Solve $\frac{2}{x - 2} - \frac{1}{x^2 + x - 6} = \frac{x}{x + 3}$.

14. A bicycle trip of 90 mi would have taken an hour less if the average speed had been increased by 3 mi/h. Find the average speed of the bicycle.

Chapter 6 Test

1. Simplify $\sqrt[3]{-0.125}$.
2. Find the real roots of $81z^4 - 1 = 0$.

Simplify.

3. $\sqrt[3]{4x^2} \cdot \sqrt[3]{4xy^2} \cdot \sqrt[3]{4y}$
4. $\sqrt[4]{\frac{64b^4c^5}{d^3}}$
5. $4\sqrt{27} + 3\sqrt{12}(2\sqrt{3} - 5\sqrt{75})$
6. $\sqrt{m^5n^3} + mn\sqrt{4m^3n} - m^2n\sqrt{16mn}$
7. $(3\sqrt{5} + \sqrt{3})^2$
8. $\frac{4}{\sqrt{2} - \sqrt{6}}$

Solve.

9. $\sqrt[3]{2x^2 - x} = 2$
10. $\sqrt{11 - s} = \sqrt{-5s} + 1$
11. Express $0.07\overline{12}$ as a common fraction.
12. Find a rational number and an irrational number between 4.1 and $\sqrt{17}$.
13. Solve $5t^2 + 125 = 0$ over the complex numbers.

Simplify.

14. $\sqrt{-27} \cdot \sqrt{-9}$
15. $\frac{7}{\sqrt{-21}}$
16. $4\sqrt{-3} + 2\sqrt{-27}$
17. $(4 + 5i) + (6 - 2i)$
18. $(6 - 3i)(4 + 3i)$
19. $\frac{2 + 3i}{4 - 5i}$

Chapter 7 Test

1. Solve $(t - 1)^2 = 7$.
2. Solve $3r^2 - 6r + 2 = 0$ by completing the square.
3. Use the quadratic formula to solve $5s^2 - 13s + 6 = 0$.
4. Without solving each equation, determine the nature of the roots.
   a. $3y^2 - 10y + 1 = 0$ **b.** $2w^2 + 6w + 10 = 0$
5. Find all real values of k for which $3t^2 - 6t + k = 0$ has two real roots.

Solve.

6. $x^4 + 5x^2 - 36 = 0$

7. $\frac{1}{\sqrt{y}} = 1 - \frac{2}{y}$

8. Find the vertex and axis of the parabola $y = 3(x + 4)^2$.
9. Find an equation of the form $y - k = a(x - h)^2$ for the parabola that has vertex $(1, -5)$ and contains $(-4, 3)$.
10. **a.** Determine whether the function $f(x) = \frac{1}{2}x^2 - 4x + 8$ has a minimum value or a maximum value.
    b. Find this value.
11. Find the domain, the range, and the zeros of $f(x) = x^2 - 12x + 36$.
12. Find a quadratic equation having $3 - \sqrt{3}$ and $3 + \sqrt{3}$ as roots.
13. Find a quadratic function $f(x) = ax^2 + bx + c$ whose graph has maximum value 25 and x-intercepts -3 and 2.

Chapter 8 Test

1. If s is directly proportional to $v + 3$ and $s = 8$ when $v = 1$, find s when $v = -2$.
2. If t varies inversely as the cube of z and directly as the square of r, and $t = 4$ when $z = 3$ and $r = 6$, find t when $z = 6$ and $r = 9$.
3. The electrical conductance of a wire varies directly as the square of its diameter and inversely as its length. The conductance of a wire 20 m long and 3 mm in diameter is 0.54 mho. If a wire of the same material has length 50 mm and diameter 5 mm, what is its conductance?
4. Divide $6b^2x^3 - 6b^2x^2 + 19bx - 10bx^2 - 15$ by $2bx^2 - 2bx + 3$.
5. Use synthetic division to divide $3x^4 - 5x^2 + 6x - 5$ by $x - 3$.
6. Find $P(\frac{1}{2})$ given that $P(x) = 2x^3 - 5x^2 + 6x - 5$.
7. Find a cubic equation with integral coefficients that has -2 and $i\sqrt{3}$ as roots.
8. Solve $x^4 - 8x^3 + 26x^2 - 40x + 21 = 0$ given that $2 - i\sqrt{3}$ is a root.
9. Solve $6x^4 - 7x^3 + 8x^2 - 7x + 2 = 0$ by first finding the rational roots.
10. Graph $P(x) = 5x^3 - 6x^2 + 20x - 24$ and estimate the real root of $5x^3 - 6x^2 + 20x - 24 = 0$ to the nearest half unit.
11. Use linear interpolation to find an approximation of $T(0.14)$ to the nearest hundredth given that $T(0.1) = 9.87$ and $T(0.2) = 10.01$.

Chapter 9 Test

1. Given the points $P(3, -5)$ and $Q(6, 1)$, find **(a)** the length PQ, **(b)** the midpoint of the segment $\overline{PQ}$, and **(c)** the point X such that P is the midpoint of $\overline{QX}$.
2. Find an equation of the circle having center (3, 3) and containing the origin.
3. Find the center and radius of the circle $x^2 + y^2 - 10x = 0$.
4. Find an equation for the set of all points in the plane equidistant from the point (2, 3) and the line $x = 4$.
5. Find the vertex, focus, directrix, and axis of the parabola $(y - 3)^2 = 7(x - 5)$.
6. Find an equation of the ellipse having foci $(-\sqrt{7}, 0)$ and $(\sqrt{7}, 0)$ and sum of focal radii 8.
7. Graph the ellipse $16x^2 + 4y^2 = 16$.
8. Graph the hyperbola $y^2 - x^2 + 9 = 0$. Show the asymptotes as dashed lines.
9. Find an equation of the ellipse having foci (2, 3) and (2, 7) and sum of focal radii 10.
10. Identify the conic $4x^2 - 8x - 3y^2 + 18y = 35$ and find its center and foci.

By sketching graphs, find the number of real solutions of each system.

11. $x^2 + y^2 = 9$
    $y^2 = x$
12. $x^2 - 4y^2 = 16$
    $x + y = 2$
13. Solve the system: $x^2 + y^2 = 25$
    $2x + y = 10$
14. Find the dimensions of a rectangle having perimeter 28 cm and a diagonal of length 10 cm.
15. Solve the system:

$$\begin{aligned} 2x - 5y + 4z &= 18 \\ x + 3y + 5z &= 7 \\ -3x - 2y + 4z &= 22 \end{aligned}$$

Chapter 10 Test

1. Simplify $\left(\frac{9}{16}\right)^{-3/2}$.
2. Write $\sqrt[3]{\frac{x^2}{y}}$ in exponential form.
3. Simplify $(\sqrt{2})^{\pi-1}(\sqrt{2})^{1-\pi}$.
4. Solve $9^{x-2} = 27^x$.
5. Suppose $f(x) = 3x - 4$ and $g(x) = \sqrt{x - 2}$. Find:
   a. $f(g(2))$ **b.** $g(f(x))$
6. Find $f^{-1}(x)$ if $f(x) = \frac{x+1}{2}$.
7. Simplify $\log_4 2\sqrt{2}$.
8. Solve: **a.** $\log_3 x = -2$
   b. $\log_5 (2x + 1) - \log_5 x = 2$
9. Given that $\log 2 = 0.301$ and $\log 5 = 0.699$, find:
   a. $\log 20$ **b.** $\log 2.5$ **c.** $\log 125$
10. Express t in terms of common logarithms: $6^{3t-1} = 5$.
11. Find $\log_7 11$ to three significant digits.
12. The population of a colony of bacteria doubles every 24 hours. How long does it take for the population to be 2.5 times its original size? Give the answer to two significant digits.
13. Write $1 + 3 \ln 2$ as a single logarithm.
14. Solve $\ln x^3 = 2$, leaving answers in terms of e.

Chapter 11 Test

1. Tell whether the sequence $-100, 20, -4, \frac{4}{5}, \ldots$ is arithmetic, geometric, or neither. Then find the next two terms in the sequence.
2. Find the first four terms of the sequence with nth term $t_n = 5 - 3n$. Then tell whether the sequence is arithmetic, geometric, or neither.
3. Find t_{20} for the arithmetic sequence $3, 7, 11, 15, \ldots$.
4. Insert three arithmetic means between 2 and 18.
5. Find a formula for the nth term of the geometric sequence $\frac{3}{2}, 6, 24, 96, \ldots$.
6. Find the geometric mean of 5 and 80.
7. Maria has taken a job with a starting salary of \$22,000 and annual raises of 5%. What will be her salary during her third year on the job?
8. Use sigma notation to write the series $81 - 54 + 36 - 24 + \cdots$.
9. Find the sum of the first fifteen terms of the arithmetic series $-7 - 1 + 5 + 11 + \cdots$.
10. Evaluate $\sum_{k=1}^{6} (-1)^{k+1}3^{k-1}$.
11. Find the sum of the infinite geometric series $1 + \frac{1}{3} + \frac{1}{9} + \frac{1}{27} + \cdots$ if it has one.
12. Expand and simplify $(2r - w)^4$.
13. Find the fourth term in the expansion of $(s - 2t)^{10}$.

Chapter 12 Test

1. Find two angles, one positive and one negative, that are coterminal with $-170°$.
2. **a.** Express $12.78°$ in degrees, minutes, and seconds.
   b. Express $120°15'36''$ in decimal degrees.
3. Find the values of the six trigonometric functions of an angle θ in standard position whose terminal side passes through $(1, 2)$.
4. If $\sin \theta = \frac{24}{25}$ and $\cos \theta < 0$, find **(a)** the quadrant of θ and **(b)** $\tan \theta$.
5. **a.** Find $\tan 140.8°$.
   b. If $\cos \theta = -0.4289$ and $0° < \theta < 180°$, find θ to the nearest tenth of a degree.
6. The angle of elevation from a boat to a hovering helicopter is $32.2°$ and the helicopter is 600 m above the water level. What is the horizontal distance, to the nearest meter, between the boat and helicopter?
7. A metal bar 12 cm long is rotated about one end so that the distance from the starting position of the other end is 20 cm from the ending position. Through what angle was the bar rotated?

Give lengths to three significant digits and angle measures to the nearest tenth of a degree.

8. Solve $\triangle ABC$, given that $b = 8$, $c = 5$, and $\angle A = 70°$.
9. Solve $\triangle ABC$, given that $\angle A = 65°$, $\angle B = 20°$, and $a = 12$.
10. Find the area of a triangle with sides of length 10 cm and 13 cm and an angle of $70°$ between those sides.

Chapter 15 Test

For Exercises 1–4, use the following distribution: 86, 72, 86, 95, 84, 78, 86, 77, 84, 92

1. Draw a stem-and-leaf plot for the distribution.
2. Find **(a)** the mode, **(b)** the median, and **(c)** the mean of the distribution.
3. Draw a box-and-whisker plot for the distribution.
4. Find, to the nearest whole number, **(a)** the variance and **(b)** the standard deviation of the distribution.
5. The mean score on a mathematics examination was 79.3 and the variance was 32.49. Assuming the scores formed a normal distribution, what fraction of the scores was above 85?
6. Draw a scatter plot for the data in the table at the right. Then describe the nature of the correlation between height and waist size.

Height (in.)	Waist Size (in.)
70	33
68	32
65	30
64	28

7. How many odd whole numbers less than 3000 can be formed using the digits 2, 3, 4, 5?
8. If you have five pictures, in how many ways can you hang two pictures side by side?
9. How many different signals can be made by displaying six pennants all at one time on a vertical flagpole? The pennants are identical except for color. Two are green, two are red, one is yellow, and one is orange.
10. How many 3-letter subsets can be formed from the letters in the set {G, H, J, K}? Specify the subsets.
11. How many combinations can be formed using the letters in the word HISTORY, taken 5 at a time?
12. A coin and a die are tossed. Specify each of the following events.
    a. The coin shows heads and the die shows a multiple of 3.
    b. The coin shows heads and the die shows a number less than 5.
13. One card is drawn at random from a 52-card deck. Find the probability of each event.
    a. It is a king or jack. **b.** It is less than 5 and more than 2. **c.** It is a club or diamond.
14. There are 4 red, 3 white, and 2 orange balls in a bag. Two are drawn at random. Let A be the event that at least one ball is white. Let B be the event that both balls are the same color.
    a. Find $P(A)$, $P(B)$, $P(A \cap B)$, and $P(A \cup B)$.
    b. Are events A and B mutually exclusive? independent?

Chapter 16 Test

1. Find the values of x and y in $\begin{bmatrix} 3x - 27 \\ 4y + 64 \end{bmatrix} = O_{2\times 1}$.

In Exercises 2–7, let $A = \begin{bmatrix} 1 & 3 & -2 \\ 6 & 5 & 0 \\ 1 & 7 & 3 \end{bmatrix}$ **and** $B = \begin{bmatrix} 1 & 6 & 3 \\ 2 & 1 & 0 \\ -5 & 2 & 6 \end{bmatrix}$.

Express each of the following as a single matrix.

2. $A + B$ **3.** $A - B$ **4.** $2A$ **5.** $-3B$ **6.** AB **7.** BA

8. Write the matrix that represents this network:

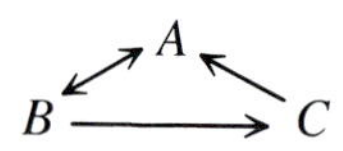

Evaluate each determinant.

9. $\begin{vmatrix} 3 & -10 \\ -5 & 3 \end{vmatrix}$

10. $\begin{vmatrix} 3 & -6 & 2 \\ 0 & 1 & 9 \\ 2 & 1 & 0 \end{vmatrix}$

11. Solve this equation for X: $\begin{bmatrix} 3 & 2 \\ -1 & 6 \end{bmatrix} X = \begin{bmatrix} 12 & -13 \\ 16 & 11 \end{bmatrix}$

12. Expand this determinant by the third column: $\begin{vmatrix} 3 & 1 & 3 \\ -1 & -3 & 0 \\ 2 & 6 & 3 \end{vmatrix}$

Use the properties of determinants to evaluate each determinant.

13. $\begin{vmatrix} 3 & 2 & 2 \\ 2 & 3 & 2 \\ 0 & 0 & 0 \end{vmatrix}$

14. $\begin{vmatrix} 1 & 0 & 2 \\ -1 & 1 & 0 \\ 3 & -1 & 5 \end{vmatrix}$

15. Solve using Cramer's rule:

$$\begin{aligned} 2x \quad\quad + z &= -2 \\ -x + 3y + 4z &= 22 \\ 2x + 3y \quad\quad &= -3 \end{aligned}$$

Cumulative Reviews

Review for Chapters 1–3

Simplify.

1. $3(7^2 - 2^2) \div (5 + 5)^2$
2. $20 + (-2) - |20 - 2|$
3. $3(x + y) - 2(x - y)$
4. Evaluate $(t + 3u)^2 + (t + 2u^2) + (t - (2u)^2)$ if $t = 5$ and $u = -2$.
5. Prove for all real numbers m and x that $m(-x) = -mx$.

Solve each equation or inequality.

6. $3(x - 2) - 4(2x + 3) = 5x$
7. $\frac{3x - 1}{4} = x$
8. $t - (2 - t) < 5t$
9. $0 \geq 2(m - 5) \geq -8$
10. $2\left|\frac{t - 1}{3}\right| + 3 \leq 5$
11. $|5 - 2d| > 3$
12. Find the solution set of $6x + 3y \leq 12$ if x and y are positive integers.
13. If a line with equation $3x - ky = 4$ has slope $\frac{5}{6}$, find the value of k.

Give an equation in standard form of the line described.

14. Has slope $-\frac{3}{4}$ and contains (4, 5)
15. Parallel to $4x + y = 3$ and contains (4, 7)
16. Perpendicular to $-2x + 3y = 7$ and y-intercept -3
17. Passes through $\left(-\frac{1}{2}, \frac{7}{9}\right)$ and $\left(\frac{1}{3}, \frac{1}{4}\right)$

Solve each system. If the system has no solution, say so.

18. $3x - y = 2x - 7$
    $2x + 5 = 2y - 9$
19. $4x + 3y = 12$
    $3x + 4y = 10$
20. $2x - 6y = 11$
    $x + 3y = 6$
21. Graph the system: $4x - 3y \geq 10$
    $2x \leq y + 6$
22. Three dozen eggs and five loaves of bread cost \$6.90; four dozen eggs and two loaves of bread cost \$5.70. Find the prices for a dozen eggs and for a loaf of bread.
23. If $g(x) = |2x - 7|$ and $h(x) = (x - 3)^2 + 5$, find $g(h(3))$ and $h(g(3))$.
24. Find an equation for the linear function f for which $f(3) = 7$ and $f(-1) = 5$.
25. If Z is a linear function such that $Z(3) = 13$ and $Z(5) = 27$, find $Z(50)$.
26. Graph the relation $\{(-2, 1), (-1, 3), (0, 0), (1, 3), (2, -3)\}$. Is the relation a function?

Review for Chapters 4–7

1. Simplify $3(2x^2y - xy + 4xy^2) - 2(xy^2 + 5xy - 3x^2y)$.
2. Multiply: **a.** $\left(\frac{4}{5}ax^2\right)^2(-10x^3a^2)^2$ **b.** $3x^2(2x - 1)(2x + 1)$
3. Find **(a)** the GCF and **(b)** the LCM of $x^2 - x$ and $x^3 - x$.
4. Factor completely: **a.** $12r^2s^2 + rs - 6$ **b.** $3t^4 - 6t^3 - 4t^2 + 8t$
5. When the product of three consecutive integers is decreased by the cube of the smallest, the result is 85. What are the integers?
6. Solve $t^2 > 5t + 6$ and graph the solution set.

Simplify.

7. $\dfrac{(5s^2t^2)^2}{4s^4(t^3)^2}$

8. $\left(\dfrac{3a}{4b}\right)^{-1}\left(\dfrac{3a}{4b}\right)^2(a^{-1}b^{-1})$

9. $\dfrac{a^4 + 2a^2 + 1}{a^3 - a^2 + a - 1}$

10. $\dfrac{4t^2 + 20t + 25}{4t^2 - 25} \div \dfrac{1}{2t^2 - 11t + 15}$

11. $\dfrac{4}{3t - 6} - \dfrac{2}{3t + 6}$

12. $\dfrac{(a + b)^{-1} - (a - b)^{-1}}{(a - b)^{-1} + (a + b)^{-1}}$

13. A chemist prepared 100 mL of 15% saline solution by using a 20% solution, an 18% solution, and 40 mL of a 10% solution. How much of the 20% and 18% solutions were used?

Solve.

14. $\dfrac{t^2 - t + 2}{2t^2 + t - 2} = \dfrac{1}{2}$

15. $5t^3 + 400 = 1025$

16. $\sqrt{4t - 10} = t - 2$

Simplify.

17. $\sqrt{\dfrac{25}{12}} - \sqrt{27}$

18. $(\sqrt{2} - 3\sqrt{2})^2$

19. $\dfrac{5 + \sqrt{2}}{2\sqrt{2} - \sqrt{2}}$

20. $\dfrac{3 + 4i}{3 - 4i}$

21. Solve $3x^2 - 18x - 5 = 4$ by completing the square.
22. Solve $3t^2 - 7t + 5 = 0$ by using the quadratic formula.
23. Find the values of k for which $4x^2 + kx + 5 = 0$ has two different real roots.
24. Graph the parabola $y = 2(x + 3)^2 + 2$. Label the vertex and axis of symmetry.
25. Find the domain, the range, and the zeros of $f(t) = 3t^2 - 12t + 17$.
26. Give the sum and product of the roots of $7r^2 - 5r + 1 = 0$.
27. Find a quadratic equation with integral coefficients having the roots $\dfrac{4 \pm 5i}{6}$.
28. If two numbers differ by seven, what is the least possible value of their product?

Review for Chapters 8–11

1. If y varies directly as the square of x and y is 100 when $x = 6$, find y when $x = 15$.
2. Divide $2x^3 + x^2 + 4x + 35$ by $2x + 5$.
3. Solve $x^4 - 7x^3 + 21x^2 - 23x - 52 = 0$ given that $2 + 3i$ is a root.
4. Find the center and radius of the circle $2x^2 + 8x + 2y^2 - 12y - 26 = 0$.
5. Find and graph an equation of the parabola with vertex $(3, 3)$ and directrix $x = 4$.
6. Graph the hyperbola $\frac{x^2}{16} - \frac{y^2}{25} = 1$. Find the coordinates of the foci.

Solve each system of equations.

7. $\begin{aligned} 4x^2 + y^2 &= 25 \\ 2x + y &= -1 \end{aligned}$

8. $\begin{aligned} y &= x^2 - 3 \\ y &= 2x \end{aligned}$

9. $\begin{aligned} 2r - s + t &= 3 \\ r + 2s - t &= 3 \\ 6r - 8s + 4t &= -2 \end{aligned}$

10. Simplify: **a.** $(16^{1/4} + 64^{1/6})^{-5/2}$ **b.** $\frac{81^{3/4}}{27^{-4/3}}$
11. Solve: **a.** $4t^{2/3} - 5 = 11$ **b.** $\log_4 x = 3$ **c.** $\log_x 243 = 5$
12. If $f(x) = 4x - 7$, find $f^{-1}(x)$ and $f^{-1}(5)$.
13. Solve $\log (2x + 8) = \log 32 + \frac{3}{4} \log 16 - \frac{1}{2} \log 64$.

Find the value of t to three significant digits.

14. $t = 4.2^{-1.5}$
15. $\sqrt{t^3} = (4.6)^2$
16. $18^t = 19$

Find t_1, t_2, t_3, t_4. Tell if each sequence is arithmetic, geometric, or neither.

17. $t_n = 10 \cdot 3^{-n+1}$
18. $t_n = 5 + \sqrt{n}$
19. $t_n = 2n + \frac{1}{2}$
20. Find two geometric means and two arithmetic means between 10 and 100.
21. Given the sequence $-8, 4, -2, 1, \ldots$, find t_7 and t_n.
22. Evaluate: **a.** $\sum_{k=2}^{7} (2k + 5)$ **b.** $\sum_{n=1}^{5} \left(\frac{2}{3}\right)^{n+1}$
23. Find the first term of an infinite geometric series with common ratio $-\frac{3}{7}$ and sum 343.
24. Expand and simplify $(2a + 2b)^4$.
25. Find and simplify the sixth term of $(3t + r)^8$.

Review for Chapters 12–14

1. Name two angles, one positive and one negative, that are coterminal with 173°.
2. Find the values of the six trigonometric functions of an angle θ in standard position whose terminal side passes through the point (1, 4).
3. An aircraft begins its runway approach. It descends 176 m for each 1000 m of horizontal distance traveled. At what angle, to the nearest tenth of a degree, is the aircraft descending?
4. In $\triangle ABC$, $\angle B = 105°$, $c = 7$, and $a = 9$. Find $\angle C$, $\angle A$, and b. Find angle measure to the nearest tenth of a degree and length to three significant digits.
5. Find the area of $\triangle ABC$ in Exercise 4.
6. Express 72° in radians.
7. Find the exact values of the six trigonometric functions of $\frac{2\pi}{3}$.
8. State the amplitude, the maximum and minimum values, and the period of $y = -3 \sin (2x) + 4$.
9. Simplify $\frac{\cos \theta \csc \theta}{\cot^2 \theta}$.
10. Prove: **a.** $\frac{\tan a - \tan b}{1 + \tan a \tan b} = \frac{\cot b - \cot a}{1 + \cot b \cot a}$ **b.** $\frac{1 + \csc a}{\cot a + \cos a} = \sec a$
11. If $\tan \theta = 3\sqrt{7}$ and $180° < \theta < 270°$, find $\cos \frac{\theta}{2}$.
12. The vector **u** has magnitude 10 and bearing 130°. The vector **v** has magnitude 20 and bearing 80°. Find the magnitude of $\mathbf{u} - \mathbf{v}$ to three significant digits and its bearing to the nearest tenth of a degree.
13. Find **(a)** a vector orthogonal to $\mathbf{v} = 3\mathbf{i} + 4\mathbf{j}$ and **(b)** a unit vector orthogonal to **v**.
14. Use De Moivre's theorem to express $(-1 + \sqrt{3}i)^{-5}$ in $x + yi$ form.
15. Find the fourth roots of -16 in $x + yi$ form.

Find the value of each of the following.

16. $\cos \left(\text{Sin}^{-1} \left(-\frac{3}{4}\right)\right)$
17. $\sin \left(\text{Tan}^{-1} \frac{1}{2} + \text{Tan}^{-1} \frac{1}{3}\right)$
18. $\text{Sin}^{-1} \left(\tan \frac{3\pi}{4}\right)$

Solve for θ, $0° \le \theta < 360°$.

19. $\sec^2 \theta - 4 = 0$
20. $2 \cos (\theta - 60°) = \sqrt{2}$
21. $\cos 2\theta(3 - 4 \sin^2 \theta) = 0$
22. $2 \sin \theta + 2 \csc \theta = 5$

Review for Chapters 15–16

For Exercises 1–4, use the following distribution:

30, 16, 24, 24, 16, 28, 14, 24, 22, 32

1. Draw a stem-and-leaf plot for the distribution.
2. Find **(a)** the mode, **(b)** the median, and **(c)** the mean of the distribution.
3. Draw a box-and-whisker plot for the distribution.
4. Find, to the nearest whole number, **(a)** the variance and **(b)** the standard deviation of the distribution.
5. In a standard normal distribution, what fraction of the data lies outside of 1.6 standard deviations of the mean?
6. In how many different ways can five different plants taken three at a time be arranged on a window sill?
7. How many distinct seven-flag messages can be sent if two flags are black, two flags are red, and three flags are green?
8. Seven points lie on a circle. How many triangles can be formed having any three of these points as vertices?
9. A bag contains two white marbles, four blue marbles, and six red marbles. A marble is drawn at random from the bag. What is the probability of each event?

   a. It is not white. **b.** It is blue or white.
10. A basketball player shoots foul shots with a 75% success rate. Find the probability of each event.

    a. The player is successful twice in a row.

    b. The player is successful in two shots out of three.
11. Use the given probabilities to tell whether A and B are independent or dependent.

    a. $P(A) = 0.25$; $P(B) = 0.20$; $P(A \cap B) = 0.05$

    b. $P(A) = 0.3$; $P(B) = 0.4$; $P(A \cap B) = 0.7$

In Exercises 12–15, let $A = \begin{bmatrix} 2 & 1 & 3 \\ -1 & 0 & 5 \\ 6 & 1 & 2 \end{bmatrix}$ **and** $B = \begin{bmatrix} 5 & -1 & 2 \\ 3 & 4 & 6 \\ -1 & -2 & -3 \end{bmatrix}$.

Express the following as a single matrix.

12. $2A + B$ **13.** $A - B$ **14.** $3B$ **15.** AB

Use the network shown for Exercises 16 and 17.

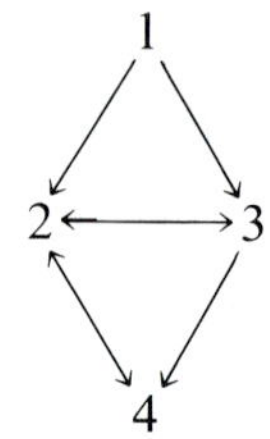

16. Write the communication matrix for the network.

17. Find the matrix that shows the number of ways a message can be sent from one point to another using at most one relay point.

Find the value of each determinant.

18. $\begin{vmatrix} 5 & -2 \\ 3 & 0 \end{vmatrix}$ **19.** $\begin{vmatrix} 6 & 3 \\ 1 & 5 \end{vmatrix}$ **20.** $\begin{vmatrix} 3 & 4 & -1 \\ 2 & 6 & 3 \\ 1 & 1 & 2 \end{vmatrix}$

21. Solve for X: $\begin{bmatrix} 3 & 4 \\ -1 & 2 \end{bmatrix} X = \begin{bmatrix} 2 & 15 \\ -4 & 5 \end{bmatrix}$

22. Solve by using Cramer's rule:
$$\begin{aligned} -2x + y - 3z &= -2 \\ x + 4y + z &= -13 \\ 3x - 4y + 2z &= 8 \end{aligned}$$

Topical Reviews

Review of Quadratics

Solve for the real roots by the most appropriate method.

1. $x^2 - 36 = 0$

2. $x^2 = 2x$

3. $(x + 3)^2 = 100$

4. $x^2 + 4x = 285$

5. $\frac{2}{2t + 1} + \frac{1}{2t - 1} = 1$

6. $(x + 2)(x - 5) = 8$

7. $2x(x + 1) = 5$

8. $(x - 3)^2 = 25(2x - 3)^2$

9. $(x^3 + 1)^2 - 4(x^3 + 1) + 3 = 0$

10. The number of diagonals that can be drawn in a polygon of n sides is $\frac{1}{2}n(n - 1) - n$. If 230 diagonals can be drawn, how many sides does the polygon have?

11. The parabola $y = 2x^2 - x - 15$ and the line $y = 3x + 1$ have two points in common. Find the points.

12. Find k if the graph of $y = 2x^2 + 2kx - k$ passes through $(2, 3)$.

13. Show that 1 is a root of $(a + b)x^2 - (a - b)x - 2b = 0$. Find the other root by using the product of the roots formula.

14. Find the coordinates of the points on the parabola $y = 2x^2 + 3x - 2$ that have equal x and y values.

15. Find a quadratic equation with integral coefficients whose roots are:
a. $3, -2$ **b.** $3 + \sqrt{2}, 3 - \sqrt{2}$ **c.** $3 + i\sqrt{2}, 3 - i\sqrt{2}$

16. The perimeter of a rectangle is 30 m. If one side is x, express the area of the rectangle in terms of x. Show that there is no value of x such that the area is 60 m^2.

17. Find k so the roots of $5x^2 - x + 1 = k$ are equal.

18. Find $P(1 + 2i)$ if $P(x) = x^2 - x$.

19. Find A, B, and C if $f(x) = Ax^2 + Bx + C$ and $f(-2) = 14$, $f(0) = -6$, and $f(2) = -2$.

20. Solve $\sqrt{2}x^2 - 3x + \sqrt{8} = 0$.

Solve.

21. $4x^2 + 4x + 1 = 0$

22. $9x^2 - 6x + 1 = 0$

23. $25x^2 + 10x + 1 = 0$

24. The difference between two numbers is 30. Their average is 10 greater than the square root of their product. Find the numbers.

25. In a right triangle the legs and the hypotenuse are consecutive integers. Show that 3, 4, 5 is the only possible solution.

26. Find two numbers whose product and sum both equal 20.

27. Find a rectangle whose perimeter and area both equal 20.

28. Find a rectangle whose perimeter and area both equal 5.

29. The area of a rectangle is 100 cm^2. If the length and width were increased by 2 cm, the area would increase to 145 cm^2. Find the dimensions of the rectangle.

30. The area of a rectangle is 100 cm^2. If the length and width were both increased by 10%, the area would increase to 121 cm^2. Find the dimensions of the rectangle.

31. Find all solutions to $\begin{vmatrix} x^2 & 4 & 9 \\ x & 2 & 3 \\ 1 & 1 & 1 \end{vmatrix} = 0.$

32. A plastic pipe can fill a tub 4 min faster than a metal pipe. If 6 plastic pipes and 6 metal pipes are used, the tub can be filled in 0.8 min. Find the time it will take one plastic pipe to fill the tub.

33. Bicycling home requires $3\frac{1}{4}$ h more time than traveling by car. The rate of the car is 65 km/h more than that of the bicycle, and the distance is 60 km. Find the rates.

34. Solve for x: $\frac{x}{m} + \frac{1}{x} + 1 + \frac{1}{m} = 0.$

35. Solve $\sqrt{x + 3} + \sqrt{x} = \frac{5}{\sqrt{x + 3}}.$

Factor completely.

36. $2x^2 - 6xy + 4y^2$

37. $6x^2 + 11x - 10$

38. $x^2 + 4xy - 5y^2$

39. $x^4 + x^3 + x + 1$

40. $x^4 - x^3 - x + 1$

41. $x^4 + x^3 - x - 1$

42. The product of two numbers is 100. If one is decreased by 5 and the other is increased by 1, the product is still 100. Find the numbers.

43. If the base of a square is increased 15 cm and its height is decreased 2 cm, the area of the rectangle formed will be twice the area of the square. Find the side of the square.

44. Two solar collectors can heat 3 m^3 of water in 20 min. It would take one collector 9 min longer to heat the water than it would the second. How long would it take each collector alone to heat the water?

45. The hypotenuse of a right triangle is 10 cm. One of the sides is 2 cm longer than the other. Find the area of the triangle.

46. A crew can row 8 km downstream and back in 4.8 h. If the rate of the current is 4 km/h, find the rate of the crew in still water.

Review of Equations and Inequalities

Solve. For Exercises 6 and 7, $0° \leq x < 360°$.

1. $(x + 1)^2 = 7$
2. $\frac{1}{x+1} + \frac{1}{x-1} = \frac{15}{8}$
3. $(3x)^{2/3} = 4$
4. $(x + 1)^{-1/3} = 2$
5. $9^x = 27$
6. $1 - \cos^2 x = \frac{1}{2}$
7. $\sin^2 x + 2 \sin x + 1 = 0$
8. $2 \text{ Tan}^{-1} x = 1$
9. $\log_x 3 = 2$
10. $\sqrt{x + 3} = 2$
11. $|2x + 1| = 3$
12. $2x^4 - x^3 - 6x^2 = 0$
13. $3^{2x} - 7 \cdot 3^x + 12 = 0$
14. $12x - 7\sqrt{x} + 1 = 0$
15. $4x^3 - 7x - 3 = 0$
16. $x^3 - 9x + 8 = 0$
17. $2x^3 + x^2 - 4x + 1 = 0$
18. $\begin{vmatrix} 3 & 3 & x \\ 4 & 4 & 4 \\ 5 & x & 5 \end{vmatrix} = 0$

Solve each system.

19. $x + y = 6$
    $x^2 - y = 6$
20. $x + y = 14$
    $\frac{1}{x} + \frac{1}{y} = \frac{7}{24}$
21. $x + \frac{1}{y} = 1$
    $y + \frac{1}{x} = 4\frac{1}{2}$
22. $\frac{x}{y} = \frac{2}{3}$
    $\frac{x+1}{y+1} = \frac{11}{16}$

Solve each system.

23. $\begin{aligned} x + 2y + 2z &= 16 \\ 2x + y + 2z &= 15 \\ 2x + 2y + z &= 14 \end{aligned}$

24. $\begin{aligned} y &= x^2 - 3x + 1 \\ x &= y + 2 \end{aligned}$

25. $\dfrac{1}{x} + \dfrac{1}{y} = 5$
$x - y = 0.3$

26. $\begin{aligned} x + y \quad &= 5 \\ x - \quad w &= 4 \\ -y + w &= -3 \end{aligned}$

27. $x^2y = 32$
$\dfrac{x}{y} = 2$

28. $\begin{aligned} 4 \sin x + y^2 &= 6 \\ 2 \sin x + 3y^2 &= 13 \end{aligned}$

29. If $0° \le x < 360°$, solve $2 \sin x = \sin 2x$.

30. Solve $\log_x 9 = \frac{2}{3}$.

Solve.

31. $2^x = 6$

32. $6^x = 2$

Solve for x, $0° \le x < 360°$.

33. $\cos \frac{1}{2}x = \frac{1}{2}$

34. $\tan 2x = \tan x$

35. $\sin (x + 90°) \sin (x - 90°) = -\frac{1}{2}$

36. $\tan (45° - x) + \cot (45° - x) = 2$

37. $\sin x \cos x = \frac{1}{4}$

38. Find x and y if $\sin (x + y) = 1$, $\tan (x - y) = 1$, and x and y are both between 0° and 360°.

39. A wire 1 m long is cut into two pieces and each piece is made into a square. If x is the length of one side of one square:
a. find the length of the side of the other square in terms of x
b. find x if the sum of the areas is 474.5 cm^2
c. find x if the sum of the areas is a minimum

40. A rectangular lot is to be fenced on three of its sides. If the length of the fence is 50 m and x is the length of one of the two equal sides:
a. find the length of the third side in terms of x
b. find x if the area of the lot is 312.5 m^2
c. find x if the area of the lot is to be as large as possible

41. The perimeter of a rectangle is 20 m. If the side is x, express the area of the rectangle in terms of x. Show that there is no value of x such that the area of the rectangle is 40 m^2.

42. The function F has the form $F(x) = Ax + \frac{B}{x}$. Find A and B if $F(3) = 8\frac{1}{2}$ and $F(1) = 1\frac{1}{2}$.

Graph the solution set of each inequality.

43. $2x + y \leq 4$

44. $x(x - 3) < 10$

45. $x^2 - 3x < 2$

46. $x^2 + y^2 < 16$

47. $y < x^2 + 2$

48. $y < \sqrt{4 - x^2}$

49. $y < x + 1$
$y < x^2 + 1$

50. $xy < 8$
$y \leq x$

51. Find $\sqrt{3 + 2\sqrt{2}}$ by solving the equation $(a + b\sqrt{2})^2 = 3 + 2\sqrt{2}$ for a and b.

52. Find $\sqrt{3 + 4i}$ by solving the equation $(a + bi)^2 = 3 + 4i$ for a and b.

Review of Operations with Numbers

Find the value of each of the following if $x = 8$ and $y = -4$.

1. $x^2 + xy + y^2$
2. $(x + y)^{1/2}$
3. $\dfrac{x^2 - y^2}{x + y}$
4. $\dfrac{x^3 + y^3}{x + y}$
5. $\dfrac{x^4 - y^4}{x - y}$
6. $\sin \pi\dfrac{x}{y}$
7. $\sin (x + y)\pi$
8. $\log_2 xy^2$
9. $\log_2 (xy)^2$
10. $2^x + 2^y$
11. $1 + \dfrac{y}{x} + \left(\dfrac{y}{x}\right)^2 + \left(\dfrac{y}{x}\right)^3 + \cdots$
12. $x^{1/2} + |xy|^{-1/2}$

Find the following if $P(x) = (x + 1)(x - 1)$.

13. $P(1 + \sqrt{3})$
14. $P(2) + P(\sqrt{3})$
15. $P(P(i))$
16. $P(1 + i)$
17. $P\left(\dfrac{1}{1 + i}\right)$
18. $P(2^x)$
19. Show that $\sqrt[3]{3} - 1$ is a root of $x^3 + 3x^2 + 3x = 2$.
20. Show that $(-1 + i\sqrt{3})^3 = 8$.
21. Simplify $\sqrt{(13 + 4\sqrt{3})(13 - 4\sqrt{3})}$.
22. If $f(x) = x + \frac{1}{x}$, find $f(2 + \sqrt{3})$.

23. If $Q(x) = \frac{\sqrt{x}}{\sqrt{x}+1} + \frac{\sqrt{x}}{\sqrt{x}-1}$, find $Q\left(\frac{1}{2}\right)$.

24. The legs of a right triangle are in the ratio 2:1. If the length of the shorter leg is x find the length of the perimeter in terms of x. Find x if the perimeter is 8 cm.

25. Simplify $(2 + \sqrt{5})^3 + (2 - \sqrt{5})^3$.

26. Show that $\sqrt{1\frac{1}{2}\sqrt{1\frac{1}{3}\sqrt{1\frac{1}{4}\sqrt{1\frac{1}{5}}}}} = \sqrt{3}$.

27. Find, to the nearest tenth, the roots of $x - \frac{1}{x} = 1$.

Simplify.

28. $(2x^2y)^3$

29. $\sqrt{16x^{16}}$

30. $\sqrt[3]{16}$

31. $8^{2/3}$

32. $\left(\frac{1}{8}\right)^{2/3}$

33. $\left(\frac{1}{8}\right)^{-2/3}$

34. $100^{1.5}$

35. $3^{\log_3 10}$

36. $\log_{10} 0.001$

37. If $A = \frac{1}{2}(2^x + 2^{-x})$ and $B = \frac{1}{2}(2^x - 2^{-x})$, find $A^2 - B^2$.

38. **a.** Express 8^4 as a power of 2.
b. Express 9^n as a power of 3.

39. **a.** Express 2^4 as a power of 8.
b. Express 3^n as a power of 9.

Simplify.

40. $\frac{2^{-1}}{2^{-1} + 2^{-2}}$

41. $\frac{2^{-1}}{2^{-1}} - \frac{2^{-1}}{3^{-1}}$

42. $\frac{2^{-1}}{2^{-1} - 3^{-1}}$

43. If $2 = 10^{0.30}$, express each as a power of 10.
a. 20 **b.** 200 **c.** 4 **d.** 8 **e.** $\sqrt{2}$ **f.** $\sqrt[3]{16}$

44. If $2^{0.2} = 1.2$, find the value of each of the following.
a. $2^{1.2}$ **b.** $2^{0.4}$ **c.** $2^{3.2}$

45. Find the value of each of the following.
a. $\sin 999°$ **b.** $\frac{\tan 0.02}{0.02}$ **c.** $\frac{\sin 0.02}{0.02}$

46. Express the product $(\cos 30° + i \sin 30°)(\cos 150° + i \sin 150°)$ in polar form.

47. Express the product $(\cos 30° + i \sin 30°)^2$ in polar form.

48. Simplify $\frac{\log_3 16}{\log_3 2}$.

49. Simplify $(\log_2 5)(\log_5 6)(\log_6 8)$.

50. How many digits does the number 2^{64} have?

51. Simplify $5^{\log_5 3 - \log_5 6}$.

Answers

Diagnostic Test

1. 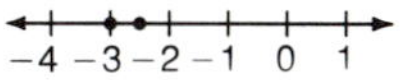

$-\frac{5}{2} > -3$ **2.** $-6, -5.7, -5.65, -4.7, |-6.2|$ **3.** 3 **4.** $\frac{5}{7}$ **5.** 2. Addition property of equality 3. Associative property of addition 4. Additive inverse 5. Additive identity 6. Multiplication property of equality 7. Associative property of multiplication 8. Multiplicative inverse 9. Multiplicative identity **6.** -28 **7.** Jerry's account; \$31.36 **8.** $-3r^3 + 2r^2 - 4$ **9.** -149 **10.** $-\frac{1}{2}$ **11.** $\frac{1}{2}$ **12.** {9}; $5(27 - 7) = 4(18 + 7)$; $5(20) = 4(25)$; $100 = 100$
13. $g = \frac{2vt - 2s}{t^2}$ **14.** $(2r + 15)$ mi **15.** 133°
16. 312 km **17.** $\{x: x > -26\}$ 

18. a. T **b.** F
19. a. $\{y: y \le -1 \text{ or } y \ge 7\}$ 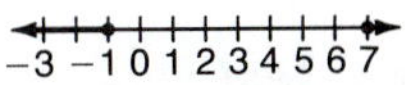

b. $\left\{x: -\frac{5}{2} < x < 6\right\}$ 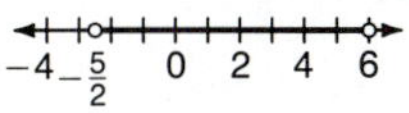

20. 22
21. a. {real numbers} 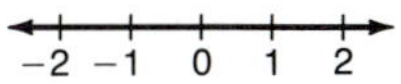

b. {2, 5} 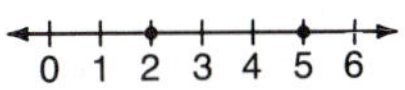

c. {−5} 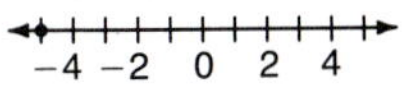

22. $|x - 3| < 2$ **23.** 1. a, b, and c are real numbers $a + c = b + c$ (Given) 2. $-c$ is a real number (Property of opposites) 3. $(a + c) + (-c) = (b + c) + (-c)$ (Addition property of equality) 4. $a + [c + (-c)] = b + [c + (-c)]$ (Associative property of addition) 5. $a + 0 = b + 0$ (Additive inverse) 6. $a = b$ (Additive identity) **24.** 1. Given 2. Second multiplicative property of order 3. Commutative property of multiplication 4. Second multiplicative property of order 5. Transitive property of order 6. Substitution **25.** $k = -12$ **26.** 6, 6, 1; 5, 5, 3; 4, 4, 5
27. (3, 0)
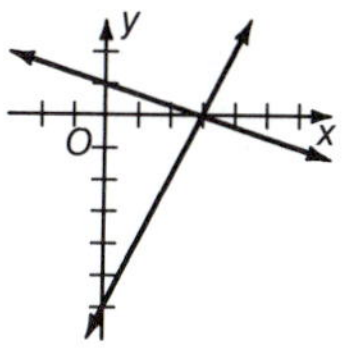

28. a. $\frac{7}{6}$ **b.** $\frac{4}{5}$ **29.** $\frac{1}{2}$
30. a. $3x - y = 8$
b. $x + y = -8$
31. $\left(\frac{2}{5}, \frac{1}{5}\right)$ **32.** bread, \$.97; peanut butter, \$1.29
33. 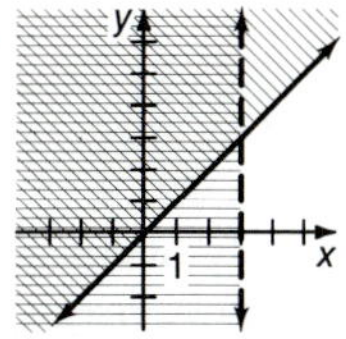

34. 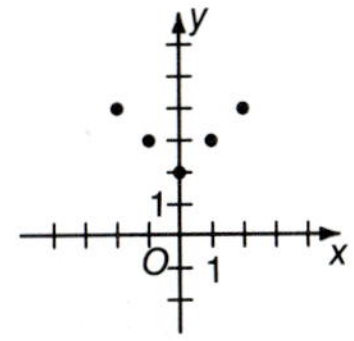

35. $f(g(0)) = -2$; $g(f(1)) = \frac{1}{5}$
36. $f(5) = 12$; $f(-5) = 8$
37. $D = \{0, 1, 2, 3, 4\}$;
See diagram at right; relation
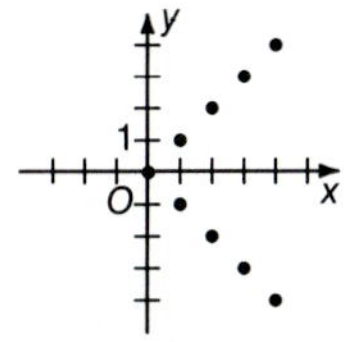

38. $13x^2y - 12xy + 15xy^2$
39. a. $-72c^{11}d^7$ **b.** $h^{2m} + h^{2m-n}$
40. a. $10r^2 - rs - 21s^2$
b. $2h^4 + 19h^3 + 38h^2 + 11h - 70$ **41. a.** $14x^2y^2$
b. $2^3 \cdot 3 \cdot 5 \cdot 7x^3y^3z^2$, or $840x^3y^3z^2$ **42. a.** $8x^2(5x - 8y)$
b. $(5r - 2s)(25r^2 + 10rs + 4s^2)$ **c.** $(y - 13)(y - 1)$
d. $(5c - 3)(c + 1)$ **e.** $(3x - 5)^2$
f. $(4u^2 + v^2)(2u + v)(2u - v)$ **43. a.** $\left\{\frac{1}{4}, \frac{1}{2}\right\}$
b. {−5, 0, 5}; 0 double root **c.** {3, 4}; 3 double root
44. 13 and 15, or −15 and −13 **45.** 50 s
46. a. $\{x: -4 < x < 0 \text{ or } x > 4\}$ 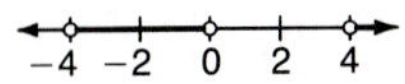

b. $\{x: 2 \le x \le 3\}$ 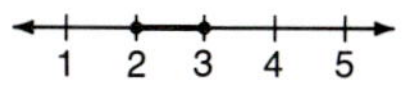

Chapter Tests

Chapter 1 Test

1. -11.7 **2.** $-1\frac{1}{5}, -\frac{3}{5}, 2\frac{2}{5}, 3\frac{4}{5}, 5$ **3.** 20 **4.** 42
5. 26 **6. a.** Associative property of addition
b. Commutative property of multiplication **c.** Additive inverse **d.** Additive identity **7.** -181 **8.** 6 **9.** $2x + 12$
10. 20 **11.** -9 **12.** $4a^3 - a^2 + 16a + 9$ **13.** {10}
14. ∅ **15.** $\frac{t}{2} - 2$ **16.** $(40 - l)$ cm **17.** \$3 and \$5
18. 2:45 P.M.

Chapter 2 Test

1. $\{x: x \ge 2\}$ **2.** $\{t: t < -11\}$ **3.** $\left\{w: -3 < w < \frac{3}{2}\right\}$

4. $\{t: t < 10\}$ **5.** 32, 34, 36 **6.** $\left\{5\frac{1}{2}, -\frac{1}{2}\right\}$

7. $\{x: 4 \le x \le 10\}$ **8.** $\left\{x: x < \frac{1}{3} \text{ or } x > 1\right\}$

9. 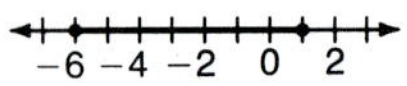

10. 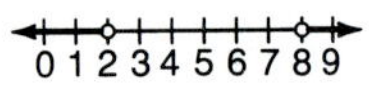

11. If an integer is not even, then it is an odd integer. **12.** 1. Given 2. Property of opposites 3. Addition property of equality 4. Associative property of addition 5. Additive inverse 6. Additive identity **13.** False. Let $a = -3$ and $b = -4$. **14.** False. Let $a = \frac{1}{4}$ and $b = \frac{1}{2}$. **15.** True **16.** False. Let $a = -2$ and $b = 1$.

Chapter 3 Test

1. $\frac{1}{2}$ **2.** (0, 7), (2, 4), (4, 1)

3. a. 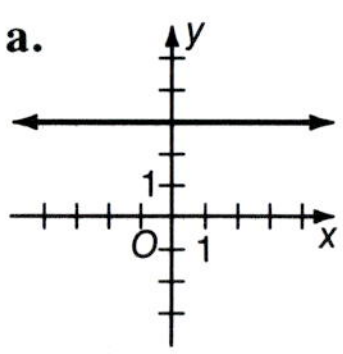

b. 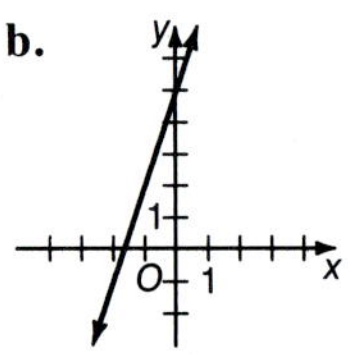

c. 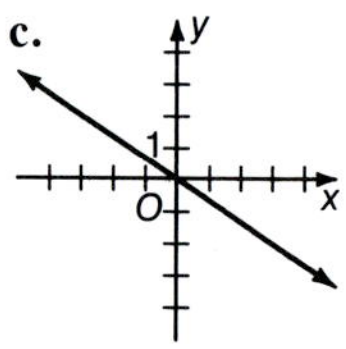

4. a. 8 **b.** $-\frac{3}{4}$ **5.** $\frac{2}{5}$ **6. a.** $2x + y = 9$ **b.** $-2x + y = 6$ **7. a.** $-3x + y = -15$ **b.** $x + 3y = -5$ **8.** $x = -1$, $y = 2$ **9.** air speed 105 km/h, wind speed 15 km/h

10. See diagram below. **11.** $R = \{-2, -1, 0, 1, 2, 3\}$

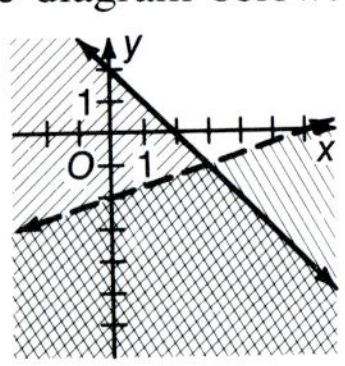

Ex. 10

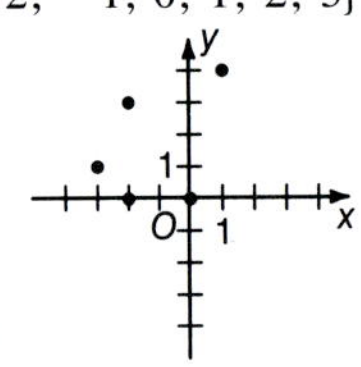

Ex. 15

12. $f(g(1)) = 3$, $g(f(1)) = 5$ **13.** $4\frac{2}{3}$ **14.** 14 cm

15. No. See diagram above.

Chapter 4 Test

1. $22 - 11x^2$ **2.** $7p^3q^2 + 12p^2q - 11pq$ **3.** $-30x^3r^5y^3$ **4.** $3w^6v - 3w^2v^5$ **5.** $6r^2 + rs - 2s^2$ **6.** $4b^4x^2 + 4b^3x^4 + x^6b^2$ **7. a.** 84 **b.** 2520 **8. a.** $5rs$ **b.** $225r^2s^3$ **9.** $(8r + 1)^2$ **10.** $(2d + e)(3d + 5)$ **11.** $(s - 2b)(s + 3b)$ **12.** $3(3x - 2)(x - 1)$ **13.** $(x^2 + 3)^2$ **14.** $a^2x(x - 10)(x - 2)$ **15.** $\left\{-5, \frac{1}{2}, -\frac{4}{3}\right\}$ **16.** $\left\{\frac{2}{3}, -\frac{1}{2}\right\}$

17. 5 cm **18.** $\{x: x \le 1 \text{ or } x \ge 12\}$ 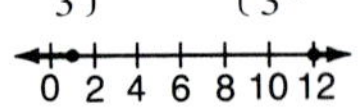

Chapter 5 Test

1. $\frac{16u^5}{3b^2}$ **2.** $\frac{y^4}{x^6}$ **3.** 3.1×10^2 **4.** $\frac{5(x - 2)}{3x - 1}$ **5. a.** {real numbers except -3} **b.** ± 2 **6.** $\frac{5t^3}{4s^2}$ **7.** $\frac{a + 5}{a - 5}$

8. $\frac{2x^2}{x^2 - y^2}$ **9.** $\frac{16(x + 2)}{3(x - 5)(x + 5)}$ **10.** $\frac{a - 2}{a - 5}$ **11.** $\{x: x \ge -4\}$

12. 3 h **13.** $\{-1, 5\}$ **14.** 15 mi/h

Chapter 6 Test

1. -0.5 **2.** $\pm\frac{1}{3}$ **3.** $4xy$ **4.** $\frac{2c}{d}|b|\sqrt[4]{4cd}$

5. $12\sqrt{3} - 414$ **6.** $-m^2n\sqrt{mn}$ **7.** $48 + 6\sqrt{15}$

8. $-(\sqrt{2} + \sqrt{6})$ **9.** $\left\{\frac{1 \pm \sqrt{65}}{4}\right\}$ **10.** $\left\{-\frac{5}{4}\right\}$ **11.** $\frac{47}{660}$

12. 4.12 and 4.1201001 . . . **13.** $\{\pm 5i\}$ **14.** $-9\sqrt{3}$

15. $-\frac{\sqrt{21}}{3}i$ **16.** $10i\sqrt{3}$ **17.** $10 + 3i$ **18.** $33 + 6i$

19. $-\frac{7}{41} + \frac{22}{41}i$

Chapter 7 Test

1. $\{1 \pm \sqrt{7}\}$ **2.** $\left\{1 \pm \frac{\sqrt{3}}{3}\right\}$ **3.** $\left\{\frac{3}{5}, 2\right\}$ **4. a.** Two real roots **b.** Complex conjugate roots **5.** $k < 3$ **6.** $\{\pm 2, \pm 3i\}$ **7.** $\{4\}$ **8.** Vertex: $(-4, 0)$; axis: $x = -4$

9. $y + 5 = \frac{8}{25}(x - 1)^2$

10. a. Minimum **b.** 0 **11.** {real numbers}; {nonnegative numbers}; 6

12. $x^2 - 6x + 6 = 0$

13. $f(x) = -4x^2 - 4x + 24$

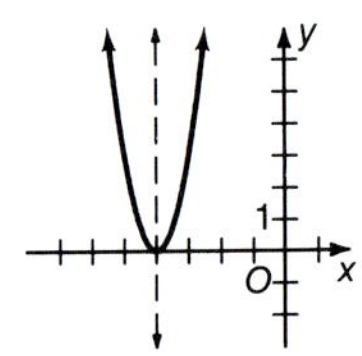

Ex. 8

Chapter 8 Test

1. 2 **2.** $\frac{9}{8}$ **3.** 600 mho **4.** $3bx - 5$

5. $3x^3 + 9x^2 + 22x + 72 + \frac{211}{x - 3}$ **6.** -3

7. $x^3 + 2x^2 + 3x + 6 = 0$

8. $\{2 - i\sqrt{3}, 2 + i\sqrt{3}, 1, 3\}$

9. $\left\{\frac{2}{3}, \frac{1}{2}, i, -i\right\}$

10. 1 (See diagram at right.)

11. 9.93

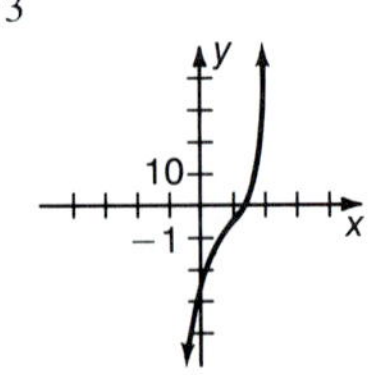

Chapter 9 Test

1. a. $3\sqrt{5}$ **b.** $(4.5, -2)$ **c.** $(0, -11)$
2. $(x-3)^2 + (y-3)^2 = 18$ **3.** Center: $(5, 0)$; radius: 5
4. $x = -\frac{1}{4}(y-3)^2 + 3$ **5.** Vertex: $(5, 3)$; focus: $\left(6\frac{3}{4}, 3\right)$; directrix: $x = 3\frac{1}{4}$; axis: $y = 3$ **6.** $\frac{x^2}{16} + \frac{y^2}{9} = 1$

7.

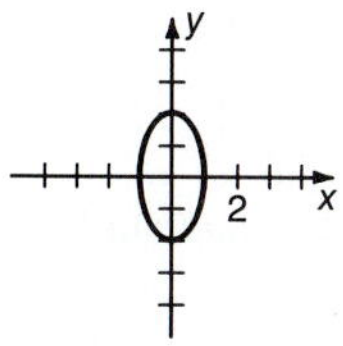

8.

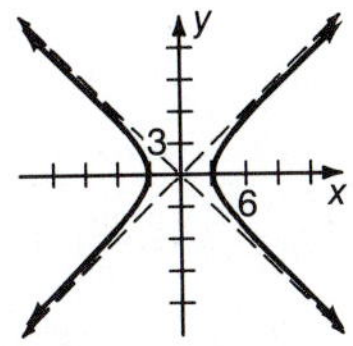

9. $\frac{(x-2)^2}{21} + \frac{(y-5)^2}{25} = 1$ **10.** Hyperbola; center: $(1, 3)$; foci: $(1 + \sqrt{7}, 3)$ and $(1 - \sqrt{7}, 3)$

11. Two solutions

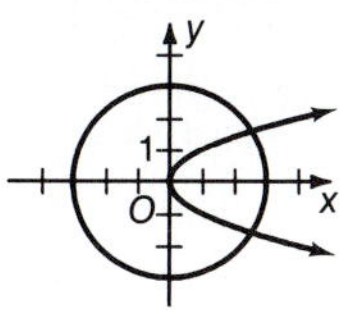

12. No solutions

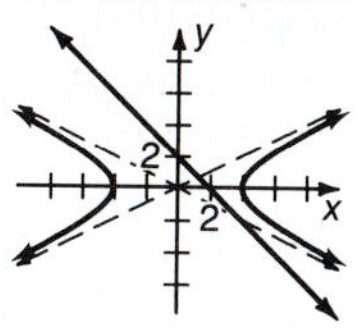

13. $(3, 4), (5, 0)$ **14.** 6 cm by 8 cm **15.** $(-2, -2, 3)$

Chapter 10 Test

1. $\frac{64}{27}$ **2.** $x^{2/3}y^{-1/3}$ **3.** 1 **4.** $\{-4\}$ **5. a.** -4
b. $g(f(x)) = \sqrt{3x-6}$ **6.** $f^{-1}(x) = 2x - 1$ **7.** $\frac{3}{4}$
8. a. $\left\{\frac{1}{9}\right\}$ **b.** $\left\{\frac{1}{23}\right\}$ **9. a.** 1.301 **b.** 0.398 **c.** 2.097
10. $t = \frac{\log 5 + \log 6}{3 \log 6}$ **11.** 1.23 **12.** 32 h **13.** $\ln 8e$
14. $\{e^{2/3}\}$

Chapter 11 Test

1. Geometric; $-\frac{4}{25}, \frac{4}{125}$ **2.** 2, −1, −4, −7; arithmetic
3. 79 **4.** 6, 10, 14 **5.** $t_n = \frac{3}{2} \cdot 4^{n-1}$ **6.** 20 **7.** \$24,255
8. $\sum_{n=1}^{\infty} 81\left(-\frac{2}{3}\right)^{n-1}$ **9.** 525 **10.** −182 **11.** $\frac{3}{2}$
12. $16r^4 - 32r^3w + 24r^2w^2 - 8rw^3 + w^4$ **13.** $-960s^7t^3$

Chapter 12 Test

1. 190°, −530° **2. a.** 12°46′48″ **b.** 120.26°
3. $\sin\theta = \frac{2\sqrt{5}}{5}$, $\cos\theta = \frac{\sqrt{5}}{5}$, $\tan\theta = 2$, $\csc\theta = \frac{\sqrt{5}}{2}$, $\sec\theta = \sqrt{5}$, $\cot\theta = \frac{1}{2}$ **4. a.** II **b.** $-\frac{24}{7}$
5. a. −0.8156 **b.** 115.4° **6.** 953 m **7.** 112.9° or 247.1°
8. $a = 7.85$, $\angle B = 73.2°$, $\angle C = 36.8°$ **9.** $\angle C = 95°$, $b = 4.53$, $c = 13.2$ **10.** 61.1 cm^2

Chapter 13 Test

1. a. $-\frac{3\pi}{4}$ **b.** 150° **2.** $\frac{5\pi}{9}$ **3.** $\sin\theta = \frac{1}{2}$, $\cos\theta = -\frac{\sqrt{3}}{2}$, $\tan\theta = -\frac{\sqrt{3}}{3}$, $\csc\theta = 2$, $\sec\theta = -\frac{2\sqrt{3}}{3}$, $\cot\theta = -\sqrt{3}$ **4.** odd
5. Amplitude: 3; maximum: 3; minimum: −3; period: 4π **6.** $y = 4\cos\left(\frac{\pi}{6}x\right)$

7.

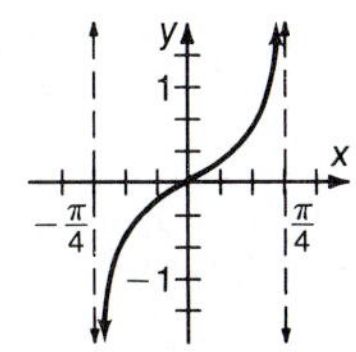

8. $\tan^2 t - \sin^2 t = \frac{\sin^2 t}{\cos^2 t} - \sin^2 t = \frac{\sin^2 t(1 - \cos^2 t)}{\cos^2 t} = \tan^2 t \sin^2 t$
9. $\sin x(\csc x - \sin x) = 1 - \sin^2 x = \cos^2 x$
10. $-\frac{\sqrt{2+\sqrt{3}}}{2}$ **11. a.** $\frac{336}{625}$ **b.** $\frac{7\sqrt{2}}{10}$ **12.** $\frac{84}{13}$

Chapter 14 Test

1. 39 km, first to second 277°, second to first 97°
2. a. 16 **b.** 106° **3.** $(3, 270°) \longleftrightarrow (0, -3)$; $(0, 4) \longleftrightarrow (4, 90°)$; $(5, 150°) \longleftrightarrow \left(\frac{-5\sqrt{3}}{2}, \frac{5}{2}\right)$; $\left(\frac{\sqrt{2}}{2}, 45°\right) \longleftrightarrow \left(\frac{1}{2}, \frac{1}{2}\right)$ **4.** $(x-1)^2 + y^2 = 4$
5. $wz = 8(\cos 270° + i \sin 270°)$; $\frac{w}{z} = 2(\cos 330° + i \sin 330°)$ **6.** i, $\frac{\sqrt{3} - i}{2}$, $\frac{-\sqrt{3} - i}{2}$
7. $-\frac{119}{169}$ **8.** $\frac{\sqrt{5}}{5}$ **9.** $210° + n \cdot 360°$, $330° + n \cdot 360°$
10. 0°, 120°, 240°

Chapter 15 Test

1.

7	2, 7, 8
8	4, 4, 6, 6, 6
9	2, 5

2. a. 86 **b.** 85 **c.** 84

3. 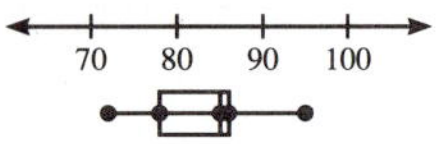

4. a. 43 b. 7 5. 0.1587

6. positive correlation 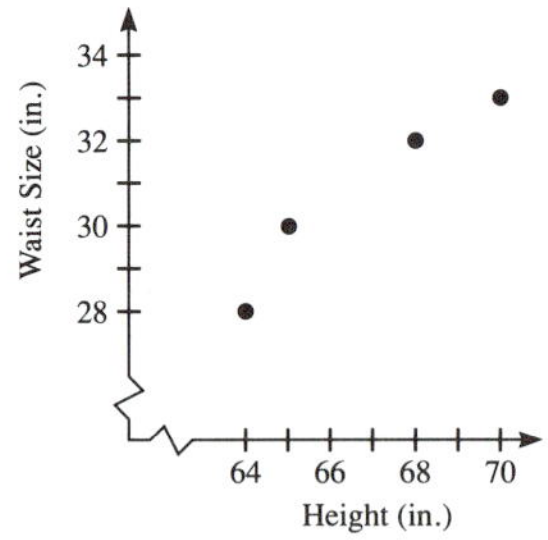

7. 74 8. 20 9. 180 10. 4; {H, J, K}, {G, J, K}, {G, H, K}, {G, H, J} 11. 21 12. a. {(H, 3), (H, 6)} b. {(H, 1), (H, 2), (H, 3), (H, 4)} 13. a. $\frac{2}{13}$ b. $\frac{2}{13}$ c. $\frac{1}{2}$ 14. a. $\frac{7}{12}, \frac{5}{18}, \frac{1}{12}, \frac{7}{9}$ b. No; no

Chapter 16 Test

1. $x = 9$, $y = -16$ 2. $\begin{bmatrix} 2 & 9 & 1 \\ 8 & 6 & 0 \\ -4 & 9 & 9 \end{bmatrix}$

3. $\begin{bmatrix} 0 & -3 & -5 \\ 4 & 4 & 0 \\ 6 & 5 & -3 \end{bmatrix}$ 4. $\begin{bmatrix} 2 & 6 & -4 \\ 12 & 10 & 0 \\ 2 & 14 & 6 \end{bmatrix}$

5. $\begin{bmatrix} -3 & -18 & -9 \\ -6 & -3 & 0 \\ 15 & -6 & -18 \end{bmatrix}$ 6. $\begin{bmatrix} 17 & 5 & -9 \\ 16 & 41 & 18 \\ 0 & 19 & 21 \end{bmatrix}$

7. $\begin{bmatrix} 40 & 54 & 7 \\ 8 & 11 & -4 \\ 13 & 37 & 28 \end{bmatrix}$ 8. $\begin{array}{c} \\ A \\ B \\ C \end{array} \begin{array}{c} A \quad B \quad C \\ \begin{bmatrix} 0 & 1 & 0 \\ 1 & 0 & 1 \\ 1 & 0 & 0 \end{bmatrix} \end{array}$

9. −41 10. −139 11. $\begin{bmatrix} 2 & -5 \\ 3 & 1 \end{bmatrix}$ 12. −24 13. 0

14. 1 15. $x = -3$, $y = 1$, $z = 4$

Cumulative Reviews

Review for Chapters 1–3

1. $\frac{27}{20}$ 2. 0 3. $x + 5y$ 4. 3

5. $m(-x) = m[(-1)x]$ Mult. prop. of −1
$= (-1)(mx)$ Comm., assoc. props. of mult.
$= -mx$ Mult. prop. of −1
$m(-x) = -mx$ Trans. prop. of =

6. $-\frac{9}{5}$ 7. −1 8. $\left\{t\colon t > -\frac{2}{3}\right\}$ 9. $\{m\colon 1 \le m \le 5\}$
10. $\{t\colon -2 \le t \le 4\}$ 11. $\{d\colon d < 1 \text{ or } d > 4\}$
12. {(1, 1), (1, 2)} 13. $\frac{18}{5}$ 14. $3x + 4y = 32$
15. $4x + y = 23$ 16. $3x + 2y = -6$
17. $114x + 180y = 83$ 18. $\{(x, y)\colon y = x + 7\}$
19. $\left\{\left(\frac{18}{7}, \frac{4}{7}\right)\right\}$ 20. $\left\{\left(\frac{23}{4}, \frac{1}{12}\right)\right\}$ 21. See diagram below.

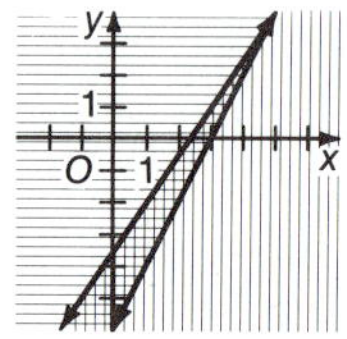

Ex. 21

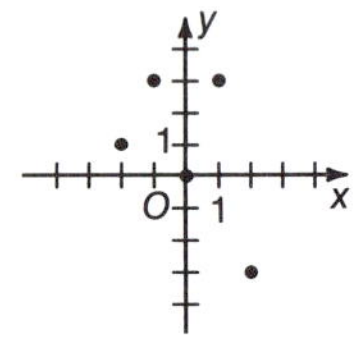

Ex. 26

22. Eggs, \$1.05; bread, \$.75 23. $g(h(3)) = 3$; $h(g(3)) = 9$
24. $y = \frac{1}{2}x + \frac{11}{2}$ 25. 342 26. Yes. See diagram above.

Review for Chapters 4–7

1. $12x^2y - 13xy + 10xy^2$ 2. a. $64a^6x^{10}$ b. $12x^4 - 3x^2$
3. a. $x^2 - x$ b. $x^3 - x$ 4. a. $(3rs - 2)(4rs + 3)$
b. $t(3t^2 - 4)(t - 2)$ 5. 5, 6, and 7
6. $\{t\colon t < -1 \text{ or } t > 6\}$ 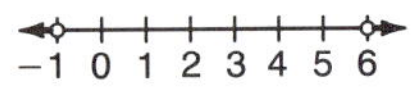

7. $\frac{25}{4t^2}$

8. $\frac{3}{4b^2}$ 9. $\frac{a^2 + 1}{a - 1}$ 10. $2t^2 - t - 15$ 11. $\frac{2t + 12}{3t^2 - 12}$

12. $-\frac{b}{a}$ 13. 50 mL of 18% solution; 10 mL of 20% solution 14. {2} 15. {5} 16. $\{4 \pm \sqrt{2}\}$ 17. $-\frac{13}{6}\sqrt{3}$

18. 8 19. $\frac{5\sqrt{2} + 2}{2}$ 20. $-\frac{7}{25} + \frac{24}{25}i$ 21. $\{3 \pm 2\sqrt{3}\}$

22. $\left\{\frac{7 \pm i\sqrt{11}}{6}\right\}$ 23. $k < -4\sqrt{5}$ or $k > 4\sqrt{5}$

24. 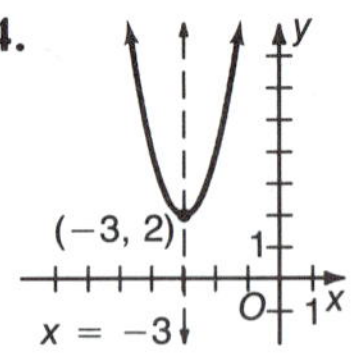

25. Domain: {real numbers}; range: $\{f(t)\colon f(t) \ge 5\}$; no zeros 26. Sum: $\frac{5}{7}$; product: $\frac{1}{7}$ 27. $36x^2 - 48x + 41 = 0$

28. $-\frac{49}{4}$

Review for Chapters 8–11

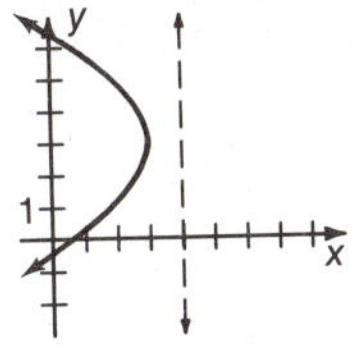

1. 625 **2.** $x^2 - 2x + 7$
3. $\{2 + 3i, 2 - 3i, 4, -1\}$
4. Center: $(-2, 3)$; radius: $\sqrt{26}$
5. $x = -\frac{1}{4}(y - 3)^2 + 3$; See diagram at right. **6.** foci: $(\pm\sqrt{41}, 0)$ (diagram at right) **7.** $\left(\frac{3}{2}, -4\right), (-2, 3)$ **8.** $(3, 6), (-1, -2)$ **9.** $(1, 3, 4)$ **10. a.** $\frac{1}{32}$ **b.** 2187 **11. a.** $\{8\}$ **b.** $\{64\}$ **c.** $\{3\}$

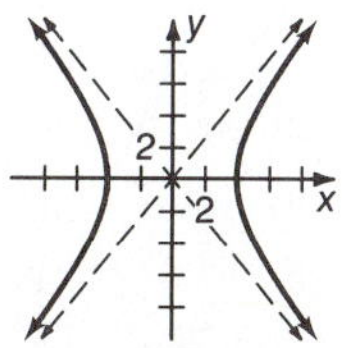

12. $f^{-1}(x) = \frac{x + 7}{4}, f^{-1}(5) = 3$ **13.** $\{12\}$ **14.** 0.116
15. 7.65 **16.** 1.02 **17.** Geometric; $10, \frac{10}{3}, \frac{10}{9}, \frac{10}{27}$
18. Neither; $6, 5 + \sqrt{2}, 5 + \sqrt{3}, 7$ **19.** Arithmetic; $\frac{5}{2}, \frac{9}{2}, \frac{13}{2}, \frac{17}{2}$ **20.** $10\sqrt[3]{10}, 10\sqrt[3]{100}$; 40, 70 **21.** $t_7 = -\frac{1}{8}$; $t_n = (-1)^n \cdot 2^{4-n}$ **22. a.** 84 **b.** $\frac{844}{729}$ **23.** 490
24. $16a^4 + 64a^3b + 96a^2b^2 + 64ab^3 + 16b^4$ **25.** $1512t^3r^5$

Review for Chapters 12–14

1. 533° and −187° **2.** $\sin\theta = \frac{4\sqrt{17}}{17}$, $\cos\theta = \frac{\sqrt{17}}{17}$, $\tan\theta = 4$, $\csc\theta = \frac{\sqrt{17}}{4}$, $\sec\theta = \sqrt{17}$, $\cot\theta = \frac{1}{4}$
3. 10.0° **4.** $b = 12.8$, $\angle C = 32.2°$, $\angle A = 42.8°$ **5.** 30.4
6. $\frac{2\pi}{5}$ **7.** $\sin\theta = \frac{\sqrt{3}}{2}$, $\cos\theta = -\frac{1}{2}$, $\tan\theta = -\sqrt{3}$, $\csc\theta = \frac{2\sqrt{3}}{3}$, $\sec\theta = -2$, $\cot\theta = -\frac{\sqrt{3}}{3}$
8. Amplitude: 3; maximum: 7; minimum: 1; period: π
9. $\tan\theta$

10. a. $\frac{\tan a - \tan b}{1 + \tan a \tan b} = \frac{\frac{1}{\tan b} - \frac{1}{\tan a}}{\frac{1}{\tan a \tan b} + 1} = \frac{\cot b - \cot a}{\cot a \cot b + 1}$
b. $\frac{1 + \csc a}{\cot a + \cos a} = \frac{\sin a + 1}{\cos a + \cos a \sin a} = \frac{(\sin a + 1)}{\cos a(\sin a + 1)} = \frac{1}{\cos a} = \sec a$ **11.** $-\frac{\sqrt{7}}{4}$ **12.** 15.6; 209.1°
13. a. $4\mathbf{i} - 3\mathbf{j}$ **b.** $\frac{4}{5}\mathbf{i} - \frac{3}{5}\mathbf{j}$ **14.** $-\frac{1}{64} + \frac{\sqrt{3}}{64}i$
15. $\sqrt{2} + \sqrt{2}i, -\sqrt{2} + \sqrt{2}i, -\sqrt{2} - \sqrt{2}i, \sqrt{2} - \sqrt{2}i$
16. $\frac{\sqrt{7}}{4}$ **17.** $\frac{\sqrt{2}}{2}$ **18.** $-\frac{\pi}{2}$ **19.** 60°, 120°, 240°, 300°
20. 15°, 105° **21.** 45°, 135°, 225°, 315°, 60°, 120°, 240°, 300° **22.** 30°, 150°

Review for Chapters 15–16

1.

1	4, 6, 6
2	2, 4, 4, 4, 8
3	0, 2

2. a. 24 **b.** 24 **c.** 23

3.

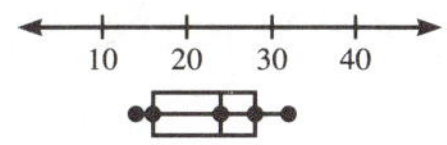

4. a. 34 **b.** 6

5. 0.1096 **6.** 60 **7.** 210 **8.** 35 **9. a.** $\frac{5}{6}$ **b.** $\frac{1}{2}$
10. a. 0.5625 **b.** 0.42 **11. a.** Independent **b.** Dependent
12. $\begin{bmatrix} 9 & 1 & 8 \\ 1 & 4 & 16 \\ 11 & 0 & 1 \end{bmatrix}$ **13.** $\begin{bmatrix} -3 & 2 & 1 \\ -4 & -4 & -1 \\ 7 & 3 & 5 \end{bmatrix}$
14. $\begin{bmatrix} 15 & -3 & 6 \\ 9 & 12 & 18 \\ -3 & -6 & -9 \end{bmatrix}$ **15.** $\begin{bmatrix} 10 & -4 & 1 \\ -10 & -9 & -17 \\ 31 & -6 & 12 \end{bmatrix}$
16. $\begin{bmatrix} 0 & 1 & 1 & 0 \\ 0 & 0 & 1 & 1 \\ 0 & 1 & 0 & 1 \\ 0 & 1 & 0 & 0 \end{bmatrix}$ **17.** $\begin{bmatrix} 0 & 2 & 2 & 2 \\ 0 & 2 & 1 & 2 \\ 0 & 2 & 1 & 2 \\ 0 & 1 & 1 & 1 \end{bmatrix}$ **18.** 6 **19.** 27 **20.** 27
21. $\begin{bmatrix} 2 & 1 \\ -1 & 3 \end{bmatrix}$ **22.** $(-2, -3, 1)$

Topical Reviews

Review of Quadratics

1. $\{\pm 6\}$ **2.** $\{0, 2\}$ **3.** $\{7, -13\}$ **4.** $\{15, -19\}$ **5.** $\left\{0, \frac{3}{2}\right\}$
6. $\{6, -3\}$ **7.** $\left\{-\frac{1}{2} \pm \frac{\sqrt{11}}{2}\right\}$ **8.** $\left\{\frac{4}{3}, \frac{18}{11}\right\}$ **9.** $\{0, \sqrt[3]{2}\}$
10. 23 **11.** $(4, 13), (-2, -5)$ **12.** $-\frac{5}{3}$
13. a. $(a + b)x^2 - (a - b)x - 2b = 0.$
When $x = 1$, then $a + b - a + b - 2b = 0$. **b.** $-\frac{2b}{a + b}$
14. $-\frac{1}{2} \pm \frac{\sqrt{5}}{2}$ **15. a.** $x^2 - x = 6$ **b.** $x^2 - 6x + 7 = 0$
c. $x^2 - 6x + 11 = 0$ **16.** $A = 15x - x^2$.
If $A = 60$ then $D < 0$, and there is no real solution.
17. $k = 0.95$ **18.** $-4 + 2i$ **19.** $A = 3, B = -4, C = -6$
20. $\left\{\frac{3\sqrt{2} \pm i\sqrt{14}}{4}\right\}$ **21.** $\left\{-\frac{1}{2}\right\}$ **22.** $\left\{\frac{1}{3}\right\}$ **23.** $\left\{-\frac{1}{5}\right\}$

24. 1.25, 31.25 **25.** Let the sides be x, $x + 1$, and $x + 2$. $x^2 + (x + 1)^2 = (x + 2)^2$; $x^2 - 2x - 3 = 0$; $(x - 3)(x + 1) = 0$. Therefore $x = 3$, and the sides are 3, 4, and 5. **26.** $10 + 4\sqrt{5}$, $10 - 4\sqrt{5}$ **27.** $5 + \sqrt{5}$, $5 - \sqrt{5}$ **28.** 6, 3 **29.** 8, 12.5 **30.** All positive real values of x and y such that $xy = 100$ **31.** 2, 3 **32.** 8 min **33.** Car: 80 km/h; bicycle: 15 km/h

34. $\{-1, -m\ (m \neq 0)\}$ **35.** $\left\{\frac{4}{7}\right\}$ **36.** $2(x - y)(x - 2y)$
37. $(2x + 5)(3x - 2)$ **38.** $(x - y)(x + 5y)$
39. $(x + 1)^2(x^2 - x + 1)$ **40.** $(x - 1)^2(x^2 + x + 1)$
41. $(x + 1)(x - 1)(x^2 + x + 1)$ **42.** 4, 25 **43.** 3 cm, 10 cm **44.** 36 min, 45 min **45.** 24 cm^2 **46.** 6 km/h

47.

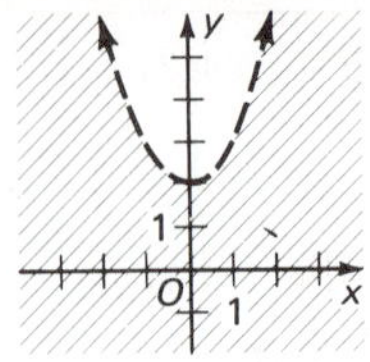

48.

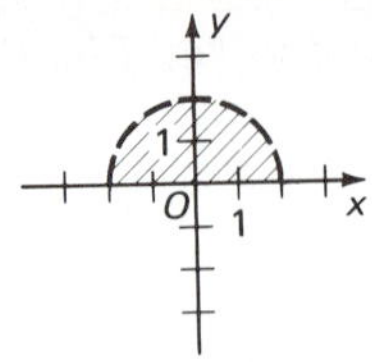

49.

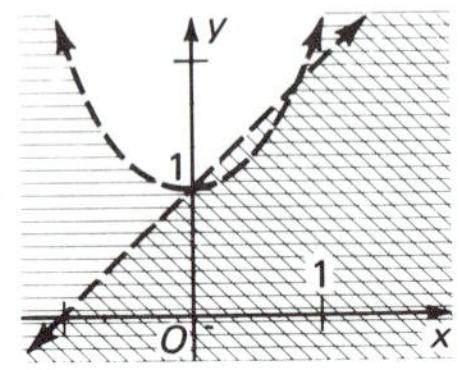

50.

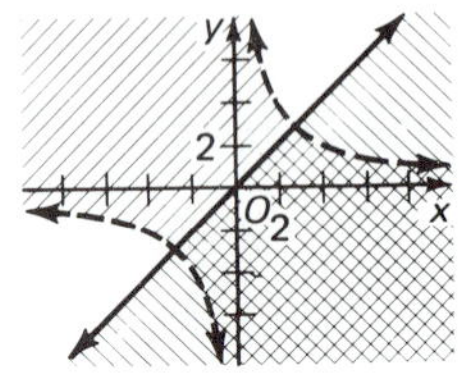

51. $\sqrt{3 + 2\sqrt{2}} = 1 + \sqrt{2}$ **52.** $\sqrt{3 + 4i} = 2 + i$

Review of Equations and Inequalities

1. $\{1 \pm \sqrt{7}\}$ **2.** $\left\{\frac{5}{3}, -\frac{3}{5}\right\}$ **3.** $\left\{\frac{8}{3}\right\}$ **4.** $\left\{-\frac{7}{8}\right\}$ **5.** $\left\{\frac{3}{2}\right\}$
6. $\{45°, 135°, 225°, 315°\}$ **7.** $\{270°\}$ **8.** $\{0.5463\}$
9. $\{\sqrt{3}\}$ **10.** $\{1\}$ **11.** $\{1, -2\}$ **12.** $\left\{0, -\frac{3}{2}, 2\right\}$
13. $\{1, \log_3 4\}$ **14.** $\left\{\frac{1}{9}, \frac{1}{16}\right\}$ **15.** $\left\{-1, \frac{3}{2}, -\frac{1}{2}\right\}$
16. $\left\{1, \frac{-1 \pm \sqrt{33}}{2}\right\}$ **17.** $\left\{1, \frac{-3 \pm \sqrt{17}}{4}\right\}$ **18.** $\{3, 5\}$
19. $\{(3, 3), (-4, 10)\}$ **20.** $\{(6, 8), (8, 6)\}$
21. $\left\{\left(\frac{2}{3}, 3\right), \left(\frac{1}{3}, \frac{3}{2}\right)\right\}$ **22.** $\{(10, 15)\}$ **23.** $\{(2, 3, 4)\}$
24. $\{(3, 1), (1, -1)\}$ **25.** $\{(0.1, -0.2), (0.6, 0.3)\}$
26. $\{(3, 2, -1)\}$ **27.** $\{(4, 2)\}$ **28.** $\left\{\left(\frac{\pi}{6} + 2k\pi, 2\right), \left(\frac{\pi}{6} + 2k\pi, -2\right), \left(\frac{5\pi}{6} + 2k\pi, 2\right), \left(\frac{5\pi}{6} + 2k\pi, -2\right)\right\}$
29. $\{0°, 180°\}$ **30.** $\{27\}$ **31.** $\{\log_2 6\}$ **32.** $\{\log_6 2\}$
33. $\{120°\}$ **34.** $\{0°, 180°\}$ **35.** $\{45°, 135°, 225°, 315°\}$
36. $\{0°, 180°\}$ **37.** $\{15°, 75°, 195°, 255°\}$ **38.** $x = 67.5°$, $y = 22.5°$ **39. a.** $\frac{1 - 4x}{4}$ **b.** 0.215 m, 0.035 m **c.** 0.125 m **40. a.** $50 - 2x$ **b.** 12.5 m **c.** 12.5 m
41. $10x - x^2$. If the area is 40 m^2, then $10x - x^2 = 40$. The value of x is not a real number, therefore there is no real solution. **42.** $A = 3$, $B = -\frac{3}{2}$

43.

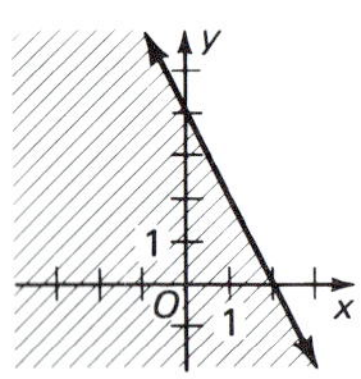

44.

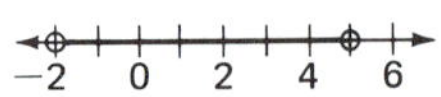

45.

46.

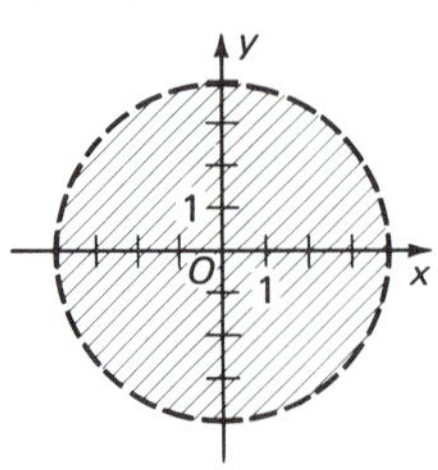

Review of Operations with Numbers

1. 48 **2.** 2 **3.** 12 **4.** 112 **5.** 320 **6.** 0 **7.** 0 **8.** 7
9. 10 **10.** $256\frac{1}{16}$ **11.** $\frac{2}{3}$ **12.** $\frac{17\sqrt{2}}{8}$ **13.** $3 + 2\sqrt{3}$
14. 5 **15.** 3 **16.** $-1 + 2i$ **17.** $-1 - \frac{1}{2}i$ **18.** $2^{2x} - 1$
19. $x^3 + 3x^2 + 3x - 2$ becomes $(2 - 3 \cdot 3^{2/3} + 3 \cdot 3^{1/3}) + (3 \cdot 3^{2/3} - 6 \cdot 3^{1/3} + 3) + (3 \cdot 3^{1/3} - 3) - 2$ which simplifies to zero. **20.** $(-1 + i\sqrt{3})^3 = (-1 + i\sqrt{3})(-2 - 2i\sqrt{3}) = 2 - 2(-1)(3) = 8$ **21.** 11
22. 4 **23.** −2 **24.** Perimeter $= x(3 + \sqrt{5})$; $x = 6 - 2\sqrt{5}$
25. 76 **26.** $\sqrt{\frac{3}{2}}\sqrt{\frac{4}{3}}\sqrt{\frac{5}{4}}\sqrt{\frac{6}{5}} = \sqrt{3}$ **27.** 1.6, −0.6
28. $8x^6y^3$ **29.** $4x^8$ **30.** $2\sqrt[3]{2}$ **31.** 4 **32.** $\frac{1}{4}$ **33.** 4
34. 1000 **35.** 10 **36.** −3 **37.** 1 **38. a.** 2^{12} **b.** 3^{2n}
39. a. $8^{4/3}$ **b.** $9^{n/2}$ **40.** $\frac{2}{3}$ **41.** $-\frac{1}{2}$ **42.** 3
43. a. $10^{1.30}$ **b.** $10^{2.30}$ **c.** $10^{0.60}$ **d.** $10^{0.90}$ **e.** $10^{0.15}$ **f.** $10^{0.40}$ **44. a.** 2.4 **b.** 1.44 **c.** 9.6 **45. a.** −0.9877 **b.** 1 **c.** 1 **46.** $-1 = \cos 180° + i \sin 180°$
47. $\frac{1}{2} + i\frac{\sqrt{3}}{2} = \cos 60° + i \sin 60°$ **48.** 4 **49.** 3 **50.** 20 digits since $2^{64} = 10^{19.264}$ (or $= 1.8 \times 10^{19}$) **51.** $\frac{1}{2}$

Strategies for Teaching

Varying the Mode of Instruction

Using Manipulatives

We commonly think of manipulatives as being most appropriate for elementary school; they are viewed as being needed less frequently at the middle school or junior high school level and rarely at the senior high level. However, concrete approaches also enhance the introduction of algebraic concepts and aid student understanding of mathematical principles when reteaching is needed. Algebra 1 students may need assistance in making the transition from arithmetic to the new world of symbols. Algebra 2 students who previously focused on learning the algorithms may need help in understanding the concepts underlying the symbolic procedures. Teachers can diminish students' anxieties by using concrete approaches to introduce or to reteach many of the topics in algebra. The goal is to have the students progress to the point where they can work entirely on the abstract level.

Manipulatives to support the teaching of algebraic concepts may be either commercially made or made by you and your students. Here are a few suggestions. See the Lesson Commentary for others.

- **Operations with integers using colored tiles or chips**
  Students gain an understanding of the operations with integers as they physically represent the problems, draw a picture of the process, record the numerical results, and arrive at the algorithms inductively by looking for patterns in their results.

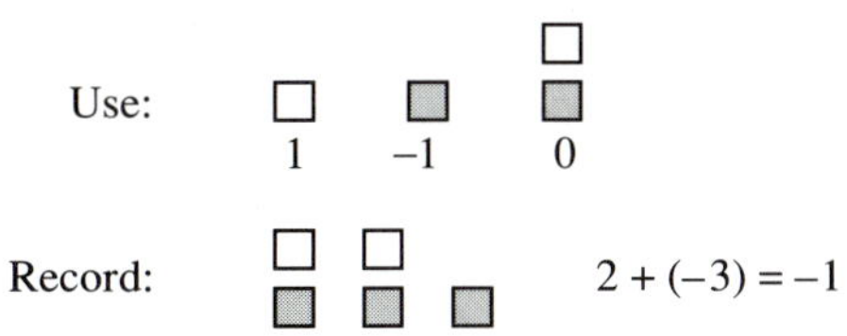

- **Operations with polynomials using algebra tiles**
  When a teacher says "x-squared" a student's mental image may be the abstract expression "x^2." It is equally valuable for the student to realize that this expression represents the area of a square having x as the length of a side. Area models are effective for multiplying binomials and factoring trinomials. (See **Explorations** on page 835.)

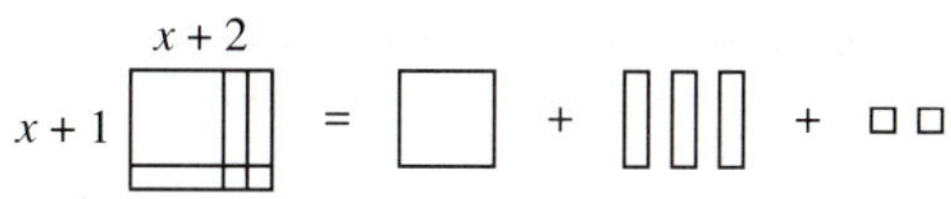

$$(x+1)(x+2) = x^2 + 3x + 2$$

- **Solving equations using a balance scale**
  An equation can be represented as a balance of unknown and known quantities, say, $x + 5 = 9$. Solving for x is represented physically by the removal of equal quantities from both sides in order to maintain the balance.

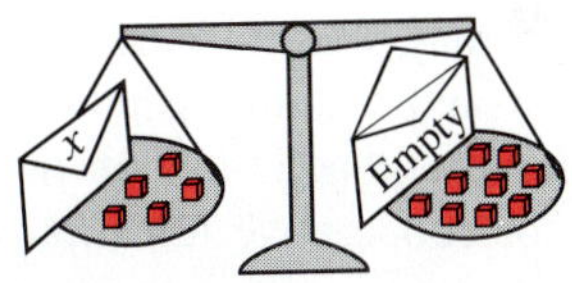

- **Studying lines and their slopes using a geoboard**
  Taking the central peg on a geoboard as (0, 0), a rubber band can "draw" the line between two points and then be stretched to show the rise and run of the slope.

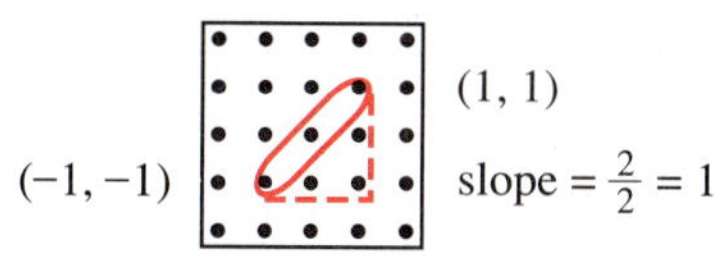

Cooperative Learning

Cooperative learning is most easily accomplished in small groups in which students see themselves as members of a team working together to achieve a specific goal. This does not mean that all the members of a group must be of equal ability or that maximum achievement is demanded of each member. It means that each member contributes according to his or her ability, that everyone's contribution is respected, and that the team does achieve its goal.

Sometimes students can understand a concept explained by another student more easily than they can follow the presentation to the class as a whole. The process of explaining and teaching is in itself a learning process. In a small group, students may find it easier to ask questions and advance ideas.

Groups involved in cooperative learning may vary in size. An even number of members is recommended (four has worked well). In setting up the groups initially, you should try to choose students who work well together. However, it is advisable to re-form the groups at intervals in order to achieve maximum interaction.

Cooperative learning can be used in several ways. You may prefer to have students work in small groups for a short time every day, or you may prefer longer but less frequent group sessions. In any case, be sure to choose material that lends itself well to group work—discovery and exploration of a concept, solution of a challenging problem, making a model, for example. You may wish to move among the groups as a counselor and resource person.

The **Explorations** activities on pages 832–847 are well-suited for cooperative learning. Also, in the Lesson Commentary for some lessons, you will find sections headed Group Activities. Many of these can be used for cooperative learning. (For example, see pages T88 and T106.) Some of them involve team activities; others call for group discussion of conclusions individually reached.

Using Technology

Competence in the use of calculators and computers is already recognized as one of the basic skills needed in many careers. Computers and related electronic devices have liberated students from a great deal of routine calculation and have vastly increased the range of information available to them.

The computer has many uses beyond that of saving students' time. Computer simulations make it possible for students to examine some concepts (function, for example) in depth and to test a variety of hypotheses. In statistics, use of computers facilitates the collection and analysis of data. The graphic presentation of data is made much easier if a graphing calculator or a computer is available.

When students become involved in writing programs, they are obliged to learn to communicate with the computer in its own language and to organize their thoughts in logical sequence. In the words of a recent report issued by the National Science Board, writing programs motivates students "to think algorithmically and develop problem solving skills."

Throughout the textbook, students will find features entitled Calculator Key-In and Computer Key-In, which can be used by students with no prior programming experience. Special sets of Computer Exercises extend some lessons for students who have had some programming experience. Suggestions for use of a calculator, a graphing calculator, or a computer appear within some of the text discussions and in directions for groups of exercises where use of these aids is particularly appropriate. In many of the **Explorations** on pages 832–847, students explore concepts using calculators and computers.

These references to the use of technology are highlighted by a special calculator or computer logo on the annotated student pages and in the side columns in this Teacher's Edition. (See, for example, pages 35 and 80.)

Communicating in Mathematics

Effective communication involves skills in four areas—listening, speaking, reading, and writing. Students need to be able to understand directions and follow them accurately, to explain processes and concepts clearly, to read with understanding, and to express their ideas clearly in writing. Much more is involved than the ability to repeat vocabulary and to manipulate symbols previously learned. Students who cannot verbalize their ideas may not be able to think clearly and to evaluate their own thinking.

The symbolism of algebra is a powerful language in itself, and students need much experience in using it if they are to realize its value as a concise way of expressing ideas that would require many words. One of the goals of mathematics education is to bring students to a point where they can use the language of algebra as a natural way of expressing mathematical ideas, and not as a foreign language.

Verbalization

Working in small groups is one way in which students can improve their communication skills. In such groups students can gain experience in explaining a concept, restating a definition, justifying a conclusion, or thinking aloud through the steps in solving a problem. Certain exercises on the pages listed in the Index under *Proof, informal,* that ask students to "explain" or to "give a convincing argument" can be used in this context. Group discussion exposes students to others' thought processes and helps them clarify their own. Try to monitor these discussions unobtrusively and to act as a coach and facilitator when help is needed.

Whole-class discussions, with you as moderator, can be valuable on occasion. The questions you raise should be aimed at encouraging independent thinking, not at eliciting specific information. An occasional brainstorming session involving a particularly difficult concept may be worth trying.

Insist that your students use the correct word for a concept or procedure. "The number in front of this thing combines with those, and the whole thing cancels." Sound familiar? Using words like *coefficient, term, factor, root, evaluate,* and *function* will keep your students aware of important and subtle differences in meaning.

Writing in Algebra

Here are a few ways to help students acquire competence in writing mathematics.

1. Have students keep a journal while working through a lesson or a section of the textbook—not as a literary exercise, but as a record of a student's false starts, frustrations, questions, and triumphs.

2. Ask students to take one of the biographies or career notes in the textbook, find additional information, and write an expansion of the article.

3. Have students write a group of word problems related to a concept being studied. Or, for a problem that a student has found particularly difficult, ask the student to try rewriting it in a way that can be more easily understood.

4. Have students prepare a list of questions for an interview of a well-known mathematician or scientist.

5. Ask students to write a set of step-by-step directions for a procedure familiar to them (not necessarily mathematical) but not well known to everyone. Other students may then evaluate these directions. Similarly, students may try writing a definition or a generalization.

6. Occasionally use five to ten minutes of class time to have students write a paragraph on an assigned topic—perhaps an explanation of a recently learned concept. This might be followed by small-group discussions.

Reading Algebra

Difficulties that students encounter in algebra are often the result of difficulties in reading. Since reading is a "learning-to-learn" skill, students become independent learners as their ability to read improves.

Clearly, successful reading calls for practice and patience on the part of both teacher and students. Consistent work on reading that is integrated into the content of the course seems to be the method most likely to lead to success. To help the learner, this textbook has been organized in a way that is easy to follow, with important material highlighted. Six major reading objectives are stressed throughout. These objectives are interrelated; it is hardly possible to teach one without the others. The Lesson Commentary suggests practical ways of accomplishing the objectives (for example, see pp. T86 and T117). We encourage you to take advantage of every opportunity throughout the year to help your students gain proficiency in the following reading skills.

1. *Reading and communicating orally*

   Since the transition from what is written to what is spoken is often difficult, students need much experience in verbalizing material that is expressed in symbolic form. You may find it helpful to have them work in small groups, reading parts of a lesson aloud and explaining the examples to one another.

2. *Reading silently*

   It is important for students to realize that the speed of silent reading depends on one's purpose—is one *skimming* for a quick preview or review of the material or is one *studying* to learn the concepts discussed in a lesson? To help students gain study skills, you may want to provide a list of questions in advance.

3. *Using symbols*

   The language of algebra is very compact because of the extensive use of symbols. Students need much help in the correct interpretation and use of symbols. By writing out a few solutions in words, they can gain an appreciation of the value of symbols as savers of time and effort.

4. *Using mathematical words*

   Algebra not only introduces students to its own specialized vocabulary, it also introduces them to new meanings for familiar words such as *power* and *variable*. Encourage your students to learn and use correct terminology and to turn to the Glossary and Index when they need help.

5. *Reading charts, graphs, and diagrams*

   These visual aids are freely used throughout the textbook. Students need to realize that they are essential parts of the text and examples that must be read along with the words. You can help students to a better understanding of charts and graphs by demonstrating how they are constructed.

6. *Reading word problems*

   Reading a problem with comprehension is the first and most essential step in its solution. You will need to work through a variety of problems with your students, applying the plan discussed on page 50 of the textbook. Help students recognize problems that are similar in structure although considerably different in wording.

Learning Strategies

One of the goals of education is to help students become independent thinkers. The purpose of this section is to provide some suggestions for you when your students need help in learning mathematics.

You can help your students by pointing out that a great deal of learning takes place by trial and error, and that the neat and orderly presentation in the textbook is the final result. When you sense that your students are having difficulty with a lesson, explain that real learning takes place through asking questions, trying guesses, starting down false paths, and trying new methods of approach.

No doubt students have often come to you with questions and statements of frustration like those that follow. The suggestions below each quotation are intended to help you respond in a way that will encourage independent thinking.

"I don't understand the book when I try to read it. What should I do?"

1. Be an active reader; as you study an example in the book, write out the solution in your notebook. In that way you'll understand why certain things are done because you'll be doing them as the authors explain them.
2. Learn the boldface words and their meanings. If a paragraph talks about factors, for example, and you don't know what a factor is, then the paragraph will not make any sense. Look up the words you don't know in the Glossary or in the Index at the back of the book. Finally, use the words when describing to someone else how you did a mathematics problem.

"When you explain a problem in class I understand it, but when I try to do my homework I just don't know how to begin."

1. If you *think* that you can do a problem, then you're ahead of the game. By saying "I can't do this problem," you're defeating yourself and giving yourself an excuse to stop trying. Your teacher believes you can do the problem; otherwise he or she wouldn't have assigned it.
2. Before your teacher or another student explains a problem in class, write the problem in your notebook. Try not to take notes while the explanation is going on, but listen carefully to get a feel for the direction of the solution and the order of the steps. When the explanation is finished, write up the solution in your notebook. If there's any confusion, or if you have forgotten any step in the solution, ask about it right then and there. Perhaps several others are stuck at the same place you are; someone had better clear up the confusion, and it might as well be you.
3. Before doing your homework it's a good idea to look at the assignment from the previous day; part of your homework is probably review.
4. Look again at the discussion preceding the exercises in the textbook and at the notes you made in class. Be sure to review the examples that are worked out in the book.
5. Here are some questions to ask yourself:
   a. What is the problem asking for? Do you understand all the words? Can you rewrite the question in another way? Have you seen a problem like this before? What does this problem have in common with others you've seen before?
   b. What is known? What are the conditions within which you must work? Can you rewrite the given items or the conditions in another way? Can you guess an answer?
6. Remember that there can be more than one way to do a mathematics problem. If the first method you try doesn't work, look for another way. Once you've solved a problem see if there are others that you can solve in the same way.

"I can't seem to organize my solutions to word problems. What do I write down first?"

1. Usually you can begin by using x or another letter to represent the quantity you are asked to find. Then see if you can express other quantities and relationships in terms of x. As the book suggests, it is often helpful to set up a chart, as in the rate-time-distance problems.
2. Read the problem carefully to find the relationships that are involved. Look for phrases such as "is twice" or "are equal to," for some form of the verb "to be" usually indicates where the equals sign in an equation will be located.
3. Sometimes it is helpful to guess an answer and then test your guess to see whether or not it is correct. The process you go through in testing your guess is usually the same process that is used to set up an equation to solve the problem.

"My friend and I work together on our homework a lot. It's helpful to have someone to work with, but most of the time she ends up doing the problem before I've even started."

It's fun to work with someone so you can share ideas and feel a sense of cooperation. Be sure to take turns in talking through the problems. When one person is talking through the problem, the other has to listen carefully to what is being said in order to understand each step. If you are doing the talking, just think out loud so your friend can hear the answers to all the questions you are asking yourself. Listen carefully if your friend makes a suggestion because she may say something that turns out to be the missing link you need.

"How should I study for the next test?"

Most studying for tests actually takes place through your daily homework assignments and class work. Don't think that the only preparation for a test should take place the night before.

Here are some study suggestions:

1. Review the meanings of the boldface words.
2. Review the examples worked out in the book by covering up each solution, writing out your solution, and then seeing if you are right.
3. Review the problems done in class that are in your notebook.
4. Do the self-test and chapter review exercises.
5. Pick out the problems from your homework that you could not do. Can you do them now?
6. Have a friend make up a problem for you to do. Then make up one for your friend.

These examples of student frustration imply that one of your major tasks is to raise students' level of confidence—especially in problem solving. Here are a few specific suggestions:

1. Before showing your students how to do a problem, spend some time discussing the steps of the five-step problem-solving plan outlined on page 50.
2. Have your students make up their own problems. You will get a sense of the sophistication of their understanding of the material and you may get some good test questions!
3. Take some time in class for your students to work together. Encourage cooperation and sharing.
4. There is often more than one way to do a problem. Help your students see these various ways. For example, one student may let x represent the time in a motion problem, and a second student may let x represent the distance.
5. Finally, give your students plenty of time in class to write in their notebooks and to think about a question before attempting to answer it.

Problem Solving Strategies

A problem solving strategy is simply a plan or technique for solving a problem. There are a number of well-known problem solving strategies that relate specifically to algebra. For example, applying the quadratic formula is a strategy for solving quadratic equations, and using the linear combination method to transform a system of linear equations into an equivalent system whose solution is obvious is a strategy for solving linear systems. One of the goals of an algebra course is to familiarize students with these standard techniques and to give students enough practice with these techniques so that they can use them confidently and successfully to solve algebra problems.

These rather specific strategies are not the only ones that students can use in solving algebra problems, however. Other, more general strategies, such as looking for a pattern or drawing a diagram, can be very effective problem solving tools. These general strategies can help not only with algebra problems but also with problems in other branches of mathematics and in other subject areas. Since these general strategies provide an *approach* to solving a problem rather than a specific method of solution, they are particularly useful for attacking a problem when the method of solution is not obvious.

For example, suppose several algebra students are confronted with a word problem that they do not know how to solve. One student might ask, "Is this problem similar to any of the types of problems I have seen before?" Another student might try to organize information in a table or a chart or a diagram. Still another might use the process of elimination to rule out what cannot be a solution. Each of these approaches can be a useful problem solving strategy.

Below is a list of general problem solving strategies and skills that are helpful in an Algebra 2 course. By using strategies such as these, students can become better problem solvers and may also grow to enjoy problem solving more.

- look for a pattern in the data
- use a table or a chart
- draw a diagram
- generalize from specific examples
- write and solve an equation, an inequality, or a system of equations or inequalities
- use the 5-step problem solving plan discussed on page 50
- apply a standard formula
- recognize a problem as a standard type
- solve a simpler, related problem
- use trial and error and the process of elimination
- reason backward
- make a deductive argument
- recognize the possibility of no solution
- simulate a problem situation

In the side columns of certain lessons throughout the textbook there are Problem Solving Strategies listed that can be used in that lesson (see, for example, pages 52 and 128). You may wish to discuss these strategies with students.

Nonroutine Problems

Learning how to solve common types of problems helps students develop their problem solving abilities. To become independent problem solvers, they must then learn how to extend familiar procedures and apply them in new situations; that is, they must learn how to solve *nonroutine* problems. There is no single strategy for solving nonroutine problems.

In order to develop their ability to solve nonroutine problems, students need frequent opportunities to extend and apply what they have learned. In this program, see the Resource Book *(Applications, Enrichment, Thinking Skills);* the Tests (*Challenge* problems); this Teacher's Edition *(Suggested Extensions);* and the student book pages listed below.

Ch.	Written Exercises	Problems	Challenges
1	p. 5: 39, 40; p. 12: 45–50, 52–54; p. 19: 25–37; p. 25: 49–56; p. 31: 36; p. 35: 29–36; p. 53: 13–18, 23		p. 48
2	p. 63: 25–33; p. 75: 29–34; p. 79: 25–33; p. 87: 15–25; p. 91: 15–18	pp. 71–72: 11–16	
3	p. 111: 34–39; p. 122: 49–57; p. 130: 33–41; p. 139: 43–45; p. 145: 44–49; p. 150: 31–33; p. 152: 11; p. 157: 37	p. 134: 18–20; p. 106: 9–14	pp. 140, 158
4	p. 170: 25–27; p. 173: 39–42; p. 176: 49–56; p. 182: 33–36; p. 192: 59; p. 197: 51–56; p. 205: 33	p. 201: 31	
5	p. 214: 35–36; p. 220: 53–58; p. 237: 40–41; pp. 240–241: 27–35	pp. 224–225: 9–15; p. 252: 21–23	p. 254: 1–4
6	p. 263: 35–38; p. 269: 67–70; p. 281: 38–39; p. 287: 36–42; p. 296: 51–56		
7	p. 314: 41–42; pp. 320–321: 38–43; p. 325: 29–32; p. 332: 35–39; p. 336: 37–44; p. 343: 39–40; pp. 344–345: 15–16		p. 325
8	p. 355: 21–25; p. 367: 30–34; p. 376: 38–39; p. 380: 23–29; pp. 384–385: 21–26; p. 389: 27–30	p. 357: 20; p. 362: 13–14	p. 363
9	p. 411: 47–50; p. 416: 35–38; p. 422: 23–27; p. 423: 34, 35; p. 435: 21; p. 438: 20; p. 448: 23–25	pp. 405–406: 9–14; p. 443: 13–14	p. 411; p. 425: 1–5
10	p. 462: 37–39; p. 467: 24–26; p. 472: 43; p. 482: 41–42; p. 491: 50–51	p. 488: 21–22	p. 492; p. 494: 1–3
11	p. 505: 31–35; p. 514: 40; p. 515: 14; p. 528: 33–38; p. 534: 27–30; p. 539: 24; p. 543: 28–29	p. 530: 13–15; pp. 534–536: 7–14	p. 516
12	p. 560: 30–34; pp. 577–578: 20–24; p. 583: 18–19	p. 579: 15–17; p. 584: 12; p. 590: 13; p. 596: 8–14	p. 601
13	p. 617: 28–32; p. 623: 29–30; p. 629: 39–48; p. 633: 29; p. 644: 34–37; p. 649: 43; p. 653: 33–35	p. 612: 11–13	p. 623; p. 635: 1–2
14	p. 670: 31–32; p. 684: 35–37; p. 688: 23; p. 696: 29		p. 647: 9–15
15	p. 712: 17–18; p. 718: 20; p. 723: 11–12; pp. 728–729; 9–13; pp. 732–733: 17–20; p. 741: 27–30		p. 733; pp. 752–753: 1–8
16	p. 773: 21–28; p. 778: 27–30; pp. 783–784: 17–18; p. 789: 19–20; p. 797: 19; p. 800: 13–15; p. 805: 17–19		

Making Connections

To appreciate the power of mathematics, students need to see the connections between the mathematics they are studying and other mathematics courses, other subject areas, and everyday and career applications. Many opportunities to explore these connections are provided in this book, as detailed below.

Connections with Earlier Math Courses

Comparing previously acquired and new concepts helps students transfer prior learning to new situations. For example, techniques for adding and subtracting rational algebraic expressions are related to similar operations with common fractions (page 235).

Connections with Geometry

Strong connections between algebra and geometry occur throughout the course in the use of geometric formulas, in the geometric modeling of algebraic concepts, in coordinate geometry, and in problem solving (e.g., pages 41–42; 124 and 444; 401–403; 46–48 and 200–201). For more examples of the integration of geometry with algebra, refer to the Index under *Geometry; Areas; Conics; Formulas, geometric;* and *Problems, geometric and three-dimensional.*

Connections with Data Analysis, Statistics, and Probability

Throughout this book, students meet techniques for dealing with data. In the illustrations facing page 1 and on pages 58 and 454, they see three different graphic presentations of data, one of them on a graphing calculator to emphasize the current use of technology in statistical work. In Lesson 5-3 (page 221) students use scientific notation and significant digits to check accuracy of data. They later interpolate and extrapolate from data tables (Lesson 8-9, page 391) and analyze data for patterns (Lesson 11-1, page 501). In many lessons, students draw and interpret graphs of data (e.g., pages 134, 152, and 159–161).

In addition to the integration of statistical work throughout, formal lessons on statistical techniques (Chapter 15) introduce students to frequency distributions, statistical measures, stem-and-leaf and box-and-whisker plots, and the normal distribution. In Lesson 15-4 (page 724) they draw scatter plots, determine correlation coefficients, and use regression lines to make predictions from data they gather (page 729). In the Application (page 752) they test hypotheses and solve problems using random sampling techniques.

Students are prepared for future coursework in statistics and probability with lessons (beginning on page 730) covering fundamental counting principles, permutations, combinations, sample spaces, and probability of mutually exclusive and independent events.

Connections with Discrete Mathematics

Topics in discrete mathematics, such as sets, algorithm analysis, math induction, functions, relations, and matrices, are listed in the Index under *Discrete Mathematics*. Some of these topics are explored in Appendix A-3 (pages 838–839).

Connections with Real-World Applications

The connections between mathematics and other curriculum areas include the physical science connection (e.g., pages 210, 350, 424–425, 672–674, and 733), the life science connection (e.g., pages 356, 500, and 535), the social science connection (e.g., pages 224, 493–494, and 779–784), and the historical connection (e.g., pages 42, 182, and 297).

Some of the innumerable applications are highlighted in the Applications features and in the problems throughout (see the Index under *Applications, Formulas,* and *Problems*).

In Biographical Notes and Career Notes students can see the contribution of others to past and contemporary society and can see that the study of mathematics opens the door to possible contributions they themselves can make.

Thinking Skills

Thinking skills are woven into the whole fabric of algebra. While such topics as proofs and problem solving may make special demands on students' thinking skills, no real understanding of any of the concepts presented in this course can take place without them.

Your students are likely to come to you with a variety of thinking skills, not always well developed or even recognized by the students themselves. In your lesson presentations you can help students improve these skills, use them more efficiently, and acquire additional skills.

The side-column notes in this Teacher's Edition point out specific thinking skills that come into play in particular lessons (for example, pp. 65 and 81). These skills are key in the study of algebra:

Recall and transfer	Applying concepts
Analysis	Interpreting
Reasoning and inferencing	
Spatial perception	Synthesis

In the Index under "Thinking Skills" you will find a list of some of the areas in which these skills are applied by students. Among these are the *recall* of methods learned earlier and their *transfer* or extension to new material; and the *application* of important concepts to the mathematical modeling of situations familiar to students.

Throughout the course, the skills of *analysis* and *interpretation* of information will be called upon. Encourage your students to be on the lookout for likenesses, differences, and patterns; and to recognize similarities between problems that appear at first glance to be quite different. In attacking a problem students will need to examine and interpret the given information and discard any irrelevant material. In working through the solution of an equation or an inequality, they will need to analyze possible values to see whether they fit the given conditions (see, for example, Examples 2 and 3 and the side-column notes on pages 102–103). In problem solving, they will need to interpret their solutions to see whether they make sense.

Reasoning comes into play whenever students consider information in an if-then format. They draw *inferences* about number relationships on the basis of observed patterns and from statistical data and graphs.

Spatial perception is required to picture all views of an object in space in order to figure the total surface area; and to understand how a cone can be cut to form circles, ellipses, parabolas, and hyperbolas.

Synthesis is employed when ideas come together to form something new; for example, when data are plotted to form a graph and commands are put together to create a computer program.

In the Resource Book there are *Thinking Skills* pages that define these critical thinking skills and provide the opportunity to practice them in the context of this algebra course.

Here are some ways to help students develop their thinking skills:

1. *Be a role model.* Talk through the steps in your solution of an exercise or a problem. By following your reasoning, students can often learn to organize their own thinking in a more logical way.
2. *Use helpful questioning techniques.* Be sure that your questions to students are directed to the way in which they have arrived at their answers rather than to their recall of specific information.
3. *Encourage active participation by students.* Take full advantage of Oral and Written Exercises that ask students to "explain," "show," "give a convincing argument." These phrases all indicate that an informal proof is expected. (See *Proof, informal,* in the Index.) As the basis for a group activity such exercises provide the opportunity for one student to present an explanation and the others to judge whether the argument is convincing.

You will need to emphasize the importance of attacking a problem analytically—of trying to see the problem as a whole and planning one's solution. Point out also that there may be a variety of ways of approaching a problem, and that there is nothing wrong with abandoning one strategy and trying another.

Error Analysis

Since mathematics builds on previously learned symbols, concepts, and skills, error patterns that are left uncorrected will impede students' progress. Of course, there are many different reasons for errors, but certain types of errors are more common than others. If you are aware of these common errors, you can help students avoid them and you can be better prepared to help students overcome them if they do occur. Throughout the book, in the side columns next to the textbook pages, common errors have been identified and suggestions for avoiding them have been provided. (See, for example, "Common Errors" on pages 22, 113, and 217.) The errors discussed in the side columns are fairly specific. However, many of them can be grouped into one or more of the following categories:

Errors in Reading and Translating

(See, for example, pp. 45 and 49.)

Students often have difficulty in translating English phrases and sentences into mathematical expressions and sentences. For example, students may translate the expression "five less than a number" as $5 - n$. Not reading word problems carefully, with concentration on their meaning, is another frequent cause of difficulty. Students may make mistakes because they do not fully understand the meanings of mathematical terms—for example, "maximum value of a function" or "angle of elevation." Words such as *or*, which have a different meaning in mathematics than in everyday speech, may cause confusion.

Failure to Understand Symbols

(See, for example, pp. 5 and 146.)

Students often do not fully understand the meanings of mathematical symbols. As a result, they may make the following errors.

$\frac{6}{0} = 0$ $\quad \frac{0}{0} = 1$ $\quad -x^2 = (-x)^2$ $\quad b^n = nb$

$2^{-2} = -4$ $\quad (a + b)^2 = a^2 + b^2$ $\quad -2 > x > 3$

$0.\overline{5} = \frac{1}{2}$ $\quad 2\sqrt{3} = \sqrt{6}$ $\quad f(4) = 4f$ $\quad \sqrt{x^2} = x$

Misunderstanding of Properties

(See, for example, pp. 39 and 195.)

Recurring errors often stem from students misapplying properties in the ways shown below.

Addition property of equality: $x + 2 = 6$
$$x = 8$$

Multiplication property of equality: $\frac{x}{2} + \frac{x}{3} = 5$
$$3x + 2x = 5$$

Division property of equality: $3x(x + 2) = 0$
$$x + 2 = 0$$

Multiplication property of order: $2x > -8$
$$x < -4$$

Distributive property: $-4(x + 1) = 9$
$$-4x + 1 = 9$$

Zero-product property: $(x - 3)(x + 2) = 14$
$$x - 3 = 14 \quad \text{or} \quad x + 2 = 14$$

Rule for simplifying fractions: $\frac{x}{\cancel{3}_1} = \frac{\cancel{6}^2}{5}$

Errors in Using Standard Forms

(See, for example, pp. 311 and 338.)

Although students may have memorized the Pythagorean theorem or the quadratic formula, they may not understand the importance of using the standard form when they apply these formulas. Consequently, they often try to work with an expression or an equation without first putting it into standard form. This means that they may substitute incorrect values in the formula $a^2 + b^2 = c^2$ or try to solve an equation by the quadratic formula before transforming it so that one side is 0. Other errors that students are liable to make are illustrated below.

$2x + 3y = 6$	$y = 1 - 3x$
slope $= 2$	slope $= 1$
	y-intercept $= -3$

$$x^2 + 1 - 2x = 0$$
$$a = 1,\ b = 1,\ c = -2$$

Use of Incorrect Formulas

(See, for example, pp. 113 and 311.)

Many errors are the result of students using formulas that are incorrect. Some of the more common "impostors" are shown below.

$p = l + w$ $\qquad C = \pi r^2$ $\qquad \text{slope} = \dfrac{x_2 - x_1}{y_2 - y_1}$

$x = -b \pm \dfrac{\sqrt{b^2 - 4ac}}{2a}$ $\qquad \text{discriminant} = \sqrt{b^2 - 4ac}$

$a^2 = b^2 + c^2 + 2bc \cos A$ $\qquad \sin \theta + \cos \theta = 1$

Errors in Simplifying

(See, for example, pp. 261 and 288.)

In addition to some of the reasons already given, students may make errors in simplifying expressions because of incorrect assumptions such as $\sqrt{a^2 - b^2} = a - b$. They may also make errors because they do not take notice of grouping symbols such as the fraction bar or because they do not follow the prescribed order of operations or because they add unlike terms. Students sometimes confuse the rules of exponents—multiplying exponents when they are multiplying and dividing exponents when they are dividing. Other times they may confuse the laws of logarithms. Students may forget to (or not realize that they must) change the sign of every term of a polynomial that they are subtracting. Simplifying fractions seems to be particularly troublesome. Thinking that $\frac{n}{n} = 0$ can lead to errors such as $6x^2 + 8x + 2 = 2(3x^2 + 4x)$.

Errors such as $\dfrac{\not{n} + 3}{\not{n}} = 3$ and $\dfrac{\overset{3}{\not{6}}\not{y}(x + 3)}{\underset{1}{\not{2}}y^2(x + 1)} = \dfrac{3(x + 3)}{2(x + 1)}$ are common.

Errors in Checking

(See, for example, pp. 190 and 277.)

Checking can help students develop self-confidence and alert them to errors. However, a check that is incorrectly performed is not useful. Students often fail to realize that it is not only helpful but necessary to check the roots of fractional and radical equations. The following checking errors may occur: Students may occasionally substitute a value for a variable such as 8 for x, get a true statement such as $4 = 4$, and conclude that the solution is 4. Students may not realize that they must check their answers with the *words* of word problems or that they must check their solutions to systems of linear equations in *all* the *original* equations. Some students may think that they should always discard negative solutions. Students may not think of checking their answers when the method involves, for example, considering whether an answer is reasonable or multiplying to check factoring.

Reteaching

Effective reteaching involves a variety of approaches and activities. Factors to be considered include the nature of the material to be retaught, the ability level of the students involved, the particular point of difficulty, and the number of students affected.

Prerequisite to reteaching is error analysis. This is discussed more fully in the article on page T55, as well as in the side-column notes headed "Common Errors." These notes not only identify frequently occurring errors; they also suggest methods of reteaching to correct them. The Cautions that appear in the student textbook are intended to anticipate students' errors and prevent their occurrence. The Mixed Review exercises at the end of every other lesson will help to minimize the need for extensive reteaching of skills.

The primary focus in reteaching needs to be on concepts that have been imperfectly understood or that students have difficulty in applying. Reteaching of concepts must be more than reiteration of what has previously been presented. In most cases a fresh approach is needed, perhaps with the aid of manipulative or visual materials. In addition to the model lesson plan that is provided for each lesson in the side columns, the Lesson Commentary (pages T79–T144) offers alternate teaching suggestions. The articles on varying the mode of instruction and communicating in mathematics under "Strategies for Teaching" (pages T47-T50) suggest new approaches that may be useful for reteaching. Note also that the student textbook itself sometimes highlights alternate methods of attacking a solution (see, for example, pages 22 and 66).

The ancillaries provide much help in the reteaching of concepts. The Study Guide and the Resource Book offer a variety of approaches to important concepts that are aimed at enhancing students' understanding. The Overhead Visuals will be helpful to students who learn best from a graphic or visual approach. The disk Algebra Plotter Plus provides software for classroom demonstration or independent use enabling users to explore concepts using computer graphing techniques. Students who are having difficulty with word problems may be directed to the problem solving pages in the Resource Book. Each lesson on these pages takes students step by step through the analysis and solution of a few problems.

Many of the errors that call for reteaching are the result of forgetting or of failure to stay with a concept or skill long enough for mastery. Built into this program are provisions for detecting and correcting errors of this type at an early stage. Students should be encouraged to use the Self-Tests, and the Chapter Summaries, Reviews, and Tests, and the Cumulative Reviews as means of detecting their own weaknesses.

Students who need additional reinforcement of skills imperfectly learned will benefit from the additional "A" exercises in the Guided Practice section in the side columns. Further materials for reteaching and reinforcing skills are provided by the Practice Masters and the practice pages in the Resource Book.

Reading References

The National Council of Teachers of Mathematics produces a number of helpful publications that give additional information about the topics discussed in the preceding pages. Here are a few suggestions:

Classroom Ideas from Research on Secondary School Mathematics, Part 1: Algebra, 1983

Curriculum and Evaluation Standards for School Mathematics, 1988

Effective Mathematics Teaching, 1987

The Ideas of Algebra, K-12, 1988

Research Within Reach: Secondary School Mathematics, 1982

Assignment Guide

The following Assignment Guide may be of help in planning the year's work. Four courses are outlined in order to allow flexibility in response to your curriculum requirements and your students' abilities:

(1) an average course covering only the topics in algebra,

(2) an average course in algebra and trigonometry,

(3) an extended algebra course,

(4) an extended course in algebra and trigonometry.

The guide is intended as a suggested schedule; you should adjust it to the needs of your students.

As each lesson in the text includes more exercises than students would normally be expected to complete, these assignments include only a portion of the exercises.

Because students' interests and backgrounds differ widely from class to class, most of the optional features, including the **Explorations** on pp. 832–847, are not listed. You will want to choose those features which best suit your individual classes. If you have access to a computer that accepts BASIC, you may wish to allow some time for your students to do the Computer Key-Ins or the Computer Exercises. See the note on page xix regarding these features.

This sample assignment explains the format of the entries in the Assignment Guide. Spiral review assignments provide a mixed review, while sequential review assignments review topics in the same order as presented in the textbook.

The Summary Time Schedule shows the number of teaching days allotted for each chapter in each of the four courses in the Assignment Guide. Semester and trimester divisions are indicated by a red rule and a blue rule, respectively.

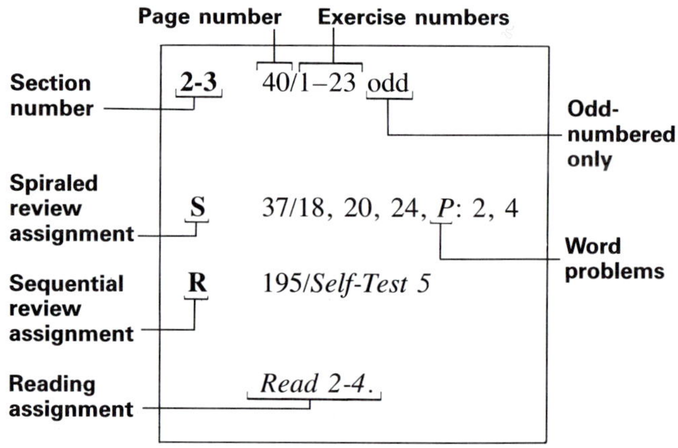

Summary Time Schedule for the Assignments

Chapter	1	2	3	4	5	6	7	8	9	10	11	12	13	14	15	16
Average Algebra Course	13	10	17	14	13	13	13	12	15	13	14	0	0	0	13	0
Average Algebra and Trigonometry Course	9	9	14	12	13	12	13	12	13	11	2	15	14	11	0	0
Extended Algebra Course	9	8	13	12	12	12	13	12	14	14	14	0	0	0	14	13
Extended Algebra and Trigonometry Course	8	6	10	8	10	11	13	12	13	13	10	11	13	12	0	10

trimester semester trimester

LESSON	AVERAGE ALGEBRA	AVERAGE ALGEBRA AND TRIGONOMETRY	EXTENDED ALGEBRA	EXTENDED ALGEBRA AND TRIGONOMETRY
1	**1-1** 4/1–39 odd **S** 5/*Mixed Review*	**1-1** 4/6–36 mult. of 3 **S** 5/*Mixed Review* **1-2** 10/3–48 mult. of 3	**1-1** 4/6–36 mult. of 3, 39, 40 **S** 5/*Mixed Review* **1-2** 10/9–54 mult. of 3	**1-1** 4/6–36 mult. of 3, 39, 40 **S** 5/*Mixed Review* **1-2** 10/9–54 mult. of 3
2	**1-2** 10/2–46 even **S** 5/34–40 even	**1-3** 17/1–23 odd, 25, 30 **S** 20/*Mixed Review* **R** 13/*Self-Test 1*	**1-3** 17/7–35 odd **S** 20/*Mixed Review* **R** 13/*Self-Test 1*	**1-3** 17/7–35 odd **S** 20/*Mixed Review* **R** 13/*Self-Test 1*
3	**1-3** 17/1–21 odd, 25–28 **S** 20/*Mixed Review* **R** 13/*Self-Test 1*	**1-4** 24/2–50 even, 51–53	**1-4** 24/2–50 even, 51, 52	**1-4** 24/12–48 mult. of 3 **1-5** 30/3–36 mult. of 3 **S** 31/*Mixed Review*
4	**1-4** 24/2-48 even, 51, 53	**1-5** 30/3–33 mult. of 3 **S** 31/*Mixed Review* **1-6** 35/4–28 mult. of 4, 29, 30 **R** 36/*Self-Test 2*	**1-5** 30/3–33 mult. of 3 **S** 31/*Mixed Review* **1-6** 35/4–28 mult. of 4, 29–31 **R** 36/*Self-Test 2*	**1-6** 35/2–28 even, 29, 30 **R** 36/*Self-Test 2* **1-7** 40/4–48 mult. of 4, 51–53 **S** 42/*Mixed Review*
5	**1-5** 30/1–33 odd **S** 31/*Mixed Review*	**1-7** 40/3–27 mult. of 3, 31–53 odd **S** 42/*Mixed Review*	**1-7** 40/3–27 mult. of 3, 31–53 odd **S** 42/*Mixed Review*	**1-8** 46/3–18 mult. of 3, 19–29 odd
6	**1-6** 35/2–28 even, 29–31 **R** 36/*Self-Test 2*	**1-8** 46/3–18 mult. of 3, 19–27 odd	**1-8** 46/3–18 mult. of 3, 19–29 odd	**1-9** 52/*P*: 2–22 even, 23 **S** 54/*Mixed Review* **R** 54/*Self-Test 3*
7	**1-7** 40/2–48 even **S** 42/*Mixed Review*	**1-9** 52/*P*: 2–22 even **S** 54/*Mixed Review* **R** 54/*Self-Test 3*	**1-9** 52/*P*: 2–22 even, 23 **S** 54/*Mixed Review* **R** 54/*Self-Test 3*	*Prepare for Chapter Test* **R** 55/*Chapter Review*
8	**1-8** 46/1–19 odd **S** 42/49–53	*Prepare for Chapter Test* **R** 55/*Chapter Review*	*Prepare for Chapter Test* **R** 55/*Chapter Review*	*Administer Chapter 1 Test* **2-1** *Read 2-1* 62/1–31 odd **S** 63/*Mixed Review*
9	**1-8** 47/16, 18, 20-27	*Administer Chapter 1 Test*	*Administer Chapter 1 Test* **2-1** *Read 2-1* 62/1–31 odd **S** 63/*Mixed Review*	**2-2** 67/5–33 odd **S** 63/32, 33 **2-3** 71/*P*: 1, 4, 7, 11, 13, 15 **S** 72/*Mixed Review* **R** 72/*Self-Test 1*
10	**1-9** 52/*P*: 1–11 **S** 54/*Mixed Review*	**2-1** 62/1–27 **S** 63/*Mixed Review*	**2-2** 67/2–34 even **S** 63/32, 33	**2-4** 75/12–34 even **2-5** 78/3–30 mult. of 3 **S** 79/*Mixed Review*

LESSON	AVERAGE ALGEBRA	AVERAGE ALGEBRA AND TRIGONOMETRY	EXTENDED ALGEBRA	EXTENDED ALGEBRA AND TRIGONOMETRY
11	**1-9** 53/*P*: 13–22 **R** 54/*Self-Test 3*	**2-2** 67/2–34 even **S** 63/28–32	**2-3** 71/*P*: 3, 6, 8, 11–16 **S** 72/*Mixed Review* **R** 72/*Self-Test 1*	**2-6** 85/3–10, 13, 15, 18, 22 **R** 79/*Self-Test 2*
12	*Prepare for Chapter Test* **R** 55/*Chapter Review*	**2-3** 71/*P*: 1, 3, 6, 8, 11–14 **S** 72/*Mixed Review* **R** 72/*Self-Test 1*	**2-4** 75/1–33 odd	**2-7** 90/1, 3, 4, 6–8, 10, 14, 15 **S** 91/*Mixed Review* **R** 92/*Self-Test 3*
13	*Administer Chapter 1 Test*	**2-4** 75/1–27 odd	**2-5** 78/2–26 even **S** 79/*Mixed Review*	*Prepare for Chapter Test* **R** 79/*Self-Test 2* **R** 92/*Self-Test 3* **R** 93/*Chapter Review*
14	**2-1** 62/1–26 **S** 63/*Mixed Review*	**2-5** 78/2–24 even **S** 79/*Mixed Review*	**2-6** 85/2–18 even **S** 79/31, 33 **R** 79/*Self-Test 2*	*Administer Chapter 2 Test* **3-1** *Read 3-1* 104/1, 3, 5, 21–35 odd 105/*P*: 2, 7–11, 13 **S** 106/*Mixed Review*
15	**2-2** 67/1–33 odd **S** 63/27–32	**2-6** 85/1–15 odd **S** 75/20–30 even **R** 79/*Self-Test 2*	**2-7** 90/1, 3, 4, 7, 8, 10–12 **S** 91/*Mixed Review* **R** 92/*Self-Test 3*	**3-2** 111/3–21 mult. of 3, 24–34 even **3-3** 116/3–48 mult. of 3 **S** 117/*Mixed Review*
16	**2-3** 71/*P*: 2–14 even **S** 72/*Mixed Review* **R** 72/*Self-Test 1*	**2-7** 90/1, 3, 4, 7, 8, 10 **S** 91/*Mixed Review* **R** 92/*Self-Test 3*	*Prepare for Chapter Test* **R** 93/*Chapter Review*	**3-4** 121/3–57 mult. of 3 **R** 123/*Self-Test 1*
17	**2-4** 75/1–25 odd **S** 71/*P*: 11, 13, 15	*Prepare for Chapter Test* **R** 93/*Chapter Review*	*Administer Chapter 2 Test* **3-1** *Read 3-1* 104/1, 3, 5, 21–35 odd 105/*P*: 2, 7–11 **S** 106/*Mixed Review* **S** 98/*P*: 2–14 even	**3-5** 129/1–37 odd **S** 130/*Mixed Review* **3-6** 132/*P*: 3, 7, 10–14
18	**2-5** 78/2–24 even **S** 79/*Mixed Review*	*Administer Chapter 2 Test* **S** 98/*P*: 1–14	**3-2** 111/2–36 even	**3-6** 133/*P*: 9, 15–18
19	**2-6** 85/1–10, 15 **R** 79/*Self-Test 2*	**3-1** 104/3–36 mult. of 3 105/*P*: 1, 3, 6–10 **S** 106/*Mixed Review*	**3-3** 116/1–47 odd **S** 117/*Mixed Review*	**3-7** 138/1–43 odd **S** 139/*Mixed Review*
20	**2-6** 87/17–22 **S** 79/13–27 odd	**3-2** 111/2–28 even	**3-4** 121/3–54 mult. of 3 **R** 123/*Self-Test 1*	**3-8** 144/3–42 mult. of 3, 48, 49 **S** 139/44, 45 **R** 140/*Self-Test 2*

LESSON	AVERAGE ALGEBRA	AVERAGE ALGEBRA AND TRIGONOMETRY	EXTENDED ALGEBRA	EXTENDED ALGEBRA AND TRIGONOMETRY
21	**2-7** 90/1, 3, 4, 6–8, 10 **S** 91/*Mixed Review* **R** 92/*Self-Test 3*	**3-3** 116/1–47 odd **S** 117/*Mixed Review*	**3-5** 129/1–37 odd **S** 130/*Mixed Review*	**3-9** 149/3–24 mult. of 3 150/*P*: 2, 4, 8–10 **S** 152/*Mixed Review*
22	*Prepare for Chapter Test* **R** 93/*Chapter Review*	**3-4** 121/2–54 even **R** 123/*Self-Test 1*	**3-6** 132/*P*: 2–16 even **S** 129/32–40 even	**3-10** 156/3–36 mult. of 3 **R** 158/*Self-Test 3*
23	*Administer Chapter 2 Test* **S** 98/*P*: 1–14	**3-5** 129/1–37 odd **S** 130/*Mixed Review*	**3-6** 133/*P*: 11–19 odd **3-7** 138/1–27 odd	*Prepare for Chapter Test* **R** 162/*Chapter Review*
24	**3-1** 104/1–37 odd	**3-6** 132/*P*: 2–14 even **S** 129/32–40 even	**3-7** 139/29–43 odd **S** 139/*Mixed Review*	*Administer Chapter 3 Test* **4-1** *Read 4-1* 170/1–27 odd **S** 170/*Mixed Review*
25	**3-1** 105/*P*: 1–11 **S** 106/*Mixed Review*	**3-6** 133/11–17 odd **3-7** 138/1–25 odd	**3-8** 144/3–42 mult. of 3 **R** 140/*Self-Test 2*	**4-2** 173/2–38 even, 39, 40
26	**3-2** 111/2–24 even **S** 105/34–38 even	**3-7** 139/27–41 odd **S** 139/*Mixed Review*	**3-9** 149/2–32 even	**4-3** 175/3–51, mult. of 3, 53, 54 **S** 176/*Mixed Review* **4-4** 181/3–30 mult. of 3, 34, 35 **R** 177/*Self-Test 1*
27	**3-3** 116/1–47 odd **S** 117/*Mixed Review*	**3-8** 144/3–42 mult. of 3 **S** 122/56, 57 **R** 140/*Self-Test 2*	**3-9** 150/27, 29, 33 150/*P*: 1–11 odd **S** 152/*Mixed Review*	**4-5** 185/1–45 odd, 49–53 odd **S** 187/*Mixed Review*
28	**3-4** 121/2–30 even **S** 111/21–35 odd	**3-9** 149/2–30 even **S** 134/*P*: 18, 19	**3-10** 156/1–31 odd **R** 158/*Self-Test 3*	**4-6** 191/3–30 mult. of 3, 32–58 even **R** 192/*Self-Test 2*
29	**3-4** 121/31–55 odd **R** 123/*Self-Test 1*	**3-9** 150/*P*: 1–9 odd **S** 152/*Mixed Review*	*Prepare for Chapter Test* **R** 162/*Chapter Review*	**4-7** 196/3–48 mult. of 3 **S** 197/*Mixed Review* **4-8** 199/*P*: 4, 8, 9, 11
30	**3-5** 129/1–31 odd **S** 130/*Mixed Review*	**3-10** 156/1–29 odd **R** 158/*Self-Test 3*	*Administer Chapter 3 Test* **R** 165/*Cumulative Review*	**4-8** 200/*P*: 13–27 odd **4-9** 204/3–30 mult. of 3 **S** 205/*Mixed Review* **R** 205/*Self-Test 3*
31	**3-5** 130/33–38 **3-6** 132/*P*: 2–10 even	*Prepare for Chapter Test* **R** 162/*Chapter Review*	**4-1** 170/1–27 odd **S** 170/*Mixed Review*	*Prepare for Chapter Test* **R** 207/*Chapter Review*

LESSON	AVERAGE ALGEBRA	AVERAGE ALGEBRA AND TRIGONOMETRY	EXTENDED ALGEBRA	EXTENDED ALGEBRA AND TRIGONOMETRY
32	**3-6** 133/*P*: 11–14, 16 **3-7** 138/1–17 odd	*Administer Chapter 3 Test* **R** 165/*Cumulative Review*	**4-2** 173/2–38 even, 39, 40	*Administer Chapter 4 Test* **5-1** *Read 5-1* 213/2–30 even, 33 **S** 215/*Mixed Review*
33	**3-7** 139/19–41 odd **S** 139/*Mixed Review*	**4-1** 170/1–25 odd **S** 170/*Mixed Review*	**4-3** 175/1–51 odd **S** 176/*Mixed Review*	**5-2** 218/1–57 odd **S** 214/34, 35
34	**3-8** 144/1–26 **R** 140/*Self-Test 2*	**4-2** 173/2–38 even	**4-4** 181/2–32 even **R** 177/*Self-Test 1*	**5-3** 223/2–30 even 224/*P*: 1–5, 13–15 **S** 225/*Mixed Review*
35	**3-8** 145/27–43 odd	**4-3** 175/1–47 odd **S** 176/*Mixed Review*	**4-5** 185/1–45 odd **S** 187/*Mixed Review*	**5-4** 228/3–42 mult. of 3 230/*Extra:* 1–11 odd **R** 226/*Self-Test 1*
36	**3-9** 149/2–30 even **S** 145/49	**4-4** 181/2–30 even **R** 177/*Self-Test 1*	**4-5** 186/32–46 even, 49–51 **4-6** 191/2–30 even	**5-5** 234/2–24 even **S** 234/*Mixed Review*
37	**3-9** 149/5, 21, 23, 27, 29 150/*P*: 1–9 odd **S** 152/*Mixed Review*	**4-5** 185/1–45 odd **S** 187/*Mixed Review*	**4-6** 191/31–50, 55–58 **R** 192/*Self-Test 2*	**5-6** 237/7–41 odd **S** 234/17–23 odd
38	**3-10** 156/1–29 odd **R** 158/*Self-Test 3*	**4-5** 186/32–46 even **4-6** 191/2–30 even	**4-7** 196/1–49 odd **S** 197/*Mixed Review*	**5-7** 239/2–28 even, 32–35 **S** 241/*Mixed Review*
39	*Prepare for Chapter Test* **R** 162/*Chapter Review*	**4-6** 191/31–54 **R** 192/*Self-Test 2*	**4-8** 199/*P*: 3–30 mult. of 3	**5-8** 245/3–24 mult. of 3 245/*P*: 7–21 odd **R** 241/*Self-Test 2*
40	*Administer Chapter 3 Test* **R** 165/*Cumulative Review*	**4-7** 196/1–49 odd **S** 197/*Mixed Review*	**4-9** 204/1–21 odd, 23, 27 **S** 205/*Mixed Review* **R** 205/*Self-Test 3*	**5-9** 249/3–30 mult. of 3 250/*P*: 3–21 mult. of 3, 22, 23 **S** 252/*Mixed Review*
41	**4-1** 170/1–25 odd **S** 170/*Mixed Review*	**4-8** 199/*P*: 2–18 even	*Prepare for Chapter Test* **R** 207/*Chapter Review*	*Prepare for Chapter Test* **R** 252/*Self-Test 3* **R** 255/*Chapter Review*
42	**4-2** 173/2–38 even	**4-9** 204/1–21 odd, 23, 27 **S** 205/*Mixed Review* **R** 205/*Self-Test 3*	*Administer Chapter 4 Test* **5-1** *Read 5-1* 213/2–30 even, 33 **S** 215/*Mixed Review*	*Administer Chapter 5 Test* **6-1** *Read 6-1* 262/1–33 odd **S** 263/*Mixed Review*

LESSON	AVERAGE ALGEBRA	AVERAGE ALGEBRA AND TRIGONOMETRY	EXTENDED ALGEBRA	EXTENDED ALGEBRA AND TRIGONOMETRY
43	**4-3** 175/1–47 odd **S** 176/*Mixed Review*	*Prepare for Chapter Test* **R** 207/*Chapter Review*	**5-2** 218/1–55 odd **S** 214/34, 35	**6-2** 267/2–66 even **S** 263/35–37
44	**4-4** 181/2–30 even **R** 177/*Self-Test 1*	*Administer Chapter 4 Test*	**5-3** 223/2–30 even 224/*P*: 1–5, 13–15 **S** 225/*Mixed Review*	**6-3** 272/1–43 odd **S** 273/*Mixed Review*
45	**4-5** 185/1–16, 18–36 even **S** 187/*Mixed Review*	**5-1** 213/2–30 even **S** 215/*Mixed Review*	**5-4** 228/3–42 mult. of 3 230/*Extra:* 1–11 odd **R** 226/*Self-Test 1*	**6-4** 275/2–44 even **S** 273/*Extra:* 1, 2
46	**4-5** 186/17–37 odd, 38–46	**5-2** 218/1–51 odd **S** 214/33, 34	**5-5** 234/2–24 even **S** 234/*Mixed Review*	**6-5** 280/1–27 odd **S** 276/43, 45, 46
47	**4-6** 191/1–30 **S** 186/49–52	**5-3** 223/2–30 even 224/*P*: 1–5, 13–15 **S** 225/*Mixed Review*	**5-6** 237/7–41 odd **S** 234/17–23 odd	**6-5** 280/29–39 odd **S** 282/*Mixed Review*
48	**4-6** 191/31–49 odd **R** 192/*Self-Test 2*	**5-4** 228/1–37 odd **R** 226/*Self-Test 1*	**5-7** 239/2–28 even **S** 241/*Mixed Review*	**6-6** 286/1–39 odd, 40 **R** 282/*Self-Test 1*
49	**4-7** 196/1–10, 12–34 even **S** 197/*Mixed Review*	**5-5** 234/2–22 even **S** 234/*Mixed Review*	**5-8** 245/3–24 mult. of 3 245/*P*: 1–13 odd **R** 241/*Self-Test 2*	**6-7** 290/2–36 even **S** 291/*Mixed Review*
50	**4-7** 196/9–35 odd, 36–40, 45, 46	**5-6** 237/1–37 odd **S** 229/28–38 even	**5-8** 246/*P*: 14–21 **S** 241/32–35	**6-7** 291/38–54 even **6-8** 295/1–21 odd
51	**4-8** 199/*P*: 2–18 even **S** 197/41–44	**5-7** 239/2–26 even **S** 241/*Mixed Review*	**5-9** 249/1–29 odd 250/*P*: 1–7 odd **S** 252/*Mixed Review*	**6-8** 295/23–55 odd 298/*Extra:* 1–17 odd
52	**4-9** 204/1–21 odd **S** 205/*Mixed Review* **R** 205/*Self-Test 3*	**5-8** 245/1–19 odd 245/*P*: 1–11 odd **R** 241/*Self-Test 2*	**5-9** 251/*P*: 9–21 odd, 22, 23 **R** 252/*Self-Test 3*	*Prepare for Chapter Test* **R** 297/*Self-Test 2* **R** 302/*Chapter Review*
53	*Prepare for Chapter Test* **R** 207/*Chapter Review*	**5-8** 245/10–20 even 245/*P*: 13–21 odd **S** 241/40, 41	*Prepare for Chapter Test* **R** 255/*Chapter Review*	*Administer Chapter 6 Test*
54	*Administer Chapter 4 Test*	**5-9** 249/1–27 odd 250/*P*: 1–7 odd **S** 252/*Mixed Review*	*Administer Chapter 5 Test*	**7-1** 309/3–21 mult. of 3, 23–41 odd **S** 310/*Mixed Review*
55	**5-1** 213/2–30 even **S** 215/*Mixed Review*	**5-9** 251/*P*: 9–19 odd **R** 252/*Self-Test 3*	**6-1** 262/1–33 odd **S** 263/*Mixed Review*	**7-2** 313/3–27 mult. of 3 314/*P*: 1–13 odd

LESSON	AVERAGE ALGEBRA	AVERAGE ALGEBRA AND TRIGONOMETRY	EXTENDED ALGEBRA	EXTENDED ALGEBRA AND TRIGONOMETRY
56	**5-2** 218/1–51 odd **S** 214/33, 34	*Prepare for Chapter Test* **R** 255/*Chapter Review*	**6-2** 267/2–66 even **S** 263/35–37	**7-2** 314/28–42 even 315/*P*: 6–14 even **S** 310/36–42 even
57	**5-3** 223/2–30 even 224/*P*: 1–5, 13–15 **S** 225/*Mixed Review*	*Administer Chapter 5 Test*	**6-3** 272/1–43 odd **S** 273/*Mixed Review*	**7-3** 320/1–39 odd, 41, 42 **S** 321/*Mixed Review* **R** 316/*Self-Test 1*
58	**5-4** 228/1–37 odd **R** 226/*Self-Test 1*	**6-1** 262/1–33 odd **S** 263/*Mixed Review*	**6-4** 275/2–42 even **S** 273/*Extra:* 1, 2	**7-4** 324/2–20 even **S** 321/43
59	**5-5** 234/2–22 even **S** 234/*Mixed Review* 230/1–7 odd	**6-2** 267/2–62 even	**6-5** 280/1–27 odd **S** 276/43–46	**7-4** 324/21–32 **R** 325/*Self-Test 2* **7-5** 331/1–25 odd
60	**5-6** 237/1–37 odd **S** 229/28–38 even	**6-3** 272/1–43 odd **S** 273/*Mixed Review*	**6-5** 280/29–39 odd **S** 282/*Mixed Review*	**7-5** 332/28–36 even, 37, 38 **S** 332/*Mixed Review*
61	**5-7** 239/2–26 even **S** 241/*Mixed Review*	**6-4** 275/2–40 even **S** 269/63–66	**6-6** 286/1–39 odd, 40 **R** 282/*Self-Test 1*	**7-6** 336/1–39 odd **S** 332/27–35 odd
62	**5-8** 245/1–19 odd 245/*P*: 1–11 odd **R** 241/*Self-Test 2*	**6-5** 280/1–23 odd **S** 276/41–44	**6-7** 290/2–36 even **S** 291/*Mixed Review*	**7-6** 336/20–42 even, 43, 44
63	**5-8** 245/10–20 even 245/*P*: 13–21 odd **S** 237/40, 41	**6-5** 280/25–37 odd **S** 282/*Mixed Review*	**6-7** 291/38–54 even **6-8** 295/1–21 odd	**7-7** 342/1–37 odd 343/*P*: 1–7 odd
64	**5-9** 249/1–27 odd 250/*P*: 1–7 odd **S** 252/*Mixed Review*	**6-6** 286/1–37 odd **R** 282/*Self-Test 1*	**6-8** 295/23–55 odd 298/*Extra:* 1–17 odd	**7-7** 343/*P*: 8–15 **S** 345/*Mixed Review*
65	**5-9** 251/*P*: 9–19 odd **R** 252/*Self-Test 3*	**6-7** 290/2–36 even **S** 291/*Mixed Review*	*Prepare for Chapter Test* **R** 297/*Self-Test 2* **R** 302/*Chapter Review*	*Prepare for Chapter Test* **R** 345/*Self-Test 3* **R** 346/*Chapter Review*
66	*Prepare for Chapter Test* **R** 255/*Chapter Review*	**6-7** 291/38–52 even **6-8** 295/1–11 odd	*Administer Chapter 6 Test*	*Administer Chapter 7 Test* **R** 348/*Cumulative Review*
67	*Administer Chapter 5 Test*	**6-8** 295/13–49 odd **S** 291/31–53 odd	**7-1** 309/3–21 mult. of 3, 23–41 odd **S** 310/*Mixed Review*	**8-1** 354/1–23 odd
68	**6-1** 262/1–33 odd **S** 263/*Mixed Review*	*Prepare for Chapter Test* **R** 297/*Self-Test 2* **R** 302/*Chapter Review*	**7-2** 313/3–27 mult. of 3 314/*P*: 1–13 odd	**8-1** 356/*P*: 1, 6, 10–17 **S** 357/*Mixed Review*

LESSON	AVERAGE ALGEBRA	AVERAGE ALGEBRA AND TRIGONOMETRY	EXTENDED ALGEBRA	EXTENDED ALGEBRA AND TRIGONOMETRY
69	**6-2** 267/2–46 even **S** 263/28–34 even	*Administer Chapter 6 Test*	**7-2** 314/28–42 even 315/*P*: 6–14 even **S** 310/36–42 even	**8-2** 360/1–10; *P*: 1, 5, 10–13 **S** 355/14–24 even
70	**6-2** 268/48–66 even **6-3** 272/1–21 odd	**7-1** 309/1–35 odd **S** 310/*Mixed Review*	**7-3** 320/1–39 odd, 41, 42 **S** 321/*Mixed Review* **R** 316/*Self-Test 1*	**8-3** 366/2–32 even **S** 367/*Mixed Review* **R** 363/*Self-Test 1*
71	**6-3** 272/23–43 odd **S** 273/*Mixed Review* **6-4** 275/1–17 odd	**7-1** 310/20–38 even **7-2** 313/1–12	**7-4** 324/2–20 even **S** 321/43	**8-4** 370/1–23 odd **S** 356/*P*: 9, 19, 20
72	**6-4** 276/19–39 odd **S** 268/45–65 odd	**7-2** 314/14–36 even *P*: 1–13 odd	**7-4** 324/21–32 **R** 325/*Self-Test 2* **7-5** 331/1–25 odd	**8-5** 375/2–38 even **S** 376/*Mixed Review*
73	**6-5** 280/2–24 even **S** 276/41–44	**7-3** 320/1–39 odd **S** 321/*Mixed Review* **R** 316/*Self-Test 1*	**7-5** 332/28–36 even, 37, 38 **S** 332/*Mixed Review*	**8-6** 380/1–23 odd, 24–27 **S** 367/29, 31, 33
74	**6-5** 280/26–36 even **S** 282/*Mixed Review*	**7-4** 324/1–12 **S** 310/37, 39	**7-6** 336/1–39 odd **S** 332/27–35 odd	**8-7** 384/2–20 even, 21–26 **S** 385/*Mixed Review* **R** 381/*Self-Test 2*
75	**6-6** 286/1–37 odd **R** 282/*Self-Test 1*	**7-4** 324/13–26 **R** 325/*Self-Test 2* **7-5** 331/1–16	**7-6** 336/20–42 even, 43, 44	**8-8** 388/1–25 odd **S** 385/27
76	**6-7** 290/2–36 even **S** 291/*Mixed Review*	**7-5** 331/17–30 **S** 332/*Mixed Review*	**7-7** 342/1–37 odd 343/*P*: 1–7 odd	**8-9** 394/1–29 odd **R** 396/*Self-Test 3*
77	**6-7** 291/38–52 even **6-8** 295/1–11 odd	**7-6** 336/1–29 odd **S** 332/31–34	**7-7** 343/*P*: 8–15 **S** 345/*Mixed Review*	*Prepare for Chapter Test* **R** 397/*Chapter Review*
78	**6-8** 295/13–49 odd **S** 291/31–53 odd	**7-6** 336/20–40 even **S** 315/*P*: 6, 10, 12	*Prepare for Chapter Test* **R** 345/*Self-Test 3* **R** 346/*Chapter Review*	*Administer Chapter 8 Test* **9-1** *Read 9-1* 404/1–31 odd
79	*Prepare for Chapter Test* **R** 297/*Self-Test 2* **R** 302/*Chapter Review*	**7-7** 342/1–33 odd 343/*P*: 1–7 odd	*Administer Chapter 7 Test* **R** 348/*Cumulative Review*	**9-1** 405/*P*: 1, 3, 5, 7–12 **S** 406/*Mixed Review* **9-2** 410/2–24 even
80	*Administer Chapter 6 Test*	**7-7** 343/*P*: 2–14 even **S** 345/*Mixed Review*	**8-1** 354/1–23 odd	**9-2** 410/26–46 even, 50 **S** 405/*P*: 13
81	**7-1** 309/1–35 odd **S** 310/*Mixed Review*	*Prepare for Chapter Test* **R** 345/*Self-Test 3* **R** 346/*Chapter Review*	**8-1** 356/*P*: 1, 6, 10–17 **S** 357/*Mixed Review*	**9-3** 415/1–35 odd **S** 417/*Mixed Review*

LESSON	AVERAGE ALGEBRA	AVERAGE ALGEBRA AND TRIGONOMETRY	EXTENDED ALGEBRA	EXTENDED ALGEBRA AND TRIGONOMETRY
82	**7-1** 310/20–38 even **7-2** 313/1–12	*Administer Chapter 7 Test* **R** 348/*Cumulative Review*	**8-2** 360/1–10; *P*: 1, 5, 10–13 **S** 355/14–24 even	**9-4** 421/1–21 odd **R** 417/*Self-Test 1*
83	**7-2** 314/14–36 even *P*: 1–13 odd	**8-1** 354/1–19 odd	**8-3** 366/2–32 even **S** 367/*Mixed Review* **R** 363/*Self-Test 1*	**9-4** 422/22–34 even **9-5** 430/2–16 even
84	**7-3** 320/1–39 odd **S** 321/*Mixed Review* **R** 316/*Self-Test 1*	**8-1** 356/*P*: 1–19 odd **S** 357/*Mixed Review*	**8-4** 370/1–23 odd **S** 356/*P*: 9, 19, 20	**9-5** 430/17–32 **S** 431/*Mixed Review*
85	**7-4** 324/1–12 **S** 310/37, 39	**8-2** 360/1–10; *P*: 1, 5, 6, 10 **S** 355/14–20 even	**8-5** 375/2–38 even **S** 376/*Mixed Review*	**9-6** 434/1–11 odd, 13–21 **S** 416/36, 37
86	**7-4** 324/13–26 **R** 325/*Self-Test 2* **7-5** 331/1–16	**8-3** 366/2–30 even **S** 367/*Mixed Review* **R** 363/*Self-Test 1*	**8-6** 380/1–23 odd, 24–27 **S** 367/29, 31, 33	**9-7** 438/2–16 even, 17–19 **S** 435/*Self-Test 2*
87	**7-5** 331/17–30 **S** 332/*Mixed Review*	**8-4** 370/1–21 odd **S** 357/*P*: 14, 16, 18	**8-7** 384/2–20 even, 21–26 **S** 385/*Mixed Review* **R** 381/*Self-Test 2*	**9-8** 441/1–25 odd 442/*P*: 1–6
88	**7-6** 336/1–29 odd **S** 332/31–34	**8-5** 375/2–36 even **S** 376/*Mixed Review*	**8-8** 388/1–25 odd **S** 385/27	**9-8** 442/*P*: 7–13 **9-9** 447/1–17 odd
89	**7-6** 336/20–40 even **S** 315/*P*: 6, 10, 12	**8-6** 380/1–23 odd, 24, 25 **S** 367/29, 31	**8-9** 394/1–29 odd **R** 396/*Self-Test 3*	**9-9** 448/19–25 odd *P*: 1–9, 11 **S** 449/*Mixed Review*
90	**7-7** 342/1–33 odd 343/*P*: 1–7 odd	**8-7** 384/2–20 even, 21–25 odd **S** 385/*Mixed Review* **R** 381/*Self-Test 2*	*Prepare for Chapter Test* **R** 397/*Chapter Review*	*Prepare for Chapter Test* **R** 450/*Self-Test 3* **R** 451/*Chapter Review*
91	**7-7** 343/*P*: 2–14 even **S** 345/*Mixed Review*	**8-8** 388/1–23 odd **S** 380/26	*Administer Chapter 8 Test*	*Administer Chapter 9 Test*
92	*Prepare for Chapter Test* **R** 345/*Self-Test 3* **R** 346/*Chapter Review*	**8-9** 394/1–25 odd **R** 396/*Self-Test 3*	**9-1** 404/1–31 odd 405/*P*: 1, 3, 5	**10-1** 458/1–49 odd **S** 458/*Mixed Review*
93	*Administer Chapter 7 Test* **R** 348/*Cumulative Review*	*Prepare for Chapter Test* **R** 397/*Chapter Review*	**9-1** 405/*P*: 7–12 **S** 406/*Mixed Review* **9-2** 410/2–24 even	**10-2** 461/1–18, 19–39 odd
94	**8-1** 354/1–19 odd	*Administer Chapter 8 Test*	**9-2** 410/26–46 even, 50 **S** 405/*P*: 13	**10-3** 466/2–26 even **S** 467/*Mixed Review*

LESSON	AVERAGE ALGEBRA	AVERAGE ALGEBRA AND TRIGONOMETRY	EXTENDED ALGEBRA	EXTENDED ALGEBRA AND TRIGONOMETRY
95	**8-1** 356/*P*: 1–19 odd **S** 357/*Mixed Review*	**9-1** 404/1–31 odd 405/*P*: 1, 3, 5	**9-3** 415/1–35 odd **S** 417/*Mixed Review*	**10-4** 470/1–22 **S** 458/44–50 even **R** 467/*Self-Test 1*
96	**8-2** 360/1–10; *P*: 1, 5, 6, 10 **S** 355/14–20 even	**9-1** 405/*P*: 7–12 **S** 406/*Mixed Review* **9-2** 410/2–14 even	**9-4** 421/1–21 odd **R** 417/*Self-Test 1*	**10-4** 471/23–36 **S** 466/19–23 odd
97	**8-3** 366/2–30 even **S** 367/*Mixed Review* **R** 363/*Self-Test 1*	**9-2** 410/16–46 even	**9-4** 422/22–34 even **9-5** 430/2–16 even	**10-5** 476/1–28 **S** 471/37–43 odd
98	**8-4** 370/1–21 odd **S** 356/*P*: 14, 16, 18	**9-3** 415/2–36 even **S** 417/*Mixed Review*	**9-5** 430/17–32 **S** 431/*Mixed Review*	**10-5** 476/29–46 **S** 477/*Mixed Review*
99	**8-5** 375/2–36 even **S** 376/*Mixed Review*	**9-4** 421/1–21 odd **R** 417/*Self-Test 1*	**9-6** 434/1–11 odd, 13–21 **S** 416/36, 37	**10-6** 481/1–29 odd, 31–34 **R** 477/*Self-Test 2*
100	**8-6** 380/1–23 odd, 24, 25 **S** 367/29, 31	**9-4** 422/16–22 even, 24–33 **9-5** 430/1–6	**9-7** 438/2–18 even **R** 435/*Self-Test 2*	**10-6** 482/35–41 **10-7** 486/*P*: 1–9 odd
101	**8-7** 384/2–20 even, 21–25 odd **S** 385/*Mixed Review* **R** 381/*Self-Test 2*	**9-5** 430/7–31 odd **S** 431/*Mixed Review*	**9-8** 441/1–25 odd 442/*P*: 1–6	**10-7** 487/*P*: 10–22 even **S** 488/*Mixed Review*
102	**8-8** 388/1–23 odd **S** 380/26	**9-7** 438/2–18 even **R** 435/*Self-Test 2*: 1–4	**9-8** 442/*P*: 7–13 **9-9** 447/1–17 odd	**10-8** 490/3–48 mult. of 3 **R** 493/*Self-Test 3*
103	**8-9** 394/1–25 odd **R** 396/*Self-Test 3*	**9-8** 441/1–19 odd 442/*P*: 1–6	**9-9** 448/19–25 odd *P*: 1–9, 11 **S** 449/*Mixed Review*	*Prepare for Chapter Test* **R** 496/*Chapter Review*
104	*Prepare for Chapter Test* **R** 397/*Chapter Review*	**9-8** 442/*P*: 7–12 **9-9** 447/1–17 odd	*Prepare for Chapter Test* **R** 450/*Self-Test 3* **R** 451/*Chapter Review*	*Administer Chapter 10 Test* **R** 498/*Mixed Problem Solving*
105	*Administer Chapter 8 Test*	**9-9** 448/19, 21 *P*: 1–9, 11 **S** 449/*Mixed Review*	*Administer Chapter 9 Test*	**11-1** 504/1–33 odd **S** 506/*Mixed Review*
106	**9-1** 404/1–31 odd 405/*P*: 1, 3, 5	*Prepare for Chapter Test* **R** 450/*Self-Test 3* **R** 451/*Chapter Review*: 1–12, 16–20	**10-1** 458/1–35 odd 458/*Mixed Review*	**11-2** 509/3–33 mult. of 3 **11-3** 513/3–39 mult. of 3 **S** 515/*Mixed Review*

LESSON	AVERAGE ALGEBRA	AVERAGE ALGEBRA AND TRIGONOMETRY	EXTENDED ALGEBRA	EXTENDED ALGEBRA AND TRIGONOMETRY
107	**9-1** 405/*P*: 7–12 **S** 406/*Mixed Review* **9-2** 410/2–14 even	*Administer Chapter 9 Test*	**10-1** 458/37–50 **10-2** 461/1–10	**11-3** 514/*P*: 1, 4, 7, 9–12 **11-4** 521/2–20 even **R** 516/*Self-Test 1*
108	**9-2** 410/16–42 even	**10-1** 458/1–49 odd **S** 458/*Mixed Review*	**10-2** 462/11–18, 19–37 odd **S** 458/26–36 even	**11-4** 521/22–34 even 523/*Extra:* 1, 3, 5 **11-5** 527/2–20 even
109	**9-3** 415/2–24 even **S** 411/43–46	**10-2** 461/1–36	**10-3** 466/2–26 even **S** 467/*Mixed Review*	**11-5** 527/22–30 even 528/*P*: 1–4, 11–13 **S** 530/*Mixed Review*
110	**9-3** 416/26–36 even **S** 417/*Mixed Review* **9-4** 421/1–6	**10-3** 466/2–24 even **S** 467/*Mixed Review*	**10-4** 470/1–22 **S** 466/19, 21, 23 **R** 467/*Self-Test 1*	**11-6** 533/1–29 odd 534/*P*: 1–9 odd
111	**9-4** 421/7–21 odd, 24–31 **R** 417/*Self-Test 1*	**10-4** 470/1–22 **S** 458/44–50 even **R** 467/*Self-Test 1*	**10-4** 471/23–33, 35, 36 **S** 462/38, 39	**11-7** 539/2–12 even, 13–22 **S** 539/*Mixed Review* **R** 536/*Self-Test 2*
112	**9-5** 430/1–16 **S** 423/32, 33	**10-4** 471/23–33, 35, 36 **S** 466/19–23 odd	**10-5** 476/1–28 **S** 471/37–43 odd	**11-8** 542/1–13, 15–27 odd **S** 535/*P*: 11, 12
113	**9-5** 430/17–24, 27, 29, 30 **S** 431/*Mixed Review*	**10-5** 476/1–32 **S** 477/*Mixed Review*	**10-5** 476/29–46 **S** 477/*Mixed Review*	*Prepare for Chapter Test* **R** 543/*Self-Test 3* **R** 545/*Chapter Review*
114	**9-7** 438/2–18 even **R** 435/*Self-Test 2:* 1–4	**10-5** 476/33–42 **10-6** 481/1–27 odd	**10-6** 481/1–29 odd, 31–34 **R** 477/*Self-Test 2*	*Administer Chapter 11 Test* **12-1** *Read 12-1* 552/3–66 mult. of 3 **S** 554/*Mixed Review*
115	**9-8** 441/1–19 odd **S** 438/17, 19	**10-6** 482/29–38 **R** 477/*Self-Test 2* **10-7** 486/*P*: 1–9 odd	**10-6** 482/35–41 **10-7** 486/*P*: 1–9 odd	**12-2** 559/1–25 odd, 26–31, 33, 34
116	**9-8** 441/12–18 even 442/*P*: 1–10	**10-7** 487/*P*: 11–19 odd **S** 488/*Mixed Review*	**10-7** 487/*P*: 10–22 even **S** 488/*Mixed Review*	**12-3** 566/2–76 even **S** 567/*Mixed Review*
117	**9-9** 447/1–21 odd **S** 449/*Mixed Review*	*Prepare for Chapter Test* **R** 493/*Self-Test 3:* 1–4 **R** 496/*Chapter Review:* 1–14	**10-8** 490/3–48 mult. of 3 **R** 493/*Self-Test 3*	**12-4** 572/2–46 even **R** 573/*Self-Test 1*
118	**9-9** 448/20, 22 *P*: 1–9, 11	*Administer Chapter 10 Test* **R** 498/*Mixed Problem Solving*	*Prepare for Chapter Test* **R** 496/*Chapter Review*	**12-5** 577/3–21 mult. of 3 578/*P*: 2–16 even **S** 579/*Mixed Review*

LESSON	AVERAGE ALGEBRA	AVERAGE ALGEBRA AND TRIGONOMETRY	EXTENDED ALGEBRA	EXTENDED ALGEBRA AND TRIGONOMETRY
119	*Prepare for Chapter Test* **R** 450/*Self-Test 3* **R** 451/*Chapter Review:* 1–12, 16–20	**11-7** 539/1–20	*Administer Chapter 10 Test* **R** 498/*Mixed Problem Solving*	**12-6** 582/2–18 even 583/*P*: 2, 7, 8, 10 **S** 579/*P*: 15
120	*Administer Chapter 9 Test*	**11-8** 542/1–13, 15–27 odd **R** 543/*Self-Test 3*	**11-1** 504/1–22	**12-7** 588/1–19 odd, 20 589/*P*: 2, 6, 8–11 **S** 590/*Mixed Review*
121	**10-1** 458/1–35 odd **S** 458/*Mixed Review*	**12-1** 552/3–66 mult. of 3 **S** 554/*Mixed Review*	**11-1** 505/23–34 **S** 506/*Mixed Review*	**12-8** 594/1–19 odd **S** 584/*P*: 9 **S** 590/*P*: 13
122	**10-1** 458/37–50 **10-2** 461/1–10	**12-2** 559/1–31 odd **S** 553/47, 49, 50, 52, 53	**11-2** 509/1–33 odd	**12-8** 595/*P*: 1, 2, 6–10, 14 **S** 577/20, 22, 23
123	**10-2** 462/11–34 **S** 458/26–36 even	**12-3** 566/1–36 **S** 559/8–26 even	**11-3** 513/3–39 mult. of 3 514/*P*: 1–9 odd	**12-9** 599/1–16 **S** 600/*Mixed Review* **R** 600/*Self-Test 2*
124	**10-3** 466/2–24 even **S** 467/*Mixed Review*	**12-3** 566/37–67 **S** 567/*Mixed Review*	**11-3** 514/*P*: 10–13 **S** 515/*Mixed Review* **11-4** 521/2–20 even	*Prepare for Chapter Test* **R** 602/*Chapter Review*
125	**10-4** 470/1–22 **S** 462/35, 36 **R** 467/*Self-Test 1*	**12-4** 572/2–46 even **S** 567/68–76 even	**11-4** 521/22–34 even **R** 516/*Self-Test 1* 517/1–7 odd	*Administer Chapter 12 Test*
126	**10-4** 471/23–33, 35, 36 **S** 466/19–23 odd	**12-5** 577/1–19 odd 578/*P*: 1, 7 **R** 573/*Self-Test 1*	**11-5** 527/2–30 even **S** 530/*Mixed Review* 523/*Extra:* 1, 3, 5	**13-1** 610/3–60 mult. of 3 611/*P*: 1–11 odd **S** 612/*Mixed Review*
127	**10-5** 476/1–28 **S** 471/37, 38	**12-5** 578/*P*: 2–14 even **S** 579/*Mixed Review*	**11-5** 527/17–29 odd 528/*P*: 1–4, 11–13 **S** 524/2, 4, 7	**13-2** 617/1–27 odd, 28–30 **S** 612/*P*: 10, 12
128	**10-5** 476/29–42 **S** 477/*Mixed Review*	**12-6** 582/1–17 odd 583/*P*: 2, 7 **S** 578/*P*: 3, 5	**11-6** 533/1–29 odd **S** 529/*P*: 9, 10, 14	**13-3** 621/2–20 even, 21–26, 29 **S** 623/*Mixed Review*
129	**10-6** 481/1–28 **R** 477/*Self-Test 2*	**12-6** 583/*P*: 1, 3, 8, 10 **12-7** 588/2–12 even	**11-6** 533/14–28 even 534/*P*: 1–9 odd	**13-4** 628/7–12, 21–26, 33–38 **S** 623/27, 28
130	**10-6** 482/29–38 **10-7** 486/*P*: 1–9 odd	**12-7** 588/9–19 odd, 20 589/*P*: 2, 3, 5, 6, 8, 9 **S** 590/*Mixed Review*	**11-7** 539/2–12 even, 13–22 **S** 539/*Mixed Review* **R** 536/*Self-Test 2*	**13-4** 629/39–44 **13-5** 633/1–16

LESSON	AVERAGE ALGEBRA	AVERAGE ALGEBRA AND TRIGONOMETRY	EXTENDED ALGEBRA	EXTENDED ALGEBRA AND TRIGONOMETRY
131	**10-7** 487/*P*: 11–19 odd **S** 488/*Mixed Review*	**12-8** 594/1–8 595/*P*: 1–3 **S** 584/*P*: 9	**11-8** 542/1–13, 15–27 odd **S** 535/*P*: 11, 12	**13-5** 633/17–28 **S** 633/*Mixed Review*
132	*Prepare for Chapter Test* **R** 493/*Self-Test 3:* 1–4 **R** 496/*Chapter Review:* 1–14	**12-8** 594/9–13, 15 595/*P*: 6, 8, 10, 14 **S** 590/*P*: 7, 10, 11	*Prepare for Chapter Test* **R** 543/*Self-Test 3* **R** 545/*Chapter Review*	**13-6** 639/1–47 odd **R** 634/*Self-Test 1*
133	*Administer Chapter 10 Test* **R** 498/*Mixed Problem Solving*	**12-9** 599/1–16 **S** 600/*Mixed Review* **R** 600/*Self-Test 2*	*Administer Chapter 11 Test* **R** 547/*Cumulative Review*	**13-6** 640/24–56 even **S** 629/47, 48
134	**11-1** 504/1–22	*Prepare for Chapter Test* **R** 602/*Chapter Review*	**15-1** 711/1–17 **S** 712/*Mixed Review,* 4–6	**13-7** 643/1–33 odd **S** 645/*Mixed Review*
135	**11-1** 505/23–34 **S** 506/*Mixed Review*	*Administer Chapter 12 Test*	**15-2** 717/1–8, 9–19 odd **S** 712/18	**13-8** 649/2–40 even **S** 644/30, 34
136	**11-2** 509/1–31 odd	**13-1** 610/3–60 mult. of 3 611/*P*: 1–7 odd **S** 612/*Mixed Review*	**15-3** 722/1–11 **S** 723/*Mixed Review*	**13-9** 652/3–24 mult. of 3, 25–32 **S** 653/*Mixed Review*
137	**11-3** 513/1–33 odd 514/*P*: 1–6 517/1–7 odd	**13-2** 617/1–27 odd **S** 611/*P*: 2, 6	**15-4** 727/1–13 **S** 718/20	*Prepare for Chapter Test* **R** 654/*Self-Test 2* **R** 655/*Chapter Review*
138	**11-3** 514/*P*: 8–12 **S** 515/*Mixed Review* **11-4** 521/1–16	**13-3** 621/2–20 even **S** 623/*Mixed Review*	**15-5** 732/1–12, 14, 16 **S** 733/*Mixed Review* **R** 729/*Self-Test 1*	*Administer Chapter 13 Test*
139	**11-4** 521/17–32 **R** 516/*Self-Test 1*	**13-4** 628/1–6, 13–20, 27–32 **S** 622/11–19 odd	**15-6** 737/1–20 **S** 723/12	**14-1** 664/1–21 odd 665/*P*: 1–9 odd **S** 665/*Mixed Review*
140	**11-5** 527/2–28 even **S** 530/*Mixed Review*	**13-4** 628/7–12, 21–26, 33–38 **S** 612/*P*: 8	**15-6** 737/21–30 **15-7** 740/2–10 even, 11–16	**14-2** 669/1–14, 15–31 odd **S** 665/*P*: 6, 8
141	**11-5** 527/17–29 odd 528/*P*: 1–6, 9, 10	**13-5** 633/1–16	**15-7** 740/17–22, 23–27 odd **S** 741/*Mixed Review*	**14-3** 678/1–28 **R** 671/*Self-Test 1*
142	**11-6** 533/1–23 odd **S** 529/*P*: 7, 8, 11	**13-5** 633/17–22 **S** 633/*Mixed Review*	**15-8** 744/1–9 odd, 10–14 **R** 742/*Self-Test 2*	**14-3** 678/29–36, *Extra:* 1–3 **S** 679/*Mixed Review*
143	**11-6** 533/14–26 even 534/*P*: 1–3, 5, 7, 9	**13-6** 639/1–37 odd **R** 634/*Self-Test 1*	**15-9** 748/1–11 **S** 750/*Mixed Review*	**14-4** 683/2–30 even, 31 **S** 670/30, 32

LESSON	AVERAGE ALGEBRA	AVERAGE ALGEBRA AND TRIGONOMETRY	EXTENDED ALGEBRA	EXTENDED ALGEBRA AND TRIGONOMETRY
144	**11-7** 539/2–12 even, 13–20 **S** 539/*Mixed Review* **R** 536/*Self-Test 2*	**13-6** 640/26–56 even	**15-10** 759/1–8 **S** 750/13, 14	**14-5** 687/1–8, 10–22 even **S** 688/*Mixed Review*
145	**11-8** 542/1–13, 15–27 odd **S** 534/*P*: 8, 10	**13-7** 643/1–29 odd **S** 645/*Mixed Review*	**15-10** 760/9–13, 16, 17 **R** 761/*Self-Test 3*	**14-6** 692/2–16 even, 18–24 **S** 687/17, 19, 21 **R** 688/*Self-Test 2*
146	*Prepare for Chapter Test* **R** 543/*Self-Test 3* **R** 545/*Chapter Review*	**13-8** 649/2–36 even **S** 644/16–20 even, 31, 32	*Prepare for Chapter Test* **R** 764/*Chapter Review*	**14-7** 695/1–17 odd, 19–26 **S** 696/*Mixed Review*
147	*Administer Chapter 11 Test* **R** 547/*Cumulative Review*	**13-9** 652/1–27 odd **S** 653/*Mixed Review*	*Administer Chapter 15 Test*	**14-8** 699/1–27 odd **S** 696/27, 28
148	**15-1** 711/1–16 **S** 712/*Mixed Review:* 4–6	*Prepare for Chapter Test* **R** 654/*Self-Test 2* **R** 655/*Chapter Review*	**16-1** 769/1–20	**14-8** 700/14–28 even, 29–45 odd **R** 700/*Self-Test 3*
149	**15-2** 717/1–8, 9–19 odd	*Administer Chapter 13 Test*	**16-2** 773/1–18 **S** 769/21, 22	*Prepare for Chapter Test* **R** 703/*Chapter Review*
150	**15-3** 722/1–10 **S** 723/*Mixed Review*	**14-1** 664/1–15 odd 665/*P*: 2–8 even **S** 665/*Mixed Review*	**16-2** 773/19–21, 23 **16-3** 777/1–7	*Administer Chapter 14 Test* *Read 16-1* 769/1–14
151	**15-4** 727/1–12	**14-2** 669/1–14, 15–29 odd **S** 665/*P*: 7, 9	**16-3** 778/8–28 even **S** 778/*Mixed Review*	**16-1** 769/15–22 **S** 769/*Mixed Review* **16-2** 773/1–23 odd
152	**15-5** 732/1–10, 11–15 odd **S** 733/*Mixed Review* **R** 729/*Self-Test 1*	**14-3** 678/1–26 **S** 679/*Mixed Review* **R** 671/*Self-Test 1*	**16-4** 782/1–10 785/*Extra:* 1, 3, 5	**16-3** 777/1–25 odd **S** 778/*Mixed Review*
153	**15-6** 737/1–20	**14-4** 683/2–30 even **S** 678/27–30	**16-4** 783/11–17 **S** 778/29, 30	**16-4** 782/1–3, 11–17 785/*Extra:* 1, 3, 5
154	**15-6** 737/21–28 **15-7** 740/1–15 odd	**14-5** 687/1–8, 9–21 odd **S** 688/*Mixed Review*	**16-5** 789/1–15 **S** 789/*Mixed Review* **R** 784/*Self-Test 1*	**16-5** 789/1–15 **S** 789/*Mixed Review* **R** 784/*Self-Test 1*
155	**15-7** 740/17–25 odd **S** 741/*Mixed Review*	**14-6** 692/1–20 **S** 687/16, 18, 20 **R** 688/*Self-Test 2*	**16-6** 792/1–18 **S** 789/17	**16-6** 792/1–18 **S** 778/27, 29

LESSON	AVERAGE ALGEBRA	AVERAGE ALGEBRA AND TRIGONOMETRY	EXTENDED ALGEBRA	EXTENDED ALGEBRA AND TRIGONOMETRY
156	**15-8** 744/1–11 odd **R** 742/*Self-Test 2*	**14-7** 695/1–22 **S** 696/*Mixed Review*	**16-7** 796/1–17 **S** 797/*Mixed Review:* 1, 3–5 **R** 793/*Self-Test 2*	**16-7** 796/1–17 **S** 797/*Mixed Review* **R** 793/*Self-Test 2*
157	**15-9** 748/1–11 **S** 750/*Mixed Review*	**14-8** 699/1–27 odd **S** 696/23–26	**16-8** 800/1–14 **S** 797/18	**16-8** 800/1–14 **S** 797/18
158	**15-10** 759/1–15 odd **R** 761/*Self-Test 3*	**14-8** 700/14–28 even, 29–42 **R** 700/*Self-Test 3*	**16-9** 804/1–13 odd, 15, 16 **S** 805/*Mixed Review:* 1–3, 5, 6 **R** 805/*Self-Test 3*	**16-9** 804/1–13 odd, 15, 16 **S** 805/*Mixed Review* **R** 805/*Self-Test 3*
159	*Prepare for Chapter Test* **R** 764/*Chapter Review*	*Prepare for Chapter Test* **R** 703/*Chapter Review*	*Prepare for Chapter Test* **R** 806/*Chapter Review*	*Prepare for Chapter Test* **R** 806/*Chapter Review*
160	*Administer Chapter 15 Test*	*Administer Chapter 14 Test*	*Administer Chapter 16 Test*	*Administer Chapter 16 Test*

Supplementary Materials Guide

For use with Lesson	Practice Masters	Tests	Study Guide (Reteaching)	Resource Book		
				Tests	Practice Exercises	Mixed Review (MR) Prob. Solving (PS) Applications (A) Enrichment (E) Technology (T) Thinking Skl. (TS)
1-1			pp. 1–2	pp. 1–3		p. 204 (A) pp. 250–256 (T)
1-2	Sheet 1		pp. 3–4		p. 100	
1-3			pp. 5–6			pp. 260–261 (TS)
1-4	Sheet 2		pp. 7–8			
1-5			pp. 9–10			
1-6	Sheet 3	Test 1	pp. 11–12		p. 101	
1-7	Sheet 4		pp. 13–14			
1-8	Sheet 5		pp. 15–16		p. 102	
1-9	Sheet 6	Test 2	pp. 17–18			pp. 190–191 (PS)
Chapter 1		Tests 3, 4		pp. 4–7	p. 103	p. 220 (E)
2-1			pp. 19–20			p. 205 (A)
2-2	Sheet 7		pp. 21–22			
2-3	Sheet 8	Test 5	pp. 23–24		p. 104	pp. 192–193 (PS)
2-4			pp. 25–26			
2-5	Sheet 9		pp. 27–28		p. 105	pp. 250–256 (T)
2-6			pp. 29–30			
2-7	Sheet 10	Test 6	pp. 31–32		p. 106	
Chapter 2		Tests 7, 8		pp. 8–11	p. 107	p. 221 (E)
3-1	Sheet 11		pp. 33–34			pp. 194–195 (PS) p. 206 (A)
3-2	Sheet 12		pp. 35–36		p. 108	p. 262 (TS)
3-3			pp. 37–38			p. 236 (T)
3-4	Sheet 13	Test 9	pp. 39–40		p. 109	pp. 236–237 (T)
3-5			pp. 41–42			p. 238 (T)
3-6	Sheet 14		pp. 43–44			pp. 196–197 (PS)
3-7	Sheet 15	Test 10	pp. 45–46		p. 110	
3-8	Sheet 16		pp. 47–48			
3-9	Sheets 17, 18		pp. 49–50			
3-10	Sheet 19	Test 11	pp. 51–52		p. 111	
Chapter 3		Tests 12, 13		pp. 12–15	p. 112	pp. 178–179 (MR) p. 222 (E)
Cum. Rev. 1–3				pp. 16–18		
4-1			pp. 53–54			p. 207 (A)
4-2			pp. 55–56			
4-3	Sheet 20		pp. 57–58		p. 113	
4-4			pp. 59–60			
4-5	Sheet 21		pp. 61–62			
4-6	Sheet 22	Test 14	pp. 63–64		p. 114	
4-7	Sheet 23		pp. 65–66			
4-8	Sheet 24		pp. 67–68			pp. 198–199 (PS)
4-9	Sheet 25	Test 15	pp. 69–70		p. 115	
Chapter 4		Tests 16, 17		pp. 19–22	p. 116	p. 223 (E)
Cum. Rev. 1–4	Sheet 26				p. 117	

For use with Lesson	Practice Masters	Tests	Study Guide (Reteaching)	Resource Book		
				Tests	Practice Exercises	Mixed Review (MR) Prob. Solving (PS) Applications (A) Enrichment (E) Technology (T) Thinking Skl. (TS)
5-1			pp. 71–72			p. 208 (A)
5-2	Sheet 27		pp. 73–74			
5-3	Sheet 28		pp. 75–76		p. 118	
5-4	Sheet 29	Test 18	pp. 77–78			
5-5			pp. 79–80			
5-6	Sheet 30		pp. 81–82			
5-7	Sheet 31		pp. 83–84		p. 119	
5-8	Sheet 32		pp. 85–86			
5-9	Sheet 33	Test 19	pp. 87–88		p. 120	pp. 200–201 (PS)
Chapter 5		Tests 20, 21		pp. 23–26	p. 121	p. 224 (E)
6-1			pp. 89–90			p. 209 (A)
6-2	Sheet 34		pp. 91–92		p. 122	
6-3			pp. 93–94			
6-4			pp. 95–96			
6-5	Sheet 35	Test 22	pp. 97–98		p. 123	
6-6	Sheet 36		pp. 99–100		p. 124	
6-7			pp. 101–102			
6-8	Sheets 37, 38	Test 23	pp. 103–104		p. 125	
Chapter 6		Tests 24, 25		pp. 27–30	p. 126	p. 225 (E)
Cum. Rev. 1–6		Test 26				
7-1			pp. 105–106			p. 210 (A)
7-2	Sheet 39		pp. 107–108		p. 127	
7-3			pp. 109–110			
7-4	Sheet 40	Test 27	pp. 111–112		p. 128	
7-5	Sheet 41		pp. 113–114			pp. 240–241 (T)
7-6	Sheet 42		pp. 115–116			pp. 263–264 (TS)
7-7	Sheet 43	Test 28	pp. 117–118		p. 129	
Chapter 7		Tests 29, 30		pp. 31–34	p. 130	pp. 180–182 (MR) p. 226 (E)
Cum. Rev. 4–7				pp. 35–37		
Cum. Rev. 1–7				pp. 38–45		
8-1			pp. 119–120			pp. 202–203 (PS) p. 211 (A)
8-2	Sheet 44		pp. 121–122		p. 131	
8-3	Sheet 45		pp. 123–124			
8-4			pp. 125–126			
8-5	Sheet 46	Test 31	pp. 127–128		p. 132	
8-6	Sheet 47		pp. 129–130			
8-7			pp. 131–132		p. 133	
8-8			pp. 133–134			p. 242 (T)
8-9	Sheet 48	Test 32	pp. 135–136		p. 134	
Chapter 8		Tests 33, 34		pp. 46–49	p. 135	p. 227 (E)
Cum. Rev. 5–8	Sheet 49					
Cum. Rev. 1–8					pp. 136–137	

For use with Lesson	Practice Masters	Tests	Study Guide (Reteaching)	Resource Book		
				Tests	Practice Exercises	Mixed Review (MR) Prob. Solving (PS) Applications (A) Enrichment (E) Technology (T) Thinking Skl. (TS)
9-1	Sheet 50		pp. 137–138			p. 212 (A)
9-2			pp. 139–140		p. 138	
9-3	Sheet 51	Test 35	pp. 141–142			
9-4			pp. 143–144			
9-5			pp. 145–146		p. 139	
9-6	Sheet 52	Test 36	pp. 147–148			p. 243 (T)
9-7			pp. 149–150			pp. 243–244 (T)
9-8	Sheet 53		pp. 151–152			
9-9	Sheet 54	Test 37	pp. 153–154		p. 140	p. 265 (TS)
Chapter 9		Tests 38, 39		pp. 50–53	p. 141	p. 228 (E)
10-1			pp. 155–156			p. 213 (A)
10-2	Sheet 55		pp. 157–158		p. 142	p. 245 (T)
10-3	Sheet 56		pp. 159–160		p. 143	p. 245 (T)
10-4			pp. 161–162			
10-5	Sheet 57	Test 40	pp. 163–164		p. 144	
10-6			pp. 165–166			
10-7			pp. 167–168			
10-8	Sheet 58	Test 41	pp. 169–170		p. 145	
Chapter 10		Tests 42, 43		pp. 54–57	p. 146	p. 229 (E)
11-1	Sheet 59		pp. 171–172			p. 214 (A)
11-2			pp. 173–174		p. 147	
11-3	Sheet 60	Test 44	pp. 175–176			
11-4			pp. 177–178			
11-5	Sheet 61		pp. 179–180		p. 148	pp. 266–267 (TS)
11-6	Sheet 62	Test 45	pp. 181–182		p. 149	
11-7			pp. 183–184			
11-8	Sheet 63	Test 46	pp. 185–186			
Chapter 11		Tests 47, 48		pp. 58–61	p. 150	p. 230 (E)
Cum. Rev. 1–11						pp. 183–185 (MR)
Cum. Rev. 7–11		Test 49				
Cum. Rev. 8–11				pp. 62–64		
12-1			pp. 187–188		p. 151	p. 215 (A)
12-2	Sheet 64		pp. 189–190		p. 152	
12-3	Sheet 65		pp. 191–192			
12-4	Sheet 66	Test 50	pp. 193–194			
12-5	Sheet 67		pp. 195–196			
12-6	Sheet 68		pp. 197–198			
12-7	Sheet 69		pp. 199–200		p. 153	
12-8	Sheet 70		pp. 201–202		p. 154	
12-9	Sheet 71	Test 51	pp. 203–204			
Chapter 12		Tests 52, 53		pp. 65–68	p. 155	p. 231 (E)
Cum. Rev. 9–12	Sheet 72				pp. 156–157	

For use with Lesson	Practice Masters	Tests	Study Guide (Reteaching)	Resource Book		
				Tests	Practice Exercises	Mixed Review (MR) Prob. Solving (PS) Applications (A) Enrichment (E) Technology (T) Thinking Skl. (TS)
13-1			pp. 205–206			p. 216 (A)
13-2	Sheet 73		pp. 207–208			
13-3			pp. 209–210		p. 158	
13-4	Sheet 74		pp. 211–212			pp. 246–247 (T)
13-5	Sheet 75	Test 54	pp. 213–214		p. 159	pp. 248–249 (T)
13-6			pp. 215–216		p. 160	
13-7	Sheet 76		pp. 217–218			
13-8			pp. 219–220			
13-9	Sheet 77	Test 55	pp. 221–222		p. 161	
Chapter 13		Tests 56, 57		pp. 69–72	p. 162	p. 232 (E)
14-1			pp. 223–224			p. 217 (A)
14-2	Sheet 78		pp. 225–226		p. 163	
14-3			pp. 227–228			
14-4			pp. 229–230			
14-5	Sheet 79	Test 58	pp. 231–232		p. 164	
14-6			pp. 233–234			
14-7	Sheet 80		pp. 235–236			
14-8	Sheet 81	Test 59	pp. 237–238		p. 165	
Chapter 14		Tests 60, 61		pp. 73–76	p. 166	p. 233 (E)
Cum. Rev. 12–14		Test 62				
15-1			pp. 239–240			p. 218 (A)
15-2	Sheet 82		pp. 241–242			p. 268 (TS)
15-3			pp. 243–244		p. 167	
15-4	Sheet 83	Test 63	pp. 245–246			
15-5			pp. 247–248		p. 168	
15-6	Sheet 84		pp. 249–250			
15-7	Sheet 85	Test 64	pp. 251–252		p. 169	
15-8			pp. 253–254			
15-9	Sheet 86		pp. 255–256			
15-10	Sheet 87	Test 65	pp. 257–258		p. 170	
Chapter 15		Tests 66, 67		pp. 77–80	p. 171	p. 234 (E)
16-1			pp. 259–260			p. 219 (A)
16-2	Sheet 88		pp. 261–262			
16-3	Sheet 89		pp. 263–264			
16-4	Sheet 90	Test 68	pp. 265–266		p. 172	
16-5			pp. 267–268			
16-6	Sheet 91		pp. 269–270		p. 173	
16-7			pp. 271–272			
16-8	Sheet 92		pp. 273–274			
16-9	Sheet 93	Test 69	pp. 275–276		p. 174	
Chapter 16		Tests 70, 71		pp. 81–84	p. 175	p. 235 (E)
Cum. Rev. 13–16	Sheet 94					
Cum. Rev. 12–16				pp. 85–87		
Cum. Rev. 9–16					pp. 176–177	
Cum. Rev. 8–16				pp. 88–91		
Cum. Rev. 1–16	Sheets 95–97			pp. 92–99		pp. 186–189 (MR)

Software

The chart correlates textbook pages in *ALGEBRA AND TRIGONOMETRY, Structure and Method, Book 2,* with the supporting menu items on the disk *Algebra Plotter Plus*. Suggestions for using the disk for classroom demonstration or independent use are found in the side-column material of this Teacher's Edition under the heading ''Using a Computer or a Graphing Calculator'' and in the booklet that accompanies the disk, *Using Algebra Plotter Plus*.

The chart also correlates the textbook with the *Computer Activities* booklet and optional disk. A more detailed correlation appears in the booklet itself.

For use with Chapter	Disk for algebra: Algebra Plotter Plus		Computer Activities
	Menu Item	**Suggested use**	
1			Activities 1–2
2			Activities 3–4
3	Line Plotter Line Quiz Inequality Plotter	pp. 110, 114, 119, 125 p. 119 pp. 135, 160	Activities 5–7
4			Activities 8–9
5	Function Plotter	p. 231	Activities 10–12
6			Activities 13–14
7	Parabola Plotter Parabola Quiz	pp. 327, 336 p. 327	Activities 5, 15–16, 37
8	Function Plotter	p. 386	Activities 17–19, 37
9	Conics Plotter Function Plotter Conics Quiz	pp. 410, 416, 421, 430, 435, 436, 441 pp. 410, 416, 421, 430, 441 p. 435	Activities 20–21, 37
10	Function Plotter	pp. 460, 465, 471	Activities 22–24, 37
11			Activities 25–26
12			Activities 27–28
13	Function Plotter Circular Function Quiz	pp. 625, 630 pp. 629, 633	Activities 29–30, 37
14			Activity 31
15	Statistics Spreadsheet Sampling Experiment	pp. 711, 716, 727 pp. 752–753	Activities 32–34
16	Matrix Reducer	p. 786	Activities 35–36

Test Generator

This software, available for Apple II and IBM computers, is for generating your own tests. There are 21 test items for every lesson of *ALGEBRA, Structure and Method, Book 2*. To generate a test, you can manually choose the items or choose by random selection, and customize the content and format of your test.

The Test Generator package includes Program Disks, Data Disks, and a Test Bank with User's Guide booklet. The Test Bank includes all the test items available in the Test Generator.

Lesson Commentary

1 Basic Concepts of Algebra

In this chapter students review basic concepts and skills of algebra studied in previous courses. This review includes real numbers and expressions, operations with real numbers, and problem solving. Emphasis is placed on dealing with real numbers symbolically and in the context of word problems. Since all of this material is subject matter that students have seen before, it might be wise to use diagnostic testing throughout the chapter to assess an appropriate pace for teaching and learning.

1-1 (pages 1–5)

Key Mathematical Ideas

- graphing real numbers on a number line
- comparing real numbers
- finding absolute value

Teaching Suggestions

It is important for students to identify various subsets of real numbers and to give examples of each type of number. The relationship among these special subsets is also important. A diagram showing the nesting relationships might help students remember the classifications.

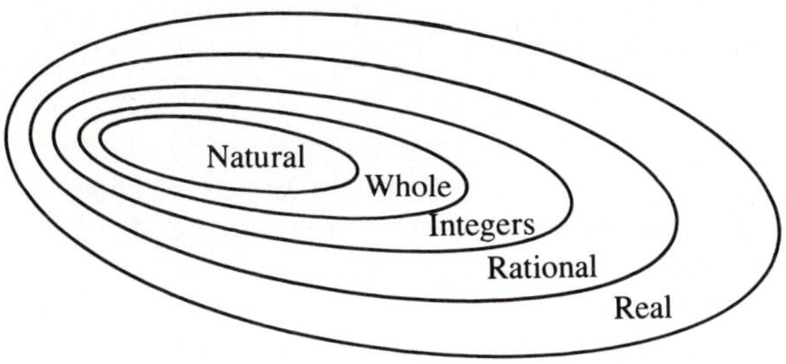

You might test student ability to compare numbers and list numbers in order by giving a list containing both positive and negative fractions and decimals.
In teaching absolute value you might also give the following definition: For any real number x, $|x| = \max\{x, -x\}$ where *max* means *maximum of*. See if students can see from this definition that the absolute value of a number is always positive or zero.
You might prepare students for problems involving absolute value to be encountered in Chapter 2 by introducing the concept of the distance from x_1 to x_2 as $d = |x_1 - x_2|$.

Suggested Extensions

Draw two number lines on the board as shown.

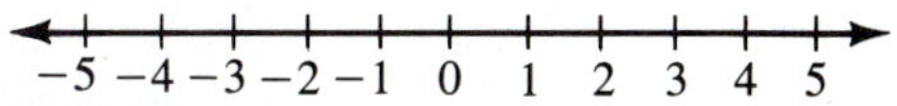

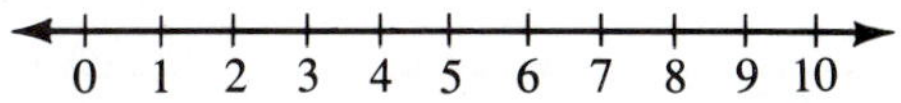

Suggest, using ordered pairs, a matching of numbers from each number line to the other number line so that the pairs are (0, 0), (1, 1), etc. Note that all the numbers on one line are not "used up." Have students try to find another matching so that *all* of the numbers on both lines are used up. (0, 0), (−1, 1), (1, 2) . . .

You might give students the following list to order from least to greatest: $10, -8.5, |-5.9|, \frac{108}{12}, -\frac{1}{5}, |6.2 - 4.8|$. $-8.5, -\frac{1}{5}, |6.2 - 4.8|, |-5.9|, \frac{108}{12}, 10$

1-2 (Pages 6–12)

Key Mathematical Ideas

- simplifying numerical expressions
- evaluating algebraic expressions

Teaching Suggestions

Be sure that the students understand all of the words and symbols in the first chart. Have the students simplify expressions that do not have grouping symbols to verify that they understand the order of operations. Have them insert the correct grouping symbols in the problem to show how the order of operations is set up and then have them change the grouping to see how many different "answers" they can get. Emphasize that evaluating an algebraic expression means finding the numerical *value* of the expression, given *values* for the variables.

You might work Oral Exercises 7–12 with students to make sure that students make the right choices in carrying out the agreement regarding order of operations. Distinguish between $-x^2$ and $(-x)^2$. In $-x^2$ the multiplication of x by x is done first. In $(-x)^2$ the additive inverse of x is taken first as indicated by the parentheses.

You might also direct student attention to the Calculator Key-In for further practice in performing the correct order of operations.

Avoid ambiguous expressions such as $10 \div 2x$. Use $10 \div (2x)$ or $(10 \div 2)x$ to be sure that the intended grouping is clear.

Suggested Extensions

Have the students use four 4's to write expressions for each integer from 1 to 10. Standard operation symbols may be used, including $\sqrt{}$.

$1 = \frac{4 + 4}{4 + 4}$ $2 = \frac{4}{4} + \frac{4}{4}$ $3 = \frac{4 + 4 + 4}{4}$

$4 = \sqrt{(4)(4)(4)} - 4$ $5 = \frac{(4)(4) + 4}{4}$ $6 = \frac{4 + 4}{4} + 4$

$7 = (4 + 4) - \frac{4}{4}$ $8 = \frac{4(4 + 4)}{4}$ $9 = 4 + 4 + \frac{4}{4}$

$10 = 4 + 4 + 4 - \sqrt{4}$

1-3 (pages 14–20)

Key Mathematical Ideas

- naming the properties of equality for real numbers
- using the properties of addition and multiplication of real numbers

Teaching Suggestions

Emphasize that the properties are the laws of algebra that *permit* or *allow* you to perform the operations needed to simplify expressions. It is important that the students not only be able to perform the operations but also be able to identify the properties used in performing those operations.

Reading Algebra

This lesson, like the preceding one, reviews many important concepts in a condensed format. It will be well worth your while to spend some time in discussing the properties reviewed on pages 14 and 15 to make sure that students have read them with understanding and that they know how to apply them.

You may wish to have the students translate the properties into words. For example, the Addition Property can be stated as follows: "If the same number is added to two equal numbers, the resulting sums will be equal."

Suggested Extensions

Have the students determine if the "clock" arithmetic defined by the following table satisfies all of the properties for addition.

+	1	2	3	4
1	2	3	4	1
2	3	4	1	2
3	4	1	2	3
4	1	2	3	4

One solution is to go through every property of addition for each numerical case. Here is a shorter solution.

1. The table only has members from {1, 2, 3, 4}. So {1, 2, 3, 4} is closed under addition.
2. Addition is commutative because the table is symmetric about the diagonal from the upper left part of the table to the lower right part of the table.
3. Addition is associative but each case must be checked.

4. The identity is 4. See the last row and the last column of the table.
5. Each member has an inverse.
   $1 + 3 = 4 \quad 3 + 1 = 4$
   $2 + 2 = 4 \quad 4 + 4 = 4$

1-4 (pages 21–25)

Key Mathematical Ideas

- adding and subtracting real numbers

Teaching Suggestions

Remind the students of the general definition for absolute value. Use the number line to illustrate the four rules of addition of real numbers. Emphasize strongly that subtraction is *defined* in terms of addition and that therefore there is no need for more rules for subtraction.

You might begin the class with a diagnostic test with questions taken from the Oral Exercises to see how much students remember from any previous course. Pay particular attention to students when they solve subtraction problems. It seems that for many students the operation of subtraction is the hardest to learn.

There are two approaches you might consider in motivating addition involving negative numbers. Ask students to imagine a person having \$5 in hand and an \$8 debt to pay or ask students to find the missing addend in $4 + (?) = 1$.

Reading Algebra

Encourage students to study the Reading Algebra feature on page 26 of the textbook. It suggests some techniques that will be helpful to students throughout the course.

Suggested Extensions

Since subtraction is just adding the opposite, it would seem that subtraction would satisfy all of the properties of addition. For example, for all a and b, $a + (-b) = (-b) + a$. However, $a - b \neq b - a$ necessarily. Have students give counterexamples to show that $a + (-b) \neq b + (-a)$ and $a + (-b) \neq (-a) + b$.
Let $a = 4$ and $b = 3$, then $4 + (-3) = 1$ and $3 + (-4) = -1$ and $(-4) + 3 = -1$.

1-5 (pages 27–32)

Key Mathematical Ideas

- multiplying real numbers

Teaching Suggestions

Emphasize the importance of the properties of -1. Correct handling of a negative sign before a product or before a sum is important. In particular, $-(ab) \neq (-a)(-b)$ but $-(a + b) = (-a) + (-b)$ and $-(a - b) = (-a) - (-b)$. Be sure that students do not confuse the properties. Ask students to try to substitute numbers for a and b in each property to see what the correct statements are. In this lesson you may want to use diagnostic testing to see if students recall correctly the rules regarding the sign of a product.

Point out the possible use of mental arithmetic in simplifying. In Example 1(b), group 3 and $-\frac{1}{3}$ to give a product of -1.

Suggested Extensions

1. Have the students prove that multiplication is distributive with respect to subtraction. For all real numbers a, b, and c,

$$a(b - c) = ab - ac.$$

STATEMENTS	REASONS
1. $a(b - c) = a(b + (-c))$	1. Def. of subtraction
2. $= ab + a(-c)$	2. Distributive prop.
3. $= ab + a((-1)c)$	3. Mult. prop. of -1
4. $= ab + (a(-1))c$	4. Assoc. prop. for mult.
5. $= ab + ((-1)a)c$	5. Comm. prop. for mult.
6. $= ab + (-1)(ac)$	6. Assoc. prop. for mult.
7. $= ab + (-(ac))$	7. Mult. prop. of -1
8. $= ab - ac$	8. Def. of subtraction
9. $\therefore a(b - c) = ab - ac$	9. Trans. prop. of eq.

Similarly $(b - c)a = ba - ca$.

2. Ask students to compute the product $[(-1)(2)](-3)$ on a number line.

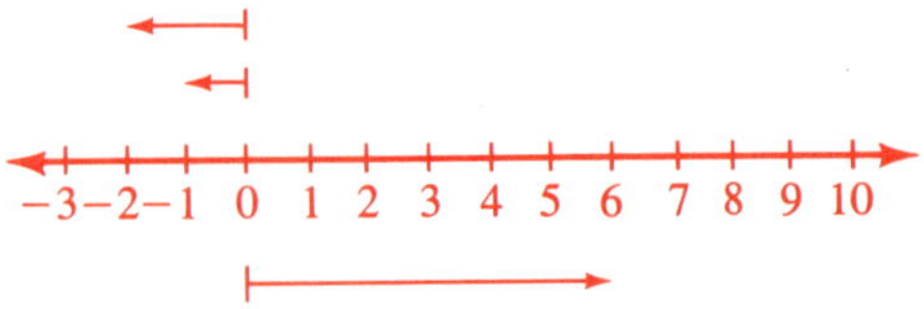

1-6 (pages 33–35)

Key Mathematical Ideas

- dividing real numbers

Teaching Suggestions

Compare the definition of division to the definition of subtraction. Since division is actually multiplication, the rules of signs remain the same. Students who have mastered multiplication should not have a difficult time mastering and reviewing division. This lesson also provides the opportunity for more practice in handling operations in the correct order.

Suggested Extensions

Have the students investigate division by zero, using a calculator. Try computing $\frac{1}{0.1}$, $\frac{1}{0.01}$, $\frac{1}{0.001}$, and so forth.

10, 100, 1000, . . . The smaller the denominator the larger the value of the fraction.

1-7 (pages 37–42)

Key Mathematical Ideas

- solving equations in one variable

Teaching Suggestions

Transformations that produce equivalent equations are used in solving any equation. They are the rules that *let* you simplify expressions. Emphasize that it is important to check the solution of an equation by substitution. Also show students what happens when solving an equation in which there is no solution and another equation in which the solution is the set of all real numbers. Give several examples. Since students sometimes have trouble relating the solving of an equation to the solving of a formula, demonstrate with several different types of formulas.

You might see how different students take alternate routes to get the same solution. Point out that sometimes there are several choices one could make to obtain a solution. Students should learn that some choices are more advantageous than other choices. Example 3 provides some early experience in handling word problems, a topic more fully discussed in Lessons 1-8 and 1-9.

Notice that the solutions to the equations are given in set notation. The notion that there may be more than one solution is emphasized by consistent references to a solution set.

Suggested Extensions

Have students solve a formula like $m = \frac{y_2 - y_1}{x_2 - x_1}$ for *each* variable without actually having values for the other variables. See also Exercises 31–42.

Example: $y_2 = m(x_2 - x_1) + y_1$

1-8 (pages 43–48)

Key Mathematical Ideas

- translating word phrases into mathematical expressions
- translating word sentences into equations

Teaching Suggestions

At this point you may need to slow the pace of the course since students often have difficulties in translating words into symbols. Algebra is a language and translating English into algebra is like translating English into a language like French or Spanish.

In preparing to solve word problems, it is important that the students first understand how to translate the written problem into algebraic form. This seems to be a problem for many uncertain and anxious students. Since there are so many different types of problems involved, some time needs to be taken to go over the "how to's" of setting up several types of problems. If students continue to have difficulty in translating words into symbols, you might try some of the Oral Exercises with numbers in place of variables first. For example, "Nine less than a number." would become "Nine less than 12.", "Nine less than 20.", and so forth.

Suggested Extensions

Give the students a group of algebraic expressions, equations, and formulas and have them write out word phrases or word sentences to match them. For example,

$2x + 5$	five more than twice a number
$3a - 14 = 7$	Seven is fourteen less than three times a number.

1-9 (pages 49–54)

Key Mathematical Ideas

- solving word problems

Teaching Suggestions

Notice that the entire lesson is devoted to examples of how to solve word problems. It is important that you demonstrate and explain carefully the steps in the plan for solving a word problem. Be sure that the students understand that the only way that they can learn to solve word problems is to do them themselves. Point out that not every problem has a solution even if the equation formed does have a solution. Encourage students to check their answers in the word problem before reporting their answers. Discuss the problem solving strategies on page T53 with the students.

Suggested Extensions

Have students identify unnecessary information in the following problems.

John and Sue drove in opposite directions from St. Louis. Sue averaged 52 mi/h and used 14 gal of gas. John averaged 57 mi/h and used 10 gal of gas. After how many hours were they 300 mi apart?
Sue used 14 gal of gas and John used 10 gal of gas.

Wilma has \$15 more than twice as much money as Sam. She has one half of her money in a savings account. Together, they have \$150. How much money does each have?
She has one half of her money in a savings account.

2 Inequalities and Proof

This chapter extends, with some modification, basic techniques for solving equations to solving inequalities. This chapter includes methods for solving inequalities involving conjunction and disjunction and methods for solving inequalities involving absolute value. Absolute value problems are solved both algebraically and graphically. Problem solving is included in this chapter as it was in Chapter 1 to reinforce and strengthen problem solving skills. The last two lessons of the chapter deal with proving theorems. Depending on the track your course is taking, you might consider these two lessons to be optional. Some of the properties of real numbers from Chapter 1 are proved to provide examples of methods of proof. Those methods are then used to extend, through proof, the theorems involving inequalities and absolute value.

2-1 (pages 59–64)

Key Mathematical Ideas

- solving inequalities in one variable

Teaching Suggestions

Show several examples of simple inequalities, such as the ones in the text, to be sure the students understand how to express the statements graphically. Point out at this time that $x < 2$ and $2 > x$ are equivalent statements. This will help later in solving inequalities. Discuss each of the properties of order, pointing out:

1. that the comparison property, although simple and logical, is a very important concept in dealing with inequalities,
2. that the transitive and addition properties of order are basically the same as the transitive and addition properties of equality,
3. that the multiplication property has one very important difference when multiplying by a negative number, and
4. that since subtraction and division are defined in terms of addition and multiplication, there is no need for a subtraction or division property of order.

Point out that the transformations that produce equivalent inequalities are the same as those for equalities and that applying them should be easy as long as the special properties for multiplication and division are remembered. Be sure that the students understand that in those inequalities that are true for all real numbers and in those that have no solution in the real numbers, transforming the inequality will result in the variable "going away" leaving a true sentence for "all real numbers" and a false sentence for "no solution."

Since many inequalities do have solutions, we begin by assuming that the inequality has a solution and proceed to find it. By logically applying the properties of inequalities we determine whether or not the assumption is justified.

Point out the different ways of writing a solution set for inequalities. In Example 1, the solution set can be written {real numbers less than -3}, but $\{x: x < -3\}$ would be more compact. On the other hand, in Example 2, the solution {real numbers} is more compact than $\{t: t \text{ is a real number}\}$.

Suggested Extensions

Although it is stated in the text that ">" may be substituted for "<" in the properties of order, many students have a hard time "seeing" this. Have the students rewrite the properties of order exactly as they are in the text except changing the "<" to ">."

2-2 (pages 65–68)

Key Mathematical Ideas

- solving combined inequalities

Teaching Suggestions

Review the meaning of "conjunction" and "disjunction." Point out that the compact form of the *and* statement corresponds to the logical order on the number line. For example, $y > -3$ and $y < 1$ should be thought of as $-3 < y$ and $y < 1$ and written as $-3 < y < 1$ in compact form. Explain that the *or* statement cannot correctly be written in compact form. For example, $x < -3$ or $x > 4$ could not logically be written $-3 > x > 4$ because x is not between -3 and 4 and the "flow" implies that $-3 > 4$. Also, show that it would not be logical to write $-3 > x < 4$ or $-3 < x > 4$.

Suggested Extensions

Ask students to give an inequality for each graph.

1. 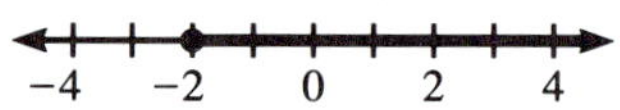

$x \geq -2$

2. 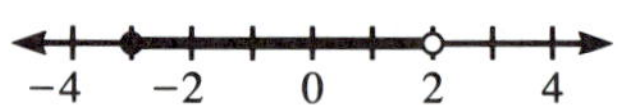

$-3 \leq x < 2$

3. 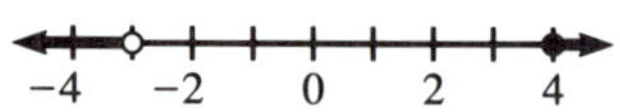

$x < -3$ or $x \geq 4$

2-3 (pages 69–72)

Key Mathematical Ideas

- solving problems with inequalities

Teaching Suggestions

Refer again to the steps in the plan for solving a word problem in Lesson 1-9. Go over the steps with the students and point out how the steps are used in solving the examples given in the text. Be sure to explain the mathematical meaning of expressions such as "at most," "at least," "smallest possible value," and "maximum number." Have the students translate several statements that include these expressions before they start on the assignment. The five-step plan for solving problems does apply, but the open sentence is an inequality in this section. Point out, therefore, that unlike problems yielding equations, problems involving inequalities might have many solutions rather than just one.

Stress to the students that the "check" step is very important in solving problems involving inequalities. This is illustrated in Example 2 where you must verify that no other set of four consecutive integers will satisfy the given conditions.

Reading Algebra

All work with word problems, of course, calls on students' reading skills as well as on their mathematical skills. The problem situations in this lesson and in many other lessons come from several different fields of activity, each of which has its own vocabulary. As pointed out in the Teaching Suggestions, you will want to make sure that your students understand the mathematical words associated with inequalities. Before assigning homework, you may also want to ask students to look quickly through the problems for other words that may be unfamiliar to them.

Suggested Extensions

In many classes the students are not required to give the solution to a problem in as much detail as shown in Examples 1 and 2. Have the students "write up" at least two different types of problems showing every step involved.

2-4 (pages 73–75)

Key Mathematical Ideas

- solving equations and inequalities involving absolute value

Teaching Suggestions

Using the simple sentences at the beginning of the lesson, explain the reason for translating $|x| = 1$ as $x = -1$ or $x = 1$, $|x| > 1$ as $x < -1$ or $x > 1$, and $|x| < 1$ as $-1 < x < 1$. Many students have a hard time "seeing" the relationships and understanding why they are used. Show a sufficient number of examples to demonstrate the different situations that can exist, especially with inequalities. Note that the student must isolate the absolute value expression on one side before changing to an equivalent conjunction or disjunction in problems like Example 3. You might write on the chalkboard the following

$$|\text{Expression}| = \text{Number}$$

gives

$$\text{Expression} = \text{Number}$$
$$\text{or}$$
$$\text{Expression} = -\text{Number}$$

to get across the idea that whatever is enclosed in the absolute value symbol is set equal to the right side or the opposite of the right side.

Suggested Extensions

Other than graphically, consider three ways to solve absolute value problems. Given: $|x + 4| = 6$

1. Intuitively, since $x + 4$ represents a number whose absolute value is 6, the number $x + 4$ equals 6 or -6. Therefore,

$$x + 4 = 6 \quad \text{or} \quad x + 4 = -6$$
$$x = 2 \quad | \quad x = -10$$

2. Algebraically, using the definition of absolute value,

$$|x + 4| = x + 4, \text{ when } x + 4 > 0 \text{ or}$$
$$|x + 4| = -(x + 4), \text{ when } x + 4 < 0.$$

Therefore: $|x + 4| = 6$

$$x + 4 = 6 \quad \text{or} \quad -(x + 4) = 6$$
$$x + 4 = -6$$
$$x = 2 \quad \text{or} \quad x = -10$$

3. Using the square root definition of absolute value ($|x| = \sqrt{x^2}$):

$$|x + 4| = \sqrt{(x + 4)^2}$$

Therefore: $|x + 4| = \sqrt{(x + 4)^2} = 6$

$$(x + 4)^2 = 6^2$$
$$x^2 + 8x + 16 = 36$$
$$x^2 + 8x - 20 = 0$$
$$(x + 10)(x - 2) = 0$$

$$x + 10 = 0 \quad \text{or} \quad x - 2 = 0$$
$$x = -10 \qquad x = 2$$

Have students try some problems using these methods. Ask students to see if they will work for inequalities involving < or >.

Reading Algebra

This lesson provides an excellent opportunity for you to point out the close relationship between diagrams and text and to emphasize to students that they should read the diagrams along with the text and the examples.

Suggested Extensions

To solve the absolute value problem, $|x + 4| = 6$, think "what numbers are 6 units from 0 on the number line?" Place $x + 4$ at those points.

$x + 4 = -6$ $\qquad$ $x + 4 = 6$

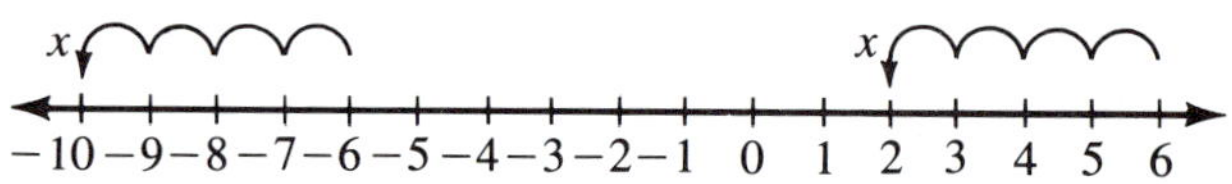

If $x + 4$ is at -6, then x is 4 more units away from 0. Therefore $x = -10$.

If $x + 4$ is at 6, then x is 4 more units closer to 0. Therefore $x = 2$.

See if this idea can be applied to inequalities.

2-5 (pages 76–79)

Key Mathematical Ideas

- graphically solving equations and inequalities involving absolute value

Teaching Suggestions

Illustrate the concept that the absolute value is used to find the distance between two points on a number line by showing several examples. Be sure that students understand that an expression like $|x - 5| = 3$ means that the distance that x is from 5 is equal to 3 and that an expression like $|x + 5| = 3$ can be changed to $|x - (-5)| = 3$ so that it will conform to the *absolute value of the difference* form that is being used.

In Oral Exercises 11–16, point out the different translations of the symbols. For example, the solution in the Sample could have used the expression "not less than 3."

2-6 (pages 81–87)

Key Mathematical Ideas

- deducing properties of real numbers from axioms

Teaching Suggestions

All of the properties of real numbers studied in Chapter 1 should be reviewed at this time. Point out that in Example 5 of Lesson 1-5, a proof was given for a property of real numbers. Since proof is an important part of the study of mathematics, every effort should be made to help the students understand the concept of mathematical proof. Many students find it difficult to generate the statements in a proof while they can fill in the reasons without much difficulty if the statements are given. Ask the students to explain why the corollary "if $ab = 0$ then $a = 0$ or $b = 0$" implies that the product of two nonzero numbers is not equal to 0.

There are many ways to help students handle proofs. Many students start by supplying reasons for steps that are already written. To help students go beyond supplying reasons to constructing their own proofs, try proof sketches. A proof sketch is a short strategic plan which focuses on the major steps of a proof.

Suggested Extensions

Use the discussion of converse to lead to a discussion of the inverse and contrapositive. Use the *if p then q* form to illustrate the inverse and contrapositive of *if* $ab = 0$*, then* $a = 0$ *or* $b = 0$.

2-7 (pages 88–91)

Key Mathematical Ideas

- proving theorems involving inequalities and absolute value

Teaching Suggestions

This lesson gives students the opportunity to use and improve their skills in proving theorems. Students should be given as much practice as possible in setting up and proving theorems. Point out the many *cases* involved in working with inequalities. All students should be able to give reasons in Exercises 1–6. By this time most students should be able to do most of the proofs in Exercises 7–16. Sometimes a ''how to'' group discussion about the proofs in an assignment helps students who are still having problems.

Suggested Extensions

Explain the concept of indirect proof which is to be used in Exercise 18. Have students see if that method would be helpful in any of the proofs in Exercises 11–16.

3 Linear Equations and Functions

The first part of this chapter deals with linear equations and their graphs. Open sentences in two variables are presented with regard to their form and their solution sets. Solution sets written as sets of ordered pairs lead to a discussion of the coordinate plane and graphing methods. Finding the slope and using the slope to find the equation of a line are main ideas in the first part of the chapter. The second part of the chapter continues with linear equations and systems of linear equations or inequalities. Transformations that produce equivalent systems are outlined. In the third part of the chapter, functions and relations are defined and compared with emphasis on linear functions and identifying relations that are also functions. Problem solving activities are given in abundance throughout the chapter.

3-1 (pages 101–106)

Key Mathematical Ideas

- solving open sentences in two variables

Teaching Suggestions

A solution of an open sentence in two variables is written as an ordered pair. Emphasize that the order is important. Point out that solutions can not only be pairs with ''nice'' integral values but can also have fractions, as is the case in Example 1. Point out also that the solutions to Example 3 are not the ordered pairs, but the numbers formed from them.

Refer students again to the five-step plan used to

solve problems. The Problems provide an opportunity to explore further restrictions placed on variables and solutions.

3-2 (pages 107–111)

Key Mathematical Ideas

- graphing linear equations in two variables

Teaching Suggestions

The first part of this lesson should be a review for students, but care should be taken to be sure they are familiar with all of the terms and with the method of plotting points, especially the *order* in ordered pairs. You may wish to use an overhead visual of a coordinate system to illustrate and discuss axes, quadrants, origin, and other terms.

Point out that not all graphs are lines and that it is important that students understand the theorem that gives the form of the equation of a line. Emphasize that in addition to the practice of finding three points to determine the graph of a linear equation, the points should be a sufficient distance apart on the coordinate plane to help avoid drawing errors.

Students are sometimes confused by equations like $y = 3$ and $x = 5$ because there is no second variable. Even though it is stated in the text, it should be repeated that a set of ordered pairs is the solution set, not just one pair. Stress that throughout the text great care is taken to distinguish a graph (a geometric object) from a solution set (an algebraic object).

Suggested Extensions

1. Graph the solution set of $x + y \leq 5$ for x and y whole numbers.

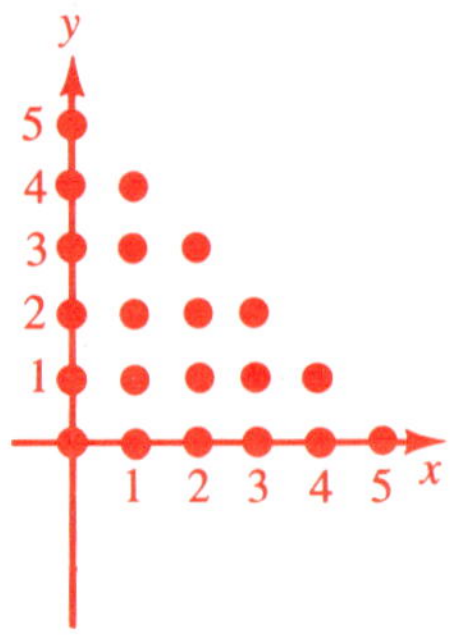

2. Graph the solution set of $y < |x|$ for x and y integers.

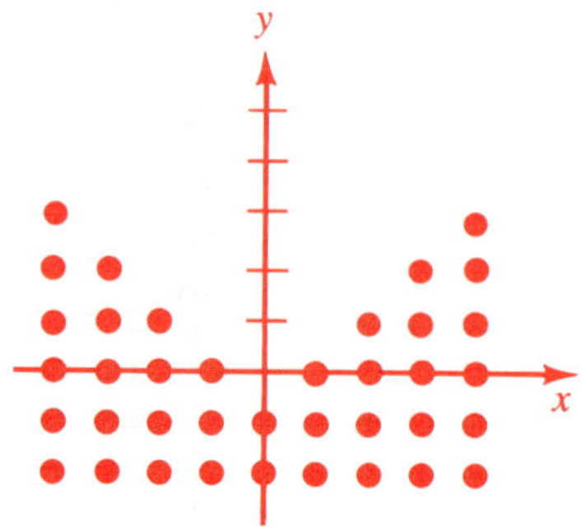

3-3 (pages 112–117)

Key Mathematical Ideas

- finding the slope of a line
- drawing a line given its slope and a point on the line

Teaching Suggestions

Emphasize the difference between "no slope" and "slope = 0." The theorem that the slope of the line $Ax + By = C$ $(B \neq 0)$ is $-\frac{A}{B}$ is very helpful in working with equations later in this chapter. Students should know both the definition of slope and the point-slope theorem. Students might find it useful to see an argument using similar triangles to show that the slope of a nonvertical line is independent of the choice of two points.

The figure at the bottom of page 114 is helpful to students in understanding what slope really means. Be sure they understand that the lines need not go through the origin even though the lines in that figure do go through the origin.

Group Activities

Students do not always observe all the details of a graph or interpret them correctly. An activity that may be helpful is to pair students so that one has a clean sheet of graph paper and the other has a graph such as the figure for Example 3, page 114. The first student tries to draw the graph as the second student describes it in every detail. When the drawing is finished, students check it and reverse roles.

Suggested Extensions

Have the students do a complete proof of the theorem on page 114. See Exercises 49–54.

Explore with students the fact that the slope of a line depends on the coordinate system established by drawing the following diagrams on the chalkboard.

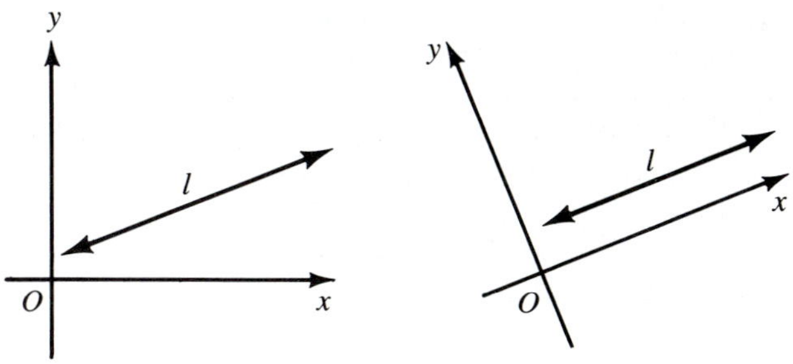

You can place a line l in many different coordinate systems. You can give a line a positive slope, a negative slope, a 0 slope or no slope depending upon how you tilt the axes.

Using Manipulatives

You may wish to review the slope of a line as related to its position on a coordinate system by constructing a large spinner attached to a reinforced piece of graph paper. Students can practice determining the slope by spinning the pointer and estimating or calculating the slope.

3-4 (pages 118–123)

Key Mathematical Ideas

- finding the equation of a line given its slope and a point on the line, two points, or its slope and the y-intercept

Teaching Suggestions

Point out that the standard form, the point-slope form, and the slope-intercept form of an equation are all important in working with equations. Students should know and be able to use all forms. Discuss the theorem regarding parallel and perpendicular lines thoroughly, referring again to the graphing lesson to illustrate what the theorem means. This theorem will also be used in the next lesson with systems of linear equations.

Suggested Extensions

Prove that the diagonals of a square are perpendicular.

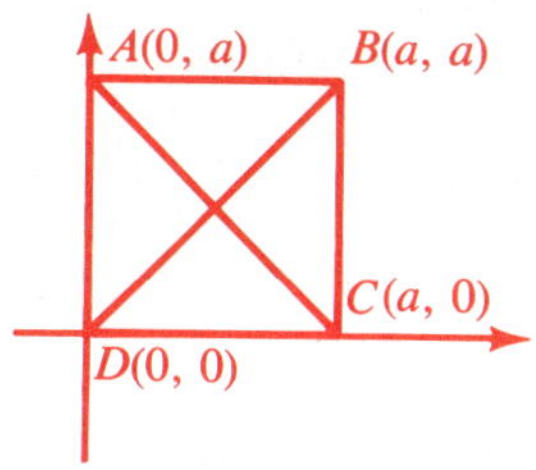

The slope of $\overline{DB} = \frac{a}{a} = 1$.

The slope of $\overline{AC} = \frac{a-0}{0-a} = -1$.

Therefore $\overline{DB}$ is perpendicular to $\overline{AC}$.

3-5 (pages 124–130)

Key Mathematical Ideas

- solving systems of linear equations in two variables

Teaching Suggestions

Point out that a system of equations is a collection of equations in conjunction. Therefore, we must find the intersection of the solution sets of the equations. Emphasize the three possible situations that can occur with a set of two linear equations in two variables. Show several examples graphically and let students graph several examples themselves. Discuss completely the transformations that produce equivalent systems. Emphasize that the final result should be of the form $x = a$ and $y = b$. Point out that both the substitution method and the linear combination method will work, but that one may be easier to use than the other on a given system.

Notice that the results that occur when a system has no solution are similar to the results for inequalities in one variable in Lesson 2-1. Discuss the terms consistent, inconsistent, and dependent.

You might pay particular attention to the possibility that different students will approach the same problem in different ways. The transformations are choices and can be selected in different orders. For example, in Example 3, x may be eliminated instead of y by multiplying the first equation by 4 and the second equation by -3, or the first by -4 and the second by 3.

Suggested Extensions

Illustrate Example 2 by graphing the following equations on the same coordinate plane.

a. $x - 2y = 5$
b. $4x + 3y = 9$
c. $x = 3$
d. $y = -1$

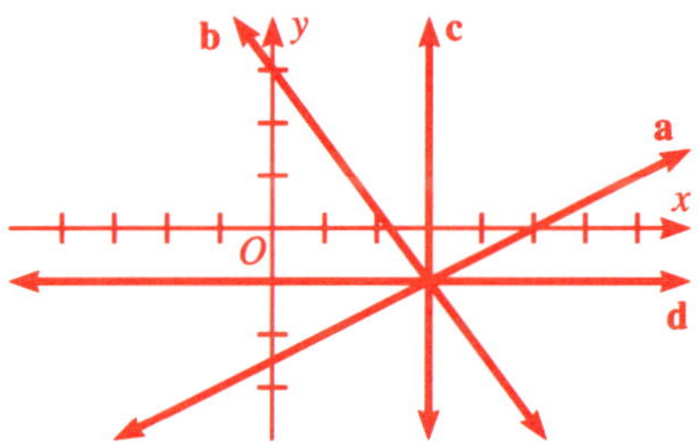

3-6 (pages 131–134)

Key Mathematical Ideas

- problem solving using systems of equation

Teaching Suggestions

Again, emphasize the five-step process of problem solving and point out that there will be two equations and two variables in each problem. Discuss some of the different types of equations and formulas that will be needed to solve the problems. Spend time letting the students discuss the problems and any difficulty they have in solving them. Remind students of the usefulness of a chart or a table in setting up a system.

Reading Algebra

If students have trouble with the word problems in this lesson, they may not be reading efficiently. Remind them that the best way to tackle a problem may often be to read it through several times with a different purpose each time. For example:

1. Skim the problem to get an overview of the situation.
2. Read carefully to discover what number you are asked to find and how this number is related to the given numbers.
3. Read to decide what formula or model you can use in setting up an equation or system of equations.
4. Don't forget to reread the problem after you have solved the equations, to determine whether your answer makes sense.

Suggested Extensions

Ask the students to make up their own problems using everyday situations relevant to their lifestyles and using names familiar to them.

3-7 (pages 135–139)

Key Mathematical Ideas

- graphing linear inequalities
- graphing solutions of systems of linear inequalities

Teaching Suggestions

You might find transparencies and overlays useful in teaching this lesson. Explain to students that the graphing of linear inequalities is very similar to graphing equations. The fact that the boundary is graphed first makes it easier to relate linear inequalities to linear equations.

The two main problems facing students are deciding whether to use a solid line ($\leq$ or $\geq$) or a dashed line ($<$ or $>$) for the boundary of the graph, and which side of the line to shade. Most students shade above the line for "greater than" and below the line for "less than." The problem is that sometimes this approach does not work. Be sure to point out situations when it does not work. Obviously, $x > 3$ is such a case. It also will not work when the direction of the inequality is changed during the solution process as in Example 2. The easiest way to check a solution is to use a test point like (0, 0), (0, 1), (1, 0), (1, 1), etc. as explained in the text. If students do not have trouble graphing single inequalities, then they should not have much trouble with systems. Most of the problems that students encounter are a result of incorrectly graphing the border line or shading the wrong side of a single inequality in the system. Sufficient examples and practice should eliminate most of the difficulties that students have.

Suggested Extensions

Ask students to read the linear programming application at the end of the chapter. Ask students to analyze the system of linear inequalities used and study the resulting graph. Then have them study the example to become familiar with an application of graphing systems of linear inequalities.

3-8 (pages 141–145)

Key Mathematical Ideas

- finding the values of functions
- graphing functions

Teaching Suggestions

Be sure the students understand not only the definition of function but also the different ways of describing a function.

1. ordered pairs
2. graph
3. mapping diagram
4. arrow notation
5. functional notation

Mention to students that great care is taken in the text to distinguish the notation f from the notation $f(x)$.

Give several examples of the composition of functions as in Example 2. Composition is studied explicitly later in the text and is useful in studying inverse functions.

Reading Algebra

In spite of the careful explanation of functional notation in the text, some students may still think that multiplication is implied by the parentheses in expressions such as $f(x)$. If students have studied probability in an earlier course, you might compare $f(x)$ with $P(x)$, the probability of x. Emphasize that multiplication is not implied in either case. A helpful activity is to list the following symbols on the chalkboard and ask students to describe the role of each pair of parentheses.

$F(t) = t^3 + 8 \qquad P(x) = \frac{1}{2}$

$8 - (6 + x) \qquad 4(x) \qquad (t^2 - 1)(t^2 + 1)$

$$\underset{\text{function}}{\underset{\uparrow}{g(x)}} = (x^3 - 1)\underset{\text{multiplication}}{\underset{\uparrow}{\,}}(x^2 + 4) + \underset{\text{grouping}}{\underset{\uparrow}{(x - 7)}}$$

Suggested Extensions

Given $f(x) = 2x + 3$, see if the students can find a function g so that $f(g(x)) = g(f(x))$.

$g(x) = \frac{x - 3}{2}$

3-9 (pages 146–152)

Key Mathematical Ideas

- finding equations for linear functions
- finding values of linear functions

Teaching Suggestions

Relate the concept of linear function to the linear equations in two variables studied in this chapter. Be sure students understand the rate of change concept used to find formulas for linear functions. Take time to discuss the problems in this lesson to be sure students can find equations for linear functions.

Suggested Extensions

Given $f(x) = 2x + 4$, find formulas for $f(x) - f(a)$ and $\frac{f(x) - f(a)}{x - a}$.

$f(x) - f(a) = 2x - 2a$

$\frac{f(x) - f(a)}{x - a} = \frac{2x - 2a}{x - a} = \frac{2(x - a)}{x - a} = 2$

3-10 (pages 153–157)

Key Mathematical Ideas

- graphing relations
- distinguishing functions from relations

Teaching Suggestions

Emphasize that a relation is *any* set of ordered pairs, while a function is a *special* set of ordered pairs. Point out that it is easy to distinguish a function when you are:

1. given a set of ordered pairs. You check the first elements.
2. given a graph. You use the vertical line test.
3. given a mapping. You check the arrows from the domain.
4. given an equation. You determine the solution set of the equation.

Suggested Extensions

Determine whether the following relations are functions for all real numbers.

a. $x + y = 5$ Yes **b.** $x^2 + y^2 = 25$ No

c. $y = x^2$ Yes **d.** $x = \sqrt{y^2}$ No

Consider the following data gathered to relate animal age A in months and mass M in kilograms. The masses of two animals were taken at each age.

A	1	1	2	2	3	3
M	1.1	1.2	1.8	1.9	2.4	2.4

Does the data viewed as a set of ordered pairs represent a function? No

Have students take the average of the masses of the animals for each age. Does the set of ordered pairs with mass replaced by average mass represent a function? Yes

4 Products and Factors of Polynomials

The chapter is divided into three groups of lessons: working with polynomials, factors of polynomials, and applications of factoring. In the first three lessons, the terms associated with polynomials are reviewed and the laws of exponents are used to develop the concepts used to find the products of polynomials. Greatest common factor and least common multiple are used to develop the methods of factoring polynomials in the next three lessons. In the last three lessons the methods developed in the chapter are used to solve polynomial equations, inequalities, and word problems.

4-1 (pages 167–170)

Key Mathematical Ideas

- simplifying polynomials
- adding and subtracting polynomials

Teaching Suggestions

Simplifying polynomials is an important skill used in solving equations. The terms in the chart should be familiar to students from a previous algebra course. Review the terms to be sure students understand them. In discussing addition and subtraction of polynomials, emphasize the combining of similar terms by showing horizontal and vertical forms to accomplish this. Remind students that in subtraction of polynomials the concept of adding the opposite of a sum is used. Remind students to change the sign of each term in the polynomial to be subtracted, not just the first term of that polynomial.

Reading Algebra

Algebra involves the manipulation of symbols, often in complex and subtle ways. In order to operate successfully with polynomials, students need a clear understanding of the meanings associated with symbols. An interesting exercise is to have students read aloud a number of expressions such as the following.

$$-8a^3b^2 \qquad (-8a^3b)^2 \qquad 8(q - \tfrac{1}{2}) - (q - 4)$$
$$P = 2l + 2w = 2(l + w) \qquad V = \tfrac{1}{3}s^2h$$

Help students to realize that there is often more than one correct way to translate a group of symbols into words. Accept any word form that makes the meaning clear. This activity can be extended by having the class write from dictation (by you or by a student) some of the expressions in the Oral Exercises.

4-2 (pages 171–173)

Key Mathematical Ideas

- using the laws of exponents in multiplying

Teaching Suggestions

Be sure that students know and understand the laws of exponents. Give several examples like those in the text. Point out the importance of the parentheses in a problem like $(2x)^3 = 8x^3$. Some students have problems with an expression like $2x^3$; these students think that it would also be $8x^3$.

Students sometimes forget to add the 1 for a first power of a variable in problems like Examples 1 and 2. Be sure that, for the first few times, they write it out as shown in Example 2.

It is important that students realize that they are using the distributive property when multiplying a polynomial by a monomial so that they can apply this later to the product of polynomials.

When students compare answers, they sometimes think that if the answers do not look the same, then one of the answers must be wrong. Remind students to look for correct answers possibly in alternate forms.

A common mistake made by students at all levels involves the use of the laws of exponents when there are no variables in the problem. They have no trouble remembering that $(a^4)(a^5) = a^9$, but they will say that $(2^4)(2^5) = 4^9$! Stop in the middle of class and have the students quickly write only the answer to a problem like the one above. This will be a check to see if they understand the concept and are thinking correctly when working problems.

4-3 (pages 174–176)

Key Mathematical Ideas

- multiplying polynomials

Teaching Suggestions

Although it is sometimes easier to use one of the other methods of multiplying two polynomials, the distributive method is a basic concept and all students should be required to use it on some problems. A slight adjustment to the problem in Example 1 makes it look more like arithmetic multiplication.

$$\begin{array}{r} x^2 + 4x - 5 \\ 2x + 3 \\ \hline + 3x^2 + 12x - 15 \\ 2x^3 + 8x^2 - 10x \qquad \\ \hline 2x^3 + 11x^2 + 2x - 15 \end{array}$$

Place like terms underneath each other (like place value). Multiply from right to left.

In most cases, students should be able to learn FOIL or a similar method and multiply two binomials mentally, just writing the answer. Students should learn to recognize the special cases of trinomial squares and difference of squares in any given situation. You might conclude the class presentation with a ten-minute mental drill on the FOIL method and on special products.

Reading Algebra

This is a good lesson for students to read silently before it is discussed in class. Some key questions that you can provide for students to answer as they read are:

- What property is used in multiplying two polynomials?
- How many ways of multiplying two polynomials are shown on page 174?
- Explain the method for multiplying two binomials.
- What name is given to the product of $(a + b)$ times $(a + b)$?

Follow the silent reading with discussion.

Suggested Extensions

Have students multiply $(a + b)(x + y)$ using the distributive property, and then have them factor. This will help them get ready for Lesson 4-6.

Ask students to write out expansions of $(a + b)$, $(a + b)^2$, and $(a + b)^3$ in a triangular array with each expansion forming the next row of the array. Ask students to determine the entries in the row for $(a + b)^4$ without actually multiplying.

$$a + b$$
$$a^2 + 2ab + b^2$$
$$a^3 + 3a^2b + 3ab^2 + b^3$$
$$a^4 + 4a^3b + 6a^2b^2 + 4ab^3 + b^4$$

4-4 (pages 179–182)

Key Mathematical Ideas

- finding the GCF and LCM of integers and monomials

Teaching Suggestions

Review several methods for factoring integers so that students can relate them to the methods they learned in other courses. Review the concepts of greatest common factor and least common multiple and give several examples of finding each expression. An illustration like the one below might help.

$4ax^2$ $-6a^2x$

$2 \cdot 2 \cdot a \cdot x \cdot x$ $-1 \cdot 2 \cdot 3 \cdot a \cdot a \cdot x$

GCF

$2ax$

(*least* power of common factors, used *once*)

$2 \cdot 2 \cdot a \cdot x \cdot x$ $-1 \cdot 2 \cdot 3 \cdot a \cdot a \cdot x$

LCM

$12a^2x^2$

(*greatest* powers of *each* factor from *both*)

4-5 (pages 183–187)

Key Mathematical Ideas

- factoring polynomials using GCF, special products, and grouping

Teaching Suggestions

Many students have trouble factoring polynomials because they forget to factor out common terms. Give several examples of this on the board. It is important that students be able to recognize the special products given in Lesson 4-3 and be able to use them in factoring. It might help to restate them in words.

Trinomial squares are the result of squaring the sum of two terms or squaring the difference of two terms.

The difference of squares is the result of multiplying the sum of two terms by the difference of the same two terms.

Sums or differences of cubes are special products that are used in several different types of problems. Students should learn these also. Again, seeing the relationships among the terms helps students learn the special products. Be sure that they do not confuse the expression $a^2 + ab + b^2$ with the expression $a^2 + 2ab + b^2$.

The factoring of four or more terms by grouping is one of the most difficult for many students. Give several examples of this type of problem and have students work a sufficient number of exercises to become familiar with the method.

Encourage students to keep trying since some students might expect to get the correct factorization on the first try.

Suggested Extensions

Have students work Exercise 59 (a), (b), and (c) and discuss their generalizations in class.

4-6 (pages 188–192)

Key Mathematical Ideas

- factoring quadratic polynomials

Teaching Suggestions

It is important in factoring quadratic polynomials that students first factor out any common factors and then look for special products. Then they should attempt to determine if the remaining polynomial can be factored further and find the factors. Many students do not like the method of listing possibilities and then checking for the correct combination. Some prefer to guess and end up trying all possibilities anyway, while others who guess are quite successful at it. It is important to point out that it is more logical, and many times quicker and easier, to list the possibilities and mentally check. After a short period of time students will learn to mentally check as they are listing the possibilities. This makes the process more efficient. Using the FOIL method in reverse as a factoring tool is another way to understand the techniques presented in this lesson.

Suggest that students learn to ask themselves questions such as "Is there a GCF of the terms other than 1?" If the answer is yes, then they should factor out this common monomial first.

Suggested Extensions

Apply the special form $a^2 - b^2 = (a + b)(a - b)$ and the fact that $2 = (\sqrt{2})^2$ to factor $x^2 - 2$.
$(x + \sqrt{2})(x - \sqrt{2})$

Group Activities

To help the class review factoring polynomials, write one polynomial each on a set of index cards. On another set write the corresponding factors. Prepare enough cards so that each student has a polynomial card and a factor card. Students, one at a time, will call out their polynomial and the student who has the factors will call out the answer. This may be adapted to smaller groups with each student having several cards. A variety of card games and scoring methods could be developed by the students.

Using Manipulatives

A manipulative approach gives an alternate method of introducing or reviewing the factoring of quadratic polynomials. The materials used in this activity may be purchased commercially or hand-made by the students. The pieces used are labeled as shown.

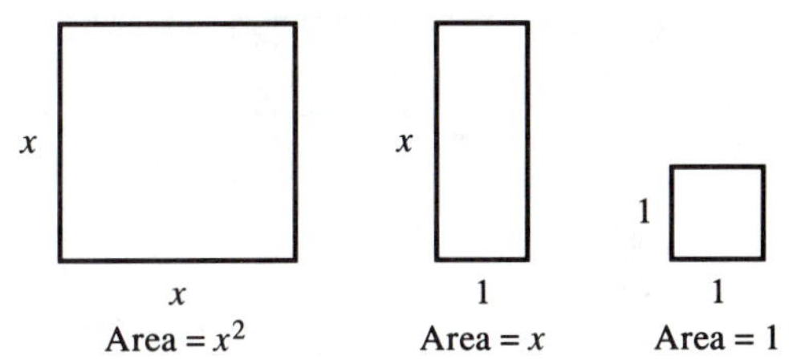

Be sure that the x-length is not a whole-number multiple of the unit length.

The desired outcome of the activity is that the student "build" a rectangle which represents a quadratic polynomial and then "read" the factors as the length and width of the rectangle. Any quadratic that is factorable yields a rectangle.

Example A: Factor $x^2 + 3x$.
The area $x^2 + 3x$ is represented by the rectangle shown on page T96. The sides of the rectangle are x and $x + 3$. Therefore, $x^2 + 3x = x(x + 3)$.

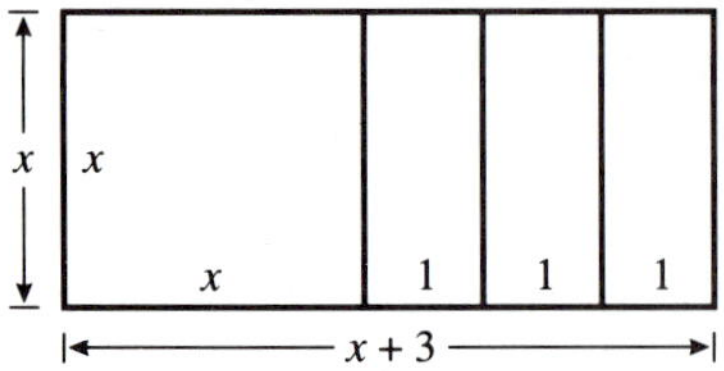

Example B: Factor $2x^2 + 5x + 2$.
The area can be represented in at least two possible ways, as shown. Notice that the representation *must* be a rectangle.

Solution 1.

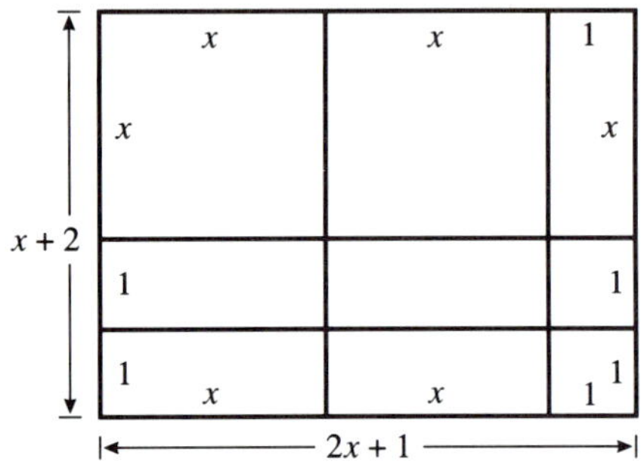

Solution 2.

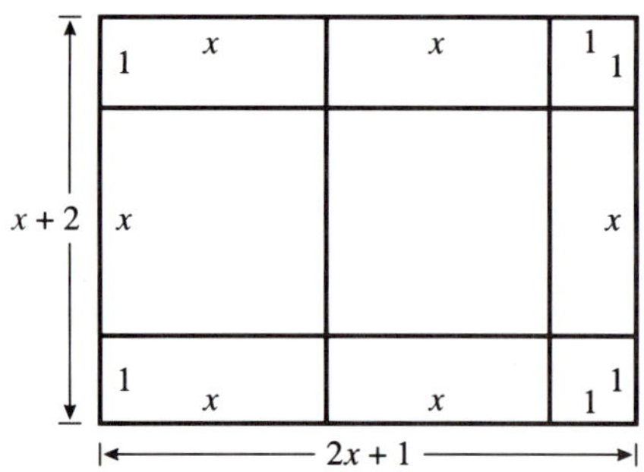

Ask the students to factor some of the exercises in the lesson using this method.

4-7 (pages 194–197)

Key Mathematical Ideas

- solving polynomial equations

Teaching Suggestions

First, make sure students make the transition from polynomials to equations. Remind them of the rules concerning solving equations. Point out that the zero-product property is very important in solving polynomial equations and that a major step is writing the equation with 0 as one side. Note that all factors, including those with a single variable, such as x, must be set equal to zero. Continue emphasizing the importance of checking the answer.

You might point out that the zero-product property allows us to write a collection of disjunctions from a single equation.

Suggested Extensions

Using the zero-product property, generate an equation of least degree for which each set is the solution set.

1. {3, 5} $x^2 - 8x + 15 = 0$
2. {0, 1, 2} $x^3 - 3x^2 + 2x = 0$

4-8 (pages 198–201)

Key Mathematical Ideas

- problem solving using polynomial equations

Teaching Suggestions

Review again the steps involved in correctly solving a word problem. Go over the examples in the lesson carefully and add other examples from the exercises so that students are as familiar as possible with the types of problems in this lesson. It is possible that students have had very little experience in working word problems, especially this type. Spend a little extra time here, if needed. Continue to remind students of the distinction between a solution of an equation and an answer to a word problem.

Suggested Extensions

Point out that in Example 1, a negative solution for width was rejected. Have students discuss situations in which solutions might have to be rejected.

4-9 (pages 202–205)

Key Mathematical Ideas

- solving polynomial inequalities

Teaching Suggestions

Review the rules for working with inequalities. Point out that the inequalities should be written with one side equal to zero and the other side factored, if possible. Since this is a familiar method, the major consideration should be getting the factors written in inequality form correctly. It is important that students understand the conjunctions and disjunctions used.

$$\text{If } ab > 0, \text{ then } a > 0 \textit{ and } b > 0, \text{ or } a < 0 \textit{ and } b < 0.$$

$$\text{If } ab < 0, \text{ then } a < 0 \textit{ and } b > 0, \text{ or } a > 0 \textit{ and } b < 0.$$

The sign graph in Example 3 is helpful if students are having trouble understanding the method of solving the inequalities.

Reading Algebra

By this time students have had a good deal of practice in reading and graphing the solution sets of equations and inequalities. As a challenge, you might reverse the process and ask students to write solution sets represented by graphs such as those below. At first they may find this difficult.

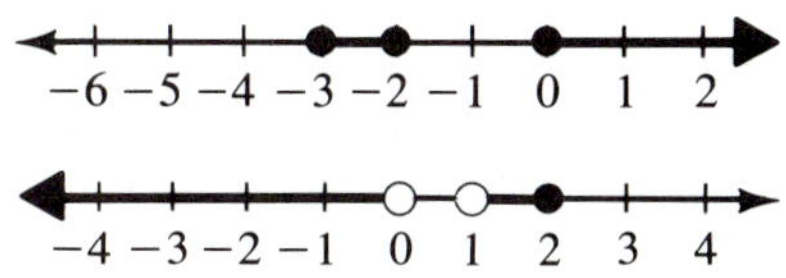

Suggested Extensions

Ask students to solve $(x - 2)(x - 3)(x - 4) < 0$ by considering all possible combinations of positive and negative factors. Then ask students to solve the same problem by dividing the number line into regions and testing points in the regions. Ask students which method is easier.

Consider $t^2 - 6t + 5 \leq 0$ and $t^2 - 6t + 5 = y$. Ask students to graph $t^2 - 6t + 5 = y$ and ask them what values of t make $y \leq 0$.

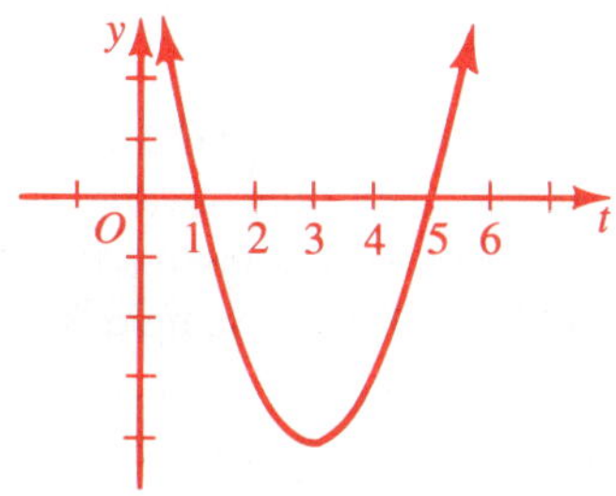

The solution set is $\{t: 1 \leq t \leq 5\}$.

Ask students to use this approach in solving some quadratic inequalities.

5 Rational Expressions

In this chapter, rational algebraic expressions are studied. The laws of exponents are reviewed and extended to include zero and negative exponents. Scientific notation is used as an application for the laws and operations using exponents. Simplifying, adding, subtracting, multiplying, and dividing rational expressions are all studied in this chapter. Once these skills are learned, they are used to solve equations with fractional coefficients and fractional equations.

5-1 (pages 211–215)

Key Mathematical Ideas

- simplifying quotients of monomials using the laws of exponents

Teaching Suggestions

Fractions are difficult even for algebra students. Be sure, through simple examples, that students understand basic skills for working with fractions. Review all of the laws of exponents and be sure that students understand their use. Emphasize that it is important for students to know when a quotient of monomials is simplified. (See the three conditions in the text.)

5-2 (pages 216–220)

Key Mathematical Ideas

- simplifying expressions with zero and negative exponents

Teaching Suggestions

Discuss with students the reasoning behind the definitions of a^0 and a^{-n}. Demonstrate each with several examples. Ask students to discuss the reasons for the agreement at the top of page 217. Discuss Sample 3 in part B of the Written Exercises and ask students why this might be an important skill to know.

Group Activities

Prepare a set of tic-tac-toe boards with answers to selected exercises from Lessons 5-1 and 5-2 in each square. Give one board to each student. Write one of the selected exercises on the chalkboard. Tell students to solve the exercise and cross out the answer if they have it on their board. The first student to have three answers in a row, as with tic-tac-toe, wins. To extend the activity, you can have the students continue until they have used all the answers on the card.

Reading Algebra

As a test for students' understanding, you may want to have them verbalize some of the Oral Exercises in this lesson. Direct them to use complete sentences in answering the exercises. For example, Oral Exercise 1 might be read as "One divided by two to the negative three power is equal to one times two to the third power, or eight." As in earlier lessons, any word form that is clear may be accepted.

Suggested Extensions

Have students try to invent a definition for 0^0 and discuss the possibilities.

Ask students to give the location of the graph of 2^n on a number line for positive integers and for negative integers. When n is positive, 2^n is to the right of 1. When n is negative, 2^n is between 0 and 1.

5-3 (pages 221–225)

Key Mathematical Ideas

- using scientific notation
- using significant digits

Teaching Suggestions

Some students might need a review of how to write a number in scientific notation. You might try the following approach: How many decimal places should the decimal point be moved to make the number between 1 and 10? Did you go left or right?

left n places? then $+ n$ for the exponent of 10
right n places? then $- n$ for the exponent of 10

See the example at the beginning of the lesson.

An important aspect of scientific notation is that it makes clear the number of significant digits in a number. Thus, by using scientific notation, one can determine the accuracy of a measurement, as discussed at the bottom of page 221.

Suggested Extensions

Ask students who have worked with scientific notation using calculators or computers to demonstrate calculations in scientific notation for the class. Ask students in physics, chemistry, or other science classes to demonstrate the use of scientific notation and significant digits in actual problems from those courses.

5-4 (pages 227–229)

Key Mathematical Ideas

- simplifying rational expressions

Teaching Suggestions

Stress the parallel between fractions and rational expressions. The idea behind simplification is the same in each case. The question that students should ask themselves is whether any further simplification can be performed.

One of the most common mistakes in simplifying rational algebraic expressions is trying to "cancel out" terms that are not factors. For example,

$$\frac{2x + 5}{2} = x + 5 \text{ (This is incorrect.)}$$

or

$$\frac{3x + 2}{2} = 3x \text{ (This is incorrect.)}$$

Emphasize to students that if an expression is to be simplified, it must first be factored. There are several types of special factoring techniques that may be used in simplifying expressions. One is "turning around" a factor by factoring out -1. For example,

$$5 - x = -1(x - 5).$$

Remind the students of the agreement about domain on page 217. Point out that the answer in Example 2 is restricted by the original expression where $x \neq \frac{1}{3}$ and $x \neq \frac{1}{2}$.

Suggested Extensions

You may wish either to discuss or to assign the material on graphing rational functions on pages 230 and 231. This will afford students an opportunity to observe the graphs of several rational functions.

5-5 (pages 232–234)

Key Mathematical Ideas

- multiplying and dividing rational expressions

Teaching Suggestions

Review the rules for multiplying and dividing fractions, using numerical fractions in some examples. Show several examples like Example 4 in the text. Require that students rewrite the problem by changing all quotients to products using the rule for division before they simplify. Emphasize to students that when simplifying, they are actually dividing numerator and denominator by a common factor. Some students tend to "oversimplify" and eliminate terms which are not factors.

Continue the parallel between fractions and rational expressions. In particular, the order of operations presented in Chapter 1 continues to apply. This will also be the case in Lesson 5-6 when addition and subtraction are discussed.

Suggested Extensions

Ask students to simplify the following expression.

$$\frac{\frac{4a^3}{5c^2 - 5c}}{\frac{c^2 - 9c - 10}{4c - 40}} \div \frac{a}{2 - 2c^2} \quad \frac{-32a^2}{5c}$$

5-6 (pages 235–237)

Key Mathematical Ideas

- adding and subtracting rational expressions

Teaching Suggestions

Illustrate the process used in adding and subtracting rational expressions using fractions from arithmetic (see the Oral Exercises). It is essential that students know how to find the least common denominator. Use Example 4 to illustrate the importance of the negative sign in handling subtraction.

You might give students a problem in which the rational expressions are not given as simplified. Ask students to find the sum of the rational expressions without simplifying them first. Then ask students to simplify the rational expressions first and then add. Ask students to compare the amount of work they did in each solution.

Suggested Extensions

For an example that illustrates some of the problems that might occur in this lesson, have students simplify the following expression.

$$5 - x - \frac{x^2 - 25}{5 - x}$$ 10

5-7 (pages 238–241)

Key Mathematical Ideas

- simplifying complex fractions

Teaching Suggestions

Discuss the two methods of simplifying complex fractions shown in the examples. You may wish to illustrate each method with several more examples. Ask students to discuss advantages and disadvantages of each method as they might be used in different situations.

Suggested Extensions

Direct student attention to the Sample above the C exercises. Discuss how every polynomial function has associated with it a rational expression that measures the rate of change of the function. This rate-of-change function is important in calculus.

5-8 (pages 242–246)

Key Mathematical Ideas

- solving equations and inequalities having fractional coefficients

Teaching Suggestions

Use Examples 1 and 2 to show students that they need never leave fractions in an equation or inequality. It is much easier to first multiply both sides of the equation by the least common denominator of the fractions. Check the way students work the multiplication. Some may keep the numerator and denominator for another step before reducing. Encourage them to mentally do the multiplication and write the resulting equation or inequality without fractions.

Encourage students to continue using the five-step plan when solving word problems. This is a good time to review all of the methods used in solving word problems and checking answers. Spend some time on the Problems that follow the Written Exercises.

Suggested Extensions

Ask students to solve $\frac{2x + 10}{x^2 - 1} < \frac{x}{x + 1}$ by the following two methods. Then discuss the methods.

Method 1 Multiply both sides of the inequality by $x^2 - 1$ to clear fractions. Solve the polynomial equation that results. (Students should beware of multiplying by an expression that may be positive or negative in value. Students must take both cases into account.)

Method 2 Write the inequality as an equivalent inequality with 0 as one side. Write the other side that results as a single rational expression. Then ask when are the signs of the numerator and denominator opposite.

By either method the solution set is $\{x: x < -2 \text{ or } x > 5 \text{ or } -1 < x < 1\}$.

5-9 (pages 247–252)

Key Mathematical Ideas

- solving fractional equations

Teaching Suggestions

Emphasize the difference between fractional equations and equations with fractional coefficients. As before, the method for solving the equations is to multiply by the least common denominator. A solid understanding of this process is required here. Point out that any time both sides of an equation are multiplied by an expression containing the variable, extraneous roots may occur. Each solution must be checked in the original equation.

Suggested Extensions

As with Lesson 5-8, any additional time should be spent working the problem solving exercises at the end of the lesson. Have exceptional students help others so that all students will increase their problem solving skills.

6 Irrational and Complex Numbers

In this chapter students are introduced to roots and principal roots of real numbers. Students learn to use various properties to simplify, add, subtract, multiply, and divide radical expressions. The theme of solving equations is continued in this chapter by consideration of radical equations. Students then learn to write rational numbers as special decimals and thereby learn to distinguish rational numbers from irrational numbers. The chapter concludes with a study of roots of negative numbers and the arithmetic of complex numbers.

It is important for students to understand what happens when the index for a radical is even and when it is odd. Emphasize the three properties on page 261. You might use Oral Exercises 1–12 to help students master the skill of writing the simplest name for a given radical.

6-1 (pages 259–263)

Key Mathematical Ideas

- finding roots of real numbers

Teaching Suggestions

Be sure that students distinguish between the solution set of $x^2 = 36$ and the number denoted by $\sqrt{36}$. In order for the number $\sqrt{36}$ to be meaningful it is necessary to give it a *single* name. That name is 6, the principal root of 36.

Point out to the students that a calculator will give the principal square root of a number. Since the principal root concept is important, it would be better to read $-\sqrt{b}$ (negative square root of b) as "the opposite of the principal square root of b."

Many students are confused as to when to write $\pm$ and when to write just $+$. The symbol $\pm$ is used when equations are solved.

6-2 (pages 264–269)

Key Mathematical Ideas

- simplifying radicals

Teaching Suggestions

After demonstrating the rules for products and quotients of radicals, point out that sums and differences do not behave the same way, as shown in the Caution on page 264. Also stress that these rules apply when indexes are the same.

You might spend some extra time making sure that students understand what it means for a radical to be in simplest form. Some students are not sure that they have obtained the right (simplest) answer.

Discuss the two theorems in this lesson and give examples.

Reading Algebra

Many of the radical expressions in this lesson may be difficult for students to verbalize, especially the generalizations involving *n*th powers and *n*th roots.

Verbalization is valuable, however, because it often gives students insight into meaning. As an exercise in oral reading, use some combinations of symbols such as the following. After each verbalization, discuss the meaning of the expression.

$$\sqrt[n]{\frac{a}{b}} = \frac{\sqrt[n]{a}}{\sqrt[n]{b}} \qquad \sqrt{|6x^3|}$$

$$\sqrt[n]{b^m} = (\sqrt[n]{b})^m \qquad \sqrt{53} \approx 7.280$$

Suggested Extensions

Ask students to prove by indirect proof that $\sqrt{a+b} \neq \sqrt{a} + \sqrt{b}$ under the assumptions that a and b are positive.

Assume that $\sqrt{a+b} = \sqrt{a} + \sqrt{b}$ and a and b are positive.

Then $a + b = a + 2\sqrt{a}\sqrt{b} + b$

or $0 = 2\sqrt{a}\sqrt{b}$.

By the zero-product property $\sqrt{a} = 0$ or $\sqrt{b} = 0$. Thus $a = 0$ or $b = 0$. This, however, contradicts the hypothesis that a and b are both positive.

Group Activities

Divide the class into groups of four or five students. The students within a group are to work together to come up with an "original" proof for Exercise 67 on page 269. If this is successful, try some of the other proofs on that page.

6-3 (pages 270–273)

Key Mathematical Ideas

- simplifying expressions involving sums of radicals

Teaching Suggestions

Stress that simplifying polynomials and simplifying radicals with sums have important features in common. The concept of like, or similar, radicals can be compared to like, or similar, terms. Stress that students should simplify radicals first before attempting to combine radicals. Students might require a great deal of practice in handling problems in this lesson.

Emphasize again that $\sqrt{a} + \sqrt{b} \neq \sqrt{a+b}$. Some students are tempted to say, for example, $\sqrt{75} - \sqrt{50} = \sqrt{25} = 5$.

Suggested Extensions

Simplify.

1. $\frac{10}{\sqrt{2}} + \frac{6}{\sqrt{3}} + \frac{12}{\sqrt{2}} + \frac{3}{\sqrt{3}}$ $11\sqrt{2} + 3\sqrt{3}$

2. $\frac{3x}{\sqrt{x}} + \sqrt{4x} + \sqrt{x^3} + \frac{5x^2}{\sqrt{x}}$

$5\sqrt{x} + 6x\sqrt{x}$ or $(5 + 6x)\sqrt{x}$

6-4 (pages 274–276)

Key Mathematical Ideas

- simplifying products and quotients of binomials that contain radicals

Teaching Suggestions

You might review the methods of multiplying binomials and then compare these methods with the process by which binomials involving radicals are multiplied. After making some comparisons stress some important differences between multiplying binomials and multiplying binomials involving radicals.

Discuss how the idea of conjugate is needed to extend the notion of rationalizing the denominator from a denominator involving one radical to a denominator involving a binomial with radicals.

6-5 (pages 277–282)

Key Mathematical Ideas

- solving equations containing radicals

Teaching Suggestions

Stress with students that the focus of this lesson is a return to equation solving. Calculations and skills learned in earlier lessons are now used to solve equations. Be sure to distinguish between radical equations and equations involving radicals. A radical equation must have a variable in a radical expression. As in other contexts the solution process might produce extraneous roots. You might ask students if they can recall other situations where extraneous roots arose. Again stress the importance of checking answers in the original exercise.

Suggested Extensions

Discuss with students the problem of no solutions. Use examples like $\sqrt{x+2} = -2$ and $-\sqrt{x-11} = 4$ to show that you can determine if there are any solutions to look for without solving the equation.

6-6 (pages 283–287)

Key Mathematical Ideas

- using the characteristics of decimal representation of real numbers

Teaching Suggestions

Review the definition of rational number with students and point out that many rational numbers are not given as quotients of integers but that they *can be* written as such. For example, $\sqrt{4}$ is rational even though there is a radical in the notation.

With increased use of calculators, it is important that the students understand that the display on their calculator is finite and that all of a repeating block or terminating sequence of numbers may not show on the display. In Example 3a, a slight adjustment avoids the "problem" of having a decimal in the numerator of the fraction. Using powers of 10, "place" the decimal in front of the repeating block in one number and behind the repeating block in the other, then subtract as shown.

$$\begin{aligned} &\text{Let } N = 0.3\overline{27} \\ \text{then} \quad &1000N = 327.\overline{27} \\ \text{and} \quad &-\ 10N = 3.\overline{27} \\ \hline &990N = 324.0 \\ &N = \frac{324}{990} = \frac{18}{55} \end{aligned}$$

Suggested Extensions

Show by the method given in Example 3 that $0.\overline{9} = 1$ and that $1.\overline{9} = 2$.

$N = 0.\overline{9}$	$N = 1.\overline{9}$
$10N = 9.\overline{9}$	$10N = 19.\overline{9}$
$9N = 9$	$9N = 18$
$N = 1$	$N = 2$

6-7 (pages 288–291)

Key Mathematical Ideas

- simplifying square roots of negative numbers

Teaching Suggestions

Discuss with students the motivation for defining a new number i. Be sure that the students understand the meaning of i. Show several examples using the definition to simplify negative square roots, possibly even showing the factoring of -1 in the radical. For example,

$$\sqrt{-9} = \sqrt{-1 \cdot 9} = \sqrt{-1} \cdot \sqrt{9} = i \cdot 3 = 3i.$$

One of the most common mistakes made by students in this area is illustrated in the Caution following Example 3. Be sure students understand why the rule does not work.

Suggested Extensions

1. Ask students to complete the following table and construct a rule for simplifying i^n, where n is an integer.

$i = i$	$i^2 = -1$	$i^3 = -i$	$i^4 = 1$
$i^5 = ?$ i	$i^6 = ?$ -1	$i^7 = ?$ $-i$	$i^8 = ?$ 1

The rule for simplifying is:
Divide n by 4, if the remainder is 0, $i^n = 1$,
if the remainder is 1, $i^n = i$,
if the remainder is 2, $i^n = -1$, and
if the remainder is 3, $i^n = -i$.

2. Ask students to prove that $\{i^n\colon n \text{ is an integer}\} = \{1, -1, i, -i\}$.

The proof that each power of i is 1, -1, i, or $-i$ is outlined in the solution of Exercise 1 above. Clearly i, -1, $-i$, and 1 are the first four powers of i. Therefore the sets are equal.

6-8 (pages 292–296)

Key Mathematical Ideas

- adding, subtracting, multiplying, and dividing complex numbers

Teaching Suggestions

Since the set of complex numbers is a new set, point out to students that new operations must be defined. The object is to make definitions that are workable and that agree with the rules of operations that govern working with real numbers. Spend time discussing multiplication of complex numbers since the definition is puzzling to many students. Point out similarities between adding two polynomials and adding two complex numbers. The idea of adding corresponding "terms" is used in addition. Unfortunately, when complex numbers are multiplied, students should not multiply corresponding "terms." Rather, stress the idea that multiplying complex numbers is like multiplying binomials. Emphasize the material between Example 3 and Example 4. The fact that the complex number system is closed is important. It is also important that students understand the "big picture" of how all of the other sets of numbers they have studied are related to the set of complex numbers.

Suggested Extensions

Ask students to read the Extra that follows this lesson and to work the exercises.

7 Quadratic Equations and Functions

The chapter begins with a discussion of completing the square and using the quadratic formula to solve quadratic equations. This is followed by a qualitative study of the nature of the roots of a quadratic equation and a method for solving equations in quadratic form. The special characteristics of the graphs of quadratic equations are studied along with their relationship to quadratic functions. The chapter concludes with a study of the interrelationships between the roots and the coefficients of quadratic equations.

7-1 (pages 307–310)

Key Mathematical Ideas

- solving quadratic equations by completing the square

Teaching Suggestions

Walk students through the steps involved in solving a quadratic equation by completing the square. You might use Example 2. Then you might focus on Example 3, particularly step 2. Some students might find Example 2 quite workable but find Example 3 difficult because of the effect of a leading coefficient other than 1.

Reading Algebra

As you walk through Example 2 with your students (see the Teaching Suggestions), you may want to mention the variety of eye movements involved. Not only do they need left-to-right and right-to-left eye movements in order to relate the directions to the successive equations in the solution, but they need vertical eye movements in order to see how each equation follows from the one that precedes it. Point out that the alignment of the equal signs and of the terms in the equations helps them to see what is happening in each step.

Suggested Extensions

Ask students to perform the process of completing the square on the general quadratic equation
$ax^2 + bx + c = 0$ where $a \neq 0$.
See Lesson 7-2.

7-2 (pages 311–315)

Key Mathematical Ideas

- solving quadratic equations by using the quadratic formula

Teaching Suggestions

Relate the discussion in this lesson to the discussion that was made in Lesson 7-1. You might point out that even though there is a formula to solve quadratic equations, the process of completing the square is still useful and will appear again in later lessons.

Point out in the solution of Example 3 that when word problems are solved and the answers are radicals, it is usually a good idea to obtain a decimal approximation. Such an approximation gives meaning to an answer and provides a quick way to check calculations.

Suggested Extensions

The following graphic solution of quadratic equations was given by the German geometer Karl Georg Christian von Staudt (1798–1867).

Given the quadratic equation $x^2 + bx + c = 0$, graph the points $M\left(-\frac{c}{b}, 0\right)$ and $N\left(-\frac{4}{b}, 2\right)$ on a rectangular coordinate system. Draw a circle with radius 1 centered at (0, 1). Draw $\overline{MN}$ and label its points of intersection with the circle as S and T. From (0, 2), draw line segments through S and T to the x-axis, intersecting the axis at $(s, 0)$ and $(t, 0)$. The values s and t are the roots.

Use von Staudt's method to solve the following.

1. $x^2 - 2x - 8 = 0$ $\{-2, 4\}$

2. $x^2 - 8x + 12 = 0$ $\{2, 6\}$

3. $x^2 + 4x - 60 = 0$ $\{-10, 6\}$

4. $x^2 + 6x - 16 = 0$ $\{-8, 2\}$

7-3 (pages 317–321)

Key Mathematical Ideas

- using the discriminant to determine the nature of the roots of a quadratic equation

Teaching Suggestions

Discuss with students the idea of a case. When students compute the discriminant, they should ask themselves in which case the equation is. The following diagram will help students keep straight what the computation implies.

$>$	2
$=$	1
$<$	0

Remind students that in this lesson the coefficients are real numbers.

The discussion of the "easiest method" might be worth some time in the classroom. Point out to students that if they solve an equation when no explicit method is chosen for them they should examine the coefficients for some direction as to which method to choose. You might also discuss $ax^2 + bx + c = 0$ when $b = 0$. Ask students which method is easiest then.

Reading Algebra

The word *discriminant* and its symbol D, introduced in this lesson, may be completely new to some of your students. To make the meaning of the word clearer, you may wish to associate it with *discriminate,* which they probably know. Basically, *discriminate* means to sort out and classify things according to their differences. Thus, we use the discriminant to help us sort out and classify the roots of a quadratic equation.

Suggested Extensions

Ask students to consider $ax^2 + bx + c = y$ rather than $ax^2 + bx + c = 0$. What does the computation of the discriminant say about x-intercepts?

If $D > 0$, then the graph has two distinct x-intercepts.
If $D = 0$, then the graph has one x-intercept.
If $D < 0$, then the graph has no x-intercepts.

Group Activities

Divide the class into "quadratic" groups of four. Within each group assign to each person the letter A, B, C, or D. These could be names instead of just letters. The group will work together to evaluate the discriminant and determine the nature of the roots of a quadratic equation. This may be done as a competitive activity or simply as a different way to do an assignment. Each person in the group is assigned a specific job. For example, A sets up the problem, putting the equation into standard form if necessary. B sets up the discriminant in the correct form with the correct numbers substituted into the formula. C works out the arithmetic and puts the answers in simplest form. D states the condition of the discriminant and states the nature of the roots based on that observation. After one or two problems, the jobs shift so that each member of the group performs each job.

7-4 (pages 322–325)

Key Mathematical Ideas

- recognizing equations in quadratic form
- solving equations in quadratic form

Teaching Suggestions

For some students this is a difficult lesson because students associate the word *quadratic* with the exponent 2 only. Without numerous examples and additional practice they have trouble identifying the quantities in quadratic form. Discuss the use of parentheses as in Example 2 to isolate quantities properly.

Some students may prefer not to use an intermediate variable, as shown in Example 1, when solving equations in quadratic form. Ask your students to try some exercises using both methods and to determine which method they prefer.

Reinforce the practice of checking answers after work is completed. Some students think that equations in quadratic form have at most two roots. Distinguish the number of roots a quadratic equation has from the number of roots an equation in quadratic form has.

Suggested Extensions

Ask students to take an alternate route in solving Example 2 by clearing the fractions first and solving for x directly.

$$\left(\frac{1}{2x}\right)^2 - 5\left(\frac{1}{2x}\right) - 6 = 0$$
$$\frac{1}{4x^2} - 5\left(\frac{1}{2x}\right) - 6 = 0$$
$$1 - 5(2x) - 6(4x^2) = 0$$
$$1 - 10x - 24x^2 = 0$$
$$(1 - 12x)(1 + 2x) = 0$$
$$12x = 1 \qquad 2x = -1$$
$$x = \frac{1}{12} \qquad x = -\frac{1}{2}$$

7-5 (pages 326–332)

Key Mathematical Ideas

- graphing equations of the form $y - k = a(x - h)^2$
- finding the vertices and axis of symmetry of $y - k = a(x - h)^2$

Teaching Suggestions

Since graphing is such an important part of advanced math courses, it is important that you emphasize the value of developing good graphing skills. Develop the various stages of the graph of a parabola of the form $y - k = a(x - h)^2$. Illustrate the physical changes caused by changing the value of a in $y = ax^2$. Then show what happens to the parabola when the equation is $y = a(x - h)^2$. Finally, illustrate all of the variations on the complete form $y - k = a(x - h)^2$. Be sure students understand the right and left shift based on h. Sometimes students get confused about the direction in which the curve opens and have a hard time correcting the problem. Emphasize the summary that is given just before Example 1. Many examples and much practice are needed for this lesson, especially if students do not have a sufficient graphing background.

You might point out that parabolas will be studied more extensively in Chapter 9. Be sure to distinguish a quadratic function from a parabola. All quadratic functions have parabolas for their graphs. However, not all parabolas are graphs of quadratic functions.

Suggested Extensions

Ask students to graph $y = x^2 - 4x + 5$ by using the techniques of this lesson.

$y = x^2 - 4x + 4 + 1$
$y - 1 = (x - 2)^2$
a parabola with vertex (2, 1); opening upward

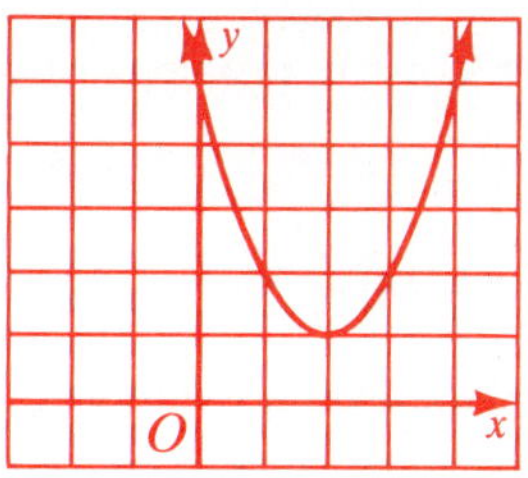

Discuss the fact that all parabolas have the same shape but vary in size with the value of a in $y = ax^2$ or $y = a(x - h)^2 + k$.

Draw on graph paper and compare the following three graphs:

a. $y = x^2$, with three squares equal to one unit and $-2 \le x \le 2$

b. $y = 3x^2$, with three squares equal to one unit and $-1 \le x \le 1$

c. $y = x^2$, with one square equal to one unit and $-3 \le x \le 3$

a. and **c.** are the same shape, dilated by a factor of 3

b. and **c.** are exactly the same

7-6 (pages 333–337)

Key Mathematical Ideas

- analyzing a quadratic function, drawing its graph, finding its maximum or minimum value

Teaching Suggestions

Carefully go over the steps of converting a quadratic function into the form $y - k = a(x - h)^2$. Relate to students that this is a ''practical'' application of completing the square. Discuss the fact that this method makes graphing of most quadratic functions much easier. Discuss with students the uses of the maximum and minimum values in a quadratic function.

Suggested Extensions

Ask students to prove that if $a > 0$ and $ax^2 + bx + c = f(x)$, then the minimum occurs at $x = -\frac{b}{2a}$.

Let $x_0 = -\frac{b}{2a} + e$ where $e \neq 0$. Then x_0 is not $-\frac{b}{2a}$.

$f(x_0) = a\left(-\frac{b}{2a} + e\right)^2 + b\left(-\frac{b}{2a} + e\right) + c =$

$a\left(\frac{b^2}{4a^2} - \frac{2be}{2a} + e^2\right) + \left(\frac{-b^2}{2a} + be\right) + c.$

(Solution continues on next page.)

After simplifying,

$f(x_0) = ae^2 + \left(-\frac{b^2 - 4ac}{4a}\right) = ae^2 + f\left(-\frac{b}{2a}\right)$

Since $a > 0$, then $ae^2 > 0$ and $f(x_0) > f\left(-\frac{b}{2a}\right)$ for all x_0 other than $-\frac{b}{2a}$. Thus the minimum occurs at $-\frac{b}{2a}$.

7-7 (pages 338–345)

Key Mathematical Ideas

- learning how the roots and coefficients of a quadratic equation are related
- writing a quadratic equation or function using information about the roots or the graph

Teaching Suggestions

Be sure that students understand the theorems on pages 338 and 339. Relate this lesson to the earlier lessons when students started with an equation and obtained its roots. Stress the importance of the theorem on page 339 in checking if two numbers are the roots of a given quadratic equation.

Graphing and reading information from graphs is an important part of mathematics. You might discuss in detail the solution to Example 3. This may help students when they work with conics in Chapter 9.

Use Example 4 to illustrate the process of using quadratic functions in problem solving.

Suggested Extensions

1. Given $x = r_1$ and $x = r_2$ write a quadratic equation in the form $ax^2 + bx + c = 0$ for which r_1 and r_2 are roots.
$(x - r_1)(x - r_2) = 0$
$\therefore x^2 - (r_1 + r_2)x + r_1r_2 = 0$

2. Given $x = r_1$, $x = r_2$, and $x = r_3$, write a cubic equation in the form $ax^3 + bx^2 + cx + d = 0$ for which r_1, r_2, and r_3 are roots.
$(x - r_1)(x - r_2)(x - r_3) = 0$
$\therefore x^3 - (r_1 + r_2 + r_3)x^2 + (r_1r_2 + r_1r_3 + r_2r_3)x - r_1r_2r_3 = 0$

3. Ask students to compare and discuss the results obtained.
The signs of coefficients alternate in each expansion. The numerical coefficients are formed from systematic combinations of the roots.

8 Variation and Polynomial Equations

In this chapter variation topics and topics in polynomial equation theory are presented. Variation is presented by way of problems in physics and other areas. Direct variation and proportion are presented first. A discussion of inverse and joint variation follows. The remainder of the chapter involves the study of polynomial equations and methods by which solutions may be obtained. Tools such as synthetic division, the remainder and factor theorems, and Descartes' rule of signs are introduced. The chapter concludes with a study of linear interpolation.

8-1 (pages 351–357)

Key Mathematical Ideas

- solving problems involving direct variation

Teaching Suggestions

Use several examples, including the Oral Exercises, to illustrate the use of direct variation in solving problems. The discussion of proportion is probably a review for most students by this time. Point out "cross-multiplying" in the transformation from proportion to equality of products.

You might point out the distinction between linear direct variation and other types of direct variation such as "varies directly with the square" or "varies directly with the square root."

Reading Algebra

If students have difficulty in setting up proportions for the problems in this lesson, suggest that they think of the relationships in terms of the word statement "y_1 is to x_1 as y_2 is to x_2" (page 352). Thus one can write for Problem 3 on page 356 "commission is to sales as commission is to sales" or "5400 is to 120,000 as x is to 145,000."

Suggested Extensions

In many cases there is not enough time for students to work as many problems as you feel are necessary. In this lesson and the next, assign as many word problems as possible in regular assignments, but also encourage students to do extra work on this type of word problem. Find some physics problems or have physics students bring some problems that apply to these two lessons.

Group Activities

Ask the students, in groups of four or five, to draw a "map" of their school. It may be of the room, building, campus, or an appropriate area surrounding the school. The "map" should be drawn using an appropriate scale so that other groups can determine measurements of items on the map using direct variation. This could be a long term project including in-class and out of class activities.

Using Manipulatives

In an extension of the group activity, the students could construct scale cut-outs of objects on their "map" and use them to suggest rearrangements, new construction, and other changes in the area.

8-2 (pages 358–362)

Key Mathematical Ideas

- solving problems involving inverse variation and joint variation

Teaching Suggestions

Continue the study of variation by comparing and contrasting inverse variation with direct variation. It is important in this lesson, as it was in Lesson 8-1, that students be comfortable with the concepts. You might consider the use of Oral Exercises 11–14. You might also use the Self-Test that follows this lesson as a form of mixed review.

You might want to talk with students about the number of variables that occur in a variation problem. It is possible for a variation problem to have three or more variables.

Reading Algebra

Point out the necessity of reading both the direction lines and the word problems carefully, so as to identify the type or types of variation involved in each case. Remind students to look for key words such as *is inversely proportional, varies directly . . . and inversely,* and *varies jointly.*

Suggested Extensions

As suggested in the last lesson, ask physics students to find examples of this type of problem to share with the class.

8-3 (pages 364–367)

Key Mathematical Ideas

- dividing one polynomial by a polynomial of equal or lower degree

Teaching Suggestions

Compare long division in arithmetic with the similar algorithm for polynomials. This reinforcement might help students retain the method. Emphasize that the terms must be arranged in decreasing order and that missing terms require a 0 as a place holder. Some students might forget to write the needed zeros.

As a way to check an answer ask students to perform multiplication.

Suggested Extensions

Ask students to solve Exercise 34 and discuss the solution in class.

8-4 (pages 368–370)

Key Mathematical Ideas

- using synthetic division to divide a polynomial by a first-degree binomial

Teaching Suggestions

Many students find it difficult to adjust to synthetic division because they are used to long division. You might spend some time on the discussion above Example 1 to help students see the advantages of using synthetic division. You might ask students to walk through Oral Exercises 1–6 and explain in words what is done at each step.

Be sure that students realize that the divisor is of the form $(x - c)$. The coefficient of x must be 1 and students must not forget that c is what follows the minus sign. Illustrate these points through the use of Examples 1 and 2.

Suggested Extensions

Have students divide $x^4 + 4x^2 + 3$ by $x^2 - 2x + 3$ using long division, and then have them try to develop a synthetic method for dividing a polynomial by a polynomial of the form $x^2 + bx + c$.

2, −3	1	0	4	0	3
		2	−3		
			4	−6	
				10	−15
	1	2	5	4	−12
	quotient			remainder	

$$(x^4 + 4x^2 + 3) \div (x^2 - 2x + 3) = x^2 + 2x + 5 + \frac{4x - 12}{x^2 - 2x + 3}$$

8-5 (pages 372–376)

Key Mathematical Ideas

- using synthetic division and the remainder theorem to find $P(c)$ for a given function P and a given number c
- using the factor theorem to determine whether $x - c$ is a factor of $P(x)$
- solving some polynomial equations

Teaching Suggestions

Take care to explain that $P(c)$ is actually the remainder when $P(x)$ is divided by $x - c$. Discuss with students the meanings of the two phrases "synthetic substitution" and "synthetic division." They are related, but they are not the same.

The term "reduced equation" is sometimes used for "depressed equation." Point out that reduced, or depressed, means that the degree of the quotient is the degree of the dividend reduced by 1. This lesson also provides the opportunity to discuss the concept of iteration. If one performs synthetic division over and over again, one can obtain more and more roots.

A common error is to forget to include the given root in the solution set. Emphasize to students that when writing the solution set of the original equation, they should include the given root.

Suggested Extensions

Discuss the use of nested forms of an equation to find $P(x)$. Consider the problem $P(x) = 3x^3 - 14x^2 - 15x + 8$ and the problem of finding $P(6)$.
Ask the students to find $P(6)$ by

a. direct substitution in $P(x)$,
b. synthetic substitution, and
c. substitution in the nested form $P(x) = ((3x - 14)x - 15)x + 8$. $P(6) = 62$

8-6 (pages 377–381)

Key Mathematical Ideas

- solving polynomial equations of degree n with real coefficients

Teaching Suggestions

The theorems in this lesson are very important in the development of methods used to solve polynomial equations. The students should know and understand the use of the fundamental theorem of algebra and the theorem that follows from it, the conjugate pair theorem, and Descartes' rule of signs. Point out that these theorems are aids that are used to help find the roots of polynomial equations. Note also that Descartes' rule of signs works only for polynomial equations with real coefficients.

Suggested Extensions

Ask students to explain why Descartes' rule of signs works only for polynomials with real coefficients.

8-7 (pages 382–385)

Key Mathematical Ideas

- using the rational root theorem to find roots of polynomials with integral coefficients

Teaching Suggestions

Many students find that using the rational root theorem to list all possible rational roots and then checking all possibilities makes the solution of a problem considerably involved. Point out that the process of elimination, along with aids from Lesson 8-6, will shorten and simplify a solution. Discuss with students the paragraph following the solution of Example 1.

Reading Algebra

Understanding the vocabulary of the rational root theorem is a prerequisite to understanding and using the theorem. You might ask students to read the theorem silently and then to pick out each mathematical word and phrase and explain it in their own words, or illustrate it by an example.

Suggested Extensions

Discuss with students Exercises 27 and 28 in Lesson 8-8 which deal with bounding roots. Ask students to prove that every polynomial equation has an upper bound and a lower bound for its roots if it has at least one real root.

Let S be the set of real roots of a polynomial equation $P(x) = 0$. The set S is nonempty by hypothesis. If S has one member r then $r - 1$ is a lower bound and $r + 1$ is an upper bound. If S has two or more members, then order them from least, r_1, to greatest, r_g. Then $r_1 - 1$ is a lower bound and $r_g + 1$ is an upper bound.

8-8 (pages 386–389)

Key Mathematical Ideas

- using the graph of a polynomial function P to approximate the real roots of $P(x) = 0$

Teaching Suggestions

Discuss the intermediate-value property in class using Example 1 to point out the steps to be taken in approximating the roots. The graph should be drawn as accurately as possible with the curve being a smooth curve, otherwise the approximation could be very inaccurate. Emphasize that the table of values is critical in finding the solution to a greater degree of accuracy, as shown in Example 2.

Students might find a discussion of the figure following Example 2 helpful because the graph and the interval where it crosses the x-axis are magnified. The more the figure is magnified the larger a small interval looks.

Suggested Extensions

The method described in Exercises 27 and 28 is very helpful at times. You may want to discuss this with students before they work the other exercises. The actual proof may be difficult for some students. Have one or two students do the proof on the chalkboard and discuss upper bound and lower bound with students.

Some students might be challenged by trying the Computer Key-In that follows this lesson.

8-9 (pages 391–395)

Key Mathematical Ideas

- using linear interpolation to approximate values from a table

Teaching Suggestions

Even though calculators are useful in obtaining trigonometric and logarithmic values, the topic of linear interpolation is still a valuable method to learn. It is a principle as well as a technique. Students use linear interpolation when studying logarithms and trigonometric functions. If they can learn the concept here, it will help in other areas. Point out that the problem should be set up exactly the same way each time, no matter what number is unknown. Establishing a pattern will help keep students from making careless errors. Discuss with students the material following Example 4. This is an important concept and they need to understand it in the general form.

Suggested Extensions

Ask students to discuss the assumptions that underlie the principle of linear interpolation. You might also discuss what would happen if one tried to devise a linear extrapolation method.

9 Analytic Geometry

This chapter begins with the distance and midpoint formulas. The Pythagorean theorem is used to derive the distance formula. In Lessons 9-2 through 9-6 the conic sections are introduced and studied. Circles, parabolas, ellipses, and hyperbolas are defined and their graphs and equations are studied. In Lessons 9-7 and 9-8 systems of nonlinear equations are solved. The chapter ends with an introduction to systems of three equations in three variables.

9-1 (pages 401–406)

Key Mathematical Ideas

- finding the distance between two points
- finding the midpoint of a line segment

Teaching Suggestions

Stress with students that the distance formula is a version of the Pythagorean theorem applied to points in a coordinate plane. Students should realize that finding the distance between two points is the same as finding the hypotenuse of a right triangle. Students should understand that the midpoint formula gives the coordinates of a point, not the distance from the midpoint to an endpoint.

Point out, using Example 3, that the figure must be represented in a general way. Coordinates must involve variables, not particular values.

You might use Oral Exercises 1-6 to give students practice in finding distance mentally.

Suggested Extensions

Have the students work Problem 8 and Problem 13. Then have them prove Problem 13 using the geometric or synthetic proof method. Provide geometry books if necessary.

From geometry the line segment through the midpoints of two sides of a triangle is parallel to the third side.

Therefore $\overline{PO} \parallel \overline{BD}$ and $\overline{MN} \parallel \overline{BD}$. Also $\overline{PM} \parallel \overline{AC}$ and $\overline{ON} \parallel \overline{AC}$. Therefore $\overline{PO} \parallel \overline{MN}$ and $\overline{PM} \parallel \overline{ON}$. Thus quadrilateral *PMNO* is a parallelogram.

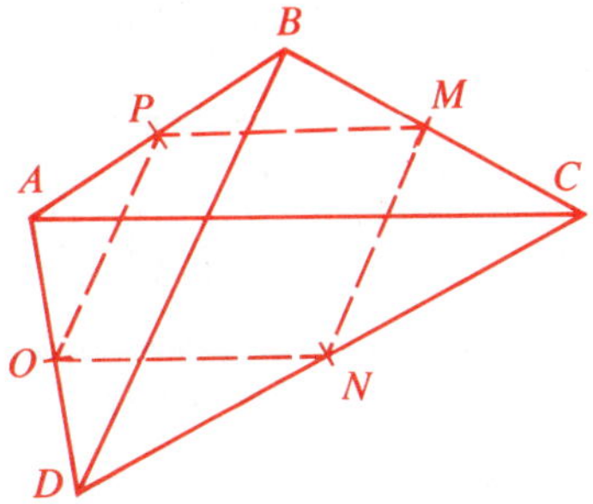

9-2 (pages 407–411)

Key Mathematical Ideas

- learning the relationship between the center and radius of a circle and its equation

Teaching Suggestions

Point out that the definition of a circle is given using distance and that the distance formula is used to derive the equation of the circle. Emphasize that the equation is of the form $(x - h)^2 + (y - k)^2 = r^2$ with minus signs in the expressions in parentheses. This use is similar to the use of minus signs in the equation of the parabola in Lesson 7-5. Continue to discuss translation in the coordinate plane. Carefully discuss Example 3a, pointing out the application of the method of completing the square.

You might discuss Example 3b with students to show that not all equations have graphs.

Suggested Extensions

The graph of the equation $x^2 + y^2 = 4$ is a circle with center (0, 0) and radius 2. By substituting values, as shown in the table, we find points on the circle.

x	y
0	± 2
1	$\pm\sqrt{3}$
-1	$\pm\sqrt{3}$
2	0
-2	0

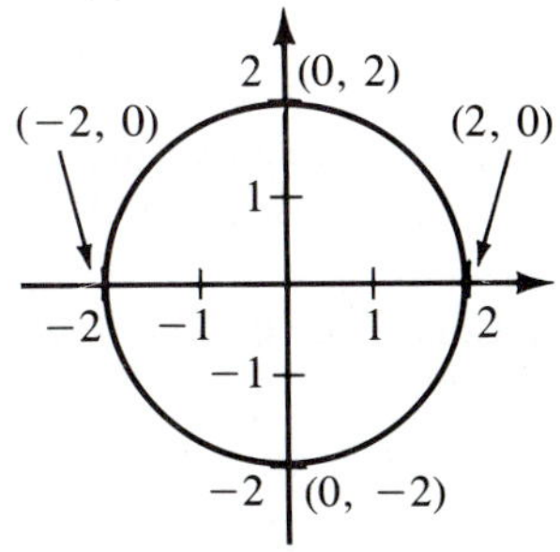

Ask students to tell whether (1, 1), (2, 3), and (−1, 0) are outside or inside the circle.

(1, 1) inside since $1^2 + 1^2 = 2 < 4$.
(2, 3) outside since $2^2 + 3^2 = 13 > 4$.
(−1, 0) inside since $(-1)^2 + 0^2 = 1 < 4$.

9-3 (pages 412–417)

Key Mathematical Ideas

- learning the relationship between the properties of a parabola and its equation

Teaching Suggestions

As with the circle, point out that the distance formula and completing the square are used in finding the equation of a parabola, given certain information. Note that the definition of a parabola is stated in terms of distance and uses the terms *directrix* and *focus*. Be sure that students know the relationships among the parabola, the directrix, the vertex, the focus, and the axis of symmetry. Students should be able to identify the properties of the parabola as illustrated in the figure at the bottom of page 412. You will probably need to work through the examples to give students a good background for working the exercises.

Reading Algebra

You may need to remind students that the diagrams in this chapter are an integral part of the exposition and that they should be read along with the text. To relate the figure at the top of page 412 to the distance formula, you may want to have some students construct the parabola on the chalkboard. Show the point (3, 0) and the line $y = 4$; then, using a piece of string with equal distances marked off on it, have students locate a number of points that are equidistant from the given point and line. When enough points have been located, they can sketch the parabola. Carefully relate the distance formula to the parabola.

Suggested Extensions

Try this paper-folding method for constructing a parabola. Draw any straight line m to be a directrix. Locate a point F not on the given line to be the focus. Fold the point F upon the directrix m, and crease the paper. Repeat this folding process from 10 to 12 times by moving F along the line m and creasing. The creases are all tangent to the parabola having F as a focus and m as a directrix. The tangents are said to "envelope" the curve and give the illusion of curvature.

9-4 (pages 418–423)

Key Mathematical Ideas

- learning the relationships among the properties of an ellipse and its equation

Teaching Suggestions

Use the illustration at the top of page 418 to show that, as with the circle and the parabola, distance is the main idea in the definition of an ellipse. Even though students should be familiar with the method by now, it is important that you discuss with them the steps in finding the equation of an ellipse. Discuss the three-part analysis of the graph of an ellipse. If students understand the three-part analysis, they should not have problems graphing in this lesson. Some students have trouble determining the major axis. The discussion just before Example 1 should help.

Stress with students that they should be able to obtain an equation from a graph and obtain the correct graph from a given equation.

Suggested Extensions

Ask students to write the equation that results from translating the graph of $\frac{x^2}{a^2} + \frac{y^2}{b^2} = 1$ h units horizontally and k units vertically.
See Lesson 9-6.

9-5 (pages 426–431)

Key Mathematical Ideas

- learning the relationships among the properties of the hyperbola and its equation

Teaching Suggestions

Since students have already found the equations for circles, parabolas, and ellipses, they should be able to find the equation for hyperbolas in the same way. Ask them to find the equation that is given in the text without looking at the steps. Emphasize the steps in analyzing the graph of the hyperbola. As with the ellipse, the steps make the task of graphing much easier. Point out the difference in the equations when the foci are on the x-axis or on the y-axis. Also, point out both the similarities and the differences of the equation of the hyperbola compared with the equation of the ellipse. Be sure students notice the "other" hyperbolas. Use the figure at the bottom of page 429 to illustrate the form of the equation and the graph.

Suggested Extensions

Ask students to give an equation for the translation of the graph of $\frac{x^2}{a^2} - \frac{y^2}{b^2} = 1$ h units horizontally and k units vertically.
See Lesson 9-6.

9-6 (pages 432–435)

Key Mathematical Ideas

- finding the equation of a conic section
- identifying a circle, ellipse, or hyperbola

Teaching Suggestions

Point out the general equations for the ellipse and hyperbola given in the text. The general equation for the circle was given in Lesson 9-2. Students should be able to transform an equation into general form and then identify it as the equation of a circle, ellipse, or hyperbola and identify its characteristics.

Suggested Extensions

1. At this point, as a review and reinforcement, use pictures or a model to show the students the conic sections as sections of a cone. Have the students discuss the conics relating what they have studied to the physical model.
2. Ask students to describe the degenerate conics $x^2 + y^2 = 0$ and $x^2 - y^2 = 0$.
3. Ask students to attempt Exercise 19 or 20 to investigate rotations.

2. The graph of $x^2 + y^2 = 0$ is the single point (0, 0). The graph of $x^2 - y^2 = 0$ is the pair of lines $x = y$ and $x = -y$.

Using Manipulatives

Divide the students into groups of two or three and provide string for them to use. Ask the students to use the description on page 418 about construction of ellipses as a guide and write a similar description for circles, parabolas, and hyperbolas.

9-7 (pages 436–438)

Key Mathematical Ideas

- using graphs to determine the number of real solutions of a quadratic system
- estimating solutions of a quadratic system

Teaching Suggestions

Since a system consisting of two quadratic equations may have 4, 3, 2, 1, or 0 real solutions, it is important that students see the possibilities. Have them draw a sketch for each of the possibilities for several different combinations of conics. Point out, as noted in the text, that the more accurate the graph, the more accurate the estimation of solutions when using the graphic method for solving a quadratic system.

9-8 (pages 439–443)

Key Mathematical Ideas

- using algebraic methods to find exact solutions of quadratic systems

Teaching Suggestions

Review the substitution and linear combination methods used in Chapter 3 to solve linear systems. Point out that the substitution method should be used if one of the equations contains a linear term. The linear combination method is more convenient with central conics. Note that a sketch of the system is not required but might help in relating to the system being solved. Be sure that the students understand that a system may have 4, 3, 2, 1, or 0 real solutions, as discussed in Lesson 9-7. Spend some time on problem solving exercises.

Group Activity

Have the students work problem 14 on page 443 as a group. Encourage discussion among students as to the best methods and procedures to use to work the problem.

9-9 (pages 444–449)

Key Mathematical Ideas

- solving systems of linear equations in three variables

Teaching Suggestions

Increasing the number of variables does not change the method of solving the system except that it may increase the number of steps. Work through the examples in the text. Students may need to see several examples of the Gaussian elimination method. Explain carefully the steps that are to be taken, using Example 3.

Suggested Extensions

Ask a group of students to study the matrix approach to transforming a system to triangular form and to explain the method to the class.

10 Exponential and Logarithmic Functions

The chapter begins by giving the definition of b^x when x is an integer, as studied in previous chapters, and extending the existing properties to include rational exponents. The meaning of b^x is extended even further to include irrational values of x so that we now have b^x defined for all real numbers x. This lets us define the exponential function. Composites and inverses of functions are discussed completely, leading up to the definition of a logarithmic function as the inverse of an exponential function. The laws of logarithms are given and applications are discussed. Then students use exponential and logarithmic functions to solve growth and decay problems. In the last lesson the natural logarithm function is introduced.

10-1 (pages 455–458)

Key Mathematical Ideas

- extending the meaning of exponents to include rational numbers

Teaching Suggestions

Emphasize the two ways that an expression with rational exponents can be written in radical form. Point out that a rational exponent is a combination of a power and a root. Show how the expression in Example 4 can be written using both exponential and radical form. Use Example 3 to show how changing to exponential form can help in simplifying expressions.

Suggested Extensions

Ask students to write $\sqrt[4]{\sqrt[3]{\sqrt{5}}}$ in exponential form.
$\sqrt[4]{\sqrt[3]{\sqrt{5}}} = 5^{1/2 \cdot 1/3 \cdot 1/4} = 5^{1/24}$

10-2 (pages 459–462)

Key Mathematical Ideas

- extending the meaning of exponents to include irrational numbers and to define exponential functions

Teaching Suggestions

Use Figures 1 through 3 to illustrate the drawing of a detailed graph of $y = 2^x$. Also use these figures to discuss with students that, as the graph is filled in, the meaning of 2^x where x is irrational becomes clearer. Define the exponential function and discuss the distinct nature of its graph.

You might pay particular attention to the steps used to solve exponential equations.

Suggested Extensions

You might direct students to the Extra that follows this lesson on page 462. Students might use a calculator to compare 2^x to x^n.

10-3 (pages 463–467)

Key Mathematical Ideas

- finding the composite of two given functions
- finding the inverse of a given function

Teaching Suggestions

In defining and illustrating the composition of functions, be sure to point out that the operations are performed from the inside out, like algebraic operations involving parentheses. $f(g(x))$ means g operates on the variable first, then f operates on that value. Point out that parts (c) and (d) of Example 2 provide an example of how to determine if two functions are inverse functions. Discuss carefully the text at the top of page 465 and the horizontal line test. You might wish to draw the graphs of $y = x^2$ and $y^2 = x$ to demonstrate the test. You might ask students to use a ruler as a physical way of applying the horizontal line test in Written Exercises 7-10.

Be sure students can successfully perform the work needed to obtain the inverse function of a function that has an inverse function. Some students become confused when they try to interchange x and y.

Suggested Extensions

Spend extra time on this lesson, if necessary, to be sure students understand and can work with composition and inversion of functions.

10-4 (pages 468–472)

Key Mathematical Ideas

- defining logarithmic functions
- learning how logarithmic functions and exponential functions are related

Teaching Suggestions

If this is the first time students have studied logarithms, it is important that you make sure they understand the concept of the inverse exponential function and the definition of logarithm based on the exponential function. Students should learn how to transform any logarithmic equation into an exponential equation.

Reading Algebra

The simple expression $\log_b N = k$ may be difficult for students to understand and read because of the unfamiliar order and position of the symbols. You may wish to keep the following diagram on the chalkboard for reference.

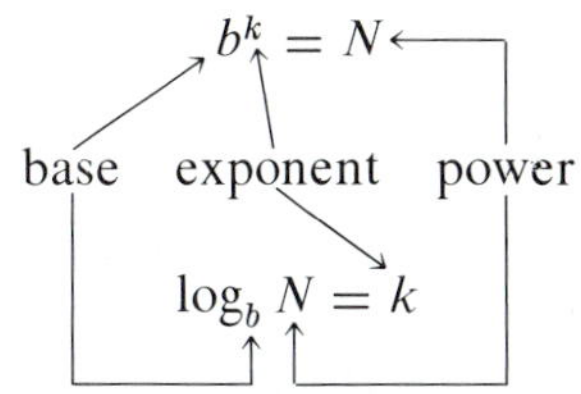

Have students answer the Oral Exercises with complete sentences until they become accustomed to the notation.

Suggested Extensions

Ask students to work Exercise 42 in the Written Exercises. Have a student write the proof on the chalkboard and explain it.

10-5 (pages 473–477)

Key Mathematical Ideas

- learning and applying the basic properties of logarithms

Teaching Suggestions

Discuss the laws of logarithms and relate them to the laws of exponents. Work through the proofs of laws

1 and 3 given in the text to be sure students understand the laws. Work several examples to show how the laws help in simplifying expressions containing logarithms and in solving logarithmic equations. Assign at least a few of Exercises 33–40 in the Written Exercises. Students need to work with logarithmic equations.

As in other lessons where equations are solved, stress the importance of checking answers.

Suggested Extensions

Ask the students to find out about the work of Napier and Briggs in relation to logarithms, and to report their findings to the class.

10-6 (pages 478–482)

Key Mathematical Ideas

- using common logarithms in solving equations

Teaching Suggestions

Even though many students have calculators that can be used to find logarithms, they should learn how to use the tables of logarithms. Stress the importance of the properties of logarithms in solving exponential equations like the equation in Example 4. Explain the change-of-base formula and give several examples.

Suggested Extensions

Ask students to compute $\log_2 12$ by using a calculator and by using a computer. This exercise should lead to a practical need for a change-of-base formula. Ask students which logarithm functions are built into the calculator and computer and which functions are not. Whether by using a calculator or computer, students must use the change-of-base formula since ln and log are usually built-in functions. The function $\log_2$ is not built-in.

10-7 (pages 483–488)

Key Mathematical Ideas

- using exponential and logarithmic functions to solve growth and decay problems

Teaching Suggestions

Help students work through the examples. Many students find this type of problem very difficult. Students should be required to work as many of these exercises as you can possibly fit into the schedule.

Suggested Extensions

Ask students to solve the equation $x = x_0 \cdot 2^{-t/h}$ for t.
Use base 2 logarithms. $t = -h \log_2 \left(\frac{x}{x_0}\right)$

10-8 (pages 489–491)

Key Mathematical Ideas

- defining and using the natural logarithm function

Teaching Suggestions

Impress upon students that if they are going to continue in the study of mathematics they will use the natural logarithm function extensively. Discuss the definition of the number e and the function ln, then work several examples with students. Have students work some of each type of problem in the Written Exercises.

Suggested Extensions

1. Ask students to work Exercise 50 and discuss the solution in class.
2. Ask students to evaluate $(1 + h)^{1/h}$ as h approaches 0 and discuss the limit.
   The limit is e.

11 Sequences and Series

The chapter starts by introducing sequences in general, giving several examples of different sequences, and defining finite and infinite sequences. More specifically, arithmetic sequences and geometric sequences are studied. Methods for finding terms of both arithmetic and geometric sequences are studied. Series are defined and sigma notation is introduced. One lesson is devoted to finding sums of arithmetic and geometric series. An important topic in advanced mathematics, the infinite geometric series, is presented and used to solve a variety of applied problems. In the last two lessons powers of binomials and the binomial theorem are studied.

11-1 (pages 501–506)

Key Mathematical Ideas

- determining whether a sequence is arithmetic, geometric, or neither
- supplying missing terms of a sequence

Teaching Suggestions

In this lesson, sequences in general are discussed. Add several examples to the ones in the text to help students understand that there are many types of sequences. An arithmetic sequence is defined using the common difference and a geometric sequence is defined using the common ratio. Compare the two types of sequences, showing similarities and differences. As is stated following Example 3, patterns are sometimes hard to find. Some students find the study of sequences difficult because they have a hard time seeing patterns. Observe students carefully to see if any are having real difficulty with the basic concepts of sequences. If so, they may need extra help and extra work on the less difficult sequences.

One way to help students understand arithmetic sequences is to relate such sequences to linear functions. Then you might relate geometric sequences to exponential functions.

As students become comfortable looking for patterns, they might find the topics in this lesson enjoyable and challenging.

Suggested Extensions

1. Ask students to continue the following sequences.
   a. O, T, T, F, F, S, S, __?__, __?__, __?__
   b. 1, 4, 9, 61, 52, 63, 94, __?__, __?__
   a. One, Two, Three, Four, Five, Six, Seven, Eight, Nine, Ten
   b. 1^2, 2^2, 3^2, 4^2, 5^2, 6^2, 7^2, 8^2, 9^2 (with digits reversed)
2. Ask students to work Exercise 27 (the Fibonacci sequence), Exercise 31 (triangular numbers), or Exercise 32 (pentagonal numbers).

Group Activities

This activity is a warm-up for the whole class when discussing sequences in general. First, a student will call out a number to begin the sequence. Students will take turns calling out a number, trying to create a sequence. The first few results will probably be 1, 2, 3, . . . , or 2, 4, 6, . . . , but as they get into the activity some students will try to make up more complicated sequences. The results of this activity could lead to a discussion of the characteristics of arithmetic and geometric sequences and of those that are neither.

11-2 (pages 507–509)

Key Mathematical Ideas

- finding a formula for the *n*th term of an arithmetic sequence
- finding specific terms of arithmetic sequences

Teaching Suggestions

Finding the nth term of an arithmetic sequence is not a difficult task with the formula given. Discuss with the students why the formula works. Point out that a single arithmetic mean, called *the arithmetic mean,* is the average, but that arithmetic means in general are not averages.

Suggested Extensions

Ask the students to work and explain Exercise 33.

Group Activities

There are many patterns in nature which are made up of the sequences discussed in this chapter. Divide the class into small groups. Ask each group to find pictures or descriptions of examples of these patterns in nature and bring them to class. Each group will then make up a folder or scrapbook illustrating the patterns they have found and describing them mathematically.

11-3 (pages 510–515)

Key Mathematical Ideas

- finding a formula for the nth term of a geometric sequence
- finding specific terms of a geometric sequence

Teaching Suggestions

Finding the nth term of a geometric sequence is not a difficult task with the formula given. Discuss with the students why the formula works. Discuss geometric means pointing out the difference between the geometric mean of two numbers and geometric means between two numbers. Another way of showing why the geometric mean of a and b is $\sqrt{ab}$ is to set up the following proportions with x, the geometric mean, in the mean position:

$$\frac{a}{x} = \frac{x}{b}$$

Then $x^2 = ab$, and $x = \sqrt{ab}$ if a and b are both positive and $-\sqrt{ab}$ if a and b are both negative.

Suggested Extensions

Ask students to prove that if a and b are positive and $a < b$ then $a < \sqrt{ab} < b$.

$a < b$. So $a^2 < ab$ and $ab < b^2$. Since a and b are positive $a < \sqrt{ab}$ and $\sqrt{ab} < b$. Therefore $a < \sqrt{ab} < b$.

11-4 (pages 518–522)

Key Mathematical Ideas

- identifying series and using sigma notation

Teaching Suggestions

Emphasize that a sequence is an ordered list of numbers while a series is the indicated sum of those same numbers. Students should not have trouble understanding the terms arithmetic and geometric as they are related to series. Introduce the sigma notation as a shortcut for writing the sum. Some students have difficulty with the notation for upper and lower limits. Use several examples to show different limits.

Suggested Extensions

Ask students to compute the double sum in Exercise 35.

11-5 (pages 525–530)

Key Mathematical Ideas

- finding sums of finite arithmetic and geometric series

Teaching Suggestions

Work out the proof of the summation formula for an arithmetic series on the chalkboard. Point out the two different ways in which the series is written and explain why that step was taken. Emphasize with students that this is not "trickery" but a legal method of deriving the formula and that they should be able to understand what has been done. After discussing the formula for an arithmetic series, see if a student can lead a discussion with the class on the proof of the formula for the sum of a geometric series.

Suggested Extensions

Write the following diagram on the chalkboard and ask students to find the sum without writing any work on paper.

$$1 + 2 + \cdots + 99 + 100$$

101

101

There are fifty 101s. The sum is $50 \times 101 = 5050$.

11-6 (pages 531–536)

Key Mathematical Ideas

- finding sums of infinite geometric series having ratios with absolute value less than one.

Teaching Suggestions

Carefully discuss the material in the text before Example 1. Many students have an idea of what an infinite series is but have difficulty with the concept of a sum of infinitely many terms. Emphasize that for an infinite geometric series to have a sum, $|r| < 1$.

Suggested Extensions

Consider the infinite geometric series $0.7 + 0.07 + 0.007 + \cdots$ as in Example 2. Since $r = 0.1$, the sum of the series is

$$S = \frac{t_1}{1 - r} = \frac{0.7}{1 - 0.1} = \frac{0.7}{0.9} = \frac{7}{9}.$$

Now have students find the sum of the infinite geometric series $0.9 + 0.09 + 0.009 + \cdots$. $\frac{9}{9}$ or 1

11-7 (pages 537–539)

Key Mathematical Ideas

- expanding powers of binomials

Teaching Suggestions

Ask students to actually multiply out the powers of $(a + b)^n$ for $n = 2$ through $n = 5$. This will help them appreciate Pascal's triangle, another method for expanding a binomial. Be sure students notice the pattern of the exponents and that they realize that if the coefficients are not equal to 1, the pattern of the coefficients of the expansion will be different. Illustrate this using Example 2.

Suggested Extensions

Ask students to write Pascal's triangle for $(a - b)^n$ for $n = 1$ through $n = 4$.

1 −1

1 −2 1

1 −3 3 −1

1 −4 6 −4 1

11-8 (pages 540–543)

Key Mathematical Ideas

- using the binomial theorem to find a term of a binomial expansion

Teaching Suggestions

Use the examples in the text to develop the concept of the binomial theorem. Some students have difficulty understanding the theorem without additional work and examples. The best way for students to learn the binomial theorem and learn how to use it is to work problems. Assign a sufficient number of exercises to assure that each student has been properly introduced to the topic.

Emphasize with students the substitution of $2x$ for a and $-y$ for b in Example 3.

Reading Algebra

An important skill in reading is the ability to read generalizations. Two tests of one's understanding of a generalization are the ability to express it in one's own words and the ability to illustrate it by means of a specific example. For any given generalization, one of these tests is likely to be much easier to use than the other. Students may gain a deeper appreciation of the clarity and compactness of mathematical symbolism if you ask them to consider expressing the binomial theorem in words.

Suggested Extensions

Ask students to work Exercise 29.

12 Triangle Trigonometry

In this chapter students are introduced to trigonometry by defining an angle using the concept of rotation. The meaning of angle measure is extended to include more than 180° and less than 0°. The six trigonometric functions of acute angles are defined in terms of the sides of a right triangle. First, values of trigonometric functions are given for 30°-60°-90° triangles and 45°-45°-90° triangles. Then the trigonometric functions of general angles are determined using reference angles. Calculators and tables are used in finding the values of trigonometric functions of angles whose values are given. After the introductory material is covered, triangles are solved; first right triangles using trigonometric functions, then general triangles using the law of sines and the law of cosines. In the last lesson, several formulas for the area of a triangle are introduced and studied.

12-1 (pages 549–554)

Key Mathematical Ideas

- using degrees to measure angles

Teaching Suggestions

In this lesson only degrees will be used. Radians will be introduced later. Changing the definition of angle and extending degree measure may cause some temporary confusion but after a few examples students should have no trouble adjusting to the change. It is very important that students learn and understand the terms and examples in this lesson because they will use them throughout the study of trigonometry. Two important terms are quadrantal angle and coterminal angle.

Suggested Extensions

1. Introduce students to another system of angle measure (grads) in which the measure of a right angle is 100. Point out that this can be considered a metric unit, since it is based on tens. Some calculators give the choice of degrees, radians, or grads.
2. Have students make up their own systems of angle measurement. Use their systems to emphasize the

arbitrary nature of units of measure. Have them convert between measures in these made-up units and degrees.

12-2 (pages 555–560)

Key Mathematical Ideas

- defining the trigonometric functions of acute angles

Teaching Suggestions

You cannot over-emphasize the importance of learning the definitions of the trigonometric functions in terms of the right triangle. Relate the discussion to the Pythagorean theorem as is done in Example 1. Point out that identities will be used extensively in later sections. A very important part of this lesson is the development of the table of values for 30°, 45°, and 60°.

Suggested Extensions

Have students investigate the linguistic origins of the terms sine, tangent, and secant.

12-3 (pages 561–567)

Key Mathematical Ideas

- defining trigonometric functions of general angles

Teaching Suggestions

Some students may be slow to accept the fact that some trigonometric functions are not defined for certain angles. Take the time to clarify this point, which is quite different from previous work with acute angles.

It is important that students understand the points made in this lesson. The signs of the trigonometric functions for each quadrant are determined by the signs of x and y. They may be remembered in table form as presented in the text or by using a memory aid like the following. The signs of the trigonometric functions in each quadrant may be remembered by using the following figure.

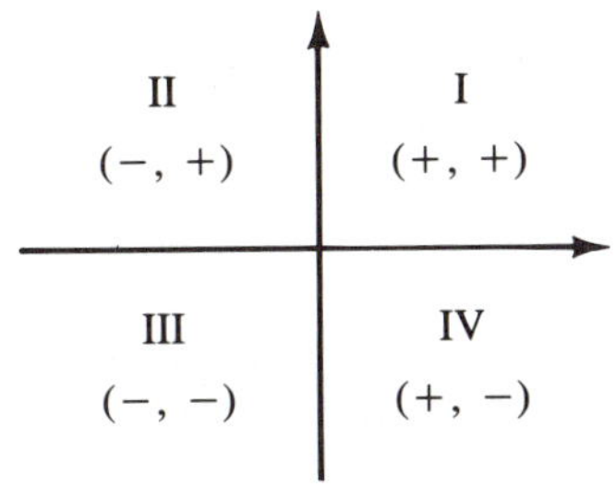

Students will appreciate reference angles when they begin using trigonometric tables; functions of any angle can be found using only angles of 90° or less. You should be aware that students using calculators will not need to use reference angles.

Example 6 makes what students may consider a surprising point: given the value of only one function and some information about the magnitude of the angle, values of all the functions can be determined.

Suggested Extensions

Have students write a formula giving the values of A for which the indicated function is not defined.

1. tan A
$A = 90° + n \cdot 180°$

2. cot A
$A = n \cdot 180°$

3. sec A
$A = 90° + n \cdot 180°$

4. csc A
$A = n \cdot 180°$

12-4 (pages 568–572)

Key Mathematical Ideas

- using calculators and tables to find values of trigonometric functions

Teaching Suggestions

Explain to students why trigonometric tables are read down for angles less than 45° and up for angles between 45° and 90°. They will appreciate this ingenious application of cofunctions. Many students will not have studied minutes before. Explain this subdivision of a degree; you may also want to include seconds in your explanation.

Students using calculators may be mystified at first about how to find cotangents, secants, and cosecants, since most inexpensive calculators have only sine, cosine, and tangent keys. Explain how to use reciprocal functions and the calculator's reciprocal key for the functions without their own keys. In Example 3 discuss, informally, the idea of inverse functions. Show students how to use their calculators to find inverse trigonometric functions. Note that there is some variation between calculators, and some students will need to be shown how to use theirs individually. Relate these inverse functions to others they have studied, such as squares and square roots, or exponents and logarithms.

Call attention to the fact that some of the trigonometric functions are decreasing functions. As in Example 2, these functions give smaller function values for greater angles. It is important to have a negative difference when interpolating with these functions, as is shown in the example.

Suggested Extensions

Have students investigate values of sin A and tan A to find over what range of values of A there is not a significant difference between the two functions. Use tables or a calculator.

Reading Algebra

Although mathematics is a discipline in which precision and clarity are among the highest values, it is nevertheless true that we sometimes find ourselves using the same symbol for totally different mathematical entities. For example, we use the symbol "$-$" for at least three distinct concepts.

1. opposite, as in $-a$
2. a negative number, as in -5.3
3. subtraction, as in $3 - 7 = -4$

The same is true of the raised $^{-1}$ symbol. Students have encountered this earlier as a negative exponent representing the reciprocal of a number. Now in Example 3 they are introduced to a new use of the same symbol, namely the inverse trigonometric functions. It is important to realize that this is not a reciprocal function at all. You may want to review the material on inverse functions in Chapter 10.

Group Activity

Students can generate their own tables of values of the sine and cosine, using graph paper. Set up a coordinate system and draw a circle with a radius of 10 spaces. Call this radius the unit of length, so that each graph paper square represents one tenth. For a table of sine values, place the given angle in standard position. Mark the point where the terminal side of the angle crosses the circle. Then since

$$\sin A = \frac{y}{r}$$

and $r = 1$, the value of sin A is simply y, which can be read from the graph.

Working in groups of three or more, have students complete three versions of the table below, one using the table in the textbook, one using a calculator, and one using the graph paper method outlined above.

A	0°	10°	20°	30°	40°
sin A					
A	50°	60°	70°	80°	90°
sin A					

Students can compare results of the three methods. They will be pleased to see that the graph paper method gives reasonable approximations. This may also reassure students that it would be a simple matter to come up with a rough value of a given trigonometric function if necessary.

12-5 (pages 574–579)

Key Mathematical Ideas

- finding the sides and angles of a right triangle

Teaching Suggestions

Explain that solving a triangle means finding the measures of its sides and angles. In a right triangle the measure of the right angle is already known and the sum of the measures of the acute angles must be 90°. Use Example 1 to illustrate that a triangle can usually be solved in more than one way. Students using tables may choose the method involving multiplication rather than division. The choice will be immaterial to those using calculators, since there is then no difference in the difficulty of the computations. Students usually find the application problems interesting, and a refreshing excursion into the "real world," which contains many problems that involve solving triangles. When discussing angles of elevation and depression, it may help to remind students of the equality of the alternate interior angles they studied in geometry. As part of the discussion of Example 2 it is a good idea to review the mechanics of solving a system of equations.

Note that Written Exercises 21–24 are very well suited for solution by calculator or computer.

Suggested Extensions

Have students measure the angle of elevation of a variety of objects such as trees, poles, and buildings. Then have them calculate each object's height using the methods of this lesson.

12-6 (pages 580–584)

Key Mathematical Ideas

- using the law of cosines

Teaching Suggestions

Stress with students that in order to use the law of cosines they must know the lengths of two sides and the measure of the angle between them.

Discuss the Pythagorean theorem as a special case of the law of cosines. Ask students to look for a pattern in the use of the variables in the three forms of the law. Each form of the law involves both a lower case letter and the capital letter corresponding to it. Ask students to complete the following equation.

$$a^2 = (\ \ )^2 + (\ \ )^2 - 2(\ \ )(\ \ ) \cos (\ \ )$$

Caution students that they will sometimes be substituting negative values for the cosine in the formula. They must be careful with signs then, since the result of the "$-2ab \cos C$" term is addition of a positive number.

When the law is expressed as a formula for $\cos C$ in terms of a, b, and c (just before Example 2), students can be shown the trigonometric analogue of the SSS property: For a given combination of a, b, and c, there is only one possible value of C between 0° and 180°. In other words, three sides uniquely determine a triangle.

Suggested Extensions

1. Using the law of cosines, have students investigate the relationships between c^2 and $a^2 + b^2$ when C is acute, right, or obtuse.
2. Ask students to derive special cases of the law of cosines for the following situations.
   a. $C = 90°$ $c^2 = a^2 + b^2$
   b. $C = 60°$ $c^2 = a^2 + b^2 - ab$
   c. a and b are the legs of an isosceles triangle. $c^2 = 2a^2(1 - \cos C)$

12-7 (pages 586–590)

Key Mathematical Ideas

- using the law of sines

Teaching Suggestions

Like the law of cosines, the law of sines applies to all triangles; furthermore, it too requires knowledge of three parts of a triangle in order to solve for the others. Although it seems that the area formulas on page 586 have no use except to enable proof of the law of sines, mention that we will see these formulas again in Lesson 12-9. Point out that for the law of sines to be used either two sides and the angle opposite one of the sides or two angles and a side must be known.

Congratulate students who recognize that SSA is not one of the triangle congruence properties. This means that given a combination of side, side, and nonincluded angle, it is possible to construct two noncongruent triangles. The preceding ideas will give students a foretaste of the ambiguous case in Lesson 12-8, as does Example 1.

Suggested Extensions

Ask students to explain how there could be two solutions in Example 1a before looking at Lesson 12-8. The equation $\sin B = 0.8571$ has two solutions for $0° < B < 180°$. The sine is positive in both the first and second quadrants.

12-8 (pages 591–596)

Key Mathematical Ideas

- solving any given triangle

Teaching Suggestions

Review the side and angle combinations listed at the top of page 591. Appeal to students' experience with these combinations in geometry. Remind students that AAS and ASA are basically the same, and that all the combinations listed except SSA are congruence properties. To help students understand why SSA is not a congruence property, see the Using Manipulatives activity below. This lesson is the most difficult of the chapter. First, students must decide whether to use the law of sines or the law of cosines. If the problem is an SSA one, students must be cautioned not to proceed too mechanically, as they may have been able to do in earlier problems. SSA problems require alertness and judgment; students must analyze the results of their calculations to decide how many solutions there are. Reassure students that SSA is the only situation in which there can be more than one solution.

Using Manipulatives

Students will understand the SSA ambiguous case better after the hands-on activity described here.

Draw an acute angle such that one side is a ray and the other side is a segment. Thus we have a fixed angle and a fixed side, but the other side is not determined.

Distribute cardboard or heavy paper strips and thumb tacks. Push the tack, point up, through the paper at the end of the segment not at the vertex. Press the end of a strip over the tack, using a pencil eraser to avoid injury. Pivot the strip until its end touches the ray of the fixed angle. Now use several longer or shorter strips, finding strips that hit the ray no times, once, or twice. In each case where the strip does touch the ray, students can see an SSA case—the strip (S), the drawn segment (S), and the drawn angle (A).

This apparatus also makes a very effective SSA demonstration for an overhead projector.

Suggested Extensions

Ask students to solve the word problems, drawing a sketch for each problem and showing all work involved in solving the problem. Have students share their solutions with the rest of the class.

12-9 (pages 597–600)

Key Mathematical Ideas

- using formulas for the areas of triangles

Teaching Suggestions

Discuss with the students each of the formulas for the area of a triangle. Work through the examples with them, explaining the use of each formula.

You might discuss with students that trigonometry can be used not only to find linear and angular measure but also area, or square measure.

Suggested Extensions

1. Since any polygonal region can be partitioned into triangles, Hero's formula can be considered a "surveyor's formula." Lay out some irregular spaces. Have students subdivide them into triangles, then use Hero's formula to find the areas of the triangles. Find the area of the entire region by summing the areas of the triangles.

2. Have students use each of the three types of triangle area formulas on page 597 to derive a formula for the area of an equilateral triangle in terms of the length of a side. If necessary, remind them that $a = b = c$ and each angle measures 60°. If a geometry book is available, ask them to confirm that the result is the same as the formula from geometry.

13 Trigonometric Graphs and Identities

This chapter extends the study of trigonometry and introduces the circular functions. To begin, radian measure is defined and conversion between radians and degrees is practiced. Using the unit circle, circular functions are defined, and their similarity to the trigonometric functions noted. As an introduction to graphing, the concepts of periodicity and symmetry are explored and amplitude, vertical and horizontal stretches, and shifts are studied. After detailed introduction to the graphs of the sine and cosine functions, the graphs of the other trigonometric functions are studied. The remainder of the chapter is devoted to a variety of trigonometric identities. The cofunction, reciprocal, and Pythagorean identities are reviewed and applied. Addition formulas for the sine and cosine are introduced, and from these double- and half-angle formulas for the sine and cosine are derived. Once the formulas for the sine and cosine are established, corresponding formulas for the tangent are derived by using the definition of the tangent.

13-1 (pages 607–612)

Key Mathematical Ideas

- using radians to measure angles

Teaching Suggestions

Discuss radian measure in general with students and develop the formula $\theta = \frac{s}{r}$, as is done in the text.

The conversion rules of $1° = \frac{\pi}{180}$ radians and $1 \text{ radian} = \frac{180}{\pi}$ degrees should be learned by students. Be sure that students understand the development and use of the formulas for circular-arc length, circular-sector area, and angular speed. You may need to work through many examples in class. Note that the formulas require that the angle measure be given in radians. Be sure students work some of the word problems in the section.

Suggested Extensions

Students can make their own radian protractors.

Trace a regular protractor on heavy paper. Use the fact that 0.2 radians ≈ 11.5° to make marks on the paper protractor. Pick up the regular protractor, and cut out the tracing. Now label the marks with the correct decimal radian measure. To help students grasp radian measure meaningfully, they should now use their radian protractors to measure given angles and to draw angles of given radian measures.

13-2 (pages 613–617)

Key Mathematical Ideas

- defining circular functions

Teaching Suggestions

Some students have trouble grasping the fact that there are circular functions that are not related to angles, even though these same students may be perfectly capable of working with the circular functions. Emphasize that the circular functions are functions of arc lengths on a unit circle and, thus, they are functions whose domains are real numbers, not angles.

Since the definitions of the circular functions depend on points on a unit circle, take time to review this circle and, particularly, the equation

$$x^2 + y^2 = 1.$$

Mention that here, as earlier, once functions of the sine and cosine are defined, the other functions are defined in terms of them.

Suggested Extensions

Some students may have discovered an ingenious way to convert between degrees and radians using their calculators. To convert from degrees to radians, set the calculator in degree mode. Press either the sine or cosine key. Now switch to radian mode, and press the inverse of that function. The result is the decimal radian equivalent of the original degree measure. A similar process converts radians to degrees. To try this method, direct students to Exercises 29–44 on page 611. Ask why the sine is better to use than the tangent (sine is defined for all angles, tangent isn't), and why calculators give incorrect results, using the sine, if the angle is not between $-90°$ and $90°$ (the inverse keys give only the principal value of the inverse function).

Using Manipulatives

The diagram on page 617 can be used as the plan for making a device for direct-reading of cosines of real numbers. On graph paper with large squares, lay out x- and y-axes and a unit circle. Glue this paper to a piece of cardboard and cut out the circle. Put the cardboard circle upright, so that it is perpendicular to the floor. (Pin it to a bulletin board, for example.) Place a strip of graph paper to form a number line with its origin at A, parallel to the y-axis. For a plumb line, cut a piece of string and tie a weight to one end. To find the cosine of a number, follow the example below.

Example $\qquad$ cos 0.65

Use a piece of light string or thread. Place the end at A. Stretch the string along the vertical number line through A, making the second endpoint of a segment 0.65 units long. Leaving the end at A fixed, wrap the string around the cardboard unit circle. Where the end of the 0.65 segment lies on the unit circle, drop the plumb line. The line will cross the x-axis; simply read the coordinate and you have the value of cos 0.65. Repeat the procedure for any other cosine. Ask students how to use this apparatus to read sines.

13-3 (pages 619–623)

Key Mathematical Ideas

- using periodicity and symmetry in graphing functions

Teaching Suggestions

The concept of a periodic function is not difficult for most students, primarily because it is so visual. Once the graph of a function is drawn, the graph shows whether the function is periodic, and, if so, what the period is. To students, periodic means that a part of the graph can be shifted right or left repeatedly to produce more of the graph. It can help students to discuss some repetitive, cyclical phenomena that they are familiar with, such as tides, seasons, plant growth, and so on.

Symmetry should be familiar from earlier studies. Review the idea of symmetry to a line and symmetry to a point, which underlie the meaning of even and odd functions. Your students may benefit from an explanation of the terms even and odd. Power functions with even exponents, like the cosine function, are symmetric to the y-axis. Those with odd exponents, like the sine, are symmetric to the origin.

Suggested Extensions

Have students work Exercise 30. Then have them find out what areas of mathematics require the use of hyperbolic sine and hyperbolic cosine functions.

13-4 (pages 624–629)

Key Mathematical Ideas

- graphing the sine, cosine, and related functions

Teaching Suggestions

Once selected values of the sine and cosine functions are obtained, this information is combined with facts established earlier about periodicity, symmetry, and oddness and evenness to sketch the full graphs of $y = \sin x$ and $y = \cos x$. Build on these basic graphs to determine the effects of constants governing shifts and stretches of the basic graphs. If students have difficulty it is likely to be with the constant b in $y = \sin bx$ or $y = \cos bx$. It is somewhat counter to student intuition that a greater value of b leads to a shrink, not a stretch, with a smaller, not larger, period. Make it clear to students that as far as the graphs of these functions are concerned the roles of a, b, and c in $y = c + a \sin bx$ and $y = c + a \cos bx$ are independent of each other, and the order in which the constants are considered in analyzing a graph does not matter. The vertical shift, vertical stretch or shrink, and change of period can be done in any order.

Suggested Extensions

Have students graph each function. Then ask them to determine the relationship among the functions.

1. $y = \sin x$ **2.** $y = \cos\left(x - \frac{\pi}{2}\right)$

3. $y = -\cos\left(x + \frac{\pi}{2}\right)$ **4.** $y = -\sin(x - \pi)$

They are all the same.

13-5 (pages 630–633)

Key Mathematical Ideas

- graphing tangent, cotangent, secant, and cosecant functions

Teaching Suggestions

Some students who have little trouble with the sine and cosine find graphing the other functions difficult. This seems to be caused by the discontinuity of these functions and the sudden swings between very large and very small values as the asymptotes are crossed. Emphasize that each value for which one of these functions is undefined corresponds to an asymptote. Visualization of the secant and cosecant graphs is improved for most students if they can see each function together with its reciprocal function, as is shown on page 632.

Reading Algebra

This is a good time to remind students of the resources available to them when they have difficulty with concepts or terms, or when they cannot recall definitions or skills from earlier work. Unfortunately, sometimes students lose their momentum in such a situation and come to a dead stop, struck mentally with the feeling "I don't know." Turn this into "I can find out." Urge students to review material when the text makes a reference to earlier work, as is done when the term asymptote is used on page 630. Remind them that this book contains a glossary and an index, real aids and time-savers in recalling or locating terms and concepts introduced earlier.

Suggested Extensions

Encourage students to try more challenging graphs such as these.

1. $y = \sin^2 x$

2. $y = |\tan x|$

3. $y = \sin x + \cos x$

1.

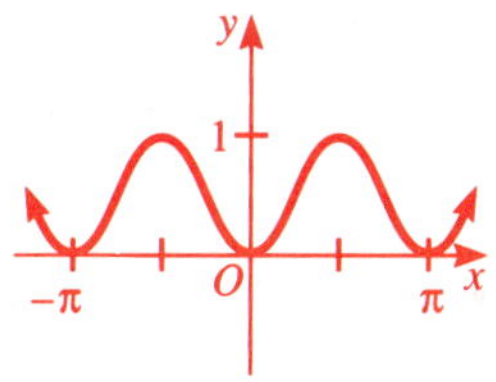

2.

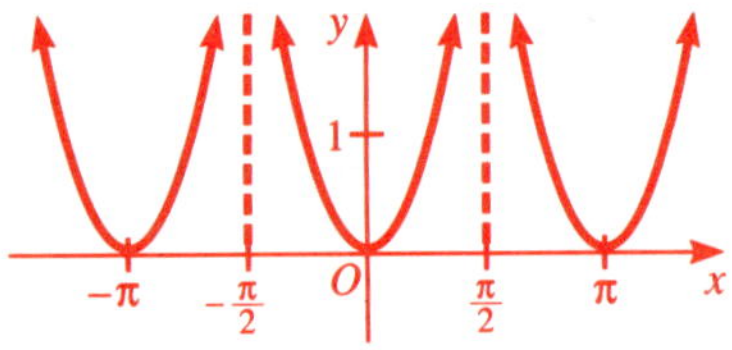

3.

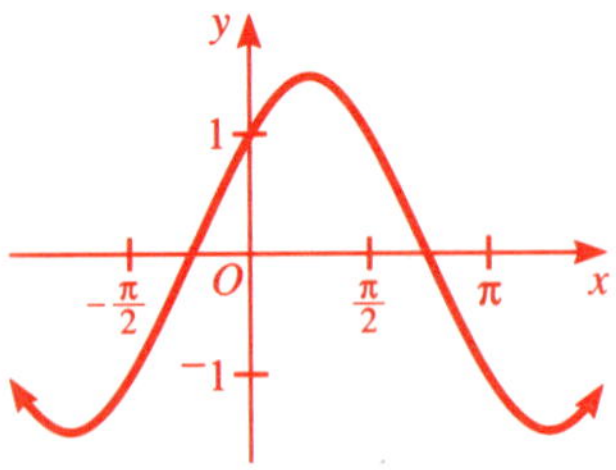

13-6 (pages 636–640)

Key Mathematical Ideas

- simplifying trigonometric expressions and proving identities

Teaching Suggestions

Review the reciprocal, cofunction, and Pythagorean identities as students have already been using these in a slightly informal way. Students are certainly familiar with the idea of "simplifying" expressions. Problems like Example 1 make it clear that the result really is "simpler." Discuss the general strategies for proving identities. Also explain the special strategies, which are techniques for trigonometric identities in particular.

Some students, either consciously or otherwise, insist on working with both sides of an identity; that is, assuming the identity is true before proof. With these students, you may want to discuss the idea of reversibility of a proof. Typically these students assume the identity and work to a conclusion that is obviously true. Thus they have proved "If $a = b$, then $c = c$" but they were asked to prove "$a = b$." They now have the outline for a correct proof if they will simply reverse the steps, assuming all steps are reversible.

Make the point that there is usually not one correct proof. Rather, expect variety in different students' proofs. Examining each other's proofs can be a valuable learning experience.

Suggested Extensions

Have students compare the steps in their proofs. Many steps will be different for different students. Have the class discuss methods used and determine if one method is as good as another. Consider efficiency as one factor in the comparison.

13-7 (pages 641–645)

Key Mathematical Ideas

- using the sum and difference formulas for the sine and cosine

Teaching Suggestions

Primary emphasis is placed on applying the formulas. They are used to find exact values of special angles, and as new tools for proving identities. Proofs of the formulas are included for more able, or interested, students. Verify that the formulas "work" by finding decimal equivalents for the results of Example 1 and comparing these to the values obtained using a calculator or table. Although the relationships in this lesson are called formulas, students should keep in mind that they are actually identities, and can be used to simplify expressions and prove other identities.

Suggested Extensions

The point made in the caution on page 641 is so familiar to teachers that sometimes it is not stressed enough. Have students show that each of the following are not identities.

1. $\sin(\alpha + \beta) \stackrel{?}{=} \sin\alpha + \sin\beta$
2. $\sin(\alpha - \beta) \stackrel{?}{=} \sin\alpha - \sin\beta$
3. $\cos(\alpha + \beta) \stackrel{?}{=} \cos\alpha + \cos\beta$
4. $\cos(\alpha - \beta) \stackrel{?}{=} \cos\alpha - \cos\beta$

$\alpha = 30°$ and $\beta = 60°$

1. $\sin 90° = 1$ but $\sin 30° + \sin 60° = \frac{1 + \sqrt{3}}{2}$
2. $\sin(-30°) = -\frac{1}{2}$ but $\sin 30° - \sin 60° = \frac{1 - \sqrt{3}}{2}$

3. $\cos(90°) = 0$ but $\cos 30° + \cos 60° = \frac{\sqrt{3}+1}{2}$

4. $\cos(-30°) = \frac{\sqrt{3}}{2}$ but $\cos 30° - \cos 60° = \frac{\sqrt{3}-1}{2}$

Group Activities

Students may enjoy playing a polar coordinate version of Tic-Tac-Toe. It goes well with three or four individual players or two teams of two players, each alternating turns.

Play on copies of the board shown below. As a student places X or O, the coordinates must be called out, giving the angle measure in radians and the distance from the center. Thus a legal move would be $\left(\frac{\pi}{3}, 3\right)$ for the red X shown. A player wins with four in a row on a given circle, four in a row on a given radius, or four in a row in a spiral.

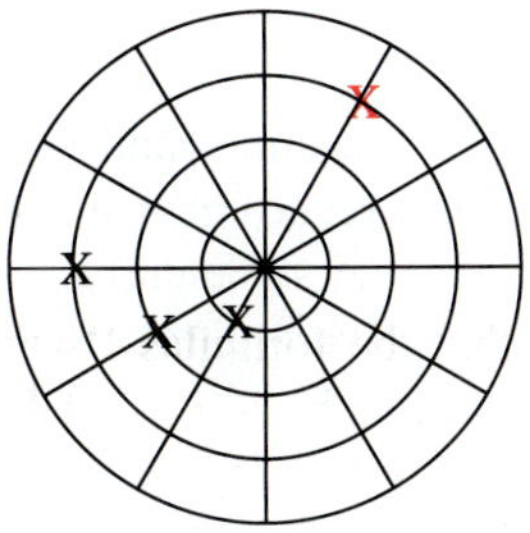

13-8 (pages 646–649)

Key Mathematical Ideas

- using double-angle and half-angle formulas for the sine and cosine

Teaching Suggestions

The proofs of the double-angle formulas are simple. Students should have no trouble seeing how they follow directly from the sum formulas in Lesson 13-7. Warn students that they must use their judgment in working with the half-angle formulas to determine the correct sign. Make sure they notice the plus-or-minus signs in the formulas.

Don't let students be daunted by the radical signs within radical signs that sometimes result from applying the formulas. In Example 2b, it will help students feel more comfortable if they evaluate the radical expression, then evaluate cos 165° directly using a calculator, to show that the results are the same.

Suggested Extensions

Evaluate sin 240° using the given formulas.

1. $\sin 240° = \sin(180° + 60°)$
2. $\sin 240° = \sin(360° - 120°)$
3. $\sin 240° = \sin(2 \cdot 120°)$
4. $\sin 240° = \sin\left(\frac{480°}{2}\right)$

All results should be $-\frac{\sqrt{3}}{2}$.

13-9 (pages 650–653)

Key Mathematical Ideas

- using addition, double-angle, and half-angle formulas for the tangent

Teaching Suggestions

These are nice derivations for students. The addition formulas follow in a straightforward way from the sine and cosine formulas, the double-angle formula from the addition formula, and the half-angle formula from the sine and cosine formulas. Students should follow the derivations readily. The formulas for the tangent complete the development. Explain to students that they could use reciprocal functions if they needed cotangents, secants, or cosecants of angle sums, double angles, or half angles.

Suggested Extensions

1. Have students write double-angle and half-angle formulas for the cotangent, secant, and cosecant.

2. Ask students to derive formulas for tan $3A$ and tan $4A$ in terms of tan A.

1. $\cot 2A = \frac{1 - \tan^2 A}{2 \tan A}$;

$\cot \frac{A}{2} = \pm\sqrt{\frac{1 + \cos A}{1 - \cos A}}$;

$\sec 2A = \frac{1}{\cos^2 A - \sin^2 A}$;

$\sec \frac{A}{2} = \pm\sqrt{\frac{2}{1 + \cos A}}$;

$\csc 2A = \frac{1}{2 \sin A \cos A}$;

$\csc \frac{A}{2} = \pm\sqrt{\frac{2}{1 - \cos A}}$

2. $\tan 3A = \frac{3 \tan A - \tan^3 A}{1 - 3 \tan^2 A}$

$$\tan 4A = \frac{2\left(\frac{\tan A}{1 - \tan^2 A}\right)}{1 - \left(\frac{2 \tan A}{1 - \tan^2 A}\right)^2}$$

14 Trigonometric Applications

Vectors are introduced, and their connection to trigonometry is shown. Plentiful applications are given. Component form, dot products, and their uses are included. Polar coordinates are defined and their use explained. Graphs of points and curves in polar coordinates are demonstrated. Conversion between polar and rectangular coordinates is presented. The polar coordinate material concludes with the development and use of DeMoivre's theorem to find nth roots. Inverse trigonometric functions, which were touched on earlier, are studied formally. Finally, trigonometric equations are solved, with both primary and general solutions given.

14-1 (pages 659–665)

Key Mathematical Ideas

- defining vector operations and applying the resultant of two vectors

Teaching Suggestions

Introduce a vector as a directed line segment with a magnitude. Use arrows to illustrate vectors and give many different vectors with different directions and different magnitudes. Be sure students understand how to move vectors so as to obtain the sum or resultant.

Point out that in this lesson multiplication is multiplication of a vector by a real number or scalar.

Be sure that students understand and can find the norm of a vector and relate the norm to the distance formula. Make sure that students can obtain a unit vector from a given vector. Stress that some unit vectors are longer than given vectors and some unit vectors are shorter than given vectors.

Stress with students the importance of drawing diagrams. Some students may need extra practice in drawing scale diagrams. You might suggest the use of graph paper.

Reading Algebra

This section is full of new vocabulary terms and new concepts. Students should be encouraged to reread the material several times, and to refer to the definitions as they read examples and work exercises. Here, as earlier, there may be a slight chance of confusion because seemingly familiar terms and symbols are used in new, specialized ways; examples are the arrow in the name of a vector, the equals sign used to mean equivalent (though not necessarily the same), and the word sum.

Suggested Extensions

Ask the students to work Problem 9 in the Problem set. Ask students to find the heading and ETA for the return trip if the conditions are the same as before and the plane leaves at 5 P.M.
317.3°; ETA of 5:55 P.M.

Using Manipulatives

Applications of vectors and resultants provide an opportunity for some interdisciplinary activities. Borrow force table apparatus from the science department. Using this equipment, students can set up three forces in balance. Then they can explore the idea of each force as the resultant of the other two, using scale drawings on graph paper or a more analytic approach with trigonometry.

14-2 (pages 666–671)

Key Mathematical Ideas

- finding vectors in component form
- applying the dot product

Teaching Suggestions

Carefully explain the concepts of x- and y-components to students. This idea is somewhat the reverse of finding the resultant. Any vector can be written as the sum of two vectors. Introduce the definition of dot product as given in the text. Then discuss with students the theorem that follows. Students should be comfortable computing the dot product by the definition and by the theorem.

Discuss with students the possible advantages and uses of the two expressions $\|\mathbf{u}\|\,\|\mathbf{v}\|\cos\theta$ and $ac + bd$ for the dot product.

You might ask students to work on the Application that follows Lesson 14-2.

Suggested Extensions

Have students look at Example 6 again. To prove that the result given is a unit vector orthogonal to **u**, have students complete the exercises.

1. Prove that $\mathbf{w} = \frac{3}{5}\mathbf{i} + \frac{4}{5}\mathbf{j}$ is a unit vector by showing that $\|\mathbf{w}\| = 1$.
$\sqrt{\left(\frac{3}{5}\right)^2 + \left(\frac{4}{5}\right)^2} = 1$

2. a. Find the measure of the angle vector **w** makes with the x-axis. 53°

b. Find the measure of the angle vector **u** makes with the x-axis. 37°

c. Find the angle between vectors **u** and **w**. 90°

3. Derive an expression for a unit vector orthogonal to a given vector $a\mathbf{i} + b\mathbf{j}$. $\frac{b\mathbf{i} - a\mathbf{j}}{\sqrt{a^2 + b^2}}$

14-3 (pages 675–679)

Key Mathematical Ideas

- defining polar coordinates and graphing polar equations

Teaching Suggestions

The basic idea of the polar coordinate system is simple, and students will have little trouble grasping it. A point is located by giving a distance and an angle. You may want to point out that coordinate systems are arbitrary; any number of different systems are possible, but there are clear advantages to both the more familiar coordinate grid and the polar system.

Polar coordinates should be introduced visually. Plot numerous examples, first showing the angle and then moving out the given distance along the terminal ray. Be sure to clarify the meaning of a negative angle (clockwise) and a negative distance (reflect through the origin). Graphs in polar coordinates are not always easy for students. They should make a table of values as in Example 3 and plot the points. Emphasize that, when sketching the curve through the points, it is important to connect the points in order, a consideration that was not especially important in graphing in rectangular coordinates. Take the time to explain why the graph in Example 3 repeats itself, forming a 3-leaf rose rather than a 4- or 2-leaf figure. (Cos 3θ runs through a full cycle of values every 120°.)

Conversion formulas between the two systems are straightforward. Students are usually interested in how fairly complicated equations in one system sometimes convert to simple equations in the other.

You may want to take advantage of the availability of special polar coordinate graph paper and chalkboards. They facilitate the teaching and learning of this topic.

Suggested Extensions

Ask students to work the Extra on page 679.

Group Activities

If you didn't use the polar coordinate tic-tac-toe game in Chapter 13, you might want to try it when introducing polar coordinates. If you did use it, this would be a good place for another round.

14-4 (pages 680–684)

Key Mathematical Ideas

- plotting complex numbers in the complex plane and using the polar form of complex numbers

Teaching Suggestions

Draw a complex plane on the chalkboard and graph several complex numbers. Relate the notions of complex number, vector, and point. Discuss how (a, b) can be looked at in a number of different ways. Be sure that students can transform a complex number to polar form.

You might explore the patterns that arise in the multiplication and division discussions before the statement of the theorem. Be sure that students understand the proof of the first part of the theorem. Then you might ask students to lead a discussion of the proof of the second part of the theorem. Students might follow Exercises 35 and 36 in constructing the proof.

Suggested Extensions

Let $z = 64(\cos 240° + i \sin 240°)$. Write each number in polar form.

1. z^2 **2.** z^3

3. $\sqrt{z}$ **4.** $\sqrt[3]{z}$

Let $z = r(\cos \theta + i \sin \theta)$.

5. Write z^2 in polar form.

6. Write z^3 in polar form.

1. $4096(\cos 480° + i \sin 480°)$
2. $262{,}144(\cos 720° + i \sin 720°)$
3. $8(\cos 120° + i \sin 120°)$
4. $4(\cos 80° + i \sin 80°)$
5. $r^2(\cos 2\theta + i \sin 2\theta)$
6. $r^3(\cos 3\theta + i \sin 3\theta)$

14-5 (pages 685–688)

Key Mathematical Ideas

- using DeMoivre's theorem

Teaching Suggestions

DeMoivre's theorem is a fascinating application of trigonometry, and students usually enjoy it. After the work in Lesson 14-4, the theorem itself is rather evident as a generalization. Students may find it almost incredible that they can find roots of imaginary numbers, which they feel came into being as roots themselves. Caution students not to stop finding the nth roots of a number until they have found n of them. Make sure they keep adding 360° to the amplitude of the original number. Students are often startled to realize that the graphs of the n nth roots of a complex number are always n points evenly spaced around a circle. Work through Example 2 thoroughly to show how these roots are found.

Suggested Extensions

1. Evaluate $(-1 - i)^3$ without using DeMoivre's theorem. Does the result verify that $-1 - i$ is a cube root of $2 - 2i$?
   $(-1 - i)^3 = 2 - 2i$; yes
2. Have students explain why the n nth roots of a number must be numbers whose graphs are the vertices of a regular n-gon.
   When you substitute consecutive integers for k in $\frac{c + k \cdot 360°}{n}$, the results differ by $\frac{360°}{n}$.

14-6 (pages 689–692)

Key Mathematical Ideas

- evaluating expressions involving the inverse sine and cosine

Teaching Suggestions

Students have already been informally introduced to inverse sines and cosines (Chapter 13). Here the treatment is more complete and formal. Make sure students understand the necessity for restricting the domains of $\sin x$ and $\cos x$ to define $\text{Sin}\, x$ and $\text{Cos}\, x$. Make sure, too, that they realize that these new, restricted functions are different from the original sine and cosine, primarily because the new functions are one-to-one and therefore invertible. Don't let students unthinkingly assume that all the trigonometric functions are restricted alike.

Some students will respond well to the formal definition of inverse functions, that is, if f and g are inverse functions then

$$f(g(x)) = x \text{ and } g(f(x)) = x.$$

This will help these students simplify expressions such as

$$\sin\left(\text{Sin}^{-1}\frac{1}{2}\right) = \frac{1}{2}$$

and

$$\text{Cos}^{-1}\left(\cos\frac{\pi}{4}\right) = \frac{\pi}{4}.$$

Suggested Extensions

1. Challenge students' understanding of one-to-one functions by asking them to give different sets of values over which the sine and cosine functions could be restricted in order to make them one-to-one.
2. Ask students to graph $y = \text{Sin}^{-1}x$ and $y = \text{Cos}^{-1}x$ in the same coordinate plane for $-1 \le x \le 1$ and show graphically that $\text{Sin}^{-1}x + \text{Cos}^{-1}x = \frac{\pi}{2}$ for all x such that $-1 \le x \le 1$.

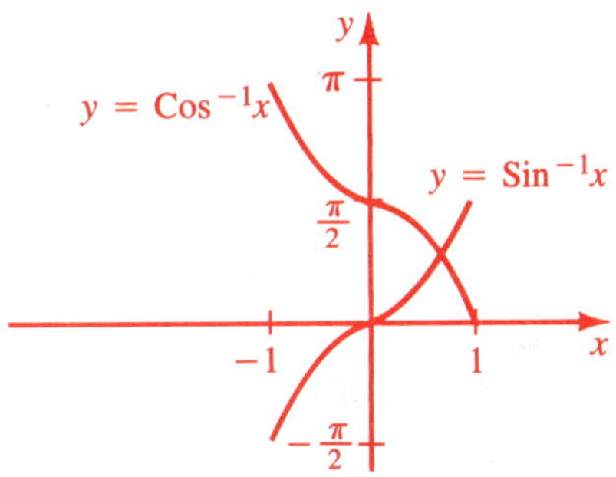

14-7 (pages 693–696)

Key Mathematical Ideas

- evaluating expressions involving all the inverse trigonometric functions

Teaching Suggestions

This lesson finishes the job that was begun in the last lesson: defining and applying the inverses of the other trigonometric functions. Keep students aware of the restrictions on the values of the inverse functions, so that they will not be tempted to give values that are not possible. When evaluating complicated expressions, have students pay attention to the outer function. If this is a sine, for example, the final answer should be a real number between -1 and 1. If the outer function is an inverse trigonometric function, then the final answer should be the measure of an angle in the restricted set of values for that function.

Suggested Extensions

Students can benefit from some explanation of how to use their calculators to evaluate expressions that include inverse trigonometric functions.

By this point students should know whether they have inverse sine, cosine, and tangent keys, or whether they must use a separate inverse key with the appropriate function key. Show with examples such as $\text{Sin}^{-1}\ 1.2$ and $\text{Cos}^{-1}\ -2$ the error messages calculators will produce.

Show the sequence of keys required to evaluate an expression such as $\tan\ (\text{Sec}^{-1}\ 2)$:

$$2 \rightarrow \frac{1}{x} \rightarrow \text{INV} \rightarrow \cos \rightarrow \tan$$

Trying to evaluate an expression such as $\text{Tan}^{-1}\ (\tan 135°)$ may help students understand the restrictions on the inverse functions better, since the value is not 135°.

14-8 (pages 697–700)

Key Mathematical Ideas

- solving trigonometric equations

Teaching Suggestions

Review using identities to rewrite the expressions in a trigonometric equation, including the addition, double-, and half-angle formulas. Review, too, the strategies given in Chapter 13 for proving identities; some of these, such as expressing everything in terms of the sine or cosine, will help here too. Students must recall their equation skills, and apply them in the new context of trigonometric equations. Warn students to look at their solutions critically. A result of $\sin x = 2$, for example, is no solution at all. Explain the difference between primary solutions (between 0° and 360°) and general solutions, which are generally infinite sets. Caution students about dividing both sides of an equation by an expression such as $\cos x$. If this is done, it implies that $\cos x$ is not zero. This may cause students to miss solutions. In this case they would need to check values of x making the expression 0 in the original equation to see if these values are solutions.

Suggested Extensions

Reverse the usual process; ask students to write a trigonometric equation with the given primary solutions. Answers may vary.

1. 0°, 30°, 180°, 330°, 360°
$(2 \cos \theta - \sqrt{3})(\cos^2 \theta - 1) = 0$

2. 45°, 90°, 225°, 270°
$(\sin^2 \theta - 1)(\cos \theta - \sin \theta) = 0$

3. 15°, 105°, 195°, 285°
$3 \tan 2\theta = \sqrt{3}$

15 Statistics and Probability

Although no prior study of statistics or probability is assumed, this chapter provides a comprehensive introduction to these topics. Summarizing data graphically is done with histograms, stem-and-leaf plots, and box-and-whisker plots. Measures of central tendency and dispersion help to analyze data. The normal distribution and correlation coefficients provide application-oriented statistics topics. As a transition to probability, the counting principles are introduced, including the counting of permutations and combinations. After sample spaces for random experiments are defined, probability itself is formally studied, frequently employing the concepts and formulas learned earlier for counting, combinations, and permutations. The chapter concludes with a discussion of mutually exclusive and independent events.

15-1 (pages 709–712)

Key Mathematical Ideas

- displaying data using frequency distributions, histograms, and stem-and-leaf plots
- computing measures of central tendency

Teaching Suggestions

Build on students' graphing experience when explaining histograms. See that students realize that intervals for a histogram are not fixed, but vary from student to student. Intervals must, however, be uniform.

Students will find stem-and-leaf plots a very simple method for representing data, with the advantage of incorporating all data into the display. Compare this to a histogram, where only the range of an interval is shown.

Students may have studied means, medians, and modes before. Make sure they realize that a distribution may have more than one mode, and that the median is not necessarily one of the actual data values.

Reading Algebra

This lesson contains many new vocabulary terms, some of which have meanings different from their common usage (''mean'' for example). Use this introductory lesson to focus on acquisition of new vocabulary, and on differentiation between several rather similar terms, such as mean, median, and mode.

Suggested Extensions

Have students provide sets of data that satisfy the given condition. Answers will vary.

1. mean = median = mode

2. mean > median

3. mean < median

15-2 (pages 713–718)

Key Mathematical Ideas

- computing measures of dispersion
- describing and comparing distributions

Teaching Suggestions

Students may have had enough experience with distributions of data to be able to appreciate the need for more than the measures of central tendency. To underscore the principle involved, have them analyze two distributions with the same mean and median but with one much more widely scattered than the other. Students should be able to see that distributions with very different characteristics can have the same mean and median. Thus, we have the need for measures of dispersion also. You may want to encourage students to use calculators (either standard or those with a statistical mode) or computers (either program their own or use commercial software). The computations required for finding standard deviation can be intimidating when done by hand.

The box-and-whisker plot will provide an appealing method for representing data. Its advantage is that it gives a visual indication of the concentration and spread of the data.

Suggested Extensions

Use the distributions in Example 5 and Written Exercises 9–14. Have students find the percent of items that lie within one and two standard deviations of the mean.

15-3 (pages 719–723)

Key Mathematical Ideas

- recognizing and analyzing normal distributions

Teaching Suggestions

Use the examples on page 719 to show that many distributions of diverse kinds of data follow the typical "bell" curve of a normal distribution. Focus on the benefits of standardizing a normal distribution, and on the properties of a standard normal deviation. Emphasize the symmetry of the curve, and show how this property is used in computations. Look ahead informally and mention the fact that, with a standard normal distribution, a probability is an area, and that, since the sum of all probabilities is 1, then the total area under the curve is also 1. Students usually enjoy making the calculations associated with standard normal distributions; it provides a stimulating challenge to their ingenuity.

Suggested Extensions

Lead a discussion of what types of information would, and would not, be expected to yield distributions that are normal. Examples that would yield such distributions are the height and the weight of sixteen-year-olds. Examples that would not are the number of math credits among Algebra 2 students and the per capita income of Americans.

15-4 (pages 724–729)

Key Mathematical Ideas

- drawing a scatter plot and determining the correlation coefficient
- using regression lines for a set of ordered pairs of data

Teaching Suggestions

Students will enjoy the conceptual aspect of their work with correlation coefficients and regression lines. This topic has a satisfying air of reality and application. Although they will be tempted to attribute high correlation to cause and effect, make sure students have read and understood the discussion at the top of page 726. The point is that there may be any number of circumstances causing high correlation. Make sure students can explain what each term in the formula for a correlation coefficient is, so that they will be able to calculate one correctly. In discussing the correlation formula, make sure students realize that r is not the slope of the regression line. Notice that the emphasis in this lesson is on using correlation and regression, not on calculating them from raw data. Your students will be able to focus more on concepts, without becoming mired in long arithmetic processes.

Suggested Extensions

1. Have students write a general equation for the regression line for a set of (x, y) pairs. Have them eliminate r from the equation, so that the only constants in the equation are means and standard deviations.

   $y = M_y + \frac{M_{xy} - M_x M_y}{(\sigma_x)^2}(x - M_x)$

2. Discuss pairs of variables similar to those in Oral Exercise 3. Look for factors that are likely to have high correlation but that are probably not linked by a cause-and-effect relationship.
   Answers will vary.

15-5 (pages 730–733)

Key Mathematical Ideas

- applying fundamental counting principles

Teaching Suggestions

The multiplication process of the first counting principle forms the backbone of this lesson. Offer a variety of diagrammatic techniques to help students apply this principle correctly. Use successive boxes to be filled with numbers of possibilities, as in Example 1. Another useful way to analyze multi-step selections is to use tree diagrams, which have an inherent clarity for most students. Make sure students understand that they add when working with mutually exclusive choices only.

Suggested Extensions

Have students apply counting principles to questions about some well-known codes.

1. Letters in Morse Code are formed from combinations of from one to four dots and dashes. How

many "letters" can be formed this way? 30

2. In binary code only ones and zeros are used to represent information. Some computers use an "eight bit byte," meaning that they can recognize a given string of eight digits consisting of ones and zeros. How many symbols can be coded using 8-bit bytes? 256
3. Have students read about the ASCII code and its relation to binary code.

15-6 (pages 734–737)

Key Mathematical Ideas

- finding the number of permutations of the elements of a set

Teaching Suggestions

In permutations, order is important. Discuss elections, where it makes a difference which office one is elected to; Zip Codes, where the same digits used in a different order identify two different locations; and telephone area codes, where the order of the given digits certainly makes a difference. Take time to introduce factorial notation, since this may be the first exposure to it for many students. Continue to use boxes to be filled with the appropriate number of selections, and tree diagrams, in leading up to a derivation of the permutation formulas.

Suggested Extensions

1. Encourage students to explore the value of $n!$ on their calculators. Those with a factorial key may find it interesting to investigate the maximum factorial their calculators can handle.
2. Students are often interested to find how rapidly the value of $n!$ grows. They might compare $n!$ to a^n, for different values of a, to see how fast they grow.
3. Ask students to explain why all values of $n!$ for $n > 4$ must end in the digit 0. For $n > 4$, $n!$ contains $2 \cdot 5$, or 10, and 10 times any integer always has a 0 in the ones place.

15-7 (pages 738–742)

Key Mathematical Ideas

- finding the combinations of a set of elements

Teaching Suggestions

Starting with the initial introduction and continuing from there, you should repeatedly emphasize the meaning of combinations and how they are different from permutations. With combinations, order does not matter. Thus for a given group of items there will be fewer combinations than permutations. Contrast appointing a three-member committee, where all members are equal, to electing officers, where it matters which office is filled by which person. Explain to students that the reason for dividing the number of permutations by $r!$ is to eliminate multiple counting of all the permutations of the same elements. Notice in Example 3b that the fundamental counting principle is employed to multiply the number of ways of choosing three teachers and the number of ways of choosing three students.

Suggested Extensions

Encourage students to explore values of ${}_nC_r$.

1. Hold n constant and compute values of ${}_nC_r$ for $r = 0, 1, 2, \ldots, n$.
2. Repeat Exercise 1 for other values of n.
3. Have students read about Pascal's triangle, and how it is related to ${}_nC_r$.
4. Express the symmetry of the rows of Pascal's triangle using ${}_nC_r$ notation.

15-8 (pages 743–744)

Key Mathematical Ideas

- specifying sample spaces and events for random experiments

Teaching Suggestions

Students often seem to have an affinity for the basic ideas of probability. The sample space of an experiment is simply all the ways it can "come out." By learning how to enumerate sample spaces now, students set themselves up for success in computing probabilities later.

Using Manipulatives

In this and following probability lessons, there are many concrete ways to illustrate and investigate probability concepts. These include dice, coins, colored disks, spinners, and lettered cards. These can be used at the introductory level for showing sample spaces and events. As students move ahead, these items can be used in more complicated experiments to compare empirical probabilities to theoretical ones and to consider expected value. Simulation is another field in which these devices can be useful.

Suggested Extensions

There are many ways to divide a given sample space into events. Have students describe several ways to separate the sample space for the experiment into mutually exclusive and exhaustive (using all elements exactly once) events. Answers will vary.

1. Drawing a card from a standard deck.
2. Selecting a state of the United States.
3. Rolling a pair of dice.
4. Choosing a letter from the alphabet.
5. Drawing a name from a hat.

15-9 (pages 745–750)

Key Mathematical Ideas

- finding the probability that an event will occur

Teaching Suggestions

This lesson draws heavily from earlier lessons in the chapter. Spend some time reviewing sample spaces and events, combinations, the specific combinations for dice and coins, and standard normal distributions, all of which appear prominently in this lesson. To describe the basic idea of probability, discuss such ordinary terms as chances, likely, probable, possible, and so on. The idea that a probability is a fraction comparing the number of successes out of the total number of possibilities is a natural one. Curious students might want to explore the idea of experiments in which the simple events do not all have equal probabilities.

Reading Algebra

Like many other topics, probability has its own notation. In some ways it should seem familiar to students, while in other ways it may be confusing at first. Make sure students understand the meaning of symbols such as $P(A)$, $P(2)$, and so on. $P(A)$, by definition, means the probability of event A. This should not be confused with the use of parentheses for multiplication. Another similar notation, that used for function values, should not cause a problem, because the probability notation is used in virtually the same way.

Suggested Extensions

Wildlife biologists sometimes use the "capture-recapture method" to estimate the population of a species in a given area. In simplest terms, consider the example of a bass population in a lake. Suppose 25 bass are caught, tagged, and released back into the lake. We assume that they spread evenly throughout the bass population. Later, bass are again caught. This time it is found that 3 bass are tagged out of 40 bass caught.

1. What is the probability that a bass is tagged, in percent? (This is based on the second catch.) 7.5%
2. What is the total population of bass in the lake? (The ratio of the 25 tagged bass to the total bass population should be the same as the probability in Exercise 1.) about 333

15-10 (pages 754–761)

Key Mathematical Ideas

- identifying mutually exclusive and independent events
- finding the probability of such events

Teaching Suggestions

The ideas of union and intersection of sets are essential to understanding mutually exclusive and independent events. Use Venn diagrams to show that $P(A \cup B) = P(A) + P(B) - P(A \cap B)$. Point out that the subtraction of the probability of the intersection is necessary to eliminate double counting. Mutually exclusive events are readily understood: $P(A \cap B) = 0$; the events cannot both happen. Independence of events is intuitively clear, but it may seem a bit strange to actually define independent events by probability, without reference to cause and effect or influence. Complementary events are a special case of mutually exclusive events. They can be useful in clarifying problem situations.

Group Activities

Students can play various games and decide if they are fair or not. If the game is not fair, that is, if the players do not have equal probabilities of winning, then they can suggest ways to make the game fair. More interesting than changing the rules of play would be changing the payoff rates to achieve fairness, while leaving the actual rules the same.

Example A

Roll two dice. You win if the total is even, I win if the total is odd.

Students can play this game with beans or chips to keep track of winnings. Since this is a fair game, after playing we expect the players to have about the same number of winnings.

Example B

Flip 4 coins. You win if 2 or 3 heads come up. I win otherwise.
After playing enough times it is clear that this is not a fair game. Your chances of winning are 10 out of 16, compared to my 6 out of 16. To make the game fair, you should get 3 points for every win, and I should get 5 points every time I win.
Other similar games can be devised and analyzed using coins, dice, spinners, and so on.

Suggested Extensions

Ask students to write a formula for $P(A \cup B \cup C)$.
$P(A) + P(B) + P(C) - P(A \cap B) - P(B \cap C) - P(A \cap C) + P(A \cap B \cap C)$

16 Matrices and Determinants

After an introduction to some basic terms and the definition of a matrix, the arithmetic operations for matrices are defined and their properties examined. Considerable attention is devoted to some of the contemporary applications of matrices. The latter lessons of the chapter build up to the major matrix skills of inverting a matrix and solving a system of equations. To reach these goals, determinants are introduced, as are methods for evaluating them, including minors and properties of determinants. Solving systems with Cramer's rule culminates this development.

16-1 (pages 767–769)

Key Mathematical Ideas

- learning and applying matrix terminology

Teaching Suggestions

Remember that this topic is totally new to most or all students. Make sure they understand that a matrix is, basically, an array of numbers in rows and columns. Check that students know a row from a column and can tell an $m \times n$ from an $n \times m$ matrix. The basic ideas of this lesson should not be difficult for students.

Suggested Extensions

1. Have students check the version of BASIC they use at home or at school to see whether and how their particular computer handles matrices.
2. Students may be interested to look into the origins of the word matrix and the use of this word in other fields.

16-2 (pages 770–773)

Key Mathematical Ideas

- finding sums and differences of matrices
- finding the product of a scalar and a matrix

Teaching Suggestions

The algorithms for matrix addition and subtraction and scalar multiplication will seem very straightforward to students. Demonstrate why matrices must have the same dimension for addition by putting several matrices of different dimensions on the chalkboard and having the students try to add or subtract them. Emphasize that the lists of properties of addition and multiplication contain the same properties that have been previously listed for real numbers.

Stress that although matrices are not numbers, we can consider them as legitimate mathematical objects for which we may define mathematical operations.

Suggested Extensions

Have students prove that scalar multiplication is distributive with respect to matrix addition.

$$a(A + B) = aA + aB$$

16-3 (pages 774–778)

Key Mathematical Ideas

- finding the product of two matrices

Teaching Suggestions

Take time to work many examples of matrix multiplication. The algorithm is somewhat complex and is difficult for some students. Show why matrices must be compatible in their dimensions. There are some major surprises for students in this lesson. Among these are that the product of nonzero matrices can be the zero matrix, and that matrix multiplication is not, in general, commutative. Some students may have some trouble getting used to these ideas.

Suggested Extensions

You may want to look ahead to upcoming work with inverses of matrices. Give students a 2×2 or 3×3 matrix A and challenge them to find a matrix B such that $AB = 1$.

16-4 (pages 779–784)

Key Mathematical Ideas

- solving problems using matrices

Teaching Suggestions

Most students will enjoy this section. Although there is new content here, this is primarily an applications section. Be sure students see how to write the matrix for a communications network. You may want to go slowly as you introduce the square of a communications or dominance matrix. There will be students who do not see the significance of the square without patient explanation, repetition, and good examples. Make sure students see the difference between A^2 and $A + A^2$, the first representing two-stage communication or dominance, the second two-stage or less.

Group Activities

Set up a communications network with its own communications matrix in the classroom. Depending on class size, first write, or have the class write, a 4×4 or 5×5 communications matrix. Distribute copies to the class. Write a secret message to be passed through the communications network. Have students pass the message around in accordance with the communications matrix. After a set time stop to assess the progress of the message and to interpret what has happened in relation to the communications matrix.

Suggested Extensions

For a matrix activity with a local flair, have students create their own dominance matrices. They can use the records of local school teams or look in the newspaper to find out which teams won and lost in the past few games.

16-5 (pages 787–789)

Key Mathematical Ideas

- finding the determinants of 2×2 and 3×3 matrices

Teaching Suggestions

Be sure your students understand that a determinant is not a matrix. As defined, a determinant is a real number associated with a square matrix. Finding the determinant of a 2×2 matrix is not difficult. Students should memorize the pattern.

Show how the evaluation process can be extended from a 2×2 determinant to a 3×3 determinant. The use of arrows and lines through entries helps students keep the right numbers in the right place. Students should also be careful in their placement of plus and minus signs.

You might mention that determinants that are 4×4 or greater cannot be evaluated by the method shown for 3×3 determinants.

Reading Algebra

Students will encounter symbols and notation in this and later lessons that need clarification. The vertical bars for a determinant should be distinguished from the more familiar, and similar, absolute value sign. The "det X" notation for the determinant of matrix X may remind students of the "f of X" notation for functions. "Det" can, actually, be thought of as a function that matches a real number with each matrix. In later lessons the notation for the inverse of a matrix is introduced. Students may find it informative to compare this with the notation for the inverse of a function.

Suggested Extensions

Ask students to find the value of each determinant and then draw a conclusion about the determinants and their values. Is the conclusion true in general?

$$\begin{vmatrix} 1 & 2 & 3 \\ 5 & 3 & 4 \\ 1 & 6 & 8 \end{vmatrix} \text{ and } \begin{vmatrix} 5 & 3 & 4 \\ 1 & 2 & 3 \\ 1 & 6 & 8 \end{vmatrix}$$

The values are 9 and -9, respectively. In general when two rows are interchanged, the sign of the value first obtained reverses.

16-6 (pages 790–792)

Key Mathematical Ideas

- solving systems of equations using inverses of matrices

Teaching Suggestions

The definition of the inverse of a matrix should remind students of multiplicative inverses for real numbers. Students will already have had enough experience to realize that determinants may equal zero; now emphasize that these are the matrices that do not have inverses. This lesson deals with the solution of systems of two equations in two variables. For solutions of more general systems, Cramer's rule is discussed in the last lesson of the chapter.

Suggested Extensions

Have the students determine what is wrong with the following solution.
Solve for A.

$$\begin{bmatrix} 3 & 2 \\ 2 & 2 \end{bmatrix} A = \begin{bmatrix} 5 & 0 \\ -1 & -1 \end{bmatrix}$$

$$\begin{bmatrix} 1 & -1 \\ -1 & \frac{3}{2} \end{bmatrix}\begin{bmatrix} 3 & 2 \\ 2 & 2 \end{bmatrix} A = \begin{bmatrix} 5 & 0 \\ -1 & -1 \end{bmatrix}\begin{bmatrix} 1 & -1 \\ -1 & \frac{3}{2} \end{bmatrix}$$

$$\begin{bmatrix} 3-2 & 2-2 \\ -3+3 & -2+3 \end{bmatrix} A = \begin{bmatrix} 5-0 & -5+0 \\ -1+1 & 1-\frac{3}{2} \end{bmatrix}$$

$$\begin{bmatrix} 1 & 0 \\ 0 & 1 \end{bmatrix} A = \begin{bmatrix} 5 & -5 \\ 0 & -\frac{1}{2} \end{bmatrix}$$

$$A = \begin{bmatrix} 5 & -5 \\ 0 & -\frac{1}{2} \end{bmatrix}$$

Matrix multiplication is not commutative.

$$\begin{bmatrix} 5 & 0 \\ -1 & -1 \end{bmatrix}\begin{bmatrix} 1 & -1 \\ -1 & \frac{3}{2} \end{bmatrix}$$

should be

$$\begin{bmatrix} 1 & -1 \\ -1 & \frac{3}{2} \end{bmatrix}\begin{bmatrix} 5 & 0 \\ -1 & -1 \end{bmatrix}.$$

16-7 (pages 794–797)

Key Mathematical Ideas

- evaluating third-order determinants using expansion by minors

Teaching Suggestions

Explain completely what a minor of an element in a determinant is and how this helps in evaluating determinants. Point out that a determinant may be expanded by minors about any row or column. The choice of row or column could be important in determining the degree of difficulty of the problem. Therefore, work Example 2 with students. Work several other examples in class.

Discuss with students that to evaluate a 4×4 determinant, they should use minors to obtain 3×3 determinants, then use minors again to obtain 2×2 determinants that are easy to evaluate.

16-8 (pages 798–800)

Key Mathematical Ideas

- using the properties of determinants to simplify the expansion of determinants by minors

Teaching Suggestions

Discuss each property with students and illustrate each property with examples. Point out that students should look closely at the determinant before beginning to choose properties by which to evaluate it. Ask students to evaluate a determinant and ask students to compare the choices they made. Even though different students choose different properties, all students should get the same answer. Stress that the better the choices, the easier the job of evaluation is.

Suggested Extensions

1. Have students examine the effect of interchanging the diagonals on the value of a determinant. The resulting determinant is the opposite of the original one.
2. It will be interesting for students to compare the methods of this lesson with the methods used in the Extra (pages 785–786) for deriving augmented matrices in triangular form.

16-9 (pages 801–805)

Key Mathematical Ideas

- solving systems of equations using determinants

Teaching Suggestions

After discussing the information in the text and working Example 1, you may want to work Example 2 before discussing Cramer's rule. It usually helps if students set up and evaluate the determinants, as shown in Example 2, before putting the values into the equations for Cramer's rule.

Algebra
and Trigonometry

Structure and Method *Book 2*

Algebra and Trigonometry

Structure and Method
Book 2

Richard G. Brown
Mary P. Dolciani
Robert H. Sorgenfrey
Robert B. Kane

Contributing Authors
Sandra K. Dawson
Barbara Nunn

Teacher Consultants
Keith Bangs
Doris Holland
Ouidasue Nash
Al Varness

McDougal Littell
A HOUGHTON MIFFLIN COMPANY
Evanston, Illinois • Boston • Dallas

Authors

Richard G. Brown, Mathematics Teacher, Phillips Exeter Academy, Exeter, New Hampshire

Mary P. Dolciani, formerly Professor of Mathematical Sciences, Hunter College of the City University of New York

Robert H. Sorgenfrey, Professor Emeritus of Mathematics, University of California, Los Angeles

Robert B. Kane, Professor of Mathematics Education, Purdue University, Lafayette, Indiana

Contributing Authors

Sandra K. Dawson, Mathematics Teacher, Glenbrook South High School, Glenview, Illinois

Barbara Nunn, Mathematics Department Chairperson, Marjory Stoneman Douglas High School, Parkland, Florida

Teacher Consultants

Keith Bangs, Mathematics Teacher, Owatonna High School, Owatonna, Minnesota

Doris Holland, Mathematics Teacher, O.D. Wyatt High School, International Baccalaureate Magnet Program, Fort Worth, Texas

Ouidasue Nash, Mathematics Teacher, Columbia High School, Columbia, South Carolina

Al Varness, former Mathematics Teacher, Lake Washington High School, Kirkland, Washington

The authors wish to thank **David L. Myers**, Computer Coordinator and Mathematics Teacher, Winsor School, Boston, Massachusetts, for writing the Portfolio Projects.

Printed in U.S.A.

ISBN: 0-395-77118-8

123456789-DC-00 99 98 97 96

Contents

1 Basic Concepts of Algebra

The Language of Algebra

1-1	Real Numbers and Their Graphs	1
1-2	Simplifying Expressions	6

Operating with Real Numbers

1-3	Basic Properties of Real Numbers	14
1-4	Sums and Differences	21
1-5	Products	27
1-6	Quotients	33

Solving Equations and Solving Problems

1-7	Solving Equations in One Variable	37
1-8	Words into Symbols	43
1-9	Problem Solving with Equations	49

- **Explorations** Exploring Irrational Numbers 832
- **Technology** Computer Exercises 5, 32 Calculator Key-In 12
- **Special Topics** Reading Algebra xiv, 26 Logical Symbols: Quantifiers 20 Biographical Note 32 Historical Note 42 Challenge 48
- **Reviews and Tests** Mixed Review Exercises 5, 20, 31, 42, 54 Self-Tests 13, 36, 54 Chapter Summary 55 Chapter Review 55 Chapter Test 57

2 Inequalities and Proof

Working with Inequalities

2-1	Solving Inequalities in One Variable	59
2-2	Solving Combined Inequalities	65
2-3	Problem Solving Using Inequalities	69

Working with Absolute Value

2-4	Absolute Value in Open Sentences	73
2-5	Solving Absolute Value Sentences Graphically	76

Proving Theorems

2-6	Theorems and Proofs	81
2-7	Theorems about Order and Absolute Value	88

- **Explorations** Exploring Inequalities 833
- **Technology** Computer Exercises 64, 75 Computer Key-In 80
- **Special Topics** Career Note / Automotive Engineer 64 Historical Note 68 Symbolic Logic: Boolean Algebra 95
- **Reviews and Tests** Mixed Review Exercises 63, 72, 79, 91 Self-Tests 72, 79, 92 Chapter Summary 92 Chapter Review 93 Chapter Test 94 Mixed Problem Solving 98 Preparing for College Entrance Exams 99

3 Linear Equations and Functions

Linear Equations and Their Graphs

3-1	Open Sentences in Two Variables	101
3-2	Graphs of Linear Equations in Two Variables	107
3-3	The Slope of a Line	112
3-4	Finding an Equation of a Line	118

Linear Systems

3-5	Systems of Linear Equations in Two Variables	124
3-6	Problem Solving: Using Systems	131
3-7	Linear Inequalities in Two Variables	135

Functions and Relations

3-8	Functions	141
3-9	Linear Functions	146
3-10	Relations	153

- **Explorations** Exploring Functions 834
- **Technology** Computer Graphing Ideas 114, 119, 160 Computer Exercises 122
- **Special Topics** Challenge 140, 158 Historical Note 152 Application / Linear Programming 159
- **Reviews and Tests** Mixed Review Exercises 106, 117, 130, 139, 152 Self-Tests 123, 140, 158 Chapter Summary 161 Chapter Review 162 Chapter Test 164 Cumulative Review 165

4 Products and Factors of Polynomials

Working with Polynomials

4-1	Polynomials	167
4-2	Using Laws of Exponents	171
4-3	Multiplying Polynomials	174

Factors of Polynomials

4-4 Using Prime Factorization 179
4-5 Factoring Polynomials 183
4-6 Factoring Quadratic Polynomials 188

Applications of Factoring

4-7 Solving Polynomial Equations 194
4-8 Problem Solving Using Polynomial Equations 198
4-9 Solving Polynomial Inequalities 202

- **Explorations** Exploring Polynomial Factors 835
- **Technology** Computer Exercises 187 Computer Key-In 193
- **Special Topics** Biographical Note 177 Reading Algebra 178 Historical Note 182
- **Reviews and Tests** Mixed Review Exercises 170, 176, 187, 197, 205 Self-Tests 177, 192, 205 Chapter Summary 206 Chapter Review 207 Chapter Test 208 Preparing for College Entrance Exams 209

5 Rational Expressions

Using the Laws of Exponents

5-1 Quotients of Monomials 211
5-2 Zero and Negative Exponents 216
5-3 Scientific Notation and Significant Digits 221

Rational Expressions

5-4 Rational Algebraic Expressions 227
5-5 Products and Quotients of Rational Expressions 232
5-6 Sums and Differences of Rational Expressions 235
5-7 Complex Fractions 238

Problem Solving Using Fractional Equations

5-8 Fractional Coefficients 242
5-9 Fractional Equations 247

- **Explorations** Exploring Continued Fractions 836
- **Technology** Computer Key-In 215 Calculator Key-In 225
- **Special Topics** Biographical Note 220 Graphing Rational Functions 230 Application / Electrical Circuits 253
- **Reviews and Tests** Mixed Review Exercises 215, 225, 234, 241, 252 Self-Tests 226, 241, 252 Chapter Summary 254 Chapter Review 255 Chapter Test 257

6 Irrational and Complex Numbers

Roots and Radicals

6-1	Roots of Real Numbers	259
6-2	Properties of Radicals	264
6-3	Sums of Radicals	270
6-4	Binomials Containing Radicals	274
6-5	Equations Containing Radicals	277

Real Numbers and Complex Numbers

6-6	Rational and Irrational Numbers	283
6-7	The Imaginary Number i	288
6-8	The Complex Numbers	292

- ***Explorations*** Exploring Radicals 837
- ***Technology*** Calculator Key-In 263, 269
- ***Special Topics*** The Irrationality of $\sqrt{2}$ 273 Historical Note 297 Conjugates and Absolute Value 298
- ***Reviews and Tests*** Mixed Review Exercises 263, 273, 282, 291 Self-Tests 282, 297 Chapter Summary 301 Chapter Review 302 Chapter Test 303 Mixed Problem Solving 304 Preparing for College Entrance Exams 305

7 Quadratic Equations and Functions

Solving Quadratic Equations

7-1	Completing the Square	307
7-2	The Quadratic Formula	311

Roots of Quadratic Equations

7-3	The Discriminant	317
7-4	Equations in Quadratic Form	322

Quadratic Functions and Their Graphs

7-5	Graphing $y - k = a(x - h)^2$	326
7-6	Quadratic Functions	333
7-7	Writing Quadratic Equations and Functions	338

- ***Explorations*** Exploring Quadratic Equations 838
- ***Technology*** Computer Graphing Ideas 327, 838 Computer Exercises 321
- ***Special Topics*** Biographical Note 316 Challenge 325 Career Note / Statistician 337
- ***Reviews and Tests*** Mixed Review Exercises 310, 321, 332, 345 Self-Tests 316, 325, 345 Chapter Summary 346 Chapter Review 346 Chapter Test 348 Cumulative Review 348

8 Variation and Polynomial Equations

Variation and Proportion

8-1 Direct Variation and Proportion 351
8-2 Inverse and Joint Variation 358

Polynomial Equations

8-3 Dividing Polynomials 364
8-4 Synthetic Division 368
8-5 The Remainder and Factor Theorems 372
8-6 Some Useful Theorems 377

Solving Polynomial Equations

8-7 Finding Rational Roots 382
8-8 Approximating Irrational Roots 386
8-9 Linear Interpolation 391

- ***Explorations*** Exploring Direct Variation 839
- ***Technology*** Computer Graphing Ideas 386, 387 Computer Key-In 389 Computer Exercises 395
- ***Special Topics*** Challenge 363 Reading Algebra 371 Historical Note 385
- ***Reviews and Tests*** Mixed Review Exercises 357, 367, 376, 385 Self-Tests 363, 381, 396 Chapter Summary 396 Chapter Review 397 Chapter Test 398 Preparing for College Entrance Exams 399

9 Analytic Geometry

Conic Sections: Circles and Parabolas

9-1 Distance and Midpoint Formulas 401
9-2 Circles 407
9-3 Parabolas 412

Conic Sections: Ellipses and Hyperbolas

9-4 Ellipses 418
9-5 Hyperbolas 426
9-6 More on Central Conics 432

Systems of Equations

9-7 The Geometry of Quadratic Systems 436
9-8 Solving Quadratic Systems 439
9-9 Systems of Linear Equations in Three Variables 444

- ***Explorations*** Exploring Circles and Ellipses 840
- ***Technology*** Computer Graphing Ideas 436, 840 Computer Exercises 406
- ***Special Topics*** Challenge 411 Career Note / Computer Graphics Artist 423 Application / Planetary Orbits 424 Historical Note 431

• *Reviews and Tests* Mixed Review Exercises 406, 417, 431, 438, 449
Self-Tests 417, 435, 450 Chapter Summary 450 Chapter Review 451
Chapter Test 453

10 Exponential and Logarithmic Functions

Exponential Functions

10-1	Rational Exponents	455
10-2	Real Number Exponents	459
10-3	Composition and Inverses of Functions	463

Logarithmic Functions

10-4	Definition of Logarithms	468
10-5	Laws of Logarithms	473

Applications

10-6	Applications of Logarithms	478
10-7	Problem Solving: Exponential Growth and Decay	483
10-8	The Natural Logarithm Function	489

• *Explorations* Exploring Powers and Roots 841

• *Technology* Computer Graphing 460, 464, 841 Calculator Key-In 492

• *Special Topics* Growth of Functions 462 Challenge 492
Application / Radiocarbon Dating 493 Historical Note 472

• *Reviews and Tests* Mixed Review Exercises 458, 467, 477, 488
Self-Tests 467, 477, 493 Chapter Summary 495 Chapter Review 496
Chapter Test 497 Mixed Problem Solving 498 Preparing for College
Entrance Exams 499

11 Sequences and Series

Sequences

11-1	Types of Sequences	501
11-2	Arithmetic Sequences	507
11-3	Geometric Sequences	510

Series

11-4	Series and Sigma Notation	518
11-5	Sums of Arithmetic and Geometric Series	525
11-6	Infinite Geometric Series	531

Binomial Expansions

11-7	Powers of Binomials	537
11-8	The General Binomial Expansion	540

• *Explorations* Exploring Pascal's Triangle 842

• *Technology* Computer Exercises 530 Calculator Key-In 543

- *Special Topics* Challenge 516 Graphing Sequences 517 Career Note / Marine Biologist 522 Induction 523 Biographical Note 544
- *Reviews and Tests* Mixed Review Exercises 506, 515, 530, 539 Self-Tests 516, 536, 543 Chapter Summary 544 Chapter Review 545 Chapter Test 546 Cumulative Review 547

12 Triangle Trigonometry

Trigonometric Functions

12-1	Angles and Degree Measure	549
12-2	Trigonometric Functions of Acute Angles	555
12-3	Trigonometric Functions of General Angles	561
12-4	Values of Trigonometric Functions	568

Triangle Trigonometry

12-5	Solving Right Triangles	574
12-6	The Law of Cosines	580
12-7	The Law of Sines	586
12-8	Solving General Triangles	591
12-9	Areas of Triangles	597

- *Explorations* Exploring Trigonometric Ratios 843
- *Technology* Calculator Key-In 554 Computer Exercises 584
- *Special Topics* Career Note / Flight Engineer 567 Historical Note 573 Reading Algebra 585 Biographical Note 601 Challenge 601
- *Reviews and Tests* Mixed Review Exercises 554, 567, 579, 590, 600 Self-Tests 572, 600 Chapter Summary 602 Chapter Review 602 Chapter Test 604 Preparing for College Entrance Exams 605

13 Trigonometric Graphs; Identities

Circular Functions and Their Graphs

13-1	Radian Measure	607
13-2	Circular Functions	613
13-3	Periodicity and Symmetry	619
13-4	Graphs of the Sine and Cosine	624
13-5	Graphs of the Other Functions	630

Trigonometric Identities

13-6	The Fundamental Identities	636
13-7	Trigonometric Addition Formulas	641
13-8	Double-Angle and Half-Angle Formulas	646
13-9	Formulas for the Tangent	650

- *Explorations* Exploring Sine Curves 844
- *Technology* Computer Graphing 624, 625, 631 Computer Exercises 645

• *Special Topics* Biographical Note 618 Historical Note 618 Challenge 623 Application / Frequencies in Music 634

• *Reviews and Tests* Mixed Review Exercises 612, 623, 633, 645, 653 Self-Tests 634, 654 Chapter Summary 654 Chapter Review 655 Chapter Test 657

14 Trigonometric Applications

Vectors

14-1 Vector Operations 659
14-2 Vectors in the Plane 666

Polar Coordinates and Complex Numbers

14-3 Polar Coordinates 675
14-4 The Geometry of Complex Numbers 680
14-5 De Moivre's Theorem 685

Inverse Functions

14-6 The Inverse Cosine and Inverse Sine 689
14-7 Other Inverse Functions 693
14-8 Trigonometric Equations 697

• *Explorations* Exploring Polar Coordinate Equations 845

• *Technology* Computer Exercises 670 Computer Key-In 701 Computer Graphing Ideas 845

• *Special Topics* Application / Force, Work, and Energy 672 Spirals 679 Biographical Note 684

• *Reviews and Tests* Mixed Review Exercises 665, 679, 688, 696 Self-Tests 671, 688, 700 Chapter Summary 702 Chapter Review 703 Chapter Test 704 Mixed Problem Solving 705 Preparing for College Entrance Exams 706 Cumulative Review 707

15 Statistics and Probability

Statistics

15-1 Presenting Statistical Data 709
15-2 Analyzing Statistical Data 713
15-3 The Normal Distribution 719
15-4 Correlation 724

Counting

15-5 Fundamental Counting Principles 730
15-6 Permutations 734
15-7 Combinations 738

Probability

15-8 Sample Spaces and Events 743
15-9 Probability 745
15-10 Mutually Exclusive and Independent Events 754

- *Explorations* Exploring Probability with Experiments 846
- *Technology* Computer Exercises 742, 750 Computer Key-In 762
- *Special Topics* Career Note / Electrician 718 Challenge 733 Reading Algebra 751 Application / Sampling 752 Random Numbers 753
- *Reviews and Tests* Mixed Review Exercises 712, 723, 733, 741, 750 Self-Tests 729, 742, 761 Chapter Summary 763 Chapter Review 764 Chapter Test 765

16 Matrices and Determinants

Matrices

16-1 Definition of Terms 767
16-2 Addition and Scalar Multiplication 770
16-3 Matrix Multiplication 774
16-4 Applications of Matrices 779

Inverses of Matrices

16-5 Determinants 787
16-6 Inverses of Matrices 790

Working with Determinants

16-7 Expansion of Determinants by Minors 794
16-8 Properties of Determinants 798
16-9 Cramer's Rule 801

- *Explorations* Exploring Matrices in Geometry 847
- *Special Topics* Augmented Matrices 785 Historical Note 793
- *Reviews and Tests* Mixed Review Exercises 769, 778, 789, 797, 805 Self-Tests 784, 793, 805 Chapter Summary 806 Chapter Review 806 Chapter Test 808 Preparing for College Entrance Exams 809

Tables 810 **Explorations** 832 **Portfolio Projects** 848

Appendix

A-1 Common Logarithms: Notation and Interpolation 856
A-2 Common Logarithms: Computation 858
A-3 Discrete Mathematics 862
A-4 Preparing for College Entrance Exams 864

Glossary 869 **Index** 879 **Answers to Selected Exercises**

Reading Your Algebra Book

An algebra book requires a different type of reading than a novel or a short story. Every sentence in a math book is full of information and logically linked to the surrounding sentences. You should read the sentences carefully and think about their meaning. As you read, remember that algebra builds upon itself; for example, the method of factoring that you'll study on page 188 will be useful to you on page 697. Be sure to read with a pencil in your hand: Do calculations, draw sketches, and take notes.

Vocabulary

You'll learn many new words in algebra. Some, such as *polynomial* and *parabola,* are mathematical in nature. Others, such as *power* and *proof,* are used in everyday speech but have different meanings when used in algebra. Important words whose meanings you'll learn are printed in **heavy type.** Also, they are listed at the beginning of each Self-Test. If you don't recall the meaning of a word, you can look it up in the Glossary or the Index at the back of the book. The Glossary will give you a definition, and the Index will give you page references for more information.

Symbols

Algebra, and mathematics in general, has its own symbolic language. You must be able to read these symbols in order to understand algebra. For example, $|x| > 2$ means "the absolute value of x is greater than 2." If you aren't sure what a symbol means, check the list of symbols on page xvi.

Diagrams

Throughout this book you'll find many diagrams. They contain information that will help you understand the concepts under discussion. Study the diagrams carefully as you read the text that accompanies them.

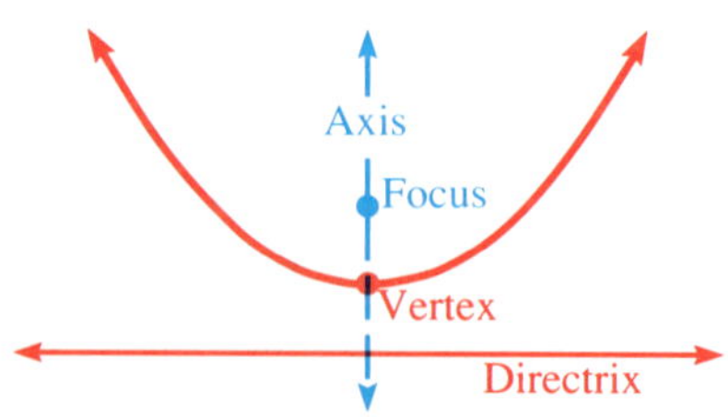

Displayed Material

Throughout this book important information is displayed in gray boxes. This information includes properties, definitions, methods, and summaries. Be sure to read and understand the material in these boxes. You should find these boxes useful when reviewing for tests and exams.

> If a is a real number and m and n are positive integers, then $\boldsymbol{a^m \cdot a^n = a^{m+n}}$.

This book also contains worked-out examples. They will help you in doing many of the exercises and problems.

Example Simplify $x^3 \cdot x^5$.

Solution $x^3 \cdot x^5 = x^{3+5} = x^8$ ***Answer***

Reading Aids

Throughout this book you will find sections called Reading Algebra. These sections deal with such topics as independent study and problem solving strategies. They will help you become a more effective reader and problem solver.

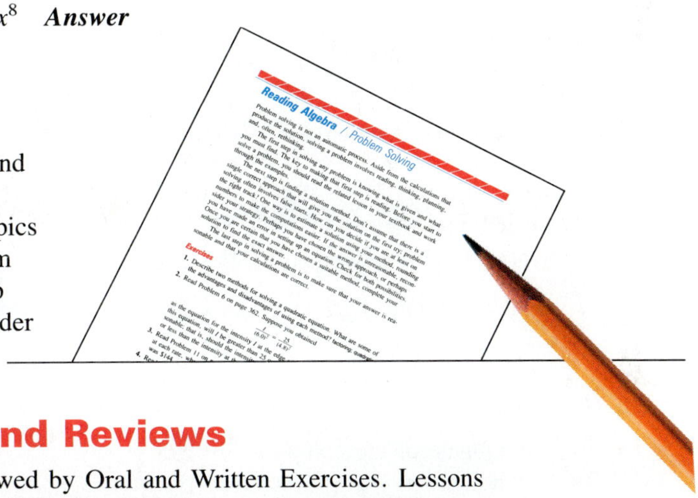

Exercises, Tests, and Reviews

Each lesson in this book is followed by Oral and Written Exercises. Lessons may also include Problems, Mixed Review Exercises, and optional Computer Exercises. Answers for all Mixed Review Exercises and for selected Written Exercises, Problems, and Computer Exercises are given at the back of this book.

Within each chapter you will find Self-Tests that you can use to check your progress. Answers for all Self-Tests are also given at the back of this book.

Each chapter concludes with a Chapter Summary that lists important ideas from the chapter, a Chapter Review in multiple-choice format, and a Chapter Test. Lesson numbers in the margins of the Review and Test indicate which lesson a group of questions covers.

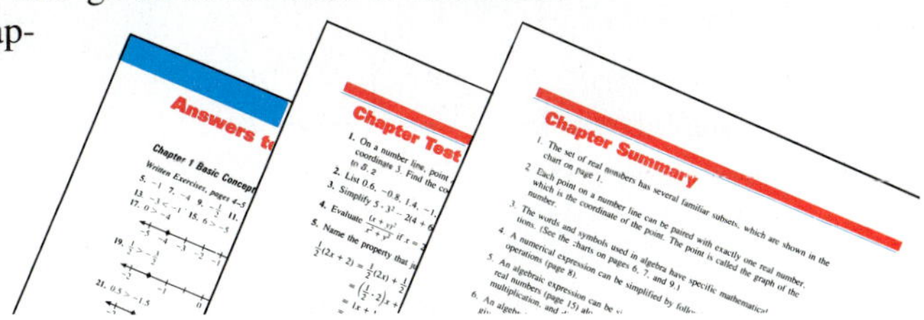

Reading Algebra/*Symbols*

		Page
$\{\ \}$	set	1
$\lvert a \rvert$	absolute value of *a*	3
$=$	equals *or* is equal to	6
$\neq$	does not equal	6
$>$	is greater than	6
$<$	is less than	6
$\leq$	is less than or equal to	6
$\geq$	is greater than or equal to	6
$\therefore$	therefore	6
a^n	the *n*th power of *a*	7
$\in$	is an element of	9
$-a$	additive inverse of *a* *or* the opposite of *a*	15
$\emptyset$	empty set or null set	38
$f(x)$	*f* of *x* *or* the value of *f* at *x*	142
$\approx$	is approximately equal to	222
$\pm$	plus-or-minus sign	259
$\sqrt[n]{b}$	*n*th root of *b*	260
i	imaginary unit ($i^2 = -1$)	288
$\bar{z}$	conjugate of the complex number *z*	298
$a^{p/q}$	*q*th root of *p*th power of *a*	455
f^{-1}	inverse function of *f*	464

		Page
$\log_b N$	logarithm of *N* to the base *b*	469
Σ	summation sign	518
!	factorial	540
°	degree	549
′	minute	550
″	second	550
$\overrightarrow{AB}$	vector *AB*	659
$\lVert \mathbf{u} \rVert$	norm of the vector **u**	661
Cos^{-1}	inverse cosine *or* arc cosine	689
σ	standard deviation	715
r	correlation coefficient	725
${}_nP_r$	number of permutations of *n* elements taken *r* at a time	735
${}_nC_r$	number of combinations of *n* elements taken *r* at a time	738
$P(E)$	probability of event *E*	745
$\cap$	intersection	754
$\cup$	union	754
$\overline{E}$	complement of event *E*	757
det	determinant	787
A^{-1}	inverse of matrix *A*	790

Greek letters: α, β, γ, θ, π, σ, ϕ, ω *alpha, beta, gamma, theta, pi, sigma, phi, omega*

Reading Algebra/*Table of Measures*

Metric Units			
Length	10 millimeters (mm)	=	1 centimeter (cm)
	100 centimeters 1000 millimeters	=	1 meter (m)
	1000 meters	=	1 kilometer (km)
Area	100 square millimeters (mm^2)	=	1 square centimeter (cm^2)
	10,000 square centimeters	=	1 square meter (m^2)
Volume	1000 cubic millimeters (mm^3)	=	1 cubic centimeter (cm^3)
	1,000,000 cubic centimeters	=	1 cubic meter (m^3)
Liquid Capacity	1000 milliliters (mL)	=	1 liter (L)
	1000 cubic centimeters	=	1 liter
Mass	1000 milligrams (mg)	=	1 gram (g)
	1000 grams	=	1 kilogram (kg)
Temperature in degrees Celsius (°C)	0°C	=	freezing point of water
	100°C	=	boiling point of water

United States Customary Units			
Length	12 inches (in.)	=	1 foot (ft)
	36 inches 3 feet	=	1 yard (yd)
	5280 feet 1760 yards	=	1 mile (mi)
Area	144 square inches ($in.^2$)	=	1 square foot (ft^2)
	9 square feet	=	1 square yard (yd^2)
Volume	1728 cubic inches ($in.^3$)	=	1 cubic foot (ft^3)
	27 cubic feet	=	1 cubic yard (yd^3)
Liquid Capacity	16 fluid ounces (fl oz)	=	1 pint (pt)
	2 pints	=	1 quart (qt)
	4 quarts	=	1 gallon (gal)
Weight	16 ounces (oz)	=	1 pound (lb)
Temperature in degrees Fahrenheit (°F)	32°F	=	freezing point of water
	212°F	=	boiling point of water

Time		
60 seconds (s)	=	1 minute (min)
60 minutes	=	1 hour (h)

Facts and Formulas from Geometry

Angle: A figure formed by two rays that have the same endpoint. To indicate that the measure of $\angle A$ is $35°$, we write $\angle A = 35°$.

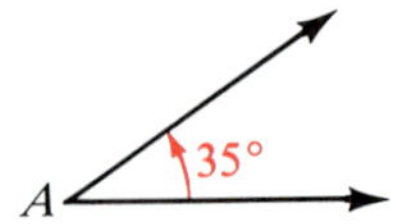

Complementary angles: two angles whose measures have the sum $90°$.

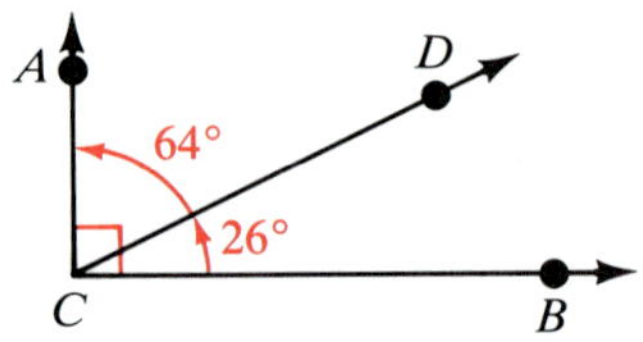

$\angle BCD + \angle DCA = 90°$

Supplementary angles: two angles whose measures have the sum $180°$.

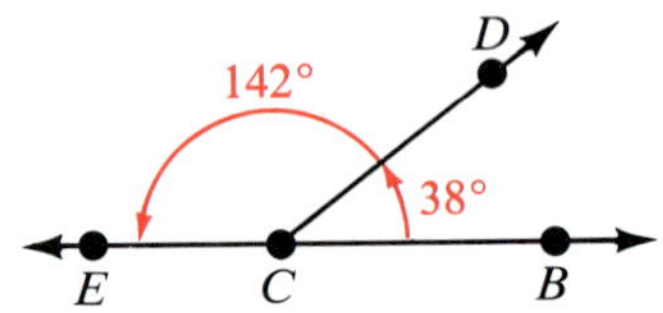

$\angle BCD + \angle DCE = 180°$

Isosceles triangle: a triangle having at least two equal sides and two equal angles. The two sides are called *legs* and the included angle is called the *vertex angle.*

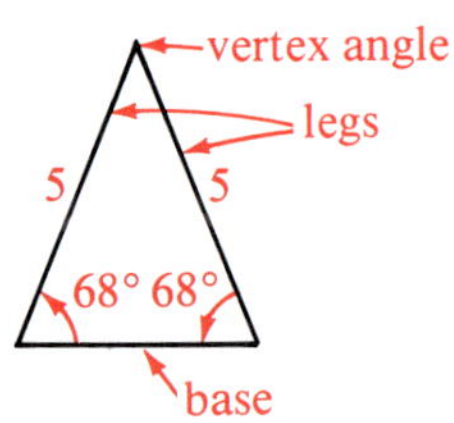

Equilateral triangle: a triangle having three equal sides. Each angle measures $60°$, so the triangle is also called *equiangular.*

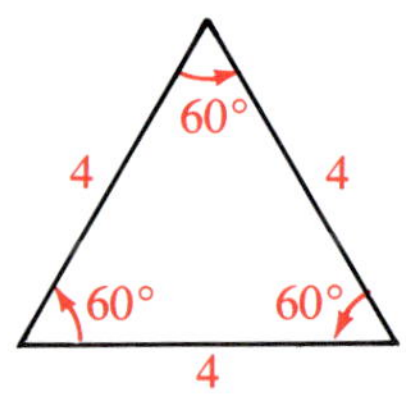

Parallelogram: a quadrilateral (a four-sided polygon) whose opposite sides are equal and parallel. The opposite angles are also equal.

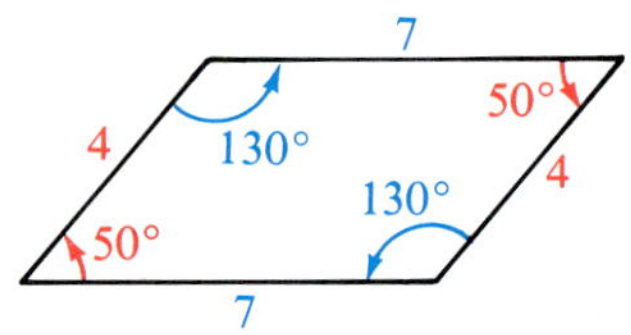

The sum of the measures of the angles of a polygon with n sides is $180(n - 2)°$. For example, the sum of the measures of the angles of a triangle is $180°$ and the sum of the measures of the angles of a quadrilateral is $360°$.

Let: A = area, P = perimeter, C = circumference, S = lateral surface area, T = total surface area, V = volume, $\pi \approx 3.1416$

Triangle

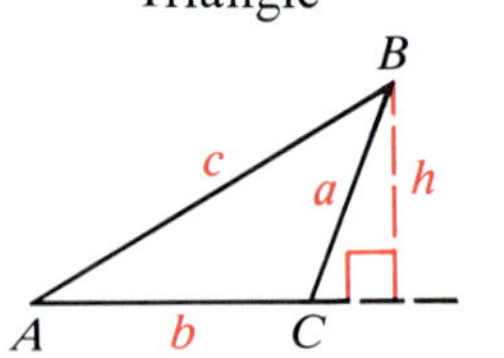

$P = a + b + c;\ A = \frac{1}{2}bh$

Rectangle

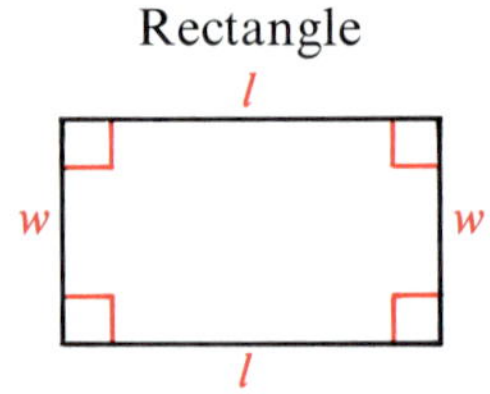

$P = 2l + 2w;\ A = lw$

Parallelogram

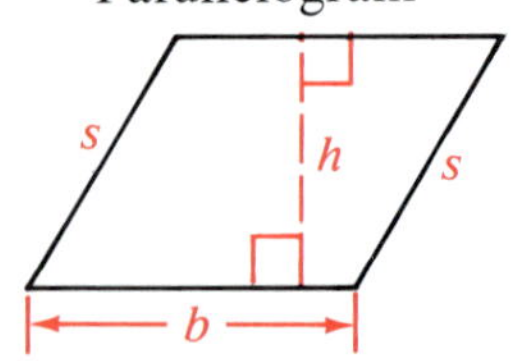

$P = 2b + 2s;\ A = bh$

Facts and Formulas from Geometry

Trapezoid

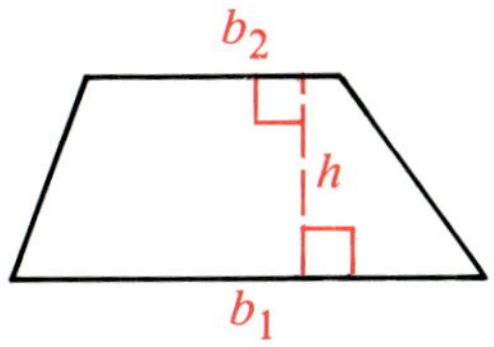

$A = \frac{1}{2}h(b_1 + b_2)$

Circle

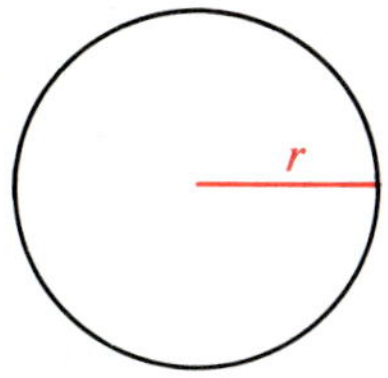

$C = 2\pi r$
$A = \pi r^2$

Rectangular Box

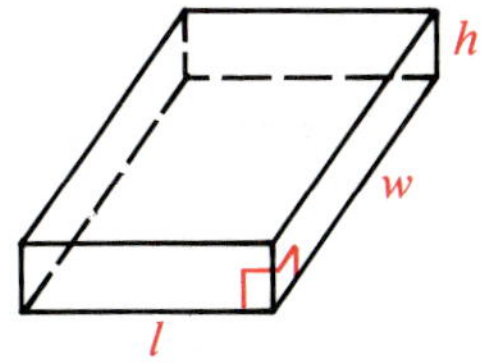

$V = lwh$
$T = 2lw + 2lh + 2wh$

Sphere

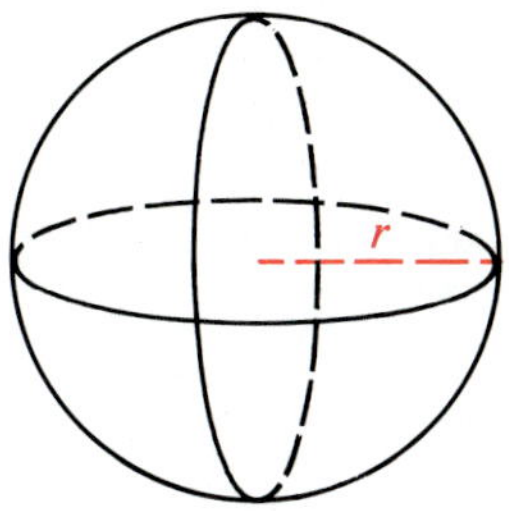

$V = \frac{4}{3}\pi r^3$

$T = 4\pi r^2$

Right Circular Cylinder

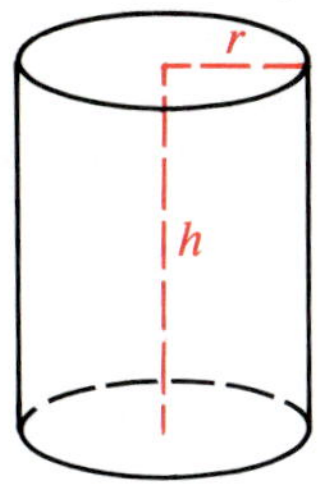

$V = \pi r^2 h$
$S = 2\pi rh$
$T = 2\pi rh + 2\pi r^2$

Right Circular Cone

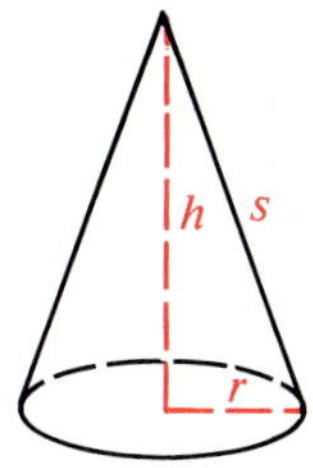

$V = \frac{1}{3}\pi r^2 h$

$S = \pi rs$
$T = \pi rs + \pi r^2$

Using Technology with This Course

There are three types of optional computer material in this text: Computer Key-In features, Computer Exercises, and suggestions for using computer graphing techniques to explore concepts and confirm results.

The Computer Key-In features can be used by students without previous programming experience. These features teach some programming in BASIC and usually include a program that students can run to explore an algebra topic covered in the chapter.

The optional Computer Exercises are designed for students who have some familiarity with programming in BASIC. Students are usually asked to write one or more programs related to the lesson just presented.

The suggestions for applying computer graphing techniques are appropriate for use with a graphing calculator or graphing software such as *Algebra Plotter Plus*.

Calculator Key-In features and certain exercise sets also suggest appropriate use of scientific calculators with this course.

1 Basic Concepts of Algebra

Objectives

1-1 To graph real numbers on a number line, to compare numbers, and to find their absolute values.

1-2 To review the methods used to simplify numerical expressions and to evaluate algebraic expressions.

1-3 To review properties of equality of real numbers and properties for adding and multiplying real numbers.

1-4 To review the rules for adding and subtracting real numbers.

1-5 To review rules for multiplying real numbers.

1-6 To review rules for dividing real numbers.

1-7 To solve certain equations in one variable.

1-8 To translate word phrases into algebraic expressions and word sentences into equations.

1-9 To solve word problems by using an equation in one variable.

Assignment Guide

See p. T58 for Key to the format of the Assignment Guide

Day	Average Algebra	Average Algebra and Trigonometry	Extended Algebra	Extended Algebra and Trigonometry
1	**1-1** 4/1–39 odd **S** 5/*Mixed Review*	**1-1** 4/6–36 mult. of 3 **S** 5/*Mixed Review* **1-2** 10/3–48 mult. of 3	**1-1** 4/6-36 mult. of 3, 39, 40 **S** 5/*Mixed Review* **1-2** 10/9–54 mult. of 3	**1-1** 4/6–36 mult. of 3, 39, 40 **S** 5/*Mixed Review* **1-2** 10/9–54 mult. of 3
2	**1-2** 10/2–46 even **S** 5/34–40 even	**1-3** 17/1–23 odd, 25, 30 **S** 20/*Mixed Review* **R** 13/*Self-Test 1*	**1-3** 17/7–35 odd **S** 20/*Mixed Review* **R** 13/*Self-Test 1*	**1-3** 17/7–35 odd **S** 20/*Mixed Review* **R** 13/*Self-Test 1*
3	**1-3** 17/1–21 odd, 25–28 **S** 20/*Mixed Review* **R** 13/*Self-Test 1*	**1-4** 24/2–50 even, 51–53	**1-4** 24/2–50 even, 51, 52	**1-4** 24/12–48 mult. of 3 **1-5** 30/3–36 mult. of 3 **S** 31/*Mixed Review*
4	**1-4** 24/2–48 even, 51, 53	**1-5** 30/3–33 mult. of 3 **S** 31/*Mixed Review* **1-6** 35/4–28 mult. of 4, 29, 30 **R** 36/*Self-Test 2*	**1-5** 30/3–33 mult. of 3 **S** 31/*Mixed Review* **1-6** 35/4–28 mult. of 4, 29–31 **R** 36/*Self-Test 2*	**1-6** 35/2–28 even, 29, 30 **R** 36/*Self-Test 2* **1-7** 40/4–48 mult. of 4, 51–53 **S** 42/*Mixed Review*
5	**1-5** 30/1–33 odd **S** 31/*Mixed Review*	**1-7** 40/3–27 mult. of 3, 31–53 odd **S** 42/*Mixed Review*	**1-7** 40/3–27 mult. of 3, 31–53 odd **S** 42/*Mixed Review*	**1-8** 46/3–18 mult. of 3, 19–29 odd
6	**1-6** 35/2–28 even, 29–31 **R** 36/*Self-Test 2*	**1-8** 46/3–18 mult. of 3, 19–27 odd	**1-8** 46/3–18 mult. of 3, 19–29 odd	**1-9** 52/*P*: 2–22 even, 23 **S** 54/*Mixed Review* **R** 54/*Self-Test 3*
7	**1-7** 40/2–48 even **S** 42/*Mixed Review*	**1-9** 52/*P*: 2–22 even **S** 54/*Mixed Review* **R** 54/*Self-Test 3*	**1-9** 52/*P*: 2–22 even, 23 **S** 54/*Mixed Review* **R** 54/*Self-Test 3*	*Prepare for Chapter Test* **R** 55/*Chapter Review*

Assignment Guide *(continued)*

Day	Average Algebra	Average Algebra and Trigonometry	Extended Algebra	Extended Algebra and Trigonometry
8	**1-8** 46/1–19 odd **S** 42/49–53	*Prepare for Chapter Test* **R** 55/*Chapter Review*	*Prepare for Chapter Test* **R** 55/*Chapter Review*	*Administer Chapter 1 Test* **2-1** *Read 2-1* 62/1–31 odd **S** 63/*Mixed Review*
9	**1-8** 47/16, 18, 20–27	*Administer Chapter 1 Test*	*Administer Chapter 1 Test* **2-1** *Read 2-1* 62/1–31 odd **S** 63/*Mixed Review*	
10	**1-9** 52/*P*: 1–11 **S** 54/*Mixed Review*			
11	**1-9** 53/*P*: 13–22 **R** 54/*Self-Test 3*			
12	*Prepare for Chapter Test* **R** 55/*Chapter Review*			
13	*Administer Chapter 1 Test*			

Supplementary Materials Guide

For Use with Lesson	Practice Masters	Tests	Study Guide (Reteaching)	Resource Book		
				Tests	Practice Exercises	Prob. Solving (PS) Applications (A) Enrichment (E) Technology (T) Thinking Skl. (TS)
1-1			pp. 1–2	pp. 1–3		p. 204 (A) pp. 250–256 (T)
1-2	Sheet 1		pp. 3–4		p. 100	
1-3			pp. 5–6			pp. 260–261 (TS)
1-4	Sheet 2		pp. 7–8			
1-5			pp. 9–10			
1-6	Sheet 3	Test 1	pp. 11–12		p. 101	
1-7	Sheet 4		pp. 13–14			
1-8	Sheet 5		pp. 15–16		p. 102	
1-9	Sheet 6	Test 2	pp. 17–18			pp. 190–191 (PS)
Chapter 1		Tests 3, 4		pp. 4–7	p. 103	p. 220 (E)

Overhead Visuals

For Use with Lesson	Visual	Title
1-1	A	Multi-Use Packet 1
1-1	B	Multi-Use Packet 2

Software

Software	Computer Activities	Test Generator
	Activities 1, 2	189 test items
For Use with Lessons	1-2, Chapter 1	all lessons

Strategies for Teaching

Communication and Problem Solving

In this chapter students review basic concepts and skills of algebra. This review includes problem solving with emphasis on real numbers in the context of word problems. It is essential for students to effectively communicate by listening, speaking, reading, and writing in order to express concepts in terms of algebraic symbols and to translate algebraic symbols into words. Therefore, students need to be able to understand directions and follow them accurately, to explain processes and concepts clearly, to read with understanding, and to express their ideas clearly in writing. Today, one of the goals of mathematics education is to bring students to a point where they can use the language of algebra as a natural way of expressing mathematical ideas.

To begin the chapter in a motivating way, see the Exploration on page 832. In this activity students explore locating irrational numbers on a number line. They make connections between the numbers, language, symbols, and graphs of real numbers.

1-2 Simplifying Expressions

Have students use the problem solving strategies "trial and error" and "process of elimination" to use four 4's to write expressions for each integer from 1 to 10. Standard operation symbols may be used including $\sqrt{\ }$. Have students read aloud each expression. This activity will also review correct use of language, symbols, and order of operations.

$1 = \frac{4+4}{4+4}$ $\quad 2 = \frac{4}{4} + \frac{4}{4}$ $\quad 3 = \frac{4+4+4}{4}$

$4 = \sqrt{(4)(4)(4)} - 4$ $\quad 5 = \frac{(4)(4)+4}{4}$ $\quad 6 = \frac{4+4}{4} + 4$

$7 = (4+4) - \frac{4}{4}$ $\quad 8 = \frac{4(4+4)}{4}$ $\quad 9 = 4 + 4 + \frac{4}{4}$

$10 = 4 + 4 + 4 - \sqrt{4}$

1-3 Basic Properties of Real Numbers

This lesson, like the preceding one, reviews many important concepts in a condensed format. It will be well worth your while to spend some time in discussing the properties reviewed on pages 14 and 15 to make sure that students have read them with understanding and that they know how to apply them.

You may wish to have the students translate the properties into words. For example, the Addition Property can be stated as follows: "If the same number is added to two equal numbers, the resulting sums will be equal."

1-4 Sums and Differences

Encourage students to study the Reading Algebra feature on page 26 of the textbook. It suggests some techniques that will be helpful to students throughout the course.

1-7 Solving Equations in One Variable

You might see how different students take alternate routes to get the same solution. When students make choices, they should learn that some choices are more advantageous than others.

1-8 Words into Symbols

In preparing to solve word problems, it is important that the students first understand how to translate written problems into algebraic form. Since there are so many different types of problems involved, some time needs to be taken to go over the "how to's" of setting up several types of problems. You might also try some of the Oral Exercises with numbers in place of variables first. For example, "Nine less than a number.'' would become ''Nine less than 12.'', ''Nine less than 20.'', and so on.

Give the students a group of algebraic expressions, equations, and formulas and have them write out word phrases or word sentences to match them. For example,

$2x + 5$	five more than twice a number
$3a - 14 = 7$	Seven is fourteen less than three times a number.

1-9 Problem Solving with Equations

In this lesson, it is important that you demonstrate and explain carefully the steps in the plan for solving a word problem. Point out that not every problem has a solution even if the equation formed does have a solution. Encourage students to check their answers in the word problem before reporting their answers. Discuss the problem solving strategies on page T53.

References to Strategies

PE: Pupil's Edition **TE:** Teacher's Edition **RB:** Resource Book

Problem Solving Strategies

PE: p. 39 (apply a standard formula); p. 48 (generalize from specific examples); pp. 49–50 (5-step problem solving plan); pp. 49, 51–52 (make a table); pp. 50–51 (checking solutions); pp. 50–52 (make a sketch); pp. 51, 53 (insufficient information); pp. 51–53 (unneeded information)
TE: pp. 45, 52
RB: pp. 190–191

Applications

PE: p. xx (data analysis); pp. 25, 43–54 (word problems); p. 39 (using a formula); pp. 39, 44, 50–51 (geometry); p. 51 (physics)
TE: pp. xx, 38, 44, 46, 50–52, 54
RB: p. 204

Nonroutine Problems

PE: p. 5 (Exs. 39, 40); p. 12 (Exs. 45–50, 52–54); p. 19 (Exs. 25–37); p. 25 (Exs. 49–56); p. 31 (Ex. 36); p. 35 (Exs. 29–36); p. 53 (Exs. 13–18, 23)
TE: pp. T82–T84 (Sugg. Extension, Lesson 1-2, 1-3, 1-5, 1-7)

Communication

PE: pp. 6, 7 (understanding definition format); p. 26 (Reading Algebra); p. 43 (translating)
TE: pp. T80, T81, 26 (Reading Algebra); pp. 43, 49 (Motivating the lesson); p. 45 (Common Errors)

Thinking Skills

PE: p. 16 (Exs. 18–19, proofs); p. 20 (logic)
RB: pp. 260–261

Explorations

PE: pp. 20 (logic), 48 (Challenge), 832 (irrational numbers)
RB: p. 220

Connections

PE: p. xx (Data Analysis); p. 20 (Logic); pp. 39, 40, 44 (Physics); pp. 39–41, 42, 44, 47 (Geometry); p. 42 (History); pp. 41, 45 (Geography)

Using Technology

PE: pp. 5, 24, 30, 32, 35, 38 (Exs.); p. 12 (Calculator Key-In)
TE: pp. 5, 10, 12, 32, 35, 38
RB: pp. 253–259
Computer Activities: pp. 1–5

Using Manipulatives/Models

PE: 832 (exploring π)
Overhead Visuals: A, B

Teaching Resources

For use in implementing the teaching strategies referenced on the previous page.

Application/Research
Resource Book, p. 204

Application—Milestones in Mathematics (For use with Chapter 1)

The list below contains important events in the history of mathematics. Use an encyclopedia, books like Howard Eves' *Great Moments in Mathematics (Before 1650)* and *Great Moments in Mathematics (After 1650)*, or other reference books to determine the date of each event. Be sure to indicate whether the date is B.C. or A.D., and bear in mind that some dates are only approximate and may vary slightly among different sources.

Once you have determined the dates of the events, you should create a mathematical time line by ordering the events along a number line marked off in centuries.

1. ______ Newton invents calculus.
2. ______ Tartaglia solves cubic equations.
3. ______ Möbius strip is discovered.
4. ______ Galois writes about group theory.
5. ______ Whitehead and Russell write *Principia Mathematica*.
6. ______ Archimedes determines formulas for the area and volume of a sphere.
7. ______ Lambert proves π is irrational.
8. ______ Oughtred invents the slide rule.
9. ______ Euclid writes *Elements*.
10. ______ Four-color map problem is solved.
11. ______ Stevin introduces decimals.
12. ______ Goldbach states famous conjecture.
13. ______ Fermat leaves last theorem.
14. ______ Descartes creates analytic geometry.
15. ______ Napier invents logarithms.
16. ______ Appollonius studies conic sections.
17. ______ Wallace introduces the symbol for infinity (∞).
18. ______ Agnesi writes *Foundations of Analysis*.
19. ______ Al-Khowarizmi uses zero.
20. ______ Argand graphs imaginary numbers.
21. ______ Rudolff introduces the radical sign.
22. ______ Riemann creates elliptic geometry.
23. ______ Eratosthenes determines the circumference of Earth.
24. ______ Lovelace describes how to program Babbage's "Analytical Engine."
25. ______ Recorde introduces the equals sign.
26. ______ Euler shows that $e^{\pi i} + 1 = 0$.
27. ______ ENIAC, the first electronic computer, is invented.
28. ______ Pascal and Fermat discuss theory of probability in their correspondence.
29. ______ Kovalevski is the first woman to earn a doctorate in mathematics.
30. ______ Pythagoreans discover irrational numbers.
31. ______ Saccheri writes *Euclid Freed of Every Flaw*.
32. ______ Gödel publishes "incompleteness" theorems.
33. ______ Leibniz invents calculus.
34. ______ Cantor creates transfinite numbers.
35. ______ Harriot introduces the inequality signs.
36. ______ Gauss determines the convergence of infinite series.

204 APPLICATIONS

Enrichment/Exploration
Resource Book, p. 220

Groups (For use with Chapter 1)

Addition and subtraction are the two basic operations on numbers. Each of these operations is called a *binary* operation since you can add or multiply only two numbers at a time.

Now consider a particular mathematical system consisting of the set of whole numbers, $\{0, 1, 2, 3, \ldots\}$, under the operation of addition.

a. Is the closure property satisfied?

b. Is the associative property satisfied?

c. Is there an identity element for that operation within the set?

d. Is the property of inverses satisfied?

If the answer to each of these questions is "Yes," then a system is said to be a *group* under the given operation. If the commutative property is also satisfied, then the system is said to be a *commutative group*.

The whole numbers are *not* a group under addition because the answer to question d is "No." No whole number (except 0) has an inverse for addition within the set.

Check to see whether each system is a group. Use questions a–d to help you decide. If you determine that a set is not a group under the given operation, tell which properties are not satisfied.

1. the set of integers, $\{\ldots -3, -2, -1, 0, 1, 2, 3, \ldots\}$, under addition ______
2. the set of integers, $\{\ldots -3, -2, -1, 0, 1, 2, 3, \ldots\}$, under multiplication ______
3. the set of natural numbers, $\{1, 2, 3, \ldots\}$, under addition ______
4. the set of natural numbers, $\{1, 2, 3, \ldots\}$, under multiplication ______
5. the set of even integers, $\{\ldots -4, -2, 0, 2, 4, \ldots\}$, under addition ______
6. the set of odd integers, $\{\ldots -3, -1, 1, 3, \ldots\}$, under addition ______
7. the finite set, $\{-1, 0, 1\}$, under multiplication ______
8. Tell which of the systems in Exercises 1–7 are commutative groups. ______

220 ENRICHMENT

Problem Solving
Resource Book, p. 190

Problem Solving with Equations (For use with Lesson 1-9)

By working through the steps in the problems below, you will gain skill in solving problems by using the five-step plan and an equation in one variable.

Problem 1 A parking meter contains $8.80 in nickels, dimes, and quarters. If the numbers of nickels, dimes, and quarters (in that order) are consecutive odd integers, how many of each coin are in the meter?

a. What does the problem ask you to find? ______

b. Complete: For each type of coin, total value = ______ × ______.

c. Let n = number of nickels. Complete this chart.

	Number	Value (cents)	Total Value
Nickels	n		
Dimes			
Quarters			

d. Which fact given in the problem is not shown in the chart?

e. Use this fact to write an equation for the problem. (*Hint:* Remember to use the same unit of measurement for all values.)

f. Solve to find the value of n. ______

g. Write your answer to the problem. ______

h. Check by finding the *numerical* values of the nickels, the dimes, the quarters, and the collection of coins.

N + D + Q = Total

______ + ______ + ______ = ______

(continued)

190 PROBLEM SOLVING

Problem Solving
Resource Book, p. 191

Problem Solving with Equations (continued)

Problem 2 At 1:00 P.M. two planes leave Rome to fly to Paris. One flies 46 mph faster than the other. One plane lands at 2:30 P.M. and the other, ten minutes later. Find the distance between Rome and Paris.

a. What does the problem ask you to find? ______

b. Recall the distance formula, distance = rate × time. Of these three quantities, which do you know? ______

c. Circle the number below that indicates the best way to start the solution of the problem.

1. Let t = flying time in hours of faster plane.
2. Let s = speed in mph of slower plane.
3. Let d = distance in miles between Rome and Paris.

d. Use your choice in part (c) to complete this chart. (*Hint:* Use fractions to express the time in hours.)

	Rate (mph)	Time (hours)	Distance (miles)
Slower plane			
Faster plane			

e. What must be true about the distances that the planes flew?

f. Write an equation for the problem. ______

g. Solve the equation and find a numerical value for each plane's speed.

Slower plane ______ Faster plane ______

h. Notice that your answers in part (g) do not solve the problem, but they enable you to do so and to check your work. Write your answer for the problem.

i. Check your results by computing and comparing the distance traveled by each plane.

Slower plane ______ Faster plane ______

PROBLEM SOLVING 191

Thinking Skills
Resource Book, p. 260

Thinking Skills (For use after Chapter 1)

Here are some thinking skills you'll be using as you continue to study algebra. Use your skills to answer the questions.

Recalling knowledge || You'll be using facts, formulas, and ideas that you've already learned in your study of mathematics.

1. Give a numerical example that illustrates the property of opposites for real numbers. ________

2. State the property of reciprocals for real numbers. ________________________

3. What is meant by the *closure* of a set of numbers for a given operation? ____________

Interpreting information || You'll be reading information about a situation that's new to you. You'll be trying to make sense of the information by looking at examples, translating the information into mathematical terms, and perhaps organizing the information in charts and tables.

4. Suppose a special form of addition, denoted by ⊕, is defined for the set {0, 1, 2, 3} as follows: If a and b are any two elements of the set, then the sum $a \oplus b$ is the *remainder* when the sum $a + b$ (that is, the "usual" sum as defined for real numbers) is divided by 4. For example, $2 \oplus 3 = 1$, because 1 is the remainder when 2 + 3, or 5, is divided by 4. Use the definition of ⊕ to complete the addition table at the right.

⊕	0	1	2	3
0	___	___	___	___
1	___	___	___	___
2	___	___	___	1
3	___	___	___	___

5. Suppose a special form of multiplication, denoted by ⊗, is defined for the set {0, 1, 2, 3} as follows: If a and b are any two elements of the set, then the product $a \otimes b$ is the *remainder* when the product ab (that is, the "usual" product as defined for real numbers) is divided by 4. For example, $2 \otimes 3 = 2$, because 2 is the remainder when 2×3, or 6, is divided by 4. Use the definition of ⊗ to complete the multiplication table at the right.

⊗	0	1	2	3
0	___	___	___	___
1	___	___	___	___
2	___	___	___	2
3	___	___	___	___

Applying concepts || You'll be recalling what you've already learned and then transferring or extending the ideas to related situations.

6. Seated around a circular table, in clockwise order, are Players A, B, C, and D, who are four people playing cards with a standard 52-card deck. Suppose Player A shuffles the cards and deals all of them, in succession from the top of the deck, to the four players in a clockwise fashion starting with Player B. If the ace of spades is the thirty-first card from the top of the deck, to which player is it dealt? __________

(continued)

Thinking Skills
Resource Book, p. 261

Thinking Skills (Chapter 1) *(continued)*

Reasoning and drawing inferences || You'll be making conjectures, gathering evidence, reaching conclusions, and defending your conclusions.

7. Based on the results of Exercises 4 and 5, is the set {0, 1, 2, 3} *closed* for ⊕ and ⊗? Explain. ________________________

8. Based on the results of Exercise 4, what number from the set {0, 1, 2, 3} is the *opposite* of 3 under ⊕? Explain. ____________________

9. Based on the results of Exercise 5, what number from the set {0, 1, 2, 3} is the *reciprocal* of 3 under ⊗? Explain. ____________________

Analysis || You'll be looking at mathematical objects in detail. You'll try to find similarities, differences, and patterns in data, in mathematical expressions, and in problem types. You'll also be looking for relationships between ideas.

10. Determine whether or not the set {0, 1, 2, 3} and the operations ⊕ and ⊗, as defined in Exercises 4 and 5, constitute a *field.* Defend your answer. ________________

Synthesis || You'll be combining ideas to generate a new idea and combining steps to generate a new result.

11. a. For the set {0, 1, 2, 3}, create a definition for a special form of subtraction, denoted by ⊖.

____
____
____

b. Use your definition to complete the subtraction table at the right.

⊖	0	1	2	3
0	___	___	___	___
1	___	___	___	___
2	___	___	___	___
3	___	___	___	___

Spatial perception || You'll be looking for symmetries in two-dimensional (2-D) and three-dimensional (3-D) figures. You'll also be picturing how 3-D objects look and visualizing movements in 2-D and 3-D space.

12. The entire four-petal "flower" shown at the right can be rotated clockwise about its center until Petal A is in Petal B's current position. Although this can be done using angles of rotation having various measures, what is the *minimum* positive angle measure of this rotation? ____________

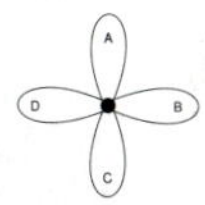

13. If you denote your answer from Exercise 12 by α, explain why it makes sense to write $2\alpha + 3\alpha = 1\alpha$. ________________

Using Technology
Resource Book, p. 253

Summary of BASIC

BASIC is one of the so-called "higher-level" languages in which programs may be written for computers. It is translated by a *compiler* or an *interpreter* inside the computer into a "machine language" that tells the computer what to do.

BASIC is essentially linear in character. A BASIC program contains *statements* that are numbered in succession. The numbering may be 1, 2, 3, ..., but it is customary to use 10, 20, 30, ... so that additional statements can be inserted if necessary. The computer follows the statements in numerical order and carries out the instruction in each statement as the computer comes to it.

The END statement

In some versions of BASIC, the last statement of a program must be an END statement. In other versions, no END statement is needed.

Besides having statements found within programs, BASIC has several *commands* that operate on programs. Since commands are not part of any program, they are not given line numbers.

The RUN command

The RUN command tells the computer to execute (carry out) a program.

The LIST command

The LIST command tells the computer to display the statements in numerical order. This is especially useful after changes have been made in a program and you want to see a "clean" copy of it.

The symbols for operations in BASIC are:

+	for addition	*	for multiplication
−	for subtraction	/	for division
( )	for grouping	↑ or ^	for raising to a power

The same order-of-operations rules are followed as in algebra.

Other symbols used in BASIC are:

=	for *is equal to*	<	for *is less than*
>	for *is greater than*	< >	for *is not equal to*
< =	for *is less than or equal to*	> =	for *is greater than or equal to*

Very large and very small numbers are expressed in E-notation, which is a version of scientific notation. For example, 1.002E + 09 means 1.002×10^9, and 0.000001 would be written as 1E − 06.

BASIC handles variables much as you do in algebra. A variable can be denoted by a single letter, a letter followed by a digit, or possibly other combinations of symbols as allowed by the various versions of BASIC. Values are given to variables, and operations are performed on them.

(continued)

Using Technology
Computer Activities, p. 3

ACTIVITY 2. *Magic Squares* (*for use with Chapter 1*)

Directions: Write all answers in the spaces provided.

PROBLEM

A magic square is a square array of positive integers, in which the sum of the elements in every row, column, and diagonal is the same. Use the computer to find magic squares with an odd number of rows (or columns).

PROGRAM •

```
10  DIM M(9, 9)
20  PRINT "HOW MANY ROWS";
30  INPUT N
40  PRINT
50  LET R = 1
60  LET C = (N + 1) / 2
70  LET M(R, C) = 1
80  FOR L = 2 TO N • N
90  IF R <> 1 THEN 170
100  IF C <> N THEN 140
110  LET R = 2
120  LET C = N
130  GOTO 260
140  LET R = N
150  LET C = C + 1
160  GOTO 260
170  IF C <> N THEN 210
180  LET R = R - 1
190  LET C = 1
200  GOTO 260
210  LET R = R - 1
220  LET C = C + 1
230  IF M(R, C) = 0 THEN 260
240  LET R = R + 2
250  LET C = C - 1
260  LET M(R, C) = L
270  NEXT L
280  FOR J = 1 TO N
290  FOR K = 1 TO N
300  PRINT M(J, K); TAB(4 • K + 1);
310  NEXT K
320  PRINT
330  NEXT J
340  PRINT
350  LET S = N • (N • N + 1) / 2
360  PRINT "THE SUM OF EACH ROW, COLUMN,"
370  PRINT "AND DIAGONAL IS  "; S; "."
380  END
```

PROGRAM CHECK

Type in the program. To test whether you entered it correctly, run the program. Enter the number 3 after the question. The computer should print:

```
8   1   6
3   5   7
4   9   2

THE SUM OF EACH ROW, COLUMN,
AND DIAGONAL IS 15.
```

(continued)

Application

Modern society seems to run on data. Numbers influence decisions, theories, and work. An individual in such a society needs to feel comfortable with numbers. To help make sense of numerical information, tools like calculators and computers are useful. The graphing calculator in the photo displays a histogram, which is a statistical graph used to present data.

One type of data that is often encountered is time. The calendar and clock in the photo indicate how the time continuum is divided and subdivided so that the passage of time can be measured. The infinite nature of time brings to mind the infinite set of real numbers, which can be depicted using a number line. The set of real numbers forms the basis for the study of algebra.

Research Activities

Represent the history of mathematics on a bulletin board using a number line marked off by centuries beginning about 600 B.C. Assign students various mathematical events and ask them to research and present brief oral or written reports. With each report the event can be added to the mathematical time line. This activity can be spread over the entire year, with reports given during the study of related chapters.

Support Materials

Resource Book p. 204

1 Basic Concepts of Algebra

The ability to understand numerical relationships and to use them in making decisions is important in many careers, as well as in daily life. Studying algebra can help develop this ability.

The Language of Algebra

1-1 Real Numbers and Their Graphs

Objective To graph real numbers on a number line, to compare numbers, and to find their absolute values.

In mathematics you have met such numbers as

$$4, -3, -\frac{5}{3}, \sqrt{3}, -4.16, \pi, \text{ and } 0.$$

These are members of the set of **real numbers.**

Some important subsets of the real numbers are shown in the chart below. Remember that when you list members of a set within braces, { }, you use three dots to show that a list continues without end.

Real Numbers

Natural numbers	{1, 2, 3, . . .}
Whole numbers	{0, 1, 2, 3, . . .}
Integers	{. . . , −3, −2, −1, 0, 1, 2, 3, . . .}
Rational numbers	numbers, such as $-\frac{5}{3}$, $\frac{7}{9}$, 0.5, and $\frac{4}{1}$, that result when an integer is divided by a nonzero integer
Irrational numbers	numbers, such as $\sqrt{3}$, $-\sqrt{7}$, and π, that are not rational

You have probably worked with number lines like the one below.

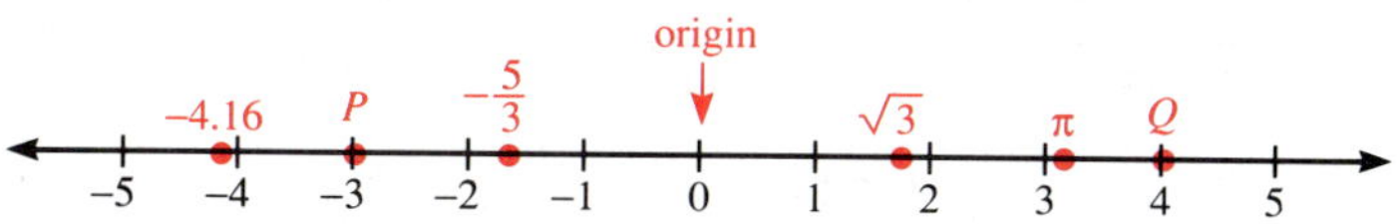

When using a number line, we take the following facts for granted.

1. Each point on a number line is paired with exactly one real number, called the **coordinate** of the point.
2. Each real number is paired with exactly one point on the line, called the **graph** of the number.

For example, on the number line above, the coordinate of point Q is 4, and the graph of -3 is point P. The graph of 0 (zero) is the **origin.**

Teaching References

Lesson Commentary, pp. T79–T83

Assignment Guide, pp. T59–T60

Supplementary Materials
- Practice Masters 1–6
- Tests 1–4
- Resource Book
  - Practice Exercises, pp. 100–103
  - Tests, pp. 4–7
  - Enrichment Activity, p. 220
  - Application, p. 204
  - Practice in Problem Solving/Word Problems, pp. 190–191
- Study Guide, pp. 1–18
- Computer Activities 1–2
- Test Generator
- Alternate Test, p. T13

Explorations, p. 832

Teaching Suggestions, p. T79

Suggested Extensions, p. T79

Warm-Up Exercises

List each set of numbers in order from least to greatest.

1. $\left\{5, -4, 0, -\frac{1}{2}\right\}$ $\quad -4, -\frac{1}{2}, 0, 5$

2. $\left\{\frac{2}{3}, 1.7, -1.6, -2\right\}$ $\quad -2, -1.6, \frac{2}{3}, 1.7$

3. $\left\{\frac{5}{8}, -1, -\frac{3}{4}, \frac{1}{2}\right\}$ $\quad -1, -\frac{3}{4}, \frac{1}{2}, \frac{5}{8}$

4. $\{3, \pi, 0, -2\}$ $\quad -2, 0, 3, \pi$

Motivating the Lesson

Point out that before computing with numbers, it is important to review some basic concepts and vocabulary about numbers. For example: What types of numbers constitute the set of real numbers? What is the *absolute value* of a number? Similarly, it is necessary to review some skills, such as comparing, ordering, and graphing real numbers, that are essential to the study of algebra.

Chalkboard Examples

1. Find the coordinate of the point two thirds of the way from A to B for the number line shown.

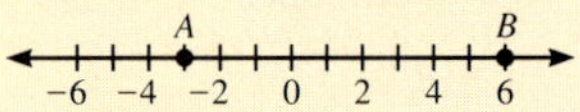

The distance from A to B is 9 units. Since $\frac{2}{3} \cdot 9 = 6$, move six units to the right from A to find the required point. The coordinate of the point is 3.

2. Graph the numbers -0.5 and -5 on a number line. Then write an inequality statement comparing them.

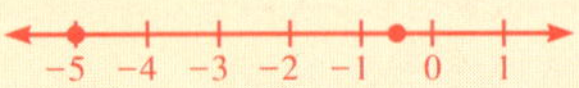

$-5 < -0.5$, or $-0.5 > -5$

Find the value of each expression.

3. $3 \cdot |-6|$ $3 \cdot 6 = 18$
4. $|-4| + |6|$ $4 + 6 = 10$

Example 1 Find the coordinate of the point one third of the way from C to D for the number line shown.

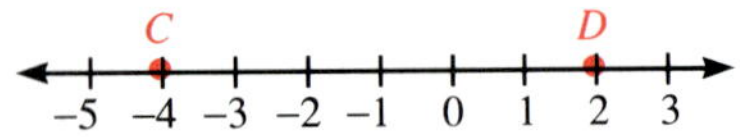

Solution The distance from C to D is 6 units. Since $\frac{1}{3} \cdot 6 = 2$, move two units to the right from C to find the required point. The coordinate of this point is -2. ***Answer***

On a horizontal number line, you can see that the larger a number is, the farther to the right its graph is. In particular, positive numbers (numbers greater than 0) are graphed to the right of the origin and negative numbers (numbers less than 0) to the left.

Notice that a number line shows the *order* of the numbers. The fact that the graph of -3 is to the left of the graph of 1 corresponds to these two equivalent statements:

In Words	In Symbols
Negative three is less than one.	$-3 < 1$
One is greater than negative three.	$1 > -3$

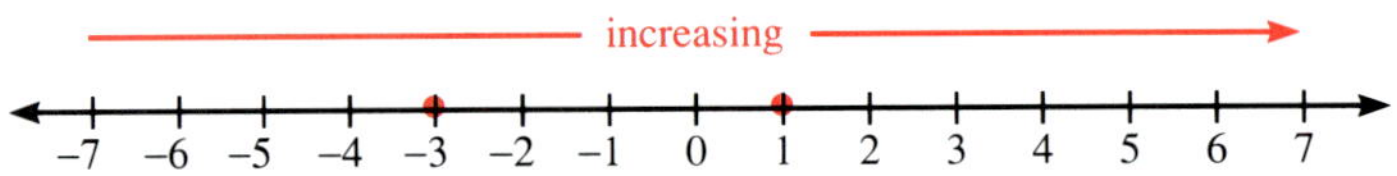

Look at the number line to see that the following statements are true.

$-4 < -1$ $\quad -4 < 0$ $\quad -1 < 3$

$-1 > -4$ $\quad 0 > -4$ $\quad 3 > -1$

Example 2 Graph the numbers -2 and -6 on a number line. Then write an inequality statement comparing them.

Solution

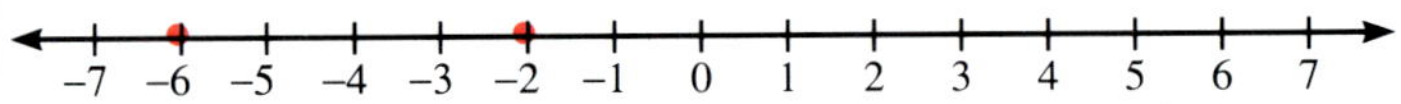

$-6 < -2$, or $-2 > -6$ ***Answer***

The diagram below illustrates that every real number has an **opposite.**

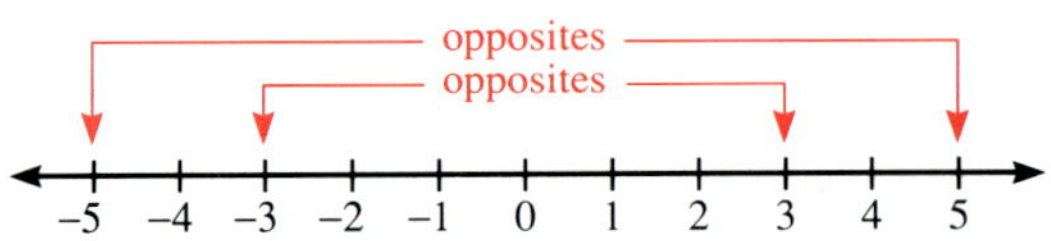

In Words	In Symbols
The opposite of three is negative three.	$-(3) = -3$
The opposite of negative five is five.	$-(-5) = 5$
The opposite of zero is zero.	$-(0) = 0$

On a number line, the graphs of opposites are on opposite sides of the origin but are the same distance from the origin. For example, the graphs of 5 and -5 are both 5 units from 0. The **absolute value** of a number is the distance between the graph of the number and the origin on a number line. Vertical bars $| \; |$ are used to denote absolute value. Thus:

$|5| = 5$ — Read: The absolute value of five is five.

$|-5| = 5$ — Read: The absolute value of negative five is five.

In general:	Examples
1. The absolute value of a positive number is the number itself.	$\|7\| = 7$
2. The absolute value of a negative number is the opposite of the number.	$\|-1\| = -(-1) = 1$
3. The absolute value of zero is zero.	$\|0\| = 0$

Example 3 Find the value of each expression.

a. $-|-4|$ **b.** $|-6| - |6|$

Solution **a.** $-|-4| = -4$ **b.** $|-6| - |6| = 6 - 6 = 0$

Oral Exercises

Estimate the coordinate of each point shown in red to the nearest half unit.

1. 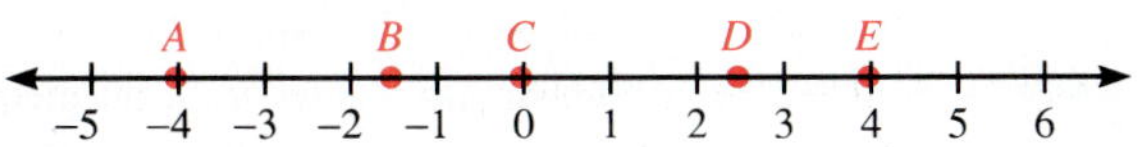

A, -4; B, $-\frac{3}{2}$; C, 0; D, $\frac{5}{2}$; E, 4

2. 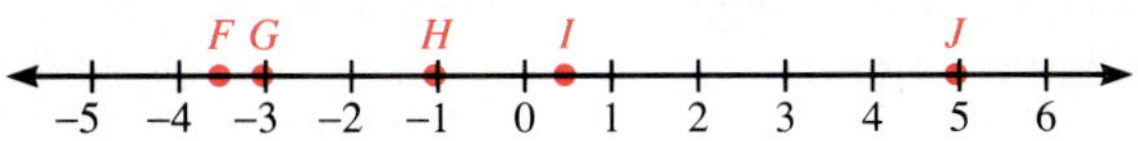

F, $-\frac{7}{2}$; G, -3; H, -1; I, $\frac{1}{2}$; J, 5

Give the opposite of each number.

3. 2 -2 **4.** 0 0 **5.** $-\sqrt{2}$ $\sqrt{2}$ **6.** $-\frac{3}{2}$ $\frac{3}{2}$

Find the value of each expression.

7. $|-7|$ 7 **8.** $|4 - 4|$ 0 **9.** $|3 - 2|$ 1 **10.** $|3| - |2|$ 1

Tell whether each statement is true or false.

11. $-4 < 1$ T **12.** $-6 > 0$ F **13.** $-2 > -7$ T **14.** $-3 < 3$ T

15. $|-4| < 1$ F **16.** $|-6| > 0$ T **17.** $|-2| > |-7|$ F **18.** $|-3| < |3|$ F

19. Is the absolute value of the sum of two numbers always equal to the sum of their absolute values? Explain. No; $|-2| + |5| = 7$ but $|-2 + 5| = 3$

List the numbers in order from least to greatest.

5. $-1, 2, -12, 21, 12$
$-12, -1, 2, 12, 21$

6. $-8.6, -8.06, -8.8, -8.08$
$-8.8, -8.6, -8.08, -8.06$

7. On a horizontal number line, point *P* has coordinate -5. Find the coordinates of the points that are 8 units from *P*.
3 and -13

Check for Understanding

Many teachers like to assess how well students have understood the explanation of a concept just presented by having students try a few exercises immediately, before a new concept is introduced. Here is a suggested use of the Oral Exercises to provide this check.

Oral Exs. 1–2: use before Example 1.
Oral Exs. 3–6: use after the discussion of opposite.
Oral Exs. 7–20: use after Example 3.

Guided Practice

After the presentation of the lesson, some teachers like to have students work some exercises under their guidance in preparation for a homework assignment. Here are some additional A exercises that can be used for this purpose.

Find the coordinate of each point. Use the number line below.

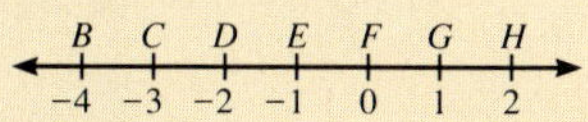

1. *H* 2

2. The point halfway between *G* and *B*. $-1\frac{1}{2}$

(continued)

Guided Practice
(continued)

3. Write in symbols: negative six is less than negative two.
$-6 < -2$

4. Write an inequality statement comparing the numbers -0.8 and -0.54.
$-0.8 < -0.54$

5. Find the value of $-|-0.3|$.
-0.3

Summarizing the Lesson

Remind students that they have learned how to compare numbers using a number line and how to find the absolute value of a number or expression. Ask students to describe what happens to the absolute values as you move from right to left on a number line.

Suggested Assignments

Average Algebra
4/1–39 odd
S 5/Mixed Review

Average Alg. and Trig.
4/6–36 mult. of 3
S 5/Mixed Review
Assign with Lesson 1-2.

Extended Algebra
4/6–36 mult. of 3, 39, 40
S 5/Mixed Review
Assign with Lesson 1-2.

Extended Alg. and Trig.
4/6–36 mult. of 3, 39, 40
S 5/Mixed Review
Assign with Lesson 1-2.

Additional Answers
Written Exercises

17. 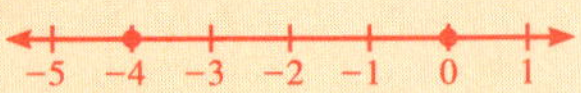

18. 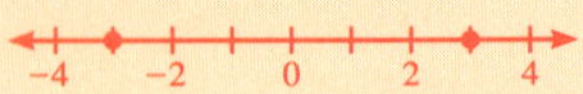

Written Exercises

Find the coordinate of each point described. Use the number line below.

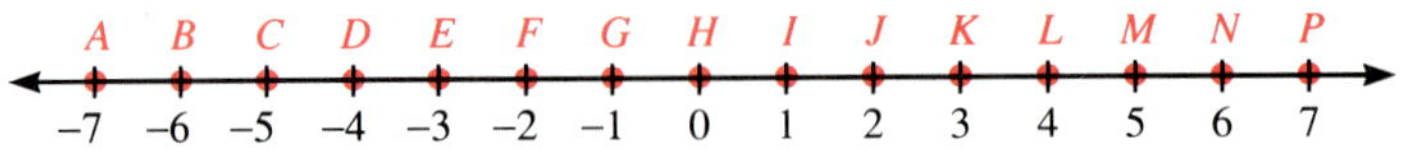

A

1. B -6
2. M 5
3. H 0
4. G -1
5. The point halfway between D and J -1
6. The point halfway between C and F $-\frac{7}{2}$
7. The point one third of the way from B to H -4
8. The point two thirds of the way from C to I -1
9. The point one fourth of the way from F to L $-\frac{1}{2}$
10. The point three fourths of the way from B to J 0

Write each statement using symbols.

Sample Negative four is less than one. **Solution** $-4 < 1$

11. Zero is greater than negative six. $0 > -6$
12. Negative two is less than two. $-2 < 2$
13. Negative three is less than negative one. $-3 < -1$
14. One is greater than negative ten. $1 > -10$
15. Six is greater than negative five. $6 > -5$
16. Negative six is less than negative five eighths. $-6 < -\frac{5}{8}$

Graph each pair of numbers on a separate number line. Then write an inequality statement comparing the numbers.

17. 0 and -4 $0 > -4$
18. 3 and -3 $3 > -3$
19. $\frac{1}{2}$ and $-\frac{3}{2}$ $\frac{1}{2} > -\frac{3}{2}$
20. $-\frac{5}{4}$ and $-\frac{3}{4}$ $-\frac{5}{4} < -\frac{3}{4}$
21. 0.5 and -1.5 $0.5 > -1.5$
22. -2.5 and -0.75 $-2.5 < -0.75$

Find the value of each expression.

23. $-|-5|$ -5
24. $2 \cdot |-5|$ 10
25. $|5| - |-2|$ 3
26. $|-5| - |2|$ 3

Arrange each list of numbers in order from least to greatest.

B

27. $4, -5, 0, -1, 2$ $-5, -1, 0, 2, 4$
28. $1, -1, -5, 3, -3$ $-5, -3, -1, 1, 3$
29. $-\frac{1}{2}, 2, -\frac{3}{2}, \frac{5}{2}, -2$ $-2, -\frac{3}{2}, -\frac{1}{2}, 2, \frac{5}{2}$
30. $-\frac{3}{4}, -\frac{1}{4}, \frac{1}{2}, -\frac{1}{2}, \frac{3}{4}$ $-\frac{3}{4}, -\frac{1}{2}, -\frac{1}{4}, \frac{1}{2}, \frac{3}{4}$
31. $-1.6, -2.6, -0.6, -1.8$ $-2.6, -1.8, -1.6, -0.6$
32. $-2.4, -1.7, -1.9, -2.0$ $-2.4, -2.0, -1.9, -1.7$

On a number line, point A has coordinate -5 and point B has coordinate 1. Find the coordinate of each point described.

33. The point 2 units to the left of B -1

34. The point 3 units to the right of A -2

35. The point 1.4 units to the right of A -3.6

36. The point 2.6 units to the left of B -1.6

37. The point $5\frac{1}{3}$ units from A and $\frac{2}{3}$ units from B $\frac{1}{3}$

38. The point $4\frac{1}{2}$ units from A and $1\frac{1}{2}$ units from B $-\frac{1}{2}$

C **39.** Each point that is twice as far from A as from B $-1, 7$

40. Each point that is half as far from A as from B $-11, -3$

Mixed Review Exercises

Tell whether each statement is true or false.

1. $\frac{1}{2} + \frac{2}{3} = \frac{3}{5}$ F

2. $3\frac{2}{5} - 2\frac{3}{10} < 1$ F

3. $\frac{4}{7} \times \frac{4}{7} > \frac{4}{7}$ F

4. $6\frac{3}{8} \div 2\frac{1}{4} = 2\frac{5}{6}$ T

5. $5.31 + 24.7 = 29.01$ F

6. $0.1 - 0.09 < 0.001$ F

7. $1.3 \times 0.1 = 0.13$ T

8. $1 \div 0.99 > 1$ T

9. $|-6| > -6$ T

10. $|-1| = -(-1)$ T

11. $|0| = 0$ T

12. $|-3| < 0$ F

Computer Exercises

The optional computer exercises in this book require familiarity with the fundamental concepts and terminology of programming in BASIC.

1. Write a program in which the user enters two numbers, and the program then prints a statement of comparison using $>$, $=$, or $<$. For example, if the user enters -3 and -5, the program should print $-3 > -5$.
2. Run the program in Exercise 1 for the following pairs of numbers.

   a. $-8, -7$ $-8 < -7$ **b.** $0.1, 0.01$ $0.1 > 0.01$ **c.** $0, 0$ $0 = 0$

   d. $3.65, 3.56$ $3.65 > 3.56$ **e.** $4, -4$ $4 > -4$ **f.** $800, 8000$ $800 < 8000$
3. Write a program in which the user enters the coordinates of two points on a number line, and the program then prints the coordinates of the point halfway between the entered points.
4. Run the program in Exercise 3 to find the coordinate of the point halfway between the two points with the given coordinates.

   a. $3, 8$ 5.5 **b.** $-1.7, 1.7$ 0 **c.** $-2.73, 5.86$ 1.565

   d. $-7.92, -2.40$ -5.16 **e.** $9.1, 9.01$ 9.055 **f.** $102, -100$ 1

19. 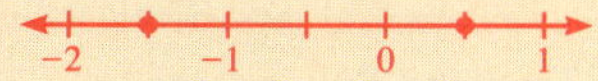

20. 

21. 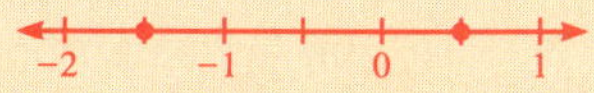

22. 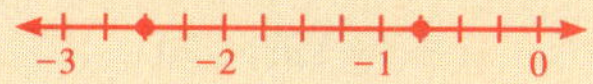

Common Errors

Some students may need to be reminded that on a horizontal number line, the numbers increase from left to right and decrease from right to left. For example, students may think that $-6 < -3$ is a false statement. Students often mistakenly guess that $|-7| + |7|$ equals 0. In reteaching, remind students that $|-7| + |7| = 7 + 7 = 14$.

Using a Computer

These exercises use a computer to compare two numbers and to find the coordinate of a point halfway between two given points.

Supplementary Materials

Study Guide	pp. 1–2
Resource Book	pp. 253–259

Teaching Suggestions, p. T80

Suggested Extensions, p. T80

Warm-Up Exercises

Perform the indicated operations.

1. $54 + 9 - 17$ 46
2. $14 + (-19) - 11$ −16
3. $-29 - (-12) + 22$ 5
4. $42 \div 7 \times 4$ 24
5. $-6 \times 12 \div 9$ −8
6. $-8 \times (-7) \times 2$ 112

Motivating the Lesson

Point out that every job and hobby has its own special vocabulary. Ask students to give examples of special terms that they use regularly in their jobs or hobbies. State that this lesson reviews some of the vocabulary that is essential to the study of algebra.

Chalkboard Examples

1. Use one of the symbols $<, =, >$ to make the statement true.
3×7 __?__ $12 + 7$
Find the value of each side.
$3 \times 7 = 21$ and $12 + 7 = 19$ so, $3 \times 7 > 12 + 7$.

Simplify.

2. $27 \div 3 - 3 \cdot 2$
$9 - 3 \cdot 2 = 9 - 6 = 3$
3. $(9 - 3)^2 \div 4 - 1$
$6^2 \div 4 - 1 = 36 \div 4 - 1 = 9 - 1 = 8$
4. $4[5 - (3^2 - 4)]$
$4[5 - (9 - 4)] = 4[5 - 5] = 4(0) = 0$

1-2 Simplifying Expressions

Objective To review the methods used to simplify numerical expressions and to evaluate algebraic expressions.

The words and symbols reviewed in the following charts are used throughout this book. Study the examples to be sure you understand the meaning of the words shown in **heavy type.**

Definition	*Examples*
Numerical expression, or **numeral:** a symbol or group of symbols used to represent a number	3×4 $5 + 5 + 2$ $24 \div 2$ 12
Value of a numerical expression: the number represented by the expression	Twelve is the value of 3×4.
Equation: a sentence formed by placing an *equals sign* = (read "equals" or "is equal to") between two expressions, called the *sides* of the equation. The equation is a true statement if both sides have the same value.	$3 \times 4 = 12$ the sides $36 \div 4 = 1 \times 9$
Inequality symbol: one of the symbols $<, >, \neq$ (read "does not equal" or "is not equal to"), $\leq$ (read "is less than or equal to"), $\geq$ (read "is greater than or equal to")	$3 + 9 < 20$ $3 \times 9 > 20$ $3 + 9 \neq 3 \times 9$ $7 \leq 1 + 6$ $8 \geq 0$
Inequality: a sentence formed by placing an inequality symbol between two expressions, called the *sides* of the inequality	$-3 > -5$ the sides $-3 \leq -0.3$

Example 1 Use one of the symbols $<$, $=$, or $>$ to make a true statement.

a. $10 - 2$ __?__ 2×4 **b.** $27 \div 3$ __?__ $3^2 + 2$

Solution Find the value of each side.

a. $10 - 2 = 8$ and
$2 \times 4 = 8$
$\therefore 10 - 2 = 2 \cdot 4$
↑
(read "therefore")

b. $27 \div 3 = 9$ and
$3^2 + 2 = 9 + 2 = 11$
$\therefore 27 \div 3 < 3^2 + 2$

As you can see in Example 1, expressions may involve sums, differences, products, quotients, and powers. The following chart reviews the meaning of these words as well as the various grouping symbols that are used to enclose parts of expressions.

Definition	***Examples***
Sum: the result of adding numbers, called the **terms** of the sum	$6 + 15 = 21$ (6 and 15: terms; 21: sum)
Difference: the result of subtracting one number from another	$8 - 6 = 2$ (2: difference)
Product: the result of multiplying numbers, called the **factors** of the product. Multiplication is sometimes shown by a raised dot · instead of a times sign ×.	$3 \times 15 = 45$ (3 and 15: factors; 45: product) $6 \cdot 7 = 42$ (6 and 7: factors; 42: product)
Quotient: the result of dividing one number by another. Division is indicated by the symbol ÷ or by a fraction bar.	$35 \div 5 = 7 = \frac{35}{5}$ (7: quotient)
Power, base, and **exponent:** A *power* is a product of equal factors. The repeated factor is the *base*. A positive *exponent* tells the number of times the base occurs as a factor.	Let the base be 3. (exponent, base) First power: $3 = 3^1$ Second power: $3 \times 3 = 3^2$ 3^2 is also called "three squared." Third power: $3 \times 3 \times 3 = 3^3$ 3^3 is also called "three cubed." Fourth power: $3 \times 3 \times 3 \times 3 = 3^4$
Grouping symbols: pairs of parentheses ( ), brackets [ ], or a bar — used to enclose part of an expression that represents a single number	$3 + (2 \times 6) = 3 + 12$ $8 \times [5 + 1] = 8 \times 6$ $\frac{4 + 18}{11} = \frac{22}{11}$

To **simplify** an expression you replace it by the simplest or most common symbol having the same value. For example,

$$4 \cdot (7 - 5) = 4 \cdot 2 = 8$$

and $$\frac{5 + 3}{5 - 3} = \frac{8}{2} = 4.$$

Evaluate the expression if $x = 7$, $y = 6$ and $z = 3$.

5. $\frac{x + z}{y} + \frac{x + y}{z}$

$\frac{7 + 3}{6} + \frac{7 + 6}{3}$

$= \frac{10}{6} + \frac{13}{3}$

$= \frac{5}{3} + \frac{13}{3} = \frac{18}{3} = 6$

Evaluate if $x = 6$ and $y = -3$.

6. $|x|^2 + 4|y|$

$|6|^2 + 4|-3| = 6^2 + 4(3) = 36 + 12 = 48$

Common Errors

Some students mistakenly think that they should always multiply before dividing and should always add before subtracting. Thus, when they attempt to simplify expressions such as $12 \div 3 \times 2$ and $8 - 3 + 1$ they get 2 and 4 respectively, instead of the correct answers 8 and 6. When reteaching, remind students that multiplication and division are done in order from left to right and that the same is true for addition and subtraction.

Check for Understanding

Here is a suggested use of the Oral Exercises to check students' understanding as you teach the lesson.

Oral Exs. 1–6: use after Example 1.
Oral Exs. 7–12: use after Example 2.
Oral Exs. 13–20: use after Example 3.
Oral Exs. 21–24: use after Example 4.

Guided Practice

Use one of the symbols <, =, > to make a true statement.

1. $2 + 2 \_\_?\_\_ 2 \cdot 2$ =
2. $2^2 + 2^2 \_\_?\_\_ 2^2 \cdot 2^2$ <
3. $\frac{9^2 - 1}{9 + 1} \_\_?\_\_ 9 - 1$ =
4. $(18 \div 6) \cdot 3 \_\_?\_\_ 18 \div (6 \div 3)$ =

Simplify.

5. $10 - 2 + 6 - 9$ 5
6. $3^2 - 4 \div 2 + 1$ 8
7. $3[9 - 2(5 - 1)]$ 3
8. $\frac{4^2(5^2 - 3^2)}{2^2}$ 64
9. Evaluate if $x = 3$, $y = 4$, and $z = 5$.
   $\frac{3xy + 4z}{7y} - \frac{10x}{5y + 2z}$ 1
10. Evaluate if $x = -2$ and $y = -3$.
   $2|y| - 3|x| + |y|^2$ 9

The following principle is used in simplifying expressions.

Substitution Principle

An expression may be replaced by another expression that has the same value.

When simplifying expressions, we agree to perform operations in the following order.

Order of Operations

1. Simplify the expression within each grouping symbol, working outward from the innermost grouping.
2. Simplify powers.
3. Perform multiplications and divisions in order from left to right.
4. Perform additions and subtractions in order from left to right.

Recall that the symbols × and · often are omitted in expressions with grouping symbols. For example, $3(5 + 2)$ means $3 \times (5 + 2)$, and $(2^3 - 1)5$ means $(2^3 - 1) \cdot 5$.

Example 2 Simplify: **a.** $2 \cdot 5 - 6 \div 3$

b. $2[4 + (2^3 - 1)]$

c. $5^2 - \frac{1 + 5}{2}$

Solution

a. $2 \cdot 5 - 6 \div 3 = 10 - \underbrace{6 \div 3}$
$= 10 - 2$
$= 8$ ***Answer***

b. $2[4 + (2^3 - 1)] = 2[4 + \underbrace{(8 - 1)}]$
$= 2\underbrace{[4 + 7]}$
$= 2[11]$
$= 22$ ***Answer***

c. $5^2 - \frac{1 + 5}{2} = 5^2 - \underbrace{\frac{6}{2}}$
$= 25 - 3$
$= 22$ ***Answer***

The chart at the top of the next page reviews the meanings of *variable* and some related terms.

Definition	Examples
Variable: a symbol, usually a letter, used to represent any member of a given set, called the **domain** or **replacement set,** of the variable	a, x, or y If the domain of x is $\{0, 1, 2, 3\}$, we write $x \in \{0, 1, 2, 3\}$ and read "x is a member of (or belongs to) the set whose members are 0, 1, 2, and 3."
Values of a variable: the members of the domain of the variable	If the domain of a is the set of positive integers, then a can have these values: 1, 2, 3, 4, . . .
Algebraic expression: a numerical expression; a variable; or a sum, difference, product, or quotient that contains one or more variables	$2^4 + 3$ x $y^2 - 2y + 6$ $\frac{a+b}{c}$ $2c^2d - \frac{4}{d}$ (Note that $2c^2d$ stands for $2 \cdot c^2 \cdot d$.)

The process of replacing each variable in an expression by a given value and simplifying the result is known as **evaluating** the expression, or **finding the value of** the expression.

Example 3 Evaluate each expression if $x = 4$ and $y = 3$.

a. $\dfrac{3x}{x^2 - 8}$ **b.** $x^2y - xy^2$

Solution

a. $$\frac{3x}{x^2-8} = \frac{3 \cdot 4}{4^2 - 8} = \frac{12}{16-8} = \frac{12}{8} = \frac{3}{2}\text{, or } 1\frac{1}{2}$$

b. $$x^2y - xy^2 = 4^2 \cdot 3 - 4 \cdot 3^2 = 16 \cdot 3 - 4 \cdot 9 = 48 - 36 = 12$$

To evaluate some expressions, you may need to find the absolute value of one or more of the terms. You learned earlier that the absolute value of a number is the distance between the graph of the number and the origin on a number line. Here is a formal definition using a variable.

Definition of Absolute Value

For each real number a,

$$|a| = \begin{cases} a & \text{if } a > 0 \\ 0 & \text{if } a = 0 \\ -a & \text{if } a < 0 \end{cases}$$

Summarizing the Lesson

In this lesson the students have reviewed several methods needed to simplify a variety of forms of numerical expressions and to evaluate algebraic expressions. Ask students to review the charts several times in the next few days.

Using a Calculator

Some calculators follow the order of operations stated in this lesson, while others do not. For this reason students should be strongly advised to read the Calculator Key-In on page 12 before evaluating any expressions with a calculator.

Example 4 Evaluate $10|x| + |y|$ if $x = 3$ and $y = -2$.

Solution
$$10|x| + |y| = 10|3| + |-2|$$
$$= 30 + 2$$
$$= 32 \quad \textit{Answer}$$

You may wish to use a calculator to evaluate some expressions. If you do, be sure to read the Calculator Key-In on page 12 first.

Oral Exercises

Use one of the symbols = or ≠ to make a true statement.

1. 3^2 _?_ $3 \cdot 3$ =
2. 2^2 _?_ $2 + 2$ =
3. $2^2 + 1$ _?_ $3 \cdot 2$ ≠
4. $\frac{2+3}{5}$ _?_ $\frac{2}{5} + \frac{3}{5}$ =
5. $\frac{1}{3+4}$ _?_ $\frac{1}{3} + \frac{1}{4}$ ≠
6. $\frac{1}{3 \cdot 2}$ _?_ $\frac{1}{3} \cdot \frac{1}{2}$ =

Tell which operation to perform first. Then simplify the expression.

7. $8 - 2 \cdot 3$ mult.; 2
8. $(8 - 2) \cdot 3$ subtr.; 18
9. $2(3 + 2)^2$ add.; 50
10. $2(3 + 2^2)$ squaring; 14
11. $2(3 + 2 \cdot 4)$ mult.; 22
12. $14 \div [(2 + 3) - 4]$ add.; 14

Evaluate each expression (a) if $x = 2$ and (b) if $x = 3$.

13. $3x + 2$ 8; 11
14. $3x - 1$ 5; 8
15. $\frac{x+1}{6}$ $\frac{1}{2}$; $\frac{2}{3}$
16. $\frac{12}{x}$ 6; 4
17. $3(x - 1)$ 3; 6
18. $2(2x - 1)$ 6; 10
19. $\frac{1}{2}x^3 + 1$ 5; $\frac{29}{2}$
20. $2x^2 - 3$ 5; 15

Evaluate each expression if $x = 3$ and $y = -2$.

21. $|x| + |y|$ 5
22. $|x| - |y|$ 1
23. $|x|^2 + |y|^2$ 13
24. $|x|^2 - |y|^2$ 5

Written Exercises

Use one of the symbols <, =, or > to make a true statement.

A

1. $5 \cdot 1$ _?_ $5 \div 1$ =
2. $1 \cdot 5$ _?_ $1 \div 5$ >
3. $\frac{3+2}{3-2}$ _?_ $\frac{4+2}{4-2}$ >
4. $\frac{3+2}{3-2}$ _?_ $\frac{6+4}{6-4}$ =
5. $3^2 \cdot 4^2$ _?_ $(3 \cdot 4)^2$ =
6. $3^2 + 4^2$ _?_ $(3 + 4)^2$ <
7. $(9 - 3) - 2$ _?_ $9 - (3 - 2)$ <
8. $(9 + 3) + 2$ _?_ $9 + (3 + 2)$ =
9. $(9 \cdot 3) \cdot 2$ _?_ $9 \cdot (3 \cdot 2)$ =
10. $(12 \div 6) \div 2$ _?_ $12 \div (6 \div 2)$ <

Simplify.

11. a. $11 - 3 + 5 - 2$ 11 **b.** $11 - (3 + 5) - 2$ 1 **c.** $11 - (3 + 5 - 2)$ 5

12. a. $12 - 5 - 2 + 3$ 8 **b.** $12 - (5 - 2) + 3$ 12 **c.** $12 - (5 - 2 + 3)$ 6

13. a. $3 \cdot 8 + 4 \cdot 5$ 44 **b.** $3 \cdot (8 + 4) \cdot 5$ 180 **c.** $3 \cdot (8 + 4 \cdot 5)$ 84

14. a. $4^2 - 6 \div 2 + 3$ 16 **b.** $(4^2 - 6) \div 2 + 3$ 8 **c.** $(4^2 - 6) \div (2 + 3)$ 2

15. $6 - [7 - (5 - 2)]$ 2

16. $14 - 2[9 - 2(5 - 3)]$ 4

17. $\frac{2^3 + 1}{2^2 - 1}$ 3

18. $\frac{3^2}{5 - (3 - 1)}$ 3

19. $\frac{1}{3}\left|\frac{1 + 7^2}{5^2}\right|$ $\frac{2}{3}$

20. $\frac{2^2(3^2 + 4^2)}{10^2}$ 1

21. $64 \div 4^2 + 3(3^2 - 1)$ 28

22. $2^2 \cdot 3^2 - (5^2 - 4^2)$ 27

23. $[3^3 - (2^3 + 2^2)] \div 5$ 3

24. $\frac{1}{10}[2(3 + 4) - 3^2]$ $\frac{1}{2}$

Evaluate each expression if $x = 3$, $y = 2$, and $z = 5$.

25. $2x^2 + x - 2$ 19

26. $3y^2 - y - 5$ 5

27. $(yz - x)^3$ 343

28. $(xz - zy)^3$ 125

29. $\frac{4z^3}{x^2 - y^2}$ 100

30. $\frac{4xyz}{z^2 - x^2}$ $\frac{15}{2}$

31. $\frac{x + z}{y} - \frac{x + y}{2z}$ $\frac{7}{2}$

32. $\frac{z^2}{x + y} - \frac{y^2}{z - x}$ 3

33. $\left(\frac{xyz}{x - y + z}\right)^4$ 625

34. $\left(\frac{z^2 - y^2 - x^2}{xy}\right)^5$ 32

35. $\frac{z^2 - (x^2 - y^2)}{3y^2z}$ $\frac{1}{3}$

36. $\frac{z^2 - y^2}{xz - 2y(z - x)}$ 3

Evaluate each expression if $a = 6$ and $b = -2$.

37. $2|a| + |b|$ 14

38. $|a| - 3|b|$ 0

39. $|a|^2 - |b|^2$ 32

40. $|a^2 - b^2|$ 32

Evaluate each expression for the given values of the variables.

B **41.** $\frac{3u^2 - 2(v - 3)^2}{2(u^2 - 1) - v^2}$; $u = 4$, $v = 5$ 8

42. $\frac{3(r^2 - s^2) + 5(r - s)^2}{(r + s)^2}$; $r = 7$, $s = 3$ 2

43. $\frac{(2r^2 + s^2)(13r - 3s)^2}{5r^2 + 4rs + s^2}$; $r = 3$, $s = 8$ 90

44. $\frac{(g^2 + h^2)(g^2 - h)^2}{(g + h)^2(3h^2 - 2gh - g^2)}$; $g = 6$, $h = 8$ $\frac{20}{3}$

Suggested Assignments

Average Algebra
10/2–46 even
S 5/34–40 even

Average Alg. and Trig.
10/3–48 mult. of 3
Assign with Lesson 1-1.

Extended Algebra
10/9–54 mult. of 3
Assign with Lesson 1-1.

Extended Alg. and Trig.
10/9–54 mult. of 3
Assign with Lesson 1-1.

Supplementary Materials

Study Guide	pp. 3–4
Practice Master	1
Computer Activity	1
Resource Book	p. 100

Additional Answers
Written Exercises

45. $18 \div 2 - 3 \cdot (2 + 1) = 0$

46. $(7 - 2) \cdot (5 - 3) + 2 = 12$

47. $[6 - (5 - 3)] \cdot 2 = 8$

48. $24 \div (3 + 1 - 2) \cdot (2 + 3) = 60$

49. $(3^2 - 2)^2 - 4 \cdot (3 + 3) = 25$

50. $[(6 + 5) \cdot 2 - 7] \div (2 \cdot 2 + 1) = 3$

Calculator Key-In Commentary

Find out how many students have calculators that follow the order of operations and how many have calculators that perform the operations in the order in which they are entered. It is important for students with the latter type to enter expressions in the manner discussed in the Key-In.

Insert grouping symbols in each expression to make a true equation.

Sample $3 + 2^2 - 10 \div 5 = 3$ **Solution** $[(3 + 2)^2 - 10] \div 5 = 3$

45. $18 \div 2 - 3 \cdot 2 + 1 = 0$

46. $7 - 2 \cdot 5 - 3 + 2 = 12$

47. $6 - 5 - 3 \cdot 2 = 8$

48. $24 \div 3 + 1 - 2 \cdot 2 + 3 = 60$

49. $3^2 - 2^2 - 4 \cdot 3 + 3 = 25$

50. $6 + 5 \cdot 2 - 7 \div 2 \cdot 2 + 1 = 3$

51. a. Find the value of $\frac{1}{4}n^2(n + 1)^2$ when $n = 2$, $n = 3$, $n = 4$, and $n = 5$. 9, 36, 100, 225

b. Evaluate $1^3 + 2^3$, $1^3 + 2^3 + 3^3$, $1^3 + 2^3 + 3^3 + 4^3$, and $1^3 + 2^3 + 3^3 + 4^3 + 5^3$. 9, 36, 100, 225

In Exercises 52–54, you may use any operation signs and any grouping symbols. 52.–54. Answers may vary.

C **52.** Use five 2's to write an expression whose value is 7. $2 + 2 + 2 + \frac{2}{2}$

53. Use five 3's to write an expression whose value is 11. $3 \cdot 3 + (3 + 3) \div 3$

54. Use four 2's to write expressions for 0, 1, 2, 3, 4, 5, and 6.
$0 = 2^2 - (2 + 2)$; $1 = (2^2 - 2) \div 2$; $2 = 2 + (2 - 2) \div 2$; $3 = \frac{2^2 + 2}{2}$
$4 = [(2 + 2) \div 2] + 2$; $5 = 2^2 + 2 \div 2$; $6 = 2 \cdot 2 \cdot 2 - 2$

Calculator Key-In

Different calculators perform a sequence of operations in different ways. You can test your calculator by entering the expression $4 + 3 \times 2$ and pressing the equals (=) key. If you get a result of 10, your calculator performs multiplication before addition. Your calculator is therefore following the order of operations as stated in Lesson 1-2. If you get 14, however, your calculator performs operations in the order in which they are entered. To get the correct result, you must enter the expression as $3 \times 2 + 4$ or, if your calculator has parentheses, as $4 + (3 \times 2)$.

Be especially careful when simplifying fractional expressions. You must remember that the fraction bar acts as a grouping symbol. For example, $\frac{4 + 6}{2}$ should be entered as $(4 + 6) \div 2$ to produce 5, the correct result. If you simply enter $\frac{4 + 6}{2}$ as $4 + 6 \div 2$, you may get 7 instead. Likewise, the expression $\frac{10}{2 + 3}$ should be entered as $10 \div (2 + 3)$ and not as $10 \div 2 + 3$.

Simplify each expression on your calculator.

1. $3 + 6 \times 7$ 45

2. $12 + 84 \div 7$ 24

3. $12 \div 3 \times 4 - 1$ 15

4. $12(10 - 5.5) - 4$ 50

5. $20 - 3.5 \cdot 4.5$ 4.25

6. $3 \cdot 12 - 5 \cdot 2.5$ 23.5

7. $\frac{3.8 + 7.7}{4.6 - 1.7}$ 3.965517

8. $(3 + 2.5)(4.5 + 5.5)$ 55

9. $\frac{10.1 + 13.7}{48.5 - 17.9}$ 0.77 . . .

10. $\frac{2.8 \times 7.3}{6.9 \times 2.5}$ 1.1849275

11. $26.54 - (3.9)^2 \div 4$ 22.7375

12. $\frac{6.1(15.8 - 9.9)}{4.3 \times 15}$ 0.55798449

Self-Test 1

Vocabulary

natural numbers (p. 1)
whole numbers (p. 1)
integers (p. 1)
rational numbers (p. 1)
irrational numbers (p. 1)
real numbers (p. 1)
coordinate (p. 1)
graph (p. 1)
origin (p. 1)
is less than (p. 2)
is greater than (p. 2)
absolute value (p. 3 and p. 9)
numerical expression (p. 6)
numeral (p. 6)
value of an expression (p. 6)
equation (p. 6)
inequality (p. 6)
sum (p. 7)
term (p. 7)
difference (p. 7)
product (p. 7)
factor (p. 7)
quotient (p. 7)
power (p. 7)
base (p. 7)
exponent (p. 7)
simplifying an expression (p. 7)
evaluating an expression (p. 9)
variable (p. 9)
domain (p. 9)
values of a variable (p. 9)
algebraic expression (p. 9)

1. Graph the following numbers on a number line: **Obj. 1-1, p. 1**

$$4,\ -2\frac{2}{3},\ 0,\ -1,\ 2.5$$

2. Use symbols to write the following statement:
"Negative two is less than negative one half." $-2 < -\frac{1}{2}$

Tell whether the statement is true or false.

3. $-3 < -5$ False

4. $1 > -2$ True

5. $|-4| = 4$ True

6. $-|7| = -7$ True

Simplify. **Obj. 1-2, p. 6**

7. $3[5(6 - 2) - 4^2]$ 12

8. $\frac{9 + 3 \cdot 5}{18 \div 6 + 3}$ 4

Evaluate each expression if $r = 5$, $s = 3$, and $t = 1$.

9. $r^2 + 2s - rst$ 16

10. $\frac{2s^2 + 3(r - t)}{r + s + 2t}$ 3

Check your answers with those at the back of the book.

Additional Answers
Self-Test 1

1. 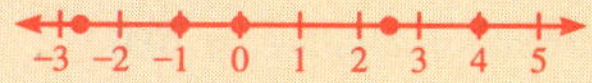

Quick Quiz

1. Graph on a number line:
$-3\frac{1}{2}, 0, 2, -1$

2. Write the numbers in Exercise 1 in order from least to greatest.
$-3\frac{1}{2}, -1, 0, 2$

3. Use symbols to write the following: "Negative two thirds is greater than negative nine."
$-\frac{2}{3} > -9$

4. Simplify $-|-6.5|$.
-6.5

5. Point P has coordinate -1 on a number line. Find the coordinate of each point that is 3 units from P.
2 and -4

Simplify.

6. $25 + 2 \cdot 3 - 12 \div [2(3)]$
29

7. $2 \cdot 3^2 - (7 - 1)$ 12

8. $\frac{5^3 - 1}{3 + 7 \cdot 4}$ 4

Evaluate if $m = 7$, $n = 3$, and $p = 1$.

9. $p^3 + mn^2$ 64

10. $\frac{2m^2 - 3n^2 + p}{m - n - p}$ 24

Teaching Suggestions, p. T80

Reading Algebra, p. T80

Suggested Extensions, p. T80

Warm-Up Exercises

True or False?

1. $3 + 5 = \frac{6}{2} + 5$ T
2. $4 \cdot 2 = 4 \cdot \frac{10}{5}$ T
3. $2 \cdot \frac{7}{2} = \frac{1}{2} \cdot 7$ F
4. $8 + (-8) = 0$ T
5. $9 \cdot 1 = 10$ F
6. $7 \cdot \frac{1}{7} = 1$ T

Motivating the Lesson

Discuss the importance of rules in everyday life. Ask what would happen if there were no rules in our schools or communities. Point out that the same need for rules exists in mathematics.

Common Errors

Students sometimes confuse the symmetric property with the commutative property. In reteaching, clarify the distinction by pointing out that the commutative property is the basis for the order in which two numbers are added or multiplied.

Operating with Real Numbers

1-3 Basic Properties of Real Numbers

Objective To review properties of equality of real numbers and properties for adding and multiplying real numbers.

When you play softball, there are definite rules that you must follow. In fact, every game has its own set of rules, and this is certainly true of the "game" of algebra. In algebra, the properties of real numbers form an important part of the rules.

The five properties below concern equality of real numbers.

Properties of Equality

For all real numbers a, b, and c:

Reflexive Property	$a = a$
Symmetric Property	If $a = b$, then $b = a$.
Transitive Property	If $a = b$ and $b = c$, then $a = c$.
Addition Property	If $a = b$, then $a + c = b + c$ and $c + a = c + b$.
Multiplication Property	If $a = b$, then $ac = bc$ and $ca = cb$.

Example 1 Name the property of equality illustrated in each statement.

a. If $x + 4 = 3$, then $2(x + 4) = 6$.
b. If $z = a + 2$ and $a + 2 = 3$, then $z = 3$.
c. If $x = y + z$, then $y + z = x$.

Solution

a. Multiplication property
b. Transitive property
c. Symmetric property

A set of numbers that has all the properties listed in the chart on the next page is called a **field.** The set of real numbers is a field. Note that when we say a **unique** real number has a certain field property we mean that *one and only one* real number has this property.

Field Properties of Real Numbers For all real numbers a, b, and c:	***Examples*** Let $a = 12$, $b = 6$, and $c = 5$.
Closure Properties $a + b$ and ab are unique real numbers.	The sum of 12 and 6 is the real number 18; the product of 12 and 6 is the real number 72.
Commutative Properties $a + b = b + a$ $ab = ba$	$12 + 6 = 6 + 12$ $12 \cdot 6 = 6 \cdot 12$
Associative Properties $(a + b) + c = a + (b + c)$ $(ab)c = a(bc)$	$(12 + 6) + 5 = 12 + (6 + 5)$ $(12 \cdot 6) \cdot 5 = 12 \cdot (6 \cdot 5)$
Identity Properties There are unique real numbers 0 and 1 $(1 \neq 0)$ such that: $a + 0 = a$ and $0 + a = a$ $a \cdot 1 = a$ and $1 \cdot a = a$	$12 + 0 = 12$ and $0 + 12 = 12$ $12 \cdot 1 = 12$ and $1 \cdot 12 = 12$
Inverse Properties **Property of Opposites:** For each a, there is a unique real number $-a$ such that: $a + (-a) = 0$ and $(-a) + a = 0$. ($-a$ is called the *opposite* or *additive inverse* of a.) **Property of Reciprocals:** For each a except 0, there is a unique real number $\frac{1}{a}$ such that: $a \cdot \frac{1}{a} = 1$ and $\frac{1}{a} \cdot a = 1$. ($\frac{1}{a}$ is called the *reciprocal* or *multiplicative inverse* of a.)	$12 + (-12) = 0$ and $-12 + 12 = 0$ $12 \cdot \frac{1}{12} = 1$ and $\frac{1}{12} \cdot 12 = 1$
Distributive Property (of multiplication with respect to addition) $a(b + c) = ab + ac$ and $(b + c)a = ba + ca$	$12(6 + 5) = 12 \cdot 6 + 12 \cdot 5$ $(6 + 5)12 = 6 \cdot 12 + 5 \cdot 12$

The next two examples illustrate that you can use the commutative and associative properties to add or to multiply numbers in any convenient order and in any groups of two.

Chalkboard Examples

Name the property that is illustrated.

1. $3 + 5 \cdot 2 = 3 + 2 \cdot 5$
   Comm. prop. of mult.
2. If $ab + c = d$, then $d = ab + c$.
   Symmetric prop. of equality
3. If $x = 7$, then $x + 3 = 10$.
   Add. prop. of equality
4. $-1 + 5 \cdot \frac{1}{5} = -1 + 1$
   Prop. of reciprocals

Simplify.

5. $21 + 4a + 18 + 6b$
   $39 + 4a + 6b$
6. $6(5a^2)(9b^2c)$
   $6 \cdot (5 \cdot 9)\, a^2b^2c$
   $6 \cdot 45\, a^2b^2c$
   $270\, a^2b^2c$
7. Name the property that justifies each step.
   a. $2(1 + x) + (-2) = [2 + 2x] + (-2)$
   Distributive prop.
   b. $= [2x + 2] + (-2)$
   Comm. prop. of add.
   c. $= 2x + [2 + (-2)]$
   Assoc. prop. of add.
   d. $= 2x + 0$
   Prop. of opposites
   e. $= 2x$
   Identity prop. of add.

Check for Understanding

Here is a suggested use of the Oral Exercises to check students' understanding as you teach the lesson.
Oral Exs. 1–8: use after the chart on page 15.
Oral Exs. 9–24: use after Example 4.

Guided Practice

Simplify.

1. $\left(5 + \frac{1}{5}\right)x$ $\frac{26}{5}x$

2. $4[a + (-a)] \div 4$ 0

Determine whether the simplification is true or false.

3. $3(x + 5) + (-6) = 12$ F

4. $\left(\frac{5}{6}d\right)\left(\frac{6}{5}e\right) = de$ T

5. State the property that justifies each step.
$y + 2(y + 1)$
a. $= y + (2y + 2)$
b. $= 1 \cdot y + (2y + 2)$
c. $= (1 \cdot y + 2y) + 2$
d. $= [(1 + 2)y] + 2$
$= 3y + 2$
(Substitution)
a. Distributive prop.
b. Identity prop. of mult.
c. Assoc. prop. of add.
d. Distributive prop.

Example 2 Simplify: **a.** $7 + 11 + 13 + 19$ **b.** $26 + 8y + 14$

Solution **a.** $7 + 11 + 13 + 19 = (7 + 13) + (11 + 19) = 20 + 30 = 50$ ***Answer***
b. $26 + 8y + 14 = (26 + 14) + 8y = 40 + 8y$ ***Answer***

Example 3 Simplify $16(4xy^2)(5z^3)$.

Solution $16(4xy^2)(5z^3) = 16 \cdot (4 \cdot 5)xy^2z^3 = 16 \cdot 20xy^2z^3 = 320xy^2z^3$ ***Answer***

The real number properties can be used to change a given expression into a simpler *equivalent expression*. Two expressions are **equivalent** when they are equal for every value of each variable they contain. *In this book, unless otherwise stated, the domain of every variable is the set of real numbers.*

Example 4 Name the property used in each step of simplifying $\frac{1}{3}(1 + 3t)$.

Solution

1.	$\frac{1}{3}(1 + 3t) = \frac{1}{3} \cdot 1 + \frac{1}{3}(3t)$	1. Distributive property
2.	$= \frac{1}{3} \cdot 1 + \left(\frac{1}{3} \cdot 3\right)t$	2. Associative property of multiplication
3.	$= \frac{1}{3} \cdot 1 + 1 \cdot t$	3. Property of reciprocals
4.	$= \frac{1}{3} + t$	4. Identity property of multiplication
5.	$\therefore \frac{1}{3}(1 + 3t) = \frac{1}{3} + t$	5. Transitive property of equality

Note that in Step 3 the property of reciprocals was used to justify the substitution of 1 for $\frac{1}{3} \cdot 3$. Of course, the substitution principle (page 8) was also used in this step. In this book we will follow the common practice of omitting many references to the substitution principle.

Oral Exercises

Give (a) the opposite and (b) the reciprocal of each number. If a number has no reciprocal, say so.

1. -2 $2; -\frac{1}{2}$
2. $\frac{1}{2}$ $-\frac{1}{2}; 2$
3. 1 $-1; 1$
4. 0 0; none
5. $-\frac{4}{3}$ $\frac{4}{3}; -\frac{3}{4}$
6. -1 $1; -1$
7. 0.1 $-0.1; 10$
8. -0.25 $0.25; -4$

Name the property illustrated in each statement.

Sample $7(2y) = (7 \cdot 2)y$ **Solution** Associative property of multiplication

9. $3 + t = t + 3$ Comm. prop. of add.
10. $5 \cdot \frac{1}{5} = 1$ Prop. of reciprocals
11. $1 \cdot (abc) = abc$ Ident. prop. of mult.
12. $-2y + 0 = -2y$ Ident. prop. of add.
13. $2\left(x + \frac{1}{2}\right) = 2x + 2 \cdot \frac{1}{2}$ Dist. prop.
14. If $x - 2 = 3$, then $(x - 2) + 2 = 3 + 2$. Add. prop. of =
15. $-1.5 + 1.5 = 0$ Prop. of opposites
16. $(a + 1)(a - 1) = (a - 1)(a + 1)$ Comm. prop. of mult.
17. If $x + 2 = 8$, then $8 = x + 2$. Symm. prop. of =
18. $\frac{1}{3}(3z^2) = \left(\frac{1}{3} \cdot 3\right)z^2$ Assoc. prop. of mult.
19. $\left(-\frac{2}{3}\right)\left(-\frac{3}{2}\right) = 1$ Prop. of reciprocals
20. $(2 + x) + 3 = (x + 2) + 3$ Comm. prop. of add.
21. $2000 + 20$ is a unique real number. Closure prop. of add.
22. If $3x = 6$ and $6 = 2y$, then $3x = 2y$. Trans. prop. of =
23. Does $a(bc) = (ab)(ac)$ for all real numbers a, b, and c? Explain. No; $2(3 \cdot 4) \neq (2 \cdot 3)(2 \cdot 4)$
24. Explain why 0 has no reciprocal. For every x, $0 \cdot x = 0$.

Written Exercises

Simplify.

A

1. $96 + 13 + 4 + 37$ 150
2. $-6 + x + 6$ x
3. $\frac{1}{3}(1 \cdot 3) + (-1)$ 0
4. $\frac{1}{3}(1 \cdot 3) + (-3 + 3)$ 1
5. $4\left(z + \frac{1}{4}\right)$ $4z + 1$
6. $\frac{1}{4}(z + 4)$ $\frac{1}{4}z + 1$
7. $1 \cdot (-t + t)$ 0
8. $\left(\frac{2}{3}a\right)\left(\frac{3}{2}b\right)$ ab
9. $2(a + 4) + (-8)$ $2a$
10. $\left(\frac{1}{3} \cdot 3\right)[p + (-1)] + 1$ p

Determine whether each simplification is true or false.

Sample 1 $7\left(2 + \frac{1}{7}x\right) - 14 = x$

Solution True, because $7\left(2 + \frac{1}{7}x\right) - 14 = 14 + x - 14 = x$.

11. $(-x + 6) + (-6 + x) = 0$ True
12. $2(x + 3) + (-3) = 2x$ False
13. $\frac{1}{2}(2t + 2) + (-2) = t$ False
14. $(7p)\frac{1}{7} + (-p) = 0$ True
15. $5(2n + 1) + (-5) = 10n$ True
16. $5(xy) = (5x)(5y)$ False

Summarizing the Lesson

Point out to students that they have learned a number of properties of real numbers. Ask students to name a property and then to explain what it means in words.

Suggested Assignments

Average Algebra
17/1–21 odd, 25–28
S 20/Mixed Review
R 13/Self-Test 1

Average Alg. and Trig.
17/1–23 odd, 25, 30
S 20/Mixed Review
R 13/Self-Test 1

Extended Algebra
17/7–35 odd
S 20/Mixed Review
R 13/Self-Test 1

Extended Alg. and Trig.
17/7–35 odd
S 20/Mixed Review
R 13/Self-Test 1

Supplementary Materials
Study Guide pp. 5–6

In Exercises 17–22, name the property used in each step of the simplification.

17. $\frac{1}{2}(1 + 2t) = \frac{1}{2} \cdot 1 + \frac{1}{2}(2t)$ **a.** ___?___ Dist. prop.
$= \frac{1}{2} \cdot 1 + \left(\frac{1}{2} \cdot 2\right)t$ **b.** ___?___ Assoc. prop. of mult.
$= \frac{1}{2} \cdot 1 + 1 \cdot t$ **c.** ___?___ Prop. of reciprocals
$= \frac{1}{2} + t$ **d.** ___?___ Ident. prop. of mult.

18. $(2 + a) + (-2) = (a + 2) + (-2)$ **a.** ___?___ Comm. prop. of add.
$= a + [2 + (-2)]$ **b.** ___?___ Assoc. prop. of add.
$= a + 0$ **c.** ___?___ Prop. of opposites
$= a$ **d.** ___?___ Ident. prop. of add.

19. $\frac{1}{5}[(n + 5) + (-n)] = \frac{1}{5}[(5 + n) + (-n)]$ **a.** ___?___ Comm. prop. of add.
$= \frac{1}{5}[5 + (n + (-n))]$ **b.** ___?___ Assoc. prop. of add.
$= \frac{1}{5}[5 + 0]$ **c.** ___?___ Prop. of opposites
$= \frac{1}{5} \cdot 5$ **d.** ___?___ Ident. prop. of add.
$= 1$ **e.** ___?___ Prop. of reciprocals

20. $3 + 4(x + 1) = 3 + (4x + 4 \cdot 1)$ **a.** ___?___ Dist. prop.
$= 3 + (4x + 4)$ **b.** ___?___ Ident. prop. of mult.
$= (4x + 4) + 3$ **c.** ___?___ Comm. prop. of add.
$= 4x + (4 + 3)$ **d.** ___?___ Assoc. prop. of add.
$= 4x + 7$ Substitution

21. $k + (k + 2) = (k + k) + 2$ **a.** ___?___ Assoc. prop. of add.
$= (1 \cdot k + 1 \cdot k) + 2$ **b.** ___?___ Ident. prop. of mult.
$= (1 + 1)k + 2$ **c.** ___?___ Dist. prop.
$= 2k + 2$ Substitution
$= 2k + 2 \cdot 1$ **d.** ___?___ Ident. prop. of mult.
$= 2(k + 1)$ **e.** ___?___ Dist. prop.

22. $x(y + 1) + (-1)x = x(y + 1) + x(-1)$ **a.** ___?___ Comm. prop. of mult.
$= x[(y + 1) + (-1)]$ **b.** ___?___ Dist. prop.
$= x[y + (1 + (-1))]$ **c.** ___?___ Assoc. prop. of add.
$= x[y + 0]$ **d.** ___?___ Prop. of opposites
$= xy$ **e.** ___?___ Ident. prop. of add.

B **23.** Show that if $3x + (-12) = 0$, then $x = 4$ by justifying each indicated step.

$3x + (-12) = 0$	Given
$[3x + (-12)] + 12 = 0 + 12$	**a.** ___?___ Add. prop. of =
$3x + [(-12) + 12] = 0 + 12$	**b.** ___?___ Assoc. prop. of add.
$3x + 0 = 0 + 12$	**c.** ___?___ Prop. of opposites
$3x = 12$	**d.** ___?___ Ident. prop. of add.
$\frac{1}{3}(3x) = \frac{1}{3} \cdot 12$	**e.** ___?___ Mult. prop. of =
$\frac{1}{3}(3x) = 4$	Substitution
$\left(\frac{1}{3} \cdot 3\right)x = 4$	**f.** ___?___ Assoc. prop. of mult.
$1 \cdot x = 4$	**g.** ___?___ Prop. of reciprocals
$x = 4$	**h.** ___?___ Ident. prop. of mult.

24. Use steps similar to those in Exercise 23 to show that if $\frac{1}{2}x + (-3) = 0$, then $x = 6$.

A set *S* of real numbers is *closed under addition* if the sum of any two members of *S* is in *S*. A set *S* is *closed under multiplication* if the product of any two members of *S* is in *S*. For each set given below tell whether the set is closed under (a) addition and (b) multiplication. If the set is not closed, give an example to show this.

Sample 2 $\{0, 1\}$

Solution **a.** Not closed under addition, since $1 + 1 = 2$ and 2 is not in the set. (Notice that "any two members" includes the same number used twice.)

b. Closed under multiplication, since every possible product ($0 \cdot 0$, $1 \cdot 1$, and $0 \cdot 1$) is in the set.

25. $\{0\}$

26. $\{-1, 0, 1\}$

27. $\{3, 6, 9, 12, \ldots\}$

28. $\{1, 3, 5, \ldots\}$

29. $\{2, 4, 6, \ldots\}$

30. $\left\{1, \frac{1}{2}, \frac{1}{3}, \frac{1}{4}, \ldots\right\}$

31. The rational numbers

32. The irrational numbers

33. The real numbers greater than -1 and less than 1

Decide whether each set is a field under the operations of addition and multiplication. If the set is not a field, name at least one field property that does not hold.

C **34.** The natural numbers No; Ident. prop. of add.

35. The integers No; Prop. of reciprocals

36. The rational numbers Yes

37. The negative rational numbers No; Ident. prop. of add.

Additional Answers Written Exercises

24. (1) $\frac{1}{2}x + (-3) = 0$ (Given)
(2) $\left[\frac{1}{2}x + (-3)\right] + 3 = 0 + 3$ (Add. prop. of =)
(3) $\frac{1}{2}x + [(-3) + 3] = 0 + 3$ (Assoc. prop. of add.)
(4) $\frac{1}{2}x + 0 = 0 + 3$ (Prop. of opposites)
(5) $\frac{1}{2}x = 3$ (Ident. prop. of add.)
(6) $2\left(\frac{1}{2}x\right) = 2 \cdot 3$ (Mult. prop. of =)
(7) $2\left(\frac{1}{2}x\right) = 6$ (Substitution)
(8) $\left(2 \cdot \frac{1}{2}\right)x = 6$ (Assoc. prop. of mult.)
(9) $1 \cdot x = 6$ (Prop. of reciprocals)
(10) $x = 6$ (Ident. prop. of mult.)

25. a. closed **b.** closed

26. a. not closed ($1 + 1 = 2$) **b.** closed

27. a. closed **b.** closed

28. a. not closed ($1 + 3 = 4$) **b.** closed

29. a. closed **b.** closed

30. a. not closed $\left(1 + \frac{1}{2} = \frac{3}{2}\right)$ **b.** closed

31. a. closed **b.** closed

32. a. not closed ($\sqrt{2} + (-\sqrt{2}) = 0$) **b.** not closed ($\sqrt{2} \cdot \sqrt{2} = 2$)

33. a. not closed $\left(\frac{3}{4} + \frac{3}{4} = \frac{3}{2}\right)$ **b.** closed

Mixed Review Exercises

Use <, =, or > to make a true statement.

1. $0.02 \underline{\ ?\ } 0.002$ >
2. $3(8 - 5) \underline{\ ?\ } 3 \cdot 8 - 5$ <
3. $17 - (8 + 5) \underline{\ ?\ } 17 - 8 + 5$ <
4. $(3 + 4)^2 \underline{\ ?\ } 3^2 + 2 \cdot 3 \cdot 4 + 4^2$ =
5. $\frac{7 + 5}{8 - 2} \underline{\ ?\ } \frac{8 + 2}{7 - 5}$ <
6. $\frac{1 + 1}{2 + 2} \underline{\ ?\ } \frac{1}{2} + \frac{1}{2}$ <

Extra / Logical Symbols: Quantifiers

Statements about the properties of real numbers often include phrases like "for all real numbers a," "for each b," or "there exists a unique real number c," that express the idea of "how many" or of quantity. Such a phrase is called a **quantifier.** In mathematical logic, quantifiers are represented by symbols, as shown in the table at the right.

Symbol	Meaning
$\forall_x$	"for all x," "for every x," "for each x," and so on
$\exists_x$	"there exists an x," "for some x," and so on
$\exists!_x$	"there is a unique x"

For examples of the use of symbolic quantifiers, study the following sentences. All variables here and in the exercises represent real numbers.

For every x, $x + 2x = 3x$.	$\forall_x\ x + 2x = 3x$
For all x, $x + 7 = 7 + x$.	$\forall_x\ x + 7 = 7 + x$
For some x, $x + 5 = 10$.	$\exists_x\ x + 5 = 10$
For all a and b, $(a + b) + 2 = a + (b + 2)$.	$\forall_a\forall_b\ (a + b) + 2 = a + (b + 2)$
For every a, there is a b for which $b = 2a$.	$\forall_a\exists_b\ b = 2a$
For every x there is a unique y such that $x + y = 0$.	$\forall_x\exists!_y\ x + y = 0$

Exercises

Express each statement using a symbol for the quantifier.

1. For all x, $2x = x + x$. $\forall_x\ 2x = x + x$
2. For some y, $y^2 = 17$. $\exists_y\ y^2 = 17$
3. For all real numbers x and all real numbers y, $x + y = y + x$. $\forall_x\forall_y\ x + y = y + x$
4. For all real numbers r, there exists a unique real number x such that $r + x = 0$. $\forall_r\exists!_x\ r + x = 0$

Tell whether each statement is true or false.

5. $\forall_a\exists_x\ x^2 = a$ F
6. $\forall_a\exists_x\ x = a^2$ T
7. $\forall_x\exists_a\ x > a$ T
8. $\forall_a\forall_b\forall_c\exists!_x\ ax + b = c$ F

1-4 Sums and Differences

Objective To review the rules for adding and subtracting real numbers.

In arithmetic you learned to add positive numbers ($4 + 7 = 11$) and to subtract a smaller positive number from a larger one ($12 - 7 = 5$). In this lesson you will review how to add and subtract *any* real numbers, positive or not.

The identity property of addition tells you how to add zero and any given number: *The sum is the given number.*

$$-6 + 0 = -6 \qquad 0 + \left(-\frac{1}{2}\right) = -\frac{1}{2}$$

The property of opposites tells you that the sum of opposites is zero.

$$-1.2 + 1.2 = 0 \qquad 4 + (-4) = 0$$

The rules covering the remaining cases can be stated in terms of absolute value. Study the examples to be sure you understand the rules.

Rules for Addition For real numbers a and b:	***Examples***
1. If a and b are negative numbers, then $a + b$ is negative and $a + b = -(\lvert a\rvert + \lvert b\rvert)$.	$-5 + (-9) = -(\lvert -5\rvert + \lvert -9\rvert)$ $= -(5 + 9)$ $= -14$
2. If a is a positive number, b is a negative number, and $\lvert a\rvert$ is greater than $\lvert b\rvert$, then $a + b$ is a positive number and $a + b = \lvert a\rvert - \lvert b\rvert$.	$9 + (-5) = \lvert 9\rvert - \lvert -5\rvert$ $= 9 - 5$ $= 4$
3. If a is a positive number, b is a negative number, and $\lvert a\rvert$ is less than $\lvert b\rvert$, then $a + b$ is a negative number and $a + b = -(\lvert b\rvert - \lvert a\rvert)$.	$5 + (-9) = -(\lvert -9\rvert - \lvert 5\rvert)$ $= -(9 - 5)$ $= -4$

Since addition of real numbers is commutative, these rules also cover sums like $-4 + 9 = 5$ and $-6 + 2 = -4$.

We say that the sign of a positive number is *plus* and the sign of a negative number is *minus,* and that a positive and a negative number have *opposite* signs. Thus, the rules for adding two nonzero numbers can be restated as follows:

If two numbers have the *same* sign, add their absolute values and then prefix their common sign.

If two numbers have *opposite* signs, subtract the lesser absolute value from the greater and then prefix the sign of the number having the greater absolute value.

Teaching Suggestions, p. T81

Reading Algebra, p. T81

Suggested Extensions, p. T81

Warm-Up Exercises

Add or subtract.

1. $18 + (-9)$ 9
2. $-14 + 6$ -8
3. $-12 + (-13)$ -25
4. $15 - 24$ -9
5. $19 - (-7)$ 26
6. $-10 - 17$ -27

Motivating the Lesson

Present the following situation to the students.

Here is a record of one contestant's responses during a televised game show.

Q	Correct	Won/Lost
1	Yes	+200
2	Yes	+400
3	No	−600
4	No	−300
5	Yes	+800
6	No	−500

Did this contestant win money, lose money, or break even in this game? Point out that today's lesson enables us to find the answer to this problem by reviewing the properties of addition and the rules for adding and subtracting real numbers.

Chalkboard Examples

Simplify.

1. $-17 + 12$ -5
2. $-17 - 12$ -29
3. $17 - (-12)$ 29
4. $-12 + 18 + (-3) + 2 + (-1)$ 4
5. $|5 - 12| - |13 - 2|$ -4

Simplify.

6. $8t + 3(4t - 9)$ $20t - 27$

Multiply.

7. $4(2x^2 - x - 1)$ $8x^2 - 4x - 4$

Find the number that makes the statement true.

8. $\underline{\ ?\ } + 20 = 12$ -8
9. $-4 - \underline{\ ?\ } = 9$ -13

Common Errors

Students are sometimes confused when the distributive property is used with a trinomial. Show students that $3(5 - x + 2y) = (3 \cdot 5) - (3 \cdot x) + (3 \cdot 2y) = 15 - 3x + 6y$.

Example 1 Simplify $-14 + 9 + (-12) + 20 + (-8)$.

Solution 1 Add the terms in order from left to right.

$$\begin{aligned} -14 + 9 + (-12) + 20 + (-8) &= \underbrace{-5 + (-12)} + 20 + (-8) \\ &= \underbrace{-17 \quad + 20} + (-8) \\ &= \underbrace{3 \quad + (-8)} \\ &= -5 \quad \textbf{\textit{Answer}} \end{aligned}$$

Solution 2 Use the commutative and associative properties to order and group the terms conveniently. For example, you may choose to group the positive terms and the negative terms.

$$\begin{aligned} -14 + 9 + (-12) + 20 + (-8) &= [-14 + (-12) + (-8)] + [9 + 20] \\ &= -34 + 29 \\ &= -5 \quad \textbf{\textit{Answer}} \end{aligned}$$

You may also use the rules shown on page 21 to subtract because subtraction of real numbers is defined in terms of addition.

Definition of Subtraction

For all real numbers a and b, $\quad \boldsymbol{a - b = a + (-b)}$.

To subtract any real number, add its opposite.

Example 2 Simplify.

a. $-7 - (-8)$ **b.** $-7 - 8$ **c.** $7 - (-8)$ **d.** $7 - 8$

Solution

a. $-7 - (-8) = -7 + 8 = 1$ **b.** $-7 - 8 = -7 + (-8) = -15$

c. $7 - (-8) = 7 + 8 = 15$ **d.** $7 - 8 = 7 + (-8) = -1$

Example 3 Simplify $10 - 8 - 15 + 6$.

Solution 1 Add the terms in order from left to right.

$$\begin{aligned} 10 - 8 - 15 + 6 &= 2 - 15 + 6 \\ &= -13 + 6 \\ &= -7 \quad \textbf{\textit{Answer}} \end{aligned}$$

Solution 2 Group the positive terms and the negative terms.

$$\begin{aligned} 10 - 8 - 15 + 6 &= 10 + (-8) + (-15) + 6 \\ &= [10 + 6] + [(-8) + (-15)] \\ &= 16 + (-23) \\ &= -7 \quad \textbf{\textit{Answer}} \end{aligned}$$

Example 4 Simplify $|14 - 5| - |5 - 19|$.

Solution $$|14 - 5| - |5 - 19| = |9| - |-14|$$
$$= 9 - 14$$
$$= -5 \quad \textit{Answer}$$

You know that multiplication is distributive with respect to addition (page 15). Since subtraction is just the addition of an opposite, multiplication is also distributive with respect to subtraction.

Distributive Property (of multiplication with respect to subtraction)

For all real numbers a, b, and c,

$$a(b - c) = ab - ac$$
and $$(b - c)a = ba - ca.$$

Example 5 Multiply $3(4x^2 - 9)$.

Solution $3(4x^2 - 9) = 3(4x^2) - 3(9) = 12x^2 - 27$ *Answer*

Example 6 Simplify $4y + 2(6y - 5)$.

Solution $$4y + 2(6y - 5) = 4y + 12y - 10$$
$$= (4 + 12)y - 10$$
$$= 16y - 10 \quad \textit{Answer}$$

In Example 6, we used the distributive property to combine the *similar terms* $4y$ and $12y$. Remember that **similar terms** contain the same variable factors. For example, $8ab$ and $2ab$ are similar terms, but $3x^2$ and $3x$ are not. Similar terms are sometimes referred to as **like terms.**

Oral Exercises

In Exercises 1–9, (a) add the numbers and (b) subtract the lower number from the upper one.

1. 3 over 7 — 10; −4

2. −2 over 5 — 3; −7

3. −9 over 4 — −5; −13

4. 4 over −7 — −3; 11

5. 8 over −3 — 5; 11

6. −5 over 5 — 0; −10

Check for Understanding

Here is a suggested use of the Oral Exercises to check students' understanding as you teach the lesson.

Oral Exs. 1–9: use after Example 2.

Oral Exs. 10–29: use after Example 6.

Guided Practice

Simplify.

1. $-12 - 18 + (5 - 8)$ -33
2. $(4 - 9 - 1) - (2 - 8)$ 0
3. $|-15 + 2| + |-12 - 9|$ 34
4. $1.05 - 7.28 - (-5 + 6.7)$ -7.93
5. $|3 - 8| - |-12 + 3|$ -4

Multiply.

6. $7(4m - 5n)$ $28m - 35n$
7. $\frac{3}{2}(8s - 6) + 2$ $12s - 7$

Simplify by combining similar terms.

8. $(8x - 4y + 7) + 3(-x - 4y + 3)$ $5x - 16y + 16$

Summarizing the Lesson

In this lesson students reviewed the rules for adding and subtracting real numbers and extended those rules to algebraic expressions.

Suggested Assignments

Average Algebra
24/2–48 even, 51, 53

Average Alg. and Trig.
24/2–50 even, 51–53

Extended Algebra
24/2–50 even, 51, 52

Extended Alg. and Trig.
24/12–48 mult. of 3

Assign with Lesson 1-5.

7. $\begin{array}{r} -5 \\ \underline{-5} \end{array}$ $-10;\ 0$

8. $\begin{array}{r} -8 \\ \underline{-3} \end{array}$ $-11;\ -5$

9. $\begin{array}{r} 0 \\ \underline{-6} \end{array}$ $-6;\ 6$

Simplify.

10. $12 + (-7)$ 5
11. $12 - 7$ 5
12. $5 - 13$ -8
13. $5 + (-13)$ -8
14. $-5 - (-13)$ 8
15. $5 - (-13)$ 18
16. $0 - 6$ -6
17. $0 - (-6)$ 6
18. $-7 - (-7)$ 0
19. $-7 - 7$ -14
20. $6 - 4 + 3$ 5
21. $8 - 3 - 2$ 3
22. $8 - (3 - 2)$ 7
23. $2 - (7 + 3) - 4$ -12
24. $3 - (8 - 2 + 7)$ -10
25. $|5| - |-6|$ -1
26. $|5 - (-6)|$ 11
27. $|-5 + (-6)|$ 11
28. Is the sum of a positive number and a negative number always, sometimes, or never a positive number? sometimes
29. Give a numerical example that shows $|a + b| \neq |a| + |b|$. $|-5 + 4| \neq |-5| + |4|$

Written Exercises

Simplify. You may wish to check your answers on a calculator.

A

1. $32 - 53$ -21
2. $-16 - 42$ -58
3. $-23 + 57$ 34
4. $17 - (-26)$ 43
5. $-15 - (-40)$ 25
6. $16 + (-50)$ -34
7. $-10.2 + 17.6$ 7.4
8. $21.2 - 32.3$ -11.1
9. $-9.6 - 13.4$ -23
10. $26.2 - (-26.2)$ 52.4
11. $0 - (-30.7)$ 30.7
12. $-15.85 - (-20.15)$ 4.3
13. $57 - 13 - 46$ -2
14. $33 - 72 + 14$ -25
15. $68 + (-42) + (-35) + 17$ 8
16. $-87 + 16 - 22 + 61$ -32
17. $5 - (2 - 9) - (4 - 11)$ 19
18. $-16 - [2 - (-61)] - [2 + (-6)]$ -75
19. $(-4 + 7 - 10) - [-6 - (-8)]$ -9
20. $(3 - 6 - 9) - [8 + (-4) - (-7)]$ -23
21. $|6 - 13| - |22 - (-6)|$ -21
22. $|-22 - 33| - |16 - 7|$ 46

Multiply.

23. $4(p - 2q)$ $4p - 8q$
24. $5(3y - 2x)$ $15y - 10x$
25. $\frac{1}{2}(2a - 4b + 6)$ $a - 2b + 3$
26. $\frac{2}{3}(6r - 9s - 27)$ $4r - 6s - 18$

Simplify by combining similar terms.

27. $8c + 2(c + 3)$ $10c + 6$
28. $5(d + 2) - 3d$ $2d + 10$
29. $7(x + 2) + 4(x - 4)$ $11x - 2$
30. $4(3 - y) + 2(1 - y)$ $14 - 6y$
31. $7p - 4q + 3q - 10p$ $-3p - q$
32. $6m - 4n + (-7)m - (-5)n$ $n - m$
33. $(-2r + s + 5) + 2(r - 3s + 2)$ $-5s + 9$
34. $(6x - 5y + 4) + 2(-2x + 3y - 2)$ $2x + y$

Find the number that makes the equation true.

Sample $-8 - \underline{\ ?\ } = -2$

Solution -6, because $-8 - (-6) = -8 + 6 = -2$

B

35. $-7 + \underline{\ ?\ } = 2$ 9
36. $\underline{\ ?\ } - 8 = -3$ 5
37. $\underline{\ ?\ } - 4 = -7$ −3
38. $-6 - \underline{\ ?\ } = -1$ −5
39. $-6 + \underline{\ ?\ } + 2 = 0$ 4
40. $-6 + \underline{\ ?\ } + 11 = -4$ −9
41. $-6.4 + \underline{\ ?\ } = 3.7$ 10.1
42. $2.6 + \underline{\ ?\ } - 8.7 = 3.2$ 9.3

Solve.

43. Temperatures on the planet Mars range from −122.2° C to 30.5° C. Find the difference between these two temperatures. 152.7°C

44. The highest point in the United States is Mt. McKinley, Alaska, at 20,320 ft above sea level, and the lowest is Bad Water, California, at 282 ft below sea level. Find the difference between these elevations. 20,602 ft
45. Lake Baikal, in Siberia, is 5315 ft deep, and its surface is 1493 ft above sea level. How many feet below sea level is the bottom of the lake at its deepest point? 3822 ft below sea level
46. A subway train was carrying 103 passengers. At the next three stops, 15 people got on and 9 got off, 27 got on and 13 got off, and 8 got on and 53 got off. How many passengers were then on the train? 78
47. A newspaper lists the closing price of a stock at $72\frac{3}{4}$ following a drop of $1\frac{7}{8}$ that day. What was the stock's opening price for the day? $74\frac{5}{8}$
48. At 7:38:16 A.M. (38 minutes and 16 seconds after 7 o'clock) the countdown on a rocket launch stood at −14:42 (14 minutes and 42 seconds before launch). There was a 37-second delay in resuming the count. At what time did the rocket take off? 7:53:35 A.M.
49. Give a numerical example to show that subtraction is not commutative. $7 - 2 \neq 2 - 7$
50. Give a numerical example to show that subtraction is not associative. $(6 - 1) - 4 \neq 6 - (1 - 4)$

Tell whether each set is closed (see page 19) under subtraction. If the set is not closed, give an example to show this.

C

51. $\{0, 1\}$ No; $0 - 1 = -1$
52. $\{-1, 0, 1\}$ No; $-1 - 1 = -2$
53. The whole numbers No; $1 - 2 = -1$
54. The negative integers No; $-2 - (-3) = 1$
55. The integers Yes
56. The irrational numbers No; $\sqrt{3} - \sqrt{3} = 0$

Supplementary Materials

Study Guide	pp. 7–8
Practice Master	2

Additional Answers
Reading Algebra

1. Solve certain equations in one variable.
2. Open sentences, solution, root, satisfy, solution set, equivalent equations, solve, empty set, null set, identity, formula, constants
3. An equation or inequality containing a variable; equations having the same solution set over a given domain.
4. See the chart on p. 37.
5. (3)
6. The glossary or the index; p. 38.
7. pp. 830–831

Reading Algebra / *Independent Study*

Your algebra textbook has been designed and written to make reading and learning algebra as easy as possible. To take full advantage of your book, begin your study of each lesson by getting an idea of the goal of the lesson. Read the lesson title and the stated objective. You may find it useful to skim the lesson first.

Heavy type indicates new words or phrases that you must understand in order to understand the lesson. It is usually very helpful to learn the definitions of these words or phrases before you begin a slow and careful reading of the lesson. When you do start to read the lesson, pay particular attention to the important information set off in boxes. If you come across any words whose meaning you do not understand, look them up in the glossary at the back of the book or in a dictionary. You may also find additional information by checking references from the index.

Be sure to work through all the examples that are included in the lesson discussions. Try doing these examples on your own before reading the solutions that are given following the examples. If you do not understand a concept, and rereading does not seem to help, make a note of the concept so that you can discuss it with your teacher.

When you have finished reading a lesson and feel that you understand what you have read, try some of the Oral and Written Exercises. They will help you determine whether you have achieved the goal set forth in the objective. Note that the answers to most odd-numbered exercises are given in the Selected Answers at the back of your book. Doing the Self-Tests and checking your answers with those at the back of the book will help you to see whether you have understood a group of lessons. The Chapter Reviews, Chapter Tests, Cumulative Reviews, and Mixed Reviews will also give you a good idea of your progress.

Exercises

Skim through Lesson 1–7 and answer the following questions.

1. What should you be able to do after studying Lesson 1–7?
2. What new words or phrases are introduced in Lesson 1–7?
3. What is an open sentence? What are equivalent equations? Find the definitions in your book.
4. Describe a transformation that produces equivalent equations.
5. Look at Example 1 on page 38. Then solve $3(4x - 9) = 5x - 6$.
6. Suppose that you have forgotten the definition of the word *identity*. Where could you look it up? On what page is this word first used?
7. Lesson 1–7 also shows you how to rewrite formulas. Where in your book could you find a list of geometric formulas?

1-5 Products

Objective To review rules for multiplying real numbers.

The following facts are useful in finding products of real numbers.

Multiplicative Property of 0

For every real number a,

$$\mathbf{a \cdot 0 = 0} \quad \text{and} \quad \mathbf{0 \cdot a = 0.}$$

Multiplicative Property of −1

For every real number a,

$$\mathbf{a(-1) = -a} \quad \text{and} \quad \mathbf{(-1)a = -a.}$$

The multiplicative property of 0 tells you that the product of any real number and 0 is 0. For example,

$$-5 \cdot 0 = 0 \quad \text{and} \quad 0\left(\frac{1}{2}\right) = 0.$$

The multiplicative property of -1 tells you that the product of any real number and -1 is the opposite of the number. For example,

$$6(-1) = -6, \quad (-1)4 = -4, \quad \text{and} \quad (-1)(-1) = -(-1) = 1.$$

You can use the multiplicative property of -1 along with the facts for positive numbers to find the product of *any* two nonzero real numbers. For example:

$$(5)(7) = 35$$
$$(-5)(-7) = (-1)(5)(-1)(7) = (-1)(-1)(5)(7) = 1(35) = 35$$
$$(5)(-7) = (5)(-1)(7) = (-1)(5)(7) = (-1)(35) = -35$$

All three products have the same absolute value, 35. When the two factors have the same sign, the product is positive; when they have opposite signs, the product is negative. Examples like these suggest the following rules.

Rules for Multiplication

1. The product of two positive numbers or two negative numbers is a positive number.
2. The product of a positive number and a negative number is a negative number.
3. The absolute value of the product of two or more numbers is the product of their absolute values.

Teaching Suggestions, p. T81

Suggested Extensions, p. T81

Warm-Up Exercises

Multiply.

1. 14×12 168
2. -13×7 −91
3. $23 \times (-5)$ −115
4. $-8 \times (-16)$ 128
5. 18×0 0

Motivating the Lesson

Pose the following problem: On your sixteenth birthday, your grandparents gave you 100 shares of stock in Skyway Airlines. When you received it, each share was worth $\$25\frac{3}{4}$. The stock's value decreased by $\frac{3}{4}$ of a point during each of the next three days. How much was your stock worth at the end of the three-day period? $2,350

Point out that today's lesson reviews the properties and rules for multiplying real numbers.

Chalkboard Examples

Simplify.

1. $(-5)(-2)(3)(-1)$ -30
2. $(-2)^5\left(\frac{1}{4}\right)$ $-32 \cdot \frac{1}{4} = -8$
3. $387(-0.2) + 13(-0.2)$
   $400(-0.2) = -80$
4. $(-18 + 7)(-18 - 7)$
   $-11(-25) = 275$
5. Multiply.
   $\left(-\frac{3}{4}\right)(-28t^2 + 12)$
   $\left(-\frac{3}{4}\right)(-28t^2) +$
   $\left(-\frac{3}{4}\right)(12) = 21t^2 - 9$
6. Give the opposite.
   a. $-5z + 6x$ **b.** $xy - 6z$
   $5z - 6x$ $-xy + 6z$

Example 1 Simplify.

a. $(-3)(-2)(-1)(4)(-5)$

b. $(-4) \cdot 3 \cdot (-2)\left(-\frac{1}{3}\right)$

c. $24(-15)(0)(13)$

Solution

a. $$(-3)(-2)(-1)(4)(-5) = \underbrace{(6)(-1)}\underbrace{(4)(-5)}$$
$$= (-6) \cdot (-20)$$
$$= 120 \quad \textit{Answer}$$

b. $$(-4) \cdot 3 \cdot (-2)\left(-\frac{1}{3}\right) = (-12)(-2)\left(-\frac{1}{3}\right)$$
$$= 24\left(-\frac{1}{3}\right)$$
$$= -8 \quad \textit{Answer}$$

c. $24(-15)(0)(13) = 0$ ***Answer***

Example 1 illustrates the following facts.

1. A product of nonzero numbers is *positive* if the number of negative factors is *even*.
2. The product is *negative* if the number of negative factors is *odd*.
3. The product is *zero* if any one of the factors is *0*.

Example 2 Name the property used in each step.

Solution

	Step	*Property*
1.	$-ab = (-1)(ab)$	1. Multiplicative property of -1
2.	$= [(-1)a]b$	2. Associative property of multiplication
3.	$= (-a)b$	3. Multiplicative property of -1
4.	$\therefore -ab = (-a)b$	4. Transitive property of equality

The last step in Example 2 states one case of the following useful fact: *The opposite of a product of real numbers is the product of the opposite of one factor and the other factors.*

Property of the Opposite of a Product

For all real numbers a and b,

$$-ab = (-a)b \quad \text{and} \quad -ab = a(-b).$$

Example 3 Simplify.

a. $2(-3x)(-5y)$

b. $(-p)(2q) + (3p)(-2q)$

Solution **a.** $2(-3x)(-5y) = 2(-3)(-5)xy = 30xy$ ***Answer***

b. $(-p)(2q) + (3p)(-2q) = -2pq + (-6pq) = -8pq$ ***Answer***

Example 4 Simplify.

a. $-2\left(3t - \frac{1}{2}\right)$

b. $16d - 5(3d - 4)$

Solution **a.** $-2\left(3t - \frac{1}{2}\right) = (-2)3t - (-2)\frac{1}{2}$

$= -6t - (-1)$

$= -6t + 1$, or $1 - 6t$ ***Answer***

b. $16d - 5(3d - 4) = 16d - 15d + 20$

$= d + 20$ ***Answer***

Example 5 Name the property used in each step.

Solution

	Step	*Property*
1.	$-(a + b) = (-1)(a + b)$	1. Multiplicative property of -1
2.	$= (-1)a + (-1)b$	2. Distributive property
3.	$= (-a) + (-b)$	3. Multiplicative property of -1
4.	$\therefore -(a + b) = (-a) + (-b)$	4. Transitive property of equality

Example 5 shows that *the opposite of a sum of real numbers is equal to the sum of the opposites of the numbers.*

Property of the Opposite of a Sum

For all real numbers a and b,

$$-(a + b) = (-a) + (-b).$$

Example 6 Give the opposite of each expression.

a. $3y + 5$ **b.** $-6x + 4$ **c.** $-xy - 5z$ **d.** $4y^3 - 8y^2 + 3y - 7$

Solution **a.** $-3y - 5$ **b.** $6x - 4$ **c.** $xy + 5z$ **d.** $-4y^3 + 8y^2 - 3y + 7$

Common Errors

Some students may confuse the multiplication rules with the addition rules and vice versa. In reteaching, compare addition and multiplication exercises. For example, show students that $-12 + (-3) = -15$ while $(-12)(-3) = 36$. Also compare $-12 + 3 = -9$ and $12 + (-3) = 9$ to $(-12)(3) = -36$ and $(12)(-3) = -36$.

Check for Understanding

Here is a suggested use of the Oral Exercises to check students' understanding as you teach the lesson.

Oral Exs. 1–6: use before Example 1.
Oral Exs. 7–12: use before Example 2.
Oral Exs. 13–20: use after Example 6.

Guided Practice

Simplify.

1. $(-5r)(-2s)(-6)$ $-60rs$
2. $(-3)^2 \cdot \frac{1}{27} \cdot (-12)$ -4
3. $-19(-12) - (-12)$ 240
4. $-2(-2+5)(-2-5)$ 42
5. $-x(-3y-4) - 3(-y-4x)$ $3xy + 16x + 3y$

Multiply.

6. $(-7)(-5a+1)$ $35a - 7$
7. $\left(-\frac{1}{4}\right)(12x^2 - 48x + 2)$ $-3x^2 + 12x - \frac{1}{2}$

Summarizing the Lesson

In this lesson students reviewed the rules for and properties of multiplying real numbers and extended these rules and properties to algebraic expressions.

Suggested Assignments

Average Algebra
30/1–33 odd
S 31/Mixed Review

Average Alg. and Trig.
30/3–33 mult. of 3
S 31/Mixed Review
Assign with Lesson 1-6.

Extended Algebra
30/3–33 mult. of 3
S 31/Mixed Review
Assign with Lesson 1-6.

Extended Alg. and Trig.
30/3–36 mult. of 3
S 31/Mixed Review
Assign with Lesson 1-4.

Oral Exercises

Tell whether the given expression represents a positive number, a negative number, or zero. Then simplify the expression.

1. $4(-3)$ neg.; -12
2. $(-2)(-5)$ pos.; 10
3. $-2 \cdot 7$ neg.; -14
4. $-[2(-3)]$ pos.; 6
5. $(-2) \cdot 3 \cdot (-5)$ pos.; 30
6. $(-6)(-1)(0)(-2)$ zero; 0

Without doing the computation, tell whether the value of each expression is positive, negative, or zero.

Sample $(-2)(3)\left(-\frac{1}{4}\right)\left(-\frac{1}{8}\right)(-12)$

Solution Positive (The expression has an even number of negative factors.)

7. $2(-6)(-13)(-37)$ neg.
8. $(-1)(-2)(-3)(-4)(-5)(-6)$ pos.
9. $2(-2)(2+2)(2-2)$ zero
10. $7\left(-\frac{1}{3}\right)\left(-\frac{2}{5}\right)(9)\left(\frac{1}{9}\right)$ pos.
11. $-(-1)^4$ neg.
12. $(-2)^{101}$ neg.

Give the opposite of each expression.

13. $5e - f$ $f - 5e$
14. $-1 + 3t$ $1 - 3t$
15. y^3 $-y^3$
16. $-3y^3 + 2y - 1$ $3y^3 - 2y + 1$

Tell whether each expression always, sometimes, or never represents a negative number.

17. $-y$ sometimes
18. $|-y|$ never
19. $y \cdot y$ never
20. $-|-(-y)|$ sometimes

Written Exercises

Simplify. You may wish to check your answers on a calculator.

A

1. $5(-2)(-7)(-3)$ -210
2. $7(-2)(6)(-1)$ 84
3. $9\left(-\frac{1}{7}\right)\left(\frac{1}{3}\right)(-28)$ 12
4. $\left(\frac{3}{4}\right)(-10)(-8)\left(\frac{1}{5}\right)$ 12
5. $(0.5)(-6)(-4)(-0.2)$ -2.4
6. $(1.5)(-3)(0.2)(-2)(-1)$ -1.8
7. $2(-5x)(-6y)$ $60xy$
8. $(-3)(-u)(-7v)$ $-21uv$
9. $(-a)(-2b)(-3c)$ $-6abc$
10. $\left(-\frac{1}{2}\right)(4r)(-s)$ $2rs$
11. $(-6-4)(-6+5)$ 10
12. $17(-13) + 17(-7)$ -340
13. $12(-1)^7(-2)^3$ 96
14. $(-5)^3\left(-\frac{1}{5}\right)^2$ -5
15. $(-1)^3(-2-2)(-6)(-4)$ 96
16. $(-9)^2(-2+2)(-5)$ 0
17. $(-2)(1 - 2x - 3x^2)$ $-2 + 4x + 6x^2$
18. $(-3)\left(2a - \frac{2}{3}\right)$ $-6a + 2$

19. $\left(-\frac{1}{2}\right)(6z^2 - 4z + 2)$ 20. $(-1)[(-t)^3 - 1]$ 21. $(-x)(3y) + (2x)(-3y)$

22. $(-a)(-b) + 6(ab)(-1)$ 23. $7k - 4(3k + 6)$ 24. $-3(p - 5) - 7p$

25. $5(x - y) - 3(x - y)$ 26. $-4(a - b) + 2(a - b)$ 27. $5(-3a^3 - 2) + 3(-2 - a^3)$

28. $4(2y^2 - 3) - 2(4 + y^2)$ 29. $t(2w - 9) - 2(2t + 7)$ 30. $-c(d + 5) + 6(2 - cd)$

Evaluate each expression for the given values of the variable.

Sample $(x + 1)(x + 3)(x - 3)$ **a.** $x = 1$ **b.** $x = -2$

Solution **a.** $(x + 1)(x + 3)(x - 3) = (1 + 1)(1 + 3)(1 - 3) = (2)(4)(-2) = -16$

b. $(x + 1)(x + 3)(x - 3) = (-2 + 1)(-2 + 3)(-2 - 3) = (-1)(1)(-5) = 5$

B 31. $x(x - 2)(x - 4)$ **a.** $x = 1$ 3 **b.** $x = 3$ −3 **c.** $x = 2$ 0

32. $t(t + 1)(t - 3)$ **a.** $t = 2$ −6 **b.** $t = -1$ 0 **c.** $t = 1$ −4

33. $y^3 - 3y^2 + 4$ **a.** $y = 2$ 0 **b.** $y = -1$ 0 **c.** $y = 3$ 4

34. $2b^3 + 3b^2 - b + 3$ **a.** $b = 2$ 29 **b.** $b = -2$ 1 **c.** $b = 0$ 3

35. Name the property used in each step.

1. $-a = -(a \cdot 1)$ 1. ___?___ Ident. prop. of mult.
2. $= a(-1)$ 2. ___?___ Prop. of opp. of a prod.
3. $= (-1)a$ 3. ___?___ Comm. prop. of mult.
4. $-a = (-1)a$ 4. ___?___ Trans. prop. of =
5. and $(-1)a = -a$ 5. ___?___ Symm. prop. of =

(This exercise is a proof of the multiplicative property of −1.)

C 36. Find a value of y for which the expression $(1 - y)(3 - y)(6 - y)$ has the given value.

a. 18 **b.** 0 **c.** −70 **d.** 120 **e.** $3\frac{1}{8}$

a. 0 **b.** 1, 3, or 6 **c.** 8 **d.** −2 **e.** 3.5

Mixed Review Exercises

Evaluate each expression if $x = 3$ and $y = -2$.

1. $x + y - 1$ 0 2. $|x| - |y|$ 1 3. $x^2 + y$ 7

4. $-|x + y|$ −1 5. $y - (x - 1)$ −4 6. $4x - y$ 14

Name the property illustrated in each statement.

7. $3 + 5 = 5 + 3$ Comm. prop. of add. 8. $-3 + 0 = -3$ Ident. prop. of add. 9. If $x = y + 1$, then $y + 1 = x$. Symm. prop. of =

10. $4 \cdot \frac{1}{4} = 1$ Prop. of reciprocals 11. $4(5 \cdot 8) = (4 \cdot 5)8$ Assoc. prop. of mult. 12. $5(2 + 7) = 5 \cdot 2 + 5 \cdot 7$ Dist. prop.

Additional Answers Written Exercises

19. $-3z^2 + 2z - 1$
20. $t^3 + 1$
21. $-9xy$
22. $-5ab$
23. $-5k - 24$
24. $-10p + 15$
25. $2x - 2y$
26. $-2a + 2b$
27. $-18a^3 - 16$
28. $6y^2 - 20$
29. $2tw - 13t - 14$
30. $-7cd - 5c + 12$

Supplementary Materials

Study Guide pp. 9–10

Using a Computer

These exercises lead to the creation of a program that will evaluate any cubic algebraic expression of the form $ax^3 + bx^2 + cx + d$.

Computer Exercises

For students with some programming experience.

1. Write a program that will evaluate the expression $5x^3 - 7x^2 + 3x + 2$ for any value of x. Use an INPUT statement to enter the values of x.

2. Run the program in Exercise 1 to find the value of the expression for each value of x.

   a. $x = 1$ 3 **b.** $x = 4.7$ 380.585 **c.** $x = -2.5$ −127.375 **d.** $x = -3.04$ −212.28352

3. Modify the program of Exercise 1 so that it will evaluate *any* expression of the form $ax^3 + bx^2 + cx + d$ for given values of x. That is, modify the program so that the user can enter the constants a, b, c, and d in the expression as well as the value to be substituted for x.

4. Run the program in Exercise 3 to evaluate $-3x^3 + 8x^2 - 6x + 11$ for each value of x.

   a. $x = 2.1$ 5.897 **b.** $x = 4.95$ −186.542125 **c.** $x = -1.06$ 29.921848 **d.** $x = -4.17$ 392.66634

Biographical Note / *Sonya Kovalevski*

When the Russian mathematician Sonya Kovalevski (1850–1891) went to Sweden to teach, a Stockholm newspaper proclaimed her "the Princess of Science." She did not reach such a position of acclaim easily, however. European universities resisted accepting women as students or faculty members during the nineteenth century.

To obtain her education, Kovalevski became the private student of Karl Weierstrass, an eminent German mathematician whose own university in Berlin would not grant her admission. Her papers on partial differential equations, Abelian integrals, and the rings of Saturn eventually qualified her for a degree from the University of Göttingen.

At the end of her five-year teaching appointment in Stockholm, Kovalevski became the first woman in modern times to hold a tenured position in mathematics. Also during this time, she won the Prix Bordin from the French Academy of Science for a paper on the rotation of solid bodies around fixed points. Her work grew out of previous work on ultraelliptic integrals, but her extension of the concept was so remarkable that the prize money was nearly doubled. In 1889 she was elected a corresponding member of the Russian Academy of Sciences.

In addition to her mathematical studies, Kovalevski published poetry, novels, plays, and essays on politics and education.

1-6 Quotients

Objective To review rules for dividing real numbers.

You know that subtraction is defined in terms of addition and opposites. Similarly, division is defined in terms of multiplication and reciprocals.

Definition of Division

The quotient a divided by b is written $\frac{a}{b}$ or $a \div b$. For every real number a and nonzero real number b,

$$\frac{a}{b} = a \cdot \frac{1}{b}, \quad \text{or} \quad a \div b = a \cdot \frac{1}{b}.$$

To divide by any nonzero number, multiply by its reciprocal. Since 0 has no reciprocal, division by 0 is not defined.

Example 1 Simplify: **a.** $12 \div 4$ **b.** $\frac{-18}{-3}$ **c.** $15 \div \left(-\frac{3}{2}\right)$

Solution **a.** $12 \div 4 = 12 \cdot \frac{1}{4} = 3$ ***Answer***

b. $\frac{-18}{-3} = (-18)\left(-\frac{1}{3}\right) = 6$ ***Answer***

c. $15 \div \left(-\frac{3}{2}\right) = 15 \cdot \left(-\frac{2}{3}\right) = -10$ ***Answer***

A number and its reciprocal have the same sign. Therefore, the sign of a quotient of real numbers follows the same rules as the sign of a product.

Rules for Division

1. The quotient of two positive numbers or two negative numbers is a positive number.
2. The quotient of two numbers when one is positive and the other negative is a negative number.

Example 2 Simplify $\frac{-9 \div (-3)}{(-1)^2(-3)}$.

Solution $\frac{-9 \div (-3)}{(-1)^2(-3)} = \frac{-9 \cdot \left(-\frac{1}{3}\right)}{(1)(-3)} = \frac{3}{-3} = -1$ ***Answer***

Teaching Suggestions, p. T82

Suggested Extensions, p. T82

Warm-Up Exercises

Divide.

1. $86 \div 4$ 21.5
2. $96 \div 8$ 12
3. $-63 \div 3$ -21
4. $115 \div (-5)$ -23
5. $-108 \div (-8)$ 13.5

Motivating the Lesson

Point out to students that addition, subtraction, multiplication, and division are used frequently in everyday life. It is important for students to know the rules and procedures for these operations if they are to function in today's and tomorrow's society.

Chalkboard Examples

Simplify.

1. $-100 \div 25 \div -\frac{1}{2}$

   $-4 \div -\frac{1}{2} = -4(-2) = 8$

2. $-35 \div (-7)(-5)$

   $5(-5) = -25$

3. $\frac{7(2-15)+1}{-(-6)^2 \div 2}$

   $\frac{7(-13)+1}{-36 \div 2} = \frac{-91+1}{-18} =$

   $\frac{-90}{-18} = 5$

(continued)

Chalkboard Examples (continued)

4. Divide: $\frac{-10z^4 + 8z^2 - 2}{-2}$
$5z^4 - 4z^2 + 1$

5. Evaluate $\frac{8 - k^3}{k - 2}$ if $k = -\frac{1}{2}$.
$\frac{8 - \left(-\frac{1}{2}\right)^3}{-\frac{1}{2} - 2} = \frac{8 - \left(-\frac{1}{8}\right)}{-\frac{5}{2}} = \frac{\frac{65}{8}}{-\frac{5}{2}} = \frac{65}{8}\left(-\frac{2}{5}\right) = -\frac{13}{4}$

Check for Understanding

Here is a suggested use of the Oral Exercises to check students' understanding as you teach the lesson.
Oral Exs. 1–12: use before Example 2.
Oral Exs. 13–19: use after Example 2.

Guided Practice

Simplify.

1. $-72 \div (-6) \div (-2)$ -6
2. $-72 \div [-6 \div (-2)]$ -24
3. $\frac{-9(11) + 43}{1 - (-3)^3}$ -2
4. $24 \div \left(-\frac{2}{3}\right)\left(-\frac{1}{4}\right) \div 27$ $\frac{1}{3}$
5. $\frac{2x^2 - 5x - 1}{-1}$
$-2x^2 + 5x + 1$
6. $\frac{18j - 36k + 5.4}{-3.6}$
$-5j + 10k - 1.5$

Notice that $\frac{a+b}{c} = (a + b) \cdot \frac{1}{c} = a \cdot \frac{1}{c} + b \cdot \frac{1}{c} = \frac{a}{c} + \frac{b}{c}$. Division, like multiplication, has these distributive properties.

For all real numbers a and b and *nonzero* real numbers c,

$$\frac{a+b}{c} = \frac{a}{c} + \frac{b}{c} \quad \text{and} \quad \frac{a-b}{c} = \frac{a}{c} - \frac{b}{c}.$$

Caution: $\frac{c}{a+b} \neq \frac{c}{a} + \frac{c}{b}$. For example, $\frac{12}{2+4} \neq \frac{12}{2} + \frac{12}{4}$, since $2 \neq 6 + 3$.

Example 3 Simplify $\frac{48 - 12x^2}{-3}$

Solution

$$\frac{48 - 12x^2}{-3} = \frac{48}{-3} - \frac{12x^2}{-3}$$
$$= -16 - (-4x^2)$$
$$= -16 + 4x^2, \text{ or } 4x^2 - 16 \quad \textit{Answer}$$

Oral Exercises

Tell whether the expression represents a positive number, a negative number, or zero. Then simplify the expression.

1. $\frac{-40}{8}$ neg.; -5
2. $\frac{-70}{-10}$ pos.; 7
3. $\frac{101}{-101}$ neg.; -1
4. $\frac{0}{-5}$ zero; 0
5. $-20 \div (-4)$ pos.; 5
6. $-8 \div (-24)$ pos.; $\frac{1}{3}$
7. $0 \div (-8)$ zero; 0
8. $19 \div (-1)$ neg.; -19
9. $\frac{2(-6)}{-3}$ pos.; 4
10. $\frac{6(-8)}{(-3)(-4)}$ neg.; -4
11. $\frac{8 \div (-4)}{-12 \div 3}$ pos.; $\frac{1}{2}$
12. $(-30) \div (15) \div (-2)$ pos.; 1

Without doing the computation, tell whether the value of the expression is positive, negative, or zero.

13. $\frac{(-23) \cdot 17 \cdot (-13)}{(41)(-31)(-29)}$ positive
14. $\frac{(-16)(-7)}{(-3)(-1)(-9)}$ negative
15. $\frac{-14 + (-26)}{(-8)(-3)(-5)}$ positive
16. $\frac{(-16)(-7)}{(-3) + (-9)}$ negative

What number makes the statement true?

17. $-12 \div (\underline{\ ?\ }) = 4$ -3
18. $-12 \div (\underline{\ ?\ }) = -4$ 3
19. $12 \div (\underline{\ ?\ }) = -4$ -3

Written Exercises

Simplify. You may wish to check your answers on a calculator.

A

1. $-21 \div (-7)$ 3
2. $-7 \div (-21)$ $\frac{1}{3}$
3. $64 \div (-4) \div (-2)$ 8
4. $64 \div [(-4) \div (-2)]$ 32
5. $-6 \div \left(-\frac{1}{3}\right) \div (-1)$ −18
6. $-\frac{1}{2} \div \left(\frac{1}{4}\right) \div (-4)$ $\frac{1}{2}$
7. $\frac{8(-18)}{3(-12)}$ 4
8. $\frac{(-3)(-4)(-2)}{(-6)(-2)}$ −2
9. $[60 \div (-5)][8 \div (-2)]$ 48
10. $[27(-2)] \div (-3)^4$ $-\frac{2}{3}$
11. $\frac{-16 \cdot 3 \div 2}{(-2)^4}$ $-\frac{3}{2}$
12. $\frac{24 \div (-3)}{4(-5) \div 2}$ $\frac{4}{5}$
13. $\frac{3^2 - 5^2}{3 + (-5)}$ 8
14. $\frac{4^2 - 5^2}{(-4) + (-5)}$ 1
15. $\frac{144 \div (-24)}{(-10) \div \left(-\frac{2}{3}\right)}$ $-\frac{2}{5}$
16. $\frac{(-12)\left(-\frac{3}{4} - \frac{1}{2}\right)}{\frac{5}{9} \div (-10)}$ −270
17. $\frac{-24\left[-18 \div \left(-\frac{2}{3}\right)\right]}{(-18)\left(-\frac{2}{3}\right)}$ −54
18. $\frac{\left[\frac{4}{9} - \left(-\frac{2}{9}\right)\right]\left[\frac{2}{3} - \left(-\frac{2}{3}\right)\right]^2}{\frac{5}{9} \div \left(-\frac{10}{3}\right)}$ $\frac{-64}{9}$
19. $\frac{9x^2 + 27}{-3}$ $-3x^2 - 9$
20. $\frac{24 - 6t^2}{2}$ $12 - 3t^2$
21. $\frac{1 - (-n)^2}{-1}$ $n^2 - 1$
22. $\frac{2n^2 - 2^2}{-2}$ $2 - n^2$
23. $\frac{36c^2 - 24c - 6}{6}$ $6c^2 - 4c - 1$
24. $\frac{-15r^3 - 5r - 5}{-5}$ $3r^3 + r + 1$

Evaluate each expression for the given values of the variable.

B

25. $\frac{y(y-3)}{y-2}$ a. $y = 1$ 2 b. $y = 4$ 2 c. $y = 3$ 0
26. $\frac{(x^2-4)(x-3)}{x+1}$ a. $x = 1$ 3 b. $x = 2$ 0 c. $x = -2$ 0
27. $\frac{(t-1)(t+1)(t-3)}{\frac{1}{2}t + 2}$ a. $t = 2$ −1 b. $t = 0$ $\frac{3}{2}$ c. $t = -2$ −15
28. $\frac{(r-3)(r)(r+3)}{(r-2)(r+2)}$ a. $r = \frac{1}{2}$ $\frac{7}{6}$ b. $r = -1$ $-\frac{8}{3}$ c. $r = 0$ 0
29. Give a numerical example to show that division is not commutative. $4 \div 2 \neq 2 \div 4$
30. Give a numerical example to show that division is not associative. $8 \div (4 \div 2) \neq (8 \div 4) \div 2$

Tell whether the given set is closed (see page 19) under division. If the set is not closed, give an example to show this.

C

31. $\{1, -1\}$ Yes
32. The rational numbers No; $\frac{1}{0}$ is not defined
33. The rational numbers without 0 Yes
34. $\left\{\ldots, \frac{1}{8}, \frac{1}{4}, \frac{1}{2}, 1, 2, 4, 8, \ldots\right\}$ Yes
35. The squares of the rational numbers without 0 Yes
36. The irrational numbers No; $\sqrt{2} \div \sqrt{2} = 1$

Summarizing the Lesson

Reinforce the objective of this lesson by asking questions such as:

- What is the definition of division?
- What are the two rules for dividing positive and negative numbers? Give an example of each rule.
- Does division have distributive properties like multiplication? If so, give an example. If not, tell why.

Using a Calculator

Some students may need to be reminded that the change-sign key for entering negative numbers is usually labeled +/−. This key is pressed *after* the number to be negated has been entered.

Suggested Assignments

Average Algebra
35/2–28 even, 29–31
R 36/Self-Test 2

Average Alg. and Trig.
35/4–28 mult. of 4, 29, 30
R 36/Self-Test 2
Assign with Lesson 1-5.

Extended Algebra
35/4–28 mult. of 4, 29–31
R 36/Self-Test 2
Assign with Lesson 1-5.

Extended Alg. and Trig.
35/2–28 even, 29, 30
R 36/Self-Test 2
Assign with Lesson 1-7.

Supplementary Material

Study Guide	pp. 11–12
Practice Master	3
Test Master	1
Resource Book	p. 101

Quick Quiz

Name the property that justifies each step in the simplification.

1. $\frac{1}{4}[(a + 4) + (-a)]$
   $= \frac{1}{4}[-a + (a + 4)]$
   Comm. prop. of add.
   $= \frac{1}{4}[(-a + a) + 4]$
   Assoc. prop. of add.
   $= \frac{1}{4}(0 + 4)$
   Prop. of opposites
   $= \frac{1}{4}(4)$
   Ident. prop. of add.
   $= 1$
   Prop. of reciprocals

Simplify.

2. $42 + (-55) + (-29) + 13$ -29
3. $|-20 + 7| - |-12 - 12|$ -11
4. $-12 - (2 - 8)$ -6
5. $-7[4 - (-1)] + (-3)^2$ -26
6. $\left(-\frac{3}{2}a^2\right)(-18)\left(\frac{5}{9}b\right)$ $15a^2b$
7. $\frac{-3 - (-3)(-2)^2}{-7 + 2(-1)}$ -1
8. $\frac{-6 \div \left(-\frac{1}{2}\right)}{-21 \div 7 \cdot 6}$ $-\frac{2}{3}$

Evaluate if $x = -4$ and $y = -3$.

9. $x^2 - 5xy + 4y^2$ -8
10. $\frac{x - y^3}{(x - y)^3}$ -23
11. Divide: $\frac{66n^4 - 3n^2 - 24n}{-6}$
    $-11n^4 + \frac{n^2}{2} + 4n$
12. Multiply: $(-20)(-3a + 2b - ab)$
    $60a - 40b + 20ab$

Self-Test 2

Vocabulary

field (p. 14)	reciprocal (p. 15)
unique (p. 14)	multiplicative inverse (p. 15)
opposite (p. 15)	equivalent expressions (p. 16)
additive inverse (p. 15)	similar terms (p. 23)

1. Name the property that justifies each step in the simplification below. **Obj. 1-3, p. 14**

$\frac{1}{2}[(x + 2) + (-x)] = \frac{1}{2}[-x + (x + 2)]$ **a.** _?_ Comm. prop. of add.

$= \frac{1}{2}[(-x + x) + 2]$ **b.** _?_ Assoc. prop. of add.

$= \frac{1}{2}(0 + 2)$ **c.** _?_ Prop. of opposites

$= \frac{1}{2} \cdot 2$ **d.** _?_ Ident. prop. of add.

$= 1$ **e.** _?_ Prop. of reciprocals

Simplify.

2. $-6 + 13 + (-5)$ 2 **Obj. 1-4, p. 21**

3. $-11 + (-8) + 9$ -10

4. $7 + (-4) - (-5)$ 8

5. $-3.2 - (4.6 - 5)$ -2.8

6. $-3(2 - 7) + 4(-6)$ -9 **Obj. 1-5, p. 27**

7. $(-6c)(4d)\left(-\frac{1}{8}\right)$ $3cd$

8. Evaluate the expression $2x^2 - 5x - 8$ if $x = -3$. 25

Simplify.

9. $24 \div (-6) \div (-2)$ 2 **Obj. 1-6, p. 33**

10. $\frac{4(-8) - 3(-4)}{(-5)(-2)}$ -2

11. $\frac{-6x^2 + 10x - 2}{-2}$ $3x^2 - 5x + 1$

12. Evaluate the expression $\frac{y + 9}{(y - 1)(3 - y)}$ if $y = -1$. -1

Check your answers with those at the back of the book.

Solving Equations and Solving Problems

1-7 Solving Equations in One Variable

Objective To solve certain equations in one variable.

An equation or inequality that contains a variable, such as

$$2t - 1 = 5 \qquad \text{and} \qquad x + 3 > 0,$$

is called an **open sentence.** Any value of the variable that makes an open sentence a true statement is called a **solution,** or **root,** of the open sentence and is said to **satisfy** it. For example,

$$3 \textit{ is a solution of } 2t - 1 = 5, \text{ because } 2 \cdot 3 - 1 = 5 \text{ is true,}$$

but 2 is *not* a solution, because $2 \cdot 2 - 1 = 5$ is false. The set of all solutions of an open sentence that belong to a given domain of the variable is the **solution set** of the sentence *over* that domain. *Unless otherwise stated, open sentences in this book are to be solved over the set of real numbers.*

Study the following two sequences of equations.

$$\begin{aligned} 7x + 12 &= 47 \\ 7x + 12 - 12 &= 47 - 12 \\ 7x &= 35 \\ \frac{7x}{7} &= \frac{35}{7} \\ x &= 5 \end{aligned} \qquad\qquad \begin{aligned} x &= 5 \\ 7x &= 7 \cdot 5 \\ 7x &= 35 \\ 7x + 12 &= 35 + 12 \\ 7x + 12 &= 47 \end{aligned}$$

The properties of real numbers guarantee that if the first statement in either sequence is true for some value of x, then the last statement in the sequence is true for that value of x. Therefore, $7x + 12 = 47$ and $x = 5$ have the same solution set, namely $\{5\}$. Equations having the same solution set over a given domain are called **equivalent equations.** To **solve** an equation you usually change, or *transform,* it into a simple equivalent equation whose solution set is easy to see.

Transformations That Produce Equivalent Equations

1. Simplifying either side of an equation.
2. Adding to (or subtracting from) each side of an equation the same number or the same expression.
3. Multiplying (or dividing) each side of an equation by the same *nonzero* number.

Teaching Suggestions, p. T82

Suggested Extensions, p. T82

Warm-Up Exercises

Evaluate the expression for the given values of the variable.

1. $5x + 4;\ x = 3$ 19

2. $8a - 3;\ a = -1$ −11

3. $-4y - 6;\ y = -2$ 2

Fill in the blank to make the statement true.

4. $5 \cdot \underline{\ ?\ } + 4 = 19$ 3

5. $8 \cdot \underline{\ ?\ } - 3 = -11$ −1

6. $-4 \cdot \underline{\ ?\ } - 6 = 2$ −2

Motivating the Lesson

Point out that in this lesson students apply their knowledge of the properties and rules of real numbers to solving equations in one variable. Emphasize the importance of solving equations in everyday life. Guide students in recalling those formulas that are particularly useful in everyday life. Some examples include formulas for area and volume, for simple interest, and for distance, rate, and time.

Chalkboard Examples

Solve.

1. $3(2x + 5) = -7(x + 9)$
   $6x + 15 = -7x - 63$
   $13x = -78$
   $x = -6$
2. $5h - 2(4 + h) = 5 + 3(1 + h)$
   $5h - 8 - 2h = 5 + 3 + 3h$
   $3h - 8 = 8 + 3h$
   $-8 = 8$ no solution
3. $\frac{1}{2}q - 7 = -15 + 2q$
   $-7 = -15 + \frac{3}{2}q$
   $8 = \frac{3}{2}q$
   $\frac{16}{3} = q$
4. The area of a trapezoid is 147 cm^2. If one base is 18 cm long and the height is 7 cm, find the length of the other base.
   Use $A = \frac{1}{2}h\,(b_1 + b_2)$.
   $147 = \frac{1}{2}(7)(18 + b_2)$
   $294 = 7(18 + b_2)$
   $42 = 18 + b_2$
   $24 = b_2$ 24 cm
5. Solve $w = \frac{az + b}{cz + d}$ for z.
   $w(cz + d) = az + b$
   $wcz + wd = az + b$
   $wcz - az = b - wd$
   $z(wc - a) = b - wd$
   $z = \frac{b - wd}{wc - a}$

Using a Calculator

Besides being used for checking solutions, calculators are especially suitable for solving problems that involve decimals, such as in Example 3.

Example 1 Solve $5(2z - 9) = 7z - 60$.

Solution

$5(2z - 9) = 7z - 60$	Copy the equation.
$10z - 45 = 7z - 60$	Simplify the left side.
$10z - 45 + 45 = 7z - 60 + 45$	Add 45 to each side.
$10z = 7z - 15$	
$10z - 7z = 7z - 15 - 7z$	Subtract $7z$ from each side.
$3z = -15$	
$\frac{3z}{3} = \frac{-15}{3}$	Divide each side by 3
$z = -5$	
Check: $5(2z - 9) = 7z - 60$	
$5[2(-5) - 9] \stackrel{?}{=} 7(-5) - 60$	Substitute -5 for z.
$5(-10 - 9) \stackrel{?}{=} -35 - 60$	
$-95 = -95$ ✓	

$\therefore$ the solution set is $\{-5\}$. ***Answer***

Example 1 illustrates how to check your work for mistakes by showing that the solution actually satisfies the *given* equation. A calculator is often helpful in checking solutions.

When solving an equation with one variable, you cannot assume that it has exactly one root. Example 2(a) shows that an equation may have *no* solutions, so that its solution set has no members. The set with no members is called the **empty set,** or the **null set,** and is denoted by $\emptyset$.

Example 2(b) shows an equation that is satisfied by *all* values of the variable. Such an equation is called an **identity.** The solution set of an identity is the set of all real numbers.

Example 2 Solve.

a. $3(2s - 3) = 6(s + 1) - 10$ **b.** $3(2s - 3) = 6(s + 1) - 15$

Solution

a. $3(2s - 3) = 6(s + 1) - 10$
$6s - 9 = 6s + 6 - 10$
$6s - 9 = 6s - 4$
$0 = 5$ False!

Since the given equation is equivalent to the *false* statement $0 = 5$, it has *no* root.

$\therefore$ the solution set is $\emptyset$. ***Answer***

b. $3(2s - 3) = 6(s + 1) - 15$
$6s - 9 = 6s + 6 - 15$
$6s - 9 = 6s - 9$

Since the given equation is equivalent to $6s - 9 = 6s - 9$, which is *true* for all values of s, it is an identity.

$\therefore$ the solution set is the set of all real numbers. ***Answer***

A **formula** is an equation that states a relationship between two or more variables. The variables usually represent physical or geometric quantities. For example, the formula $h = -16t^2 + vt$ gives the height h (in feet) of a launched object t seconds after firing with initial velocity v (in ft/s). Given values for all but one of the variables in a formula, you can find the value of the remaining variable.

Example 3 A model rocket launched with initial velocity v reaches a height of 40 ft after 2.5 s. Find v.

Solution Substitute the given values of the variables.

$$h = -16t^2 + vt$$
$$40 = -16(2.5)^2 + v(2.5)$$
$$40 = -100 + 2.5v$$
$$140 = 2.5v$$
$$v = 56 \text{ (ft/s)} \quad \textbf{\textit{Answer}}$$

When you solve a formula or equation for a certain variable, you can think of all the other variables as **constants,** that is, as fixed numbers. Then solve by the usual methods.

Example 4 The volume V of a pyramid with height h and square base with sides s is given by the formula $V = \frac{1}{3}s^2h$. Solve this formula for h.

Solution You need to express h in terms of the other variables. That is, you need to get h alone on one side of the equation.

$$V = \frac{1}{3}s^2h$$
$$3(V) = 3\left(\frac{1}{3}s^2h\right)$$
$$3V = s^2h$$
$$\frac{3V}{s^2} = \frac{s^2h}{s^2} \quad \text{(Assume } s \neq 0.\text{)}$$
$$h = \frac{3V}{s^2} \quad \textbf{\textit{Answer}}$$

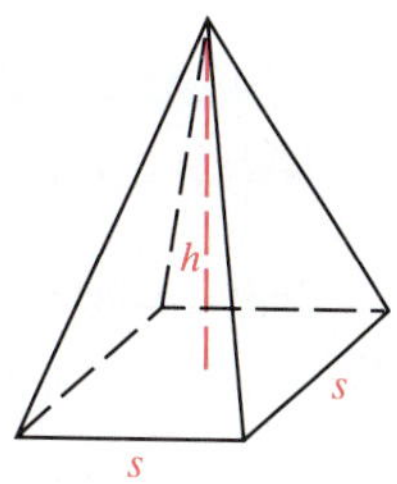

Check for Understanding

Here is a suggested use of the Oral Exercises to check students' understanding as you teach the lesson.

Oral Exs. 1–8: use before Example 1.
Oral Exs. 9–10: use before Example 3.
Oral Exs. 11–12: use after Example 4.

Common Errors

When solving an equation like $6n = 2n - 42 - 3n$ students may need to be reminded to first simplify each side and then use the addition property of equality. Students who do not fully understand the addition property of equality may begin solving by adding $3n$ to all terms with the variable n in them:

$6n = 2n - 42 - 3n$
$6n + 3n = 2n + 3n - 42 - 3n + 3n$
$9n = 5n - 42$

Guided Practice

Solve.

1. $-7.5(x - 3) = -30$ 7
2. $\frac{2}{3}m - 1 = m + 12$ -39
3. $3p - (p - 9) = 2(p + 5)$ no solution
4. $\frac{6}{5}k - \frac{k}{3} = -39$ -45
5. Tell whether each number is a solution of the equation $2z(z - 4)(z + 3) = 0$.
   4 Y 3 N 0 Y

Summarizing the Lesson

Point out that in this review of solving equations many important terms were studied. Write statements such as the following on the chalkboard and ask whether each is true or false. For each statement that is false, ask students to give a correct statement.

- An open sentence can only have an equals sign such as $7x = 49$. False
- A formula is an open sentence that expresses a relationship between two or more variables. True

Suggested Assignments

Average Algebra
40/2–48 even
S 42/Mixed Review

Average Alg. and Trig.
40/3–27 mult. of 3, 31–53 odd
S 42/Mixed Review

Extended Algebra
40/3–27 mult. of 3, 31–53 odd
S 42/Mixed Review

Extended Alg. and Trig.
40/4–48 mult. of 4, 51–53
S 42/Mixed Review
Assign with Lesson 1-6.

Oral Exercises

Explain how you transform the first equation to obtain the second equation.

Sample a. $2t + 5 = 11;\ 2t = 6$ b. $3x - 2 - x = 16;\ 2x - 2 = 16$

Solution a. Subtract 5 from both sides. b. Simplify the left side of the equation.

1. $6z = 42;\ z = 7$ Div. both sides by 6.
2. $4x - 3 = 7;\ 4x = 10$ Add 3 to both sides.
3. $r = 2(r - 1) + 5;\ r = 2r + 3$ Simplify the right side.
4. $\frac{1}{2}(t - 5) = 3;\ t - 5 = 6$ Mult. both sides by 2.
5. $2x - 7 = x + 1;\ 2x = x + 8$ Add 7 to both sides.
6. $5y - 1 = 3y;\ 2y - 1 = 0$ Subtr. $3y$ from both sides.
7. $\frac{t - 5}{3} = 2t;\ t - 5 = 6t$ Mult. both sides by 3.
8. $3x - 9 = 12;\ x - 3 = 4$ Div. both sides by 3.

Each formula below has been transformed into an equivalent one. Name the transformation used at each step.

9. Area of a triangle:

$$A = \frac{1}{2}bh$$

a. $2A = bh$ a. ___?___ Mult. both sides by 2.

b. $\frac{2A}{h} = b$ b. ___?___ Div. both sides by h.

10. Kinetic energy:

$$K = \frac{1}{2}mv^2$$

a. $2K = mv^2$ a. ___?___ Mult. both sides by 2.

b. $\frac{2K}{v^2} = m$ b. ___?___ Div. both sides by v^2.

11. Explain how you would solve the equation $A = \frac{x + y}{2}$ for y. Mult. both sides by 2; subtr. x from both sides.

12. a. Solve $2x = 6$. {3}
 b. Solve $0 \cdot 2x = 0 \cdot 6$. {real numbers}
 c. If you multiply each side of an equation by 0, do you get an equivalent equation? No
 d. Explain why it is not possible to divide each side of an equation by 0. Division by 0 is not defined.

Written Exercises

Solve. Check your work when there is a single solution.

A

1. $3x - 4 = 5$ {3}
2. $4z + 11 = 3$ {−2}
3. $\frac{2}{3}t - 8 = 0$ {12}
4. $15 - \frac{1}{8}d = -1$ {128}
5. $5r = 18 + 2r$ {6}
6. $24 - 2y = 6y$ {3}
7. $3(t - 1) = -(t - 5)$ {2}
8. $2(x - 3) = x + 3$ {9}

9. $2k - 1 = k - 5 + 3k$ {2}

10. $3(2z - 1) = 6z + 5$ ∅

11. $-(5 - x) = x + 3$ ∅

12. $\frac{1}{3}(s - 2) = s + 4$ {−7}

13. $1.5(u + 2) = 7.5$ {3}

14. $1.6(2t - 1) = 14.4$ {5}

15. $0.2(x - 5) = x + 5$ {−7.5}

16. $0.3(2r - 3) = 0.2r + 0.9$ {4.5}

17. $3(x - 2) - x = 2(2x + 1)$ {−4}

18. $\frac{2}{5}(x - 2) = x + 4$ {−8}

19. $2z - (1 - z) = 11 - z$ {3}

20. $3(1 - t) + 5 = 3(1 + t) - 7$ {2}

21. $2(5t - 3) - t = 3(3t - 2)$ {real numbers}

22. $3(5z - 1) + 5(3z + 2) = 7$ {0}

23. $\frac{6x - 2(x - 4)}{3} = 8$ {4}

24. $\frac{3y - 2(y - 1)}{6} = -1$ {−8}

Tell whether each number at the right of the given equation is a solution of the equation.

25. $x(x - 3)(x + 2) = 0;\ -2,\ -3$ Yes; No

26. $z(z + 1)(z - 2) = 0;\ 0,\ -1$ Yes; Yes

27. $z^3 - 4z^2 + z + 6 = 0;\ 2,\ 3$ Yes; Yes

28. $u^3 - 7u - 6 = 0;\ 1,\ -2$ No; Yes

29. $\frac{x + 12}{x - 4} = x - 3;\ -2,\ 0$ No; Yes

30. $\frac{2y}{2y - 1} = \frac{y + 2}{y + 1};\ 0,\ 2$ No; Yes

Solve the equation for the given variable.

31. $2x - 5y = 10$ for x $x = \frac{5}{2}y + 5$

32. $A = \frac{1}{2}bh$ for h $h = \frac{2A}{b}$

33. $I = prt$ for p $p = \frac{I}{rt}$

34. $C = 2\pi r$ for r $r = \frac{C}{2\pi}$

35. $y = mx + b$ for x $x = \frac{y - b}{m}$

36. $ax + by = c$ for y $y = \frac{c - ax}{b}$

37. $P = 2l + 2w$ for w $w = \frac{1}{2}P - l$

38. $P = 2(l + w)$ for l $l = \frac{1}{2}P - w$

B 39. $a(x - b) = c + ab$ for x $x = 2b + \frac{c}{a}$

40. $5cy - d = 4d - cy$ for y $y = \frac{5d}{6c}$

41. $S = -\frac{1}{2}gt^2 + vt$ for v $v = \frac{S}{t} + \frac{1}{2}gt$

42. $C = \frac{5}{9}(F - 32)$ for F $F = \frac{9}{5}C + 32$

In each formula, substitute the given values of the variables. Then find the value of the remaining variable, which is printed in red.

43. Volume of a cylinder: $V = \pi r^2 h;\ V = 128,\ r = 8$ $h = \frac{2}{\pi}$

44. Volume of a cone: $V = \frac{1}{3}\pi r^2 h;\ V = 48,\ r = 4$ $h = \frac{9}{\pi}$

45. Amount at simple interest: $A = P(1 + rt);\ A = 168,\ P = 150,\ r = 0.08$ $t = 1.5$

46. Distance an object falls: $d = \frac{v^2}{2g};\ d = 1000,\ v = 140$ $g = 9.8$

47. Area of a trapezoid: $A = \frac{h}{2}(b_1 + b_2);\ A = 100,\ h = 5,\ b_2 = 12$ $b_1 = 28$

Supplementary Materials

Study Guide	pp. 13–14
Practice Master	4

48. Total area of a cylinder: $A = 2\pi r(r + h)$; $A = 80\pi$, $r = 5$ $h = 3$

49. Total area of a cone: $A = \pi r(s + r)$; $A = 100\pi$, $r = 5$ $s = 15$

50. Sum of a geometric series: $S = \dfrac{a(1 - r^n)}{1 - r}$; $S = 80$, $r = 3$, $n = 4$ $a = 2$

C **51–53.** Solve the formulas in Exercises 48–50 for the variables printed in red. (Ignore the given values of the variables.)

51. $h = \dfrac{A}{2\pi r} - r$ **52.** $s = \dfrac{A}{\pi r} - r$ **53.** $a = \dfrac{S(1 - r)}{1 - r^n}$

Mixed Review Exercises

Simplify.

1. $7 - (3 - 4)$ 8

2. $-2(-5 + 8)$ −6

3. $(4 - 9)^2$ 25

4. $|-6 + 2|$ 4

5. $(-8)\left(\frac{1}{2}\right)(-3)(-1)$ −12

6. $\dfrac{-3 + 11}{-1 - 3}$ −2

7. $3x - 2(x + 4)$ $x - 8$

8. $(-4a)(5b) + (-3a)(-8b)$ $4ab$

9. $(-x)^2(-y)^3$ $-x^2y^3$

10. $2(c - 3d) + 5(2d - c)$ $-3c + 4d$

11. $-3 + \frac{1}{2}(6 - 10x) + 5x$ 0

12. $\dfrac{1 - m}{-1}$ $m - 1$

Historical Note / *Word Problems*

Consider the following problems.

1. A number added to $\frac{1}{7}$ of the number is 19. What is the number? $16\frac{5}{8}$

2. A dog chasing a rabbit, which has a head start of 150 feet, jumps 9 feet every time the rabbit jumps 7 feet. In how many jumps will the dog catch up with the rabbit? 75 jumps

3. If B gives A 7 denars, then A will have 5 times as much money as B. If instead A gives B 5 denars, then B will have 7 times as much as A. How much money has each? A: $7\frac{2}{17}$ denars; B: $9\frac{14}{17}$ denars

Students of mathematics have been solving problems similar to these for thousands of years. Problem 1 above appears in the Rhind Papyrus, written about 1650 B.C.; Problem 2 appears in a Latin problem collection compiled about A.D. 775; and Problem 3 is from a book by the Italian mathematician Fibonacci dating from A.D. 1202.

The mathematical symbols we use to represent quantities and relationships in word problems have their history too. For example, the plus sign (+) was used for addition as early as 1514; the equals sign (=) we know today originated with Robert Recorde in 1557; and the dot (·) was adopted as a symbol for multiplication by Leibniz in the seventeenth century. With these symbols you can write Problem 1 as

$$\text{Number} + \frac{1}{7} \cdot \text{Number} = 19.$$

1-8 Words into Symbols

Objective To translate word phrases into algebraic expressions and word sentences into equations.

To use algebra as a problem-solving tool, you often must translate word phrases into algebraic expressions.

Example 1 Represent each word phrase by an algebraic expression. Use n for the variable.

a. A number decreased by 2

b. Five more than three times a number

c. The difference between a number and its square

d. The sum of twice a number and 6

e. Twice the sum of a number and 6

Solution **a.** $n - 2$ **b.** $3n + 5$ **c.** $n - n^2$

d. $2n + 6$ **e.** $2(n + 6)$

Notice that the answer to part (c) of Example 1 is $n - n^2$ and *not* $n^2 - n$. In this book, when we say "the difference between x and y," we mean $x - y$. Similarly, "the quotient of x and y" means $\frac{x}{y}$, or $x \div y$.

Example 2 Ann is biking at r mi/h. Use the variable r to represent each word phrase by an algebraic expression.

a. Ann's speed if she bikes 5 mi/h slower

b. Ann's speed if she bikes 3 mi/h faster

c. The average of Ann's and Juan's speeds if Juan bikes at 10 mi/h

Solution **a.** $r - 5$ **b.** $r + 3$ **c.** $\frac{r + 10}{2}$

In certain applications, like the one in Example 3, facts and formulas from geometry are needed.

Teaching Suggestions, p. T82

Suggested Extensions, p. T83

Warm-Up Exercises

List the following sets of numbers.

1. Three consecutive integers, starting with 4. 4, 5, 6
2. Three consecutive odd integers, starting with 7. 7, 9, 11
3. The next consecutive even integer after 8. 10
4. The next consecutive odd integer after 11. 13

Motivating the Lesson

Ask if anyone speaks or understands a language other than English. Then ask those students to translate "2 more than 5" into that language. Point out that algebra has a language of its own and that in this lesson students will translate word phrases into algebraic expressions.

Chalkboard Examples

Express answers in simplest form.

1. a. 7 less than a number $n - 7$
   b. Three times a number and 8 $3n + 8$
2. Aline left Vista and drove for t hours at 40 mi/h toward Centerville. If Vista and Centerville are 200 mi apart, how many more miles must Aline drive?
   $200 - 40t$ (miles)
3. Find the measure of the third angle of a triangle if one angle has a measure $x°$ and the measure of the second angle is 15° less than half the measure of the first angle.
   $180 - x - \left(\frac{1}{2}x - 15\right) =$
   $180 - x - \frac{1}{2}x + 15 =$
   $\left(195 - \frac{3}{2}x\right)^{\circ}$

Use the formula $d = rt$ to write an equation for the following problem.

4. A car leaves town traveling at 40 mi/h. Two hours later, a second car leaves the same town, on the same road, traveling at 60 mi/h. In how many hours will the second car pass the first car?
   $40(t + 2) = 60t$
   $t = 4$
5. What is the sum of three consecutive even integers if m is the middle integer?
   $(m - 2) + m + (m + 2)$, or $3m$

Example 3 The base of an isosceles triangle has length b cm and each base angle measures $a°$.

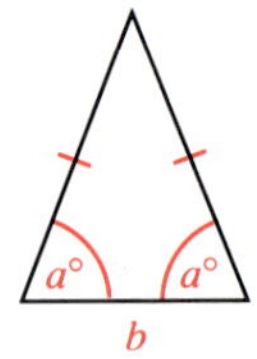

a. Find the measure of the vertex angle in terms of a.

b. If the perimeter is 120 cm, find the length of one of the legs in terms of b.

Solution a. The sum of the measures of the angles of a triangle is 180°.

$\therefore$ the measure of the vertex angle is $180° - (a + a)°$, or $(180 - 2a)°$. ***Answer***

b. The legs of an isosceles triangle are equal in length. Call this length s. Then,

$$s + s + b = 120$$
$$2s + b = 120$$
$$s = \frac{1}{2}(120 - b)$$

$\therefore$ the length of each leg is $\frac{1}{2}(120 - b)$ cm. ***Answer***

Formulas from science and technology are used in many applications. For example, to describe **uniform motion,** that is, motion at a constant speed, you use the formula

$$\text{distance} = \text{rate} \times \text{time, or } d = rt.$$

Example 4 A helicopter left Midcity airport at noon and flew east at 110 km/h. One hour later a light plane left Midcity flying west at 320 km/h. How far apart were the aircraft x hours after noon? Express your answer in terms of x.

Solution The table and the diagram show the given information. Note that the time for the plane is represented by $x - 1$ because the plane left one hour *later* than the helicopter.

	Rate r (km/h)	× Time t (h)	= Distance d (km)
Helicopter	110	x	$110x$
Plane	320	$x - 1$	$320(x - 1)$

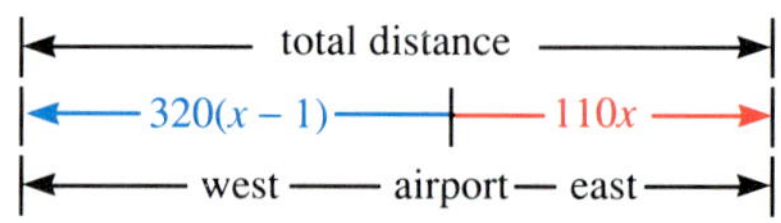

total distance $= 320(x - 1) + 110x = 430x - 320$

$\therefore$ the aircraft are $(430x - 320)$ km apart. ***Answer***

Use the chart below to review the meaning of consecutive numbers.

Consecutive Numbers

Integers	$\ldots, -3, -2, -1, 0, 1, 2, 3, \ldots$	$n-1, n, n+1, n+2$ are four consecutive integers if n is an integer.
Even Integers (multiples of 2)	$\ldots, -6, -4, -2, 0, 2, 4, 6, \ldots$	$n-2, n, n+2, n+4$ are four consecutive even integers if n is even
Odd Integers	$\ldots, -5, -3, -1, 1, 3, 5, \ldots$	and four consecutive odd integers if n is odd.

Example 5 What is the sum of five consecutive odd integers if:

a. the middle one is m? **b.** the next to largest is x?

Solution **a.** $(m-4)+(m-2)+m+(m+2)+(m+4)$, or $5m$

b. $(x-6)+(x-4)+(x-2)+x+(x+2)$, or $5x-10$

The next example shows how you can choose a variable to represent an unknown number and then write an equation to describe a given situation.

Example 6 A state legislature has 45 people. The number of men is six less than twice the number of women.

a. Choose a variable to represent the number of women.
b. Write an expression for the number of men in terms of that variable.
c. Write an equation that describes the situation.

Solution **a.** Let w = the number of women.

b. Then $2w - 6$ = the number of men.

c. The sum of the number of women and the number of men is 45.
$\therefore w + (2w - 6) = 45$, or $3w - 6 = 45$ ***Answer***

Oral Exercises

Represent each word phrase by an algebraic expression.

1. Five more than a number $x + 5$
2. Ten less than a number $x - 10$
3. One less than twice a number $2x - 1$
4. The difference between a number and six $x - 6$
5. Seven more than half a number $\frac{1}{2}x + 7$
6. The sum of a number and its reciprocal $x + \frac{1}{x}$
7. One more than the square of a number $x^2 + 1$
8. The square of one more than a number $(x + 1)^2$

Check for Understanding

Here is a suggested use of the Oral Exercises to check students' understanding as you teach the lesson.
Oral Exs. 1–10: use after Example 2.
Oral Exs. 11–20: use after Example 6.

Common Errors

A troublesome phrase for some students is one such as "Eight less than a number." Many students write $8 - x$ or $8 < x$ rather than $x - 8$. Point out that to have eight less than a number we first must have a number to subtract 8 from. Stress that $8 < x$ means "eight is less than a number."

Problem Solving Strategies

Some students may find it helpful to use the strategy of *drawing a diagram,* or that of *making a chart,* to write an algebraic expression or equation for a given situation. You may wish to point out that many of the situations presented in the exercises are standard types of problems that students will encounter throughout their study of algebra. Stress that being able to *recognize standard types of problems* is another strategy that they will find useful.

Guided Practice

Give answers in simplest form.

1. The perimeter of a rectangle is 8.4 cm. If the length is x cm, find the width.
$\frac{1}{2}(8.4 - 2x) = 4.2 - x$(cm)

2. The measure of $\angle 1$ is $y°$. Find the sum of the measures of $\angle 1$, a complement of $\angle 1$, and a supplement of $\angle 1$.
$y + (90 - y) + (180 - y) = (270 - y)°$

3. Patrick's father is f years old. Find Patrick's age next year if this year his age is 3 years more than half his father's age.
$\left(\frac{1}{2}f + 3\right) + 1$
$= \frac{1}{2}f + 4$ (years old)

4. How many hours does it take to produce 1000 pencils if t pencils are produced per minute?
t pencils per minute $= 60t$ pencils per hour
$\frac{60t \text{ pencils}}{1 \text{ hour}} = \frac{1000 \text{ pencils}}{x \text{ hours}}$
$x = \frac{1000}{60t} = \frac{50}{3t}$ (hours)

5. The Yees drove k kilometers at 60 km/h. On the return trip the weather improved and their average speed was 80 km/h. Find the total traveling time.
$\frac{k}{60} + \frac{k}{80} = \frac{4k + 3k}{240} =$
$\frac{7k}{240}$ (hours)

9. The reciprocal of two more than a number $\frac{1}{x+2}$
10. Two more than the reciprocal of a number $\frac{1}{x} + 2$

In Exercises 11–20, express each answer in simplest form in terms of the given variable.

11. What is the area of a rectangular desk top that is w cm wide and twice as long as it is wide? $2w^2$ cm²
12. How long is a rug that is w meters wide and covers an area of 20 m²? $\frac{20}{w}$ m
13. How far will a car traveling at 90 km/h go in:
    a. h hours? $90h$ km **b.** $h + 2$ hours? $90(h + 2)$ km
14. How long will it take a car to go 200 km traveling at:
    a. v km/h? $\frac{200}{v}$ h **b.** $v - 5$ km/h? $\frac{200}{v-5}$ h
15. The vertex angle of an isosceles triangle measures $v°$. What is the measure of each base angle? $\frac{1}{2}(180 - v)°$
16. What is the sum of the measures of a supplement and a complement of an angle with measure $x°$? $(270 - 2x)°$
17. What is the area of an isosceles right triangle if one of its legs is r cm long? $\frac{1}{2}r^2$ cm²
18. Yoneko is y inches tall. Her height and the heights of her two taller sisters are consecutive odd numbers. What is the average of their heights? $y + 2$ in.
19. What is the cost *in cents* of one baseball if baseballs are priced at d dollars per dozen? $\frac{25d}{3}$ cents
20. What is the cost *in dollars* of 20 pounds of potatoes priced at c cents per pound? $\frac{c}{5}$ dollars

Written Exercises

Express each answer in simplest form in terms of the given variable.

A

1. A rectangular garden that is w ft wide is enclosed by 120 ft of fencing. How long is the garden? $(60 - w)$ ft
2. The perimeter of an isosceles triangle is 300 cm, and its base is b cm long. How long is each leg? $\frac{300 - b}{2}$ cm
3. In a basketball game, one team's score is two points less than half the other team's score, which is x. What is the difference in the scores? $\frac{1}{2}x + 2$
4. Dan has y cassette tapes. If he had 15 more tapes, he would have half the number his brother has. How many tapes does his brother have? $2y + 30$
5. The length and the width of a rectangle are consecutive even integers, and the length is l cm. Find **(a)** the area and **(b)** the perimeter of the rectangle.
a. $l(l - 2)$ cm² b. $(4l - 4)$ cm

6. The length, width, and height of a rectangular box are consecutive integers, and the largest dimension is k cm. Find the volume V of the box. (*Hint:* $V = lwh$.) $k(k-1)(k-2)$ cm^3

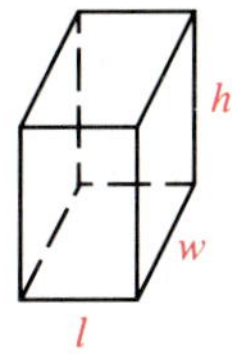

7. Two jets leave an airport at noon, one flying north at r mi/h, and the other flying south at twice that speed. After 3 h, how far apart are the planes? $9r$ mi
8. A bus traveled for 2 h at r mi/h, then decreased the speed by 10 mi/h and traveled for 1 more hour. How far did the bus go? $(3r - 10)$ mi
9. One angle of a quadrilateral has measure $a°$. Find the average of the measures of the other three angles. (*Hint:* The sum of the measures of the angles of a quadrilateral is 360°.) $\left(120 - \frac{a}{3}\right)°$
10. An angle has measure $x°$. Find the average of the measures of a complement and a supplement of the angle. $(135 - x)°$

In Exercises 11 and 12, use the fact that the volume V of a pyramid is given by $V = \frac{1}{3}Bh$, where B is the area of the base and h is the height.

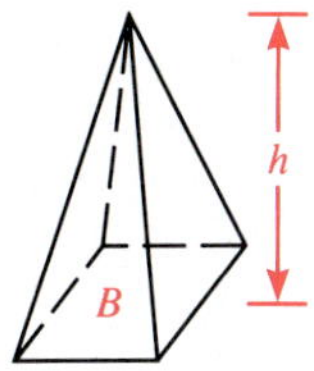

11. A pyramid has height x cm and a *square base* whose edges are 3 cm less than twice the height. Find the volume. $\frac{1}{3}x(2x-3)^2$ cm^3
12. A pyramid has a *rectangular base*. Find the volume if the length and width of the base and the height are three consecutive odd integers and x is the largest integer. $\frac{1}{3}x(x-2)(x-4)$
13. The Drama Club sold t students' tickets at \$1.50 each and 100 fewer adults' tickets at \$2.50 each. How much money did the club collect? $(4t - 250)$ dollars
14. Jorge bought s 40-cent stamps and three times as many 25-cent stamps. How many dollars did he spend? $1.15s$ dollars
15. Jessica's bank contains 18 quarters and dimes, of which q are quarters. Find the total value of the coins in dollars. $(0.15q + 1.80)$ dollars
16. The length of a rectangular field is 45 ft greater than its width, w ft. How much fencing is needed to enclose the field and divide it into two parts as shown? $(5w + 90)$ ft

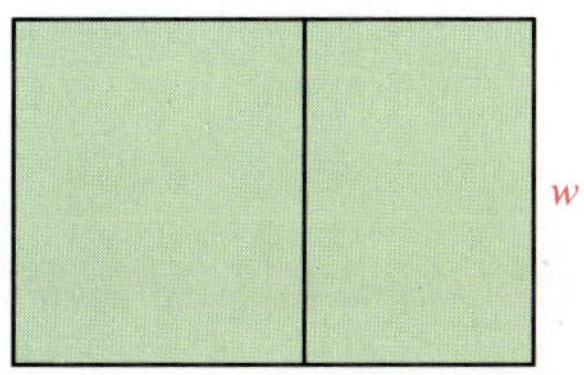

In Exercises 17–29, choose a variable to represent an unknown number, and then write an equation to describe the given situation.

17. A quadrilateral has perimeter 60 cm, and the lengths of its sides (in centimeters) are consecutive odd numbers. $x + (x+2) + (x+4) + (x+6) = 60$

6. Find the difference between the largest of six consecutive odd integers and the sum of the three smallest if the smallest is n.
The numbers are n, $n+2$, $n+4$, $n+6$, $n+8$, and $n+10$. $(n+10) - [n + (n+2) + n + 4)] = (n+10) - (3n+6) = n + 10 - 3n - 6 = -2n + 4$

Summarizing the Lesson

In this lesson students have taken the first step in learning to solve word problems. The translating of word phrases and sentences into algebraic expressions and equations is an important part of problem solving.

Suggested Assignments

Average Algebra
Day 1: 46/1–19 odd
S 42/49–53
Day 2: 47/16, 18, 20–27

Average Alg. and Trig.
46/3–18 mult. of 3, 19–27 odd

Extended Algebra
46/3–18 mult. of 3, 19–29 odd

Extended Alg. and Trig
46/3–18 mult. of 3, 19–29 odd

Supplementary Materials

Study Guide	pp. 15–16
Practice Master	5
Resource Book	p. 102

18. The numerator of a fraction is 6 less than its denominator, and the value of the fraction is $\frac{3}{5}$. $\frac{d-6}{d} = \frac{3}{5}$

B **19.** Adam purchased a shirt at regular price. Later, when the shirts were on sale, he purchased two more at \$2 off the regular price. He spent a total of \$41 for the three shirts. $x + 2(x - 2) = 41$

20. Lupe swims two fewer laps than Mary. If both added seven laps to their daily swims, the sum of their laps would be three times as many as Mary now swims. $[(m - 2) + 7] + (m + 7) = 3m$

21. Max has twice as much money as Katy, who has \$12 more than Greg. All three together have \$124. $g + (g + 12) + 2(g + 12) = 124$

22. Paula's purse contains twice as many dimes as quarters and three fewer nickels than dimes. The total of these coins is \$3.15. $0.25q + 0.10(2q) + 0.05(2q - 3) = 3.15$

23. A car and a truck left Elton at noon and traveled in opposite directions. The truck's speed is two thirds of the car's speed, and the vehicles are 140 mi apart at 2 P.M. $2c + 2\left(\frac{2}{3}c\right) = 140$

24. Tom and Tina set out on their bikes at noon and travel toward each other, meeting at 2:30 P.M. Tina's speed is 4 mi/h faster than Tom's speed, and their starting points are 50 mi apart. $2.5r + 2.5(r + 4) = 50$

25. In quadrilateral $ABCD$, the measure of $\angle A$ exceeds the measure of $\angle B$ by 20°. Also, the measure of $\angle D$ is twice the measure of $\angle B$ and half the measure of $\angle C$. $(b + 20) + b + 2b + 4b = 360$

26. In an equilateral triangle, the length of one side is 20 cm more than one third the length of another. $s = \frac{1}{3}s + 20$

27. A grocer mixed cashews and almonds to produce 20 kg of mixed nuts worth \$7.80/kg. Cashews are worth \$7/kg, and almonds are worth \$9/kg. $7c + 9(20 - c) = 7.80(20)$

28. When the Kims went to the ball game, they bought two adults' tickets and three children's tickets. A child's ticket cost \$1.50 less than an adult's, and the family's average price per ticket was \$3.35. $\frac{2a + 3(a - 1.5)}{5} = 3.35$

C **29.** Kevin drove 320 mi to a mountain resort. His return trip took 20 min longer because his speed returning was 4 mi/h slower than his speed going. $\frac{320}{r} + \frac{1}{3} = \frac{320}{r - 4}$

Challenge

Every positive integer can be written as a sum of cubes of positive integers. For example, 10 can be written as a sum of three cubes, and 44 can be written as a sum of four cubes: $10 = 2^3 + 1^3 + 1^3$ and $44 = 3^3 + 2^3 + 2^3 + 1^3$. Find a positive integer that can be written as a sum of nine cubes, but that cannot be written as the sum of any smaller number of cubes.
Answers may vary. $23 = 2^3 + 2^3 + 1^3 + 1^3 + 1^3 + 1^3 + 1^3 + 1^3 + 1^3$

1-9 Problem Solving with Equations

Objective To solve word problems by using an equation in one variable.

''Word problems'' describe relationships among numbers. If you can translate the relationship described in a problem into an equation, then you can solve the problem by solving the equation. Example 1 illustrates a step-by-step method for solving word problems.

Example 1 Tickets to the Civic Center Auditorium for a rock concert were $19 for seats near the front and $14 for rear seats. There were 525 more rear seats sold than front seats, and sales for all tickets totaled $31,770. How many of each kind of ticket were sold?

Solution

Step 1 The problem asks for the number of each type of ticket sold.

Step 2 Let x = the number of front-seat tickets sold.
Then $x + 525$ = the number of rear-seat tickets sold.

	Price ×	Number =	Sales
Front	19	x	$19x$
Rear	14	$x + 525$	$14(x + 525)$
	Total Ticket Sales		31,770

Step 3 Front-seat Sales + Rear-seat Sales = Total Sales

$$19x \quad + \quad 14(x + 525) \quad = \quad 31{,}700$$

Step 4

$$19x + 14x + 7350 = 31{,}770$$
$$33x + 7350 = 31{,}770$$
$$33x = 24{,}420$$
$$x = 740 \quad \text{(Front)}$$
$$x + 525 = 1265 \quad \text{(Rear)}$$

Step 5 Is the number of rear-seat tickets 525 more than the number of front-seat tickets?

$$1265 \stackrel{?}{=} 740 + 525$$
$$1265 = 1265 \quad \checkmark$$

Is the sum of the ticket sales of front seats and rear seats $31,770?

$$740(19) + 1265(14) \stackrel{?}{=} 31{,}770$$
$$14{,}060 + 17{,}710 \stackrel{?}{=} 31{,}770$$
$$31{,}770 = 31{,}770 \quad \checkmark$$

∴ 740 front-seat tickets and 1265 rear-seat tickets were sold. ***Answer***

Teaching Suggestions, p. T83

Suggested Extensions, p. T83

Warm-Up Exercises

Students should review the "Facts and Formulas from Geometry," Table 7 on page 830. Point out that many of these terms and formulas are used in problem-solving exercises.

Motivating the Lesson

Point out to students that the translation of words into equations is one of the major steps in solving a word problem. Tell them that they already have the basic tools that they need: the vocabulary and the algebraic symbols. Tell students not to be discouraged: even a faulty first step is a good first step. The topic of today's lesson is getting started.

Common Errors

Stress the importance of Step 2 in the problem-solving plan. Point out the advantage of making a table to gather and sort the information in the problem accurately. Guide students to realize that if they do not organize the data carefully in Step 2, the equation that they write in Step 3 may be incorrect.

Chalkboard Examples

Solve.

1. Each of the base angles of an isosceles triangle has a measure that is 30° less than the measure of the vertex angle. Find the measure of each angle.

   S.1 What does the problem ask for?

   The measures of the three angles.

   What is true of the base angles of an isosceles triangle?

   Their measures are equal.

   S.2 Let $x°$ = measure of the vertex angle. Then $(x - 30)°$ = measure of each base angle.

   S.3 What do you know about the measures of the angles of a triangle?

   Their sum is 180°.

   Write an equation.

   $x + (x - 30) + (x - 30) = 180$

   S.4 Solve to find the value of x.

   $3x - 60 = 180$;

   $3x = 240$; $x = 80$

   Measure of the vertex angle = 80°

   Measure of each base angle = 80 − 30, or 50°

   S.5 Check: Do the angles have the required sum?

   $80 + 50 + 50 = 180$; yes

The step-by-step method of Example 1 is part of this general strategy.

Plan for Solving a Word Problem

Step 1 Read the problem carefully a few times. Decide what numbers are asked for and what information is given. Making a sketch may be helpful.

Step 2 Choose a variable and use it with the given facts to represent the number(s) described in the problem. Labeling your sketch or arranging the given information in a chart may help.

Step 3 Reread the problem. Then write an equation that represents relationships among the numbers in the problem.

Step 4 Solve the equation and find the required numbers.

Step 5 Check your results with the original statement of the problem. Give the answer.

Example 2 Melissa, a city planner, had two pieces of wire of equal length. She shaped one piece into a square to represent a new building site and the other into an isosceles triangle to represent a nearby park. The base of the isosceles triangle is 4 cm shorter than a side of the square, and each leg is 9 cm longer than a side of the square. How long was each piece of wire?

Solution

Step 1 The problem asks for the length of each piece of wire. This length is the perimeter of the square and the perimeter of the triangle. Sketch each figure.

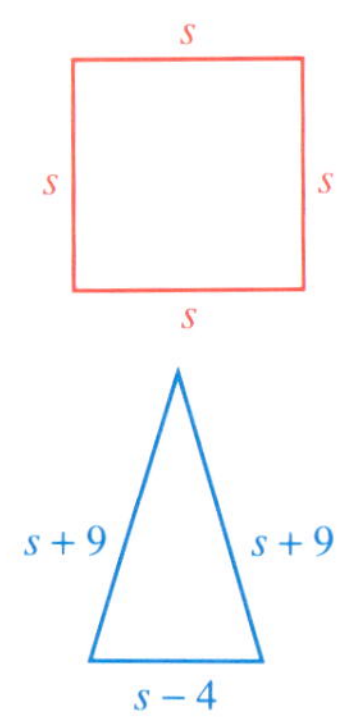

Step 2 Let s = the length of a side of the square in centimeters. (Note that the variable does not have to represent a number that is asked for.) Then the base of the triangle has length $s - 4$, and each leg has length $s + 9$.

Perimeter of square $= 4s$

Perimeter of triangle $= (s - 4) + (s + 9) + (s + 9)$

Step 3 The pieces of wire were equally long.

Perimeter of square = Perimeter of triangle

$$4s = (s - 4) + (s + 9) + (s + 9)$$

Step 4

$$4s = 3s + 14$$
$$s = 14$$

Perimeter of square $= 4s = 56$ (cm)

Step 5 ∴ each piece of wire was 56 cm long. (The check is left for you.) ***Answer***

Example 3 Find the length of each piece of wire in Example 2 if the base of the isosceles triangle is 6 cm shorter than a side of the square, and each leg is 3 cm longer than a side of the square.

Solution Modify Step 2 of Example 2 to obtain the following equation in Step 3:

Step 3 Perimeter of square = Perimeter of triangle

$$4s = (s - 6) + (s + 3) + (s + 3)$$

Step 4

$$4s = (s - 6) + (s + 3) + (s + 3)$$
$$4s = 3s$$
$$s = 0$$

Step 5 $\therefore$ the side of the square is 0.

But there is *no* square with sides 0 units long. Therefore, the problem has no solution. ***Answer***

As Example 3 shows, *not every word problem has a solution.* Some problems without solutions have contradictory facts like the one in Example 3; others do not give enough facts to provide a solution. Of course, *sometimes a problem will give you more information than you need to find the solution.*

Example 4 At noon a cargo plane leaves McHare Airport and heads east at 180 mi/h. Its destination is Jamesville, 500 mi away. At 1:00 P.M. a jet takes off from McHare and flies east after the cargo plane at 450 mi/h. At what time will the jet overtake the cargo plane?

Solution

Step 1 The problem asks for the time at which the jet overtakes the cargo plane.

Step 2 Let t = the number of hours *after noon* when the jet overtakes the cargo plane.

Make a chart and a sketch.

(Solution continues on the next page.)

2. Adults' tickets for the art museum cost \$1.50 more than children's tickets. One day 200 adults' tickets and 160 children's tickets were sold at the museum. The income from the adults' tickets was \$30 less than twice the income from the children's tickets. How much does each type of ticket cost?

S.1 The problem asks for the cost of an adult's ticket and the cost of a child's ticket.

S.2 Let c = cost in dollars of a child's ticket.

Then $c + 1.5$ = cost of adult's ticket. Income from children's tickets = $160c$; income from adults' tickets = $200(c + 1.5) = 200c + 300$.

S.3 Income from adults' = 2(income from children's) − 30; $200c + 300 = 2(160c) - 30$

S.4 $200c + 300 = 320c - 30$

$330 = 120c$

$c = \frac{330}{120}$

$= 2.75$

A child's ticket costs \$2.75. An adult's ticket costs \$2.75 + \$1.50, or \$4.25.

S.5 Income (children's) = 160(2.75) = \$440

income (adults') = 200(4.25) = \$850

$850 = 2(440) - 30$

$850 = 850$ ✓

Problem Solving Strategies

Point out to students that the *five-step plan* is used to solve the problem in each Example. You may also wish to review the other strategies on page T53 to be certain that students are familiar with them.

Guided Practice

Solve.

1. Judy has \$12 more than Barbara. Together they have \$58. How much money does each girl have? Judy \$35, Barbara \$23
2. A newspaper ad costs \$12.50 plus 30¢ for each word after the first twelve. Yolanda paid \$17 for her ad. How many words were in it? 27 words
3. A truck left Midville at 2:00 P.M. for Carmine, traveling at 40 mi/h. At 3:15 P.M., a car left Carmine for Midville at 55 mi/h. The truck and car passed each other exactly 2 h before the car reached Midville. How far is it from Midville to Carmine? 192.5 mi
4. The lengths of the sides of a quadrilateral are consecutive odd integers. The perimeter of the quadrilateral is 160 cm. Find the length of the longest side of the quadrilateral. 43 cm

	Rate (mi/h)	× Time = (h)	Distance (mi)
Cargo plane	180	t	$180t$
Jet	450	$t - 1$	$450(t - 1)$

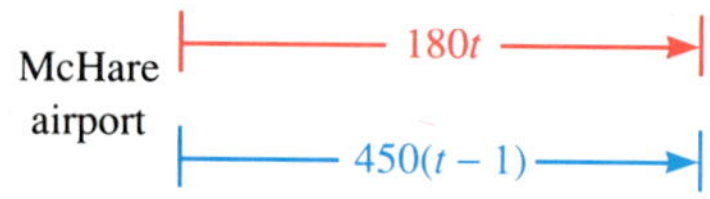

Step 3 The jet will overtake the cargo plane when:

Cargo plane's distance = Jet's distance

$$180t = 450(t - 1)$$

Step 4

$$180t = 450t - 450$$
$$450 = 270t$$
$$t = 1\frac{2}{3} \quad \text{(hours after noon)}$$

Step 5 ∴ the jet overtakes the cargo plane $1\frac{2}{3}$ h after noon, or 1:40 P.M.

The check is left for you. ***Answer***

The information concerning the cargo plane's destination 500 mi away was not needed in the solution of this example.

Problems

A

1. Amy has \$8 less than Maria. Together they have \$30. How much money does each girl have? Amy: \$11; Maria: \$19
2. Jim's weekly pay is two thirds of Alicia's. Together they earn \$600 per week. What is each person's weekly pay? Alicia: \$360; Jim: \$240
3. A music dealer ran a sale of records and tapes. Records were reduced to \$7 each and tapes to \$7.50 each. The dealer sold 60 more records than tapes for a total sale of \$2160. How many records did the dealer sell? 180
4. At the homecoming football game, the Senior Class officers sold slices of pizza for \$.75 each and hamburgers for \$1.35 each. They sold 40 more slices of pizza than hamburgers, and sales totaled \$292.50. How many slices of pizza did they sell? 165
5. The perimeter of a certain basketball court is 266 ft, and its length is 35 ft more than its width. Find the dimensions of the court. 84 ft × 49 ft
6. If one side of a square is increased by 8 cm and an adjacent side decreased by 2 cm, a rectangle is formed whose perimeter is 40 cm. Find the length of a side of the square. 7 cm
7. The measure of a supplement of an angle is 12° greater than three times the measure of a complement. Find the measure of the angle. 51°
8. The degree measures of the angles of a pentagon are consecutive even integers. Find the measure of the largest angle. (*Hint:* The sum of the measures of the angles of a pentagon is 540°.) 112°

9. At 10:30 A.M. two planes leave Houston, one flying east at 560 km/h and the other flying west at 640 km/h. At what time will they be 2100 km apart? 12:15 P.M.

10. In a walkathon to raise money for a charity, Elisa walked a certain distance at 5 mi/h and then jogged twice that distance at 8 mi/h. Her total time walking and jogging was 2 h and 15 min. How many miles long was the walkathon? $\left(\textit{Hint: } \text{time} = \frac{\text{distance}}{\text{rate}}.\right)$ 15 mi

11. A jar contains 40 coins consisting of dimes and quarters and having a total value of \$4.90. How many of each kind of coin are there? 34 dimes, 6 quarters

12. Larry has an annual return of \$213 from \$3000 invested at simple interest, some at 5% and the rest at 8%. How much is invested at each rate? (*Hint:* Interest earned = Amount invested × Rate of interest.) \$900 at 5%, \$2100 at 8%

In Problems 13–18, find the solution if possible. If there is not enough information to solve the problem or if it has no solution, say so. If extra information is given, identify it.

13. Two planes leave Wichita at noon. One plane flies east 30 mi/h faster than the other plane, which is flying west. At what time will they be 1200 mi apart? not enough information

14. A collection of 30 coins worth \$5.50 consists of nickels, dimes, and quarters. There are twice as many dimes as nickels. How many quarters are there? 18 quarters

15. John drove part of a 260 km trip at 80 km/h and the rest at 100 km/h. Find the distance he traveled at 80 km/h if his total driving time was 2 h and 30 min. no solution

16. A triangle has perimeter 29 cm. The sides have lengths, in centimeters, that are consecutive odd integers. What is the length of the longest side? no solution

B 17. Jan invested \$1200 at a certain simple interest rate and \$2200 at a rate 3% higher. Her annual earnings were \$253. Find the two interest rates if she earned \$121 more on the larger investment than on the smaller. 17. $5\frac{1}{2}$% and $8\frac{1}{2}$%; extra information

18. A school cafeteria sells milk at 25 cents per carton and salads at 45 cents each. One week the total sales for these items were \$132.50. How many salads were sold that week? No solution; not enough information

Solve.

19. At noon a train leaves Bridgton heading east at 90 mi/h to Cogsville, 450 mi away. At 12:15 P.M. a train leaves Cogsville heading west to Bridgton at 100 mi/h. At what time will they pass each other? 2:30 P.M.

20. In a bicycle race, Lionel gives Robert a 500 m advantage. Also, Lionel agrees to start 15 min after Robert. If Lionel bikes at 17 km/h and Robert at 14 km/h, how long will it take Lionel after he starts biking to overtake Robert? $1\frac{1}{3}$ h

Summarizing the Lesson

Remind students that the objective of this lesson was to solve a problem by writing and solving an equation. Point out that a five-step plan for solving a word problem was introduced. Ask volunteers to state the five steps.

Suggested Assignments

Average Algebra
Day 1: 52/*P*: 1–11
S 54/Mixed Review
Day 2: 53/*P*: 13–22
R 54/Self-Test 3

Average Alg. and Trig.
52/*P*: 2–22 even
S 54/Mixed Review
R 54/Self-Test 3

Extended Algebra
52/*P*: 2–22 even, 23
S 54/Mixed Review
R 54/Self-Test 3

Extended Alg. and Trig.
52/*P*: 2–22 even, 23
S 54/Mixed Review
R 54/Self-Test 3

Supplementary Materials

Study Guide	pp. 17–18
Practice Master	6
Test Master	2
Computer Activity	2

21. A pollution control device reduces the rate of emission of an air pollutant in a car's exhaust by 56 ppm (parts per million) per hour. When the device is installed, it will take the car 10 h to emit the same amount of pollutant that it formerly did in 3 h. What was the original rate of emission of the pollutant in the car's exhaust? 80 ppm/h

22. A grocer wants to mix peanuts and cashews to produce 20 lb of mixed nuts worth \$6.20/lb. How many pounds of each kind of nut should she use if peanuts cost \$4.80/lb and cashews cost \$8/lb? peanuts: 11.25 lb, cashews: 8.75 lb

C 23. In a two-candidate election 1401 votes were cast. If 30 voters had switched their votes from the winner to the loser, the loser would have won by 5 votes. How many votes did each candidate actually receive? winner: 728 votes, loser: 673 votes

Mixed Review Exercises

Solve.

1. $3x + 2 = -4$ {−2}
2. $4(2 - y) = y - 7$ {3}
3. $1 - 5t = t + 7$ {−1}

Evaluate each expression if $c = -4$ and $d = 6$.

4. $c^2 - d^2$ −20
5. $(c - d)^2$ 100
6. $|2c - 3d|$ 26
7. $\frac{c + d}{-2}$ −1
8. $\frac{cd}{8}$ −3
9. $\frac{d - c}{5}$ 2

Self-Test 3

Vocabulary open sentence (p. 37), solution (p. 37), root (p. 37), solve (p. 37), solution set (p. 37), equivalent equations (p. 37), empty set (p. 38), identity (p. 38), formula (p. 39), constant (p. 39)

Solve.

1. $3x - 8 = 7$ {5}
2. $4(1 - x) = 2(x - 4)$ {2} **Obj. 1-7, p. 37**
3. Solve the formula $m = \frac{1}{2}(a + b)$ for b. $b = 2m - a$
4. Using n for the variable, translate this word phrase into an algebraic expression: Twice the sum of a number and its square. $2(n + n^2)$ **Obj. 1-8, p. 43**
5. Express your answer in terms of x: What is the perimeter of a rectangle that is x cm wide and 5 cm longer than it is wide? $4x + 10$
6. Two cars, heading toward each other on a divided highway, are 250 mi apart. If one car travels 45 mi/h and the other 10 mi/h faster, in how many hours will the cars pass each other? $2\frac{1}{2}$ h **Obj. 1-9, p. 49**

Quick Quiz

Solve.

1. $4m + 6 = 14$ 2
2. $2(2x + 1) = 8x + 4$ $-\frac{1}{2}$
3. Solve for x. $y - b = \frac{b}{a}(x - a)$ $x = \frac{ay}{b}$
4. Using c for the variable, translate this word phrase into an algebraic expression: Three times the difference of a number and its cube. $3(c - c^3)$
5. Express your answer in terms of y. What is the area of a rectangle that is y cm wide and 6 cm longer than it is wide? $y(y + 6)$
6. Two motorcycles, heading towards each other on a divided highway, are 392 mi apart. If one motorcycle travels 52 mi/h and the other 8 mi/h faster, in how many hours will they pass each other? 3.5 h

Chapter Summary

1. The set of real numbers has several familiar subsets, which are shown in the chart on page 1.
2. Each point on a number line can be paired with exactly one real number, which is the coordinate of the point. The point is called the graph of the number.
3. The words and symbols used in algebra have specific mathematical definitions. (See the charts on pages 6, 7, and 9.)
4. A numerical expression can be simplified by following the rules for order of operations (page 8).
5. An algebraic expression can be simplified by using the field properties of real numbers (page 15) along with the definitions of addition, subtraction, multiplication, and division.
6. An algebraic expression can be evaluated by replacing each variable by a given value and simplifying the result.
7. An equation can be solved by applying the transformations on page 37. The solution set of an equation is the set of values of the variable that make the equation a true statement.
8. Word problems can be solved algebraically after translating the given information into an equation. Refer to the five-step problem-solving method presented on page 50.

Chapter Review

Write the letter of the correct answer.

1. On a number line, the coordinate of point A is -3, and the coordinate of point B is 7. Find the coordinate of the point halfway between A and B. **1-1**
a. 4 **b.** 3 **c.** 2 **d.** 1

2. Simplify $|-5| - |5|$.
a. -10 **b.** 0 **c.** 10 **d.** -5

3. Simplify $\frac{2 \cdot 5^2 + 1}{2 \cdot 3^2 - 1}$. **1-2**
a. 2 **b.** $\frac{13}{4}$ **c.** $\frac{101}{35}$ **d.** 3

4. Evaluate $x(y - 1)^2$ if $x = 4$ and $y = 3$.
a. 16 **b.** 12 **c.** 64 **d.** 32

Supplementary Materials

Test Masters	3, 4
Resource Book	p. 103

5. Simplify $\frac{1}{2}(2a + 1)$. **1-3**

(**a.**) $a + \frac{1}{2}$ **b.** $\frac{1}{2}a + 1$ **c.** $\frac{1}{2}a + \frac{1}{2}$ **d.** $a + 1$

6. Name the property that is illustrated by this statement:

$$2a + (3b + c) = (3b + c) + 2a.$$

a. Symmetric property of equality **b.** Associative property of addition

c. Addition property of equality (**d.**) Commutative property of addition

7. Simplify $(-14 + 8) - (4 - 11)$. **1-4**

a. -13 **b.** 13 (**c.**) 1 **d.** -1

8. Simplify $8m - 6n - m + 2n$.

a. $3mn$ (**b.**) $7m - 4n$ **c.** $4n - 7m$ **d.** $8 - 4n$

9. Simplify $3(2x - y) - 4(x - 2y)$. **1-5**

a. $2x - 11y$ **b.** $10x - 11y$ **c.** $2x - 5y$ (**d.**) $2x + 5y$

10. Simplify $(-a)^2 + a^2$.

a. $-2a^2$ **b.** $2a$ **c.** 0 (**d.**) $2a^2$

11. Simplify $6(3 - 7) \div (-2)^3 \div (-1)$. **1-6**

a. 3 **b.** 4 (**c.**) -3 **d.** -4

12. Simplify $\frac{24 - 20x}{-4}$.

(**a.**) $5x - 6$ **b.** $5x + 6$ **c.** $-5x - 6$ **d.** $-5x + 6$

13. Solve $2(3 - x) = 3x + 1$. **1-7**

a. $\frac{5}{4}$ (**b.**) 1 **c.** -1 **d.** 5

14. Solve $A = 2\pi r(r + h)$ for h.

a. $\frac{A - r}{2\pi r}$ **b.** $\frac{2\pi r}{A} - r$ **c.** $\frac{2\pi r}{A - r}$ (**d.**) $\frac{A}{2\pi r} - r$

15. The hourly wages of three workers are consecutive even integers. If the highest-paid worker's hourly wage is x, what is the sum of all three workers' hourly wages? **1-8**

a. $3x$ **b.** $3x + 6$ (**c.**) $3x - 6$ **d.** $3x - 3$

16. The perimeter of a rectangle is 24 cm. If the length is y cm, what is the width?

(**a.**) $12 - y$ **b.** $24 - y$ **c.** $24 - \frac{y}{2}$ **d.** $12 - \frac{y}{2}$

17. The school book store sold 8 more pencils than pens one day. The cost of a pencil is \$.05, and the cost of a pen is \$.20. If the day's sales of pens and pencils totaled \$8.90, how many pencils were sold? **1-9**

(**a.**) 42 **b.** 34 **c.** 26 **d.** 50

18. A woman drove part of a 185 mi trip at 50 mi/h and the rest at 55 mi/h. Find the distance she traveled at 50 mi/h if her total driving time was 3 h and 30 min.

a. 110 mi **b.** 100 mi (**c.**) 75 mi **d.** 150 mi

Chapter Test

Alternate Test p. T13

1. On a number line, point A has coordinate -4, and point B has coordinate 5. Find the coordinate of the point two thirds of the way from A to B. 2 1-1

2. List 0.6, -0.8, 1.4, -1, and -0.2 in order from least to greatest. $-1, -0.8, -0.2, 0.6, 1.4$

3. Simplify $5 \cdot 3^2 - 2(4 + 6)$. 25 1-2

4. Evaluate $\frac{(x + y)^2}{x^2 + y^2}$ if $x = 2$ and $y = 4$. $\frac{9}{5}$

5. Name the property that justifies each lettered step. 1-3

$$\frac{1}{2}(2x + 2) = \frac{1}{2}(2x) + \frac{1}{2} \cdot 2$$ **a.** _?_ Dist. prop.

$$= \left(\frac{1}{2} \cdot 2\right)x + \frac{1}{2} \cdot 2$$ **b.** _?_ Assoc. prop. of mult.

$$= 1x + 1$$ **c.** _?_ Prop. of reciprocals

$$= x + 1$$ **d.** _?_ Ident. prop. of mult.

Simplify.

6. $|8 - 13| - |-7 - (-3) + 1|$ 2

7. $8m - 4 - 9m + 7$ $3 - m$ 1-4

8. $-(-1)^3(-2)^2$ 4

9. $5(3x - 2y) - 4(6y - x)$ $19x - 34y$ 1-5

Evaluate if $x = -3$.

10. $x(4 - x)(x + 1)$ 42

11. $7 - 2x - x^2$ 4

Simplify.

12. $\left|\frac{(-6)^2 - 3(-4)}{-8 + 2}\right|$ 8

13. $\frac{9 - 12x^2}{-3}$ $4x^2 - 3$ 1-6

Solve.

14. $\frac{3}{2}x - 7 = 2x + 3$ $\{-20\}$

15. $6(5x - 4) = 7(4x + 5) - 19$ $\{20\}$ 1-7

Express your answer in terms of the given variable.

16. A car's gas tank held x gallons of gas before a trip. The trip consumed three quarters of the gas in the tank, but 10 gallons of gas were added after the trip. How much gas is now in the tank? $\left(\frac{1}{4}x + 10\right)$ gallons 1-8

Solve.

17. A man invested \$6000, part of it at 5% simple interest and the rest at 7% simple interest. If his annual interest income is \$372, how much did he invest at each rate? \$2400 at 5%, \$3600 at 7% 1-9

2 Inequalities and Proof

Objectives

2-1 To solve simple inequalities in one variable.

2-2 To solve conjunctions and disjunctions.

2-3 To solve word problems by using inequalities in one variable.

2-4 To solve open sentences involving absolute value.

2-5 To use number lines to obtain quick solutions to certain equations and inequalities involving absolute value.

2-6 To use axioms, definitions, and theorems to prove some properties of real numbers.

2-7 To prove theorems about inequalities and absolute value.

Assignment Guide

See p. T58 for Key to the format of the Assignment Guide

Day	Average Algebra	Average Algebra and Trigonometry	Extended Algebra	Extended Algebra and Trigonometry
1	**2-1** 62/1–26 **S** 63/*Mixed Review*	**2-1** 62/1–27 **S** 63/*Mixed Review*	*Administer Chapter 1 Test* **2-1** *Read 2-1* 62/1–31 odd **S** 63/*Mixed Review*	*Administer Chapter 1 Test* **2-1** *Read 2-1* 62/1–31 odd **S** 63/*Mixed Review*
2	**2-2** 67/1–33 odd **S** 63/27–32	**2-2** 67/2–34 even **S** 63/28–32	**2-2** 67/2–34 even **S** 63/32, 33	**2-2** 67/5–33 odd **S** 63/32, 33 **2-3** 71/*P*: 1, 4, 7, 11, 13, 15 **S** 72/*Mixed Review* **R** 72/*Self-Test 1*
3	**2-3** 71/*P*: 2–14 even **S** 72/*Mixed Review* **R** 72/*Self-Test 1*	**2-3** 71/*P*: 1, 3, 6, 8, 11–14 **S** 72/*Mixed Review* **R** 72/*Self-Test 1*	**2-3** 71/*P*: 3, 6, 8, 11–16 **S** 72/*Mixed Review* **R** 72/*Self-Test 1*	**2-4** 75/12–34 even **2-5** 78/3–30 mult. of 3 **S** 79/*Mixed Review*
4	**2-4** 75/1–25 odd **S** 71/*P*: 11, 13, 15	**2-4** 75/1–27 odd	**2-4** 75/1–33 odd	**2-6** 85/3–10, 13, 15, 18, 22 **R** 79/*Self-Test 2*
5	**2-5** 78/2–24 even **S** 79/*Mixed Review*	**2-5** 78/2–24 even **S** 79/*Mixed Review*	**2-5** 78/2–26 even **S** 79/*Mixed Review*	**2-7** 90/1, 3, 4, 6-8, 10, 14, 15 **S** 91/*Mixed Review* **R** 92/*Self-Test 3*
6	**2-6** 85/1–10, 15 **R** 79/*Self-Test 2*	**2-6** 85/1–15 odd **S** 75/20–30 even **R** 79/*Self-Test 2*	**2-6** 85/2–18 even **S** 79/31, 33 **R** 79/*Self-Test 2*	*Prepare for Chapter Test* **R** 79/*Self-Test 2* **R** 92/*Self-Test 3* **R** 93/*Chapter Review*
7	**2-6** 87/17–22 **S** 79/13–27 odd	**2-7** 90/1, 3, 4, 7, 8, 10 **S** 91/*Mixed Review* **R** 92/*Self-Test 3*	**2-7** 90/1, 3, 4, 7, 8 10–12 **S** 91/*Mixed Review* **R** 92/*Self-Test 3*	*Administer Chapter 2 Test* **3-1** *Read 3-1* 104/1, 3, 5, 21–35 odd 105/*P*: 2, 7–11, 13 **S** 106/*Mixed Review*

Assignment Guide *(continued)*

Day	Average Algebra	Average Algebra and Trigonometry	Extended Algebra	Extended Algebra and Trigonometry
8	**2-7** 90/1, 3, 4, 6–8, 10 **S** 91/*Mixed Review* **R** 92/*Self-Test 3*	*Prepare for Chapter Test* **R** 93/*Chapter Review*	*Prepare for Chapter Test* **R** 93/*Chapter Review*	
9	*Prepare for Chapter Test* **R** 93/*Chapter Review*	*Administer Chapter 2 Test* **S** 98/*P*: 1–14	*Administer Chapter 2 Test* **3-1** *Read 3-1* 104/1, 3, 5, 21–35 odd 105/*P*: 2, 7–11 **S** 106/*Mixed Review* **S** 98/*P*: 2–14 even	
10	*Administer Chapter 2 Test* **S** 98/*P*: 1–14			

Supplementary Materials Guide

For Use with Lesson	Practice Masters	Tests	Study Guide (Reteaching)	Resource Book: Tests	Resource Book: Practice Exercises	Resource Book: Prob. Solving (PS) Applications (A) Enrichment (E) Technology (T)
2-1			pp. 19–20			p. 205 (A)
2-2	Sheet 7		pp. 21–22			
2-3	Sheet 8	Test 5	pp. 23–24		p. 104	pp. 192–193 (PS)
2-4			pp. 25–26			
2-5	Sheet 9		pp. 27–28		p. 105	pp. 250–256 (T)
2-6			pp. 29–30			
2-7	Sheet 10	Test 6	pp. 31–32		p. 106	
Chapter 2		Tests 7, 8		pp. 8–11	p. 107	p. 221 (E)

Overhead Visuals

For Use with Lessons	Visual	Title
2-1, 2-2, 2-3, 2-4, 2-5	A	Multi-Use Packet 1

Software

Software	Computer Activities	Test Generator
	Activities 3, 4	147 test items
For Use with Lessons	Chapter 2, 2-5	all lessons

Strategies for Teaching

Communication

Today it is recognized that strong reading skills are essential for the mastery of algebraic techniques. Students need to gain experience in explaining concepts, restating definitions, justifying conclusions, and thinking aloud through the steps in solving problems. Certain exercises that ask students to "explain" or to "give a convincing argument" can be used in this context. Insist that students use the correct word for a concept or a procedure. Using words like *coefficient, term, factor, root, evaluate,* and *function* will keep students aware of important and subtle differences in meaning. As a writing exercise to help students with logical progression, have them write a set of step-by-step directions for a procedure familiar to them (not necessarily mathematics) but not well known to everyone. Other students may then evaluate these directions. Exercises such as these may help students when they attempt to write proofs.

See the Exploration on page 833 for an activity in which students explore statements about inequalities.

2-3 Problem Solving Using Inequalities

Be sure to explain the mathematical meaning of expressions such as "at most," "at least," "smallest possible value," and "maximum number." Have the students translate several statements that include these expressions before they start on the assignment.

All work with word problems calls on students' reading skills as well as on their mathematical skills. The problem situations in this lesson and in many other lessons come from several different fields of activity, each of which has its own vocabulary. You will want to make sure that your students understand the mathematical words associated with inequalities.

Before assigning homework, you may also want to ask students to look quickly through the problems for other words that may be unfamiliar to them.

In many classes the students are not required to give the solution to a problem in as much detail as shown in Examples 1 and 2. Have the students "write up" at least two different types of problems showing every step involved.

2-5 Solving Absolute Value Sentences Graphically

This lesson provides an excellent opportunity for you to point out the close relationship between diagrams and text and to emphasize to students that they should read the diagrams along with the text and examples.

In Oral Exercises 11–16, point out the different translations of the symbols. For example, the solution in the Sample could have used the expression "not less than 3."

2-6 Theorems and Proofs

All of the properties of real numbers studied in Chapter 1 should be reviewed at this time. Point out that in Example 5 of Lesson 1-5, a proof was given for a property of real numbers. Since proof is an important part of the study of mathematics, every effort should be made to help the students understand the concept of mathematical proof. Many students find it difficult to generate the statements in a proof while they can fill in the reasons without much difficulty if the statements are given. Ask the students to explain why the corollary "if $ab = 0$ then $a = 0$ or $b = 0$" implies that the product of two nonzero numbers is not equal to 0.

There are many ways to help students handle proofs. Many students start by supplying reasons for steps that are already written. To help students go beyond supplying reasons to constructing their own proofs, try proof sketches. A proof sketch is a short strategic plan which focuses on the major steps of a proof.

Use the discussion of converse to lead to a discussion of the inverse and contrapositive. Use the *if p then q* form to illustrate the inverse and contrapositive of *if ab = 0, then a = 0 or b = 0.*

2-7 Theorems about Order and Absolute Value

This lesson gives students the opportunity to use and improve their skills in proving theorems. Students should be given as much practice as possible in setting up and proving theorems. Point out the many *cases* involved in working with inequalities. All students should be able to give reasons in Exercises 1–6. By this time most students should be able to do most of the proofs in Exercises 7–16. Sometimes a "how to" group discussion about the proofs in an assignment helps students who are still having problems.

References to Strategies

PE: Pupil's Edition **TE:** Teacher's Edition **RB:** Resource Book

Problem Solving Strategies

PE: pp. 69–70 (checking solutions)
TE: p. 69
RB: pp. 192–193

Applications

PE: p. 58 (art), pp. 65, 69, 71–72 (word problems)
TE: pp. 58, 65, 69–72
RB: p. 205

Nonroutine Problems

PE: p. 63 (Exs. 25–33); pp. 71–72 (Probs. 11–16); p. 75 (Exs. 29–34); p. 79 (Exs. 25–33); p. 87 (Exs. 15–25); p. 91 (Exs. 15–18)
TE: pp. T85, T87 (Sugg. Extensions, Lessons 2-3, 2-4, 2-6)

Communication

PE: p. 62 (Ex. 17, discussion/convincing argument); p. 63 (Ex. 33, discussion); p. 70 (translation); p. 87 (Exs. 24–25, convincing argument)
TE: pp. 61, T86 (Reading Algebra); pp. 81, 88 (Warm-up Exs.)

Thinking Skills

PE: pp. 81–91 (proofs)
TE: pp. 65, 73, 81

Explorations

PE: pp. 95–97 (Boolean algebra); p. 833 (inequalities)

Connections

PE: p. 58 (Data Analysis); p. 64 (Engineering); p. 68 (History); pp. 95–97 (Logic); pp. 71–72 (Geometry)

Using Technology

PE: pp. 64, 75 (Exs.); p. 80 (Computer Key-In)
TE: pp. 64, 80
RB: pp. 250–256
Computer Activities: pp. 6–10

Using Manipulatives/Models

Overhead Visuals: A

Teaching Resources

For use in implementing the teaching strategies referenced on the previous page.

Application
Resource Book, p. 205

Application—Inequalities and Electric Circuits (For use with Chapter 2)

When two simple inequalities are joined by "and" or "or," a compound inequality is formed. When "and" is used, as in "$x > 3$ and $x < 7$," the compound inequality is called a *conjunction.* When "or" is used, as in "$x < 3$ or $x > 7$," the compound inequality is called a *disjunction.* The graphs of a conjunction and a disjunction are shown below.

Conjunction: $x > 3$ and $x < 7$ Disjunction: $x < 3$ or $x > 7$

You can use diagrams of electric circuits with on-off switches to illustrate compound inequalities. From the diagram of the *series* circuit at the right, you can see that *both* switches must be closed in order for current to flow through the circuit. Since a conjunction is true if and only if *both* simple inequalities are true, a conjunction can be represented by a series circuit.

1. Explain why a *parallel* circuit, shown in the diagram at the right, can be used to represent a disjunction.

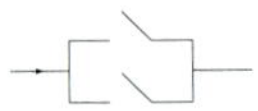

2. The first column of the following table contains the graphs of compound inequalities. For each graph, write an open sentence in the second column. Then match the statement with one of the circuit diagrams at the bottom of the page and write the letter of the diagram in the third column.

Graph *Open sentence* *Circuit*

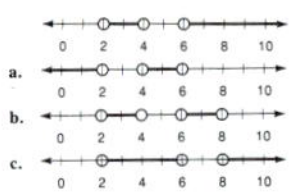

($x > 2$ and $x < 4$) or $x > 6$ D

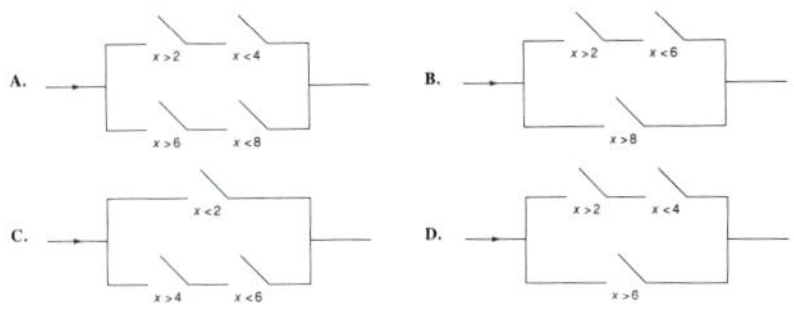

APPLICATIONS 205

Enrichment
Resource Book, p. 221

More About Truth Tables (For use with Chapter 2)

In the textbook, *conjunction* and *disjunction* were defined both verbally and by means of truth tables. Since both parts of a conjunction must be true for the conjunction to be true, it follows that

the negation of a conjunction is obtained by negating *either* part of the conjunction.

This is proved by the following truth table. (If you have not yet studied the EXTRA topic on pages 93–95, you should study it now before continuing this page.)

p	q	$\sim p$	$\sim q$	$(p \wedge q)$	$\sim(p \wedge q)$	$\sim p \vee \sim q$
T	T	F	F	T	F	F
T	F	F	T	F	T	T
F	T	T	F	F	T	T
F	F	T	T	F	T	T

As you know, the first two columns show the four possible true-false combinations of statements p and q. The entries in all the other columns are obtained by applying the definitions of conjunction, disjunction, and negation. When you compare the last two columns, you can see that they have the same array of truth values. Thus, we have proved that

$$\sim(p \wedge q) \leftrightarrow \sim p \vee \sim q.$$

Prove the following statements by means of truth tables. As the first step in each proof, rewrite the statement using statement variables and symbols of symbolic logic.

1. In order to negate a disjunction, it is necessary to negate both statements that make up the disjunction.

2. To negate the negation of a statement is the same as to affirm the statement.

ENRICHMENT 221

Problem Solving
Resource Book, p. 192

Problem Solving with Inequalities (For use with Lesson 2-3)

By working through the steps in the problems below, you will gain skill in solving problems involving inequalities.

Problem 1 A company offers its salespersons a monthly salary of \$1400 plus a 10% commission on all sales over \$5000. Makoto, a salesman for the company, wants to earn at least \$2000 this month. His sales for the month must therefore be at least how much?

a. What does the problem ask you to find? ______

b. Let s = Makoto's sales for the month. Give an expression in terms of s for:

the amount of sales to which the 10% commission is applied ______

the dollar amount of his commission ______

the amount he earns for the month ______

c. Replace each phrase with an algebraic expression or symbol.

Makoto's earnings for the month should be at least \$2000.

______ ______ ______

d. Solve the inequality. (Remember to multiply both side of the inequality by 10 to remove the decimal.)

e. Write your answer to the problem. ______

f. Check your results with the statements of the problem.

At least how much must Makoto's sales for the month be? ______

By how much do these sales exceed \$5000? ______

What would be the dollar amount of Makoto's commission on sales over \$5000? ______

What would be his total earnings for the month? ______

Are his earnings at least \$2000? ______

(continued)

192 PROBLEM SOLVING

Problem Solving
Resource Book, p. 193

Problem Solving with Inequalities *(continued)*

Problem 2 Denise bought 14 tropical fish at Aquaworld. Some were neons at 39¢ each and the rest were fantails at 59¢ each. If Denise spent no more than \$6 for the fish, find the greatest possible number of fantails she bought.

a. What does the problem ask you to find? ______

b. Let f = number of fantails purchased. Give an expression in terms of f for:

the number of neons purchased ______

the cost in cents for the fantails ______

the cost in cents for the neons ______

c. Use the fact that Denise spent no more than \$6 to write an inequality for the problem.

d. Solve the inequality. ______

e. Write your answer to the problem. (*Remember:* You cannot buy part of a fish!) ______

f. Check your results with the statements of the problem.

Did Denise buy 14 fish in all? ______

At most, what did the fantails cost? ______

What did the remaining fish (neons) cost? ______

Is the total cost less than \$6.00? ______

PROBLEM SOLVING 193

Problem Solving
Study Guide, p. 23

2–3 *Problem Solving Using Inequalities*

Objective: To solve word problems by using inequalities in one variable.

Symbols
$x \ge a$ (x is at least a, or x is no less than a.)
$x \le b$ (x is at most b, or x is no greater than b.)
$a < x < b$ (x is between a and b.)
$a \le x \le b$ (x is a, b, or between a and b.)

Example 1 A video store charges \$19.99 for a lifetime membership. Members pay \$2.00 to rent a movie, while nonmembers pay \$2.25. At least how many movies would a member have to rent in order to pay less, overall, than a nonmember?

Solution

Step 1 The problem asks for the least number of movies a member would have to rent for rental costs to be less than those for a nonmember.

Step 2 Let m = the number of movies.
Then $19.99 + 2.00m$ = the amount a member would pay for m movies; and $2.25m$ = the amount a nonmember would pay for m movies.

Step 3 Rental for a member is less than rental for a nonmember.
$$19.99 + 2.00m < 2.25m$$

Step 4 Multiply both sides of the inequality in Step 3 by 100 to clear the decimals.
$$1999 + 200m < 225m$$
$$-25m < -1999$$
$$m > 79.96$$

Interpret the result: Since the number of movies must be a whole number, $m \ge 80$.

Step 5 *Check:* Has a member who has rented 80 movies paid less than a nonmember who has rented 80 movies? Compare.
$$19.99 + 2.00(80) \overset{?}{<} 2.25(80)$$
$$179.99 < 180 \checkmark$$

Check: Is 80 the least number of movies a member must rent to pay less, overall, than a nonmember? Try 79.
$$19.99 + 2.00(79) \overset{?}{>} 2.25(79)$$
$$177.99 > 177.75 \checkmark$$

∴ a member must rent at least 80 movies to pay less, overall, than a nonmember.

Solve.

1. The owners of a skating rink sell discount cards for \$15.00 that are worth \$.50 off the regular admission price of \$3.00. The discount cards are good for six months. At least how many times would you have to go skating in order to pay less, overall, with the discount card than without it?
2. A summer recreation department charges \$45.00 for a season ticket to the town pool. Admission to the pool for one day is \$1.75. How many days would you have to go swimming at the regular price in order to spend at least the cost of a season ticket?

Study Guide, ALGEBRA AND TRIGONOMETRY, Structure and Method, Book 2

23

Problem Solving
Study Guide, p. 24

2–3 *Problem Solving Using Inequalities* (continued)

Example 2 Two sides of a triangle are consecutive *even* integers. The other side is 65 cm. If the perimeter is between 215 cm and 230 cm, what are the possible lengths for the first two sides?

Solution

Step 1 The problem asks you to find all possible lengths for the first two sides if the other side is 65 cm and the perimeter is between 215 cm and 230 cm. Draw a sketch.

Step 2 Let s = the length of the first side. Then $s + 2$ = the length of the second side.

Step 3 The perimeter is between 215 cm and 230 cm.
$$215 < s + (s + 2) + 65 < 230$$

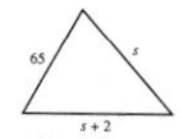

Step 4
$$215 < 2s + 67 < 230$$
$$148 < 2s < 163$$
$$74 < s < 81.5$$

Interpret the result: Since s must be an even integer, there are only three possible values for s: 76, 78, and 80. There are three pairs of consecutive even integers that meet the requirements of the problem: 76, 78; 78, 80; and 80, 82.

Step 5 To check, you must verify that the perimeter is between 215 cm and 230 cm in all three cases, and that neither the "next smaller" nor the "next larger" pair of even consecutive integers is a solution. The work is left for you.

∴ the lengths, in centimeters, of the first two sides are: 76, 78; 78, 80; or 80, 82.

Solve.

3. The length of a rectangle is 3 cm more than twice its width. Find the largest possible width if the perimeter is at most 66 cm.
4. The length of a leg of an isosceles triangle is twice the length of the base. What is the minimum length of the base if the perimeter is at least 20 cm?
5. Find all sets of four consecutive integers whose sum is between 95 and 105.
6. Ellen's first three test scores were consecutive *odd* integers. Her fourth score was 83. She had a B− average (between 80 and 82, inclusive) for the four tests. What was her lowest test score?

Mixed Review Exercises

Solve each open sentence and graph each solution set that is not empty.

1. $2x - 3 > -7$
2. $0.5p \le 1$ and $p + 1 \ge -2$
3. $3(n - 1) > 2 - 3(1 - n)$
4. $-2y < -8$ or $1 + 2y < 3$
5. $5 - 2d < 3$
6. $-2 > 1 - 2x > -6$

Evaluate if $a = -2$ and $b = 5$.

7. $|a - b|$
8. $|a| - |b|$
9. $|ab|$
10. $|b| - |a|$

24
Study Guide, ALGEBRA AND TRIGONOMETRY, Structure and Method, Book 2

Using Technology
Computer Activities, p. 6

ACTIVITY 3. *String Variables in BASIC* (*for use with Chapter 2*)

Directions: Write all answers in the spaces provided.

PROBLEM

What are string variables and how are they used?

PROGRAM ■ ★ ●

```
10 LET A$ = "YOU HAVE  "
20 LET B$ = "  AND  "
30 PRINT "NAME SOMETHING THAT IS YOURS";
40 INPUT M$
50 PRINT "WHAT ELSE IS YOURS";
60 INPUT N$
70 LET S$ = A$ + M$ + B$ + N$ + "."
80 PRINT
90 PRINT S$
100 END
```

PROGRAM CHECK

Type in the program. To test whether you entered it correctly, run the program. After the first question, enter A VIOLIN. After the second question, enter A PET CATFISH. The computer should print:

YOU HAVE A VIOLIN AND A PET CATFISH.

ANALYSIS

In BASIC, a *string* is a sequence of characters (numbers, letters, blanks, and other keyboard symbols). A *string variable* is a variable whose "value" is a string. The symbol \$ must be the last character of a string variable. In the program, the string variable A\$ has the value YOU HAVE, and the string variable B\$ has the value AND.

Strings can be combined to form larger strings. In line 70, A\$, M\$, B\$, N\$, and the one-character string consisting of a period are combined in that order to form S\$.

USING THE PROGRAM

Run the program for each sentence shown. In response to the questions, enter the strings that will produce the given sentence. Write your inputs in the spaces provided.

1. YOU HAVE A PET CATFISH AND A VIOLIN.

   First input ______________ Second input ______________

2. YOU HAVE TWO BABY GUPPIES AND AN ERASER.

   First input ______________ Second input ______________

3. YOU HAVE POISE AND CONFIDENCE.

   First input ______________ Second input ______________

(*continued*)

6
Computer Activities, ALGEBRA AND TRIGONOMETRY, Structure and Method, Book 2

Using Technology
Computer Activities, p. 7

(*Activity 3 continued*)

EXTENSIONS

1. You can use an instruction like

```
200 A$ = A$ + X$
```

to add the contents of X\$ to the contents of A\$. We can use this idea to build a sentence. Delete your original program from the computer's memory. Then type in the following program.

■ ★ ●

```
10 LET S$ = "IN THE NEXT 10 YEARS YOU WOULD LIKE TO  "
20 PRINT "WHAT WOULD YOU LIKE TO DO"
30 PRINT "IN THE NEXT 10 YEARS";
40 INPUT X$
50 LET A$ = A$ + X$
60 LET P$ = A$ + "."
70 PRINT
80 PRINT S$; P$
90 PRINT
100 PRINT "WHAT ELSE WOULD YOU LIKE TO DO";
110 INPUT X$
120 PRINT
130 IF X$ = "NOTHING" THEN 160
140 LET A$ = A$ + "  AND  "
150 GOTO 50
160 END
```

Run the program for the given list of responses. Answer each question with the next response.

STUDY PHYSICS
ATTEND COLLEGE
VISIT ITALY
FINISH THIS ASSIGNMENT
NOTHING

2. a. Delete the previous program from the computer's memory. Then complete line 40 of the following program so that it builds and prints a string, T\$, consisting of a single word repeated 15 times.

★ ●

```
10 PRINT "WHAT IS THE WORD YOU WANT REPEATED";
20 INPUT Y$
30 FOR I = 1 TO 15
40 LET ______________________________
50 NEXT I
60 PRINT
70 PRINT T$
80 END
```

   b. Type in the program and run it using the word READ.

Computer Activities, ALGEBRA AND TRIGONOMETRY, Structure and Method, Book 2

7

Application

Pottery has been used throughout the ages for both practical purposes and decorative purposes. All pottery is basically clay that has been baked or "fired." The firing removes water from the clay and also causes a chemical change that makes the clay rigid.

When clay is mixed with other substances, different types of pottery are produced. The three basic types are earthenware, stoneware, and porcelain, and each must be fired within a specific range of temperatures.

The firing of pottery is done in an oven called a kiln. Because kilns produce temperatures that are much too high to be measured with thermometers, pyrometric (or Seger) cones are used. A Seger cone is a small clay pyramid to which a flux that melts at a known temperature has been added.

Research Activities

Students can research the temperatures for firing various types of pottery. They can also research how Seger cones are used to measure kiln temperatures.

If the art department in your school uses a kiln, suggest that students determine what its temperature range is and how the temperature is measured.

References

Kenny, John B. *The Complete Book of Pottery Making.* Radnor, PA: Chilton Book Company, 1976.

2 Inequalities and Proof

Earthenware pottery is a mixture of clay, flint, and feldspar. Depending on how these elements are mixed, pottery must be heated to a temperature between 1000°C and 1150°C.

Working with Inequalities

2-1 Solving Inequalities in One Variable

Objective To solve simple inequalities in one variable.

The inequality

$$x > -2$$

is satisfied by every real number greater than -2. The graph of this inequality is shown in red on the number line below. Note the use of the open circle to show that -2 is not a solution.

To find solutions of more complicated inequalities like

$$5x + 17 < 2$$

and

$$5(3 - t) < 7 - t$$

you use methods similar to those used to solve equations. These methods are based on the properties of order for real numbers stated below. Recall that a real number c is called *positive* if $c > 0$ and *negative* if $c < 0$.

Properties of Order

Let a, b, and c be any real numbers.

Comparison Property

Exactly one of the following statements is true:

$$a < b, \quad a = b, \quad \text{or } a > b.$$

Transitive Property

If $a < b$ and $b < c$, then $a < c$.

Addition Property

If $a < b$, then $a + c < b + c$.

Multiplication Property

1. If $a < b$ and c is *positive,* then $ac < bc$.
2. If $a < b$ and c is *negative,* then $ac > bc$.

Teaching References

Lesson Commentary, pp. T83–T87

Assignment Guide, pp. T59–T61

Supplementary Materials

- Practice Masters 7–10
- Tests 5–8
- Resource Book
  - Practice Exercises, pp. 104–107
  - Tests, pp. 8–11
  - Enrichment Activity, p. 221
  - Application, p. 205
  - Practice in Problem Solving/Word Problems, pp. 192–193
- Study Guide, pp. 19–32
- Computer Activities 3–4
- Test Generator
- Alternate Test, p. T14

Explorations p. 833

Teaching Suggestions p. T84

Suggested Extensions p. T84

Warm-Up Exercises

Solve.

1. $x + 3 = 5$ $\{2\}$

2. $3a - 4 = 11$ $\{5\}$

3. $4z = 2(3 + 2z)$ $\emptyset$

4. $5(2r + 3) = 2(r - 3) + r$ $\{-3\}$

5. $\frac{7x}{4} = 5x - 2$ $\left\{\frac{8}{13}\right\}$

Motivating the Lesson

Explain to students that the temperature in a freezer must be below 32° F or the food will thaw. This relationship can be shown by the inequality $t < 32$, where t is the freezer temperature.

Since subtraction is defined in terms of addition, the addition property of order applies to subtraction. Similarly, the multiplication properties apply to division. Furthermore, since the statement $b > a$ has the same meaning as $a < b$, the properties of order hold true if $<$ and $>$ are interchanged throughout.

When you multiply or divide both sides of an inequality by a *negative* number, you must *reverse* the direction of the inequality. For example:

$$5 < 8, \quad \text{but} \quad (-2)5 > (-2)8 \quad (\text{that is, } -10 > -16)$$

$$1 > -4, \quad \text{but} \quad \frac{1}{-2} < \frac{-4}{-2} \quad \left(\text{that is, } -\frac{1}{2} < 2\right)$$

You solve an inequality by transforming it into an inequality whose solution set is easy to see. Transformations that produce **equivalent inequalities,** that is, inequalities with the same solution set, are listed in the chart below.

Transformations that Produce Equivalent Inequalities

1. Simplifying either side of an inequality.
2. Adding to (or subtracting from) each side of an inequality the same number or the same expression.
3. Multiplying (or dividing) each side of an inequality by the same *positive* number.
4. Multiplying (or dividing) each side of an inequality by the same *negative* number and *reversing* the direction of the inequality.

Notice how these transformations are used in the following example.

Example 1 Solve each inequality and graph its solution set.

a. $5x + 17 < 2$ **b.** $5(3 - t) < 7 - t$

Solution

a. $5x + 17 - 17 < 2 - 17$

$5x < -15$

$\frac{5x}{5} < \frac{-15}{5}$

$x < -3$

$\therefore$ the solution set consists of all real numbers less than -3.

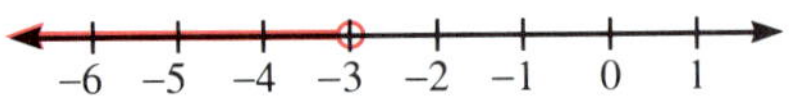

b. $15 - 5t < 7 - t$

$15 - 5t + t < 7 - t + t$

$15 - 4t < 7$

$-4t < -8$

$\frac{-4t}{-4} > \frac{-8}{-4}$

$t > 2$

$\therefore$ the solution set consists of all real numbers greater than 2.

Chalkboard Examples

Solve and graph.

1. $3x + 16 < 7$
   $3x < -9$
   $x < -3$
   $\{x: x < -3\}$

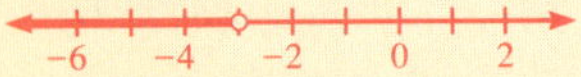

2. $11 - \frac{3}{2}r > -4$
   $-\frac{3}{2}r > -15$
   $r < 10$
   $\{r: r < 10\}$

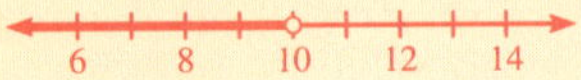

3. $3(4z - 1) > 2(6z - 5)$
   $12z - 3 > 12z - 10$
   $-3 > -10$
   {real numbers}

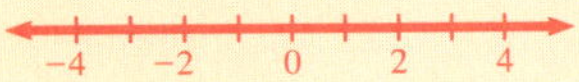

4. $3 - 2(q - 1) < 5$
   $3 - 2q + 2 < 5$
   $-2q < 0$
   $q > 0$
   $\{q: q > 0\}$

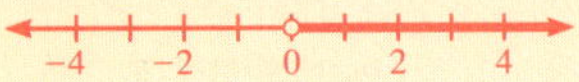

5. $-(10 - r) < r - 15$
   $-10 + r < r - 15$
   $-10 < -15$
   $\emptyset$

Check for Understanding

Here is a suggested use of the Oral Exercises to check students' understanding as you teach the lesson.

Oral Exs. 1–10: use after Example 1.

Oral Exs. 11–17: use after Example 2.

To quickly check whether the direction of the inequality in your answer is correct, choose a test point and see if it satisfies the given inequality. For example, in part (a) of Example 1, substituting 0 for x gives $17 < 2$, which is false. So 0 is not in the solution set and the direction of the inequality $x < -3$ is correct.

The solution set of part (a) of Example 1 can be written as $\{x: x < -3\}$, which is read, "the set of all x such that x is less than -3." Similarly, the solution set of part (b) can be written $\{t: t > 2\}$. This notation will be used throughout the rest of this book.

Some inequalities are true for all real numbers, and others have no solution, as Example 2 illustrates.

Example 2 Solve each inequality and graph its solution set.

a. $4x > 2(3 + 2x)$ **b.** $2t < -\frac{1}{2}(-6 - 4t)$

Solution

a. $4x > 2(3 + 2x)$

$4x > 6 + 4x$

$0 > 6$

Since the equivalent inequality $0 > 6$ is *false*, the given inequality is false and has no solution.

$\therefore$ the solution set is $\emptyset$, and there is no graph.

b. $2t < -\frac{1}{2}(-6 - 4t)$

$2t < 3 + 2t$

$0 < 3$

Since $0 < 3$ is *true*, the given inequality is true for all values of t.

$\therefore$ the solution set is {real numbers}.

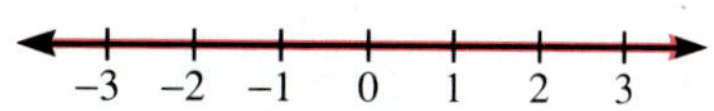

Oral Exercises

Explain how to transform the first inequality into the second inequality.

Sample $-3t > 15$; $t < -5$

Solution Divide each side by -3 and reverse the direction of the inequality.

1. $x - 3 < 4$; $x < 7$ Add 3.
2. $3s > 6$; $s > 2$ Div. by 3.
3. $-\frac{u}{4} > 3$; $u < -12$ Mult. by -4; reverse sign.
4. $k + 2 < 1$; $k < -1$ Subtr. 2.
5. $2(y + 1) + 3 < y$; $2y + 5 < y$ Simplify left side.
6. $10 > -5y$; $-2 < y$ Div. by -5; reverse sign.
7. $\frac{3x + 1}{2} < 5$; $3x + 1 < 10$ Mult. by 2.
8. $5t < 3(t - 2) + 1$; $5t < 3t - 5$ Simplify right side.
9. $z - 3 > 3z - 7$; $-3 > 2z - 7$ Subtr. z.
10. $4x - 7 > x - 2$; $4x > x + 5$ Add 7.

Reading Algebra

Before graphing, students should be encouraged to read, either aloud or mentally, their answers with the variable read first. By reading $-3 < x$ as "x is greater than negative 3," students may more easily see that the graph is to the right, and not the left, of -3.

Common Errors

The most common error students make while solving inequalities is to forget to switch the inequality symbol when multiplying or dividing by a negative number. When reteaching, stress this concept by pointing out transformation 4 on page 60.

Guided Practice

Solve each inequality.

1. $\frac{4}{5}k - 8 > 8 - \frac{4}{5}k$
   $\{k: k > 10\}$
2. $-\frac{1}{2}a - 19 > -15$
   $\{a: a < -8\}$
3. $8 - (2b - 3) < 2b - 3$
   $\left\{b: b > \frac{7}{2}\right\}$
4. $p - (7 - p) > 2p + 1$
   $\emptyset$
5. $5 - 2(-z - 4) > z - (z - 1)$
   $\{z: z > -6\}$
6. $2.7(3v - 8) > 7.2(8v - 3)$
   $\{v: v < 0\}$

Summarizing the Lesson

In this lesson we have applied the properties of order and the methods of transforming inequalities to solve inequalities. Stress to students that they must understand the transformations that are listed on page 60.

Suggested Assignments

Average Algebra
62/1–26
S 63/Mixed Review
Average Alg. and Trig.
62/1–27
S 63/Mixed Review
Extended Algebra
62/1–31 odd
S 63/Mixed Review
Extended Alg. and Trig.
62/1–31 odd
S 63/Mixed Review

Additional Answers Written Exercises

1. 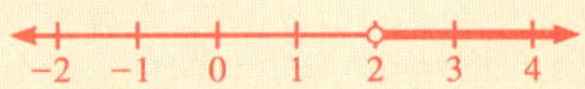

2. 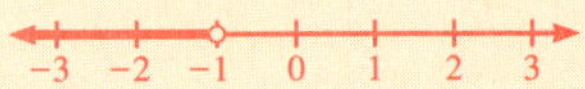

3. 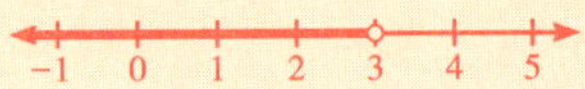

4. 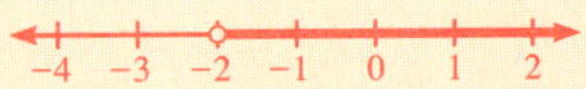

5. 

6. 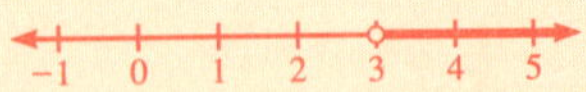

7. 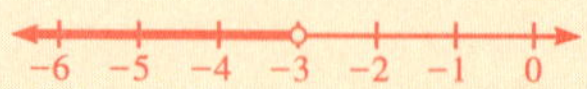

8. 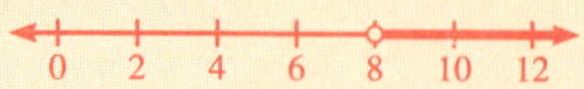

9. 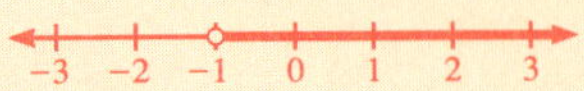

10. 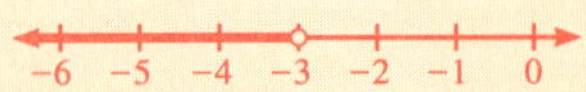

11. 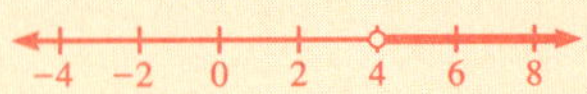

12. 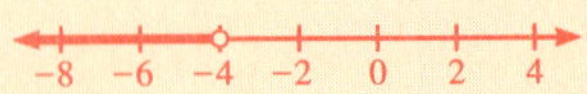

Match each inequality with the graph of its solution set.

11. $-6x > 3$ g
12. $4 - x > 2$ a
13. $1 - x < x$ e
14. $2(x - 1) < 2x - 3$ c
15. $2(x + 1) + 1 > 0$ b
16. $2(x - 1) > 2x - 3$ f

a. 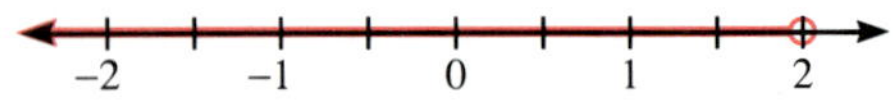

b. 

c. 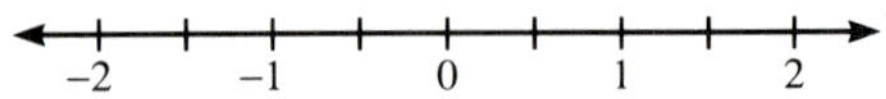

d. 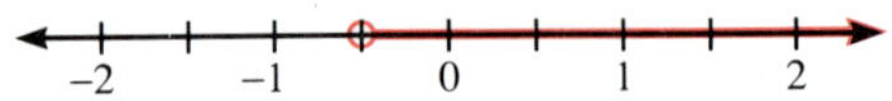

e. 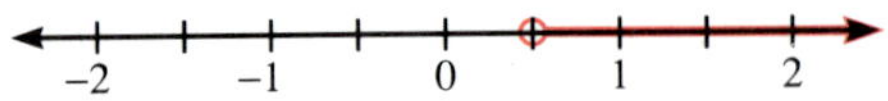

f. 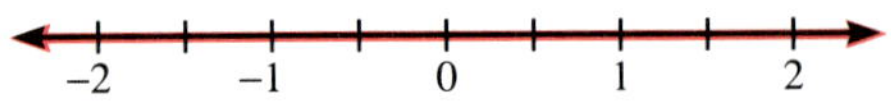

g. 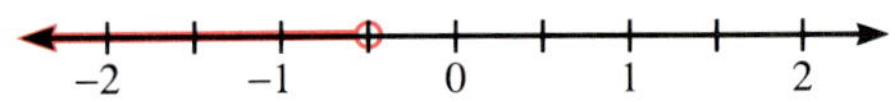

h. 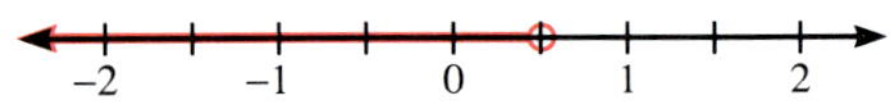

17. a. Explain the meaning of this statement: *If* $a < 0$, *then* $a^2 > 0$.
b. Decide whether the statement in (a) is true or false. Give a convincing argument to support your answer.
a. The square of a negative number is positive.
b. True, since $a^2 = a \cdot a$ and the product of two negative numbers is positive.

Written Exercises

Solve each inequality and graph each solution set that is not empty.

Sample $1 - 2t > -5$

Solution $\{t\colon t < 3\}$

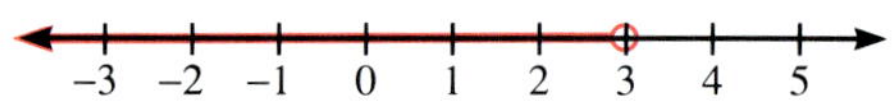

A

1. $x - 7 > -5$ $\{x\colon x > 2\}$
2. $y + 4 < 3$ $\{y\colon y < -1\}$
3. $2t < 6$ $\{t\colon t < 3\}$
4. $3u > -6$ $\{u\colon u > -2\}$
5. $-5x < 10$ $\{x\colon x > -2\}$
6. $-12 > -4y$ $\{y\colon y > 3\}$
7. $-\frac{t}{2} > \frac{3}{2}$ $\{t\colon t < -3\}$
8. $-\frac{3}{4}k < -6$ $\{k\colon k > 8\}$
9. $3s - 1 > -4$ $\{s\colon s > -1\}$
10. $2r + 5 < -1$ $\{r\colon r < -3\}$
11. $y < 7y - 24$ $\{y\colon y > 4\}$
12. $3t > 6t + 12$ $\{t\colon t < -4\}$
13. $2 - h < 4 + h$ $\{h\colon h > -1\}$
14. $1 + 2x < 2(x - 1)$ $\emptyset$
15. $5(2u + 3) > 2(u - 3) + u$ $\{u\colon u > -3\}$
16. $3(x - 2) - 2 < x - 5$ $\left\{x\colon x < \frac{3}{2}\right\}$

17. $5(x - 7) + 2(1 - x) > 3(x - 11)$ ∅

18. $4s + 3(2 - 3s) < 5(2 - s)$ {real numbers}

19. $7y - 2(y - 4) > 6 - (2 - y)$ $\{y: y > -1\}$

20. $4(2 - x) - 3(1 + x) < 5(1 - x)$ $\{x: x > 0\}$

Solve.

B 21. $k - 3(2 - 4k) < 7 - (8k - 9 + k)$ $\{k: k < 1\}$

22. $\frac{2}{3}t - (2 - 3t) < 5t + 2(1 - t)$ $\{t: t < 6\}$

23. $4(y + 2) - 9y > y - 3(2y + 1) - 1$ {real numbers}

24. $4[5x - (3x - 7)] < 2(4x - 5)$ ∅

Tell whether each statement is true for all real numbers. If you think it is not, give a numerical example to support your answer.

25. If $a < b$, then $a - c < b - c$. True

26. If $a < b$, then $a - b < 0$. True

27. If $a < b$, then $a^2 < b^2$. False; $-1 < 0$, but $(-1)^2 > 0^2$.

28. If $a < b$, then $a^3 < b^3$. True

29. If $a < b$ and $c < d$, then $a + c < b + d$. True

30. If $a < b$ and $c < d$, then $a - c < b - d$. False; $3 < 4$ and $0 < 2$, but $(3 - 0) > (4 - 2)$.

C 31. If $a \neq b$, then $a^2 + b^2 > 2ab$. True

32. If $a > 0$ and $a \neq 1$, then $a + \frac{1}{a} > 2$. True

33. The following statement is true:

If $a > 0$ and $a < 1$, then $a^2 < a$.

Write a short paragraph explaining the meaning of this statement to someone who has studied arithmetic but not algebra.

33. If you multiply any number between 0 and 1 with itself, the result is smaller than the original number.

Mixed Review Exercises

Simplify.

1. $(8t - 5) - (5 - 8t)$ $16t - 10$

2. $(-4)(5)(-1)(-3)$ -60

3. $(-2)^4(-p)^3$ $-16p^3$

4. $\frac{5^2 - 7^2}{5 - 7}$ 12

5. $\frac{8cd - 6}{-2}$ $-4cd + 3$

6. $\frac{6(2 + 3)}{6 \cdot 2 + 3}$ 2

7. $|3(-4) - 2|$ 14

8. $-2(5 - 8)^3$ 54

9. $2a^2 - 5a - (a^2 - 7)$ $a^2 - 5a + 7$

10. $6x - y + 2x - 4y$ $8x - 5y$

11. $|-4| - |-9|$ -5

12. $4(3 - m) - (2m + 1)$ $11 - 6m$

13. 

15. 

16. 

18. 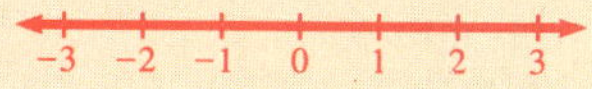

19. 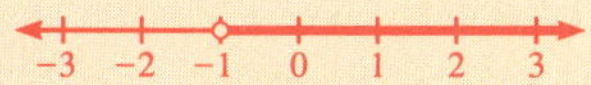

20. 

Supplementary Materials

Study Guide pp. 19–20

Using a Computer

These exercises use the computer to test for integral solutions to a simple inequality over a given domain. This is similar to a trial and error approach for a student not using a computer.

Computer Exercises

For students with some programming experience.

1. Write a computer program that will print all solutions of the inequality $ax + b < c$, where a, b, and c are entered by the user. In this program the domain of x will be a set of consecutive integers, where the smallest and largest members of the set will also be entered by the user. (*Hint:* Use a FOR . . . NEXT loop to test each integer from smallest to largest.)

2. Run the program in Exercise 1 to find solutions of the inequality $3x - 7 < 17$ for the following domains. **a.** $\{1, 2, 3, 4, 5, 6\}$ **b.** $\{5, 6, 7\}$ **c.** $\emptyset$

   a. $\{1, 2, \ldots, 6\}$ **b.** $\{5, 6, \ldots, 12\}$ **c.** $\{10, 11, \ldots, 20\}$

3. Modify the program in Exercise 1 so that if there are no solutions in the given domain, the computer will print a message to this effect.

4. Run the program in Exercise 3 to find solutions of the inequality $5x + 3 < 12$ for the following domains.

   a. $\{6, 7, \ldots, 12\}$ **b.** $\{-10, -9, \ldots, 6\}$ **c.** $\{1, 2, \ldots, 6\}$

   a. No solutions **b.** $\{-10, -9, -8, -7, -6, -5, -4, -3, -2, -1, 0, 1\}$ **c.** $\{1\}$

Career Note / *Automotive Engineer*

When you read that automobile manufacturers have increased the fuel efficiency of cars, you are actually reading about the work of automotive engineers. Automotive engineering is a specialized field of mechanical engineering, which is concerned with the use, production, and transmission of mechanical power.

The role of an automotive engineer, as with all engineers, is to apply the theories and principles of science and mathematics to the solution of practical problems. Automotive engineers in particular are responsible for designing automobiles that deliver efficient and economical performance. Evaluating the overall cost, reliability, and safety of automobiles is also part of an automotive engineer's work.

To assist them in their work, automotive engineers rely upon calculators, computer simulations, and other engineers.

Computer-aided design systems have become especially important for the production and analysis of automotive engine-and-body designs.

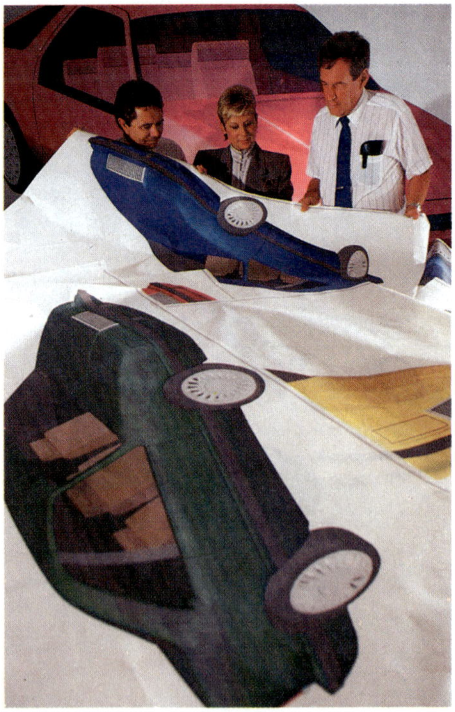

2-2 Solving Combined Inequalities

Objective To solve conjunctions and disjunctions.

The general admission price at the Cinema V theater is \$4.50. Children 12 years of age or under and adults who are at least 65 are charged only half price. Therefore, you must pay the full price of \$4.50 if your age, a, satisfies both of the inequalities $a > 12$ and $a < 65$. The combined inequality "$a > 12$ and $a < 65$" is an example of a *conjunction*.

A sentence formed by joining two sentences with the word *and* is called a **conjunction.** A conjunction is true when *both* sentences are true.

Example 1 Graph the solution set of the conjunction $x > -2$ *and* $x < 3$.

Solution The conjunction is true for all values of x between -2 and 3. The graph is shown below. Notice that the numbers *between* -2 and 3 include neither -2 nor 3.

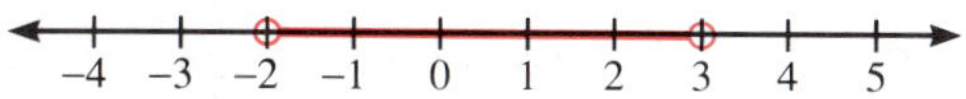

The conjunction "$x > a$ and $x < b$" is usually written as

$$a < x < b,$$

which is read "x is greater than a and less than b." Using this notation, the solution set of Example 1 is written $\{x: -2 < x < 3\}$.

A sentence formed by joining two sentences with the word *or* is called a **disjunction.** A disjunction is true when *at least one* of the sentences is true.

You learned earlier that "$x \le 2$" means "x is less than or equal to 2" and therefore represents the disjunction

$$x < 2 \quad or \quad x = 2.$$

In the graph of this disjunction, shown below, a solid red dot has been used to show that 2 is included in the solution set.

Teaching Suggestions p. T84

Suggested Extensions p. T84

Warm-Up Exercises

Write an inequality for each graph. Use x as the variable.

1. 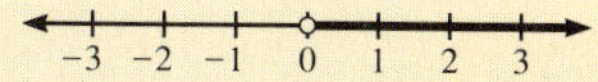
$x > 0$

2. 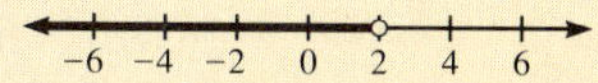
$x < 2$

3. 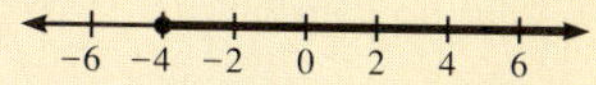
$x \ge -4$

4. 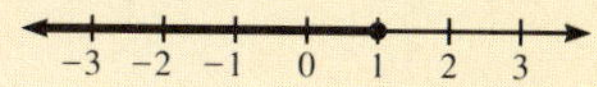
$x \le 1$

5. 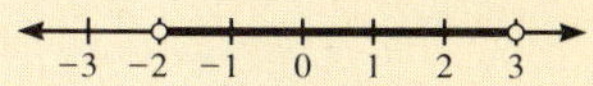
$x > -2$ and $x < 3$

Motivating the Lesson

Point out that many real world quantities can be defined by combined inequalities. For example, a multilane highway may have a minimum speed of 45 mi/h and a maximum speed of 60 mi/h. This can be represented by $45 \le x \le 60$.

Common Errors

Students may make the mistake of writing a disjunction like $x < -3$ or $x > 1$ as $-3 > x > 1$. When reteaching, show students that $-3 > x > 1$ is not correct since -3 is *not* greater than 1.

Thinking Skills

Students should *recall* the methods of graphing simple inequalities from Lesson 2-1. These methods will aid in the graphing of conjunctions and disjunctions.

The chart below summarizes how to solve conjunctions and disjunctions of open sentences in one variable.

Conjunction: Find the values of the variable for which *both* sentences are true.

Disjunction: Find the values of the variable for which *at least one* of the sentences is true.

Example 2 Solve $3 < 2x + 5 \leq 15$ and graph its solution set.

Solution 1 You can first rewrite the inequality using *and*. Then solve both inequalities.

$$\begin{array}{ccc} 3 < 2x + 5 & \text{and} & 2x + 5 \leq 15 \\ -2 < 2x & \downarrow & 2x \leq 10 \\ -1 < x & \text{and} & x \leq 5 \end{array}$$

$\therefore$ the solution set is $\{x\colon -1 < x \leq 5\}$. ***Answer***

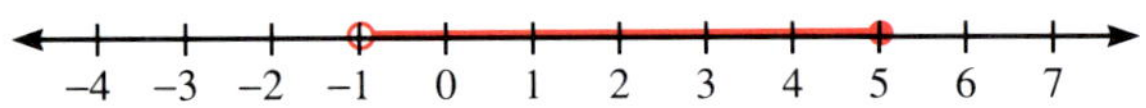

Solution 2 You can solve the inequality using this shortened method that involves operating on all three parts of the inequality at the same time.

$$\begin{array}{ll} 3 < 2x + 5 \leq 15 & \text{Subtract 5 from all three parts.} \\ -2 < \quad 2x \quad \leq 10 & \text{Divide all three parts by 2.} \\ -1 < \quad x \quad \leq 5 & \end{array}$$

$\therefore$ the solution set is $\{x\colon -1 < x \leq 5\}$. ***Answer***

Example 3 Solve the disjunction $7 - 2y \leq 1$ *or* $3y + 10 < 4 - y$ and graph its solution set.

Solution

$$\begin{array}{ccc} 7 - 2y \leq 1 & \text{or} & 3y + 10 < 4 - y \\ -2y \leq -6 & \downarrow & 4y < -6 \\ y \geq 3 & \text{or} & y < -\frac{3}{2} \end{array}$$

$\therefore$ the solution set is $\left\{y\colon y \geq 3 \text{ or } y < -\frac{3}{2}\right\}$. ***Answer***

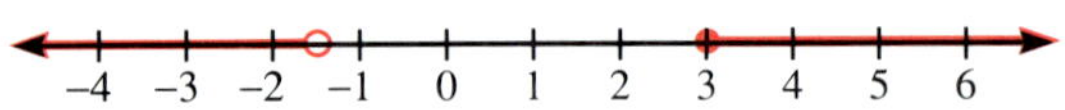

Chalkboard Examples

Solve each inequality and draw its graph.

1. $-2 \leq -x + 3 < 4$
$-5 \leq -x < 1$
$5 \geq x > -1$
$\{x\colon -1 < x \leq 5\}$

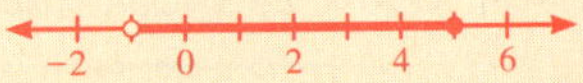

2. $c + 7 < 4$ or $7 - c < 1$
$c < -3$ or $-c < -6$
$c < -3$ or $c > 6$
$\{c\colon c < -3 \text{ or } c > 6\}$

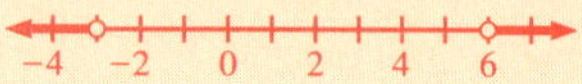

3. $-4 > \frac{3}{4}m - 7 > -7$
$-4 > \frac{3}{4}m - 7$ and
$\frac{3}{4}m - 7 > -7$
$3 > \frac{3}{4}m$ and $\frac{3}{4}m > 0$
$4 > m$ and $m > 0$
$\{m\colon 0 < m < 4\}$

Solve.

4. $-(k - 1) > 4$ and $-(1 - k) < 4$
$-k + 1 > 4$ and $-1 + k < 4$
$-k > 3$ and $k < 5$
$k < -3$ and $k < 5$
$\{k\colon k < -3\}$

5. $2f + 1 < -9$ or $0.1f + 0.5 \geq 0$
$2f < -10$ or $0.1f \geq -0.5$
$f < -5$ or $f \geq -5$
{real numbers}

Check for Understanding

Here is a suggested use of the Oral Exercises to check students' understanding as you teach the lesson.

Oral Exs. 1–6: use after Example 2.

Oral Exs. 7–14: use after Example 3.

Oral Exercises

Tell whether each conjunction or disjunction is true or false.

1. $-5 < -3$ and $3 < 5$ True
2. $-1 < 2$ and $-1 < -2$ False
3. $-6 > -2$ or $-6 > 2$ False
4. $-6 > -3$ or $-3 < 6$ True
5. $-7 < -5$ or $7 < -5$ True
6. $-4 < -6$ and $-4 < 6$ False

Match each graph with one of the open sentences in a–h.

7. 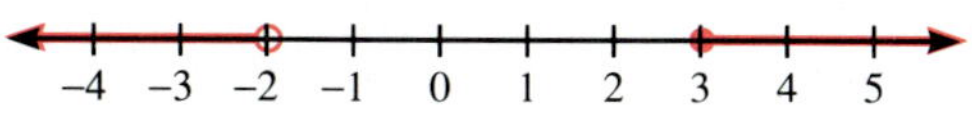 g

8. 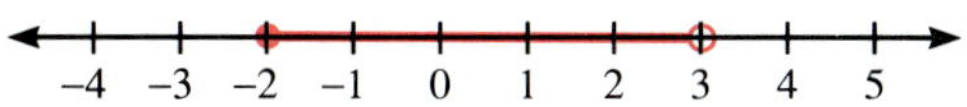 f

9. 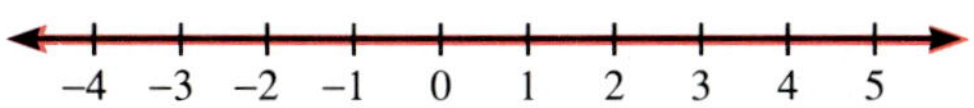 d

10. 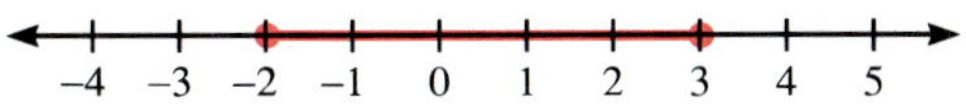 a

11.  c

12. 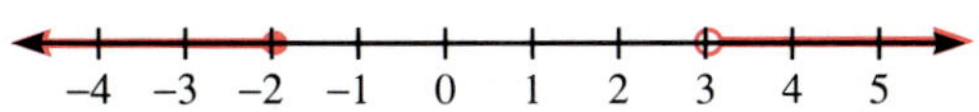 h, e

a. $-2 \le x \le 3$
b. $-2 < x < 3$
c. $x < 0$ or $x > 0$
d. $x \ge -2$ or $x \le 3$
e. $x > 3$ or $x \le -2$
f. $-2 \le x < 3$
g. $x < -2$ or $x \ge 3$
h. $x \le -2$ or $x > 3$

13. The inequality $x \ne 2$ is equivalent to the combined inequality $x < 2$ __?__ $x > 2$. or

14. Explain why it would be incorrect to write $-2 < x < -5$.
No number is both larger than -2 and smaller than -5.

Written Exercises

Solve each conjunction or disjunction and graph each solution set that is not empty.

A

1. $3 \le x < 5$ $\{x: 3 \le x < 5\}$
2. $z > -1$ and $z < 3$ $\{z: -1 < z < 3\}$
3. $t < 1$ or $t \ge 3$ $\{t: t < 1 \text{ or } t \ge 3\}$
4. $p > 1$ or $p < 1$ $\{p: p \ne 1\}$
5. $y \ge -1$ and $y \ge 3$ $\{y: y \ge 3\}$
6. $y \ge -1$ or $y \ge 3$ $\{y: y \ge -1\}$
7. $t > 0$ or $t < 2$ {real numbers}
8. $w < 0$ and $w \ge 4$ $\emptyset$
9. $0 \le x - 2 < 3$ $\{x: 2 \le x < 5\}$
10. $2 > y + 2 \ge 0$ $\{y: -2 \le y < 0\}$
11. $-1 > 2r - 5 > -9$ $\{r: -2 < r < 2\}$
12. $-1 \le 3z + 2 \le 8$ $\{z: -1 \le z \le 2\}$
13. $2z - 1 \le 5$ or $3z - 5 > 10$ $\{z: z \le 3 \text{ or } z > 5\}$
14. $3k + 7 < 1$ or $2k - 3 > 1$ $\{k: k < -2 \text{ or } k > 2\}$

Guided Practice

Solve each open sentence.

1. $-9 < w - 2 < -5$
$\{w: -7 < w < -3\}$
2. $8 + m < 1$ or $8 - m < 1$
$\{m: m < -7 \text{ or } m > 7\}$
3. $1 < 10 - 3y < 16$
$\{y: -2 < y < 3\}$
4. $\frac{1}{2}s \le -3$ or $-\frac{1}{2}s \le -4$
$\{s: s \le -6 \text{ or } s \ge 8\}$
5. $-2 > z - 2 > 2$ $\emptyset$

Summarizing the Lesson

This lesson discussed the concepts of conjunction and disjunction. Methods for solving these combined inequalities were presented in Examples 2 and 3. Stress to students that the graph of a conjunction is the intersection of the two graphs of its separate parts, while the graph of a disjunction is the union of the two graphs.

Suggested Assignments

Average Algebra
67/1–33 odd
S 63/27–32

Average Alg. and Trig.
67/2–34 even
S 63/28–32

Extended Algebra
67/2–34 even
S 63/32, 33

Extended Alg. and Trig.
67/5–33 odd
S 63/32, 33
Assign with Lesson 2-3.

Additional Answers Written Exercises

2. 

4. 

(continued)

Additional Answers Written Exercises
(continued)

6. 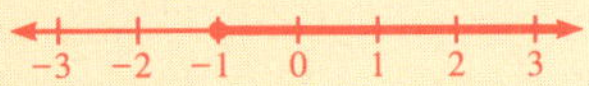

10. 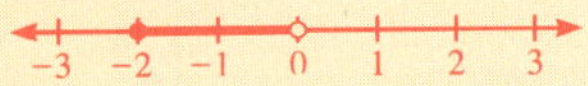

12. 

14. 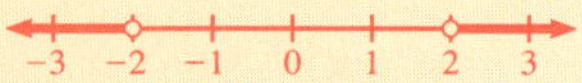

16. 

18. 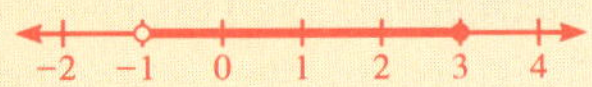

20. 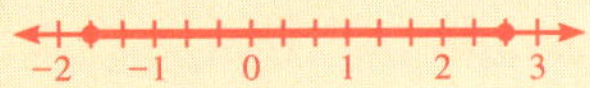

22. 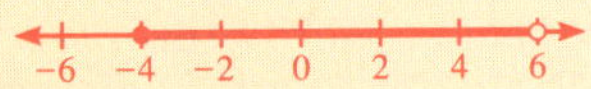

26. 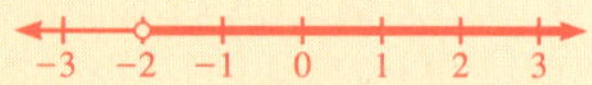

28. 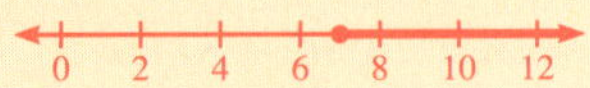

Supplementary Materials

Study Guide pp. 21–22
Practice Master 7

15. $2t + 7 \geq 13$ or $5t - 4 < 6$ $\{t: t \geq 3 \text{ or } t < 2\}$

16. $2x + 3 > 1$ or $5x - 9 \leq 6$ {real numbers}

17. $2t + 7 \geq 13$ and $5t - 4 < 6$ $\emptyset$

18. $2x + 3 > 1$ and $5x - 9 \leq 6$ $\{x: -1 < x \leq 3\}$

19. $-5 < 1 - 2k < 3$ $\{k: -1 < k < 3\}$

20. $-6 \leq 2 - 3m \leq 7$ $\left\{m: -\frac{5}{3} \leq m \leq \frac{8}{3}\right\}$

21. $-3 < 2 - \frac{d}{3} \leq -1$ $\{d: 9 \leq d < 15\}$

22. $3 \geq 1 - \frac{n}{2} > -2$ $\{n: -4 \leq n < 6\}$

B **23.** $7q - 1 > q + 11$ or $-11q > -33$ {real numbers}

24. $5n - 1 > 0$ and $4n + 2 < 0$ $\emptyset$

25. $x - 7 < 3x - 5 < x + 11$ $\{x: -1 < x < 8\}$

26. $3y + 5 \geq 2y + 1 > y - 1$ $\{y: y > -2\}$

27. $-\frac{3}{4}m \geq m - 1$ or $-\frac{3}{4}m < m + 1$ {real numbers}

28. $3z + 7 \leq 4z$ and $3z + 7 > -4z$ $\{z: z \geq 7\}$

Solve.

29. $-3 \leq -2(t - 3) < 6$ $\left\{t: 0 < t \leq \frac{9}{2}\right\}$

30. $-5 < 2(2 - s) + 1 \leq 9$ $\{s: -2 \leq s < 5\}$

31. $\frac{t}{4} + 2 < t + 3$ and $t - 3 > \frac{t}{2} - 4$ $\left\{t: t > -\frac{4}{3}\right\}$

32. $\frac{r - 3}{6} \leq r - 1$ or $\frac{r - 6}{3} \leq r + 4$ $\{r: r \geq -9\}$

C **33.** $0 < 1 - x \leq 3$ or $-1 \leq 2x - 3 \leq 5$ $\{x: -2 \leq x \leq 4\}$

34. $1 < -(2s + 1) < 5$ or $1 < 2s - 1 < 5$ $\{s: -3 < s < -1 \text{ or } 1 < s < 3\}$

35. $2 < \frac{y + 6}{2} < 5$ and $(4 - y > 5$ or $4 + y > 7)$ $\{y: -2 < y < -1 \text{ or } 3 < y < 4\}$

36. $\left(x \leq \frac{x + 4}{3} + 2 \text{ or } x \geq 2x - 1\right)$ and $1 \leq \frac{x - 1}{2} \leq 3$ $\{x: 3 \leq x \leq 5\}$

Historical Note / *Linkages*

The problem of constructing a linkage made of hinged rods that will draw a straight line has practical as well as theoretical interest. James Watt, the inventor of the steam engine, tried to construct such a device to guide the motion of the engine's piston. Watt's linkage, however, drew only an approximation of a straight line.

In 1864 a French army officer named Peaucellier solved the problem with the linkage shown at the right, in which one pivot moves in a perfectly straight line. Peaucellier was awarded the mechanical prize of the Institute of France in 1873.

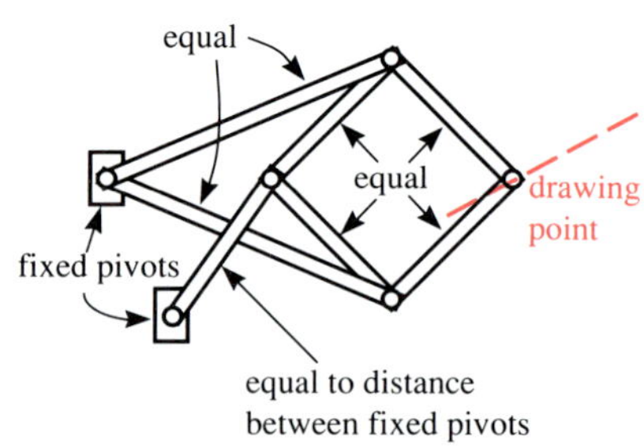

2-3 Problem Solving Using Inequalities

Objective To solve word problems by using inequalities in one variable.

Sometimes solving a word problem involves using an inequality.

Example 1 A bus is to be chartered for the senior class trip. The basic fare is $9.50 per passenger. If more than 20 people go, everyone's fare is reduced by $.30 for each passenger over this number (20). *At least* how many people must go to make the fare less than $7.50 per passenger?

Solution

Step 1 The problem asks for the least number of passengers needed to make the fare for each less than $7.50.

Step 2 Let n = the number of passengers.

Then $n - 20$ = the number of passengers over 20;

$0.30(n - 20)$ = the amount each passenger's fare is reduced; and

$9.50 - 0.30(n - 20)$ = the reduced fare per passenger.

Step 3 Reduced fare per passenger is less than $7.50.

$$9.50 - 0.30(n - 20) < 7.50$$

Step 4 Multiplying both sides of the inequality in Step 3 by 10 clears the decimals and gives this equivalent inequality to solve:

$$95 - 3(n - 20) < 75$$
$$95 - 3n + 60 < 75$$
$$-3n < -80$$
$$n > \frac{-80}{-3}$$
$$n > 26\frac{2}{3}$$

Interpret the result: Since the number of passengers must be an integer, $n \geq 27$.

Step 5 *Check:* Is the reduced fare less than $7.50? If at least 27 people go, the fare per passenger is reduced by $0.30(27 - 20) = 2.10$.
Then the reduced fare is $9.50 - 2.10 = 7.40 < 7.50$.

Is 27 the least number of passengers? Try 26. If 26 people go, the fare per passenger is reduced by $0.30(26 - 20) = 1.80$.
Then the reduced fare is $9.50 - 1.80 = 7.70 > 7.50$.

$\therefore$ at least 27 passengers must go. ***Answer***

Teaching Suggestions p. T85

Suggested Extensions p. T85

Warm-Up Exercises

Ask students to review the five-step plan. Have them look at Example 1 on page 49 and compare it to Example 1 on page 69.

Motivating the Lesson

Point out that there are many instances where inequalities can be used to represent actual data. For example, the statement "there were more than 20,000 people at the concert" can be represented as $x > 20{,}000$.

Chalkboard Examples

1. A store is charged $5.50 each for calculators and a delivery charge of $25 for the order. If the store sells the calculators for $8 each, how many calculators must be ordered and sold to produce a profit of at least $80?

S.1 The problem asks for the number of calculators to produce the given profit.

S.2 Let n = required number of calculators.
Income = $8n$ dollars
Cost = $5.5n + 25$ dollars

S.3 Profit = income − cost
$8n - (5.5n + 25) \geq 80$

S.4 $2.5n \geq 105$
$n \geq 42$

S.5
$8(42) - (5.5 \cdot 42 + 25) \overset{?}{\geq} 80$
$336 - 256 \geq 80$ ✓
$\therefore$ at least 42 calculators must be ordered and sold.

Chalkboard Examples

(continued)

2. The Perez family has 100 more shares of stock B than of stock A. The current price per share of stock A is \$18.50 and of stock B is \$24.75. At most how many shares of each do they have if the average price per share is greater than \$22?

S.1 The problem asks for the maximum number of shares of each stock.

S.2 Let a = number of shares of stock A. Then $a + 100$ = number of shares of stock B.

S.3

$$\frac{18.5a + 24.75(a + 100)}{a + (a + 100)} > 22$$

S.4

$$43.25a + 2475 > 44a + 2200$$
$$275 > 0.75a$$
$$366\tfrac{2}{3} > a$$

At most they have 366 shares of A and 466 shares of B.

S.5 Is the average price per share greater than \$22?

$$\frac{18.5(366) + 24.75(466)}{366 + 466} =$$
$$\frac{18{,}304.5}{832} > 22 \quad \checkmark$$

∴ at most they have 366 shares of stock A and 466 shares of stock B.

Guided Practice

1. Bert's bank contains twice as many nickels as quarters and 3 more dimes than nickels. If the coins are worth more than \$5, at least how many dimes does he have? 21 dimes

Certain phrases can be translated into mathematical terms using inequalities. Here is a list of the more common ones.

Phrase	Translation
x is at least a. x is no less than a.	$x \geq a$
x is at most b. x is no greater than b.	$x \leq b$
x is between a and b. x is between a and b, inclusive.	$a < x < b$ $a \leq x \leq b$

Example 2 Find all sets of 4 consecutive integers whose sum is between 10 and 20.

Solution

Step 1 The problem asks for 4 consecutive integers; their sum must be greater than 10 and less than 20.

Step 2 Let n = the first of these integers.
Then the other three are $n + 1$, $n + 2$, and $n + 3$.

Step 3

$$10 < \text{the sum} < 20$$
$$10 < n + (n + 1) + (n + 2) + (n + 3) < 20$$

Step 4

$$10 < 4n + 6 < 20$$
$$4 < 4n < 14$$
$$1 < n < \frac{14}{4}$$

Interpret the result: Since n is an integer, there are only two values possible for n: 2 and 3. There are two sets of consecutive integers that fulfill the requirements of the problem:

$$\{2, 3, 4, 5\} \text{ and } \{3, 4, 5, 6\}$$

Step 5 *Check:* Is the sum between 10 and 20?

For $\{2, 3, 4, 5\}$: $10 < 2 + 3 + 4 + 5 < 20$
$10 < 14 < 20$ ✓

For $\{3, 4, 5, 6\}$: $10 < 3 + 4 + 5 + 6 < 20$
$10 < 18 < 20$ ✓

To complete the check, you must show that any other set of four consecutive integers will not satisfy the requirements. In fact, you need only eliminate the set of "next greater" integers, $\{4, 5, 6, 7\}$, and the set of "next smaller" integers, $\{1, 2, 3, 4\}$. That work is left for you to do.

∴ the required sets are $\{2, 3, 4, 5\}$ and $\{3, 4, 5, 6\}$. ***Answer***

Problems

Solve.

A 1. For the Hawks' 80 basketball games next year, you can buy separate tickets for each game at \$9 each, or you can buy a season ticket for \$580. At most how many games could you attend at the \$9 price before spending more than the cost of a season ticket? At most 64 games

2. The usual toll charge to use the Bingham tunnel is 50 cents. If you purchase a special sticker for \$5.50, the toll is only 35 cents. At least how many trips through the tunnel are needed before the sticker costs less than paying for each trip separately? At least 37 trips

3. The length of a rectangle is 5 cm more than twice its width. Find the largest possible width if the perimeter is at most 64 cm. 9 cm

4. The lengths of the legs of an isosceles triangle are integers. The base is half as long as each leg. What are the possible lengths of the legs if the perimeter is between 6 units and 16 units? 3, 4, 5, or 6 units

5. Find all sets of three consecutive *odd* integers whose sum is between 20 and 30. {5, 7, 9}, {7, 9, 11}

6. Find all sets of three consecutive *even* integers whose sum is between 25 and 45. {8, 10, 12}, {10, 12, 14}, {12, 14, 16}

7. Jeannie's scores on her first four tests were 80, 65, 87, and 75. What will she have to score on her next test to obtain an average of at least 80 for the term? 93

8. Jim's second test score was 8 points higher than his first score. His third score was 88. He had a B average (between 80 and 89, inclusive) for the three tests. What can you conclude about his first test score?

8. First score was between 72 and 85.5, inclusive.

9. The sides $\overline{AB}$ and $\overline{AD}$ of a square are extended 10 cm and 6 cm, respectively, to form sides $\overline{AE}$ and $\overline{AF}$ of a rectangle. At most how long is the side of the square if the perimeter of the rectangle is at least twice the perimeter of the square? At most 8 cm

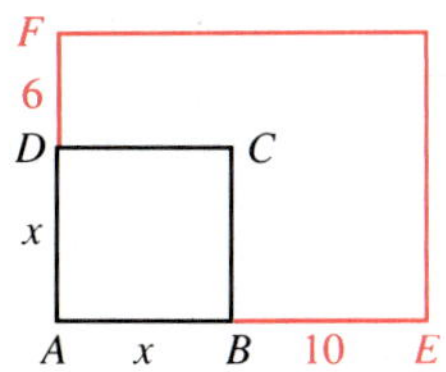

10. The three sides of an equilateral triangle are increased by 20 cm, 30 cm, and 40 cm, respectively. The perimeter of the resulting triangle is between twice and three times the perimeter of the original triangle. What can you conclude about the length of a side of the original triangle?
The length of a side is between 15 cm and 30 cm.

B 11. The telephone company offers two types of service. With Plan A, you can make an unlimited number of local calls per month for \$18.50. With Plan B, you pay \$6.50 monthly, plus 10 cents for each min of calls after the first 40 min. At least how many min would you have to use the telephone each month to make Plan A the better option? At least 160 min

2. A regular hexagon, a square, and an equilateral triangle all have equal sides. If the sum of the perimeters of the square and the triangle is no more than 18 cm less than twice the perimeter of the hexagon, what is the minimum length of each side? 3.6 cm

3. Find all the sets of five consecutive even integers whose sum is between 225 and 250.
{42, 44, 46, 48, 50} and {44, 46, 48, 50, 52}

Summarizing the Lesson

In this lesson the problem solving skills studied in Chapter 1 and the methods of solution for inequalities studied in Lessons 2-1 and 2-2 were used to solve word problems involving inequalities.

Suggested Assignments

Average Algebra
71/*P*: 2–14 even
S 72/Mixed Review
R 72/Self-Test 1

Average Alg. and Trig.
71/*P*: 1, 3, 6, 8, 11–14
S 72/Mixed Review
R 72/Self-Test 1

Extended Algebra
71/*P*: 3, 6, 8, 11–16
S 72/Mixed Review
R 72/Self-Test 1

Extended Alg. and Trig.
71/*P*: 1, 4, 7, 11, 13, 15
S 72/Mixed Review
R 72/Self-Test 1
Assign with Lesson 2-2.

Supplementary Materials

Study Guide	pp. 23–24
Practice Master	8
Test Master	5
Resource Book	p. 104

Quick Quiz

Solve each inequality and graph the solution set.

1. $-\left(\frac{3}{2}x + 18\right) \leq 6$
   $\{x: x \geq -16\}$

2. $3(y - 7) < y + 5$
   $\{y: y < 13\}$

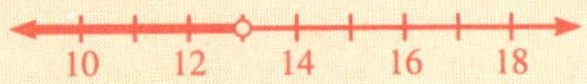

3. $3(8 - 6t) \geq -2(5 + 9t)$
   {real numbers}

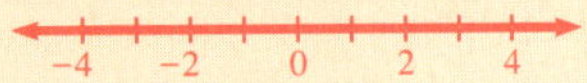

4. $9 < \frac{m}{2} - 3 < 11$
   $\{m: 24 < m < 28\}$

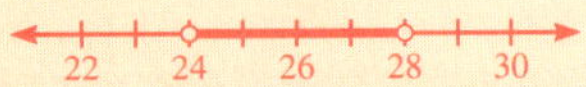

5. $-7 - 5z \geq -3(z + 1)$
   $\{z: z \leq -2\}$

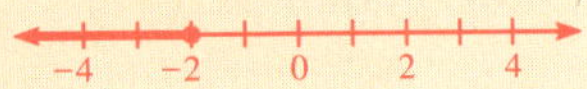

6. $-j + 1 \geq -4$ or $-\frac{2}{3}j < -6$
   $\{j: j \leq 5 \text{ or } j > 9\}$

7. Tickets for the school concert were $1.50 each for students and $3 each for others. Five hundred tickets were sold in all and receipts exceeded $1000. What can you conclude about the number of student tickets sold? At most, 333 student tickets were sold.

12. During the first 20 mi of a 50 mi bicycle race, Roger's average speed was 16 mi/h. What must his average speed be during the remainder of the race if he is to finish the race in less than 2.5h? More than 24 mi/h

13. A subway train makes six stops of equal length during its 21 km run. The train is actually moving for 20 min of the trip. At most how long can the train remain at each station if the average speed for the trip, including stops, is to be at least 36 km/h? At most $2\frac{1}{2}$ min

14. The length of a rectangular sheet of paper was twice its width. After 1 cm was trimmed from each edge of the sheet, the perimeter was at most 1 m. Find the largest possible dimensions of the trimmed sheet. 16 cm × 34 cm

C 15. A swimming pool is 5 m longer than it is wide and is surrounded by a deck 2 m wide. The area of the pool and deck together is at least 140 m^2 greater than the area of the pool alone. What can you conclude about the dimensions of the pool? Width: at least 13 m; length: at least 18 m

16. Find all triples of consecutive integers such that 11 times the largest of the integers is at least 46 more than the product of the other two. {4, 5, 6}, {5, 6, 7}, {6, 7, 8}

Mixed Review Exercises

Solve each open sentence and graph each solution set that is not empty.

1. $3x - 2 \geq -8$ $\{x: x \geq -2\}$
2. $\frac{1}{2}t \leq -2$ or $t - 4 \geq -3$ $\{t: t \leq -4 \text{ or } t \geq 1\}$
3. $2(m - 2) > 4 - 3(1 - m)$ $\{m: m < -5\}$
4. $-2y < -6$ and $y + 3 \leq 1$ $\emptyset$
5. $7 - 4d < 3$ $\{d: d > 1\}$
6. $-1 < 5 - 2p < 5$ $\{p: 0 < p < 3\}$

Evaluate if $x = -7$ and $y = 3$.

7. $|x + y|$ 4
8. $|x| + |y|$ 10
9. $|xy|$ 21
10. $|x| \cdot |y|$ 21

Self-Test 1

Vocabulary equivalent inequalities (p. 60) disjunction (p. 65) conjunction (p. 65)

Solve each inequality and graph the solution set.

1. $6m - 13 \leq 5$ 2. $2y + 9 < 5(y + 3)$ 3. $8 - 3x > -7$ **Obj. 2-1, p. 59**

4. $\frac{n-3}{2} \geq n - 1$ 5. $1 < 4a + 5 < 9$ 6. $c - 3 > 2c$ or $\frac{c}{3} \geq 1$ **Obj. 2-2, p. 65**

7. Bill's scores on his first two tests were 75 and 82. What will he have to score on his next test to obtain an average of at least 80? 83 **Obj. 2-3, p. 69**

Working with Absolute Value

2-4 Absolute Value in Open Sentences

Objective To solve open sentences involving absolute value.

If you think of the absolute value of a real number x (see page 3) as the distance between the graph of x and the origin on a number line, you can see why the sentences below are equivalent.

Sentence	Equivalent Sentence	Graph
$\|x\| = 1$ Distance between x and 0 equals 1.	$x = -1$ or $x = 1$	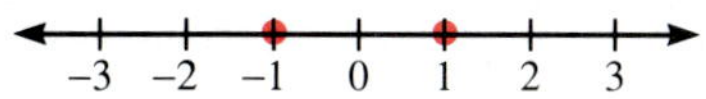
$\|x\| > 1$ Distance between x and 0 is greater than 1.	$x < -1$ or $x > 1$	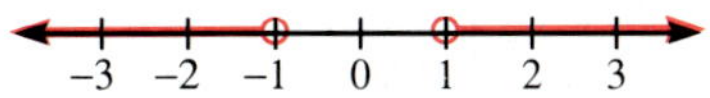
$\|x\| < 1$ Distance between x and 0 is less than 1.	$-1 < x < 1$	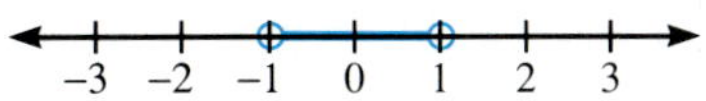

You can often solve an open sentence involving absolute value by first writing an equivalent disjunction or conjunction.

Example 1 Solve $|3x - 2| = 8$.

Solution $|3x - 2| = 8$ is equivalent to this disjunction:

$$3x - 2 = -8 \quad \text{or} \quad 3x - 2 = 8$$
$$3x = -6 \qquad\qquad 3x = 10$$
$$x = -2 \quad \text{or} \quad x = \frac{10}{3}$$

$\therefore$ the solution set is $\left\{-2, \frac{10}{3}\right\}$. ***Answer***

Example 2 Solve $|3 - 2t| < 5$.

Solution $|3 - 2t| < 5$ is equivalent to this conjunction:

$$-5 < 3 - 2t < 5$$
$$-8 < -2t < 2$$
$$4 > t > -1$$

$\therefore$ the solution set is $\{t\colon -1 < t < 4\}$. ***Answer***

Teaching Suggestions p. T85

Suggested Extensions p. T85

Warm-Up Exercises

State whether the inequality is true for the values given.

1. $|x| < 2$ $\{0, -3, 3\}$ T F F
2. $|x - 1| > 8$ $\{0, 8, -8\}$ F F T
3. $|2x - 1| < 10$ $\{0, 2, -5\}$ T T F
4. $|3x + 2| - 5 \le 8$ $\{0, 1, 4\}$ T T F

Motivating the Lesson

This lesson extends the concept of absolute value beyond the definition. Methods for solving equations and inequalities involving absolute value and graphing the solution sets are presented.

Thinking Skills

Before attempting to solve an inequality involving absolute value, students should *analyze* the problem to see if it is equivalent to a conjunction or a disjunction.

Common Errors

Some students mistakenly think that $|x - 7| = 5$ is equivalent to $x - 7 = 5$. When reteaching, emphasize that $|x - 7| = 5$ means that $(x - 7)$ is 5 units from the origin on a number line and thus $x - 7 = -5$ or $x - 7 = 5$.

Supplementary Materials

Study Guide pp. 25–26

Chalkboard Examples

Solve, and graph the solution set.

1. $|2s + 3| = 5$
$2s + 3 = 5$ or $2s + 3 = -5$
$2s = 2$ or $2s = -8$
$s = 1$ or $s = -4$
$\{1, -4\}$

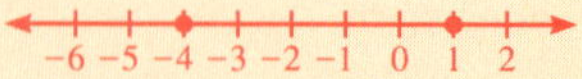

2. $|3 - k| < 2$
$-2 < 3 - k < 2$
$-5 < -k < -1$
$5 > k > 1$
$\{k: 1 < k < 5\}$

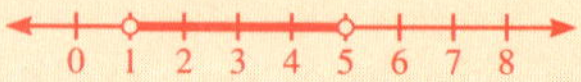

3. $\frac{3}{2} \leq \left|2 + \frac{1}{2}x\right|$
$\left|2 + \frac{1}{2}x\right| \geq \frac{3}{2}$ is equivalent to $2 + \frac{1}{2}x \leq -\frac{3}{2}$ or
$2 + \frac{1}{2}x \geq \frac{3}{2}$
$\frac{1}{2}x \leq -\frac{7}{2}$ or $\frac{1}{2}x \geq -\frac{1}{2}$
$x \leq -7$ or $x \geq -1$
$\{x: x \leq -7$ or $x \geq -1\}$

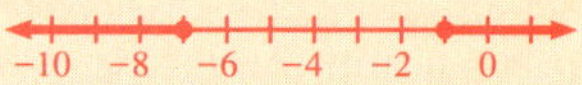

Check for Understanding

Here is a suggested use of the Oral Exercises to check students' understanding as you teach the lesson.
Oral Exs. 1–11: use after Example 3.

Guided Practice

Solve and graph the solution set.

1. $|2w - 5| > 7$
$\{w: w > 6$ or $w < -1\}$

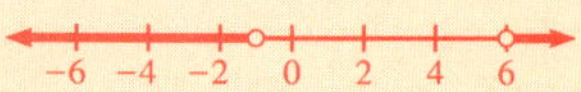

2. $|6 + x| \leq 4$
$\{x: -10 \leq x \leq -2\}$

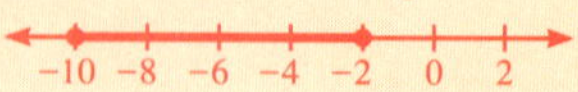

Example 3 Solve $|2z - 1| + 3 \geq 8$ and graph its solution set.

Solution First transform the inequality to an equivalent inequality in which the expression involving absolute value is alone on one side.

$$|2z - 1| + 3 \geq 8$$
$$|2z - 1| + 3 - 3 \geq 8 - 3$$
$$|2z - 1| \geq 5$$

The last inequality is equivalent to this disjunction:

$2z - 1 \leq -5$	or	$2z - 1 \geq 5$
$2z \leq -4$	↓	$2z \geq 6$
$z \leq -2$	or	$z \geq 3$

$\therefore$ the solution set is $\{z: z \leq -2$ or $z \geq 3\}$. ***Answer***

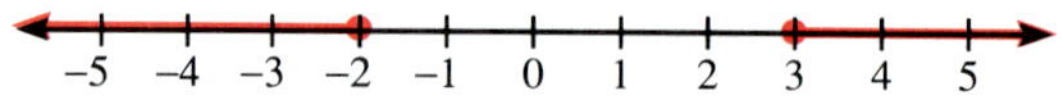

You can help guard against errors by testing one value from each region of the graph. Substitute values in the original inequality $|2z - 1| + 3 \geq 8$.

Try $z = -4$:	$|2(-4) - 1| + 3 = |-9| + 3 = 12 \geq 8$	True	✓
Try $z = 0$:	$|2 \cdot 0 - 1| + 3 = |-1| + 3 = 4 \geq 8$	False	✓
Try $z = 4$:	$|2 \cdot 4 - 1| + 3 = |7| + 3 = 10 \geq 8$	True	✓

You can tell at a glance that an inequality such as $|x - 3| \geq -2$ is true for all real numbers x, because the absolute value of every real number is nonnegative. On the other hand, an inequality such as $|t + 5| < -1$ has $\emptyset$ as its solution set (why?).

Oral Exercises

Express each open sentence as an equivalent conjunction or disjunction without absolute value.

Sample 1 $|3t - 1| > 2$ **Solution** $3t - 1 < -2$ or $3t - 1 > 2$

1. $|x| \leq 3$
2. $|t| = 2$
3. $|z| > 0$
4. $|y - 3| \leq 2$
5. $|s + 3| = 3$
6. $|2x - 3| \geq 1$
7. $|3t - 1| \leq 2$
8. $|5 - 2z| < 3$

Express each conjunction or disjunction as an equivalent open sentence involving absolute value.

Sample 2 $-1 \leq x - 2 \leq 1$ **Solution** $|x - 2| \leq 1$

9. $u = -3$ or $u = 3$ $|u| = 3$
10. $t \geq -3$ and $t \leq 3$ $|t| \leq 3$
11. $3 > 4(x - 1) > -3$ $|4(x - 1)| < 3$

Written Exercises

A **1–8.** Graph the solution set of each open sentence in Oral Exercises 1–8.

Solve and graph the solution set.

9. $|2t + 5| < 3$ $\{t: -4 < t < -1\}$

10. $|3x + 2| > 4$ $\left\{x: x < -2 \text{ or } x > \frac{2}{3}\right\}$

11. $|2u - 5| = 0$ $\left\{\frac{5}{2}\right\}$

12. $8 = |5y + 2|$ $\left\{-2, \frac{6}{5}\right\}$

13. $\left|1 - \frac{x}{3}\right| \geq \frac{2}{3}$ $\{x: x \leq 1 \text{ or } x \geq 5\}$

14. $\left|1 - \frac{p}{2}\right| \leq 2$ $\{p: -2 \leq p \leq 6\}$

15. $0 \leq |4u - 7|$ {real numbers}

16. $|3r - 12| > 0$ $\{r: r \neq 4\}$

17. $\left|\frac{t - 2}{4}\right| \leq \frac{1}{2}$ $\{t: 0 \leq t \leq 4\}$

18. $1 > |2 - 0.8n|$ $\left\{n: \frac{5}{4} < n < \frac{15}{4}\right\}$

Solve.

B **19.** $|x + 5| - 3 = 1$ $\{-1, -9\}$

20. $|2t - 3| + 2 = 5$ $\{0, 3\}$

21. $|2u - 1| + 3 \leq 6$ $\{u: -1 \leq u \leq 2\}$

22. $4 - |3k + 1| < 2$ $\left\{k: k < -1 \text{ or } k > \frac{1}{3}\right\}$

23. $7 - 3|4d - 7| \geq 4$ $\left\{d: \frac{3}{2} \leq d \leq 2\right\}$

24. $6 + 5|2r - 3| \geq 4$ {real numbers}

25. $4 + 2\left|\frac{3t - 5}{2}\right| > 5$ $\left\{t: t < \frac{4}{3} \text{ or } t > 2\right\}$

26. $2\left|\frac{2t - 5}{3}\right| - 3 \geq 5$ $\left\{t: t \leq -\frac{7}{2} \text{ or } t \geq \frac{17}{2}\right\}$

27. $7 + 5|c| \leq 1 - 3|c|$ $\emptyset$

28. $\frac{1}{2}|d| + 5 \geq 2|d| - 13$ $\{d: -12 \leq d \leq 12\}$

Graph the solution set of each open sentence.

C **29.** $2 < |w| < 4$

30. $1 \leq |s - 2| \leq 3$

31. $1 \leq |2x + 1| < 3$

32. $0 < |2 - r| \leq 2$

Solve.

33. $|2x| \leq |x - 3|$ $\{x: -3 \leq x \leq 1\}$

34. $|t| > |2t - 6|$ $\{t: 2 < t < 6\}$

Computer Exercises

For students with some programming experience.

1. Write a program to list all integers x in the interval $-50 \leq x \leq 50$ that are solutions of an open sentence of the form $a < |cx + d| < b$. The values of a, b, c, and d are to be entered by the user. If no integers in the given interval satisfy the inequality, have the output state this. You will need to use the BASIC function ABS in your program.

2. Use the program in Exercise 1 to find the integer solutions of each open sentence. **a.** $x = -3, -2, -1, 0, 1, 2, 15, 16, 17, 18, 19$

a. $17 < |3x - 25| < 35$ **b.** $1 < |18x + 120| < 100$

b. $x = -12, -11, -10, -9, -8, -7, -6, -5, -4, -3, -2$

Summarizing the Lesson

The concepts of conjunction and disjunction were used in this lesson to solve equations and inequalities involving absolute value.

Suggested Assignments

Average Algebra
75/1–25 odd
S 71/*P*: 11, 13, 15

Average Alg. and Trig.
75/1–27 odd

Extended Algebra
75/1–33 odd

Extended Alg. and Trig.
75/12–34 even

Assign with Lesson 2-5.

Additional Answers
Oral Exercises

1. $-3 \leq x \leq 3$
2. $t = 2$ or $t = -2$
3. $z > 0$ or $z < 0$
4. $-2 \leq y - 3 \leq 2$
5. $s + 3 = 3$ or $s + 3 = -3$
6. $2x - 3 \geq 1$ or $2x - 3 \leq -1$
7. $-2 \leq 3t - 1 \leq 2$
8. $-3 < 5 - 2z < 3$

Additional Answers
Written Exercises

2. 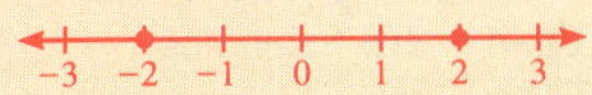

4. 

6. 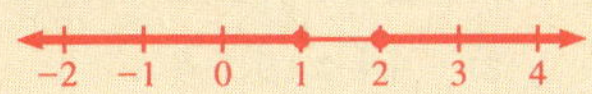

8. 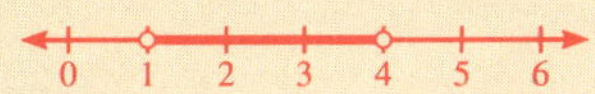

10. 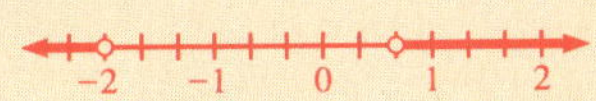

(continued on p. 98)

Teaching Suggestions p. T86

Reading Algebra p. T86

Suggested Extensions p. T86

Warm-Up Exercises

Graph the solution set of each open sentence.

1. $|x| = 3$

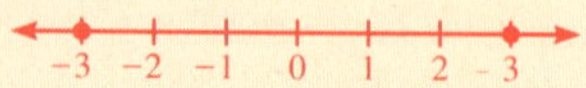

2. $|x| \le 2$

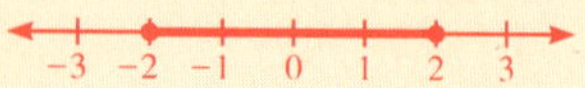

3. $|x| > 1$

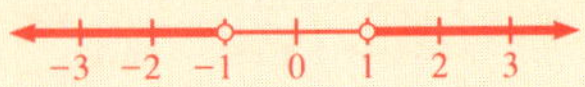

4. $-5 < x < 1$

5. $x > 2$ or $x < -1$

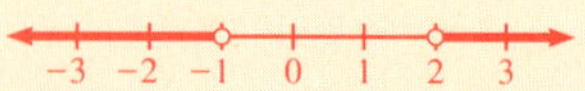

Motivating the Lesson

It is often helpful, and in some cases easier, to solve absolute value equations and inequalities by using graphs. The solution set may be easily seen from the graph. Also, graphing techniques are useful for solving more complicated problems.

Chalkboard Examples

1. Find the distance between the graphs of (a) 2 and -7, (b) $-\frac{3}{4}$ and $\frac{5}{4}$, and (c) -14 and -41.
   a. $|2 - (-7)| = |9| = 9$
   b. $\left|-\frac{3}{4} - \frac{5}{4}\right| = \left|-\frac{8}{4}\right| = |-2| = 2$
   c. $|-14 - (-41)| = |27| = 27$

2-5 Solving Absolute Value Sentences Graphically

Objective To use number lines to obtain quick solutions to certain equations and inequalities involving absolute value.

You know that on a number line the distance between the graph of a real number x and the origin is $|x|$. *The distance on the number line between the graphs of real numbers a and b is* $|a - b|$, *or equivalently,* $|b - a|$. You can use this fact to solve many open sentences almost at sight.

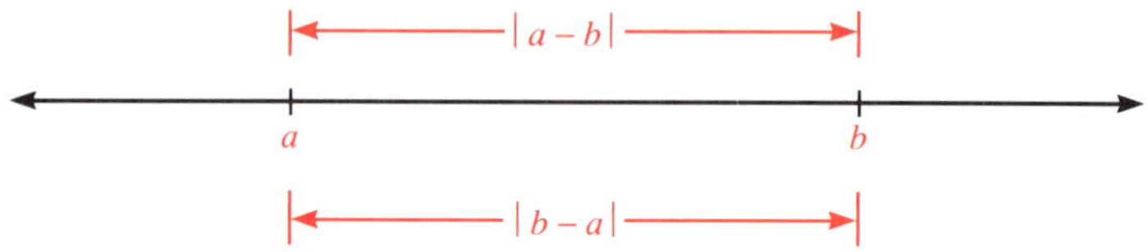

Example 1 Find the distance between the graphs of each pair of numbers.

a. 8 and 17 **b.** -11 and -6 **c.** -9 and 12

Solution

a. $|8 - 17| = |-9| = 9$ ***Answer***

b. $|-11 - (-6)| = |-11 + 6| = |-5| = 5$ ***Answer***

c. $|-9 - 12| = |-21| = 21$ ***Answer***

Example 2 Solve $|x - 3| = 2$.

Solution To satisfy $|x - 3| = 2$, x must be a number whose distance from 3 is 2 units. So, to find x, start at 3 and move 2 units in each direction on a number line.

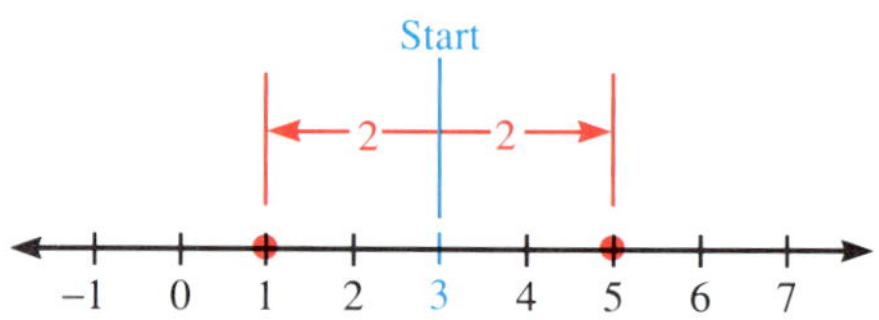

You arrive at 1 and 5 as the values of x.

$\therefore$ the solution set is $\{1, 5\}$. ***Answer***

Example 3 Solve $|y + 2| \le 3$.

Solution $y + 2 = y - (-2)$

Therefore, $|y + 2| \le 3$ is equivalent to $|y - (-2)| \le 3$.

So the distance between y and -2 must be 3 units or less. To find y, start at -2 and move 3 units in each direction on a number line.

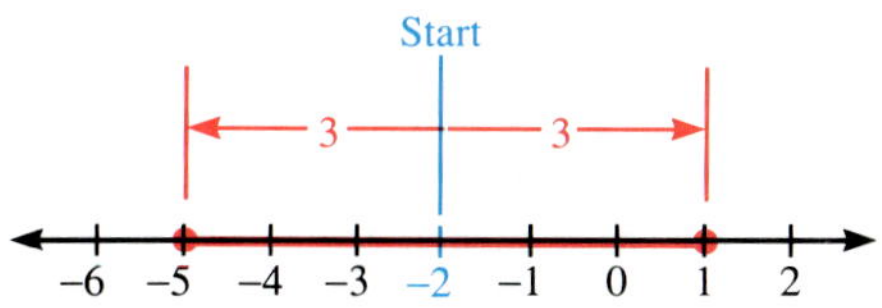

The numbers up to and including 1 and the numbers down to and including -5 will satisfy the inequality.

$\therefore$ the solution set is $\{y: -5 \le y \le 1\}$. ***Answer***

Certain equations and inequalities, such as the ones in Examples 1–3, lend themselves more easily to a graphic solution. With these types, the expression involving absolute value is of the form $|x - \text{constant}|$. When the expression is more complicated, as in Example 4, a graphic method can also be applied, though not as easily. For these types, you might prefer to use the algebraic method learned in the previous lesson.

Example 4 Solve $|5 - 2t| > 3$.

Solution Use the facts that $|a - b| = |b - a|$ and $|ab| = |a| \cdot |b|$ (page 27) to rewrite $|5 - 2t|$ this way:

$$|5 - 2t| = |2t - 5| = \left|2 \cdot \left(t - \frac{5}{2}\right)\right|$$
$$= |2| \cdot \left|t - \frac{5}{2}\right|$$
$$= 2\left|t - \frac{5}{2}\right|$$

Therefore, the given inequality

$$|5 - 2t| > 3$$

is equivalent to $\quad 2\left|t - \frac{5}{2}\right| > 3$, or $\left|t - \frac{5}{2}\right| > \frac{3}{2}$

To find t, start at $\frac{5}{2}$ and move *more than* $\frac{3}{2}$ units to the right and to the left.

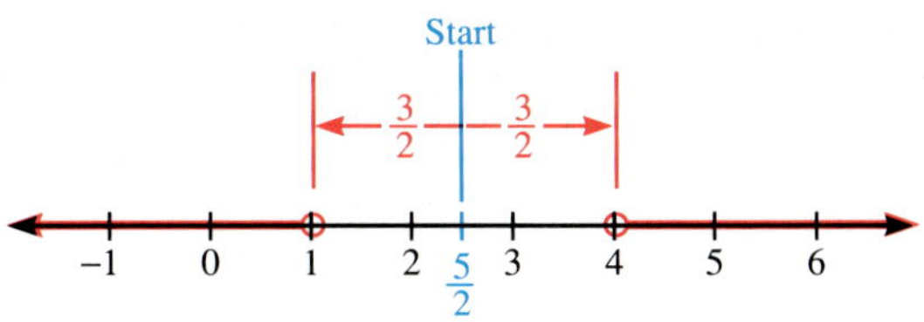

$\therefore$ the solution set is $\{t: t < 1 \text{ or } t > 4\}$. ***Answer***

Use graphs to solve each open sentence.

2. $|5 - x| = 2$
Note that $|5 - x| = |x - 5|$. The distance between x and 5 is 2.

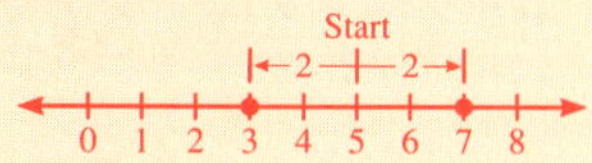

$\{3, 7\}$

3. $|b + 5| > 3$
Since $|b - (-5)| > 3$, the distance between b and -5 must be greater than 3.

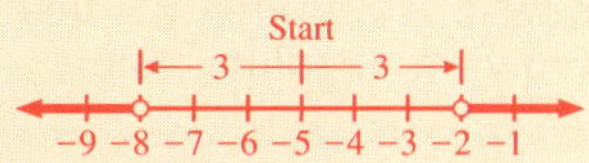

$\{b: b < -8 \text{ or } b > -2\}$

4. $|2n + 5| \le 3$

$$\left|2\left(n + \frac{5}{2}\right)\right| \le 3$$
$$|2| \cdot \left|n + \frac{5}{2}\right| \le 3$$
$$2 \cdot \left|n + \frac{5}{2}\right| \le 3$$
$$\left|n - \left(-\frac{5}{2}\right)\right| \le \frac{3}{2}$$

The distance between n and $-\frac{5}{2}$ is at most $\frac{3}{2}$.

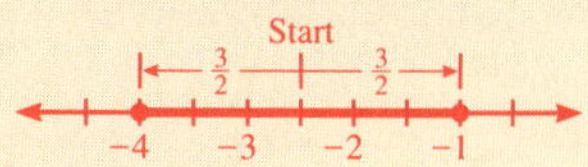

$\{n: -4 \le n \le -1\}$

Check for Understanding

Oral Exs. 1–10: use after Example 2.
Oral Exs. 11–16: use after Example 3.
Oral Exs. 17–24: use after Example 4.

Supplementary Materials

Study Guide	pp. 27–28
Practice Master	9
Computer Activities	3, 4
Resource Book	p. 105

Guided Practice

Solve graphically.

1. $|-6 + 3g| = 6$

2. $|3n + 24| > 0$

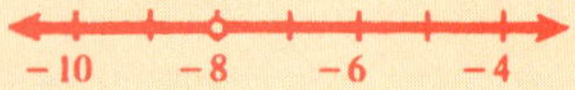

3. $|2t - 3| \leq 5$

Summarizing the Lesson

In this lesson graphing techniques were used to solve equations and inequalities involving absolute value.

Additional Answers
Oral Exercises

11. The distance between t and 0 is at most 5.
12. The distance between r and 2 is greater than 1.
13. The distance between 3 and s is 4.
14. The distance between x and -1 is at most 4.
15. The distance between u and -5 is at least 3.
16. The distance between y and 2 is less than 2.

Suggested Assignments

Average Algebra
78/2–24 even
S 79/Mixed Review

Average Alg. and Trig.
78/2–24 even
S 79/Mixed Review

Extended Algebra
78/2–26 even
S 79/Mixed Review

Extended Alg. and Trig.
78/3–30 mult. of 3
S 79/Mixed Review
Assign with Lesson 2-4.

Oral Exercises

Find the distance between the graphs of each pair of numbers.

1. 3 and -4 7
2. -6 and -1 5
3. 12 and 5 7
4. -7 and 1 8
5. 11 and 8 3
6. -11 and 8 19
7. 11 and -8 19
8. -11 and -8 3
9. t and $4t$ if **(a)** t is positive, and **(b)** t is negative. **a.** $3t$ **b.** $-3t$
10. $-2k$ and $2k$ if **(a)** k is positive, and **(b)** k is negative. **a.** $4k$ **b.** $-4k$

Translate each open sentence into a statement involving distance.

Sample $|z + 2| \geq 3$ **Solution** The distance between z and -2 is at least 3.

11. $|t| \leq 5$
12. $|r - 2| > 1$
13. $|3 - s| = 4$
14. $|x + 1| \leq 4$
15. $|u + 5| \geq 3$
16. $|2 - y| < 2$

Translate each statement into an open sentence using absolute value.

Sample The numbers whose distance from -3 is at most 2 **Solution** $|x - (-3)| \leq 2$, or $|x + 3| \leq 2$

17. The numbers whose distance from 2 is less than 5 $|x - 2| < 5$
18. The numbers whose distance from -3 is at least 3 $|x + 3| \geq 3$
19. The numbers whose distance from -1 is not more than 3 $|x + 1| \leq 3$
20. The numbers whose distance from 4 is equal to 4 $|x - 4| = 4$
21. The numbers whose distance from $-\frac{2}{3}$ is not less than 5 $\left|x + \frac{2}{3}\right| \geq 5$
22. The numbers whose distance from 0 is at most 6 $|x| \leq 6$
23. The numbers whose distance from -1.2 is greater than 4.8 $|x + 1.2| > 4.8$
24. The numbers whose distance from a is not more than h $|x - a| \leq h$

Written Exercises

Solve each open sentence graphically.

A

1. $|w| = 4$
2. $|x - 3| = 2$
3. $|t| < 3$
4. $|y| > 4$
5. $|u| \geq 2$
6. $|p| \leq 5$
7. $|y - 2| < 3$
8. $|k - 4| > 1$
9. $\left|t - \frac{3}{2}\right| \leq \frac{5}{2}$
10. $\left|x - \frac{1}{2}\right| > \frac{3}{2}$
11. $|r + 2| > 5$
12. $|w + 2| \leq 2$

Solve each open sentence. Use whichever method you prefer.

B 13. $|2x - 1| = 3$ 14. $\left|\frac{1}{4}y + 1\right| = \frac{1}{2}$ 15. $|2p + 5| \geq 3$

16. $|3k - 2| < 4$ 17. $\left|\frac{1}{3}t - 1\right| \leq \frac{2}{3}$ 18. $\left|\frac{y}{4} + \frac{1}{2}\right| > \frac{3}{4}$

19. $3 - |2x - 3| > 1$ 20. $7 - |3y - 2| \leq 1$ 21. $|9 + 3f| < 4$

22. $|4 + 2y| \geq 3$ 23. $|1.2 + 0.4t| < 2$ 24. $|1 - 0.3x| \geq 1.5$

Solve for x in terms of the other variables. Assume that a, b, and c are positive numbers.

C 25. $|x - a| \geq c$ 26. $|x + a| \leq c$ 27. $|x - c| < c$

28. $|x + c| > c$ 29. $|a + bx| \geq c$ 30. $|a - bx| \leq c$

31. $a - |bx| < c$ (Assume $a - c \geq 0$.) 32. $a + |bx| > c$ (Assume $c - a \geq 0$.) 33. $b|x + a| < c$

Mixed Review Exercises

Solve each open sentence.

1. $1 \leq 3x + 4 \leq 13$ $\{x: -1 \leq x \leq 3\}$
2. $|5 - 2n| = 3$ $\{1, 4\}$
3. $4(2c - 3) > 7c - 9$ $\{c: c > 3\}$
4. $|6y + 6| > 0$ $\{y: y \neq -1\}$
5. $|w| + 4 \geq 6$ $\{w: w \leq -2 \text{ or } w \geq 2\}$
6. $p + 2 < -1$ or $-4p \leq -8$ $\{p: p < -3 \text{ or } p \geq 2\}$
7. $5z + 11 \leq 1$ $\{z: z \leq -2\}$
8. $|t + 4| = 0$ $\{-4\}$
9. $\frac{1}{2}m > -1$ and $5 - m > 1$ $\{m: -2 < m < 4\}$

Self-Test 2

1. Express the following conjunction as an equivalent open sentence involving absolute value: $3x + 2 \geq -4$ and $3x + 2 \leq 4$. $|3x + 2| \leq 4$ **Obj. 2-4, p. 73**

Solve and graph each solution set.

2. $|2x + 1| > 3$ $\{x: x < -2 \text{ or } x > 1\}$
3. $5 - |3 - y| \geq 4$ $\{y: 2 \leq y \leq 4\}$
4. Translate the following statement into an open sentence using absolute value and the variable x: The numbers whose distance from -2 is at least 5 units. $|x + 2| \geq 5$ **Obj. 2-5, p. 76**

Solve each open sentence graphically.

5. $|m - 4| \leq 2$ $\{m: 2 \leq m \leq 6\}$
6. $\left|n + \frac{3}{2}\right| > \frac{1}{2}$ $\{n: n < -2 \text{ or } n > -1\}$

Check your answers with those at the back of the book.

Additional Answers Written Exercises

1. $\{-4, 4\}$ 2. $\{1, 5\}$
3. $\{t: -3 < t < 3\}$
4. $\{y: y < -4 \text{ or } y > 4\}$
5. $\{u: u \leq -2 \text{ or } u \geq 2\}$
6. $\{p: -5 \leq p \leq 5\}$
7. $\{y: -1 < y < 5\}$
8. $\{k: k < 3 \text{ or } k > 5\}$
9. $\{t: -1 \leq t \leq 4\}$
10. $\{x: x < -1 \text{ or } x > 2\}$
11. $\{r: r < -7 \text{ or } r > 3\}$
12. $\{w: -4 \leq w \leq 0\}$
13. $\{-1, 2\}$ 14. $\{-6, -2\}$
15. $\{p: p \leq -4 \text{ or } p \geq -1\}$
16. $\left\{k: -\frac{2}{3} < k < 2\right\}$
17. $\{t: 1 \leq t \leq 5\}$
18. $\{y: y < -5 \text{ or } y > 1\}$
19. $\left\{x: \frac{1}{2} < x < \frac{5}{2}\right\}$
20. $\left\{y: y \leq -\frac{4}{3} \text{ or } y \geq \frac{8}{3}\right\}$
21. $\left\{f: -\frac{13}{3} < f < -\frac{5}{3}\right\}$
22. $\left\{y: y \leq -\frac{7}{2} \text{ or } y \geq -\frac{1}{2}\right\}$
23. $\{t: -8 < t < 2\}$
24. $\left\{x: x \leq -\frac{5}{3} \text{ or } x \geq \frac{25}{3}\right\}$
25. $\{x: x \leq a - c \text{ or } x \geq a + c\}$
26. $\{x: -a - c < x < -a + c\}$
27. $\{x: 0 < x < 2c\}$
28. $\{x: x < -2c \text{ or } x > 0\}$

(continued)

Quick Quiz on p. 98

Additional Answers Written Exercises
(continued)

29. $\left\{x: x \le \frac{-a-c}{b} \text{ or } x \ge \frac{-a+c}{b}\right\}$

30. $\left\{x: \frac{a-c}{b} \le x \le \frac{a+c}{b}\right\}$

31. $\left\{x: x < -\frac{a-c}{b} \text{ or } x > \frac{a-c}{b}\right\}$

32. $\left\{x: x < -\frac{c-a}{b} \text{ or } x > \frac{c-a}{b}\right\}$

33. $\left\{x: -a - \frac{c}{b} < x < -a + \frac{c}{b}\right\}$

Computer Key-In Commentary

There are many versions of BASIC. They are similar enough so that, generally, programs in this book run on all machines. Any minor modifications are noted in the side column. See the Resource Book for a summary of the BASIC language.

Line 40 has several variables in one INPUT statement. When typing items of data, the user must enter the numbers and strings, separated by commas, in exactly the same order in which variables are listed. If a user tries to store an inequality symbol in a numeric variable, a message may be printed and the user will need to re-enter the data.

Computer Key-In

In Lesson 2-4 you solved inequalities involving absolute value. The program below solves inequalities of the form $|ax + b| < c$ or $|ax + b| > c$ when the user enters the values of a, b, and c, and the inequality sign. The program uses the following four basic solutions.

For $|ax + b| < c$, where $c > 0$:

If $a > 0$, $\frac{-c-b}{a} < x$ and $x < \frac{c-b}{a}$. If $a < 0$, $\frac{c-b}{a} < x$ and $x < \frac{-c-b}{a}$.

For $|ax + b| > c$, where $c \ge 0$:

If $a > 0$, $x < \frac{-c-b}{a}$ or $x > \frac{c-b}{a}$. If $a < 0$, $x < \frac{c-b}{a}$ or $x > \frac{-c-b}{a}$.

The program also handles two special cases, which you will be asked about in Exercises 7 and 8 below.

```
10 PRINT "THIS PROGRAM WILL SOLVE AN INEQUALITY"
20 PRINT "OF THE FORM ABS (A * X + B) SIGN C,"
30 PRINT "WITH A NOT ZERO. THE SIGN IS < OR >."
40 INPUT "ENTER A,B,C, AND THE SIGN: ";A,B,C,S$
50 IF S$ = "<" AND C < = 0 THEN 60
55 PRINT "SOLUTION SET CONSISTS OF X SUCH THAT"
60 IF S$ = "<" AND A > 0 AND C > 0 THEN PRINT
   (-C - B) / A;" < X AND X < "; (C - B) / A
70 IF S$ = "<" AND A < 0 AND C > 0 THEN PRINT
   (C - B) / A;" < X AND X < "; (-C - B) / A
80 IF S$ = ">" AND A > 0 AND C > = 0 THEN PRINT
   "X < ";(-C - B) / A;" OR X > ";(C - B) / A
90 IF S$ = ">" AND A < 0 AND C > = 0 THEN PRINT
   "X < "; (C - B) / A;" OR X > ";(-C - B) / A
100 IF S$ = ">" AND C < 0 THEN PRINT
   "X IS ANY REAL NUMBER."
110 IF S$ = "<" AND C < = 0 THEN PRINT
   "THERE IS NO SOLUTION."
120 END
```

Exercises

Run the program to solve each inequality. Be sure each inequality is in one of the forms $|ax + b| < c$ or $|ax + b| > c$.

1. $|-4x + 5| < 11$ $-1.5 < x < 4$
2. $|2x + 9| > 1$ $x < -5$ or $x > -4$
3. $|5x - 1| + 4 > 6$ $x < -0.2$ or $x > 0.6$
4. $3|5 - x| > 12$ $x < 1$ or $x > 9$
5. $|2x + 3| - 6 < 7$ $-8 < x < 5$
6. $-4|x - 1| > -8$ $-1 < x < 3$

Run the program to solve each inequality. Explain the result.

7. $|3x + 4| < -1$ No solution
8. $|-2x + 5| > -3$ Any real number

Proving Theorems

2-6 Theorems and Proofs

Objective To use axioms, definitions, and theorems to prove some properties of real numbers.

You have used many properties of real numbers in the previous lessons of this book. It can be shown that if just a few of these properties are accepted as true statements, then all the other properties will necessarily follow.

Statements that we assume to be true are called **axioms** (or **postulates**). The axioms that we accept include:

The substitution principle (page 8)
The properties of equality (page 14)
The field properties of real numbers (page 15)
The properties of order (page 59)

At this time, you should review all the properties and the definitions included up to this point.

Example 1 shows how you can reason from a **hypothesis** (a statement that is given or assumed to be true) to a **conclusion** (a statement that follows logically from the assumptions). In the example, each step of reasoning from the hypothesis "a, b, and c are real numbers and $a + c = b + c$" to the conclusion "$a = b$" is justified by a given fact or an axiom.

Example 1 Show that for all real numbers a, b, and c, if $a + c = b + c$, then $a = b$.

Solution

Statements	Reasons
1. a, b, and c are real numbers; $a + c = b + c$.	1. Hypothesis (or Given)
2. $-c$ is a real number.	2. Property of opposites
3. $(a + c) + (-c) = (b + c) + (-c)$	3. Addition property of equality
4. $a + [c + (-c)] = b + [c + (-c)]$	4. Associative property of addition
5. $a + 0 = b + 0$	5. Property of opposites
6. $\therefore a = b$	6. Identity property of addition

This form of logical reasoning from hypothesis to conclusion is called a **proof.** A statement that can be proved is called a **theorem.**

A theorem that can be proved easily from another is called a **corollary.** You can use the commutative property of addition to prove the following corollary of the theorem proved in Example 1.

Teaching Suggestions p. T86

Suggested Extensions p. T87

Warm-Up Exercises

Have students review all the properties in Chapter 1 and in the first five lessons of Chapter 2. Have students translate the properties into words as they review.

Motivating the Lesson

The concept of proof is a natural idea in our everyday lives. Many times when we make a statement, we are asked to prove it. It is important that students learn the concept of what it means to prove a mathematical statement. This lesson provides an experience with simple proof.

Thinking Skills

Skill and confidence in proving theorems is gained through practice. A student must be able to *think logically* to proceed from one step to another, as well as be able to *analyze* each step so as to provide a reason for it.

Chalkboard Examples

1. Give reasons for the steps in the proof.
   If $a \neq 0$, $b \neq 0$, and $\frac{1}{a} = \frac{1}{b}$, then $a = b$.
   a. $a \neq 0, b \neq 0, \frac{1}{a} = \frac{1}{b}$
   Given
   b. $ab \cdot \frac{1}{a} = ab \cdot \frac{1}{b}$
   Mult. Prop. of Eq.
   c. $ba \cdot \frac{1}{a} = ab \cdot \frac{1}{b}$
   Comm. Prop. of Mult.
   d. $b\left(a \cdot \frac{1}{a}\right) = a\left(b \cdot \frac{1}{b}\right)$
   Assoc. Prop. of Mult.
   e. $b \cdot 1 = a \cdot 1$
   Prop. of Reciprocals
   f. $b = a$
   Identity Prop. for Mult.
   g. $\therefore a = b$
   Symmetric Prop. of Eq.
2. Prove: If $a \neq 0$, then $\frac{b+a}{a} = \frac{b}{a} + 1$.
   1. $a \neq 0$ (Given)
   2. $\frac{b+a}{a} = (b+a)\frac{1}{a}$ (Def. of Div.)
   3. $= b \cdot \frac{1}{a} + a \cdot \frac{1}{a}$ (Distrib. Prop.)
   4. $= b \cdot \frac{1}{a} + 1$ (Prop. of Reciprocals)
   5. $= \frac{b}{a} + 1$ (Def. of Div.)
   6. $\therefore \frac{b+a}{a} = \frac{b}{a} + 1$ (Trans. Prop. of Eq.)

Corollary: If $c + a = c + b$, then $a = b$.

The theorem of Example 1 and its corollary make up the *cancellation property of addition*.

Cancellation Property of Addition

For all real numbers a, b, and c:

If $a + c = b + c$, then $a = b$.
If $c + a = c + b$, then $a = b$.

A proof of the following property is outlined in Exercise 14.

Cancellation Property of Multiplication

For all real numbers a and b and *nonzero* real numbers c:

If $ac = bc$, then $a = b$.
If $ca = cb$, then $a = b$.

If you interchange the hypothesis and conclusion of an if-then statement, you get the **converse** of the statement. The converse of a true statement is not necessarily true. For example, the converse of "If $a = 1$, then $a^2 = a$" is "If $a^2 = a$, then $a = 1$." This converse is false, as shown by the *counterexample* $0^2 = 0$, but $0 \neq 1$. (In algebra, a **counterexample** is a single numerical example that makes a statement false.)

After a theorem has been proved, it may be used along with axioms and definitions in other proofs. For example, the cancellation property of addition is used in the following proof.

Example 2 Prove the *multiplicative property of 0* (page 27):
For every real number a, $a \cdot 0 = 0$ and $0 \cdot a = 0$.

Proof

Statements	Reasons
1. a is a real number.	1. Given
2. $0 + 0 = 0$	2. Identity property of addition
3. $a \cdot (0 + 0) = a \cdot 0$	3. Multiplication property of equality
4. $a \cdot 0 + a \cdot 0 = a \cdot 0$	4. Distributive property
5. $a \cdot 0 + a \cdot 0 = 0 + a \cdot 0$	5. Identity property of addition
6. $\therefore a \cdot 0 = 0$	6. Cancellation property of addition
7. and $0 \cdot a = 0$	7. Commutative property of multiplication

The multiplicative property of 0 can be used to prove an important property, shown at the top of the next page.

Zero-Product Property

For all real numbers a and b:

$$ab = 0 \text{ if and only if } a = 0 \text{ or } b = 0.$$

An "if and only if" statement such as the one above is equivalent to two "if-then" statements that are converses of each other:

If $ab = 0$, then $a = 0$ or $b = 0$.

If $a = 0$ or $b = 0$, then $ab = 0$.

The second statement follows directly as a corollary to the multiplicative property of 0. A proof of the first statement is outlined in Exercise 11.

The property of opposites (page 15) states that the opposite of a real number *exists* and that it is *unique* (exactly one exists). These facts are used in the two following proofs.

Example 3 Prove: For every real number a, $-(-a) = a$.

Proof

Statements	Reasons
1. a is a real number.	1. Given
2. $-a + a = 0$	2. Property of opposites (existence of $-a$)
3. $-a + [-(-a)] = 0$	3. Property of opposites (existence of $-(-a)$)
4. $\therefore -(-a) = a$	4. Steps 2 and 3 and property of opposites (uniqueness)

Example 4 Prove the property of the *opposite of a product* (page 28): For all real numbers a and b, $-ab = (-a)b$ and $-ab = a(-b)$.

Proof

Statements	Reasons
1. a and b are real numbers.	1. Given
2. $ab + (-a)b = [a + (-a)]b$	2. Distributive property
3. $ab + (-a)b = 0 \cdot b$	3. Property of opposites
4. $ab + (-a)b = 0$	4. Multiplicative property of 0
5. $ab + (-ab) = 0$	5. Property of opposites (existence of $-ab$)
6. $\therefore -ab = (-a)b$	6. Steps 4 and 5 and property of opposites (uniqueness)

(The proof of the second part is outlined in Exercise 5.)

Guided Practice

Give reasons for each step in the proof.
$(a - b) + b = a$

a. $(a - b) + b = [a + (-b)] + b$
Def. of Subtr.

b. $= a + (-b + b)$
Assoc. Prop. for Add.

c. $= a + 0$
Prop. of Opp.

d. $= a$
Identity Prop. for Add.

e. $\therefore (a - b) + b = a$
Trans. Prop. of Eq.

Check for Understanding

Here is a suggested use of the Oral Exercises to check students' understanding as you teach the lesson.
Oral Exs. 1–6: use before Example 2.
Oral Exs. 7–8: use after Example 4.

Summarizing the Lesson

Students studied and proved several properties of real numbers in this lesson. The method of proof and terms, such as theorem, corollary, converse, and counter-example, were presented.

Suggested Assignments

Average Algebra
Day 1: 85/1–10, 15
R 79/Self-Test 2
Day 2: 87/17–22
S 79/13–27 odd

Average Alg. and Trig.
85/1–15 odd
S 75/20–30 even
R 79/Self-Test 2

Extended Algebra
85/2–18 even
S 79/31, 33
R 79/Self-Test 2

Extended Alg. and Trig.
85/3–10, 13, 15, 18, 22
R 79/Self-Test 2

Additional Answers
Oral Exercises

1. a. H: $x = 1$; C: $|x| = 1$.
   b. If $|x| = 1$, then $x = 1$.
   c. False; $|-1| = 1$, but $-1 \neq 1$.
2. a. H: $|x - 1| = 0$; C: $x = 1$.
   b. If $x = 1$, then $|x - 1| = 0$. c. True
3. a. H: an integer is divisible by 2; C: it is even.
   b. If an integer is even, then it is divisible by 2.
   c. True
4. a. H: $y > 0$; C: $y^2 > 0$.
   b. If $y^2 > 0$, then $y > 0$.
   c. False, $(-1)^2 > 0$, but -1 is not greater than 0.
5. a. H: $0 < x < 1$; C: $|x| < 1$.
   b. If $|x| < 1$, then $0 < x < 1$.
   c. False; $\left|-\frac{1}{2}\right| < 1$, but $-\frac{1}{2}$ is not between 0 and 1.
6. a. H: a triangle is isosceles; C: two of its angles are congruent.
   b. If two angles of a triangle are congruent, then the triangle is isosceles. c. True

A corollary to the theorem of Example 4 is given in the next example.

Example 5 Prove the *multiplicative property of −1* (page 27):
For every real number a, $a(-1) = -a$ and $(-1)a = -a$.

Proof

Statements	Reasons
1. a is a real number.	1. Given
2. $-a$ is a real number.	2. Property of opposites
3. $-a = -(a \cdot 1)$	3. Identity property of multiplication
4. $-a = a(-1)$	4. Property of the opposite of a product
5. $\therefore a(-1) = -a$	5. Symmetric property of equality
6. $a(-1) = (-1)a$	6. Commutative property of multiplication
7. $\therefore (-1)a = -a$	7. Substitution

Example 6 Prove the property of the *opposite of a sum* (page 29):
For all real numbers a and b, $-(a + b) = (-a) + (-b)$.

Proof

Statements	Reasons
1. a and b are real numbers.	1. Given
2. $-(a + b) = (-1)(a + b)$	2. Multiplicative property of -1
3. $-(a + b) = (-1)a + (-1)b$	3. Distributive property
4. $\therefore -(a + b) = (-a) + (-b)$	4. Multiplicative property of -1

Oral Exercises

For each statement, (a) identify the hypothesis and the conclusion, (b) give the converse of the statement, and (c) tell if the converse is true or false, and if false, give a counterexample.

1. If $x = 1$, then $|x| = 1$.
2. If $|x - 1| = 0$, then $x = 1$.
3. An integer is even if it is divisible by 2.
4. $y^2 > 0$ if $y > 0$.
5. If $0 < x < 1$, then $|x| < 1$.
6. If a triangle is isosceles, then two of its angles are congruent.

Give reasons for the steps shown in each proof. The domain of each variable is the set of real numbers unless otherwise stated.

7. If $a + a = a$, then $a = 0$.

Proof

1. $a + a = a$ Given
2. $-a$ is a real number. Prop. of opp.
3. $(a + a) + (-a) = a + (-a)$ Add. prop. of eq.
4. $a + [a + (-a)] = a + (-a)$ Assoc. prop. of add.
5. $a + 0 = 0$ Prop. of opp.
6. $\therefore a = 0$ Ident. prop. of add.

8. If $b \neq 0$, then $\frac{ab}{b} = a$.

Proof

1. $b \neq 0$ Given
2. $\frac{1}{b}$ is a real number. Prop. of recip.
3. $\frac{ab}{b} = (ab) \cdot \frac{1}{b}$ Def. of div.
4. $\frac{ab}{b} = a\left(b \cdot \frac{1}{b}\right)$ Assoc. prop. of mult.
5. $\frac{ab}{b} = a \cdot 1$ Prop. of recip.
6. $\therefore \frac{ab}{b} = a$ Ident. prop. of mult.

Written Exercises

In Exercises 1–4, answers may vary.

Give a counterexample to show that each statement is false. The domain of each variable is the set of real numbers.

A **1.** If $a^2 = b^2$, then $a = b$. $(-1)^2 = 1^2$, but $-1 \neq 1$

2. If $b < a$, then $a - b < 0$. $1 < 2$, but $2 - 1 \not< 0$

3. $|a - b| = |a| - |b|$ $|0 - 1| \neq |0| - |1|$

4. $|-a| < 0$ $|-(-1)| \not< 0$

Give reasons for the steps shown in each proof. You may use the axioms listed at the top of page 81, definitions, theorems proved in the examples of this lesson, and the results of earlier exercises as reasons. The domain of each variable is the set of real numbers unless otherwise stated.

5. The opposite of a product: $-ab = a(-b)$.

Proof

1. $-ab = -ba$ Comm. prop. of mult.
2. $-ab = (-b)a$ Proved in Example 4 (See Example 4.)
3. $\therefore -ab = a(-b)$ Comm. prop. of mult.

6. The product of opposites: $(-a)(-b) = ab$.

Proof

1. $(-a)(-b) = -[a(-b)]$ Opp. of a prod.
2. $(-a)(-b) = -[-(ab)]$ Opp. of a prod.
3. $\therefore (-a)(-b) = ab$ Proved in Example 3 (See Example 3.)

7. $(a + b) - b = a$.

Proof

1. $(a + b) - b = (a + b) + (-b)$ Def. of subtr.
2. $(a + b) - b = a + [b + (-b)]$ Assoc. prop. of add.
3. $(a + b) - b = a + 0$ Prop. of opp.
4. $\therefore (a + b) - b = a$ Ident. prop. of add.

8. If $x + c = 0$, then $x = -c$.

Proof

1. $x + c = 0$ Given
2. $(x + c) + (-c) = 0 + (-c)$ Add. prop.
3. $x + [c + (-c)] = 0 + (-c)$ Assoc. prop.
4. $x + 0 = 0 + (-c)$ Prop. of opp.
5. $\therefore x = -c$ Ident. prop. of add.

Additional Answers
Written Exercises

15. 1. $a \neq 0$ Given
2. $\frac{1}{a} \cdot a = 1$ Prop. of recip.
3. $\frac{1}{a} \cdot \frac{1}{\frac{1}{a}} = 1$ Prop. of recip.
4. $\therefore a = \frac{1}{\frac{1}{a}}$ Steps 2 and 3 and prop. of recip. (uniqueness)

16. 1. $b \neq 0$ Given
2. $\frac{-a}{b} = (-a) \cdot \frac{1}{b}$ Def. of div.
3. $\frac{-a}{b} = -\left(a \cdot \frac{1}{b}\right)$ Opp. of a prod.
4. $\therefore \frac{-a}{b} = -\frac{a}{b}$ Def. of div.

17. 1. $b \neq 0$ Given
2. $(-b)\left(-\frac{1}{b}\right) = b \cdot \frac{1}{b}$ Prod. of opp. (Ex. 6)
3. $(-b)\left(-\frac{1}{b}\right) = 1$ Prop. of recip.
4. $\therefore -\frac{1}{b} = \frac{1}{-b}$ Prop. of recip. (uniqueness of recip. of $-b$)

18. 1. $b \neq 0$ Given
2. $\frac{a}{-b} = a\left(\frac{1}{-b}\right)$ Def. of div.
3. $\frac{a}{-b} = a\left(-\frac{1}{b}\right)$ Proved in Ex. 17
4. $\frac{a}{-b} = -\left(a \cdot \frac{1}{b}\right)$ Opp. of a prod.
5. $\therefore \frac{a}{-b} = -\frac{a}{b}$ Def. of div.

Additional Answers Written Exercises

(continued)

19. 1. $b \neq 0$
Given
2. $\frac{-a}{-b} = (-a)\left(\frac{1}{-b}\right)$
Def. of div.
3. $\frac{-a}{-b} = (-a)\left(-\frac{1}{b}\right)$
Proved in Ex. 17
4. $\frac{-a}{-b} = a \cdot \frac{1}{b}$
Prod. of opp. (Ex. 6)
5. $\therefore \frac{-a}{-b} = \frac{a}{b}$
Def. of div.

20. 1. $a \neq 0$; $b \neq 0$
Given
2. $(ab)\left(\frac{1}{a} \cdot \frac{1}{b}\right) = \left(a \cdot \frac{1}{a}\right)\left(b \cdot \frac{1}{b}\right)$
Assoc. and comm. prop. of mult.
3. $(ab)\left(\frac{1}{a} \cdot \frac{1}{b}\right) = 1 \cdot 1$
Prop. of reciprocals
4. $(ab)\left(\frac{1}{a} \cdot \frac{1}{b}\right) = 1$
Ident. prop. of mult.
5. $\therefore \frac{1}{a} \cdot \frac{1}{b} = \frac{1}{ab}$
Prop. of reciprocals (uniqueness of reciprocal of ab)

21. 1. $b \neq 0$; $d \neq 0$
Given
2. $\frac{a}{b} \cdot \frac{c}{d} = \left(a \cdot \frac{1}{b}\right)\left(c \cdot \frac{1}{d}\right)$
Def. of div.
3. $\frac{a}{b} \cdot \frac{c}{d} = (ac)\left(\frac{1}{b} \cdot \frac{1}{d}\right)$
Assoc. and comm. prop. of mult.
4. $\frac{a}{b} \cdot \frac{c}{d} = (ac)\left(\frac{1}{bd}\right)$
Proved in Ex. 20
5. $\therefore \frac{a}{b} \cdot \frac{c}{d} = \frac{ac}{bd}$
Def. of div.

9. If $u \neq 0$ and $u^2 = u$, then $u = 1$.

Proof

1. $u \neq 0$; $u^2 = u$ Given
2. $u \cdot u = u$ Def. of u^2
3. $\frac{1}{u}$ is a real number. Prop. of recip.
4. $(u \cdot u) \cdot \frac{1}{u} = u \cdot \frac{1}{u}$ Mult. prop. of eq.
5. $u \cdot \left(u \cdot \frac{1}{u}\right) = u \cdot \frac{1}{u}$ Assoc. prop. of mult.
6. $u \cdot 1 = 1$ Prop. of recip.
7. $\therefore u = 1$ Ident. prop. of mult.

10. If $c \neq 0$ and $cx = 1$, then $x = \frac{1}{c}$.

Proof

1. $c \neq 0$; $cx = 1$ Given
2. $\frac{1}{c}$ is a real number. Prop. of recip.
3. $\frac{1}{c}(cx) = \frac{1}{c} \cdot 1$ Mult. prop. of eq.
4. $\left(\frac{1}{c} \cdot c\right)x = \frac{1}{c} \cdot 1$ Assoc. prop. of mult.
5. $1 \cdot x = \frac{1}{c} \cdot 1$ Prop. of recip.
6. $\therefore x = \frac{1}{c}$ Ident. prop. of mult.

11. Zero-product property: If $ab = 0$, then $a = 0$ or $b = 0$.

Proof

1. $ab = 0$ Given
2. Suppose $a \neq 0$. Then $\frac{1}{a}$ is a real number. (Note that if $a = 0$, the given statement is obviously true.) Prop of recip.
3. $\frac{1}{a}(ab) = \frac{1}{a} \cdot 0$ Mult. prop. of eq.
4. $\frac{1}{a}(ab) = 0$ Mult. prop. of 0
5. $\left(\frac{1}{a} \cdot a\right)b = 0$ Assoc. prop. of mult.
6. $1 \cdot b = 0$ Prop. of recip.
7. $b = 0$ Ident. prop. of mult.
8. $\therefore$ if $a \neq 0$, then $b = 0$. Steps 2–7

12. Multiplication is distributive with respect to subtraction: $a(b - c) = ab - ac$.

Proof

1. $a(b - c) = a[b + (-c)]$ Def. of subtr.
2. $a(b - c) = ab + a(-c)$ Dist. prop.
3. $a(b - c) = ab + (-ac)$ (*Hint:* Use Exercise 5.) Opp. of a prod.
4. $\therefore a(b - c) = ab - ac$ Def. of subtr.

13. Division is distributive with respect to subtraction: If $c \neq 0$, then $\frac{a - b}{c} = \frac{a}{c} - \frac{b}{c}$.

Proof

1. $c \neq 0$; $\frac{1}{c}$ is a real number. Given; prop. of recip.
2. $\frac{a - b}{c} = (a - b) \cdot \frac{1}{c}$ Def. of div.
3. $\frac{a - b}{c} = a \cdot \frac{1}{c} - b \cdot \frac{1}{c}$ (*Hint:* Use Exercise 12.) Distr. prop. of mult. with respect to subtr.
4. $\therefore \frac{a - b}{c} = \frac{a}{c} - \frac{b}{c}$ Def. of div.

14. Cancellation property of multiplication: If $c \neq 0$ and $ca = cb$, then $a = b$.

Proof

1. $c \neq 0;\ ca = cb$ Given
2. $\frac{1}{c}$ is a real number. Prop. of recip.
3. $\frac{1}{c}(ca) = \frac{1}{c}(cb)$ Mult. prop. of eq.
4. $\left(\frac{1}{c} \cdot c\right)a = \left(\frac{1}{c} \cdot c\right)b$ Assoc. prop. of mult.
5. $1 \cdot a = 1 \cdot b$ Prop. of recip.
6. $\therefore a = b$ Ident. prop. of mult.

B 15. Follow the steps outlined in Example 3 (with multiplication replacing addition) to prove that if $a \neq 0$, then $\frac{1}{\frac{1}{a}} = a$.

16. Prove: If $b \neq 0$, then $\frac{-a}{b} = -\frac{a}{b}$.
(*Hint:* Use the definition of division and Example 4.)

17. Prove: If $b \neq 0$, then $\frac{1}{-b} = -\frac{1}{b}$.
$\left(\textit{Hint: } \text{Show that } -\frac{1}{b} \text{ is the reciprocal of } -b.\right)$

18. Prove: If $b \neq 0$, then $\frac{a}{-b} = -\frac{a}{b}$. (*Hint:* Use Exercises 17 and 5.)

19. Prove: If $b \neq 0$, then $\frac{-a}{-b} = \frac{a}{b}$. (*Hint:* Use Exercises 17 and 6.)

20. Prove: If $a \neq 0$ and $b \neq 0$, then $\frac{1}{a} \cdot \frac{1}{b} = \frac{1}{ab}$.
$\left(\textit{Hint: } \text{Show that } \frac{1}{a} \cdot \frac{1}{b} \text{ is the reciprocal of } ab.\right)$

21. Prove: If $b \neq 0$ and $d \neq 0$, then $\frac{a}{b} \cdot \frac{c}{d} = \frac{ac}{bd}$.
(*Hint:* Use Exercise 20.)

C 22. Prove: If $c \neq 0$ and $d \neq 0$, then $\frac{1}{\frac{c}{d}} = \frac{d}{c}$.

23. Prove: If $c \neq 0$ and $d \neq 0$, then $\frac{a}{b} \div \frac{c}{d} = \frac{ad}{bc}$.

24. Use the zero-product property to give a convincing argument why 0 has no reciprocal.

25. Use the zero-product property to give a convincing argument why the product of two nonzero real numbers is never equal to 0.

22. 1. $c \neq 0;\ d \neq 0$
Given
2. $\frac{c}{d} \cdot \frac{d}{c} = \frac{cd}{dc}$
Proved in Ex. 21
3. $\frac{c}{d} \cdot \frac{d}{c} = (cd)\left(\frac{1}{cd}\right)$
Comm. prop. of mult.; def. of div.
4. $\frac{c}{d} \cdot \frac{d}{c} = 1$
Prop. of recip.
5. $\therefore \frac{1}{\frac{c}{d}} = \frac{d}{c}$
Prop. of recip. (uniqueness of recip.)

23. 1. $c \neq 0;\ d \neq 0$
Given
2. $\frac{a}{b} \div \frac{c}{d} = \frac{a}{b} \cdot \frac{1}{\frac{c}{d}}$
Def. of div.
3. $\frac{a}{b} \div \frac{c}{d} = \frac{a}{b} \cdot \frac{d}{c}$
Proved in Ex. 22
4. $\therefore \frac{a}{b} \div \frac{c}{d} = \frac{ad}{bc}$
Proved in Ex. 21

24. If $a = 0$ and a has a reciprocal (call it b), then $ab = 1$. But by the zero-product property, $ab = 0$ if $a = 0$.

25. Call the numbers a and b. If $ab = 0$, then at least one of a and b must be 0 by the zero-product property. But both a and b are given to be nonzero.

Supplementary Materials

Study Guide pp. 29–30

2-7 Theorems about Order and Absolute Value

Objective To prove theorems about inequalities and absolute value.

The properties of order (page 59) play an important part in proving the theorems of this section.

Example 1 Prove: If a is a real number and $a \neq 0$, then $a^2 > 0$.

Proof

Statements	Reasons
1. a is a real number; $a \neq 0$	1. Given
2. Either $a > 0$ or $a < 0$.	2. Comparison property of order
3. *Case 1:* $a > 0$ Multiply this inequality by the *positive* number a: $\underbrace{a \cdot a} > \underbrace{a \cdot 0}$	3. First multiplication property of order (with $<$ replaced by $>$)
4. $\therefore\ a^2 > 0$	4. Definition of a^2 (page 7) and multiplicative property of 0
5. *Case 2:* $a < 0$ Multiply this inequality by the *negative* number a: $\underbrace{a \cdot a} > \underbrace{a \cdot 0}$	5. Second multiplication property of order
6. $\therefore\ a^2 > 0$	6. See the reasons in Step 4.

The rules for multiplication are proved as theorems in Examples 2 and 3.

Example 2 Prove for all real numbers a and b:

a. If $a > 0$ and $b > 0$, then $ab > 0$.
b. If $a < 0$ and $b < 0$, then $ab > 0$.
c. If $a > 0$ and $b < 0$, then $ab < 0$.

Proof We will prove parts (**b**) and (**c**). The proof of (**a**) is left as Exercise 9.

*Proof of part (**b**)*

Statements	Reasons
1. $a < 0$ and $b < 0$	1. Given
2. Multiply both sides of $b < 0$ by the *negative* number a: $ab > a \cdot 0$	2. Second multiplication property of order
3. $\therefore\ ab > 0$	3. Multiplicative property of 0

Teaching Suggestions P. T87

Suggested Extensions P. T87

Warm-Up Exercises

Have students review the properties of order on page 59. Tell students to think about what the properties mean, using numbers for the variables if necessary, rather than just reading the words and symbols.

Motivating the Lesson

This lesson continues the study of proof and extends it to cover proofs about inequalities and absolute value. Lessons 2-6 and 2-7 not only provide practice in proving theorems, but also give properties which will be used throughout the course.

Chalkboard Examples

1. Supply reasons for the steps in the proof.
If $a > 1$, then $a^2 > 1$.
a. $a > 1$ Given
b. $a \cdot a > a \cdot 1$
Mult. Prop. of Order
c. $a^2 > a \cdot 1$
Def. of a^2
d. $a^2 > a$
Identity Prop. for Mult.
e. $a^2 > 1$
Trans. Prop. of Order

2. Prove: If $x > y$, then $x - y > 0$.
1. $x > y$ (Given)
2. $x + (-y) > y + (-y)$ (Add. Prop. of Order)
3. $x + (-y) > 0$ (Prop. of Opposites)
4. $x - y > 0$ (Def. of Subtr.)

Proof of part (c)

Statements	Reasons
1. $a > 0$ and $b < 0$	1. Given
2. Multiply both sides of $b < 0$ by the *positive* number a: $ab < a \cdot 0$	2. First multiplication property of order
3. $\therefore ab < 0$	3. Multiplicative property of 0

Recall that the **absolute value** of a real number x is defined as follows:

$$|x| = x \quad \text{if } x \geq 0$$
$$|x| = -x \text{ if } x < 0$$

Example 3 Prove: For all real numbers a and b, $|ab| = |a| \cdot |b|$.

Proof If either a or b is 0, $|ab| = |a| \cdot |b|$ because both products are 0. There are three remaining cases to prove. Case 1: $a > 0$ and $b > 0$; Case 2: $a < 0$ and $b < 0$; and Case 3: $a > 0$ and $b < 0$. Cases 2 and 3 follow. Case 1 is left as Exercise 10.

Case 2: $a < 0$ and $b < 0$

Statements	Reasons
1. $a < 0$; $b < 0$	1. Given
2. $ab > 0$	2. Example 2, part (b)
3. $\|ab\| = ab$	3. Definition of absolute value
4. $\|a\| = -a$, $\|b\| = -b$	4. Definition of absolute value
5. $\|a\| \cdot \|b\| = (-a)(-b)$	5. Multiplication property of equality
6. $\|a\| \cdot \|b\| = ab$	6. Exercise 6, page 85
7. $\therefore \|ab\| = \|a\| \cdot \|b\|$	7. Substitution principle (Steps 3 and 6)

Case 3: $a > 0$ and $b < 0$

Statements	Reasons
1. $a > 0$; $b < 0$	1. Given
2. $ab < 0$	2. Example 2, part (c)
3. $\|ab\| = -ab$	3. Definition of absolute value
4. $\|a\| = a$, $\|b\| = -b$	4. Definition of absolute value
5. $\|a\| \cdot \|b\| = a(-b)$	5. Multiplication property of equality
6. $\|a\| \cdot \|b\| = -ab$	6. Property of the opposite of a product
7. $\therefore \|ab\| = \|a\| \cdot \|b\|$	7. Substitution principle (Steps 3 and 6)

Check for Understanding

Here is a suggested use of the Oral Exercises to check students' understanding as you teach the lesson.
Oral Exs. 1–2: use after Example 2.
Oral Ex. 3: use after Example 3.

Guided Practice

Give the reasons.

1. $|a| \geq a$

a. $a < 0$, $a = 0$, or $a > 0$
Comparison Prop. of Order

case 1

b. If $a < 0$, $|a| = -a$.
Def. of $|a|$

c. If $a < 0$, then $-a > 0$.
2nd mult. prop. of order

d. If $a < 0$, then $|a| > 0$.
Subst. Prop.

e. If $a < 0$, then $|a| > a$.
Trans. Prop. of Order

case 2

f. If $a \geq 0$, then $|a| = a$.
Def of $|a|$

g. If $a \geq 0$, then $|a| \geq a$.
Def of $\geq$

2. $a^2 \geq 0$

a. $a < 0$, $a = 0$, or $a > 0$
Comparison Prop. of Order

case 1

b. If $a \neq 0$, $a^2 \geq 0$.
Example, page 88 and def. of $\geq$

case 2

c. If $a = 0$, $a \cdot a = 0 \cdot a$.
Mult. Prop. of Eq.

d. If $a = 0$, $a^2 \geq 0$.
Def. of a^2, def. of $\geq$, and Mult. Prop. of 0

Summarizing the Lesson

In this lesson we have extended the proof of theorems to include those involving inequalities and absolute value.

Suggested Assignments

Average Algebra
90/1, 3, 4, 6–8, 10
S 91/Mixed Review
R 92/Self-Test 3

Average Alg. and Trig.
90/1, 3, 4, 7, 8, 10
S 91/Mixed Review
R 92/Self-Test 3

Extended Algebra
90/1, 3, 4, 7, 8, 10–12
S 91/Mixed Review
R 92/Self-Test 3

Extended Alg. and Trig.
90/1, 3, 4, 6–8, 10, 14, 15
S 91/Mixed Review
R 92/Self-Test 3

Additional Answers Written Exercises

8. 1. $a < 1$ and $a > 0$ Given
2. $a \cdot a < a \cdot 1$ First mult. prop. of order
3. $a \cdot a < a$ Ident. prop. of mult.
4. $\therefore a^2 < a$ Def. of a^2

10. 1. $a > 0$ and $b > 0$ Given
2. $ab > 0$ Proved in Ex. 9
3. $|ab| = ab$ Def. of abs. value
4. $|a| = a$, $|b| = b$ Def. of abs. value
5. $|a| \cdot |b| = ab$ Mult. prop. of eq.
6. $\therefore |ab| = |a| \cdot |b|$ Subst. prin. (Steps 3 and 5)

Oral Exercises

Give reasons for the steps shown in each proof.

1. Prove: For all real numbers a and b, if $a < b$, then $-a > -b$.

Proof

1. $a < b$ Given
2. $(-1)a > (-1)b$ Second mult. prop. of order
3. $\therefore -a > -b$ Mult. prop. of -1

2. Prove: If a is a real number and $a < 0$, then $-a > 0$.

Proof

1. $a < 0$ Given
2. $a + (-a) < 0 + (-a)$ Add. prop. of order
3. $0 < 0 + (-a)$ Prop. of opp.
4. $\therefore 0 < -a$, or $-a > 0$ Ident. prop. of add.

3. In Example 3, why is there no need to consider the case $a < 0$ and $b > 0$? The possibility of a and b having opposite signs is covered in Case 3.

Written Exercises

Give reasons for the steps shown in the proof of each theorem. The domain of each variable is the set of real numbers unless otherwise stated.

A

1. If $a > b$, then $-a < -b$.

Proof

1. $a > b$ Given
2. $(-1)a < (-1)b$ Second mult. prop. of order
3. $\therefore -a < -b$ Mult. prop. of -1

2. If $a > b$, then $a - c > b - c$.

Proof

1. $a > b$ Given
2. $a + (-c) > b + (-c)$ Add. prop. of order
3. $\therefore a - c > b - c$ Def. of subtr.

3. If $a < b$ and $c < d$, then $a + c < b + d$.

Proof

1. $a < b$ Given
2. $a + c < b + c$ Add. prop. of order
3. $c < d$ Given
4. $b + c < b + d$ Add. prop. of order
5. $\therefore a + c < b + d$ Trans. prop. of order (Steps 2 and 4)

4. If $0 < a < b$, then $a^2 < b^2$.

Proof

1. $0 < a < b$ Given
2. $a \cdot a < a \cdot b$ First mult. prop. of order
3. $a \cdot b < b \cdot b$ First mult. prop. of order
4. $a \cdot a < b \cdot b$ Trans. prop. of order (Steps 2 and 3)
5. $\therefore a^2 < b^2$ Def. of a^2 and b^2

5. If $a < b < 0$, then $a^2 > b^2$.

Proof

1. $a < b$ and $a < 0$ Given
2. $a \cdot a > a \cdot b$ Second mult. prop. of order
3. $a < b$ and $b < 0$ Given
4. $a \cdot b > b \cdot b$ Second mult. prop. of order
5. $a \cdot a > b \cdot b$ Trans. prop. of order (Steps 2 and 4)
6. $\therefore a^2 > b^2$ Def. of a^2 and b^2

6. If $0 < a < b$ and $0 < c < d$, then $ac < bd$.

Proof

1. $a < b$ and $c > 0$ Given
2. $ac < bc$ First mult. prop. of order
3. $c < d$ and $b > 0$ Given
4. $bc < bd$ First mult. prop. of order
5. $\therefore ac < bd$ Trans. prop. of order (Steps 2 and 4)

7. Prove: If $a > 0$, then $-a < 0$. (*Hint:* See Oral Exercise 2.)

8. Prove: If $0 < a < 1$, then $a^2 < a$.

9. Prove: If $a > 0$ and $b > 0$, then $ab > 0$. (Part (a), Example 2)

10. Prove: If $a > 0$ and $b > 0$, then $|ab| = |a|\,|b|$. (Case 1, Example 3)

In Exercises 11–16, assume that if $a > 0$, then $\frac{1}{a} > 0$, and if $a < 0$, then $\frac{1}{a} < 0$.

B 11. Prove: If $a > 0$ and $b > 0$, then $\frac{a}{b} > 0$.

12. Prove: If $a < 0$ and $b < 0$, then $\frac{a}{b} > 0$.

13. Prove: If $a > 0$ and $b < 0$, then $\frac{a}{b} < 0$.

14. Prove: If $a < 0$ and $b > 0$, then $\frac{a}{b} < 0$.

15. Prove: If $a > 0$, $b > 0$, and $\frac{1}{a} > \frac{1}{b}$, then $a < b$.

16. Prove: If $0 < a < b$, then $\frac{1}{a} > \frac{1}{b}$. $\left(\textit{Hint:}\text{ Multiply by } \frac{1}{ab}.\right)$

C 17. Prove: If $a < b$, then $a < \frac{a+b}{2} < b$. This proof shows that the average of two numbers lies between them.

18. Prove: For every $a > 0$, if $a > \frac{1}{a}$, then $a > 1$.
$\Big($*Hint:* Give an *indirect proof*. That is, show that the assumptions $a = 1$ and $a < 1$ each lead to a contradiction of the hypothesis that $a > \frac{1}{a}$.$\Big)$

Mixed Review Exercises

Tell whether each statement is true for all real numbers.

1. If $a > b$, then $b < a$. True
2. If $a > b$ and $c < 0$, then $ac < bc$. True
3. If $ac = bc$ and $c \neq 0$, then $a = b$. True

Solve each open sentence and graph each solution set that is not empty.

4. $|x - 3| = 1$ $\{2, 4\}$
5. $4d + 5 \geq 1$ $\{d: d \geq -1\}$
6. $-1 < 2 - y < 3$ $\{y: -1 < y < 3\}$
7. $6r + 13 = 25$ $\{2\}$
8. $|n| + 7 < 5$ $\emptyset$
9. $-2k > 8$ or $k - 4 \geq 0$ $\{k: k < -4 \text{ or } k \geq 4\}$

12. 1. $a < 0$ and $b < 0$
Given
2. $a \cdot \frac{1}{b} > 0 \cdot \frac{1}{b}$
Second mult. prop. of order (assuming $\frac{1}{b} < 0$)
3. $a \cdot \frac{1}{b} > 0$
Mult. prop. of 0
4. $\therefore \frac{a}{b} > 0$
Def. of div.

14. 1. $a < 0$ and $b > 0$
Given
2. $a \cdot \frac{1}{b} < 0 \cdot \frac{1}{b}$
First mult. prop. of order (assuming $\frac{1}{b} > 0$)
3. $a \cdot \frac{1}{b} < 0$
Mult. prop. of 0
4. $\therefore \frac{a}{b} < 0$
Def. of div.

16. 1. $a > 0$, $b > 0$, and $a < b$
Given
2. $ab > 0$
Proved in Ex. 9
3. $\frac{1}{ab} \cdot a < \frac{1}{ab} \cdot b$
First mult. prop. of order (assuming $\frac{1}{ab} > 0$)
4. $\left(\frac{1}{a} \cdot \frac{1}{b}\right) \cdot a < \left(\frac{1}{a} \cdot \frac{1}{b}\right) \cdot b$
Ex. 20, page 87
5. $\left(\frac{1}{a} \cdot a\right) \cdot \frac{1}{b} < \frac{1}{a} \cdot \left(\frac{1}{b} \cdot b\right)$
Assoc. and comm. prop. of mult.
6. $1 \cdot \frac{1}{b} < \frac{1}{a} \cdot 1$
Prop. of recip.
7. $\therefore \frac{1}{b} < \frac{1}{a}$, or $\frac{1}{a} > \frac{1}{b}$
Ident. prop. of mult.
(continued on p. 99)

Supplementary Materials

Study Guide	pp. 31–32
Practice Master	10
Test Master	6
Resource Book	p. 106

Additional Answers Self-Test 3

2. 1. $ab = c;\ b \neq 0$ Given
2. $\frac{1}{b}$ is a real number. Prop. of reciprocals
3. $(ab) \cdot \frac{1}{b} = c \cdot \frac{1}{b}$ Mult. prop. of eq.
4. $a\left(b \cdot \frac{1}{b}\right) = c \cdot \frac{1}{b}$ Assoc. prop. of mult.
5. $a(1) = c \cdot \frac{1}{b}$ Prop. of reciprocals
6. $a = c \cdot \frac{1}{b}$ Ident. prop. of mult.
7. $\therefore a = \frac{c}{b}$ Def. of div.

3. 1. $a > 1$ Given
2. $a > 0$ Trans. prop. of order (using the fact $1 > 0$)
3. $a \cdot a > a \cdot 1$ First mult. prop. of order
4. $a \cdot a > a$ Ident. prop. of mult.
5. $\therefore a^2 > a$ Def. of a^2

Quick Quiz

Give reasons for the steps in the proofs.

1. If $ab = 0$ and $a \neq 0$, then $b = 0$.
a. $ab = 0$ and $a \neq 0$ Given

Self-Test 3

Vocabulary axiom (p. 81), hypothesis (p. 81), conclusion (p. 81), proof (p. 81), theorem (p. 81), corollary (p. 81), converse (p. 82), counterexample (p. 82)

1. Give reasons for the steps shown in the following proof. If $a + b = c$, then $a = c - b$. **Obj. 2-6, p. 81**

Proof

1. $a + b = c$ Given
2. $-b$ is a real number. Prop. of opp.
3. $(a + b) + (-b) = c + (-b)$ Add. prop. of eq.
4. $a + [b + (-b)] = c + (-b)$ Assoc. prop. of add.
5. $a + 0 = c + (-b)$ Prop. of opp.
6. $a = c + (-b)$ Ident. prop. of add.
7. $\therefore a = c - b$ Def. of subtr.

2. Prove: If $ab = c$ and $b \neq 0$, then $a = \frac{c}{b}$.

3. Prove: If $a > 1$, then $a^2 > a$. **Obj. 2-7, p. 88**

Chapter Summary

1. Inequalities can be solved by using the properties of order on page 59 and the transformations on page 60.
2. If an inequality is a conjunction, it can be solved by finding the values of the variable for which both sentences are true. If an inequality is a disjunction, it can be solved by finding the values of the variable for which at least one of the sentences is true.
3. Some word problems can be solved algebraically by translating the given information into an inequality and then solving the inequality.
4. Equations and inequalities that involve absolute value can be solved algebraically by writing an equivalent conjunction or disjunction.
5. Equations and inequalities that involve absolute value can be solved geometrically by using this fact: On a number line the distance between the graphs of two numbers is the absolute value of the difference between the numbers.
6. Axioms are statements assumed to be true. Using the axioms for real numbers (referred to on page 81), other properties of real numbers can be proved as theorems. Every step in the proof of a theorem can be justified by either an axiom, a definition, a given fact, or a theorem.

Chapter Review

Give the letter of the correct answer.

Solve.

1. $-\frac{m}{2} < -2$ 2-1

a. $m > 1$ **b.** $m > 4$ **c.** $m < 4$ **d.** $m > -4$

2. $3(n - 1) > 5n + 7$

a. $n > -5$ **b.** $n < -4$ **c.** $n > -4$ **d.** $n < -5$

3. $-3 < 4c + 5 \leq 1$ 2-2

a. $-2 < c \leq -1$ **b.** $-2 < c \leq 1$

c. $2 < c \leq -1$ **d.** $2 < c \leq 1$

4. Graph the solution set of the following disjunction:
$4 - w \leq 3$ or $w + 5 < 3$

a. 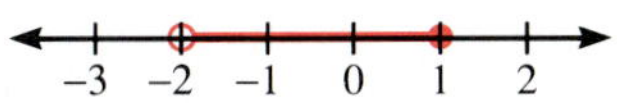

b. 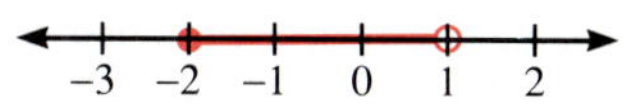

c. 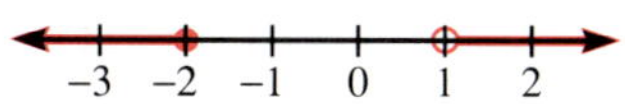

d. 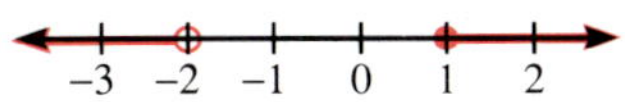

Solve.

5. At the video store, you can rent tapes for \$1.50 each per day, or you can get unlimited rentals for a monthly fee of \$20. At most how many tapes could you rent at \$1.50 before spending more than the cost for a month's unlimited rentals? 2-3

a. 12 **b.** 13 **c.** 14 **d.** 15

6. $\left|5 - \frac{x}{3}\right| = 7$ 2-4

a. $\{-6, 36\}$ **b.** $\{-6, -36\}$ **c.** $\{6, -36\}$ **d.** $\{6, 36\}$

7. $|2y + 9| < 13$

a. $-2 < y < 11$ **b.** $y > 2$ **c.** $-11 < y < 2$ **d.** $y < -11$ or $y > 2$

8. $|4 - h| \geq 5$

a. $h \leq -1$ or $h \geq 9$ **b.** $h \leq -1$

c. $h \leq -9$ or $h \geq 1$ **d.** $-1 \leq h \leq 9$

9. Describe the graph of the solution set of $|k + 3| < 2$. 2-5

a. Points at least 2 units from 3. **b.** Points less than 2 units from -3.

c. Points more than 3 units from 2. **d.** Points at most 2 units from -3.

10. A theorem that can be proved easily from another theorem is a(n) __?__. 2-6

a. axiom **b.** corollary **c.** property **d.** theorem

11. The converse of a true statement is __?__.

a. a corollary **b.** never true **c.** always true **d.** sometimes true

b. $\frac{1}{a}$ is a real number.
Prop. of Reciprocals

c. $\frac{1}{a} \cdot ab = \frac{1}{a} \cdot 0$
Mult. Prop. of Eq.

d. $\frac{1}{a} \cdot ab = 0$
Mult. Prop. of 0

e. $\left(\frac{1}{a} \cdot a\right)b = 0$
Assoc. Prop. for Mult.

f. $1 \cdot b = 0$
Prop. of Reciprocals

g. $b = 0$
Identity Prop. for Mult.

2. If $a < 0$ and $b < a$, then $b^2 > a^2$.

a. $a < 0$ and $b < a$
Given

b. $a \cdot b > a \cdot a$
Mult. Prop. of Order

c. $ab > a^2$
Def of a^2

d. $b < 0$
Trans. Prop. of Order

e. $b \cdot b > b \cdot a$
Mult. Prop. of Order

f. $b^2 > ab$
Def. of b^2; Comm. Prop. of Mult.

g. $b^2 > a^2$
Trans. Prop. of Order

3. Prove:
$a + [-(a + b)] = -b$.

1. $a + [-(a + b)]$
$= a + [(-1)(a + b)]$
(Mult. Prop. of -1)
2. $= a + [-1(a) + (-1)b]$
(Distrib. Prop.)
3. $= a + [-a + (-b)]$
(Mult. Prop. of -1)
4. $= [a + (-a)] + (-b)$
(Assoc. Prop. of Add.)
$= 0 + (-b)$
(Prop. of Opposites)
5. $= -b$
(Identity Prop. of Add.)

Alternate Test p. T14

Additional Answers Chapter Test

9.

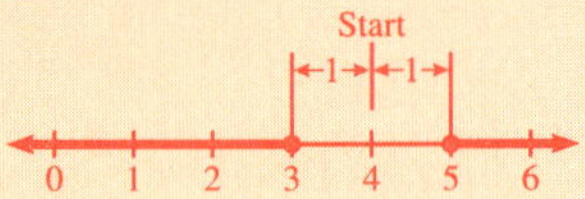

$\{w: w \le 3 \text{ or } w \ge 5\}$

10.

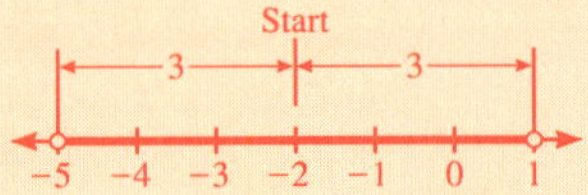

$\{h: -5 < h < 1\}$

13. False; $1 > -2$, but $|1|$ is not greater than $|-2|$.

14. False; $-1 < 2$ and $-2 > -3$, but $(-1)(-3)$ is not less than $(2)(-2)$.

15. False; $-2 < -3$ and $-1 < 0$, but $(-2)(-1)$ is not less than $(3)(-1)$.

16. False; $-2 < -1 < 0$, but $(-2)^2$ is not less than $(-1)^2$.

Supplementary Materials

Test Masters	7, 8
Resource Book	p. 107

Supply the conclusion that makes each statement true.

12. For real numbers a and b, $ab = 0$ if and only if __?__.

a. $a = 0$ **b.** $b = 0$ **(c.)** $a = 0$ or $b = 0$ **d.** $a = 0$ and $b = 0$

13. For real numbers a, b, and c, if $a > b$ and $c < 0$, then __?__. **2-7**

a. $\frac{a}{c} > \frac{b}{c}$ **b.** $ac > bc$ **(c.)** $\frac{a}{c} < \frac{b}{c}$ **d.** $a + c < b + c$

Chapter Test

Solve.

1. $5x - 9 > 6x$ $\{x: x < -9\}$ **2.** $3(2y + 1) < 2(y - 3) + 1$ $\{y: y < -2\}$ **2-1**

3. $-3 < 4 - m < 6$ $\{m: -2 < m < 7\}$ **4.** $-2n \ge 8$ or $n + 3 < 7$ $\{n: n < 4\}$ **2-2**

5. The dimensions of a rectangle are consecutive odd integers. Find the smallest such rectangle with a perimeter of at least 35 cm. 9 cm × 11 cm **2-3**

Solve.

6. $|9 - 2k| = 5$ $\{2, 7\}$ **7.** $\left|\frac{c}{3} + 1\right| > 2$ $\{c: c < -9 \text{ or } c > 3\}$ **8.** $|4f + 3| \le 5$ $\left\{f: -2 \le f \le \frac{1}{2}\right\}$ **2-4**

Solve each sentence graphically.

9. $|w - 4| \ge 1$ **10.** $|h + 2| < 3$ **2-5**

11. Write the converse of the following statement and tell whether the converse is true or false: "If $x > 1$, then $x^2 > x$." If $x^2 > x$, then $x > 1$. False **2-6**

12. Supply reasons for the steps in the proof of the following statement: For all real numbers a and nonzero real numbers b, if $ab = b$, then $a = 1$.

1. a is a real number; b is a nonzero real number; $ab = b$	1. __?__ Given
2. $\frac{1}{b}$ is a real number.	2. __?__ Prop. of recip.
3. $(ab) \cdot \frac{1}{b} = b \cdot \frac{1}{b}$	3. __?__ Mult. prop. of eq.
4. $a\left(b \cdot \frac{1}{b}\right) = b \cdot \frac{1}{b}$	4. __?__ Assoc. prop. of mult.
5. $a(1) = 1$	5. __?__ Prop. of recip.
6. $a = 1$	6. __?__ Ident. prop. of mult.

Tell whether the statement is true for all real numbers. If it is not, give a counterexample.

13. If $a > b$, then $|a| > |b|$. False **14.** If $a < b$ and $c > d$, then $ad < bc$. False **2-7**

15. If $a < b$ and $c < 0$, then $ac < bc$. False **16.** If $a < b < 0$, then $a^2 < b^2$. False

Extra / Symbolic Logic: Boolean Algebra

Most people think of algebra as the study of operations with numbers and variables. Another kind of algebra that is used in the design of electronic digital computers involves operations with logical statements. In fact, complex electrical circuits can be analyzed using logical statements similar to those that you will work with in Exercises 1–18. This "algebra of logic" is called **Boolean algebra** in honor of its originator, George Boole (British, 1815–1864).

In Boolean algebra you use letters such as p, q, r, s, and so on, to stand for statements. For example, you might let p represent the statement "3 is an odd integer," and q, the statement "5 is less than 3." In this case, the statement p has the truth value T (the statement is true), whereas q has the truth value F (the statement is false).

The table below shows the operations that are used in Boolean algebra to produce compound statements from any given statements p and q.

Operation	In words	In symbols
Conjunction	p and q	$p \wedge q$
Disjunction	p or q	$p \vee q$
Conditional	If p, then q	$p \rightarrow q$
Equivalence	p if and only if q	$p \leftrightarrow q$
Negation	not p	$\sim p$

Note that in ordinary English the word *or* is sometimes used in the exclusive sense, to mean that just one of two alternatives occurs. In mathematical usage, however, the word *or* generally is used in the inclusive sense. As you can see in the table below, the disjunction p or q is true if only p is true, if only q is true, or if both are true. The rules for assigning truth values to compound statements are shown in the following five *truth tables*.

The conjunction $p \wedge q$ is true whenever *both* p and q are true.

Conjunction

p	q	$p \wedge q$
T	T	T
T	F	F
F	T	F
F	F	F

The disjunction $p \vee q$ is true whenever *at least one* of the statements is true.

Disjunction

p	q	$p \vee q$
T	T	T
T	F	T
F	T	T
F	F	F

Additional Answers
Exercises

10–12. $p \rightarrow q$ is false only when p is true and q is false.

10. q is false; p is true; $q \rightarrow p$ is true.

11. p is true; $\sim q$ is true; $p \wedge \sim q$ is true.

12. p is true; q is false; $p \vee q$ is true; $(p \vee q) \wedge p$ is true.

13.

p	q	$p \vee q$	$q \vee p$
T	T	T	T
T	F	T	T
F	T	T	T
F	F	F	F

Logically equivalent

14.

q	$\sim q$	$q \vee \sim q$
T	F	T
F	T	T

Not logically equivalent

15.

p	q	$p \vee q$	$(p \vee q) \rightarrow p$
T	T	T	T
T	F	T	T
F	T	T	F
F	F	F	T

Not a tautology

16.

p	q	$\sim q$	$q \wedge \sim q$	$(q \wedge \sim q) \rightarrow p$
T	T	F	F	T
T	F	T	F	T
F	T	F	F	T
F	F	T	F	T

Tautology

17.

p	q	r	$p \vee q$
T	T	T	T
T	T	F	T
T	F	T	T
T	F	F	T
F	T	T	T
F	T	F	T
F	F	T	F
F	F	F	F

(continued)

The conditional $p \rightarrow q$ is false only when p is true and q is false.

Conditional

p	q	$p \rightarrow q$
T	T	T
T	F	F
F	T	T
F	F	T

The equivalence $p \leftrightarrow q$ is true whenever p and q have the same truth value.

Equivalence

p	q	$p \leftrightarrow q$
T	T	T
T	F	F
F	T	F
F	F	T

The negation $\sim p$ is the denial of p. Therefore, it is reasonable to agree that $\sim p$ is false when p is true, and true when p is false.

Negation

p	$\sim p$
T	F
F	T

Example 1 Let r stand for "$1 > 2$," and s for "$2 < 4$." Read each of the following statements. Then by referring to the truth tables above and on page 95, give the truth value of the statement and a reason for your answer.

a. $r \wedge s$ **b.** $r \vee s$ **c.** $r \rightarrow s$ **d.** $r \leftrightarrow s$ **e.** $\sim r$

Solution

a. $r \wedge s$: $1 > 2$ and $2 < 4$. F; the truth value of r is F.

b. $r \vee s$: $1 > 2$ or $2 < 4$. T; the truth value of s is T.

c. $r \rightarrow s$: If $1 > 2$, then $2 < 4$. T; the truth value of r is F, and that of s is T.

d. $r \leftrightarrow s$: $1 > 2$ if and only if $2 < 4$. F; r and s have different truth values.

e. $\sim r$: not $(1 > 2)$, that is $1 \leq 2$. T; the truth value of r is F.

Two compound statements are *logically equivalent* if they have the same truth value for each combination of truth values of the individual statements.

Example 2 Construct a truth table for $q \vee \sim p$ and $p \rightarrow q$ for all possible truth values of p and q. Show that $q \vee \sim p$ and $p \rightarrow q$ are logically equivalent.

Solution

p	q	$\sim p$	$q \vee \sim p$	$p \rightarrow q$
T	T	F	T	T
T	F	F	F	F
F	T	T	T	T
F	F	T	T	T

Column 4 was constructed by using disjunction with columns 2 and 3.

Column 5 was constructed by using the conditional with columns 1 and 2.

Since columns 4 and 5 match, $q \vee \sim p$ and $p \rightarrow q$ are logically equivalent.

A compound statement that is true for all truth values of its component statements is called a **tautology.** For example, $p \leftrightarrow p$ is a tautology.

Exercises

In Exercises 1–9 assume that r and p are true statements and that q is a false statement. Determine the truth value of each statement.

1. $q \rightarrow r$ True
2. $\sim r \wedge p$ False
3. $p \vee \sim q$ True
4. $r \wedge (p \vee q)$ True
5. $p \vee (q \wedge r)$ True
6. $(p \vee r) \rightarrow q$ False
7. $\sim p \rightarrow (q \vee \sim r)$ True
8. $r \rightarrow (q \rightarrow r)$ True
9. $[r \wedge (p \vee q)] \rightarrow [(r \wedge p) \vee (r \wedge q)]$ True

Given that $p \rightarrow q$ is false, show that each statement in Exercises 10–12 is true.

10. $q \rightarrow p$
11. $p \wedge \sim q$
12. $(p \vee q) \wedge p$

Construct a truth table and determine if the two statements in each exercise are logically equivalent.

13. $p \vee q,\ q \vee p$
14. $q \vee \sim q,\ q$

Construct a truth table for each statement and determine whether the statement is a tautology.

15. $(p \vee q) \rightarrow p$
16. $(q \wedge \sim q) \rightarrow p$
17. $r \wedge (p \vee q) \leftrightarrow (r \wedge p) \vee (r \wedge q)$
18. $[(p \rightarrow q) \wedge (q \rightarrow r)] \rightarrow (p \rightarrow r)$

$r \wedge (p \vee q)$	$r \wedge p$	$r \wedge q$
T	T	T
F	F	F
T	T	F
F	F	F
T	F	T
F	F	F
F	F	F
F	F	F

$(r \wedge p) \vee (r \wedge q)$	$r \wedge (p \vee q) \leftrightarrow (r \wedge p) \vee (r \wedge q)$
T	T
F	T
T	T
F	T
T	T
F	T
F	T
F	T

Tautology

18.

p	q	r	$p \rightarrow q$
T	T	T	T
T	T	F	T
T	F	T	F
T	F	F	F
F	T	T	T
F	T	F	T
F	F	T	T
F	F	F	T

$q \rightarrow r$	$(p \rightarrow q) \wedge (q \rightarrow r)$
T	T
F	F
T	F
T	F
T	T
F	F
T	T
T	T

$p \rightarrow r$	$[(p \rightarrow q) \wedge (q \rightarrow r)] \rightarrow (p \rightarrow r)$
T	T
F	T
T	T
F	T
T	T
T	T
T	T
T	T

Tautology

Additional Answers
Written Exercises

(continued from p. 75)

12. 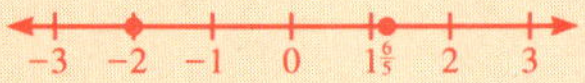

14. 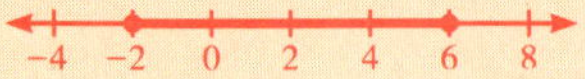

16. 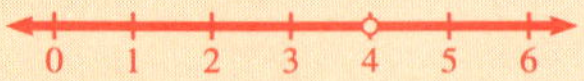

18. 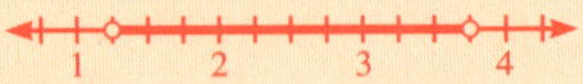

30. 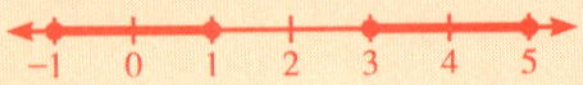

32. 

Quick Quiz from p. 79

1. Express the following conjunction as an equivalent open sentence involving absolute value: $4x - 2 \geq -6$ and $4x - 2 \leq 6$
   $|4x - 2| \leq 6$

Solve and graph the solution set.

2. $|3h + 7| > 1$
   $\{h\colon h < -\frac{8}{3} \text{ or } h > -2\}$
   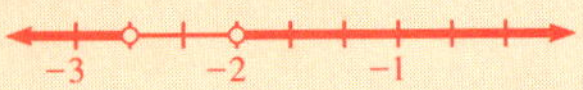
3. $|2 - x| - 4 < -1$
   $\{x\colon -1 < x < 5\}$
   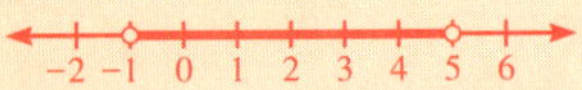
4. Translate the statement into an open sentence involving absolute value. The distance between d and 2 is greater than 4.
   $|d - 2| > 4$

Solve the open sentence graphically.

5. $|d + 4| \geq 2$
   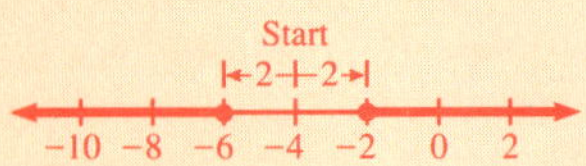
   $\{d\colon d \geq -2 \text{ or } d \leq -6\}$

Mixed Problem Solving

Solve each problem that has a solution. If a problem has no solution, explain why.

A

1. A child has 40 coins worth more than \$6.00. If the coins are dimes and quarters only, what can you conclude about the number of quarters? There are at least 14 quarters.
2. The width and length of a rectangle are consecutive odd integers. If the width is w, find the perimeter, in simplest form, in terms of w. $4w + 4$
3. Percy works 8 hours more per week than Selena. Selena, however, earns \$2 more per hour than Percy, who earns \$6 per hour. If their weekly pay is the same, how many hours does Percy work per week? 32 h
4. In Sioux Falls, South Dakota, the average low temperatures in January and July are $-16°$ C and $17°$ C, respectively. Find the difference between these temperatures. 33° C
5. Marcus invested \$4000 at simple annual interest, some at 5% and the rest at 8%. If he received \$272 in interest for one year, how much did he invest at each rate? \$1600 at 5%; \$2400 at 8%
6. Megan bought four times as many pencils as erasers. Pencils cost \$.04 each, and erasers cost \$.19 each. If Megan spent \$2.10 in all, how many pencils did she buy? 24 pencils
7. Four tennis balls cost d dollars. How many tennis balls can you buy for \$8? Give your answer in terms of d. $\frac{32}{d}$
8. Drew's second test score was 6 points higher than his first score. If his average was less than 80, what can you conclude about his first score? 8. His first score was less than 77.
9. The measure of a supplement of an angle is four times the measure of a complement. Find the measure of the angle. 60°
10. Helen has \$57 more in her checking account than in her savings account. If she has \$239 altogether, how much is in savings? \$91

B

11. At a health food store, dried apple slices that cost \$1.80/lb are mixed with dried banana slices that cost \$2.10/lb. If the mixture weighs 5 lb and costs \$1.92/lb, how many pounds of apple slices were used? 3 lb
12. The Petersons drove to a friend's house at 80 km/h. On the return trip, rain reduced their speed by 10 km/h. If their total driving time was 3 h, find the total distance driven. 224 km
13. The width of a rectangle is 1 cm more than half the length. When both the length and the width are increased by 1 cm, the area increases by 20 cm^2. Find the original dimensions of the rectangle. 7 cm × 12 cm
14. At 1:00 P.M. Sue left her home and began walking at 6 km/h toward Sandy's house. Fifteen minutes later, Sandy left her home and walked at 8 km/h toward Sue's house. If Sue lives 5 km from Sandy, at what time did they meet? Who walked farther? 1:30 P.M.; Sue

Preparing for College Entrance Exams

Strategy for Success

If there is a penalty for incorrect answers, random guessing may not be to your advantage. However, if you can eliminate several answers in a given question and then take a guess, your chance of selecting the correct answer is improved and it may be worthwhile to guess.

Decide which is the best of the choices given and write the corresponding letter on your answer sheet.

1. If $a > b$ and $c < 0$, which of the following *must* be true: C

I. $ab > bc$ II. $a^2 > b^2$ III. $\frac{a+c}{c} < \frac{b+c}{c}$

(A) I only **(B)** II only **(C)** III only **(D)** I, II, and III **(E)** II and III only

2. Which of the following open sentences is true for all real numbers x? D

(A) $3[x - 5(2 - x)] = 2(9x + 15)$ **(B)** $|7 - x| > 0$

(C) $(7x - 1) \div \frac{1}{2} = 2 - 14x$ **(D)** $-\frac{5}{2}(-6 + 4x) = 10\left(\frac{3}{2} - x\right)$

(E) $x + 3 < 3 - x$

3. Lucy spent \$2.36 on 15¢ and 22¢ stamps. If she bought twice as many 22¢ stamps as 15¢ stamps, how many 22¢ stamps did she buy? C

(A) 2 **(B)** 4 **(C)** 8 **(D)** 5 **(E)** 10

4. Evaluate $x^3 - 3x^2 \div (4 + x)$ if $x = -2$. D

(A) 2 **(B)** -2 **(C)** -10 **(D)** -14 **(E)** -6

5. Solve $A = \frac{h}{2}(b_1 + b_2)$ for b_2. D

(A) $\frac{A}{2h} - b_1$ **(B)** $A - \frac{hb_1}{2}$ **(C)** $\frac{2A - b_1}{h}$ **(D)** $\frac{2A}{h} - b_1$ **(E)** $\frac{2A}{hb_1}$

6. How many sets of four consecutive integers are there such that half the sum of the first three is greater than the fourth, and the sum of the last three is at least four times the first? D

(A) none **(B)** one **(C)** two **(D)** three **(E)** four

7. The graph at the right shows the solution set to which inequality? A

(A) $|2t + 3| < 1$ **(B)** $|2t + 3| > 1$

(C) $|3t + 2| < 1$ **(D)** $|3t + 2| \geq -1$

(E) $\frac{1}{2} \geq \left|t + \frac{3}{2}\right|$

Additional Answers
Written Exercises

(continued from p. 91)

18. Part 1:

1. $a > 0$ and $a > \frac{1}{a}$ Given
2. Assume $a = 1$. Assumption for indirect proof
3. $\frac{1}{a} \cdot a = \frac{1}{a} \cdot 1$ Mult. prop. of eq.
4. $1 = \frac{1}{a}$ Prop. of recip.; ident. prop. of mult.
5. $\therefore a = \frac{1}{a}$ Trans. prop. of eq. (Steps 2 and 4) But $a = \frac{1}{a}$ contradicts the given, $a > \frac{1}{a}$. So $a \neq 1$.

Part 2:

1. $a > 0$ and $a > \frac{1}{a}$ Given
2. Assume $a < 1$. Assumption for indirect proof
3. $\frac{1}{a} \cdot a < \frac{1}{a} \cdot 1$ First mult. prop. of order (assuming $\frac{1}{a} > 0$)
4. $1 < \frac{1}{a}$ Prop. of recip.; ident. prop. of mult.
5. $\therefore a < \frac{1}{a}$ Trans. prop. of order (Steps 2 and 4) But $a < \frac{1}{a}$ contradicts the given, $a > \frac{1}{a}$. So a is not less than 1. Since $a \neq 1$, (part 1) and a is not less than 1 (part 2), $a > 1$ by the comparison prop. of order.

3 Linear Equations and Functions

Objectives

3-1 To find solutions of open sentences in two variables and to solve problems involving open sentences in two variables.

3-2 To graph a linear equation in two variables.

3-3 To find the slope of a line and to graph a line given its slope and a point on it.

3-4 To find an equation of a line given its slope and a point on the line, or two points, or its slope and the y-intercept.

3-5 To solve systems of linear equations in two variables.

3-6 To use systems of equations to solve problems.

3-7 To graph linear inequalities in two variables and systems of such inequalities.

3-8 To find values of functions and to graph functions.

3-9 To find equations of linear functions and to apply properties of linear functions.

3-10 To graph relations and to determine when relations are also functions.

Assignment Guide

See p. T58 for Key to the format of the Assignment Guide

Day	Average Algebra	Average Algebra and Trigonometry	Extended Algebra	Extended Algebra and Trigonometry
1	**3-1** 104/1–37 odd	**3-1** 104/3–36 mult. of 3 105/*P*: 1, 3, 6–10 **S** 106/*Mixed Review*	*Administer Chapter 2 Test* **3-1** *Read 3-1* 104/1, 3, 5, 21–35 odd 105/*P*: 2, 7–11 **S** 106/*Mixed Review* **S** 98/*P*: 2–14 even	*Administer Chapter 2 Test* **3-1** *Read 3-1* 104/1, 3, 5, 21–35 odd 105/*P*: 2, 7–11, 13 **S** 106/*Mixed Review*
2	**3-1** 105/*P*: 1–11 **S** 106/*Mixed Review*	**3-2** 111/2–28 even	**3-2** 111/2–36 even	**3-2** 111/3–21 mult. of 3, 24–34 even **3-3** 116/3–48 mult. 3 **S** 117/*Mixed Review*
3	**3-2** 111/2–24 even **S** 105/34–38 even	**3-3** 116/1–47 odd **S** 117/*Mixed Review*	**3-3** 116/1–47 odd **S** 117/*Mixed Review*	**3-4** 121/3–57 mult. 3 **R** 123/*Self-Test 1*
4	**3-3** 116/1–47 odd **S** 117/*Mixed Review*	**3-4** 121/2–54 even **R** 123/*Self-Test 1*	**3-4** 121/3–54 mult. of 3 **R** 123/*Self-Test 1*	**3-5** 129/1–37 odd **S** 130/*Mixed Review* **3-6** 132/*P*: 3, 7, 10–14
5	**3-4** 121/2–30 even **S** 111/21–35 odd	**3-5** 129/1–37 odd **S** 130/*Mixed Review*	**3-5** 129/1–37 odd **S** 130/*Mixed Review*	**3-6** 133/*P*: 9, 15–18
6	**3-4** 121/31–55 odd **R** 123/*Self-Test 1*	**3-6** 132/*P*: 2–14 even **S** 129/32–40 even	**3-6** 132/*P*: 2–16 even **S** 129/32–40 even	**3-7** 138/1–43 odd **S** 139/*Mixed Review*
7	**3-5** 129/1–31 odd **S** 130/*Mixed Review*	**3-6** 133/11–17 odd **3-7** 138/1–25 odd	**3-6** 133/*P*: 11–19 odd **3-7** 138/1–27 odd	**3-8** 144/3–42 mult. of 3, 48, 49 **S** 139/44, 45 **R** 140/*Self-Test 2*
8	**3-5** 130/33–38 **3-6** 132/*P*: 2–10 even	**3-7** 139/27–41 odd **S** 139/*Mixed Review*	**3-7** 139/29–43 odd **S** 139/*Mixed Review*	**3-9** 149/3–24 mult. 3 150/*P*: 2, 4, 8–10 **S** 152/*Mixed Review*

9	**3-6** 133/*P*: 11–14, 16 **3-7** 138/1–17 odd	**3-8** 144/3–42 mult. 3 **S** 122/56, 57 **R** 140/*Self-Test 2*	**3-8** 144/3–42 mult. 3 **R** 140/*Self-Test 2*	**3-10** 156/3–36 mult. 3 **R** 158/*Self-Test 3*
10	**3-7** 139/19–41 odd **S** 139/*Mixed Review*	**3-9** 149/2–30 even **S** 134/*P*: 18, 19	**3-9** 149/2–32 even	**R** 162/*Chapter Review*
11	**3-8** 144/1–26 **R** 140/*Self-Test 2*	**3-9** 150/*P*: 1–9 odd **S** 152/*Mixed Review*	**3-9** 150/27, 29, 33 150/*P*: 1–11 odd **S** 152/*Mixed Review*	*Administer Chapter 3 Test* **4-1** 170/1–27 odd **S** 170/*Mixed Review*
12	**3-8** 145/27–43 odd	**3-10** 156/1–29 odd **R** 158/*Self-Test 3*	**3-10** 156/1–31 odd **R** 158/*Self-Test 3*	
13	**3-9** 149/2–30 even **S** 145/49	**R** 162/*Chapter Review*	**R** 162/*Chapter Review*	
14	**3-9** 149/5,21,23,27,29 150/*P*: 1–9 odd **S** 152/*Mixed Review*	*Administer Chapter 3 Test* **R** 165/*Cumulative Review*	*Administer Chapter 3 Test* **R** 165/*Cumulative Review*	
15	**3-10** 156/1–29 odd **R** 158/*Self-Test 3*			
16	**R** 162/*Ch. Review*			
17	*Administer Chapter 3 Test* **R** 165/*Cum. Review*			

Supplementary Materials Guide

For Use with Lesson	Practice Masters	Tests	Study Guide (Reteaching)	Resource Book		
				Tests	Practice	Other
3-1	Sheet 11		pp. 33–34			pp. 194–195 (Prob. Solv.) p. 206 (Applications)
3-2	Sheet 12		pp. 35–36		p. 108	p. 262 (Thinking Skl.)
3-3			pp. 37–38			p. 236 (Technology)
3-4	Sheet 13	Test 9	pp. 39–40		p. 109	pp. 236–237 (Tech.)
3-5			pp. 41–42			p. 238 (Technology)
3-6	Sheet 14		pp. 43–44			pp. 196–197 (Prob. Solv.)
3-7	Sheet 15	Test 10	pp. 45–46		p. 110	
3-8	Sheet 16		pp. 47–48			
3-9	Sheets 17, 18		pp. 49–50			
3-10	Sheet 19	Test 11	pp. 51–52		p. 111	
Chapter 3		Tests 12, 13		pp. 12–15	p. 112	pp. 178–179 (Mix. Rev.) p. 222 (Enrichment)
Cum. Rev. 1–3				pp. 16–18		

Overhead Visuals

For Use with Lessons	Visual
3-2, 3-3, 3-5, 3-7	A
3-2, 3-3, 3-5, 3-7	B
3-4, 3-5	1
3-7	2

Software

Algebra Plotter Plus	Using Algebra Plotter Plus	Computer Activities
Line Plotter, Line Quiz, Function Plotter, Inequality Plotter, Absolute Value Plotter	Scripted Demo, pp. 26–26 Enrichment, pp. 35–37 Activity Sheets, pp. 50–52	Activities 5–7
3-2, 3-3, 3-4, 3-5, 3-7, 3-8, 3-10	3-2, 3-3, 3-4, 3-5, 3-7, 3-8, 3-10	3-3, 3-4, Chapter 3

Strategies for Teaching

Problem Solving and Using Manipulatives

Today there are recognized a number of well-known problem solving strategies that relate to algebra. For example, using the linear combination method to transform a system of linear equations into an equivalent system whose solution is obvious is a strategy for solving linear systems. Other, more general strategies, such as looking for a pattern or drawing a diagram, can be very effective problem solving tools. Since these general strategies provide an *approach* to solving a problem rather than a specific method of solution, they are particularly useful for attacking a problem when the method of solution is not obvious. Here is a list of some of the strategies students may use in this chapter: checking solutions; using a table or a chart; drawing a diagram; writing and solving equations, inequalities, systems of equations or inequalities; using the 5-step problem solving plan.

Manipulatives are also useful when learning concepts or working through problems and exercises. Algebra 2 students who previously focused on learning the algorithms may need help understanding the concepts underlying the symbolic procedures. The goal is to have the students progress to the point where they can work entirely on the abstract level.

See the Exploration on page 834 for an activity in which students explore properties and graphs of step functions before formally learning about them in Lesson 3-8. One of the functions they explore is a real-life application of problem solving.

3-3 The Slope of a Line

A geoboard can be used to show lines and their slopes. Taking the central peg on a geoboard as (0, 0), a rubber band can "draw" the line between two points and then be stretched to show the rise and run of the slope.

(−1, −1)

(1, 1)

slope = $\frac{2}{2}$ = 1

Students might find it useful to see an argument using similar triangles to show that the slope of a nonvertical line is independent of the choice of two points. This example may be simulated using similar triangles cut from cardboard so the students may manipulate them until the concept is clear.

You may wish to review the slope of a line as related to its position on a coordinate system by constructing a large spinner attached to a reinforced piece of graph paper. Students can practice determining the slope by spinning the pointer and estimating or calculating the slope.

3-5 Systems of Linear Equations in Two Variables

You might pay particular attention to the possibility that different students will approach the same problem in different ways. The transformations are choices and can be selected in different orders. For example, in Example 3, x may be eliminated instead of y.

3-6 Problem Solving: Using Systems

Again, emphasize the five-step process of problem solving and point out that there will be two equations and two variables in each problem. Discuss some of the different types of equations and formulas that will be needed to solve the problems. Spend time letting the students discuss the problems and any difficulty they have in solving them. Remind students of the usefulness of a chart or a table in setting up a system.

3-7 Linear Inequalities in Two Variables

The two main problems facing students are deciding whether to use a solid line ($\leq$ or $\geq$) or a dashed line ($<$ or $>$) for the boundary of the graph, and which side of the line to shade. Most students shade above the line for "greater than" and below the line for "less than." The problem is that sometimes this approach does not work. Be sure to point out situations when it does not work. Obviously, $x > 3$ is such a case. It also will not work when the direction of the inequality is changed during the solution process as in Example 2. The easiest way to check a solution is to use a test point like (0, 0), (0, 1), (1, 0), (1, 1), etc., as explained in the text.

References to Strategies

PE: Pupil's Edition **TE:** Teacher's Edition **RB:** Resource Book

Problem Solving Strategies

PE: p. 132 (make a table)
TE: pp. 128, 131, 146
RB: pp. 194–197

Applications

PE: p. 100 (architecture); p. 102 (banking); pp. 105–106 (money); pp. 105–106, 116–117, 131–134, 140, 147–148, 150–152, 158 (word problems); pp. 159–161 (linear programming)
TE: pp. 100, 102–103, 132–133, 135, 140, 146–148, 158
RB: p. 206

Nonroutine Problems

PE: p. 106 (Probs. 9–14); p. 111 (Exs. 34–39); p. 112 (Exs. 49–57); p. 130 (Exs. 33–41); p. 134 (Probs. 18–20); p. 139 (Exs. 43–45); p. 140 (Challenge); p. 145 (Exs. 44–49); p. 150 (Exs. 31–33); p. 152 (Ex. 11); p. 157 (Ex. 37); p. 158 (Challenge)
TE: pp. T89, T91–T92 (Sugg. Extensions, Lessons 3-3, 3-4, 3-8, 3-9)

Communication

PE: p. 117 (Exs. 49–54, convincing argument); p. 157 (Ex. 37, convincing argument)
TE: pp. T90–T91 (Reading Algebra)

Thinking Skills

RB: p. 262

Explorations

PE: pp. 140, 158 (Challenge); pp. 159–160 (linear programming); p. 834 (functions)

Connections

PE: p. 100 (Architecture); pp. 102–103, 105–106 (Money); p. 124 (Geometry); p. 134 (Data Analysis); pp. 147–148 (Mathematical modeling); pp. 131–151 (Physics); p. 152 (History); pp. 159–161 (Linear programming)

Using Technology

PE: pp. 111, 114, 119, 122, 138, 161 (Exs.)
TE: pp. 110, 114, 119, 122, 125, 135, 160
RB: pp. 236–239
Computer Activities: pp. 11–19
Using Algebra Plotter Plus: pp. 26–27, 35–37, 50–52

Using Manipulatives/Models

TE: p. T89 (Lesson 3-3)
Overhead Visuals: A, B, 1, 2

Cooperative Learning

TE: p. T88 (Lesson 3-3)

Teaching Resources

For use in implementing the teaching strategies referenced on the previous page.

Application
Resource Book, p. 206

Application—Slope (For use with Chapter 3)

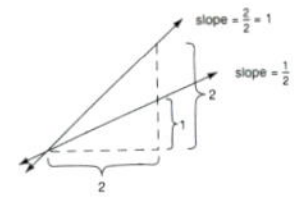

The *slope* of a line is defined to be the ratio of vertical displacement, called *rise*, to horizontal displacement, called *run*. For a given run, a greater rise results in a greater slope. Slope is a measure of the steepness of a line. Builders use slope when constructing roofs, stairs, and ramps.

1. Using a ruler to measure the appropriate lengths, find the indicated slopes.

   a. Slope of a roof $\left(\frac{\text{rise}}{\text{run}}\right)$: ______

   b. Slope of a flight of stairs $\left(\frac{\text{rise}}{\text{tread}}\right)$: ______

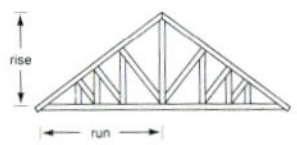

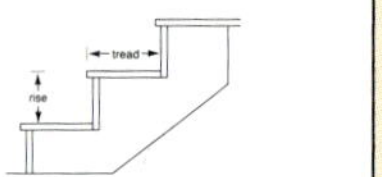

2. To meet federal guidelines, the slope of a ramp for the handicapped should be no more than a one-inch rise for a one-foot run. In the space at the right, make a scale drawing of the side view of a ramp that meets these guidelines.

3. a. Suppose the three steps pictured in part (b) of Exercise 1 rise a total of 1 ft 10 in. If a ramp is to replace these steps, find the amount of *run* needed for the ramp to meet federal guidelines. ______

   b. What is the distance traveled by a person using the ramp of part (a)? (Use the Pythagorean theorem: $a^2 + b^2 = c^2$, where a and b are the legs of a right triangle and c is the hypotenuse.) ______

4. a. Measure the rise and tread of a variety of flights of stairs, including stairs in your home, school, and other public buildings. Use the table at the right to record your measurements.

Location of stairs	*Rise*	*Tread*
______	______	______
______	______	______
______	______	______

   b. Find the sum of the rise and tread for each flight of stairs. Into what range do the sums fall? ______

   c. Find the slope of each flight of stairs. Into what range do the slopes fall? ______

206 APPLICATIONS

Enrichment/Nonroutine Problems
Resource Book, p. 222

Surprising Systems (For use with Chapter 3)

Carlos needed practice in solving simultaneous equations so he asked Roz to make up a problem for him. Roz, somewhat in a hurry, made all the coefficients consecutive integers.

$$x + 2y = 3$$
$$4x + 5y = 6$$

Carlos solved this system and found that the answer is $x = -1$ and $y = 2$. He then asked for another problem, so Roz made up the following:

$$3x + 4y = 5$$
$$8x + 9y = 10$$

After some work, Carlos came up with $x = -1$ and $y = 2$. "Come on Roz," said Carlos. "Make up one that's going to be different."

Roz thought that the next system would surely do the job.

$$55x + 56y = 57$$
$$13x + 12y = 11$$

To everyone's surprise, Carlos got $x = -1$ and $y = 2$ once more. What's going on here?

1. On the given set of axes graph the following lines:

$$x + 2y = 3$$
$$4x + 5y = 6$$
$$3x + 4y = 5$$

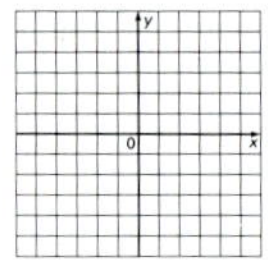

2. If the coefficients A, B, and C of the equation $Ax + By = C$ are three consecutive integers, show that the line passes through the point $(-1, 2)$. (Let $A = n$, $B = n + 1$, and $C = n + 2$ for increasing consecutive integers. Then let $A = n$, $B = n - 1$, and $C = n - 2$ for decreasing consecutive integers.)

222 ENRICHMENT

Thinking Skills
Resource Book, p. 262

Thinking Skills (For use after Chapter 3)

Interpreting information

1. When the x- and y-coordinates of a set of points are given in terms of a third variable, such as t, a system of *parametric* equations results. (The variable t is called the *parameter* of the system.) Complete the table below for this system of parametric equations: $x = 6 - 2t$, $y = -1 + t$

	t	x	y
a.	0	______	______
b.	1	______	______
c.	2	______	______
d.	3	______	______
e.	4	______	______
f.	5	______	______
g.	6	______	______

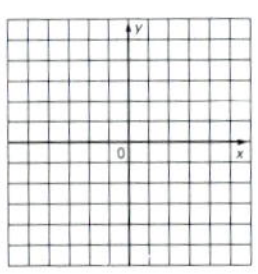

2. Using the axes at the right above, graph the ordered pairs (x, y) from the table in Exercise 1. What do you notice about these points? ______

3. The system of parametric equations given in Exercise 1 can be rewritten as a single equation relating x and y directly if the parameter t is eliminated. To do so, solve one of the parametric equations for t and substitute this expression for t in the other parametric equation. Write the result: ______ How does this equation confirm your observation in Exercise 2? ______

Applying concepts

4. Suppose the system of parametric equations given in Exercise 1 represents the position (x, y) of an ant on a tabletop at time t. (*Note:* The tabletop is bounded by the lines $x = \pm 6$ and $y = \pm 6$.) Using the axes at the right, draw the complete path of the ant if t takes on all real values from $t = 0$ to $t = 6$.

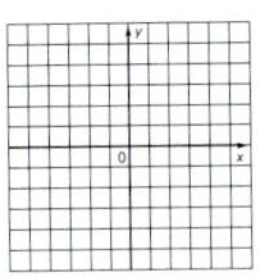

5. Suppose a second ant on the tabletop of Exercise 4 follows a path determined by this system of parametric equations: $x = -4 + t$, $y = -6 + 2t$

   a. Using the axes at the right, draw the complete path of the second ant.

   b. At what point do the paths of the two ants cross? ______

   c. Although the paths of the ants cross, the ants never actually meet. Explain. ______

262 THINKING SKILLS

Using Technology
Resource Book, p. 239

Using a Computer or a Graphing Calculator

To complete these activities, you should use a computer with graphics software (such as ALGEBRA PLOTTER PLUS) or a graphing calculator.

Linear Programming (For use with Application/Linear Programming)

1. Suppose the feasible region for a linear programming problem is given by:

$$x + y \leq 9$$
$$x + 2y \leq 16$$
$$4x + y \leq 24$$
$$x \geq 0$$
$$y \geq 0$$

A computer or a graphing calculator can help you determine the vertices of the feasible region. If you have the ALGEBRA PLOTTER PLUS disk, you can use the "Inequality Plotter" program to graph the first three inequalities. Note that the last two inequalities do not really need to be graphed, since they simply imply that the feasible region lies entirely in the first quadrant. You may therefore want to rescale—using 0 to 12 on the x-axis and 0 to 10 on the y-axis, for example—before graphing the first three inequalities. The region that is shaded the darkest is the feasible region.

An alternate approach to graphing the inequalities is to graph the related equations instead. Of course, you must determine the feasible region for yourself by deciding on which side of each line the feasible region lies.

Once the feasible region has been graphed, you should notice that it has five vertices. List them: ______, ______, ______, ______, ______

2. For the feasible region of Exercise 1, suppose you are given the objective function $P = 3x + 2y$. To determine the minimum or maximum value of P in the feasible region, you only need to check the values of P at the vertices. For example, one of the vertices in your list for Exercise 1 should have been the origin, as shown in the table below. Complete the table for the other vertices from your list.

	x	y	$3x + 2y = P$
	0	0	$3(0) + 2(0) = 0$
a.	____	____	________
b.	____	____	________
c.	____	____	________
d.	____	____	________

   e. Minimum value of P: ______ f. Maximum value of P: ______

3. Use the feasible region of Exercise 1 to find the minimum and maximum values of the objective function $P = 4x + 5y$.

   a. Minimum value of P: ______ b. Maximum value of P: ______

USING TECHNOLOGY 239

Using Technology/Exploration

Using Algebra Plotter Plus, p. 50

Exploring Slopes — Algebra Plotter Plus | Book 2: Lesson 3-3

Use the Line Plotter program of Algebra Plotter Plus.

1. Graph these equations: $y = \frac{1}{2}x$, $y = 1x$, $y = 2x$

   The coefficient of x in each equation is called the *slope* of the line. What do you notice about the steepness of the lines as the slopes get larger?

   ____

2. Notice that the lines in Exercise 1 all pass through the origin. Find the point on each line with x-coordinate 1 and complete the table below.

	Equation of line	*Slope*	*Common point*	*Point with x-coordinate 1*
a.	$y = \frac{1}{2}x$	____	(0, 0)	(1, ____)
b.	$y = 1x$	____	(0, 0)	(1, ____)
c.	$y = 2x$	____	(0, 0)	(1, ____)

   d. Describe the relationship between the point with x-coordinate 1 and the slope of any line that passes through the origin.

   ____

   ____

3. Graph these equations: $y = 1x + 1$, $y = 0x + 1$, $y = -1x + 1$

   a. What happens to the lines as the slope changes from positive to negative?

   ____

   ____

   b. What point do the lines have in common? (____, ____)

4. Try to complete the table below before graphing the given equations.

	Equation of line	*Slope*	*Common point*	*Point with x-coordinate 1*
a.	$y = -\frac{1}{2}x + 1$	____	(____, ____)	(1, ____)
b.	$y = -1x + 1$	____	(____, ____)	(1, ____)
c.	$y = -2x + 1$	____	(____, ____)	(1, ____)

   Now examine the graphs of the given equations to confirm the entries in the table.

5. Use the "Line Quiz" program of Algebra Plotter Plus to see if you can correctly determine the equations of random lines drawn by the computer.

50 USING ALGEBRA PLOTTER PLUS

Using Technology/Exploration

Using Algebra Plotter Plus, p. 52

Estimating Solutions of Linear Systems — Algebra Plotter Plus | Book 2: Lesson 3-5

Use the Line Plotter program of Algebra Plotter Plus. Select EQUATIONS and use the right-arrow key to move to the standard form.

1. Graph this system: $3x + 2y = 11$, $2x - 3y = 3$

   Write the coordinates of the point of intersection: (____, ____)

   This is called a *solution* of the system. Check the solution by substituting the x- and y-coordinates of the point of intersection into each equation of the system. You should get a true statement each time.

2. Graph this system: $8x + y = 2$, $20x + 5y = 2$

   a. The coordinates of the point of intersection for the system are *not* integers. Estimate the coordinates from the graph: (_____, _____)

   b. To obtain a better approximation of the solution, select ZOOM and move the cross hairs near the point of intersection. Press <RETURN> or <ENTER> and use the arrow keys to change the size of the rectangle that indicates the portion of the graph to be magnified. Then press <RETURN> or <ENTER> and the computer will automatically draw the magnified graph. Now give a better estimate of the coordinates of the point of intersection: (_____, _____)

3. Graph this system: $2x - 4y = 7$, $3x + 5y = 9$

   Find the solution of the system to the nearest tenth: (_____, _____)
   (*Note*: If you used ZOOM in the previous exercise, you may first need to return the scale to the standard values. Select SCALE and use arrow keys to move to the "Standard Scale" option, then highlight "Yes.")

4. Graph this system: $9x - 5y = 13$, $-7x + 3y = 6$

   Find the solution of the system to the nearest tenth: (_____, _____)
   (*Note*: If the point of intersection is off the screen, you will need to rescale to see where the lines intersect. Select SCALE and redefine the minimum and maximum x-values and y-values before using DRAW.)

5. Select TABLE to see a table of values for the equations in Exercise 4. Explain how this table helps you find the point of intersection *without* looking at the graph.

   ____

   ____

52 USING ALGEBRA PLOTTER PLUS

Using Models

Overhead Visual 1, Sheets 1 and 3

PARALLEL AND PERPENDICULAR LINES

$y = \frac{3}{2}x - 6$

Slope $\frac{3}{2}$

$y = -\frac{2}{3}x - \frac{5}{3}$

Slope $-\frac{2}{3}$

$\left(\frac{3}{2}\right)\left(-\frac{2}{3}\right) = -1$

Because the slopes are negative reciprocals of each other, the lines are perpendicular.

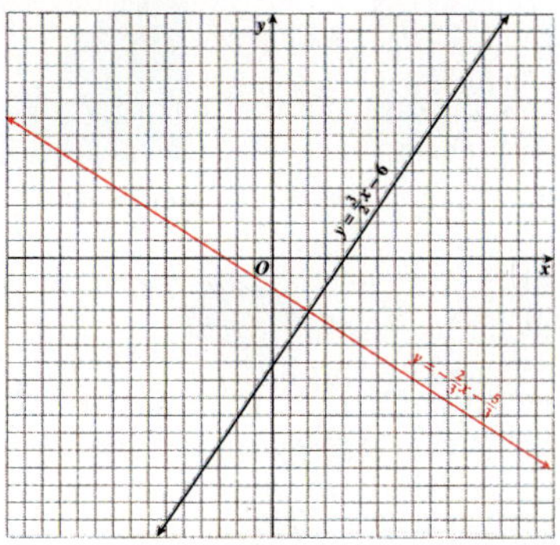

VISUAL 1

Using Models

Overhead Visual 2, Sheets 1 and 2

LINEAR INEQUALITIES IN TWO VARIABLES

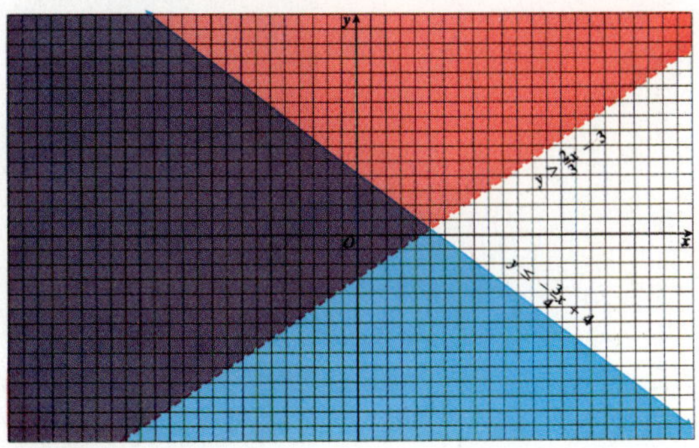

VISUAL 2

Application

Because slope is a measure of steepness, people use slope whenever surfaces are not level. Builders, for instance, consider slope when constructing the roof of a house. Slope is also important to highway engineers. The curves of a highway need to be banked to help vehicles grip the road.

One group of people whose lives are affected by slope are wheelchair users. Current federal guidelines require new buildings constructed with federal funds to be accessible to disabled people. In particular, the maximum slope allowed for a sidewalk is 5%, and a wheelchair ramp can be no steeper than a one-inch rise for a one-foot run.

Group Activities

Student can look for local buildings with ramps and measure their slopes. Students can also look for buildings without ramps and determine the lengths of the ramps needed to replace any existing steps.

Support Materials

Resource Book p. 206

3 Linear Equations and Functions

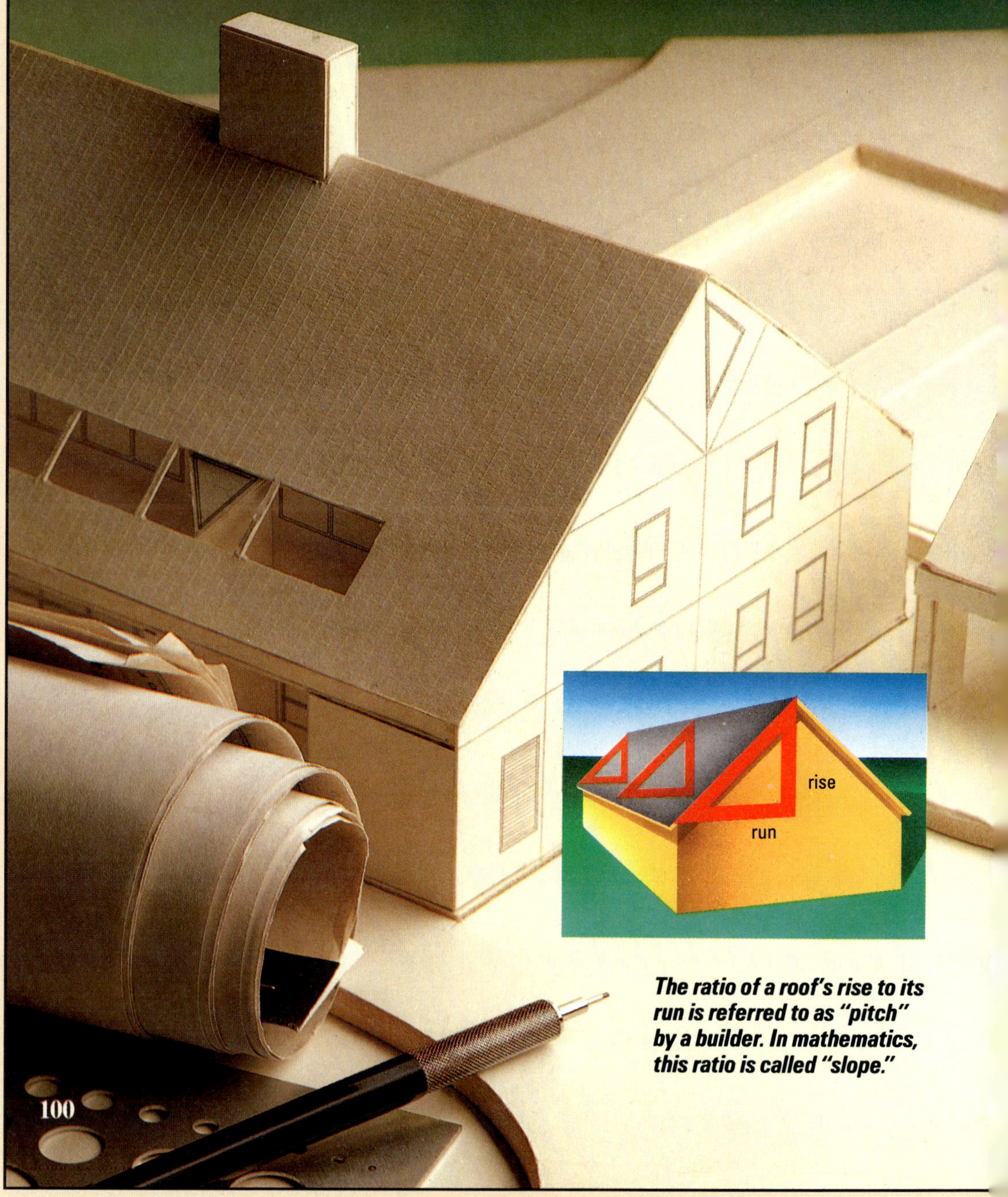

The ratio of a roof's rise to its run is referred to as "pitch" by a builder. In mathematics, this ratio is called "slope."

Linear Equations and Their Graphs

3-1 Open Sentences in Two Variables

Objective To find solutions of open sentences in two variables and to solve problems involving open sentences in two variables.

Equations and inequalities such as

$$9x + 2y = 15, \quad y = x^2 - 4, \quad \text{and} \quad 2x - y \geq 6$$

are called **open sentences in two variables.**

A **solution** of an open sentence in the two variables x and y is a pair of numbers, one a value of x and the other a value of y, that together make the sentence a true statement. We usually write such a solution as an **ordered pair** in which the value of x is listed first and value of y is listed second. For example,

$(1, 3)$ is a solution of $9x + 2y = 15$ because $9(1) + 2(3) = 15$.

However, $(3, 1)$ is *not* a solution because $9(3) + 2(1) \neq 15$. A solution of an open sentence is said to *satisfy* the sentence. The set of all solutions is called the **solution set** of the open sentence. Finding the solution set is called **solving** the open sentence.

Example 1 Solve the equation $9x + 2y = 15$ if the domain of x is $\{-1, 0, 1, 2, 3\}$.

Solution 1 Solve the equation for y.

$$y = \frac{15 - 9x}{2}$$

Then replace x with each value in its domain and find the corresponding value of y. The last column of the table lists the five solutions. Solving the equation over other domains will produce different solutions.

x	$\frac{15 - 9x}{2}$	y	Solution
-1	$\frac{15 - 9(-1)}{2}$	12	$(-1, 12)$
0	$\frac{15 - 9(0)}{2}$	$\frac{15}{2}$	$\left(0, \frac{15}{2}\right)$
1	$\frac{15 - 9(1)}{2}$	3	$(1, 3)$
2	$\frac{15 - 9(2)}{2}$	$-\frac{3}{2}$	$\left(2, -\frac{3}{2}\right)$
3	$\frac{15 - 9(3)}{2}$	-6	$(3, -6)$

$\therefore$ the solution set is $\left\{(-1, 12), \left(0, \frac{15}{2}\right), (1, 3), \left(2, -\frac{3}{2}\right), (3, -6)\right\}$. ***Answer***

Teaching References

Lesson Commentary, pp. T87–T92
Assignment Guide, pp. T60–T62
Supplementary Materials
- Practice Masters 11–19
- Tests 9–13
- Resource Book
  - Practice Exercises, pp. 108–112
  - Tests, pp. 12–18
  - Enrichment Activity, p. 222
  - Application, p. 206
  - Mixed Review, pp. 178–179
  - Practice in Problem Solving/Word Problems, pp. 194–197
- Study Guide, pp. 33–52
- Computer Activities, 5–7
- Test Generator
- Disk for Algebra

Alternate Test, p. T15
Cumulative Review, p. T29

Teaching Suggestions p. T87

Warm-Up Exercises

Evaluate if $x \in \{-2, -1, 0, 1, 2\}$.

1. $5x - 1$ $\{-11, -6, -1, 4, 9\}$
2. $\frac{7}{2}x$ $\{-7, -\frac{7}{2}, 0, \frac{7}{2}, 7\}$
3. $3(x + 1)$ $\{-3, 0, 3, 6, 9\}$
4. $\frac{1}{2}(5x - 8)$ $\{-9, -\frac{13}{2}, -4, -\frac{3}{2}, 1\}$

Motivating the Lesson

Point out to students that many real life situations require solving sentences in two variables. For example, the president of the junior class may need to determine the number of adult and student tickets that must be sold to a benefit basketball game in order for the class to raise money needed for a project.

Chalkboard Examples

1. Find five solutions of $4x + 3y = 24$.
   $3y = 24 - 4x$
   $y = \frac{24 - 4x}{3}$

x	y	Sol.
0	$\frac{24}{3} = 8$	(0, 8)
6	$\frac{24-24}{3} = 0$	(6, 0)
3	$\frac{24-12}{3} = 4$	(3, 4)
−3	$\frac{24+12}{3} = 12$	(−3, 12)
$-\frac{3}{4}$	$\frac{24+3}{3} = 9$	$\left(-\frac{3}{4}, 9\right)$

2. Roberto has $22. He buys some notebooks costing $2 each and some binders costing $5 each. If Roberto spends all $22 how many of each does he buy?
   Let n = number of notebooks and b = number of binders. Both n and b must be whole numbers.
   $2n + 5b = 22$
   $n = \frac{22 - 5b}{2}$
   If b is odd, then $22 - 5b$ is odd and n is not a whole number.

Solution 2 Substitute each value in the domain of x in the given equation $9x + 2y = 15$. Then solve for y.

$x = -1$: $9(-1) + 2y = 15$; $2y = 24$; $y = 12$ *Solution* $(-1, 12)$

$x = 0$: $9(0) + 2y = 15$; $2y = 15$; $y = \frac{15}{2}$ *Solution* $\left(0, \frac{15}{2}\right)$

When you substitute the other values in the domain of x in the given equation, you will find the complete solution set:

$\left\{(-1, 12), \left(0, \frac{15}{2}\right), (1, 3), \left(2, -\frac{3}{2}\right), (3, -6)\right\}$. ***Answer***

Open sentences in two variables can be used to solve certain word problems.

Example 2 A customer asks a bank teller for $390 in traveler's checks, some worth $50 and some worth $20. Find all possibilities for the number of each type of check the customer could receive.

Solution

Step 1 The problem asks for the number of $50 checks and the number of $20 checks whose total value is $390.

Step 2 Let f = the number of $50 checks and let t = the number of $20 checks.

Step 3 The total value of the $50 checks is $50f$ dollars and the total value of the $20 checks is $20t$ dollars. Write an equation that expresses the total value of the checks in dollars.

$$50f + 20t = 390$$

Step 4 Solve the equation for one variable, say t, in terms of the other. First divide both sides by 10 to obtain this simpler equation:

$$5f + 2t = 39$$
$$t = \frac{39 - 5f}{2}$$

Remember: The number of each type of check *must* be a whole number. If f is even, then $5f$ is even and $39 - 5f$ is odd. If $39 - 5f$ is odd, t is not a whole number. Also, if f exceeds 7, t is negative. Therefore, replace f with 1, 3, 5, and 7 as shown in the table.

f	t	(f, t)
1	17	(1, 17)
3	12	(3, 12)
5	7	(5, 7)
7	2	(7, 2)

∴ the solution set of $50f + 20t = 390$ over the domain of the whole numbers is $\{(1, 17), (3, 12), (5, 7), (7, 2)\}$.

Step 5 The check is left for you.

$\therefore$ the customer can receive: 1 \$50 check and 17 \$20 checks
or 3 \$50 checks and 12 \$20 checks
or 5 \$50 checks and 7 \$20 checks
or 7 \$50 checks and 2 \$20 checks. ***Answer***

Example 3 Find all positive two-digit odd numbers with this property: When the digits are interchanged, the result exceeds the original number by more than 36.

Solution

Step 1 The problem asks for all positive two-digit odd numbers such that when the digits are interchanged, the new number is greater than the original number plus 36. Recall that every positive integer can be written in expanded form. For example:

$$73 = 7 \cdot 10 + 3$$

(3: units' digit; 7: tens' digit)

Step 2 Let u = the units' digit and let t = the tens' digit. Then the original number is $10t + u$. The number with the digits interchanged is $10u + t$.

Step 3 Write an inequality relating t and u.

new number → $10u + t$; is greater than → $>$; original number → $(10t + u)$; plus → $+$; 36 → 36

Step 4 Solve the inequality:

$$10u + t > 10t + u + 36$$
$$9u > 9t + 36$$

Divide both sides by 9:

$$u > t + 4$$

Since the original number has two digits, the tens' digit is not zero, so $t \neq 0$ and the replacement set for t is $\{1, 2, 3, 4, 5, 6, 7, 8, 9\}$. Since the number is odd, the replacement set for u is $\{1, 3, 5, 7, 9\}$.

t	$t+4$	$u > t+4$	(t, u)
1	5	7, 9	(1, 7), (1, 9)
2	6	7, 9	(2, 7), (2, 9)
3	7	9	(3, 9)
4	8	9	(4, 9)

If $t > 4$, then $u > 9$ and there are no more possibilities.

Step 5 The check is left for you.

$\therefore$ the numbers are 17, 19, 27, 29, 39, and 49. ***Answer***

$\therefore$ replace b by 0, 2, 4 as shown below.

b	n	(b, n)
0	11	(0, 11)
2	6	(2, 6)
4	1	(4, 1)

If $b > 4$, then $n < 0$.
$\therefore$ Roberto buys 0 binders and 11 notebooks or 2 binders and 6 notebooks or 4 binders and 1 notebook.

3. A two-digit number is such that the tens digit is at least 5 more than twice the units digit. Find all positive integers with this property.
Let u = units digit and t = tens digit. The number is $10t + u$ where $t \in \{1, 2, 3, \cdots, 9\}$ and $u \in \{0, 1, 2, \cdots, 9\}$.
$t \geq 5 + 2u$

u	$5 + 2u$	t
0	5	5, 6, 7, 8, 9
1	7	7, 8, 9
2	9	9

If $u > 2$, then $t > 9$.
$\therefore$ the numbers are 50, 60, 70, 80, 90, 71, 81, 91, 92

Check for Understanding

Here is a suggested use of the Oral Exercises to check students' understanding as you teach the lesson.
Oral Exs. 1–10: use after Example 1.
Oral Ex. 11: use after Example 2.

Guided Practice

Solve. The domain of x is $\{1, 0, -2\}$.

1. $3x + 2y = 8$
$\left\{\left(1, \frac{5}{2}\right), (0, 4), (-2, 7)\right\}$

2. $-3x + \frac{1}{2}y = 6$
$\{(1, 18), (0, 12), (-2, 0)\}$

Complete.

3. $2x + 3y = 12$
(0, _?_), (_?_, 0), (4, _?_) 4; 6; $\frac{4}{3}$

4. $\frac{1}{3}x - 2y = 6$
(_?_, 0), (6, _?_), (0, _?_) 18; −2; −3

Find k so that the ordered pair satisfies the equation.

5. $x + 2y = k$; (2, 1) 4
6. $kx + 3y = 12$; (3, −3) 7

Summarizing the Lesson

In this lesson students learned to solve open sentences in two variables over a given domain and to apply this to solving problems. Ask students to discuss the importance of order in the solutions to open sentences in two variables.

Suggested Assignments

Average Algebra
Day 1: 104/1–37 odd
Day 2: 105/*P*: 1–11
S 106/Mixed Review

Average Alg. and Trig.
104/3–36 mult. of 3
105/*P*: 1, 3, 6–10
S 106/Mixed Review

Extended Algebra
104/1, 3, 5, 21–35 odd
105/*P*: 2, 7–11
S 106/Mixed Review

Extended Alg. and Trig.
104/1, 3, 5, 21–35 odd
105/*P*: 2, 7–11, 13
S 106/Mixed Review

Oral Exercises

Tell whether each ordered pair is a solution of the open sentence.

1. $3x + y = 1$ (1, 0), (0, 1), (1, −2), (1, −4) no, yes, yes, no
2. $2x - 3y = 5$ (−1, 1), (1, −1), (2, −3), (−5, −5) no, yes, no, yes
3. $7x - 2y = 8$ (1, −1), (2, 3), (0, −4), (4, 10) no, yes, yes, yes
4. $3x + 5y = 4$ (−2, −2), (−2, 2), (8, 4), (7, −3) no, yes, no, no
5. $5x - 2y > 6$ (2, 0), (0, −3), (1, −1), (1, 1) yes, no, yes, no
6. $4x - 3y \le 2$ (1, 2), (2, 1), (1, −1), (0, −1) yes, no, no, no

7.–10. Answers may vary. Examples are given.

Give three ordered pairs of integers that satisfy the open sentence.

7. $x - y = 2$ (1, −1), (0, −2), (2, 0)
8. $5x + y = 6$ (0, 6), (1, 1), (2, −4)
9. $2x + 3y > 10$ (2, 5), (0, 4), (6, 0)
10. $3x + 5y \le 8$ (0, 0), (1, 1), (2, 0)
11. Make up a word problem that could be solved using this equation: $10d + 25q = 160$. Answers may vary. Example: A collection of dimes and quarters is worth \$1.60. Find all possibilities for the number of each type of coin.

1. $\left\{(-1, 3), \left(0, \frac{7}{3}\right), (2, 1)\right\}$ 2. $\left\{(-1, 2), \left(0, \frac{3}{2}\right), \left(2, \frac{1}{2}\right)\right\}$ 3. $\left\{\left(-1, \frac{1}{2}\right), (0, 0), (2, -1)\right\}$

Written Exercises

4. $\{(-1, -5), (0, -3), (2, 1)\}$ 5. $\left\{(-1, -1), \left(0, -\frac{5}{9}\right), \left(2, \frac{1}{3}\right)\right\}$
6. $\{(-1, -18), (0, -6), (2, 18)\}$

Solve each equation if the domain of x is $\{-1, 0, 2\}$.

A

1. $2x + 3y = 7$
2. $3x + 6y = 9$
3. $-x - 2y = 0$
4. $-2x + y = -3$
5. $4x - 9y = 5$
6. $6x - \frac{1}{2}y = 3$

7–12. Solve each equation in Exercises 1–6 if the domain of x is $\{-2, 1, 3\}$.

Complete each ordered pair to form a solution of the equation.

Sample $x + 2y = 8$; (0, _?_), (_?_, 0) **Solution** (0, 4), (8, 0)

13. $3x + 2y = 12$ (0, _?_), (_?_, 0), (2, _?_) 6, 4, 3
14. $4x + 3y = 8$ (0, _?_), (_?_, 0), (5, _?_) $\frac{8}{3}$, 2, −4
15. $5x - 2y = 7$ (0, _?_), (_?_, 0), (−3, _?_) $-\frac{7}{2}$, $\frac{7}{5}$, −11
16. $x + 6y = -9$ (0, _?_), (_?_, 0), (−3, _?_) $-\frac{3}{2}$, −9, −1
17. $2x - 2y = 3$ (1, _?_), $\left(\frac{1}{2}, \_?\_\right)$, $\left(\_?\_, \frac{1}{2}\right)$ $-\frac{1}{2}$, −1, 2
18. $3x + 5y = 3$ (1, _?_), $\left(-\frac{2}{3}, \_?\_\right)$, $\left(\_?\_, \frac{7}{5}\right)$ 0, 1, $-\frac{4}{3}$
19. $\frac{1}{2}x - 2y = 1$ (1, _?_), (6, _?_), (_?_, 0) $-\frac{1}{4}$, 1, 2
20. $x + \frac{1}{3}y = 2$ (1, _?_), (_?_, 6), $\left(\frac{1}{3}, \_?\_\right)$ 3, 0, 5

Find the value of k so that the ordered pair satisfies the equation.

Sample $x + 2y = k;\ (3, 1)$

Solution
$$x + 2y = k$$
$$3 + 2(1) = k$$
$$k = 5$$

21. $2x + y = k;\ (2, 1)$ 5
22. $3x - y = k;\ (1, -3)$ 6
23. $3x - ky = 4;\ (2, -1)$ −2
24. $kx + 3y = 7;\ (-1, 3)$ 2
25. $kx + 2y = k;\ (3, 3)$ −3
26. $6x - ky = k;\ (2, 2)$ 4

Solve each equation if each variable represents a *whole number*.

B 27. $x + y = 4$ {(0, 4), (1, 3), (2, 2), (3, 1), (4, 0)}
28. $2x + y = 6$ {(0, 6), (1, 4), (2, 2), (3, 0)}
29. $4x + y = 15$ {(0, 15), (1, 11), (2, 7), (3, 3)}
30. $x + 5y = 24$
31. $2x + 3y = 18$ {(0, 6), (3, 4), (6, 2), (9, 0)}
32. $5x + 2y = 30$ {(0, 15), (2, 10), (4, 5), (6, 0)}

30. {(4, 4), (9, 3), (14, 2), (19, 1), (24, 0)}

Solve each open sentence if each variable represents a *positive integer*.

33. $x + y < 5$ {(1, 1), (1, 2), (1, 3), (2, 1), (2, 2), (3, 1)}
34. $2x + y < 6$ {(1, 1), (1, 2), (1, 3), (2, 1)}
35. $2x + 3y \le 12$
36. $3x + 5y \le 19$
37. $x + y^2 = 10$ {(1, 3), (6, 2), (9, 1)}
38. $x^2 + 2y < 11$ {(1, 1), (1, 2), (1, 3), (1, 4), (2, 1), (2, 2), (2, 3)}

In Exercises 39–40, the digits of a positive two-digit integer N are interchanged to form an integer K.

C 39. Show that $N - K$ is an integral multiple of 9.

40. Show that $N + K$ is an integral multiple of 11.

35. {(1, 1), (1, 2), (1, 3), (2, 1), (2, 2), (3, 1), (3, 2), (4, 1)}
36. {(1, 1), (1, 2), (1, 3), (2, 1), (2, 2), (3, 1), (3, 2), (4, 1)}

Problems

In each problem (a) choose two variables to represent the numbers asked for, (b) write an open sentence relating the variables, and (c) solve the open sentence and give the answer to the problem. (Include solutions in which one of the variables is zero.)

A 1. A bank teller needs to pay out \$75 using \$5 bills and \$20 bills. Find all possibilities for the number of each type of bill the teller could use.

2. Bruce, an appliance salesman, earns a commission of \$50 for each washing machine he sells and \$100 for each refrigerator. Last month he earned \$500 in commissions. Find all possibilities for the number of each kind of appliance he could have sold.

3. Luis has 95 cents in dimes and quarters. Find all possibilities for the number of each type of coin he could have.

4. Kimberly has \$1.95 in dimes and quarters. Find all possibilities for the number of each type of coin she could have.

Additional Answers Written Exercises

7. $\left\{\left(-2, \frac{11}{3}\right), \left(1, \frac{5}{3}\right), \left(3, \frac{1}{3}\right)\right\}$
8. $\left\{\left(-2, \frac{5}{2}\right), (1, 1), (3, 0)\right\}$
9. $\left\{(-2, 1), \left(1, -\frac{1}{2}\right), \left(3, -\frac{3}{2}\right)\right\}$
10. $\{(-2, -7), (1, -1), (3, 3)\}$
11. $\left\{\left(-2, -\frac{13}{9}\right), \left(1, -\frac{1}{9}\right), \left(3, \frac{7}{9}\right)\right\}$
12. $\{(-2, -30), (1, 6), (3, 30)\}$

Additional Answers Problems

1. b. $5f + 20t = 75$
   c. 15 \$5 bills and 0 \$20 bills; 11 \$5 bills and 1 \$20 bill; 7 \$5 bills and 2 \$20 bills; or 3 \$5 bills and 3 \$20 bills
2. b. $50w + 100r = 500$
   c. 10 washers and 0 refrig.; 8 washers and 1 refrig.; 6 washers and 2 refrig.; 4 washers and 3 refrig.; 2 washers and 4 refrig.; or 0 washers and 5 refrig.
3. b. $10d + 25q = 95$
   c. 7 dimes and 1 quarter; or 2 dimes and 3 quarters
4. b. $10d + 25q = 195$
   c. 17 dimes and 1 quarter; 12 dimes and 3 quarters; 7 dimes and 5 quarters; or 2 dimes and 7 quarters

(continued)

Additional Answers Problems

5. b. $3x + y = 19$
   c. 4 cm, 4 cm, 4 cm and 7 cm; 5 cm, 5 cm, 5 cm, and 4 cm; or 6 cm, 6 cm, 6 cm, and 1 cm
6. b. $2x + y = 15$
   c. 4 m, 4 m, and 7 m; 5 m each; 6 m, 6 m, and 3 m; or 7 m, 7 m, and 1 m
7. b. $5n + 25q + 10(q + 3) = 200$
   c. 34 nick., 0 quar., 3 dimes; 27 nick., 1 quar., 4 dimes; 20 nick., 2 quar., 5 dimes; 13 nick., 3 quar., 6 dimes; 6 nick., 4 quar., 7 dimes
8. b. $5n + 25q + 2n + 10(q + 4) = 201$
   c. 23 nick., 0 quar., 46 pen., 4 dimes; 18 nick., 1 quar., 36 pen., 5 dimes; 13 nick., 2 quar., 26 pen., 6 dimes; 8 nick., 3 quar., 16 pen., 7 dimes; 3 nick., 4 quar., 6 pen., 8 dimes

Additional Answers Mixed Review Exercises

9. 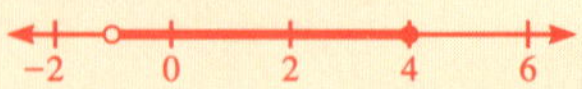

10. 

11. 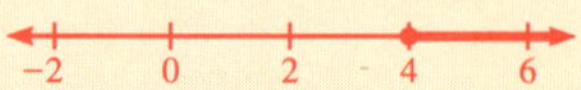

Supplementary Materials

Study Guide	pp. 33–34
Practice Master	11

5. A certain quadrilateral has three sides of equal length and its perimeter is 19 cm. Find all integral possibilities for the lengths of the sides in centimeters. (*Hint:* The sum of the lengths of any three sides of a quadrilateral must exceed the length of the fourth side.)

6. An isosceles triangle has perimeter 15m. Find all integral possibilities for the lengths of the sides in meters. (*Hint:* The sum of the lengths of any two sides of a triangle must exceed the third side.)

B

7. A box contains nickels, dimes, and quarters worth \$2.00. Find all possibilities for the number of each coin if there are three more dimes than quarters.

8. A bag contains twice as many pennies as nickels and four more dimes than quarters. Find all possibilities for the number of each coin if their total value is \$2.01.

In Exercises 9–12, the digits of a positive two-digit integer N are interchanged to form an integer K. Find all possibilities for N under the conditions described.

9. N is odd and exceeds K by more than 18. 41, 51, 61, 63, 71, 73, 81, 83, 85, 91, 93, 95

10. The average of N, K, and 35 is 30. 14, 23, 32, 41, 50

C

11. The sum of K and twice N is less than 60. 10, 11, 12, 13, 20, 21

12. N is even and exceeds K by more than 50. 60, 70, 80, 82, 90, 92

Solve.

13. A 15-member special committee met three times. Twice as many members were present at the third meeting as at the first, and the average attendance was 9 people. Find all possibilities for the number of people present at each meeting. 4, 15, and 8; 5, 12, and 10; 6, 9, and 12; 7, 6, and 14

14. A stick of wood is to be cut into three unequal pieces. The first piece is shorter than the second piece, and the second piece is shorter than the third. If the stick is 24 cm in length and the length of each piece is an even integer, what are the possibilities for the lengths of the pieces?
2 cm, 4 cm, and 18 cm; 2 cm, 6 cm, and 16 cm; 2 cm, 8 cm, and 14 cm; 2 cm, 10 cm, and 12 cm; 4 cm, 6 cm, and 14 cm; 4 cm, 8 cm, and 12 cm; 6 cm, 8 cm, and 10 cm

Mixed Review Exercises

Evaluate each expression if $x = -3$ and $y = 4$.

1. $2x - 5y$ −26
2. $-x^2y$ −36
3. $|x - y|$ 7
4. $(x - 2)(y + 1)$ −25
5. $\frac{3x + 1}{y}$ −2
6. $|xy|$ 12
7. $x + 3y$ 9
8. $\frac{x - y}{x + y}$ −7

Solve each open sentence and graph each solution set that is not empty.

9. $-7 < 2y - 5 \le 3$ $\{y: -1 < y \le 4\}$
10. $|3 - m| > 1$ $\{m: m < 2 \text{ or } m > 4\}$
11. $3n + 7 \le 8n - 13$ $\{n: n \ge 4\}$

3-2 Graphs of Linear Equations in Two Variables

Objective To graph a linear equation in two variables.

Solutions of open sentences in x and y can be graphed in an ***xy*-coordinate plane.** To set up a **plane rectangular coordinate system,** draw two number lines, or **axes,** meeting at right angles at a point O, the **origin.** The horizontal axis is called the ***x*-axis,** and the vertical axis, the ***y*-axis.** The axes divide the plane into four **quadrants.**

With each ordered pair of numbers, you can associate a unique point in the plane. To associate a point with the pair $(-4, -3)$, imagine drawing a vertical line through -4 on the x-axis and a horizontal line through -3 on the y-axis, as shown below. These lines intersect at a point P, the **graph** of $(-4, -3)$. Locating a point in this way is called **graphing** the ordered pair, or **plotting** the point.

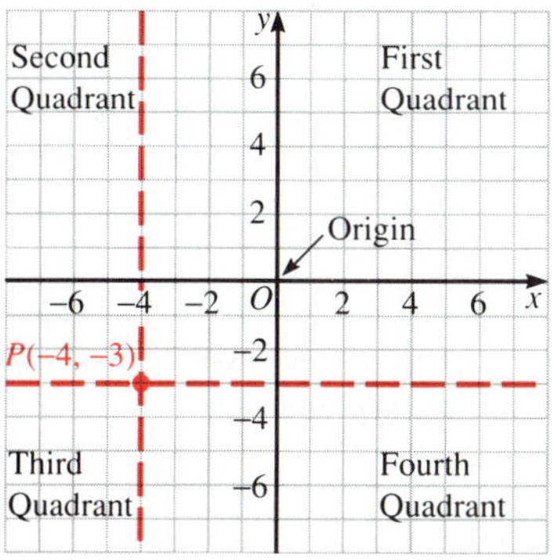

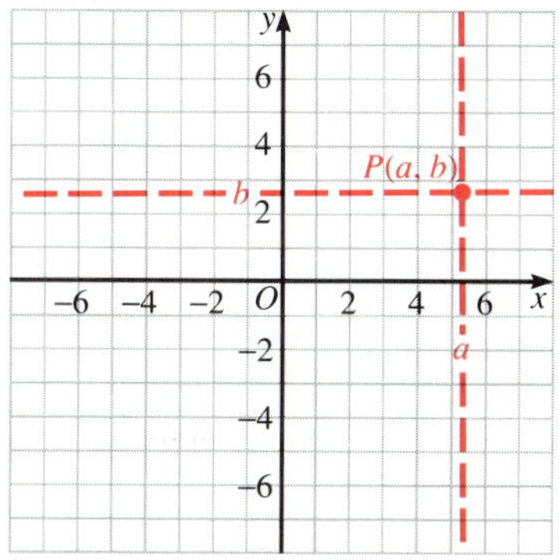

By reversing the process just described, you can associate with each point P in the plane a unique ordered pair (a, b) of real numbers, called the **coordinates** of P (see the diagram above). The first coordinate, a, is called the ***x*-coordinate,** or **abscissa,** of P; the second coordinate, b, is called the ***y*-coordinate,** or **ordinate** of P.

This **one-to-one correspondence** between ordered pairs of real numbers and points of the plane can be summarized as follows:

1. There is exactly one point in the coordinate plane associated with each ordered pair of real numbers.
2. There is exactly one ordered pair of real numbers associated with each point in the coordinate plane.

A rectangular coordinate system is sometimes called a **Cartesian coordinate system** in honor of René Descartes (1596–1650), the French mathematician who introduced coordinates.

Teaching Suggestions p. T88

Suggested Extensions p. T88

Warm Up Exercises

Solve for y.

1. $3x + 2y = 7$ $y = \frac{7 - 3x}{2}$
2. $4x - 6y = 12$ $y = \frac{2x - 6}{3}$
3. $\frac{1}{2}x + 3y = 9$ $y = \frac{18 - x}{6}$
4. $5x - 6y = -4$ $y = \frac{5x + 4}{6}$
5. $-2x - 2y = -2$ $y = 1 - x$

Motivating the Lesson

Point out to students that state maps have a number-letter coordinate system to help locate cities and towns. Lesson 3-2 discusses plotting points and graphing linear equations in two variables.

Chalkboard Examples

1. Graph (−1, −3), (3, 3), (0, −4), (3, −3), and (4, 0) on the same coordinate plane.

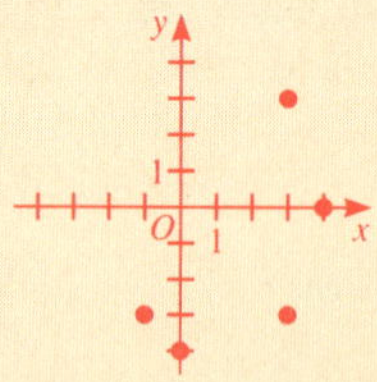

Draw the graph of each equation.

2. $x + y = 7$
   $y = 7 - x$

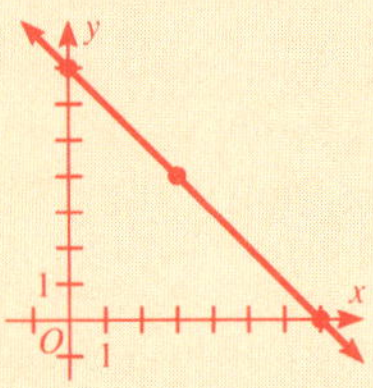

x	y
0	7
7	0
3	4

3. $y - 2x = 6$
   If $x = 0$, $y = 6$
   If $y = 0$, $x = -3$

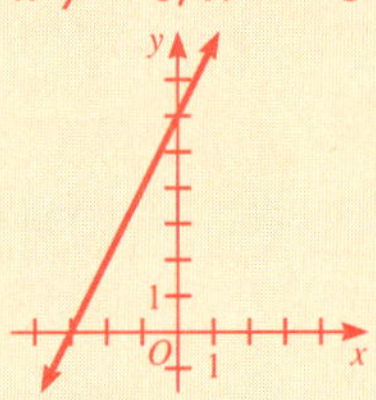

4. Graph in a coordinate plane.

   a. $y = -2$ b. $x = \frac{3}{2}$

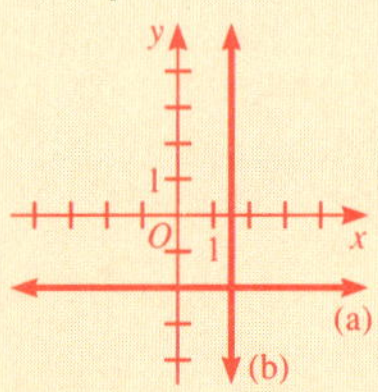

Example 1 Graph the ordered pairs (4, 1), (0, 2), (−2, 4), (2, −4), and (−4, 0) in the same coordinate plane.

Solution

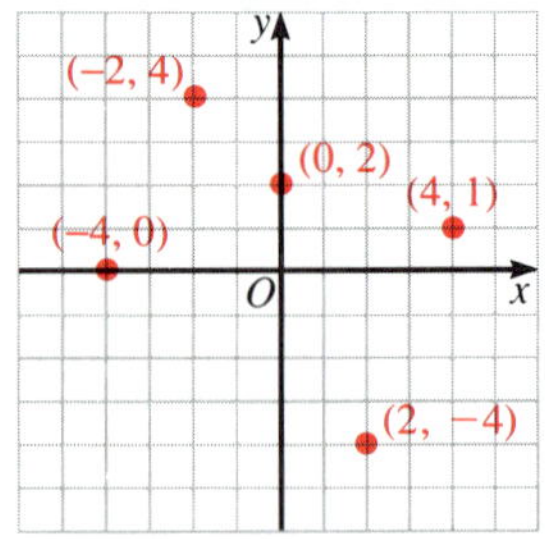

Example 2 Find and graph five solutions of $3x + 2y = 4$.

Solution Solve for one variable, say y, $\quad y = \frac{4 - 3x}{2}$

in terms of the other. Choose convenient values of x and find the corresponding values of y. Then graph the resulting ordered pairs.

x	$\frac{4 - 3x}{2}$	y
−2	$\frac{4 - 3(-2)}{2}$	5
−1	$\frac{4 - 3(-1)}{2}$	$\frac{7}{2}$
0	$\frac{4 - 3(0)}{2}$	2
1	$\frac{4 - 3(1)}{2}$	$\frac{1}{2}$
2	$\frac{4 - 3(2)}{2}$	−1

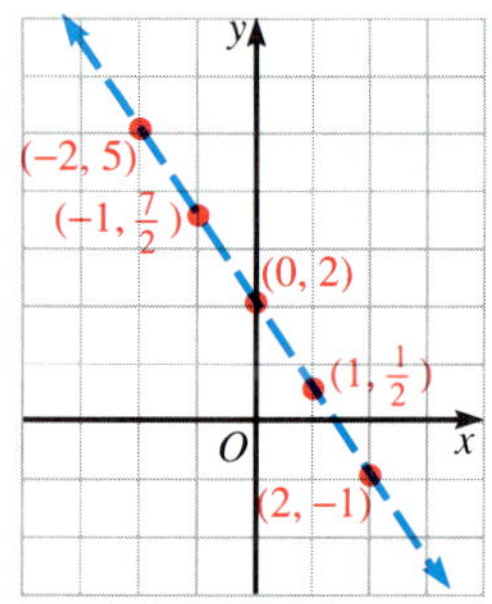

The **graph** of an open sentence in two variables is the set of *all* points in the coordinate plane whose coordinates satisfy the open sentence. Example 2 suggests that if we were to graph *all* the ordered pairs satisfying $3x + 2y = 4$, we would obtain the line shown in blue. The following theorem, which is proved in more advanced courses, tells us that the graph of the equation $3x + 2y = 4$ is in fact a line.

Theorem

The graph of every equation of the form

$$Ax + By = C \quad (A \text{ and } B \text{ not both } 0)$$

is a line. Conversely, every line in the coordinate plane is the graph of an equation of this form.

Because of this property, any equation that can be expressed in the form $Ax + By = C$ (A and B not both 0) is called a **linear equation in two variables.** The equation $3y + 8 = -5x$ is linear since it can be written as $5x + 3y = -8$. The following equations are *not* linear:

$$2x + 3y^2 = 4, \quad xy = 2, \quad 2x + \frac{3}{y} = 5$$

Although you need only two points to determine the graph of a linear equation, it is a good practice to plot a third point as a check. Points where the graph crosses the axes are often easy to find and plot.

Example 3 Graph $2x - 3y = -9$.

Solution The graph crosses the y-axis at a point whose x-coordinate is 0. The graph crosses the x-axis at a point whose y-coordinate is 0.

Let $x = 0$.	Let $y = 0$.
$2(0) - 3y = -9$	$2x - 3(0) = -9$
$-3y = -9$	$2x = -9$
$y = 3$	$x = -\frac{9}{2}$
Solution $(0, 3)$	*Solution* $\left(-\frac{9}{2}, 0\right)$

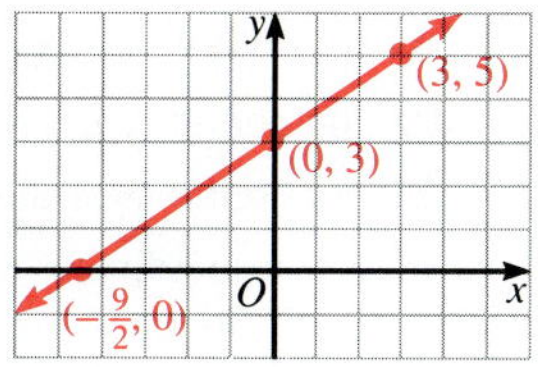

The graph is the line through the points with coordinates $(0, 3)$ and $\left(-\frac{9}{2}, 0\right)$. As a check, note that $(3, 5)$ is a solution of $2x - 3y = -9$ and its graph lies on the line.

Example 4 Graph in a coordinate plane.

a. $y = 4$ **b.** $x = -\frac{5}{2}$

Solution **a.** The equation $y = 4$ can be written as

$$0 \cdot x + 1 \cdot y = 4.$$

The graph consists of all points having y-coordinate 4 and is therefore a horizontal line.

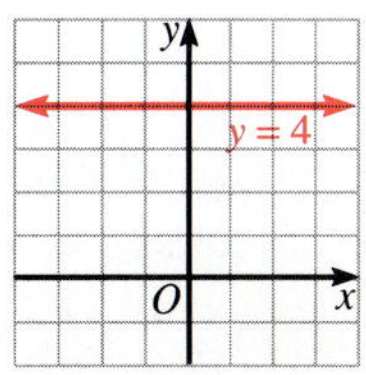

b. The equation $x = -\frac{5}{2}$ can be written as

$$1 \cdot x + 0 \cdot y = -\frac{5}{2}.$$

The graph consists of all points having x-coordinate $-\frac{5}{2}$ and is therefore a vertical line.

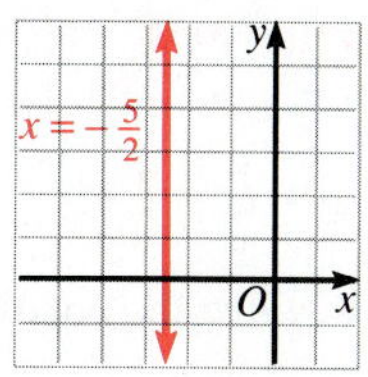

Check for Understanding

Here is a suggested use of the Oral Exercises to check students' understanding as you teach the lesson.
Oral Exs. 1–16: use after Example 1.
Oral Exs. 17–31: use after Example 4.

Guided Practice

1. Graph $(0, -2)$, $(-3, 0)$, $(-5, 5)$, and $(2, -1)$.

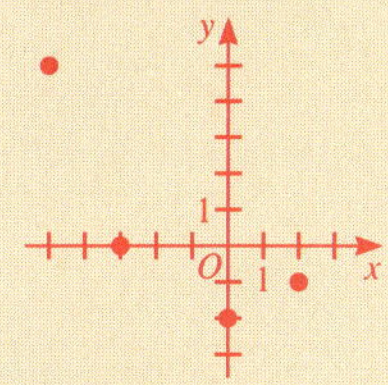

Graph each equation.

2. $y - x = 3$

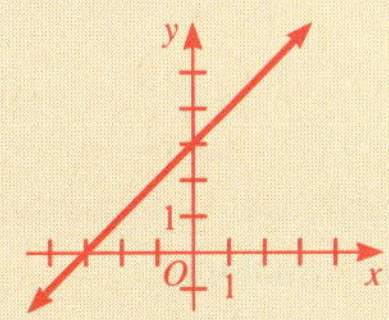

3. $2x + y = 5$

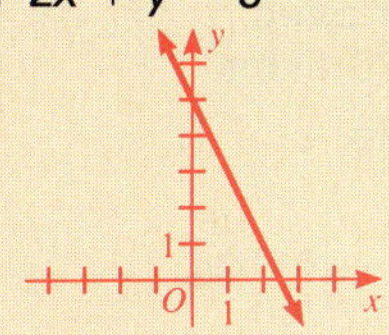

4. $x + 3y = 9$

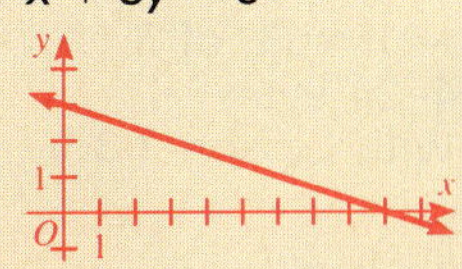

Supplementary Materials

Study Guide	pp. 35–36
Practice Master	12
Resource Book	p. 108

Summarizing the Lesson

Students have learned to graph linear equations in two variables. Ask students to explain what occurs when $A = 0$ and when $B = 0$ in the equation $Ax + By = C$.

Additional Answers
Oral Exercises

1. $A(3, 0)$; x-axis
2. $B(-4, 4)$; II
3. $C(-3, -3)$; III
4. $D(0, 5)$; y-axis
5. $E(5, 2)$; I
6. $F(3, -4)$; IV
7. $G(-2, 3)$; II
8. $H(-5, 0)$; x-axis
9. $I(4, 4)$; I
10. $J(-1, -4)$; III
11. $K(0, -2)$; y-axis
12. $L(0, 0)$; origin, x-axis and y-axis

Suggested Assignments

Average Algebra
111/2–24 even
S 105/34–38 even

Average Alg. and Trig.
111/2–28 even

Extended Algebra
111/2–36 even

Extended Alg. and Trig.
111/3–21 mult. of 3, 24–34 even
Assign with Lesson 3-3.

Using a Computer or a Graphing Calculator

For Written Exercises 5-22, once students have drawn their own graphs, they can use a computer or graphing calculator to check their work.

Support Materials
Disk for Algebra
Menu Item: Line Plotter

Example 4 illustrates the fact that the graph of $Ax + By = C$ is a horizontal line if $A = 0$, and is a vertical line if $B = 0$.

From now on, we will follow the common practice of using the simpler phrase "the point (a, b)" instead of "the point whose coordinates are (a, b)." Also, the phrase "the line $Ax + By = C$" will mean "the line whose equation is $Ax + By = C$."

Oral Exercises

1–12. Give the coordinates of each of the points A through L, and name the quadrant or axis that contains the point.

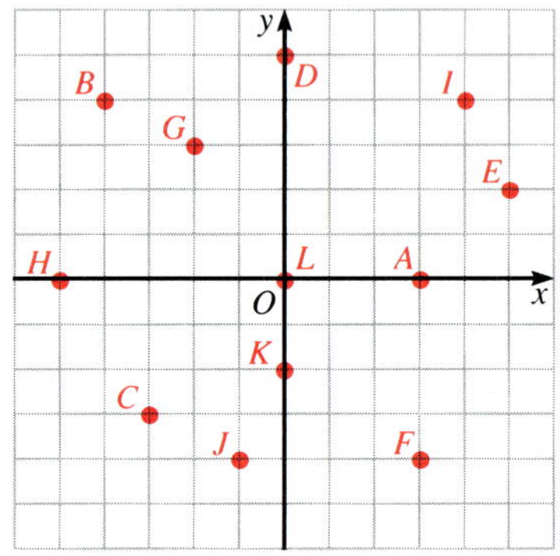

Name the quadrant of the point described.

13. Its coordinates are both negative. III

14. Its coordinates are both positive. I

15. Its x-coordinate is negative and its y-coordinate is positive. II

16. Its x-coordinate is positive and its y-coordinate is negative. IV

In Exercises 17–24, (a) tell whether or not the equation is linear, and (b) if it is linear, tell whether its graph is a horizontal line, a vertical line, or neither.

17. $x + y = 2$ yes; neither

18. $2x^2 + y = 3$ no

19. $x - 3 = 0$ yes; vertical

20. $2x + 3y + 4 = 0$ yes; neither

21. $x^2 - y^2 = 4$ no

22. $2y + 3 = 0$ yes; horizontal

23. $\frac{x}{2} - \frac{y}{3} = 1$ yes; neither

24. $\frac{2}{x} - \frac{3}{y} = 1$ no

Give the coordinates of the points where each line crosses the coordinate axes.

Sample $2x + 3y = 6$ **Solution** $(0, 2)$ and $(3, 0)$

25. $x + 2y = 4$ (0, 2), (4, 0)

26. $x + y = 3$ (0, 3), (3, 0)

27. $x - y = 2$ (0, −2), (2, 0)

28. $2x - y = 6$ (0, −6), (3, 0)

29. $\frac{x}{2} + \frac{y}{3} = 1$ (0, 3), (2, 0)

30. $\frac{x}{5} + \frac{y}{-2} = 1$ (0, −2), (5, 0)

31. a. Give the coordinates of the point where the line $x = -2$ crosses the x-axis. (−2, 0) **b.** The line is vertical and therefore parallel to the y-axis.
b. Explain why the line $x = -2$ does not cross the y-axis.

Written Exercises

For each exercise, graph the ordered pairs in the same coordinate plane.

A

1. (1, 2), (0, 3), (3, −2), (−3, 2), (−4, 0)
2. (−3, −3), (0, −3), (−2, 4), (3, 0), (1, 4)
3. (6, −4), (3, −2), (0, 0), (−3, 2), $\left(\frac{3}{2}, -1\right)$
4. (3, 3), (1, 1), $\left(-\frac{3}{2}, -\frac{3}{2}\right)$, (0, 0), (−2, −2)

Graph each equation. You may wish to verify your graphs on a computer or a graphing calculator.

5. $x - y = 4$
6. $x + y = 3$
7. $2y - 3 = 0$
8. $2x + 1 = 0$
9. $x + 2y = 4$
10. $2x - y = 6$
11. $3x - 2y + 18 = 0$
12. $2x + 3y + 12 = 0$
13. $x - 3y = 0$
14. $x = 2 - 2y$
15. $2x + 5y = 15$
16. $2x - 3y = 0$
17. $3x - 2y = 7$
18. $3x + 2y - 9 = 0$

B

19. $y = \frac{1}{2}x - 2$
20. $\frac{x}{3} + y = 1$
21. $y - 2 = x - y + 2$
22. $x + y = 2(x + y + 3)$

Find the value of *k* so that the point *P* lies on the line *L*.

23. $P(2, 1)$, L: $3x + ky = 8$ 2
24. $P(2, 3)$, L: $kx - 2y + k = 0$ 2
25. $P(2, 2)$, L: $kx + (k + 1)y = 2$ 0
26. $P(k, -2)$, L: $3x + 2y = k$ 2

Graph each pair of equations in the same coordinate plane. Find the coordinates of the point where the graphs intersect. Then show by substitution that the coordinates satisfy both equations.

27. $2x + 5y = 0$; $2x + y = 8$
28. $x - 2y = -4$; $3x + 2y = 12$
29. $3x + 2y - 1 = 0$; $x - 2y + 13 = 0$
30. $3x + y = -6$; $3x - 5y = 12$

Graph each equation.

31. $|x| = 2$
32. $|y| = 3$
33. $y = |x|$

C

34. $|y| = |x|$
35. $y = |x - 1|$
36. $y = |x| - 1$
37. $y = |x| - x$
38. $y = x - |x|$
39. $y = |x| + x$

Additional Answers
Written Exercises

2.
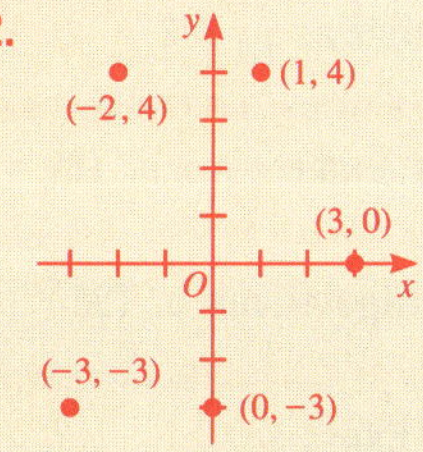

4.
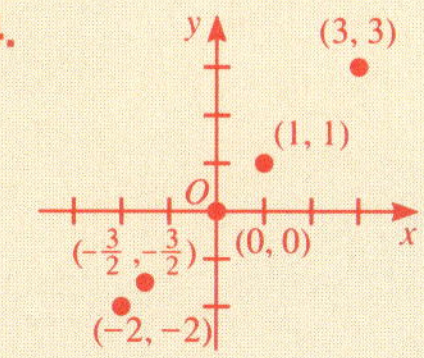

6.
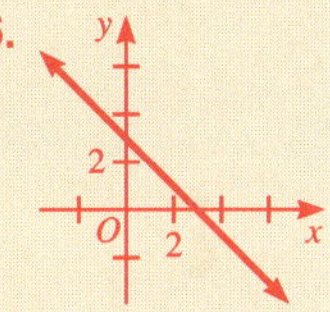

8.
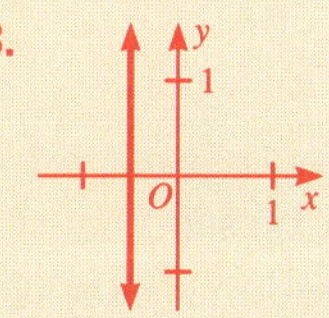

10.
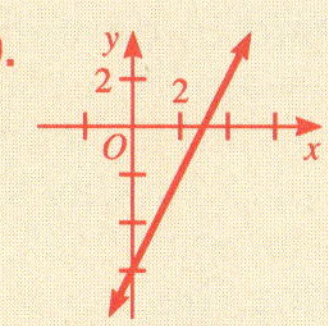

12.
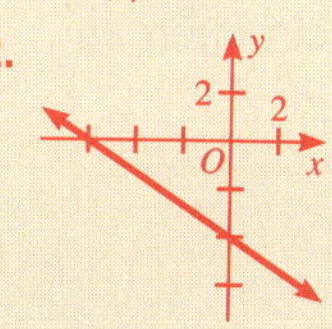

14.
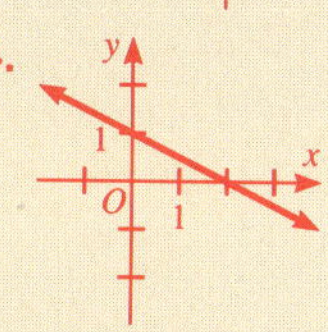

(continued on p. 158)

Teaching Suggestions p. T88

Group Activities p. T88

Suggested Extensions p. T89

Using Manipulatives p. T89

Warm-Up Exercises

Simplify.

1. $\frac{7-3}{4-1}$ $\frac{4}{3}$
2. $\frac{-2-5}{4-(-1)}$ $-\frac{7}{5}$
3. $\frac{3-(-6)}{1-3}$ $-\frac{9}{2}$
4. $\frac{4-(-4)}{3-2}$ 8
5. $\frac{-5-(-5)}{2-6}$ 0
6. $\frac{8-7}{3-3}$ undef.

Motivating the Lesson

Explain to students that it is sometimes hard to describe how steep a ramp or a roof is without drawing a picture. Slope, the topic of today's lesson, is a mathematical way to describe the steepness of objects such as ramps and roofs.

3-3 The Slope of a Line

Objective To find the slope of a line and to graph a line given its slope and a point on it.

A jet plane can take off at an angle that will lessen the noise on the ground. The path of the plane shown is a line that rises 150 ft for each 500 ft traveled horizontally. The steepness, or *slope,* of the path is

$$\frac{\text{rise}}{\text{run}} = \frac{150}{500} = \frac{3}{10}, \text{ or } 0.30.$$

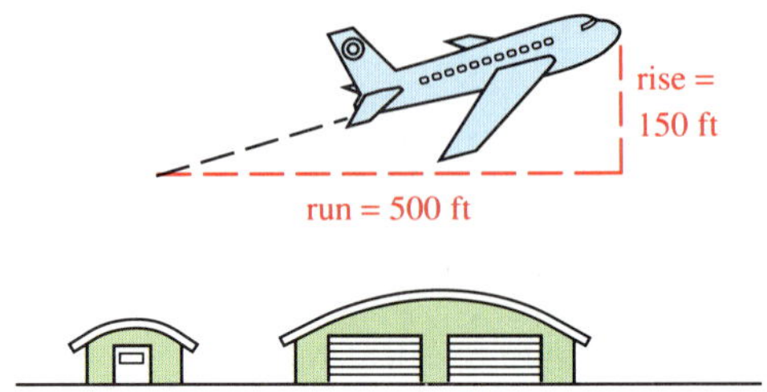

You can use this same method to measure the steepness, or *slope,* of a nonvertical line L in a coordinate plane. First choose any two distinct points (x_1, y_1) and (x_2, y_2) on L. (See Figure 1. You read (x_1, y_1) as "x one, y one" or "x sub one, y sub one.") Then,

$$\textbf{slope} \text{ of line } L = \frac{\text{rise}}{\text{run}} = \frac{y_2 - y_1}{x_2 - x_1} \quad (x_2 \neq x_1).$$

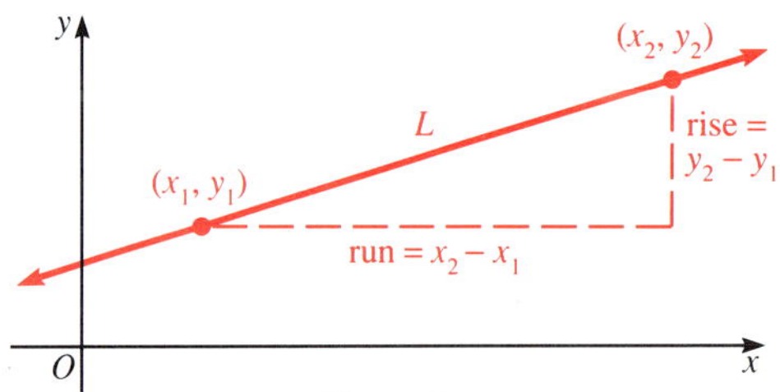

Figure 1

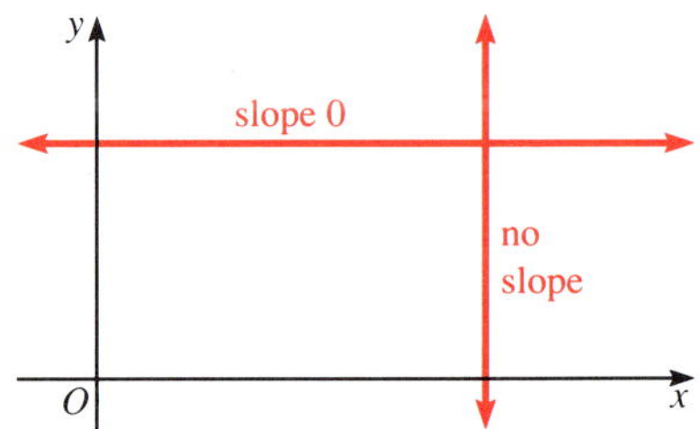

Figure 2

Look at Figure 2. If line L is vertical, then x_2 equals x_1, $x_2 - x_1 = 0$, and $\frac{y_2 - y_1}{x_2 - x_1}$ does not represent a real number. *Vertical lines have no slope.* If L is horizontal, then $y_2 = y_1$ and $y_2 - y_1 = 0$. *Horizontal lines have slope 0.*

Example 1 Find the slope of the line containing the given points.

a. $(-2, 3)$ and $(4, -1)$ **b.** $(3, -4)$ and $(3, 5)$

Solution **a.** slope $= \frac{-1-3}{4-(-2)} = \frac{-4}{6} = -\frac{2}{3}$ ***Answer***

b. slope $= \frac{5-(-4)}{3-3} = \frac{9}{0}$; the line is vertical and has no slope. ***Answer***

When you know an equation of a line, an easy way to find the slope is to use the following theorem.

Theorem

The slope of the line $Ax + By = C$ $(B \neq 0)$ is $-\frac{A}{B}$.

Proof Let (x_1, y_1) and (x_2, y_2) be any two different points of the line $Ax + By = C$. Since $B \neq 0$ (given), the line is nonvertical and $x_2 \neq x_1$. Therefore, $x_2 - x_1 \neq 0$ and the slope of the line is $\frac{y_2 - y_1}{x_2 - x_1}$.

Since (x_1, y_1) and (x_2, y_2) are on the line, the coordinates of these points must satisfy $Ax + By = C$.

$$Ax_1 + By_1 = C$$
$$Ax_2 + By_2 = C$$

Subtracting the first equation from the second gives:

$$(Ax_2 + By_2) - (Ax_1 + By_1) = C - C$$
$$Ax_2 - Ax_1 + By_2 - By_1 = 0$$
$$A(x_2 - x_1) + B(y_2 - y_1) = 0$$
$$B(y_2 - y_1) = -A(x_2 - x_1)$$

Dividing both sides by $B(x_2 - x_1)$ completes the proof by showing that the slope is the constant $-\frac{A}{B}$:

$$\frac{y_2 - y_1}{x_2 - x_1} = -\frac{A}{B}. \text{ (Recall that } B \neq 0 \text{ and } x_2 - x_1 \neq 0.)$$

When you solve $Ax + By = C$ for y, you obtain $y = -\frac{A}{B}x + \frac{C}{B}$. Thus, *the slope of a line is the numerical factor, or* ***coefficient,*** *of the x-term when the equation of the line is solved for y.*

Example 2 Find the slope of the line $5x - 2y = 20$.

Solution 1 Solve the given equation for y.

$$-2y = -5x + 20$$
$$y = \frac{5}{2}x - 10$$

The slope is the coefficient of x, which is $\frac{5}{2}$. ***Answer***

Solution 2 Since $A = 5$ and $B = -2$, slope $= -\frac{A}{B} = -\frac{5}{-2} = \frac{5}{2}$. ***Answer***

Chalkboard Examples

Give the slope of the line through the given points. If the line is vertical, so state.

1. (3, 5), (4, 2)
   $\frac{5-2}{3-4} = \frac{3}{-1} = -3$
2. (−5, 2), (8, 2)
   $\frac{2-2}{8-(-5)} = \frac{0}{13} = 0$
3. (5, −4), (−4, 5)
   $\frac{5-(-4)}{-4-5} = \frac{9}{-9} = -1$
4. (2, 4), (2, 7)
   $\frac{7-4}{2-2} = \frac{3}{0}$; vertical
5. (−1, 4), (5, 6)
   $\frac{6-4}{5-(-1)} = \frac{2}{6} = \frac{1}{3}$

Give the slope of each line.

6. $x - y = 2$
   $x = y + 2$; $x - 2 = y$; 1
7. $y + 3x = 1$
   $y = -3x + 1$; -3
8. $2x + 3y = 5$
   $3y = -2x + 5$;
   $y = -\frac{2}{3}x + \frac{5}{3}$; $-\frac{2}{3}$

Common Errors

Given two points, students may make the error of using the reciprocal of the definition of slope to calculate slope. Point out to students that the change in y is the numerator in the slope formula even though they may have become accustomed to considering x first.

Check for Understanding

Here is a suggested use of the Oral Exercises to check students' understanding as you teach the lesson.

Oral Exs. 1–6: use before Example 1.
Oral Exs. 7–12: use after Example 1.
Oral Exs. 13–22: use after Example 2.
Oral Exs. 23–26: use after Example 3.

Guided Practice

Give the slope of each line through the given points. If the line is vertical, so state.

1. (5, 2), (−5, 2) 0

2. (6, 1), (−4, 6) $-\frac{1}{2}$

3. (2, 4), (2, 9) vertical

4. (8, 9), (12, 13) 1

Give the slope of each line.

5. $x + y = 0$ −1

6. $8x + 2y = 4$ −4

7. $x + y = x - 7$ 0

8. $ax + ay = 7$ $(a \neq 0)$ −1

Using a Computer or a Graphing Calculator

You may want to have students discover the meaning of slope for themselves by using a computer or graphing calculator to graph $y = mx + 1$ for various values of m such as 2, 1, $\frac{1}{2}$, 0, $-\frac{1}{2}$, −1, and −2.

Support Materials
Disk for Algebra
Menu Item: Line Plotter
Resource Book p. 236

The following theorem states that there is exactly one line passing through a given point and having a given slope. A proof of the theorem is outlined in Exercises 49–54.

Theorem

Let $P(x_1, y_1)$ be a point and m a real number. There is one and only one line L through P having slope m. An equation of L is

$$y - y_1 = m(x - x_1).$$

Example 3 Graph the line through the point $P(2, 1)$ having slope $m = -\frac{3}{4}$.

Solution Since

$$m = \frac{\text{rise}}{\text{run}} = -\frac{3}{4} = \frac{-3}{4},$$

start at $P(2, 1)$ and move 3 units down and 4 units to the right to reach the point $Q(6, -2)$. Draw the line through P and Q. As a check, you should also find a third point on the line, such as $(-2, 4)$.

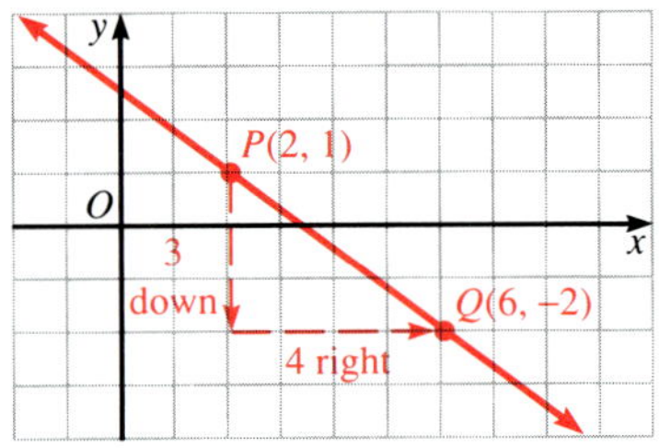

Answer

You can explore the meaning of slope by drawing graphs such as those shown at the right. A computer or a graphing calculator may be helpful. The diagram shows the graphs of the equation

$$y - y_1 = m(x - x_1)$$

when $(x_1, y_1) = (0, 0)$ for various values of m.

Notice that as you move from left to right along a line having slope m, the line:

rises if m is positive;
is horizontal if $m = 0$;
falls if m is negative.

The larger $|m|$ is, the steeper the line is.

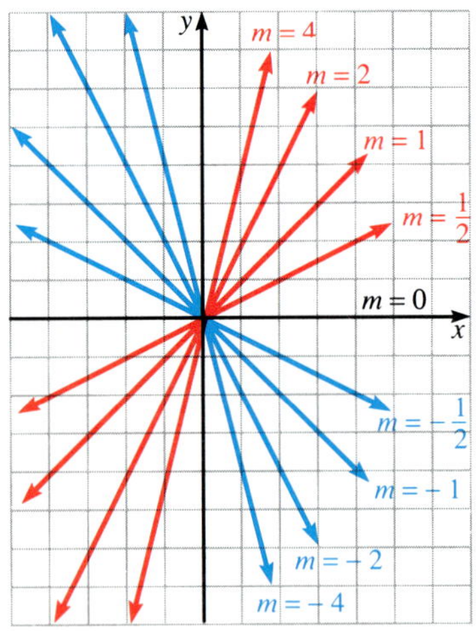

Oral Exercises

1. $-\frac{1}{2}$ 2. 0 3. $\frac{1}{3}$ 4. 2 5. −4 6. $-\frac{2}{3}$

Give the slope of each line shown.

1. 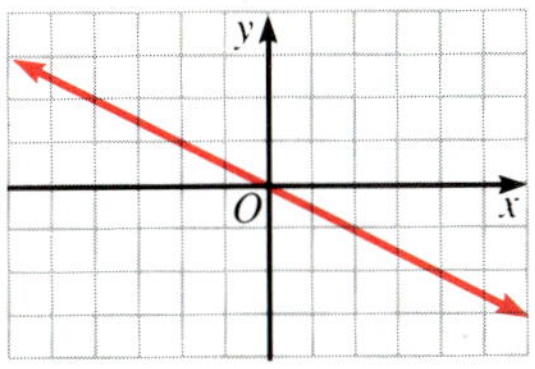

2. 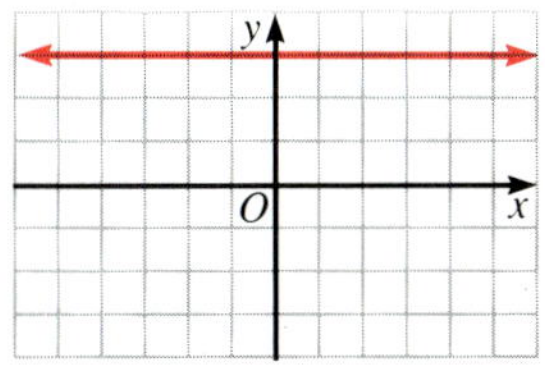

3. 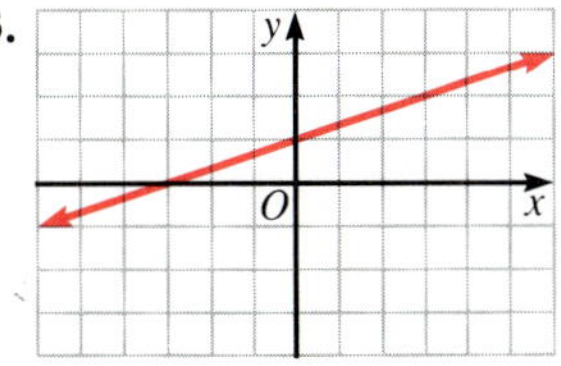

4. 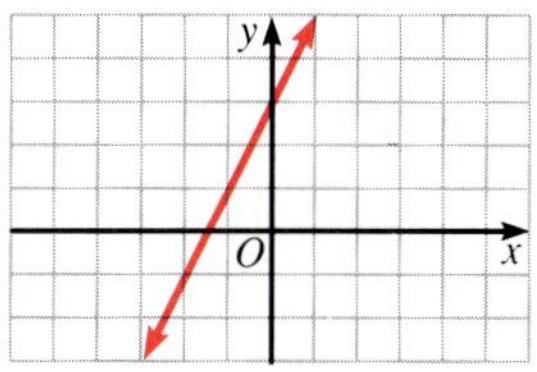

5. 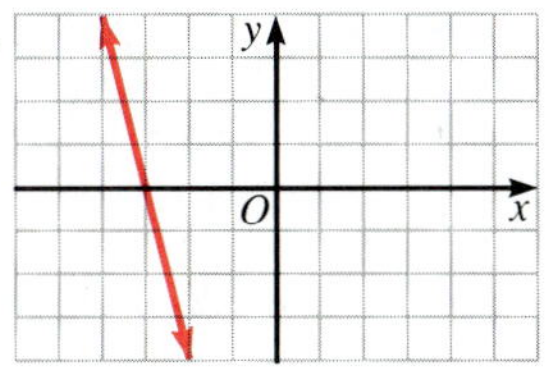

6. 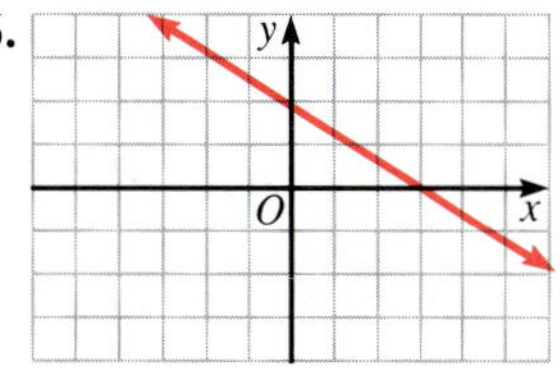

Give the slope of the line containing the given points.

7. (0, 0), (−2, 6) −3
8. (4, −2), (0, 0) $-\frac{1}{2}$
9. (3, 0), (0, 9) −3
10. (3, 2), (−1, 2) 0
11. (3, 1), (2, 2) −1
12. (1, 1), (−1, −1) 1

Give the slope of each line.

13. $y = 2x + 3$ 2
14. $y = 3x - 1$ 3
15. $y = -x + 2$ −1
16. $y + 2x = 4$ −2
17. $y = 0$ 0
18. $y = \frac{1}{2}x$ $\frac{1}{2}$

Determine whether the points whose coordinates are given in the table lie on a line. If they do, give the slope of the line.

Sample

change in x	x	y	change in y
	−2	5	
3			−2
	1	3	
3			−2
	4	1	
6			−4
	10	−3	

Solution Each change of 3 in x produces a change of −2 in y.
∴ the points lie on a line because the slope between any two points is a constant, $-\frac{2}{3}$.

19. yes; −2

x	y
1	9
2	7
3	5
4	3

20. no

x	y
0	3
2	0
6	3
8	6

21. yes; $\frac{2}{5}$

x	y
−4	2
1	4
6	6
11	8

22. yes; 2

x	y
0	1
1	3
3	7
7	15

Summarizing the Lesson

Students have learned how to find the slope of a line and how to graph a line if the slope and a point on the line are given. Ask students to draw ten lines through the point (−2, 3) having the following slopes: 4, −1, −4, 2, −2, $-\frac{1}{2}$, no slope, 0, 1, $\frac{1}{2}$.

Suggested Assignments

Average Algebra
116/1–47 odd
S 117/Mixed Review

Average Alg. and Trig.
116/1–47 odd
S 117/Mixed Review

Extended Alg. and Trig.
116/1–47 odd
S 117/Mixed Review

Extended Alg. and Trig.
116/3–48 mult. of 3
S 117/Mixed Review
Assign with Lesson 3-2.

Additional Answers
Written Exercises

26.–36. Points chosen may vary. Examples are given.

26.

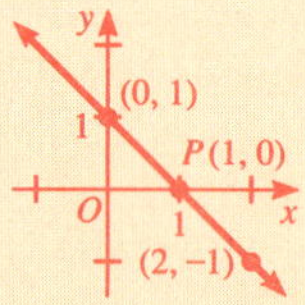

28.

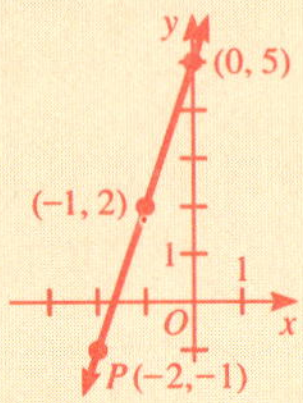

30.

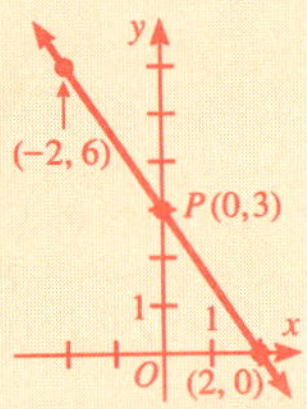

32.

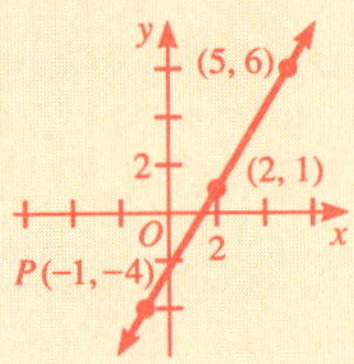

34.

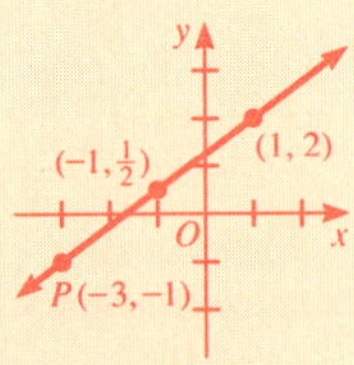

36.

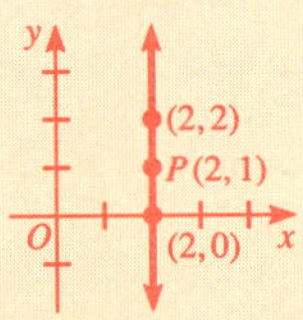

What do you know about the value(s) of A and B if the line $Ax + By = C$ has the property described?

23. The line is horizontal. $A = 0$

24. The line is vertical. $B = 0$

25. The line rises from left to right. A and B have opposite signs.

26. The line falls from left to right. A and B have the same sign.

Written Exercises

Find the slope of the line containing the given points. If the line has no slope, write "vertical."

A

1. (3, 1), (5, 5) 2

2. (4, 3), (0, 1) $\frac{1}{2}$

3. (4, −1), (−2, 3) $-\frac{2}{3}$

4. (−5, −2), (5, −2) 0

5. (3, −4), (3, −2) vertical

6. (3, −1), (−3, 1) $-\frac{1}{3}$

7. $\left(\frac{3}{2}, -3\right), \left(\frac{1}{2}, -7\right)$ 4

8. $\left(\frac{1}{2}, -2\right)$, (0, −4) 4

9. (6, −5), (−4, 3) $-\frac{4}{5}$

10. (0.5, 2.4), (1.5, −1.6) −4

11. $(a, b), (b, a)$ $(a \neq b)$ −1

12. $(a, b), (-b, -a)$ $(a \neq -b)$ 1

Find the slope of each line.

13. $x + y = 7$ −1

14. $x - y + 1 = 0$ 1

15. $2x + 4y = 5$ $-\frac{1}{2}$

16. $4x - 3y = 3$ $\frac{4}{3}$

17. $3x - 3y = 5$ 1

18. $4y - 5 = 6x$ $\frac{3}{2}$

19. $x = 3y + 2$ $\frac{1}{3}$

20. $2(1 - y) = x$ $-\frac{1}{2}$

21. $\frac{1}{2}x + \frac{1}{3}y = 1$ $-\frac{3}{2}$

22. $\frac{1}{4}x - \frac{1}{2}y = 1$ $\frac{1}{2}$

23. $\frac{x}{-1} + \frac{y}{6} = 1$ 6

24. $\frac{x}{3} - \frac{y}{-5} = 1$ $-\frac{5}{3}$

Graph the line through point P having slope m. Find the coordinates of two other points on the line.

25. $P(0, 2)$, $m = 1$

26. $P(1, 0)$, $m = -1$

27. $P(3, -1)$, $m = -2$

28. $P(-2, -1)$, $m = 3$

29. $P(2, -3)$, $m = \frac{1}{2}$

30. $P(0, 3)$, $m = -\frac{3}{2}$

B

31. $P(-2, 3)$, $m = -\frac{2}{3}$

32. $P(-1, -4)$, $m = \frac{5}{3}$

33. $P(0, 0)$, $m = 0.25$

34. $P(-3, -1)$, $m = 0.75$

35. $P(2, 1)$, $m = 0$

36. $P(2, 1)$, no slope

Solve.

37. A ramp to provide handicapped people access to a certain building is to be constructed with a slope of 5%. If the entrance to the building is 3 ft above ground level, how long should the base of the ramp be? 60 ft

Solve.

38. A jetliner covered a horizontal distance of 5 mi while following a flight path with slope 0.25. How much altitude did it gain? 1.25 mi

Find the value of k so that the given line has slope m.

39. $kx - 3y = 7$, $m = 2$ 6

40. $6x + ky = 10$, $m = -2$ 3

41. $(k + 3)x - 3y = 1$, $m = k$ $\frac{3}{2}$

42. $(k + 1)x + 2y = 6$, $m = k - 2$ 1

Find the value of k so that the line through the given points has slope m.

43. $(2k, 3)$, $(1, k)$; $m = 2$ 1

44. $(k, k + 1)$, $(3, 2)$; $m = 3$ 4

45. $(k + 1, k - 1)$, $(k, -k)$; $m = k + 1$ 2

46. $(k + 1, 3 + 2k)$, $(k - 1, 1 - k)$; $m = k$ −2

C **47.** $(3, k)$, $(-1, |k|)$; $m = -2$ −4

48. $(1, k)$, $(5, |k|)$; $m = 3$ −6

Exercises 49–54 outline a proof of the theorem stated on page 114.

49. Show that the graph of the equation $y - y_1 = m(x - x_1)$ is a line L by expressing the equation in the form $Ax + By = C$. **49.–50.** See below.

50. Explain how you know L passes through $P(x_1, y_1)$.

51. Explain how you know L has slope m. slope $= -\frac{A}{B} = -\frac{m}{-1} = m$

Exercises 49–51 showed that there is *at least one* line through P having slope m. Now let L' be *any* line through P having slope m.

52. Let $Q(x', y')$ be any point of L' different from $P(x_1, y_1)$. Explain why $\frac{y' - y_1}{x' - x_1} = m$. Since $P(x_1, y_1)$ and $Q(x', y')$ are points on a line with slope m, $\frac{y' - y_1}{x' - x_1} = m$.

53. From Exercise 52, $y' - y_1 = m(x' - x_1)$. Explain why this shows that $Q(x', y')$ is on L. The coordinates of $Q(x', y')$ satisfy the equation of L.

54. Explain why L' and L must be the same line. Through any two points there is exactly one line.

Therefore, there is *only one* line through P having slope m.

49. $y - y_1 = m(x - x_1)$; $y - y_1 = mx - mx_1$; $mx - y = mx_1 - y_1$; $A = m$; $B = -1$; $C = mx_1 - y_1$

50. The coordinates of $P(x_1, y_1)$ satisfy the equation in Ex. 49 above; $m(x_1) - y_1 = mx_1 - y_1$

Mixed Review Exercises

Complete the ordered pair to form a solution of the given equation.

1. $2x + y = 5$; $(4, \underline{\ ?\ })$ −3

2. $x - 3y = 7$; $(\underline{\ ?\ }, -2)$ 1

3. $-x + 4y = 9$; $(-1, \underline{\ ?\ })$ 2

4. $5x + 2y = -8$; $(\underline{\ ?\ }, 1)$ −2

Graph each equation.

5. $-x + y = 2$

6. $y = -4$

7. $2x + 3y = 6$

8. $y = -x$

9. $x + 3 = 0$

10. $y = \frac{1}{2}x - 3$

Additional Answers
Mixed Review Exercises

5.
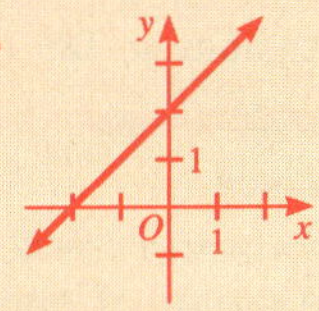

6.
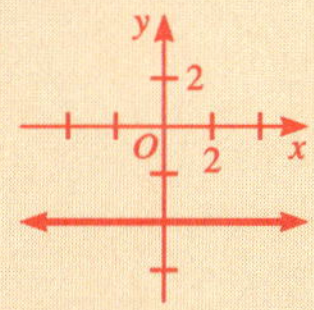

7.
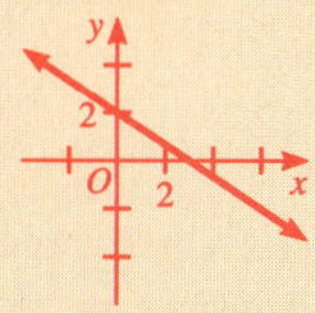

8.
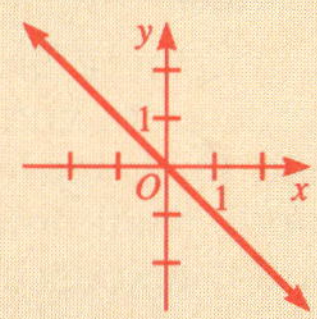

9.
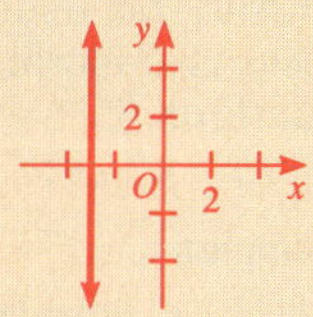

10.
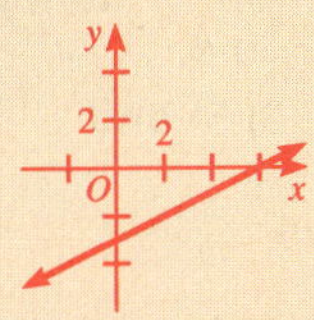

Supplementary Materials

Study Guide	pp. 37–38
Computer Activity	5

Teaching Suggestions p. T89

Suggested Extensions p. T89

Warm-Up Exercises

Solve each equation for y. Then find the slope of each line.

1. $3x + 4y = 7$
   $y = -\frac{3}{4}x + \frac{7}{4}; -\frac{3}{4}$
2. $-2x - 6y = 2$
   $y = -\frac{1}{3}x - \frac{1}{3}; -\frac{1}{3}$
3. $8x - 4y = 12$
   $y = 2x - 3; 2$

Motivating the Lesson

Tell students that there are times when architects and engineers need to describe certain situations mathematically. Lesson 3-4 discusses methods that may be used to write equations describing these situations.

Chalkboard Examples

Find an equation in standard form for each line.

1. through $(-3, 5)$ and $(-6, 4)$
   $m = \frac{5 - 4}{-3 - (-6)} = \frac{1}{3}$
   $y - 5 = \frac{1}{3}(x + 3)$
   $3y - 15 = x + 3$
   $x - 3y = -18$
2. through $(4, 3)$ with slope 2
   $y - 3 = 2(x - 4)$;
   $y - 3 = 2x - 8$;
   $2x - y = 5$
3. y-intercept -4 with slope $\frac{1}{2}$ $y = \frac{1}{2}x - 4$; $2y = x - 8$; $x - 2y = 8$

3-4 Finding an Equation of a Line

Objective To find an equation of a line given its slope and a point on the line, or two points, or its slope and the y-intercept.

You know that the graph of $Ax + By = C$ is a line in an xy-plane. Equations of lines are usually given in this **standard form,** with A, B, and C integers. However, when you are asked to *find* the equation of a certain line, the two other forms discussed in this lesson may be more useful.

By the theorem on page 114, the line containing the point (x_1, y_1) and having slope m has the equation $y - y_1 = m(x - x_1)$.

The equation

$$y - y_1 = m(x - x_1)$$

is called the **point-slope form** of the equation of a line.

Example 1 (a point and a slope given) Find an equation in standard form of the line containing the point $(4, -3)$ and having slope $-\frac{2}{5}$.

Solution Use the point-slope form with $(x_1, y_1) = (4, -3)$ and $m = -\frac{2}{5}$.

$$y - y_1 = m(x - x_1)$$

$$y - (-3) = -\frac{2}{5}(x - 4) \qquad \text{Multiply both sides by 5.}$$

$$5(y + 3) = -2(x - 4)$$

$$5y + 15 = -2x + 8$$

$$2x + 5y = -7 \quad \textit{Answer}$$

Check: $2(4) + 5(-3) = -7$; $m = -\frac{A}{B} = -\frac{2}{5}$ ✓

Example 2 (two points given) Find an equation in standard form of the line containing $(1, -2)$ and $(5, 1)$.

Solution First find the slope of the line: $m = \frac{1 - (-2)}{5 - 1} = \frac{3}{4}$.

Then use the point-slope form with either given point.

$y - y_1 = m(x - x_1)$	$y - y_1 = m(x - x_1)$
$y - (-2) = \frac{3}{4}(x - 1)$	$y - 1 = \frac{3}{4}(x - 5)$
$4y + 8 = 3x - 3$	$4y - 4 = 3x - 15$
$3x - 4y = 11$	$3x - 4y = 11$ ***Answer***

You should check the answer in Example 2 by verifying that $(1, -2)$ and $(5, 1)$ are both solutions of $3x - 4y = 11$.

A line (or any curve) has **y-intercept** b if it intersects the y-axis at the point $(0, b)$. A line has **x-intercept** a if it intersects the x-axis at the point $(a, 0)$. If you let $(x, y) = (0, b)$ in the point-slope form, then you get $y - b = m(x - 0)$, or $y = mx + b$.

If b is the y-intercept of a line, the equation

$$y = mx + b$$

is called the **slope-intercept** form of the equation of the line.

Example 3 (slope and y-intercept given) Find an equation in standard form of the line having slope -2 and y-intercept $\frac{4}{3}$.

Solution Use $y = mx + b$ with $m = -2$ and $b = \frac{4}{3}$.

$$y = -2x + \frac{4}{3} \qquad \text{Multiply both sides by 3.}$$

$$3y = -6x + 4$$

$$6x + 3y = 4 \quad \textbf{\textit{Answer}}$$

You can explore the relationships between the slopes and the graphs of two equations by graphing $y = mx + b$ for various values of m and b. A computer or a graphing calculator may be helpful. If you graph $y = 3x + b$ and $y = -\frac{1}{3}x + b$ for various values of b on the same axes, your results should suggest the following theorem. This theorem tells how slopes indicate when graphs of equations are parallel or perpendicular lines. (A proof of part (1) is outlined in Exercises 42 and 43, page 130, and of part (2) in Problem 14, page 406.)

Theorem

Let L_1 and L_2 be two different lines, with slopes m_1 and m_2 respectively.

1. L_1 and L_2 are *parallel* if and only if $m_1 = m_2$.
2. L_1 and L_2 are *perpendicular* if and only if $m_1 m_2 = -1$.

By this theorem, the lines $y = 3x - 2$ and $3x - y = 5$ are parallel because the slope of each is 3. The lines $y = 3x - 2$ and $-3y = x + 1$ are perpendicular because their slopes are 3 and $-\frac{1}{3}$, respectively, and $(3)(-\frac{1}{3}) = -1$.

4. parallel to $y = 3x$, through $(5, 10)$
$y - 10 = 3(x - 5)$;
$y - 10 = 3x - 15$;
$3x - y = 5$

5. perpendicular to $y + 2x = 6$ with the same y-intercept
$y + 2x = 6$ in slope-intercept form is $y = -2x + 6$; slope of perpendicular line is $\frac{1}{2}$;
$y = \frac{1}{2}x + 6$;
$2y = x + 12$;
$x - 2y = -12$

Using a Computer or a Graphing Calculator

You may want to have students discover for themselves the relationship between the slopes of parallel or perpendicular lines. They can use a computer or graphing calculator to draw pairs of lines with slopes that are equal or negative reciprocals.

You may also want to have students find an equation for a given linear graph. The program "Line Quiz" on the Disk for Algebra will draw a random line and let students visually experiment to find the corresponding equation.

Support Materials
Disk for Algebra
Menu Items: Line Plotter
Line Quiz
Resource Book pp. 236–237

Supplementary Materials

Study Guide	pp. 39–40
Practice Master	13
Test Master	9
Computer Activity	6
Resource Book	p. 109
Overhead Visual	1

Check for Understanding

Here is a suggested use of the Oral Exercises to check students' understanding as you teach the lesson.
Oral Exs. 1–8: use before Example 1.
Oral Exs. 9–14: use before Example 3.
Oral Exs. 15–25: use after Example 4.

Guided Practice

Give the standard form of the equation for each line.

1. through (5, 1), slope 2
   $2x - y = 9$
2. through (−1, 3), no slope
   $x + 0y = -1$ or $x = -1$
3. through (5, −2), slope $\frac{1}{3}$
   $x - 3y = 11$
4. slope $-\frac{2}{3}$, y-intercept 7
   $2x + 3y = 21$
5. through (1, 3) and (3, 7)
   $2x - y = -1$
6. through (4, 2) and (8, 2)
   $0x + y = 2$ or $y = 2$

Give the standard form of the equation of the line through point P that is:

7. perpendicular to L. $P(1, 2)$; L: $x + y = 4$
   $x - y = -1$
8. parallel to L. $P(-3, 2)$; L: $x - 4 = 0$
   $x = -3$

Summarizing the Lesson

In this lesson students learned the point-slope form, the slope-intercept form, and the standard form of the equation of a line. Ask students to write the equation $x = 3 + 4y$ in standard form, in slope-intercept form, and in point-slope form using the point (7, 1).

Because the equation $m_1m_2 = -1$ can be written as $m_2 = -\frac{1}{m_1}$ and also as $m_1 = -\frac{1}{m_2}$, we say that nonvertical lines are perpendicular if and only if their slopes are *negative reciprocals* of each other.

Example 4 Let P be the point (4, 1) and L the line $2x + 5y = -10$.

a. Find an equation in standard form of the line L_1 through P and parallel to L.

b. Find an equation in standard form of the line L_2 through P and perpendicular to L.

Solution Solve the equation of line L for y:

$$2x + 5y = -10;\ y = -\frac{2}{5}x - 2$$

$\therefore$ the slope of L is $-\frac{2}{5}$.

a. To find an equation of L_1, use the point-slope form with $(x_1, y_1) = (4, 1)$ and $m = -\frac{2}{5}$:

$$y - 1 = -\frac{2}{5}(x - 4)$$
$$5y - 5 = -2x + 8$$
$$L_1\text{: } 2x + 5y = 13$$

b. The slope of L_2 is the negative reciprocal of $-\frac{2}{5}$, which is $\frac{5}{2}$. Use the point-slope form with $(x_1, y_1) = (4, 1)$ and $m = \frac{5}{2}$:

$$y - 1 = \frac{5}{2}(x - 4)$$
$$2y - 2 = 5x - 20$$
$$L_2\text{: } 5x - 2y = 18$$

Oral Exercises

Give the slope and y-intercept of each line.

1. $y = 3x + 2$ — $m = 3$; $b = 2$
2. $y = x - 1$ — $m = 1$; $b = -1$
3. $y = 3 - x$ — $m = -1$; $b = 3$
4. $y = 2(x + 1)$ — $m = 2$; $b = 2$
5. $2y = x + 6$ — $m = \frac{1}{2}$; $b = 3$
6. $x + 2y = 4$ — $m = -\frac{1}{2}$; $b = 2$
7. $6x - 2y = 1$ — $m = 3$; $b = -\frac{1}{2}$
8. $2x + y + 5 = 0$ — $m = -2$; $b = -5$

Give an equation of the line described in point-slope or slope-intercept form.

9. Has slope 2 and y-intercept 3 — $y = 2x + 3$
10. Passes through (2, 1) with slope 1 — $y - 1 = 1(x - 2)$
11. Contains (1, 3) and has slope 2 — $y - 3 = 2(x - 1)$
12. Has y-intercept −2 and slope −1 — $y = -1x - 2$
13. Has slope −2 and x-intercept 3 — $y - 0 = -2(x - 3)$
14. Has slope $\frac{2}{3}$ and passes through (0, 0) — $y = \frac{2}{3}x$

Give the slope of any line that is (a) parallel to, and (b) perpendicular to, the graph of the given equation.

15. $y = -2x + 6$
16. $y = 5(x + 1)$
17. $x = 2y + 5$
18. $x - 3y = 3$

 15. a. −2 b. $\frac{1}{2}$ 16. a. 5 b. $-\frac{1}{5}$ 17. a. $\frac{1}{2}$ b. −2 18. a. $\frac{1}{3}$ b. −3

Tell whether the given lines are parallel, perpendicular, or neither.

19. $y = x + 2$ parallel
$x - y = 3$

20. $y = 2x - 4$ perpendicular
$y = 3 - \frac{1}{2}x$

21. $2x + y = 1$ neither
$x + 2y = 3$

22. $x - 2y = 1$ parallel
$2x = 4y + 3$

23. $2x + 3y = 6$ perpendicular
$3x - 2y = 4$

24. $3x - y = 6$ neither
$x - 3y = 6$

25. Explain why the lines $Ax + By = C_1$ and $Ax + By = C_2$ $(C_1 \neq C_2)$ are parallel. The lines are different ($C_1 \neq C_2$) and have the same slope $\left(-\frac{A}{B}\right)$.

Written Exercises

5. $x - 2y = 1$ 6. $2x - 3y = 1$
7. $x - 5y = 19$ 8. $3x + 2y = 12$

Find an equation in standard form of the line containing point *P* and having slope *m*.

A

1. $P(2, 3)$, $m = 1$ $x - y = -1$
2. $P(2, 1)$, $m = -1$ $x + y = 3$
3. $P(5, 0)$, $m = -2$ $2x + y = 10$
4. $P(-1, 4)$, $m = 0$ $y = 4$
5. $P(-3, -2)$, $m = \frac{1}{2}$
6. $P(2, 1)$, $m = \frac{2}{3}$
7. $P(4, -3)$, $m = \frac{1}{5}$
8. $P(0, 6)$, $m = -\frac{3}{2}$
9. $P(-2, -1)$, $m = 0$ $y = -1$
10. $P(-3, 3)$, $m = -\frac{4}{3}$ $4x + 3y = -3$
11. $P(-2, 4)$, $m = 0.4$ $2x - 5y = -24$
12. $P(4, 0)$, $m = -0.6$ $3x + 5y = 12$

Find an equation in standard form of the line having slope *m* and *y*-intercept *b*.

13. $m = -1$, $b = 2$ $x + y = 2$
14. $m = 1$, $b = -3$ $x - y = 3$
15. $m = \frac{1}{2}$, $b = \frac{3}{2}$ $x - 2y = -3$
16. $m = -\frac{3}{4}$, $b = -\frac{5}{4}$ $3x + 4y = -5$
17. $m = 1.2$, $b = -0.6$ $6x - 5y = 3$
18. $m = -0.8$, $b = 1.4$ $4x + 5y = 7$

Find an equation in standard form of the line containing the given points.

19. $(0, 0)$, $(5, -2)$ $2x + 5y = 0$
20. $(0, 0)$, $(-3, 1)$ $x + 3y = 0$
21. $(3, -2)$, $(-2, 3)$ $x + y = 1$
22. $(3, -2)$, $(2, -3)$ $x - y = 5$
23. $(3, -2)$, $(-3, 2)$ $2x + 3y = 0$
24. $(3, -2)$, $(-3, -2)$ $y = -2$
25. $(-2, 3)$, $(-2, -3)$ $x = -2$
26. $(4, -5)$, $(1, -4)$ $x + 3y = -11$
27. $\left(-3, \frac{1}{2}\right)$, $\left(3, \frac{1}{2}\right)$ $y = \frac{1}{2}$
28. $\left(\frac{3}{2}, -\frac{1}{2}\right)$, $\left(-\frac{1}{2}, \frac{5}{2}\right)$ $6x + 4y = 7$
29. $\left(\frac{2}{3}, -\frac{1}{2}\right)$, $\left(\frac{1}{6}, -1\right)$ $6x - 6y = 7$
30. $\left(\frac{3}{4}, \frac{5}{4}\right)$, $\left(-\frac{1}{4}, \frac{1}{2}\right)$ $12x - 16y = -11$

Find equations in standard form of the lines through point *P* that are (a) parallel to, and (b) perpendicular to, line *L*.

31. $P(0, 3)$; L: $x + y = 5$
32. $P(0, -2)$; L: $y = x - 3$
33. $P(0, -4)$; L: $2y = x$
34. $P(0, 1)$; L: $3x + 2y = 1$
35. $P(2, 0)$; L: $x + 2y = 3$
36. $P(-1, 2)$; L: $x - 3y = -2$
37. $P(-4, 1)$; L: $y + 2 = 0$
38. $P(-1, -2)$; L: $x - 3 = 0$

Suggested Assignments

Average Algebra
Day 1: 121/2–30 even
S 111/21–35 odd
Day 2: 121/31–55 odd
R 123/Self-Test 1

Average Alg. and Trig.
121/2–54 even
R 123/Self-Test 1

Extended Algebra
121/3–54 mult. of 3
R 123/Self-Test 1

Extended Alg. and Trig.
121/3–57 mult. of 3
R 123/Self-Test 1

Additional Answers
Written Exercises

31. a. $x + y = 3$
b. $x - y = -3$

32. a. $x - y = 2$
b. $x + y = -2$

33. a. $x - 2y = 8$
b. $2x + y = -4$

34. a. $3x + 2y = 2$
b. $2x - 3y = -3$

35. a. $x + 2y = 2$
b. $2x - y = 4$

36. a. $x - 3y = -7$
b. $3x + y = -1$

37. a. $y = 1$ b. $x = -4$

38. a. $x = -1$ b. $y = -2$

55. The slope of $L_1 = -\frac{A_1}{B_1}$; the slope of $L_2 = -\frac{A_2}{B_2}$

a. The lines are parallel if and only if $-\frac{A_1}{B_1} = -\frac{A_2}{B_2}$; $-A_1B_2 = -A_2B_1$; $A_1B_2 = A_2B_1$.

b. The lines are perpendicular if and only if $\left(-\frac{A_1}{B_1}\right)\left(-\frac{A_2}{B_2}\right) = -1$; $A_1A_2 = -B_1B_2$; $A_1A_2 + B_1B_2 = 0$

(continued)

Additional Answers Written Exercises
(continued)

56. The slope is $\frac{y_2 - y_1}{x_2 - x_1} = m$. Substitution into the point-slope form $y - y_1 = m(x - x_1)$ gives $y - y_1 = \frac{y_2 - y_1}{x_2 - x_1}(x - x_1)$.

57. The line passes through $(a, 0)$ and $(0, b)$. Assume $a \neq 0$. The slope is $\frac{b - 0}{0 - a} = -\frac{b}{a}$. Using the slope-intercept form, $y = -\frac{b}{a}x + b$. Dividing by b yields $\frac{y}{b} = -\frac{x}{a} + 1$; $\frac{x}{a} + \frac{y}{b} = 1$. (If $a = 0$ or $b = 0$, this form cannot be used.)

Using a Computer

These exercises ask students to write programs that will find the slope of a line and the slope-intercept equation of a line. Ask students with programming experience to demonstrate the programs. Allow other students to use the programs to work selected problems.

Additional Answers Self-Test 1

3.
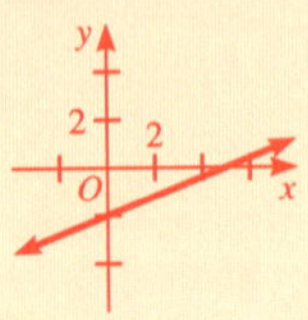

4.
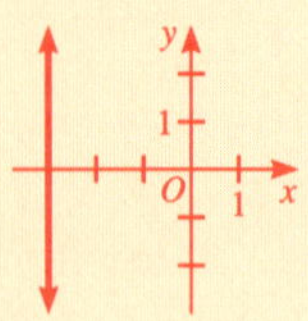

Find an equation in standard form for the line described.

B

39. Passing through the points (1, 4) and (−3, 4) $y = 4$

40. Passing through the points (−2, 3) and (−2, 6) $x = -2$

41. Passing through the origin and with no slope $x = 0$

42. Passing through the origin and with slope 0 $y = 0$

43. Having y-intercept 6 and parallel to the x-axis $y = 6$

44. Having x-intercept −4 and parallel to the y-axis $x = -4$

45. Having x-intercept 4 and y-intercept 3 $3x + 4y = 12$

46. Having x-intercept −3 and y-intercept −1 $x + 3y = -3$

47. Through $P(-2, 1)$ and parallel to the line containing (1, 4) and (2, 3) $x + y = -1$

48. Through $Q(-3, 2)$ and parallel to the line containing (2, 3) and (1, −2) $5x - y = -17$

In Exercises 49–54 the vertices of a quadrilateral *ABCD* are given. Determine whether or not *ABCD* is a parallelogram, and if it is, whether or not it is a rectangle. (*Hint:* Check to see if the lines containing opposite sides are parallel. If they are, check to see if the lines containing two adjacent sides are perpendicular.) **49., 50.** Parallelogram; not a rectangle

49. $A(2, 0)$, $B(-2, 2)$, $C(2, 8)$, $D(6, 6)$

50. $A(-2, 3)$, $B(1, 4)$, $C(3, -1)$, $D(0, -2)$

51. $A(0, -2)$, $B(6, 0)$, $C(5, 3)$, $D(-1, 1)$

52. $A(4, -2)$, $B(5, 1)$, $C(0, 3)$, $D(-2, 0)$

53. $A(7, 1)$, $B(1, 4)$, $C(-2, -1)$, $D(6, -5)$

54. $A(5, 1)$, $B(-1, 5)$, $C(-3, 2)$, $D(3, -2)$

51., 54. Parallelogram; rectangle **52., 53.** Not a parallelogram

C

55. Consider two different nonvertical lines.

$$L_1\text{: } A_1x + B_1y = C_1 \quad \text{and} \quad L_2\text{: } A_2x + B_2y = C_2$$

a. Show that L_1 and L_2 are parallel if and only if $A_1B_2 = A_2B_1$.
b. Show that L_1 and L_2 are perpendicular if and only if $A_1A_2 + B_1B_2 = 0$.

56. Show that an equation of the line containing the points (x_1, y_1) and (x_2, y_2) when $x_1 \neq x_2$ is $y - y_1 = \frac{y_2 - y_1}{x_2 - x_1}(x - x_1)$. This is called the **two-point form.**

57. Show that an equation of the line having x-intercept a and y-intercept b is $\frac{x}{a} + \frac{y}{b} = 1$. This is called the **intercept form.**

Computer Exercises

For students with some programming experience.

1. Write a program to find the slope of a line, given two distinct points on the line. If the line has no slope, then the program should print that the line is vertical.

2. Run the program in Exercise 1 to find the slope of the line through each pair of points or to identify the line as vertical. **b.** −2.28571429

a. (3, −1) and (5, 8) 4.5 **b.** (−5, 7) and (2, −9) **c.** (7, −3) and (7, 10) vertical

3. Write a program to find the slope-intercept equation of a line, given the slope and a point on the line.

4. Run the program in Exercise 3 to find the slope-intercept equation of a line with slope 1.72 and containing the point $(-5, 8.6)$. $y = 1.72x + 17.2$

5. Write a program based on the programs in Exercises 1 and 3 to find the slope-intercept equation of a line given two distinct points on the line. Allow for the possibility that the line may be vertical.

6. Run the program in Exercise 5 to find the slope-intercept equation of the line containing each pair of points in Exercise 2.
a. $y = 4.5x - 14.5$ **b.** $y = -2.28571429x - 4.42857143$ **c.** $x = 7$

Self-Test 1

Vocabulary

open sentence in two variables (p. 101)
ordered pair (p. 101)
solution set (p. 101)
xy-coordinate plane (p. 107)
plane rectangular coordinate system (p. 107)
axes (p. 107)
origin (p. 107)
quadrant (p. 107)
graphing an ordered pair (p. 107)
plotting a point (p. 107)
coordinates (p. 107)
x-coordinate (p. 107)
abscissa (p. 107)
y-coordinate (p. 107)
ordinate (p. 107)
Cartesian coordinate system (p. 107)
graph of an open sentence (p. 108)
linear equation (p. 109)
slope (p. 112)
standard form (p. 118)
point-slope form (p. 118)
y-intercept (p. 119)
x-intercept (p. 119)
slope-intercept form (p. 119)

$\left\{\left(-3, -\frac{21}{2}\right), \left(-1, -\frac{13}{2}\right), \left(2, -\frac{1}{2}\right)\right\}$

1. Solve $4x - 2y = 9$ if the domain of x is $\{-3, -1, 2\}$. **Obj. 3-1, p. 101**

2. At a yard sale, Toni bought some jazz records at \$4 each and some classical records at \$6 each. If her purchases totaled \$46, find all possibilities for the number of each type of record that she bought. 10 jazz, 1 classical; 7 jazz, 3 classical; 4 jazz, 5 classical; 1 jazz, 7 classical

3. Graph $2x - 5y = 10$. 4. Graph $x = -3$. **Obj. 3-2, p. 107**

5. Find the slope of line $9x - 6y = 5$. $\frac{3}{2}$ **Obj. 3-3, p. 112**

6. Find the slope of the line containing the points $(-2, 6)$ and $(2, 1)$. $-\frac{5}{4}$

7. Find the slope of the line $2y - 12 = 0$. 0

8. Find an equation in point-slope form of the line containing $(2, 4)$ and $(-4, 3)$. $y - 4 = \frac{1}{6}(x - 2)$ or $y - 3 = \frac{1}{6}(x + 4)$ **Obj. 3-4, p. 118**

9. Find an equation in standard form of the line through $(-3, 2)$ and parallel to the line $2x - y = 1$. $2x - y = -8$

10. Find an equation in standard form of the line through $(5, 1)$ that is perpendicular to the line $x = 4$. $y = 1$

Check your answers with those at the back of the book.

Quick Quiz

1. Solve $3y - 2x = 10$ if the domain of x is $\left\{0, -5, \frac{5}{2}\right\}$.
$\left\{\left(0, \frac{10}{3}\right), (-5, 0), \left(\frac{5}{2}, 5\right)\right\}$

2. A collection of dimes and quarters has a value of \$1.35. List all possible combinations of dimes and quarters.
{(1 dime, 5 quarters), (6 dimes, 3 quarters), (11 dimes, 1 quarter)}

Graph each line.

3. $y = 5$

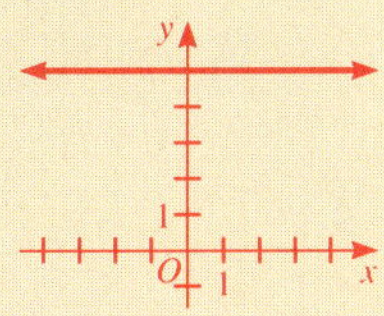

4. $2x + 3y = 12$

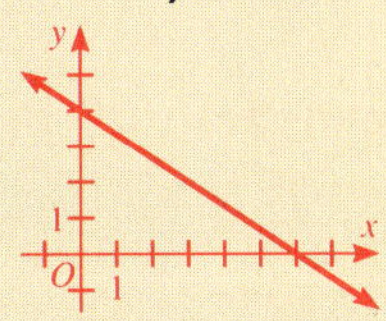

5. Find the slope of $5x - 7y = 12$. $\frac{5}{7}$

6. Find the slope of the line through $(2, -5)$ and $(-4, -2)$. $-\frac{1}{2}$

7. Give the slope-intercept form for the equation of the line through $(-7, 3)$ and $(5, -3)$. $y = -\frac{1}{2}x - \frac{1}{2}$

8. Give the standard form for the equation of the line through $(1, 2)$ and perpendicular to $y - 2x = 3$.
$x + 2y = 5$

Teaching Suggestions p. T89

Suggested Extensions p. T90

Warm-Up Exercises

Graph each pair of equations in the same coordinate plane.

1. $x + y = 6$
$x - y = 4$

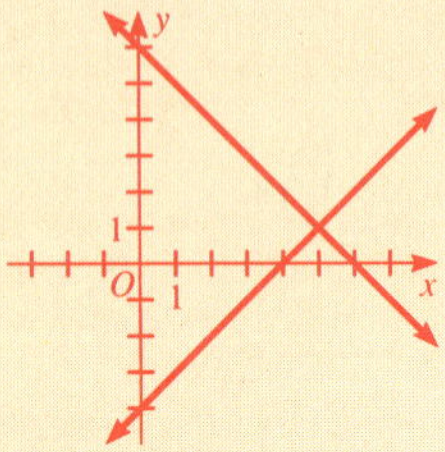

2. $2x - y = 4$
$2x + 3y + 12 = 0$

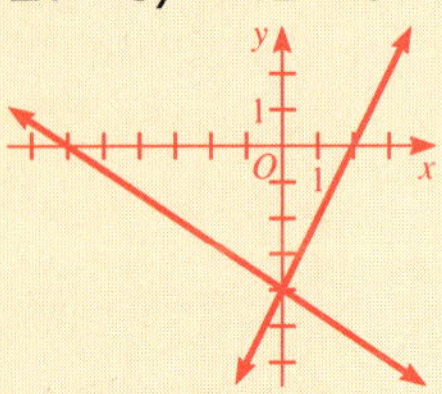

3. $3x + 2y = 6$
$y = -\frac{3}{2}x - 1$

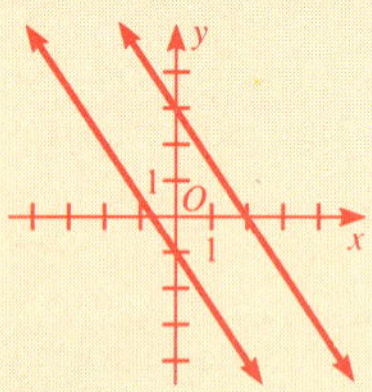

4. $\frac{1}{2}x - 2y = 4$
$4y - x + 8 = 0$

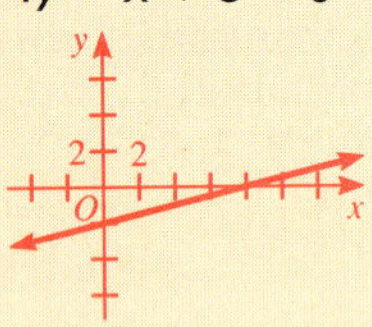

Linear Systems

3-5 Systems of Linear Equations in Two Variables

Objective To solve systems of linear equations in two variables.

A set of linear equations in the same two variables is called a **system of linear equations,** or a **linear system.** Any ordered pair of numbers that is a solution of each equation in the system is called a **solution** of the system, or a **simultaneous solution** of the equations.

When you graph a linear system with two equations in the same coordinate plane, the result is one of three types of graphs, as illustrated below.

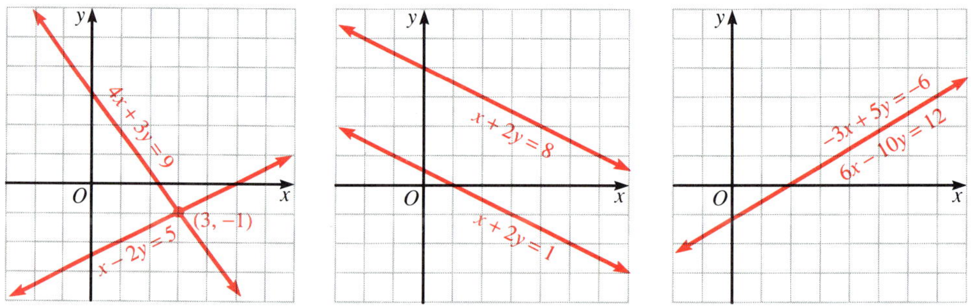

(a) Intersecting lines
$x - 2y = 5$
$4x + 3y = 9$

(b) Parallel lines
$x + 2y = 1$
$y = -\frac{1}{2}x + 4$

(c) Coinciding lines
$-3x + 5y = -6$
$6x - 10y = 12$

Algebraically, these geometric relationships mean:

(a) When the graphs intersect in only one point, the system has one solution.

(b) When the graphs are parallel, the system has no solution.

(c) When the graphs coincide (that is, are the same line), every ordered pair that satisfies one equation also satisfies the other equation. The solution set for the system contains infinitely many ordered pairs.

We can extend the definition of equivalent equations to systems. **Equivalent systems** are systems that have the same solution set. To solve a system of equations, you transform the system into an equivalent system whose solution is easily seen. Transformations that produce equivalent systems are listed in the chart at the top of the next page.

Transformations that Produce Equivalent Systems

1. Replacing an equation by an equivalent equation. (Recall that multiplying each side of an equation by the same nonzero number produces an equivalent equation.)
2. Substituting for one variable in any equation an equivalent expression for that variable obtained from another equation in the system. (This expression may be the value of the variable, if known.)
3. Replacing any equation by the sum of that equation and another equation in the system. (Recall that to add two equations, you add their left sides, add their right sides, and equate the results.)

Example 1 Solve this system: $4x + 3y = 5$
$2x - 5y = -17$

Solution At each step replace the system by an equivalent one, using one of the transformations listed in the chart.

1. Multiply the second given equation by -2. (Transformation 1) $$\left.\begin{aligned} 4x + 3y &= 5 \\ -4x + 10y &= 34 \end{aligned}\right\} (1)$$
2. Replace the second equation in (1) by the sum of the two equations. (Transformation 3) $$\left.\begin{aligned} 4x + 3y &= 5 \\ 13y &= 39 \end{aligned}\right\} (2)$$
3. Solve the second equation in (2) for y. (Transformation 1) $$\left.\begin{aligned} 4x + 3y &= 5 \\ y &= 3 \end{aligned}\right\} (3)$$
4. Substitute 3 for y in the *first* equation in (3) (Transformation 2) $$\left.\begin{aligned} 4x + 3(3) &= 5 \\ y &= 3 \end{aligned}\right\} (4)$$
5. Solve the first equation in (4) for x. (Transformation 1) $$\left.\begin{aligned} x &= -1 \\ y &= 3 \end{aligned}\right\} (5)$$

Check that the solution $(-1, 3)$ satisfies *both of the original equations:*

$$4x + 3y = 5 \qquad\qquad 2x - 5y = -17$$
$$4(-1) + 3(3) \stackrel{?}{=} 5 \qquad\qquad 2(-1) - 5(3) \stackrel{?}{=} -17$$
$$5 = 5 \ \checkmark \qquad\qquad -17 = -17 \ \checkmark$$

$\therefore$ the solution of the system is $(-1, 3)$. ***Answer***

When you add two equations, the result is a **linear combination** of the equations. To solve a system of equations, a useful goal is to obtain a linear combination that has fewer variables than the given system.

Solution 1 of Example 2 illustrates the linear-combination method for solving a linear system and shows you an efficient way to set up your work. You do not need to write out each system in the chain of equivalent systems as was done in Example 1.

Motivating the Lesson

Tell students that streets and roads illustrate the relationships between various lines in a plane. For example, roads may intersect or be parallel, and some roads may be known by more than one name.

Thinking Skills

Solving systems of linear equations in two variables requires the application of several skills learned previously, as well as the thinking skills of *interpreting, applying,* and *analyzing.* This is evident in Oral Exercises 13–17.

Using a Computer or a Graphing Calculator

For demonstration purposes, you may want to use the program "Line Plotter" on the Disk for Algebra when solving systems of linear equations graphically.

You may also want to have students use a computer or graphing calculator to estimate visually the solution of a linear system, as in Written Exercises 13–16 on page 129.

Support Materials
Disk for Algebra
Menu Item: Line Plotter
Resource Book p. 238

Chalkboard Examples

1. Is (5, 2) a solution of the system $2x + 3y = 16$, $3x - 7y = -1$?
 $2(5) + 3(2) = 10 + 6 = 16$; $3(5) - 7(2) = 15 - 14 = 1$; no
2. Are the following systems equivalent? $x = 5$, $y = 3x$ and $y - x = 10$, $y + x = 20$
 The solution to the first system is (5, 15). The solution to the second sytem is (5, 15). yes
3. Solve by the substitution method. $2x + 3y = 13$, $x - y = 9$
 $x = 9 + y$; $2(9 + y) + 3y = 13$; $18 + 2y + 3y = 13$; $18 + 5y = 13$; $5y = -5$; $y = -1$. $x - (-1) = 9$; $x + 1 = 9$; $x = 8$ (8, −1)
4. Solve using the linear combination method.
 $3x + 2y = 4$
 $2x - 5y = -29$
 $5(3x + 2y) = 20$
 $2(2x - 5y) = -58$
 $15x + 10y = 20$
 $4x - 10y = -58$
 $19x = -38$
 $x = -2$
 $3(-2) + 2y = 4$; $-6 + 2y = 4$; $2y = 10$; $y = 5$; (−2, 5)

Example 2 Solve this system:
$$x - 2y = 5$$
$$4x + 3y = 9$$

Solution 1 (Linear-Combination Method)

Find a linear combination that eliminates x.

1. Multiply the first equation by −4 so that the coefficients of x will be *opposites*.
 $-4x + 8y = -20$
 $4x + 3y = 9$
2. Add the equations in Step 1 and solve the resulting equation for y.
 $0x + 11y = -11$
 $11y = -11$
 $y = -1$
3. Substitute −1 for y in either of the original equations to find x.
 $x - 2(-1) = 5$
 $x = 3$
4. Check that (3, −1) satisfies both equations.
 $3 - 2(-1) = 5$ ✓
 $4(3) + 3(-1) = 9$ ✓

∴ the solution of the system is (3, −1). ***Answer***

This algebraic solution of the system verifies the graphic solution shown in diagram (a) on page 124.

Solution 2 (Substitution Method)

1. Express x in terms of y in the first equation (since the coefficient of x is 1).
 $x - 2y = 5$
 $x = 2y + 5$
2. Substitute $2y + 5$ for x in the second equation. Solve for y.
 $4x + 3y = 9$
 $4(2y + 5) + 3y = 9$
 $8y + 20 + 3y = 9$
 $11y = -11$
 $y = -1$
3. Substitute −1 for y in either of the original equations to find x.
 $4x + 3(-1) = 9$
 $4x = 12$
 $x = 3$
4. The check is the same as Step 4 above.

∴ the solution of the system is (3, −1). ***Answer***

Example 3 Solve this system:
$$3x - 4y = 14$$
$$4x + 10y = 11$$

Solution (Linear-Combination Method)

Find a linear combination that eliminates y.

1. Multiply the first equation by 10 and the second equation by 4 so that the coefficients of y will be *opposites*.
 $30x - 40y = 140$
 $16x + 40y = 44$

2. Add the equations in Step 1 and solve the resulting equation for x.

$$46x + 0y = 184$$
$$46x = 184$$
$$x = 4$$

3. Substitute 4 for x in either of the original equations to find y.

$$3(4) - 4y = 14$$
$$-4y = 2$$
$$y = -\frac{1}{2}$$

4. Check that $\left(4, -\frac{1}{2}\right)$ satisfies both equations.

$$3(4) - 4\left(-\frac{1}{2}\right) = 14 \ \checkmark$$
$$4(4) + 10\left(-\frac{1}{2}\right) = 11 \ \checkmark$$

$\therefore$ the solution is $\left(4, -\frac{1}{2}\right)$. ***Answer***

When the graphs of the equations of a system are parallel (see diagram (b) on page 124), the system has no solution. If you try to solve such a system algebraically, both variables are eliminated and the result is a false statement.

Example 4 Solve this system: $x + 2y = 1$

$$y = -\frac{1}{2}x + 4$$

Solution Substitute the expression for y from the second equation into the first equation:

$$x + 2y = 1$$
$$x + 2\left(-\frac{1}{2}x + 4\right) = 1$$
$$x - x + 8 = 1$$
$$8 = 1 \quad \leftarrow \text{False!}$$

Since $8 = 1$ is false, the system has no solution. ***Answer***

When the graphs of the equations of a system coincide (see diagram (c) on page 124), the system has infinitely many solutions. If you try to solve such a system algebraically, you obtain an identity and all solutions of one equation are solutions of the other.

Example 5 Solve this system:

$$-3x + 5y = -6$$
$$6x - 10y = 12$$

Solution Multiply the first equation by 2, and then add the equations:

$$\begin{array}{r} -6x + 10y = -12 \\ 6x - 10y = 12 \\ \hline 0 = 0 \end{array}$$

(Solution continues on the next page.)

Check for Understanding

Here is a suggested use of the Oral Exercises to check students' understanding as you teach the lesson.

Oral Exs. 1–6: use before Example 1.

Oral Exs. 7–16: use after Example 3.

Guided Practice

Solve.

1. $2x + y = 7$
   $3x - y = 8$ (3, 1)
2. $4x - y = 9$
   $3x - 5y = 11$ (2, −1)
3. $a + b = 9$
   $a - b = 13$ (11, −2)
4. Graph both equations in the same coordinate plane. Estimate the solution to nearest half unit.
   $2x + y = 7$; $x - 2y = 7$

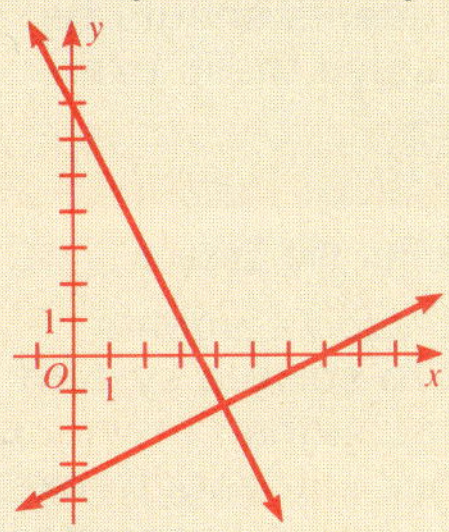

$\left(4, -1\frac{1}{2}\right)$

Solve each system. If no solution exists or if the solution set is infinite, say so.

5. $x + y = -8$
   $x - y = 8$ (0, −8)
6. $2x - y = 7$
   $4x - 2y = 9$
   no solution
7. $3x - 2y = x + 8$
   $y = x - 4$
   infinite

Common Errors

When solving a system of two dependent equations, some students conclude that there is no solution, rather than correctly concluding that there are infinitely many solutions. Emphasize that if students try to find a solution and obtain a contradiction, then the system has no solution. If after trying to find a solution, students obtain an identity, then the system has infinitely many solutions.

Problem Solving Strategies

Exercises 33–38 present a type of problem that is sometimes difficult for students. By applying the strategy of *solving a simpler problem* (as given in the hint), students should be able to solve these exercises.

Summarizing the Lesson

In this lesson students learned to solve a system of linear equations in two variables. Ask students to explain the terms "consistent," "inconsistent," and "dependent" as they relate to systems of linear equations.

Since $0 = 0$ is an identity, the solution set of the system is the same as the solution set of either of the equations, namely, the infinite set $\{(x, y): -3x + 5y = -6\}$. ***Answer***

Three of the solutions are $\left(0, -\frac{6}{5}\right)$, $(2,0)$, and $(7, 3)$.

If a system of equations has at least one solution, the equations in the system are called **consistent.** This was the case in Examples 1, 2, 3, and 5. If the system has no solution, the equations are called **inconsistent** and their graphs are parallel lines. This was the case in Example 4. If a consistent system has infinitely many solutions, then the equations are called **dependent** and their graphs are coinciding lines. This was the case in Example 5.

Another method for solving linear systems, called *Cramer's Rule,* will be presented in Chapter 16. The method is especially useful when using a computer to solve linear systems.

Oral Exercises

Tell whether the given ordered pair is a solution of the system.

1. $(2, 1)$ yes
   $x + 2y = 4$
   $3x - 2y = 4$
2. $(4, 3)$ yes
   $3x - 2y = 6$
   $4x - 3y = 7$
3. $(-3, 2)$ no
   $x + 3y = 3$
   $3x - 2y = 13$
4. $(3, -5)$ yes
   $3x + 2y = -1$
   $2x + y = 1$
5. $(-4, -2)$ yes
   $x - 3y = 2$
   $2x - 5y = 2$
6. $(-3, 4)$ no
   $2x + 3y = 6$
   $3x - 2y = -1$

Explain how you would form a linear combination of the given equations to eliminate the variable printed in red.

Sample $5x + 2y = 2$
$2x + 3y = 7$; y

Solution Multiply the first equation by 3 and the second by -2; then add.

7. $2x + y = 1$
   $2x + 3y = 7$; x
   Mult. 1st eqn. by -1; then add
8. $2x - 3y = 7$
   $3x + y = 5$; y
   Mult. 2nd eqn. by 3; then add
9. $5x - 6y = 9$
   $2x - 3y = 3$; y
   Mult. 2nd eqn. by -2; then add
10. $4x - 3y = 6$
    $2x - 5y = -4$; x
    Mult. 2nd eqn. by -2; then add
11. $2x - 3y = 3$
    $5x + 2y = 17$; y
    Mult. 1st eqn. by 2 and 2nd eqn. by 3; then add
12. $4x + 5y = 3$
    $3x + 2y = 4$; x
    Mult. 1st eqn. by 3 and 2nd eqn. by -4; then add

Use *intersect, are parallel,* or *coincide* to make a true statement.

13. If two lines have the same slope and different y-intercepts, then the lines __?__. are parallel
14. If one equation can be obtained from another equation by multiplying both sides by the same nonzero number, then the graphs __?__. coincide

Use *intersect, are parallel,* or *coincide* to make a true statement.

15. If two lines have different slopes and the same x-intercept, then the lines __?__. intersect

16. If two lines have more than one point in common, then the lines __?__. coincide

17. Why do you think the linear-combination method is more efficient than substitution in solving the system in Example 3?
The linear-combination method avoids fractional coefficients.

Written Exercises

Solve each system.

1. $(-1, 3)$ 2. $(2, -1)$ 3. $(3, 1)$ 4. $(3, 2)$ 5. $(3, 1)$ 6. $(2, -1)$

A **1–6.** The systems in Oral Exercises 7–12.

7. $2x - 7y = 10$, $5x - 6y = 2$ $(-2, -2)$

8. $5x + 6y + 8 = 0$, $3x - 2y + 16 = 0$ $(-4, 2)$

9. $3x - 2y + 2 = 0$, $x + 3y = 14$ $(2, 4)$

10. $8x - 3y = 3$, $3x - 2y + 5 = 0$ $(3, 7)$

11. $6u + 5v = -2$, $2u + 3v = 6$ $\left(-\frac{9}{2}, 5\right)$

12. $3p + 2q = -2$, $9p - q = -6$ $\left(-\frac{2}{3}, 0\right)$

Graph both equations of each system in the same coordinate plane. Then estimate the solution (that is, the coordinates of the point of intersection) to the nearest half unit.

13. $y = -x + 3$, $y = x - 4$

14. $2x + y = -2$, $2x - 3y = 15$

15. $y = 5x$, $x + y = 10$

16. $3x + 5y = 15$, $x - y = 4$

Solve each system. If the system has an infinite solution set, specify it and give three solutions. If the system has no solution, say so.

17. $3x + 3y = 6$, $5x - 6y = 15$ $\left(\frac{27}{11}, -\frac{5}{11}\right)$

18. $3x - 2y = 6$, $5x + 3y + 9 = 0$ $(0, -3)$

19. $x - y = 2x - 2$, $x + y = 2y - 2$ $(0, 2)$

20. $6x = 4y + 5$, $6y = 9x - 5$ No sol.

21. $2p - 5q = 14$, $p + \frac{3}{2}q = 5$ $\left(\frac{23}{4}, -\frac{1}{2}\right)$

22. $d = 2 - 6c$, $\frac{1}{2}d - c = 1$ $(0, 2)$

23. $x + y = 3x - 1$, $x - y = 1 - x$

23. $\{(x, y)\colon 2x - y = 1\}$; $(0, -1)$, $\left(\frac{1}{2}, 0\right)$, $(2, 3)$

24. $2x + y = 2 - x$, $x + 2y = 2 + y$ $(0, 2)$

25. $3y = x - 2y - 1$, $3x = 2x - y - 1$ $\left(-\frac{2}{3}, -\frac{1}{3}\right)$

26. $x + y = 4(y + 2)$, $x - y = 2(y + 4)$

26. $\{(x, y)\colon x - 3y = 8\}$; $\left(0, -\frac{8}{3}\right)$, $(8, 0)$, $(2, -2)$

27. $2x - 3y = 2 - x$, $3x - 2y = -2 + y$ No sol.

28. $2(y - x) = 5 + 2x$, $2(y + x) = 5 - 2y$ $\left(-\frac{1}{2}, \frac{3}{2}\right)$

Write each equation of the system in slope-intercept form. By comparing slopes and y-intercepts, determine whether the equations are consistent or inconsistent.

B **29.** $3x = 4y - 4$, $4y = 3x - 3$

30. $3x = 4y + 8$, $3y = 4x + 8$

31. $3x - 4y = 12$, $4x - 3y = 12$

32. $3x - 6y = 9$, $4x - 3y = 12$

Suggested Assignments

Average Algebra
Day 1: 129/1–31 odd
S 130/Mixed Review
Day 2: 130/33–38
Assign with Lesson 3-6.

Average Alg. and Trig.
129/1–37 odd
S 130/Mixed Review

Extended Algebra
129/1–37 odd
S 130/Mixed Review

Extended Alg. and Trig.
129/1–37 odd
S 130/Mixed Review
Assign with Lesson 3-6.

Additional Answers Written Exercises

13.
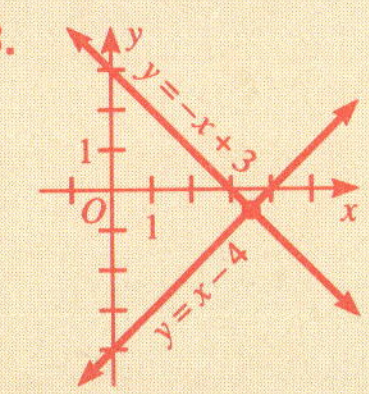

14.
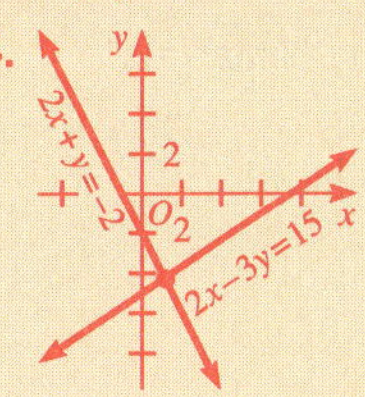

15.
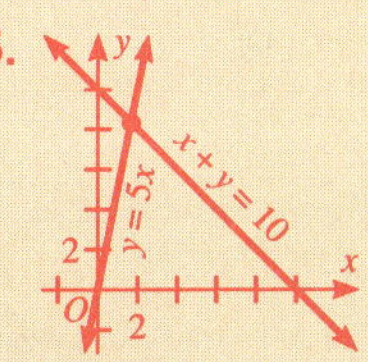

16.
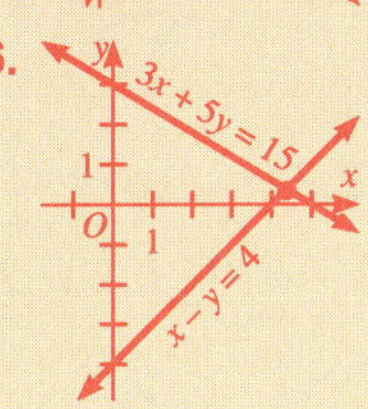

Additional Answers Written Exercises

42. Suppose the equations have a common solution; substitution for y yields $mx + b_1 = mx + b_2$, or $b_1 = b_2$; but this contradicts the assumption that $b_1 \neq b_2$; so the equations have no common solution.

43. $y = m_1x + b_1$ and $y = m_2x + b_2$. Eliminating y: $0 = x(m_1 - m_2) + b_1 - b_2$, $x = \frac{b_2 - b_1}{m_1 - m_2}$. Since $m_1 \neq m_2$, $x \in R$. $y = m_1\left(\frac{b_2 - b_1}{m_1 - m_2}\right) + b_1$ $= \frac{m_1b_2 - m_1b_1 + b_1m_1 - b_1m_2}{m_1 - m_2}$ $= \frac{m_1b_2 - m_2b_1}{m_1 - m_2}$, $y \in R$. $\left(\frac{b_2 - b_1}{m_1 - m_2}, \frac{m_1b_2 - m_2b_1}{m_1 - m_2}\right)$

Supplementary Materials

Study Guide	pp. 41–42
Overhead Visual	1

Solve each system. Note that the equations in Exercises 33–38 are not linear in the original variables. $\left(\textit{Hint:}\ \text{Let } x = \frac{1}{u} \text{ and } y = \frac{1}{v}\right.$**. Rewrite the system in terms of x and y and solve. Then use x and y to find u and v.**$\left.\right)$

33. $\frac{6}{u} + \frac{3}{v} = 2$, $\frac{2}{u} - \frac{9}{v} = 4$ (2, −3)

34. $\frac{6}{u} + \frac{5}{v} = 1$, $\frac{3}{u} - \frac{10}{v} = 3$ (3, −5)

35. $\frac{2}{u} - \frac{3}{v} + 2 = 0$, $\frac{4}{u} + \frac{3}{v} + 1 = 0$ (−2, 3)

36. $\frac{3}{u} + \frac{4}{v} = 1$, $\frac{6}{u} - \frac{2}{v} = 1$ (5, 10)

37. $\frac{3}{u} - \frac{4}{v} = 4$, $\frac{5}{u} - \frac{6}{v} = 7$ $\left(\frac{1}{2}, 2\right)$

38. $\frac{4}{u} + \frac{3}{v} = 3$, $\frac{6}{u} + \frac{5}{v} = 4$ $\left(\frac{2}{3}, -1\right)$

In Exercises 39–41, *a*, *b*, *c*, *d*, *e*, and *f* represent constants. Solve each system. What conditions must be placed on the constants in each system to be certain that the system has a unique solution?

C **39.** $ax + y = b$, $bx + y = a$ $(-1, a + b)$; $a \neq b$

40. $ax - by = c$, $bx + ay = d$ See below.

41. $ax + by = e$, $cx + dy = f$ See below.

Exercises 42 and 43 outline a proof of the first part of the theorem stated on page 119.

42. Proof of the "if" part: If $m_1 = m_2 = m$, then L_1 and L_2 have equations of the form L_1: $y = mx + b_1$ and L_2: $y = mx + b_2$ where $b_1 \neq b_2$. Show that these equations have no common solution.

43. The "only if" part is equivalent to: If $m_1 \neq m_2$, then L_1 and L_2 are not parallel. Show that if $m_1 \neq m_2$, then the equations L_1: $y = m_1x + b_1$ and L_2: $y = m_2x + b_2$ have a common solution.

40. $\left(\frac{ac + bd}{a^2 + b^2}, \frac{ad - bc}{a^2 + b^2}\right)$; a, b not both 0 **41.** $\left(\frac{de - bf}{ad - bc}, \frac{af - ce}{ad - bc}\right)$; $ad \neq bc$

Mixed Review Exercises

For the line containing the given points, find (a) the slope and (b) an equation in standard form.

1. (−2, 0), (0, 1) a. $\frac{1}{2}$ b. $x - 2y = -2$

2. (1, −2), (3, 4) a. 3 b. $3x - y = 5$

3. (−6, −4), (3, 2) a. $\frac{2}{3}$ b. $2x - 3y = 0$

4. (1, −1), (5, −1) a. 0 b. $y = -1$

5. (−3, 2), (1, −2) a. −1 b. $x + y = -1$

6. (0, 0), (−3, 4) a. $-\frac{4}{3}$ b. $4x + 3y = 0$

Find the slope and *y*-intercept of each line.

7. $x - y = 4$ $m = 1, b = -4$

8. $5x + 3y = 6$ $m = -\frac{5}{3}, b = 2$

9. $-2x + y = 1$ $m = 2, b = 1$

10. $y - 5 = 0$ $m = 0, b = 5$

11. $y = 3x + 2$ $m = 3, b = 2$

12. $x = 4 - 6y$ $m = -\frac{1}{6}, b = \frac{2}{3}$

3-6 Problem Solving: Using Systems

Objective To use systems of equations to solve problems.

Sometimes you can solve a problem involving two unknown quantities most easily by using a system of equations. The problem solving plan given in Lesson 1-9 is useful here. However, you must set up and solve a system of equations rather than a single equation.

Example 1 To use a certain computer data base, the charge is \$30/h during the day and \$10.50/h at night. If a research company paid \$411 for 28 h of use, find the number of hours charged at the daytime rate and at the nighttime rate.

Solution

Step 1 The problem asks for the number of hours charged at each rate.

Step 2 Let d = the number of hours of use at the daytime rate.
Let n = the number of hours of use at the nighttime rate.

Step 3 Set up a system of two equations.

Total number of hours charged is 28. $\longrightarrow$ $d + n = 28$
Total amount charged is 411. $\longrightarrow$ $30d + 10.50n = 411$

Step 4 Solve the system. Express d in terms of n in the first equation and substitute in the second equation.

$$d = 28 - n$$

$$\begin{aligned} 30(28 - n) + 10.50n &= 411 \\ 840 - 30n + 10.50n &= 411 \\ -19.50n &= -429 \\ n &= 22 \end{aligned}$$

Since $n = 22$, $d = 28 - 22 = 6$. The solution is $(6, 22)$.

Step 5 The check is left for you.
$\therefore$ there were 6 h charged at the daytime rate and 22 h at the nighttime rate. ***Answer***

The following terms are used in connection with aircraft flight.

air speed the speed of an aircraft in still air
wind speed the speed of the wind
tail wind a wind blowing in the same direction as the path of the aircraft
head wind a wind blowing in the direction opposite to the path of the aircraft
ground speed the speed of the aircraft relative to the ground

With a tail wind: ground speed = air speed + wind speed.
With a head wind: ground speed = air speed − wind speed.

Teaching Suggestions p. T90

Reading Algebra p. T90

Suggested Extensions p. T90

Warm-Up Exercises

Have students turn to page 50 and review the five-step plan for solving a word problem.

Motivating the Lesson

Tell students that there are a variety of problems that can be solved using systems of equations. These problems may involve money, speed, geometry, or business decisions.

Chalkboard Examples

1. If 8 pens and 7 pencils cost $3.37 while 5 pens and 11 pencils cost $3.10, how much does each pen and each pencil cost?
Let x = cost of a pen and y = cost of a pencil.
$8x + 7y = 337$;
$5x + 11y = 310$
$11(8x + 7y) = 11 \cdot 337$
$-7(5x + 11y) = (-7)310$
$88x + 77y = 3707$
$-35x - 77y = -2170$
$53x = 1537$
$x = 29$
$5(29) + 11y = 310$;
$145 + 11y = 310$;
$11y = 165$; $y = 15$
$\therefore$ pens cost $.29 each and pencils cost $.15 each.

2. Loren's marble jar contains plain marbles and colored marbles. If there are 32 more plain marbles than colored marbles and there are 180 marbles in the jar, how many colored marbles does Loren have?
Let p = number of plain marbles and c = number of colored marbles.
$p = c + 32$
$p + c = 180$
$(c + 32) + c = 180$;
$2c + 32 = 180$;
$2c = 148$; $c = 74$;
$\therefore$ Loren has 74 colored marbles.

Example 2 To measure the speed of the jet stream (a high-speed, high-altitude, west-to-east wind), a weather-service plane flew 1800 km with the jet stream as a tail wind and then back again. The eastbound flight took 2 h, and the westbound one took 3 h 20 min. Find the speed of the jet stream and the air speed of the plane.

Solution

Step 1 The problem asks for the plane's air speed and the speed of the wind.

Step 2 Let p = the air speed in km/h and w = the wind speed in km/h.

Step 3 Using the fact that rate × time = distance, construct a table.

	Ground speed km/h	Time h	Distance km
Eastbound	$p + w$	2	$2(p + w)$
Westbound	$p - w$	$3\frac{1}{3}$, or $\frac{10}{3}$	$\frac{10}{3}(p - w)$

The distance east is 1800 km.

$$2(p + w) = 1800$$

The distance west is 1800 km.

$$\frac{10}{3}(p - w) = 1800$$

Step 4 Solve the system.

$$2(p + w) = 1800 \text{ is equivalent to } p + w = 900$$
$$\frac{10}{3}(p - w) = 1800 \text{ is equivalent to } p - w = 540$$

Using the linear-combination method, you will find that $p = 720$ and $w = 900 - 720 = 180$. The solution is (720, 180).

Step 5 The check is left for you.
$\therefore$ the air speed of the plane is 720 km/h and the speed of the jet stream is 180 km/h. ***Answer***

Problems

A

1. Kerry asked a bank teller to cash a $390 check using $20 bills and $50 bills. If the teller gave her a total of 15 bills, how many of each type of bill did she receive? 12 $20 bills, 3 $50 bills

2. Tickets for the homecoming dance cost $20 for a single ticket or $35 for a couple. Ticket sales totaled $2280, and 128 people attended. How many tickets of each type were sold? 16 single's tickets, 56 couple's tickets

3. Find the measures of the angles of an isosceles triangle if the measure of the vertex angle is 40° less than the sum of the measures of the base angles. 55°, 55°, 70°

4. Two isosceles triangles have the same base length. The legs of one of the triangles are twice as long as the legs of the other. Find the lengths of the sides of the triangles if their perimeters are 23 cm and 41 cm. 4. small triangle: 5 cm, 9 cm, 9 cm; large triangle: 5 cm, 18 cm, 18 cm

5. On Friday, the With-It Clothiers sold some jeans at $25 a pair and some shirts at $18 each. Receipts for the day totaled $441. On Saturday the store priced both items at $20, sold exactly the same number of each item, and had receipts of $420. How many pairs of jeans and how many shirts were sold each day? 9 pairs of jeans, 12 shirts

6. A grain-storage warehouse has a total of 30 bins. Some hold 20 tons of grain each, and the rest hold 15 tons each. How many of each type of bin are there if the capacity of the warehouse is 510 tons? 12 20-ton bins, 18 15-ton bins

7. With a tail wind, a helicopter traveled 300 mi in 1 h 40 min. The return trip against the same wind took 20 min longer. Find the wind speed and also the air speed of the helicopter. air speed, 165 mi/h wind speed, 15 mi/h

8. With a head wind, a plane traveled 1000 mi in 4 h. With the same wind as a tail wind, the return trip took 3 h 20 min. Find the plane's air speed and the wind speed. 275 mi/h; 25 mi/h

9. An overseas phone call is charged at one rate (a fixed amount) for the first minute and at a different rate for each additional minute. If a 7 min call costs $10, and a 4 min call costs $6.40, find each rate. first minute, $2.80; add. minute, $1.20

10. A caterer's total cost for catering a party includes a fixed cost, which is the same for every party. In addition the caterer charges a certain amount for each guest. If it costs $300 to serve 25 guests and $420 to serve 40 guests, find the fixed cost and the cost per guest. fixed cost, $100 cost per guest, $8

11. A financial planner wants to invest $8000, some in stocks earning 15% annually and the rest in bonds earning 6% annually. How much should be invested at each rate to get a return of $930 annually from the two investments? $5000 at 15%, $3000 at 6%

12. For a recent job, a plumber earned $28/h, and the plumber's apprentice earned $15/h. The plumber worked 3 hours more than the apprentice. If together they were paid $213, how much did each earn? (*Hint:* First write an expression for the number of hours each worked on the job.) plumber, $168; apprentice, $45

B 13. Marcia flew her ultralight plane to a nearby town against a head wind of 15 km/h in 2 h 20 min. The return trip under the same wind conditions took 1 h 24 min. Find the plane's air speed and the distance to the nearby town. air speed, 60 km/h; distance, 105 km

Guided Practice

1. Ann can paddle her canoe 15 mi downstream in 2 h. The return trip upstream takes 3 times as long. What is the rate of the current? 2.5 mi/h
2. A grocer mixed nuts worth $4 per kilogram with raisins worth $3.25 per kilogram to make 15 kg of a mixture worth $3.60 per kilogram. How many kilograms of nuts were used? 7 kg

Summarizing the Lesson

In this lesson students applied the methods of solving systems of equations to problem solving. Ask students to tell why the five-step problem solving method is especially important in working this type of problem.

Suggested Assignments

Average Algebra
Day 1: 132/*P*: 2–10 even
Assign with Lesson 3-5.
Day 2: 133/*P*: 11–14, 16
Assign with Lesson 3-7.

Average Alg. and Trig.
Day 1: 132/*P*: 2–14 even
S 129/32–40 even
Day 2: 133/11–17 odd
Assign with Lesson 3-7.

Extended Algebra
Day 1: 132/*P*: 2–16 even
S 129/32–40 even
Day 2: 133/*P*: 11–19 odd
Assign with Lesson 3-7.

Extended Alg. and Trig.
Day 1: 132/*P*: 3, 7, 10–14
Assign with Lesson 3-5.
Day 2: 133/*P*: 9, 15–18

Additional Answers Problems

18. a.

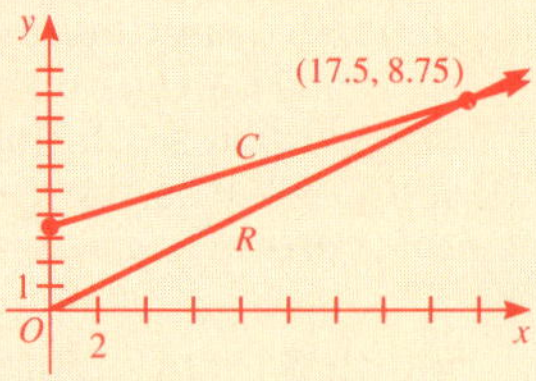

19. a.

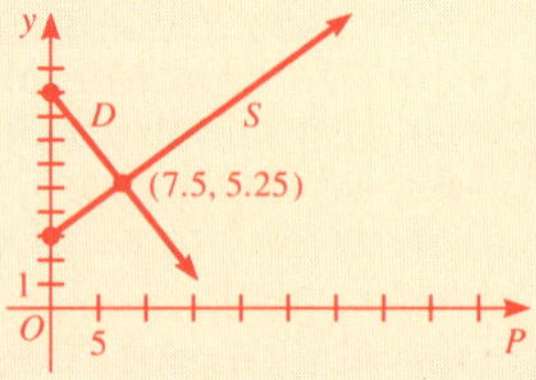

Supplementary Materials

Study Guide	pp. 43–44
Practice Master	14

14. A plane whose air speed is 150 mi/h flew from Abbot to Blair in 2 h with a tail wind. On the return trip against the same wind, the plane was still 60 mi from Abbot after two hours. Find the wind speed and the distance between Abbot and Blair. wind speed 15 mi/h, distance 330 mi

15. If a particle starting with initial speed v_0 has constant acceleration a, then its speed after t seconds is given by

$$v = v_0 + at.$$

Find v_0 and a if $v = 28$ m/s when $t = 4$ s and $v = 43$ m/s when $t = 7$ s. $v_0 = 8$ m/s, $a = 5$ m/s^2

16. While training for a biathlon race, Kevin covered a total distance of 9 km by swimming for 45 min and running for 20 min. The next day he swam for 30 min and ran for 40 min, covering 14 km. Find his rates (in km/h) for swimming and running. (Assume that these rates are constant.) swimming, 4 km/h; running, 18 km/h

17. Davis Rent-A-Car charges a fixed amount per weekly rental plus a charge for each mile driven. A one-week trip of 520 miles costs \$250, and a two-week trip of 800 miles costs \$440. Find the weekly charge and the charge for each mile driven. weekly charge, \$120; charge per mile, \$.25

18. In a certain mill the cost C in thousands of dollars of producing x tons of steel is given by

$$C = 0.3x + 3.5.$$

The revenue R in thousands of dollars from selling x tons is given by

$$R = 0.5x.$$

a. Graph each equation in the same coordinate plane. (Let the vertical axis represent cost or revenue in thousands of dollars.)

b. Find the point at which $R = C$ (the *break-even point*). $x = 17.5$

c. What does the difference $R - C$ measure for all x greater than the x-coordinate of the break-even point? $R - C$ measures profit.

19. The *supply* S of a commodity (in thousands of units), if the price is p dollars per unit, is given by the equation

$$S = 0.3p + 3.$$

The *demand* D for that commodity (in thousands of units), if the price is p dollars per unit, is given by the equation

$$D = -0.5p + 9.$$

The value of p for which $S = D$ is called the *equilibrium price*.

a. Graph both equations in the same coordinate plane. (Let the horizontal axis represent price and the vertical axis represent supply or demand in thousands of units.)

b. Determine the equilibrium price. \$7.50

c. What will happen at prices greater than this price? Supply is greater than demand.

C **20.** Two miles upstream from his starting point, a canoeist passed a log floating in the river's current. After paddling upstream for one more hour, he paddled back and reached his starting point just as the log arrived. Find the speed of the current. 1 mi/h

3-7 Linear Inequalities in Two Variables

Objective To graph linear inequalities in two variables and systems of such inequalities.

When you replace the equals sign in a linear equation in two variables by one of the symbols $>$, $<$, $\geq$, or $\leq$, you obtain a **linear inequality in two variables.** For example,

$$y > \frac{1}{2}x + 1, \qquad 2x - 3y < 12, \qquad \text{and} \qquad y \geq x$$

are linear inequalities in the two variables x and y.

A **solution** of an inequality in two variables is an ordered pair of numbers that satisfies the inequality. For example, the pair (4, 5) is a solution of $y > \frac{1}{2}x + 1$ since $5 > \frac{1}{2}(4) + 1$. There are infinitely many ordered pairs that satisfy this inequality, and so its graph contains infinitely many points.

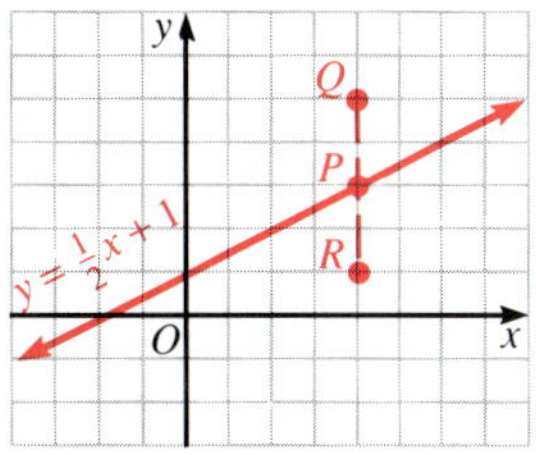

When graphing a linear inequality, it is helpful first to graph the **associated equation** of the inequality. The associated equation of $y > \frac{1}{2}x + 1$ is $y = \frac{1}{2}x + 1$. Its graph is shown at the right. Every point $P(x,y)$ on the line $y = \frac{1}{2}x + 1$ has y-coordinate equal to $\frac{1}{2}x + 1$. It is not difficult to see that if a point Q is above P, its y-coordinate is greater than $\frac{1}{2}x + 1$. Similarly, if a point R is below P, its y-coordinate is less than $\frac{1}{2}x + 1$. Therefore, the graph of $y > \frac{1}{2}x + 1$ is the set of all points *above* the line $y = \frac{1}{2}x + 1$, and the graph of $y < \frac{1}{2}x + 1$ is the set of all points *below* the line $y = \frac{1}{2}x + 1$.

The graphs of $y > \frac{1}{2}x + 1$ and $y < \frac{1}{2}x + 1$ are shown as shaded regions in Figure 1. Each graph is a **half-plane** with the line $y = \frac{1}{2}x + 1$ as its **boundary.** We show each boundary as a dashed line because it is not included in the graph. We call each graph an **open half-plane.**

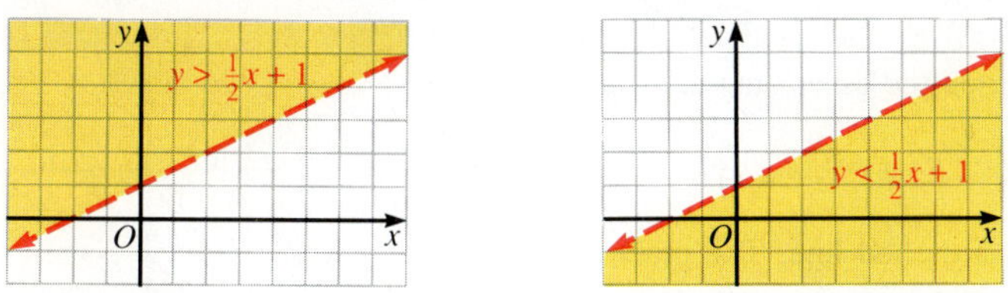

Figure 1

Teaching Suggestions p. T90

Suggested Extensions p. T91

Warm-Up Exercises

Solve each inequality. Graph the solution set.

1. $3x + 17 < 2$ $x < -5$

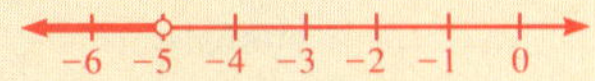

2. $2x - 12 > 14$ $x > 13$

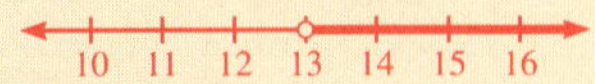

3. $7t \leq 4t - 3$ $t \leq -1$

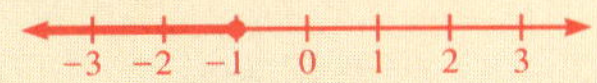

4. $5 > 4 + 3y$ $y < \frac{1}{3}$

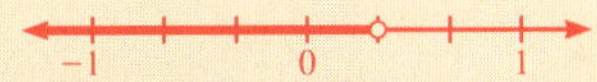

Motivating the Lesson

Tell students that you can indicate an area on a map or on an aerial photograph by drawing the area's boundary lines and shading that part of the map or photo. This illustrates an application of the topic of Lesson 3-7.

Using a Computer or a Graphing Calculator

For demonstration purposes, you may want to use the program "Inequality Plotter" on the Disk for Algebra when graphing linear inequalities.

Also, once students have drawn their own graphs in Written Exercises 1–34 on pages 138–139, you may want to have them check their work by using a computer or graphing calculator.

Support Materials
Disk for Algebra
Menu Item: Inequality Plotter

Chalkboard Examples

1. Graph $3x - 2y \le 8$.

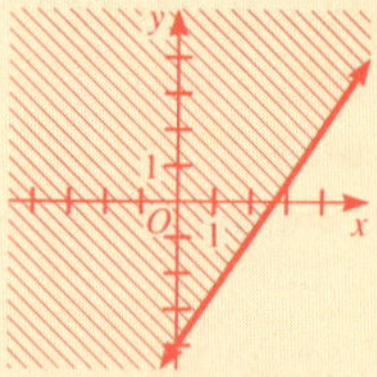

2. Graph the system:
   $2x + y > 1$
   $x - y \le 3$

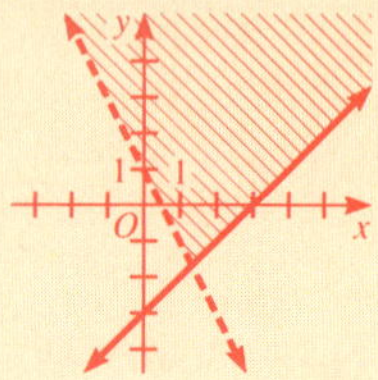

3. Graph the system:
   $x < 6$
   $x + y > 4$
   $y < -\frac{1}{2}x + 4$

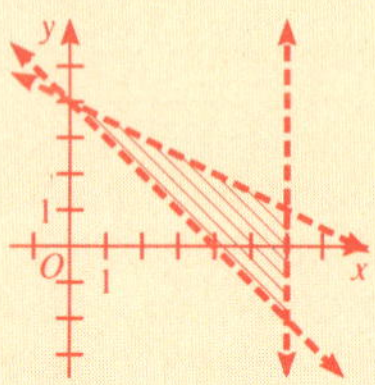

Common Errors

After transforming an inequality into an equivalent inequality having *y* as one side and graphing the associated equation, students often have difficulty deciding which half-plane to shade. Point out that if *y* is the left side of the inequality then the symbol > indicates shading above the line and the symbol < indicates shading below the line. If *x* is the left side of the inequality then the symbol > indicates shading to the right of the line and the symbol < indicates shading to the left of the line.

The graphs of $y \ge \frac{1}{2}x + 1$ and $y \le \frac{1}{2}x + 1$ are shown in Figure 2. Since the boundary line $y = \frac{1}{2}x + 1$ is included in each graph, we show it as a solid line. We call a half-plane together with its boundary a **closed half-plane.**

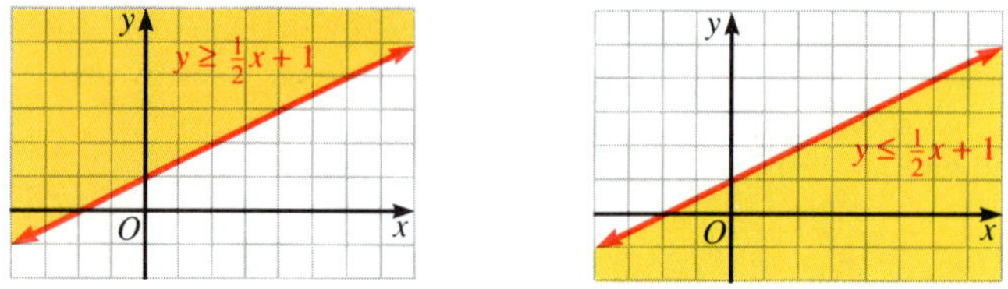

Figure 2

Example 1 Graph $3x + 2y + 4 \ge 0$.

Solution Transform the inequality into an equivalent inequality with *y* alone on one side.

$$3x + 2y + 4 \ge 0$$
$$2y \ge -3x - 4$$
$$y \ge -\frac{3}{2}x - 2$$

Graph the associated equation

$$y = -\frac{3}{2}x - 2$$

as a solid line. The graph of the inequality includes this boundary line and the half-plane above it.

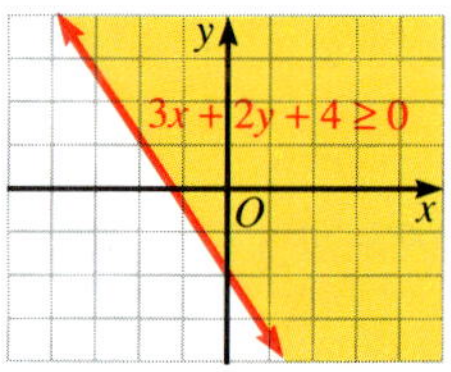

Example 2 Graph $5x - 4y > 10$.

Solution

$$5x - 4y > 10$$
$$-4y > -5x + 10$$
$$y < \frac{5}{4}x - \frac{5}{2}$$

Graph the associated equation

$$y = \frac{5}{4}x - \frac{5}{2}$$

as a dashed line. The graph of the inequality is the open half-plane below this boundary line.

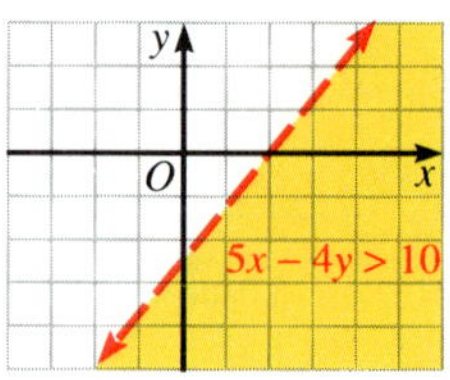

You can tell whether you have shaded the correct half-plane by testing a point that is not on the boundary. The origin is a convenient point to use. In Example 1, the ordered pair (0, 0) *satisfies* the given inequality ($4 \geq 0$ is true), and the point (0, 0) is *in* the shaded region. In Example 2, (0, 0) *does not satisfy* the given inequality ($0 > 10$ is false), and the point (0, 0) is *not in* the shaded region.

Example 3 Graph $x < 2$ in a coordinate plane.

Solution The graph of $x = 2$ is a vertical line. A point $P(x, y)$ belongs to the graph of the inequality if and only if the x-coordinate of P is less than 2 (that is, if and only if P is to the left of the boundary line $x = 2$).

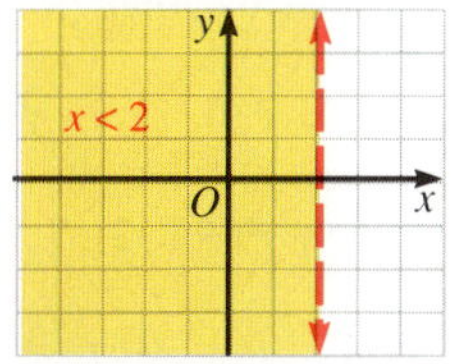

The graph of a **system of inequalities** consists of points satisfying all of the inequalities in the system. This is the region where the graphs of the individual inequalities overlap.

Example 4 Graph this system:

$$x + y \leq 6$$
$$2y > x$$

Solution Graph $x + y \leq 6$, showing the boundary as a solid line. In the same coordinate plane graph $2y > x$, showing the boundary as a dashed line. The graph of the system is the doubly-shaded region in the figure at the right.

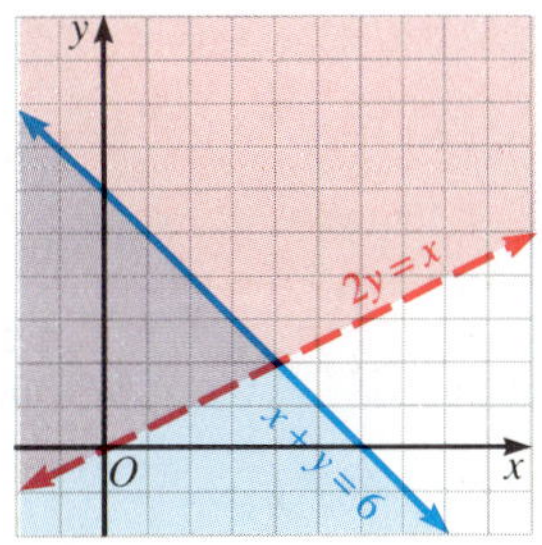

Example 5 Graph this system:

$$x > 0$$
$$y > 0$$
$$3x + 2y \leq 12$$
$$x - 2y \geq -4$$

Solution The inequalities $x > 0$ and $y > 0$ tell you that the graph of the system is contained in the first quadrant. The graph of the system is the shaded region in the first quadrant below both of the lines $3x + 2y = 12$ and $x - 2y = -4$.

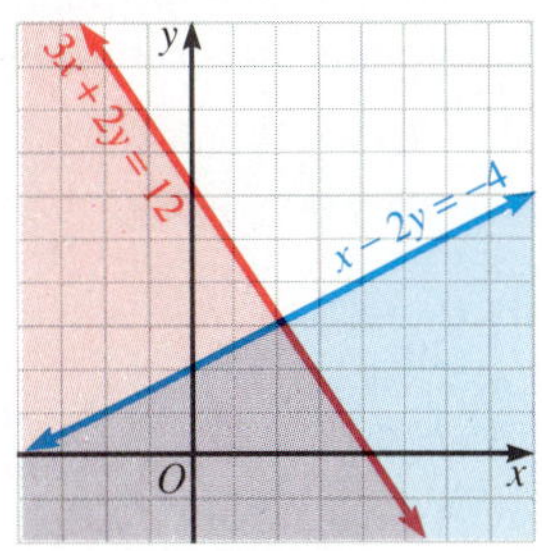

Check for Understanding

Here is a suggested use of the Oral Exercises to check students' understanding as you teach the lesson.

Oral Exs. 1–6: use before Example 1.
Oral Exs. 7–12: use after Example 3.
Oral Exs. 13–16: use after Example 3.

Guided Practice

Graph each inequality or system of inequalities in a coordinate plane.

1. $x - 3 \leq 0$

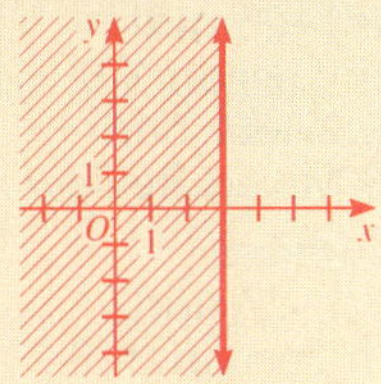

2. $y + 1 \geq 0$

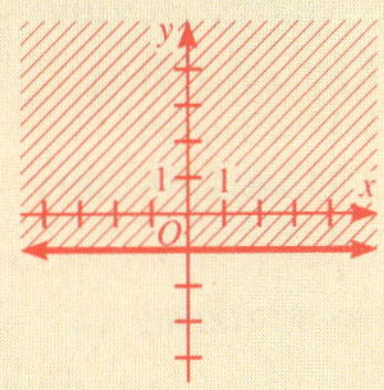

3. $3x - 2y < 6$

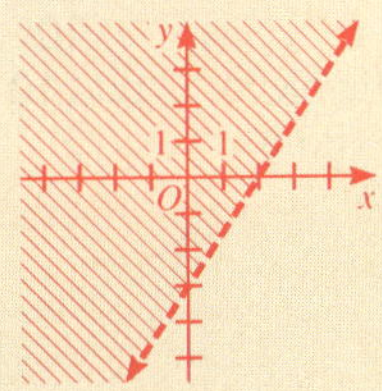

4. $x \geq 0$; $y \leq 0$
$x + y > -1$

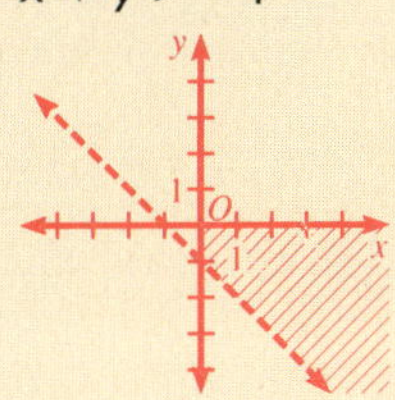

Summarizing the Lesson

Students have learned to graph linear inequalities and systems of inequalities. Ask students to explain the terms "open half-plane" and "closed half-plane" using inequalities to illustrate the differences.

Suggested Assignments

Average Algebra
Day 1: 138/1–17 odd
Assign with Lesson 3-6.
Day 2: 139/19–41 odd
S 139/Mixed Review

Average Alg. and Trig.
Day 1: 138/1–25 odd
Assign with Lesson 3-6.
Day 2: 139/27–41 odd
S 139/Mixed Review

Extended Algebra
Day 1: 138/1–27 odd
Assign with Lesson 3–6.
Day 2: 139/29-43 odd
S 139/Mixed Review

Extended Alg. and Trig.
138/1–43 odd
S 139/Mixed Review

Supplementary Material

Study Guide	pp. 45–46
Practice Master	15
Test Master	10
Resource Book	p. 110
Overhead Visual	2

Additional Answers Written Exercises

2. 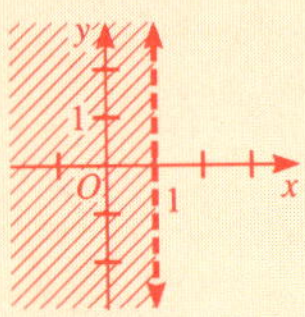

4. 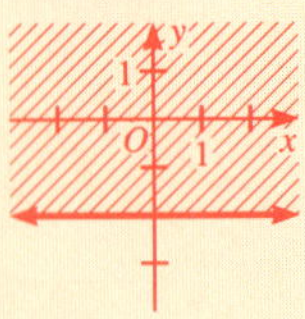

Oral Exercises

Tell whether each ordered pair satisfies the given inequality.

1. $(3, -4), (-3, 4); x + y > 0$ no, yes
2. $(3, -1), (1, -1); x - y < 2$ no, no
3. $(2, 1), (1, 2); 2x - y \leq 3$ yes, yes
4. $(-2, -1), (-1, -2); 3x - y \geq -1$ no, yes
5. $(3, -2), (2, -3); 3x + 2y < 0$ no, no
6. $(5, 1), (-3, -1); x - 4y + 1 > 0$ yes, yes

Tell whether point P lies above, on, or below the line L.

7. $P(0, 2); L: y = 2x + 1$ above
8. $P(2, 2); L: y = \frac{3}{2}x - 1$ on
9. $P(-1, 2); L: y = -\frac{1}{2}x + 2$ below
10. $P(-3, 2); L: y = \frac{1}{2}x + 3$ above
11. $P(4, 5); L: y = \frac{1}{2}x + 3$ on
12. $P(-5, -3); L: y = -\frac{1}{2}x - 1$ below

Give the letter of the inequality whose graph is shown.

13. a. $x + 1 \geq 0$
    b. $y \geq 1$
    (c.) $x - 1 \geq 0$
    d. $x > 1$

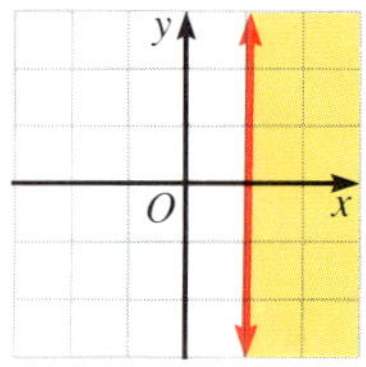

14. a. $y \leq 2$
    (b.) $y - 2 < 0$
    c. $y + 2 < 0$
    d. $x < 2$

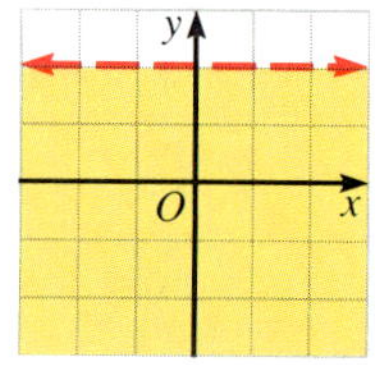

15. (a.) $y > -x$
    b. $x - y > 0$
    c. $x + y \geq 0$
    d. $y > x$

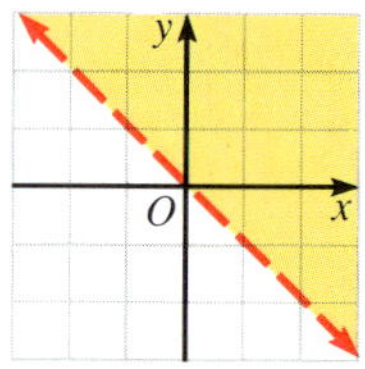

16. a. $x + y \geq 0$
    b. $x + y > 0$
    (c.) $x - y \geq 0$
    d. $y - x \geq 0$

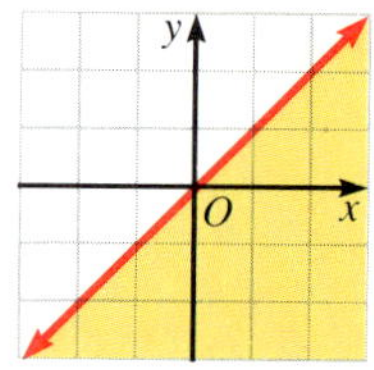

Written Exercises

Graph each inequality in a coordinate plane. You may wish to check your graphs on a computer or a graphing calculator.

A

1. $y - 2 \geq 0$
2. $x - 1 < 0$
3. $x + 1 < 0$
4. $y + 2 \geq 0$
5. $x < y$
6. $x + y < 0$
7. $x \leq 2y$
8. $x + 2y \geq 0$
9. $x + y > 1$
10. $x - y < 1$
11. $2x - y \leq 2$
12. $x + 2y \leq 2$
13. $3x + 2y < 6$
14. $2x - 3y < 6$
15. $3x \geq 2(y - 1)$
16. $y \leq \frac{1}{2}(4 - x)$
17. $y \geq \frac{1}{3}(2x + 3)$
18. $2(y - 1) > 3(x + 1)$

Graph each system of inequalities.

19. $y \le 2$; $y \ge x$

20. $x + y > 0$; $y + 2 > 0$

21. $x + y > 0$; $x - y \le 0$

22. $x - y \le 0$; $x + y < 0$

23. $y < x + 2$; $1 - y < 0$

24. $y + x \le 2$; $y - x \le 2$

25. $x - y \le 1$; $y - x \le 1$

26. $x + y > 0$; $x + y < 2$

27. $x + 2y < 2$; $x - 2y > 2$

28. $x - 2y + 2 \ge 0$; $x + 2y - 2 \ge 0$

29. $3x + 2y \le 6$; $2x + 3y \ge 6$

30. $y - 3x < 3$; $3y - x > 3$

B **31.** $x \ge 0$; $y \ge 0$; $x + 2y \le 4$

32. $y \ge 0$; $y \le x$; $x + y \le 3$

33. $y - x > 0$; $y + x > 0$; $y - 2 < 0$

34. $y - x < 3$; $y + x < 3$; $y - 1 > 0$

35. $x \ge 0$; $y \ge 0$; $x + y \le 6$; $x - y \le 2$

36. $x \ge 0$; $y \ge 0$; $x - y + 2 > 0$; $2x - y < 2$

Graph each inequality or system of inequalities.

Sample $|x + 1| < 2$

Solution The given inequality is equivalent to

$$-2 < x + 1 < 2$$

so $x > -3$ and $x < 1$.

The graph is the shaded region between the vertical lines $x = -3$ and $x = 1$.

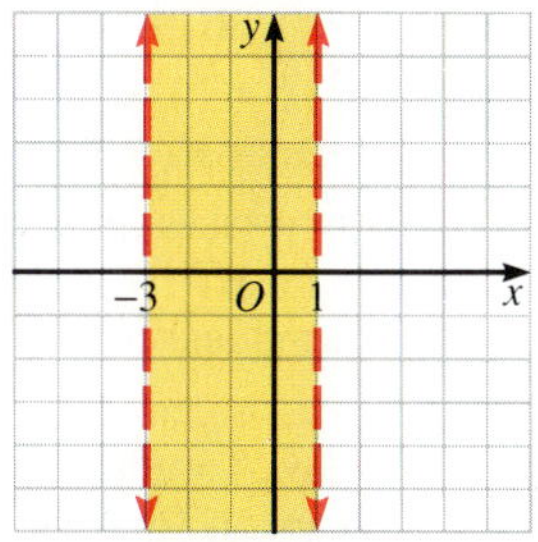

37. $0 \le y \le 3$

38. $1 < x < 3$

39. $|x - 2| < 2$

40. $|y - 2| \le 1$

41. $|x| \le 2$; $|y| \le 3$

42. $|x - 1| < 1$; $|y - 1| < 1$

C **43.** $|x| < |y|$

44. $|y| \le |x|$

45. $|x| + |y| \le 2$

Mixed Review Exercises

Each system has a unique solution; find it.

1. $2x + y = -3$; $x + 3y = 1$ (−2, 1)

2. $3x - y = 4$; $x + 2y = -8$ (0, −4)

3. $5x + 2y = 7$; $x - y = 7$ (3, −4)

Find an equation in standard form of the line through P having the given slope. Then graph the line.

4. $P(-1, 3)$, $m = 1$ $x - y = -4$

5. $P(2, 3)$, vertical $x = 2$

6. $P(0, 0)$, $m = -4$ $4x + y = 0$

7. $P(-6, -2)$, $m = \frac{2}{3}$

8. $P(-5, 7)$, $m = 0$ $y = 7$

9. $P(0, -3)$, $m = -\frac{1}{2}$

7. $2x - 3y = -6$ **9.** $x + 2y = -6$

6.
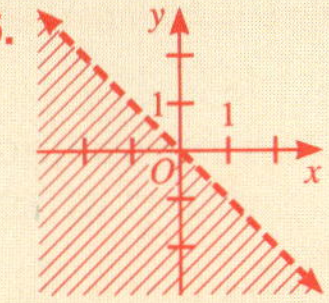

8.
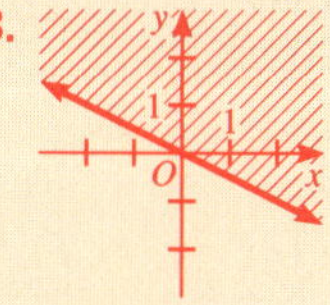

10.
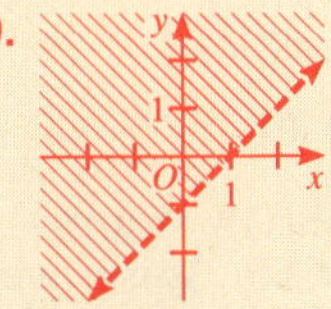

12.
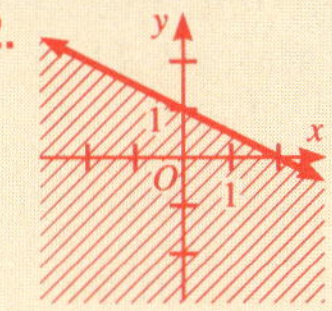

14.
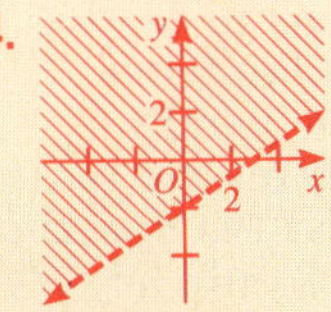

16.
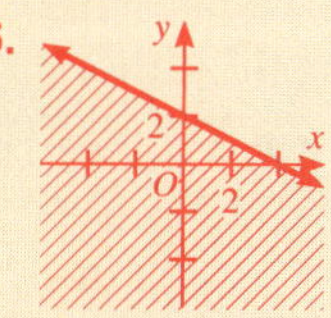

18.
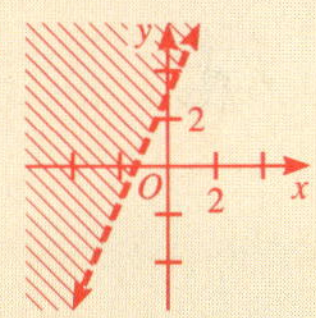

(continued on p. 160)

Quick Quiz

Solve. If there is no solution or an infinite number of solutions, say so.

1. $2x + y = 13$
   $3x + y = 20$ $(7, -1)$
2. $y + 17 = 5x$
   $y - 5x = 9$ no solution
3. $5x + 6y = 12$
   $6(3 - y) = 6 + 5x$ infinite
4. $3x + 2y = 10$
   $5x + 6y = 26$ $\left(1, \frac{7}{2}\right)$
5. Sue bought 25 stamps costing \$4.55. Some were 15¢ stamps and the rest were 25¢ stamps. How many of each did she buy? 8 25¢ stamps, 17 15¢ stamps
6. A small plane can fly from Athens to Madrid in 9.5 h with the wind. Against the wind the same 1425 mi trip takes 12.5 h. Find the speed of the plane in still air. 132 mi/h

Graph.

7. $8x - 3y \leq 12$

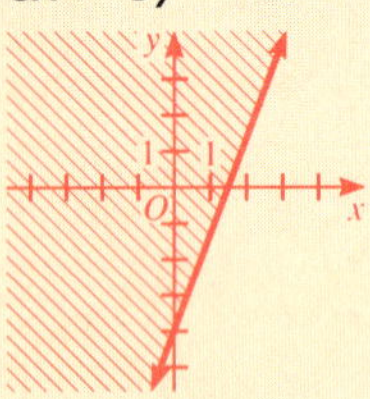

8. $y - x < 6$
   $x < 2$
   $y + x > -3$

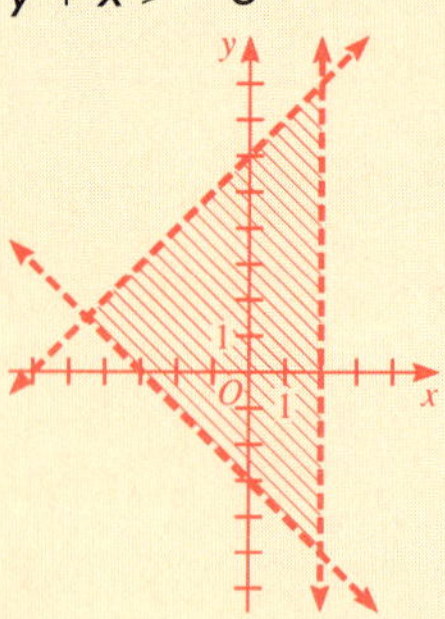

Self-Test 2

Vocabulary

system of linear equations (p. 124)
linear system (p. 124)
solution of a system (p. 124)
simultaneous solution (p. 124)
equivalent systems (p. 124)
linear combination (p. 125)
consistent (p. 128)
inconsistent (p. 128)
dependent (p. 128)
linear inequality in two variables (p. 135)
associated equation (p. 135)
open half-plane (p. 135)
boundary (p. 135)
closed half-plane (p. 136)
system of inequalities in two variables (p. 137)

Solve each system. If the system has an infinite solution set, specify it and give three solutions. If the system has no solution, say so.

Obj. 3-5, p. 124

1. $4x - y = 12$
   $3x + 4y = 66$ $(6, 12)$
2. $6x + 3y = 5$
   $4x + 2y = 9$ No sol.
3. $4x = 10 - 3y$
   $12x = 30 - 9y$ $\{(x, y): 4x + 3y = 10\}$; $\left(0, \frac{10}{3}\right)$, $\left(\frac{5}{2}, 0\right)$, $(1, 2)$
4. $8x + 3y = 5$
   $5x + 4y = 18$ $(-2, 7)$

Obj. 3-6, p. 131

5. If it takes an airplane 3 h to fly 360 mi with the wind and 4 h to make the return trip against the wind, find the speed of the wind and the air speed of the airplane. wind speed, 15 mi/h; air speed, 105 mi/h
6. Last summer Bret earned \$32 a day and Sandra earned \$36 a day. Together they earned a total of \$3560. How much did each earn, if Bret worked 5 days more than Sandra? Bret, \$1760; Sandra, \$1800

7. Graph $2x + 3y < 12$.
8. Graph $5x - 2y \geq -10$.

Obj. 3-7, p. 135

9. Graph $y > 0$ in a coordinate plane.
10. Graph this system: $x + 2y \leq 6$
    $x - 3y \geq 4$

Check your answers with those at the back of the book.

Challenge

Suppose that you want to rank four people (Alicia, Beth, Carlos, and Dave) in order from youngest to oldest. No two of them have the same age. They won't tell you their ages, but they will answer questions like "Dave, are you older than Beth?" What is the least number of such questions with which you can be *sure* of finding out the order of their ages? Describe your method for finding the order. 5

Functions and Relations

3-8 Functions

Objective To find values of functions and to graph functions.

The post office provides an everyday example of a *function:* The cost of mailing a letter is a function of its weight. Here there is a correspondence, cost to weight, given by a definite rule. The mathematical concept of function involves such a correspondence between two sets.

The **mapping diagram** at the right pictures a correspondence between two sets, D and R. A rule that defines this particular correspondence is:

To each whole number, assign
the number of letters in its name.

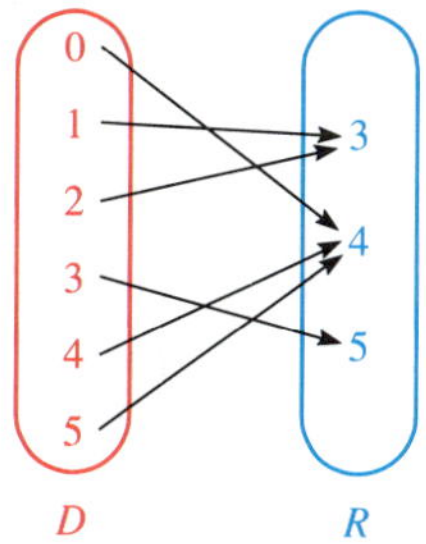

For example, 5 is assigned to 3 because there are 5 letters in the word "three." Such a pairing is called a *mapping,* or a *function.*

In Lesson 3-10, you will learn a general definition of function stated in terms of ordered pairs. Here you will work with a definition that emphasizes a correspondence between two sets.

A **function** is a correspondence between two sets, D and R, that assigns to each member of D exactly one member of R.

The set D is the **domain** of the function. The **range** of the function is the set of all members of R assigned to at least one member of D. *In this book, domains and ranges of functions will be sets of real numbers, unless otherwise stated.*

To define a function, you can give a rule describing the correspondence and then specify the domain. For example, if the function with the rule

"double the number"

has the set of integers as its domain, it assigns 4 to 2, 2 to 1, 0 to 0, -2 to -1, -4 to -2, and so on, as illustrated at the right. The range of this function is the set of even integers.

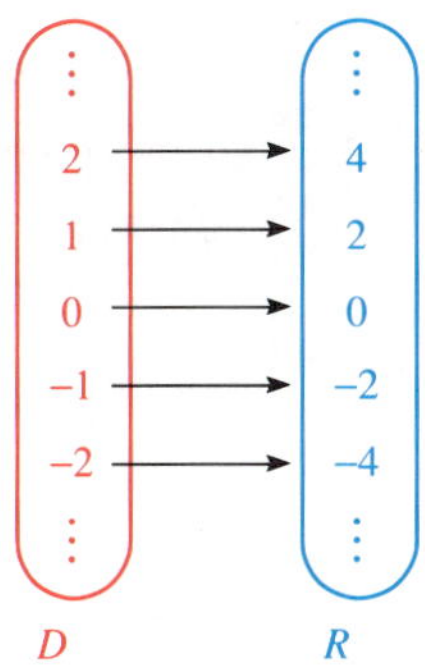

Explorations p. 834

Teaching Suggestions p. T91

Reading Algebra p. T91

Suggested Extensions p. T91

Warm-Up Exercises

Complete the table.

x	$4 - 2x$	(x, y)
-1	$4 - 2(-1)$	$(-1, 6)$
0	$4 - 2(0)$	$(0, 4)$
1	$4 - 2(1)$	$(1, 2)$
2	$4 - 2(2)$	$(2, 0)$
3	$4 - 2(3)$	$(3, -2)$

Motivating the Lesson

Tell students that when they match their numerical grade with a letter grade, they are using the concept of a mapping or a function.

Chalkboard Examples

1. Give the range of and graph $f: x \to -\frac{1}{2}x + 2$ if the domain is $\{0, 1, 2, 3, 4\}$.

x	$-\frac{1}{2}x + 2$	(x, y)
0	$0 + 2 = 2$	$(0, 2)$
1	$-\frac{1}{2} + 2 = \frac{3}{2}$	$\left(1, \frac{3}{2}\right)$
2	$-1 + 2 = 1$	$(2, 1)$
3	$-\frac{3}{2} + 2 = \frac{1}{2}$	$\left(3, \frac{1}{2}\right)$
4	$-2 + 2 = 0$	$(4, 0)$

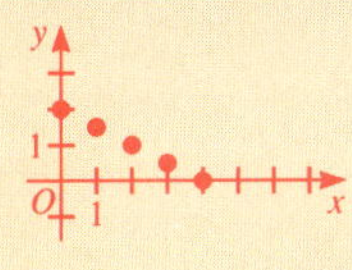

(continued)

Chalkboard Examples
(continued)

2. Given $f: x \to 5x - 1$ and $g: x \to 2 + x^2$, with both f and g having the real numbers as their domain, find the following.
   a. $f(10)$ b. $g(7)$
   c. $g(f(3))$ d. $f(g(3))$
   a. $5(10) - 1 = 50 - 1 = 49$
   b. $2 + 7^2 = 2 + 49 = 51$
   c. $f(3) = 5(3) - 1 = 14$; $g(14) = 2 + 14^2 = 2 + 196 = 198$; $g(f(3)) = 198$
   d. $g(3) = 2 + 3^2 = 2 + 9 = 11$; $f(11) = 5(11) - 1 = 55 - 1 = 54$; $f(g(3)) = 54$

For each domain, state the range of $f: x \to |x - 3|$.

3. $D = \{-2, -1, 0, 1, 2\}$
   $R = \{5, 4, 3, 2, 1\}$
4. $D = \{x: x > 3\}$
   $R = \{f(x): f(x) > 0\}$
5. $D = \{\text{real numbers}\}$
   $R = \{\text{nonnegative real numbers}\}$

Check for Understanding

Here is a suggested use of the Oral Exercises to check students' understanding as you teach the lesson.
Oral Exs. 1–10: use after Example 2.
Oral Exs. 11–17: use after Example 4.

The letters f, g, F, G, and ϕ (Greek phi) are often used to name functions. If the rule of a function is given by an algebraic expression, you can describe it using "arrow notation." For example, if g is the name of the doubling function, you can write

$$g: x \to 2x.$$

This is read "g, the function that assigns to x the number $2x$."

The **graph of a function** f is the set of all points (x, y) such that x is in the domain of f and the rule of the function assigns y to x.

Example 1 Given $f: x \to 4x - x^2$ with domain $D = \{1, 2, 3, 4, 5\}$.
a. Find the range of f. **b.** Graph f.

Solution

a. Make a table showing the numbers that the rule of f assigns to each member of the domain.

x	$4x - x^2$	(x, y)
1	$4 \cdot 1 - 1^2 = 3$	$(1, 3)$
2	$4 \cdot 2 - 2^2 = 4$	$(2, 4)$
3	$4 \cdot 3 - 3^2 = 3$	$(3, 3)$
4	$4 \cdot 4 - 4^2 = 0$	$(4, 0)$
5	$4 \cdot 5 - 5^2 = -5$	$(5, -5)$

From the second column, $R = \{-5, 0, 3, 4\}$. ***Answer***

b. Graph the ordered pairs from the table.

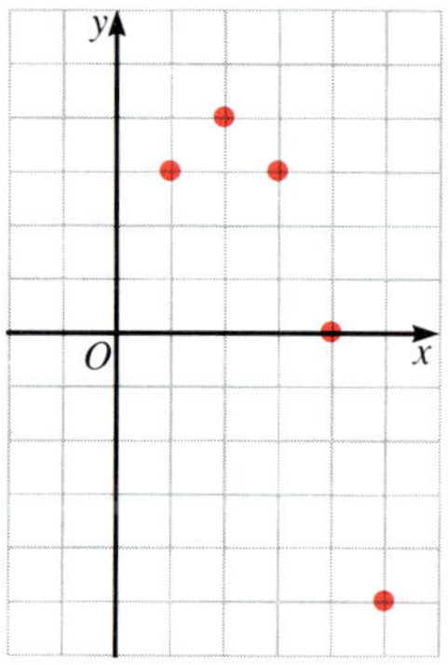

The **values** of a function are the members of its range. Thus, in Example 1, f assigns to 2 the *value* 4. To indicate this you use **functional notation** and write $f(2) = 4$, read "the value of f at 2 is 4" or "f of 2 is 4." In general, for any function f,

$$f(x) \text{ denotes the value of } f \text{ at } x.$$

Caution: $f(2)$ is *not* the product of f and 2.

Example 2 Given $f: x \to 2x - 1$ and $g: x \to x^2 + 1$, find the indicated values.
a. $f(2)$ **b.** $g(2)$ **c.** $f(a)$ **d.** $f(g(2))$ **e.** $g(f(2))$

Solution
a. $f(2) = 2 \cdot 2 - 1 = 3$
b. $g(2) = 2^2 + 1 = 5$
c. $f(a) = 2a - 1$
d. Use (b): $f(g(2)) = f(5) = 2 \cdot 5 - 1 = 9$
e. Use (a): $g(f(2)) = g(3) = 3^2 + 1 = 10$

Often a function f is defined simply by giving an equation for $f(x)$ and specifying the domain, as in Example 3.

Example 3 Given $f(x) = |x - 1|$ with domain $D = \{-2, -1, 0, 1, 2\}$.
a. Find the range of f. **b.** Graph f.

Solution

a.

x	$\lvert x - 1\rvert = f(x)$	(x, y)
-2	$\lvert -2 - 1\rvert = 3$	$(-2, 3)$
-1	$\lvert -1 - 1\rvert = 2$	$(-1, 2)$
0	$\lvert 0 - 1\rvert = 1$	$(0, 1)$
1	$\lvert 1 - 1\rvert = 0$	$(1, 0)$
2	$\lvert 2 - 1\rvert = 1$	$(2, 1)$

b.

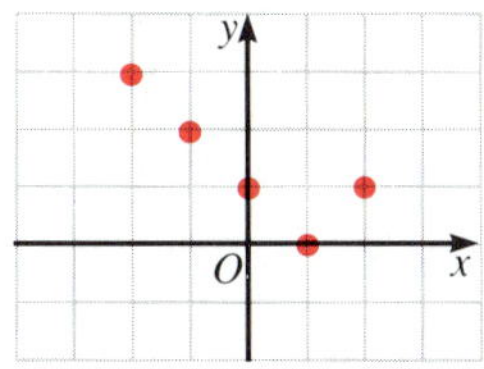

From the second column, $R = \{3, 2, 1, 0\}$. ***Answer***

Sometimes a function is defined by an equation and its domain is not specified. *For such functions, take as the domain the set of those real numbers for which the equation produces real numbers.*

Example 4 Give the domain of each function.

a. $f(x) = \dfrac{1}{x^2 + 4}$ **b.** $g(x) = \dfrac{1}{x(x + 2)}$ **c.** $h(x) = \sqrt{2 - x}$

Solution **a.** $D = \{\text{all real numbers}\}$

b. The expression $\dfrac{1}{x(x + 2)}$ is not defined when $x = 0$ and $x = -2$.
$\therefore D = \{\text{all real numbers except } 0 \text{ and } -2\}$

c. The expression $\sqrt{2 - x}$ is not defined when $2 - x < 0$, or $x > 2$.
$\therefore D = \{x\colon x \le 2\}$

Oral Exercises

For each function, find the indicated values.

1. $g\colon x \to x + 2$ **a.** $g(0)$ 2 **b.** $g(-2)$ 0 **c.** $g(-1)$ 1 **d.** $g(2)$ 4

2. $f\colon x \to 1 - x$ **a.** $f(-2)$ 3 **b.** $f(1)$ 0 **c.** $f(0)$ 1 **d.** $f(2)$ −1

3. $G\colon x \to 1 - 2x$ **a.** $G(-2)$ 5 **b.** $G(-1)$ 3 **c.** $G(0)$ 1 **d.** $G(3)$ −5

4. $F\colon x \to 3x - 2$ **a.** $F(1)$ 1 **b.** $F(0)$ −2 **c.** $F(-1)$ −5 **d.** $F(2)$ 4

5. $h\colon x \to 1 - x^2$ **a.** $h(1)$ 0 **b.** $h(-1)$ 0 **c.** $h(2)$ −3 **d.** $h(-2)$ −3

Guided Practice

State the range and draw the graph of each function.

1. $Q\colon x \to 8 - \frac{1}{2}x$;
$D = \{-4, -2, 0, 2, 4\}$
$R = \{6, 7, 8, 9, 10\}$

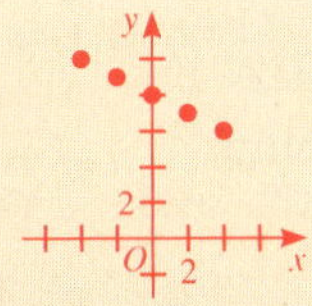

2. $G\colon x \to 5 - x^2$;
$D = \{0, 1, 2, 3\}$
$R = \{-4, 1, 4, 5\}$

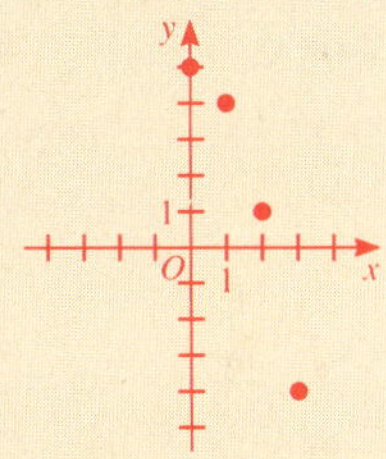

3. $H\colon x \to |2x + 1|$;
$D = \{-2, -1, 0, 1, 2\}$
$R = \{1, 3, 5\}$

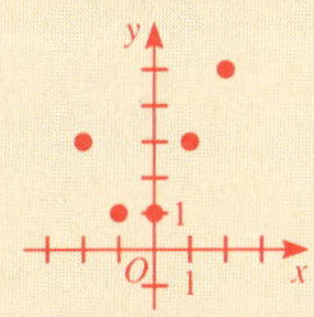

Summarizing the Lesson

In this lesson students studied functions, their notation, and their graphs. Ask students to write the equation $y = 3x + 4$ using function notation and arrow notation.

Suggested Assignments

Average Algebra
Day 1: 144/1–26
R 140/Self-Test 2
Day 2: 145/27–43 odd

Average Alg. and Trig.
144/3–42 mult. of 3
S 122/56, 57
R 140/Self-Test 2

Extended Algebra
144/3–42 mult. of 3
R 140/Self-Test 2

Extended Alg. and Trig.
144/3–42 mult. of 3, 48, 49
S 139/44, 45
R 140/Self-Test 2

Additional Answers Written Exercises

14. 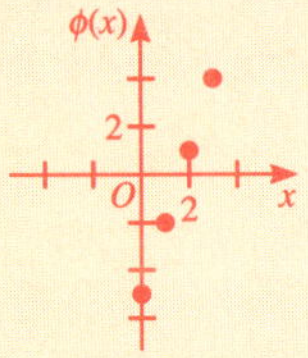

16. 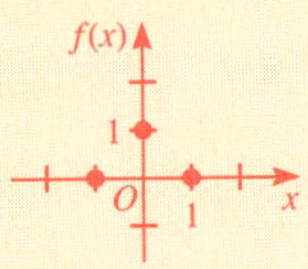

18. 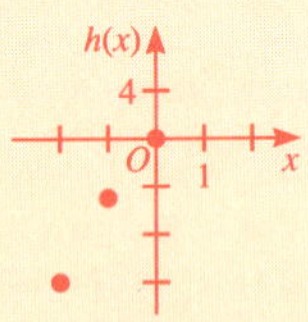

20. 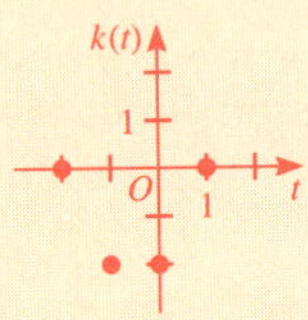

For each function, find the indicated values.

6. $\phi: x \to x - x^2$	**a.** $\phi(0)$ 0	**b.** $\phi(-1)$ −2	**c.** $\phi(2)$ −2	**d.** $\phi(-2)$ −6
7. $k: x \to \lvert 2 - x \rvert$	**a.** $k(0)$ 2	**b.** $k(-2)$ 4	**c.** $k(2)$ 0	**d.** $k(3)$ 1
8. $H: x \to 2 - \lvert x \rvert$	**a.** $H(0)$ 2	**b.** $H(-2)$ 0	**c.** $H(2)$ 0	**d.** $H(-3)$ −1
9. $s: x \to \dfrac{10}{x^2 + 1}$	**a.** $s(0)$ 10	**b.** $s(1)$ 5	**c.** $s(2)$ 2	**d.** $s(3)$ 1
10. $h: x \to \dfrac{1 - x}{x^3}$	**a.** $h(-1)$ −2	**b.** $h(0)$ undefined	**c.** $h(1)$ 0	**d.** $h(2)$ $-\frac{1}{8}$

11. $D = \{-2, -1, 0, 1, 2, 3, 4\}$; $R = \{1, 3\}$

Give the domain and range of the function whose complete graph is shown in red.

11. 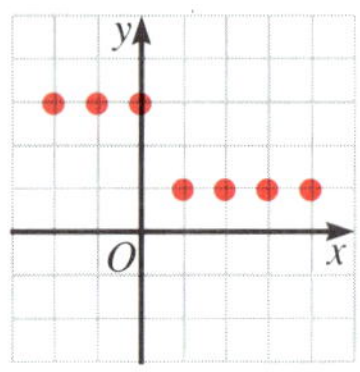

12. 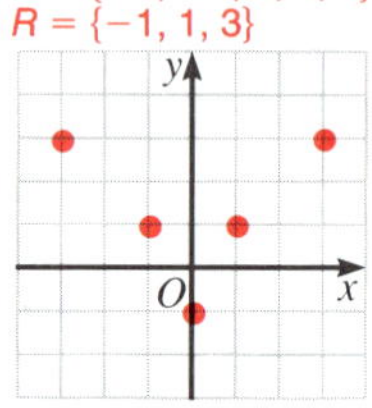

$D = \{-3, -1, 0, 1, 3\}$
$R = \{-1, 1, 3\}$

13. 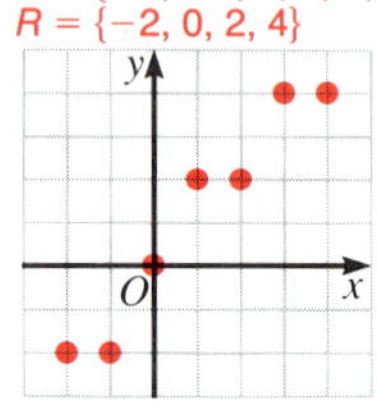

$D = \{-2, -1, 0, 1, 2, 3, 4\}$
$R = \{-2, 0, 2, 4\}$

14. 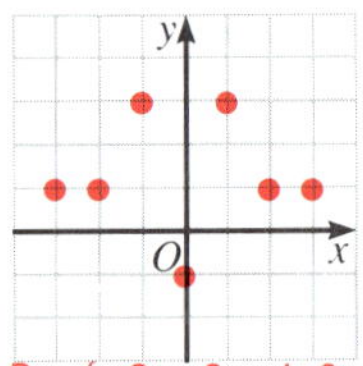

$D = \{-3, -2, -1, 0, 1, 2\ 3\}$
$R = \{-1, 1, 3\}$

15. 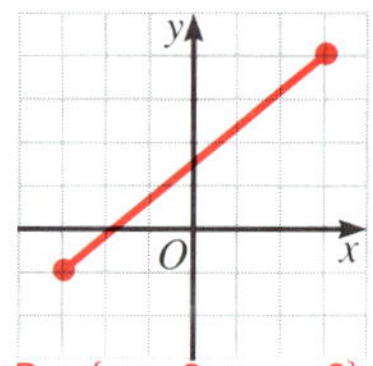

$D = \{x: -3 \le x \le 3\}$
$R = \{y: -1 \le y \le 4\}$

16. 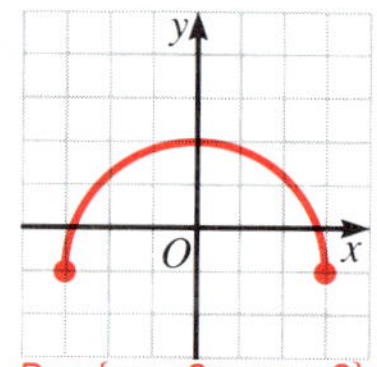

$D = \{x: -3 \le x \le 3\}$
$R = \{y: -1 \le y \le 2\}$

17. Suppose that the domain of a function has n members.

a. What is the least number of members in its range? 1

b. What is the greatest number of members in its range? n

Written Exercises

5. $\{-2, 0, 4\}$ 6. $\{-12, -5, 0\}$ 7. $\{0, 1, 4\}$ 8. $\{-2, 0\}$ 9. $\{0, 12\}$ 10. $\{-4, 0, 2\}$

In Exercises 1–12, the domains D and rules of some functions are given. Find the range of each function.

A

1. $F: x \to 3 - 2x$; $D = \{0, 1, 2, 3\}$ $\{-3, -1, 1, 3\}$

2. $\phi: x \to 3x - 5$; $D = \{0, 1, 2, 3\}$ $\{-5, -2, 1, 4\}$

3. $g: x \to x^2 - 2$; $D = \{-2, 0, 2\}$ $\{-2, 2\}$

4. $f: x \to 1 - x^2$; $D = \{-1, 0, 1\}$ $\{0, 1\}$

5. $f: x \to x^2 - 3x$; $D = \{-1, 0, 1, 2, 3\}$

6. $h: x \to 4x - x^2$; $D = \{-1, -2, 0\}$

7. $G: x \to x^2 - 4x + 4$; $D = \{0, 1, 2, 3\}$

8. $k: t \to t^2 + t - 2$; $D = \{-2, -1, 0, 1\}$

9. $g: x \to x^4 - x^2$; $D = \{-2, 0, 2\}$

10. $H: z \to z^2 - z^3$; $D = \{-1, 0, 1, 2\}$

11. $m: z \to 1 - \lvert z \rvert$; $D = \{-2, -1, 0, 2\}$ $\{-1, 0, 1\}$

12. $r: t \to \lvert 1 - t \rvert$; $D = \{-2, -1, 0, 1, 2\}$ $\{0, 1, 2, 3\}$

13–24. Graph the functions in Exercises 1–12.

Complete the equation for the function pictured in each mapping diagram.

B **25.** $f(x) = \underline{\ ?\ } \cdot x + 1$ 2

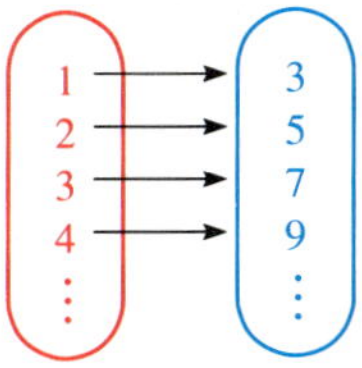

26. $g(x) = x \cdot \underline{\ ?\ }$ x

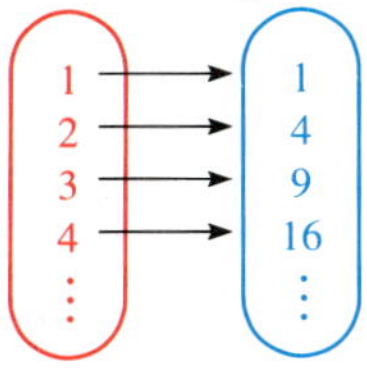

27. $h(x) = \underline{\ ?\ }^{x}$ 2

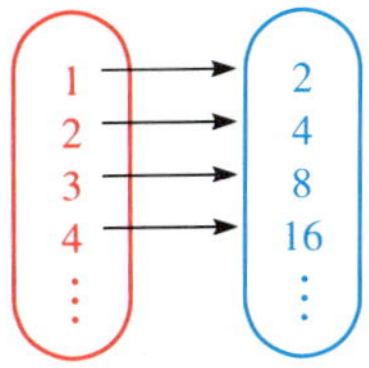

28. $F(x) = \underline{\ ?\ } \cdot x + 5$ −2

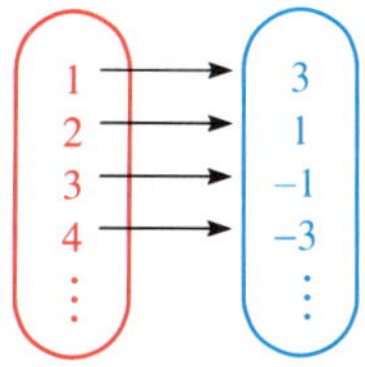

Give the domain of each function.

29. $f(x) = x^3 - 2$ {real numbers}

30. $g(x) = \frac{2}{x+3}$ $\{x: x \neq -3\}$

31. $h(x) = \frac{2}{x^2+3}$ {real numbers}

32. $F(x) = \sqrt{2x-1}$ $\left\{x: x \geq \frac{1}{2}\right\}$

33. $f(x) = \frac{1}{x^2 + 5x + 6}$ $\{x: x \neq -2 \text{ and } x \neq -3\}$

34. $G(x) = \frac{2}{(x-1)(x+2)}$ $\{x: x \neq 1 \text{ and } x \neq -2\}$

Let $f(x) = x^2 - 1$ and $g(x) = 1 - 2x$. Find the indicated values.

35. $f(g(1))$ and $g(f(1))$ 0, 1

36. $f(g(-1))$ and $g(f(-1))$ 8, 1

37. $f(g(2))$ and $g(f(2))$ 8, −5

38. $f(g(-2))$ and $g(f(-2))$ 24, −5

39. $f(f(2))$ and $f(2f(1))$ 8, −1

40. $g(g(2))$ and $g(2g(1))$ 7, 5

C **41.** $f\left(\frac{1}{g(3)}\right)$ $-\frac{24}{25}$

42. $g\left(\frac{1}{f(3)}\right)$ $\frac{3}{4}$

43. $g(a+1) - g(a)$ −2

44. $f(a+1) - f(a)$ $2a + 1$

45. $\frac{f(x+h) - f(x)}{h}$ $2x + h$

46. $\frac{g(x+h) - g(x)}{h}$ −2

47. If f is a function such that $f(a+b) = f(a) + f(b)$, for all real numbers a and b, then show that the following statements are true.

a. $f(0) = 0$ (*Hint:* Let $a = b = 0$.)

b. $f(2a) = 2f(a)$ (*Hint:* Let $b = a$.)

c. $f(-a) = -f(a)$

48. The **greatest-integer function** is defined for all real numbers x as follows:

$$g(x) = \text{the greatest integer } \textit{not exceeding } x$$

For example: $g(2) = 2$, $g\left(2\frac{1}{2}\right) = 2$, $g(3) = 3$, $g(\pi) = 3$, and $g\left(-2\frac{1}{2}\right) = -3$. Graph the greatest-integer function for $-3 \leq x \leq 3$.

49. The **signum function** is defined for all real numbers x as follows:

$$s(x) = \begin{cases} 1 \text{ if } x > 0 \\ 0 \text{ if } x = 0 \\ -1 \text{ if } x < 0 \end{cases}$$

Graph the signum function for $3 \leq x \leq 3$. How does the signum function differ from the function $f(x) = \frac{x}{|x|}$?

22.

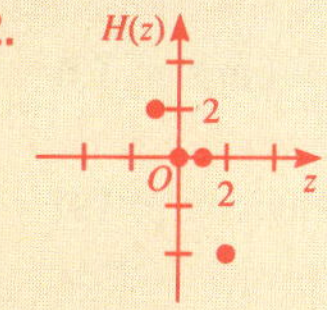

24.

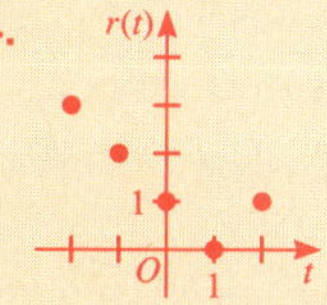

47. a. $f(0) = f(0+0) = f(0) + f(0)$. Since $f(0) = f(0) + f(0)$, $f(0)$ is an additive identity, i.e. $f(0) = 0$.

b. $f(2a) = f(a+a) = f(a) + f(a) = 2f(a)$

c. $f(a + (-a)) = f(a) + f(-a)$; $f(0) = f(a) + f(-a)$; from part (a), $f(0) = 0$; $0 = f(a) + f(-a)$; $f(-a) = -f(a)$

Supplementary Materials

Study Guide	pp. 47–48
Practice Master	16

Teaching Suggestions p. T91

Suggested Extensions p. T92

Warm-Up Exercises

Find the indicated values.

1. F: $x \rightarrow 2x - 1$; $F(14)$, $F(0)$, $F(-8)$ 27; −1; −17
2. G: $x \rightarrow 3 - 4x$; $G(-3)$, $G(2)$, $G(0)$ 15; −5; 3
3. h: $x \rightarrow x^2 - x$: $h(0)$, $h(1)$, $h(-5)$ 0; 0; 30
4. g: $x \rightarrow 2x - 3$; $g\left(\frac{1}{2}\right)$, $g(0)$, $g(-3)$ −2; −3; −9
5. f: $x \rightarrow \frac{x+2}{4}$; $f(0)$, $f(1)$, $f(-2)$ $\frac{1}{2}$; $\frac{3}{4}$; 0

Motivating the Lesson

Tell students that some real life situations can be represented mathematically by linear equations. Examples are given on pages 147 and 148.

Problem Solving Strategies

Many problem solving strategies can be used to solve problems involving linear functions. Among them are the *five-step plan, drawing a diagram,* applying a *standard formula,* and *generalizing* from a specific example.

Common Errors

Some students may need to be reminded that $f(x)$ denotes the value of f at x. For example, $f(3) = 5$ indicates that f assigns to 3 the value 5. Also show that $f(3) = 5$ can be written as the ordered pair (3, 5).

3-9 Linear Functions

Objective To find equations of linear functions and to apply properties of linear functions.

The graph of the function $f(x) = 2x - 3$ is the set of all points (x, y) such that $y = 2x - 3$. This graph is the line with slope 2 and y-intercept -3, as shown at the right. A function such as f, which has a straight line for its graph, is called a *linear function*.

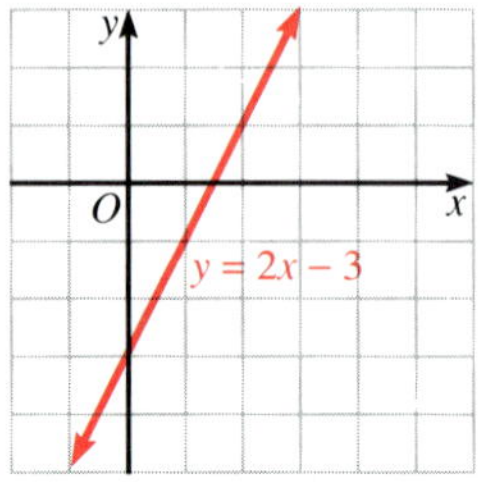

A **linear function** is a function f that can be defined by

$$f(x) = mx + b$$

where x, m, and b are any real numbers. The graph of f is the graph of $y = mx + b$, a line with slope m and y-intercept b.

If $f(x) = mx + b$ and $m = 0$, then $f(x) = b$ for all x, and f is a **constant function.** Its graph is the horizontal line $y = b$.

For any linear function f that is not a constant function, a change in the value of x produces a change in the value of $f(x)$. For a linear function $f(x) = mx + b$, the *rate of change of $f(x)$ with respect to x* is the constant m. (See Exercise 32.) The tables below illustrate this fact.

$f(x) = 2x - 3$, $m = 2$

x	$f(x)$
0	−3
1	−1
2	1
3	3
4	5

(Braces: x from 0 to 1: 1; $f(x)$ from −3 to −1: 2; x from 1 to 4: 3; $f(x)$ from −1 to 5: 6)

When x increases by 1, $f(x)$ changes by $2 \cdot 1$, or 2. When x increases by 3, $f(x)$ changes by $2 \cdot 3$, or 6.

$f(x) = -5x + 7$, $m = -5$

x	$f(x)$
0	7
1	2
2	−3
3	−8
4	−13

(Braces: x from 0 to 1: 1; $f(x)$ from 7 to 2: −5; x from 1 to 4: 3; $f(x)$ from 2 to −13: −15)

When x increases by 1, $f(x)$ changes by $-5 \cdot 1$, or -5. When x increases by 3, $f(x)$ changes by $-5 \cdot 3$, or -15.

Notice that changing x by an amount a changes $f(x)$ by a corresponding amount $m \cdot a$. In general, to find the change in $f(x)$ you multiply the change in x by the rate of change m. If m is positive, the change in $f(x)$ is an increase; if m is negative, the change in $f(x)$ is a decrease.

When you know the change in $f(x)$ produced by some change in x, you can find the rate of change m by using this formula:

$$\text{rate of change } m = \frac{\text{change in } f(x)}{\text{change in } x}$$

Example 1 Find equations of the linear functions f and g using the given information.

a. $f(4) = 1$ and $f(8) = 7$

b. $g(0) = 5$ and an increase of 4 units in x causes a decrease of 12 units in $g(x)$.

Solution **a.** Since f is linear, let $f(x) = mx + b$. First find the value of m.

$$\text{rate of change } m = \frac{\text{change in } f(x)}{\text{change in } x}$$

$$m = \frac{f(8) - f(4)}{8 - 4} = \frac{7 - 1}{4} = \frac{3}{2}$$

Thus, $m = \frac{3}{2}$ and $f(x) = \frac{3}{2}x + b$. Now find the value of b.

Since $f(4) = 1$, $\quad 1 = \frac{3}{2}(4) + b$

$$1 = 6 + b$$

$$-5 = b$$

$\therefore f(x) = \frac{3}{2}x - 5$ ***Answer***

b. Since g is linear, let $g(x) = mx + b$. First find m.

$$\text{rate of change } m = \frac{\text{change in } g(x)}{\text{change in } x} = \frac{-12}{4} = -3$$

Thus, $m = -3$ and $g(x) = -3x + b$. Now find b.

Since $g(0) = 5$, $\quad 5 = -3(0) + b$

$$5 = b$$

$\therefore g(x) = -3x + 5$ ***Answer***

Many real-life situations can be described by linear functions. Here are some examples:

1. A manufacturer of television sets spends \$6000 per day in fixed costs and \$250 for each set produced. The total cost C for producing x sets a day is given by

$$C(x) = 6000 + 250x.$$

Notice that the domain of the linear function C is restricted to positive integers, since x represents a number of television sets.

Chalkboard Examples

1. Use the given facts to find a formula for the linear function f: the slope of the graph of f is $\frac{1}{3}$ and $f(6) = 0$.
Since f is linear, $f(x) = mx + b$. $m = \frac{1}{3}$, so $f(x) = \frac{1}{3}x + b$. Since $f(6) = 0$, $0 = \frac{1}{3}(6) + b$ or $0 = 2 + b$. Thus, $b = -2$.
$\therefore f(x) = \frac{1}{3}x - 2$.

2. On an endurance test a person scored 140 on a scale of 0 to 175. Find a function rule and use it to convert this score to a scale of 0 to 100.
Let $S(x)$ be the score on a scale of 0 to 100 when x is the score on a scale of 0 to 175. Then $S(0) = 0$ and $S(175) = 100$.
$m = \frac{S(175) - S(0)}{175 - 0} = \frac{100}{175} = \frac{4}{7}$.
$\therefore S(x) = \frac{4}{7}x + b$.
Since $S(0) = 0$,
$0 = \frac{4}{7}(0) + b$ and
$b = 0$.
$\therefore S(x) = \frac{4}{7}x$ and a score of 140 converts to a score of 80.

Check for Understanding

Here is a suggested use of the Oral Exercises to check students' understanding as you teach the lesson.
Oral Exs. 1–8: use after Example 1.

Guided Practice

Give a linear function in the form $f(x) = mx + b$ for each of the following.

1. $m = -3$, $f(0) = 12$
   $f(x) = -3x + 12$
2. $m = 3$, $f(0) = 9$
   $f(x) = 3x + 9$
3. $m = 4$, $f(2) = 7$
   $f(x) = 4x - 1$
4. $f(0) = 3$, $f(4) = 1$
   $f(x) = -\frac{1}{2}x + 3$
5. $f(3) = 12$, $f(5) = -2$
   $f(x) = -7x + 33$
6. $f(3) = 2$, $f(12) = 4$
   $f(x) = \frac{2}{9}x + \frac{4}{3}$
7. Write a linear function to describe the following and then use it to answer the question.
   A piece of machinery loses $15 in value every month. If its value after one year is $2250, what was its initial value?
   $f(x) = 2430 - 15x$; $2430

Summarizing the Lesson

Students have learned to find equations of linear functions and to use them to solve problems. Ask students to give further examples of everyday situations that can be described by linear equations.

2. A computer purchased new for $3600 loses $600 in value each year (straight-line depreciation). Its remaining worth, or book value B, y years after purchase is given by

$$B(y) = 3600 - 600y.$$

Notice that the domain of B is restricted to $\{y: 0 \le y \le 6\}$, since after 6 years the book value has reached $0.

3. A hiker climbing Mount Jefferson left the 4000-foot level at 6 A.M. and gained elevation at the rate of 800 ft/h. The hiker's elevation E at any time t hours after 6 A.M. is given by

$$E(t) = 4000 + 800t.$$

Here t is restricted also, because the hiker's elevation cannot exceed the height of the mountain.

It is customary to call functions such as C, B, and E linear functions even though the domain of each is not the set of all real numbers.

In Example 2, an equation of a linear function is found using a method different from that shown in Example 1.

Example 2 At noon a pump started emptying an oil storage tank at a constant rate. At 2 P.M. there were 1680 barrels left in the tank. At 5 P.M. there were 600 barrels left.

a. Write a linear function to describe this situation.
b. How many barrels were in the tank at noon?
c. What time will the tank be empty?

Solution

a. Let $f(t)$ = the number of barrels of petroleum in the tank t hours after noon. Then, since f is linear, $f(t) = mt + b$.

To find the values of m and b, use the facts that $f(2) = 1680$ and $f(5) = 600$ to write a system of linear equations:

$$f(2) = 2m + b = 1680$$
$$f(5) = 5m + b = 600$$

Solve this system to obtain $m = -360$ and $b = 2400$. Therefore,

$$f(t) = -360t + 2400. \quad \textit{Answer}$$

b. At noon, $t = 0$, and

$$f(0) = -360(0) + 2400 = 2400.$$

$\therefore$ there were 2400 barrels in the tank at noon. ***Answer***

c. The tank will be empty when $f(t) = 0$.

$$0 = -360t + 2400$$
$$360t = 2400$$
$$t = \frac{2400}{360} = 6\frac{2}{3} \text{ (hours after noon)}$$

$\therefore$ the tank will be empty at 6:40 P.M. ***Answer***

Oral Exercises

Each table is a table of values of a *linear* function. Supply the missing entries.

Sample

x	$f(x)$
0	3
3	5
6	?
12	?

(Braces beside the x column: 3, 3, 6; beside the $f(x)$ column: 2)

Solution

The rate of change $m = \frac{2}{3}$. This means an increase of 3 in x (from 3 to 6) causes an increase of $3 \cdot \frac{2}{3}$, or 2, in $f(x)$: from 5 to 7. An increase of 6 in x (from 6 to 12) causes an increase of $6 \cdot \frac{2}{3}$, or 4, in $f(x)$: from 7 to 11. $\therefore$ the missing entries are 7 and 11.

1.

x	$f(x)$
1	1
2	4
3	7 ?
4	10 ?

2.

x	$f(x)$
0	3
1	1
2	−1?
3	−3?

3.

x	$f(x)$
0	5
2	4
4	3 ?
6	2 ?

4.

x	$f(x)$
2	−1
4	1
6	3 ?
8	5 ?

5.

x	$f(x)$
0	2 ?
1	2
3	2 ?
6	2

6.

x	$f(x)$
0	−4 ?
−1	−1
−2	2
−3	5 ?

7.

x	$f(x)$
2	−5
4	−2 ?
6	1 ?
8	4

8.

x	$f(x)$
1	6
−1	2 ?
−3	−2 ?
−5	−6

Written Exercises

Find an equation of the linear function f using the given information.

A

1. $m = 2$, $b = 3$ $f(x) = 2x + 3$
2. $m = -1$, $b = \frac{1}{2}$ $f(x) = -x + \frac{1}{2}$
3. $m = 3$, $f(0) = 1$ $f(x) = 3x + 1$
4. $m = -\frac{3}{2}$, $f(0) = -\frac{1}{2}$ $f(x) = -\frac{3}{2}x - \frac{1}{2}$
5. $f(0) = -2$, slope of graph $= \frac{1}{2}$ $f(x) = \frac{1}{2}x - 2$
6. $f(0) = 1$, slope of graph $= -1$ $f(x) = -x + 1$
7. $f(0) = 1$; $f(x)$ increases by 6 when x increases by 3. $f(x) = 2x + 1$
8. $f(0) = -1$; $f(x)$ decreases by 3 when x increases by 1. $f(x) = -3x - 1$
9. $m = 2$, $f(1) = 5$ $f(x) = 2x + 3$
10. $m = -1$, $f(2) = 3$ $f(x) = -x + 5$
11. $m = -\frac{3}{2}$, $f(4) = -1$ $f(x) = -\frac{3}{2}x + 5$
12. $m = \frac{2}{3}$, $f(6) = -2$ $f(x) = \frac{2}{3}x - 6$

Suggested Assignments

Average Algebra
Day 1: 149/2–30 even
S 145/49
Day 2: 149/5, 21, 23, 27, 29
150/*P*: 1–9 odd
S 152/Mixed Review

Average Alg. and Trig.
Day 1: 149/2–30 even
S 134/18, 19
Day 2: 150/*P*: 1–9 odd
S 152/Mixed Review

Extended Algebra
Day 1: 149/2–32 even
Day 2: 150/27, 29, 33
150/*P*: 1–11 odd
S 152/Mixed Review

Extended Alg. and Trig.
149/3–24 mult. of 3
150/*P*: 2, 4, 8–10
S 152/Mixed Review

Additional Answers Written Exercises

32. $\frac{f(x_2) - f(x_1)}{x_2 - x_1} = \frac{(mx_2 + b) - (mx_1 + b)}{x_2 - x_1} = \frac{mx_2 - mx_1}{x_2 - x_1} = \frac{m(x_2 - x_1)}{x_2 - x_1} = m$

33. a. If $m > 0$, then $m = \frac{f(x_2) - f(x_1)}{x_2 - x_1} > 0$; if $x_2 > x_1$, then $x_2 - x_1 > 0$ and $f(x_2) - f(x_1) > 0$; so $f(x_2) > f(x_1)$, and f is increasing.

b. If $m < 0$, then $m = \frac{f(x_2) - f(x_1)}{x_2 - x_1} < 0$; if $x_2 > x_1$, then $x_2 - x_1 > 0$ and $f(x_2) - f(x_1) < 0$; so $f(x_2) < f(x_1)$, and f is decreasing.

Supplementary Materials

Study Guide pp. 49–50
Practice Masters 17, 18

13. $f(0) = 1$, $f(3) = 7$ $f(x) = 2x + 1$

14. $f(0) = -2$, $f(2) = 4$ $f(x) = 3x - 2$

15. $f(0) = -3$, $f(3) = -3$ $f(x) = -3$

16. $f(0) = 0$, $f(-2) = 4$ $f(x) = -2x$

17. $f(1) = 2$, $f(2) = 5$ $f(x) = 3x - 1$

18. $f(2) = 6$, $f(4) = 0$ $f(x) = -3x + 12$

19. $f(-2) = 3$, $f(2) = -3$ $f(x) = -\frac{3}{2}x$

20. $f(-1) = 2$, $f(2) = 2$ $f(x) = 2$

21. $f(-1) = 1$, $f(1) = -2$ $f(x) = -\frac{3}{2}x - \frac{1}{2}$

22. $f(3) = 4$, $f(6) = -3$ $f(x) = -\frac{7}{3}x + 11$

Complete the table, given that g is a linear function. (*Hint:* See the Sample on page 149.)

B **23.**

x	$g(x)$
3	4
1	−2
0	−5 ?
−1 ?	−8

24.

x	$g(x)$
2	1
4	4
0 ?	−2
−2	−5?

25.

x	$g(x)$
−4	4
0 ?	2
2	1
3	$\frac{1}{2}$?

26.

x	$g(x)$
5 ?	−3
2	−1
1	$-\frac{1}{3}$?
−1	1

In Exercises 27–30, f is a linear function.

27. If $f(6) = 7$ and $f(3) = 2$, find $f(-3)$ and $f(10)$. $f(-3) = -8$; $f(10) = \frac{41}{3}$

28. If $f(2) = 10$ and $f(10) = -2$, find $f(-10)$ and $f(100)$. $f(-10) = 28$; $f(100) = -137$

29. If $f(1) = 2$ and $f(7) = -6$, find $f(10)$ and $f(20)$. $f(10) = -10$; $f(20) = -\frac{70}{3}$

30. If $f(5) = -5$ and $f(-25) = -25$, find $f(2)$ and $f(50)$. $f(2) = -7$; $f(50) = 25$

C **31.** If $f(x) = mx + b$ and $k \neq 0$, find $\frac{f(x + k) - f(x)}{k}$. m

32. If $f(x) = mx + b$ and $x_1 \neq x_2$, show that $\frac{f(x_2) - f(x_1)}{x_2 - x_1} = m$.

33. A function f is *increasing* if $f(x_2) > f(x_1)$ whenever $x_2 > x_1$. A function f is *decreasing* if $f(x_2) < f(x_1)$ whenever $x_2 > x_1$. Show that if $f(x) = mx + b$, then **(a)** f is increasing if $m > 0$, and **(b)** f is decreasing if $m < 0$.

Problems

Solve Problems 1–11. Assume that each situation can be described by a linear function.

A **1.** A photocopying machine purchased new for \$4500 loses \$900 in value each year.

a. Find the book value of the machine after 18 months. \$3150

b. When will the book value be \$1200? 3 years, 8 months

2. The charge for a one-day rental of a car from Shurtz Car Rental Agency is \$24 plus 15 cents for each mile driven.

a. If Jill drives 85 mi in one day, how much is the charge? \$36.75

b. If a one-day rental cost Jill \$42, how far did she drive? 120 mi

3. A plumber charged \$110 for a three-hour job and \$160 for a five-hour job. At this rate, how much would he charge for an eight-hour job? \$235

4. It costs Ace Electronics Company \$1900 to manufacture 10 VCRs and \$2200 to manufacture 16 VCRs. At this rate, what would be the cost of manufacturing 25 VCRs? \$2650

5. Allied Airlines charges \$90 for a ticket to fly between two cities 260 mi apart and \$150 for a ticket to fly between two cities 500 mi apart. At this rate, what would it cost for a trip between two cities 1000 mi apart? \$275

6. A load of 8 kg attached to the bottom of a coil spring stretches the spring to a length of 76 cm, and a load of 14 kg stretches it to a length of 85 cm. Find the natural (unstretched) length of the spring. 64 cm

7. Fifteen days after Alan began a diet he weighed 176 lb. After 45 days he weighed 170 lb.
   a. How much did he weigh at the beginning of the diet? 179 lb
   b. At this rate, when will he weigh 165 lb? 70 days

8. A climber left base camp at 5 A.M. to ascend a 7400-meter peak. The climber gained altitude at the rate of 240 m/h and at 8 A.M. was at the 6500-meter level.
   a. Find the elevation of the base camp. 5780 m
   b. At what time did the climber reach the summit? 11:45 A.M.

B 9. The table below shows the freezing and boiling points of water in degrees Celsius (°C) and degrees Fahrenheit (°F).

	Celsius	Fahrenheit
freezing	0	32
boiling	100	212

   a. Find a linear function that gives °C in terms of °F. $c(f) = \frac{5}{9}f - \frac{160}{9}$
   b. What °C temperature corresponds to 98.6° F? 37° C
   c. At what temperature do the two scales give the same reading? $-40°$

10. On a 50 question true-false test, a student received 3 points for each correct answer. For each incorrect answer, the student lost 1 point. For an unanswered question, no points were added or deducted.
   a. Find the best score possible. 150
   b. Find the worst score possible. -50
   c. Find the linear function that converts the raw score described above to a score from 0 to 100. $f(x) = \frac{1}{2}x + 25$
   d. What score from 0 to 100 corresponds to a test result of 38 correct, 4 incorrect, and 8 omitted? 80

Additional Answers
Problems

11. c.

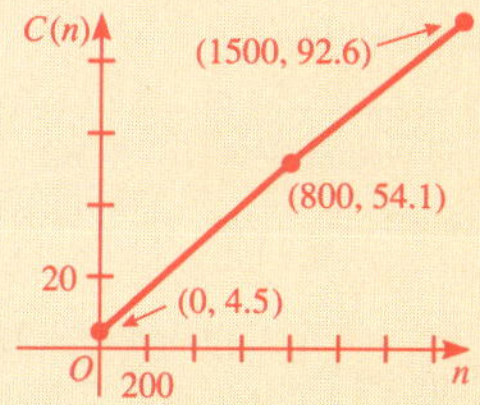

Additional Answers
Mixed Review Exercises

1.

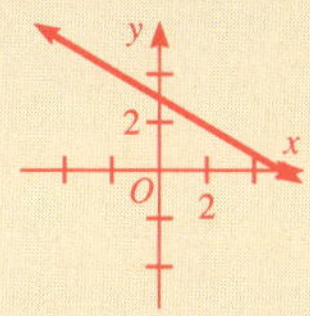

2.

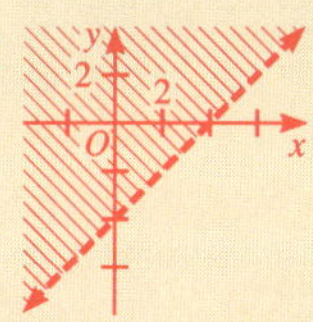

(continued)

Additional Answers
Mixed Review Exercises
(continued)

3.
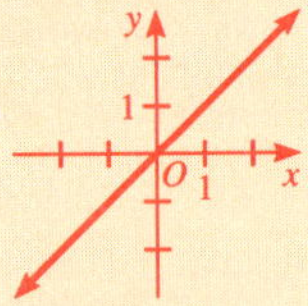

4.
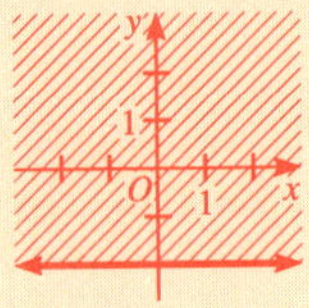

5.
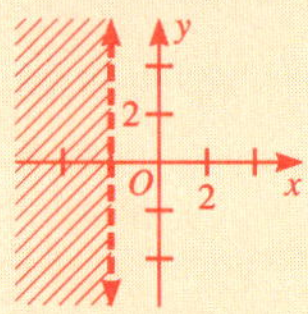

6.
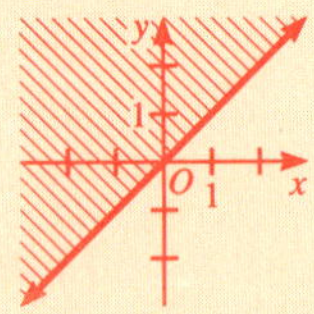

7.
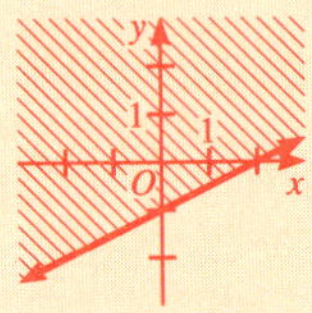

8.
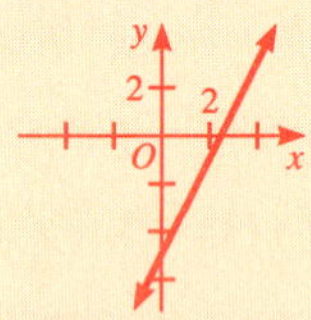

9.
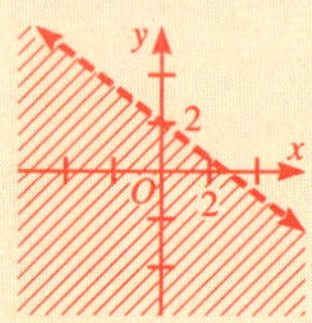

C **11.** South Coast Utilities makes the following monthly charges for residential electrical services:

Service Charge $4.50	
Energy Charge:	
First 800 kW · h	⟶ $0.062 per kW · h
All over 800 kW · h	⟶ $0.055 per kW · h

Express the monthly charge C in terms of the number n of kilowatt-hours used, for **(a)** $0 \le n \le 800$ and **(b)** $n > 800$. Then **(c)** draw a graph of $C(n)$ for $0 \le c \le 1500$. **a.** $C(n) = 4.5 + 0.062n$ **b.** $C(n) = 10.1 + 0.055n$

Mixed Review Exercises

Graph each open sentence.

1. $2x + 3y = 9$

2. $x - y < 4$

3. $y = x$

4. $y + 2 \ge 0$

5. $x + 2 < 0$

6. $y \ge x$

7. $y \ge \frac{1}{2}x - 1$

8. $2x - y = 5$

9. $3x + 4y < 8$

Find the range of each function with domain D.

10. $f: x \to 3x + 2$, $D = \{-2, -1, 0\}$ $\{-4, -1, 2\}$

11. $g: x \to 1 - x^2$, $D = \{-1, 0, 1\}$ $\{0, 1\}$

12. $F: x \to |x - 4|$, $D = \{-3, -2, -1\}$ $\{5, 6, 7\}$

13. $G: x \to 8 - 5x$, $D = \{1, 2, 3\}$ $\{-7, -2, 3\}$

Historical Note / *Functions*

In the seventeenth century, two different concepts of function developed. One was *geometric:* Mathematicians studied curves defined by geometric conditions (like the conic sections you will study in Chapter 9) and explored geometric properties of graphs such as their slopes and maximum and minimum points.

Another approach dealt with functions that could be defined by a formula, such as a polynomial in the variable x. The methods of calculus were used to analyze these and other basic functions, such as the exponential, logarithmic, and trigonometric functions you will meet in Chapters 10 and 12–14. It was Leibniz, a pioneer of this *analytic* approach, who introduced the term *function*.

As mathematicians became interested in a greater variety of functions, a more formal definition was needed. In 1837 the German mathematician Dirichlet proposed the following definition: y is a single-valued function of the variable x when a definite value of y corresponds to each value of x, no matter in what form this correspondence is specified.

Most modern treatments of functions use a definition based on ordered pairs, like the one stated on page 153.

3-10 Relations

Objective To graph relations and to determine when relations are also functions.

The mapping diagram, the table, and the graph below illustrate a correspondence between the members of the set $D = \{1, 2, 3\}$ and those of the set $R = \{0, 1, 2, 3, 4\}$.

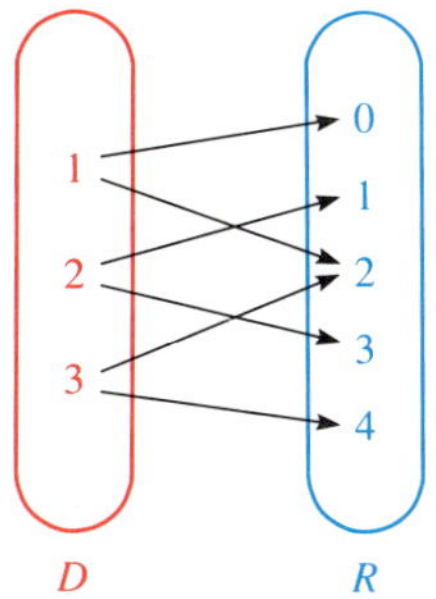

D	R
1	0
1	2
2	1
2	3
3	2
3	4

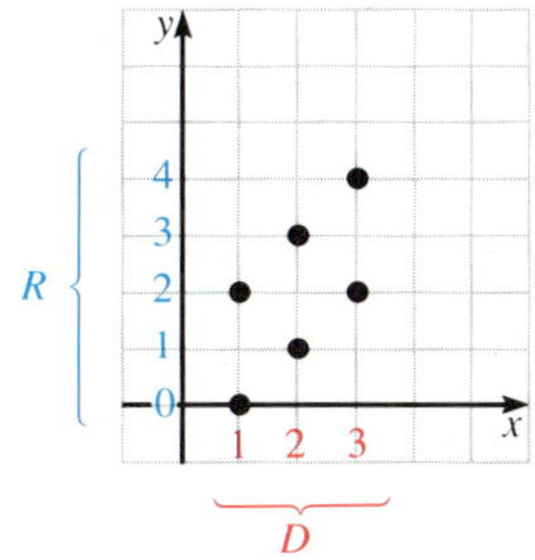

A rule that defines this particular correspondence is: "to each x in D assign a number y if and only if $|x - y| = 1$." Another way of describing this correspondence is to list the ordered pairs determined by the rule:

$$\{(1, 0), (1, 2), (2, 1), (2, 3), (3, 2), (3, 4)\}$$

A more efficient way of describing this set is to use *set notation:*

$$\{(x, y): |x - y| = 1 \text{ and } x \in D\}$$

This is read "the set of all ordered pairs (x, y) such that $|x - y| = 1$ and x is in D."

The correspondence just discussed does *not* assign exactly one member of R to each member of D. For example, both 0 and 2 are assigned to 1. Therefore, this correspondence is *not* a function. It is, however, a *relation*.

A **relation** is any set of ordered pairs. The set of first coordinates in the ordered pairs is the **domain** of the relation, and the set of second coordinates is the **range.**

A function can now be defined as a special kind of relation.

A **function** is a relation in which different ordered pairs have *different first coordinates*.

Teaching Suggestions p. T92

Suggested Extensions p. T92

Warm-Up Exercises

Find the range of each function with the given domain.

1. $f(x) = 3x + 1$ $\{-2, -1, 0, 1, 2\}$ $\{-5, -2, 1, 4, 7\}$
2. $g(x) = 2x$ $\{-3, -1, 0, 1, 3\}$ $\{-6, -2, 0, 2, 6\}$
3. $h(x) = 1 - x$ $\{-2, -1, 0, 1, 2\}$ $\{3, 2, 1, 0, -1\}$
4. $G: x = |x|$ $\{-6, -3, 0, 3, 6\}$ $\{6, 3, 0\}$

Motivating the Lesson

Tell students that some games of strategy are based on a grid system and use the concept of ordered pairs. For example, in a game in which you try to sink your opponent's "ship," the resulting pattern of "shots" could be the graph of a relation.

Common Errors

Some students have trouble determining whether a given relation is a function when two or more elements of the domain are assigned to one element of the range. When reteaching, tell students to consider each element of the domain separately and to determine whether there is a unique assignment for that element.

Chalkboard Examples

State the domain and the range and whether the relation is a function.

1. $\{(1, 2), (2, 2), (3, 1), (4, 1)\}$
   $D = \{1, 2, 3, 4\}$,
   $R = \{1, 2\}$, function

2. 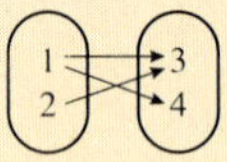

   $D = \{1, 2\}$; $R = \{3, 4\}$; not a function

3. 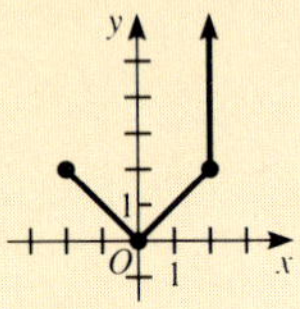

   $D = \{x: -2 \le x \le 2\}$;
   $R = \{y: y \ge 0\}$; not a function

4. $\{(x, y): |y| = |x|\}$
   $D = \{\text{real numbers}\}$;
   $R = \{\text{real numbers}\}$; not a function

5. $\left\{(x, y): y = \frac{1}{|x|}\right\}$
   $D = \{x: x \ne 0\}$,
   $R = \{y: y > 0\}$, function

Check for Understanding

Here is a suggested use of the Oral Exercises to check students' understanding as you teach the lesson.

Oral Exs. 1–18: use before Example 2.

Oral Exs. 19–24: use after Example 2.

Note that the definition just given does not refer to any rule. The set of ordered pairs $\{(1, 5), (3, 2), (5, 2)\}$ is a function even though no rule is given. The set of ordered pairs $\{(1, 5), (2, 3), (2, 5)\}$ is a relation but *not* a function, since $(2, 3)$ and $(2, 5)$ are different pairs with the *same* first coordinate.

When a relation is defined by an open sentence in two variables and a domain and range are not specified, *include in the domain and range those real numbers and only those real numbers for which the open sentence is true*. The relation is the solution set of the open sentence. The graph of the relation is the graph of the open sentence.

Example 1 Graph each relation, and then tell whether it is a function.

a. $\left\{(1, 1), (2, 1), \left(2, \frac{1}{2}\right), \left(3, \frac{1}{3}\right), \left(4, \frac{1}{4}\right)\right\}$

b. $\{(x, y): |x| + y = 3\}$

c. $\{(x, y): y > |x|\}$

Solution

a. This relation is *not* a function. The ordered pairs $(2, 1)$ and $\left(2, \frac{1}{2}\right)$ have the same first coordinate. ***Answer***

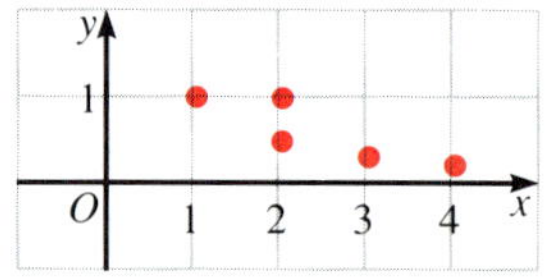

b. The open sentence $|x| + y = 3$ is equivalent to:

$$y = 3 - |x| = \begin{cases} 3 - x & \text{if } x \ge 0 \\ 3 + x & \text{if } x < 0 \end{cases}$$

This relation *is* a function, since each value of x is paired with exactly one value of y. ***Answer***

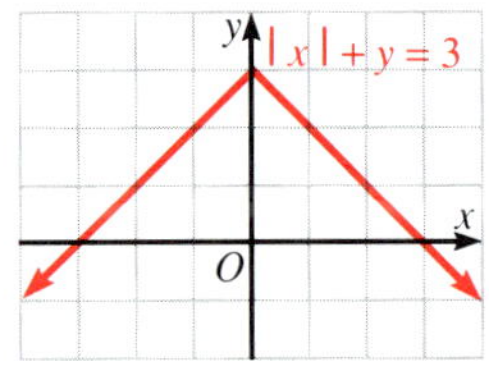

c. $$y = |x| = \begin{cases} x & \text{if } x \ge 0 \\ -x & \text{if } x < 0 \end{cases}$$

The boundary of $y > |x|$ is the V-shaped region shown at the right. The relation is *not* a function, since it includes different pairs with the same first coordinates, for example, $(1, 2)$ and $(1, 3)$. ***Answer***

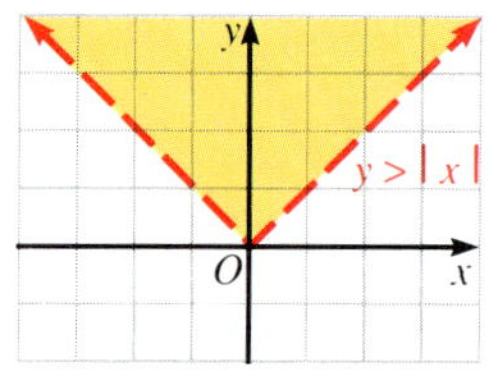

The graphs in Example 1 suggest a convenient way of telling whether a relation is a function. It is called the *vertical-line test*.

Vertical-Line Test

A relation is a function if and only if no vertical line intersects its graph more than once.

Example 2 In the relation $\{(x, y): |x| + |y| = 3 \text{ and } |x| \leq 2\}$, x and y are integers. Find the domain D of the relation and draw its graph. Is the relation a function?

Solution Since x is an integer and $|x| \leq 2$, the domain $D = \{-2, -1, 0, 1, 2\}$.

The relation $|x| + |y| = 3$
is equivalent to $|y| = 3 - |x|$.

Here is a table of values.

x	$\lvert y\rvert = 3 - \lvert x\rvert$	y
-2	1	1 or -1
-1	2	2 or -2
0	3	3 or -3
1	2	2 or -2
2	1	1 or -1

Here is a graph.

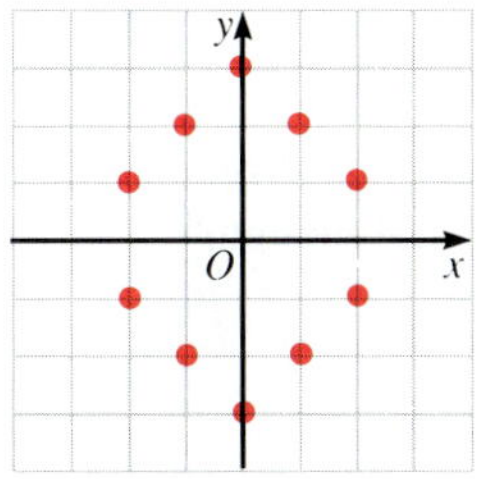

Since the vertical line $x = 1$ intersects the graph in the two points $(1, 2)$ and $(1, -2)$, the relation is *not* a function. ***Answer***

Oral Exercises

Give the domain and range of each relation. Is the relation a function?

Sample $\{(3, 2), (1, 4), (5, 2), (7, 6)\}$

Solution $D = \{1, 3, 5, 7\}$, $R = \{2, 4, 6\}$; yes

3. $D = \{-1, 0, 1\}$, $R = \{-1, 0, 1\}$; no
4. $D = \{1, 2, 3, 4\}$, $R = \{1, 2, 3, 4\}$; yes

1. $\{(1, 3), (2, 5), (3, 5), (4, 3)\}$ — $D = \{1, 2, 3, 4\}$, $R = \{3, 5\}$; yes

2. $\{(2, 4), (1, 3), (0, 2), (1, 1)\}$ — $D = \{2, 1, 0\}$, $R = \{1, 2, 3, 4\}$; no

3. $\{(-1, 0), (1, 0), (0, -1), (-1, 1)\}$

4. $\{(4, 1), (3, 2), (2, 3), (1, 4)\}$

5. $\{(a, b), (b, a)\}$
$D = \{a, b\}$, $R = \{a, b\}$; yes

6. $\{(a, a + 1), (a, a - 1)\}$
$D = \{a\}$, $R = \{a - 1, a + 1\}$; no

List the ordered pairs in the relation pictured in each diagram. Is the relation a function?

7.

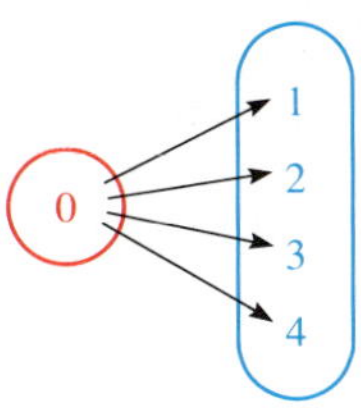

8.

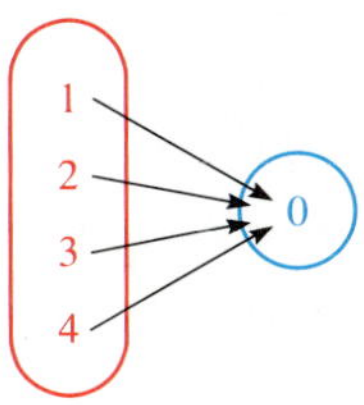

9.

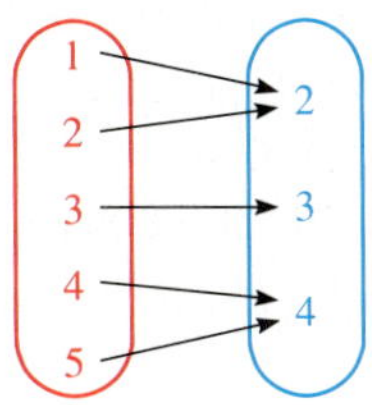

10.

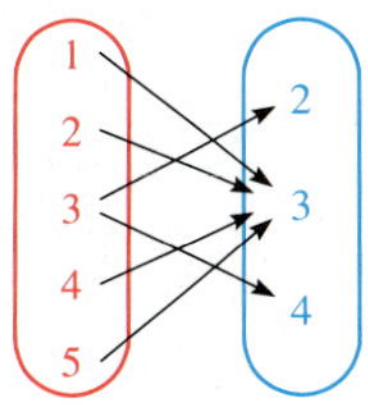

Guided Practice

Graph each relation and state if it is a function.

1. $\{(2, 1), (1, -2), (1, 2)\}$

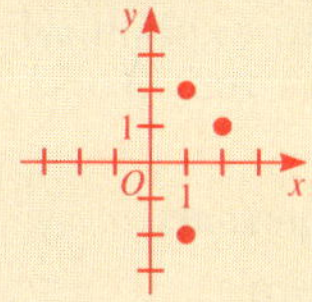

not a function

2. $\{(|-2|, 1), (1, |-2|), (1, 2)\}$

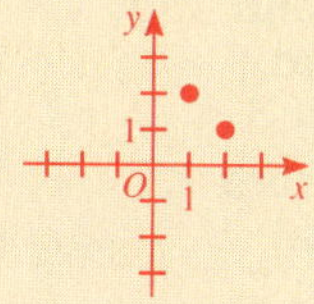

function

Graph each relation if x and y are integers. State which are functions.

3. $\{(x, y): |x| + |y| < 3\}$

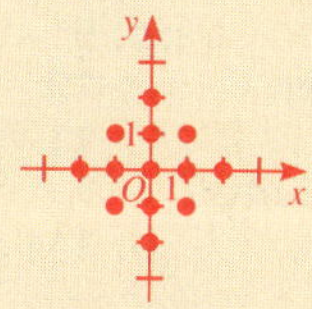

not a function

4. $\{(x, y): x^2 + y^2 < 6\}$

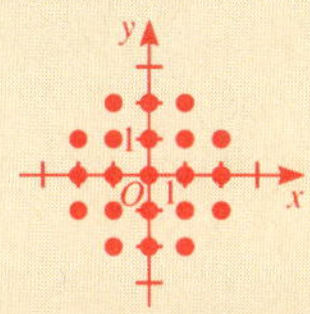

not a function

Summarizing the Lesson

In this lesson students learned to graph relations and to determine whether or not relations are functions. Ask students to explain why the vertical line test works.

Suggested Assignments

Average Algebra
156/1–29 odd
R 158/Self-Test 3

Average Alg. and Trig.
156/1–29 odd
R 158/Self-Test 3

Extended Algebra
156/1–31 odd
R 158/Self-Test 3

Extended Alg. and Trig.
156/3–36 mult. of 3
R 158/Self-Test 3

Supplementary Materials

Study Guide	pp. 51–52
Practice Master	19
Test Master	11
Computer Activity	7
Resource Book	p. 111

Additional Answers
Oral Exercises

7. $\{(0, 1), (0, 2), (0, 3), (0, 4)\}$; no

8. $\{(1, 0), (2, 0), (3, 0), (4, 0)\}$; yes

9. $\{(1, 2), (2, 2), (3, 3), (4, 4), (5, 4)\}$; yes

10. $\{(1, 3), (2, 3), (3, 2), (3, 4), (4, 3), (5, 3)\}$; no

11. $\{(-3, 1), (-2, 2), (-1, 3), (0, 0), (1, -3), (2, -2), (3, -1)\}$; yes

12. $\{(-2, 4), (-1, 4), (0, 2), (0, -2), (1, -4), (2, -4)\}$; no

List the ordered pairs in the relation pictured in each graph. Is the relation a function?

11. 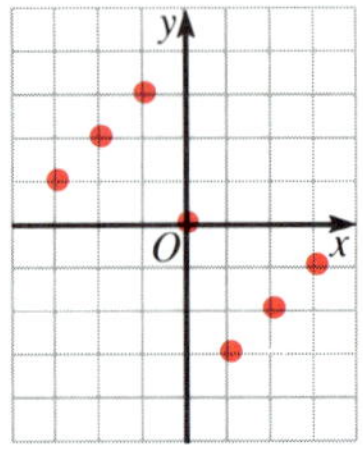

12. 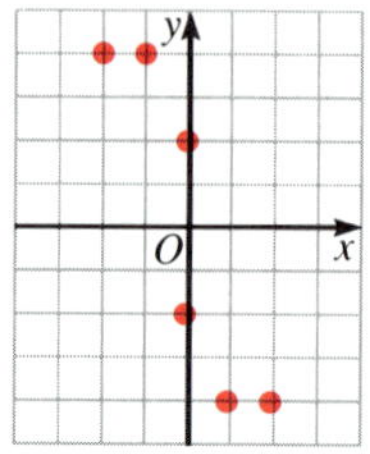

13. 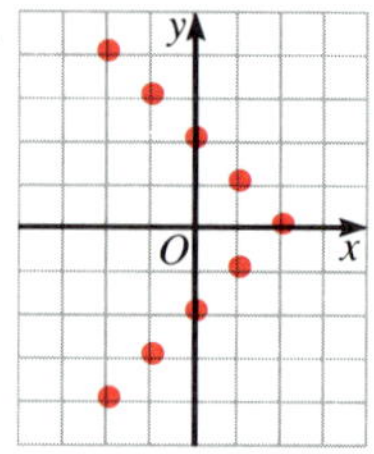

14. 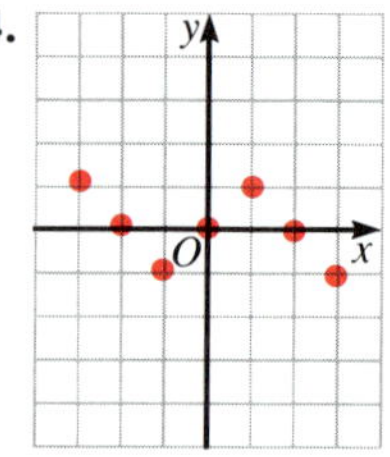

Give the domain and range of each relation graphed below. Is the relation a function?

15. 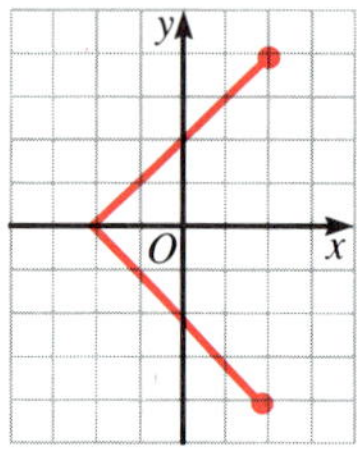

16. 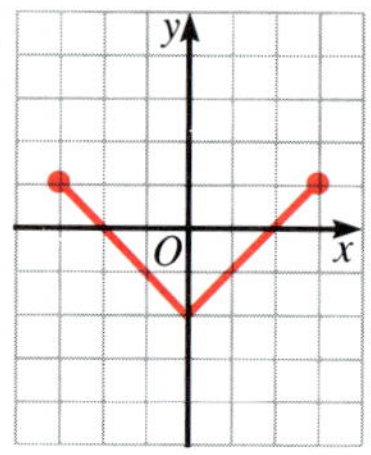

17. 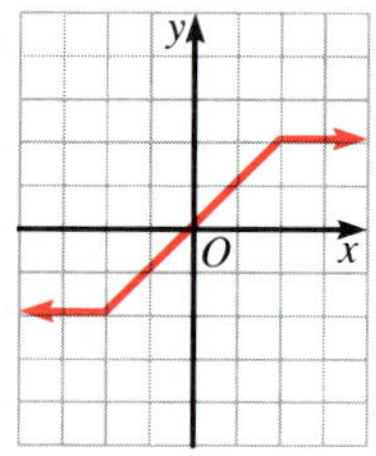

18. 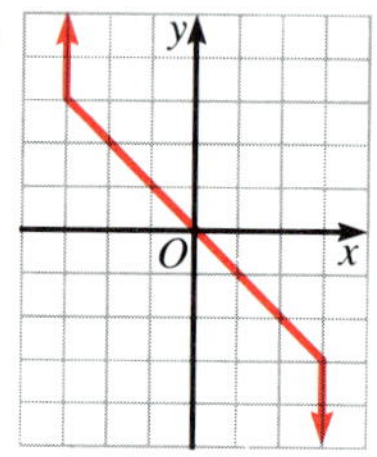

15.–18. See below.

Give the domain of each relation.

19. $\left\{(x, y)\colon y = \frac{3}{x-5}\right\}$ $\{x\colon x \neq 5\}$

20. $\left\{(x, y)\colon y < \frac{1}{x^2-4}\right\}$ $\{x\colon x \neq 2, x \neq -2\}$

21. $\left\{(x, y)\colon y = \frac{x}{|x-7|}\right\}$ $\{x\colon x \neq 7\}$

22. $\{(x, y)\colon |x| + |y| = 1\}$ $\{x\colon -1 \leq x \leq 1\}$

True or false?

23. All relations are functions. False

24. All functions are relations. True

15. $D = \{x\colon -2 \leq x \leq 2\}$, $R = \{y\colon -4 \leq y \leq 4\}$; no
16. $D = \{x\colon -3 \leq x \leq 3\}$, $R = \{y\colon -2 \leq y \leq 1\}$; yes
17. $D = \{\text{real numbers}\}$, $R = \{y\colon -2 \leq y \leq 2\}$; yes
18. $D = \{x\colon -3 \leq x \leq 3\}$, $R = \{\text{real numbers}\}$; no

Written Exercises

Graph each relation. Then tell whether it is a function. If it is not a function, draw a vertical line that intersects the graph more than once.

A

1. $\{(1, 2), (2, 0), (1, 1)\}$ not a function

2. $\{(-1, 2), (0, 1), (1, 2)\}$ function

3. $\{(-2, 1), (|-2|, 0), (-1, |-2|), (|-1|, -2)\}$ function

4. $\{(2, 1), (1, -1), (0, 2), (2, 0)\}$ not a function

5. $\{(|-3|, 2), (3, -2), (-2, -1), (2, |-1|)\}$ not a function

6. $\{(1, 2), (2, -1), (-1, 1), (1, -1), (0, 1)\}$ not a function

10. $D = \{-2, -1, 0, 1, 2\}$; yes

In each relation in Exercises 7–20, x and y are *integers*. Find the domain of the relation and draw its graph. Is the relation a function?

$D = \{-2, -1, 0, 1, 2\}$; no

7. $\{(x, y): |x| = |y| \text{ and } |x| \le 1\}$ $D = \{-1, 0, 1\}$; no
8. $\{(x, y): |x| = |y| \text{ and } |y| \le 2\}$
9. $\{(x, y): |y| = x \text{ and } x \le 3\}$ $D = \{0, 1, 2, 3\}$; no
10. $\{(x, y): y + |x| = 0 \text{ and } y > -3\}$
11. $\{(x, y): |x| + |y| = 2\}$ $D = \{-2, -1, 0, 1, 2\}$; no
12. $\{(x, y): |x| + |y| \le 1\}$ $D = \{-1, 0, 1\}$; no
13. $\{(x, y): |x + y| = 0 \text{ and } |y| \le 2\}$
14. $\{(x, y): |x| \le y \text{ and } y \le 2\}$
15. $\{(x, y): |y| = 2 \text{ and } |x| \le 2\}$ $D = \{-2, -1, 0, 1, 2\}$; no
16. $\{(x, y): |x| = 2 \text{ and } |y| \le 2\}$ $D = \{-2, 2\}$; no

13. $D = \{-2, -1, 0, 1, 2\}$; yes
14. $D = \{-2, -1, 0, 1, 2\}$; no

B

17. $\{(x, y): |xy| = 2\}$ $D = \{-2, -1, 1, 2\}$; no
18. $\{(x, y): x|y| = 2\}$ $D = \{1, 2\}$; no
19. $\{(x, y): |x|y = 2\}$ $D = \{-2, -1, 1, 2\}$; yes
20. $\{(x, y): |xy| \le 2\}$ $D = \{\text{integers}\}$; no

Two functions are *equal* if they consist of the same ordered pairs. In Exercises 21–24, f and g have the same domain D and the same range. Determine whether f and g are equal.

21. $D = \{-1, 0, 1, 2, 3\}$ $f: x \to 2x + 1$ $g: x \to 5 - 2x$ $f \ne g$
22. $D = \{x: x \ge -1\}$ $f: x \to x^2 + x$ $g: x \to x|1 + x|$ $f = g$
23. $D = \{-2, -1, 1, 2\}$ $f: x \to 4 - x^2$ $g: x \to 6 - 3|x|$ $f = g$
24. $D = \{-2, -1, 0, 1, 2\}$ $f: x \to |x| + 2x$ $g: x \to |x| - 2x$ $f \ne g$

In Exercises 25–36, each open sentence defines a relation, and x and y are real numbers. Graph each relation and tell whether it is a function.

25. $\{(x, y): y = -|x|\}$ yes
26. $\{(x, y): |y| = 2\}$ no
27. $\{(x, y): y = 1 - |x|\}$ yes
28. $\{(x, y): y = 1 + |x|\}$ yes
29. $\{(x, y): y = |1 - x|\}$ yes
30. $\{(x, y): y = |1 + x|\}$ yes

C

31. $\{(x, y): |y| = 1 - x\}$ no
32. $\{(x, y): |y| = 1 + x\}$ no
33. $\{(x, y): |y| \le 1 - |x|\}$ no
34. $\{(x, y): |y| \ge 1 + |x|\}$ no
35. $\{(x, y): x^2 = 4y^2\}$ no
36. $\{(x, y): x^2 = 9 - y^2\}$ no

37. A *definition of "ordered pair"*: The **pair** "a, b" is the set $\{a, b\}$ (which is the same as $\{b, a\}$). We can define **ordered pair** in terms of the simpler concept of pair: The ordered pair "a, b," written (a, b), is the pair having as its two members a and the pair "a, b." Thus,

$$(a, b) = \{a, \{a, b\}\} \quad \text{or} \quad (a, b) = \{\{b, a\}, a\}.$$

For example, in $\{\{3, 5\}, 5\}$, the pair member $\{3, 5\}$ tells you that the pair is either (3, 5) or (5, 3), while the lone member 5 tells you that the first coordinate is 5, so that $\{\{3, 5\}, 5\} = (5, 3)$.

$\{\{5, 3\}, 5\}, \{5, \{5, 3\}\}, \{5, \{3, 5\}\}$

a. Express the ordered pair (5, 3) in terms of pairs in three other ways.

b. What ordered pair is $\{4, \{1, 4\}\}$? (4, 1)

c. Write a definition of the ordered triple (a, b, c) in terms of **pairs**. $\{a, \{\{a, b\}, \{a, b, c\}\}\}$

13. $\{(-2, 4), (-2, -4), (-1, 3), (-1, -3), (0, 2), (0, -2), (1, 1), (1, -1), (2, 0)\}$; no

14. $\{(-3, 1), (-2, 0), (-1, -1), (0, 0), (1, 1), (2, 0), (3, -1)\}$; yes

Additional Answers
Written Exercises

2.
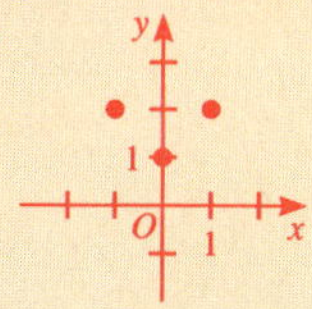

4.
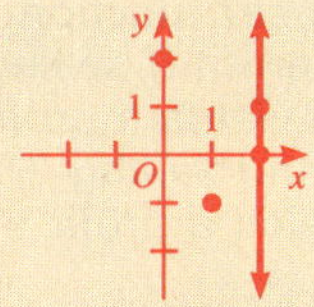

6.
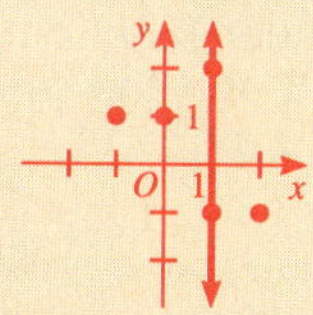

8.
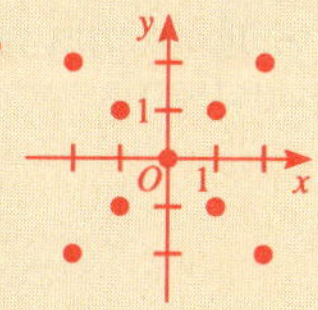

10.
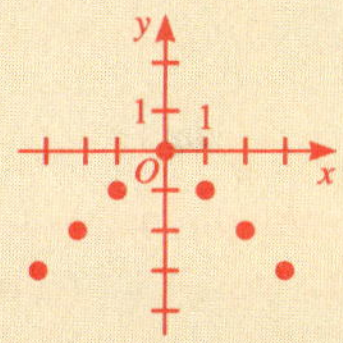

12.
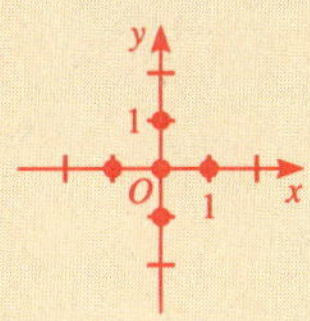

14.
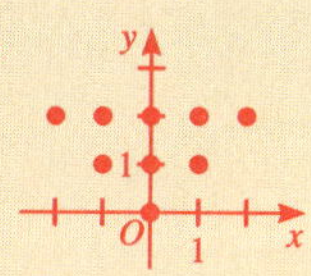

(continued on p. 162)

Self-Test 3

Vocabulary

mapping diagram (p. 141)
function (p. 141, 153)
domain (p. 141)
range (p. 141)
graph of a function (p. 142)
values of a function (p. 142)
functional notation (p. 142)
linear function (p. 146)
constant function (p. 146)
relation (p. 153)
domain of a relation (p. 153)
range of a relation (p. 153)

1. If $f: x \rightarrow x^2 - x + 1$ has domain $D = \{-2, -1, 0, 1, 2\}$, what is the range of f? $R = \{1, 3, 7\}$ **Obj. 3-8, p. 141**
2. Give the domain of $g(x) = \frac{3x + 4}{2x - 5}$. $D = \left\{x: x \neq \frac{5}{2}\right\}$
3. Graph $f(x) = |x - 2|$ if the domain $D = \{0, 1, 2, 3, 4\}$.
4. Find an equation of the linear function f if $f(3) = -1$ and $f(x)$ decreases by 4 when x increases by 6. $f(x) = -\frac{2}{3}x + 1$ **Obj. 3-9, p. 146**
5. Find an equation of a linear function f having slope -2 if $f(1) = 3$. $f(x) = -2x + 5$
6. Write an equation of a linear function f that describes this situation: A bank's total monthly charge for maintaining a checking account is \$.20 for each check written plus an \$8.50 service charge. $f(x) = 0.2x + 8.5$
7. The cost of printing the school newspaper is \$500 for 800 copies and \$620 for 1200 copies. If the printing cost is a linear function of the copies printed, find the cost of printing 1500 copies. \$710
8. Graph the relation $\{(-2, 0), (-1, -1), (0, -2), (2, 0), (1, 1), (0, 2)\}$. Is the relation a function? no **Obj. 3-10, p. 153**
9. If x and y are integers, find the domain of the relation $\{(x, y): y = 2 - |x| \text{ and } y \geq 0\}$ and draw its graph. Is the relation a function? $D = \{-2, -1, 0, 1, 2\}$; yes

Check your answers with those at the back of the book.

Challenge

Using three different weights, you want to be able to balance an object with any integer weight from 1 gram to 13 grams on an equal-arm balance. You can use one, two, or all three of your weights, and one or two of them can go in the same pan with the object you are balancing. What should your three weights weigh? 1g, 3g, and 9g

Quick Quiz

1. Find the range of $g: x \rightarrow 2x - x^2$ if the domain is $\{-1, 0, 1, 2\}$. $\{-3, 0, 1\}$
2. Graph $f(x) = 4 - |x|$ if the domain is $\{-2, -1, 0, 1, 2\}$.

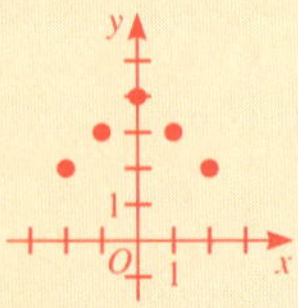

3. Find a linear function to express profit if selling 15 items yields a profit of \$125 and selling 21 items yields a profit of \$167. $P(x) = 7x + 20$
4. Find a linear function g if $g(2) = 8$ and the graph of g has a slope of -3. $g(x) = -3x + 14$
5. Graph $\{(x, y): x = |y + 1|\}$ if the range is $\{-2, -1, 0, 1\}$. Is this relation a function?

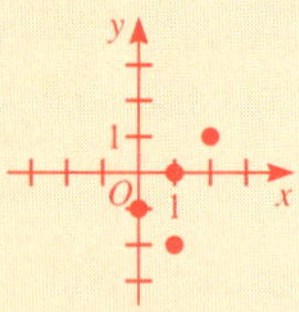

not a function

Additional Answers Written Exercises

(continued from p. 111)

16.

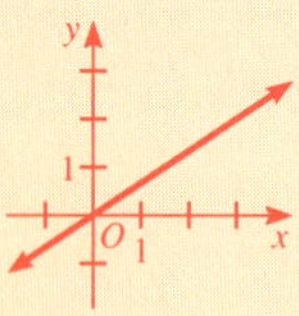

18.

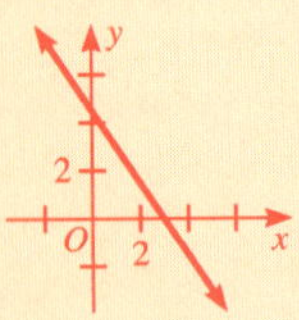

Application / *Linear Programming*

A farmer wants to maximize the revenue resulting from the production and harvesting of corn and wheat. For every 100 bushels of corn he produces he receives \$265 and for every 100 bushels of wheat he produces he receives \$365. However, his production is restricted by the availability of land, capital, and labor. For example, he needs one acre to produce 100 bushels of corn and three acres to produce 100 bushels of wheat, and he can plant a total of 100 acres with these crops. The following table shows the availability of resources and the restrictions placed upon them.

	Input requirements per 100 bushels of		
	Corn	Wheat	Available Material
Land (acre)	1	3	100
Capital (\$)	120	90	9000
Aug. labor (h)	1	2	200
Sept. labor (h)	1	6	160
Value of output of 100 bushels (\$)	265	365	

The farmer can formulate the problem of maximizing revenue mathematically.

Let x = the number of hundreds of bushels of corn produced.
Let y = the number of hundreds of bushels of wheat produced.

If R denotes the total return in dollars, then $R = 265x + 365y$.

The complete problem facing the farmer is to maximize $R = 265x + 365y$ subject to

$$\begin{aligned} x + 3y &\le 100 \\ 120x + 90y &\le 9000 \\ x + 2y &\le 200 \\ x + 6y &\le 160 \\ x &\ge 0 \\ y &\ge 0 \end{aligned}$$

Notice that the set of restrictions forms a system of linear inequalities. Each linear inequality is called a **constraint,** and the expression to be maximized is called the **objective function.** The problem of maximizing (or minimizing) a linear function subject to a set of linear constraints is called a problem in

Additional Answers Written Exercises

(continued from p. 111)

20. 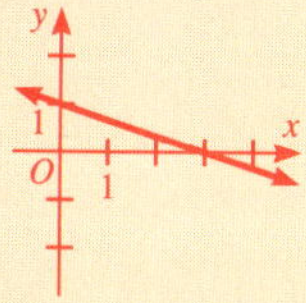

22. 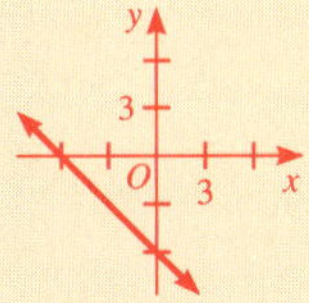

28. 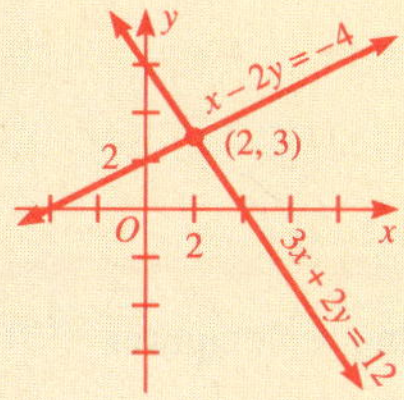

Graphs intersect at (2, 3); 2 − 2(3) = −4; 3(2) + 2(3) = 12

30. 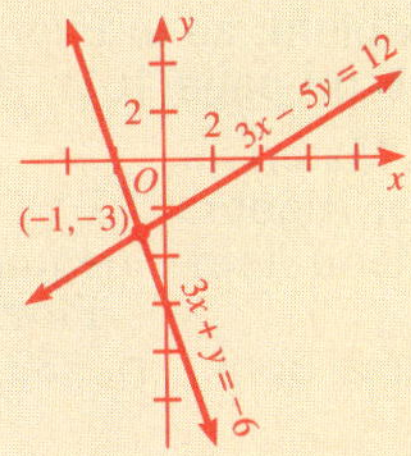

Graphs intersect at (−1, −3); 3(−1) − 5(−3) = 12; 3(−1) + (−3) = −6

32. 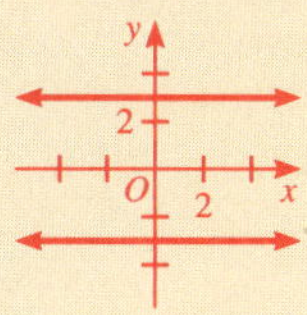

34. 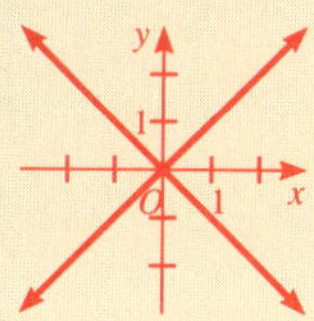

(continued)

linear programming. Decision makers in many organizations formulate and solve linear programming problems as part of their management activities.

The graph of the solution set of the system of constraints for a linear programming problem is called the **feasible region.** The figure below shows the feasible region for the farmer's problem. A computer may be helpful in finding the feasible region, or you can obtain it by hand.

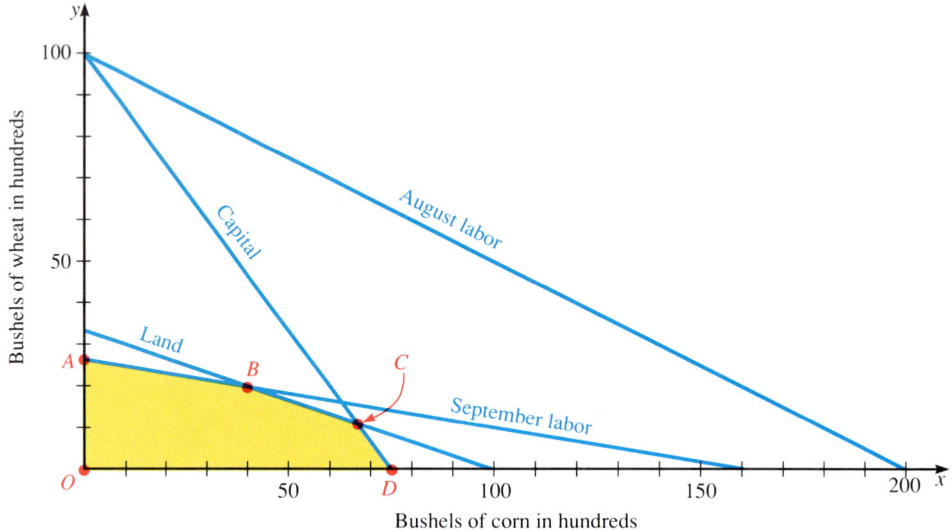

The farmer needs to determine which point in the feasible region gives the maximum value of R. Although there are infinitely many points in this region, the following theorem from linear programming (which is proved in more advanced courses) makes the list of points to check manageable.

The maximum and minimum values of the objective function occur at vertices of the feasible region.

By solving $x + 3y = 100$ and $x + 6y = 160$ simultaneously, point B, $(40, 20)$, is obtained. Each of the remaining vertices O, A, C, and D is obtained by solving a system of two equations associated with a pair of constraints. The table shows the vertices and the corresponding values of R.

Vertex	$265x + 365y = R$
$O(0, 0)$	$265(0) + 365(0) = 0$
$A\left(0, 26\frac{2}{3}\right)$	$265(0) + 365\left(26\frac{2}{3}\right) = 9733.33$
$B(40, 20)$	$265(40) + 365(20) = 17900$
$C\left(66\frac{2}{3}, 11\frac{1}{9}\right)$	$265\left(66\frac{2}{3}\right) + 365\left(11\frac{1}{9}\right) = 21722.22$
$D(75, 0)$	$265(75) + 365(0) = 19875$

A maximum revenue of \$21,722.22 is achieved if the farmer produces $66\frac{2}{3} \times 100 = 6667$ bushels of corn and $11\frac{1}{9} \times 100 = 1111$ bushels of wheat. Thus he should plant $1 \times 66\frac{2}{3}$, or $66\frac{2}{3}$, acres of corn and $3 \times 11\frac{1}{9}$, or $33\frac{1}{3}$ acres of wheat.

Additional Answers Written Exercises
(continued)

36.

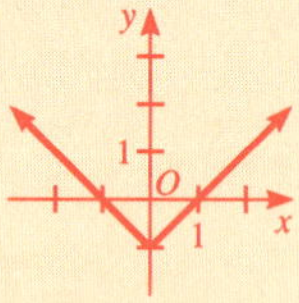

38.

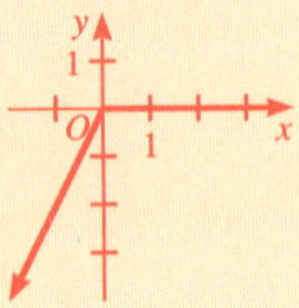

Using a Computer

The program "Inequality Plotter" can be used to find the feasible region when solving a linear programming problem, as in Exercises 3–5 on page 161.

Support Materials
Disk for Algebra
Menu Item: Inequality Plotter
Resource Book p. 239

Additional Answers Written Exercises
(continued from p. 139)

20.

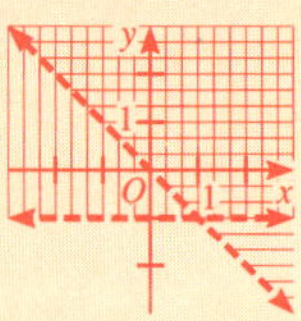

22.

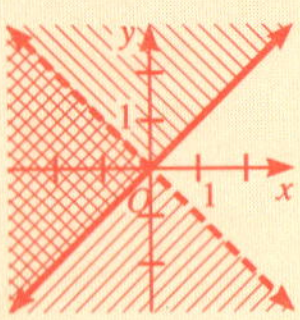

Exercises (A computer may be helpful in Exercises 3–5.)

1. A linear programming problem gives (0, 0), (0, 90), (50, 80), (70, 20), and (70, 0) as consecutive vertices of the feasible region. Find the maximum and minimum values of the objective function $P = 3x + 4y$ in the feasible region. max. value = 470; min. value = 0

2. A linear programming problem gives (10, 0), (10, 40), (30, 50), and (40, 20) as the consecutive vertices of the feasible region. Find the maximum and minimum values of the objective function $P = 13x - 5y$ in the feasible region. max. value = 420; min. value = −70

3. Find the maximum and minimum values of the objective function $S = 10x + 3y$ in the region which is the solution set of the system $0 \le x \le 10$, $y \ge 5$, and $y \le -0.3x + 10$. max. value = 121; min. value = 15

4. Find the maximum and minimum values of the objective function $S = 6x - 2y$ in the region which is the solution set of the system $0 \le y \le 10$, $y \ge -2x + 4$, and $y \le -2x + 12$. max. value = 36; min. value = −38

5. A nutrition center sells health food to mountain-climbing teams. The Trailblazer mix package contains one pound of corn cereal mixed with four pounds of wheat cereal and sells for \$9.75. The Frontier mix package contains two pounds of corn cereal mixed with three pounds of wheat cereal and sells for \$9.50. The center has available 60 pounds of corn cereal and 120 pounds of wheat cereal. How many packages of each mix should the center sell to maximize its income?
12 packages of Trailblazer and 24 packages of Frontier

Chapter Summary

1. An equation that can be expressed in the form
$$Ax + By = C$$
(A and B not both zero) is called a *linear equation* because its graph in the coordinate plane is a straight line. The line is horizontal if $A = 0$ and vertical if $B = 0$.

2. The slope, m, of a nonvertical line L is given by
$$m = \frac{y_2 - y_1}{x_2 - x_1}$$
where (x_1, y_1) and (x_2, y_2) are two points of L. An equation of L is
$$y - y_1 = m(x - x_1) \qquad \text{(point-slope form)},$$
and if L has y-intercept b, it has the equation
$$y = mx + b \qquad \text{(slope-intercept form)}.$$

3. If the slopes of the nonvertical lines L_1 and L_2 are m_1 and m_2, then:
(a) L_1 and L_2 are parallel if and only if $m_1 = m_2$.
(b) L_1 and L_2 are perpendicular if and only if $m_1 m_2 = -1$.

24.
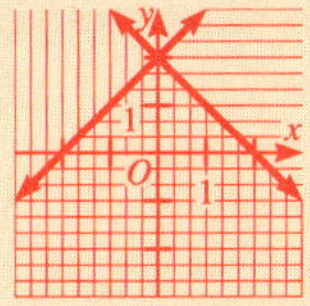

26.
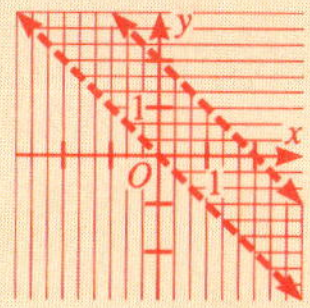

28.
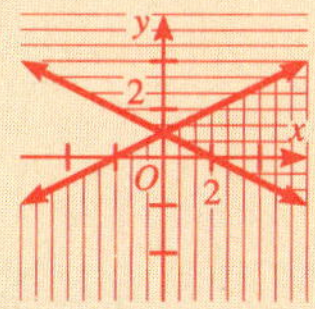

30.
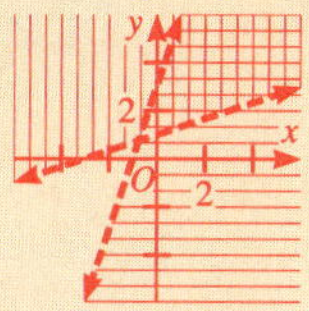

32.
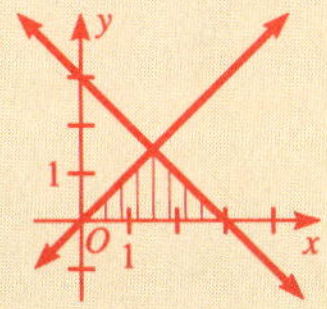

34.

36.
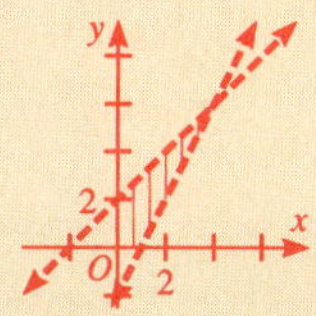

(continued)

Additional Answers Written Exercises

(continued)

38. 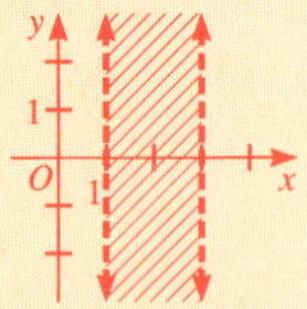

40. 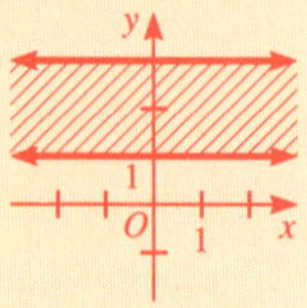

42. 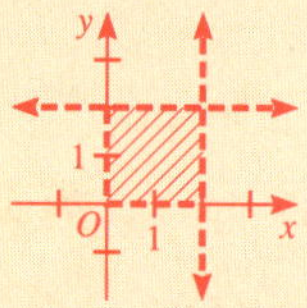

44. 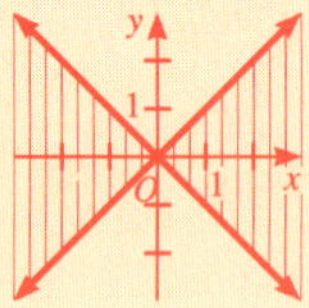

Additional Answers Written Exercises

(continued from p. 157)

16. 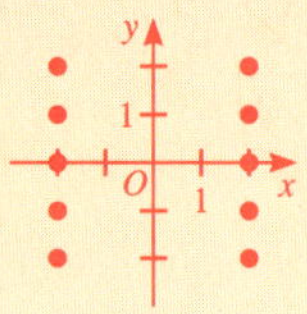

18. 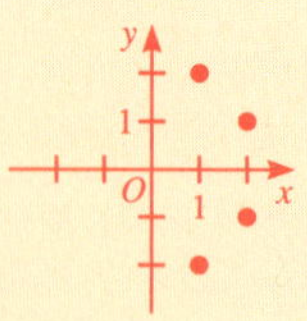

4. A solution of a system of linear equations in two variables is an ordered pair that satisfies each of the equations. The number of solutions of such a system is one, none, or infinitely many according to whether the graphs of the equations intersect in one point, are parallel, or coincide.
5. Systems of equations can be solved algebraically by the *substitution method* or the *linear-combination method*.
6. The graph of a linear inequality in two variables is an open or a closed half-plane. The graph of the solution set of a system of such inequalities is the intersection of the solution sets of the individual inequalities.
7. A *function* is a correspondence that assigns to each member of its *domain* exactly one member of its *range*. The *value* assigned by the function f to the number x is denoted by $f(x)$. The *graph* of a function f is the set of all points having coordinates (x, y).
8. A *linear function* is a function f that can be defined by $f(x) = mx + b$ where x, m, and b are any real numbers. The graph of f is the graph of $y = mx + b$, a line with slope m and y-intercept b.
9. A set of ordered pairs is called a *relation*. A function is a relation in which different ordered pairs have different first coordinates.

Chapter Review

Give the letter of the correct answer.

1. Find the solution set of $x + 2y = 4$ if each variable represents a whole number. 3-1

a. $\{(2, 1)\}$ **b.** $\{(2, 0), (1, 2), (0, 4)\}$

c. $\{(0, 2), (2, 1), (4, 0)\}$ **d.** $\{(0, 2), \left(1, \frac{3}{2}\right), (2, 1), \left(3, \frac{1}{2}\right), (4, 0)\}$

2. Frank has \$1.85 in dimes and quarters. Find all possibilities for the number of dimes that he has.

a. 1, 3, 5, or 7 **b.** 7 or fewer

c. 1, 6, 11, or 16 **d.** 18 or fewer

3. Graph the equation $2x - y = 1$. 3-2

a. 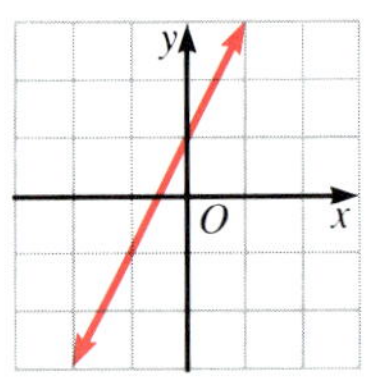

b. 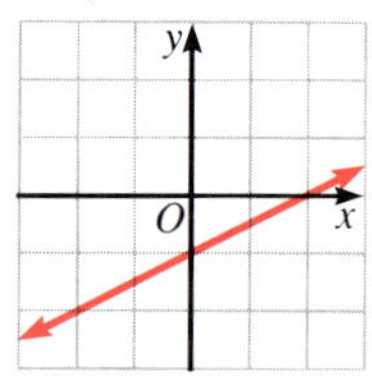

c. 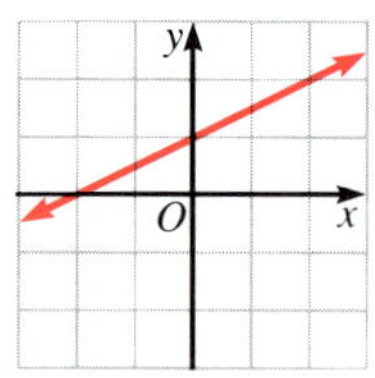

d. 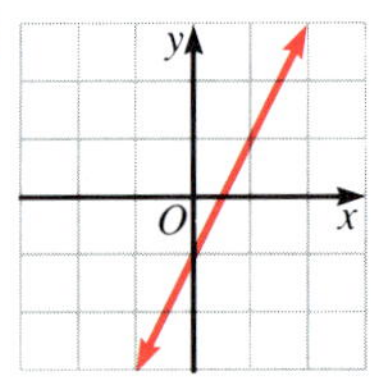

4. Determine k so that the line $3x + ky = 7$ will pass through the point $P(-3, -2)$.

a. 1 **b.** 8 **c.** -8 **d.** $-\frac{13}{3}$

5. Find the slope of the line containing the points $(-4, 3)$ and $(2, -5)$. 3-3

a. $\frac{3}{4}$ **b.** $-\frac{4}{3}$ **c.** $\frac{4}{3}$ **d.** $-\frac{3}{4}$

6. Find the slope of the line $2x - 5y = 8$.

a. $-\frac{8}{5}$ **b.** $\frac{8}{5}$ **c.** $\frac{2}{5}$ **d.** $-\frac{2}{5}$

Find an equation in standard form for each line.

7. The line through $(-1, 2)$ having slope $-\frac{3}{4}$. 3-4

a. $3x + 4y = 5$ **b.** $-3x + 4y = 11$ **c.** $4x + 3y = 2$ **d.** $4x + 3y = 5$

8. The line through $(3, -2)$ and parallel to $4x - y = 5$.

a. $x + 4y = -5$ **b.** $4x - y = 14$ **c.** $4x + y = 10$ **d.** $x - 4y = 11$

9. The line through $(4, 1)$ and perpendicular to $3x + y = 5$.

a. $x + 3y = 7$ **b.** $3x + y = 13$ **c.** $x - 3y = 1$ **d.** $3x - y = 11$

10. Solve this system: $\begin{aligned} 4x + 3y &= 7 \\ -2x + y &= 9 \end{aligned}$ 3-5

a. $(1, 1)$ **b.** $(-4, 1)$ **c.** $(-2, 5)$ **d.** $(5, -2)$

11. Find an equation of the form $y = ax^2 + b$ whose graph passes through the points $(-1, 1)$ and $(2, 7)$.

a. $y = 2x^2 + 1$ **b.** $y = -2x^2 - 1$ **c.** $y = -2x^2 + 1$ **d.** $y = 2x^2 - 1$

12. Two shirts and one tie cost \$42, but one shirt and two ties cost \$39. What does one shirt cost? 3-6

a. \$12 **b.** \$13 **c.** \$14 **d.** \$15

13. Find the ordered pair that does *not* satisfy this system: $\begin{aligned} 2x + y &\le 3 \\ x - 3y &> 4 \end{aligned}$ 3-7

a. $(-1, -3)$ **b.** $(1, 1)$ **c.** $(0, -2)$ **d.** $(2, -1)$

14. If $f(x) = 3 - x^2$ and $g(x) = x + 2$, find $f(g(1))$. 3-8

a. -6 **b.** 0 **c.** 2 **d.** 4

15. The function $g: x \rightarrow x + |x|$ has domain $\{-2, -1, 0, 1, 2\}$. Find the range of g.

a. $\{0, 1, 2\}$ **b.** $\{-4, -2, 0\}$ **c.** $\{-2, 0, 2, 4\}$ **d.** $\{0, 2, 4\}$

16. The function f is linear. If $f(2) = 2$ and $f(-6) = 6$, find $f(6)$. 3-9

a. 0 **b.** -6 **c.** 6 **d.** 1

17. A student's test score was 684 on a scale of 200–800. Convert this score to a scale of 0–100.

a. 80 **b.** $80\frac{1}{3}$ **c.** $80\frac{2}{3}$ **d.** 81

18. Which relation graphed below is a function? 3-10

a.
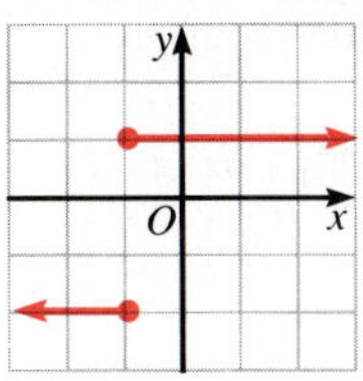

b.
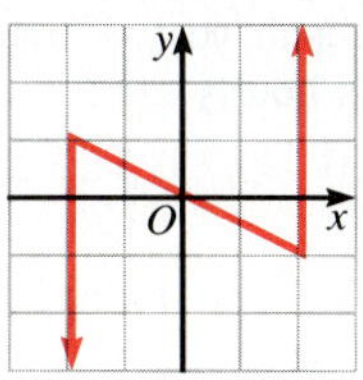

c.
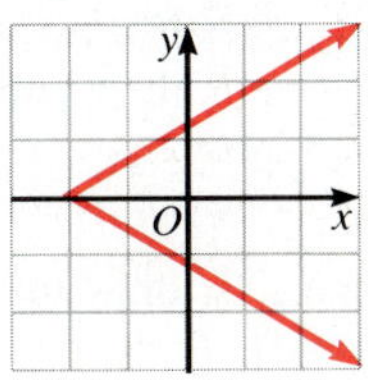

d.
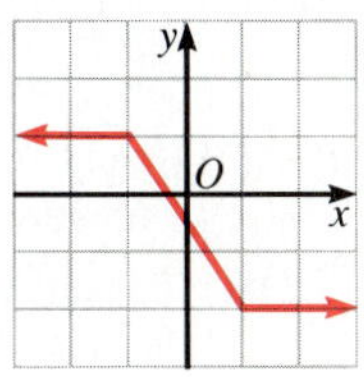

20.
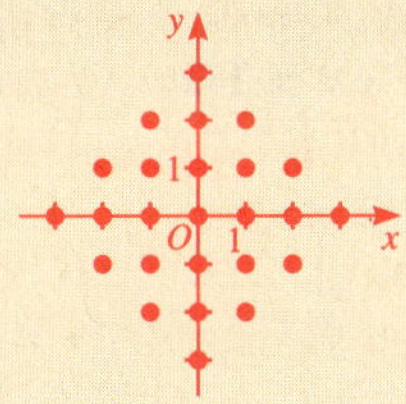

26.
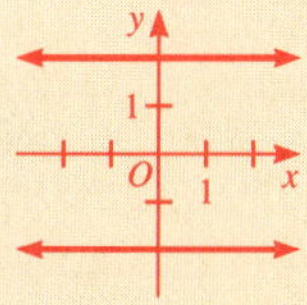

28.
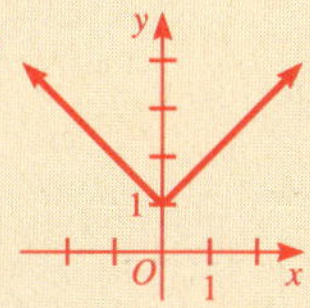

30.
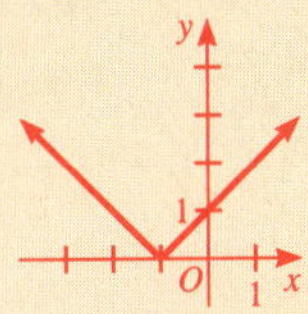

32.
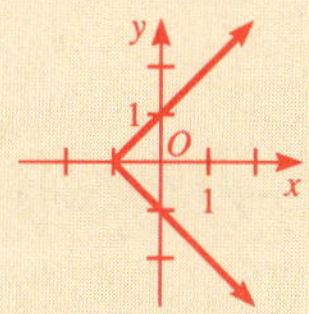

34.
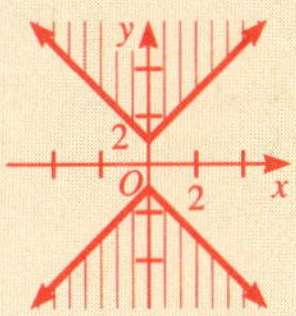

36.
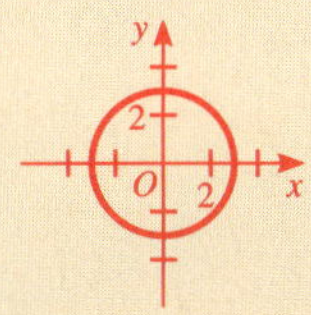

Supplementary Materials

Test Masters	12, 13
Resource Book	pp. 16–18, 112

Alternate Test p. T15

Additional Answers Chapter Test

3. a.
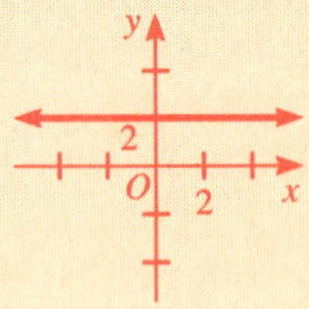

b.
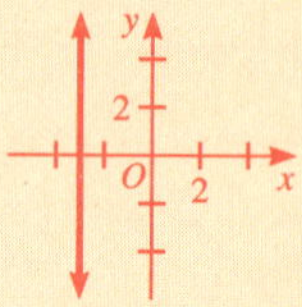

c.
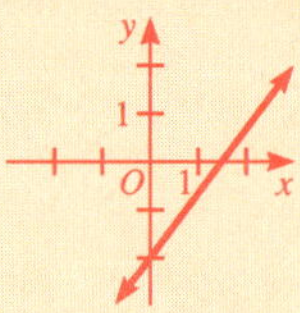

d.
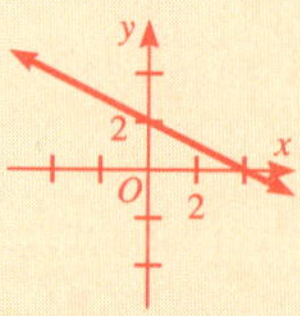

10.

15.
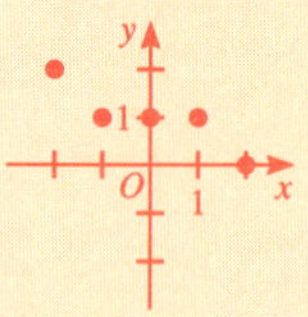

16.
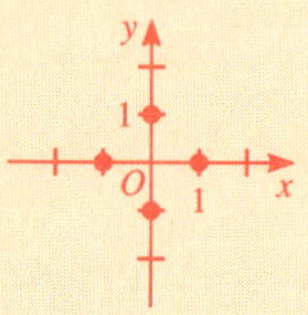

Chapter Test

1. Find the value of k so that $(-2, 1)$ satisfies $kx + 6y = k$. 2 **3-1**

2. Find the solution set of $3x + 2y = 12$ if each variable represents a positive integer. $\{(2, 3)\}$

3. Graph each equation. **3-2**
a. $y - 2 = 0$
b. $x = -3$
c. $4x - 3y = 6$
d. $y = -\frac{1}{2}x + 2$

4. Find the slope of each line described. **3-3**
a. Passes through $(-7, 1)$ and $(3, -4)$. $-\frac{1}{2}$
b. Has equation $y = -\frac{1}{3}x + 7$. $-\frac{1}{3}$
c. Has equation $2x - 5y = 3$. $\frac{2}{5}$

5. Find the value of k so that $(k + 2)x - 3y = 1$ has slope 2. 4

6. Find an equation in standard form for each line described. **3-4**
a. Through $(-2, 4)$ and slope -3. $3x + y = -2$
b. Through $(-5, -1)$ and $(3, 3)$. $x - 2y = -3$
c. Having x-intercept 4 and y-intercept -6. $3x - 2y = 12$

7. Find equations in standard form for the lines through $(3, -1)$ that are **(a)** parallel to and **(b)** perpendicular to the line $4x - 3y = 7$. **a.** $4x - 3y = 15$ **b.** $3x + 4y = 5$

8. Solve this system: $5x - 4y = 13$ **3-5**
$2x + 3y = -4$ $(1, -2)$

9. Two cement blocks and three bricks weigh 102 lb, as do one cement block and ten bricks. What does one brick weigh? 6 lb **3-6**

10. Graph this system: $2x + y \geq 2$ **3-7**
$x - y < 1$

11. The domain of the function $f: x \rightarrow (x - 2)^2$ is $D = \{0, 1, 2, 3, 4\}$. Find the range of f. $R = \{0, 1, 4\}$ **3-8**

12. If $f(x) = |2x + 1|$ and $g(x) = x - 4$, find $f(g(1))$ and $g(f(1))$. $f(g(1)) = 5$; $g(f(1)) = -1$

13. Find an equation of the linear function f if $f(1) = -1$ and $f(4) = 8$. $f(x) = 3x - 4$ **3-9**

14. A refrigerator repairman charges a fixed amount for a service call in addition to his hourly rate. If a two-hour repair costs \$50 and a four-hour repair costs \$74, what is the repairman's hourly rate? \$12

15. Graph the relation $\{(-2, 2), (-1, 1), (0, 1), (1, 1), (2, 0)\}$. Is the relation a function? yes **3-10**

16. If x and y are integers, draw the graph of the relation $\{(x, y): |x| + |y| = 1\}$. Is the relation a function? no

Cumulative Review (Chapters 1–3)

Simplify.

1. $-1(4-7)^2+6$ -3
2. $2-|(-3)(4)+(-1)^2|$ -9
3. $2(x+y)-3(y-x)$ $5x-y$
4. Evaluate $\frac{|a|+|b|}{|a+b|}$ if $a=-6$ and $b=2$. 2
5. On a 5-hour trip, Thomas drove part of the time at 30 mi/h and the rest at 55 mi/h. If he traveled a distance of 250 mi, how much time did he spend driving at 55 mi/h? 4 h

Solve each open sentence and graph each solution set.

6. $3(2x-1)=4x+7$ $\{5\}$
7. $13-7y\ge 34$ $\{y: y\le -3\}$
8. $5<3w+8<14$ $\{w: -1<w<2\}$
9. $|2m+3|=5$ $\{-4, 1\}$
10. $\left|\frac{c-3}{2}\right|<1$ $\{c: 1<c<5\}$
11. $6-|n|\ge 4$ $\{n: -2\le n\le 2\}$
12. Prove: If a and b are real numbers, then $a+(b-a)=b$.
13. Solve $3x-2y=1$ if the domain of x is $\{-1, 0, 1\}$. $\left\{(-1,-2), \left(0,-\frac{1}{2}\right), (1,1)\right\}$
14. Find the slope of the line $8x-6y=5$. $\frac{4}{3}$
15. Find the slope of the line $3y-9=0$. 0

Find an equation in standard form for the line described.

16. Passes through $(4, -3)$ with slope $-\frac{1}{2}$ $x+2y=-2$
17. Passes through $(-2, 4)$ and $\left(-3, \frac{1}{2}\right)$ $7x-2y=-22$
18. Has x-intercept 6 and y-intercept 4 $2x+3y=12$
19. Parallel to the line $5x-4y=7$ and contains $(-1, -2)$ $5x-4y=3$
20. Perpendicular to the line $2x+y=5$ and contains $(3, 2)$ $x-2y=-1$

Solve each system. If the system has no solution, say so.

21. $3x-4y=10$, $2x+3y=1$ $(2, -1)$
22. $y=\frac{1}{2}x-1$, $x-2y=4$ No sol.
23. $x=7y-4$, $y=7x+4$ $\left(-\frac{1}{2}, \frac{1}{2}\right)$
24. Graph this system: $2x-y<3$, $x+y\ge 0$
25. Admission to a county fair costs \$20 for two adults and one child and \$16 for one adult and two children. What is the cost for one adult? \$8
26. If $f(x)=4-x^2$ and $g(x)=|2x-7|$, find $f(g(2))$ and $g(f(2))$. $f(g(2))=-5$; $g(f(2))=7$
27. If $h(x)$ is a linear function such that $h(-3)=2$ and $h(1)=-4$, find $h(7)$. -13
28. Graph the relation $\{(-2, 0), (-1, 1), (0, 2), (2, 0), (1, -1), (0, -2)\}$. Is the relation a function? no

Cumulative Review p. T29

Additional Answers
Cumulative Review

6. 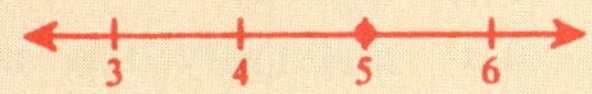

7. 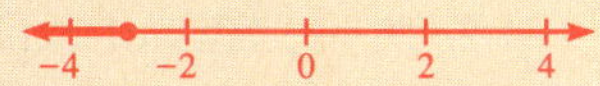

8. 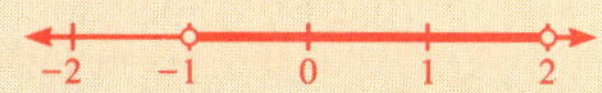

9. 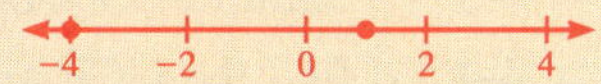

10. 

11. 

12.
1. $a+(b-a)=a+(b+(-a))$ Def. of subtr.
2. $a+(b-a)=a+((-a)+b)$ Comm. prop. of add.
3. $a+(b-a)=(a+(-a))+b$ Assoc. prop. of add.
4. $a+(b-a)=0+b$ Prop. of opposites
5. $a+(b-a)=b$ Ident. prop. of add.

24. 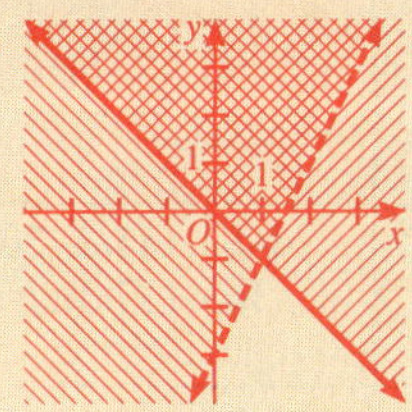

28. 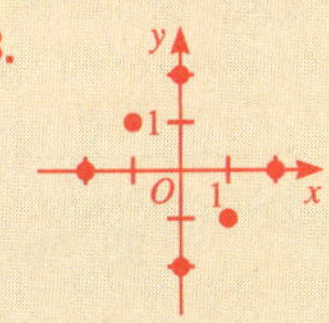

4 Products and Factors of Polynomials

Objectives

4-1 To simplify, add, and subtract polynomials.

4-2 To use laws of exponents to multiply a polynomial by a monomial.

4-3 To multiply polynomials.

4-4 To find the GCF and LCM of integers and monomials.

4-5 To factor polynomials by using the GCF, by recognizing special products, and by grouping terms.

4-6 To factor quadratic polynomials.

4-7 To solve polynomial equations.

4-8 To solve problems using polynomial equations.

4-9 To solve polynomial inequalities.

Assignment Guide

See p. T58 for Key to the format of the Assignment Guide

Day	Average Algebra	Average Algebra and Trigonometry	Extended Algebra	Extended Algebra and Trigonometry
1	**4-1** 170/1–25 odd **S** 170/*Mixed Review*	**4-1** 170/1–25 odd **S** 170/*Mixed Review*	**4-1** 170/1–27 odd **S** 170/*Mixed Review*	*Administer Chapter 3 Test* **4-1** *Read 4-1* 170/1–27 odd **S** 170/*Mixed Review*
2	**4-2** 173/2–40 even	**4-2** 173/2–40 even	**4-2** 173/2–40 even, 39	**4-2** 173/2–40 even, 39
3	**4-3** 175/1–47 odd **S** 176/*Mixed Review*	**4-3** 175/1–47 odd **S** 176/*Mixed Review*	**4-3** 175/1–51 odd **S** 176/*Mixed Review*	**4-3** 175/3–51, mult. of 3, 53, 54 **S** 176/*Mixed Review* **4-4** 181/3–30 mult. of 3, 34, 35 **R** 177/*Self-Test 1*
4	**4-4** 181/2–30 even **R** 177/*Self-Test 1*	**4-4** 181/2–30 even **R** 177/*Self-Test 1*	**4-4** 181/2–32 even **R** 177/*Self-Test 1*	**4-5** 185/1–45 odd, 49–53 odd **S** 187/*Mixed Review*
5	**4-5** 185/1–16, 18–36 even **S** 187/*Mixed Review*	**4-5** 185/1–45 odd **S** 187/*Mixed Review*	**4-5** 185/1–45 odd **S** 187/*Mixed Review*	**4-6** 191/3–30 mult. of 3, 32–58 even **R** 192/*Self-Test 2*
6	**4-5** 186/17–37 odd, 38–46	**4-5** 186/32–46 even **4-6** 191/2–30 even	**4-5** 186/32–46 even, 49–51 **4-6** 191/2–30 even	**4-7** 196/3–48 mult. of 3 **S** 197/*Mixed Review* **4-8** 199/*P*: 4, 8, 9, 11
7	**4-6** 191/1–30 **S** 186/49–52	**4-6** 191/31–54 **R** 192/*Self-Test 2*	**4-6** 191/31–50, 55–58 **R** 192/*Self-Test 2*	**4-8** 200/*P*: 13–27 odd **4-9** 204/3–30 mult. of 3 **S** 205/*Mixed Review* **R** 205/*Self-Test 3*
8	**4-6** 191/31–49 odd **R** 192/*Self-Test 2*	**4-7** 196/1–49 odd **S** 197/*Mixed Review*	**4-7** 196/1–49 odd **S** 197/*Mixed Review*	*Prepare for Chapter Test* **R** 207/*Chapter Review*
9	**4-7** 196/1–10, 12–34 even **S** 197/*Mixed Review*	**4-8** 199/*P*: 2–18 even	**4-8** 199/*P*: 3–30 mult. of 3	*Administer Chapter 4 Test* **5-1** *Read 5-1* 213/2–30 even, 33 **S** 215/*Mixed Review*

Assignment Guide *(continued)*

Day	Average Algebra	Average Algebra and Trigonometry	Extended Algebra	Extended Algebra and Trigonometry
10	**4-7** 196/9–35 odd, 36–40, 45, 46	**4-9** 204/1–21 odd, 23, 27 **S** 205/*Mixed Review* **R** 205/*Self-Test 3*	**4-9** 204/1–21 odd, 23, 27 **S** 205/*Mixed Review* **R** 205/*Self-Test 3*	
11	**4-8** 199/*P*: 2–18 even **S** 197/41–44	*Prepare for Chapter Test* **R** 207/*Chapter Review*	*Prepare for Chapter Test* **R** 207/*Chapter Review*	
12	**4-9** 204/1–21 odd **S** 205/*Mixed Review* **R** 205/*Self-Test 3*	*Administer Chapter 4 Test*	*Administer Chapter 4 Test* **5-1** *Read 5-1* 213/2–30 even, 33 **S** 215/*Mixed Review*	
13	*Prepare for Chapter Test* **R** 207/*Chapter Review*			
14	*Administer Chapter 4 Test*			

Supplementary Materials Guide

For Use with Lesson	Practice Masters	Tests	Study Guide (Reteaching)	Resource Book: Tests	Resource Book: Practice Exercises	Resource Book: Prob. Solving (PS) Applications (A) Enrichment (E)
4-1			pp. 53–54			p. 207 (A)
4-2			pp. 55–56			
4-3	Sheet 20		pp. 57–58		p. 113	
4-4			pp. 59–60			
4-5	Sheet 21		pp. 61–62			
4-6	Sheet 22	Test 14	pp. 63–64		p. 114	
4-7	Sheet 23		pp. 65–66			
4-8	Sheet 24		pp. 67–68			pp. 198–199 (PS)
4-9	Sheet 25	Test 15	pp. 69–70		p. 115	
Chapter 4		Tests 16, 17		pp. 19–22	p. 116	p. 223 (E)
Cum. Rev. 1–4	Sheet 26				p. 117	

Overhead Visuals

For Use with Lesson	Visual	Title
4-3	3	Multiplying Polynomials—Geometric Model

Software

Software	Algebra Plotter Plus	Using Algebra Plotter Plus	Computer Activities	Test Generator
	Function Plotter	Enrichment, p. 37	Activities 8, 9	189 test items
For Use with Lessons	4-7	4-7	Chapter 4, 4-6	all lessons

Strategies for Teaching

Communication and Using Manipulatives

The language of algebra is very compact because of the extensive use of symbols. Students need help in the correct interpretation and use of symbols. By writing out a few solutions in words, they can gain an appreciation of the value of symbols as savers of time and effort.

It is important for students to realize that the speed of silent reading depends on one's purpose—is one *skimming* for a quick preview or review of the material or is one *studying* to learn the concepts discussed in a lesson? To help students gain study skills, you may want to provide a list of questions in advance.

When a teacher says "x-squared" a student's mental image may be the abstract expression "x^2." It is equally valuable for the student to realize that this expression represents the area of a square having x as the length of a side. Area models and manipulatives are effective for multiplying binomials and factoring trinomials.

See the Exploration on page 835 for an activity in which students use manipulatives to factor polynomials.

4-1 Polynomials

Algebra involves the use of symbols, often in complex and subtle ways. In order to operate successfully with polynomials, students need a clear understanding of the meanings associated with symbols. An interesting exercise is to have students read aloud a number of expressions such as the following.

$$-8a^3b^2 \qquad (-8a^3b)^2 \qquad 8(q - \tfrac{1}{2}) - (q - 4)$$
$$P = 2l + 2w = 2(l + w) \qquad V = \tfrac{1}{3}s^2h$$

Help students to realize that there is often more than one correct way to translate a group of symbols into words. Accept any word form that makes the meaning clear. This activity can be extended by having the class write from dictation (by you or by a student) some of the expressions in the Oral Exercises.

4-3 Multiplying Polynomials

This is a good lesson for students to read silently before it is discussed in class. Some key questions to provide for students to answer as they read are:

- What property is used to multiply two polynomials?
- How many ways of multiplying two polynomials are shown on page 174?
- Explain the method for multiplying two binomials.
- What name is given to the product of $(a + b)$ times $(a + b)$?

4-6 Factoring Quadratic Polynomials

A manipulative approach gives an alternate method of introducing or reviewing the factoring of quadratic polynomials. The materials used in this activity may be purchased commercially or hand-made by the students. The pieces used are labeled as shown.

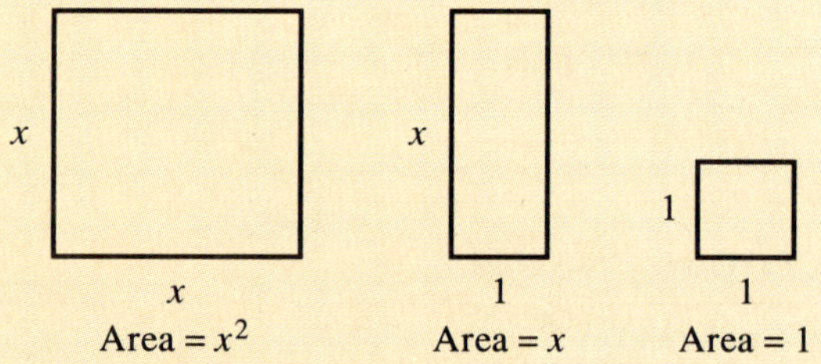

Be sure that the x-length is not a whole-number multiple of the unit length.

The desired outcome of the activity is that the students "build" a rectangle which represents a quadratic polynomial and then "read" the factors as the length and width of the rectangle. Any quadratic that is factorable yields a rectangle.

Example: Factor $x^2 + 3x$.

The area $x^2 + 3x$ is represented by the rectangle shown below. The sides of the rectangle are x and $x + 3$. Therefore, $x^2 + 3 = x(x + 3)$.

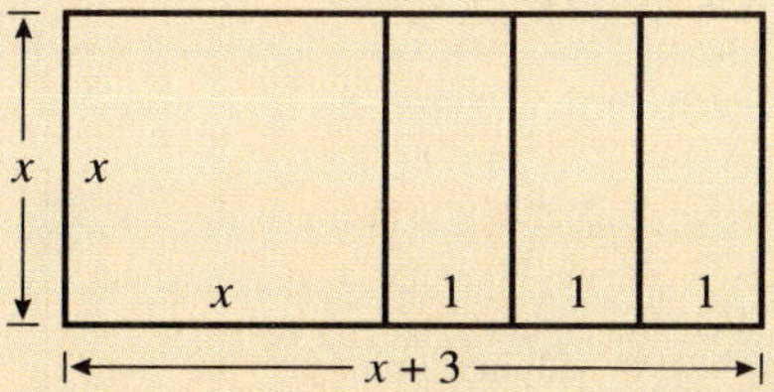

Use the Exploration on page 835, and then ask the students to factor some of the exercises in the lesson using this method.

4-9 Solving Polynomial Inequalities

By this time students have had a good deal of practice in reading and graphing the solution sets of equations and inequalities. As a challenge, you might reverse the process and ask students to write solution sets represented by graphs such as those below.

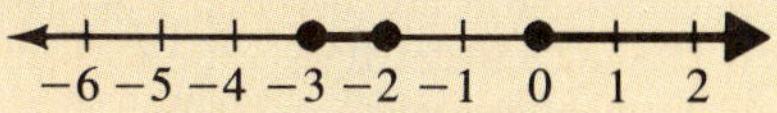

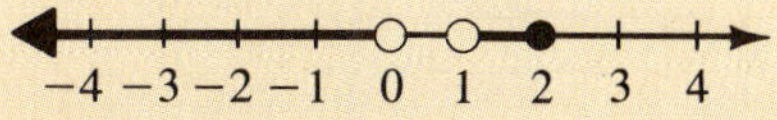

References to Strategies

PE: Pupil's Edition **TE:** Teacher's Edition **RB:** Resource Book

Problem Solving Strategies

PE: pp. 175, 183–184 (recognize a pattern); p. 198 (make a sketch); p. 199 (check solutions)
TE: pp. 179, 189, 199
RB: pp. 198–199

Applications

PE: p. 183 (physics); pp. 198–201 (word problems)
TE: pp. 198–199, 207
RB: p. 207

Nonroutine Problems

PE: p. 170 (Exs. 25–27); p. 173 (Exs. 39–42); p. 176 (Exs. 49–56); p. 182 (Exs. 33–36); p. 192 (Ex. 59); p. 197 (Exs. 51–56); p. 201 (Prob. 31); p. 205 (Ex. 33)
TE: pp. T94, T96, T97 (Sugg. Extensions, Lessons 4-4, 4-7, 4-8, 4-9)

Communication

PE: pp. 181–182, 201, 204 (convincing argument); p. 178 (Reading Algebra)
TE: pp. T93–T94, T97 (Reading Algebra); p. 167 (Warm-up Exs.)

Thinking Skills

PE: p. 182 (Exs. 35, 36)

Explorations

PE: p. 835 (polynomial factors)

Connections

PE: pp. 198–201 (Geometry); p. 166 (Light); pp. 177, 182 (History); pp. 199, 201 (Physics)

Using Technology

PE: p. 187 (Exs.); p. 193 (Computer Key-In)
TE: pp. 187, 193
Computer Activities: pp. 20–24
Using Algebra Plotter Plus: p. 37

Using Manipulatives/Models

PE: 835 (exploring polynomial factors)
TE: p. T95 (Lesson 4-6)
Overhead Visuals: 3

Cooperative Learning

TE: p. T95 (Lesson 4-6)

Teaching Resources

For use in implementing the teaching strategies referenced on the previous page.

Application
Resource Book, p. 207

Application—Motion of a Projectile (For use with Chapter 4)

To determine the path of an object projected at an angle θ with the ground, it is convenient to use two equations:

$$x = v_x t$$

and

$$y = v_y t - 16t^2$$

where v_x and v_y represent the horizontal and vertical components of the initial speed v (measured in feet per second) and (x, y) is the ordered pair for the position (measured horizontally and vertically in feet) of the object at time t (measured in seconds).

Suppose a ball is thrown from ground level with initial horizontal and vertical speeds $v_x = 48$ ft/s and $v_y = 64$ ft/s.

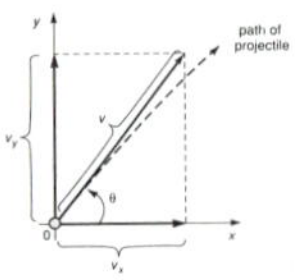

1. Write the two equations that locate the ball at time t.

$x =$ ______________ $y =$ ______________

2. Use the equations from Exercise 1 to complete the table below. Then use the table to sketch the path of the ball while in flight.

	t	x	y
a.	0	____	____
b.	0.5	____	____
c.	1	____	____
d.	1.5	____	____
e.	2	____	____
f.	2.5	____	____
g.	3	____	____
h.	3.5	____	____
i.	4	____	____

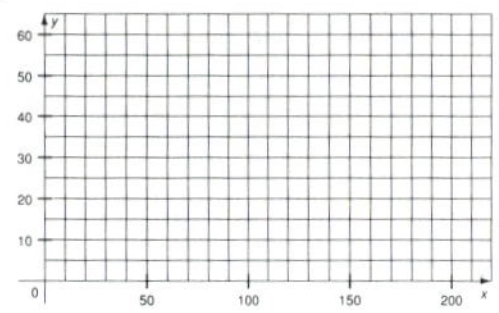

3. Use the table and graph of Exercise 2 to answer the following questions.

a. How long is the ball in the air? ______________

b. The ball reaches what maximum height? ______________

c. What horizontal distance does the ball travel? ______________

d. How can you obtain the facts in parts (a)–(c) from the equations in Exercise 1 directly (that is, without using a table or a graph)?

____

4. Using the right triangle relationship among v, v_x, and v_y, find the initial speed v. ______________

5. Solve the first equation in Exercise 1 for t and then substitute into the second equation to write an equation for y in terms of x: $y =$ ______________

Resource Book, ALGEBRA, Structure and Method, Book 2

APPLICATIONS 207

Enrichment
Resource Book, p. 223

The Absolute Value Function (For use with Chapter 4)

At the right is the graph of $y = |x|$.

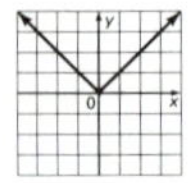

Graph each equation using the given axes.

1. $y = |x - 2|$

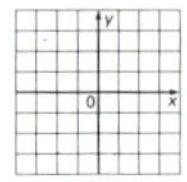

2. $y = |x| - 2$

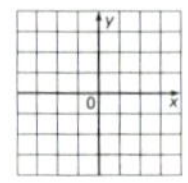

3. $y = ||x| - 2|$

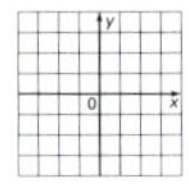

4. $|y| = |x|$

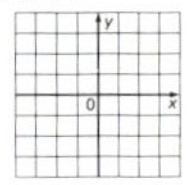

5. $|y| = 2 + |x|$

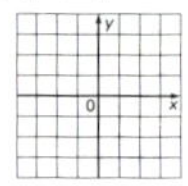

6. $|y| = |2 + x|$

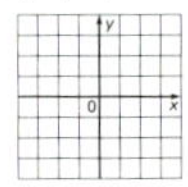

7. $y = |x + 1| + |x - 1|$

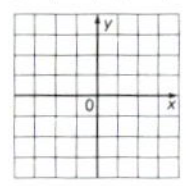

8. $y = \frac{|x|}{x}$

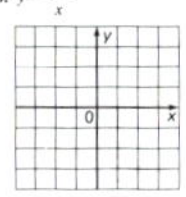

Resource Book, ALGEBRA, Structure and Method, Book 2

ENRICHMENT 223

Problem Solving
Resource Book, p. 198

Using Polynomial Equations to Solve Problems (For use with Lessons 4-8 and 7-2)

By working through the steps in the problems below, you will gain skill in solving problems by using polynomial equations.

Problem 1 The sum of the squares of two consecutive multiples of 4 is 976. Find the multiples.

a. What are the first three positive multiples of 4? ______________

b. If n is a multiple of 4, what is the next greater multiple of 4? ______________

c. Express the first sentence of the problem as an equation.

____

d. Express the equation in simplified form.

____

e. Solve the equation.

____

f. Answer the question. (*Hint:* There are two sets of answers.)

______________ and ______________

Problem 2 The length of a photograph is 6 inches less than twice the width. The photograph is mounted in a frame that is 3 inches wide on all sides. If the area of the framed picture is 270 square inches, find the dimensions of the unframed photograph.

a. Let w = the width of the unframed photo. Write an expression for the length of the unframed photo. ______________

b. Draw a sketch of the facts and label the dimensions in terms of w.

c. Write the length and the width of the *framed* photo in terms of w.

length: ______________ width: ______________

d. Use the facts of the problem to write an equation for the problem.

____

(continued)

198 PROBLEM SOLVING
Resource Book, ALGEBRA, Structure and Method, Book 2

Problem Solving
Resource Book, p. 199

Using Polynomial Equations to Solve Problems (continued)

e. Write your equation in simplified form.

____

f. Solve to find the value(s) of w.

____

g. Write your answer to the problem.

____

h. Find the numerical areas of the unframed photo and the framed photo.

______________ and ______________

Do your answers agree with the facts of the problem?

Resource Book, ALGEBRA, Structure and Method, Book 2

PROBLEM SOLVING 199

Problem Solving
Study Guide, p. 67

4–8 Problem Solving Using Polynomial Equations

Objective: To solve problems using polynomial equations.

Vocabulary

Mathematical model An equation that represents a real-life problem.

CAUTION After solving the mathematical model (equation), you should always check your answer against the conditions in the problem because an answer that satisfies the equation may not be reasonable in real life.

Example 1 Find two consecutive even integers such that the sum of their squares is 100.

Solution

Step 1 You are asked to find two consecutive even integers whose squares add up to 100.

Step 2 Let n = first consecutive even integer.
Then $n + 2$ = second consecutive even integer.

Step 3 Write an equation that shows the sum of squares of the integers is 100.

$$n^2 + (n + 2)^2 = 100$$

Step 4 Solve the equation.

$n^2 + n^2 + 4n + 4 = 100$	Square $n + 2$.
$2n^2 + 4n - 96 = 0$	Make one side 0.
$n^2 + 2n - 48 = 0$	Divide both sides by 2.
$(n + 8)(n - 6) = 0$	Factor the polynomial.
$n + 8 = 0$ or $n - 6 = 0$	Use the zero-product property.
$n = -8$ or $n = 6$	

Step 5 Check each possible solution.

If $n = -8$, then $n + 2 = -6$.
sum of squares $= (-8)^2 + (-6)^2 = 64 + 36 = 100$

If $n = 6$, then $n + 2 = 8$.
sum of squares $= (6)^2 + (8)^2 = 36 + 64 = 100$

$\therefore$ there are two correct answers: -8 and -6 or 6 and 8.

Solve each problem. If there are two correct answers, give both of them.

1. Find two consecutive integers such that the sum of their squares is 113.
2. Find two consecutive even integers such that the sum of their squares is 340.
3. Find two consecutive odd integers whose product is 195.
4. The sum of a number and its square is 42. Find the number.

Problem Solving
Study Guide, p. 68

4–8 Problem Solving Using Polynomial Equations *(continued)*

Example 2 A rectangular residential lot with area 7475 m^2 is 50 m longer than it is wide. Find the dimensions of the lot.

Solution

Step 1 You are asked to find the width and length of the lot. Draw a diagram.

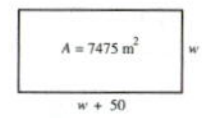

Step 2 Let w = width of the lot in meters.
Then $w + 50$ = length of the lot.
Label your diagram.

Step 3 Write an equation that shows the area of the lot is 7475 m^2.

$$\text{Length} \times \text{Width} = \text{Area}$$
$$(w + 50)w = 7475$$

Step 4 Solve the equation.

$w^2 + 50w = 7475$	
$w^2 + 50w - 7475 = 0$	Make one side 0.
$(w + 115)(w - 65) = 0$	Factor the polynomial.
$w + 115 = 0$ or $w - 65 = 0$	Use the zero-product property.
$w = -115$ or $w = 65$	

Step 5 Although $w = -115$ is a solution of $(w + 50)w = 7475$, it must be rejected because width cannot be negative. The other solution, $w = 65$, gives a lot that is 65 m wide and 115 m long. Check these dimensions:

$$\text{Lot's area} = (115)(65) = 7475 \checkmark$$

$\therefore$ the dimensions are 115 m by 65 m.

Solve each problem. If there are two correct answers, give both of them.

5. A rectangle is 5 cm longer than it is wide, and its area is 176 m^2. Find its dimensions.
6. An entry hall with an area of 72 ft^2 is 14 ft longer than it is wide. Find the dimensions of the hall.
7. The height of a triangle is 3 cm less than the length of its base, and its area is 27 cm^2. Find the height. (Area of a triangle $= \frac{1}{2} \times$ Base $\times$ Height.)
8. The area of a right triangle is 96 m^2. The length of one leg is 8 m less than twice the length of the other. Find the length of each leg.
9. A rectangular lot has perimeter 78 ft and area 350 ft^2. Find the dimensions of the lot.
10. The hypotenuse of a right triangle is 13 in. long. One leg is 7 in. longer than the other leg. Find the length of each leg.

Using Manipulatives/Models
Overhead Visual 3, Sheets 1–3

MULTIPLYING POLYNOMIALS—GEOMETRIC MODEL

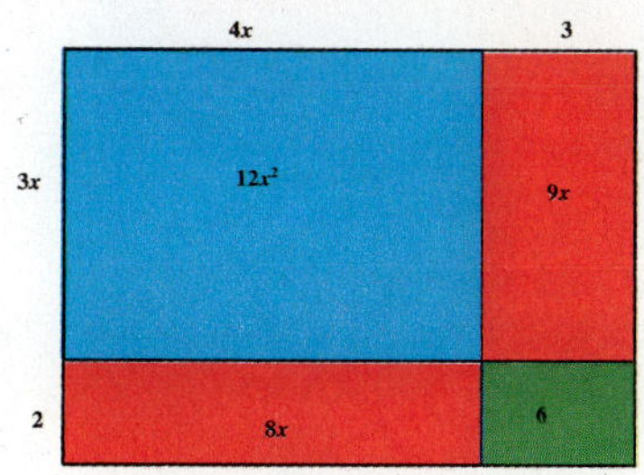

$$(3x + 2)(4x + 3) = 3x \cdot 4x + 3x \cdot 3 + 2 \cdot 4x + 2 \cdot 3$$
$$= 12x^2 + 9x + 8x + 6$$
$$= 12x^2 + 17x + 6$$

VISUAL 3

Application

White light is a blending of all the colors in the color spectrum. When a beam of white light passes through a prism, the prism bends the light. Colors with shorter wave lengths, such as red and orange, bend more, while colors with longer wave lengths, such as blue and violet, bend less. The colors in white light thereby become separated and visible.

The color spectrum has its most dramatic form in a rainbow. If the sun shines immediately after a shower, a rainbow may appear in the part of the sky opposite the sun. The rainbow is a result of raindrops acting as tiny prisms and mirrors. Sunlight entering a raindrop is separated into its component colors, each of which is reflected from the raindrop at a different angle. Those reflected colors form the bands of the rainbow.

Research Activities

Interested students can investigate the mathematics of a rainbow. Questions to consider include: Why is a rainbow an arc? What are the key angles involved in seeing a rainbow?

References

Austin, Joe Dan, and F. Barry Dunning. "Applications: Mathematics of the Rainbow." *Mathematics Teacher* 81 (September 1988): 484–488.

4 Products and Factors of Polynomials

Just as a prism separates white light into its component colors, you can often factor a polynomial into a product of "simpler" polynomials.

166

Working with Polynomials

4-1 Polynomials

Objective To simplify, add, and subtract polynomials.

Here is a review of some terms you'll use when working with polynomials.

Definition	*Examples*
Constant: a number	$-2, \frac{3}{5}, 0$
Monomial: a constant, a variable, or a product of a constant and one or more variables	$-7, u, \frac{1}{3}m^2, -s^2t^3, 6xy^3$
Coefficient (or **numerical coefficient**): the constant (or numerical) factor in a monomial	The coefficient of $3m^2$ is 3. The coefficient of u is 1. The coefficient of $-s^2t^3$ is -1.
Degree of a variable in a monomial: the number of times the variable occurs as a factor in the monomial	$6xy^3$ The degree of x is 1. The degree of y is 3.
Degree of a monomial: the sum of the degrees of the variables in the monomial. A nonzero constant has degree 0. The constant 0 has *no degree*.	$6xy^3$ has degree $1 + 3$, or 4. $-s^2t^3$ has degree $2 + 3$, or 5. u has degree 1. -7 has degree 0.
Similar (or **like**) **monomials:** monomials that are identical or that differ only in their coefficients	$-s^2t^3$ and $2s^2t^3$ are similar. $6xy^3$ and $6x^3y$ are *not* similar.
Polynomial: a monomial or a sum of monomials. The monomials in a polynomial are called the **terms** of the polynomial.	$x^2 + (-4)x + 5$, or $x^2 - 4x + 5$ The terms are x^2, $-4x$, and 5.
Simplified polynomial: a polynomial in which no two terms are similar. The terms are usually arranged in order of decreasing degree of one of the variables.	$2x^3 - 5 + 4x + x^3$ is *not* simplified, but $3x^3 + 4x - 5$ is simplified.
Degree of a polynomial: the greatest of the degrees of its terms after it has been simplified	The degrees of the terms of $x^4 - 2x^2y^3 + 6y - 11$ are, in order, 4, 5, 1, and 0. The polynomial has degree 5.

Teaching References

Lesson Commentary, pp. T92–T97

Assignment Guide, pp. T61–T63

Supplementary Materials
- Practice Masters 20–26
- Tests 14–17
- Resource Book
  - Practice Exercises, pp. 113–117
  - Tests, pp. 19–22
  - Enrichment Activity, p. 223
  - Application, p. 207
  - Practice in Problem Solving/Word Problems, pp. 198–199
- Study Guide, pp. 53–70
- Computer Activities 8–9
- Test Generator

Alternate Test, p. T16

Teaching Suggestions p. T93

Reading Algebra p. T93

Warm-Up Exercises

Ask students to read the definitions on page 167. For each term listed, have the students write an example different from those given.

Motivating the Lesson

Tell students that a number like 376 can be written in expanded form as $3 \cdot 10^2 + 7 \cdot 10 + 6$. If x is used in place of 10, you get $3x^2 + 7x + 6$. This is an example of a polynomial, the topic of today's lesson.

Chalkboard Examples

Simplify. Then give the degree of the polynomial.

1. $3x - x^2 - 9 + 2x^2 + 2 - 5x$
   $(-x^2 + 2x^2) + (3x - 5x) + (-9 + 2) = x^2 - 2x - 7$; degree 2
2. $x^2y^2 - 8xy^4 + 2x^4y + 5x^3y^3 + 2xy^4$
   $2x^4y + 5x^3y^3 + x^2y^2 + (-8xy^4 + 2xy^4) = 2x^4y + 5x^3y^3 + x^2y^2 - 6xy^4$; degree 6
3. Add $-3x^2 - 4x + 9$ and $2x^3 - 4x^2 - 5$.
   $$\begin{array}{r} -3x^2 - 4x + 9 \\ \underline{2x^3 - 4x^2 \qquad - 5} \\ 2x^3 - 7x^2 - 4x + 4 \end{array}$$
4. Subtract $3x^2 + 4x - 5$ from $2x^2 - 9x + 7$.
   $(2x^2 - 9x + 7) - (3x^2 + 4x - 5) = 2x^2 - 3x^2 - 9x - 4x + 7 + 5 = -x^2 - 13x + 12$

Common Errors

When subtracting polynomials, some students may take the opposite of just the *first* term of the polynomial being subtracted. Emphasize that the opposite of *every* term of the polynomial being subtracted must be taken.

Check for Understanding

Here is a suggested use of the Oral Exercises to check students' understanding as you teach the lesson.

Oral Exs. 1–8: use before Example 1.
Oral Exs. 9–14: use after Example 1.
Oral Exs. 15–18: use after Example 3.

When simplifying a polynomial, you use many of the properties of real numbers discussed in Chapter 1. By using the commutative and associative properties of addition, you can order and group the terms of a polynomial in any way. Similar terms are usually grouped together and then combined. For example,

$$\begin{aligned} 4s - 3s^2t - s + 5s^2t - 7 &= (-3s^2t + 5s^2t) + (4s - s) - 7 \\ &= 2s^2t + 3s - 7 \end{aligned}$$

Example 1 Simplify, arranging terms in order of decreasing degree of x. Then write the degree of the polynomial.

a. $x - 3x^2 + 8 + x^2 - 2 + 4x$
b. $x^3y^3 - 6xy^4 + 2x^3y - x^3y^3 + 3xy^4 - 4x^2y$

Solution **a.**
$$\begin{aligned} x - 3x^2 + 8 + x^2 - 2 + 4x &= (-3x^2 + x^2) + (x + 4x) + (8 - 2) \\ &= -2x^2 + 5x + 6 \end{aligned}$$

The degrees of the terms are, in order, 2, 1, and 0.
∴ the degree of the polynomial is 2. ***Answer***

b. $x^3y^3 - 6xy^4 + 2x^3y - x^3y^3 + 3xy^4 - 4x^2y =$
$(x^3y^3 - x^3y^3) + 2x^3y - 4x^2y + (-6xy^4 + 3xy^4) =$
$0 + 2x^3y - 4x^2y - 3xy^4 =$
$2x^3y - 4x^2y - 3xy^4$

The degrees of the terms are, in order, 4, 3, and 5.
∴ the degree of the polynomial is 5. ***Answer***

Adding and Subtracting Polynomials

To add two or more polynomials, write their sum and then simplify by combining similar terms.

To subtract one polynomial from another, add the opposite of each term of the polynomial you're subtracting.

Example 2 Add $2x^2 - 3x + 5$ and $x^3 - 5x^2 + 2x - 5$.

Solution 1 $(2x^2 - 3x + 5) + (x^3 - 5x^2 + 2x - 5) =$
$x^3 + [2x^2 + (-5x^2)] + (-3x + 2x) + [5 + (-5)] =$
$x^3 - 3x^2 - x$ ***Answer***

Solution 2 You can also add vertically.

$$\begin{array}{r} 2x^2 - 3x + 5 \\ \underline{x^3 - 5x^2 + 2x - 5} \\ x^3 - 3x^2 - x \phantom{{}+5} \end{array} \quad \textbf{\textit{Answer}}$$

Example 3 Subtract $2x^2 - 3x + 5$ from $x^3 - 5x^2 + 2x - 5$.

Solution

$$(x^3 - 5x^2 + 2x - 5) - (2x^2 - 3x + 5) =$$
$$(x^3 - 5x^2 + 2x - 5) + (-2x^2 + 3x - 5) =$$
$$x^3 + [-5x^2 + (-2x^2)] + (2x + 3x) + [-5 + (-5)] =$$
$$x^3 - 7x^2 + 5x - 10$$

Answer

Example 4 Simplify $x(2y - 3) + 4(x + 2y) - 3y(x - 1)$.

Solution Use the distributive property first. Then combine similar terms.

$$x(2y - 3) + 4(x + 2y) - 3y(x - 1) = 2xy - 3x + 4x + 8y - 3xy + 3y$$
$$= x - xy + 11y$$

Answer

Oral Exercises

Give (a) the degree and (b) the coefficient of each monomial.

1. $3x^4$ a. 4 b. 3
2. $-t$ a. 1 b. −1
3. xy^2z^2 a. 5 b. 1
4. 2^3 a. 0 b. 8

Name the similar monomials.

5. $3t,\quad -t^4,\quad 4t^2,\quad -\frac{1}{2}t,\quad 6,\quad 3t^3,\quad -1,\quad \frac{3}{4}t^2,\quad -\frac{1}{2}t^4,\quad 16t^2$

6. $uv^2,\quad -3u^2v,\quad -u^3,\quad \frac{3}{2}uv^2,\quad uv,\quad 4,\quad 2vu,\quad v^3,\quad u^2v,\quad -7$

7. $6xy,\quad -4xz,\quad yz,\quad 4zx,\quad -zxy,\quad -xy,\quad 2yzx,\quad \frac{xz}{2},\quad \frac{2yx}{3}$

8. $p^2q,\quad -3pq^3,\quad 2qp,\quad 6qp^2,\quad \frac{pq}{2},\quad 4pq^3,\quad -8p^2q,\quad p^2q^3,\quad q^3p$

Read each polynomial with terms in order of decreasing degree of *x*. Then give the degree of the polynomial.

9. $11x - 7 + 2x^2$ $2x^2 + 11x - 7$; 2
10. $1 - x^3 + 2x - 4x^2$ $-x^3 - 4x^2 + 2x + 1$; 3
11. $1 - 3x^2 + 5x^4 - 7x^6$ $-7x^6 + 5x^4 - 3x^2 + 1$; 6
12. $7xz - 5z^2 + 3x^2$ $3x^2 + 7xz - 5z^2$; 2
13. $ux - u^3x^2 + ux^3 - u^2$ $ux^3 - u^3x^2 + ux - u^2$; 5
14. $xy^3 - 2x^3y + 4x^2y^2 - y^4$ $-2x^3y + 4x^2y^2 + xy^3 - y^4$; 4

In Exercises 15–18, (a) add the polynomials and (b) subtract the second polynomial from the first.

15. $x + 7,\ 3x + 2$ a. $4x + 9$ b. $-2x + 5$
16. $3t + 5,\ 4t - 7$ a. $7t - 2$ b. $-t + 12$
17. $3r - 5,\ r^2 - 2r + 6$ a. $r^2 + r + 1$ b. $-r^2 + 5r - 11$
18. $u^2 - 6u + 7,\ u^2 + 6u - 8$ a. $2u^2 - 1$ b. $-12u + 15$

Guided Practice

Simplify. Then give the degree of the polynomial.

1. $-3x^2 + 5 - x + 4x^2 - 9$ $x^2 - x - 4$; 2
2. $4xz^4 - 2x^2z^2 - 5x^3z + x^2z^2$ $-5x^3z - x^2z^2 + 4xz^4$; 5

Add the polynomials.

3. $8r - 7,\ -3r - 4$ $5r - 11$
4. $2a^2b^2 - 4a^4 - 4b^4,\ 7a^4 + 6b^4 - 5a^2b^2$ $3a^4 - 3a^2b^2 + 2b^4$

Subtract the second polynomial from the first.

5. $6v + 4,\ -2v - 5$ $8v + 9$
6. $3p^2 - 4pq + 7q^2,\ 8p^2 - 12q^2$ $-5p^2 - 4pq + 19q^2$

Summarizing the Lesson

In this lesson students have reviewed the terms needed for working with polynomials and learned the methods of adding and subtracting polynomials. Ask students to discuss the merits of using the vertical format or the horizontal format.

Additional Answers
Oral Exercises

5. $3t$ and $-\frac{1}{2}t$; $-t^4$ and $-\frac{1}{2}t^4$; $4t^2$, $\frac{3}{4}t^2$, and $16t^2$; 6 and −1
6. uv^2 and $\frac{3}{2}uv^2$; $-3u^2v$ and u^2v; uv and $2vu$; 4 and −7
7. $6xy$, $-xy$, and $\frac{2yx}{3}$; $-4xz$, $4zx$, and $\frac{xz}{2}$; $-zxy$ and $2yzx$
8. p^2q, $6qp^2$, and $-8p^2q$; $-3pq^3$, $4pq^3$, and q^3p; $2qp$ and $\frac{pq}{2}$

Suggested Assignments

Average Algebra
170/1–25 odd
S 170/Mixed Review

Average Alg. and Trig.
170/1–25 odd
S 170/Mixed Review

Extended Algebra
170/1–27 odd
S 170/Mixed Review

Extended Alg. and Trig.
170/1–27 odd
S 170/Mixed Review

Additional Answers Written Exercises

11. a. $2t^2 - 3t - 13$ b. $-13t - 1$

12. a. $3n^2 - n + 6$ b. $n^2 - n + 4$

13. a. $5v^3 + v^2 - 1$ b. $5v^3 - v^2 - 4v + 3$

14. a. $-2w^2$ b. $2w^3 + 2w - 2$

Supplementary Materials

Study Guide pp. 53–54

Written Exercises

4. $5x^2 + 3x - 7$; 2 5. $9x^2y^2 - 3x^2 + 5xy^2$; 4 6. $2x^3y + 4x^2y^3 - 3xy^2$; 5

Simplify, arranging terms in order of decreasing degree of *x*. Then write the degree of the polynomial.

A

1. $2 - x^2 + 3x + 2x^2 - 5x$ $x^2 - 2x + 2$; 2
2. $x^3 - 4x + 7x^2 + 3 + 2x$ $x^3 + 7x^2 - 2x + 3$; 3
3. $-x^2 + 3x^3 - 3x + x^2 + 2x$ $3x^3 - x$; 3
4. $2x^3 - 7 + 5x^2 - x^3 + 3x - x^3$
5. $x^2y^2 - x^2 + 8x^2y^2 + 5xy^2 - 2x^2$
6. $4x^2y^3 - xy^2 + 2x^3y - 2xy^2$
7. $4x^2yz^3 - xyz + 2x^2yz^3 + 5x^3y^2z^2$ $5x^3y^2z^2 + 6x^2yz^3 - xyz$; 7
8. $7xy^2z^3 - 4xy^2z^3 + 2x^2yz^2 - 3xy^2z^3$ $2x^2yz^2$; 5

In Exercises 9–16, (a) add the polynomials and (b) subtract the second polynomial from the first.

9. $5m - 4,\ 2m + 3$ a. $7m - 1$ b. $3m - 7$
10. $3u + 7,\ u - 8$ a. $4u - 1$ b. $2u + 15$
11. $t^2 - 8t - 7,\ t^2 + 5t - 6$
12. $2n^2 - n + 5,\ n^2 + 1$
13. $5v^3 - 2v + 1,\ v^2 + 2v - 2$
14. $w^3 - w^2 + w - 1,\ 1 - w - w^2 - w^3$
15. $3x^2 - 2xy + 4y^2,\ 2x^2 + 3y^2$ a. $5x^2 - 2xy + 7y^2$ b. $x^2 - 2xy + y^2$
16. $4a^2 + 3ab - b^2,\ b^2 - 2ab$ a. $4a^2 + ab$ b. $4a^2 + 5ab - 2b^2$

Simplify.

B

17. $3(x^2 - 2x + 4) + 2(5x^2 - 7)$ $13x^2 - 6x - 2$
18. $4(3y^2 - 2y) + 3(y^2 + 5y - 1)$ $15y^2 + 7y - 3$
19. $2(4m^2 + 3) - 7(m^2 - 2) + 1$ $m^2 + 21$
20. $5(2n^2 - 3) - 2(5n^2 + 2) - 6$ -25
21. $4a(x - y) + 3a(x + y) + ay$ $7ax$
22. $2d(3m + n) - 5d(m - 4n) - 10dm$ $-9dm + 22dn$
23. $3[2p^2 - q(3p + 4q)] - 2[4q^2 - 3p(p - 2q)]$ $12p^2 - 21pq - 20q^2$
24. $4[2a(3a - b) + 3ab] + 5[3b(a + 2b) - 4ab]$ $24a^2 - ab + 30b^2$

Find values of *a*, *b*, *c*, and *d* that make the equation true.

C

25. $(4t^3 - at^2 - 2bt + 5) - (ct^3 + 2t^2 - 6t + 3) = t^3 - 2t + d$ $a = -2, b = 4, c = 3, d = 2$
26. $(ax^3 - 3x^2 + 2bx - 2) - (2x^3 - cx^2 - 5x - 4d) = x^2 + x - 6$ $a = 2, b = -2, c = 4, d = -1$
27. $(x^2 + ax + 2b) + (x^2 - 2bx + 3a) = cx^2 - 7x + 3$ $a = -1, b = 3, c = 2$

Mixed Review Exercises

Find an equation of the form $f(x) = mx + b$ for the linear function *f*.

1. $m = -2;\ f(4) = -6$ $f(x) = -2x + 2$
2. $f(0) = 5;\ f(2) = 7$ $f(x) = x + 5$
3. $f(-1) = 3;\ f(3) = 1$ $f(x) = -\frac{1}{2}x + \frac{5}{2}$

For the line containing the given points, find (a) the slope and (b) an equation in standard form.

4. $(0, 4), (4, 0)$ a. -1 b. $x + y = 4$
5. $(-3, 1), (5, 3)$ a. $\frac{1}{4}$ b. $x - 4y = -7$
6. $(0, 0), (-3, 2)$ a. $-\frac{2}{3}$ b. $2x + 3y = 0$
7. $(6, -5), (-2, 1)$
8. $(-4, 7), (1, 9)$
9. $(-2, 5), (3, 5)$ a. 0 b. $y = 5$

 7. a. $-\frac{3}{4}$ b. $3x + 4y = -2$ 8. a. $\frac{2}{5}$ b. $2x - 5y = -43$

4-2 Using Laws of Exponents

Objective To use laws of exponents to multiply a polynomial by a monomial.

In this lesson you'll review the laws of exponents that are used when multiplying polynomials.

Laws of Exponents

Let a and b be real numbers and m and n be positive integers. Then:

1. $a^m \cdot a^n = a^{m+n}$ **2.** $(ab)^m = a^m b^m$ **3.** $(a^m)^n = a^{mn}$

Without the first law you would have to simplify a product of powers, such as $c^5 \cdot c^3$, by counting factors:

$$c^5 \cdot c^3 = \underbrace{\overbrace{(c \cdot c \cdot c \cdot c \cdot c)}^{5\ factors}\overbrace{(c \cdot c \cdot c)}^{3\ factors}}_{8\ factors} = c^8.$$

But using the first law you can write

$$c^5 \cdot c^3 = c^{5+3} = c^8.$$

Using the second law you can simplify a power of a product:

$$(2x)^3 = 2^3x^3 = 8x^3,$$

and using the third law you can simplify a power of a power:

$$(10^3)^2 = 10^{3 \cdot 2} = 10^6.$$

We'll prove the law $(ab)^m = a^m b^m$ by counting factors; the proofs of the other laws are left as Exercises 39 and 40.

Statements	Reasons
1. $(ab)^m = \overbrace{(ab)(ab)\ .\ .\ .\ (ab)}^{m\ factors\ of\ ab}$	1. Definition of a power
2. $= \underbrace{(a \cdot a \cdot\ .\ .\ .\ \cdot a)}_{m\ factors\ of\ a}\underbrace{(b \cdot b \cdot\ .\ .\ .\ \cdot b)}_{m\ factors\ of\ b}$	2. Commutative and associative properties of multiplication
3. $= a^m b^m$	3. Definition of a power

You can use the laws of exponents along with the commutative and associative properties of multiplication to simplify products and powers like the ones in Examples 1 and 2 on the next page.

Teaching Suggestions p. T93

Warm-Up Exercises

Simplify.

1. 2^5 32 **2.** $2^2 \cdot 2^3$ 32

3. $(2 \cdot 3)^2$ 36 **4.** $2^2 \cdot 3^2$ 36

5. 2^6 64 **6.** $(2^2)^3$ 64

Motivating the Lesson

Tell students that when an exponent is a positive integer, exponentiation can be thought of as "repeated multiplication." For example, the power x^4 is just an abbreviated form of the product $x \cdot x \cdot x \cdot x$. Today's lesson will use this idea to develop three basic laws of exponents.

Chalkboard Examples

Simplify.

1. $2x^4 \cdot x^3$ $2x^7$

2. $3x \cdot 4x^5$
$3 \cdot 4 \cdot x \cdot x^5 = 12x^6$

3. $(2a^2)^3$
$(2a^2)(2a^2)(2a^2) = 8a^6$

4. $(ab^4)^2$ $(ab^4)(ab^4) = a^2b^8$

5. $(m^r)^2$ $(m^r)(m^r) = m^{2r}$

6. $c^x \cdot c^{5-x}$ $c^{x+5-x} = c^5$

7. $6y^2(y^2 + y - 2)$
$6y^2(y^2) + 6y^2(y) + 6y^2(-2) =$
$6y^4 + 6y^3 - 12y^2$

Common Errors

Some students have no trouble simplifying $(x^2)^4 \cdot x^5$, but may not correctly simplify $(x^m)^n \cdot x^p$. If students use the laws of exponents incorrectly when dealing with variable exponents, show them a similar example with constant exponents when reteaching.

Check for Understanding

Here is a suggested use of the Oral Exercises to check students' understanding as you teach the lesson.
Oral Exs. 1–18: use before Example 1.
Oral Exs. 19–27: use after Example 4.

Guided Practice

Simplify.

1. $5x \cdot 3x^2 \cdot 4x^4$ $60x^7$
2. $(-r^2)(r^4)$ $-r^6$
3. $(3x^2yz^3)^2$ $9x^4y^2z^6$
4. $2pq(p^2 - 3pq + 2q^2)$ $2p^3q - 6p^2q^2 + 4pq^3$
5. $a^{x-4}(a^{x+2})^2$ a^{3x}

Summarizing the Lesson

Students have studied how to multiply a polynomial by a monomial using the laws of exponents. Ask the students to copy the three laws of exponents given in this lesson and then write in sentence form what each law means.

Suggested Assignments

Average Algebra
173/2–38 even

Average Alg. and Trig.
173/2–38 even

Extended Algebra
173/2–38 even, 39, 40

Extended Alg. and Trig.
173/2–38 even, 39, 40

Additional Answers
Oral Exercises

22. $4y^4z - 4y^2z^3$
23. $-3s^5 - 6s^3t^3$
24. $-2x^4y^2 + 2x^2y^4$

Example 1 Simplify: **a.** $(-3x^2y^3)(4xy^2)$ **b.** $(st^4)^3$ **c.** $(-x^3)^2$

Solution **a.** Use the first law of exponents and the fact that $x = x^1$.

$$(-3x^2y^3)(4xy^2) = (-3 \cdot 4)(x^2 \cdot x)(y^3 \cdot y^2)$$
$$= -12x^3y^5 \quad \textbf{Answer}$$

b. Use the second law of exponents. Then apply the third law.

$$(st^4)^3 = s^3(t^4)^3 = s^3t^{12} \quad \textbf{Answer}$$

c. Use the fact that $-x^3 = (-1)x^3$ and the second law of exponents.

$$(-x^3)^2 = [(-1)x^3]^2$$
$$= (-1)^2(x^3)^2 = x^6 \quad \textbf{Answer}$$

Example 2 Simplify: **a.** $u \cdot (u^2)^3 \cdot u^5$ **b.** $(3xy^2z^3)^3$

Solution **a.** $u(u^2)^3u^5 = u^1 \cdot u^6 \cdot u^5$

$$= u^{1+6+5} = u^{12} \quad \textbf{Answer}$$

b. $(3xy^2z^3)^3 = 3^3x^3(y^2)^3(z^3)^3 = 27x^3y^6z^9$ ***Answer***

To multiply a polynomial by a monomial, use the distributive property.

Example 3 Simplify $3t^2(t^3 - 2t^2 + t - 4)$.

Solution

$$3t^2(t^3 - 2t^2 + t - 4) = (3t^2)t^3 - (3t^2)(2t^2) + (3t^2)t - (3t^2)4$$
$$= 3t^5 - 6t^4 + 3t^3 - 12t^2 \quad \textbf{Answer}$$

Example 4 Simplify. Assume that variable exponents represent positive integers.

a. $(a^2)^k(a^k)^3$ **b.** $x^{m-n}(x^{m+n} + x^n)$

Solution **a.** $(a^2)^k(a^k)^3 = a^{2k} \cdot a^{3k} = a^{5k}$ ***Answer***

b.

$$x^{m-n}(x^{m+n} + x^n) = x^{m-n} \cdot x^{m+n} + x^{m-n} \cdot x^n$$
$$= x^{(m-n)+(m+n)} + x^{(m-n)+n}$$
$$= x^{2m} + x^m \quad \textbf{Answer}$$

Oral Exercises

Simplify. Assume that variable exponents represent positive integers.

1. $y^3 \cdot y^2$ y^5
2. $c \cdot c^3$ c^4
3. $(a^2)^3$ a^6
4. $(2k)^4$ $16k^4$
5. $(-x^3)^2$ x^6
6. $(-x^2)^3$ $-x^6$
7. $z^3 \cdot z^n$ z^{3+n}
8. $(z^3)^n$ z^{3n}
9. $(p^3q)^2$ p^6q^2
10. $b^3 \cdot b^2 \cdot b$ b^6
11. $r^k \cdot r^k$ r^{2k}
12. $(r^k)^k$ r^{k^2}

If the statement is true, give the law of exponents that justifies it. If the statement is false, give a counterexample. Counterexamples will vary.

13. $(ab)^2 = ab^2$ F
14. $a^3 \cdot a^2 = a^6$ F
15. $a^3 \cdot a^3 = a^6$ T; 1
16. $(a^3)^3 = a^9$ T; 3
17. $(a^3)^4 = a^7$ F
18. $(a^2b)^2 = a^2b^2$ F

Simplify. Assume that variable exponents represent positive integers.

19. $t^2(t^2 - 3)$ $t^4 - 3t^2$
20. $uv(u + v)$ $u^2v + uv^2$
21. $p^2(p - 2q)$ $p^3 - 2p^2q$
22. $4y^2z(y^2 - z^2)$
23. $-3s(s^4 + 2s^2t^3)$
24. $-2xy^2(x^3 - xy^2)$
25. $u^3 \cdot u^{m-3}$ u^m
26. $a^{n-1} \cdot a^{n+1}$ a^{2n}
27. $x^4(x^{k-2})^2$ x^{2k}

Written Exercises

Simplify. Assume that variable exponents represent positive integers.

A

1. $3z^2 \cdot 2z^3$ $6z^5$
2. $5r^2 \cdot r^4$ $5r^6$
3. $(-t^4)^3$ $-t^{12}$
4. $(-t^3)^4$ t^{12}
5. $(3x^2y)(xy^2)$ $3x^3y^3$
6. $(4p^2q)(p^2q^3)$ $4p^4q^4$
7. $(-2u^2)(uv^3)(-u^2v^2)$ $2u^5v^5$
8. $(r^2s)(-3rs^3)(2rs)$ $-6r^4s^5$
9. $(4a^3b^2)^2$ $16a^6b^4$
10. $(2c^2d^3)^3$ $8c^6d^9$
11. $(-3pq^4r^2)^3$ $-27p^3q^{12}r^6$
12. $(-x^2yz^3)^4$ $x^8y^4z^{12}$
13. $(-z^3)(-z)^3$ z^6
14. $(-c)^2(-c^4)$ $-c^6$
15. $(s^2t)^3(st^3)^2$ s^8t^9
16. $(2x^2y^3)^3(3x^3y)^2$ $72x^{12}y^{11}$
17. $3y(y^3 - 2y^2 + 3)$ $3y^4 - 6y^3 + 9y$
18. $x^2(x - 2x^2 + 3x^3)$ $x^3 - 2x^4 + 3x^5$
19. $rs^2(r^2 - 2rs - s^2)$ $r^3s^2 - 2r^2s^3 - rs^4$
20. $p^2q^3(p^2 - 4q)$ $p^4q^3 - 4p^2q^4$
21. $z^{n-2} \cdot z^{n+2}$ z^{2n}
22. $t^4 \cdot t^{k-4}$ t^k
23. $x^{m-1} \cdot x \cdot x^m$ x^{2m}
24. $y^{p+2} \cdot y^p \cdot y^{p-2}$ y^{3p}
25. $r^{h-2}(r^{h+1})^2$ r^{3h}
26. $s^3(s^{2k-1})^3$ s^{6k}

B

27. $t(t^{n-1} + t^n + t^{n+1})$ $t^n + t^{n+1} + t^{n+2}$
28. $x^2(x^k - x^{k-1} + x^{k-2})$ $x^{k+2} - x^{k+1} + x^k$
29. $p^n(p^{m-n+1} + p^{m-n})$ $p^{m+1} + p^m$
30. $s^{2n}(s^{2m-n} - s^{m-2n})$ $s^{2m+n} - s^m$
31. $z^{m-n}(z^{n+m} - z^{n-m} + z^n)$ $z^{2m} + z^m - 1$
32. $x^{h+k}(x^{2h-k} - x^{h-2k} + x^k)$ $x^{3h} - x^{2h-k} + x^{h+2k}$
33. $(t^m)^n(t^n)^{n-m}$ t^{n^2}
34. $(y^{h-k})^h(y^{h+k})^k$ $y^{h^2+k^2}$

In Exercises 35–38, solve for *n*.

35. $3^{5n} = 3^5(3^{2n})^2$ {5}
36. $(2^{3n})^2 = (2^n)^3 \cdot 2^{n+6}$ {3}
37. $3 \cdot 9^{2n} = (3^{n+1})^3$ {2}
38. $4^{n+3} \cdot 16^n = 8^{3n}$ {2}

C

39. Prove the first law of exponents.
40. Prove the third law of exponents.
41. Prove that for positive integers m, n, and r, $((a^m)^n)^r = a^{mnr}$.
42. Prove that for positive integers m and n, $(a^m)^n = (a^n)^m$.

Additional Answers
Written Exercises

39. 1. $a^m \cdot a^n = \underbrace{(a \cdot a \cdot \ldots \cdot a)}_{m \text{ factors}}\underbrace{(a \cdot a \cdot \ldots \cdot a)}_{n \text{ factors}}$
Definition of a power
2. $= \underbrace{a \cdot a \cdot \ldots \cdot a}_{m+n \text{ factors}}$
Assoc. prop. for mult.
3. $= a^{m+n}$
Definition of a power
4. $\therefore a^m \cdot a^n = a^{m+n}$
Transitive prop. of eq.

40. 1. $(a^m)^n = \underbrace{a^m \cdot a^m \cdot \ldots \cdot a^m}_{n \text{ factors}}$
Definition of a power
2. $= a^{\overbrace{m+m+\cdots+m}^{n \text{ addends}}}$
First law of exponents
3. $= a^{mn}$
Definition of mult.
4. $\therefore (a^m)^n = a^{mn}$
Transitive prop. of eq.

41. 1. $((a^m)^n)^r = (a^{mn})^r$
Third law of exponents
2. $= a^{(mn)r}$
Third law of exponents
3. $= a^{mnr}$
Assoc. prop. for mult.
4. $\therefore ((a^m)^n)^r = a^{mnr}$
Transitive prop. of eq.

42. 1. $(a^m)^n = a^{mn}$
Third law of exponents
2. $= a^{nm}$
Comm. prop. for mult.
3. $= (a^n)^m$
Third law of exponents
4. $\therefore (a^m)^n = (a^n)^m$
Transitive prop. of eq.

Supplementary Materials
Study Guide pp. 55–56

Teaching Suggestions p. T93

Reading Algebra p. T94

Suggested Extensions p. T94

Warm-Up Exercises

Simplify.

1. $(3x^2)(-x)(x^3)$ $-3x^6$
2. $(xz)(2xyz^2)$ $2x^2yz^3$
3. $4y(2y^2 - 3y + 1)$ $8y^3 - 12y^2 + 4y$
4. $(2m^2n)(m + n)$ $2m^3n + 2m^2n^2$
5. $t^2(t + 1) + 2t(t^2 - t)$ $t^3 + t^2 + 2t^3 - 2t^2 = 3t^3 - t^2$

Motivating the Lesson

Tell students that the statement "My Dear Aunt Sally" is a way of remembering the order of operations. Today's lesson, which deals with multiplying polynomials, includes a memory aid called FOIL to help students multiply two binomials.

Chalkboard Examples

Multiply.

1. $(2a - 5)(3a + 4)$ $6a^2 + 8a - 15a - 20 = 6a^2 - 7a - 20$
2. $(3c + 2)^2$ $(3c + 2)(3c + 2) = 9c^2 + 2(3c)(2) + 4 = 9c^2 + 12c + 4$
3. $(4k + 7)(4k - 7)$ $16k^2 - 28k + 28k - 49 = 16k^2 - 49$
4. $(3x + 1)(x^2 + 2x - 6)$ $3x^3 + 6x^2 - 18x + x^2 + 2x - 6 = 3x^3 + 7x^2 - 16x - 6$

4-3 Multiplying Polynomials

Objective To multiply polynomials.

To multiply two polynomials, you use the distributive property: Multiply each term of one polynomial by each term of the other and add the resulting monomials.

Example 1 Multiply: $(2x + 3)(x^2 + 4x - 5)$

Solution 1 You can arrange the work horizontally.

$$\begin{aligned}(2x + 3)(x^2 + 4x - 5) &= 2x(x^2 + 4x - 5) + 3(x^2 + 4x - 5)\\ &= 2x^3 + 8x^2 - 10x + 3x^2 + 12x - 15\\ &= 2x^3 + 11x^2 + 2x - 15 \quad \textbf{\textit{Answer}}\end{aligned}$$

Solution 2 You can arrange the work vertically.

$$\begin{array}{rrrrl} & x^2 & + 4x & - 5 & \\ & 2x & + 3 & & \\ \hline 2x^3 + & 8x^2 & - 10x & & \leftarrow \text{This is } 2x(x^2 + 4x - 5). \\ & 3x^2 & + 12x & - 15 & \leftarrow \text{This is } 3(x^2 + 4x - 5). \\ \hline 2x^3 + & 11x^2 & + 2x & - 15 & \textbf{\textit{Answer}} \end{array}$$

In Example 1, the factors of the product $(2x + 3)(x^2 + 4x - 5)$ are a *binomial* and a *trinomial*, respectively. A **binomial** is a polynomial that has two terms. A **trinomial** is a polynomial that has three terms.

When multiplying two binomials, you can save time if you learn to find the product mentally. Study the method illustrated in Example 2.

Example 2 Multiply: $(2a - b)(3a + 5b)$

Solution

$$(2a - b)(3a + 5b) = 6a^2 + 7ab - 5b^2$$

1. Multiply the *first* terms of the binomials.
   $(2a)(3a)$ → $6a^2$
2. Multiply the *outer* terms of the binomials.
   Multiply the *inner* terms of the binomials.
   Then add the products.
   $(2a)(5b) + (3a)(-b)$ → $7ab$
3. Multiply the *last* terms of the binomials.
   $(-b)(5b)$ → $-5b^2$

The method used in Example 2 is sometimes called the FOIL method.

The word FOIL reminds you to multiply the *F*irst, *O*uter, *I*nner, and *L*ast terms when multiplying two binomials.

Certain special products occur so frequently that their patterns should be memorized.

Special Product	*Pattern and Example*
$(a + b)^2 = a^2 + 2ab + b^2$	$(\text{first} + \text{second})^2 = (\text{first})^2 + 2(\text{first})(\text{second}) + (\text{second})^2$ $(4s + 3t)^2 = (4s)^2 + 2(4s)(3t) + (3t)^2$ $= 16s^2 + 24st + 9t^2$
$(a - b)^2 = a^2 - 2ab + b^2$	$(\text{first} - \text{second})^2 = (\text{first})^2 - 2(\text{first})(\text{second}) + (\text{second})^2$ $(3x - 5)^2 = (3x)^2 - 2(3x)(5) + 5^2$ $= 9x^2 - 30x + 25$
$(a + b)(a - b) = a^2 - b^2$	$(\text{first} + \text{second})(\text{first} - \text{second}) = (\text{first})^2 - (\text{second})^2$ $(2p + 3q)(2p - 3q) = (2p)^2 - (3q)^2$ $= 4p^2 - 9q^2$

Oral Exercises

4. $2n^2 + 13n - 24$ 5. $6 - 5p - 6p^2$ 6. $3 - 14w + 8w^2$

Use the FOIL method to find each product mentally.

1. $(x + 6)(x + 4)$ $x^2 + 10x + 24$
2. $(y + 7)(y - 3)$ $y^2 + 4y - 21$
3. $(3m + 1)(2m + 5)$ $6m^2 + 17m + 5$
4. $(2n - 3)(n + 8)$
5. $(2 - 3p)(3 + 2p)$
6. $(1 - 4w)(3 - 2w)$
7. $(3x + y)(x + 2y)$ $3x^2 + 7xy + 2y^2$
8. $(c + d)(2c - 5d)$ $2c^2 - 3cd - 5d^2$
9. $(2p - q)(3p - q)$ $6p^2 - 5pq + q^2$

Find each special product.

10. $(x + 3)^2$ $x^2 + 6x + 9$
11. $(2y + 5)^2$ $4y^2 + 20y + 25$
12. $(3m + n)^2$ $9m^2 + 6mn + n^2$
13. $(a - 2)^2$ $a^2 - 4a + 4$
14. $(4z - 3)^2$ $16z^2 - 24z + 9$
15. $(p - 2q)^2$
16. $(k + 1)(k - 1)$ $k^2 - 1$
17. $(3r + 5)(3r - 5)$ $9r^2 - 25$
18. $(4c + d)(4c - d)$ $16c^2 - d^2$

15. $p^2 - 4pq + 4q^2$

Written Exercises

6. $16k^2 - 40k + 25$ 9. $14t^2 - 3t - 2$

Multiply.

A

1. $(3v + 1)(2v - 5)$ $6v^2 - 13v - 5$
2. $(2x - 3)(3x + 2)$ $6x^2 - 5x - 6$
3. $(4z + 3)(3z - 4)$ $12z^2 - 7z - 12$
4. $(r - 4)(3r - 2)$ $3r^2 - 14r + 8$
5. $(3x + 10)^2$ $9x^2 + 60x + 100$
6. $(4k - 5)^2$
7. $(5y - 2)(5y + 2)$ $25y^2 - 4$
8. $(2s + 7)(2s - 7)$ $4s^2 - 49$
9. $(7t + 2)(2t - 1)$
10. $(5z + 6)(6z - 5)$ $30z^2 + 11z - 30$
11. $(9t + 1)(1 - 9t)$ $1 - 81t^2$
12. $(9 - 5t)(5t - 9)$ $-25t^2 + 90t - 81$

Check for Understanding

Here is a suggested use of the Oral Exercises to check students' understanding as you teach the lesson.

Oral Exs. 1–9: use after Example 2.

Oral Exs. 10–18: use after the table of special products on page 175.

Guided Practice

Multiply.

1. $(3y - 2)(y + 4)$ $3y^2 + 10y - 8$
2. $(5k - 3)(7k - 2)$ $35k^2 - 31k + 6$
3. $(6 - 5m)^2$ $36 - 60m + 25m^2$
4. $(2a + 3b)(3a - 2b)$ $6a^2 + 5ab - 6b^2$
5. $(4t - 1)(1 + 4t)$ $16t^2 - 1$
6. $(a - b)(3a + b)$ $3a^2 - 2ab - b^2$
7. $(q^2 - p)^2$ $q^4 - 2pq^2 + p^2$
8. $r^2(r + 1)(r - 1)$ $r^4 - r^2$
9. $(x^2 - 2xy + y^2)(x + y)$ $x^3 - x^2y - xy^2 + y^3$
10. $(3x + y + 1) \times (3x - y - 1)$ $9x^2 - y^2 - 2y - 1$

Summarizing the Lesson

Point out that in this lesson, the methods for multiplying polynomials are based on the distributive property. The FOIL and special-product methods help students find products of binomials quickly.

Suggested Assignments

Average Algebra
175/1–47 odd
S 176/Mixed Review

Average Alg. and Trig.
175/1–47 odd
S 176/Mixed Review

Extended Algebra
175/1–51 odd
S 176/Mixed Review

Extended Alg. and Trig.
175/3–51 mult. of 3, 53, 54
S 176/Mixed Review
Assign with Lesson 4-4.

Additional Answers Written Exercises

33. $a^4 - 4a^2b^2 - ab^3 - 2b^4$

34. $3s^4 + 2s^3t - 9s^2t^2 + 4t^4$

Common Errors

Some students think that $(a + b)^2 = a^2 + b^2$. When reteaching, show the students that $(a + b)^2 = (a + b)(a + b) = a^2 + ab + ab + b^2 = a^2 + 2ab + b^2$. A similar error may occur with $(a - b)^2$.

Supplementary Materials

Study Guide	pp. 57–58
Practice Master	20
Resource Book	p. 113
Overhead Visual	3

Multiply. Assume that variable exponents represent positive integers.

13. $(x - 2y)(3x + 4y)$ $3x^2 - 2xy - 8y^2$

14. $(5h - 3k)(h - 2k)$ $5h^2 - 13hk + 6k^2$

15. $(2p + 3q)(3p - 2q)$ $6p^2 + 5pq - 6q^2$

16. $(10r - 3s)(r + 2s)$ $10r^2 + 17rs - 6s^2$

17. $(x^2 - 3)(x^2 + 3)$ $x^4 - 9$

18. $(p^2 - 2q^2)(p^2 + 2q^2)$ $p^4 - 4q^4$

19. $(s^3 + t^3)^2$ $s^6 + 2s^3t^3 + t^6$

20. $(2z^2 - 5)^2$ $4z^4 - 20z^2 + 25$

21. $t(t - 2)(t + 1)$ $t^3 - t^2 - 2t$

22. $x^2(x - 3)(x + 3)$ $x^4 - 9x^2$

23. $xy(x - y)^2$ $x^3y - 2x^2y^2 + xy^3$

24. $mn(m - n)(m - 2n)$ $m^3n - 3m^2n^2 + 2mn^3$

25. $(2c + 1)(c^2 - 3c + 2)$ $2c^3 - 5c^2 + c + 2$

26. $(t - 3)(2t^2 - t + 2)$ $2t^3 - 7t^2 + 5t - 6$

27. $(x^2 + 3x - 5)(x + 2)$ $x^3 + 5x^2 + x - 10$

28. $(z^2 - 2z + 4)(z + 3)$ $z^3 + z^2 - 2z + 12$

29. $(y^4 - 3y^2 + 1)(y^2 - 2)$ $y^6 - 5y^4 + 7y^2 - 2$

30. $(3 - k^2)(2 - k^2 - k^4)$ $k^6 - 2k^4 - 5k^2 + 6$

B **31.** $(x^2 - x + 2)(x^2 + x - 1)$ $x^4 + 3x - 2$

32. $(y^2 - 2y + 1)(y^2 + y + 1)$ $y^4 - y^3 - y + 1$

33. $(a + 2b)(a^3 - 2a^2b - b^3)$

34. $(3s + 2t)(s^3 - 3st^2 + 2t^3)$

35. $(p^n - 1)^2$ $p^{2n} - 2p^n + 1$

36. $(x^{2n} - y^n)^2$ $x^{4n} - 2x^{2n}y^n + y^{2n}$

37. $(r^n - s^n)(r^n + 2s^n)$ $r^{2n} + r^ns^n - 2s^{2n}$

38. $(x^n + 1)(x^n - 1)$ $x^{2n} - 1$

39. $(a - b)^3$ $a^3 - 3a^2b + 3ab^2 - b^3$

40. $(a + b)^3$ $a^3 + 3a^2b + 3ab^2 + b^3$

41. $(a + b)(a^2 - ab + b^2)$ $a^3 + b^3$

42. $(a - b)(a^2 + ab + b^2)$ $a^3 - b^3$

43. $(a - b)(a^3 + a^2b + ab^2 + b^3)$ $a^4 - b^4$

44. $(a + b)(a^3 - a^2b + ab^2 - b^3)$ $a^4 - b^4$

45. $(x + y)(x - y)(x^2 + y^2)$ $x^4 - y^4$

46. $(x + y)^2(x - y)^2$ $x^4 - 2x^2y^2 + y^4$

47. $(x^2 + 2x + 2)(x^2 - 2x + 2)$ $x^4 + 4$

48. $(x^2 - 4x + 8)(x^2 + 4x + 8)$ $x^4 + 64$

Without actually finding the product, determine how many terms the simplified product has.

C **49.** $(u + v)(x + y)(u - v)(x - y)$ 4

50. $(u + v + w)(x + y + z)$ 9

51. $(u + v + w)(u + v - w)$ 4

52. $(x - y + z)(x + y + z)$ 4

Find the value or values of k that make the equation true.

53. $(x + 2k)(x - 3k) = x^2 + 2x - 24$ -2

54. $(2x + k)(x - 2k) = 2x^2 + 9x - 18$ -3

55. $(2x - k)(3x + 2k) = 6x^2 + kx - 32$ $4, -4$

56. $(3kx + 2)^2 = 81x^2 + 12kx + 4$ $3, -3$

Mixed Review Exercises

Simplify.

1. $(3x^2 - 7x + 9) - (x^2 + 4x - 1)$ $2x^2 - 11x + 10$

2. $(a^2b^3)^3$ a^6b^9

3. $(4m^2n)(-3mn^3)$ $-12m^3n^4$

4. $5(2y - 1) - 3(y + 2)$ $7y - 11$

5. $2c(d - 3) + 3d(c + 4)$ $5cd - 6c + 12d$

6. $4p(2p^2 - p + 5)$ $8p^3 - 4p^2 + 20p$

7. $(-u^2)^4(-u)^3$ $-u^{11}$

8. $(9z^3 - 4z) + (5z^2 - 8)$ $9z^3 + 5z^2 - 4z - 8$

9. $(y^2 - 5y + 9) - (9 + 5y + y^2)$ $-10y$

10. $\left(-\frac{1}{2}x^2\right)(-4x^4)$ $2x^6$

Self-Test 1

Vocabulary

constant (p. 167)
monomial (p. 167)
coefficient (p. 167)
degree of a variable (p. 167)
degree of a monomial (p. 167)
similar monomials (p. 167)
polynomial (p. 167)
simplified polynomial (p. 167)
degree of a polynomial (p. 167)
binomial (p. 174)
trinomial (p. 174)

Simplify. Assume that variable exponents represent positive integers.

1. $(x^2 + 3x - 2) + (1 - 2x)$ $x^2 + x - 1$ **2.** $3(t^2 - 2) - t(3t - 1)$ $t - 6$ **Obj. 4-1, p. 167**

3. $(3p^2)(-2pq^2)^2$ $12p^4q^4$ **4.** $t^2 \cdot t^{n+1} \cdot t^{n-3}$ t^{2n} **Obj. 4-2, p. 171**

5. $(4a - 5b)(2a + 3b)$ $8a^2 + 2ab - 15b^2$ **6.** $(2y - 1)(y^2 + 2y + 3)$ **Obj. 4-3, p. 174**

7. $(3x - 2)^2$ $9x^2 - 12x + 4$ **8.** $(4c - 3)(4c + 3)$ $16c^2 - 9$

6. $2y^3 + 3y^2 + 4y - 3$

Check your answers with those at the back of the book.

Biographical Note / *Ernest Everett Just*

How the biologist Ernest Everett Just (1883–1941) felt about life affected how he studied it. He saw living things as more than a mass of cells and a chain of chemical interactions. He was therefore very careful about how he conducted his laboratory work.

Isolated cells, unless handled properly, can easily die or grow unnaturally. At the Marine Biology Laboratories in Woods Hole, Massachusetts, Just showed how to keep cells normal and healthy in a laboratory.

Just was particularly interested in the function of the cell wall. Describing the cell wall as more than just a "dam against the outside world," he saw it as a living, interacting part of the cell. One result of his work with cells was an important original book, *The Biology of the Cell Surface*.

Besides spending the summers between 1912 and 1929 working at Woods Hole, Just taught and performed research at Howard University in Washington, D.C. By applying new experimental techniques to the study of cell development, Just became an important force in the advancement of genetics and embryology.

Quick Quiz

Simplify.

1. $(2x^2 + 3x - 4) - (x^2 + 5x - 1)$
$x^2 - 2x - 3$

2. $3x^2(2x^3)^2$ $12x^8$

3. $3y^2(2y^3 - y + 1)$
$6y^5 - 3y^3 + 3y^2$

4. $(2t + 1)(2t - 2)$
$4t^2 - 2t - 2$

5. $(6q^2 + 7)(q^2 - q + 1)$
$6q^4 - 6q^3 + 13q^2 - 7q + 7$

Reading Algebra / Symbols

At times you may think that the use of symbols only complicates algebra and that it's not worth the effort to master the skill of reading symbols. To see why this is not the case, try to write in simple words an understandable version of the law $a^m \cdot a^n = a^{m+n}$. The task is difficult even if you are allowed to use variables. The use of symbols permits the presentation of ideas in a clear, concise, and easy-to-remember fashion.

When you come across an unfamiliar symbol, be sure you learn what it means. If necessary, check the list on page xvi, which includes each symbol in your textbook, its meaning, and a reference to the page on which it is introduced. Practice translating back and forth between expressions using the symbol and equivalent expressions in words.

The exercises below deal with two symbols related to divisibility. Let a, b, and n be integers with $n \neq 0$. We say that *a is congruent to b* (mod n) if n divides the difference $a - b$. (Recall that one integer is said to *divide* another if and only if the second integer is an integral multiple of the first.) The symbol $\equiv$ stands for "is congruent to," and the symbol $|$ stands for "divides." Therefore, $a \equiv b \pmod{n}$ if $n|(a - b)$. For example, $18 \equiv 3 \pmod{5}$ since $5|(18 - 3)$.

Exercises

1. Rewrite each expression in words and tell whether it is true or false.
   a. $2|8$ **b.** $4|4$ **c.** $2|1$ **d.** $2|0$ **e.** $0|3$
2. What does the symbol $\neq$ mean? What do you think the symbol $\nmid$ means? does not equal; does not divide
3. Tell whether each statement is true or false.
   a. $2 \nmid 5$ True **b.** $n \nmid (n + 1)$ if $n \geq 2$ True **c.** $5 \nmid 0$ False
4. If $a|b$ and $b|c$, is it necessarily true that $a|c$? Yes
5. If $m|n$ and $n|m$, is it necessarily true that $m = n$? No
6. Write the statement that 17 is congruent to 2 (mod 5) using the congruency symbol $\equiv$. $17 \equiv 2 \pmod{5}$
7. Write $19 \equiv 4 \pmod{5}$ using the divisibility symbol $|$. $5|(19 - 4)$
8. If $k|s$ and $k|t$, is it necessarily true that $s \equiv t \pmod{k}$? Yes
9. Rewrite each expression in words and tell whether it is true or false.
   a. $100 \equiv 25 \pmod{5}$ **b.** $-3 \equiv 3 \pmod{5}$
   c. $6 \equiv 1 \pmod{5}$ **d.** $0 \equiv 5 \pmod{5}$
10. What do we call the set of integers n that satisfy $n \equiv 0 \pmod{2}$? even integers
11. What do we call the set of integers n that satisfy $n \equiv 1 \pmod{2}$? odd integers
12. Find a value of n that makes the statement true. Answers may vary.
   a. $11 \equiv 27 \pmod{n}$ 2 **b.** $6 \equiv n \pmod{2}$ 2

Additional Answers
Exercises

1. a. Two divides eight; true
   b. Four divides four; true
   c. Two divides one; false
   d. Two divides zero; true
   e. Zero divides three; false
9. a. One hundred is congruent to twenty-five (mod five); true.
   b. Negative three is congruent to three (mod five); false.
   c. Six is congruent to one (mod five); true.
   d. Zero is congruent to five (mod five); true.

Factors of Polynomials

4-4 Using Prime Factorization

Objective To find the GCF and LCM of integers and monomials.

To **factor** a number over a set of numbers, you write it as a product of numbers chosen from that set, called the **factor set.** For example, the number 14 can be factored over the integers in the following ways:

$$(1)(14), \quad (-1)(-14), \quad (2)(7), \quad \text{and} \quad (-2)(-7).$$

In this book, integers will be factored over the set of integers unless some other set is specified.

The adjective corresponding to the word "integer" is *integral.* Because $2 \cdot 7 = 14$, you say that 2 and 7 are *integral factors* of 14 and that 14 is an *integral multiple* of 2 and of 7.

A **prime number,** or **prime,** is an integer greater than 1 whose only positive integral factors are itself and 1. The first ten primes are

$$2, 3, 5, 7, 11, 13, 17, 19, 23, \text{ and } 29.$$

If the factor set is restricted to the set of primes, then 14 has only one factorization, $2 \cdot 7$. This is called the *prime factorization* of 14. To find the **prime factorization** of a positive integer, you write the integer as a product of primes. If a prime factor occurs more than once, use an exponent.

Example 1 Find the prime factorization of 936.

Solution Here is a systematic way to find the prime factorization of a large number: Try the primes, in order, as factors. Use each repeatedly until it is no longer a factor. Then try the next prime.

$$\begin{aligned} 936 &= 2 \cdot 468 \\ &= 2 \cdot 2 \cdot 234 \\ &= 2 \cdot 2 \cdot 2 \cdot 117 \\ &= 2 \cdot 2 \cdot 2 \cdot 3 \cdot 39 \\ &= 2 \cdot 2 \cdot 2 \cdot 3 \cdot 3 \cdot 13 \end{aligned}$$

$\therefore 936 = 2^3 \cdot 3^2 \cdot 13$ ***Answer***

The **greatest common factor (GCF)** of two or more integers is the greatest integer that is a factor of each. The **least common multiple (LCM)** of two or more integers is the least positive integer having each as a factor. When given two or more integers, you can use their prime factorizations to find their GCF and LCM.

Teaching Suggestions p. T94

Warm-Up Exercises

State whether or not each number is prime. If not prime, write the number as a product of primes.

1. 37 Prime
2. 35 Not prime; $5 \cdot 7$
3. 43 Prime
4. 42 Not prime; $2 \cdot 3 \cdot 7$

Motivating the Lesson

Tell students that when scientists analyze a chemical compound, they break it down into its constituent elements. Likewise, mathematicians "break" a number into a product of primes. This process is called factoring, the topic of today's lesson.

Problem Solving Strategies

Students may wish to *use a diagram,* like the "factor tree" shown below, to find the prime factorization of an integer.

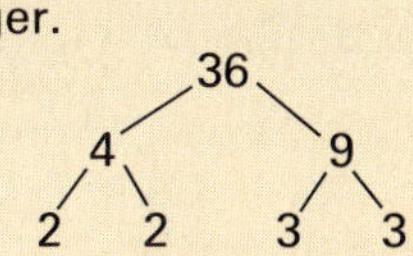

$\therefore 36 = 2^2 \cdot 3^2$

Chalkboard Examples

1. Give the prime factorization of 924.
   $924 = 2 \cdot 2 \cdot 3 \cdot 7 \cdot 11$
   $= 2^2 \cdot 3 \cdot 7 \cdot 11$

Find the GCF and the LCM.

2. 20, 28
   $20 = 2 \cdot 2 \cdot 5 = 2^2 \cdot 5$
   $28 = 2 \cdot 2 \cdot 7 = 2^2 \cdot 7$
   $\text{GCF} = 2^2 = 4,$
   $\text{LCM} = 2^2 \cdot 5 \cdot 7 = 140$
3. 100, 120, 90
   $100 = 2 \cdot 2 \cdot 5 \cdot 5$
   $= 2^2 \cdot 5^2$
   $120 = 2 \cdot 2 \cdot 2 \cdot 3 \cdot 5$
   $= 2^3 \cdot 3 \cdot 5$
   $90 = 2 \cdot 3 \cdot 3 \cdot 5$
   $= 2 \cdot 3^2 \cdot 5$
   $\text{GCF} = 2 \cdot 5 = 10,$
   $\text{LCM} = 2^3 \cdot 3^2 \cdot 5^2$
   $= 1800$
4. $8ax^2$, $12a^2x$
   $8ax^2 = 2 \cdot 2 \cdot 2 \cdot ax^2$
   $= 2^3ax^2$
   $12a^2x = 2 \cdot 2 \cdot 3a^2x$
   $= 2^2 \cdot 3a^2x$
   $\text{GCF} = 2^2 \cdot a \cdot x = 4ax,$
   $\text{LCM} = 2^3 \cdot 3 \cdot a^2 \cdot x^2$
   $= 24a^2x^2$
5. $32x^3y^3$, $120xy^4$, $42x^2y^3$
   $32x^3y^3 = 2 \cdot 2 \cdot 2 \cdot 2 \cdot 2x^3y^3 = 2^5x^3y^3$
   $120xy^4 = 2 \cdot 2 \cdot 2 \cdot 3 \cdot 5xy^4 = 2^3 \cdot 3 \cdot 5xy^4$
   $42x^2y^3 = 2 \cdot 3 \cdot 7x^2y^3$
   $\text{GCF} = 2 \cdot x \cdot y^3 = 2xy^3,$
   $\text{LCM} = 2^5 \cdot 3 \cdot 5 \cdot 7 \cdot x^3y^4 = 3360x^3y^4$

Check for Understanding

Oral Exs. 1–12: use after Example 1.
Oral Exs. 13–16: use after Example 2.
Oral Exs. 17–24: use after Example 3.

Example 2 Find **(a)** the GCF and **(b)** the LCM of the following numbers:

72, 108, and 126.

Solution The prime factorizations are

$$72 = 2^3 \cdot 3^2, \quad 108 = 2^2 \cdot 3^3, \quad \text{and} \quad 126 = 2 \cdot 3^2 \cdot 7.$$

a. To find the GCF, take the *least* power of each *common* prime factor.

$\therefore$ GCF $= 2 \cdot 3^2 = 18$

b. To find the LCM, take the *greatest* power of *each* prime factor.

$\therefore$ LCM $= 2^3 \cdot 3^3 \cdot 7 = 1512$

Factoring monomials is very similar to factoring integers. For example, $6xy^2$ is a *factor* of $18x^3y^2$ because $18x^3y^2 = 6xy^2 \cdot 3x^2$. Moreover, $18x^3y^2$ is a *multiple* of $6xy^2$ and of $3x^2$.

In this book, monomials with integral coefficients will be factored over the set of monomials with integral coefficients unless some other factor set is specified.

The **greatest common factor (GCF)** of two or more monomials is the common factor that has the greatest degree and the greatest numerical coefficient. The **least common multiple (LCM)** of two or more monomials is the common multiple that has the least degree and the least positive numerical coefficient. Here is an example.

The GCF of $9a^2x$ and $-6ax^3$ is $3ax$.

The LCM of $9a^2x$ and $-6ax^3$ is $18a^2x^3$.

Although numerical coefficients of monomials may be negative, the numerical coefficients of the GCF and LCM of monomials are always positive.

Example 3 Find **(a)** the GCF and **(b)** the LCM of the following monomials:

$48u^2v^2$ and $60uv^3w$.

Solution The prime factorizations of the coefficients are

$$48 = 2^4 \cdot 3 \quad \text{and} \quad 60 = 2^2 \cdot 3 \cdot 5.$$

a. The GCF of 48 and 60 is

$2^2 \cdot 3 = 12$.

Compare the powers of each variable occurring in *both* monomials. Use the power with the *least* exponent.

Compare u^2 and u. Use u
Compare v^2 and v^3. Use v^2.

$\therefore$ GCF $= 12uv^2$

b. The LCM of 48 and 60 is

$2^4 \cdot 3 \cdot 5 = 240$.

Compare the powers of each variable occurring in *either* monomial. Use the power with the *greatest* exponent.

Compare u^2 and u. Use u^2.
Compare v^3 and v^2. Use v^3.
Use w.

$\therefore$ LCM $= 240u^2v^3w$

Oral Exercises

Find all positive factors of each integer.

1. 12 1, 2, 3, 4, 6, 12
2. 15 1, 3, 5, 15
3. 26 1, 2, 13, 26
4. 35 1, 5, 7, 35

Find the prime factorization of each integer.

5. 10 $2 \cdot 5$
6. 33 $3 \cdot 11$
7. 25 5^2
8. 50 $2 \cdot 5^2$
9. 20 $2^2 \cdot 5$
10. 49 7^2
11. 98 $2 \cdot 7^2$
12. 18 $2 \cdot 3^2$

Find the GCF and LCM of each pair of integers.

13. 8 and 12 4; 24
14. 15 and 20 5; 60
15. 9 and 11 1; 99
16. 13 and 26 13; 26

Find the GCF and LCM of the given monomials.

17. a^2b and ab^2 ab; a^2b^2
18. $2x^3$ and x^2y x^2; $2x^3y$
19. $4st^2$ and $6rs^2t$ $2st$; $12rs^2t^2$
20. $4uv^2w$ and $-2u^2v^2$ $2uv^2$; $4u^2v^2w$
21. $15x^2y$, $-25xyz^2$, and $20xy^2$ $5xy$; $300x^2y^2z^2$
22. $12ar^2s^3$, $8a^2r^2s$, and ar^3 ar^2; $24a^2r^3s^3$
23. There are fifteen primes less than 50. Ten of them are given on page 179. What are the other five? 31, 37, 41, 43, 47
24. Give a convincing argument that justifies the following statement: If a is a factor of b and b is a factor of c, then a is a factor of c.
Let $b = k_1a$ and $c = k_2b$. Then $c = k_2(k_1a) = k_2k_1a$.

Written Exercises

Find the prime factorization of each integer.

A

1. 140 $2^2 \cdot 5 \cdot 7$
2. 198 $2 \cdot 3^2 \cdot 11$
3. 89 prime
4. 756 $2^2 \cdot 3^3 \cdot 7$
5. 441 $3^2 \cdot 7^2$
6. 203 $7 \cdot 29$
7. 2548 $2^2 \cdot 7^2 \cdot 13$
8. 3861 $3^3 \cdot 11 \cdot 13$

Find (a) the GCF and (b) the LCM of the following monomials.

9. 20, 35 5; 140
10. 45, 75 15; 225
11. -48, 108 12; 432
12. 315, -525 105; 1575
13. 84, -56, 140 28; 840
14. 168, 280, 196 28; 5880
15. 3, 5, 7, 9 1; 315
16. 30, 35, 36, 42 1; 1260
17. $9p^3q$, $15p^2$ $3p^2$; $45p^3q$
18. $49x^3$, $35x^2y$ $7x^2$; $245x^3y$
19. $68xy^2z$, $51y^2z^2$ $17y^2z$; $204xy^2z^2$
20. $52r^2s$, $78rs^2t$ $26rs$; $156r^2s^2t$
21. $110h^3k^2r$, $-88h^2k^2r^2$ $22h^2k^2r$; $440h^3k^2r^2$
22. $98a^2b^2c$, $-70abc^2$ $14abc$; $490a^2b^2c^2$
23. $14ab$, $14bc$, $21ac$ 7; $42abc$
24. $22xy^2z^2$, $33x^2yz^2$, $44x^2yz$ $11xyz$; $132x^2y^2z^2$
25. $26p^3q^2r^2$, $39p^2q^3r^2$, $78p^2q^2r^3$ $13p^2q^2r^2$; $78p^3q^3r^3$
26. $200a^3b^2c$, $300a^2bc^3$, $400ab^3c^2$ $100abc$; $1200a^3b^3c^3$

Guided Practice

Give the prime factorization.

1. 420 $2^2 \cdot 3 \cdot 5 \cdot 7$
2. 130 $2 \cdot 5 \cdot 13$
3. 490 $2 \cdot 5 \cdot 7^2$

Find the GCF and LCM.

4. 96, 54 6; 864
5. 8, 12, 30 2; 120
6. $8ab^2c^3$, $16a^3b^2c$ $8ab^2c$; $16a^3b^2c^3$
7. $14b^2$, $35b^3$ $7b^2$; $70b^3$
8. $8x^5$, x^3, $7x^2$ x^2; $56x^5$
9. $132xy^3$, $44x^2$, $99xy^4$ $11x$; $396x^2y^4$

Summarizing the Lesson

Students have studied prime factorization and its relationship to finding the greatest common factor and the least common multiple. Ask students to describe, in their own words, the process for finding the GCF and LCM of two monomials.

Suggested Assignments

Average Algebra
181/2–30 even
R 177/Self-Test 1

Average Alg. and Trig.
181/2–30 even
R 177/Self-Test 1

Extended Algebra
181/2–32 even
R 177/Self-Test 1

Extended Alg. and Trig.
181/3–30 mult. of 3, 34, 35
R 177/Self-Test 1
Assign with Lesson 4-3.

Supplementary Materials

Study Guide pp. 59–60

Additional Answers
Written Exercises

33. The LCM of two numbers is the product of the greatest power of each prime factor of both numbers. If the numbers are relatively prime, they have no factors in common and the LCM is the product of the two numbers.

34. If the integers are divided by their GCF, the resulting quotients have no common factors other than 1 and are relatively prime.

35. False; 2 and 3 are relatively prime, as are 3 and 4, but 2 and 4 are not relatively prime since their GCF is 2.

36. Let h, k, m, and n be integers such that $hm - kn = 1$ and suppose that m and n are not relatively prime. Then there is an integer $x > 1$ and integers y and z such that $m = xy$ and $n = xz$. Then, $hxy - kxz = 1$ or $x(hy - kz) = 1$ and $hy - kz$ is the reciprocal of x. Since h, y, k, and z are integers, so is $hy - kz$. There are only two integers whose reciprocals are integers, -1 and 1. This contradicts the fact that $x > 1$. Thus, m and n are relatively prime.

B

27. The GCF of 84 and another integer is 42, and their LCM is 252. Find the other integer. 126

28. The GCF of $12x^2y^3$ and another monomial is $6xy^3$, and their LCM is $36x^2y^4$. Find the other monomial. $18xy^4$

A positive integer n is *perfect* if it is the sum of all its positive factors except n itself.

29. Show that 496 is perfect. $1 + 2 + 4 + 8 + 16 + 31 + 62 + 124 + 248 = 496$

30. There are two perfect integers less than 30. Find them. 6 and 28

In Exercises 31 and 32, p and q represent different prime numbers.

31. List all the positive factors of each product.
a. pq **b.** p^2q^2 **c.** p^3q^3

31. a. $1, p, q, pq$
b. $1, p, q, p^2, q^2, pq, p^2q, pq^2, p^2q^2$
c. $1, p, q, p^2, q^2, pq, p^3, p^2q, pq^2, q^3, p^3q, p^2q^2, pq^3, p^3q^2, p^2q^3, p^3q^3$

32. How many positive factors does p^mq^m have? $(m + 1)^2$

Two integers are *relatively prime* if their greatest common factor is 1. Give a convincing argument to justify each statement.

33. The LCM of two relatively prime integers is their product.

34. If each of two integers is divided by their GCF, the two quotients are relatively prime.

C

35. Prove or disprove: If a and b are relatively prime and b and c are relatively prime, then a and c are relatively prime.

36. Prove that m and n are relatively prime if there are integers h and k such that $hm - kn = 1$. (The converse of this statement is also true. See the Euclidean Algorithm, page 215.)

Historical Note / *Prime Numbers*

The Greek mathematician Euclid, author of the *Elements,* is best known for his work in geometry. In fact his famous book still provides the basis for most modern geometry courses. The book, however, is not entirely devoted to geometry. Euclid also stated and proved many algebraic theorems. One of the most important theorems states that there are *infinitely many* primes. Euclid's reasoning, considered today a model of elegant simplicity, was as follows:

> Suppose there were only finitely many prime numbers: $2, 3, 5, \ldots, p$. Let $Q = 2 \cdot 3 \cdot 5 \cdot \ldots \cdot p$, the product of all the existing primes. Then $Q + 1$ could *not* be prime, since $Q + 1 > p$ and p is the largest in the finite list of primes. But if $Q + 1$ were composite (that is, not a prime), it would have to be divisible by one of the primes $2, 3, 5, \ldots, p$. This is impossible, since each of these primes divides Q and therefore leaves a remainder of 1 when divided into $Q + 1$. Therefore, there must exist infinitely many primes, since the assumption that there are only a finite number of them leads to a contradiction.

4-5 Factoring Polynomials

Objective To factor polynomials by using the GCF, by recognizing special products, and by grouping terms.

Many real life situations can be described by using polynomials. For example, the height of a baseball hit directly upward with initial speed 96 ft/s will be about $96t - 16t^2$ feet after t seconds. This polynomial can be written as $16t(6 - t)$ by factoring, as you'll soon see in this lesson. The factored polynomial can then be used to determine how long the ball is in the air and what its maximum height is, as you'll see later in this chapter.

To **factor** a polynomial you express it as a product of other polynomials taken from a specified factor set. *In this chapter the factor set is the set of polynomials having integral coefficients.*

The first step in factoring a polynomial is to find its **greatest monomial factor,** that is, the GCF of its terms. If this factor is other than 1, "factor it out." In other words, write the given polynomial as the product of its greatest monomial factor and a polynomial whose greatest monomial factor is 1.

Example 1 Factor each polynomial.

a. $2x^4 - 4x^3 + 8x^2$ **b.** $10ab^3 - 15a^2b^2$

Solution

a. The GCF of $2x^4$, $-4x^3$, and $8x^2$ is $2x^2$.

$\therefore 2x^4 - 4x^3 + 8x^2 = 2x^2(x^2 - 2x + 4)$ ***Answer***

b. The GCF of $10ab^3$ and $-15a^2b^2$ is $5ab^2$.

$\therefore 10ab^3 - 15a^2b^2 = 5ab^2(2b - 3a)$ ***Answer***

When a polynomial has 1 as its greatest monomial factor, you may still be able to factor it by recognizing it as one of the special products listed on page 175. The polynomials $a^2 + 2ab + b^2$ and $a^2 - 2ab + b^2$, which are the result of squaring $a + b$ and $a - b$, respectively, are called **perfect square trinomials.** Also, the polynomial $a^2 - b^2$, which is the product of $a + b$ and $a - b$, is called a **difference of squares.**

Perfect Square Trinomials	**Difference of Squares**
$a^2 + 2ab + b^2 = (a + b)^2$	$a^2 - b^2 = (a + b)(a - b)$
$a^2 - 2ab + b^2 = (a - b)^2$	

Explorations p. 835

Teaching Suggestions p. T94

Suggested Extensions p. T95

Warm-Up Exercises

Multiply.

1. $3x^2(2x - 5)$ $6x^3 - 15x^2$
2. $(x - 2)(3x + 4)$ $3x^2 - 2x - 8$
3. $(5x + 1)^2$ $25x^2 + 10x + 1$
4. $(6x + 7)(6x - 7)$ $36x^2 - 49$

Motivating the Lesson

Students already know how to multiply polynomials, but some, given the product $3x^2 - 2x - 8$, may not know how to find its factors. Tell students that they will look at several special cases of reversing the multiplication process to find factors.

Chalkboard Examples

Factor.

1. $3x^3 - 15x^2$ $3x^2(x - 5)$
2. $y^2 - 49$ $(y + 7)(y - 7)$
3. $a^3 + 8$ $a^3 + 2^3 = (a + 2)(a^2 - 2a + 4)$
4. $xy + 5x + 3y + 15$ $x(y + 5) + 3(y + 5) = (x + 3)(y + 5)$
5. $4m^2 + 4m + 1$ $(2m + 1)(2m + 1) = (2m + 1)^2$

Common Errors

Some students may incorrectly factor examples of the form $a^3 + b^3$ as $(a + b)(a + b)(a + b)$. When reteaching have the students find the product of their incorrect answer and then show them that the correct solution is $(a + b)(a^2 - ab + b^2)$.

Check for Understanding

Here is a suggested use of the Oral Exercises to check students' understanding as you teach the lesson.
Oral Exs. 1–21: use after Example 5.

Guided Practice

Factor.

1. $24x^2y - 40xy$ $8xy(3x - 5)$
2. $x^2 + 14x + 49$ $(x + 7)^2$
3. $81 - 4a^2$ $(9 - 2a)(9 + 2a)$
4. $5a^4 - 5a$ $5a(a - 1)(a^2 + a + 1)$
5. $b^3 - 10b^2 + 25b$ $b(b - 5)^2$
6. $xy + 3y + 2x + 6$ $(y + 2)(x + 3)$
7. $10q^2 - 5q + 2qt - t$ $(5q + t)(2q - 1)$
8. $16s^4 - 1$ $(4s^2 + 1) \cdot (2s + 1)(2s - 1)$
9. $2u^4 + 54u$ $2u(u + 3)(u^2 - 3u + 9)$
10. $4m^8 - 12m^4p^3 + 9p^6$ $(2m^4 - 3p^3)^2$

Summarizing the Lesson

In this lesson students have seen how to factor polynomials by identifying special cases. Ask students to give the factorization of each of the following special cases: $a^2 + 2ab + b^2$, $a^2 - 2ab + b^2$, $a^2 - b^2$, $a^3 + b^3$, and $a^3 - b^3$.

Example 2 Factor each polynomial.

a. $z^2 + 6z + 9$ **b.** $4s^2 - 4st + t^2$ **c.** $25x^2 - 16a^2$

Solution

a. $z^2 + 6z + 9 = z^2 + 2 \cdot z \cdot 3 + 3^2$ ← perfect square trinomial
$= (z + 3)^2$ ***Answer***

b. $4s^2 - 4st + t^2 = (2s)^2 - 2(2s)t + t^2$ ← perfect square trinomial
$= (2s - t)^2$ ***Answer***

c. $25x^2 - 16a^2 = (5x)^2 - (4a)^2$ ← difference of squares
$= (5x + 4a)(5x - 4a)$ ***Answer***

Sometimes you need to use more than one method to factor a given polynomial. Remember that the first step is to look for and factor out the GCF of the terms.

Example 3 Factor $3x^5 - 48x$.

Solution The GCF of the terms is $3x$.

$3x^5 - 48x = 3x(x^4 - 16)$
$= 3x[(x^2)^2 - 4^2]$ ← difference of squares
$= 3x(x^2 + 4)(x^2 - 4)$
$= 3x(x^2 + 4)(x^2 - 2^2)$ ← difference of squares
$= 3x(x^2 + 4)(x + 2)(x - 2)$

The following patterns are useful when factoring polynomials that are sums or differences of cubes.

Sum and Difference of Cubes

$$\mathbf{a^3 + b^3 = (a + b)(a^2 - ab + b^2)}$$
$$\mathbf{a^3 - b^3 = (a - b)(a^2 + ab + b^2)}$$

Example 4 Factor each polynomial.

a. $y^3 - 1$ **b.** $8u^3 + v^3$

Solution

a. $y^3 - 1 = y^3 - 1^3 = (y - 1)(y^2 + y + 1)$ ***Answer***

b. $8u^3 + v^3 = (2u)^3 + v^3 = (2u + v)(4u^2 - 2uv + v^2)$ ***Answer***

When a polynomial is not a special product, you may be able to factor it by rearranging and grouping its terms. This method can be particularly helpful when there are four or more terms, as Example 5 on the next page shows.

Example 5 Factor each polynomial.

a. $3xy - 4 - 6x + 2y$ **b.** $s^2 - 4t^2 - 4s + 4$

Solution **a.** The first and third terms have a common factor of $3x$, and the second and fourth terms have a common factor of 2.

$$3xy - 4 - 6x + 2y = (3xy - 6x) + (2y - 4)$$
$$= 3x(y - 2) + 2(y - 2)$$
$$= (3x + 2)(y - 2) \quad \textbf{\textit{Answer}}$$

b. The first, third, and fourth terms form a perfect square trinomial.

$$s^2 - 4t^2 - 4s + 4 = (s^2 - 4s + 4) - 4t^2$$
$$= (s - 2)^2 - 4t^2$$
$$= (s - 2)^2 - (2t)^2 \longleftarrow \text{difference of squares}$$
$$= [(s - 2) + 2t][(s - 2) - 2t]$$
$$= (s + 2t - 2)(s - 2t - 2) \quad \textbf{\textit{Answer}}$$

Don't be discouraged if your first attempt at factoring doesn't work. In Example 5(b) you might write $s^2 - 4t^2 - 4s + 4 = (s + 2t)(s - 2t) - 4(s - 1)$, but that would not lead to a factorization, so try a different grouping.

Oral Exercises

3. $3ay(3y - 5a)$ 6. $(3t - 1)(3t + 1)$ 9. $(h + 3)^2$ 12. $(2s + 1)^2$ 15. $t(t - 6)^2$

Factor each polynomial.

1. $3z^2 + 6z$ $3z(z + 2)$
2. $11x^2 - 33x^3$ $11x^2(1 - 3x)$
3. $9ay^2 - 15a^2y$
4. $18r^2s^3 + 12r^4s$ $6r^2s(3s^2 + 2r^2)$
5. $x^2 - 16$ $(x - 4)(x + 4)$
6. $9t^2 - 1$
7. $4a^2 - z^2$ $(2a - z)(2a + z)$
8. $x^2 - 8x + 16$ $(x - 4)^2$
9. $h^2 + 6h + 9$
10. $u^4 - 4v^2$ $(u^2 - 2v)(u^2 + 2v)$
11. $25y^2 - 10y + 1$ $(5y - 1)^2$
12. $4s^2 + 4s + 1$
13. $x^3 - x$ $x(x - 1)(x + 1)$
14. $4hk^2 + 16h^2k$ $4hk(k + 4h)$
15. $t^3 - 12t^2 + 36t$
16. $xy^3 - 2xy^2 + xy$ $xy(y - 1)^2$
17. $ax + bx + a + b$ $(a + b)(x + 1)$
18. $ax - bx + a - b$ $(a - b)(x + 1)$

For what value(s) of k will the polynomial be a perfect square trinomial?

19. $x^2 - 10x + k$ 25
20. $y^2 + ky + 9$ 6, −6
21. $100t^2 + kt + 1$ 20, −20

Written Exercises

Factor each polynomial.

A

1. $16x^3 - 64x^2$ $16x^2(x - 4)$
2. $6x^2y^2 + 8x^3y$ $2x^2y(3y + 4x)$
3. $t^2 + 18t + 81$ $(t + 9)^2$
4. $z^2 - 12z + 36$ $(z - 6)^2$

Suggested Assignments

Average Algebra
Day 1: 185/1–16, 18–36 even
S 187/Mixed Review
Day 2: 186/17–37 odd, 38–46

Average Alg. and Trig.
Day 1: 185/1–45 odd
S 187/Mixed Review
Day 2: 186/32–46 even
Assign with Lesson 4-6.

Extended Algebra
Day 1: 185/1–45 odd
S 187/Mixed Review
Day 2: 186/32–46 even, 49–51
Assign with Lesson 4–6.

Extended Alg. and Trig.
185/1–45 odd, 49–53 odd
S 187/Mixed Review

Supplementary Materials

Study Guide	pp. 61–62
Practice Master	21

Factor each polynomial.

5. $16k^2 - 1$ $(4k-1)(4k+1)$
6. $121x^2 - 1$ $(11x-1)(11x+1)$
7. $4y^2 + 20y + 25$ $(2y+5)^2$
8. $9s^2 - 24s + 16$ $(3s-4)^2$
9. $16x^2 - 25$ $(4x-5)(4x+5)$
10. $4h^2 - 81$ $(2h-9)(2h+9)$
11. $121s^2 - 66st + 9t^2$ $(11s-3t)^2$
12. $16x^2 + 40xy + 25y^2$ $(4x+5y)^2$
13. $36p^2 - 49q^2$ $(6p-7q)(6p+7q)$
14. $9x^4 - 16z^2$ $(3x^2-4z)(3x^2+4z)$
15. $st^2 - s$ $s(t-1)(t+1)$
16. $p^3q - pq$ $pq(p-1)(p+1)$
17. $t^3 - 27$ $(t-3)(t^2+3t+9)$
18. $8p^3 + 1$ $(2p+1)(4p^2-2p+1)$
19. $16r^4s + 2rs^4$ $2rs(2r+s)(4r^2-2rs+s^2)$
20. $3x^2y^4 - 81x^2y$ $3x^2y(y-3)(y^2+3y+9)$
21. $x(y-3) + 2(y-3)$ $(x+2)(y-3)$
22. $u(v-1) - 2(v-1)$ $(u-2)(v-1)$
23. $x(y-3) + 2(3-y)$ $(x-2)(y-3)$
24. $u(v-1) - 2(1-v)$ $(u+2)(v-1)$
25. $pq - 2q + 2p - 4$ $(q+2)(p-2)$
26. $xy - 2y - x + 2$ $(y-1)(x-2)$
27. $ab - 2 - 2b + a$ $(a-2)(b+1)$
28. $4ab + 1 - 2a - 2b$ $(2a-1)(2b-1)$
29. $x^2 - 6x + 9 - 4y^2$ $(x-3+2y)(x-3-2y)$
30. $z^2 + 2z + 1 - w^2$ $(z+1-w)(z+1+w)$
31. $u^2 - v^2 + 2v - 1$ $(u-v+1)(u+v-1)$
32. $x^2 - y^2 - 4y - 4$ $(x-y-2)(x+y+2)$

B

33. $x^4 - 2x^2y + y^2$ $(x^2-y)^2$
34. $4u^4v^2 + 4u^2v + 1$ $(2u^2v+1)^2$
35. $a^6 + b^3$ $(a^2+b)(a^4-a^2b+b^2)$
36. $250x^2 - 2x^5$ $2x^2(5-x)(25+5x+x^2)$
37. $16s^4 - 81$ $(2s-3)(2s+3)(4s^2+9)$
38. $p^4 - q^4$ $(p-q)(p+q)(p^2+q^2)$
39. $x^6 - y^6$ $(x-y)(x^2+xy+y^2)(x+y)(x^2-xy+y^2)$
40. $64 - z^6$
41. $u^2 - v^2 - 2u - 2v$ $(u+v)(u-v-2)$
42. $a^2 - b^2 + a - b$ $(a-b)(a+b+1)$
43. $(p+q)^3 - (p-q)^3$ $2q(3p^2+q^2)$
44. $(x+y)^3 + (x-y)^3$ $2x(x^2+3y^2)$
45. $s^3 + t^3 + s^2t + st^2$ $(s+t)(s^2+t^2)$
46. $u^3 - v^3 - u^2v + uv^2$ $(u-v)(u^2+v^2)$

C

47. $(a+b)^6 - (a-b)^6$ $4ab(3a^2+b^2)(a^2+3b^2)$
48. $(a+b)^4 - (a-b)^4$ $8ab(a^2+b^2)$

40. $(2-z)(4+2z+z^2)(2+z)(4-2z+z^2)$

Factor each polynomial. In Exercises 49–54, assume that *n* represents a positive integer.

49. $x^{2n} - 1$ $(x^n-1)(x^n+1)$
50. $x^{2n} - 2x^n + 1$ $(x^n-1)^2$
51. $x^{2n} + 2x^ny^n + y^{2n}$ $(x^n+y^n)^2$
52. $x^{3n} + y^{3n}$ $(x^n+y^n)(x^{2n}-x^ny^n+y^{2n})$
53. $x^{4n} - x^{2n}y^{2n}$ $x^{2n}(x^n-y^n)(x^n+y^n)$
54. $x^{4n} - 2x^{3n}y^n + x^{2n}y^{2n}$ $x^{2n}(x^n-y^n)^2$

55. $x^4 + x^2 + 1$ (*Hint:* Add and subtract x^2.) $(x^2-x+1)(x^2+x+1)$
56. $x^4 + 4$ (*Hint:* Add and subtract $4x^2$.) $(x^2-2x+2)(x^2+2x+2)$
57. $x^4 + x^2y^2 + y^4$ (*Hint:* See Exercise 55.) $(x^2+y^2-xy)(x^2+y^2+xy)$
58. $x^4 + 4y^4$ (*Hint:* See Exercise 56.) $(x^2+2y^2-2xy)(x^2+2y^2+2xy)$
59. a. Show that $x^4 - y^4 = (x-y)(x+y)(x^2+y^2)$.
    b. Show that $x^8 - y^8 = (x-y)(x+y)(x^2+y^2)(x^4+y^4)$.
    c. Write a general pattern for the factorization of $x^n - y^n$ where n is a power of 2.

Additional Answers Written Exercises

59. a. $x^4 - y^4 = (x^2-y^2) \times (x^2+y^2) = (x-y) \times (x+y)(x^2+y^2)$
    b. $x^8 - y^8 = (x^4-y^4) \times (x^4+y^4) = (x^2-y^2) \times (x^2+y^2)(x^4+y^4) = (x-y)(x+y) \times (x^2+y^2)(x^4+y^4)$
    c. $x^n - y^n = (x-y) \times (x+y)(x^2+y^2) \times (x^4+y^4)\cdots(x^{n/2}+y^{n/2})$

Mixed Review Exercises

Write as a simplified polynomial.

1. $(x - 3)^2$ $x^2 - 6x + 9$
2. $(2y + 3)(y - 4)$ $2y^2 - 5y - 12$
3. $m^2n(2m - 3n)$ $2m^3n - 3m^2n^2$
4. $(3c - 4) - (5 - c)$ $4c - 9$
5. $(a^2 + 2)(a - 6)$ $a^3 - 6a^2 + 2a - 12$
6. $(-w)^2(2w)^3$ $8w^5$

Find (a) the GCF and (b) the LCM of the following monomials.

7. $18x$, $24x^3$ $6x$; $72x^3$
8. $30a^2b^4$, $75a^3b^3$ $15a^2b^3$; $150a^3b^4$
9. $63mp^2$, $42mp^3$ $21mp^2$; $126mp^3$

Computer Exercises

For students with some programming experience.

How can a computer determine whether a given positive integer is a perfect square? One way is to use a loop like this:

```
LET R=0
FOR I=1 TO N
IF I*I=N THEN LET R=I
NEXT I
```

If the given integer N is a perfect square of some integer I, then the value of I will be stored in R by the time the computer completes the FOR . . . NEXT loop; otherwise the value of R will be 0.

You can use loops like the one above to determine whether the polynomial $Ax^2 + Bxy + Cy^2$ with $A > 0$ and $C > 0$ is a perfect square trinomial.

Exercises

1. Write a program to determine whether $Ax^2 + Bxy + Cy^2$ is a perfect square trinomial for given integers A, B, and C. (Assume $A > 0$ and $C > 0$.) If the polynomial is a perfect square trinomial, the program should print its factorization.
2. Use the program in Exercise 1 to test each polynomial and, if the polynomial is a perfect square trinomial, print its factorization.
   a. $x^2 - 56xy + 784y^2$
   b. $9x^2 + 144xy + 576y^2$
   c. $64x^2 + 56xy + 49y^2$
   d. $625x^2 - 1050xy + 441y^2$
3. Modify the program in Exercise 1 to determine whether the polynomial $Ax^2 + Bxy + Cy^2$ is a perfect square trinomial or a difference of squares. (Assume $A > 0$.) If the polynomial is factorable, print the factorization.
4. Use the program in Exercise 3 to test each polynomial and, if it is factorable, print the factorization.
   a. $144x^2 - 48xy + 4y^2$
   b. $144x^2 - 4y^2$
   c. $144x^2 + 48xy + 4y^2$
   d. $144x^2 + 4y^2$

Using a Computer

These exercises involve factoring perfect square trinomials and differences of squares. Ask students with programming experience to write the programs described in the exercises and then to demonstrate them to the class.

Additional Answers Computer Exercises

2. a. $(x - 28y)(x - 28y)$
   b. $(3x + 24y)(3x + 24y)$
   c. not a perfect square
   d. $(25x - 21y) \times (25x - 21y)$
4. a. $(12x - 2y)(12x - 2y)$
   b. $(12x + 2y)(12x - 2y)$
   c. $(12x + 2y)(12x + 2y)$
   d. not a perfect square or diff. of squares

Explorations p. 835

Teaching Suggestions p. T95

Suggested Extensions p. T95

Group Activity p. T95

Using Manipulatives p. T95

Warm-Up Exercises

Factor.

1. $4x^3 - 12x$ $4x(x^2 - 3)$
2. $27x^3 - 8y^3$ $(3x - 2y)(9x^2 + 6xy + 4y^2)$
3. $9c^2 + 30c + 25$ $(3c + 5)^2$
4. $4m^2 - 121n^2$ $(2m + 11n)(2m - 11n)$
5. $18p^2 - 48pq + 32q^2$ $2(3p - 4q)^2$

Motivating the Lesson

Tell students that trial and error can be used to solve problems. In today's lesson on factoring quadratic polynomials, a type of "controlled" trial and error is used to determine the correct factors.

4-6 Factoring Quadratic Polynomials

Objective To factor quadratic polynomials.

Polynomials of the form

$$ax^2 + bx + c \quad (a \neq 0)$$

are called **quadratic** or **second-degree polynomials.** The term ax^2 is the **quadratic term,** bx is the **linear term,** and c is the **constant term.**

A **quadratic trinomial** is a quadratic polynomial for which a, b, and c are all nonzero integers. In Lesson 4-5 you factored a special type of quadratic trinomial, the perfect square trinomial. In this lesson you'll review how to factor quadratic trinomials that are not necessarily perfect squares.

If the quadratic trinomial $ax^2 + bx + c$ can be factored into the product $(px + q)(rx + s)$ where p, q, r, and s are integers, then

$$\begin{aligned} ax^2 + bx + c &= (px + q)(rx + s) \\ &= prx^2 + (ps + qr)x + qs \end{aligned}$$

Setting corresponding coefficients equal gives

$$a = pr, \qquad b = ps + qr, \qquad \text{and} \qquad c = qs.$$

These equations suggest a way to find p, q, r, and s; it is illustrated in the following examples.

Example 1 Factor $x^2 + 2x - 15$.

Solution

1. Since the coefficient of x^2 is 1, $pr = 1$. Therefore, let both p and r equal 1 so that the factorization begins $(x \quad)(x \quad)$.
2. Since the constant term is -15, $qs = -15$. Therefore, q and s must have opposite signs. There are four possibilities:

   $(-1)(15)$, $\quad (1)(-15)$, $\quad (-3)(5)$, $\quad$ and $\quad (3)(-5)$.
3. Since the coefficient of x is 2, $ps + qr = 2$. But p and r are both 1, so $ps + qr = s + q = 2$. Therefore, choose the pair of factors having sum 2: -3 and 5.

$\therefore x^2 + 2x - 15 = (x - 3)(x + 5)$ ***Answer***

Example 2 Factor $15t^2 - 16t + 4$.

Solution

1. $pr = 15$. The positive factors of 15 are 1, 3, 5, and 15, so the factorization begins in one of these ways:

   $(t \quad)(15t \quad)$ $\quad$ or $\quad$ $(3t \quad)(5t \quad)$.
2. $qs = 4$. Since the product qs is positive, its factors are both positive or both negative. But the coefficient of x is negative, so choose negative factors of 4. Therefore, the factorization ends as

   $(\quad -1)(\quad -4)$, $\quad (\quad -2)(\quad -2)$, $\quad$ or $\quad (\quad -4)(\quad -1)$.

3. There are six possible factorizations to check:

Trial Factorization	Linear Term of Product
$(t-1)(15t-4)$	$-19t$
$(t-2)(15t-2)$	$-32t$
$(t-4)(15t-1)$	$-61t$
$(3t-1)(5t-4)$	$-17t$
$(3t-2)(5t-2)$	$-16t$
$(3t-4)(5t-1)$	$-23t$

Since the required linear term is $-16t$, the fifth factorization is the correct one.

$\therefore 15t^2 - 16t + 4 = (3t-2)(5t-2)$ ***Answer***

Example 3 Factor $3 - 2z - z^2$.

Solution

$3 - 2z - z^2 = -z^2 - 2z + 3$	Write the terms in order of decreasing degree of z.
$= -1(z^2 + 2z - 3)$	Factor out -1.
$= -(z+3)(z-1)$	Factor $z^2 + 2z - 3$.

$\therefore 3 - 2z - z^2 = -(z+3)(z-1)$ ***Answer***

There are quadratic polynomials with integral coefficients that cannot be factored over the set of polynomials with integral coefficients.

Example 4 Factor $x^2 + 4x - 3$.

Solution Since 1 is the coefficient of the quadratic term, you need factors of -3 whose sum is 4. The only factorizations of -3 are $(-1)(3)$ and $(1)(-3)$. Neither pair of factors has sum 4. Therefore $x^2 + 4x - 3$ cannot be factored over the set of polynomials with integral coefficients.

A polynomial that has more than one term and cannot be expressed as a product of polynomials of lower degree taken from a given factor set is said to be **irreducible** over that set. Example 1 shows that $x^2 + 2x - 15$ is reducible, while Example 4 shows that $x^2 + 4x - 3$ is irreducible. An irreducible polynomial with integral coefficients is **prime** if the greatest common factor of its coefficients is 1. Therefore $x^2 + 4x - 3$ is prime, but $2x^2 + 8x - 6$ is *not* prime since 2 can be factored out.

A polynomial is **factored completely** when it is written as a product of factors and each factor is either a monomial, a prime polynomial, or a power of a prime polynomial.

Problem Solving Strategies

The method of *trial and error* is used throughout the factoring process, as shown in Example 2.

Chalkboard Examples

Factor. If prime, say so.

1. $x^2 - 2x - 8$
   $(x-4)(x+2)$
2. $y^2 + 3y + 4$ prime
3. $r^2 - 7rs + 12s^2$
   $(r-3s)(r-4s)$
4. $2m^2 - m - 1$
   $(2m+1)(m-1)$
5. Find the GCF of $6y^2 - 6y$ and $4y^4 - 24y^3 + 20y^2$.
   $6y^2 - 6y = 6y(y-1)$
   $4y^4 - 24y^3 + 20y^2 = 4y^2(y-5)(y-1)$
   $\therefore$ GCF $= 2y(y-1)$

Common Errors

Students often factor correctly, but do not factor completely. After factoring, have students check each factor to determine if it is prime. For example, if students factor $6x^2 + 3xy - 9y^2$ and give $(2x + 3y)(3x - 3y)$ for the answer, have them check each factor so that the complete solution $3(2x + 3y)(x - y)$ is obtained.

Check for Understanding

Here is a suggested use of the Oral Exercises to check students' understanding as you teach the lesson.

Oral Exs. 1–8: use after Example 3.

Oral Exs. 9–16: use after Example 5.

Guided Practice

Factor. If prime, say so.

1. $t^2 + 4t + 12$ prime
2. $m^2 - 2m - 35$ $(m - 7)(m + 5)$
3. $x^2 - 3xy - 28y^2$ $(x - 7y)(x + 4y)$
4. $a^2 - 14a + 15$ prime
5. $y^2 - 14y + 13$ $(y - 13)(y - 1)$
6. $6c^2 - 5c + 1$ $(3c - 1)(2c - 1)$
7. $2n^2 + 9n - 18$ $(2n - 3)(n + 6)$
8. $3z^2 + 10z + 3$ $(3z + 1)(z + 3)$
9. $5r^2 - 10r + 7$ prime
10. $13p^2 + p - 12$ $(13p - 12)(p + 1)$

Example 5 Factor $3x^6 - 48x^2$ completely.

Solution

$3x^6 - 48x^2 = 3x^2(x^4 - 16)$ — Factor out $3x^2$.

$= 3x^2(x^2 + 4)(x^2 - 4)$ — Factor $x^4 - 16$.

$= 3x^2(x^2 + 4)(x + 2)(x - 2)$ — Factor $x^2 - 4$.

The polynomial is now factored completely since $x^2 + 4$, $x + 2$, and $x - 2$ are prime polynomials.

$\therefore 3x^6 - 48x^2 = 3x^2(x^2 + 4)(x + 2)(x - 2)$ ***Answer***

The **greatest common factor (GCF)** of two or more polynomials is the common factor having the greatest degree and the greatest constant factor. The **least common multiple (LCM)** of two or more polynomials is the common multiple having least degree and least positive constant factor. To find the GCF and LCM of two or more polynomials, first write the complete factorization of each.

Example 6 Find **(a)** the GCF and **(b)** the LCM of the following polynomials:

$$12p^3q + 12p^2q^2 + 3pq^3 \text{ and } 12p^4 - 6p^3q - 6p^2q^2.$$

Solution Factor the polynomials completely.

$$12p^3q + 12p^2q^2 + 3pq^3 = 3pq(4p^2 + 4pq + q^2) = 3pq(2p + q)^2$$

$$12p^4 - 6p^3q - 6p^2q^2 = 6p^2(2p^2 - pq - q^2) = 6p^2(2p + q)(p - q)$$

a. The GCF of the monomial factors is $3p$.
The GCF of the binomial factors is $2p + q$, because each binomial factor is taken the *least* number of times that it occurs in *both* factorizations.

$\therefore$ the GCF of the polynomials is $3p(2p + q)$. ***Answer***

b. The LCM of the monomial factors is $6p^2q$.
The LCM of the binomial factors is $(2p + q)^2(p - q)$, because each binomial factor is taken the *greatest* number of times that it occurs in *either* factorization.

$\therefore$ the LCM of the polynomials is $6p^2q(2p + q)^2(p - q)$. ***Answer***

Oral Exercises

Tell which factorization is correct.

1. $x^2 - 8x - 48$ $(x + 4)(x - 12)$
   $(x - 4)(x + 12)$ or $(x + 4)(x - 12)$

2. $4x^2 - 4x - 15$ $(2x - 5)(2x + 3)$
   $(2x - 5)(2x + 3)$ or $(4x + 5)(x - 3)$

Factor each polynomial.

3. $x^2 + 3x + 2$ $(x + 2)(x + 1)$
4. $t^2 + 11t + 10$ $(t + 10)(t + 1)$
5. $u^2 + 7u - 8$ $(u - 1)(u + 8)$
6. $y^2 + 4y - 5$ $(y + 5)(y - 1)$
7. $z^2 + z - 12$ $(z + 4)(z - 3)$
8. $r^2 - r - 2$ $(r - 2)(r + 1)$

State whether or not each polynomial is prime.

9. $4x + 9$ prime
10. $x^2 + 3x$ not prime
11. $x^2 - 2x + 1$ not prime
12. $x^2 + 1$ prime
13. $x^4 + x^2$ not prime
14. $x^3 + 1$ not prime
15. One factor of $x^2 + 3x - 10$ is $x - 2$. What is the other? $x + 5$
16. One factor of $2x^2 + 5x - 12$ is $x + 4$. What is the other? $2x - 3$

Written Exercises

Factor completely. If the polynomial is prime, say so.

A

1. $x^2 - 9x + 8$ $(x - 8)(x - 1)$
2. $t^2 + 9t + 14$ $(t + 7)(t + 2)$
3. $z^2 - 11z + 18$ $(z - 9)(z - 2)$
4. $u^2 - 10u + 9$ $(u - 9)(u - 1)$
5. $r^2 + 12r + 20$ $(r + 2)(r + 10)$
6. $y^2 - 5y + 6$ $(y - 3)(y - 2)$
7. $p^2 - 8p + 9$ prime
8. $h^2 - 10h + 24$ $(h - 4)(h - 6)$
9. $s^2 - 20s + 36$ $(s - 2)(s - 18)$
10. $z^2 - 9z + 12$ prime
11. $x^2 + x - 12$ $(x + 4)(x - 3)$
12. $t^2 + 2t - 15$ $(t + 5)(t - 3)$
13. $t^2 - 2t - 35$ $(t - 7)(t + 5)$
14. $s^2 - 6s - 27$ $(s + 3)(s - 9)$
15. $3z^2 + 4z + 1$ $(3z + 1)(z + 1)$
16. $5v^2 + 4v - 1$ $(5v - 1)(v + 1)$
17. $8 + 2s - s^2$ $-(s - 4)(s + 2)$
18. $21 - 4x - x^2$ $-(x + 7)(x - 3)$
19. $x^2 - xy - 30y^2$ $(x - 6y)(x + 5y)$
20. $p^2 + 2pq - 24q^2$ $(p + 6q)(p - 4q)$
21. $u^2 - 8uv - 12v^2$ prime
22. $h^2 - 8hk - 15k^2$ prime
23. $2t^2 + 5t - 3$ $(2t - 1)(t + 3)$
24. $3x^2 - 8x + 5$ $(3x - 5)(x - 1)$
25. $3p^2 - 7p - 6$ $(p - 3)(3p + 2)$
26. $4r^2 + 8r + 3$ $(2r + 1)(2r + 3)$
27. $6x^2 - 7xy - 3y^2$ $(2x - 3y)(3x + y)$
28. $6s^2 + st - 5t^2$ $(s + t)(6s - 5t)$
29. $2h^2 + 7hk - 15k^2$ $(2h - 3k)(h + 5k)$
30. $2u^2 + uv - 21v^2$ $(2u + 7v)(u - 3v)$

B

31. $6x^2 + 7x - 10$ $(6x - 5)(x + 2)$
32. $4y^2 - 17y + 15$ $(y - 3)(4y - 5)$
33. $4t^2 - 9t + 6$ prime
34. $25u^2 - 20u + 4$ $(5u - 2)^2$
35. $12p^2 - 32pq - 5q^2$ prime
36. $4r^2 + 16rs - 10s^2$ $2(2r^2 + 8rs - 5s^2)$
37. $4x^3 + 8x^2y - 5xy^2$ $x(2x - y)(2x + 5y)$
38. $4x^2 + 3xy - 15y^2$ prime
39. $4pq^4 - 32pq$ $4pq(q - 2)(q^2 + 2q + 4)$
40. $81uv^3 + 3u^4$ $3u(3v + u)(9v^2 - 3uv + u^2)$
41. $r^4 - 16s^4$ $(r - 2s)(r + 2s)(r^2 + 4s^2)$
42. $x^6 - 64y^6$
43. $x^4 - 3x^2 - 4$ $(x - 2)(x + 2)(x^2 + 1)$
44. $z^4 - 10z^2 + 9$ $(z - 3)(z + 3)(z - 1)(z + 1)$

42. $(x - 2y)(x^2 + 2xy + 4y^2)(x + 2y)(x^2 - 2xy + 4y^2)$

Summarizing the Lesson

Students have extended their study of factoring to include quadratic polynomials that are not necessarily special cases. Ask students to tell how to determine whether a polynomial is prime.

Suggested Assignments

Average Algebra
Day 1: 191/1–30
S 186/49–52
Day 2: 191/31–49 odd
R 192/Self-Test 2

Average Alg. and Trig.
Day 1: 191/2–30 even
Assign with Lesson 4-5.
Day 2: 191/31–54
R 192/Self-Test 2

Extended Algebra
Day 1: 191/2–30 even
Assign with Lesson 4-5.
Day 2: 191/31–50, 55–58
R 192/Self-Test 2

Extended Alg. and Trig.
191/3–30 mult. of 3, 32–58 even
R 192/Self-Test 2

Supplementary Materials

Study Guide	pp. 63–64
Practice Master	22
Test Master	14
Computer Activity	9
Resource Book	p. 114

Additional Answers Written Exercises

45. $x - 2$; $(x - 1)(x - 2)^2$

46. $x + 4$; $(x - 4)(x + 4)(x - 2)$

47. $t(t + 1)$; $t(t + 1)(t - 3)(t + 4)$

48. $y(y - 2)$; $y(y - 2)(y + 2)(y^2 + 2y + 4)$

49. $p - q$; $(p - q)(p^2 + pq + q^2)$

50. 1; $(x + y)(x^2 - xy + y^2)(x^2 + y^2)$

51. Suppose $x^2 + x + k$ is not prime. Then there are nonzero integers r and s such that $rs = k$ and $r + s = 1$. Since k is positive, r and s are both negative or both positive. The sum of two negative numbers is negative, so r and s must both be positive. But the sum of two positive integers is greater than or equal to 2. Thus, there can be no such integers r and s, and $x^2 + x + k$ is prime.

52. $x^2 + (k + 1)x + k = (x + k)(x + 1)$ for all positive integers k.

54. $(a^n - b^n) \cdot (a^{2n} + a^nb^n + b^{2n}) \cdot (a^n + b^n) \cdot (a^{2n} - a^nb^n + b^{2n})$

55. $(x^n - 2)(x^n + 2) \cdot (x^n - 1)(x^n + 1)$

56. $(x^n - 1)^2(x^n + 1)^2(x^{2n} + 1)^2$

57. $4(x + 2)(x^2 - 3x + 1)$

58. $8x(x - 2)(x + 2)$

Quick Quiz

1. Find the prime factorization of 3528. $2^3 \cdot 3^2 \cdot 7^2$.

Find (a) the GCF and (b) the LCM of the given polynomials.

45. $x^2 - 3x + 2;\ x^2 - 4x + 4$

46. $x^2 - 16;\ x^2 + 2x - 8$

47. $t^3 - 2t^2 - 3t;\ t^3 + 5t^2 + 4t$

48. $y^3 - 4y;\ y^4 - 8y$

49. $p - q;\ p^3 - q^3$

50. $x^3 + y^3;\ x^2 + y^2;\ x + y$

C **51.** Show that $x^2 + x + k$ is prime for all positive integers k.

52. Show that $x^2 + (k + 1)x + k$ is reducible for all positive integers k.

Factor completely. Assume that *n* represents a positive integer.

53. $a^{4n} - b^{4n}$ $(a^n - b^n)(a^n + b^n)(a^{2n} + b^{2n})$

54. $a^{6n} - b^{6n}$

55. $x^{4n} - 5x^{2n} + 4$

56. $x^{8n} - 2x^{4n} + 1$

57. $(x^2 - 2x + 3)^2 - (x^2 - 4x - 1)^2$

58. $(x^2 + 2x - 4)^2 - (x^2 - 2x - 4)^2$

59. One factor of $x^3 - 3x^2 + 3x - 2$ is $x - 2$. What is the other? $x^2 - x + 1$

Self-Test 2

Vocabulary

factor, factor set (p. 179)
prime number, prime (p. 179)
prime factorization (p. 179)
greatest common factor (GCF) of integers (p. 179)
least common multiple (LCM) of integers (p. 179)
greatest common factor (GCF) of monomials (p. 180)
least common multiple (LCM) of monomials (p. 180)
greatest monomial factor (p. 183)
perfect square trinomial (p. 183)
difference of squares (p. 183)
sum of cubes (p. 184)
difference of cubes (p. 184)
quadratic polynomial (p. 188)
quadratic term (p. 188)
linear term (p. 188)
constant term (p. 188)
quadratic trinomial (p. 188)
irreducible polynomial (p. 189)
prime polynomial (p. 189)
factor completely (p. 189)
greatest common factor (GCF) of polynomials (p. 190)
least common multiple (LCM) of polynomials (p. 190)

Find the prime factorization of each integer.

1. 990 $2 \cdot 3^2 \cdot 5 \cdot 11$

2. 3000 $2^3 \cdot 3 \cdot 5^3$ **Obj. 4-4, p. 179**

Find (a) the GCF and (b) the LCM of the following.

3. $36a^2b^3,\ 54abc^2$ $18ab$; $108a^2b^3c^2$

4. $6p^3q^2r,\ 8p^2q^2r^2,\ 4p^3q^3$ $2p^2q^2$; $24p^3q^3r^2$

Factor completely.

5. $27x^3 - 12x$ $3x(3x - 2)(3x + 2)$

6. $9t^2 + 6t + 1$ $(3t + 1)^2$ **Obj. 4-5, p. 183**

7. $125s^3 - 8t^3$ $(5s - 2t)(25s^2 + 10st + 4t^2)$

8. $ab - a + b - 1$ $(a + 1)(b - 1)$

9. $6t^2 + 4t - 2$ $2(3t - 1)(t + 1)$

10. $4z^2 + z - 14$ $(4z - 7)(z + 2)$ **Obj. 4-6, p. 188**

Computer Key-In

In Lesson 4–6 you learned how to factor $ax^2 + bx + c$ where a, b, and c are integers and $a \neq 0$. The following program uses a similar method to factor $Ax^2 + Bx + C$ where $A \neq 0$ and $C \neq 0$. Although the process of complete factorization includes factoring out the GCF of the terms of a polynomial, this program does *not* factor out the GCF of A, B, and C. (See the Computer Key-In in Chapter 5 for a program that will find the GCF of two integers.)

```
10 PRINT "THIS PROGRAM FINDS FACTORS OF"
20 PRINT "A*X^2+B*X+C."
30 INPUT "ENTER A, B, AND C (A, C NONZERO): "; A, B, C
40 IF A=0 OR C=0 THEN 30
50 LET FLAG=0
60 FOR M=1 TO SQR (ABS (A))
70 IF INT (A/M)<>A/M OR FLAG<>0 THEN 170
80 LET P=M
90 LET R=A/M
100 FOR N=1 TO ABS (C)
110 IF INT (C/N)<>C/N OR FLAG<>0 THEN 160
120 LET Q=N
130 LET S=C/N
140 IF P*S+Q*R=B THEN LET FLAG=1
150 IF P*S+Q*R=-B THEN LET FLAG=2
160 NEXT N
170 NEXT M
180 IF FLAG=0 THEN
        PRINT A; "*X^2+ "; B; " *X+ "; C; " IS IRREDUCIBLE."
190 IF FLAG=1 THEN
        PRINT "("; P; " *X+ "; Q; ")*("; R; " *X+ "; S; ")"
200 IF FLAG=2 THEN
        PRINT "("; P; " *X+ ";-Q; ")*("; R; " *X+ "; -S; ")"
210 END
```

For each possible pair of factors P and R of A, the program tests each possible pair of factors Q and S of C. If $PS + QR = B$ in line 140, the program will print the factorization $(Px + Q)(Rx + S)$. If $PS + QR = -B$ in line 150, then reversing the signs of Q and S will yield a factorization. If all pairs of factors of A and C are checked and $PS + QR$ is neither B nor $-B$, then the computer will print that the given trinomial is irreducible.

Exercises

Run the program to display a factorization, if possible, for each quadratic trinomial.

1. $12x^2 - 13x - 120$ $(3x + 8)(4x - 15)$
2. $-15x^2 + 22x + 48$ $(3x - 8)(-5x - 6)$
3. $6x^2 - 55x + 56$ $(x - 8)(6x - 7)$
4. $x^2 + 5x - 4$ irreducible
5. $2x^2 + 8x - 10$ $(x - 1)(2x + 10)$
6. $4x^2 - 25$ $(2x + 5)(2x - 5)$

2. Find the GCF and LCM.

a. $12x^2y^3z$, $36x^3y$ $12x^2y$; $36x^3y^3z$

b. $15a^4b^2$, $10ab^5$, $3a^3b^3$ ab^2; $30a^4b^5$

Factor completely.

3. $27x^3 + 1$ $(3x + 1)(9x^2 - 3x + 1)$

4. $y^2 - 7y + 12$ $(y - 4)(y - 3)$

5. $8n^3 + 8n^2 + 2n$ $2n(2n + 1)^2$

6. $s^3 - 49s$ $s(s - 7)(s + 7)$

7. $a^2 - 7a - 15$ prime

8. $2r^2 - 3r - 5$ $(2r - 5)(r + 1)$

9. $12t^2 + 13tv + 3v^2$ $(3t + v)(4t + 3v)$

10. $10r^2 + 11r - 6$ $(5r - 2)(2r + 3)$

Computer Key-In Commentary

The conjunctions AND and OR can be used to combine conditions in an IF . . . THEN statement. With the conjunction OR (line 40), the action will be executed if either condition is true.

The statement in line 40 is an example of error-trapping. Its purpose is to test the input values to make certain a user did not inadvertently type inappropriate values. It is good programming practice to include checks such as this.

In line 70, the INT (greatest integer) function is used to determine if the given quotient is an integer. If INT(A/M) <> A/M, then M does not divide A evenly and is therefore not a factor. Line 110 similarly tests for factors of C.

Teaching Suggestions p. T96

Suggested Extensions p. T96

Warm-Up Exercises

Rewrite each equation so that one side equals zero.

1. $x^2 = 2x + 6$
   $x^2 - 2x - 6 = 0$
2. $12 = -2x^2 + 9x$
   $2x^2 - 9x + 12 = 0$
3. $x + 2 = 3x^2$
   $3x^2 - x - 2 = 0$

Multiply.

4. $(y + 2)(y^2 - 3)$
   $y^3 + 2y^2 - 3y - 6$
5. $4x(x + 2)^2$
   $4x^3 + 16x^2 + 16x$

Motivating the Lesson

Tell students that their ability to factor polynomials will be used as a base for solving polynomial equations, the topic of today's lesson.

Chalkboard Examples

Solve. Identify double roots.

1. $(2c - 1)(c + 4)^2 = 0$
   $(2c - 1)(c + 4)(c + 4) = 0$
   $c = \frac{1}{2}, c = -4, c = -4$
   $\left\{\frac{1}{2}, -4 \text{ (dbl. root)}\right\}$
2. $(x - 2)^2(x^2 - 9) = 0$
   $(x - 2)(x - 2)(x - 3)(x + 3) = 0$
   $x = 2, x = 2, x = 3, x = -3$
   $\{2 \text{ (dbl. root)}, 3, -3\}$
3. $y^2 + 7y = 18$
   $y^2 + 7y - 18 = 0$
   $(y - 2)(y + 9) = 0$
   $y = 2, y = -9$ $\{2, -9\}$
4. $3r^2 = 10r + 8$
   $3r^2 - 10r - 8 = 0$
   $(r - 4)(3r + 2) = 0$
   $r = 4, r = -\frac{2}{3}$ $\left\{4, -\frac{2}{3}\right\}$

Applications of Factoring

4-7 Solving Polynomial Equations

Objective To solve polynomial equations.

The factoring that you have learned to do is a necessary step in solving polynomial equations. A **polynomial equation** is an equation that is equivalent to one with a polynomial as one side and 0 as the other. Both $x^2 - 5x - 24 = 0$ and $x^2 = 5x + 24$ are polynomial equations.

A **root,** or **solution,** of a polynomial equation is a value of the variable that satisfies the equation. The roots of $x^2 - 5x - 24 = 0$ are -3 and 8, because $(-3)^2 - 5(-3) - 24 = 0$ and $8^2 - 5(8) - 24 = 0$.

When a polynomial equation in x is written with 0 as one side, you can solve the equation if you can factor the polynomial on the other side into *linear factors* of the form $ax + b$ $(a \neq 0)$. The basis of this method is the *zero-product property* (page 83):

$$ab = 0 \text{ if and only if } a = 0 \text{ or } b = 0.$$

The zero-product property can be extended to any number of factors. Therefore, a product of real numbers is zero if and only if at least one of its factors is zero.

To use the *zero-product property* to solve a polynomial equation, you need to

1. write the equation with 0 as one side,
2. factor the other side of the equation, and
3. solve the equation obtained by setting each factor equal to 0.

Example 1 Solve $(x - 5)(x + 2) = 0$.

Solution The given equation already has 0 as one side, and the other side is already factored. Simply set each factor equal to 0 and solve:

$$x - 5 = 0 \quad \text{or} \quad x + 2 = 0$$
$$x = 5 \quad \text{or} \quad x = -2$$

Check $x = 5$:

$$(5 - 5)(5 + 2) \stackrel{?}{=} 0$$
$$0 \cdot 7 \stackrel{?}{=} 0$$
$$0 = 0 \ \checkmark$$

Check $x = -2$:

$$(-2 - 5)(-2 + 2) \stackrel{?}{=} 0$$
$$-7 \cdot 0 \stackrel{?}{=} 0$$
$$0 = 0 \ \checkmark$$

$\therefore$ the solution set is $\{5, -2\}$. ***Answer***

Example 2 Solve $x^2 = x + 30$.

Solution

1. $x^2 - x - 30 = 0$ Make one side 0.
2. $(x + 5)(x - 6) = 0$ Factor the polynomial.
3. $x + 5 = 0$ or $x - 6 = 0$ Use the zero-product property.
   $x = -5$ or $x = 6$

Check: Substitute -5 and 6 into $x^2 = x + 30$ to verify that they are solutions.

$\therefore$ the solution set is $\{-5, 6\}$. ***Answer***

Example 3 Solve $3x^3 = 4x(2x - 1)$.

Solution

1. $3x^3 = 8x^2 - 4x$
   $3x^3 - 8x^2 + 4x = 0$ Make one side 0.
2. $x(3x^2 - 8x + 4) = 0$ Factor the polynomial.
   $x(3x - 2)(x - 2) = 0$
3. $x = 0$ or $3x - 2 = 0$ or $x - 2 = 0$ Use the zero-product property.
   $x = 0$ or $x = \frac{2}{3}$ or $x = 2$

$\therefore$ the solution set is $\left\{0, \frac{2}{3}, 2\right\}$. The check is left for you. ***Answer***

As you have seen, the zero-product property is used to solve polynomial equations. You can also use it to find *zeros* of polynomial functions. A number r is a **zero** of a function f if $f(r) = 0$. For example, 3 and -3 are zeros of $g(x) = x^2 - 9$ because $g(3) = 3^2 - 9 = 0$ and $g(-3) = (-3)^2 - 9 = 0$.

Example 4 Find the zeros of $f(x) = (x - 4)^3 - 4(3x - 16)$.

Solution

1. $f(x) = (x - 4)^3 - 4(3x - 16)$ To find the zeros of f, write $f(x) = 0$.
   $(x - 4)^3 - 4(3x - 16) = 0$
2. $(x^3 - 12x^2 + 48x - 64) - 12x + 64 = 0$ Simplify.
   $x^3 - 12x^2 + 36x = 0$
   $x(x^2 - 12x + 36) = 0$ Factor the polynomial.
   $x(x - 6)^2 = 0$
3. $x = 0$ or $x - 6 = 0$ or $x - 6 = 0$ Use the zero-product property.
   $x = 0$ or $x = 6$ or $x = 6$

Check: $f(0) = (0 - 4)^3 - 4(3 \cdot 0 - 16) = -64 + 64 = 0$ ✓
$f(6) = (6 - 4)^3 - 4(3 \cdot 6 - 16) = 8 - 8 = 0$ ✓

$\therefore$ the zeros of f are 0 and 6. ***Answer***

Common Errors

When solving polynomial equations, caution students against factoring before they write the equation with zero on one side. For example, some students solve the equation $x^2 - 5x + 6 = 2$ by the following incorrect method: $(x - 3)(x - 2) = 2$; $x = 3$ or $x = 2$. These students may not understand the zero-product property and so reteaching this property may be in order.

Check for Understanding

Here is a suggested use of the Oral Exercises to check students' understanding as you teach the lesson.
Oral Exs. 1–10: use after Example 3.
Oral Exs. 11–18: use after Example 4.

Guided Practice

Solve. Identify all double roots.

1. $q(2q + 6) = 0$ $\{0, -3\}$
2. $s^2(2s - 7)^2 = 0$ $\left\{0 \text{ (dbl. root)}, \frac{7}{2} \text{ (dbl. root)}\right\}$
3. $t^2 + 7t + 6 = 0$ $\{-1, -6\}$
4. $x^2 + 25 = 10x$ $\{5 \text{ (dbl. root)}\}$
5. $u^4 = 36u^2$ $\{0 \text{ (dbl. root)}, 6, -6\}$
6. $15y^2 + 1 = 8y$ $\left\{\frac{1}{5}, \frac{1}{3}\right\}$
7. $12 + 4m = m^2$ $\{6, -2\}$
8. $(a + 3)(a - 3) = 40$ $\{7, -7\}$
9. $(c - 6)^2 = c$ $\{4, 9\}$
10. $x^2(x - 1) = 6x$ $\{0, 3, -2\}$

Summarizing the Lesson

In this lesson students have seen how factoring can be used to solve polynomial equations. Ask students to explain the importance of the zero-product property in solving polynomial equations.

Suggested Assignments

Average Algebra
Day 1: 196/1–10, 12–34 even
S 197/Mixed Review
Day 2: 196/9–35 odd, 36–40, 45–46

Average Alg. and Trig.
196/1–49 odd
S 197/Mixed Review

Extended Algebra
196/1–49 odd
S 197/Mixed Review

Extended Alg. and Trig.
196/3–48 mult. of 3
S 197/Mixed Review
Assign with Lesson 4-8.

From Example 4 you can see that $f(x) = x(x-6)^2$. Since $x - 6$ occurs as a factor of f twice, the number 6 is called a **double zero** of the function f and a **double root** of the equation $f(x) = 0$. In general, the zeros and roots arising from repeated factors are called **multiple zeros** of functions and **multiple roots** of equations.

Oral Exercises

Solve.

1. $(x-3)(x-7) = 0$ $\{3, 7\}$
2. $(y+2)(y-5) = 0$ $\{-2, 5\}$
3. $(u-1)(u+2)(u-3) = 0$ $\{1, -2, 3\}$
4. $t(t-3)(t+4) = 0$ $\{0, 3, -4\}$
5. $y^2 - 2y = 0$ $\{0, 2\}$
6. $s^2 - 5s = 0$ $\{0, 5\}$
7. $x^2 - 4x + 4 = 0$ $\{2\}$
8. $z^2 + 6z + 9 = 0$ $\{-3\}$
9. $t^2 - 25 = 0$ $\{5, -5\}$
10. $4x^2 - 1 = 0$ $\left\{\frac{1}{2}, -\frac{1}{2}\right\}$

Find all zeros of the given function. Identify all double zeros.

11. $f(x) = 3x - 12$ 4
12. $g(y) = y^2 - 16$ 4, −4
13. $h(t) = (2t-3)(t+2)$ $\frac{3}{2}$, −2
14. $k(z) = z(z-3)(z+1)$ 0, 3, −1
15. $f(s) = s(s-2)^2$ 0, 2 (d.z.)
16. $g(u) = u^2(u-2)$ 0 (d.z.), 2
17. $g(x) = x(x^2 - 2x + 1)$ 0, 1 (d.z.)
18. $f(t) = (t-2)^2(t-3)^2$ 2 (d.z.), 3 (d.z.)

Written Exercises

Solve. Identify all double roots.

A

1. $(x-1)(x-4) = 0$ $\{1, 4\}$
2. $(t+2)(t-5) = 0$ $\{-2, 5\}$
3. $t(t+1)(t-2) = 0$ $\{0, -1, 2\}$
4. $z^2(2z-1) = 0$ $\left\{0 \text{ (d.r.)}, \frac{1}{2}\right\}$
5. $(s-1)^2(s-3)^2 = 0$ $\{1 \text{ (d.r.)}, 3 \text{ (d.r.)}\}$
6. $y(y+1)^2(y-2) = 0$ $\{0, -1 \text{ (d.r.)}, 2\}$
7. $z^2 + 3 = 4z$ $\{1, 3\}$
8. $x^2 - 12 = 4x$ $\{-2, 6\}$
9. $t^3 - t = 0$ $\{-1, 0, 1\}$
10. $s^3 - s^2 = 0$ $\{0 \text{ (d.r.)}, 1\}$
11. $x^3 + 4x = 4x^2$ $\{0, 2 \text{ (d.r.)}\}$
12. $y^3 + 6y^2 = 27y$ $\{-9, 0, 3\}$
13. $3r^2 = 4r - 1$ $\left\{\frac{1}{3}, 1\right\}$
14. $6x^2 = 1 - x$ $\left\{-\frac{1}{2}, \frac{1}{3}\right\}$
15. $2y^2 + y = 6$ $\left\{-2, \frac{3}{2}\right\}$
16. $10t^2 - 9t = 1$ $\left\{-\frac{1}{10}, 1\right\}$
17. $6 - 7u = 3u^2$ $\left\{-3, \frac{2}{3}\right\}$
18. $5s - 1 = 6s^2$ $\left\{\frac{1}{3}, \frac{1}{2}\right\}$
19. $6(x+12) = x^2$ $\{-6, 12\}$
20. $(u+3)(u-3) = 8u$ $\{-1, 9\}$
21. $(y-4)^2 = 2y$ $\{2, 8\}$
22. $x = (x-6)^2$ $\{4, 9\}$
23. $3t(t+1) = 4(t+1)$ $\left\{-1, \frac{4}{3}\right\}$
24. $2(r^2+1) = 5r$ $\left\{\frac{1}{2}, 2\right\}$
25. $(x-1)(x^2+x-2) = 0$ $\{-2, 1 \text{ (d.r.)}\}$
26. $(x+2)(x^2-4) = 0$ $\{-2 \text{ (d.r.)}, 2\}$
27. $y^2(y-3)(y^2-9) = 0$ $\{-3, 0 \text{ (d.r.)}, 3 \text{ (d.r.)}\}$
28. $(x^2-1)(x^2+3x+2) = 0$ $\{-2, -1 \text{ (d.r.)}, 1\}$

Solve. Identify all double roots.

B **29.** $x^4 - 2x^2 + 1 = 0$ {−1 (d.r.), 1 (d.r.)}

30. $y^4 - 5y^2 + 4 = 0$ {−2, −1, 1, 2}

31. $t^6 + 9t^2 = 10t^4$

32. $x^2(x^4 + 16) = 8x^4$

33. $(u - 2)^3 - 3u + 8 = 0$ {0, 3 (d.r.)}

34. $(s - 6)^3 = 27(s - 8)$ {0, 9 (d.r.)}

31. {−3, −1, 0 (d.r.), 1, 3}

32. {−2 (d.r.), 0 (d.r.), 2 (d.r.)}

Multiply both sides of the given equation by a number that will eliminate decimals or fractions. Then solve.

35. $0.3x^2 + 0.2x - 0.1 = 0$ $\left\{-1, \frac{1}{3}\right\}$

36. $0.2x^2 - 1.1x + 0.5 = 0$ $\left\{\frac{1}{2}, 5\right\}$

37. $\frac{1}{6}x^2 + \frac{1}{2}x - \frac{2}{3} = 0$ {1, −4}

38. $\frac{1}{3}x^2 - \frac{1}{2}x - \frac{1}{3} = 0$ $\left\{-\frac{1}{2}, 2\right\}$

Solve. Identify all multiple roots.

39. $(x - 1)^3 - (x - 1)^2 = 0$ {1 (d.r.), 2}

40. $(t - 2)^3 - (t - 2) = 0$ {1, 2, 3}

41. $(x^2 - 4)^3 = 0$ {−2 (t.r.), 2 (t.r.)}

42. $(x^2 - 3x + 2)^3 = 0$ {1 (t.r.), 2 (t.r.)}

43. $(x^2 + 1)^2 - 4(x^2 + 1) + 4 = 0$ {−1 (d.r.), 1 (d.r.)}

44. $(x^2 - 3)^2 + (x^2 - 3) - 2 = 0$ {−2, −1, 1, 2}

Find all zeros of *f*. Identify all multiple zeros.

45. $f(x) = (x - 4)^3 - 4(x - 4)$ 2, 4, 6

46. $f(t) = (t - 2)^3 - (t - 2)$ 1, 2, 3

47. $f(z) = 9z^4 - 12z^3 + 4z^2$ 0 (d.z.), $\frac{2}{3}$ (d.z.)

48. $f(s) = 4s^4 - 17s^2 + 4$ $-\frac{1}{2}, \frac{1}{2}, -2, 2$

49. $f(x) = (x - 1)^4 - 4(x - 1)^3$ 1 (t.z.), 5

50. $f(y) = (y^2 - 9)^3$ −3 (t.z.), 3 (t.z.)

In Exercises 51–54, *a* and *b* are constants. Solve each equation for *x*.

C **51.** $a^2x^2 - b^2 = 0\ (a \neq 0)$ $\left\{-\frac{b}{a}, \frac{b}{a}\right\}$

52. $x^2 + ax - bx - ab = 0$ {−a, b}

53. $x^3 + ax^2 = x + a$ {−a, 1, −1}

54. $x^3 - bx^2 - 4x + 4b = 0$ {b, 2, −2}

55. Under what conditions placed on x and y is $(x + y)^2 = x^2 + y^2$ true? $x = 0$ or $y = 0$

56. Under what conditions placed on x and y is $(x + y)^3 = x^3 + y^3$ true? $x = 0$, $y = 0$, or $x = -y$

Mixed Review Exercises

Factor completely. If the polynomial is prime, say so.

1. $x^2 - 6x + 9$ $(x - 3)^2$

2. $3y^2 - 5y + 8$ prime

3. $8mn - 10n + 12m - 15$ $(2n + 3)(4m - 5)$

4. $16a^2 - 25b^2$ $(4a - 5b)(4a + 5b)$

5. $12x + 24x^3 - 34x^2$ $2x(4x - 3)(3x - 2)$

6. $8u^3 + 1$ $(2u + 1)(4u^2 - 2u + 1)$

Express as a simplified polynomial.

7. $z^2(z - 1)$ $z^3 - z^2$

8. $(4a + 5)^2$ $16a^2 + 40a + 25$

9. $(3m - 7) + 2(m + 4)$ $5m + 1$

10. $(p^2 + 2)(p^2 - 3)$ $p^4 - p^2 - 6$

11. $x - (x + 1)$ -1

12. $(-3c^2d)^3$ $-27c^6d^3$

Supplementary Materials

Study Guide	pp. 65–66
Practice Master	23

Teaching Suggestions p. T96

Suggested Extensions p. T96

Warm-Up Exercises

Review the plan for solving a word problem on page 50.

Motivating the Lesson

Tell students that just as engineers build physical models to test theories and designs, mathematicians build mathematical models to solve real life problems. In this lesson we will look at problems that lead to models involving polynomial equations.

Chalkboard Examples

1. Two numbers differ by 5 and have a product of 84. Find the numbers.
   Let x = larger number.
   Then $x - 5$ = smaller number.
   $x(x - 5) = 84$
   $x^2 - 5x - 84 = 0$
   $(x + 7)(x - 12) = 0$
   $x = -7$ or $x = 12$
   $\therefore$ the numbers are -7, -12 or 7, 12.
2. The sum of two numbers is 9. The sum of their squares is 101. Find the numbers.
   Let x = one number.
   Then $9 - x$ = the other.
   $x^2 + (9 - x)^2 = 101$
   $x^2 + 81 - 18x + x^2 = 101$
   $2x^2 - 18x - 20 = 0$
   $x^2 - 9x - 10 = 0$
   $(x - 10)(x + 1) = 0$
   $x = 10$ or $x = -1$
   $\therefore$ the numbers are 10 and -1.

4-8 Problem Solving Using Polynomial Equations

Objective To solve problems using polynomial equations.

An equation that represents a real life problem is called a **mathematical model.** The plan for solving word problems outlined in Lesson 1-9 can be used to develop a mathematical model for a given problem. After solving the equation, you should always check your answer against the conditions stated in the original problem. Checking your answer just in the equation is not sufficient, because an acceptable answer in mathematics may not be reasonable in real life.

Example 1 A graphic artist is designing a poster that consists of a rectangular print with a uniform border. The print is to be twice as tall as it is wide, and the border is to be 3 in. wide. If the area of the poster is to be 680 in.^{2}, find the dimensions of the print.

Solution

Step 1 You are asked to find the width and height of the print. A border 3 in. wide surrounds the print. Draw a diagram.

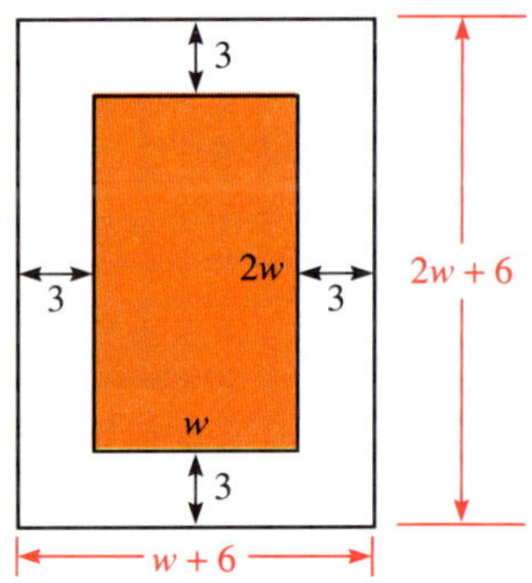

Step 2 Let w = width of the print. Then $2w$ = height of the print. Since the dimensions of the poster are 6 in. greater than the dimensions of the print, $w + 6$ = width of the poster and $2w + 6$ = height of the poster. Label your diagram.

Step 3 Write an equation that shows the area of the poster is 680 in.^{2}:

Length × Width = Area

$$(2w + 6)(w + 6) = 680$$

Step 4 Solve the equation.

$2(w + 3)(w + 6) = 680$	Factor $2w + 6$.
$(w + 3)(w + 6) = 340$	Divide both sides by 2.
$w^2 + 9w + 18 = 340$	
$w^2 + 9w - 322 = 0$	Make one side 0.
$(w - 14)(w + 23) = 0$	Factor the polynomial.
$w - 14 = 0$ or $w + 23 = 0$	Use the zero-product property.
$w = 14$ or $w = -23$	

Step 5 Although $w = -23$ is a solution of $(2w + 6)(w + 6) = 680$, it must be rejected because width cannot be negative. The other solution, $w = 14$, results in a print that is 14 in. wide and 28 in. tall. Check these two dimensions:

$$\text{Print's height} = 28 = 2 \cdot 14 = \text{twice print's width} \quad \checkmark$$

$$\text{Poster's area} = (14 + 6)(28 + 6) = 20 \cdot 34 = 680 \quad \checkmark$$

$\therefore$ the print is 14 in. by 28 in. ***Answer***

Vertical motion affected only by gravity leads to a mathematical model that is a polynomial equation. When a projectile is launched vertically upward with an initial speed v, its height h above the launch point t seconds later is given approximately by:

$h = vt - 4.9t^2$ if the distance is measured in meters,

and $h = vt - 16t^2$ if the distance is measured in feet.

Example 2 A batter hits a baseball directly upward with speed 96 ft/s.

a. How long is the ball in the air before being caught by the catcher?

b. How high does the ball go?

Solution **a.** Use the formula $h = 96t - 16t^2$. When the ball returns to the level at which it is hit, the height h is 0.

$$0 = 96t - 16t^2$$

$$0 = 16t(6 - t)$$

$$t = 0 \quad \text{or} \quad t = 6$$

$\therefore$ the ball is in the air 6 seconds. ***Answer***

b. The ball reaches its maximum height halfway between the time when it is hit ($t = 0$) and the time when it is caught ($t = 6$).

$$\text{If } t = \frac{6}{2} = 3,\ h = 96(3) - 16(3)^2 = 144.$$

$\therefore$ the maximum height of the ball above the level of the bat is 144 ft. ***Answer***

Problems

Solve each problem. If there are two correct answers, give both of them.

A

1. The sum of a number and its square is 72. Find the number. -9 or 8
2. Find a number that is 56 less than its square. -7 or 8
3. Find two consecutive odd integers whose product is 143. -13 and -11 or 11 and 13
4. Find two consecutive odd integers the sum of whose squares is 130. -9 and -7 or 7 and 9

3. The length of a rectangle is 1 m less than twice the width. If the area is 55 m², find the perimeter. Let w = width in meters. Then $2w - 1$ = length in meters.

$$w(2w - 1) = 55$$

$$2w^2 - w - 55 = 0$$

$$(2w - 11)(w + 5) = 0$$

$$w = \frac{11}{2} \text{ or } w = -5 \text{ (reject)}$$

The width is 5.5 m and the length is 10 m. The perimeter $= 2w + 2l = 2(5.5) + 2(10) = 11 + 20 = 31$.

$\therefore$ the perimeter is 31 m.

Problem Solving Strategies

The *five-step plan* and *formulas* are used in this lesson to solve problems involving polynomial equations.

Guided Practice

1. The sum of the squares of two consecutive even numbers is 100. Find the numbers. 6, 8 or -6, -8
2. The shorter leg of a right triangle is 17 cm less than the longer leg. If the hypotenuse is 25 cm long, find the perimeter of the triangle. 56 cm
3. The product of three consecutive integers is 21 more than the cube of the smallest integer. Find the integers. -3, -2, -1
4. The perimeter of a rectangle is 18 cm. The area is 14 cm². Find the dimensions. 2 cm by 7 cm

Summarizing the Lesson

In this lesson the study of problem solving was continued by applying the skill of solving polynomial equations to problems. Ask the students to state the five steps of the plan for solving a word problem.

Suggested Assignments

Average Algebra
199/*P*: 2–18 even
S 197/41–44

Average Alg. and Trig.
199/*P*: 2–18 even

Extended Algebra
199/*P*: 3–30 mult. of 3

Extended Alg. and Trig.
Day 1: 199/ *P*: 4, 8, 9, 11
Assign with Lesson 4-7.
Day 2: 200/*P*: 13–27 odd
Assign with Lesson 4-9.

Supplementary Materials

Study Guide	pp. 67–68
Practice Master	24

5. A rectangle is 4 cm longer than it is wide, and its area is 117 cm^2. Find its dimensions. 9 cm by 13 cm

6. A rectangular garden has perimeter 66 ft and area 216 ft^2. Find the dimensions of the garden. 9 ft by 24 ft

7. The area of a right triangle is 44 m^2. Find the lengths of its legs if one of the legs is 3 m longer than the other. 8 m and 11 m

8. Two ships leave port, one sailing east and the other south. Some time later they are 17 mi apart, with the eastbound ship 7 mi farther from port than the southbound ship. How far is each from port? 8 mi and 15 mi

9. The top of a 15-foot ladder is 3 ft farther up a wall than the foot of the ladder is from the bottom of the wall. How far is the foot of the ladder from the bottom of the wall? 9 ft

10. The height of a triangle is 7 cm greater than the length of its base, and its area is 15 cm^2. Find the height. 10 cm

11. The hypotenuse of a right triangle is 25 m long. The length of one leg is 10 m less than twice the other. Find the length of each leg. 15 m and 20 m

12. The side of a large tent is in the shape of an isosceles triangle whose area is 54 ft^2 and whose base is 6 ft shorter than twice its height. Find the height and the base of the side of the tent. 9 ft high and 12 ft long base

13. A rectangle is 15 cm wide and 18 cm long. If both dimensions are decreased by the same amount, the area of the new rectangle formed is 116 cm^2 less than the area of the original. Find the dimensions of the new rectangle. 11 cm by 14 cm

14. A rectangle is twice as long as it is wide. If its length is increased by 4 cm and its width is decreased by 3 cm, the new rectangle formed has an area of 100 cm^2. Find the dimensions of the original rectangle. 8 cm by 16 cm

15. A projectile is launched upward from ground level with an initial speed of 98 m/s. How high will it go? When will it return to the ground? 490 m; 20 s

16. A ball is thrown directly upward from ground level with an initial speed of 80 ft/s. How high will it go? When will it return to the ground? 100 ft; 5 s

B **17.** A signal flare is fired upward from ground level with an initial speed of 294 m/s. A balloonist cruising at a height of 2450 m sees it pass on the way up. How long will it be before the flare passes the balloonist again on the way down? 40 s

18. Luis wanted to throw an apple to Kim, who was on a balcony 40 ft above him, so he tossed it upward with an initial speed of 56 ft/s. Kim missed it on the way up, but then caught it on the way down. How long was the apple in the air? 2.5 s

19. A rocket is moving vertically upward with speed 245 m/s when its fuel runs out. How much farther will it travel upward before starting to fall back to the ground? 3062.5 m

20. A ball is thrown upward from the top of a 98-meter tower with an initial speed of 39.2 m/s. How much later will it hit the ground? (*Hint:* If h is the height of the ball above the top of the tower, then $h = -98$ when the ball hits the ground.) 10 s

21. The distance d, in feet, required for a vehicle traveling at r mi/h to come to a stop is given approximately by $d = 0.05r^2 + r$. If an automobile in a 55 mi/h speed zone required 240 ft to stop, was its speed within the legal limit? No

22. The cost C of manufacturing n calculators per day at a certain plant is given by $C = n(20 - 0.01n) + 100$. The size of the plant limits the maximum output to 500 calculators per day. If the company plans to invest $5200 per day in manufacturing costs, how many calculators per day can it manufacture? 300 calculators

23. When 0.5 cm was planed off each of the six faces of a wooden cube, its volume decreased by 169 cm^3. Find its new volume. 343 cm^3

24. A farmer plans to use 21 m of fencing to enclose a rectangular pen having area 55 m^2. Only three sides of the pen need fencing because part of an existing wall will form the fourth side. Find the dimensions of the pen. 5 m by 11 m or 5.5 m by 10 m

25. The width, length, and diagonal of a rectangle are consecutive even integers. Find the integers. 6, 8, 10

26. Give a convincing argument why the width, length, and diagonal of a rectangle cannot be consecutive odd integers.

27. A rectangular corner lot, originally twice as long as it was wide, lost a 2-meter strip along two adjacent sides due to street widening. Its new area is 684 m^2. Find its new dimensions. 18 m by 38 m

28. A decorator plans to place a rug in a room 9 m by 12 m in such a way that a uniform strip of flooring around the rug will remain uncovered. If the rug is to cover half the floor space, what should the dimensions of the rug be? 6 m by 9 m

29. A garden plot 5 m by 15 m has one of its longer sides next to a wall. The area of the plot is to be doubled by digging up a strip of uniform width along the other three sides. How wide should the border be? 2.5 m

30. A rancher plans to use 160 yd of fencing to enclose a rectangular corral and to divide it into two parts by a fence parallel to the shorter sides of the corral. Find the dimensions of the corral if its area is 1000 yd^2. 20 yd by 50 yd

C 31. The general formula for the height h after t seconds of a projectile launched upward from ground level with initial speed v is $h = vt - \frac{1}{2}gt^2$, where g is the gravitational constant (approximately 9.8 m/s^2 or 32 ft/s^2). Show that the greatest height of the projectile is $\frac{v^2}{2g}$.

Additional Answers
Problems

26. Let n, $n + 2$, and $n + 4$ be three consecutive odd integers. Assume that n, $n + 2$, and $n + 4$ are the two sides and the diagonal respectively of a rectangle. By the Pythagorean theorem, $n^2 + (n + 2)^2 = (n + 4)^2$ or $n^2 + n^2 + 4n + 4 = n^2 + 8n + 16$.
$n^2 + 4n + 4 = 8n + 16$
$n^2 - 4n - 12 = 0$
$(n - 6)(n + 2) = 0$
$n = 6$ or $n = -2$ (reject)
But n is supposed to be odd.
$\therefore$ the width, length, and diagonal of a rectangle cannot be consecutive odd integers.

31. The projectile is at ground level when $h = 0$. $h = vt - \frac{1}{2}gt^2 = 0$ when $t = 0$ or $t = \frac{2v}{g}$. The total time of the flight is $\frac{2v}{g}$. The greatest height is reached halfway through the flight at $t = \frac{1}{2}\left(\frac{2v}{g}\right) = \frac{v}{g}$. $h = v\left(\frac{v}{g}\right) - \frac{1}{2}g\left(\frac{v}{g}\right)^2 = \frac{v^2}{g} - \left(\frac{1}{2}\right)\frac{v^2}{g}$. The greatest height is $\frac{v^2}{2g}$.

Teaching Suggestions p. T97

Reading Algebra p. T97

Suggested Extensions p. T97

Warm-Up Exercises

Find and graph the solution set.

1. $4m \le 12$ or $m + 7 > 12$ $\{m\colon m \le 3 \text{ or } m > 5\}$

2. $|3 + 4x| > 9$ $\{x\colon x > \frac{3}{2}$ or $x < -3\}$

3. $|p - 1| \le 3$ $\{p\colon -2 \le p \le 4\}$

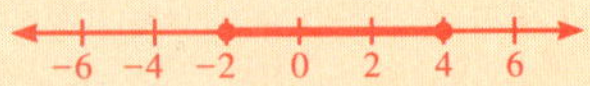

Motivating the Lesson

Tell students that just as a roller coaster has places where it rises and places where it falls, a polynomial has intervals where it is positive and intervals where it is negative. Finding these intervals is the topic of today's lesson.

Chalkboard Examples

Find and graph the solution set.

1. $y^2 - 3y \le 0$ $y(y - 3) \le 0$ For $y(y - 3) = 0$, $y = 0$ or $y = 3$. For $y(y - 3) < 0$, the factors must have opposite signs. Thus, $y < 0$ and $y > 3$ (no solution), or $y > 0$ and $y < 3$ which gives $0 < y < 3$. $\{y\colon 0 \le y \le 3\}$

4-9 Solving Polynomial Inequalities

Objective To solve polynomial inequalities.

A **polynomial inequality** is an inequality that is equivalent to an inequality with a polynomial as one side and 0 as the other side. Both $x^2 - x - 6 > 0$ and $x^2 > x + 6$ are polynomial inequalities.

You can often solve a polynomial inequality that has 0 as one side by factoring the polynomial into linear factors and applying one of the following facts:

$ab > 0$ if and only if a and b have the same sign;

$ab < 0$ if and only if a and b have opposite signs.

Example 1 Find and graph the solution set of $x^2 - 1 > x + 5$.

Solution

$$x^2 - 1 > x + 5$$

$x^2 - x - 6 > 0$ — Make one side 0.

$(x + 2)(x - 3) > 0$ — Factor the polynomial.

The product is positive if and only if both factors have the *same sign*.

Both factors positive			or	**Both factors negative**		
$x + 2 > 0$	and	$x - 3 > 0$		$x + 2 < 0$	and	$x - 3 < 0$
$x > -2$	and	$x > 3$		$x < -2$	and	$x < 3$
The solution set of this conjunction is $\{x\colon x > 3\}$.				The solution set of this conjunction is $\{x\colon x < -2\}$.		

$\therefore$ the solution set of the given inequality is $\{x\colon x > 3 \text{ or } x < -2\}$.

The graph of the solution set is given below.

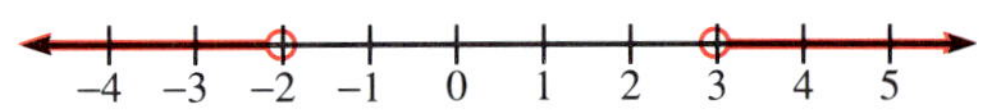

Example 2 Find and graph the solution set of $3t < 4 - t^2$.

Solution

$$3t < 4 - t^2$$

$t^2 + 3t - 4 < 0$ — Make one side 0.

$(t - 1)(t + 4) < 0$ — Factor the polynomial.

The product is negative if and only if the factors have opposite signs.

			or			
$t - 1 > 0$	and	$t + 4 < 0$		$t - 1 < 0$	and	$t + 4 > 0$
$t > 1$	and	$t < -4$		$t < 1$	and	$t > -4$
no solution				$\{t\colon -4 < t < 1\}$		

$\therefore$ the solution set of the given inequality is $\{t: -4 < t < 1\}$. The graph of the solution set is shown below.

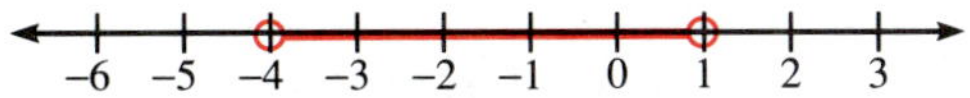

After solving a polynomial inequality, you can make a partial check of your solution by choosing *test points* on the number line. For instance, with the answer in Example 2, first check a number to the left of the endpoint -4, say -5: -5 does *not* satisfy the inequality. Then check a number between -4 and 1, say 0: 0 *does* satisfy the inequality. Finally, check a number to the right of the endpoint 1, say 2: 2 does *not* satisfy the inequality.

In Example 3 a **sign graph** is used to help find and graph the solution set of a polynomial inequality. This visual method is particularly helpful when the polynomial has three or more factors.

Example 3 Find and graph the solution set of $9x \le x^3$.

Solution

$$9x \le x^3$$

$$x^3 - 9x \ge 0 \qquad \text{Make one side 0.}$$

$$x(x^2 - 9) \ge 0 \qquad \text{Factor the polynomial.}$$

$$x(x + 3)(x - 3) \ge 0$$

To draw a sign graph, draw a number line for each factor and label the part where the factor is positive, where it is zero, and where it is negative.

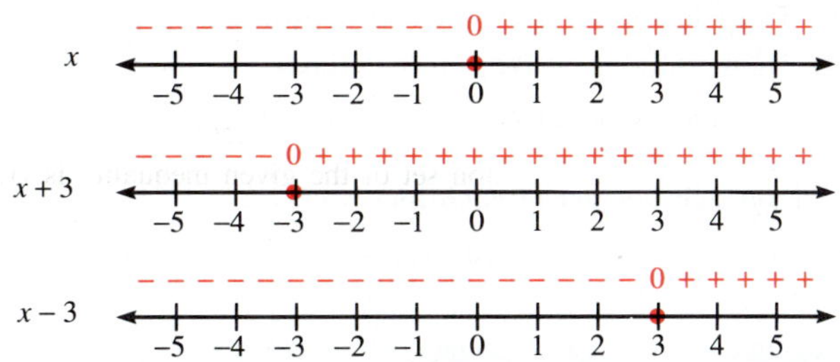

The sign of the product $x(x + 3)(x - 3)$ is determined by the number of negative factors. When exactly one or exactly three factors are negative, the product is negative. When no factors or exactly two factors are negative, the product is positive.

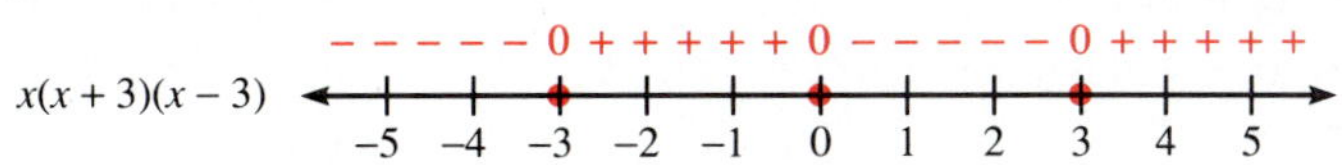

$\therefore$ the solution set of the given inequality is $\{x: -3 \le x \le 0 \text{ or } x \ge 3\}$. The graph of the solution set is shown below.

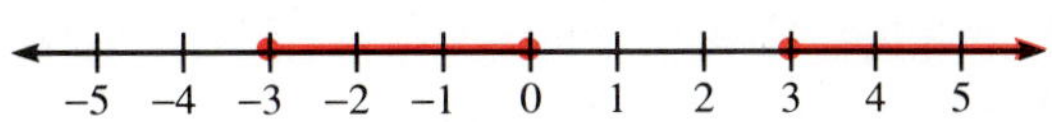

2. $y^3 - 16y \ge 0$

$y(y^2 - 16) \ge 0$

$y(y - 4)(y + 4) \ge 0$

Using a sign graph:

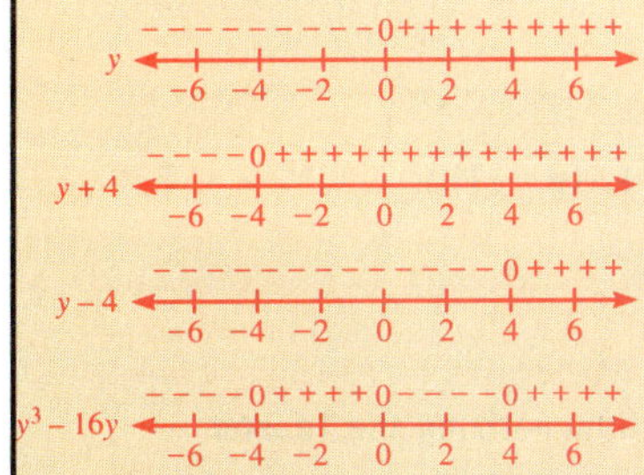

The product is negative when one factor is negative or when three factors are negative. The product is positive when no factors or exactly two factors are negative. The product is zero when any one of the factors is zero, that is, at 0, -4, and 4.

$\{y: y \ge 4 \text{ or } -4 \le y \le 0\}$.

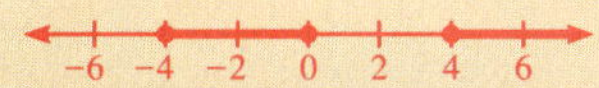

Check for Understanding

Here is a suggested use of the Oral Exercises to check students' understanding as you teach the lesson.

Oral Exs. 1–6: use after Example 2.

Oral Exs. 7–16: use after Example 3.

Guided Practice

Find and graph the solution set.

1. $(x - 4)(x + 2) < 0$

$\{x: -2 < x < 4\}$

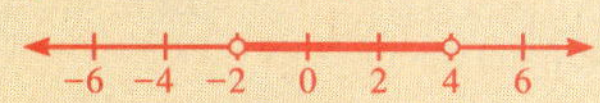

2. $36 < x^2$

$\{x: x < -6 \text{ or } x > 6\}$

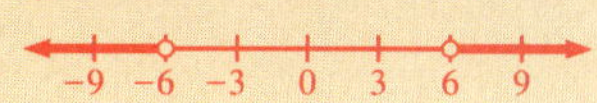

(continued)

Oral Exercises

Match each inequality with the graph of its solution set.

1. $(x-2)(x+3)<0$ e
2. $(x-2)(x+3)>0$ c
3. $(x+2)(x-3)<0$ a
4. $(x+2)(x-3)\geq 0$ f
5. $(x+2)(x-3)>0$ d
6. $(x-2)(x+3)\leq 0$ b

a. 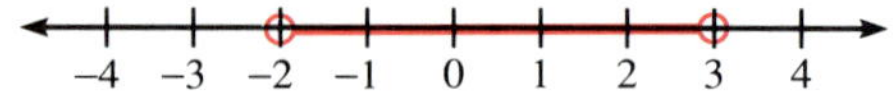

b. 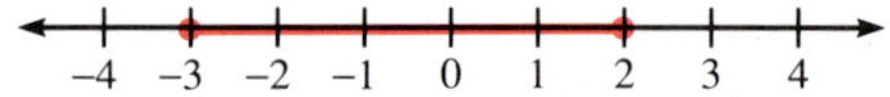

c. 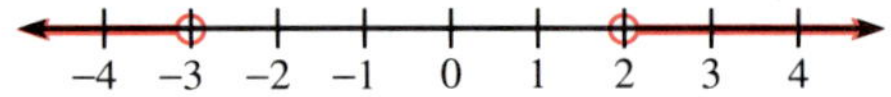

d. 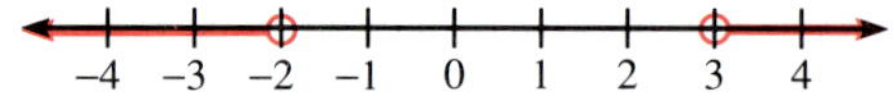

e. 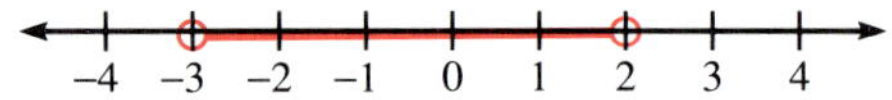

f. 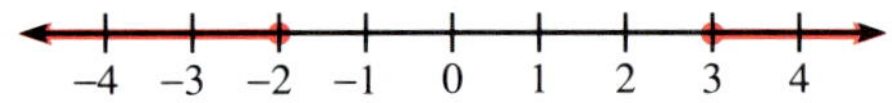

Justify the following statements.

7. The solution set of $x^2\leq 0$ is $\{0\}$.
8. The solution set of $x^2>0$ is {real numbers except 0}.
9. The solution set of $x^2+4\leq 0$ is $\emptyset$.
10. The solution set of $x^2+1>0$ is {real numbers}.
11. The solution set of $x(x^2+1)>0$ is $\{x\colon x>0\}$.
12. The solution set of $(x-2)(x^2+1)<0$ is $\{x\colon x<2\}$.

Give the solution set of each inequality.

13. $x(x+1)<0$ $\{x\colon -1<x<0\}$
14. $x(x-3)>0$ $\{x\colon x<0 \text{ or } x>3\}$
15. $x(x+2)\geq 0$ $\{x\colon x\leq -2 \text{ or } x\geq 0\}$
16. $x(x-4)\leq 0$ $\{x\colon 0\leq x\leq 4\}$

Written Exercises

Find and graph the solution set of each inequality.

A

1. $(x-2)(x-5)<0$ $\{x\colon 2<x<5\}$
2. $(t+1)(t-2)>0$ $\{t\colon t<-1 \text{ or } t>2\}$
3. $x^2-4\leq 0$ $\{x\colon -2\leq x\leq 2\}$
4. $x^2\leq 9$ $\{x\colon -3\leq x\leq 3\}$
5. $z^2-4z>0$ $\{z\colon z<0 \text{ or } z>4\}$
6. $x^2+2x\geq 0$ $\{x\colon x\leq -2 \text{ or } x\geq 0\}$
7. $4y^2\geq 36$ $\{y\colon y\leq -3 \text{ or } y\geq 3\}$
8. $3s^2<48$ $\{s\colon -4<s<4\}$
9. $x^2-5x+4<0$ $\{x\colon 1<x<4\}$
10. $z^2-z-6>0$ $\{z\colon z<-2 \text{ or } z>3\}$
11. $t^2>9(t-2)$ $\{t\colon t<3 \text{ or } t>6\}$
12. $r^2\leq 2(r+4)$ $\{r\colon -2\leq r\leq 4\}$
13. $4z(z-1)\leq 15$ $\left\{z\colon -\frac{3}{2}\leq z\leq \frac{5}{2}\right\}$
14. $4x(x+1)\geq 3$ $\left\{x\colon x\leq -\frac{3}{2} \text{ or } x\geq \frac{1}{2}\right\}$

Guided Practice
(continued)

3. $4k^2-8k>0$
$\{k\colon k<0 \text{ or } k>2\}$
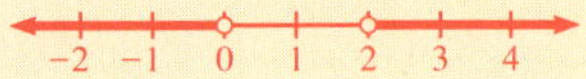

4. $(y-1)(y+1)<y+1$
$\{y\colon -1<y<2\}$
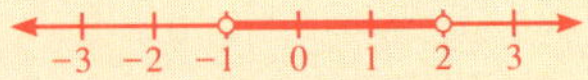

Summarizing the Lesson

Students have solved polynomial inequalities by determining where their factors take on positive, negative, and zero values. Ask students to explain why a sign graph is useful when solving polynomial inequalities.

Additional Answers
Oral Exercises

7. For every real number x, $x^2\geq 0$, thus the solution set of $x^2\leq 0$ is $\{0\}$.
8. For every real number x, $x^2\geq 0$, thus the solution set of $x^2>0$ is {real numbers except 0}.
9. For every real number x, $x^2\geq 0$, thus $x^2+4>0$.
10. For every real number x, $x^2\geq 0$, thus $x^2+1>0$.
11. $x(x^2+1)>0$; then, $x>0$ and $x^2+1>0$ or $x<0$ and $x^2+1<0$; for every real number x, $x^2+1>0$; thus the solution set is $\{x\colon x>0\}$
12. $(x-2)(x^2+1)<0$; then $x-2>0$ and $x^2+1<0$ or $x-2<0$ and $x^2+1>0$; for every real number x, $x^2+1>0$; thus $x-2<0$ and the solution set is $\{x\colon x<2\}$

15. $t^2 + 16 \geq 8t$ {real numbers}

16. $x^2 + 9 \leq 6x$ {3}

17. $12 + s - s^2 \geq 0$ $\{s: -3 \leq s \leq 4\}$

18. $9 - 3t - 2t^2 > 0$ $\left\{t: -3 < t < \frac{3}{2}\right\}$

B 19. $x^3 - 16x > 0$ $\{x: -4 < x < 0 \text{ or } x > 4\}$

20. $t^3 < 9t^2$ $\{t: t \neq 0 \text{ and } t < 9\}$

21. $y^3 + y^2 < 6y$ $\{y: y < -3 \text{ or } 0 < y < 2\}$

22. $z^3 + 7z^2 + 10z > 0$ $\{z: -5 < z < -2 \text{ or } z > 0\}$

Find and graph the solution set of each inequality.

23. $(x^2 - 4x)(x^2 - 4) < 0$

24. $(x^2 + 4)(x^2 + 4x) > 0$

25. $x^4 > 3x^2 + 4$

26. $x^4 - 18 < 7x^2$

C 27. $x^4 + 9 \leq 10x^2$

28. $x^4 + 100 \geq 29x^2$

29. $(x^2 - x - 6)(x^2 - 2x + 1) > 0$

30. $(x^2 + x - 2)(x^2 - 4x + 4) \leq 0$

31. $x^2(x^2 + 4) \leq 4x^3$

32. $x^2(x^2 + 9) > 6x^3$

33. Under what conditions placed on the real numbers p and q will $(x + p)(x + q) < 0$ have a nonempty solution set? For all real numbers p and q such that $p \neq q$.

Mixed Review Exercises

Solve.

1. $(2x + 3)(x - 1) = 0$ $\left\{-\frac{3}{2}, 1\right\}$
2. $3p - 7 = 8$ {5}
3. $5 - |b| = 1$ {−4, 4}
4. $z(z - 1) = z(z + 1)$ {0}
5. $|2u - 5| = 3$ {1, 4}
6. $2a^2 = 7a + 4$ $\left\{-\frac{1}{2}, 4\right\}$
7. $|8 - t| = 0$ {8}
8. $(n - 1)^2 = 1$ {0, 2}
9. $(3c + 1)^2 = 0$ $\left\{-\frac{1}{3}\right\}$
10. $m^2 = m$ {0, 1}
11. $y^2(4 - y) = 0$ {0, 4}
12. $4(3d - 5) = 2(5d + 1)$ {11}

Self-Test 3

Vocabulary polynomial equation (p. 194)
root (p. 194)
zero of a function (p. 195)
double zero (p. 196)
double root (p. 196)
multiple zero (p. 196)
multiple root (p. 196)
mathematical model (p. 198)
polynomial inequality (p. 202)
sign graph (p. 203)

Solve each equation.

1. $x^2 - 11x - 42 = 0$ {−3, 14} 2. $3t^2 = 2 - t$ $\left\{-1, \frac{2}{3}\right\}$ **Obj. 4-7, p. 194**

3. A 10 m by 20 m pool is surrounded by a deck of uniform width. The area of the deck is 216 m^2. How wide is the deck? 3 m **Obj. 4-8, p. 198**

Find and graph the solution set of each inequality.

4. $2x^2 + 7x - 15 > 0$ 5. $(x - 2)^2 \leq x$ **Obj. 4-9, p. 202**

Check your answers with those at the back of the book.

Suggested Assignments

Average Algebra
204/1–21 odd
S 205/Mixed Review
R 205/Self-Test 3

Average Alg. and Trig.
204/1–21 odd, 23, 27
S 205/Mixed Review
R 205/Self-Test 3

Extended Algebra
204/1–21 odd, 23, 27
S 205/Mixed Review
R 205/Self-Test 3

Extended Alg. and Trig.
204/3–30 mult. of 3
S 205/Mixed Review
R 205/Self-Test 3
Assign with Lesson 4-8.

Supplementary Materials

Study Guide	pp. 69–70
Practice Master	25
Test Master	15
Computer Activity	8
Resource Book	p. 115

Additional Answers Written Exercises

2. 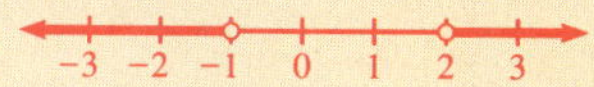

4. 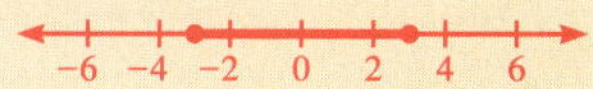

6. 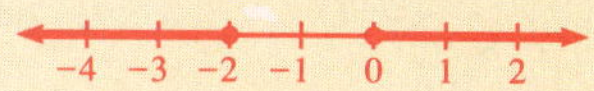

8. 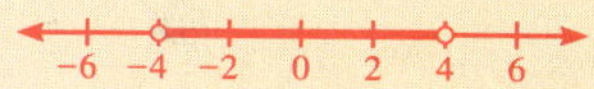

10. 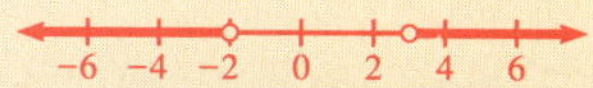

12. 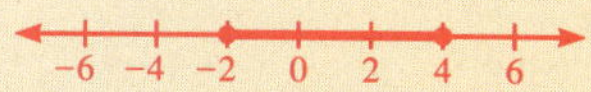

14. 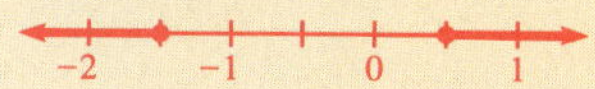

16. 

(continued on p. 206)

Quick Quiz on p. 207.

Additional Answers
Written Exercises
(continued)

18. 

20. 

22. 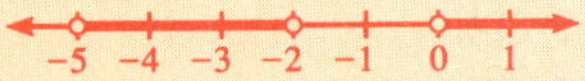

23. $\{x: -2 < x < 0$ or $2 < x < 4\}$

24. $\{x: x < -4$ or $x > 0\}$

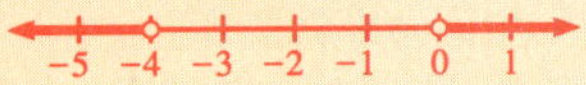

25. $\{x: x < -2$ or $x > 2\}$

26. $\{x: -3 < x < 3\}$

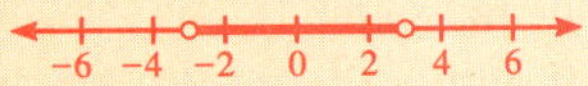

27. $\{x: -3 \le x \le -1$ or $1 \le x \le 3\}$

28. $\{x: x \le -5$ or $-2 \le x \le 2$ or $x \ge 5\}$

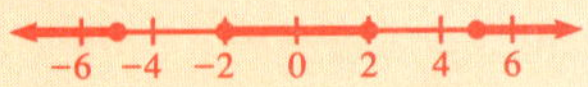

29. $\{x: x < -2$ or $x > 3\}$

30. $\{x: -2 \le x \le 1$ or $x = 2\}$

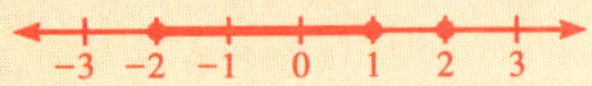

31. $\{0, 2\}$

32. $\{x: x \ne 0$ and $x \ne 3\}$

Chapter Summary

1. A *monomial* is a constant, a variable, or a product of a constant and one or more variables. A *polynomial* is a monomial or a sum of monomials. A *simplified* polynomial has no two terms similar. The terms of a simplified polynomial are usually arranged in decreasing degree of one of the variables.

2. If a and b are real numbers and m and n are positive integers, then:

   Law 1 $\quad a^m \cdot a^n = a^{m+n}$

   Law 2 $\quad (ab)^m = a^m b^m$

   Law 3 $\quad (a^m)^n = a^{mn}$

   These *laws of exponents* are useful in simplifying polynomials.

3. To find the product of two polynomials, multiply each term of one polynomial by each term of the other and simplify the result.

4. A *prime* is an integer greater than 1 whose only positive integral factors are itself and 1. To find the *prime factorization* of an integer, write it as the product of primes.

5. Use prime factorization to determine the *greatest common factor* (GCF) or *least common multiple* (LCM) of two or more integers. To find the GCF, take the least power of each common prime factor in the factorization. To find the LCM, take the greatest power of each prime factor in the factorizations.

6. The GCF of two or more monomials is the factor of each that has the greatest degree and greatest coefficient. The *greatest monomial factor* of a polynomial is the GCF of its terms.

7. The LCM of two or more monomials is the multiple of each that has the least degree and least positive coefficient.

8. The following strategies are useful in factoring polynomials:

   Factor out the greatest monomial factor.

   Look for special products such as a perfect square trinomial, a difference of squares, and a sum or difference of cubes.

   Rearrange and group terms.

   Factor quadratic trinomials into products of linear factors.

9. Many *polynomial equations* can be solved by factoring and using the *zero-product property*. Many *polynomial inequalities* can be solved by factoring and determining the signs of the factors. If the polynomial has three or more factors, it may be helpful to use a *sign graph* to determine the solution.

Chapter Review

Write the letter of the correct answer.

1. Simplify $(x^3 - 3x^2 - 2x + 5) - (x^2 - 2x + 2)$. **4-1**

a. $-x^2 + 7$ **b.** $x^3 - 2x^2 - 4x + 3$ **c.** (circled) $x^3 - 4x^2 + 3$ **d.** $x^3 - 4x^2 - 4x + 3$

2. Simplify $p(p - 2q) + 3(pq - q^2)$.

a. $p^2 + pq - q^2$ **b.** (circled) $p^2 + pq - 3q^2$ **c.** $p^2 - 2q + 3pq - q^2$ **d.** $p^2 - 5pq - 3q^2$

3. Simplify $(-x^3)^2x^4$. **4-2**

a. $-x^{10}$ **b.** x^{24} **c.** $-x^9$ **d.** (circled) x^{10}

4. Simplify $xy^2(x + y) - x^2y(y - x)$.

a. $xy^3 - x^3y$ **b.** 0 **c.** (circled) $xy^3 + x^3y$ **d.** $2x^2y^2$

5. Express $(3t - 4)(2t + 3)$ as a simplified polynomial. **4-3**

a. (circled) $6t^2 + t - 12$ **b.** $6t^2 - t + 12$ **c.** $6t^2 + 11t - 12$ **d.** $6t^2 - 12$

6. Express $ax(x - a)(x + a)$ as a simplified polynomial.

a. $ax^3 - 2a^2x^2 + a^3x$ **b.** $ax^3 + 2a^2x^2 + a^3x$ **c.** (circled) $ax^3 - a^3x$ **d.** $-a^4x^4$

7. Find the prime factorization of 1350. **4-4**

a. $2^3 \cdot 3 \cdot 5^2$ **b.** $2^2 \cdot 3^4 \cdot 5$ **c.** (circled) $2 \cdot 3^3 \cdot 5^2$ **d.** $2^2 \cdot 3^2 \cdot 5^2$

8. Find the GCF of $54s^2t^3$, $90s^3t^2$, and $108s^4t$.

a. (circled) $18s^2t$ **b.** $9s^2t$ **c.** $540s^4t^3$ **d.** $36st$

9. Find the LCM of $9a^2b^2c$, $15a^2b^3$, and $6a^3b^2$.

a. $45abc$ **b.** $90a^2b^2$ **c.** $180a^3b^3c$ **d.** (circled) $90a^3b^3c$

10. Factor $4t^2 + 4t + 1$ completely. **4-5**

a. $4t(t + 1) + 1$ **b.** $4(t^2 + t + 1)$ **c.** (circled) $(2t + 1)^2$ **d.** $(2t - 1)^2$

11. Factor $3x^3 - 27x$ completely.

a. $3x(x - 3)^2$ **b.** (circled) $3x(x + 3)(x - 3)$ **c.** $3(x - 3)(x^2 + 3x + 9)$ **d.** $3(x + 3)(x^2 - 3x + 9)$

12. Factor $x^2 - x - a - a^2$ completely.

a. (circled) $(x + a)(x - a - 1)$ **b.** $x(x - 1) - a(1 + a)$ **c.** $(x - a)(x + a - 1)$ **d.** $(x - a)(x + a + 1)$

13. Factor $3x^2 - 4x - 4$ completely. **4-6**

a. $(3x - 2)(x + 2)$ **b.** (circled) $(3x + 2)(x - 2)$ **c.** $(3x - 4)(x + 1)$ **d.** $(3x - 1)(x + 4)$

14. Factor $x^4 - 3x^2 - 4$ completely.

a. $(x^2 + 4)(x + 1)(x - 1)$ **b.** $(x + 1)(x - 1)(x + 2)(x - 2)$ **c.** (circled) $(x^2 + 1)(x - 2)(x + 2)$ **d.** $(x^2 + 1)(x^2 + 4)$

15. Solve the equation $6x^2 = 5x - 1$. **4-7**

a. $\{2, 3\}$ **b.** $\left\{0, \frac{1}{5}\right\}$ **c.** (circled) $\left\{\frac{1}{2}, \frac{1}{3}\right\}$ **d.** $\{2, 3\}$

Quick Quiz

Solve.

1. $x^2 - 5x + 6 = 0$
$\{3, 2\}$

2. $2y^2 = 7 + 5y$ $\left\{-1, \frac{7}{2}\right\}$

3. The base of a triangle is 1 cm more than twice the height. The area is 18 cm^2. Find the base.
9 cm

Find and graph the solution set.

4. $4x^2 - 25 \le 0$
$\left\{x: -\frac{5}{2} \le x \le \frac{5}{2}\right\}$

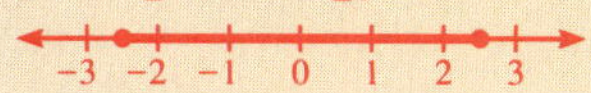

5. $2q^2 > 3 - 5q$
$\left\{q: q < -3 \text{ or } q > \frac{1}{2}\right\}$

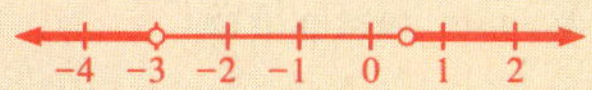

16. Find the zeros of the function $f(x) = x^3 - 4x$.
a. 0 and 2 **b.** 2 and -2 **(c.)** 0, 2, and -2 **d.** 0 and 4

17. A farmer plans to use 25 m of fencing to enclose a rectangular pen that has area 63 m^2. Only three sides of the pen need fencing because part of an existing wall will be used for one of the longer sides. Find the dimensions of the pen. 4-8
(a.) 3.5 m by 18 m **b.** 4.5 m by 16 m **c.** 7 m by 9 m **d.** 6.3 m by 10 m

18. Find the solution set of $x^2 + 2x - 8 > 0$. 4-9
a. $\{x: -4 < x < 2\}$ **(b.)** $\{x: x < -4 \text{ or } x > 2\}$
c. $\{x: -2 < x < 4\}$ **d.** $\{x: x < -2 \text{ or } x > 4\}$

19. Graph the solution set of $x^3 + x^2 < 2x$.

(a.) 

b. 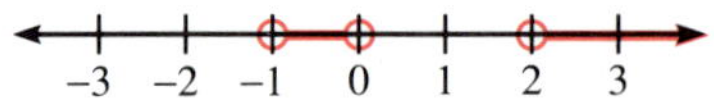

c. 

d. 

Chapter Test

Simplify.

1. $-3(x^2 - 2) + 2(3 - 2x^2)$ $-7x^2 + 12$ 4-1
2. $(2xy^2 + 2x^3 - x^2y) - (-2x^2y + 2xy^2 - y^3)$ $2x^3 + x^2y + y^3$
3. $(-2p^2q^3)^2(-p^2q)^3$ $-4p^{10}q^9$
4. $3xy^2(x^2 - 2xy)$ $3x^3y^2 - 6x^2y^3$ 4-2
5. $(2x - 3a)(x + 2a)$ $2x^2 + ax - 6a^2$
6. $rs(r + 2s)(2s - r)$ $4rs^3 - r^3s$ 4-3

Find (a) the GCF and (b) the LCM of the following.

7. 315 and 882 63; 4410
8. $12a^2b^2$, $8ab^3$, and $16b^4$ $4b^2$; $48a^2b^4$ 4-4

Factor each polynomial completely.

9. $x^4 - 8x^2 + 16$ $(x - 2)^2(x + 2)^2$
10. $xy + 2 - 2x - y$ $(x - 1)(y - 2)$ 4-5
11. $5s^2 - 16s + 12$ $(5s - 6)(s - 2)$
12. $6a^3 - 7a^2b - 2ab^2$ $a(6a^2 - 7ab - 2b^2)$ 4-6

Solve. Identify double roots.

13. $3t^2 - 10t + 3 = 0$ $\left\{\frac{1}{3}, 3\right\}$
14. $x^4 = 8(x^2 - 2)$ {−2 (d.r.), 2 (d.r.)} 4-7
15. The minute hand of a clock is 3 cm longer than the hour hand. If the distance between the tips of the hands at nine o'clock is 15 cm, how long is the minute hand? (*Hint:* Draw a diagram.) 12 cm 4-8

Find and graph the solution set of each inequality.

16. $x^2 \le x + 12$ $\{x: -3 \le x \le 4\}$
17. $x^3 > 4x$ $\{x: -2 < x < 0 \text{ or } x > 2\}$ 4-9

Alternate Test p. T16

Supplementary Materials

Practice Master	26
Test Masters	16, 17
Resource Book	pp. 116–117

Additional Answers
Chapter Test

16. 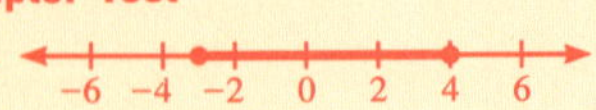

17. 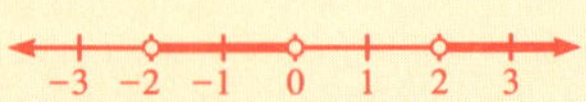

Preparing for College Entrance Exams

Strategy for Success

Use your time wisely. Solve the easy problems first and skip the more difficult ones. Do not spend too much time on one problem. When you come to the end of the test, return to the problems that were skipped. If time remains, check your work.

Decide which is the best of the choices given and write the corresponding letter on your answer sheet.

1. The lengths of the three sides of a right triangle are consecutive multiples of three. What is the area of the triangle? B
(A) 108 **(B)** 54 **(C)** 36 **(D)** 90 **(E)** 45

2. A line contains the points $(2, -7)$ and $(-4, 2)$. Which statement is *not* true? C
(A) The line has y-intercept -4.
(B) The line contains $\left(1, -\frac{11}{2}\right)$.
(C) The line is parallel to the line $3x - 2y = 6$.
(D) The line is perpendicular to the line $y = \frac{2}{3}x + 4$.
(E) No portion of the graph of the line lies in the first quadrant.

3. The graph of the solution set of the system $2(x + y) = 3(x - 1)$
$x - 3y = 3 - y$
is best described as: C
(A) a point **(B)** no point **(C)** a line
(D) a pair of parallel lines **(E)** a pair of intersecting lines

4. The Simon brothers rowed 18 km upstream in 3.75 h. The return trip with the same current took only 2.5 h. What was the speed of the current? B
(A) 1 km/h **(B)** 1.2 km/h **(C)** 4.5 km/h **(D)** 6 km/h **(E)** 6.5 km/h

5. Which expression is equivalent to $x^{n+5} \cdot x^{n-5}$? A
(A) x^{2n} **(B)** x^{10} **(C)** x^{n^2-25} **(D)** $(x^2)^{n^2-25}$ **(E)** x^{n^2}

6. Which polynomial is prime? D
(A) $4x^2 - 17x + 15$ **(B)** $4x^4 + 13x$ **(C)** $8x^6 + 27$
(D) $8x^2 + 7x - 16$ **(E)** $x^4 - 4$

7. Which inequality is false for every real number? D
(A) $r^2 - 4r \geq 5$ **(B)** $s^3 < 1$ **(C)** $t^3 + 9t \leq 0$
(D) $9u^2 + 49 < 42u$ **(E)** $(3z - 1)(z - 2) \leq 6$

5 Rational Expressions

Objectives

5-1 To simplify quotients using the laws of exponents.

5-2 To simplify expressions involving the exponent zero and negative integral exponents.

5-3 To use scientific notation and significant digits.

5-4 To simplify rational algebraic expressions.

5-5 To multiply and divide rational expressions.

5-6 To add and subtract rational expressions.

5-7 To simplify complex fractions.

5-8 To solve equations and inequalities having fractional coefficients.

5-9 To solve and use fractional equations.

Assignment Guide

See p. T58 for Key to the format of the Assignment Guide

Day	Average Algebra	Average Algebra and Trigonometry	Extended Algebra	Extended Algebra and Trigonometry
1	**5-1** 213/2–30 even **S** 215/*Mixed Review*	**5-1** 213/2–30 even **S** 215/*Mixed Review*	*Administer Chapter 4 Test* **5-1** *Read 5-1* 213/2–30 even, 33 **S** 215/*Mixed Review*	*Administer Chapter 4 Test* **5-1** *Read 5-1* 213/2–30 even, 33 **S** 215/*Mixed Review*
2	**5-2** 218/1–51 odd **S** 214/33, 34	**5-2** 218/1–51 odd **S** 214/33, 34	**5-2** 218/1–55 odd **S** 214/34, 35	**5-2** 218/1–57 odd **S** 214/34, 35
3	**5-3** 223/2–30 even 224/*P*: 1–5, 13–15 **S** 225/*Mixed Review*	**5-3** 223/2–30 even 224/*P*: 1–5, 13–15 **S** 225/*Mixed Review*	**5-3** 223/2–30 even 224/*P*: 1–5, 13–15 **S** 225/*Mixed Review*	**5-3** 223/2–30 even 224/*P*: 1–5, 13–15 **S** 225/*Mixed Review*
4	**5-4** 228/1–37 odd **R** 226/*Self-Test 1*	**5-4** 228/1–37 odd **R** 226/*Self-Test 1*	**5-4** 228/3–42 mult. of 3 230/*Extra:* 1–11 odd **R** 226/*Self-Test 1*	**5-4** 228/3–42 mult. of 3 230/*Extra:* 1–11 odd **R** 226/*Self-Test 1*
5	**5-5** 234/2–22 even **S** 234/*Mixed Review*	**5-5** 234/2–22 even **S** 234/*Mixed Review*	**5-5** 234/2–24 even **S** 234/*Mixed Review*	**5-5** 234/2–24 even **S** 234/*Mixed Review*
6	**5-6** 237/1–37 odd **S** 229/28–38 even	**5-6** 237/1–37 odd **S** 229/28–38 even	**5-6** 237/7–41 odd **S** 234/17–23 odd	**5-6** 237/7–41 odd **S** 234/17–23 odd
7	**5-7** 239/2–26 even **S** 241/*Mixed Review*	**5-7** 239/2–26 even **S** 241/*Mixed Review*	**5-7** 239/2–28 even **S** 241/*Mixed Review*	**5-7** 239/2–28 even, 32–35 **S** 241/*Mixed Review*
8	**5-8** 245/1–19 odd 245/*P*: 1–11 odd **R** 241/*Self-Test 2*	**5-8** 245/1–19 odd 245/*P*: 1–11 odd **R** 241/*Self-Test 2*	**5-8** 245/3–24 mult. of 3 245/*P*: 1–13 odd **R** 241/*Self-Test 2*	**5-8** 245/3–24 mult. of 3 245/*P*: 7–21 odd **R** 241/*Self-Test 2*
9	**5-8** 245/10–20 even 245/*P*: 13–21 odd **S** 237/40, 41	**5-8** 245/10–20 even 245/*P*: 13–21 odd **S** 241/40, 41	**5-8** 246/*P*: 14–21 **S** 241/32–35	**5-9** 249/3–30 mult. of 3 250/*P*: 3–21 mult. of 3, 22, 23 **S** 252/*Mixed Review*

Assignment Guide *(continued)*

Day	Average Algebra	Average Algebra and Trigonometry	Extended Algebra	Extended Algebra and Trigonometry
10	**5-9** 249/1–27 odd 250/*P*: 1–7 odd **S** 252/*Mixed Review*	**5-9** 249/1–27 odd 250/*P*: 1–7 odd **S** 252/*Mixed Review*	**5-9** 249/1–29 odd 250/*P*: 1–7 odd **S** 252/*Mixed Review*	*Prepare for Chapter Test* **R** 252/*Self-Test 3* **R** 255/*Chapter Review*
11	**5-9** 251/*P*: 9–19 odd **R** 252/*Self-Test 3*	**5-9** 251/*P*: 9–19 odd **R** 252/*Self-Test 3*	**5-9** 251/*P*: 9–21 odd, 22, 23 **R** 252/*Self-Test 3*	*Administer Chapter 5 Test* **6-1** *Read 6-1* 262/1–33 odd **S** 263/*Mixed Review*
12	*Prepare for Chapter Test* **R** 255/*Chapter Review*	*Prepare for Chapter Test* **R** 255/*Chapter Review*	*Prepare for Chapter Test* **R** 255/*Chapter Review*	
13	*Administer Chapter 5 Test*	*Administer Chapter 5 Test*	*Administer Chapter 5 Test*	

Supplementary Materials Guide

For Use with Lesson	Practice Masters	Tests	Study Guide (Reteaching)	Resource Book		
				Tests	Practice Exercises	Prob. Solving (PS) Applications (A) Enrichment (E)
5-1			pp. 71–72			p. 208 (A)
5-2	Sheet 27		pp. 73–74			
5-3	Sheet 28		pp. 75–76		p. 118	
5-4	Sheet 29	Test 18	pp. 77–78			
5-5			pp. 79–80			
5-6	Sheet 30		pp. 81–82			
5-7	Sheet 31		pp. 83–84		p. 119	
5-8	Sheet 32		pp. 85–86			
5-9	Sheet 33	Test 19	pp. 87–88		p. 120	pp. 200–201 (PS)
Chapter 5		Tests 20, 21		pp. 23–26	p. 121	p. 224 (E)

Software

Software	Computer Activities	Test Generator
	Activities 10–12	189 test items
For Use with Lessons	5-3, 5-8, 5-9	all lessons

Strategies for Teaching

Using Technology and Using Manipulatives

Competence in the use of calculators and computers is already recognized as one of the basic skills needed in many careers. Computers and related electronic devices have liberated students from a great deal of routine calculation and have vastly increased the range of information available to them.

The computer has many uses beyond that of saving students' time. Computer simulations make it possible for students to examine some concepts (function, for example) in depth and to test a variety of hypotheses. In statistics, use of computers facilitates the collection and analysis of data. The graphic presentation of data is made much easier if a graphing calculator or a computer is available.

When students become involved in writing programs, they are obliged to learn to communicate with the computer in its own language and to organize their thoughts in logical sequence. In the words of a recent report issued by the National Science Board, writing programs motivates students "to think algorithmically and develop problem solving skills."

Throughout this chapter, students will find features entitled Calculator Key-In and Computer Key-In, which can be used by students with no prior programming experience. Special sets of Computer Exercises extend some lessons for students who have had some programming experience. Suggestions for use of a calculator, a graphing calculator, or a computer appear within some of the text discussions and in directions for groups of exercises where use of these aids is particularly appropriate.

See the Exploration on page 836 for an activity in which students use a calculator to explore and make conjectures about continued fractions.

Again, manipulatives are always useful for students to take concrete approaches to enhance algebraic concepts and to aid student understanding of mathematical principles.

5-2 Zero and Negative Exponents

Students can fold pieces of paper or draw on graph paper to answer these questions.

1. Is it possible to divide a square into two smaller squares so that the whole square is used?
2. Is it possible to divide a square into three smaller squares?
3. Is it possible to divide a square into four smaller squares?
4. The figure shows that it is possible to divide a square into seventeen smaller squares. Write all the possible numbers of squares between one and twenty inclusive into which a large square can be divided. You can use graph paper as an aid in drawing the squares.

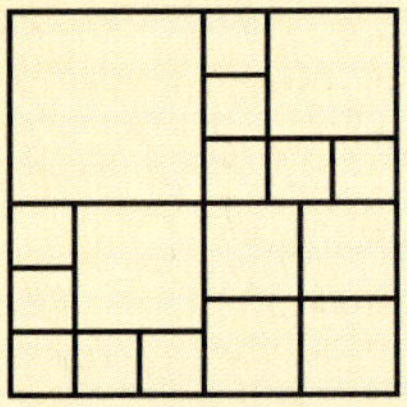

5-3 Scientific Notation and Significant Digits

Ask students who have worked with scientific notation using calculators or computers to demonstrate calculations in scientific notation for the rest of the class. Ask students in physics, chemistry, or other science classes to demonstrate the use of scientific notation and significant digits in actual problems from those courses.

5-4 Rational Algebraic Expressions

Students can use a computer or graphing calculator to verify their graphs for Exercises 1–12. Although asymptotes will not be shown, students should be able to infer where the asymptotes occur.

Using ALGEBRA PLOTTER PLUS

CHOOSE FUNCTION PLOTTER. Go to EQUATIONS and enter the equation, letting y equal the rule for the function. DRAW the graph.

References to Strategies

PE: Pupil's Edition **TE:** Teacher's Edition **RB:** Resource Book

Problem Solving Strategies

PE: pp. 244, 248–249 (make a table)
TE: pp. 224, 242
RB: pp. 200–201

Applications

PE: p. 210 (light); p. 221 (data analysis); pp. 224–225, 243–246, 248–254 (word problems); pp. 253–254 (electrical circuits)
TE: pp. 210, 243, 252
RB: p. 208

Nonroutine Problems

PE: p. 214 (Exs. 35, 36); p. 220 (Exs. 53–58); pp. 224–225 (Probs. 9–15); p. 237 (Exs. 40–41); pp. 240–241 (Exs. 27–35); p. 252 (Probs. 21–23); p. 254 (Challenge 1–4)
TE: pp. T99–T100 (Sugg. Extensions, Lessons 5-5, 5-8)
RB: p. 224

Communication

TE: p. T98 (Reading Algebra and Sugg. Extension, Lesson 5-2)

Thinking Skills

PE: p. 214 (Exs. 35); p. 220 (Exs. 53–58); p. 237 (Ex. 39, proofs)
TE: pp. 212, 221, 247

Explorations

PE: pp. 230–231 (graphing rational functions); pp. 253–254 (electrical circuits); p. 836 (continued fractions)

Connections

PE: p. 210 (Light); p. 220 (History); p. 221 (Data Analysis); pp. 224–225 (Astronomy); p. 244 (Chemistry); pp. 253–254 (Electricity); pp. 246, 248–251, 253–254 (Physics)

Using Technology

PE: p. 215 (Computer Key-In); pp. 225–226 (Calculator Key-In); pp. 222, 224, 231, 240
TE: pp. 215, 222, 225, 231, 240
Computer Activities: pp. 25–33

Cooperative Learning

TE: p. T98 (Lesson 5-2)

Teaching Resources

For use in implementing the teaching strategies referenced on the previous page.

Application
Resource Book, p. 208

Application—Astronomical Calculations (For use with Chapter 5)

1. Light travels at a speed of 1.86×10^5 mi/s in space. You can calculate the time it takes sunlight to reach one of the planets in our solar system by using the formula $\frac{\text{distance}}{\text{rate}} = \text{time}$. For example, since Earth is 9.30×10^7 mi from the sun, sunlight would take $\frac{9.30 \times 10^7}{1.86 \times 10^5} \approx 500$ s, or 8 min 20 s, to reach Earth.

   Perform similar calculations to complete the table below.

	Planet	*Distance (in miles) of planet from sun*	*Time (in hours, minutes, seconds) sunlight takes to reach planet*
a.	Mercury	3.60×10^7	________
b.	Venus	6.72×10^7	________
c.	Mars	1.42×10^8	________
d.	Jupiter	4.84×10^8	________
e.	Saturn	8.85×10^8	________
f.	Uranus	1.78×10^9	________
g.	Neptune	2.79×10^9	________
h.	Pluto	3.66×10^9	________

2. The light leaving the sun at any given moment spreads out over a sphere with ever-increasing radius. By the time the light reaches the orbit of one of the planets, the sphere is so large that the planet receives only a small fraction of the available light. For example, since Earth is 9.30×10^7 mi from the sun, the surface area of the sphere of sunlight, given by $A = 4\pi r^2$, is at that distance from the sun $4(3.14)(9.30 \times 10^7)^2 \approx 1.09 \times 10^{17}$ mi^2. Also, since the radius of Earth is 3.96×10^3 mi, the cross-sectional area of Earth, given by $A = \pi r^2$, is $(3.14)(3.96 \times 10^3)^2 \approx 4.92 \times 10^7$ mi^2. Therefore, Earth receives only $\frac{4.92 \times 10^7}{1.09 \times 10^{17}} \approx 4.51 \times 10^{-10}$ (which is less than one billionth) of the available sunlight. Perform similar calculations to complete the table below.

	Planet	*Distance (in miles) of planet from sun*	*Radius (in miles) of planet*	*Fraction of sunlight reaching planet*
a.	Mercury	3.60×10^7	1.52×10^3	________
b.	Venus	6.72×10^7	3.76×10^3	________
c.	Mars	1.42×10^8	2.10×10^3	________
d.	Jupiter	4.83×10^8	4.44×10^4	________
e.	Saturn	8.86×10^8	3.73×10^4	________
f.	Uranus	1.78×10^9	1.58×10^4	________
g.	Neptune	2.79×10^9	1.51×10^3	________
h.	Pluto	3.66×10^9	9.50×10^2	________

208 APPLICATIONS

Enrichment/Nonroutine Problems
Resource Book, p. 224

Squares and Such (For use with Chapter 5)

1. Is it possible to divide a square into two smaller squares so that the whole square is used? ________
2. Is it possible to divide a square into three smaller squares? ________
3. Is it possible to divide a square into four smaller squares? ________
4. The figure at the right shows that it is possible to divide a square into seventeen smaller squares.
   Write all the possible numbers of squares between one and twenty inclusive into which a large square can be divided. You can use the grid below as an aid in drawing the squares.

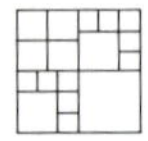

____

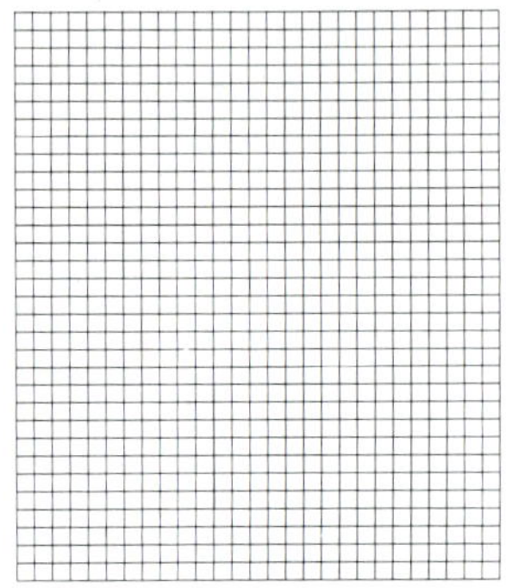

224 ENRICHMENT

Problem Solving
Resource Book, p. 200

Using Fractional Equations to Solve Problems (For use with Lesson 5-9)

By working through the steps in the problems below, you will gain skill in solving problems involving fractional equations.

Problem 1 At 9:00 A.M. June began working on the daily reports. At 10:00 A.M. Rosa joined her and the two completed the reports at 11:00 A.M. If June, a trainee, needs 2 h more to prepare the reports than Rosa needs, how long would June need to do the job alone?

a. What does the problem ask you to find? ________________________

____

b. Let j = the number of hours June would need to prepare the reports alone. Write an expression for the number of hours Rosa would need to do the job alone.

____

c. How many hours did each person work?

June ____________ Rosa ____________

d. What fraction of the job did June handle? Answer in terms of j. (*Hint:* If she can do the whole job in j hours, then each hour she handles $\frac{1}{j}$ of the job.) ____________

e. Write an equation, using 1 to represent the whole job. ________________

f. Solve your equation. Show your work.

g. Are both roots of the equation solutions to the problem? Explain. ________________

____

h. Write your answer to the problem. ________________________

i. Check by finding the numerical part of the job done by each person:

June: ____________ Rosa: ____________

Is the sum of the parts equal to 1? ____________

(continued)

200 PROBLEM SOLVING

Problem Solving
Resource Book, p. 201

Using Fractional Equations to Solve Problems _(continued)_

Problem 2 The numerator of a fraction is 4 greater than the denominator. This fraction is added to the fraction obtained by adding 3 to both the numerator and the denominator. If the resulting sum is equal to 4, find the original fraction.

a. Let d = the denominator of original fraction.
Complete the chart.

	numerator	denominator
original fraction		
new fraction		

b. The problem tells us that

original fraction + new fraction = 4.

Write this equation algebraically.

____

c. Multiply both sides by the LCD of the fractions and simplify the equation.

____

d. Solve to find the value(s) for d. ________________

e. Write the answer to the problem. ________________

f. For each solution, check that the sum of the original fraction and the new fraction is 4.

____

PROBLEM SOLVING 201

Reteaching/Practice
Study Guide, p. 87

5–9 *Fractional Equations*

Objective: To solve and use fractional equations.

Vocabulary

Fractional equation An equation in which a variable occurs in a denominator.

Extraneous root A root of a transformed equation that is not a root of the original equation.

CAUTION Since multiplying an equation by a polynomial may produce extraneous roots, you must *always check* each root of the new equation in the *original* equation.

Example Solve $\frac{12}{t^2 - 4} - \frac{3}{t - 2} = -1$.

Solution

$$\frac{12}{(t + 2)(t - 2)} - \frac{3}{t - 2} = -1 \qquad \text{The LCD is } (t + 2)(t - 2).$$

$$(t + 2)(t - 2)\left[\frac{12}{(t - 2)(t + 2)} - \frac{3}{t - 2}\right] = (t + 2)(t - 2)(-1) \qquad \left\{\begin{array}{l}\text{Multiply both}\\ \text{sides by the LCD.}\end{array}\right.$$

$$12 - 3(t + 2) = -1(t^2 - 4)$$

$$12 - 3t - 6 = -t^2 + 4$$

$$t^2 - 3t + 2 = 0$$

$$(t - 2)(t - 1) = 0 \longrightarrow t - 2 = 0 \text{ or } t - 1 = 0$$

$$t = 2 \text{ or } t = 1$$

Check the possible solutions in the *original* equation.

When $t = 2$: $\frac{12}{2^2 - 4} - \frac{3}{2 - 2} \stackrel{?}{=} -1$

$\frac{12}{0} - \frac{3}{0} \stackrel{?}{=} -1$

not defined

2 is an extraneous root.

When $t = 1$: $\frac{12}{1^2 - 4} - \frac{3}{1 - 2} \stackrel{?}{=} -1$

$\frac{12}{-3} - \frac{3}{-1} \stackrel{?}{=} -1$

$-4 + 3 = -1$ ✓

1 is a root of the original equation.

$\therefore$ the solution set is $\{1\}$.

Solve and check. If an equation has no solution, say so.

1. $\frac{3}{y} - \frac{1}{2y} = \frac{5}{4}$
2. $\frac{4}{3z} + \frac{2}{z} = \frac{5}{6}$
3. $\frac{6}{x + 1} = \frac{3}{x - 2}$
4. $\frac{12}{a} = \frac{4}{a - 4}$
5. $\frac{3}{r - 3} + 9 = \frac{r}{r - 3}$
6. $\frac{12}{n} = \frac{12}{n + 1} + 1$
7. $\frac{6p}{2p - 1} - 3 = \frac{3}{p}$
8. $\frac{7}{k - 3} - \frac{3}{k - 4} = \frac{1}{2}$
9. $\frac{9}{m + 5} - \frac{1}{m - 5} = \frac{3m}{m^2 - 25}$
10. $\frac{60}{d^2 - 36} + 1 = \frac{5}{d - 6}$
11. $\frac{2}{b^2 - 2b} - \frac{1}{b} = \frac{1}{3}$
12. $\frac{5}{x - 2} + \frac{x^2 - 4}{x^2 + 3x - 10} = \frac{x}{x + 5}$

Study Guide, ALGEBRA AND TRIGONOMETRY, Structure and Method, Book 2

87

Problem Solving
Study Guide, p. 88

5–9 *Fractional Equations* *(continued)*

Vocabulary

Work rate The fractional part of a job done in a given unit of time.
Example: Lenny can paint a room in 3 h. His work rate is $\frac{1}{3}$ job per hour.

Special rate formulas work rate × time = work done rate × time = distance

Complete each table and solve.

13. Stan can load his truck in 24 min. If Chris helps him, it takes 15 min to load the truck. How long does it take Chris alone?

Let x = the time it takes Chris alone.

	Work rate ×	Time =	Work done
Stan	?	15	?
Chris	$\frac{1}{x}$	15	?

Stan's part of job + Chris' part of job = Whole job
? + ? = 1

14. Bonnie can complete her paper route in 45 min. When her sister Jean helps her it takes them 18 min to complete the route. How long would it take Jean alone?

Let x = the time it takes Jean alone.

	Work rate ×	Time =	Work done
Bonnie	?	18	?
Jean	$\frac{1}{x}$	18	?

15. An express train travels 150 km in the same time that a freight train travels 100 km. The average speed of the freight train is 20 km/h less than that of the express train. Find the speed of each train.

Use the fact that time = $\frac{\text{distance}}{\text{rate}}$.

	Distance	Rate	Time
Express	?	r	?
Freight	?	$r - 20$	?

time for express train = time for freight train

16. Helen can ride 15 km on her bicycle in the same time it takes her to walk 6 km. If her rate riding is 6 km/h faster than her rate walking, how fast does she walk?

	Distance	Rate	Time
Riding	?	?	?
Walking	?	r	?

Mixed Review Exercises

Simplify.

1. $\frac{x^2 - 4}{2 - x}$
2. $\frac{72m^2n^3}{27mn^4}$
3. $\frac{1 + a^{-1}}{a^{-2} - 1}$
4. $\frac{k^2 - k - 6}{k^2 - 2k - 8}$

88
Study Guide, ALGEBRA AND TRIGONOMETRY, Structure and Method, Book 2

Using Technology/Exploration
Computer Activities, p. 25

ACTIVITY 10. *Powers of Numbers Near 1* *(for use with Section 5-3)*

Directions: Write all answers in the spaces provided.

PROBLEM

Use the computer to investigate the value of 0.5^a for large values of a.

PROGRAM

```
10  PRINT "WHAT IS THE BASE";
20  INPUT X
30  PRINT "WHAT IS THE LARGEST EXPONENT";
40  INPUT N
50  PRINT
60  PRINT "A", X; "↑ A"
70  PRINT
80  FOR A = 1 TO N
90  PRINT A, X ↑ A
100  NEXT A
110  END
```

PROGRAM CHECK

Type in the program. To test whether you entered it correctly, run the program. Enter .5 after the first question and 8 after the second question. The computer should print:

A	.5 ↑ A
1	.5
2	.25
3	.125
4	.0625
5	.03125
6	.015625
7	7.8125E-03
8	3.90625E-03

ANALYSIS

Your display shows the values of 0.5^a for $a = 1, 2, 3, \ldots, 8$. You can see that as a gets larger, the value of 0.5^a gets smaller. Your computer may have printed 0.5^8 as

3.90625E−03,

which is scientific notation in BASIC. In BASIC:

$$3.90625\text{E}-03 = 3.90625 \times 10^{-3} = 0.00390625$$

(continued)

Computer Activities, ALGEBRA AND TRIGONOMETRY, Structure and Method, Book 2

25

Using Technology/Exploration
Computer Activities, p. 26

(Activity 10 continued)

USING THE PROGRAM

In Exercises 1-12, modify line 80 where indicated. Then run the program to compute values of x^a for the given value of x. You must input the base, x, and the largest value of a, which is n.

Write down the value of x^n (the last value of x^a) to the nearest thousandth. Use standard (not scientific) notation.

Look at the display and determine whether x^a increases or decreases as a increases. Write your answer in the right-hand column.

Change in line 80	x		x^n	*x^a increases or decreases*
1. 80 FOR A = 1 TO N STEP 2	0.6	17	______	______
2. *No change*	0.7	17	______	______
3. *No change*	0.8	17	______	______
4. 80 FOR A = 1 TO N STEP 4	0.9	33	______	______
5. 80 FOR A = 1 TO N STEP 15	0.95	91	______	______
6. 80 FOR A = 1 TO N STEP 50	0.99	401	______	______
7. *No change*	1.01	401	______	______
8. 80 FOR A = 1 TO N STEP 15	1.05	91	______	______
9. 80 FOR A = 1 TO N STEP 4	1.1	33	______	______
10. 80 FOR A = 1 TO N STEP 2	1.2	17	______	______
11. *No change*	1.3	17	______	______
12. *No change*	1.4	17	______	______

Fill in the blank with "increases" or "decreases," whichever makes the sentence correct.

13. (Refer to Exercises 1-6.) For a given value of x between 0 and 1, x^a ______ as a increases.
14. (Refer to Exercises 7-12.) For a given value of x greater than 1, x^a ______ as a increases.

EXTENSION

1. Delete lines 10, 70, 80, 90, and 100 of the program and change lines 20 and 60 as follows:

```
20  LET X = .99
60  PRINT X; " ↑ "; N; " = "; X ↑ N
```

Run the program as many times as is necessary to find the smallest integral value of n for which $0.99^n < 0.0001$. You will need to experiment with a different value of n each time you run the program.

n = ______

2. Repeat Extension 1, replacing .99 with .999 in line 20. n = ______

26
Computer Activities, ALGEBRA AND TRIGONOMETRY, Structure and Method, Book 2

Application

The illumination (E) of a surface depends upon the intensity (I) of the light source and the distance (d) between the surface and the light source. The formula that expresses this relationship is

$$E = \frac{I}{d^2}.$$

In practical terms, consider the illumination of a book by a light bulb: To increase the illumination, either the intensity of the light bulb must be increased or the distance between the book and the light bulb must be decreased. Notice that doubling the intensity doubles the illumination; halving the distance, however, *quadruples* the illumination.

The relationship that exists between a book and a light bulb also applies to Earth and the sun. By the time sunlight has reached Earth's orbit, it has spread out over a sphere of radius 93,000,000 miles. Since the surface area of this sphere is about 1.1×10^{17} mi^2 and the cross-sectional area of Earth (whose radius is 4000 mi) is about 5.0×10^7 mi^2, the fraction of sunlight intercepted by Earth is about $\frac{5.0 \times 10^7}{1.1 \times 10^{17}}$, or 4.5×10^{-10}. This is less than one billionth of the available sunlight.

Group Activities

For each planet in the solar system, students can use its radius and its distance from the sun to determine the fraction of sunlight intercepted by the planet.

Support Materials

Resource Book p. 208

5 Rational Expressions

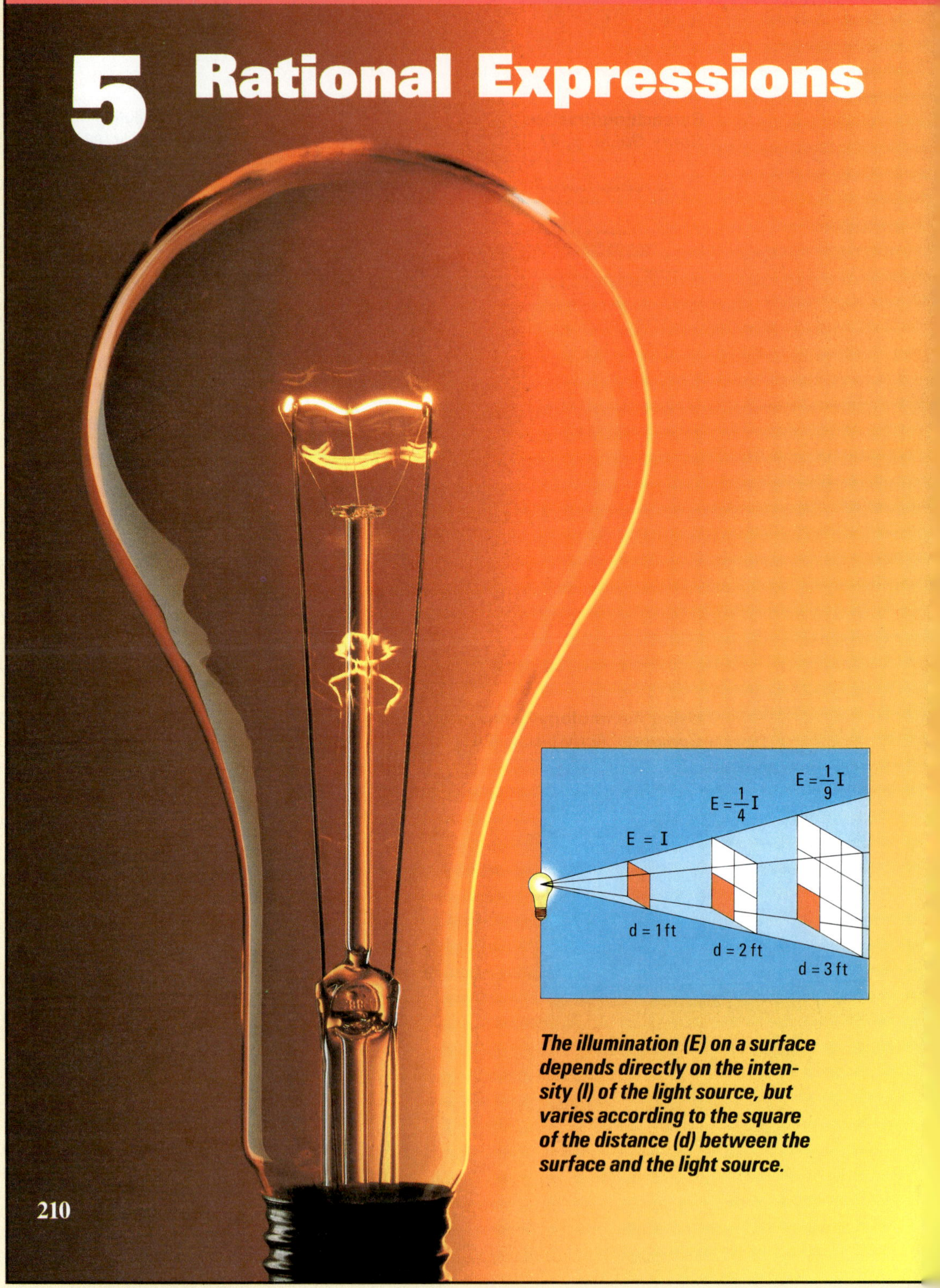

The illumination (E) on a surface depends directly on the intensity (I) of the light source, but varies according to the square of the distance (d) between the surface and the light source.

210

Using the Laws of Exponents

5-1 Quotients of Monomials

Objective To simplify quotients using the laws of exponents.

When you multiply fractions, you use the *multiplication rule for fractions*. For example.

$$\frac{3}{5} \cdot \frac{4}{7} = \frac{3 \cdot 4}{5 \cdot 7} = \frac{12}{35}.$$

Multiplication Rule for Fractions

Let p, q, r, and s be real numbers with $q \neq 0$ and $s \neq 0$. Then

$$\frac{p}{q} \cdot \frac{r}{s} = \frac{pr}{qs}.$$

A proof of this rule is asked for in Exercise 33.

Because equality is symmetric, the rule can be rewritten as $\frac{pr}{qs} = \frac{p}{q} \cdot \frac{r}{s}$. If $r = s$, you can replace s by r, obtaining $\frac{pr}{qr} = \frac{p}{q} \cdot \frac{r}{r} = \frac{p}{q} \cdot 1 = \frac{p}{q}$. This proves the following *rule for simplifying fractions*.

Let p, q, and r be real numbers with $q \neq 0$ and $r \neq 0$. Then

$$\frac{pr}{qr} = \frac{p}{q}.$$

Example 1 Simplify: **a.** $\frac{30}{48}$ **b.** $\frac{9xy^3}{15x^2y^2}$

Solution Find the GCF of the numerator and denominator. Then use the rule $\frac{pr}{qr} = \frac{p}{q}$ to simplify the fractions.

a. $\frac{30}{48} = \frac{5 \cdot 6}{8 \cdot 6} = \frac{5}{8}$ (GCF: 6)

b. $\frac{9xy^3}{15x^2y^2} = \frac{3y \cdot 3xy^2}{5x \cdot 3xy^2} = \frac{3y}{5x}$ (GCF: $3xy^2$)

Teaching References

Lesson Commentary, pp. T97–T101
Assignment Guide, pp. T62–T64
Supplementary Materials
- Practice Masters 27–33
- Tests 18–21
- Resource Book
  - Practice Exercises, pp. 118–121
  - Tests, pp. 23–26
  - Enrichment Activity, p. 224
  - Application, p. 208
  - Practice in Problem Solving/Word Problems, pp. 200–201
- Study Guide, pp. 71–88
- Computer Activities 10–12
- Test Generator
- Disk for Algebra

Alternate Test, p. T17

Teaching Suggestions, p. T98

Warm-Up Exercises

Write each fraction in simplest form.

1. $\frac{24}{36}$ $\frac{2}{3}$
2. $\frac{6}{9}$ $\frac{2}{3}$
3. $\frac{6}{8}$ $\frac{3}{4}$
4. $\frac{21}{24}$ $\frac{7}{8}$
5. $\frac{51}{57}$ $\frac{17}{19}$

Common Errors

Some students may incorrectly think that $\frac{x^2}{x^2} = 0$. When reteaching, remind students that any nonzero number divided by itself yields a quotient of one.

Motivating the Lesson

Remind students that they know how to multiply, divide, and simplify fractions. In this lesson students will use the rules for multiplying, dividing, and simplifying fractions and the laws of exponents to work with fractions composed of monomials.

Thinking Skills

In this lesson students should *recall* their earlier work with simple fractions. Many of the procedures for working with algebraic fractions are similar to those for working with simple fractions.

Chalkboard Examples

Simplify.

1. $\frac{9x^3y^2}{3x^2y^5}$ $\frac{9}{3} \cdot \frac{x^3}{x^2} \cdot \frac{y^2}{y^5} =$
$3 \cdot x \cdot \frac{1}{y^3} = \frac{3x}{y^3}$

2. $\frac{4a^3b}{18ab^4}$ $\frac{4}{18} \cdot \frac{a^3}{a} \cdot \frac{b}{b^4} =$
$\frac{2}{9} \cdot a^2 \cdot \frac{1}{b^3} = \frac{2a^2}{9b^3}$

3. $\frac{3c^4}{2d^2} \cdot \frac{d^5}{6c^2}$ $\frac{3}{2 \cdot 6} \cdot \frac{c^4}{c^2} \cdot$
$\frac{d^5}{d^2} = \frac{1}{4} \cdot c^2 \cdot d^3 = \frac{c^2d^3}{4}$

4. $\left(\frac{3r}{5s^3}\right)^2$ $\frac{3r}{5s^3} \cdot \frac{3r}{5s^3} = \frac{3 \cdot 3}{5 \cdot 5} \cdot$
$\frac{r \cdot r}{s^3 \cdot s^3} = \frac{9}{25} \cdot \frac{r^2}{s^6} = \frac{9r^2}{25s^6}$

5. $\frac{4x}{3y^7}\left(\frac{y^4}{2x^2}\right)^3$ $\frac{4x}{3y^7} \cdot \frac{y^{12}}{8x^6} =$
$\frac{4}{3 \cdot 8} \cdot \frac{x}{x^6} \cdot \frac{y^{12}}{y^7} = \frac{y^5}{6x^5}$

6. $\frac{(-2u^3v^2)^2}{12u^2v^4}$
$\frac{(-2u^3v^2)(-2u^3v^2)}{12u^2v^4} = \frac{4u^6v^4}{12u^2v^4} =$
$\frac{4}{12} \cdot \frac{u^6}{u^2} \cdot \frac{v^4}{v^4} = \frac{u^4}{3}$

In Lesson 4-2, you learned the first three laws of exponents listed below. The additional laws, Laws 4 and 5, cover quotients involving exponents.

The Laws of Exponents

Let m and n be positive integers and a and b be real numbers, with $a \neq 0$ and $b \neq 0$ when they are divisors. Then:

1. $a^m \cdot a^n = a^{m+n}$ **2.** $(ab)^m = a^m b^m$ **3.** $(a^m)^n = a^{mn}$

4a. If $m > n$, $\frac{a^m}{a^n} = a^{m-n}$ **4b.** If $n > m$, $\frac{a^m}{a^n} = \frac{1}{a^{n-m}}$

5. $\left(\frac{a}{b}\right)^m = \frac{a^m}{b^m}$

Here is a concise proof of Law 4(b). We omit many steps that use only the basic properties of real numbers. For example, no reasons are given for the fact that $n = (n - m) + m$.

Statement	*Reason*
1. $a^n = a^{(n-m)+m} = a^{n-m} \cdot a^m$	Law 1 of Exponents
2. $\frac{a^m}{a^n} = \frac{1 \cdot a^m}{a^{n-m} \cdot a^m}$	Substitution
3. $\frac{a^m}{a^n} = \frac{1}{a^{n-m}}$	Rule for simplifying fractions

Proofs of Laws 4(a) and 5 are left as Exercises 34 and 35.

Notice how Laws 4(a), 4(b), and 5 are used in Example 2.

Example 2 Simplify: **a.** $\frac{2^8}{2^5}$ **b.** $\frac{5x^3}{x^7}$ **c.** $\left(\frac{t^3}{3}\right)^2$

Solution **a.** $\frac{2^8}{2^5} = 2^{8-5}$
$= 2^3$
$= 8$ ***Answer***

b. $\frac{5x^3}{x^7} = \frac{5}{x^{7-3}}$
$= \frac{5}{x^4}$ ***Answer***

c. $\left(\frac{t^3}{3}\right)^2 = \frac{(t^3)^2}{3^2}$
$= \frac{t^6}{9}$ ***Answer***

A quotient of monomials having integral coefficients is *simplified* when:

1. the integral coefficients are relatively prime, that is, have no common factor except 1 and −1;
2. each base appears only once; and
3. there are no "powers of powers" [such as $(a^3)^4$].

Example 3 Simplify: **a.** $\frac{24s^4t^3}{32st^5}$ **b.** $\frac{5x}{4y^2}\left(\frac{2y}{x^2}\right)^3$

Solution **a.** *Method 1* Think of each variable separately and use Law 4(a) or 4(b).

$$\frac{24s^4t^3}{32st^5} = \frac{24}{32} \cdot \frac{s^4}{s^1} \cdot \frac{t^3}{t^5} = \frac{3}{4} \cdot \frac{s^3}{1} \cdot \frac{1}{t^2} = \frac{3s^3}{4t^2}$$

Method 2 As in Example 1, find the GCF of the numerator and the denominator. Then simplify the fraction.

$$\frac{24s^4t^3}{32st^5} = \frac{3s^3 \cdot 8st^3}{4t^2 \cdot 8st^3} = \frac{3s^3}{4t^2}$$

b. $\frac{5x}{4y^2}\left(\frac{2y}{x^2}\right)^3 = \frac{5x}{4y^2} \cdot \frac{8y^3}{x^6} = \frac{5 \cdot 8}{4} \cdot \frac{x}{x^6} \cdot \frac{y^3}{y^2} = \frac{10y}{x^5}$

(This solution used Laws 5 and 3 and then Method 1.)

Oral Exercises

Simplify. Assume that no denominator equals 0 and that variables in exponents represent positive integers.

1. $\frac{6^5}{6^3}$ 36
2. $\frac{3^5}{3^2}$ 27
3. $\frac{(-2)^7}{(-2)^4}$ -8
4. $\frac{(-5)^2}{(-5)^4}$ $\frac{1}{25}$
5. $\frac{t^8}{t^5}$ t^3
6. $\frac{x^4}{x^7}$ $\frac{1}{x^3}$
7. $\frac{y^n}{y^{2n}}$ $\frac{1}{y^n}$
8. $\frac{s^{k+2}}{s^k}$ s^2
9. $\frac{6x^3y^2}{3x^2y}$ $2xy$
10. $\frac{4ab^4}{12ab^2}$ $\frac{b^2}{3}$
11. $\frac{(x^2)^3}{x^5}$ x
12. $\frac{(pq)^2}{pq^2}$ p

Written Exercises

Simplify. Assume that no denominator equals 0.

A

1. $\frac{18x^3}{6x}$ $3x^2$
2. $\frac{5t^3}{15t^5}$ $\frac{1}{3t^2}$
3. $\frac{-12p^3q}{4p^2q^2}$ $-\frac{3p}{q}$
4. $\frac{30x^2y^3}{-6x^3y^2}$ $-\frac{5y}{x}$
5. $\frac{-15u^5v^3}{-25u^4v^2}$ $\frac{3uv}{5}$
6. $\frac{48x^5y^5}{32x^4y^6}$ $\frac{3x}{2y}$
7. $\left(\frac{3r}{s^2}\right)^3$ $\frac{27r^3}{s^6}$
8. $\left(\frac{2x^2}{-y}\right)^4$ $\frac{16x^8}{y^4}$
9. $\frac{3s}{t^2} \cdot \frac{s^2}{t}$ $\frac{3s^3}{t^3}$
10. $\frac{2u}{v^2} \cdot \frac{3u}{2v^2}$ $\frac{3u^2}{v^4}$
11. $\frac{3x^2}{y^2} \cdot \frac{3y}{6x}$ $\frac{3x}{2y}$
12. $\frac{xy^2}{2} \cdot \frac{6x}{y^2}$ $3x^2$
13. $\frac{rs^2t^3}{r^3s^2t}$ $\frac{t^2}{r^2}$
14. $\frac{a^2b^3c}{a^3bc^2}$ $\frac{b^2}{ac}$
15. $\frac{u^2}{v}\left(\frac{3v}{u^2}\right)^2$ $\frac{9v}{u^2}$
16. $\frac{2x^2}{y^3}\left(\frac{-y^3}{2x^2}\right)^2$ $\frac{y^3}{2x^2}$
17. $\frac{(4r^2s^2)^2}{(4r^2s)^2}$ s^2
18. $\frac{(2hk^3)^3}{(-h^2k^2)^2}$ $\frac{8k^5}{h}$
19. $\frac{(xyz^2)^2}{(x^2yz)^2}$ $\frac{z^2}{x^2}$
20. $\frac{(pq^2r^3)^3}{(p^3qr^2)^2}$ $\frac{q^4r^5}{p^3}$

Check for Understanding

Oral Exs. 1–12: use after Example 3.

Guided Practice

Simplify. Assume that no denominator equals zero.

1. $\frac{18x^5}{6x^2}$ $3x^3$
2. $-\frac{35p^4q}{14pq^4}$ $-\frac{5p^3}{2q^3}$
3. $\left(\frac{5a}{b^3}\right)^3$ $\frac{125a^3}{b^9}$
4. $\frac{a^5}{b} \cdot \frac{3a^2}{2b^5}$ $\frac{3a^7}{2b^6}$
5. $\frac{(-2m^3n)^3}{(3mn^2)^2}$ $-\frac{8m^7}{9n}$
6. $\frac{2x^2y^2}{(2xy^2)^2}$ $\frac{1}{2y^2}$
7. $\frac{5k^2}{p^3} \cdot \left(\frac{p^2}{10k}\right)^2$ $\frac{p}{20}$
8. $\frac{3(2r^2)^2}{9} \cdot \frac{6}{r^4}$ 8

Summarizing the Lesson

In this lesson students learned how to simplify quotients of monomials using the laws of exponents and the rule for simplifying fractions. Ask students to state in words the five laws of exponents.

Suggested Assignments

Average Algebra
213/2–30 even
S 215/Mixed Review

Average Alg. and Trig.
213/2–30 even
S 215/Mixed Review

Extended Algebra
213/2–30 even, 33
S 215/Mixed Review

Extended Alg. and Trig.
213/2–30 even, 33
S 215/Mixed Review

Additional Answers Written Exercises

35. 1. $m > n$, or $m - n > 0$ Given
2. $a^{m-n} \cdot a^n = a^{m-n+n} = a^m$ Law 1 of exponents
3. $a^{m-n} = \frac{a^m}{a^n}$ Div. prop. of eq.

36. 1. $ma + na = (m + n)a$; yes
2. $m(a + b) = ma + mb$; yes
3. $n(m(a)) = (mn)a$; yes
4. a. If $m > n$, $ma - na = (m - n)a$; yes
b. If $n > m$, $ma - na = 1 - (n - m)a = 1 - (na - ma) = 1 - na + ma = ma - na + 1$; no
5. $m(a - b) = ma - mb$; yes

Supplementary Materials

Study Guide pp. 71–72

B **21.** $\left(\frac{2y^2}{3}\right)^2 \cdot \frac{3x}{y^4}$ $\frac{4x}{3}$ **22.** $\frac{4x^2}{yz^2}\left(\frac{z}{2x}\right)^3$ $\frac{z}{2xy}$ **23.** $\left(\frac{c^3}{d^4}\right)^2\left(\frac{-cd}{h}\right)^3$ $-\frac{c^9}{d^5h^3}$ **24.** $\left(\frac{-4a^2}{3b}\right)^2\left(\frac{-b}{2a}\right)^3$ $-\frac{2ab}{9}$

In Exercises 25–30, assume that no denominator equals 0 and that *m* and *n* are integers greater than 1.

25. $\frac{a^{2m}b^{2m+1}}{(a^2b^2)^m}$ b **26.** $\frac{x^{n+1}y^n}{x^ny^{n-1}}$ xy **27.** $\frac{(pq)^n}{pq^n}$ p^{n-1}

28. $\frac{(z^n)^3}{z^nz^3}$ z^{2n-3} **29.** $\frac{t^{n+1}t^{n-1}}{t^n}$ t^n **30.** $\frac{a^{n-1}b^{2n}}{a^{n+1}(b^2)^{n-1}}$ $\frac{b^2}{a^2}$

31. The multiplication rule for fractions can be rewritten in the form $(p \div q) \cdot (r \div s) = (p \cdot r) \div (q \cdot s)$. Replace $\div$ by $-$ and $\cdot$ by $+$. Is the resulting statement true? Yes

32. **a.** Rewrite the rule $\frac{pr}{qr} = \frac{p}{q}$ using $\cdot$ for multiplication and $\div$ for division. $(p \cdot r) \div (q \cdot r) = p \div q$

b. In your answer to (a) replace $\cdot$ by $+$ and $\div$ by $-$. Is the resulting statement true? $(p + r) - (q + r) = p - q$; Yes

33. Supply reasons for the steps in the following proof of the multiplication rule for fractions.

1. $\frac{p}{q} \cdot \frac{r}{s} = \left(p \cdot \frac{1}{q}\right)\left(r \cdot \frac{1}{s}\right)$ ___?___ Def. of division
2. $= pr\left(\frac{1}{q} \cdot \frac{1}{s}\right)$ ___?___ and ___?___ Comm. prop. and Assoc. prop. of mult.
3. $= pr \cdot \frac{1}{qs}$ ___?___ Prop. of the reciprocal of a product
4. $= \frac{pr}{qs}$ ___?___ Def. of division

$\therefore \frac{p}{q} \cdot \frac{r}{s} = \frac{pr}{qs}$

34. Supply reasons for the steps in the following proof of Exponent Law 5.

1. $\left(\frac{a}{b}\right)^m = \underbrace{\frac{a}{b} \cdot \frac{a}{b} \cdot \cdots \cdot \frac{a}{b}}_{m \text{ factors}}$ ___?___ Def. of exponent
2. $= \frac{a \cdot a \cdot \cdots \cdot a}{\underbrace{b \cdot b \cdot \cdots \cdot b}_{m \text{ factors}}}$ ___?___ Mult. rule for fractions

$\therefore \left(\frac{a}{b}\right)^m = \frac{a^m}{b^m}$

C **35.** Give a concise proof of Exponent Law 4a.

36. In each of the laws of exponents, replace $\cdot$ by $+$, $\div$ by $-$, and, for example, a^m by ma. Is the resulting statement true? (*Example:* $a^m \cdot a^n = a^{m+n}$ becomes $ma + na = (m + n)a$, which is true.)

Mixed Review Exercises

Find the solution set of each inequality.

1. $x^2 + 2x - 8 > 0$	**2.** $4y + 9 \le 1$	**3.** $w^2 < 4$
4. $5 - 3p > -7$	**5.** $\lvert 2t + 5 \rvert \le 3$	**6.** $2a^2 < a + 3$
7. $-3 < 2c - 1 < 5$	**8.** $m(m - 1) \ge 2$	**9.** $\lvert d + 2 \rvert > 1$
10. $g \le g^2$	**11.** $n^2 + 2 \le 3n$	**12.** $-2u < -6$ or $u - 1 \le 1$

Computer Key-In

If you divide a by b, you obtain a quotient q and a remainder r less than b. For example, $192 \div 84$ gives

$$192 = 84 \cdot 2 + 24$$
$$a = b \cdot q + r.$$

The *Euclidean Algorithm* uses repeated division to find the GCF of two positive integers a and b. To find the GCF of 192 and 84 by the Euclidean Algorithm you would proceed as follows.

$$192 = 84 \cdot 2 + 24$$
$$84 = 24 \cdot 3 + 12$$
$$24 = 12 \cdot 2 + 0$$

The last nonzero remainder, 12, is the GCF of 192 and 84.

Here is a program that uses the Euclidean Algorithm to find GCF's.

```
10 PRINT "FIND THE GCF OF TWO POSITIVE INTEGERS."
20 INPUT "ENTER THE NUMBERS (LARGER FIRST): "; A, B
30 PRINT "THE GCF OF "; A; " AND "; B; " IS ";
40 IF INT (A/B)=A/B THEN 90
50 LET R=A-B*INT (A/B)
60 LET A=B
70 LET B=R
80 GOTO 40
90 PRINT B
95 END
```

Exercises

1. Run the program to find the GCF of each pair of numbers.

a. (432, 528) 48 **b.** (2592, 4682) 2 **c.** (999, 259) 37 **d.** (30,030, 13,122) 6

2. Run the program to find the GCF of the numerator and the denominator of each fraction. Write each fraction in simplest form.

a. $\frac{87}{1653}$ $\frac{1}{19}$ **b.** $\frac{11{,}776}{15{,}872}$ $\frac{23}{31}$ **c.** $\frac{2772}{5940}$ $\frac{7}{15}$ **d.** $\frac{2048}{40{,}387}$ $\frac{2048}{40{,}387}$

Additional Answers
Mixed Review

1. $\{x: x < -4 \text{ or } x > 2\}$
2. $\{y: y \le -2\}$
3. $\{w: -2 < w < 2\}$
4. $\{p: p < 4\}$
5. $\{t: -4 \le t \le -1\}$
6. $\left\{a: -1 < a < \frac{3}{2}\right\}$
7. $\{c: -1 < c < 3\}$
8. $\{m: m \le -1 \text{ or } m \ge 2\}$
9. $\{d: d < -3 \text{ or } d > -1\}$
10. $\{g: g \le 0 \text{ or } g \ge 1\}$
11. $\{n: 1 \le n \le 2\}$
12. $\{u: u \le 2 \text{ or } u > 3\}$

Computer Key-In Commentary

In the program that follows, the statement in line 50 is derived from this equation: $a = b \cdot q + r$. This may not be apparent to your students. If they solve the equation for r to get $a - b \cdot q = r$ and recognize that q is indeed the integer value of $\frac{a}{b}$, then line 50 becomes clear.

Your students may wonder why the message in line 30 is not incorporated into the answer of line 90. The values of the original numbers A and B are "lost" when they are replaced by the new values in lines 60 and 70. Unless B is the GCF of A, by the time the computer executes line 90, the values of A and B have changed.

Teaching Suggestions p. T98

Group Activities p. T98

Reading Algebra p. T98

Suggested Extensions p. T98

Warm-Up Exercises

Simplify.

1. $\frac{a^2}{a^3}$ $\frac{1}{a}$ 2. $\frac{c^4}{c^{10}}$ $\frac{1}{c^6}$

3. $\frac{z^3}{z^5}$ $\frac{1}{z^2}$ 4. $\frac{m}{m^2}$ $\frac{1}{m}$

5. $\frac{4ab^4}{6a^3b^4}$ $\frac{2}{3a^2}$

Motivating the Lesson

Some students may be familiar with scientific notation (Lesson 5-3). Negative and zero exponents are used extensively in this notation for the size and mass of extremely small objects.

Chalkboard Examples

Express in simplest form without negative exponents.

1. 458×10^{-4} $458 \times \frac{1}{10^4}$

$= \frac{458}{10,000} = 0.0458$

2. $\frac{6^{-1}x^{-2}z^3}{5^0x^{-1}z^{-2}}$ $\left(\frac{1}{6}\right)x^{-2-(-1)}z^{3-(-2)}$

$= \frac{1}{6}x^{-1}z^5 = \frac{z^5}{6x}$

3. $\frac{4m^{-2}}{n}\left(\frac{m^3}{3n^2}\right)^{-1}$

$\frac{4m^{-2}}{n} \cdot \frac{m^{-3}}{3^{-1}n^{-2}} =$

$\frac{4 \cdot 3^1 \cdot m^{-2+(-3)}}{n^{1+(-2)}} =$

$\frac{12m^{-5}}{n^{-1}} = \frac{12n}{m^5}$

5-2 Zero and Negative Exponents

Objective To simplify expressions involving the exponent zero and negative integral exponents.

The definition of *power* can be extended so that any integer may be used as an exponent, not just a positive integer. The power a^n when n is 0 or n is a negative integer is defined in such a way that the laws of exponents stated on page 212 continue to hold.

If Law 1, $a^m \cdot a^n = a^{m+n}$, is to hold when $n = 0$, then this statement must be true:

$$a^m \cdot a^0 = a^{m+0} = a^m$$

If $a \neq 0$, you can divide both sides of the resulting equation $a^m \cdot a^0 = a^m$ by a^m to obtain

$$a^0 = \frac{a^m}{a^m} = 1.$$

If Law 1 is to hold for negative exponents, then this statement must be true: If $-n$ is a negative integer and $a \neq 0$,

$$a^{-n} \cdot a^n = a^{-n+n} = a^0 = 1.$$

Since $a^{-n} \cdot a^n = 1$, a^{-n} must be the reciprocal of a^n:

$$a^{-n} = \frac{1}{a^n}$$

The discussion above leads to these definitions.

If n is a positive integer and $a \neq 0$:

$$\mathbf{a^0 = 1} \qquad \mathbf{a^{-n} = \frac{1}{a^n}}$$

The expression 0^0 is not defined.

Example 1 Write in simplest form without negative or zero exponents. Assume that the variables are nonzero.

a. 10^{-2} **b.** $-3c^{-2}$ **c.** $2^{-1}a^0b^{-3}$ **d.** $(3x)^{-2}$

Solution

a. $10^{-2} = \frac{1}{10^2} = \frac{1}{100}$

b. $-3c^{-2} = -3 \cdot \frac{1}{c^2} = -\frac{3}{c^2}$

c. $2^{-1}a^0b^{-3} = \frac{1}{2} \cdot 1 \cdot \frac{1}{b^3} = \frac{1}{2b^3}$

d. $(3x)^{-2} = \frac{1}{(3x)^2} = \frac{1}{9x^2}$

Since $a^{-n} = \frac{1}{a^n}$ and division by zero is not defined, bases of powers with negative exponents must not be zero. To simplify matters we make the following agreement, to hold throughout the rest of this book.

The domains of all variables in any algebraic expression are automatically restricted so that denominators of fractions and bases of powers with negative or zero exponents will not be zero.

For example, in the expression

$$\frac{3a^0b^{-2}}{x^2 - 4},$$

you may assume that $a \neq 0$, $b \neq 0$, $x \neq 2$, and $x \neq -2$ even though these restrictions are not stated.

It can be shown that all the laws of exponents hold even if some of the exponents are negative or zero. For example, Law 3 now implies that

$$(a^3)^{-2} = a^{3(-2)} = a^{-6}.$$

This statement can be justified with the help of Law 3 for *positive* exponents:

$$(a^3)^{-2} = \frac{1}{(a^3)^2} = \frac{1}{a^6} = a^{-6}$$

You can now see that Laws 4(a) and 4(b) give the same results. For example,

by Law 4(a): $\frac{a^2}{a^7} = a^{2-7} = a^{-5} = \frac{1}{a^5}$

by Law 4(b): $\frac{a^2}{a^7} = \frac{1}{a^{7-2}} = \frac{1}{a^5}.$

Example 2 Write in simplest form without negative or zero exponents.

a. $10^{-2} \times 10^{-3}$ **b.** $(-3 \cdot 5^{-1})^{-2}$ **c.** $\frac{3^0x^{-3}y}{2x^{-1}y^{-2}}$

Solution **a.** $10^{-2} \times 10^{-3} = 10^{-2+(-3)} = 10^{-5} = \frac{1}{10^5} = \frac{1}{100{,}000}$ ***Answer***

b. $(-3 \cdot 5^{-1})^{-2} = (-3)^{-2}(5^{-1})^{-2}$

$= \frac{1}{(-3)^2} \cdot 5^2$

$= \frac{25}{9}$ ***Answer***

c. $\frac{3^0x^{-3}y}{2x^{-1}y^{-2}} = \frac{1 \cdot x^{-3-(-1)}y^{1-(-2)}}{2}$

$= \frac{1x^{-2}y^3}{2}$

$= \frac{y^3}{2x^2}$ ***Answer***

Express without using fractions.

4. $\frac{3}{100{,}000}$

$\frac{3}{10^5} = 3 \times 10^{-5}$

5. $\frac{2a}{t^3b}$ $2at^{-3}b^{-1}$

6. $\frac{8x^2y}{xy^3}$ $8x^2yx^{-1}y^{-3} =$

$8x^{2+(-1)}y^{1+(-3)} = 8xy^{-2}$

Common Errors

Some students mistakenly think that $\left(-\frac{1}{3}\right)^{-3} = \frac{1}{27}$. When reteaching, emphasize the definition of negative exponents. Show students that $\left(-\frac{1}{3}\right)^{-3} = \frac{1}{\left(-\frac{1}{3}\right)^3} = \frac{1}{-\frac{1}{27}} = -27.$

Check for Understanding

Here is a suggested use of the Oral Exercises to check students' understanding as you teach the lesson.
Oral Exs. 1–16: use after Example 4.

Guided Practice

Express the following without exponents.

1. $2^{-2} \cdot 3^2$ $\frac{9}{4}$
2. $(-5^{-1})^{-2}$ 25
3. $\left(-\frac{2}{3}\right)^{-2}$ $\frac{9}{4}$
4. 8.15×10^{-2} 0.0815
5. $\left[\left(\frac{1}{2}\right)^{-2} \cdot \left(\frac{5}{7}\right)\right]^{-2}$ $\frac{49}{400}$

Write without using fractions.

6. $\frac{5}{1000}$ 5×10^{-3}
7. $\frac{8y^2}{-2x^3}$ $-4x^{-3}y^2$

Express without zero or negative exponents.

8. $\frac{3x^{-1}}{y^{-2}}$ $\frac{3y^2}{x}$
9. $\frac{(3a^3)^{-2}}{a^{-3}b^{-3}}$ $\frac{b^3}{9a^3}$
10. $\left(\frac{x^2}{y}\right)^{-1}\left(\frac{x}{y^2}\right)^2$ $\frac{1}{y^3}$
11. $\frac{(a^{-2}b)^{-1}}{(ab^2)^{-2}}$ a^4b^3
12. $\frac{r^{-1}}{s^{-2}}\left(\frac{r^{-1}}{s^2}\right)^{-1}$ s^4

Summarizing the Lesson

Students learned to simplify expressions involving zero and negative exponents. Ask students to explain why 0^0 is undefined and to explain the agreement made at the top of page 217.

Since a^{-n} and a^n are reciprocals, $\frac{1}{a^{-n}} = a^n$. Example 3 uses this fact.

Example 3 Write in simplest form without negative or zero exponents.

a. $\left(\frac{2}{3}\right)^{-3}$ **b.** $\left(\frac{2x^{-2}}{5y^3}\right)^{-1}$

Solution **a.** $\left(\frac{2}{3}\right)^{-3} = \frac{2^{-3}}{3^{-3}} = 2^{-3} \cdot \frac{1}{3^{-3}} = \frac{1}{2^3} \cdot 3^3 = \frac{3^3}{2^3} = \frac{27}{8}$ ***Answer***

b. $\left(\frac{2x^{-2}}{5y^3}\right)^{-1} = \frac{(2x^{-2})^{-1}}{(5y^3)^{-1}} = \frac{2^{-1}x^2}{5^{-1}y^{-3}} = \frac{5x^2y^3}{2}$ ***Answer***

Negative exponents are introduced in order to write expressions without using fractions. The form of the answer to Example 4(a) below is called *scientific notation*. This notation will be discussed in Lesson 5-3.

Example 4 Write without using fractions: **a.** $\frac{3}{10{,}000}$ **b.** $\frac{5x^2}{yz^3}$

Solution **a.** $\frac{3}{10{,}000} = \frac{3}{10^4} = 3 \times 10^{-4}$ **b.** $\frac{5x^2}{yz^3} = 5x^2y^{-1}z^{-3}$

Oral Exercises

Express in simplest form without negative or zero exponents.

1. $\frac{1}{2^{-3}}$ 8
2. $\frac{5}{3^{-1}}$ 15
3. $10^7 \cdot 10^{-5}$ 100
4. $(3^{-1})^{-2}$ 9
5. $c^{-3} \cdot c^5$ c^2
6. $a^{-4} \cdot a^3$ $\frac{1}{a}$
7. $(y^{-1})^{-2}$ y^2
8. $y^{-1} \cdot y^{-2}$ $\frac{1}{y^3}$
9. $(5x)^0$ 1
10. $5x^0$ 5
11. p^2q^{-2} $\frac{p^2}{q^2}$
12. $(p^2q)^{-2}$ $\frac{1}{p^4q^2}$
13. $\left(\frac{a}{b}\right)^{-2}$ $\frac{b^2}{a^2}$
14. $\frac{a}{2^{-1}b^{-2}}$ $2ab^2$
15. $\frac{x^{-3}}{y^{-3}}$ $\frac{y^3}{x^3}$
16. $\left(\frac{r^{-1}}{s^{-1}}\right)^{-1}$ $\frac{r}{s}$

Written Exercises

Write in simplest form without negative or zero exponents.

A

1. $3 \cdot 5^{-1}$ $\frac{3}{5}$
2. $(3 \cdot 5)^{-1}$ $\frac{1}{15}$
3. $(-3^{-1})^{-2}$ 9
4. $(-2^{-2})^{-1}$ -4
5. $(2^{-2} \cdot 3^{-1} \cdot 5^0)^{-1}$ 12
6. $5^{-1}(3^{-2} \cdot 2^{-3})^0$ $\frac{1}{5}$
7. $2\left(\frac{2}{5}\right)^{-2}$ $\frac{25}{2}$
8. $\left(\frac{3}{4}\right)^{-1}\left(\frac{4}{3}\right)^{-2}$ $\frac{3}{4}$

Write without using fractions.

9. $\frac{7}{10{,}000}$ 7×10^{-4} 10. $\frac{3}{1000}$ 3×10^{-3} 11. $\frac{6x^2}{y^3}$ $6x^2y^{-3}$ 12. $\frac{x^2}{yz^4}$ $x^2y^{-1}z^{-4}$

Express as a decimal numeral. A calculator may be helpful.

Sample 1 $(-5 \cdot 3^{-2})^{-2} = (-5)^{-2} \cdot 3^4 = \frac{81}{25} = 3.24$

13. 596×10^{-2} 5.96 14. 238×10^{-3} 0.238 15. 7.2×10^{-3} 0.0072 16. 1.45×10^{-2} 0.0145

17. $\left(-\frac{5}{3}\right)^{-3}$ −0.216 18. $\left(-\frac{2}{5}\right)^{-2}$ 6.25 19. $(3^{-1} \cdot 2^2)^{-2}$ 0.5625 20. $(-5^{-1} \cdot 2^{-2})^2$ 0.0025

Write in simplest form without negative or zero exponents.

21. $\frac{3x^{-2}}{y^{-1}}$ $\frac{3y}{x^2}$ 22. $\frac{p^{-1}q^{-2}}{p^{-3}}$ $\frac{p^2}{q^2}$ 23. $\frac{s^{-2}t^{-3}}{s^{-1}t^0}$ $\frac{1}{st^3}$

24. $\frac{6xy^{-1}}{-2x^{-2}y^{-1}}$ $-3x^3$ 25. $\left(\frac{u^{-2}}{v}\right)^{-1}$ u^2v 26. $\left(\frac{2}{h^2k^{-3}}\right)^{-2}$ $\frac{h^4}{4k^6}$

27. $(2x^{-2}y^2)^{-2}$ $\frac{x^4}{4y^4}$ 28. $\frac{(3x^{-2}y)^{-1}}{(2xy^{-2})^0}$ $\frac{x^2}{3y}$ 29. $3x^2(3xy^{-1})^{-2}$ $\frac{y^2}{3}$

30. $5t(s^{-1}t^{-2})^{-2}$ $5s^2t^5$ 31. $\frac{(2x^{-1})^{-2}}{2(y^{-1})^{-2}}$ $\frac{x^2}{8y^2}$ 32. $\left(\frac{2pq^{-1}}{4q^2}\right)^{-1}$ $\frac{2q^3}{p}$

33. $\left(\frac{x}{y^2}\right)^{-1}\left(\frac{x^{-2}}{y}\right)^2$ $\frac{1}{x^5}$ 34. $\left(\frac{3}{t^2}\right)^{-1}\left(\frac{t}{3}\right)^{-2}$ 3 35. $\left(\frac{p^{-2}q^{-1}}{p^{-1}q^{-2}}\right)^{-1}$ $\frac{p}{q}$

36. $\frac{(ax^2)^{-1}}{a^{-2}x^{-2}}$ a 37. $\left(\frac{x^2}{y^{-1}}\right)^{-2}\left(\frac{y^2}{x^{-1}}\right)^2$ $\frac{y^2}{x^2}$ 38. $\frac{r^{-2}}{s^2}\left(\frac{1}{rs}\right)^{-2}$ 1

39. $\left(\frac{u}{v^{-1}}\right)^0\left(\frac{u^{-1}}{v^2}\right)^2(uv^2)^{-1}$ $\frac{1}{u^3v^6}$ 40. $\left(\frac{a^0}{b}\right)^{-2}\left(\frac{a}{b^{-2}}\right)^{-2}$ $\frac{1}{a^2b^2}$

41. $4x^3y^{-6} + (x^{-1}y^2)^{-3}$ $\frac{5x^3}{y^6}$ 42. $\left(\frac{u^2}{v}\right)^2 + (-u^{-2}v)^{-2}$ $\frac{2u^4}{v^2}$

Show by counterexample that the given expressions are *not* equivalent.

Sample 2 $(x + y)^{-2}$ and $x^{-2} + y^{-2}$

Solution Let $x = 1$ and $y = 2$. Then:

$$(x + y)^{-2} = (1 + 2)^{-2} = 3^{-2} = \frac{1}{9} \qquad x^{-2} + y^{-2} = 1^{-2} + 2^{-2} = 1 + \frac{1}{4} = \frac{5}{4} \qquad \left(\frac{1}{9} \neq \frac{5}{4}\right)$$

B 43. $(x + y)^{-1}$ and $x^{-1} + y^{-1}$ 44. $(xy)^{-1}$ and $\frac{x}{y}$

45. xy^{-1} and $\frac{1}{xy}$ 46. $(1 - x)^{-2}$ and $1 - x^{-2}$

Suggested Assignments

Average Algebra
218/1–51 odd
S 214/33, 34

Average Alg. and Trig.
218/1–51 odd
S 214/33, 34

Extended Algebra
218/1–55 odd
S 214/34, 35

Extended Alg. and Trig.
218/1–57 odd
S 214/34, 35

Supplementary Materials

Study Guide pp. 73–74
Practice Master 27

Additional Answers Written Exercises

43–46. Answers may vary; examples are given.

43. Let $x = 1$ and $y = 2$; $(x + y)^{-1} = (1 + 2)^{-1} = 3^{-1} = \frac{1}{3}$; $x^{-1} + y^{-1} = 1 + \frac{1}{2} = \frac{3}{2}$; $\frac{1}{3} \neq \frac{3}{2}$

44. Let $x = 2$ and $y = 3$; $(xy)^{-1} = (2 \cdot 3)^{-1} = \frac{1}{6}$; $\frac{x}{y} = \frac{2}{3}$; $\frac{1}{6} \neq \frac{2}{3}$

45. Let $x = 2$ and $y = 3$; $xy^{-1} = 2 \cdot 3^{-1} = \frac{2}{3}$; $\frac{1}{xy} = \frac{1}{2 \cdot 3} = \frac{1}{6}$; $\frac{2}{3} \neq \frac{1}{6}$

46. Let $x = -1$; $(1 - x)^{-2} = \frac{1}{(1 - x)^2} = \frac{1}{(1 - (-1))^2} = \frac{1}{2^2} = \frac{1}{4}$; $1 - x^{-2} = 1 - \frac{1}{x^2} = 1 - \frac{1}{(-1)^2} = 1 - 1 = 0$; $\frac{1}{4} \neq 0$

Additional Answers
Written Exercises

53. 1. m and n are pos. integers; $m > n$
Given

2. $a^m \cdot a^{-n} = a^m \cdot \frac{1}{a^n}$
Def. of neg. exp.

3. $= \frac{a^m}{a^n}$
Def. of div.

4. $= a^{m-n}$
Law 4a of pos. exp.

5. $= a^{m+(-n)}$
Def. of subtr.

6. $\therefore a^m a^{-n} = a^{m+(-n)}$
Trans. prop. of eq.

54. 1. $(a^m)^{-n} = \frac{1}{(a^m)^n}$
Def. of neg. exp.

2. $= \frac{1}{a^{nm}}$
Subst.; Law 3 of pos. exp.

3. $= a^{-(mn)}$
Def. of neg. exp.

4. $= a^{m(-n)}$
Opp. of a prod. is prod. of opp. of one factor and the other factors.

5. $\therefore (a^m)^{-n} = a^{m(-n)}$
Trans. prop. of eq.

(continued on p. 254)

In Exercises 47–52, replace the (__?__) by a polynomial to make a true statement.

Sample 3 $x^{-3} + 2x^{-2} - 3x^{-1} = x^{-3}(\underline{\ ?\ })$

Solution $x^{-3} + 2x^{-2} - 3x^{-1} = x^{-3}(1 + 2x - 3x^2)$

47. $x^{-1} - 4x^{-2} + 2x^{-3} = x^{-3}(\underline{\ ?\ })$ $x^2 - 4x + 2$

48. $2x^{-2} + x^{-1} - 3 = x^{-2}(\underline{\ ?\ })$ $2 + x - 3x^2$

49. $4 - 5x^{-1} + x^{-2} = x^{-2}(\underline{\ ?\ })$ $4x^2 - 5x + 1$

50. $x^{-1} - 9x^{-3} = x^{-3}(\underline{\ ?\ })$ $x^2 - 9$

C **51.** $x^2(x-1)^{-2} - 4(x-1)^{-1} = (x-1)^{-2}(\underline{\ ?\ })$ $x^2 - 4x + 4$

52. $(x^2+4)^{-1} - 5(x^2+4)^{-2} = (x^2+4)^{-2}(\underline{\ ?\ })$ $x^2 - 1$

In Exercises 53–58, use the exponent laws for positive exponents to prove the forms of the laws given below. In each case, *m* and *n* are positive integers with *m* > *n*.

53. $a^m a^{-n} = a^{m+(-n)}$

54. $(a^m)^{-n} = a^{m(-n)}$

55. $a^{-m}a^n = a^{(-m)+n}$

56. $\frac{a^{-m}}{a^n} = a^{(-m)-n}$

57. $\frac{a^m}{a^{-n}} = a^{m-(-n)}$

58. $\frac{a^{-m}}{a^{-n}} = a^{(-m)-(-n)}$

Biographical Note / *Ch'in Chiu-Shao*

Ch'in Chiu-Shao (ca. 1202–1261) has been called one of the greatest mathematicians of his time. This achievement is particularly remarkable because Ch'in did not devote his life to mathematics. He was accomplished in many other fields and held a series of bureaucratic positions in Chinese provinces.

Ch'in's mathematical reputation rests on one celebrated treatise, *Shu-shu chiu-chang* ("Mathematical Treatise in Nine Sections"), which appeared in 1247. The treatise covers topics ranging from indeterminate analysis to military matters and surveying. Ch'in included a version of the Chinese remainder theorem, which used algorithms to solve problems.

His interest in indeterminate analysis led Ch'in into related fields. He wrote down the earliest explanation of how Chinese calendar experts calculated astronomical data according to the timing of the winter solstice. He also introduced techniques for solving equations, finding sums of arithmetic series, and solving linear systems. His use of the zero symbol is a milestone in Chinese mathematics.

5-3 Scientific Notation and Significant Digits

Objective To use scientific notation and significant digits.

In scientific work, you meet very large and very small numbers. For example, the distance that light travels in a year is about 5,680,000,000,000 mi. The time that it takes a signal to travel from one component of a supercomputer to another might be 0.0000000024 s. The wavelength of ultraviolet light is about 0.00000136 cm. To make such numbers easier to work with, you can write them in *scientific notation*:

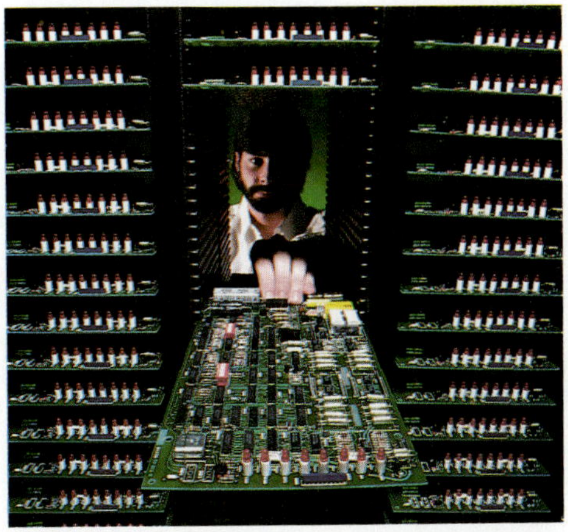

$$5{,}680{,}000{,}000{,}000 = 5.68 \times 10^{12}$$
$$0.0000000024 = 2.4 \times 10^{-9}$$
$$0.00000136 = 1.36 \times 10^{-6}$$

In **scientific notation,** a number is expressed in the form

$$m \times 10^n$$

where $1 \le m < 10$, and n is an integer.

The digits in the factor m should all be *significant*. A **significant digit** of a number written in decimal form is any nonzero digit or any zero that has a purpose other than placing the decimal point. In the following examples the significant digits are printed in red.

4006 0.4050 320.0 0.00203

In scientific notation

$4006 = 4.006 \times 10^3$, $0.4050 = 4.050 \times 10^{-1}$,
$320.0 = 3.200 \times 10^2$, $0.00203 = 2.03 \times 10^{-3}$.

For a number such as 2300 it is not clear which, if any, of the zeros are significant. Scientific notation eliminates this problem. Writing 2300 as

2.3×10^3, 2.30×10^3, or 2.300×10^3

indicates, respectively, that none, one, or two of the zeros are significant.

Most measurements are approximate, and the more significant digits that are given in such an approximation, the more *accurate* it is. For example, if a length is given as 2.30×10^3 cm, the measurement has three significant digits. Since the 0 is significant, you know that the length is *not* 2.31×10^3 cm or 2.29×10^3 cm, but it might be 2.302×10^3 cm.

Teaching Suggestions p. T98

Suggested Extensions p. T98

Warm-Up Exercises

Write without using fractions.

1. $\frac{9}{1000}$ 9×10^{-3}
2. $\frac{3}{100}$ 3×10^{-2}
3. $\frac{1.7}{10^2}$ 1.7×10^{-2}

Express as a decimal.

4. 3.05×10^{-2} 0.0305
5. 4.57×10^3 4570

Motivating the Lesson

Tell students that the study of physics and chemistry involves working with very large and very small numbers. The topic of today's lesson, scientific notation and significant digits, is directly applicable to that work.

Thinking Skills

Before beginning a multiplication or division involving approximations, students should *analyze* the problem to see how many digits the least accurate approximation has. The product or quotient should have the same number of digits.

Chalkboard Examples

Express in scientific notation.

1. 7,920,000 7.92×10^6
2. 0.000041 4.1×10^{-5}

Express in decimal form.

3. 2.7×10^3 2700
4. 6.18×10^{-7} 0.000000618
5. Find a one-significant-digit estimate of x, where

$$x = \frac{3.25 \times 12{,}120 \times 0.039}{0.00186}$$

$$3.25 = 3.25 \times 10^0$$
$$12{,}120 = 1.212 \times 10^4$$
$$0.039 = 3.9 \times 10^{-2}$$
$$0.00186 = 1.86 \times 10^{-3}$$
$$x = \frac{3.25 \times 1.212 \times 3.9}{1.86} \times 10^{0+4+(-2)-(-3)}$$
$$\approx \frac{3 \times 1 \times 4}{2} \times 10^{0+4+(-2)-(-3)}$$
$$= 6 \times 10^5$$
$$\therefore x \approx 6 \times 10^5, \text{ or } 600{,}000.$$

Check for Understanding

Here is a suggested use of the Oral Exercises to check students' understanding as you teach the lesson.
Oral Exs. 1–8: use before Example 1.
Oral Exs. 9–12: use after Example 2.

Using a Calculator

As an introduction to scientific notation on a calculator, you might want to have students enter the number 1 and repeatedly multiply or divide by 10 until the calculator's display switches from standard to scientific notation.

Calculator use is recommended for solving the problems on pages 224–225.

When you multiply or divide measurements, their accuracy limits the accuracy of the answer. *When approximations are multiplied or divided, the product or quotient should have the same number of digits as the least accurate factor.*

Example 1 Write each number in scientific notation. If the number is an integer and ends in zeros, assume that the zeros are not significant.

a. 75,040 **b.** 0.0000702 **c.** 3,050,000

Solution

a. $75{,}040 = 7.504 \times 10^4$ decimal point moved 4 places to the left

b. $0.0000702 = 7.02 \times 10^{-5}$ decimal point moved 5 places to the right

c. $3{,}050{,}000 = 3.05 \times 10^6$ decimal point moved 6 places to the left

Numbers expressed in scientific notation are easy to compare.

Example 2

a. $6.84 \times 10^{-2} > 4.96 \times 10^{-2}$ because $6.84 > 4.96$.

b. $8.7 \times 10^4 < 5.3 \times 10^5$ because $4 < 5$.

c. $5.25 \times 10^{-2} > 7.50 \times 10^{-3}$ because $-2 > -3$.

Example 3 Find a one-significant-digit estimate of x, where

$$x = \frac{5260 \times 0.0682 \times 86.1}{0.420}.$$

Solution

$$x = \frac{5.26 \times 10^3 \times 6.82 \times 10^{-2} \times 8.61 \times 10^1}{4.2 \times 10^{-1}}$$ Write each number in scientific notation.

$$\approx \frac{5 \times 10^3 \times 7 \times 10^{-2} \times 9 \times 10^1}{4 \times 10^{-1}}$$ Round each decimal to a whole number.

$$= \frac{5 \times 7 \times 9}{4} \times 10^{3+(-2)+1-(-1)}$$ Compute, and give the result to one significant digit.

$$= \frac{315}{4} \times 10^3$$

$$\approx 80 \times 10^3 = 8 \times 10^4$$

$$\therefore x \approx 8 \times 10^4, \text{ or } 80{,}000$$

Round off errors may cause the digit obtained in the process shown in Example 3 to be incorrect. (In the example, $x = 73{,}500$ to three significant digits.) Nevertheless, such estimates do describe the "size" of the answer.

Before using a calculator in the exercises, see the Calculator Key-In on page 225 to determine how your calculator handles numbers that are too large or small or too long for the display.

Oral Exercises

Tell how many significant digits each number has.

1. 40.05 4
2. 10,608 5
3. 0.050 2
4. 0.0250 3
5. 4.0×10^3 2
6. 7.06×10^{-2} 3
7. 3.600×10^2 4
8. 4.8×10^6 2

Replace the __?__ with < or > to make a true statement.

9. 4.60×10^3 __?__ 3.75×10^3 >
10. 2.60×10^3 __?__ 6.75×10^4 <
11. 7.73×10^{-3} __?__ 2.50×10^{-2} <
12. 2.78×10^{-4} __?__ 2.78×10^{-6} >

Written Exercises

Write each number in scientific notation. If the number is an integer and ends in zeros, assume that the zeros are not significant.

8. 4.60×10^{-5}

A
1. 7500 7.5×10^3
2. 106,000 1.06×10^5
3. 0.608 6.08×10^{-1}
4. 0.0038 3.8×10^{-3}
5. 10.05 1.005×10^1
6. 762.20 7.6220×10^2
7. 0.0320 3.20×10^{-2}
8. 0.0000460
9. 655×10^3 6.55×10^5
10. 0.025×10^{-2} 2.5×10^{-4}
11. 0.560×10^{-3} 5.60×10^{-4}
12. 4775×10^{-2} 4.775×10^1

Write each number in decimal form.

13. 5×10^3 5000
14. 10^{-4} 0.0001
15. 4.3×10^{-3} 0.0043
16. 10^7 10,000,000
17. 6.75×10^4 67,500
18. 6.20×10^{-2} 0.0620
19. 7.50×10^{-3} 0.00750
20. 4.000×10^2 400.0

Replace the __?__ with < or > to make a true statement.

21. $(2 \times 10^3)^2$ __?__ 5×10^5 >
22. 1.6×10^{18} __?__ $(1.2 \times 10^9)^2$ >
23. $(0.005)^3$ __?__ 2×10^{-8} >
24. $(2 \times 10^3)^{-2}$ __?__ $(2 \times 10^{-3})^2$ <

Find a one-significant-digit estimate of each of the following.

B
25. $\frac{478 \times 0.230}{0.0281 \times 32.4}$ 100
26. $\frac{0.525 \times 7820}{22.5 \times 0.00475}$ 40,000
27. $\frac{73.1 \times (0.493)^2}{0.620 \times (32.6)^2}$ 0.03
28. $\frac{0.000212 \times (588)^2}{57.7 \times (0.0620)^2}$ 300

Simplify, assuming that the factors are approximations. Give answers in scientific notation with the same number of significant digits as in the least accurate factor.

29. $\frac{(8 \times 10^6)(2 \times 10^{-2})}{4 \times 10^2}$ 4×10^2
30. $\frac{(7.5 \times 10^6)(5.0 \times 10^{-1})}{1.5 \times 10^8}$ 2.5×10^{-2}
31. $\frac{(8.4 \times 10^{15})(1.5 \times 10^{-5})}{(4.0 \times 10^4)(1.2 \times 10^3)}$ 2.6×10^3

Guided Practice

Express in decimal form.

1. 3×10^4 30,000
2. 3.5×10^{-2} 0.035

Express in scientific notation.

3. 17,000 1.7×10^4
4. 0.709 7.09×10^{-1}
5. 0.0040 4.0×10^{-3}
6. 10,800,000 1.08×10^7
7. 3400 3.4×10^3
8. 0.0023 2.3×10^{-3}

Replace the __?__ with < or >.

9. $(3 \times 10^2)^3$ __?__ 2.4×10^5 >
10. $(2 \times 10^2)^{-3}$ __?__ $(2 \times 10^{-2})^3$ <

Summarizing the Lesson

In this lesson students learned some applications of and methods for using scientific notation and significant digits. Ask students to explain an answer, such as 3.7×10^0, that involves a zero exponent.

Suggested Assignments

Average Algebra
223/2–30 even
224/*P*: 1–5, 13–15
S 225/Mixed Review

Average Alg. and Trig.
223/2–30 even
224/*P*: 1–5, 13–15
S 225/Mixed Review

Extended Algebra
223/2–30 even
224/*P*: 1–5, 13–15
S 225/Mixed Review

Extended Alg. and Trig.
223/2–30 even
224/*P*: 1–5, 13–15
S 225/Mixed Review

Problem Solving Strategies

Some problems may be solved by finding *information in charts.* This procedure is shown by Problems 1–5 and 9–11.

Supplementary Materials

Study Guide	pp. 75–76
Practice Master	28
Computer Activity	10
Resource Book	p. 118

Problems

In these problems, all given data are approximate. In Problems 1–5, give answers reflecting the situation in 1987 using the following table. A calculator may be helpful.

	1987 Population	Land Area (km^2)		
U.S.A.	2.44×10^8	9.36×10^6	1. 244,000,000	9,360,000
China	1.06×10^9	9.60×10^6	1,060,000,000	9,600,000
Italy	5.74×10^7	3.01×10^5	57,400,000	301,000
World	5.03×10^9	1.49×10^8	5,030,000,000	149,000,000

A

1. Write the table above using decimal notation.
2. What percent of the world's land area has **(a)** the United States, **(b)** China, and **(c)** Italy? a. 6.28% b. 6.44% c. 0.202%
3. What percent of the world's population has **(a)** the United States, **(b)** China, and **(c)** Italy? a. 4.85% b. 21.1% c. 1.14%
4. Find the population density (persons per km^2) of **(a)** the United States, **(b)** China, **(c)** Italy, and **(d)** the world. a. 26.1 persons per km^2 b. 110 persons per km^2 c. 191 persons per km^2 d. 33.8 persons per km^2
5. In 1987 the national debt of the United States was about 2.50×10^{12} dollars. How much debt is this per person? $10,200
6. How many nanoseconds (1 nanosecond = 10^{-9} s) does it take a computer signal to travel 60 cm at a rate of 2.4×10^{10} cm/s? 2.5 nanoseconds
7. The designer of a minicomputer wants to limit to 4.0 nanoseconds (see Problem 6) the time that it takes signals to travel from one component to another. What restriction does this put on the distance between components? no greater than 96 cm
8. The estimated masses of a proton and an electron are 1.67×10^{-24} g and 9.11×10^{-28} g, respectively. Find the ratio of the mass of the proton to the mass of the electron. 1830 to 1

B

9. The astronomical unit (AU), the light year, and the parsec are units of distance used in astronomy. Copy and complete the following table showing how these units are related.

	AU	parsec	light year	
One AU equals	1	?	1.58×10^{-5}	4.85×10^{-6}
One parsec equals	?	1	?	2.06×10^5; 3.26
One light year equals	?	0.307	1	6.33×10^4

10. Given that 1 light year = 9.46×10^{12} km, express 1 AU and 1 parsec in kilometers. (See Problem 9.) 1.49×10^8 km; 3.08×10^{13} km

11. The mean distances from the sun to the planets Mercury and Neptune are 5.79×10^7 km and 4.51×10^9 km, respectively. Express these distances in astronomical units. (See Problems 9 and 10.) 0.389 AU; 30.3 AU

12. A steady current of 1 ampere flowing through a solution of silver nitrate will deposit 1.12×10^{-3} g of silver in one second. How much silver will be deposited by a current of 12.0 amperes in one hour? 48.4 g

13. An electric current of one ampere corresponds to a flow of 6.2×10^{18} electrons per second past any point of the circuit. How many electrons flow through the filament of a 100 watt, 115 volt light bulb during one hour? (*Note:* Watts = amperes × volts) 1.9×10^{22}

14. If you were to receive as wages 1 cent for the first day, 2 cents for the second day, 4 cents for the third day, and so on, with each day's wages twice the previous day's, about how many dollars would you receive on the 31st day? Use the approximation $2^{10} \approx 10^3$ and express your answer in decimal form. $10,000,000

C 15. The masses of the sun and Earth are 2.0×10^{30} kg and 6.0×10^{24} kg, respectively, and their radii are 7.0×10^8 m and 6.4×10^6 m. Show that the average density of Earth is about four times that of the sun. (*Note:* Density = mass ÷ volume)

Mixed Review Exercises

Simplify.

1. $\frac{18x^3y}{27x^2y^4}$ $\frac{2x}{3y^3}$
2. $\left(\frac{3a}{2b}\right)^2$ $\frac{9a^2}{4b^2}$
3. $\frac{4m}{n} \cdot \frac{m^2}{5n}$ $\frac{4m^3}{5n^2}$
4. $\frac{(c^2d)^3}{(cd^2)^3}$ $\frac{c^3}{d^3}$
5. $\frac{2p^2}{q} \cdot \frac{q^3}{2p^3}$ $\frac{q^2}{p}$
6. $\frac{(-w)^2}{-w^2}$ -1
7. $\frac{14z^5}{10z^5}$ $\frac{7}{5}$
8. $\frac{u^4}{v}\left(\frac{v^2}{u^3}\right)^2$ $\frac{v^3}{u^2}$

Express in simplest form without negative or zero exponents.

9. $(x^{-3}y^2)^{-1}$ $\frac{x^3}{y^2}$
10. $\left(\frac{c^{-1}}{d}\right)^{-2}$ c^2d^2
11. $(6m^2)(2m)^{-2}$ $\frac{3}{2}$
12. $\frac{a^{-4}b}{a^2b^{-3}}$ $\frac{b^4}{a^6}$

Calculator Key-In

Some calculators will indicate an error when the result of a computation has too many digits to be shown in full on the display. Others will display such a number in scientific notation, with the exponent of 10 shown at the right. Most scientific calculators have a key that allows the user to select scientific notation for all numbers displayed and to enter numbers in scientific notation.

Additional Answers Problems

15. volume of sun = $\frac{4}{3}\pi(7.0 \times 10^8)^3 = \frac{4}{3}\pi(3.43 \times 10^{26})$; density of sun = $\frac{2.0 \times 10^{30}}{\frac{4}{3}\pi(3.43 \times 10^{26})}$; volume of Earth = $\frac{4}{3}\pi(6.4 \times 10^6)^3 = \frac{4}{3}\pi(2.62 \times 10^{20})$; density of Earth = $\frac{6.0 \times 10^{24}}{\frac{4}{3}\pi(2.62 \times 10^{20})}$; $\frac{\text{density of Earth}}{\text{density of sun}} = \frac{6.0 \times 10^{24}}{\frac{4}{3}\pi(2.62 \times 10^{20})} \div \frac{2.0 \times 10^{30}}{\frac{4}{3}\pi(3.43 \times 10^{26})} = 3.9 \times 10^0 \approx 4$

Calculator Key-In Commentary

Ask students to read the manuals for their calculators and then to work the problems. Students with the same kind of calculator might work together. The class may wish to discuss the differences in the way their calculators work with very large or very small numbers.

When the result of a computation has more significant digits than the display can show, the number may be *truncated* (the extra digits are discarded) or *rounded* (the last place shown is rounded to the correct value).

You can find answers to some of the exercises below without a calculator, but doing them on your calculator will help you find out how your calculator handles numbers that are too large or small or too long for the display.

Evaluate on your calculator.

1. $20{,}000 \times 50{,}000$ 1×10^9

2. $352{,}129 \times 416{,}187$ 1.4655×10^{11}

3. $(0.00001)^2$ 1×10^{-10}

Find the value of each expression to four significant digits. Give your answers in scientific notation.

4. $(4.325 \times 10^5)(9.817 \times 10^{12})$ 4.246×10^{18}

5. $(1.589 \times 10^3)(6.805 \times 10^{-8})$ 1.081×10^{-4}

6. $2.563 \div 493{,}412$ 5.194×10^{-6}

7. $\dfrac{(4.303 \times 10^6)(2.115 \times 10^3)}{9.563 \times 10^8}$ 9.517×10^0

8. a. Find $\frac{2}{3}$ as a decimal on your calculator. 0.66666666 or 0.66666667

b. From part (a), tell whether your calculator truncates or rounds.
If the last digit is 6, the calculator truncates.
If the last digit is 7, the calculator rounds.

Self-Test 1

Vocabulary scientific notation (p. 221) significant digit (p. 221)

Simplify.

1. $\dfrac{24s^3t^2}{16st^3}$ $\dfrac{3s^2}{2t}$

2. $\left(\dfrac{2x^2y}{w^2}\right)^2\left(\dfrac{y^2w}{x^2}\right)^3$ $\dfrac{4y^8}{wx^2}$

Obj. 5-1, p. 211

Write in simplest form without negative or zero exponents.

3. $\dfrac{p^{-2}qr^{-1}}{p^0q^{-2}r^{-3}}$ $\dfrac{q^3r^2}{p^2}$

4. $\left(\dfrac{x^2y^{-1}}{z}\right)^{-2}\left(\dfrac{x^2y^2}{z^{-3}}\right)$ $\dfrac{y^4z^5}{x^2}$

Obj. 5-2, p. 216

Replace the _?_ by a polynomial to make a true statement.

5. $x^{-1} - 4x^{-3} = x^{-3}$ (_?_) $x^2 - 4$

Write each number in scientific notation.

6. 482.60 4.8260×10^2

7. 0.000210 2.10×10^{-4}

Obj. 5-3, p. 221

Write each number in decimal form.

8. 3.609×10^3 3609

9. 5.400×10^{-2} 0.05400

10. Give a one-significant-digit estimate of $\dfrac{683 \times 0.536}{0.00392}$. 90,000

Quick Quiz

Simplify.

1. $\dfrac{6x^5y^3}{20xy^4}$ $\dfrac{3x^4}{10y}$

2. $\left(\dfrac{2a^2b}{c^3}\right)^3 \cdot \left(\dfrac{c^2}{6a}\right)^2$ $\dfrac{2a^4b^3}{9c^5}$

Express without zero or negative exponents.

3. $\dfrac{(4r^2)^0(3s^{-2})^2}{(2sr^{-2})^{-2}}$ $\dfrac{36}{r^4s^2}$

4. $\dfrac{8a^{-2}bc^{-1}}{4a^{-1}b^{-1}c^3}$ $\dfrac{2b^2}{ac^4}$

Express in scientific notation.

5. 0.000101 1.01×10^{-4}

6. 302.511 3.02511×10^2

Express in decimal form.

7. 8.04×10^5 804,000

8. 8.04×10^{-3} 0.00804

9. Express with one significant digit: $\dfrac{12.01(3.1 \times 10^3)}{5.07 \times 10^{-2}}$ 7×10^5

Rational Expressions

5-4 Rational Algebraic Expressions

Objective To simplify rational algebraic expressions.

You know that a *rational number* is one that can be expressed as a quotient of integers. Similarly, a **rational algebraic expression,** or **rational expression,** is one that can be expressed as a quotient of polynomials. Some examples of rational expressions are

$$\frac{4xy^2}{7y}, \quad \frac{x^2-3x-4}{x^2-1}, \quad \text{and} \quad x(x^2-4)^{-1} = \frac{x}{x^2-4}.$$

A rational expression is *simplified,* or in *simplest form,* when it is expressed as a quotient of polynomials whose greatest common factor is 1. (Recall Lesson 4-6.) To simplify a rational expression you factor the numerator and denominator and then look for common factors.

Example 1 Simplify $\frac{x^2-2x}{x^2-4}$.

Solution $\frac{x^2-2x}{x^2-4} = \frac{x(x-2)}{(x+2)(x-2)}$ Factor the numerator and the denominator.

$= \frac{x}{x+2}$ Simplify.

Answer

Factors of the numerator and denominator may be opposites of each other, as in Example 2.

Example 2 Simplify $(3x-5x^2-2x^3)(6x^2-5x+1)^{-1}$.

Solution

$(3x-5x^2-2x^3)(6x^2-5x+1)^{-1} = \frac{3x-5x^2-2x^3}{6x^2-5x+1}$ Write as a fraction.

$= \frac{x(3+x)(1-2x)}{(2x-1)(3x-1)}$ Factor.

$= \frac{x(3+x)(-1)(2x-1)}{(2x-1)(3x-1)}$ Replace $1-2x$ by $(-1)(2x-1)$.

$= \frac{-x(3+x)}{3x-1}$, or $-\frac{x(x+3)}{3x-1}$ ***Answer***

Teaching Suggestions p. T99

Suggested Extensions p. T99

Warm-Up Exercises

Simplify.

1. $\frac{2\cdot3\cdot8}{6\cdot5\cdot4}$ $\frac{2}{5}$
2. $\frac{3x^2yz^4}{7xy^5z}$ $\frac{3xz^3}{7y^4}$
3. $\frac{8x^{-1}yz}{4xy^{-2}z}$ $\frac{2y^3}{x^2}$

Motivating the Lesson

Explain to students that while a rational algebraic expression may look very complicated, the procedure for simplifying it is very simple. Lesson 5-4 explains this procedure.

Chalkboard Examples

Simplify.

1. $\frac{5a^2+4a-1}{5a^2-10a-15}$

   $\frac{(5a-1)(a+1)}{5(a-3)(a+1)} = \frac{5a-1}{5(a-3)}$

2. $(z^4-5z^3+6z^2)\times(9z-z^3)^{-1}$

   $\frac{z^4-5z^3+6z^2}{9z-z^3} =$

   $\frac{z^2(z-2)(z-3)}{(-1)z(z-3)(3+z)} = -\frac{z(z-2)}{z+3}$

3. A rational function g is defined by:

   $g(x) = \frac{(x+4)(x-1)}{2x(x+3)(x-3)}$

   a. Find the domain of g.
   b. Find the zeros of g.

   a. $2x(x+3)(x-3) \neq 0$
   $x \neq 0, 3, \text{ or } -3$
   The domain of g consists of all real numbers except 0, 3, and −3.
   b. $g(x) = 0$ if and only if $(x+4)(x-1) = 0$.
   The zeros of g are −4 and 1.

Check for Understanding

Here is a suggested use of the Oral Exercises to check students' understanding as you teach the lesson.
Oral Exs. 1–6: use after Example 2.
Oral Exs. 7–15: use after Example 3.

Guided Practice

Simplify.

1. $\frac{8x^2 - 6x}{4x^2}$ $\frac{4x-3}{2x}$
2. $\frac{x^3 + 6x^2 + 9x}{x^3 - 9x}$ $\frac{x+3}{x-3}$
3. $\frac{(y-2)^2}{y^2 - 4}$ $\frac{y-2}{y+2}$
4. $(x^2 + x - 6)(x + 3)^{-2}$ $\frac{x-2}{x+3}$
5. $\frac{x^2 - y^2}{x^2 + 3xy + 2y^2}$ $\frac{x-y}{x+2y}$
6. $\frac{(k+1)(k^2-1)}{(k-1)(k+1)^2}$ 1

Give (a) the domain and (b) the zeros of *f*.

7. $f(x) = \frac{x^3 + 27}{x^2 + 9}$
   a. all real numbers
   b. -3
8. $f(x) = \frac{3x^3 - x^2 - 2x}{x(x+1)^2}$
   a. $\{x: x \neq 0, x \neq -1\}$
   b. $1, -\frac{2}{3}$

Summarizing the Lesson

In this lesson students learned the procedure for simplifying rational algebraic expressions and how to find the domain and zeros of a rational function. Ask students to explain where the graph of a rational function is when it has no zero. As a hint, tell students to look at the examples in the Extra on pages 230 and 231.

A function that is defined by a simplified rational expression in one variable is called a **rational function.**

Example 3 Let $f(x) = \frac{2x^2 - 7x + 3}{x^3 + x^2 - 2x}$.

a. Find the domain of *f*.

b. Find the zeros of *f*, if any.

Solution

$$f(x) = \frac{2x^2 - 7x + 3}{x^3 + x^2 - 2x} = \frac{(2x-1)(x-3)}{x(x-1)(x+2)}$$

The zero-product property is used for both (a) and (b).

a. According to the agreement made on page 217, the domain of *f* consists of all real numbers except 0, 1, and -2.

b. $f(x) = 0$ if and only if $(2x - 1)(x - 3) = 0$. Therefore, the zeros of *f* are $\frac{1}{2}$ and 3.

Oral Exercises

Simplify.

1. $\frac{5x^2y^2}{xy^3}$ $\frac{5x}{y}$
2. $\frac{5x^2y^3}{3xy^2}$ $\frac{5xy}{3}$
3. $\frac{x(x-2)}{x(x+2)}$ $\frac{x-2}{x+2}$
4. $\frac{(x-1)(x+2)}{2+x}$ $x-1$
5. $\frac{x^2 - x}{x - 1}$ x
6. $\frac{2-x}{x-2}$ -1

Find (a) the domain of each rational function *f* and (b) the zeros of *f*, if any.

7. $f(x) = \frac{x-2}{x}$
8. $f(x) = \frac{x}{x-2}$
9. $f(x) = \frac{x+1}{x-1}$
10. $f(x) = \frac{x^2+1}{x+1}$
11. $f(x) = \frac{x+3}{x^2+9}$
12. $f(x) = \frac{(x-1)(x-3)}{x-2}$
13. $f(x) = \frac{x^2+1}{x^2-1}$
14. $f(x) = \frac{x^2-4}{x^2+4}$
15. $f(x) = \frac{x-4}{x^2-4}$

Written Exercises

Simplify.

A

1. $\frac{5x^2 - 15x}{10x^2}$ $\frac{x-3}{2x}$
2. $\frac{3t^4 - 9t^3}{6t^2}$ $\frac{t(t-3)}{2}$
3. $\frac{u^2 - u - 2}{u^2 + u}$ $\frac{u-2}{u}$
4. $\frac{z^3 - 4z}{z^2 - 4z + 4}$ $\frac{z(z+2)}{z-2}$

5. $(p - q)(q - p)^{-1}$ -1

6. $(r^2 - rs)(r^2 - s^2)^{-1}$ $\frac{r}{r+s}$

7. $\frac{s^2 - t^2}{(t - s)^2}$ $\frac{s+t}{s-t}$

8. $\frac{(a - x)^2}{x^2 - a^2}$ $\frac{x-a}{x+a}$

9. $\frac{x^2 - 5x + 6}{x^2 - 7x + 12}$ $\frac{x-2}{x-4}$

10. $\frac{2t^2 + 5t - 3}{2t^2 + 7t + 3}$ $\frac{2t-1}{2t+1}$

11. $\frac{6y^2 - 5y + 1}{1 - y - 6y^2}$ $\frac{1-2y}{1+2y}$

12. $\frac{9 - 4z^2}{6z^2 - 5z - 6}$ $-\frac{3+2z}{3z+2}$

13. $(r^2 - 5r + 4)(r - 4)^{-2}$ $\frac{r-1}{r-4}$

14. $(p^2 + 4p - 5)(p - 1)^{-2}$ $\frac{p+5}{p-1}$

15. $\frac{x^2 + 2x - 8}{(2 - x)(4 + x)}$ -1

16. $\frac{(t + 1)(t^2 - 1)}{(t - 1)(t + 1)^2}$ 1

17. $(z^4 - 1)(z^4 - z^2)^{-1}$ $\frac{z^2+1}{z^2}$

18. $(1 - r^3)(1 - r)^{-3}$ $\frac{1+r+r^2}{(1-r)^2}$

19. $\frac{(x^2 - a^2)^2}{(x - a)^2}$ $(x+a)^2$

20. $\frac{t^4 - c^4}{(t + c)^2(t^2 + c^2)}$ $\frac{t-c}{t+c}$

Find (a) the domain of each function and (b) its zeros, if any.

21. $f(t) = \frac{t^2 - 9}{t^2 - 9t}$

22. $g(x) = \frac{x^3 + 2x}{x^2 - 4}$

23. $F(x) = (x^4 - 16)(x^3 - 1)^{-1}$

24. $h(y) = (y^3 - 8)(y + 2)^{-3}$

25. $g(t) = \frac{2t^2 + 3t - 9}{t^3 - 4t}$

26. $G(s) = \frac{4s^2 + 15s - 4}{(2s - 1)^2}$

B 27. $f(x) = \frac{x^3 - 2x^2 + x - 2}{x^4 + x^2 - 2}$

28. $h(t) = \frac{t^3 + 4t^2 - t - 4}{t^3 - t^2 + t - 1}$

Simplify.

29. $\frac{x^3 + x^2 - x - 1}{x^3 - x^2 - x + 1}$ $\frac{x+1}{x-1}$

30. $\frac{t^4 - 1}{t^3 + t^2 + t + 1}$ $t - 1$

31. $\frac{x^3 - x^2y + xy^2 - y^3}{x^4 - y^4}$ $\frac{1}{x+y}$

32. $\frac{x^4 - 2x^2y^2 + y^4}{x^4 - x^3y - xy^3 + y^4}$ $\frac{(x+y)^2}{x^2+xy+y^2}$

33. $\frac{s^4 - t^4}{s^4 - 2s^2t^2 + t^4}$ $\frac{s^2+t^2}{(s+t)(s-t)}$

34. $\frac{u^4 - v^4}{u^4 + 2u^2v^2 + v^4}$ $\frac{(u+v)(u-v)}{u^2+v^2}$

35. $\frac{x^4 + x^3y - xy^3 - y^4}{x^4 - y^4}$ $\frac{x^2+xy+y^2}{x^2+y^2}$

36. $\frac{x^2 - y^2 - 4x + 4y}{x^2 - y^2 + 4x - 4y}$ $\frac{x+y-4}{x+y+4}$

37. $\frac{x^2 - y^2 - 4y - 4}{x^2 - y^2 - 4x + 4}$ $\frac{x+y+2}{x+y-2}$

38. $\frac{ax + by - bx - ay}{ax - by + bx - ay}$ $\frac{a-b}{a+b}$

C 39. $\frac{x^2 - y^2 - z^2 - 2yz}{x^2 - y^2 + z^2 - 2xz}$ $\frac{x+y+z}{x+y-z}$

40. $\frac{x^2 + y^2 - z^2 - 2xy}{x^2 - y^2 + z^2 - 2xz}$ $\frac{x-y+z}{x+y-z}$

41. $\frac{x^4 + x^2y^2 + y^4}{x^3 + y^3}$ (*Hint:* See Exercise 57, p. 186.) $\frac{x^2+xy+y^2}{x+y}$

42. $\frac{x^{2n} + 2x^ny^n - 3y^{2n}}{x^{2n} + 5x^ny^n + 6y^{2n}}$ $\frac{x^n-y^n}{x^n+2y^n}$

43. $\frac{x^{2n} - 2x^ny^n + y^{2n}}{x^{2n} + 3x^ny^n - 4y^{2n}}$ $\frac{x^n-y^n}{x^n+4y^n}$

Suggested Assignments

Average Algebra
228/1–37 odd
R 226/Self-Test 1

Average Alg. and Trig.
228/1–37 odd
R 226/Self-Test 1

Extended Algebra
228/3–42 mult. of 3
230/1–11 odd
R 226/Self-Test 1

Extended Alg. and Trig.
228/3–42 mult. of 3
230/1–11 odd
R 226/Self-Test 1

Supplementary Materials

Study Guide	pp. 77–78
Practice Master	29
Test Master	18

Additional Answers
Oral Exercises

7. a. reals except 0 b. 2
8. a. reals except 2 b. 0
9. a. reals except 1 b. -1
10. a. reals except -1 b. no zeros
11. a. all reals b. -3
12. a. reals except 2 b. 1 and 3
13. a. reals except 1 and -1 b. no zeros
14. a. all reals b. 2 and -2
15. a. reals except 2 and -2 b. 4

Additional Answers
Written Exercises

21. a. reals except 0 and 9 b. 3 and -3
22. a. reals except 2 and -2 b. 0
23. a. reals except 1 b. 2 and -2

(continued on p. 256)

Additional Answers
Exercises

1.

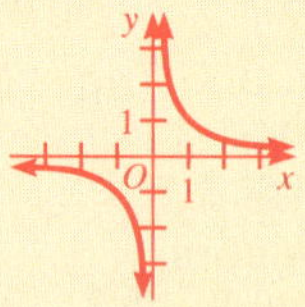

2.

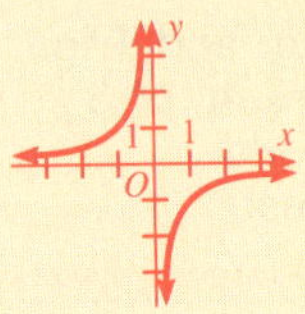

3.

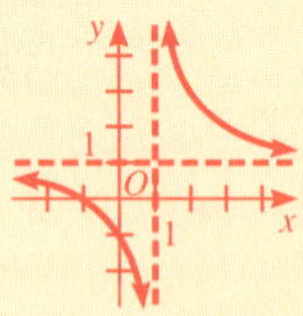

4.

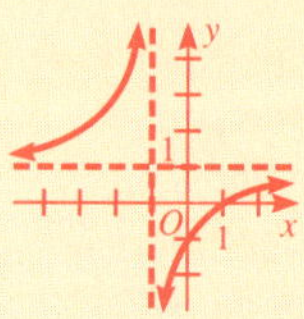

5.

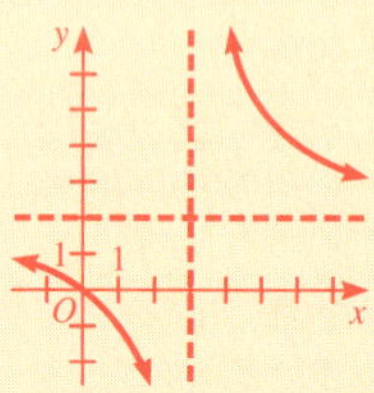

6.

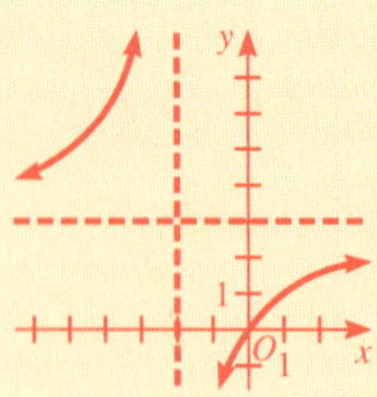

Graphing Rational Functions

To graph a rational function, $f(x) = \frac{x}{x-2}$ for example, you need to plot points (x, y) for which $y = \frac{x}{x-2}$. As you will see, the graph of a rational function can approach, without intersecting, certain lines called *asymptotes*.

Example 1 Graph $f(x) = \frac{x}{x-2}$.

Solution The domain of f consists of all real numbers except 2. Using a calculator, prepare two tables of x-y values: one for $x < 2$ and one for $x > 2$.

$x < 2$

x	$y = \frac{x}{x-2}$
-2	$\frac{-2}{-2-2} = 0.5$
-1	$\frac{-1}{-1-2} \approx 0.33$
0	$\frac{0}{0-2} = 0$
1	$\frac{1}{1-2} = -1$
1.5	$\frac{1.5}{1.5-2} = -3$
1.9	$\frac{1.9}{1.9-2} = -19$
1.99	$\frac{1.99}{1.99-2} = -199$

As x approaches 2 from the left, y becomes increasingly small.

$x > 2$

x	$y = \frac{x}{x-2}$
6	$\frac{6}{6-2} = 1.5$
5	$\frac{5}{5-2} \approx 1.67$
4	$\frac{4}{4-2} = 2$
3	$\frac{3}{3-2} = 3$
2.5	$\frac{2.5}{2.5-2} = 5$
2.1	$\frac{2.1}{2.1-2} = 21$
2.01	$\frac{2.01}{2.01-2} = 201$

As x approaches 2 from the right, y becomes increasingly large.

Some of the points from the two tables are plotted on the grid at the right. Notice that for values of x near 2, the points on the graph of f "follow" the vertical line $x = 2$. The line $x = 2$ is an asymptote of the graph of f.

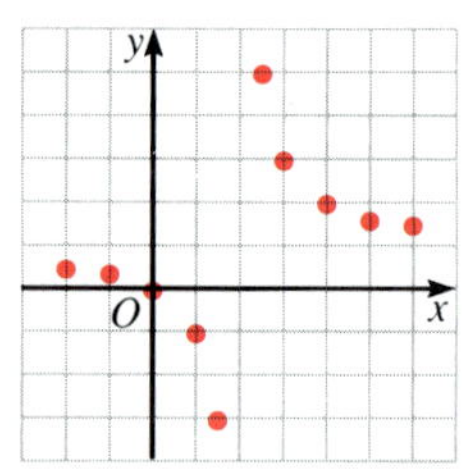

The graph of f has another asymptote. As $|x|$ gets increasingly large, $f(x)$ approaches 1:

$$f(12) = 1.2, \quad f(102) = 1.02, \quad f(1002) = 1.002;$$
$$f(-8) = 0.8, \quad f(-98) = 0.98, \quad f(-998) = 0.998.$$

Therefore the horizontal line $y = 1$ is also an asymptote of the graph of f.

To obtain the complete graph of f, draw the asymptotes $x = 2$ and $y = 1$ as dashed lines and connect the points already plotted with a smooth curve, as shown at the right.

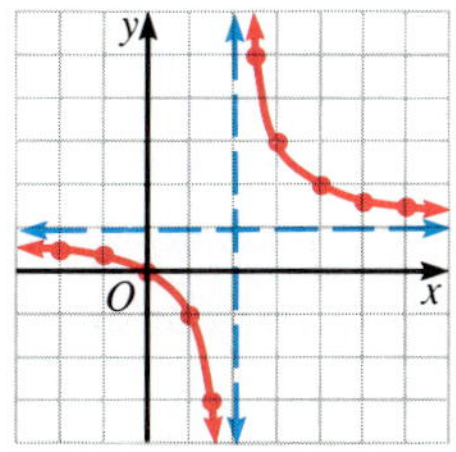

Example 2 Graph $f(x) = \dfrac{1}{(x + 2)^2}$.

Solution The domain of f consists of all real numbers except -2. Because $f(x)$ gets increasingly large as x approaches -2, the line $x = -2$ is a vertical asymptote of the graph of f.

As $|x|$ gets increasingly large, $f(x)$ approaches 0. Therefore the x-axis is a horizontal asymptote of the graph of f.

A table of x-y values and the graph of f are shown below.

x	$y = \dfrac{1}{(x + 2)^2}$
-4	0.25
-3	1
-2.5	4
-1.5	4
-1	1
0	0.25

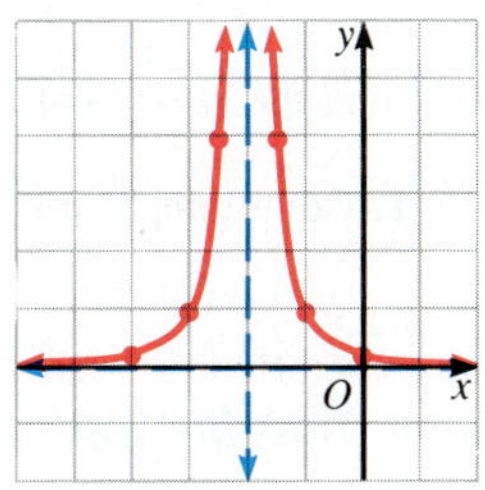

Exercises

Graph each function. Show any asymptotes as dashed lines. You may wish to check your graphs on a computer or a graphing calculator.

A

1. $f(x) = \dfrac{1}{x}$ **2.** $g(x) = -\dfrac{1}{x}$ **3.** $h(x) = \dfrac{x + 1}{x - 1}$

4. $f(x) = \dfrac{x - 1}{x + 1}$ **5.** $g(x) = \dfrac{2x}{x - 3}$ **6.** $h(x) = \dfrac{3x}{x + 2}$

B

7. $f(x) = \dfrac{1}{x^2}$ **8.** $g(x) = \dfrac{1}{(x - 1)^2}$ **9.** $h(x) = \dfrac{x^2 + 1}{x^2}$

10. $f(x) = \dfrac{1 - x^2}{x^2}$ **11.** $g(x) = \dfrac{10}{x^2 + 1}$ **12.** $h(x) = \dfrac{10x^2}{x^2 + 1}$

7.

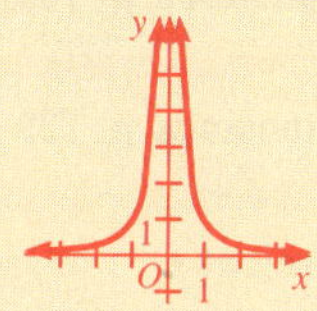

8.

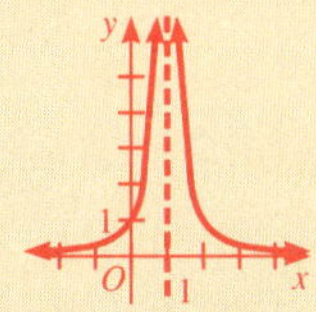

9.

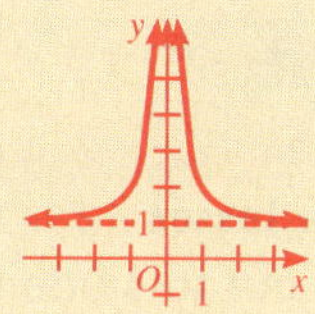

10.

(continued on p. 256)

Using a Computer or a Graphing Calculator

Once students have drawn their own graphs for Exercises 1–12, you may want to have them check their work using a computer or graphing calculator. Although asymptotes will not be shown, students should be able to infer where they occur.

Support Materials
Disk for Algebra
Menu Item: Function Plotter

Teaching Suggestions p. T99

Suggested Extensions p. T99

Warm-Up Exercises

Perform the indicated operations.

1. $\frac{2x}{3z} \cdot \frac{4x}{5y}$ $\frac{8x^2}{15yz}$
2. $\frac{2a^2}{9b} \cdot \frac{1}{3a}$ $\frac{2a}{27b}$
3. $\frac{4z^3}{7y^2} \cdot \frac{9y^2}{10z^2}$ $\frac{18z}{35}$
4. $\frac{5c^4}{9a^2} \div \frac{2a}{3c^7}$ $\frac{5c^{11}}{6a^3}$
5. $\frac{3x^3}{4b} \div \frac{x^3}{4b}$ 3

Motivating the Lesson

Again point out to students that while multiplying or dividing rational expressions may look complicated the procedures used are the same as for simple fractions.

Chalkboard Examples

Simplify.

1. $\frac{b^3}{xy} \cdot \frac{2x^2y}{4ab}$ $\frac{b^2x}{2a}$
2. $\frac{a^3 - 6a^2}{a^2 + 2a - 3} \cdot \frac{a^2 - 2a + 1}{a^3}$

$\frac{a^2(a-6)}{(a+3)(a-1)} \cdot \frac{(a-1)(a-1)}{a^3} = \frac{(a-6)(a-1)}{a(a+3)}$

3. $\frac{1 - x^2}{3x + 4} \div \frac{x^2 + x - 2}{9x^2 + 24x + 16}$

$\frac{(1+x)(1-x)}{3x+4} \cdot \frac{(3x+4)(3x+4)}{(x+2)(x-1)}$

$= \frac{(-1)(x-1)(x+1)}{3x+4} \cdot \frac{(3x+4)(3x+4)}{(x+2)(x-1)}$

$= -\frac{(x+1)(3x+4)}{(x+2)}$

5-5 Products and Quotients of Rational Expressions

Objective To multiply and divide rational expressions.

To find the product of two or more rational expressions, you use the multiplication rule for fractions: $\frac{a}{b} \cdot \frac{c}{d} = \frac{ac}{bd}$. Products of rational expressions should always be expressed in simplest form.

Example 1 Simplify $\frac{3x^2 - 6x}{x^2 - 6x + 9} \cdot \frac{x^2 - x - 6}{x^2 - 4}$.

Solution

$$\frac{3x^2 - 6x}{x^2 - 6x + 9} \cdot \frac{x^2 - x - 6}{x^2 - 4} = \frac{3x(x-2)}{(x-3)(x-3)} \cdot \frac{(x+2)(x-3)}{(x+2)(x-2)}$$

$$= \frac{3x(x-2)(x-3)}{(x-3)(x-3)(x-2)}$$

$$= \frac{3x}{x-3} \quad \textit{Answer}$$

By the definition of division (page 33), a quotient is the product of the dividend and the reciprocal of the divisor. The reciprocal of $\frac{r}{s}$ is $\frac{s}{r}$ because $\frac{r}{s} \cdot \frac{s}{r} = 1$. Combining these facts gives the following rule.

Division Rule for Fractions

Let p, q, r, and s be real numbers with $q \neq 0$, $r \neq 0$, and $s \neq 0$. Then

$$\frac{p}{q} \div \frac{r}{s} = \frac{p}{q} \cdot \frac{s}{r}.$$

Example 2 Simplify.

a. $\frac{14}{15} \div \frac{7}{5}$ b. $\frac{6xy}{a^2} \div \frac{3y}{a^3x}$

Solution

a. $\frac{14}{15} \div \frac{7}{5} = \frac{14}{15} \cdot \frac{5}{7} = \frac{14 \cdot 5}{15 \cdot 7} = \frac{2}{3}$ ***Answer***

b. $\frac{6xy}{a^2} \div \frac{3y}{a^3x} = \frac{6xy}{a^2} \cdot \frac{a^3x}{3y} = \frac{6a^3x^2y}{3a^2y} = 2ax^2$ ***Answer***

When multiplying rational expressions, you can divide out factors common to a numerator and a denominator *before* you write the product as a single fraction. For instance, for the product in Example 2(b),

$$\frac{\overset{2}{\cancel{6}}x\cancel{y}}{\underset{1}{\cancel{a^2}}} \cdot \frac{\overset{a}{a^{\cancel{3}}}x}{\underset{1}{\cancel{3}}\underset{1}{\cancel{y}}} = 2ax^2.$$

Sometimes you may find it helpful to rewrite a quotient using the ÷ sign.

Example 3 Simplify $\dfrac{\dfrac{a^2 - 4ab + 3b^2}{a + 2b}}{a^2 - ab - 6b^2}$.

Solution

$$\frac{\dfrac{a^2 - 4ab + 3b^2}{a + 2b}}{a^2 - ab - 6b^2} = \frac{a^2 - 4ab + 3b^2}{a + 2b} \div \frac{a^2 - ab - 6b^2}{1}$$

$$= \frac{a^2 - 4ab + 3b^2}{a + 2b} \cdot \frac{1}{a^2 - ab - 6b^2}$$

$$= \frac{(a - b)\cancel{(a - 3b)}}{a + 2b} \cdot \frac{1}{(a + 2b)\cancel{(a - 3b)}}$$

$$= \frac{a - b}{(a + 2b)^2} \quad \textbf{Answer}$$

When you simplify expressions, perform multiplication and division in order from left to right, as in Example 4.

Example 4 Simplify $\dfrac{6p^2q}{r} \div \dfrac{3pq^2}{r} \cdot \dfrac{2q^2}{pr}$.

Solution

$$\frac{6p^2q}{r} \div \frac{3pq^2}{r} \cdot \frac{2q^2}{pr} = \left(\frac{6p^2q}{r} \div \frac{3pq^2}{r}\right) \cdot \frac{2q^2}{pr}$$

$$= \frac{6p^2q}{r} \cdot \frac{r}{3pq^2} \cdot \frac{2q^2}{pr}$$

$$= \frac{4q}{r} \quad \textbf{Answer}$$

Oral Exercises

Simplify.

1. $\frac{4}{5} \cdot 2$ $\frac{8}{5}$
2. $\frac{4}{5} \div 2$ $\frac{2}{5}$
3. $\frac{4}{5} \div \frac{1}{2}$ $\frac{8}{5}$
4. $\frac{4}{5} \cdot \frac{1}{2}$ $\frac{2}{5}$
5. $\frac{a^2}{b} \div \frac{a}{b^2}$ ab
6. $\frac{a^2}{b} \cdot \frac{a}{b^2}$ $\frac{a^3}{b^3}$
7. $\frac{3s}{4t} \cdot \frac{2t}{s^2}$ $\frac{3}{2s}$
8. $\frac{3s}{4t} \div \frac{2t}{s^2}$ $\frac{3s^3}{8t^2}$

4. $\dfrac{\dfrac{y^2 - 5py + 6p^2}{y + 2p}}{y^2 - 4p^2}$

$\dfrac{y^2 - 5py + 6p^2}{y + 2p} \div (y^2 - 4p^2)$
$= \dfrac{(y - 2p)(y - 3p)}{y + 2p} \cdot \dfrac{1}{(y + 2p)(y - 2p)} = \dfrac{y - 3p}{(y + 2p)^2}$

5. $\dfrac{r^4s}{t} \cdot \dfrac{t^3}{r^2s} \div \dfrac{rs}{t}$

$\left(\dfrac{r^4s}{t} \cdot \dfrac{t^3}{r^2s}\right) \div \dfrac{rs}{t} =$
$r^2t^2 \cdot \dfrac{t}{rs} = \dfrac{rt^3}{s}$

Check for Understanding

Here is a suggested use of the Oral Exercises to check students' understanding as you teach the lesson.
Oral Exs. 1–8: use after Example 4.

Guided Practice

Simplify.

1. $\frac{21}{10} \div \frac{9}{10} \cdot \left(-\frac{4}{7}\right)$ $-\frac{4}{3}$
2. $\frac{5y^3}{-7} \cdot \frac{-14}{25y} \div \left(\frac{2}{3y}\right)^0$ $\frac{2y^2}{5}$
3. $\frac{bq^2}{c} \div \frac{q}{c^2} \div \frac{q}{bc}$ b^2c^2
4. $\frac{x^2 - 1}{x + 2} \div \frac{x + 1}{x - 2} \cdot \frac{x^2 - 4}{x - 1}$ $(x - 2)^2$
5. $\left(\frac{2x^2 - x - 1}{2x - 1}\right)^{-1}\left(\frac{4x^2 - 1}{x^2 - x}\right)^{-1}$ $\frac{x}{(2x + 1)^2}$

Summarizing the Lesson

In this lesson students learned to multiply and divide rational expressions using the multiplication and division rules for fractions. Ask students to explain, in words, the procedures for multiplying and dividing rational expressions.

Common Errors

Some students may need to be reminded that dividing by a nonzero number is equivalent to multiplying by the reciprocal of the number. Students may mistakenly think that $\frac{x}{y^2} \div \frac{y}{x}$ equals $\frac{1}{y}$, rather than $\frac{x^2}{y^3}$.

Suggested Assignments

Average Algebra
234/2–22 even
S 234/Mixed Review
230/1–7 odd

Average Alg. and Trig.
234/2–22 even
S 234/Mixed Review

Extended Algebra
234/2–24 even
S 234/Mixed Review

Extended Alg. and Trig.
234/2–24 even
S 234/Mixed Review

Supplementary Materials

Study Guide pp. 79–80

Written Exercises

11. $\frac{x}{(x-1)(x-2)}$ 12. $\frac{t-1}{t+1}$

Simplify. Write answers without negative or zero exponents.

A

1. $\frac{-10}{21} \div \frac{15}{28}$ $-\frac{8}{9}$
2. $-\frac{26}{25} \div \frac{39}{20}$ $-\frac{8}{15}$
3. $\frac{5x^3}{-3} \cdot \frac{-6}{10x^2}$ x
4. $\frac{22z^2}{15} \cdot \frac{-5}{11z^3}$ $-\frac{2}{3z}$
5. $\frac{8t^2}{3} \div \frac{2t}{9}$ $12t$
6. $\frac{28x}{25} \div \frac{21x^3}{15}$ $\frac{4}{5x^2}$
7. $\frac{x^2}{4} \cdot \left(\frac{xy}{6}\right)^{-1} \cdot \frac{2y^2}{x}$ $3y$
8. $2uv \div \frac{2u^2}{v} \div \frac{2v^2}{u}$ $\frac{1}{2}$
9. $\frac{4rs^2}{45} \div \frac{8s}{27r} \div \frac{9rs}{10}$ $\frac{r}{3}$
10. $\frac{7x^2}{9y} \div \frac{4x}{15y^2} \cdot \left(\frac{35}{6xy}\right)^{-1}$ $\frac{x^2y^2}{2}$
11. $\frac{x(x-1)}{(x-2)^2} \div \frac{(x-1)^2}{x-2}$
12. $\frac{t-2}{t+3} \cdot \frac{t^2+2t-3}{t^2-t-2}$
13. $\frac{4u^2-1}{u^2-4} \cdot \frac{u-2}{2u-1}$ $\frac{2u+1}{u+2}$
14. $\frac{x^2}{x-1} \cdot \frac{x+1}{x+2} \div \frac{x}{(x-1)(x+2)}$ $x(x+1)$
15. $\frac{x^2-4}{2x^2-5x+2} \div \frac{2x^2-3x-2}{4x^2-1}$ $\frac{x+2}{x-2}$
16. $\frac{3x^2-8x+4}{9x^2-4} \div \frac{3x^2-5x-2}{9x^2-3x-2}$ $\frac{3x-2}{3x+2}$

B

17. $\dfrac{\frac{p^4-q^4}{(p+q)^2}}{p^2+q^2}$ $\frac{p-q}{p+q}$
18. $\dfrac{\frac{x^2-y^2}{x+y}}{x^4-y^4}$ $\frac{1}{(x^2+y^2)(x+y)}$
19. $\frac{u^2v}{u+v} \div (u+v) \cdot \frac{u^2+2uv+v^2}{uv^2-u^2v}$ $\frac{u}{v-u}$
20. $\frac{x^2+3ax}{3a-x} \cdot \frac{x^2-4ax+3a^2}{a^2-x^2} \div \frac{x+3a}{x+a}$ x
21. $\frac{3x^2+xy-2y^2}{3x^2-xy-2y^2} \div \frac{3x^2+7xy-6y^2}{3x^2-2xy-y^2} \div \frac{3x+y}{3x+2y}$ $\frac{x+y}{x+3y}$
22. $\frac{r^2+4rs+3s^2}{r^2+5rs+6s^2} \cdot (r+2s)^{-1} \div \frac{r+s}{r^2+4rs+4s^2}$ 1
23. $(a^4+2a^2b^2+b^4) \div (a^4-b^4) \cdot (a-b)$ $\frac{a^2+b^2}{a+b}$

C

24. $(u^4+u^2v^2+v^4) \div (u^6-v^6) \cdot (u^2-v^2)$ 1

Mixed Review Exercises

Express in scientific notation.

1. 0.00054 5.4×10^{-4}
2. 63,400,000 6.34×10^7
3. 0.1 1×10^{-1}
4. 3281 3.281×10^3

Simplify.

5. $\frac{x^2+x-6}{x^2-x-2}$ $\frac{x+3}{x+1}$
6. $\frac{15y^2}{15y^3-10y^2}$ $\frac{3}{3y-2}$
7. $\frac{16c^3d^2}{24cd^5}$ $\frac{2c^2}{3d^3}$
8. $\frac{a^2-1}{a^2+2a+1}$ $\frac{a-1}{a+1}$
9. $\frac{3m+12}{2m+8}$ $\frac{3}{2}$
10. $\frac{(6z^2)^2}{(4z^3)^3}$ $\frac{9}{16z^5}$
11. $\frac{4t^2+4}{8t^2-8}$ $\frac{t^2+1}{2(t+1)(t-1)}$
12. $(u^3+1)(u+1)^{-1}$ u^2-u+1

5-6 Sums and Differences of Rational Expressions

Objective To add and subtract rational expressions.

The definition of division and the distributive property can be used to prove the following rules for adding and subtracting fractions with *equal* denominators.

$$\frac{a}{c} + \frac{b}{c} = \frac{a+b}{c} \qquad \frac{a}{c} - \frac{b}{c} = \frac{a-b}{c}$$

These rules can be extended to more than two terms.

Example 1 Simplify: **a.** $\frac{4}{15} + \frac{13}{15} - \frac{7}{15}$ **b.** $\frac{3x^2-8}{x+2} - \frac{x^2-9}{x+2}$

Solution **a.** $\frac{4}{15} + \frac{13}{15} - \frac{7}{15} = \frac{4+13-7}{15} = \frac{10}{15} = \frac{2}{3}$ ***Answer***

b. $\frac{3x^2-8}{x+2} - \frac{x^2-9}{x+2} = \frac{3x^2-8-x^2+9}{x+2}$

$= \frac{2x^2+1}{x+2}$ ***Answer***

To add or subtract fractions with *different* denominators, use this rule.

Rule for Adding and Subtracting Fractions

1. Find the **least common denominator (LCD)** of the fractions (that is, the least common multiple (LCM) of their denominators).
2. Express each fraction as an equivalent fraction with the LCD as denominator.
3. Add or subtract and then simplify the result.

Example 2 Simplify $\frac{25}{42} + \frac{11}{18} - 2$.

Solution The LCM of $42 = 2 \cdot 3 \cdot 7$ and $18 = 2 \cdot 3^2$ is $2 \cdot 3^2 \cdot 7 = 126$. So the LCD is 126.

$$\frac{25}{42} + \frac{11}{18} - 2 = \frac{25 \cdot 3}{42 \cdot 3} + \frac{11 \cdot 7}{18 \cdot 7} - \frac{2 \cdot 126}{1 \cdot 126}$$

$$= \frac{75}{126} + \frac{77}{126} - \frac{252}{126}$$

$$= -\frac{100}{126} = -\frac{50}{63}$$ ***Answer***

Teaching Suggestions p. T99

Suggested Extensions p. T100

Warm-Up Exercises

Find the least common multiple.

1. 21, 9 63
2. $3a^2$, $4b^2$ $12a^2b^2$
3. $8x^2$, $2xy$, $3y^2$ $24x^2y^2$
4. $x + 2$, $x - 3$ $(x + 2)(x - 3)$

Motivating the Lesson

Once again, point out to students that the procedures involved in this lesson are ones that they are already familiar with from their work with simple fractions.

Chalkboard Examples

Simplify.

1. $\frac{1}{6a^2} + \frac{1}{3b^2} - \frac{2}{ab}$

$\text{LCM} = 6a^2b^2$

$\frac{1}{6a^2} = \frac{b^2}{6a^2b^2}$;

$\frac{1}{3b^2} = \frac{2a^2}{6a^2b^2}$;

$\frac{2}{ab} = \frac{12ab}{6a^2b^2}$;

$\frac{b^2}{6a^2b^2} + \frac{2a^2}{6a^2b^2} - \frac{12ab}{6a^2b^2} = \frac{b^2+2a^2-12ab}{6a^2b^2}$

2. $\frac{2}{y^2+1} + \frac{5}{y}$

$\text{LCM} = y(y^2+1)$

$\frac{2}{y^2+1} = \frac{2y}{y(y^2+1)}$;

$\frac{5}{y} = \frac{5(y^2+1)}{y(y^2+1)}$;

$\frac{2y}{y(y^2+1)} + \frac{5(y^2+1)}{y(y^2+1)} = \frac{2y+5y^2+5}{y(y^2+1)}$

Check for Understanding

Here is a suggested use of the Oral Exercises to check students' understanding as you teach the lesson.
Oral Exs. 1–8: use after Example 3.

Guided Practice

Simplify.

1. $\frac{1}{2} + \frac{1}{6} - \frac{1}{3}$ $\frac{1}{3}$
2. $2 - \frac{3}{10} + \frac{5}{14}$ $\frac{72}{35}$
3. $\frac{5}{6k} - \frac{4}{3k}$ $-\frac{1}{2k}$
4. $\frac{a-1}{a^2} + \frac{a+1}{2a}$ $\frac{a^2+3a-2}{2a^2}$
5. $\frac{1}{a^2b^3} - \frac{1}{2a^3b}$ $\frac{2a-b^2}{2a^3b^3}$
6. $2c^{-1}d + 2cd^{-1}$ $\frac{2(d^2+c^2)}{cd}$
7. $\frac{2}{3xy} - \frac{1}{3y} - \frac{1}{3x}$ $\frac{2-x-y}{3xy}$
8. $\frac{a}{b-a} + \frac{b}{a-b}$ -1
9. $\frac{5}{s^2-s} + \frac{1}{s^2-1}$ $\frac{6s+5}{s(s-1)(s+1)}$

Summarizing the Lesson

In this lesson students used the rules for adding and subtracting fractions to add and subtract rational expressions. Ask the students to show that $\frac{a+b}{b} \neq \frac{a+\not b}{\not b} = a$

Supplementary Materials

Study Guide	pp. 81–82
Practice Master	30
Test Master	19

Recall (page 190) that the least common multiple (LCM) of two or more polynomials is the common multiple having the least degree and least positive common factor.

Example 3 Simplify $\frac{1}{6a^2} - \frac{1}{2ab} + \frac{3}{8b^2}$.

Solution The LCM of $6a^2$, $2ab$, and $8b^2$ is $24a^2b^2$. So the LCD is $24a^2b^2$.

$$\begin{aligned}\frac{1}{6a^2} - \frac{1}{2ab} + \frac{3}{8b^2} &= \frac{1 \cdot 4b^2}{6a^2 \cdot 4b^2} - \frac{1 \cdot 12ab}{2ab \cdot 12ab} + \frac{3 \cdot 3a^2}{8b^2 \cdot 3a^2} \\ &= \frac{4b^2}{24a^2b^2} - \frac{12ab}{24a^2b^2} + \frac{9a^2}{24a^2b^2} \\ &= \frac{4b^2 - 12ab + 9a^2}{24a^2b^2} \\ &= \frac{(2b-3a)^2}{24a^2b^2} \quad \textbf{Answer}\end{aligned}$$

When you add fractions in which the denominators are trinomials, factoring these trinomials will often shorten the work.

Example 4 Simplify $\frac{3}{x^2+x-6} - \frac{2}{x^2-3x+2}$.

Solution $x^2 + x - 6 = (x-2)(x+3)$; $x^2 - 3x + 2 = (x-1)(x-2)$
So the LCD is $(x-1)(x-2)(x+3)$.

$$\begin{aligned}\frac{3}{x^2+x-6} - \frac{2}{x^2-3x+2} &= \frac{3}{(x-2)(x+3)} - \frac{2}{(x-1)(x-2)} \\ &= \frac{3(x-1)}{(x-1)(x-2)(x+3)} - \frac{2(x+3)}{(x-1)(x-2)(x+3)} \\ &= \frac{3x-3-2x-6}{(x-1)(x-2)(x+3)} \\ &= \frac{x-9}{(x-1)(x-2)(x+3)} \quad \textbf{Answer}\end{aligned}$$

Oral Exercises

Simplify.

1. $\frac{3}{5} + \frac{1}{5}$ $\frac{4}{5}$
2. $\frac{5}{7} - \frac{2}{7}$ $\frac{3}{7}$
3. $\frac{5}{6} - \frac{1}{6}$ $\frac{2}{3}$
4. $\frac{3}{8} + \frac{1}{4}$ $\frac{5}{8}$
5. $\frac{1}{3} + \frac{1}{3a}$ $\frac{a+1}{3a}$
6. $\frac{2}{x} - \frac{1}{2x}$ $\frac{3}{2x}$
7. $\frac{1}{x} + \frac{1}{y}$ $\frac{y+x}{xy}$
8. $z + \frac{1}{z}$ $\frac{z^2+1}{z}$

Written Exercises

Simplify.

A

1. $\frac{5}{16} - \frac{12}{16} + \frac{3}{16}$ $-\frac{1}{4}$
2. $\frac{7}{10} + \frac{11}{20} - \frac{9}{20}$ $\frac{4}{5}$
3. $\frac{1}{2} - \frac{3}{7} + \frac{5}{14}$ $\frac{3}{7}$
4. $\frac{2}{3} - \frac{3}{5} + \frac{4}{15}$ $\frac{1}{3}$
5. $\frac{5}{12} + \frac{5}{18} - \frac{5}{36}$ $\frac{5}{9}$
6. $\frac{7}{12} - 1 + \frac{19}{20}$ $\frac{8}{15}$
7. $\frac{t+2}{3} + \frac{t-4}{6}$ $\frac{t}{2}$
8. $\frac{x+3}{5} - \frac{2x+1}{10}$ $\frac{1}{2}$
9. $\frac{z-1}{z} + \frac{z+1}{z^2}$ $\frac{z^2+1}{z^2}$
10. $\frac{x+2}{x^2} + \frac{x-2}{2x}$ $\frac{x^2+4}{2x^2}$
11. $\frac{t-4}{2t} - \frac{t-6}{3t}$ $\frac{1}{6}$
12. $\frac{2x+5}{4x^2} + \frac{2x-5}{10x}$ $\frac{4x^2+25}{20x^2}$
13. $\frac{1}{2pq^4} + \frac{2}{p^3q^2}$ $\frac{p^2+4q^2}{2p^3q^4}$
14. $\frac{1}{xy^3} - \frac{1}{x^3y}$ $\frac{(x+y)(x-y)}{x^3y^3}$
15. $ab^{-1} - a^{-1}b$ $\frac{(a+b)(a-b)}{ab}$
16. $4s^{-2} - (4t)^{-2}$ $\frac{(8t+s)(8t-s)}{16s^2t^2}$
17. $\frac{a}{bc} + \frac{b}{ac} + \frac{c}{ab}$ $\frac{a^2+b^2+c^2}{abc}$
18. $\frac{1}{yz} + \frac{1}{zx} + \frac{1}{xy}$ $\frac{x+y+z}{xyz}$
19. $\frac{y-z}{yz} - \frac{z-x}{zx} - \frac{x-y}{xy}$ $\frac{2(y-z)}{yz}$
20. $\frac{1}{x^2} + \frac{1}{xy} + \frac{1}{4y^2}$ $\frac{(x+2y)^2}{4x^2y^2}$
21. $\frac{1}{z-4} - \frac{1}{z+4}$ $\frac{8}{(z+4)(z-4)}$
22. $\frac{x}{x-1} - \frac{1}{x+1}$ $\frac{x^2+1}{(x+1)(x-1)}$
23. $\frac{1}{t^2+t} + \frac{1}{t^2-t}$ $\frac{2}{(t+1)(t-1)}$
24. $\frac{1}{u^2-2u} - \frac{1}{u^2-4}$ $\frac{2}{u(u+2)(u-2)}$
25. $\frac{1}{x^2-1} - \frac{1}{(x-1)^2}$ $-\frac{2}{(x+1)(x-1)^2}$
26. $\frac{1}{y^2-y-2} + \frac{1}{y^2+y}$ $\frac{2(y-1)}{y(y+1)(y-2)}$
27. $\frac{1}{s^2+2s+1} - \frac{1}{s^2-1}$ $-\frac{2}{(s+1)^2(s-1)}$
28. $\frac{1}{p^2-2p+1} - \frac{1}{p^2+p-2}$ $\frac{3}{(p+2)(p-1)^2}$

B

29. $\frac{a+b}{a-b} + \frac{a-b}{a+b} + \frac{b-a}{a-b} + \frac{b-a}{a+b}$ $\frac{2b}{a-b}$
30. $\frac{a+b}{a-b} + \frac{a-b}{a+b} - \frac{b-a}{a-b} + \frac{b-a}{a+b}$ $\frac{2a}{a-b}$
31. $(x-y)^{-1} - (x+y)^{-1}$ $\frac{2y}{(x+y)(x-y)}$
32. $(x-y)^{-2} - (x+y)^{-2}$ $\frac{4xy}{(x+y)^2(x-y)^2}$
33. $\frac{3}{x^2-5x+6} + \frac{2}{x^2-4}$
34. $\frac{1}{4t^2-4t+1} + \frac{1}{4t^2-1}$
35. $\frac{1}{2u^2-3uv+v^2} + \frac{1}{4u^2-v^2}$
36. $\frac{3}{4x^2-12xy+9y^2} + \frac{1}{2xy-3y^2}$
37. $\frac{x}{x-a} - \frac{x^2+a^2}{x^2-a^2} + \frac{a}{x+a}$
38. $\frac{3u}{2u-v} - \frac{2u}{2u+v} + \frac{2v^2}{4u^2-v^2}$
39. Prove: **a.** $\frac{a}{c} + \frac{b}{c} = \frac{a+b}{c}$ **b.** $\frac{a}{c} - \frac{b}{c} = \frac{a-b}{c}$

Find constants A and B that make the equation true.

C

40. $\frac{2x-9}{x^2-x-6} = \frac{A}{x-3} + \frac{B}{x+2}$ $A = -\frac{3}{5};\ B = \frac{13}{5}$
41. $\frac{x-7}{x^2+x-6} = \frac{A}{x+3} + \frac{B}{x-2}$ $A = 2;\ B = -1$

Suggested Assignments

Average Algebra
237/1–37 odd
S 229/28–38 even

Average Alg. and Trig.
237/1–37 odd
S 229/28–38 even

Extended Algebra
237/7–41 odd
S 234/17–23 odd

Extended Alg. and Trig.
237/7–41 odd
S 234/17–23 odd

Additional Answers Written Exercises

33. $\frac{5x}{(x+2)(x-2)(x-3)}$

34. $\frac{4t}{(2t+1)(2t-1)^2}$

35. $\frac{3u}{(2u+v)(2u-v)(u-v)}$

36. $\frac{2x}{y(2x-3y)^2}$

37. $\frac{2a}{x+a}$

38. $\frac{u+2v}{2u-v}$

39. a.
1. $\frac{a}{c} + \frac{b}{c} = a \cdot \frac{1}{c} + b \cdot \frac{1}{c}$ Def. of division
2. $= (a+b)\frac{1}{c}$ Dist. prop.
3. $= \frac{a+b}{c}$ Def. of division
4. $\therefore \frac{a}{c} + \frac{b}{c} = \frac{a+b}{c}$ Trans. prop.

b.
1. $\frac{a}{c} - \frac{b}{c} = a \cdot \frac{1}{c} - b \cdot \frac{1}{c}$ Def. of division
2. $= (a-b)\frac{1}{c}$ Dist. prop.
3. $= \frac{a-b}{c}$ Def. of division
4. $\therefore \frac{a}{c} - \frac{b}{c} = \frac{a-b}{c}$ Trans. prop.

Explorations p. 836

Teaching Suggestions p. T100

Suggested Extensions p. T100

Warm-Up Exercises

Simplify.

1. $\frac{5}{7} \cdot \frac{3}{4}$ $\frac{15}{28}$
2. $12\left(\frac{2}{3} + \frac{1}{4}\right)$ 11
3. $2\left(\frac{x}{2} + \frac{y}{2}\right)$ $x + y$
4. $\frac{x+1}{2} \div \frac{x+1}{3}$ $\frac{3}{2}$
5. $2\left(2 + \frac{y}{2}\right)$ $4 + y$

Motivating the Lesson

Tell students that some real-life activities, such as tuning up a car, are really a series of smaller, simpler activities. One method of simplifying complex fractions makes use of the idea of working on smaller, simpler parts.

Chalkboard Examples

Simplify.

1. $\dfrac{\frac{3}{4} - \frac{1}{8}}{2 + \frac{1}{12}}$

Method 1

$\left(\frac{3}{4} - \frac{1}{8}\right) \div \left(2 + \frac{1}{12}\right)$

$= \frac{5}{8} \div \frac{25}{12}$

$= \frac{5}{8} \cdot \frac{12}{25} = \frac{3}{10}$

Method 2

$\dfrac{\left(\frac{3}{4} - \frac{1}{8}\right) \cdot 24}{\left(2 + \frac{1}{12}\right) \cdot 24} =$

$\frac{18 - 3}{48 + 2} = \frac{15}{50} = \frac{3}{10}$

5-7 Complex Fractions

Objective To simplify complex fractions.

A fraction is a **complex fraction** if its numerator or denominator (or both) has one or more fractions or powers with negative exponents. For example,

$$\dfrac{\frac{7}{12} - \frac{1}{6}}{2 + \frac{4}{9}} \quad \text{and} \quad \dfrac{x^{-1} - y^{-1}}{x^{-2} - y^{-2}}$$

are complex fractions. Here are two methods for simplifying such fractions.

Method 1 Simplify the numerator and denominator separately; then divide.

Method 2 Multiply the numerator and denominator by the LCD of all the fractions appearing in the numerator and denominator.

Example 1 Simplify $\dfrac{\frac{7}{12} - \frac{1}{6}}{2 + \frac{4}{9}}$.

Solution

Method 1

$$\dfrac{\frac{7}{12} - \frac{1}{6}}{2 + \frac{4}{9}} = \dfrac{\frac{7-2}{12}}{\frac{18+4}{9}} = \dfrac{\frac{5}{12}}{\frac{22}{9}} = \frac{5}{12} \cdot \frac{9}{22} = \frac{15}{88}$$

Method 2

$$\dfrac{\frac{7}{12} - \frac{1}{6}}{2 + \frac{4}{9}} = \dfrac{\left(\frac{7}{12} - \frac{1}{6}\right) \cdot 36}{\left(2 + \frac{4}{9}\right) \cdot 36} = \frac{21 - 6}{72 + 16} = \frac{15}{88}$$

Example 2 Simplify $\dfrac{z - \frac{1}{z}}{1 - \frac{1}{z}}$.

Solution

Method 1

$$\dfrac{z - \frac{1}{z}}{1 - \frac{1}{z}} = \dfrac{\frac{z^2 - 1}{z}}{\frac{z - 1}{z}}$$

$$= \frac{z^2 - 1}{z} \cdot \frac{z}{z - 1}$$

$$= \frac{\cancel{z}\cancel{(z - 1)}(z + 1)}{\cancel{z}\cancel{(z - 1)}}$$

$= z + 1$ ***Answer***

Method 2

$$\dfrac{z - \frac{1}{z}}{1 - \frac{1}{z}} = \dfrac{\left(z - \frac{1}{z}\right) \cdot z}{\left(1 - \frac{1}{z}\right) \cdot z}$$

$$= \frac{z^2 - 1}{z - 1}$$

$$= \frac{\cancel{(z - 1)}(z + 1)}{\cancel{z - 1}}$$

$= z + 1$ ***Answer***

When the numerator or denominator of a complex fraction has powers with negative exponents, you should first rewrite the powers using positive exponents. Then simplify the fraction using either of the methods shown in Example 3.

Example 3 Simplify $\frac{a^{-1} - x^{-1}}{a^{-2} - x^{-2}}$.

Solution

Method 1

$$\frac{a^{-1} - x^{-1}}{a^{-2} - x^{-2}} = \left(\frac{1}{a} - \frac{1}{x}\right) \div \left(\frac{1}{a^2} - \frac{1}{x^2}\right)$$
$$= \frac{x-a}{ax} \div \frac{x^2 - a^2}{a^2x^2}$$
$$= \frac{x-a}{ax} \cdot \frac{a^2x^2}{x^2 - a^2}$$
$$= \frac{a^2x^2(x-a)}{ax(x+a)(x-a)}$$
$$= \frac{ax}{x+a}$$
Answer

Method 2

$$\frac{a^{-1} - x^{-1}}{a^{-2} - x^{-2}} = \frac{\left(\frac{1}{a} - \frac{1}{x}\right) \cdot a^2x^2}{\left(\frac{1}{a^2} - \frac{1}{x^2}\right) \cdot a^2x^2}$$
$$= \frac{ax^2 - a^2x}{x^2 - a^2}$$
$$= \frac{ax(x-a)}{(x+a)(x-a)}$$
$$= \frac{ax}{x+a}$$
Answer

Written Exercises

Simplify.

A

1. $\frac{1 - \frac{1}{3}}{\frac{1}{2} - \frac{1}{6}}$ 2
2. $\frac{\frac{1}{2} + \frac{1}{3}}{1 - \frac{1}{6}}$ 1
3. $\frac{1 - \frac{4}{5}}{\frac{1}{4} - \frac{1}{5}}$ 4
4. $\frac{\frac{2}{3} - \frac{5}{6}}{\frac{1}{3} + \frac{2}{9}}$ $-\frac{3}{10}$
5. $\frac{x + 1}{1 + \frac{1}{x}}$ x
6. $\frac{z - \frac{1}{z}}{1 - \frac{1}{z}}$ $z + 1$
7. $\frac{a - b}{a^{-1} - b^{-1}}$ $-ab$
8. $\frac{1 - xy^{-1}}{x^{-1} - y^{-1}}$ x
9. $\frac{u^{-2} - v^{-2}}{u^{-1} - v^{-1}}$ $\frac{v + u}{uv}$
10. $\frac{a^{-2} - b^{-2}}{a^{-1} + b^{-1}}$ $\frac{b - a}{ab}$
11. $\frac{\frac{1}{x^2} - \frac{1}{y^2}}{\frac{1}{x^2} + \frac{2}{xy} + \frac{1}{y^2}}$ $\frac{y - x}{y + x}$
12. $\frac{\frac{1}{p^2} - \frac{1}{q^2}}{\frac{2}{p^2} - \frac{1}{pq} - \frac{1}{q^2}}$ $\frac{q + p}{2q + p}$
13. $\frac{h + h^{-2}}{1 + h^{-1}}$ $\frac{h^2 - h + 1}{h}$
14. $\frac{x^{-2} - x^2}{x^{-1} - x}$ $\frac{1 + x^2}{x}$
15. $\frac{s^2 - t^{-2}}{s - t^{-1}}$ $\frac{st + 1}{t}$

2. $\frac{r^{-3} + s^{-3}}{r^{-1} + s^{-1}}$

Method 1

$\left(\frac{1}{r^3} + \frac{1}{s^3}\right) \div \left(\frac{1}{r} + \frac{1}{s}\right)$

$= \frac{s^3 + r^3}{r^3s^3} \cdot \frac{rs}{s + r}$

$= \frac{(s + r)(s^2 - rs + r^2)rs}{r^3s^3(s + r)}$

$= \frac{s^2 - rs + r^2}{r^2s^2}$

Method 2

$\frac{(r^{-3} + s^{-3}) \cdot r^3s^3}{(r^{-1} + s^{-1}) \cdot r^3s^3}$

$= \frac{s^3 + r^3}{r^2s^3 + r^3s^2}$

$= \frac{(s + r)(s^2 - rs + r^2)}{r^2s^2(s + r)}$

$= \frac{s^2 - rs + r^2}{r^2s^2}$

3. $\frac{1 + \frac{3}{x} + \frac{2}{x^2}}{1 + \frac{2}{x}}$

Method 1

$\left(1 + \frac{3}{x} + \frac{2}{x^2}\right) \div \left(1 + \frac{2}{x}\right)$

$= \frac{x^2 + 3x + 2}{x^2} \cdot \frac{x}{x + 2}$

$= \frac{(x + 1)(x + 2)}{x^2} \cdot \frac{x}{x + 2}$

$= \frac{x + 1}{x}$

Method 2

$\frac{\left(1 + \frac{3}{x} + \frac{2}{x^2}\right) \cdot x^2}{\left(1 + \frac{2}{x}\right) \cdot x^2}$

$\frac{x^2 + 3x + 2}{x^2 + 2x} =$

$\frac{(x + 1)(x + 2)}{x(x + 2)} =$

$\frac{x + 1}{x}$

Common Errors

Students are sometimes confused by examples such as $\frac{x}{\frac{1}{x}+\frac{1}{y}}$. They may remember that dividing by a nonzero number is equivalent to multiplying by its reciprocal and incorrectly think the above example is equivalent to $x \cdot \left(\frac{x}{1}+\frac{y}{1}\right)$. When reteaching, point out that if this rule is used they must multiply by the reciprocal of the entire divisor:

$$\frac{x}{\frac{1}{x}+\frac{1}{y}} = \frac{x}{\frac{y+x}{xy}} = \frac{x^2y}{y+x}.$$

Guided Practice

Simplify.

1. $\dfrac{2-\frac{1}{2}}{1-\frac{1}{4}}$ 2

2. $\dfrac{1-a}{a^{-1}-1}$ a

3. $\dfrac{\frac{1}{x^2}-\frac{1}{y^2}}{\frac{x}{y}+\frac{y}{x}+2}$ $\dfrac{y-x}{xy(x+y)}$

4. $\dfrac{k+\frac{1}{k-2}}{\frac{k^2}{k-2}+1}$ $\dfrac{k-1}{k+2}$

Summarizing the Lesson

In this lesson students learned two methods of simplifying complex fractions. Ask students which of the two methods they prefer.

Using a Calculator

Point out to students that they can keep the number of key presses to a minimum by using the reciprocal key when evaluating the expressions in Exercise 27.

Simplify.

16. $\dfrac{\frac{1}{x}-\frac{1}{y}}{\frac{y}{x}-\frac{x}{y}}$ $\dfrac{1}{y+x}$

17. $\dfrac{\frac{2}{y+2}-1}{\frac{1}{y+2}+1}$ $-\dfrac{y}{y+3}$

18. $\dfrac{1+\frac{1}{t-1}}{1-\frac{1}{t+1}}$ $\dfrac{t+1}{t-1}$

B

19. $\dfrac{\frac{1}{a+1}+\frac{1}{a-1}}{\frac{1}{a+1}-\frac{1}{a-1}}$ $-a$

20. $\dfrac{\frac{1}{x}+\frac{1}{x+1}}{\frac{1}{x}-\frac{1}{x+1}}$ $2x+1$

21. $\dfrac{1+\frac{1}{x-1}}{1+\frac{1}{x^2-1}}$ $\dfrac{x+1}{x}$

22. $\dfrac{\frac{1}{1-t}-\frac{1}{t}}{\frac{1}{1+t}-\frac{1}{t}}$ $\dfrac{(2t-1)(t+1)}{t-1}$

23. $\dfrac{\frac{a}{b}-\frac{a-b}{a+b}}{\frac{a}{b}+\frac{a+b}{a-b}}$ $\dfrac{a-b}{a+b}$

24. $\dfrac{\frac{u+v}{u-v}-\frac{u-v}{u+v}}{\frac{u+v}{u-v}+\frac{u-v}{u+v}}$ $\dfrac{2uv}{u^2+v^2}$

25. $\dfrac{1-\frac{2-\frac{1}{x}}{x}}{1-\frac{1}{x}}$ $\dfrac{x-1}{x}$

26. $\dfrac{u+\frac{1}{1+\frac{1}{u}}}{\frac{1}{u+1}}$ u^2+2u

27. Evaluate to three decimal places. A calculator may be helpful.

a. $1+\dfrac{1}{2}$ 1.500

b. $1+\dfrac{1}{2+\frac{1}{2}}$ 1.400

c. $1+\dfrac{1}{2+\frac{1}{2+\frac{1}{2}}}$ 1.417

d. $1+\dfrac{1}{2+\frac{1}{2+\frac{1}{2+\frac{1}{2}}}}$ 1.414

(The farther this process is carried out, the closer the results will be to $\sqrt{2} = 1.41421\ldots.$)

In Exercises 28–31, express $\dfrac{f(x+h)-f(x)}{h}$ as a single simplified fraction. (These exercises might be met in calculus.)

Sample $f(x) = \dfrac{1}{1-x}$

Solution

$$\frac{f(x+h)-f(x)}{h} = \frac{\frac{1}{1-(x+h)}-\frac{1}{1-x}}{h}$$

$$= \frac{(1-x)-(1-x-h)}{h(1-x-h)(1-x)}$$

$$= \frac{1}{(1-x-h)(1-x)}$$

C

28. $f(x) = \dfrac{1}{x}$ $-\dfrac{1}{x(x+h)}$

29. $f(x) = \dfrac{1}{x+1}$ $-\dfrac{1}{(x+1)(x+h+1)}$

30. $f(x) = \dfrac{1-x}{x}$ $-\dfrac{1}{x(x+h)}$

31. $f(x) = \dfrac{1}{x^2}$ $-\dfrac{2x+h}{x^2(x+h)^2}$

In Exercises 32–35, express $f(f(x))$ as a single simplified fraction.

32. $f(x) = \frac{1}{x+1}$ $\frac{x+1}{x+2}$

33. $f(x) = \frac{x}{x+1}$ $\frac{x}{2x+1}$

34. $f(x) = \frac{1+x}{1-x}$ $-\frac{1}{x}$

35. $f(x) = (1-x)^{-1}$ $\frac{x-1}{x}$

Mixed Review Exercises

Simplify.

1. $\frac{1}{x^2y} + \frac{1}{xy^2}$ $\frac{y+x}{x^2y^2}$

2. $\frac{8}{a^2-3a} \cdot \frac{a^2-9}{6a}$ $\frac{4(a+3)}{3a^2}$

3. $\frac{36u^2}{25v} \div \frac{27u^4}{10v^3}$ $\frac{8v^2}{15u^2}$

4. $\frac{t}{t-2} - \frac{1}{t+2}$ $\frac{t^2+t+2}{(t+2)(t-2)}$

5. $-\frac{8a^2b^3}{16a^3b^2}$ $-\frac{b}{2a}$

6. $\frac{x}{2y} + \frac{y}{2x}$ $\frac{x^2+y^2}{2xy}$

7. $\frac{x-2}{4} - \frac{2-x}{8}$ $\frac{3x-6}{8}$

8. $\frac{2y-x}{x-2y}$ -1

9. $\frac{(a-2)^2}{b^2} \div \frac{6a-12}{4b}$ $\frac{2(a-2)}{3b}$

Find the unique solution of each system. Check your answer by using substitution.

10. $4x - y = 14$ $(3, -2)$
$5x + 3y = 9$

11. $2x + 5y = 17$ $(1, 3)$
$x - 2y = -5$

12. $x - 4y = -4$ $\left(-2, \frac{1}{2}\right)$
$2x + 10y = 1$

Self-Test 2

Vocabulary rational algebraic expression (p. 227); rational expression (p. 227); rational function (p. 228); least common denominator (p. 235); complex fraction (p. 238)

Simplify.

1. $\frac{t^3-t}{t^3-t^2}$ $\frac{t+1}{t}$

2. $\frac{x^3-x^2-4x+4}{x^3-4x}$ $\frac{x-1}{x}$ **Obj. 5-4, p. 227**

3. $\frac{2p^4r^3}{3s^2} \cdot \frac{p^3r^2}{6s^3}$ $\frac{p^7r^5}{9s^5}$

4. $\frac{2u^2+3uv-2v^2}{2u^2-3uv-2v^2} \cdot \frac{2u+v}{2u-v}$ $\frac{u+2v}{u-2v}$ **Obj. 5-5, p. 232**

5. $\frac{1}{a^2} - \frac{2}{ab} + \frac{1}{b^2}$ $\frac{(b-a)^2}{a^2b^2}$

6. $\frac{1}{4x^2-4x+1} - \frac{1}{4x^2-1}$ $\frac{2}{(2x+1)(2x-1)^2}$ **Obj. 5-6, p. 235**

7. $\frac{\frac{2}{3}-\frac{1}{4}}{1-\frac{1}{6}}$ $\frac{1}{2}$

8. $\frac{1-hk^{-1}}{h^{-1}-k^{-1}}$ h **Obj. 5-7, p. 238**

Check your answers with those at the back of the book.

Suggested Assignments

Average Algebra
239/2–26 even
S 241/Mixed Review

Average Alg. and Trig.
239/2–26 even
S 241/Mixed Review

Extended Algebra
239/2–28 even
S 241/Mixed Review

Extended Alg. and Trig.
239/2–28 even, 32–35
S 241/Mixed Review

Supplementary Materials

Study Guide	pp. 83–84
Practice Master	31
Resource Book	p. 119

Quick Quiz

Simplify.

1. $\frac{8x^3-50x}{4x^2-6x-10}$ $\frac{x(2x+5)}{x+1}$

2. $\frac{14b^2x}{3y^2} \div \frac{35b^3x}{6y}$ $\frac{4}{5by}$

3. $(4k^2+4k+1) \div (2k^2+k)$ $\frac{2k+1}{k}$

4. $\frac{2}{a} + \frac{3}{a^2}$ $\frac{2a+3}{a^2}$

5. $\frac{\frac{3}{2}+\frac{5}{6}}{4-\frac{1}{2}}$ $\frac{2}{3}$

6. $\frac{a}{b^2} - \frac{2}{ab} + \frac{b}{a^2}$ $\frac{a^3-2ab+b^3}{a^2b^2}$

7. $\frac{4}{3-3z^2} - \frac{2}{z^2+5z+4}$ $-\frac{10}{3(z-1)(z+4)}$

8. $\frac{x^{-1}+y}{\frac{y}{x}+(xy)^{-1}}$ $\frac{y(xy+1)}{y^2+1}$

Teaching Suggestions p. T100

Suggested Extensions p. T100

Warm-Up Exercises

Solve.

1. $2x = 5x + 6$ -2
2. $5a + 6 < 2a - 9$ $\{a: a < -5\}$
3. $2x^2 + 2x + 9 = x^2 - 3x + 3$ $\{-2, -3\}$
4. $(x + 2)(x - 3) \geq 0$ $\{x: x \leq -2 \text{ or } x \geq 3\}$
5. $(x + 3)^2 < 0$ no soln.

Motivating the Lesson

Tell students that many problems in science and medicine involve mixtures. Problems of this type, as well as others, can be solved by using equations with fractional coefficients, the topic of today's lesson.

Problem Solving Strategies

Students should learn to *recognize each problem* in the problem set *as a standard type* (mixture, working together, etc.). They should then apply the method necessary to solve that type of problem, such as using a chart or the five-step plan.

Chalkboard Examples

Solve.

1. $\frac{x^2}{9} - \frac{x}{6} - \frac{1}{2} = 0$

$18 \cdot \left(\frac{x^2}{9} - \frac{x}{6} - \frac{1}{2}\right) = 18 \cdot 0$

$2x^2 - 3x - 9 = 0$

$(2x + 3)(x - 3) = 0$

$x = -\frac{3}{2}$ or $x = 3$

$\left\{-\frac{3}{2}, 3\right\}$

Problem Solving Using Fractional Equations

5-8 Fractional Coefficients

Objective To solve equations and inequalities having fractional coefficients.

To solve an equation or an inequality having fractional coefficients, it is helpful to multiply both sides of the open sentence by the least common denominator of the fractions.

Example 1 Solve $\frac{x^2}{2} = \frac{2x}{15} + \frac{1}{10}$.

Solution

$$\frac{x^2}{2} = \frac{2x}{15} + \frac{1}{10}$$

$$30 \cdot \frac{x^2}{2} = 30\left(\frac{2x}{15} + \frac{1}{10}\right)$$ Multiply both sides by 30, the LCM of 2, 15, and 10.

$$15x^2 = 4x + 3$$ Use the distributive property.

$$15x^2 - 4x - 3 = 0$$

$$(3x + 1)(5x - 3) = 0$$

$$x = -\frac{1}{3} \quad \text{or} \quad x = \frac{3}{5}$$

$\therefore$ the solution set is $\left\{-\frac{1}{3}, \frac{3}{5}\right\}$. ***Answer***

Example 2 Solve $\frac{x}{8} - \frac{x-2}{3} \geq \frac{x+1}{6} - 1$.

Solution

$$\frac{x}{8} - \frac{x-2}{3} \geq \frac{x+1}{6} - 1$$

$$24\left(\frac{x}{8} - \frac{x-2}{3}\right) \geq 24\left(\frac{x+1}{6} - 1\right)$$ Multiply both sides by 24, the LCM of 8, 3, and 6.

$$3x - 8x + 16 \geq 4x + 4 - 24$$ Use the distributive property.

$$-9x \geq -36$$

$$x \leq 4$$ Reverse the inequality.

$\therefore$ the solution set is $\{x: x \leq 4\}$. ***Answer***

Example 3 Crane A can unload the container ship in 10 h, and crane B can unload it in 14 h. Crane A started to unload the ship at noon and was joined by crane B at 2 P.M. At what time was the unloading job of the ship completed?

Solution

Step 1 Essentially, the problem asks for the number of hours crane A worked.

Step 2 Let t = the number of hours crane A worked.

Then $t - 2$ = the number of hours crane B worked.

The part of the job crane A can do in 1 hour $= \frac{1}{10}$.

Thus, the part of the job crane A can do in t hours is $t \cdot \frac{1}{10} = \frac{t}{10}$.

Similarly, the part of the job crane B can do in $t - 2$ hours is $(t - 2) \cdot \frac{1}{14} = \frac{t - 2}{14}$.

Step 3

Part done by crane A	plus	part done by crane B	is	the whole job.
$\frac{t}{10}$	$+$	$\frac{t-2}{14}$	$=$	1

Step 4 $\frac{t}{10} + \frac{t-2}{14} = 1$ Multiply both sides by the LCM, 70.

$$7t + 5t - 10 = 70$$
$$12t = 80$$
$$t = \frac{80}{12}$$
$$t = \frac{20}{3} = 6\frac{2}{3}$$

Step 5 The check is left for you.

$\therefore$ the unloading job was completed $6\frac{2}{3}$ h after noon, or at 6:40 P.M.

Recall that *percent* means hundredths. For example, 60% of 400 means 0.60×400, or 240. Fractions often enter problems through the use of percents, as in Example 4 at the top of the next page.

2. $\frac{x - 14}{4} \geq \frac{5x}{12} - 3$

$12\left(\frac{x-14}{4}\right) \geq 12\left(\frac{5x}{12} - 3\right)$

$3x - 42 \geq 5x - 36$

$-42 \geq 2x - 36$

$-6 \geq 2x$

$-3 \geq x$, or

$x \leq -3$

$\{x: x \leq -3\}$

3. The German Club collected the same amount from each member participating in a trip to a music festival. When 6 members could not attend, each of the remaining members were charged \$2 more to cover the total cost of \$360. How many members went on the excursion?

Let m = number of members going on the excursion. Then $m + 6$ = original number of members.

$\frac{360}{m} = \frac{360}{m+6} + 2$

$m(m + 6)\left(\frac{360}{m}\right) =$

$m(m + 6)\left[\frac{360}{m+6} + 2\right]$

$360(m + 6) =$

$360m + 2m(m + 6)$

$360m + 2160 =$

$360m + 2m^2 + 12m$

$2m^2 + 12m - 2160 = 0$

$m^2 + 6m - 1080 = 0$

$(m + 36)(m - 30) = 0$

$m = -36$ or $m = 30$

-36 cannot be a solution.

$\therefore$ 30 members went on the excursion.

Common Errors

After learning to solve equations and inequalities having fractional coefficients, students sometimes incorrectly simplify addition problems by multiplying by the LCM. For example, students may conclude that $\frac{1}{6x} + \frac{1}{8y}$ is equivalent to $24xy \cdot \frac{1}{6x} + 24xy \cdot \frac{1}{8y} = 4y + 3x$. When reteaching, stress that this property only applies to equations and inequalities.

Check for Understanding

Here is a suggested use of the Oral Exercises to check students' understanding as you teach the lesson.
Oral Exs. 1–9: use after Example 2.

Guided Practice

Solve.

1. $\frac{m}{5} - \frac{1}{2} = \frac{3}{10}$ 4
2. $\frac{x+1}{3} + \frac{x-1}{6} = \frac{2x+1}{2}$ $-\frac{2}{3}$
3. $\frac{2}{9} - \frac{4y}{3} \le 2$ $y \ge -\frac{4}{3}$
4. $\frac{2a+3}{4} - \frac{a}{6} < a + 1$ $a > -\frac{3}{8}$
5. $\frac{s^2}{2} - \frac{4s}{3} = \frac{1}{2}$ $\left\{3, -\frac{1}{3}\right\}$
6. $\frac{x(x+5)}{12} = \frac{x+8}{4}$ $\{4, -6\}$
7. $\frac{b+2}{3} - \frac{b-2}{7} \ge 2$ $b \ge \frac{11}{2}$
8. $\frac{k}{5} - \frac{k}{4} = \frac{2k}{3} - \frac{3k}{2}$ 0

Example 4 A nurse wishes to obtain 800 mL of a 7% solution of boric acid by mixing 4% and 12% solutions. How much of each should be used?

Solution

Step 1 The problem asks for a number of mL of 4% and 12% solutions of boric acid to be used.

Step 2 Let x = number of mL of 4% solution.

Then $800 - x$ = number of mL of 12% solution.

Show the known facts in a table.

	mL of solution	× % boric acid	= mL of boric acid
4% solution	x	4%	$0.04x$
12% solution	$800 - x$	12%	$0.12(800 - x)$
Final mixture	800	7%	0.07(800)

Step 3 From the last column of the table, write an equation relating the 4% and 12% solutions to the final mixture.

$$0.04x + 0.12(800 - x) = 0.07(800)$$

or

$$4x + 12(800 - x) = 7(800)$$

To clear decimals, multiply both sides by 100.

Step 4

$$4x + 9600 - 12x = 5600$$
$$-8x = -4000$$
$$x = 500 \quad \text{(4\% solution)}$$
$$800 - x = 300 \quad \text{(12\% solution)}$$

Step 5 The check is left for you.

∴ the nurse should mix 500 mL of the 4% solution and 300 mL of the 12% solution. ***Answer***

Oral Exercises

Find an open sentence with integers as coefficients that is equivalent to the given open sentence.

3. $2t - 3 > 6$
6. $2t + 5 = 9$

1. $\frac{x}{2} - 1 = \frac{5}{6}$ $3x - 6 = 5$
2. $\frac{y}{6} + 2 = \frac{2}{3}$ $y + 12 = 4$
3. $\frac{t}{3} - \frac{1}{2} > 1$
4. $\frac{x}{3} - \frac{2}{5} \le 2$ $5x - 6 \le 30$
5. $u = \frac{2}{5} - \frac{u}{2}$ $10u = 4 - 5u$
6. $\frac{t}{3} + \frac{5}{6} = \frac{3}{2}$
7. $2.7 + 0.5s = 1.2$ $27 + 5s = 12$
8. $0.14x - 0.06 = 0.15$ $14x - 6 = 15$
9. $\frac{z}{3} - \frac{3}{4} = \frac{5}{6}$ $4z - 9 = 10$

Written Exercises

5. $\{z: z \le 3\}$ 6. $\{t: t \le -6\}$
7. $\{r: r > -2\}$ 8. $\{x: x \ge -7\}$

Solve each open sentence.

A

1. $\frac{x}{9} + \frac{1}{6} = \frac{2}{3}$ $\left\{\frac{9}{2}\right\}$
2. $\frac{3u}{5} - \frac{5}{6} = \frac{u}{10}$ $\left\{\frac{5}{3}\right\}$
3. $\frac{2t-1}{6} = \frac{t+2}{4} + \frac{1}{3}$ $\{12\}$
4. $\frac{s-2}{2} - \frac{s-1}{5} = \frac{1}{4}$ $\left\{\frac{7}{2}\right\}$
5. $\frac{z}{4} - \frac{z-1}{6} \le \frac{5}{12}$
6. $\frac{t}{4} - \frac{t+2}{3} + \frac{1}{6} \ge 0$
7. $\frac{r-2}{8} < \frac{3r+1}{6} + \frac{1}{3}$
8. $\frac{x-5}{9} \le \frac{2x}{15} - \frac{2}{5}$
9. $\frac{y^2}{4} - \frac{3y}{2} + 2 = 0$ $\{2, 4\}$
10. $\frac{t^2}{6} - \frac{t}{2} - \frac{2}{3} = 0$ $\{4, -1\}$
11. $\frac{z^2}{3} - \frac{z}{6} = 1$ $\left\{-\frac{3}{2}, 2\right\}$
12. $\frac{v^2}{6} + \frac{1}{8} = \frac{v}{3}$ $\left\{\frac{3}{2}, \frac{1}{2}\right\}$
13. $\frac{x^2}{9} + \frac{x-1}{10} = 0$ $\left\{-\frac{3}{2}, \frac{3}{5}\right\}$
14. $\frac{t(t-1)}{3} = \frac{t+1}{2}$ $\left\{-\frac{1}{2}, 3\right\}$
15. $\frac{y(2y-1)}{2} = \frac{1-y}{3}$ $\left\{\frac{2}{3}, -\frac{1}{2}\right\}$
16. $\frac{w(w-1)}{3} + \frac{1}{2} = \frac{w+1}{4}$ $\left\{\frac{3}{4}, 1\right\}$

B

17. $\frac{x(x+1)}{5} - \frac{x+1}{6} = \frac{1}{3}$ $\left\{\frac{3}{2}, -\frac{5}{3}\right\}$
18. $\frac{2t(3t+1)}{5} - \frac{t+1}{2} = \frac{1}{10}$ $\left\{-\frac{2}{3}, \frac{3}{4}\right\}$
19. $\frac{y(2y-1)}{4} + \frac{3}{10} = \frac{y(y+2)}{5}$ $\left\{\frac{2}{3}, \frac{3}{2}\right\}$
20. $\frac{u(u-1)}{2} + \frac{1}{3} = \frac{u(2-u)}{4}$ $\left\{\frac{2}{3}\right\}$
21. $\frac{y^2+4}{6} + \frac{y+1}{3} < \frac{3}{2}$ $\{y: -3 < y < 1\}$
22. $\frac{z^2+1}{6} \ge \frac{z+2}{3}$ $\{z: z \le -1 \text{ or } z \ge 3\}$
23. $\frac{t^2}{6} + \frac{t-2}{4} \ge \frac{t+1}{3}$ $\left\{t: t \le -2 \text{ or } t \ge \frac{5}{2}\right\}$
24. $\frac{u^2}{15} \le \frac{u+2}{6} - \frac{2}{5}$ $\left\{u: \frac{1}{2} \le u \le 2\right\}$

Problems

In Problems 1–4, find the number described.

A

1. $\frac{5}{16}$ of $\frac{4}{5}$ of the number is 15. 60
2. 12 is $\frac{3}{5}$ of $\frac{10}{21}$ of the number. 42
3. 30 is 20% of 30% of the number. 500
4. 75% of 60% of the number is 36. 80
5. Pump A can unload the *Lunar Petro* in 30 h and pump B can unload it in 24 h. Because of an approaching storm, both pumps were used. How long did they take to empty the ship? $13\frac{1}{3}$ h
6. An old conveyor belt takes 21 h to move one day's coal output from the mine to a rail line. A new belt can do it in 15 h. How long does it take when both are used at the same time? $8\frac{3}{4}$ h
7. How much pure antifreeze must be added to 12 L of a 40% solution of antifreeze to obtain a 60% solution? 6 L
8. How much water must be evaporated from a 300 L tank of a 2% salt solution to obtain a 5% solution? 180 L

Summarizing the Lesson

In this lesson students learned to solve equations and inequalities having fractional coefficients by using equivalent equations with integral coefficients. Ask students to list the types of problems which may result in equations with fractional coefficients.

Suggested Assignments

Average Algebra
Day 1: 245/1–19 odd
245/*P*: 1–11 odd
R 241/Self-Test 2
Day 2: 245/10–20 even
245/*P*: 13–21 odd
S 237/40, 41

Average Alg. and Trig.
Day 1: 245/1–19 odd
245/*P*: 1–11 odd
R 241/Self-Test 2
Day 2: 245/10–20 even
245/*P*: 13–21 odd
S 241/40, 41

Extended Algebra
Day 1: 245/3–24 mult. of 3
245/*P*: 1–13 odd
R 241/Self-Test 2
Day 2: 246/*P*: 14–21
S 241/32–35

Extended Alg. and Trig.
245/3–24 mult. of 3
245/*P*: 7–21 odd
R 241/Self-Test 2

Supplementary Materials

Study Guide	pp. 85–86
Practice Master	32
Computer Activity	11

9. The river boat *Delta Duchess* paddled upstream at 12 km/h, stopped for 2 h of sightseeing, and paddled back at 18 km/h. How far upstream did the boat travel if the total time for the trip, including the stop, was 7 h? 36 km

10. Pam jogged up a hill at 6 km/h and then jogged back down at 10 km/h. How many kilometers did she travel in all if her total jogging time was 1 h 20 min? 10 km

11. The Computer Club invested \$2200, part at 4.5% interest and the rest at 7%. The total annual interest earned was \$144. How much was invested at each rate? \$400 at 4.5%; \$1800 at 7%

12. Lina Chen invested \$24,000, part at 8% and the rest at 7.2%. How much did she invest at each rate if her income from the 8% investment is two thirds that of the 7.2% investment? \$9000 at 8%; \$15,000 at 7.2%

B **13.** A pharmacist wishes to make 1.8 L of a 10% solution of boric acid by mixing 7.5% and 12% solutions. How much of each type of solution should be used? 0.8 L of 7.5%; 1 L of 12%

14. How much of an 18% solution of sulfuric acid should be added to 360 mL of a 10% solution to obtain a 15% solution? 600 mL

15. The county's new asphalt paving machine can surface 1 km of highway in 10 h. A much older machine can surface 1 km in 18 h. How long will it take them to surface 21 km of highway if they start at opposite ends and work day and night? 135 h

16. Pipes A and B can fill a storage tank in 8 h and 12 h, respectively. With the tank empty, pipe A was turned on at noon, and then pipe B was turned on at 1:30 P.M. At what time was the tank full? 5:24 P.M.

17. Sharon drove for part of a 150 km trip at 45 km/h and the rest of the trip at 75 km/h. How far did she drive at each speed if the entire trip took her 2 h 40 min? 75 km at 45 km/h; 75 km at 75 km/h

18. An elevator went from the bottom to the top of a tower at an average speed of 4 m/s, remained at the top for 90 s, and then returned to the bottom at 5 m/s. If the total elapsed time was $4\frac{1}{2}$ min, how high is the tower? 400 m

C **19.** A commercial jet can fly from San Francisco to Dallas in 3 h. A private jet can make the same trip in $3\frac{1}{2}$ h. If the two planes leave San Francisco at noon, after how many hours is the private jet twice as far from Dallas as the commercial jet? $2\frac{5}{8}$ h

20. A car radiator is filled with 5 L of a 25% antifreeze solution. How many liters must be drawn off and replaced by a 75% antifreeze solution to leave the radiator filled with a 55% antifreeze solution? 3 L

21. The rail line between two cities consists of two segments, one 96 km longer than the other. A passenger train averages 60 km/h over the shorter segment, 120 km/h over the longer, and 100 km/h for the entire trip. How far apart are the cities? 160 km

5-9 Fractional Equations

Objective To solve and use fractional equations.

An equation in which a variable occurs in a denominator, such as

$$\frac{3}{x^2 - 7x + 10} + 2 = \frac{x-4}{x-5},$$

is called a **fractional equation.** When you solve a fractional equation by multiplying both sides by the LCD of the fractions, the resulting equation is *not* always equivalent to the original one.

Example 1 Solve $\dfrac{3}{x^2 - 7x + 10} + 2 = \dfrac{x-4}{x-5}$.

Solution $x^2 - 7x + 10 = (x-2)(x-5)$. So the LCD is $(x-2)(x-5)$.

$$(x-2)(x-5)\left[\frac{3}{x^2-7x+10} + 2\right] = (x-2)(x-5)\left[\frac{x-4}{x-5}\right]$$

$$3 + 2(x-2)(x-5) = (x-2)(x-4)$$

$$3 + 2x^2 - 14x + 20 = (x-2)(x-4)$$

$$2x^2 - 14x + 23 = x^2 - 6x + 8$$

$$x^2 - 8x + 15 = 0$$

$$(x-3)(x-5) = 0$$

$$x = 3 \quad \text{or} \quad x = 5$$

Check: When x equals 2 or 5, a denominator in the original equation has a value of 0. Therefore, 2 and 5 are not in the domain of x and cannot be solutions. Since 5 is not permissible, check 3 in the original equation:

$$\frac{3}{3^2 - 7 \cdot 3 + 10} + 2 \stackrel{?}{=} \frac{3-4}{3-5}$$

$$-\frac{3}{2} + 2 \stackrel{?}{=} \frac{-1}{-2}$$

$$\frac{1}{2} = \frac{1}{2} \quad \checkmark$$

$\therefore$ the solution set is $\{3\}$. ***Answer***

In Example 1, the equation obtained by multiplying the given equation by $(x-2)(x-5)$ has the *extraneous root* 5. An **extraneous root** is a root of the transformed equation that is *not* a root of the original equation.

Caution: If you transform an equation by multiplying by a polynomial, always check each root of the new equation in the *original* one.

Teaching Suggestions p. T100

Suggested Extensions p. T101

Warm-Up Exercises

Solve each open sentence.

1. $3x + \frac{1}{2} = \frac{2}{3}$ $\frac{1}{18}$
2. $\frac{a}{9} + \frac{5}{7} = \frac{2a}{3}$ $\frac{9}{7}$
3. $\frac{v-5}{8} = 12$ 101
4. $\frac{t^2+t-2}{5} = \frac{4}{5}$ $\{-3, 2\}$
5. $\frac{(u+5)(u-7)}{3} = \frac{1}{3}(u+5)$ $\{-5, 8\}$

Motivating the Lesson

Tell students that sometimes they may want to know how long it will take a person or a thing to do a task, or how fast a certain object is traveling. Problems of this type can be solved by fractional equations.

Thinking Skills

Students should realize that every time a fractional equation is solved by multiplying both sides by the LCD there is the possibility of extraneous roots. Students should *interpret* the answers to check if they make sense. For example, the time of −4 h generated in Example 2 makes no sense.

Chalkboard Examples

Solve.

1. $\frac{z+3}{3-z} + \frac{11z+3}{z^2-9} = \frac{1-5z}{z+3}$

$(z^2-9)\left(-\frac{z+3}{z-3}\right) +$

$(z^2-9)\left(\frac{11z+3}{z^2-9}\right) =$

$(z^2-9)\left(\frac{1-5z}{z+3}\right)$

$-(z+3)(z+3) + 11z + 3 = (z-3)(1-5z)$

$-z^2 - 6z - 9 + 11z + 3 = -5z^2 + 16z - 3$

$4z^2 - 11z - 3 = 0$

$(4z+1)(z-3) = 0$

$z = -\frac{1}{4}$ or $z = 3$

3 is excluded from the domain and cannot be a root.

$\frac{-\frac{1}{4}+3}{3+\frac{1}{4}} + \frac{11\left(-\frac{1}{4}\right)+3}{\left(-\frac{1}{4}\right)^2-9} = \frac{1-5\left(-\frac{1}{4}\right)}{-\frac{1}{4}+3}$; $\frac{9}{11} = \frac{9}{11}$ ✓

$\left\{-\frac{1}{4}\right\}$

2. $\frac{10}{x^2-4} = 1 - \frac{1}{x+2}$

$(x^2-4)\left(\frac{10}{x^2-4}\right) =$

$(x^2-4)\left(1 - \frac{1}{x+2}\right)$

$10 = x^2 - 4 - (x-2)$

$10 = x^2 - x - 2$

$x^2 - x - 12 = 0$

$(x-4)(x+3) = 0$

$x = 4$ or $x = -3$

$\frac{10}{4^2-4} = 1 - \frac{1}{4+2}$;

$\frac{5}{6} = \frac{5}{6}$ ✓

$\frac{10}{(-3)^2-4} =$

$1 - \frac{1}{(-3)+2}$; $2 = 2$ ✓

$\{4, -3\}$

Example 2 One pump can empty the town swimming pool in 7 h less time than a smaller second pump can. Together they can empty the pool in 12 h. How much time would it take the larger pump alone to empty it?

Solution

Step 1 The problem asks for the number of hours for the larger pump to empty the pool.

Step 2 Let t = the number of hours for the larger pump to empty the pool.

Then $t + 7$ = the number of hours for the smaller pump to empty the pool.

Now write an expression for the part of the pool that is emptied by each pump in 1 h and the part emptied by each in 12 h.

$\frac{1}{t}$ = part of pool emptied by larger pump in 1 h

$\frac{12}{t}$ = part of pool emptied by larger pump in 12 h

$\frac{1}{t+7}$ = part of pool emptied by smaller pump in 1 h

$\frac{12}{t+7}$ = part of pool emptied by smaller pump in 12 h

Step 3 The sum of the parts emptied by each pump in 12 h is 1.

$$\frac{12}{t} + \frac{12}{t+7} = 1$$

Step 4

$$12(t+7) + 12t = t(t+7)$$
$$t^2 - 17t - 84 = 0$$
$$(t+4)(t-21) = 0$$
$$t = -4 \quad \text{or} \quad t = 21$$

Since the time t cannot be negative, reject $t = -4$ as an answer.

Step 5 *Check* $t = 21$: If the larger pump can empty the pool in 21 h, then the smaller can empty it in 21 + 7, or 28 h. Therefore, the parts of the pool emptied by the pumps in 12 h are $\frac{12}{21} = \frac{4}{7}$ and $\frac{12}{28} = \frac{3}{7}$.

$$\frac{4}{7} + \frac{3}{7} = 1 \quad ✓$$

∴ the larger pump can empty the pool in 21 h.

Answer

Example 3 Because of strong headwinds, an airplane's ground speed (see page 131) for the first half of a 2000 km trip averaged only 600 km/h. What must its ground speed be for the rest of the trip if it is to average 720 km/h for the entire trip?

Solution Let $r =$ the ground speed in km/h that the airplane must travel for the second half of its trip. Use the fact that $time = \frac{distance}{rate}$ and make a table.

	Distance (km)	Rate (km/h)	Time (h)
First half	1000	600	$\frac{1000}{600} = \frac{5}{3}$
Second half	1000	r	$\frac{1000}{r}$
Entire trip	2000	720	$\frac{2000}{720} = \frac{25}{9}$

Time for first half of trip plus Time for second half of trip equals Time for entire trip

$$\frac{5}{3} + \frac{1000}{r} = \frac{25}{9}$$

Multiply both sides by $9r$: $15r + 9000 = 25r$

$r = 900$

The check is left for you.

$\therefore$ the ground speed for the second half must be 900 km/h. ***Answer***

Written Exercises

Solve and check. If an equation has no solution, say so.

A

1. $\frac{3}{t} - \frac{1}{3t} = \frac{2}{3}$ {4}
2. $\frac{2}{x} + \frac{1}{4} = \frac{1}{x}$ {−4}
3. $\frac{1}{x} = \frac{2}{x-3}$ {−3}
4. $\frac{3}{u+2} = \frac{1}{u-2}$ {4}
5. $\frac{2}{s+3} - \frac{1}{s-3} = 0$ {9}
6. $\frac{5}{2r+1} - \frac{3}{2r-1} = 0$ {2}
7. $\frac{t}{t-1} = \frac{t+2}{t}$ {2}
8. $\frac{x-3}{x} = \frac{x-4}{x-2}$ {6}
9. $\frac{6t^2 - t - 1}{3(t^2+1)} = 2$ {−7}
10. $\frac{(y+1)^2}{(y-3)^2} = 1$ {1}
11. $\frac{x}{x+3} + \frac{1}{x-3} = 1$ {6}
12. $\frac{1}{s} + \frac{s}{s+2} = 1$ {2}
13. $\frac{1}{y-2} + \frac{1}{y+2} = \frac{4}{y^2-4}$ no sol.
14. $\frac{k}{k+1} + \frac{k}{k-2} = 2$ {−4}
15. $\frac{3}{x+1} - \frac{1}{x-2} = \frac{1}{x^2-x-2}$ {4}
16. $\frac{1}{t-1} + \frac{1}{t+2} = \frac{3}{t^2+t-2}$ no sol.
17. $\frac{6}{t} - \frac{2}{t-1} = 1$ {2, 3}
18. $\frac{1}{x+1} - \frac{1}{x+2} = \frac{1}{2}$ {0, −3}

Guided Practice

Solve.

1. $\frac{1}{x} - \frac{1}{2x} = 3$ $\frac{1}{6}$
2. $\frac{5}{y-1} - \frac{4}{y+1} = 0$ −9
3. $\frac{a}{a+5} = \frac{a-6}{a}$ −30
4. $\frac{2x+3}{x-1} - \frac{2x-3}{x+1} = \frac{10}{x^2-1}$ no sol.
5. $\frac{1}{x} + \frac{4}{x+1} = 6$ $\left\{\frac{1}{3}, -\frac{1}{2}\right\}$
6. $\frac{1}{a-3} + \frac{2}{a^2-9} = \frac{5}{a+3}$ 5

Summarizing the Lesson

In this lesson students learned how to solve fractional equations and to check for extraneous roots. Ask students to identify errors that might be made when working with problems involving fractional equations.

Suggested Assignments

Day 1: 249/1–27 odd
250/*P*: 1–7 odd
S 252/Mixed Review
Day 2: 251/*P*: 9–19 odd
R 252/Self-Test 3

Average Alg. and Trig.
Day 1: 249/1–27 odd
250/*P*: 1–7 odd
S 252/Mixed Review
Day 2: 251/*P*: 9–19 odd
R 252/Self-Test 3

Extended Algebra
Day 1: 249/1–29 odd
250/*P*: 1–7 odd
S 252/Mixed Review
Day 2: 251/*P*: 9–21 odd, 22, 23
R 252/Self-Test 3

Extended Alg. and Trig.
249/3–30 mult. of 3
250/*P*: 3–21 mult. of 3, 22, 23
S 252/Mixed Review

Supplementary Materials

Study Guide	pp. 87–88
Practice Master	33
Computer Activity	12
Resource Book	p. 120

B **19.** $\frac{u}{u-2} + \frac{30}{u+2} = 9$ {1, 3}

20. $\frac{1}{s+3} + \frac{1}{s-5} = \frac{1-s}{s+3}$ {1, 3}

21. $\frac{2}{x-1} - \frac{x}{x+3} = \frac{6}{x^2+2x-3}$ {0, 3}

22. $\frac{x}{x+3} + \frac{1}{x-1} = \frac{4}{x^2+2x-3}$ {−1}

23. $\frac{5}{u^2+u-6} = 2 - \frac{u-3}{u-2}$ {−4}

24. $\frac{y}{y-2} - \frac{2}{y+3} = \frac{10}{y^2+y-6}$ no sol.

25. $\frac{t}{t-1} = \frac{1}{t+2} + \frac{3}{t^2+t-2}$ no sol.

26. $\frac{9}{t^2-2t-8} + \frac{t}{t+2} = 2$ {5, −5}

27. $\left(\frac{x-3}{x+1}\right)^2 = 2 \cdot \frac{x-3}{x+1} + 3$ {1, −3}

28. $\left(\frac{t+3}{t-1}\right)^2 = 2 + \frac{t+3}{t-1}$ {−1, 5}

C **29.** $\dfrac{\frac{1}{x^2} - x^2}{\frac{1}{x} + x} = \frac{3}{2}$ $\left\{-2, \frac{1}{2}\right\}$

30. $\left(\frac{1}{x} - \frac{x}{2}\right)\left(\frac{2}{x} - x\right) = \frac{1}{2}$ {−2, 2, −1, 1}

Problems

Solve.

A **1.** Find two positive numbers that differ by 8 and whose reciprocals differ by $\frac{1}{6}$. 4 and 12

2. Find two numbers whose sum is 25 and the sum of whose reciprocals is $\frac{1}{6}$. 15 and 10

3. The reciprocal of half a number increased by half the reciprocal of the number is $\frac{1}{2}$. Find the number. 5

4. The reciprocal of one third of a number decreased by one third of the reciprocal of the number is $\frac{1}{3}$. Find the number. 8

5. A town's old street sweeper can clean the streets in 60 h. The old sweeper together with a new sweeper can clean the streets in 15 h. How long would it take the new sweeper to do the job alone? 20 h

6. The intake pipe can fill a certain tank in 6 h when the outlet pipe is closed, but with the outlet pipe open it takes 9 h. How long would it take the outlet pipe to empty a full tank? 18 h

7. During 60 mi of city driving, Jenna averaged 15 mi/gal. She then drove 140 mi on an expressway and averaged 25 mi/gal for the entire 200 mi. Find the average fuel consumption on the expressway. 35 mi/gal

8. Helped by a strong jet stream, a Los Angeles-to-Boston plane flew 10% faster than usual and made the 4400 km trip in 30 min less time than usual. At what speed does the plane usually fly? 800 km/h

9. The excursion boat *Holiday* travels 35 km upstream and then back again in 4 h 48 min. If the speed of the *Holiday* in still water is 15 km/h, what is the speed of the current? 2.5 km/h

10. Tim paddled his kayak 12 km upstream against a 3 km/h current and back again in 5 h 20 min. In that time how far could he have paddled in still water? 32 km

11. Members of the Computer Club were assessed equal amounts to raise $1200 to buy some software. When 8 new members joined, the per-member assessment was reduced by $7.50. What was the new size of the club? 40

12. Members of the Ski Club contributed equally to obtain $1800 for a holiday trip. When 6 members found that they could not go, their contributions were refunded and each remaining member then had to pay $10 more to raise the $1800. How many went on the trip? 30

13. To measure the speed of the jet stream, a weather plane left its base at noon and flew 800 km directly against the stream with an air speed of 750 km/h. It then returned directly to its base, arriving at 2:24 P.M. What was the speed of the jet stream? 250 km/h

14. When Ace Airlines changed to planes that flew 100 km/h faster than its old ones, the time of its 2800 km Dallas-Seattle flight was reduced by 30 min. Find the speed of the new planes. 800 km/h

B 15. Elvin drove halfway from Ashton to Dover at 40 mi/h and the rest of the way at 60 mi/h. What was his average speed for the whole trip? (*Hint:* Let the distance for the whole trip be, say, 100 mi.) 48 mi/h

16. Elizabeth drove the first half of a trip at 36 mi/h. At what speed should she cover the remaining half in order to average 45 mi/h for the whole trip? (See the hint in Exercise 15.) 60 mi/h

17. A train averaged 120 km/h for the first two thirds of a trip and 100 km/h for the whole trip. Find its average speed for the last third of the trip. 75 km/h

18. Because of traffic Maria could average only 40 km/h for the first 20% of her trip, but she averaged 75 km/h for the whole trip. What was her average speed for the last 80% of her trip? 96 km/h

19. Pipe A can fill a tank in 5 h. Pipe B can fill it in 2 h less time than it takes pipe C, a drainpipe, to empty the tank. With all three pipes open, it takes 3 h to fill the tank. How long would it take pipe C to empty it? 5 h

20. An elevator went from the bottom to the top of a 240 m tower, remained there for 12 s, and returned to the bottom in an elapsed time of 2 min. If the elevator traveled 1 m/s faster on the way down, find its speed going up. 4 m/s

A number x is the **harmonic mean** of a and b if $\frac{1}{x}$ is the average of $\frac{1}{a}$ and $\frac{1}{b}$.

C **21.** The harmonic mean of a number and 5 is 8. Find the number. 20

22. Find two positive numbers that differ by 12 and have harmonic mean 5. 3 and 15

23. Suppose that a vehicle averages u km/h for the first half of a trip and v km/h for the second half. Show that its average speed for the whole trip is the harmonic mean of u and v. Let d = length of trip. Let s = average speed of trip. Then time first half + time second half = total time. $\frac{\frac{1}{2}d}{u} + \frac{\frac{1}{2}d}{v} = \frac{d}{s}$; $\frac{\frac{1}{2}}{u} + \frac{\frac{1}{2}}{v} = \frac{1}{s}$; $\frac{1}{2}\left(\frac{1}{u} + \frac{1}{v}\right) = \frac{1}{s}$; $\therefore$ s is the harmonic mean.

Mixed Review Exercises

Simplify.

1. $\frac{x^2 - 1}{1 - x}$ $-x - 1$

2. $\frac{8t^2 - 2}{8t + 4}$ $\frac{2t - 1}{2}$

3. $\frac{32m^5n}{24m^2n^2}$ $\frac{4m^3}{3n}$

4. $\frac{(4z^3)^{-2}}{(2z^4)^{-1}}$ $\frac{1}{8z^2}$

5. $\frac{a^{-1} + b^{-2}}{a^{-2} + b^{-1}}$ $\frac{a(b^2 + a)}{b(b + a^2)}$

6. $\frac{c^2 - c - 12}{c^2 - 2c - 8}$ $\frac{c + 3}{c + 2}$

Solve.

7. $\frac{2x - 1}{3} = \frac{3x + 2}{6}$ $\{4\}$

8. $3 - \frac{1}{2}x \le 4$ $\{x: x \ge -2\}$

9. $\frac{x^2}{6} = \frac{x}{3} + \frac{5}{2}$ $\{-3, 5\}$

10. $\frac{3x}{4} + 1 > \frac{x - 1}{2}$ $\{x: x > -6\}$

11. $\frac{3}{2}x - 1 = 5$ $\{4\}$

12. $\frac{1}{2}x^2 + \frac{7}{3}x - 4 < 0$ $\left\{x: -6 < x < \frac{4}{3}\right\}$

Self-Test 3

Vocabulary fractional equation (p. 247) extraneous root (p. 247)

Solve.

1. $\frac{3t - 2}{8} = \frac{t - 1}{12} + 1$ $\{4\}$

2. $\frac{y^2}{3} + \frac{y}{15} - \frac{2}{5} = 0$ $\left\{-\frac{6}{5}, 1\right\}$ **Obj. 5-8, p. 242**

3. One computer can process a payroll in 6 h. An older computer can do the same job in 7 h 30 min. How long would it take both computers to do the processing? $3\frac{1}{3}$ h

Solve and check. If an equation has no solution, say so.

4. $\frac{x}{x - 1} = \frac{x - 2}{x - 3}$ no sol.

5. $\frac{z}{z - 2} - \frac{1}{z + 3} = \frac{10}{z^2 + z - 6}$ $\{-4\}$ **Obj. 5-9, p. 247**

6. A train averaged 80 km/h for the first half of its trip. How fast must it travel for the second half of the trip in order to average 96 km/h for the whole trip? 120 km/h

Compare your answers with those at the back of the book.

Quick Quiz

Solve.

1. $\frac{x^2}{3} = \frac{x}{6} + \frac{1}{2}$ $\left\{-1, \frac{3}{2}\right\}$

2. $\frac{k + 2}{k - 1} = \frac{2k + 1}{k - 1} + 1$ no sol.

3. $\frac{4}{m} - \frac{2}{m + 1} = \frac{2m}{m + 1}$ 2

4. Two numbers differ by 6 and their reciprocals differ by $\frac{8}{15}$. Find the numbers. $\frac{3}{2}, \frac{15}{2}$ or $-\frac{3}{2}, -\frac{15}{2}$

5. At noon a pump is turned on to fill an empty pool. Normally the pump would fill the pool in 12 hours, but at 2 P.M. a valve is accidentally opened that could drain a full pool in 20 hours. If the valve remains open, at what time will the pool be full? 3 P.M. the next day

Application / *Electrical Circuits*

Figure (a) pictures an electrical circuit containing an E-volt (V) battery, an R-ohm (Ω) resistance, and an ammeter for measuring the current I in amperes (A). Ohm's law states that $I = \dfrac{E}{R}$.

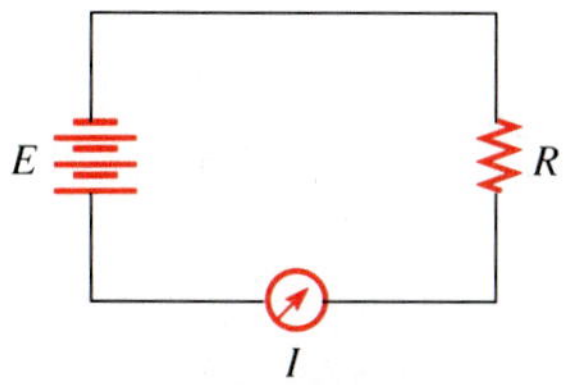

(a)

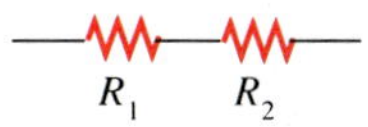

$$R_c = R_1 + R_2$$

(b)

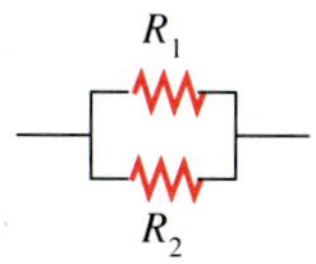

$$\frac{1}{R_c} = \frac{1}{R_1} + \frac{1}{R_2}$$

(c)

Figures (b) and (c) show resistances R_1 and R_2 in series and in parallel, respectively, and also formulas for finding the total resistance R_c of the combination.

Example Find the quantity labeled X in the following circuit.

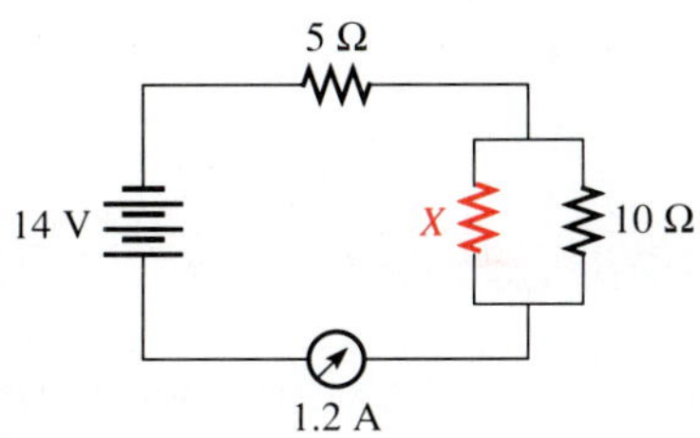

Solution The resistance of the 10 ohm–X ohm combination is R_c, where

$$\frac{1}{R_c} = \frac{1}{10} + \frac{1}{X} = \frac{X + 10}{10X}.$$

$$\therefore R_c = \frac{10X}{X + 10}.$$

The total resistance is

$$R = 5 + R_c$$
$$= 5 + \frac{10X}{X + 10}.$$

From $I = \dfrac{E}{R}$, or $IR = E$, you have

$$1.2\left(5 + \frac{10X}{X + 10}\right) = 14$$

Solving the equation, you find that $X = 20\ \Omega$. ***Answer***

253

Additional Answers
Written Exercises

(continued from p. 220)

55. 1. m and n are pos. integers; $m > n$
Given

2. $a^{-m}a^{n} = \frac{1}{a^{m}} \cdot a^{n}$
Def. of neg. exp.

3. $= \frac{a^{n}}{a^{m}}$
Def. of div.

4. $= \frac{1}{a^{m-n}}$
Law 4b of pos. exp.

5. $= a^{-(m-n)}$
Def. of neg. exp.

6. $= a^{-(m+(-n))}$
Def. of subtr.

7. $= a^{(-m)+[-(-n)]}$
Prop. of the opp. of a sum

8. $= a^{(-m)+n}$
For every real number x, $-(-x) = x$.

9. $\therefore a^{-m}a^{n} = a^{(-m)+n}$
Trans. prop. of eq.

56. 1. $\frac{a^{-m}}{a^{n}} = a^{-m} \cdot \frac{1}{a^{n}}$
Def. of div.

2. $= \frac{1}{a^{m}} \cdot \frac{1}{a^{n}}$
Def. of neg. exp.

3. $= \frac{1}{a^{m}a^{n}}$
Rule for mult. fractions

4. $= \frac{1}{a^{m+n}}$
Law 1 of pos. exp.

5. $= a^{-(m+n)}$
Def. of neg. exp.

6. $= a^{(-m)+(-n)}$
Prop. of the opposite of a sum

7. $= a^{(-m)-n}$
Def. of subtr.

8. $\therefore \frac{a^{-m}}{a^{n}} = a^{(-m)-n}$
Trans. prop. of eq.

Exercises

Find the quantity labeled X in each of the following circuits.

1. 1.5 A

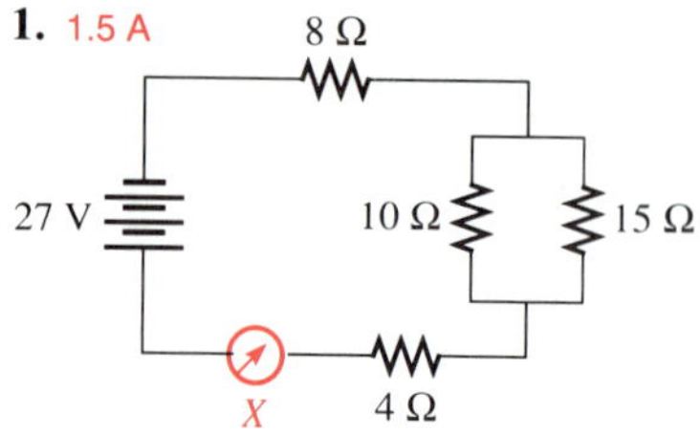

2. 12.5 V

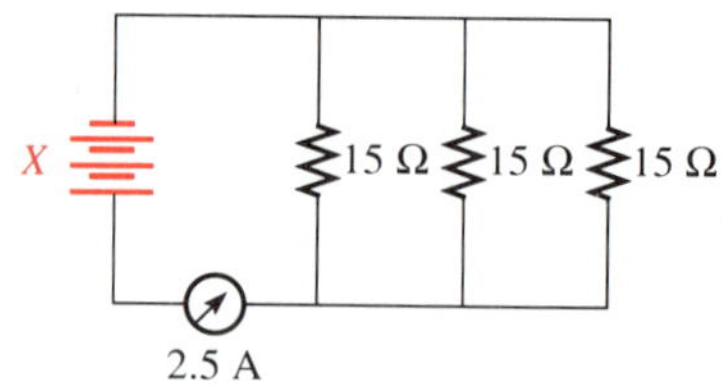

3. 10 Ω

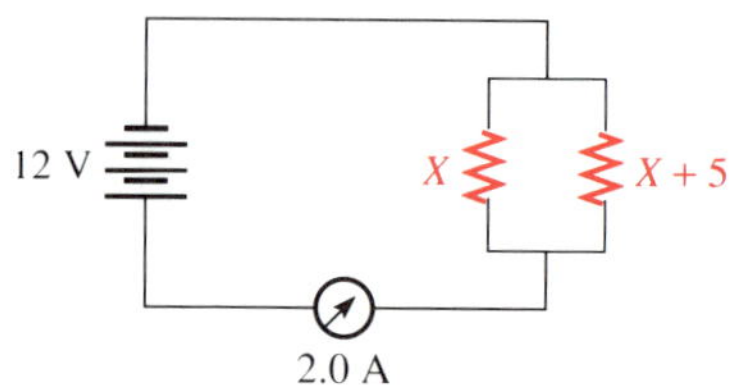

4. 22 Ω

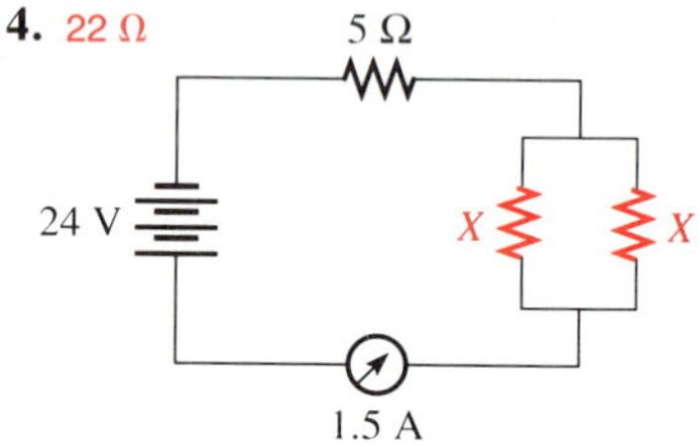

Chapter Summary

1. The definitions

$$a^0 = 1 \quad \text{and} \quad a^{-n} = \frac{1}{a^n}$$

permit zero and negative integers as exponents. The laws of exponents stated on page 212 continue to hold.

2. In *scientific notation* a positive number is expressed in the form

$$m \times 10^n$$

where $1 \le m < 10$ and n is an integer. The digits of m should all be *significant*.

3. A quotient of polynomials is called a *rational expression*. Such an expression is simplified if its numerator and denominator have no common factors other than 1.

4. To multiply and divide rational expressions, you can use the product and quotient rules for fractions:

$$\frac{p}{q} \cdot \frac{r}{s} = \frac{pr}{qs}$$

$$\text{and} \qquad \frac{p}{q} \div \frac{r}{s} = \frac{p}{q} \cdot \frac{s}{r}$$

5. To add or subtract rational expressions, find their *least common denominator* (LCD), express each as an equivalent fraction having the LCD as denominator, and then add or subtract.

6. To simplify a *complex fraction,* use either of these methods:

   Simplify the numerator and denominator separately; then divide.

   Multiply the numerator and denominator by the LCD of all the fractions appearing in the numerator and the denominator.

7. To solve an equation or an inequality having fractional coefficients, multiply both sides by the LCD of the fractions. To solve a *fractional equation* (one having a variable in the denominator), use the same method, but always check the results for any extraneous roots that may have been introduced.

Chapter Review

Give the letter of the correct answer.

1. Simplify $\frac{18x^3y}{12x^2y^4}$. 5-1

a. $\frac{2}{3xy^3}$ (**b.**) $\frac{3x}{2y^3}$ **c.** $\frac{2xy^3}{3}$ **d.** $\frac{3y^3}{2x}$

2. Simplify $\frac{(a^2bc^3)^3}{(-ab^3c^2)^2}$.

a. $-\frac{a^4c^5}{b^3}$ **b.** $a^4b^3c^5$ **c.** $-\frac{ac}{b^2}$ (**d.**) $\frac{a^4c^5}{b^3}$

3. Simplify $\frac{(p^2q)^{-1}}{p^2q^{-1}}$. 5-2

(**a.**) $\frac{1}{p^4}$ **b.** q^2 **c.** $\frac{q^2}{p^4}$ **d.** 1

4. Simplify $\left(\frac{m^2}{n}\right)^{-2} \cdot \left(\frac{m^0}{n^{-1}}\right)^3$.

a. $\frac{n^5}{m}$ **b.** m^4n **c.** $\frac{1}{m^4n}$ (**d.**) $\frac{n^5}{m^4}$

5. Express 0.000543 in scientific notation. 5-3

a. 5.43×10^{-3} (**b.**) 5.43×10^{-4} **c.** 54.3×10^{-5} **d.** 543×10^{-6}

6. Find a one-significant-digit estimate for $(0.0593)(2045) \div 0.32$.

a. 5×10^3 (**b.**) 4×10^2 **c.** 2×10^{-1} **d.** 3×10

7. Simplify $\frac{y^2 - 4}{y^2 + y - 6}$. 5-4

a. $\frac{2}{y+3}$ **b.** $\frac{y-2}{y-3}$ **c.** $\frac{y-4}{y-6}$ (**d.**) $\frac{y+2}{y+3}$

57. 1. $\frac{a^m}{a^{-n}} = a^m \div \frac{1}{a^n}$
Def. of neg. exp.
2. $= a^m \cdot \frac{1}{\frac{1}{a^n}}$
Def. of div.
3. $= a^m \cdot a^n$
For every nonzero real number x, $\frac{1}{\frac{1}{x}} = x$.
4. $= a^{m+n}$
Law 1 of pos. exp.
5. $= a^{m+[-(-n)]}$
For every real number x, $-(-x) = x$
6. $= a^{m-(-n)}$
Def. of subtr.
7. $\therefore \frac{a^m}{a^{-n}} = a^{m-(-n)}$
Trans. prop. of eq.

58. 1. $\frac{a^{-m}}{a^{-n}} = a^{-m} \cdot \frac{1}{a^{-n}}$
Def. of div.
2. $= \frac{1}{a^m} \cdot a^n$
Def. of neg. exp.
3. $= \frac{a^n}{a^m}$
Def. of div.
4. $= \frac{1}{a^{m-n}}$
Law 4b of pos. exp.
5. $= a^{-(m-n)}$
Def. of neg. exp.
6. $= a^{-[m+(-n)]}$
Def. of subtr.
7. $= a^{(-m)+[-(-n)]}$
Prop. of the opp. of a sum
8. $= a^{(-m)-(-n)}$
Def. of subtr.
9. $\therefore \frac{a^{-m}}{a^{-n}} = a^{(-m)-(-n)}$
Trans. prop. of eq.

Additional Answers Written Exercises

(continued from p. 229)

24. **a.** reals except -2
b. 2

25. **a.** reals except -2, 0, and 2
b. $\frac{3}{2}$ and -3

26. **a.** reals except $\frac{1}{2}$
b. $\frac{1}{4}$ and -4

27. **a.** reals except 1 and -1
b. 2

28. **a.** reals except 1
b. -4 and -1

Additional Answers Written Exercises

(continued from p. 231)

11.

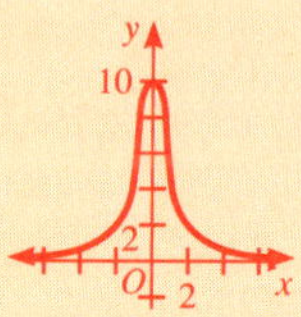

12.

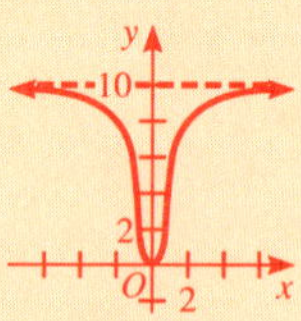

8. Find the zeros of $f(x) = \dfrac{2x^2 - x - 1}{2x^2 + 3x - 2}$.

a. $-\frac{1}{2}, 1$ **b.** $\frac{1}{2}, -2$ **c.** $\frac{1}{2}, -1$ **d.** $-\frac{1}{2}, 2$

9. Simplify $\dfrac{u^2 + 3u}{u^2 + 2u - 3} \cdot \dfrac{u^3 - u}{u^2 - u - 2}$. 5-5

a. $\dfrac{u^2}{u - 2}$ **b.** $u - 2$ **c.** $\dfrac{1}{u - 2}$ **d.** $\dfrac{u}{u - 2}$

10. Simplify $\dfrac{m^2}{n^3} \cdot \dfrac{n}{m^4} \div \dfrac{1}{mn}$.

a. mn **b.** m^3n^3 **c.** $\dfrac{1}{mn}$ **d.** $\dfrac{1}{m^3n^3}$

11. Simplify $\dfrac{a + 2}{2a} + \dfrac{1 - a}{a^2}$. 5-6

a. $\dfrac{a^2 + 2}{2a^3}$ **b.** $\dfrac{a^2 + 2}{2a^2}$ **c.** $\dfrac{a + 2}{2a}$ **d.** $\dfrac{a + 1}{a}$

12. Simplify $\dfrac{x - 1}{x + 1} - \dfrac{x^2 - x - 6}{x^2 + 4x + 3}$.

a. $\dfrac{1}{x + 1}$ **b.** $\dfrac{2x^2 + x - 9}{(x + 1)(x + 3)}$

c. $\dfrac{3}{x + 3}$ **d.** $\dfrac{x - 9}{(x + 1)(x + 3)}$

13. Simplify $\dfrac{y - \dfrac{2}{y + 1}}{1 - \dfrac{2}{y + 1}}$. 5-7

a. $y + 1$ **b.** $\dfrac{y}{y - 1}$ **c.** $\dfrac{y - 2}{y + 1}$ **d.** $y + 2$

14. Solve $\dfrac{x + 1}{6} < x - \dfrac{3x - 2}{4}$. 5-8

a. $\{x: x > -4\}$ **b.** $\{x: x < 4\}$ **c.** $\{x: x < -4\}$ **d.** $\{x: x > 4\}$

15. Solve $\dfrac{x(x - 2)}{6} = \dfrac{x + 1}{8}$.

a. $\{3\}$ **b.** $\left\{-3, \frac{1}{4}\right\}$ **c.** $\left\{-\frac{1}{4}, 3\right\}$ **d.** $\{-1, 0, 2\}$

16. How much of an 8% saline solution should be added to 600 mL of a 3% solution to produce a 5% solution?

a. 200 mL **b.** 300 mL **c.** 400 mL **d.** 500 mL

17. Solve $\dfrac{2}{x + 2} + \dfrac{x^2}{x^2 - 4} = \dfrac{1}{x - 2}$. 5-9

a. $\{-3, 2\}$ **b.** $\{-3\}$ **c.** $\{-2, 3\}$ **d.** $\{3\}$

18. Bill rowed 30 km upstream against a 2 km/h current and back again in a total of 8 h. How fast can he row in still water?

a. 7 km/h **b.** 8 km/h **c.** 9 km/h **d.** 10 km/h

Chapter Test

Simplify. Use only positive exponents in the answer.

1. $\dfrac{(3xy^3)^2}{6x^5y^4}$ $\dfrac{3y^2}{2x^3}$

2. $\dfrac{m^4}{4n^3}\left(\dfrac{2n}{m^3}\right)^3$ $\dfrac{2}{m^5}$ 5-1

3. $pq^4(p^{-1}q^2)^{-2}$ p^3

4. $\left(\dfrac{u^2}{v}\right)^{-2}\left(\dfrac{v^3}{u}\right)^{-1}$ $\dfrac{1}{u^3v}$ 5-2

Evaluate. Give each answer in scientific notation with two significant digits.

5. $(2.1 \times 10^3)(1.6 \times 10^{-5})$ 3.4×10^{-2}

6. $\dfrac{5.2 \times 10^2}{1.2 \times 10^{-4}}$ 4.3×10^6 5-3

Find (a) the domain and (b) the zeros of each function.

7. $f(x) = \dfrac{3x - 2}{x^2 - x - 12}$
   a. reals except 4 and -3 **b.** $\frac{2}{3}$

8. $f(x) = \dfrac{x^2 - x - 12}{3x - 2}$ 5-4
   a. reals except $\frac{2}{3}$ **b.** 4 and -3

Simplify.

9. $\dfrac{6x^3}{y^2} \div \dfrac{3xy}{2}$ $\dfrac{4x^2}{y^3}$

10. $\dfrac{u^2 - 9}{u + 2} \div \dfrac{u + 3}{u^2 - 4}$ $(u - 3)(u - 2)$ 5-5

11. $\dfrac{p - 2}{2p} + \dfrac{p + 3}{3p}$ $\dfrac{5}{6}$

12. $\dfrac{1}{s - 1} - \dfrac{2}{s^2 - 1}$ $\dfrac{1}{s + 1}$ 5-6

13. $\dfrac{ab^{-1} + 1}{a^{-1}b + 1}$ $\dfrac{a}{b}$

14. $\dfrac{\dfrac{x + y}{x - y}}{\dfrac{1}{x} - \dfrac{1}{y}}$ $-\dfrac{xy(x + y)}{(x - y)^2}$ 5-7

Solve.

15. $\dfrac{x + 7}{8} - \dfrac{1}{2} \ge \dfrac{x}{4}$ $\{x\colon x \le 3\}$

16. $\dfrac{x^2}{9} = \dfrac{x + 2}{2}$ $\left\{-\frac{3}{2}, 6\right\}$ 5-8

17. One crew can detassel a field of corn in 6 days. Another crew can do the same job in 4 days. If the slower crew works alone for the first two days of detasseling and then is joined by the faster crew, how long will it take the two crews to finish? $1\frac{3}{5}$ days

18. Solve $\dfrac{3}{x^2 - 2x - 8} + \dfrac{1}{x + 2} = \dfrac{4}{x^2 - 16}$. $\{-3\}$ 5-9

19. The members of a computer club were to be assessed equally to raise \$400 for the purchase of software. When four more persons joined the club, the per-member assessment was reduced by \$5.00. What was the new size of the club membership? 20

Alternate Test p. T17

Supplementary Materials

Test Masters	20, 21
Resource Book	p. 121

6 Irrational and Complex Numbers

Objectives

6-1 To find roots of real numbers.

6-2 To simplify expressions involving radicals.

6-3 To simplify expressions involving sums of radicals.

6-4 To simplify products and quotients of binomials that contain radicals.

6-5 To solve equations containing radicals.

6-6 To find and use decimal representations of real numbers.

6-7 To use the number *i* to simplify square roots of negative numbers.

6-8 To add, subtract, multiply, and divide complex numbers.

Assignment Guide

See p. T58 for Key to the format of the Assignment Guide

Day	Average Algebra	Average Algebra and Trigonometry	Extended Algebra	Extended Algebra and Trigonometry
1	**6-1** 262/1–33 odd **S** 263/*Mixed Review*	**6-1** 262/1–33 odd **S** 263/*Mixed Review*	**6-1** 262/1–33 odd **S** 263/*Mixed Review*	*Administer Chapter 5 Test* **6-1** *Read 6-1* 262/1–33 odd **S** 263/*Mixed Review*
2	**6-2** 267/2–46 even **S** 263/28–34 even	**6-2** 267/2–62 even	**6-2** 267/2–66 even **S** 263/35–37	**6-2** 267/2–66 even **S** 263/35–37
3	**6-2** 268/48–66 even **6-3** 272/1–21 odd	**6-3** 272/1–43 odd **S** 273/*Mixed Review*	**6-3** 272/1–43 odd **S** 273/*Mixed Review*	**6-3** 272/1–43 odd **S** 273/*Mixed Review*
4	**6-3** 272/23–43 odd **S** 273/*Mixed Review* **6-4** 275/1–17 odd	**6-4** 275/2–40 even **S** 269/63–66	**6-4** 275/2–42 even **S** 273/*Extra:* 1, 2	**6-4** 275/2–44 even **S** 273/*Extra:* 1, 2
5	**6-4** 276/19–39 odd **S** 268/45–65 odd	**6-5** 280/1–23 odd **S** 276/41–44	**6-5** 280/1–27 odd **S** 276/43–46	**6-5** 280/1–27 odd **S** 276/43, 45, 46
6	**6-5** 280/2–24 even **S** 276/41–44	**6-5** 280/25–37 odd **S** 282/*Mixed Review*	**6-5** 280/29–39 odd **S** 282/*Mixed Review*	**6-5** 280/29–39 odd **S** 282/*Mixed Review*
7	**6-5** 280/26–36 even **S** 282/*Mixed Review*	**6-6** 286/1–37 odd **R** 282/*Self-Test 1*	**6-6** 286/1–39 odd, 40 **R** 282/*Self-Test 1*	**6-6** 286/1–39 odd, 40 **R** 282/*Self-Test 1*
8	**6-6** 286/1–37 odd **R** 282/*Self-Test 1*	**6-7** 290/2–36 even **S** 291/*Mixed Review*	**6-7** 290/2–36 even **S** 291/*Mixed Review*	**6-7** 290/2–36 even **S** 291/*Mixed Review*
9	**6-7** 290/2–36 even **S** 291/*Mixed Review*	**6-7** 291/38–52 even **6-8** 295/1–11 odd	**6-7** 291/38–54 even **6-8** 295/1–21 odd	**6-7** 291/38–54 even **6-8** 295/1–21 odd
10	**6-7** 291/38–52 even **6-8** 295/1–11 odd	**6-8** 295/13–49 odd **S** 291/31–53 odd	**6-8** 295/23–55 odd 298/*Extra:* 1–17 odd	**6-8** 295/23–55 odd 298/*Extra:* 1–17 odd
11	**6-8** 295/13–49 odd **S** 291/31–53 odd	*Prepare for Chapter Test* **R** 297/*Self-Test 2* **R** 302/*Chapter Review*	*Prepare for Chapter Test* **R** 297/*Self-Test 2* **R** 302/*Chapter Review*	*Prepare for Chapter Test* **R** 297/*Self-Test 2* **R** 302/*Chapter Review*

Assignment Guide *(continued)*

Day	Average Algebra	Average Algebra and Trigonometry	Extended Algebra	Extended Algebra and Trigonometry
12	*Prepare for Chapter Test* **R** 297/*Self-Test 2* **R** 302/*Chapter Review*	*Administer Chapter 6 Test*	*Administer Chapter 6 Test*	*Administer Chapter 6 Test*
13	*Administer Chapter 6 Test*			

Supplementary Materials Guide

For Use with Lesson	Practice Masters	Tests	Study Guide (Reteaching)	Resource Book: Tests	Resource Book: Practice Exercises	Resource Book: Applications (A) Enrichment (E)
6-1			pp. 89–90			p. 209 (A)
6-2	Sheet 34		pp. 91–92		p. 122	
6-3			pp. 93–94			
6-4			pp. 95–96			
6-5	Sheet 35	Test 22	pp. 97–98		p. 123	
6-6	Sheet 36		pp. 99–100		p. 124	
6-7			pp. 101–102			
6-8	Sheets 37, 38	Test 23	pp. 103–104		p. 125	
Chapter 6		Tests 24, 25		pp. 27–30	p. 126	p. 225 (E)
Cum. Rev. 1–6		Test 26				

Overhead Visuals

For Use with Lessons	Visual	Title
6-6, 6-7, 6-8	B	Multi-Use Packet 2

Software

Software	Algebra Plotter Plus	Using Algebra Plotter Plus	Computer Activities	Test Generator
	Function Plotter	Enrichment, p. 37	Activities 13, 14	168 test items
For Use with Lessons	6-1	6-1	6-5, 6-8	all lessons

Strategies for Teaching

Using Technology

Graphing calculators and computers are useful in this chapter for graphing equations and finding the solutions of equations as shown in the following example. Computers can also help students verify their answers and graphs as shown below.

See the Exploration on page 837 for an activity in which students use a calculator to explore some properties of radicals. Use this activity before Lesson 6-3 so that students can construct their own thinking before seeing the rules presented in the lesson.

6-1 Roots of Real Numbers

Using *ALGEBRA PLOTTER PLUS*

You can have students approximate the *n*th root of a number a by using the Function Plotter program to graph $y = x^n - a$. For example, have students graph $y = x^3 + 4$. By zooming in on the point where the graph crosses the x-axis, students can obtain approximations of the point's x-coordinate, which is $\sqrt[3]{-4}$. In the diagram at the top of the next column, the approximation is -1.58740.

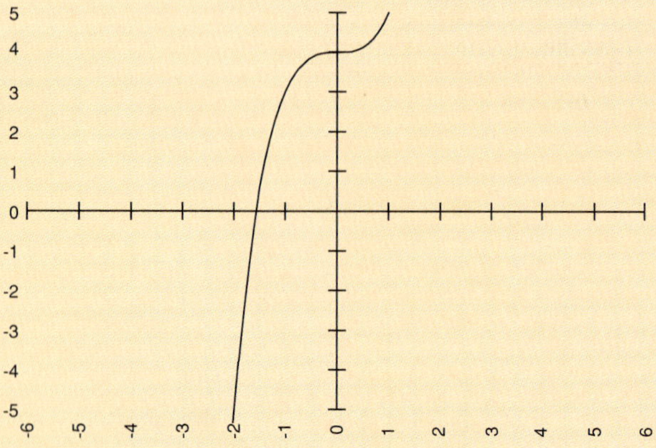

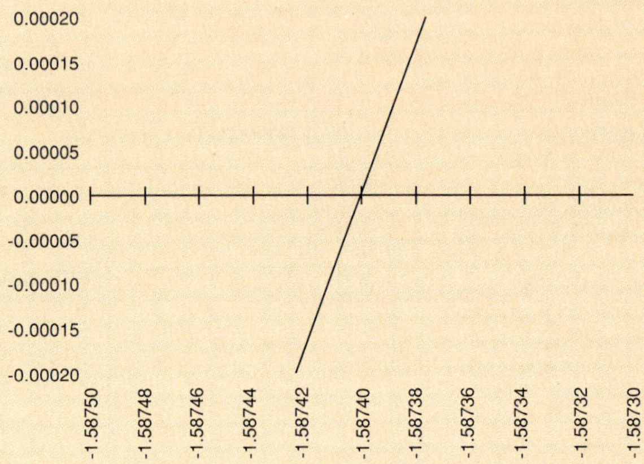

You might wish to point out (or have students discover) that when n is even, the graph of $y = x^n - a$ (where a is a positive real number) crosses the x-axis at two points (producing a positive and a negative *n*th root), and when n is odd, the graph of $y = x^n - a$ (where a is any real number) crosses the x-axis at only one point (producing a single *n*th root).

6-5 Equations Containing Radicals

Have students use a computer or graphing calculator to verify their answers for Written Exercises 1–18 and 25–32 on page 280. For example, to verify that 2 is the solution $\sqrt{x + 2} = x$ (Exercise 13), students should rewrite the equation as $\sqrt{x + 2} - x = 0$, graph the equation $y = \sqrt{x + 2} - x$, and then determine the values of x for which the graph intersects the x-axis. Alternatively, you might point out to students that the two sides of a radical equation can be graphed and solved as a system. For example, to solve $\sqrt{x + 2} = x$, students can graph $y = \sqrt{x + 2}$ and $y = x$ together and then determine the x-coordinate of each point of intersection.

Using *ALGEBRA PLOTTER PLUS*

Choose FUNCTION PLOTTER. Go to EQUATIONS and enter the equation(s). For example, enter $\sqrt{x + 2}$ as SQR(X + 2), and $\sqrt[3]{3m + 1}$ as (3X + 1) ~ 3. (Notice the use of ~ to indicate a root. Square roots can be entered using ~ 2 instead of SQR.) DRAW the graph(s).

References to Strategies

PE: Pupil's Edition **TE:** Teacher's Edition **RB:** Resource Book

Problem Solving Strategies

TE: pp. 261, 277; p. T103 (Teaching Sugg., Lesson 6-5)

Applications

PE: p. 258 (electricity); p. 277 (physics); pp. 280–281 (word problems)
TE: pp. 258, 277
RB: p. 209

Nonroutine Problems

PE: p. 263 (Exs. 35–38); p. 269 (Exs. 67–70); p. 281 (Exs. 38, 39); p. 287 (Exs. 36–42); p. 296 (Exs. 51–56)
TE: pp. T102–T104 (Sugg. Extensions, Lessons 6-2, 6-5, 6-6, 6-7)

Communication

PE: p. 273 (writing indirect proof); p. 287 (Exs. 34, 38–42, convincing argument)
TE: p. T101 (Reading Algebra); p. T104 (indirect proof)

Thinking Skills

PE: p. 269 (Exs. 67–70); p. 276 (Ex. 47); p. 296 (Exs. 53–56, proofs)
TE: pp. 270, 284

Explorations

PE: p. 273 (irrationality); pp. 298–300 (conjugates, absolute value); p. 837 (radicals)

Connections

PE: p. 258 (Electricity); pp. 273, 276, 281 (Geometry); pp. 292–296 (Arithmetic); p. 297 (History)

Using Technology

PE: pp. 263, 269 (Calculator Key-In); pp. 267–268, 285, 837
TE: pp. 258, 260, 263, 267, 269, 284, T103
Computer Activities: pp. 34–37
Using Algebra Plotter Plus: p. 37

Using Manipulatives/Models

Overhead Visuals: B

Cooperative Learning

TE: p. T102 (Lesson 6-2)

Teaching Resources

For use in implementing the teaching strategies referenced on the previous page.

Application
Resource Book, p. 209

Application—The Mandelbrot Set (For use with Chapter 6)

The figure at the right is a computer-generated graph of a set of complex numbers called the *Mandelbrot set*. This set is an example of a *fractal*, which has the unusual property of being self-similar at various levels of magnification. Self-similarity makes fractals of interest both to mathematicians and to artists.

The Mandelbrot set is based on the function $f(z) = z^2 + c$, where c is a complex constant. For a given value of c, if the sequence

$$f(0), f(f(0)), f(f(f(0))), \dots$$

is bounded (that is, if every term of the sequence, when graphed in the complex plane, stays within a finite region of the plane), then c belongs to the set.

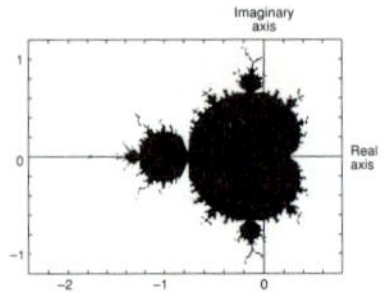

1. Enter a (real) number on a calculator and repeatedly press the squaring key (usually labeled x^2). What happens if your number has absolute value
   a. greater than 1? ____________
   b. equal to 1? ____________
   c. between 0 and 1? ____________

2. Applying a function repeatedly as you did with the squaring function in Exercise 1 is called *iteration*. Iteration can also be done with complex numbers. Use the fact that $(a + bi)^2 = a^2 - b^2 + 2abi$ to find the first three iterations of squaring $1 + 2i$:
   a. $(1 + 2i)^2 =$ ____________
   b. $(\_\_\_\_\_\_)^2 =$ ____________
   c. $(\_\_\_\_\_\_)^2 =$ ____________

3. The Mandelbrot set is based on iterations of the function $f(z) = z^2 + c$, where c is a complex constant. Suppose $c = 1 - i$. Then $f(0) = 1 - i$, and $f(f(0)) = f(1 - i) = (1 - i)^2 + 1 - i = (1 - 1 - 2i) + 1 - i = 1 - 3i$. Find the next two iterations:
   a. $f(f(f(0))) = f(1 - 3i) = (1 - 3i)^2 + 1 - i =$ ____________
   b. $f(f(f(f(0)))) = f(\_\_\_\_\_\_) = (\_\_\_\_\_\_)^2 + 1 - i =$ ____________

4. The Mandelbrot set consists of all complex numbers c for which the sequence of iterations of $f(z) = z^2 + c$ evaluated at $z = 0$ is bounded. To check boundedness, you must compute the *magnitude* of each number in the sequence. The magnitude of the complex number $a + bi$ is denoted by $|a + bi|$ and defined as $\sqrt{a^2 + b^2}$. (Interpreted geometrically, the magnitude of $a + bi$ gives the distance between the origin of the complex plane and the point (a, b), which is the graph of $a + bi$ in the complex plane.) Thus, if the numbers $|f(0)|, |f(f(0))|, |f(f(f(0)))|, \dots$ head toward infinity (that is, grow increasingly large), c is *not* in the Mandelbrot set. If, however, the numbers show stability (that is, stay at or near one or more fixed values), then c *is* in the Mandelbrot set. Determine whether or not the following values of c are in the Mandelbrot set.
   a. $c = 0.5 + 0.5i$: ____________ b. $c = 0 + 1i$: ____________

Resource Book, ALGEBRA, Structure and Method, Book 2

APPLICATIONS 209

Enrichment/Reasoning
Resource Book, p. 225

Squares and Square Roots (For use with Chapter 6)

1. It is conjectured that if the difference between two numbers is even, then the product of the two numbers can be written as the difference of two squares. What do you think?

   If you think the conjecture is true, try to prove it. Otherwise, give an example to show that the conjecture is false.

2. It is conjectured that the product of two numbers each of which is the sum of two squares is a number that is also the sum of two squares. What do you think? Try to prove the conjecture or give an example to show that it is false.

3. Give an example to show that if $c = a + b$, then $\sqrt{c} = \sqrt{a} + \sqrt{b}$ is false.

4. It is known that it is impossible to find two integers such that the square of one is exactly twice the square of the other. But you can find two integers such that the square of one is 1 less than twice the square of the other; for example, $7^2 = 2(5^2) - 1$. Can you find another example?

5. Find the value of the following product of square roots.

   $$\sqrt{1\tfrac{1}{2}} \cdot \sqrt{1\tfrac{1}{3}} \cdot \sqrt{1\tfrac{1}{4}} \cdot \sqrt{1\tfrac{1}{5}} \cdot \sqrt{1\tfrac{1}{6}} \cdot \sqrt{1\tfrac{1}{7}} \cdot \sqrt{1\tfrac{1}{8}} \cdot \sqrt{1\tfrac{1}{9}} \cdot \sqrt{1\tfrac{1}{10}} \cdot \sqrt{1\tfrac{1}{11}}$$

6. If $f(x) = \sqrt{3x + 2} + \sqrt{3 - 2x}$, for what values of x are there real values for $f(x)$?

7. Show that the following expression is a perfect square no matter what values are used for X and Y.

   $$(XY)^2 + (X^2 + Y^2)(X + Y)^2$$

8. Find a relation between the product of four consecutive integers and a special set of numbers. Try to prove this relation.

9. Make a conjecture about the difference of the squares of two odd integers. Consider the largest number by which the difference is divisible. Test your conjecture with several examples. Then try to prove it.

Resource Book, ALGEBRA, Structure and Method, Book 2

ENRICHMENT 225

Reteaching/Practice
Study Guide, p. 99

6–6 Rational and Irrational Numbers

Objective: To find and use decimal representations of real numbers.

Vocabulary

Rational number A number that can be written as a fraction with an integer in the numerator and a nonzero integer in the denominator. The decimal equivalent of such a number either terminates (ends) or repeats. Examples:

$\frac{3}{4} = 0.75$ (a *terminating*, or *finite*, *decimal*) $\frac{5}{3} = 1.\overline{6}$ (a *repeating*, or *infinite*, *decimal*)

Irrational number A real number that is not rational. The decimal equivalent of such a number is infinite and nonrepeating. Examples: π $6\sqrt{2}$ 0.525225222...

Completeness property of real numbers Every real number has a decimal representation, and every decimal represents a real number.

CAUTION Since calculators have finite displays, calculators can give only a *rational approximation* of the decimal representation of an irrational number.

Example 1 Classify each real number as either rational or irrational.

a. $\sqrt{3}$ b. $\sqrt{\frac{9}{16}}$ c. 3.712712...

Solution

a. Since $\sqrt{3} = 1.7320508\dots$, an infinite, nonrepeating decimal, $\sqrt{3}$ is *irrational*. (Note: $\sqrt{3}$ can't be written as the quotient of two integers.)

b. Since $\sqrt{\frac{9}{16}} = \frac{3}{4} = 0.75$, a terminating decimal, $\sqrt{\frac{9}{16}}$ is *rational*. (Note: $\frac{3}{4}$ is a fraction with integers in the numerator and denominator.)

c. Since 3.712712... (or $3.\overline{712}$) is a repeating decimal, it is *rational*.

Classify each real number or expression as either rational or irrational.

1. a. $1.\overline{48}$ b. 1.481481148111... 2. a. $\sqrt{3} + \sqrt{27}$ b. $\sqrt{3} \cdot \sqrt{27}$

3. a. $\pi - \frac{1}{\pi}$ b. $\pi \div \pi$ 4. a. $\sqrt{\frac{25}{64}}$ b. $\sqrt{\frac{1}{10}}$

Example 2 Write as a decimal: a. $\frac{9}{8}$ b. $\frac{3}{11}$

Solution a. $\frac{9}{8} = 9 \div 8 = 1.125$ b. $\frac{3}{11} = 3 \div 11 = 0.2727\dots = 0.\overline{27}$

Write each fraction as a terminating or repeating decimal.

5. $\frac{2}{7}$ 6. $\frac{7}{9}$ 7. $\frac{15}{16}$ 8. $\frac{32}{5}$

Study Guide, ALGEBRA AND TRIGONOMETRY, Structure and Method, Book 2

99

Reteaching/Practice
Study Guide, p. 100

6–6 Rational and Irrational Numbers (continued)

Example 3 Write each terminating decimal as a common fraction in lowest terms.

a. 5.635 b. 0.0048

Solution

a. The decimal place is thousandths. $5.635 = \frac{5635}{1000} = \frac{1127}{200}$

b. The decimal place is ten-thousandths. $0.0048 = \frac{48}{10{,}000} = \frac{3}{625}$

Example 4 Write each repeating decimal as a common fraction in lowest terms.

a. $0.2\overline{13}$ b. $5.76\overline{576}$

Solution In each case let N be the number. Multiply the given number by 10^n where n is the number of digits in the block of repeating digits. Then subtract N. This will eliminate the repeating block. Finally, solve for N.

a.
$$\begin{aligned} N &= 0.2\overline{13} \\ 100N &= 21.3\overline{13} \\ -\quad N &= 0.2\overline{13} \\ \hline 99N &= 21.1 \\ N &= \frac{21.1}{99} = \frac{211}{990} \end{aligned}$$
$\therefore 0.2\overline{13} = \frac{211}{990}$

b.
$$\begin{aligned} N &= 5.76\overline{576} \\ 1000N &= 5765.76\overline{576} \\ -\quad N &= 5.76\overline{576} \\ \hline 999N &= 5760 \\ N &= \frac{5760}{999} = \frac{640}{111} \end{aligned}$$
$\therefore 5.76\overline{576} = \frac{640}{111}$

Write each decimal as a common fraction in lowest terms.

9. 2.005 10. 3.74 11. 5.0125 12. $0.8\overline{7}$ 13. $3.\overline{72}$ 14. $4.62\overline{462}$

Example 5 Find a rational number r and an irrational number s between $\sqrt{5}$ and $\sqrt{6}$.

Solution Begin by finding the decimal approximations for the radicals:

$\sqrt{5} \approx 2.236068\dots$ and $\sqrt{6} \approx 2.4494897\dots$

Then find one finite or repeating decimal between them, and one infinite, nonrepeating decimal between them. Two possible answers are:

$r = 2.\overline{37}$ and $s = 2.301401501\dots$

Find (a) a rational and (b) an irrational number between each pair.

15. 0.7 and 0.8 16. $\sqrt{7}$ and $\sqrt{8}$ 17. $\sqrt{8}$ and 3

18. 10^{-4} and 10^{-3} 19. $2\frac{1}{12}$ and $2\frac{1}{11}$ 20. $1.\overline{96}$ and $1.\overline{9}$

100
Study Guide, ALGEBRA AND TRIGONOMETRY, Structure and Method, Book 2

Using Technology
Computer Activities, p. 34

ACTIVITY 13. *Radical Equations* *(for use with Section 6-5)*

Directions: Write all answers in the spaces provided.

PROBLEM

Use the computer to help solve the radical equation $\sqrt{2x-6} = \sqrt{4x+5} - 3$.

PROGRAM

```
10  PRINT "WHAT ARE P, Q, R, S, T";
20  INPUT P, Q, R, S, T
30  LET M = P - R
40  LET N = Q - S - T • T
50  LET A = M • M
60  LET B = 2 • M • N - 4 • R • T • T
70  LET C = N • N - 4 • S • T • T
80  LET D = B • B - 4 • A • C
90  IF D < 0 THEN 220
100  PRINT
110  PRINT "POSSIBLE SOLUTIONS ARE:"
120  IF A = 0 THEN 190
130  LET X1 = (-B + SQR(D)) / (2 • A)
140  LET X2 = (-B - SQR(D)) / (2 • A)
150  LET X1 = INT(100 • X1 + .5) / 100
160  LET X2 = INT(100 • X2 + .5) / 100
170  PRINT "X  =  "; X1; "  OR X  =  "; X2
180  GOTO 230
190  LET X = -C / B
200  PRINT "X  =  "; X
210  GOTO 230
220  PRINT "NO REAL SOLUTION"
230  END
```

PROGRAM CHECK

Type in the program. To test whether you entered it correctly, run the program. After the question, enter 2, −6, 4, 5, −3. The computer should print:

```
POSSIBLE SOLUTIONS ARE:
X = 11 OR X = 5
```

USING THE PROGRAM

The program finds possible solutions to equations of the form

$$\sqrt{px+q} = \sqrt{rx+s} + t.$$

For each equation, list the values of p, q, r, s, and t. Then run the program and determine, by substitution, which of the possible solutions is in fact a solution of the given equation.

	P	*Q*	*R*	*S*	*T*	*Possible solutions*	*Solutions*
1. $\sqrt{3x+1} = \sqrt{6x+1} - 2$	____	____	____	____	____	________	________
2. $\sqrt{2x-1} = \sqrt{x-1} + 1$	____	____	____	____	____	________	________

(continued)

Using Technology
Computer Activities, p. 35

(Activity 13 continued)

	P	*Q*	*R*	*S*	*T*	*Possible solutions*	*Solutions*
3. $\sqrt{x+4} = \sqrt{2x-1} - 6$	____	____	____	____	____	________	________
4. $\sqrt{2x+1} = \sqrt{x} + 5$	____	____	____	____	____	________	________
5. $\sqrt{3x+4} = \sqrt{x} - 6$	____	____	____	____	____	________	________
6. $\sqrt{x+6} = \sqrt{x-2} + 2$	____	____	____	____	____	________	________
7. $\sqrt{2x+5} = \sqrt{x+3}$	____	____	____	____	____	________	________
8. $\sqrt{-x+2} = \sqrt{3x+4} + 5$	____	____	____	____	____	________	________

Before running the program for each equation below, rewrite the equation in the form $\sqrt{px+q} = \sqrt{rx+s} + t$.

	Rewritten equation	*Possible solutions*	*Solutions*
9. $\sqrt{2x-5} + 2 = \sqrt{4x-3}$	____________________	________	________
10. $\sqrt{4x} - \sqrt{2x+7} = 1$	____________________	________	________
11. $\sqrt{2x+1} - \sqrt{6x-1} = 0$	____________________	________	________
12. $1 = \sqrt{y} - \sqrt{y+2}$	____________________	________	________
13. $\sqrt{-2-3x} - \sqrt{-\frac{x}{2}} = 1$	____________________	________	________

ANALYSIS

To solve the equation $\sqrt{px+q} = \sqrt{rx+s} + t$ for x, first square both sides and rearrange terms as follows:

$$(\sqrt{px+q})^2 = (\sqrt{rx+s} + t)^2$$
$$px + q = rx + s + 2t\sqrt{rx+s} + t^2$$
$$(p-r)x + q - s - t^2 = 2t\sqrt{rx+s}$$

To simplify further manipulation, let $m = p - r$ and $n = q - s - t^2$ (lines 30 and 40 of the program). Then you obtain

$$mx + n = 2t\sqrt{rx+s}.$$

Now, by squaring both sides again and expressing the result in the standard form for a quadratic equation, you get

$$m^2x^2 + (2mn - 4rt^2)x + n^2 - 4st^2 = 0.$$

By letting

$$A = m^2, \quad B = 2mn - 4rt^2, \quad \text{and} \quad C = n^2 - 4st^2$$

(lines 50-70), you obtain

$$Ax^2 + Bx + C = 0.$$

This equation is solved, using the quadratic formula, in lines 130 and 140.

Using Technology
Using Algebra Plotter Plus, p. 37

Book 2, Lesson 3-10 • Solving Absolute Value Inequalities

The graphs of all equations having the form $y = a|bx + c| + d$ are V-shaped curves. Students can use such graphs, produced by a computer, to solve absolute-value equations and inequalities in a single variable.

For example, to solve $|x + 3| - 1 = 0$, students could use the Absolute Value Plotter to graph the equation $y = |x + 3| - 1$ and then determine the values of x for which the graph crosses the x-axis (that is, where $y = 0$).

Similarly, to solve $|2x - 1| < 3$, students should rewrite the inequality as $|2x - 1| - 3 < 0$, graph the equation $y = |2x - 1| - 3$, and then determine the values of x for which the graph lies below the x-axis (that is, where $y < 0$). Use this method to solve the A- or B-level Written Exercises for Lessons 2-4 and 2-5.

Book 2, Lesson 4-7 • Solving Polynomial Equations by Graphing

Have students use the Function Plotter program to check their solutions of Written Exercises 1–50 on pages 196–197. For example, to find the solutions of $6x^2 = 1 - x$ (Exercise 14), students should rewrite the equation as $6x^2 + x - 1 = 0$, graph the equation $y = 6x^2 + x - 1$, and then determine the values of x at which the graph intersects the x-axis (that is, where $y = 0$).

You might point out to students what happens when an equation has a root that occurs more than once. If the multiplicity of a root (the number of times the root occurs) is odd, the graph *crosses* the x-axis at that root. But if the multiplicity of a root is even, the graph is *tangent* to the x-axis at that root.

Book 2, Lesson 6-1 • Graphing to Find *n*th Roots of Real Numbers

You can have students approximate the nth root of a number a by using the Function Plotter program to graph $y = x^n - a$. For example, have students graph $y = x^3 + 4$. By zooming in on the point where the graph crosses the x-axis, students can obtain approximations of the point's x-coordinate, which is $\sqrt[3]{-4}$. In the diagram at the right below, the approximation is −1.58740.

You might wish to point out (or have students discover) that when n is even, the graph of $y = x^n - a$ (where a is a positive real number) crosses the x-axis at two points (producing a positive and a negative nth root), and when n is odd, the graph of $y = x^n - a$ (where a is any real number) crosses the x-axis at only one point (producing a single nth root).

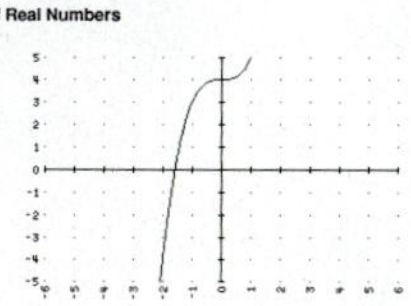

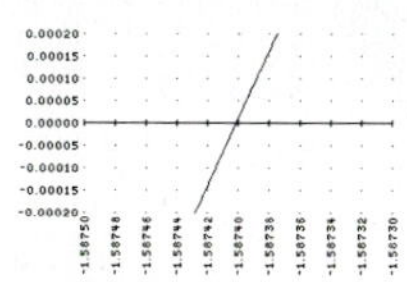

Using Models
Overhead Visual B

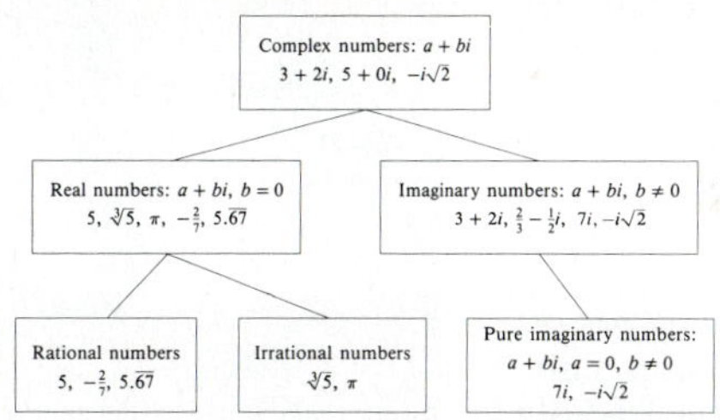

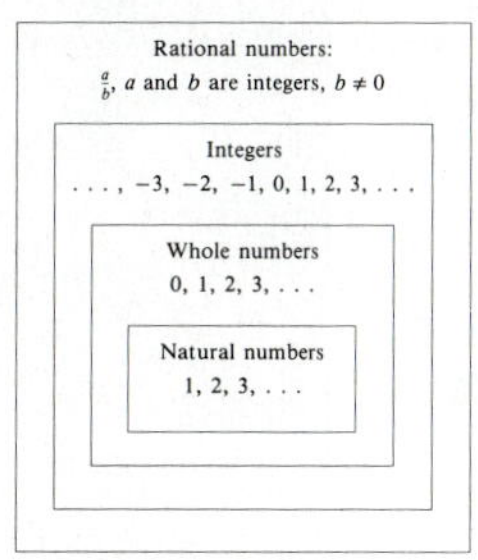

6 Irrational and Complex Numbers

The flow of an electrical current through a circuit can be described using a formula that includes the imaginary number i, defined as $\sqrt{-1}$.

258

Application

When an electrical current (I) flows through a circuit with a resistor (R), an inductor (L), and a capacitor (C), the electromotive force (E) is given by $E = E_0 e^{i\omega t}$ at time t. In the formula, ω is a constant that depends upon the values of R, L, and C; E_0 is the electromotive force at $t = 0$; e is an irrational number approximately equal to 2.718; and i is the imaginary number defined as $\sqrt{-1}$.

The imaginary number i also plays a central role in fractal geometry. Fractal geometry utilizes complex numbers, which have the form $a + bi$ where a and b are real. Just as real numbers can be associated with points on a line, complex numbers can be associated with points in a plane. By using a "complex plane" to graph an infinite sequence of complex numbers obtained through an iterative process, a fractal is created. Some fractals have a striking similarity to certain irregular forms found in nature, such as coastlines.

Group Activities
Students can investigate fractals and their applications. Students with programming experience can generate their own fractals.

References

Gleick, James. *Chaos: Making a New Science.* New York: Viking Press, 1987.

Peitgen, Heinz-Otto, and Dietmar Saupe, eds. *The Science of Fractal Images.* New York: Springer-Verlag, 1988.

Support Materials
Resource Book p. 209

Roots and Radicals

6-1 Roots of Real Numbers

Objective To find roots of real numbers.

A **square root** of a number b is a solution of the equation $x^2 = b$.
Every positive number b has two square roots, denoted $\sqrt{b}$ and $-\sqrt{b}$.

For example, the square roots of 25 are $\sqrt{25} = 5$ and $-\sqrt{25} = -5$, since $5^2 = 25$ and $(-5)^2 = 25$. The *positive* square root, $\sqrt{b}$, is called the **principal square root** of b. The principal square root of 25 is $\sqrt{25} = 5$.

Example 1 Simplify: **a.** $\sqrt{9}$ **b.** $-\sqrt{9}$ **c.** $\sqrt{\frac{1}{9}}$ **d.** $\sqrt{0.09}$

Solution **a.** 3 **b.** -3 **c.** $\frac{1}{3}$ **d.** 0.3

Since the square of a real number is never negative, the equation $x^2 = b$ has no real-number solution if $b < 0$. Therefore, *a negative number b has no real square roots.*

Example 2 Find the real roots of each equation. If there are none, say so.
a. $x^2 = 9$ **b.** $x^2 + 4 = 0$ **c.** $5x^2 = 15$

Solution

a. $x^2 = 9$
$x = \pm\sqrt{9} = \pm 3$
$\therefore$ the roots are 3 and -3.

b. $x^2 + 4 = 0$
$x^2 = -4$
$\therefore$ there are no real roots.

c. $5x^2 = 15$
$x^2 = 3$
$x = \pm\sqrt{3}$
$\therefore$ the roots are $\sqrt{3}$ and $-\sqrt{3}$.

In part (a) of Example 2, notice that the two solutions are written $\pm\sqrt{9}$, or ± 3. You read $\pm\sqrt{9}$ as "plus or minus the square root of 9."

Caution: Do not confuse solving $x^2 = 9$, as in Example 2(a), with simplifying $\sqrt{9}$, as in Example 1(a). It is correct to write $x = \pm\sqrt{9} = \pm 3$ in solving $x^2 = 9$. However, it is *incorrect* to write $\sqrt{9} = \pm 3$ in simplifying $\sqrt{9}$, since $\sqrt{9}$ represents only the *positive* square root of 9.

Teaching References

Lesson Commentary, pp. T101–T104
Assignment Guide, pp. T62–T65
Supplementary Materials
- Practice Masters 34–38
- Tests 22–26
- Resource Book
  - Practice Exercises, pp. 122–126
  - Tests, pp. 27–30
  - Enrichment Activity, p. 225
  - Application, p. 209
- Study Guide, pp. 89–104
- Computer Activities 13–14
- Test Generator
- Disk for Algebra

Alternate Test, p. T18

Teaching Suggestions p. T101

Warm-Up Exercises

Use a calculator to find the following to four decimal places.

1. $\sqrt{2}$ 1.4142
2. $\sqrt{17}$ 4.1231
3. $\sqrt{23}$ 4.7958
4. $\sqrt{31}$ 5.5678
5. $\sqrt{40}$ 6.3246
6. $\sqrt{47}$ 6.8557
7. $\sqrt{53}$ 7.2801
8. $\sqrt{69}$ 8.3066
9. $\sqrt{75}$ 8.6603
10. $\sqrt{88}$ 9.3808

Motivating the Lesson

Students should be familiar with the radical symbol from earlier math courses. Tell students that this lesson will extend the concept of root to higher orders, such as cube roots and fourth roots.

Chalkboard Examples

Evaluate each radical.

1. $\sqrt[3]{125}$ 5
2. $\sqrt[5]{-243}$ -3
3. $\sqrt[4]{0.0081}$ 0.3

Find the real roots of each equation. If there are none, say so.

4. $x^2 = 0.49$ $x = \pm 0.7$
5. $3x^2 = 432$
   $x^2 = 144$
   $x = \pm 12$
6. $5y^2 + 45 = 0$
   $5y^2 = -45$
   $y^2 = -9$
   no real roots
7. $4z^2 - 24 = 0$
   $4z^2 = 24$
   $z^2 = 6$
   $z = \pm\sqrt{6}$

A **cube root** of a number b is a solution of the equation $x^3 = b$.

Every number b, whether positive, negative, or zero, has exactly one real cube root, denoted $\sqrt[3]{b}$.

Example 3 illustrates that the cube root of a positive number is positive and the cube root of a negative number is negative.

Example 3 Simplify.

a. $\sqrt[3]{8}$ **b.** $\sqrt[3]{-27}$ **c.** $\sqrt[3]{10^6}$ **d.** $\sqrt[3]{a^9}$

Solution

a. Since $2^3 = 8$, $\sqrt[3]{8} = 2$.
b. Since $(-3)^3 = -27$, $\sqrt[3]{-27} = -3$.
c. Since $(10^2)^3 = 10^6$, $\sqrt[3]{10^6} = 10^2$, or 100.
d. Since $(a^3)^3 = a^9$, $\sqrt[3]{a^9} = a^3$.

Instead of defining the fourth, fifth, and higher roots of a number b separately, we will give a general definition of the nth root of b (where n is a positive integer).

1. An ***n*th root** of b is a solution of the equation $x^n = b$.
2. a. If n is even and $b > 0$, there are two real nth roots of b.
   The **principal** (or positive) ***n*th root** of b is denoted $\sqrt[n]{b}$.
   The other nth root of b is denoted $-\sqrt[n]{b}$.

   b. If n is even and $b = 0$, there is one nth root: $\sqrt[n]{0} = 0$.

   c. If n is even and $b < 0$, there is no real nth root of b.
3. If n is odd, there is exactly one real nth root of b, whether b is positive, negative, or zero.

Example 4 Simplify.

a. $\sqrt[4]{81}$ **b.** $\sqrt[5]{32}$ **c.** $\sqrt[5]{-32}$ **d.** $\sqrt[6]{-1}$

Solution

a. Since $3^4 = 81$, $\sqrt[4]{81} = 3$.
b. Since $2^5 = 32$, $\sqrt[5]{32} = 2$.
c. Since $(-2)^5 = -32$, $\sqrt[5]{-32} = -2$.
d. Since $x^6 = -1$ has no solution, $\sqrt[6]{-1}$ does not represent a real number. (See part 2(c) in the chart above.)

The symbol $\sqrt[n]{b}$ is called a **radical.** Each part of a radical is given a name, as indicated below.

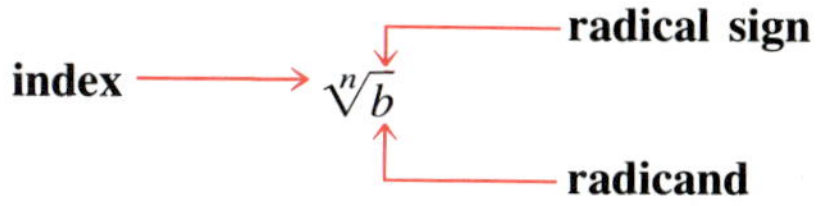

The index must be a positive integer. For a square root, the index 2 is usually omitted: $\sqrt{b}$ has the same meaning as $\sqrt[2]{b}$.

Here are some properties that will help you in simplifying radicals.

Properties of Radicals	***Examples***
1. $(\sqrt[n]{b})^n = b$, because $\sqrt[n]{b}$ satisfies the equation $x^n = b$.	$(\sqrt{7})^2 = 7$ $(\sqrt[3]{-5})^3 = -5$
2. $\sqrt[n]{b^n} = b$ if n is odd.	$\sqrt[3]{10^3} = 10$ $\sqrt[5]{x^5} = x$
3. $\sqrt[n]{b^n} = \lvert b\rvert$ if n is even, because the principal nth root is always nonnegative for even values of n.	$\sqrt{(-3)^2} = \lvert -3\rvert = 3$ $\sqrt{(x-1)^2} = \lvert x-1\rvert = \begin{cases} x-1 & \text{if } x \geq 1 \\ 1-x & \text{if } x < 1 \end{cases}$

Oral Exercises

Simplify each expression. If the expression does not represent a real number, say so.

1. $\sqrt{36}$ 6
2. $\sqrt[3]{-8}$ -2
3. $\sqrt{-\frac{1}{4}}$ Not real
4. $-\sqrt{\frac{1}{25}}$ $-\frac{1}{5}$
5. $-\sqrt[4]{16}$ -2
6. $(\sqrt[3]{5})^3$ 5
7. $(\sqrt{2})^2$ 2
8. $(-\sqrt{11})^2$ 11
9. $(-\sqrt[3]{13})^3$ -13
10. $\sqrt[4]{(-7)^4}$ 7
11. $\sqrt[5]{(-6)^5}$ -6
12. $\sqrt[6]{0}$ 0

Give the real roots of each equation. If an equation has no solution, say so.

13. $x^2 = 49$ ± 7
14. $x^2 = 1$ ± 1
15. $x^2 = -1$ No solution
16. $x^2 - 6 = 0$ $\pm\sqrt{6}$
17. $3x^2 = 27$ ± 3
18. $5x^2 = 20$ ± 2

Tell whether each equation is true for all real values of the variable.

19. $\sqrt{w^2} = \lvert w\rvert$ T
20. $\sqrt{y^4} = y^2$ T
21. $(\sqrt[4]{y})^4 = \lvert y\rvert$ T
22. $\sqrt[3]{x^3} = \lvert x\rvert$ F
23. $(\sqrt[4]{\lvert y\rvert})^4 = y$ F
24. $(\sqrt[5]{z})^5 = z$ T

Common Errors

Students often simplify $\sqrt{a^2 + b^2}$ incorrectly as $a + b$. When reteaching, remind students to check their results by squaring. Also, illustrate this concept with a numerical example such as $\sqrt{5^2 + 5^2} \neq 5 + 5 = 10$.

Check for Understanding

Here is a suggested use of the Oral Exercises to check students' understanding as you teach the lesson.

Oral Exs. 1–12: use after Example 4.

Oral Exs. 13–18: use after Example 2.

Oral Exs. 19–24: use after discussing the table of properties of radicals.

Guided Practice

Evaluate. If not real, say so.

1. $-\sqrt{49}$ -7
2. $\sqrt{-49}$ not real
3. $\sqrt[3]{-343}$ -7
4. $\sqrt[3]{343}$ 7
5. $\sqrt{0.09}$ 0.3
6. $\sqrt[4]{-1}$ not real
7. $\sqrt[3]{x^3}$ x
8. $\sqrt[4]{(-2)^4}$ 2

Solve.

9. $2x^2 = 2$ $\{1, -1\}$
10. $3x^3 + 81 = 0$ -3
11. $\sqrt{x^2} = 7$ $\{7, -7\}$
12. $-\sqrt{(2-k)^2} = 2k - 2$ 0

Summarizing the Lesson

In this lesson students learned how to find roots of real numbers and how to use roots to solve equations. Ask students to explain the properties of radicals listed on page 261.

Suggested Assignments

Average Algebra
262/1–33 odd
S 263/Mixed Review

Average Alg. and Trig.
262/1–33 odd
S 263/Mixed Review

Extended Algebra
262/1–33 odd
S 263/Mixed Review

Extended Alg. and Trig.
262/1–33 odd
S 263/Mixed Review

Supplementary Materials

Study Guide pp. 89–90

Tell whether each statement is true for all real values of the variable.

25. $\sqrt[3]{x}$ is always a real number. T

26. $\sqrt[6]{x}$ is always a real number. F

27. $\sqrt{x + 1}$ is always a real number. F

28. $\sqrt[3]{x + 1}$ is always a real number. T

29. $\sqrt{x^2 + 1}$ is always a real number. T

30. $\sqrt[4]{x^2 + 1}$ is always a real number. T

31. For every real number x, $\sqrt[4]{x^4} = |x|$ and $\sqrt[5]{x^5} = x$. Using a numerical example, explain why the absolute value symbol must be used in the first case but not in the second.
Answers may vary. For example, $\sqrt[4]{(-1)^4} = 1 = |-1|$ and $\sqrt[5]{(-1)^5} = -1$.

Written Exercises

Simplify each expression. If the expression does not represent a real number, say so.

A

1. a. $\sqrt{16}$ 4 b. $-\sqrt{16}$ −4 c. $\sqrt{-16}$ Not real d. $\sqrt[4]{16}$ 2

2. a. $\sqrt{64}$ 8 b. $\sqrt{-64}$ Not real c. $\sqrt[3]{64}$ 4 d. $\sqrt[3]{-64}$ −4

3. a. $\sqrt{81}$ 9 b. $-\sqrt{81}$ −9 c. $\sqrt{-81}$ Not real d. $\sqrt[4]{81}$ 3

4. a. $\sqrt{144}$ 12 b. $\sqrt{-144}$ Not real c. $-\sqrt{144}$ −12 d. $\sqrt[4]{-144}$ Not real

5. a. $\sqrt{0.01}$ 0.1 b. $\sqrt{-0.01}$ Not real c. $\sqrt[3]{0.001}$ 0.1 d. $\sqrt[3]{-0.001}$ −0.1

6. a. $\sqrt{0.04}$ 0.2 b. $-\sqrt{0.04}$ −0.2 c. $\sqrt{0.0004}$ 0.02 d. $\sqrt{-0.0004}$ Not real

7. a. $\sqrt{7^2}$ 7 b. $\sqrt[3]{7^3}$ 7 c. $\sqrt[4]{(-7)^4}$ 7 d. $\sqrt[5]{(-7)^5}$ −7

8. a. $\sqrt{5^2}$ 5 b. $\sqrt{-5^2}$ Not real c. $\sqrt[3]{(-5)^3}$ −5 d. $\sqrt[3]{-5^3}$ −5

9. a. $\sqrt{\frac{1}{64}}$ $\frac{1}{8}$ b. $\frac{1}{\sqrt{64}}$ $\frac{1}{8}$ c. $\sqrt[3]{-\frac{1}{64}}$ $-\frac{1}{4}$ d. $-\frac{1}{\sqrt[3]{64}}$ $-\frac{1}{4}$

10. a. $\sqrt{\frac{1}{16}}$ $\frac{1}{4}$ b. $\sqrt{\frac{81}{16}}$ $\frac{9}{4}$ c. $\sqrt[4]{\frac{1}{16}}$ $\frac{1}{2}$ d. $\sqrt[4]{\frac{81}{16}}$ $\frac{3}{2}$

11. a. $\sqrt{10^2}$ 10 b. $\sqrt{10^4}$ 10^2 c. $\sqrt{10^6}$ 10^3 d. $\sqrt{10^{20}}$ 10^{10}

12. a. $\sqrt[3]{10^{-3}}$ $\frac{1}{10}$ b. $\sqrt[3]{10^{-6}}$ $\frac{1}{10^2}$ c. $\sqrt[3]{10^{-9}}$ $\frac{1}{10^3}$ d. $\sqrt[3]{10^{-30}}$ $\frac{1}{10^{10}}$

13. a. $\sqrt{a^2}$ $|a|$ b. $\sqrt{a^4}$ a^2 c. $\sqrt[3]{a^6}$ a^2 d. $\sqrt[6]{a^6}$ $|a|$

14. a. $\sqrt{-a^2}$ Not real b. $\sqrt{(-a)^2}$ $|a|$ c. $\sqrt[4]{a^4}$ $|a|$ d. $\sqrt{a^6}$ $|a^3|$

Find the real roots of each equation. If there are none, say so.

15. $x^2 = 144$ ± 12

16. $y^2 = 0$ 0

17. $x^2 + 9 = 0$ None

18. $y^2 - 7 = 0$ $\pm \sqrt{7}$

19. $9x^2 = 4$ $\pm \frac{2}{3}$

20. $25y^2 = -16$ None

21. $16y^2 = 25$ $\pm \frac{5}{4}$

22. $9x^2 - 81 = 0$ ± 3

23. $4 - 16x^2 = 0$ $\pm \frac{1}{2}$

24. $0 = 4 + 16x^2$ None

25. $81 - 9x^2 = 0$ ± 3

26. $25y^2 + 16 = 17$ $\pm \frac{1}{5}$

For what values of the variable is each equation true?

Sample $\sqrt{(x-3)^2} = x - 3$

Solution $\sqrt{(x-3)^2} = |x-3| = x - 3$ if $x \geq 3$
(Note that if $x < 3$, $|x - 3| = -(x-3) = 3 - x$.)

B **27.** $\sqrt{(x+5)^2} = x + 5$ $x \geq -5$
28. $\sqrt{(a-2)^2} = 2 - a$ $a \leq 2$
29. $\sqrt[3]{(n+1)^3} = n + 1$ All reals
30. $\sqrt{y^4} = y^2$ All reals
31. $\sqrt{c^2 + 4c + 4} = |c + 2|$ All reals
32. $\sqrt{b^2 - 2b + 1} = b - 1$ $b \geq 1$

For what values of x does each expression represent a real number?

33. a. $\sqrt{x+1}$ $x \geq -1$ **b.** $\sqrt{x-1}$ $x \geq 1$ **c.** $\sqrt[3]{x-1}$ All reals **d.** $\sqrt{x^2 - 1}$ $x \leq -1$ or $x \geq 1$

34. a. $\sqrt{4-x}$ $x \leq 4$ **b.** $\sqrt{4-x^2}$ $-2 \leq x \leq 2$ **c.** $\sqrt[3]{4-x^2}$ All reals **d.** $\sqrt{4+x^2}$ All reals

C **35.** $\sqrt{x^3 - 9x}$ $-3 \leq x \leq 0$ or $x \geq 3$
36. $\sqrt{16x - x^2}$ $0 \leq x \leq 16$
37. $\sqrt{\sqrt{x} - x}$ $0 \leq x \leq 1$
38. $\sqrt{x - \sqrt{x}}$ $x \geq 1$

Mixed Review Exercises

Solve. If an equation has no solution, say so.

1. $\frac{x}{x-1} = \frac{2x}{x+2}$ $\{0, 4\}$
2. $-3y = y^2$ $\{-3, 0\}$
3. $\frac{3n-1}{4} + \frac{n+1}{2} = 4$ $\{3\}$
4. $|5 - 2w| = 1$ $\{2, 3\}$
5. $\frac{d}{d+2} + \frac{3}{d} = 2$ $\{-3, 2\}$
6. $\frac{u^2}{3} = \frac{u+1}{6}$ $\left\{-\frac{1}{2}, 1\right\}$
7. $3(2m - 1) = 4m + 7$ $\{5\}$
8. $2p^2 = 12 - 5p$ $\left\{-4, \frac{3}{2}\right\}$
9. $3k + 6 = 3(k + 4)$ No solution
10. $\frac{1}{f} + \frac{1}{f-1} = \frac{1}{f^2 - f}$ No solution

Calculator Key-In

Use the square root key, usually labeled $\sqrt{x}$, to simplify each radical. (In Exercises 5 and 6, use scientific notation.)

1. $\sqrt{11{,}025}$ 105
2. $\sqrt{1{,}010{,}025}$ 1005
3. $\sqrt{0.9801}$ 0.99
4. $\sqrt{0.998001}$ 0.999
5. $\sqrt{40{,}000{,}000{,}000}$ 2×10^5
6. $\sqrt{0.000000000025}$ 5×10^{-6}

7. The expressions $\sqrt{2 - \sqrt{3}}$ and $\frac{\sqrt{6} - \sqrt{2}}{2}$ represent the same real number. Use your calculator to confirm this fact. Both expressions equal 0.51763809 on a calculator.

Calculator Key-In Commentary

Notice the number patterns in Exercises 1–2 and Exercises 3–4. Calculators can be very useful in finding number patterns. Students may use their calculators to explore other similar patterns.

Explorations p. 837

Teaching Suggestions p. T101

Reading Algebra p. T101

Suggested Extensions p. T102

Group Activities p. T102

Warm-Up Exercises

Find the prime factorization.

1. 18 $3^2 \cdot 2$
2. 75 $5^2 \cdot 3$
3. 32 2^5
4. 220 $2^2 \cdot 5 \cdot 11$
5. 176 $2^4 \cdot 11$

Motivating the Lesson

Students may be familiar with the properties presented in this lesson with regard to square roots. This lesson will now extend these properties to higher order roots.

6-2 Properties of Radicals

Objective To simplify expressions involving radicals.

Since $\sqrt{4 \cdot 9} = \sqrt{36} = 6$ and $\sqrt{4} \cdot \sqrt{9} = 2 \cdot 3 = 6$, you can see that

$$\sqrt{4 \cdot 9} = 6 = \sqrt{4} \cdot \sqrt{9}.$$

Similarly,

$$\sqrt{\frac{4}{9}} = \frac{2}{3} = \frac{\sqrt{4}}{\sqrt{9}}.$$

These examples illustrate two important properties of radicals, stated in the chart below.

Product and Quotient Properties of Radicals

If $\sqrt[n]{a}$ and $\sqrt[n]{b}$ are real numbers, then:

$$\textbf{1. } \sqrt[n]{ab} = \sqrt[n]{a} \cdot \sqrt[n]{b}$$

$$\textbf{2. } \sqrt[n]{\frac{a}{b}} = \frac{\sqrt[n]{a}}{\sqrt[n]{b}}$$

Caution: Do *not* assume that $\sqrt[n]{a+b} = \sqrt[n]{a} + \sqrt[n]{b}$ or $\sqrt[n]{a-b} = \sqrt[n]{a} - \sqrt[n]{b}$. For example, $\sqrt{9+16} \neq \sqrt{9} + \sqrt{16}$, and $\sqrt{4-x^2}$ is not equivalent to $2 - x$.

You are asked to prove the product and quotient properties of radicals in Exercises 67 and 68. As the following examples illustrate, these properties can be used to simplify expressions involving radicals.

Example 1 Simplify.

a. $\sqrt{98}$ **b.** $\sqrt[3]{25} \cdot \sqrt[3]{10}$ **c.** $\sqrt[3]{\frac{81}{8}}$ **d.** $\frac{\sqrt{60}}{\sqrt{5}}$

Solution

a. $\sqrt{98} = \sqrt{49 \cdot 2}$
$= \sqrt{49} \cdot \sqrt{2}$
$= 7\sqrt{2}$ ***Answer***

b. $\sqrt[3]{25} \cdot \sqrt[3]{10} = \sqrt[3]{250} = \sqrt[3]{125 \cdot 2}$
$= \sqrt[3]{125} \cdot \sqrt[3]{2}$
$= 5\sqrt[3]{2}$ ***Answer***

c. $\sqrt[3]{\frac{81}{8}} = \frac{\sqrt[3]{81}}{\sqrt[3]{8}}$
$= \frac{\sqrt[3]{27 \cdot 3}}{2}$
$= \frac{\sqrt[3]{27} \cdot \sqrt[3]{3}}{2}$
$= \frac{3\sqrt[3]{3}}{2}$ ***Answer***

d. $\frac{\sqrt{60}}{\sqrt{5}} = \sqrt{\frac{60}{5}}$
$= \sqrt{12}$
$= \sqrt{4 \cdot 3}$
$= \sqrt{4} \cdot \sqrt{3}$
$= 2\sqrt{3}$ ***Answer***

Example 2 Simplify. Assume that each radical represents a real number.

a. $\sqrt{2a^2b}$ b. $\sqrt{36w^3}$ c. $\sqrt[3]{\frac{x^5}{y^3}}$ d. $\sqrt{4a^2 + 4b^2}$

Solution

a. $\sqrt{2a^2b} = \sqrt{a^2} \cdot \sqrt{2b}$

$= |a|\sqrt{2b}$ ***Answer***

b. $\sqrt{36w^3} = \sqrt{36w^2 \cdot w}$

$= \sqrt{36w^2} \cdot \sqrt{w}$

$= 6w\sqrt{w}$ ***Answer***

(Since $\sqrt{36w^3}$ is real, w is nonnegative, and you can write $6w\sqrt{w}$ rather than $6|w|\sqrt{w}$.)

c. $\sqrt[3]{\frac{x^5}{y^3}} = \frac{\sqrt[3]{x^5}}{\sqrt[3]{y^3}} = \frac{\sqrt[3]{x^3 \cdot x^2}}{y} = \frac{\sqrt[3]{x^3} \cdot \sqrt[3]{x^2}}{y} = \frac{x\sqrt[3]{x^2}}{y}$ ***Answer***

d. $\sqrt{4a^2 + 4b^2} = \sqrt{4(a^2 + b^2)}$

$= \sqrt{4} \cdot \sqrt{a^2 + b^2}$

$= 2\sqrt{a^2 + b^2}$ ***Answer***

The next example illustrates a process known as **rationalizing the denominator,** in which a perfect square, cube, or other power is created in the denominator so that the radical expression can be written without a fraction in the radicand or a radical in the denominator.

Example 3 Simplify: a. $\sqrt{\frac{5}{3}}$ b. $\frac{4}{\sqrt[3]{c}}$

Solution

a. $\sqrt{\frac{5}{3}} = \sqrt{\frac{5}{3} \cdot \frac{3}{3}}$ ⟵ Numerator and denominator are multiplied by 3 so that the denominator will be a perfect square.

$= \sqrt{\frac{15}{9}}$

$= \frac{\sqrt{15}}{\sqrt{9}}$

$= \frac{\sqrt{15}}{3}$ ***Answer***

b. $\frac{4}{\sqrt[3]{c}} = \frac{4}{\sqrt[3]{c}} \cdot \frac{\sqrt[3]{c^2}}{\sqrt[3]{c^2}}$ ⟵ Numerator and denominator are multiplied by $\sqrt[3]{c^2}$ so that the radicand in the denominator will be a perfect cube.

$= \frac{4\sqrt[3]{c^2}}{\sqrt[3]{c^3}}$

$= \frac{4\sqrt[3]{c^2}}{c}$ ***Answer***

Chalkboard Examples

Simplify.

1. $\sqrt[5]{64}$ $\sqrt[5]{32 \cdot 2} = \sqrt[5]{32} \cdot \sqrt[5]{2} = 2\sqrt[5]{2}$
2. $\sqrt{18m^3}$ $\sqrt{9m^2 \cdot 2m} = \sqrt{9m^2} \cdot \sqrt{2m} = 3m\sqrt{2m}$
3. $\sqrt[3]{\frac{81}{8}}$ $\frac{\sqrt[3]{81}}{\sqrt[3]{8}} = \frac{\sqrt[3]{27 \cdot 3}}{\sqrt[3]{8}} = \frac{\sqrt[3]{27} \cdot \sqrt[3]{3}}{2} = \frac{3\sqrt[3]{3}}{2}$
4. $\sqrt[3]{\frac{5}{4}}$ $\sqrt[3]{\frac{5}{4} \cdot \frac{2}{2}} = \sqrt[3]{\frac{10}{8}} = \frac{\sqrt[3]{10}}{\sqrt[3]{8}} = \frac{\sqrt[3]{10}}{2}$
5. $5\sqrt{24} \cdot 3\sqrt{8}$ $5 \cdot 3\sqrt{24} \cdot \sqrt{8} = 15 \cdot \sqrt{192} = 15 \cdot \sqrt{64 \cdot 3} = 15 \cdot \sqrt{64} \cdot \sqrt{3} = 15 \cdot 8 \cdot \sqrt{3} = 120\sqrt{3}$
6. $\frac{12}{\sqrt{5a}}$ $\frac{12}{\sqrt{5a}} \cdot \frac{\sqrt{5a}}{\sqrt{5a}} = \frac{12\sqrt{5a}}{\sqrt{25a^2}} = \frac{12\sqrt{5a}}{5a}$

Find a value correct to the nearest hundredth.

7. $\sqrt[4]{169}$ $\sqrt{\sqrt{169}} = \sqrt{13} \approx 3.61$
8. $\sqrt[3]{(2600)^2}$ $(\sqrt[3]{2600})^2 = (\sqrt[3]{1000} \cdot \sqrt[3]{2.6})^2 \approx (10 \cdot 1.375)^2 \approx 100(1.8906) = 189.06$

Common Errors

Some students may mistakenly think that $\sqrt{x^9} = x^3$. When reteaching, explain that they are to find the square root of the entire expression, not the exponent. Instruct students to look for two identical factors whose product is x^9 and remind them that when multiplying expressions with the same base the exponents are added.

Check for Understanding

Here is a suggested use of the Oral Exercises to check students' understanding as you teach the lesson.

Oral Exs. 1–4: use after Example 1.

Oral Exs. 5–20: use after Example 3.

Oral Exs. 21–23: use after Example 5.

Guided Practice

Simplify.

1. $\sqrt{75}$ $5\sqrt{3}$
2. $\sqrt{\frac{8}{25}}$ $\frac{2}{5}\sqrt{2}$
3. $\frac{15}{\sqrt{3}}$ $5\sqrt{3}$
4. $9\sqrt{\frac{5}{27}}$ $\sqrt{15}$
5. $\sqrt[3]{\frac{1}{4}}$ $\frac{\sqrt[3]{2}}{2}$
6. $\sqrt[4]{162}$ $3\sqrt[4]{2}$
7. $\sqrt[5]{\frac{3}{32}}$ $\frac{\sqrt[5]{3}}{2}$
8. $\sqrt[3]{0.064}$ 0.4

Simplify. Then express to the nearest hundredth.

9. $\sqrt{3}\cdot\sqrt{15}$ $3\sqrt{5}$; 6.71
10. $\sqrt[3]{\frac{2}{9}}\cdot 3\sqrt[3]{4}$ $2\sqrt[3]{3}$; 2.88
11. $(-4\sqrt{6})^3$ $-384\sqrt{6}$; -940.60
12. $\frac{10\sqrt{5}}{5\sqrt{2}}$ $\sqrt{10}$; 3.16

The process of rationalizing denominators was very important before the invention of calculators. You can see why if you try to evaluate $\sqrt{\frac{5}{3}}$ using a table of square roots (see Table 1, page 810) instead of a calculator: $\frac{\sqrt{15}}{3}$ is easier to evaluate than $\sqrt{\frac{5}{3}}$ $\left(\text{or even } \frac{\sqrt{5}}{\sqrt{3}}\right)$.

Although rationalizing denominators is now less important for numerical work than it used to be, it is often used in mathematics classes as a convenient way to compare answers. This is also the reason that answers involving radicals are given in *simplest radical form*.

An expression containing *n*th roots is in **simplest radical form** if:

1. no radicand contains a factor (other than 1) that is a perfect *n*th power, and
2. every denominator has been rationalized, so that no radicand is a fraction and no radical is in a denominator.

All the answers in Examples 1, 2, and 3 are in simplest form.

You can use a calculator or a table of roots to obtain a decimal approximation of some radicals. Table 1 and Table 2 on pages 810–811 are provided for this purpose.

Example 4 Give a decimal approximation to the nearest hundredth for each radical.

a. $\sqrt{53}$ **b.** $\sqrt[3]{7100}$

Solution **a.** Use Table 1. Find 5.3 and look under $\sqrt{10N}$.

$\sqrt{53} \approx 7.28$ ***Answer***

b. Use Table 2.

$\sqrt[3]{7100} = \sqrt[3]{1000}\cdot\sqrt[3]{7.1} \approx 10(1.922) = 19.22$ ***Answer***

The following two theorems are often useful in numerical work with radicals. You are asked to prove these theorems in Exercises 69 and 70.

Theorem 1. If each radical represents a real number, then $\sqrt[nq]{b} = \sqrt[n]{\sqrt[q]{b}}$.

Theorem 2. If $\sqrt[n]{b}$ represents a real number, then $\sqrt[n]{b^m} = (\sqrt[n]{b})^m$.

The first of these theorems allows you to transform a radical with a large index into two radicals with smaller indexes. The second allows you to transform a root of a power into a power of a root.

Example 5 Give a decimal approximation to the nearest hundredth for each radical.

a. $\sqrt[4]{100}$ b. $\sqrt[3]{170^2}$

Solution a. $\sqrt[4]{100} = \sqrt{\sqrt{100}} = \sqrt{10} \approx 3.16$

b. $\sqrt[3]{170^2} = (\sqrt[3]{170})^2 \approx (5.540)^2 \approx 30.69$

To investigate how you can use your calculator to find roots of numbers, try the Calculator Key-Ins on pages 263 and 269.

Oral Exercises

Tell whether each radical is in simplest form. Explain your answer.

1. $\sqrt{\frac{1}{3}}$ No
2. $\frac{\sqrt{6}}{5}$ Yes
3. $\sqrt{72}$ No
4. $\frac{2}{\sqrt[3]{4}}$ No

Simplify.

5. $\sqrt{75}$ $5\sqrt{3}$
6. $\sqrt{24}$ $2\sqrt{6}$
7. $\sqrt{600}$ $10\sqrt{6}$
8. $\sqrt{98}$ $7\sqrt{2}$
9. $\sqrt{\frac{3}{16}}$ $\frac{\sqrt{3}}{4}$
10. $\frac{1}{\sqrt{5}}$ $\frac{\sqrt{5}}{5}$
11. $\frac{2}{\sqrt{2}}$ $\sqrt{2}$
12. $\sqrt{\frac{2}{3}}$ $\frac{\sqrt{6}}{3}$
13. $\sqrt{2} \cdot \sqrt{6}$ $2\sqrt{3}$
14. $\sqrt{10} \cdot \sqrt{15}$ $5\sqrt{6}$
15. $\frac{\sqrt{27}}{\sqrt{3}}$ 3
16. $\frac{\sqrt{56}}{\sqrt{7}}$ $2\sqrt{2}$
17. $\sqrt[3]{40}$ $2\sqrt[3]{5}$
18. $\sqrt[3]{\frac{16}{27}}$ $\frac{2\sqrt[3]{2}}{3}$
19. $\sqrt[3]{8} \cdot \sqrt[3]{-8}$ -4
20. $\frac{\sqrt[3]{48}}{\sqrt[3]{2}}$ $2\sqrt[3]{3}$
21. Can you apply the product property of radicals to $\sqrt{6} \cdot \sqrt[3]{5}$? Explain your answer. No; the indexes are not the same.
22. Does $\sqrt{x^2 + y^2} = x + y$? Explain. No; $(x + y)^2 = x^2 + 2xy + y^2$.
23. Does $\sqrt{4 - x^2} = 2 - x$? Explain. No; $(2 - x)^2 = 4 - 4x + x^2$.

Written Exercises

Simplify.

A

1. $\sqrt{52}$ $2\sqrt{13}$
2. $\sqrt{125}$ $5\sqrt{5}$
3. $\sqrt{162}$ $9\sqrt{2}$
4. $\sqrt{363}$ $11\sqrt{3}$
5. $\sqrt{196}$ 14
6. $\sqrt{324}$ 18
7. $\sqrt{\frac{8}{9}}$ $\frac{2\sqrt{2}}{3}$
8. $\sqrt{\frac{50}{49}}$ $\frac{5\sqrt{2}}{7}$
9. $\sqrt{\frac{4}{3}}$ $\frac{2\sqrt{3}}{3}$

Summarizing the Lesson

In this lesson students learned to use the product and quotient properties of radicals to simplify expressions involving radicals. Ask students to discuss the differences in rationalizing a denominator with a square root and with higher roots.

Using a Calculator

A calculator will be especially helpful for Written Exercises 33–38 on page 268.

Suggested Assignments

Average Algebra
Day 1: 267/2–46 even
S 263/28–34 even
Day 2: 268/48–66 even
Assign with Lesson 6-3.

Average Alg. and Trig.
267/2–62 even

Extended Algebra
267/2–66 even
S 263/35–37

Extended Alg. and Trig.
267/2–66 even
S 263/35–37

Supplementary Materials

Study Guide	pp. 91–92
Practice Master	34
Resource Book	p. 122

Additional Answers
Written Exercises

67. a. $(\sqrt[n]{a}\cdot\sqrt[n]{b})^n = (\sqrt[n]{a})^n\cdot(\sqrt[n]{b})^n = ab$

b. If n is odd, $\sqrt[n]{a}\cdot\sqrt[n]{b}$ is the only root and therefore the principal root. If n is even, $\sqrt[n]{a}$ and $\sqrt[n]{b}$ are both nonnegative; thus $\sqrt[n]{a}\cdot\sqrt[n]{b}$ is the nonnegative nth root of ab.

68. $\left(\frac{\sqrt[n]{a}}{\sqrt[n]{b}}\right)^n = \frac{(\sqrt[n]{a})^n}{(\sqrt[n]{b})^n} = \frac{a}{b}$.

Thus $\frac{\sqrt[n]{a}}{\sqrt[n]{b}}$ is an nth root of $\frac{a}{b}$. If n is odd, $\frac{\sqrt[n]{a}}{\sqrt[n]{b}}$ is the only nth root of $\frac{a}{b}$. If n is even, $\sqrt[n]{a}$ and $\sqrt[n]{b}$ are both nonnegative, and so is $\frac{\sqrt[n]{a}}{\sqrt[n]{b}}$.

69. 1. $(\sqrt[n]{(\sqrt[q]{b})})^{nq} = \left[(\sqrt[n]{(\sqrt[q]{b})})^n\right]^q$
(Third law of exponents, p. 171)
2. $= (\sqrt[q]{b})^q$
(Property 1, p. 261)
3. $= b$
(Property 1, p. 261)

Thus, $\sqrt[n]{\sqrt[q]{b}}$ is an nqth root of b. If $b < 0$, then n and q are odd and there is only one root. If $b \geq 0$, then $\sqrt[q]{b}$ is nonnegative and so is $\sqrt[n]{\sqrt[q]{b}}$.

Simplify.

10. $\sqrt{\frac{9}{5}}$ $\frac{3\sqrt{5}}{5}$
11. $\frac{4}{\sqrt{2}}$ $2\sqrt{2}$
12. $\frac{6}{\sqrt{3}}$ $2\sqrt{3}$
13. $\frac{\sqrt{270}}{\sqrt{6}}$ $3\sqrt{5}$
14. $\frac{\sqrt{96}}{\sqrt{3}}$ $4\sqrt{2}$
15. $\sqrt{30}\cdot\sqrt{42}$ $6\sqrt{35}$
16. $\sqrt{35}\cdot\sqrt{21}$ $7\sqrt{15}$
17. $\sqrt{6}\cdot\sqrt{\frac{2}{3}}$ 2
18. $\sqrt{15}\cdot\sqrt{\frac{3}{5}}$ 3
19. $\sqrt[3]{250}$ $5\sqrt[3]{2}$
20. $\sqrt[3]{135}$ $3\sqrt[3]{5}$
21. $\sqrt[3]{\frac{5}{4}}$ $\frac{\sqrt[3]{10}}{2}$
22. $\sqrt[3]{\frac{2}{9}}$ $\frac{\sqrt[3]{6}}{3}$
23. $\frac{9\sqrt{2}}{\sqrt{18}}$ 3
24. $\frac{4\sqrt{3}}{\sqrt{12}}$ 2
25. $(2\sqrt{7})^2$ 28
26. $(3\sqrt{6})^2$ 54
27. $\sqrt[3]{45}\cdot\sqrt[3]{12}$ $3\sqrt[3]{20}$
28. $\sqrt[3]{20}\cdot\sqrt[3]{14}$ $2\sqrt[3]{35}$
29. $\frac{\sqrt[3]{60}}{\sqrt[3]{36}}$ $\frac{\sqrt[3]{45}}{3}$
30. $\frac{\sqrt[3]{175}}{\sqrt[3]{50}}$ $\frac{\sqrt[3]{28}}{2}$

31. a. $\sqrt{32}$ $4\sqrt{2}$ **b.** $\sqrt[3]{32}$ $2\sqrt[3]{4}$ **c.** $\sqrt[4]{32}$ $2\sqrt[4]{2}$ **d.** $\sqrt[5]{32}$ 2
32. a. $\sqrt{\frac{3}{8}}$ $\frac{\sqrt{6}}{4}$ **b.** $\sqrt[3]{\frac{3}{8}}$ $\frac{\sqrt[3]{3}}{2}$ **c.** $\sqrt[4]{\frac{3}{8}}$ $\frac{\sqrt[4]{6}}{2}$ **d.** $\sqrt[5]{\frac{3}{8}}$ $\frac{\sqrt[5]{12}}{2}$

Give a decimal approximation to the nearest hundredth for each radical. Use a calculator or Tables 1 and 2 on pages 810–811.

33. $\sqrt{39}$ 6.24
34. $\sqrt{870}$ 29.50
35. $\sqrt[3]{450}$ 7.66
36. $\sqrt[3]{7300}$ 19.40
37. $\sqrt[4]{144}$ 3.46
38. $\sqrt[4]{900}$ 5.48

Simplify. Assume that each radical represents a real number.

39. $\sqrt{18x^2}$ $3|x|\sqrt{2}$
40. $\sqrt{12x^5}$ $2x^2\sqrt{3x}$
41. $\sqrt[3]{375a^5}$ $5a\sqrt[3]{3a^2}$
42. $\sqrt[3]{16c^4}$ $2c\sqrt[3]{2c}$
43. $\sqrt{\frac{x^2}{y^3}}$ $\frac{|x|\sqrt{y}}{y^2}$
44. $\sqrt{\frac{y^2}{x^5}}$ $\frac{|y|\sqrt{x}}{x^3}$
45. $\sqrt[3]{\frac{27a}{4b^4}}$ $\frac{3\sqrt[3]{2ab^2}}{2b^2}$
46. $\sqrt[3]{\frac{8c}{9d^5}}$ $\frac{2\sqrt[3]{3cd}}{3d^2}$

B
47. $\sqrt{16a + 16b}$ $4\sqrt{a+b}$
48. $\sqrt{9a^2 - 9b^2}$ $3\sqrt{a^2-b^2}$
49. $\sqrt{2a^2 + 4a + 2}$ $|a+1|\sqrt{2}$
50. $\sqrt{3x^2 - 12x + 12}$ $|x-2|\sqrt{3}$

Evaluate the following radicals if $x = 4$, $y = 3$, and $z = 8$.

51. $\sqrt{x^{-1}y^{-2}}$ $\frac{1}{6}$
52. $\sqrt[3]{(xz)^{-1}}$ $\frac{\sqrt[3]{2}}{4}$
53. $\sqrt[3]{x^{-1} + z^{-1}}$ $\frac{\sqrt[3]{3}}{2}$
54. $\sqrt{x^{-2} + y^{-2}}$ $\frac{5}{12}$
55. $\sqrt[3]{(y - x)^{-1}}$ -1
56. $\sqrt{(x + z)^{-1}}$ $\frac{\sqrt{3}}{6}$
57. $\sqrt[6]{(xy)^z}$ $12\sqrt[3]{12}$
58. $\sqrt[4]{xy^z}$ $9\sqrt{2}$

Simplify, giving answers with no negative exponents. Assume that each radical represents a real number.

59. $\sqrt{27x^3y^{-2}}$ $\frac{3x\sqrt{3x}}{|y|}$
60. $\sqrt[3]{32a^{-3}b^4}$ $\frac{2b\sqrt[3]{4b}}{a}$
61. $\sqrt{3a} \cdot \sqrt{15a^{-2}}$ $\frac{3\sqrt{5a}}{a}$
62. $\sqrt{2x^{-3}} \cdot \sqrt{6x}$ $\frac{2\sqrt{3}}{x}$
63. $\sqrt[3]{6y^{-4}} \cdot \sqrt[3]{9y^2}$ $\frac{3\sqrt[3]{2y}}{y}$
64. $\sqrt[3]{6c} \cdot \sqrt[3]{8c^{-5}}$ $\frac{2\sqrt[3]{6c^2}}{c^2}$
65. $\frac{\sqrt{2x^{-1}}}{\sqrt{8x}}$ $\frac{1}{2x}$
66. $\frac{\sqrt{27x}}{\sqrt{3x^{-1}}}$ $3x$

C 67. The purpose of this exercise is to prove the product property of radicals.

a. Using the exponent law $(xy)^n = x^n \cdot y^n$, show that $(\sqrt[n]{a} \cdot \sqrt[n]{b})^n = ab$.

b. Part (a) shows that $\sqrt[n]{a} \cdot \sqrt[n]{b}$ is a solution of $x^n = ab$ and is therefore an nth root of ab. Prove that it is the principal nth root of ab.

68. Prove the quotient property of radicals. Follow the method suggested in Exercise 67.

69. Prove the first theorem on page 266. (*Hint:* Show that $\sqrt[n]{\sqrt[q]{b}}$ is an nqth root of b by justifying these statements:

$$(\sqrt[n]{\sqrt[q]{b}})^{nq} = [(\sqrt[n]{\sqrt[q]{b}})^n]^q = (\sqrt[q]{b})^q = b$$

Then show that $\sqrt[n]{\sqrt[q]{b}}$ is the principal nqth root by considering the cases $b \geq 0$ and $b < 0$. Note that if $b < 0$, then q and n are both odd.)

70. Prove the second theorem on page 266. (*Hint:* Show that $(\sqrt[n]{b})^m$ is an nth root of b^m by justifying these statements:

$$[(\sqrt[n]{b})^m]^n = [(\sqrt[n]{b})^n]^m = b^m$$

Then show that it is the principal nth root by considering the cases n even and n odd.)

Additional Answers Written Exercises

70. 1. $[(\sqrt[n]{b})^m]^n = (\sqrt[n]{b})^{mn}$ (Third law of exponents, p. 171)
2. $= (\sqrt[n]{b})^{nm}$ (Comm. prop. of mult.)
3. $= [(\sqrt[n]{b})^n]^m$ (Third law of exponents, p. 171)
4. $= b^m$ (Property 1, p. 261)

Thus $(\sqrt[n]{b})^m$ is an nth root of b^m. If n is odd, there is only one nth root of b^m. If n is even, then $\sqrt[n]{b}$ is nonnegative and so is $(\sqrt[n]{b})^m$.

Calculator Key-In

Example 5(a) in Lesson 6-2 illustrates a method for finding the fourth root of a number using square roots. Use your calculator's square root key to find a decimal approximation for each radical.

1. $\sqrt[4]{897}$ 5.47
2. $\sqrt[4]{523.72}$ 4.78
3. $\sqrt[8]{0.002}$ 0.46
4. $\sqrt[16]{8.3 \times 10^{-80}}$ 1.14×10^{-5}
5. Can you use the square root key to find $\sqrt[10]{1024}$? Explain. No; 10 is not a power of 2.

You can evaluate the nth root of a positive number for any n if your calculator has an $\sqrt[x]{y}$ key. (Some calculators instead have an $x^{1/y}$ key or a combination of INV and y^x keys.) Investigate how your calculator works by finding $\sqrt[3]{8}$, which you know is 2. Then find a decimal approximation to the nearest hundredth for each radical.

6. $\sqrt[3]{2463}$ 13.50
7. $\sqrt[3]{479.552}$ 7.83
8. $\sqrt[6]{0.518}$ 0.90
9. $\sqrt[20]{4.5 \times 10^{40}}$ 107.81
10. What value does $\sqrt[n]{n}$ approach as n gets very large? 1

Calculator Key-In Commentary

Although it is possible to take the nth root of a negative number if n is odd, some calculators display an error message. If that is the case, tell students to use the fact that $\sqrt[n]{-x} = -\sqrt[n]{x}$ for $x > 0$ and n odd.

Teaching Suggestions p. T102

Suggested Extensions p. T102

Warm-Up Exercises

Simplify.

1. $3a + 4b - 8a + 9$ $-5a + 4b + 9$
2. $5(x + 3) - 4(x - 9)$ $x + 51$
3. $3x^2 + 5x + 4 - 9x^2 + 7x - 2$ $-6x^2 + 12x + 2$
4. $3x^2 + (x + 2)^2 - 4x - 4$ $4x^2$
5. $4z - (5z + 1) - (2z - 5)$ $-3z + 4$

Motivating the Lesson

Tell students that calculator values of most expressions involving radicals are not exact. In some situations there may be a need for the value to be expressed exactly in radical form. Methods for doing this are the topic of today's lesson.

Thinking Skills

The method of combining like terms to simplify algebraic expressions is now *transferred* as a method for simplifying radical expressions.

Common Errors

When simplifying expressions such as $\sqrt{50} + \sqrt{98}$, students may need to be reminded that like radicals are necessary before addition can be performed. When reteaching, compare $\sqrt{50} + \sqrt{98} = 5\sqrt{2} + 7\sqrt{2} = 12\sqrt{2}$ to $5x + 7x = 12x$.

6-3 Sums of Radicals

Objective To simplify expressions involving sums of radicals.

Two radicals with the same index and radicand are called *like radicals*. You can apply the distributive property to add or subtract like radicals in the same way as like terms.

Combine like terms:

$$5x + y - 4x = (5x - 4x) + y$$
$$= x + y$$

Combine like radicals:

$$5\sqrt{2} + \sqrt[3]{2} - 4\sqrt{2} = (5\sqrt{2} - 4\sqrt{2}) + \sqrt[3]{2}$$
$$= \sqrt{2} + \sqrt[3]{2}$$

You cannot combine $\sqrt{2}$ and $\sqrt[3]{2}$ because they are *not* like radicals; they have different indexes. Example 1 shows that sometimes you can combine unlike radicals if you first simplify them so that they become like radicals.

Example 1 Simplify.

a. $\sqrt{8} + \sqrt{98}$ **b.** $\sqrt[3]{81} - \sqrt[3]{24}$ **c.** $\sqrt{\frac{32}{3}} + \sqrt{\frac{2}{3}}$

Solution

a. $\sqrt{8} + \sqrt{98} = \sqrt{4 \cdot 2} + \sqrt{49 \cdot 2}$
$= 2\sqrt{2} + 7\sqrt{2}$
$= 9\sqrt{2}$ ***Answer***

b. $\sqrt[3]{81} - \sqrt[3]{24} = \sqrt[3]{27 \cdot 3} - \sqrt[3]{8 \cdot 3}$
$= 3\sqrt[3]{3} - 2\sqrt[3]{3}$
$= \sqrt[3]{3}$ ***Answer***

c. $\sqrt{\frac{32}{3}} + \sqrt{\frac{2}{3}} = \sqrt{\frac{32 \cdot 3}{3 \cdot 3}} + \sqrt{\frac{2 \cdot 3}{3 \cdot 3}}$
$= \sqrt{\frac{96}{9}} + \sqrt{\frac{6}{9}}$
$= \frac{\sqrt{96}}{\sqrt{9}} + \frac{\sqrt{6}}{\sqrt{9}}$
$= \frac{\sqrt{16 \cdot 6}}{3} + \frac{\sqrt{6}}{3}$
$= \frac{4\sqrt{6}}{3} + \frac{\sqrt{6}}{3}$
$= \frac{5\sqrt{6}}{3}$ ***Answer***

Example 2 Simplify.

a. $\sqrt{6}(\sqrt{2} + \sqrt{3})$ **b.** $\dfrac{\sqrt{21} + \sqrt{15}}{\sqrt{3}}$

Solution **a.** $\sqrt{6}(\sqrt{2} + \sqrt{3}) = \sqrt{6} \cdot \sqrt{2} + \sqrt{6} \cdot \sqrt{3}$

$= \sqrt{12} + \sqrt{18}$

$= \sqrt{4 \cdot 3} + \sqrt{9 \cdot 2}$

$= 2\sqrt{3} + 3\sqrt{2}$ ***Answer***

b. $\dfrac{\sqrt{21} + \sqrt{15}}{\sqrt{3}} = \dfrac{\sqrt{21}}{\sqrt{3}} + \dfrac{\sqrt{15}}{\sqrt{3}}$

$= \sqrt{\dfrac{21}{3}} + \sqrt{\dfrac{15}{3}}$

$= \sqrt{7} + \sqrt{5}$ ***Answer***

Example 3 Simplify. Assume that each radical represents a real number.

a. $\sqrt{12x^5} - x\sqrt{3x^3} + 5x^2\sqrt{3x}$ **b.** $\sqrt{6y} - \dfrac{\sqrt{3y}}{\sqrt{2}}$

Solution **a.** $\sqrt{12x^5} - x\sqrt{3x^3} + 5x^2\sqrt{3x} = \sqrt{4x^4 \cdot 3x} - x\sqrt{x^2 \cdot 3x} + 5x^2\sqrt{3x}$

$= 2x^2\sqrt{3x} - x^2\sqrt{3x} + 5x^2\sqrt{3x}$

$= 6x^2\sqrt{3x}$ ***Answer***

b. $\sqrt{6y} - \dfrac{\sqrt{3y}}{\sqrt{2}} = \sqrt{6y} - \dfrac{\sqrt{3y}}{\sqrt{2}} \cdot \dfrac{\sqrt{2}}{\sqrt{2}}$

$= \sqrt{6y} - \dfrac{\sqrt{6y}}{2}$

$= \dfrac{2\sqrt{6y} - \sqrt{6y}}{2}$

$= \dfrac{\sqrt{6y}}{2}$ ***Answer***

Oral Exercises

Simplify. If no simplification is possible, say so.

1. $\sqrt{5} + \sqrt{10}$ Not possible **2.** $\sqrt{7} + \sqrt[3]{7}$ Not possible **3.** $\sqrt{5} + \sqrt{2} + 2\sqrt{5}$ $3\sqrt{5} + \sqrt{2}$

4. $\sqrt{18} + 2\sqrt{3}$ $3\sqrt{2} + 2\sqrt{3}$ **5.** $\sqrt[3]{16} - \sqrt[3]{2}$ $\sqrt[3]{2}$ **6.** $\sqrt{2} + \sqrt{\dfrac{1}{2}}$ $\dfrac{3\sqrt{2}}{2}$

7. $\sqrt[3]{3} - \sqrt[3]{2}$ Not possible **8.** $\sqrt{12} - \sqrt{27}$ $-\sqrt{3}$ **9.** $\sqrt{3} + \sqrt{30} + \sqrt{300}$ $11\sqrt{3} + \sqrt{30}$

10. $\sqrt{3}(\sqrt{3} + \sqrt{6})$ $3 + 3\sqrt{2}$ **11.** $\sqrt{2}(2 - \sqrt{2})$ $2\sqrt{2} - 2$ **12.** $\sqrt{\dfrac{3}{4}} + \sqrt{\dfrac{27}{4}}$ $2\sqrt{3}$

Chalkboard Examples

Simplify.

1. $2\sqrt{50} - \sqrt{245} + 3\sqrt{125}$
$2 \cdot 5\sqrt{2} - 7\sqrt{5} + 3 \cdot 5\sqrt{5} = 10\sqrt{2} - 7\sqrt{5} + 15\sqrt{5} = 10\sqrt{2} + 8\sqrt{5}$

2. $2\sqrt[3]{9} + \sqrt[3]{81} - 2\sqrt[3]{\dfrac{1}{3}}$
$2\sqrt[3]{9} + 3\sqrt[3]{3} - 2 \cdot \dfrac{1}{\sqrt[3]{3}} \cdot \dfrac{\sqrt[3]{9}}{\sqrt[3]{9}} = 2\sqrt[3]{9} + 3\sqrt[3]{3} - \dfrac{2}{3}\sqrt[3]{9} = \dfrac{4}{3}\sqrt[3]{9} + 3\sqrt[3]{3}$

3. $-4\sqrt{6}(3\sqrt{2} - 5\sqrt{3}) + 5\sqrt{48}$ $-12\sqrt{12} + 20\sqrt{18} + 5\sqrt{48} = -24\sqrt{3} + 60\sqrt{2} + 20\sqrt{3} = -4\sqrt{3} + 60\sqrt{2}$

4. $\dfrac{3\sqrt{40} + 2\sqrt{24}}{\sqrt{8}} - \sqrt{5}$
$\dfrac{3\sqrt{40}}{\sqrt{8}} + \dfrac{2\sqrt{24}}{\sqrt{8}} - \sqrt{5} = 3\sqrt{5} + 2\sqrt{3} - \sqrt{5} = 2\sqrt{5} + 2\sqrt{3}$

Check for Understanding

Here is a suggested use of the Oral Exercises to check students' understanding as you teach the lesson.
Oral Exs. 1–15: use after Example 2.
Oral Ex. 16: use after Example 3.

Guided Practice

Simplify.

1. $\sqrt{27} + \sqrt{12} - \sqrt{75}$ 0
2. $\sqrt{8} + \sqrt{12} + \sqrt{6}$ $2\sqrt{2} + 2\sqrt{3} + \sqrt{6}$
3. $\sqrt[3]{250} - \sqrt[3]{54} + \sqrt[3]{4}$ $2\sqrt[3]{2} + \sqrt[3]{4}$
4. $\sqrt{\frac{4}{3}} + \frac{\sqrt{6}}{\sqrt{2}} - \frac{4\sqrt{2}}{\sqrt{6}}$ $\frac{\sqrt{3}}{3}$
5. $\sqrt{6}(\sqrt{3} - \sqrt{2})$ $3\sqrt{2} - 2\sqrt{3}$
6. $\frac{\sqrt{10} + \sqrt{250}}{\sqrt{2}}$ $6\sqrt{5}$
7. $(\sqrt{98} + \sqrt{28})\sqrt{14}$ $14\sqrt{7} + 14\sqrt{2}$

Summarizing the Lesson

In this lesson students applied the concept of combining like algebraic terms to combining like radicals. Ask students to explain why the directions for Exercises 35–42 say to assume that each radical represents a real number.

Suggested Assignments

Average Algebra
Day 1: 272/1–21 odd
Assign with Lesson 6-2.
Day 2: 272/23–43 odd
S 273/Mixed Review
Assign with Lesson 6-4.

Average Alg. and Trig.
272/1–43 odd
S 273/Mixed Review

Extended Algebra
272/1–43 odd
S 273/Mixed Review

Extended Alg. and Trig.
272/1–43 odd
S 273/Mixed Review

Simplify.

13. $\frac{\sqrt{8} + \sqrt{18}}{\sqrt{2}}$ 5

14. $\frac{\sqrt{15} + \sqrt{35}}{\sqrt{5}}$ $\sqrt{3} + \sqrt{7}$

15. $\frac{8\sqrt{6} - 2\sqrt{3}}{2\sqrt{3}}$ $4\sqrt{2} - 1$

16. Is $\sqrt{x} + \sqrt{x}$ equivalent to $\sqrt{2x}$? Explain. No; $\sqrt{x} + \sqrt{x} = 2\sqrt{x}$, but $\sqrt{2x} = \sqrt{2}\sqrt{x}$.

Written Exercises

6. Not possible 7. $6 + 7\sqrt{6}$ 8. $5 + 6\sqrt{5}$ 9. $\sqrt{2} + 3\sqrt{7}$ 10. $3\sqrt{2} - \sqrt{6}$ 11. $5\sqrt[3]{2} + 2\sqrt[3]{5}$ 12. $5\sqrt[3]{3} - 2\sqrt[3]{7}$

Simplify. If no simplification is possible, say so.

A

1. $\sqrt{50} + \sqrt{18}$ $8\sqrt{2}$
2. $\sqrt{45} - \sqrt{20}$ $\sqrt{5}$
3. $3\sqrt{12} - \sqrt{48}$ $2\sqrt{3}$
4. $\sqrt{27} + 2\sqrt{75}$ $13\sqrt{3}$
5. $5\sqrt{2} - 2\sqrt{5}$ Not possible
6. $7\sqrt{3} - 3\sqrt{7}$
7. $\sqrt{6} + \sqrt{36} + \sqrt{216}$
8. $\sqrt{5} + \sqrt{25} + \sqrt{125}$
9. $\sqrt{50} + \sqrt{63} - \sqrt{32}$
10. $\sqrt{18} + \sqrt{24} - \sqrt{54}$
11. $\sqrt[3]{54} + \sqrt[3]{40} + \sqrt[3]{16}$
12. $\sqrt[3]{24} - \sqrt[3]{56} + \sqrt[3]{81}$
13. $\sqrt{\frac{27}{5}} - \sqrt{\frac{3}{5}}$ $\frac{2\sqrt{15}}{5}$
14. $\sqrt{\frac{75}{2}} - \sqrt{\frac{3}{2}}$ $2\sqrt{6}$
15. $\sqrt{\frac{2}{3}} + \sqrt{\frac{3}{2}}$ $\frac{5\sqrt{6}}{6}$
16. $\sqrt{\frac{5}{2}} + \sqrt{\frac{2}{5}}$ $\frac{7\sqrt{10}}{10}$
17. $\sqrt[3]{4} + \sqrt[3]{\frac{1}{2}}$ $\frac{3\sqrt[3]{4}}{2}$
18. $\sqrt[3]{16} - \sqrt[3]{\frac{1}{4}}$ $\frac{3\sqrt[3]{2}}{2}$
19. $\sqrt{2}(\sqrt{8} + \sqrt{10})$ $4 + 2\sqrt{5}$
20. $\sqrt{3}(\sqrt{12} - \sqrt{24})$ $6 - 6\sqrt{2}$
21. $\sqrt{15}(\sqrt{3} + 2\sqrt{5})$ $3\sqrt{5} + 10\sqrt{3}$
22. $\sqrt{7}(3\sqrt{14} - \sqrt{21})$ $21\sqrt{2} - 7\sqrt{3}$
23. $2\sqrt{3}(\sqrt{48} - 5\sqrt{12})$ -36
24. $3\sqrt{5}(\sqrt{5} + 2\sqrt{75})$ $15 + 30\sqrt{15}$
25. $\frac{\sqrt{6} - \sqrt{24}}{\sqrt{2}}$ $-\sqrt{3}$
26. $\frac{\sqrt{18} - \sqrt{6}}{\sqrt{3}}$ $\sqrt{6} - \sqrt{2}$
27. $\frac{4\sqrt{300} - \sqrt{108}}{\sqrt{12}}$ 17
28. $\frac{\sqrt{40} - 2\sqrt{5}}{\sqrt{10}}$ $2 - \sqrt{2}$
29. $\sqrt{\frac{2}{3}}\left(\sqrt{\frac{27}{2}} - \frac{3}{\sqrt{2}}\right)$ $3 - \sqrt{3}$
30. $\sqrt{\frac{3}{8}}\left(\sqrt{\frac{3}{4}} + \frac{2}{\sqrt{3}}\right)$ $\frac{7\sqrt{2}}{8}$

B

31. $\sqrt[3]{5}(\sqrt[3]{200} - \sqrt[3]{16})$ $10 - 2\sqrt[3]{10}$
32. $\sqrt[3]{40}(\sqrt[3]{25} + 2\sqrt[3]{5})$ $10 + 4\sqrt[3]{25}$
33. $\frac{\sqrt[3]{18} + 3\sqrt[3]{54}}{\sqrt[3]{3}}$ $\sqrt[3]{6} + 3\sqrt[3]{18}$
34. $\frac{\sqrt[3]{320} + \sqrt[3]{1250}}{2\sqrt[3]{5}}$ $2 + \frac{5\sqrt[3]{2}}{2}$

Simplify. Assume that each radical represents a real number.

35. $\sqrt{8x^3} - x\sqrt{18x}$ $-x\sqrt{2x}$
36. $y^2\sqrt{45y} + 2y\sqrt{5y^3}$ $5y^2\sqrt{5y}$
37. $\sqrt{p^3r} + \sqrt{pr^3}$ $(p + r)\sqrt{pr}$
38. $\sqrt{2a^2b^4} + \sqrt{8a^2b^4}$ $3|a|b^2\sqrt{2}$
39. $\sqrt{10a} - \frac{\sqrt{5a}}{\sqrt{2}} + \sqrt{\frac{2a}{5}}$ $\frac{7\sqrt{10a}}{10}$
40. $\sqrt{6x} + \frac{\sqrt{2x}}{\sqrt{3}} - \sqrt{\frac{3x}{2}}$ $\frac{5\sqrt{6x}}{6}$
41. $\sqrt{6w}(\sqrt{3w} + \sqrt{2w^3})$ $3w\sqrt{2} + 2w^2\sqrt{3}$
42. $\sqrt{10t}(\sqrt{2t^5} - \sqrt{5t})$ $2t^3\sqrt{5} - 5t\sqrt{2}$

Use the Pythagorean theorem to find x in each right triangle shown.

43. 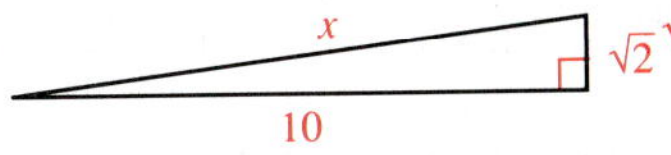

$\sqrt{102}$

44. 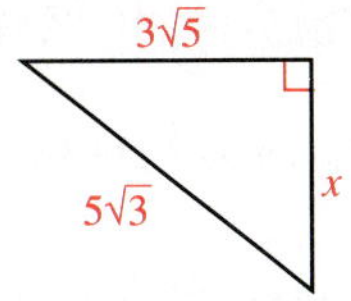

$\sqrt{30}$

Mixed Review Exercises

Simplify.

1. $\sqrt{28x^3}$ $2|x|\sqrt{7x}$

2. $\sqrt{\frac{81}{49}}$ $\frac{9}{7}$

3. $\frac{a^3x^2y}{a^2x^3y^2}$ $\frac{a}{xy}$

4. $\sqrt{\frac{45x}{32y^4}}$ $\frac{3\sqrt{10x}}{8y^2}$

5. $\frac{x^2-5x+6}{2-x}$ $3-x$

6. $\sqrt[3]{-135}$ $-3\sqrt[3]{5}$

7. $(3x+1)(3x-1)$ $9x^2-1$

8. $\sqrt{\frac{1}{x}}$ $\frac{\sqrt{x}}{|x|}$

9. $\sqrt[3]{8x^4y^6}$ $2xy^2\sqrt[3]{x}$

Extra / The Irrationality of $\sqrt{2}$

The proof that the number $\sqrt{2}$ is irrational, outlined below, is usually attributed to Pythagoras. The proof is *indirect;* that is, we first assume that $\sqrt{2}$ is rational and then show that this assumption leads to a contradiction. Here is the proof:

Suppose that $\sqrt{2}$ is rational. Then we can write $\sqrt{2}$ as a fraction and reduce this fraction to lowest terms. That is,

$$\sqrt{2} = \frac{a}{b}$$

where a and b are integers having no common factor except 1. Then $\frac{a^2}{b^2} = 2$ and $a^2 = 2b^2$, so a^2 is even. Since a^2 is even, a is even, and we can write $a = 2c$ for some integer c. Then $a^2 = 4c^2$, so $2b^2 = 4c^2$ and $b^2 = 2c^2$. Therefore b^2 is even and b is even. Since a and b are both even, they have the common factor 2, which contradicts our assumption that $\frac{a}{b}$ is in lowest terms. Therefore $\sqrt{2}$ *cannot* be written as a quotient of integers; that is, $\sqrt{2}$ is irrational.

Exercises

1. Prove that $\sqrt{3}$ is irrational.

2. Prove that $\sqrt{p}$, where p is a prime number, is irrational.

Supplementary Materials

Study Guide pp. 93–94

Additional Answers
Extra

1. Suppose $\sqrt{3}$ is rational. Then $\sqrt{3} = \frac{a}{b}$ for some integers a and b. Assume $\frac{a}{b}$ is in lowest terms. Then $3 = \frac{a^2}{b^2}$, and $3b^2 = a^2$. Since 3 divides $3b^2$, 3 divides a^2, and so 3 divides a. Hence $a = 3c$ for some integer c. Therefore $3b^2 = (3c)^2 = 9c^2$, and $b^2 = 3c^2$. Thus 3 divides b^2 and 3 divides b. But this is impossible since a and b have no common factor other than 1. $\therefore \sqrt{3}$ is irrational.

2. Suppose $\sqrt{p}$ is rational. Then $\sqrt{p} = \frac{a}{b}$ for some integers a and b. Assume $\frac{a}{b}$ is in lowest terms. Then $p = \frac{a^2}{b^2}$ and $pb^2 = a^2$. Since p divides pb^2, p divides a^2, and (since p is prime) p divides a. Hence $a = pc$ for some integer c. Therefore $pb^2 = (pc)^2 = p^2c^2$, and $b^2 = pc^2$. Thus p divides b^2 and p divides b. But this is impossible since a and b have no common factor other than 1. $\therefore \sqrt{p}$ is irrational.

Teaching Suggestions p. T102

Warm-Up Exercises

Simplify.

1. $(2a + 1)(3a - 4)$ $6a^2 - 5a - 4$
2. $(m - 4)(3m - 8)$ $3m^2 - 20m + 32$
3. $(7z + 2)(5z + 3)$ $35z^2 + 31z + 6$
4. $(2x + 3)(2x - 3)$ $4x^2 - 9$
5. $(4p + 5)^2$ $16p^2 + 40p + 25$

Motivating the Lesson

Remind students that they learned to multiply binomials by recognizing special products and by using the FOIL method. These techniques can also be applied to products and quotients of binomials containing radicals, the topic of today's lesson.

Chalkboard Examples

Simplify.

1. $(5 - \sqrt{3})(3 + 2\sqrt{3})$
   $15 + 7\sqrt{3} - 2 \cdot 3 = 9 + 7\sqrt{3}$
2. $(4\sqrt{2} + \sqrt{3})^2$
   $32 + 8\sqrt{6} + 3 = 35 + 8\sqrt{6}$
3. $\frac{2 + \sqrt{3}}{1 + \sqrt{3}}$
   $\frac{2 + \sqrt{3}}{1 + \sqrt{3}} \cdot \frac{1 - \sqrt{3}}{1 - \sqrt{3}} = \frac{2 - \sqrt{3} - 3}{1 - 3} = \frac{-1 - \sqrt{3}}{-2} = \frac{1 + \sqrt{3}}{2}$
4. $\frac{1}{\sqrt{5} - 1}$
   $\frac{1}{\sqrt{5} - 1} \cdot \frac{\sqrt{5} + 1}{\sqrt{5} + 1} = \frac{\sqrt{5} + 1}{5 - 1} = \frac{\sqrt{5} + 1}{4}$

6-4 Binomials Containing Radicals

Objective To simplify products and quotients of binomials that contain radicals.

You can multiply binomials containing radicals just as you would multiply any binomials.

Example 1 Simplify.

a. $(4 + \sqrt{7})(3 + 2\sqrt{7})$ **b.** $(2\sqrt{3} - \sqrt{6})^2$

c. $(4\sqrt{5} + 3\sqrt{2})(4\sqrt{5} - 3\sqrt{2})$

Solution **a.** Recall how you simplify $(4 + x)(3 + 2x)$:

$$(4 + x)(3 + 2x) = 12 + 8x + 3x + 2x^2$$
$$= 12 + 11x + 2x^2$$
$$\therefore\ (4 + \sqrt{7})(3 + 2\sqrt{7}) = 12 + 8\sqrt{7} + 3\sqrt{7} + 2(\sqrt{7})^2$$
$$= 12 + 11\sqrt{7} + 14$$
$$= 26 + 11\sqrt{7} \quad \textit{Answer}$$

b. Recall this pattern:

$$(a - b)^2 = a^2 - 2ab + b^2$$
$$\therefore\ (2\sqrt{3} - \sqrt{6})^2 = (2\sqrt{3})^2 - 2 \cdot 2\sqrt{3} \cdot \sqrt{6} + (\sqrt{6})^2$$
$$= 12 - 4\sqrt{18} + 6$$
$$= 12 - 12\sqrt{2} + 6$$
$$= 18 - 12\sqrt{2} \quad \textit{Answer}$$

c. Recall this pattern:

$$(a + b)(a - b) = a^2 - b^2$$
$$\therefore\ (4\sqrt{5} + 3\sqrt{2})(4\sqrt{5} - 3\sqrt{2}) = (4\sqrt{5})^2 - (3\sqrt{2})^2$$
$$= 80 - 18 = 62 \quad \textit{Answer}$$

In part (c) of Example 1, notice that the product of $4\sqrt{5} + 3\sqrt{2}$ and $4\sqrt{5} - 3\sqrt{2}$ is an integer. Expressions of the form $a\sqrt{b} + c\sqrt{d}$ and $a\sqrt{b} - c\sqrt{d}$ are called **conjugates.** Since

$$(a\sqrt{b} + c\sqrt{d})(a\sqrt{b} - c\sqrt{d}) = (a\sqrt{b})^2 - (c\sqrt{d})^2$$
$$= a^2b - c^2d,$$

the product of conjugates is always an integer when a, b, c, and d are integers.

Conjugates can be used to rationalize any denominator containing a binomial radical expression, as Examples 2 and 3 illustrate.

Example 2 Simplify $\frac{3+\sqrt{5}}{3-\sqrt{5}}$.

Solution

$$\frac{3+\sqrt{5}}{3-\sqrt{5}} = \frac{3+\sqrt{5}}{3-\sqrt{5}} \cdot \frac{3+\sqrt{5}}{3+\sqrt{5}}$$ ← Multiply numerator and denominator by the conjugate of the denominator.

$$= \frac{9+6\sqrt{5}+5}{9-5}$$

$$= \frac{14+6\sqrt{5}}{4} = \frac{7+3\sqrt{5}}{2} \quad \textit{Answer}$$

Example 3 If $f(x) = \frac{x+1}{x}$, find $f(\sqrt{3}+2)$.

Solution

$$f(\sqrt{3}+2) = \frac{(\sqrt{3}+2)+1}{\sqrt{3}+2}$$

$$= \frac{\sqrt{3}+3}{\sqrt{3}+2} \cdot \frac{\sqrt{3}-2}{\sqrt{3}-2}$$

$$= \frac{3+3\sqrt{3}-2\sqrt{3}-6}{3-4} = \frac{-3+\sqrt{3}}{-1} = 3-\sqrt{3} \quad \textit{Answer}$$

Oral Exercises

Simplify. Assume that each radical represents a real number.

1. a. $(x+y)(x-y)$ x^2-y^2 b. $(\sqrt{5}+\sqrt{2})(\sqrt{5}-\sqrt{2})$ 3 c. $(2+\sqrt{3})(2-\sqrt{3})$ 1
2. a. $(x+y)^2$ $x^2+2xy+y^2$ b. $(\sqrt{5}+\sqrt{2})^2$ $7+2\sqrt{10}$ c. $(2+\sqrt{3})^2$ $7+4\sqrt{3}$
3. a. $(x-y)^2$ $x^2-2xy+y^2$ b. $(\sqrt{5}-\sqrt{2})^2$ $7-2\sqrt{10}$ c. $(2-\sqrt{3})^2$ $7-4\sqrt{3}$
4. a. $\frac{1}{2+\sqrt{3}} \cdot \frac{2-\sqrt{3}}{2-\sqrt{3}}$ $2-\sqrt{3}$ b. $\frac{1}{x+\sqrt{y}} \cdot \frac{x-\sqrt{y}}{x-\sqrt{y}}$ $\frac{x-\sqrt{y}}{x^2-y}$
5. a. $\frac{1}{\sqrt{5}-\sqrt{2}} \cdot \frac{\sqrt{5}+\sqrt{2}}{\sqrt{5}+\sqrt{2}}$ $\frac{\sqrt{5}+\sqrt{2}}{3}$ b. $\frac{1}{\sqrt{x}-\sqrt{y}} \cdot \frac{\sqrt{x}+\sqrt{y}}{\sqrt{x}+\sqrt{y}}$ $\frac{\sqrt{x}+\sqrt{y}}{x-y}$

Written Exercises

Simplify.

A

1. $(3+\sqrt{7})(3-\sqrt{7})$ 2
2. $(5+\sqrt{2})(5-\sqrt{2})$ 23
3. $(\sqrt{7}+1)^2$ $8+2\sqrt{7}$
4. $(\sqrt{5}+2)^2$ $9+4\sqrt{5}$
5. $(1+\sqrt{2})(3+\sqrt{2})$ $5+4\sqrt{2}$
6. $(6-\sqrt{3})(4+\sqrt{3})$ $21+2\sqrt{3}$
7. $\frac{1}{4-\sqrt{3}}$ $\frac{4+\sqrt{3}}{13}$
8. $\frac{1}{6+\sqrt{3}}$ $\frac{6-\sqrt{3}}{33}$
9. $(\sqrt{7}-\sqrt{2})^2$ $9-2\sqrt{14}$
10. $(3\sqrt{11}-\sqrt{10})^2$ $109-6\sqrt{110}$
11. $(3+4\sqrt{3})(2-\sqrt{3})$ $-6+5\sqrt{3}$
12. $(5-\sqrt{2})(3-2\sqrt{2})$ $19-13\sqrt{2}$

Common Errors

Students often simplify $(\sqrt{7}+\sqrt{2})^2$ incorrectly as $7+2$, or 9. When reteaching, remind students that squaring requires them to use the base as a factor twice.

Check for Understanding

Oral Exs. 1–3: use after Example 1.
Oral Exs. 4–5: use after Example 2.

Guided Practice

Simplify.

1. $2\sqrt{2}(\sqrt{6}-\sqrt{3})$ $4\sqrt{3}-2\sqrt{6}$
2. $(4+\sqrt{3})(4-\sqrt{3})$ 13
3. $(3\sqrt{5}-5\sqrt{3}) \cdot (3\sqrt{5}+5\sqrt{3})$ -30
4. $\frac{\sqrt{2}+3}{3} \cdot \frac{\sqrt{2}-3}{2}$ $-\frac{7}{6}$
5. $(3-\sqrt{5})^2$ $14-6\sqrt{5}$
6. $(2\sqrt{3}+1)(3\sqrt{3}-2)$ $16-\sqrt{3}$
7. $\frac{1}{4-\sqrt{15}}$ $4+\sqrt{15}$
8. $\frac{\sqrt{3}}{\sqrt{2}+\sqrt{3}}$ $3-\sqrt{6}$
9. $\frac{3}{\sqrt{2}+1}$ $3\sqrt{2}-3$
10. $\frac{\sqrt{5}+\sqrt{2}}{\sqrt{5}-\sqrt{2}}$ $\frac{7+2\sqrt{10}}{3}$

Summarizing the Lesson

Products of binomials containing radicals can be simplified using the FOIL method. Quotients of binomials containing radicals can be simplified by multiplying numerator and denominator by the conjugate of the denominator. Ask students to explain why conjugates are important in simplifying quotients.

Suggested Assignments

Average Algebra
Day 1: 275/1–17 odd
Assign with Lesson 6-3.
Day 2: 276/19–39 odd
S 268/45–65 odd

Average Alg. and Trig.
275/2–40 even
S 269/63–66

Extended Algebra
275/2–42 even
S 273/Extra: 1, 2

Extended Alg. and Trig.
275/2–44 even
S 273/Extra: 1, 2

Additional Answers Written Exercises

35. $(3+\sqrt{5})^2 - 6(3+\sqrt{5}) + 4 = 9 + 6\sqrt{5} + 5 - 18 - 6\sqrt{5} + 4 = 0$; $(3-\sqrt{5})^2 - 6(3-\sqrt{5}) + 4 = 9 - 6\sqrt{5} + 5 - 18 + 6\sqrt{5} + 4 = 0$

36. $2\left(1+\frac{\sqrt{2}}{2}\right)^2 - 4\left(1+\frac{\sqrt{2}}{2}\right) + 1 = 2\left(1+\sqrt{2}+\frac{1}{2}\right) - 4 - 2\sqrt{2} + 1 = 2 + 2\sqrt{2} + 1 - 4 - 2\sqrt{2} + 1 = 0$; $2\left(1-\frac{\sqrt{2}}{2}\right)^2 - 4\left(1-\frac{\sqrt{2}}{2}\right) + 1 = 2\left(1-\sqrt{2}+\frac{1}{2}\right) - 4 + 2\sqrt{2} + 1 = 2 - 2\sqrt{2} + 1 - 4 + 2\sqrt{2} + 1 = 0$

38. The reciprocal of $\frac{\sqrt{5}+1}{2}$ is $\frac{2}{\sqrt{5}+1} = \frac{2}{\sqrt{5}+1}\cdot\frac{\sqrt{5}-1}{\sqrt{5}-1} = \frac{2(\sqrt{5}-1)}{4} = \frac{\sqrt{5}-1}{2}$, the conjugate of $\frac{\sqrt{5}+1}{2}$

Supplementary Materials

Study Guide pp. 95–96

13. $\frac{3}{\sqrt{5}+\sqrt{2}}$ $\sqrt{5}-\sqrt{2}$

14. $\frac{10}{2\sqrt{3}-\sqrt{7}}$ $4\sqrt{3}+2\sqrt{7}$

15. $(\sqrt{11}-\sqrt{7})(\sqrt{11}+\sqrt{7})$ 4

16. $(\sqrt{13}-\sqrt{3})(\sqrt{13}+\sqrt{3})$ 10

17. $(5+\sqrt{3})(8-2\sqrt{3})$ $34-2\sqrt{3}$

18. $(3+2\sqrt{6})(4-5\sqrt{6})$ $-48-7\sqrt{6}$

19. $\frac{\sqrt{15}}{\sqrt{3}+\sqrt{5}}$ $\frac{5\sqrt{3}-3\sqrt{5}}{2}$

20. $\frac{\sqrt{6}}{\sqrt{2}+\sqrt{3}}$ $3\sqrt{2}-2\sqrt{3}$

21. $(2\sqrt{5}+\sqrt{7})^2$ $27+4\sqrt{35}$

22. $(3\sqrt{2}+\sqrt{6})^2$ $24+12\sqrt{3}$

23. $(2\sqrt{3}+\sqrt{5})(2\sqrt{3}-\sqrt{5})$ 7

24. $(3\sqrt{7}-2\sqrt{5})(3\sqrt{7}+2\sqrt{5})$ 43

25. $(\sqrt{6}-\sqrt{15})^2$ $21-6\sqrt{10}$

26. $(2\sqrt{5}-\sqrt{10})^2$ $30-20\sqrt{2}$

27. $\frac{\sqrt{5}+\sqrt{3}}{2}\cdot\frac{\sqrt{5}-\sqrt{3}}{2}$ $\frac{1}{2}$

28. $\frac{2\sqrt{7}+1}{3}\cdot\frac{2\sqrt{7}-1}{3}$ 3

29. $(5\sqrt{6}+3\sqrt{2})(2\sqrt{6}-4\sqrt{3})$ $60-60\sqrt{2}+12\sqrt{3}-12\sqrt{6}$

30. $(3\sqrt{5}+2\sqrt{15})(4\sqrt{3}-3\sqrt{15})$ $12\sqrt{15}-45\sqrt{3}+24\sqrt{5}-90$

31. $\frac{\sqrt{5}+1}{\sqrt{5}-3}$ $-2-\sqrt{5}$

32. $\frac{2\sqrt{7}-\sqrt{3}}{\sqrt{7}+\sqrt{3}}$ $\frac{17-3\sqrt{21}}{4}$

B **33.** If $f(x) = \frac{x}{x+1}$, find $f(1-\sqrt{2})$. $-\frac{\sqrt{2}}{2}$

34. If $g(x) = \frac{x^2}{x-1}$, find $g(1+\sqrt{2})$. $\frac{3\sqrt{2}+4}{2}$

35. Show by substitution that $3+\sqrt{5}$ and $3-\sqrt{5}$ are roots of $x^2-6x+4=0$.

36. Show by substitution that $1+\frac{\sqrt{2}}{2}$ and $1-\frac{\sqrt{2}}{2}$ are roots of $2x^2-4x+1=0$.

37. a. What is the conjugate of $2\sqrt{5}-3\sqrt{2}$? $2\sqrt{5}+3\sqrt{2}$
b. What is the reciprocal of the conjugate of $2\sqrt{5}-3\sqrt{2}$? $\frac{2\sqrt{5}-3\sqrt{2}}{2}$
c. What is the conjugate of the reciprocal of $2\sqrt{5}-3\sqrt{2}$? $\frac{2\sqrt{5}-3\sqrt{2}}{2}$

38. Show that the reciprocal of $\frac{\sqrt{5}+1}{2}$ is also the conjugate of $\frac{\sqrt{5}+1}{2}$.

Use the Pythagorean theorem to find x.

39.

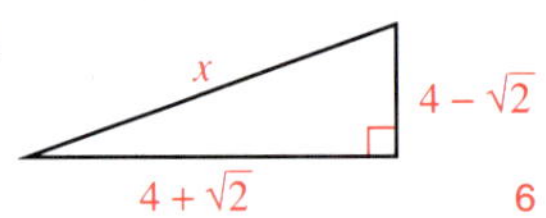

40.

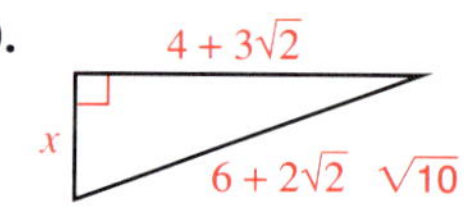

Simplify. Assume that each radical represents a real number.

41. $(\sqrt{n+1}+\sqrt{n})(\sqrt{n+1}-\sqrt{n})$ 1

42. $(b+\sqrt{b})^2-(b-\sqrt{b})^2$ $4b\sqrt{b}$

43. $\frac{\sqrt{w}}{\sqrt{w}+1}+\frac{\sqrt{w}}{\sqrt{w}-1}$ $\frac{2w}{w-1}$

44. $\sqrt{1-y^2}+\frac{y^2}{\sqrt{1-y^2}}$ $\frac{\sqrt{1-y^2}}{1-y^2}$

C **45.** $\frac{\sqrt{a-\sqrt{a}}\cdot\sqrt{a+\sqrt{a}}}{\sqrt{a-1}}$ $\sqrt{a}$

46. $\frac{\sqrt{x}}{\sqrt{x}+\sqrt{y}}+\frac{\sqrt{y}}{\sqrt{x}-\sqrt{y}}$ $\frac{x+y}{x-y}$

47. Prove that if a and b are rational numbers, $(a+\sqrt{b})^3+(a-\sqrt{b})^3$ is also rational.

6-5 Equations Containing Radicals

Objective To solve equations containing radicals.

By measuring a car's skid marks, investigators can determine how fast the car was traveling when the brakes were applied. On dry pavement, the speed s in miles per hour is given approximately by

$$s = \sqrt{22d},$$

where d is the distance in feet that the car travels as it brakes to a complete stop. For example, if the length of the skid marks is $d = 130$ ft, then the car's speed was

$$s = \sqrt{22 \cdot 130} \approx 53.5 \text{ (mi/h)}.$$

The formula $s = \sqrt{22d}$ can also be used to find a car's braking distance for a given speed. For example, at a speed s of 40 mi/h,

$40 = \sqrt{22d}$	Substitute 40 for s.
$40^2 = (\sqrt{22d})^2$	Square both sides.
$1600 = 22d$	Solve for d.
$d = \frac{1600}{22} \approx 73 \text{ (ft)}$	

Therefore, a speed of 40 mi/h requires a braking distance of 73 ft.

An equation like $40 = \sqrt{22d}$, which contains a radical with a variable in the radicand, is called a **radical equation.** To solve a radical equation involving square roots, start by isolating the radical term on one side of the equation. Then square both sides of the equation. If cube roots are involved instead of square roots, then cube both sides of the equation.

Example 1 Solve: **a.** $\sqrt{2x - 1} = 3$ **b.** $2\sqrt[3]{x} - 1 = 3$

Solution

a.

$\sqrt{2x-1} = 3$	The radical term is isolated.	*Check:*
$(\sqrt{2x-1})^2 = 3^2$	Square both sides.	$\sqrt{2 \cdot 5 - 1} \stackrel{?}{=} 3$
$2x - 1 = 9$	Solve for x.	$3 = 3$ ✓
$x = 5$		$\therefore$ the solution set is $\{5\}$.

b.

$2\sqrt[3]{x} - 1 = 3$	Isolate the radical term.	*Check:*
$2\sqrt[3]{x} = 4$		$2\sqrt[3]{8} - 1 \stackrel{?}{=} 3$
$(2\sqrt[3]{x})^3 = 4^3$	Cube both sides.	$3 = 3$ ✓
$8x = 64$	Solve for x.	$\therefore$ the solution set is $\{8\}$.
$x = 8$		

Teaching Suggestions p. T103

Suggested Extensions p. T103

Warm-Up Exercises

Solve and check for extraneous roots.

1. $4x + 3 = 15$ $\{3\}$
2. $2x^2 - 3x - 5 = 0$ $\left\{-1, \frac{5}{2}\right\}$
3. $(x - 6)^2 = x$ $\{4, 9\}$
4. $\frac{x}{x-2} = \frac{x+3}{x}$ $\{6\}$
5. $2 - \frac{5}{x^2 - x - 6} = \frac{x+3}{x+2}$ $\{4\}$

Motivating the Lesson

The time t, in seconds, that a free-falling object takes to travel s feet is given by the formula $t = \frac{\sqrt{s}}{4}$. This is an example of a radical equation, the topic of today's lesson.

Common Errors

Some students may fail to check for extraneous roots when given a radical equation to solve. When reteaching, emphasize the checking of all solutions.

Chalkboard Examples

Solve.

1. $\sqrt{2t-1} - 3 = 0$
   $\sqrt{2t-1} = 3$
   $2t - 1 = 9$
   $2t = 10$
   $t = 5$
   $\{5\}$
2. $\sqrt{7x-12} = x$
   $7x - 12 = x^2$
   $x^2 - 7x + 12 = 0$
   $(x-3)(x-4) = 0$
   $x = 3$ or $x = 4$
   $\{3, 4\}$
3. $3\sqrt{x+1} - x = 1$
   $3\sqrt{x+1} = x + 1$
   $9(x+1) = x^2 + 2x + 1$
   $9x + 9 = x^2 + 2x + 1$
   $x^2 - 7x - 8 = 0$
   $(x-8)(x+1) = 0$
   $x = 8$ or $x = -1$
   $3\sqrt{8+1} - 8 =$
   $3\sqrt{9} - 8 = 1$ ✓
   $3\sqrt{-1+1} - (-1) =$
   $3\sqrt{0} + 1 = 1$ ✓
   $\{8, -1\}$
4. $\sqrt{2x-2} - \sqrt{x+6} = 1$
   $\sqrt{2x-2} = \sqrt{x+6} + 1$
   $2x - 2 = x + 6 + 2\sqrt{x+6} + 1$
   $x - 9 = 2\sqrt{x+6}$
   $x^2 - 18x + 81 = 4(x+6)$
   $x^2 - 18x + 81 = 4x + 24$
   $x^2 - 22x + 57 = 0$
   $(x-3)(x-19) = 0$
   $x = 3$ or $x = 19$
   $\sqrt{6-2} - \sqrt{3+6} =$
   $\sqrt{4} - \sqrt{9} = -1 \neq 1$
   $\sqrt{38-2} - \sqrt{19+6} =$
   $\sqrt{36} - \sqrt{25} = 6 - 5 =$
   1 ✓
   $\{19\}$

Example 2 Solve $3x - 5\sqrt{x} = 2$.

Solution

$3x - 2 = 5\sqrt{x}$	Isolate the radical term.
$9x^2 - 12x + 4 = 25x$	Square both sides.
$9x^2 - 37x + 4 = 0$	Solve for x.
$(x-4)(9x-1) = 0$	
$x = 4 \quad \text{or} \quad x = \frac{1}{9}$	

Check: Check each possible solution in the *original* equation.

$3x - 5\sqrt{x} = 2$	$3x - 5\sqrt{x} = 2$
$3 \cdot 4 - 5\sqrt{4} \stackrel{?}{=} 2$	$3 \cdot \frac{1}{9} - 5\sqrt{\frac{1}{9}} \stackrel{?}{=} 2$
$12 - 10 \stackrel{?}{=} 2$	$\frac{1}{3} - \frac{5}{3} \stackrel{?}{=} 2$
$2 = 2$ ✓	$-\frac{4}{3} \neq 2$
The root 4 checks.	The root $\frac{1}{9}$ does not check.

$\therefore$ the solution set is $\{4\}$. ***Answer***

Example 2 shows that when you square both sides of a radical equation, the resulting equation may have a solution that is not a solution of the original equation. Such a solution is called an **extraneous root.** When you square the equation in the first step of this example, you get $(3x-2)^2 = (5\sqrt{x})^2$, which is equivalent to $(3x-2) = \pm 5\sqrt{x}$. The root 4 satisfies $3x - 2 = +5\sqrt{x}$ and the extraneous root $\frac{1}{9}$ satisfies $3x - 2 = -5\sqrt{x}$.

Caution: Since you may get an extraneous root when you solve a radical equation, you must not forget to check your answer in the *original* equation.

If a radical equation has more than one term with a variable in a radical, it may be necessary to square both sides of the equation more than once.

Example 3 Solve $\sqrt{2x+5} = 2\sqrt{2x} + 1$.

Solution

$(\sqrt{2x+5})^2 = (2\sqrt{2x} + 1)^2$	Square both sides.
$2x + 5 = 8x + 4\sqrt{2x} + 1$	Isolate the radical term.
$4 - 6x = 4\sqrt{2x}$	Square both sides.
$16 - 48x + 36x^2 = 32x$	
$36x^2 - 80x + 16 = 0$	Divide both sides by 4.
$9x^2 - 20x + 4 = 0$	Solve for x.
$(9x-2)(x-2) = 0$	
$x = \frac{2}{9} \quad \text{or} \quad x = 2$	

Check:

$$\sqrt{2 \cdot \frac{2}{9} + 5} \stackrel{?}{=} 2\sqrt{2 \cdot \frac{2}{9}} + 1$$

$$\sqrt{\frac{49}{9}} \stackrel{?}{=} 2 \cdot \frac{2}{3} + 1$$

$$\frac{7}{3} = \frac{7}{3} \checkmark$$

$\therefore$ the solution set is $\left\{\frac{2}{9}\right\}$. ***Answer***

Check:

$$\sqrt{2 \cdot 2 + 5} \stackrel{?}{=} 2\sqrt{2 \cdot 2} + 1$$

$$\sqrt{9} \stackrel{?}{=} 2\sqrt{4} + 1$$

$$3 \neq 5$$

Consider these two similar-looking equations:

(a) $3x = 2 + 5\sqrt{x}$ (b) $3x = 2 + x\sqrt{5}$

Although both equations contain radicals, only equation (a) is a radical equation. Equation (b) is a *linear* equation having the form $ax = c + bx$, where the coefficient b in this case is a radical. Such linear equations can be solved without squaring both sides, as Example 4 illustrates.

Example 4 Solve $3x = 2 + x\sqrt{5}$ without squaring both sides.

Solution

$$3x = 2 + x\sqrt{5}$$
$$3x - x\sqrt{5} = 2$$
$$(3 - \sqrt{5})x = 2$$
$$x = \frac{2}{3 - \sqrt{5}}$$
$$= \frac{2}{3 - \sqrt{5}} \cdot \frac{3 + \sqrt{5}}{3 + \sqrt{5}}$$
$$= \frac{2(3 + \sqrt{5})}{4} = \frac{3 + \sqrt{5}}{2}$$

$\therefore$ the solution set is $\left\{\frac{3 + \sqrt{5}}{2}\right\}$. ***Answer***

Oral Exercises

Solve. If an equation has no real solution, say so.

1. $\sqrt{x} = 5$ {25}
2. $\sqrt{2x} = 4$ {8}
3. $\sqrt{x - 1} = 3$ {10}
4. $\sqrt{x} - 7 = 0$ {49}
5. $\sqrt{x} + 6 = 0$ No sol.
6. $2\sqrt{x} - 3 = 1$ {4}
7. $\sqrt[3]{x - 3} = 2$ {11}
8. $\sqrt{x} + \sqrt{x + 2} = 0$ No sol.
9. $\sqrt[3]{x} + 5 = 3$ {−8}
10. Tell whether each equation is a radical equation or a linear equation.
    a. $x\sqrt{2} = 3$ linear b. $2\sqrt{x} = 3$ radical c. $x\sqrt{3} + x\sqrt{2} = 1$ linear

Check for Understanding

Here is a suggested use of the Oral Exercises to check students' understanding as you teach the lesson.
Oral Exs. 1–9: use after Example 3.
Oral Exs. 10–12: use after Example 4.

Guided Practice

Solve.

1. $\sqrt{2x-1} = 7$ {25}
2. $\sqrt[3]{x^2+2x} = 2$ {−4, 2}
3. $\sqrt{x-10} = \sqrt{2x}$ ∅
4. $x = \sqrt{20-x}$ {4}
5. $x\sqrt{2} = 8$ $\{4\sqrt{2}\}$
6. $2\sqrt{x} = 8$ {16}
7. $3x = \sqrt{7-x} - 3$ $\left\{-\frac{1}{9}\right\}$
8. $3x + x\sqrt{5} = 2$ $\left\{\frac{3-\sqrt{5}}{2}\right\}$
9. $3x + 5\sqrt{x} = 2$ $\left\{\frac{1}{9}\right\}$
10. $2x = \sqrt{7x+2}$ {2}

Summarizing the Lesson

In this lesson students learned to solve a radical equation by isolating the radical term and then raising both sides of the equation to an appropriate power. This technique may need to be applied more than once if the equation contains more than one radical term. Ask students to explain why they must check their solutions in the original equation.

11. Solve each of the equations in Exercise 10. **a.** $\left\{\frac{3\sqrt{2}}{2}\right\}$ **b.** $\left\{\frac{9}{4}\right\}$ **c.** $\{\sqrt{3}-\sqrt{2}\}$

12. Explain why $\sqrt{x}$ should be isolated in an equation like $\sqrt{x} + 4 = 7$ *before* both sides are squared. Otherwise a radical term remains on the left side.

Written Exercises

Solve. If an equation has no real solution, say so.

A

1. $\sqrt{4x-3} = 5$ {7}
2. $\sqrt{3n+1} = 7$ {16}
3. $3\sqrt{t} - 5 = 13$ {36}
4. $7 + 4\sqrt{a} = 3$ No sol.
5. $\sqrt{2x^2-7} = 5$ {−4, 4}
6. $\sqrt{5y^2+1} = 9$ {−4, 4}
7. $\sqrt[3]{3m+1} = 4$ {21}
8. $\sqrt[3]{2w-5} = 3$ {16}
9. $\sqrt[3]{2d} + 5 = 3$ {−4}
10. $7 - \sqrt[3]{9c} = 4$ {3}
11. $2\sqrt[3]{x} = \sqrt[3]{x^2}$ {0, 8}
12. $\frac{\sqrt[3]{x}}{2} = \sqrt[3]{x-7}$ {8}
13. $\sqrt{x+2} = x$ {2}
14. $\sqrt{2n+3} = n$ {3}
15. $\sqrt{t-2} + t = 4$ {3}
16. $5 + \sqrt{a+7} = a$ {9}
17. $\sqrt{2x+5} - 1 = x$ {2}
18. $\sqrt{3n+10} - 4 = n$ {−3, −2}

In Exercises 19–24, a radical equation and a linear equation are given. Solve the radical equation by squaring both sides. Solve the linear equation without squaring both sides.

19. a. $5\sqrt{x} = 10$ {4}
    b. $x\sqrt{5} = 10$ $\{2\sqrt{5}\}$
20. a. $3\sqrt{x} = 12$ {16}
    b. $x\sqrt{3} = 12$ $\{4\sqrt{3}\}$
21. a. $5 + 2\sqrt{x} = 7$
    b. $5 + x\sqrt{2} = 7$
    a. {1} b. $\{\sqrt{2}\}$
22. a. $2 + 3\sqrt{x} = 8$ {4}
    b. $2 + x\sqrt{3} = 8$ $\{2\sqrt{3}\}$
23. a. $x = 3 + 2\sqrt{x}$ {9}
    b. $x = 3 + x\sqrt{2}$ $\{-3-3\sqrt{2}\}$
24. a. $3x = 7\sqrt{x} - 2$
    b. $3x = x\sqrt{7} - 2$
    a. $\left\{\frac{1}{9}, 4\right\}$ b. $\{-3-\sqrt{7}\}$

Solve. If an equation has no real solution, say so.

B

25. $\sqrt{y} + \sqrt{y+5} = 5$ {4}
26. $\sqrt{x-7} + \sqrt{x} = 7$ {16}
27. $\sqrt{2n-5} - \sqrt{3n+4} = 2$ No sol.
28. $\sqrt{3a-2} - \sqrt{2a-3} = 1$ {2, 6}
29. $\sqrt{3b-2} - \sqrt{2b+5} = 1$ {22}
30. $\sqrt{5y-1} - \sqrt{7y+9} = 2$ No sol.
31. $\sqrt{x} + \sqrt{3} = \sqrt{x+3}$ {0}
32. $\sqrt{n+6} - \sqrt{n} = \sqrt{6}$ {0}
33. If you are near the top of a tall building on a clear day, how far can you see? If a building is h ft high, then the distance d (in miles) to the earth's horizon is approximately

$$d = \sqrt{\frac{3}{2}h}.$$

    a. The observatory of a tall building in Chicago is 607 ft high. What is the distance to the horizon from this observatory? 30 mi
    b. Solve the formula for h. $h = \frac{2}{3}d^2$

34. If a pendulum is l cm long, then the time T (in seconds) that it takes the pendulum to swing back and forth once is given by $T = 2\pi\sqrt{\frac{l}{g}}$, where $\pi \approx 3.14$ and $g \approx 980$.

a. Find the value of T if the pendulum is 20 cm long. 0.9 s

b. Solve the formula for l in terms of T, g, and π. $l = \frac{gT^2}{4\pi^2}$

35. The diagram shows an isosceles right triangle.

a. If its perimeter is 10, find x. $10 - 5\sqrt{2}$

b. If its area is 12, find x. $2\sqrt{6}$

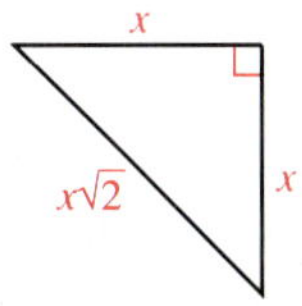

36. The diagram shows a 30°-60°-90° right triangle.

a. If the perimeter of the triangle is 18, find x. $9 - 3\sqrt{3}$

b. If the area of the triangle is 24, show that $x = 4\sqrt[4]{3}$.

Area $= \frac{1}{2}x^2\sqrt{3} = 24$; $x^2 = \frac{48}{\sqrt{3}} = 16\sqrt{3}$; $x = \sqrt{16\sqrt{3}} = 4\sqrt[4]{3}$

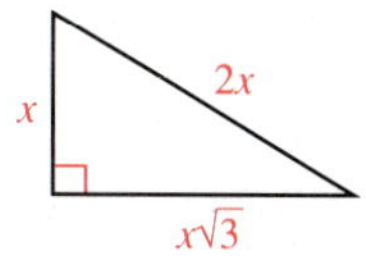

37. PQR is an isosceles triangle with base $\overline{QR}$ 10 cm long. The length of altitude $\overline{PN}$ is x cm.

a. Show that each leg of the triangle has length $\sqrt{25 + x^2}$.

b. If the perimeter of the triangle is three times the length of $\overline{PN}$, find x. **a.** $QN = NR = 5$ cm; $(PQ)^2 = (QN)^2 + (PN)^2 = 5^2 + x^2 = 25 + x^2$; $PQ = \sqrt{25 + x^2}$; since $PR = PQ$, $PR = \sqrt{25 + x^2}$

b. 12 cm

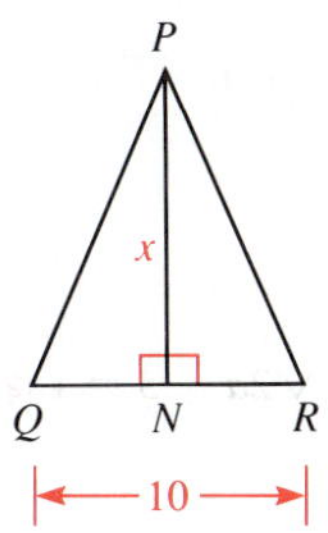

Ex. 37

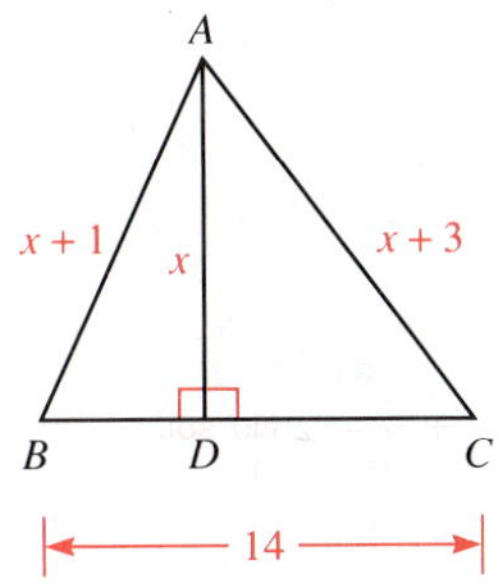

Ex. 38

C 38. Use the dimensions given in the figure to find x, the length of altitude $\overline{AD}$. 12

39. The eight edges of the square pyramid shown each have length x.

a. Show that the volume of the pyramid is $\frac{x^3\sqrt{2}}{6}$.

$\left(\text{Volume} = \frac{1}{3} \times \text{base area} \times \text{height}\right)$

b. If the volume is 9, find x. $3\sqrt[6]{2}$

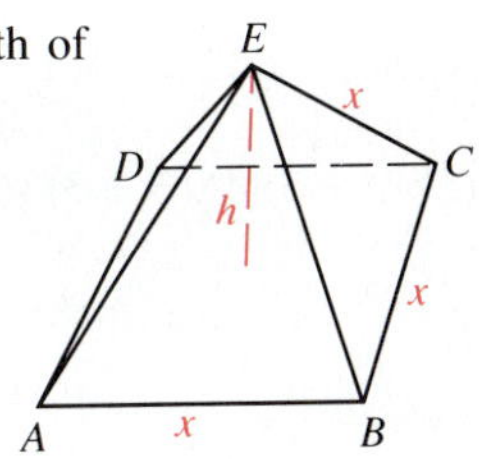

Suggested Assignments

Average Algebra
Day 1: 280/2–24 even
S 276/41–44
Day 2: 280/26–36 even
S 282/Mixed Review

Average Alg. and Trig.
Day 1: 280/1–23 odd
S 276/41–44
Day 2: 280/25–37 odd
S 282/Mixed Review

Extended Algebra
Day 1: 280/1–27 odd
S 276/43–46
Day 2: 280/29–39 odd
S 282/Mixed Review

Extended Alg. and Trig.
Day 1: 280/1–27 odd
S 276/43, 45, 46
Day 2: 280/29–39 odd
S 282/Mixed Review

Supplementary Materials

Study Guide	pp. 97–98
Practice Master	35
Test Master	22
Computer Activity	13
Resource Book	p. 123

Additional Answers Written Exercises

39. a. $V = \frac{1}{3}Bh = \frac{1}{3}x^2h$.
Let F be the foot of the altitude; note $FC = \frac{x\sqrt{2}}{2}$, since it is half the diagonal of a square of side x.
Then $(EF)^2 + (FC)^2 = (EC)^2$;
$h^2 + \left(\frac{x\sqrt{2}}{2}\right)^2 = x^2$;
$h^2 + \frac{x^2}{2} = x^2$; $h^2 = \frac{x^2}{2}$;
$h = \frac{x}{\sqrt{2}} = \frac{x\sqrt{2}}{2}$.
$V = \frac{1}{3}x^2\left(\frac{x\sqrt{2}}{2}\right) = \frac{x^3\sqrt{2}}{6}$.

Mixed Review Exercises

Simplify.

1. $\dfrac{\sqrt{35}+\sqrt{14}}{\sqrt{7}}$ $\sqrt{5}+\sqrt{2}$
2. $\sqrt{6}(2\sqrt{3}-3\sqrt{2})$ $6\sqrt{2}-6\sqrt{3}$
3. $\dfrac{a^3-b^3}{a-b}$ a^2+ab+b^2
4. $\dfrac{2}{4-\sqrt{6}}$ $\dfrac{4+\sqrt{6}}{5}$
5. $(m^2n)^3(-mn^3)^2$ m^8n^9
6. $(\sqrt{5}+\sqrt{10})^2$ $15+10\sqrt{2}$
7. $(1-x)^{-2}(x-1)$ $\dfrac{1}{x-1}$
8. $\sqrt{75}-\sqrt{12}+\sqrt{27}$ $6\sqrt{3}$
9. $(3\sqrt{6}+\sqrt{2})(\sqrt{6}-4\sqrt{3})$ $18-36\sqrt{2}+2\sqrt{3}-4\sqrt{6}$
10. $\dfrac{3\sqrt{32}}{2\sqrt{27}}$ $\dfrac{2\sqrt{6}}{3}$
11. $(5y+2)(y^2-3y+4)$ $5y^3-13y^2+14y+8$
12. $\sqrt[3]{40}+\sqrt[3]{625}-\sqrt[3]{135}$ $4\sqrt[3]{5}$

Self-Test 1

Vocabulary

square root (p. 259)	radical sign (p. 261)
principal square root (p. 259)	index of a radical (p. 261)
cube root (p. 260)	rationalizing the denominator (p. 265)
*n*th root (p. 260)	simplest radical form (p. 266)
principal *n*th root (p. 260)	conjugates (p. 274)
radical (p. 261)	radical equation (p. 277)
radicand (p. 261)	extraneous root (p. 278)

Simplify.

1. $\sqrt{\dfrac{49}{121}}$ $\dfrac{7}{11}$
2. $\sqrt[3]{-0.125}$ -0.5
3. $\sqrt[4]{(-13)^4}$ 13

Obj. 6-1, p. 259

4. For what values of x is $\sqrt{(x-4)^2}=4-x$ true? $x \le 4$

Simplify.

5. $\sqrt{10}\cdot\sqrt{15}$ $5\sqrt{6}$
6. $\sqrt{\dfrac{4a^4}{9b^3}}$ $\dfrac{2a^2\sqrt{b}}{3b^2}$
7. $\sqrt{2x^2-12x+18}$ $|x-3|\sqrt{2}$

Obj. 6-2, p. 264

8. $\sqrt[3]{250}-\sqrt[3]{128}+\sqrt[3]{16}$ $3\sqrt[3]{2}$
9. $\sqrt{\dfrac{3}{5}}\left(\sqrt{\dfrac{20}{3}}-\dfrac{\sqrt{15}}{6}\right)$ $\dfrac{3}{2}$

Obj. 6-3, p. 270

10. $(2\sqrt{3}+5)(4\sqrt{3}-3)$ $9+14\sqrt{3}$
11. $\dfrac{4}{3\sqrt{2}-4}$ $6\sqrt{2}+8$

Obj. 6-4, p. 274

Solve.

12. $2\sqrt{x}+3=x$ $\{9\}$
13. $1+\sqrt{x-2}=\sqrt{x+3}$ $\{6\}$

Obj. 6-5, p. 277

Check your answers with those at the back of the book.

Quick Quiz

1. Evaluate $\sqrt[3]{0.064}+\sqrt[3]{-8}+\sqrt{(-2)^2}$ 0.4
2. For what values of x is $\sqrt[3]{9-\sqrt{4-x}}$ a real number? $\{x : x \le 4\}$

Simplify.

3. $\sqrt{6}\cdot\sqrt{50}$ $10\sqrt{3}$
4. $\sqrt{x^2+4x+4}$ $|x+2|$
5. $\sqrt[3]{\dfrac{y^3}{4}}$ $\dfrac{y\sqrt[3]{2}}{2}$
6. $2\sqrt{3}(2-\sqrt{3})^2$ $14\sqrt{3}-24$
7. $\dfrac{7}{4-\sqrt{2}}$ $\dfrac{4+\sqrt{2}}{2}$

Solve.

8. $3x-x\sqrt{7}=6$ $\{9+3\sqrt{7}\}$
9. $2x+1=\sqrt{4-7x}$ $\left\{\dfrac{1}{4}\right\}$
10. $\sqrt{x+3}-\sqrt{x}=1$ $\{1\}$

Real Numbers and Complex Numbers

6-6 Rational and Irrational Numbers

Objective To find and use decimal representations of real numbers.

The set of real numbers includes both rational numbers, such as $\frac{37}{16}$ and $\frac{19}{22}$, and irrational numbers, such as $\sqrt{2}$ and π. One way to learn about real numbers is to look at their decimal representations. We accept as an axiom the following property of real numbers.

Completeness Property of Real Numbers

Every real number has a decimal representation, and every decimal represents a real number.

Recall that a **rational number** is any number that can be expressed as the ratio, or quotient, of two integers. To find the decimal representation of a rational number, you can use the division process.

Example 1 Find a decimal representation for each rational number.

a. $\frac{37}{16}$ **b.** $\frac{19}{22}$

Solution **a.** $\frac{37}{16} = 37 \div 16$

$$\begin{array}{r} 2.3125 \\ 16\overline{)37.0000} \\ \underline{32} \\ 5\,0 \\ \underline{4\,8} \\ 20 \\ \underline{16} \\ 40 \\ \underline{32} \\ 80 \\ \underline{80} \\ 0 \end{array}$$

Division terminates

$\therefore \frac{37}{16} = 2.3125$ ***Answer***

b. $\frac{19}{22} = 19 \div 22$

$$\begin{array}{r} 0.86363 \\ 22\overline{)19.00000} \\ \underline{17\,6} \\ 1\,40 \\ \underline{1\,32} \\ 80 \\ \underline{66} \\ 140 \\ \underline{132} \\ 80 \\ \underline{66} \\ 14 \end{array}$$

Recurring remainders

$\therefore \frac{19}{22} = 0.8636363\ldots$ ***Answer***

Teaching Suggestions p. T103

Suggested Extensions p. T103

Warm-Up Exercises

Write as a fraction.

1. 0.7 $\frac{7}{10}$
2. 0.25 $\frac{1}{4}$
3. 0.004 $\frac{1}{250}$
4. 1.5 $\frac{3}{2}$
5. 2.35 $\frac{47}{20}$

Write as a decimal.

6. $\frac{3}{10}$ 0.3
7. $\frac{7}{8}$ 0.875
8. $\frac{23}{50}$ 0.46
9. $\frac{4}{9}$ $0.\overline{4}$
10. $\frac{2}{3}$ $0.\overline{6}$

Motivating the Lesson

Suppose a spinner had the digits 0 through 9 on it. If you created an infinitely large decimal whose digits were randomly obtained by using the spinner, the decimal would represent an irrational number, one of the topics of today's lesson.

Thinking Skills

The conversion of $\frac{37}{16}$ and $\frac{19}{22}$ to decimals, as in Example 1, can be *analyzed* as follows. When an integer p is divided by a positive integer q, the remainder at each step is either 0 or a positive integer less than q. Therefore, when the division process reaches the last nonzero digit of p and only zeros are left in the dividend, something must happen within q - 1 steps: Either 0 occurs as a remainder, and the division process ends; or one of the nonzero remainders recurs, and the division process produces a repeating sequence of remainders with a repeating block of digits in the quotient.

Using a Calculator

Students should be aware that the value of π displayed on their calculators is a rational approximation of an irrational number.

Notice that the representation for $\frac{37}{16}$ is a **terminating decimal.** The representation for $\frac{19}{22}$, however, is a nonterminating **repeating decimal.** Terminating decimals are also called *finite decimals*. Nonterminating decimals are called *infinite decimals*. When you write a repeating decimal, you can use a bar to indicate the block of digits that repeats. For example,

$$2.\overline{45} = 2.454545\ldots$$

and

$$0.8\overline{63} = 0.8636363\ldots$$

The conversion of $\frac{37}{16}$ and $\frac{19}{22}$ to decimals illustrates the first fact stated in the chart below.

1. The decimal representation of any rational number is either terminating or repeating.
2. Every terminating or repeating decimal represents a rational number. (In other words, every terminating or repeating decimal can be written in the form $\frac{p}{q}$, where p and q are integers and $q \neq 0$.)

Exercise 42 asks you to prove the second fact stated above.

Example 2 Write each terminating decimal as a common fraction in lowest terms.

a. 2.571 **b.** 0.0036

Solution

a. $2.571 = 2\frac{571}{1000} = \frac{2571}{1000}$ ***Answer***

b. $0.0036 = \frac{36}{10{,}000} = \frac{9}{2500}$ ***Answer***

Example 3 Write each repeating decimal as a common fraction in lowest terms.

a. $0.3\overline{27}$ **b.** $1.89\overline{189}$

Solution In each case let N be the number. Multiply the given number by 10^n where n is the number of digits in the block of repeating digits. Then subtract N.

a. Let $N = 0.3\overline{27}$

$$100N = 32.7\overline{27}$$
$$\text{Subtract } N = 0.3\overline{27}$$
$$99N = 32.4$$
$$N = \frac{32.4}{99} = \frac{324}{990} = \frac{18}{55}$$

$\therefore\ 0.3\overline{27} = \frac{18}{55}$ ***Answer***

b. Let $N = 1.89\overline{189}$

$$1000N = 1891.89\overline{189}$$
$$\text{Subtract } N = 1.89\overline{189}$$
$$999N = 1890$$
$$N = \frac{1890}{999} = \frac{70}{37}$$

$\therefore\ 1.89\overline{189} = \frac{70}{37}$ ***Answer***

An **irrational number** is a real number that is not rational. Therefore, the decimal representation of an irrational number is neither terminating nor repeating. For example,

$$\pi = 3.14159265\ldots \qquad \text{and} \qquad \sqrt{3} = 1.7320508\ldots$$

are irrational.

1. The decimal representation of any irrational number is infinite and nonrepeating.
2. Every infinite and nonrepeating decimal represents an irrational number.

This fact is significant when you use a calculator to obtain the decimal representation of an irrational number. Because the calculator's display is finite, you can get only a *rational approximation* of the irrational number.

Example 4 Classify each real number as either rational or irrational.

a. $\sqrt{2}$ **b.** $\sqrt{\frac{4}{9}}$

c. 2.030303 . . . **d.** 2.030030003 . . .

Solution

a. Since $\sqrt{2} = 1.41421356\ldots$ (an infinite, nonrepeating decimal), $\sqrt{2}$ is irrational. (For a proof of the irrationality of $\sqrt{2}$, see the Extra on page 273.)

b. Since $\sqrt{\frac{4}{9}} = \frac{2}{3}$, $\sqrt{\frac{4}{9}}$ is rational.

c. Since 2.030303 . . . (or $2.\overline{03}$) is a repeating decimal, 2.030303 . . . is rational.

d. Since 2.030030003 . . . is infinite and nonrepeating, 2.030030003 . . . is irrational.

A set S of real numbers is **dense** if between any two numbers in the set there is a member of S. Both the set of rational numbers and the set of irrational numbers are dense.

Example 5 Find a rational number r and an irrational number s between 1.51287 and 1.51288.

Solution There are an infinite number of possible answers for r and s.

For example, $r = 1.512871111111111111\ldots$

and $s = 1.512871011011101110\ldots$

Notice that r is repeating and is therefore rational and that s is nonrepeating and is therefore irrational.

Chalkboard Examples

Write as a finite or repeating decimal.

1. $\frac{17}{80}$

$$\begin{array}{r} 0.2125 \\ 80\overline{)17.0000} \\ \underline{16\ 0} \\ 1\ 00 \\ \underline{80} \\ 200 \\ \underline{160} \\ 400 \\ \underline{400} \\ 0 \end{array}$$

$\therefore \frac{17}{80} = 0.2125$

2. $\frac{13}{33}$

$$\begin{array}{r} 0.3939 \\ 33\overline{)13.0000} \\ \underline{9\ 9} \\ 3\ 10 \\ \underline{2\ 97} \\ 130 \\ \underline{99} \\ 310 \\ \underline{297} \\ 13 \end{array}$$

$\therefore \frac{13}{33} = 0.\overline{39}$

3. Write as a common fraction: $0.\overline{348}$

Let $N = 0.\overline{348}$

$$\begin{array}{rl} 1000N = & 348.\overline{348} \\ -\quad N = & 0.\overline{348} \\ \hline 999N = & 348 \end{array}$$

$N = \frac{348}{999} = \frac{116}{333}$

4. Find a rational number r and an irrational number s between $\frac{1}{5}$ and $\frac{1}{4}$. Since $\frac{1}{5}$ and $\frac{1}{4}$ are rational, the number halfway between them is rational, so $r = \frac{9}{40}$. Since $\frac{1}{5} = 0.2$ and $\frac{1}{4} = 0.25$, $s = 0.212112111\cdots$.

Check for Understanding

Here is a suggested use of the Oral Exercises to check students' understanding as you teach the lesson.
Oral Exs. 1–16: use after Example 4.

Guided Practice

1. Classify as rational or irrational.
   a. $\sqrt[3]{16}$ irr.
   b. $\sqrt[3]{27}$ rat.
   c. 5π irr.
   d. $\frac{7.23}{3.14}$ rat.
2. Write as a decimal.
   a. $\frac{5}{8}$ 0.625
   b. $\frac{5}{11}$ $0.\overline{45}$
   c. $\frac{5}{6}$ $0.8\overline{3}$
   d. $\frac{5}{7}$ $0.\overline{714285}$
3. Write as a common fraction in lowest terms.
   a. $0.\overline{8}$ $\frac{8}{9}$
   b. 0.08 $\frac{2}{25}$
   c. $0.\overline{08}$ $\frac{8}{99}$
   d. $0.0\overline{8}$ $\frac{4}{45}$

Summarizing the Lesson

In this lesson students learned to distinguish between rational and irrational numbers by examining their decimal representations. Ask students to explain the meaning of the term "dense" as it applies to this lesson.

Supplementary Materials

Study Guide	pp. 99–100
Practice Master	36
Resource Book	p. 124

Oral Exercises R = rational; Ir = irrational

Tell whether each number is rational or irrational.

1. $\frac{15}{34}$ R
2. -7 R
3. 9.26 R
4. $\sqrt{0.01}$ R
5. $\sqrt[3]{8}$ R
6. $\sqrt[3]{9}$ Ir
7. $\sqrt{\frac{16}{25}}$ R
8. $\frac{\sqrt{2}}{2}$ Ir
9. π Ir
10. $\frac{\pi}{\pi + 1}$ Ir
11. $(2\sqrt{3})^2$ R
12. $(2\sqrt{3})^3$ Ir

Tell whether or not each number is real. If it is real, tell whether it is rational or irrational.

13. $\sqrt{1000}$ Real, irrational
14. $\sqrt{-1000}$ Not real
15. $\sqrt{(-1000)^2}$ Real, rational
16. $\sqrt[3]{-1000}$ Real, rational

Written Exercises R = rational; Ir = irrational

Classify each real number or expression as either rational or irrational.

A

1. a. $\sqrt{49}$ R b. $\sqrt{50}$ Ir
2. a. π Ir b. $\frac{22}{7}$ R
3. a. $\pi + \frac{1}{\pi}$ Ir b. $\pi \cdot \frac{1}{\pi}$ R
4. a. $\sqrt{\frac{1}{2}} + \sqrt{\frac{1}{8}}$ Ir b. $\sqrt{\frac{1}{2}} \cdot \sqrt{\frac{1}{8}}$ R

Classify each real number as either rational or irrational.

5. a. 1.23 R b. $1.\overline{23}$ R c. 1.2345678910111213 . . . Ir
6. a. 75.4682 R b. $75.\overline{4682}$ R c. 75.468244682444682 . . . Ir

Write each fraction as a terminating or repeating decimal.

7. $\frac{5}{8}$ 0.625
8. $\frac{5}{11}$ $0.\overline{45}$
9. $\frac{13}{7}$ $1.\overline{857142}$
10. $\frac{13}{4}$ 3.25

Write each decimal as a common fraction in lowest terms.

11. 5.06 $\frac{253}{50}$
12. 3.004 $\frac{751}{250}$
13. 4.72 $\frac{118}{25}$
14. 0.1375 $\frac{11}{80}$
15. $0.\overline{4}$ $\frac{4}{9}$
16. $0.\overline{5}$ $\frac{5}{9}$
17. $0.8\overline{3}$ $\frac{5}{6}$
18. $0.08\overline{3}$ $\frac{1}{12}$
19. $2.\overline{36}$ $\frac{26}{11}$
20. $1.\overline{27}$ $\frac{14}{11}$
21. $3.\overline{033}$ $\frac{1010}{333}$
22. $1.\overline{101}$ $\frac{1100}{999}$

Find (a) a rational, and (b) an irrational, number between each given pair.

23. 0.1 and 0.2
24. 0.3725 and 0.3726
25. $\sqrt{6}$ and $\sqrt{7}$
26. 3 and π
27. $\sqrt{15}$ and 4
28. 10^{-9} and 10^{-8}
29. $\frac{7}{8}$ and $\frac{8}{9}$
30. $3\frac{1}{7}$ and $3\frac{1}{6}$
31. $0.\overline{4}$ and $0.\overline{45}$

B 32. a. The *arithmetic mean* of two numbers x and y is the number $\frac{x + y}{2}$. If x and y are rational numbers, what can you conclude about their arithmetic mean? The arithmetic mean is rational.

b. The *geometric mean* of two positive numbers x and y is the number $\sqrt{xy}$. If x and y are positive rational numbers, can you conclude that their geometric mean is also rational? Explain. No; if the product xy is not a perfect square, $\sqrt{xy}$ is irrational.

33. a. Find two irrational numbers x and y such that $\frac{x + y}{2}$ is rational.

b. Find two irrational numbers a and b such that $\frac{a + b}{2}$ is irrational.

34. If $x < y$, then $x < \frac{x + y}{2} < y$. Use this fact to give a convincing argument for the following: Between any two rational numbers there are infinitely many rational numbers.

35. Each of these numbers is between $\sqrt{2}$ and $\sqrt{3}$. Which are irrational? (1) and (4)

(1) $\sqrt{2\frac{1}{2}}$ **(2)** $\sqrt{2\frac{1}{4}}$ **(3)** $1.\overline{4}$ **(4)** 1.456789101112 . . .

36. a. A student who used long division to find the repeating decimal for $\frac{1}{17}$ claimed that there were 20 digits in the repeating block of digits. Explain why this is incorrect. There can be no more than 16 digits.

b. Find the repeating decimal for $\frac{1}{17}$. (Be patient!) $0.\overline{0588235294117647}$

37. The following unit fractions all have finite decimals: $\frac{1}{2}$, $\frac{1}{4}$, $\frac{1}{5}$, $\frac{1}{8}$, $\frac{1}{25}$, $\frac{1}{40}$. Name four other unit fractions with finite decimals. Answers may vary. For example: $\frac{1}{10}$, $\frac{1}{16}$, $\frac{1}{20}$, $\frac{1}{50}$

38. Suppose x and y are rational numbers. Prove that $x + y$ is also rational. $\left(\textit{Hint:}\text{ Let } x = \frac{a}{b} \text{ and } y = \frac{c}{d} \text{ where } a, b, c, \text{ and } d \text{ are integers. Show that } x + y \text{ can be written as a quotient of integers.}\right)$

39. a. Prove that the product of two rational numbers is rational. (See the hint for Exercise 38.)

b. What can you say about the product of two irrational numbers?

C 40. Suppose x is rational and z is irrational. Prove that $x + z$ is irrational. $\left(\textit{Hint:}\text{ Let } x = \frac{a}{b} \text{ where } a \text{ and } b \text{ are integers, and use an indirect proof by assuming that } x + z \text{ is a rational number } \frac{c}{d}.\right)$

41. What can you say about the product of a nonzero rational number and an irrational number? Prove your answer by using an indirect proof.

42. Show that each repeating decimal represents a rational number as follows: Let N = the number. Let p = the number of digits in the repeating block of N. Justify the following statements.

a. $N \cdot 10^p$ has the same repeating block as N.

b. $N \cdot 10^p - N$ is a terminating decimal.

c. $N(10^p - 1)$ is rational. **d.** $\frac{N(10^p - 1)}{10^p - 1}$ is rational.

Suggested Assignments

Average Algebra
286/1–37 odd
R 282/Self-Test 1

Average Alg. and Trig.
286/1–37 odd
R 282/Self-Test 1

Extended Algebra
286/1–39 odd, 40
R 282/Self-Test 1

Extended Alg. and Trig.
286/1–39 odd, 40
R 282/Self-Test 1

Additional Answers Written Exercises

23–31. Answers may vary; examples are given.

23. a. 0.15 **b.** $\frac{\sqrt{2}}{10}$

24. a. 0.37255
b. 0.372515115111 . . .

25. a. 2.5 **b.** $\sqrt{6.5}$

26. a. 3.1 **b.** $\frac{3 + \pi}{2}$

27. a. 3.9 **b.** $\sqrt{15.5}$

28. a. 5×10^{-9}
b. 0.00000000101001000 . . .

29. a. $\frac{127}{144}$
b. 0.87515115111 . . .

30. a. $3\frac{13}{24}$ **b.** $\pi + 0.01$

31. a. 0.45 **b.** $\frac{\sqrt{5}}{5}$

33. Answers may vary; examples are given.
a. $x = \sqrt{2}$ and $y = -\sqrt{2}$
b. $x = \sqrt{2}$ and $y = \sqrt{3}$

34. It is possible to create an infinite sequence of rational numbers between x and y by successively finding the arithmetic mean of x and the previous mean.

(continued on p. 302)

Teaching Suggestions p. T103

Suggested Extensions p. T103

Warm-Up Exercises

Simplify.

1. $\sqrt{24}$ $2\sqrt{6}$
2. $\sqrt{18}\cdot\sqrt{6}$ $6\sqrt{3}$
3. $\frac{5}{\sqrt{10}}$ $\frac{\sqrt{10}}{2}$
4. $\sqrt{35x}\cdot\sqrt{7x^3}$ $7x^2\sqrt{5}$
5. $\sqrt{8x}+\sqrt{50x}$ $7\sqrt{2x}$

Motivating the Lesson

As in other human endeavors, mathematics has grown as a result of imaginative thinking. An example of this fact is the number i, defined as $\sqrt{-1}$. This "imaginary" number gave mathematicians the power to solve certain equations that have no real solutions.

Common Errors

Often students will simplify $\sqrt{-4}\cdot\sqrt{-9}$ incorrectly as $\sqrt{36}$. These students must be instructed that the rule $\sqrt{a}\cdot\sqrt{b}=\sqrt{ab}$ does not hold if a and b are both negative. When reteaching, emphasize that the students must express the radicals in pure imaginary form before simplifying.

6-7 The Imaginary Number *i*

Objective To use the number i to simplify square roots of negative numbers.

Throughout the history of mathematics, new numbers have been invented to extend the existing number system. For example, the invention of negative numbers made it possible to solve an equation like $x + 3 = 2$. Similarly, the invention of irrational numbers made an equation like $x^2 = 2$ solvable.

Now consider the equation $x^2 = -1$. This equation has no real solution, because the square of a real number is never negative. To solve this equation the *imaginary number i* was invented.

Definition of *i*

$$i = \sqrt{-1} \quad \text{and} \quad i^2 = -1.$$

The equation $x^2 = -1$ can now be solved: $x = \pm\sqrt{-1} = \pm i$. Because i and $-i$ are solutions of $x^2 = -1$, both i and $-i$ are square roots of -1.

The number i can be used to find square roots of other negative numbers. For example, since

$$(2i)^2 = 2^2 \cdot i^2 = 4(-1) = -4,$$
$$2i = \sqrt{-4}.$$

Similarly, $$3i = \sqrt{-9}.$$

If r is a positive real number, then $\sqrt{-r} = i\sqrt{r}$.

Example 1 Simplify: **a.** $\sqrt{-5}$ **b.** $\sqrt{-25}$ **c.** $\sqrt{-50}$

Solution **a.** $\sqrt{-5} = i\sqrt{5}$ **b.** $\sqrt{-25} = 5i$ **c.** $\sqrt{-50} = i\sqrt{50} = 5i\sqrt{2}$

Numbers like $i\sqrt{5}$ and $5i$, which have the form bi (where b is a nonzero real number), are called **pure imaginary numbers.** Example 2 illustrates how you can add and subtract these numbers.

Example 2 Simplify.

a. $\sqrt{-16} - \sqrt{-49}$ **b.** $i\sqrt{2} + 3i\sqrt{2}$

Solution **a.** $\sqrt{-16} - \sqrt{-49} = 4i - 7i$
$= -3i$ ***Answer***

b. $i\sqrt{2} + 3i\sqrt{2} = (1 + 3)i\sqrt{2}$
$= 4i\sqrt{2}$ ***Answer***

Examples 3 and 4 illustrate how you can multiply and divide pure imaginary numbers.

Example 3 Simplify.

a. $\sqrt{-4} \cdot \sqrt{-25}$

b. $i\sqrt{2} \cdot i\sqrt{3}$

Solution

a. $\sqrt{-4} \cdot \sqrt{-25} = i\sqrt{4} \cdot i\sqrt{25}$
$= 2i \cdot 5i$
$= 10i^2 = -10$ ***Answer***

b. $i\sqrt{2} \cdot i\sqrt{3} = i^2\sqrt{6}$
$= (-1)\sqrt{6}$
$= -\sqrt{6}$ ***Answer***

Caution: The rule $\sqrt{a} \cdot \sqrt{b} = \sqrt{ab}$ does *not* hold if a and b are both negative. Notice that if you tried to use this rule in part (a) in Example 3 you would *incorrectly* get $\sqrt{-4} \cdot \sqrt{-25} = \sqrt{100} = 10$. Therefore, whenever you are simplifying expressions involving square roots of negative numbers, you must first express the radicals as pure imaginary numbers.

Example 4 Simplify.

a. $\frac{2}{3i}$

b. $\frac{6}{\sqrt{-2}}$

Solution

a. $\frac{2}{3i} = \frac{2}{3i} \cdot \frac{i}{i}$
$= \frac{2i}{3i^2}$
$= \frac{2i}{3(-1)}$
$= -\frac{2}{3}i$ ***Answer***

b. $\frac{6}{\sqrt{-2}} = \frac{6}{i\sqrt{2}}$
$= \frac{6}{i\sqrt{2}} \cdot \frac{i\sqrt{2}}{i\sqrt{2}}$
$= \frac{6i\sqrt{2}}{2i^2}$
$= \frac{6i\sqrt{2}}{2(-1)}$
$= -3i\sqrt{2}$ ***Answer***

Example 5 Simplify. Assume that each variable represents a positive real number.

a. $\sqrt{-9x^2} + \sqrt{-x^2}$

b. $\sqrt{-6y} \cdot \sqrt{-2y}$

Solution

a. $\sqrt{-9x^2} + \sqrt{-x^2} = 3xi + xi$
$= 4xi$ ***Answer***

b. $\sqrt{-6y} \cdot \sqrt{-2y} = i\sqrt{6y} \cdot i\sqrt{2y}$
$= i^2\sqrt{12y^2}$
$= (-1)\sqrt{4y^2 \cdot 3}$
$= -2y\sqrt{3}$ ***Answer***

Chalkboard Examples

Simplify.

1. $\sqrt{-32}$
$i\sqrt{32} = i\sqrt{16 \cdot 2}$
$= 4i\sqrt{2}$

2. $5\sqrt{-6} - \sqrt{-24}$
$5i\sqrt{6} - 2i\sqrt{6} = 3i\sqrt{6}$

3. $\sqrt{-3} \cdot \sqrt{-27}$
$i\sqrt{3} \cdot i\sqrt{27} =$
$i^2\sqrt{3} \cdot 3\sqrt{3} =$
$(-1)(3)(\sqrt{3})^2 =$
$-3 \cdot 3 = -9$

4. $\sqrt{-5} \cdot \sqrt{-75}$
$i\sqrt{5} \cdot i\sqrt{75} =$
$i^2 \cdot \sqrt{5} \cdot 5\sqrt{3} =$
$(-1)(5)(\sqrt{5})(\sqrt{3}) =$
$-5\sqrt{15}$

5. $\frac{4\sqrt{-18}}{\sqrt{-12}}$
$\frac{12i\sqrt{2}}{2i\sqrt{3}} = \frac{6\sqrt{2}}{\sqrt{3}} \cdot \frac{\sqrt{3}}{\sqrt{3}} =$
$\frac{6\sqrt{6}}{3} = 2\sqrt{6}$

6. $\frac{15}{\sqrt{-32}}$
$\frac{15}{4i\sqrt{2}} =$
$-i \cdot \frac{15}{4\sqrt{2}} \cdot \frac{\sqrt{2}}{\sqrt{2}} = -\frac{15i\sqrt{2}}{8}$

Solve.

7. $4x^2 + 25 = -23$
$4x^2 = -48$
$x^2 = -12$
$x = \pm i\sqrt{12}$
$x = \pm 2i\sqrt{3}$

Check for Understanding

Here is a suggested use of the Oral Exercises to check students' understanding as you teach the lesson.
Oral Exs. 1–6: use after Example 1.
Oral Exs. 7–15: use after Example 3.
Oral Exs. 16–22: use after Example 6.

Guided Practice

Simplify.

1. $3\sqrt{-4}$ $6i$
2. $\sqrt{-45}$ $3i\sqrt{5}$
3. $-\sqrt{-27}$ $-3i\sqrt{3}$
4. $(5i)^2$ -25
5. $-(i\sqrt{2})^2$ 2
6. $(-i\sqrt{7})^2$ -7
7. $(2i)^3$ $-8i$
8. $\frac{4}{3i}$ $-\frac{4}{3}i$
9. $\frac{\sqrt{-7}}{2i}$ $\frac{\sqrt{7}}{2}$
10. $\sqrt{-8} + \sqrt{-18}$ $5i\sqrt{2}$
11. $\sqrt{-4} \cdot \sqrt{-9}$ -6
12. $i\sqrt{4} \cdot \sqrt{-4}$ -4

Solve.

13. $x^2 + 144 = 0$ $\{12i, -12i\}$
14. $4x^2 = x^2 - 18$ $\{i\sqrt{6}, -i\sqrt{6}\}$

Summarizing the Lesson

In this lesson students learned to use the number i in simplifying expressions containing the square root of a negative number. Ask students to explain why i is not a real number.

Using pure imaginary numbers, you can solve simple quadratic equations that have no real solutions.

Example 6 Solve $2x^2 + 19 = 3$.

Solution

$$2x^2 + 19 = 3$$
$$2x^2 = -16$$
$$x^2 = -8$$
$$x = \pm\sqrt{-8} = \pm 2i\sqrt{2}$$

$\therefore$ the solution set is $\{2i\sqrt{2}, -2i\sqrt{2}\}$. ***Answer***

Oral Exercises

Simplify.

1. $\sqrt{-64}$ $8i$
2. $\sqrt{-100}$ $10i$
3. $\sqrt{-11}$ $i\sqrt{11}$
4. $-\sqrt{-9}$ $-3i$
5. $\sqrt{-12}$ $2i\sqrt{3}$
6. $-\sqrt{-18}$ $-3i\sqrt{2}$
7. $3i^2$ -3
8. $(3i)^2$ -9
9. $(-4i)^2$ -16
10. $2i \cdot 6i$ -12
11. $-i \cdot 8i$ 8
12. $(i\sqrt{5})^2$ -5
13. $\sqrt{-3} \cdot \sqrt{-3}$ -3
14. $\sqrt{11} \cdot \sqrt{-11}$ $11i$
15. $-\sqrt{2} \cdot \sqrt{-2}$ $-2i$

Solve.

16. $x^2 = -16$ $\{\pm 4i\}$
17. $-y^2 = 11$ $\{\pm i\sqrt{11}\}$
18. $3w^2 = -12$ $\{\pm 2i\}$
19. $-2n^2 = 14$ $\{\pm i\sqrt{7}\}$
20. $6 - t^2 = 7$ $\{\pm i\}$
21. $8 + z^2 = 3$ $\{\pm i\sqrt{5}\}$
22. What is the reciprocal of i (in simplified form)? $-i$

Written Exercises

Simplify.

A

1. $\sqrt{-81}$ $9i$
2. $\sqrt{-121}$ $11i$
3. $-4\sqrt{-36}$ $-24i$
4. $-2\sqrt{-144}$ $-24i$
5. $\sqrt{-20}$ $2i\sqrt{5}$
6. $\sqrt{-75}$ $5i\sqrt{3}$
7. $3\sqrt{-8}$ $6i\sqrt{2}$
8. $5\sqrt{-27}$ $15i\sqrt{3}$
9. $2i \cdot 3i$ -6
10. $5i \cdot 3i$ -15
11. $\sqrt{7} \cdot \sqrt{-7}$ $7i$
12. $\sqrt{-6} \cdot \sqrt{2}$ $2i\sqrt{3}$
13. $\sqrt{-5} \cdot \sqrt{-10}$ $-5\sqrt{2}$
14. $\sqrt{-3} \cdot \sqrt{-6}$ $-3\sqrt{2}$
15. $(7i)^2$ -49
16. $(8i)^2$ -64
17. $(-i)^2$ -1
18. $(-5i)^2$ -25
19. $(i\sqrt{2})^2$ -2
20. $(3i\sqrt{5})^2$ -45
21. $(-i\sqrt{3})^2$ -3
22. $(-3i\sqrt{6})^2$ -54
23. $-\frac{2}{i}$ $2i$
24. $\frac{8}{3i}$ $-\frac{8i}{3}$

Simplify.

25. $\frac{1}{\sqrt{-5}}$ $-\frac{i\sqrt{5}}{5}$

26. $\frac{4}{\sqrt{-4}}$ $-2i$

27. $\frac{\sqrt{18}}{2i\sqrt{6}}$ $-\frac{i\sqrt{3}}{2}$

28. $\frac{\sqrt{28}}{4i\sqrt{7}}$ $-\frac{i}{2}$

29. $\frac{\sqrt{60}}{\sqrt{-15}}$ $-2i$

30. $-\frac{\sqrt{12}}{\sqrt{-18}}$ $\frac{i\sqrt{6}}{3}$

Solve.

31. $x^2 + 144 = 0$ $\{\pm 12i\}$

32. $y^2 + 400 = 0$ $\{\pm 20i\}$

33. $2w^2 = -98$ $\{\pm 7i\}$

34. $5t^2 = -20$ $\{\pm 2i\}$

35. $3u^2 + 40 = 4$ $\{\pm 2i\sqrt{3}\}$

36. $4z^2 + 39 = 7$ $\{\pm 2i\sqrt{2}\}$

Simplify.

37. *a.* $\sqrt{-25} + \sqrt{-36}$ $11i$ *b.* $\sqrt{-25} \cdot \sqrt{-36}$ -30

38. *a.* $\sqrt{-3} + \sqrt{-27}$ $4i\sqrt{3}$ *b.* $\sqrt{-3} \cdot \sqrt{-27}$ -9

39. *a.* $3\sqrt{-2} - \sqrt{-50}$ $-2i\sqrt{2}$ *b.* $3\sqrt{-2} \cdot (-\sqrt{-50})$ 30

40. *a.* $2\sqrt{-24} - \sqrt{-54}$ $i\sqrt{6}$ *b.* $2\sqrt{-24} \cdot (-\sqrt{-54})$ 72

41. *a.* $i\sqrt{18} + \sqrt{-8}$ $5i\sqrt{2}$ *b.* $i\sqrt{18} \cdot \sqrt{-8}$ -12

42. *a.* $i\sqrt{-98} - \sqrt{98}$ $-14\sqrt{2}$ *b.* $i\sqrt{-98} \cdot (-\sqrt{98})$ 98

Simplify. Assume that each variable represents a positive number.

B 43. $\sqrt{-12a} \cdot \sqrt{-3a}$ $-6a$

44. $-\sqrt{18c} \cdot \sqrt{-2c^3}$ $-6c^2i$

45. $\sqrt{-\frac{r}{5}} \cdot \sqrt{-\frac{20}{r}}$ -2

46. $\sqrt{-\frac{t^5}{2}} \cdot \sqrt{-\frac{2}{t^3}}$ $-t$

47. $\sqrt{-3c^2} + \sqrt{-27c^2} - \sqrt{-45c^2}$ $4ic\sqrt{3} - 3ic\sqrt{5}$

48. $\sqrt{-2t^5} + \sqrt{-8t^5} - \sqrt{-18t^5}$ 0

49. $\sqrt{-4r^3} + \sqrt{-64r^3} - 4r\sqrt{-16r}$ $-6ri\sqrt{r}$

50. $\sqrt{-25a^3} - \sqrt{-225a^3} + 20a\sqrt{-a}$ $10ai\sqrt{a}$

51. $\sqrt{-x^5} + x\sqrt{-25x^2} - x^2\sqrt{-25x}$ $5x^2i - 4x^2i\sqrt{x}$

52. $yi\sqrt{-16y^2} + \sqrt{16y^4} - y^2i\sqrt{-9}$ $3y^2$

53. Simplify i^n for $n = 2, 3, 4, \ldots, 12$. What pattern do you see?

54. Simplify: **a.** i^{100} 1 **b.** i^{101} i **c.** i^{102} -1 **d.** i^{103} $-i$

Mixed Review Exercises

Solve. If an equation has no real solutions, say so.

1. $\sqrt{2x - 3} = 5$ $\{14\}$

2. $15 - 2n = n^2$ $\{-5, 3\}$

3. $\sqrt{y^2 + 12} = 2y$ $\{2\}$

4. $\frac{3y - 4}{5} = \frac{y + 1}{2}$ $\{13\}$

5. $\frac{1}{n} + \frac{2}{n - 2} = \frac{4}{n(n - 2)}$ No sol.

6. $2\sqrt[3]{x} + 9 = 5$ $\{-8\}$

7. $y = \sqrt{5y - 6}$ $\{2, 3\}$

8. $5|n| - 7 = 3$ $\{-2, 2\}$

9. $x = 2 + \sqrt{x + 4}$ $\{5\}$

Classify each real number as either rational or irrational. R = rational; Ir = irrational

10. $\sqrt[3]{-125}$ R

11. $3.7\overline{82}$ R

12. $\sqrt[5]{12}$ Ir

13. $0.121121112\ldots$ Ir

Suggested Assignments

Average Algebra
Day 1: 290/2–36 even
S 291/Mixed Review
Day 2: 291/38–52 even
Assign with Lesson 6-8.

Average Alg. and Trig.
Day 1: 290/2–36 even
S 291/Mixed Review
Day 2: 291/38–52 even
Assign with Lesson 6-8.

Extended Algebra
Day 1: 290/2–36 even
S 291/Mixed Review
Day 2: 291/38–54 even
Assign with Lesson 6-8.

Extended Alg. and Trig.
Day 1: 290/2–36 even
S 291/Mixed Review
Day 2: 291/38–54 even
Assign with Lesson 6-8.

Supplementary Materials

Study Guide pp. 101–102

Additional Answers Written Exercises

53. $i^2 = -1$; $i^3 = -i$; $i^4 = 1$; $i^5 = i$; $i^6 = -1$; $i^7 = -i$; $i^8 = 1$; $i^9 = i$; $i^{10} = -1$; $i^{11} = -i$; $i^{12} = 1$. If $n = 4k + r$ (k and r integers), then $i^n = i^r$.

Teaching Suggestions p. T104

Suggested Extensions p. T104

Warm-Up Exercises

Simplify.

1. $\sqrt{-36}$ $6i$
2. $\sqrt{-12} + \sqrt{-27}$ $5i\sqrt{3}$
3. $\sqrt{18} + \sqrt{50}$ $8\sqrt{2}$
4. $(4i)^2$ -16
5. $\sqrt{32} + \sqrt{-9} \cdot \sqrt{-2}$ $\sqrt{2}$

Motivating the Lesson

Until students encountered pure imaginary numbers, they worked exclusively with real numbers. Today students see how these two types of numbers can be combined to form still other types of numbers.

Common Errors

When subtracting one complex number from another, students may change the sign of only the first term of the complex number that is being subtracted. When reteaching, remind students that they must add the opposite of each term of the complex number that is being subtracted.

When squaring a complex number, students may only square each term. When reteaching, remind the students that $(a + bi)^2 = (a + bi) \cdot (a + bi)$ and that the distributive property must be used to simplify this expression further.

6-8 The Complex Numbers

Objective To add, subtract, multiply, and divide complex numbers.

In this lesson, you will see how the real numbers can be extended to form a number system that includes the imaginary number i. This extended system, called the *complex numbers,* has become an important mathematical tool for physicists and engineers working in fields such as electricity and magnetism, optics, and hydrodynamics.

In the preceding lesson you worked with pure imaginary numbers, which have the form bi $(b \neq 0)$. When pure imaginary and real numbers are combined, **imaginary numbers** of the form $a + bi$ $(b \neq 0)$ are the result. The real numbers and the imaginary numbers together form the set of *complex numbers*.

A **complex number** is a number of the form $a + bi$, where a and b are real numbers. The chart below shows the relationship between the various sets of numbers that make up the complex numbers. In the chart, a and b are real numbers.

- Complex numbers: $a + bi$ — $3 + 2i,\ 5 + 0i,\ -i\sqrt{2}$
  - Real numbers: $a + bi,\ b = 0$ — $5,\ \sqrt[3]{5},\ \pi,\ -\frac{2}{7},\ 5.\overline{67}$
    - Rational numbers — $5,\ -\frac{2}{7},\ 5.\overline{67}$
    - Irrational numbers — $\sqrt[3]{5},\ \pi$
  - Imaginary numbers: $a + bi,\ b \neq 0$ — $3 + 2i,\ \frac{2}{3} - \frac{1}{2}i,\ 7i,\ -i\sqrt{2}$
    - Pure imaginary numbers: $a + bi,\ a = 0,\ b \neq 0$ — $7i,\ -i\sqrt{2}$

The number a is called the **real part** of the complex number $a + bi$, and b (not bi) is called the **imaginary part.**

Two complex numbers are equal if and only if they have equal real parts and equal imaginary parts. For example, if $x + yi = 5 - i$, then $x = 5$ and $y = -1$.

Equality of Complex Numbers

$$a + bi = c + di \quad \text{if and only if} \quad a = c \text{ and } b = d.$$

The sum and the product of two complex numbers are defined as follows.

Sum of Complex Numbers

$$(a + bi) + (c + di) = (a + c) + (b + d)i$$

Product of Complex Numbers

$$(a + bi)(c + di) = (ac - bd) + (ad + bc)i$$

Notice that you add or subtract complex numbers by combining their real parts and their imaginary parts separately.

Example 1 Simplify.

a. $(3 + 6i) + (4 - 2i)$ **b.** $(3 + 6i) - (4 - 2i)$

Solution **a.** $(3 + 6i) + (4 - 2i) = (3 + 4) + (6 - 2)i = 7 + 4i$

b. $(3 + 6i) - (4 - 2i) = (3 - 4) + (6 + 2)i = -1 + 8i$

The definition of multiplication given above is equivalent to the following rule for multiplying two complex numbers: *Multiply two complex numbers as you would multiply any two binomials and use the fact that $i^2 = -1$.*

Example 2 Simplify.

a. $(3 + 4i)(5 + 2i)$ **b.** $(3 + 4i)^2$ **c.** $(3 + 4i)(3 - 4i)$

Solution **a.** $(3 + 4i)(5 + 2i) = 15 + 20i + 6i + 8i^2$
$= 15 + 26i + 8(-1)$
$= 7 + 26i$ ***Answer***

b. $(3 + 4i)^2 = 9 + 24i + 16i^2$
$= 9 + 24i - 16$
$= -7 + 24i$ ***Answer***

c. $(3 + 4i)(3 - 4i) = 9 - 16i^2$
$= 9 + 16 = 25$ ***Answer***

Chalkboard Examples

Simplify.

1. $(4 - 2i) + (3 - 3i) + (-4 + 7i)$
$(4 + 3 - 4) + (-2i - 3i + 7i) = 3 + 2i$

2. $(3 - 2i) - (2 - 3i)$
$(3 - 2) + (-2i + 3i) = 1 + i$

3. $(2 + 5i)(3 - 2i)$
$6 + 15i - 4i - 10i^2 = 6 + 11i + 10 = 16 + 11i$

4. $3i(2i - 1)^2$
$3i(4i^2 - 4i + 1) = 3i(-3 - 4i) = -9i - 12i^2 = 12 - 9i$

5. $\frac{4 - 3i}{2 - 4i}$
$\frac{4 - 3i}{2 - 4i} \cdot \frac{2 + 4i}{2 + 4i} = \frac{8 + 10i - 12i^2}{4 - 16i^2} = \frac{20 + 10i}{20} = 1 + \frac{1}{2}i$

6. Find the reciprocal of $4 + i$.
$\frac{1}{4 + i} = \frac{1}{4 + i} \cdot \frac{4 - i}{4 - i} = \frac{4 - i}{16 - i^2} = \frac{4 - i}{17} = \frac{4}{17} - \frac{1}{17}i$

Check for Understanding

Here is a suggested use of the Oral Exercises to check students' understanding as you teach the lesson.
Oral Exs. 1–12: use after Example 2.
Oral Exs. 13–20: use after discussing the chart on page 292.

Guided Practice

Simplify.

1. $(12 - 2i) - (8 - 3i)$ $4 + i$
2. $3i(2 - 3i)$ $9 + 6i$
3. $(6 + i)(6 - i)$ 37
4. $(-2 - 2i)(3 - i)$ $-8 - 4i$
5. $(3 + 2i)^2$ $5 + 12i$
6. $(\sqrt{3} - 2i)^2$ $-1 - 4i\sqrt{3}$
7. $(\sqrt{3} + \sqrt{-2}) \cdot (\sqrt{3} - \sqrt{-2})$ 5
8. $\frac{-5}{2 + i}$ $-2 + i$
9. $\frac{5i}{3 - 4i}$ $-\frac{4}{5} + \frac{3}{5}i$
10. $\frac{2 + i}{6 + 2i}$ $\frac{7}{20} + \frac{1}{20}i$

The numbers $3 + 4i$ and $3 - 4i$ in part (c) of Example 2 are examples of *conjugates*. In general, the complex numbers $a + bi$ and $a - bi$ are **complex conjugates,** and their product is the real number $a^2 + b^2$. This fact can be used to simplify the quotient of two imaginary numbers.

Example 3 Simplify $\frac{5 - i}{2 + 3i}$.

Solution

$$\frac{5 - i}{2 + 3i} = \frac{5 - i}{2 + 3i} \cdot \frac{2 - 3i}{2 - 3i}$$ ← Multiply numerator and denominator by the conjugate of the denominator.

$$= \frac{10 - 2i - 15i + 3i^2}{4 - 9i^2}$$

$$= \frac{10 - 17i + 3(-1)}{4 - 9(-1)}$$

$$= \frac{7 - 17i}{13} = \frac{7}{13} - \frac{17}{13}i$$ ***Answer***

You usually express the result of a computation with complex numbers in the form $a + bi$, as was done in Example 3. This form emphasizes that the answer is another complex number.

The examples of this lesson show that the sum, difference, product, or quotient of two complex numbers is another complex number. A power or root of a complex number is also a complex number. In short, the complex numbers form a closed number system.

Example 4 illustrates that the reciprocal of a complex number is another complex number.

Example 4 Find the reciprocal of $3 - i$.

Solution

$$\frac{1}{3 - i} = \frac{1}{3 - i} \cdot \frac{3 + i}{3 + i}$$

$$= \frac{3 + i}{9 - i^2}$$

$$= \frac{3 + i}{10} = \frac{3}{10} + \frac{1}{10}i$$ ***Answer***

In addition to being closed for all operations, the complex number system has many other properties that the real number system has: Addition and multiplication are commutative and associative, and the distributive property holds. (See Exercises 53–56.) However, the complex numbers cannot be ordered along a number line like the real numbers can. Although you cannot say that one complex number is greater than another, you can measure the "size" of a complex number by using absolute value. See the Extra on page 298 for more on absolute values of complex numbers and conjugates.

Oral Exercises

Simplify.

1. $(3 + 2i) + (5 + 7i)$ $8 + 9i$
2. $(9 + 4i) + (7 - 5i)$ $16 - i$
3. $(-4 + 3i) - (6 - i)$ $-10 + 4i$
4. $(9 - 2i) - (6 - 3i)$ $3 + i$
5. $2i(5 - i)$ $2 + 10i$
6. $3i(7 + 2i)$ $-6 + 21i$
7. $(6 + i)(6 - i)$ 37
8. $(3 + 2i)(3 - 2i)$ 13
9. $(2 + i)(2 - i)$ 5
10. $(-1 + 3i)(-1 - 3i)$ 10
11. $(1 + i)^2$ $2i$
12. $(4 - i)^2$ $15 - 8i$

Study the chart on page 292 and tell in which category each of the following numbers belongs.

13. $5 - 9i$ Imaginary
14. $-3i$ Pure imaginary
15. $5.\overline{67}$ Rational
16. $\sqrt{27}$ Irrational
17. $\sqrt[3]{27}$ Rational
18. 3^{-2} Rational
19. $7 + i\sqrt{3}$ Imaginary
20. $0.121121112\ldots$ Irrational

Written Exercises

Simplify.

A

1. $(9 + 2i) + (1 - 7i)$ $10 - 5i$
2. $(3 - 4i) + (-5 - 2i)$ $-2 - 6i$
3. $(5 - 7i) - (8 + 2i)$ $-3 - 9i$
4. $(6 - 4i) - (-4 + i)$ $10 - 5i$
5. $3(-2 + i) - 4(3 - 2i)$ $-18 + 11i$
6. $2(-1 + 6i) - 3(2 + 5i)$ $-8 - 3i$
7. $i(3 + 4i)$ $-4 + 3i$
8. $3i(5 - 6i)$ $18 + 15i$
9. $-4i(-2 + i)$ $4 + 8i$
10. $-2i(1 - 3i)$ $-6 - 2i$
11. $(3 - i)(3 + i)$ 10
12. $(4 + i)(4 - i)$ 17
13. $(3 + 7i)(3 - 7i)$ 58
14. $(-5 + 3i)(2 - 3i)$ $-1 + 21i$
15. $(-4 + i)(8 + 5i)$ $-37 - 12i$
16. $(3 - 7i)(2 + 4i)$ $34 - 2i$
17. $(-2 + 5i)(1 + 3i)$ $-17 - i$
18. $(3 + 4i)(-2 + 3i)$ $-18 + i$
19. $(4 + i\sqrt{5})(4 - i\sqrt{5})$ 21
20. $(2 - i\sqrt{3})(2 + i\sqrt{3})$ 7
21. $(2 - 4i)^2$ $-12 - 16i$
22. $(6 - 7i)^2$ $-13 - 84i$
23. $(-1 + i\sqrt{3})^2$ $-2 - 2i\sqrt{3}$
24. $(3 + i\sqrt{5})^2$ $4 + 6i\sqrt{5}$
25. $(3 + 2i)^2(3 - 2i)^2$ 169
26. $(2 - 3i)^2(2 + 3i)^2$ 169
27. $(\sqrt{2} - \sqrt{-5})(\sqrt{2} + \sqrt{-5})$ 7
28. $(\sqrt{3} + \sqrt{-7})(\sqrt{3} - \sqrt{-7})$ 10

Simplify each quotient.

32. $\frac{1}{5} - \frac{7}{5}i$ 36. $\frac{2}{7} - \frac{3\sqrt{5}}{7}i$

29. $\dfrac{5}{3 + 4i}$ $\frac{3}{5} - \frac{4}{5}i$
30. $\dfrac{15}{2 - i}$ $6 + 3i$
31. $\dfrac{2}{3 - i}$ $\frac{3}{5} + \frac{1}{5}i$
32. $\dfrac{10}{1 + 7i}$
33. $\dfrac{-1 - 2i}{-1 + 2i}$ $-\frac{3}{5} + \frac{4}{5}i$
34. $\dfrac{5 + i}{5 - i}$ $\frac{12}{13} + \frac{5}{13}i$
35. $\dfrac{6 - i\sqrt{2}}{6 + i\sqrt{2}}$ $\frac{17}{19} - \frac{6\sqrt{2}}{19}i$
36. $\dfrac{-3 + i\sqrt{5}}{-3 - i\sqrt{5}}$

Summarizing the Lesson

In this lesson students learned how to perform arithmetic operations using complex numbers. Using the chart on page 292, ask the students to give additional examples for each type of number shown.

Suggested Assignments

Average Algebra
Day 1: 295/1–11 odd
Assign with Lesson 6-7.
Day 2: 295/13–49 odd
S 291/31–53 odd

Average Alg. and Trig.
Day 1: 295/1–11 odd
Assign with Lesson 6-7.
Day 2: 295/13–49 odd
S 291/31–53 odd

Extended Algebra
Day 1: 295/1–21 odd
Assign with Lesson 6-7.
Day 2: 295/23–55 odd
298/Extra: 1–17 odd

Extended Alg. and Trig.
Day 1: 295/1–21 odd
Assign with Lesson 6-7.
Day 2: 295/23–55 odd
298/Extra: 1–17 odd

Supplementary Materials

Study Guide	pp. 103–104
Practice Masters	37, 38
Test Master	23
Computer Activity	14
Resource Book	p. 125

Additional Answers Written Exercises

47. a. $\left(\frac{\sqrt{2}}{2} + \frac{\sqrt{2}}{2}i\right)^2 = \frac{1}{2} + \frac{1}{2}i + \frac{1}{2}i + \frac{1}{2}i^2 = i$

48. a. $(-1 - i\sqrt{3})^3 = (-1 - i\sqrt{3}) \cdot (-1 - i\sqrt{3})^2 = (-1 - i\sqrt{3})(-2 + 2i\sqrt{3}) = 2 - 2i\sqrt{3} + 2i\sqrt{3} - 6i^2 = 8$

53–56. Let $a + bi$, $c + di$, and $e + fi$ be any three complex numbers where a, b, c, d, e, and f are real numbers.

53. (1) $[(a + bi) + (c + di)] + (e + fi) = [(a + c) + (b + d)i] + (e + fi)$ (Def. of complex add.);
(2) $[(a + bi) + (c + di)] + (e + fi) = [(a + c) + e] + [(b + d) + f]i$ (Def. of complex add.);
(3) $[(a + bi) + (c + di)] + (e + fi) = [a + (c + e)] + [b + (d + f)]i$ (Assoc. prop. of real add.);
(4) $[(a + bi) + (c + di)] + (e + fi) = (a + bi) + [(c + e) + (d + f)i]$ (Def. of complex add.);
(5) $[(a + bi) + (c + di)] + (e + fi) = (a + bi) + [(c + di) + (e + fi)]$ (Def. of complex add.)

Find the reciprocal of each complex number.

37. $2 + 3i$ $\frac{2}{13} - \frac{3}{13}i$ **38.** $1 - 4i$ $\frac{1}{17} + \frac{4}{17}i$ **39.** $-\sqrt{3} + i\sqrt{6}$ $-\frac{\sqrt{3}}{9} - \frac{\sqrt{6}}{9}i$ **40.** $-\sqrt{5} - i\sqrt{2}$ $-\frac{\sqrt{5}}{7} + \frac{\sqrt{2}}{7}i$

41. If $f(x) = x + \frac{1}{x}$, find $f(1 + 3i)$. $\frac{11}{10} + \frac{27}{10}i$

42. If $g(x) = x - \frac{1}{x}$, find $g(2 - i)$. $\frac{8}{5} - \frac{6}{5}i$

43. If $g(x) = \frac{x - 1}{x + 1}$, find $g(1 + i\sqrt{3})$. $\frac{3}{7} + \frac{2\sqrt{3}}{7}i$

44. If $f(x) = \frac{2 + x}{2 - x}$, find $f(1 - i)$. $1 - 2i$

B **45.** Show by substitution that $2 + i$ is a solution of the equation $x^2 - 4x + 5 = 0$. $(2 + i)^2 - 4(2 + i) + 5 = 3 + 4i - 8 - 4i + 5 = 0$

46. Show by substitution that $1 - 3i$ is a solution of the equation $x^2 - 2x + 10 = 0$. $(1 - 3i)^2 - 2(1 - 3i) + 10 = -8 - 6i - 2 + 6i + 10 = 0$

47. a. Show that $\frac{\sqrt{2}}{2} + \frac{\sqrt{2}}{2}i$ is a square root of i. $\left(\textit{Hint:} \text{ Square } \frac{\sqrt{2}}{2} + \frac{\sqrt{2}}{2}i.\right)$ $\left(\frac{\sqrt{2}}{2} + \frac{\sqrt{2}}{2}i\right)^2 = i$

b. Can you find another square root of i? $-\frac{\sqrt{2}}{2} - \frac{\sqrt{2}}{2}i$

48. a. Show that $-1 - i\sqrt{3}$ is a cube root of 8. (*Hint:* Cube $-1 - i\sqrt{3}$.) $(-1 - i\sqrt{3})^3 = 8$

b. Decide whether or not $-1 + i\sqrt{3}$ is a cube root of 8. Yes

49. Show that the conjugates $\frac{3}{5} + \frac{4}{5}i$ and $\frac{3}{5} - \frac{4}{5}i$ are also reciprocals. $\left(\frac{3}{5} + \frac{4}{5}i\right)\left(\frac{3}{5} - \frac{4}{5}i\right) = 1$

50. Show that the conjugates $\frac{1}{2} + \frac{\sqrt{3}}{2}i$ and $\frac{1}{2} - \frac{\sqrt{3}}{2}i$ are also reciprocals. $\left(\frac{1}{2} + \frac{\sqrt{3}}{2}i\right)\left(\frac{1}{2} - \frac{\sqrt{3}}{2}i\right) = 1$

C **51.** The numbers in Exercises 49 and 50 are both conjugates and reciprocals. Find another pair of such numbers. Answers may vary. Example: $\frac{1}{3} + \frac{2\sqrt{2}}{3}i$ and $\frac{1}{3} - \frac{2\sqrt{2}}{3}i$

52. What condition must be placed on a and b if the reciprocal of $a + bi$ is to be $a - bi$? $a^2 + b^2 = 1$

In Exercises 53–56, prove each property of complex numbers.

Sample The addition of complex numbers is commutative.

Solution Let $a + bi$ and $c + di$ be any two complex numbers where a, b, c, and d are real numbers.

$(a + bi) + (c + di) = (a + c) + (b + d)i$ Def. of complex add.
$= (c + a) + (d + b)i$ Comm. prop. of real add.
$= (c + di) + (a + bi)$ Def. of complex add.

53. The addition of complex numbers is associative.

54. The multiplication of complex numbers is commutative.

55. The multiplication of complex numbers is associative.

56. In the complex number system, multiplication is distributive over addition.

Self-Test 2

Vocabulary rational number (p. 283)
terminating decimal (p. 284)
repeating decimal (p. 284)
irrational number (p. 285)
dense (p. 285)
pure imaginary number (p. 288)
imaginary number (p. 292)
complex number (p. 292)
real part (p. 292)
imaginary part (p. 292)
complex conjugate (p. 294)

1. Find the decimal representation of $\frac{5}{13}$. $0.\overline{384615}$ **Obj. 6-6, p. 283**

2. Express $0.1\overline{72}$ as a common fraction in lowest terms. $\frac{19}{110}$

3. Find a rational number and an irrational number between 10 and $\sqrt{101}$. Answers may vary. Examples: 10.02; 10.010110111 . . .

Simplify. **Obj. 6-7, p. 288**

4. $(-3i)^2$ -9

5. $\sqrt{-28} + \sqrt{-63}$ $5i\sqrt{7}$

6. $\sqrt{-6} \cdot \sqrt{-10}$ $-2\sqrt{15}$

7. $-\frac{1}{2i}$ $\frac{i}{2}$

8. Solve $2y^2 + 90 = 0$. $\{\pm 3i\sqrt{5}\}$

Simplify. **Obj. 6-8, p. 292**

9. $(5 + 3i) - (2 - i)$ $3 + 4i$

10. $(1 - 2i)^2$ $-3 - 4i$

11. $(4 - i\sqrt{3})(-5 + 2i\sqrt{3})$ $-14 + 13i\sqrt{3}$

12. $\frac{2 + 3i}{1 - 4i}$ $-\frac{10}{17} + \frac{11}{17}i$

Historical Note / *Complex Numbers*

As early as the mid-sixteenth century, mathematicians like Girolamo Cardano and Rafael Bombelli began working with square roots of negative numbers. They used these non-real numbers in a formula that produced *real* solutions of cubic equations. Although the Swiss mathematician Leonhard Euler introduced the symbol i to represent $\sqrt{-1}$ about 1750, most mathematicians shunned "imaginary" numbers. Attitudes changed 50 years later when the great German mathematician Karl Friedrich Gauss used complex numbers in his doctoral thesis to prove this geometric fact: A regular polygon can be constructed with straightedge and compass if its number of sides is a prime number of the form $2^m + 1$, where m is a positive integer. (The numbers 3, 5, and 17 are the first three such primes, so regular triangles, pentagons, and 17-gons can be constructed with straightedge and compass.)

Gauss' accomplishment finally made complex numbers acceptable and paved the way for their widespread use in modern mathematics and physics.

54. (1) $(a + bi)(c + di) = (ac - bd) + (ad + bc)i$ (Def. of complex mult.); (2) $(a + bi)(c + di) = (ca - db) + (da + cb)i$ (Comm. prop. of real mult.); (3) $(a + bi)(c + di) = (c + di)(a + bi)$ (Def. of complex mult.)

55. (1) $[(a + bi)(c + di)](e + fi) = [(ac - bd) + (ad + bc)i](e + fi)$ (Def. of complex mult.); (2) $[(a + bi)(c + di)](e + fi) = [(ac - bd)e - (ad + bc)f] + [(ad + bc)e + (ac - bd)f]i$ (Def. of complex mult.); (3) $[(a + bi)(c + di)](e + fi) = [ace - bde - adf - bcf] + [ade + bce + acf - bdf]i$ (Dist. prop. of reals); (4) $[(a + bi)(c + di)](e + fi) = [a(ce - df) - b(de + cf)] + [a(de + cf) + b(ce - df)]i$ (Dist. prop. of reals); (5) $[(a + bi)(c + di)](e + fi) = (a + bi)[(ce - df) + (de + cf)i]$ (Def. of complex mult.); (6) $[(a + bi)(c + di)](e + fi) = (a + bi)[(c + di)(e + fi)]$ (Def. of complex mult.)

(continued)

(Quick Quiz on p. 299)

Additional Answers Written Exercises

(continued)

56. (1) $(a + bi)[(c + di) + (e + fi)] = (a + bi)[(c + e) + (d + f)i]$ (Def. of complex add.);

(2) $(a + bi)[(c + di) + (e + fi)] = [a(c + e) - b(d + f)] + [a(d + f) + b(c + e)]i$ (Def. of complex mult.);

(3) $(a + bi)[(c + di) + (e + fi)] = [ac + ae - bd - bf] + [ad + af + bc + be]i$ (Dist. prop. of reals);

(4) $(a + bi)[(c + di) + (e + fi)] = [(ac - bd) + (ae - bf)] + [(ad + bc) + (af + be)]i$ (Assoc. and Comm. prop. of real add.);

(5) $(a + bi)[(c + di) + (e + fi)] = [(ac - bd) + (ad + bc)i] + [(ae - bf) + (af + be)i]$ (Def. of complex add.);

(6) $(a + bi)[(c + di) + (e + fi)] = (a + bi)(c + di) + (a + bi)(e + fi)$ (Def. of complex mult.)

Extra / Conjugates and Absolute Value

When complex numbers are being discussed, they are often represented by single letters. For example, let

$$z = x + yi.$$

The *complex conjugate* of z is denoted by $\bar{z}$ (read "z bar"). That is,

$$\bar{z} = x - yi.$$

The following theorem summarizes the properties of conjugates.

Theorem 1

Let w and z be complex numbers.

a. $\overline{w + z} = \bar{w} + \bar{z}$ **b.** $\overline{w - z} = \bar{w} - \bar{z}$

c. $\overline{w \cdot z} = \bar{w} \cdot \bar{z}$ **d.** $\overline{\left(\frac{w}{z}\right)} = \frac{\bar{w}}{\bar{z}}$ $(z \neq 0)$

Proof of part (c): Let $w = u + vi$ and $z = x + yi$. Then:

$$\begin{aligned}
\overline{w \cdot z} &= \overline{(u + vi)(x + yi)} \\
&= \overline{(ux - vy) + (uy + vx)i} \\
&= (ux - vy) - (uy + vx)i \\
\bar{w} \cdot \bar{z} &= (u - vi)(x - yi) \\
&= [ux - (-v)(-y)] + [u(-y) + (-v)x]i \\
&= (ux - vy) - (uy + vx)i \\
\therefore \overline{w \cdot z} &= \bar{w} \cdot \bar{z}
\end{aligned}$$

Proofs of the other parts of Theorem 1 are left as Exercises 16 and 17.

To measure the "size" of the complex number $z = x + yi$, you can use its **absolute value,** defined by

$$|z| = \sqrt{x^2 + y^2}.$$

Example Find $|z|$ for each complex number z.

a. $z = 4 - 3i$ **b.** $z = -6i$ **c.** $z = -2$

Solution

a. $|z| = \sqrt{4^2 + (-3)^2} = \sqrt{25} = 5$

b. $|z| = \sqrt{0^2 + (-6)^2} = \sqrt{36} = 6$

c. $|z| = \sqrt{(-2)^2 + 0^2} = \sqrt{4} = 2$

Notice that if z is real, as in part (c) of Example 1, the definition of absolute value for *complex* numbers agrees with the definition of absolute value for *real* numbers found in Chapter 1.

The following theorem states a useful relationship between absolute value and conjugates.

Theorem 2

If z is a complex number, then $|z|^2 = z \cdot \bar{z}$.

Proof: Let $z = x + yi$. Then

$$|z|^2 = (\sqrt{x^2 + y^2})^2 = x^2 + y^2,$$

and $$z \cdot \bar{z} = (x + yi)(x - yi) = x^2 + y^2.$$

$$\therefore |z|^2 = z \cdot \bar{z}$$

If you divide both sides of $|z|^2 = z \cdot \bar{z}$ by $|z|^2 \cdot z$, you obtain this convenient expression for the reciprocal of a nonzero complex number:

$$\frac{1}{z} = \frac{\bar{z}}{|z|^2}$$

Example 2 Use the fact that $\frac{1}{z} = \frac{\bar{z}}{|z|^2}$ to find $\frac{1}{z}$ for each z given in Example 1.

Solution

a. Since $|z|^2 = 25$, $\frac{1}{z} = \frac{\bar{z}}{|z|^2} = \frac{4 + 3i}{25} = \frac{4}{25} + \frac{3}{25}i$.

b. Since $|z|^2 = 36$, $\frac{1}{z} = \frac{\bar{z}}{|z|^2} = \frac{6i}{36} = \frac{1}{6}i$.

c. Since $|z|^2 = 4$, $\frac{1}{z} = \frac{\bar{z}}{|z|^2} = \frac{-2}{4} = -\frac{1}{2}$.

Some properties of absolute value are stated in the following theorem.

Theorem 3

Let w and z be complex numbers.

a. $|w \cdot z| = |w| \cdot |z|$

b. $\left|\frac{w}{z}\right| = \frac{|w|}{|z|}$ $(z \neq 0)$

c. $|w + z| \leq |w| + |z|$

The proof of part (a) is given on the next page. Proofs of parts (b) and (c) are left as Exercises 20–22.

Quick Quiz

1. Write $0.20\overline{3}$ as a fraction. $\frac{61}{300}$
2. Write $\frac{5}{13}$ as a decimal. $0.\overline{384615}$

Simplify.

3. $\sqrt{-16} + \sqrt{-9}$ $7i$
4. $i\sqrt{8}(\sqrt{-24})$ $-8\sqrt{3}$
5. $(2i)^5$ $32i$
6. Solve $\frac{x^2}{5} + 5 = 0$. $\{5i, -5i\}$

Simplify.

7. $(2i + 3) - (4 - i)$ $-1 + 3i$
8. $(2 + i)^2(2 - i)$ $10 + 5i$
9. $\frac{8 + 2i}{1 - i}$ $3 + 5i$

Additional Answers
Extra

13. Let $z = u + vi$; $\bar{z} = u - vi$. $|\bar{z}| = \sqrt{u^2 + (-v)^2} = \sqrt{u^2 + v^2} = |z|$
14. Let $z = u + vi$. $z + \bar{z} = (u + vi) + (u - vi) = 2u$

(continued)

Additional Answers Extra

(continued)

15. a. Let $z = u + vi$. $z - \bar{z} = (u + vi) - (u - vi) = 2vi$; $v = \frac{z - \bar{z}}{2i} \cdot \frac{i}{i} = \frac{(z - \bar{z})i}{-2} = \frac{(\bar{z} - z)i}{2}$

b. Let $z = u + vi$. If $z = \bar{z}$, then $u + vi = u - vi$; $v = 0$. If $v = 0$, then $z = u + 0 = u$ and $\bar{z} = u - 0 = u$; $z = \bar{z}$.

16. Let $w = u + vi$, $z = x + yi$. $\overline{w + z} = \overline{(u + x) + (v + y)i} = (u + x) - (v + y)i$. $\bar{w} + \bar{z} = (u - vi) + (x - yi) = (u + x) + (-v - y)i = (u + x) - (v + y)i$.

17. Let $w = u + vi$, $z = x + yi$. $\overline{w - z} = \overline{(u + vi) - (x + yi)} = \overline{(u - x) + (v - y)i} = (u - x) + (y - v)i$. $\bar{w} - \bar{z} = (u - vi) - (x - yi) = (u - x) + (-v + y)i = (u - x) + (y - v)i$.

18. Let $z = u + vi$. $\overline{\left(\frac{1}{z}\right)} = \overline{\left(\frac{1}{u + vi} \cdot \frac{u - vi}{u - vi}\right)} = \overline{\left(\frac{u - vi}{u^2 + v^2}\right)} = \frac{u}{u^2 + v^2} + \frac{v}{u^2 + v^2}i$; $\frac{1}{\bar{z}} = \frac{1}{u - vi} \cdot \frac{u + vi}{u + vi} = \frac{u + vi}{u^2 + v^2} = \frac{u}{u^2 + v^2} + \frac{v}{u^2 + v^2}i$.

Proof of part (a): By Theorems 1 and 2 and the commutative and associative properties:

$$|w \cdot z|^2 = (w \cdot z)(\overline{w \cdot z}) = w \cdot \bar{w} \cdot z \cdot \bar{z} = |w|^2 \cdot |z|^2$$

Since $|wz|$, $|w|$, and $|z|$ are all nonnegative, when you take the principal square root of each side of $|w \cdot z|^2 = |w|^2 \cdot |z|^2$, you obtain

$$|w \cdot z| = |w| \cdot |z|.$$

Exercises

7. $-\frac{4}{25} - \frac{3}{25}i$ **8.** $\frac{12}{169} + \frac{5}{169}i$ **9.** $-\frac{1}{2}i$ **10.** $-\frac{1}{3}$ **11.** $\frac{\sqrt{2}}{2} - \frac{\sqrt{2}}{2}i$ **12.** $\frac{\sqrt{3}}{2} + \frac{1}{2}i$

Find $|z|$ for each complex number z.

A

1. $z = -4 + 3i$ 5
2. $z = 12 - 5i$ 13
3. $z = 2i$ 2
4. $z = -3$ 3
5. $z = \frac{\sqrt{2}}{2} + \frac{\sqrt{2}}{2}i$ 1
6. $z = \frac{\sqrt{3}}{2} - \frac{1}{2}i$ 1

7–12. Use the fact that $\frac{1}{z} = \frac{\bar{z}}{|z|^2}$ to find $\frac{1}{z}$ for each complex number z in Exercises 1–6.

13. Show that $|\bar{z}| = |z|$.

14. Show that $z + \bar{z}$ is twice the real part of z.

15. a. Express the imaginary part of z in terms of z and $\bar{z}$.

b. Show that z is real if and only if $z = \bar{z}$.

16. Prove part (a) of Theorem 1.

17. Prove part (b) of Theorem 1.

B

18. Prove that $\overline{\left(\frac{1}{z}\right)} = \frac{1}{\bar{z}}$.

19. Prove part (d) of Theorem 1 by combining part (c) and Exercise 18. $\left(\textit{Hint: } \frac{w}{z} = w \cdot \frac{1}{z}.\right)$

20. Prove part (b) of Theorem 3. $\left(\textit{Hint: }\right.$ Apply Theorem 2 to $\left|\frac{w}{z}\right|^2$. Then use part (d) of Theorem 1.$\left.\right)$

Exercises 21–22 outline a proof of part (c) of Theorem 3.

C

21. Show that $w \cdot \bar{z} + \bar{w} \cdot z \le 2|w||z|$.
(*Hint:* Show that $w \cdot \bar{z} + \bar{w} \cdot z = w \cdot \bar{z} + \overline{w \cdot \bar{z}}$ and therefore is twice the real part of $w \cdot \bar{z}$. How does the real part of a complex number compare in size with the absolute value of the number?)

22. Show that $|w + z| \le |w| + |z|$ by using the theorems of this Extra and the preceding exercises to justify these steps:

$$\begin{aligned} |w + z|^2 &= (w + z)(\overline{w + z}) \\ &= (w + z)(\bar{w} + \bar{z}) \\ &= w \cdot \bar{w} + w \cdot \bar{z} + \bar{w} \cdot z + z \cdot \bar{z} \\ &\le |w|^2 + 2|w||z| + |z|^2 \\ &\le (|w| + |z|)^2 \\ \therefore\ |w + z| &\le |w| + |z| \end{aligned}$$

Chapter Summary

1. A solution of the equation $x^n = b$ is called an *n*th root of *b*. The radical $\sqrt[n]{b}$ denotes the *principal* *n*th root of *b*.

$$(\sqrt[n]{b})^n = b \qquad \sqrt[n]{b^n} = \begin{cases} b & \text{if } n \text{ is odd} \\ |b| & \text{if } n \text{ is even} \end{cases}$$

2. In working with radicals you can use the following properties when the radicals involved represent real numbers.

$$\sqrt[n]{ab} = \sqrt[n]{a} \cdot \sqrt[n]{b} \qquad \sqrt[n]{\frac{a}{b}} = \frac{\sqrt[n]{a}}{\sqrt[n]{b}}$$

$$\sqrt[nq]{b} = \sqrt[n]{\sqrt[q]{b}} \qquad \sqrt[n]{b^m} = (\sqrt[n]{b})^m$$

3. An expression containing *n*th roots is in *simplest radical form* if:
   (a) no radicand contains a factor (other than 1) that is a perfect *n*th power, and
   (b) every denominator has been rationalized, so that no radicand is a fraction and no radical is in a denominator.
4. Radical expressions with the same index and radicand can be added or subtracted in the same way as like terms.
5. To solve a *radical equation* involving square roots, first isolate the radical term and then square both sides of the equation. Repeat this process if necessary. Be sure to check all possible solutions in the original equation to determine if any extraneous roots were introduced.
6. A real number that can be expressed as a quotient of two integers is *rational*. Otherwise it is *irrational*.
7. (a) Every terminating or repeating decimal represents a rational number, and any rational number can be expressed as a terminating or repeating decimal.
   (b) Every infinite nonrepeating decimal represents an irrational number, and any irrational number can be expressed as a nonrepeating decimal.
8. A *complex number* is a number of the form $a + bi$, where a and b are real numbers and $i = \sqrt{-1}$. The number a is called the *real part* of the complex number, and b is called the *imaginary part*. The *complex conjugate* of $a + bi$ is $a - bi$. The product of these complex conjugates is the real number $a^2 + b^2$.
9. To simplify an expression containing square roots of negative numbers, first rewrite each radical using i.
10. Complex numbers can be added or subtracted by combining their real parts and their imaginary parts separately. Multiply two complex numbers as you would multiply two binomials, and use the fact that $i^2 = -1$. To simplify a quotient, multiply both the numerator and denominator by the conjugate of the denominator.

19. $\overline{\left(\frac{w}{z}\right)} = \overline{\left(w \cdot \frac{1}{z}\right)} = \overline{w}\overline{\left(\frac{1}{z}\right)} = \overline{w} \cdot \frac{1}{\overline{z}} = \frac{\overline{w}}{\overline{z}}$.

20. $\left|\frac{w}{z}\right|^2 = \frac{w}{z}\overline{\left(\frac{w}{z}\right)} = \frac{w}{z} \cdot \frac{\overline{w}}{\overline{z}} = \frac{|w|^2}{|z|^2}$. Since $\left|\frac{w}{z}\right|$, $|w|$, $|z| \ge 0$, taking principal square roots gives $\left|\frac{w}{z}\right| = \frac{|w|}{|z|}$.

21. $w\overline{z} + \overline{w}z = w\overline{z} + \overline{w\overline{z}}$. Let $w\overline{z} = u + vi$. $w\overline{z} + \overline{w}z = (u + vi) + (u - vi) = 2u$. $u \le |w\overline{z}|$. $w\overline{z} + \overline{w}z = 2u \le 2|w\overline{z}| = 2|w|\,|\overline{z}| = 2|w|\,|z|$.

22.
$|w + z|^2 = (w + z)\overline{(w + z)}$
(If z is a complex number, $|z|^2 = z\overline{z}$.)
$= (w + z)(\overline{w} + \overline{z})$
(If w and z are complex numbers, $\overline{w + z} = \overline{w} + \overline{z}$.)
$= w\overline{w} + w\overline{z} + \overline{w}z + z\overline{z}$
(Distributive prop.)
$\le w\overline{w} + 2|w|\,|z| + z\overline{z}$
(Exercise 21, above)
$\le |w|^2 + 2|w|\,|z| + |z|^2$
(If z is a complex number, $|z|^2 = z\overline{z}$.)
$\le (|w| + |z|)^2$
(Distributive prop.)
$\therefore |w + z| \le |w| + |z|$
(If $a, b \ge 0$ and $a^2 < b^2$, then $a < b$.)

Additional Answers
Written Exercises
(continued from p. 287)

38. Let $x = \frac{a}{b}$ and $y = \frac{c}{d}$ where a, b, c, and d are integers and $bd \neq 0$. Then $x + y = \frac{a}{b} + \frac{c}{d} = \frac{ad + bc}{bd}$. Since $ad + bc$ and bd are integers, $x + y$ is rational.

39. a. Let $x = \frac{a}{b}$ and $y = \frac{c}{d}$ where a, b, c, and d are integers and $bd \neq 0$. Then $xy = \frac{ac}{bd}$, which is the ratio of two integers.
b. The product may be rational ($\sqrt{2} \cdot \sqrt{2} = 2$) or irrational ($\sqrt{2} \cdot \sqrt{3} = \sqrt{6}$).

40. Let $x = \frac{a}{b}$ where a and b are integers and $b \neq 0$. Assume $x + z$ is rational; that is, $x + z = \frac{c}{d}$ where c and d are integers and $d \neq 0$. Then $z = (x + z) - x = \frac{c}{d} - \frac{a}{b} = \frac{bc - ad}{bd}$. Since $bc - ad$ and bd are integers, z is rational. But since this contradicts the fact that z is irrational, $x + z$ must be irrational.

Chapter Review

Write the letter of the correct answer.

1. Simplify $\sqrt{x^2}$. **6-1**
a. $-x$ **b.** x **c.** $|x|$ **d.** $-|x|$

2. Find the real roots of the equation $2 - y^2 = 0$.
a. $\sqrt{2}$ **b.** $\pm i\sqrt{2}$ **c.** $\pm\sqrt{2}$ **d.** no real roots

3. Simplify $\sqrt{12x^4y^7}$. Assume that the radical represents a real number. **6-2**
a. $x^2y^3\sqrt{12y}$ **b.** $2x^2y^3\sqrt{3y}$
c. $6x^2y^3\sqrt{y}$ **d.** $x^2\sqrt{12y^7}$

4. Simplify $\sqrt{18} \cdot \sqrt{\frac{3}{10}}$.
a. $\frac{3\sqrt{15}}{5}$ **b.** $15\sqrt{3}$
c. $\frac{\sqrt{15}}{5}$ **d.** $\frac{\sqrt{15}}{5}$

5. Simplify $\sqrt{6}\left(\frac{\sqrt{2}}{2} + \sqrt{3}\right) - \sqrt{8}$. **6-3**
a. $\sqrt{3} + \sqrt{2}$ **b.** $\sqrt{3} + 5\sqrt{2}$
c. $\sqrt{5}$ **d.** $2\sqrt{3} + \sqrt{2}$

6. Simplify $\sqrt{72x^3} - 5x\sqrt{2x}$. Assume that each radical represents a real number.
a. $\sqrt{2x}$ **b.** $2x\sqrt{x}$
c. $x^2\sqrt{2x}$ **d.** $x\sqrt{2x}$

7. Simplify $(2\sqrt{6} - \sqrt{3})^2$. **6-4**
a. 21 **b.** $27 - 6\sqrt{2}$
c. $27 - 12\sqrt{2}$ **d.** $15 - 12\sqrt{2}$

8. Simplify $\frac{6}{3 + 2\sqrt{3}}$.
a. $6 - 4\sqrt{3}$ **b.** $6 + 4\sqrt{3}$ **c.** $-6 + 4\sqrt{3}$ **d.** $-6 - 4\sqrt{3}$

9. Solve $3 - 2\sqrt{x} = 7$. **6-5**
a. $\{-2\}$ **b.** $\{4\}$ **c.** $\{-4\}$ **d.** $\emptyset$

10. Solve $2 - y = \sqrt{y + 4}$.
a. $\{0\}$ **b.** $\{0, 5\}$ **c.** $\{5\}$ **d.** $\emptyset$

11. Express $0.\overline{675}$ as a common fraction in lowest terms. **6-6**
a. $\frac{27}{40}$ **b.** $\frac{25}{37}$ **c.** $\frac{67}{99}$ **d.** $\frac{2}{3}$

12. Which of the following is *not* a rational number?
a. $\frac{407}{528}$ **b.** $\sqrt[3]{-8}$ **c.** $0.\overline{34}$ **d.** $1.626626662\ldots$

13. Simplify $3i\sqrt{2} \cdot \sqrt{-12}$. 6-7

(a.) $-6\sqrt{6}$ b. $6i\sqrt{6}$ c. $-6i\sqrt{6}$ d. $6\sqrt{6}$

14. Solve $3x^2 + 14 = 8$.

a. $\{\pm 2i\}$ b. $\{\pm\sqrt{2}\}$ (c.) $\{\pm i\sqrt{2}\}$ d. $\{\pm 2\}$

15. Simplify $(2 + 3i)(1 - i)$. 6-8

a. $-1 + i$ (b.) $5 + i$ c. $-1 + 5i$ d. -1

16. Simplify $\frac{3 - 4i}{-2 + i}$.

a. $-\frac{2}{3} + \frac{5}{3}i$ b. -2 c. $-\frac{2}{3} + \frac{11}{3}i$ (d.) $-2 + i$

Chapter Test

Simplify.

1. $\sqrt[3]{(-8)^3}$ -8
2. $(\sqrt{49})^2$ 49 6-1
3. $\sqrt[3]{-64}$ -4
4. $\sqrt[4]{81}$ 3
5. Find the real roots of the equation $16x^2 = 81$. $\left\{\pm \frac{9}{4}\right\}$
6. Find the real roots of the equation $2x^3 = -16$. $\{-2\}$

Simplify. Assume that each radical represents a real number.

7. $\frac{\sqrt{15} \cdot \sqrt{42}}{\sqrt{35}}$ $3\sqrt{2}$
8. $\sqrt{\frac{75a^6}{32b^3}}$ $\frac{5|a|^3\sqrt{6b}}{8b^2}$ 6-2
9. $13\sqrt{5} - \sqrt{10}(3\sqrt{2} + 4\sqrt{5})$ $7\sqrt{5} - 20\sqrt{2}$
10. $\frac{\sqrt{5x}}{\sqrt{3}} + \sqrt{\frac{3x}{5}}$ $\frac{8\sqrt{15x}}{15}$ 6-3
11. $(5 + 2\sqrt{6})(8 - 3\sqrt{6})$ $4 + \sqrt{6}$
12. $\frac{2}{3 - \sqrt{5}}$ $\frac{3 + \sqrt{5}}{2}$ 6-4

Solve.

13. $\sqrt[3]{4x - 1} = 3$ $\{7\}$
14. $\sqrt{3y + 4} = 2 + \sqrt{y + 2}$ $\{7\}$ 6-5
15. Express $0.7\overline{45}$ as a common fraction in lowest terms. $\frac{41}{55}$ 6-6
16. Find a rational number and an irrational number between 1.999 and 2. Answers may vary. 1.9995; 1.99909999099999 . . .

Simplify.

17. $\sqrt{-45} \cdot \sqrt{-20}$ -30
18. $2i\sqrt{3} - \sqrt{-75}$ $-3i\sqrt{3}$ 6-7
19. $(-3 + 4i) + (5 - i)$ $2 + 3i$
20. $(7 - 3i) - (-2 + 5i)$ $9 - 8i$ 6-8
21. $(4 - 6i)(3 + 8i)$ $60 + 14i$
22. $\frac{6 + i}{3 - 2i}$ $\frac{16}{13} + \frac{15}{13}i$

41. The product is irrational. Let z be an irrational number. Let $x = \frac{a}{b}$ where a and b are nonzero integers. Assume xz is rational, that is, $xz = \frac{c}{d}$ where c and d are integers and $d \neq 0$. Then $z = (xz) \div x = \frac{c}{d} \div \frac{a}{b} = \frac{bc}{ad}$, which is rational. But since this contradicts the fact that z is irrational, xz must be irrational.

42. a. Multiplying N by 10^p shifts the decimal point p digits to the right but does not affect the digits' pattern.
b. The repeating blocks in $N \cdot 10^p$ and N subtract to zero, yielding a terminating decimal.
c. All terminating decimals are rational; $N(10^p - 1) = N \cdot 10^p - N$
d. $\frac{N(10^p - 1)}{10^p - 1}$ is the quotient of rational numbers and is therefore rational.

Alternate Test p. T18

Supplementary Materials

Test Masters 24, 25, 26
Resource Book p. 126

Mixed Problem Solving

Solve each problem that has a solution. If a problem has no solution, explain why.

A

1. For \$1 you can buy either five lemons or four limes. If you pay \$2.10 for some lemons and limes, find all possibilities for the number of each fruit bought. 3 lemons, 6 limes; 8 lemons, 2 limes
2. Cheryl has 20 coins worth \$3.15. If the coins are dimes and quarters only, how many of each type of coin does she have? no solution; $q = 7\frac{2}{3}$ is not an integer.
3. A season pass to the city pool costs \$34. If daily passes cost \$1.25 each, at most how many daily passes can be purchased before their cost exceeds that of the season pass? 27
4. A micron is a millionth of a meter. Some bacteria are about 0.2 micron in diameter. Use scientific notation to express this diameter in meters. 2×10^{-7} m
5. The Khos invested \$7000, part at 6% and the rest at 10% simple annual interest. If they had reversed the amounts invested at each rate, they would have earned \$120 less in annual interest. How much was invested at each rate? \$2000 at 6% and \$5000 at 10%
6. As a salesman, George earns a monthly base salary of \$1500 plus a 2% commission on his sales. If he wants an income of at least \$2000 each month, what must his minimum monthly sales total be? at least \$25,000
7. The numerator of a fraction is three more than twice the denominator. If the fraction is equal to $\frac{9}{4}$, find the fraction. $\frac{27}{12}$

B

8. A boat traveled 36 km downstream in 2 h. The return trip upstream took 3 h. Find the rate of the current. 3 km/h
9. Find the dimensions of a rectangle whose perimeter is 18 cm and whose area is 18 cm^2. length 6 cm; width 3 cm
10. A company's monthly sales of \$50,000 result in \$6000 profit, while monthly sales of \$70,000 result in \$9000 profit. If profit is considered as a linear function of sales, what minimum amount of sales is needed to avoid a loss? at least \$10,000
11. Marie can prepare the salads at the Lunch Inn in 2 h. Bob needs 3 h to prepare the salads. At 10:00 A.M. Marie began the job. At 11:00 A.M. Bob joined her. At what time did they finish preparing the salads? 11:36 A.M.
12. The lengths of the sides of a right triangle are consecutive multiples of 3. Find the length of the hypotenuse. 15

Preparing for College Entrance Exams

Strategy for Success

Prepare for the test by becoming familiar with the format of the test, including the directions, explanations, and types of questions. Examine and complete some sample tests well ahead of the test date.

Decide which is the best of the choices given and write the corresponding letter on your answer sheet.

1. If $y^2 - y - 6 = 0$, then $y =$

(A) $1, -6$ **(B)** $-2, 3$ **(C)** $-1, 6$ **(D)** $2, 3$ **(E)** $-3, 2$

2. If $\frac{1}{1-x} + \frac{1}{1+x} = 4$, then $x^4 =$

(A) $\frac{\sqrt{2}}{2}$ **(B)** $\frac{1}{4}$ **(C)** $\frac{1}{2}$ **(D)** $\pm\frac{\sqrt{2}}{2}$ **(E)** $\pm\frac{1}{2}$

3. $\sqrt[3]{-54} - \sqrt[3]{250} + \sqrt[3]{32} =$

(A) $-6\sqrt[3]{2}$ **(B)** $-12\sqrt[3]{2}$ **(C)** $2\sqrt[3]{2}(\sqrt[3]{2} + 4)$ **(D)** $2(\sqrt[3]{4} - 4\sqrt[3]{2})$ **(E)** $2(\sqrt[3]{4} - \sqrt[3]{2})$

4. What are the zeros of $\frac{r^4 - 13r^2 + 36}{r^3 + r^2 - 6r}$?

(A) $0, \pm 2, 3$ **(B)** $0, 3$ **(C)** $2, \pm 3$ **(D)** $-2, 3$ **(E)** $0, \pm 2, \pm 3$

5. How many solutions does $\sqrt{\frac{1}{2}x + 5} = \frac{1}{2}x - 1$ have?

(A) 0 **(B)** 1 **(C)** 2 **(D)** 3 **(E)** cannot be determined

6. $\sqrt{-36}\left(\sqrt{-81} - \frac{\sqrt{8}}{\sqrt{-18}}\right) =$

(A) -58 **(B)** 58 **(C)** $58i$ **(D)** -50 **(E)** $-50i$

7. $\frac{\left(\frac{1}{3}\right)^{-2} - (-3)^{-1}}{\left(\frac{1}{2}\right)^{-1} - (-2)^{-2}} =$

(A) $\frac{112}{27}$ **(B)** $\frac{16}{9}$ **(C)** $\frac{16}{3}$ **(D)** $\frac{104}{21}$ **(E)** $-\frac{8}{9}$

8. What is the reciprocal of $\sqrt{2} + \sqrt{3} - \sqrt{5}$?

(A) $\sqrt{2} + \sqrt{3} + \sqrt{5}$ **(B)** $\frac{2\sqrt{3} + 3\sqrt{2} + \sqrt{30}}{12}$ **(C)** $\frac{15\sqrt{2} + 10\sqrt{3} - 6\sqrt{5}}{30}$

(D) $\sqrt{5} - \sqrt{2} - \sqrt{3}$ **(E)** $\frac{2\sqrt{3} + 3\sqrt{2} + \sqrt{30}}{6}$

305

7 Quadratic Equations and Functions

Objectives

7-1 To solve quadratic equations by completing the square.

7-2 To solve quadratic equations by using the quadratic formula.

7-3 To determine the nature of the roots of a quadratic equation by using its discriminant.

7-4 To recognize and solve equations in quadratic form.

7-5 To graph parabolas whose equations have the form $y - k = a(x - h)^2$ and to find the vertices and axes of symmetry.

7-6 To analyze a quadratic function, draw its graph, and find its minimum or maximum value.

7-7 To learn the relationship between the roots and coefficients of a quadratic equation.

To write a quadratic equation or function using information about the roots or the graph.

Assignment Guide

See p. T58 for Key to the format of the Assignment Guide

Day	Average Algebra	Average Algebra and Trigonometry	Extended Algebra	Extended Algebra and Trigonometry
1	**7-1** 309/1–35 odd **S** 310/*Mixed Review*	**7-1** 309/1–35 odd **S** 310/*Mixed Review*	**7-1** 309/3–21 mult. of 3, 23–41 odd **S** 310/*Mixed Review*	**7-1** 309/3–21 mult. of 3, 23–41 odd **S** 310/*Mixed Review*
2	**7-1** 310/20–38 even **7-2** 313/1–12	**7-1** 310/20–38 even **7-2** 313/1–12	**7-2** 313/3–27 mult. of 3 314/*P*: 1–13 odd	**7-2** 313/3–27 mult. of 3 314/*P*: 1–13 odd
3	**7-2** 314/14–36 even *P*: 1–13 odd	**7-2** 314/14–36 even *P*: 1–13 odd	**7-2** 314/28–42 even 315/*P*: 6–14 even **S** 310/36–42 even	**7-2** 314/28–42 even 315/*P*: 6–14 even **S** 310/36–42 even
4	**7-3** 320/1–39 odd **S** 321/*Mixed Review* **R** 316/*Self-Test 1*	**7-3** 320/1–39 odd **S** 321/*Mixed Review* **R** 316/*Self-Test 1*	**7-3** 320/1–41 odd, 42 **S** 321/*Mixed Review* **R** 316/*Self-Test 1*	**7-3** 320/1–41 odd, 42 **S** 321/*Mixed Review* **R** 316/*Self-Test 1*
5	**7-4** 324/1–12 **S** 310/37, 39	**7-4** 324/1–12 **S** 310/37, 39	**7-4** 324/2–20 even **S** 321/43	**7-4** 324/2–20 even **S** 321/43
6	**7-4** 324/13–26 **R** 325/*Self-Test 2* **7-5** 331/1–16	**7-4** 324/13–26 **R** 325/*Self-Test 2* **7-5** 331/1–16	**7-4** 324/21–32 **R** 325/*Self-Test 2* **7-5** 331/1–25 odd	**7-4** 324/21–32 **R** 325/*Self-Test 2* **7-5** 331/1–25 odd
7	**7-5** 331/17–30 **S** 332/*Mixed Review*	**7-5** 331/17–30 **S** 332/*Mixed Review*	**7-5** 332/28–36 even, 37, 38 **S** 332/*Mixed Review*	**7-5** 332/28–36 even, 37, 38 **S** 332/*Mixed Review*
8	**7-6** 336/1–29 odd **S** 332/31–34	**7-6** 336/1–29 odd **S** 332/31–34	**7-6** 336/1–39 odd **S** 332/27–35 odd	**7-6** 336/1–39 odd **S** 332/27–35 odd
9	**7-6** 336/20–40 even **S** 315/*P*: 6, 10, 12	**7-6** 336/20–40 even **S** 315/*P*: 6, 10, 12	**7-6** 336/20–42 even, 43, 44	**7-6** 336/20–42 even, 43, 44
10	**7-7** 342/1–33 odd 343/*P*: 1–7 odd	**7-7** 342/1–33 odd 343/*P*: 1–7 odd	**7-7** 342/1–37 odd 343/*P*: 1–7 odd	**7-7** 342/1–37 odd 343/*P*: 1–7 odd

Assignment Guide (continued)

Day	Average Algebra	Average Algebra and Trigonometry	Extended Algebra	Extended Algebra and Trigonometry
11	**7-7** 343/*P*: 2–14 even **S** 345/*Mixed Review*	**7-7** 343/*P*: 2–14 even **S** 345/*Mixed Review*	**7-7** 343/*P*: 8–15 **S** 345/*Mixed Review*	**7-7** 343/*P*: 8–15 **S** 345/*Mixed Review*
12	*Prepare for Chapter Test* **R** 345/*Self-Test 3* **R** 346/*Chapter Review*	*Prepare for Chapter Test* **R** 345/*Self-Test 3* **R** 346/*Chapter Review*	*Prepare for Chapter Test* **R** 345/*Self-Test 3* **R** 346/*Chapter Review*	*Prepare for Chapter Test* **R** 345/*Self-Test 3* **R** 346/*Chapter Review*
13	*Administer Chapter 7 Test* **R** 348/*Cumulative Review*	*Administer Chapter 7 Test* **R** 348/*Cumulative Review*	*Administer Chapter 7 Test* **R** 348/*Cumulative Review*	*Administer Chapter 7 Test* **R** 348/*Cumulative Review*

Supplementary Materials Guide

For Use with Lesson	Practice Masters	Tests	Study Guide (Reteaching)	Resource Book		
				Tests	Practice Exercises	Mixed Review (MR) Applications (A) Enrichment (E) Technology (T) Thinking Skl. (TS)
7-1			pp. 105–106			p. 210 (A)
7-2	Sheet 39		pp. 107–108		p. 127	
7-3			pp. 109–110			
7-4	Sheet 40	Test 27	pp. 111–112		p. 128	
7-5	Sheet 41		pp. 113–114			pp. 240–241 (T)
7-6	Sheet 42		pp. 115–116			pp. 263–264 (TS)
7-7	Sheet 43	Test 28	pp. 117–118		p. 129	
Chapter 7		Tests 29, 30		pp. 31–34	p. 130	pp. 180–182 (MR) p. 226 (E)
Cum. Rev. 4–7				pp. 35–37		
Cum. Rev. 1–7				pp. 38–45		

Overhead Visuals

For Use with Lesson	Visual	Title
7-5	B	Multi-Use Packet 2
7-5	4	Parabolas

Software

Software	Algebra Plotter Plus	Using Algebra Plotter Plus	Computer Activities	Test Generator
	Function Plotter, Parabola Plotter, Parabola Quiz	Scripted Demo, pp. 27–30 Activity Sheet, p. 53	Activities 5, 15, 16, 37	147 test items
For Use with Lessons	7-5, 7-6	7-5, 7-6	7-3, 7-5, Chapter 7	all lessons

Using Technology

Students may have already used computers in Chapters 5 and 6 to facilitate finding roots of equations and graphing equations. The use of technology can be continued in this chapter by using graphing calculators and computers when students wish to investigate how changes in equations affect graphs and when they wish to verify their graphs.

See the Exploration on page 838 for an activity in which students use a graphing calculator or a computer with graphing software to explore the graphs of quadratic equations.

7-3 The Discriminant

Students can explore the nature of the roots of a quadratic equation by examining cases in which the discriminant is greater than, less than, or equal to zero. The following activity is a good introduction to the definition of the discriminant.

Students can use a computer with graphing software or a graphing calculator. It would be suitable for students to work together in small groups as they explore the concept of the discriminant. Or, this activity can be done as a demonstration for the class.

On the board, write the roots of the quadratic equation, $ax^2 + bx + c = 0$ in the form

$$r_1 = \frac{-b + \sqrt{D}}{2a} \quad \text{and} \quad r_2 = \frac{-b - \sqrt{D}}{2a},$$

where $D = b^2 - 4ac$.

Now choose values for a, b, and c such that $D > 0$. For example, let $a = 1$, $b = 6$, and $c = 5$ so that the quadratic equation is $x^2 + 6x + 5 = 0$. Compute the value of D. Have the students note whether it is greater than, less than, or equal to zero. Now graph, or have the students graph, the quadratic equation and find the roots of the equation. How many times does the graph cross the x-axis? Are the roots real or imaginary? Are they equal?

Repeat this exercise for $x^2 + 8x + 16 = 0$, and then for $x^2 + 6x + 10 = 0$. Note the values of the discriminants, graph the equations, and then find the roots of the equations. How many times does each graph cross the x-axis? Are the roots real or imaginary? Are they equal?

Have the students generalize the results to determine the nature of the roots and the graph of quadratic equations by examining the discriminant.

7-5 Graphing $y - k = a(x - h)^2$

Students can use a computer or graphing calculator to investigate how changes in an equation of a parabola affect the coordinates of its vertex, as suggested in the lesson on page 327. A worksheet for this exploration is given on pages 240–241 of the *Resource Book*.

Using *ALGEBRA PLOTTER PLUS*

Choose PARABOLA PLOTTER. Go to EQUATIONS and enter the equations of the parabolas, using X ∧ 2 for x^2. DRAW the graphs.

Students can use a computer or graphing calculator to verify their graphs for Written Exercises 1–18 on page 331 and Exercises 33 and 34 on page 332.

7-7 Writing Quadratic Equations and Functions

Students can use a computer or graphing calculator to verify their answers for Written Exercises 7–16 and 25–36 on page 342 by graphing their equations and inspecting the graphs for the given characteristics. For example, the answer for Exercise 7 is $x^2 - 7x + 10 = 0$, so students can graph $y = x^2 - 7x + 10$ and observe that the x-intercepts of the graph are 2 and 5 (the given roots).

References to Strategies

PE: Pupil's Edition **TE:** Teacher's Edition **RB:** Resource Book

Problem Solving Strategies

PE: p. 312 (make a sketch); p. 313 (check solutions)
TE: p. 322
RB: pp. 198–199

Applications

PE: p. 306 (physics); pp. 312–316, 340–341, 343–345 (word problems)
TE: pp. 306, 312, 316, 340, 345
RB: p. 210

Nonroutine Problems

PE: p. 314 (Exs. 41, 42); pp. 320–321 (Exs. 38–43); p. 325 (Exs. 29–32, Challenge); p. 332 (Exs. 35–39); p. 336 (Exs. 37–44); p. 343 (Exs. 39, 40); pp. 344–345 (Exs. 15, 16)
TE: pp. T105–T108 (Sugg. Extensions, Lessons 7-2, 7-3, 7-6, 7-7)

Communication

PE: p. 321 (Exs. 43c, convincing argument); pp. 327–328 (translating); p. 332 (Exs. 35–36, convincing argument)
TE: pp. T105–T106 (Reading Algebra)

Thinking Skills

PE: p. 314 (Exs. 42, proof)
RB: pp. 263–264

Explorations

PE: p. 325 (Challenge); p. 838 (quadratic equations)

Connections

PE: pp. 312–315, 330–331, 343–345 (Geometry); p. 316 (History); p. 337 (Statistics); pp. 340–341 (Business); p. 344 (Physics)

Using Technology

PE: pp. 312–314, 316, 319–320, 321, 324, 327, 331, 336, 838
TE: pp. 313, 319, 321, 324, 327, 336
RB: pp. 240–241
Computer Activities: pp. 11–13, 38–43, 100–102
Using Algebra Plotter Plus: pp. 27–30, 53

Using Manipulatives/Models

Overhead Visuals: B, 4

Cooperative Learning

TE: p. 306; p. T106 (Lesson 7-3)

Teaching Resources

For use in implementing the teaching strategies referenced on the previous page.

Application
Resource Book, p. 210

Application—Maximizing Income (For use with Chapter 7)

Suppose you're in charge of a concession stand that sells hot dogs for $1.00 each at school football games. At that price, the average number of hot dogs sold is 100. To increase the income from hot dog sales, you want to raise the price, but a survey of customers indicates that for each $.10 increase in the price of a hot dog, the average number of hot dogs sold would drop by 5.

1. Complete the table below.

	Number of $.10 increases	*Price of a hot dog*	*Number of hot dogs sold*	*Total income*
	0	$1.00	100	$100.00
a.	1	$1.10	95	______
b.	2	$1.20	______	______
c.	3	______	85	______
d.	4	______	______	______
e.	5	______	______	______
f.	6	______	______	______
g.	7	______	______	______
h.	8	______	______	______
i.	9	______	______	______
j.	10	______	______	______

2. Describe how the total income changes as the price of a hot dog varies from $1.00 to $2.00.

____

3. Let x represent the number of $.10 increases in the price of a hot dog, and let y represent the total income from hot dog sales. For example, $x = 0$ and $y = 100$ initially. Use the table of Exercise 1 to plot the ordered pairs (x, y) on the set of axes at the right. Notice that the first ordered pair, (0, 100), has already been plotted.

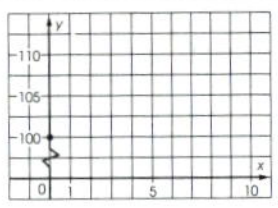

4. Let x represent the number of $.10 increases in the price of a hot dog.

a. Write an expression for the price of a hot dog in terms of x: ______

b. Write an expression for the number of hot dogs sold in terms of x: ______

c. Using your answers to parts (a) and (b), write an expression for the total income from hot dog sales: ______

d. Based on the table of Exercise 1 and the graph of Exercise 3, what value of x maximizes the expression for total income in part (c)? ______

210 APPLICATIONS

Enrichment/Exploration
Resource Book, p. 226

Factoring by Fours (For use with Chapter 7)

1. Factor each of the following quadratic expressions.

$x^2 + 5x + 6$ ______ $x^2 + 5x - 6$ ______ $x^2 - 5x + 6$ ______ $x^2 - 5x - 6$ ______

2. The graph of one of the following parabolas is shown.

$y = x^2 + 5x + 6$
$y = x^2 + 5x - 6$
$y = x^2 - 5x + 6$
$y = x^2 - 5x - 6$

Note that the scales on the x- and y-axes are not the same. Determine the scales on the axes and draw the remaining three parabolas on the same set of axes. Clearly indicate the intercepts on each axis.

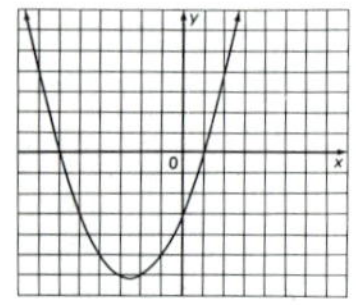

3. Repeat Exercise 2 for the following parabolas, one of which is drawn.

$y = x^2 + 10x + 24$
$y = x^2 + 10x - 24$
$y = x^2 - 10x + 24$
$y = x^2 - 10x - 24$

How do the graphs and the scales compare with those in Exercise 2?

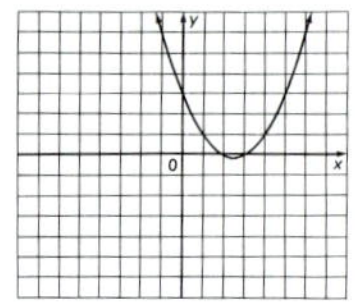

4. Repeat Exercise 2 for the following parabolas. One of them is drawn. Then compare graphs and scales with those in Exercise 2.

$y = 2x^2 + 5x + 3$
$y = 2x^2 + 5x - 3$
$y = 2x^2 - 5x + 3$
$y = 2x^2 - 5x - 3$

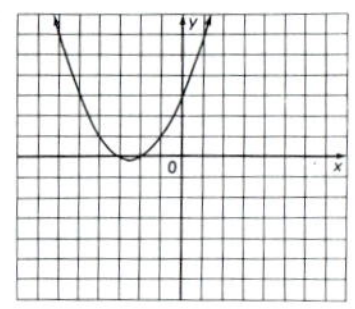

226 ENRICHMENT

Thinking Skills
Resource Book, p. 263

Thinking Skills (For use after Chapter 7)

Interpreting information

1. a. The x-coordinate of a point where the graph of a function crosses the x-axis is called a *zero* of the function. Why is the word "zero" appropriate in this context?

____

b. The graph of a quadratic function is a parabola. How does the graph show that a quadratic function can have at most two real zeros?

____

2. a. When the graph of a quadratic function is a parabola that opens upward, why is the y-coordinate of the parabola's vertex called the *minimum value* of the function?

____

b. When the graph of a quadratic function is a parabola that opens downward, why is the y-coordinate of the parabola's vertex called the *maximum value* of the function?

____

Analysis

3. Examine the quadratic functions and their graphs at the right. The functions have the form $f(x) = x^2 - 2x + c$, where $c = 2, 1$, or 0. How does the constant c affect the graph of f?

____

____

$f(x) = x^2 - 2x + 2$

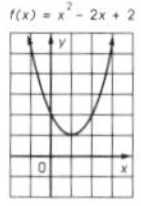

$f(x) = x^2 - 2x + 1$

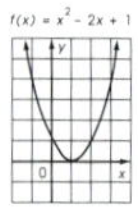

$f(x) = x^2 - 2x$

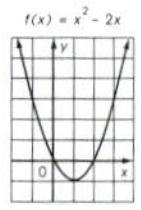

4. Use the graphs at the right above to complete the table below.

	Function	*Number of real zeros*	*Minimum value*
a.	$f(x) = x^2 - 2x + 2$	______	______
b.	$f(x) = x^2 - 2x + 1$	______	______
c.	$f(x) = x^2 - 2x$	______	______

Applying concepts

5. Complete the table of values at the right for the *cubic* function $f(x) = x^3 - 3x$.

	x	$x^3 - 3x = y$
a.	−3	______
b.	−2	______
c.	−1	______
d.	0	______
e.	1	______
f.	2	______
g.	3	______

(continued)

THINKING SKILLS 263

Thinking Skills
Resource Book, p. 264

Thinking Skills (Chapter 7) ***(continued)***

6. a. In the table of Exercise 5, what happens to the y-values as the positive x-values become larger and larger? ______

b. In the table of Exercise 5, what happens to the y-values as the negative x-values become smaller and smaller? ______

c. How would the results of parts (a) and (b) differ if the given function were quadratic instead of cubic? ______

7. Using the table of Exercise 5 and the axes at the right, sketch the graph of $f(x) = x^3 - 3x$.

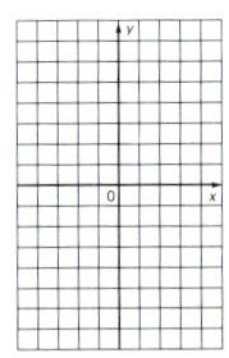

8. How many real zeros does $f(x) = x^3 - 3x$ have? ______

9. Notice that the graph of $f(x) = x^3 - 3x$ reaches a high point at $x = -1$ and a low point at $x = 1$. The y-coordinates of these points are not an *absolute* maximum and an *absolute* minimum of f, however. They are instead a *relative* maximum and a *relative* minimum. Explain the difference between "absolute" and "relative" in this context.

____

____

10. a. For what values of the constant c does the function $f(x) = x^3 - 3x + c$ have only two distinct real zeros? (*Hint:* This occurs when the graph of f is tangent to the x-axis.) ______

b. For what values of the constant c does f have only one real zero? ______

Spatial perception

11. When a curve opens upward, it is said to have *positive concavity.* When a curve opens downward, it is said to have *negative concavity.* While the concavity of the graph of a quadratic function is strictly positive or strictly negative, the graph of a cubic function changes concavity. Examine the graph of $f(x) = x^3 - 3x$ in Exercise 7.

a. For what values of x does the graph of f have positive concavity? ______

b. For what values of x does the graph of f have negative concavity? ______

c. A point where the graph of a function changes concavity is called a *point of inflection.* The graph of f has one point of inflection. What are its coordinates? ______

12. A quadratic function's graph, which is a parabola, is symmetric with respect to the vertical line through the vertex of the parabola. What symmetry does the graph of a cubic function have? (*Hint:* Look at the graph in Exercise 7, and compare such pairs of points as (a) and (g), (b) and (f) from the table in Exercise 5.)

____

264 THINKING SKILLS

Using Technology
Resource Book, p. 240

Using a Computer or a Graphing Calculator

To complete these activities, you should use a computer with graphics software (such as ALGEBRA PLOTTER PLUS) or a graphing calculator.

Translations of Parabolas (For use with Lesson 7-5)

1. Graph these equations: $y = (x - 2)^2$, $y = -2(x - 1)^2$, $y = \frac{1}{2}(x - 3)^2$

   Each graph, called a *parabola*, has a highest or lowest point, called a *vertex*. For example, the graph of $y = x^2$ has its vertex at the origin. Complete the table below.

	Equation of parabola	*Coordinates of vertex*
a.	$y = (x - 2)^2$	(? , ?) ______
b.	$y = -2(x - 1)^2$	(? , ?) ______
c.	$y = \frac{1}{2}(x - 3)^2$	(? , ?) ______

   d. What do you notice about the x-coordinate of each vertex?

   ____

   e. What do you notice about the y-coordinate of each vertex? ______________

2. Graph these equations: $y = (x + 2)^2$, $y = -2(x + 1)^2$, $y = \frac{1}{2}(x + 3)^2$

   Complete the table below.

	Equation of parabola	*Coordinates of vertex*
a.	$y = (x + 2)^2$	(? , ?) ______
b.	$y = -2(x + 1)^2$	(? , ?) ______
c.	$y = \frac{1}{2}(x + 3)^2$	(? , ?) ______

   d. What do you notice about the x-coordinate of each vertex?

   ____

   e. What do you notice about the y-coordinate of each vertex? ______________

3. Try to complete the table below before graphing the given equations.

	Equation of parabola	*Coordinates of vertex*
a.	$y = -(x + 3)^2$	(? , ?) ______
b.	$y = 3(x - 2)^2$	(? , ?) ______
c.	$y = -\frac{2}{3}(x + 1)^2$	(? , ?) ______

   Now examine the graphs of the given equations to confirm the entries in the table.

(continued)

Resource Book, ALGEBRA, Structure and Method, Book 2

Using Technology
Resource Book, p. 241

Translations of Parabolas *(continued)*

4. Graph these equations: $y - 2 = x^2$, $y + 1 = -2x^2$, $y - 3 = \frac{1}{2}x^2$

   Complete the table below.

	Equation of parabola	*Coordinates of vertex*
a.	$y - 2 = x^2$	(? , ?) ______
b.	$y + 1 = -2x^2$	(? , ?) ______
c.	$y - 3 = \frac{1}{2}x^2$	(? , ?) ______

   d. What do you notice about the x-coordinate of each vertex? ______________

   e. What do you notice about the y-coordinate of each vertex?

   ____

5. Try to complete the table below before graphing the given equations.

	Equation of parabola	*Coordinates of vertex*
a.	$y + 3 = -x^2$	(? , ?) ______
b.	$y - 2 = 3x^2$	(? , ?) ______
c.	$y + 1 = -\frac{2}{3}x^2$	(? , ?) ______

   Now examine the graphs of the given equations to confirm the entries in the table.

6. Try to complete the table below before graphing the given equations.

	Equation of parabola	*Coordinates of vertex*
a.	$y - 2 = (x - 3)^2$	(? , ?) ______
b.	$y - 1 = -3(x + 2)^2$	(? , ?) ______
c.	$y + 3 = \frac{2}{3}(x + 1)^2$	(? , ?) ______

   Now examine the graphs of the given equations to confirm the entries in the table.

7. Write equations of three different parabolas with vertex $(2, -1)$.

   a. ______________________________

   b. ______________________________

   c. ______________________________

   Graph the equations to confirm that the parabolas have the correct vertex.

8. If you have the ALGEBRA PLOTTER PLUS disk, use the "Parabola Quiz" program to see if you can correctly determine the equations of random parabolas drawn by the computer.

Resource Book, ALGEBRA, Structure and Method, Book 2

Using Models
Overhead Visual 4, Sheets 1 and 2

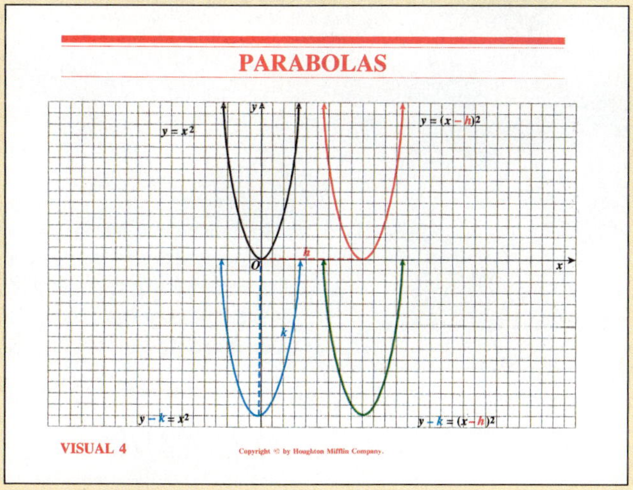

Using Models
Overhead Visual 4, Sheets 1 and 3

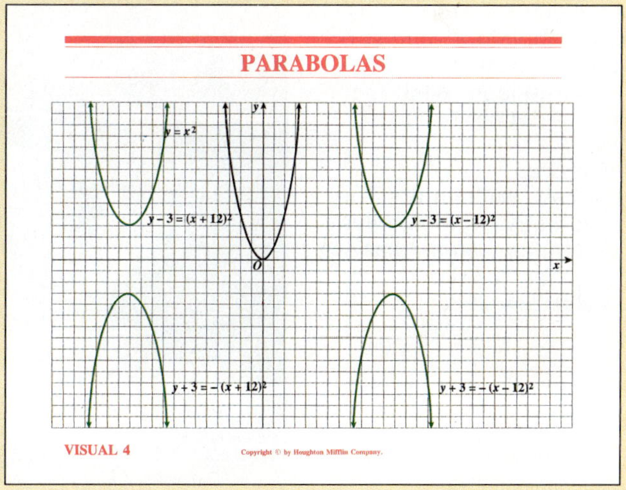

Application

When a parabola is rotated about its axis, a three-dimensional surface called a *paraboloid* results. Because of its reflection properties, a paraboloid is often used in transmitting and receiving light rays, sound waves, radio waves, and so forth. When a paraboloid is used as a transmitter, the outgoing waves, which emanate from a point called the focus, bounce off the paraboloid's surface and leave the paraboloid traveling in one direction. The situation is reversed when a paraboloid is used as a receiver.

Group Activities

Divide the class into groups and have each group investigate a particular instance of a paraboloid being used as a transmitter or a receiver. Besides flashlights and headlights, some examples of paraboloids are satellite dishes, solar furnaces, and radio telescopes.

Each group should answer such questions as: Is the focus a source of energy or a collection point? What is being sent or received? Is the paraboloid a solid surface or a mesh surface? How large is it? What factors determine its size?

References

Parzynski, William R. "The Geometry of Microwave Antennas." *Mathematics Teacher* 77 (May 1984):294–296.

7 Quadratic Equations and Functions

Headlights and flashlights use a curved mirror to direct light from a single bulb into parallel rays. The curve of the mirror, in cross section, is a parabola.

306

Solving Quadratic Equations

7-1 Completing the Square

Objective To solve quadratic equations by completing the square.

Here are three general types of polynomial equations in one variable ($a \neq 0$):

linear: $ax + b = 0$

quadratic: $\boldsymbol{ax^2 + bx + c = 0}$

cubic: $ax^3 + bx^2 + cx + d = 0$

Earlier in this book you learned how to solve all linear equations and some quadratic equations. In this chapter you'll learn methods for solving *all* types of quadratic equations. Cubic equations will be considered in the next chapter.

In Chapter 6 you learned to solve a simple quadratic equation $x^2 = r$ by writing the equivalent equation $x = \pm\sqrt{r}$. Example 1 illustrates how a similar method can be used to solve quadratic equations of the form $(px + q)^2 = r$.

Example 1 Solve.

a. $(x - 3)^2 = 7$ **b.** $(2x - 3)^2 = 7$ **c.** $(x + 5)^2 = -4$

Solution

a. $(x - 3)^2 = 7$

$$x - 3 = \pm\sqrt{7}$$

$$x = 3 \pm \sqrt{7}$$

$\therefore$ the solution set is $\{3 + \sqrt{7}, 3 - \sqrt{7}\}$. ***Answer***

b. $(2x - 3)^2 = 7$

$$2x - 3 = \pm\sqrt{7}$$

$$2x = 3 \pm \sqrt{7}$$

$$x = \frac{3 \pm \sqrt{7}}{2}$$

$\therefore$ the solution set is $\left\{\frac{3 + \sqrt{7}}{2}, \frac{3 - \sqrt{7}}{2}\right\}$. ***Answer***

c. $(x + 5)^2 = -4$

$$x + 5 = \pm\sqrt{-4} = \pm 2i$$

$$x = -5 \pm 2i$$

$\therefore$ the solution set is $\{-5 + 2i, -5 - 2i\}$. ***Answer***

To solve a quadratic equation of the form $ax^2 + bx + c = 0$, you can transform the equation into the form $(x + q)^2 = r$, which can be solved by the method of Example 1.

Teaching References

Lesson Commentary, pp. T104–T108

Assignment Guide, pp. T63–T66

Supplementary Materials
- Practice Masters 39–43
- Tests 27–30
- Resource Book
  - Practice Exercises, pp. 127–130
  - Tests, pp. 31–45
  - Enrichment Activity, p. 226
  - Application, p. 210
  - Mixed Review, pp. 180–182
- Study Guide, pp. 105–118
- Computer Activities 5, 15–16
- Disk for Algebra

Alternate Test, p. T19

Cumulative Review, p. T30

Teaching Suggestions p. T104

Reading Algebra p. T105

Suggested Extensions p. T105

Warm-Up Exercises

Solve.

1. $x^2 - 6x + 9 = 0$ $\{3\}$
2. $t^2 - 5t + 4 = 0$ $\{1, 4\}$
3. $m^2 + 10 = -7m$ $\{-5, -2\}$
4. $4z^2 = 16z$ $\{0, 4\}$
5. $2x^2 - 17x + 15 = 0$ $\left\{\frac{15}{2}, 1\right\}$
6. $x^2 + 121 = 0$ $\{\pm 11i\}$
7. $3x^2 = -48$ $\{\pm 4i\}$

Motivating the Lesson

Tell students that quadratic equations are used in many areas of science and industry and that it is important that they understand the methods used to solve these equations. The method of completing the square is the topic of today's lesson.

Chalkboard Examples

Solve.

1. $(x+3)^2 = 5$

$x + 3 = \pm\sqrt{5}$

$x = -3 \pm \sqrt{5}$

$\{-3+\sqrt{5}, -3-\sqrt{5}\}$

2. $x^2 - 4x + 4 = -2$

$(x-2)^2 = -2$

$x - 2 = \pm i\sqrt{2}$

$x = 2 \pm i\sqrt{2}$

$\{2+i\sqrt{2}, 2-i\sqrt{2}\}$

Solve. Complete the square.

3. $a^2 - 5a + 3 = 0$

$a^2 - 5a = -3$

$a^2 - 5a + \frac{25}{4} = -3 + \frac{25}{4}$

$\left(a - \frac{5}{2}\right)^2 = \frac{13}{4}$

$a - \frac{5}{2} = \pm\frac{\sqrt{13}}{2}$

$a = \frac{5}{2} \pm \frac{\sqrt{13}}{2} = \frac{5 \pm \sqrt{13}}{2}$

$\left\{\frac{5+\sqrt{13}}{2}, \frac{5-\sqrt{13}}{2}\right\}$

4. $7x^2 - 8x + 3 = 0$

$7x^2 - 8x = -3$

$x^2 - \frac{8}{7}x = -\frac{3}{7}$

$x^2 - \frac{8}{7}x + \frac{16}{49} = -\frac{3}{7} + \frac{16}{49}$

$\left(x - \frac{4}{7}\right)^2 = -\frac{5}{49}$

$x - \frac{4}{7} = \pm i\frac{\sqrt{5}}{7}$

$x = \frac{4 \pm i\sqrt{5}}{7}$

$\left\{\frac{4+i\sqrt{5}}{7}, \frac{4-i\sqrt{5}}{7}\right\}$

The method of changing the form of a quadratic equation so that the left side is a trinomial square is called **completing the square.**

Example 2 Solve $x^2 - 6x - 3 = 0$.

Solution $x^2 - 6x - 3 = 0$

Step 1 Add 3 to both sides to get 3 alone on one side. $x^2 - 6x = 3$

Step 2 Check the coefficient of x^2. Since it is 1, you are ready to complete the square. $1x^2 - 6x = 3$

Step 3 Add the square of half the coefficient of x to both sides. Thus, add $\left(\frac{-6}{2}\right)^2$, or 9, to both sides. $x^2 - 6x + 9 = 3 + 9$

Step 4 Factor the left side as the square of a binomial. $(x-3)^2 = 12$

Step 5 Solve as in Example 1. $x - 3 = \pm\sqrt{12}$

$x = 3 \pm 2\sqrt{3}$

$\therefore$ the solution set is $\{3 + 2\sqrt{3}, 3 - 2\sqrt{3}\}$. ***Answer***

Solving $ax^2 + bx + c = 0$ by Completing the Square

Step 1 Transform the equation so that the constant term c is alone on the right side.

Step 2 If a, the coefficient of the second-degree term, is not equal to 1, then divide both sides by a.

Step 3 Add the square of half the coefficient of the first-degree term, $\left(\frac{b}{2a}\right)^2$, to *both sides*. (Completing the square)

Step 4 Factor the left side as the square of a binomial.

Step 5 Complete the solution using the fact that $(x+q)^2 = r$ is equivalent to $x + q = \pm\sqrt{r}$.

Example 3 Solve $2y^2 + 2y + 5 = 0$.

Solution $2y^2 + 2y + 5 = 0$

Step 1 $2y^2 + 2y = -5$

Step 2 $y^2 + y = -\frac{5}{2}$ Since $a = 2$, divide both sides by 2.

Step 3 $y^2 + y + \left(\frac{1}{2}\right)^2 = -\frac{5}{2} + \left(\frac{1}{2}\right)^2$ ⟵ Complete the square.

$y^2 + y + \frac{1}{4} = -\frac{5}{2} + \frac{1}{4}$

Step 4 $\left(y + \frac{1}{2}\right)^2 = -\frac{9}{4}$

Step 5 $y + \frac{1}{2} = \pm\sqrt{-\frac{9}{4}} = \pm\frac{3}{2}i$

$y = -\frac{1}{2} \pm \frac{3}{2}i$

$\therefore$ the solution set is $\left\{-\frac{1}{2} + \frac{3}{2}i, -\frac{1}{2} - \frac{3}{2}i\right\}$. ***Answer***

Oral Exercises

What two numbers are represented by each of the following?

1. 4 ± 1 5, 3
2. -3 ± 5 2, −8
3. $\frac{1}{2} \pm \frac{3}{2}$ 2, −1
4. $-2 \pm \frac{1}{4}$ $-\frac{7}{4}, -\frac{9}{4}$

5. $11 \pm \sqrt{2}$ $11 + \sqrt{2}, 11 - \sqrt{2}$
6. $-6 \pm \sqrt{6}$ $-6 + \sqrt{6}, -6 - \sqrt{6}$
7. $6 \pm 3i$ $6 + 3i, 6 - 3i$
8. $\frac{4}{3} \pm \frac{i\sqrt{2}}{3}$ $\frac{4}{3} + \frac{\sqrt{2}}{3}i, \frac{4}{3} - \frac{\sqrt{2}}{3}i$

Give the first step in the solution of each equation.

9. a. $x^2 = 5$ $x = \pm\sqrt{5}$ **b.** $(x - 3)^2 = 5$ $x - 3 = \pm\sqrt{5}$ **c.** $(x + 3)^2 = 5$ $x + 3 = \pm\sqrt{5}$ **d.** $(2x + 3)^2 = 5$ $2x + 3 = \pm\sqrt{5}$

10. a. $y^2 = -1$ $y = \pm\sqrt{-1}$ **b.** $(y + 2)^2 = -1$ $y + 2 = \pm\sqrt{-1}$ **c.** $(y - 2)^2 = -1$ $y - 2 = \pm\sqrt{-1}$ **d.** $(6y - 2)^2 = -1$ $6y - 2 = \pm\sqrt{-1}$

What must be added to make the given expression a trinomial square?

11. $x^2 + 2x +$ ___?___ 1
12. $m^2 - 4m +$ ___?___ 4
13. $y^2 + 8y +$ ___?___ 16

14. $a^2 + 12a +$ ___?___ 36
15. $t^2 - t +$ ___?___ $\frac{1}{4}$
16. $y^2 - 3y +$ ___?___ $\frac{9}{4}$

17. $z^2 - 7z +$ ___?___ $\frac{49}{4}$
18. $y^2 + \frac{2}{5}y +$ ___?___ $\frac{1}{25}$
19. $r^2 - \frac{1}{2}r +$ ___?___ $\frac{1}{16}$

20. Solve $t^2 - 6t = 0$ by completing the square and by factoring. Which method seems easier? $t = 0, 6$; Answers may vary; factoring

Written Exercises

Solve.

A

1. a. $x^2 = 3$ $\{\pm\sqrt{3}\}$ **b.** $(x - 1)^2 = 3$ $\{1 \pm \sqrt{3}\}$ **c.** $(2x - 1)^2 = 3$ $\left\{\frac{1 \pm \sqrt{3}}{2}\right\}$

2. a. $x^2 = 6$ $\{\pm\sqrt{6}\}$ **b.** $(x + 4)^2 = 6$ $\{-4 \pm \sqrt{6}\}$ **c.** $(3x + 4)^2 = 6$ $\left\{\frac{-4 \pm \sqrt{6}}{3}\right\}$

3. a. $y^2 = 16$ $\{\pm 4\}$ **b.** $(y + 7)^2 = 16$ $\{-3, -11\}$ **c.** $(3y + 7)^2 = 16$ $\left\{-1, -\frac{11}{3}\right\}$

4. a. $y^2 = 49$ $\{\pm 7\}$ **b.** $(y - 8)^2 = 49$ $\{15, 1\}$ **c.** $(5y - 8)^2 = 49$ $\left\{3, \frac{1}{5}\right\}$

Common Errors

When solving equations by completing the square, students may forget to divide both sides by the coefficient of the second degree term when it is not equal to one. When reteaching, recommend that the five steps on page 308 be followed completely.

Check for Understanding

Here is a suggested use of the Oral Exercises to check students' understanding as you teach the lesson.
Oral Exs. 1–8: use before Example 1.
Oral Exs. 9–10: use after Example 1.
Oral Exs. 11–19: use after Example 2.
Oral Ex. 20: use before Example 3.

Guided Practice

Solve.

1. $m^2 = 50$ $\{\pm 5\sqrt{2}\}$
2. $(p + 2)^2 = -50$ $\{-2 \pm 5i\sqrt{2}\}$
3. $(f - 12)^2 = -18$ $\{12 \pm 3i\sqrt{2}\}$
4. $4(2x - 3)^2 = -100$ $\left\{\frac{3 \pm 5i}{2}\right\}$
5. $\frac{3(5 - x)^2 + 1}{4} = 7$ $\{2, 8\}$
6. $x^2 + 6x + 13 = 0$ $\{-3 \pm 2i\}$
7. $y^2 - 14y - 1 = 0$ $\{7 \pm 5\sqrt{2}\}$
8. $9x^2 = 36x - 35$ $\left\{\frac{5}{3}, \frac{7}{3}\right\}$
9. $m^2 + m = 2$ $\{1, -2\}$

Summarizing the Lesson

In this lesson students learned the method of completing the square. Ask students to discuss when it is best to use completing the square and when it is best to use factoring to solve quadratic equations. Also ask students if they can recall any other methods for solving quadratic equations.

Suggested Assignments

Average Algebra
Day 1: 309/1–35 odd
S 310/Mixed Review
Day 2: 310/20–38 even
Assign with Lesson 7-2.

Average Alg. and Trig.
Day 1: 309/1–35 odd
S 310/Mixed Review
Day 2: 310/20–38 even
Assign with Lesson 7-2.

Extended Algebra
309/3–21 mult. of 3, 23–41 odd
S 310/Mixed Review

Extended Alg. and Trig.
309/3–21 mult. of 3, 23–41 odd
S 310/Mixed Review

Supplementary Materials

Study Guide pp. 105–106

Solve.

5. a. $x^2 = -4$ $\{\pm 2i\}$ b. $(x + 7)^2 = -4$ $\{-7 \pm 2i\}$ c. $(2x + 7)^2 = -4$ $\left\{-\frac{7}{2} \pm i\right\}$
6. a. $z^2 = -5$ $\{\pm i\sqrt{5}\}$ b. $(z - 3)^2 = -5$ $\{3 \pm i\sqrt{5}\}$ c. $(5z - 3)^2 = -5$ $\left\{\frac{3}{5} \pm \frac{\sqrt{5}}{5}i\right\}$
7. $(y - 7)^2 = 12$ $\{7 \pm 2\sqrt{3}\}$
8. $(t - 3)^2 = 8$ $\{3 \pm 2\sqrt{2}\}$
9. $\left(\frac{1}{3}n + 2\right)^2 = 18$ $\{-6 \pm 9\sqrt{2}\}$
10. $\left(\frac{1}{2}t - 12\right)^2 = 50$ $\{24 \pm 10\sqrt{2}\}$
11. $3(y - 7)^2 = -12$ $\{7 \pm 2i\}$
12. $\frac{(x - 5)^2}{3} = -8$ $\{5 \pm 2i\sqrt{6}\}$

Solve by completing the square. 15. $\{-3 \pm \sqrt{11}\}$ 18. $\{5 \pm i\sqrt{5}\}$ 21. $\left\{-1 \pm \frac{\sqrt{2}}{2}\right\}$ 24. $\left\{2 \pm \frac{\sqrt{22}}{2}\right\}$

13. $x^2 - 2x - 5 = 0$ $\{1 \pm \sqrt{6}\}$
14. $x^2 - 4x + 2 = 0$ $\{2 \pm \sqrt{2}\}$
15. $y^2 + 6y - 2 = 0$
16. $y^2 + 8y + 6 = 0$ $\{-4 \pm \sqrt{10}\}$
17. $p^2 + 20p + 200 = 0$ $\{-10 \pm 10i\}$
18. $k^2 - 10k + 30 = 0$
19. $x^2 - 1 = 4x$ $\{2 \pm \sqrt{5}\}$
20. $t^2 + 8 = 4t$ $\{2 \pm 2i\}$
21. $2t^2 + 4t + 1 = 0$
22. $3n^2 + 12n + 1 = 0$ $\left\{-2 \pm \frac{\sqrt{33}}{3}\right\}$
23. $5n^2 + 100 = 30n$ $\{3 \pm i\sqrt{11}\}$
24. $2n^2 - 8n - 3 = 0$
25. $x^2 - x - 1 = 0$ $\left\{\frac{1}{2} \pm \frac{\sqrt{5}}{2}\right\}$
26. $y^2 - 3y - 5 = 0$ $\left\{\frac{3}{2} \pm \frac{\sqrt{29}}{2}\right\}$
27. $3k^2 + 5k + 2 = 0$ $\left\{-\frac{2}{3}, -1\right\}$

B

28. $\frac{1}{2}x^2 - 3x = 2$ $\{3 \pm \sqrt{13}\}$
29. $\frac{y^2}{4} - \frac{y}{2} + 1 = 0$ $\{1 \pm i\sqrt{3}\}$
30. $0.1x^2 - 0.6x + 9 = 0$ $\{3 \pm 9i\}$
31. $0.6x^2 + 2 = 2.4x$ $\left\{2 \pm \frac{\sqrt{6}}{3}\right\}$
32. $7x(1 - x) = 5(x - 2)$ $\left\{\frac{1}{7} \pm \frac{\sqrt{71}}{7}\right\}$
33. $2x(x - 4) = 3(1 - x)$ $\left\{-\frac{1}{2}, 3\right\}$

Solve each equation. Be sure to check for extraneous roots by substituting your answers into the original equation.

34. $\frac{1}{x + 1} + \frac{1}{x - 1} = 1$ $\{1 \pm \sqrt{2}\}$
35. $\frac{1}{y + 2} + \frac{1}{y + 6} = 1$ $\{-3 \pm \sqrt{5}\}$
36. $\sqrt{x + 3} = 2x$ $\{1\}$
37. $\frac{x}{x - 1} - \frac{x}{x + 1} = 3 + \frac{2x^2}{1 - x^2}$ $\{3\}$
38. $\frac{x + 2}{x - 2} + \frac{x - 2}{x + 2} = \frac{8 - 4x}{x^2 - 4}$ $\{0\}$
39. $\sqrt{x - 4} - \frac{2}{\sqrt{x - 4}} = 1$ $\{8\}$

Solve by completing the square. Assume that *a*, *b*, and *c* are nonzero constants.

C

40. $x^2 + x + c = 0$ $\left\{-\frac{1}{2} \pm \frac{\sqrt{1 - 4c}}{2}\right\}$
41. $x^2 + bx + c = 0$ $\left\{-\frac{b}{2} \pm \frac{\sqrt{b^2 - 4c}}{2}\right\}$
42. $ax^2 + bx + c = 0$ $\left\{-\frac{b}{2a} \pm \frac{\sqrt{b^2 - 4ac}}{2a}\right\}$

Mixed Review Exercises

Simplify.

1. $\sqrt{-50}$ $5i\sqrt{2}$
2. $(3 - 2i)(4 + i)$ $14 - 5i$
3. $(3\sqrt{2})^3$ $54\sqrt{2}$
4. $\sqrt{98} - \sqrt{8}$ $5\sqrt{2}$
5. $(2i\sqrt{6})^2$ -24
6. $2(8 - 3i) + (-4 + 5i)$ $12 - i$
7. $\sqrt{\frac{2}{5}} + \sqrt{\frac{5}{2}}$ $\frac{7\sqrt{10}}{10}$
8. $\frac{5x^2y^3}{-15xy^4}$ $-\frac{x}{3y}$
9. $\frac{x^2 - 5x + 6}{x^2 - 9}$ $\frac{x - 2}{x + 3}$

7-2 The Quadratic Formula

Objective To solve quadratic equations by using the quadratic formula.

Quadratic equations are used in many applications, so it is useful to have a formula that gives their solutions directly from the coefficients. You can derive this formula by applying the method of completing the square to the general quadratic equation:

$$ax^2 + bx + c = 0$$

$$x^2 + \frac{b}{a}x + \frac{c}{a} = 0 \text{ (Recall that } a \neq 0.)$$

$$x^2 + \frac{b}{a}x = -\frac{c}{a}$$

$$x^2 + \frac{b}{a}x + \left(\frac{b}{2a}\right)^2 = -\frac{c}{a} + \left(\frac{b}{2a}\right)^2 \quad \text{Complete the square.}$$

$$\left(x + \frac{b}{2a}\right)^2 = \frac{-4ac + b^2}{4a^2} \quad \text{Factor the left side.}$$

$$x + \frac{b}{2a} = \pm\sqrt{\frac{b^2 - 4ac}{4a^2}} = \pm\frac{\sqrt{b^2 - 4ac}}{2a}$$

$$x = -\frac{b}{2a} \pm \frac{\sqrt{b^2 - 4ac}}{2a} = \frac{-b \pm \sqrt{b^2 - 4ac}}{2a}$$

The Quadratic Formula

The solutions of the quadratic equation $ax^2 + bx + c = 0$ $(a \neq 0)$ are given by the formula

$$x = \frac{-b \pm \sqrt{b^2 - 4ac}}{2a}.$$

Example 1 Solve $3x^2 + x - 1 = 0$.

Solution For the equation $3x^2 + 1x - 1 = 0$,

$a = 3$, $b = 1$, and $c = -1$

Substitute these values in the quadratic formula.

$$x = \frac{-b \pm \sqrt{b^2 - 4ac}}{2a}$$

$$x = \frac{-(1) \pm \sqrt{(1)^2 - 4(3)(-1)}}{2(3)}$$

$$= \frac{-1 \pm \sqrt{1 - (-12)}}{6} = \frac{-1 \pm \sqrt{13}}{6}$$

$\therefore$ the solution set is $\left\{\frac{-1 + \sqrt{13}}{6}, \frac{-1 - \sqrt{13}}{6}\right\}$. ***Answer***

Teaching Suggestions p. T105

Suggested Extensions p. T105

Warm-Up Exercises

Find x if $a = 3$, $b = 4$, and $c = 5$.

1. $x = \frac{a + b + c}{3}$ $\{4\}$
2. $ax + b = c$ $\left\{\frac{1}{3}\right\}$
3. $\frac{x}{b} = \frac{c}{a}$ $\left\{\frac{20}{3}\right\}$
4. $x^2 + ax - b = 0$ $\{1, -4\}$
5. $x = \sqrt{b^2 - 4ac}$ $\{2i\sqrt{11}\}$

Motivating the Lesson

Tell students that the quadratic formula is one of the most widely known and used methods of solving quadratic equations. Point out that the formula can be used to solve any quadratic equation and that it is important that students remember it.

Thinking Skills

The derivation of the quadratic formula in this lesson requires students to *apply* the method of completing the square to the general quadratic equation. Also, for students to use the formula, they must *transform* quadratic equations into the form $ax^2 + bx + c = 0$.

Common Errors

When using the quadratic formula to solve an equation, students may forget that the equation must be in the form $ax^2 + bx + c = 0$ $(a \neq 0)$ before the formula can be applied. When reteaching, emphasize this fact.

The solutions just obtained are exact and expressed in simplest radical form. In applications you may want approximate solutions to the nearest hundredth. Since $\sqrt{13} \approx 3.606$ (from a calculator or Table 1, page 810),

$$x \approx \frac{-1 + 3.606}{6} \quad \text{or} \quad x \approx \frac{-1 - 3.606}{6}$$

$$x \approx 0.43 \quad \text{or} \quad x \approx -0.77$$

$\therefore$ the solution set is $\{0.43, -0.77\}$. ***Answer***

Example 2 Solve $5y^2 = 6y - 3$.

Solution First rewrite the equation in the form $ax^2 + bx + c = 0$:

$$5y^2 - 6y + 3 = 0$$

Then substitute 5 for a, -6 for b, and 3 for c in the quadratic formula.

$$y = \frac{-b \pm \sqrt{b^2 - 4ac}}{2a}$$

$$y = \frac{-(-6) \pm \sqrt{(-6)^2 - 4(5)(3)}}{2(5)}$$

$$= \frac{6 \pm \sqrt{-24}}{10}$$

$$= \frac{6 \pm 2i\sqrt{6}}{10}$$

$$= \frac{3 \pm i\sqrt{6}}{5}$$

$\therefore$ the solution set is $\left\{\frac{3 + i\sqrt{6}}{5}, \frac{3 - i\sqrt{6}}{5}\right\}$. ***Answer***

In Examples 1 and 2, the coefficients a, b, and c were integers. Keep in mind, however, that the quadratic formula can be used to solve *any* quadratic equation, whether the coefficients are fractions, decimals, irrational numbers, or imaginary numbers. (See Exercises 29–40, page 314.)

Example 3 A swimming pool 6 m wide and 10 m long is to be surrounded by a walk of uniform width. The area of the walk happens to equal the area of the pool. What is the width of the walk?

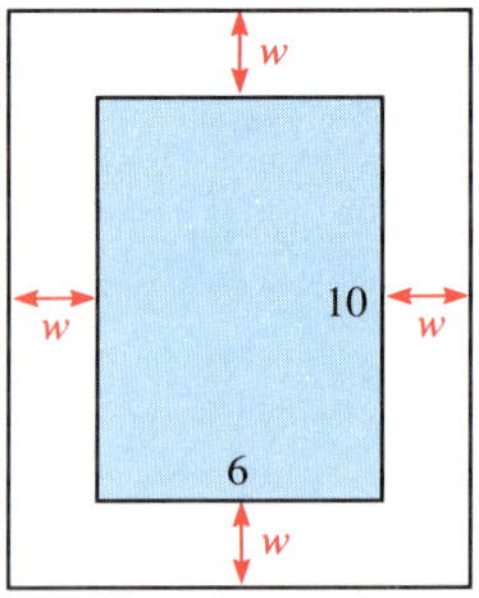

Solution 1. Make a sketch.
Let w = the width of the walk.
Then the dimensions of the pool plus the walkway are $10 + 2w$ by $6 + 2w$.

Chalkboard Examples

Solve by using the quadratic formula.

1. $x^2 - 5x - 14 = 0$
$x = \frac{5 \pm \sqrt{25 + 56}}{2}$
$= \frac{5 \pm \sqrt{81}}{2} = \frac{5 \pm 9}{2}$
$x = 7$ or $x = -2$
$\{7, -2\}$

2. $5x^2 + 4x - 2 = 0$
$x = \frac{-4 \pm \sqrt{16 + 40}}{10}$
$= \frac{-4 \pm \sqrt{56}}{10} = \frac{-4 \pm 2\sqrt{14}}{10}$
$= \frac{-2 \pm \sqrt{14}}{5}$
$\left\{\frac{-2 + \sqrt{14}}{5}, \frac{-2 - \sqrt{14}}{5}\right\}$

3. $4m^2 - 5m + 3 = 0$
$m = \frac{5 \pm \sqrt{25 - 48}}{8}$
$= \frac{5 \pm \sqrt{-23}}{8} = \frac{5 \pm i\sqrt{23}}{8}$
$\left\{\frac{5 + i\sqrt{23}}{8}, \frac{5 - i\sqrt{23}}{8}\right\}$

4. The volume of a rectangular jewelry box 4 cm in height is 136 cm³. If the perimeter of the base is 24 cm, find the dimensions of the base.
$V = lwh$, $136 = lw(4)$,
$34 = lw$, $l = \frac{34}{w}$;
$P = 2l + 2w$, $24 = 2l + 2w$; $12 = l + w$,
$12 = \frac{34}{w} + w$
$12w = 34 + w^2$
$w^2 - 12w + 34 = 0$
$w = \frac{12 \pm \sqrt{144 - 136}}{2}$
$= \frac{12 \pm \sqrt{8}}{2} = \frac{12 \pm 2\sqrt{2}}{2}$
$= 6 \pm \sqrt{2}$
$\therefore$ the dimensions of the base are $(6 - \sqrt{2})$ cm and $(6 + \sqrt{2})$ cm.
$\therefore$ the box is about 4.6 cm by 7.4 cm.

2. The area of the pool is $10 \cdot 6 = 60$ (m^2).
The area of the walk can be determined by subtracting the pool's area from the total area covered by both the pool and the walk.

$$\text{Area of walk} = \text{Total area} - \text{Area of pool}$$
$$= (10 + 2w)(6 + 2w) - 60$$

3.
$$\text{Area of walk} = \text{Area of pool}$$
$$(10 + 2w)(6 + 2w) - 60 = 60$$
$$(60 + 32w + 4w^2) - 60 = 60$$
$$4w^2 + 32w = 60$$
$$4w^2 + 32w - 60 = 0 \longleftarrow \text{Divide both sides by 4.}$$
$$w^2 + 8w - 15 = 0$$

4. By completing the square or using the quadratic formula you will find that
$$w = -4 \pm \sqrt{31}.$$
Since $-4 - \sqrt{31} \approx -9.57$, you must reject this root because the width of the walk cannot be negative.
∴ the width of the walk is $(-4 + \sqrt{31})$ m, or approximately 1.57 m. ***Answer***

5. *Check:* A calculator is helpful in checking approximate solutions.
$$\text{Area of walk} = \text{Area of pool}$$
$$(10 + 2w)(6 + 2w) - 60 \stackrel{?}{\approx} 60$$
$$(10 + 2 \cdot 1.57)(6 + 2 \cdot 1.57) - 60 \stackrel{?}{\approx} 60$$
$$(13.14)(9.14) - 60 \stackrel{?}{\approx} 60$$
$$120.10 - 60 \stackrel{?}{\approx} 60 \quad \checkmark$$

Oral Exercises

Give the values that you would substitute for *a*, *b*, and *c* in the quadratic formula.

1. $2x^2 - 3x + 7 = 0$ ($a = 2, b = -3, c = 7$)
2. $3x^2 + 7x - 2 = 0$ ($a = 3, b = 7, c = -2$)
3. $5 - 7x - 4x^2 = 0$ ($a = -4, b = -7, c = 5$)
4. $x^2 = 4 - 2x$ ($a = 1, b = 2, c = -4$)
5. $x^2 - x\sqrt{5} + 1 = 0$ ($a = 1, b = -\sqrt{5}, c = 1$)
6. $x(x - 2) = 9$ ($a = 1, b = -2, c = -9$)

Written Exercises

Solve each equation. Give answers involving radicals in simplest radical form.

A
1. $x^2 + 6x + 4 = 0$ $\{-3 \pm \sqrt{5}\}$
2. $v^2 + 3v - 5 = 0$ $\left\{\frac{-3 \pm \sqrt{29}}{2}\right\}$
3. $y^2 - 4y + 13 = 0$ $\{2 \pm 3i\}$
4. $t^2 + 6t + 6 = 0$ $\{-3 \pm \sqrt{3}\}$
5. $5k^2 + 3k - 2 = 0$ $\left\{\frac{2}{5}, -1\right\}$
6. $2p^2 - 3p - 2 = 0$ $\left\{2, -\frac{1}{2}\right\}$

Using a Calculator

When the solutions of a quadratic equation involve radicals, students may not be aware of the magnitude of the solutions. Encourage students to use calculators to obtain decimal approximations, as is done in Example 1. Such use is suggested for Exercises 19–24 on page 314, Problems 1–15 on pages 314–315, and Exercises 5–6 from Self-Test 1.

Also encourage students to use calculators to check their approximate solutions, as is done in Example 3. Point out that approximate solutions will not satisfy a given equation exactly but should come reasonably close to doing so.

Check for Understanding

Oral Exs. 1–6: use after Example 2.

Guided Practice

Solve.

1. $x^2 + 6x + 2 = 0$ $\{-3 \pm \sqrt{7}\}$
2. $5t^2 + 8 = -12t$ $\left\{\frac{-6 \pm 2i}{5}\right\}$

Solve. Approximate the answers to the nearest hundredth.

3. $3r^2 - 6r - 7 = 0$ $\{2.83, -0.83\}$
4. $3x(x + 2) + 2.5 = 0$ $\{-0.59, -1.41\}$

Solve (a) by factoring and (b) by the quadratic formula.

5. $3a^2 - 2a - 1 = 0$ $\left\{-\frac{1}{3}, 1\right\}$
6. $4c^2 + 4c - 15 = 0$ $\left\{\frac{3}{2}, -\frac{5}{2}\right\}$

Summarizing the Lesson

In this lesson students learned to use the quadratic formula to solve equations. Ask students to discuss the relative ease of using factoring, completing the square, and the quadratic formula.

Suggested Assignments

Average Algebra
Day 1: 313/1–12
Assign with Lesson 7-1.
Day 2: 314/14–36 even
P: 1–13 odd

Average Alg. and Trig.
Day 1: 313/1–12
Assign with Lesson 7-1.
Day 2: 314/14–36 even
P: 1–13 odd

Extended Algebra
Day 1: 313/3–27 mult. of 3
314/*P*: 1–13 odd
Day 2: 314/28–42 even
315/*P*: 6–14 even
S 310/36–42 even

Extended Alg. and Trig.
Day 1: 313/3–27 mult. of 3
314/*P*: 1–13 odd
Day 2: 314/28–42 even
315/*P*: 6–14 even
S 310/36–42 even

Additional Answers Written Exercises

42. The roots of $ax^2 + bx + c = 0$ are $x_1 = \frac{-b + \sqrt{b^2 - 4ac}}{2a}$ and $x_2 = \frac{-b - \sqrt{b^2 - 4ac}}{2a}$. If x_1 and x_2 are reciprocals, $x_1x_2 = 1$; $\frac{-b + \sqrt{b^2 - 4ac}}{2a} \times \frac{-b - \sqrt{b^2 - 4ac}}{2a} = 1$; $\frac{b^2 - (b^2 - 4ac)}{4a^2} = 1$; $\frac{4ac}{4a^2} = 1$; $a = c$.

Solve each equation after rewriting it in the form $ax^2 + bx + c = 0$. Give answers involving radicals in simplest radical form.

7. $5r^2 + 8 = -12r$ $\left\{-\frac{6}{5} \pm \frac{2}{5}i\right\}$
8. $2w^2 + 4w = -3$ $\left\{-1 \pm \frac{\sqrt{2}}{2}i\right\}$
9. $3y^2 = 1 - y$ $\left\{\frac{-1 \pm \sqrt{13}}{6}\right\}$
10. $8x = 1 - x^2$ $\{-4 \pm \sqrt{17}\}$
11. $2x(x + 1) = 7$ $\left\{\frac{-1 \pm \sqrt{15}}{2}\right\}$
12. $5 = 4r(2r + 3)$
13. $(3n - 5)(2n - 2) = 6$
14. $(2x + 1)(2x - 1) = 4x$
15. $\frac{w^2}{2} - w = \frac{3}{4}$
16. $\frac{t^2}{2} + 1 = \frac{t}{5}$ $\left\{\frac{1}{5} \pm \frac{7}{5}i\right\}$
17. $\frac{2m^2 + 16}{5} = 2m$ $\left\{\frac{5}{2} \pm \frac{\sqrt{7}}{2}i\right\}$
18. $\frac{4 - 2y^2}{7} = 2y$ $\left\{\frac{-7 \pm \sqrt{57}}{2}\right\}$

12. $\left\{\frac{-3 \pm \sqrt{19}}{4}\right\}$ **13.** $\left\{\frac{4 \pm \sqrt{10}}{3}\right\}$ **14.** $\left\{\frac{1 \pm \sqrt{2}}{2}\right\}$ **15.** $\left\{\frac{2 \pm \sqrt{10}}{2}\right\}$

Solve each equation and approximate solutions to the nearest hundredth. A calculator may be helpful.

19. $2n^2 - 4n = 8$ $\{-1.24, 3.24\}$
20. $2x^2 - 3x = 7$ $\{-1.27, 2.77\}$
21. $3t^2 - 6t - 7 = 0$ $\{-0.83, 2.83\}$
22. $4x(x + 1) = 2.75$ $\{-1.47, 0.47\}$
23. $3x(x + 2) = -2.5$ $\{-1.41, -0.59\}$
24. $2t(t - 4) = -3$ $\{0.42, 3.58\}$

Solve each equation (a) by factoring and (b) by using the quadratic formula.

25. $5x^2 - 45 = 0$ $\{\pm 3\}$
26. $3y^2 - 48 = 0$ $\{\pm 4\}$
27. $3x^2 - 6x + 3 = 0$ $\{1\}$
28. $4y^2 + 4y - 15 = 0$ $\left\{-\frac{5}{2}, \frac{3}{2}\right\}$

Solve each equation. Give answers involving radicals in simplest radical form.

B **29.** $x^2 - x\sqrt{2} - 1 = 0$ $\left\{\frac{\sqrt{2} \pm \sqrt{6}}{2}\right\}$
30. $x^2 - x\sqrt{5} - 1 = 0$ $\left\{\frac{\sqrt{5} \pm 3}{2}\right\}$
31. $t^2 - 2t\sqrt{2} + 1 = 0$ $\{\sqrt{2} \pm 1\}$
32. $u^2 + 2u\sqrt{3} - 3 = 0$ $\{-\sqrt{3} \pm \sqrt{6}\}$
33. $\sqrt{2}x^2 + 5x + 2\sqrt{2} = 0$ $\left\{-2\sqrt{2}, -\frac{\sqrt{2}}{2}\right\}$
34. $\sqrt{3}x^2 - 2x + 2\sqrt{3} = 0$ $\left\{\frac{\sqrt{3}}{3} \pm \frac{\sqrt{15}}{3}i\right\}$
35. $z^2 + iz + 2 = 0$ $\{-2i, i\}$
36. $z^2 + 2iz - 1 = 0$ $\{-i\}$
37. $z^2 - (3 + 2i)z + (1 + 3i) = 0$ $\{2 + i, 1 + i\}$
38. $iz^2 + (2 - 3i)z - (3 + i) = 0$ $\{i, 3 + i\}$
39. $\frac{2w + i}{w - i} = \frac{3w + 4i}{w + 3i}$ $\{-i, 7i\}$
40. $\frac{1}{2z + i} + \frac{1}{2z - i} = \frac{4}{z + 2i}$ $\left\{-\frac{1}{3}i, i\right\}$
41. Show that the solutions of $3x^2 - 2x + 3 = 0$ are reciprocals.
The solutions are $x_1 = \frac{1}{3} + \frac{2\sqrt{2}}{3}i$ and $x_2 = \frac{1}{3} - \frac{2\sqrt{2}}{3}i$; $x_1x_2 = 1$, so x_1 and x_2 are reciprocals.

C **42.** Prove that if the roots of $ax^2 + bx + c = 0$ $(a \neq 0)$ are reciprocals, then $a = c$.

Problems

Solve each problem. Approximate any answers involving radicals to the nearest hundredth. A calculator may be helpful.

A **1.** Each side of a square is 4 m long. When each side is increased by x m, the area is doubled. Find the value of x. 1.66

2. A rectangle is 6 cm long and 5 cm wide. When each dimension is increased by x cm, the area is tripled. Find the value of x. 4

3. A positive real number is 1 more than its reciprocal. Find the number. 1.62

4. Two positive real numbers have a sum of 5 and product of 5. Find the numbers. 1.38, 3.62

5. A rectangular field with area 5000 m^2 is enclosed by 300 m of fencing. Find the dimensions of the field. 50 m by 100 m

6. A rectangular animal pen with area 1200 m^2 has one side along a barn. The other three sides are enclosed by 100 m of fencing. Find the dimensions of the pen. 20 m by 60 m; 30 m by 40 m

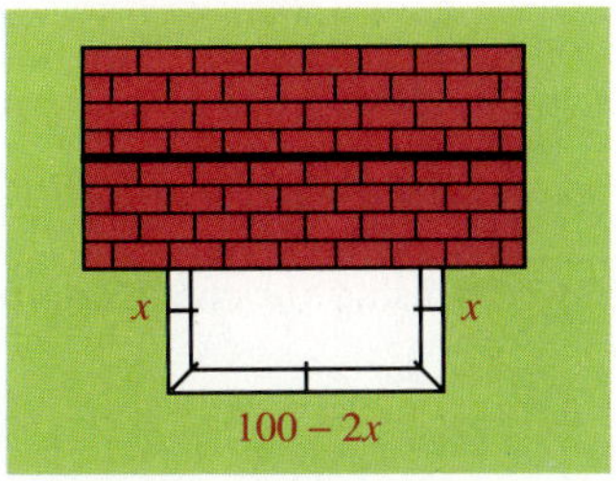

7. A walkway of uniform width has area 72 m^2 and surrounds a swimming pool that is 8 m wide and 10 m long. Find the width of the walkway. 1.68 m

8. A 5 in. by 7 in. photograph is surrounded by a frame of uniform width. The area of the frame equals the area of the photograph. Find the width of the frame. 1.21 in.

9. When mineral deposits formed a coating 1 mm thick on the inside of a pipe, the area through which fluid can flow was reduced by 20%. Find the original inside diameter of the pipe.
(*Remember:* Area of circle = πr^2 and diameter = $2r$.) 18.94 mm

10. The area of the trapezoid shown below is 90 square units. Find the value of x. 5.31

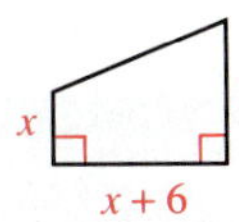

Ex. 10

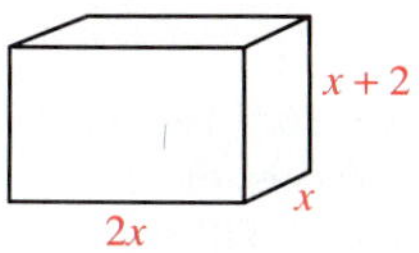

Ex. 11

B 11. The total surface area of the rectangular solid shown is 36 m^2. Find the value of x. 1.39

12. In a *golden rectangle* the ratio of the length to the width equals the ratio of the length plus width to the length. Find the value of this *golden ratio*. (Do not approximate the answer.) $\frac{1+\sqrt{5}}{2}$

13. A box with height $(x + 5)$ cm has a square base with side x cm. A second box with height $(x + 2)$ cm has a square base with side $(x + 1)$ cm. If the two boxes have the same volume, find the value of x. 5.37

14. A box with a square base and no lid is to be made from a square piece of metal by cutting squares from the corners and folding up the sides. The cut-off squares are 5 cm on a side. If the volume of the box is 100 cm^3, find the dimensions of the original piece of metal. 14.47 cm on a side

15. A hydrofoil made a round trip of 144 km in 4 h. Because of head winds, the average speed on returning was 15 km/h less than the average speed going out. Find the two speeds. Going out, 45 km/h; returning, 30 km/h

Supplementary Materials

Study Guide	pp. 107–108
Practice Master	39
Resource Book	p. 127

Quick Quiz

Solve by completing the square.

1. $y^2 - 6y - 9 = 0$
   $\{3 \pm 3\sqrt{2}\}$
2. $2a^2 + 6a + 7 = 2$
   $\left\{-\frac{3}{2} \pm \frac{1}{2}i\right\}$

Solve by the quadratic formula.

3. $x^2 - 2x = 4$
   $\{1 \pm \sqrt{5}\}$
4. $4k^2 + k + 1 = 0$
   $\left\{\frac{-1 \pm i\sqrt{15}}{8}\right\}$

Solve. Approximate answers to the nearest hundredth.

5. A rectangle is twice as long as it is wide. If the length and width are both increased by 5 cm, the resulting rectangle has an area of 50 cm^2. Find the dimensions of the original rectangle.
   1.40 cm by 2.81 cm
6. Two numbers differ by 2. Their reciprocals add to 3. Find the numbers.
   2.39 and 0.39 or 0.28 and −1.72

Self-Test 1

Vocabulary quadratic equation (p. 307) | completing the square (p. 308) | quadratic formula (p. 311)

Solve by completing the square.

1. $x^2 - 6x + 2 = 0$ $\{3 \pm \sqrt{7}\}$ **2.** $3y^2 + 9y = -2$ $\left\{\frac{-9 \pm \sqrt{57}}{6}\right\}$ **Obj. 7-1, p. 307**

Solve by using the quadratic formula.

3. $w^2 - 5w = 3$ $\left\{\frac{5 \pm \sqrt{37}}{2}\right\}$ **4.** $4z^2 + 2z + 1 = 0$ $\left\{-\frac{1}{4} \pm \frac{\sqrt{3}}{4}i\right\}$ **Obj. 7-2, p. 311**

Solve each problem. Approximate answers involving radicals to the nearest hundredth. A calculator may be helpful.

5. Find the dimensions of a rectangle whose perimeter is 10 cm and whose area is 3 cm^2. 0.70 cm by 4.30 cm

6. A sidewalk of uniform width has area 180 ft^2 and surrounds a flower bed that is 11 ft wide and 13 ft long. Find the width of the sidewalk. 3 ft

Check your answers with those at the back of the book.

Biographical Note / *Charles Steinmetz*

Charles Proteus Steinmetz (1865–1923) was an electrical engineer and mathematician who combined technical skill with theoretical insight. He showed that sophisticated mathematics could help to solve problems in the design of motors and transformers, and he pioneered the use of complex numbers in the analysis of alternating current circuits.

Born in Germany, Steinmetz immigrated to the United States at the age of 24 and was soon working for an electric company, where he spent most of his career. Even though he suffered from a deformed spine, Steinmetz had a tremendous capacity for work. In his laboratory he devised and improved designs for arc-lamp electrodes and generators and studied the effects of transient currents like those produced by lightning. Steinmetz also researched solar energy, electrical networks, electrification of railways, synthetic production of protein, and electric cars. His 195 patents are proof of the range of his invention.

Roots of Quadratic Equations

7-3 The Discriminant

Objective To determine the nature of the roots of a quadratic equation by using its discriminant.

Using the quadratic formula, you can write the roots of the quadratic equation

$$ax^2 + bx + c = 0$$

in the form

$$r_1 = \frac{-b + \sqrt{D}}{2a} \quad \text{and} \quad r_2 = \frac{-b - \sqrt{D}}{2a},$$

where

$$D = b^2 - 4ac.$$

The following examples illustrate how D is related to the nature of the roots.

Example 1 Solve.

a. $x^2 + 6x - 2 = 0$ **b.** $3x^2 - 4\sqrt{3}x + 4 = 0$ **c.** $x^2 - 6x + 10 = 0$

Solution

a. $x^2 + 6x - 2 = 0$

$D = 6^2 - 4(1)(-2) = 36 + 8 = 44$ — D is positive.

$r_1 = \frac{-6 + \sqrt{44}}{2} = -3 + \sqrt{11}$

$r_2 = \frac{-6 - \sqrt{44}}{2} = -3 - \sqrt{11}$

The roots are real and unequal.

b. $3x^2 - 4\sqrt{3}x + 4 = 0$

$D = (-4\sqrt{3})^2 - 4(3)(4) = 48 - 48 = 0$ — D is zero.

$r_1 = \frac{-(-4\sqrt{3}) + \sqrt{0}}{2 \cdot 3} = \frac{2\sqrt{3}}{3}$

$r_2 = \frac{-(-4\sqrt{3}) - \sqrt{0}}{2 \cdot 3} = \frac{2\sqrt{3}}{3}$

The roots are real and equal. We say there is a *double root*.

c. $x^2 - 6x + 10 = 0$

$D = (-6)^2 - 4(1)(10) = 36 - 40 = -4$ — D is negative.

$r_1 = \frac{-(-6) + \sqrt{-4}}{2} = 3 + i$

$r_2 = \frac{-(-6) - \sqrt{-4}}{2} = 3 - i$

The roots are imaginary conjugates.

Teaching Suggestions p. T105

Reading Algebra p. T106

Suggested Extensions p. T106

Group Activities p. T106

Warm-Up Exercises

Solve by the quadratic formula.

1. $6x^2 + 7x = 20$ $\left\{\frac{4}{3}, -\frac{5}{2}\right\}$
2. $x^2 - 8x - 10 = 0$ $\{4 \pm \sqrt{26}\}$
3. $10 = x - 9x^2$ $\left\{\frac{1 \pm i\sqrt{359}}{18}\right\}$

Motivating the Lesson

Point out to students that it may be possible to determine the nature or characteristics of an object by studying its parts or observing its behavior. The discriminant, $b^2 - 4ac$, allows one to determine the nature of the roots of a quadratic equation without finding the actual roots.

Chalkboard Examples

Without solving the equation, determine the nature of its roots.

1. $x^2 - 8x + 5 = 0$
   $b^2 - 4ac = (-8)^2 - 4(1)(5) = 64 - 20 = 44 > 0$
   two different real roots
2. $x^2 + 10x + 25 = 0$
   $b^2 - 4ac = (10)^2 - 4(1)(25) = 100 - 100 = 0$
   one real double root

(continued)

Chalkboard Examples

(continued)

3. $x^2 - 4x + 13 = 0$
   $b^2 - 4ac = (-4)^2 - 4(1)(13)$
   $= 16 - 52$
   $= -36 < 0$
   imaginary conjugates

4. $x^2 + \frac{7}{3}x + \frac{2}{3} = 0$
   $3x^2 + 7x + 2 = 0$
   $b^2 - 4ac = (7)^2 - 4(3)(2)$
   $= 49 - 24 = 25$
   real, rational roots

5. $x^2 + 4\sqrt{2}x - 8 = 0$
   $b^2 - 4ac = (4\sqrt{2})^2 - 4(1)(-8)$
   $= 32 + 32 = 64$
   The roots are real but may not be rational. The equation does not have integral coefficients.

Check for Understanding

Oral Exs. 1–10: use after Example 2.

Guided Practice

Determine the nature of the roots.

1. $y^2 - 3y - 1 = 0$
   two real irrational roots
2. $3a^2 - 10a + 11 = 0$
   two imaginary conjugate roots
3. $5x^2 + 2\sqrt{10}x + 2 = 0$
   one double real root
4. $2.5k^2 = 1 - \sqrt{6}k$
   two real roots
5. $3b^2 - 14b = 24$
   two rational roots

Solve by whatever method you prefer.

6. $x^2 = 3x + 28$ $\{-4, 7\}$
7. $0.5k^2 - \sqrt{2}k - 1 = 0$
   $\{\sqrt{2} + 2, \sqrt{2} - 2\}$
8. $(3y - 5)^2 = 75$ $\left\{\frac{5 \pm 5\sqrt{3}}{3}\right\}$
9. $x^2 - 4ax + a^2 = 0$
   $(a \geq 0)$ $\{2a \pm a\sqrt{3}\}$

The Nature of the Roots of a Quadratic Equation

Let $ax^2 + bx + c = 0$ be a quadratic equation with *real* coefficients.

1. If $b^2 - 4ac > 0$, then there are two unequal real roots.
2. If $b^2 - 4ac = 0$, then there is a real double root.
3. If $b^2 - 4ac < 0$, then there are two conjugate imaginary roots.

Since the value $D = b^2 - 4ac$ "discriminates" among the three cases, D is called the **discriminant** of the quadratic equation.

The discriminant also shows you whether a quadratic equation with integral coefficients has rational roots. If $b^2 - 4ac$ is a perfect square, then

$$x = \frac{-b \pm \sqrt{\text{perfect square}}}{2a}$$

$$x = \frac{-b \pm \text{integer}}{2a}$$

Since $\frac{-b \pm \text{integer}}{2a}$ is always a quotient of integers, the roots are rational.

Test for Rational Roots

If a quadratic equation has integral coefficients and its discriminant is a perfect square, then the equation has rational roots.

If a quadratic equation can be transformed into an equivalent equation that meets this test, then it has rational roots.

Example 2 Without solving each equation, determine the nature of its roots.

a. $3x^2 - 7x + 5 = 0$ **b.** $2x^2 - 13x + 15 = 0$

c. $x^2 + 2\sqrt{3}x - 1 = 0$ **d.** $\frac{x^2}{4} - \frac{5}{2}x + \frac{25}{4} = 0$

Solution

a. $D = (-7)^2 - 4(3)(5) = 49 - 60 = -11$ (negative)
$\therefore$ the roots are imaginary conjugates. ***Answer***

b. $D = (-13)^2 - 4(2)(15) = 169 - 120 = 49$ (positive, perfect square)
The given equation has integral coefficients.
$\therefore$ the roots are real, unequal, and rational. ***Answer***

c. $D = (2\sqrt{3})^2 - 4(1)(-1) = 12 + 4 = 16$ (positive, perfect square)
The given equation does *not* have integral coefficients.
$\therefore$ the roots are real and unequal. ***Answer***

d. The given equation is equivalent to $x^2 - 10x + 25 = 0$, which has integral coefficients.
$D = (-10)^2 - 4(1)(25) = 100 - 100 = 0$ (perfect square)
$\therefore$ there is a rational double root. ***Answer***

You now have three methods of solving quadratic equations:

1. Using the quadratic formula
2. Completing the square
3. Factoring

Using the quadratic formula: For real-life applications, computing the quadratic formula on a calculator to find exact rational roots or to approximate irrational roots may be the most efficient method.

If a calculator is not available, the quadratic formula can be computed by hand to solve *any* quadratic equation.

Completing the square: If the equation has the form

$$x^2 + (\text{even number}) \cdot x + \text{constant} = 0,$$

then completing the square may be the easiest method.

Factoring: If $b^2 - 4ac$ is a perfect square, then you can solve by factoring.

Example 3 The equation $kx^2 + 12x + 9k = 0$ has different roots for different values of k. Find the values of k for which the equation has the following:

a. a real double root **b.** two different real roots **c.** imaginary roots

Solution The discriminant $D = 12^2 - 4(k)(9k) = 144 - 36k^2$.

a. Equation has a double root if $D = 0$:

$$144 - 36k^2 = 0$$
$$-36k^2 = -144$$
$$k^2 = 4$$
$$k = \pm 2$$

b. Equation has real roots if $D > 0$:

$$144 - 36k^2 > 0$$
$$-36k^2 > -144$$
$$k^2 < 4$$
$$|k| < 2, \text{ or}$$
$$-2 < k < 2$$

c. Equation has imaginary roots if $D < 0$:

$$144 - 36k^2 < 0$$
$$-36k^2 < -144$$
$$k^2 > 4$$
$$|k| > 2, \text{ or}$$
$$k > 2 \text{ or } k < -2$$

Caution: When solving $k^2 = 4$, it is correct to write $k = 2$ or $k = -2$. However, when solving $k^2 < 4$, it is *incorrect* to write $k < 2$ or $k < -2$. To see why, consider $k = -3$. This value satisfies $k < -2$, but *not* $k^2 < 4$.

Oral Exercises

Give the value of the discriminant and determine the nature of the roots of each equation. If the equation is factorable, say so.

1. $x^2 + 6x + 3 = 0$ 24
2. $x^2 + 6x + 5 = 0$ 16
3. $x^2 + 8x + 16 = 0$ 0
4. $x^2 + 6x + 10 = 0$ −4
5. $t^2 - 5t - 5 = 0$ 45
6. $3y^2 - 4y + 2 = 0$ −8
7. $5x^2 + \sqrt{5}x - 1 = 0$ 25
8. $k^2 - 4\sqrt{2}k + 4 = 0$ 16
9. $\sqrt{5}x^2 - 6x + \sqrt{5} = 0$ 16
10. Explain why the roots of $ax^2 + bx + c = 0$ are unequal when $b^2 - 4ac > 0$.

Using a Calculator

Point out to students that while exact solutions are generally preferred within mathematics, approximate solutions (in decimal form) are preferred outside mathematics. Calculators, of course, are a convenient means of obtaining decimal values for solutions of real-life problems.

Summarizing the Lesson

In this lesson students learned to evaluate the discriminant and to use it to determine the nature of the roots of a quadratic equation. Ask students to state the three cases of the discriminant and to give the nature of the roots for each case.

Additional Answers
Oral Exercises

1. Real, unequal, irrational
2. Real, unequal, rational; factorable
3. Real, equal, rational; factorable
4. Imaginary, conjugate
5. Real, unequal, irrational
6. Imaginary, conjugate
7. Real, unequal, irrational
8. Real, unequal, irrational
9. Real, unequal, irrational
10. If the roots were equal, then $\sqrt{b^2 - 4ac} = -\sqrt{b^2 - 4ac}$, so $b^2 - 4ac = 0$. But $b^2 - 4ac > 0$.

Suggested Assignments

Average Algebra
320/1–39 odd
R 316/Self-Test 1
S 321/Mixed Review

Average Alg. and Trig.
320/1–39 odd
R 316/Self-Test 1
S 321/Mixed Review

Extended Algebra
320/1–39 odd, 41, 42
R 316/Self-Test 1
S 321/Mixed Review

Extended Alg. and Trig.
320/1–39 odd, 41, 42
R 316/Self-Test 1
S 321/Mixed Review

Additional Answers Written Exercises

1. Real, unequal, irrational
2. Real, unequal, rational
3. Imaginary, conjugate
4. Real, unequal, irrational
5. Real, unequal, irrational
6. Imaginary, conjugate
7. Imaginary, conjugate
8. Real, double, rational
9. Real, unequal, rational
10. Real, unequal, rational
11. Real, unequal, irrational
12. Real, double, irrational

31. a. $k = 2$ b. $k < 2$ c. $k > 2$

32. a. $k = -3$ b. $k > -3$ c. $k < -3$

33. a. $k = \pm 2$ b. $-2 < k < 2$ c. $k < -2$ or $k > 2$

34. a. $k = \pm 1$ b. $-1 < k < 1$ c. $k < -1$ or $k > 1$

35. a. $k = \pm 2$ b. $-2 < k < 2$ c. $k < -2$ or $k > 2$

36. a. $k = \pm 2$ b. $k < -2$ or $k > 2$ c. $-2 < k < 2$

Written Exercises

Without solving each equation, determine the nature of its roots. A calculator may be helpful.

A

1. $x^2 + 3x - 9 = 0$
2. $x^2 - 4x - 5 = 0$
3. $t^2 + 8t + 20 = 0$
4. $3m^2 - 8m - 5 = 0$
5. $2y^2 - 9y + 3 = 0$
6. $5t^2 - 4t + 3 = 0$
7. $z^2 + \frac{5}{4} = z$
8. $\frac{r^2}{4} + 1 = r$
9. $d^2 + \frac{7}{3}d = 2$
10. $\sqrt{3}x^2 - 4\sqrt{3} = 0$
11. $2u^2 - 5u + \sqrt{8} = 0$
12. $7y^2 + 2 = 2\sqrt{14}y$

Solve each equation using whichever method seems easiest to you.

13. $x^2 - 6x + 5 = 0$ $\{1, 5\}$
14. $y^2 + 2y - 24 = 0$ $\{-6, 4\}$
15. $9x^2 - 12x + 4 = 0$ $\left\{\frac{2}{3}\right\}$
16. $4y^2 + 12y + 9 = 0$ $\left\{-\frac{3}{2}\right\}$
17. $5(x + 7)^2 = 0$ $\{-7\}$
18. $5(x + 7)^2 = 25$ $\{-7 \pm \sqrt{5}\}$
19. $y^2 - 2y = 99$ $\{-9, 11\}$
20. $10 = 6t - t^2$ $\{3 \pm i\}$
21. $3(x - 2)^2 = 18$ $\{2 \pm \sqrt{6}\}$
22. $x^2 + 4x - 396 = 0$ $\{-22, 18\}$
23. $(2x + 5)(x - 3) = 0$ $\left\{-\frac{5}{2}, 3\right\}$
24. $(2x + 5)(x - 3) = 6$ $\left\{-3, \frac{7}{2}\right\}$

B

25. $2(y - 1)^2 = y^2$ $\{2 \pm \sqrt{2}\}$
26. $2w^2 = 3(w - 2)^2$ $\{6 \pm 2\sqrt{6}\}$
27. $\frac{x}{3} + \frac{3}{x} = 1$ $\left\{\frac{3}{2} \pm \frac{3\sqrt{3}}{2}i\right\}$
28. $\frac{x + 1}{x} - \frac{x}{x + 1} = 2$ $\left\{\pm\frac{\sqrt{2}}{2}\right\}$
29. $\frac{y + 1}{y - 1} - \frac{y}{3} = \frac{2}{y - 1}$ $\{3\}$
30. $\frac{t}{t - 2} + \frac{2t}{t - 1} = 6$ $\left\{\frac{4}{3}, 3\right\}$

Find the value(s) of *k* for which each equation has the following:

a. a real double root **b. two different real roots** **c. imaginary roots**

31. $2x^2 + 4x + k = 0$
32. $3x^2 - 6x - k = 0$
33. $k^2x^2 - 8x + 4 = 0$
34. $9x^2 - 6x + k^2 = 0$
35. $kx^2 - 4x + k = 0$
36. $3x^2 - 6kx + 12 = 0$

37. Find the value(s) of k for which the expression $16x^2 + 8x + 2k$ is a perfect square. $\frac{1}{2}$

38. Find three values of k for which the expression $3x^2 - 6x + k$ is factorable.
38. Answers may vary. For example: 0, 3, −9

39. Cliff solved the equation $2x^2 - 5x + 7 = 0$ and noticed that the sum of the roots was $\frac{5}{2}$ and the product of the roots was $\frac{7}{2}$.
 a. Solve the equation and verify Cliff's results.
 b. Find the sum and product of the roots of $2x^2 + 13x + 11 = 0$. Sum, $-\frac{13}{2}$; product, $\frac{11}{2}$
 c. On the basis of your answers to parts (a) and (b), predict the sum and the product of the roots of the equation $ax^2 + bx + c = 0$. Sum, $-\frac{b}{a}$; product, $\frac{c}{a}$

40. a. Find the roots of $9x^2 - 6x + 2 = 0$. $\frac{1}{3} + \frac{1}{3}i, \frac{1}{3} - \frac{1}{3}i$
 b. Verify that the sum of the roots is $\frac{6}{9}$ and the product of the roots is $\frac{2}{9}$.
 c. Predict the sum and the product of the roots of the equation $3x^2 + 7x + 8 = 0$. Sum, $-\frac{7}{3}$; product, $\frac{8}{3}$

C 41. Show that the roots of $\frac{1}{x+r} = \frac{1}{x} + \frac{1}{r}$ are always imaginary if r is a nonzero real number.

42. Show that the roots of $x^2 + kx + k = 1$ are always rational if k is an integer.

43. Consider the equation $iz^2 - 3z - 2i = 0$.
   a. Show that the discriminant is 1. $D = (-3)^2 - 4(-2i)(i) = 9 + 8i^2 = 9 - 8 = 1$
   b. The roots of the given equation are imaginary. Find these roots. $-2i, -i$
   c. Explain why parts (a) and (b) do not contradict the rule that a quadratic with positive discriminant has real roots.
   The rule applies only to quadratic equations with real coefficients.

Mixed Review Exercises

Solve each equation over the complex numbers.

1. $(x-2)^2 = 18$ $\{2 \pm 3\sqrt{2}\}$
2. $y^2 = 3y + 4$ $\{-1, 4\}$
3. $2w(w-3) = -1$ $\left\{\frac{3 \pm \sqrt{7}}{2}\right\}$
4. $u^2 + 5 = u$ $\left\{\frac{1}{2} \pm \frac{\sqrt{19}}{2}i\right\}$
5. $6v^2 - 13v - 28 = 0$ $\left\{-\frac{4}{3}, \frac{7}{2}\right\}$
6. $3t^2 = 1 - 4t$ $\left\{\frac{-2 \pm \sqrt{7}}{3}\right\}$

Express in simplest form without negative exponents. Assume that all radicals represent real numbers.

7. $(x^3)^2$ x^6
8. $\sqrt{72x^3}$ $6x\sqrt{2x}$
9. $(x^{-1})^2$ $\frac{1}{x^2}$
10. $(x^2y)(xy^3)$ x^3y^4
11. $(\sqrt{x})^{-3}$ $\frac{\sqrt{x}}{x^2}$
12. $\frac{x^{-2}y}{(xy)^{-1}}$ $\frac{y^2}{x}$
13. $(\sqrt{5x})^2$ $5x$
14. $\left(\frac{x^2}{y^3}\right)^{-2}$ $\frac{y^6}{x^4}$

Computer Exercises

For students with some programming experience.

2. a. real, irrational b. real, rational c. imaginary

1. Write a program in which the user enters the integer coefficients a, b, and c of any quadratic equation $ax^2 + bx + c = 0$, and the program then prints the nature of the roots (imaginary roots, a double real root, rational real roots, or irrational real roots). Use X * X to represent x^2.
2. Run the program in (1) to find the nature of the roots of each equation.
   a. $2x^2 + 3x - 1 = 0$ b. $x^2 - 6x = 16$ c. $5x^2 - 7x + 12 = 0$
3. Modify the program in (1) so that the computer finds and prints the values of the roots when they are real.
4. Run the program in (3) to print the roots of each equation.
   a. $3x^2 - 8x - 35 = 0$ b. $x^2 + 5x - 1 = 0$ c. $x^2 = x + 21$
5. Modify the program in (3) so that the computer finds the roots of any quadratic, whether the roots are real or imaginary. Have the computer print imaginary roots in the form $a + bi$. You will need to have the computer print the symbol "i."
6. Run the program in (5) to print the roots of each equation.
   a. $x^2 + 2x + 8 = 0$ b. $4x^2 - 5x - 11 = 0$ c. $2x^2 - 6x + 7 = 0$

Supplementary Materials

Study Guide	pp. 109–110
Test Master	27
Computer Activity	15

Using a Computer

These exercises create a computer program that determines the nature of the roots of a quadratic equation and then finds the roots. Ask students who have programming experience to demonstrate the program to the class, explaining the relationship of the parts of the program to Lessons 7-1 and 7-2.

Additional Answers
Computer Exercises

4. a. 5, −2.33333334
   b. 0.192582404, −5.1925824
   c. 5.10977223, −4.10977223
6. a. $-1 \pm 2.64575131i$
   b. 2.39718086, −1.14718086
   c. $1.5 \pm 1.11803399i$

Teaching Suggestions p. T106

Suggested Extensions p. T106

Warm-Up Exercises

Multiply.

1. $(x + a)(x - a)$ $x^2 - a^2$
2. $(\sqrt{3} + 2)(\sqrt{3} - 2)$ -1

Simplify.

3. $3x + 4y + 7x + 3 - y$
   $10x + 3y + 3$
4. $3\sqrt{2} + 4\sqrt{5} + 7\sqrt{2} + 3 - \sqrt{5}$
   $10\sqrt{2} + 3\sqrt{5} + 3$
5. $(3x - 2)^2 - 5(3x - 2) - 6$
   $9x^2 - 27x + 8$

Motivating the Lesson

Tell students that many seemingly difficult equations can be solved by using the methods of quadratic equations. Today's lesson will illustrate this topic.

Problem Solving Strategies

The strategy of *solving a simpler problem* is evident in the worked-out examples.

Thinking Skills

To solve equations such as those in Examples 3 and 4, students must *analyze* the equation to determine in what expression it is quadratic.

Chalkboard Examples

Solve.

1. $x - 5\sqrt{x} + 6 = 0$
   Let $z = \sqrt{x}$ to obtain
   $z^2 - 5z + 6 = 0$
   $(z - 2)(z - 3) = 0$
   $z = 2$ or $z = 3$
   $\sqrt{x} = 2$ or $\sqrt{x} = 3$
   $x = 4$ or $x = 9$
   $\{4, 9\}$

7-4 Equations in Quadratic Form

Objective To recognize and solve equations in quadratic form.

Study the three equations below and note their similarity.

$x^2 - 5x - 6 = 0$ This equation is quadratic in x.

$(3x - 2)^2 - 5(3x - 2) - 6 = 0$ This equation is quadratic in $3x - 2$.

$\left(\frac{1}{2x}\right)^2 - 5\left(\frac{1}{2x}\right) - 6 = 0$ This equation is quadratic in $\frac{1}{2x}$.

Each of the equations above has the same *quadratic form:*

$$(\text{function of } x)^2 - 5 \cdot (\text{function of } x) - 6 = 0$$

In general, an equation is in **quadratic form** if it can be written as

$$a[f(x)]^2 + b[f(x)] + c = 0,$$

where $a \neq 0$ and $f(x)$ is some function of x. When solving an equation in quadratic form, it is often helpful to replace $f(x)$ by a single variable.

Example 1 Solve $(3x - 2)^2 - 5(3x - 2) - 6 = 0$.

Solution Let $z = 3x - 2$. Then $z^2 = (3x - 2)^2$ and the equation becomes quadratic in z.

$$z^2 - 5z - 6 = 0$$
$$(z + 1)(z - 6) = 0$$

$z = -1$ or $z = 6$

$3x - 2 = -1$ $3x - 2 = 6$ ⟵ Remember $z = 3x - 2$

$x = \frac{1}{3}$ $x = \frac{8}{3}$

$\therefore$ the solution set is $\left\{\frac{1}{3}, \frac{8}{3}\right\}$. ***Answer***

The check is left for you.

Example 2 Solve $\left(\frac{1}{2x}\right)^2 - 5\left(\frac{1}{2x}\right) - 6 = 0$.

Solution Let $z = \frac{1}{2x}$. Then $z^2 = \left(\frac{1}{2x}\right)^2$ and the equation becomes quadratic in z.

$$z^2 - 5z - 6 = 0$$
$$(z + 1)(z - 6) = 0$$

$z = -1$ or $z = 6$

$\frac{1}{2x} = -1$ $\frac{1}{2x} = 6$ ⟵ Remember $z = \frac{1}{2x}$

$x = -\frac{1}{2}$ $x = \frac{1}{12}$

$\therefore$ the solution set is $\left\{-\frac{1}{2}, \frac{1}{12}\right\}$. ***Answer***

Example 3 illustrates that the quadratic form of an equation can sometimes be less obvious than in Examples 1 and 2.

Example 3 Solve $3x + 4\sqrt{x} - 2 = 0$. Approximate real solutions to the nearest hundredth.

Solution Let $z = \sqrt{x}$. Then $z^2 = x$ and the equation becomes

$$3z^2 + 4z - 2 = 0.$$

By using the quadratic formula, you will find that

$$z = \frac{-2 \pm \sqrt{10}}{3} \approx \frac{-2 \pm 3.162}{3}.$$

$z = \sqrt{x} \approx \frac{-2 + 3.162}{3}$ or $z = \sqrt{x} \approx \frac{-2 - 3.162}{3}$

$\sqrt{x} \approx 0.387$

$x \approx (0.387)^2$

$x \approx 0.15$

$\sqrt{x} \approx -1.72$

No solution, because the principal square root is always nonnegative.

$\therefore$ the solution set is $\{0.15\}$. ***Answer***

Example 4 Solve $x^4 + 7x^2 - 18 = 0$.

Solution The given equation is quadratic in x^2. You can either do all your work in terms of x or let $z = x^2$. Both methods are shown below.

Method 1

$(x^2)^2 + 7(x^2) - 18 = 0$

$(x^2 + 9)(x^2 - 2) = 0$

$x^2 = -9$ or $x^2 = 2$

$x = \pm 3i$ or $x = \pm\sqrt{2}$

Method 2

Let $z = x^2$. Then:

$z^2 + 7z - 18 = 0$

$(z + 9)(z - 2) = 0$

$z = -9$ or $z = 2$

$x^2 = -9$ or $x^2 = 2$

$x = \pm 3i$ or $x = \pm\sqrt{2}$

$\therefore$ the solution set is $\{\sqrt{2}, -\sqrt{2}, 3i, -3i\}$. ***Answer***

Oral Exercises

1. $z = x + 2$; $z^2 - 5z - 14 = 0$ 2. $z = 3x + 4$; $z^2 + 6z - 16 = 0$

For each equation, substitute z for some expression in x to make the equation easier to solve. Then give the resulting quadratic equation in z.

1. $(x + 2)^2 - 5(x + 2) - 14 = 0$

2. $(3x + 4)^2 + 6(3x + 4) - 16 = 0$

3. $\left(\frac{1}{3x}\right)^2 - 4\left(\frac{1}{3x}\right) - 5 = 0$ $z = \frac{1}{3x}$; $z^2 - 4z - 5 = 0$

4. $\left(\frac{2}{x}\right)^2 + 5\left(\frac{2}{x}\right) - 24 = 0$ $z = \frac{2}{x}$; $z^2 + 5z - 24 = 0$

2. $2x + 2\sqrt{2x} - 15 = 0$

Let $z = \sqrt{2x}$ to obtain

$z^2 + 2z - 15 = 0$

$(z + 5)(z - 3) = 0$

$z = -5$ or $z = 3$

$\sqrt{2x} = -5$ or $\sqrt{2x} = 3$

no solution $\quad 2x = 9$

$x = \frac{9}{2}$

$\left\{\frac{9}{2}\right\}$

3. $\frac{6}{u} - \frac{17}{\sqrt{u}} = -5$

Let $z = \frac{1}{\sqrt{u}}$ to obtain

$6z^2 - 17z + 5 = 0$

$(3z - 1)(2z - 5) = 0$

$z = \frac{1}{3}$ or $z = \frac{5}{2}$

$\frac{1}{\sqrt{u}} = \frac{1}{3}$ or $\frac{1}{\sqrt{u}} = \frac{5}{2}$

$\sqrt{u} = 3$ or $\sqrt{u} = \frac{2}{5}$

$u = 9$ or $u = \frac{4}{25}$

$\left\{9, \frac{4}{25}\right\}$

Check for Understanding

Here is a suggested use of the Oral Exercises to check students' understanding as you teach the lesson.

Oral Exs. 1–12: use after Example 4.

Guided Practice

Solve.

1. $x^4 - 9x^2 + 8 = 0$ $\{1, -1, 2\sqrt{2}, -2\sqrt{2}\}$
2. $y - 5 = 4\sqrt{y}$ $\{25\}$
3. $4 + \frac{1}{a^2} = \frac{3\sqrt{2}}{a}$ $\left\{\frac{\sqrt{2}}{2}, \frac{\sqrt{2}}{4}\right\}$
4. $\left(\frac{2b+1}{b}\right)^2 = \frac{2b+1}{b} + 6$ $\left\{1, -\frac{1}{4}\right\}$
5. $4x^4 + 7x^2 - 36 = 0$ $\left\{\frac{3}{2}, -\frac{3}{2}, 2i, -2i\right\}$
6. $\frac{1}{r} + 4 = \frac{5}{\sqrt{r}}$ $\left\{\frac{1}{16}, 1\right\}$
7. $5a + 6\sqrt{a} + 1 = 0$ no sol.

Summarizing the Lesson

In this lesson students learned that certain equations can be put into quadratic form and solved as if they were quadratic equations.

Using a Calculator

When an equation is quadratic in x^2, $\sqrt{x}$, or $|x|$, students must recognize and eliminate any negative values obtained from the quadratic formula. By using a calculator to find decimal approximations for these values, the sign of each value becomes apparent.

5. $x^4 - 10x^2 + 9 = 0$ $z = x^2;\ z^2 - 10z + 9 = 0$
6. $x^6 - 9x^3 + 8 = 0$ $z = x^3;\ z^2 - 9z + 8 = 0$
7. $2x - 3\sqrt{x} + 1 = 0$ $z = \sqrt{x};\ 2z^2 - 3z + 1 = 0$
8. $7|x|^2 - 20|x| - 3 = 0$ $z = |x|;\ 7z^2 - 20z - 3 = 0$
9. $(x^2 - 5)^2 - (x^2 - 5) - 2 = 0$
10. $5\left(\frac{x+2}{x}\right)^2 - 3\left(\frac{x+2}{x}\right) = 2$
11. $2x^{-2} + 3x^{-1} - 2 = 0$ $z = x^{-1};\ 2z^2 + 3z - 2 = 0$
12. $3x^{-2} - 4x^{-1} + 1 = 0$ $z = x^{-1};\ 3z^2 - 4z + 1 = 0$

9. $z = x^2 - 5;\ z^2 - z - 2 = 0$
10. $z = \frac{x+2}{x};\ 5z^2 - 3z = 2$

Written Exercises

Solve each equation.

A

1. a. $(x+3)^2 - 5(x+3) + 4 = 0$ $\{1, -2\}$
   b. $(2x-1)^2 - 5(2x-1) + 4 = 0$ $\left\{1, \frac{5}{2}\right\}$
   c. $x^4 - 5x^2 + 4 = 0$ $\{\pm1, \pm2\}$
2. a. $(x-7)^2 - 13(x-7) + 36 = 0$ $\{11, 16\}$
   b. $(1-3x)^2 - 13(1-3x) + 36 = 0$ $\left\{-1, -\frac{8}{3}\right\}$
   c. $x^4 - 13x^2 + 36 = 0$ $\{\pm2, \pm3\}$
3. a. $2\left(\frac{1}{2y}\right)^2 + 5\left(\frac{1}{2y}\right) - 3 = 0$ $\left\{-\frac{1}{6}, 1\right\}$
   b. $2(y^2-4)^2 + 5(y^2-4) - 3 = 0$ $\left\{\pm\frac{3\sqrt{2}}{2}, \pm1\right\}$
   c. $2y^{-2} + 5y^{-1} - 3 = 0$ $\left\{-\frac{1}{3}, 2\right\}$
4. a. $3\left(\frac{w}{6}\right)^2 - 8\left(\frac{w}{6}\right) + 4 = 0$ $\{4, 12\}$
   b. $3(w^2-2)^2 - 8(w^2-2) + 4 = 0$ $\left\{\pm2, \pm\frac{2\sqrt{6}}{3}\right\}$
   c. $3w^{-2} - 8w^{-1} + 4 = 0$ $\left\{\frac{1}{2}, \frac{3}{2}\right\}$
5. a. $x^4 - 3x^2 - 4 = 0$ $\{\pm i, \pm2\}$
   b. $x - 3\sqrt{x} - 4 = 0$ $\{16\}$
6. a. $x^4 + 7x^2 - 8 = 0$ $\{\pm2i\sqrt{2}, \pm1\}$
   b. $x + 7\sqrt{x} - 8 = 0$ $\{1\}$
7. a. $x^4 + 5x^2 - 36 = 0$ $\{\pm3i, \pm2\}$
   b. $x^{-4} + 5x^{-2} - 36 = 0$ $\left\{\pm\frac{1}{3}i, \pm\frac{1}{2}\right\}$
8. a. $2x^4 + 7x^2 - 4 = 0$ $\left\{\pm2i, \pm\frac{\sqrt{2}}{2}\right\}$
   b. $2(x-1)^4 + 7(x-1)^2 - 4 = 0$ $\left\{1 \pm 2i, 1 \pm \frac{\sqrt{2}}{2}\right\}$
9. $x - 11\sqrt{x} + 30 = 0$ $\{25, 36\}$
10. $x + 4\sqrt{x} - 21 = 0$ $\{9\}$
11. $(x^2-1)^2 - 11(x^2-1) + 24 = 0$ $\{\pm2, \pm3\}$
12. $(x^2+3)^2 - 6(x^2+3) = 7$ $\{\pm2i, \pm2\}$

B

13. $\left(\frac{1+x}{2}\right)^2 - 3\left(\frac{1+x}{2}\right) = 18$ $\{-7, 11\}$
14. $\left(\frac{1}{x-1}\right)^2 - \frac{1}{x-1} = 2$ $\left\{0, \frac{3}{2}\right\}$
15. $\frac{2}{y} + \frac{1}{\sqrt{y}} = 1$ $\{4\}$
16. $\frac{3}{n} - \frac{7}{\sqrt{n}} - 6 = 0$ $\left\{\frac{1}{9}\right\}$

Solve each equation (a) by squaring both sides of the equation and then (b) by using a quadratic form. Be sure to check whether either method gives an extraneous root.

17. $x - 10 = 3\sqrt{x}$ $\{25\}$
18. $x + 3 = 4\sqrt{x}$ $\{1, 9\}$
19. $2y - 2 = 3\sqrt{y}$ $\{4\}$
20. $2t + 3 = 5\sqrt{t}$ $\left\{1, \frac{9}{4}\right\}$

Solve each equation. Approximate real solutions to the nearest hundredth. A calculator may be helpful.

21. $3|x|^2 = 7|x| + 5$ $\{\pm2.91\}$
22. $|x| \cdot (|x| - 3) = 1$ $\{\pm3.30\}$
23. $3x + 6\sqrt{x} - 2 = 0$ $\{0.08\}$
24. $\frac{2}{x} - \frac{5}{\sqrt{x}} = 1$ $\{0.14\}$
25. $(x^2 - 3x)^2 - 3(x^2 - 3x) = 10$ $\{-1.19, 1, 2, 4.19\}$
26. $(x^2 + 6x)^2 + 9(x^2 + 6x) + 20 = 0$ $\{-5.24, -5, -1, -0.76\}$

27. Suppose you know the fourth-degree equation $4x^4 - 73x^2 + 144 = 0$ has roots $\pm\frac{3}{2}$ and ± 4. Write a fourth-degree equation that has roots $\pm\frac{2}{3}$ and $\pm\frac{1}{4}$. (*Hint:* See Exercise 7.) $144x^4 - 73x^2 + 4 = 0$

28. Suppose you know the fourth-degree equation $4x^4 - 55x^2 + 39 = 0$ has roots $\pm\sqrt{13}$ and $\pm\frac{\sqrt{3}}{2}$. Write a fourth-degree equation that has roots $1 \pm \sqrt{13}$ and $1 \pm \frac{\sqrt{3}}{2}$. (*Hint:* See Exercise 8.) $4(x-1)^4 - 55(x-1)^2 + 39 = 0$

Solve.

C 29. $\sqrt{x+6} - 6\sqrt[4]{x+6} + 8 = 0$ $\{10, 250\}$

30. $\sqrt{x-2} + \sqrt{\sqrt{x-2}} = 2$ $\{3\}$

31. $(s^2 - 9)(s^4 - 3s^2 - 2) = 2(s^2 - 9)$ $\{\pm 2, \pm 3, \pm i\}$

32. $(\sqrt{y} - 3)(y - \sqrt{y} - 1) = \sqrt{y} - 3$ $\{4, 9\}$

Self-Test 2

Vocabulary discriminant (p. 318) quadratic form (p. 322)

Without solving each equation, determine the nature of its roots. **Obj. 7-3, p. 317**

1. $2t^2 - 6t + 5 = 0$ Imaginary, conjugate
2. $7y^2 - 9y + 2 = 0$ Real, unequal, rational
3. $2x^2 - 6x - 5 = 0$ Real, unequal, irrational
4. $3y^2 - 9y = 0$ Real, unequal, rational

Solve each equation using whichever method seems easiest to you.

5. $4(x-2)^2 = 20$ $\{2 \pm \sqrt{5}\}$
6. $w^2 - 4 = 3w$ $\{-1, 4\}$
7. $(x-3)(x-5) = 9$ $\{4 \pm \sqrt{10}\}$
8. Find all real values of k for which $3t^2 - 4t + k = 0$ has imaginary roots. $k > \frac{4}{3}$

Solve. **Obj. 7-4, p. 322**

9. $x^4 + x^2 - 12 = 0$ $\{\pm 2i, \pm\sqrt{3}\}$
10. $3x + 2\sqrt{3x} - 8 = 0$ $\left\{\frac{4}{3}\right\}$

Challenge

Find three different positive integers a, b, and c, such that the sum

$$\frac{1}{a} + \frac{1}{b} + \frac{1}{c}$$

is an integer. $a = 2, b = 3, c = 6$; $\frac{1}{2} + \frac{1}{3} + \frac{1}{6} = 1$

Suggested Assignments

Average Algebra
Day 1: 324/1–12
S 310/37, 39
Day 2: 324/13–26
R 325/Self-Test 2
Assign with Lesson 7-5.

Average Alg. and Trig.
Day 1: 324/1–12
S 310/37, 39
Day 2: 324/13–26
R 325/Self-Test 2
Assign with Lesson 7-5.

Extended Algebra
Day 1: 324/2–20 even
S 321/43
Day 2: 324/21–32
R 325/Self-Test 2
Assign with Lesson 7-5.

Extended Alg. and Trig.
Day 1: 324/2–20 even
S 321/43
Day 2: 324/21–32
R 325/Self-Test 2
Assign with Lesson 7-5.

Supplementary Materials

Study Guide	pp. 111–112
Practice Master	40
Resource Book	p. 128

Quick Quiz

Without solving, state the nature of the roots.

1. $3m^2 + m - 7 = 0$ two real irrational roots
2. $r^2 - 5r + 10 = 0$ two complex conjugate roots
3. $8x^2 + \frac{1}{2} = 4x$ one rational double root
4. For what values of k will $3x^2 + kx + 12 = 0$ have a double root? 12, −12

Explorations p. 838

Teaching Suggestions p. T107

Suggested Extensions p. T107

Warm-Up Exercises

Complete the table for each equation.

1. $3x + 4y = 12$

x	y	Answer
0	?	3
4	?	0
$\frac{20}{3}$	?	−2

2. $x - 2y = 8$

x	y	Answer
0	?	−4
8	?	0
−2	?	−5

3. $y = |x|$

x	y	Answer
−3	?	3
0	?	0
3	?	3

4. Graph $y = |x|$.

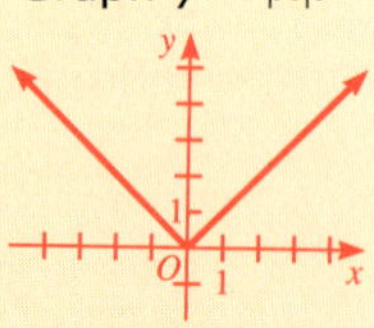

Motivating the Lesson

Tell students that a projectile, such as a ball thrown into the air, follows a parabolic path that can be written as a quadratic equation and graphed. Graphing parabolas is the topic of today's lesson.

Quadratic Functions and Their Graphs

7-5 Graphing $y - k = a(x - h)^2$

Objective To graph parabolas whose equations have the form $y - k = a(x - h)^2$ and to find the vertices and axes of symmetry.

To graph the equation

$$y = x^2$$

first make a table of pairs of values of x and y that satisfy the equation. Then plot the graph of each ordered pair of coordinates, as shown in Figure 1. If you plotted more points, you would see that they all lie on the smooth curve shown in Figure 2. This curve, called a **parabola,** is the graph of $y = x^2$.

x	y
−3	9
−2	4
−1	1
0	0
1	1
2	4
3	9

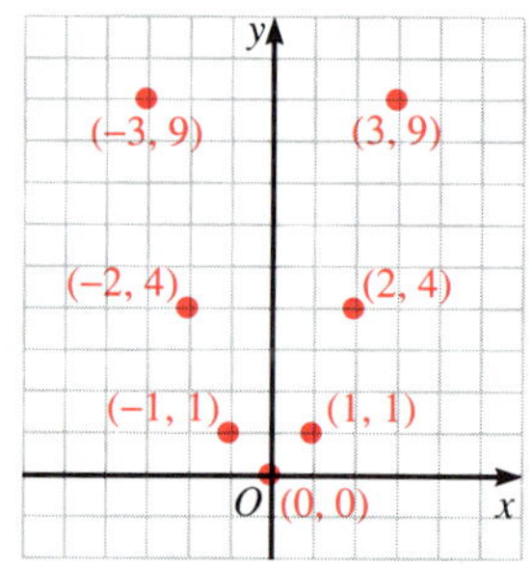

Figure 1

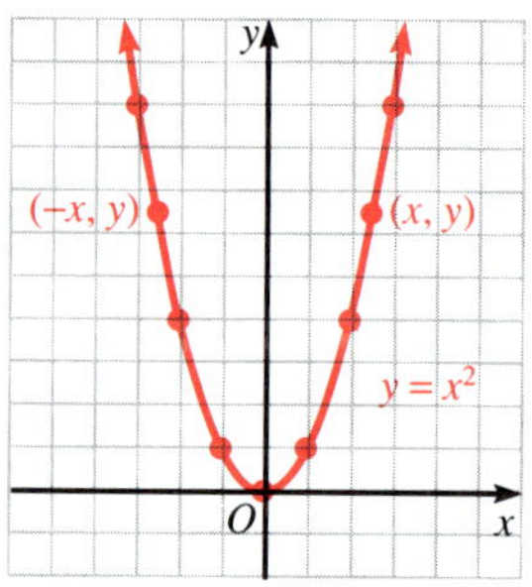

Figure 2

In Figure 2 notice that if the point (x, y) is on the parabola, then $(-x, y)$, its "mirror image" across the y-axis, is also on the parabola. This property can also be seen in the table where the coordinate pairs connected by red lines are mirror images. Because of this property, the y-axis is called the **axis of symmetry,** or simply the **axis,** of the parabola. The **vertex** of a parabola is the point where the parabola crosses its axis. In the case of $y = x^2$, the vertex is the origin.

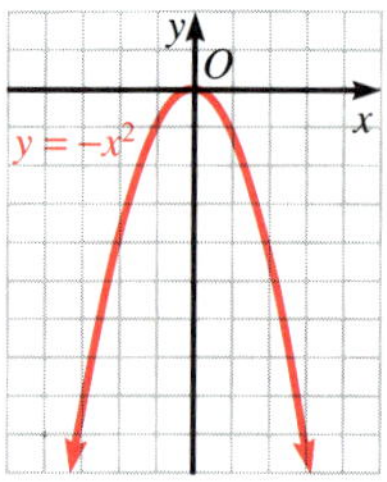

Figure 3

The graph of $y = -x^2$, shown in Figure 3 at the right, is a mirror image or *congruent copy* of the graph of $y = x^2$. If the graph of $y = x^2$ is reflected across the x-axis, then the result is the graph of $y = -x^2$.

Figures 4 and 5 on the next page show the effect of the value of a on the graph of an equation of the form $y = ax^2$.

The graph of $y = ax^2$ opens upward if $a > 0$ and downward if $a < 0$.
The larger $|a|$ is, the "narrower" the graph is.

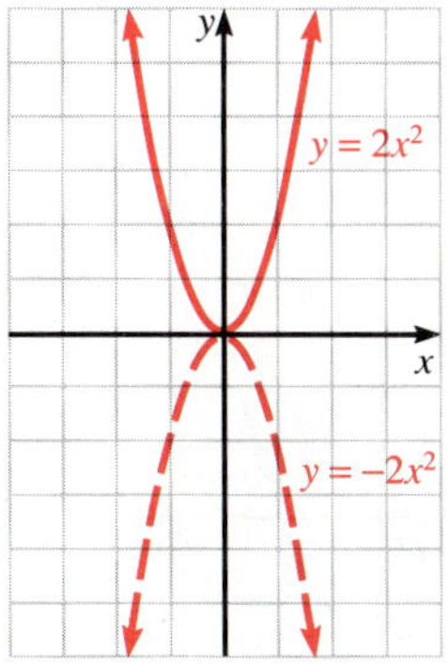

Figure 4

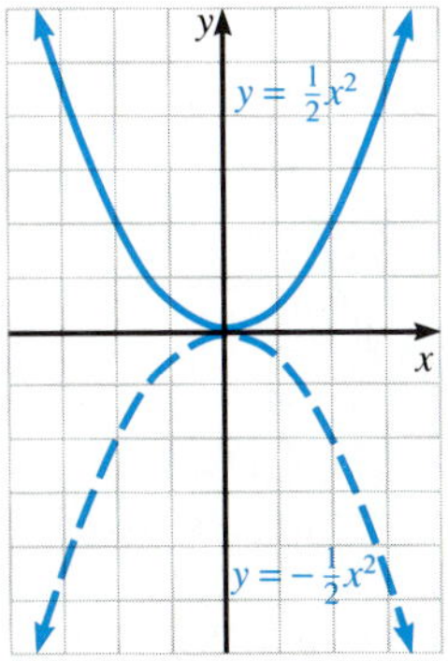

Figure 5

By graphing pairs of quadratic equations on the same axes, such as those in Figures 6–11 that follow, you may investigate the methods for graphing parabolas described in this lesson. A computer or a graphing calculator may be helpful.

Figures 6 and 7 below illustrate the following method for graphing a parabola whose equation has the form $y = a(x - h)^2$.

To graph $y = a(x - h)^2$, slide the graph of $y = ax^2$ horizontally h units.
If $h > 0$, slide it to the right; if $h < 0$, slide it to the left.
The graph has vertex $(h, 0)$ and its axis is the line $x = h$.

Slide right 3.

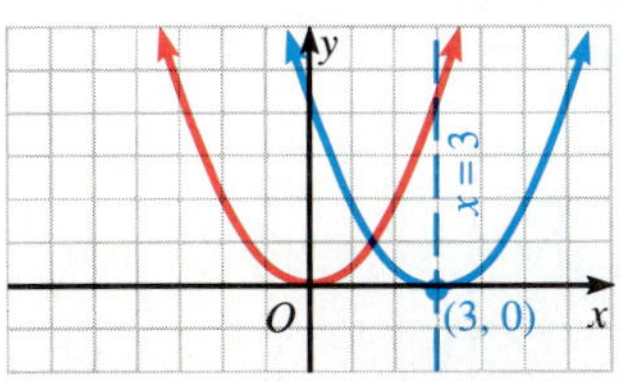

$y = \frac{1}{2}(x - 0)^2$, or $y = \frac{1}{2}x^2$

$y = \frac{1}{2}(x - 3)^2$

Figure 6

Slide left 3.

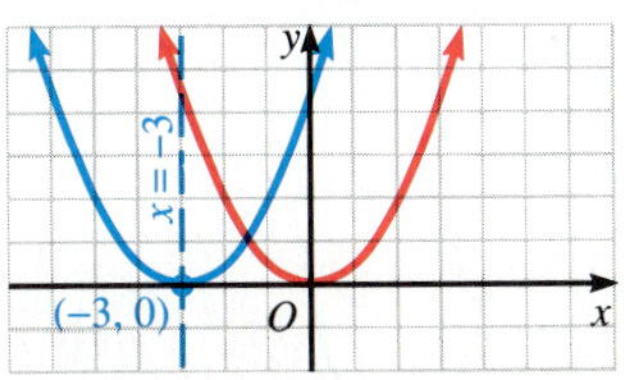

$y = \frac{1}{2}x^2$

$y = \frac{1}{2}(x - (-3))^2$, or $y = \frac{1}{2}(x + 3)^2$

Figure 7

Figures 8 and 9 on the next page illustrate a method for graphing a parabola whose equation has the form $y - k = ax^2$.

Common Errors

Some students may calculate a few ordered pairs incorrectly, plot their corresponding points, and sketch a graph without using their knowledge of equations of the form $y - k = a(x - h)^2$ to check their graph. When reteaching, have the students state whether the parabola opens upward or downward and state the coordinates of the vertex as an aid before graphing.

Using a Computer or a Graphing Calculator

Vertical and horizontal translations of the graph of $y = ax^2$ can be demonstrated by you and investigated by the students using a computer or graphing calculator. Students can also use a computer or graphing calculator to check their graphs for Written Exercises 1–18 on page 331.

You may wish to have students determine an equation for a given parabola. The program "Parabola Quiz" on the Disk for Algebra will draw a random parabola and let students visually experiment to find the corresponding equation.

Support Materials
Disk for Algebra
Menu Items: Parabola Plotter
Parabola Quiz
Resource Book pp. 240–241

Chalkboard Examples

Graph.

1. $y + 2 = (x + 3)^2$
   $a = 1$; parabola opens upward. $h = -3$, $k = -2$; the vertex is $(-3, -2)$; the axis is $x = -3$.

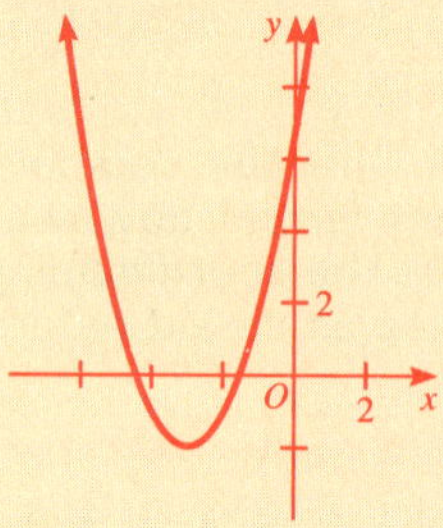

2. $y - 3 = -(x + 1)^2$
   $a = -1$; parabola opens downward. $h = -1$, $k = 3$; the vertex is $(-1, 3)$; the axis is $x = -1$.

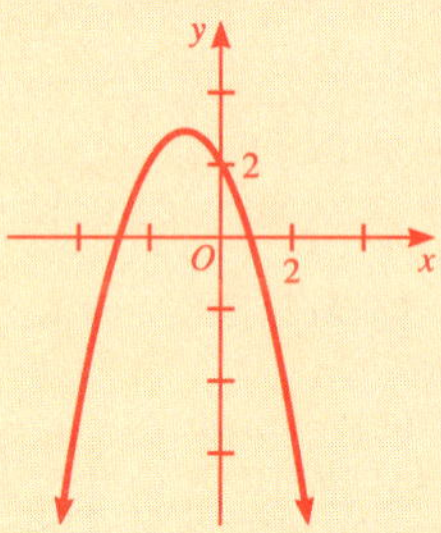

3. Find an equation of the parabola having vertex $(-1, -2)$ and containing the point $(2, -5)$.
   $y - k = a(x - h)^2$
   $y + 2 = a(x + 1)^2$
   $-5 + 2 = a(2 + 1)^2$
   $-3 = 9a$
   $-\frac{1}{3} = a$
   $\therefore$ an equation is
   $y + 2 = -\frac{1}{3}(x + 1)^2$.

To graph $y - k = ax^2$, slide the graph of $y = ax^2$ vertically k units.

If $k > 0$, slide it upward; if $k < 0$, slide it downward.

The graph has vertex $(0, k)$ and its axis is the line $x = 0$ (the y-axis).

Slide up 3.

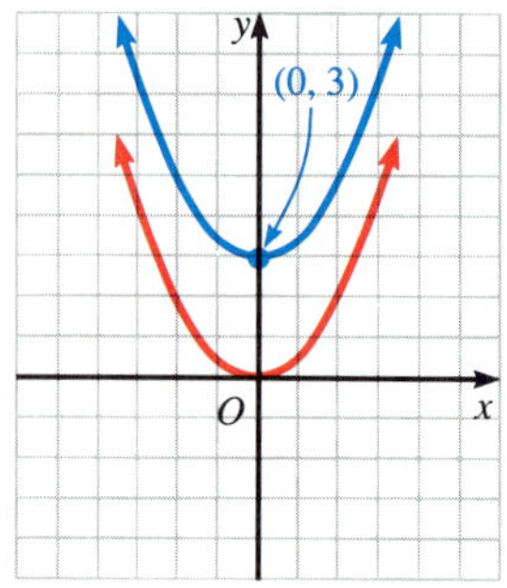

$y - 0 = \frac{1}{2}x^2$, or $y = \frac{1}{2}x^2$

$y - 3 = \frac{1}{2}x^2$

Figure 8

Slide down 3.

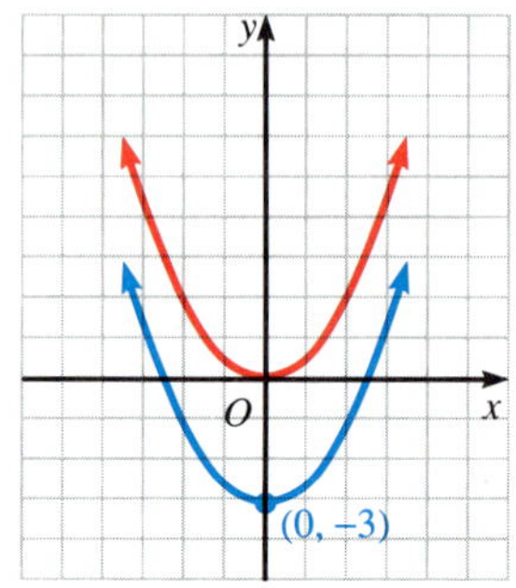

$y = \frac{1}{2}x^2$

$y - (-3) = \frac{1}{2}x^2$, or $y + 3 = \frac{1}{2}x^2$

Figure 9

Figures 10 and 11 below illustrate the following method for graphing a parabola whose equation has the form $y - k = a(x - h)^2$.

To graph $y - k = a(x - h)^2$,

slide the graph of $y = ax^2$ horizontally h units and vertically k units.

The graph has vertex (h, k) and its axis is the line $x = h$.

Slide right 4
and up 2.

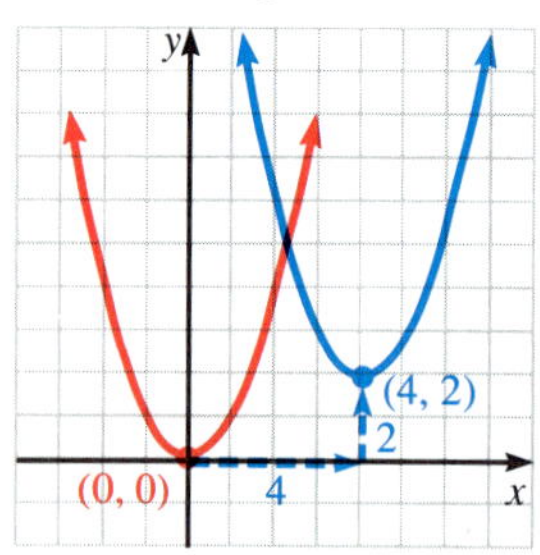

$y = x^2$ $\quad y - 2 = (x - 4)^2$

Figure 10

Slide left 2
and up 3.

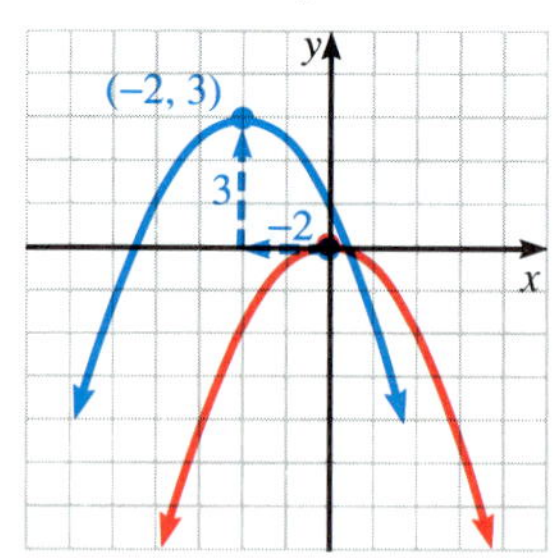

$y = -\frac{1}{2}x^2$ $\quad y - 3 = -\frac{1}{2}(x + 2)^2$

Figure 11

Example 1 Graph $y - 5 = -(x + 2)^2$. Label the vertex and axis.

Solution Since $a = -1$, the parabola opens downward.
Since $h = -2$ and $k = 5$, the vertex is $(-2, 5)$.
The axis of symmetry is the line $x = -2$.

Calculate a few convenient ordered pairs and plot the corresponding points. Also plot their images by reflection across the axis $x = -2$. Now draw the parabola by connecting the points with a smooth curve.

x	y
0	1
−1	4
−2	5

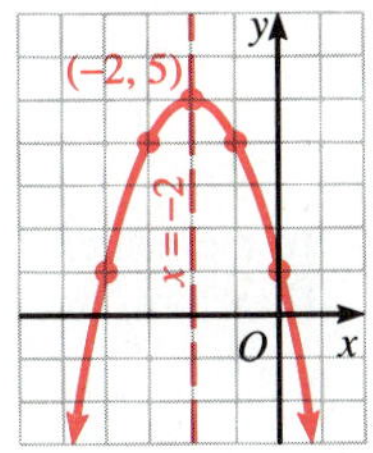

When graphing an equation in the coordinate plane, it is usually helpful to know the *intercepts* of the graph. The y-coordinate of a point where a graph crosses the y-axis is called the *y-intercept*. The x-coordinate of a point where a graph crosses the x-axis is called an *x-intercept*.

A parabola may have no x-intercepts, one x-intercept, or two x-intercepts, as illustrated below.

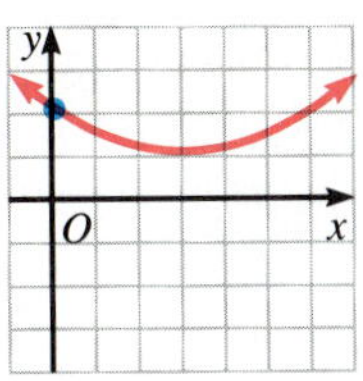

no x-intercepts
y-intercept 2

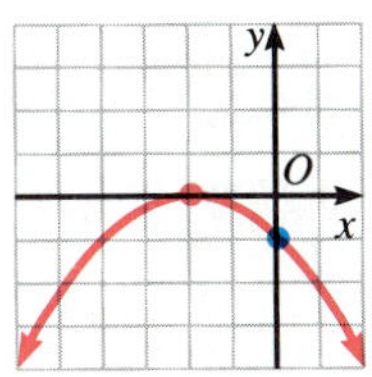

x-intercept −2
y-intercept −1

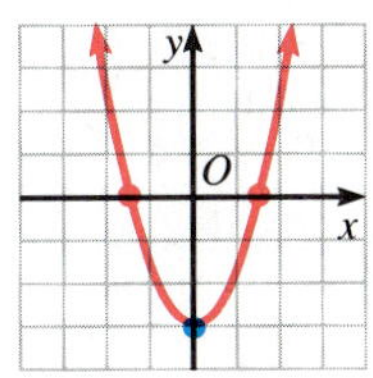

x-intercepts 1.5 and −1.5
y-intercept −3

To find the y-intercept of a parabola, set x equal to zero in the equation of the parabola and solve for y. To find the x-intercepts of a parabola, set y equal to zero in the equation and solve the resulting quadratic equation for x. If the roots are real they are the x-intercepts. If the roots are imaginary, the graph has no x-intercepts.

Example 2 Graph $y + 6 = 2(x + 1)^2$. Label the vertex and axis. Find all intercepts.

Solution 1. Since $a = 2$, the parabola opens upward.
Since $h = -1$ and $k = -6$ the vertex is $(-1, -6)$.
The axis of symmetry is the line $x = -1$.

(Solution continues on the next page.)

4. Find an equation of the parabola having vertex $(2, -3)$ and y-intercept 9.

$y - k = a(x - h)^2$
$y + 3 = a(x - 2)^2$
$9 + 3 = a(0 - 2)^2$
$12 = 4a$
$3 = a$
$\therefore$ an equation is
$y + 3 = 3(x - 2)^2$.

Check for Understanding

Here is a suggested use of the Oral Exercises to check students' understanding as you teach the lesson.
Oral Exs. 1–8: use before Example 1.
Oral Exs. 9–17: use after Example 1.

Guided Practice

State the coordinates of the vertex and the y-intercept of each parabola.

1. $y = (x + 1)^2$ vertex: $(-1, 0)$; y-int.: $(0, 1)$
2. $y - 3 = -(x + 2)^2$ vertex: $(-2, 3)$; y-int.: $(0, -1)$
3. $y + 2x^2 = 1$ vertex: $(0, 1)$; y-int.: $(0, 1)$
4. $4y = -(x + 4)^2 - 8$ vertex: $(-4, -2)$; y-int.: $(0, -6)$

Give an equation for each parabola.

5. vertex: $(-3, 5)$; $a = 2$ $y - 5 = 2(x + 3)^2$
6. vertex: $(-1, -2)$; $a = -2$ $y + 2 = -2(x + 1)^2$
7. vertex: $(-5, 0)$; $|a| = 3$; opens downward $y = -3(x + 5)^2$
8. vertex: $(-1, -2)$; $|a| = 3$; opens upward $y + 2 = 3(x + 1)^2$

Summarizing the Lesson

In this lesson students studied the equation $y - k = a(x - h)^2$ and how a, h, and k affect the shape and location of the parabola associated with the equation. Ask students to describe in words what effect changes in a, h, and k have on the parabola.

2. To find the y-intercept, set $x = 0$ and solve for y.

$$y + 6 = 2(0 + 1)^2$$
$$y + 6 = 2$$
$$y = -4 \longleftarrow y\text{-intercept}$$

Therefore, the graph crosses the y-axis at $(0, -4)$. Since $(0, -4)$ is on the graph, so is its mirror image across the axis of symmetry, $(-2, -4)$.

3. To find any x-intercepts, set $y = 0$.

$$0 + 6 = 2(x + 1)^2$$
$$3 = (x + 1)^2$$
$$\pm\sqrt{3} = x + 1$$
$$x = -1 \pm \sqrt{3}$$

To the nearest tenth, $\sqrt{3} \approx 1.7$.
Therefore, the x-intercepts are:

$x \approx -1 - 1.7$ or $x \approx -1 + 1.7$
$x \approx -2.7$ or $x \approx 0.7$

4. Plot the vertex $(-1, -6)$ and the intercepts. Then complete the curve using symmetry.

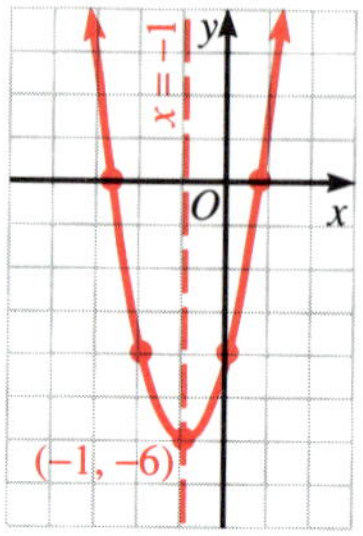

Example 3 Find an equation $y - k = a(x - h)^2$ of the parabola having vertex $(1, -2)$ and containing the point $(3, 6)$.

Solution Substitute $(1, -2)$ for (h, k) in the equation $y - k = a(x - h)^2$.

$$y - (-2) = a(x - 1)^2$$
$$y + 2 = a(x - 1)^2$$

Since the parabola contains the point $(3, 6)$, the coordinates of this point must satisfy the equation.

$$6 + 2 = a(3 - 1)^2$$
$$8 = 4a$$
$$a = 2$$

$\therefore$ an equation of the parabola is $y + 2 = 2(x - 1)^2$. ***Answer***

Oral Exercises

Match each equation with one of the graphs at the top of the next page.

1. $y = 3x^2$ c
2. $y = -\frac{1}{3}x^2$ e
3. $y = 3(x + 2)^2$ h
4. $y + 1 = 3(x + 2)^2$ b
5. $y - 1 = \frac{1}{3}(x + 2)^2$ a
6. $y + 1 = -3(x - 2)^2$ g
7. $y = -\frac{1}{3}(x - 2)^2$ f
8. $y - 1 = -\frac{1}{3}(x - 2)^2$ d

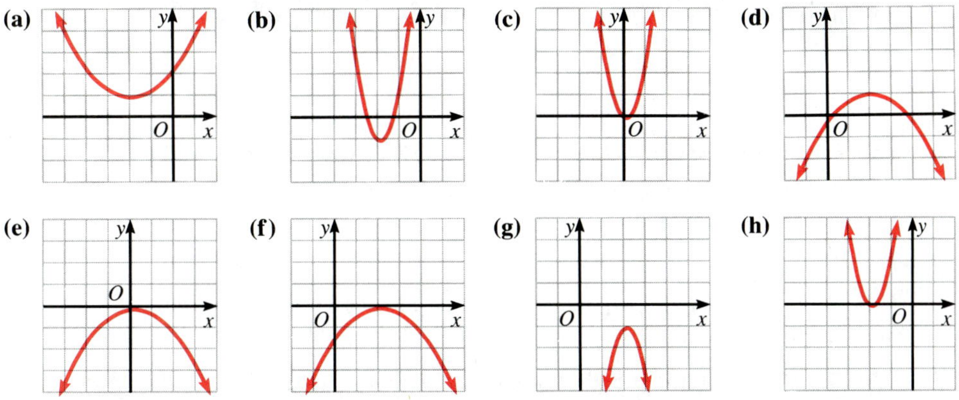

Tell: (a) whether the parabola opens upward or downward, (b) the coordinates of the vertex, and (c) the equation of the axis of symmetry.

9. $y = -3x^2$ a. down b. (0, 0) c. $x = 0$

10. $y + 1 = x^2$ a. up b. (0, −1) c. $x = 0$

11. $y = -3(x - 1)^2$ a. down b. (1, 0) c. $x = 1$

12. $y - 4 = 2(x - 5)^2$ a. up b. (5, 4) c. $x = 5$

13. $y + 3 = -(x - 2)^2$ a. down b. (2, −3) c. $x = 2$

14. $y + 2 = 5(x + 1)^2$ a. up b. (−1, −2) c. $x = -1$

Tell how to shift the graph of $y = 2x^2$ to obtain the graph of each equation.

15. $y = 2(x - 1)^2$ Right 1

16. $y + 4 = 2x^2$ Down 4

17. $y + 4 = 2(x - 1)^2$ Right 1, down 4

Written Exercises

Graph each equation. Label the vertex and axis of symmetry. Find all intercepts. You may wish to check your graphs on a computer or a graphing calculator.

A

1. $y = -3x^2$

2. $y = \frac{1}{3}x^2$

3. $y - 4 = -x^2$

4. $y + 1 = \frac{1}{2}x^2$

5. $y = (x - 3)^2$

6. $y = -(x + 3)^2$

7. $y + 2 = (x - 1)^2$

8. $y - 5 = -(x + 2)^2$

9. $y - 3 = 2(x + 1)^2$

10. $y - 8 = -2(x - 3)^2$

11. $y + 8 = \frac{1}{2}(x + 1)^2$

12. $y + 3 = \frac{1}{3}(x - 6)^2$

In Exercises 13–18, graph the given equations on the same axes.

13. $y = x^2$
$y = (x - 2)^2$
$y = (x + 3)^2$

14. $y = 2x^2$
$y = 2(x - 1)^2$
$y = 2(x + 3)^2$

15. $y = -x^2$
$y + 3 = -x^2$
$y - 4 = -x^2$

16. $y = \frac{1}{2}x^2$
$y - 2 = \frac{1}{2}(x - 1)^2$

17. $y = -2x^2$
$y + 1 = -2(x - 2)^2$

18. $y = -\frac{1}{3}x^2$
$y - 3 = -\frac{1}{3}(x + 6)^2$

Suggested Assignments

Average Algebra
Day 1: 331/1–16
Assign with Lesson 7-4.
Day 2: 331/17–30
S 332/Mixed Review

Average Alg. and Trig.
Day 1: 331/1–16
Assign with Lesson 7-4.
Day 2: 331/17–30
S 332/Mixed Review

Extended Algebra
Day 1: 331/1–25 odd
Assign with Lesson 7-4.
Day 2: 332/28–36 even, 37, 38
S 332/Mixed Review

Extended Alg. and Trig.
Day 1: 331/1–25 odd
Assign with Lesson 7-4.
Day 2: 332/28–36 even, 37, 38
S 332/Mixed Review

Supplementary Materials

Study Guide	pp. 113–114
Practice Master	41
Overhead Visual	4

Additional Answers Written Exercises

2.

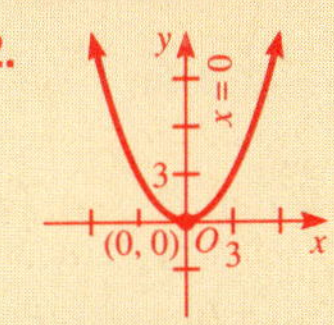

4.

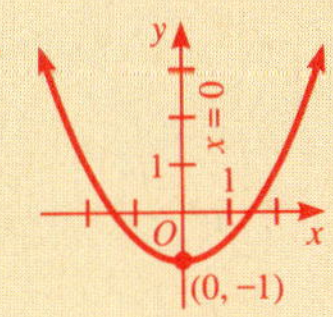

6.

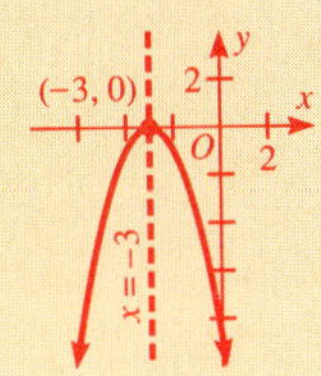

(continued)

Additional Answers Written Exercises

(continued)

8.
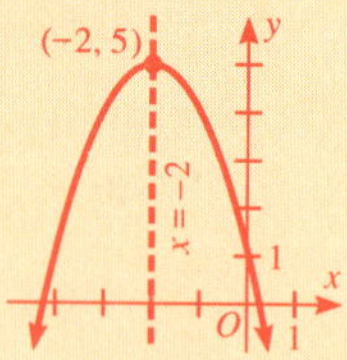

10.
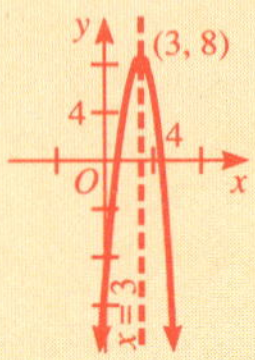

12.
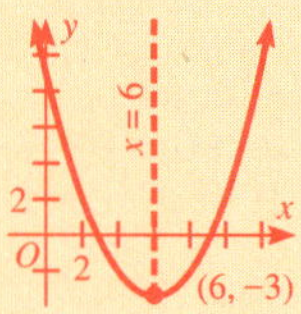

14.
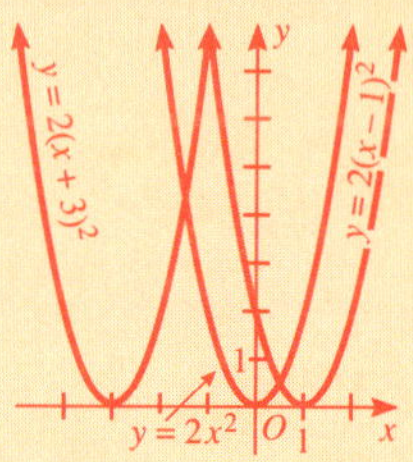

16.
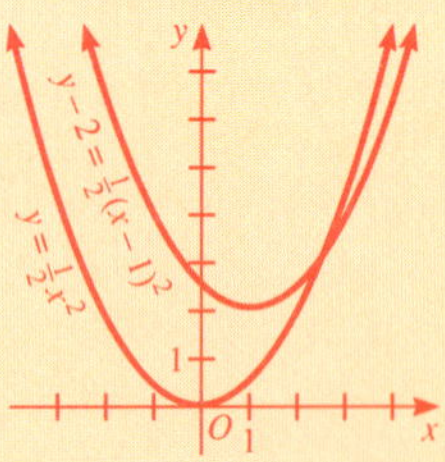

18.
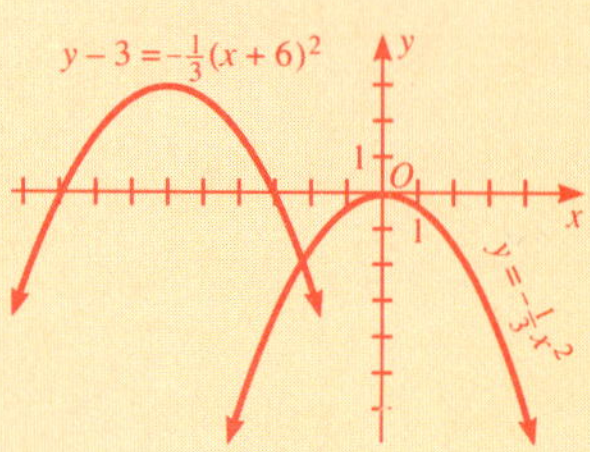

(continued on p. 346)

Find an equation $y - k = a(x - h)^2$ for each parabola described.

19. Vertex $(4, -3)$; contains $(2, -1)$

20. Vertex $(4, 5)$; contains $(5, 3)$

21. Vertex $(0, 0)$; contains $(-3, 3)$

22. Vertex $(-3, 6)$; contains the origin

B

23. Vertex $(3, 5)$; y-intercept 2

24. Vertex $(-2, 6)$; y-intercept -2

25. Vertex $(4, 2)$; one x-intercept is 3

26. Vertex $(-3, 4)$; one x-intercept is -1

27. If the parabola $y - 8 = 3(x - h)^2$ has the line $x = 5$ as its axis, find h. 5

28. If the parabola $y - k = -3(x - 1)^2$ passes through the origin, find k. 3

29. If the parabola $y - 3 = a(x - 1)^2$ passes through the point $(2, 5)$, find a. 2

30. If the parabola $y + 5 = a(x + 2)^2$ has y-intercept 4, find a. $\frac{9}{4}$

Find values of a and k for which each parabola contains the given points.

31. $y - k = a(x - 1)^2$; $(3, 11)$, $(1, 3)$ $a = 2$, $k = 3$

32. $y - k = a(x + 3)^2$; $(-5, 1)$, $(1, 7)$ $a = \frac{1}{2}$, $k = -1$

C

33. Graph $y + 3 = |x + 5|$.

34. Graph $y - 1 = |x + 4|$.

35. Prove algebraically that if the point $(h + r, s)$ is on the parabola $y - k = (x - h)^2$, then so is the point $(h - r, s)$. What fact about a parabola is established by this proof? The parabola is symmetric about the line $x = h$.

36. **a.** Using integral values of y from -3 to 3, make a table of values for the relation $x = y^2$.

b. Use the table in part (a) to draw the graph of $x = y^2$.

c. Is $x = y^2$ a function? Explain. No; two values of y correspond to every positive value of x.

37. Make a table of values for each relation using integer values of y in the given range. Then draw the graph of the relation over that range.

a. $x = -y^2$; -3 to 3

b. $x = y^2 + 2$; -3 to 3

c. $x = (y + 1)^2$; -4 to 2

d. $x = (y + 1)^2 + 2$; -4 to 2

e. $x = -\frac{1}{2}(y - 2)^2 + 3$; -1 to 5

38. Find the coordinates of the vertex and the equation of the axis of symmetry for each parabola in Exercise 37.

39. **a.** Find a general equation that represents all parabolas that have a horizontal axis. $x - h = a(y - k)^2$

b. How many x-intercepts can a parabola with a horizontal axis have? How many y-intercepts? 1 x-intercept; 0, 1, or 2 y-intercepts

Mixed Review Exercises

Without solving each equation, determine the nature of its roots.

1. $5x^2 - 3x + 2 = 0$
Imaginary, conjugate

2. $16y^2 - 24y + 9 = 0$
Real, double, rational

3. $t^2 + t - 1 = 0$
Real, unequal, irrational

Solve each equation over the complex numbers.

4. $t^4 - 5t^2 + 4 = 0$ $\{\pm 1, \pm 2\}$

5. $4w^2 - 4w + 5 = 0$ $\left\{\frac{1}{2} \pm i\right\}$

6. $x + \sqrt{x} - 6 = 0$ $\{4\}$

7-6 Quadratic Functions

Objective To analyze a quadratic function, draw its graph, and find its minimum or maximum value.

A **quadratic function** is a function that can be written in either of two forms.

$$\textit{General form:}\quad f(x) = ax^2 + bx + c \quad (a \neq 0)$$
$$\textit{Completed-square form:}\quad f(x) = a(x - h)^2 + k \quad (a \neq 0)$$

The graph of a quadratic function in either form is a parabola.

Example 1 Graph $f(x) = 2(x - 3)^2 + 1$.

Solution When graphing a function, it is often helpful to replace $f(x)$ with y.

$$f(x) = 2(x - 3)^2 + 1$$
$$y = 2(x - 3)^2 + 1$$

Then, $y - 1 = 2(x - 3)^2$.

This last equation is the equation of a parabola with vertex (3, 1) and axis $x = 3$. ***Answer***

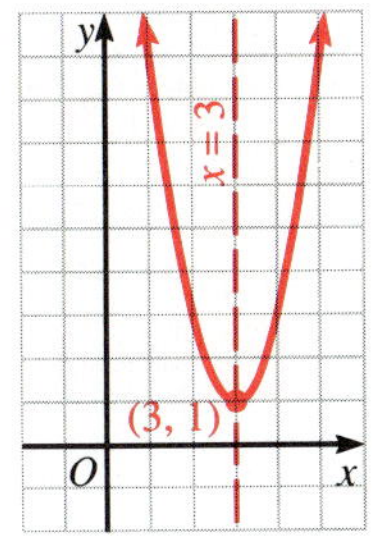

Example 2 **a.** Show that the graph of $f(x) = 3x^2 - 6x + 1$ is a parabola.
b. Find the vertex and graph the parabola.

Solution **a.** You want to rewrite the equation in the form $y - k = a(x - h)^2$.

Step 1 Replace $f(x)$ with y. $\quad y = 3x^2 - 6x + 1$

Step 2 Subtract the constant term from both sides. $\quad y - 1 = 3x^2 - 6x$

Step 3 Factor so that the coefficient of x^2 becomes 1. $\quad y - 1 = 3(x^2 - 2x)$

Step 4 Complete the square in x. $\quad y - 1 + 3 \cdot 1 = 3(x^2 - 2x + 1)$
$\quad y + 2 = 3(x - 1)^2$

$\therefore$ since the last equation is in the form $y - k = a(x - h)^2$, the graph is a parabola. ***Answer***

b. Since $h = 1$ and $k = -2$, the vertex is $(1, -2)$. Since $a = 3$, a is positive and the parabola opens upward. The graph is shown at the right. ***Answer***

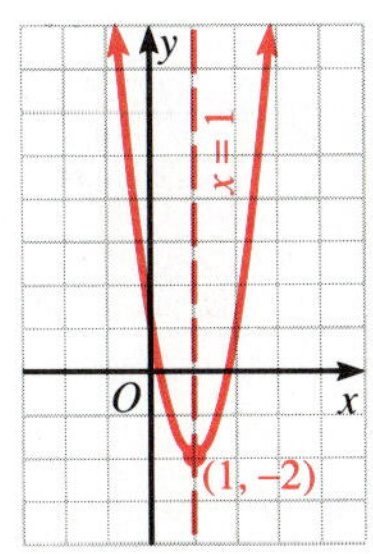

Teaching Suggestions p. T107

Suggested Extensions p. T107

Warm-Up Exercises

Solve for y. Write the answer in the form $y = ax^2 + bx + c$.

1. $y + 6 = 2(x + 1)^2$
   $y = 2x^2 + 4x - 4$
2. $y - 1 = \frac{1}{3}(x + 3)^2$
   $y = \frac{1}{3}x^2 + 2x + 4$
3. $y + 1 = -2(x - 2)^2$
   $y = -2x^2 + 8x - 9$
4. $y - 3 = -1(x + 1)^2$
   $y = -x^2 - 2x + 2$
5. $y - k = a(x - h)^2$
   $y = ax^2 - 2ahx + ah^2 + k$

Motivating the Lesson

Determining extremes is an important use of mathematics in the real world. Businesses need to minimize costs, factories need to maximize output, and so forth. Quadratic functions, which can occur in such applications, will be analyzed in today's lesson for their extreme values.

Common Errors

Students attempting to rewrite an equation of the form $y = ax^2 + bx + c$ $(a \neq 0, a \neq 1)$ into the form $y - k = a(x - h)^2$ may forget to multiply the number needed to complete the square by a before adding it to the left side. When reteaching, caution students about this error by emphasizing Step 4 in Example 2.

The completing-the-square method illustrated in Example 2 can be used (see Exercises 41 and 42) to show that the quadratic function $y = ax^2 + bx + c$ is equivalent to

$$y - \left(-\frac{b^2 - 4ac}{4a}\right) = a\left(x - \frac{-b}{2a}\right)^2.$$
$$\downarrow \qquad\qquad\qquad\qquad \downarrow$$
$$y - k \qquad = \quad a(x - h)^2$$

Therefore, the graph of $y = ax^2 + bx + c$ is a parabola having the vertex $(h, k) = \left(-\frac{b}{2a}, -\frac{b^2 - 4ac}{4a}\right)$. You do not have to remember the expression for k, because you can easily find it by substituting $-\frac{b}{2a}$ for x in the equation of the function. Example 3 illustrates.

Example 3 Given $g(x) = 6 + 6x - 3x^2$, find the following.

a. The vertex of the graph of g. **b.** The domain of g.
c. The range of g. **d.** The zeros of g.

Solution **a.** Since $a = -3$ and $b = 6$, the x-coordinate of the vertex is

$$x = -\frac{b}{2a} = -\frac{6}{2(-3)} = 1.$$

Then the corresponding y-coordinate is

$$g(1) = 6 + 6(1) - 3(1)^2 = 9.$$

$\therefore$ the vertex is $(1, 9)$. ***Answer***

b. g is defined for all real values of x.

$\therefore$ the domain $D = \{\text{real numbers}\}$. ***Answer***

c. Since a is negative, the parabola opens downward and the vertex $(1, 9)$ is the highest point of the graph. This means 9 is the *maximum value* of the function.

$\therefore$ the range $R = \{y: y \le 9\}$. ***Answer***

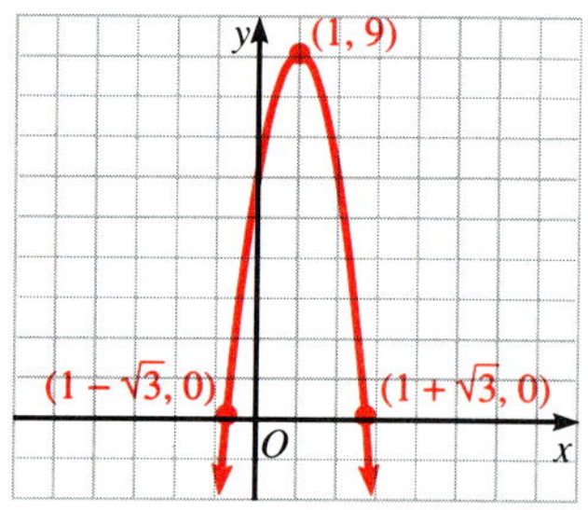

d. Recall (page 195) that the zeros of g are the roots of $g(x) = 0$.

$$g(x) = 6 + 6x - 3x^2$$
$$0 = -3x^2 + 6x + 6$$
$$x = \frac{-6 \pm \sqrt{6^2 - 4 \cdot (-3) \cdot 6}}{2 \cdot (-3)}$$
$$x = 1 \pm \sqrt{3}$$

$\therefore$ the zeros of g are $1 + \sqrt{3}$ and $1 - \sqrt{3}$. ***Answer***

Note that the zeros of g are the x-intercepts of the graph of g.

Chalkboard Examples

1. **a.** Show that the graph of $g(x) = 3x^2 - 12x + 7$ is a parabola.
   b. Find the vertex and sketch the parabola.

   a. $y = 3x^2 - 12x + 7$
   $y - 7 = 3(x^2 - 4x)$
   $y - 7 + 3 \cdot 4 = 3(x^2 - 4x + 4)$
   $y + 5 = 3(x - 2)^2$
   parabola

   b. The vertex is $(2, -5)$.

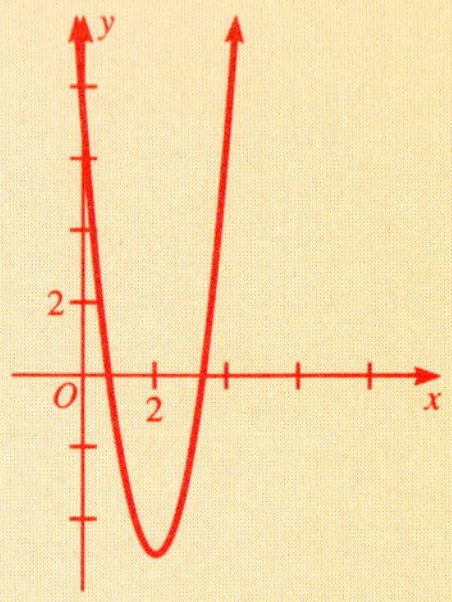

2. **a.** Find the vertex of the graph of $f(x) = 3 + 4x - x^2$.
   b. Find the domain, range, and the zeros of f.

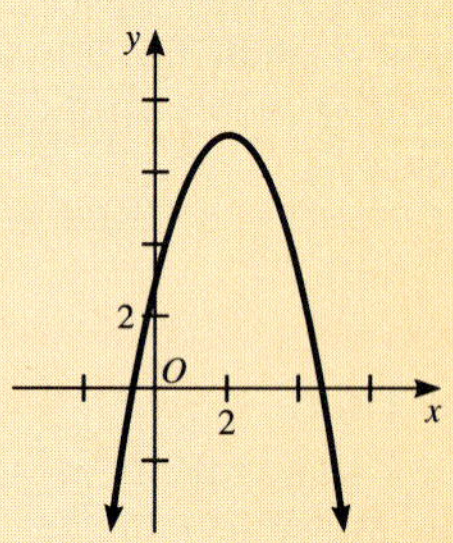

   a. $a = -1$ and $b = 4$;
   $-\frac{b}{2a} = 2$ and $f(2) = 7$.
   The vertex is $(2, 7)$.

   b. The domain consists of all real values of x. Since $a < 0$, the vertex is the highest point. The range of f consists of all real numbers less than or equal to 7.

Example 3, part (c), illustrated one of the following important facts about a quadratic function.

Let $f(x) = ax^2 + bx + c,\ a \neq 0$.

If $a < 0$, f has a **maximum value.**

If $a > 0$, f has a **minimum value.**

The graph of f is a parabola. This maximum or minimum value of f is the y-coordinate when $x = -\frac{b}{2a}$, at the vertex of the graph.

Example 4 Given $f(x) = \frac{1}{2}x^2 + 3x - \frac{7}{4}$, find the following.

a. The maximum or minimum value of f.

b. The vertex of the graph of f.

Solution **a.** Since $a = \frac{1}{2}$, $a > 0$ and f has a minimum value.

This minimum occurs when $x = -\frac{b}{2a} = -\frac{3}{2 \cdot \frac{1}{2}} = -3$

$$f(-3) = \frac{1}{2}(-3)^2 + 3(-3) - \frac{7}{4}$$
$$= \frac{9}{2} - 9 - \frac{7}{4}$$
$$= -\frac{25}{4}$$

$\therefore$ the minimum value of f is $-\frac{25}{4}$. ***Answer***

b. The vertex is $\left(-3, -\frac{25}{4}\right)$. ***Answer***

Oral Exercises

Tell whether each function has a maximum value or a minimum value. Give the value of x associated with this maximum or minimum.

1. $f(x) = x^2 - 5$ min.; 0

2. $f(x) = 5 - x^2$ max.; 0

3. $f(x) = x^2 - 2x$ min.; 1

4. $f(x) = 2x - x^2$ max.; 1

5. $f(x) = (x - 5)^2$ min.; 5

6. $f(x) = (5 - x)^2$ min.; 5

7. Mentally draw the graph of $f(x) = (x - 1)^2 + 5$.

a. Does the graph have 0, 1, or 2 x-intercepts? 0

b. How many zeros does f have? 0

$0 = 3 + 4x - x^2$

$x = \frac{-4 \pm \sqrt{16 + 12}}{-2}$

$x = 2 \pm \sqrt{7}$

$\therefore$ the zeros of f are $2 + \sqrt{7}$ and $2 - \sqrt{7}$.

3. Let $f(x) = 4x^2 + 8x - 1$. Determine whether f has a minimum or a maximum value and find this value.
$a = 4$ and $b = 8$. Since $a > 0$, f has a minimum value.
$-\frac{b}{2a} = -1$; $f(-1) = -5$
$\therefore$ the minimum value of f is -5.

Check for Understanding

Here is a suggested use of the Oral Exercises to check students' understanding as you teach the lesson.
Oral Exs. 1–7: use after Example 4.

Guided Practice

Show that the graph of each function is a parabola.

1. $f(x) = x^2 - 8x + 1$
$y + 15 = (x - 4)^2$

2. $f(x) = 2x^2 + 12x + 21$
$y - 3 = 2(x + 3)^2$

3. $f(x) = 12 - 3x - \frac{1}{2}x^2$
$y - \frac{33}{2} = -\frac{1}{2}(x + 3)^2$

Give the minimum or maximum value of each function. Then give the domain, range, and zeros.

4. $f(x) = 2x^2 - 12x$
-18; $D = \{\text{real numbers}\}$; $R = \{f(x): f(x) \geq -18\}$; 0, 6

5. $f(x) = 3x^2 + 6x + 7$
4; $D = \{\text{real numbers}\}$; $R = \{f(x): f(x) \geq 4\}$; no zeros

Summarizing the Lesson

In this lesson students used their understanding of parabolas to analyze and graph quadratic functions. Ask the students to describe, in general terms, how to find the minimum or maximum value of a quadratic function.

Suggested Assignments

Average Algebra
Day 1: 336/1–29 odd
S 332/31–34
Day 2: 336/20–40 even
S 315/*P*: 6, 10, 12

Average Alg. and Trig.
Day 1: 336/1–29 odd
S 332/31–34
Day 2: 336/20–40 even
S 315/*P*: 6, 10, 12

Extended Algebra
Day 1: 336/1–39 odd
S 332/27–35 odd
Day 2: 336/20–42 even, 43, 44

Extended Alg. and Trig.
Day 1: 336/1–39 odd
S 332/27–35 odd
Day 2: 336/20–42 even, 43, 44

Supplementary Materials

Study Guide	pp. 115–116
Practice Master	42
Computer Activities	5, 16

Using a Computer or a Graphing Calculator

Students will find a computer or graphing calculator helpful in checking their graphs for Written Exercises 1–18.

Support Materials
Disk for Algebra
Menu Item: Parabola Plotter

Written Exercises

Graph each function. Follow the method of Example 1. You may wish to check your graphs on a computer or a graphing calculator.

A

1. $f(x) = x^2 - 1$ **2.** $f(x) = 4 - x^2$ **3.** $f(x) = (x - 1)^2 + 3$

4. $f(x) = 2(x + 3)^2 + 5$ **5.** $f(x) = 4 - 2(x - 1)^2$ **6.** $f(x) = 6 - \frac{1}{2}(x + 4)^2$

Graph each function. Follow the method of Example 2. You may wish to check your graphs on a computer or a graphing calculator.

7. $h(x) = x^2 + 2x - 3$ **8.** $g(x) = x^2 + 4x + 1$ **9.** $f(x) = 2x^2 - 4x + 1$

10. $g(x) = 3x^2 - 6x + 5$ **11.** $f(x) = 5 + 2x - x^2$ **12.** $h(x) = 6 - 6x - x^2$

13. $f(x) = 3x^2 - 12x$ **14.** $h(x) = 4x - 2x^2$ **15.** $g(x) = 2x^2 - 12x + 18$

16. $h(x) = 4x^2 + 4x - 5$ **17.** $g(x) = \frac{1}{2}x^2 + x + \frac{1}{2}$ **18.** $f(x) = \frac{1}{4}x^2 - 2x + \frac{5}{2}$

Find the maximum value or minimum value of each function. Then give the coordinates of the vertex of its graph.

19. $g(x) = 2x^2 + 8x$ min., -8; $(-2, -8)$

20. $f(x) = x^2 + 8x + 6$ min., -10; $(-4, -10)$

21. $g(x) = 2x^2 - 6x + 2$ min., $-\frac{5}{2}$; $\left(\frac{3}{2}, -\frac{5}{2}\right)$

22. $f(x) = 3x^2 - 5x - 2$ min., $-\frac{49}{12}$; $\left(\frac{5}{6}, -\frac{49}{12}\right)$

23. $h(x) = (2 - x)(5 + x)$ max., $\frac{49}{4}$; $\left(-\frac{3}{2}, \frac{49}{4}\right)$

24. $h(x) = (2x - 5)(2x + 3)$ min., -16; $\left(\frac{1}{2}, -16\right)$

For each function find (a) the vertex of its graph, (b) its domain, (c) its range, and (d) its zeros.

25. $f(x) = x^2 - 4x - 3$ **26.** $g(x) = 6x - 3x^2$ **27.** $f(x) = 8 - 2x - x^2$

28. $F(x) = 9 - 8x - x^2$ **29.** $f(x) = x^2 - 2x - 5$ **30.** $G(x) = 2x^2 + 8x + 5$

B

31. $h(x) = 2(x - 7)(x + 5)$ **32.** $g(x) = \frac{1}{2}(6 + x)(4 + x)$ **33.** $f(x) = 9 - (x + 6)^2$

34. $F(x) = 2(x - 5)^2 - 8$ **35.** $G(x) = 2 - \frac{1}{2}(x + 1)^2$ **36.** $H(x) = \frac{1}{2}(3x - 2)(3x - 4)$

In Exercises 37–39, a parabola represented by an equation $y = ax^2 + bx + c$ is given. Study each graph and then tell whether the value of each expression that follows is *positive, negative,* or *zero:*

a. $b^2 - 4ac$ **b.** c **c.** a **d.** b

37. a. neg. b. pos. c. pos. d. zero

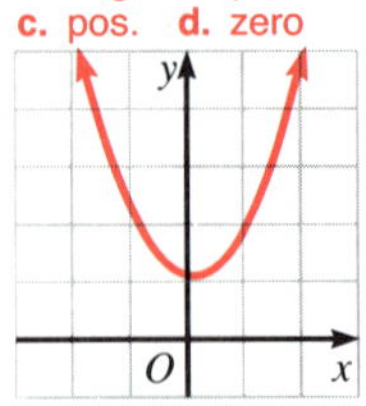

38. a. pos. b. zero c. neg. d. pos.

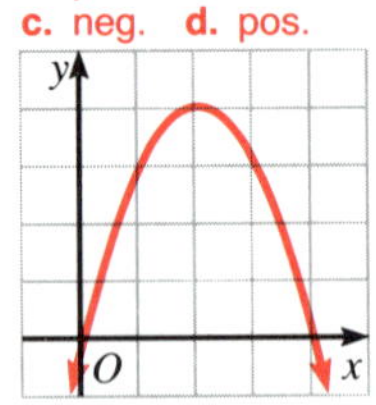

39. a. zero b. pos. c. pos. d. pos.

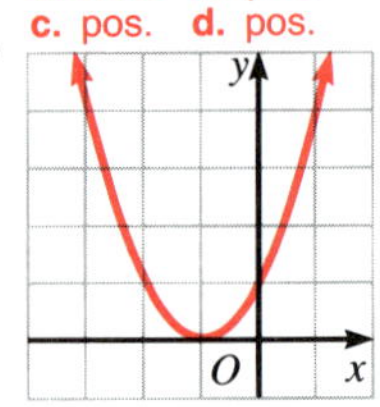

40. Tell how many times the parabola $y = x^2 - 6x + 10$ intersects:

a. the x-axis 0
b. the line $y = 1$ 1
c. the line $y = 2$ 2

41. By completing the square, rewrite the equation $y = x^2 + bx + c$ in the form $y - k = a(x - h)^2$. $y - \frac{4c - b^2}{4} = \left(x - \frac{-b}{2}\right)^2$

42. Repeat Exercise 41 using the equation $y = ax^2 + bx + c$. $y - \frac{4ac - b^2}{4a} = a\left(x - \frac{-b}{2a}\right)^2$

C **43.** A function f is an **even function**

$$\text{if } f(-x) = f(x) \text{ for every } x.$$

Show that $f(x) = ax^2 + bx + c$ is an even function if and only if $b = 0$.

44. A function f is an **odd function**

$$\text{if } f(-x) = -f(x) \text{ for every } x.$$

Show that there are no odd quadratic functions. That is, show that if $f(x) = ax^2 + bx + c$, then $f(-x) \neq -f(x)$.

Career Note / *Statistician*

Every day, people's opinions are being polled by means of surveys—opinions about political candidates, television favorites, consumer goods, and taste preferences, for example. Statisticians are involved in designing such surveys and experiments, implementing them, and interpreting their numerical results. Statistical methods are applied to a wide variety of situations, from predicting human behavior to controlling the quality of a manufactured product.

One of the most important characteristics of statistical work is the way in which accurate information about a large group of people can be obtained by surveying a small part of the group. Television rating services, for example, can determine the size of a national network's audience by asking only a few thousand households, rather than all viewers, what programs are watched. To achieve accurate results, however, statisticians must carefully choose where and how to get their data and determine the type and size of the sample group.

Because statistics are used so widely, statisticians generally combine their knowledge of mathematics with another field, such as business or psychology.

Statisticians also find a background in computer science helpful, since computers are frequently used as tools in statistical applications.

Additional Answers
Written Exercises

2.
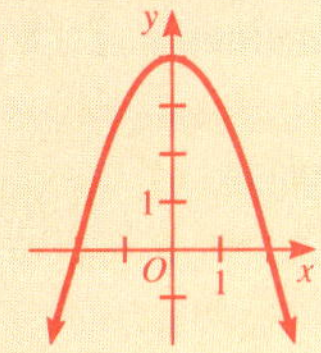

4.
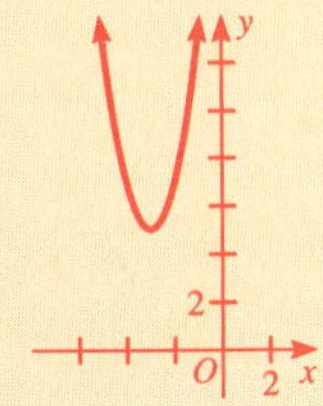

6.
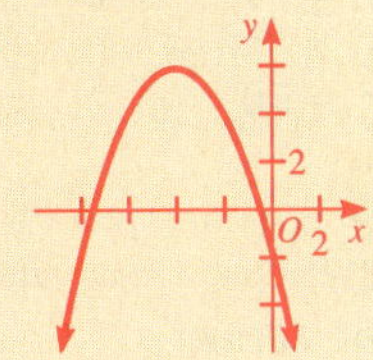

8.
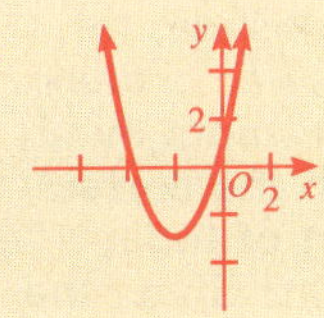

10.
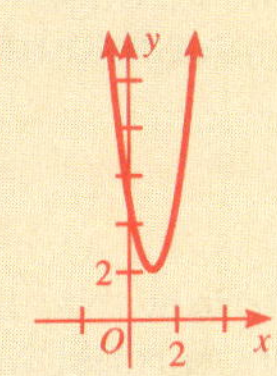

12.
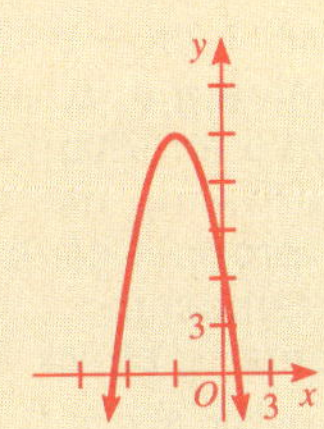

(continued on p. 348)

Teaching Suggestions p. T108

Suggested Extensions p. T108

Warm-Up Exercises

Solve.

1. $y^2 + 5y + 4 = 0$ $\{-4, -1\}$
2. $3r^2 - 8r + 5 = 0$ $\left\{1, \frac{5}{3}\right\}$
3. $x^2 - 3x - 7 = 0$ $\left\{\frac{3 \pm \sqrt{37}}{2}\right\}$
4. $3(x + 5)^2 = 0$ $\{-5\}$
5. $(2x - 3)^2 = 50$ $\left\{\frac{3 \pm 5\sqrt{2}}{2}\right\}$

Motivating the Lesson

In some game-show formats, the answer is given and the contestants must come up with the correct question. Part of today's lesson is like that: Given the roots of a quadratic equation, students must find the equation.

Common Errors

Students may forget that a quadratic equation must be in the form $ax^2 + bx + c = 0$ before determining *a, b,* and *c,* and thus give an incorrect answer when finding the sum and the product of the roots. When reteaching, remind students to check for the form $ax^2 + bx + c = 0$ before applying the theorem for the roots of quadratic equations.

7-7 Writing Quadratic Equations and Functions

Objective To learn the relationship between the roots and coefficients of a quadratic equation. To write a quadratic equation or function using information about the roots or the graph.

So far in this chapter, you have learned to find roots of quadratic equations and to graph quadratic functions. In this lesson, you will reverse the process and learn how to *write* a quadratic equation or function using given information about the roots or the graph. These new problems are closely related to ones you've already studied. For example, consider the familiar problem of solving the quadratic equation $2x^2 - 6x - 8 = 0$:

$$\begin{aligned} 2x^2 - 6x - 8 &= 0 \\ 2(x^2 - 3x - 4) &= 0 \\ 2(x + 1)(x - 4) &= 0 \\ x = -1 \quad \text{or} \quad x &= 4 \end{aligned}$$

Now suppose you are given that -1 and 4 are the roots of a quadratic equation. You can reverse the steps above and reason that the equation is

$$\begin{aligned} (x + 1)(x - 4) &= 0 \\ x^2 - 3x - 4 &= 0 \\ \text{or} \qquad a(x^2 - 3x - 4) &= 0 \qquad (a \neq 0). \end{aligned}$$

This idea can be generalized as follows: If r_1 and r_2 are roots of a quadratic equation, then

$$(x - r_1)(x - r_2) = 0.$$

Multiplying the binomials on the left side gives

$$x^2 - (r_1 + r_2)x + r_1r_2 = 0.$$

Theorem

A quadratic equation with roots r_1 and r_2 is

$$x^2 - (r_1 + r_2)x + r_1r_2 = 0$$

or

$$a[x^2 - (r_1 + r_2)x + r_1r_2] = 0.$$

The equation just given is equivalent to

$$a[x^2 - (\text{sum of roots})x + (\text{product of roots})] = 0.$$

The theorem above provides an easy method of writing a quadratic equation from the sum and product of its roots, as Example 1 illustrates.

Example 1 Find a quadratic equation with roots $\frac{2+i}{3}$ and $\frac{2-i}{3}$.

Solution Sum of roots $= \frac{2+i}{3} + \frac{2-i}{3} = \frac{4}{3}$

Product of roots $= \frac{2+i}{3} \cdot \frac{2-i}{3} = \frac{4-i^2}{9} = \frac{5}{9}$

$$x^2 - (\text{sum of roots})x + (\text{product of roots}) = 0$$

$$x^2 - \frac{4}{3}x + \frac{5}{9} = 0.$$

To clear fractions, multiply both sides of the equation by 9.

$\therefore$ the equation is $9x^2 - 12x + 5 = 0$. ***Answer***

The next theorem gives you an easy way to find the sum and product of the roots of a quadratic equation of the form $ax^2 + bx + c = 0$.

Theorem

If r_1 and r_2 are the roots of a quadratic equation $ax^2 + bx + c = 0$, then

$$r_1 + r_2 = \text{sum of roots} = -\frac{b}{a} \quad \text{and} \quad r_1r_2 = \text{product of roots} = \frac{c}{a}.$$

Proof By the theorem on page 338, an equation with roots r_1 and r_2 is

$$a[x^2 - (r_1 + r_2)x + r_1r_2] = 0.$$

Factoring a from the left side of the equation $ax^2 + bx + c = 0$ gives this equivalent equation:

$$a\left[x^2 - \left(-\frac{b}{a}\right)x + \frac{c}{a}\right] = 0$$

Setting the corresponding coefficients equal to each other gives

$$r_1 + r_2 = -\frac{b}{a} \quad \text{and} \quad r_1r_2 = \frac{c}{a}.$$

This theorem is often used to check solutions of a quadratic equation.

Example 2 Find the roots of $2x^2 + 9x + 5 = 0$. Check your answer by using the theorem about the sum and product of the roots.

Solution $2x^2 + 9x + 5 = 0$

$$x = \frac{-9 \pm \sqrt{9^2 - 4(2)(5)}}{2(2)} = \frac{-9 \pm \sqrt{41}}{4}$$

Thus, $r_1 = \frac{-9 + \sqrt{41}}{4}$ and $r_2 = \frac{-9 - \sqrt{41}}{4}$.

(Solution continues on the next page.)

Chalkboard Examples

1. Find a quadratic equation with roots $\frac{3+i}{2}$ and $\frac{3-i}{2}$.

   $r_1 + r_2 = 3$; $r_1r_2 = \frac{10}{4} = \frac{5}{2}$

   $x^2 - 3x + \frac{5}{2} = 0$

   $2x^2 - 6x + 5 = 0$

2. Find the roots of $2y^2 + y - 5 = 0$. Check your answer by using the sum and product of the roots.

   $y = \frac{-1 \pm \sqrt{1 + 40}}{4}$

   $= \frac{-1 \pm \sqrt{41}}{4}$

   $\therefore r_1 = \frac{-1 + \sqrt{41}}{4}$ and

   $r_2 = \frac{-1 - \sqrt{41}}{4}$

   $r_1 + r_2 = \frac{-1 + \sqrt{41}}{4} + \frac{-1 - \sqrt{41}}{4}$

   $= \frac{-2}{4} = -\frac{1}{2}$

   $= -\frac{b}{a}$

   $r_1r_2 = \left(\frac{-1 + \sqrt{41}}{4}\right) \cdot \left(\frac{-1 - \sqrt{41}}{4}\right)$

   $= \frac{1 - 41}{16} = -\frac{40}{16}$

   $= -\frac{5}{2} = \frac{c}{a}$

(continued)

Chalkboard Examples
(continued)

3. Find a quadratic function with minimum value −1 whose graph has x-intercepts 1 and 3.
The x-intercepts of the graph of f are the roots of $f(x) = 0$. By the theorem on page 338, $f(x) = a(x^2 - 4x + 3)$. The x-coordinate of the vertex of the graph of f is $\frac{1+3}{2}$, or 2. Since $f(2) = -1$,
$a[2^2 - 4(2) + 3] = -1$
$-a = -1$
$a = 1$
$\therefore f(x) = x^2 - 4x + 3$

4. A rectangular dog kennel is to be constructed alongside a house with 60 m of fencing. If the house serves as one side of the kennel, determine the greatest possible area that can be enclosed.
Let x = width of kennel
$60 - 2x$ = length of kennel
$x(60 - 2x)$ = area of enclosure
$A(x) = x(60 - 2x)$
$= 60x - 2x^2$
Maximum value is when
$x = -\frac{b}{2a} = -\frac{60}{2(-2)}$
$= 15$ (m).
$\therefore$ maximum area is $15(60 - 2 \cdot 15) = 450$ (m^2).

Check: $r_1 + r_2 = \frac{-9 + \sqrt{41}}{4} + \frac{-9 - \sqrt{41}}{4} = -\frac{9}{2} = -\frac{b}{a}$ √

$r_1 r_2 = \frac{-9 + \sqrt{41}}{4} \cdot \frac{-9 - \sqrt{41}}{4} = \frac{81 - 41}{16} = \frac{5}{2} = \frac{c}{a}$ √

$\therefore \frac{-9 + \sqrt{41}}{4}$ and $\frac{-9 - \sqrt{41}}{4}$ are the roots. ***Answer***

Example 3 shows how to write the equation of a quadratic function using information about its graph.

Example 3 Find a quadratic function $f(x) = ax^2 + bx + c$ such that the maximum value of f is 8 and the graph of f has x-intercepts 3 and 7.

Solution Recall that the x-intercepts of the graph of f are also the roots of the related quadratic equation $f(x) = 0$. Therefore, you can use the theorem on page 338 and write this equation:

$$f(x) = a(x^2 - 10x + 21)$$

Since the maximum value of f is 8, the y-coordinate of the vertex of the graph is 8. To find the x-coordinate of the vertex, note that the axis of symmetry intersects the x-axis at a point midway between the x-intercepts. This means the x-coordinate of the vertex is simply the *average* of the x-intercepts:

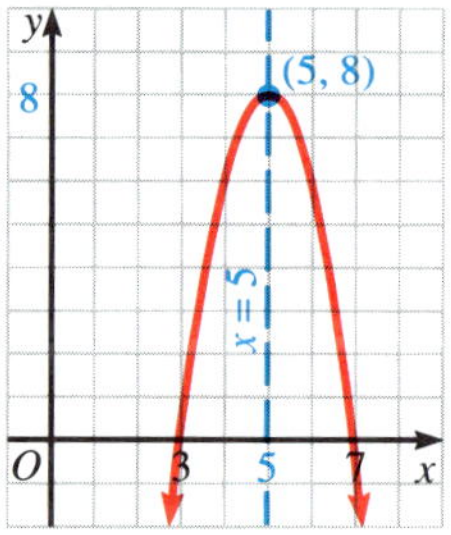

$$x = \frac{3 + 7}{2} = 5$$

Therefore, the vertex is (5, 8). Since the vertex lies on the graph of f, you can substitute its coordinates in the equation of f to find a.

$$f(x) = a(x^2 - 10x + 21)$$
$$8 = a(25 - 50 + 21)$$
$$8 = -4a$$
$$a = -2$$

$\therefore$ the equation is $f(x) = -2(x^2 - 10x + 21)$. ***Answer***

There are many situations in which knowing when maximum or minimum values occur is useful. For example, a business may want to maximize a profit or minimize a loss. If the profit (or loss) can be described as a quadratic function, then you can find the maximum value (or minimum value). Example 4 illustrates this idea.

Example 4 An automobile dealership sells only 35 new cars per month when the markup over factory price is \$3000. Marketing research indicates that for each \$200 decrease in the markup, the dealership can expect to sell an additional 5 cars per month. What should the markup be per car in order to maximize total monthly profits? What will the total profit be?

Solution

Step 1 The problem asks for the markup needed per car to produce maximum monthly profits. It also asks for the maximum monthly profit.

Step 2 Let n = the number of \$200 decreases in markup per car. Making a chart like the one below is often helpful.

Number of \$200 decreases	Markup per car	Number of car sales	Total monthly profit
1	$3000 - 200(1) = 2800$	$35 + 5(1) = 40$	$(2800)(40) = 112{,}000$
2	$3000 - 200(2) = 2600$	$35 + 5(2) = 45$	$(2600)(45) = 117{,}000$
⋮	⋮	⋮	⋮
n	$3000 - 200n$	$35 + 5n$	$(3000 - 200n)(35 + 5n)$

Step 3 Express the total profit P as a function of n.

Total profit = (Markup per car)(Number of car sales)

$$P(n) = (3000 - 200n)(35 + 5n)$$
$$P(n) = 105{,}000 + 8000n - 1000n^2$$

Step 4 This function has its maximum value when

$$n = -\frac{b}{2a} = -\frac{8000}{2(-1000)} = 4.$$

Then the markup per car is $3000 - 200(4) = 2200$, and the maximum profit is

$$P(4) = 105{,}000 + 8000(4) - 1000(4)^2 = 121{,}000$$

Step 5 The check is left for you.

∴ the markup per car should be \$2200 to obtain maximum monthly profit of \$121,000. ***Answer***

Oral Exercises

Give the sum and product of the roots of each equation.

1. $x^2 - 3x + 6 = 0$ 3; 6
2. $x^2 - 6x + 8 = 0$ 6; 8
3. $2y^2 - y - 6 = 0$ $\frac{1}{2}$; -3
4. $2t^2 - 6t + 1 = 0$ 3; $\frac{1}{2}$
5. $2x^2 + 2 = 3x$ $\frac{3}{2}$; 1
6. $3u^2 - 4 = 0$ 0; $-\frac{4}{3}$

Check for Understanding

Here is a suggested use of the Oral Exercises to check students' understanding as you teach the lesson.
Oral Exs. 1–15: use after discussing the theorem on page 339.

Guided Practice

Find a quadratic equation with integral coefficients having the given roots.

1. $3, -5$
$x^2 + 2x - 15 = 0$

2. $\frac{1}{2}, -\frac{5}{2}$
$4x^2 + 8x - 5 = 0$

3. $1 + \sqrt{5}, 1 - \sqrt{5}$
$x^2 - 2x - 4 = 0$

4. $\frac{2 + \sqrt{5}}{3}, \frac{2 - \sqrt{5}}{3}$
$9x^2 - 12x - 1 = 0$

5. $8 + i, 8 - i$
$x^2 - 16x + 65 = 0$

6. $-3 - i, -3 + i$
$x^2 + 6x + 10 = 0$

7. $2 + i\sqrt{3}, 2 - i\sqrt{3}$
$x^2 - 4x + 7 = 0$

8. $\frac{3 + i\sqrt{2}}{2}, \frac{3 - i\sqrt{2}}{2}$
$4x^2 - 12x + 11 = 0$

Find a quadratic function for each parabola described.

9. maximum value 4, x-intercepts -5 and -1
$f(x) = -x^2 - 6x - 5$

10. minimum value -25, x-intercepts -2 and 3
$f(x) = 4x^2 - 4x - 24$

Summarizing the Lesson

In this lesson students learned that the sum and product of the roots of a quadratic equation are related to the coefficients and that a quadratic equation can be written using information about the roots. Ask students to explain how the theorem on page 339 helps them check their solutions of quadratic equations.

Suggested Assignments

Average Algebra
Day 1: 342/1–33 odd
343/*P*: 1–7 odd
Day 2: 343/*P*: 2–14 even
S 345/Mixed Review

Average Alg. and Trig.
Day 1: 342/1–33 odd
343/*P*: 1–7 odd
Day 2: 343/*P*: 2–14 even
S 345/Mixed Review

Extended Algebra
Day 1: 342/1–37 odd
343/*P*: 1–7 odd
Day 2: 343/*P*: 8–15
S 345/Mixed Review

Extended Alg. and Trig.
Day 1: 342/1–37 odd
343/*P*: 1–7 odd
Day 2: 343/*P*: 8–15
S 345/Mixed Review

Supplementary Materials

Study Guide	pp. 117–118
Practice Master	43
Test Master	28
Resource Book	p. 129

Find a quadratic equation whose roots have the given sum and product.
Answers may vary.

7. $r_1 + r_2 = 3,\ r_1r_2 = 4$ $x^2 - 3x + 4 = 0$

8. $r_1 + r_2 = -2,\ r_1r_2 = -2$ $x^2 + 2x - 2 = 0$

9. $r_1 + r_2 = 0,\ r_1r_2 = -1$ $x^2 - 1 = 0$

10. $r_1 + r_2 = -4,\ r_1r_2 = 0$ $x^2 + 4x = 0$

Find a quadratic equation with the given roots. Answers may vary.

11. $3, -2$ $x^2 - x - 6 = 0$

12. $1, 9$ $x^2 - 10x + 9 = 0$

13. $0, 7$ $x^2 - 7x = 0$

14. $\sqrt{5}, -\sqrt{5}$ $x^2 - 5 = 0$

15. Explain why 2 and $\frac{3}{2}$ cannot be the roots of $2y^2 - 7y + 3 = 0$.
The product of the roots should be $\frac{3}{2}$, but $2 \cdot \frac{3}{2} = 3$.

Written Exercises

13. $x^2 - 2x - 2 = 0$ **14.** $x^2 - 4x - 3 = 0$ **15.** $9x^2 - 6x - 1 = 0$ **16.** $16x^2 + 16x - 1 = 0$ **19.** $x^2 - 6x + 10 = 0$ **20.** $x^2 - 8x + 20 = 0$ **21.** $x^2 - 10x + 27 = 0$

A

1–6. Solve the equations in Oral Exercises 1–6. Check each answer by using the theorem about the sum and product of the roots.

1. $\left\{\frac{3}{2} \pm \frac{\sqrt{15}}{2}i\right\}$ **2.** $\{2, 4\}$ **3.** $\left\{-\frac{3}{2}, 2\right\}$ **4.** $\left\{\frac{3 \pm \sqrt{7}}{2}\right\}$ **5.** $\left\{\frac{3}{4} \pm \frac{\sqrt{7}}{4}i\right\}$ **6.** $\left\{\pm\frac{2\sqrt{3}}{3}\right\}$

Find a quadratic equation with *integral coefficients* having the given roots.
Answers may vary.

7. $2, 5$ $x^2 - 7x + 10 = 0$

8. $-3, 1$ $x^2 + 2x - 3 = 0$

9. $-2, \frac{5}{2}$ $2x^2 - x - 10 = 0$

10. $\frac{3}{2}, -\frac{1}{2}$ $4x^2 - 4x - 3 = 0$

11. $-\sqrt{3}, \sqrt{3}$ $x^2 - 3 = 0$

12. $-\frac{\sqrt{5}}{2}, \frac{\sqrt{5}}{2}$ $4x^2 - 5 = 0$

13. $1 + \sqrt{3}, 1 - \sqrt{3}$

14. $2 + \sqrt{7}, 2 - \sqrt{7}$

15. $\frac{1 + \sqrt{2}}{3}, \frac{1 - \sqrt{2}}{3}$

16. $\frac{-2 + \sqrt{5}}{4}, \frac{-2 - \sqrt{5}}{4}$

17. $i\sqrt{5}, -i\sqrt{5}$ $x^2 + 5 = 0$

18. $2i\sqrt{2}, -2i\sqrt{2}$ $x^2 + 8 = 0$

19. $3 + i, 3 - i$

20. $4 + 2i, 4 - 2i$

21. $5 + i\sqrt{2}, 5 - i\sqrt{2}$

22. $-2 + i\sqrt{7}, -2 - i\sqrt{7}$ $x^2 + 4x + 11 = 0$

23. $\frac{1 - i\sqrt{5}}{4}, \frac{1 + i\sqrt{5}}{4}$ $8x^2 - 4x + 3 = 0$

24. $\frac{2 + i\sqrt{3}}{2}, \frac{2 - i\sqrt{3}}{2}$ $4x^2 - 8x + 7 = 0$

Find a quadratic function $f(x) = ax^2 + bx + c$ for each parabola described.

25. maximum value 10
x-intercepts 1 and 3 $f(x) = -10x^2 + 40x - 30$

26. maximum value 6
x-intercepts -2 and 4 $f(x) = -\frac{2}{3}x^2 + \frac{4}{3}x + \frac{16}{3}$

27. minimum value -8
x-intercepts 0 and 8 $f(x) = \frac{1}{2}x^2 - 4x$

28. minimum value -5
x-intercepts -2 and 3 $f(x) = \frac{4}{5}x^2 - \frac{4}{5}x - \frac{24}{5}$

B

29. vertex $(2, 12)$; x-intercepts -4 and 8 $f(x) = -\frac{1}{3}x^2 + \frac{4}{3}x + \frac{32}{3}$

30. vertex $(-1, -10)$; x-intercepts -6 and 4 $f(x) = \frac{2}{5}x^2 + \frac{4}{5}x - \frac{48}{5}$

31. minimum value -6; zeros of f are -1 and 5 $f(x) = \frac{2}{3}x^2 - \frac{8}{3}x - \frac{10}{3}$
(*Hint:* Recall that the zeros of f are the *roots* of $f(x) = 0$.)

32. maximum value 9; zeros of f are -6 and 0 $f(x) = -x^2 - 6x$

33. maximum value 6 when $x = -2$; one zero is 1 $f(x) = -\frac{2}{3}x^2 - \frac{8}{3}x + \frac{10}{3}$

34. minimum value -4 when $x = 3$; one zero is 6 $f(x) = \frac{4}{9}x^2 - \frac{8}{3}x$

35. y-intercept 2; x-intercepts 1 and 5 $f(x) = \frac{2}{5}x^2 - \frac{12}{5}x + 2$

36. range: $\{y: y \le 9\}$; x-intercepts -2 and 4 $f(x) = -x^2 + 2x + 8$

37. **a.** Show that $3 + i\sqrt{2}$ is a root of $x^2 - 6x + 11 = 0$.
b. Find another root. **a.** $(3 + i\sqrt{2})^2 - 6(3 + i\sqrt{2}) + 11 = 7 + 6i\sqrt{2} - 18 - 6i\sqrt{2} + 11 = 0$
b. $3 - i\sqrt{2}$

38. **a.** Show that $\frac{5 - i\sqrt{2}}{4}$ is a root of $16x^2 - 40x + 27 = 0$.
b. Find another root. **a.** $16\left(\frac{5 - i\sqrt{2}}{4}\right)^2 - 40\left(\frac{5 - i\sqrt{2}}{4}\right) + 27 = 23 - 10i\sqrt{2} - 50 + 10i\sqrt{2} + 27 = 0$
b. $\frac{5 + i\sqrt{2}}{4}$

C 39. Show that the roots of $ax^2 + bx + c = 0$ are reciprocals of each other if and only if $a = c$. The roots are reciprocals if and only if the product of the roots is 1; that is, $\frac{c}{a} = 1$, or $a = c$.

40. Show that the sum of the reciprocals of the roots of $ax^2 + bx + c = 0$ is $-\frac{b}{c}$.

Let the roots be r_1 and r_2. Then $\frac{1}{r_1} + \frac{1}{r_2} = \frac{r_1 + r_2}{r_1 r_2} = \frac{-\frac{b}{a}}{\frac{c}{a}} = -\frac{b}{c}$.

Problems

A 1. The sum of two numbers is 20. If one number is x, then the other number is __?__. Their product is $p(x) =$ __?__. Find the maximum value of p. $20 - x$; $-x^2 + 20x$; 100

2. The difference of two numbers is 8. If the smaller number is x, then the other number is __?__. Their product is $p(x) =$ __?__. Find the minimum value of p. $x + 8$; $x^2 + 8x$; -16

3. The sum of two numbers is 40. Find their greatest possible product. 400

4. Find two numbers such that their sum is 20 and the sum of their squares is as small as possible. 10, 10

5. A rectangle has a perimeter of 100 cm. Find the greatest possible area for the rectangle. (*Hint:* Draw a diagram and label the length and width.) 625 cm^2

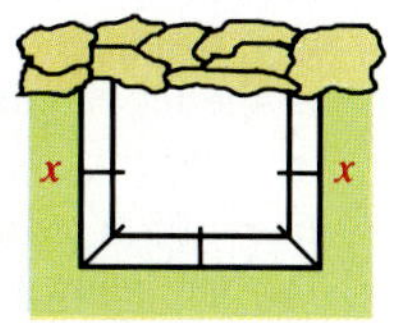

6. A rectangular pen is made with 100 m of fencing on three sides. The fourth side is a stone wall. Find the greatest possible area of such an enclosure. 1250 m^2

7. There are currently 20 members in a school's Ski Club and the dues are \$8 per member. To encourage the recruiting of new members, the club treasurer suggests that for each *new* member recruited, the dues for *all* members be reduced by 10 cents.
   a. How much membership money will the club have if it recruits 10 new members? 15 new members? n new members? \$210; \$227.50; $(20 + n)(8 - 0.10n)$
   b. For what value of n will the club's total membership money be a maximum? 30

8. In Problem 7, suppose that the Ski Club has 24 members instead of 20. How many new members should be recruited in order to maximize the total membership money? What will this total be? 28; \$270.40

9. A charter company will provide a plane for a fare of \$60 each for 20 or fewer passengers. For each passenger in excess of 20, the fare is decreased \$2 per person for everyone. What number of passengers will produce the greatest revenue for the company? 25 passengers

Additional Answers
Mixed Review Exercises

1.

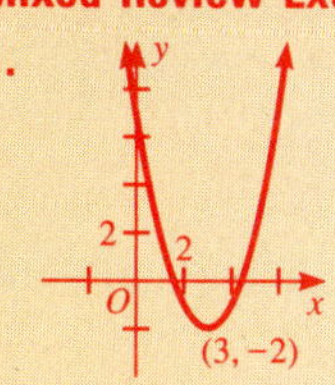

2.

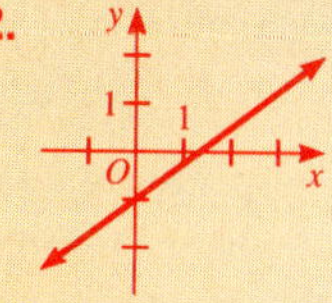

(continued)

Additional Answers
Mixed Review Exercises
(continued)

3.
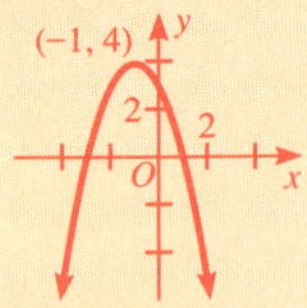

4.
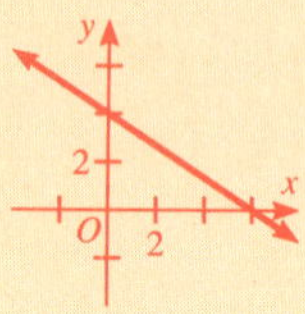

5.
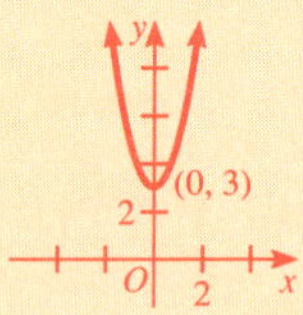

6.
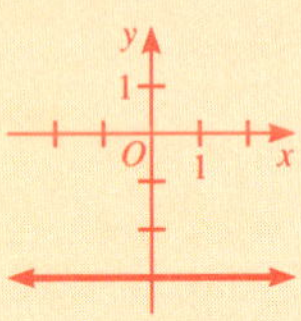

7. $D = \{\text{real numbers}\}$; zeros: ± 3
8. $D = \{\text{real numbers}\}$; zero: $\frac{8}{3}$
9. $D = \{\text{real numbers}\}$; zeros: $-3 \pm \sqrt{11}$
10. $D = \{\text{real numbers}\}$ no zeros
11. $D = \{\text{real numbers}\}$; no real zeros
12. $D = \left\{x: x \neq \frac{5}{2}\right\}$; zero: $-\frac{1}{4}$

B **10.** A ferry service transporting passengers to an island charges a fare of \$10 and carries 300 persons per day. The manager estimates that the company will lose 15 passengers for each increase of \$1 in the fare. Find the fare that yields the greatest income. \$15

11. In a 120 volt electrical circuit having a resistance of 16 ohms, the available power P in watts is a function of I, the amount of current flowing in amperes. If $P = 120I - 16I^2$, how many amperes will produce the maximum power in the circuit? What will this maximum power be? 3.75 amps; 225 watts

12. A ball is thrown vertically upward with an initial speed of 80 ft/s. Its height after t seconds is given by $h = 80t - 16t^2$.

a. How high does the ball go? 100 ft

b. When does the ball hit the ground? 5 s after being thrown

13. A rectangular field is to be enclosed by a fence and divided into two parts by another fence. Find the maximum area that can be enclosed and separated in this way with 800 m of fencing. $26{,}666\frac{2}{3}$ m²

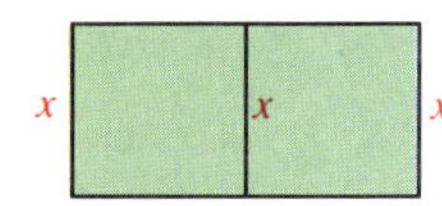

14. A parking lot is to be formed by fencing in a rectangular plot of land except for an entrance 12 m wide along one of the sides. Find the dimensions of the lot of greatest area if 300 m of fencing is to be used. 78 m by 78 m

15. A football kicker tries to make a field goal with the ball on the 25 yard line (105 ft from the goal posts). If you imagine a coordinate system with the ball being kicked at $x = 0$ and the goal post at $x = 105$, then the equation of the ball's path is $y = \frac{4}{3}x - \frac{x^2}{90}$, where y is the ball's height in feet.

a. What is the maximum height of the ball? 40 ft

b. Will the ball clear the goal-post crossbar that is 10 ft above the ground? Yes

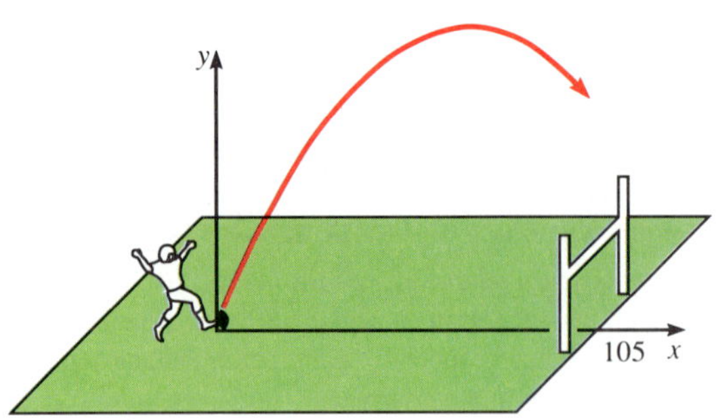

C **16.** A rectangle is to be inscribed in an isosceles triangle of height 8 and base 10, as shown. Find the greatest area of such a rectangle. (*Hint:* $\triangle APQ$ is similar to $\triangle ABC$. Use this fact to express y in terms of x.) 20

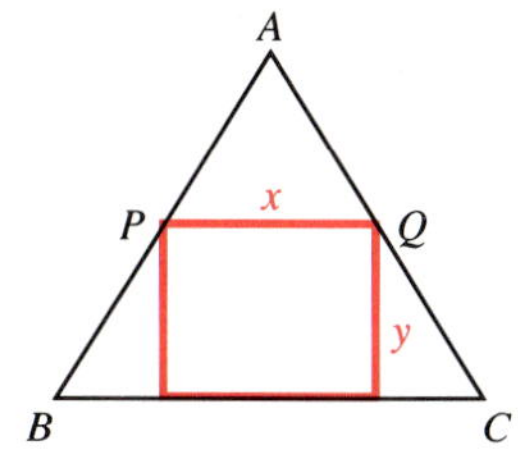

Mixed Review Exercises

Sketch the graph of each equation.

1. $y + 2 = (x - 3)^2$
2. $y = \frac{3}{4}x - 1$
3. $y = 3 - 2x - x^2$
4. $2x + 3y = 12$
5. $y = 2x^2 + 3$
6. $y = -3$

Find the domain and zeros of each function. If there are no zeros, say so.

7. $f(x) = 9 - x^2$
8. $g(x) = 3x - 8$
9. $h(x) = x^2 + 6x - 2$
10. $F(x) = 2$
11. $G(x) = 2x^2 - 3x + 4$
12. $H(x) = \frac{4x + 1}{2x - 5}$

Self-Test 3

Vocabulary parabola (p. 326) vertex (p. 326) axis of symmetry (p. 326) quadratic function (p. 333)

1. Find the vertex and axis of the parabola $y - 4 = -2(x + 3)^2$. $(-3, 4)$; $x = -3$ **Obj. 7-5, p. 326**
2. Find all intercepts of the parabola $y - 2 = -\frac{1}{2}(x - 4)^2$. $(0, -6)$, $(6, 0)$, $(2, 0)$
3. Find an equation $y - k = a(x - h)^2$ of the parabola having vertex $(-2, 1)$ and containing the point $(-3, 4)$. $y - 1 = 3(x + 2)^2$
4. Graph the function $f(x) = 2x^2 + 12x + 17$. **Obj. 7-6, p. 333**
5. Find the domain, range, and zeros of the function $g(x) = x^2 - 6x + 3$. D = {real numbers}; $R = \{y: y \geq\ \ 6\}$; zeros: $3 \pm \sqrt{6}$
6. Find a quadratic equation with integral coefficients having roots $\frac{1 + \sqrt{3}}{2}$ and $\frac{1 - \sqrt{3}}{2}$. Answers may vary. $2x^2 - 2x - 1 = 0$ **Obj. 7-7, p. 338**
7. A ball is thrown vertically upward with an initial speed of 48 ft/s. Its height, in feet, after t seconds is given by $h = 48t - 16t^2$. How high does the ball go? 36 ft

Check your answers with those at the back of the book.

Quick Quiz

1. Find the vertex, axis of symmetry, and y-intercept of $y - 3 = -\frac{1}{2}(x + 2)^2$.
   vertex: $(-2, 3)$
   axis: $x = -2$
   y-int.: $(0, 1)$
2. Find an equation of the form $y - k = a(x - h)^2$ for the parabola with vertex $(1, -2)$ and passing through $(2, 1)$.
   $y + 2 = 3(x - 1)^2$
3. Let $f(x) = 10 + 6x - 2x^2$. Determine whether f has a minimum or maximum value and find this value.
   Maximum; $\frac{29}{2}$

Find a quadratic equation with integral coefficients having the given roots.

4. $\frac{5}{3}, -\frac{1}{3}$
   $9x^2 - 12x - 5 = 0$
5. $2 - i\sqrt{5}, 2 + i\sqrt{5}$
   $x^2 - 4x + 9 = 0$
6. A plane with 125 seats may be chartered for \$250 per person plus a fee of \$6 *per person* for each empty seat. What number of passengers will provide the maximum income? 83

Additional Answers
Self-Test 3

4. $y = 2x^2 + 12x + 17$;
   $y + 1 = 2(x + 3)^2$

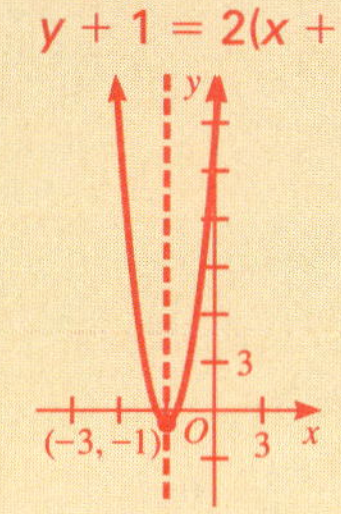

Additional Answers Written Exercises

(continued from p. 332)

19. $y + 3 = \frac{1}{2}(x - 4)^2$

20. $y - 5 = -2(x - 4)^2$

21. $y = \frac{1}{3}x^2$

22. $y - 6 = -\frac{2}{3}(x + 3)^2$

23. $y - 5 = -\frac{1}{3}(x - 3)^2$

24. $y - 6 = -2(x + 2)^2$

25. $y - 2 = -2(x - 4)^2$

26. $y - 4 = -1(x + 3)^2$

34.

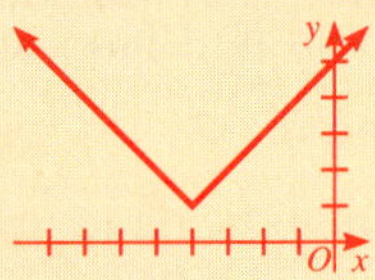

35. Let $x = h + r$ and $y = s$; since (x,y) is on the parabola, $s - k = a(h + r - h)^2$; $s - k = ar^2$; $s - k = a(-r)^2$; $s - k = a(h - r - h)^2$; since $(h - r, s)$ satisfies the equation of the parabola, it is also on the graph.

36. a.

x	9	4	1	0
y	± 3	± 2	± 1	0

b.

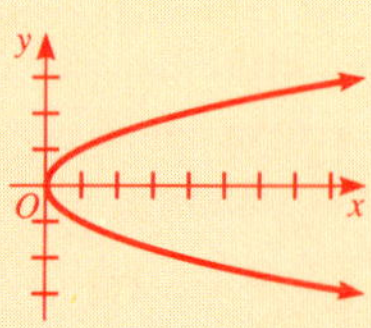

37. a.

x	-9	-4	-1	0
y	± 3	± 2	± 1	0

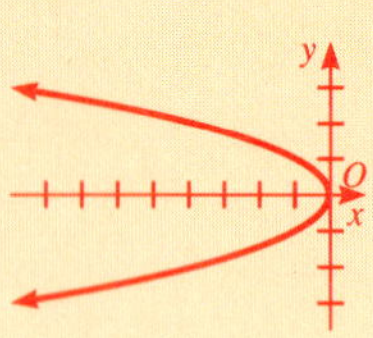

Chapter Summary

1. The general quadratic equation, $ax^2 + bx + c = 0$ $(a \neq 0)$, can be solved either by *completing the square* or by using the *quadratic formula:*

$$x = \frac{-b \pm \sqrt{b^2 - 4ac}}{2a}$$

2. The *discriminant,* $b^2 - 4ac$, can be used to determine the nature and number of the roots of a quadratic equation: real or imaginary; rational or irrational; one root or two roots.
3. An equation in *quadratic form,* $a[f(x)]^2 + b[f(x)] + c = 0$, can usually be solved by first replacing the expression for $f(x)$ by a single variable such as z and solving the resulting equation for z. Then to find x, substitute for z the expression for $f(x)$.
4. A parabola with vertex (h, k) and a vertical axis of symmetry has an equation of the form

$$y - k = a(x - h)^2.$$

If $a > 0$, the parabola opens upward; if $a < 0$, the parabola opens downward. The axis has the equation $x = h$.

5. The graph of the *quadratic function* $f(x) = ax^2 + bx + c$ (or $f(x) = a(x - h)^2 + k$) is a parabola. The *minimum* or *maximum* value of a quadratic function is the y-coordinate of its vertex. This value is a minimum if $a > 0$, a maximum if $a < 0$.
6. The sum of the roots of a quadratic equation is $-\frac{b}{a}$; the product of the roots is $\frac{c}{a}$.

Chapter Review

Write the letter of the correct answer.

1. Solve $2(x - 1)^2 = -16$. 7-1

a. $\{1 + 2i, 1 - 2i\}$ **b.** $\{1 + 2\sqrt{2}, 1 - 2\sqrt{2}\}$
(**c.**) $\{1 + 2i\sqrt{2}, 1 - 2i\sqrt{2}\}$ **d.** $\{1 - i\sqrt{2}, 1 + i\sqrt{2}\}$

2. If $3y^2 - 18y + d$ is a trinomial square, find d.

a. 81 (**b.**) 27 **c.** 6 **d.** 108

3. Find the roots of $2x^2 = 3x - 1$. 7-2

(**a.**) $\frac{1}{2}, 1$ **b.** $\frac{-3 \pm \sqrt{41}}{4}$ **c.** $-\frac{1}{2}, -1$ **d.** $\frac{3 \pm \sqrt{41}}{4}$

4. A vegetable garden measures 8 ft by 12 ft. By what equal amount must each dimension be increased if the area is to be doubled?

a. 2 ft (**b.**) 4 ft **c.** 6 ft **d.** 8 ft

5. Describe the roots of $9y^2 - 6y + 5 = 0$. 7-3
 (a.) imaginary conjugates b. real, equal, rational
 c. real, unequal, rational d. real, unequal, irrational

6. Find the real values of k for which the equation $3y^2 - 2y + k = 0$ has two real roots.
 a. $k > \frac{1}{3}$ b. $k > -\frac{1}{3}$ c. $k = \frac{1}{3}$ (d.) $k < \frac{1}{3}$

7. Solve $(2x + 1)^2 - 5(2x + 1) + 6 = 0$. 7-4
 (a.) $\left\{1, \frac{1}{2}\right\}$ b. $\left\{-1, \frac{5}{2}\right\}$ c. $\left\{-2, -\frac{3}{2}\right\}$ d. $\left\{0, -\frac{7}{2}\right\}$

8. Find the roots of $y^{-2} - 2y^{-1} + 2 = 0$.
 (a.) $\frac{1 \pm i}{2}$ b. $1 \pm i$ c. $-1 \pm i$ d. $\frac{-1 \pm i}{2}$

9. Find the vertex of the parabola $y + 3 = -(x - 2)^2$. 7-5
 a. $(-3, 2)$ b. $(2, 3)$ (c.) $(2, -3)$ d. $(3, -2)$

10. Find an equation of the parabola having vertex $(-1, 4)$ and containing the point $(1, 2)$.
 a. $y + 4 = \frac{1}{2}(x - 1)^2$ b. $y - 4 = \frac{1}{2}(x + 1)^2$
 c. $y + 4 = -\frac{1}{2}(x - 1)^2$ (d.) $y - 4 = -\frac{1}{2}(x + 1)^2$

11. Find the minimum value of $f(x) = 2x^2 - 6x + 5$. 7-6
 a. $\frac{3}{2}$ (b.) $\frac{1}{2}$ c. $-\frac{3}{2}$ d. $\frac{37}{2}$

12. Which curve is the graph of $g(x) = 4x - x^2$?

a.
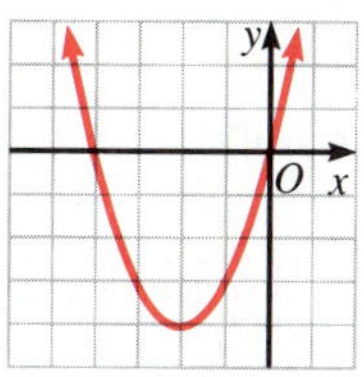
b.
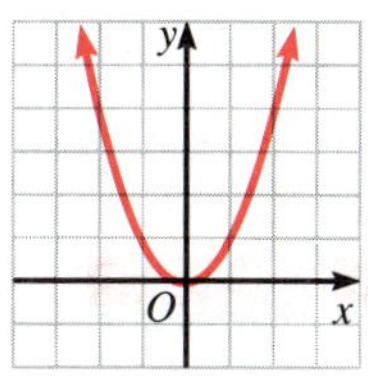
(c.)
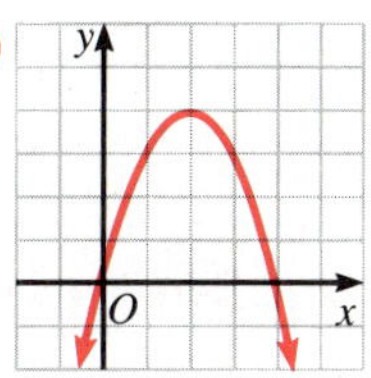
d.
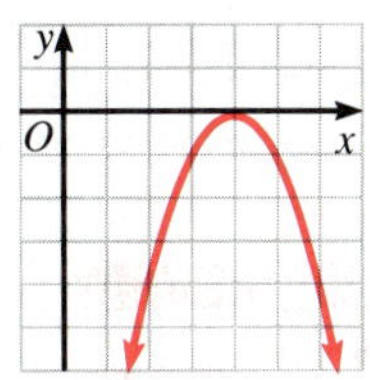

13. Find the range of the function $f(x) = 1 - 6x - 3x^2$.
 (a.) $\{y: y \leq 4\}$ b. $\{y: y \geq 4\}$ c. $\{y: y \leq -8\}$ d. $\{y: y \geq -8\}$

14. Find a quadratic equation with integral coefficients having the roots $\frac{1 + i\sqrt{2}}{2}$ and $\frac{1 \quad i\sqrt{2}}{2}$. 7-7
 a. $4x^2 + 4x + 3 = 0$ b. $4x^2 - 4x - 3 = 0$
 (c.) $4x^2 - 4x + 3 = 0$ d. $4x^2 + 4x - 3 = 0$

15. Find a quadratic function for the parabola having minimum value -2 and x-intercepts 1 and -3.
 a. $f(x) = 2x^2 + 4x - 6$ b. $f(x) = 6 - 4x - 2x^2$
 c. $f(x) = \frac{3}{2} - x - \frac{1}{2}x^2$ (d.) $f(x) = \frac{1}{2}x^2 + x - \frac{3}{2}$

b.

x	11	6	3	2
y	±3	±2	±1	0

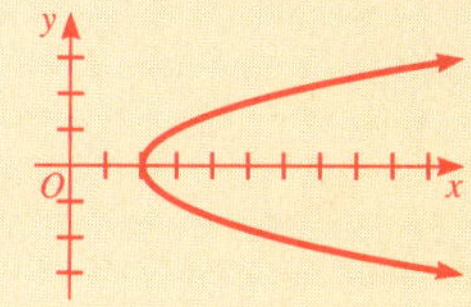

c.

x	9	4	1	0	1	4	9
y	−4	−3	−2	−1	0	1	2

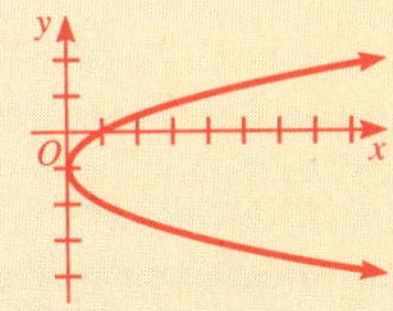

d.

x	11	6	3	2	3	6	11
y	−4	−3	−2	−1	0	1	2

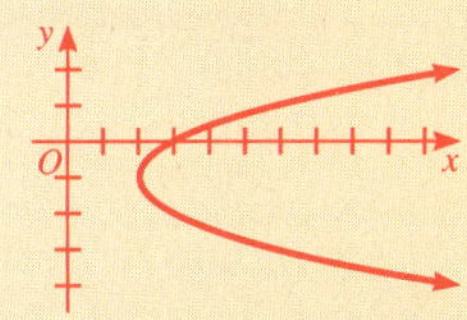

e.

x	$-\frac{3}{2}$	1	$\frac{5}{2}$	3	$\frac{5}{2}$	1	$-\frac{3}{2}$
y	−1	0	1	2	3	4	5

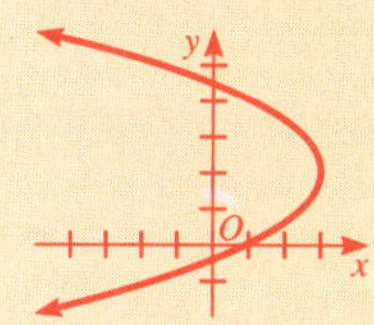

38. a. $(0, 0)$; $y = 0$
b. $(2, 0)$; $y = 0$
c. $(0, -1)$; $y = -1$
d. $(2, -1)$; $y = -1$
e. $(3, 2)$; $y = 2$

Supplementary Materials

Test Masters	29, 30
Resource Book	p. 130

Additional Answers
Chapter Test

8. 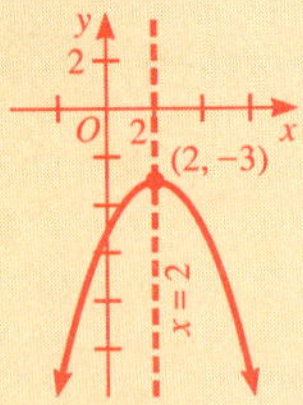

10. 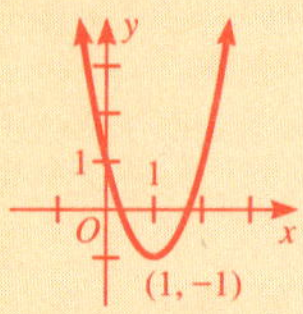

Alternate Test p. T19

Cumulative Review p. T30

Additional Answers
Cumulative Review

35. 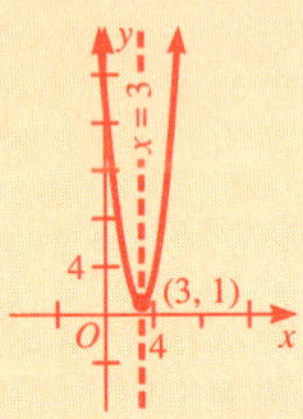

Additional Answers
Written Exercises
(continued from p. 337)

14. 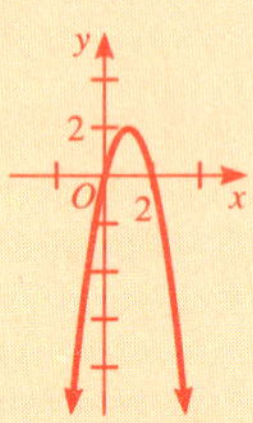

16. 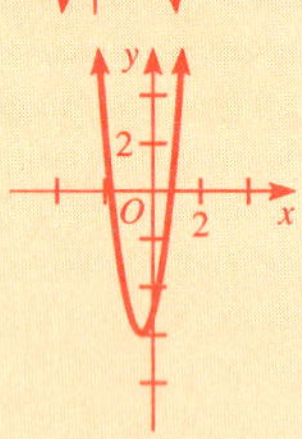

Chapter Test

1. Solve $(3x + 1)^2 = 8$. $\left\{\frac{-1 \pm 2\sqrt{2}}{3}\right\}$ **7-1**
2. Solve by completing the square: $2x^2 + 6x + 3 = 0$. $\left\{\frac{-3 \pm \sqrt{3}}{2}\right\}$
3. Use the quadratic formula to solve $4x^2 - 3x + 2 = 0$. $\left\{\frac{3}{8} \pm \frac{\sqrt{23}}{8}i\right\}$ **7-2**
4. Two positive real numbers have a sum of 7 and a product of 11. Find the numbers. $\frac{7 + \sqrt{5}}{2}$ and $\frac{7 - \sqrt{5}}{2}$
5. Without solving the equation, determine the nature of its roots. **7-3**
   a. $5x^2 + 7x + 2 = 0$ unequal, rational **b.** $3x^2 - 4x + 2 = 0$ imaginary conjugates
6. Find all real values of k for which $2x^2 + kx + 3 = 0$ has a double root. $k = \pm 2\sqrt{6}$
7. Solve each equation over the complex numbers. **7-4**
   a. $x^4 + x^2 - 12 = 0$ **b.** $x^{-2} - 2x^{-1} - 1 = 0$
   a. $\{\pm\sqrt{3}, \pm 2i\}$ **b.** $\{-1 \pm \sqrt{2}\}$
8. Graph the parabola $y + 3 = -\frac{1}{2}(x - 2)^2$. Label the vertex and axis of symmetry. **7-5**
9. Find an equation in the form $y - k = a(x - h)^2$ for the parabola having vertex $(-2, 5)$ and containing the point $(2, 9)$. $y - 5 = \frac{1}{4}(x + 2)^2$
10. Graph the function $f(x) = 2x^2 - 4x + 1$ after rewriting its equation in the form $y - k = a(x - h)^2$. **7-6**
    11. $D = \{\text{real numbers}\}$, $R = \{y: y \geq -5\}$, zeros are $3 \pm \sqrt{5}$
11. Find the domain, range, and zeros of $g(x) = x^2 - 6x + 4$.
12. Find a quadratic equation with *integral* coefficients having roots $\frac{1 + \sqrt{3}}{4}$ and $\frac{1 - \sqrt{3}}{4}$. $8x^2 - 4x - 1 = 0$ **7-7**
13. Find a quadratic function $f(x) = ax^2 + bx + c$ having minimum value -9 and zeros $\frac{1}{2}$ and $-\frac{5}{2}$. $f(x) = 4x^2 + 8x - 5$
14. Find the dimensions of the rectangle of greatest area whose perimeter is 20 cm. 5 cm × 5 cm

Cumulative Review *(Chapters 4–7)*

Simplify.

1. $2(3x^2 - 4x + 1) - (5x^2 + x - 6)$ $x^2 - 9x + 8$
2. $(4a^2b)(-3a^3b^2)$ $-12a^5b^3$
3. $(-x^2yz^3)^4$ $x^8y^4z^{12}$
4. $(4y - 3)(2y + 5)$ $8y^2 + 14y - 15$

Factor completely.

5. $2x^2 - 11x + 12$ $(2x - 3)(x - 4)$
6. $3x^2 - 2x + 6xy - 4y$ $(x + 2y)(3x - 2)$
7. $16a^2 - 9b^2$ $(4a - 3b)(4a + 3b)$
8. $3y^4 - 6y^3 - 9y^2$ $3y^2(y - 3)(y + 1)$

Solve.

9. $(x - 4)(x + 1) = 0$ $\{-1, 4\}$

10. $x^2 - 10x + 25 = 0$ $\{5\}$

11. $x^3 = 9x$ $\{-3, 0, 3\}$

12. $y^2 + y < 20$ $\{y: -5 < y < 4\}$

13. Find two consecutive odd integers whose product is 99. 9 and 11

14. One leg of a right triangle is 8 cm less than the hypotenuse, while the other leg is only 1 cm less than the hypotenuse. Find the length of the hypotenuse. 13 cm

Simplify.

15. $\dfrac{12x^3y}{(4xy^2)^2}$ $\dfrac{3x}{4y^3}$

16. $\left(\dfrac{x^3}{y^2}\right)^{-1} \cdot \left(\dfrac{x^2}{y}\right)^{-2}$ $\dfrac{y^4}{x^7}$

17. $\dfrac{a^2 - a}{a^2 + a - 2}$ $\dfrac{a}{a + 2}$

18. $\dfrac{18x^2}{x^2 - 2x - 8} \cdot \dfrac{x^2 - x - 6}{12x} \div \dfrac{3x - 9}{x - 4}$ $\dfrac{x}{2}$

19. $\dfrac{1}{2y + 4} + \dfrac{1}{y^2 + 2y}$ $\dfrac{1}{2y}$

20. $\dfrac{1 - m^{-2}}{m^{-1} - 1}$ $-\dfrac{m + 1}{m}$

Solve.

21. $\dfrac{x}{x + 2} - \dfrac{1}{x + 1} = \dfrac{2}{x^2 + 3x + 2}$ $\{2\}$

22. $y + \sqrt{y - 1} = 7$ $\{5\}$

23. Mr. Ramirez invested \$4000, part at 5% simple annual interest and the rest at 8%. If his total interest income for a year is \$284, how much did he invest at each rate? \$1200 at 5%, \$2800 at 8%

Simplify.

24. $\sqrt{80}$ $4\sqrt{5}$

25. $\sqrt{24} - \sqrt{\dfrac{3}{2}}$ $\dfrac{3\sqrt{6}}{2}$

26. $(7 - \sqrt{5})(3 + 4\sqrt{5})$ $1 + 25\sqrt{5}$

27. $\dfrac{2}{\sqrt{7} - \sqrt{3}}$ $\dfrac{\sqrt{7} + \sqrt{3}}{2}$

28. $\sqrt{-8} \cdot \sqrt{-6}$ $-4\sqrt{3}$

29. $\dfrac{3 - 2i}{3 + 2i}$ $\dfrac{5}{13} - \dfrac{12}{13}i$

30. Write $0.1\overline{54}$ as a common fraction in lowest terms. $\dfrac{17}{110}$

31. Solve by completing the square: $4x^2 - 12x + 7 = 0$. $\left\{\dfrac{3 \pm \sqrt{2}}{2}\right\}$

32. Solve by using the quadratic formula: $9y^2 + 12y + 5 = 0$. $\left\{-\dfrac{2}{3} \pm \dfrac{1}{3}i\right\}$

33. Find the real values of k for which $kx^2 - 6x + 3 = 0$ has two imaginary roots. $k > 3$

34. Solve: $(4x - 3)^2 - 6(4x - 3) + 5 = 0$. $\{1, 2\}$

35. Graph $y - 1 = 2(x - 3)^2$. Label the vertex and axis of symmetry.

36. Find the domain, range, and zeros of $f(x) = x^2 - 8x + 9$. (See below.)

37. If two numbers differ by 6, then what is the least possible value of their product? -9

36. $D = \{\text{real numbers}\}$, $R = \{y: y \geq -7\}$, zeros are $4 \pm \sqrt{7}$

18. 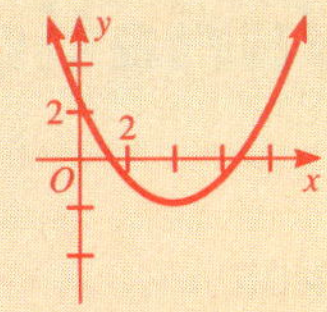

25. a. $(2, -7)$ b. {real numbers} c. $\{y: y \geq -7\}$ d. $2 \pm \sqrt{7}$

26. a. $(1, 3)$ b. {real numbers} c. $\{y: y \leq 3\}$ d. 0, 2

27. a. $(-1, 9)$ b. {real numbers} c. $\{y: y \leq 9\}$ d. $-4, 2$

28. a. $(-4, 25)$ b. {real numbers} c. $\{y: y \leq 25\}$ d. $-9, 1$

29. a. $(1, -6)$ b. {real numbers} c. $\{y: y \geq -6\}$ d. $1 \pm \sqrt{6}$

30. a. $(-2, -3)$ b. {real numbers} c. $\{y: y \geq -3\}$ d. $\dfrac{-4 \pm \sqrt{6}}{2}$

31. a. $(1, -72)$ b. {real numbers} c. $\{y: y \geq -72\}$ d. $-5, 7$

32. a. $\left(-5, -\dfrac{1}{2}\right)$ b. {real numbers} c. $\left\{y: y \geq -\dfrac{1}{2}\right\}$ d. $-6, -4$

33. a. $(-6, 9)$ b. {real numbers} c. $\{y: y \leq 9\}$ d. $-9, -3$

34. a. $(5, -8)$ b. {real numbers} c. $\{y: y \geq -8\}$ d. 3, 7

35. a. $(-1, 2)$ b. {real numbers} c. $\{y: y \leq 2\}$ d. $-3, 1$

36. a. $\left(1, -\dfrac{1}{2}\right)$ b. {real numbers} c. $\left\{y: y \geq -\dfrac{1}{2}\right\}$ d. $\dfrac{2}{3}, \dfrac{4}{3}$

8 Variation and Polynomial Equations

Objectives

8-1 To solve problems involving direct variation.

8-2 To solve problems involving inverse and joint variation.

8-3 To divide one polynomial by another polynomial.

8-4 To use synthetic division to divide a polynomial by a first-degree binomial.

8-5 To use the remainder and factor theorems to find factors of polynomials and to solve polynomial equations.

8-6 To find or solve a polynomial equation with real coefficients and positive degree n by using these facts:

1. There are exactly n roots.
2. The imaginary roots occur in conjugate pairs.
3. Descartes' rule of signs gives information about the numbers of positive and negative real roots.

8-7 To find rational roots of polynomial equations with integral coefficients.

8-8 To approximate the real roots of a polynomial equation $P(X) = 0$ by using the graph of $y = P(X)$.

8-9 To use linear interpolation to find values not listed in a given table of data.

Assignment Guide

See p. T58 for Key to the format of the Assignment Guide

Day	Average Algebra	Average Algebra and Trigonometry	Extended Algebra	Extended Algebra and Trigonometry
1	**8-1** 354/1–19 odd	**8-1** 354/1–19 odd	**8-1** 354/1–23 odd	**8-1** 354/1–23 odd
2	**8-1** 356/*P*: 1–19 odd **S** 357/*Mixed Review*	**8-1** 356/*P*: 1–19 odd **S** 357/*Mixed Review*	**8-1** 356/*P*: 1, 6, 10–17 **S** 357/*Mixed Review*	**8-1** 356/*P*: 1, 6, 10–17 **S** 357/*Mixed Review*
3	**8-2** 360/1–10; *P*: 1, 5, 6, 10 **S** 355/14–20 even	**8-2** 360/1–10; *P*: 1, 5, 6, 10 **S** 355/14–20 even	**8-2** 360/1–10; *P*: 1, 5, 10–13 **S** 355/14–24 even	**8-2** 360/1–10; *P*: 1, 5, 10–13 **S** 355/14–24 even
4	**8-3** 366/2–30 even **S** 367/*Mixed Review* **R** 363/*Self-Test 1*	**8-3** 366/2–30 even **S** 367/*Mixed Review* **R** 363/*Self-Test 1*	**8-3** 366/2–32 even **S** 367/*Mixed Review* **R** 363/*Self-Test 1*	**8-3** 366/2–32 even **S** 367/*Mixed Review* **R** 363/*Self-Test 1*
5	**8-4** 370/1–21 odd **S** 357/*P*: 14, 16, 18	**8-4** 370/1–21 odd **S** 357/*P*: 14, 16, 18	**8-4** 370/1–23 odd **S** 356/*P*: 9, 19, 20	**8-4** 370/1–23 odd **S** 356/*P*: 9, 19, 20
6	**8-5** 375/2–36 even **S** 376/*Mixed Review*	**8-5** 375/2–36 even **S** 376/*Mixed Review*	**8-5** 375/2–38 even **S** 376/*Mixed Review*	**8-5** 375/2–38 even **S** 376/*Mixed Review*
7	**8-6** 380/1–23 odd, 24, 25 **S** 367/29, 31	**8-6** 380/1–23 odd, 24, 25 **S** 367/29, 31	**8-6** 380/1–23 odd, 24–27 **S** 367/29, 31, 33	**8-6** 380/1–23 odd, 24–27 **S** 367/29, 31, 33
8	**8-7** 384/2–20 even, 21–25 odd **S** 385/*Mixed Review* **R** 381/*Self-Test 2*	**8-7** 384/2–20 even, 21–25 odd **S** 385/*Mixed Review* **R** 381/*Self-Test 2*	**8-7** 384/2–20 even, 21–26 **S** 385/*Mixed Review* **R** 381/*Self-Test 2*	**8-7** 384/2–20 even, 21–26 **S** 385/*Mixed Review* **R** 381/*Self-Test 2*
9	**8-8** 388/1–23 odd **S** 380/26	**8-8** 388/1–23 odd **S** 380/26	**8-8** 388/1–25 odd **S** 385/27	**8-8** 388/1–25 odd **S** 385/27

Assignment Guide *(continued)*

Day	Average Algebra	Average Algebra and Trigonometry	Extended Algebra	Extended Algebra and Trigonometry
10	**8-9** 394/1–25 odd **R** 396/*Self-Test 3*	**8-9** 394/1–25 odd **R** 396/*Self-Test 3*	**8-9** 394/1–29 odd **R** 396/*Self-Test 3*	**8-9** 394/1–29 odd **R** 396/*Self-Test 3*
11	*Prepare for Chapter Test* **R** 397/*Chapter Review*	*Prepare for Chapter Test* **R** 397/*Chapter Review*	*Prepare for Chapter Test* **R** 397/*Chapter Review*	*Prepare for Chapter Test* **R** 397/*Chapter Review*
12	*Administer Chapter 8 Test*	*Administer Chapter 8 Test*	*Administer Chapter 8 Test*	*Administer Chapter 8 Test* **9-1** *Read 9-1* 404/1–31 odd

Supplementary Materials Guide

For Use with Lesson	Practice Masters	Tests	Study Guide (Reteaching)	Resource Book		
				Tests	Practice Exercises	Prob. Solving (PS) Applications (A) Enrichment (E) Technology (T)
8-1			pp. 119–120			pp. 202–203 (PS) p. 211 (A)
8-2	Sheet 44		pp. 121–122		p. 131	
8-3	Sheet 45		pp. 123–124			
8-4			pp. 125–126			
8-5	Sheet 46	Test 31	pp. 127–128		p. 132	
8-6	Sheet 47		pp. 129–130			
8-7			pp. 131–132		p. 133	
8-8			pp. 133–134			p. 242 (T)
8-9	Sheet 48	Test 32	pp. 135–136		p. 134	
Chapter 8		Tests 33, 34		pp. 46–49	p. 135	p. 227 (E)
Cum. Rev. 5–8	Sheet 49					
Cum. Rev. 1–8					pp. 136–137	

Software

Software	Algebra Plotter Plus	Using Algebra Plotter Plus	Computer Activities	Test Generator
	Statistics Spreadsheet Function Plotter	Enrichment, pp. 38–39 Activity Sheet, p. 54	Activities 17–19, 37	189 test items
For Use with Lessons	8-1, 8-8	8-1, 8-8	8-2, 8-7, 8-8, Chapter 8	all lessons

Strategies for Teaching

Cooperative Learning and Communication

Cooperative learning is most easily accomplished in small groups in which students see themselves as members of a team working together to achieve a specific goal. This does not mean that all the members of a group must be of equal ability or that maximum achievement is demanded of each member. It means that each member contributes according to his or her ability, that everyone's contribution is respected, and that the team does achieve its goal.

Sometimes students can understand a concept explained by another student more easily than they can follow the presentation to the class as a whole. The process of explaining and teaching is in itself a learning process. In a small group, students may find it easier to ask questions and advance ideas. Be sure to choose material that lends itself well to group work—discovery and exploration of a concept, solution of a challenging problem, making a model, for example. You may wish to move among the groups as a counselor and resource person.

See the Exploration on page 839 for a motivating activity in which students use springs and weights to explore direct variation. This is an ideal cooperative learning activity for groups of students to complete even before beginning the chapter.

Strong language skills will also help students to succeed when working with concepts in this chapter. Algebra not only introduces students to its own specialized vocabulary, it also introduces them to new meanings for familiar words like *power* and *variable*.

8-1 Direct Variation and Proportion

If students have difficulty in setting up proportions for the problems in this lesson, the difficulty may be in their communication skills. Suggest that they think of the relationships in terms of the word statement "y_1 is to x_1 as y_2 is to x_2" (page 352). Thus one can write for Problem 3 on page 356 "commission is to sales as commission is to sales" or "5400 is to 120,000 as x is to 145,000."

This lesson is also suitable for a cooperative learning activity. Ask the students, in groups of four or five, to draw a "map" of their school. It may be of the room, building, campus, or an appropriate area surrounding the school. The "map" should be drawn using an appropriate scale so that other groups can determine measurements of items on the map using direct variation. This could be a long term project including in-class and out of class activities.

As an extension of this activity, have the groups construct scale cut-outs of objects on their "map" and use them to suggest rearrangements, new construction, and other changes in the area.

8-2 Inverse and Joint Variation

Point out the necessity of reading both the direction lines and the word problems carefully, so as to identify the type or types of variation involved in each case. Remind students to look for key words such as *is inversely proportional, varies directly . . . and inversely,* and *varies jointly*.

8-4 Synthetic Division

Many students find it difficult to adjust to synthetic division because they are used to long division. You might spend some time on the discussion above Example 1 to help students see the advantages of using synthetic division. You might ask students to walk through Oral Exercises 1–6 and explain in words what is done at each step.

Be sure that students realize that the divisor is of the form $(x - c)$. The coefficient of x must be 1 and c is what follows the minus sign.

8-6 Some Useful Theorems

Ask students to explain why Descartes' rule of signs works only for polynomials with real coefficients.

8-7 Finding Rational Roots

Understanding the vocabulary of the rational root theorem is a prerequisite to understanding and using the theorem. You might ask students to read the theorem silently and then to pick out each mathematical word and phrase and explain it in their own words, or illustrate it by an example.

8-9 Linear Interpolation

Ask students to discuss the assumptions that underlie the principle of linear interpolation. You might also discuss what would happen if one tried to devise a linear extrapolation method.

References to Strategies

PE: Pupil's Edition **TE:** Teacher's Edition **RB:** Resource Book

Problem Solving Strategies

PE: p. 371 (estimate answers, check solutions, 5-step plan)
TE: p. 387
RB: pp. 202–203

Applications

PE: pp. 350–351, 353, 361–363 (physics); pp. 356–359, 391–394 (word problems)
TE: pp. 350, 352, 359–360, 363, 394
RB: p. 211

Nonroutine Problems

PE: p. 355 (Exs. 21–25); p. 357 (Prob. 20); p. 362 (Probs. 13, 14); p. 363 (Challenge); p. 367 (Exs. 30–34); p. 376 (Exs. 38, 39); p. 380 (Exs. 23–29); pp. 384–385 (Exs. 21–26); p. 389 (Exs. 27–30)
TE: pp. T110–T111 (Sugg. Extensions, Lessons 8-3, 8-4, 8-6, 8-7)

Communication

PE: pp. 367 (Ex. 34, convincing argument); p. 379 (Exs. 7–9, convincing argument); p. 389 (Ex. 28, convincing argument); p. 371 (Reading Algebra)
TE: pp. T109, T111 (Reading Algebra); pp. T111–T112 (Sugg. Extensions, Lessons 8-6, 8-7, 8-8, 8-9)

Thinking Skills

PE: p. 355 (Exs. 13–22); p. 385 (Ex. 28, proofs)
TE: p. 387

Explorations

PE: p. 363 (Challenge); p. 839 (direct variation)

Connections

PE: p. 368 (Arithmetic); p. 350 (Astronomy); pp. 357, 362 (Geometry); p. 385 (History); pp. 351–354, 356–357, 359–362 (Physics); pp. 391–395 (Statistics)

Using Technology

PE: pp. 386–388, 839; pp. 389–390 (Computer Key-In); p. 395 (Exs.)
TE: pp. 350, 386, 389, 395
RB: p. 242
Computer Activities: pp. 44–51, 100–102
Using Algebra Plotter Plus: pp. 38–39, 54

Using Manipulatives/Models

TE: p. T109 (Lesson 8-1)

Cooperative Learning

TE: p. T109 (Lesson 8-1)

Teaching Resources

For use in implementing the teaching strategies referenced on the previous page.

Application
Resource Book, p. 211

Application—Kepler's Laws of Planetary Motion (For use with Chapter 8)

Johannes Kepler (1571–1630) was an astronomer and mathematician who, after many years of analyzing astronomical observations and doing involved calculations, discovered three basic laws of planetary motion.

1. Kepler's first law of planetary motion states that each planet moves in an elliptical orbit with the sun at one focus of the ellipse, as shown at the right. A planet is therefore closest to the sun at one point, called the *perihelion*, and farthest from the sun at another point, called the *aphelion*. Locate the perihelion in the given drawing and label it P; then locate the aphelion and label it A.

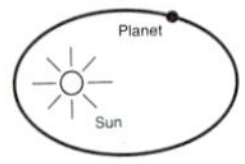

2. Kepler's second law of planetary motion states that an imaginary line connecting the sun and a planet sweeps out equals areas within the planet's elliptical orbit in equal periods of time, as shown at the right. Based on this law, what conclusion can you draw about the planet's orbital speed near its perihelion as compared with its orbital speed near its aphelion?

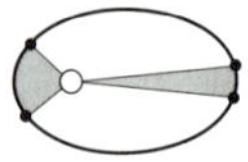

____

3. Kepler's third law of planetary motion states that the square of a planet's period, or orbit time, P is directly proportional to the cube of its mean distance d from the sun. That is, $P^2 = kd^3$, where k is a constant that depends on the units of measurement.

   a. If the orbit time of Earth is 365.25 days and its mean distance from the sun is 93,000,000 mi, what is the value of k in Kepler's third law? $k =$ ______________

   b. A convenient unit of measure in the solar system is an *astronomical unit* (AU), where 1 AU is the mean distance of Earth from the sun (about 93,000,000 mi). What is the value of k in Kepler's third law if the orbit time of Earth is 1 year and its mean distance from the sun is 1 AU? $k =$ ______________

6. If a planet's period P is measured in Earth years and its mean distance from the sun d is measured in AU, Kepler's third law simply becomes $P^2 = d^3$. Use this law to complete the following table for our solar system, rounding P to the nearest hundredth and d to the nearest thousandth.

	Planet	*Period (in years)*	*Mean distance (in AU)*
a.	Mercury	______	0.387
b.	Venus	______	0.723
c.	Earth	______	______
d.	Mars	1.88	______
e.	Jupiter	11.86	______
f.	Saturn	______	9.538
g.	Uranus	______	19.182
h.	Neptune	164.79	______
i.	Pluto	246.37	______

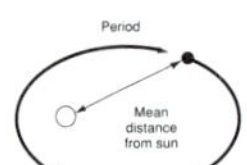

Resource Book, ALGEBRA, Structure and Method, Book 2

APPLICATIONS 211

Enrichment/Reasoning
Resource Book, p. 227

Factoring (For use with Chapter 8)

If $x = 1$ the expression $x^3 + x - 2$ will equal zero. Therefore, using the Factor Theorem, we can say that the expression has $x - 1$ as a factor. You can verify that $x^3 + x - 2 = (x - 1)(x^2 + x + 2)$.

1. Verify that $a - b$, $b - c$, and $c - a$ are factors of the expression $a^2(b - c) + b^2(c - a) + c^2(a - b)$ by letting $a = b$, $b = c$, and $c = a$.

2. Verify that $a - d$ is a factor of the expression:
   $(a + b)^2 + (a + c)^2 - (c + d)^2 - (b + d)^2$
   Can you find another factor of this expression?

3. Verify that $b - a - c$, $c - a - b$, and $a - b - c$ are all factors of the expression:
   $a^2(a - b - c) + b^2(b - a - c) + c^2(c - a - b) + 2abc$

4. Verify that $x + 3$ and $y + 2x + 1$ are factors of the expression:
   $2x^2 + xy + 7x + 3y + 3$

5. Verify that $ax - by$ and $bx + ay$ are factors of the expression:
   $ab(x^2 - y^2) + xy(a^2 - b^2)$

Resource Book, ALGEBRA, Structure and Method, Book 2

ENRICHMENT 227

Problem Solving
Resource Book, p. 202

Direct Variation and Proportion (For use with Lesson 8-1)

By working through the steps in the problems below, you will gain skill in solving problems involving direct variation.

Problem 1 The average gain in the masses of rats fed an experimental diet was directly proportional to the number of weeks that the experiment was run. If the average mass gained after 5 weeks was 71 g, find the average gain in mass after 3 weeks.

a. What ratio must be constant? Give your answer in words. ______________________________

b. Let g = the gain in mass in grams and w = the number of weeks. Write an algebraic equation that includes k, the constant of proportionality.

____

c. Write an algebraic equation in the form of a proportion.

____

d. Use your equation from part (b) or your equation from part (c) to write an equation with exactly one unknown.

____

e. Solve your equation and answer the question.

____

f. Check to see if the two values of the ratio identified in part (a) are equal.

____

(continued)

202 PROBLEM SOLVING
Resource Book, ALGEBRA, Structure and Method, Book 2

Problem Solving
Resource Book, p. 203

Direct Variation and Proportion *(continued)*

Problem 2 The population density in North America is about 42 people per square mile. The population density in Asia is about 162 people per square mile. How many more people would there be in a square region 4 miles on each side in Asia than in an equivalent region in North America? (Assume uniform distribution of people.)

a. What is the area of a square region 4 miles on each side? ______________

b. Write a proportion relating the population density in North America and the unknown population in the square region.

____

c. Solve your proportion. ______________

d. Write a proportion relating the population density in Asia and the unknown population in the square region.

____

e. Solve your proportion. ______________

f. Write the answer to the problem. ______________

Resource Book, ALGEBRA, Structure and Method, Book 2

PROBLEM SOLVING 203

Using Technology/Exploration
Resource Book, p. 242

Using a Computer or a Graphing Calculator

To complete these activities, you should use a computer with graphics software (such as ALGEBRA PLOTTER PLUS) or a graphing calculator.

Approximating Roots of Polynomial Equations **(For use with Lesson 8–8)**

1. Graph this equation: $y = x^3 - 2x + 3$

The x-coordinate of any point where the graph crosses the x-axis is a *root* of the equation $x^3 - 2x + 3 = 0$. As you can see from the graph, the equation has only one real root. To approximate this root, you can look at a table of values. The root must lie between consecutive integral x-values for which the corresponding y-values have opposite sign. Write down these values:

	x	y
a.	____	____
b.	____	____

c. Of the two x-values listed above, which is the better approximation of the root? ______

d. Justify your answer to part (c): ______

2. You can obtain more accurate approximations of the root of the equation in Exercise 1 by magnifying the graph in the vicinity of the root and examining the corresponding table of values. (For a discussion of rescaling and zooming, see Resource Book page 238.) Follow this procedure to approximate the root to the nearest hundredth: ______

3. Graph this equation: $y = x^4 + 3x^2 - x - 4$

a. How many real roots does the equation $x^4 + 3x^2 - x - 4 = 0$ have? ______

b. Approximate each root to the nearest hundredth: ______

4. Graph this equation: $y = 81x^4 - 81x^3 + 9x^2 + 4$

Notice that the graph is *tangent* to the x-axis (that is, the graph touches the x-axis without crossing it).

a. The procedure, described in Exercise 1, for using a table of values to approximate a root does not apply when a graph is tangent to the x-axis. Describe an alternate procedure for using a table of values to approximate a root that occurs at a point of tangency: ______

b. Use your procedure from part (a) to approximate the root of $81x^4 - 81x^3 + 9x^2 + 4 = 0$ to the nearest hundredth: ______

5. Graph this equation: $y = 147x^3 - 77x^2 - 15x + 9$

(*Note:* To see the graph better, you might initially make the x-axis scale -2 to 2 and the y-axis scale -10 to 10.) Approximate all roots of the equation $147x^3 - 77x^2 - 15x + 9 = 0$ to the nearest hundredth: ______

242 USING TECHNOLOGY

Resource Book, ALGEBRA, Structure and Method, Book 2

Using Technology/Exploration
Using Algebra Plotter Plus, p. 38

Book 2, Lesson 8-1 · Curve Fitting

As a means of introducing students to the concept of curve fitting, have the students use the Statistics Spreadsheet program to determine the relationship between x and y for each data set below. Each set involves a direct variation of the form $y = k \cdot f(x)$ where k is the constant of variation and f is one of the following functions: $f(x) = x$, $f(x) = x^2$, $f(x) = \sqrt{x}$.

To determine which function f applies to a given set, students should separately plot the points (x, y), (x^2, y), and $(\sqrt{x}, y)$. On only one of these graphs will the plotted points lie on a straight line that passes through the origin. This identifies the function of x to which y is directly proportional, and the slope of the line is the constant of proportionality. (For example, if the points (x^2, y) lie on a line of slope 2, then the relationship between x and y is given by $y = 2x^2$.)

Select Statistics Spreadsheet. Select ENTER DATA. Relabel "Column A" as "X," "Column B" as "X ^ 2," "Column C" as "SQR(X)," and "Column D" as "Y."

Enter the data for an exercise. (Students will need to calculate the values for the columns x^2 and $\sqrt{x}$.)

Select GRAPH. Draw the SCATTER plots for the x-y, x^2-y, and $\sqrt{x}$-y pairs.

When the line of best fit that is drawn with each scatter plot passes through all the plotted points (as well as the origin), you have direct variation.

Press the space bar to see an equation of the line of best fit.

For each of the following data sets, find an equation relating x and y.

1.

x	y
1	0.5
4	8
9	40.5

$[y = 0.5x^2]$

2.

x	y
1	1.6
4	6.4
9	14.4

$[y = 1.6x]$

3.

x	y
1	2.4
4	4.8
9	7.2

$[y = 2.4\sqrt{x}]$

4.

x	y
1	3.7
4	14.8
9	33.3

$[y = 3.7x]$

Book 2, Lesson 8-8 · Approximations with Greater Accuracy

Students can use the Function Plotter program to approximate the real zeros of the polynomial functions given in Written Exercises 1–10 on page 388 to a higher degree of accuracy than called for in the directions (for example, to the nearest thousandth). [See the answers on the next page.]

Select Function Plotter. Select EQUATIONS and enter the function. DRAW the graph. Select ZOOM and choose one of the intersections of the graph with the x-axis.

By repeated ZOOMing, you can obtain better approximations of the root.

To approximate another root, select SCALE and set it to STANDARD SCALE. Then choose a different intersection of the graph with the x-axis and repeat the ZOOM procedure.

38

USING ALGEBRA PLOTTER PLUS

Reteaching/Practice
Study Guide, p. 119

8 Variation and Polynomial Equations

8–1 Direct Variation and Proportion

Objective: To solve problems involving direct variation.

Vocabulary

Direct variation A linear function defined by an equation of the form $y = mx$ $(m \neq 0)$. The constant m in the equation is called the *constant of variation* or *constant of proportionality*. We say that *y varies directly as x* because if x increases, y also increases, and if x decreases, y also decreases.

Proportion An equality of ratios. A proportion can be written in the form $\frac{y_1}{x_1} = \frac{y_2}{x_2}$, or $y_1 : x_1 = y_2 : x_2$, where the ordered pairs (x_1, y_1) and (x_2, y_2) are solutions of a direct variation, and $\frac{y_1}{x_1} = \frac{y_2}{x_2} = m$ (x_1 and $x_2 \neq 0$).

In a direct variation, y is often said to be *directly proportional* to x.

Means and extremes In the proportion $y_1 : x_1 = y_2 : x_2$, the means are the numbers x_1 and y_2, and the extremes are y_1 and x_2. In a proportion, *the product of the extremes equals the product of the means*, or $y_1 x_2 = x_1 y_2$.

Example 1 If y varies directly as x, and $y = 12$ when $x = 20$, find y when $x = 50$.

Solution First find m and write an equation of the direct variation.

$y = mx$ — Start with the general equation.
$12 = m(20)$ — Substitute $y = 12$ and $x = 20$.
$m = \frac{12}{20} = \frac{3}{5}$ — Solve for m.

$\therefore y = \frac{3}{5}x$ is an equation of the direct variation.

You can find y when $x = 50$ by substituting 50 for x in this equation.

$$y = \frac{3}{5}(50) = 30$$

Example 2 If a is directly proportional to $b + 3$, and $a = 6$ when $b = 15$, find b when $a = 7$.

Solution Since a is directly proportional to $b + 3$, you can write a proportion.

$\frac{a_1}{b_1 + 3} = \frac{a_2}{b_2 + 3}$ — Set up a proportion.
$\frac{6}{15 + 3} = \frac{7}{b_2 + 3}$ — Substitute values for variables.
$6(b_2 + 3) = 18(7)$ — Multiply the extremes, and the means.
$b_2 + 3 = 126 \div 6 = 21$ — Solve for b_2.
$b_2 = 18$

Study Guide, ALGEBRA AND TRIGONOMETRY, Structure and Method, Book 2

119

Reteaching/Practice
Study Guide, p. 120

8–1 Direct Variation and Proportion *(continued)*

CAUTION It doesn't matter whether you solve direct variation problems by finding the equation first or by using the proportion method. However, if you decide to use the proportion method, be sure that the values in the numerators represent the same variable.

Solve.

1. If y varies directly as x, and $y = 6$ when $x = 4$, find y when $x = 12$.
2. If a is directly proportional to b, and $a = 25$ when $b = 35$, find b when $a = 40$.
3. If w varies directly as z, and $w = 4.5$ when $z = 3$, find z when $w = 1.5$.
4. If p is directly proportional to q^3, and $p = 3$ when $q = 2$, find p when $q = 4$.
5. If r varies directly as $s + 1$, and $r = 4$ when $s = 5$, find r when $s = 8$.
6. If a varies directly as $3b + 2$, and $a = 10$ when $b = 6$, find b when $a = 7$.

Example 3 If a car travels 70 km in 2 hours, how far can it travel in 4.5 hours, traveling at the same rate of speed?

Solution Let d be the required distance in kilometers.

Since the ratio $\frac{distance}{time} = rate$ is constant, a proportion can be written.

$$\frac{d_1}{t_1} = \frac{d_2}{t_2} \longrightarrow \frac{70}{2} = \frac{d}{4.5}$$
$$2d = 70(4.5)$$
$$d = 157.5$$

$\therefore$ the distance the car will travel is 157.5 km.

Solve.

7. If the sales tax on a $38 purchase is $2.85, what will the tax be on an $84 purchase?
8. A survey showed that 52 out of 234 people questioned preferred hot cereal to cold cereal. In a school population of 1800, how many people are likely to prefer hot cereal?
9. A real estate agent received a commission of $2232 on a piece of land that sold for $124,000. At this rate, what commission will the agent receive for a piece of land that sold for $160,000?

Mixed Review Exercises

Solve each equation over the real numbers.

1. $4x^2 - 7x + 2 = 0$ **2.** $\frac{a - 1}{a + 5} = 1 - \frac{3}{a}$ **3.** $\frac{3y^2}{8} + \frac{y}{4} = 1$ **4.** $|2a - 8| = 6$

5. $\sqrt{2m + 15} = m$ **6.** $4q^{-2} + 7q^{-1} = 2$ **7.** $6n^2 = 7n$ **8.** $(2x - 5)^2 = 18$

120

Study Guide, ALGEBRA AND TRIGONOMETRY, Structure and Method, Book 2

Application

Johannes Kepler (1571–1630) spent years analyzing tables of astronomical observations and doing calculations to account for the movements of the planets. Astronomers of his day believed that the planets moved in circular orbits, but Kepler showed that the orbits must in fact be elliptical. This was the first of his three laws of planetary motion. The second law states that the speed of a planet varies so that a line joining it with the sun sweeps equal areas in equal times. The third law is given in the caption.

Research Activities

Students can check newspapers to determine which planets are visible at night. Encourage students to locate these planets.

Students can also make use of astronomy computer programs that plot a planet's position on a star map or give the time of a planet's rising and setting.

References

Burgess, Eric. *Celestial BASIC: Astronomy on Your Computer.* Alameda, CA: Sybex, 1982.

Duffet-Smith, Peter. *Practical Astronomy with Your Calculator.* 3d ed. New York: Cambridge University Press, 1980.

Support Materials

Resource Book p. 211

8 Variation and Polynomial Equations

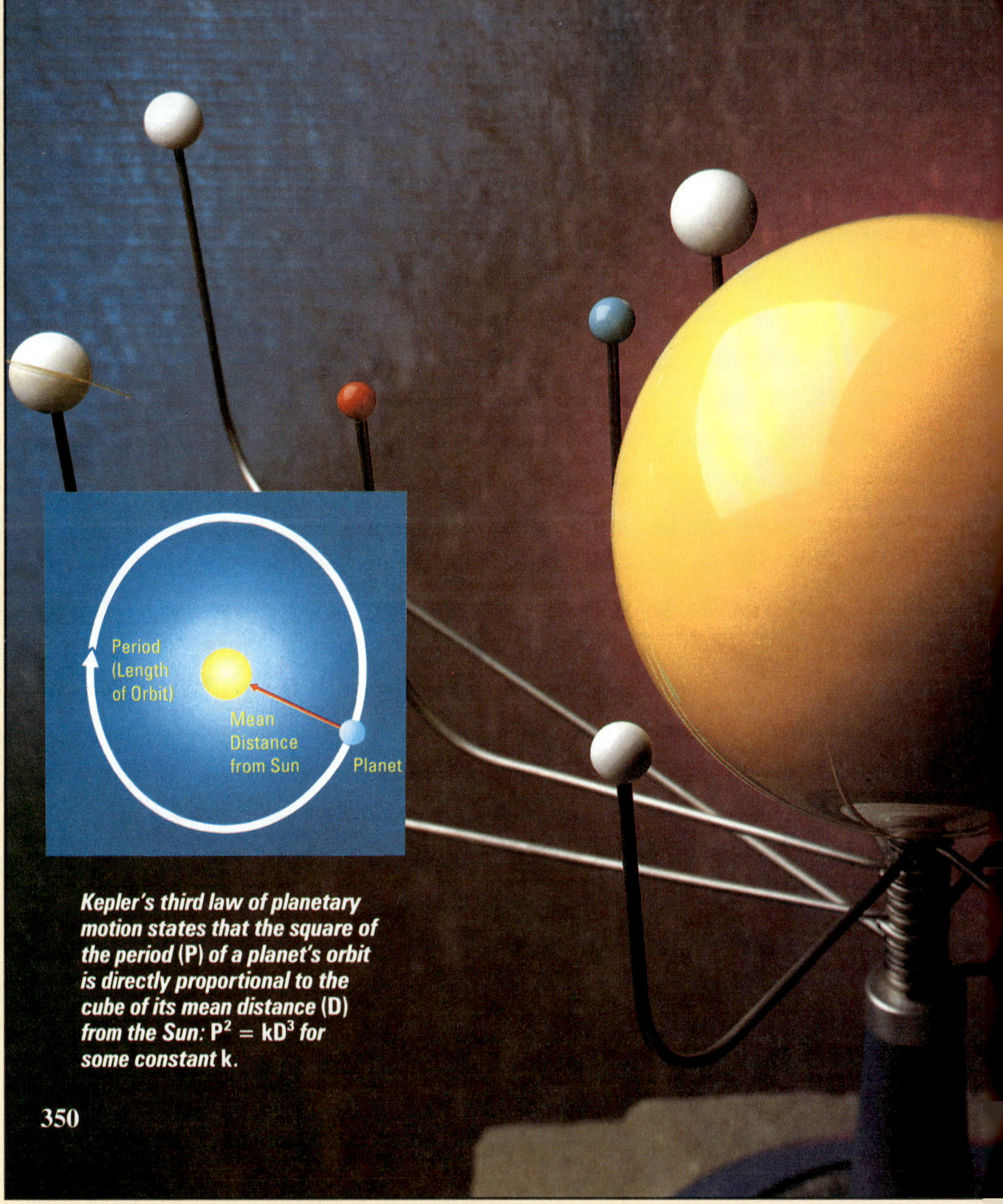

***Kepler's third law of planetary motion states that the square of the period* (P) *of a planet's orbit is directly proportional to the cube of its mean distance* (D) *from the Sun:* $P^2 = kD^3$ *for some constant* k.**

350

Variation and Proportion

8-1 Direct Variation and Proportion

Objective To solve problems involving direct variation.

The water pressure, y, on a scuba diver increases with the diver's depth x meters beneath the surface, as shown in the table at the right. (Here pressure is measured in kilopascals (kPa).) Notice that for each ordered pair (x, y), the ratio of pressure to depth is constant: $\frac{y}{x} = 9.8$; and therefore, $y = 9.8x$. Because of this, we say that the pressure *varies directly* as the depth.

Depth x (m)	Pressure y (kPa)	$\frac{y}{x}$ (kPa/m)
3	29.4	9.8
6	58.8	9.8
9	88.2	9.8
12	117.6	9.8

A linear function defined by an equation of the form

$$y = mx \qquad (m \neq 0)$$

is called a **direct variation,** and we say that *y varies directly as x.* The constant m is called the **constant of variation.**

Example 1 The stretch in a loaded spring varies directly as the load it supports (within the spring's elastic limit). A load of 8 kg stretches a certain spring 9.6 cm.

a. Find the constant of variation (the stiffness constant) and the equation of the direct variation.

b. What load would stretch the spring 6 cm?

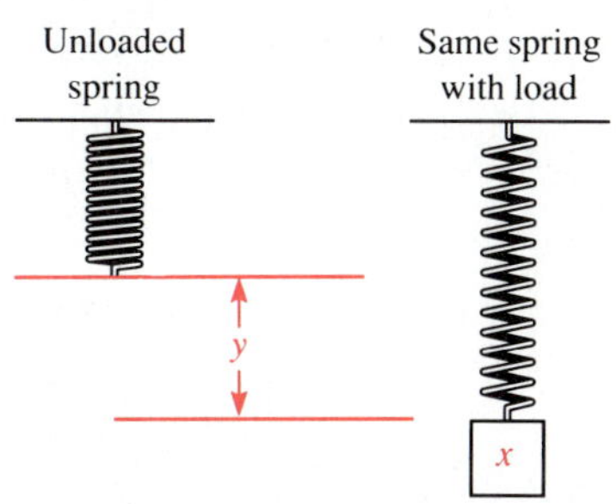

Solution Let x = the load in kilograms and y = the resulting stretch in centimeters. Since y varies directly as x, $y = mx$ where m is the constant of variation.

a. Find m. Since $y = 9.6$ when $x = 8$, substitute these values of y and x in $y = mx$ and solve for m:

$$9.6 = m(8)$$
$$m = 9.6 \div 8 = 1.2$$

$\therefore$ the constant of variation is 1.2, and the equation is $y = 1.2x$.

(Solution continues on the next page.)

Teaching References

Lesson Commentary, pp. T108–T112

Assignment Guide, pp. T64–T66

Supplementary Materials
- Practice Masters 44–48
- Tests 31–34
- Resource Book
  - Practice Exercises, pp. 131–137
  - Tests, pp. 46–49
  - Enrichment Activity, p. 227
  - Application, p. 211
  - Practice in Problem Solving/Word Problems, pp. 202–203
- Study Guide, pp. 119–136
- Computer Activities 17–19
- Test Generator
- Disk for Algebra

Alternate Test, p. T20

Explorations p. 839

Teaching Suggestions p. T108

Reading Algebra p. T109

Suggested Extensions p. T109

Group Activities p. T109

Using Manipulatives p. T109

Warm-Up Exercises

1. If $y = 4x$ and $x = 7$, then $y =$ __?__. 28
2. If $y = \frac{3}{5}x$ and $y = 6$, then $x =$ __?__. 10
3. If $\frac{y}{x} = 6$ and $x = 2$, then $y =$ __?__. 12
4. If $2x = 3y$ and $y = 4$, then $x =$ __?__. 6
5. If $y = mx$ and $x = 12$ and $y = 9$, then $m =$ __?__. $\frac{3}{4}$

Motivating the Lesson

Tell students that cartographers use satellite photographs to update maps. The process of matching the scale of the photograph to the scale of the map involves direct variation, the topic of today's lesson.

Chalkboard Examples

1. If y varies directly as x, and $y = 25$ when $x = 15$, find x when $y = 40$.
   $y = mx$
   $25 = m(15)$
   $\frac{5}{3} = m$
   $40 = \frac{5}{3}x$
   $x = \frac{3}{5}(40)$
   $x = 24$
2. If a is directly proportional to b^3, *and* $a = 10$ when $b = 2$, find a when $b = 4$.
   $\frac{a_1}{b_1^3} = \frac{a_2}{b_2^3}$
   $\frac{10}{(2)^3} = \frac{a_2}{(4)^3}$
   $8a_2 = (10)(64)$
   $a_2 = 80$
3. The amount of hydrogen produced when sodium is added to water varies directly as the amount of sodium added. If 92 g of sodium produces 4 g of hydrogen, find the amount of sodium needed to produce 10 g of hydrogen.
   $h = ks$
   $4 = k(92)$
   $k = \frac{4}{92} = \frac{1}{23}$
   $10 = \frac{1}{23}s$
   $230 = s$
   $\therefore$ 230 g of sodium are needed.

b. Use the equation $y = 1.2x$ to find x when $y = 6$:

$$6 = 1.2x$$
$$x = 6 \div 1.2 = 5$$

$\therefore$ the load is 5 kg. ***Answer***

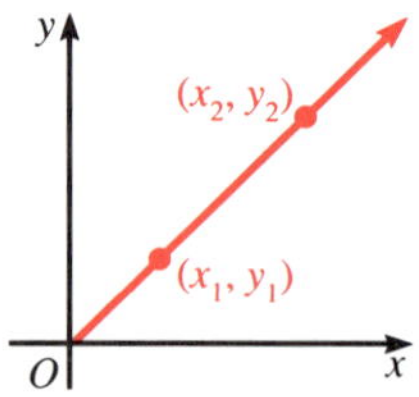

The graph of $y = mx$ is a line through the origin with slope m. In most applications of direct variation, the variables are positive, so the graph of the variation is a half-line, as shown at the right. If (x_1, y_1) and (x_2, y_2) are ordered pairs of the variation $y = mx$, and neither x_1 nor x_2 is zero, then $\frac{y_1}{x_1} = m$ and $\frac{y_2}{x_2} = m$. Therefore,

$$\frac{y_1}{x_1} = \frac{y_2}{x_2}.$$

Such an equality of ratios is called a **proportion.** For this reason, in a direct variation y is often said to be **directly proportional** to x, and m is called the **constant of proportionality.** The proportion is sometimes written

$$y_1 : x_1 = y_2 : x_2$$

and is read "y_1 is to x_1 as y_2 is to x_2." The numbers x_1 and y_2 are called the **means,** and y_1 and x_2 the **extremes** of the proportion.

means
$$y_1 : x_1 = y_2 : x_2$$
extremes

Multiplying both sides of $\frac{y_1}{x_1} = \frac{y_2}{x_2}$ by x_1x_2 gives the result

$$y_1x_2 = x_1y_2.$$

Therefore, in any proportion, *the product of the extremes equals the product of the means.*

Example 2 If y varies directly as x, and $y = 15$ when $x = 24$, find x when $y = 25$.

Solution 1 First find m and write an equation of the direct variation.

$$y = mx$$
$$15 = m(24)$$
$$m = \frac{15}{24} = \frac{5}{8}$$

$\therefore$ an equation of the direct variation is $y = \frac{5}{8}x$.

Then, to find x when $y = 25$, substitute in $y = \frac{5}{8}x$.

$$25 = \frac{5}{8}x$$
$$x = 40 \quad \textbf{\textit{Answer}}$$

Solution 2 Since "y varies directly as x" means that y is directly proportional to x, a proportion can be used.

$$\frac{y_1}{x_1} = \frac{y_2}{x_2}$$
$$\frac{15}{24} = \frac{25}{x}$$
$$15x = 24 \cdot 25$$
$$x = \frac{24 \cdot 25}{15} = 40 \quad \textit{Answer}$$

Example 3 The electrical resistance in ohms (Ω) of a wire varies directly as its length. If a wire 110 cm long has a resistance of 7.5 Ω, what length wire will have a resistance of 12 Ω?

Solution Let l be the required length in centimeters. Since the ratio $\frac{\text{resistance}}{\text{length}}$ is constant, you can write this proportion:

$$\frac{7.5}{110} = \frac{12}{l}$$
$$7.5l = 12(110)$$
$$l = 176$$

$\therefore$ the wire's length is 176 cm. ***Answer***

In many applications direct variations are nonlinear functions. One quantity may vary directly as a power of another (that is, $y = kx^n$ for some constant k) or as a root of another (that is, $y = k\sqrt[n]{x}$ for some constant k).

Example 4 The period of a pendulum (the time for a complete back-and-forth trip) is directly proportional to the square root of the length of the pendulum. If a pendulum 64 cm long has period 1.6 seconds, what is the period of a pendulum 1 m long?

Solution Let p = the period of the pendulum, l = the length of the pendulum, and k = the constant of proportionality. Then $p = k\sqrt{l}$, or $\frac{p}{\sqrt{l}} = k$. Therefore, you can use this proportion:

$$\frac{p_1}{\sqrt{l_1}} = \frac{p_2}{\sqrt{l_2}}$$
$$\frac{1.6}{\sqrt{64}} = \frac{p_2}{\sqrt{100}} \qquad \text{Use 100 cm for 1 m.}$$
$$\frac{1.6}{8} = \frac{p_2}{10}$$
$$8p_2 = 16, \text{ so } p_2 = 2$$

$\therefore$ the period of the pendulum is 2 s. ***Answer***

Check for Undertstanding

Oral Exs. 1–4: use before Example 2.
Oral Exs. 5–12: use after Example 4.

Guided Practice

Solve.

1. If y varies directly as x, and $y = 15$ when $x = 10$, find y when $x = 14$. 21
2. If m is directly proportional to n and $m = 4.8$ when $n = 3$, find m when $n = 12.5$. 20
3. If y is directly proportional to x^2 and $y = 12$ when $x = 4$, find y when $x = 6$. 27
4. If a varies directly as $b + 5$ and $a = 2$ when $b = 1$, find a when $b = 4$. 3

Summarizing the Lesson

In this lesson students learned to solve problems involving direct variation. Ask students to describe two methods of solving such problems. (Refer to Example 2.)

Suggested Assignments

Average Algebra
Day 1: 354/1–19 odd
Day 2: 356/*P*: 1–19 odd
S 357/Mixed Review

Average Alg. and Trig.
Day 1: 354/1–19 odd
Day 2: 356/*P*: 1–19 odd
S 357/Mixed Review

Extended Algebra
Day 1: 354/1–23 odd
Day 2: 356/*P*: 1, 6, 10–17
S 357/Mixed Review

Extended Alg. and Trig.
Day 1: 354/1–23 odd
Day 2: 356/*P*: 1, 6, 10–17
S 357/Mixed Review

Supplementary Materials

Study Guide pp. 119–120

Additional Answers Written Exercises

13. 1. $\frac{a}{b} = \frac{c}{d}$; a, b, c, and d nonzero
Given
2. $\frac{bd}{1}\left(\frac{a}{b}\right) = \frac{bd}{1}\left(\frac{c}{d}\right)$
Mult. prop. of eq.
3. $\frac{bda}{b} = \frac{bdc}{d}$
Mult. rule for fractions
4. $da = bc$
Rule for simplifying fractions
5. $ad = bc$
Comm. prop. for mult.

14. 1. $\frac{a}{b} = \frac{c}{d}$; a, b, c, and d nonzero
Given
2. $\frac{b}{c}\left(\frac{a}{b}\right) = \frac{b}{c}\left(\frac{c}{d}\right)$
Mult. prop. of eq.
3. $\frac{ba}{cb} = \frac{bc}{cd}$
Mult. rule for fractions
4. $\frac{a}{c} = \frac{b}{d}$
Rule for simplifying fractions

15. 1. $\frac{a}{b} = \frac{c}{d}$; a, b, c, and d nonzero
Given
2. $\frac{d}{a}\left(\frac{a}{b}\right) = \frac{d}{a}\left(\frac{c}{d}\right)$
Mult. prop. of eq.
3. $\frac{da}{ab} = \frac{dc}{ad}$
Mult. rule for fractions
4. $\frac{d}{b} = \frac{c}{a}$
Rule for simplifying fractions

Oral Exercises

In Exercises 1–4, several ordered pairs of a function are given in a table. Tell whether the function is a direct variation. If it is, give the constant of variation.

1.

x	y
0.5	3
2	12
7	42

Yes; 6

2.

w	z
8	6
12	9
15	12

No

3.

s	t
4	10
10	25
12	30

Yes; 2.5

4.

t	d
$\frac{3}{2}$	$\frac{3}{4}$
6	3
$\frac{32}{3}$	$\frac{16}{3}$

Yes; $\frac{1}{2}$

For each direct variation, give an equation that relates the two variables and includes the value of the constant of variation.

5. The circumference (C) of a circle varies directly as the radius (r). $C = 2\pi r$

6. The perimeter (P) of a square is directly proportional to the length (s) of a side. $P = 4s$

7. The area (A) of a square is directly proportional to the square of the length (s) of a side. $A = 1s^2$

8. The area (A) of a circle varies directly as the square of the radius (r). $A = \pi r^2$

9. The momentum (M) of a moving 1-kg mass is directly proportional to the speed (v) of the mass in m/s. (Constant of proportionality = 1) $M = 1v$

10. The kinetic energy (E) of a moving 1-kg mass varies directly as the square of the speed (v) of the mass in m/s. (Constant of variation = $\frac{1}{2}$) $E = \frac{1}{2}v^2$

11. The distance (d) a car travels at 90 km/h is directly proportional to the time (t) traveled. $d = 90t$

12. A shop is selling all its merchandise at a 20% discount. The sale price (s) varies directly as the original price (p). $s = 0.8p$

Written Exercises

Solve.

A **1.** If y varies directly as x, and $y = 6$ when $x = 15$, find y when $x = 25$. 10

2. If s is directly proportional to t, and $s = 40$ when $t = 15$, find t when $s = 64$. 24

3. If p is directly proportional to q, and $p = 9$ when $q = 7.5$, find q when $p = 24$. 20

4. If a varies directly as b, and $a = 75$ when $b = 40$, find a when $b = 12$. 22.5

5. If s varies directly as r^2, and $s = 12$ when $r = 2$, find s when $r = 5$. 75

6. If y is directly proportional to $\sqrt{x}$, and $y = 25$ when $x = 3$, find x when $y = 100$. 48

7. If p is directly proportional to $r - 2$, and $p = 20$ when $r = 6$, find p when $r = 12$. 50

8. If w varies directly as $2x - 1$, and $w = 9$ when $x = 2$, find x when $w = 15$. 3

If a, b, and c are positive, and $\frac{a}{b} = \frac{b}{c}$, then b is called the *mean proportional*, or *geometric mean*, between a and c. Find the mean proportional between each pair of numbers.

9. 2 and 18 6

10. 3 and 27 9

11. 8 and 9 $6\sqrt{2}$

12. 5 and 15 $5\sqrt{3}$

Let a, b, c and d be positive. Prove the following properties of the proportion $\frac{a}{b} = \frac{c}{d}$. (*Hint:* For Exercises 13–16, multiply both sides of $\frac{a}{b} = \frac{c}{d}$ by an expression that will produce the given equation.)

B 13. $ad = bc$

14. $\frac{a}{c} = \frac{b}{d}$

15. $\frac{d}{b} = \frac{c}{a}$

16. $\frac{d}{c} = \frac{b}{a}$

17. $\frac{a+b}{b} = \frac{c+d}{d}$ (*Hint:* Add 1 to each side of $\frac{a}{b} = \frac{c}{d}$.)

18. $\frac{a-b}{b} = \frac{c-d}{d}$ (*Hint:* Add -1 to each side of $\frac{a}{b} = \frac{c}{d}$.)

19. If $c \neq d$, then $\frac{a-b}{c-d} = \frac{a}{c}$. (*Hint:* Use Exercises 14 and 18.)

20. If $a \neq b$ and $c \neq d$, then $\frac{a+b}{a-b} = \frac{c+d}{c-d}$. (*Hint:* Use Exercises 17 and 18.)

C 21. Prove that if g is a direct variation over the real numbers, then for every pair of real numbers a and c,

$$g(a + c) = g(a) + g(c).$$

22. Prove that if f is a linear function but not a direct variation and not the constant function 0, then for every pair of real numbers a and c

$$f(a + c) \neq f(a) + f(c).$$

In Exercises 23–25, x, y, and z are positive numbers such that y varies directly as x, and z varies directly as x.

23. Show that $y + z$ varies directly as x.

24. Show that $\sqrt{yz}$ varies directly as x.

25. Show that yz varies directly as $y^2 + z^2$.

16. 1. $\frac{a}{b} = \frac{c}{d}$; a, b, c, and d nonzero
Given
2. $\frac{db}{ac}\left(\frac{a}{b}\right) = \frac{db}{ac}\left(\frac{c}{d}\right)$
Mult. prop. of eq.
3. $\frac{dba}{acb} = \frac{dbc}{acd}$
Mult. rule for fractions
4. $\frac{d}{c} = \frac{b}{a}$
Rule for simplifying fractions

17. 1. $\frac{a}{b} = \frac{c}{d}$; a, b, c, and d nonzero
Given
2. $\frac{a}{b} + 1 = \frac{c}{d} + 1$
Add. prop. of eq.
3. $\frac{a}{b} + b \cdot \frac{1}{b} = \frac{c}{d} + d \cdot \frac{1}{d}$
Prop. of reciprocals
4. $\frac{a}{b} + \frac{b}{b} = \frac{c}{d} + \frac{d}{d}$
Def. of division
5. $\frac{a+b}{b} = \frac{c+d}{d}$
Add. rule for fractions

18. 1. $\frac{a}{b} = \frac{c}{d}$; a, b, c, and d nonzero
Given
2. $\frac{a}{b} - 1 = \frac{c}{d} - 1$
Add. prop. of eq.; def. of subtr.
3. $\frac{a}{b} - b \cdot \frac{1}{b} = \frac{c}{d} - d \cdot \frac{1}{d}$
Prop. of reciprocals
4. $\frac{a}{b} - \frac{b}{b} = \frac{c}{d} - \frac{d}{d}$
Def. of division
5. $\frac{a-b}{b} = \frac{c-d}{d}$
Subtr. rule for fractions

(continued)

Additional Answers Written Exercises
(continued)

19. 1. $\frac{a}{b} = \frac{c}{d}$; a, b, c, and d nonzero; $c \neq d$
Given
2. $\frac{a-b}{b} = \frac{c-d}{d}$
Ex. 18, above
3. $\frac{a-b}{c-d} = \frac{b}{d}$
Ex. 14, above
4. $\frac{a}{c} = \frac{b}{d}$
Ex. 14, above
5. $\frac{a-b}{c-d} = \frac{a}{c}$
Substitution

20. 1. $\frac{a}{b} = \frac{c}{d}$; a, b, c, and d nonzero; $a \neq b$ and $c \neq d$
Given
2. $\frac{a+b}{b} = \frac{c+d}{d}$ and $\frac{a-b}{b} = \frac{c-d}{d}$
Ex. 17 and Ex. 18, above
3. $\frac{a+b}{c+d} = \frac{b}{d}$ and $\frac{a-b}{c-d} = \frac{b}{d}$
Ex. 14, above
4. $\frac{a+b}{c+d} = \frac{a-b}{c-d}$
Substitution
5. $\frac{a+b}{a-b} = \frac{c+d}{c-d}$
Ex. 14, above

21. 1. g is a direct variation over the real numbers.
Given
2. There is a real number m such that for every real number x, $g(x) = mx$.
Def. of direct variation

Problems

Solve.

A

1. Refer to the table on page 351 that shows water pressure at various depths. At what depth would the pressure be 147 kPa? 15 m
2. If the sales tax on a \$60 purchase is \$3.90, what would it be on a \$280 purchase? \$18.20
3. A real estate agent made a commission of \$5400 on a house that sold at \$120,000. At this rate, what commission will the agent make on a house that sells for \$145,000? \$6525
4. The acceleration of an object varies directly as the force acting on it. If a force of 240 newtons causes an acceleration of 150 m/s^2, what force will cause an acceleration of 100 m/s^2? 160 newtons
5. On a certain map, a field 280 ft long is represented by a 5 in. by 8 in. rectangle. How wide is the field? 175 ft
6. At the Gourmet Grocery Mart, a 575 g can of green beans costs 39¢ and a 810 g can of the same product costs 52¢. Which is the better buy? 810 g can
7. To estimate the number of flamingos in the Lake Manyara flock, the Kenyan Wildlife Service banded 600 birds. At a later time, they caught 1000 birds and found that 48 were banded. About how large is the entire flock? 12,500

8. A public-opinion poll found that of a sample of 450 voters, 252 favored a school bond measure. If 20,000 persons vote, about how many are likely to vote for the bond measure? 11,200
9. The stretch in a loaded spring varies directly as the load it supports. A load of 15 kg stretches a certain spring 3.6 cm. What load would stretch the spring 6 cm? 25 kg
10. The speed of an object falling from rest in a vacuum is directly proportional to the time it has fallen. After an object has fallen for 1.5 s, its speed is 14.7 m/s. What is its speed after it has fallen 5 s? 49 m/s

B

11. *Absolute temperature* is measured in degrees Kelvin (° K). A degree on the Kelvin scale is the same size as a Celsius degree, but 0° K is −273° C. The volume of a fixed amount of gas kept at constant pressure varies directly as its absolute temperature. If the gas occupies 100 L at −13° C, what is its volume at 26° C? (*Hint:* First convert the Celsius temperatures to absolute temperatures.) 115 L

12. *Newton's law of cooling* states that the rate at which an object cools varies directly as the difference between its temperature and the temperature of the surrounding air. At the moment a steel plate at 270° C is placed in air that is at 20° C, its rate of cooling is 50° C per minute. How fast is it cooling when its temperature is 100° C? 16° C/min

13. A load of 12 kg stretches a spring to a total length of 15 cm, and a load of 30 kg stretches it to a length of 18 cm. Find the natural (unstretched) length of the spring. 13 cm

14. The centrifugal force acting on an object moving in a circle is directly proportional to the square of the speed of the object. If the force is 2240 newtons when the object is moving at 8 m/s, what is the force when the object is moving at 12 m/s? 5040 newtons

15. The distance an object falls from rest varies directly as the square of the time it has fallen. If the object fell 4 ft during the first half second, how far did it fall during the next two seconds? 96 ft

16. The speed of an object falling from rest is directly proportional to the square root of the distance the object has fallen. When an object has fallen 36 ft, its speed is 48 ft/s. How much farther must it fall before its speed is 80 ft/s? 64 ft

17. The power developed by an electric current varies directly as the square of the magnitude of the current. If a current of 0.5 amperes develops 100 watts of power, what current will develop 1.6 kilowatts of power? 2 amperes

18. The average number of red cells in a milliliter (mL) of human blood is 5×10^6. For a certain test, a laboratory technician diluted one part of a blood sample with 199 parts of salt water. How many red cells would the technician expect to find in 6 mL of the diluted solution? 1.5×10^5

19. The volume of a sphere is directly proportional to the cube of its diameter and is 288π cm^3 when the diameter is 12 cm. Find the constant of proportionality in terms of π. Then write the formula for the volume of a sphere in terms of its diameter. $\frac{\pi}{6}$; $V = \frac{\pi}{6}d^3$

C 20. The kinetic energy of a moving object varies directly as the square of its speed. By how much would a car traveling at v_0 km/h need to increase its speed in order to double its kinetic energy?
It would need to multiply its speed by $\sqrt{2}$, or add $(\sqrt{2} - 1)v_0$ to its speed.

Mixed Review Exercises

Solve each equation over the real numbers.

6. $\left\{-4, \frac{1}{2}\right\}$
9. $\left\{\frac{1 \pm 2\sqrt{3}}{3}\right\}$

1. $3x^2 - 5x + 1 = 0$ $\left\{\frac{5 \pm \sqrt{13}}{6}\right\}$
2. $\frac{y-1}{y+2} = 1 - \frac{2}{y}$ $\{4\}$
3. $2m^2 = 5m$ $\left\{0, \frac{5}{2}\right\}$
4. $\left|\frac{a}{2} + 1\right| = 3$ $\{-8, 4\}$
5. $3(2 - c) = c + 4$ $\left\{\frac{1}{2}\right\}$
6. $\frac{w^2}{2} + \frac{7w}{4} = 1$
7. $p^{-2} - 2p^{-1} - 1 = 0$ $\{-1 \pm \sqrt{2}\}$
8. $\sqrt{5n + 6} = n$ $\{6\}$
9. $(3k - 1)^2 = 12$

3. $g(a + c) = m(a + c)$; $g(a) + g(c) = ma + mc$
Application of Step 2
4. $m(a + c) = ma + mc$
Dist. prop.
5. $g(a + c) = g(a) + g(c)$
Substitution

22. 1. For every real number x, $f(x) = mx + b$, $m \neq 0$, $b \neq 0$.
Def. of linear function; f is not a direct variation, and f is not the constant function 0.
2. $f(a + c) - [f(a) + f(c)] = m(a + c) + b - (ma + b + mc + b) = ma + mc + b - ma - mc - 2b = -b$
Substitution; dist. prop.
3. $b \neq 0$ and $-b \neq 0$
Step 1
4. $f(a + c) \neq f(a) + f(c)$
For real numbers x and y, $x = y$ if and only if $x - y = 0$.

23. $y = kx$ and $z = cx$ where k and c are constants.
$y + z = kx + cx$
$y + z = (k + c)x$
Since $k + c$ is a constant, $y + z$ varies directly as x.

24. $y = kx$ and $z = cx$ where k and c are constants.
$yz = (kx)(cx)$
$yz = kcx^2$
$\sqrt{yz} = \sqrt{kcx^2}$
$\sqrt{yz} = x\sqrt{kc}$
Since $\sqrt{kc}$ is a constant, $\sqrt{yz}$ varies directly as x.

(continued on p. 371)

Teaching Suggestions p. T109

Reading Algebra p. T109

Suggested Extensions p. T109

Warm-Up Exercises

If $x = 10$ and $y = 4$, find k.

1. $y = kx$ $\frac{2}{5}$
2. $xy = k$ 40

If $x = 6$, $y = 9$, and $z = 12$, find k.

3. $z = kx\sqrt{y}$ $\frac{2}{3}$
4. $z = k\frac{x^2}{y}$ 3

Motivating the Lesson

Tell students that if the area of a rectangle is to remain constant, an increase in the length of the rectangle must be accompanied by a decrease in the width. This is an example of inverse variation, one of the topics of today's lesson.

Chalkboard Examples

1. If y varies inversely as q, and $y = 22$ when $q = 6$, find q when $y = 15$.

$$y = \frac{k}{q}$$
$$22 = \frac{k}{6}$$
$$132 = k$$
$$15 = \frac{132}{q}$$
$$q = \frac{132}{15} = \frac{44}{5}, \text{ or } 8.8$$

2. If r varies jointly as s and t and inversely as u, and $r = 18$ when $s = 2$, $t = 3$, and $u = 4$, find s when $r = 6$, $t = 2$, and $u = 4$.

8-2 Inverse and Joint Variation

Objective To solve problems involving inverse and joint variation.

You know that the distance d traveled in time t at speed r is given by

$$rt = d.$$

If the distance d is held constant and r increases, the time t decreases. For example, the greater the average speed of a sprinter, the less time it takes the sprinter to run 100 m. We say that the time *varies inversely* as the speed. In the case of the sprinter, the formula above becomes

$$rt = 100 \quad \text{or} \quad t = \frac{100}{r}$$

where t is measured in seconds and r in meters per second.

A function defined by an equation of the form

$$xy = k \quad \text{or} \quad y = \frac{k}{x} \quad (x \neq 0, k \neq 0)$$

is called an **inverse variation,** and we say that y *varies inversely as* x, or y *is inversely proportional to* x. The constant k is called the **constant of variation,** or the **constant of proportionality.**

In most applications of inverse variation, x and y are positive. In such cases the larger x is, the smaller y is, as shown by the graph of $xy = k$ in the diagram at the right. (You will study this type of curve in Lesson 9-5.)

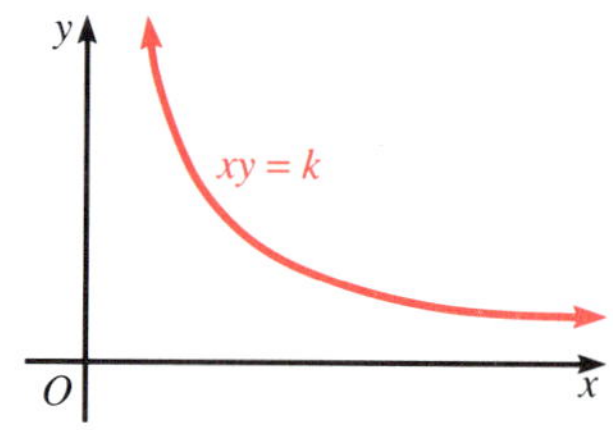

Example 1 If y is inversely proportional to x, and $y = 6$ when $x = 5$, find x when $y = 12$.

Solution First find k and write an equation of the inverse variation.

$$xy = k$$
$$(5)(6) = k$$
$$k = 30$$

$\therefore$ an equation of the inverse variation is $xy = 30$.

Then, to find x when $y = 12$, substitute in $xy = 30$.

$$x(12) = 30$$
$$x = 2.5 \quad \textit{Answer}$$

When a quantity varies directly as the product of two or more other quantities, the variation is called **joint variation.** For example, if z varies jointly as x and the square of y, then $z = kxy^2$ for some nonzero constant k. In this case, we also say that z is **jointly proportional** to x and y^2.

Example 2 If z varies jointly as x and the square root of y, and $z = 6$ when $x = 3$ and $y = 16$, find z when $x = 7$ and $y = 4$.

Solution First find k and write an equation of the joint variation.

$$z = kx\sqrt{y}$$
$$6 = k(3)\sqrt{16}$$
$$6 = 12k$$
$$k = \frac{1}{2}$$

$\therefore$ an equation of the joint variation is $z = \frac{1}{2}x\sqrt{y}$.

Then, to find z when $x = 7$ and $y = 4$, substitute in $z = \frac{1}{2}x\sqrt{y}$.

$$z = \frac{1}{2}(7)\sqrt{4}$$
$$z = 7 \quad \textbf{\textit{Answer}}$$

Several important physical laws combine joint and inverse variation. For example, Newton's law of gravitation states that the force of attraction F between two spherical bodies varies jointly as their masses, m_1 and m_2, and inversely as the square of the distance r between their centers. That is,

$$F = k\frac{m_1m_2}{r^2}.$$

Example 3 The electrical resistance of a wire varies directly as its length and inversely as the square of its diameter. One hundred meters of a wire with diameter 6 mm has resistance 12 ohms (Ω). Eighty meters of a second wire of the same material has resistance 15 Ω. Find the diameter of the second wire.

Solution Let R = the resistance of the wire in ohms, d = the diameter of the wire in millimeters, and l = its length in meters. Then R varies directly as l and inversely as d^2, that is,

$$R = k\frac{l}{d^2}.$$

For the first wire, $R = 12$, $d = 6$, and $l = 100$.

$$12 = k \cdot \frac{100}{6^2}$$
$$k = \frac{12 \cdot 6^2}{100}$$

(Solution continues on the next page.)

$$r = k\frac{st}{u}$$
$$18 = \frac{k(2)(3)}{4}$$
$$k = 12$$
$$6 = \frac{(12)s(2)}{4}$$
$$s = 1$$

3. The time required to travel a given distance is inversely proportional to the speed of travel. If a trip can be made in 3.6 h at a speed of 70 km/h, how long will it take to make the same trip at 90 km/h?
Let s = speed and t = time.

$$t = \frac{k}{s}$$
$$3.6 = \frac{k}{70}$$
$$k = 252$$
$$t = \frac{252}{90} = 2.8$$

$\therefore$ the trip will take 2.8 h at 90 km/h.

4. For cones of equal volume, the height is inversely proportional to the square of the radius of the base. If the radius of a cone is 10 cm and its height is 15 cm, find the radius of a second cone of equal volume with a height of 60 cm.
Let r = radius and h = height.

$$h = \frac{k}{r^2}$$
$$15 = \frac{k}{100}$$
$$k = 1500$$
$$60 = \frac{1500}{r^2}$$
$$r^2 = \frac{1500}{60} = 25$$
$$r = 5$$

$\therefore$ the radius is 5 cm.

Check for Understanding

Here is a suggested use of the Oral Exercises to check students' understanding as you teach the lesson.
Oral Exs. 1–4: use before Example 2.
Oral Exs. 5–14: use after Example 2.

Guided Practice

1. If c is inversely proportional to d, and $c = 2$ when $d = 3.6$, find:
   a. c when $d = 4.5$. 1.6
   b. d when $c = 6.4$. 1.125
2. Suppose r varies jointly as t and s and inversely as the square of v. When $t = 3$, $s = 18$, and $v = 5$, $r = 3.78$. Find r when $t = 4$, $s = 12$, and $v = 4$. 5.25
3. The surface area of a cylinder varies jointly as the radius and the sum of the radius and the height. A cylinder with height 8 cm and radius 4 cm has a surface area of 96 π cm^2. Find the surface area of a cylinder with radius 3 cm and height 10 cm. 78 π cm^2

Summarizing the Lesson

In this lesson students learned to solve problems involving inverse and joint variation. Ask students to contrast inverse variation with direct variation.

For the second wire, $R = 15$ and $l = 80$.

$$15 = k \cdot \frac{80}{d^2}$$

$$15 = \frac{12 \cdot 6^2}{100} \cdot \frac{80}{d^2} \qquad \text{Use the value of } k \text{ obtained for the first wire.}$$

$$d^2 = \frac{12 \cdot 6^2 \cdot 80}{100 \cdot 15} = 23.04$$

$$d = 4.8$$

$\therefore$ the diameter of the second wire is 4.8 mm. ***Answer***

Oral Exercises

In Exercises 1–4, k is a constant and the other letters are variables. Express in words each equation as a variation.

Sample $E = k\frac{p}{r}$ ***Solution*** E varies directly as p and inversely as r.

1. $P = kst$
2. $z = kqr^2$
3. $h = k\frac{s}{t^2}$
4. $M = k\frac{jm^2}{r}$

In Exercises 5–10, give an equation of the variation described. Use k for the constant of variation.

5. V varies directly as the cube of r. $V = kr^3$
6. i is inversely proportional to the square of j. $i = \frac{k}{j^2}$
7. h is directly proportional to t and inversely proportional to z. $h = k\frac{t}{z}$
8. A is jointly proportional to l and w. $A = klw$
9. w varies jointly as x and y and inversely as z. $w = k\frac{xy}{z}$
10. m varies directly as n and inversely as z. $m = k\frac{n}{z}$

In Exercises 11–14, z varies jointly as x and y. Tell what happens to z in each of the situations described.

11. x and y are both doubled. z is quadrupled.
12. x and y are both halved. z is quartered.
13. x is halved and y is quadrupled. z is doubled.
14. x is doubled and y is quadrupled. z is multiplied by 8.

Written Exercises

A

1. If y varies inversely as x, and $y = 3$ when $x = 6$, find x when $y = 18$. 1
2. If z is inversely proportional to r, and $z = 32$ when $r = 1.5$, find r when $z = 8$. 6

3. If w is inversely proportional to the square of v, and $w = 3$ when $v = 6$, find w when $v = 3$. 12

4. If p varies inversely as the square root of q, and $p = 12$ when $q = 36$, find p when $q = 16$. 18

5. If z is jointly proportional to x and y, and $z = 18$ when $x = 0.4$ and $y = 3$, find z when $x = 1.2$ and $y = 2$. 36

6. If w is jointly proportional to u and v, and $w = 24$ when $u = 0.8$ and $v = 5$, for what value of u will $w = 18$ when $v = 2$? 1.5

B 7. If s varies directly as r and inversely as t, and $s = 10$ when $r = 5$ and $t = 3$, for what value of t will $s = 3$ when $r = 4$? 8

8. Suppose that r varies directly as p and inversely as q^2, and that $r = 27$ when $p = 3$ and $q = 2$. Find r when $p = 2$ and $q = 3$. 8

9. Suppose that z varies jointly as u and v and inversely as w, and that $z = 0.8$ when $u = 8$, $v = 6$ and $w = 5$. Find z when $u = 3$, $v = 10$ and $w = 5$. 0.5

10. Suppose that w varies directly as z^2 and inversely as xy, and that $w = 10$ when $x = 15$, $y = 2$, and $z = 5$. Find z when $w = 2$, $x = 8$, and $y = 27$. 6

Problems

A 1. The frequency of a radio signal varies inversely as the wave length. A signal of frequency 1200 kilohertz (kHz), which might be the frequency of an AM radio station, has wave length 250 m. What frequency has a signal of wave length 400 m? 750 kHz

2. By *Ohm's law,* the current flowing in a wire is inversely proportional to the resistance of the wire. If the current is 5 amperes (A) when the resistance is 24 ohms (Ω), for what resistance will the current be 8 A? 15 Ω

3. The heat loss through a glass window varies jointly as the area of the window and the difference between the inside and outside temperatures. If the loss through a window with area 3 m^2 is 720 BTU when the temperature difference is 15° C, what is the heat loss through a window with area 4.5 m^2 when the temperature difference is 12° C? 864 BTU

4. The conductance of a wire varies directly as the square of the wire's diameter and inversely as its length. Fifty meters of wire with diameter 2 mm has conductance 0.12 mho (*mho,* which is "ohm" spelled backwards, is a unit of conductance). If a wire of the same material has length 75 m and diameter 2.5 mm, what is its conductance? 0.125 mho

In Problems 5 and 6, use the fact that the intensity of light, measured in *lux,* is inversely proportional to the square of the distance between the light source and the object illuminated.

5. A light meter 7.5 m from a light source registers 24 lux. What intensity would it register 15 m from the light source? 6 lux

Additional Answers
Oral Exercises

1. P varies jointly as s and t.
2. z varies jointly as q and the square of r.
3. h varies directly as s and inversely as the square of t.
4. M varies jointly as j and the square of m and inversely as r.

Suggested Assignments

Average Algebra
360/1–10; *P*: 1, 5, 6, 10
S 355/14–20 even

Average Alg. and Trig.
360/1–10; *P*: 1, 5, 6, 10
S 355/14–20 even

Extended Algebra
360/1–10; *P*: 1, 5, 10–13
S 355/14–24 even

Extended Alg. and Trig.
360/1–10; *P*: 1, 5, 10–13
S 355/14–24 even

Supplementary Materials

Study Guide	pp. 121–122
Practice Master	44
Computer Activity	17
Resource Book	p. 131

6. A light hangs 4.8 ft above the center of a circular table 7.2 ft in diameter. If the illumination is 25 lux at the center of the table, what is it at the edge of the table? 16 lux

7. A sprocket gear 8 in. in diameter is connected to a gear 3 in. in diameter. How fast does the smaller gear rotate when the larger one rotates at 216 r/min (revolutions per minute)? If the smaller gear is attached to the 28-inch-diameter rear wheel of a bicycle, how fast does the wheel rotate? (*Hint:* r/min · circumference of gear = k.)

8. The volume of a cone varies jointly as the height and the square of the radius of the base. A cone of height 8 cm and base diameter 9 cm has volume 54π. Find the constant of variation and a general formula for the volume of a cone.

7. 576 r/min; about 61.7 r/min **8.** $\frac{\pi}{3}$; $V = \frac{\pi}{3}r^2h$

B 9. The stretch in a wire under a given tension varies directly as the length of the wire and inversely as the square of its diameter. A wire having length 2 m and diameter 1.5 mm stretches 1.2 mm. If a second wire of the same material (and under the same tension) has length 3 m and diameter 2.0 mm, find the amount of stretch. 1.0125 mm

10. The volume of a given mass of a gas varies directly as its absolute temperature and inversely as its pressure. (See Problem 11, page 356.) At 2° C and 99 kPa, 4 g of helium occupies 23.1 L. What is the volume of the helium at 27° C and 121 kPa? About 20.6 L

In Problems 11 and 12, use the fact that the load a beam with a rectangular cross section can support is jointly proportional to the beam's width and the square of its depth and inversely proportional to its length.

11. A beam 3 cm wide and 5 cm deep can support a load of 630 kg. What load can it support when turned on its side? 378 kg

12. One beam is half as wide, twice as long, and three times as deep as a second beam. Find the ratio of the loads they can support. 9:4

C 13. By one of Kepler's laws (proved by Newton), the square of the period of a satellite in circular orbit is directly proportional to the cube of the radius of the orbit. A satellite in orbit 9600 km above the surface of the Earth has period 5.6 h. How high is a satellite in a "stationary" orbit that has period 24 h? (Use 6400 km as the radius of Earth.) About 35,800 km

14. Part of a monument consists of two iron spheres whose centers are 3 m apart. Each sphere has radius 1 m and mass 3200 kg. How many times greater is the gravitational force of Earth on each of them than between them? (*Hint:* Use Newton's law of gravitation (see page 359). Also use 6400 km as the radius of Earth and 6×10^{24} kg as the mass of Earth.) About 4×10^8 times

Self-Test 1

Vocabulary

direct variation (p. 351)	extremes (p. 352)
y varies directly as x (p. 351)	inverse variation (p. 358)
constant of (direct) variation (p. 351)	y varies inversely as x (p. 358)
proportion (p. 352)	inversely proportional (p. 358)
directly proportional (p. 352)	constant of (inverse) variation (p. 358)
constant of proportionality (p. 352)	joint variation (p. 359)
means (p. 352)	jointly proportional (p. 359)

Obj. 8-1, p. 351

1. Suppose that y varies directly as x, and $y = 8$ when $x = 56$.
   a. Find the constant of variation. $\frac{1}{7}$
   b. Find y when $x = 21$. 3
2. If w is directly proportional to z, and $z = 81$ when $w = 45$, find z when $w = 120$. 216
3. The water pressure on a diver is directly proportional to the diver's depth below the surface. If the pressure at 21 m is 205.8 kPa, what is the pressure at 7 m? 68.6 kPa

Obj. 8-2, p. 358

4. Suppose that y varies inversely as x, and $y = 45$ when $x = \frac{1}{3}$.
   a. Find the constant of variation. 15
   b. Find x when $y = 30$. $\frac{1}{2}$
5. If w is inversely proportional to the square of t, and $w = 1$ when $t = 6$, find w when $t = 2$. 9
6. Suppose y varies jointly as r and s^2 and inversely as t, and $y = 120$ when $r = 5$, $s = 3$, and $t = 2$. Find r when $y = 80$, $s = 1$, and $t = 6$. 90
7. Two meshed gears have 32 and 50 teeth, respectively. If the speeds of the gears are inversely proportional to the numbers of teeth, at what speed should the second gear be driven so that the first gear will run at 1250 revolutions per minute? 800 r/min

Check your answers with those at the back of the book.

Challenge

A group of friends went to a restaurant for lunch. They had agreed to split the bill equally. However, when the bill arrived two of them discovered that they had left their money at home. The others in the group then agreed to make up the difference, which resulted in each one having to pay an extra \$1.30. If the total bill was \$78.00, how many people were in the group? 12

Quick Quiz

1. If y varies directly as t and $y = 15$ when $t = 18$, find:
   a. the constant of variation. $\frac{5}{6}$
   b. y when $t = 24$. 20
2. If P is inversely proportional to the cube of Q, and $P = 13.5$ when $Q = 2$, find P when $Q = 3$. 4
3. The time needed to complete one lap of an auto test track is inversely proportional to the speed. If a lap takes 5 min at 30 mi/h, how fast must a car travel to complete a lap in 3 min? 50 mi/h
4. Suppose A varies jointly as P and $\sqrt{Q}$ and inversely as B. If $A = 100$ when $P = 20$, $Q = 9$, and $B = 3$, find:
   a. B when $A = 140$, $P = 7$, and $Q = 4$. $\frac{1}{2}$
   b. Q when $A = 200$, $B = 5$, and $P = 8$. 625
5. The centripetal force on an object, when that object is swung in a circle, varies jointly as the mass of the object and the square of its velocity and inversely as the radius of the circle. What would be the effect on the force if the mass, the velocity, and the radius were all doubled? The force would be multiplied by 4.

Teaching Suggestions p. T109

Suggested Extensions p. T110

Warm-Up Exercises

Simplify.

1. $\frac{1-x}{-1}$ $x - 1$
2. $\frac{8y^2 - 6y + 10}{2}$
   $4y^2 - 3y + 5$
3. $\frac{10n^2 - 15n - 5}{-5}$
   $-2n^2 + 3n + 1$
4. $\frac{12m^3 - 30m^2 + 24m}{6m}$
   $2m^2 - 5m + 4$
5. $\frac{9a^4 - 12a^3 - 15a^2}{-3a^2}$
   $-3a^2 + 4a + 5$

Motivating the Lesson

Tell students that the process of dividing polynomials, while it may look complicated, is directly related to long division of whole numbers that they are already familiar with.

Polynomial Equations

8-3 Dividing Polynomials

Objective To divide one polynomial by another polynomial.

Polynomial division is similar to the division process you used in arithmetic to write a fraction as a mixed number. For example:

$$\frac{19}{5} = 3 + \frac{4}{5}, \quad \text{or} \quad 3\frac{4}{5}$$

$$\frac{3x+4}{x} = 3 + \frac{4}{x}$$

These two equations illustrate the **division algorithm:**

$$\frac{\text{Dividend}}{\text{Divisor}} = \text{Quotient} + \frac{\text{Remainder}}{\text{Divisor}}$$

or

$$\text{Dividend} = \text{Quotient} \times \text{Divisor} + \text{Remainder}$$

When you cannot perform a division mentally, you can use long division, illustrated below for $\frac{873}{14}$.

$$\begin{array}{r l}
62 & \\
14\overline{)873} & \\
\underline{84} & \leftarrow \text{subtract } 6 \times 14 \\
33 & \\
\underline{28} & \leftarrow \text{subtract } 2 \times 14 \\
5 &
\end{array}$$

$$\therefore \frac{873}{14} = 62\frac{5}{14}$$

Check:

$873 \stackrel{?}{=} 62 \times 14 + 5$

$873 \stackrel{?}{=} 868 + 5$

$873 = 873$ ✓

You can use a similar long-division process to divide one polynomial by another, as Example 1 illustrates.

Example 1 Divide: $\frac{x^3 - 5x^2 + 4x - 2}{x - 2}$

Solution

$$\begin{array}{r r l}
 & x^2 - 3x \;\; - 2 & \leftarrow \text{quotient} \\
\text{divisor} \rightarrow x - 2 & \overline{)x^3 - 5x^2 + 4x - 2} & \leftarrow \text{dividend} \\
 & \underline{x^3 - 2x^2} \qquad\qquad & \leftarrow \text{subtract } x^2(x-2) \\
 & -3x^2 + 4x \qquad & \\
 & \underline{-3x^2 + 6x} \qquad & \leftarrow \text{subtract } -3x(x-2) \\
 & -2x - 2 & \\
 & \underline{-2x + 4} & \leftarrow \text{subtract } -2(x-2) \\
 & -6 & \leftarrow \text{remainder}
\end{array}$$

$$\therefore \underbrace{\frac{x^3 - 5x^2 + 4x - 2}{x - 2}} = \underbrace{x^2 - 3x - 2} + \underbrace{\frac{-6}{x - 2}}$$

$$\frac{\text{Dividend}}{\text{Divisor}} = \text{Quotient} + \frac{\text{Remainder}}{\text{Divisor}}$$

Check: To check the result, use this form of the division algorithm:

Quotient × Divisor + Remainder = Dividend

$$(x^2 - 3x - 2)(x - 2) + (-6) \stackrel{?}{=} x^3 - 5x^2 + 4x - 2$$
$$x^3 - 3x^2 - 2x - 2x^2 + 6x + 4 - 6 \stackrel{?}{=} x^3 - 5x^2 + 4x - 2$$
$$x^3 - 5x^2 + 4x - 2 = x^3 - 5x^2 + 4x - 2 \quad \checkmark$$

Notice that the division process ends when the remainder is 0 or is of lower degree than the divisor.

The next example illustrates the two points mentioned in the following caution.

Caution: Before using long division, always arrange the terms of both dividend and divisor in order of decreasing degree of the same variable. Be sure to insert any "missing" terms by using 0 as a coefficient.

Example 2 Divide: $\dfrac{3x^4 - 9a^2x^2 + 4a^4 - 2ax^3}{ax - 2a^2 + 3x^2}$

Solution Arrange the terms of the dividend and divisor in decreasing degree of x.

$$\begin{array}{r}
x^2 - ax - 2a^2 \qquad\qquad\qquad \\
3x^2 + ax - 2a^2\overline{\big)3x^4 - 2ax^3 - 9a^2x^2 + 0a^3x + 4a^4} \\
\underline{3x^4 + ax^3 - 2a^2x^2} \qquad\qquad\qquad \\
-3ax^3 - 7a^2x^2 + 0a^3x \qquad \\
\underline{-3ax^3 - a^2x^2 + 2a^3x} \qquad \\
-6a^2x^2 - 2a^3x + 4a^4 \\
\underline{-6a^2x^2 - 2a^3x + 4a^4} \\
0
\end{array}$$

$\therefore$ the quotient is $x^2 - ax - 2a^2$. ***Answer***

Check: Since the remainder is 0, you only have to show that

Quotient × Divisor = Dividend.

$$\begin{array}{r}
x^2 - ax - 2a^2 \qquad\qquad \\
\underline{3x^2 + ax - 2a^2} \qquad\qquad \\
3x^4 - 3ax^3 - 6a^2x^2 \qquad\qquad \\
ax^3 - a^2x^2 - 2a^3x \qquad \\
\underline{- 2a^2x^2 + 2a^3x + 4a^4} \\
3x^4 - 2ax^3 - 9a^2x^2 \qquad\qquad + 4a^4 \quad \checkmark
\end{array}$$

Chalkboard Examples

Divide the first polynomial by the second.

1. $16x^5 + 10x^3 + 4x^2;\ 2x^3$

$$\frac{16x^5 + 10x^3 + 4x^2}{2x^3} = \frac{16x^5}{2x^3} + \frac{10x^3}{2x^3} + \frac{4x^2}{2x^3} = 8x^2 + 5 + \frac{2}{x}$$

2. $x^2 + 6x - 10;\ x + 2$

$$\begin{array}{r}
x + 4 \\
x + 2\overline{\big)x^2 + 6x - 10} \\
\underline{x^2 + 2x} \qquad \\
4x - 10 \\
\underline{4x + 8} \\
-18
\end{array}$$

$$x + 4 - \frac{18}{x + 2}$$

3. $3x^3 + x^2 - 4x;\ x - 1$

$$\begin{array}{r}
3x^2 + 4x \\
x - 1\overline{\big)3x^3 + x^2 - 4x} \\
\underline{3x^3 - 3x^2} \qquad \\
4x^2 - 4x \\
\underline{4x^2 - 4x} \\
0
\end{array}$$

$$3x^2 + 4x$$

4. $2a^3 + a - 3;\ a + 2$

$$\begin{array}{r}
2a^2 - 4a + 9 \\
a + 2\overline{\big)2a^3 + 0a^2 + a - 3} \\
\underline{2a^3 + 4a^2} \qquad\qquad \\
-4a^2 + a \qquad \\
\underline{-4a^2 - 8a} \qquad \\
9a - 3 \\
\underline{9a + 18} \\
-21
\end{array}$$

$$2a^2 - 4a + 9 - \frac{21}{a + 2}$$

Common Errors

Students having trouble with polynomial long division may not line up like terms between each partial product and the dividend. They may also forget that each partial product is *subtracted* from the dividend. When reteaching, remind students of these facts.

Check for Understanding

Here is a suggested use of the Oral Exercises to check students' understanding as you teach the lesson.
Oral Exs. 1–12: use after Example 1.

Guided Practice

Find the quotient and the remainder.

1. $\frac{x^2 - 4x + 3}{x + 2}$ $x - 6;\ 15$
2. $\frac{6 - 7p + 3p^2}{2 - p}$ $-3p + 1;\ 4$
3. $\frac{8r^3 - 10r^2 - r - 4}{2r + 1}$ $4r^2 - 7r + 3;\ -7$
4. $\frac{4x^5 + 4x^2 - 2x - 2}{2x^2 + 1}$ $2x^3 - x + 2;\ -x - 4$

Summarizing the Lesson

In this lesson students learned to divide polynomials using long division. Ask students to describe how they can check their answers when doing polynomial long division.

Suggested Assignments

Average Algebra
366/2–30 even
S 367/Mixed Review
R 363/Self-Test 1

Average Alg. and Trig.
366/2–30 even
S 367/Mixed Review
R 363/Self-Test 1

Extended Algebra
366/2–32 even
S 367/Mixed Review
R 363/Self-Test 1

Extended Alg. and Trig.
366/2–32 even
S 367/Mixed Review
R 363/Self-Test 1

Oral Exercises

Give the quotient and remainder in each division.

1. $\frac{3x + 5}{x}$ $q = 3;\ r = 5$
2. $\frac{x^2 - 2}{x}$ $q = x;\ r = -2$
3. $\frac{2x^2 + 1}{x^2}$ $q = 2;\ r = 1$
4. $\frac{x^2 + 2x}{x}$ $q = x + 2;\ r = 0$
5. $\frac{x^2 + 3x + 5}{x}$ $q = x + 3;\ r = 5$
6. $\frac{2x^2 - x - 3}{x}$ $q = 2x - 1;\ r = -3$

Sample $\frac{x + 1}{x + 4} = \frac{(x + 4) - 3}{x + 4} = \frac{x + 4}{x + 4} + \frac{-3}{x + 4} = 1 + \frac{-3}{x + 4}$

$\therefore$ the quotient is 1 and the remainder is -3. ***Answer***

7. $\frac{x + 4}{x + 1}$ $q = 1;\ r = 3$
8. $\frac{x - 2}{x - 3}$ $q = 1;\ r = 1$
9. $\frac{x}{x + 2}$ $q = 1;\ r = -2$
10. $\frac{x - 3}{x + 2}$ $q = 1;\ r = -5$
11. $\frac{x^2 + 5x + 6}{x + 5}$ $q = x;\ r = 6$
12. $\frac{x^2 - 3x + 2}{x - 3}$ $q = x;\ r = 2$

Written Exercises

3. $x + 4 + \frac{-2}{2 - x}$
6. $2u + 3 + \frac{8}{3u - 1}$

Divide.

A

1. $\frac{x^2 + 3x - 4}{x + 2}$ $x + 1 + \frac{-6}{x + 2}$
2. $\frac{x^2 - x + 3}{x + 1}$ $x - 2 + \frac{5}{x + 1}$
3. $\frac{6 - 2x - x^2}{2 - x}$
4. $\frac{9z - z^2}{z - 3}$ $-z + 6 + \frac{18}{z - 3}$
5. $\frac{4t^2 - 4t + 1}{2t + 1}$ $2t - 3 + \frac{4}{2t + 1}$
6. $\frac{6u^2 + 7u + 5}{3u - 1}$
7. $\frac{x^3 - x^2 - 10x + 10}{x - 3}$ $x^2 + 2x - 4 + \frac{-2}{x - 3}$
8. $\frac{z^3 + 3z^2 - 13z + 6}{z - 2}$ $z^2 + 5z - 3$
9. $\frac{2s^3 - 29s + 13}{s + 4}$ $2s^2 - 8s + 3 + \frac{1}{s + 4}$
10. $\frac{15y^3 + y^2 - 21y}{5y - 3}$ $3y^2 + 2y - 3 + \frac{-9}{5y - 3}$
11. $\frac{6t^3 + t^2 + 7t + 10}{3t + 2}$ $2t^2 - t + 3 + \frac{4}{3t + 2}$
12. $\frac{6x^3 - 5x^2 + 15x - 5}{2x^2 - x + 3}$ $3x - 1 + \frac{5x - 2}{2x^2 - x + 3}$
13. $\frac{15z^3 - z^2 - 11z - 3}{3z^2 - 2z - 1}$ $5z + 3$
14. $\frac{4u^4 - 4u^3 - 5u^2 - 9u - 1}{2u - 1}$ $2u^3 - u^2 - 3u - 6 + \frac{-7}{2u - 1}$
15. $\frac{6x^4 - 3x^3 + 5x^2 + 2x - 6}{3x^2 - 2}$ $2x^2 - x + 3$
16. $\frac{4t^4 - 2t^3 - 3t - 9}{2t^2 - t - 3}$ $2t^2 + 3$
17. $\frac{9u^4 + 6u^3 + 4u + 4}{3u^2 + 2u + 2}$ $3u^2 - 2 + \frac{8u + 8}{3u^2 + 2u + 2}$
18. $\frac{2x^4 - 3x^2 + 7x - 8}{x^2 + x - 3}$ $2x^2 - 2x + 5 + \frac{-4x + 7}{x^2 + x - 3}$

B

19. $\frac{x^3 + a^3 + 4a^2x + 4ax^2}{x + 2a}$ $x^2 + 2ax + \frac{a^3}{x + 2a}$
20. $\frac{2p^3 - 2q^3 + pq^2 - p^2q}{2p + q}$ $p^2 - pq + q^2 + \frac{-3q^3}{2p + q}$
21. $\frac{6t^4 + ct^3 - c^3t + c^4}{2t^2 + ct + c^2}$ $3t^2 - ct - c^2 + \frac{c^3t + 2c^4}{2t^2 + ct + c^2}$
22. $\frac{2x^4 - x^3y + x^2y^2 + 4xy^3 - 4y^4}{2x^2 + xy - 2y^2}$ $x^2 - xy + 2y^2$

23. $\dfrac{x^6 + x^5 + x^3 + x + 1}{x^4 - x^2 + 1}$ $x^2 + x + 1 + \dfrac{2x^3}{x^4 - x^2 + 1}$

24. $\dfrac{x^5 - x^4 + x^3 - x^2 + x - 1}{x^2 - x + 1}$ $x^3 - 1$

25. $\dfrac{x^4 + a^4}{x^2 + a^2}$ $x^2 - a^2 + \dfrac{2a^4}{x^2 + a^2}$

26. $\dfrac{x^5 - a^5}{x - a}$ $x^4 + ax^3 + a^2x^2 + a^3x + a^4$

27. $\dfrac{x^6 - a^6}{x^2 + ax + a^2}$ $x^4 - ax^3 + a^3x - a^4$

28. $\dfrac{x^8 - a^8}{x^3 - ax^2 + a^2x - a^3}$ $x^5 + ax^4 + a^4x + a^5$

29. When the polynomial $P(x)$ is divided by $x - 3$, the quotient is $x^2 + 2x + 6$ and the remainder is 8. Find $P(x)$. $x^3 - x^2 - 10$

C

30. When $x^3 - 7x + 4$ is divided by the polynomial $D(x)$, the quotient is $x^2 - 3x + 2$ and the remainder is -2. Find $D(x)$. $x + 3$

31. Find k so that when $x^3 + kx^2 - kx + 1$ is divided by $x - 2$, the remainder is 0. $-\frac{9}{2}$

32. Find k so that when $x^3 + kx^2 + k^2x + 14$ is divided by $x + 2$, the remainder is 0. -1 or 3

33. When $3x^2 - 5x + c$ is divided by $x + k$, the quotient is $3x + 1$ and the remainder is 3. Find c and k. $c = 1; k = -2$

34. Here is a complete statement of the division algorithm for polynomials:

Let $P(x)$ and $D(x)$ be polynomials with $D(x)$ not equal to the constant 0. Then there are *unique* polynomials $Q(x)$ and $R(x)$ such that

$$\frac{P(x)}{D(x)} = Q(x) + \frac{R(x)}{D(x)}$$

and either R is the constant 0, or the degree of R is less than the degree of D.

Verify that

$$\frac{x^3 + 2x^2 - 5x - 1}{x - 2} = x^2 + 4x + 3 + \frac{5}{x - 2}$$

and that

$$\frac{x^3 + 2x^2 - 5x - 1}{x - 2} = x^2 + 4x + 2 + \frac{x + 3}{x - 2}.$$

Explain why this does not violate the uniqueness part of the algorithm.

Mixed Review Exercises

In Exercises 1–4, assume that y varies directly as x.

1. If $y = 8$ when $x = 6$, find y when $x = 9$. 12

2. If $y = 4$ when $x = \frac{1}{2}$, find y when $x = 2$. 16

3. If $y = 3$ when $x = \sqrt{2}$, find y when $x = \sqrt{6}$. $3\sqrt{3}$

4. If $y = 3.2$ when $x = 0.2$, find y when $x = 1.6$. 25.6

5–8. Solve Exercises 1–4 assuming that y varies inversely as x. **5.** $\frac{16}{3}$ **6.** 1 **7.** $\sqrt{3}$ **8.** 0.4

Additional Answers Written Exercises

34. $\dfrac{P(x)}{D(x)} = Q(x) + \dfrac{R(x)}{D(x)}$ if and only if $D(x) \neq 0$ and $P(x) = Q(x) \cdot D(x) + R(x)$; $(x^2 + 4x + 3)(x - 2) + 5 = x^3 + 2x^2 - 5x - 1$; $(x^2 + 4x + 2)(x - 2) + x + 3 = x^3 + 2x^2 - 5x - 1$; In the second of the given equations, the degree of R is not less than the degree of D. Rewriting the right side of the equation to produce a remainder with degree less than that of D produces the following:

$x^2 + 4x + 2 + \dfrac{x + 3}{x - 2} =$

$x^2 + 4x + 2 + 1 + \dfrac{5}{x - 2} =$

$x^2 + 4x + 3 + \dfrac{5}{x - 2}$

Supplementary Materials

Study Guide	pp. 123–124
Practice Master	45

Teaching Suggestions p. T110

Suggested Extensions p. T110

Warm-Up Exercises

Divide.

1. $\dfrac{x^2 + 2x - 1}{x + 3}$

   $x - 1 + \dfrac{2}{x + 3}$

2. $\dfrac{2 + 5x - x^2}{x - 3}$

   $-x + 2 + \dfrac{8}{x - 3}$

3. $\dfrac{4x - x^2}{2 - x}$

   $x - 2 + \dfrac{4}{2 - x}$

4. $\dfrac{x^4 - 2x^3 - 2x + 1}{x - 2}$

   $x^3 - 2 - \dfrac{3}{x - 2}$

5. $\dfrac{6x^3 - 5x^2 + 2x - 1}{2x + 1}$

   $3x^2 - 4x + 3 + \dfrac{-4}{2x + 1}$

Motivating the Lesson

In arithmetic, students may have learned a short cut for long division when the divisor is a one-digit whole number. Today students will learn a similar short cut for polynomial long division when the divisor is a first-degree binomial. This short-cut is called synthetic division.

Chalkboard Examples

Divide by synthetic division.

1. $2x^3 + 9x^2 + x - 12$ by $x + 4$

$$\begin{array}{r|rrrr} -4 & 2 & 9 & 1 & -12 \\ & & -8 & -4 & 12 \\ \hline & 2 & 1 & -3 & 0 \end{array}$$

$2x^2 + x - 3$

8-4 Synthetic Division

Objective To use synthetic division to divide a polynomial by a first-degree binomial.

Synthetic division is an efficient way to divide a polynomial by a binomial of the form $x - c$. The process is illustrated below with the division of $2x^3 - 10x^2 + 9x + 15$ by $x - 3$. In this example, $c = 3$.

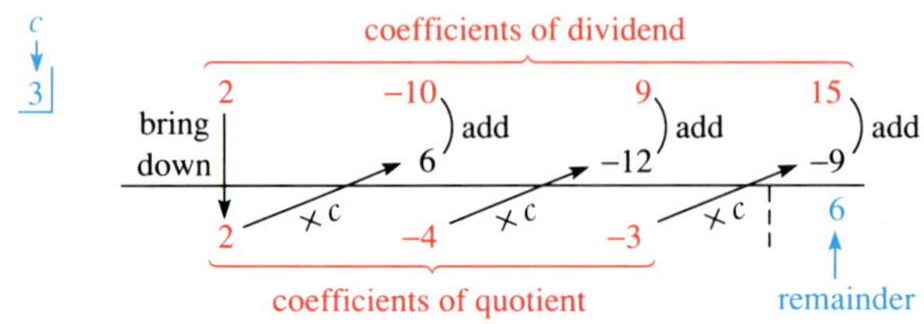

Therefore, $\dfrac{2x^3 - 10x^2 + 9x + 15}{x - 3} = 2x^2 - 4x - 3 + \dfrac{6}{x - 3}$

or $\qquad 2x^3 - 10x^2 + 9x + 15 = (2x^2 - 4x - 3)(x - 3) + 6.$

Synthetic division is derived from ordinary long division by using only the coefficients of the polynomials involved. Compare the two methods:

$$\begin{array}{r} 2x^2 - 4x - 3 \\ x - 3\overline{)2x^3 - 10x^2 + 9x + 15} \\ \underline{2x^3 - 6x^2} \qquad\qquad \\ -4x^2 + 9x \qquad \\ \underline{-4x^2 + 12x} \qquad \\ -3x + 15 \\ \underline{-3x + 9} \\ 6 \end{array}$$

$$\begin{array}{r|rrrr} 3 & 2 & -10 & 9 & 15 \\ & & 6 & -12 & -9 \\ \hline & 2 & -4 & -3 & 6 \end{array}$$

Example 1 Use synthetic division to divide $x^4 - 2x^3 + 13x - 6$ by $x + 2$.

Solution Since $x + 2 = x - (-2)$, you use -2 for c. Insert the missing term, $0x^2$, as you would in long division. Note that the coefficient of x^4 is 1.

$$\begin{array}{r|rrrrr} -2 & 1 & -2 & 0 & 13 & -6 \\ & & -2 & 8 & -16 & 6 \\ \hline & 1 & -4 & 8 & -3 & 0 \end{array}$$

The quotient is $x^3 - 4x^2 + 8x - 3$, and the remainder is 0.

$\therefore \dfrac{x^4 - 2x^3 + 13x - 6}{x + 2} = x^3 - 4x^2 + 8x - 3$ ***Answer***

Example 1 illustrates the use of synthetic division when the coefficient of x in the divisor is 1. In fact, synthetic division can be used when the divisor is *any* first-degree binomial.

Example 2 Use synthetic division to divide $6x^3 + 7x^2 + x + 1$ by $2x + 3$.

Solution $2x + 3 = 2\left(x + \frac{3}{2}\right)$

$$\therefore \frac{6x^3 + 7x^2 + x + 1}{2x + 3} = \frac{1}{2}\left(\frac{6x^3 + 7x^2 + x + 1}{x + \frac{3}{2}}\right)$$

Now use synthetic division to divide $6x^3 + 7x^2 + x + 1$ by $x + \frac{3}{2}$.

$$\begin{array}{r|rrrr} -\frac{3}{2} & 6 & 7 & 1 & 1 \\ & & -9 & 3 & -6 \\ \hline & 6 & -2 & 4 & -5 \end{array}$$

$$\therefore \frac{6x^3 + 7x^2 + x + 1}{2x + 3} = \frac{1}{2}\left(6x^2 - 2x + 4 + \frac{-5}{x + \frac{3}{2}}\right)$$

$$= 3x^2 - x + 2 + \frac{-5}{2x + 3} \quad \textit{Answer}$$

Oral Exercises

For each synthetic division shown below, express as a polynomial (a) the divisor, (b) the dividend, (c) the quotient, and (d) the remainder. Use x as the variable.

1. $\begin{array}{r|rrrr} 2 & 1 & -3 & 5 & -4 \\ & & 2 & -2 & 6 \\ \hline & 1 & -1 & 3 & 2 \end{array}$

2. $\begin{array}{r|rrrr} -3 & 1 & 1 & -4 & 3 \\ & & -3 & 6 & -6 \\ \hline & 1 & -2 & 2 & -3 \end{array}$

3. $\begin{array}{r|rrrrr} -1 & 5 & 1 & 0 & 6 & 2 \\ & & -5 & 4 & -4 & -2 \\ \hline & 5 & -4 & 4 & 2 & 0 \end{array}$

4. $\begin{array}{r|rrrrr} -2 & 3 & 5 & -2 & -1 & 1 \\ & & -6 & 2 & 0 & 2 \\ \hline & 3 & -1 & 0 & -1 & 3 \end{array}$

5. $\begin{array}{r|rrrrr} 3 & 1 & -3 & -1 & 5 & -3 \\ & & 3 & 0 & -3 & 6 \\ \hline & 1 & 0 & -1 & 2 & 3 \end{array}$

6. $\begin{array}{r|rrrrr} 1 & 2 & -1 & 3 & 0 & -4 \\ & & 2 & 1 & 4 & 4 \\ \hline & 2 & 1 & 4 & 4 & 0 \end{array}$

2. $8x^3 + 2x^2 - 49$ by $4x - 7$

$4x - 7 = 4\left(x - \frac{7}{4}\right)$

$\therefore \frac{8x^3 + 2x^2 - 49}{4x - 7} = \frac{1}{4}\left(\frac{8x^3 + 2x^2 - 49}{x - \frac{7}{4}}\right)$

$$\begin{array}{r|rrrr} \frac{7}{4} & 8 & 2 & 0 & -49 \\ & & 14 & 28 & 49 \\ \hline & 8 & 16 & 28 & 0 \end{array}$$

$\therefore$ the quotient is $\frac{1}{4}(8x^2 + 16x + 28) = 2x^2 + 4x + 7$

Check for Understanding

Oral Exs. 1–6: use after Example 1.

Additional Answers
Oral Exercises

1. a. $x - 2$
   b. $x^3 - 3x^2 + 5x - 4$
   c. $x^2 - x + 3$
   d. 2
2. a. $x + 3$
   b. $x^3 + x^2 - 4x + 3$
   c. $x^2 - 2x + 2$
   d. -3
3. a. $x + 1$
   b. $5x^4 + x^3 + 6x + 2$
   c. $5x^3 - 4x^2 + 4x + 2$
   d. 0
4. a. $x + 2$
   b. $3x^4 + 5x^3 - 2x^2 - x + 1$
   c. $3x^3 - x^2 - 1$
   d. 3
5. a. $x - 3$
   b. $x^4 - 3x^3 - x^2 + 5x - 3$
   c. $x^3 - x + 2$
   d. 3
6. a. $x - 1$
   b. $2x^4 - x^3 + 3x^2 - 4$
   c. $2x^3 + x^2 + 4x + 4$
   d. 0

Guided Practice

Use synthetic division to find the quotient and the remainder.

1. $\frac{2x^3 - 3x^2 + 4x + 5}{x - 1}$

$2x^2 - x + 3; 8$

2. $\frac{2t^3 + t^2 - 24t - 10}{t - 3}$

$2t^2 + 7t - 3; -19$

3. $\frac{2y^3 - 9y + 3}{y + 2}$

$2y^2 - 4y - 1; 5$

4. $\frac{z^7 + 128}{z + 2}$ $z^6 - 2z^5 + 4z^4 - 8z^3 + 16z^2 - 32z + 64; 0$

5. $\frac{x^3 + 3x^2 - 2x + 3i}{x - i}$

$x^2 + (3 + i)x - 3 + 3i; -3$

Summarizing the Lesson

In this lesson students learned to use synthetic division for dividing a polynomial by a first-degree binomial. Ask students to describe, in general terms, the steps to be taken when using synthetic division.

Suggested Assignments

Average Algebra
370/1–21 odd
S 356/*P*: 14, 16, 18

Average Alg. and Trig.
370/1–21 odd
S 357/*P*: 14, 16, 18

Extended Algebra
370/1–23 odd
S 356/*P*: 9, 19, 20

Extended Alg. and Trig.
370/1–23 odd
S 356/*P*: 9, 19, 20

Supplementary Materials

Study Guide pp. 125–126

Written Exercises

In Exercises 1–18, divide using synthetic division.

A

1. $\frac{3x^3 - 5x^2 + x - 2}{x - 2}$ $3x^2 + x + 3 + \frac{4}{x - 2}$
2. $\frac{2x^3 - 4x^2 - 7x + 5}{x - 3}$ $2x^2 + 2x - 1 + \frac{2}{x - 3}$
3. $\frac{x^3 + 3x^2 - 2x - 6}{x + 3}$ $x^2 - 2$
4. $\frac{3x^3 - 2x^2 + x + 4}{x + 1}$ $3x^2 - 5x + 6 + \frac{-2}{x + 1}$
5. $\frac{t^4 + 5t^3 - 2t - 7}{t + 5}$ $t^3 - 2 + \frac{3}{t + 5}$
6. $\frac{2u^4 - 5u^3 - 12u^2 + 2u - 8}{u - 4}$ $2u^3 + 3u^2 + 2$
7. $\frac{2s^4 - 7s^3 + 7s + 6}{s - 3}$ $2s^3 - s^2 - 3s - 2$
8. $\frac{y^4 - 4y^2 + y + 4}{y + 2}$ $y^3 - 2y^2 + 1 + \frac{2}{y + 2}$
9. $\frac{x^5 - 1}{x - 1}$ $x^4 + x^3 + x^2 + x + 1$
10. $\frac{x^6 - 1}{x + 1}$ $x^5 - x^4 + x^3 - x^2 + x - 1$
11. $\frac{2x^4 + x^3 - x - 2}{x + 1}$ $2x^3 - x^2 + x - 2$
12. $\frac{3x^6 - 2x^5 - x^4 - x^2 - 2x + 3}{x - 1}$ $3x^5 + x^4 - x - 3$

B

13. $\frac{2x^3 - 3x^2 + 4x - 2}{2x + 1}$ $x^2 - 2x + 3 + \frac{-5}{2x + 1}$
14. $\frac{4x^3 + 2x^2 - 4x + 3}{2x + 3}$ $2x^2 - 2x + 1$
15. $\frac{6t^4 + 5t^3 - 10t + 4}{3t - 2}$ $2t^3 + 3t^2 + 2t - 2$
16. $\frac{5s^4 - 3s^3 + 10s + 2}{5s - 3}$ $s^3 + 2 + \frac{8}{5s - 3}$

Sample $\frac{2z^3 - z^2 + 8z - 4}{z + 2i}$

Solution

$-2i$	2	-1	8	-4
		$-4i$	$-8 + 2i$	4
	2	$-1 - 4i$	$2i$	0

$\therefore$ the quotient is $2z^2 - (1 + 4i)z + 2i$, and the remainder is 0.

17. $\frac{z^3 - 2z^2 + 4z - 5}{z - 2i}$ $z^2 + (-2 + 2i)z - 4i + \frac{3}{z - 2i}$
18. $\frac{z^3 + 3z^2 - 2z + 3}{z - i}$ $z^2 + (3 + i)z + (-3 + 3i) + \frac{-3i}{z - i}$

Find the polynomial $Q(x)$ and the constant R.

19. $2x^3 + 5x^2 + 4 = (x + 3)Q(x) + R$ $Q(x) = 2x^2 - x + 3; R = -5$
20. $x^4 - 5x^3 + 2x - 5 = (x - 5)Q(x) + R$ $Q(x) = x^3 + 2; R = 5$
21. $2z^3 - 3z^2 + 8z - 10 = (z + 2i)Q(z) + R$ $Q(z) = 2z^2 - (3 + 4i)z + 6i; R = 2$
22. $z^3 - 2z^2 - 3z + 10 = (z - 2 + i)Q(z) + R$ $Q(z) = z^2 - iz - (4 + 2i); R = 0$

Determine k so that the first polynomial is a factor of the second.

C

23. $x + 2; 2x^3 + 3x^2 + k$ 4
24. $x - 2; x^4 - 2x^3 + kx + 6$ -3

Reading Algebra / Problem Solving

Problem solving is not an automatic process. Aside from the calculations that produce the solution, solving a problem involves reading, thinking, planning, and, often, rethinking.

The first step in solving any problem is knowing what is given and what you must find. The key to making that first step is reading. Before you start to solve a problem, you should read the related lesson in your textbook and work through the examples.

The next step is finding a solution method. Don't assume that there is a single correct approach that will give you the solution on the first try; problem solving often involves false starts. How can you decide if you are at least on the right track? One way is to estimate a solution using your method, rounding numbers to make the computations easier. If the answer is unreasonable, reconsider your strategy. Perhaps you have chosen the wrong approach, or perhaps you have made an error in setting up an equation. Check for both possibilities. Once you are certain that you have chosen a suitable method, complete your solution to find the exact answer.

The last step in solving a problem is to make sure that your answer is reasonable and that your calculations are correct.

Exercises

1. Describe two methods for solving a quadratic equation. What are some of the advantages and disadvantages of using each method? factoring; quadratic formula
2. Read Problem 6 on page 362. Suppose you obtained

$$\frac{I}{(6.0)^2} = \frac{25}{(4.8)^2}$$

as the equation for the intensity I at the edge of the table. According to this equation, will I be greater than 25 or less than 25? Is this answer reasonable; that is, should the intensity at the edge of the table be greater than or less than the intensity at the center? Is the equation correct? $I > 25$; Not reasonable Equation is incorrect.

3. Read Problem 11 on page 246. If an *equal* amount of money was invested at each rate, what would the total annual yield be? Since the actual yield was \$144, which account has more money? \$126.50; 7% account
4. Read Problem 20 on page 134. Is the current greater than 2 mi/h or less than 2 mi/h? less than
5. Working together, Tom and Sue can complete a task in 3 h. Working alone, Sue takes 5 h.
   a. If they worked at the *same* rate, how long would each person take working alone? 6 h
   b. Is Sue's actual time greater than or less than your answer to (a)? What, then, can you conclude about Tom's time working alone? less than; Tom's time is greater than Sue's
   c. Find Tom's time working alone, and check it against your conclusion from (b). 7.5 h

Additional Answers
Written Exercises

(continued from p. 357)

25. $y = kx$ and $z = cx$ where k and c are constants.

$yz = (kx)(cx) = kcx^2$

$y^2 + z^2 = k^2x^2 + c^2x^2 = (k^2 + c^2)x^2$

Since $x^2 = \left(\frac{1}{kc}\right)yz$ and

$x^2 = \left(\frac{1}{k^2 + c^2}\right)(y^2 + z^2)$,

$\left(\frac{1}{kc}\right)yz = \left(\frac{1}{k^2 + c^2}\right)(y^2 + z^2)$

or $yz = \left(\frac{kc}{k^2 + c^2}\right)(y^2 + z^2)$.

Since $\frac{kc}{k^2 + c^2}$ is a constant, yz varies directly as $y^2 + z^2$.

Teaching Suggestions p. T110

Suggested Extensions p. T110

Warm-Up Exercises

Divide using synthetic division.

1. $\frac{2x^3 - 7x^2 + 6x - 3}{x - 3}$
   $2x^2 - x + 3 + \frac{6}{x-3}$
2. $\frac{x^3 + 6x^2 + 5x - 10}{x + 2}$
   $x^2 + 4x - 3 + \frac{-4}{x+2}$
3. $\frac{x^6 - 1}{x + 1}$
   $x^5 - x^4 + x^3 - x^2 + x - 1$

Evaluate for the given value of x.

4. $2x^3 - 7x^2 + 6x - 3$; $x = 3$ 6
5. $x^3 + 6x^2 + 5x - 10$; $x = -2$ −4
6. $x^6 - 1$; $x = -1$ 0

Motivating the Lesson

Tell students that not only is synthetic division a short cut for dividing a polynomial by a first-degree binomial, but it is also an efficient means of finding values of polynomial functions, as today's lesson will show.

8-5 The Remainder and Factor Theorems

Objective To use the remainder and factor theorems to find factors of polynomials and to solve polynomial equations.

You can evaluate the polynomial $P(x) = 2x^3 - 7x^2 + 5x - 1$ when $x = 3$ by substituting 3 for x:

$$P(3) = 2 \cdot 3^3 - 7 \cdot 3^2 + 5 \cdot 3 - 1 = 54 - 63 + 15 - 1 = 5$$

To find the remainder when $P(x)$ is divided by $x - 3$, use synthetic division:

$$\begin{array}{r|rrrr} 3 & 2 & -7 & 5 & -1 \\ & & 6 & -3 & 6 \\ \hline & 2 & -1 & 2 & 5 \end{array}$$

Notice that the remainder when $P(x)$ is divided by $x - 3$ is equal to $P(3)$. To see that this did not just happen by chance, rewrite $P(x)$ using the division algorithm with $x - 3$ as divisor:

$$\begin{aligned} \text{Dividend} &= \text{Quotient} \times \text{Divisor} + \text{Remainder} \\ P(x) &= (2x^2 - x + 2)(x - 3) + 5 \\ P(3) &= (2 \cdot 3^2 - 3 + 2)(3 - 3) + 5 \\ &= (2 \cdot 3^2 - 3 + 2)(0) + 5 \\ &= 0 + 5 \\ &= 5 \end{aligned}$$

In general, if $Q(x)$ is the quotient and R is the remainder when the polynomial $P(x)$ is divided by $x - c$, then $P(c) = R$. You can reason as follows:

$$\begin{aligned} P(x) &= Q(x) \cdot (x - c) + R \\ P(c) &= Q(c) \cdot (c - c) + R \\ &= Q(c) \cdot (0) + R \\ &= 0 + R \\ &= R \end{aligned}$$

We have just proved the *remainder theorem.*

Remainder Theorem

Let $P(x)$ be a polynomial of positive degree n. Then for any number c,

$$P(x) = Q(x) \cdot (x - c) + P(c),$$

where $Q(x)$ is a polynomial of degree $n - 1$.

The remainder theorem is *not* restricted to polynomials with real coefficients. It holds for any polynomial $P(x)$ with complex coefficients and any complex number c.

Because synthetic division provides a convenient way to find values of polynomials, the process is sometimes called **synthetic substitution.**

Example 1 Use synthetic substitution to find the value $P(-4)$ for the polynomial $P(x) = x^4 - 14x^2 + 5x - 3$.

Solution By the remainder theorem, $P(-4)$ is equal to the remainder when $P(x)$ is divided by $x + 4$. Since $x + 4 = x - (-4)$, $c = -4$.

$$\begin{array}{r|rrrrr} c \rightarrow -4 & 1 & 0 & -14 & 5 & -3 \\ & & -4 & 16 & -8 & 12 \\ \hline & 1 & -4 & 2 & -3 & 9 \leftarrow P(c) \end{array}$$

$\therefore P(-4) = 9$ ***Answer***

Check: You can check the answer by directly calculating $P(-4)$.

$$P(-4) = (-4)^4 - 14(-4)^2 + 5(-4) - 3$$
$$= 256 - 224 - 20 - 3 = 9 \quad \checkmark$$

The following theorem is a corollary of the remainder theorem.

Factor Theorem

The polynomial $P(x)$ has $x - r$ as a factor if and only if r is a root of the equation $P(x) = 0$.

Proof If r is a root of $P(x) = 0$, then by the definition of root, $P(r) = 0$. By the remainder theorem,

$$P(x) = Q(x) \cdot (x - r) + P(r) = Q(x) \cdot (x - r) + 0 = Q(x) \cdot (x - r).$$

Therefore, $x - r$ is a factor of $P(x)$.

Conversely, if $x - r$ is a factor of $P(x)$, then $P(x) = Q(x) \cdot (x - r)$ for some polynomial $Q(x)$, and thus

$$P(r) = Q(r) \cdot (r - r) = Q(r) \cdot 0 = 0.$$

Therefore r is a root of $P(x) = 0$.

Example 2 Determine whether $x + 1$ is a factor of $P(x) = x^{12} - 3x^8 - 4x - 2$.

Solution By the factor theorem, if $P(-1) = 0$, then $x + 1$ is a factor.

$$P(-1) = (-1)^{12} - 3(-1)^8 - 4(-1) - 2 = 1 - 3 + 4 - 2 = 0$$

$\therefore x + 1$ is a factor of $P(x)$. ***Answer***

Chalkboard Examples

1. Use synthetic division to find the value of $P(x) = x^3 + 2x^2 - x + 4$ at -5.

$$\begin{array}{r|rrrr} -5 & 1 & 2 & -1 & 4 \\ & & -5 & 15 & -70 \\ \hline & 1 & -3 & 14 & -66 \end{array}$$

$P(-5) = -66$

2. Is $x - 2$ a factor of $P(x) = x^4 - 4x^3 + 5x^2 + 4x - 12$?

$$\begin{array}{r|rrrrr} 2 & 1 & -4 & 5 & 4 & -12 \\ & & 2 & -4 & 2 & 12 \\ \hline & 1 & -2 & 1 & 6 & 0 \end{array}$$

Since $P(2) = 0$, $x - 2$ is a factor of $P(x)$.

3. Find a polynomial whose roots are 0 and -1 (double root).

$P(x)$ must have $(x - 0)$ and $(x + 1)^2$ as factors.
$\therefore P(x) = x(x + 1)^2 = x^3 + 2x^2 + x$.

4. Solve $3x^3 - x^2 - 18x + 6 = 0$, given that $\frac{1}{3}$ is a root.

$$\begin{array}{r|rrrr} \frac{1}{3} & 3 & -1 & -18 & 6 \\ & & 1 & 0 & -6 \\ \hline & 3 & 0 & -18 & 0 \end{array}$$

$\therefore P(x) = \left(x - \frac{1}{3}\right)(3x^2 - 18)$.

The depressed equation is $3x^2 - 18 = 0$.

$x^2 = 6$
$x = \pm\sqrt{6}$

$\left\{\frac{1}{3}, \sqrt{6}, -\sqrt{6}\right\}$

Check for Understanding

Here is a suggested use of the Oral Exercises to check students' understanding as you teach the lesson.
Oral Exs. 1–6: use after Example 1.

Guided Practice

Use synthetic substitution to find the following.

1. $P(2)$ if $P(x) = x^3 - 3x^2 + x - 1$ -3
2. $P(-1)$ if $P(x) = x^5 - x^3 + x^2 - 2$ -1

Is the first polynomial a factor of the second?

3. $a + 3$; $a^5 + 3a^4 - 2a^3 - 6a^2 + a + 3$ yes
4. $b - 4$; $b^3 - 5b^2 - 2b - 8$ no
5. $k + \sqrt{3}$; $k^3 + 2k^2 - 3k - 6$ yes

Solve each equation given the indicated root.

6. $x^3 - 2x^2 - x + 2 = 0$; -1 $\{-1, 1, 2\}$
7. $2y^3 - 5y^2 - 4y + 3$; 3 $\left\{3, -1, \frac{1}{2}\right\}$
8. $2k^3 + \sqrt{3}k^2 - 15k + 6\sqrt{3}$; $\sqrt{3}$ $\left\{\sqrt{3}, -2\sqrt{3}, \frac{\sqrt{3}}{2}\right\}$
9. $2m^3 - 5m^2 - 13m - 5$; $-\frac{1}{2}$ $\left\{-\frac{1}{2}, \frac{3+\sqrt{29}}{2}, \frac{3-\sqrt{29}}{2}\right\}$

Example 3 Find a polynomial equation with integral coefficients that has 1, -2, and $\frac{3}{2}$ as roots.

Solution By the factor theorem, the required polynomial must have factors $(x - 1)$, $(x - (-2))$, and $\left(x - \frac{3}{2}\right)$. To obtain integers, you can use $(2x - 3)$ in place of $\left(x - \frac{3}{2}\right)$, since both produce the root $\frac{3}{2}$. Therefore, a solution is

$$(x - 1)(x + 2)(2x - 3) = 0, \quad \text{or}$$
$$2x^3 - x^2 - 7x + 6 = 0 \quad \textbf{\textit{Answer}}$$

In Example 3 many polynomial equations with the given roots are possible. If a is any nonzero constant, then $a(x - 1)(x + 2)(2x - 3) = 0$ has the given roots 1, -2, and $\frac{3}{2}$.

When a given root is a multiple root, the corresponding factor of the polynomial will be to a power greater than 1. For example, an equation with the root 0 and the double root 3 is $x(x - 3)^2 = 0$, or $x^3 - 6x^2 + 9x = 0$.

Suppose that you are given one root r of the nth degree polynomial equation $P(x) = 0$. By the theorems of this lesson,

$$P(x) = Q(x) \cdot (x - r),$$

where $Q(x)$ is a polynomial of degree $n - 1$. You can find the remaining roots of $P(x) = 0$ by solving the **depressed equation**

$$Q(x) = 0$$

because any root of $Q(x)$ is also a root of $P(x)$.

Example 4 Solve $x^3 + x + 10 = 0$ given that -2 is a root.

Solution To find the depressed equation, divide $P(x) = x^3 + x + 10$ by $x - (-2)$ or $x + 2$.

-2	1	0	1	10
		-2	4	-10
	1	-2	5	0

So $x^3 + x + 10 = (x + 2)(x^2 - 2x + 5)$ and the *depressed equation* is
$$x^2 - 2x + 5 = 0.$$

To solve the depressed equation, use the quadratic formula:
$$x = \frac{2 \pm \sqrt{4 - 20}}{2} = 1 \pm 2i$$

$\therefore$ the solution set is $\{-2, 1 + 2i, 1 - 2i\}$. ***Answer***

Oral Exercises

Use the remainder theorem to find the remainder when $P(x)$ is divided by each of the two given binomials.

Sample $P(x) = x^3 + x^2 + 3x + 3$ **a.** $x - 1$ **b.** $x + 2$

Solution **a.** $P(1) = 1^3 + 1^2 + 3(1) + 3 = 8$; the remainder is 8.
b. $P(-2) = (-2)^3 + (-2)^2 + 3(-2) + 3 = -7$; the remainder is -7.

1. $P(x) = x^3 - 3x^2 - 2x + 2$
a. $x - 1$ −2 **b.** $x + 1$ 0

2. $P(x) = x^3 - x^2 - x - 2$
a. $x - 2$ 0 **b.** $x + 1$ −3

3. $P(x) = x^3 - 2x^2 - x + 2$
a. $x - 1$ 0 **b.** $x - 2$ 0

4. $P(x) = x^4 - x^3 + x^2 - x + 1$
a. $x - 1$ 1 **b.** $x + 1$ 5

5. $P(x) = x^6 - x^4 - x^2 + 1$
a. $x + 1$ 0 **b.** $x - 1$ 0

6. $P(x) = x^4 - 4x^2 + 2x - 4$
a. $x + 2$ −8 **b.** $x - 2$ 0

Written Exercises

Use synthetic substitution to find $P(c)$ for the given polynomial $P(x)$ and the given number c.

A

1. $P(x) = x^3 - 2x^2 - 5x - 7$; $c = 4$ 5

2. $P(x) = x^3 + 4x^2 - 8x - 6$; $c = -5$ 9

3. $P(x) = 2 - 5x + 3x^2 + 2x^3$; $c = -3$ −10

4. $P(x) = 1 - 7x - 4x^2 + x^3$; $c = 6$ 31

5. $P(x) = 4x^3 - 4x^2 + 5x + 1$; $c = \frac{3}{2}$ 13

6. $P(x) = 6x^3 - x^2 + 4x + 3$; $c = -\frac{1}{3}$ $\frac{4}{3}$

7. $P(x) = 2x^4 - x^3 + x - 2$; $c = -\frac{3}{2}$ 10

8. $P(x) = 2x^4 - 3x^3 + 3x^2 + 1$; $c = -\frac{1}{2}$ $\frac{9}{4}$

Use the factor theorem to determine whether the binomial is a factor of the given polynomial.

9. $x + 1$; $P(x) = x^7 - x^5 + x^3 - x$ Yes

10. $t + 1$; $P(t) = t^5 + t^4 + t^3 + t^2 + t + 1$ Yes

11. $s + 1$; $P(s) = s^6 - s^5 - s + 1$ No

12. $z + 2$; $P(z) = z^5 + 2z^4 + z^3 + 2z^2 + z + 2$ Yes

13. $x - \sqrt{3}$; $P(x) = x^3 - 2x^2 - 3x + 6$ Yes

14. $t + \sqrt{2}$; $P(t) = t^5 + t^4 + 4t + 4$ No

15. $z - i$; $P(z) = z^7 + z^6 + z^5 + z^4 + z^3 + z^2 + z + 1$ Yes

16. $z + 2i$; $P(z) = z^3 + z^2 + 4z + 4$ Yes

In Exercises 17–20, a root of the equation is given. Solve the equation.

17. $x^3 + 3x^2 - 3x - 9 = 0$; -3 $\{-3, \pm\sqrt{3}\}$

18. $2x^3 + 9x^2 + 7x - 6 = 0$; -2 $\{-3, -2, \frac{1}{2}\}$

19. $t^3 - 11t + 20 = 0$; -4 $\{-4, 2 \pm i\}$

20. $2z^3 + z^2 - 8z + 3 = 0$; $\frac{3}{2}$ $\{\frac{3}{2}, -1 \pm \sqrt{2}\}$

Summarizing the Lesson

In this lesson students learned to use synthetic division, along with the remainder and factor theorems, to find factors of polynomials and to solve polynomial equations. Ask students to state the remainder and factor theorems in words.

Suggested Assignments

Average Algebra
375/2–36 even
S 376/Mixed Review

Average Alg. and Trig.
375/2–36 even
S 376/Mixed Review

Extended Algebra
375/2–38 even
S 376/Mixed Review

Extended Alg. and Trig.
375/2–38 even
S 376/Mixed Review

Supplementary Materials

Study Guide	pp. 127–128
Practice Master	46
Test Master	31
Resource Book	p. 132

Additional Answers
Written Exercises

21. $x^3 - 7x + 6 = 0$
22. $x^3 + 3x^2 - 4x - 12 = 0$
23. $x^4 + 2x^3 - 5x^2 - 6x = 0$
24. $x^3 + 2x^2 + x + 2 = 0$
25. $2x^3 - 3x^2 - 11x + 6 = 0$
26. $2x^3 - 3x^2 - 5x + 6 = 0$
27. $x^4 - x^3 - 12x^2 - 4x + 16 = 0$
28. $x^3 - 5x^2 + 11x - 15 = 0$
33.

-1	1	0	2	8	5
		-1	1	-3	-5
-1	1	-1	3	5	0
		-1	2	-5	
	1	-2	5	0	

34.

$\frac{3}{2}$	4	-12	13	-12	9
		6	-9	6	-9
$\frac{3}{2}$	4	-6	4	-6	0
		6	0	6	
	4	0	4	0	

38. $x - a - b = x - (a + b)$ is a factor of $x^3 - a^3 - b^3 - 3ab(a + b)$ if $a + b$ is a root of $x^3 - a^3 - b^3 - 3ab(a + b) = 0$; $(a + b)^3 - a^3 - b^3 - 3ab(a + b) = a^3 + 3a^2b + 3ab^2 + b^3 - a^3 - b^3 - 3a^2b - 3ab^2 = 0$.

39. $x(x - a)$ is a factor of $ax^n - a^nx$ if 0 and a are roots of $ax^n - a^nx = 0$; $a(0^n) - a^n(0) = 0 - 0 = 0$; for every positive integer n and every nonzero number a, $a(a^n) - a^n(a) = a^{n+1} - a^{n+1} = 0$.

Find a polynomial equation with integral coefficients that has the given numbers as roots.

21. 1, 2, -3 **22.** -2, 2, -3 **23.** 0, -1, 2, -3 **24.** -2, $-i$, i

B **25.** $\frac{1}{2}$, -2, 3 **26.** 1, $-\frac{3}{2}$, 2 **27.** -2 (double), 1, 4 **28.** 3, $1 + 2i$, $1 - 2i$

Solve each equation given the two indicated roots. (*Hint:* Perform the process illustrated in Example 4 twice.) 29. $\{-1, 4, -2, 2\}$

29. $x^4 - 3x^3 - 8x^2 + 12x + 16 = 0$; -1, 4

30. $2x^4 - 5x^3 - 11x^2 + 20x + 12$; -2, 3 $\left\{-2, 3, 2, -\frac{1}{2}\right\}$

31. $3x^4 + 5x^3 - 7x^2 - 3x + 2$; 1, $-\frac{2}{3}$ $\left\{1, -\frac{2}{3}, -1 \pm \sqrt{2}\right\}$

32. $2x^4 - 3x^3 - 3x - 2 = 0$; 2, $-\frac{1}{2}$ $\left\{2, -\frac{1}{2}, \pm i\right\}$

Show that r is a double root of $P(x) = 0$ by verifying that $x - r$ is a factor of both $P(x)$ and $P(x) \div (x - r)$.

33. $P(x) = x^4 + 2x^2 + 8x + 5$; $r = -1$

34. $P(x) = 4x^4 - 12x^3 + 13x^2 - 12x + 9$; $r = \frac{3}{2}$

Find the number of times r is a root of $P(x) = 0$.

35. $P(x) = x^4 + 4x^3 - 16x - 16$; $r = -2$ 3

36. $P(x) = x^5 + 3x^4 + 2x^3 - 2x^2 - 3x - 1$; $r = -1$ 4

37. When $x^3 - 5x^2 + 4x + 5$ is divided by $x - c$, the quotient is $x^2 - 3x - 2$ and the remainder is 1. Find c. 2

C **38.** Show that $x - a - b$ is a factor of $x^3 - a^3 - b^3 - 3ab(a + b)$.

39. Show that for every nonzero number a and integer $n > 1$, $x(x - a)$ is a factor of $ax^n - a^nx$.

Mixed Review Exercises

Divide.

1. $\frac{x^2 + 3x - 1}{x + 2}$ $x + 1 + \frac{-3}{x + 2}$

2. $\frac{y^3 - 1}{y^2 + y + 1}$ $y - 1$

3. $\frac{4u^2 - 8u + 3}{2u - 1}$ $2u - 3$

4. $\frac{3w^4 - 4w^2 - 5}{w^2 - 2}$ $3w^2 + 2 + \frac{-1}{w^2 - 2}$

5. $\frac{a^5 - 4a^3 + 3a}{a + 1}$ $a^4 - a^3 - 3a^2 + 3a$

6. $\frac{3c^3 + 8c^2 - 8}{3c^2 + 2c - 4}$ $c + 2$

Without solving, determine whether the roots of each equation are rational, irrational, or imaginary.

7. $2x^2 - 3x + 2 = 0$ Imag. **8.** $2x^2 - 3x + 1 = 0$ Rat. **9.** $2x^2 - 3x - 1 = 0$ Irrat.

8-6 Some Useful Theorems

Objective To find or solve a polynomial equation with real coefficients and positive degree n by using these facts:
1. There are exactly n roots.
2. The imaginary roots occur in conjugate pairs.
3. Descartes' rule of signs gives information about the numbers of positive and negative real roots.

The following equations can be solved using methods you already know. Each is a polynomial equation of the form $P(x) = 0$, and the **degree of the polynomial equation** is the degree of $P(x)$.

Equation	Roots	Degree of equation Number of roots
$x + 1 = 0$	-1	1
$x^2 - 4x + 13 = 0$	$2 + 3i,\ 2 - 3i$	2
$x^3 + 4x^2 + 4x = 0$	$0,\ -2,\ -2$	3
$x^4 - 10x^2 + 9 = 0$	$1,\ -1,\ 3,\ -3$	4
$x^5 - 16x = 0$	$0,\ 2,\ -2,\ 2i,\ -2i$	5

Notice that in each case the degree of the equation and the number of roots of the equation are equal, provided that you count multiple roots as many times as each appears. This relationship is true for any polynomial equation.

Theorem

Every polynomial equation with complex coefficients and positive degree n has exactly n roots.

The theorem above follows from the *Fundamental Theorem of Algebra,* proved by the German mathematician Karl Friedrich Gauss (1777–1855):

> Every polynomial equation with complex coefficients and positive degree has at least one complex root.

In the examples given above, the imaginary roots occurred in conjugate pairs. For example, $x^2 - 4x + 13 = 0$ has roots $2 + 3i$ and $2 - 3i$. This is true of every polynomial equation with *real* coefficients.

Conjugate Root Theorem

If a polynomial equation with *real* coefficients has $a + bi$ as a root (a and b real, $b \neq 0$), then $a - bi$ is also a root.

The proof of this theorem is given in Exercise 29, page 381.

Teaching Suggestions p. T111

Suggested Extensions p. T111

Warm-Up Exercises

Solve using factoring, completing the square, or the quadratic formula.

1. $x^2 + 6x + 5 = 0$ $\{-5, -1\}$
2. $x^2 + 6x + 4 = 0$ $\{-3 \pm \sqrt{5}\}$
3. $2x^2 - 6x + 5 = 0$ $\left\{\frac{3 \pm i}{2}\right\}$
4. $x^3 + 6x^2 + 5x = 0$ $\{-5, -1, 0\}$
5. $x^4 + 6x^2 + 5 = 0$ $\{\pm i\sqrt{5}, \pm i\}$

Motivating the Lesson

Although students have already learned a number of techniques for solving polynomial equations, all these techniques have their limitations. Today students will learn theorems that will extend their ability to solve polynomial equations.

Common Errors

When applying Descartes' rule of signs, students may need to be reminded to ignore terms having coefficent 0.

Chalkboard Examples

1. Give the degree and the number of roots of:
   a. $x^2 + 8x + 16 = 0$
      degree = 2;
      number of roots = 2
   b. $x^3 + 125 = 0$
      degree = 3;
      number of roots = 3

(continued)

Chalkboard Examples

(continued)

2. Solve $P(x) = x^4 - 6x^3 + 6x^2 + 24x - 40 = 0$, given that $3 + i$ is a root.
   $P(x) = [x - (3 + i)] \cdot [x - (3 - i)]Q(x) = (x^2 - 6x + 10)Q(x)$
   $Q(x) = P(x) \div (x^2 - 6x + 10) = x^2 - 4$
   The depressed equation is $x^2 - 4 = 0$.
   $(x - 2)(x + 2) = 0$
   $x = 2$ or $x = -2$
   $\{3 + i, 3 - i, 2, -2\}$

3. List the possibilities for the nature of the roots (positive real, negative real, and imaginary) of $P(x) = 0$ for the polynomial $P(x) = 5x^4 - 3x^2 + 6x - 10$.
   $P(x)$ has 3 variations in sign. Thus $P(x) = 0$ has 3 or 1 positive root(s).
   $P(-x)$ has 1 variation in sign. Thus $P(x) = 0$ has 1 negative root.

pos. real	neg. real	imag.
3	1	0
1	1	2

Problem Solving Strategies

Organized lists are useful both in this lesson when using Descartes' rule of signs and in the next lesson when using the rational root theorem.

Check for Understanding

Oral Exs. 1–9: use after discussing the conjugate root theorem.

Oral Exs. 10–15: use after Example 3.

Example 1 Find a cubic equation with integral coefficients that has 2 and $3 - i$ as roots.

Solution The third root must be $3 + i$, the conjugate of $3 - i$.
An equation with these three roots is:

$$(x - 2)[x - (3 - i)][x - (3 + i)] = 0$$
$$(x - 2)[x^2 - (3 + i)x - (3 - i)x + (9 - i^2)] = 0$$
$$(x - 2)(x^2 - 6x + 10) = 0$$
$$x^3 - 8x^2 + 22x - 20 = 0 \quad \textit{Answer}$$

Example 2 Solve $x^4 - 12x - 5 = 0$ given that $-1 + 2i$ is a root.

Solution Let $P(x) = x^4 - 12x - 5$.

1. If $-1 + 2i$ is a root of $P(x) = 0$, then $-1 - 2i$ is also a root. Therefore, $x - (-1 + 2i)$ and $x - (-1 - 2i)$ are factors of $P(x)$. Their product,
$$[x - (-1 + 2i)][x - (-1 - 2i)] = x^2 + 2x + 5,$$
is also a factor of $P(x)$; that is, $P(x) = (x^2 + 2x + 5) \cdot Q(x)$.
2. Using long division, $Q(x) = P(x) \div (x^2 + 2x + 5) = x^2 - 2x - 1$.
3. The resulting equation is $x^2 - 2x - 1 = 0$, and its roots are $1 + \sqrt{2}$ and $1 - \sqrt{2}$.

$\therefore$ the solution set of the given equation is $\{-1 + 2i, -1 - 2i, 1 + \sqrt{2}, 1 - \sqrt{2}\}$. ***Answer***

The next theorem, **Descartes' rule of signs,** gives you information about the number of real roots of a polynomial equation with *real* coefficients. Consider a simplified polynomial, $P(x)$, with terms arranged in decreasing degree of x. Whenever the coefficients of two adjacent terms have opposite signs, we say that $P(x)$ has a **variation in sign.** For example,

$$\underset{1}{x^6 - 2x^4} - \underset{2}{5x^2 + 3x} - \underset{3}{\;6}$$

has three variations in sign. (Notice that you ignore any "missing" terms.)

Descartes' Rule of Signs

Let $P(x)$ be a simplified polynomial with real coefficients and terms arranged in decreasing degree of x.

1. The number of positive real roots of $P(x) = 0$ equals the number of variations of sign of $P(x)$ or is fewer than this number by an even integer.
2. The number of negative real roots of $P(x) = 0$ equals the number of variations of sign of $P(-x)$ or is fewer than this number by an even integer.

Sometimes Descartes' rule gives complete information about the number of roots of various kinds. For example, $P(x) = x^4 - 12x - 5$ has one variation in sign, so that $P(x) = 0$ has exactly one positive real root. Since the polynomial $P(-x) = (-x)^4 - 12(-x) - 5 = x^4 + 12x - 5$ also has one variation in sign, $P(x) = 0$ has exactly one negative real root. Because $P(x) = 0$ has four roots in all, two of them must be imaginary. This is what was found in Example 2.

Usually, instead of giving complete information, Descartes' rule leaves us with several possibilities.

Example 3 List the possibilities for the nature of the roots (positive real, negative real, and imaginary) for the equation $P(x) = 0$, where

$$P(x) = x^5 + x^4 - 3x^2 + 4x + 6.$$

Solution

1. $P(x)$ has two variations in sign, so by part (1) of Descartes' rule the number of positive real roots of $P(x) = 0$ is 2 or 0.
2. $P(-x) = (-x)^5 + (-x)^4 - 3(-x)^2 + 4(-x) + 6$
   $= -x^5 + x^4 - 3x^2 - 4x + 6$

   $P(-x)$ has three variations in sign, so by part (2) of Descartes' rule the number of negative real roots of $P(x) = 0$ is 3 or 1.
3. Since $P(x) = 0$ has five roots in all, there are only four possibilities.

Number of positive real roots	Number of negative real roots	Number of imaginary roots
2	3	0
2	1	2
0	3	2
0	1	4

Oral Exercises

In Exercises 1–6, a polynomial equation with real coefficients has the given root(s). What other root(s) must it have?

1. $2 - 3i$ $2 + 3i$ **2.** $2i - 1$ $-1 - 2i$ **3.** $-i$, $2i$ i, $-2i$

4. $3i$, $-\frac{1}{3}i$ $-3i$, $\frac{1}{3}i$ **5.** $i + 1$, $i - 1$ $1 - i$, $-1 - i$ **6.** $2 - i$, $-2 + i$ $2 + i$, $-2 - i$

In Exercises 7 and 8, explain why each statement is true.

7. Every third-degree polynomial equation with real coefficients has at least one real root.

8. Every fifth-degree polynomial equation with real coefficients has at least one real root.

9. What general statement is suggested by Exercises 7 and 8?

Guided Practice

Find a cubic equation with real coefficients that has the following roots.

1. -1, $3 - i$ $x^3 - 5x^2 + 4x + 10 = 0$

2. $\sqrt{2}$, $i\sqrt{5}$ $x^3 - \sqrt{2}x^2 + 5x - 5\sqrt{2} = 0$

3. Find the remaining roots of $x^4 + x^3 + 6x^2 - 14x - 20 = 0$ if two roots are 2 and $-1 + 3i$. $-1 - 3i$, -1

List the possibilities for the nature of the roots of each equation.

4. $2x^4 - 3x - 1 = 0$
1 pos., 1 neg., 2 imag.

5. $x^3 + x + 5 = 0$
1 neg., 2 imag.

Summarizing the Lesson

In this lesson students learned to find and solve polynomial equations using several important theorems, such as the conjugate root theorem and Descartes' rule of signs. Ask students to state Descartes' rule of signs and explain what information it provides, as well as the limitations of that information.

Additional Answers Oral Exercises

7. Every third-degree equation has three roots; since complex roots occur in pairs, at least one root must be real.

8. Every fifth-degree equation has five roots; since complex roots occur in pairs, at least one root must be real.

9. Every polynomial equation of odd degree has at least one real root.

Suggested Assignments

Average Algebra
380/1–23 odd, 24, 25
S 367/29, 31

Average Alg. and Trig.
380/1–23 odd, 24, 25
S 367/29, 31

Extended Algebra
380/1–23 odd, 24–27
S 367/29, 31, 33

Extended Alg. and Trig.
380/1–23 odd, 24–27
S 367/29, 31, 33

Supplementary Materials

Study Guide pp. 129–130
Practice Master 47

Additional Answers
Written Exercises

23. a. $(-1+i)^3 + 2(-1+i)^2 + (-1+i) - 1 + i = 2 + 2i - 4i - 1 + i - 1 + i = 0$; $-1+i$ is a root.

b. $(-1-i)^3 + 2(-1-i)^2 + (-1-i) - 1 + i = 2 - 2i + 4i - 1 - i - 1 + i = 2i$; $-1-i$ is not a root.

24. a. $(2-i)^3 - 2(2-i)^2 + 2(2-i) + 5i = 2 - 11i - 6 + 8i + 4 - 2i + 5i = 0$; $2-i$ is a root.

b. $(2+i)^3 - 2(2+i)^2 + 2(2+i) + 5i = 2 + 11i - 6 - 8i + 4 + 2i + 5i = 10i$; $2+i$ is not a root.

26.

No. of pos. real roots	No. of neg. real roots	No. of imag. roots
3	0	0
2	1	0
1	2	0
1	0	2
0	1	2
0	3	0

Use Descartes' rule of signs to give the possibilities for the numbers of positive and negative real roots.

10. $x^3 - 3x^2 - 6 = 0$ 1 pos.; 0 neg.

11. $x^3 + x^2 + 2 = 0$ 0 pos.; 1 neg.

12. $x^4 - 2x^3 - 4 = 0$ 1 pos.; 1 neg.

13. $x^4 - 6x + 3 = 0$ 2 or 0 pos.; 0 neg.

14. $x^5 - x^2 + 3x - 1 = 0$ 3 or 1 pos.; 0 neg.

15. $x^6 - 10 = 0$ 1 pos.; 1 neg.

Written Exercises

Find a cubic equation with integral coefficients that has the given numbers as roots.

A

1. $-1, 5i$ $x^3 + x^2 + 25x + 25 = 0$

2. $3, i\sqrt{2}$ $x^3 - 3x^2 + 2x - 6 = 0$

3. $-2, -1 + i$ $x^3 + 4x^2 + 6x + 4 = 0$

4. $1, 2 - 3i$ $x^3 - 5x^2 + 17x - 13 = 0$

In Exercises 5–8, all but one of the equation's roots are given. Find the remaining root. Check your answer by substituting it for *x* in the equation.

5. $x^3 - 3x^2 + 4x - 12 = 0$; 3 and $2i$ $-2i$

6. $x^3 - 2x + 4 = 0$; -2 and $1 - i$ $1 + i$

7. $x^4 - 2x^3 + 4x^2 + 2x - 5 = 0$; 1, -1, and $1 - 2i$ $1 + 2i$

8. $x^4 - 3x^3 + 4x^2 - 6x + 4 = 0$; 1, 2, and $i\sqrt{2}$ $-i\sqrt{2}$

In Exercises 9–12, a root of the equation is given. Solve the equation.

9. $x^3 + x - 10 = 0$; $-1 + 2i$ $\{-1 \pm 2i, 2\}$

10. $2x^3 - x^2 + 10x - 5 = 0$; $i\sqrt{5}$ $\left\{\pm i\sqrt{5}, \frac{1}{2}\right\}$

11. $x^4 - 6x^3 + 60x - 100 = 0$; $3 + i$ $\{3 \pm i, \pm\sqrt{10}\}$

12. $x^4 - 5x^2 - 10x - 6 = 0$; $-1 + i$ $\{-1 \pm i, -1, 3\}$

List the possibilities for the nature of the roots of each equation.
Answers are given in this order: pos. real, neg. real, imaginary.

B

13. $x^4 + 3x^2 - 4 = 0$ 1, 1, 2

14. $x^4 - x + 3 = 0$ 2, 0, 2; or 0, 0, 4

15. $x^4 + 2x^3 + x^2 + 1 = 0$ 0, 2, 2; or 0, 0, 4

16. $x^4 - 3x^3 + 5x^2 - 2x + 5 = 0$ (See below.)

17. $x^5 - x^3 - x - 2 = 0$ 1, 2, 2; or 1, 0, 4

18. $x^5 - x^4 + 2x - 3 = 0$ 3, 0, 2; or 1, 0, 4

19. $x^5 - x^3 - x^2 + x - 2 = 0$ 3, 2, 0; 1, 2, 2; 3, 0, 2; or 1, 0, 4

20. $x^6 + x^5 + x^4 + 3x - 2 = 0$

16. 4, 0, 0; 2, 0, 2; or 0, 0, 4 **20.** 1, 3, 2; or 1, 1, 4

Find a fourth-degree polynomial equation with integral coefficients that has the given numbers as roots.

21. $2i, 1 - i$ $x^4 - 2x^3 + 6x^2 - 8x + 8 = 0$

22. $1 + i, 2 + i$ $x^4 - 6x^3 + 15x^2 - 18x + 10 = 0$

For the equations in Exercises 23 and 24, show that (a) the given number is a root, and (b) its conjugate is *not* a root.

23. $x^3 + 2x^2 + x - 1 + i = 0$; $-1 + i$

24. $x^3 - 2x^2 + 2x + 5i = 0$; $2 - i$

25. Explain why Exercises 23 and 24 do *not* contradict the conjugate root theorem. The theorem holds for equations with *real* coefficients.

26. There can be six possibilities for the nature of the roots of a cubic equation with real coefficients. List them in a table.

In Exercises 27 and 28, find one real root of the equation by inspection. Then use Descartes' rule to show that there are no other real roots.

C **27.** $x^5 - x^4 + 2x^3 - 2x^2 + 3x - 3 = 0$ 1 **28.** $x^5 + x^4 + x^3 + x^2 + 2x + 2 = 0$ −1

29. A proof of the conjugate root theorem is outlined below. Let $P(x)$ be a polynomial with real coefficients, and let $a + bi$ $(b \neq 0)$ be an imaginary root of $P(x) = 0$. Justify each of the following statements.

1. If $S(x) = [x - (a + bi)][x - (a - bi)] = x^2 - 2ax + (a^2 + b^2)$, then $S(a + bi) = 0$ and $S(a - bi) = 0$.
2. There are real numbers c and d and a polynomial $Q(x)$ for which $P(x) = Q(x)S(x) + cx + d$.
3. Setting $x = a + bi$, you obtain $0 = Q(a + bi) \cdot 0 + c(a + bi) + d$.
4. $c(a + bi) + d = 0$, or $(ac + d) + bci = 0$
5. $bc = 0$, and since $b \neq 0$, $c = 0$
6. $ac + d = 0$, and since $c = 0$, $d = 0$
7. $P(x) = Q(x)S(x)$
8. $P(a - bi) = Q(a - bi)S(a - bi) = Q(a - bi) \cdot 0 = 0$

Self-Test 2

Vocabulary

division algorithm (p. 364)
synthetic division (p. 368)
remainder theorem (p. 372)
synthetic substitution (p. 373)
factor theorem (p. 373)
depressed equation (p. 374)
degree of a polynomial equation (p. 377)
conjugate-root theorem (p. 377)
Descartes' rule of signs (p. 378)
variation in sign (p. 378)

1. Divide: $\dfrac{x^4 + x^3 - 5x^2 + 13x - 6}{x^2 + 3x - 2}$ $x^2 - 2x + 3$ **Obj. 8-3, p. 364**

2. Use synthetic division to divide $2x^4 - 3x^3 - 5x + 4$ by $x - 2$. $2x^3 + x^2 + 2x - 1 + \frac{2}{x-2}$ **Obj. 8-4, p. 368**

3. If $P(x) = 2x^3 - 4x^2 + x - 5$, use synthetic substitution to find $P(3)$. 16 **Obj. 8-5, p. 372**

4. Determine whether $x + 1$ is a factor of $x^7 + x^4 + x + 1$. Yes

5. Find a cubic equation with integral coefficients that has $\frac{1}{2}$, −1, and 3 as roots. $2x^3 - 5x^2 - 4x + 3 = 0$

6. Consider the equation $2x^4 - 9x^3 + 13x^2 - x - 5 = 0$. **Obj. 8-6, p. 377**
a. What are the possibilities for the numbers of positive and negative real roots? 3 or 1 pos.; 1 neg.
b. Given that one root is $2 + i$, solve the equation. $\left\{2 \pm i, 1, -\frac{1}{2}\right\}$

7. Find a cubic equation with integral coefficients that has −2 and $1 + 3i$ as roots. $x^3 + 6x + 20 = 0$

Check your answers with those at the back of the book.

27. The polynomial can be factored as $(x - 1) \cdot (x^4 + 2x^2 + 3) = 0$; by Descartes' rule the second factor has neither positive nor negative real roots.

28. The polynomial can be factored as $(x + 1) \cdot (x^4 + x^2 + 2) = 0$; by Descartes' rule the second factor has neither positive nor negative real roots.

(continued on p. 390)

Quick Quiz

1. Divide $50 - 35x + 3x^3 - x^4$ by $4 - 2x - x^2$. $x^2 - 5x + 14 + \dfrac{13x - 6}{4 - 2x - x^2}$

2. Divide by synthetic division: $\dfrac{2x^4 + 5x^3 + 4x^2 - 9}{x + 2}$
$2x^3 + x^2 + 2x - 4 - \dfrac{1}{x + 2}$

3. Use synthetic substitution to find $P(-2)$ if $P(x) = 5x^3 + 8x^2 + 2x - 2$. −14

4. Is $x - 1$ a factor of $x^{15} - 2x + 1$? yes

5. Find a fourth-degree polynomial equation with integral coefficients that has $1 - 2i$ as a double root.
$x^4 - 4x^3 + 14x^2 - 20x + 25 = 0$

6. a. List the possibilities for the nature of the roots of $x^4 - 2x^3 - 6x^2 + 22x - k$ if $k > 0$. 3 pos. and 1 neg. or 1 pos., 1 neg., and 2 imag.
b. If $k = 15$ in the equation in part (a) and one root is $2 - i$, find the other three roots.
$2 + i$, 1, −3

Teaching Suggestions p. T111

Reading Algebra p. T111

Suggested Extensions p. T111

Warm-Up Exercises

List how many are possible: positive real roots, negative real roots, imaginary roots.

1. $y^4 - y^3 + y^2 - y + 1 = 0$ 4, 0, 0; 2, 0, 2; 0, 0, 4
2. $a^4 - 6a^2 - 8 = 0$ 1, 1, 2
3. $m^5 - 6m^2 + 3m - 2 = 0$ 3, 0, 2; 1, 0, 4

A root of the equation is given. Solve the equation.

4. $x^3 - 4x^2 + 3x - 12 = 0$; $i\sqrt{3}$ $\{\pm i\sqrt{3}, 4\}$
5. $x^4 - 4x^3 + 20x - 25 = 0$; $2 + i$ $\{2 \pm i, \pm\sqrt{5}\}$

Motivating the Lesson

Ask students to suppose they had to solve a given polynomial equation by direct substitution of randomly chosen numbers. Tell students that even if they limited themselves to rational numbers, their task would be hopeless, since there are infinitely many such numbers. Today's lesson presents a theorem that will produce a *finite* list of possible rational solutions, thereby making the task of solving the given equation manageable.

Chalkboard Examples

1. List all possible rational roots of $4x^4 + 3x^2 - 1 = 0$. $\pm 1, \pm\frac{1}{2}, \pm\frac{1}{4}$

Solving Polynomial Equations

8-7 Finding Rational Roots

Objective To find rational roots of polynomial equations with integral coefficients.

The theorems in Lesson 8-6 give you information about the nature of the roots of polynomial equations but not about how to find them. The following theorem (proved in Exercises 27 and 28, page 385) tells you how to find rational roots of any polynomial equation with integral coefficients. Recall that every rational number can be written as the quotient of relatively prime integers, that is, integers whose GCF is 1.

Rational Root Theorem

Suppose that a polynomial equation with integral coefficients has the root $\frac{h}{k}$, where h and k are relatively prime integers. Then h must be a factor of the constant term of the polynomial and k must be a factor of the coefficient of the highest-degree term.

For example, any rational root of

$$6x^3 + 8x^2 - 7x - 3 = 0$$

must have a numerator that is a factor of -3 (namely, ± 1 or ± 3) and a denominator that is a factor of 6 (namely, ± 1, ± 2, ± 3, or ± 6). There are, therefore, only twelve *possible* rational roots:

$$\pm\frac{1}{1}, \pm\frac{3}{1}, \pm\frac{1}{2}, \pm\frac{3}{2}, \pm\frac{1}{3}, \pm\frac{1}{6}$$

You can verify that the only root among the twelve possibilities is $-\frac{1}{3}$.

Example 1 Solve the equation $2x^4 + 3x^3 - 7x^2 + 3x - 9 = 0$ by first finding any rational roots.

Solution The only possible rational roots are

$$\pm 1, \pm 3, \pm 9, \pm\frac{1}{2}, \pm\frac{3}{2}, \text{ and } \pm\frac{9}{2}.$$

Since computing with integers is easier, try the integral possibilities first. When you substitute 1 and -1 in the equation, you find that neither is a root. To try other possibilities, it is best to use synthetic substitution.

$$\begin{array}{r|rrrrr} 3 & 2 & 3 & -7 & 3 & -9 \\ & & 6 & 27 & 60 & 189 \\ \hline & 2 & 9 & 20 & 63 & 180 \end{array} \leftarrow 3 \text{ is } \textit{not} \text{ a root.}$$

$$\begin{array}{r|rrrrr} -3 & 2 & 3 & -7 & 3 & -9 \\ & & -6 & 9 & -6 & 9 \\ \hline & 2 & -3 & 2 & -3 & 0 \end{array} \leftarrow -3 \text{ } \textit{is} \text{ a root.}$$

The depressed equation $2x^3 - 3x^2 + 2x - 3 = 0$ has

$$\pm 1, \pm 3, \pm \frac{1}{2}, \text{ and } \pm \frac{3}{2}$$

as possible rational roots. The possibilities 1, -1, and 3 were eliminated using the original equation. Since -3 may be a multiple root, it remains a possibility, along with $\pm\frac{1}{2}$ and $\pm\frac{3}{2}$. Try each of these until another root is found.

$$\begin{array}{r|rrrr} -3 & 2 & -3 & 2 & -3 \\ & & -6 & 27 & -87 \\ \hline & 2 & -9 & 29 & -90 \end{array} \qquad \begin{array}{r|rrrr} \frac{1}{2} & 2 & -3 & 2 & -3 \\ & & 1 & -1 & \frac{1}{2} \\ \hline & 2 & -2 & 1 & -\frac{5}{2} \end{array}$$

$$\begin{array}{r|rrrr} -\frac{1}{2} & 2 & -3 & 2 & -3 \\ & & -1 & 2 & -2 \\ \hline & 2 & -4 & 4 & -5 \end{array} \qquad \begin{array}{r|rrrr} \frac{3}{2} & 2 & -3 & 2 & -3 \\ & & 3 & 0 & 3 \\ \hline & 2 & 0 & 2 & 0 \end{array}$$

The synthetic substitution shows that $\frac{3}{2}$ is another root. The second depressed equation is $2x^2 + 2 = 0$ (or $x^2 + 1 = 0$). From your study of quadratic equations you know that it has roots $\pm i$.

$\therefore$ the original equation has solution set $\{-3, \frac{3}{2}, i, -i\}$. ***Answer***

Sometimes you can eliminate several possibilities by using Descartes' rule of signs. In Example 1, the given equation $2x^4 + 3x^3 - 7x^2 + 3x - 9 = 0$ has only one negative root. Therefore, after finding that -3 is a root, you could have eliminated the other negative possibilities, leaving only $\frac{1}{2}$ and $\frac{3}{2}$ to try.

The rational root theorem can be used to prove that certain numbers are irrational.

Example 2 Show that $\sqrt[3]{9}$ is irrational.

Solution $\sqrt[3]{9}$ is a root of the equation $x^3 - 9 = 0$. The only possible rational roots of this equation are ± 1, ± 3, and ± 9. None of these six numbers satisfies the equation. Since $x^3 - 9 = 0$ has no rational roots, $\sqrt[3]{9}$ is irrational.

2. If the equation in Exercise 1 has one or more rational roots, solve it completely.

$$\begin{array}{r|rrrrr} \frac{1}{2} & 4 & 0 & 3 & 0 & -1 \\ & & 2 & 1 & 2 & 1 \\ \hline -\frac{1}{2} & 4 & 2 & 4 & 2 & 0 \\ & & -2 & 0 & -2 & \\ \hline & 4 & 0 & 4 & 0 & \end{array}$$

$\therefore \frac{1}{2}$ and $-\frac{1}{2}$ are roots, and the depressed equation is

$4x^2 + 4 = 0.$

$x^2 = -1$

$x = \pm i$

$\left\{\frac{1}{2}, -\frac{1}{2}, i, -i\right\}$

3. Show that $\sqrt[4]{3}$ is irrational. $\sqrt[4]{3}$ is a root of the equation $x^4 - 3 = 0$. The only possible rational roots are ± 1 and ± 3. None of these four numbers satisfies the equation. Since the equation has no rational roots, $\sqrt[4]{3}$ is irrational.

Check for Understanding

Here is a suggested use of the Oral Exercises to check students' understanding as you teach the lesson.

Oral Exs. 1–12: use after discussing the rational root theorem on page 382.

Oral Exs. 13–15: use after Example 1.

Guided Practice

1. Find any rational roots of $2x^3 - 5x^2 + 6x - 2 = 0$. If the equation has at least one rational root, solve it completely. $\left\{\frac{1}{2}, 1 \pm i\right\}$

(continued)

Guided Practice

(continued)

2. Show that $\sqrt{7}$ is irrational. $\sqrt{7}$ is a root of $x^2 - 7 = 0$. The possible rational roots are ±1 and ±7. None of these satisfies the equation. Since $x^2 - 7 = 0$ has no rational roots, $\sqrt{7}$ is irrational.
3. Verify that $\sqrt{3} - \sqrt{2}$ is a root of $x^4 - 10x^2 + 1 = 0$. Show that the number is irrational. $(\sqrt{3} - \sqrt{2})^4 - 10(\sqrt{3} - \sqrt{2})^2 + 1 = 49 - 20\sqrt{6} - 10(5 - 2\sqrt{6}) + 1 = 0$. The possible rational roots are ±1. Since $\sqrt{3} - \sqrt{2} \neq 1$ and $\sqrt{3} - \sqrt{2} \neq -1$, $\sqrt{3} - \sqrt{2}$ is irrational.

Summarizing the Lesson

In this lesson students learned how to use the rational root theorem to find rational roots of polynomial equations and to show that certain numbers are irrational. Ask students to state the rational root theorem.

Suggested Assignments

Average Algebra
384/2–20 even, 21–25 odd
S 385/Mixed Review
R 381/Self-Test 2

Average Alg. and Trig.
384/2–20 even, 21–25 odd
S 385/Mixed Review
R 381/Self-Test 2

Extended Algebra
384/2–20 even, 21–26
S 385/Mixed Review
R 381/Self-Test 2

Extended Alg. and Trig.
384/2–20 even, 21–26
S 385/Mixed Review
R 381/Self-Test 2

Oral Exercises

List the possible rational roots of each equation.

1. $x^3 - 7x + 6 = 0$
2. $x^3 + x^2 - 4x + 4 = 0$
3. $x^3 - 3x^2 + 2x - 8 = 0$
4. $x^4 + 3x^3 - x^2 - 9x - 6 = 0$
5. $x^4 - 2x^2 - 16x - 15 = 0$
6. $x^4 + 3x^2 - 8x + 10 = 0$
7. $2x^3 + 7x^2 + 6x - 5 = 0$
8. $2x^3 - 5x^2 - 11x - 4 = 0$
9. $2x^3 - 11x^2 + 16x - 6 = 0$
10. $4x^4 + 4x^3 + 17x^2 + 16x + 4 = 0$
11. $3x^4 + 4x^3 - x^2 + 4x - 4 = 0$
12. $6x^4 - 7x^3 + 8x^2 - 7x + 2 = 0$

Tell how you know that the following equations have no rational roots.

13. $x^3 - x + 1 = 0$
14. $x^3 + x^2 - x + 1 = 0$
15. $x^4 + 2x + 2 = 0$
16. $x^3 + 3x - 3 = 0$

Written Exercises

A

1–12. For each equation in Oral Exercises 1–12, find any rational roots. If the equation has at least one rational root, solve it completely.

Use the method of Example 2 to show that the following numbers are irrational.

13. $\sqrt{3}$
14. $\sqrt{6}$
15. $\sqrt[3]{-4}$
16. $\sqrt[3]{2}$
17. $\sqrt[4]{8}$
18. $\sqrt[5]{-9}$

Verify that the given number is a root of the equation and use this fact to show that the number is irrational.

19. $\sqrt{3} + \sqrt{2}$; $x^4 - 10x^2 + 1 = 0$
20. $\sqrt{5} - \sqrt{3}$; $x^4 - 16x^2 + 4 = 0$

B

21. Explain how to use the rational root theorem with an equation that has rational coefficients.

Use the method you described in Exercise 21 to solve the following equations.

22. $\frac{1}{3}x^3 - \frac{1}{2}x^2 + \frac{1}{3}x + \frac{1}{3} = 0$ $\left\{-\frac{1}{2}, 1 \pm i\right\}$
23. $\frac{1}{3}x^3 + \frac{1}{2}x^2 + \frac{2}{3}x + 1 = 0$ $\left\{-\frac{3}{2}, \pm i\sqrt{2}\right\}$
24. $2.0x^3 - 0.8x^2 + 0.5x - 0.2 = 0$ $\left\{\frac{2}{5}, \pm\frac{1}{2}i\right\}$
25. $0.2x^3 - 0.5x^2 + 0.8x - 2.0 = 0$ $\left\{\frac{5}{2}, \pm 2i\right\}$

C

26. Show that if k is an integer and $\sqrt[n]{k}$ is rational, then k is the nth power of an integer. (*Hint:* Consider the possible rational roots of $x^n - k = 0$.)

The proof of the rational root theorem is essentially the same whatever the degree of the polynomial. The proof for degree 3 is given in Exercises 27 and 28. Let $\frac{h}{k}$ (h and k relatively prime integers) be a root of the polynomial equation $ax^3 + bx^2 + cx + d = 0$, where a, b, c, and d are integers. We must show that h is a factor of d and k is a factor of a.

27. Justify the following statements.

1. $a\left(\frac{h}{k}\right)^3 + b\left(\frac{h}{k}\right)^2 + c\left(\frac{h}{k}\right) + d = 0$ $\frac{k}{h}$ is a root of the equation.
2. $ah^3 + bh^2k + chk^2 + dk^3 = 0$ Mult. prop. of eq.
3. $dk^3 = -ah^3 - bh^2k - chk^2$ Add. prop. of eq.
4. $\frac{dk^3}{h} = -ah^2 - bhk - ck^2$ Div. prop. of eq.
5. $\frac{dk^3}{h}$ is an integer. Closure properties for integers

The proof is completed as follows: Since $\frac{dk^3}{h}$ is an integer, all the prime factors of h must divide out with the prime factors of dk^3. But h and k have no common prime factor. Therefore h must be a factor of d.

28. Modify the proof in Exercise 27 to show that k is a factor of a.

Mixed Review Exercises

Solve each equation completely. In Exercises 7–10, one root is given.

1. $4(2 - m) + 5 = 3m - 8$ $\{3\}$

2. $3u^2 - 4u + 2 = 0$ $\left\{\frac{2}{3} \pm \frac{\sqrt{2}}{3}i\right\}$

3. $\sqrt{2r + 1} = r - 1$ $\{4\}$

4. $\frac{1}{w} + \frac{w}{w - 1} = \frac{1}{w - w^2}$ $\{-1\}$

5. $2v^{-2} - 5v^{-1} - 12 = 0$ $\left\{-\frac{2}{3}, \frac{1}{4}\right\}$

6. $n^4 - 2n^2 - 8 = 0$ $\{\pm 2, \pm i\sqrt{2}\}$

7. $x^3 - 2x^2 + x - 2 = 0;\ i$ $\{\pm i, 2\}$

8. $c^3 + c^2 - 7c - 3 = 0;\ -3$ $\{-3, 1 \pm \sqrt{2}\}$

9. $a^3 - 3a^2 + a + 5 = 0;\ 2 - i$ $\{2 \pm i, -1\}$

10. $y^4 - 3y^3 - 2y^2 + 10y - 12 = 0;\ 1 + i$ $\{1 \pm i, -2, 3\}$

Historical Note / *Roots of Polynomial Equations*

Although the Fundamental Theorem of Algebra guarantees the existence of roots for any polynomial equation, it does not provide a method for finding them. Of course, the quadratic formula gives roots of any quadratic equation, and there are similar but more complicated formulas for the roots of cubic and fourth-degree equations. But in 1824 the Norwegian mathematician Niels Henrik Abel (1802–1829) showed that for equations of degree greater than four no such general formulas exist. Numerical methods like those you are studying in this chapter, therefore, are often used to find approximate values for the roots of higher-degree polynomial equations.

Supplementary Materials

Study Guide pp. 131–132
Computer Activity 18
Resource Book p. 133

Additional Answers
Oral Exercises

1. $\pm 1, \pm 2, \pm 3, \pm 6$
2. $\pm 1, \pm 2, \pm 4$
3. $\pm 1, \pm 2, \pm 4, \pm 8$
4. $\pm 1, \pm 2, \pm 3, \pm 6$
5. $\pm 1, \pm 3, \pm 5, \pm 15$
6. $\pm 1, \pm 2, \pm 5, \pm 10$
7. $\pm 1, \pm 5, \pm\frac{1}{2}, \pm\frac{5}{2}$
8. $\pm 1, \pm 2, \pm 4, \pm\frac{1}{2}$
9. $\pm 1, \pm 2, \pm 3, \pm 6, \pm\frac{1}{2}, \pm\frac{3}{2}$
10. $\pm 1, \pm 2, \pm 4, \pm\frac{1}{2}, \pm\frac{1}{4}$
11. $\pm 1, \pm 2, \pm 4, \pm\frac{1}{3}, \pm\frac{2}{3}, \pm\frac{4}{3}$
12. $\pm 1, \pm 2, \pm\frac{1}{2}, \pm\frac{1}{3}, \pm\frac{2}{3}, \pm\frac{1}{6}$
13. The possible rational roots are ± 1; neither satisfies the equation.
14. The possible rational roots are ± 1; neither satisfies the equation.
15. The possible rational roots are $\pm 1, \pm 2$; the equation has no positive roots; neither -1 nor -2 satisfies the equation.
16. The possible rational roots are $\pm 1, \pm 3$; the equation has no negative roots; neither 1 nor 3 satisfies the equation.

Additional Answers
Written Exercises (on p. 396)

Teaching Suggestions p. T111

Suggested Extensions p. T111

Warm-Up Exercises

For each polynomial function, find $P(1)$ and $P(2)$.

1. $P(x) = x^2 - 3x + 2$ 0; 0
2. $P(x) = 2x^3 + x^2 - 5x - 1$ −3; 9
3. $P(x) = 5 - 6x + 3x^2 - x^3$ 1; −3
4. $P(x) = x^4 - 8x^2 + 16$ 9; 0
5. $P(x) = 2x^5 - 3x^3 + x^2 + x - 8$ −7; 38

Motivating the Lesson

Tell students that although the rational root theorem is useful in solving polynomial equations, it gives no information about irrational roots. Instead, we rely upon graphs and the intermediate-value theorem to help approximate irrational roots.

Using a Computer or a Graphing Calculator

Due to the "wiggles" in the graphs of higher-degree polynomial functions, students may find graphing such functions difficult. A computer or graphing calculator is helpful in these cases.

When the roots of a polynomial equation must be approximated, the zoom or scaling features on a computer or graphing calculator can be used for increasing the accuracy of the approximations. Such use is suggested for Written Exercises 1–20 on page 388.

Support Materials
Disk for Algebra
Menu Item: Function Plotter
Resource Book p. 242

8-8 Approximating Irrational Roots

Objective To approximate the real roots of a polynomial equation $P(x) = 0$ by using the graph of $y = P(x)$.

Using methods discussed in previous lessons, you can show that the equation

$$x^3 - 2x^2 - 4x + 2 = 0$$

has either one or three real roots, none of which are rational. In this lesson you will learn a method for approximating the irrational roots of such equations. Because the real *roots* of the given equation are the same as the *x-intercepts* of the graph of $y = x^3 - 2x^2 - 4x + 2$, the graph can give us more information about these roots.

Example 1 **a.** Graph the equation $y = x^3 - 2x^2 - 4x + 2$.

b. Use the graph to estimate to the nearest half unit the roots of the equation $x^3 - 2x^2 - 4x + 2 = 0$.

Solution **a.** You may wish to use a computer or a graphing calculator if one is available. Otherwise make a table of values, plot the corresponding points, and join the points with a smooth unbroken curve. (Use a smaller scale on the y-axis than on the x-axis to accommodate the greater range of y-values.)

$y = x^3 - 2x^2 - 4x + 2$

x	y
−2	−6
−1	3
0	2
1	−3
2	−6
3	−1
4	18

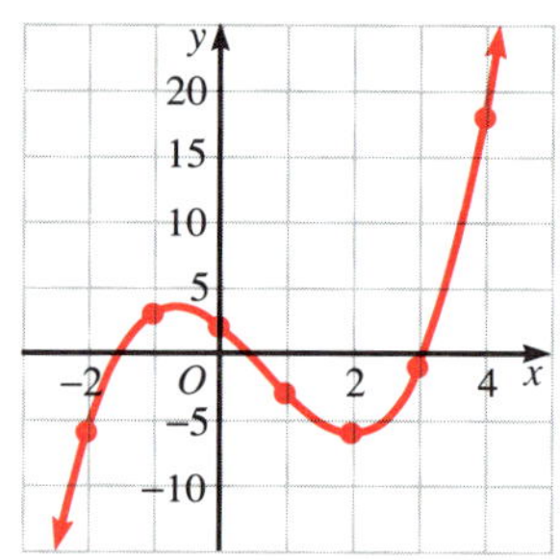

b. The roots of the equation are the x-coordinates of the points where the graph intersects the x-axis. To the nearest half unit, the roots are −1.5, 0.5, and 3. ***Answer***

The method of Example 1 can be described in terms of the polynomial *function* $P(x) = x^3 - 2x^2 - 4x + 2$. The *zeros* of P are the *roots* of the polynomial *equation* $P(x) = 0$. When you draw the graph of the function P as an unbroken curve, you assume that as x goes from a to b, P takes on all values between $P(a)$ and $P(b)$. For example, since $P(-2) = -6$ and $P(-1) = 3$, and 0 is between −6 and 3, $P(x)$ must equal 0 for some x between −2 and −1.

This property, known as the *intermediate-value theorem,* is stated formally below.

Intermediate-Value Theorem

If P is a polynomial function with real coefficients, and m is any number between $P(a)$ and $P(b)$, then there is at least one number c between a and b for which $P(c) = m$.

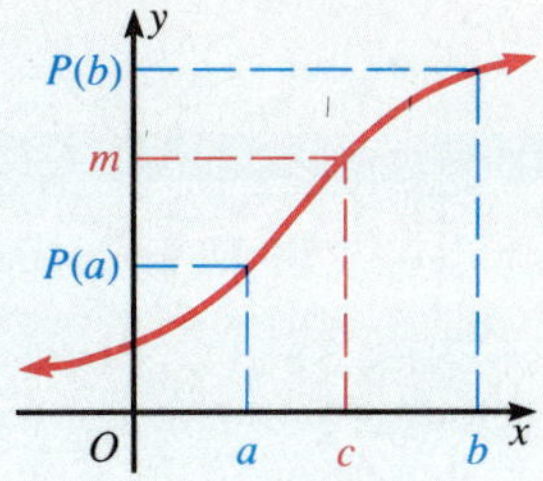

Example 2 Approximate to the nearest tenth the real zero of the function $P(x) = x^3 - 2x^2 + x - 5$.

Solution By Descartes' rule $P(x) = 0$ has no negative roots, so you can make a table using nonnegative values of x. The table at the left below shows that $P(2) < 0$ and $P(3) > 0$. Therefore, for some r between 2 and 3, $P(r) = 0$.

x	$P(x)$
0	−5
1	−5
2	−3
3	7

$r \rightarrow$ (between 2 and 3) $\leftarrow 0$

x	$P(x)$
2.1	−2.459
2.2	−1.832
2.3	−1.113
2.4	−0.296
2.5	0.625

$r \rightarrow$ (between 2.4 and 2.5) $\leftarrow 0$

The table of values at the right above, which is more easily generated using a computer or a calculator, shows that r is between 2.4 and 2.5. Since $P(2.4)$ is closer to 0 than $P(2.5)$, you may assume that r is closer to 2.4. Therefore, to the nearest tenth, the real zero of P is 2.4. ***Answer***

The graphs below show the "magnification" process used in Example 2. You can obtain the same result on a computer or graphing calculator by changing the scaling of the graph.

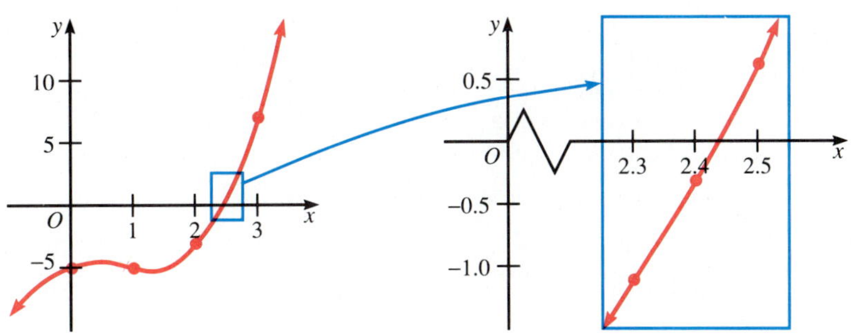

Chalkboard Examples

1. Use a graph to estimate to the nearest half unit the roots of the equation $x^3 - 9x + 9 = 0$.

x	y
−4	−19
−3	9
−2	19
−1	17
0	9
1	1
2	−1
3	9

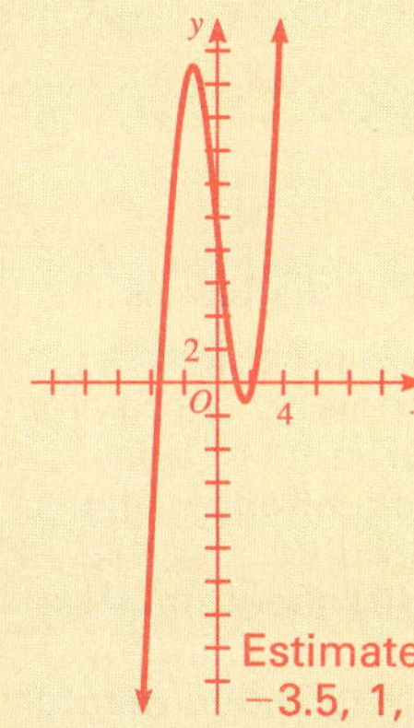

Estimated roots: −3.5, 1, 2

2. Approximate to the nearest tenth the positive zero of the function $P(x) = x^3 - x^2 - 5x - 4$.

x	$P(x)$
0	−4
1	−9
2	−10
3	−1
4	24

x	$P(x)$
3.0	−1
3.1	0.681

$P(3.1)$ is closer to 0 than $P(3)$. The positive zero is 3.1 to the nearest tenth.

Thinking Skills

Given sufficient exposure to the graphs of higher-order polynomial functions, students should be able to *visualize* the general shape of such curves before plotting.

Check for Understanding

Oral Exs. 1–4: use after discussing the intermediate-value theorem.

Oral Exs. 5–8: use after Example 2.

Guided Practice

Approximate each real root of $P(x) = 0$ to the nearest half unit.

1. $P(x) = x^3 - 11$ 2
2. $P(x) = -x^4 + 10x + 4$ −0.5, 2
3. $P(x) = x^3 - 8x^2 + 10x + 2$ 0, 2, 6.5
4. $P(x) = x^4 - x^3 + x^2 - 3x + 1$ 0.5, 1.5

Summarizing the Lesson

In this lesson students learned how to approximate irrational roots of polynomial equations using graphs and the intermediate-value theorem. Ask students to explain how the intermediate-value theorem can be used to approximate irrational roots.

Suggested Assignments

Average Algebra
388/1–23 odd
S 380/26

Average Alg. and Trig.
388/1–23 odd
S 380/26

Extended Algebra
388/1–25 odd
S 385/27

Extended Alg. and Trig.
388/1–25 odd
S 385/27

Supplementary Materials

Study Guide pp. 133–134
Computer Activity 19

The "magnification" process shown on the previous page can be repeated if a better approximation is desired. For example, another "magnification" shows that the zero asked for in Example 2 is about 2.43.

Oral Exercises

In Exercises 1 and 2, a table of values for a polynomial function P is given. In each case find one or more pairs of numbers between which roots of $P(x) = 0$ must lie.

1.

x	−3	−2	−1	0	1	2	3	4
$P(x)$	6	−4	3	5	4	1	−3	−10

−3, −2; −2, −1; 2, 3

−0.3, −0.4; −0.7, −0.8

2.

x	−0.1	−0.2	−0.3	−0.4	−0.5	−0.6	−0.7	−0.8	−0.9
$P(x)$	−2.64	−1.25	−0.72	0.27	1.31	0.98	0.14	−0.55	−1.28

Suppose P is a polynomial function for which $P(1) = 6$, $P(2) = 1$, and $P(3) = -5$. Tell whether each statement is true or false.

3. P has no zeros between 1 and 2. F
4. P has exactly one zero between 2 and 3. F

Locate between consecutive integers each real root of each equation.

5. $x^3 - x^2 - 5 = 0$ 2, 3
6. $x^3 + x + 3 = 0$ −2, −1
7. $x^3 - x + 1 = 0$ −2, −1
8. $x^4 - 2x^3 - x + 1 = 0$ 0, 1; 2, 3

Written Exercises

For each polynomial P, draw a graph to approximate to the nearest half unit the real root(s) of the equation $P(x) = 0$. You may wish to use a computer or a graphing calculator.

A

1. $P(x) = x^3 - 16$ 2.5
2. $P(x) = x^3 + 11$ −2
3. $P(x) = x^3 + x^2 + 1$ −1.5
4. $P(x) = x^3 + 2x - 8$ 1.5
5. $P(x) = x^3 - x^2 - x - 3$ 2
6. $P(x) = x^3 - 3x^2 + 3x + 2$ −0.5
7. $P(x) = x^3 - 3x - 1$ −1.5, −0.5, 2
8. $P(x) = x^3 - 4x + 1$ −2, 0.5, 2
9. $P(x) = x^4 - 3x^3 + 5$ 1.5, 3
10. $P(x) = x^4 + x^3 - 10x^2 - 4x + 10$ −3.5, −1, 1, 2.5

11–16. In Exercises 1–6 above, each polynomial equation $P(x) = 0$ has one real root. Approximate it to the nearest tenth.
11. 2.5 **12.** −2.2 **13.** −1.5 **14.** 1.7 **15.** 2.1 **16.** −0.4

B **17–20.** In Exercises 7–10 above, each polynomial equation $P(x) = 0$ has several real roots. Approximate each root to the nearest tenth.

 17. −1.5, −0.3, 1.9 **18.** −2.1, 0.3, 1.9 **19.** 1.5, 2.8 **20.** −3.4, −1.2, 0.9, 2.7

For each polynomial P, (a) verify that m is between $P(a)$ and $P(b)$, and (b) find a number c between a and b such that $P(c) = m$.

21. $P(x) = x^2 - 2x + 3;\ a = 1;\ b = 4;\ m = 6$

22. $P(x) = x^2 + x - 5;\ a = -5;\ b = -1;\ m = 7$

23. $P(x) = 2x^2 + 5x - 4;\ a = -2;\ b = 2;\ m = -1$

24. $P(x) = x^3 + x^2 - 2x - 5;\ a = -3;\ b = 2;\ m = -5$

C **25.** Show that $P(x) = x^3 - 3x + 1$ has a zero between 1.5 and 1.6 and approximate it to the nearest hundredth. 1.53

26. Show that $P(x) = x^3 + x^2 - 3x + 2$ has a zero between -2.6 and -2.5 and approximate it to the nearest hundredth. -2.51

27. This exercise develops a method for finding an *upper bound* for the positive roots of the polynomial equation $P(x) = 0$. Assume that the coefficient of the highest-degree term, called the *leading coefficient,* of $P(x)$ is positive. Let $P(x) = (x - m)Q(x) + P(m)$, where m is a positive number. Show that if $P(m)$ and the coefficients of $Q(x)$ are all nonnegative, then $P(x) = 0$ has no roots greater than m. Therefore m is an upper bound for the roots. (*Hint:* For all $x > m$, $x - m > 0$ and $Q(x) > 0$.)

28. Explain how to use Exercise 27 to find a *lower bound* for the negative roots of $P(x) = 0$. (*Hint:* Both $P(-x) = 0$ and $-P(-x) = 0$ have the absolute values of the negative roots of $P(x) = 0$ as their positive roots, and one of them has a positive leading coefficient.)

Use Exercises 27 and 28 to find the *least* integral upper bound and *greatest* integral lower bound for the roots of the given equations.

29. $2x^3 - 5x^2 - 8x - 9 = 0$ 4; -2

30. $2x^4 - 6x^2 - 3x - 9 = 0$ 3; -2

Computer Key-In

In Example 2 of Lesson 8-8 you found that a zero of $P(x) = x^3 - 2x^2 + x - 5$ was between 2 and 3 by observing that $P(2) = -3 < 0$ and $P(3) = 7 > 0$ and by using the intermediate-value theorem. Then, dividing the interval from 2 to 3 into tenths and observing that $P(2.4) < 0$ and $P(2.5) > 0$, you found that the zero was between 2.4 and 2.5.

By continuing to subdivide intervals containing the zero and by using the intermediate-value theorem, you could approximate the zero more accurately.

The program given on the next page can be used to search for real zeros of a polynomial function. The program is based on subdividing an interval and using the intermediate-value theorem. Lines 20–100 accept data about the given polynomial and the interval to be searched. In lines 140–170 the polynomial is evaluated by means of synthetic substitution.

The program then tests whether the current x-value is a zero (line 180) or whether a zero lies between the current x-value and its predecessor (line 190). The intermediate-value theorem is used in line 190, where the program checks whether the corresponding y-values have opposite sign. If the product of the

Additional Answers Written Exercises

2.

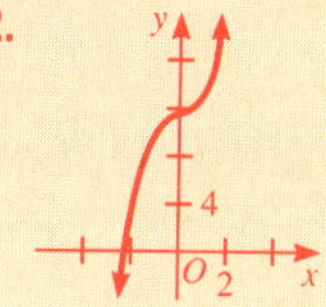

4.

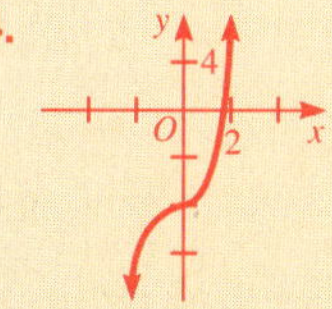

6.

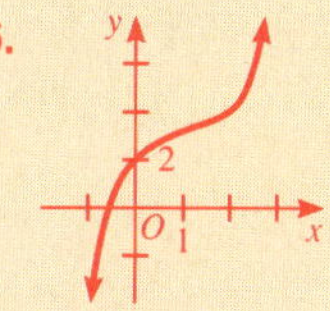

8.

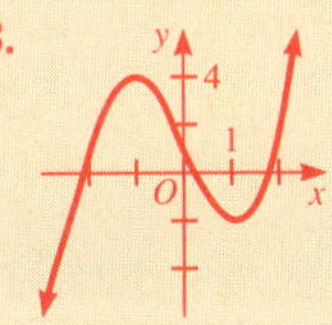

10.

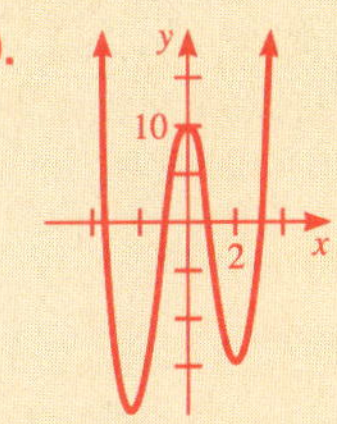

Computer Key-In Commentary

A computer is ideally suited for the tedious task of searching for the real zeros of a polynomial function. The ability of the program given here to locate zeros depends upon the interval being searched and the subdivision size, both of which the user controls.

y-values is negative, then they have opposite signs, and there must be a zero between the two subdivision points that produced those y-values.

```
10 PRINT "THIS PROGRAM WILL SEARCH FOR REAL ZEROS OF A
   POLYNOMIAL FUNCTION."
20 INPUT "ENTER THE DEGREE OF THE POLYNOMIAL: "; D
30 DIM C(D)
40 FOR J=D TO 1 STEP-1
50 PRINT "ENTER THE COEFFICIENT OF X^"; J; ": ";
60 INPUT C(J)
70 NEXT J
80 INPUT "ENTER THE CONSTANT TERM: "; C(0)
90 INPUT "ENTER THE ENDPOINTS OF THE INTERVAL TO BE
   SEARCHED: "; X1, X2
100 INPUT "ENTER THE SUBDIVISION SIZE: "; I
110 LET F=0
120 LET Y1=0
130 FOR X=X1 TO X2 STEP I
140 LET Y=C(D)
150 FOR J=D-1 TO 0 STEP -1
160 LET Y=Y*X+C(J)
170 NEXT J
180 IF Y=0 THEN PRINT X; " IS A ZERO."
190 IF Y*Y1<0 THEN PRINT "THERE IS A ZERO BETWEEN ";
    X-I; " AND "; X; "."
200 IF Y=0 OR Y*Y1 < 0 THEN LET F=1
210 LET Y1=Y
220 NEXT X
230 IF F=0 THEN PRINT "NO ZEROS FOUND."
240 END
```

Exercises

4. No zeros are found. The zeros are $\frac{1}{7}$ and $\frac{1}{6}$, both of which are between 0.1 and 0.2.

1. The function $P(x) = x^3 - 4x^2 + 2x + 4$ has three real zeros.
   a. Approximate these zeros to the nearest tenth. (*Hint:* Use -3 and 3 as the left and right endpoints. Use a subdivision size of 0.01.) $x \approx -0.7, 2.0, 2.7$
   b. Approximate the largest of the zeros to the nearest hundredth. $x \approx 2.73$
2. The function $P(x) = x^5 - 14x^4 + 38x^3 + 32x^2 - 118x - 20$ has five real zeros. Approximate each zero to the nearest tenth as follows:
   a. Run the program for the interval from $x = -3$ to $x = 3$. What zeros did you locate? $x \approx -1.6, -0.2, 2.2$
   b. Run the program again with an interval large enough to locate the remaining zeros. $x \approx 3.6, 10.0$
3. Find $\sqrt[5]{2}$ to three decimal places. (*Hint:* Use $P(x) = x^5 - 2$.) 1.149
4. The function $P(x) = 42x^2 - 13x + 1$ has two real zeros between $x = 0$ and $x = 1$. What happens when you use a subdivision size of 0.1 to locate these zeros? Explain why this happens. (See above.)

Additional Answers Written Exercises

(continued from p. 381)

29. 1. $S(a + bi) = [(a + bi) - (a + bi)] \cdot [(a + bi) - (a - bi)] = 0(2bi) = 0$
$S(a - bi) = [(a - bi) - (a + bi)] \cdot [(a - bi) - (a - bi)] = (-2bi)0 = 0$

2. Division algorithm; since the divisor $S(x)$ is quadratic, the remainder has the form $cx + d$, which is linear if $c \neq 0$ or constant if $c = 0$.

3. Since $a + bi$ is a root of $P(x) = 0$, $P(a + bi) = 0$; also, part (1) showed that $S(a + bi) = 0$.

4. From part (3), $0 = 0 + c(a + bi) + d$; so $c(a + bi) + d = 0$, $ac + bci + d = 0$, and $(ac + d) + bci = 0$.

5. If $(ac + d) + bci = 0 + 0i$, the imaginary parts must be equal; so $bc = 0$; since $b \neq 0$ is given, $c = 0$ by the zero-product property.

6. If $(ac + d) + bci = 0 + 0i$, the real parts must be equal; so $ac + d = 0$; since $c = 0$ by part (5), $a \cdot 0 + d = 0$; so $d = 0$.

7. From part (2), $P(x) = Q(x)S(x) + cx + d$ since $c = d = 0$ by parts (5) and (6), $P(x) = Q(x)S(x)$.

8. Setting $x = a - bi$ in part (7), you obtain $P(a - bi) = Q(a - bi)S(a - bi)$; since $S(a - bi) = 0$ by part (1), $P(a - bi) = Q(a - bi) \cdot 0 = 0$.

8-9 Linear Interpolation

Objective To use linear interpolation to find values not listed in a given table of data.

Newspapers, magazines, and journals often display numerical information in tables. Look, for instance, at the Census Bureau data shown in the table at the right.

To approximate values not given in the table, you can use **linear interpolation.** For example, to find the population of the United States in 1953, you reason as follows: 1953 is $\frac{3}{10}$ of the way from 1950 to 1960; therefore the number p that is $\frac{3}{10}$ of the way from 151 to 179 is an acceptable approximation of the 1953 population (in millions). To find p, add $\frac{3}{10}$ of the difference between 179 and 151 to 151:

$$p = 151 + \frac{3}{10}(179 - 151) \approx 151 + 8 = 159$$

Therefore, the 1953 population was about 159 million.

Example 1 shows how the work done above can be arranged in a way that makes it easy to set up a proportion.

Year	U.S. population (millions)
1900	76
1910	92
1920	106
1930	123
1940	132
1950	151
1960	179
1970	203
1980	227
1990	243

Example 1 Find the approximate population of the United States in 1953.

Solution

		Year	Population		
10	3	1950	151	d	28
		1953	p		
		1960	179		

$$\frac{d}{28} = \frac{3}{10};\ d = \frac{3}{10} \cdot 28 \approx 8$$

$$p = 151 + d \approx 151 + 8 = 159$$

$\therefore$ the population in 1953 was about 159 million. ***Answer***

Given a population size, the process used above can be reversed to find the corresponding year. This process is called **inverse interpolation.**

Example 2 Refer to the table above. Use inverse interpolation to find approximately in which year the population was 140 million.

Solution

		Year	Population		
10	c	1940	132	8	19
		y	140		
		1950	151		

$$\frac{c}{10} = \frac{8}{19};\ c = \frac{80}{19} \approx 4$$

$$y = 1940 + c \approx 1940 + 4 = 1944$$

$\therefore$ the population was 140 million in about 1944. ***Answer***

Teaching Suggestions p. T112

Suggested Extensions p. T112

Warm-Up Exercises

Solve each proportion.

1. $\frac{x}{10} = \frac{4}{35}$ $\left\{\frac{8}{7}\right\}$
2. $\frac{x}{140} = \frac{7}{10}$ {98}
3. $\frac{x}{0.1} = \frac{3.2}{0.4}$ {0.8}
4. $\frac{x}{4.8} = \frac{0.7}{1.6}$ {2.1}
5. $\frac{x}{250} = \frac{240}{1000}$ {60}

Motivating the Lesson

Tell students that in most real world applications involving two variables, the relationship between the variables comes not in the form of an equation, but in the form of a finite number of data pairs. Determining other values of the variables from the known ones can be done using linear interpolation, the topic of today's lesson.

Chalkboard Examples

The table below give entries for a function $F(x)$.

x	$F(x)$
0.01	4.3
0.02	5.0
0.03	5.6
0.04	6.2
0.05	6.7

Use linear interpolation or inverse interpolation to approximate the following.

1. x if $F(x) = 6.0$

		x	$F(x)$		
0.01	c	0.03	5.6	0.4	0.6
		x	6.0		
		0.04	6.2		

$c = \frac{0.4}{0.6}(0.01) \approx 0.007$

$\therefore x = 0.03 + c$
≈ 0.037

2. $F(x)$ if $x = 0.017$

		x	$F(x)$		
0.01	0.007	0.01	4.3	d	0.7
		0.017	y		
		0.02	5.0		

$d = \frac{0.007}{0.01}(0.7) = 0.49$

$\therefore y = 4.3 + 0.49$
$= 4.79$

In the table used in Examples 1 and 2, the population values increased as the years increased. In some tables the entries in one column (or row) decrease as the entries in the other increase.

Example 3 The table below gives the density of dry air at various altitudes.

Altitude (m)	0	500	1000	1500	2000	2500	3000	3500
Density (kg/m^3)	1.225	1.167	1.112	1.058	1.007	0.957	0.909	0.863

a. Approximate the density at 2300 m.

b. At about what altitude is the density 1.025 kg/m^3?

Solution **a.**

		Altitude	Density		
500	300	2000	1.007	d	0.050
		2300	y		
		2500	0.957		

$\frac{d}{0.050} = \frac{300}{500} = \frac{3}{5};$

$d = \frac{3}{5} \cdot 0.050 = 0.030$

$y = 1.007 - d$
$\approx 1.007 - 0.030 = 0.977$

$\therefore$ the density at 2300 m is about 0.977 kg/m^3. ***Answer***

b.

		Altitude	Density		
500	c	1500	1.058	0.033	0.051
		x	1.025		
		2000	1.007		

$\frac{c}{500} = \frac{0.033}{0.051} = \frac{11}{17};$

$c = \frac{11}{17} \cdot 500 \approx 324$

$x = 1500 + c$
$\approx 1500 + 324 = 1824$

$\therefore$ the density is 1.025 km/m^3 at about 1824 m. ***Answer***

You can use linear interpolation to approximate the zeros of functions.

Example 4 Approximate to the nearest hundredth the real zero of the function $P(x) = x^3 - 2x^2 + x - 5$.

Solution First use the method of Example 2, page 387, to locate the zero between consecutive tenths. In that example you found that $P(2.4) = -0.296$ and $P(2.5) = 0.625$. Now use linear interpolation.

		x	$P(x)$		
0.1	c	2.4	-0.296	0.296	0.921
		r	0		
		2.5	0.625		

$\frac{c}{0.1} = \frac{0.296}{0.921} \approx 0.32;$

$c \approx 0.1 \cdot 0.32 \approx 0.03$

$r = 2.4 + c$
$\approx 2.4 + 0.03 = 2.43$

$\therefore$ the zero of P is about 2.43. ***Answer***

To illustrate how linear interpolation works, suppose $P_1(x_1, f(x_1))$ and $P_2(x_2, f(x_2))$ are two points on the graph of a function f. Linear interpolation essentially replaces the graph of f between these points by the line segment $\overline{P_1P_2}$. (See the figure below, where the graph of f is shown in red and the approximation line in blue.)

For any x between x_1 and x_2, linear interpolation approximates $f(x)$ by $f(x_1) + d$. The amount of error in this approximation depends upon the vertical distance between the graph of f and the approximation line. The less the graph of f "strays" from the approximation line, the smaller the error.

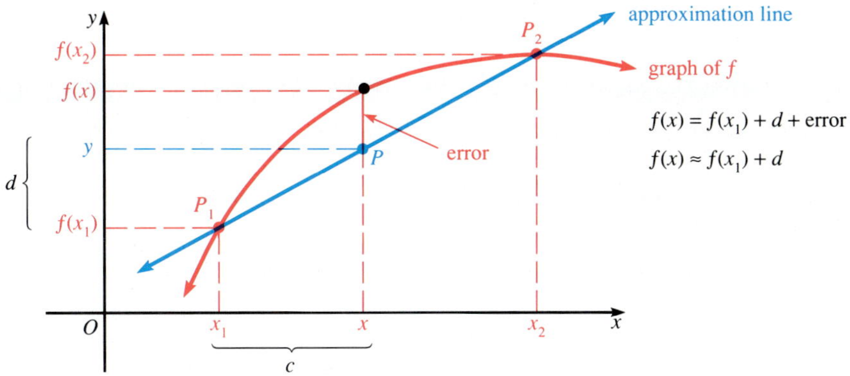

Using P_1 and P you can see that the slope of the approximation line is $\frac{d}{c}$; using P_1 and P_2 the slope is $\frac{f(x_2) - f(x_1)}{x_2 - x_1}$. Since the slope of a line is constant, you can write $\frac{f(x_2) - f(x_1)}{x_2 - x_1} = \frac{d}{c}$, which is equivalent to

$$\frac{c}{x_2 - x_1} = \frac{d}{f(x_2) - f(x_1)}.$$

This proportion is the one used in linear interpolation.

		x	$f(x)$		
$x_2 - x_1$	c	x_1	$f(x_1)$	d	$f(x_2) - f(x_1)$
		x	y		
		x_2	$f(x_2)$		

Oral Exercises

1. Explain why, in part (a) of Example 3, d is *subtracted* from 1.007 in order to approximate the density at 2300 m. As altitude increases, density decreases.

2. Use linear interpolation to estimate the indicated function value.
   a. $f(1) = 0$, $f(5) = 100$; $f(2) = \underline{\ ?\ }$ 25
   b. $g(-1) = 10$, $g(1) = -10$; $g(0.5) = \underline{\ ?\ }$ −5

3. Use linear interpolation to approximate to the nearest hundredth a zero of a function f given that $f(-8.1) = -16.2$ and $f(-8.2) = 3.8$.

		x	$f(x)$		
-0.1	c	-8.1	-16.2	16.2	20
		r	0		
		-8.2	3.8		

$\frac{c}{-0.1} = \frac{16.2}{20}$
$c \approx -0.08$
$r = -8.1 + (-0.08)$
≈ -8.18
$\therefore$ a zero of f is approximately -8.18.

Check for Understanding

Here is a suggested use of the Oral Exercises to check students' understanding as you teach the lesson.
Oral Exs. 1–2: use after Example 3.
Oral Ex. 3: use after the geometric discussion of linear interpolation on page 393.

Guided Practice

Use the following tax table.

Income ($)	Tax ($)
6000	81
8000	378
10,000	698
12,000	1058
16,000	1840
20,000	2739
25,000	4050

What would be the approximate tax on each of the following incomes?

1. $9000 $538

2. $15,000 $1645

3. $21,200 $3054

What approximate income would have the following tax on it?

4. $878 $11,000

5. $628 $9562

6. $3000 $20,995

Summarizing the Lesson

In this lesson students learned to use linear interpolation to find missing values in a table of data. Ask students to discuss the geometric concept underlying linear interpolation.

Suggested Assignments

Average Algebra
394/1–25 odd
R 396/Self-Test 3

Average Alg. and Trig.
394/1–25 odd
R 396/Self-Test 3

Extended Algebra
394/1–29 odd
R 396/Self-Test 3

Extended Alg. and Trig.
394/1–29 odd
R 396/Self-Test 3

3. This exercise shows that the error in using linear interpolation may be large: Use linear interpolation to approximate $f(x) = x^2$ at $x = 1$ with $x_1 = 0$ and $x_2 = 2$, as follows.

a. $f(0) =$ __?__ and $f(2) =$ __?__. 0; 4

b. What linear function g has the same values as f at 0 and 2? $g(x) = 2x$

c. What is the value of this linear function at $x = 1$? 2

d. By how much does the functional value found in part (c) differ from $f(1)$? 1

e. How could you choose x_1 and x_2 to make the approximation for $f(1)$ closer to the true value? Choose x_1 and x_2 closer to 1.

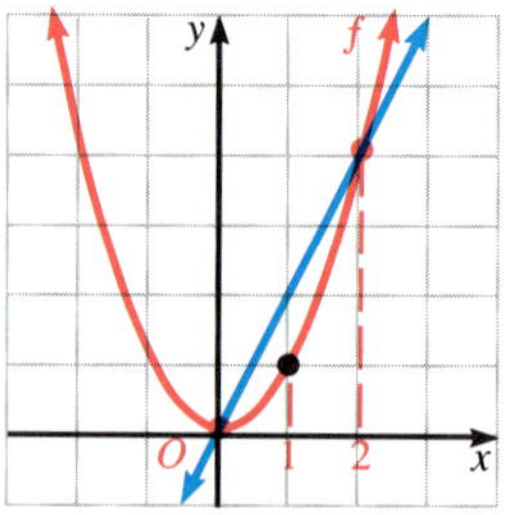

Written Exercises

Throughout these exercises, use linear interpolation. In Exercises 1–8, use the table on page 391.

Approximate the population of the United States in each year.

A **1.** 1915 99 million **2.** 1963 186 million **3.** 1968 198 million **4.** 1976 217 million

Approximate the year in which the United States had each population (in millions).

5. 100 1916 **6.** 115 1925 **7.** 170 1957 **8.** 220 1977

Use the table in Example 3 on page 392 to find an approximation of the density of air at each altitude (in meters).

9. 1200 1.090 kg/m^3 **10.** 3200 0.8906 kg/m^3 **11.** 400 1.179 kg/m^3 **12.** 320 1.188 kg/m^3

Use the table in Example 3 on page 392 to find an approximation of the altitude at which air has the given density (in kg/m^3).

13. 1.200 216 m **14.** 1.030 1775 m **15.** 1.000 2070 m **16.** 0.930 2781 m

In Exercises 17–26, use Tables 1 and 2 on pages 810 and 811 to find an approximation of each square root and cube root.

Sample $\sqrt[3]{726}$

Solution Use the column headed $\sqrt[3]{100N}$ with $N = 7.2$ and $N = 7.3$.

	x	$\sqrt[3]{x}$	
10 [6 [	720	8.963	] d] 0.041
	726	$\sqrt[3]{726}$	
	730	9.004	

$$d = \frac{6}{10} \cdot 0.041 \approx 0.024$$
$$\sqrt[3]{726} \approx 8.963 + 0.024 = 8.987 \quad \textit{Answer}$$

B 17. $\sqrt{7.63}$ 2.762 18. $\sqrt{4.87}$ 2.207 19. $\sqrt{67.8}$ 8.234 20. $\sqrt{24.2}$ 4.919

21. $\sqrt[3]{5.26}$ 1.739 22. $\sqrt[3]{526}$ 8.072 23. $\sqrt[3]{26.8}$ 2.992 24. $\sqrt[3]{84.3}$ 4.385

25. $\sqrt{5280}$ (*Hint:* $5280 = 10 \times 5.28 \times 10^2$) 72.66

26. $\sqrt[3]{62300}$ (*Hint:* $62300 = 10 \times 6.23 \times 10^3$) 39.64

In Exercises 27–30, (a) locate the real zero of each polynomial function between consecutive tenths, and (b) approximate the zero to the nearest hundredth using linear interpolation.

27. $P(x) = x^3 + 2x - 5$ 1.3, 1.4; 1.33

28. $P(x) = x^3 - 2x^2 - 3$ 2.4, 2.5; 2.48

29. $P(x) = x^3 + x^2 + 9$ −2.5, −2.4; −2.47

30. $P(x) = x^3 + 2x + 8$ −1.7, −1.6; −1.67

Computer Exercises

For students with some programming experience.

1. Write a program that will carry out linear interpolation between two given functional values, $y_1 = f(x_1)$ and $y_2 = f(x_2)$. You will enter the numbers for x_1, y_1, x_2, y_2, and x (where $x_1 < x < x_2$). Then have the program compute an estimate for $y = f(x)$ using linear interpolation.

For Exercises 2 and 3, refer to the population table on page 391.

2.b. 107.5 million; it is greater than the table value by 1.5 million.

2. a. Run the program in Exercise 1 to estimate the population (in millions) for the years 1916, 1949, and 1972. 100.4; 149.1; 207.8
   b. Run the program with $x_1 = 1910$ and $x_2 = 1930$ to estimate the population in the year 1920. How does this approximate value compare to the exact value for the 1920 population given in the table?

Instead of estimating a value *between* two known values, it is possible to use the same process involved in linear interpolation to estimate a value that is either *less than* or *greater than* the two known values. This process is called **linear extrapolation.**

3. Run the program in Exercise 1 to estimate the population (in millions) for the years 1890, 1895, 1995, and 2000. 60, 68, 251, 259
4. Modify the program in Exercise 1 so that you can enter a rule for the function f. You may want to use the DEF FN statement so that you can change the function by retyping one line. After finding an estimate of $y = f(x)$, have the program compute the exact value of $y = f(x)$ and print it for comparison.
5. Run the program in Exercise 4 for each given function and each given value of x_1, x_2, and x.
   a. $f(x) = x^3$; $x_1 = 2$; $x_2 = 3$, and $x = 2.8$ $f(2.8) = 23.2$; exact value = 21.952
   b. $f(x) = 3x - 7$; $x_1 = 5$, $x_2 = 10$, and $x = 7$ $f(7) = 14$; exact value = 14
   c. $f(x) = \frac{1}{x}$; $x_1 = 10$, $x_2 = 20$, and $x = 13.71$ $f(13.71) = 0.08145$; exact value = 0.0729394603
   d. $f(x) = x^5 + 0.75x^3 - x^2 + 8$; $x_1 = 14$, $x_2 = 15$, and $x = 14.3$ $f(14.3) = 606292.576$; exact value = 599967.756

Supplementary Materials

Study Guide	pp. 135–136
Practice Master	48
Test Master	32
Resource Book	p. 134

Using a Computer

Ask students who have programming experience to work the computer exercises and demonstrate the results to the class. Note that these exercises introduce the concept of linear extrapolation.

Quick Quiz

1. List the possible rational roots for $4x^3 - 3x^2 + 8x + 3 = 0$. $\pm 3, \pm 1, \pm\frac{3}{2}, \pm\frac{1}{2}, \pm\frac{3}{4}, \pm\frac{1}{4}$
2. Approximate the real root of $x^3 - 7 = 0$ to the nearest tenth. 1.9
3. Solve $3x^4 + 2x^3 - 4x^2 + 7x - 2 = 0$. $\left\{\frac{1}{3}, -2, \frac{1 \pm i\sqrt{3}}{2}\right\}$
4. If $\sqrt{1.5} \approx 1.225$ and $\sqrt{1.7} \approx 1.304$, approximate $\sqrt{1.62}$. 1.272
5. Use the following table.

Age in days	2	7	15
Length (cm)	8.6	10.4	13.2

a. Approximate the length of a 4-day-old gerbil. 9.3 cm

b. At what age will a gerbil be 12 cm long? 11.6 days

Additional Answers Written Exercises

(continued from p. 385)

19. $(\sqrt{3} + \sqrt{2})^4 - 10(\sqrt{3} + \sqrt{2})^2 + 1 = 49 + 20\sqrt{6} - 10(5 + 2\sqrt{6}) + 1 = 49 + 20\sqrt{6} - 50 - 20\sqrt{6} + 1 = 0$; the only possible rational roots of $x^4 - 10x^2 + 1 = 0$ are ± 1, neither of which satisfy the equation; $\sqrt{3} + \sqrt{2}$ is irrational.

20. $(\sqrt{5} - \sqrt{3})^4 - 16(\sqrt{5} - \sqrt{3})^2 + 4 = 124 - 32\sqrt{15} - 16(8 - 2\sqrt{15}) + 4 = 124 - 32\sqrt{15} - 128 + 32\sqrt{15} + 4 = 0$;

Self-Test 3

Vocabulary rational root theorem (p. 382); intermediate-value theorem (p. 387); linear interpolation (p. 391); inverse interpolation (p. 391)

1. Explain why $x^3 + 2x^2 - x + 3 = 0$ has no rational roots. The only possible rational roots, ± 1 and ± 3, do not satisfy the equation. **Obj. 8-7, p. 382**
2. Solve the equation $2x^4 - x^3 + 7x^2 - 4x - 4 = 0$ completely. $\left\{-\frac{1}{2}, 1, \pm 2i\right\}$
3. Approximate to the nearest tenth the real zero of the polynomial function $P(x) = x^3 + 5$. -1.7 **Obj. 8-8, p. 386**
4. Given that $\sqrt{5.3} \approx 2.302$ and $\sqrt{5.4} \approx 2.324$, use linear interpolation to find an approximation of $\sqrt{5.36}$. 2.315 **Obj. 8-9, p. 391**
5. The temperature of a beaker of water is read every 0.1 min as it is heated, with the results shown in the following table.

Time in minutes	0.0	0.1	0.2	0.3	0.4
Temperature (° C)	30.2	31.5	32.7	34.0	35.1

a. Estimate the temperature of the water 0.24 min after the heating began. 33.2° C

b. Estimate how long it took the temperature to rise to 32.1° C. 0.15 min

Check your answers with those at the back of the book.

Chapter Summary

1. If $y = mx$, then y is said to *vary directly* as x or to be *directly proportional* to x. If $y_1 = mx_1$ and $y_2 = mx_2$, then the *proportion* $\frac{y_1}{x_1} = \frac{y_2}{x_2}$ holds. If $y = \frac{k}{x}$, then y *varies inversely* as x, and if $z = kxy$, then z *varies jointly* as x and y.

2. To divide one polynomial by another, find the quotient and remainder using the *division algorithm:*

$$\frac{\text{Dividend}}{\text{Divisor}} = \text{Quotient} + \frac{\text{Remainder}}{\text{Divisor}}$$

Synthetic division can be used instead of long division if the divisor is a first-degree binomial.

3. The *remainder theorem* states that when a polynomial $P(x)$ is divided by $x - c$, the remainder is $P(c)$, that is,

$$P(x) = Q(x) \cdot (x - c) + P(c).$$

A corollary is the *factor theorem:* The polynomial $P(x)$ has $x - r$ as a factor if and only if r is a root of the equation $P(x) = 0$.

4. Every polynomial equation $P(x) = 0$ with complex coefficients and of positive degree n has exactly n roots. By the *conjugate root theorem,* if $P(x)$ has *real* coefficients and $a + bi$ (a and b real, $b \neq 0$) is a root of $P(x) = 0$, then $a - bi$ is also a root. *Descartes' rule of signs* gives information about the numbers of positive and negative roots.

5. By the *rational root theorem,* if a polynomial equation $P(x) = 0$ with integral coefficients has the *rational* root $\frac{h}{k}$ (h and k relatively prime), then h is a factor of the constant term of $P(x)$ and k is a factor of the coefficient of the highest-degree term.

6. The *irrational* roots of a polynomial equation $P(x) = 0$ (P having real coefficients) can be approximated using the *intermediate-value theorem*. If P is a polynomial function with real coefficients, and m is any number between $P(a)$ and $P(b)$, then there is at least one number c between a and b for which $P(c) = m$.

7. *Linear interpolation* can be used to approximate values not given in a table.

Chapter Review

Write the letter of the correct answer.

1. If t varies directly as s and $t = 21$ when $s = 12$, find t when $s = 28$. 8-1
a. 9 **b.** 16 **c.** 49 **d.** 54

2. In a pre-election poll of 480 voters, 260 favored candidate Harrison. How many of an anticipated 12,000 votes is Harrison likely to get?
a. 6000 **b.** 6500 **c.** 7000 **d.** 7500

3. Suppose z varies directly as the square of x and inversely as y. If $z = 8$ when $x = 4$ and $y = 6$, find z when $x = 6$ and $y = 12$. 8-2
a. $\frac{3}{2}$ **b.** 6 **c.** $\frac{64}{9}$ **d.** 9

4. The kinetic energy of an object varies jointly as the mass and the square of the speed. The kinetic energy of an object with mass 3 kg and speed 4 m/s is 24 joules. Find the kinetic energy of an object with mass 4 kg and speed 3 m/s.
a. 30 joules **b.** 24 joules **c.** 18 joules **d.** 12 joules

5. Divide $3x^3 - x^2 - 4x + 1$ by $x^2 - 2$. 8-3
a. $3x - 1 + \frac{2x + 1}{x^2 - 2}$ **b.** $3x - 1 + \frac{2x - 1}{x^2 - 2}$
c. $3x + 1 + \frac{2x - 1}{x^2 - 2}$ **d.** $3x + 1 + \frac{2x + 1}{x^2 - 2}$

6. Use synthetic division to find the remainder when $x^4 + 4x^3 - 5x + 3$ is divided by $x + 2$. 8-4
a. 21 **b.** 41 **c.** -3 **d.** 17

the only possible rational roots of $x^4 - 16x^2 + 4 = 0$ are ±1, ±2, and ±4, none of which satisfy the equation; $\sqrt{5} - \sqrt{3}$ is irrational.

21. Multiply both sides by the LCD of the coefficients to produce an equivalent equation with integral coefficients. Then use the rational root theorem.

26. The possible rational roots of $x^n - k = 0$ are the integral factors of k. Since $\sqrt[n]{k}$ is rational and $\sqrt[n]{k}$ is a root of $x^n - k = 0$, then $\sqrt[n]{k}$ is an integral factor of k and there is an integer z such that $\sqrt[n]{k} \cdot z = k$; $\sqrt[n]{k} = \frac{k}{z}$ (an integer); $k = \left(\frac{k}{z}\right)^n$

28. 1. $a\left(\frac{h}{k}\right)^3 + b\left(\frac{h}{k}\right)^2 + c\left(\frac{h}{k}\right) + d = 0$ $\left(\frac{h}{k}\text{ is a root.}\right)$
2. $ah^3 + bh^2k + chk^2 + dk^3 = 0$ (Mult. prop. of eq.)
3. $ah^3 = -bh^2k - chk^2 - dk^3$ (Add. prop. of eq.)
4. $\frac{ah^3}{k} = -bh^2 - chk - dk^2$ (Div. prop. of eq.)
5. $\frac{ah^3}{k}$ is an integer. (Int. closure prop.)
6. k is a factor of a. (All prime factors of k divide out with prime factors of ah^3, and h and k are relatively prime.)

7. For what value of k will $x + 1$ be a factor of $x^{13} - 2x^7 + 3x + k$?
a. -6 b. -2 (c.) 2 d. 6

8. Find $P(2)$ if $P(x) = x^5 - 2x^4 + 3x^2 - 1$. 8-5
(a.) 11 b. 5 c. -23 d. -55

9. Solve $x^4 - 4x^3 + 4x^2 - 9 = 0$ given that $1 + i\sqrt{2}$ is a root. 8-6
a. $\{1 + i\sqrt{2}, 1 - i\sqrt{2}\}$ b. $\{1 + i\sqrt{2}, 1 - i\sqrt{2}, -3, -1\}$
(c.) $\{1 + i\sqrt{2}, 1 - i\sqrt{2}, -1, 3\}$ d. $\{1 + i\sqrt{2}, 1 - i\sqrt{2}, 1, -3\}$

10. Find a cubic equation with integral coefficients that has -2 and $3 - i$ as roots.
a. $x^3 + 8x^2 + 22x + 20 = 0$ b. $x^3 - 8x^2 + 22x - 20 = 0$
c. $x^3 + 4x^2 - 2x - 20 = 0$ (d.) $x^3 - 4x^2 - 2x + 20 = 0$

11. Use the rational root theorem to determine which of the following could *not* be a root of $6x^3 + 23x^2 - 6x - 8 = 0$. 8-7
a. -4 b. $\frac{2}{3}$ (c.) $\frac{3}{4}$ d. $-\frac{1}{2}$

12. Approximate to the nearest tenth the real zero of $P(x) = x^3 - 4$. 8-8
a. 1.5 (b.) 1.6 c. 1.7 d. 1.8

13. Use linear interpolation to find an approximation of $f(2.64)$ to the nearest hundredth given that $f(2.6) = 4.97$ and $f(2.7) = 5.12$. 8-9
a. 5.01 b. 5.02 (c.) 5.03 d. 5.04

Chapter Test

1. The distance an object falls from rest is proportional to the square of the length of time it has fallen. If an object falls 64 ft in 2 s, how far will it fall in 3 s? 144 ft 8-1

2. The volume of a cone varies jointly as the height and the square of the base radius. A cone with height 6 cm and base radius 4 cm has volume 32π cm^3. What is the volume of a cone with height 4 cm and base radius 6 cm? 48π cm^3 8-2

3. Divide $x^4 - 3x^3 + 6x - 5$ by $x^2 - 2x + 2$. $x^2 - x - 4 + \frac{3}{x^2 - 2x + 2}$ 8-3

4. Use synthetic division to divide $x^4 + 2x^3 + 4x + 5$ by $x + 2$. $x^3 + 4 + \frac{-3}{x + 2}$ 8-4

5. Find $P\left(\frac{1}{2}\right)$ given that $P(x) = 6x^3 - x^2 + 3x + 5$. 7 8-5

6. Solve $x^4 - 2x^3 + 6x - 9 = 0$ completely given that $1 - i\sqrt{2}$ is a root. $\{1 \pm i\sqrt{2}, \pm\sqrt{3}\}$ 8-6

7. Solve $2x^4 + 3x^3 - 11x^2 + 2x + 4 = 0$ completely by first finding any rational roots. $\left\{-\frac{1}{2}, 1, -1 \pm \sqrt{5}\right\}$ 8-7

8. Graph $P(x) = x^3 - x^2 - 2$ and estimate the real root of $x^3 - x^2 - 2 = 0$ to the nearest half unit. 1.5 8-8

9. Use linear interpolation to find an approximation of $f(-1.76)$ to the nearest hundredth given that $f(-1.7) = 4.63$ and $f(-1.8) = 4.45$. 4.52 8-9

Additional Answers Chapter Test

8.

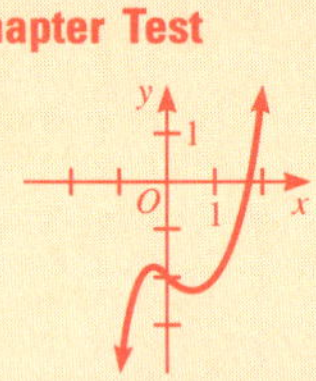

Alternate Test p. T20

Supplementary Materials

Test Masters 33, 34
Resource Book pp. 135–137

Preparing for College Entrance Exams

Strategy for Success

You are not expected to solve the test problems mentally. Do your calculations in the test booklet and cross out inappropriate choices. If you decide to skip a particular problem, mark it in your test booklet so you can easily find it later. Do not put any stray marks on your answer sheet.

Decide which is the best of the choices given and write the corresponding letter on your answer sheet.

1. How many real roots does $2(x^2 + 1)^2 + (x^2 + 1) - 3 = 0$ have? B

(A) 0 **(B)** 1 **(C)** 2 **(D)** 3 **(E)** 4

2. The roots of $1.5y^2 - 5y + 2 = 0$: C

(A) are imaginary **(B)** are rational **(C)** are irrational
(D) have the sum $-\frac{10}{3}$ **(E)** have the product $-\frac{4}{3}$

3. Find $P(1 - i)$ given that $P(x) = 2x^3 - x^2 + 3x + 1$. E

(A) $2 + 9i$ **(B)** 0 **(C)** $1 + i$ **(D)** $-1 + i$ **(E)** $-5i$

4. Given that $\sqrt{44} \approx 6.633$ and $\sqrt{45} \approx 6.708$, use linear interpolation to find an approximate value for $\sqrt{44.2}$. D

(A) 6.652 **(B)** 6.618 **(C)** 6.693 **(D)** 6.648 **(E)** 6.723

5. Suppose z varies directly as x and inversely as y. If $z = 12$ when $x = 4$ and $y = 5$, find z when $x = 6$ and $y = 45$. B

(A) $\frac{1}{9}$ **(B)** 2 **(C)** 112.5 **(D)** 15 **(E)** 72

6. Use the rational root theorem to determine which number is *not* a possible root of $6t^3 + t^2 - 31t + 10 = 0$. E

(A) $-\frac{2}{3}$ **(B)** $\frac{5}{2}$ **(C)** 1 **(D)** -2 **(E)** 3

7. Find a quadratic equation whose roots have the sum -2 and the product $-\frac{3}{2}$. C

(A) $2x^2 + 3x - 4 = 0$ **(B)** $2x^2 - 4x + 3 = 0$ **(C)** $2x^2 + 4x - 3 = 0$
(D) $2x^2 - 4x - 3 = 0$ **(E)** $2x^2 - 3x + 4 = 0$

8. Find the maximum value of g if $g(x) = 7 - 8x - 2x^2$. C

(A) 7 **(B)** -2 **(C)** 15 **(D)** -17 **(E)** 13

9. The volume of a cylinder varies jointly as the height and the square of the base radius. If a cylinder's height is doubled and its base radius is halved, then its volume: E

(A) is quadrupled **(B)** is tripled **(C)** is doubled
(D) remains the same **(E)** is halved

9 Analytic Geometry

Objectives

9-1 To find the distance between any two points and the midpoint of the line segment joining them.

9-2 To learn the relationship between the center and radius of a circle and the equation of the circle.

9-3 To learn the relationship between the focus, directrix, vertex, and axis of a parabola and the equation of the parabola.

9-4 To learn the relationship between the center, foci, and intercepts of an ellipse and the equation of the ellipse.

9-5 To learn the relationship between the foci, intercepts, and asymptotes of a hyperbola and the equation of the hyperbola.

9-6 To find an equation of a conic section with center not at the origin and to identify a conic as a circle, ellipse, or hyperbola.

9-7 To use graphs to determine the number of real solutions of a quadratic system and to estimate the solutions.

9-8 To use algebraic methods to find exact solutions of quadratic systems.

9-9 To solve systems of linear equations in three variables.

Assignment Guide

See p. T58 for Key to the format of the Assignment Guide

Day	Average Algebra	Average Algebra and Trigonometry	Extended Algebra	Extended Algebra and Trigonometry
1	**9-1** 404/1–31 odd 405/*P*: 1, 3, 5	**9-1** 404/1–31 odd 405/*P*: 1, 3, 5	**9-1** 404/1–31 odd 405/*P*: 1, 3, 5	*Administer Chapter 8 Test* **9-1** *Read 9-1* 404/1–31 odd
2	**9-1** 405/*P*: 7–12 **S** 406/*Mixed Review* **9-2** 410/2–14 even	**9-1** 405/*P*: 7–12 **S** 406/*Mixed Review* **9-2** 410/2–14 even	**9-1** 405/*P*: 7–12 **S** 406/*Mixed Review* **9-2** 410/2–24 even	**9-1** 405/1–7 odd, 12 **S** 406/*Mixed Review* **9-2** 410/2–24 even
3	**9-2** 410/16–42 even	**9-2** 410/16–46 even	**9-2** 410/26–46 even, 50 **S** 405/*P*: 13	**9-2** 410/26–46 even, 50 **S** 405/*P*: 13
4	**9-3** 415/2–24 even **S** 411/43–46	**9-3** 415/2–36 even **S** 417/*Mixed Review*	**9-3** 415/1–35 odd **S** 417/*Mixed Review*	**9-3** 415/1–35 odd **S** 417/*Mixed Review*
5	**9-3** 416/26–36 even **S** 417/*Mixed Review* **9-4** 421/1–6	**9-4** 421/1–21 odd **R** 417/*Self-Test 1*	**9-4** 421/1–21 odd **R** 417/*Self-Test 1*	**9-4** 421/1–21 odd **R** 417/*Self-Test 1*
6	**9-4** 421/7–21 odd, 24–31 **R** 417/*Self-Test 1*	**9-4** 422/16–22 even, 24–33 **9-5** 430/1–6	**9-4** 422/22–34 even **9-5** 430/2–16 even	**9-4** 422/22–34 even **9-5** 430/2–16 even
7	**9-5** 430/1–16 **S** 423/32, 33	**9-5** 430/7–31 odd **S** 431/*Mixed Review*	**9-5** 430/17–32 **S** 431/*Mixed Review*	**9-5** 430/17–32 **S** 431/*Mixed Review*
8	**9-5** 430/17–24, 27, 29, 30 **S** 431/*Mixed Review*	**9-7** 438/2–18 even **R** 435/*Self-Test 2* 1–4	**9-6** 434/1–11 odd, 13–21 **S** 416/36, 37	**9-6** 434/1–11 odd, 13–21 **S** 416/36, 37
9	**9-7** 438/2–18 even **R** 435/*Self-Test 2* 1–4	**9-8** 441/1–19 odd 442/*P*: 1–6	**9-7** 438/2–18 even **R** 435/*Self-Test 2*	**9-7** 438/2–16 even, 17–19 **S** 435/*Self-Test 2*

10	**9-8** 441/1–19 odd **S** 438/17, 19	**9-8** 442/*P*: 7–12 **9-9** 447/1–17 odd	**9-8** 441/1–25 odd 442/*P*: 1–6	**9-8** 441/1–25 odd 442/*P*: 1–6
11	**9-8** 441/12–18 even 442/*P*: 1–10	**9-9** 448/19, 21 *P*: 1–9, 11 **S** 449/*Mixed Review*	**9-8** 442/*P*: 7–13 **9-9** 447/1–17 odd	**9-8** 442/*P*: 7–13 **9-9** 447/1–17 odd
12	**9-9** 447/1–21 odd **S** 449/*Mixed Review*	**R** 450/*Self-Test 3* **R** 451/*Ch. Review:* 1–12, 16–20	**9-9** 448/19–25 odd *P*: 1–9, 11 **S** 449/*Mixed Review*	**9-9** 448/19–25 odd *P*: 1–9, 11 **S** 449/*Mixed Review*
13	**9-9** 448/20, 22 *P*: 1–9, 11	*Administer Chapter 9 Test*	**R** 450/*Self-Test 3* **R** 451/*Ch. Review*	**R** 450/*Self-Test 3* **R** 451/*Ch. Review*
14	**R** 450/*Self-Test 3* **R** 451/*Ch. Review:* 1–12, 16–20		*Administer Chapter 9 Test*	*Administer Chapter 9 Test*
15	*Administer Chapter 9 Test*			

Supplementary Materials Guide

For Use with Lesson	Practice Masters	Tests	Study Guide (Reteaching)	Resource Book		
				Tests	Practice Exercises	Applications (A) Enrichment (E) Technology (T) Thinking Skl. (TS)
9-1	Sheet 50		pp. 137–138			p. 212 (A)
9-2			pp. 139–140		p. 138	
9-3	Sheet 51	Test 35	pp. 141–142			
9-4			pp. 143–144			
9-5			pp. 145–146		p. 139	
9-6	Sheet 52	Test 36	pp. 147–148			p. 243 (T)
9-7			pp. 149–150			pp. 243–244 (T)
9-8	Sheet 53		pp. 151–152			
9-9	Sheet 54	Test 37	pp. 153–154		p. 140	p. 265 (TS)
Chapter 9		Tests 38, 39		pp. 50–53	p. 141	p. 228 (E)

Overhead Visuals

For Use with Lessons	Visual	Title
9-2	B	Multi-Use Packet 2
9-2, 9-3, 9-4, 9-5	5	Conic Sections
9-4, 9-6	6	Ellipses
9-5, 9-6	7	Hyperbolas
9-9	8	Intersecting Planes

Software

	Algebra Plotter Plus	Using Algebra Plotter Plus	Computer Activities	Test Generator
Software	Parabola Plotter, Conics Plotter, Conics Quiz, Function Plotter	Enrichment, p. 39 Activity Sheets, pp. 55, 56	Activities 20, 21, 37	189 test items
For Use with Lessons	9-2, 9-3, 9-4, 9-5, 9-6, 9-7, 9-8	9-3, 9-6, 9-7	9-1, 9-9, Chapter 9	all lessons

Strategies for Teaching

Making Connections and Using Manipulatives

Strong connections between algebra and geometry occur throughout this chapter—in the use of geometric formulas, in the geometric modeling of algebraic concepts, in coordinate geometry, and in problem solving as on pages 401–403. For additional examples of the integration of geometry with algebra in this book, refer to the Index under *Geometry; Areas; Conics; Formulas, geometric;* and *Problems, geometric and three-dimensional.*

Manipulatives and models are helpful in this chapter to reinforce geometric concepts such as conic sections—ellipses, parabolas, circles, and hyperbolas. These concrete approaches enhance the introduction of analytic geometry and can diminish students' anxieties. The goal is to have student progress to the point where they can work entirely on the abstract level.

9-1 Distance and Midpoint Formulas

Have the students work Problem 8 and Problem 13. Then have them prove Problem 13 using the geometric or synthetic proof method. Provide geometry books if necessary.

From geometry the line segment through the midpoints of two sides of a triangle is parallel to the third side. Therefore $\overline{PO} \parallel \overline{BD}$ and $\overline{MN} \parallel \overline{BD}$. Also $\overline{PM} \parallel \overline{AC}$ and $\overline{ON} \parallel \overline{AC}$. Therefore $\overline{PO} \parallel \overline{MN}$ and $\overline{PM} \parallel \overline{ON}$. Thus quadrilateral *PMNO* is a parallelogram.

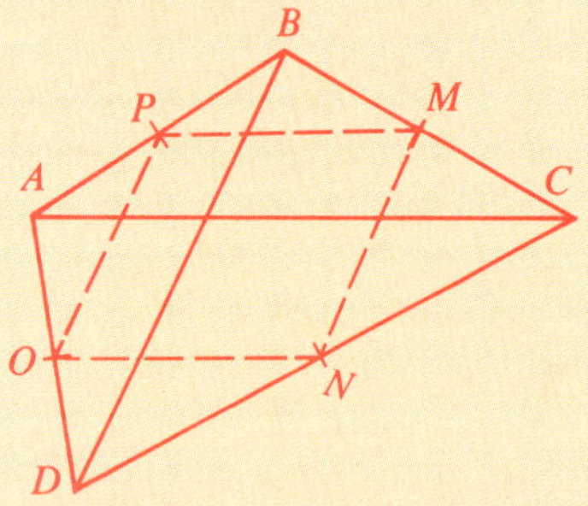

9-2 Circles

The graph of the equation $x^2 + y^2 = 4$ is a circle with center (0, 0) and radius 2. By substituting values, as shown in the table, we find points on the circle.

x	y
0	± 2
1	$\pm\sqrt{3}$
−1	$\pm\sqrt{3}$
2	0
−2	0

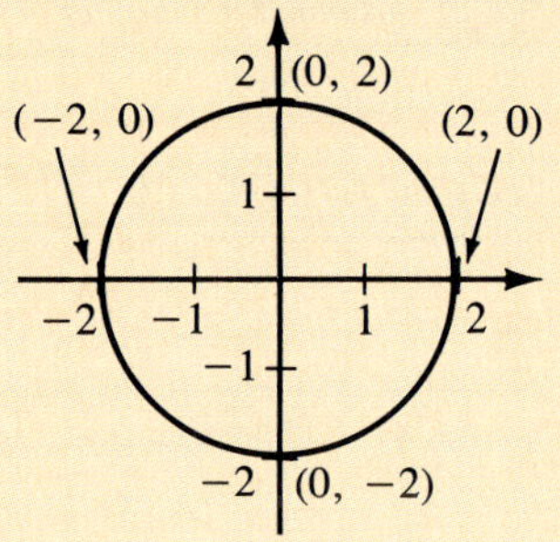

Ask students to tell whether (1, 1), (2, 3), and (−1, 0) are outside or inside the circle.

(1, 1) inside since $1^2 + 1^2 = 2 < 4$.
(2, 3) outside since $2^2 + 3^2 = 13 > 4$.
(−1, 0) inside since $(-1)^2 + 0^2 = 1 < 4$.

9-3 Parabolas

Try this paper-folding method for constructing a parabola. Draw any straight line *m* to be a directrix. Locate a point *F* not on the given line to be the focus. Fold the point *F* upon the directrix *m*, and crease the paper. Repeat this folding process from 10 to 12 times by moving *F* along the line *m* and creasing. The creases are all tangent to the parabola having *F* as a focus and *m* as a directrix. The tangents are said to "envelope" the curve and give the illusion of curvature.

9-6 More on Central Conics

1. As a review and reinforcement, use pictures or a model to show the students the conic sections as sections of a cone. Have the students discuss the conics relating what they have studied to the physical model.
2. Ask students to describe the degenerate conics $x^2 + y^2 = 0$ and $x^2 - y^2 = 0$.
(The graph of $x^2 + y^2 = 0$ is the single point (0, 0). The graph of $x^2 - y^2 = 0$ is the pair of lines $x = y$ and $x = -y$.)
3. Ask students to attempt Exercise 19 or 20 to investigate rotations.

See the Exploration on page 840 for an activity in which students use graphing software to explore the graphs of various circles and ellipses.

Divide the students into groups of two or three and provide string for them to use. Ask the students to use the description on page 418 about construction of ellipses as a guide and write a similar description for circles, parabolas, and hyperbolas.

References to Strategies

PE: Pupil's Edition **TE:** Teacher's Edition **RB:** Resource Book

Problem Solving Strategies

TE: p. 402

Applications

PE: pp. 405–406, 411–412, 422, 442–443 (geometry); pp. 424–426, 436 (physics); p. 426 (navigation); pp. 448–449 (word problems)
TE: p. 400
RB: p. 212

Nonroutine Problems

PE: pp. 405–406 (Probs. 9–14); p. 411 (Exs. 47–50, Challenge); p. 416 (Exs. 35–38); p. 422 (Exs. 23–27); p. 423 (Exs. 34, 35); p. 425 (Challenge 1–5); p. 435 (Ex. 21); p. 438 (Ex. 20); p. 443 (Probs. 13, 14)
TE: pp. T113–T114 (Sugg. Extensions, Lessons 9-1, 9-4, 9-5)

Communication

PE: p. 416 (Exs. 35, 37, 38, convincing argument)
TE: p. T113 (Reading Algebra); p. T116 (Sugg. Extension, Lesson 9-9)

Thinking Skills

PE: pp. 403, 405–406 (proofs)
TE: p. 433
RB: p. 265

Explorations

PE: p. 411 (Challenge); p. 840 (circles and ellipses)

Connections

PE: pp. 401–453 (Analytic Geometry); p. 423 (Art); pp. 424–425, 436 (Astronomy); p. 426 (Navigation); p. 431 (History); p. 443 (Physics); pp. 448–449 (Money)

Using Technology

PE: p. 406 (Exs.); pp. 410, 416, 419, 421, 423, 427, 430, 435–437, 441, 840
TE: pp. 406, 410, 416, 418, 421, 426, 430, 435–436, 441
RB: pp. 243–244
Computer Activities: pp. 52–56, 100–102
Using Algebra Plotter Plus: pp. 39, 55, 56

Using Manipulatives/Models

TE: pp. T114–T116 (Lessons 9-3, 9-6, 9-9)
Overhead Visuals: B, 5, 6, 7, 8

Cooperative Learning

TE: p. 400; p. T115 (Lesson 9-8)

Teaching Resources

For use in implementing the teaching strategies referenced on the previous page.

Application
Resource Book, p. 212

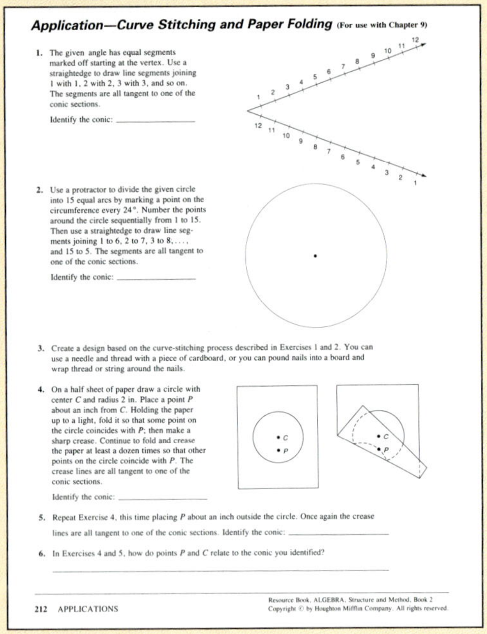

Application—Curve Stitching and Paper Folding (For use with Chapter 9)

1. The given angle has equal segments marked off starting at the vertex. Use a straightedge to draw line segments joining 1 with 1, 2 with 2, 3 with 3, and so on. The segments are all tangent to one of the conic sections.

   Identify the conic: ______

2. Use a protractor to divide the given circle into 15 equal arcs by marking a point on the circumference every 24°. Number the points around the circle sequentially from 1 to 15. Then use a straightedge to draw line segments joining 1 to 6, 2 to 7, 3 to 8, . . . , and 15 to 5. The segments are all tangent to one of the conic sections.

   Identify the conic: ______

3. Create a design based on the curve-stitching process described in Exercises 1 and 2. You can use a needle and thread with a piece of cardboard, or you can pound nails into a board and wrap thread or string around the nails.

4. On a half sheet of paper draw a circle with center C and radius 2 in. Place a point P about an inch from C. Holding the paper up to a light, fold it so that some point on the circle coincides with P; then make a sharp crease. Continue to fold and crease the paper at least a dozen times so that other points on the circle coincide with P. The crease lines are all tangent to one of the conic sections.

   Identify the conic: ______

5. Repeat Exercise 4, this time placing P about an inch outside the circle. Once again the crease lines are all tangent to one of the conic sections. Identify the conic: ______

6. In Exercises 4 and 5, how do points P and C relate to the conic you identified?

   ______

212 APPLICATIONS

Resource Book, ALGEBRA, Structure and Method, Book 2

Enrichment
Resource Book, p. 228

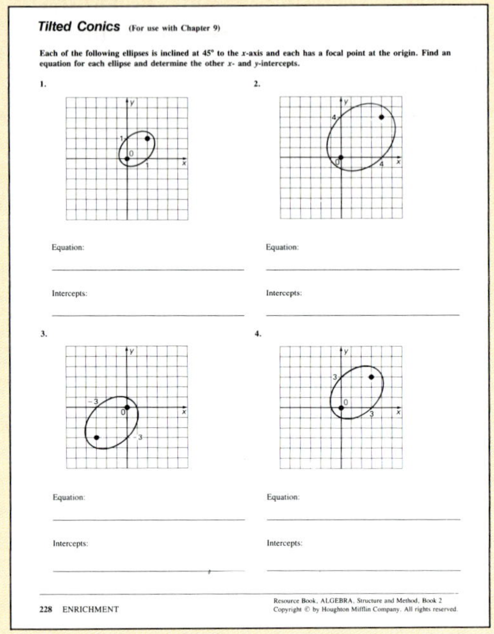

Tilted Conics (For use with Chapter 9)

Each of the following ellipses is inclined at 45° to the x-axis and each has a focal point at the origin. Find an equation for each ellipse and determine the other x- and y-intercepts.

1. Equation: ______ Intercepts: ______

2. Equation: ______ Intercepts: ______

3. Equation: ______ Intercepts: ______

4. Equation: ______ Intercepts: ______

228 ENRICHMENT

Resource Book, ALGEBRA, Structure and Method, Book 2

Thinking Skills
Resource Book, p. 265

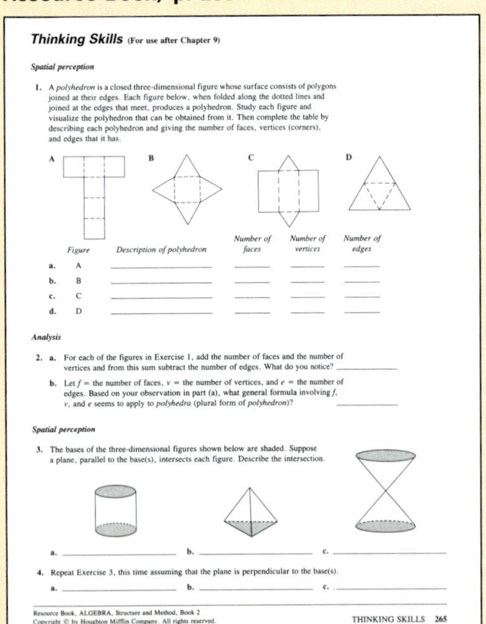

Thinking Skills (For use after Chapter 9)

Spatial perception

1. A *polyhedron* is a closed three-dimensional figure whose surface consists of polygons joined at their edges. Each figure below, when folded along the dotted lines and joined at the edges that meet, produces a polyhedron. Study each figure and visualize the polyhedron that can be obtained from it. Then complete the table by describing each polyhedron and giving the number of faces, vertices (corners), and edges that it has.

	Figure	*Description of polyhedron*	*Number of faces*	*Number of vertices*	*Number of edges*
a.	A	______	______	______	______
b.	B	______	______	______	______
c.	C	______	______	______	______
d.	D	______	______	______	______

Analysis

2. a. For each of the figures in Exercise 1, add the number of faces and the number of vertices and from this sum subtract the number of edges. What do you notice? ______

   b. Let f = the number of faces, v = the number of vertices, and e = the number of edges. Based on your observation in part (a), what general formula involving f, v, and e seems to apply to *polyhedra* (plural form of *polyhedron*)? ______

Spatial perception

3. The bases of the three-dimensional figures shown below are shaded. Suppose a plane, parallel to the base(s), intersects each figure. Describe the intersection.

   a. ______ b. ______ c. ______

4. Repeat Exercise 3, this time assuming that the plane is perpendicular to the base(s).

   a. ______ b. ______ c. ______

Resource Book, ALGEBRA, Structure and Method, Book 2

THINKING SKILLS 265

Using Technology
Resource Book, p. 243

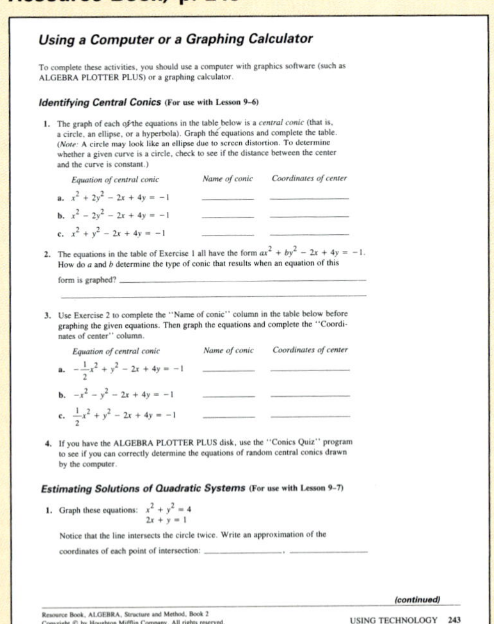

Using a Computer or a Graphing Calculator

To complete these activities, you should use a computer with graphics software (such as ALGEBRA PLOTTER PLUS) or a graphing calculator.

Identifying Central Conics (For use with Lesson 9-6)

1. The graph of each of the equations in the table below is a *central conic* (that is, a circle, an ellipse, or a hyperbola). Graph the equations and complete the table. (*Note:* A circle may look like an ellipse due to screen distortion. To determine whether a given curve is a circle, check to see if the distance between the center and the curve is constant.)

	Equation of central conic	*Name of conic*	*Coordinates of center*
a.	$x^2 + 2y^2 - 2x + 4y = -1$	______	______
b.	$x^2 - 2y^2 - 2x + 4y = -1$	______	______
c.	$x^2 + y^2 - 2x + 4y = -1$	______	______

2. The equations in the table of Exercise 1 all have the form $ax^2 + by^2 - 2x + 4y = -1$. How do a and b determine the type of conic that results when an equation of this form is graphed? ______

3. Use Exercise 2 to complete the "Name of conic" column in the table below before graphing the given equations. Then graph the equations and complete the "Coordinates of center" column.

	Equation of central conic	*Name of conic*	*Coordinates of center*
a.	$-\frac{1}{2}x^2 + y^2 - 2x + 4y = -1$	______	______
b.	$-x^2 - y^2 - 2x + 4y = -1$	______	______
c.	$\frac{1}{2}x^2 + y^2 - 2x + 4y = -1$	______	______

4. If you have the ALGEBRA PLOTTER PLUS disk, use the "Conics Quiz" program to see if you can correctly determine the equations of random central conics drawn by the computer.

Estimating Solutions of Quadratic Systems (For use with Lesson 9-7)

1. Graph these equations: $x^2 + y^2 = 4$
   $2x + y = 1$

   Notice that the line intersects the circle twice. Write an approximation of the coordinates of each point of intersection: ______, ______

(continued)

Resource Book, ALGEBRA, Structure and Method, Book 2

USING TECHNOLOGY 243

Using Models
Overhead Visual 5, Sheets 1 and 4

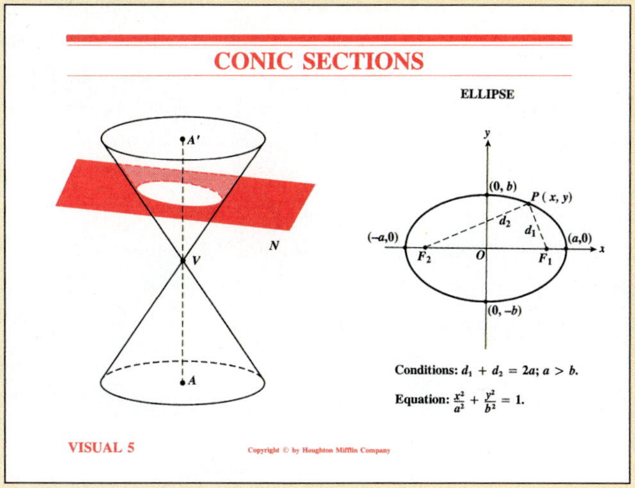

Using Models
Overhead Visual 6, Sheets 1 and 4

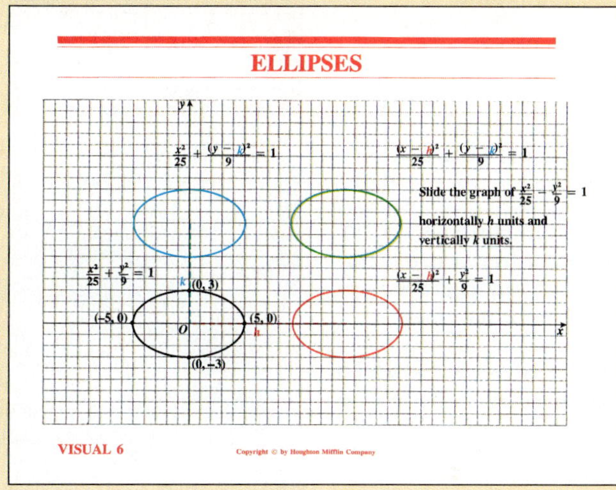

Using Models
Overhead Visual 8, Sheet 4

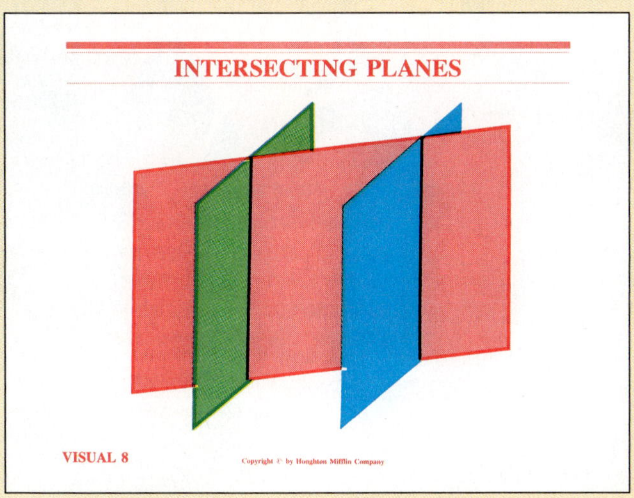

Using Models
Overhead Visual 8, Sheet 5

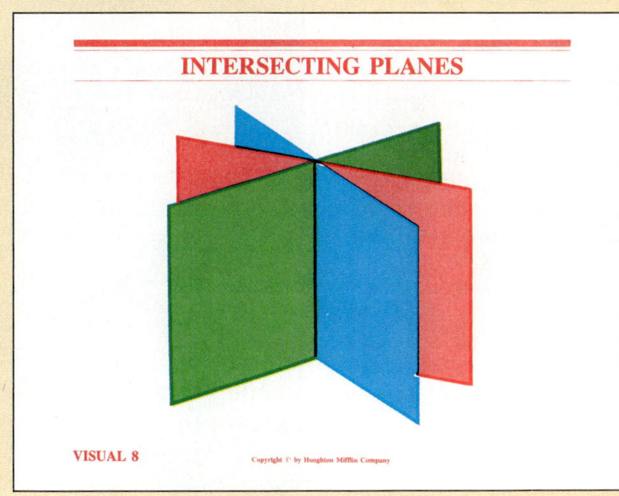

Application

When a plane intersects a double cone but does not pass through the common vertex, the type of conic section that results depends on the angle that the plane makes with the double cone's axis of symmetry. If this angle is denoted by θ and the angle between the axis of symmetry and either cone is α, then the intersection is a circle when $\theta = 90°$, an ellipse when $\alpha < \theta < 90°$, a parabola when $\theta = \alpha$, and a hyperbola when $0° \leq \theta \leq \alpha$.

Special cases occur when the plane passes through the common vertex. These "degenerate conics" include a point (when $\alpha < \theta \leq 90°$), a line (when $\theta = \alpha$), and a pair of intersecting lines (when $0° \leq \theta < \alpha$).

Group Activities

Divide the class into four groups and assign to each group one of the conic sections. Ask the groups to investigate the following: the derivation of the names of the conics, the focus-directrix definitions of the conics, the work of Apollonius (c. 262–190 B.C.) and Dandelin (1794–1847), curve-stitching and paper-folding methods involving tangents to the conics, and practical applications of the conics. Each group should then make a presentation to the class at an appropriate time during the study of this chapter.

Support Materials

Resource Book p. 212

9 Analytic Geometry

Circles, ellipses, parabolas, and hyperbolas are called "conics" because each of these curves is the intersection of a plane and a double cone.

400

Conic Sections: Circles and Parabolas

9-1 Distance and Midpoint Formulas

Objective To find the distance between any two points and the midpoint of the line segment joining them.

The distance between any two points P and Q is written as PQ. If P and Q are points on a number line, PQ is the absolute value of the difference between their coordinates. Since $|a - b| = |b - a|$, you can subtract the coordinates in either order. For the points P and Q shown,

$PQ = |2 - (-3)| = |5| = 5$

or $PQ = |-3 - 2| = |-5| = 5.$

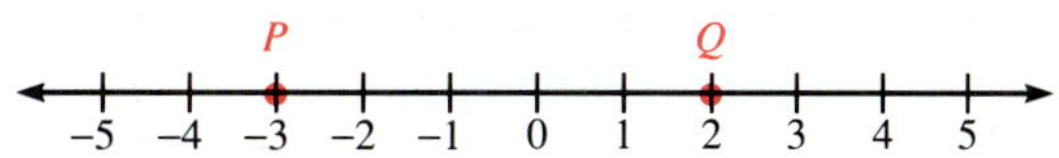

A similar method can be used to find the distance between two points *in a coordinate plane* if the points lie on the same horizontal or vertical line. For example, the distances between the points shown in the diagram at the right are

$$AB = |4 - (-2)| = 6$$
$$CD = |-1 - 6| = 7$$
$$QR = |y_2 - y_1| = |y_1 - y_2|$$
$$PQ = |x_2 - x_1| = |x_1 - x_2|.$$

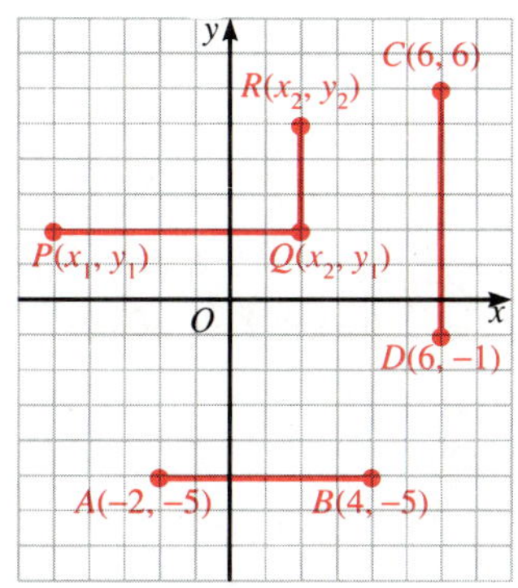

You can use the Pythagorean theorem to get a general formula for finding the distance between *any* two points in a coordinate plane.

Pythagorean Theorem

If the length of the hypotenuse of a right triangle is c, and the lengths of the other two sides (legs) are a and b, then

$$c^2 = a^2 + b^2.$$

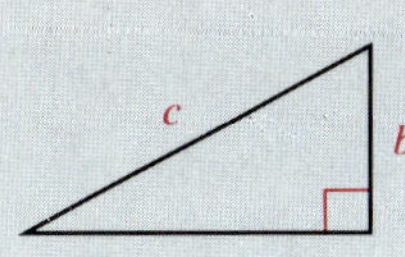

The converse of this theorem is true as well: If the lengths a, b, and c of the sides of a triangle satisfy the equation $c^2 = a^2 + b^2$, then the triangle is a right triangle and c is the length of the hypotenuse.

Teaching References

Lesson Commentary, pp. T112–T116

Assignment Guide, pp. T65–T69

Supplementary Materials
- Practice Masters 50–54
- Tests 35–39
- Resource Book
  - Practice Exercises, pp. 138–141
  - Tests, pp. 50–53
  - Enrichment Activity, p. 228
  - Application, p. 212
- Study Guide, pp. 137–154
- Computer Activities 20–21
- Test Generator
- Disk for Algebra

Alternate Test, p. T21

Teaching Suggestions p. T112

Suggested Extensions p. T113

Warm-Up Exercises

Simplify.

1. $-5 + 2$ -3
2. $\frac{7 + (-3)}{2}$ 2
3. $(-1 - (-4))^2$ 9
4. $|-2 + (-6)|$ 8
5. $(3 - (-5))^2 - (-4 - 2)^2$ 28

Motivating the Lesson

Tell the students that many of the theorems that they proved in geometry using two-column proof can also be proved algebraically. Such algebraic proofs often use the distance and midpoint formulas, which are the topics of today's lesson.

Problem Solving Strategies

Deduction is used in this lesson to establish the distance formula which is then used in subsequent lessons to obtain general forms for equations of circles, parabolas, ellipses, and hyperbolas.

Chalkboard Examples

1. Find the distance between $P_1(6, 2)$ and $P_2(0, -2)$.
$P_1P_2 =$
$\sqrt{(6-0)^2 + (2+2)^2} =$
$\sqrt{36+16} = \sqrt{52} = 2\sqrt{13}$

2. Find the midpoint of the line segment joining (4, 1) and (1, −5).
$\left(\frac{4+1}{2}, \frac{1+(-5)}{2}\right) =$
$\left(\frac{5}{2}, \frac{-4}{2}\right) = \left(\frac{5}{2}, -2\right)$

3. Find an equation for the perpendicular bisector of $\overline{AB}$ for the points $A(3, 5)$ and $B(9, -3)$. Let $P(x, y)$ be any point on the perpendicular bisector.
$AP = BP$
$\sqrt{(3-x)^2 + (5-y)^2} =$
$\sqrt{(x-9)^2 + (y+3)^2}$
$9 - 6x + x^2 + 25 - 10y +$
$y^2 = x^2 - 18x + 81 + y^2 +$
$6y + 9$
$12x - 16y - 56 = 0$
$3x - 4y - 14 = 0$

Check for Understanding

Here is a suggested use of the Oral Exercises to check students' understanding as you teach the lesson.
Oral Exs. 1–9: use after Example 1.
Oral Exs. 10–15: use after Example 2.

To find a general formula for distance, let $P_1(x_1, y_1)$ and $P_2(x_2, y_2)$ be any two points not on the same horizontal or vertical line. Then P_1P_2 is the length of the hypotenuse of the right triangle having vertices P_1, P_2, and $Q(x_2, y_1)$. The lengths of the legs are

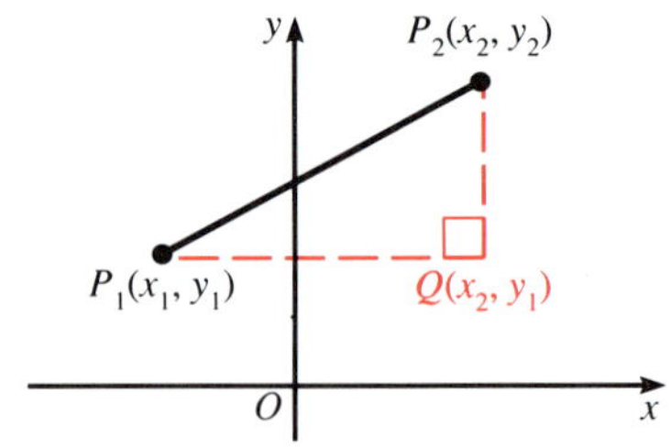

$$P_1Q = |x_2 - x_1| \quad \text{and} \quad P_2Q = |y_2 - y_1|.$$

By the Pythagorean theorem,

$$\begin{aligned}(P_1P_2)^2 &= (P_1Q)^2 + (P_2Q)^2 \\ &= |x_2 - x_1|^2 + |y_2 - y_1|^2 \\ &= (x_2 - x_1)^2 + (y_2 - y_1)^2 && \longleftarrow \text{Remember, } |a|^2 = a^2. \\ P_1P_2 &= \sqrt{(x_2 - x_1)^2 + (y_2 - y_1)^2} && \longleftarrow \text{Taking the principal square root of each side.}\end{aligned}$$

The Distance Formula

The distance between the points $P_1(x_1, y_1)$ and $P_2(x_2, y_2)$ is

$$\mathbf{P_1P_2 = \sqrt{(x_2 - x_1)^2 + (y_2 - y_1)^2}.}$$

Example 1 Find the distance between $P_1(-2, -1)$ and $P_2(-4, 3)$.

Solution

$$\begin{aligned}P_1P_2 &= \sqrt{(-4-(-2))^2 + (3-(-1))^2} \\ &= \sqrt{(-2)^2 + 4^2} \\ &= \sqrt{4+16} \\ &= \sqrt{20} \\ &= 2\sqrt{5} \quad \textbf{\textit{Answer}}\end{aligned}$$

The distance formula can be used to prove the next result (see Problem 11).

The Midpoint Formula

The midpoint of the line segment joining $P_1(x_1, y_1)$ and $P_2(x_2, y_2)$ is

$$M\left(\frac{x_1 + x_2}{2}, \frac{y_1 + y_2}{2}\right).$$

Example 2 Find the midpoint of the line segment joining (4, −6) and (−3, 2).

Solution $\left(\frac{x_1 + x_2}{2}, \frac{y_1 + y_2}{2}\right) = \left(\frac{4 + (-3)}{2}, \frac{-6+2}{2}\right)$, or $\left(\frac{1}{2}, -2\right)$.
Answer

Example 3 Prove that the midpoint of the hypotenuse of any right triangle is equidistant from the three vertices.

Solution Any right triangle can be placed so that its vertices are $O(0, 0)$, $P(a, 0)$, and $Q(0, b)$, as shown. By the distance formula or the Pythagorean theorem, $PQ = \sqrt{a^2 + b^2}$.

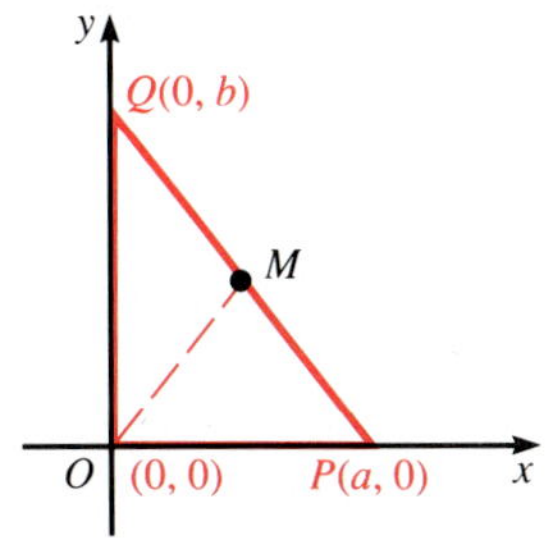

Let point M be the midpoint of $\overline{PQ}$. Then $MP = MQ = \frac{1}{2}\sqrt{a^2 + b^2}$. The coordinates of M are $\left(\frac{a+0}{2}, \frac{0+b}{2}\right)$, or $\left(\frac{a}{2}, \frac{b}{2}\right)$.

By the distance formula,

$$MO = \sqrt{\left(\frac{a}{2} - 0\right)^2 + \left(\frac{b}{2} - 0\right)^2}$$
$$= \sqrt{\frac{a^2}{4} + \frac{b^2}{4}} = \sqrt{\frac{a^2 + b^2}{4}} = \frac{1}{2}\sqrt{a^2 + b^2}$$

$\therefore MO = MP = MQ$

Example 4 Find an equation of the perpendicular bisector of $\overline{AB}$ for the points $A(-2, 1)$ and $B(1, -3)$.

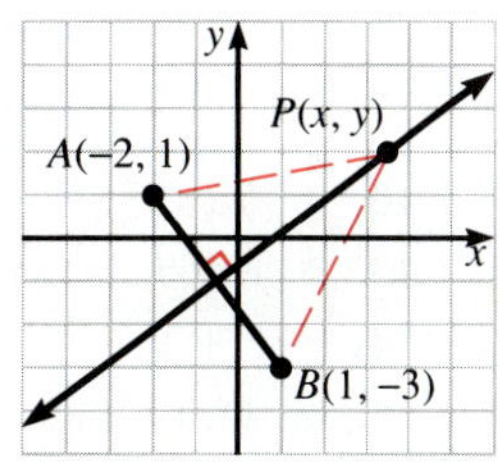

Solution 1 Make a sketch of the points $A(-2, 1)$ and $B(1, -3)$, as shown. Let $P(x, y)$ be any point on the perpendicular bisector. Since any point on the perpendicular bisector is equidistant from A and B,

$$AP = BP$$
$$\sqrt{(x+2)^2 + (y-1)^2} = \sqrt{(x-1)^2 + (y+3)^2}$$

Square both sides and multiply out the binomials.

$$x^2 + 4x + 4 + y^2 - 2y + 1 = x^2 - 2x + 1 + y^2 + 6y + 9$$
$$6x - 8y - 5 = 0 \quad \textbf{\textit{Answer}}$$

Solution 2 The midpoint of $\overline{AB}$ is $\left(-\frac{1}{2}, -1\right)$, and the slope of the line containing A and B is $-\frac{4}{3}$. Then the perpendicular bisector is the line passing through $\left(-\frac{1}{2}, -1\right)$ having slope which is the negative reciprocal of $-\frac{4}{3}$, namely $\frac{3}{4}$. The point-slope form of the equation of this line is $y + 1 = \frac{3}{4}\left(x + \frac{1}{2}\right)$, which is equivalent to $6x - 8y - 5 = 0$.

Guided Practice

Find (a) the distance between the two points and (b) the midpoint of the segment joining the points.

1. $(-3, -1)$, $(-7, 5)$
   a. $2\sqrt{13}$ **b.** $(-5, 2)$
2. $\left(\frac{1}{2}, 0\right)$, $(3, -6)$
   a. $6\frac{1}{2}$ **b.** $\left(\frac{7}{4}, -3\right)$
3. $(\sqrt{2} + 1, \sqrt{2})$, $(\sqrt{2} - 1, 2\sqrt{2})$
   a. $\sqrt{6}$ **b.** $\left(\sqrt{2}, \frac{3\sqrt{2}}{2}\right)$

Summarizing the Lesson

In this lesson students learned to use the distance formula and the midpoint formula. Ask the students to state the two formulas.

Suggested Assignments

Average Algebra
Day 1: 404/1–31 odd
405/*P*: 1, 3, 5
Day 2: 405/*P*: 7–12
S 406/Mixed Review
Assign with Lesson 9-2.

Average Alg. and Trig.
Day 1: 404/1–31 odd
405/*P*: 1, 3, 5
Day 2: 405/*P*: 7–12
S 406/Mixed Review
Assign with Lesson 9-2.

Extended Algebra
Day 1: 404/1–31 odd
405/*P*: 1, 3, 5
Day 2: 405/*P*: 7–12
S 406/Mixed Review
Assign with Lesson 9-2.

Extended Alg. and Trig.
Day 1: 404/1–31 odd
Day 2: 405/*P*: 1, 3, 5, 7–12
S 406/Mixed Review
Assign with Lesson 9-2.

Supplementary Materials

Study Guide	pp. 137–138
Practice Master	50
Computer Activity	20

Additional Answers
Problems

7. $(CD)^2 = (EF)^2 = 25$; $(DE)^2 = (CF)^2 = 34$; $(CD)^2 + (DE)^2 + (EF)^2 + (CF)^2 = 25 + 34 + 25 + 34 = 118$; $(CE)^2 = 89$; $(DF)^2 = 29$; $(CE)^2 + (DF)^2 = 89 + 29 = 118$; $\therefore (CD)^2 + (DE)^2 + (EF)^2 + (CF)^2 = (CE)^2 + (DF)^2$

8. Let $M(0, 5)$, $N(4, 4)$, $P(3, -1)$, and $Q(-1, 0)$ be the midpoints of $\overline{AB}$, $\overline{BC}$, $\overline{CD}$, and $\overline{AD}$, respectively; slope of $\overline{MN} = -\frac{1}{4} =$ slope of $\overline{PQ}$; slope of $\overline{NP} = 5 =$ slope of $\overline{MQ}$; since $\overline{MN} \parallel \overline{PQ}$ and $\overline{NP} \parallel \overline{MQ}$, $MNPQ$ is a parallelogram.

9. Let M, N, and P be the midpts. of $\overline{AB}$, $\overline{BC}$, and $\overline{AC}$, respectively; $M = \left(\frac{a}{2}, 0\right)$, $N = \left(\frac{a+b}{2}, \frac{c}{2}\right)$, $P = \left(\frac{b}{2}, \frac{c}{2}\right)$; $MN = \frac{\sqrt{b^2 + c^2}}{2}$; $AC = \sqrt{b^2 + c^2}$; slope of $\overline{MN} = \frac{c}{b} =$ slope of $\overline{AC}$; $\overline{MN} \parallel \overline{AC}$ and $MN = \frac{1}{2}AC$; it can be shown similarly that $\overline{MP} \parallel \overline{BC}$ and $MP = \frac{1}{2}BC$ and that $\overline{PN} \parallel \overline{AB}$ and $PN = \frac{1}{2}AB$.

Oral Exercises

Find the distance between each pair of points.

1. $(8, 4), (8, -2)$ 6
2. $(6, -5), (0, -5)$ 6
3. $(4, -1), (-2, -1)$ 6
4. $(2, -7), (2, -1)$ 6
5. $(0, 0), (4, 3)$ 5
6. $(0, -8), (6, 0)$ 10
7. $(2, 3), (0, 0)$ $\sqrt{13}$
8. $(7, 0), (0, -2)$ $\sqrt{53}$
9. $(3, 1), (2, 4)$ $\sqrt{10}$

Find the midpoint of the line segment joining each pair of points.

10. $(7, 5), (-1, -3)$ $(3, 1)$
11. $(3, -8), (5, -9)$ $\left(4, -\frac{17}{2}\right)$
12. $(-3, 6), (2, 2)$ $\left(-\frac{1}{2}, 4\right)$
13. $(5, -1), (-6, -2)$ $\left(-\frac{1}{2}, -\frac{3}{2}\right)$
14. $(2.5, 0.5), (6.5, 3)$ $(4.5, 1.75)$
15. $(6, -4.5), (-3, -1.5)$ $(1.5, -3)$

Written Exercises

3. $\sqrt{74}$; $\left(-\frac{5}{2}, \frac{5}{2}\right)$ 6. $2\sqrt{13}$; $(-1, -1)$ 9. $11\sqrt{2}$; $\left(\frac{11}{2}, \frac{11}{2}\right)$

Find (a) the distance between each pair of points and (b) the midpoint of the line segment joining the points. Express all radicals in simplest form.

A

1. $(13, 6), (0, 6)$ 13; $\left(\frac{13}{2}, 6\right)$
2. $(0, 8), (-6, 0)$ 10; $(-3, 4)$
3. $(0, 6), (-5, -1)$
4. $(9, 1), (2, -1)$ $\sqrt{53}$; $\left(\frac{11}{2}, 0\right)$
5. $(3, 2), (5, 6)$ $2\sqrt{5}$; $(4, 4)$
6. $(-4, -3), (2, 1)$
7. $(2, 2), \left(\frac{1}{3}, -2\right)$ $\frac{13}{3}$; $\left(\frac{7}{6}, 0\right)$
8. $\left(\frac{1}{2}, -1\right), (-1, 1)$ $\frac{5}{2}$; $\left(-\frac{1}{4}, 0\right)$
9. $(0, 0), (11, 11)$
10. $(0, 0), (5, 5)$ $5\sqrt{2}$; $\left(\frac{5}{2}, \frac{5}{2}\right)$
11. $(\sqrt{2}, 1), (-\sqrt{2}, 0)$ 3; $\left(0, \frac{1}{2}\right)$
12. $(5, \sqrt{5}), (3, -\sqrt{5})$ $2\sqrt{6}$; $(4, 0)$
13. $(1 + \sqrt{5}, 2 + \sqrt{3}), (1 - \sqrt{5}, -2 + \sqrt{3})$ 6; $(1, \sqrt{3})$
14. $(\sqrt{6} + 1, \sqrt{3} - \sqrt{2}), (\sqrt{6} - 1, \sqrt{3} + \sqrt{2})$ $2\sqrt{3}$; $(\sqrt{6}, \sqrt{3})$
15. $(a, b), (0, b)$ $|a|$; $\left(\frac{a}{2}, b\right)$
16. $(-a, b), (2a, 4b)$ $3\sqrt{a^2 + b^2}$; $\left(\frac{a}{2}, \frac{5b}{2}\right)$
17. $(a + b, a - b), (b - a, b + a)$ $2\sqrt{a^2 + b^2}$; (b, a)
18. $(a, \sqrt{ab}), (b, -\sqrt{ab})$ $|a + b|$; $\left(\frac{a + b}{2}, 0\right)$

Find the coordinates of Q given that M is the midpoint of $\overline{PQ}$.

Sample $P(-6, 2)$, $M(-1, 1)$

Solution Let Q be the point (x, y). Then use the midpoint formula.

$$\frac{x_1 + x_2}{2} = -1 \quad \text{and} \quad \frac{y_1 + y_2}{2} = 1$$

$$\frac{-6 + x}{2} = -1 \qquad \frac{2 + y}{2} = 1$$

$$-6 + x = -2 \qquad 2 + y = 2$$

$$x = 4 \qquad y = 0$$

$\therefore Q$ is the point $(4, 0)$. ***Answer***

B

19. $P(0, 0)$, $M(3, 5)$ $(6, 10)$
20. $P(-4, 3)$, $M(0, 0)$ $(4, -3)$
21. $P(-4, 0)$, $M(3, 3)$ $(10, 6)$
22. $P(6, -2)$, $M(0, 5)$ $(-6, 12)$
23. $P(h, k)$, $M(0, 0)$ $(-h, -k)$
24. $P(0, 0)$, $M(h, k)$ $(2h, 2k)$

The vertices A, B, and C of a triangle are given. For each triangle, determine (a) whether it is isosceles, and (b) whether it is a right triangle. If it is a right triangle, find its area.

25. $A(-2, 2)$, $B(2, 1)$, $C(1, -3)$ yes; yes; $\frac{17}{2}$

26. $A(4, -5)$, $B(-2, -8)$, $C(-8, 4)$ no; yes; 45

27. $A(5, -4)$, $B(-3, 4)$, $C(-2, -3)$ yes; no

28. $A(2, -3)$, $B(2, 3)$, $C(5, 0)$ yes; yes; 9

Use the distance formula to determine whether the given points are collinear. (*Hint:* Find the distance between each pair of points. If one of these distances is the sum of the other two, the three points are collinear.)

29. $(1, 2)$, $(7, 4)$, $(-2, 1)$ collinear

30. $(5, 0)$, $(-7, 3)$, $(1, 1)$ collinear

31. $(-5, -2)$, $(-2, 1)$, $(1, 3)$ not collinear

32. $(1, 5)$, $(2, 0)$, $(4, -10)$ collinear

Problems

Find an equation of the perpendicular bisector of $\overline{AB}$.

A

1. $A(-3, 0)$, $B(0, 5)$ $3x + 5y - 8 = 0$

2. $A(2, 1)$, $B(-2, 3)$ $2x - y + 2 = 0$

3. $A(8, -3)$, $B(-2, 5)$ $5x - 4y - 11 = 0$

4. $A(-9, -3)$, $B(1, -7)$ $5x - 2y + 10 = 0$

5. Find the points on the coordinate axes that are equidistant from the points A and B in Problem 1. $\left(\frac{8}{3}, 0\right)$, $\left(0, \frac{8}{5}\right)$

6. Which of the following represents the distance from any point (x, y) to the origin $(0, 0)$? a

a. $\sqrt{x^2 + y^2}$ b. $\sqrt{x} + \sqrt{y}$ c. $x + y$

7. The sum of the squares of the lengths of the sides of any parallelogram equals the sum of the squares of the lengths of its diagonals. Verify this for the parallelogram with vertices $C(0, 0)$, $D(5, 0)$, $E(8, 5)$, and $F(3, 5)$.

8. The midpoints of the sides of any quadrilateral are the vertices of a parallelogram. Verify this for the quadrilateral with vertices $A(-2, 3)$, $B(2, 7)$, $C(6, 1)$, and $D(0, -3)$.

B

9. Prove: The segment joining the midpoints of two sides of a triangle is parallel to, and half as long as, the third side. [*Hint:* Any triangle can be placed so that its vertices are $A(0, 0)$, $B(a, 0)$, and $C(b, c)$.]

10. Suppose that the midpoints of the sides of a triangle are given. Explain how to draw the triangle.

11. Use the distance formula to prove the midpoint formula. (That is, show that $MP_1 = MP_2 = \frac{1}{2}P_1P_2$.)

C

12. Prove the fact stated in Problem 7. [*Hint:* Any parallelogram can be placed so that its vertices are $A(0, 0)$, $B(a, 0)$, $C(a + b, c)$, and $D(b, c)$.]

13. Prove the fact stated in Problem 8. [*Hint:* Any quadrilateral can be placed so that its vertices are $A(0, 0)$, $B(2a, 0)$, $C(2b, 2c)$, and $D(2d, 2e)$.]

10. Methods may vary; example: Let M, N, and P be the given points. Draw $\overline{NP}$ and construct line j through $M \parallel \overline{NP}$. Draw $\overline{MP}$ and construct line k through $N \parallel \overline{MP}$, intersecting j at A. Extend $\overrightarrow{AN}$ through N to B so that $NB = AN$; extend $\overrightarrow{AM}$ through M to C so that $MC = AM$. Draw $\overline{BC}$. $\triangle ABC$ is the triangle.

11. $MP_1 = \sqrt{\frac{(x_1 - x_2)^2}{4} + \frac{(y_1 - y_2)^2}{4}} = \frac{\sqrt{(x_1 - x_2)^2 + (y_1 - y_2)^2}}{2} = \frac{1}{2}P_1P_2$; $MP_2 = \sqrt{\frac{(x_2 - x_1)^2}{4} + \frac{(y_2 - y_1)^2}{4}} = \frac{\sqrt{(x_1 - x_2)^2 + (y_1 - y_2)^2}}{2} = \frac{1}{2}P_1P_2$

12. $(AB)^2 = (CD)^2 = a^2$; $(BC)^2 = (AD)^2 = b^2 + c^2$; $(AB)^2 + (BC)^2 + (CD)^2 + (AD)^2 = 2a^2 + 2b^2 + 2c^2$; $(AC)^2 = (a + b)^2 + c^2$; $(BD)^2 = (a - b)^2 + c^2$; $(AC)^2 + (BD)^2 = 2a^2 + 2b^2 + 2c^2$; $\therefore (AB)^2 + (BC)^2 + (CD)^2 + (AD)^2 = (AC)^2 + (BD)^2$

13. Let M, N, P, and Q be the midpoints of $\overline{AB}$, $\overline{BC}$, $\overline{CD}$, and $\overline{AD}$, respectively; $M = (a, 0)$, $N = (a + b, c)$, $P = (b + d, c + e)$, and $Q = (d, e)$; slope of $\overline{MN} = \frac{c}{b} =$ slope of $\overline{PQ}$; slope of $\overline{PN} = \frac{e}{d - a} =$ slope of $\overline{QM}$; since $\overline{MN} \parallel \overline{PQ}$ and $\overline{PN} \parallel \overline{QM}$, $MNPQ$ is a parallelogram.

Additional Answers Problems

14. a. Since L_1 and L_2 intersect at (r, s), $s = m_1 r + b_1$ and $s = m_2 r + b_2$. Let $T_1 = (x_1, y_1)$ and $T_2 = (x_2, y_2)$. Since T_1 and T_2 are on $x = r + 1$, $x_1 = r + 1$ and $x_2 = r + 1$; $y_1 = m_1 x_1 + b_1 = m_1(r + 1) + b_1 = m_1 r + m_1 + b_1 = s + m_1$ and $y_2 = m_2 x_2 + b_2 = m_2(r + 1) + b_2 = m_2 r + m_2 + b_2 = s + m_2$; $T_1 = (r + 1, s + m_1)$ and $T_2 = (r + 1, s + m_2)$

b. $T_1T_2 = \sqrt{(m_1 - m_2)^2} = |m_1 - m_2|$; $PT_1 = \sqrt{1 + m_1^2}$; $PT_2 = \sqrt{1 + m_2^2}$

c. $(T_1T_2)^2 = (m_1 - m_2)^2 = m_1^2 - 2m_1m_2 + m_2^2 = m_1^2 + 2 + m_2^2 = m_1^2 + 1 + m_2^2 + 1 = (PT_1)^2 + (PT_2)^2$

d. Since $\triangle T_1PT_2$ has a right angle at P, $(T_1T_2)^2 = (PT_1)^2 + (PT_2)^2$; $(m_1 - m_2)^2 = 1 + m_1^2 + 1 + m_2^2$; $m_1^2 - 2m_1m_2 + m_2^2 = 2 + m_1^2 + m_2^2$; $-2m_1m_2 = 2$; $m_1m_2 = -1$

Using a Computer

The exercises ask students to write programs that incorporate the distance formula. Ask students with programming experience to complete the exercises and demonstrate their programs to the class.

14. This problem outlines a proof of part (2) of the theorem on page 119.

Let $P(r, s)$ be the point of intersection of

L_1: $y = m_1x + b_1$ and L_2: $y = m_2x + b_2$,

and let T_1 and T_2 be the points where the line $x = r + 1$ intersects L_1 and L_2, respectively.

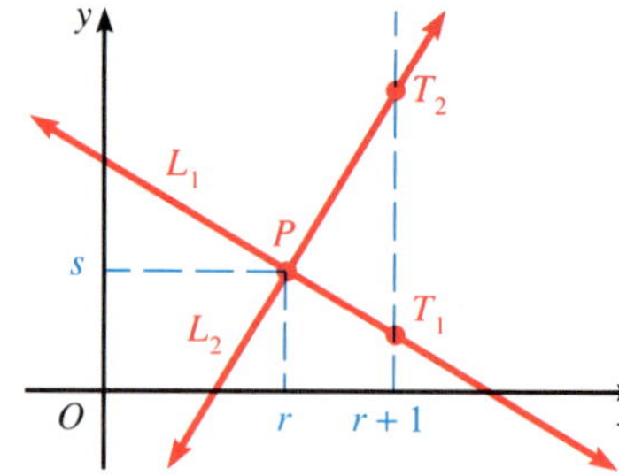

a. Show that $T_1 = (r + 1, s + m_1)$ and $T_2 = (r + 1, s + m_2)$.

b. Find the distances T_1T_2, PT_1, and PT_2.

c. Proof of the "if" part: You are given that $m_1m_2 = -1$. Use this fact and part (b) to show that $(T_1T_2)^2 = (PT_1)^2 + (PT_2)^2$. It follows from the converse of the Pythagorean theorem that triangle T_1PT_2 has a right angle at P. Therefore, L_1 and L_2 are perpendicular.

d. Proof of the "only if" part: You are given that L_1 and L_2 are perpendicular, so that triangle T_1PT_2 has a right angle at P. Apply the Pythagorean theorem and part (b). Then simplify to obtain $m_1m_2 = -1$.

Mixed Review Exercises

Graph each equation.

1. $x - 2y = 4$ **2.** $y = 3 - x^2$ **3.** $y = x^2 + 4x + 5$

Find the value of each function if $x = -3$.

4. $f(x) = -x + 5$ 8 **5.** $g(x) = \frac{x - 2}{3x - 1}$ $\frac{1}{2}$ **6.** $h(x) = x^5 - 7x^3 - 13x + 11$ −4

Computer Exercises

For students with some programming experience.

1. Write a program to compute the distance between two given points.

2. Run your program to find the distance between the given points.

a. (0, 0), (5, 5) 7.07106782 **b.** (−4, 7), (8, −7) 18.4390889 **c.** (100, 85), (242, −173) 294.49618

3. Modify the program to accept the coordinates of 3 distinct points and compute the distance between each pair of points. If the 3 points are collinear, the program should print a message stating this; and if they are not collinear, the program should print the perimeter of the resulting triangle. (*Hint:* Since the computer approximates the square roots used in finding distances, consider A, B, and C collinear if $|AB - (BC + AC)|$, $|AC - (BC + AB)|$, or $|BC - (AC + AB)| < 0.00001$.)

4. Run the program in Exercise 3 for each set of points.

a. (4, 7), (5, 9), (6, 18) 22.471793 **b.** (74, −18), (100, 36), (120, 0) 150.512173

9-2 Circles

Objective To learn the relationship between the center and radius of a circle and the equation of the circle.

Analytic geometry uses algebra to investigate geometric figures. In Lessons 9-2 through 9-6 you will study plane curves having second-degree equations. These curves are called **conic sections,** or simply **conics,** because they can be obtained by slicing a double cone with a plane (see page 400).

Of these conics you are probably most familiar with the circle. A **circle** is the set of all points in a plane that are a fixed distance, called the **radius,** from a fixed point, called the **center.**

The diagram shows a circle having center $C(2, -3)$ and radius 6. To find the equation of the circle, use the distance formula. A point $P(x, y)$ is on the circle if and only if the distance between P and the center $C(2, -3)$ equals 6.

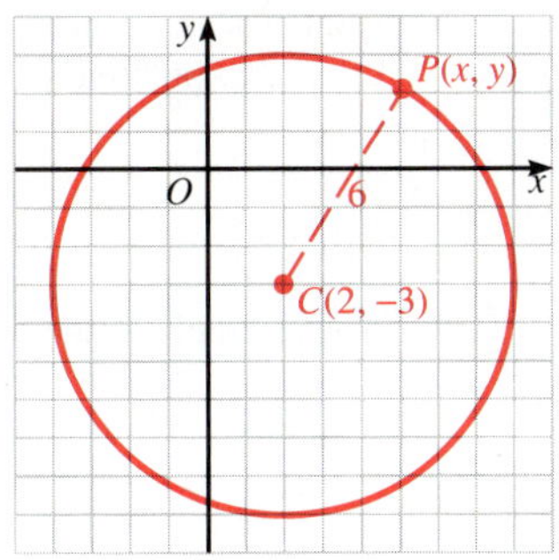

$$CP = 6$$
$$\sqrt{(x-2)^2 + (y+3)^2} = 6$$
Thus $$(x-2)^2 + (y+3)^2 = 36$$

Every circle has an equation of this form.

Equation of a Circle

The circle with center (h, k) and radius r has the equation

$$(x-h)^2 + (y-k)^2 = r^2.$$

Example 1 Find an equation of the circle with center $(-2, 5)$ and radius 3.

Solution Substitute $h = -2$, $k = 5$, and $r = 3$ in this equation:

$$(x-h)^2 + (y-k)^2 = r^2.$$
$$[x-(-2)]^2 + (y-5)^2 = 3^2$$
$$(x+2)^2 + (y-5)^2 = 9 \quad \textbf{\textit{Answer}}$$

If the center of a circle is at the origin, then (h, k) is $(0, 0)$ and the equation of the circle is

$$(x-0)^2 + (y-0)^2 = r^2$$
or $$x^2 + y^2 = r^2.$$

Teaching Suggestions p. T113

Suggested Extensions p. T113

Warm-Up Exercises

Give the constant that makes each of the following a perfect square trinomial.

1. $x^2 + 8x + \underline{\ ?\ }$ 16
2. $y^2 - 36y + \underline{\ ?\ }$ 324
3. $x^2 - x + \underline{\ ?\ }$ $\frac{1}{4}$
4. $y^2 + 3y + \underline{\ ?\ }$ $\frac{9}{4}$
5. $4x^2 + 4x + \underline{\ ?\ }$ 1
6. $9y^2 - 12y + \underline{\ ?\ }$ 4

Motivating the Lesson

Tell students that circles can be found everywhere: the steering wheel of a car, the face of a clock, a class ring. Exploring the relationship between a circle and its equation is the topic of today's lesson.

Chalkboard Examples

1. Find an equation of the circle with center $(3, -4)$ and radius 6.
   $(x-3)^2 + [y-(-4)]^2 = 6^2$
   $(x-3)^2 + (y+4)^2 = 36$
2. Graph $x^2 + y^2 + 8x - 2y + 22 = 0$.
   $x^2 + 8x + 16 + y^2 - 2y + 1 = -22 + 16 + 1$;
   $(x+4)^2 + (y-1)^2 = -5$
   no solution
   Equation has no graph.
3. Find the center and radius of the circle $x^2 + y^2 - 6x + 2y + 6 = 0$.
   $x^2 - 6x + 9 + y^2 + 2y + 1 = -6 + 9 + 1$; $(x-3)^2 + (y+1)^2 = 4$
   center: $(3, -1)$, radius: 2

(continued)

Chalkboard Examples
(continued)

4. The center of a circle of radius 5 lies in the first quadrant. Find an equation of the circle if it is tangent to the x-axis at $(1, 0)$.
   The center of the circle lies on the line $x = 1$ and must be 5 units from $(1, 0)$. Thus the center is $(1, 5)$. The equation is $(x - 1)^2 + (y - 5)^2 = 25$.

Check for Understanding

Oral Exs. 1–6: use before Example 1.
Oral Exs. 7–15: use after Example 1.

Common Errors

Students may need a review of completing the square before being able to determine if the graph of a given equation is a circle. Completing the square when the coefficient of the second degree term is not equal to one may be especially troublesome for some students.

Guided Practice

1. Find an equation of the circle with center $(-1, 3)$ and radius 2.
   $(x + 1)^2 + (y - 3)^2 = 4$
2. Graph $x^2 + y^2 - 4y - 5 = 0$.
   $x^2 + (y^2 - 4y + 4) = 5 + 4$
   $x^2 + (y - 2)^2 = 9$
   center: $(0, 2)$, radius: 3

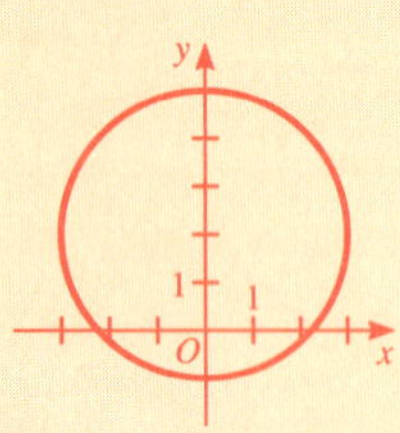

The graph of the equation $x^2 + y^2 = 9$ is a circle with center at the origin and radius 3. If you slide every point of this circle to the right 2 units and up 5 units, the equation of the *translated* circle is

$$(x - 2)^2 + (y - 5)^2 = 9.$$

Sliding a graph to a new position in the coordinate plane without changing its shape is called a **translation.**

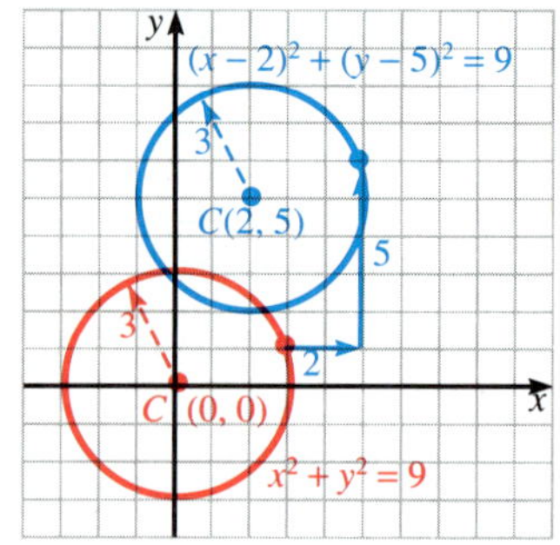

In general, replacing x by $x - h$ and y by $y - k$ in an equation slides the corresponding graph h units horizontally and k units vertically. Slide:

right if h is positive,	left if h is negative,
up if k is positive,	down if k is negative.

Example 2 Graph $(x - 2)^2 + (y + 6)^2 = 4$.

Solution 1 The graph of the equation $x^2 + y^2 = 4$ is a circle with center at the origin and radius 2. Slide this circle 2 units to the right and then 6 units down to get the graph of $(x - 2)^2 + [y - (-6)]^2 = 4$, or $(x - 2)^2 + (y + 6)^2 = 4$.

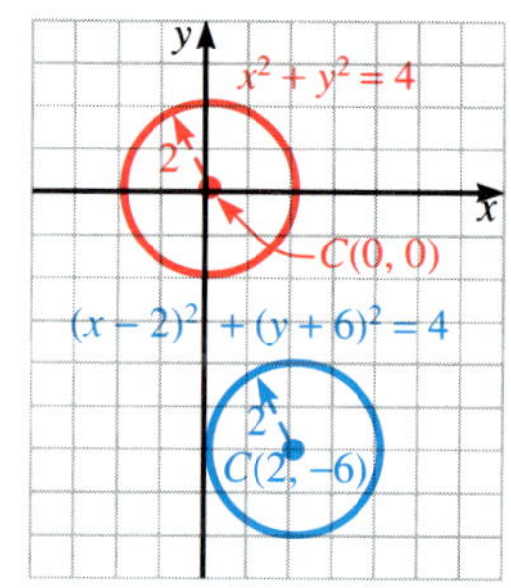

Solution 2 Rewrite the given equation in the form $(x - h)^2 + (y - k)^2 = r^2$:

$$(x - 2)^2 + [y - (-6)]^2 = 2^2$$

The graph is a circle. The center (h, k) is $(2, -6)$ and the radius $r = 2$.

Example 3 If the graph of the given equation is a circle, find its center and radius. If the equation has no graph, say so.

a. $x^2 + y^2 + 10x - 4y + 21 = 0$ **b.** $x^2 + y^2 - 8x + 6y + 30 = 0$

Solution **a.** If the graph is a circle, the equation can be written in the form $(x - h)^2 + (y - k)^2 = r^2$. Complete the square twice, once using the terms in x, and once using the terms in y.

$$x^2 + y^2 + 10x - 4y + 21 = 0$$

$$(x^2 + 10x + \underline{\ ?\ }) + (y^2 - 4y + \underline{\ ?\ }) = -21 \qquad \text{Rearrange terms.}$$

$$(x^2 + 10x + 25) + (y^2 - 4y + 4) = -21 + 25 + 4 \qquad \text{Add 25 and 4.}$$

$$(x + 5)^2 + (y - 2)^2 = 8$$

$\therefore$ the center is $(-5, 2)$ and the radius is $2\sqrt{2}$. ***Answer***

b.

$$x^2 + y^2 - 8x + 6y + 30 = 0$$
$$(x^2 - 8x + 16) + (y^2 + 6y + 9) = -30 + 16 + 9$$
$$(x - 4)^2 + (y + 3)^2 = -5$$

Since the square of any number is positive and the sum of two positive numbers is positive, *no* ordered pair satisfies the equation. This equation has no graph.
Answer

Recall this fact from geometry: Let L be the line tangent to a given circle at a point P. Then the line perpendicular to L at P passes through the center of the circle.

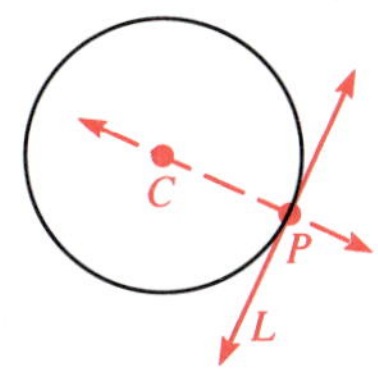

Example 4 Find an equation of a circle of radius 3 that has its center in the first quadrant and is tangent to the y-axis at (0, 2).

Solution First make a sketch using the given information. From the geometric fact stated above, the center of the circle lies on the line $y = 2$ and must be 3 units from (0, 2), which is on the circle. Thus the center is (3, 2).

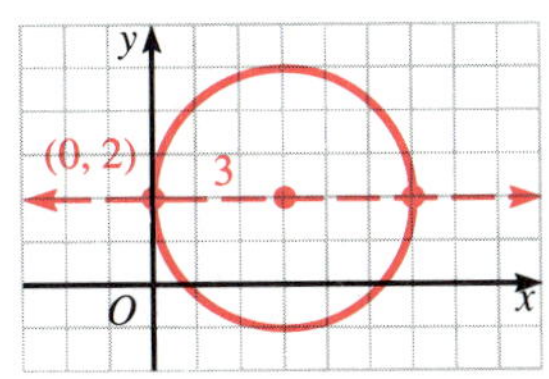

The equation is

$$(x - 3)^2 + (y - 2)^2 = 3^2.$$
Answer

Oral Exercises

Give the center and radius of each circle.

1. $x^2 + y^2 = 49$ (0, 0); 7
2. $x^2 + y^2 = 1$ (0, 0); 1
3. $(x - 2)^2 + (y - 4)^2 = 36$ (2, 4); 6
4. $(x - 5)^2 + y^2 = 25$ (5, 0); 5
5. $x^2 + (y + 5)^2 = \frac{1}{9}$ (0, −5); $\frac{1}{3}$
6. $(x + 4)^2 + (y - 3)^2 = 7$ (−4, 3); $\sqrt{7}$

Give an equation of the circle with the indicated center and radius.

7. (0, 0); 4 $x^2 + y^2 = 16$
8. (0, 0); $\sqrt{2}$ $x^2 + y^2 = 2$
9. (1, 1); 3 $(x - 1)^2 + (y - 1)^2 = 9$
10. (3, 2); 5 $(x - 3)^2 + (y - 2)^2 = 25$
11. (4, 0); 1 $(x - 4)^2 + y^2 = 1$
12. (−1, 0); 4 $(x + 1)^2 + y^2 = 16$
13. (−3, −3); 3 $(x + 3)^2 + (y + 3)^2 = 9$
14. (−3, 4); $\sqrt{8}$ $(x + 3)^2 + (y - 4)^2 = 8$
15. (1, −1); $\frac{1}{2}$ $(x - 1)^2 + (y + 1)^2 = \frac{1}{4}$

If the graph of the given equation is a circle, find its center and radius; if the equation has no graph, say so.

3. $x^2 + y^2 + 2y = 8$
(0, −1); 3
4. $x^2 + y^2 + 4x + 10y = 7$
(−2, −5); 6
5. $x^2 + y^2 - 2x + 2y + 8 = 0$
no graph

Summarizing the Lesson

In this lesson students learned to write the equation of a circle and to relate that equation to the center and radius of the circle. Ask the students to state the equation of a circle with center (h, k) and radius r.

Suggested Assignments

Average Algebra
Day 1: 410/2–14 even
Assign with Lesson 9-1.
Day 2: 410/16–42 even

Average Alg. and Trig.
Day 1: 410/2–14 even
Assign with Lesson 9-1.
Day 2: 410/16–46 even

Extended Algebra
Day 1: 410/2–24 even
Assign with Lesson 9-1.
Day 2: 410/26–46 even, 50
S 405/*P*: 13

Extended Alg. and Trig.
Day 1: 410/2–24 even
Assign with Lesson 9-1.
Day 2: 410/26–46 even, 50
S 405/*P*: 13

Supplementary Materials

Study Guide	pp. 139–140
Resource Book	p. 138
Overhead Visual	5

Using a Computer or a Graphing Calculator

Students may wish to use a computer or a graphing calculator to check their graphs for Exs. 9–14 and 43–46.

Point out to students using a graphing calculator that circles must be graphed in two parts. For example, to graph $x^2 + y^2 = 1$, students must graph the semicircles $y = \pm\sqrt{1 - x^2}$ separately.

Support Materials
Disk for Algebra
Menu Items: Conics Plotter
Function Plotter

Additional Answers Written Exercises

10.

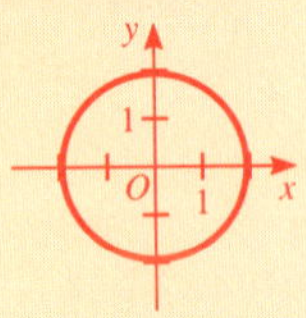

12.

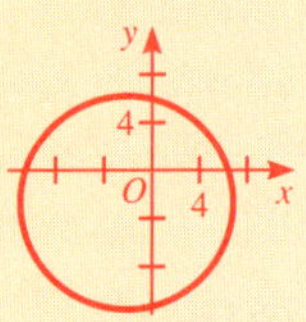

14.

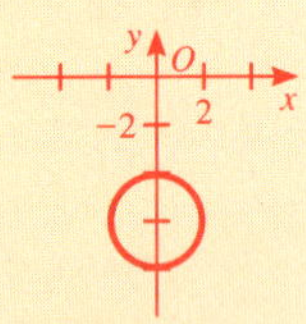

26.

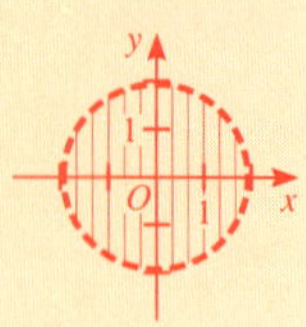

Written Exercises

Find an equation of the circle with the given center and radius.

A

1. $(3, 0); 3$ $(x - 3)^2 + y^2 = 9$
2. $(0, -1); 1$ $x^2 + (y + 1)^2 = 1$
3. $(2, -5); 8$ $(x - 2)^2 + (y + 5)^2 = 64$
4. $(-3, 1); 5$ $(x + 3)^2 + (y - 1)^2 = 25$
5. $(0, 0); 12$ $x^2 + y^2 = 144$
6. $(-4, -2); 10$ $(x + 4)^2 + (y + 2)^2 = 100$
7. $(6, 1); \sqrt{2}$ $(x - 6)^2 + (y - 1)^2 = 2$
8. $(-5, 3); \frac{1}{6}$ $(x + 5)^2 + (y - 3)^2 = \frac{1}{36}$

Graph each equation. You may wish to check your graphs on a computer or a graphing calculator.

9. $x^2 + y^2 = 25$
10. $x^2 + y^2 = 4$
11. $(x - 4)^2 + (y - 5)^2 = 1$
12. $(x + 2)^2 + (y + 3)^2 = 81$
13. $(x - 3)^2 + y^2 = 36$
14. $x^2 + (y + 6)^2 = 4$

If the graph of the given equation is a circle, find its center and radius. If the equation has no graph, say so.

15. $x^2 + y^2 - 16 = 0$ $(0, 0); 4$
16. $x^2 + y^2 - 81 = 0$ $(0, 0); 9$
17. $x^2 + y^2 = -8y$ $(0, -4); 4$
18. $x^2 + y^2 - 6x = 0$ $(3, 0); 3$
19. $x^2 + y^2 - 4x + 2y - 4 = 0$ $(2, -1); 3$
20. $x^2 + y^2 + 10x - 4y + 20 = 0$ $(-5, 2); 3$
21. $x^2 + y^2 + 8x + 2y + 18 = 0$ no graph
22. $x^2 + y^2 + 12x - 6y = 0$ $(-6, 3); 3\sqrt{5}$
23. $x^2 + y^2 + 3x - 4y = 0$ $\left(-\frac{3}{2}, 2\right); \frac{5}{2}$
24. $x^2 + y^2 - 5y + 4 = 0$ $\left(0, \frac{5}{2}\right); \frac{3}{2}$

Graph each inequality.

Sample $x^2 + y^2 - 8x + 2y + 1 < 0$

Solution Rewrite the inequality by completing the squares in x and y. The inequality becomes $(x - 4)^2 + (y + 1)^2 < 16$.

The graph of the equation $(x - 4)^2 + (y + 1)^2 = 16$ is the circle with center $(4, -1)$ and radius 4. Sketch this circle using dashed lines to indicate that points on the circle do not satisfy the inequality. To determine whether the graph is the set of points inside or outside the circle, substitute a test point, say $(3, 0)$, in the inequality. Since $(3, 0)$ satisfies the inequality, shade the region inside the circle for the graph.

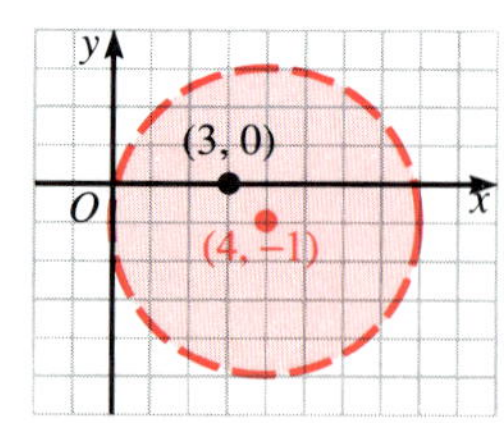

B

25. $x^2 + y^2 \geq 1$
26. $x^2 + y^2 < 4$
27. $x^2 + y^2 - 4y > 0$
28. $x^2 + y^2 \geq 2y$
29. $x^2 + y^2 + 6x - 6y + 9 \leq 0$
30. $x^2 + y^2 + 4x - 10y < 7$

Find the center and radius of each circle. (***Hint:*** **First divide both sides by the coefficient of the second-degree terms.)**

31. $4x^2 + 4y^2 - 16x - 24y + 36 = 0$ $(2, 3);\ 2$

32. $9x^2 + 9y^2 + 6x + 18y + 9 = 0$ $\left(-\frac{1}{3}, -1\right);\ \frac{1}{3}$

33. $16x^2 + 16y^2 - 32x + 8y = 0$ $\left(1, -\frac{1}{4}\right);\ \frac{\sqrt{17}}{4}$

34. $3x^2 + 3y^2 - 6x + 24y + 24 = 0$ $(1, -4);\ 3$

Find an equation of the circle described. (A sketch may be helpful.)

35. Center (0, 5); passes through (0, 0). $x^2 + (y - 5)^2 = 25$

36. Center (−2, 0); passes through (2, 0). $(x + 2)^2 + y^2 = 16$

37. A diameter has endpoints (2, 5) and (0, 3). $(x - 1)^2 + (y - 4)^2 = 2$

38. Center in quadrant two; radius 3; tangent to y-axis at (0, 4). $(x + 3)^2 + (y - 4)^2 = 9$

39. Center on line $y - 4 = 0$; tangent to x-axis at (−2, 0). $(x + 2)^2 + (y - 4)^2 = 16$

40. Center on line $x + y = 4$; tangent to both coordinate axes. $(x - 2)^2 + (y - 2)^2 = 4$

41. Center in quadrant four; tangent to the lines $x = 1$, $x = 9$, and $y = 0$. $(x - 5)^2 + (y + 4)^2 = 16$

42. Tangent to both coordinate axes and the line $x = -8$. (Two answers) $(x + 4)^2 + (y - 4)^2 = 16$ or $(x + 4)^2 + (y + 4)^2 = 16$

Graph each semicircle. Recall that $\sqrt{a} \geq 0$.

C **43.** $y = \sqrt{25 - x^2}$

44. $y = -\sqrt{25 - x^2}$

45. $y = \sqrt{4x - x^2}$

46. $x = \sqrt{2y - y^2}$

47. A boat sails so that it is always twice as far from one buoy as from a second buoy 3 mi from the first one. Describe the path of the boat. (*Hint:* Introduce a coordinate system so that one buoy is at (0, 0) and the other is at (3, 0).)

48. A ladder 6 m long leaning against a wall slips to the ground. Describe the path followed by the midpoint of the ladder. (*Hint:* See the figure at the right.)

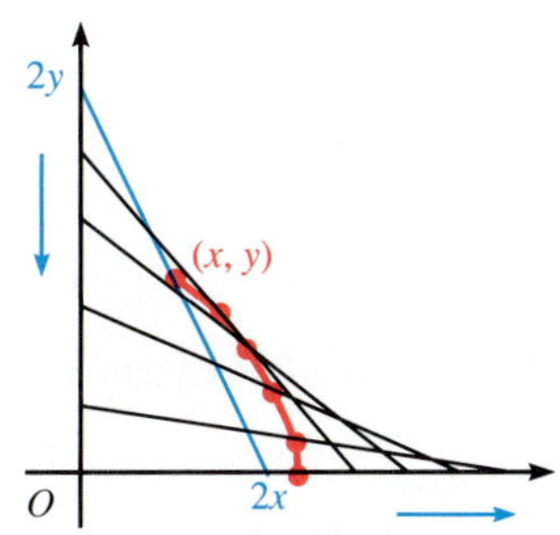

49. Find an equation of the circle of radius 4 that is tangent to both branches of the graph of $y = |x|$. $x^2 + (y - 4\sqrt{2})^2 = 16$

50. Use analytic geometry to prove that an angle inscribed in a semicircle is a right angle. (*Hint:* Let the semicircle be the top half of the circle $x^2 + y^2 = r^2$ and $P(x, y)$ be any point on the semicircle. Use the slopes of $\overline{PA}$ and $\overline{PB}$ to show that they are perpendicular.)

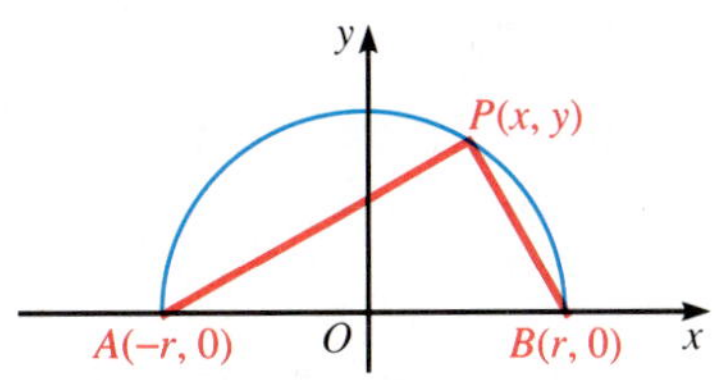

Challenge

Answers may vary. Example: The circle with equation $(x - 4)^2 + \left(y - \frac{5}{3}\right)^2 = \frac{169}{9}$ has lattice points (0, 0), (8, 0), and (4, 6).

Points with integer coordinates are called *lattice points*. For example, (0, 0) and (−2, 5) are lattice points, but $\left(\frac{1}{2}, 3\right)$ is not. The circle $x^2 + y^2 = 1$ passes through the 4 lattice points (1, 0), (−1, 0), (0, 1), and (0, −1). Find an equation of a circle that passes through exactly 3 lattice points.

28.

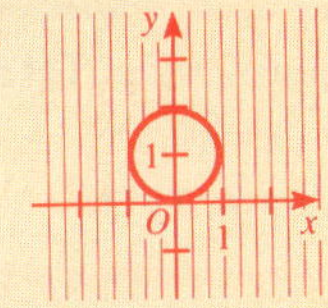

30.

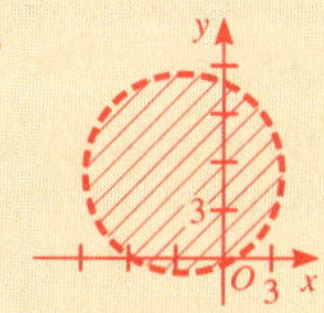

44.

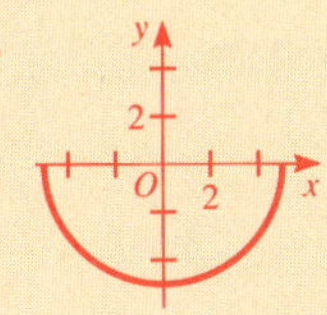

46.

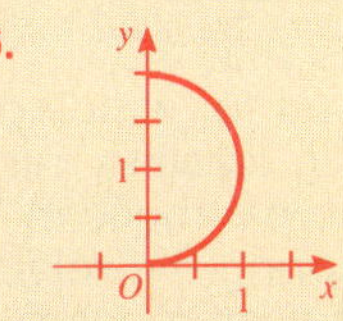

48. The path is the first-quadrant portion of the circle with center at the origin and radius 3 m.

50. Let $P(x, y)$ be any point on the semicircle; slope of $\overline{PA} = \frac{y - 0}{x - (-r)} = \frac{y}{x + r}$; slope of $\overline{PB} = \frac{y - 0}{x - r} = \frac{y}{x - r}$; $\frac{y}{x + r} \cdot \frac{y}{x - r} = \frac{y^2}{x^2 - r^2} = \frac{y^2}{-(r^2 - x^2)} = \frac{y^2}{-y^2} = -1$; since the product of the slopes of $\overline{PA}$ and $\overline{PB}$ is −1, $\overline{PA} \perp \overline{PB}$ and $\angle P$ is a right angle.

Teaching Suggestions p. T113

Reading Algebra p. T113

Suggested Extensions p. T114

Warm-Up Exercises

Determine the distance between the given point and line. Draw a sketch if necessary.

1. (3, 4); x-axis 4
2. $(-1, 2)$; y-axis 1
3. $(-2, 3)$; $x = 1$ 3
4. $(5, -4)$; $y = -2$ 2
5. $(1, -3)$; $x = -4$ 5

Motivating the Lesson

Tell students that the "dishes" used for sending and receiving satellite transmissions are parabolic in shape. Exploring the relationship between a parabola and its equation is the topic of today's lesson.

Chalkboard Examples

1. A parabola has focus $(4, -3)$ and directrix $y = 1$. Find its vertex. Since the vertex is midway between the focus and directrix, the vertex is $(4, -1)$.
2. Find an equation of the parabola having focus (1, 0) and directrix $y = 3$.
$\sqrt{(x-1)^2 + (y-0)^2} = \sqrt{(x-x)^2 + (y-3)^2}$
$(x-1)^2 + y^2 = (y-3)^2$
$(x-1)^2 = -6y + 9$
$y = -\frac{1}{6}(x-1)^2 + \frac{3}{2}$

9-3 Parabolas

Objective To learn the relationship between the focus, directrix, vertex, and axis of a parabola and the equation of the parabola.

Suppose that you draw the line $y = 4$ and plot the point $F(3, 0)$. Then plot several points P that appear to be the same distance from the line $y = 4$ as they are from the point F.

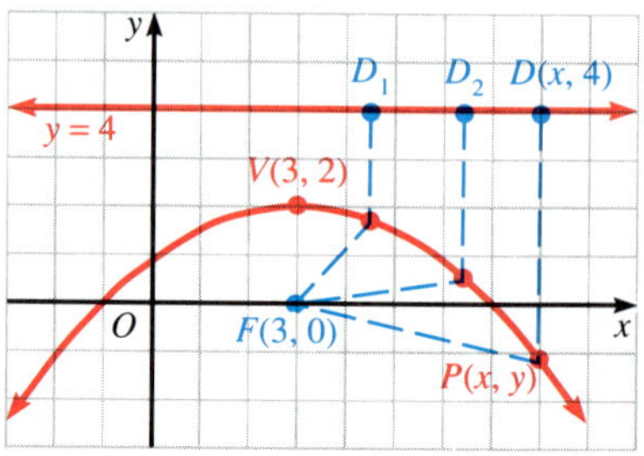

In the diagram, the distance from point P to the line is measured along the perpendicular $\overline{PD}$. To find an equation of the path of P, you use the distance formula.

$$PD = PF$$
$$\sqrt{(x-x)^2 + (y-4)^2} = \sqrt{(x-3)^2 + (y-0)^2}$$
$$(y-4)^2 = (x-3)^2 + y^2$$
$$-8y + 16 = (x-3)^2$$
$$-8(y-2) = (x-3)^2$$
$$y - 2 = -\frac{1}{8}(x-3)^2$$

The last equation is of the form $y - k = a(x - h)^2$. In Lesson 7-5, you learned that the graph of such an equation is a parabola with vertex (h, k) and axis $x = h$. Therefore, the graph of the set of points P is a parabola with vertex (3, 2) and axis $x = 3$. The following general definition of a parabola is stated in terms of distance.

A **parabola** is the set of all points equidistant from a fixed line, called the **directrix,** and a fixed point not on the line, called the **focus.**

The important features of a parabola are shown in the diagram below. Notice that the vertex is midway between the focus and the directrix.

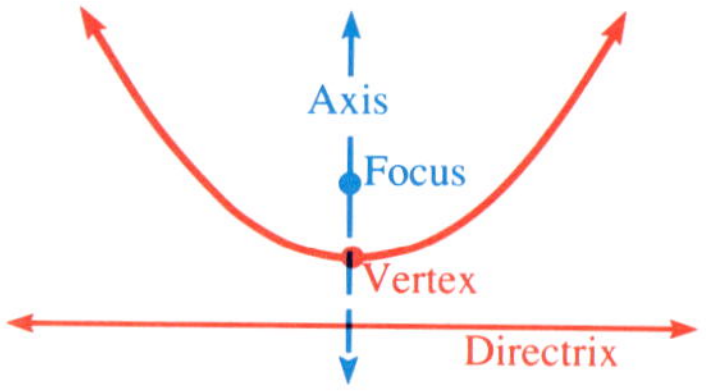

Example 1 The vertex of a parabola is $(-5, 1)$ and the directrix is the line $y = -2$. Find the focus of the parabola.

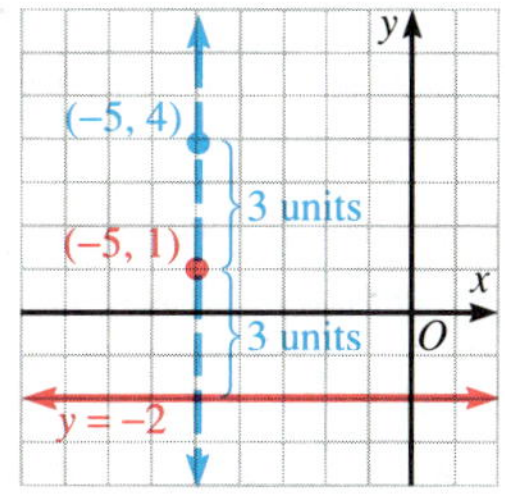

Solution It is helpful to make a sketch. The vertex is 3 units above the directrix. Since the vertex is midway between the focus and the directrix, the focus is $(-5, 4)$. ***Answer***

Example 2 Find an equation of the parabola having the point $F(0, -2)$ as focus and the line $x = 3$ as directrix. Draw the curve and label the vertex V, the focus F, the directrix, and the axis of symmetry.

Solution From the definition, $P(x, y)$ is on the parabola if and only if $PD = PF$, where PD is the perpendicular distance from P to the directrix.

$$PD = PF$$
$$\sqrt{(x-3)^2 + (y-y)^2} = \sqrt{(x-0)^2 + (y-(-2))^2}$$
$$(x-3)^2 = x^2 + (y+2)^2$$
$$x^2 - 6x + 9 - x^2 = (y+2)^2$$
$$-6\left(x - \frac{3}{2}\right) = (y+2)^2$$
$$x - \frac{3}{2} = -\frac{1}{6}(y+2)^2$$

To plot a few points, choose convenient values of y and compute the corresponding values of x.

x	y
$-\frac{9}{2}$	-8
0	-5
$\frac{3}{2}$	-2
0	1
$-\frac{9}{2}$	4

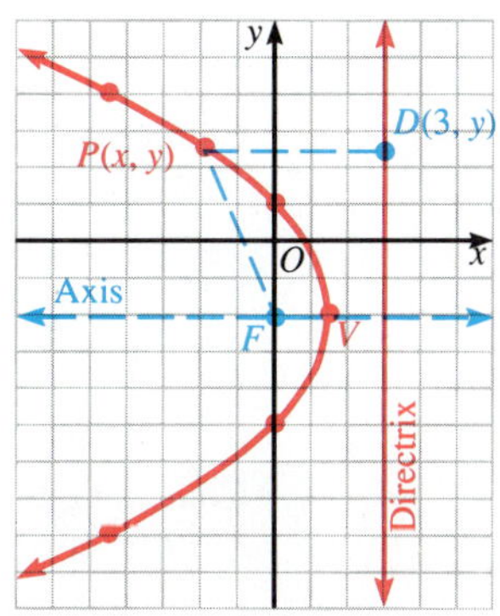

Notice that the parabola in Example 2 has a horizontal axis and an equation of the form $x - h = a(y - k)^2$ where the point (h, k) is the vertex. This is similar to the equation of a parabola that has a vertical axis, except that the roles of x and y are reversed.

3. Find the axis, vertex, focus, and directrix of the parabola
$x^2 - 2x - 2y = 0$.
Sketch the parabola.
$x^2 - 2x + 1 = 2y + 1$
$(x-1)^2 = 2\left(y + \frac{1}{2}\right)$; $h = 1$, $k = -\frac{1}{2}$, and $c = \frac{1}{2}$.
axis: $x = 1$
vertex: $\left(1, -\frac{1}{2}\right)$
focus: $(1, 0)$
directrix: $y = -1$

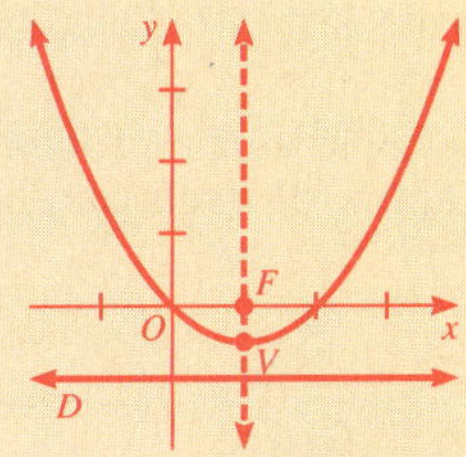

4. Find an equation of the parabola that has vertex $(3, 0)$ and focus $(3, -3)$.

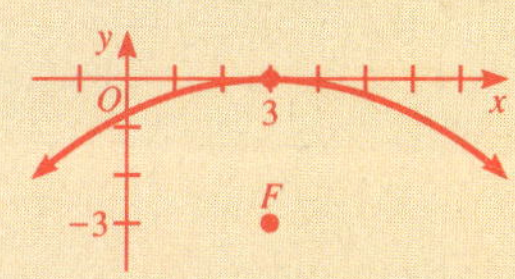

The distance from the vertex to the focus is 3, so $|c| = 3$. Since the vertex is above the focus, the parabola opens downward, so $c < 0$. The equation is: $(x-3)^2 = -12y$

Check for Understanding

Here is a suggested use of the Oral Exercises to check students' understanding as you teach the lesson.
Oral Exs. 1–6: use after Example 1.
Oral Exs. 7–12: use before Example 3.

Guided Practice

1. A parabola has vertex $(-2, 3)$ and focus $(-2, 1)$. Find its directrix. $y = 5$

Find an equation for the parabola with the given features.

2. Vertex $(0, 0)$; focus $(0, -2)$ $x^2 = -8y$
3. Focus $(0, 0)$; directrix $x = -6$ $y^2 = 12(x + 3)$
4. Vertex $(-1, 1)$, directrix $y = -5$ $(x + 1)^2 = 24(y - 1)$

Find the axis, vertex, focus and directrix of each parabola. Graph the parabola.

5. $y^2 + 2x = 0$ $y = 0$; $(0, 0)$; $\left(-\frac{1}{2}, 0\right)$; $x = \frac{1}{2}$

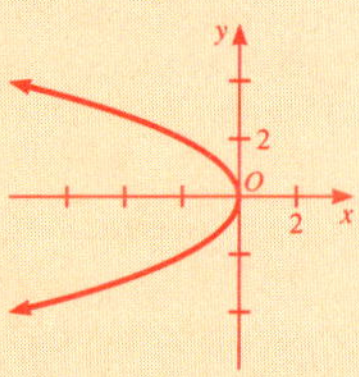

6. $x^2 - 6x - 4y + 5 = 0$ $x = 3$; $(3, -1)$; $(3, 0)$; $y = -2$

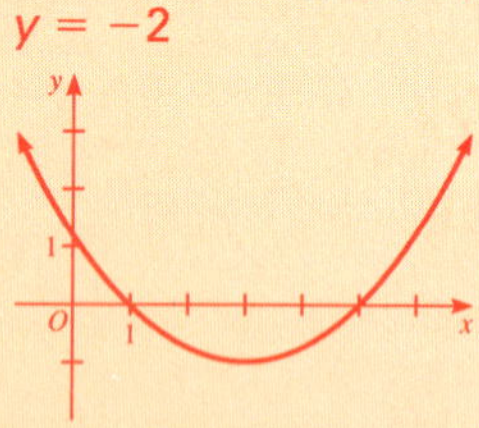

Summarizing the Lesson

In this lesson students learned to write the equation of a parabola and to relate that equation to the focus, directrix, vertex, and axis of the parabola. Ask the students to give the general form of the equation of a parabola (a) with vertical axis and (b) with horizontal axis.

If the distance between the vertex and the focus of a parabola is $|c|$, then it can be shown that $a = \frac{1}{4c}$ in the equation of the parabola (see Exercise 36).

The parabola whose equation is

$$\mathbf{y - k = a(x - h)^2}, \text{ where } a = \frac{1}{4c},$$

opens upward if $a > 0$, downward if $a < 0$;
has vertex $V(h, k)$,
focus $F(h, k + c)$,
directrix $y = k - c$,
and axis of symmetry $x = h$.

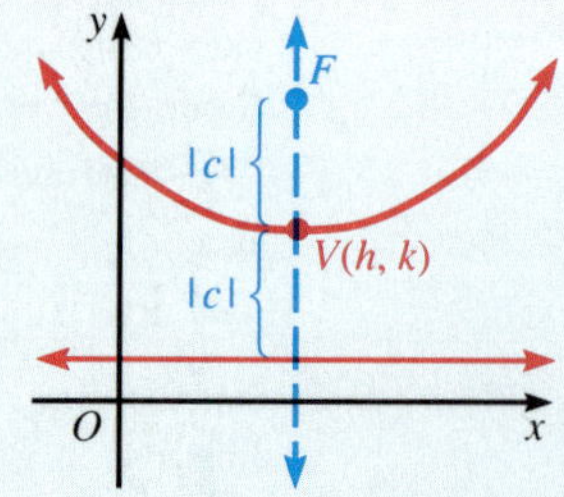

The parabola whose equation is

$$\mathbf{x - h = a(y - k)^2}, \text{ where } a = \frac{1}{4c},$$

opens to the right if $a > 0$, to the left if $a < 0$;
has vertex $V(h, k)$,
focus $F(h + c, k)$,
directrix $x = h - c$,
and axis of symmetry $y = k$.

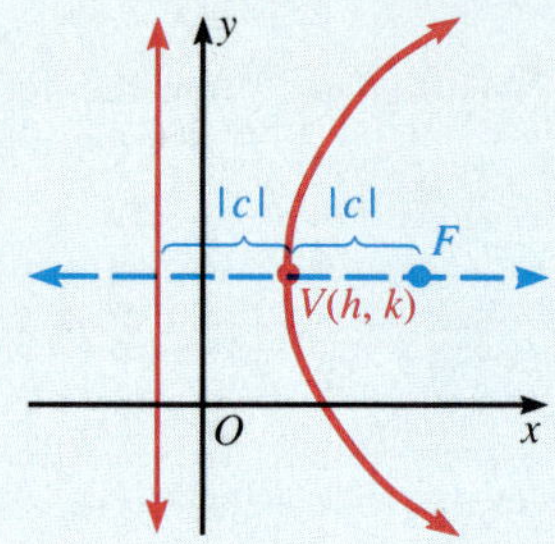

Example 3 Find the vertex, focus, directrix, and axis of symmetry of the parabola $y^2 - 12x - 2y + 25 = 0$. Then graph the parabola.

Solution Complete the square using the terms in y:

$$y^2 - 12x - 2y + 25 = 0$$
$$y^2 - 2y = 12x - 25$$
$$y^2 - 2y + 1 = 12x - 25 + 1$$
$$(y - 1)^2 = 12(x - 2)$$
$$x - 2 = \frac{1}{12}(y - 1)^2$$

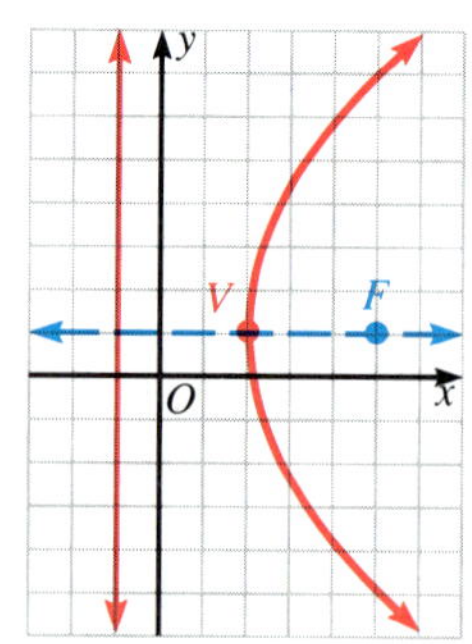

Comparing this equation with $x - h = a(y - k)^2$, you can see that $a = \frac{1}{12}$, $h = 2$, and $k = 1$.

Since $a = \frac{1}{12} = \frac{1}{4c}$, $c = 3$.

Thus, the parabola opens to the right (since $a > 0$).

The vertex is $V(2, 1)$, the focus is $F(5, 1)$, the directrix is $x = -1$, and the axis of symmetry is $y = 1$. The graph is shown above. ***Answer***

Example 4 Find an equation of the parabola that has vertex (4, 2) and directrix $y = 5$.

Solution The distance from the vertex to the directrix is 3, so $|c| = 3$. Since the directrix is above the vertex, the parabola opens downward. Therefore the squared term is the term with x, and c is negative. If $c = -3$, then $a = -\frac{1}{12}$. Thus the equation is

$y - 2 = -\frac{1}{12}(x - 4)^2$. ***Answer***

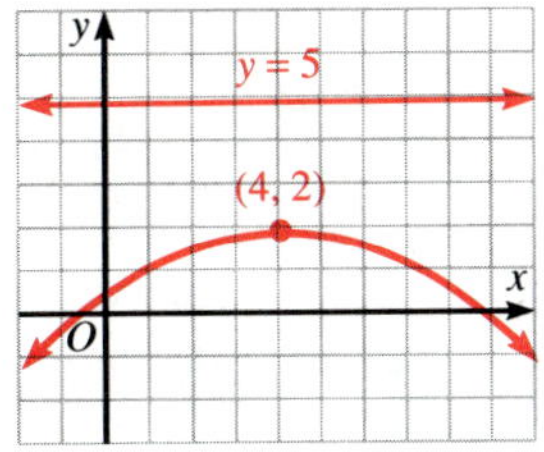

Oral Exercises

In the following exercises, V denotes the vertex of a parabola, F the focus, and D the directrix. Two of these are given. Find the third.

1. $V(0, 0)$, $F(-3, 0)$ D: x = 3
2. $V(0, 0)$, D: $x = -2$ F(2, 0)
3. $F(4, 6)$, D: $x = 0$ V(2, 6)
4. D: $y = -1$, $F(4, -4)$ $V\left(4, -\frac{5}{2}\right)$
5. D: $y = -2$, $V(3, -1)$ F(3, 0)
6. $F(-2, 5)$, $V(-2, 1)$ D: y = −3

For each parabola, find (a) $V(h, k)$, (b) $|c|$, the distance between its vertex and focus, and (c) the direction in which it opens (up, down, left, right).

7. $y = \frac{1}{8}x^2$ (0, 0); 2; up
8. $x = -\frac{1}{12}y^2$ (0, 0); 3; left
9. $x = -(y - 2)^2$ (0, −2); $\frac{1}{4}$; left
10. $y = (x + 5)^2$ (−5, 0); $\frac{1}{4}$; up
11. $x + 3 = \frac{1}{4}(y - 1)^2$ (−3, 1); 1; right
12. $y + 4 = -\frac{1}{36}(x - 4)^2$ (4, −4); 9; down

Written Exercises

In the following exercises, V denotes the vertex of a parabola, F the focus, and D the directrix. Two of these are given. Find the third.

A

1. $V(4, 2)$, D: $y = -3$ F(4, 7)
2. D: $y = 2$, $V(2, 4)$ F(2, 6)
3. $V(0, 2)$, $F(0, 0)$ D: y = 4
4. $F(-3, -1)$, $V(1, -1)$ D: x = 5
5. $F(1, -2)$, $V(1, -5)$ D: y = −8
6. D: $x = -2$, $F(2, 0)$ V(0, 0)

Find an equation of the parabola described. Then graph the parabola.

7. Focus (0, 0); directrix $y = 4$
8. Focus (0, 0); directrix $x = 4$
9. Vertex (0, 0); focus (0, −4)
10. Vertex (0, 0); focus (3, 0)
11. Vertex (0, 0); directrix $x + 1 = 0$
12. Vertex (0, 0); directrix $y = -4$
13. Focus (0, 2); directrix $x = 2$
14. Focus (−2, 0); directrix $y = 3$
15. Focus (3, 4); vertex (3, 2)
16. Focus (−2, 1); vertex (−3, 1)

Suggested Assignments

Average Algebra
Day 1: 415/2–24 even
S 411/43–46
Day 2: 416/26–36 even
S 417/Mixed Review
Assign with Lesson 9-4.

Average Alg. and Trig.
415/2–36 even
S 417/Mixed Review

Extended Algebra
415/1–35 odd
S 417/Mixed Review

Extended Alg. and Trig.
415/1–35 odd
S 417/Mixed Review

Supplementary Materials

Study Guide	pp. 141–142
Practice Master	51
Test Master	35
Overhead Visual	5

Additional Answers Written Exercises

7. $y - 2 = -\frac{1}{8}x^2$
8. $x - 2 = -\frac{1}{8}y^2$
9. $y = -\frac{1}{16}x^2$
10. $x = \frac{1}{12}y^2$
11. $x = \frac{1}{4}y^2$
12. $y = \frac{1}{16}x^2$
13. $x - 1 = -\frac{1}{4}(y - 2)^2$
14. $y - \frac{3}{2} = -\frac{1}{6}(x + 2)^2$
15. $y - 2 = \frac{1}{8}(x - 3)^2$
16. $x + 3 = \frac{1}{4}(y - 1)^2$

Using a Computer or a Graphing Calculator

Once students have drawn their own graphs for Exercises 17–26, you may wish to have them check their work using a computer or a graphing calculator. The same can be done for Exercises 31–34.

Support Materials
Disk for Algebra
Menu Items: Conics Plotter
Function Plotter

Additional Answers Written Exercises

21. $(-2, 1)$; $(-2, -1)$; $y = 3$; $x = -2$

22. $(2, -3)$; $(0, -3)$; $x = 4$; $y = -3$

23. $\left(-\frac{3}{2}, 3\right)$; $\left(\frac{1}{2}, 3\right)$; $x = -\frac{7}{2}$; $y = 3$

24. $(3, 1)$; $\left(3, -\frac{3}{2}\right)$; $y = \frac{7}{2}$; $x = 3$

25. $(-5, -2)$; $\left(-5, -\frac{3}{2}\right)$; $y = -\frac{5}{2}$; $x = -5$

26. $(4, 1)$; $\left(\frac{13}{4}, 1\right)$; $x = \frac{19}{4}$; $y = 1$

28.

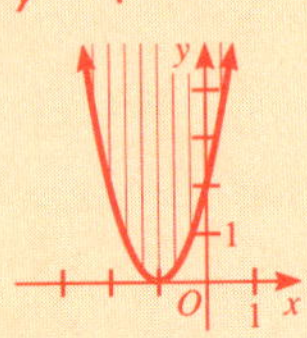

30.

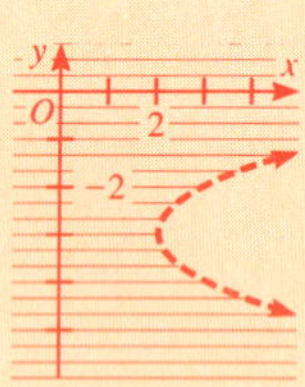

(continued on p. 425)

Find the vertex, focus, directrix, and axis of symmetry of each parabola. Then graph the parabola. You may wish to check your graphs on a computer or a graphing calculator.

17. $y = \frac{1}{6}x^2$ $(0, 0)$; $\left(0, \frac{3}{2}\right)$; $y = -\frac{3}{2}$; $x = 0$

18. $6x + y^2 = 0$ $(0, 0)$; $\left(-\frac{3}{2}, 0\right)$; $x = \frac{3}{2}$; $y = 0$

19. $4x = y^2 - 4y$ $(-1, 2)$; $(0, 2)$; $x = -2$; $y = 2$

20. $x^2 = y + 2x$ $(1, -1)$; $\left(1, -\frac{3}{4}\right)$; $y = -\frac{5}{4}$; $x = 1$

B **21.** $x^2 + 8y + 4x - 4 = 0$

22. $y^2 + 6y + 8x - 7 = 0$

23. $y^2 - 8x - 6y - 3 = 0$

24. $x^2 - 6x + 10y - 1 = 0$

25. $x^2 + 10x - 2y + 21 = 0$

26. $y^2 + 3x - 2y - 11 = 0$

Graph each inequality.

27. $y < (x - 3)^2$

28. $y \geq 2(x + 1)^2$

29. $y + 4 \geq (x + 2)^2$

30. $x - 11 < y^2 + 6y$

Graph each equation. Each graph is half of a parabola, since $\sqrt{a} \geq 0$.

31. $y = \sqrt{x}$

32. $y = \sqrt{x - 2}$

33. $y = -\sqrt{1 - x}$

34. $y = -\sqrt{3 - x}$

C **35.** The line through the focus of a parabola perpendicular to its axis intersects the parabola in two points, P and Q. ($\overline{PQ}$ is called the *latus rectum* of the parabola.) Explain why the length of $\overline{PQ}$ is twice the distance from the focus to the directrix.

36. A parabola having vertex (h, k) and a vertical axis has as focus the point $(h, k + c)$ and as directrix the line $y = k - c$. Show that an equation of the parabola is $y - k = \frac{1}{4c}(x - h)^2$.

37. F is the focus of the parabolic arc shown below. Line L is parallel to the directrix, D. Explain why the sum $FP + PQ$ is the same for all positions of P. (*Hint:* Extend $\overline{PQ}$ so that it intersects the directrix.)

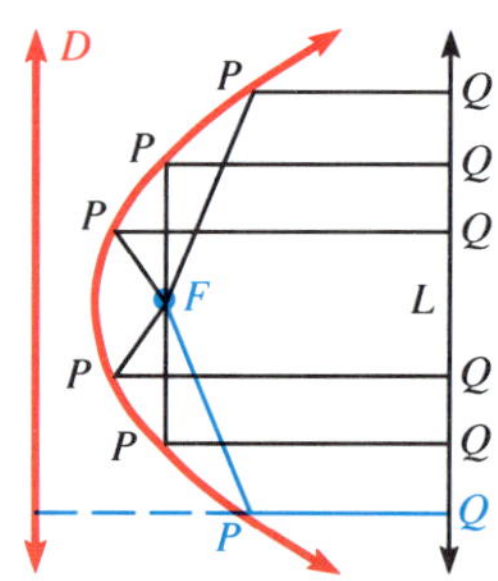

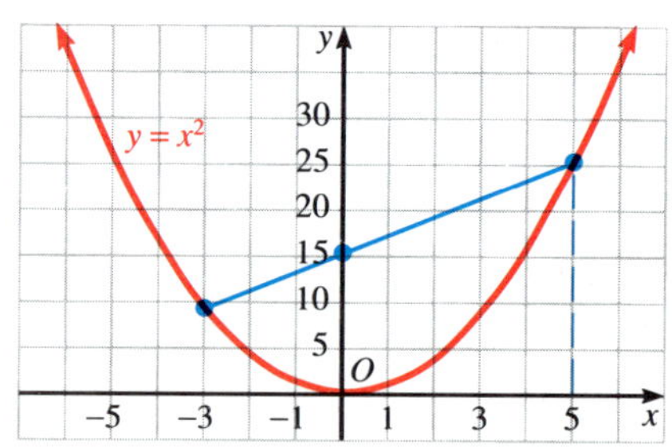

38. A *multiplication machine:* Draw the parabola $y = x^2$. Given any two positive numbers a and b, locate the points of the parabola having the x-coordinates $-a$ and b. Explain why the line joining these points has y-intercept ab. (*Hint:* Find an equation of the line through $(-a, a^2)$ and (b, b^2). Use the diagram above as a model.)

Mixed Review Exercises

For the given points P and Q, find (a) the slope of $\overline{PQ}$, (b) the distance PQ, and (c) the midpoint of $\overline{PQ}$. Express radicals in simplest form.

1. $P(-3, 2)$, $Q(5, 2)$ 0; 8; (1, 2)

2. $P(1, -6)$, $Q(3, 2)$ 4; $2\sqrt{17}$; (2, −2)

3. $P(4, 3)$, $Q(-3, 4)$ $-\frac{1}{7}$; $5\sqrt{2}$; $\left(\frac{1}{2}, \frac{7}{2}\right)$

Find an equation for each figure described.

4. The line having slope $\frac{2}{3}$ and passing through $(-2, -3)$. $2x - 3y = 5$

5. The perpendicular bisector of the segment with endpoints $(7, -2)$ and $(1, -8)$. $x + y = -1$

6. The circle with center $(-1, 3)$ and radius 4. $(x + 1)^2 + (y - 3)^2 = 16$

Self-Test 1

Vocabulary

Pythagorean theorem (p. 401)
distance formula (p. 402)
midpoint formula (p. 402)
conic section (p. 407)
conic (p. 407)
circle (p. 407)
radius (p. 407)
center (p. 407)
translation (p. 408)
parabola (p. 412)
directrix (p. 412)
focus (of a parabola) (p. 412)
vertex (p. 412)
axis of symmetry (p. 412)

Obj. 9-1, p. 401

1. Find the distance between $(3, -6)$ and $(4, 2)$. $\sqrt{65}$

2. Find the midpoint of the line segment having endpoints $(-3, 5)$ and $(8, 11)$. $\left(\frac{5}{2}, 8\right)$

3. Determine whether the three points $A(2, 2)$, $B(8, 6)$, and $C(11, 10)$ are collinear by comparing the lengths of $\overline{AB}$, $\overline{BC}$, and $\overline{AC}$. not collinear

Obj. 9-2, p. 407

4. Find an equation of the circle with center $(5, -4)$ and radius 7. $(x - 5)^2 + (y + 4)^2 = 49$

5. Find the center and radius of the circle whose equation is $x^2 + y^2 + 12x - 4y + 32 = 0$. $(-6, 2)$; $2\sqrt{2}$

Obj. 9-3, p. 412

6. Find an equation of the parabola having focus $(0, 3)$ and vertex $(-4, 3)$. $x + 4 = \frac{1}{16}(y - 3)^2$

7. Find the vertex, focus, directrix, and axis of the parabola $x^2 + 10x + 16y - 7 = 0$. $(-5, 2)$; $(-5, -2)$; $y = 6$; $x = -5$

8. Graph $(x + 1) = \frac{1}{8}(y + 3)^2$. Label the vertex, focus, directrix, and axis.

Check your answers with those at the back of the book.

Quick Quiz

1. Find (a) the distance between P and Q and (b) the midpoint of $\overline{PQ}$ given $P(-3, 7)$ and $Q(1, -1)$.
a. $4\sqrt{5}$ **b.** $(-1, 3)$

2. Given $A(-4, -5)$, $B(3, -4)$, and $C(2, 3)$, show that $\triangle ABC$ is an isosceles right triangle.
$AB = BC = 5\sqrt{2}$;
$(AB)^2 + (BC)^2 = 50 + 50 = 100 = (AC)^2$

3. Find the center and the radius of the circle with equation $x^2 + y^2 + 6x - 8y - 119 = 0$.
$(-3, 4)$; 12

4. Find an equation of a circle whose diameter has endpoints $(0, 3)$ and $(4, -3)$.
$(x - 2)^2 + y^2 = 13$

5. Find an equation of the parabola with directrix $y = 6$ and focus $(0, 2)$.
$x^2 = -8(y - 4)$

6. Find the vertex, focus, directrix, and axis of the parabola $y^2 + 8y - 12x + 4 = 0$.
$(-1, -4)$; $(2, -4)$; $x = -4$; $y = -4$

7. Graph $(x - 1)^2 = -8(y + 1)$. Label the vertex, focus, directrix, and axis.

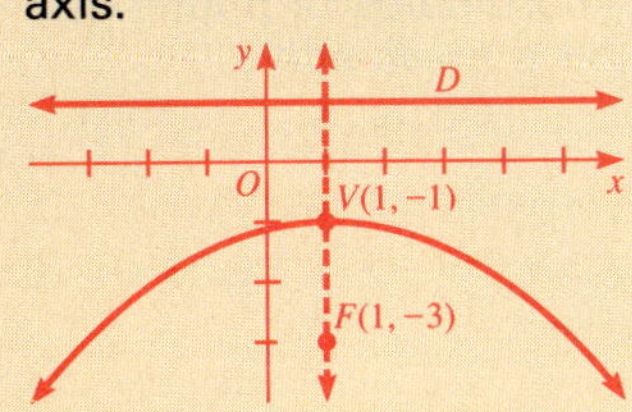

Teaching Suggestions p. T114

Suggested Extensions p. T114

Warm-Up Exercises

State the following in symbols.

1. Pythagorean theorem $a^2 + b^2 = c^2$
2. Distance formula $P_1P_2 = \sqrt{(x_2 - x_1)^2 + (y_2 - y_1)^2}$
3. Midpoint formula $M\left(\frac{x_1 + x_2}{2}, \frac{y_1 + y_2}{2}\right)$
4. Equation of a circle $(x - h)^2 + (y - k)^2 = r^2$
5. Equation of a parabola (two forms) $y - k = a(x - h)^2$ $x - h = a(y - k)^2$

Motivating the Lesson

Tell students that although early astronomers thought the planets moved in circular orbits, modern astronomers know that the orbits are in fact elliptical. The relationship between an ellipse and its equation is the topic of today's lesson.

Using a Calculator

Encourage students to use a calculator when finding the coordinates of points that lie on a given conic.

Conic Sections: Ellipses and Hyperbolas

9-4 Ellipses

Objective To learn the relationship between the center, foci, and intercepts of an ellipse and the equation of the ellipse.

The diagram at the right shows a piece of string whose ends are fastened at the points F_1 and F_2. If you use a pencil to keep the string tight, the point P of the pencil will move along a path that is an oval curve called an *ellipse*. Because the sum of the distances from the pencil to the points F_1 and F_2 is the length of the string, this sum, $PF_1 + PF_2$, is the same for all positions of P.

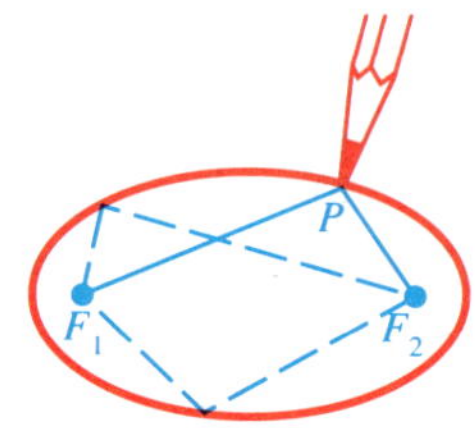

An **ellipse** is the set of all points P in the plane such that the sum of the distances from P to two fixed points is a given constant.

Each of the fixed points is called a **focus** (plural: *foci*) of the ellipse. If the foci are F_1 and F_2, then for each point P of the ellipse, PF_1 and PF_2 are the **focal radii** of P.

To find an equation of the ellipse having foci $F_1(-4, 0)$ and $F_2(4, 0)$ and sum of focal radii 10, use the distance formula. A point $P(x, y)$ is on the ellipse if and only if the following is true:

$$\begin{aligned} PF_1 \quad + \quad PF_2 &= 10 \\ \sqrt{(x+4)^2 + y^2} + \sqrt{(x-4)^2 + y^2} &= 10 \\ \sqrt{(x+4)^2 + y^2} &= 10 - \sqrt{(x-4)^2 + y^2} \end{aligned}$$

Square both sides and then simplify:

$$\begin{aligned} (x+4)^2 + y^2 &= 100 - 20\sqrt{(x-4)^2 + y^2} + (x-4)^2 + y^2 \\ 20\sqrt{(x-4)^2 + y^2} &= 100 - 16x \\ 5\sqrt{(x-4)^2 + y^2} &= 25 - 4x \end{aligned}$$

Square both sides again:

$$\begin{aligned} 25(x^2 - 8x + 16 + y^2) &= 625 - 200x + 16x^2 \\ 9x^2 + 25y^2 &= 225 \end{aligned}$$

Divide both sides by 225:

$$\frac{x^2}{25} + \frac{y^2}{9} = 1$$

The following analysis will help you graph the equation $\frac{x^2}{25} + \frac{y^2}{9} = 1$.

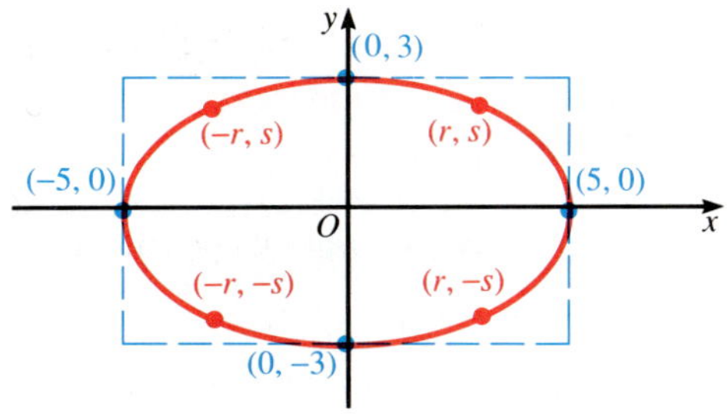

1. The graph has four *intercepts*.
   If $y = 0$, then
   $$\frac{x^2}{25} + \frac{0^2}{9} = 1.$$
   $\therefore x^2 = 25$, or $x = \pm 5$.
   If $x = 0$, then
   $$\frac{0^2}{25} + \frac{y^2}{9} = 1.$$
   $\therefore y^2 = 9$, or $y = \pm 3$.
2. The *extent* of the graph is limited.
   Solving the equation for each variable in terms of the other gives
   $$y = \pm\frac{3}{5}\sqrt{25 - x^2} \quad \text{and} \quad x = \pm\frac{5}{3}\sqrt{9 - y^2}.$$
   In the first equation, y is a real number if and only if $|x| \leq 5$. In the second equation, x is a real number if and only if $|y| \leq 3$. Thus the extent of the graph is limited to the dashed rectangle shown above.
3. The graph is *symmetric about the x-axis*: If (r, s) is on the graph, so is $(r, -s)$. The graph is *symmetric about the y-axis*: If (r, s) is on the graph, so is $(-r, s)$. Thus the graph is also *symmetric about the origin:* If (r, s) is on the graph, then so is $(-r, -s)$. See the diagram above.

Because of the symmetry described, you need to choose only values of x in the first quadrant from $x = 0$ to $x = 5$. A calculator may be helpful to set up a table of values. You can then draw the complete graph.

x	$y = 0.6\sqrt{25 - x^2}$
0	3
1	2.9
2	2.7
3	2.4
4	1.8
5	0

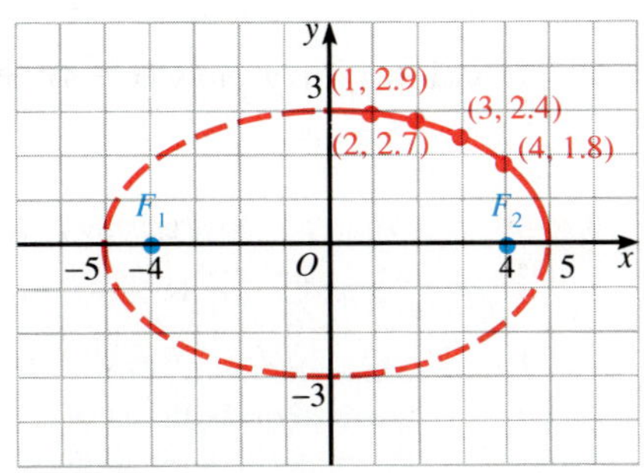

The **center** of an ellipse is the midpoint of the line segment joining its foci. The **major axis** of an ellipse is the chord passing through its foci. The **minor axis** is the chord containing the center and perpendicular to the major axis.

Chalkboard Examples

1. Give the x- and y-intercepts and find the foci of the ellipse $9x^2 + 4y^2 = 144$.
   $\frac{9x^2}{144} + \frac{4y^2}{144} = 1$
   $\frac{x^2}{16} + \frac{y^2}{36} = 1$
   x-intercepts are 4 and -4.
   y-intercepts are 6 and -6.
   $a^2 = 36$, $b^2 = 16$,
   $c^2 = a^2 - b^2 = 20$, and
   $c = 2\sqrt{5}$.
   Foci are $(0, 2\sqrt{5})$ and $(0, -2\sqrt{5})$.
2. Graph the ellipse $x^2 + 4y^2 - 16 = 0$ and find the foci.
   $\frac{x^2}{16} + \frac{y^2}{4} = 1$
   The major axis is horizontal; the x-intercepts are 4 and -4; and the y-intercepts are 2 and -2.

   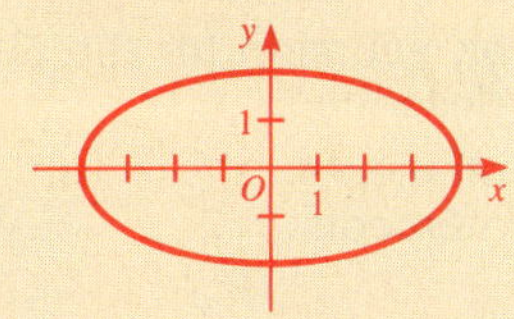

   $a^2 = 16$, $b^2 = 4$,
   $c^2 = a^2 - b^2 = 12$, and
   $c = 2\sqrt{3}$.
   Foci are $(-2\sqrt{3}, 0)$ and $(2\sqrt{3}, 0)$.
3. Find an equation for the ellipse having foci at $(0, -3)$ and $(0, 3)$ and sum of focal radii equal to 10.
   The distance from each focus to the center is 3. Thus $c = 3$, $2a = 10$, $a = 5$.
   $b^2 = a^2 - c^2$
   $= 25 - 9 = 16$.
   Foci are on y-axis.
   $\therefore$ the equation is
   $\frac{x^2}{16} + \frac{y^2}{25} = 1$.

Check for Understanding

Oral Exs. 1–6: use before Example 1.

Guided Practice

Find **(a)** the x-intercepts, **(b)** the y-intercepts, and **(c)** the foci of each ellipse.

1. $\frac{x^2}{25} + y^2 = 1$ **a.** ± 5 **b.** ± 1 **c.** $(\pm 2\sqrt{6}, 0)$

2. $9x^2 + y^2 = 36$ **a.** ± 2 **b.** ± 6 **c.** $(0, \pm 4\sqrt{2})$

3. Graph $16x^2 + 25y^2 = 400$ and show the foci.

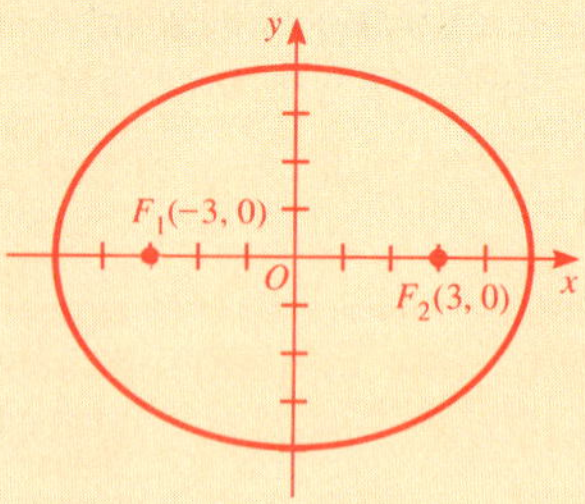

Give an equation for the ellipse described.

4. x-intercepts: ± 1; y-intercepts: ± 2 $\frac{x^2}{1} + \frac{y^2}{4} = 1$

5. foci: $(0, \pm 5)$; y-intercepts: ± 13 $\frac{x^2}{144} + \frac{y^2}{169} = 1$

Summarizing the Lesson

In this lesson students learned to write the equation of an ellipse and relate that equation to the center, foci, and intercepts. Ask students to write the general form of the equation of an ellipse centered at the origin and having (a) a horizontal major axis and (b) a vertical major axis.

An ellipse has a *quadratic equation in two variables*. A general form for the equation of an ellipse with foci on the x-axis and equidistant from the origin is shown at the left in the chart. An ellipse may of course have its foci on the y-axis. The corresponding equation is shown at the right.

The ellipse having center (0, 0), foci $(-c, 0)$ and $(c, 0)$, and sum of focal radii $2a$ has the equation

$$\frac{x^2}{a^2} + \frac{y^2}{b^2} = 1$$

where $b^2 = a^2 - c^2$.

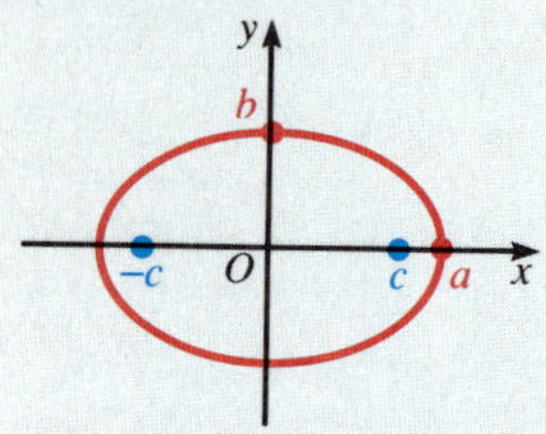

The ellipse having center (0, 0), foci $(0, -c)$ and $(0, c)$, and sum of focal radii $2a$ has the equation

$$\frac{x^2}{b^2} + \frac{y^2}{a^2} = 1$$

where $b^2 = a^2 - c^2$.

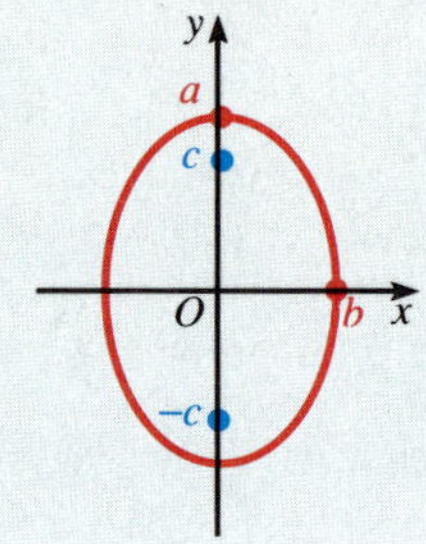

For an ellipse having an equation with one of the forms above, the major axis is a segment of the x-axis or the y-axis. Note that the major axis is horizontal if the x^2-term has the larger denominator and vertical if the y^2-term has the larger denominator. Since the larger of the two denominators is a^2, the length of the major axis is always $2a$. The length of the minor axis is $2b$.

The distance from the center to either focus is $|c|$. If you know the values of a and b and the coordinates of the center, you can find the coordinates of each focus by solving the equation $b^2 = a^2 - c^2$ for c (see Exercise 34).

Example 1 Graph the ellipse $4x^2 + y^2 = 64$ and find its foci.

Solution $4x^2 + y^2 = 64$ ⟵ Divide both sides by 64.

$$\frac{x^2}{16} + \frac{y^2}{64} = 1$$

Since the denominator of the y^2-term is the larger, the major axis is vertical. The y-intercepts are 8 and -8. The x-intercepts are 4 and -4. Solve the equation for y and make a short table of first-quadrant points.

$$y = \pm 2\sqrt{16 - x^2}$$

x	y
0	8
1	7.7
2	6.9
3	5.3
4	0

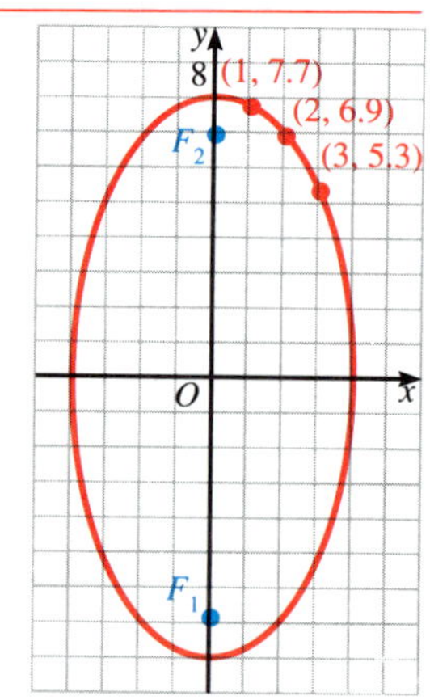

Use the relationship $c^2 = a^2 - b^2$ to find the foci. Since $a^2 = 64$ and $b^2 = 16$, $c^2 = 48$ and $c = \sqrt{48} = 4\sqrt{3} \approx 6.9$. Thus the foci of the ellipse are $F_1(0, -4\sqrt{3})$ and $F_2(0, 4\sqrt{3})$. ***Answer***

Example 2 Find an equation of an ellipse having x-intercepts $\sqrt{2}$ and $-\sqrt{2}$ and y-intercepts 3 and -3.

Solution Since $3^2 > (\sqrt{2})^2$, the y^2-term has the larger denominator, and so the major axis is vertical. The center is (0, 0), $a^2 = 9$, and $b^2 = 2$.

$\therefore$ an equation is $\frac{x^2}{2} + \frac{y^2}{9} = 1$. ***Answer***

Example 3 Find an equation of an ellipse having foci $(-3, 0)$ and $(3, 0)$ and sum of focal radii equal to 12.

Solution The center of the ellipse is (0, 0). The distance from each focus to the center is 3, so $c = 3$. The sum of the focal radii is $2a$. Since $2a = 12$, $a = 6$ and $a^2 = 36$. So $b^2 = a^2 - c^2 = 36 - 9 = 27$. Since the foci are on the x-axis, the major axis is horizontal.

$\therefore$ an equation is $\frac{x^2}{36} + \frac{y^2}{27} = 1$. ***Answer***

Oral Exercises

Give the x- and y-intercepts of each ellipse and tell on which of the coordinate axes its foci lie.

1. $\frac{x^2}{9} + \frac{y^2}{4} = 1$ ±3; ±2; *x*-axis
2. $\frac{x^2}{16} + \frac{y^2}{25} = 1$ ±4; ±5; *y*-axis
3. $x^2 + 9y^2 = 36$ ±6; ±2; *x*-axis
4. $x^2 + 4y^2 = 16$ ±4; ±2; *x*-axis
5. $3x^2 + y^2 = 9$ $\pm\sqrt{3}$; ±3; *y*-axis
6. $2x^2 + 3y^2 = 6$ $\pm\sqrt{3}$; $\pm\sqrt{2}$; *x*-axis

Written Exercises

1. $(-\sqrt{5}, 0), (\sqrt{5}, 0)$ 2. $(0, -3), (0, 3)$ 3. $(-4\sqrt{2}, 0), (4\sqrt{2}, 0)$ 4. $(-2\sqrt{3}, 0), (2\sqrt{3}, 0)$ 5. $(0, -\sqrt{6}), (0, \sqrt{6})$ 6. $(-1, 0), (1, 0)$

Graph each ellipse and find its foci. You may wish to check your graphs on a computer or a graphing calculator.

A

1–6. Use the equations in Oral Exercises 1–6.

7. $5x^2 + y^2 = 25$ $(0, -2\sqrt{5}), (0, 2\sqrt{5})$
8. $x^2 + 25y^2 = 100$ $(-4\sqrt{6}, 0), (4\sqrt{6}, 0)$
9. $3x^2 + y^2 = 12$ $(0, -2\sqrt{2}), (0, 2\sqrt{2})$
10. $2x^2 + y^2 = 8$ $(0, -2), (0, 2)$
11. $4x^2 + 3y^2 = 48$ $(0, -2), (0, 2)$
12. $5x^2 + 9y^2 = 45$ $(-2, 0), (2, 0)$
13. $x^2 + 9y^2 = 1$ $\left(-\frac{2\sqrt{2}}{3}, 0\right), \left(\frac{2\sqrt{2}}{3}, 0\right)$
14. $9x^2 + 4y^2 = 9$ $\left(0, -\frac{\sqrt{5}}{2}\right), \left(0, \frac{\sqrt{5}}{2}\right)$

Suggested Assignments

Average Algebra
Day 1: 421/1–6
Assign with Lesson 9-3.
Day 2: 421/7–21 odd, 24–31
R 417/Self-Test 1

Average Alg. and Trig.
Day 1: 421/1–21 odd
R 417/Self-Test 1
Day 2: 422/16–22 even, 24–33
Assign with Lesson 9-5.

Extended Algebra
Day 1: 421/1–21 odd
R 417/Self-Test 1
Day 2: 422/22–34 even
Assign with Lesson 9-5.

Extended Alg. and Trig.
Day 1: 421/1–21 odd
R 417/Self-Test 1
Day 2: 422/22–34 even
Assign with Lesson 9-5.

Supplementary Materials

Study Guide pp. 143–144
Overhead Visuals 5, 6

Using a Computer or a Graphing Calculator

Once students have drawn their own graphs for Written Exercises 1–14, you may wish to have them check their work using a computer or a graphing calculator. The same can be done for Written Exercises 32–33.

Support Materials
Disk for Algebra
Menu Items: Conics Plotter
Function Plotter

Additional Answers Written Exercises

2.

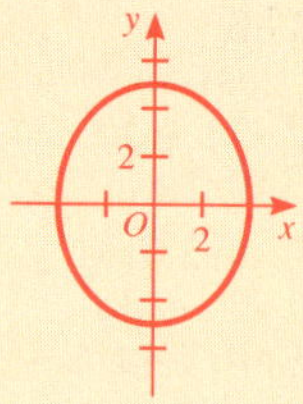

4.

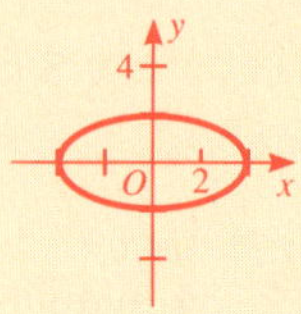

6.

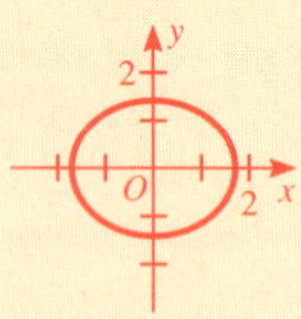

8.

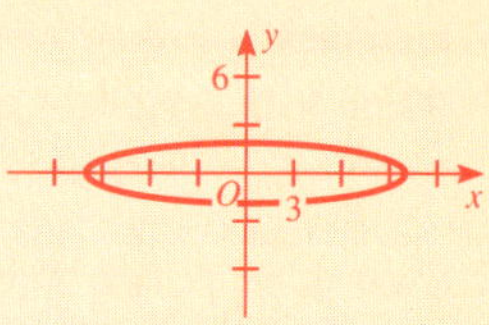

10.

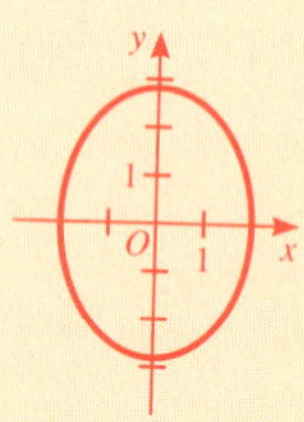

12.

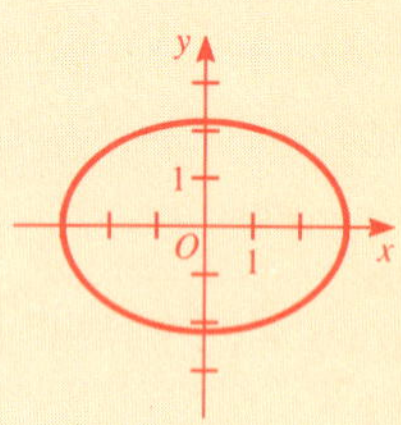

Find an equation of an ellipse having the given intercepts.

15. x-intercepts: ± 5 $\frac{x^2}{25} + \frac{y^2}{4} = 1$
y-intercepts: ± 2

16. x-intercepts: ± 3 $\frac{x^2}{9} + \frac{y^2}{16} = 1$
y-intercepts: ± 4

17. x-intercepts: ± 2 $\frac{x^2}{4} + \frac{y^2}{2} = 1$
y-intercepts: $\pm\sqrt{2}$

18. x-intercepts: $\pm\sqrt{6}$ $\frac{x^2}{6} + \frac{y^2}{12} = 1$
y-intercepts: $\pm 2\sqrt{3}$

Find an equation of the ellipse having the given points as foci and the given number as sum of focal radii.

B **19.** $(-6, 0), (6, 0)$; 18 $\frac{x^2}{81} + \frac{y^2}{45} = 1$

20. $(0, -5), (0, 5)$; 20 $\frac{x^2}{75} + \frac{y^2}{100} = 1$

21. $(0, -4), (0, 4)$; 24 $\frac{x^2}{128} + \frac{y^2}{144} = 1$

22. $(-9, 0), (9, 0)$; 30 $\frac{x^2}{225} + \frac{y^2}{144} = 1$

23. Describe a way of constructing the foci of a given ellipse using a compass. Give a convincing argument to justify the method. (*Hint:* See the figure at the left below.) Let $B = (0, b)$; draw the circle with center B and radius a, intersecting the major axis at F_1 and F_2; $F_1B + F_2B = a + a = 2a$; F_1 and F_2 are the foci of the ellipse.

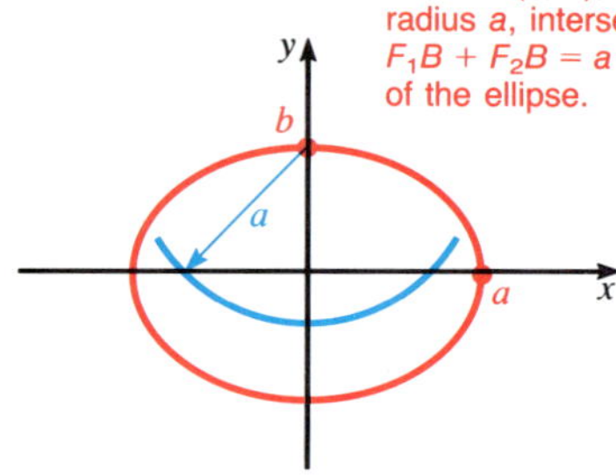

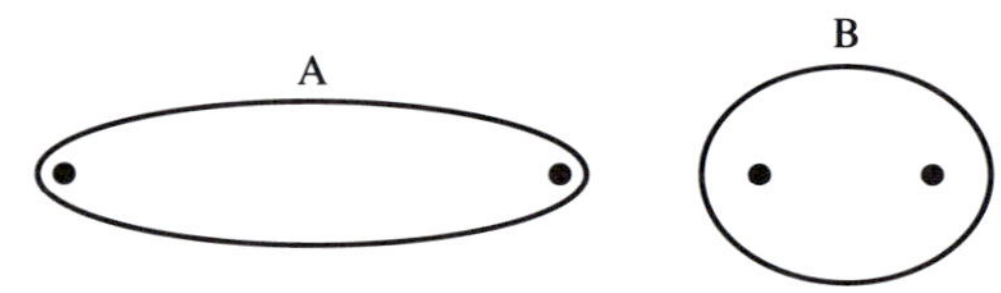

24. An indicator of the shape of the ellipse $\frac{x^2}{a^2} + \frac{y^2}{b^2} = 1$ is its *eccentricity* e, defined by $e = \frac{c}{a}$, where $c = \sqrt{a^2 - b^2}$.

a. Explain why is e a number between 0 and 1. Since $0 < \sqrt{a^2 - b^2} < a$, $0 < c < a$; $0 < \frac{c}{a} < 1$; $0 < e < 1$

b. Which of the ellipses shown above has the greater eccentricity? A

25. Suppose that $a = b$ in the equation $\frac{x^2}{a^2} + \frac{y^2}{b^2} = 1$.

a. What is the value of c? 0

b. What is the value of e? 0

c. Where are the two foci? At the center

d. What special ellipse will the graph be? A circle

26. Describe the graph of $\frac{x^2}{a^2} + \frac{y^2}{b^2} = 0$. The origin, (0, 0)

27. The arch of a bridge is in the form of half an ellipse, with the major axis horizontal. The span of the bridge (the length of the major axis) is 12 m and the height of the arch above the water at the center is 4 m. How high above the water is the arch at a point on the water 2 m from one of the ends of the arch? $\frac{4\sqrt{5}}{3}$ m ($\approx$3 m)

Graph each inequality.

28. $4x^2 + 9y^2 < 36$

29. $4x^2 + y^2 \geq 16$

30. $5x^2 + 4y^2 \geq 20$

31. $3x^2 + 16y^2 \leq 48$

Graph each equation. Each graph is a semi-ellipse, since $\sqrt{a}$ is always nonnegative.

32. $y = 3\sqrt{1 - x^2}$

33. $y = \frac{1}{2}\sqrt{4 - x^2}$

C **34.** For the ellipse $\frac{x^2}{a^2} + \frac{y^2}{b^2} = 1$ with foci at $(-c, 0)$ and $(c, 0)$, show that $b^2 = a^2 - c^2$.

35. Use the definition to find an equation of the ellipse with foci at $(-c, 0)$ and $(c, 0)$ and sum of focal radii equal to $2a$. At the appropriate point let $b^2 = a^2 - c^2$ to simplify the equation.

Career Note / *Computer Graphics Artist*

In movies and on television you have probably seen believable pictures of things that do not exist and cannot be photographed. These pictures are the product of computer graphics. The computer graphics artist gives the computer a complete description of each object that is to be in a picture. The object's shape may be built up from flat polygons or from geometric solids such as spheres. The location of each object in the scene is given in three-dimensional coordinates so that objects farther from the viewer will appear smaller.

A computer program compares the boundaries of objects and decides which surfaces are visible to the viewer. In order to give realistic texture to images, the graphics artist uses formulas describing how different materials reflect light. Even using powerful computers, it may take the computer graphics artist hours to make one second of movie film.

The computer has become a valuable tool for graphic artists in many other fields. For example, computer-generated shapes and color patterns are now used in the design of magazine illustrations, posters, packaging, and textiles.

14.

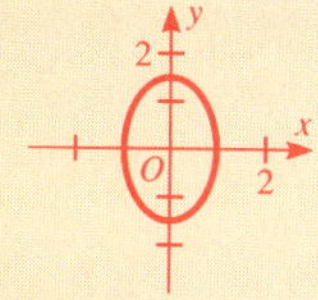

28.

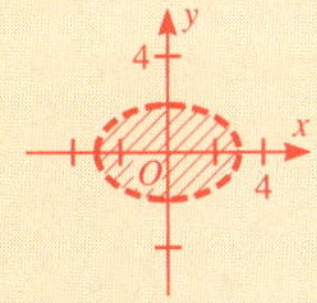

30.

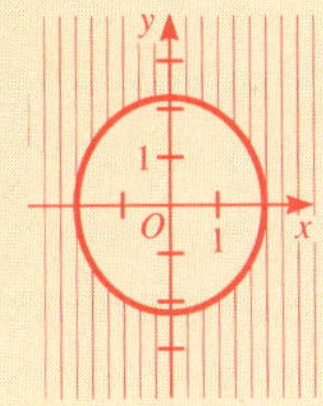

32.

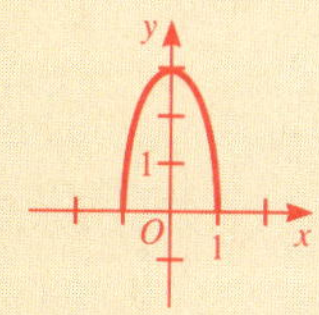

34. Let $F_1 = (-c, 0)$, $F_2 = (c, 0)$, and $P = (0, b)$; $F_1P + F_2P = 2a$; $\sqrt{(-c - 0)^2 + (0 - b)^2} + \sqrt{(c - 0)^2 + (0 - b)^2} = 2a$; $2\sqrt{c^2 + b^2} = 2a$; $c^2 + b^2 = a^2$; $b^2 = a^2 - c^2$

(continued)

Additional Answers
Written Exercises
(continued)

35. Let $P(x, y)$ be any point on the ellipse. Then $\sqrt{(x + c)^2 + (y - 0)^2} + \sqrt{(x - c)^2 + (y - 0)^2} = 2a$; $\sqrt{(x + c)^2 + y^2} = 2a - \sqrt{(x - c)^2 + y^2}$; $(x + c)^2 + y^2 = 4a^2 - 4a\sqrt{(x - c)^2 + y^2} + (x - c)^2 + y^2$; $(x + c)^2 - (x - c)^2 = 4a^2 - 4a\sqrt{(x - c)^2 + y^2}$; $4cx = 4a^2 - 4a\sqrt{(x - c)^2 + y^2}$; $cx - a^2 = -a\sqrt{(x - c)^2 + y^2}$; $c^2x^2 - 2a^2cx + a^4 = a^2(x - c)^2 + a^2y^2$; $c^2x^2 - 2a^2cx + a^4 = a^2x^2 - 2a^2cx + a^2c^2 + a^2y^2$; $a^2(a^2 - c^2) = (a^2 - c^2)x^2 + a^2y^2$; let $b^2 = a^2 - c^2$; $a^2b^2 = b^2x^2 + a^2y^2$; $1 = \frac{x^2}{a^2} + \frac{y^2}{b^2}$.

Additional Answers
Application

2.

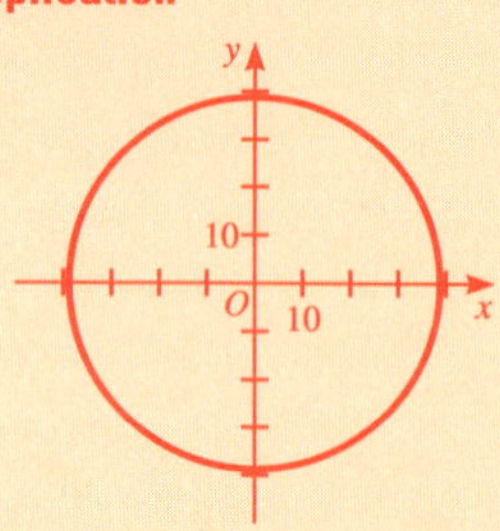

Application / *Planetary Orbits*

In 1609 the German astronomer Johann Kepler proposed two of his three famous laws of planetary motion (all of which were later confirmed by Newton's laws). In one of these, Kepler stated that the orbits of the planets in our solar system are ellipses with the sun at one focus. This discovery was all the more remarkable because it represented the first practical application of the geometry of the ellipse, which had been extensively studied by Greek mathematicians about 2000 years earlier.

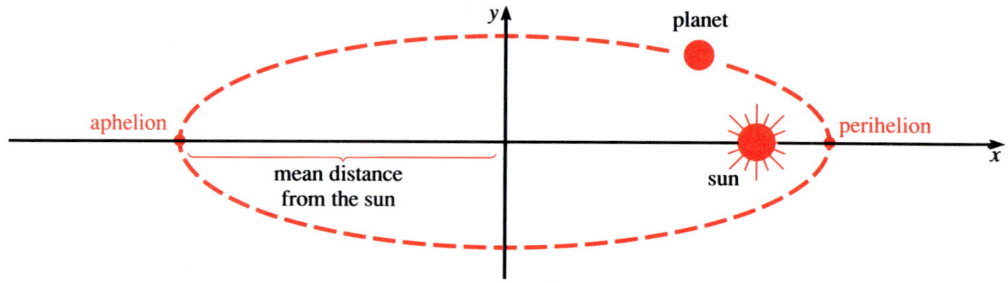

The greater the eccentricity of an ellipse, the more elongated it is (see Exercise 24 on page 422). The eccentricity of the orbit shown above has been exaggerated for clarity. The planetary orbits are all more nearly circular.

As the diagram above indicates, a planet's *mean distance from the sun,* a figure given in many astronomical tables, is defined to be half the length of the major axis of the orbit. This is the mean, or average, of the planet's maximum and minimum distances from the sun. The positions at which a planet achieves these distances are called *aphelion* and *perihelion,* respectively.

Much of the computation involving distances within the solar system is carried out using the *astronomical unit* (AU), defined to be Earth's mean distance from the sun, or about 150 million km, as a unit of measure. The table below gives the eccentricities of the orbits of the planets and their mean distances from the sun.

Planet	Mean Distance (in AU)	Eccentricity
Mercury	0.387	0.206
Venus	0.723	0.007
Earth	1.000	0.017
Mars	1.524	0.093
Jupiter	5.203	0.049
Saturn	9.539	0.056
Uranus	19.182	0.047
Neptune	30.058	0.009
Pluto	39.44	0.250

Suppose that you choose a coordinate system so that the major axis of the orbit lies on the x-axis and the center of the ellipse is at the origin. Then using the information from the table, you can find the Cartesian equation of the orbit of a planet, such as Mercury, in the standard form

$$\frac{x^2}{a^2} + \frac{y^2}{b^2} = 1.$$

First note that for Mercury a = mean distance from the sun = 0.387. Then use the fact that eccentricity $= \frac{c}{a} = 0.206$. To find the value of c:

$$\frac{c}{0.387} = 0.206, \text{ so } c = 0.080.$$

Now use the relationship $b^2 = a^2 - c^2$, which holds for any ellipse, to find the value of b:

$$\begin{aligned} b &= \text{half the length of the minor axis} \\ &= 0.379. \end{aligned}$$

Thus the equation for the orbit of Mercury is

$$\frac{x^2}{(0.387)^2} + \frac{y^2}{(0.379)^2} = 1.$$

Exercises

5. ≈1.78 : 1; since the ratio is much larger, the orbit is much more elongated than that of Earth

1. Find the maximum and minimum distances from Earth to the sun in AU and in km. max: 1.017 AU ≈ 153 million km; min: 0.983 AU ≈ 147 million km
2. What is the planet with the least circular orbit? Find the lengths of the major and minor axes for this orbit. What is the ratio of these lengths? Sketch the graph. Pluto; major axis: 78.88 AU; minor axis: 76.38 AU; ≈1.03 : 1
3. Find the ratio of the length of the major axis of Earth's orbit to the length of the minor axis. ≈1.0001 : 1
4. Using the equation for the orbit of Mercury, find how far the planet is from the sun in AU when Mercury is on the y-axis of the coordinate system. ≈0.387 AU
5. The asteroid Icarus has an elliptical orbit around the sun. Its mean distance from the sun is 1.078 AU and the eccentricity of its orbit is 0.827. Find the ratio of the length of the major axis of this orbit to the length of the minor axis. Contrast this ratio with the same ratio for Earth (Exercise 3). (See above.)

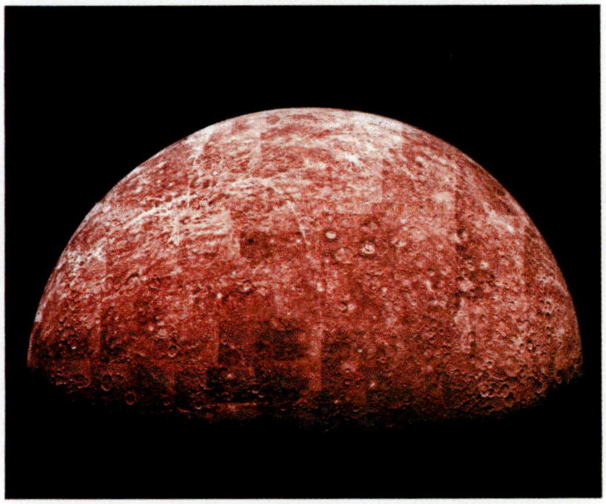

Additional Answers
Written Exercises
(continued from p. 416)

32.

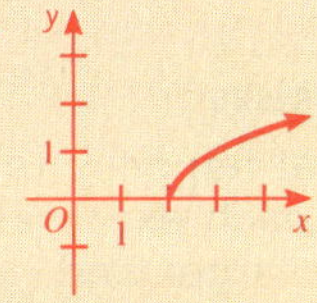

34.

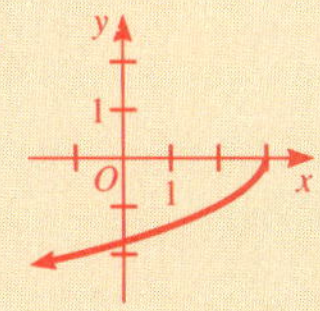

Teaching Suggestions p. T114

Suggested Extensions p. T114

Warm-Up Exercises

Describe what happens to the value of each expression as x is given larger and larger values.

1. $\frac{1}{x}$ Approaches 0
2. $\frac{1}{x^2}$ Approaches 0
3. $1 - \frac{1}{x^2}$ Approaches 1
4. $\sqrt{1 - \frac{1}{x^2}}$ Approaches 1

Motivating the Lesson

Tell students that a "small" change in the definition of an ellipse (that is, replacing the word sum by difference) produces a completely different curve, called a hyperbola. Exploring the relationship between a hyperbola and its equation is the topic of today's lesson.

Using a Calculator

Encourage students to use a calculator when finding the coordinates of points that lie on a given conic.

9-5 Hyperbolas

Objective To learn the relationship between the foci, intercepts, and asymptotes of a hyperbola and the equation of the hyperbola.

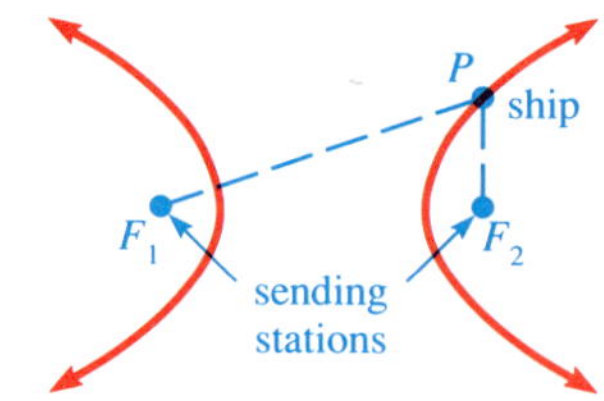

A ship's navigator can plot the location of a ship by using a radio navigation system called LORAN (LOng RAnge Navigation). In this system, a pair of radio sending stations continually transmit signals at different time intervals. The navigator measures the difference in time between the two signals received and then uses this information to plot the path of the ship. Its curved path is called a *hyperbola*. The signals from another pair of stations are used to plot a second hyperbolic path. The intersection of the two hyperbolas determines the exact position of the ship. In the diagram above, the sending stations are represented by two fixed points, F_1 and F_2, and the ship by the point P. The ship, or point P, moves along a path that maintains a constant difference between the distances from point P to the fixed points. In other words, the difference $PF_1 - PF_2$ is the same for all positions of P.

A **hyperbola** is the set of all points P in the plane such that the difference between the distances from P to two fixed points is a given constant.

Each of the fixed points is a **focus.** If P is a point on the hyperbola and F_1 and F_2 are the foci, then PF_1 and PF_2 are the **focal radii.** As you will see, hyperbolas are curves with two pieces, or *branches*.

To find an equation of the hyperbola having foci $F_1(-5, 0)$ and $F_2(5, 0)$ and difference of focal radii 6, use the definition and write:

$$|PF_1 - PF_2| = 6, \quad \text{or} \quad PF_1 - PF_2 = \pm 6.$$

Now apply the distance formula:

$$\sqrt{(x+5)^2 + y^2} - \sqrt{(x-5)^2 + y^2} = \pm 6.$$

Next eliminate the radicals and simplify.

$$\begin{aligned}
\sqrt{(x+5)^2 + y^2} &= \sqrt{(x-5)^2 + y^2} \pm 6 \\
(x+5)^2 + y^2 &= (x-5)^2 + y^2 \pm 12\sqrt{(x-5)^2 + y^2} + 36 \\
20x - 36 &= \pm 12\sqrt{(x-5)^2 + y^2} \\
5x - 9 &= \pm 3\sqrt{(x-5)^2 + y^2} \\
25x^2 - 90x + 81 &= 9[(x-5)^2 + y^2] \\
16x^2 - 9y^2 &= 144
\end{aligned}$$

Divide both sides by 144:

$$\frac{x^2}{9} - \frac{y^2}{16} = 1$$

Before graphing this equation, you analyze it using similar steps as for ellipses (page 419).

1. The x-intercepts, 3 and -3, are the values of x for which $y = 0$. When $x = 0$, the equation becomes $-\frac{y^2}{16} = 1$. Since no real value of y satisfies this equation, there are no y-intercepts.
2. Determine the extent of the graph.

$$y = \pm\frac{4}{3}\sqrt{x^2 - 9} \qquad x = \pm\frac{3}{4}\sqrt{y^2 + 16}$$

In the first equation, y is real if and only if $|x| \geq 3$. Thus, no part of the graph lies between the lines $x = -3$ and $x = 3$. In the second equation, x is real for all values of y, so no values of y are excluded.

3. It is easy to see that the graph is symmetric with respect to the x-axis, the y-axis, and the origin.
4. Notice that $y = \pm\frac{4}{3}\sqrt{x^2 - 9} = \pm\frac{4}{3}x\sqrt{1 - \frac{9}{x^2}}$.

When $|x|$ is very large, $\sqrt{1 - \frac{9}{x^2}}$ is very near 1, and therefore $y \approx \pm\frac{4}{3}x$. Thus as $|x|$ becomes large the graph approaches the lines

$$y = \frac{4}{3}x \quad \text{and} \quad y = -\frac{4}{3}x.$$

These lines are the *asymptotes* of the hyperbola $\frac{x^2}{3^2} - \frac{y^2}{4^2} = 1$. These asymptotes are also the diagonals of the rectangle formed by the lines $x = 3$, $x = -3$, $y = 4$, and $y = -4$. Asymptotes are useful guides in drawing a hyperbola, as you can see in the diagram below.

Using the facts above, you can construct a table of first-quadrant points and then draw the complete graph. A calculator may be helpful.

Since $\frac{x^2}{9} - \frac{y^2}{16} = 1$,

$$y = \pm\frac{4}{3}\sqrt{x^2 - 9}.$$

x	y
3	0
4	3.5
5	5.3
7	8.4

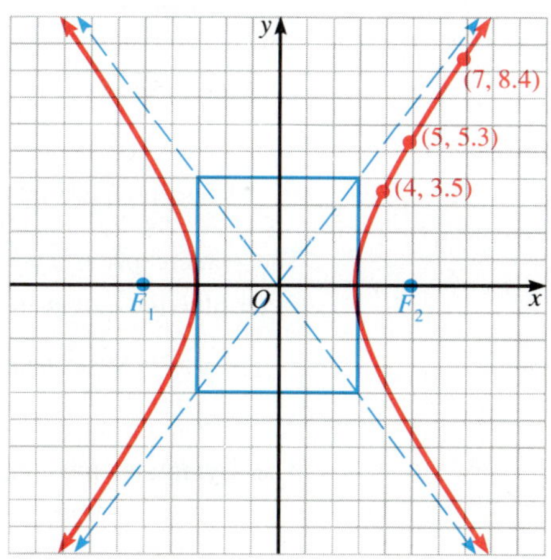

The graph is shown at the right. Notice that the graph consists of two *branches*.

Chalkboard Examples

For each hyperbola, sketch the graph, showing asymptotes as dashed lines. Find its foci.

1. $16x^2 = 9y^2 + 144$

$\frac{x^2}{9} - \frac{y^2}{16} = 1$

x-intercepts are 3 and -3; no y-intercepts. The asymptotes have equations $y = \frac{4}{3}x$ and $y = -\frac{4}{3}x$.

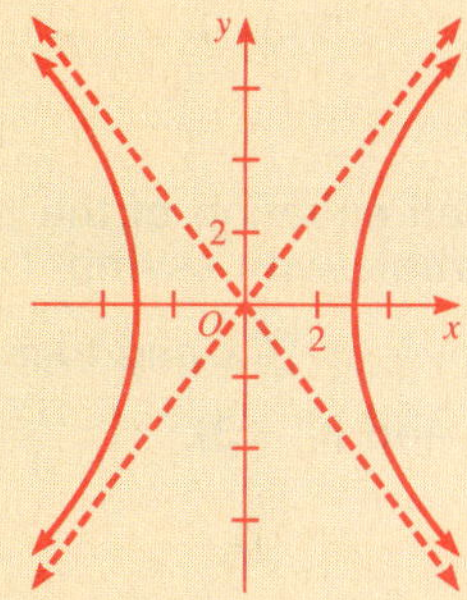

$c^2 = 3^2 + 4^2 = 25$;
$c = 5$
The foci are $(5, 0)$ and $(-5, 0)$.

2. $y^2 = 9x^2 + 9$

$\frac{y^2}{9} - \frac{x^2}{1} = 1$

y-intercepts are 3 and -3; no x-intercepts. The asymptotes have equations $y = 3x$ and $y = -3x$.

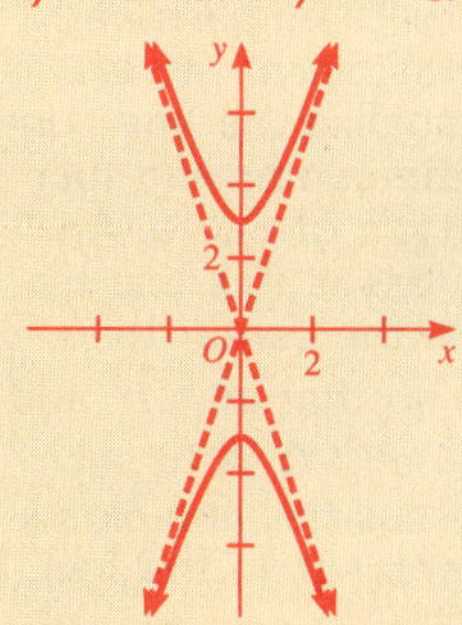

$c^2 = 1^2 + 3^2 = 10$;
$c = \sqrt{10}$
The foci are $(0, \sqrt{10})$ and $(0, -\sqrt{10})$.

(continued)

Chalkboard Examples (continued)

3. Find an equation of the hyperbola having foci (0, 3) and (0, −3) and difference of focal radii equal to 2.
The distance from each focus to the center is 3, so $c = 3$. The difference of focal radii is 2, so $a = 1$. $b^2 = c^2 - a^2 = 8$. Foci are on y-axis.
$\therefore$ the equation is
$$\frac{y^2}{1} - \frac{x^2}{8} = 1.$$

4. Find an equation of the hyperbola with asymptotes $y = \pm\frac{\sqrt{5}}{2}x$ and foci (0, 3) and (0, −3).
The y^2-term is positive because the foci are on the y-axis. $a = \sqrt{5}$ and $b = 2$, so $a^2 = 5$ and $b^2 = 4$.
$\therefore$ the equation is
$$\frac{y^2}{5} - \frac{x^2}{4} = 1.$$

Common Errors

Students often get confused about the relationship between a^2, b^2, and c^2 when working with ellipses and hyperbolas. Have the students visualize the line containing the center and foci of each type of curve. An ellipse crosses this line farther from the center than the foci, so that $a^2 > c^2$ and $b^2 = a^2 - c^2 > 0$. A hyperbola crosses the line closer to the center than the foci, so that $c^2 > a^2$ and $b^2 = c^2 - a^2 > 0$.

The **center** of a hyperbola is the midpoint of the line segment joining its foci. A hyperbola with foci on the x-axis or y-axis and equidistant from the origin has a *quadratic equation in two variables* with one of the two forms shown below. Notice that in each case a^2 is the denominator of the *positive* squared term.

The hyperbola with center (0, 0), foci $(-c, 0)$ and $(c, 0)$, and difference of focal radii $2a$ has the equation
$$\frac{x^2}{a^2} - \frac{y^2}{b^2} = 1, \text{ where } b^2 = c^2 - a^2.$$
The equations of the asymptotes are $y = \frac{b}{a}x$ and $y = -\frac{b}{a}x$.

If the foci are on the y-axis at $(0, -c)$ and $(0, c)$, then the equation is
$$\frac{y^2}{a^2} - \frac{x^2}{b^2} = 1, \text{ where } b^2 = c^2 - a^2.$$
In this case the equations of the asymptotes are $y = \frac{a}{b}x$ and $y = -\frac{a}{b}x$.

For a hyperbola having an equation with one of the forms shown above, the foci are either on the x-axis or the y-axis. If the x^2-term is positive, the line containing the center and foci is horizontal. If the y^2-term is positive, the line containing the center and foci is vertical.

The denominator of the positive squared term is a^2. The difference of the focal radii is $2a$. If you know the values of a and b and the coordinates of the center, you can find the coordinates of the foci by solving $b^2 = c^2 - a^2$ for c.

Example 1 Graph the hyperbola $y^2 = 3x^2 + 12$, showing its asymptotes as dashed lines. Find its foci.

Solution

$$y^2 = 3x^2 + 12$$
$$y^2 - 3x^2 = 12$$
$$\frac{y^2}{12} - \frac{x^2}{4} = 1$$

The y-intercepts are $\pm\sqrt{12} \approx 3.5$. There are no x-intercepts.

Determine the extent of the graph:
$$y = \pm\sqrt{3x^2 + 12} \quad \text{and} \quad x = \pm\frac{\sqrt{3}}{3}\sqrt{y^2 - 12}$$
$\therefore$ *x is real if and only if* $|y| \geq \sqrt{12}$.

The asymptotes have equations
$$y = \pm\frac{\sqrt{12}}{\sqrt{4}}x, \text{ or } y = \sqrt{3}x \text{ and } y = -\sqrt{3}x.$$

Make a short table of first-quadrant points and sketch the complete graph using symmetry.

$$y = \pm\sqrt{3x^2 + 12}$$

x	y
0	3.5
1	3.9
2	4.9
3	6.2

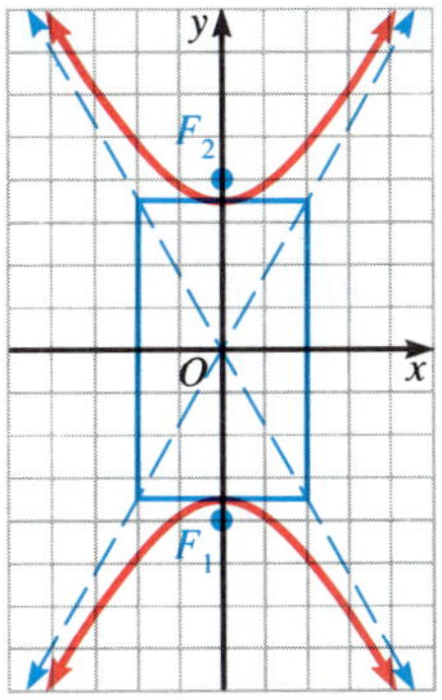

The graph is shown at the right.

Since $c^2 = a^2 + b^2 = 12 + 4 = 16$, $c = 4$.

$\therefore$ the foci are $F_1(0, -4)$ and $F_2(0, 4)$. ***Answer***

Example 2 Find an equation of the hyperbola having foci at (3, 0) and (−3, 0) and difference of focal radii equal to 4.

Solution The center of the hyperbola is (0, 0). The distance from each focus to the center is 3, so $c = 3$. The difference of the focal radii is $2a = 4$, and so $a = 2$. Since $b^2 = c^2 - a^2$, $b^2 = 3^2 - 2^2 = 5$. Since the foci are on the x-axis, an equation is $\frac{x^2}{4} - \frac{y^2}{5} = 1$. ***Answer***

Example 3 Find an equation of the hyperbola with asymptotes $y = \frac{3}{4}x$ and $y = -\frac{3}{4}x$ and foci (5, 0) and (−5, 0).

Solution The center of the hyperbola is (0, 0). The x^2-term is positive because the foci are on the x-axis. Since the equations of the asymptotes are $y = \pm\frac{3}{4}x$, $b = 3$ and $a = 4$. Thus $b^2 = 9$ and $a^2 = 16$. An equation of the hyperbola is $\frac{x^2}{16} - \frac{y^2}{9} = 1$. ***Answer***

In Lesson 8-2, you learned that an inverse variation has an equation of the form

$$xy = k \quad (k \neq 0).$$

It can be shown that the graph of such an equation is a hyperbola (see Exercises 34 and 35 on page 431). The coordinate axes are the asymptotes. If $k > 0$, the branches of the hyperbola are in the first and third quadrants. If $k < 0$, the branches are in the second and fourth quadrants. The graph of $xy = 9$ is shown at the right.

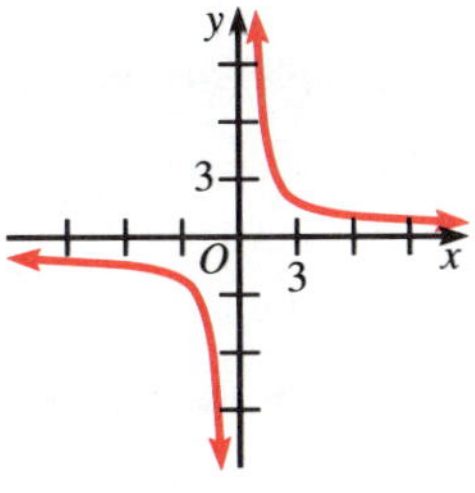

Check for Understanding

Here is a suggested use of the Oral Exercises to check students' understanding as you teach the lesson.

Oral Exs. 1–6: use before Example 1.

Oral Exs. 7–9: use at the end of the lesson.

Guided Practice

Give the intercepts of the hyperbola, the equations of the asymptotes, and the coordinates of the foci.

1. $\frac{x^2}{9} - \frac{y^2}{4} = 1$ $(\pm 3, 0)$; $y = \pm\frac{2}{3}x$; $(\pm\sqrt{13}, 0)$
2. $y^2 - 4x^2 = 4$ $(0, \pm 2)$; $y = \pm 2x$; $(0, \pm\sqrt{5})$
3. Graph $\frac{x^2}{64} - \frac{y^2}{36} = 1$.

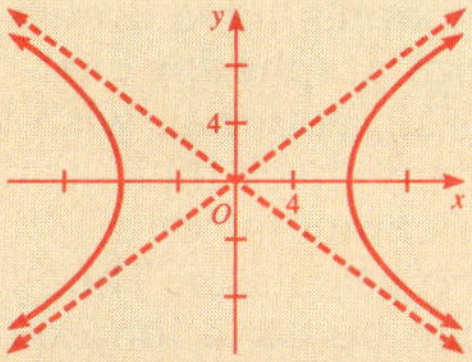

4. Find an equation of the hyperbola with foci (0, 13) and (0, −13) and with 24 as the difference of the focal radii.
$\frac{y^2}{144} - \frac{x^2}{25} = 1$

Summarizing the Lesson

In this lesson students learned to write the equation of a hyperbola and relate that equation to the foci, intercepts, and asymptotes of the hyperbola. Ask the students to state the general form of the equation of a hyperbola centered at the origin and with foci (a) on the x-axis and (b) on the y-axis.

Suggested Assignments

Average Algebra
Day 1: 430/1–16
S 423/32, 33
Day 2: 430/17–24, 27, 29, 30
S 431/Mixed Review

Average Alg. and Trig.
Day 1: 430/1–6
Assign with Lesson 9-4.
Day 2: 430/7–31 odd
S 431/Mixed Review

Extended Algebra
Day 1: 430/2–16 even
Assign with Lesson 9-4.
Day 2: 430/17–32
S 431/Mixed Review

Extended Alg. and Trig.
Day 1: 430/2–16 even
Assign with Lesson 9-4.
Day 2: 430/17–32
S 431/Mixed Review

Using a Computer or a Graphing Calculator

Once students have drawn their own graphs for Exercises 1–12, you may wish to have them check their work using a computer or a graphing calculator. The same can be done for Exercises 25–28.

Support Materials
Disk for Algebra
Menu Items: Conics Plotter
Function Plotter

Additional Answers
Oral Exercises

1. x-int.: ± 5; $y = \pm\frac{4}{5}x$; x-axis
2. y-int.: ± 1; $y = \pm\frac{1}{3}x$; y-axis
3. y-int.: ± 1; $y = \pm\frac{1}{5}x$; y-axis
4. x-int.: ± 2; $y = \pm 2x$; x-axis
5. x-int.: ± 2; $y = \pm\frac{5}{2}x$; x-axis

Oral Exercises

Give the intercepts of the hyperbola and the equations of its asymptotes. Tell on which of the coordinate axes its foci lie.

1. $\frac{x^2}{25} - \frac{y^2}{16} = 1$
2. $\frac{y^2}{1} - \frac{x^2}{9} = 1$
3. $x^2 - 25y^2 + 25 = 0$
4. $4x^2 - y^2 = 16$
5. $25x^2 - 4y^2 = 100$
6. $4x^2 - 9y^2 + 36 = 0$

Describe the graph of each equation.

7. $xy = k$, when $k = 0$
8. $y = \frac{1}{x}$
9. $\frac{x^2}{a^2} - \frac{y^2}{b^2} = 0$

Written Exercises

1. b. $(\pm\sqrt{41}, 0)$ 2. b. $(0, \pm\sqrt{10})$ 3. b. $(0, \pm\sqrt{26})$
4. b. $(\pm 2\sqrt{5}, 0)$ 5. b. $(\pm\sqrt{29}, 0)$ 6. b. $(0, \pm\sqrt{13})$

In Exercises 1–12, (a) graph each hyperbola, showing its asymptotes as dashed lines; (b) find the coordinates of the foci. You may wish to check your graphs on a computer or a graphing calculator.

A

1–6. Use the equations in Oral Exercises 1–6.

7. $x^2 = 9y^2 - 81$ $(0, \pm 3\sqrt{10})$
8. $y^2 = 5x^2 + 25$ $(0, \pm\sqrt{30})$
9. $75x^2 - 100y^2 = 7500$ $(\pm 5\sqrt{7}, 0)$
10. $25x^2 - 144y^2 = 3600$ $(\pm 13, 0)$
11. $4x^2 - y^2 + 1 = 0$ $\left(0, \pm\frac{\sqrt{5}}{2}\right)$
12. $16x^2 - 4y^2 + 64 = 0$ $(0, \pm 2\sqrt{5})$

Find an equation of the hyperbola described.

13. Foci $(0, -8)$ and $(0, 8)$; difference of focal radii 10. $\frac{y^2}{25} - \frac{x^2}{39} = 1$
14. Foci $(-4, 0)$ and $(4, 0)$; difference of focal radii 4. $\frac{x^2}{4} - \frac{y^2}{12} = 1$
15. Asymptotes $y = \frac{3}{2}x$ and $y = -\frac{3}{2}x$; foci $(0, -\sqrt{13})$ and $(0, \sqrt{13})$. $\frac{y^2}{9} - \frac{x^2}{4} = 1$
16. Asymptotes $y = \frac{\sqrt{2}}{2}x$ and $y = -\frac{\sqrt{2}}{2}x$; foci $(0, -\sqrt{6})$ and $(0, \sqrt{6})$. $\frac{y^2}{2} - \frac{x^2}{4} = 1$

B

17. Asymptotes $y = 3x$ and $y = -3x$; y-intercepts 3 and -3. $\frac{y^2}{9} - \frac{x^2}{1} = 1$
18. Asymptotes $y = x$ and $y = -x$; foci $(-4, 0)$ and $(4, 0)$. $\frac{x^2}{8} - \frac{y^2}{8} = 1$

Graph each inequality.

19. $y^2 - x^2 > 4$
20. $y^2 \le x^2 - 4$
21. $4x^2 \le y^2 + 16$
22. $9x^2 > 4y^2 - 36$

The hyperbolas $\frac{x^2}{p^2} - \frac{y^2}{q^2} = 1$ and $\frac{y^2}{q^2} - \frac{x^2}{p^2} = 1$ are *conjugates* of each other. Graph the following conjugate hyperbolas on the same coordinate axes.

23. $\frac{x^2}{9} - \frac{y^2}{4} = 1$; $\frac{y^2}{4} - \frac{x^2}{9} = 1$
24. $\frac{x^2}{1} - \frac{y^2}{4} = 1$; $\frac{y^2}{4} - \frac{x^2}{1} = 1$

Graph each equation. Each graph is half of a hyperbola, since $\sqrt{a}$ is nonnegative.

25. $y = \sqrt{x^2 + 16}$

26. $y = \sqrt{x^2 - 1}$

27. $y = \sqrt{x^2 - 16}$

28. $y = \sqrt{x^2 + 1}$

29. The statement "traveling 200 miles at x mi/h for y hours" can be described by the equation $xy = 200$. Consider the restrictions on x and y and then graph this equation. What does the graph tell you about the relationship between x and y? x and y vary inversely.

Use the definition to find an equation of the hyperbola having the given points as foci and the given number as difference of focal radii.

30. $(0, -5)$, $(0, 5)$; 4 $\frac{y^2}{4} - \frac{x^2}{21} = 1$

31. $(-3, 0)$, $(3, 0)$; 2 $\frac{x^2}{1} - \frac{y^2}{8} = 1$

C **32.** $(-c, 0)$, $(c, 0)$; $2a$ $\frac{x^2}{a^2} - \frac{y^2}{c^2 - a^2} = 1$

33. $(0, -c)$, $(0, c)$; $2a$ $\frac{y^2}{a^2} - \frac{x^2}{c^2 - a^2} = 1$

34. (a, a), $(-a, -a)$; $2a$ $xy = \frac{a^2}{2}$

35. $(-a, a)$, $(a, -a)$; $2a$ $xy = -\frac{a^2}{2}$

Mixed Review Exercises

Graph each equation.

1. $x^2 + 4y^2 = 16$

2. $x^2 + y^2 - 2x + 4y + 1 = 0$

3. $3x - 4y = 6$

4. $2x^2 - 4x + y + 5 = 0$

Draw the graph of each relation and tell whether or not it is a function.

5. $\{(x, y): y = 2x\}$ yes

6. $\{(x, y): x = 2y\}$ yes

7. $\{(x, y): x = y^2\}$ no

8. $\{(x, y): y = x^2\}$ yes

9. $\{(x, y): y = |x|\}$ yes

10. $\{(x, y): x = |y|\}$ no

Historical Note / *The Area of a Parabolic Section*

About 240 B.C. the great Greek mathematician Archimedes discovered a formula for the area of any section of a parabola. By summing the areas of a sequence of smaller and smaller triangles that fill the section more and more completely, Archimedes was able to show that the area of any section is $\frac{2}{3}bh$, where b is the length of the base (the chord that cuts off the section) and h is the height (the length of a perpendicular segment to the base from the point on the parabola where a tangent parallel to the base touches the parabola). Therefore, the area of the section is $\frac{4}{3}$ as large as the area of the largest triangle that can be inscribed in it.

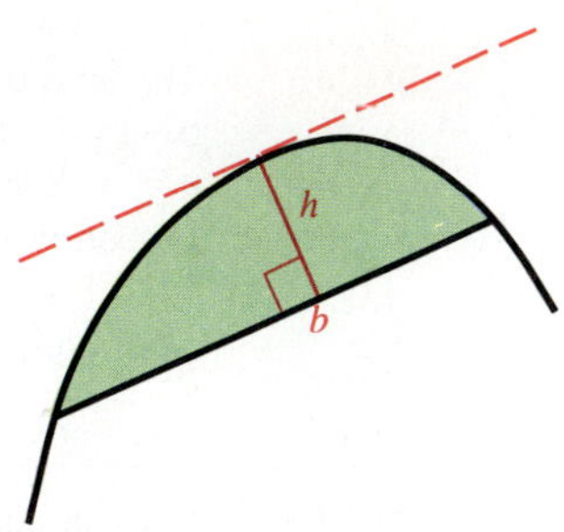

6. y-int.: ± 2; $y = \pm\frac{2}{3}x$; y-axis

7. The x- and y-axes

8. A hyperbola with branches in the first and third quadrants and with the coordinate axes as asymptotes

9. Two lines having slopes $\pm\frac{b}{a}$ and intersecting at the origin

Supplementary Materials

Study Guide	pp. 145–146
Resource Book	p. 139
Overhead Visuals	5, 7

Additional Answers
Written Exercises

2. a.

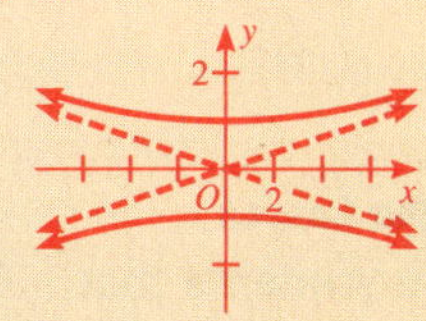

4. a.

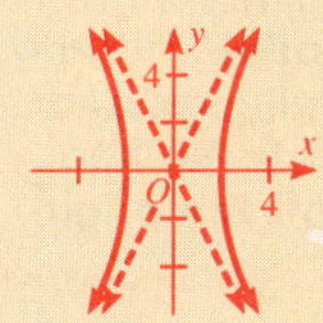

6. a.

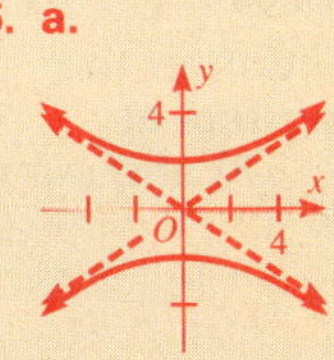

8. a.

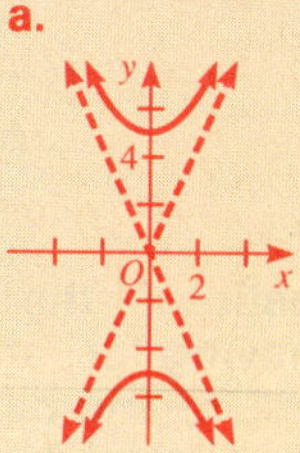

(continued on p. 443)

Explorations p. 840

Teaching Suggestions p. T115

Suggested Extensions p. T115

Using Manipulatives p. T115

Warm-Up Exercises

Identify each of the following as an equation of a circle, parabola, ellipse, or hyperbola.

1. $y = (x + 5)^2$ parabola
2. $\frac{y^2}{25} - \frac{x^2}{4} = 1$ hyperbola
3. $\frac{x^2}{4} + \frac{y^2}{12} = 1$ ellipse
4. $(x - 3)^2 + (y + 2)^2 = 4$ circle
5. $xy = 9$ hyperbola

Motivating the Lesson

Remind the students that they have already dealt with circles having centers *not* at the origin. In today's lesson the centers of ellipses and hyperbolas will be shifted, or translated, off the origin.

Chalkboard Examples

1. Find an equation of the ellipse having foci at (2, 1) and (2, −3) and sum of focal radii 6. Center is at (2, −1); distance from center to each focus is 2. Thus $a = 3$, $c = 2$, and $b^2 = a^2 - c^2 = 9 - 4 = 5$. The equation is $\frac{(x-2)^2}{5} + \frac{(y+1)^2}{9} = 1$.
2. Find an equation of the hyperbola having foci at (−6, 5) and (2, 5) and difference of focal radii 4.

9-6 More on Central Conics

Objective To find an equation of a conic section with center not at the origin and to identify a conic as a circle, ellipse, or hyperbola.

Circles, ellipses, and hyperbolas are called **central conics** because they have centers. Now you'll study central conics with centers not at the origin.

As you learned in Lesson 9-2, you can translate a graph centered at the origin by sliding every point of the graph h units horizontally and k units vertically. Replacing x by $x - h$ and y by $y - k$ in the equation of a central conic with center at the origin gives the equation of the same conic with center now at (h, k). Using this fact, the formulas of Lessons 9-4 and 9-5 can be rewritten in the following more general forms. The constants a and b have the same meaning here as in those lessons.

Ellipses with Center (h, k)

Horizontal major axis:

$$\frac{(x-h)^2}{a^2} + \frac{(y-k)^2}{b^2} = 1$$

Foci at $(h - c, k)$ and $(h + c, k)$, where $c^2 = a^2 - b^2$.

Vertical major axis:

$$\frac{(x-h)^2}{b^2} + \frac{(y-k)^2}{a^2} = 1$$

Foci at $(h, k - c)$ and $(h, k + c)$, where $c^2 = a^2 - b^2$.

Hyperbolas with Center (h, k)

Horizontal major axis:

$$\frac{(x-h)^2}{a^2} - \frac{(y-k)^2}{b^2} = 1$$

Foci at $(h - c, k)$ and $(h + c, k)$, where $c^2 = a^2 + b^2$.

Vertical major axis:

$$\frac{(y-k)^2}{a^2} - \frac{(x-h)^2}{b^2} = 1$$

Foci at $(h, k - c)$ and $(h, k + c)$, where $c^2 = a^2 + b^2$.

Example 1 Find an equation of the ellipse having foci (−3, 4) and (9, 4) and sum of focal radii 14.

Solution The sum of the focal radii, $2a$, is 14. So $a = 7$. The center is halfway between the foci, at (3, 4). The distance from the center to each focus is 6, so $c = 6$. Substituting 7 for a and 6 for c in the equation $b^2 = a^2 - c^2$ (from Lesson 9-4) gives $b^2 = 49 - 36 = 13$.

$\therefore$ an equation of the ellipse is $\frac{(x-3)^2}{49} + \frac{(y-4)^2}{13} = 1$. ***Answer***

Example 2 Find an equation of the hyperbola having foci (−3, −2) and (−3, 8) and difference of focal radii 8.

Solution The difference of the focal radii is $8 = 2a$, so $a = 4$. The center is halfway between the foci, at $(-3, 3)$. The distance from the center to each focus is 5, so $c = 5$. Substituting 4 for a and 5 for c in the equation $b^2 = c^2 - a^2$ gives $b^2 = 25 - 16 = 9$. The y^2-term is positive because the line containing the foci is vertical.

$\therefore$ an equation of the hyperbola is $\frac{(y-3)^2}{16} - \frac{(x+3)^2}{9} = 1$. ***Answer***

When the equation of a central conic is given in the form

$$Ax^2 + By^2 + Cx + Dy + E = 0,$$

you can complete the square in x and y to identify the conic and find its center and its foci, as shown in Example 3.

Example 3 Identify the conic $x^2 - 4y^2 - 2x - 16y - 11 = 0$. Find its center and foci. Then draw its graph.

Solution Complete the square using the x-terms and then using the y-terms.

$$(x^2 - 2x + \underline{\ ?\ }) + (-4y^2 - 16y + \underline{\ ?\ }) = 11 + \underline{\ ?\ } + \underline{\ ?\ }$$
$$(x^2 - 2x + 1) - 4(y^2 + 4y + 4) = 11 + 1 - 16$$
$$(x-1)^2 - 4(y+2)^2 = -4$$

Divide both sides by -4.

$$\frac{(y+2)^2}{1} - \frac{(x-1)^2}{4} = 1$$

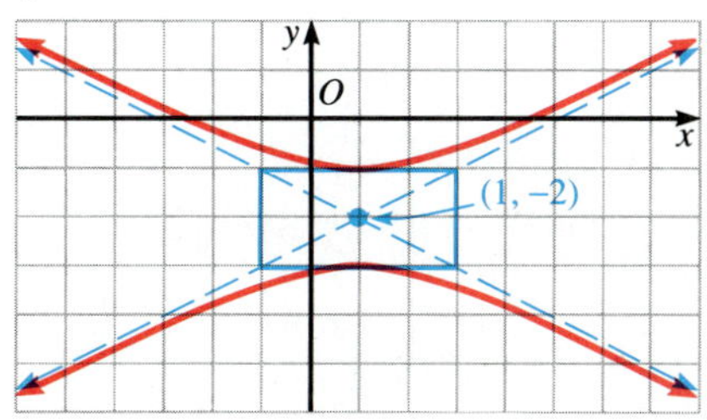

Comparing this equation with the general forms, you can see that the conic is a hyperbola with center $(1, -2)$. The line containing the center and foci is vertical. Since $a^2 = 1$ and $b^2 = 4$, $c^2 = a^2 + b^2 = 5$, and $c = \sqrt{5}$. Thus, the foci are the points $(1, -2 + \sqrt{5})$ and $(1, -2 - \sqrt{5})$, or approximately $(1, 0.2)$ and $(1, -4.2)$. The graph is shown above. ***Answer***

If you apply the vertical-line test (see page 154) to the graph of the hyperbola in Example 3, you will see that the graph does *not* represent a function. *No central conic can be a function.*

Oral Exercises

Identify each conic and give its center and foci.

1. $\frac{(x+3)^2}{16} + \frac{(y-5)^2}{12} = 1$ ellipse; $(-3, 5)$; $(-5, 5)$, $(-1, 5)$

2. $\frac{(x+7)^2}{9} - \frac{(y+1)^2}{16} = 1$ hyperbola; $(-7, -1)$; $(-12, -1)$, $(-2, -1)$

Center is at $(-2, 5)$; distance from each focus to center is 4. Thus $a = 2$, $c = 4$, and $b^2 = c^2 - a^2 = 12$. Line containing center is horizontal. The equation is $\frac{(x+2)^2}{4} - \frac{(y-5)^2}{12} = 1$.

3. Identify the conic $4x^2 + 9y^2 - 16x + 18y - 11 = 0$. Find its center and foci. Then draw its graph.

$(4x^2 - 16x) + (9y^2 + 18y) = 11$

$4(x^2 - 4x + 4) + 9(y^2 + 2y + 1) = 11 + 16 + 9$

$4(x-2)^2 + 9(y+1)^2 = 36$

$\frac{(x-2)^2}{9} + \frac{(y+1)^2}{4} = 1$

The conic is an ellipse with center $(2, -1)$. Since $a^2 = 9$ and $b^2 = 4$, $c^2 = a^2 - b^2 = 5$ and $c = \sqrt{5}$. The foci are $(2 + \sqrt{5}, -1)$ and $(2 - \sqrt{5}, -1)$.

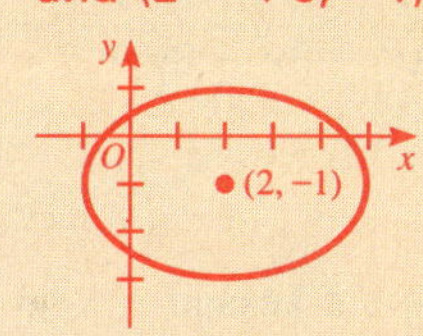

Thinking Skills

Conics can be identified directly from their equations without first completing the square in x and y. The *analysis* of the general equation $Ax^2 + By^2 + Cx + Dy + E = 0$ is outlined in Written Exercise 21 on page 435.

Check for Understanding

Here is a suggested use of the Oral Exercises to check students' understanding as you teach the lesson.
Oral Exs. 1–14: use before Example 1.

Guided Practice

Find an equation of the ellipse having the given foci and sum of focal radii.

1. (0, 2), (8, 2); 10
   $\frac{(x-4)^2}{25} + \frac{(y-2)^2}{9} = 1$
2. (−6, 1), (−2, 1); 6
   $\frac{(x+4)^2}{9} + \frac{(y-1)^2}{5} = 1$

Find an equation of the hyperbola having the given foci and difference of focal radii.

3. (0, 1), (6, 1); 4
   $\frac{(x-3)^2}{4} - \frac{(y-1)^2}{5} = 1$
4. (2, −3), (2, 7); 8
   $\frac{(y-2)^2}{16} - \frac{(x-2)^2}{9} = 1$

Identify each conic. Find its center and foci (if any).

5. $x^2 - y^2 - 4x + 2y + 2 = 0$
   hyperbola; center (2, 1); foci $(2 \pm \sqrt{2}, 1)$
6. $4x^2 + 4y^2 - 8x - 8y + 7 = 0$
   circle; center (1, 1)

Summarizing the Lesson

In this lesson students learned to write the equation of a conic not centered at the origin and to identify a conic by its equation. Ask the students to discuss how a translation affects the equation of a conic section.

Suggested Assignments

Extended Algebra
434/1–11 odd, 13–21
S 416/36, 37

Extended Alg. and Trig.
434/1–11 odd, 13–21
S 416/36, 37

The given conic is to be translated so that its new center is at the given point. What will its new equation be?

3. $\frac{x^2}{25} + \frac{y^2}{4} = 1$; (0, −5) $\frac{x^2}{25} + \frac{(y+5)^2}{4} = 1$
4. $\frac{x^2}{16} - \frac{y^2}{1} = 1$; (−5, 0) $\frac{(x+5)^2}{16} - \frac{y^2}{1} = 1$
5. $x^2 - y^2 = 49$; (−4, 3) $(x+4)^2 - (y-3)^2 = 49$
6. $x^2 + y^2 = 9$; (−1, 1) $(x+1)^2 + (y-1)^2 = 9$
7. $4x^2 + y^2 = 16$; (1, −4) $4(x-1)^2 + (y+4)^2 = 16$
8. $4x^2 - 9y^2 = 36$; (3, −2) $4(x-3)^2 - 9(y+2)^2 = 36$

Match each equation with its graph.

9. $\frac{(x+1)^2}{1} + \frac{(y-2)^2}{4} = 1$ **e**
10. $\frac{(x+1)^2}{1} - \frac{(y-2)^2}{4} = 1$ **a**
11. $x + 1 = (y-2)^2$ **d**
12. $(x+1)^2 + (y-2)^2 = 1$ **b**
13. $\frac{(y-2)^2}{1} - \frac{(x+1)^2}{4} = 1$ **f**
14. $\frac{(x+1)^2}{4} + \frac{(y-2)^2}{1} = 1$ **c**

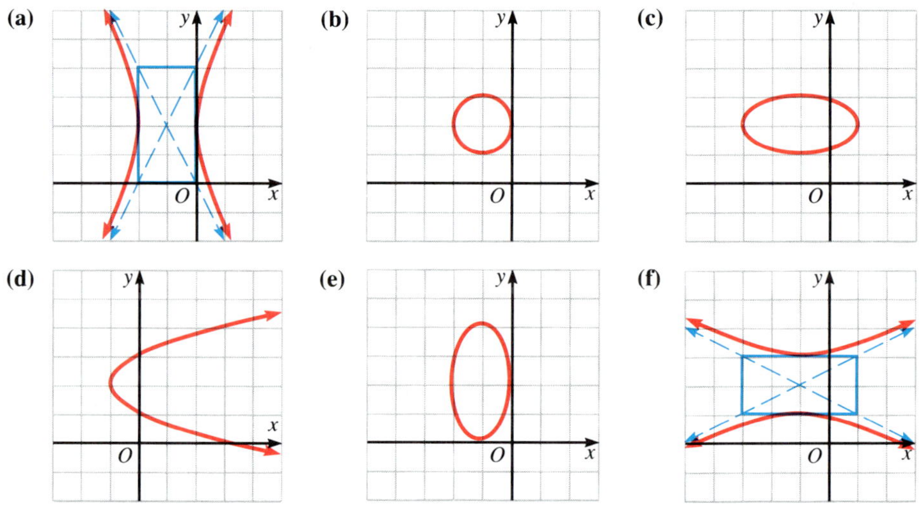

Written Exercises

Find an equation of the ellipse having the given foci and sum of focal radii.

A
1. (6, 0), (6, 6); 10
2. (0, 0), (0, 8); 12
3. (−3, −3), (−3, 3); 8
4. (−5, 1), (3, 1); 16
5. (−2, −3), (6, −3); 10
6. (−10, 2), (−2, 2); 14

Find an equation of the hyperbola having the given foci and difference of focal radii.

7. (0, −2), (8, −2); 2
8. (0, 4), (0, 10); 4
9. (3, −8), (3, −2); 4
10. (−5, 3), (9, 3); 6
11. (5, −9), (5, −1); 6
12. (−4, −4), (4, −4); 6

Identify each conic. Find its center and its foci (if any). Then draw its graph. You may wish to check your graphs on a computer or graphing calculator.

13. $x^2 - 4y^2 - 2x - 24y - 39 = 0$

14. $x^2 + 9y^2 + 2x - 18y + 1 = 0$

15. $x^2 + y^2 - 6x - 16y + 57 = 0$

16. $9x^2 - y^2 - 18x - 6y - 9 = 0$

17. $9x^2 + 25y^2 + 36x - 150y + 36 = 0$

18. $16x^2 - 9y^2 + 64x + 18y + 199 = 0$

B **19.** Use the definition of an ellipse to find an equation of the ellipse having foci (1, 1) and (−1, −1) and sum of focal radii 3. $20x^2 + 20y^2 - 32xy - 9 = 0$

20. Use the definition of a hyperbola to find an equation of the hyperbola having foci (−1, 1) and (1, −1) and difference of focal radii 2. $xy = -\frac{1}{2}$

C **21.** Every conic section has an equation of the form

$$Ax^2 + By^2 + Cx + Dy + E = 0$$

where A and B are not both zero. Let $A = 1$, $C = 2$, $D = -8$, and $E = 1$. Graph the resulting quadratic equation in two variables for each given value of B. Then identify the graph.

a. $B = 0$ parabola **b.** $B = 1$ circle **c.** $B = 4$ ellipse **d.** $B = -4$ hyperbola

e. Analyze the different equations you graphed in parts (a)–(d). What is the relationship between the coefficients A and B for which the general equation gives a circle? a parabola? an ellipse? a hyperbola?
circle: $A = B$; parabola: $AB = 0$; ellipse: $AB > 0$ and $A \neq B$; hyperbola: $AB < 0$

Self-Test 2

Vocabulary

ellipse (p. 418)
focus (of an ellipse)(p. 418)
focal radii (of an ellipse) (p. 418)
symmetric about the x-axis (p. 419)
symmetric about the y-axis (p. 419)
symmetric about the origin (p. 419)
center (of an ellipse) (p. 419)
major axis (p. 419)
minor axis (p. 419)
hyperbola (p. 426)
focal radii (of a hyperbola) (p. 426)
asymptotes (p. 427)
central conic (p. 432)

1. Graph $16x^2 + 9y^2 = 144$. **Obj. 9-4, p. 418**

2. Find an equation of the ellipse having foci (3, 0) and (−3, 0) and sum of focal radii 8. $\frac{x^2}{16} + \frac{y^2}{7} = 1$

3. Graph $25y^2 - x^2 = 25$, showing the asymptotes as dashed lines. **Obj. 9-5, p. 426**

4. Find an equation of the hyperbola having foci (0, 4) and (0, −4) and difference of focal radii 4. $\frac{y^2}{4} - \frac{x^2}{12} = 1$

5. Identify and graph $4x^2 - 25y^2 - 24x + 50y - 89 = 0$. hyperbola **Obj. 9-6, p. 432**

6. Find an equation of the ellipse having foci (5, 2) and (−5, 2) and sum of focal radii 12. $\frac{x^2}{36} + \frac{(y-2)^2}{11} = 1$

Check your answers with those at the back of the book.

Supplementary Materials

Study Guide	pp. 147–148
Practice Master	52
Test Master	36
Overhead Visuals	6, 7

Using a Computer or a Graphing Calculator

Once students have drawn their own graphs for Written Exercises 13–18, you may wish to have them check their work using a computer or a graphing calculator. The same can be done for Written Exercise 21.

You may also wish to have students determine an equation for a given conic. The program "Conics Quiz" on the Disk for Algebra will draw a random conic and let students visually experiment to find the corresponding equation.

Support Materials
Disk for Algebra
Menu Items: Conics Plotter
Conics Quiz
Resource Book p. 243

Additional Answers Written Exercises

1. $\frac{(x-6)^2}{16} + \frac{(y-3)^2}{25} = 1$
2. $\frac{x^2}{20} + \frac{(y-4)^2}{36} = 1$
3. $\frac{(x+3)^2}{7} + \frac{y^2}{16} = 1$
4. $\frac{(x+1)^2}{64} + \frac{(y-1)^2}{48} = 1$
5. $\frac{(x-2)^2}{25} + \frac{(y+3)^2}{9} = 1$
6. $\frac{(x+6)^2}{49} + \frac{(y-2)^2}{33} = 1$

(continued on p. 449)

Quick Quiz on p. 451

Systems of Equations

9-7 The Geometry of Quadratic Systems

Objective To use graphs to determine the number of real solutions of a quadratic system and to estimate the solutions.

You know that a solution of a linear system can be found geometrically by locating the point where the graphs of the system's equations intersect. You can also use graphs to determine the real solutions of a *quadratic system*. A system containing only quadratic equations or a combination of linear and quadratic equations in the same two variables is called a **quadratic system.**

If you have a computer or a graphing calculator, you may wish to explore the geometry of some quadratic systems. For example, a system of one quadratic and one linear equation may have 2, 1, or 0 real solutions, as illustrated below.

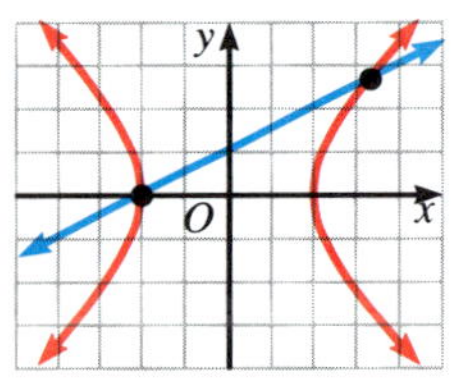

$x^2 - y^2 = 4$
$x - 2y + 2 = 0$
Two solutions

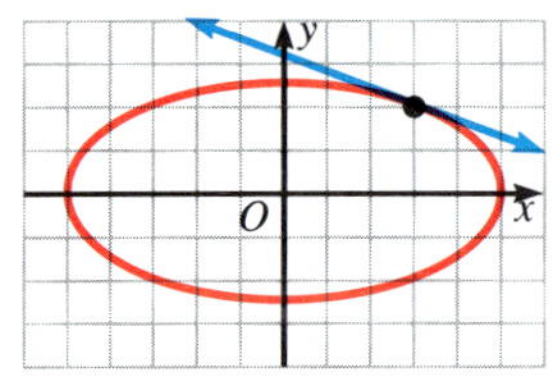

$x^2 + 4y^2 = 25$
$3x + 8y = 25$
One solution

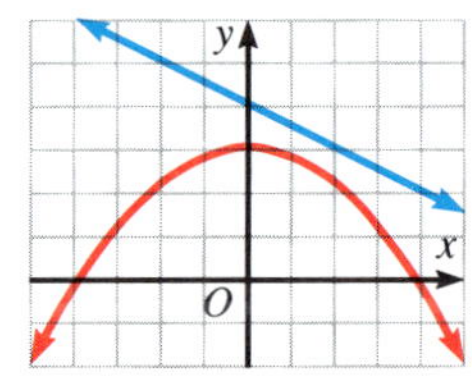

$x^2 + 5y = 15$
$x + 2y = 8$
No solution

A system of two quadratic equations may have 4, 3, 2, 1, or 0 real solutions. For example, a circle and an ellipse can have four points of intersection and their equations would have four common real solutions (see Oral Exercise 9).

Finding the points of intersection of quadratic graphs can be useful in astronomy. Like planetary orbits, the orbits of many comets are elliptical. To observe a comet closely, astronomers are naturally interested in where its orbit crosses that of the earth's.

You can use graphs not only to find the number of real solutions of a quadratic system, but also to estimate these solutions. Example 1 at the top of the next page illustrates.

Teaching Suggestions p. T115

Warm-Up Exercises

Give a system of linear equations having the indicated number of solutions. Answers may vary.

1. One $x + y = 1$; $x - y = 1$
2. Zero $x + y = 1$; $x + y = 2$
3. Infinitely many $x + y = 1$; $2x + 2y = 2$

Motivating the Lesson

Tell the students that one use of the space shuttle is to intercept and repair satellites. This situation could be illustrated by the intersection of two graphs representing the paths of the shuttle and a satellite. The graphic solution of systems of quadratic equations is the topic of today's lesson.

Using a Computer or a Graphing Calculator

You may want to use the program "Conics Plotter" on the Disk for Algebra when graphing quadratic systems. The program's zoom feature is helpful when estimating solutions of such systems.

You may also wish to have students use a computer or a graphing calculator when checking their graphs for Exercises 1–9.

Support Materials
Disk for Algebra
Menu Item: Conics Plotter
Resource Book pp. 243–244

Check for Understanding

Oral Exs. 1–12: use after the example.

Example Graph each system and estimate the real solutions.

a. $25x^2 + 4y^2 = 100$
$5x = y^2 - 15$

b. $x^2 + 6y = 30$
$xy + 6 = 0$

Solution Graph the equations of each system and estimate the coordinates of their points of intersection.

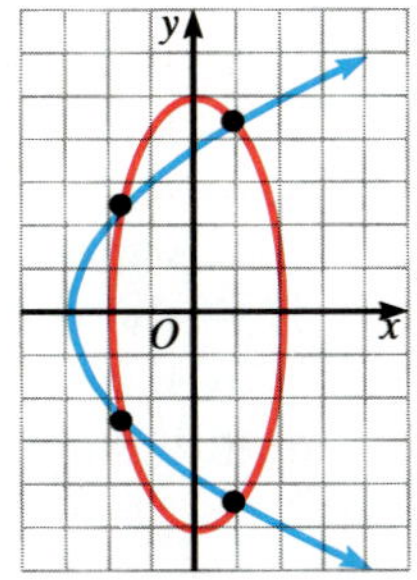

a. $25x^2 + 4y^2 = 100$
$5x = y^2 - 15$

There are four real solutions. Estimates of the solutions are (0.9, 4.4), (0.9, −4.4), (−1.7, 2.5), and (−1.7, −2.5).

b. $x^2 + 6y = 30$
$xy + 6 = 0$

There are three real solutions. Estimates of the solutions are (6, −1), (−1.3, 4.7), and (−4.7, 1.3).

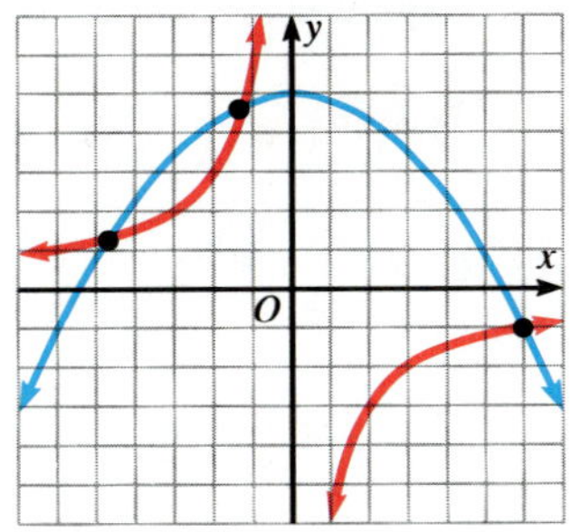

Actually, (6, −1) is an exact solution of the system in part (b) of the example, as the following check shows.

$(6)^2 + 6(-1) = 30$ ✓
$(6)(-1) + 6 = 0$ ✓

The other solutions in the example are approximations. If you have a calculator, you might check to see how accurate they are.

Oral Exercises

Mentally sketch the graph of each equation and give the *number* of real solutions the system has.

1. $x^2 + y^2 = 4$
   $y = x$
   circle and line: 2
2. $x^2 + y^2 = 4$
   $x + y = 8$
   circle and line: 0
3. $x^2 + y^2 = 4$
   $y = 2$
   circle and line: 1
4. $y = x^2$
   $y = 2x$
   parabola and line: 2
5. $x^2 + y^2 = 1$
   $y = x^2$
   circle and parabola: 2
6. $x^2 - y^2 = 1$
   $x^2 + y^2 = 9$
   hyperbola and circle: 4
7. $x^2 - y^2 = 4$
   $x + y = 2$
   hyperbola and line: 1
8. $x^2 + y^2 = 9$
   $\frac{x^2}{3} + \frac{y^2}{4} = 1$
   circle and ellipse: 0
9. By drawing rough sketches, illustrate that a circle and an ellipse can have the following possible points of intersection.
   a. four   b. three   c. two   d. one   e. zero

Can a line intersect a conic of the specified type in a single point without being tangent to it? If so, explain how.

10. A parabola
    10. Yes; the axis of symmetry intersects the parabola at a single point but is not tangent to it.
11. A hyperbola
    11. Yes; a line parallel to one of the asymptotes (but not itself an asymptote) intersects one branch at a single point.
12. An ellipse
    No

Chalkboard Examples

Use graphs to estimate the solutions of each system.

1. $x^2 + y^2 = 25$
   $2x - y = 2$
   (3, 4) and (−1.4, −4.8)

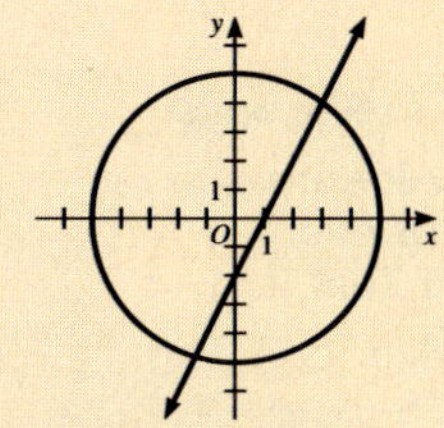

2. $x^2 - y^2 = 16$
   $\frac{x^2}{9} + \frac{y^2}{4} = 1$
   no solutions

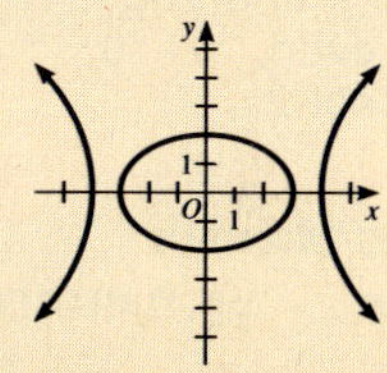

Guided Practice

Graph each system to determine the number of real solutions.

1. $x^2 - y^2 = 16$
   $y - 2x = 0$ 0

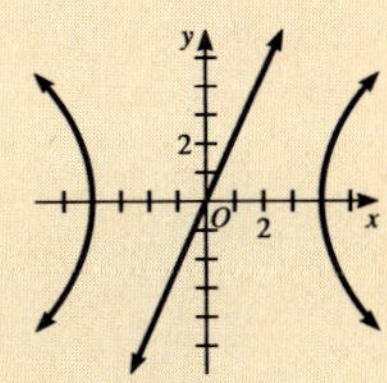

2. $x^2 + y^2 = 16$
   $y + 2x^2 = 0$ 2

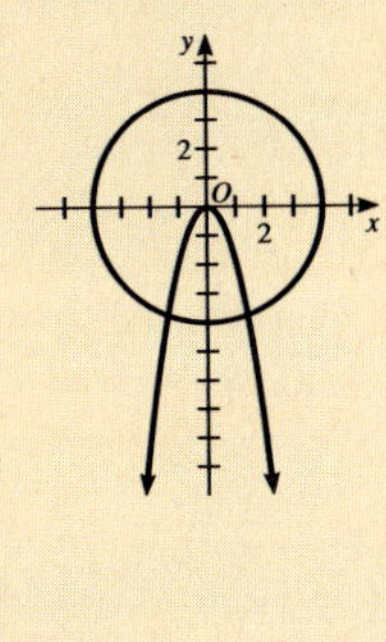

Written Exercises

By sketching graphs, find the *number* of real solutions the system has.

A

1. $4x^2 + 9y^2 = 36$ two
   $2x + 3y = 6$
2. $16x^2 + 4y^2 = 64$ two
   $3x + 4y = 12$
3. $x^2 - 9y = 0$ zero
   $x - 2y = 2$
4. $4x^2 + 25y^2 = 100$ zero
   $x^2 + y^2 = 64$
5. $x^2 - y^2 = 4$ one
   $x + y = 4$
6. $y^2 - x^2 = 4$ four
   $y = x^2 - 3$
7. $x^2 + y^2 = 25$ four
   $xy = 10$
8. $9x^2 + 25y^2 = 225$ two
   $y^2 = 2x + 4$
9. $x^2 - 4y^2 = 4$ three
   $6y^2 - x = 2$

10–15. Estimate the real solutions (if any) of the systems of Exercises 4–9 to the nearest half unit.

Graph each system of inequalities.

Sample $x^2 + 4y^2 \le 4$
$x^2 \ge y^2 + 1$

Solution

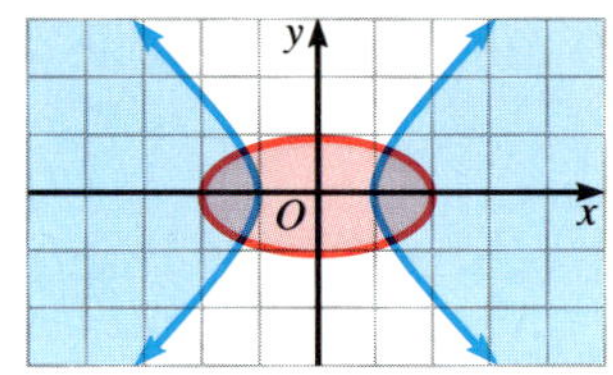

B

16. $y \ge x + 3$
    $y \le x^2$
17. $x^2 + 4y^2 \le 16$
    $x^2 \le y^2 + 4$
18. $4x^2 + y^2 \le 16$
    $x^2 + 4y^2 \le 16$
19. $x \ge -y^2$
    $x^2 + y^2 \le 49$

C

20. Use parts (a), (b), and (c) below to show that no isosceles triangle has perimeter 4 and area 1.
    a. Write equations for the area and perimeter. (*Hint:* Let base = $2x$ and height = y.) Area: $xy = 1$; perimeter: $2x + 2\sqrt{x^2 + y^2} = 4$
    b. Simplify the perimeter equation to eliminate radicals. $y^2 = -4x + 4$
    c. Graph the resulting system and use the graph to reach the required conclusion. Since the graphs of $xy = 1$ and $y^2 = -4x + 4$ intersect only for negative values of x and y, there is no isosceles triangle with perimeter 4 and area 1.

Mixed Review Exercises

Find an equation for each figure described.

1. Parabola with focus (2, 3) and directrix $y = -1$. $y - 1 = \frac{1}{8}(x - 2)^2$
2. Ellipse with x-intercepts $\pm\sqrt{6}$ and y-intercepts ± 2. $\frac{x^2}{6} + \frac{y^2}{4} = 1$
3. Hyperbola with foci (0, −3) and (0, 3) and difference of focal radii 4. $\frac{y^2}{4} - \frac{x^2}{5} = 1$
4. The circle having center (−3, 4) and passing through the origin. $(x + 3)^2 + (y - 4)^2 = 25$

Find the unique solution for each system.

5. $y = 3x + 2$ {(1, 5)}
   $2x - y = -3$
6. $x - 3y = 5$ {(2, −1)}
   $4x + y = 7$
7. $5x + 2y = 9$ {(3, −3)}
   $4x - 3y = 21$

Summarizing the Lesson

In this lesson students solved systems of quadratic equations graphically. Ask students to state the number of possible real solutions of a quadratic system.

Suggested Assignments

Average Algebra
438/2–18 even
R 435/Self-Test 2: 1–4

Average Alg. and Trig.
438/2–18 even
R 435/Self-Test 2: 1–4

Extended Algebra
438/2–18 even
R 435/Self-Test 2

Extended Alg. and Trig.
438/2–16 even, 17–19
S 435/Self-Test 2

Supplementary Materials

Study Guide pp. 149–150

Using a Calculator

You may wish to have students use a calculator to check the accuracy of their estimated solutions for Written Exercises 10–15.

Additional Answers Written Exercises

2.

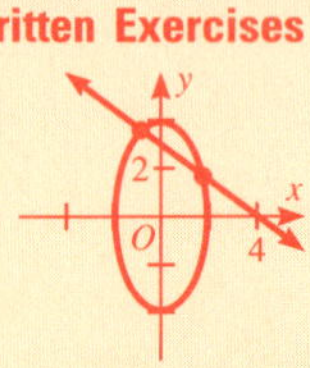

4.

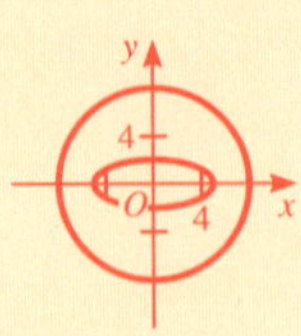

(continued on p. 452)

9-8 Solving Quadratic Systems

Objective To use algebraic methods to find exact solutions of quadratic systems.

The substitution and linear-combination methods that you used in Chapter 3 to solve linear systems can also be used to solve quadratic systems. Although it is possible for the solutions of quadratic systems to be complex numbers, in this lesson we will consider only real solutions.

Example 1 Solve this system: $4x^2 + y^2 = 25$
$2x + y = -1$

Solution (Substitution Method) Solve the linear equation for one of the variables. If you solve for y, you can avoid fractions.

$$y = -2x - 1$$

Substitute $-2x - 1$ for y in the quadratic equation and solve the resulting equation.

$$4x^2 + (-2x - 1)^2 = 25$$
$$4x^2 + 4x^2 + 4x + 1 = 25$$
$$8x^2 + 4x - 24 = 0$$
$$2x^2 + x - 6 = 0$$
$$(2x - 3)(x + 2) = 0$$
$$x = \frac{3}{2} \quad \text{or} \quad x = -2$$

Substitute $\frac{3}{2}$ and -2 for x in $y = -2x - 1$ to find the y-values.

$y = -2\left(\frac{3}{2}\right) - 1 = -4$

$\therefore \left(\frac{3}{2}, -4\right)$ is a solution.

$y = -2(-2) - 1 = 3$

$\therefore (-2, 3)$ is a solution.

Checking the ordered pairs $\left(\frac{3}{2}, -4\right)$ and $(-2, 3)$ in both *given* equations is left for you.

The solution set is $\left\{\left(\frac{3}{2}, -4\right), (-2, 3)\right\}$. ***Answer***

The equations are graphed in the diagram at the right. As you can see, the graphs have two points of intersection.

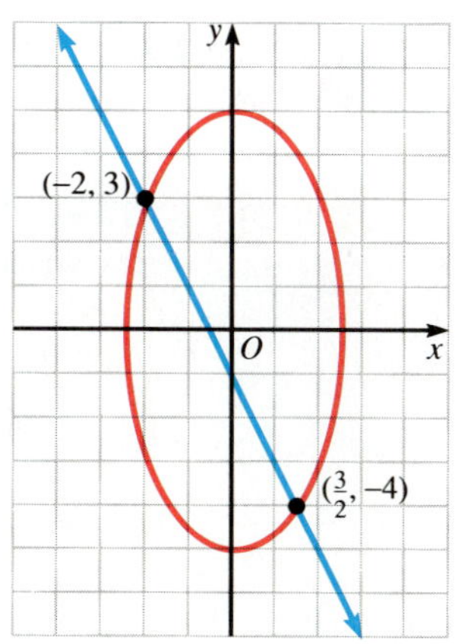

Teaching Suggestions p. T115

Group Activities p. T115

Warm-Up Exercises

Solve this system as indicated:

$x + 3y = 9$
$2x - y = 4$

1. Solve graphically. (3, 2)

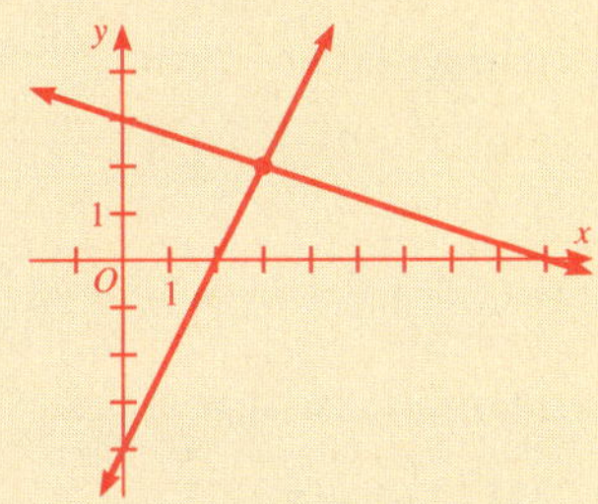

2. Solve using substitution.
$x = -3y + 9$
$2(-3y + 9) - y = 4$
$-6y + 18 - y = 4$
$-7y = -14$
$y = 2$
$x = -3(2) + 9 = 3$
(3, 2)

3. Solve using linear combinations.
$x + 3y = 9$
$6x - 3y = 12$
$7x = 21$
$x = 3$
$3 + 3y = 9$
$3y = 6$
$y = 2$
(3, 2)

Motivating the Lesson

Remind students that solving a quadratic system by graphing usually results in approximate solutions. To obtain exact solutions, the algebraic methods of substitution and linear combinations, which are the topics of today's lesson, must be used.

Chalkboard Examples

Solve each system.

1. $x^2 - 9y^2 = 9$
   $x + 6y = 3$
   (Substitution Method)
   $x = -6y + 3$
   $(-6y + 3)^2 - 9y^2 = 9$
   $36y^2 - 36y + 9 - 9y^2 = 9$
   $27y^2 - 36y = 0$
   $9y(3y - 4) = 0$
   $y = 0$ or $y = \frac{4}{3}$
   $x + 6(0) = 3$
   $x = 3$
   $x + 6\left(\frac{4}{3}\right) = 3$
   $x + 8 = 3$
   $x = -5$
   $(3, 0)$ and $\left(-5, \frac{4}{3}\right)$

2. $4x^2 - y^2 = 5$
   $xy - 3 = 0$
   (Substitution Method)
   $y = \frac{3}{x}$
   $4x^2 - \left(\frac{3}{x}\right)^2 = 5$
   $4x^4 - 5x^2 - 9 = 0$
   $(x^2 + 1)(4x^2 - 9) = 0$
   $x^2 = -1$ or $x^2 = \frac{9}{4}$
   no solution | $x = \pm\frac{3}{2}$
   $\left(\frac{3}{2}\right)y - 3 = 0$
   $y = 2$
   $\left(-\frac{3}{2}\right)y - 3 = 0$
   $y = -2$
   $\left(\frac{3}{2}, 2\right)$ and $\left(-\frac{3}{2}, -2\right)$

Example 2 Solve this system: $x^2 - y^2 = 12$
$xy = 8$

Solution (Substitution Method) Solve the second equation for y.

$$y = \frac{8}{x} \qquad (x \neq 0)$$

Substitute $\frac{8}{x}$ for y in the first equation and solve the resulting equation.

$$x^2 - \left(\frac{8}{x}\right)^2 = 12$$
$$x^2 - \frac{64}{x^2} = 12$$
$$x^4 - 64 = 12x^2$$
$$x^4 - 12x^2 - 64 = 0 \qquad \text{This equation is quadratic in } x^2.$$
$$(x^2 - 16)(x^2 + 4) = 0$$
$$x^2 - 16 = 0 \qquad x^2 + 4 \text{ has no real solutions.}$$
$$x = 4 \quad \text{or} \quad x = -4$$

Substitute the x-values in $y = \frac{8}{x}$.

If $x = 4$, $y = \frac{8}{4} = 2$.

If $x = -4$, $y = \frac{8}{-4} = -2$.

The solution set is $\{(4, 2), (-4, -2)\}$. ***Answer***

The graph is shown at the right.

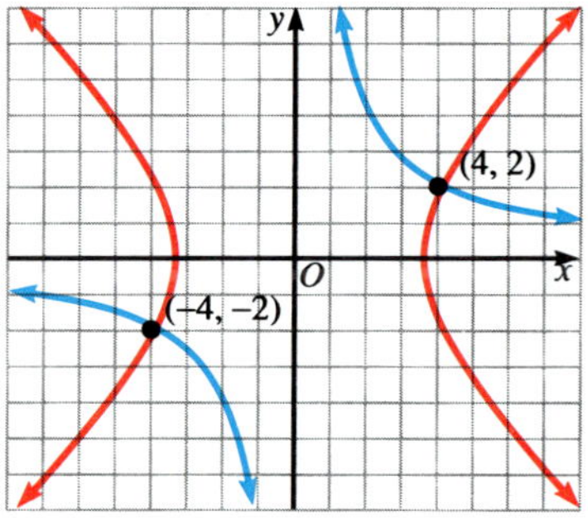

Example 3 Solve this system: $x^2 + 2y^2 = 23$
$2x^2 - y^2 = 1$

Solution (Linear-Combination Method) Multiply the second equation by 2 and add the two equations. Then solve the resulting equation.

$$x^2 + 2y^2 = 23$$
$$4x^2 - 2y^2 = 2$$
$$5x^2 = 25$$
$$x^2 = 5$$
$$x = \sqrt{5} \quad \text{or} \quad x = -\sqrt{5}$$

Substitute $\sqrt{5}$ and $-\sqrt{5}$ for x in $x^2 + 2y^2 = 23$ to find the corresponding values of y.

If $x = \sqrt{5}$:

$$(\sqrt{5})^2 + 2y^2 = 23$$
$$2y^2 = 18$$
$$y^2 = 9$$
$$y = 3 \quad \text{or} \quad y = -3$$

If $x = -\sqrt{5}$:

$$(-\sqrt{5})^2 + 2y^2 = 23$$
$$2y^2 = 18$$
$$y^2 = 9$$
$$y = 3 \quad \text{or} \quad y = -3$$

$\therefore$ the solution set is $\{(\sqrt{5}, 3), (\sqrt{5}, -3), (-\sqrt{5}, 3), (-\sqrt{5}, -3)\}$.

As the preceding examples show, substitution is usually the more appropriate method for solving a system consisting of a linear and a quadratic equation. When a system's equations are both quadratic, either the substitution or the linear-combination method may be used.

Oral Exercises

Which of the given ordered pairs are solutions of the system?

1. $xy + 5 = 0$; $2x + y + 3 = 0$ — (5, −1), (−5, 1), (1, −5), (−1, 5) (1, −5)
2. $x^2 + y^2 = 20$; $x + y + 2 = 0$ — (4, 2), (−4, 2), (2, 4), (−2, 4) (−4, 2)
3. $x^2 + 4y^2 = 25$; $y^2 - x = 1$ — (3, 2), (−3, 2), (−3, −2), (3, −2) (3, 2), (3, −2)

Written Exercises

6. $\left\{\left(-\frac{39}{5}, -\frac{2}{5}\right), (5, 6)\right\}$ 9. {(−43, 9), (5, −3)}
11. {(4, 3), (4, −3), (−4, 3), (−4, −3)} 12. No real solution

Find the real solutions, if any, of each system. You may wish to check your answers visually on a computer or a graphing calculator.

A

1. $x^2 - y = 5$; $2x + y = 3$ {(−4, 11), (2, −1)}
2. $x = y^2 - 9$; $x - 4y - 12 = 0$ {(40, 7), (0, −3)}
3. $y = x^2$; $x^2 + y^2 = 12$ $\{(\sqrt{3}, 3), (-\sqrt{3}, 3)\}$
4. $y^2 = 2x$; $x^2 + y^2 = 8$ {(2, 2), (2, −2)}
5. $x^2 - y^2 = 15$; $x + y = 1$ {(8, −7)}
6. $x^2 + y^2 = 61$; $x - 2y + 7 = 0$
7. $xy = 8$; $x + y = 6$ {(2, 4), (4, 2)}
8. $xy + 6 = 0$; $x - y = 5$ {(3, −2), (2, −3)}
9. $2y^2 + 3x - 33$; $x + 4y + 7 = 0$
10. $4x^2 - y^2 + 12 = 0$; $x + y = 3$ {(−1, 4)}
11. $x^2 + y^2 = 25$; $x^2 - y^2 = 7$
12. $8x^2 + y^2 = 25$; $8x^2 - y^2 = 39$

13. $\{(\sqrt{21}, 2), (\sqrt{21}, -2), (-\sqrt{21}, 2), (-\sqrt{21}, -2)\}$ 14. {(2, 2), (2, −2), (−2, 2), (−2, −2)}

B

13. $2x^2 - 3y^2 = 30$; $x^2 + y^2 = 25$
14. $x^2 + 2y^2 = 12$; $3x^2 - y^2 = 8$
15. $5x^2 + 3y^2 = 7$; $3x^2 - 7y^2 = 13$
16. $9x^2 + 9y^2 = 1$; $x = y^2 + 1$ No real solution
17. $x^2 + y^2 = 13$; $xy + 6 = 0$
18. $2x^2 - y^2 = 7$; $xy = 3$

15. No real solution

17. {(3, −2), (−3, 2), (2, −3), (−2, 3)} 18. $\left\{\left(\frac{3\sqrt{2}}{2}, \sqrt{2}\right), \left(-\frac{3\sqrt{2}}{2}, -\sqrt{2}\right)\right\}$

3. $x^2 + y^2 = 7$
$4x^2 + 3y^2 = 24$
(Linear-Combination Method)
$$-3x^2 - 3y^2 = -21$$
$$4x^2 + 3y^2 = 24$$
$$x^2 = 3$$
$$x = \pm\sqrt{3}$$
$$(\pm\sqrt{3})^2 + y^2 = 7$$
$$y^2 = 4$$
$$y = \pm 2$$
$(\sqrt{3}, 2)$, $(\sqrt{3}, -2)$, $(-\sqrt{3}, 2)$, and $(-\sqrt{3}, -2)$

Check for Understanding

Here is a suggested use of the Oral Exercises to check students' understanding as you teach the lesson.
Oral Exs. 1–3: use before Example 1.

Guided Practice

Find the real solutions, if any, of each system.

1. $x^2 + y^2 = 16$; $x^2 - y^2 = 20$ no sol.
2. $4x^2 + y^2 = 25$; $8x - 3y = 25$ {(2, −3)}
3. $x^2 + y^2 = 24$; $x^2 + 5y = 0$ $\{(\sqrt{15}, -3), (-\sqrt{15}, -3)\}$
4. $xy + 4 = 0$; $x - y = 5$ {(1, −4), (4, −1)}

Using a Computer or a Graphing Calculator

You may wish to have students confirm their solutions to Exercises 1–20 using a computer or a graphing calculator.

Support Materials
Disk for Algebra
Menu Items: Conics Plotter
Function Plotter

Summarizing the Lesson

In this lesson students learned to solve quadratic systems algebraically. Ask the students to answer these questions: could Example 2 on page 440 have been solved using the linear-combination method? (No) Could Example 3 on page 440 have been solved using the substitution method? (Yes)

Suggested Assignments

Average Algebra
Day 1: 441/1–19 odd
S 438/17, 19
Day 2: 441/12–18 even
442/*P*: 1–10

Average Alg. and Trig.
Day 1: 441/1–19 odd
442/*P*: 1–6
Day 2: 442/*P*: 7–12
Assign with Lesson 9-9.

Extended Algebra
Day 1: 441/1–25 odd
442/*P*: 1–6
Day 2: 442/*P*: 7–13
Assign with Lesson 9-9.

Extended Alg. and Trig.
Day 1: 441/1–25 odd
442/*P*: 1–6
Day 2: 442/*P*: 7–13
Assign with Lesson 9-9.

Supplementary Materials

Study Guide pp. 151–152
Practice Master 53

Exercises 19 and 20 lead to cubic equations. Use the methods of Lesson 8-7 to find their real roots.

19. $y^2 = x + 7$ {(2, 3), (−3, −2), (−6, −1)}
$xy = 6$

20. $y = x^2 - 1$ {(−2, 3)}
$xy + 6 = 0$

Find the square roots of each complex number.

Sample $12 + 16i$

Solution Let a square root of $12 + 16i$ be $x + yi$, where x and y are real. Then $(x + yi)^2 = 12 + 16i$, or $(x^2 - y^2) + 2xyi = 12 + 16i$. Equating the real and imaginary parts gives this system:

$$x^2 - y^2 = 12$$
$$2xy = 16$$

This system is equivalent to the one in Example 2, page 440, and thus has the solutions (4, 2) and (−4, −2). Therefore, the square roots of $12 + 16i$ are $4 + 2i$ and $-4 - 2i$.

C **21.** $3 + 4i$ $2 + i$, $-2 - i$ **22.** $7 - 24i$ $4 - 3i$, $-4 + 3i$ **23.** $5 + 12i$ $3 + 2i$, $-3 - 2i$

24. $-5 - 12i$ $2 - 3i$, $-2 + 3i$ **25.** $-7 + 24i$ $3 + 4i$, $-3 - 4i$ **26.** $12 - 16i$ $4 - 2i$, $-4 + 2i$

Problems

Solve.

A **1.** The sum of two numbers is 16, and the sum of their squares is 146. Find the numbers. 5 and 11

2. The product of two numbers is 1, and the difference of their squares is $\frac{15}{4}$. Find the numbers. 2 and $\frac{1}{2}$; −2 and $-\frac{1}{2}$

3. The fence around a rectangular piece of property is 156 m long. If the area of the property is 1505 m^2, find the dimensions of the property. 35 m by 43 m

4. An ellipse with center at the origin and horizontal major axis is to fit snugly inside a rectangle that has its longer sides horizontal. The area of the rectangle is 12 square units, and the perimeter is 14 units. Find an equation for the ellipse. $\frac{x^2}{4} + \frac{4y^2}{9} = 1$

5. Find the dimensions of a rectangle having perimeter 34 ft and a diagonal of length 13 ft. 5 ft by 12 ft

6. Find the length of the legs of a right triangle having perimeter 56 m if the hypotenuse is 25 m. 7 m and 24 m

7. The product of a two-digit number and its tens digit is 285. The units digit is two more than the tens digit. Find the original number. 57

8. Find the dimensions of a rectangle that has area 10 and a diagonal of length 5. Leave your answer in terms of radicals. $\sqrt{5}$ by $2\sqrt{5}$

B **9.** A rectangular plot of land having area 1350 m^2 is to be enclosed and divided into two parts, as shown. Find the dimensions of the plot if the total length of fencing used is 180 m. 30 m by 45 m

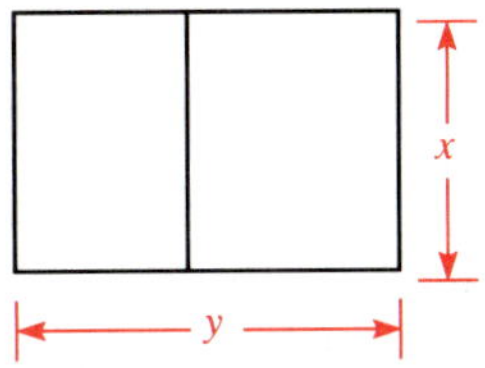

Ex. 9

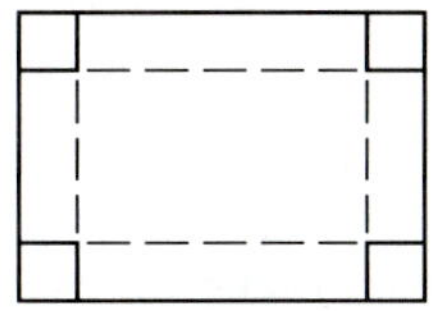

Ex. 10

10. Four squares, each with sides 4 cm long, are cut from the corners of a rectangular piece of cardboard having area 560 cm^2. The flaps are then bent up to form an open-topped box having volume 960 cm^3. Find the dimensions of the original piece of cardboard. 20 cm by 28 cm

11. Two people part company and walk along perpendicular paths. One person walks 1 km/h faster than the other. They are 6 km apart after one hour. Find the rate at which each person walks. Give your answers to the nearest tenth. 3.7 km/h and 4.7 km/h

12. A 20 m ladder and a 15 m ladder were leaned against a building. The bottom of the longer ladder was 7 m farther from the building than the bottom of the shorter ladder, but both ladders reached the same distance up the building. Find the distance. 12 m

C **13.** Find the point on the circle with equation $x^2 + y^2 = 1$ that is closest to the point (4, 3). $\left(\frac{4}{5}, \frac{3}{5}\right)$

14. From the top of a vertical canyon wall 300 m high, a person throws a rock with a speed of 20 m/s toward the top of the opposite side of the canyon, which is at the same elevation, but is 510 m away. According to the laws of physics, the rock follows a path described approximately by the equation

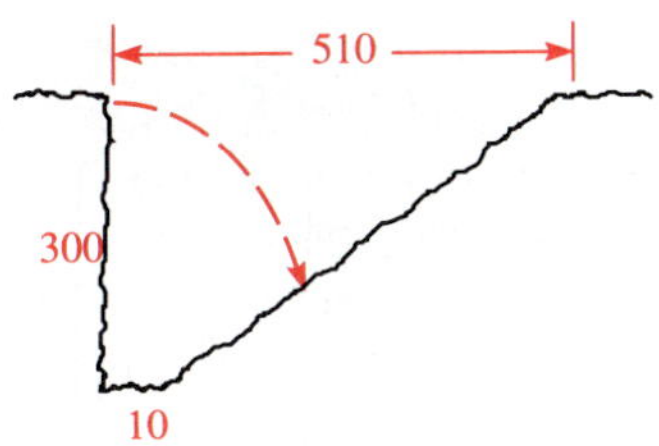

$$y = 300 - \frac{x^2}{80}$$

where y is the rock's height in meters above the canyon floor and x is the rock's horizontal distance in meters to the right of the vertical canyon wall. The opposite side of the canyon is steeply sloping and becomes level at a point 10 m from the base of the vertical wall. When the rock strikes the sloping side of the canyon, how far is it from where it was thrown? Express your answer rounded to the nearest tenth of a meter. (*Hint:* Introduce a coordinate system whose origin is the base of the vertical wall. Write an equation for the sloping side, and find a common solution of this equation and the equation of the rock's path.) 262.4 m

Additional Answers
Written Exercises

(continued from p. 431)

10. a.

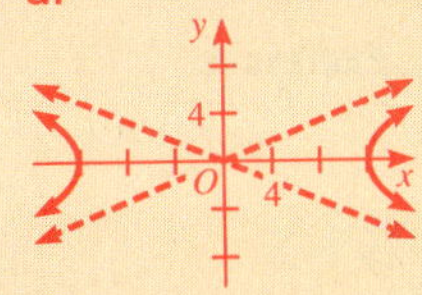

12. a.

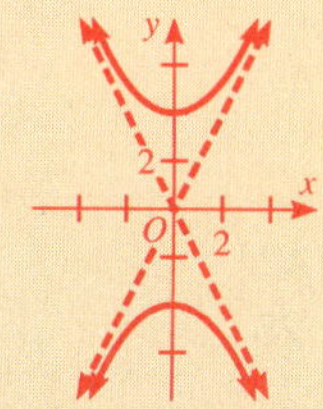

20.

22.

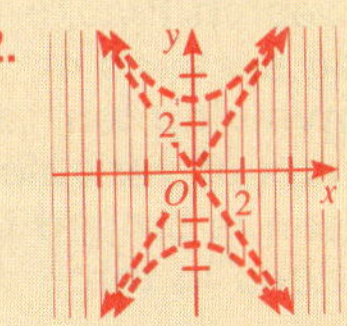

24.

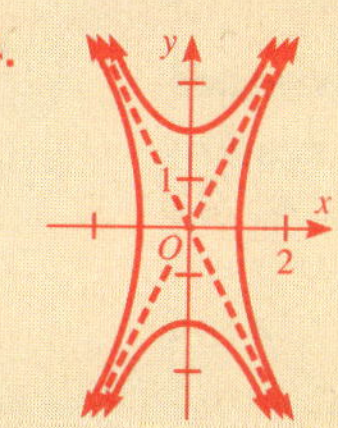

26.

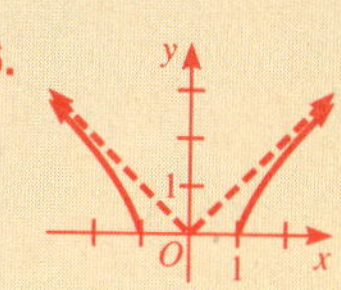

28.

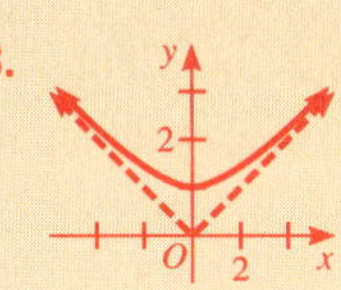

Teaching Suggestions p. T116

Suggested Extensions p. T116

Warm-Up Exercises

Find the real solutions, if any.

1. $3x - 2y = 8$
   $x + 4y = -2$ $\{(2, -1)\}$
2. $y^2 = x + 4$
   $x + 3y = 6$ $\{(0, 2), (21, -5)\}$
3. $xy = 10$
   $2x + y = 12$ $\{(5, 2), (1, 10)\}$
4. $x^2 + y^2 = 25$
   $x^2 - y^2 = 7$ $\{(4, 3), (4, -3), (-4, 3), (-4, -3)\}$

Motivating the Lesson

Tell students that there are problems in engineering and scientific research that require solving systems of equations involving tens or even hundreds of variables. Today's lesson is limited to linear equations in three variables.

Chalkboard Examples

Solve each system.

1. $3x + y + 2z = 8$
   $2y - z = 7$
   $z = 3$

$2y - 3 = 7$
$2y = 10$
$y = 5$
$3x + 5 + 2(3) = 8$
$3x = -3$
$x = -1$
$(-1, 5, 3)$

9-9 Systems of Linear Equations in Three Variables

Objective To solve systems of linear equations in three variables.

Just as the ordered pair $(x, y) = (2, -4)$ is a solution of the equation $3x + 2y = -2$, so the **ordered triple** $(x, y, z) = (2, -4, 3)$ is a solution of the equation

$$3x + 2y - z = -5$$

because $3(2) + 2(-4) - 3 = -5$ is a true statement.

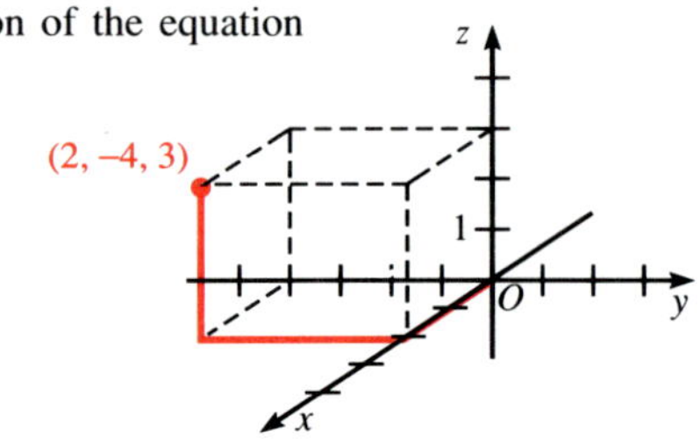

Any equation of the form $Ax + By + Cz = D$, where A, B, C, and D are real numbers with A, B, and C not all zero, is called a **linear equation in three variables.** Although such equations are called "linear," their graphs are planes, not lines.

Solutions of equations in x, y, and z can be graphed in a three-dimensional coordinate system having an x-axis, a y-axis, and a z-axis that meet at right angles at a common point O, the origin. The graph of the *ordered triple* $(2, -4, 3)$ is shown in the diagram above.

To graph a *system* of three linear equations in three variables, you draw the three graphs in the same coordinate system. Although any two of the planes may intersect, coincide, or be parallel, the graph of the system's solution consists only of the points that *all three* planes have in common. Some of the ways that three planes can intersect are shown below.

(a)

(b)

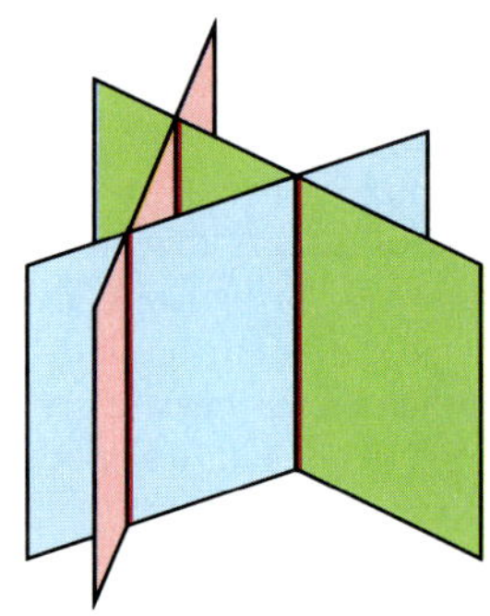

(c)

(d)

Algebraically, these geometrical relationships mean:

(a) When the three planes intersect in only *one* point, the system has *one solution*.

(b)–(c) When the three planes do *not* contain a common point, the system has *no solution*. (This is also true when the three planes are parallel.)

(d) When the three planes intersect in a common line, the system has *infinitely many solutions*. (This is also true when the three planes coincide.)

You can solve linear systems algebraically by substitution or linear combinations, as you did in Lesson 3-5.

Example 1 Solve this system:

$$\begin{aligned} 2x + 3y + 2z &= 13 \\ 2y + z &= 1 \\ z &= 3 \end{aligned}$$

Solution (Substitution Method) The third equation gives the value of z: $z = 3$. Substitute 3 for z in the second equation and solve for y.

$$\begin{aligned} 2y + z &= 1 \\ 2y + 3 &= 1 \\ 2y &= -2 \\ y &= -1 \end{aligned}$$

Substitute 3 for z and -1 for y in the first equation and solve for x.

$$\begin{aligned} 2x + 3y + 2z &= 13 \\ 2x + 3(-1) + 2(3) &= 13 \\ 2x - 3 + 6 &= 13 \\ 2x &= 10 \\ x &= 5 \end{aligned}$$

The check is left for you.

$\therefore$ the solution of the system is $(5, -1, 3)$. ***Answer***

Because its shape suggests a triangle, the system in Example 1 is called a **triangular system.** Since triangular systems are easy to solve, a good way to solve any linear system is to transform it into an equivalent triangular system and then to proceed as in Example 1. This method is called *Gaussian elimination* in honor of Karl Friedrich Gauss (1777–1855), a very famous mathematician.

$$\begin{aligned} 2x + 3y + 2z &= 13 \\ 2y + z &= 1 \\ z &= 3 \end{aligned}$$

Example 2 Solve this system:

$$\left.\begin{aligned} x + y - 2z &= 7 \\ -x + 4y + 3z &= 2 \\ 2x - 3y + 2z &= -2 \end{aligned}\right\} \quad (1)$$

2. $x + y + z = 5$
$2x + z = 8$
$3y + 2z = 5$

$x + y + z = 5$
$-2y - z = -2$
$3y + 2z = 5$

$x + y + z = 5$
$-2y - z = -2$
$z = 4$

$-2y - 4 = -2$
$-2y = 2$
$y = -1$

$x - 1 + 4 = 5$
$x = 2$

$(2, -1, 4)$

3. $x - y - z = 1$
$y - 2z = 0$
$x + 2y - 7z = 1$

$x - y - z = 1$
$x + 2y - 7z = 1$
$y - 2z = 0$

$x - y - z = 1$
$3y - 6z = 0$
$y - 2z = 0$

$x - y - z = 1$
$3y - 6z = 0$
$0 = 0$

$3y - 6z = 0$
$y = 2z$

$x - 2z - z = 1$
$x = 3z + 1$

$(3z + 1, 2z, z)$ for any number z

Common Errors

Transforming a system of three linear equations in three variables with no zero coefficients into triangular form can be troublesome for some students. When reteaching, remind students to first transform two of the given equations into equations consisting of the same two variables and then transform one of the equations in two variables into an equation consisting of one variable.

Check for Understanding

Here is a suggested use of the Oral Exercises to check students' understanding as you teach the lesson.

Oral Exs. 1–8: use after the initial discussion of an ordered triple.

Oral Exs. 9–12: use after Example 2.

Guided Practice

Solve each system.

1. $x + 2y + 2z = 15$
   $2y + z = 18$
   $z = 2$
   $(-5, 8, 2)$
2. $2x + y + 3z = 9$
   $x - 2y + z = 8$
   $-4x + 3y + 2z = -4$
   $(1, -2, 3)$
3. $x + y + 3z = 7$
   $2x - 4y + z = 7$
   $3x + 5y - 4z = -6$
   $\left(\frac{3}{2}, -\frac{1}{2}, 2\right)$
4. $x - 2y = -(9 + z)$
   $2(x + y) = 5 + 3z$
   $3y - x = 7 - 4z$
   $(-2, 3, -1)$
5. A pen and a pencil cost 30¢; two pencils and an eraser cost 39¢; a pencil, an eraser, and two pens cost 63¢. How much is each item? pencil, 12¢; eraser, 15¢; pen, 18¢

Summarizing the Lesson

In this lesson students learned to solve systems of linear equations in three variables using the triangular-system method. Ask the students to describe the geometric interpretation of a system of three linear equations having (a) one solution, (b) no solution, and (c) infinitely many solutions.

Solution (Triangular-System Method)

In (1), add the first equation to the second. Then multiply the first equation by -2 and add the result to the third equation.

$$\left.\begin{aligned} x + y - 2z &= 7 \\ 5y + z &= 9 \\ -5y + 6z &= -16 \end{aligned}\right\} \quad (2)$$

In (2), add the second and third equations together.

$$\left.\begin{aligned} x + y - 2z &= 7 \\ 5y + z &= 9 \\ 7z &= -7 \end{aligned}\right\} \quad (3)$$

When you solve the triangular system in (3) you get $x = 3$, $y = 2$, and $z = -1$. The solution is $(3, 2, -1)$. ***Answer***

The system in Example 2 has a unique solution. Example 3 shows a system that has infinitely many solutions.

Example 3 Solve the system:

$$\left.\begin{aligned} x - 2y - z &= -1 \\ 2x - y + z &= 1 \\ x + 4y + 5z &= 5 \end{aligned}\right\} \quad (1)$$

Solution (Triangular-System Method)

In (1), multiply the first equation by -2 and add it to the second equation. Then subtract the third equation from the first. Here is the resulting system:

$$\left.\begin{aligned} x - 2y - z &= -1 \\ 3y + 3z &= 3 \\ -6y - 6z &= -6 \end{aligned}\right\} \quad (2)$$

In (2), add twice the second equation to the third.

$$\left.\begin{aligned} x - 2y - z &= -1 \\ 3y + 3z &= 3 \\ 0 &= 0 \end{aligned}\right\} \quad (3)$$

Solve $3y + 3z = 3$ for y and substitute for y in the first equation in (3)

$$\begin{aligned} 3y + 3z &= 3 \\ y + z &= 1 \\ y &= 1 - z \end{aligned}$$

$$\begin{aligned} x - 2(1 - z) - z &= -1 \\ x - 2 + 2z - z &= -1 \\ x &= 1 - z \end{aligned}$$

You can check that, for any number z, the triple $(1 - z, 1 - z, z)$ is a solution of the given system. If, in this general expression, you assign z the values 0, 2, and -3, you obtain the particular solutions $(1, 1, 0)$, $(-1, -1, 2)$, and $(4, 4, -3)$, respectively. ***Answer***

Recall from Chapter 3 that a linear system having infinitely many solutions, such as the one in Example 3, is said to be *dependent*. If the second equation in Example 3 had been $x - 2y - z = 2$, then the system would have had no solution, since the first two equations would have had no common solution. Such a system is said to be *inconsistent*.

Other methods can be used to solve systems of equations. One such method involves using numbers in a rectangular array called a **matrix.** You will learn about matrices in Chapter 16.

Oral Exercises

Which of the ordered triples, if any, is a solution of the equation?

1. $x + y - z = 7$; $(4, -2, 5)$, $(3, 0, -4)$ (3, 0, −4)
2. $2x + y - z = 0$; $(-1, 2, 3)$, $(0, 2, 2)$ (0, 2, 2)
3. $4x - y + 2z = 8$; $(1, 2, -3)$, $(2, -3, -1)$ none
4. $x - 3y + 4z = 6$; $(-2, 0, 2)$, $(2, -2, 0)$ (−2, 0, 2)

Find two ordered triples that satisfy the equation. Answers may vary.

5. $x + y + z = 7$ (7, 0, 0), (2, 3, 2)
6. $x + y + z = -5$ (0, 0, −5), (−1, −2, −2)
7. $y + z = 6$ (0, 2, 4), (0, 4, 2)
8. $3x + z = 0$ (0, 0, 0), (1, 0, −3)

Tell how to combine the given equations to obtain an equation in which the coefficient of *x* is 0. What is the new coefficient of *y*? Answers may vary.

9. $x + 3y - 2z = 5$; $-x + y + 4z = 6$ Add; 4
10. $x + 2y - z = 2$; $2x + 5y - 4z = 20$ Subtract twice the first equation from the second; 1
11. $3x + y - 2z = 6$; $2x + 2y + z = 4$ Subtract twice the first equation from 3 times the second; 4
12. $7x - y + 4z = 3$; $-2x + 3y - z = 5$ Add twice the first equation to 7 times the second; 19

Written Exercises

7. {(−1, 0, −4)} 8. {(−2, 1, −1)} 9. {(−1, −3, 5)} 10. {(−3, 1, −2)} 11. {(1, 2, 2)} 12. {(3, −3, 4)}

Solve each system.

A

1. $x + y - 3z = 10$; $y + z = 12$; $z = -2$ {(−10, 14, −2)}
2. $x + 2y + 3z = -1$; $3y + 2z = -1$; $z = -2$ {(3, 1, −2)}
3. $x + 2y - z = 3$; $3y + z = -10$; $-2z = 8$ {(3, −2, −4)}
4. $5x - 3y + 4z = 4$; $4y - 3z = 10$; $3z = -6$ {(3, 1, −2)}
5. $2z + 3y - x = 0$; $4y - x = -4$; $-2x = 8$ {(−4, −2, 1)}
6. $z + 3x + y = 3$; $2x + 3y = 10$; $2y = 8$ {(−1, 4, 2)}
7. $2x - y - z = 2$; $x - 5y + 3z = -13$; $-2x - 2y + z = -2$
8. $x + 3y + 2z = -1$; $-3x - 2y + z = 3$; $2x - y + 3z = -8$
9. $2x + y + 3z = 10$; $x - 2y + z = 10$; $-4x + 3y + 2z = 5$
10. $2x + 5y + 2z = -5$; $-3x + 3y + 5z = 2$; $x + 4y - z = 3$
11. $2x + 3y = 6 + z$; $x - 2y = -1 - z$; $3x + y = -1 + 3z$
12. $2x - y = 3z - 3$; $3x + 2y = z - 1$; $x + 3y = z - 10$

Suggested Assignments

Average Algebra
Day 1: 447/1–21 odd
S 449/Mixed Review
Day 2: 448/20, 22; *P*: 1–9, 11

Average Alg. and Trig.
Day 1: 447/1–17 odd
Assign with Lesson 9-8.
Day 2: 448/19, 21; *P*: 1–9, 11
S 449/Mixed Review

Extended Algebra
Day 1: 447/1–17 odd
Assign with Lesson 9-8.
Day 2: 448/19–25 odd
P: 1–9, 11
S 449/Mixed Review

Extended Algebra
Day 1: 447/1–17 odd
Assign with Lesson 9-8.
Day 2: 448/19–25 odd
P: 1–9, 11
S 449/Mixed Review

Supplementary Materials

Study Guide	pp. 153–154
Practice Master	54
Test Master	37
Computer Activity	21
Resource Book	p. 140
Overhead Visual	8

Additional Answers Written Exercises

21. $x - 2y - z = -1$
$2x - y + z = 3$
$x + 4y + 5z = 5$
Subtract twice the first equation from the second and subtract the first equation from the third.
$x - 2y - z = -1$
$3y + 3z = 5$
$6y + 6z = 6$
Subtract twice the second equation from the third.
$x - 2y - z = -1$
$3y + 3z = 5$
$0 = -4$
Since $0 \neq -4$, the system is inconsistent.

22. $x + 2y - 2z = 3$
$x + 3y - 4z = 6$
$4x + 5y - 2z = 6$
Subtract the first equation from the second and subtract 4 times the first equation from the third.
$x + 2y - 2z = 3$
$y - 2z = 3$
$-3y + 6z = -6$
Add 3 times the second equation to the third.
$x + 2y - 2z = 3$
$y - 2z = 3$
$0 = 3$
Since $0 \neq 3$, the system is inconsistent.

B

13. $\{(1, 3, 2)\}$
$$\begin{aligned} 3a - 2b + 2c &= 1 \\ 2a + 5b - 5c &= 7 \\ 4a - 3b + c &= -3 \end{aligned}$$

14. $\{(5, -5, 3)\}$
$$\begin{aligned} 2x + 2y + z &= 3 \\ 3x + 2y - 2z &= -1 \\ 5x - 2y - 6z &= 17 \end{aligned}$$

15. $\{(-1, 2, 3)\}$
$$\begin{aligned} 3u + 2v + w &= 4 \\ 5u + 3v - w &= -2 \\ 2u + w &= 1 \end{aligned}$$

16.
$$\begin{aligned} a + 3b + 4c &= 6 \\ a - 2b + c &= 10 \\ 2a + 3b - c &= 0 \end{aligned}$$
$\{(4, -2, 2)\}$

17.
$$\begin{aligned} \frac{1}{x} - \frac{2}{y} + \frac{3}{z} &= -3 \\ \frac{2}{x} - \frac{3}{y} - \frac{1}{z} &= 7 \\ \frac{3}{x} + \frac{1}{y} - \frac{2}{z} &= 6 \end{aligned}$$
$\left\{\left(1, -1, -\frac{1}{2}\right)\right\}$

18.
$$\begin{aligned} \frac{1}{r} + \frac{3}{s} - \frac{2}{t} &= 1 \\ \frac{2}{r} + \frac{3}{s} - \frac{4}{t} &= 1 \\ \frac{1}{r} - \frac{6}{s} - \frac{6}{t} &= 0 \end{aligned}$$
$\{(-1, 3, -2)\}$

Show that each of the following systems is dependent by (a) finding a general expression for all of its solutions and (b) using this expression to find three particular solutions.

19.
$$\begin{aligned} x - 2y + 3z &= 1 \\ x + y - 3z &= 7 \\ 3x - 4y + 5z &= 7 \end{aligned}$$
a. $\{(5 + z, 2 + 2z, z)\}$
b. Examples: (5, 2, 0), (8, 8, 3), (3, −2, −2)

20.
$$\begin{aligned} x + 2y - 2z &= 3 \\ x + 3y - 4z &= 6 \\ 4x + 5y - 2z &= 3 \end{aligned}$$
a. $\{(-3 - 2z, 3 + 2z, z)\}$
b. Examples: (−5, 5, 1), (−3, 3, 0), (−9, 9, 3)

Show that each of the following systems is inconsistent.

C

21.
$$\begin{aligned} x - 2y - z &= -1 \\ 2x - y + z &= 3 \\ x + 4y + 5z &= 5 \end{aligned}$$

22.
$$\begin{aligned} x + 2y - 2z &= 3 \\ x + 3y - 4z &= 6 \\ 4x + 5y - 2z &= 6 \end{aligned}$$

The following systems are said to be *homogeneous* because the constant terms are all 0. Each of them has (0, 0, 0) as an obvious solution. Determine whether there are any other solutions, and, if there are, find a general expression for them.

23.
$$\begin{aligned} x - 2z &= 0 \\ 2x - y + z &= 0 \\ x - y + 3z &= 0 \end{aligned}$$
$\{(2z, 5z, z)\}$

24.
$$\begin{aligned} x + 2y - 2z &= 0 \\ 2x + 5y + 2z &= 0 \\ 3x + 4y - 2z &= 0 \end{aligned}$$
(0, 0, 0) is the only solution.

25.
$$\begin{aligned} x - 2y + 4z &= 0 \\ 3x - y + 2z &= 0 \\ x + 3y - 6z &= 0 \end{aligned}$$
$\{(0, 2z, z)\}$

Problems

Solve.

A

1. Barbara has nickels, dimes, and quarters worth \$2.35 in her purse. The number of dimes is three less than the sum of the number of nickels and quarters. How many of each type of coin does she have if there are 19 coins in all? 6 nickels, 8 dimes, 5 quarters

2. David paid \$9.50 for some 15¢, 25¢, and 45¢ stamps. He bought 38 stamps in all. The number of 25¢ stamps was 8 more than twice the number of 45¢ stamps. How many stamps of each type did David buy? 12 15¢ stamps, 20 25¢ stamps, 6 45¢ stamps

3. When three large diamonds are weighed in pairs, the masses of the pairs are found to be 6 carats, 10 carats, and 12 carats. Find the mass of each diamond. 2 carats, 4 carats, 8 carats

4. A chemist has three samples of unknown mass. Each pair of the samples is balanced against the third sample and one of three known masses. Find the unknown masses if the known ones are 9 g, 15 g, and 27 g. 12 g, 18 g, 21 g

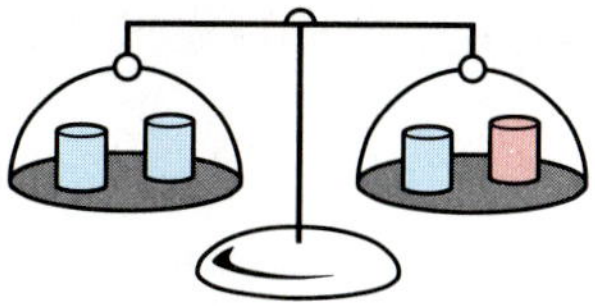

5. In a certain triangle, the measure of the largest angle is 20° more than twice the measure of the smallest angle. Five times the measure of the smallest angle equals the sum of the measures of the other two angles. Find the measures of the three angles. (*Hint:* Remember that the sum of the measures of the angles of a triangle is 180°.) 30°, 70°, 80°

6. The sum of the length, width, and height of a rectangular box is 75 cm. The length is twice the sum of the width and height, and twice the width exceeds the height by 5 cm. Find the dimensions of the box. 50 cm by 10 cm by 15 cm

B 7. The Charity Ball Committee sold \$10 patron tickets, \$5 sponsor tickets, and \$2.50 donor tickets. The number of donor tickets sold was 24 more than the number of sponsor and patron tickets combined. There were 326 tickets sold, and the receipts totaled \$1432.50. How many tickets of each type were sold? 48 patrons, 103 sponsors, 175 donors

8. The sum of the digits of a 3-digit number is 21. The sum of the ten's and hundred's digits is 3 less than twice the unit's digit. The number is increased by 198 if the digits are reversed. Find the original number. 678

A parabola with vertical axis passes through the given points. Find its equation. (*Hint:* Substitute the coordinates of the given points in turn in $y = ax^2 + bx + c$. Then solve the resulting system for a, b, and c.)

9. $(-1, 4)$, $(2, 7)$, $(1, 0)$ $y = 3x^2 - 2x - 1$

10. $(1, 3)$, $(2, 10)$, $(-2, -6)$ $y = x^2 + 4x - 2$

A circle passes through the given points. Find its equation. (*Hint:* Use the equation $x^2 + y^2 + Cx + Dy + E = 0$.)

11. $(2, -1)$, $(4, -3)$, $(0, -3)$ $x^2 + y^2 - 4x + 6y + 9 = 0$

12. $(-6, -3)$, $(1, 4)$, $(2, 3)$ $x^2 + y^2 + 4x - 21 = 0$

Mixed Review Exercises

11. $\frac{y^4}{x^5}$

Find the real solutions of each system.

1. $x^2 - y^2 = 1$
$x^2 + y^2 = 3$
$\{(\sqrt{2}, 1), (\sqrt{2}, -1), (-\sqrt{2}, 1), (-\sqrt{2}, -1)\}$

2. $4x - y = 3$ $\{(2, 5)\}$
$x + 3y = 17$

3. $y^2 = x + 1$
$2x + y = 4$
$\left\{\left(\frac{5}{4}, \frac{3}{2}\right), (3, -2)\right\}$

Simplify.

4. $(3x^{-1}y^2)^{-2}$ $\frac{x^2}{9y^4}$

5. $3\sqrt{16x^3}$ $12x\sqrt{x}$

6. $(4x^3y)(-3x^2y^4)$ $-12x^5y^5$

7. $(2x\sqrt{3})^2$ $12x^2$

8. $\frac{10xy^{-2}}{6x^{-1}y^3}$ $\frac{5x^2}{3y^5}$

9. $\frac{18x^3y}{30x^2y^4}$ $\frac{3x}{5y^3}$

10. $\sqrt{\frac{27x^4}{4}}$ $\frac{3x^2\sqrt{3}}{2}$

11. $\left(\frac{x}{y^2}\right)^{-1} \cdot \left(\frac{y}{x^2}\right)^2$

Additional Answers Written Exercises

(continued from p. 435)

7. $\frac{(x-4)^2}{1} - \frac{(y+2)^2}{15} = 1$

8. $\frac{(y-7)^2}{4} - \frac{x^2}{5} = 1$

9. $\frac{(y+5)^2}{4} - \frac{(x-3)^2}{5} = 1$

10. $\frac{(x-2)^2}{9} - \frac{(y-3)^2}{40} = 1$

11. $\frac{(y+5)^2}{9} - \frac{(x-5)^2}{7} = 1$

12. $\frac{x^2}{9} - \frac{(y+4)^2}{7} = 1$

13. hyperbola; center: $(1, -3)$; foci: $(1 - \sqrt{5}, -3)$, $(1 + \sqrt{5}, -3)$

14. ellipse; center: $(-1, 1)$; foci: $(-1 - 2\sqrt{2}, 1)$, $(-1 + 2\sqrt{2}, 1)$

15. circle; center: $(3, 8)$

16. hyperbola; center: $(1, -3)$; foci: $(1 - \sqrt{10}, -3)$, $(1 + \sqrt{10}, -3)$

17. ellipse; center: $(-2, 3)$; foci: $(-6, 3)$, $(2, 3)$

18. hyperbola; center: $(-2, 1)$ foci: $(-2, -4)$, $(-2, 6)$

Quick Quiz

For each system, (a) graph the equations, (b) determine the number of real solutions, and (c) estimate each real solution to the nearest half unit.

1. $9x^2 + 4y^2 = 36$
   $x - 2y = 2$

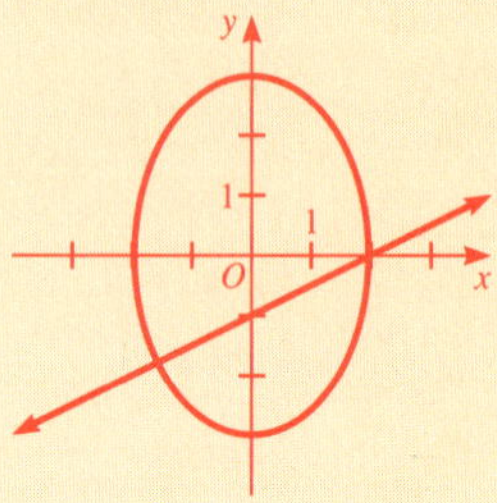

$(2, 0), (-1.5, -2.0)$

2. $x^2 - y^2 = 4$
   $x + 2 = y^2$

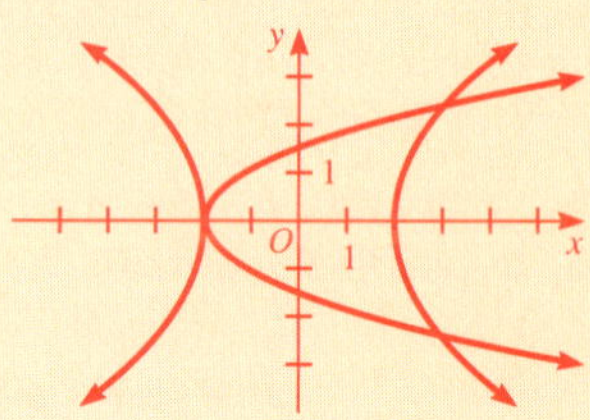

$(-2, 0); (3, 2.0), (3, -2.0)$

Solve each system algebraically.

3. $5y^2 - x^2 = 4$
   $2y = x + 3$
   $\{(-1, 1), (-29, -13)\}$

4. $5x^2 - 3y^2 = 5$
   $x^2 + 3y^2 = 91$
   $\{(4, 5), (4, -5), (-4, -5), (-4, 5)\}$

5. $y = 2x^2 - 3x - 3$
   $5x - y = 11$
   $\{(2, -1)\}$

6. $x + 2y + 3z = 2$
   $2x + 3y = 1$
   $x = 5$
   $\{(5, -3, 1)\}$

7. $3x + y + 2z = 5$
   $x - y + 4z = 12$
   $2x + y - 2z = -3$
   $\left\{\left(2, -4, \frac{3}{2}\right)\right\}$

Self-Test 3

Vocabulary quadratic system (p. 436); ordered triple (p. 444); triangular system (p. 445); linear equation in three variables (p. 444); matrix (p. 447)

For each system, (a) graph the equations, (b) determine the number of real solutions, and (c) estimate each real solution to the nearest half-unit.

1. $4x^2 - 9y^2 = 36$
$x^2 + y^2 = 25$

2. $x^2 = 4y$
$xy = 8$

3. $9y^2 + 16x^2 = 144$
$x^2 + y^2 = 9$ **Obj. 9-7, p. 436**

Solve each system algebraically.

4. $25x^2 + 16y^2 = 400$
$x^2 + y^2 = 25$ $\{(0, 5), (0, -5)\}$

5. $3x^2 - 2y^2 = 1$
$y = x - 2$ $\{(-9, -11), (1, -1)\}$ **Obj. 9-8, p. 439**

6. $3x + 2y - z = 1$
$4y + 3z = 5$
$z = 3$ $\{(2, -1, 3)\}$

7. $2x - y + z = 11$
$x + 2y + 3z = 8$
$3x - 4y - 5z = -2$ $\{(2, -3, 4)\}$ **Obj. 9-9, p. 444**

Check your answers with those at the back of the book.

Chapter Summary

1. The *distance* between the points $P_1(x_1, y_1)$ and $P_2(x_2, y_2)$ is given by
$$P_1P_2 = \sqrt{(x_2 - x_1)^2 + (y_2 - y_1)^2}.$$
The *midpoint* of the line segment P_1P_2 is
$$\left(\frac{x_1 + x_2}{2}, \frac{y_1 + y_2}{2}\right).$$

2. The *circle* with center (h, k) and radius r has the equation
$$(x - h)^2 + (y - k)^2 = r^2.$$

3. A *parabola* with vertex (h, k) has an equation of the form
$$y - k = a(x - h)^2 \text{ if its axis is vertical,}$$
$$x - h = a(y - k)^2 \text{ if its axis is horizontal,}$$
where $a = \frac{1}{4c}$. The distance between its vertex and focus is $|c|$.

4. The *ellipse* with center $(0, 0)$, foci $(-c, 0)$ and $(c, 0)$, and sum of focal radii $2a$ has the equation
$$\frac{x^2}{a^2} + \frac{y^2}{b^2} = 1, \quad \text{where } b^2 = a^2 - c^2.$$
If the foci are $(0, -c)$ and $(0, c)$, then the equation has the form
$$\frac{x^2}{b^2} + \frac{y^2}{a^2} = 1, \quad \text{where } b^2 = a^2 - c^2.$$

5. The *hyperbola* with center (0, 0), foci $(-c, 0)$ and $(c, 0)$, and difference of focal radii $2a$ has the equation

$$\frac{x^2}{a^2} - \frac{y^2}{b^2} = 1, \quad \text{where } b^2 = c^2 - a^2.$$

If the foci are $(0, c)$ and $(0, -c)$, then the equation has the form

$$\frac{y^2}{a^2} - \frac{x^2}{b^2} = 1, \quad \text{where } b^2 = c^2 - a^2.$$

6. If an ellipse (or hyperbola) having foci on a horizontal or vertical line has its center at (h, k), then its equation can be obtained from the one given in 4 (or 5) by replacing x by $x - h$ and y by $y - k$.

7. When the equation of a central conic is given in the form

$$Ax^2 + By^2 + Cx + Dy + E = 0,$$

you can complete the squares in x and y to identify the conic and find its center and its foci.

8. Graphical methods can be used to determine the number of real solutions of a quadratic system and to estimate these solutions. To obtain exact solutions, the *substitution method* or the *linear-combination method* often can be used.

9. A good way to solve any system of linear equations in three variables algebraically is to transform it into an equivalent *triangular system* by using linear combinations and then to solve the triangular system by substitution.

Chapter Review

Write the letter of the correct answer.

Exercises 1–3 refer to the points $P(-4, 3)$ and $Q(-6, -1)$.

1. Find the distance PQ. 9-1
a. 12 **b.** 6 **c.** $2\sqrt{26}$ **(d.)** $2\sqrt{5}$

2. Find the midpoint of $\overline{PQ}$.
a. (5, 1) **(b.)** $(-5, 1)$ **c.** $(-1, 1)$ **d.** $(-5, -1)$

3. If Q is the midpoint of $\overline{PM}$, find M.
a. $(-5, 1)$ **b.** $(-2, -4)$ **(c.)** $(-8, -5)$ **d.** (2, 4)

4. Find an equation of the circle having center $(4, -3)$ and radius 5. 9-2
a. $x^2 + y^2 - 8x + 6y - 25 = 0$ **(b.)** $x^2 + y^2 - 8x + 6y = 0$
c. $x^2 + y^2 + 8x - 6y = 0$ **d.** $x^2 + y^2 - 4x + 3y = 0$

5. Find the center and radius of the circle $x^2 + y^2 - 8x + 4y + 12 = 0$.
(a.) $(4, -2)$; $2\sqrt{2}$ **b.** $(-4, 2)$; $2\sqrt{2}$
c. $(4, -2)$; $2\sqrt{3}$ **d.** $(-4, 2)$; $2\sqrt{3}$

Quick Quiz for p. 435

1. Graph $9x^2 + y^2 = 9$ and find the foci.
$(0, 2\sqrt{2})$; $(0, -2\sqrt{2})$

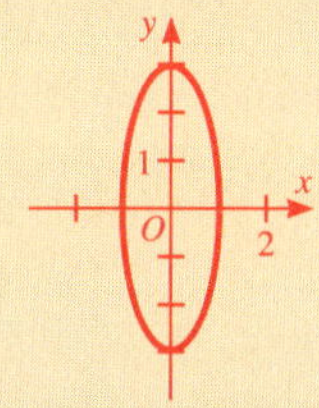

2. Find an equation of the ellipse with x-intercepts ± 5 and y-intercepts $\pm 2\sqrt{3}$. $\frac{x^2}{25} + \frac{y^2}{12} = 1$

3. Graph $25x^2 - 4y^2 = 100$.

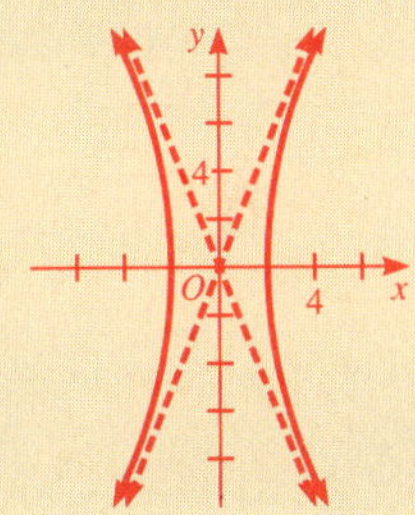

4. Find an equation of the hyperbola having foci (4, 0) and $(-4, 0)$ and difference of focal radii 6.
$\frac{x^2}{9} - \frac{y^2}{7} = 1$

5. Find an equation of the ellipse with foci (6, 2) and $(-10, 2)$ and with 20 as the sum of the focal radii.
$\frac{(x + 2)^2}{100} + \frac{(y - 2)^2}{36} = 1$

Identify each conic.

6. $4x^2 + y^2 - 8x + 6y - 23 = 0$
$\frac{(x - 1)^2}{9} + \frac{(y + 3)^2}{36} = 1$; ellipse

7. $9x^2 - 16y^2 - 32y + 128 = 0$
$\frac{(y + 1)^2}{9} - \frac{x^2}{16} = 1$; hyperbola

Additional Answers
Written Exercises
(continued from p. 438)

6.
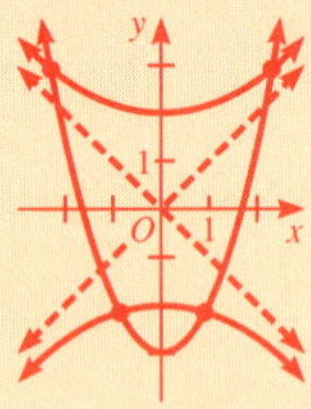

8.
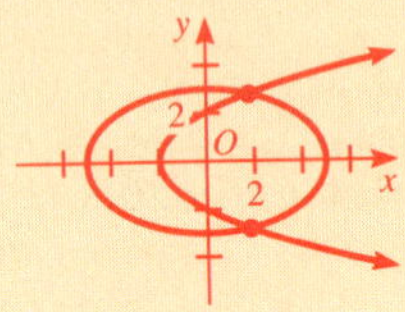

10. none

12. $\{(2.5, 3), (-2.5, 3), (1, -2), (-1, -2)\}$

14. $\{(2, 3), (2, -3)\}$

16.
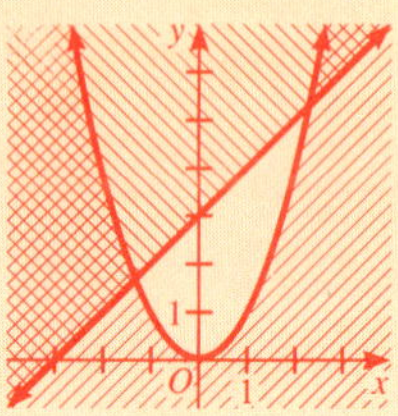

18.
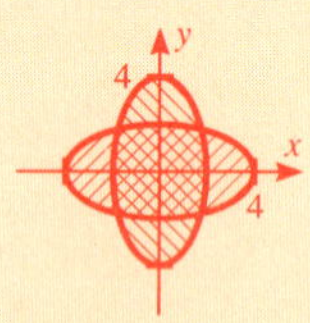

20. c.
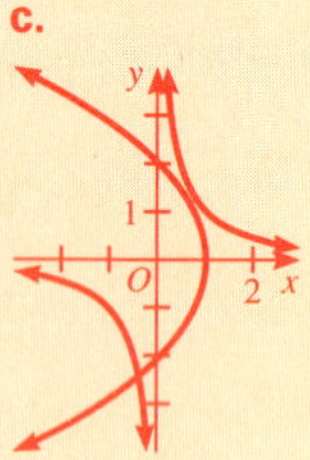

6. Find an equation of the circle shown at the right.
a. $x^2 + y^2 - 8x + 8y + 57 = 0$
b. $x^2 + y^2 - 8x + 8y + 7 = 0$
c. $x^2 + y^2 + 8x - 8y + 57 = 0$
d. $x^2 + y^2 + 8x - 8y + 7 = 0$

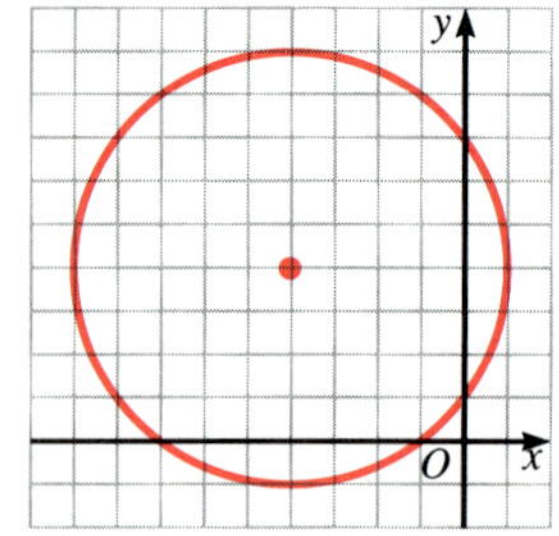

7. Find an equation of the parabola having vertex (0, 0) and directrix $y = 2$. 9-3
a. $y^2 = -4x$
b. $x^2 = -4y$
c. $x^2 = -8y$
d. $x^2 = 4 - 4y$

8. Find the vertex of the parabola $4x = y^2 - 4y$.
a. $(1, -2)$
b. $(-1, 2)$
c. $(0, -2)$
d. $(0, 4)$

9. Find an equation of the parabola shown at the right.
a. $x^2 = -y + 3$
b. $y^2 = -x - 3$
c. $y^2 = -x + 3$
d. $x^2 = -y - 3$

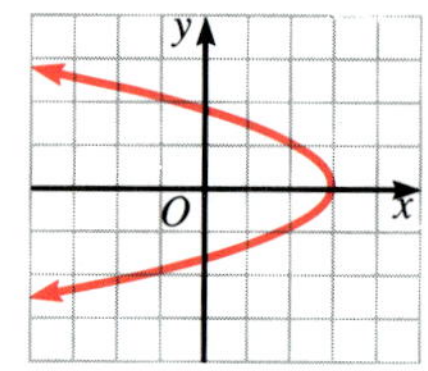

10. Find an equation of the ellipse having foci 9-4
(0, −2) and (0, 2) and sum of focal radii 8.
a. $\frac{x^2}{4} + \frac{y^2}{16} = 1$
b. $\frac{x^2}{16} + \frac{y^2}{20} = 1$
c. $\frac{x^2}{12} + \frac{y^2}{16} = 1$
d. $\frac{x^2}{16} + \frac{y^2}{12} = 1$

11. Find a focus of the ellipse $\frac{x^2}{4} + \frac{y^2}{20} = 1$.
a. $(0, 4)$
b. $(4, 0)$
c. $(0, 2\sqrt{6})$
d. $(2\sqrt{6}, 0)$

12. Find a focus of the hyperbola $8x^2 - 3y^2 = 48$. 9-5
a. $(0, \sqrt{22})$
b. $(\sqrt{22}, 0)$
c. $(0, \sqrt{10})$
d. $(\sqrt{10}, 0)$

13. Find an equation of the hyperbola shown at the right.
a. $xy = 4$
b. $xy = 2$
c. $x^2 - y^2 = 4$
d. $y^2 - x^2 = 4$

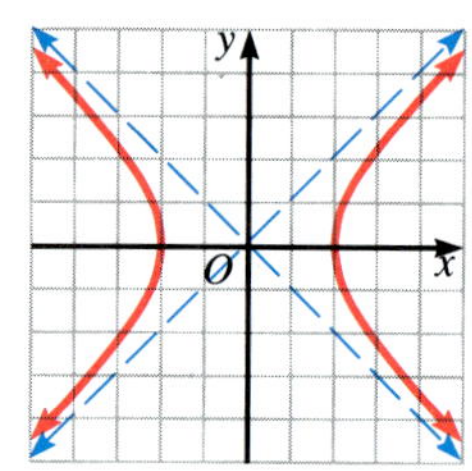

14. Find the center of the conic 9-6
$9x^2 - y^2 - 18x + 4y - 31 = 0$.
a. $(1, 2)$
b. $(3, 2)$
c. $(1, -2)$
d. $(-1, -2)$

15. Find an equation of the ellipse having foci (1, 0) and (3, 0) and sum of focal radii 4.
a. $x^2 + 9y^2 = 9$
b. $4x^2 + 9y^2 = 36$
c. $4x^2 + 3y^2 = 12y$
d. $3x^2 + 4y^2 = 12x$

16. How many solutions does this system have? 9-7
$$x^2 - 4y^2 = 4$$
$$x + y^2 = 1$$
a. 0
b. 1
c. 2
d. 3
e. 4

17. Find the real solutions of the system: 9-8
$$x^2 + y^2 = 17$$
$$x^2 - 2y = 9$$
a. $\{(1, -4), (-1, -4)\}$
b. $\{(-\sqrt{13}, 2), (\sqrt{13}, 2)\}$
c. $\{(\sqrt{13}, -2), (-\sqrt{13}, -2)\}$
d. $\{(1, -4), (-1, -4), (\sqrt{13}, 2), (-\sqrt{13}, 2)\}$

18. Find the dimensions of a rectangle having perimeter 14 m and a diagonal of length 5 m.

a. 3 m by 5 m **b.** 4 m by 5 m **(c.)** 3 m by 4 m **d.** 2 m by 3 m

19. Solve this system: 9-9

$$x + y + z = 2$$
$$x - 2y - z = 2$$
$$3x + 2y + z = 2$$

a. $\{(-1, -2, 5)\}$ **b.** $\{(1, 2, -1)\}$ **(c.)** $\{(1, -2, 3)\}$ **d.** $\{(1, -3, 4)\}$

20. Sara has \$36 in \$1, \$5, and \$10 bills. She has the same number of \$5 bills as \$10 bills, and she has 10 bills in all. How many bills of each denomination does she have?

a. 8 \$1 bills, 1 \$5 bill, 1 \$10 bill **(b.)** 6 \$1 bills, 2 \$5 bills, 2 \$10 bills
c. 4 \$1 bills, 3 \$5 bills, 3 \$10 bills **d.** 2 \$1 bills, 4 \$5 bills, 4 \$10 bills

Chapter Test

1. a. $2\sqrt{41}$ b. $(-1, 1)$ c. $(11, -14)$ 2. $(x - 6)^2 + (y + 3)^2 = 116$
3. $C(4, -1)$; $r = \sqrt{6}$ 5. $V(1, -3)$, $F(3, -3)$, D: $x = -1$

1. Given the points $P(3, -4)$ and $Q(-5, 6)$, find **(a)** the length PQ, **(b)** the midpoint of $\overline{PQ}$, and **(c)** the point X such that P is the midpoint of $\overline{QX}$. 9-1

2. Find an equation of the circle with center $(6, -3)$ that passes through $(-4, 1)$. 9-2

3. Find the center and radius of the circle $x^2 + y^2 - 8x + 2y + 11 = 0$.

4. Find an equation for the set of all points equidistant from the point $(2, 0)$ and the line $y = -4$. $y + 2 = \frac{1}{8}(x - 2)^2$ 9-3

5. Find the vertex, focus, and directrix of the parabola $x - 1 = \frac{1}{8}(y + 3)^2$.

6. Find an equation of the ellipse having foci $(\sqrt{5}, 0)$ and $(-\sqrt{5}, 0)$ and sum of focal radii 6. $\frac{x^2}{9} + \frac{y^2}{4} = 1$ 9-4

7. Graph the ellipse $x^2 + 4y^2 = 16$.

8. Graph the hyperbola $x^2 - y^2 + 4 = 0$, showing the asymptotes as dashed lines. 9-5

9. Find an equation of the ellipse having foci $(0, 0)$ and $(4, 0)$ and sum of focal radii 8. $\frac{(x-2)^2}{16} + \frac{y^2}{12} = 1$ 9-6

10. Identify the conic $x^2 - 4y^2 - 4x - 8y - 4 = 0$ and find its center and foci.
Hyperbola; center: $(2, -1)$; foci: $(2 + \sqrt{5}, -1)$, $(2 - \sqrt{5}, -1)$

By sketching graphs, find the *number* of real solutions of each system.

11. $4x^2 + 9y^2 = 36$ three
$x^2 = y + 2$

12. $4x^2 - y^2 = 4$ two 9-7
$xy = -9$

13. Solve this system: $x^2 - 4y^2 = 4$ $\{(6, 2\sqrt{2}), (6, -2\sqrt{2}), (-2, 0)\}$ 9-8
$y^2 = x + 2$

14. Find the dimensions of a rectangle whose area is 36 m² and whose diagonal is $5\sqrt{3}$ m long. $3\sqrt{3}$ m by $4\sqrt{3}$ m

15. Solve this system: $\{(-1, 5, -2)\}$ 9-9

$$2x - y - 3z = -1$$
$$2x - y + z = -9$$
$$x + 2y - 4z = 17$$

Additional Answers Chapter Test

7.
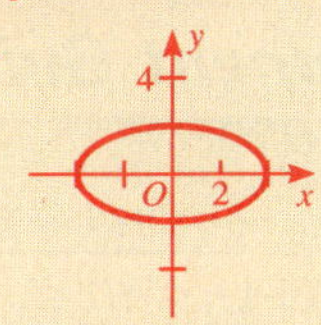

8.
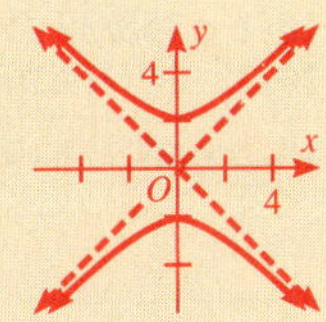

11.
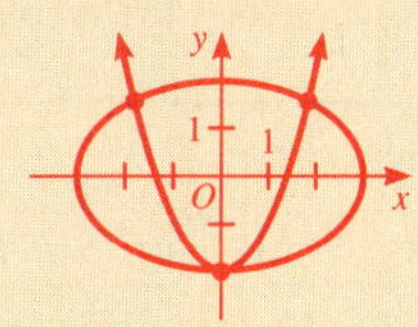

12.
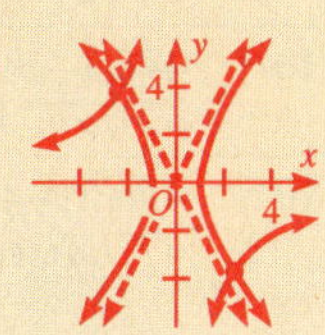

Alternate Test p. T21

Supplementary Materials

Test Masters	38, 39
Resource Book	p. 141

10 Exponential and Logarithmic Functions

Objectives

10-1 To extend the meaning of exponents to include rational numbers.

10-2 To extend the meaning of exponents to include irrational numbers and to define exponential functions.

10-3 To find the composite of two given functions and to find the inverse of a given function.

10-4 To define logarithmic functions and to learn how they are related to exponential functions.

10-5 To learn and apply the basic properties of logarithms.

10-6 To use common logarithms to solve equations involving powers and to evaluate logarithms with any given base.

10-7 To use exponential and logarithmic functions to solve growth and decay problems.

10-8 To define and use the natural logarithm function.

Assignment Guide

See p. T58 for Key to the format of the Assignment Guide

Day	Average Algebra	Average Algebra and Trigonometry	Extended Algebra	Extended Algebra and Trigonometry
1	**10-1** 458/1–35 odd **S** 458/*Mixed Review*	**10-1** 458/1–49 odd **S** 458/*Mixed Review*	**10-1** 458/1–35 odd 458/*Mixed Review*	**10-1** 458/1–49 odd **S** 458/*Mixed Review*
2	**10-1** 458/37–50 **10-2** 461/1–10	**10-2** 461/1–36	**10-1** 458/37–50 **10-2** 461/1–10	**10-2** 461/1–18, 19–39 odd
3	**10-2** 462/11–34 **S** 458/26–36 even	**10-3** 466/2–24 even **S** 467/*Mixed Review*	**10-2** 462/11–18, 19–37 odd **S** 458/26–36 even	**10-3** 466/2–26 even **S** 467/*Mixed Review*
4	**10-3** 466/2–24 even **S** 467/*Mixed Review*	**10-4** 470/1–22 **S** 458/44–50 even **R** 467/*Self-Test 1*	**10-3** 466/2–26 even **S** 467/*Mixed Review*	**10-4** 470/1–22 **S** 458/44–50 even **R** 467/*Self-Test 1*
5	**10-4** 470/1–22 **S** 462/35, 36 **R** 467/*Self-Test 1*	**10-4** 471/23–33, 35, 36 **S** 466/19–23 odd	**10-4** 470/1–22 **S** 466/19, 21, 23 **R** 467/*Self-Test 1*	**10-4** 471/23–36 **S** 466/19–23 odd
6	**10-4** 471/23–33, 35, 36 **S** 466/19–23 odd	**10-5** 476/1–32 **S** 477/*Mixed Review*	**10-4** 471/23–33, 35, 36 **S** 462/38, 39	**10-5** 476/1–28 **S** 471/37–43 odd
7	**10-5** 476/1–28 **S** 471/37, 38	**10-5** 476/33–42 **10-6** 481/1–27 odd	**10-5** 476/1–28 **S** 471/37–43 odd	**10-5** 476/29–46 **S** 477/*Mixed Review*
8	**10-5** 476/29–42 **S** 477/*Mixed Review*	**10-6** 482/29–38 **R** 477/*Self-Test 2* **10-7** 486/*P*: 1–9 odd	**10-5** 476/29–46 **S** 477/*Mixed Review*	**10-6** 481/1–29 odd, 31–34 **R** 477/*Self-Test 2*
9	**10-6** 481/1–28 **R** 477/*Self-Test 2*	**10-7** 487/*P*: 11–19 odd **S** 488/*Mixed Review*	**10-6** 481/1–29 odd, 31–34 **R** 477/*Self-Test 2*	**10-6** 482/35–41 **10-7** 486/*P*: 1–9 odd
10	**10-6** 482/29–38 **10-7** 486/*P*: 1–9 odd	*Prepare for Chapter Test* **R** 493/*Self-Test 3:* 1–4 **R** 496/*Chapter Review:* 1–14	**10-6** 482/35–41 **10-7** 486/*P*: 1–9 odd	**10-7** 487/*P*: 10–22 even **S** 488/*Mixed Review*

Assignment Guide *(continued)*

Day	Average Algebra	Average Algebra and Trigonometry	Extended Algebra	Extended Algebra and Trigonometry
11	**10-7** 487/*P*: 11–19 odd **S** 488/*Mixed Review*	*Administer Chapter 10 Test* **R** 498/*Mixed Problem Solving*	**10-7** 487/*P*: 10–22 even **S** 488/*Mixed Review*	**10-8** 490/3–48 mult. of 3 **R** 493/*Self-Test 3*
12	*Prepare for Chapter Test* **R** 493/*Self-Test 3*: 1–4 **R** 496/*Chapter Review*: 1–14		**10-8** 490/3–48 mult. of 3 **R** 493/*Self-Test 3*	*Prepare for Chapter Test* **R** 496/*Chapter Review*
13	*Administer Chapter 10 Test* **R** 498/*Mixed Problem Solving*		*Prepare for Chapter Test* **R** 496/*Chapter Review*	*Administer Chapter 10 Test* **R** 498/*Mixed Problem Solving*
14			*Administer Chapter 10 Test* **R** 498/*Mixed Problem Solving*	

Supplementary Materials Guide

For Use with Lesson	Practice Masters	Tests	Study Guide (Reteaching)	Resource Book: Tests	Resource Book: Practice Exercises	Resource Book: Applications (A) Enrichment (E) Technology (T)
10-1			pp. 155–156			p. 213 (A)
10-2	Sheet 55		pp. 157–158		p. 142	p. 245 (T)
10-3	Sheet 56		pp. 159–160		p. 143	p. 245 (T)
10-4			pp. 161–162			
10-5	Sheet 57	Test 40	pp. 163–164		p. 144	
10-6			pp. 165–166			
10-7			pp. 167–168			
10-8	Sheet 58	Test 41	pp. 169–170		p. 145	
Chapter 10		Tests 42, 43		pp. 54–57	p. 146	p. 229 (E)

Overhead Visuals

For Use with Lessons	Visual	Title
10-1	B	Multi-Use Packet 2
10-2, 10-3, 10-4	9	Functions and Inverses

Software

Software	Algebra Plotter Plus	Using Algebra Plotter Plus	Computer Activities	Test Generator
	Function Plotter, Statistics Spreadsheet	Enrichment, pp. 39–40 Activity Sheets, pp. 57–59	Activities 22–24, 37	168 test items
For Use with Lessons	10-2, 10-3, 10-4	10-2, 10-3, 10-4	10-3, 10-6, 10-7, Chapter 10	all lessons

Strategies for Teaching

Communication and Using Technology

Difficulties that students encounter in algebra are often the result of difficulties in reading. Since reading is a "learning-to-learn" skill, students become independent learners as their ability to read improves.

Clearly, successful reading calls for practice and patience on the part of both teacher and students. Consistent work on reading that is integrated into the content of the course seems to be the method most likely to lead to success.

Since the transition from what is written to what is spoken is often difficult, students need much experience in verbalizing material that is expressed in symbolic form. You may find it helpful to have them work in small groups, reading parts of a lesson aloud and explaining the examples to one another.

Working in small groups is one way in which students can improve their communication skills. In such groups students can gain experience in explaining a concept, restating a definition, justifying a conclusion, or thinking aloud through the steps in solving a problem. Certain exercises on the pages listed in the Index under *Proof, informal,* that ask students to "explain" or to "give a convincing argument" can be used in this context. Group discussion exposes students to others' thought processes and helps them clarify their own. Try to monitor these discussions unobtrusively and to act as a coach and facilitator when help is needed.

The language of algebra is very compact because of the extensive use of symbols. Students need much help in the correct interpretation and use of symbols. By writing out a few solutions in words, they can gain an appreciation of the value of symbols as savers of time and effort.

Technology is also an important aspect when working with exponential and logarithmic functions. Before calculators and computers, computation was quite involved. The Scottish mathematician John Napier found a way to do these computations with the use of logarithms. Their use as a computational aid is now outdated due to calculators and computers. Therefore, competence in the use of calculators and computers is now considered one of the basic skills needed by students in order to use mathematics efficiently. Computers and related electronic devices have liberated students from a great deal of routine calculations and have vastly increased the range of information available to them.

10-2 Real Number Exponents

You might direct students to the Extra that follows this lesson on page 462. Students might use a calculator to compare 2^x to x^n.

See the Exploration on page 841 for an activity in which students use a graphing calculator or a computer with graphing software to investigate powers and roots.

10-4 Definition of Logarithms

The simple expression $\log_b N = k$ may be difficult for students to understand and read because of the unfamiliar order and position of the symbols. You may wish to keep the following diagram on the chalkboard for reference.

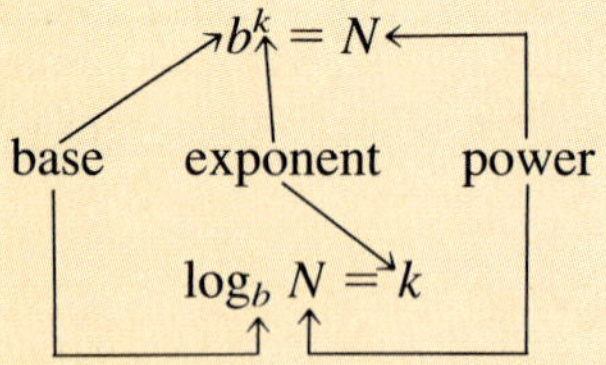

Have students answer the Oral Exercises with complete sentences until they become accustomed to the notation.

Ask students to work Exercise 42 in the Written Exercises. Have a student write the proof on the chalkboard and explain it.

10-5 Laws of Logarithms

Ask the students to find out about the work of Napier and Briggs in relation to logarithms, and to report their findings to the class.

10-6 Applications of Logarithms

Ask students to compute $\log_2 12$ by using a calculator and by using a computer. This exercise should lead to a practical need for a change-of-base formula. Ask students which logarithm functions are built into the calculator and computer and which functions are not. Whether by using a calculator or computer, students must use the change-of-base formula since ln and log are usually built-in functions. The function $\log_2$ is not built-in.

References to Strategies

PE: Pupil's Edition **TE:** Teacher's Edition **RB:** Resource Book

Problem Solving Strategies

TE: p. 483

Applications

PE: pp. 454, 493–494 (biology); p. 472 (physics); p. 492 (business); pp. 484–488 (word problems); 478–482, 489–491 (logarithms); 483–484 (compound interest); 483–488 (exponential growth and decay)
TE: pp. 454, 483–486
RB: p. 213

Nonroutine Problems

PE: p. 462 (Exs. 37–39); p. 467 (Exs. 24–26); p. 472 (Ex. 43); p. 482 (Exs. 41, 42); p. 488 (Probs. 21, 22); p. 491 (Exs. 50, 51); p. 492 (Challenge); p. 494 (Challenge 1–3)
TE: pp. T117–T118 (Sugg. Extensions, Lessons 10-2, 10-4, 10-6, 10-7, 10-8)

Communication

PE: p. 482 (Ex. 41, convincing argument)
TE: p. T117 (Reading Algebra and Sugg. Extension, Lesson 10-4); p. T118 (Sugg. Extension, Lesson 10-5)

Thinking Skills

PE: p. 482 (Ex. 42, formula derivation)

Explorations

PE: p. 462 (growth of functions); p. 492 (Challenge); p. 841 (powers and roots)

Connections

PE: pp. 454, 484–488 (Biology); p. 472 (Physics); p. 472 (History); pp. 483–484 (Money); pp. 493–494 (Geology)

Using Technology

PE: pp. 459–460, 462, 464, 466, 471, 478–482, 485–489, 841; p. 492 (Calculator Key-In)
TE: pp. 459–460, 465, 471, 480, 485, 489, 493, T118
RB: p. 245
Computer Activities: pp. 57–65, 100–102
Using Algebra Plotter Plus: pp. 39–40, 57–59

Using Manipulatives/Models

Overhead Visuals: B, 9

Teaching Resources

For use in implementing the teaching strategies referenced on the previous page.

Application
Resource Book, p. 213

Application—Population Growth Curves (For use with Chapter 10)

1. a. As the table below indicates, the population of the United States has grown steadily throughout U.S. history. Using the table and the given axes, sketch the population growth curve for the United States.

Year	*Population*
1800	5 million
1850	23 million
1900	76 million
1950	151 mil!ion

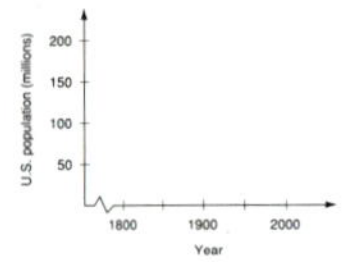

b. The curve that comes closest to fitting the data in part (a) is given by the equation $P = 6(1.023)^t$, where P is the population in millions and t is the number of years since 1800. Assuming the U.S. population continues to follow this exponential growth model, use the equation to estimate the population of the United States in the year 2000: ________ million

c. Would you expect the U.S. population to increase exponentially indefinitely?

Explain why or why not: ________________________________

2. a. The table below is based on experimental data involving the growth of a population of yeast cells placed in a closed environment with a limited food supply. Using the table and the given axes, sketch the population growth curve for the yeast cells.

Time (hours)	*Number of cells*
0	6
2	18
4	49
6	121
8	236
10	345
12	409
14	435
16	445
18	448
20	449

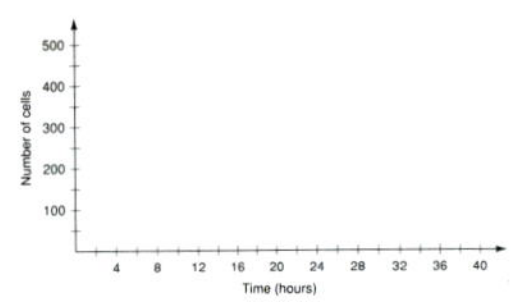

b. Notice that the first part of the growth curve for the yeast cells is similar to that for the United States in Exercise 1, but as time goes on the graph levels out. If you were a scientist studying the yeast population, how would you account for this S-shaped (*sigmoid*) curve? ________________

b. What do you predict will happen to the curve in part (a) when the food supply for the yeast cells becomes insufficient to support the population? ________________

d. Illustrate your answer to part (c) by continuing the curve of part (a) beyond 20 hours.

Enrichment
Resource Book, p. 229

Inverse Functions (For use with Chapter 10)

If you put on your socks and then your shoes in the morning, at night you would first take off your shoes, then your socks.

Morning: socks on → shoes on
Night: shoes off → socks off

The night operation is the *inverse* of the morning operation.

Look at the function $f(x) = 2x + 7$. To evaluate it, you would multiply an x-value by 2. Then you would add 7.

Function: Multiply by 2. → Add 7.

The inverse operation would undo these steps.

Inverse: Subtract 7. → Divide by 2.

If $f(x) = 2x + 7$, then $f^{-1}(x) = \frac{x - 7}{2}$.

Find the inverse of each function.

1. $f(x) = 3x + 5$
2. $f(x) = 2x - 1$
3. $f(x) = \frac{x + 6}{3}$
4. $f(x) = \frac{3x + 1}{7}$
5. $f(x) = \frac{x - 1}{4}$
6. $f(x) = \frac{5x + 12}{3}$

What operation is the inverse of raising 2 to a power? It is the base 2 logarithm operation.

If $f(x) = 2^x$, then $f^{-1}(x) = \log_2 x$.

Find the inverse of each function.

7. $f(x) = 3^x$
8. $f(x) = \log_{10} x$
9. $f(x) = \log_{\frac{1}{2}} x$

How would you find the inverse of $f(x) = \log_2 (3x + 2)$? Think of the function as built up with the following steps.

f: multiply x by 3 → add 2 → take the base 2 logarithm

Then the inverse function is built up as follows:

f^{-1}: raise 2 to the power of x → subtract 2 → divide by 3

Therefore if $f(x) = \log_2 (3x + 2)$, then $f^{-1}(x) = \frac{2^x - 2}{3}$. Note that $f(2) = \log_2 8 = 3$ and $f^{-1}(3) = \frac{8 - 2}{3} = 2$.

Find the inverse of each function.

10. $f(x) = \log_3 (2x - 1)$
11. $f(x) = 3(2 + 5^x)$
12. $f(x) = \log_4 (2^x + 1)$
13. $f(x) = 2^{\log_3(x+1)}$

ANSWERS

1. ________
2. ________
3. ________
4. ________
5. ________
6. ________
7. ________
8. ________
9. ________
10. ________
11. ________
12. ________
13. ________

Using Technology
Resource Book, p. 245

Using a Computer or a Graphing Calculator

To complete these activities, you should use a computer with graphics software (such as ALGEBRA PLOTTER PLUS) or a graphing calculator.

Exponential Functions (For use with Lesson 10-2)

1. Graph these equations: $y = 2^x$, $y = 3^x$, $y = 4^x$
   a. Which graph lies above the other two for $x < 0$? ________
   b. Which graph lies above the other two for $x > 0$? ________
2. Graph these equations: $y = \left(\frac{2}{3}\right)^x$, $y = \left(\frac{3}{2}\right)^x$
   a. Which graph is *increasing* from left to right? ________
   b. Which graph is *decreasing* from left to right? ________
3. a. Predict whether the graph of $y = 1.2^x$ is increasing or decreasing: ________
   b. Predict whether the graph of $y = 0.8^x$ is increasing or decreasing: ________
   Now graph the equations to confirm your predictions.
4. Generalize the results of Exercises 2 and 3: Under what conditions is the graph of $y = b^x$ increasing? decreasing? ________________

Functions and Their Inverses (For use with Lesson 10-3)

1. Graph these equations: $y = 3x + 2$, $y = \frac{x - 2}{3}$, $y = x$

   The functions $f(x) = 3x + 2$ and $g(x) = \frac{x - 2}{3}$ are *inverses*, since each one "undoes" what the other does. For example, $f(1) = 5$ and $g(5) = 1$. When you compare the graphs of f and g, what geometric relationship involving the line $y = x$ do you see?

   ____

2. Graph this equation: $y = \frac{1}{2}x^3$

   Based on your answer to Exercise 2, sketch the graph of the inverse of $f(x) = \frac{1}{2}x^3$ on the axes at the right.

   Then graph $y = \sqrt[3]{2x}$ to confirm your sketch.

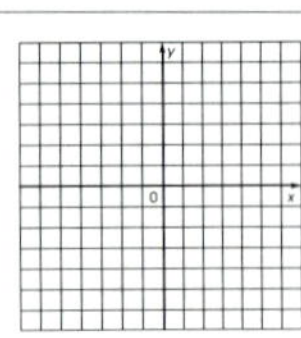

Using Technology
Using Algebra Plotter Plus, p. 59

Functions and Their Inverses

Algebra Plotter Plus | Book 2: Lesson 10-3

Use the Function Plotter program of Algebra Plotter Plus. To minimize distortion, you may wish to set the SCALE for each exercise so the graph displays the intervals $-6 \le x \le 6$ and $-4 \le y \le 4$.

1. Graph these equations: $y = 3x + 2$; $y = \frac{(x - 2)}{3}$; $y = x$

   The functions $f(x) = 3x + 2$ and $g(x) = \frac{(x-2)}{3}$ are *inverses*, since each one "undoes" what the other does. For example, $f(1) = 5$ and $g(5) = 1$. When you compare the graphs of f and g, what geometric relationship involving the line $y = x$ do you see?

   ____

2. Graph these equations: $y = (x - 2)^3$; $y = \sqrt[3]{x} + 2$; $y = x$

   You can enter the new equations by typing: y = (x – 2)^3 and y = x^(1/3) + 2. What do you notice about the graphs of these inverse functions?

   ____

3. Graph this equation: $y = \frac{1}{2}x^3$

   Based on your answers to Exercises 1 and 2, sketch the graph of the inverse of $f(x) = \frac{1}{2}x^3$. Then graph $y = \sqrt[3]{2x}$ by entering y = (2x)^(1/3) to confirm your sketch.

4. Graph this equation: $y = x^2 - 4$

   Sketch the graph of the inverse of $f(x) = x^2 - 4$. Then enter y = SQR(x + 4) and y = –SQR(x + 4) to confirm your sketch. Why do you have to tell the Function Plotter to graph two parts to get the inverse of f?

   ____

Using Technology
Using Algebra Plotter Plus, p. 57

Exponential Growth | Algebra Plotter Plus | Book 2: Lesson 10-2

Use the Function Plotter program of Algebra Plotter Plus.

1. Graph these equations: $y = 2^x$, $y = 3^x$, $y = 4^x$
   a. Which graph lies above the other two for $x < 0$? _____
   b. Which graph lies above the other two for $x > 0$? _____
2. Graph these equations: $y = (\frac{2}{3})^x$, $y = (\frac{3}{2})^x$
   a. Which graph is *increasing* from left to right? _____
   b. Which graph is *decreasing* from left to right? _____
3. Based on your answer to Exercise 2, complete the following.
   a. Predict whether the graph of $y = 1.2^x$ is increasing or decreasing. _________
   b. Predict whether the graph of $y = 0.8^x$ is increasing or decreasing. _________
   Now graph the equations to confirm your predictions.
4. Generalize the results of Exercises 2 and 3. Assume $b > 0$.
   a. Under what conditions does the graph of $y = b^x$ increase?
   ____
   b. Under what conditions does the graph of $y = b^x$ decrease?
   ____
   c. When would the graph of $y = b^x$ neither increase nor decrease?
   ____
5. The following table shows the population y of a colony of bacteria over time x in hours. Assume the population grows exponentially.
   a. Complete the table by finding y when $x = 4$, 5, and 6.

x	0	1	2	3	4	5	6
y	5	10	20	40			

   b. Use the pattern in the table to write an equation for y in terms of x.
   ____
   c. Find y when $x = 10$: _____ when $x = 12$: _____.
   d. Use the computer to graph the equation in (b). Set the SCALE so that the graph will display the interval $0 \le x \le 12$. What interval on the y-axis must be used to show the function values? _______________
   e. Set the SCALE to display the interval $0 \le x \le 4$. From the screen display, estimate the population when $x = 2.5$ hours. _____
   Check your estimate by selecting TABLE.
   f. After how many hours was the population one-half of the maximum population shown on the screen? _____

(continued)

Using Technology
Using Algebra Plotter Plus, p. 58

Exponential Growth *(continued)*

6. A rare population of tortoises now numbering 38 increases at the rate of 4% per year. Round your answers to the nearest whole tortoise.
   a. Find the population one, two, and three years from now. _________
   b. The equation $y = 38(1.04)^x$ gives the population y after x years. Do your calculations in (a) and (b) agree with this equation? _____
   c. Use the computer to graph the equation in (b) for the interval $0 \le x \le 10$.
   d. Estimate the population to the nearest whole tortoise after 9 years. _____
   e Approximately how many years will it take for the population to reach 51? _____
   f. Approximately how many years will it take for the population to double from 38 to 76? _____
   g. Approximate the number of years it will take for the population to double from 76 to 152. _____ Is your answer the same as (f)? _____
7. Use the computer to graph $y = 10(1.06)^x$. From the screen display, estimate the time it will take for y to change from 10 to 20. _____
8. Repeat Exercise 7 for the function $y = 10(1.04)^x$. _____
9. The growth function of a certain population is $y = 10(1.05)^x$.
   a. Predict the time it takes for the population to double. _____
   b. Use the computer to graph the function in (a) to check your prediction.

Some populations decline over time. In Exercises 10 and 11, y represents population and x represents time in years.

10. An initial population of 24 decreases at the rate of 3% per year.
   a. Find the population after 1 year. (*Hint:* If at the start, the population is 100%, then after 1 year, the population is 97% of the original population.) _____
   b. Find the population after two years. That is, find 97% of 97% of 24.
   _____
   c. Write an equation of the form $y = ab^x$ to describe the population as a function of time. _____
   d. Use the computer to graph the function. Does the screen display indicate that the population is declining? _____
11. An initial population of 40 decreases at the rate of 4% per year.
   a. Write an equation to give the population as a function of time. _____
   b. Use the computer to graph the function over the interval $0 \le x \le 20$.
   c. After how many years will the population be half its original size? _____
   d. Graph the function over the interval $20 \le x \le 40$. Estimate the population after 40 years. _____

Using Models
Overhead Visual 9, Sheets 1 and 2

FUNCTIONS AND INVERSES

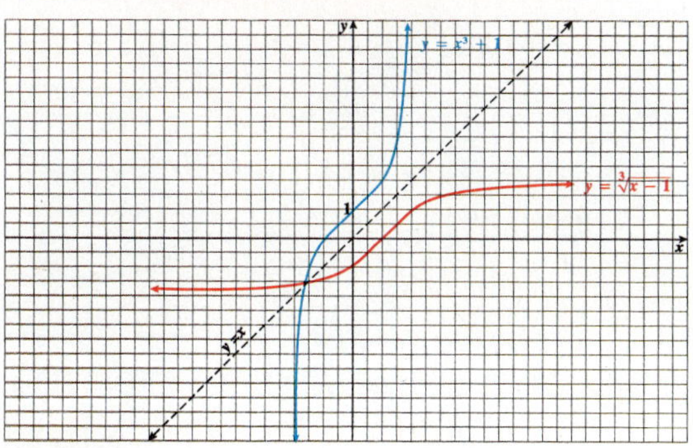

VISUAL 9

Using Models
Overhead Visual 9, Sheets 3 and 4

FUNCTIONS AND INVERSES

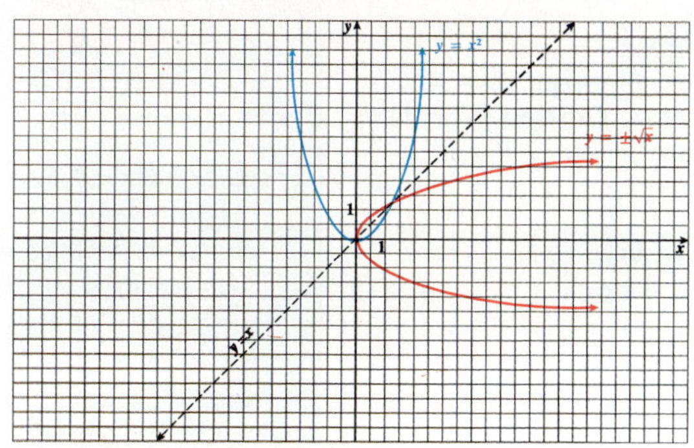

VISUAL 9

Application

Under such ideal conditions as a hospitable climate, an ample food supply, and no natural enemies, a population can grow exponentially. This means that the rate of population growth at any time is proportional to the current population size; that is, the larger the population gets, the faster it grows.

In most natural settings, a population may grow exponentially for a while, but overcrowding, disease, and other factors often intervene to check the growth. In terms of a graph, the population growth curve (like the one shown with the photo) tends to level off at some point rather than continue its upward climb.

Research Activities

Have students research and report on the successes and failures of introducing wildlife to new environments. For example, students can investigate the consequences of introducing rabbits to Australia, gray squirrels to England, muskrats to Europe, and ring-necked pheasants, sparrows, starlings, carp, killer bees, European boars, and Russian aphids to the United States.

References

Laycock, George. *The Alien Animals.* Garden City, NY: Natural History Press, 1966.

Support Materials

Resource book p. 213

10 Exponential and Logarithmic Functions

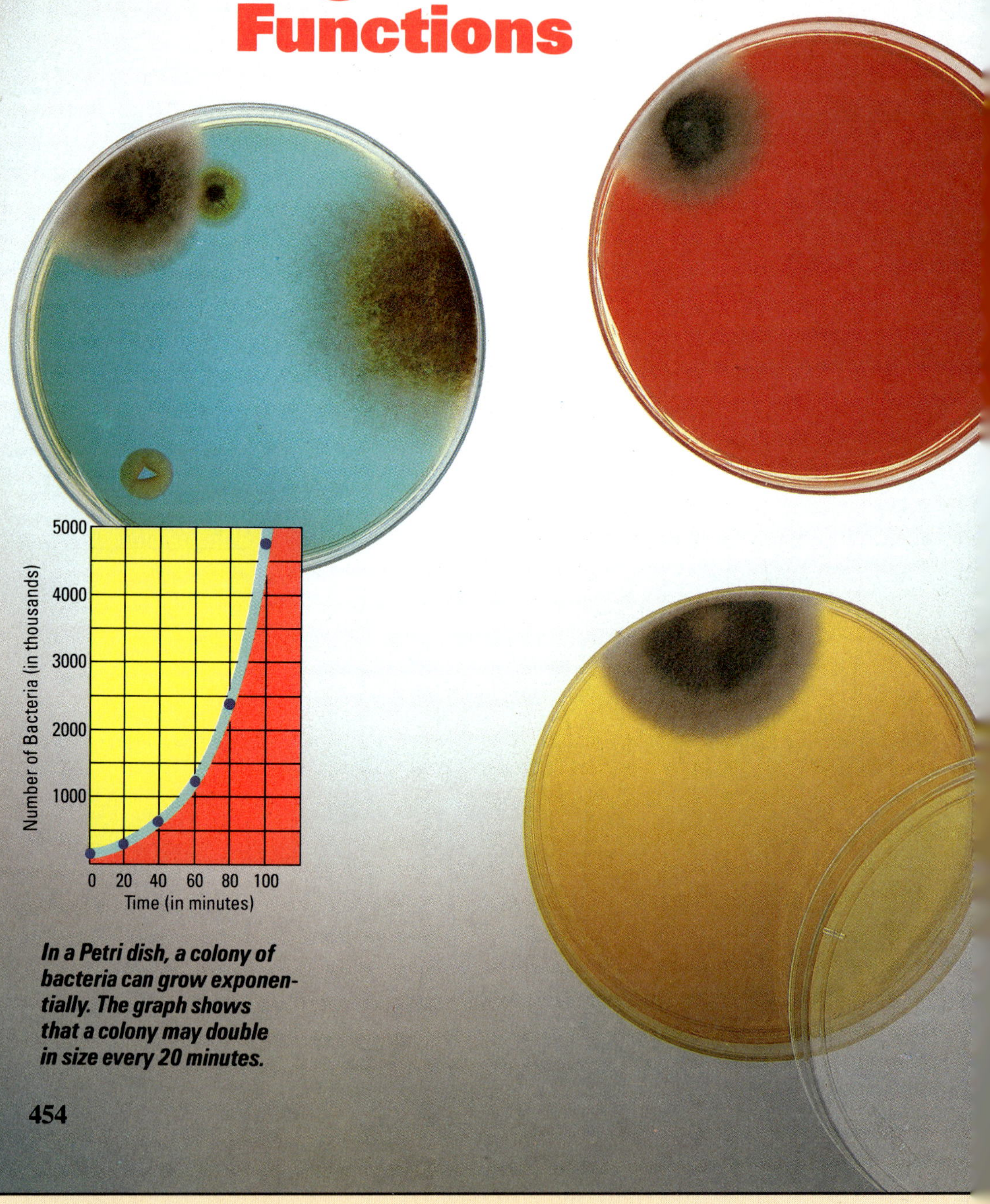

In a Petri dish, a colony of bacteria can grow exponentially. The graph shows that a colony may double in size every 20 minutes.

Exponential Functions

10-1 Rational Exponents

Objective To extend the meaning of exponents to include rational numbers.

You know the meaning of the power b^x when x is an integer. For example,

$$5^2 = 25, \quad 5^{-2} = \frac{1}{25}, \quad \text{and} \quad 5^0 = 1.$$

Now you'll learn the meaning of b^x when x is any *rational number*.

Definition of $b^{1/2}$: (read as "b to the one-half power" or as "b to the one half") If the laws of exponents on page 212 are to hold, then

$$(b^{1/2})^2 = b^{(1/2)2} = b^1 = b.$$

Since the square of $b^{1/2}$ is b, *$b^{1/2}$ is defined to be $\sqrt{b}$.*

Definition of $b^{1/3}$: *$b^{1/3}$ is defined to be $\sqrt[3]{b}$*, since its cube is b:

$$(b^{1/3})^3 = b^{(1/3)3} = b^1 = b$$

Definition of $b^{2/3}$: Using the law $(a^m)^n = a^{mn}$, you can express $b^{2/3}$ in either of two ways:

$$b^{2/3} = (b^{1/3})^2 = (\sqrt[3]{b})^2 \quad \text{or} \quad b^{2/3} = (b^2)^{1/3} = \sqrt[3]{b^2}$$

Therefore, *$b^{2/3}$ is defined to be either of the equivalent expressions $(\sqrt[3]{b})^2$ or $\sqrt[3]{b^2}$.*

Definition of $b^{p/q}$

If p and q are integers with $q > 0$, and b is a positive real number, then

$$b^{p/q} = (\sqrt[q]{b})^p = \sqrt[q]{b^p}.$$

Example 1 Simplify $16^{3/4}$.

Solution 1 $16^{3/4} = (\sqrt[4]{16})^3 = 2^3 = 8$ ***Answer***

Solution 2 $16^{3/4} = \sqrt[4]{16^3} = \sqrt[4]{4096} = 8$ ***Answer***

Notice that computing the root first is easier. This is usually the case.

Teaching References

Lesson Commentary, pp. T116–T118

Assignment Guide, pp. T66–T70

Supplementary Materials
- Practice Masters 55–58
- Tests 40–43
- Resource Book
  - Practice Exercises, pp. 142–146
  - Tests, pp. 54–57
  - Enrichment Activity, p. 229
  - Application, p. 213
- Study Guide, pp. 155–170
- Computer Activities 22–24
- Test Generator
- Disk for Algebra

Alternate Test, p. T22

Teaching Suggestions p. T116

Suggested Extensions p. T116

Warm-Up Exercises

Simplify.

1. $a^2 \cdot a^3$ a^5
2. $(xy^3)(x^2y)$ x^3y^4
3. $(ab^2)^3$ a^3b^6
4. $\frac{m^6}{m^2}$ m^4
5. $\frac{n^3}{n^5}$ $\frac{1}{n^2}$
6. $\left(\frac{x^3}{y^4}\right)^2$ $\frac{x^6}{y^8}$

Motivating the Lesson

Tell students that until now the laws of exponents they have used since Chapter 5 covered only integral exponents. In this lesson the laws will be extended to cover fractional exponents as well.

Chalkboard Examples

Evaluate.

1. $81^{3/2}$ $(\sqrt{81})^3 = 9^3 = 729$
2. $8^{-5/3}$
   $(\sqrt[3]{8})^{-5} = 2^{-5} = \frac{1}{2^5} = \frac{1}{32}$
3. $9^{1.5}$
   $(9)^{3/2} = (\sqrt{9})^3 = 3^3 = 27$
4. $\left(\frac{1}{\sqrt[3]{81}}\right)^{-3/4}$
   $(81^{-1/3})^{-3/4} = 81^{1/4} = \sqrt[4]{81} = 3$
5. $\sqrt{\frac{1}{3}} \cdot \sqrt[3]{9}$
   $\frac{1}{\sqrt{3}} \cdot \sqrt[3]{3^2} = 3^{-1/2} \cdot 3^{2/3} = 3^{-3/6+4/6} = 3^{1/6} = \sqrt[6]{3}$
6. Write $\sqrt[3]{9} \cdot \sqrt{3}$ in exponential form and in simplest radical form.
   $3^{2/3} \cdot 3^{1/2} = 3^{7/6}$
   $3^{7/6} = 3\sqrt[6]{3}$

Solve.

7. $6x^{2/3} = 54$
   $x^{2/3} = 9$
   $(x^{2/3})^{3/2} = 9^{3/2}$
   $x = 27$
   $\therefore$ the solution set is $\{27\}$.
8. $(t-4)^{2/5} - 3 = 1$
   $(t-4)^{2/5} = 4$
   $[(t-4)^{2/5}]^{5/2} = 4^{5/2}$
   $t - 4 = 32$
   $t = 36$
   $\therefore$ the solution set is $\{36\}$.

Example 2 Simplify: **a.** $25^{-3/2}$ **b.** $9^{2.5}$

Solution

a. $25^{-3/2} = \frac{1}{25^{3/2}}$

$= \frac{1}{(\sqrt{25})^3}$

$= \frac{1}{125}$ ***Answer***

b. $9^{2.5} = 9^{5/2}$

$= (\sqrt{9})^5$

$= 3^5$

$= 243$ ***Answer***

The laws of exponents on page 212 also hold for powers with rational exponents. These laws can be used with the definition of rational exponents to simplify many radical expressions. The first step is to write the expression in **exponential form,** that is, as a power or a product of powers.

Example 3 **a.** Write $\sqrt[3]{\frac{a^5b^3}{c^2}}$ in exponential form. **b.** Simplify $\left(\frac{1}{\sqrt[3]{4}}\right)^{-3/2}$

Solution

a. $\sqrt[3]{\frac{a^5b^3}{c^2}} = \left(\frac{a^5b^3}{c^2}\right)^{1/3}$

$= \frac{a^{5/3}b^{3/3}}{c^{2/3}}$

$= a^{5/3}bc^{-2/3}$ ***Answer***

b. $\left(\frac{1}{\sqrt[3]{4}}\right)^{-3/2} = (4^{-1/3})^{-3/2}$

$= 4^{(-1/3)(-3/2)}$

$= 4^{1/2}$

$= 2$ ***Answer***

Example 4 Write $\sqrt{8} \cdot \sqrt[3]{4}$ in exponential form and in simplest radical form.

Solution

$\sqrt{8} \cdot \sqrt[3]{4} = \sqrt{2^3} \cdot \sqrt[3]{2^2}$

$= 2^{3/2} \cdot 2^{2/3}$

$= 2^{13/6}$ ⟵ in exponential form ***Answer***

$= 2^2 \cdot 2^{1/6}$

$= 4\sqrt[6]{2}$ ⟵ in simplest radical form ***Answer***

Example 5 Solve: **a.** $(x-1)^{3/2} = 8$ **b.** $5x^{-1/3} = 20$

Solution

a. $(x-1)^{3/2} = 8$

Raise both sides to the $\frac{2}{3}$ power. $[(x-1)^{3/2}]^{2/3} = 8^{2/3}$

$x - 1 = 4$

$x = 5$

$\therefore$ the solution set is $\{5\}$.

Check: $(5-1)^{3/2} = 4^{3/2} = 8$ ✓

b. $5x^{-1/3} = 20$

$x^{-1/3} = 4$

$(x^{-1/3})^{-3} = 4^{-3}$ Raise both sides to the -3 power.

$x = \frac{1}{64}$

$\therefore$ the solution set is $\left\{\frac{1}{64}\right\}$.

Check: $5\left(\frac{1}{64}\right)^{-1/3} = 5 \cdot 4 = 20$ ✓

Oral Exercises

Simplify.

1. a. $9^{1/2}$ 3 b. $9^{-1/2}$ $\frac{1}{3}$
2. a. $8^{1/3}$ 2 b. $8^{-1/3}$ $\frac{1}{2}$
3. a. $8^{2/3}$ 4 b. $8^{-2/3}$ $\frac{1}{4}$
4. a. $16^{1/4}$ 2 b. $16^{-1/4}$ $\frac{1}{2}$
5. $27^{-1/3}$ $\frac{1}{3}$
6. $27^{2/3}$ 9
7. $4^{3/2}$ 8
8. $25^{-1/2}$ $\frac{1}{5}$

Give the letter of the correct answer. Assume x and y are positive numbers.

9. $x^{1/2}$ equals: a. $\frac{x}{2}$ b. $\frac{2}{x}$ (c.) $\sqrt{x}$
10. $x^{2/3}$ equals: (a.) $(\sqrt[3]{x})^2$ b. $\sqrt{x^3}$ c. $\frac{2\sqrt{x}}{3}$
11. $2x^{-1/2}$ equals: a. $\frac{1}{\sqrt{2x}}$ b. $\frac{1}{2\sqrt{x}}$ (c.) $\frac{2}{\sqrt{x}}$
12. $(2x)^{-1/2}$ equals: (a.) $\frac{1}{\sqrt{2x}}$ b. $\frac{1}{2\sqrt{x}}$ c. $\frac{\sqrt{2}}{x}$
13. $-8x^{-1/3}$ equals: a. $\frac{-2}{\sqrt[3]{x}}$ (b.) $\frac{-8}{\sqrt[3]{x}}$ c. $\frac{-1}{2\sqrt[3]{x}}$
14. $\sqrt[4]{x^3y^{-4}}$ equals: (a.) $\frac{x^{3/4}}{y}$ b. $\frac{x^{3/4}}{y^{-1}}$ c. $\frac{3x}{4y}$
15. $\sqrt[3]{\frac{x^2}{8y^{-1}}}$ equals: a. $\frac{x^{2/3}y}{8}$ (b.) $\frac{x^{2/3}y^{1/3}}{2}$ c. $\frac{x^{2/3}}{2y}$
16. $8 \cdot 2^{1/2}$ equals: (a.) $2^{7/2}$ b. $16^{1/2}$ c. $16\sqrt{2}$

Tell whether each equation is true or false.

17. a. $9^{1/2} + 4^{1/2} = (9 + 4)^{1/2}$ F
    b. $9^{1/2} \cdot 4^{1/2} = (9 \cdot 4)^{1/2}$ T
18. a. $2^{1/3} + 4^{1/3} = (2 + 4)^{1/3}$ F
    b. $2^{1/3} \cdot 4^{1/3} = (2 \cdot 4)^{1/3}$ T
19. a. $\left(\frac{1}{a} + \frac{1}{b}\right)^{-1} = a + b$ F
    b. $\left(\frac{1}{a} \cdot \frac{1}{b}\right)^{-1} = ab$ T
20. a. $(\sqrt{a} + \sqrt{b})^2 = a + b$ F
    b. $(\sqrt{a} \cdot \sqrt{b})^2 = ab$ T
21. a. $(a^{-1} + b^{-1})^{-2} = a^2 + b^2$ F
    b. $(a^{-1} \cdot b^{-1})^{-2} = a^2b^2$ T
22. a. $(x^{1/3} + y^{1/3})^6 = x^2 + y^2$ F
    b. $(x^{1/3} \cdot y^{1/3})^6 = x^2y^2$ T

Give the power to which you would raise both sides of each equation in order to solve the equation.

23. $x^{1/2} = 9$ 2
24. $x^{2/3} = 4$ $\frac{3}{2}$
25. $x^{-1/3} = 2$ -3
26. $x^{-3/4} = 8^{-1}$ $-\frac{4}{3}$

Tell what steps you would use to solve each equation.

27. $3x^{1/4} = 6$
28. $5x^{-3/2} = 40$
29. $(x - 3)^{-2} = \frac{1}{4}$
30. $(5x)^{-1/2} = 3$

Check for Understanding

Here is a suggested use of the Oral Exercises to check students' understanding as you teach the lesson.
Oral Exs. 1–8: use after Example 2.
Oral Exs. 9–16: use after Example 3.
Oral Exs. 17–22: use after Example 4.
Oral Exs. 23–30: use after Example 5.

Guided Practice

Evaluate.

1. $49^{-1/2}$ $\frac{1}{7}$ 2. $8^{5/3}$ 32

Write in exponential form.

3. $\sqrt{ab^3}$ $a^{1/2}b^{3/2}$
4. $\frac{1}{\sqrt[3]{xy^5}}$ $x^{-1/3}y^{-5/3}$

Express in simplest form.

5. $\sqrt[6]{5^5} \div \sqrt[3]{25}$ $\sqrt[6]{5}$
6. $\sqrt[3]{16} \cdot \sqrt{8}$ $4\sqrt[6]{32}$

Summarizing the Lesson

In this lesson students learned the meaning of rational exponents and how to simplify expressions and solve equations where rational exponents appear. Ask students to explain why $b^{p/q}$ can be defined as either $(\sqrt[q]{b})^p$ or $\sqrt[q]{b^p}$.

Additional Answers Oral Exercises

27. Divide by 3; raise to the fourth power.
28. Divide by 5; raise to the $-\frac{2}{3}$ power.
29. Raise to the $-\frac{1}{2}$ power; add 3.
30. Raise to the -2 power; divide by 5.

Suggested Assignments

Average Algebra
Day 1: 458/1–35 odd
S 458/Mixed Review
Day 2: 458/37–50
Assign with Lesson 10-2.

Average Alg. and Trig.
458/1–49 odd
S 458/Mixed Review

Extended Algebra
Day 1: 458/1–35 odd
S 458/Mixed Review
Day 2: 458/37–50
Assign with Lesson 10-2.

Extended Alg. and Trig.
458/1–49 odd
S 458/Mixed Review

Common Errors

When evaluating an expression such as $27^{2/3}$, students may square 27 and then have difficulty finding the cube root of the result. Remind students that in some cases finding the *n*th root first is easier.

Students may understand that $a^{1/7} \cdot a^{2/7} = a^{3/7}$ but fail to see that $4^{1/7} \cdot 4^{2/7} = 4^{3/7}$. Students may think that the result should be $16^{3/7}$. When reteaching, compare the two examples and remind students that the variable *a* represents an unknown number and that the number could be 4.

Supplementary Materials

Study Guide pp. 155–156

Written Exercises

Simplify.

A

1. $81^{1/2}$ 9
2. $27^{1/3}$ 3
3. $49^{-1/2}$ $\frac{1}{7}$
4. $32^{-1/5}$ $\frac{1}{2}$
5. $4^{3/2}$ 8
6. $27^{2/3}$ 9
7. $16^{3/4}$ 8
8. $25^{-3/2}$ $\frac{1}{125}$
9. $(-125)^{-1/3}$ $-\frac{1}{5}$
10. $(-32)^{-3/5}$ $-\frac{1}{8}$
11. $4^{-0.5}$ $\frac{1}{2}$
12. $\left(\frac{4}{9}\right)^{-1.5}$ $\frac{27}{8}$
13. $-8^{2/3}$ -4
14. $-9^{3/2}$ -27
15. $(5^{1/3})^{-3}$ $\frac{1}{5}$
16. $(7^{-2/3})^3$ $\frac{1}{49}$
17. $(16^{-5})^{1/20}$ $\frac{1}{2}$
18. $(27^4)^{-1/12}$ $\frac{1}{3}$
19. $(9^{1/2} + 16^{1/2})^2$ 49
20. $(8^{2/3} - 8^{1/3})^3$ 8

Write in exponential form.

21. $\sqrt{x^3y^5}$ $x^{3/2}y^{5/2}$
22. $\sqrt[3]{p^4q}$ $p^{4/3}q^{1/3}$
23. $\sqrt{a^{-2}b^3}$ $a^{-1}b^{3/2}$
24. $\sqrt[3]{x^6y^{-4}}$ $x^2y^{-4/3}$
25. $(\sqrt{a^{-2}b})^5$ $a^{-5}b^{5/2}$
26. $\sqrt[3]{8b^6c^{-4}}$ $2b^2c^{-4/3}$
27. $\sqrt[4]{\dfrac{16^3 \cdot a^{-2}}{b^6}}$ $8a^{-1/2}b^{-3/2}$
28. $\dfrac{1}{\sqrt[4]{p^4q^{-8}}}$ $p^{-1}q^2$

Express in simplest radical form.

29. $\sqrt[3]{4} \cdot \sqrt[3]{4}$ $2\sqrt[3]{2}$
30. $\sqrt{8} \cdot \sqrt[6]{8}$ 4
31. $\dfrac{\sqrt[3]{4}}{\sqrt[6]{2}}$ $\sqrt{2}$
32. $\dfrac{\sqrt[5]{27^3}}{\sqrt[5]{9^2}}$ 3
33. $\sqrt[10]{32} \div \sqrt[8]{4}$ $\sqrt[4]{2}$
34. $\sqrt[6]{8^3} \div \sqrt[6]{4^2}$ $\sqrt[6]{32}$
35. $\sqrt[4]{27} \cdot \sqrt[8]{9}$ 3
36. $\sqrt[4]{128} \cdot \sqrt[8]{256}$ $4\sqrt[4]{8}$

Simplify each expression. Give answers in exponential form.

B

37. $\sqrt{x} \cdot \sqrt[3]{x} \cdot \sqrt[6]{x}$ x
38. $\sqrt[3]{a^2} \cdot \sqrt[3]{a^4}$ a^2
39. $\sqrt[4]{x} \cdot \sqrt[6]{x} \div \sqrt[3]{x}$ $x^{1/12}$
40. $((b^{1/2})^{-2/3})^{3/4}$ $b^{-1/4}$
41. $a^{1/2}(a^{3/2} - 2a^{1/2})$ $a^2 - 2a$
42. $(x^{3/2} - 2x^{5/2}) \div x^{1/2}$ $x - 2x^2$

Solve each equation.

43. a. $a^{3/4} = 8$ $\{16\}$
 b. $(3x + 1)^{3/4} = 8$ $\{5\}$
44. a. $y^{-1/2} = 6$ a. $\left\{\frac{1}{36}\right\}$
 b. $(3y)^{-1/2} = 6$ b. $\left\{\frac{1}{108}\right\}$
45. a. $2y^{-1/2} = 10$ a. $\left\{\frac{1}{25}\right\}$
 b. $(2y)^{-1/2} = 10$ b. $\left\{\frac{1}{200}\right\}$
46. a. $(9t)^{-2/3} = 4$ a. $\left\{\frac{1}{72}\right\}$
 b. $9t^{-2/3} = 4$ b. $\left\{\frac{27}{8}\right\}$
47. $(8 - y)^{1/3} = 4$ $\{-56\}$
48. $(3n - 1)^{-2/3} = \frac{1}{4}$ $\{3\}$
49. $(x^2 + 4)^{2/3} = 25$ $\{\pm 11\}$
50. $(x^2 + 9)^{1/2} = 5$ $\{\pm 4\}$

Mixed Review Exercises

Express in simplest form without negative exponents.

1. $\dfrac{x}{x-1} - \dfrac{2}{x^2-1}$ $\dfrac{x+2}{x+1}$
2. $\sqrt{\dfrac{8x}{y^3}}$ $\dfrac{2\sqrt{2xy}}{y^2}$
3. $(a^2 + 5) - (4 + a - a^2)$ $2a^2 - a + 1$
4. $(-2c^2)(-2c)^2$ $-8c^4$
5. $\left(\dfrac{a}{b^2}\right)^{-2} \cdot \left(\dfrac{b}{a^3}\right)^{-1}$ ab^3
6. $(2n + 3)(n - 4)$ $2n^2 - 5n - 12$

10-2 Real Number Exponents

Objective To extend the meaning of exponents to include irrational numbers and to define exponential functions.

You have learned the meaning of the power b^x when x is a rational number. Now let's consider the meaning of b^x when x is an irrational number, as in the expression $2^{\sqrt{3}}$.

The graph of $y = 2^x$ for the rational values of x in the table below is shown in Figure 1. In the table irrational values of y are rounded to the nearest tenth.

x	-3	-2	-1	-0.5	0	0.5	1	1.5	2
$y = 2^x$	0.125	0.25	0.5	0.7	1	1.4	2	2.8	4

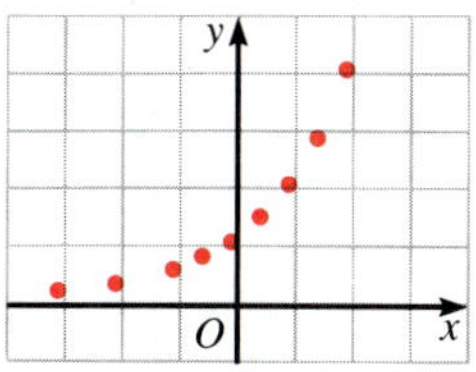

Figure 1

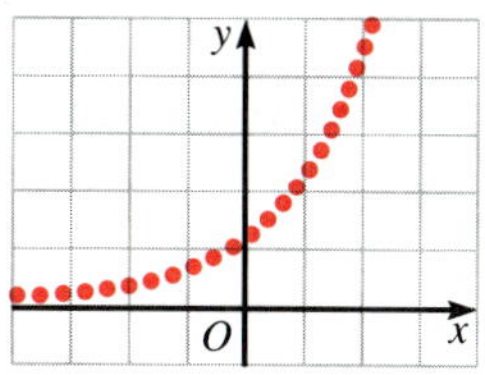

Figure 2

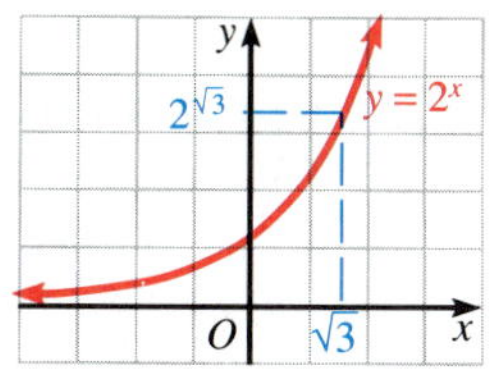

Figure 3

Figure 2 illustrates what happens when you plot many more points of $y = 2^x$ for rational values of x. If the complete graph is to be the smooth curve shown in Figure 3, there must be a point on the curve for every real value of x. In order for there to be no "breaks" in the curve, there must be a point where $x = \sqrt{3}$ and $y = 2^{\sqrt{3}}$.

To find a decimal approximation of $2^{\sqrt{3}}$, look at a sequence of rational numbers x that more and more closely approximate $\sqrt{3}$. Then consider the corresponding sequence of powers 2^x.

sequence of exponents x:	1,	1.7,	1.73,	1.732
sequence of powers 2^x:	2^1,	$2^{1.7}$,	$2^{1.73}$,	$2^{1.732}$
	$\downarrow$	$\downarrow$	$\downarrow$	$\downarrow$
approximations:	2,	3.2490,	3.3173,	3.3219

As the value of the exponent x gets closer and closer to $\sqrt{3}$, the power 2^x gets closer and closer to a definite positive real number, which is defined to be $2^{\sqrt{3}}$. A scientific calculator can approximate $2^{\sqrt{3}}$ to eight or more significant digits: $2^{\sqrt{3}} \approx 3.3219971$.

The method just discussed can be used to define b^x where $b > 0$ and x is *any real number*. It can be shown in more advanced courses that the laws of exponents also hold for these powers.

Explorations p. 841

Teaching Suggestions p. T116

Suggested Extensions p. T117

Warm-Up Exercises

Simplify.

1. 16^2 256 **2.** 16^{-1} $\frac{1}{16}$

3. 16^0 1 **4.** $16^{-1/2}$ $\frac{1}{4}$

5. $16^{3/4}$ 8 **6.** $16^{3/2}$ 64

Motivating the Lesson

Ask students whether they think 16 can be raised to the π power. Point out that 16 can certainly be raised to any *rational* approximation of π. This idea is the basis for dealing with irrational exponents, the topic of today's lesson.

Using a Calculator

You might wish to have students use their calculators to convert the sequence 2^3, $2^{3.1}$, $2^{3.14}$ . . . to decimal form. Then ask the students what number this sequence approaches. (2^{π})

Chalkboard Examples

Simplify.

1. $2^{4-\sqrt{5}} \cdot 2^{1+\sqrt{5}}$
   $2^{(4-\sqrt{5})+(1+\sqrt{5})} =$
   $2^5 = 32$
2. $(5^{2\sqrt{3}})^{\sqrt{3}}$
   $5^{2\sqrt{3}\cdot\sqrt{3}} = 5^6 = 15{,}625$
3. $\dfrac{3^{2+\sqrt{7}}}{3^{2-\sqrt{7}}}$
   $3^{(2+\sqrt{7})-(2-\sqrt{7})} = 3^{2\sqrt{7}}$

Solve.

4. $4^{2x} = \left(\frac{1}{2}\right)^{1-2x}$
   $(2^2)^{2x} = (2^{-1})^{1-2x}$
   $2^{4x} = 2^{-1+2x}$
   $4x = -1 + 2x$
   $2x = -1$
   $x = -\frac{1}{2}$
   $\therefore$ the solution set is $\left\{-\frac{1}{2}\right\}$.
5. $5^{3x} = \left(\frac{1}{25}\right)^{x-1}$
   $5^{3x} = (5^{-2})^{x-1}$
   $5^{3x} = 5^{-2x+2}$
   $3x = -2x + 2$
   $5x = 2$
   $x = \frac{2}{5}$
   $\therefore$ the solution set is $\left\{\frac{2}{5}\right\}$.

Using a Computer or a Graphing Calculator

The graphs of exponential functions can be demonstrated by you and explored by the students using a computer or graphing calculator. Students can also use a computer or graphing calculator to check their graphs for Written Exercises 33–36 on page 462.

Support Materials
Disk for Algebra
Menu Item: Function Plotter
Resource Book p. 245

Example 1 Simplify: **a.** $\dfrac{6^{\sqrt{2}}}{6^{-\sqrt{2}}}$ **b.** $4^{\pi} \cdot 2^{3-2\pi}$ **c.** $(3^{2\sqrt{2}})^{\sqrt{2}}$

Solution

a. $\dfrac{6^{\sqrt{2}}}{6^{-\sqrt{2}}} = 6^{\sqrt{2}-(-\sqrt{2})}$
$= 6^{2\sqrt{2}}$, or $36^{\sqrt{2}}$ ***Answer***

b. $4^{\pi} \cdot 2^{3-2\pi} = (2^2)^{\pi} \cdot 2^{3-2\pi}$
$= 2^{2\pi} \cdot 2^{3-2\pi}$
$= 2^{2\pi+(3-2\pi)}$
$= 2^3 = 8$ ***Answer***

c. $(3^{2\sqrt{2}})^{\sqrt{2}} = 3^{2\sqrt{2}\cdot\sqrt{2}} = 3^4 = 81$ ***Answer***

Now that the meaning of exponents has been extended to include all real numbers, *exponential functions* can be defined.

If $b > 0$ and $b \neq 1$, the function defined by $y = b^x$ is called the **exponential function** with base b.

A computer or a graphing calculator may be helpful in exploring the graphs of $y = b^x$ for various values of b.

If $b > 1$, the graph has the general shape shown in red in the diagram. It rises continuously as x increases.

If $0 < b < 1$, the graph has the general shape shown in blue. It falls continuously as x increases.

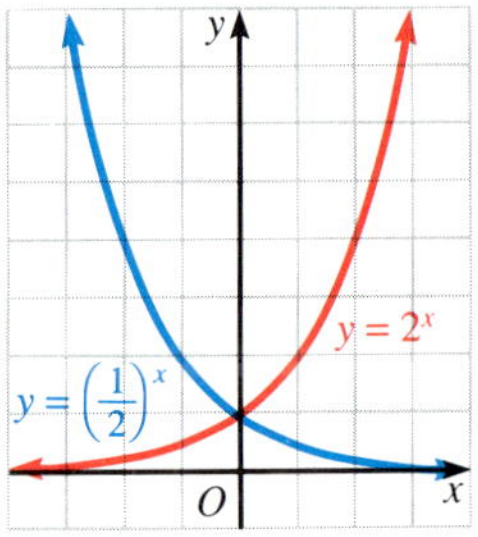

The graph of every exponential function has 1 for its y-intercept (since $b^0 = 1$) and the x-axis as an asymptote.

For any exponential function,

$$b^p = b^q \text{ if and only if } p = q.$$

This condition expresses the fact that every exponential function is *one-to-one*. A function f is called **one-to-one** if for every p and q in the domain of f, $f(p) = f(q)$ if and only if $p = q$. This is equivalent to saying that any horizontal line intersects the graph of a one-to-one function in at most one point.

In the diagram you can see that any horizontal line intersects the graph of $y = 2^x$ in at most one point. This is also true of the graph of $y = (\frac{1}{2})^x$.

The fact that every exponential function is one-to-one allows you to solve certain kinds of *exponential equations*. An **exponential equation** is an equation in which a variable appears in an exponent.

Solving Exponential Equations

Step 1 Express each side of the equation as a power of the same base.

Step 2 Set the exponents equal and then solve.

Step 3 Check the answer.

Example 2 Solve: **a.** $8^x = \frac{1}{4}$ **b.** $5^{4-t} = 25^{t-1}$

Solution

a. *Step 1* $(2^3)^x = 2^{-2}$

$2^{3x} = 2^{-2}$

Step 2 $3x = -2$

$x = -\frac{2}{3}$

Step 3 *Check:* $8^{-2/3} = \frac{1}{8^{2/3}} = \frac{1}{4}$ ✓

$\therefore$ the solution set is $\left\{-\frac{2}{3}\right\}$. ***Answer***

b. *Step 1* $5^{4-t} = (5^2)^{t-1}$

$5^{4-t} = 5^{2t-2}$

Step 2 $4 - t = 2t - 2$

$-3t = -6$

$t = 2$

Step 3 The check is left to you.

$\therefore$ the solution set is $\{2\}$.

Answer

Oral Exercises

Which number of the given pair is greater?

1. $7^{\sqrt{2}}, 7^{1.5}$ $7^{1.5}$
2. $5^{\pi}, 5^{3.14}$ 5^{π}
3. $4^{2\pi}, 16^3$ $4^{2\pi}$
4. $6^{-\sqrt{2}}, 6^{-\sqrt{3}}$ $6^{-\sqrt{2}}$

Simplify.

5. $3^{1+\pi} \cdot 3^{1-\pi}$ 9
6. $8^{\sqrt{2}} \cdot 8^{-\sqrt{2}}$ 1
7. $\frac{7^{3\sqrt{2}}}{7^{2\sqrt{2}}}$ $7^{\sqrt{2}}$
8. $\frac{9^{\sqrt{10}}}{3^{\sqrt{10}}}$ $3^{\sqrt{10}}$

Solve.

9. $3^x = \frac{1}{9}$ $\{-2\}$
10. $25^x = 125$ $\left\{\frac{3}{2}\right\}$
11. $4^x = \frac{1}{8}$ $\left\{-\frac{3}{2}\right\}$
12. $36^x = \sqrt{6}$ $\left\{\frac{1}{4}\right\}$

Written Exercises

Simplify.

A 1. **a.** $3^{\sqrt{2}} \cdot 3^{\sqrt{2}}$ $9^{\sqrt{2}}$ **b.** $(3^{\sqrt{2}})^2$ $9^{\sqrt{2}}$ **c.** $(3^{\sqrt{2}})^{\sqrt{2}}$ 9 **d.** $\frac{3^{\sqrt{2}+2}}{3^{\sqrt{2}-2}}$ 81

2. **a.** $7^{\sqrt{3}} \cdot 7^{\sqrt{2}}$ $7^{\sqrt{3}+\sqrt{2}}$ **b.** $(7^{\sqrt{3}})^2$ $49^{\sqrt{3}}$ **c.** $(7^{\sqrt{3}})^{\sqrt{2}}$ $7^{\sqrt{6}}$ **d.** $\frac{7^{\sqrt{3}+2}}{49}$ $7^{\sqrt{3}}$

Simplify.

3. $(10^{\pi})^2$ 100^{π}
4. $(5^{-\pi})^{-1}$ 5^{π}
5. $\sqrt{6^{2\pi}}$ 6^{π}
6. $\sqrt[3]{4^{6\pi}}$ 16^{π}
7. $\frac{10^{\sqrt{3}-2}}{10^{\sqrt{3}+2}}$ $\frac{1}{10,000}$
8. $\frac{6^{\sqrt{2}} \cdot 6^{\sqrt{8}}}{6^{3\sqrt{2}}}$ 1
9. $[(\sqrt{2})^{\pi}]^0$ 1
10. $(\sqrt{3})^{\sqrt{2}}(\sqrt{3})^{-\sqrt{2}}$ 1

Common Errors

Students may mistakenly think that $\frac{36^{2.4}}{6^{2.8}}$ equals $6^{-0.4}$. When reteaching, remind students that to use the law of exponents for division, the numerator and denominator must have the same base. Thus, $\frac{36^{2.4}}{6^{2.8}} = \frac{(6^2)^{2.4}}{6^{2.8}} = \frac{6^{4.8}}{6^{2.8}} = 6^2 = 36$.

Check for Understanding

Here is a suggested use for the Oral Exercises to check students' understanding as you teach the lesson.

Oral Exs. 1–4: use before Example 1.

Oral Exs. 5–8: use after Example 1.

Oral Exs. 9–12: use after Example 2.

Guided Practice

Simplify.

1. $(4^{\sqrt{3}})^{\sqrt{3}}$ 64
2. $(5^{\sqrt{3}+1}) \div (5^0 \cdot 5^{\sqrt{3}})$ 5
3. $\frac{7^{\sqrt{2}-1}}{7^{\sqrt{2}-2}}$ 7
4. $\frac{6^{\sqrt{12}} \cdot 6^{\sqrt{27}}}{6^{\sqrt{3}}}$ $6^{4\sqrt{3}}$
5. $\frac{(\sqrt{2}+\sqrt{3})^{2+\pi}}{(\sqrt{2}+\sqrt{3})^{\pi}}$ $(\sqrt{2}+\sqrt{3})^2 = 5 + 2\sqrt{6}$

Solve.

6. $\sqrt[3]{32} = 2^x$ $\left\{\frac{5}{3}\right\}$
7. $9^{x-3} = 27$ $\left\{\frac{9}{2}\right\}$
8. $4^{x-2} = 64^x$ $\{-1\}$

Simplify.

11. $(2^{\sqrt{2}})^{-1/\sqrt{2}}$ $\frac{1}{2}$
12. $(\sqrt{2}^{\sqrt{2}})^{\sqrt{2}}$ 2
13. $8^{1.2} \cdot 2^{-3.6}$ 1
14. $\frac{25^{2.4}}{5^{5.8}}$ $\frac{1}{5}$
15. $\frac{(1 + \sqrt{3})^{\pi - 1}}{(1 + \sqrt{3})^{\pi + 1}}$ $\frac{2 - \sqrt{3}}{2}$
16. $\frac{(\sqrt{2} - 1)^{2 + \pi}}{(\sqrt{2} - 1)^{\pi}}$ $3 - 2\sqrt{2}$
17. $\sqrt[4]{\frac{9^{1-\pi}}{9^{1+\pi}}}$ $\frac{1}{3^{\pi}}$
18. $\sqrt{\frac{2^{\sqrt{3}+3}}{8}}$ $2^{\sqrt{3}/2}$

Solve. If an equation has no solution, say so. 25. No sol. 26. $\left\{-\frac{5}{9}\right\}$

19. $3^x = \frac{1}{27}$ $\{-3\}$
20. $5^x = \sqrt{125}$ $\left\{\frac{3}{2}\right\}$
21. $8^{2+x} = 2$ $\left\{-\frac{5}{3}\right\}$
22. $4^{1-x} = 8$ $\left\{-\frac{1}{2}\right\}$
23. $27^{2x-1} = 3$ $\left\{\frac{2}{3}\right\}$
24. $49^{x-2} = 7\sqrt{7}$ $\left\{\frac{11}{4}\right\}$
25. $4^{2x+5} = 16^{x+1}$
26. $3^{-(x+5)} = 9^{4x}$
27. $25^{2x} = 5^{x+6}$ $\{2\}$
28. $6^{x+1} = 36^{x-1}$ $\{3\}$
29. $10^{x-1} = 100^{4-x}$ $\{3\}$

B

30. $3^{2x} - 6 \cdot 3^x + 9 = 0$ $\{1\}$
31. $4^{2x} - 63 \cdot 4^x - 64 = 0$ $\{3\}$
32. $3^{2x} - 10 \cdot 3^x + 9 = 0$ $\{0, 2\}$

Graph each pair of equations in the same coordinate system. You may wish to verify your graphs on a computer or a graphing calculator.

33. $y = 3^x$ and $y = 3^{x-2}$
34. $y = 3^x - 2$ and $y = 3^x + 2$
35. $y = 2^x$ and $y - 3 = 2^{x-4}$
36. $y = 2^{-x}$ and $y + 3 = 2^{-(x+1)}$

Solve each equation.

C

37. $2^{(2/3)x+1} - 3 \cdot 2^{(1/3)x} - 20 = 0$ $\{6\}$
38. $2^{2x-1} - 3 \cdot 2^{x-1} + 1 = 0$ $\{0, 1\}$
39. $2^x - 3^x = 0$ $\{0\}$

Extra / *Growth of Functions*

A function f is *increasing* for positive x if $f(x_2) > f(x_1)$ whenever $x_2 > x_1 > 0$. To compare the rate at which two increasing functions grow with x, you look at the ratio $\frac{f(x)}{g(x)}$. The function f will *grow faster* than g if you can make $\frac{f(x)}{g(x)}$ larger than any positive number by taking x large enough.

An important fact about exponential functions is that the exponential function b^x with $b > 1$ grows faster than any polynomial. For example, 2^x grows faster than the polynomial function x^{100}. When $x = 10$ the ratio $\frac{2^x}{x^{100}}$ is very small (approximately 10^{-97}!), but as x increases, the numerator catches up. For $x = 996$ the ratio is about 1, for $x = 1000$ it is about 10, and by taking x still larger we can make the ratio as large as we want. Even an exponential function with a base very close to 1, such as 1.0001^x, will eventually overtake any power x^n, no matter how large n is.

Summarizing the Lesson

In this lesson students learned the meaning of real number exponents and how to solve certain exponential equations. Ask students to describe the difference between the graphs of $y = \left(\frac{2}{3}\right)^x$ and $y = \left(\frac{3}{2}\right)^x$.

Suggested Assignments

Average Algebra
Day 1: 461/1–10
Assign with Lesson 10-1.
Day 2: 462/11–34
S 458/26–36 even

Average Alg. and Trig.
461/1–36

Extended Algebra
Day 1: 461/1–10
Assign with Lesson 10-1.
Day 2: 462/11–18, 19–37 odd
S 458/26–36 even

Extended Alg. and Trig.
461/1–18, 19–39 odd

Supplementary Materials

Study Guide	pp. 157–158
Practice Master	55
Resource Book	p. 142
Overhead Visual	9

Additional Answers
Written Exercises

34.

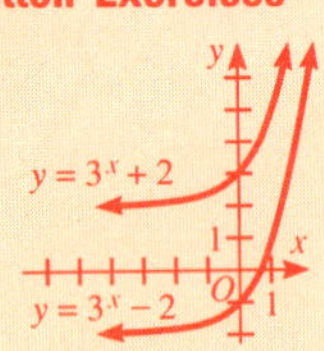

36.

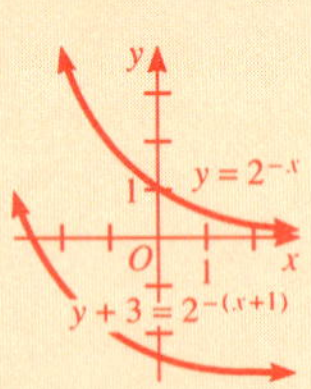

10-3 Composition and Inverses of Functions

Objective To find the composite of two given functions and to find the inverse of a given function.

Consider the squaring function $f(x) = x^2$ and the doubling function $g(x) = 2x$. As the diagram below shows, these two functions can be combined to produce a new function whose value at x is $f(g(x))$, read "f of g of x."

	g is the doubling function		f is the squaring function	
3	$\longrightarrow$	$g(3) = 6$	$\longrightarrow$	$f(g(3)) = f(6) = 36$
5	$\longrightarrow$	$g(5) = 10$	$\longrightarrow$	$f(g(5)) = f(10) = 100$
x	$\longrightarrow$	$g(x) = 2x$	$\longrightarrow$	$f(g(x)) = f(2x) = (2x)^2$

Notice that $f(g(x))$ is evaluated by working from the innermost parentheses to the outside. You begin with x, then g doubles x, and then f squares the result.

$$f(g(x)) = f(2x) = (2x)^2$$

The function whose value at x is $f(g(x))$ is called the **composite** of the functions f and g. The operation that combines f and g to produce their composite is called **composition.**

Example 1 If $f(x) = 3x - 5$ and $g(x) = \sqrt{x}$, find the following.

a. $f(g(4))$ **b.** $g(f(4))$
c. $f(g(x))$ **d.** $g(f(x))$

Solution

a. Since $g(4) = \sqrt{4} = 2$, $f(g(4)) = 3 \cdot 2 - 5 = 1$.
b. Since $f(4) = 3 \cdot 4 - 5 = 7$, $g(f(4)) = \sqrt{7}$.
c. $f(g(x)) = f(\sqrt{x}) = 3\sqrt{x} - 5$
d. $g(f(x)) = g(3x - 5) = \sqrt{3x - 5}$

Notice in Example 1 that $f(g(x)) \neq g(f(x))$.

The function $I(x) = x$ is called the **identity function.** It behaves like the multiplicative identity 1.

$$a \cdot 1 = a \quad \text{for all numbers } a$$
$$f(I(x)) = f(x) \quad \text{for all functions } f$$

For the two functions f and g defined in the example at the top of the next page, the composites $f(g(x))$ and $g(f(x))$ are both equal to the identity function.

Teaching Suggestions p. T117

Suggested Extensions p. T117

Warm-Up Exercises

Given the functions $f(x) = |x - 1|$ and $g(x) = 3x + 2$, find:

1. $f(0)$ 1
2. $g(-1)$ -1
3. $f(2) + g(2)$ 9
4. $f(-3) \cdot g(1)$ 20
5. $g(4) - f(7)$ 8
6. $g(-2) \div f(-1)$ -2

Motivating the Lesson

Tell students that just as numbers can be combined through arithmetic operations, functions can be combined through a process known as composition. An identity function and inverse functions are associated with composition just as there are identities and inverses associated with addition and multiplication.

Chalkboard Examples

If $f(x) = \sqrt{2x}$ and $g(x) = x + 5$, find the following.

1. $f(g(3))$
   $f(g(3)) = f(8) = \sqrt{16} = 4$
2. $g(f(3))$
   $g(f(3)) = g(\sqrt{6}) = \sqrt{6} + 5$
3. $f(g(x))$
   $f(g(x)) = f(x + 5) = \sqrt{2(x + 5)} = \sqrt{2x + 10}$
4. $g(f(x))$
   $g(f(x)) = g(\sqrt{2x}) = \sqrt{2x} + 5$

(continued)

Chalkboard Examples (continued)

5. Show that $f(x) = \frac{1}{3}x - 1$ and $g(x) = 3x + 3$ are inverses.
$f(g(x)) = f(3x + 3) = \frac{1}{3}(3x + 3) - 1 = x$
$g(f(x)) = g\left(\frac{1}{3}x - 1\right) = 3\left(\frac{1}{3}x - 1\right) + 3 = x$

6. Let $g(x) = |x|$. Does g have an inverse function? Explain.
g does not have an inverse function since g fails the horizontal-line test.

Common Errors

When finding a function's inverse, students may forget to determine first whether the inverse exists. When reteaching, remind students to check for a one-to-one function *before* proceeding to interchange x and y.

Check for Understanding

Here is a suggested use of the Oral Exercises to check students' understanding as you teach the lesson.
Oral Exs. 1–4: use after Example 1.
Oral Exs. 5–7: use after Example 3.

Example 2 If $f(x) = \frac{x + 4}{2}$ and $g(x) = 2x - 4$, find the following.

a. $g(1)$ and $f(g(1))$ **b.** $f(-3)$ and $g(f(-3))$ **c.** $f(g(x))$ **d.** $g(f(x))$

Solution

a. $g(1) = -2$ and $f(g(1)) = f(-2) = 1$. ***Answer***
Notice that $g\colon 1 \to -2$ and $f\colon -2 \to 1$.

b. $f(-3) = \frac{1}{2}$ and $g(f(-3)) = g\left(\frac{1}{2}\right) = -3$. ***Answer***

Notice that $f\colon -3 \to \frac{1}{2}$ and $g\colon \frac{1}{2} \to -3$.

Parts (a) and (b) suggest that the functions f and g "undo each other." Parts (c) and (d) prove that this is so for any number x.

c. $f(g(x)) = f(2x - 4) = \frac{2x - 4 + 4}{2} = x$ ***Answer***

d. $g(f(x)) = g\left(\frac{x + 4}{2}\right) = 2\left(\frac{x + 4}{2}\right) - 4 = x$ ***Answer***

In multiplication, two numbers whose product is the identity 1, such as 2 and 2^{-1}, are called inverses. Similarly, two functions whose composite is the identity I, such as f and g in Example 2, are called *inverse functions*.

Inverse Functions

The functions f and g are **inverse functions** if

$f(g(x)) = x$ for all x in the domain of g

and $g(f(x)) = x$ for all x in the domain of f.

The inverse of a function f is usually denoted f^{-1}, read "f inverse."

Caution: The superscript -1 in f^{-1} is *not* an exponent. The symbol $f^{-1}(x)$ denotes the value of f inverse at x; it does *not* mean $\frac{1}{f(x)}$.

Suppose that two functions f and g are inverses and that $f(a) = b$, so that (a, b) is on the graph of f. Then $g(b) = g(f(a)) = a$, and the point (b, a) must be on the graph of g. This means every point (a, b) on the graph of f corresponds to a point (b, a) on the graph of g. Therefore, the graphs are *mirror images* of each other with respect to the line $y = x$. You can verify this by drawing graphs of inverse functions on the same axes. A computer or a graphing calculator may be helpful. The diagram shows the inverse functions of Example 2.

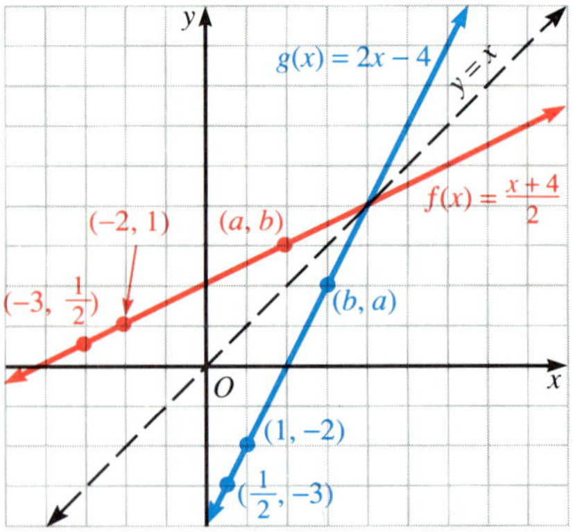

Some functions do not have inverse functions. If the reflection of the graph of a function f is itself to be the graph of a function, the graph of f must *not* contain two different points with the same y-coordinate. Therefore, a function has an inverse function if and only if it is one-to-one.

As you learned in Lesson 10-2, every horizontal line intersects the graph of a one-to-one function in at most one point. Therefore, you can use the *horizontal-line test* to tell whether a given function has an inverse function.

Horizontal-Line Test

A function has an inverse function if and only if every horizontal line intersects the graph of the function in *at most* one point.

If a function has an inverse function, you can find it by writing y for $f(x)$, interchanging x and y, and solving for y. Example 3 illustrates.

Example 3 Let $f(x) = x^3 - 1$.

a. Graph f and determine whether f has an inverse function. If so, graph f^{-1} by reflecting f across the line $y = x$.

b. Find $f^{-1}(x)$.

Solution

a. The graph of f is shown in red at the right. The graph of f passes the horizontal-line test, so f has an inverse. The graph of f^{-1}, shown in blue, is the reflection of the graph of f across the line $y = x$.

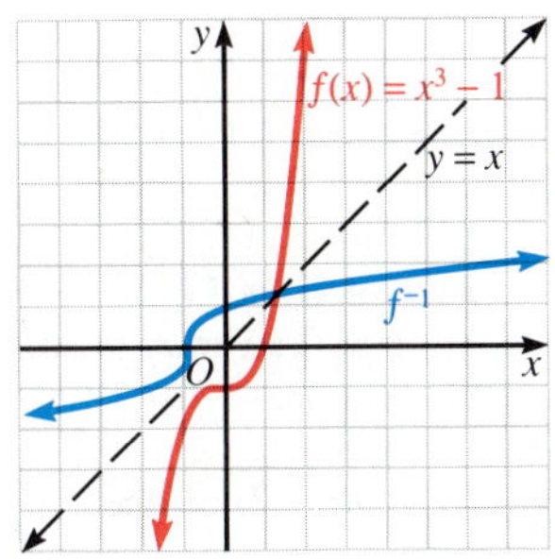

b. Replace $f(x)$ by y: $y = x^3 - 1$

Interchange x and y: $x = y^3 - 1$

Solve for y: $y^3 = x + 1$

$y = \sqrt[3]{x + 1}$

$f^{-1}(x) = \sqrt[3]{x + 1}$ ***Answer***

Oral Exercises

Suppose $f(x) = 3x$, $g(x) = x + 1$, and $h(x) = x^2 + 2$. Find the following.

1. a. $f(g(3))$ 12 b. $f(g(0))$ 3 c. $f(g(-6))$ −15 d. $f(g(x))$ $3x + 3$
2. a. $g(f(4))$ 13 b. $g(f(5))$ 16 c. $g(f(-6))$ −17 d. $g(f(x))$ $3x + 1$
3. a. $f(h(2))$ 18 b. $h(f(2))$ 38 c. $f(h(x))$ $3x^2 + 6$ d. $h(f(x))$ $9x^2 + 2$
4. a. $g(h(3))$ 12 b. $h(g(3))$ 18 c. $g(h(x))$ $x^2 + 3$ d. $h(g(x))$ $x^2 + 2x + 3$
5. Find $f^{-1}(x)$. $\frac{x}{3}$
6. Find $g^{-1}(x)$. $x - 1$
7. Does h^{-1} exist? Why or why not?

7. Yes; $h^{-1}(x) = \pm\sqrt{x - 2}$; h^{-1} is not a function.

Guided Practice

Let $f(x) = 2x + 1$, $g(x) = \sqrt{x + 1}$, and $h(x) = x + 3$. Find:

1. $f(g(8))$ 7
2. $h(f(-2))$ 0
3. $h(h(5))$ 11
4. $f(h(x))$ $2x + 7$
5. $h^{-1}(x)$ $x - 3$
6. $g(f(x))$ $\sqrt{2x + 2}$
7. Find $f^{-1}(x)$. Sketch f and f^{-1} in the same coordinate plane.
   $f^{-1}(x) = \frac{1}{2}x - \frac{1}{2}$

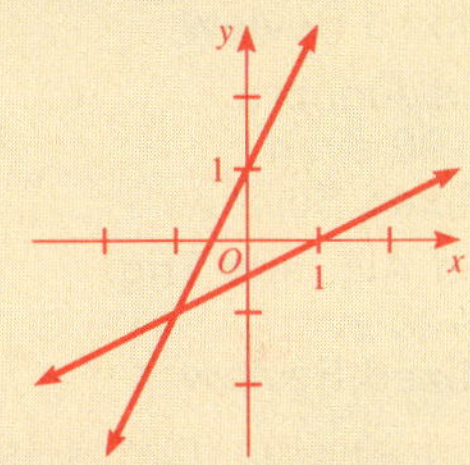

8. If $j(x) = -\frac{6}{x}$, find $j^{-1}(x)$.
   $-\frac{6}{x}$

Using a Computer or a Graphing Calculator

The fact that the graphs of such inverse functions as $f(x) = 3x + 2$ and $g(x) = \frac{x - 2}{3}$ are reflections with respect to the line $y = x$ can be demonstrated by you and investigated by the students using a computer or graphing calculator.

Students will find a computer or graphing calculator useful when checking their graphs for Written Exercises 11–22 on page 466.

Support Materials
Disk for Algebra
Menu Item: Function Plotter
Resource Book p. 245

Summarizing the Lesson

In this lesson students learned how to find the composite of two functions and how to find the inverse of a function. Ask the students to state the horizontal-line test and explain why it works.

Suggested Assignments

Average Algebra
466/2–24 even
S 467/Mixed Review

Average Alg. and Trig.
466/2–24 even
S 467/Mixed Review

Extended Algebra
466/2–26 even
S 467/Mixed Review

Extended Alg. and Trig.
466/2–26 even
S 467/Mixed Review

Supplementary Materials

Study Guide	pp. 159–160
Practice Master	56
Computer Activity	22
Resource Book	p. 143
Overhead Visual	9

Additional Answers Written Exercises

8.

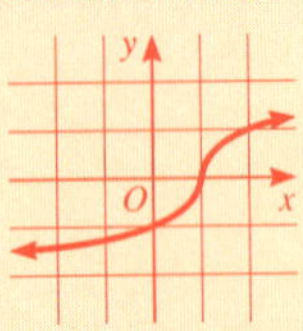

10.

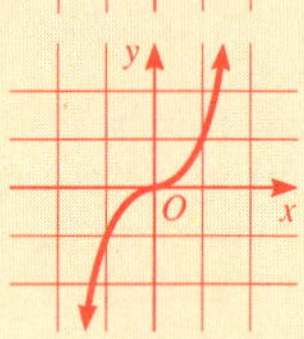

12.

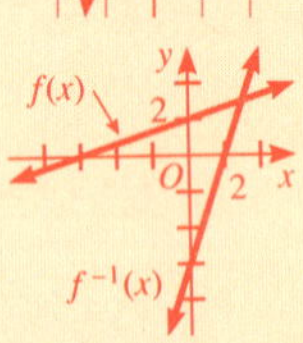

(continued on p. 494)

Written Exercises

2. d. $\frac{x}{2} - 3$ 3. d. $\frac{\sqrt{x}}{2}$ 4. d. $\frac{x}{4}$ 5. d. $\sqrt[4]{x}$

Suppose $f(x) = \frac{x}{2}$, $g(x) = x - 3$, and $h(x) = \sqrt{x}$. Find a real-number value or an expression in x for each of the following. If no real value can be found, say so.

A

1. a. $f(g(8))$ $\frac{5}{2}$ b. $f(g(-5))$ -4 c. $f(g(0))$ $-\frac{3}{2}$ d. $f(g(x))$ $\frac{x-3}{2}$
2. a. $g(f(8))$ 1 b. $g(f(-5))$ $-\frac{11}{2}$ c. $g(f(0))$ -3 d. $g(f(x))$
3. a. $f(h(9))$ $\frac{3}{2}$ b. $f(h(4))$ 1 c. $f(h(-4))$ not real d. $f(h(x))$
4. a. $h(f(32))$ 4 b. $h(f(16))$ $2\sqrt{2}$ c. $h(f(x))$ $\frac{\sqrt{2x}}{2}$ d. $f(f(x))$
5. a. $h(g(12))$ 3 b. $h(g(2))$ not real c. $h(g(x))$ $\sqrt{x-3}$ d. $h(h(x))$
6. a. $g(h(9))$ 0 b. $g(h(\sqrt{3}))$ $\sqrt[4]{3} - 3$ c. $g(h(x))$ $\sqrt{x} - 3$ d. $g(g(x))$ $x - 6$

Use the horizontal-line test to determine whether each function f has an inverse function. If so, draw a rough sketch of f^{-1} by reflecting f across $y = x$.

7. no

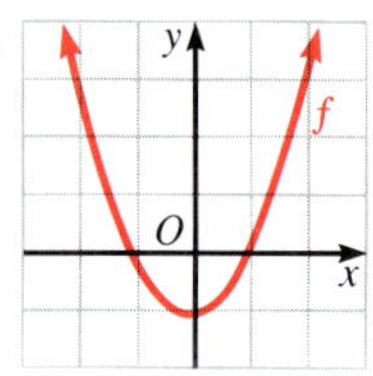

8. yes

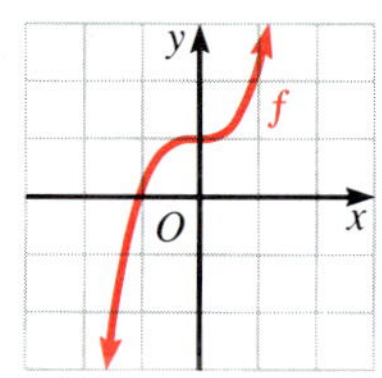

9. no

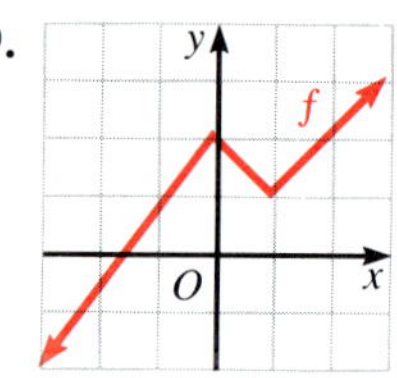

10. yes

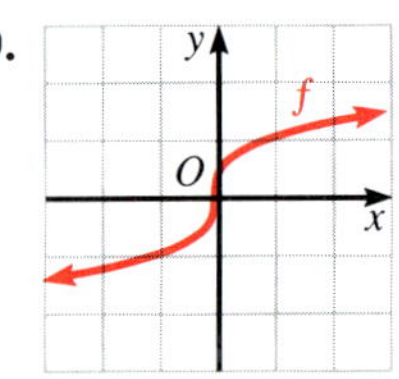

In Exercises 11–14, find $f^{-1}(x)$. Then graph f and f^{-1} in the same coordinate system. You may wish to verify your graphs on a computer or a graphing calculator.

11. $f(x) = 2x - 3$ $\frac{x+3}{2}$ 12. $f(x) = \frac{x+6}{3}$ $3x - 6$ 13. $f(x) = x^3$ $\sqrt[3]{x}$ 14. $f(x) = \frac{12}{x}$ $\frac{12}{x}$

In Exercises 15–22, graph g and use the horizontal-line test to determine if g has an inverse function. If so, find $g^{-1}(x)$. If g has no inverse, say so. You may wish to verify your graphs on a computer or a graphing calculator.

15. $g(x) = \left(\frac{8}{x}\right)^3$ $\frac{8}{\sqrt[3]{x}}$ 16. $g(x) = \sqrt[3]{2x}$ $\frac{x^3}{2}$ 17. $g(x) = x^4$ no inverse 18. $g(x) = |x|$ no inverse

B

19. $g(x) = x^2 - x$ no inverse 20. $g(x) = x^3 + 2$ $\sqrt[3]{x-2}$ 21. $g(x) = \sqrt{x^2}$ no inverse 22. $g(x) = (2x + 3)^5$ $\frac{\sqrt[5]{x} - 3}{2}$

23. a. Draw the graph of $f(x) = 2^x$ by making a table of values and carefully plotting several points.
b. Draw the graph of f^{-1} on the same coordinate system by reflecting the graph of f in the line $y = x$.
c. Find $f^{-1}(2)$, $f^{-1}(4)$, $f^{-1}(8)$, and $f^{-1}(\frac{1}{2})$.
d. Give the domain and range of f and f^{-1}.

24. The squaring function $f(x) = x^2$ has no inverse function. However, the square root function $g(x) = \sqrt{x}$ *does* have an inverse function. Find this inverse and explain why it is not the same as $f(x) = x^2$. [*Hint:* Compare the domains of f and g^{-1}.] $\{(x, x^2): x \geq 0\}$; it is the function $f(x) = x^2$ with domain restricted to nonnegative numbers.

C **25.** Let $f(x) = mx + b$. For what values of m and b will $f(x) = f^{-1}(x)$ for all x? $m = 1, b = 0$; or $m = -1$

26. Prove that the graphs of a linear function and its inverse function are never perpendicular. Let $f(x) = mx + b$; $f^{-1}(x) = \frac{1}{m}x - \frac{b}{m}$; if the graphs of f and f^{-1} are $\perp$, then $m\left(\frac{1}{m}\right) = -1$; since $1 \neq -1$, the graphs of a linear function and its inverse are not $\perp$.

Mixed Review Exercises

Simplify.

1. $25^{-3/2}$ $\frac{1}{125}$

2. $\sqrt{18} + \sqrt{32}$ $7\sqrt{2}$

3. $(2 - 3i)^2$ $-5 - 12i$

4. $(3^{\sqrt{2}})^{2\sqrt{2}}$ 81

5. $\sqrt{-48} \cdot \sqrt{-12}$ -24

6. $\frac{2^{\pi-2}}{2^{\pi+2}}$ $\frac{1}{16}$

7. $(-27)^{2/3}$ 9

8. $\frac{\sqrt{2} - 3}{1 + \sqrt{2}}$ $5 - 4\sqrt{2}$

9. $\frac{5^2 - 1}{1 - 3^2}$ -3

10. $16^{1.75}$ 128

11. $\frac{4 + 3i}{2 - i}$ $1 + 2i$

12. $4^{1.5} \cdot 2^{-2}$ 2

Self-Test 1

Vocabulary

exponential form (p. 456)
exponential function (p. 460)
one-to-one function (p. 460)
exponential equation (p. 460)
composite (p. 463)
composition (p. 463)
identity function (p. 463)
inverse functions (p. 464)
horizontal-line test (p. 465)

1. Write in exponential form. **Obj. 10-1, p. 455**

a. $\sqrt[6]{64x^4y^{-2}}$ $2x^{2/3}y^{-1/3}$ **b.** $\sqrt[3]{6^{1/2}} \cdot \sqrt[6]{6^{-3}}$ $6^{-1/3}$

2. Write in simplest radical form.

a. $\left(\frac{25}{\sqrt[3]{20}}\right)^{3/2}$ $\frac{25\sqrt{5}}{2}$ **b.** $\sqrt[3]{x^2y^4} \cdot \sqrt{x^3y}$ $x^2y\sqrt[6]{xy^5}$

3. Solve $(x - 2)^{3/5} = 8$. $\{34\}$

4. Simplify. **Obj. 10-2, p. 459**

a. $4^{3/\sqrt{2}} \div 16^{2\sqrt{2}}$ $2^{-5\sqrt{2}}$ **b.** $\frac{16^{\sqrt{5}}}{4^{\sqrt{5}}} \cdot 32^{-2\sqrt{5}}$ $2^{-8\sqrt{5}}$

5. Solve $7^{x-1} = 49^{4-x}$. $\{3\}$

6. If $f(x) = 6x + 1$ and $g(x) = \sqrt{x}$, find the following. **Obj. 10-3, p. 463**

a. $f(g(4))$ 13 **b.** $g(f(4))$ 5 **c.** $f(g(x))$ $6\sqrt{x} + 1$ **d.** $g(f(x))$ $\sqrt{6x + 1}$

7. Show that $f(x) = 3x - 7$ and $g(x) = \frac{x + 7}{3}$ are inverse functions.

Additional Answers Self-Test 1

7. For every x in the domain of g, $f(g(x)) = 3\left(\frac{x + 7}{3}\right) - 7 = x + 7 - 7 = x$; for every x in the domain of f, $g(f(x)) = \frac{(3x - 7) + 7}{3} = \frac{3x}{3} = x$.

Quick Quiz

Write in exponential form.

1. $\sqrt[3]{3x^5}$ $3^{1/3}x^{5/3}$

2. $\sqrt{5^{-1}} \cdot \sqrt[3]{5^2}$ $5^{1/6}$

Write in simplest radical form.

3. $\left(\frac{\sqrt[3]{5}}{8}\right)^{2/3}$ $\frac{\sqrt[9]{25}}{4}$

4. $(4x^{-1})^{-3/2}$ $\frac{1}{8}x\sqrt{x}$

5. Solve $(y + 1)^{2/3} = 9$. $\{26\}$

Simplify.

6. $\frac{2^{\pi} \cdot 2^{-\pi}}{2^{-5}}$ 32

7. $8^{\sqrt{2}} \div 2^{\sqrt{8}}$ $2^{\sqrt{2}}$

Solve.

8. $27^{x+1} = 81$ $\left\{\frac{1}{3}\right\}$

9. $\left(\frac{1}{10}\right)^y = 100^{y+3}$ $\{-2\}$

10. If $f(x) = x^2$ and $g(x) = 1 - 2x$, find:

a. $f(g(2))$ 9

b. $g(f(x))$ $1 - 2x^2$

c. $f(g(x))$ $(1 - 2x)^2$

d. $g^{-1}(x)$ $\frac{1 - x}{2}$

11. Let $h(x) = x^4 - 1$. Show that h has no inverse function. Since $h(1) = 0$ and $h(-1) = 0$, the line $y = 0$ intersects the graph of h in two points. $\therefore$ h has no inverse function.

Teaching Suggestions p. T117

Reading Algebra p. T117

Suggested Extensions p. T117

Warm-Up Exercises

Write each of the following as a power of 2.

1. 8 2^3 **2.** $2\sqrt{2}$ $2^{3/2}$

3. $\frac{\sqrt{2}}{2}$ $2^{-1/2}$ **4.** 1 2^0

5. $\sqrt[3]{4}$ $2^{2/3}$ **6.** 0.0625 2^{-4}

Motivating the Lesson

Tell students that logarithms, the topic of today's lesson, represented a milestone in the history of mathematics, because they greatly facilitated computational work, much as the abacus did centuries before. Although their use as a computational aid is now outdated due to calculators and computers, logarithmic functions continue to play an important role in mathematics and science.

Chalkboard Examples

1. Express $\log_3 81 = 4$ in exponential form.
   $3^4 = 81$
2. Express $5^3 = 125$ in logarithmic form.
   $\log_5 125 = 3$

Simplify.

3. $\log_2 0.125$
   Let $\log_2 0.125 = x$
   $2^x = 0.125$
   $2^x = \frac{1}{8}$
   $2^x = 2^{-3}$
   $x = -3$
   $\therefore \log_2 0.125 = -3$

Logarithmic Functions

10-4 Definition of Logarithms

Objective To define logarithmic functions and to learn how they are related to exponential functions.

The graph of $f(x) = 2^x$ is shown in Figure 1. Since the function passes the horizontal-line test, it has an inverse function. The graph of f^{-1} can be found by reflecting the graph of f across the line $y = x$, as shown in Figure 2.

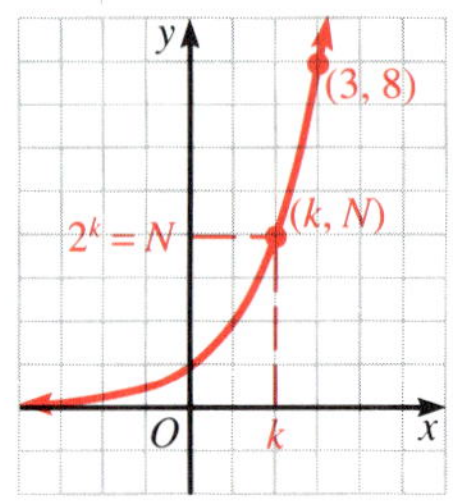

$f(x) = 2^x$

Figure 1

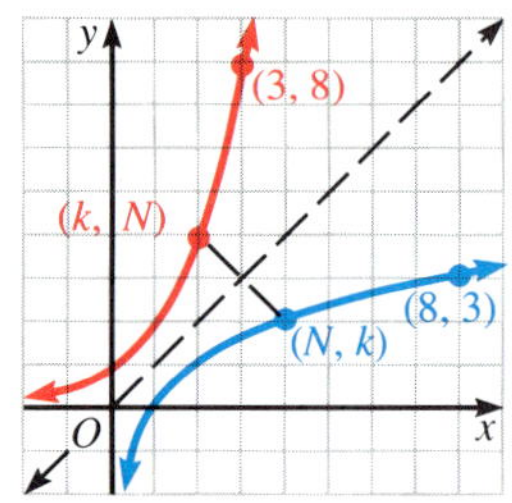

Figure 2

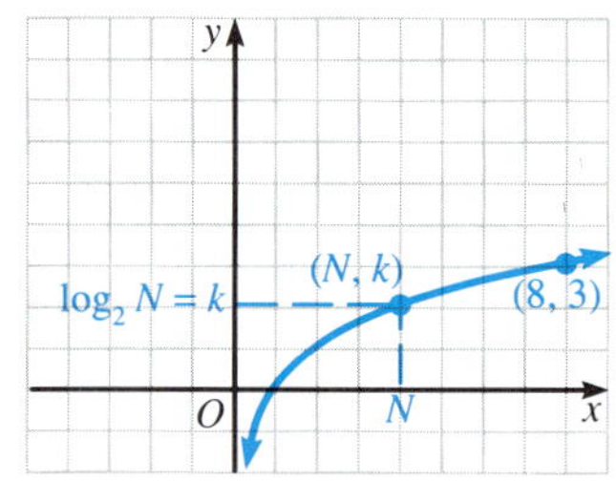

$f^{-1}(x) = \log_2 x$

Figure 3

The inverse of f, the exponential function with base 2, is called the **logarithmic function** with base 2, and is denoted by $f^{-1}(x) = \log_2 x$. (See Figure 3.)

$\log_2 x$ is read "the base 2 logarithm of x" or "log base 2 of x."

Notice that (3, 8) is on the graph of $f(x) = 2^x$ and (8, 3) is on the graph of $f^{-1}(x) = \log_2 x$. Since these functions are inverses, it follows that $2^3 = 8$ means $\log_2 8 = 3$. This and other examples are given below.

Exponential Form		*Logarithmic Form*
$2^3 = 8$	means	$\log_2 8 = 3$
$2^4 = 16$	means	$\log_2 16 = 4$
$2^0 = 1$	means	$\log_2 1 = 0$
$2^{-1} = \frac{1}{2}$	means	$\log_2 \frac{1}{2} = -1$
$2^k = N$	means	$\log_2 N = k$

A base other than 2 can be used with logarithmic functions. In fact the base can be any positive number except 1, since every power of 1 is 1.

Definition of Logarithm

If b and N are positive numbers ($b \neq 1$),

$$\log_b N = k \text{ if and only if } b^k = N.$$

Every positive number has a unique logarithm with base b. Since the base b exponential function is one-to-one, its inverse is one-to-one. Therefore,

$$\log_b M = \log_b N \text{ if and only if } M = N.$$

Several other properties of logarithms are easy to verify. Since the statements $\log_b N = k$ and $b^k = N$ are equivalent, by substitution

$$\log_b b^k = k \qquad \text{and} \qquad b^{\log_b N} = N.$$

Since $b = b^1$ and $1 = b^0$, it follows from the definition of a logarithm that

$$\log_b b = 1 \qquad \text{and} \qquad \log_b 1 = 0.$$

Example 1 Write each equation in exponential form.

a. $\log_6 36 = 2$ **b.** $\log_2 2 = 1$ **c.** $\log_{10} (0.001) = -3$

Solution **a.** $6^2 = 36$ **b.** $2^1 = 2$ **c.** $10^{-3} = 0.001$

Example 2 Write each equation in logarithmic form.

a. $6^0 = 1$ **b.** $8^{-2/3} = \frac{1}{4}$ **c.** $5^{3/2} = 5\sqrt{5}$

Solution **a.** $\log_6 1 = 0$ **b.** $\log_8 \frac{1}{4} = -\frac{2}{3}$ **c.** $\log_5 (5\sqrt{5}) = \frac{3}{2}$

Example 3 Simplify each logarithm.

a. $\log_5 25$ **b.** $\log_2 8\sqrt{2}$ **c.** $2^{\log_2 7}$

Solution **a.** Since $5^2 = 25$, $\log_5 25 = 2$. ***Answer***

b. Let $\log_2 8\sqrt{2} = x$.

Then: $2^x = 8\sqrt{2}$

$2^x = 2^3 \cdot 2^{1/2}$

$2^x = 2^{7/2}$ Set the exponents equal.

$x = \frac{7}{2}$

$\therefore \log_2 8\sqrt{2} = \frac{7}{2}$ ***Answer***

c. Since $b^{\log_b N} = N$, $2^{\log_2 7} = 7$. ***Answer***

4. $\log_{1/9} \sqrt[4]{3}$

Let $\log_{1/9} \sqrt[4]{3} = x$

Then $\left(\frac{1}{9}\right)^x = \sqrt[4]{3}$

$(3^{-2})^x = 3^{1/4}$

$3^{-2x} = 3^{1/4}$

$-2x = \frac{1}{4}$

$x = -\frac{1}{8}$

$\therefore \log_{1/9} \sqrt[4]{3} = -\frac{1}{8}$

Solve.

5. $\log_4 2 = x$

$4^x = 2$

$2^{2x} = 2^1$

$2x = 1$

$x = \frac{1}{2}$

$\therefore$ the solution set is $\left\{\frac{1}{2}\right\}$.

6. $\log_x 5 = -1$

$x^{-1} = 5$

$(x^{-1})^{-1} = 5^{-1}$

$x = \frac{1}{5}$

$\therefore$ the solution set is $\left\{\frac{1}{5}\right\}$.

Common Errors

Some students find changing from logarithmic form to exponential form or exponential form to logarithmic form confusing. When reteaching, remind students that a logarithm of a number is an exponent.

Check for Understanding

Here is a suggested use of the Oral Exercises to check students' understanding as you teach the lesson.

Oral Exs. 1–4: use after Example 1.

Oral Exs. 5–8: use after Example 2.

Oral Exs. 9–16: use after Example 3.

Oral Exs. 17–20: use after Example 4.

Guided Practice

Simplify.

1. $\log_5 625$ 4
2. $\log_4\left(\frac{1}{64}\right)$ -3
3. $\log_2 4\sqrt{2}$ $\frac{5}{2}$
4. $\log_6 \sqrt[3]{36}$ $\frac{2}{3}$
5. $\log_4 \frac{1}{8}$ $-\frac{3}{2}$
6. $\log_{1/3} \sqrt[3]{81}$ $-\frac{4}{3}$

Solve.

7. $\log_9 x = \frac{3}{2}$ {27}
8. $\log_{1/2} x = -2$ {4}

Summarizing the Lesson

In this lesson students learned the definition of logarithmic functions and their relationship to exponential functions. Ask the students to state the definition of logarithm and give several examples to illustrate.

Suggested Assignments

Average Algebra

Day 1: 470/1–22
S 462/35, 36
R 467/Self-Test 1

Day 2: 471/23–33, 35, 36
S 466/19–23 odd

Example 4 Solve each equation.

a. $\log_4 x = 3$

b. $\log_x 81 = 4$

Solution

a. Rewrite in exponential form.

$$4^3 = x$$
$$64 = x$$

$\therefore$ the solution set is {64}. ***Answer***

b. Rewrite in exponential form.

$$x^4 = 81$$

Since $81 = 3^4$,

$$x^4 = 3^4$$
$$x = 3$$

$\therefore$ the solution set is {3}. ***Answer***

Oral Exercises

Express in exponential form.

1. $\log_2 32 = 5$ $2^5 = 32$
2. $\log_3 9 = 2$ $3^2 = 9$
3. $\log_7 \sqrt{7} = \frac{1}{2}$ $7^{1/2} = \sqrt{7}$
4. $\log_3 \frac{1}{81} = -4$ $3^{-4} = \frac{1}{81}$

Express in logarithmic form.

5. $4^3 = 64$ $\log_4 64 = 3$
6. $9^{3/2} = 27$ $\log_9 27 = \frac{3}{2}$
7. $10^{-2} = 0.01$ $\log_{10} 0.01 = -2$
8. $16^{-3/4} = \frac{1}{8}$ $\log_{16} \frac{1}{8} = -\frac{3}{4}$

Simplify.

9. $\log_6 36$ 2
10. $\log_2 16$ 4
11. $\log_{10} 100$ 2
12. $\log_3 \frac{1}{9}$ -2
13. $\log_2 2\sqrt{2}$ $\frac{3}{2}$
14. $\log_7 1$ 0
15. $4^{\log_4 16}$ 16
16. $\log_6 (6^5)$ 5
17. $\log_2 10$ lies between the consecutive integers __?__ and __?__. 3; 4
18. $\log_{10} 101$ lies between the consecutive integers __?__ and __?__. 2; 3
19. $\log_3 \frac{1}{50}$ lies between the consecutive integers __?__ and __?__. -4; -3
20. If you solve $\log_2 x^2 = \log_2 100$, then the solution set is {?, ?}. -10, 10

Written Exercises

Simplify each logarithm.

A

1. $\log_5 125$ 3
2. $\log_4 16$ 2
3. $\log_3 81$ 4
4. $\log_6 6$ 1
5. $\log_3 1$ 0
6. $\log_8 4$ $\frac{2}{3}$
7. $\log_5 \frac{1}{25}$ -2
8. $\log_2 \frac{1}{8}$ -3
9. $\log_6 6\sqrt{6}$ $\frac{3}{2}$
10. $\log_5 25\sqrt{5}$ $\frac{5}{2}$
11. $\log_4 \sqrt{2}$ $\frac{1}{4}$
12. $\log_{27} \sqrt{3}$ $\frac{1}{6}$

13. $\log_7 \sqrt[3]{49}$ $\frac{2}{3}$

14. $\log_3 \sqrt[5]{9}$ $\frac{2}{5}$

15. $\log_{1/2} 8$ -3

16. $\log_{1/3} 27$ -3

17. $\log_2 \sqrt[3]{\frac{1}{4}}$ $-\frac{2}{3}$

18. $\log_{10} \frac{1}{\sqrt{1000}}$ $-\frac{3}{2}$

Solve for *x*.

19. $\log_7 x = 2$ $\{49\}$

20. $\log_6 x = 3$ $\{216\}$

21. $\log_9 x = -\frac{1}{2}$ $\left\{\frac{1}{3}\right\}$

22. $\log_6 x = 2.5$ $\{36\sqrt{6}\}$

23. $\log_4 x = -\frac{3}{2}$ $\left\{\frac{1}{8}\right\}$

24. $\log_{1/9} x = -\frac{1}{2}$ $\{3\}$

B 25. $\log_x 27 = \frac{3}{2}$ $\{9\}$

26. $\log_x 64 = 6$ $\{2\}$

27. $\log_x 7 = -\frac{1}{2}$ $\left\{\frac{1}{49}\right\}$

28. $\log_x 7 = 1$ $\{7\}$

29. $\log_x 1 = 0$ $\{x: x > 0$ and $x \neq 1\}$

30. $\log_x 2 = 0$ No solution

31. a. Show that $\log_2 8 + \log_2 4 = \log_2 32$ by simplifying the three logarithms.

b. Show that $\log_9 3 + \log_9 27 = \log_9 81$ by simplifying the three logarithms.

c. State a generalization based on parts (a) and (b).

32. a. Simplify $\log_2 8$ and $\log_8 2$. $3; \frac{1}{3}$

b. Simplify $\log_3 \sqrt{3}$ and $\log_{\sqrt{3}} 3$. $\frac{1}{2}; 2$

c. State a generalization based on parts (a) and (b). c. For positive numbers a and b where $a \neq 1$ and $b \neq 1$, $\log_b a = \frac{1}{\log_a b}$.

33. a. If $f(x) = 6^x$, then $f^{-1}(x) =$ __?__. $\log_6 x$

b. Find $f^{-1}(36)$ and $f^{-1}\left(\frac{1}{\sqrt{6}}\right)$. $2; -\frac{1}{2}$

c. Give the domain and range of f and f^{-1}.

34. a. If $g(x) = \log_4 x$, then $g^{-1}(x) =$ __?__. 4^x

b. Find $g^{-1}(2)$ and $g^{-1}\left(-\frac{3}{2}\right)$. $16; \frac{1}{8}$

c. Give the domain and range of g and g^{-1}.

Sketch the graph of each function and its inverse function in the same coordinate system. Label at least three points on each graph. You may wish to check your graphs on a computer or a graphing calculator.

35. $f(x) = 6^x$

36. $g(x) = \log_4 x$

37. $h(x) = \log_{10} x$

38. $k(x) = \left(\frac{1}{2}\right)^x$

Solve.

C 39. $\log_5 (\log_3 x) = 0$ $\{3\}$

40. $\log_4 (\log_3 (\log_2 x)) = 0$ $\{8\}$

41. If $0 < b < 1$ and $0 < u < 1$, is $\log_b u$ positive or negative? positive

42. If $1 < a < b$, is $\log_b (\log_b a)$ positive or negative? negative

Average Alg. and Trig.
Day 1: 470/1–22
S 458/44–50 even
R 467/Self-Test 1
Day 2: 471/23–33, 35, 36
S 466/19, 21, 23

Extended Algebra
Day 1: 470/1–22
S 466/19, 21, 23
R 467/Self-Test 1
Day 2: 471/23–33, 35, 36
S 462/38, 39

Extended Alg. and Trig.
Day 1: 470/1–22
S 458/44–50 even
R 467/Self-Test 1
Day 2: 471/23–36
S 466/19–23 odd

Supplementary Materials

Study Guide pp. 161–162
Overhead Visual 9

Using a Computer or a Graphing Calculator

Once students have drawn their own graphs for Written Exercises 35–38, you may wish to have them check their work using a computer or graphing calculator.

Support Materials
Disk for Algebra
Menu Item: Function Plotter

Additional Answers Written Exercises

31. a. $\log_2 8 + \log_2 4 = \log_2 2^3 + \log_2 2^2 = 3 + 2 = 5 = \log_2 2^5 = \log_2 32$

b. $\log_9 3 + \log_9 27 = \log_9 9^{1/2} + \log_9 9^{3/2} = \frac{1}{2} + \frac{3}{2} = 2 = \log_9 9^2 = \log_9 81$

c. For all positive numbers a, b, and c, $a \neq 1$, $\log_a b + \log_a c = \log_a bc$

Additional Answers Written Exercises

(continued)

33. c. domain of f: {real numbers}
range of f: $\{y: y > 0\}$
domain of f^{-1}: $\{x: x > 0\}$
range of f^{-1}: {real numbers}

34. c. domain of g: $\{x: x > 0\}$
range of g: {real numbers}
domain of g^{-1}: {real numbers}
range of g^{-1}: $\{y: y > 0\}$

35.

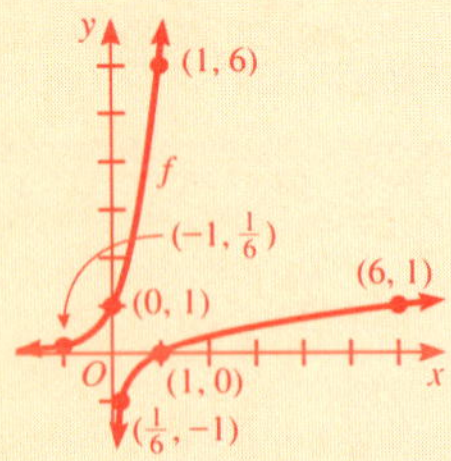

36.

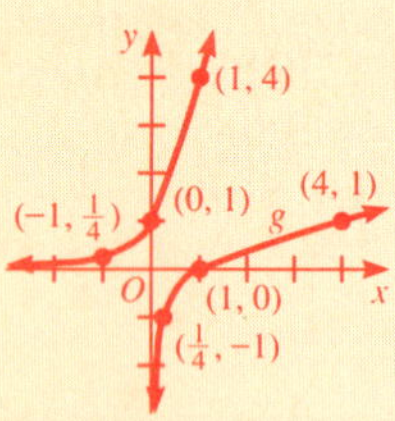

37.

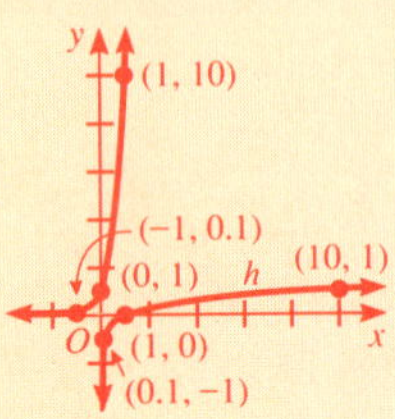

38.

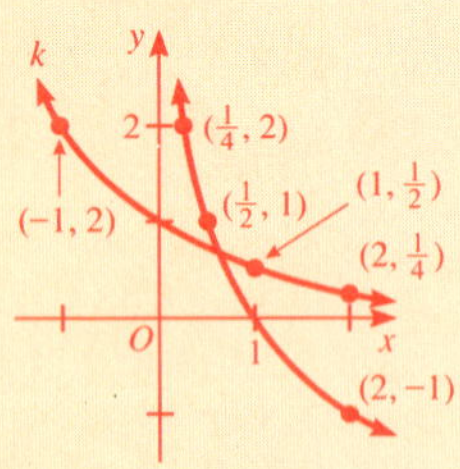

43. The decibel (dB) is used to measure the loudness of sound. Sound that is just barely audible has a decibel level of 0 and an intensity level of 10^{-12} W/m² (watts per square meter). Sound painful to the ear has a decibel level of 130 and an intensity level of 10 W/m². The formula $D = 10 \log_{10}\left(\frac{I}{I_0}\right)$ gives the decibel level of a sound whose intensity is I, where I_0 is the intensity of barely audible sound. Some decibel levels and intensities are shown in the table below. The intensity is given in watts per square meter.

	Intensity	Decibels
Barely audible sound	10^{-12}	0
Quiet conversation	10^{-9}	30
Heavy traffic	10^{-3}	90
Jet plane (at 20 m)	10^{2}	140

a. The intensity of thunder is 10^{12} times the intensity of barely audible sound. What is the decibel level of thunder? 120 dB

b. The decibel levels of a subway train entering a station and normal conversation are 100 dB and 60 dB, respectively. How many times as intense as normal conversation is the noise of the subway train? 10^4

Historical Note / *Logarithms*

The late sixteenth century was a time of great mathematical activity, particularly in astronomy and navigation. Because of the involved computations that these fields required, there was a need for a method to simplify arithmetic calculation and make multiplication and division easier. The Scottish mathematician John Napier found a way to turn multiplication into addition and division into subtraction, and in 1614 he published a work demonstrating the use of logarithms.

In the nineteenth century, the great French mathematician Laplace said of Napier that by lessening the work of computation his invention doubled the life of the astronomer. Scientists and mathematicians continued to use logarithms until calculators and computers provided more efficient tools.

Napier's logarithms are related to what we today call *natural logarithms*. A year after Napier published his work, the English mathematician Henry Briggs became interested in Napier's invention. It was Briggs who suggested that 10 be the logarithmic base and who introduced the terms *characteristic* and *mantissa*. He compiled an extensive table of logarithms giving 14-place approximations for logarithms of numbers from 1 to 20,000 and 90,000 to 100,000.

10-5 Laws of Logarithms

Objective To learn and apply the basic properties of logarithms.

The laws of exponents can be used to derive the *laws of logarithms*.

Laws of Logarithms

Let b be the base of a logarithmic function ($b > 0$, $b \neq 1$). Let M and N be positive numbers.

1. $\log_b MN = \log_b M + \log_b N$
2. $\log_b \frac{M}{N} = \log_b M - \log_b N$
3. $\log_b M^k = k \log_b M$

Laws 1 and 3 are proved below. The proof of Law 2 is left as Exercise 43.

Proof: Let $\log_b M = x$ and $\log_b N = y$. Therefore, $b^x = M$ and $b^y = N$.

Law 1	*Law 3*
$MN = b^x b^y = b^{x+y}$	$M^k = (b^x)^k = b^{kx}$
$\log_b MN = \log_b b^{x+y}$	$\log_b M^k = \log_b b^{kx}$
$\log_b MN = x + y$	$\log_b M^k = kx$
$\therefore \log_b MN = \log_b M + \log_b N$	$\therefore \log_b M^k = k \log_b M$

Example 1 Express $\log_6 M^2N^3$ in terms of $\log_6 M$ and $\log_6 N$.

Solution $\log_6 M^2N^3 = \log_6 M^2 + \log_6 N^3$ Use Law 1.

$= 2 \log_6 M + 3 \log_6 N$ Use Law 3.

Example 2 Express $\log_2 \sqrt{\frac{M}{N^5}}$ in terms of $\log_2 M$ and $\log_2 N$.

Solution $\log_2 \sqrt{\frac{M}{N^5}} = \log_2 \left(\frac{M}{N^5}\right)^{1/2}$

$= \frac{1}{2} \log_2 \left(\frac{M}{N^5}\right)$ Use Law 3.

$= \frac{1}{2}(\log_2 M - \log_2 N^5)$ Use Law 2.

$= \frac{1}{2}(\log_2 M - 5 \log_2 N)$ Use Law 3.

Teaching Suggestions p. T117

Suggested Extensions p. T118

Warm-Up Exercises

Complete each of the following laws of exponents.

1. $a^m \cdot a^n =$? a^{m+n}
2. $(ab)^m =$? $a^m b^m$
3. $(a^m)^n =$? a^{mn}
4. $\frac{a^m}{a^n} =$? a^{m-n}
5. $\left(\frac{a}{b}\right)^m =$? $\frac{a^m}{b^m}$

Motivating the Lesson

Tell students that logarithms have amazing properties: They can change multiplication to addition, division to subtraction, and exponentiation to multiplication. These properties are stated more formally in the laws of logarithms, the topic of today's lesson.

Chalkboard Examples

Express in terms of $\log_{10} a$ and $\log_{10} b$.

1. $\log_{10} (a^2\sqrt{b})$

   $\log_{10} (a^2 b^{1/2})$
   $= \log_{10} a^2 + \log_{10} b^{1/2}$
   $= 2 \log_{10} a + \frac{1}{2} \log_{10} b$

2. $\log_{10} \frac{a^5}{\sqrt{b^3}}$

   $\log_{10} \frac{a^5}{b^{3/2}} =$
   $5 \log_{10} a - \frac{3}{2} \log_{10} b$

(continued)

Chalkboard Examples (continued)

3. Express $3 \log_6 m - \frac{\log_6 n}{2}$ as a single logarithm.

$3 \log_6 m - \frac{\log_6 n}{2} =$

$\log_6 m^3 - \log_6 n^{1/2} =$

$\log_6 \left(\frac{m^3}{\sqrt{n}}\right)$

Given that $\log_2 3 = 1.585$ and $\log_2 5 = 2.322$, find the following.

4. $\log_2 81$

$81 = 3^4$

$\log_2 81 = 4 \log_2 3 =$

$4(1.585) = 6.340$

5. $\log_2 6$

$6 = 2 \cdot 3$

$\log_2 6 = \log_2 2 + \log_2 3 =$

$1 + 1.585 = 2.585$

Solve.

6. $\log_5 x = 3 \log_5 2 + \log_5 7$

$\log_5 x = \log_5 2^3 + \log_5 7$

$\log_5 x = \log_5 (8 \cdot 7)$

$\log_5 x = \log_5 56$

$x = 56$

$\therefore$ the solution set is $\{56\}$.

7. $\log_7 (x + 1) + \log_7 (x - 5) = 1$

$\log_7 (x + 1)(x - 5) = 1$

$(x + 1)(x - 5) = 7^1$

$x^2 - 4x - 5 = 7$

$x^2 - 4x - 12 = 0$

$(x + 2)(x - 6) = 0$

$x = -2$ or $x = 6$

If $x = -2$, $\log_7 (x + 1)$ and $\log_7 (x - 5)$ are not defined.

$\log_7 (6 + 1) + \log_7 (6 - 5) = 1$

$\log_7 7 + \log_7 1 = 1$

$1 + 0 = 1$ ✓

$\therefore$ the solution set is $\{6\}$.

The process illustrated in Examples 1 and 2 can also be reversed.

Example 3 Express as a single logarithm.

a. $\log_{10} p + 3 \log_{10} q$ **b.** $4 \log_{10} p - 2 \log_{10} q$

Solution

a. $\log_{10} p + 3 \log_{10} q = \log_{10} p + \log_{10} q^3$ Use Law 3.

$= \log_{10} pq^3$ Use Law 1.

b. $4 \log_{10} p - 2 \log_{10} q = \log_{10} p^4 - \log_{10} q^2$ Use Law 3.

$= \log_{10} \frac{p^4}{q^2}$ Use Law 2.

You can often use the laws of logarithms to compute the numerical value of a logarithm from given logarithms.

Example 4 If $\log_{10} 2 = 0.30$ and $\log_{10} 3 = 0.48$, find the following.

a. $\log_{10} 18$ **b.** $\log_{10} \frac{20}{3}$ **c.** $\log_{10} 5$ **d.** $\log_{10} \left(\frac{1}{\sqrt[3]{2}}\right)$

Solution

a. Since $18 = 2 \cdot 3^2$,

$$\log_{10} 18 = \log_{10} 2 + 2 \log_{10} 3$$
$$= 0.30 + 2(0.48)$$
$$= 1.26 \quad \textbf{\textit{Answer}}$$

b. Since $\frac{20}{3} = \frac{2 \cdot 10}{3}$,

$$\log_{10} \frac{20}{3} = \log_{10} 2 + \log_{10} 10 - \log_{10} 3$$
$$= 0.30 + 1 - 0.48$$
$$= 0.82 \quad \textbf{\textit{Answer}}$$

c. Since $5 = \frac{10}{2}$,

$$\log_{10} 5 = \log_{10} 10 - \log_{10} 2$$
$$= 1 - 0.30$$
$$= 0.70 \quad \textbf{\textit{Answer}}$$

d. Since $\frac{1}{\sqrt[3]{2}} = 2^{-1/3}$,

$$\log_{10} \left(\frac{1}{\sqrt[3]{2}}\right) = -\frac{1}{3} \log_{10} 2$$
$$= -\frac{1}{3}(0.30)$$
$$= -0.10 \quad \textbf{\textit{Answer}}$$

You can use the laws and properties of logarithms to solve logarithmic equations. *Because the logarithm of a variable expression is defined only if the expression is positive, be sure to check the answers you obtain.*

Example 5 Solve $\log_3 x + \log_3 (x - 6) = 3$ for x.

Solution

$$\log_3 x + \log_3 (x - 6) = 3$$
$$\log_3 x(x - 6) = 3 \qquad \text{Use Law 1.}$$
$$\log_3 (x^2 - 6x) = 3$$

Change to exponential form: $x^2 - 6x = 3^3$

$$x^2 - 6x - 27 = 0$$
$$(x + 3)(x - 9) = 0$$
$$x = -3 \quad \text{or} \quad x = 9$$

Check: If $x = -3$, $\log_3 x$ is not defined. So -3 is *not* a root.

If $x = 9$, $\log_3 x + \log_3 (x - 6) = \log_3 9 + \log_3 3$

$$= \log_3 (9 \cdot 3)$$
$$= \log_3 27$$
$$= 3 \ \checkmark$$

$\therefore$ the solution set is $\{9\}$. ***Answer***

Oral Exercises

Express each logarithm in terms of $\log_3 M$ and $\log_3 N$.

1. $\log_3 M^4$ $4 \log_3 M$
2. $\log_3 N^6$ $6 \log_3 N$
3. $\log_3 M^4N$ $4 \log_3 M + \log_3 N$
4. $\log_3 \left(\frac{M}{N^3}\right)$ $\log_3 M - 3 \log_3 N$
5. $\log_3 \left(\frac{1}{M}\right)$ $-\log_3 M$
6. $\log_3 \sqrt{M}$ $\frac{1}{2} \log_3 M$
7. $\log_3 \sqrt[3]{N^2}$ $\frac{2}{3} \log_3 N$
8. $\log_3 \left(\frac{1}{N\sqrt{N}}\right)$ $-\frac{3}{2} \log_3 N$

Express as a logarithm of a single number or expression.

9. $\log_a 3 + \log_a 4$ $\log_a 12$
10. $\log_a 7 - \log_a 5$ $\log_a \frac{7}{5}$
11. $4 \log_a 2$ $\log_a 16$
12. $2 \log_a 9$ $\log_a 81$
13. $\frac{1}{2} \log_a 36$ $\log_a 6$
14. $-\log_a \frac{1}{6}$ $\log_a 6$
15. $\log_b 3 + \log_b 5 + \log_b 2$ $\log_b 30$
16. $\log_b 6 + \log_b 5 - \log_b 2$ $\log_b 15$
17. $2 \log_b p + \log_b q$ $\log_b p^2q$
18. $\log_b x - 3 \log_b y$ $\log_b \frac{x}{y^3}$
19. $\frac{1}{2} \log_b r + \frac{1}{2} \log_b s$ $\log_b \sqrt{rs}$
20. $\frac{1}{2}(\log_b x - \log_b y)$ $\log_b \sqrt{\frac{x}{y}}$

Let $c = \log_3 10$ and $d = \log_3 5$. Express the following in terms of c and d.

21. $\log_3 50$ $c + d$
22. $\log_3 500$ $2c + d$
23. $\log_3 250$ $c + 2d$
24. $\log_3 2$ $c - d$

Common Errors

Students may mistakenly think that $\log (M + N)$ is equivalent to $\log M \cdot \log N$. Inform students that $\log (M + N)$ is in simple form and that they are incorrectly using the laws of logarithms.

When solving logarithmic equations, students may forget to check the answers obtained, and as a result may give incorrect values for roots. When reteaching, stress checking each answer.

Check for Understanding

Oral Exs. 1–8: use after Example 2.
Oral Exs. 9–20: use after Example 3.
Oral Exs. 21–24: use after Example 4.

Guided Practice

Express in terms of $\log_3 C$ and $\log_3 D$.

1. $\log_3 C^2D^3$ $2 \log_3 C + 3 \log_3 D$
2. $\log_3 \left(\frac{\sqrt{C}}{D}\right)^3$ $\frac{3}{2} \log_3 C - 3 \log_3 D$

If $\log_{10} 4 = 0.60$ and $\log_{10} 3 = 0.48$, find:

3. $\log_{10} 36$ 1.56
4. $\log_{10} \sqrt[3]{16}$ 0.40
5. $\log_{10} \frac{1}{3000}$ -3.48

Express as a single logarithm.

7. $4 \log M - \log N$ $\log \left(\frac{M^4}{N}\right)$
8. $\log_2 A + 1 + \frac{\log_2 B}{2}$ $\log_2 2A\sqrt{B}$

Summarizing the Lesson

In this lesson students learned the laws of logarithms, their relationship to the laws of exponents, and how to use them when simplifying logarithmic expressions and solving logarithmic equations. Ask the students to state the three laws of logarithms studied in this lesson.

Suggested Assignments

Average Algebra
Day 1: 476/1–28
S 471/37, 38
Day 2: 476/29–42
S 477/Mixed Review

Average Alg. and Trig.
Day 1: 476/1–32
S 477/Mixed Review
Day 2: 476/33–42
Assign with Lesson 10-6.

Extended Algebra
Day 1: 476/1–28
S 471/37–43 odd
Day 2: 476/29–46
S 477/Mixed Review

Extended Alg. and Trig.
Day 1: 476/1–28
S 471/37–43 odd
Day 2: 476/29–46
S 477/Mixed Review

Supplementary Materials

Study Guide	pp. 163–164
Practice Master	57
Test Master	40
Resource Book	p. 144

Written Exercises

1. $6 \log_2 M + 3 \log_2 N$ 2. $4 \log_2 M + 4 \log_2 N$ 3. $\log_2 M + \frac{1}{2} \log_2 N$ $\frac{2}{3} \log_2 M + \frac{1}{3} \log_2 N$

Express each logarithm in terms of $\log_2 M$ and $\log_2 N$.

A

1. $\log_2 M^6N^3$ **2.** $\log_2 (MN)^4$ **3.** $\log_2 M\sqrt{N}$ **4.** $\log_2 \sqrt[3]{M^2N}$

5. $\log_2 \frac{M^4}{N^3}$ $4 \log_2 M - 3 \log_2 N$

6. $\log_2 \left(\frac{M}{N}\right)^7$ $7 \log_2 M - 7 \log_2 N$

7. $\log_2 \sqrt{\frac{M}{N^3}}$ $\frac{1}{2} \log_2 M - \frac{3}{2} \log_2 N$

8. $\log_2 \frac{1}{MN}$ $-\log_2 M - \log_2 N$

If $\log_{10} 9 = 0.95$ and $\log_{10} 2 = 0.30$ (accurate to two decimal places), find the following.

9. $\log_{10} 81$ 1.90 **10.** $\log_{10} \frac{9}{2}$ 0.65 **11.** $\log_{10} \sqrt{2}$ 0.15 **12.** $\log_{10} 3$ 0.48

13. $\log_{10} 8$ 0.90 **14.** $\log_{10} 36$ 1.55 **15.** $\log_{10} \frac{20}{9}$ 0.35 **16.** $\log_{10} 900$ 2.95

17. $\log_{10} \frac{1}{9}$ -0.95 **18.** $\log_{10} \frac{1}{2000}$ -3.30 **19.** $\log_{10} \sqrt[3]{\frac{2}{9}}$ -0.22 **20.** $\log_{10} 162$ 2.20

Express as a logarithm of a single number or expression.

21. $5 \log_4 p + \log_4 q$ $\log_4 p^5q$

22. $\log_{10} x - 4 \log_{10} y$ $\log_{10} \frac{x}{y^4}$

23. $4 \log_3 A - \frac{1}{2} \log_3 B$ $\log_3 \frac{A^4}{\sqrt{B}}$

24. $\log_5 M + \frac{1}{4} \log_5 N$ $\log_5 M\sqrt[4]{N}$

B

25. $\log_2 M + \log_2 N + 3$ $\log_2 8MN$

26. $\log_5 x - \log_5 y + 2$ $\log_5 \frac{25x}{y}$

27. $1 - 3 \log_5 x$ $\log_5 \frac{5}{x^3}$

28. $\frac{1 + \log_9 x}{2}$ $\log_9 3\sqrt{x}$

Simplify.

29. $2 \log_{10} 5 + \log_{10} 4$ 2

30. $2 \log_3 6 - \log_3 4$ 2

31. $\log_4 40 - \log_4 5$ $\frac{3}{2}$

32. $\log_4 3 - \log_4 48$ -2

Solve each equation.

33. $\log_a x = 2 \log_a 3 + \log_a 5$ {45}

34. $\log_a x = \frac{3}{2} \log_a 9 + \log_a 2$ {54}

35. $\log_b (x + 3) = \log_b 8 - \log_b 2$ {1}

36. $\log_b (x^2 + 7) = \frac{2}{3} \log_b 64$ {±3}

37. $\log_a x - \log_a (x - 5) = \log_a 6$ {6}

38. $\log_a (3x + 5) - \log_a (x - 5) = \log_a 8$ {9}

39. $\log_2 (x^2 - 9) = 4$ {±5}

40. $\log_3 (x + 2) + \log_3 6 = 3$ $\left\{\frac{5}{2}\right\}$

41. If $f(x) = \log_2 x$ and $g(x) = 4^x$, find: **a.** $f(g(3))$ 6 **b.** $g\left(f\left(\frac{1}{2}\right)\right)$ $\frac{1}{4}$ **c.** $f(g^{-1}(16))$ 1

42. If $f(x) = 3^x$ and $g(x) = \log_9 x$, find: **a.** $f(f(-1))$ $\sqrt[3]{3}$ **b.** $g\left(g\left(\frac{1}{81}\right)\right)$ undefined **c.** $f^{-1}(g(9))$ 0

43. Prove the second law of logarithms.

44. Simplify $4^{\log_2 (2^{\log_2 5})}$. 25

Solve each equation.

C **45.** $\log_5(\log_3 x) = 0$ {3}

46. $\log_2(\log_4 x) = 1$ {16}

47. $\log_6(x+1) + \log_6 x = 1$ {2}

48. $\log_{10}(x+6) + \log_{10}(x-6) = 2$ $\{2\sqrt{34}\}$

49. $\frac{1}{2}\log_a(x+2) + \frac{1}{2}\log_a(x-2) = \frac{2}{3}\log_a 27$ $\{\sqrt{85}\}$

50. $2\log_3 x - \log_3(x-2) = 2$ {3, 6}

51. $\log_b(x-1) + \log_b(x+2) = \log_b(8-2x)$ {2}

Mixed Review Exercises

Solve.

1. $\log_2 x = -\frac{1}{2}$ $\left\{\frac{\sqrt{2}}{2}\right\}$

2. $x^3 - 7x + 6 = 0$ {−3, 1, 2}

3. $2^{x+3} = 4^{x-1}$ {5}

4. $\sqrt{x} + 2 = x$ {4}

5. $3^x = \frac{\sqrt{3}}{9}$ $\left\{-\frac{3}{2}\right\}$

6. $3(2x-1) = 5x + 4$ {7}

7. $\frac{x}{x+1} - 1 = \frac{1}{x}$ $\left\{-\frac{1}{2}\right\}$

8. $\log_x 8 = \frac{3}{2}$ {4}

9. $(x-2)^2 = 5$ $\{2 \pm \sqrt{5}\}$

If $f(x) = x^2 - 1$ and $g(x) = \sqrt{x+1}$, find each of the following.

10. $f(-2)$ 3

11. $g(3)$ 2

12. $f(g(1))$ 1

13. $g(f(-1))$ 1

Self-Test 2

Vocabulary logarithmic function (p. 468) logarithm (p. 469) laws of logarithms (p. 473)

1. Write in exponential form. **Obj. 10-4 p. 468**

a. $\log_3 81 = 4$ $3^4 = 81$

b. $\log_6 216 = 3$ $6^3 = 216$

2. Write in logarithmic form.

a. $5^4 = 625$ $\log_5 625 = 4$

b. $25^{3/2} = 125$ $\log_{25} 125 = \frac{3}{2}$

3. Evaluate each expression.

a. $\log_2 4^{3/2}$ 3

b. $4^{\log_4 12}$ 12

4. Solve $\log_b 27 = 3$. {3}

5. Write $\log_2(M^5N^6)^{1/3}$ in terms of $\log_2 M$ and $\log_2 N$. $\frac{5}{3}\log_2 M + 2\log_2 N$ **Obj. 10-5 p. 473**

6. If $\log_{10} 5 = 0.70$, find $\log_{10} 0.04$. −1.40

7. Solve $\log_a x + \log_a(x-2) = \log_a 3$. {3}

Check your answers with those at the back of the book.

Additional Answers Written Exercises

43. Let $\log_b M = x$ and $\log_b N = y$. Therefore, $b^x = M$ and $b^y = N$.

$\frac{M}{N} = \frac{b^x}{b^y} = b^{x-y}$

$\log_b \frac{M}{N} = \log_b b^{x-y}$

$\log_b \frac{M}{N} = x - y$

$\therefore \log_b \frac{M}{N} = \log_b M - \log_b N$

Quick Quiz

1. Express $\log_4 2 = \frac{1}{2}$ in exponential form. $4^{1/2} = 2$

2. Express $2^9 = 512$ in logarithmic form. $\log_2 512 = 9$

Evaluate.

3. $\log_2 8\sqrt{2}$ $\frac{7}{2}$

4. $\log_{1/6} 216$ −3

Solve.

5. $\log_8 x = -\frac{1}{3}$ $\left\{\frac{1}{2}\right\}$

6. $\log_x 7 = 3$ $\{\sqrt[3]{7}\}$

7. Express $\log_4\left(\frac{A^2}{\sqrt{B}}\right)$ in terms of $\log_4 A$ and $\log_4 B$. $2\log_4 A - \frac{1}{2}\log_4 B$

8. Evaluate $\log_{10} 240$ if $\log_{10} 3 = 0.48$ and $\log_{10} 4 = 0.60$. 2.38

9. Evaluate $\log_2 56 - \log_2 3.5$. 4

10. Solve $\log_a x + \log_a(2x+3) = \log_a 2$. $\left\{\frac{1}{2}\right\}$

Teaching Suggestions p. T118

Suggested Extensions p. T118

Warm-Up Exercises

Complete each of the following laws of logarithms.

1. $\log_b MN = \underline{\ ?\ }$
   $\log_b M + \log_b N$
2. $\log_b \frac{M}{N} = \underline{\ ?\ }$
   $\log_b M - \log_b N$
3. $\log_b M^k = \underline{\ ?\ }$
   $k \log_b M$

Write as a logarithm of a single expression.

4. $3 \log_4 R + \log_4 S$
   $\log_4 R^3 S$
5. $\log_2 X + \log_2 Y - 4$
   $\log_2 \frac{XY}{16}$

Motivating the Lesson

Ask students if any of them have ever seen a slide rule. Tell them that until computers and calculators became commonplace, everyone relied upon slide rules (which were based on logarithms) and tables of logarithms to do complex calculations. In this lesson students will see why logarithms make computational work easier.

Chalkboard Examples

1. Use a calculator or Table 3 to find the following logarithms.
   a. log 4.6 0.6628
   b. log 283,000 5.4518
   c. log 0.0065
      $0.0065 = 6.5 \times 10^{-3}$
      $0.8129 + (-3) = -2.1871$

Applications

10-6 Applications of Logarithms

Objective To use common logarithms to solve equations involving powers and to evaluate logarithms with any given base.

Logarithms were invented to simplify difficult calculations. Because of the decimal nature of our number system, it is easiest to work with base 10 logarithms. These are called **common logarithms.** When common logarithms are used in calculations, the base 10 is usually not written. For example, log 6 means $\log_{10} 6$.

Most scientific calculators have a key marked "log" that gives the common logarithm of a number. You also can find the common logarithm of a number by using Table 3 on page 812.

Although calculators and computers have replaced logarithms for doing heavy computational work, logarithms still provide the best, or even the only, method of solution for many types of problems. Some of these will be shown later in this lesson and in the next.

If you have a calculator, you should practice finding the logarithms of a few numbers and then skip ahead to the paragraph preceding Example 2.

Without a calculator, you will need to use the table of logarithms (Table 3), a portion of which is shown below.

N	0	1	2	3	4	5	6	7	8	9
20	3010	3032	3054	3075	3096	3118	3139	3160	3181	3201
21	3222	3243	3263	3284	3304	3324	3345	3365	3385	3404
22	3424	3444	3464	3483	3502	3522	3541	3560	3579	3598
23	3617	3636	3655	3674	3692	3711	3729	3747	3766	3784
24	3802	3820	3838	3856	3874	3892	3909	3927	3945	3962

The table gives the logarithms of numbers between 1 and 10 rounded to four decimal places. The decimal point is omitted. To find an approximation for $\log x$ for $1 < x < 10$, find the first two digits of x in the column headed N and the third digit in the row to the right of N. For example, to find log 2.36, look for 23 under N and move across row 23 to the column headed 6, where you find 3729. Therefore, $\log 2.36 \approx 0.3729$.

To find an approximation for the logarithm of a positive number greater than 10 or less than 1, write the number in scientific notation (see page 221) and use Law 1: $\log MN = \log M + \log N$. For example,

$$\begin{aligned} \log 236 &= \log (2.36 \times 10^2) \\ &= \log 2.36 + \log 10^2 && \text{Use Law 1.} \\ &= \log 2.36 + 2 \\ &\approx 0.3729 + 2, \text{ or } 2.3729 && \text{Use Table 3.} \end{aligned}$$

The example shows that the common logarithm of a number can be written as the sum of an integer, called the **characteristic,** and a nonnegative number less than 1, called the **mantissa,** which can be found in a table of logarithms.

Example 1 Find each logarithm. Use Table 3.

a. log 3.8 **b.** log 97,500 **c.** log 0.000542

Solution

a. Find 38 under N; then read across to the column headed 0.

$\therefore$ log 3.8 = 0.5798 ***Answer***

b. Write 97,500 in scientific notation and use laws of logarithms.

$$\log 97{,}500 = \log (9.75 \times 10^4) = \log 9.75 + \log 10^4$$

The mantissa for 9.75 is 0.9890. The characteristic is 4.

$\therefore$ $\log 97{,}500 = 0.9890 + 4 = 4.9890$ ***Answer***

c. Write 0.000542 in scientific notation and use laws of logarithms.

$$\log 0.000542 = \log (5.42 \times 10^{-4}) = \log 5.42 + \log 10^{-4}$$

The mantissa for 5.42 is 0.7340. The characteristic is −4.

$\therefore$ $\log 0.000542 = 0.7340 + (-4) = -3.2660$ ***Answer***

The logarithms in the tables, and therefore the answers, are approximations. However, it is a common practice to use =, as in Example 1, rather than ≈.

If $\log y = a$, then the number y is sometimes called the **antilogarithm** of a. The value of y can be found by using a calculator or log tables.

Example 2 Find y to three significant digits if:

a. $\log y = 0.8995$ **b.** $\log y = 2.4825$

Solution 1 Using a Calculator

On some calculators you find antilogarithms by using the inverse function key with the logarithmic function key. On many others, you can use the 10^x key as shown.

a. If $\log_{10} y = 0.8995$, then $y = 10^{0.8995} = 7.93$. ***Answer***

b. $y = 10^{2.4825} = 304$ ***Answer***

Solution 2 Using Tables

a. In the body of Table 3 find the value closest to 0.8995, that is, 0.8993. This is the entry in the row labeled 79 and in the column labeled 3.

$\therefore$ $y = 7.93$ ***Answer***

b. First find the number that has 0.4825 as its mantissa. The table entry with mantissa closest to 0.4825 is 3.04.

$$\begin{aligned}\log y = 2.4825 &= 0.4825 + 2\\ &= \log 3.04 + \log 10^2\\ &= \log (3.04 \times 10^2)\\ \log y &= \log 304\end{aligned}$$

$\therefore$ $y = 304$ ***Answer***

Find x to three significant digits.

2. $\log x = 3.7135$

$$\begin{aligned}\log x &= 0.7135 + 3\\ &= \log 5.17 + \log 10^3\\ &= \log (5.17 \times 10^3)\\ &= \log 5170\end{aligned}$$

$\therefore x = 5170$

3. $\log x = -2.4535$

$$\begin{aligned}\log x &= 3 - 2.4535 + (-3)\\ &= 0.5465 + (-3)\\ &= \log 3.52 + \log 10^{-3}\\ &= \log (3.52 \times 10^{-3})\\ &= \log 0.00352\end{aligned}$$

$\therefore x = 0.00352$

4. Find the value of $\sqrt[4]{38.2}$ to three significant digits.

Let $x = \sqrt[4]{38.2}$

$$\begin{aligned}\log x &= \frac{1}{4}\log 38.2\\ &= \frac{1}{4}(1.5821)\\ &= 0.3955\end{aligned}$$

$\therefore x = 2.49$

5. Solve $2^{3x} = 7$ **(a)** in calculation-ready form and **(b)** as a decimal with three significant digits.

a. $\log 2^{3x} = \log 7$

$3x \log 2 = \log 7$

$$x = \frac{\log 7}{3 \log 2}$$

b. $\log 7 = 0.8451$

$\log 2 = 0.3010$

$$x = \frac{0.8451}{3(0.3010)} = 0.936$$

6. Find $\log_3 5$.

$$\log_3 5 = \frac{\log_{10} 5}{\log_{10} 3} = \frac{0.6990}{0.4771} = 1.47$$

Using a Calculator

Since students are less likely to make errors using a calculator than reading a table of logarithms, encourage them to use calculators with this lesson. Students need to familiarize themselves with the key presses needed to find logarithms and antilogarithms on their calculators.

Since the characteristic gives the "size" of a number (for example, a number in the hundreds has characteristic 2 and a number in the thousands has characteristic 3), you might suggest that students use the characteristic as a means of checking the reasonableness of results obtained from a calculator.

Check for Understanding

Here is a suggested use of the Oral Exercises to check students' understanding as you teach the lesson.

Oral Exs. 1–8: use after Example 1.
Oral Exs. 9–11: use after Example 2.
Oral Exs. 12–16: use after Example 4.

Guided Practice

Use a calculator or Table 3 to find the value of each expression to three significant digits.

1. $(1.62)^8$ 47.4
2. $\sqrt[4]{(12.3)^3}$ 6.57

Find x to three significant digits.

3. $\log x = 2.9149$ 822
4. $\log x = -1.5575$ 0.0277

In the previous example the answers were rounded to three significant digits. For most problems the answers you find using Table 3 will be accurate to three significant digits.

Example 3 Find the value of each expression to three significant digits.

a. $\sqrt[5]{493}$ **b.** $(0.173)^6$

Solution 1 Using a Calculator

a. Since $\sqrt[5]{493} = 493^{1/5} = 493^{0.2}$, use the power key, y^x.

$y^x = 493^{0.2} = 3.46$ ***Answer***

b. $y^x = (0.173)^6 = 0.0000268$ ***Answer***

Solution 2 Using Tables

a. Let $x = \sqrt[5]{493}$.

$$\log x = \frac{1}{5}\log 493$$
$$\log x = \frac{1}{5}(2.6928)$$
$$\log x = 0.5386$$
$$\therefore\ x = 3.46$$ ***Answer***

b. $(0.173)^6 = (1.73 \times 10^{-1})^6 = (1.73)^6 \times 10^{-6}$

Let $x = (1.73)^6$. Then:

$$\log x = 6 \log 1.73$$
$$\log x = 1.428$$
$$x = 26.8$$
$$\therefore (0.173)^6 = 26.8 \times 10^{-6} = 0.0000268$$ ***Answer***

Whether or not you use a calculator, you need the properties of logarithms to solve the exponential equation given in Example 4. The solution to Example 4 is given in *calculation-ready form,* where the next step is to obtain a decimal approximation using a calculator or a table.

Example 4 Solve $3^{2x} = 5$.

a. Give the solution in calculation-ready form.
b. Give the solution as a decimal with three significant digits.

Solution

a.

$3^{2x} = 5$ Take logarithms of both sides.

$\log 3^{2x} = \log 5$

$2x \log 3 = \log 5$ Use laws of logarithms to simplify.

$x = \dfrac{\log 5}{2 \log 3}$ Solve for x.

$\therefore$ the solution in calculation-ready form is $\dfrac{\log 5}{2 \log 3}$. ***Answer***

b. Find log 5 and log 3 with a calculator or table.

$$x = \frac{0.6990}{2(0.4771)} = 0.733$$

$\therefore$ to three significant digits, the solution is 0.733. ***Answer***

If you know the base b logarithm of a number and wish to find its base a logarithm, you can use the following formula:

Change-of-Base Formula

$$\log_a x = \frac{\log_b x}{\log_b a}$$

The proof of this formula is left as Exercise 42 on page 482.

Example 5 Find $\log_4 7$.

Solution Use the change-of-base formula letting $b = 10$:

$$\log_4 7 = \frac{\log_{10} 7}{\log_{10} 4} = \frac{0.8451}{0.6021} = 1.404 \quad \textit{Answer}$$

Oral Exercises

Use a calculator or Table 3 to find each logarithm.

1. log 4.05 0.6075
2. log 40.5 1.6075
3. log 405 2.6075
4. log 0.405 −0.3925
5. log 8.36 0.9222
6. log 83.6 1.9222
7. log 0.836 −0.0778
8. log 0.0836 −1.0778

Use a calculator or Table 3 to find x to three significant digits.

9. $\log x = 0.6072$ 4.05
10. $\log x = 0.9212$ 8.34
11. $\log x = 1.9212$ 83.4
12. $\log x = 0.9212 - 1$ 0.834

Solve for x in calculation-ready form.

13. $2^x = 7$ $\frac{\log 7}{\log 2}$
14. $9^x = 8$ $\frac{\log 8}{\log 9}$
15. $8^{2x} = 3$ $\frac{\log 3}{2 \log 8}$
16. $3^{-x} = 7$ $-\frac{\log 7}{\log 3}$

Written Exercises

Use a calculator or Table 3 to find the value of each expression to three significant digits.

A

1. $(1.06)^{10}$ 1.79
2. $(10.6)^{10}$ 1.79×10^{10}
3. $(0.38)^5$ 0.00792
4. $(347)^{1.5}$ 6460
5. $(12.7)^{5/2}$ 575
6. $\sqrt[6]{786}$ 3.04
7. $\sqrt[5]{(81.2)^4}$ 33.7
8. $\sqrt[3]{(412)^2}$ 55.4

Use a calculator or Table 3 to find x to three significant digits.

9. $\log x = 0.8531$ 7.13
10. $\log x = 0.4065$ 2.55
11. $\log x = 2.84$ 692
12. $\log x = 1.605$ 40.3
13. $\log x = -1.8$ 0.0158
14. $\log x = -2.91$ 0.00123

Give the solution in calculation-ready form. Then give the solution to three significant digits.

5. $4^x = 50$ $\frac{\log 50}{\log 4}$; 2.82
6. $1.5^{-x} = 0.5$ $-\frac{\log 0.5}{\log 1.5}$; 1.71
7. Solve $9^x = 3\sqrt{3}$ *without* using a calculator or logarithms. $\left\{\frac{3}{4}\right\}$
8. Solve $x^{3/4} = 80$ to three significant digits. {345}

Summarizing the Lesson

In this lesson students learned to use common logarithms to solve equations and to change the base of a logarithmic expression using the change-of-base formula. Ask the students to state the change-of-base formula and illustrate its use.

Suggested Assignments

Average Algebra
Day 1: 481/1–28
R 477/Self-Test 2
Day 2: 482/29–38
Assign with Lesson 10-7.

Average Alg. and Trig.
Day 1: 481/1–27 odd
Assign with Lesson 10-5.
Day 2: 482/29–38
R 477/Self-Test 2
Assign with Lesson 10-7.

Extended Algebra
Day 1: 481/1–29 odd, 31–34
R 477/Self-Test 2
Day 2: 482/35–41
Assign with Lesson 10-7.

Extended Alg. and Trig.
Day 1: 481/1–29 odd, 31–34
R 477/Self-Test 2
Day 2: 482/35–41
Assign with Lesson 10-7.

Additional Answers Written Exercises

15. b. 3.10 **16. b.** 1.43
17. b. 2.34 **18. b.** 35.0
19. b. −0.473 **20. b.** 1.39
21. b. 1.63 **22. b.** −0.196
42. Let $\log_a x = y$, so that $x = a^y$. Then $\log_b x = \log_b a^y$, $\log_b x = y \log_b a$, and $y = \frac{\log_b x}{\log_b a}$. Since $y = \log_a x$, $\log_a x = \frac{\log_b x}{\log_b a}$.

Supplementary Materials

Study Guide pp. 165–166
Computer Activity 23

Solve each equation.

a. Give the solution in calculation-ready form.
b. Give the solution to three significant digits.

15. $3^x = 30$ **a.** $\frac{\log 30}{\log 3}$ **16.** $5^t = 10$ **a.** $\frac{1}{\log 5}$ **17.** $5.6^x = 56$ **a.** $\frac{\log 56}{\log 5.6}$ **18.** $(1.02)^x = 2$ **a.** $\frac{\log 2}{\log 1.02}$

19. $30^{-x} = 5$ **a.** $-\frac{\log 5}{\log 30}$ **20.** $12^{2x} = 1000$ **a.** $\frac{3}{2 \log 12}$ **21.** $3.5^{2t} = 60$ **a.** $\frac{\log 60}{2 \log 3.5}$ **22.** $\frac{4^{2-t}}{3} = 7$ **a.** $2 - \frac{\log 21}{\log 4}$

Solve each equation *without* using a calculator or logarithms. (See Example 2 of Lesson 10-2, page 461.)

23. $4^x = 8\sqrt{2}$ $\left\{\frac{7}{4}\right\}$ **24.** $3^x = \sqrt[5]{9}$ $\left\{\frac{2}{5}\right\}$ **25.** $125^x = 25\sqrt{5}$ $\left\{\frac{5}{6}\right\}$ **26.** $8^x = 16\sqrt[3]{2}$ $\left\{\frac{13}{9}\right\}$

Unlike Exercises 15–26, Exercises 27–34 are *not* exponential equations. Solve each equation using a calculator or logarithms. Give answers to three significant digits.

Sample Solve $x^{2/3} = 12$.

Solution 1. Raise each side to the $\frac{3}{2}$ power. You find that $x = 12^{3/2}$.

2. To simplify $12^{3/2}$.
 (a) Use a calculator and the y^x key to obtain $x = 41.6$, or
 (b) Use logarithms to obtain $\log x = \frac{3}{2} \log 12 = 1.619$. Then find the antilogarithm: $x = 41.6$. ***Answer***

27. $x^{2/5} = 34$ {6740} **28.** $x^{2/3} = 50$ {354} **29.** $\sqrt[3]{x^4} = 60$ {21.6} **30.** $\sqrt[5]{x^3} = 900$ {83,900}

B **31.** $2x^5 = 100$ {2.19} **32.** $\frac{\sqrt[5]{x}}{9} = 7$ $\{9.92 \times 10^8\}$ **33.** $(3y - 1)^6 = 80$ {1.03} **34.** $\sqrt[3]{4t + 3} = 8.15$ {135}

Find each logarithm. Use the change-of-base formula.

35. $\log_2 9$ 3.17 **36.** $\log_6 8$ 1.16 **37.** $\log_3 40$ 3.36 **38.** $\log_7 \frac{1}{2}$ −0.356

Solve for x to three significant digits.

39. $3^{2x} - 7 \cdot 3^x + 10 = 0$ {0.631, 1.46} **40.** $3^{2x} - 7 \cdot 3^x + 12 = 0$ {1, 1.26}

41. a. Simplify $\log_7 49$ and $\log_{49} 7$. 2; $\frac{1}{2}$
b. Simplify $\log_2 8$ and $\log_8 2$. 3; $\frac{1}{3}$
c. How are $\log_b a$ and $\log_a b$ related? Give a convincing argument to justify your answer. They are reciprocals; by the change-of-base formula, $\log_b a = \frac{\log_a a}{\log_a b} = \frac{1}{\log_a b}$.

C **42.** Derive the change-of-base formula. (*Hint:* Let $\log_a x = y$, so that $x = a^y$.)

10-7 Problem Solving: Exponential Growth and Decay

Objective To use exponential and logarithmic functions to solve growth and decay problems.

Suppose an investment of P dollars earns 8% interest compounded annually. Then the value of the investment will be multiplied by 1.08 each year because:

$$\begin{aligned}\text{Value at end of year} &= 100\%(\text{value at beginning of year}) + \text{interest earned}\\ &= 1.00(\text{value at beginning of year}) + 0.08(\text{value at beginning of year})\\ &= 1.08(\text{value at beginning of year})\end{aligned}$$

From this equation you can obtain the following table of values.

Time in years	0	1	2	3	...	t
Value in dollars	P	$P(1.08)$	$P(1.08)^2$	$P(1.08)^3$	...	$P(1.08)^t$
	×1.08	×1.08	×1.08			

Because the value of the investment in t years is given by the exponential function $A = P(1.08)^t$, we say that the investment has **exponential growth** and that the annual *growth rate* is 8%.

If the 8% annual interest is compounded quarterly (that is, compounded four times per year), the quarterly growth rate is 2%. Then

$$\text{value at end of } q \text{ quarters} = P(1.02)^q$$

and

$$\text{value at end of } t \text{ years} = P(1.02)^{4t}.$$

This result can be generalized as follows.

Compound Interest Formula

If an amount P (called the *principal*) is invested at an annual interest rate r compounded n times a year, then in t years the investment will grow to an amount A given by

$$A = P\left(1 + \frac{r}{n}\right)^{nt}$$

where r is expressed as a decimal.

Example 1 How long will it take an investment of \$1000 to triple in value if it is invested at an annual rate of 12% compounded quarterly?

(Solution is on the next page.)

Teaching Suggestions p. T118

Suggested Extensions p. T118

Warm-Up Exercises

Complete.

1. $\frac{\log_b x}{\log_b a} = \underline{\ ?\ }$ $\log_a x$

Solve for x in calculation-ready form.

2. $2^x = 9$ $\frac{\log 9}{\log 2}$
3. $3^{-x} = 8$ $-\frac{\log 8}{\log 3}$
4. $5^{2x} = 3$ $\frac{\log 3}{2 \log 5}$

Motivating the Lesson

Tell students that examples of exponential growth and decay can be found in the activities of people and the phenomena of nature. An investment grows exponentially when interest is compounded, and a radioactive substance like carbon-14 decays exponentially. These and other such examples will be considered in today's lesson.

Problem Solving Strategies

Students should be able to *recognize a problem type* and use the correct exponential growth or decay formula to solve it.

Solution Use the compound interest formula. Let $P = 1000$, $A = 3000$, $r = 0.12$, and $n = 4$.

$$3000 = 1000\left(1 + \frac{0.12}{4}\right)^{4t}$$
$$3 = (1.03)^{4t}$$
$$\log 3 = \log (1.03)^{4t}$$
$$\log 3 = 4t \log 1.03$$
$$t = \frac{\log 3}{4 \log 1.03} = \frac{0.4771}{4(0.0128)}$$
$$t = 9.3 \text{ (years)}$$

Since the interest is compounded *quarterly*, you cannot give 9.3 as an answer. Since 9.3 is closer to $9\frac{1}{4}$ than $9\frac{1}{2}$, the investment will triple in approximately $9\frac{1}{4}$ years. ***Answer***

Investing money is just one example of exponential growth. Another example is the growth of a population. If a population now has N_0 people and is growing at a rate of 1.4% per year, then in t years the population will have $N_0(1.014)^t$ people.

A growth rate of 1.4% per year approximately doubles the population every 50 years, since $N_0(1.014)^{50} \approx 2N_0$. From this equation you can see that $1.014 \approx 2^{1/50}$. Therefore, the number N of people in the population t years from now can be written in either of two ways:

$N = N_0(1.014)^t$ (The population grows 1.4% per year.)

$N = N_0 \cdot 2^{t/50}$ (The population doubles every 50 years.)

Doubling-Time Growth Formula

If a population of size N_0 doubles every d years (or hours, or days, or any other unit of time), then the number N in the population at time t is given by

$$\mathbf{N = N_0 \cdot 2^{t/d}}.$$

Example 2 A certain bacteria population doubles in size every 12 hours. By how much will it grow in 2 days?

Solution Use the doubling-time growth formula. Let $t = 48$ hours (2 days) and let the doubling time $d = 12$ hours.

$$N = N_0 \cdot 2^{t/d}$$
$$N = N_0 \cdot 2^{48/12} = N_0 \cdot 2^4 = 16N_0$$

$\therefore$ the population grows by a factor of 16 in 2 days. ***Answer***

Chalkboard Examples

1. How many dollars must be invested at 16%, compounded quarterly, to yield \$10,000 at the end of 5 years?

$$A = P\left(1 + \frac{r}{n}\right)^{nt}$$
$$10{,}000 = P\left(1 + \frac{0.16}{4}\right)^{4(5)}$$
$$10{,}000 = P(1.04)^{20}$$
$$\log 10{,}000 = \log [P(1.04)^{20}]$$
$$4 = \log P + 20 \log 1.04$$
$$\log P = 4 - 20(0.0170)$$
$$\log P = 3.660$$
$P = 4570$ to three significant digits

$\therefore$ \$4570 must be invested at 16%.

2. The population of a city was 3,500,000 on April 1, 1980 and 5,000,000 on April 1, 1990. Assuming exponential growth, find the month and year that the population will be 6,500,000.

Let n = the population in millions, then $n = 3.5 \cdot 2^{kt}$, where t is the time from April 1, 1980.

$$5.0 = 3.5 \cdot 2^{k(10)}$$
$$1.429 = 2^{10k}$$
$$k = \frac{\log 1.429}{10 \log 2} = 0.05146$$
$$2^{(0.05146)t} = \frac{6.5}{3.5} = 1.8571$$
$$\log 2^{(0.05146)t} = \log 1.8571$$
$$(0.05146)t \cdot \log 2 = \log 1.8571$$
$$t = \frac{\log 1.8571}{(0.05146) \log 2}$$
$$t = \frac{0.2688}{(0.05146)(0.3010)}$$
$$t = \frac{0.2688}{0.01549} = 17.353$$

17 yrs + 12(0.353) mos = 17 yrs + 4.2 mos

$\therefore$ the population will reach 6,500,000 in August 1997.

While investments and populations grow exponentially, the value of a car decreases exponentially, or experiences **exponential decay.** If a car loses 25% of its value each year, then its value N at the end of a year is

$$N = 0.75 \times (\text{value at beginning of year}).$$

Therefore, after t years a car initially valued at N_0 dollars will have a value of

$$N = N_0(0.75)^t.$$

One frequently studied example of exponential decay is the decay of a radioactive substance. Over a period of time, some of the substance will change to a different element. The *half-life* of the radioactive substance is the amount of time it takes until exactly half of the original substance remains unchanged. For example, the half-life of radioactive radium is 1600 years. Therefore, every 1600 years half of the radium decays.

Time in years	0	1600	3200	4800	. . .	t
Amount left	N_0	$N_0 \cdot \frac{1}{2}$	$N_0 \cdot \left(\frac{1}{2}\right)^2$	$N_0 \cdot \left(\frac{1}{2}\right)^3$	. . .	$N_0 \cdot \left(\frac{1}{2}\right)^{t/1600}$

The half-life decay formula is very similar to the doubling-time growth formula.

Half-Life Decay Formula

If an amount N_0 has a half-life h, then the amount remaining at time t is

$$N = N_0\left(\frac{1}{2}\right)^{t/h}.$$

Example 3 The half-life of carbon-14 (C-14) is 5730 years. How much of a 10.0 mg sample will remain after 4500 years?

Solution Use the half-life decay formula. Let $N_0 = 10.0$, $h = 5730$, and $t = 4500$.

$$N = N_0\left(\frac{1}{2}\right)^{t/h}$$

$$N = 10.0\left(\frac{1}{2}\right)^{4500/5730}$$

Using a calculator: $N = 5.80$ (to three significant digits)

If a calculator is not available, take logarithms of both sides:

$$\log N = \log 10.0 + \frac{4500}{5730} \log 0.5$$

$$\log N = 1 + (0.7853)(-0.3010)$$

$$\log N = 0.7636$$

$$N = 5.80$$

$\therefore$ 5.80 mg of the original C-14 remains after 4500 years. ***Answer***

3. The half-life of radioactive radon gas is 3.8 days. How much of 100 mg of the gas will be left after 7 days?

$$x = x_0 \cdot 2^{-t/h}$$
$$= 100 \cdot 2^{-7/3.8}$$
$$= 100 \cdot 2^{-1.8421}$$
$$\log x = \log(100 \cdot 2^{-1.8421})$$
$$= \log 100 - 1.8421 \log 2$$
$$= 2 - 1.8421(0.3010)$$
$$= 1.4455$$
$$x = 27.9$$

After 7 days, 27.9 mg of radon gas is left.

Using a Calculator

When evaluating an exponential expression like $\left(\frac{1}{2}\right)^{4500/5730}$, students should evaluate the exponent first and store the result in memory. They can then enter the base, press the exponential key (usually labeled y^x or x^y), recall the exponent from memory, and press the equals key.

You may wish to encourage students to use calculators for Problems 1–22 on pages 486–488.

Check for Understanding

Here is a suggested use of the Oral Exercises to check students' understanding as you teach the lesson.
Oral Exs. 1–2: use after Example 1.
Oral Ex. 3: use after Example 2.
Oral Ex. 4: use after Example 3.

Guided Practice

Solve. Give money answers in dollars and cents and all other answers to three significant digits.

1. One isotope of chromium has a half-life of 23 h. How long does it take 50 g to decay to 40 g? 7.40 h
2. The population of Waterville increased 12% during 4 years. How many years are required for the population to double its initial value? 24.5 years
3. How much will a $4000 investment be worth after 5 years if it is invested at 8% interest compounded quarterly? $5943.79

Summarizing the Lesson

In this lesson students learned to use exponential and logarithmic functions to solve growth and decay problems. Write each of the formulas presented in this lesson on the board and ask students to identify it.

Oral Exercises

Complete each table.

1.

Item	Cost now	Annual growth rate	Cost in t years	
Movie ticket	$5	10%	?	$5(1.1)^t$
Sweater	$40	5%	?	$40(1.05)^t$
Sneakers	? $60	? 7%	$60(1.07)^t$	

2.

Item	Cost now	Annual decay rate	Value in t years	
Car	$10,000	30%	?	$10{,}000(0.70)^t$
Bike	$300	20%	?	$300(0.80)^t$
Skis	? $250	? 5%	$250(0.95)^t$	

3.

Population size now	Doubling time	Size in t years	
1000	9 years	?	$1000 \cdot 2^{t/9}$
3 million	25 years	?	$(3 \times 10^6)2^{t/25}$
? 7500	? 12 years	$7500 \cdot 2^{t/12}$	

4.

Amount of radioactive substance now	Half-life in days	Amount of substance in t days	
80	8	?	$80\left(\frac{1}{2}\right)^{t/8}$
1200	30	?	$1200\left(\frac{1}{2}\right)^{t/30}$
? 500	? 3	$500\left(\frac{1}{2}\right)^{t/3}$	

Problems

3. a. $1125.51 b. $1266.77 c. $1425.76 d. $3262.04
4. a. $1126.83 b. $1269.73 c. $1430.77 d. $3300.39

Solve. Give money answers in dollars and cents and all other answers to three significant digits. A calculator may be helpful.

A

1. One thousand dollars is invested at 12% interest compounded annually. Determine how much the investment is worth after:
   a. 1 year $1120.00 **b.** 2 years $1254.40 **c.** 3 years $1404.93 **d.** 10 years $3105.85
2. One hundred dollars is invested at 7.2% interest compounded annually. Determine how much the investment is worth after:
   a. 1 year $107.20 **b.** 5 years $141.57 **c.** 10 years $200.42 **d.** 20 years $401.69
   e. Use your answers to parts (a)–(d) to estimate the doubling time for the investment. 10 years
3. Redo Exercise 1 assuming that the interest is compounded quarterly.
4. Redo Exercise 1 assuming that the interest is compounded monthly.

5. The value of a new \$12,500 automobile decreases 20% per year. Find its value after:

a. 1 year **b.** 2 years
c. 3 years **d.** 10 years

6. The value of a new \$3500 sailboat decreases 10% per year. Find its value after:

a. 1 year **b.** 5 years
c. 10 years **d.** 20 years

7. A certain population of bacteria doubles every 3 weeks. The number of bacteria in the population is now N_0. Find its size in:

a. 6 weeks $4N_0$ **b.** 15 weeks $32N_0$ **c.** W weeks $(2^{W/3})N_0$

8. A culture of yeast doubles in size every 20 min. The size of the culture is now N_0. Find its size in:

a. 1 hour $8N_0$ **b.** 12 hours $2^{36}N_0$, or $(6.87 \times 10^{10})N_0$ **c.** 1 day $2^{72}N_0$, or $(4.72 \times 10^{21})N_0$

9. The half-life of carbon-14 is approximately 6000 years. Determine how much of 100 kg of this substance will remain after:

a. 12,000 years 25 kg **b.** 24,000 years 6.25 kg **c.** y years $100\left(\frac{1}{2}\right)^{y/6000}$ kg

10. The radioactive gas radon has a half-life of approximately $3\frac{1}{2}$ days. About how much of a 100 g sample will remain after 1 week? 25 g

B 11. How long will it take you to double your money if you invest it at a rate of 8% compounded annually? 9.01, or approximately 9 years

12. How long will it take you to triple your money if you invest it at a rate of 6% compounded annually? 18.9, or approximately 19 years

13. Savings institutions sometimes use the term "effective yield" to describe the interest rate compounded *annually* (once a year) that is equivalent to a given rate. The effective yield can be found by computing the value of \$1 at the end of the year at the given rate. For example, at 8% compounded quarterly, the value A of \$1 at the end of a year is

$$A = 1\left(1 + \frac{0.08}{4}\right)^4 = (1.02)^4.$$

Evaluating A directly (without using logs), you find that $A = 1.0824$ to four decimal places. Therefore, the original \$1 has grown in value by \$0.0824, an increase of 8.24%. The effective yield is 8.24% per year. Find the effective yield of 12% compounded quarterly. 12.6%

14. Bank A offers 6% interest compounded monthly. Bank B offers 6.1% compounded quarterly. Which bank pays more interest per year? B

15. One of the many by-products of uranium is radioactive plutonium, which has a half-life of approximately 25,000 years. If some of this plutonium is encased in concrete, what fraction of it will remain after 100 years? after a million years? $0.997; 9.09 \times 10^{-13}$

Suggested Assignments

Average Algebra
Day 1: 486/*P*: 1–9 odd
Assign with Lesson 10-6.
Day 2: 487/*P*: 11–19 odd
S 488/Mixed Review

Average Alg. and Trig.
Day 1: 486/*P*: 1–9 odd
Assign with Lesson 10-6.
Day 2: 487/*P*: 11–19 odd
S 488/Mixed Review

Extended Algebra
Day 1: 486/*P*: 1–9 odd
Assign with Lesson 10-6.
Day 2: 487/*P*: 10–22 even
S 488/Mixed Review

Extended Alg. and Trig.
Day 1: 486/*P*: 1–9 odd
Assign with Lesson 10-6.
Day 2: 487/*P*: 10–22 even
S 488/Mixed Review

Supplementary Materials

Study Guide pp. 167–168
Computer Activity 24

Additional Answers Problems

5. a. \$10,000.00
b. \$8000.00
c. \$6400.00
d. \$1342.18

6. a. \$3150.00
b. \$2066.72
c. \$1220.37
d. \$425.52

21. d. Compounding produces interest on the interest the investment earns. The more frequently interest is compounded, the more additional interest is earned. However, the additional interest is small compared to the amount invested.

16. Sugar is put into a large quantity of water and the mixture is stirred. After 2 minutes 50% of the sugar has dissolved. How much longer will it take until 90% of the sugar has dissolved? (Use the half-life decay formula.) 4.64 min

Sample A baseball card increased in value from \$50 to \$500 in 15 years. Find its average annual rate of appreciation.

Solution Let r be the annual rate of appreciation.

Then $$50(1 + r)^{15} = 500$$
$$(1 + r)^{15} = 10$$
$$1 + r = 10^{1/15} \approx 1.17$$
$$\therefore r \approx 0.17 = 17\%$$

17. A gold coin appreciated in value from \$100 to \$238 in 8 years. Find the average annual rate of appreciation. 11.4%

18. Ten years ago Michael paid \$250 for a rare 1823 stamp. Its current value is \$1000. Find the average annual rate of growth. 14.9%

19. A new car that cost \$12,000 decreased in value to \$4000 in 5 years. Find the average annual rate of depreciation. 19.7%

20. A tractor that 4 years ago cost \$8000, now is worth only \$3200. Find the average annual rate of depreciation. 20.5%

C 21. One million dollars is invested at 6.4% interest. Find the value of the investment after one year if the interest is compounded:

a. quarterly \$1,065,552.45 **b.** daily \$1,066,086.39 **c.** hourly \$1,066,091.81

d. Why do you think your answers to parts (a)–(c) are so close in value?

22. A colony of bacteria has 6.5×10^6 members at 8 A.M. and 9.75×10^6 members at 10:30 A.M. Find its population at noon. 1.24×10^7

Mixed Review Exercises

If the domain of each function is $D = \{1, 2, 4\}$, find the range.

1. $f(x) = \log_4 x$ $\left\{0, \frac{1}{2}, 1\right\}$
2. $g(x) = |x - 3|$ $\{1, 2\}$
3. $h(x) = (\sqrt{2})^x$ $\{\sqrt{2}, 2, 4\}$
4. $F(x) = x^2 - 3x + 2$ $\{0, 6\}$
5. $G(x) = \frac{x}{x + 2}$ $\left\{\frac{1}{3}, \frac{1}{2}, \frac{2}{3}\right\}$
6. $H(x) = x^{1/2}$ $\{1, \sqrt{2}, 2\}$

Simplify.

7. $\log_3 12 - 2 \log_3 2$ 1
8. $(4^{1/3})^{3/2}$ 2
9. $\frac{2^{-1} - 2^{-2}}{1 + 2^{-1}}$ $\frac{1}{6}$
10. $\frac{8^{0.5}}{2^{1.5}}$ 1
11. $\sqrt{\left(\frac{1}{3}\right)^2 + \left(\frac{1}{4}\right)^2}$ $\frac{5}{12}$
12. $\log_6 24 + 2 \log_6 3$ 3

10-8 The Natural Logarithm Function

Objective To define and use the natural logarithm function.

In advanced work in science and mathematics, the most important logarithm function is the **natural logarithm function.** Its base is the irrational number e, which has the approximate value 2.71828. The **natural logarithm** of x is sometimes denoted by $\log_e x$, but more often by $\ln x$.

The number e is defined to be the limiting value of $\left(1 + \frac{1}{n}\right)^n$ as n becomes larger and larger. Using a calculator, you can make a table and show that as n gets very large, the value of this expression approaches a number a bit larger than 2.718.

n	100	1000	10,000	100,000
$\left(1 + \frac{1}{n}\right)^n$	2.70481	2.71692	2.71815	2.71827

The exercises of this lesson are just like exercises in previous lessons, except that the symbol $\ln x$ is used instead of $\log_b x$. Example 1 illustrates.

Example 1

Working with base 2 logs

1. If $\log_2 x = 5$, then $x = 2^5$.
2. If $2^x = 7$, then $x = \log_2 7$.
3. $\log_2 2^5 = 5$ and $2^{\log_2 7} = 7$

Working with base e logs

1. If $\ln x = 5$, then $x = e^5$.
2. If $e^x = 7$, then $x = \ln 7$.
3. $\ln e^5 = 5$ and $e^{\ln 7} = 7$

Example 2

a. Simplify $\ln \frac{1}{e^2}$.

b. Write as a single logarithm: $2 \ln 5 + \ln 4 - 3$.

Solution

a. $\ln \frac{1}{e^2} = \ln e^{-2} = -2 \ln e = -2 \cdot 1 = -2$ ***Answer***

b. $2 \ln 5 + \ln 4 - 3 = \ln 5^2 + \ln 4 - \ln e^3$

$= \ln \frac{5^2 \cdot 4}{e^3} = \ln \frac{100}{e^3}$ ***Answer***

Example 3 Solve: **a.** $\ln x = 2$ **b.** $\ln \frac{1}{x} = 2$

Solution

a. $x = e^2$ ***Answer***

b. $\frac{1}{x} = e^2$

$x = \frac{1}{e^2}$, or e^{-2} ***Answer***

Teaching Suggestions p. T118

Suggested Extensions p. T118

Warm-Up Exercises

Complete.

1. If $\log x = 2.5$, then $x =$ _?_. $10^{2.5}$
2. If $10^x = 9$, then $x =$ _?_. log 9
3. $\log 10^{-1.7} =$ _?_. -1.7
4. $10^{\log 5} =$ _?_. 5
5. $\log_2 6 = \frac{\log 6}{?}$. log 2

Motivating the Lesson

Tell students that although they are now accustomed to using the common logarithm, there is another type of logarithm that is used more frequently in advanced mathematical and scientific work. This logarithm, called the natural logarithm, is the topic of today's lesson.

Using a Calculator

You might have students use a calculator to confirm and extend the entries in the table preceding Example 1.

Chalkboard Examples

Find the following.

1. $\ln e^{-2}$
   $\ln e^{-2} = -2$
2. $\ln 1$
   $\ln 1 = 0$
3. $e^{\ln 5}$
   $e^{\ln 5} = 5$

Solve for x.

4. $\ln x = \frac{1}{3}$
   $x = e^{1/3}$
5. $e^{3x} = 27$
   $\ln e^{3x} = \ln 27$
   $3x = \ln 27$
   $x = \frac{1}{3}\ln 27$
   $x = \ln 27^{1/3} = \ln 3$
6. Express $\ln 5 - \ln 4 + \ln 12$ as a single logarithm.
   $\ln 5 - \ln 4 + \ln 12$
   $= (\ln 5 - \ln 4) + \ln 12$
   $= \ln \frac{5}{4} + \ln 12$
   $= \ln\left(\frac{5}{4} \cdot 12\right)$
   $= \ln 15$

Check for Understanding

Here is a suggested use of the Oral Exercises to check students' understanding as you teach the lesson.

Oral Exs. 1–6: use after Example 1.
Oral Exs. 7–9: use after Example 2.
Oral Exs. 10–15: use after Example 3.
Oral Ex. 16: use after Example 4.

Example 4 Solve $e^{2x} = 9$.

Solution 1

$$e^{2x} = 9$$
$$2x = \ln 9 \quad \text{Rewrite in logarithmic form.}$$
$$x = \frac{1}{2}\ln 9$$
$$x = \ln 9^{1/2} = \ln 3 \quad \textbf{Answer}$$

Solution 2

$$e^{2x} = 9$$
$$\ln e^{2x} = \ln 9 \quad \text{Take the natural log of both sides of the equation.}$$
$$2x \cdot \ln e = \ln 9$$
$$2x \cdot 1 = \ln 9$$
$$x = \frac{1}{2}\ln 9 = \ln 9^{1/2} = \ln 3 \quad \textbf{Answer}$$

Oral Exercises

Give each equation in exponential form.

1. $\ln 4 = 1.39$ $e^{1.39} = 4$
2. $\ln \frac{1}{4} = -1.39$ $e^{-1.39} = \frac{1}{4}$
3. $\ln e = 1$ $e^1 = e$

Give each equation in logarithmic form.

4. $e^2 = 7.39$ $\ln 7.39 = 2$
5. $e^{-2} = 0.14$ $\ln 0.14 = -2$
6. $e^{1/5} = 1.22$ $\ln 1.22 = \frac{1}{5}$

Simplify.

7. $\ln \frac{1}{e}$ -1
8. $\ln e^{12}$ 12
9. $\ln \sqrt{e}$ $\frac{1}{2}$

Solve.

10. $\ln x = 5$ $\{e^5\}$
11. $\ln x = \frac{1}{3}$ $\{e^{1/3}\}$
12. $e^x = 3$ $\{\ln 3\}$
13. $x = \ln e^{3/4}$ $\left\{\frac{3}{4}\right\}$
14. $e^{\ln x} = 10$ $\{10\}$
15. $\ln x = -\frac{1}{2}$ $\{e^{-1/2}\}$
16. Approximate to three decimal places the value of $\left(1 + \frac{1}{5000}\right)^{5000}$. 2.718

Written Exercises

Write each equation in exponential form.

A

1. $\ln 8 = 2.08$ $e^{2.08} = 8$
2. $\ln 100 = 4.61$ $e^{4.61} = 100$
3. $\ln \frac{1}{3} = -1.10$ $e^{-1.10} = \frac{1}{3}$
4. $\ln \frac{1}{e^2} = -2$ $e^{-2} = \frac{1}{e^2}$

Write each equation in logarithmic form.

5. $e^3 = 20.1$ $\ln 20.1 = 3$
6. $e^7 = 1097$ $\ln 1097 = 7$
7. $e^{1/2} = 1.65$ $\ln 1.65 = \frac{1}{2}$
8. $\sqrt[3]{e} = 1.40$ $\ln 1.40 = \frac{1}{3}$

Simplify. If the expression is undefined, say so.

9. $\ln e^2$ 2
10. $\ln e^{10}$ 10
11. $\ln \frac{1}{e^3}$ -3
12. $\ln \frac{1}{\sqrt{e}}$ $-\frac{1}{2}$
13. $\ln 1$ 0
14. $\ln 0$ undefined
15. $e^{\ln 5}$ 5
16. $e^{\ln 0.5}$ 0.5

Write as a single logarithm.

17. $\ln 3 + \ln 4$ $\ln 12$
18. $\ln 8 - \ln 2$ $\ln 4$
19. $2 \ln 3 - \ln 5$ $\ln \frac{9}{5}$
20. $\ln 7 + \frac{1}{2} \ln 9$ $\ln 21$
21. $\frac{1}{3} \ln 8 + \ln 5 + 3$ $\ln 10e^3$
22. $4 \ln 2 - \ln 3 - 1$ $\ln \frac{16}{3e}$

Solve for x. Leave answers in terms of e.

23. $\ln x = 3$ $\{e^3\}$
24. $\ln \frac{1}{x} = 2$ $\{e^{-2}\}$
25. $\ln (x - 4) = -1$ $\{4 + e^{-1}\}$
26. $\ln |x| = 1$ $\{\pm e\}$
27. $\ln x^2 = 9$ $\{\pm e^{9/2}\}$
28. $\ln \sqrt{x} = 3$ $\{e^6\}$

Solve for x. Leave answers in terms of natural logarithms.

29. $e^x = 2$ $\{\ln 2\}$
30. $e^{-x} = 3$ $\left\{\ln \frac{1}{3}\right\}$
31. $e^{2x} = 25$ $\{\ln 5\}$
32. $e^{3x} = 8$ $\{\ln 2\}$
33. $e^{x-2} = 2$ $\{2 + \ln 2\}$
34. $\frac{1}{e^x} = 7$ $\left\{\ln \frac{1}{7}\right\}$

Solve. Leave answers in terms of e or natural logarithms.

35. $\sqrt{e^x} = 3$ $\{\ln 9\}$
36. $e^{-2x} = 0.2$ $\left\{\frac{1}{2} \ln 5\right\}$
37. $(e^x)^5 = 1000$ $\left\{\frac{3}{5} \ln 10\right\}$
38. $3e^{2x} + 2 = 50$ $\{\ln 4\}$
39. $\ln (\ln x) = 0$ $\{e\}$
40. $|\ln x| = 1$ $\{e, e^{-1}\}$
41. $\ln x + \ln (x + 3) = \ln 10$ $\{2\}$
42. $2 \ln x = \ln (x + 1)$ $\left\{\frac{1 + \sqrt{5}}{2}\right\}$
43. $e^{2x} - 7e^x + 12 = 0$ $\{\ln 3, \ln 4\}$

Give the domain and range of each function.

B 44. $f(x) = \ln x$
45. $f(x) = \ln |x|$
46. $f(x) = \ln x^2$
47. $f(x) = \ln (x - 5)$

48. Graph $y = \ln x$ and $y = e^x$ in the same coordinate system.
49. Graph $y = 2^x$, $y = e^x$, and $y = 3^x$ in the same coordinate system.

C 50. Express in terms of e the approximate value of each expression when n is very large.
   a. $\left(1 + \frac{1}{n}\right)^{5n}$ e^5
   b. $\left(1 + \frac{2}{n}\right)^n$ e^2
   c. $\left(\frac{n}{n + 1}\right)^{2n}$ e^{-2}

51. Refer to the compound interest formula on page 483.
   a. Show that if interest is compounded daily, so that n is quite large, then $A \approx Pe^{rt}$.
   b. Use part (a) to find the amount of money that you would have after 1 year if you invest \$1000 at 6% interest compounded daily. $A \approx \$1061.84$

Guided Practice

1. Give $\ln 2 = 0.69$ in exponential form.
   $e^{0.69} = 2$
2. Give $\sqrt[6]{e^4} = 1.95$ in logarithmic form.
   $\ln 1.95 = \frac{2}{3}$
3. Simplify $e^{\ln e^2}$. e^2
4. Write as a single logarithm.
   a. $\ln 7 - \ln 2 + \ln 8$
   $\ln 28$
   b. $\frac{1}{2} \ln 4 - 1$ $\ln \frac{2}{e}$

Solve for x. Leave answers in terms of e.

5. $\ln x = \frac{1}{2}$ $x = \sqrt{e}$
6. $\ln (x + 2) = 4$
   $x = e^4 - 2$

Solve for x. Leave answers in terms of natural logarithms.

7. $e^{2x} = 5$ $x = \ln \sqrt{5}$
8. $e^{x+1} = 7$ $x = -1 + \ln 7$

Summarizing the Lesson

In this lesson students learned the definition of natural logarithm and how to use the natural logarithm function. Ask the students to state what the base of the natural logarithm is and how it is defined.

Suggested Assignments

Extended Algebra
490/3–48 mult. of 3
R 493/Self-Test 3

Extended Alg. and Trig.
490/3–48 mult. of 3
R 493/Self-Test 3

Additional Answers
Written Exercises

44. D: $\{x: x > 0\}$
R: {real numbers}

45. D: $\{x: x \neq 0\}$
R: {real numbers}

46. D: $\{x: x \neq 0\}$
R: {real numbers}

47. D: $\{x: x > 5\}$
R: {real numbers}

48.

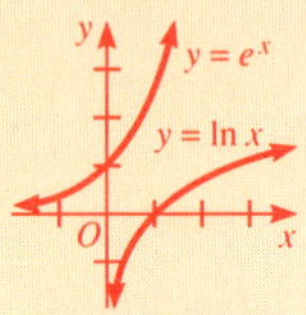

49.

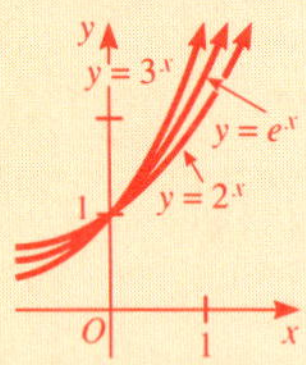

51. a. $A = P\left(1 + \frac{r}{n}\right)^{nt}$
$= P\left(1 + \frac{1}{\frac{n}{r}}\right)^{r(n/r)t}$
$= P\left[\left(1 + \frac{1}{\frac{n}{r}}\right)^{n/r}\right]^{rt}$
$\approx Pe^{rt}$, because $\left(1 + \frac{1}{\frac{n}{r}}\right)^{n/r}$ is approximately equal to e for large n (and constant r) just as $\left(1 + \frac{1}{n}\right)^{n}$ is.

Supplementary Materials

Study Guide	pp. 169–170
Practice Master	58
Test Master	41
Resource Book	p. 145

Calculator Key-In

If your calculator has an e^x or y^x key, or a natural logarithm ($\ln x$) and an inverse function key, you can evaluate powers of e.

1. If you invest P dollars at an annual interest rate r compounded daily, then the value of the investment t years later will be about Pe^{rt}. (Note that the rate is expressed as a decimal. Therefore, if the rate is 6%, then $r = 0.06$.) Find the value of a \$1000 investment at 6% interest compounded daily after:

a. 1 year \$1061.84 **b.** 2 years \$1127.50 **c.** 10 years \$1822.12

2. Which investment is worth more? B

Investment A: \$1000 at 8% compounded daily after 10 years

Investment B: \$2000 at 8% compounded daily after 5 years

3. a. Complete the following table for $y = e^{-x^2}$.

x	0	0.1	0.2	0.3	0.4	0.5	0.6	0.7	0.8	0.9	1.0
y	1	0.99	0.96	0.91	0.85	0.78	0.70	0.61	0.53	0.44	0.37

b. Use the table to sketch the graph of $y = e^{-x^2}$. (*Hint:* The graph is symmetric about the y-axis.) This graph is closely related to the so-called bell-shaped curve so important in statistics.

4. a. Evaluate $\left(1 + \frac{1}{n}\right)^n$ when n is a million. 2.718280469

b. Compare your answer with the following approximation for e: 2.718281828. They are equal to 5 decimal places.

5. The notation $n!$ (read "n factorial") represents the product of the integers from 1 to n.

$$n! = 1 \cdot 2 \cdot 3 \cdot \ \ldots \ \cdot n$$

A very important number in mathematics is approximated by the sum

$$1 + \frac{1}{1!} + \frac{1}{2!} + \frac{1}{3!} + \frac{1}{4!} + \frac{1}{5!} + \frac{1}{6!}.$$

Find this sum and tell what important number it approximates.
$1 + 1 + \frac{1}{2} + \frac{1}{6} + \frac{1}{24} + \frac{1}{120} + \frac{1}{720} \approx 2.718056 \approx e$

Challenge

A friend is thinking of an integer between 1 and 1,000,000. You are to guess the number by asking your friend questions that require only yes-or-no answers. What is the minimum number of questions that you must ask to be sure of guessing the number? How should you guess?

Self-Test 3

Vocabulary

common logarithms (p. 478)	compound interest formula (p. 483)
characteristic (p. 479)	doubling-time formula (p. 484)
mantissa (p. 479)	exponential decay (p. 485)
antilogarithm (p. 479)	half-life (p. 485)
calculation-ready form (p. 480)	half-life decay formula (p. 485)
change-of-base formula (p. 481)	natural logarithm function (p. 489)
exponential growth (p. 483)	natural logarithm (p. 489)

Solve. Give answers to three significant digits.

1. $x^{2/3} = 75$ {650} **2.** $2^{x-2} = 14$ {5.81} **3.** $10^{2x} = 5^{x+2}$ {1.07} **Obj. 10-6, p. 478**

4. You invest \$2000 in a bank offering 10% interest compounded quarterly. Find the value of your investment after a total of five years' growth. \$3277.23 **Obj. 10-7, p. 483**

5. If $e^{3x+2} = 5$, find x in terms of natural logarithms. $\ln\left(\frac{5}{e^2}\right)^{1/3}$ $\left(\text{or } \frac{\ln 5 - 2}{3}\right)$ **Obj. 10-8, p. 489**

Check your answers with those at the back of the book.

Application / *Radiocarbon Dating*

When archaeologists uncover a piece of bone or wood from the site of a dig, they are often able to date the time when the animal or tree was alive by using radiocarbon dating. Most of the carbon in Earth's atmosphere is the isotope C-12, but a small amount is the radioactive isotope C-14. The C-14 decays with a half-life of 5730 years, but the ratio of C-14 to C-12 in the atmosphere remains approximately constant because the C-14 is restored by cosmic ray bombardment of atoms in the atmosphere. The carbon taken in and used by growing plants and animals contains the same fraction of C-14 as the atmosphere. When an organism dies, the amount of C-12 it contains remains the same, but the C-14 decays. Archaeologists therefore use the half-life decay formula,

$$N = N_0 \cdot \left(\frac{1}{2}\right)^{t/h}, \qquad \text{or } \frac{N}{N_0} = \left(\frac{1}{2}\right)^{t/h},$$

where N is the amount of C-14 in the sample discovered, N_0 is the amount of C-14 in the sample when it was alive, and h is the half-life of C-14, 5730 years.

Calculator Key-In Commentary

Students can learn more about the number e by completing the calculator exercises.

Additional Answers Calculator Key-In

3. b.

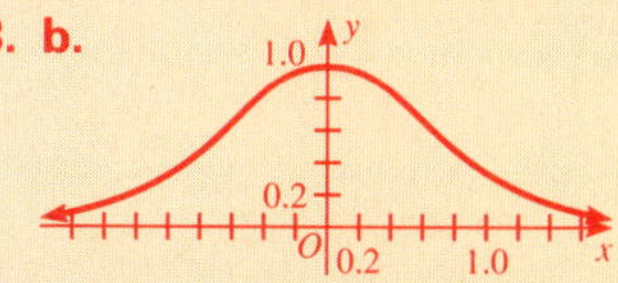

Additional Answers Challenge

20; each question should eliminate half the numbers. For example, the first question should be "Is the number between 1 and 500,000, inclusive?" (If the answer is yes, then ask "Is the number between 1 and 250,000?" Otherwise ask "Is the number between 500,000 and 750,000?") Continue asking such questions until the number is reached.

Quick Quiz

Solve to three significant digits.

1. $1.53^x = 53.1$ $x = 9.34$

2. $x^{5/3} = 50$ $x = 10.5$

3. $(x + 2)^7 = 32$ $x = -0.359$

4. How long does it take an investment to double in value if it is invested at 12% compounded monthly? 5 years, 10 months

5. Evaluate.

a. $\ln \frac{1}{\sqrt[3]{e^2}}$ $-\frac{2}{3}$

b. $-\ln 1$ 0

6. Solve: $\ln x + \ln 3 = \ln (x + 4)$ $x = 2$

Additional Answers
Written Exercises
(continued from p. 466)

14.

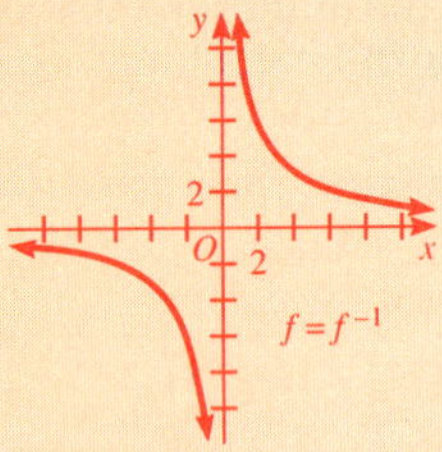

16.

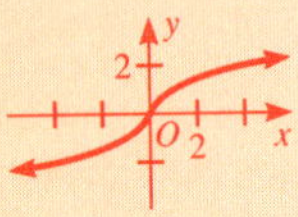

18.

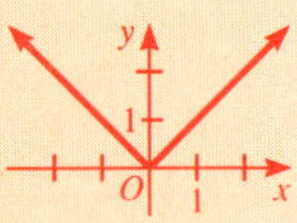

20.

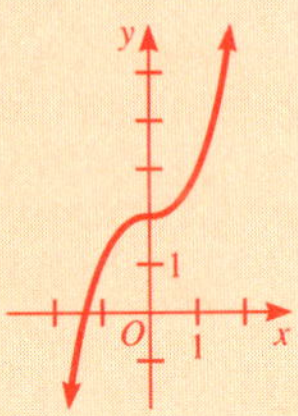

22.

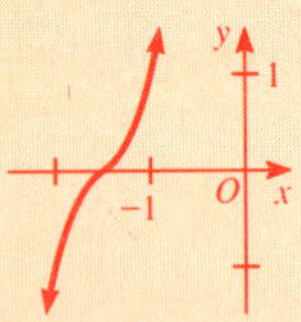

The quantity N is found by measuring the radioactivity of the sample. N_0 is not known, but it can be estimated by finding the total amount of carbon in the sample and assuming that the atmospheric C-14 to C-12 ratio was the same when the sample was alive as it is now, about 1 to 10^{12}. The age t of the sample can then be computed. For example, if the amount N of C-14 is $\frac{1}{8}$ of the estimated amount N_0 that was in the sample when it was alive, then

$$\frac{N}{N_0} = \frac{1}{8} = \left(\frac{1}{2}\right)^{t/5730}$$

$$2^{-3} = 2^{-t/5730}$$

$$-3 = -\frac{t}{5730}$$

$$t = 3 \cdot 5730 = 17{,}190$$

Therefore, the age of the sample is approximately 17,000 years.

The dates found by radiocarbon dating can often be checked against other dating methods. For example, *dendrochronology* (establishing a chronology by correlating sequences of growth rings in wood) has dated some very old samples of long-lived tree species. A piece of wood from a bristle-cone pine has been dated in this way to 6000 B.C. Radiocarbon dating of this sample gave a date of 5500 B.C., showing that 8000 years ago the atmospheric C-14 to C-12 ratio was greater than it is now. (Can you see why a greater ratio then gives a younger date?) Estimates of such atmospheric changes made by comparing tree-ring and radiocarbon dates can be used to obtain more accurate radiocarbon dates. With such corrections these dates are usually accurate to within two centuries for dates up to 10,000 years ago.

Exercises

1. The amount of C-14 in a sample of wood is 45% of the amount that would be found in a living sample with the same total carbon content. How many years have passed since the sample was part of a living tree? Give your answer to two significant digits. 6600

2. A scientist finds that a bone sample contains C-14 and C-12 in the ratio 0.08 to 10^{12}.

a. What is the ratio of C-14 in the sample now (N) to C-14 in the sample when it was part of a living animal (N_0)? Assume that the atmospheric C-14 fraction was the same then as it is now. 0.08 : 1

b. How many years have passed since the animal was living? Give your answer to two significant digits. 21,000

3. Using tree-ring chronology it is found that 400 years have passed since a given wood sample was part of a living tree.

a. What percent of the original C-14 can a scientist expect to find today? 95.3%

b. What percent would a scientist 200 years from now expect to find? 93.0%

Give your answers to the nearest tenth of a percent.

Chapter Summary

1. The meaning of rational exponents is defined as follows:

$$b^{p/q} = (\sqrt[q]{b})^p = \sqrt[q]{b^p}$$

   Rational exponents can be used to define powers with real exponents, and the laws of exponents hold for real exponents.

2. The function defined by $y = b^x$ is called the *exponential function* with base b, where $b > 0$ and $b \neq 1$.
3. The function whose value at x is $f(g(x))$ is called the *composite* of f and g. The operation that combines f and g to produce their composite is called *composition*.
4. *Inverse functions* are functions that "undo" one another. The functions f and g are inverse functions if

$$f(g(x)) = x \text{ for all } x \text{ in the domain of } g$$

and

$$g(f(x)) = x \text{ for all } x \text{ in the domain of } f.$$

5. The function $y = \log_b x$ is called the *logarithmic function* with base b, where $b > 0$ and $b \neq 1$. This function is the inverse of the exponential function with base b.
6. *Logarithms* are defined by the relationship

$$\log_b N = k \text{ if and only if } b^k = N,$$

   where b and N are positive numbers and $b \neq 1$.

7. In working with logarithms, you can use the *laws of logarithms*.

   Law 1 $\quad \log_b MN = \log_b M + \log_b N$

   Law 2 $\quad \log_b \frac{M}{N} = \log_b M - \log_b N$

   Law 3 $\quad \log_b M^k = k \log_b M$

8. There are certain types of equations for which logarithms provide the best, or even the only, method of solution. Logarithms are used in studying exponential equations and their applications to such occurrences as compound interest, population growth, and radioactive decay. Here are some related formulas:

   Compound interest formula: $A = P\left(1 + \frac{r}{n}\right)^{nt}$ (See page 483.)

   Doubling-time growth formula: $N = N_0 \cdot 2^{t/d}$ (See page 484.)

   Half-life decay formula: $N = N_0 \cdot \left(\frac{1}{2}\right)^{t/h}$ (See page 485.)

9. An important function used in science and mathematics is the *natural logarithm function* denoted by $\ln x$ or $\log_e x$. Its base is the irrational number e, which is approximately 2.71828.

23. a.–b.

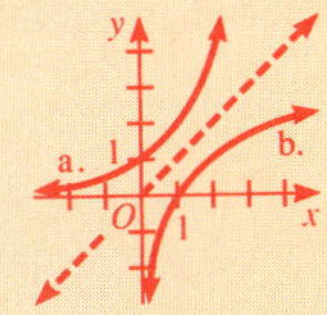

c. $f^{-1}(2) = 1$;
$f^{-1}(4) = 2$;
$f^{-1}(8) = 3$;
$f^{-1}\left(\frac{1}{2}\right) = -1$

d. domain of f: {real numbers}
range of f: $\{y: y > 0\}$
domain of f^{-1}: $\{x: x > 0\}$
range of f^{-1}: {real numbers}

Chapter Review

Write the letter of the correct answer.

1. For positive x and y, which expression is equivalent to $(x^{1/2} - y^{1/2})^2$? 10-1

a. $x - y$ **b.** $x + 2\sqrt{xy} + y$ **c.** $x - 2\sqrt{xy} + y$

2. Simplify $\frac{2}{\sqrt[6]{8}}$.

a. $\frac{1}{\sqrt{2}}$ **b.** $\frac{1}{2}$ **c.** $\sqrt{2}$ **d.** $\sqrt[3]{4}$

3. Solve $25^{x+2} = 5^{3x-3}$. 10-2

a. 2 **b.** 3 **c.** 4 **d.** 7

4. Simplify $\left(\frac{2^{1+\sqrt{2}}}{2^{1-\sqrt{2}}}\right)^{\sqrt{2}}$.

a. 4 **b.** $2^{\sqrt{2}}$ **c.** 8 **d.** 16

5. Which of the following could be the graph of $y = \left(\frac{1}{3}\right)^x$?

a.

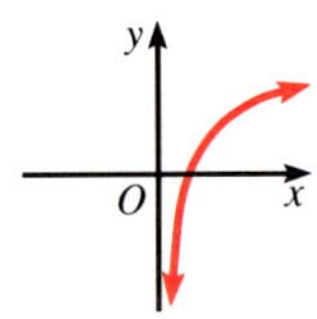

b.

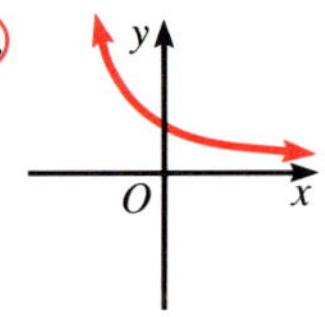

c.

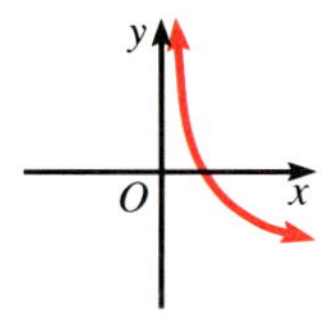

d.

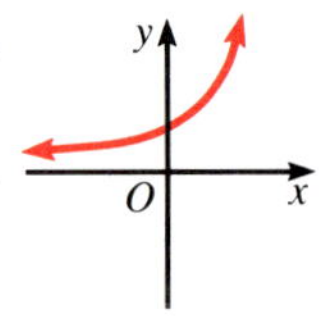

6. If $f(x) = 3x - 2$ and $g(x) = 2x - 1$, then $g(f(x))$ is equal to: 10-3

a. $6x - 4$ **b.** $6x - 5$ **c.** $6x^2 - x - 2$ **d.** $6x - 3$

7. Are the functions f and g inverse functions if $f(x) = \frac{5}{3}x + 1$ and $g(x) = \frac{3(x-1)}{5}$?

a. Yes **b.** No

8. If $f(x) = 9x - 7$, find the function g that is the inverse of f.

a. $g(x) = 7x + 9$ **b.** $g(x) = \frac{1}{9}x + 7$ **c.** $g(x) = \frac{x+7}{9}$

9. Simplify $\log_4 64$. 10-4

a. $\frac{1}{3}$ **b.** 3 **c.** 16 **d.** 4

10. Between which two numbers is $\log_5 150$?

a. 0 and 1 **b.** 1 and 2 **c.** 2 and 3 **d.** 3 and 4

11. Simplify $\log_6\left(\frac{36}{6^{-10}}\right)$. 10-5

a. 12 **b.** -8 **c.** -20 **d.** -5

12. If $\log_b z = \frac{1}{3}\log_b x + \log_b y$, write z in terms of x and y.

a. $y\sqrt[3]{x}$ **b.** $(x + y)^{1/3}$ **c.** $(xy)^{1/3}$ **d.** $\frac{x}{3} + y$

13. Solve $10^{5t} = 2$. 10-6

a. $\frac{2}{5}$ **b.** $\frac{\log_{10} 2}{\log_{10} 5}$ **c.** $5 \log_{10} 2$ **(d.)** $\frac{1}{5} \log_{10} 2$

14. If 40 mg of a radioactive substance decays to 5 mg in 12 min, find the half-life, in minutes, of the substance. 10-7

a. 2 min **(b.)** 4 min **c.** 6 min **d.** 8 min

15. Simplify $\ln \frac{1}{e^3}$. 10-8

a. 3 **(b.)** -3 **c.** 0 **d.** 1

16. Solve $e^{2x-1} = 3$.

a. $\ln 3 + 1$ **b.** $\frac{\ln 3}{2x - 1}$ **(c.)** $\frac{\ln 3 + 1}{2}$ **d.** $1 - \ln 3$

Chapter Test

1. Find the value of x. 10-1

a. $\left(\frac{1}{16}\right)^{-3/4} = x$ 8 **b.** $27^x = 81$ $\frac{4}{3}$

2. Simplify.

a. $\sqrt[3]{\sqrt{125y^6}}$ $|y|\sqrt{5}$ **b.** $(64^{2/3} + 27^{2/3})^{3/2}$ 125

3. Solve $4^{x-2} = 8^{\pi+1} \div 8^{\pi-1}$. {5} 10-2

4. Suppose $f(x) = 2x - 1$ and $g(x) = x^2 + 4$. Find: 10-3

a. $f(g(-2))$ 15 **b.** $g(f(x))$ $4x^2 - 4x + 5$

5. Suppose that $f(x) = \sqrt[3]{x - 1}$ and $g(x) = x^3 + 1$. Show that f and g are inverse functions.

6. a. Write in logarithmic form using the base 2: $32^{3/5} = 8$. $\log_2 8 = 3$ 10-4

b. Write in exponential form: $\log_{16}\left(\frac{1}{64}\right) = -\frac{3}{2}$. $16^{-3/2} = \frac{1}{64}$

7. Simplify: **a.** $6^{\log_6 3}$ 3 **b.** $\log_3 27^{\sqrt{2}}$ $3\sqrt{2}$

8. Solve: **a.** $\log_3 x = \log_3 12 + \log_3 2 - \log_3 6$ {4} 10-5

b. $\log_4 (x - 6) + \log_4 x = 2$ {8}

9. If $\log_{10} 2 = 0.301$ and $\log_{10} 3 = 0.477$, find the following.

a. $\log_{10} 8$ 0.903 **b.** $\log_{10} 12$ 1.079 **c.** $\log_{10} 15$ 1.176

10. Express t in terms of common logarithms: $5^{3t} = 2$. $\frac{\log 2}{3 \log 5}$ 10-6

11. The population of a certain colony of bacteria doubles every 5 hours. How long will it take for the population to triple? Give the answer to two significant digits. 7.9 h 10-7

12. Write x in terms of e: $\ln x^2 = 8$. $\pm e^4$ 10-8

13. Write as a single natural logarithm: $\frac{1}{2} - \ln 7$. $\ln \frac{\sqrt{e}}{7}$

Additional Answer
Chapter Test

5. $f(g(x)) = \sqrt[3]{(x^3 + 1) - 1} = \sqrt[3]{x^3} = x$;
$g(f(x)) = (\sqrt[3]{x - 1})^3 + 1 = (x - 1) + 1 = x$

Alternate Test p. T22

Supplementary Materials

Test Masters	42, 43
Resource Book	p. 146

Mixed Problem Solving

Solve each problem that has a solution. If a problem has no solution, explain why.

A

1. How much iodine should be added to 50 mL of a 10% iodine solution to obtain a 25% solution? 10 mL
2. When the reciprocal of a real number is subtracted from 1, the difference is equal to the number. Find the number. The equation $1 - \frac{1}{x} = x$ has no real solution.
3. The number of representatives to Congress from California is equal to the sum of the numbers of representatives from New York and North Carolina. Together the three states have 90 representatives. The number from New York is 23 more than the number from North Carolina. How many representatives does each state have? CA: 45; NY: 34; NC: 11
4. When two meshed gears revolve, their speeds vary inversely as the numbers of teeth they have. A gear with 36 teeth runs at 200 r/min (revolutions/min). Find the speed of a 60-tooth gear meshed with it. 120 r/min
5. The difference of two numbers is 6. Find their minimum product. −9
6. A rectangular garden has perimeter 44 m. A walkway 2 m wide surrounds the garden. The combined area of the garden and walkway is 224 m^2. Find the dimensions of the garden. 10 m × 12 m
7. When Susan had annual incomes of \$15,000 and \$18,000, she saved \$600 and \$900, respectively. If the amount saved is a linear function of the amount earned, how much would she save when her income becomes \$24,000? \$1500
8. The *eagle* is a 10-dollar gold United States coin minted from 1795 to 1933. Find all the possible ways a person could receive \$25 using eagles and quarter-eagles. 2 eagles, 2 quarter-eagles; 1 eagle, 6 quarter-eagles; 10 quarter-eagles

B

9. The area of a circle is directly proportional to the square of its circumference. A circle with circumference 14π has area 49π. Find a general formula relating the circumference and the area of a circle. $A = \frac{C^2}{4\pi}$
10. Find three consecutive integers such that the square of the sum of the first two is 80 more than the sum of the squares of the last two. 7, 8, 9; −6, −5, −4
11. Two water purification plants can process a day's water supply in 5 h. Operating alone, the smaller plant would need 4 h more than the larger one would. How long would each need to do the job alone? Larger plant: 8.4 h; smaller: 12.4 h
12. The value of a certain car decreases at a rate of 12% annually. About how long does it take for the car to be worth half its original value? 5.4 years
13. If Rachel jogs for awhile at 8 km/h and then walks at 5 km/h, she travels 23 km. If she reverses the amount of time spent jogging and walking, she travels 6 km farther. Find her total traveling time. 4 h

Preparing for College Entrance Exams

Strategy for Success

If you skip a question, don't forget to leave the line for that answer blank on your answer sheet. To ensure an accurate pairing between questions and answers, every so often compare the numbering in your test booklet with the one on your answer sheet.

Decide which is the best of the choices given and write the corresponding letter on your answer sheet.

1. Which equation represents the set of all points that are equidistant from $(-7, 4)$ and $(3, 6)$? C

(A) $x - 5y = -27$ **(B)** $5x - y = -15$ **(C)** $5x + y = -5$
(D) $5x - y = 15$ **(E)** $x + 5y = 23$

2. The graph of $9x^2 - 18 = y^2$ is D

(A) a parabola **(B)** a circle **(C)** an ellipse
(D) a hyperbola **(E)** a point

3. Which expression below is equivalent to $-3 \log_8 4$? E

(A) $\frac{\sqrt[5]{32^4}}{32}$ **(B)** $\log_5 \sqrt{5} - \log_5 5\sqrt{5}$ **(C)** $\log 1 - \log 0.01$
(D) $\sqrt{-4}$ **(E)** $\ln \frac{1}{e^2}$

4. An ellipse with equation $\frac{x^2}{16} + \frac{y^2}{36} = 1$ has which point as a focus? B

(A) $(0, -2\sqrt{3})$ **(B)** $(0, 2\sqrt{5})$ **(C)** $(-2\sqrt{5}, 0)$
(D) $(2\sqrt{13}, 0)$ **(E)** $(0, 6)$

5. If $f(x) = \frac{4}{x} - 1$, find $f^{-1}(7)$ if possible. A

(A) $\frac{1}{2}$ **(B)** $-\frac{3}{7}$ **(C)** 4 **(D)** 0 **(E)** f^{-1} does not exist.

6. Solve $27^{2t-1} = 81^{t+2}$. E

(A) 3 **(B)** -3 **(C)** $-\frac{1}{2}$ **(D)** -2 **(E)** $\frac{11}{2}$

7. Find the number of real solutions of this system: $x^2 + 16y^2 = 25$ E
$xy - 3 = 0$

(A) 0 **(B)** 1 **(C)** 2 **(D)** 3 **(E)** 4

8. Find the z-coordinate of the solution of this system: C

$$\begin{aligned} 5x - 3y + z &= 5 \\ 4x + 3y - 2z &= -4 \\ 2x - 3y - 7z &= 13 \end{aligned}$$

(A) 0 **(B)** 1 **(C)** -1 **(D)** 2 **(E)** -2

11 Sequences and Series

Objectives

11-1 To determine whether a sequence is arithmetic, geometric, or neither and to supply missing terms of a sequence.

11-2 To find a formula for the *n*th term of an arithmetic sequence and to find specified terms of arithmetic sequences.

11-3 To find a formula for the *n*th term of a geometric sequence and to find specified terms of geometric sequences.

11-4 To identify series and to use sigma notation.

11-5 To find sums of finite arithmetic and geometric series.

11-6 To find sums of infinite geometric series having ratios with absolute value less than one.

11-7 To expand powers of binomials.

11-8 To use the binomial theorem to find a particular term of a binomial expansion.

Assignment Guide

See p. T58 for Key to the format of the Assignment Guide

Day	Average Algebra	Average Algebra and Trigonometry	Extended Algebra	Extended Algebra and Trigonometry
1	**11-1** 504/1–22	**11-7** 539/1–20	**11-1** 504/1–22	**11-1** 504/1–33 odd **S** 506/*Mixed Review*
2	**11-1** 505/23–24 **S** 506/*Mixed Review*	**11-8** 542/1–13, 15–27 odd **R** 543/*Self-Test 3*	**11-1** 505/23–34 **S** 506/*Mixed Review*	**11-2** 509/3–33 mult. 3 **11-3** 513/3–39 mult. 3 **S** 515/*Mixed Review*
3	**11-2** 509/1–31 odd		**11-2** 509/1–33 odd	**11-3** 514/*P*: 1, 4, 7, 9–12 **11-4** 521/2–20 even **R** 516/*Self-Test 1*
4	**11-3** 513/1–33 odd 514/*P*: 1–6		**11-3** 513/3–39 mult. of 3 514/*P*: 1–9 odd	**11-4** 521/22–34 even 523/*Extra:* 1, 3, 5 **11-5** 527/2–20 even
5	**11-3** 514/*P*: 8–12 **S** 515/*Mixed Review* **11-4** 521/1–16		**11-3** 514/*P*: 10–13 **S** 515/*Mixed Review* **11-4** 521/2–20 even	**11-5** 527/22–30 even 528/*P*: 1–4, 11–13 **S** 530/*Mixed Review*
6	**11-4** 521/17–32 **R** 516/*Self-Test 1*		**11-4** 521/22–34 even **R** 516/*Self-Test 1* 523/*Extra:* 1, 3, 5	**11-6** 533/1–29 odd 534/*P*: 1–9 odd
7	**11-5** 527/2–28 even **S** 530/*Mixed Review*		**11-5** 527/2–30 even **S** 530/*Mixed Review*	**11-7** 539/2–12 even, 13–22 **S** 539/*Mixed Review* **R** 536/*Self-Test 2*
8	**11-5** 527/17–29 odd 528/*P*: 1–6, 9, 10		**11-5** 527/17–29 odd 528/*P*: 1–4, 11–13 **S** 524/2, 4, 7	**11-8** 542/1–13, 15–27 odd **S** 535/*P*: 11, 12
9	**11-6** 533/1–23 odd **S** 529/*P*: 7, 8, 11		**11-6** 533/1–29 odd **S** 529/*P*: 9, 10, 14	*Prepare for Chapter Test* **R** 543/*Self-Test 3* **R** 545/*Ch. Review*

Assignment Guide *(continued)*

Day	Average Algebra	Average Algebra and Trigonometry	Extended Algebra	Extended Algebra and Trigonometry
10	**11-6** 533/14–26 even 534/*P*: 1–3, 5, 7, 9		**11-6** 533/14–28 even 534/*P*: 1–9 odd	*Administer Chapter 11 Test* **12-1** *Read 12-1* 552/3–66 mult. 3 **S** 554/*Mixed Review*
11	**11-7** 539/2–12 even, 13–20 **S** 539/*Mixed Review* **R** 536/*Self-Test 2*		**11-7** 539/2–12 even, 13–22 **S** 539/*Mixed Review* **R** 536/*Self-Test 2*	
12	**11-8** 542/1–13, 15–27 odd **S** 534/*P*: 8, 10		**11-8** 542/1–13, 15–27 odd **S** 535/*P*: 11, 12	
13	*Prepare for Chapter Test* **R** 543/*Self-Test 3* **R** 545/*Ch. Review*		*Prepare for Chapter Test* **R** 543/*Self-Test 3* **R** 545/*Ch. Review*	
14	*Administer Chapter 11 Test* **R** 547/*Cumulative Review*		*Administer Chapter 11 Test* **R** 547/*Cumulative Review*	

Supplementary Materials Guide

For Use with Lesson	Practice Masters	Tests	Study Guide (Reteaching)	Resource Book: Tests	Resource Book: Practice Exercises	Resource Book: Mixed Review (MR) Applications (A) Enrichment (E) Thinking Skl. (TS)
11-1	Sheet 59		pp. 171–172			p. 214 (A)
11-2			pp. 173–174		p. 147	
11-3	Sheet 60	Test 44	pp. 175–176			
11-4			pp. 177–178			
11-5	Sheet 61		pp. 179–180		p. 148	pp. 266–267 (TS)
11-6	Sheet 62	Test 45	pp. 181–182		p. 149	
11-7			pp. 183–184			
11-8	Sheet 63	Test 46	pp. 185–186			
Chapter 11		Tests 47, 48		pp. 58–61	p. 150	p. 230 (E)
Cum. Rev. 1–11						pp. 183–185 (MR)
Cum. Rev. 7–11		Test 49				
Cum. Rev. 8–11				pp. 62–64		

Overhead Visuals

For Use with Lesson	Visual	Title
11-7	B	Multi-Use Packet 2

Software

Software	Computer Activities	Test Generator
	Activities 25, 26	168 test items
For Use with Lessons	11-6, 11-8	all lessons

Strategies for Teaching

Thinking Skills and Cooperative Learning

Thinking skills are woven into the whole fabric of algebra. While such topics as proofs and problem solving may make special demands on students' thinking skills, no real understanding of any of the concepts presented in this course can take place without them.

Your students are likely to come to you with a variety of thinking skills, not always well developed or even recognized by the students themselves. In your lesson presentations you can help students improve these skills, use them more efficiently, and acquire additional skills.

In the Index under "Thinking Skills" you will find a list of some of the areas in which these skills are applied by students.

You will need to emphasize the importance of attacking a problem analytically—of trying to see the problem as a whole and planning one's solution. Point out also that there may be a variety of ways of approaching a problem, and that there is nothing wrong with abandoning one strategy and trying another.

In business and industry, leaders have recognized that the ability to work with others in thinking and solving problems is a characteristic of successful and productive members of the work force. In a school setting, research shows that sometimes students can understand a concept explained by another student more easily than they can follow the presentation to the class as a whole. The process of explaining and teaching is in itself a learning process, and students may find it easier to ask questions, advance ideas, and test hypotheses in a small group.

See the Exploration on page 842 for an activity in which students explore Pascal's triangle. This is an ideal cooperative learning activity for students to do before learning about binomial expansions in Lesson 11-7.

11-1 Types of Sequences

As is stated following Example 3, patterns are sometimes hard to find. Some students find the study of sequences difficult because they have a hard time seeing patterns. Observe students carefully to see if any are having real difficulty with the basic concepts of sequences. If so, they may need extra help and extra work on the less difficult sequences.

One way to help students understand arithmetic sequences is to relate such sequences to linear functions. Then you might relate geometric sequences to exponential functions.

As students become comfortable looking for patterns, they might find the topics in this lesson enjoyable and challenging.

1. Ask students to continue the following sequences.
   a. O, T, T, F, F, S, S, ?, ?, ?
   b. 1, 4, 9, 61, 52, 63, 94, ?, ?
   a. One, Two, Three, Four, Five, Six, Seven, Eight, Nine, Ten
   b. 1^2, 2^2, 3^2, 4^2, 5^2, 6^2, 7^2, 8^2, 9^2 (with digits reversed)
2. Ask students to work Exercise 27 (the Fibonacci sequence), Exercise 31 (triangular numbers), or Exercise 32 (pentagonal numbers).

This activity is a warm-up for the whole class when discussing sequences in general. First, a student will call out a number to begin the sequence. Students will take turns calling out a number, trying to create a sequence. The first few results will probably be 1, 2, 3, . . . , or 2, 4, 6, . . . , but as they get into the activity some students will try to make up more complicated sequences. The results of this activity could lead to a discussion of the characteristics of arithmetic and geometric sequences and of those that are neither.

11-3 Geometric Sequences

There are many patterns in nature which are made up of the sequences discussed in this chapter. Divide the class into small groups. Ask each group to find pictures or descriptions of examples of these patterns in nature and bring them to class. Each group will then make up a folder or scrapbook illustrating the patterns they have found and describing them mathematically.

11-5 Sums of Arithmetic and Geometric Series

Write the following diagram on the chalkboard and ask students to find the sum without writing any work on paper.

$$1 + 2 + \cdots + 99 + 100$$

101

101

There are fifty 101s. The sum is $50 \times 101 = 5050$.

References to Strategies

PE: Pupil's Edition **TE:** Teacher's Edition **RB:** Resource Book

Problem Solving Strategies

PE: pp. 501–503, 505–506 (recognize patterns)
TE: pp. 504, 518; pp. T119, T121 (Sugg. Extensions)

Applications

PE: pp. 500–501 (biology); p. 512 (banking); pp. 514–516, 528–530 (word problems)
TE: pp. 500, 507, 510, 527
RB: p. 214

Nonroutine Problems

PE: p. 505 (Exs. 31–35); p. 514 (Ex. 40); p. 515 (Ex. 14); p. 516 (Challenge); p. 528 (Exs. 33–38); p. 530 (Probs. 13–15); p. 534 (Exs. 27–30); p. 534–536 (Probs. 7–14); p. 539 (Ex. 24); p. 543 (Exs. 28, 29)
TE: pp. T119–T121 (Sugg. Extensions, Lessons 11-1, 11-2, 11-3, 11-5, 11-6)

Communication

PE: p. 530 (Exs. 14, 15); p. 534 (Ex. 27, convincing argument)
TE: p. T120 (Sugg. Extensions, Lessons 11-2, 11-3); p. T120 (Teaching Sugg., Lesson 11-4); p. T112 (Reading Algebra); p. 518 (translating)

Thinking Skills

PE: pp. 523–542 (induction)
TE: pp. T119, T121 (Sugg. Extensions, Lessons 11-1, 11-5, 11-6)
RB: pp. 266–267

Explorations

PE: p. 517 (graphing sequences); pp. 523–524 (induction); p. 842 (Pascal's Triangle)

Connections

PE: pp. 500, 522 (Biology); pp. 500, 506, 534–536 (Geometry); p. 544 (History)

Using Technology

PE: p. 543 (Calculator Key-In); pp. 514, 528–530, 534–536
TE: pp. 514, 529, 530, 535, 543
Computer Activities: pp. 66–70

Using Manipulatives/Models

Overhead Visuals: B

Cooperative Learning

TE: pp. T119–T120 (Lessons 11-1, 11-3)

Teaching Resources

For use in implementing the teaching strategies referenced on the previous page.

Application
***Resource Book,* p. 214**

Application—The Golden Ratio (For use with Chapter 11)

From the Parthenon of ancient Greece to the paperback book of modern times, the shape of one particular rectangle has been considered to be the most pleasing to the eye. The steps for constructing the rectangle are as follows:

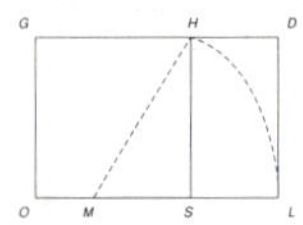

Step 1 Construct a square labeled *GOSH*.

Step 2 Locate the midpoint of $\overline{OS}$ and label it *M*.

Step 3 Draw an arc intersecting the extension of $\overline{OS}$ at *L*.

Step 4 Construct a perpendicular to $\overline{OL}$ at *L*, and extend $\overline{GH}$ to intersect the perpendicular at *D*. Figure *GOLD* is called a "golden rectangle."

1. Follow these steps to find the ratio $\frac{OL}{LD}$ of the sides of a "golden rectangle":

a. Suppose *GOSH* is a unit square. Then $OS = SH =$ ________ and $OM = MS =$ ________.

b. Use the Pythagorean theorem to find the exact length of $\overline{MH}$: $MH = ML =$ ________.

c. Since $OL = OM + ML$, $OL =$ ________.

d. Therefore, the "golden ratio" $\frac{OL}{LD} =$ ________. To the nearest thousandth, this value is ________.

2. Just as the symbols π and e are used for the special irrational numbers 3.141 . . . and 2.718 . . . , the ratio you found in part (d) of Exercise 1 is important enough to be given a special letter, the Greek letter phi (ϕ). The "golden ratio," $\phi = \frac{1 + \sqrt{5}}{2} = 1.618\ldots$, shows up in some surprising places!

a. Each term (after the first two) of the Fibonacci sequence 1, 1, 2, 3, 5, 8, 13, 21, . . . is generated by adding the two previous terms. For example, $1 + 1 = 2$; $1 + 2 = 3$; $2 + 3 = 5$; and so on. Find a decimal approximation of each ratio of consecutive terms in the Fibonacci sequence:

$\frac{1}{1} =$ ______; $\frac{2}{1} =$ ______; $\frac{3}{2} =$ ______; $\frac{5}{3} =$ ______; $\frac{8}{5} =$ ______; $\frac{13}{8} =$ ______;

$\frac{21}{13}$ ______; $\frac{34}{21} =$ ______; $\frac{55}{34} =$ ______; $\frac{89}{55} =$ ______; $\frac{144}{89} =$ ______.

What number do the ratios of consecutive terms of the Fibonacci sequence seem to approach? ________

b. You can simplify the infinitely continued fraction $1 + \cfrac{1}{1 + \cfrac{1}{1 + \ldots}}$

by letting x equal this fraction and noting that $x = 1 + \frac{1}{x}$. Now solve the equation for x: ________

3. a. Find the next two terms in this sequence: $1, \phi, 1 + \phi, 1 + 2\phi, 2 + 3\phi, 3 + 5\phi, 5 + 8\phi,$ ______, ______

b. How is the sequence in part (a) similar to the Fibonacci sequence described in part (a) of Exercise 2? ________

c. In part (b) of Exercise 2, you discovered that ϕ is a solution of the quadratic equation $x^2 = x + 1$. Therefore, $\phi^2 = 1 + \phi$. Use this fact to show that the sequence $1, \phi, \phi^2, \phi^3, \phi^4, \ldots$ and the one from part (a) are exactly the same.

214 APPLICATIONS

Resource Book, ALGEBRA, Structure and Method, Book 2

Enrichment/Nonroutine Problems
***Resource Book,* p. 230**

Sequences of Moves (For use with Chapter 11)

1. Gwen's house is one mile from her school. She starts walking to school one day, goes half the distance from home to her school and stops. She turns around, heads home, goes half the distance from the point where she turned around and her home, and stops. She turns around, heads toward school, goes half the distance from where she turned around and her school, and stops. She turns around, heads toward home, goes half the distance from where she turned around and her home, and stops. She keeps doing this for a long time! What will her limiting positions be? You can get a feel for her motion by trying out a similar walk in your classroom.

____

2. An ant starts at the origin and walks along the x-axis a distance of 12 units. It then turns left and walks parallel to the y-axis a distance of 6 units. It then turns right and walks parallel to the x-axis a distance of 3 units. It then turns left and walks parallel to the y-axis a distance of 1.5 units. If this sequence of moves is continued in this manner, what will the ant's limiting position be?

____

3. Suppose the ant in problem 2 kept turning left instead of alternating left and right. What would be the limiting position now?

____

4. Assume that the process of shading regions illustrated at right is continued indefinitely. Find the limiting value of the total area of the shaded regions if the largest square is 1 cm by 1 cm.

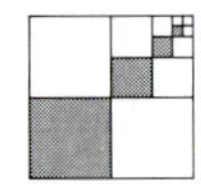

____

5. Take the unit interval $0 \leq x \leq 1$ and erase the middle third. This leaves two lines each of length one-third. Erase the middle third of each of the two remaining lines. Then erase the middle third of each of the four remaining lines, and keep up this process.

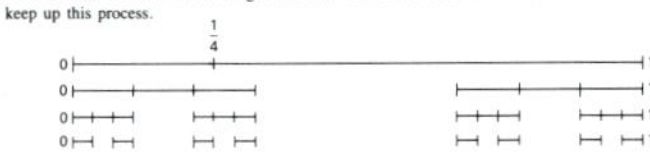

Will the point $\frac{1}{4}$ ever get erased? ________

230 ENRICHMENT

Resource Book, ALGEBRA, Structure and Method, Book 2

Thinking Skills
***Resource Book,* p. 266**

Thinking Skills (For use after Chapter 11)

Analysis

1. A *polygonal* number can be represented by dots that are arranged in the shape of a polygon. For each set of polygonal numbers given below, study the pattern of the dots and then draw the fourth member of the set.

a. Triangular numbers:

b. Square numbers:

c. Pentagonal numbers:

d. Hexagonal numbers:

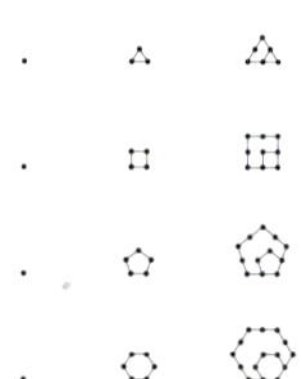

2. After analyzing the patterns in the drawings of Exercise 1, complete the table below.

	P_1^N	P_2^N	P_3^N	P_4^N	P_5^N
a. Triangular numbers ($N = 3$):	1	3	______	______	______
b. Square numbers ($N = 4$):	1	4	______	______	______
c. Pentagonal numbers ($N = 5$):	1	5	______	______	______
d. Hexagonal numbers ($N = 6$):	1	6	______	______	______

3. The nth triangular number is given by the formula $P_n^3 = \frac{n(n + 1)}{2}$. For example, the second triangular number is $P_2^3 = \frac{2(2 + 1)}{2} = 3$. Find a formula, in terms only of n, for:

a. the nth square number: $P_n^4 =$ ________

b. the nth pentagonal number: $P_n^5 =$ ________

c. the nth hexagonal number: $P_n^6 =$ ________

(continued)

266 THINKING SKILLS

Resource Book, ALGEBRA, Structure and Method, Book 2

Thinking Skills
***Resource Book,* p. 267**

Thinking Skills (Chapter 11) *(continued)*

4. a. Look down the columns of the table in Exercise 2. What patterns do you see?

____

b. Using your answer for part (a), predict the values of the first five heptagonal numbers.

	P_1^N	P_2^N	P_3^N	P_4^N	P_5^N
Heptagonal numbers ($N = 7$):	______	______	______	______	______

c. Use part (d) of Exercise 2 and part (b) of this exercise to complete the following statements.

$P_1^7 = P_1^6 +$ ______; $P_2^7 = P_2^6 +$ ______; $P_3^7 = P_3^6 +$ ______; $P_4^7 = P_4^6 +$ ______; $P_5^7 = P_5^6 +$ ______

d. What is the significance of the numbers you wrote in part (c)?

____

e. Using your answer from part (d), complete the following: $P_n^7 = P_n^6 +$ ________

f. Substituting your answer from part (c) of Exercise 3 for P_n^6 in the equation from part (e) of this exercise, find a formula for P_n^7 in terms only of n: $P_n^7 =$ ________

5. Based on your work in Exercise 4, find a formula for the nth octagonal number in terms only of n: $P_n^8 =$ ________

Synthesis

6. Use the table at the right to collect the formulas for P_n^3, P_n^4, P_n^5, P_n^6, P_n^7, and P_n^8, from Exercise 3, part (f) of Exercise 4, and Exercise 5. After examining the pattern in the formulas, try to determine a formula for P_n^N in terms only of N and n. (*Hint:* Any polynomials that appear in these formulas should be written in factored form. Also, you may need to rewrite the formulas for P_n^4, P_n^6, and P_n^8 as fractions with a denominator of 2 so that they look more like the formulas for P_n^3, P_n^5, and P_n^7.)

$P_n^N =$ ________

Type of number	*Formula for* P_n^N
Triangular ($N = 3$)	________
Square ($N = 4$)	________
Pentagonal ($N = 5$)	________
Hexagonal ($N = 6$)	________
Heptagonal ($N = 7$)	________
Octagonal ($N = 8$)	________

Resource Book, ALGEBRA, Structure and Method, Book 2

THINKING SKILLS 267

Reteaching/Practice
Study Guide, p. 183

11–7 Powers of Binomials

Objective: To expand powers of binomials.

Vocabulary

Binomial expansion The sum of terms that results from multiplying out a power of a binomial. Example: The expansion of $(a + b)^2$ is $a^2 + 2ab + b^2$.

Pascal's triangle A triangular array of numbers representing the coefficients of the expansion of $(a + b)^n$, where n is a nonnegative integer.

	Row
1	0
1 1	1
1 2 1	2
1 3 3 1	3
1 4 6 4 1	4
1 5 10 (10) (5) 1	5
1 6 15 20 (15) 6 1	6

The triangle has 1's at the beginning and end of each row. Each of the other numbers is the sum of the two numbers above it. Example: 15 = 10 + 5

Example 1 Expand $(a + b)^5$. Use row 5 of Pascal's triangle.

Solution

The coefficients are:	1	5	10	10	5	1
The powers of a are:	a^5	a^4	a^3	a^2	a^1	a^0
The powers of b are:	b^0	b^1	b^2	b^3	b^4	b^5

Combining this information, you get:

$$(a + b)^5 = a^5 + 5a^4b + 10a^3b^2 + 10a^2b^3 + 5ab^4 + b^5$$

Example 2 Expand and simplify $(2x - y^2)^7$.

Solution First find row 7 of Pascal's triangle. (Remember: Except for the 1's at the ends of a given row, each number in the row is the sum of the two numbers above it in the triangle.)

	Row
1 6 15 20 15 6 1	6
1 7 21 35 35 21 7 1	7

Now expand $(a + b)^7$:

$$(a + b)^7 = a^7 + 7a^6b + 21a^5b^2 + 35a^4b^3 + 35a^3b^4 + 21a^2b^5 + 7ab^6 + b^7$$

Replace a by $2x$ and b by $-y^2$ and simplify:

$$(2x - y^2)^7 = (2x)^7 + 7(2x)^6(-y^2) + 21(2x)^5(-y^2)^2 + 35(2x)^4(-y^2)^3 + 35(2x)^3(-y^2)^4 + 21(2x)^2(-y^2)^5 + 7(2x)(-y^2)^6 + (-y^2)^7$$

$$= 128x^7 - 448x^6y^2 + 672x^5y^4 - 560x^4y^6 + 280x^3y^8 - 84x^2y^{10} + 14xy^{12} - y^{14}$$

Reteaching/Practice
Study Guide, p. 184

11–7 Powers of Binomials *(continued)*

Expand and simplify each expression.

1. $(x + y)^4$
2. $(x + y)^5$
3. $(x - y)^3$
4. $(x + 2)^5$
5. $(y - 3)^4$
6. $(2a + 1)^3$
7. $(3a + 2)^5$
8. $(x^2 - 1)^7$
9. $(\sqrt{x} + \sqrt{y})^4$
10. $(x^2 - y^3)^8$
11. $(x + \sqrt{2})^6$
12. $(x - x^{-1})^5$

Example 3 The first three terms in the expansion of $(a + b)^{15}$ are $a^{15} + 15a^{14}b + 105a^{13}b^2$. Write the last three terms.

Solution For any row of Pascal's triangle, notice that the pattern of the coefficients is symmetric. That is, the first and last coefficients are equal, the second and next-to-last coefficients are equal, and so on. So if the coefficients of the first three terms in the expansion of $(a + b)^{15}$ are 1, 15, and 105, then the coefficients of the last three terms must be 105, 15, and 1. Also, the powers of a are decreasing to 0 and the powers of b are increasing to 15.
$\therefore$ the last three terms are $105a^2b^{13} + 15ab^{14} + b^{15}$.

In Exercises 13–17, use the symmetry of the coefficients.

13. The first three terms of $(x + y)^{18}$ are $x^{18} + 18x^{17}y + 153x^{16}y^2$. Write the last three terms.
14. The first three terms of $(x - y)^{24}$ are $x^{24} - 24x^{23}y + 276x^{22}y^2$. Write the last three terms.
15. The first three terms of $(x + y)^{12}$ are $x^{12} + 12x^{11}y + 66x^{10}y^2$. Write the last three terms.
16. The ninth term of $(x + 1)^{18}$ is $43{,}758x^{10}$. Write the eleventh term. (*Hint*: The tenth term is the middle term.)
17. The tenth term of $(x - y)^{19}$ is $-92{,}378x^{10}y^9$. Write the eleventh term.

Mixed Review Exercises

Tell whether each sequence is arithmetic, geometric, or neither. Then find a formula for the nth term of the sequence.

1. $\frac{3}{2}, \frac{6}{5}, \frac{9}{8}, \frac{12}{11}, \ldots$
2. 24, 12, 6, 3, . . .
3. −6, −1, 4, 9, . . .
4. $\frac{2}{3}$, 2, 6, 18, . . .
5. 1, 4, 9, 16, . . .
6. 100, 92, 84, 76, . . .

Find the sum of each series.

7. $8 + 11 + 14 + \cdots + 59$
8. $4 + \frac{8}{3} + \frac{16}{9} + \cdots$
9. $\sum_{n=1}^{20} (2n + 5)$
10. $\sum_{n=1}^{8} 3(2^{-n})$

Using Technology
Computer Activities, p. 69

ACTIVITY 26. *Pascal's Triangle* *(for use with Section 11-8)*

Directions: Write all answers in the spaces provided.

PROBLEM

Use the computer to find the seventh row of Pascal's triangle.

PROGRAM ◆

```
10 PRINT "WHAT ROW DO YOU WANT";
20 INPUT R
30 LET N = R - 1
40 LET C = 1
50 PRINT
60 FOR K = 1 TO R
70 PRINT C; "  ";
80 LET C = (N - (K - 1)) * C / K
90 NEXT K
100 PRINT
110 END
```

PROGRAM CHECK

Type in the program. To test whether you entered it correctly, run the program. After the question, enter 7. The computer should print:

1 6 15 20 15 6 1

The numbers displayed are the coefficients of the expansion of $(a + b)^6$.

ANALYSIS

The letter c represents, in turn, each number in the seventh row of Pascal's triangle. Each value of c is used to compute the next value of c in line 80.

USING THE PROGRAM

Run the program to determine the numbers in the given row of Pascal's triangle.

1. 9th row ____________________
2. 12th row ____________________
3. 14th row ____________________
4. 17th row ____________________
5. How many numbers are there in the 109th row? __________

(continued)

Using Models
Overhead Visual B

Pascal's Triangle

1	Row 0
1 1	1
1 2 1	2
1 3 3 1	3
1 4 6 4 1	4
1 5 10 10 5 1	5
1 6 15 20 15 6 1	6
1 7 21 35 35 21 7 1	7
1 8 28 56 70 56 28 8 1	8
1 9 36 84 126 126 84 36 9 1	9
1 10 45 120 210 252 210 120 45 10 1	10

Row n: $\frac{n!}{(n-k)!\,k!}$ for $k = 0, 1, 2, \ldots, n$

Binomial Expansions

$$(a + b)^0 = 1$$
$$(a + b)^1 = 1a^1 + 1b^1$$
$$(a + b)^2 = 1a^2 + 2a^1b^1 + 1b^2$$
$$(a + b)^3 = 1a^3 + 3a^2b^1 + 3a^1b^2 + 1b^3$$
$$(a + b)^4 = 1a^4 + 4a^3b^1 + 6a^2b^2 + 4a^1b^3 + 1b^4$$
$$(a + b)^5 = 1a^5 + 5a^4b^1 + 10a^3b^2 + 10a^2b^3 + 5a^1b^4 + 1b^5$$
$$(a + b)^6 = 1a^6 + 6a^5b^1 + 15a^4b^2 + 20a^3b^3 + 15a^2b^4 + 6a^1b^5 + 1b^6$$
$$(a + b)^7 = 1a^7 + 7a^6b^1 + 21a^5b^2 + 35a^4b^3 + 35a^3b^4 + 21a^2b^5 + 7a^1b^6 + 1b^7$$

$$(a + b)^n = \sum_{k=0}^{n} \frac{n!}{(n-k)!\,k!} a^{n-k}b^k$$

Application

Look at the drawing that accompanies the photo. If the two smallest squares have sides of length 1, then the next larger square has sides of length 2. By following the squares in their spiral sequence and noting the length of each square's sides, you obtain the sequence 1, 1, 2, 3, 5, 8, 13, ... Notice that each term in the sequence (after the initial 1's) is the sum of the two preceding terms. This is the Fibonacci sequence.

By taking the ratios of consecutive terms in the Fibonacci sequence, you obtain $\frac{1}{1}, \frac{2}{1}, \frac{3}{2}, \frac{5}{3}, \frac{8}{5}, \frac{13}{8}, \ldots$. The terms of this sequence approach a number called the *golden ratio,* denoted by ϕ and having an approximate decimal value of 1.618. A rectangle for which the ratio of length to width is ϕ is called a *golden rectangle.*

Research Activities
Students can investigate and report on various aspects of the Fibonacci sequence and the golden ratio. Some topics are: the occurrence of Fibonacci numbers in nature, the geometric construction of a golden rectangle; golden rectangles in art and architecture, and sequences involving ϕ.

References

Ghyka, Matila. *The Geometry of Art and Life.* New York: Dover, 1977.

Huntley, H. E. *The Divine Proportion: A Study in Mathematical Beauty.* New York: Dover, 1970.

Support Materials
Resource Book p. 214

11 Sequences and Series

The spiral form of a nautilus shell can be approximated by drawing quarter circles within squares positioned in a spiral sequence.

500

Sequences

11-1 Types of Sequences

Objective To determine whether a sequence is arithmetic, geometric, or neither and to supply missing terms of a sequence.

Sequences of numbers are very common in the world around us. Here are some examples:

(1)	Detroit's noontime temperatures (in °F) for a particularly cold week	18°, 15°, 17°, 14°, 0°, −2°, −5°
(2)	Value (in dollars and cents) of an investment at three-month intervals	507.50, 515.11, 522.84, 530.68, . . .
(3)	Total distance (in miles) traveled by a jet after each hour of a transcontinental flight	500, 1100, 1700, 2300, 2900
(4)	Number of bacteria at each stage in the growth of a colony	1, 2, 4, 8, 16, . . .

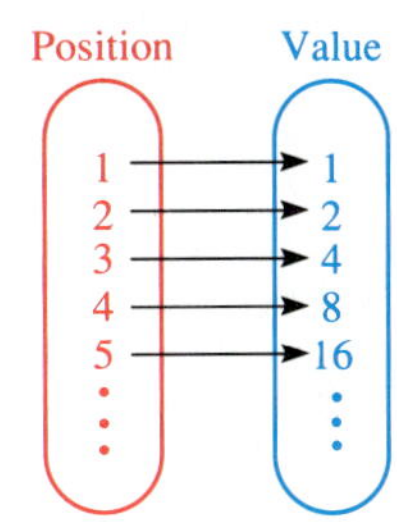

Because the members of a sequence are arranged in a definite order, you can set up a correspondence between the members and their positions in the list. A correspondence for sequence (4) is shown in the mapping diagram at the right.

A **sequence** can be defined as a function whose domain consists of consecutive positive integers. Each corresponding value is called a **term** of the sequence. A sequence is **finite** if it has a limited number of terms and **infinite** if it does not.

The terms of sequence (1) above seem to have no pattern in their occurrence. On the other hand, there are definite patterns in sequences (2)–(4). If you know what the pattern is, then you can find terms of the sequence that are not shown. For example, in sequence (3), the *difference* between each term and the term before it is always 600. If the terms of this sequence were to continue, then the next term would be 2900 + 600, or 3500. Sequence (3) is an example of an *arithmetic sequence*.

A sequence in which the difference between any two successive terms is constant is called an **arithmetic sequence,** or **arithmetic progression.** The constant difference is called the **common difference** and is usually denoted by d.

Teaching References

Lesson Commentary, pp. T119–T122

Assignment Guide, pp. T67–T71

Supplementary Materials
- Practice Masters 59–63
- Tests 44–49
- Resource Book
  - Practice Exercises, pp. 147–150
  - Tests, pp. 58–61
  - Enrichment Activity, p. 230
  - Application, p. 214
  - Mixed Review, pp. 183–185
- Study Guide, pp. 171–186
- Computer Activities 25–26
- Test Generator

Alternate Test, p. T23

Cumulative Review, p. T31

Teaching Suggestions p. T119

Suggested Extensions p. T119

Group Activities p. T119

Warm-Up Exercises

Give the next two numbers in the list.

1. 1, 2, 3, 4, _, _ 5, 6

2. 2, 4, 6, 8, _, _ 10, 12

3. 2, 4, 8, 16, _, _ 32, 64

4. 1, 4, 9, 16, _, _ 25, 36

5. 1, −1, 1, −1, _, _ 1, −1

Motivating the Lesson

Point out to students that lists of numbers can be found everywhere: the dates printed on a calendar, the won-lost columns found in the sports section of a newspaper, the test scores recorded in a teacher's grade book, and so on. Such lists of numbers are known in mathematics as sequences, the topic of today's lesson.

Chalkboard Examples

1. For each arithmetic sequence, find the common difference and the next two terms.
   a. 9, 7, 5, 3, . . .
   $d = 7 - 9 = -2$
   The next two terms of the sequence are $3 + (-2) = 1$ and $1 + (-2) = -1$.
   b. 3, 4.5, 6, 7.5, . . .
   $d = 4.5 - 3 = 1.5$
   The next two terms of the sequence are $7.5 + 1.5 = 9$ and $9 + 1.5 = 10.5$.
2. For each geometric sequence, find the common ratio and the next two terms.
   a. 40, 20, 10, . . .
   $r = \frac{20}{40} = \frac{1}{2}$
   The next two terms are $10 \cdot \frac{1}{2} = 5$ and $5 \cdot \frac{1}{2} = \frac{5}{2}$.
   b. $\sqrt{6}, 6, 6\sqrt{6}, \ldots$
   $r = \frac{6}{\sqrt{6}} = \sqrt{6}$
   The next two terms are $6\sqrt{6} \cdot \sqrt{6} = 36$ and $36 \cdot \sqrt{6} = 36\sqrt{6}$.

Example 1 For each arithmetic sequence, find the common difference and the next two terms of the sequence.

a. 4, 7, 10, 13, 16, . . . **b.** 50, 45, 40, 35, 30, . . .

Solution **a.** The common difference is found by subtracting any term from the term that follows it.

$$d = 7 - 4 = 3$$

The next two terms are $16 + 3 = 19$ and $19 + 3 = 22$. ***Answer***

b. The common difference is

$$d = 45 - 50 = -5.$$

The next two terms are $30 + (-5) = 25$ and $25 + (-5) = 20$. ***Answer***

Look at the pattern in sequence (4): 1, 2, 4, 8, 16, ... The sequence is not arithmetic because the difference between successive terms is *not* constant. However, the *ratio* of successive terms *is* constant.

$$\frac{2}{1} = \frac{4}{2} = \frac{8}{4} = \frac{16}{8} = \ldots$$

Since the ratio is always 2, every term is twice the preceding term. So the next term of this sequence would be $16 \cdot 2$, or 32. Sequence (4) is an example of a *geometric sequence*.

A sequence in which the ratio of every pair of successive terms is constant is called a **geometric sequence,** or **geometric progression.** The constant ratio is called the **common ratio** and is usually denoted by r.

Notice that sequence (2) on the preceding page is almost a geometric sequence. Each pair of successive terms has a ratio approximately equal to 1.015. In fact, the sequence was obtained by rounding the terms of a true geometric sequence to the nearest hundredth.

Example 2 For each geometric sequence, find the common ratio and the next two terms of the sequence.

a. 2, 6, 18, 54, . . . **b.** 80, −40, 20, −10, . . .

Solution **a.** Find the common ratio by dividing any term by the term before it:

$$r = \frac{6}{2} = 3$$

The next two terms are $54 \cdot 3 = 162$ and $162 \cdot 3 = 486$. ***Answer***

b. The common ratio is

$$r = \frac{-40}{80} = -\frac{1}{2}.$$

The next two terms are $-10 \cdot \left(-\frac{1}{2}\right) = 5$ and $5 \cdot \left(-\frac{1}{2}\right) = -\frac{5}{2}$. ***Answer***

The nth term of a sequence is usually denoted by the symbol t_n. For example, in the sequence 2, 6, 18, 54, . . . , the first term is $t_1 = 2$, the second term is $t_2 = 6$, and so on. If you know a formula for the nth term of a sequence in terms of n, then you can find *any* term of the sequence.

Example 3 Using the given formula for the nth term, find t_1, t_2, t_3, and t_4. Then tell whether the sequence is arithmetic, geometric, or neither.

a. $t_n = 5 + 4n$ **b.** $t_n = n^2$ **c.** $t_n = 3 \cdot 2^n$

Solution

a.

n	$t_n = 5 + 4n$
1	$t_1 = 5 + 4(1) = 9$
2	$t_2 = 5 + 4(2) = 13$
3	$t_3 = 5 + 4(3) = 17$
4	$t_4 = 5 + 4(4) = 21$

The sequence 9, 13, 17, 21, . . . is arithmetic since the common difference is 4. ***Answer***

b.

n	$t_n = n^2$
1	$t_1 = 1^2 = 1$
2	$t_2 = 2^2 = 4$
3	$t_3 = 3^2 = 9$
4	$t_4 = 4^2 = 16$

The sequence 1, 4, 9, 16, . . . is neither arithmetic nor geometric. ***Answer***

c.

n	$t_n = 3 \cdot 2^n$
1	$t_1 = 3 \cdot 2^1 = 6$
2	$t_2 = 3 \cdot 2^2 = 12$
3	$t_3 = 3 \cdot 2^3 = 24$
4	$t_4 = 3 \cdot 2^4 = 48$

The sequence 6, 12, 24, 48, . . . is geometric since the common ratio is 2. ***Answer***

Sometimes it is possible to find a formula for the nth term of a sequence if a sufficient number of terms are given to indicate a pattern. (This will be discussed in Lessons 11-2 and 11-3.) Of course, when you use a pattern to guess a formula for t_n, you are making the assumption that the pattern continues throughout the sequence.

When it is difficult to find a pattern in a sequence of numbers, you may be able to find a pattern in the differences between terms.

Example 4 Find the next term in the sequence 2, 6, 12, 20, 30, . . . by using the pattern in the differences between terms.

Solution

sequence	2		6		12		20		30		?
differences		4		6		8		10		?	

Although the original sequence is neither arithmetic nor geometric, the *differences* form an arithmetic sequence with common difference 2. So the next difference will be 12, and the next term in the original sequence will be 30 + 12, or 42. ***Answer***

Using the given formula for the nth term, find t_1, t_2, t_3, and t_4. Then tell whether the sequence is arithmetic, geometric, or neither.

3. $\frac{1}{2}n - 1$

$n \quad t_n = \frac{1}{2}n - 1$

$1 \quad t_1 = \frac{1}{2}(1) - 1 = -\frac{1}{2}$

$2 \quad t_2 = \frac{1}{2}(2) - 1 = 0$

$3 \quad t_3 = \frac{1}{2}(3) - 1 = \frac{1}{2}$

$4 \quad t_4 = \frac{1}{2}(4) - 1 = 1$

arithmetic; the common difference is $\frac{1}{2}$.

4. $t_n = n^2 - 1$

$n \quad t_n = n^2 - 1$

$1 \quad t_1 = (1)^2 - 1 = 0$

$2 \quad t_2 = (2)^2 - 1 = 3$

$3 \quad t_3 = (3)^2 - 1 = 8$

$4 \quad t_4 = (4)^2 - 1 = 15$

neither

5. $t_n = \frac{2^{n-1}}{4}$

$n \quad t_n = \frac{2^{n-1}}{4}$

$1 \quad t_1 = \frac{2^{1-1}}{4} = \frac{1}{4}$

$2 \quad t_2 = \frac{2^{2-1}}{4} = \frac{1}{2}$

$3 \quad t_3 = \frac{2^{3-1}}{4} = 1$

$4 \quad t_4 = \frac{2^{4-1}}{4} = 2$

geometric; the common ratio is 2.

6. Find the next term in the sequence 1, 4, 8, 13, 19, . . . by using the pattern in the differences between terms.
The differences between terms form the following arithmetic sequence: 3, 4, 5, 6, ... The next term in this sequence of differences is 7, so the next term in the given sequence is 19 + 7, or 26.

Problem Solving Strategies

The students must *look for patterns* in a given sequence in order to determine whether it is arithmetic, geometric, or neither.

Check for Understanding

Here is a suggested use of the Oral Exercises to check students' understanding as you teach the lesson.
Oral Exs. 1–4: use after Example 1.
Oral Exs. 5–8: use after Example 2.
Oral Exs. 9–12: use after Example 3.
Oral Exs. 13–14: use after Example 4.

Guided Practice

Tell whether the sequence is arithmetic, geometric, or neither. Then supply the missing terms.

1. $\frac{3}{5}, \frac{5}{7}, \frac{7}{9}, \frac{9}{11}$, ___?___, ___?___
neither; $\frac{11}{13}, \frac{13}{15}$
2. $\frac{1}{2}\cdot 1, \frac{1}{2}\cdot 2, \frac{1}{2}\cdot 3, \frac{1}{2}\cdot 4$, ___?___, ___?___
arithmetic; $\frac{5}{2}$, 3
3. 0.2, 0.02, 0.002, 0.0002, ___?___, ___?___
geometric; 2×10^{-5}, 2×10^{-6}
4. $\log_2 2, \log_3 4, \log_4 6, \log_5 8$, ___?___, ___?___
neither; $\log_6 10, \log_7 12$

Oral Exercises

Give the common difference and supply the missing terms for each arithmetic sequence.

1. 3, 7, 11, 15, ___?___, ___?___ 4; 19, 23
2. 21, 15, 9, 3, ___?___, ___?___ −6; −3, −9
3. 7, 10, ___?___, 16, 19, ___?___ 3; 13, 22
4. ___?___, ___?___, 25, 50, 75, 100 25; −25, 0

Give the common ratio and supply the missing terms for each geometric sequence.

5. 3, 6, 12, 24, ___?___, ___?___ 2; 48, 96
6. 1, −2, 4, −8, ___?___, ___?___ −2; 16, −32
7. $\frac{1}{100}, \frac{1}{10}$, ___?___, 10, 100, ___?___ 10; 1, 1000
8. ___?___, ___?___, $\frac{1}{3}, \frac{1}{9}, \frac{1}{27}, \frac{1}{81}$ $\frac{1}{3}$; 3, 1

Tell the first four terms of the sequence with the given formula. Then tell whether the sequence is arithmetic, geometric, or neither.

9. $t_n = 1 - 2n$ −1, −3, −5, −7; arith.
10. $t_n = \frac{1}{n+1}$ $\frac{1}{2}, \frac{1}{3}, \frac{1}{4}, \frac{1}{5}$; neither
11. $t_n = 3^n$ 3, 9, 27, 81; geom.
12. $t_n = n^2 - 1$ 0, 3, 8, 15; neither

Give the next two terms of each sequence by using the pattern in the differences between terms.

13. 8, 9, 11, 14, ___?___, ___?___ 18, 23
14. 5, 7, 11, 17, ___?___, ___?___ 25, 35

Written Exercises

Tell whether each sequence is arithmetic, geometric, or neither. Then supply the missing terms of the sequence. A = arithmetic, G = geometric, N = neither

A

1. 20, 17, 14, 11, ___?___, ___?___ A; 8, 5
2. 5, 9, 13, 17, ___?___, ___?___ A; 21, 25
3. 1, 5, 25, 125, ___?___, ___?___ G; 625, 3125
4. 256, 64, 16, 4, ___?___, ___?___ G; 1, $\frac{1}{4}$
5. 18, 22, 26, ___?___, 34, ___?___ A; 30, 38
6. 4, ___?___, −4, −8, −12, ___?___ A; 0, −16
7. $1, \frac{1}{4}, \frac{1}{9}, \frac{1}{16}$, ___?___, ___?___ N; $\frac{1}{25}, \frac{1}{36}$
8. 32, −16, 8, −4, ___?___, ___?___ G; 2, −1
9. $4^{1/2}, 4^{3/2}, 4^{5/2}, 4^{7/2}$, ___?___, ___?___ G; $4^{9/2}, 4^{11/2}$
10. $\frac{1}{12}, \frac{2}{13}, \frac{3}{14}, \frac{4}{15}$, ___?___, ___?___ N; $\frac{5}{16}, \frac{6}{17}$

Find the first four terms of the sequence with the given formula. Then tell whether the sequence is arithmetic, geometric, or neither.

11. $t_n = 4n + 3$ 7, 11, 15, 19; A
12. $t_n = 2n + 1$ 3, 5, 7, 9; A
13. $t_n = 3^{n-1}$ 1, 3, 9, 27; G
14. $t_n = 2 \cdot 3^n$ 6, 18, 54, 162; G
15. $t_n = \frac{(-2)^n}{8}$
16. $t_n = 13 - 4n$ 9, 5, 1, −3; A
17. $t_n = \log(n+1)$ log 2, log 3, log 4, log 5; N
18. $t_n = \log 10^n$ 1, 2, 3, 4; A

 15. $-\frac{1}{4}, \frac{1}{2}$, −1, 2; G

19. **a.** What type of sequence is $-3, -1, 1, 3, \ldots$? A

b. What type of sequence is $2^{-3}, 2^{-1}, 2^{1}, 2^{3}, \ldots$? G

20. **a.** What type of sequence is $1, 4, 16, 64, \ldots$? G

b. What type of sequence is $\log_2 1, \log_2 4, \log_2 16, \log_2 64, \ldots$? A

Find the next two terms of each sequence by using the pattern in the differences between terms.

B **21.** 2, 4, 8, 14, 22, . . . 32, 44

22. −3, 1, 9, 21, 37, . . . 57, 81

23. 60, 48, 38, 30, 24, . . . 20, 18

24. 24, 23, 21, 17, 9, . . . −7, −39

25. 1, 3, 7, 15, 31, . . . 63, 127

26. 0, 1, 4, 13, 40, . . . 121, 364

27. 1, 1, 2, 3, 5, 8, 13, . . . 21, 34

28. 1, 2, 6, 15, 31, . . . 56, 92

(The well-known sequence in Exercise 27 is called the *Fibonacci sequence*. Each term is the sum of the two terms before it.)

29. 1, 3, 6, 11, 19, 31, . . . 48, 71

30. 5, 7, 10, 16, 27, 45, . . . 72, 110

[*Hint:* For Exercises 29 and 30, look at the *second* differences (that is, the differences of the differences between terms).]

31. A *triangular number* can be represented by dots that are arranged in the shape of an equilateral triangle, as shown below. The first four triangular numbers are given.

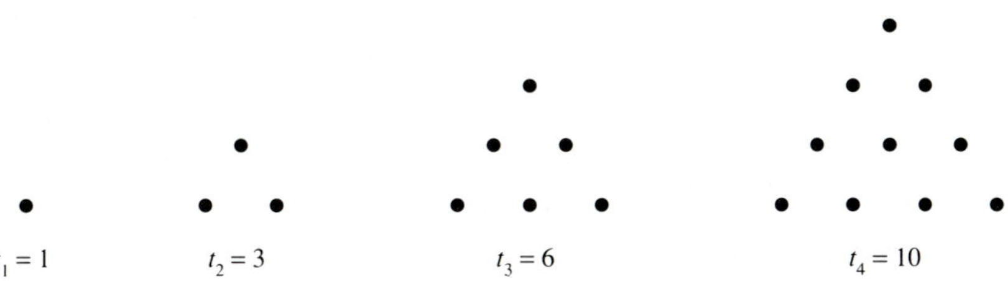

a. Find the next two triangular numbers. 15, 21

b. Find the tenth triangular number without actually drawing a diagram. 55

32. A *pentagonal number* can be represented by dots that are arranged in the shape of a pentagon, as shown below. The first four pentagonal numbers are given.

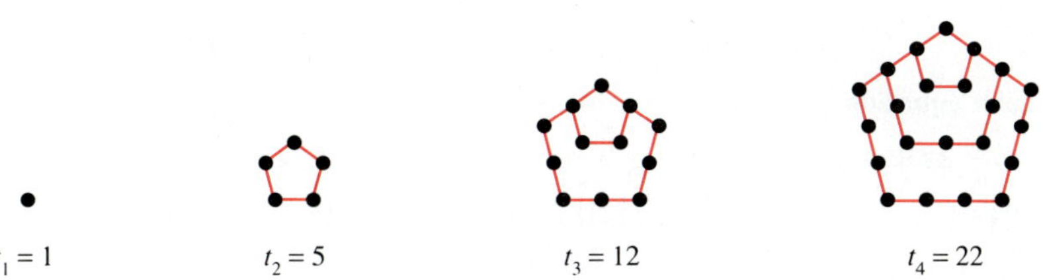

a. Find the next two pentagonal numbers. 35, 51

b. Find the tenth pentagonal number without drawing a diagram. 145

Find the first four terms of the sequence. Tell whether the sequence is arithmetic, geometric, or neither.

5. $t_n = -5 \cdot 2^{n-1}$
$-5, -10, -20, -40$; geometric

6. $t_n = \frac{1}{2}(n - 7)$
$-3, -\frac{5}{2}, -2, -\frac{3}{2}$; arithmetic

7. $t_n = n\sqrt{n}$
$1, 2\sqrt{2}, 3\sqrt{3}, 8$; neither

8. $t_n = 3^n + 1$
4, 10, 28, 82; neither

Summarizing the Lesson

In this lesson students learned to identify arithmetic sequences, geometric sequences, and those which are neither. Students also learned methods for finding the missing terms of a sequence. Ask students to give the definitions of arithmetic and geometric sequences.

Suggested Assignments

Average Algebra
Day 1: 504/1–22
Day 2: 505/23–34
S 506/Mixed Review

Extended Algebra
Day 1: 504/1–22
Day 2: 505/23–34
S 506/Mixed Review

Extended Alg. and Trig.
504/1–33 odd
S 506/Mixed Review

Supplementary Materials

Study Guide	pp. 171–172
Practice Master	59

Additional Answers
Mixed Review Exercises

1.
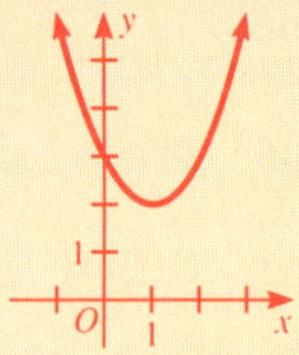

2.
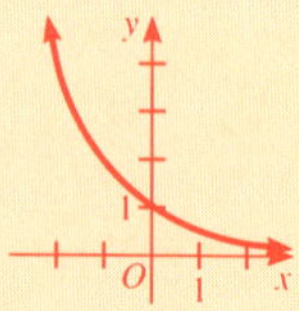

3.
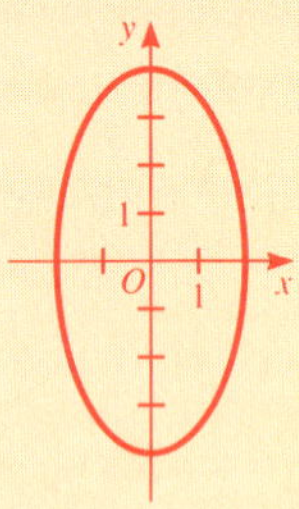

4.
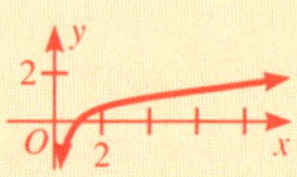

5.
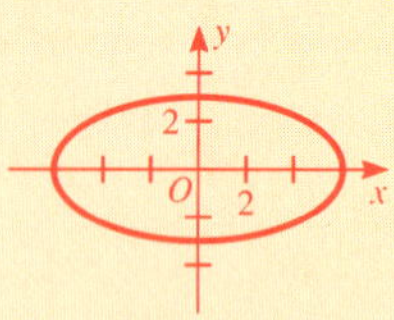

6.
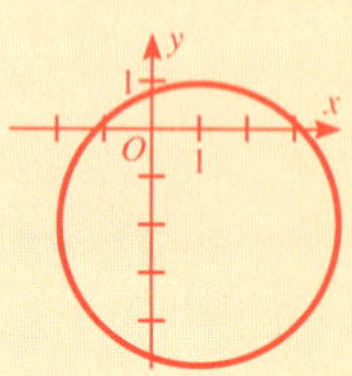

33. A diagonal of a polygon is a line segment joining any two nonadjacent vertices of the polygon. If n is the number of sides of a polygon ($n \geq 3$), then let t_n be the number of its diagonals.

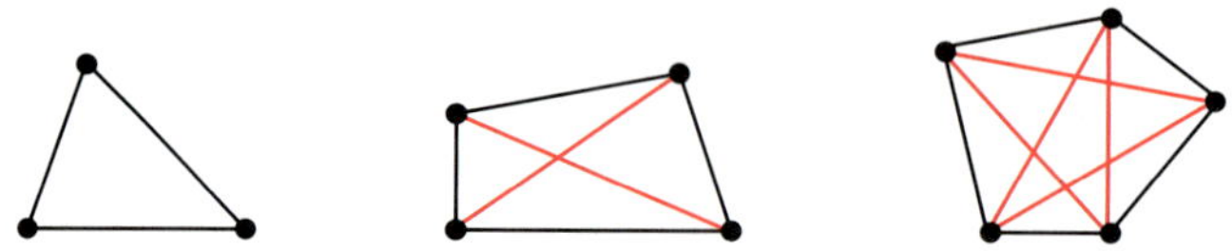

$t_3 = 0$ $t_4 = 2$ $t_5 = 5$

a. Draw a hexagon and find t_6. 9 **b.** Find t_{10} without drawing a decagon. 35

34. If n chords of a circle are drawn, then let t_n be the *maximum* number of non-overlapping regions that can be formed within the circle, as shown.

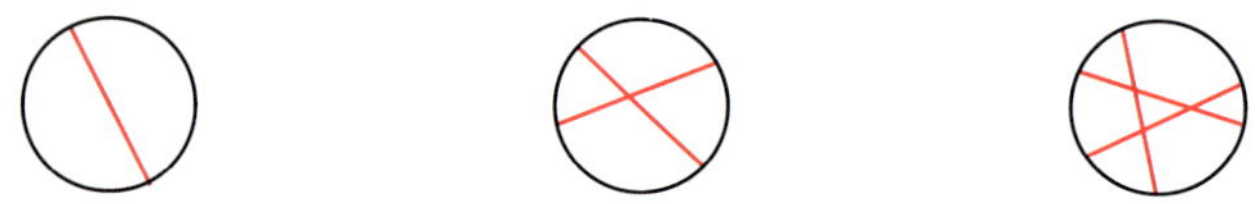

1 chord $t_1 = 2$ regions; 2 chords $t_2 = 4$ regions; 3 chords $t_3 = 7$ regions

a. Use drawings to find t_4 and t_5. 11, 16 **b.** Find t_6 without using a drawing. 22

C **35.** Let t_n be the *maximum* number of non-overlapping regions that can be formed when n points of a circle are connected, as shown.

1 point $t_1 = 1$ region; 2 points $t_2 = 2$ regions; 3 points $t_3 = 4$ regions; 4 points $t_4 = 8$ regions

a. Make careful drawings and determine t_5 and t_6 by counting the regions formed. Is the resulting sequence geometric? 16, 31; No

b. Use the pattern in the *third* differences (that is, the differences of the differences of the differences between terms) to guess the value of t_7. 57

Mixed Review Exercises

Graph each equation.

1. $y = x^2 - 2x + 3$ **2.** $y = 2^{-x}$ **3.** $4x^2 + y^2 = 16$

4. $y = \log_4 x$ **5.** $x^2 + 4y^2 = 36$ **6.** $(x - 1)^2 + (y + 2)^2 = 9$

Solve each system.

7. $2x + y = 1$, $x - y = 5$ (2, −3)

8. $y = 2x^2 - 1$, $x + y = 2$ (1, 1), $\left(-\frac{3}{2}, \frac{7}{2}\right)$

9. $x^2 + y^2 = 5$, $x^2 - y^2 = 3$ (2, 1), (−2, 1), (2, −1), (−2, −1)

11-2 Arithmetic Sequences

Objective To find a formula for the nth term of an arithmetic sequence and to find specified terms of arithmetic sequences.

How would you find the hundredth term (t_{100}) of the arithmetic sequence

$$3, 7, 11, 15, \ldots ?$$

One way is to consider the number of terms in the sequence and the number of differences between the terms.

sequence	3	7	11	15	. . .	t_{100}
differences	4	4	4	4	. . .	4

99 differences

As you can see from the diagram, the number of differences is always one less than the number of terms. Therefore, to find t_{100}, start with the first term 3 and add the common difference 4 ninety-nine times.

$$t_{100} = 3 + 99 \cdot 4$$
$$= 399$$

Example 1 Find a formula for the nth term of the sequence 5, 8, 11, 14, ...

Solution The first term is $t_1 = 5$. The common difference is $d = 8 - 5 = 3$. Since there are n terms, start with 5 and add the difference $n - 1$ times.

$$t_n = 5 + (n - 1)3$$
$$= 2 + 3n \quad \textit{Answer}$$

The method used in Example 1 can be generalized to *any* arithmetic sequence.

In an arithmetic sequence with first term t_1 and common difference d, the nth (or *general*) term is given by

$$\mathbf{t_n = t_1 + (n - 1)d.}$$

Example 2 Find t_{17} for the following arithmetic sequence:

$$5, 8, 11, 14, \ldots$$

Solution Use the formula $t_n = 2 + 3n$ from Example 1.

$$t_{17} = 2 + 3(17)$$
$$= 53 \quad \textit{Answer}$$

Teaching Suggestions p. T120

Suggested Extensions p. T120

Warm-Up Exercises

Simplify.

1. $-7 + (6 - 1)3$ 8
2. $4 + (n - 1)(-2)$ $6 - 2n$

Evaluate if $n = 50$.

3. $-3 + 4n$ 197
4. $100 - \frac{1}{2}n$ 75
5. Solve the system:
   $6 = x + y$
   $14 = x + 5y$ (4, 2)

Motivating the Lesson

Pose the following to the students: An employer agrees to raise an employee's current $1500 monthly salary by $25 a month until it reaches $2000. How many months will this take?

Although this problem can be solved by listing the sequence of salaries term by term, it can be solved more efficiently using the techniques of today's lesson.

Chalkboard Examples

1. Find the nth term of the arithmetic sequence 7, 11, 15, 19, ...
   The first term is $t_1 = 7$.
   The common difference is $d = 4$. Since there are n terms, start with 7 and add the difference $(n - 1)$ times.
   $t_n = 7 + (n - 1)4$
   $t_n = 3 + 4n$
2. Find t_{11} for the arithmetic sequence of Chalkboard Example 1.
   $t_{11} = 3 + 4(11) = 47$

(continued)

Chalkboard Examples (continued)

3. Find t_{20} for the arithmetic sequence in which $t_2 = 13$ and $t_5 = 22$.
Substitute 13 for t_2 and 22 for t_5 in the formula $t_n = t_1 + (n - 1)d$ to obtain a system of linear equations in t_1 and d.
$t_2 = t_1 + (2 - 1)d \rightarrow 13 = t_1 + d$
$t_5 = t_1 + (5 - 1)d \rightarrow 22 = t_1 + 4d$
Solve the first equation for t_1: $t_1 = 13 - d$. Substitute into the second equation:
$22 = (13 - d) + 4d$
$9 = 3d;\ d = 3$
Then $t_1 = 13 - d = 10$. Using $t_1 = 10$ and $d = 3$ in the formula $t_n = t_1 + (n - 1)d$, find t_{20}:
$t_{20} = 10 + (19)3 = 67$

4. Find the arithmetic mean of −7.8 and 3.6.
$\frac{-7.8 + 3.6}{2} = \frac{-4.2}{2} = -2.1$

5. Insert three arithmetic means between 18 and 54.
Outline the sequence: 18, _?_, _?_, _?_, 54
Substitute $t_5 = 54$ and $t_1 = 18$ in the formula $t_n = t_1 + (n - 1)d$ to find d.
$54 = 18 + (5 - 1)d$
$36 = 4d;\ d = 9$
The three means are obtained by adding 9 to successive terms: 27, 36, 45

Check for Understanding

Oral Exs. 1–3: use after Example 1.
Oral Exs. 4–6: use after Example 2.

Example 3 Find t_{23} for the arithmetic sequence in which $t_2 = 4$ and $t_5 = 22$.

Solution Substitute 4 for t_2 and 22 for t_5 in the formula $t_n = t_1 + (n - 1)d$ to obtain a system of linear equations in t_1 and d.

$$t_2 = t_1 + (2 - 1)d \longrightarrow 4 = t_1 + d$$
$$t_5 = t_1 + (5 - 1)d \longrightarrow 22 = t_1 + 4d$$

Solve the first equation for t_1: $t_1 = 4 - d$.
Substitute $4 - d$ for t_1 in the second equation and solve for d:

$$22 = (4 - d) + 4d$$
$$22 = 4 + 3d$$
$$d = 6$$

Then $t_1 = 4 - d = -2$. Now using $t_1 = -2$ and $d = 6$ in the formula $t_n = t_1 + (n - 1)d$, find t_{23}:

$$t_{23} = t_1 + (23 - 1)d$$
$$= -2 + 22 \cdot 6$$
$$= 130 \quad \textit{Answer}$$

The terms between two given terms of an arithmetic sequence are called **arithmetic means** of the given terms.

10, 13, 16, 19, 22	10, 14, 18, 22	10, 16, 22
Three arithmetic means between 10 and 22	Two arithmetic means between 10 and 22	*The* arithmetic mean of 10 and 22

A single arithmetic mean between two numbers is called *the* **arithmetic mean** of the numbers. The arithmetic mean, or *average,* of two numbers a and b is $\frac{a + b}{2}$.

Example 4 **a.** Find the arithmetic mean of 4 and 9.
b. Insert four arithmetic means between 15 and 50.

Solution **a.** The arithmetic mean is the average of 4 and 9.

$$\frac{4 + 9}{2} = \frac{13}{2} = 6.5 \quad \textit{Answer}$$

b. Outline the sequence: 15, _?_, _?_, _?_, _?_, 50
In this sequence, 15 is the *first* term and 50 is the *sixth* term. So to find d substitute 15 for t_1 and 50 for t_6 in the formula $t_n = t_1 + (n - 1)d$.

$$50 = 15 + (6 - 1)d$$
$$35 = 5d$$
$$d = 7$$

The four means are obtained by adding 7 to successive terms:
15, 22, 29, 36, 43, 50. ***Answer***

Oral Exercises

Give a formula for the *n*th term of each arithmetic sequence.

1. $-1, 0, 1, 2, \ldots$ $t_n = n - 2$
2. $2, 4, 6, 8, \ldots$ $t_n = 2n$
3. $0, 3, 6, 9, \ldots$ $t_n = 3n - 3$

For each arithmetic sequence, find the value of t_{11}.

4. $4, 7, 10, 13, \ldots$ 34
5. $10, 5, 0, -5, \ldots$ -40
6. $3, 13, 23, 33, \ldots$ 103

Written Exercises

Find a formula for the *n*th term of each arithmetic sequence.

A

1. $24, 32, 40, 48, \ldots$ $t_n = 8n + 16$
2. $30, 20, 10, 0, \ldots$ $t_n = 40 - 10n$
3. $-3, -10, -17, -24, \ldots$ $t_n = 4 - 7n$
4. $-6, -1, 4, 9, \ldots$ $t_n = 5n - 11$
5. $7, 11, 15, 19, \ldots$ $t_n = 4n + 3$
6. $13, 4, -5, -14, \ldots$ $t_n = 22 - 9n$

Find the specified term of each arithmetic sequence.

7. $4, 9, 14, 19, \ldots; t_{21}$ 104
8. $3, 11, 19, \ldots; t_{31}$ 243
9. $100, 98, 96, \ldots; t_{25}$ 52
10. $3, 3.5, 4, 4.5, \ldots; t_{101}$ 53
11. $-2, -11, -20, \ldots; t_{101}$ -902
12. $17, 7, -3, \ldots; t_{1000}$ -9973
13. $t_1 = 5, t_3 = 20; t_{12}$ 87.5
14. $t_2 = 7, t_4 = 8; t_1$ 6.5
15. $t_5 = 24, t_9 = 40; t_1$ 8
16. $t_8 = 60, t_{12} = 48; t_{40}$ -36
17. $t_7 = -19, t_{10} = -28; t_{21}$ -61
18. $t_{10} = 41, t_{15} = 61; t_3$ 13

Find the arithmetic mean of each pair of numbers.

19. $-3, 7$ 2
20. $2.3, 9.1$ 5.7
21. $\frac{4}{5}, \frac{11}{5}$ $\frac{3}{2}$
22. $-\sqrt{2}, 3\sqrt{2}$ $\sqrt{2}$

Insert (a) two, (b) three, and (c) four arithmetic means between each pair.

B

23. $-27, 33$
24. $15, 45$
25. $11, 35$
26. $0, 20$
27. How many terms are in the sequence $18, 24, 30, \ldots, 618$? 101
28. How many terms are in the sequence $44, 36, 28, \ldots, -380$? 54
29. How many multiples of 3 are there between 100 and 1000? 300
30. How many numbers between 50 and 500 are divisible by 7? 64

Find the position *n* of the term in red in each arithmetic sequence.

31. $25, 33, 41, \ldots, 145, \ldots$ 16
32. $40, 37, 34, \ldots, -29, \ldots$ 24

C

33. Show that the sequence $a, \frac{a+b}{2}, b$ is arithmetic.

Guided Practice

1. Find the formula for the *n*th term of the arithmetic sequence $-1, 0, 1, 2, 3, \ldots$; $t_n = n - 2$

Find the specified term of each arithmetic sequence.

2. $2, 5, 8, \ldots; t_{17}$ 50
3. $t_2 = 9, t_5 = 21; t_{41}$ 165
4. Find the arithmetic mean of $\frac{1}{2}$ and $\frac{1}{3}$. $\frac{5}{12}$

Summarizing the Lesson

In this lesson students learned to find a specified term of an arithmetic sequence. Ask students to state the formula for the *n*th term of an arithmetic sequence.

Suggested Assignments

Average Algebra
509/1–31 odd

Extended Algebra
509/1–33 odd

Extended Alg. and Trig.
509/3–33 mult. of 3
Assign with Lesson 11-3.

Supplementary Materials

Study Guide	pp. 173–174
Test Master	44
Resource Book	p. 147

Additional Answers
Written Exercises

23. a. $-7, 13$
b. $-12, 3, 18$
c. $-15, -3, 9, 21$

24. a. 25, 35
b. 22.5, 30, 37.5
c. 21, 27, 33, 39

25. a. 19, 27 b. 17, 23, 29
c. 15.8, 20.6, 25.4, 30.2

26. a. $\frac{20}{3}, \frac{40}{3}$ b. 5, 10, 15
c. 4, 8, 12, 16

Teaching Suggestions p. T120

Suggested Extensions p. T120

Group Activities p. T120

Warm-Up Exercises

Evaluate if $n = 4$.

1. $3 \cdot 5^{n-1}$ 375
2. $8\left(-\frac{3}{2}\right)^{n-1}$ -27

Find the common ratio for each geometric sequence.

3. 3, 12, 48, . . . 4
4. 12, 18, 27, . . . $\frac{3}{2}$
5. -2, 10, -50, . . . -5

Motivating the Lesson

Pose the following problem to the students:
Suppose an aunt decides to give her newborn nephew a penny on his first birthday and double the gift on subsequent birthdays. How much will the nephew receive on his 21st birthday? The problem can be solved by finding a formula for the nth term of the geometric sequence \$.01, \$.02, \$.04, . . . and letting $n = 21$. This approach will be taken in today's lesson, just as it was for arithmetic sequences in the previous lesson.

11-3 Geometric Sequences

Objective To find a formula for the nth term of a geometric sequence and to find specified terms of geometric sequences.

The method for finding the nth term of a geometric sequence is similar to the method used for arithmetic sequences. For example, to find the tenth term of the geometric sequences 5, 10, 20, 40, . . . , look at the number of times the first term 5 is multiplied by the common ratio 2 to produce t_{10}.

sequence	5	10	20	40	. . .	t_{10}
ratios	2	2	2	2	. . .	2

9 ratios

As the diagram shows, the number of ratios is always one less than the number of terms. Therefore, the tenth term is the product of 5 and the ninth power of 2.

$$\begin{aligned} t_{10} &= 5 \cdot 2^9 \\ &= 2560 \end{aligned}$$

Example 1 Find a formula for the nth term of the sequence 3, -12, 48, -192, ...

Solution The first term is $t_1 = 3$. The common ratio is $r = \frac{-12}{3} = -4$.

Since there are n terms, start with 3 and multiply by the ratio $n - 1$ times:

$$t_n = 3(-4)^{n-1} \quad \textbf{\textit{Answer}}$$

The method used in Example 1 can be generalized to *any* geometric sequence.

In a geometric sequence with first term t_1 and common ratio r, the nth (or *general*) term is given by

$$\boldsymbol{t_n = t_1 \cdot r^{n-1}}.$$

Example 2 Find t_6 for the following geometric sequence:

3, -12, 48, . . .

Solution Use the formula $t_n = 3(-4)^{n-1}$ from Example 1.

$$\begin{aligned} t_6 &= 3(-4)^{6-1} \\ &= 3(-4)^5 \\ &= 3(-1024) = -3072 \quad \textbf{\textit{Answer}} \end{aligned}$$

Example 3 Find t_7 for the geometric sequence in which $t_2 = 24$ and $t_5 = 3$.

Solution Substitute 24 for t_2 and 3 for t_5 in the formula $t_n = t_1 \cdot r^{n-1}$ to obtain a system of equations in t_1 and r.

$$t_2 = t_1 \cdot r^{2-1} \longrightarrow 24 = t_1 \cdot r$$
$$t_5 = t_1 \cdot r^{5-1} \longrightarrow 3 = t_1 \cdot r^4$$

Solve the first equation for t_1: $t_1 = \frac{24}{r}$. Substitute this expression for t_1 in the second equation and solve for r:

$$3 = \frac{24}{r} \cdot r^4$$
$$\frac{1}{8} = r^3, \quad \text{so } r = \frac{1}{2}$$

Substitute $\frac{1}{2}$ for r in the first equation:

$$24 = t_1 \cdot \frac{1}{2}$$
$$48 = t_1$$

Using $t_1 = 48$ and $r = \frac{1}{2}$ in the formula $t_n = t_1 \cdot r^{n-1}$, find t_7:

$$t_7 = 48 \cdot \left(\frac{1}{2}\right)^{7-1}$$
$$= 48 \cdot \frac{1}{64} = \frac{3}{4} \quad \textit{Answer}$$

Geometric means are the terms between two given terms of a geometric sequence.

1, 2, 4, 8, 16	1, −2, 4, −8, 16	1, 4, 16
Three geometric means between 1 and 16	Three geometric means between 1 and 16	*The* geometric mean of 1 and 16

Because 1, 4, 16 and 1, −4, 16 are both geometric sequences, you might think that 4 or −4 is the geometric mean of 1 and 16. However, so that the geometric mean of two numbers will be unique, it is a common practice to consider *the* **geometric mean** of a and b to be $\sqrt{ab}$ if a and b are positive and $-\sqrt{ab}$ if a and b are negative.

Example 4 **a.** Find the geometric mean of 4 and 9.

b. Insert three geometric means between $\frac{1}{2}$ and $\frac{1}{162}$.

Solution **a.** The geometric mean of 4 and 9 is $\sqrt{4 \cdot 9} = \sqrt{36}$, or 6. ***Answer***

b. Outline the sequence: $\frac{1}{2}$, _?_, _?_, _?_, $\frac{1}{162}$

(Solution continues on the next page.)

Chalkboard Examples

1. Find a formula for the nth term of the geometric sequence 81, 54, 36, 24, ...

The first term is $t_1 = 81$. The common ratio is $\frac{54}{81} = \frac{2}{3}$.

$\therefore t_n = 81\left(\frac{2}{3}\right)^{n-1}$

2. Find t_6 for the geometric sequence in Chalkboard Example 1.

$t_6 = 81\left(\frac{2}{3}\right)^{6-1} = 81\left(\frac{32}{243}\right) = \frac{32}{3}$

3. Find t_7 for the geometric sequence in which $t_2 = 6$ and $t_5 = 750$.

Substitute 6 for t_2 and 750 for t_5 in the formula $t_n = t_1 \cdot r^{n-1}$.

$t_2 = t_1 \cdot r^{2-1} \rightarrow 6 = t_1 \cdot r$

$t_5 = t_1 \cdot r^{5-1} \rightarrow 750 = t_1 \cdot r^4$

Solve the first equation for t_1: $t_1 = \frac{6}{r}$. Substitute this expression for t_1 in the second equation and solve for r:

$750 = \frac{6}{r} \cdot r^4$

$750 = 6 \cdot r^3$

$125 = r^3$

$r = 5$

Then $t_1 = \frac{6}{5}$.

Using $t_1 = \frac{6}{5}$ and $r = 5$ in the formula $t_n = t_1 \cdot r^{n-1}$ find t_7:

$t_7 = \frac{6}{5} \cdot 5^{7-1}$

$= \frac{6}{5} \cdot 5^6$

$= \frac{6}{5}(15{,}625)$

$= 18{,}750$

(continued)

Chalkboard Examples (continued)

6. Find the geometric mean of -2 and -18.
$-\sqrt{(-2)(-18)} = -\sqrt{36} = -6$

7. Insert two geometric means between $\frac{1}{2}$ and $\frac{1}{54}$.
Outline the sequence:
$\frac{1}{2}, \underline{\ ?\ }, \underline{\ ?\ }, \frac{1}{54}$
Substitute $t_4 = \frac{1}{54}$ and $t_1 = \frac{1}{2}$ in the formula $t_n = t_1 \cdot r^{n-1}$ to find r.
$\frac{1}{54} = \frac{1}{2} \cdot r^{4-1}$
$\frac{1}{27} = r^3$
$\frac{1}{3} = r$
The required means are obtained by multiplying consecutive terms by $\frac{1}{3}$:
$\frac{1}{6}, \frac{1}{18}$

Check for Understanding

Here is a suggested use of the Oral Exercises to check students' understanding as you teach the lesson.
Oral Exs. 1–4: use after Example 1.
Oral Exs. 5–7: use after Example 2.
Oral Ex. 8: use after Example 4.

To find r, substitute $\frac{1}{162}$ for t_5 and $\frac{1}{2}$ for t_1 in the formula $t_n = t \cdot r^{n-1}$:

$$\frac{1}{162} = \frac{1}{2} \cdot r^{5-1}$$
$$\frac{1}{81} = r^4$$
$$\pm\frac{1}{3} = r$$

If $r = \frac{1}{3}$, then the geometric sequence is $\frac{1}{2}, \frac{1}{6}, \frac{1}{18}, \frac{1}{54}, \frac{1}{162}$.

If $r = -\frac{1}{3}$, then the geometric sequence is $\frac{1}{2}, -\frac{1}{6}, \frac{1}{18}, -\frac{1}{54}, \frac{1}{162}$. ***Answer***

Example 5 George has taken a job with a starting salary of \$20,000. Find his salary during his fourth year on the job if he receives annual raises of:

a. \$1100 **b.** 5%

Solution **a.** With annual raises of \$1100, George's yearly salaries form an *arithmetic* sequence with $t_1 = 20{,}000$ and $d = 1100$.
The fourth term in the sequence is:

$$\begin{aligned} t_4 &= 20{,}000 + (4-1)1100 \\ &= 20{,}000 + 3300 \\ &= 23{,}300 \end{aligned}$$

$\therefore$ George's fourth-year salary will be \$23,300. ***Answer***

b. With annual raises of 5%, George's yearly salaries form a *geometric* sequence with $t_1 = 20{,}000$ and $r = 1.05$. (Notice that the common ratio is 1.05 and not 0.05, because a 5% raise means that for a salary S one year, the salary will be $S + 0.05S$, or $1.05S$, the next year.)
The fourth term in the sequence is:

$$\begin{aligned} t_4 &= 20{,}000(1.05)^{4-1} \\ &= 20{,}000(1.157625) \\ &= 23{,}152.50 \end{aligned}$$

$\therefore$ George's fourth-year salary will be \$23,152.50. ***Answer***

Oral Exercises

Give a formula for the *n*th term of each geometric sequence.

1. 1000, 200, 40, 8, . . . $t_n = 1000\left(\frac{1}{5}\right)^{n-1}$
2. 3, −6, 12, −24, . . . $t_n = 3(-2)^{n-1}$
3. 4, −12, 36, −108, . . . $t_n = 4(-3)^{n-1}$
4. 125, 25, 5, 1, . . . $t_n = 125\left(\frac{1}{5}\right)^{n-1}$

For each geometric sequence, find the value of t_{11} (in exponential form).

5. 1, 4, 16, 64, . . . 4^{10} **6.** 5, 10, 20, 40, . . . $5 \cdot 2^{10}$ **7.** 18, −6, 2, $-\frac{2}{3}$, . . . $18 \cdot \left(-\frac{1}{3}\right)^{10}$

8. Give the geometric mean of each pair of numbers.

a. 1, 4 2 **b.** −3, −27 −9 **c.** 3, 12 6

Written Exercises

Find a formula for the nth term of each geometric sequence.

A

1. 2, 6, 18, 54, . . . $t_n = 2 \cdot 3^{n-1}$

2. 500, 100, 20, 4, . . . $t_n = 500\left(\frac{1}{5}\right)^{n-1}$

3. 1, $\sqrt{2}$, 2, $2\sqrt{2}$, . . . $t_n = \sqrt{2}^{n-1}$

4. 8, 12, 18, 27, . . . $t_n = 8\left(\frac{3}{2}\right)^{n-1}$

5. 64, −48, 36, −27, . . . $t_n = 64\left(-\frac{3}{4}\right)^{n-1}$

6. −1, 0.1, −0.01, 0.001, . . . $t_n = -\left(-\frac{1}{10}\right)^{n-1}$

Find the specified term of each geometric sequence.

7. 2, 6, 18, 54, . . . ; t_{10} 39,366

8. 5, 10, 20, 40, . . . ; t_{12} 10,240

9. 320, 80, 20, 5, . . . ; t_8 $\frac{5}{256}$

10. 1, −3, 9, −27, . . . ; t_8 −2187

11. 40, −20, 10, −5, . . . ; t_{11} $\frac{5}{128}$

12. −10, 50, −250, 1250, . . . ; t_9 −3,906,250

13. $t_2 = 18$, $t_3 = 12$; t_5 $\frac{16}{3}$

14. $t_3 = -12$, $t_6 = 96$; t_9 −768

15. $t_1 = 5$, $t_3 = 80$; t_6 5120

16. $t_2 = 8$, $t_4 = 72$; t_1 $\frac{8}{3}$

17. $y, y^3, y^5, \ldots$; t_{20} y^{39}

18. $ab^2, a^2b^5, a^3b^8, \ldots$; t_{25} $a^{25}b^{74}$

Find the geometric mean of each pair of numbers.

19. 2, 8 4

20. $\frac{1}{12}, \frac{1}{18}$ $\frac{\sqrt{6}}{36}$

21. $\sqrt{3}, 3\sqrt{3}$ 3

22. −18, −36 $-18\sqrt{2}$

Insert the given number of geometric means between the pairs of numbers.

23. Three; 5, 80 10, 20, 40

24. Two; −4, 108 12, −36

25. Four; 1, 2 $\sqrt[5]{2}, \sqrt[5]{4}, \sqrt[5]{8}, \sqrt[5]{16}$

26. Three; $\frac{1}{5}, \frac{5}{4}$ $\frac{\sqrt{10}}{10}, \frac{1}{2}, \frac{\sqrt{10}}{4}$

Tell whether each sequence is arithmetic or geometric. Then find a formula for the nth term. A = arithmetic, G = geometric

B

27. The sequence of positive even integers A; $t_n = 2n$

28. The sequence of odd integers greater than two A; $t_n = 2n + 1$

29. 25, 33, 41, 49, . . . A; $t_n = 25 + 8(n - 1)$

30. −17, −11, −5, 1, . . . A; $t_n = -17 + 6(n - 1)$

31. 200, −100, 50, −25 G; $t_n = 200\left(-\frac{1}{2}\right)^{n-1}$

32. $e^x, e^{2x}, e^{3x}, e^{4x}, \ldots$ G; $t_n = e^{nx}$

33. $2a + 1, 3a + 3, 4a + 5, \ldots$ A; $t_n = (n + 1)a + (2n - 1)$

34. $\frac{a^2}{2}, \frac{a^4}{4}, \frac{a^6}{8}, \ldots$ G; $t_n = \left(\frac{a^2}{2}\right)^n$

35. The sequence of negative even integers that are multiples of 5 A; $t_n = -10n$

36. The sequence of positive integers that give a remainder of 1 when divided by 4 A; $t_n = 1 + 4(n - 1)$

Guided Practice

1. Find the formula for the nth term of the geometric sequence 2, −2, 2, −2, 2, ...
$t_n = 2(-1)^{n-1}$

Find the specified term of each geometric sequence.

2. 3, −6, 12, −24, . . . ; t_{10}
−1536
3. 15, 3, 0.6, . . . ; t_7
0.00096
4. $t_3 = 24$, $t_5 = 2400$; t_1 0.24
5. $k^2, k^5, k^8, \ldots$; t_{20} k^{59}
6. Insert two geometric means between 2 and −128.
−8, 32

Summarizing the Lesson

In this lesson students learned to find a specified term of a geometric sequence. Ask students to state the formula for the nth term of a geometric sequence.

Suggested Assignments

Average Algebra
Day 1: 513/1–33 odd
514/*P*: 1–6
517/1–7 odd
Day 2: 514/*P*: 8–12
S 515/Mixed Review
Assign with Lesson 11-4.

Extended Algebra
Day 1: 513/3–39 mult. of 3
514/*P*: 1–9 odd
Day 2: 514/*P*: 10–13
S 515/Mixed Review
Assign with Lesson 11-4.

Extended Alg. and Trig.
Day 1: 513/3–39 mult. of 3
S 515/Mixed Review
Assign with Lesson 11-2.
Day 2: 514/*P*: 1, 4, 7, 9–12
Assign with Lesson 11-4.

Supplementary Materials

Study Guide pp. 175–176
Practice Master 60

Additional Answers Written Exercises

39. $\frac{a}{\sqrt{ab}} = \frac{a}{\sqrt{ab}} \cdot \frac{\sqrt{ab}}{\sqrt{ab}} = \frac{a\sqrt{ab}}{ab} = \frac{\sqrt{ab}}{b}$

$\frac{\sqrt{ab}}{b} = \frac{a}{\sqrt{ab}}$

Since the ratio of every pair of successive terms is constant, the sequence is geometric.

Using a Calculator

Remind students to use the power key, usually labeled x^y or y^x, when finding a specified term of a geometric sequence, as in Problems 2–4, 7–8, and 11–13.

Find a formula for the *n*th term of each sequence. The sequences are neither arithmetic nor geometric. (*Hint:* Analyze the patterns in the numerators and denominators separately.)

37. $\frac{2}{1}, \frac{3}{4}, \frac{4}{9}, \frac{5}{16}, \ldots$ $t_n = \frac{n+1}{n^2}$

38. $\frac{1}{2}, \frac{2}{5}, \frac{3}{8}, \frac{4}{11}, \ldots$ $t_n = \frac{n}{3n-1}$

C **39.** Show that the sequence a, $\sqrt{ab}$, b (a, b positive) is geometric.

40. The first three numbers of the sequence 8, x, y, 36 form an arithmetic sequence, but the last three numbers form a geometric sequence. Find all possible values of x and y. $x = 16$, $y = 24$ or $x = 1$, $y = -6$

Problems

The following problems involve arithmetic and geometric sequences. A calculator may be helpful in solving these problems.

A **1.** Allysa has taken a job with a starting salary of \$17,600 and annual raises of \$850. What will be her salary during her fifth year on the job? \$21,000

2. Frank has taken a job with a starting salary of \$15,000 and annual raises of 4%. What will be his salary during his third year on the job? \$16,224

3. The cost of an annual subscription to a magazine is \$20 this year. The cost is expected to rise by 10% each year. What will be the cost 6 years from now? \$35.43

4. A new pair of running shoes costs \$70 now. Assuming an annual 8% price increase, find the price 4 years from now. \$95.23

5. A carpenter is building a staircase from the first floor to the second floor of a house. The distance between floors is 3.3 m. Each step rises 22 cm. Not counting either floor itself, how many steps will there be? 14

6. The width of a pyramid decreases by 1.57 m for each successive 1 m of height. If the width at a height of 1 m is 229.22 m, what is the width at a height of 86 m? 95.77 m

7. A culture of yeast doubles in size every 4 hours. If the yeast population is estimated to be 3 million now, what will it be one day from now? 192 million

8. An advertisement for a mutual fund claims that people who invested in the fund 5 years ago have doubled their money. If the fund's future performance is similar to its past performance, how much would a \$2000 investment be worth in 40 years? \$512,000

B **9.** A pile of bricks has 85 bricks in the bottom row, 79 bricks in the second row, 73 in the third row, and so on until there is only 1 brick in the top row.

a. How many bricks are in the 12th row? 19

b. How many rows are there in all? 15

10. A projectile fired vertically upward rises 15,840 ft the first second, 15,808 ft the following second, and 15,776 ft the third second.

a. How many feet does it rise the 45th second? 14,432 ft

b. How many feet and in what direction does it move during the first second of the tenth minute after it has been fired? 1440 ft downward

11. A new $12,000 automobile decreases in value by 25% each year. What is its value 7 years from now? $1601.87

12. A house purchased last year for $80,000 is now worth $96,000. Assuming that the value of the house continues to appreciate (increase) at the same rate each year, find the value 2 years from now. $138,240

13. Job A has a starting salary of $12,000 with annual increases of $800. Job B has a starting salary of $11,000 with annual increases of 10%. Which job will pay more after 3 years? after 5 years? B; B

C **14.** There are 12 steps in the chromatic scale from the A below middle C to the next higher A. This scale can be played by playing the white and black keys of a piano in order from left to right. The frequency of a note is given in hertz (Hz). For example, the A below middle C has a frequency of 220 Hz, which means the piano string vibrates 220 times per second. Notice that the frequency of the A above middle C is twice as great, 440 Hz. If the frequencies of the notes in the chromatic scale form a geometric sequence, show that the common ratio is $2^{1/12}$ and then find the frequency of middle C. $t_{13} = 440$, $t_1 = 220$; ratio $= x \cdot 440 = 220x^{12}$; $2 = x^{12}$; $x = 2^{1/12}$

Frequency of middle C is 261.6 Hz.

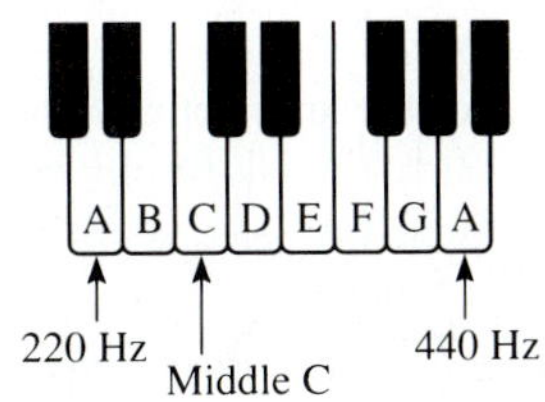

Mixed Review Exercises

Solve each inequality and graph the solution set.

1. $|2x - 5| > 9$ $\{x: x < -2 \text{ or } x > 7\}$
2. $x^2 + x \le 6$ $\{x: -3 \le x \le 2\}$
3. $2 \le \frac{x}{2} + 4 < 5$ $\{x: -4 \le x < 2\}$
4. $x^3 > 4x$ $\{x: -2 < x < 0 \text{ or } x > 2\}$
5. $5 - 3x \ge -1$ $\{x: x \le 2\}$
6. $|x + 3| < 2$ $\{x: -5 < x < -1\}$

Find an equation for each figure described.

7. The line containing $(-2, 3)$ and $(1, 0)$ $x + y = 1$
8. The parabola with focus $(0, -1)$ and directrix $y = 1$ $y = -\frac{1}{4}x^2$
9. The ellipse with x-intercepts ± 2 and y-intercepts $\pm\sqrt{3}$ $\frac{x^2}{4} + \frac{y^2}{3} = 1$

Additional Answers Mixed Review Exercises

1. 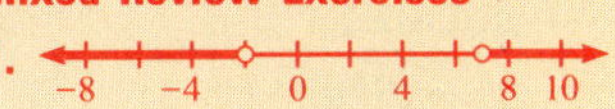
2. 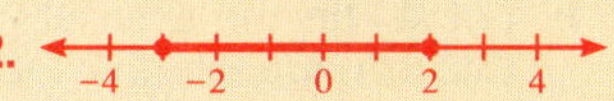
3. 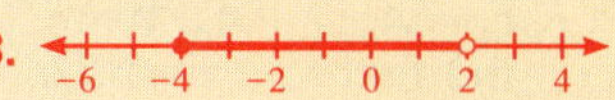
4. 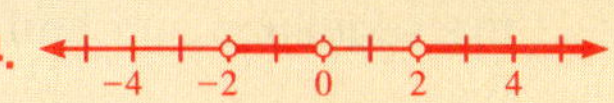
5. 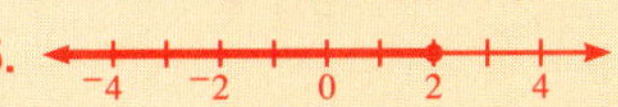
6. 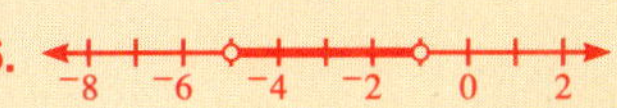

Quick Quiz

1. Tell whether the sequence is arithmetic, geometric, or neither. Supply the missing terms.
   a. $10\frac{1}{8}, 6\frac{3}{4}, 4\frac{1}{2}$, ___?___, ___?___
   geometric; 3, 2
   b. $-1, -11, -111$, ___?___, ___?___
   neither; -1111, $-11{,}111$
   c. 2, 16, 30, ___?___, ___?___
   arithmetic; 44, 58
2. For the arithmetic sequence $-12, -5, 2, 9, \ldots$, find (a) a formula for nth term and (b) t_{23}. $t_n = 7n - 19$; 142
3. Insert (a) one and (b) two arithmetic means between -7 and 2. -2.5; $-4, -1$
4. For the geometric sequence 3, 6, 12, 24, . . . , find (a) a formula for the nth term and (b) t_{12}. $t_n = 3 \cdot 2^{n-1}$; 6144
5. Insert (a) one and (b) two geometric means between 3 and 81. $9\sqrt{3}$; 9, 27
6. One square has sides of length 20 cm. A second square has sides of length 25 cm. Each of the succeeding squares has sides that are 5 cm longer than the preceding square.
   a. Find the perimeters of the fifth and the tenth squares.
   160 cm, 260 cm
   b. Let A_n be the area of the nth square and let $t_n = A_{n+1} - A_n$. Write the first four terms of the sequence and find t_{10}. 225, 275, 325, 375; 675

Self-Test 1

Vocabulary

sequence (p. 501)
terms of a sequence (p. 501)
finite sequence (p. 501)
infinite sequence (p. 501)
arithmetic sequence (p. 501)
arithmetic progression (p. 501)
common difference (p. 501)
arithmetic means (p. 508)
the arithmetic mean (p. 508)
average (p. 508)
geometric sequence (p. 502)
geometric progression (p. 502)
common ratio (p. 502)
geometric means (p. 511)
the geometric mean (p. 511)

1. Tell whether each sequence is arithmetic, geometric, or neither. Then supply the missing terms of the sequence. **Obj. 11-1, p. 501**
   A = arithmetic, G = geometric, N = neither
   a. $\frac{2}{3}, 1, \frac{3}{2}, \frac{9}{4}$, ___?___, ___?___ G; $\frac{27}{8}, \frac{81}{16}$
   b. $5, 1, -3, -7$, ___?___, ___?___ A; $-11, -15$
   c. $\frac{1}{3}, \frac{2}{5}, \frac{3}{7}, \frac{4}{9}$, ___?___, ___?___ N; $\frac{5}{11}, \frac{6}{13}$
   d. 1, 1.3, 1.6, 1.9, ___?___, ___?___ A; 2.2, 2.5
2. For the arithmetic sequence 36, 29, 22, 15, . . . , find **(a)** a formula for the nth term and **(b)** t_{19}. $t_n = 43 - 7n$; -90 **Obj. 11-2, p. 507**
3. Insert **(a)** one and **(b)** two arithmetic means between -2 and 10. 4; 2, 6
4. For the geometric sequence $48, -24, 12, -6, \ldots$, find **(a)** a formula for the nth term and **(b)** t_{10}. $t_n = 48\left(-\frac{1}{2}\right)^{n-1}$; $-\frac{3}{32}$ **Obj. 11-3, p. 510**
5. Insert **(a)** one and **(b)** two geometric means between 2 and 16. $4\sqrt{2}$; 4, 8
6. During a week-long sale, a store reduced its prices on sale items by 10% each day. A coat that was priced \$80 on the first day of the sale was sold on the fourth day. At what price was the coat sold? \$58.32

Check your answers with those at the back of the book.

Challenge

The two arithmetic sequences given below have infinitely many terms in common. Find three of these common terms.

2, 14, 26, . . .

and 1, 8, 15, . . . 50, 134, 218

Graphing Sequences

The table and graphs below show that the arithmetic sequence $t_n = 2n - 1$ is related to the linear function $f(x) = 2x - 1$. The only difference is that $f(x)$ is defined for all real numbers x, while t_n is defined only for positive integers n.

n	1	2	3	4
$t_n = 2n - 1$	1	3	5	7

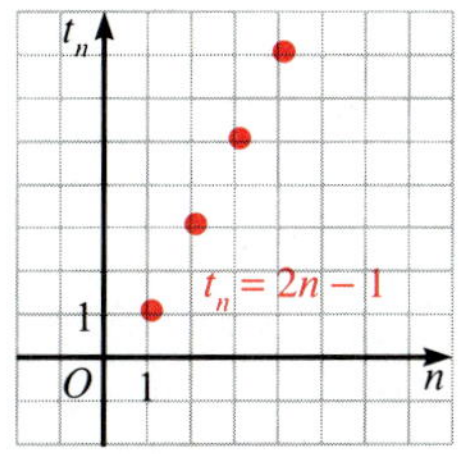

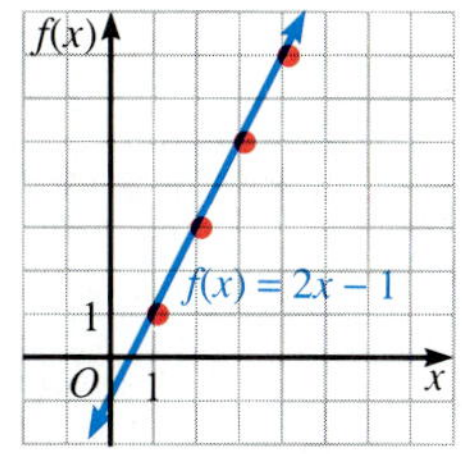

In general, the arithmetic sequence $t_n = t_1 + (n - 1)d$ can be rewritten as $t_n = dn + (t_1 - d)$. In this form you can see that the slope of the graph is the common difference of the sequence.

$$\left.\begin{aligned} t_n &= dn + (t_1 - d) \\ f(x) &= mx + b \end{aligned}\right\} \quad \text{common difference } d = \text{slope } m$$

Just as arithmetic sequences and linear functions are related, geometric sequences and exponential functions are also related. A comparison of the sequence $t_n = 16\left(\frac{1}{2}\right)^n$ and the function $f(x) = 16\left(\frac{1}{2}\right)^x$ is shown below.

n	1	2	3	4
$t_n = 16\left(\frac{1}{2}\right)^n$	8	4	2	1

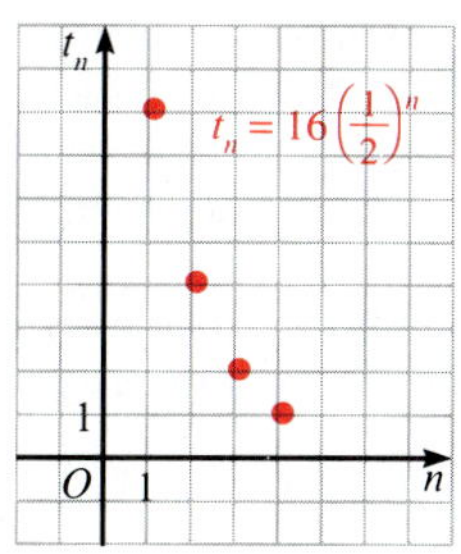

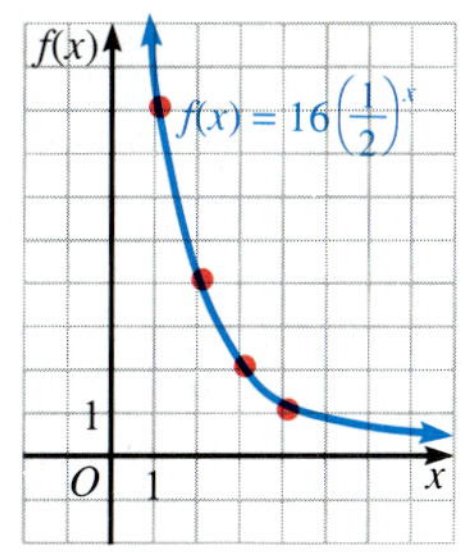

Exercises

Give the slope of the graph of each arithmetic sequence.

1. 3, 9, 15, . . . 6 **2.** 9, 5, 1, . . . −4 **3.** $t_n = 7 - 2n$ −2 **4.** $t_n = 3 + 5(n - 1)$ 5

Give the domain of the function $f(x)$ and of the sequence t_n. Sketch the graph of each.

5. $f(x) = 3x + 5$; $t_n = 3n + 5$

6. $f(x) = 8 - 2x$; $t_n = 8 - 2n$

7. $f(x) = 2^x$; $t_n = 2^n$

8. $f(x) = 9\left(\frac{2}{3}\right)^x$; $t_n = 9\left(\frac{2}{3}\right)^n$

Additional Answers
Extra

6. D = {real numbers}

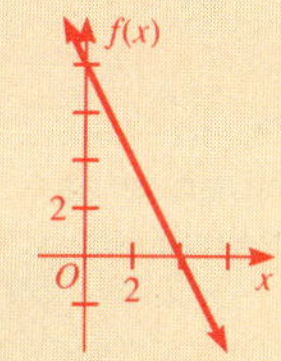

D = {positive integers}

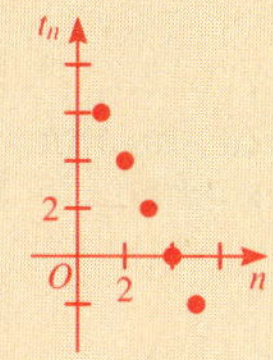

8. D = {real numbers}

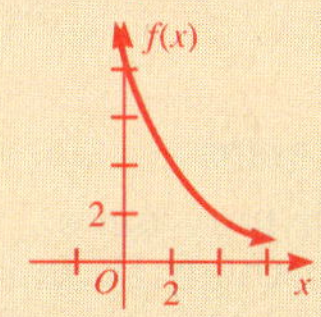

D = {positive integers}

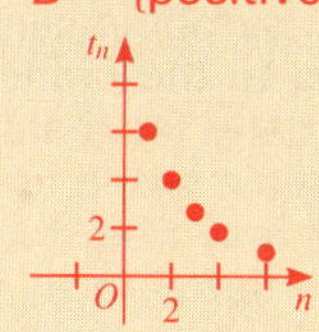

Teaching Suggestions p. T120

Suggested Extensions p. T120

Warm-Up Exercises

Identify each sequence as arithmetic, geometric, or neither.

1. 7, 9, 11, 13, ... A
2. 2, 6, 18, 54, ... G
3. $\frac{2}{3}, \frac{3}{4}, \frac{4}{5}, \frac{5}{6}, \ldots$ N

Give a formula for the nth term of each sequence.

4. 13, 10, 7, 4, ... $t_n = 16 - 3n$
5. 64, 48, 36, 27, ... $t_n = 64\left(\frac{3}{4}\right)^{n-1}$

Motivating the Lesson

Ask students if they have ever flipped through an unfamiliar math text and wondered what the strange symbols in the book meant. Although such symbols can be intimidating to the uninitiated, they greatly facilitate a mathematician's work. In today's lesson a symbol for the sum of the terms in a sequence will be introduced.

Problem Solving Strategies

When using sigma notation to write a series that is neither arithmetic nor geometric, the student must *look for patterns* in the terms of the series to determine the correct expression to be used.

Series

11-4 Series and Sigma Notation

Objective To identify series and to use sigma notation.

When the terms of a sequence are added together, the resulting expression is called a **series.** Here are some examples:

Finite sequence	3, 7, 11, 15, 19
Related finite series	3 + 7 + 11 + 15 + 19
Infinite sequence	$\frac{1}{2}, \frac{1}{4}, \frac{1}{8}, \frac{1}{16}, \ldots$
Related infinite series	$\frac{1}{2} + \frac{1}{4} + \frac{1}{8} + \frac{1}{16} + \cdots$

Many of the words used to describe sequences are also used to describe series. For example, the finite series above is an arithmetic series with first term $t_1 = 3$, last term $t_5 = 19$, and general term $t_n = 4n - 1$. An **arithmetic series** is a series whose related sequence is arithmetic. Similarly, the infinite series above is a **geometric series** because its related sequence is geometric.

A series can be written in an abbreviated form by using the Greek letter Σ *(sigma),* called the **summation sign.** For instance, to abbreviate the writing of the series

$$2 + 4 + 6 + 8 + \cdots + 100$$

first notice that the general term of the series is $2n$. The series begins with the term for $n = 1$ and ends with the term for $n = 50$. Using sigma notation you can write this series as $\sum_{n=1}^{50} 2n$, which is read "the sum of $2n$ for values of n from 1 to 50."

$$\sum_{n=1}^{50} 2n = 2 \cdot 1 + 2 \cdot 2 + 2 \cdot 3 + 2 \cdot 4 + \cdots + 2 \cdot 50$$
$$= 2 + 4 + 6 + 8 + \cdots + 100$$

Similarly, the sigma expression below represents "the sum of n^2 for values of n from 1 to 10."

$$\sum_{n=1}^{10} n^2 = 1^2 + 2^2 + 3^2 + \cdots + 10^2$$

In sigma notation the general term, n^2, is called the **summand,** and the letter n is called the **index.**

Any letter can be used as the index. For example, replacing the index n by k in the series just given does not change the series:

$$\sum_{k=1}^{10} k^2 = 1^2 + 2^2 + 3^2 + \cdots + 10^2$$

Example 1 Write the series $\sum_{j=1}^{20} (-1)^j(j + 2)$ in expanded form.

Solution

$$\sum_{j=1}^{20} (-1)^j(j + 2) = (-1)^1(1 + 2) + (-1)^2(2 + 2) + (-1)^3(3 + 2) + \cdots + (-1)^{20}(20 + 2)$$

$$= -3 + 4 - 5 + \cdots + 22 \quad \textit{Answer}$$

The first and last values of the index are called the **limits of summation.** In Example 1, the *lower limit* is 1 and the *upper limit* is 20. If a series is infinite, then the symbol ∞ is used for the upper limit to indicate that the summation does not end. For example,

$$\sum_{k=1}^{\infty} \frac{1}{2^{k-1}}$$

is read "the sum of $\frac{1}{2^{k-1}}$ for values of k from 1 to infinity."

$$\sum_{k=1}^{\infty} \frac{1}{2^{k-1}} = \frac{1}{2^{1-1}} + \frac{1}{2^{2-1}} + \frac{1}{2^{3-1}} + \frac{1}{2^{4-1}} + \cdots$$

$$= 1 + \frac{1}{2} + \frac{1}{4} + \frac{1}{8} + \cdots$$

By changing the lower limit of summation from 1 to 0, this infinite geometric series can be rewritten with a simpler summand.

$$\sum_{k=0}^{\infty} \frac{1}{2^k} = \frac{1}{2^0} + \frac{1}{2^1} + \frac{1}{2^2} + \frac{1}{2^3} + \cdots$$

$$= 1 + \frac{1}{2} + \frac{1}{4} + \frac{1}{8} + \cdots$$

As you will see in Lesson 11-6, this infinite series has a finite sum of 2.

Example 2 Use sigma notation to write the series $10 + 15 + 20 + \cdots + 100$.

Solution 1 By inspection,

$$10 + 15 + 20 + \cdots + 100 = 5 \cdot 2 + 5 \cdot 3 + 5 \cdot 4 + \cdots + 5 \cdot 20.$$

$\therefore$ the series is $\sum_{k=2}^{20} 5k$. ***Answer***

Chalkboard Examples

1. Write the series $\sum_{i=1}^{5} (3i + 1)$ in expanded form.
   $\sum_{i=1}^{5} (3i + 1) = 3(1) + 1 + 3(2) + 1 + 3(3) + 1 + 3(4) + 1 + 3(5) + 1 = 4 + 7 + 10 + 13 + 16$
2. Use sigma notation to write the series $\frac{1}{3} + \frac{2}{3} + 1 + \frac{4}{3} + \frac{5}{3} + 2$.
   The terms are the first six positive multiples of $\frac{1}{3}$.
   $\therefore$ the series is $\sum_{k=1}^{6} \frac{k}{3}$.
3. Use sigma notation to write the series $-\frac{7}{1} + \frac{7}{4} - \frac{7}{7} + \cdots$
   Since the series is neither arithmetic nor geometric, you need to look for patterns. The terms alternate in sign and the denominators form an arithmetic sequence. The general term is $(-1)^n \cdot \frac{7}{3n - 2}$. The series is $\sum_{n=1}^{\infty} \frac{7(-1)^n}{3n - 2}$.

Check for Understanding

Oral Exs. 1–10: use before Example 2.
Oral Exs. 11–14: use before Example 3.

Guided Practice

Write in expanded form.

1. $\sum_{j=3}^{7} (j + 7)$
   $10 + 11 + 12 + 13 + 14$
2. $\sum_{n=1}^{4} (8 - 3n)$
   $5 + 2 - 1 - 4$
3. $\sum_{k=2}^{100} \frac{(-1)^k}{2k}$
   $\frac{1}{4} - \frac{1}{6} + \frac{1}{8} - \frac{1}{10} + \cdots + \frac{1}{200}$
4. $\sum_{n=0}^{\infty} \frac{|2 - n|}{2^n}$
   $2 + \frac{1}{2} + 0 + \frac{1}{8} + \frac{2}{16} + \cdots$

Write using sigma notation. Answers may vary.

5. $7 + 14 + 21 + \cdots + 700$
   $\sum_{n=1}^{100} 7n$
6. $\frac{1}{5} + \frac{1}{7} + \frac{1}{9} + \frac{1}{11} + \cdots + \frac{1}{25}$
   $\sum_{j=1}^{11} \frac{1}{2j + 3}$
7. $5 + 10 + 20 + 40 + \cdots$
   $\sum_{n=0}^{\infty} 5 \cdot 2^n$
8. $1 - \frac{1}{3} + \frac{1}{9} - \frac{1}{27} + \frac{1}{81}$
   $\sum_{n=0}^{4} \left(-\frac{1}{3}\right)^n$

Solution 2 Since the series is arithmetic with common difference 5, the nth term is:

$$t_n = 10 + (n - 1)5$$
$$= 5n + 5, \text{ or } 5(n + 1)$$

Now find n such that the last term is 100.

$$t_n = 5n + 5$$
$$100 = 5n + 5$$
$$n = 19$$

$\therefore$ the series is $\sum_{n=1}^{19} 5(n + 1)$. ***Answer***

In the two solutions of Example 2, notice that the expressions $\sum_{k=2}^{20} 5k$ and $\sum_{n=1}^{19} 5(n + 1)$ represent the same series.

Example 3 Use sigma notation to write the series $\frac{5}{2} - \frac{5}{4} + \frac{5}{6} - \frac{5}{8} + \cdots$.

Solution Since the infinite series is neither arithmetic nor geometric, you need to look for patterns:

(1) The numerators are all equal to 5.

(2) Since the denominators are consecutive even integers, a general expression for the denominators is $2n$.

(3) To make the terms of a series alternate in sign, choose $n = 1$ for the lower limit of the summand and include in the expression for the summand one of the following factors:

$(-1)^n$ makes *odd-numbered* terms *negative;*
$(-1)^{n+1}$ makes *even-numbered* terms *negative*.

So the expression for the summand is $(-1)^{n+1}\left(\frac{5}{2n}\right)$.

$\therefore$ the series is $\sum_{n=1}^{\infty} (-1)^{n+1}\left(\frac{5}{2n}\right)$. ***Answer***

Oral Exercises

Exercises 1–4 refer to the series $\sum_{m=2}^{5} (4m^2 - 3)$.

1. What is the index? m
2. What is the summand? $4m^2 - 3$
3. What are the limits of summation? 2, 5
4. What are the first and last terms? 13, 97

For each series, read each symbol aloud and give the expanded form.

5. $\sum_{i=1}^{3} i$ 1 + 2 + 3 6. $\sum_{j=2}^{4} (5 - j)$ 3 + 2 + 1 7. $\sum_{n=1}^{5} 4n$ 4 + 8 + 12 + 16 + 20

8. $\sum_{k=1}^{4} k^3$ 1 + 8 + 27 + 64 9. $\sum_{z=5}^{\infty} \frac{1}{z-1}$ $\frac{1}{4} + \frac{1}{5} + \frac{1}{6} + \frac{1}{7} + \cdots$ 10. $\sum_{j=0}^{6} (-1)^j$ 1 + (−1) + 1 + (−1) + 1 + (−1) + 1

Tell whether or not each pair of series is the same.

11. $\sum_{k=0}^{8} \frac{1}{k+3}, \sum_{j=3}^{11} \frac{1}{j}$ Same 12. $\sum_{n=0}^{4} (-1)^n(n+1)^2, \sum_{j=1}^{5} (-1)^j(j^2)$ Different

13. $\sum_{k=0}^{4} \frac{3k+3}{k+6}, \sum_{i=1}^{3} \frac{3i}{i+5}$ Different 14. $\sum_{k=0}^{\infty} \frac{k+1}{3^k}, \sum_{n=1}^{\infty} \frac{n}{3^{n-1}}$ Same

Written Exercises

Write each series in expanded form.

A 1. $\sum_{n=1}^{6} (n + 10)$ 2. $\sum_{k=1}^{8} 3k$ 3. $\sum_{n=1}^{6} 2^n$ 4. $\sum_{n=4}^{10} (3n - 2)$

5. $\sum_{j=0}^{5} \frac{(-1)^j}{j+1}$ 6. $\sum_{k=0}^{3} 4^{-k}$ 7. $\sum_{n=3}^{8} |5 - n|$ 8. $\sum_{k=1}^{4} (-k)^{k+1}$

Write each series using sigma notation. 12. $\sum_{n=1}^{24} 3 \cdot 4^n$

9. $2 + 4 + 6 + \cdots + 1000$ $\sum_{n=1}^{500} 2n$ 10. $5 + 10 + 15 + \cdots + 250$ $\sum_{n=1}^{50} 5n$

11. $1^3 + 2^3 + 3^3 + \cdots + 20^3$ $\sum_{n=1}^{20} n^3$ 12. $3 \cdot 4 + 3 \cdot 4^2 + 3 \cdot 4^3 + \cdots + 3 \cdot 4^{24}$

13. $\frac{1}{2} + \frac{2}{3} + \frac{3}{4} + \cdots + \frac{99}{100}$ $\sum_{n=1}^{99} \frac{n}{n+1}$ 14. $\frac{1}{5^3} + \frac{1}{5^4} + \frac{1}{5^5} + \cdots + \frac{1}{5^{15}}$ $\sum_{n=1}^{13} 5^{-(n+2)}$

15. $1 + 3 + 5 + \cdots + 199$ $\sum_{n=1}^{100} (2n - 1)$ 16. $3 + 7 + 11 + 15 + \cdots + 399$ $\sum_{n=1}^{100} (4n - 1)$

17. $1 + 2 + 4 + 8 + \cdots + 64$ $\sum_{n=0}^{6} 2^n$ 18. $\sqrt{7} + \sqrt{14} + \sqrt{21} + \cdots + \sqrt{77}$ $\sum_{n=1}^{11} \sqrt{7n}$

19. $1 + \frac{1}{2} + \frac{1}{3} + \frac{1}{4} + \cdots$ $\sum_{n=1}^{\infty} \frac{1}{n}$ 20. $1 + \frac{1}{3} + \frac{1}{9} + \frac{1}{27} + \cdots$ $\sum_{n=0}^{\infty} \frac{1}{3^n}$

B 21. $-9 + 3 - 1 + \frac{1}{3} + \cdots$ $\sum_{n=1}^{\infty} 27\left(-\frac{1}{3}\right)^n$ 22. $8 - 4 + 2 - 1 + \cdots$ $\sum_{n=0}^{\infty} 8\left(-\frac{1}{2}\right)^n$

23. $6 - 12 + 24 - 48$ $\sum_{n=0}^{3} 6(-2)^n$ 24. $2 + 5 + 10 + 17 + 26 + 37$ $\sum_{n=1}^{6} (n^2 + 1)$

25. $1 + \frac{1}{4} + \frac{1}{9} + \frac{1}{16} + \cdots$ $\sum_{n=1}^{\infty} \frac{1}{n^2}$ 26. $1 - \frac{1}{3} + \frac{1}{5} - \frac{1}{7} + \cdots$ $\sum_{n=0}^{\infty} \frac{(-1)^n}{2n+1}$

27. The series consisting of positive three-digit integers divisible by 5

28. The series consisting of positive two-digit integers ending in 2

27. $\sum_{n=20}^{199} 5n$ 28. $\sum_{n=1}^{9} (10n + 2)$

Summarizing the Lesson

In this lesson students learned the definition of an arithmetic series and a geometric series and how to use sigma notation. Ask students what $\sum_{k=1}^{100} k$ represents. Also ask students how to write "the sum of the powers of 2 from 2 to 64" using sigma notation.

Suggested Assignments

Average Algebra
Day 1: 521/1–16
Assign with Lesson 11-3.
Day 2: 521/17–32
R 516/Self-Test 1

Extended Algebra
Day 1: 521/2–20 even
Assign with Lesson 11-3.
Day 2: 521/22–34 even
R 516/Self-Test 1
517/1–7 odd

Extended Alg. and Trig.
Day 1: 521/2–20 even
R 516/Self-Test 1
Assign with Lesson 11-3.
Day 2: 521/22–34 even
523/*Extra*: 1, 3, 5
Assign with Lesson 11-5.

Additional Answers Written Exercises

1. 11 + 12 + 13 + 14 + 15 + 16
2. 3 + 6 + 9 + 12 + 15 + 18 + 21 + 24
3. 2 + 4 + 8 + 16 + 32 + 64
4. 10 + 13 + 16 + 19 + 22 + 25 + 28
5. $1 - \frac{1}{2} + \frac{1}{3} - \frac{1}{4} + \frac{1}{5} - \frac{1}{6}$
6. $1 + \frac{1}{4} + \frac{1}{16} + \frac{1}{64}$
7. 2 + 1 + 0 + 1 + 2 + 3
8. 1 − 8 + 81 − 1024

Common Errors

Since pattern recognition is difficult for some students, they may have trouble writing a given series using sigma notation. When reteaching, have these students examine the sigma notation given in Oral Exercises 5–10 and Written Exercises 1–8 and then compare these to the corresponding expanded forms.

Supplementary Materials

Study Guide pp. 177–178

29. An infinite geometric series with first term 1 and common ratio $\frac{1}{4}$ $\sum_{n=0}^{\infty} \left(\frac{1}{4}\right)^n$

30. An arithmetic series consisting of eleven terms with first term 8 and last term 508 $\sum_{n=1}^{11} (50n - 42)$

Determine the missing summand so that each statement is true.

31. $\sum_{k=5}^{8} \frac{k}{k+4} = \sum_{j=1}^{4} \underline{\ ?\ }$ $\frac{j+4}{j+8}$

32. $\sum_{k=6}^{9} \frac{k+1}{k-1} = \sum_{j=7}^{10} \underline{\ ?\ }$ $\frac{j}{j-2}$

C **33.** Show that $\sum_{k=1}^{4} k \log 5 = \log (5^{10})$. $\sum_{k=1}^{4} k \log 5 = \log 5 + 2 \log 5 + 3 \log 5 + 4 \log 5 = \log 5\,(1 + 2 + 3 + 4) = 10 \log 5 = \log 5^{10}$

34. Write $\log (1 \cdot 2 \cdot 3 \cdot 4 \cdot 5 \cdot 6)$ using sigma notation. $\sum_{n=1}^{6} \log n$

35. Find the value of $\sum_{k=1}^{6} \left(\sum_{j=1}^{k} 1\right)$. 21

Career Note / *Marine Biologist*

Although nearly seventy percent of the earth's surface is covered by water, we know much less about what life exists and how it exists in the sea than we do about life on land. Scientists who study the variety and activity of life in the sea are called marine biologists.

Some marine biologists engage in basic research. They seek and describe new species of plants and animals living in the sea, and they study the relationships between the sealife and its environment. Other marine biologists undertake applied research. They investigate how sealife can meet human needs, especially in providing new sources of food and pharmaceuticals.

Marine biologists are as committed to conserving the resources of the sea as to utilizing them. Conservation requires knowing a species' population, and this is done through random sampling. A sample of the population is caught, tagged, and then released. Later on when a second sample is taken, some of those caught will already be tagged. The ratio of those that are found tagged

to the size of the second sample should approximately equal the ratio of the size of the original sample to the entire population. From this proportion marine biologists can estimate the population of the species.

Extra / Induction

How could you prove that

$$\frac{1}{1 \cdot 2} + \frac{1}{2 \cdot 3} + \frac{1}{3 \cdot 4} + \cdots + \frac{1}{n(n+1)} = \frac{n}{n+1}$$

is true for all positive integers n? Since the series is neither arithmetic nor geometric, there is no formula that can be used to find the sum. Instead, you can think of the general statement as a *sequence of statements*.

(1) $$\frac{1}{1 \cdot 2} = \frac{1}{2}$$

(2) $$\frac{1}{1 \cdot 2} + \frac{1}{2 \cdot 3} = \frac{2}{3}$$

(3) $$\frac{1}{1 \cdot 2} + \frac{1}{2 \cdot 3} + \frac{1}{3 \cdot 4} = \frac{3}{4}$$

$\vdots$

(n) $$\frac{1}{1 \cdot 2} + \frac{1}{2 \cdot 3} + \cdots + \frac{1}{n(n+1)} = \frac{n}{n+1}$$

It is easy to verify that statements (1), (2), and (3) above are true. However, this does not *prove* that statement (n) is true for all positive integers n. One way to prove statement (n) is to use a method called the *principle of mathematical induction*.

Principle of Mathematical Induction

To show that a sequence of statements is true for all positive integers n:

1. Show that statement (1) is true.
2. Show that if statement (k) is true, then statement ($k + 1$) is also true.

Using mathematical induction to prove a sequence of statements can be compared to knocking down an infinite row of dominoes. The first part of such a proof shows that the first domino falls. The second part shows that if any one domino falls, then the next in line also falls. Taken together, the two parts of the proof guarantee that all the dominoes fall.

Example Show that statement (n) above is true for all positive integers n.

Solution 1. Statement (1) is $\frac{1}{1 \cdot 2} = \frac{1}{2}$. This is obviously true.

2. Assume that statement (k) is true:

$$\frac{1}{1 \cdot 2} + \frac{1}{2 \cdot 3} + \cdots + \frac{1}{k(k+1)} = \frac{k}{k+1}$$

(Solution continues on next page.)

Prove that statement $(k + 1)$ is true:

$$\frac{1}{1 \cdot 2} + \frac{1}{2 \cdot 3} + \cdots + \frac{1}{k(k+1)} + \frac{1}{(k+1)(k+2)} = \frac{k+1}{k+2}$$

Proof:

Assuming that

$$\frac{1}{1 \cdot 2} + \frac{1}{2 \cdot 3} + \cdots + \frac{1}{k(k+1)} = \frac{k}{k+1}$$

is true, add $\frac{1}{(k+1)(k+2)}$ to both sides and then simplify the right side of the equation:

$$\begin{aligned}\frac{1}{1 \cdot 2} + \frac{1}{2 \cdot 3} + \cdots + \frac{1}{k(k+1)} + \frac{1}{(k+1)(k+2)} &= \frac{k}{k+1} + \frac{1}{(k+1)(k+2)} \\ &= \frac{k(k+2)+1}{(k+1)(k+2)} \\ &= \frac{k^2+2k+1}{(k+1)(k+2)} \\ &= \frac{(k+1)^2}{(k+1)(k+2)} \\ &= \frac{k+1}{k+2} \quad \checkmark\end{aligned}$$

Exercises

Use the principle of mathematical induction to prove that each statement is true for all positive integers *n*. See solution key for proofs.

1. $\frac{1}{1 \cdot 3} + \frac{1}{3 \cdot 5} + \frac{1}{5 \cdot 7} + \cdots + \frac{1}{(2n-1)(2n+1)} = \frac{n}{2n+1}$
2. $\frac{1}{1 \cdot 4} + \frac{1}{4 \cdot 7} + \frac{1}{7 \cdot 10} + \cdots + \frac{1}{(3n-2)(3n+1)} = \frac{n}{3n+1}$
3. $1^2 + 2^2 + 3^2 + \cdots + n^2 = \frac{n(n+1)(2n+1)}{6}$
4. $1 \cdot 2 + 2 \cdot 3 + 3 \cdot 4 + \cdots + n(n+1) = \frac{n(n+1)(n+2)}{3}$
5. If, in a room of n people, every person shakes hands once with every other person, there will be $\frac{n^2-n}{2}$ handshakes.
6. A convex polygon with n sides has $\frac{n^2-3n}{2}$ diagonals. (*Note:* The first step in the proof is to show that the statement is true for $n = 3$, since a polygon must have at least three sides.)
7. $n^3 + 2n$ is a multiple of 3. (This means $n^3 + 2n = 3m$ where m is some integer.)
8. $1^3 + 2^3 + 3^3 + \cdots + n^3 = (1 + 2 + 3 + \cdots + n)^2$

11-5 Sums of Arithmetic and Geometric Series

Objective To find sums of finite arithmetic and geometric series.

In the preceding lesson you worked with series such as

$$2 + 4 + 6 + 8 + \cdots + 100$$

but did not compute the sums. If a series consists of just a few terms, then the sum is easy to compute. For example, the sum of the first three terms of the series above, which we denote by S_3, is

$$S_3 = 2 + 4 + 6$$
$$= 12$$

The sum of the first n terms of a series is denoted by S_n. If a series is arithmetic or geometric, there are ways to find its sum without actually adding all of the terms. Theorems 1 and 2 state methods of finding such sums.

Theorem 1 The sum of the first n terms of an arithmetic series is

$$S_n = \frac{n(t_1 + t_n)}{2}.$$

Proof Write the series for S_n twice, the second time with the order reversed. Then add the two equations, term by term:

$$S_n = t_1 + (t_1 + d) + (t_1 + 2d) + \cdots + (t_n - d) + t_n$$
$$S_n = t_n + (t_n - d) + (t_n - 2d) + \cdots + (t_1 + d) + t_1$$
$$2S_n = \underbrace{(t_1 + t_n) + (t_1 + t_n) + (t_1 + t_n) + \cdots + (t_1 + t_n) + (t_1 + t_n)}_{n \text{ addends}}$$

$$2S_n = n(t_1 + t_n)$$
$$S_n = \frac{n(t_1 + t_n)}{2}$$

Example 1 Find the sum of the first 40 terms of the arithmetic series $2 + 5 + 8 + 11 + \cdots$.

Solution First find the 40th term: $t_{40} = t_1 + (n - 1)d$
$= 2 + (39)3 = 119$

Then use Theorem 1: $S_n = \frac{n(t_1 + t_n)}{2}$

$$S_{40} = \frac{40(2 + 119)}{2} = 2420 \quad \textit{Answer}$$

Teaching Suggestions p. T120

Suggested Extensions p. T121

Warm-Up Exercises

Write each series in expanded form.

1. $\sum_{k=1}^{5} 3k + 1$
   $4 + 7 + 10 + 13 + 16$
2. $\sum_{n=0}^{4} \left(\frac{2}{3}\right)^n$
   $1 + \frac{2}{3} + \frac{4}{9} + \frac{8}{27} + \frac{16}{81}$
3. $\sum_{i=1}^{5} (-1)^i i$
   $-1 + 2 - 3 + 4 - 5$

Write each series using sigma notation.

4. $8 + 16 + 24 + 32 + \cdots + 400$
   $\sum_{n=1}^{50} 8n$
5. $\frac{1}{4} + \frac{1}{2} + 1 + 2 + \cdots + 128$
   $\sum_{n=1}^{10} \frac{1}{8}(2^n)$

Motivating the Lesson

Tell students that when Karl Friedrich Gauss, a famous German mathematician, was a child, his teacher asked his class to find the sum of the integers from 1 to 100. Gauss astounded his teacher by getting the correct answer in an incredibly short time. Today's lesson will reveal the secret to Gauss's accomplishment.

Chalkboard Examples

1. Find the sum of the first twenty terms of the arithmetic series $8 + 5 + 2 + \cdots$.
   Find the twentieth term. $t_{20} = t_1 + (n - 1)d = 8 + (19)(-3) = -49$
   Then use $S_n = \frac{n(t_1 + t_n)}{2}$.
   $S_{20} = \frac{20(8 - 49)}{2} = -410$
2. Evaluate $\sum_{j=1}^{16} (7 - 2j)$.
   Find S_{16} for the series $5 + 3 + 1 + \cdots + (-25)$.
   $S_{16} = \frac{16[5 + (-25)]}{2} = -160$
3. Evaluate $\sum_{j=1}^{5} \left(-\frac{2}{3}\right)^j$.
   Find S_5 for the series $-\frac{2}{3} + \frac{4}{9} - \frac{8}{27} + \cdots$ where the common ratio is $-\frac{2}{3}$.
   $S_n = \frac{t_1(1 - r^n)}{1 - r}$
   $S_5 = \frac{-\frac{2}{3}\left(1 - \left(-\frac{2}{3}\right)^5\right)}{1 - \left(-\frac{2}{3}\right)} = \frac{-\frac{2}{3}\left(1 + \frac{32}{243}\right)}{1 + \frac{2}{3}} = -\frac{110}{243}$

Check for Understanding

Here is a suggested use of the Oral Exercises to check students' understanding as you teach the lesson.
Oral Exs. 1–4: use after Example 3.

Example 2 Evaluate $\sum_{k=1}^{20} (5k + 2)$.

Solution Evaluating $\sum_{k=1}^{20} (5k + 2)$ means finding S_{20} for this arithmetic series:

$$7 + 12 + 17 + \cdots + 102$$

$$\therefore \quad S_{20} = \frac{20(7 + 102)}{2} = 1090 \quad \textbf{Answer}$$

Theorem 2 The sum of the first n terms of a geometric series with common ratio r is

$$S_n = \frac{t_1(1 - r^n)}{1 - r}. \quad (r \neq 1)$$

Proof Multiply the series for S_n by the common ratio r. Then subtract the new series from the original one, as shown below:

$$\begin{aligned} S_n &= t_1 + t_1r + t_1r^2 + \cdots + t_1r^{n-2} + t_1r^{n-1} \\ rS_n &= \phantom{t_1 +{}} t_1r + t_1r^2 + \cdots + t_1r^{n-2} + t_1r^{n-1} + t_1r^n \\ \hline S_n - rS_n &= t_1 + 0 + 0 + \cdots + 0 + 0 - t_1r^n \\ S_n - rS_n &= t_1 - t_1r^n \\ S_n(1 - r) &= t_1(1 - r^n) \end{aligned}$$

Since $r \neq 1$, divide both sides by $1 - r$:

$$S_n = \frac{t_1(1 - r^n)}{1 - r}$$

Note that the formula just given is not defined for $r = 1$. (If $r = 1$, the geometric series $t_1 + t_1r + t_1r^2 + \cdots + t_1r^{n-1}$ consists of n terms equal to t_1, and the sum is nt_1.)

Example 3 Evaluate $\sum_{n=1}^{10} 3(-2)^{n-1}$.

Solution You are finding S_{10} for the series $3 - 6 + 12 - 24 + \cdots$, whose common ratio is -2.

$$S_n = \frac{t_1(1 - r^n)}{1 - r}$$

$$S_{10} = \frac{3[1 - (-2)^{10}]}{1 - (-2)}$$

$$= \frac{3(1 - 1024)}{3} = -1023 \quad \textbf{Answer}$$

Oral Exercises

Find S_5 for each series.

1. $8 + 12 + 16 + 20 + \cdots$ 80
2. $3 - 6 + 12 - 24 + \cdots$ 33
3. $1 + \frac{1}{2} + \frac{1}{4} + \frac{1}{8} + \cdots$ $\frac{31}{16}$
4. $7 + 5 + 3 + 1 + \cdots$ 15

Written Exercises

Find the sum of each arithmetic series.

A

1. $n = 20,\ t_1 = 5,\ t_{20} = 62$ 670
2. $n = 100,\ t_1 = 17,\ t_{100} = 215$ 11,600
3. $n = 40,\ t_1 = -12,\ t_{40} = 183$ 3420
4. $n = 50,\ t_1 = 187,\ t_{50} = 40$ 5675
5. $\sum_{k=1}^{100} 5k$ 25,250
6. $\sum_{n=1}^{24} (2n - 1)$ 576
7. $\sum_{j=1}^{50} (3j + 2)$ 3925
8. $\sum_{m=10}^{20} (30 - m)$ 165
9. The first 100 terms of the series $4 + 7 + 10 + 13 + \cdots$ 15,250
10. The first 100 terms of the series $100 + 98 + 96 + 94 + \cdots$ 100
11. $11 + 15 + 19 + \cdots + 83$ 893
12. $50 + 48 + 46 + \cdots + 10$ 630

Find the sum of each geometric series.

13. $n = 8,\ r = 2,\ t_1 = 1$ 255
14. $n = 10,\ r = -2,\ t_1 = 1$ -341
15. $n = 10,\ r = -3,\ t_1 = 2$ $-29{,}524$
16. $n = 12,\ r = 3,\ t_1 = \frac{1}{9}$ $29{,}524\frac{4}{9}$
17. $\sum_{k=1}^{12} 2^{-k}$ $\frac{4095}{4096}$
18. $\sum_{j=1}^{10} \left(-\frac{1}{2}\right)^j$ $-\frac{341}{1024}$
19. Find S_{10} if the series $24 + 12 + \cdots$ is **(a)** arithmetic; **(b)** geometric.
20. Find S_{20} if the series $1 + 1.1 + \cdots$ is **(a)** arithmetic; **(b)** geometric.

19. (a) -300 (b) $\frac{3069}{64}$ 20. (a) 39 (b) 57.275

Find the sum of each of the following.

B

21. The first 20 positive integers ending in 3 1960
22. The positive two-digit integers ending in 4 486
23. The positive three-digit odd integers 247,500
24. The positive three-digit integers divisible by 6 82,350
25. **a.** The first twenty multiples of 2, starting with 2 420
    b. The first twenty powers of 2, starting with 2^1 2,097,150
26. **a.** The first ten powers of 5 12,207,030
    b. The first ten powers of 5 that are perfect squares $\approx 9.93 \times 10^{13}$

Guided Practice

Find the sum of each series.

1. Arithmetic with $n = 10$, $t_1 = 3$, and $t_{10} = 300$ 1515
2. S_{20} for $3 + 11 + 19 + 27 + \cdots$ 1580
3. $50 + 33 + 16 + \cdots + (-86)$ -162
4. Geometric with $n = 10$, $r = 2$, and $t_1 = \frac{1}{4}$ 255.75
5. S_8 for $48 - 24 + 12 - 6 + \cdots$ $31\frac{7}{8}$
6. $\sum_{k=1}^{100} (2k - 51)$ 5000
7. $\sum_{n=1}^{9} 640\left(-\frac{3}{2}\right)^n$ $-15{,}146.25$
8. A hose is coiled so that each complete loop is 6 inches longer than the previous loop. If the innermost loop is 18 inches long and there are 15 loops in all, how long is the hose? 900 inches or 75 feet
9. The Thompsons paid \$300 for electricity during 1985. If their costs increased 10% each year thereafter, how much did they pay for electricity between 1985 and 1990, inclusive? \$2314.68

Summarizing the Lesson

Students learned to find the sums of finite arithmetic and geometric series using the formulas derived in the lesson. Ask students to state the formula for the sum of the first n terms of (a) an arithmetic series and (b) a geometric series.

Suggested Assignments

Average Algebra
Day 1: 527/2–28 even
S 530/Mixed Review
Day 2: 527/17–29 odd
528/*P*: 1–6, 9, 10

Extended Algebra
Day 1: 527/2–30 even
S 530/Mixed Review
523/*Extra:* 1, 3, 5
Day 2: 527/17–29 odd
528/*P*: 1–4, 11–13
S 524/2, 4, 7

Extended Alg. and Trig.
Day 1: 527/2–20 even
Assign with Lesson 11-4.
Day 2: 527/22–30 even
528/*P*: 1–4, 11–13
S 530/Mixed Review

Supplementary Materials

Study Guide	pp. 179–180
Practice Master	61
Test Master	45
Resource Book	p. 148

Find the sum of each of the following.

27. The positive two-digit integers that are *not* divisible by 3 3240

28. The positive two-digit integers that are *not* divisible by 5 3960

29. a. Show that $\sum_{k=1}^{n} 2^{k-1} = 2^n - 1$. $\sum_{k=1}^{n} 2^{k-1} = \frac{2^0(1-2^n)}{(1-2)} = \frac{1-2^n}{-1} = 2^n - 1$

b. How many terms are needed before the sum exceeds one million? 20

30. If n is an even number, show that the sum of series (A) is three times the sum of series (B).

(A) $1 + \frac{1}{2} + \frac{1}{4} + \cdots + \left(\frac{1}{2}\right)^{n-1}$ (B) $1 - \frac{1}{2} + \frac{1}{4} - \cdots + \left(-\frac{1}{2}\right)^{n-1}$

31. The sum of a series is given by $S_n = 8(3^n - 1)$. Find the first three terms of the related sequence. Then identify the series as arithmetic, geometric, or neither. (*Hint:* $S_1 = t_1$.) 16, 48, 144; geometric

32. If $S_n = 2n^2 + 5n$, find t_n. $4n + 3$

33. Guess the approximate value of the following expressions. Check your guess with a calculator. Answers will vary.

a. $\left(\frac{1}{2}\right)^{20}$ 0.00000095 **b.** $(0.9)^{100}$ 0.00002656 **c.** $(-0.8)^{50}$ 0.00001427

Use the results of Exercise 33 to estimate the answers for Exercise 34.

C **34.** If $|r| < 1$ and n is a very large integer, find the approximate value of:

a. r^n 0 **b.** S_n where $S_n = \frac{t_1(1 - r^n)}{1 - r}$ $\frac{t_1}{1-r}$

The following series are neither arithmetic nor geometric, but by analyzing their patterns, you can find their sums. Find the sum of each series.

35. $2 - 4 + 6 - 8 + 10 - \cdots - 100$ -50

36. $1 + 2 + 4 + 5 + 7 + 8 + \cdots + 95 + 97 + 98$ 3267

37. $\sum_{k=1}^{20} (2^k + k)$ 2,097,360 **38.** $\sum_{n=1}^{12} (2^n - 1)$ 8178

Problems

Solve. A calculator may be helpful.

A **1.** The front row of a theater has 25 seats. Each of the other rows has two more seats than the row before it. How many seats are there altogether in the first 20 rows? 880

2. Kristen is given a test consisting of 15 questions. The first question is worth five points, and each question after the first is worth three points more than the question before it. What is the maximum score that Kristen can obtain? 390

3. Every hour a clock chimes as many times as the hour. How many times does it chime from 1 A.M. through midnight, inclusive? 156

4. A ship's clock strikes every half hour of a 4-hour period. After the first half hour it strikes "one bell," after the second half hour it strikes "two bells," and so on until it strikes "eight bells" at the end of the 4-hour period. It then begins a new 4-hour period. How many strikes of the bell occur in one day? 216

5. Kurt can trace his ancestors back through 10 generations. He counts his parents as the first generation back, his four grandparents as the second generation back, and so on. How many ancestors does he have in these 10 generations? 2046

6. The Kemps have rented a house for the past six years. During their first year of renting, they paid \$600 per month. If their rent was increased by 5% for each year after that, what is the total amount of money they have paid in rent over the past six years? \$48,973.77

B

7. You have won a contest sponsored by a local radio station. If you are given the choice of the two payment plans listed below, which plan will pay you more? How much more? Plan B; \$58.83 more

   Plan A: \$1 on the first day, \$2 on the second day, \$3 on the third day, and so on for two weeks.

   Plan B: 1¢ on the first day, 2¢ on the second day, 4¢ on the third day, and so on for two weeks.

8. Refer to Problem 7 and find how much each plan would pay you if the payments extended over three weeks instead of two. A: \$231; B: \$20,971.51

9. Which of the two jobs described below will pay you the higher salary during the fifth year of employment? Which will pay you the greater total amount for all five years? B; A

   Job A: Make \$20,000 the first year with annual raises of \$1500.

   Job B: Make \$18,000 the first year with annual raises of 10%.

10. Refer to the two jobs described in Problem 9. Which job will pay you the greater total amount over a 10-year period? How much more? B; \$19,373.65

11. On the first day of each year, the Ortegas invest \$1000 at 6% interest compounded annually. Find the value of their investment after 10 years. \$13,971.64

12. The 1560 members of the Western Environmental Society have a method of quickly notifying members. The president and treasurer each call 3 members (round 1), each of whom then calls 3 other members (round 2), each of whom then calls 3 more members (round 3), and so on. How many rounds of calls are needed before everyone is contacted? 6

[Teacher's edition margin notes]

Additional Answers Written Exercises

30. Note that each series has n terms, from $1 = 2^0$ to $\pm\left(\frac{1}{2}\right)^{n-1}$.

$$S_A = \frac{1 \cdot \left(1 - \left(\frac{1}{2}\right)^n\right)}{1 - \frac{1}{2}} = 2\left(1 - \left(\frac{1}{2}\right)^n\right);$$

$$S_B = \frac{1 \cdot \left(1 - \left(-\frac{1}{2}\right)^n\right)}{1 - \left(-\frac{1}{2}\right)} = \frac{2}{3}\left(1 - \left(-\frac{1}{2}\right)^n\right)$$

Now if n is even, $\left(-\frac{1}{2}\right)^n = \left(\frac{1}{2}\right)^n$, and we see $S_A = 3S_B$.

Using a Calculator

Because the formulas for finding the sums of finite arithmetic and geometric series are fairly complex, you may wish to encourage students to use a calculator when solving Problems 1–15.

Additional Answers
Problems

14. A perfect square;
$T_n + T_{n+1}$
$= \frac{n(n+1)}{2} + \frac{(n+1)(n+2)}{2}$
$= \frac{n^2 + n + n^2 + 3n + 2}{2}$
$= \frac{2n^2 + 4n + 2}{2}$
$= n^2 + 2n + 1 = (n+1)^2.$

15. The square of an odd integer;
$1 + 8T_n$
$= 1 + 8\left(\frac{n(n+1)}{2}\right)$
$= 1 + 4n^2 + 4n$
$= (2n+1)^2.$

Using a Computer

Students who have programming experience are asked to write a program that will compute sums of series without relying upon any summation formulas.

Note that part (b) of Exercise 2 could be used as a springboard for Lesson 11-6. By letting the upper limit of summation get larger and larger, students will see that 2 seems to be the sum of the infinite geometric series $\frac{2}{3} + \frac{4}{9} + \frac{8}{27} + \cdots$.

13. The number $T_n = 1 + 2 + 3 + 4 + \cdots + n$ is called a *triangular number* because it is possible to represent the number by dots arranged as an equilateral triangle.

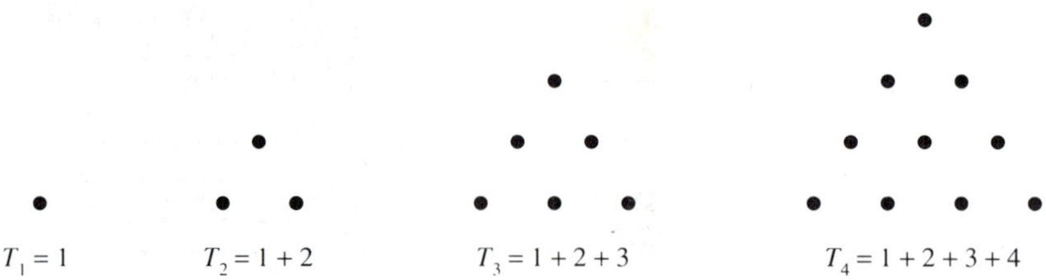

a. Find a formula for T_n in terms of n. $T_n = \frac{n(n+1)}{2}$
b. Write your formula using sigma notation. $\sum_{k=1}^{n} k = \frac{n(n+1)}{2}$

C **14.** Add any two consecutive triangular numbers. What special kind of number is obtained? Make a conjecture and then prove it.

15. Almost 2000 years ago, Plutarch observed that 1 more than eight times a triangular number always gives a special kind of number. What is this kind of number? Give a convincing argument to support your answer.

Mixed Review Exercises

Evaluate if $x = -2$ and $y = 8$.

1. $y^{1/x}$ $\frac{\sqrt{2}}{4}$ **2.** $2x^2 - y$ 0 **3.** $x^{-1} \cdot y^{2/3}$ -2

4. $\sqrt{-xy}$ 4 **5.** $\frac{y}{x^3}$ -1 **6.** $\sqrt{x^2 + y^2}$ $2\sqrt{17}$

7. $|xy|$ 16 **8.** $\log_4\left(-\frac{x}{y}\right)$ -1 **9.** $\frac{y - x^2}{x}$ -2

10. Find the slope of the line $3x - 2y = 6$. $\frac{3}{2}$

11. Find the radius of the circle $x^2 + y^2 - 4x + 6y - 3 = 0$. 4

12. Find the x-intercepts of the hyperbola $4x^2 - 9y^2 = 36$. (3, 0) and (−3, 0)

Computer Exercises

For students with some programming experience.

1. Write a program that will find the sum of a series when the user enters the limits of summation and the general term of the series. (*Hint:* You can have the user enter the general term by retyping a numbered program line before each run.)

2. Use the program in Exercise 1 to find the sum of each series.

a. $\sum_{k=10}^{20} (3k + 1)$ 506 **b.** $\sum_{k=1}^{30} \left(\frac{2}{3}\right)^k$ 1.99998957 **c.** $\sum_{k=2}^{100} \frac{1}{k^2 - 1}$ 0.740049503

11-6 Infinite Geometric Series

Objective To find sums of infinite geometric series having ratios with absolute value less than one.

Consider the infinite geometric series

$$\frac{1}{2} + \frac{1}{4} + \frac{1}{8} + \frac{1}{16} + \cdots$$

and several of the following *partial sums* for this series.

$$S_1 = \frac{1}{2}$$

$$S_2 = \frac{1}{2} + \frac{1}{4} = \frac{3}{4}$$

$$S_3 = \frac{1}{2} + \frac{1}{4} + \frac{1}{8} = \frac{7}{8}$$

$$S_4 = \frac{1}{2} + \frac{1}{4} + \frac{1}{8} + \frac{1}{16} = \frac{15}{16}$$

$$\vdots$$

$$S_{10} = \frac{1}{2} + \frac{1}{4} + \frac{1}{8} + \frac{1}{16} + \cdots + \frac{1}{1024} = \frac{1023}{1024}$$

You can see that as more terms are added, the sum gets closer and closer to the value 1. In fact, this can be proved as follows:

1. First, find a formula for S_n, the sum of the first n terms.

$$S_n = \frac{t_1(1 - r^n)}{1 - r} = \frac{\frac{1}{2}[1 - (\frac{1}{2})^n]}{1 - \frac{1}{2}} = \frac{\frac{1}{2}[1 - (\frac{1}{2})^n]}{\frac{1}{2}} = 1 - \left(\frac{1}{2}\right)^n$$

This formula tells you that S_n is always less than 1 for any n.

2. Next, notice that as n becomes increasingly large, $\left(\frac{1}{2}\right)^n$ gets closer and closer to 0. This means S_n gets closer and closer to 1:

$$S_n = 1 - \left(\frac{1}{2}\right)^n \approx 1 - 0 = 1$$

If enough terms are added (making n sufficiently large), the sum will approximate 1 as closely as you may require. Therefore, we say that the sum of the infinite geometric series $\frac{1}{2} + \frac{1}{4} + \frac{1}{8} + \frac{1}{16} + \cdots$ is 1.

Some infinite geometric series do not have a sum. For example, the partial sums of the series

$$1 + 10 + 100 + 1000 + \cdots$$

continue to become larger and larger. The sums do not approach, or "home in" on, a fixed value.

Teaching Suggestions p. T121

Suggested Extensions p. T121

Warm-Up Exercises

Find the sum of each arithmetic series.

1. $1 + 3 + 5 + \cdots + 49$ 625

2. $\sum_{n=1}^{50} (39 - 2n)$ -600

Find the sum of each geometric series.

3. $2 + 4 + 8 + \cdots + 1024$ 2046

4. $n = 20$, $r = -1$, $t_1 = 1$ 0

Motivating the Lesson

Ask the students if they think it is possible to add up infinitely many numbers and get a finite sum. Today's lesson will show that this can happen.

Chalkboard Examples

Find the sum of the geometric series.

1. $\frac{9}{10} - \frac{9}{100} + \frac{9}{1000} - \cdots$

$r = -\frac{1}{10}$ and so $|r| < 1$.

$$S = \frac{t_1}{1 - r} = \frac{\frac{9}{10}}{1 - \left(-\frac{1}{10}\right)} = \frac{9}{11}$$

2. Express $0.36\overline{36}$ as a common fraction.

$0.36\overline{36}$ can be written as the infinite series $0.36 + 0.0036 + 0.000036 + \cdots$. Since $r = 0.01$, the sum of the series is:

$$S = \frac{t_1}{1 - r} = \frac{0.36}{1 - 0.01} = \frac{0.36}{0.99} = \frac{4}{11}$$

Common Errors

Students may forget to check whether an infinite geometric series has a sum before applying the formula. Have these students find each common ratio and state this ratio before solving Written Exercises 1–12.

Check for Understanding

Here is a suggested use of the Oral Exercises to check students' understanding as you teach the lesson.
Oral Exs. 1–10: use after Example 1.

Guided Practice

Find the sum of each geometric series, if possible. If the series has no sum, say so.

1. $64 + 48 + 36 + \cdots$ 256
2. $80 - 40 + 20 - \cdots$ $\frac{160}{3}$
3. $6 - 6\sqrt{3} + 18 - \cdots$ no sum
4. $6 - 2\sqrt{3} + 2 - \cdots$ $9 - 3\sqrt{3}$
5. $\frac{2}{5} + \frac{4}{25} + \frac{8}{125} + \cdots$ $\frac{2}{3}$
6. $\sum_{n=1}^{\infty} (4 \times 10^{-n})$ $\frac{4}{9}$
7. $\sum_{n=1}^{\infty} -2(1.1)^{n+1}$ no sum
8. For the geometric series $1 + 0.2 + 0.04 + 0.008 + \cdots$, find S_1, S_2, S_3, S_4, and S_5. Use these sums to approximate S. Then use the formula to find S.
$S_1 = 1$, $S_2 = 1.2$, $S_3 = 1.24$, $S_4 = 1.248$, $S_5 = 1.2496$, $S = 1.25$

In order for an infinite geometric series to have a sum, the common ratio r must be between -1 and 1. Then as n increases, r^n gets closer and closer to 0. The fact that r^n approximates 0 for large n is used to prove the following theorem.

Theorem

An infinite geometric series with common ratio r has a sum S if $|r| < 1$. This sum is

$$S = \frac{t_1}{1 - r}.$$

Proof

1. The sum of the first n terms is $S_n = \frac{t_1(1 - r^n)}{1 - r}$.
2. As n becomes larger and larger, r^n gets closer and closer to 0 since $|r| < 1$. Then

$$S_n = \frac{t_1(1 - r^n)}{1 - r} \approx \frac{t_1(1 - 0)}{1 - r} = \frac{t_1}{1 - r}.$$

Therefore, the sum S is given by

$$S = \frac{t_1}{1 - r}.$$

Example 1 Find the sum of each infinite geometric series. If the series has no sum, say so.

a. $8 - 4 + 2 - 1 + \cdots$ **b.** $8 + 12 + 18 + 27 + \cdots$

Solution **a.** Since $r = -\frac{1}{2}$, $|r| < 1$ and the series has a sum.

$$S = \frac{t_1}{1 - r} = \frac{8}{1 - (-\frac{1}{2})} = \frac{8}{\frac{3}{2}} = \frac{16}{3} \quad \textit{Answer}$$

b. Since $r = \frac{3}{2}$, $|r| \not< 1$. $\therefore$ the series has no sum. ***Answer***

The next example shows that any infinite repeating decimal can be expressed as an infinite geometric series.

Example 2 Write 0.121212 . . . as a common fraction.

Solution The infinite repeating decimal can be written as the infinite series

$$0.12 + 0.0012 + 0.000012 + \cdots.$$

Since $r = 0.01$, $|r| < 1$ and the series has a sum.

$$S = \frac{t_1}{1 - r} = \frac{0.12}{1 - 0.01} = \frac{0.12}{0.99} = \frac{4}{33} \quad \textit{Answer}$$

Oral Exercises

Give the common ratio of each geometric series and tell whether the series has a sum.

1. $27 + 9 + 3 + \cdots$ $\frac{1}{3}$; yes
2. $27 - 9 + 3 - \cdots$ $-\frac{1}{3}$; yes
3. $\frac{2}{3} + 1 + \frac{3}{2} + \cdots$ $\frac{3}{2}$; no
4. $\frac{9}{4} + \frac{3}{2} + 1 + \cdots$ $\frac{2}{3}$; yes
5. $1 + \sqrt{3} + 3 + \cdots$ $\sqrt{3}$; no
6. $8^{-1} - 8^{-2} + 8^{-3} - \cdots$ $-\frac{1}{8}$; yes
7. $\sum_{n=0}^{\infty} 3\left(\frac{4}{5}\right)^n$ $\frac{4}{5}$; yes
8. $\sum_{n=1}^{\infty} 2(1.1)^n$ 1.1; no
9. $\sum_{n=1}^{\infty} (-0.35)^n$ -0.35; yes
10. $\sum_{n=1}^{\infty} \frac{1}{10^n}$ $\frac{1}{10}$; yes

Written Exercises

For each geometric series, find the sum. If the series has no sum, say so.

A

1. $24 + 12 + 6 + 3 + \cdots$ 48
2. $24 - 12 + 6 - 3 + \cdots$ 16
3. $27 - 18 + 12 - 8 + \cdots$ $\frac{81}{5} = 16.2$
4. $27 + 18 + 12 + 8 + \cdots$ 81
5. $256 + 320 + 400 + 500 + \cdots$ no sum
6. $500 + 400 + 320 + 256 + \cdots$ 2500
7. $3 + 4 + 5\frac{1}{3} + 7\frac{1}{9} + \cdots$ no sum
8. $\frac{1}{2} - \frac{1}{3} + \frac{2}{9} - \frac{4}{27} + \cdots$ $\frac{3}{10}$
9. $3\sqrt{3} - 3 + \sqrt{3} - 1 + \cdots$ $\frac{9}{2}(\sqrt{3} - 1)$
10. $4^{-1/2} + 4^{-3/2} + 4^{-5/2} + 4^{-7/2} + \cdots$ $\frac{2}{3}$
11. $\sum_{n=0}^{\infty} 3\left(\frac{1}{4}\right)^n$ 4
12. $\sum_{n=1}^{\infty} \frac{2^n}{5^n}$ $\frac{2}{3}$

For each geometric series find S_1, S_2, S_3, S_4, and S_5. Use these sums to approximate S. Then use the formula to find S. Compare your approximation with the value obtained from the formula.

13. $8 + 2 + \frac{1}{2} + \frac{1}{8} + \cdots$
14. $9 + \frac{9}{10} + \frac{9}{100} + \frac{9}{1000} + \cdots$
15. $\frac{3}{4} - \frac{3}{8} + \frac{3}{16} - \frac{3}{32} + \cdots$
16. $\frac{1}{3} + \frac{1}{9} + \frac{1}{27} + \frac{1}{81} + \cdots$

Write each repeating decimal as a common fraction. Use the method of Example 2.

17. 0.3333 . . . $\frac{1}{3}$
18. 0.4444 . . . $\frac{4}{9}$
19. 3.12312312 . . . $\frac{1040}{333}$
20. 0.363636 . . . $\frac{4}{11}$
21. 0.49999 . . . $\frac{1}{2}$
22. 1.045045045 . . . $\frac{116}{111}$

Write the first three terms of the infinite geometric series satisfying the given condition.

B

23. $t_1 = 8$, $S = 12$ 8, $\frac{8}{3}$, $\frac{8}{9}$
24. $t_1 = 40$, $S = 200$ 40, 32, 25.6
25. $r = -\frac{1}{3}$, $S = 30$ 40, $-\frac{40}{3}$, $\frac{40}{9}$
26. $r = \frac{2}{5}$, $S = 125$ 75, 30, 12

Write each repeating decimal as a common fraction. Use the method of Example 2.

9. 0.7777 . . . $\frac{7}{9}$
10. 0.4545 . . . $\frac{5}{11}$
11. 1.321321 . . . $\frac{440}{333}$

Summarizing the Lesson

In this lesson students learned to find the sums of infinite geometric series having ratios with absolute value less than one. Ask students to state the formula presented in this lesson.

Suggested Assignments

Average Algebra
Day 1: 533/1–23 odd
S 529/*P*: 7, 8, 11
Day 2: 533/14–26 even
534/*P*: 1–3, 5, 7, 9

Extended Algebra
Day 1: 533/1–29 odd
S 529/*P*: 9, 10, 14
Day 2: 533/14–28 even
534/*P*: 1–9 odd

Extended Alg. and Trig.
533/1–29 odd
534/*P*: 1–9 odd

Supplementary Materials

Study Guide	pp. 181–182
Practice Master	62
Computer Activity	25

Additional Answers Written Exercises

13. $S_1 = 8$, $S_2 = 10$, $S_3 = \frac{21}{2}$, $S_4 = \frac{85}{8}$, $S_5 = \frac{341}{32}$; $S = \frac{32}{3}$

14. $S_1 = 9$, $S_2 = 9.9$, $S_3 = 9.99$, $S_4 = 9.999$, $S_5 = 9.9999$; $S = 10$

(continued)

Additional Answers Written Exercises

(continued)

15. $S_1 = \frac{3}{4}$, $S_2 = \frac{3}{8}$, $S_3 = \frac{9}{16}$, $S_4 = \frac{15}{32}$, $S_5 = \frac{33}{64}$; $S = \frac{1}{2}$

16. $S_1 = \frac{1}{3}$, $S_2 = \frac{4}{9}$, $S_3 = \frac{13}{27}$, $S_4 = \frac{40}{81}$, $S_5 = \frac{121}{243}$; $S = \frac{1}{2}$

27. If $S = 4$ for a geometric series, then $4 = \frac{t_1}{1 - r}$, so $1 - r = \frac{t_1}{4}$ or $r = 1 - \frac{t_1}{4}$. Now $-1 < r < 1$, so $-1 < 1 - \frac{t_1}{4} < 1$. Solving this inequality, we get $8 > t_1 > 0$. t_1 cannot equal 10.

27. Explain why there is no infinite geometric series with first term 10 and sum 4.

28. There are two infinite geometric series with $t_1 = 100$ and $t_3 = 1$. Find the sum of each. $\frac{1000}{9} = 111.111\ldots$ and $\frac{1000}{11} = 90.9090\ldots$

C **29. a.** Find the sum of the series $1 + \frac{2}{3} + \frac{4}{9} + \frac{8}{27} + \cdots$. 3

b. How many terms of this series must be added before their sum is within 0.00001 of the sum found in part (a)? (*Hint:* Consider $S - S_n$.) 32

30. If $|x| < 1$, find the sum S of the series $1 + 2x + 3x^2 + 4x^3 + \cdots$. $\frac{1}{(1 - x)^2}$ (*Hint:* Consider $S - xS$.)

Problems

Solve. A calculator may be helpful.

A **1.** A child on a swing is given a big push. She travels 12 ft on the first back-and-forth swing but only $\frac{5}{6}$ as far on each successive back-and-forth swing. How far does she travel before the swing stops? 72 ft

2. Suppose an indecisive man starts out from home and walks 1 mi east, then $\frac{1}{2}$ mi west, then $\frac{1}{4}$ mi east, then $\frac{1}{8}$ mi west, and so on. Relative to his home, approximately where would he end up? $\frac{2}{3}$ mi east

3. A side of a square is 12 cm. The midpoints of its sides are joined to form an inscribed square, and this process is continued as shown in the diagram. Find the sum of the perimeters of the squares if this process is continued without end. $48(2 + \sqrt{2})$ cm

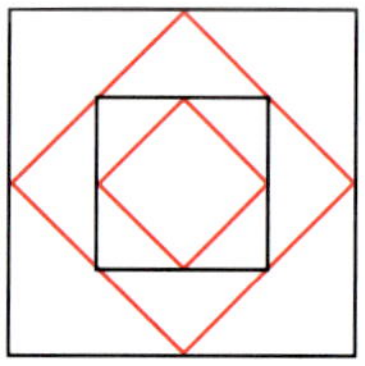

4. A side of an equilateral triangle is 10 cm. The midpoints of its sides are joined to form an inscribed equilateral triangle, and this process is continued. Find the sum of the perimeters of the triangles if the process is continued without end. 60 cm

5. Find the sum of the areas of the squares in Problem 3. 288 cm^2

6. Find the sum of the areas of the triangles in Problem 4. $\frac{100\sqrt{3}}{3}$ cm^2

B **7.** The diagram shows a superball that rebounds 95% of the distance it falls. This ball is thrown 12 m in the air (so that the initial up-and-down distance traveled is 24 m). What is the total vertical distance traveled by the ball before it stops bouncing? 480 m

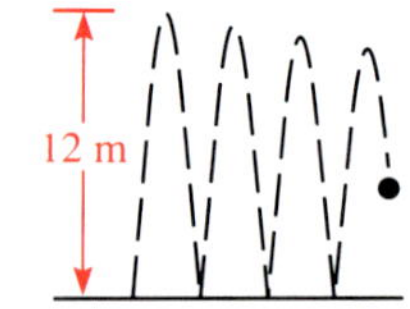

8. Refer to Problem 7 and suppose that the ball is dropped from a height of 10 m. Make a diagram of the bouncing ball's path and note that it differs from the diagram above because the ball does not *begin* at ground level. What is the total vertical distance traveled by the ball before it stops bouncing? 390 m

9. A morning glory vine grows more slowly as it gets longer. Suppose that its growth during the week after it emerges from the soil is 50 in. and that during each succeeding week it grows $\frac{4}{5}$ as much as it did the previous week.

a. How much does the vine grow during its 26th week? 0.18889 in.

b. What is its total growth for a 26-week growing period? 249.24 in.

c. Assuming that the vine could grow forever, what would be the vine's total growth? 250 in.

(*Note:* Since the answer to part (c) is easier to calculate than the answer to part (b) and since the two answers are nearly the same, the sum of a finite geometric series with many terms can be estimated by the sum of the corresponding infinite series.)

10. A piledriver pounds a steel column into the earth. On its first drive, the column penetrates 1.5 m into the earth and on each succeeding drive it moves 92% as far as it did on its previous drive.

a. How far is the column driven on the 60th drive? 0.01095 m

b. What is the approximate total distance that the column moves in 60 drives? (See the note after Problem 9(c).) 18.75 m

C 11. The Koch curve is constructed as follows: Start with a square (see Figure 1). Construct a square on the middle third of each side of Figure 1 (see Figure 2). Then construct a square on the middle third of each side of Figure 2 (see Figure 3). Continue forming new figures by constructing a square on the middle third of each side of the previous figure.

Figure 1

Figure 2

Figure 3

a. Complete the table below.

	Fig. 1	Fig. 2	Fig. 3	Fig. 4	. . .	Fig. n	
Number of new squares	1	4	? 20	? 100	. . .	?	$4 \cdot 5^{n-2}$
Area of each new square	1	$\frac{1}{9}$	? $\frac{1}{81}$	? $\frac{1}{729}$	. . .	?	$\left(\frac{1}{9}\right)^{n-1}$
Total new area	1	$\frac{4}{9}$	? $\frac{20}{81}$	? $\frac{100}{729}$	. . .	?	$\frac{4 \cdot 5^{n-2}}{9^{n-1}} = \frac{4}{9}\left(\frac{5}{9}\right)^{n-2}$

b. Find the total area bounded by the Koch curve. (*Hint:* Add the entries in the last row of the table. Notice that the entries *after* the first one form a geometric series.) 2

12. Show that the Koch curve described in Problem 11 has no perimeter (that is, the perimeter is of unbounded length).

Using a Calculator

Students will find a calculator particularly helpful when solving Problems 9–10.

Additional Answers Problems

12. Two sides of each new square get added to the original perimeter. This sequence of additions to the perimeter begins with: $4 \cdot 1$, $4 \cdot 2 \cdot \frac{1}{3}$, $20 \cdot 2 \cdot \frac{1}{9}$, $100 \cdot 2 \cdot \frac{1}{27}$, and so on, and we see that this is a geometric series (after the first term) with $r = \frac{5}{3} > 1$. Therefore its sum is infinite.

14. After the first stage, each side from the present figure becomes 4 sides in the next, each having $\frac{1}{3}$ the previous length. Thus the perimeter increases by a ratio of $\frac{4}{3} > 1$, and grows without bound.

13. A snowflake curve is constructed as follows: Start with an equilateral triangle (see Figure 1). Construct an equilateral triangle on the middle third of each side of Figure 1 (see Figure 2). Then construct an equilateral triangle on the middle third of each side of Figure 2 (see Figure 3). Continue forming new figures by constructing an equilateral triangle on the middle third of each side of the previous figure. If the first equilateral triangle has sides of length 1, what is the total area bounded by the snowflake curve? $\frac{2\sqrt{3}}{5}$

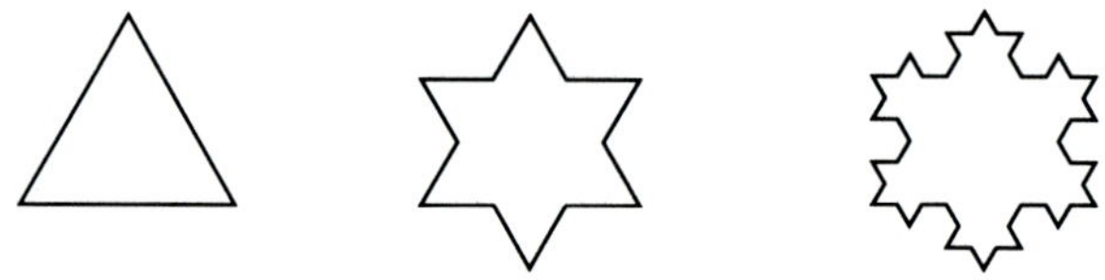

Figure 1 **Figure 2** **Figure 3**

14. Show that the snowflake curve described in Problem 13 has no perimeter (that is, the perimeter is of unbounded length).

Self-Test 2

Vocabulary

series (p. 518)	summation sign (p. 518)
arithmetic series (p. 518)	summand (p. 518)
geometric series (p. 518)	index (p. 518)
sigma (p. 518)	limits of summation (p. 519)

1. Identify each series as arithmetic, geometric, or neither. Then rewrite the series using sigma notation. **Obj. 11-4, p. 518**

   a. $27 - 18 + 12 - 8$ Geometric; $\sum_{n=0}^{3} 27\left(-\frac{2}{3}\right)^n$

   b. $\frac{1}{2} + \frac{2}{3} + \frac{3}{4} + \cdots + \frac{15}{16}$ Neither; $\sum_{n=1}^{15} \frac{n}{n+1}$

   c. $20 + 17 + 14 + \cdots + 2$ c. Arithmetic; $\sum_{n=1}^{7} (23 - 3n)$

   d. $1 + 4 + 9 + 16 + \cdots$ Neither; $\sum_{n=1}^{\infty} n^2$

2. Find the sum of each arithmetic series. **Obj. 11-5, p. 525**

   a. $n = 30,\ t_1 = 7,\ t_{30} = 65$ 1080

   b. $\sum_{k=1}^{15} (45 - 6k)$ -45

3. Find the sum of each geometric series.

   a. $n = 6,\ r = 2,\ t_1 = 5$ 315

   b. $\sum_{k=1}^{4} \left(\frac{2}{3}\right)^k$ $\frac{130}{81}$

4. Find the sum of each infinite geometric series. If the series has no sum, say so. **Obj. 11-6, p. 531**

   a. $\frac{25}{8} + \frac{5}{2} + 2 + \frac{8}{5} + \cdots$ $\frac{125}{8}$

   b. $1 - 1.1 + 1.21 - 1.331 + \cdots$ No sum

Check your answers with those at the back of the book.

Quick Quiz

1. Identify each series as arithmetic, geometric, or neither. Rewrite the series using sigma notation.

   a. $\frac{5}{1} + \frac{7}{2} + \frac{9}{3} + \cdots + \frac{63}{30}$
   Neither; for example, $\sum_{n=1}^{30} \frac{2n+3}{n}$

   b. $27 + 9 + 3 + 1 + \cdots$
   Geometric; for example, $\sum_{n=1}^{\infty} 3^{4-n}$

   c. The series consisting of positive two-digit multiples of 3
   Arithmetic; for example, $\sum_{n=1}^{30} (3n + 9)$

2. Find the sum of each series.

   a. $\sum_{k=0}^{50} (3k - 1)$ 3774

   b. $\sum_{j=1}^{9} (-3)^j$ $-14{,}763$

   c. An arithmetic series of 40 terms with $t_1 = 5$ and $t_{30} = 63$ 1760

3. For each series, find the sum or state that it does not exist.

   a. $\frac{1}{25} - \frac{1}{20} + \frac{1}{16} - \cdots$
   no sum

   b. $25 - 20 + 16 - \cdots$
   $\frac{125}{9}$

4. Express $0.\overline{296}$ as a common fraction by writing it as an infinite series. $\frac{8}{27}$

Binomial Expansions

11-7 Powers of Binomials

Objective To expand powers of binomials.

When you multiply out a power of a binomial, the resulting sum of terms is called a **binomial expansion.** An interesting pattern is formed by the coefficients of the expansion of $(a + b)^n$ where n is a nonnegative integer.

$$(a + b)^0 = 1$$
$$(a + b)^1 = 1a^1 + 1b^1$$
$$(a + b)^2 = 1a^2 + 2ab + 1b^2$$
$$(a + b)^3 = 1a^3 + 3a^2b + 3ab^2 + 1b^3$$
$$(a + b)^4 = 1a^4 + 4a^3b + 6a^2b^2 + 4ab^3 + 1b^4$$
$$(a + b)^5 = 1a^5 + 5a^4b + 10a^3b^2 + 10a^2b^3 + 5ab^4 + 1b^5$$

The pattern of coefficients suggests the triangular array of numbers, shown at the right, called **Pascal's triangle.** The triangle has 1's at the beginning and end of each row. Each of the other numbers is the sum of the two numbers above it. For example, 15 = 10 + 5

	Row
1	Row 0
1 1	1
1 2 1	2
1 3 3 1	3
1 4 6 4 1	4
1 5 10 10 5 1	5
1 6 15 20 15 6 1	6

Because the top row of the triangle corresponds to $(a + b)^0$, it is convenient to call it row 0. The next row is row 1, followed by row 2, and so on. Then for any positive integer n, the coefficients of the expansion of $(a + b)^n$ are the elements of row n of Pascal's triangle.

Example 1 Expand $(a + b)^6$. Use row 6 of Pascal's triangle.

Solution From row 6 of Pascal's triangle, you find that the coefficients are

1, 6, 15, 20, 15, 6, 1.

The powers of a are a^6, a^5, a^4, a^3, a^2, a^1, a^0.

The powers of b are b^0, b^1, b^2, b^3, b^4, b^5, b^6.

Combining this information, you get

$(a + b)^6 = a^6 + 6a^5b + 15a^4b^2 + 20a^3b^3 + 15a^2b^4 + 6ab^5 + b^6$. ***Answer***

Explorations p. 842

Teaching Suggestions p. T121

Suggested Extensions p. T121

Warm-Up Exercises

Multiply.

1. $(a + b)^2$ $a^2 + 2ab + b^2$
2. $(a - b)^2$ $a^2 - 2ab + b^2$
3. $(3x + 4y)^2$ $9x^2 + 24xy + 16y^2$
4. $(7x - 2y)^2$ $49x^2 - 28xy + 4y^2$
5. $(a + b)^3$ $a^3 + 3a^2b + 3ab^2 + b^3$

Motivating the Lesson

Tell students that Blaise Pascal, a French mathematician, showed his mathematical talent early. By age 17, he had written an essay in which he used projective geometry to examine the conic sections. Pascal is also famous for his work in probability theory, where he used a special triangular array of numbers.

Chalkboard Examples

1. Find the first three terms of the expansion of $(2x - y)^6$.
   Row 6 of Pascal's triangle is:
   1 6 15 20 15 6 1
   Now expand $(2x - y)^6$:
   $(2x)^6 + 6(2x)^5(-y)^1 + 15(2x)^4(-y)^2 + \cdots$
   The first three terms are:
   $64x^6 - 192x^5y + 240x^4y^2$
2. The first three terms of $(x + y)^{17}$ are $x^{17} + 17x^{16}y + 136x^{15}y^2$. Write the last three terms.
   $136x^2y^{15} + 17xy^{16} + y^{17}$

Check for Understanding

Oral Exs. 1–2: use before Example 1.
Oral Exs. 3–5: use after Example 1.

Guided Practice

Expand and simplify each expression.

1. $(a-b)^4$
   $a^4 - 4a^3b + 6a^2b^2 - 4ab^3 + b^4$
2. $(x-3)^3$
   $x^3 - 9x^2 + 27x - 27$
3. $(2x+3y)^5$
   $32x^5 + 240x^4y + 720x^3y^2 + 1080x^2y^3 + 810xy^4 + 243y^5$
4. The first three terms of $(x+y)^{15}$ are $x^{15} + 15x^{14}y + 105x^{13}y^2$. Write the last three terms.
   $105x^2y^{13} + 15xy^{14} + y^{15}$

Summarizing the Lesson

In this lesson students learned to expand powers of binomials using Pascal's triangle. Ask the students to look at Pascal's triangle on page 537 and state what numbers should appear in rows 8, 9, and 10.

Suggested Assignments

Average Algebra
539/2–12 even, 13–20
R 536/Self-Test 2
S 539/Mixed Review

Average Alg. and Trig.
539/1–20

Extended Algebra
539/2–12 even, 13–22
R 536/Self-Test 2
S 539/Mixed Review

Extended Alg. and Trig.
539/2–12 even, 13–22
R 536/Self-Test 2
S 539/Mixed Review

Notice that the exponents of a in Example 1 decrease from 6 to 0 and the exponents of b increase from 0 to 6. There are seven terms, and the sum of the exponents in each term is equal to 6. When the binomial contains a negative term, the signs in the expansion will alternate, as Example 2 shows.

Example 2 Find the first four terms in the expansion of $(x-2y)^7$.

Solution First find row 7 of Pascal's triangle:

	1	6	15	20	15	6	1	Row 6
1	7	21	35	35	21	7	1	Row 7

Now expand $(a+b)^7$.

$$(a+b)^7 = a^7 + 7a^6b + 21a^5b^2 + 35a^4b^3 + \cdots$$

Then replace a by x and b by $(-2y)$.

$$(x-2y)^7 = x^7 + 7x^6(-2y) + 21x^5(-2y)^2 + 35x^4(-2y)^3 + \cdots$$

The first four terms are: $x^7 - 14x^6y + 84x^5y^2 - 280x^4y^3$. ***Answer***

The pattern of coefficients given in Pascal's triangle is symmetric. For example, in the expansion of $(a+b)^5$ the coefficients of the last three terms are the same as the coefficients of the first three in reverse order. And in the expansion of $(a+b)^6$ the coefficients to the right of the middle term, $20a^3b^3$, are the same as those to the left. This symmetry is useful in writing expansions and finding particular terms, as shown in Example 3.

Example 3 The first three terms in the expansion of $(a+b)^{20}$ are $a^{20} + 20a^{19}b + 190a^{18}b^2$. Write the last three terms.

Solution $190a^2b^{18} + 20ab^{19} + b^{20}$ ***Answer***

Oral Exercises

Row 8 of Pascal's triangle is 1, 8, 28, 56, 70, 56, 28, 8, 1. Use this row to complete Exercises 1 and 2.

1. Give the first four terms in the expansion of $(a+b)^8$. $a^8 + 8a^7b + 28a^6b^2 + 56a^5b^3$
2. What is row 9 of Pascal's triangle? 1, 9, 36, 84, 126, 126, 84, 36, 9, 1
3. Give the number of terms in the expansion of each of the following.
   a. $(m-n)^1$ 2 **b.** $(2x+y)^6$ 7 **c.** $(3p^2-1)^9$ 10
4. Give the first term in the expansion of each of the following.
   a. $(a+b)^{10}$ a^{10} **b.** $(3m+n)^3$ $27m^3$ **c.** $(x^3-y^2)^5$ x^{15}
5. Give the last term in the expansion of each of the following.
   a. $(a-b)^7$ $-b^7$ **b.** $(p^2-2)^4$ 16 **c.** $(m+n^2)^8$ n^{16}

Written Exercises

Expand and simplify each expression.

A **1.** $(x + y)^3$ **2.** $(x + y)^4$ **3.** $(c - d)^5$ **4.** $(p - q)^6$

5. $(a + 1)^8$ **6.** $(x - 2)^3$ **7.** $(3t + 4)^5$ **8.** $(2y - 3)^4$

9. $(x^2 - 1)^6$ **10.** $(\sqrt{x} + 3)^4$ **11.** $(p^2 + q^3)^3$ **12.** $\left(a + \frac{1}{a}\right)^5$

In Exercises 13–16, use the symmetry of the coefficients.

13. The first three terms of $(x + y)^{17}$ are $x^{17} + 17x^{16}y + 136x^{15}y^2$. Write the last three terms. $136x^2y^{15} + 17xy^{16} + y^{17}$

14. The first three terms of $(p - q)^{22}$ are $p^{22} - 22p^{21}q + 231p^{20}q^2$. Write the last three terms. $231p^2q^{20} - 22pq^{21} + q^{22}$

15. The tenth term of $(x + 1)^{20}$ is $167{,}960x^{11}$. Write the twelfth term. (*Hint:* The eleventh term is the middle term.) $167{,}960x^9$

16. The eighth term of $(p - q)^{15}$ is $-6435p^8q^7$. Write the ninth term. $6435p^7q^8$

Expand and simplify each expression.

B **17.** $(a + b)^7 + (a - b)^7$ **18.** $(a + b)^8 - (a - b)^8$

19. $(a^2 + 2ab + b^2)^3$ **20.** $(x - 1)^4(x + 1)^4$

21. Use the first three terms of a binomial expansion to approximate $(2.1)^7$. [*Hint:* $(2.1)^7 = (2 + 0.1)^7$.] 179.52

22. Use the first four terms of a binomial expansion to approximate $(9.8)^4$. [*Hint:* $(9.8)^4 = (10 - 0.2)^4$.] 9223.68

C **23.** **a.** Find the sum of the entries in each of the first 7 rows of Pascal's triangle. 1, 2, 4, 8, 16, 32, 64
b. Guess a formula for the sum of the entries in the nth row. 2^n

24. Find a formula for the sum of *all* entries in rows 0 through n of Pascal's triangle. $2^{n+1} - 1$

Mixed Review Exercises

Tell whether each sequence is arithmetic, geometric, or neither. Then find a formula for the nth term of the sequence.

1. $\frac{2}{3}, \frac{4}{5}, \frac{6}{7}, \ldots$ Neither; $\frac{2n}{2n+1}$

2. $\frac{9}{16}, \frac{3}{4}, 1, \ldots$ Geometric; $\frac{9}{16}\left(\frac{4}{3}\right)^{n-1}$

3. $7, 3, -1, \ldots$ Arithmetic; $11 - 4n$

4. $-24, 12, -6, \ldots$ Geometric; $-24\left(-\frac{1}{2}\right)^{n-1}$

Find the sum of each series.

5. $5 + 9 + 13 + \cdots + 45$ 275 **6.** $6 + \frac{9}{2} + \frac{27}{8} + \cdots$ 24 **7.** $\sum_{n=1}^{6} 3 \cdot 2^n$ 378

Supplementary Materials
Study Guide pp. 183–184
Resource Book p. 149

Additional Answers
Written Exercises

1. $x^3 + 3x^2y + 3xy^2 + y^3$
2. $x^4 + 4x^3y + 6x^2y^2 + 4xy^3 + y^4$
3. $c^5 - 5c^4d + 10c^3d^2 - 10c^2d^3 + 5cd^4 - d^5$
4. $p^6 - 6p^5q + 15p^4q^2 - 20p^3q^3 + 15p^2q^4 - 6pq^5 + q^6$
5. $a^8 + 8a^7 + 28a^6 + 56a^5 + 70a^4 + 56a^3 + 28a^2 + 8a + 1$
6. $x^3 - 6x^2 + 12x - 8$
7. $243t^5 + 1620t^4 + 4320t^3 + 5760t^2 + 3840t + 1024$
8. $16y^4 - 96y^3 + 216y^2 - 216y + 81$
9. $x^{12} - 6x^{10} + 15x^8 - 20x^6 + 15x^4 - 6x^2 + 1$
10. $x^2 + 12x\sqrt{x} + 54x + 108\sqrt{x} + 81$
11. $p^6 + 3p^4q^3 + 3p^2q^6 + q^9$
12. $a^5 + 5a^3 + 10a + \frac{10}{a} + \frac{5}{a^3} + \frac{1}{a^5}$
17. $2a^7 + 42a^5b^2 + 70a^3b^4 + 14ab^6$
18. $16a^7b + 112a^5b^3 + 112a^3b^5 + 16ab^7$
19. $a^6 + 6a^5b + 15a^4b^2 + 20a^3b^3 + 15a^2b^4 + 6ab^5 + b^6$
20. $x^8 - 4x^6 + 6x^4 - 4x^2 + 1$

Teaching Suggestions p. T121

Reading Algebra p. T122

Suggested Extensions p. T122

Warm-Up Exercises

Write rows 7, 8, 9, and 10 of Pascal's triangle.

Row 7: 1, 7, 21, 35, 35, 21, 7, 1
Row 8: 1, 8, 28, 56, 70, 56, 28, 8, 1
Row 9: 1, 9, 36, 84, 126, 126, 84, 36, 9, 1
Row 10: 1, 10, 45, 120, 210, 252, 210, 120, 45, 10, 1

Motivating the Lesson

Point out to the students that generating the rows of Pascal's triangle (beyond the first few) is a tedious task to do by hand. Fortunately, there is a way to find any particular entry in any given row of the triangle *without* generating any preceding entries. Today's lesson on the binomial theorem will show how this is done.

Chalkboard Examples

1. The first three terms in the expansion of $(x - y)^{22}$ are $x^{22} - 22x^{21}y + 231x^{20}y^2$. Find the next two terms.
$-\frac{22 \cdot 21 \cdot 20}{1 \cdot 2 \cdot 3}x^{19}y^3 = -1540x^{19}y^3$
$\frac{22 \cdot 21 \cdot 20 \cdot 19}{1 \cdot 2 \cdot 3 \cdot 4}x^{18}y^4 = 7315x^{18}y^4$

11-8 The General Binomial Expansion

Objective To use the binomial theorem to find a particular term of a binomial expansion.

If you used Pascal's triangle to find the coefficients in the expansion of $(a + b)^{20}$, you would need to compute all the rows up through row 20. Fortunately, there is another way to find these coefficients. The following example shows how each coefficient depends on the power of the binomial and the position of the term in the expansion.

Example 1 The first four terms in the expansion of $(a + b)^{20}$ are given below. Find the next two terms.

$$(a + b)^{20} = a^{20} + \frac{20}{1}a^{19}b + \frac{20 \cdot 19}{1 \cdot 2}a^{18}b^2 + \frac{20 \cdot 19 \cdot 18}{1 \cdot 2 \cdot 3}a^{17}b^3 + \cdots$$

Solution The coefficient of the term containing b^k has k factors in the numerator beginning with the power 20 and decreasing, and k factors in the denominator beginning with 1 and increasing. So the next two terms are

$$\frac{20 \cdot 19 \cdot 18 \cdot 17}{1 \cdot 2 \cdot 3 \cdot 4}a^{16}b^4 \quad \text{and} \quad \frac{20 \cdot 19 \cdot 18 \cdot 17 \cdot 16}{1 \cdot 2 \cdot 3 \cdot 4 \cdot 5}a^{15}b^5. \quad \textit{Answer}$$

Example 2 Find the first five terms in the expansion of $(a - b)^{12}$. Simplify each coefficient.

Solution The terms containing odd powers of b will have a minus sign. The expansion begins

$$a^{12} - \frac{12}{1}a^{11}b + \frac{12 \cdot 11}{1 \cdot 2}a^{10}b^2 - \frac{12 \cdot 11 \cdot 10}{1 \cdot 2 \cdot 3}a^9b^3 + \frac{12 \cdot 11 \cdot 10 \cdot 9}{1 \cdot 2 \cdot 3 \cdot 4}a^8b^4 - \cdots.$$

When the fractions are simplified, each coefficient is an integer. The first five terms are

$$a^{12} - 12a^{11}b + 66a^{10}b^2 - 220a^9b^3 + 495a^8b^4. \quad \textit{Answer}$$

The coefficients in the expansions of $(a + b)^n$ can be written more simply using **factorial notation.** The symbol $r!$ (read "r factorial") is defined as follows:

$$r! = r(r - 1)(r - 2) \cdot \cdots \cdot 3 \cdot 2 \cdot 1 \text{ if } r \text{ is a positive integer}$$
$$0! = 1$$

For example, $4! = 4 \cdot 3 \cdot 2 \cdot 1 = 24$ and $5! = 5 \cdot 4 \cdot 3 \cdot 2 \cdot 1 = 120$.

Notice what happens when factorials are divided.

$$\frac{12!}{8!} = \frac{12 \cdot 11 \cdot 10 \cdot 9 \cdot 8 \cdot 7 \cdot 6 \cdot 5 \cdot 4 \cdot 3 \cdot 2 \cdot 1}{8 \cdot 7 \cdot 6 \cdot 5 \cdot 4 \cdot 3 \cdot 2 \cdot 1} = 12 \cdot 11 \cdot 10 \cdot 9$$

Therefore, you can write the fifth term of the expansion in Example 2 as follows:

$$\frac{12 \cdot 11 \cdot 10 \cdot 9}{1 \cdot 2 \cdot 3 \cdot 4}a^8b^4 = \frac{12! \div 8!}{4!}a^8b^4 = \frac{12!}{8!4!}a^8b^4.$$

The terms found in Examples 1 and 2 illustrate the general formula for the expansion of $(a + b)^n$ given by the *binomial theorem,* which is proved in more advanced courses by using the principle of mathematical induction. (See the Extra on page 523.)

The Binomial Theorem

If n is a positive integer, then

$$(a + b)^n = a^n + \frac{n}{1}a^{n-1}b + \frac{n(n-1)}{1 \cdot 2}a^{n-2}b^2 + \frac{n(n-1)(n-2)}{1 \cdot 2 \cdot 3}a^{n-3}b^3 + \cdots + b^n$$

$$= a^n + \frac{n!}{(n-1)!1!}a^{n-1}b + \frac{n!}{(n-2)!2!}a^{n-2}b^2 + \frac{n!}{(n-3)!3!}a^{n-3}b^3 + \cdots + b^n$$

The $(k + 1)$st term of $(a + b)^n$ is $\dfrac{n!}{(n-k)!k!}a^{n-k}b^k$.

Example 3 Find and simplify the seventh term in the expansion of $(2x - y)^{10}$.

Solution First find the seventh term in the expansion of $(a + b)^{10}$. This is the term containing b^6.

$$\frac{10!}{4!6!}a^4b^6 = \frac{10 \cdot \overset{3}{\cancel{9}} \cdot \cancel{8} \cdot 7}{\cancel{4} \cdot \cancel{3} \cdot \cancel{2} \cdot 1}a^4b^6 = 210a^4b^6$$

Now replace a by $2x$ and b by $-y$.

$$210(2x)^4(-y)^6 = 210 \cdot 16x^4 \cdot y^6$$
$$= 3360x^4y^6 \quad \textbf{Answer}$$

Because 0! is defined as 1, the formula for the $(k + 1)$st term in the expansion of $(a + b)^n$ remains valid for the first and last terms.

First term $(k = 0)$: $\dfrac{n!}{(n-0)!0!}a^{n-0}b^0 = \dfrac{n!}{n!}a^n \cdot 1 = a^n$

Last term $(k = n)$: $\dfrac{n!}{(n-n)!n!}a^{n-n}b^n = \dfrac{n!}{n!}1 \cdot b^n = b^n$

2. Find the first four terms in the expansion of $(a + b)^{17}$. Simplify each coefficient.
$a^{17} + \frac{17}{1}a^{16}b + \frac{17 \cdot 16}{1 \cdot 2}a^{15}b^2 + \frac{17 \cdot 16 \cdot 15}{1 \cdot 2 \cdot 3}a^{14}b^3 = a^{17} + 17a^{16}b + 136a^{15}b^2 + 680a^{14}b^3$
3. Find and simplify the fourth term of $(x + y)^{12}$.
$\frac{12 \cdot 11 \cdot 10}{1 \cdot 2 \cdot 3}x^9y^3 = 220x^9y^3$

Common Errors

Students may mistakenly think that $\frac{6!}{3!2!}$ is equivalent to $\frac{2!}{2!} = 1$. When reteaching, show students that $\frac{6!}{3!} \neq 2!$ by showing that $\frac{6!}{3!} = \frac{6 \cdot 5 \cdot 4 \cdot 3 \cdot 2 \cdot 1}{3 \cdot 2 \cdot 1} = 6 \cdot 5 \cdot 4 = 120$. Thus $\frac{6!}{3!2!} = \frac{120}{2} = 60$.

Check for Understanding

Here is a suggested use of the Oral Exercises to check students' understanding as you teach the lesson.
Oral Exs. 1–8: use before the binomial theorem on page 541.
Oral Exs. 9–10: use after Example 3.

Guided Practice

1. Evaluate.

   a. $\frac{20!}{18!}$ 380 b. $\frac{8!}{5!3!}$ 56

   c. $\frac{10!}{2!8!}$ 45 d. $\frac{(x+1)!}{(x-1)!}$ $x(x+1)$

Write the first four terms in the expansion of each of the following.

2. $(a - x)^{18}$ $a^{18} - 18a^{17}x + 153a^{16}x^2 - 816a^{15}x^3$
3. $(1 + 2x)^{16}$ $1 + 32x + 480x^2 + 4480x^3$
4. Find the term containing x^5 in $(x + y)^{12}$. $792x^5y^7$
5. Find the term containing y^3 in $(2x - y)^8$. $-1792x^5y^3$
6. Find the fifth term of $(1 + y^2)^6$. $15y^8$

Summarizing the Lesson

In this lesson students learned about factorial notation and the binomial theorem. Students then learned to use the binomial theorem to find a particular term of a binomial expansion. Ask students to state the $(k + 1)$st term in the expansion of $(a + b)^n$.

Suggested Assignments

Average Algebra
542/1–13, 15–27 odd
S 534/*P*: 8, 10
R 543/Self-Test 3

Average Alg. and Trig.
542/1–13, 15–27 odd
R 543/Self-Test 3

This means you can write the binomial theorem in a compact form using sigma notation:

$$(a + b)^n = \sum_{k=0}^{n} \frac{n!}{(n-k)!k!} a^{n-k}b^k$$

Oral Exercises

Evaluate.

1. $3!$ 6 **2.** $2!$ 2 **3.** $\frac{5!}{4!}$ 5 **4.** $\frac{11!}{9!}$ 110

5. $\frac{4!}{3!1!}$ 4 **6.** $\frac{5!}{2!3!}$ 10 **7.** $\frac{6!}{4!2!}$ 15 **8.** $\frac{4!}{4!0!}$ 1

9. Give the coefficient of the third term in the expansion of each of the following.
a. $(a + b)^{10}$ 45 **b.** $(c - d)^8$ 28 **c.** $(x + 2y)^5$ 40

10. Give **(a)** the power of a and **(b)** the power of b for the given term in the expansion of $(a + b)^{20}$.
a. The fifth term (a) 16 (b) 4 **b.** The eleventh term (a) 10 (b) 10

Written Exercises

Evaluate.

A

1. $6!$ 720 **2.** $7!$ 5040 **3.** $\frac{5!}{3!}$ 20 **4.** $\frac{8!}{7!}$ 8

5. $\frac{100!}{99!}$ 100 **6.** $\frac{20!}{18!}$ 380 **7.** $\frac{10!}{7!3!}$ 120 **8.** $\frac{10!}{4!6!}$ 210

9. $\frac{(n+1)!}{n!}$ $n + 1$ **10.** $\frac{n!}{(n-2)!}$ $n^2 - n$ **11.** $\frac{(n+1)!}{(n-1)!2!}$ $\frac{n^2 + n}{2}$ **12.** $\frac{n!}{3!(n-3)!}$ 12. $\frac{n(n-1)(n-2)}{6}$

Write the first four terms in the expansion of each of the following.

13. a. $(a + b)^{14}$ **b.** $(a - b)^{14}$
14. a. $(a + b)^{12}$ **b.** $(a - b)^{12}$
15. a. $(a + b)^{19}$ **b.** $(x - 2y)^{19}$
16. a. $(a + b)^{21}$ **b.** $(c - d^2)^{21}$

Find and simplify the specified term in each expansion.

17. The term containing b^5 in $(a + b)^{20}$ $15{,}504a^{15}b^5$
18. The term containing b^{13} in $(a - b)^{15}$ $-105a^2b^{13}$
19. The eleventh term of $(s - t)^{14}$ $1001s^4t^{10}$
20. The eighteenth term of $(s + t)^{18}$ $18st^{17}$

B

21. The term containing a^4 in $(a + 2b)^9$ $4032a^4b^5$
22. The term containing x^2 in $(2x - y)^7$ $-84x^2y^5$
23. The middle term of $(c^2 - 2d)^8$ $1120c^8d^4$
24. The term containing b^9 in $(a + 3b^3)^5$ $270a^2b^9$

Decide whether each statement is true or false. If the statement is false, give a counterexample.

25. $(a \cdot b)! = a!b!$ False **26.** $(n^2)! = (n!)^2$ False **27.** $(2n)! = 2^n(n!)$ False

C 28. Find the term in the expansion of $(y - y^{-2})^{12}$ that does not contain y. the fifth term: 495

29. Assume that the binomial theorem continues to hold when n is a positive rational number that is not an integer. In general, the coefficients in the expansion will not be integers, and instead of terminating, the expansion will be an infinite series:

$$(a + b)^n = a^n + \frac{n}{1}a^{n-1}b + \frac{n(n-1)}{1 \cdot 2}a^{n-2}b^2 + \cdots$$

a. Write the first four terms in the expansion of $(1 + x)^{1/2}$. $1 + \frac{1}{2}x - \frac{1}{8}x^2 + \frac{1}{16}x^3$

b. Use your answer in (a) with $x = 1$ to find a numerical estimate of $\sqrt{2}$. 1.4375

Calculator Key-In

1. Use the factorial key on your calculator to find 10!. 3,628,800
2. What is the largest positive integer n for which $n!$ can be displayed exactly (that is, every digit is shown) on your calculator? Answers will vary.
3. If your calculator displays numbers in scientific notation, you will be able to find approximate values of $n!$ when the exact integer value of $n!$ is too long to be displayed.
   a. Find 20! to four significant digits. 2.433×10^{18}
   b. Find the coefficient of the middle term of $(a + b)^{36}$ to four significant digits. 9.075×10^9
4. What is the largest positive integer n for which $n!$ can be displayed in scientific notation on your calculator? Answers will vary.

Self-Test 3

Vocabulary binomial expansion (p. 537) factorial notation (p. 540) Pascal's triangle (p. 537) binomial theorem (p. 541)

1. $x^5 - 10x^4 + 40x^3 - 80x^2 + 80x - 32$ 2. $256a^4 + 768a^3 + 864a^2 + 432a + 81$

Expand and simplify each expression.

1. $(x - 2)^5$ **2.** $(4a + 3)^4$ **Obj. 11-7, p. 537**

3. Write the first three terms in the expansion of $(2x + 3y)^{10}$. **Obj. 11-8, p. 540**
$1024x^{10} + 15{,}360x^9y + 103{,}680x^8y^2$

Find and simplify the specified term in each expansion.

4. The fourth term in $(a^2 - b)^7$ $-35a^8b^3$

5. The term containing x^8 in $(x - 2y)^{11}$ $-1320x^8y^3$

Check your answers with those at the back of the book.

Extended Algebra
542/1–13, 15–27 odd
S 535/*P*: 11, 12
R 543/Self-Test 3

Extended Alg. and Trig.
542/1–13, 15–27 odd
S 535/*P*: 11, 12
R 543/Self-Test 3

Supplementary Materials

Study Guide	pp. 185–186
Practice Master	63
Test Master	46
Computer Activity	26

Additional Answers
Written Exercises

13. a. $a^{14} + 14a^{13}b + 91a^{12}b^2 + 364a^{11}b^3$
b. $a^{14} - 14a^{13}b + 91a^{12}b^2 - 364a^{11}b^3$

14. a. $a^{12} + 12a^{11}b + 66a^{10}b^2 + 220a^9b^3$
b. $a^{12} - 12a^{11}b + 66a^{10}b^2 - 220a^9b^3$

15. a. $a^{19} + 19a^{18}b + 171a^{17}b^2 + 969a^{16}b^3$
b. $x^{19} - 38x^{18}y + 684x^{17}y^2 - 7752x^{16}y^3$

16. a. $a^{21} + 21a^{20}b + 210a^{19}b^2 + 1330a^{18}b^3$
b. $c^{21} - 21c^{20}d^2 + 210c^{19}d^4 - 1330c^{18}d^6$

25. False; $6! = 720 = (2 \cdot 3)! \neq 2!3! = 12$

26. False; $n = 2$, $(2^2)! = 4! = 24 \neq (2!)^2 = 4$

27. False; $(2 \cdot 4)! = 40{,}320 \neq 2^4(4!) = 384$

Calculator Key-In Commentary

In these exercises students are asked to use the factorial key on their calculators. Students should come to realize that factorials "grow" very fast.

Quick Quiz on p. 544

Quick Quiz

Expand and simplify.

1. $(y - 2)^5$ $y^5 - 10y^4 + 40y^3 - 80y^2 + 80y - 32$
2. $\left(2a + \frac{1}{2b}\right)^4$ $16a^4 + \frac{16a^3}{b} + \frac{6a^2}{b^2} + \frac{a}{b^3} + \frac{1}{16b^4}$
3. Which term of $(x + 1)^{19}$ has the same coefficient as the coefficient of the fifth term? the sixteenth term
4. Write the first four terms of $(x^2 - y)^{13}$. $x^{26} - 13x^{24}y + 78x^{22}y^2 - 286x^{20}y^3$

Find and simplify the specified term in the expansion.

5. The fourth term of $(x + 1)^{30}$ $4060x^{27}$
6. The term containing y^2 in $(3y - 1)^{16}$ $1080y^2$

Biographical Note / *Alice Hamilton*

Alice Hamilton (1869–1970), Harvard's first female professor, thought factory workers should have a safe environment in which to work. As a physician and specialist in industrial toxicology, she was dismayed by the high death rates in American industry in the early part of this century. Before laws protecting workers could be passed, someone had to gather data about the number of deaths and fight for better working conditions. Hamilton performed this service for the pottery, lead, and munitions industries.

In her zeal to find poisonous dusts, she explored mine shafts and climbed dangerous catwalks. Her commitment to preventing lead and benzene poisoning led to some of the first industrial disease laws in the United States.

Chapter Summary

1. Sequences can be *arithmetic, geometric,* or neither. If there is a *common difference d* between any two successive terms, then the sequence is arithmetic. If the *ratio* of successive terms is constant, then the sequence is geometric.
2. In any arithmetic sequence with first term t_1 and common difference d, the nth term is given by $t_n = t_1 + (n - 1)d$.

   In any geometric sequence with first term t_1 and common ratio r, the nth term is given by $t_n = t_1 \cdot r^{n-1}$.
3. The *arithmetic mean* of two numbers a and b is $\frac{a+b}{2}$. The *geometric mean* is $\sqrt{ab}$ (if $a, b > 0$) or $-\sqrt{ab}$ (if $a, b < 0$).
4. A *series* is an expression in which the terms of a sequence are added together. If the related sequence is arithmetic or geometric, then the series is classified respectively as arithmetic or geometric.

   The Greek letter Σ (sigma) can be used to represent the sum of a series in an abbreviated form. The upper and lower limits of summation are the first and last values of the *index*. If a series is infinite, then the symbol ∞ is used for the upper limit.

5. The sum of the first n terms of an *arithmetic series* is given by

$$S_n = \frac{n(t_1 + t_n)}{2}.$$

6. The sum of the first n terms of a *geometric series* with common ratio r is given by

$$S_n = \frac{t_1(1 - r^n)}{1 - r}, \quad r \neq 1.$$

7. An *infinite geometric series* with common ratio r has a sum if $|r| < 1$. This sum is

$$S = \frac{t_1}{1 - r}.$$

8. When $(a + b)^n$ is multiplied out, the resulting sum of terms is a *binomial expansion*. For any positive integer n, the coefficients of the expansion of $(a + b)^n$ are the elements of row n of Pascal's triangle. The *binomial theorem* can be used to expand $(a + b)^n$.

Chapter Review

Write the letter of the correct answer.

1. The sequence 56, 28, 14, 7, . . . is __?__. **11-1**
a. arithmetic **(b.)** geometric **c.** neither

2. Find the next term in the sequence 8, 5, 2, −1, ...
a. −2 **b.** −3 **(c.)** −4 **d.** −5

3. For the arithmetic sequence −4, −1.5, 1, 3.5, . . . , find t_{21}. **11-2**
a. 43.5 **(b.)** 46 **c.** 48.5 **d.** 51

4. Insert two arithmetic means between −2 and 13.
a. 1, 10 **b.** 2, 9 **(c.)** 3, 8 **d.** 4, 7

5. For the geometric sequence $\frac{81}{8}, \frac{27}{4}, \frac{9}{2}, 3, \ldots$, find t_9. **11-3**
a. $\frac{8}{9}$ **b.** $\frac{16}{27}$ **(c.)** $\frac{32}{81}$ **d.** $\frac{64}{243}$

6. Find the geometric mean of 10 and 40.
a. 15 **(b.)** 20 **c.** 25 **d.** 30

7. Write the series $\sum_{k=0}^{3} 2^{k-1}$ in expanded form. **11-4**
(a.) $\frac{1}{2} + 1 + 2 + 4$ **b.** $1 + 2 + 4 + 8$
c. $-2 + 0 + 2 + 4$ **d.** $0 + 2 + 4 + 6$

8. Write the series $47 + 41 + 35 + \cdots + 5$ using sigma notation.
a. $\sum_{n=0}^{8} (53 - 6n)$ **(b.)** $\sum_{n=1}^{8} (53 - 6n)$
c. $\sum_{n=1}^{7} (47 - 6n)$ **d.** $\sum_{n=0}^{8} (47 - 6n)$

9. Find the sum of the arithmetic series $\sum_{n=1}^{25} (3n + 2)$. 11-5

a. 984 b. 1000 (c.) 1025 d. 2050

10. Find the sum of the geometric series with $n = 7$, $r = -2$, and $t_1 = 5$.

a. -105 (b.) 215 c. 315 d. 635

11. Find the sum of the infinite geometric series $64 + 48 + 36 + 27 + \cdots$ if it has one. 11-6

a. $\frac{256}{7}$ b. $\frac{256}{3}$ (c.) 256 d. no sum

12. Expand and simplify $(x^2 - 2)^4$. 11-7

a. $x^8 + 8x^6 + 24x^4 + 32x^2 + 16$ b. $x^8 + 4x^6 + 6x^4 + 4x^2 + 1$

(c.) $x^8 - 8x^6 + 24x^4 - 32x^2 + 16$ d. $x^4 - 8x^3 + 24x^2 - 32x + 16$

13. Find the fourth term in the expansion of $(3x - y^2)^6$. 11-8

(a.) $-540x^3y^6$ b. $540x^3y^6$ c. $-540x^3y^3$ d. $-27x^3y^6$

Chapter Test

Alternate Test p. T23

Supplementary Materials

Test Masters 47, 48, 49
Resource Book p. 150

1. Tell whether the sequence $\frac{2}{3}, \frac{3}{5}, \frac{4}{7}, \frac{5}{9}, \ldots$ is arithmetic, geometric, or neither. Then find the next term in the sequence. Neither; $\frac{6}{11}$ 11-1
2. Find a formula for the nth term of the arithmetic sequence $-7, -3, 1, 5, \ldots$. $4n - 11$ 11-2
3. Insert three arithmetic means between 1 and 47. 12.5, 24, 35.5
4. Find the eighth term of the geometric sequence $11, -22, 44, -88, \ldots$. -1408 11-3
5. Find the geometric mean of $\frac{3}{2}$ and $\frac{8}{3}$. 2
6. Write the series $\frac{1}{2} - \frac{1}{4} + \frac{1}{6} - \frac{1}{8} + \cdots$ using sigma notation. $\sum_{n=1}^{\infty} \frac{(-1)^{n-1}}{2n}$ 11-4
7. Find the sum of the first 20 terms of the arithmetic series $53 + 46 + 39 + 32 + \cdots$. -270 11-5
8. Find the sum of the geometric series $\sum_{n=1}^{5} 2 \cdot 3^{n-1}$. 242
9. Find the sum of the infinite geometric series $250 + 150 + 90 + 54 + \cdots$ if it has one. 625 11-6
10. Expand and simplify $(a^2 + b)^8$. (See below.) 11-7
11. Find the 12th term in the expansion of $(2x - y)^{13}$. $-312x^2y^{11}$ 11-8

10. $a^{16} + 8a^{14}b + 28a^{12}b^2 + 56a^{10}b^3 + 70a^8b^4 + 56a^6b^5 + 28a^4b^6 + 8a^2b^7 + b^8$

Cumulative Review *(Chapters 8–11)*

1. Suppose z varies directly as x and inversely as the square of y. If $z = 2$ when $x = 36$ and $y = 3$, find z when $x = 24$ and $y = 2$. 3
2. Divide $8x^3 - 12x + 11$ by $2x + 3$. $4x^2 - 6x + 3 + \frac{2}{2x+3}$
3. Solve $x^4 - 2x^3 + 4x^2 + 30x - 13 = 0$ completely given that $2 + 3i$ is a root. $\{2 \pm 3i, -1 \pm \sqrt{2}\}$
4. Find the center and radius of the circle $x^2 + y^2 - 6x + 8y + 21 = 0$. $C(3, -4)$; $r = 2$
5. Find and graph an equation of the parabola with vertex (4, 0) and directrix $x = 0$. $x = \frac{1}{16}y^2 + 4$
6. Graph the hyperbola $\frac{x^2}{16} - \frac{y^2}{4} = 1$ and find the coordinates of its foci. $(2\sqrt{5}, 0)$ and $(-2\sqrt{5}, 0)$

Solve each system of equations.

7. $2x^2 + y^2 = 6$, $x - y = 3$ (1, −2)
8. $y = x^2 - 1$, $y^2 - x^2 = 5$ (2, 3) (−2, 3)
9. $x - y - z = 5$, $2x - y + z = 3$, $x + 2y - 2z = 4$ (2, −1, −2)
10. $5x - 3y + z = -6$, $x + 3y - z = 0$, $4x + y + z = 7$ (−1, 3, 8)
11. Evaluate $\frac{64^{2/3}}{64^{-1/2}}$. 128
12. If $f(x) = x^2 + 2$ and $g(x) = 2x - 3$, find the following:
    a. $f(g(1))$ 3 **b.** $g(f(1))$ 3

Solve.

13. $9^{x+2} = 27$ $-\frac{1}{2}$
14. $4^{x-2} = \frac{1}{8}$ $\frac{1}{2}$
15. $\log_9 x = 1.5$ 27
16. $\log_x 16 = -2$ $\frac{1}{4}$
17. Give the solution of $5^x = 40$ to three significant digits. 2.29
18. Evaluate $e^{2 \ln 3}$. 9
19. Tell whether the sequence 108, −72, 48, −32, . . . is arithmetic, geometric, or neither. Then find a formula for the nth term. Geometric; $t_n = 108\left(-\frac{2}{3}\right)^{n-1}$
20. Tell whether the series $\sum_{n=1}^{11} (35 - 3n)$ is arithmetic, geometric, or neither. Then find its sum. Arithmetic; 187
21. Expand and simplify $(x - y^2)^4$. $x^4 - 4x^3y^2 + 6x^2y^4 - 4xy^6 + y^8$
22. Find and simplify the twelfth term in the expansion of $(2x + y)^{13}$. $312x^2y^{11}$

Additional Answers Cumulative Review

5.
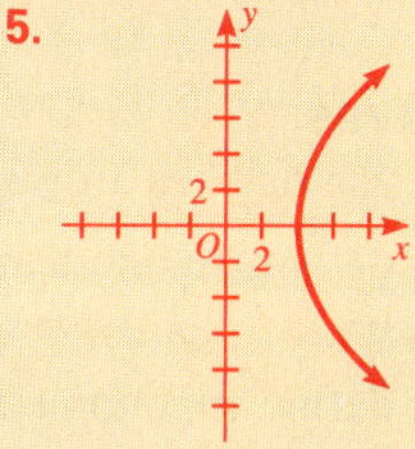

6.
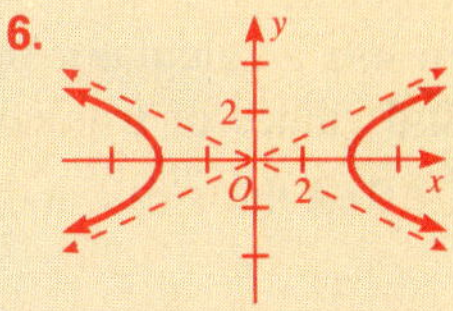

12 Triangle Trigonometry

Objectives

12-1 To use degrees to measure angles.

12-2 To define trigonometric functions of acute angles.

12-3 To define trigonometric functions of general angles.

12-4 To use a calculator or trigonometric tables to find values of trigonometric functions.

12-5 To find the sides and angles of a right triangle.

12-6 To use the law of cosines to find sides and angles of triangles.

12-7 To use the law of sines to find sides and angles of triangles.

12-8 To solve any given triangle.

12-9 To apply triangle area formulas.

Assignment Guide

See p. T58 for Key to the format of the Assignment Guide

Day	Average Algebra	Average Algebra and Trigonometry	Extended Algebra	Extended Algebra and Trigonometry
1		**12-1** 552/3–66 mult. of 3 **S** 554/*Mixed Review*		*Administer Chapter 11 Test* **12-1** *Read 12-1* 552/3–66 mult. of 3 **S** 554/*Mixed Review*
2		**12-2** 559/1–31 odd **S** 553/47, 49, 50, 52, 53		**12-2** 559/1–25 odd, 26–31, 33, 34
3		**12-3** 566/1–36 **S** 559/8–26 even		**12-3** 566/2–76 even **S** 567/*Mixed Review*
4		**12-3** 566/37–67 **S** 567/*Mixed Review*		**12-4** 572/2–46 even **R** 573/*Self-Test 1*
5		**12-4** 572/2–46 even **S** 567/68–76 even		**12-5** 577/3–21 mult. of 3 578/*P*: 2–16 even **S** 579/*Mixed Review*
6		**12-5** 577/1–19 odd 578/*P*: 1, 7 **R** 573/*Self-Test 1*		**12-6** 582/2–18 even 583/*P*: 2, 7, 8, 10 **S** 579/*P*: 15
7		**12-5** 578/*P*: 2–14 even **S** 579/*Mixed Review*		**12-7** 588/1–19 odd, 20 589/*P*: 2, 6, 8–11 **S** 590/*Mixed Review*
8		**12-6** 582/1–17 odd 583/*P*: 2, 7 **S** 578/*P*: 3, 5		**12-8** 594/1–19 odd **S** 584/*P*: 9 **S** 590/*P*: 13
9		**12-6** 583/*P*: 1, 3, 8, 10 **12-7** 588/2–12 even		**12-8** 595/*P*: 1, 2, 6–10, 14 **S** 577/20, 22, 23

Assignment Guide (continued)

Day	Average Algebra	Average Algebra and Trigonometry	Extended Algebra	Extended Algebra and Trigonometry
10		**12-7** 588/9–19 odd, 20 589/*P*: 2, 3, 5, 6, 8, 9 **S** 590/*Mixed Review*		**12-9** 599/1–16 **S** 600/*Mixed Review* **R** 600/*Self-Test 2*
11		**12-8** 594/1–8 595/*P*: 1–3 **S** 584/*P*: 9		*Prepare for Chapter Test* **R** 602/*Chapter Review*
12		**12-8** 594/9–13, 15 595/*P*: 6, 8, 10, 14 **S** 590/*P*: 7, 10, 11		*Administer Chapter 12 Test*
13		**12-9** 599/1–16 **S** 600/*Mixed Review* **R** 600/*Self-Test 2*		
14		*Prepare for Chapter Test* **R** 602/*Chapter Review*		
15		*Administer Chapter 12 Test*		

Supplementary Materials Guide

For Use with Lesson	Practice Masters	Tests	Study Guide (Reteaching)	Resource Book		
				Tests	Practice Exercises	Applications (A) Enrichment (E)
12-1			pp. 187–188		p. 151	p. 215 (A)
12-2	Sheet 64		pp. 189–190		p. 152	
12-3	Sheet 65		pp. 191–192			
12-4	Sheet 66	Test 50	pp. 193–194			
12-5	Sheet 67		pp. 195–196			
12-6	Sheet 68		pp. 197–198			
12-7	Sheet 69		pp. 199–200		p. 153	
12-8	Sheet 70		pp. 201–202		p. 154	
12-9	Sheet 71	Test 51	pp. 203–204			
Chapter 12		Tests 52, 53		pp. 65–68	p. 155	p. 231 (E)
Cum. Rev. 9–12	Sheet 72				pp. 156–157	

Overhead Visuals

For Use with Lessons	Visual	Title
12-1, 12-2, 12-3	10	Angles

Software

Software	Computer Activities	Test Generator
	Activities 27, 28	189 test items
For Use with Lessons	12-5, 12-8	all lessons

Strategies for Teaching

Communication and Using Manipulatives

Students often have difficulty in translating English phrases and sentences into mathematical expressions and sentences. Not reading word problems carefully, with concentration on their meaning, is another frequent cause of difficulty. Students may make mistakes because they do not fully understand the meanings of mathematical terms or symbols.

Manipulatives and models are very useful in this chapter for students to experiment with rotating angles, the unit circle, the trigonometric functions sine, cosine, and tangent, and the SSA ambiguous case.

12-1 Angles and Degree Measure

Display Overhead Visual 10, Sheets 1 and 3. Align the green rays with the positive x-axis.

1. Rotate one green ray to many varied positions, including all quadrants and both axes. Ask students to identify the quadrant or axis. Ask students to estimate the angle measure.
2. Rotate a green ray counterclockwise to a particular position, such as in the second quadrant. Ask students to describe other rotations, both clockwise and counterclockwise, that move the other green ray to the same terminal ray.
3. Give a positive and a negative angle that are coterminal with the given angle. Answers may vary. Samples are given. a. 150° 510°, –210° b. –100° 260°, –460°

Project onto a chalkboard so that additional marks can be made.

4. Align the green rays with the positive x-axis. Rotate a green ray to the 30° position. Drop a perpendicular from the intersection of the green ray and the circle to the x-axis. If the radius of the circle is 1, what are the lengths of the legs of the triangle? $\frac{1}{2}$ and $\frac{\sqrt{3}}{2}$
5. Use the lengths from Question 4. What are the sine, cosine, and tangent of 30°? $\frac{1}{2}, \frac{\sqrt{3}}{2}, \frac{\sqrt{3}}{3}$

   This kind of derivation can be repeated for other special angles.

Align the green ray at the positive x-axis.

6. As you rotate a green ray to various angles, ask students to state the sign of the sine, cosine, and tangent at that position.

12-2 Trigonometric Functions of Acute Angles

Have students investigate the linguistic origins of the terms sine, tangent, and secant.

See the Exploration on page 843 for an activity in which students explore trigonometric ratios.

12-4 Values of Trigonometric Functions

Although mathematics is a discipline in which precision and clarity are among the highest values, it is nevertheless true that we sometimes find ourselves using the same symbol for totally different mathematical entities. For example, we use the symbol "–" for at least three distinct concepts.

1. opposite, as in $-a$
2. a negative number, as in –5.3
3. subtraction, as in $3 - 7 = -4$

The same is true of the raised $^{-1}$ symbol. Students have encountered this earlier as a negative exponent representing the reciprocal of a number. Now in Example 3 they are introduced to a new use of the same symbol, namely the inverse trigonometric functions. It is important to realize that this is not a reciprocal function at all.

12-8 Solving General Triangles

Students will understand the SSA ambiguous case better after the hands-on activity described here.

Draw an acute angle such that one side is a ray and the other side is a segment. We have a fixed angle and a fixed side, but the other side is not determined.

Distribute cardboard or heavy paper strips and thumb tacks. Push the tack, point up, through the paper at the end of the segment not at the vertex. Press the end of a strip over the tack, using a pencil eraser to avoid injury. Pivot the strip until its end touches the ray of the fixed angle. Now use several longer or shorter strips, finding strips that hit the ray no times, once, or twice. In each case where the strip does touch the ray, students can see the SSA case—the strip (S), the drawn segment (S), and the drawn angle (A).

This apparatus also makes a very effective SSA demonstration for an overhead projector.

References to Strategies

PE: Pupil's Edition **TE:** Teacher's Edition **RB:** Resource Book

Problem Solving Strategies

PE: pp. 576, 585 (make a sketch)
TE: p. 556; p. T126 (Sugg. Extension, Lesson 12-8)

Applications

PE: p. 548 (astronomy); pp. 575, 578–579, 589–590, 595–597, 601 (word problems)
TE: pp. 548, 549, 568, 575, 587, 591, 594, T125
RB: p. 215

Nonroutine Problems

PE: p. 560 (Exs. 30–34); pp. 577–578 (Exs. 20–24); p. 579 (Probs. 15–17); p. 583 (Exs. 18, 19); p. 584 (Prob. 12); p. 590 (Prob. 13); p. 596 (Probs. 8–14); p. 601 (Challenge)
TE: pp. T122, T125–T127 (Lessons 12-1, 12-6, 12-7, 12-9)

Communication

PE: p. 600 (convincing argument); p. 585 (Reading Algebra)
TE: pp. T123–T124 (Reading Algebra); p. T126 (Sugg. Extension, Lesson 12-7)

Thinking Skills

PE: p. 560 (Exs. 30–31, proofs)
TE: p. 580

Explorations

PE: p. 601 (Challenge); p. 843 (trigonometric ratios)

Connections

PE: p. 548 (Astronomy); pp. 555–572, 578–579, 583–584, 589–590, 595–596, 599 (Trigonometry); pp. 563, 586, 591–600 (Geometry); p. 567 (Engineering); pp. 573, 601 (History); p. 579 (Physics)

Using Technology

PE: p. 550; p. 554 (Calculator Key-In); pp. 568–572, 575, 577–583, 584, 588–590, 594–596, 599–600 (Exs.)
TE: pp. 554, 576, 581, 584, 588, 595, 598, T124
Computer Activities: pp. 71–76

Using Manipulatives/Models

TE: p. T126 (Lesson 12-8)
Overhead Visuals: 10

Cooperative Learning

TE: p. T124 (Lesson 12-4)

Teaching Resources

For use in implementing the teaching strategies referenced on the previous page.

Application
Resource Book, p. 215

Application—Using Trigonometry to Determine Height
(For use with Chapter 12)

Right triangle trigonometry and a homemade instrument for measuring angles can be used to find the heights of trees, flagpoles, antennas, buildings, and other objects whose heights are difficult to measure directly.

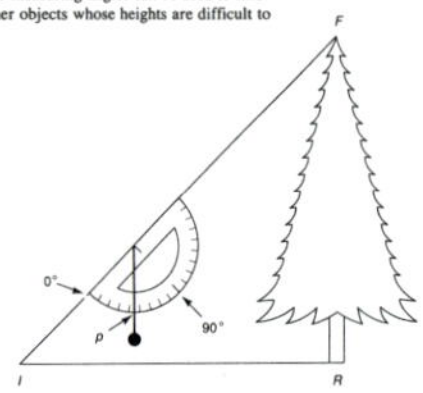

1. A large protractor (chalkboard size) has a weight attached to a string that hangs from the center of the protractor as shown. Suppose that when you sight along the straight base of the protractor to the top of a tree, the string crosses the protractor at p.
   a. What is the relationship between the value p obtained from the protractor and $\angle FIR$, the angle of elevation to the top of the tree? ____________
   b. What assumption must you make about $\angle FRI$ if the height of the tree is to be found using the equation $\tan FIR = \frac{FR}{IR}$? ____________
   c. Suppose $p = 50°$ and $IR = 31$ ft. To the nearest foot, how tall is the tree? ____________

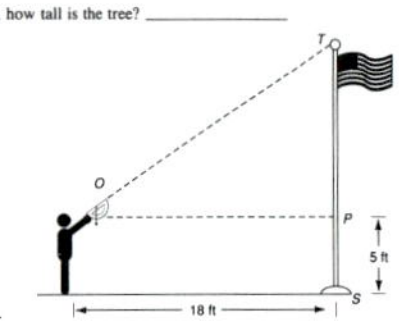

2. Both Sue and Bob calculated the height of a flagpole by standing 18 ft from the base of the pole and measuring the angle of elevation to the top. In each case, their eyes were 5 ft above the ground. Since it is difficult to measure angles exactly using the homemade weighted protractor, Sue measured $\angle TOP$ as 65°, while Bob measured it as 62°. To the nearest foot, find the height of the flagpole (TS) using
   a. Sue's measurement: ____________
   b. Bob's measurement: ____________
   c. By how much do the two heights differ? ____________

3. Working with a partner, select an object whose height is to be determined, such as a nearby tree or flagpole. You and your partner should use a weighted protractor as described in Exercise 1 to measure the angle of elevation.
   a. Standing at various distances from the object, measure the angle of elevation and determine the object's height to the nearest foot. Use the table at the right to record your results.
   b. Based on the results of part (a), what do you think is a good approximation of the object's height? Explain. ____________

Distance from object	*Angle of elevation*	*Approximate height of object*
______	______	______
______	______	______
______	______	______
______	______	______
______	______	______

Resource Book, ALGEBRA, Structure and Method, Book 2

APPLICATIONS 215

Enrichment/Connection
Resource Book, p. 231

Testing a Formula (For use with Chapter 12)

If a convex quadrilateral has sides a, b, c, and d units long, then the following formula is said to give the area of the quadrilateral.

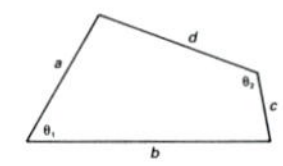

$A = \sqrt{(s-a)(s-b)(s-c)(s-d) - abcd \cdot \cos^2 \theta}$

where $s = \frac{1}{2}(a+b+c+d)$ and $\theta = \frac{1}{2}(\theta_1 + \theta_2)$. The angles θ_1 and θ_2 are any two opposite angles of the quadrilateral.

For each figure, find the area (a) by using the formula and (b) by using another technique.

1. The quadrilateral is a square with sides 8.
   a. ____________ b. ____________
2. The quadrilateral is a rectangle with sides 8 and 10.
   a. ____________ b. ____________
3. The quadrilateral is a parallelogram with sides 6 and 15 and an acute angle of 60°.
   a. ____________ b. ____________
4. The quadrilateral is a rhombus with sides 6 and acute angle 45°.
   a. ____________ b. ____________
5. The quadrilateral is a trapezoid with sides 10, 10, 10, and 20.
   a. ____________ b. ____________
6. The quadrilateral has side $a = 0$, and the other sides are all 6 units long. In other words, the "quadrilateral" is a triangle.
   a. ____________ b. ____________
7. The quadrilateral is composed of two right triangles as shown in the figure.

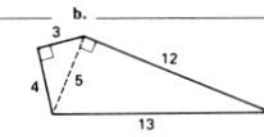

   a. ____________ b. ____________

Resource Book, ALGEBRA, Structure and Method, Book 2

ENRICHMENT 231

Communication
Study Guide, p. 187

12 Triangle Trigonometry

12–1 Angles and Degree Measure

Objective: To use degrees to measure angles.

Vocabulary

Degree A measure of rotation. One complete revolution measures 360 degrees.

Angle A figure formed by two rays that have the same endpoint.

Directed angle An angle generated by the rotation of a ray (the *initial side*) onto another ray (the *terminal side*).

Positive angle An angle generated by a counterclockwise rotation.

Negative angle An angle generated by a clockwise rotation.

Standard position An angle is in standard position when its initial side coincides with the positive x-axis.

Quadrantal angle An angle whose terminal side lies on a coordinate axis.

Coterminal angles Two angles whose terminal sides coincide when the angles are in standard position.

Symbols $1°$ (1 degree $= \frac{1}{360}$ revolution) $1'$ (1 minute $= \frac{1}{60}°$) $1''$ (1 second $= \frac{1}{3600}°$)

Example 1 a. Sketch $-315°$ in standard position. Indicate its rotation by a curved arrow and classify the angle by the quadrant containing its terminal side.
b. Sketch $\frac{3}{4}$ of a counterclockwise revolution and find its measure.

Solution a.

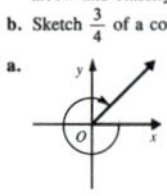

Quadrant I

b.

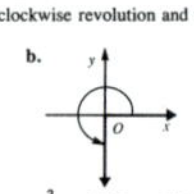

$\frac{3}{4} \times 360° = 270°$

Sketch each angle in standard position. Indicate its rotation by a curved arrow. Classify each angle by its quadrant. If the angle is a quadrantal angle, say so.

1. 60°	**2.** −60°	**3.** 120°	**4.** −120°	**5.** 200°
6. −200°	**7.** 330°	**8.** −390°	**9.** 450°	**10.** −820°

Sketch in standard position the angle described and find its measure.

11. $\frac{1}{6}$ of a counterclockwise revolution
12. $\frac{2}{3}$ of a clockwise revolution
13. $\frac{3}{4}$ of a counterclockwise revolution
14. $1\frac{1}{8}$ of a clockwise revolution

Study Guide, ALGEBRA AND TRIGONOMETRY, Structure and Method, Book 2

187

Communication
Study Guide, p. 188

12–1 Angles and Degree Measure (continued)

Example 2 a. Write a formula for the measure of all angles coterminal with a 60° angle.
b. Use the formula to find two positive angles and two negative angles that are coterminal with a 60° angle.

Solution a. $60° + n \cdot 360°$, where n is an integer.
b. To find the measure of the angles, let $n = 1, 2, -1, -2$.
$60° + 1(360°) = 420°$ $60° + 2(360°) = 780°$
$60° + (-1)(360°) = -300°$ $60° + (-2)(360°) = -660°$

For Exercises 15–22: (a) Write a formula for the measures of all angles coterminal with the given angle. (b) Use the formula to find two angles (one positive and one negative) that are coterminal with the given angle.

15. 40°	**16.** 120°	**17.** −60°	**18.** −235°
19. 450°	**20.** 210°	**21.** 900°	**22.** −720°

Example 3 a. Express 28° 12′ 38″ in decimal degrees.
b. Express 48.21° in degrees, minutes, and seconds to the nearest second.

Solution Use these facts: $1' = \left(\frac{1}{60}\right)°$ and $1'' = \left(\frac{1}{60}\right)' = \left(\frac{1}{3600}\right)°$

a. $28° 12' 38'' = 28° + \left(\frac{12}{60}\right)° + \left(\frac{38}{3600}\right)° = 28° + 0.2° + 0.011° = 28.211°$

b. $48.21° = 48° + (0.21 \times 60)' = 48° + 12.6' = 48° + 12' + (0.6 \times 60)''$
$= 48° + 12' + 36'' = 48° 12' 36''$

Express in decimal degrees to the nearest hundredth of a degree.

23. 16° 40′	**24.** 49° 27′	**25.** 83° 52′	**26.** 28° 45′

Express in degrees, minutes, and seconds to the nearest second.

27. 26.52°	**28.** 39.45°	**29.** 85.39°	**30.** 55.73°

Mixed Review Exercises

Find the third term in the expansion of each binomial.

1. $(x - 3y)^4$ **2.** $(x + \sqrt{2}y)^6$ **3.** $(x^3 + y^2)^{10}$

Simplify.

4. $\sqrt{-5} \cdot \sqrt{-20}$ **5.** $\log_2 28 - \log_2 7$ **6.** $(9^{3/4} - 9^{5/4})^2$
7. $\ln \sqrt{e}$ **8.** $(5 + 2i)(3 - 4i)$ **9.** $\sqrt{27} - \sqrt{12} + \sqrt{48}$

188
Study Guide, ALGEBRA AND TRIGONOMETRY, Structure and Method, Book 2

Using Technology/Exploration
Computer Activities, p. 71

ACTIVITY 27. ***Areas of Polygons and Circles*** *(for use with Section 12-5)*

Directions: Write all answers in the spaces provided.

PROBLEM

Use the computer to investigate the areas of regular polygons inscribed in a unit circle.

PROGRAM

```
10  LET R = 1
20  LET P1 = 3.14159
30  LET C = P1 * R ↑ 2
40  PRINT "START WITH HOW MANY SIDES";
50  INPUT S
60  PRINT
70  PRINT "NO. OF", "AREA OF", "AREA OF"
80  PRINT "SIDES", "POLYGON", "CIRCLE"
90  PRINT
100 FOR N = S TO S + 2
110 LET A = (360 / N) * P1 / 180
120 LET T = .5 * R * R * SIN(A)
130 LET P = N * T
140 LET P = INT(1000 * P + .5) / 1000
150 PRINT N, P, C
160 NEXT N
170 END
```

PROGRAM CHECK

Type in the program. To test whether you entered it correctly, run the program. After the question, enter 3. The computer should print:

NO. OF SIDES	AREA OF POLYGON	AREA OF CIRCLE
3	1.299	3.14159
4	2	3.14159
5	2.378	3.14159

(continued)

71

Using Technology/Exploration
Computer Activities, p. 72

(Activity 27 continued)

USING THE PROGRAM

1. State whether the area of the inscribed polygon increases or decreases as the number of sides increases. ________

2. Change line 100 to

```
100 FOR N = S TO S + 10
```

and run the program again starting with 3 sides. Use the output to calculate, for the given number of sides, the difference between the area of the circle and the area of the polygon. Round off your answer to the nearest thousandth.

Number of sides	*Difference*
4	________
8	________
12	________

3. Run the program again starting with 16 sides. What is the least number of sides for which the tenths' place digit of the area of the polygon is the same as the tenths' place digit of the area of the circle? ________

4. Use the program to find the least number of sides of the inscribed polygon for which the hundredths' place digit of the areas of the polygon and the circle is the same. ________

EXTENSION

Modify lines 110, 120, and 130 so that the program will compute the area of regular polygons *circumscribed* about a unit circle.

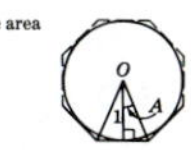


```
• 110 LET A = 360 / (2 * N) * P1 / 180
  120 LET T = .5 * R * R * TAN(A)
  130 LET P = 2 * N * T
```

Run the program starting with 3 sides.

1. State whether the area of the circumscribed polygon increases or decreases as the number of sides increases. ________

2. Use the output to calculate, for the given number of sides, the difference between the area of the circle and the area of the circumscribed polygon. Round off your answer to the nearest thousandth.

Number of sides	*Difference*
4	________
8	________
12	________

(continued)

72

Using Models
Overhead Visual 10, Sheets 1 and 2

ANGLES

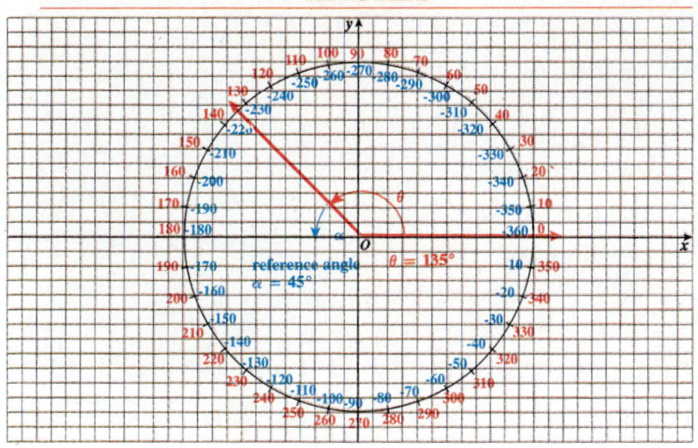

VISUAL 10

Application

When a star is viewed from Earth and then viewed again six months later (so that Earth is at opposite points in its orbit), the star shifts its apparent position against the background of more distant stars. This effect, called *parallax,* is the basis for determining the distance of the star from Earth.

Astronomers also use parallax to find the height of a meteor before it disintegrates in Earth's atmosphere. Two cameras are positioned at least three miles apart and aimed directly overhead. With coordinated exposures of 10–15 minutes on both cameras, the same meteor appears on both photographs but the star background will be shifted. This parallax can then be used to determine the meteor's height.

Research Activities

Students might want to look up the dates of the major annual meteor showers and participate in a local astronomy group's observations. Students who are particularly interested in measuring the heights of meteors can send a stamped, self-addressed envelope to one of the organizations whose addresses are given at the end of Philip Bagnall's article, "Dark Skies for the Perseid Meteor Shower," in the August, 1988 issue of *Astronomy* magazine.

References

Hoff, Darrell B.; Linda J. Kelsey; and John S. Neff. "Height of a Meteor." In *Activities in Astronomy* (2nd ed.), 66–68. Dubuque, IA: Kendall/Hunt Publishing Company, 1984.

12 Triangle Trigonometry

By finding angle measurements between our sun and a star on different occasions, an astronomer can use trigonometry to determine how far the star is from Earth.

548

Trigonometric Functions

12-1 Angles and Degree Measure

Objective To use degrees to measure angles.

Trigonometry, which means "triangle measurement," is a branch of mathematics that was originally used for surveying. Today trigonometry is used in astronomy, navigation, architecture, and many other fields where measurement is important.

In this chapter angle measurements will be given in *degrees*. Recall that one **degree** ($1°$) is $\frac{1}{360}$ of a complete revolution. In geometry most angles you work with have measures between $0°$ and $180°$. In trigonometry you will also use angles with measures greater than $180°$ and less than $0°$.

You can generate any angle by using a rotation that moves one of the rays forming the angle, called the **initial side,** onto the other ray, called the **terminal side.** As shown in the figure below, *counterclockwise* rotations produce **positive angles** and *clockwise* rotations produce **negative angles.** An angle generated in this manner is called a **directed angle.**

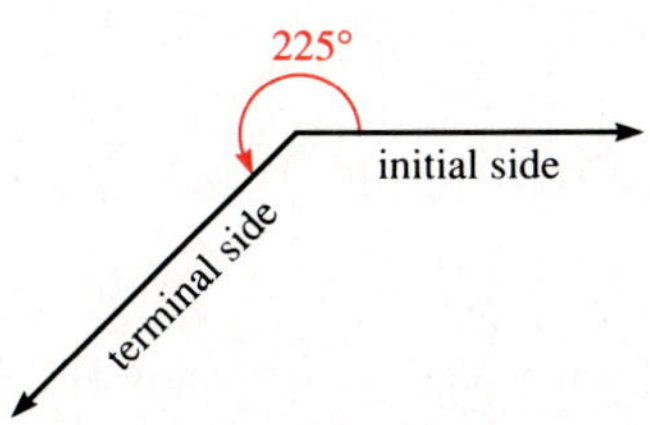

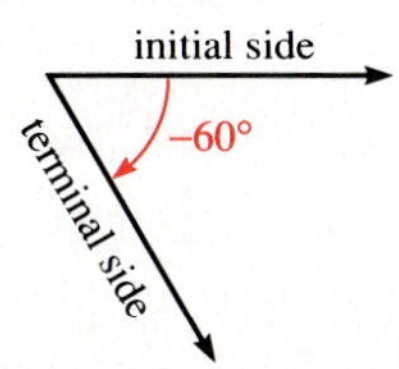

An angle is in **standard position** when its initial side coincides with the positive x-axis. As shown in the diagrams below, angles in standard position can be classified according to the quadrant in which their terminal side lies. If the terminal side lies on a coordinate axis, the angle is called a **quadrantal angle.**

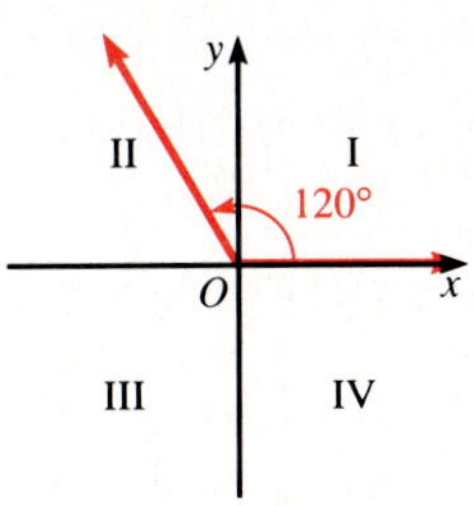

Second-quadrant angle

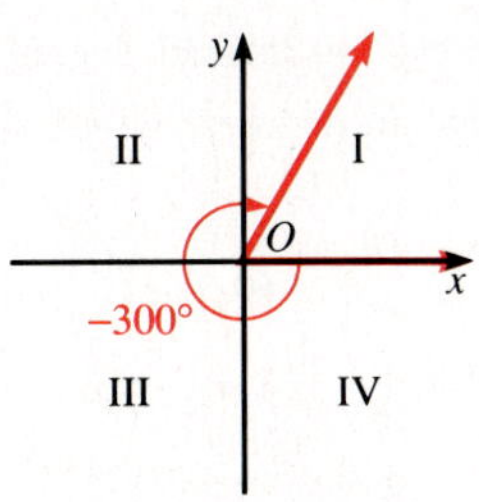

First-quadrant angle

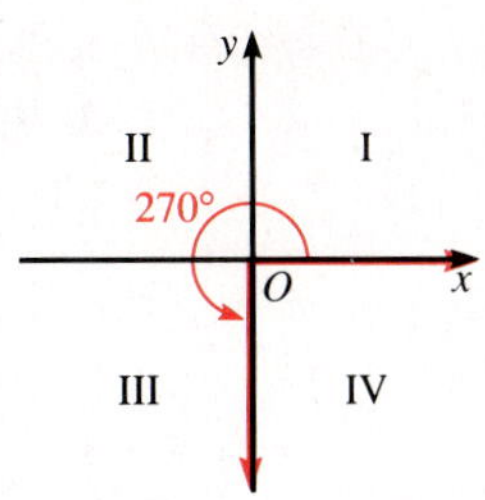

Quadrantal angle

Teaching References

Lesson Commentary, pp. T122–T127

Assignment Guide, pp. T68–T70

Supplementary Materials
- Practice Masters 67–72
- Tests 50–53
- Resource Book
  - Practice Exercises, pp. 151–155
  - Tests, pp. 65–68
  - Enrichment Activity, p. 231
  - Application, p. 215
- Study Guide, pp. 187–204
- Computer Activities 27–28
- Test Generator

Alternate Test, p. T24

Teaching Suggestions p. T122

Suggested Extensions p. T122

Warm-Up Exercises

Find the measure of the angle formed by the hands of a clock at the given time.

1. 6:00 180°
2. 9:00 90°
3. 2:00 60°
4. 11:00 30°
5. 4:30 45°

Motivating the Lesson

As the Earth rotates, a given point on the surface passes through an angle of 15° each hour. To measure rotations over more than 12 hours requires the use of angles greater than 180°. Angles of any measure are the topic of today's lesson.

Chalkboard Examples

1. Give a formula for the measures of all angles coterminal with a 70° angle. Use the formula to find measures of four specific angles coterminal with a 70° angle.
   Use the formula $70° + n \cdot 360°$, where n is an integer.
   Let $n = 1, 2, -1$, and -2.
   $70° + (1)360° = 430°$
   $70° + (2)360° = 790°$
   $70° + (-1)360° = -290°$
   $70° + (-2)360° = -650°$
2. **a.** Express 64°9′ in decimal degrees.
   b. Express 53.721° in degrees, minutes, and seconds.
   a. $1' = \left(\frac{1}{60}\right)^{\circ}$
   $64°9' = 64° + \left(\frac{9}{60}\right)^{\circ} = 64.15°$
   b. $53.721° = 53° + (0.721 \times 60)' = 53° + 43.26' = 53° + 43' + (0.26 \times 60)'' = 53° + 43' + 16'' = 53°43'16''$

Check for Understanding

Here is a suggested use of the Oral Exercises to check students' understanding as you teach the lesson.
Oral Exs. 1–18: use after discussing page 549.
Oral Exs. 19–26: use after Example 1.

Two angles in standard position are **coterminal** if their terminal sides coincide. The diagrams below show several angles coterminal with a 150° angle.

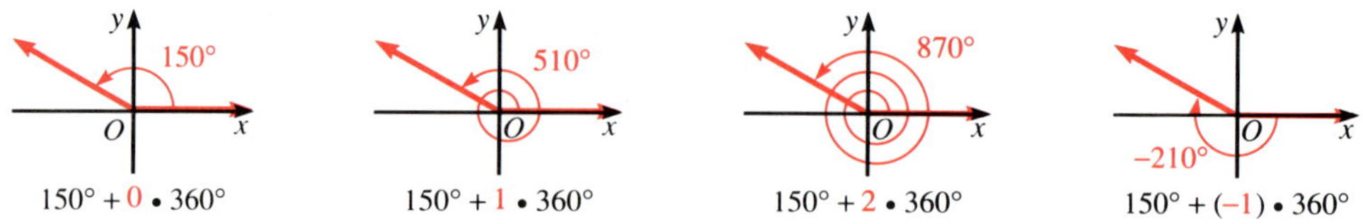

The measures of all angles coterminal with a 150° angle are given by the expression

$$150° + n \cdot 360°, \text{ where } n \text{ is an integer.}$$

In the diagrams above, n has the values 0, 1, 2, and -1, respectively.

Example 1 **a.** Write a formula for the measures of all angles coterminal with a 30° angle.
b. Use the formula to find two positive angles and two negative angles that are coterminal with a 30° angle.

Solution **a.** $30° + n \cdot 360°$, where n is an integer.
b. To find the measures of the angles, let n equal 1, 2, -1, and -2.

$30° + (1)360° = 390°$
$30° + (2)360° = 750°$
$30° + (-1)360° = -330°$
$30° + (-2)360° = -690°$

To measure angles more precisely than to the nearest degree, you can use either decimal degrees or degrees, minutes, and seconds. One **minute** (1′) is $\frac{1}{60}$ of 1°, and one **second** (1″) is $\frac{1}{60}$ of 1′.

Many calculators will convert decimal degrees to degrees, minutes, and seconds, and vice versa. (See the Calculator Key-In on page 554.) If you do not use a calculator, you can use the methods shown in Example 2.

Example 2 **a.** Express 14°36′54″ in decimal degrees.
b. Express 72.568° in degrees, minutes, and seconds to the nearest second.

Solution **a.** Use these facts: $1' = \left(\frac{1}{60}\right)^{\circ}$ and $1'' = \left(\frac{1}{60}\right)' = \left(\frac{1}{3600}\right)^{\circ}$.

$$14°36'54'' = 14° + \left(\frac{36}{60}\right)^{\circ} + \left(\frac{54}{3600}\right)^{\circ}$$
$$= 14° + 0.6° + 0.015°$$
$$= 14.615° \quad \textit{Answer}$$

b. $72.568° = 72° + (0.568 \times 60)'$
$= 72° + 34.08'$
$= 72° + 34' + (0.08 \times 60)''$
$= 72° + 34' + 4.8''$
$\approx 72°34'5''$
Answer

We often use capital letters, as in $\angle A$, or Greek letters such as θ (theta), ϕ (phi), and α (alpha) to name angles. (A list of Greek letters used in this book appears on page xvi.) To denote the measure of an angle, we may write, for example, $\theta = 20°$, $\angle A = 20°$, $m\angle A = 20°$, or $\angle A = 20$. In this book we will use $\theta = 20°$ or $\angle A = 20°$ to denote an angle with a measure of 20°. Also, when we say that 120° is a second-quadrant angle, we mean that a 120° *angle* in standard position has its terminal side in the second quadrant.

Oral Exercises

1. The measure of an angle in quadrant I is between 0° and 90°. What can you say about the measure of an angle in quadrant II? in quadrant III? in quadrant IV? between 90° and 180°, between 180° and 270°, between 270° and 360°
2. Give the measures of the quadrantal angles from 0° to 360°, inclusive. 0°, 90°, 180°, 270°, 360°

Estimate the measure of angle θ.

3. 130°

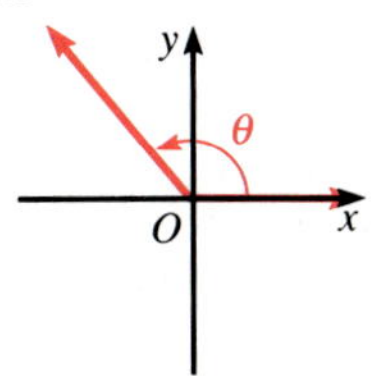

4. 330°

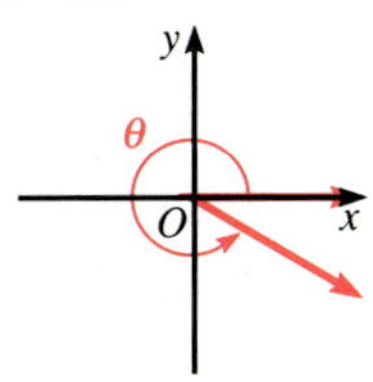

5. 240°

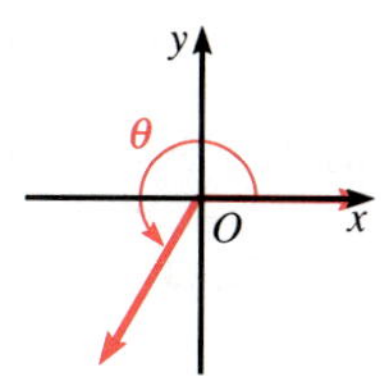

6. −315°

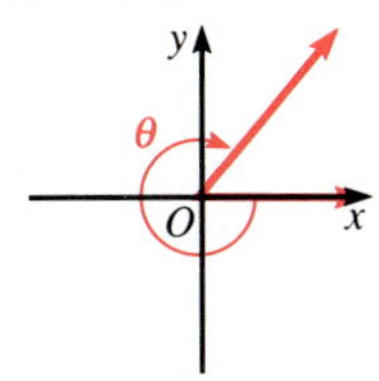

7. −135°

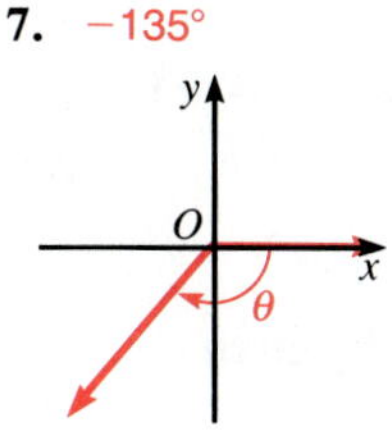

8. 480°

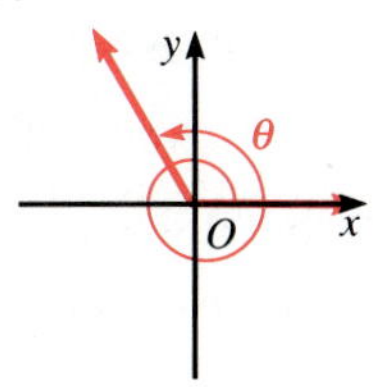

9. 380°

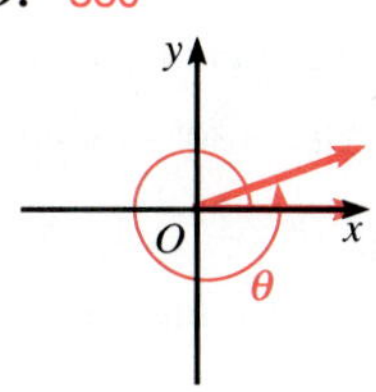

10. −30°

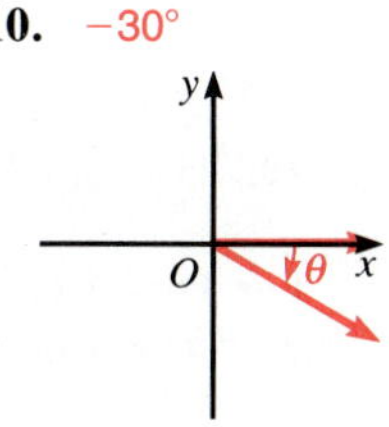

Name the quadrant of each angle.

11. 140° II **12.** 215° III **13.** −45° IV **14.** −150° III
15. −315° I **16.** 440° I **17.** −400° IV **18.** 500° II

Guided Practice

Sketch each angle in standard position. Indicate its rotation by a curved arrow. Classify the angle by its quadrant. If the angle is a quadrantal angle, say so.

1. 210° III

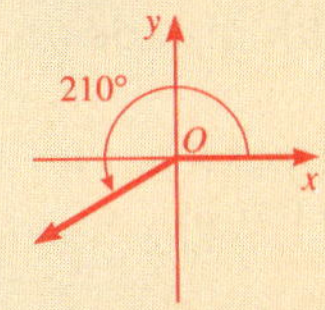

2. −180° Quad.

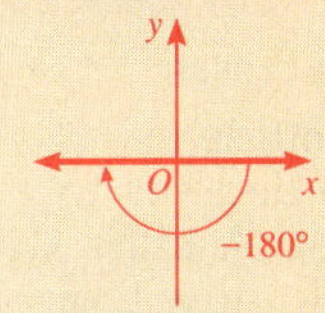

3. Sketch in standard position the angle that is $\frac{1}{3}$ of a clockwise revolution. Then find the measure of the angle. −120°

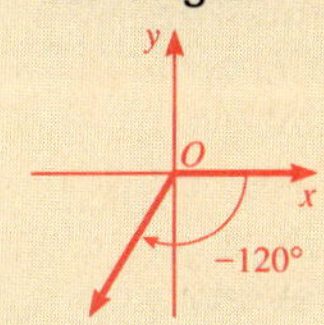

4. Sketch $90° + n \cdot 180°$ for the given values of n.
a. $n = 1$ **b.** $n = -1$

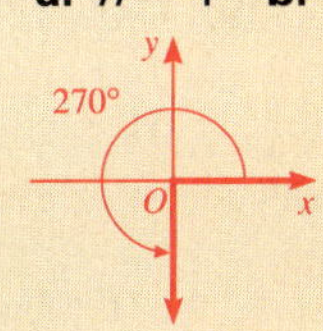

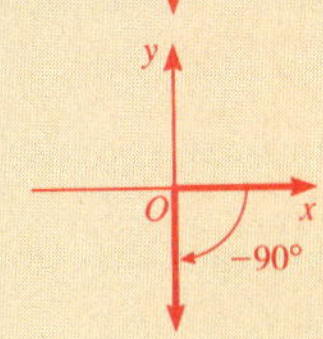

(continued)

Guided Practice

(continued)

5. Write a formula for all angles coterminal with a 45° angle.
$45° + n \cdot 360°$
6. Express 7°24′18″ in decimal degrees to the nearest hundredth of a degree. 7.41°
7. Express in degrees, minutes, and seconds.
a. 5.45° 5°27′
b. 59.72° 59°43′12″

Summarizing the Lesson

In this lesson students learned to use degrees to measure angles. Ask the students to explain how they can tell from the measures of two angles whether the angles are coterminal.

Suggested Assignments

Average Alg. and Trig.
552/3–66 mult. of 3
S 554/Mixed Review

Extended Alg. and Trig.
552/3–66 mult. of 3
S 554/Mixed Review

Supplementary Materials

Study Guide	pp. 187–188
Resource Book	p. 151
Overhead Visual	10

Additional Answers Written Exercises

2. a.

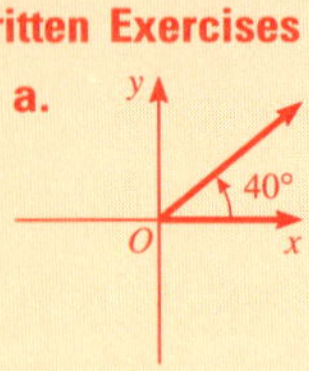

b.

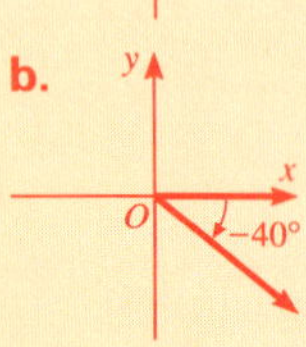

Give two angles, one positive and one negative, that are coterminal with the given angle. Answers may vary. Examples are given.

19. 50° 410°, −310° **20.** 100° 460°, −260° **21.** 120° 480°, −240° **22.** 40° 400°, −320°
23. −45° 315°, −405° **24.** −210° 150°, −570° **25.** 390° 30°, −330° **26.** 480° 120°, −240°

Written Exercises

Sketch each angle in standard position. Indicate its rotation by a curved arrow. Classify each angle by its quadrant. If the angle is a quadrantal angle, say so. (See the diagrams at the bottom of page 549.)

A

1. a. 135° II **b.** −135° III **2. a.** 40° I **b.** −40° IV
3. a. 300° IV **b.** −300° I **4. a.** 240° III **b.** −240° II
5. −270° Quad. **6.** 315° IV **7.** 290° IV **8.** −90° Quad.
9. 495° II **10.** 1080° Quad. **11.** −810° Quad. **12.** 750° I

Sketch in standard position the angle described and then find its measure.

Sample 1 $\frac{2}{5}$ of a clockwise rotation

Solution $\frac{2}{5} \times (-360) = -144$
$\therefore$ the measure is $-144°$.

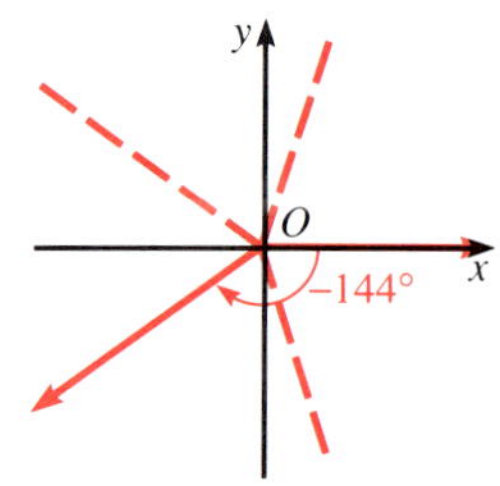

13. $\frac{2}{3}$ of a counterclockwise revolution 240°

14. $\frac{3}{8}$ of a counterclockwise revolution 135°

15. $\frac{3}{4}$ of a clockwise revolution −270°

16. $\frac{1}{6}$ of a clockwise revolution −60°

17. $1\frac{3}{5}$ counterclockwise revolutions 576°

18. $2\frac{1}{3}$ counterclockwise revolutions 840°

Sketch each angle in standard position when $n = 0$, $n = 1$, $n = 2$, and $n = -1$.

19. $45° + n \cdot 360°$ **20.** $-30° + n \cdot 360°$ **21.** $60° + n \cdot 180°$ **22.** $-30° + n \cdot 180°$

For Exercises 23–30:

a. Write a formula for the measures of all angles coterminal with the given angle.
b. Use the formula to find two angles, one positive and one negative, that are coterminal with the given angle. Answers may vary. Examples are given.

23. 35° 395°, −325° **24.** 140° 500°, −220° **25.** −100° 260°, −460° **26.** −210° 150°, −570°
27. 520° 160°, −200° **28.** 355° 715°, −5° **29.** 1000° 280°, −80° **30.** −3605° 355°, −5°

Express in decimal degrees to the nearest tenth of a degree.

31. 15°30′ 15.5° **32.** 47°36′ 47.6° **33.** 72°50′ 72.8° **34.** 51°20′ 51.3°

Express in decimal degrees to the nearest hundredth of a degree.

35. 25°45′ 25.75° **36.** 33°15′ 33.25° **37.** 45°18′20″ 45.31° **38.** 0°42′30″ 0.71°

Express in degrees and minutes to the nearest minute.

39. 25.4° 25°24′ **40.** 63.6° 63°36′ **41.** 44.9° 44°54′ **42.** 27.1° 27°6′

Express in degrees, minutes, and seconds to the nearest second.

43. 34.41° 34°24′36″ **44.** 18.27° 18°16′12″ **45.** 23.67° 23°40′12″ **46.** 58.83° 58°49′48″

The terminal side of an angle in standard position passes through the given point. Draw the angle and use a protractor to estimate its measure. ($\sqrt{3} \approx 1.73$)

Sample 2 $(-1, \sqrt{3})$

Solution

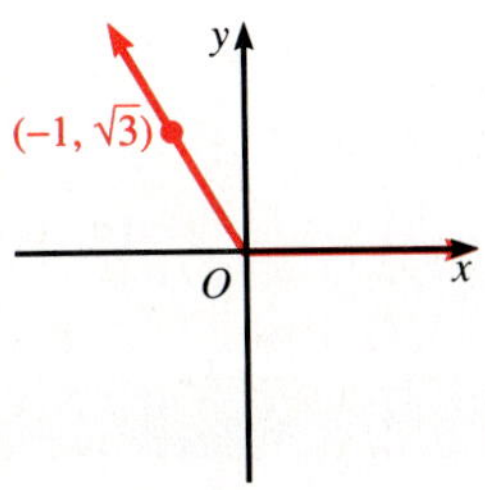

The measure is about 120°. ***Answer***

47. (4, −4) 315° **48.** (3, 4) 53° **49.** (4, 3) 37° **50.** (−4, 4) 135°

51. $(\sqrt{3}, 1)$ 30° **52.** $(-1, \sqrt{3})$ 120° **53.** $(-1, -\sqrt{3})$ 240° **54.** $(\sqrt{3}, -1)$ 330°

Find a first-quadrant angle θ, $0° < \theta < 90°$, for which an angle four times as large as θ will be in the given quadrant.

Sample 3 Find a first-quadrant angle θ, $0° < \theta < 90°$, for which an angle four times as large will be in the second quadrant.

Solution You want 4θ to be in quadrant II: $90° < 4\theta < 180°$

$$\therefore 22.5° < \theta < 45°$$

Since any angle θ between 22.5° and 45° is in quadrant I, and 4θ is in quadrant II, you can choose, for example, 30°. ***Answer***

B **55.** the first quadrant 55. Any angle between 0° and 22.5° **56.** the third quadrant 56. Any angle between 45° and 67.5° **57.** the fourth quadrant 57. Any angle between 67.5° and 90°

4. a.

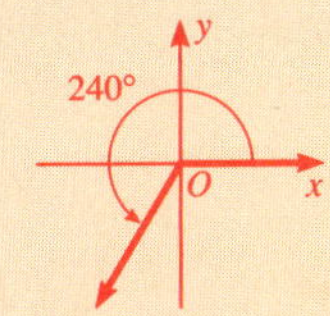

b.

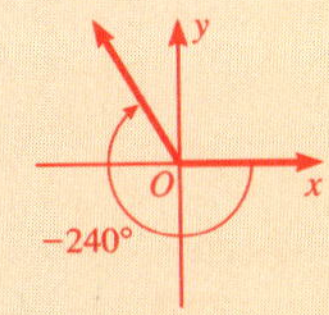

6.

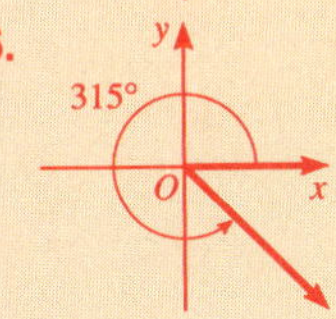

8.

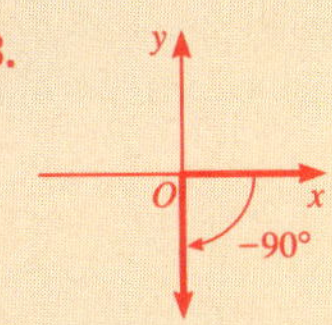

10.

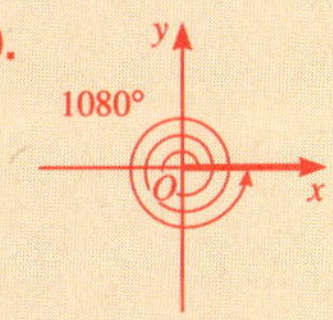

12.

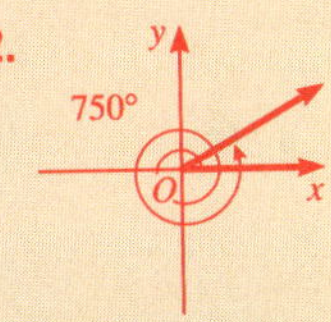

14.

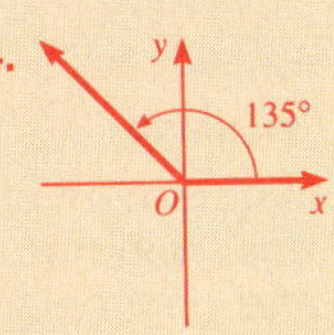

16.

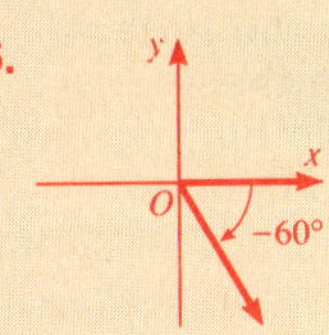

(continued)

Additional Answers Written Exercises

(continued)

18. 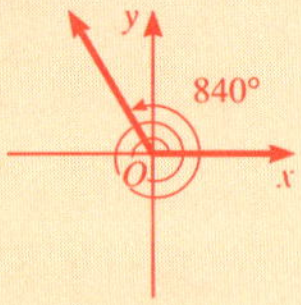

20. 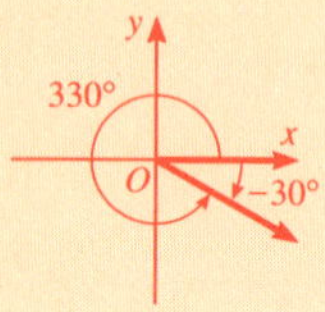

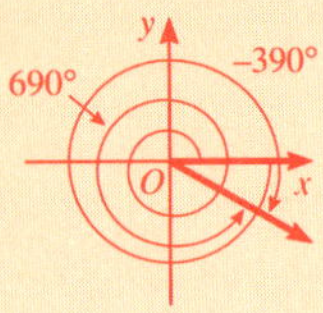

22. 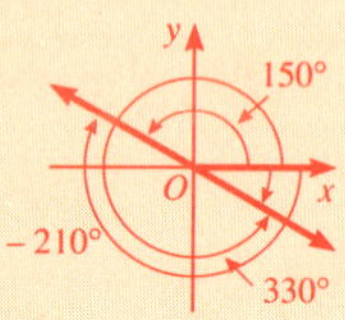

23. a. $35° + n \cdot 360°$
24. a. $140° + n \cdot 360°$
25. a. $-100° + n \cdot 360°$
26. a. $-210° + n \cdot 360°$
27. a. $520° + n \cdot 360°$
28. a. $355° + n \cdot 360°$
29. a. $1000° + n \cdot 360°$
30. a. $-3605° + n \cdot 360°$

Using a Calculator

Different calculators convert between degrees, minutes, and seconds and decimal degrees in various ways. If students take the time to experiment in these exercises, they will be prepared for any future degree/minute/second and decimal degree conversions.

Find a first-quadrant angle θ, $0° < \theta < 90°$, for which an angle six times as large as θ will be in the given quadrant.

58. the second quadrant 58. Any angle between 15° and 30°
59. the third quadrant 59. Any angle between 30° and 45°
60. the fourth quadrant 60. Any angle between 45° and 60°

Find an angle θ for which an angle half as large as θ will be in the given quadrant.

61. the first quadrant 61. Any angle between 0° and 180°
62. the second quadrant 62. Any angle between 180° and 360°
63. the third quadrant 63. Any angle between 360° and 540°

Find an angle θ for which an angle one fifth as large as θ will be in the given quadrant.

64. the first quadrant 64. Any angle between 0° and 450°
65. the second quadrant 65. Any angle between 450° and 900°
66. the third quadrant 66. Any angle between 900° and 1350°

Mixed Review Exercises

Find the third term in the expansion of each binomial.

1. $(x + 2y)^5$ $40x^3y^2$
2. $(a - 3b)^8$ $252a^6b^2$
3. $(m^2 + n^3)^{13}$ $78m^{22}n^6$

Simplify.

4. $\sqrt{-3} \cdot \sqrt{-12}$ -6
5. $\log_2 14 - \log_2 7$ 1
6. $(2^{3/4} - 2^{1/4})^2$ $3\sqrt{2} - 4$
7. $\ln \sqrt[3]{e^2}$ $\frac{2}{3}$
8. $(3 - 2i)(4 + i)$ $14 - 5i$
9. $\sqrt{20} + \sqrt{45} - \sqrt{125}$ 0
10. $\dfrac{-6(3 - 5)}{(-7 + 9)^2}$ 3
11. $\sqrt{\dfrac{98}{27}}$ $\dfrac{7\sqrt{6}}{9}$
12. $\log_3 3\sqrt{3}$ $\frac{3}{2}$

Calculator Key-In

Many scientific calculators have keys that convert between degrees, minutes, and seconds and decimal degrees. On some calculators there is a conversion key (usually labeled ° ′ ″) that allows you to convert, say, 32°45′10″ to decimal degrees as follows: enter 32, press the key; enter 45, press the key again; enter 10, press the key a third time. The result is 32.7527778°.

To convert from decimal degrees to degrees, minutes, and seconds, you just enter the decimal degrees and press the inverse conversion key. (If there is no such key, press the inverse key and then the conversion key.)

On other calculators you may be required to enter 32°45′10″ differently and the keys may be labeled with "DMS" for degrees-minutes-seconds. Test your calculator using 32°45′10″ and its decimal equivalent 32.7527778° before doing the following exercises.

Convert each measure to degrees, minutes, and seconds or to decimal degrees.

1. 58.6° 58°36′
2. 29.7° 29°42′
3. 86.43° 86°25′48″
4. 108.26° 108°15′36″
5. 36°25′36″ $36.42\overline{6}°$
6. 45°11′19″ $45.1886\overline{1}°$
7. 73°52′25″ $73.8736\overline{1}°$
8. 115°42′51″ $115.7141\overline{6}°$

12-2 Trigonometric Functions of Acute Angles

Objective To define trigonometric functions of acute angles.

To define the trigonometric functions of an acute angle θ, first place θ in standard position as shown in the diagram below. Next, choose any point $P(x, y)$, other than the origin, on the terminal side of θ and let r be the distance OP. Then the following definitions can be made:

The **sine** of θ, written $\sin \theta$, is equal to $\frac{y}{r}$.

The **cosine** of θ, written $\cos \theta$, is equal to $\frac{x}{r}$.

The **tangent** of θ, written $\tan \theta$, is equal to $\frac{y}{x}$.

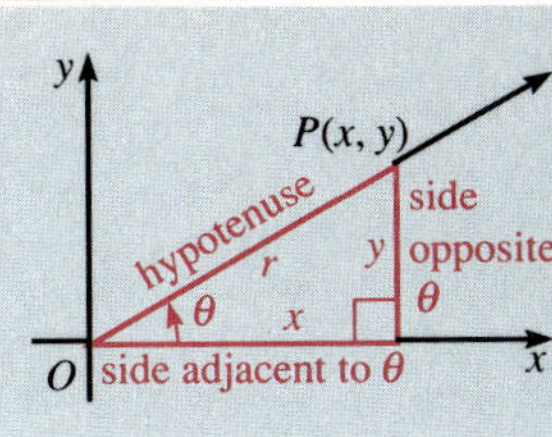

The ratios $\frac{y}{r}$, $\frac{x}{r}$, and $\frac{y}{x}$ depend only on θ and not on the choice of P. (See Exercise 32.) Therefore, the sine, cosine, and tangent are functions of θ. Notice that x, y, and r are the lengths of the legs and hypotenuse, respectively, of right triangle OMP. The definitions above can be restated in terms of these lengths, for any right triangle containing the acute angle θ.

$$\sin \theta = \frac{y}{r} = \frac{\text{length of side opposite } \theta}{\text{length of the hypotenuse}}$$

$$\cos \theta = \frac{x}{r} = \frac{\text{length of side adjacent to } \theta}{\text{length of the hypotenuse}}$$

$$\tan \theta = \frac{y}{x} = \frac{\text{length of side opposite } \theta}{\text{length of side adjacent to } \theta}$$

Example 1 An acute angle θ is in standard position and its terminal side passes through $P(4, 5)$. Find $\sin \theta$, $\cos \theta$, and $\tan \theta$.

Solution Here $x = 4$ and $y = 5$. To find r, use the Pythagorean theorem (page 401):

$$r^2 = 4^2 + 5^2 = 16 + 25 = 41, \quad \text{so} \quad r = \sqrt{41}$$

$$\therefore \sin \theta = \frac{y}{r} = \frac{5}{\sqrt{41}} = \frac{5\sqrt{41}}{41}; \ \cos \theta = \frac{x}{r} = \frac{4}{\sqrt{41}} = \frac{4\sqrt{41}}{41}; \ \tan \theta = \frac{5}{4}$$

Explorations p. 843

Teaching Suggestions p. T123

Suggested Extensions p. T123

Warm-Up Exercises

$\triangle MNP$ is a right $\triangle$ with right $\angle N$. $\triangle JKL \cong \triangle MNP$.

1. Name the hypotenuse of $\triangle MNP$. $\overline{MP}$
2. Name the hypotenuse of $\triangle JKL$. $\overline{JL}$
3. Name the sides that include $\angle P$. $\overline{PM}$, $\overline{PN}$
4. Name the side opposite $\angle J$. $\overline{KL}$

Name a ratio that is equal to the given ratio.

5. $JL:KL$ $MP:NP$
6. $MN:NP$ $JK:KL$

Motivating the Lesson

Tell students that if they were flying a kite and they knew the amount of string let out and the angle that the string made with the ground, they could use a trigonometric function like cosine to determine the kite's height above the ground.

Common Errors

When students first encounter the definitions of the trigonometric functions, some may think that the values of the functions of a given angle depend not only on the measure of the angle, but also on the point used. When reteaching, use similar triangles to help convince these students that the values of the trigonometric functions depend only on the measure of the angle.

Problem Solving Strategies

Problems in trigonometry are usually easier to understand and solve if students *make a drawing or sketch* of the problem situations. As students read Example 1, they should be encouraged to sketch the angle in standard position, with P labeled. This makes it clear how the Pythagorean theorem applies and gives more meaning to the rote application of the definition.

Chalkboard Examples

1. When an acute angle θ is in standard position, its terminal side passes through (2, 4). Find $\sin\theta$, $\cos\theta$, and $\tan\theta$.
   $r = \sqrt{2^2 + 4^2} = \sqrt{20} = 2\sqrt{5}$
   $\sin\theta = \frac{4}{2\sqrt{5}} = \frac{2\sqrt{5}}{5}$
   $\cos\theta = \frac{2}{2\sqrt{5}} = \frac{\sqrt{5}}{5}$
   $\tan\theta = \frac{4}{2} = 2$
2. Find the values of the six trigonometric functions of an angle θ in standard position whose terminal side passes through the point (3, 4).
   $r^2 = x^2 + y^2 = 3^2 + 4^2 = 25. \therefore r = 5$
   $\sin\theta = \frac{y}{r} = \frac{4}{5}$
   $\cos\theta = \frac{x}{r} = \frac{3}{5}$
   $\tan\theta = \frac{y}{x} = \frac{4}{3}$
   $\cot\theta = \frac{x}{y} = \frac{3}{4}$
   $\sec\theta = \frac{r}{x} = \frac{5}{3}$
   $\csc\theta = \frac{r}{y} = \frac{5}{4}$

The three remaining trigonometric functions are **reciprocal functions** of those functions already defined.

The **cotangent** of θ, written $\cot\theta$, is equal to $\frac{x}{y}$, or $\frac{1}{\tan\theta}$.

The **secant** of θ, written $\sec\theta$, is equal to $\frac{r}{x}$, or $\frac{1}{\cos\theta}$.

The **cosecant** of θ, written $\csc\theta$, is equal to $\frac{r}{y}$, or $\frac{1}{\sin\theta}$.

Example 2 Find the values of the six trigonometric functions of an angle θ in standard position whose terminal side passes through (5, 12).

Solution Sketch θ in standard position.
By the Pythagorean theorem:

$$r^2 = x^2 + y^2 = 5^2 + 12^2 = 169$$
$$r = 13$$

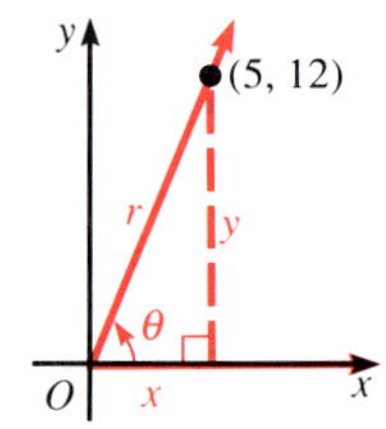

$$\sin\theta = \frac{y}{r} = \frac{12}{13} \qquad \csc\theta = \frac{r}{y} = \frac{13}{12}$$
$$\cos\theta = \frac{x}{r} = \frac{5}{13} \qquad \sec\theta = \frac{r}{x} = \frac{13}{5}$$
$$\tan\theta = \frac{y}{x} = \frac{12}{5} \qquad \cot\theta = \frac{x}{y} = \frac{5}{12}$$

An equation involving trigonometric functions of an angle θ that is true for all values of θ is a *trigonometric identity*. Using the diagram above and the definitions of this lesson, notice that

$$\frac{\sin\theta}{\cos\theta} = \frac{\frac{y}{r}}{\frac{x}{r}} = \frac{y}{x} = \tan\theta \quad \text{and} \quad \frac{\cos\theta}{\sin\theta} = \frac{\frac{x}{r}}{\frac{y}{r}} = \frac{x}{y} = \cot\theta.$$

Therefore, the following equations are identities:

$$\tan\theta = \frac{\sin\theta}{\cos\theta} \qquad \cot\theta = \frac{\cos\theta}{\sin\theta}$$

The next identity is called a *Pythagorean identity*. Notice that we write $\sin^2\theta$ for $(\sin\theta)^2$ and $\cos^2\theta$ for $(\cos\theta)^2$.

$$\sin^2\theta + \cos^2\theta = 1$$

To prove this identity, you use the Pythagorean relationship $x^2 + y^2 = r^2$.

$$\sin^2\theta + \cos^2\theta = \left(\frac{y}{r}\right)^2 + \left(\frac{x}{r}\right)^2 = \frac{y^2}{r^2} + \frac{x^2}{r^2} = \frac{y^2 + x^2}{r^2} = \frac{r^2}{r^2} = 1$$

Example 3 Find $\cos\theta$ and $\tan\theta$ if θ is an acute angle and $\sin\theta = \frac{1}{3}$.

Solution 1 Use the identity $\sin^2\theta + \cos^2\theta = 1$.

$$\left(\frac{1}{3}\right)^2 + \cos^2\theta = 1$$

$$\cos^2\theta = 1 - \left(\frac{1}{3}\right)^2 = 1 - \frac{1}{9} = \frac{8}{9}$$

$$\cos\theta = \sqrt{\frac{8}{9}} = \frac{2\sqrt{2}}{3} \quad \textbf{\textit{Answer}}$$

$$\tan\theta = \frac{\sin\theta}{\cos\theta} = \frac{\frac{1}{3}}{\frac{2\sqrt{2}}{3}} = \frac{1}{2\sqrt{2}} = \frac{\sqrt{2}}{4} \quad \textbf{\textit{Answer}}$$

Solution 2 Make a diagram showing

$$\sin\theta = \frac{\text{length of side opposite } \theta}{\text{length of the hypotenuse}} = \frac{1}{3}.$$

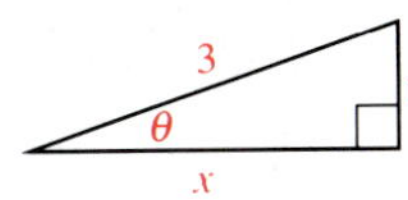

Then find x: $x^2 = 3^2 - 1^2 = 8$

$$x = 2\sqrt{2}$$

Then $\cos\theta = \frac{2\sqrt{2}}{3}$ and $\tan\theta = \frac{1}{2\sqrt{2}} = \frac{\sqrt{2}}{4}$. ***Answer***

The sine and cosine are called **cofunctions.** Other pairs of cofunctions are the tangent and cotangent and the secant and cosecant. There are six cofunction identities that can be derived from right triangle ABC.

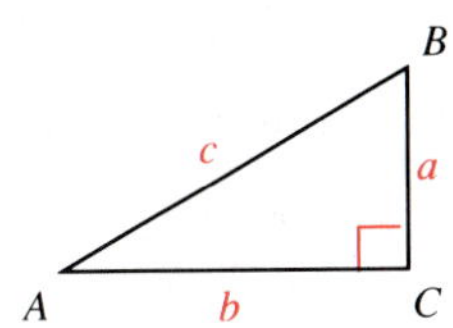

$\sin A = \frac{a}{c} = \cos B$ $\quad$ $\cos A = \frac{b}{c} = \sin B$

$\tan A = \frac{a}{b} = \cot B$ $\quad$ $\cot A = \frac{b}{a} = \tan B$

$\sec A = \frac{c}{b} = \csc B$ $\quad$ $\csc A = \frac{c}{a} = \sec B$

Notice that $\angle A$ and $\angle B$ are complementary; that is, $m\angle A + m\angle B = 90°$. Notice also that:

$\sin\theta = \cos(90° - \theta)$	$\cos\theta = \sin(90° - \theta)$
$\tan\theta = \cot(90° - \theta)$	$\cot\theta = \tan(90° - \theta)$
$\sec\theta = \csc(90° - \theta)$	$\csc\theta = \sec(90° - \theta)$

Thus, any trigonometric function of an acute angle is equal to the cofunction of the complement of the angle. Since 30° angles and 60° angles are complementary, they illustrate the cofunction relationship. For example, $\tan 60° = \cot 30°$.

3. Find the following if φ is an acute angle and $\sin\varphi = \frac{5}{13}$.

a. $\cos\varphi$ **b.** $\tan\varphi$

a. Use $\sin^2\varphi + \cos^2\varphi = 1$.

$\left(\frac{5}{13}\right)^2 + \cos^2\varphi = 1$

$\cos^2\varphi = 1 - \left(\frac{5}{13}\right)^2 = \frac{144}{169}$

$\cos\varphi = \frac{12}{13}$

b. $\tan\varphi = \frac{\sin\varphi}{\cos\varphi} = \frac{\frac{5}{13}}{\frac{12}{13}} = \frac{5}{12}$

4. The length of one side of a right triangle is given in the diagram below. Find the lengths of the other two sides.

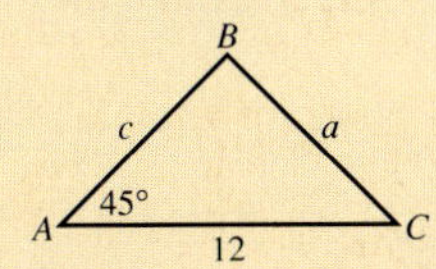

$\sin 45° = \frac{BC}{AC} = \frac{a}{12}$

$\frac{\sqrt{2}}{2} = \frac{a}{12}$

$12\sqrt{2} = 2a$

$a = 6\sqrt{2}$

$\cos 45° = \frac{AB}{AC} = \frac{c}{12}$

$\frac{\sqrt{2}}{2} = \frac{c}{12}$

$6\sqrt{2} = c$

5. Use the diagram to find the measure of $\angle X$. 30°

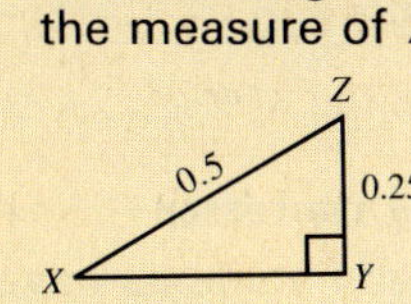

Check for Understanding

Oral Exs. 1–4: use after Example 2.
Oral Exs. 5–8: use after the boxed cofunction identities on page 557.
Oral Exs. 9–11: use after Example 4.
Oral Ex. 12: use after Example 3.

Guided Practice

Find the values of the six trigonometric functions of angle θ.

1.

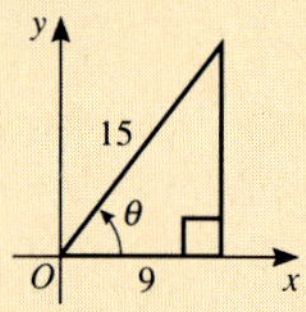

$\sin\theta = \frac{4}{5}$, $\cos\theta = \frac{3}{5}$, $\tan\theta = \frac{4}{3}$, $\csc\theta = \frac{5}{4}$, $\sec\theta = \frac{5}{3}$, $\cot\theta = \frac{3}{4}$

2. $P(12, 5)$

$\sin\theta = \frac{5}{13}$, $\cos\theta = \frac{12}{13}$, $\tan\theta = \frac{5}{12}$, $\csc\theta = \frac{13}{5}$, $\sec\theta = \frac{13}{12}$, $\cot\theta = \frac{12}{5}$

3. If θ is an acute angle and $\sin\theta = \frac{4}{5}$, find $\cos\theta$ and $\tan\theta$. $\frac{3}{5}$; $\frac{4}{3}$

4. Find the measure of acute angle ϕ if $\cot\phi = \tan 20°$. 70°

Summarizing the Lesson

In this lesson students have learned the definitions of the six trigonometric functions of acute angles. Ask the students to state these definitions.

You can read the trigonometric functions of 30°, 45°, and 60° from the triangles shown below. To recall these values, you may find it easier to draw the triangles rather than to memorize the table.

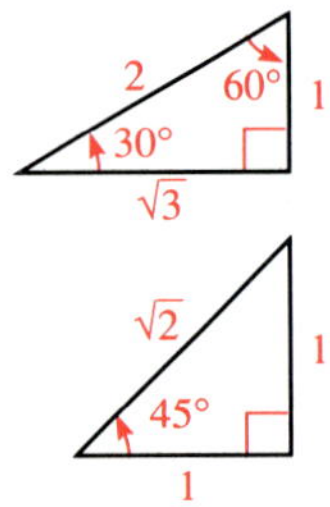

θ	$\sin\theta$	$\cos\theta$	$\tan\theta$	$\csc\theta$	$\sec\theta$	$\cot\theta$
30°	$\frac{1}{2}$	$\frac{\sqrt{3}}{2}$	$\frac{\sqrt{3}}{3}$	2	$\frac{2\sqrt{3}}{3}$	$\sqrt{3}$
45°	$\frac{\sqrt{2}}{2}$	$\frac{\sqrt{2}}{2}$	1	$\sqrt{2}$	$\sqrt{2}$	1
60°	$\frac{\sqrt{3}}{2}$	$\frac{1}{2}$	$\sqrt{3}$	$\frac{2\sqrt{3}}{3}$	2	$\frac{\sqrt{3}}{3}$

Example 4 Use the diagram at the right. Find the lengths of side $\overline{BC}$ and side $\overline{AB}$.

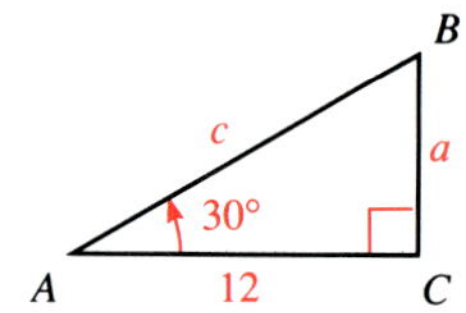

Solution

$$\tan 30° = \frac{BC}{AC} \qquad \cos 30° = \frac{AC}{AB}$$

$$\frac{\sqrt{3}}{3} = \frac{a}{12} \qquad \frac{\sqrt{3}}{2} = \frac{12}{c}$$

$$3a = 12\sqrt{3} \qquad \sqrt{3}c = 24$$

$$a = 4\sqrt{3} \qquad c = \frac{24}{\sqrt{3}} = 8\sqrt{3}$$

Example 5 Use the diagram at the right to find the measure of $\angle A$.

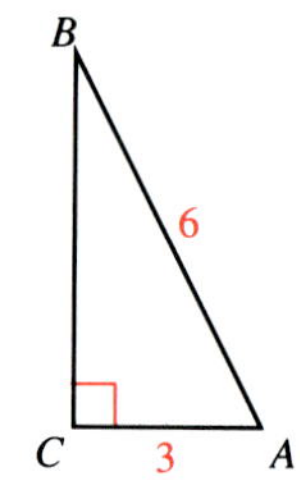

Solution

$$\cos A = \frac{AC}{AB} = \frac{3}{6} = \frac{1}{2}$$

$$\cos 60° = \frac{1}{2}$$

$\therefore\ \angle A = 60°$. ***Answer***

Oral Exercises

Give the values of the six trigonometric functions of θ.

1.

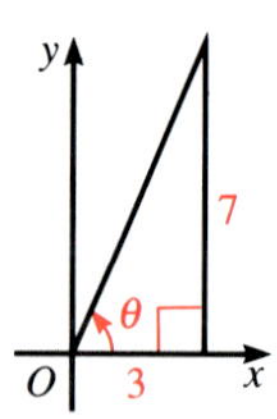

2.

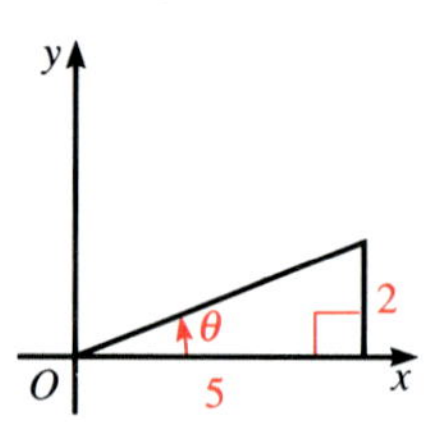

3.

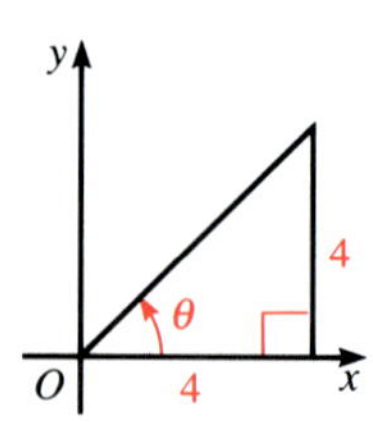

4.

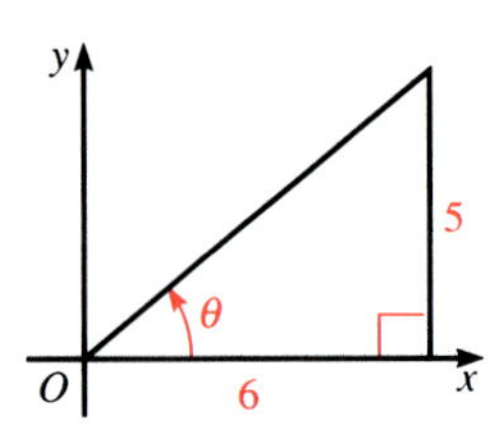

Use the cofunction identities to find the measure of the acute angle θ.

5. $\sin \theta = \cos 25°$ 65°　　**6.** $\tan \theta = \cot 70°$ 20°

7. $\sec \theta = \csc 15°$ 75°　　**8.** $\cos \theta = \sin 45°$ 45°

The length of one side of a right triangle is given. Give the lengths of the other two sides.

9.

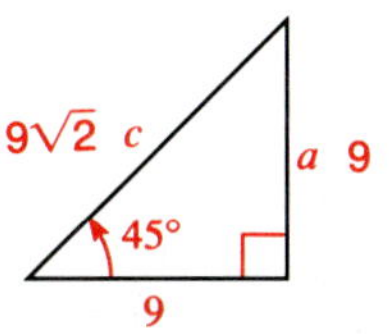

10.

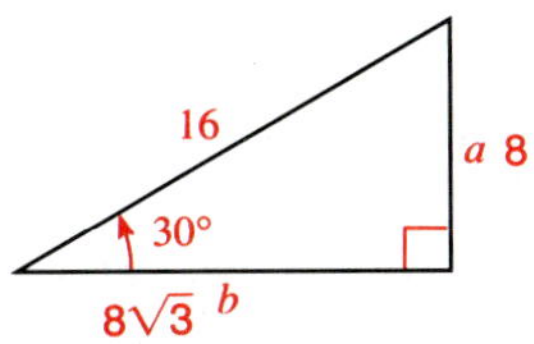

11.

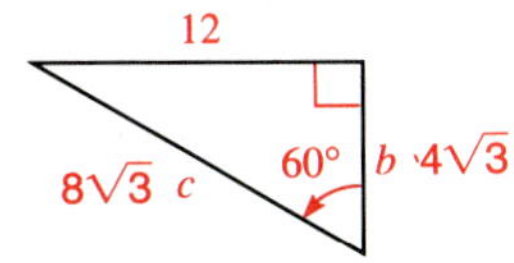

12. If $\sin \theta = \frac{2}{3}$ for an acute angle θ, state two ways to find $\cos \theta$ and $\tan \theta$.

Written Exercises

Find the values of the six trigonometric functions of angle θ.

A

1.

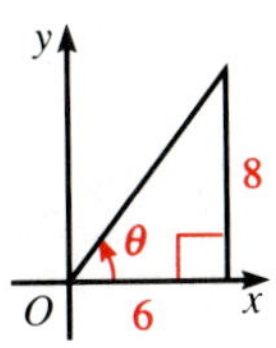

2.

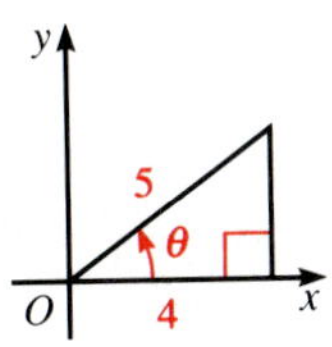

3.

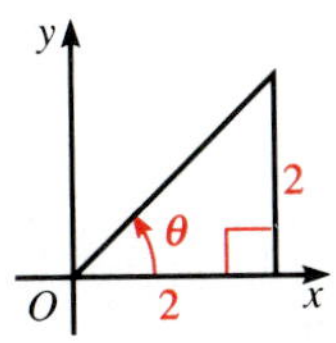

4.

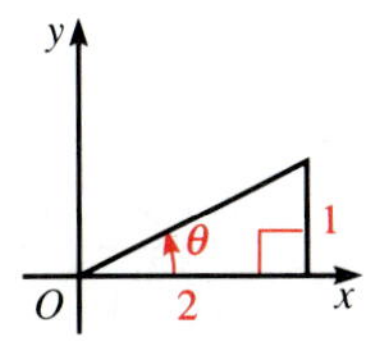

Find the values of the six trigonometric functions of an angle θ in standard position whose terminal side passes through point P.

5. $P(8, 15)$　　**6.** $P(3, 4)$　　**7.** $P(3, 1)$　　**8.** $P(5, 5)$

Complete the table. In each case, θ is an acute angle.

	9.	10.	11.	12.	13.	14.	15.
$\sin \theta$	$\frac{3}{5}$	? $\frac{\sqrt{3}}{2}$	? $\frac{1}{2}$	$\frac{\sqrt{2}}{3}$	$\frac{\sqrt{161}}{15}$	? $\frac{12}{13}$	? $\frac{\sqrt{15}}{4}$
$\cos \theta$	? $\frac{4}{5}$	$\frac{1}{2}$	$\frac{\sqrt{3}}{2}$	? $\frac{\sqrt{7}}{3}$	? $\frac{8}{15}$	$\frac{5}{13}$	$\frac{1}{4}$
$\tan \theta$	? $\frac{3}{4}$	? $\sqrt{3}$	? $\frac{\sqrt{3}}{3}$	? $\frac{\sqrt{14}}{7}$	? $\frac{\sqrt{161}}{8}$	? $\frac{12}{5}$	? $\sqrt{15}$

Use the cofunction identities to find the measure of the acute angle ϕ.

16. $\sin \phi = \cos 50°$ 40°　　**17.** $\cos \phi = \sin 40°$ 50°　　**18.** $\tan \phi = \cot 17°$ 73°　　**19.** $\sec \phi = \csc 80°$ 10°

Suggested Assignments

Average Alg. and Trig.
559/1–31 odd
S 553/47, 49, 50, 52, 53

Extended Alg. and Trig.
559/1–25 odd, 26–31, 33, 34

Additional Answers
Oral Exercises

1. $\sin \theta = \frac{7\sqrt{58}}{58}$, $\cos \theta = \frac{3\sqrt{58}}{58}$, $\tan \theta = \frac{7}{3}$, $\csc \theta = \frac{\sqrt{58}}{7}$, $\sec \theta = \frac{\sqrt{58}}{3}$, $\cot \theta = \frac{3}{7}$

2. $\sin \theta = \frac{2\sqrt{29}}{29}$, $\cos \theta = \frac{5\sqrt{29}}{29}$, $\tan \theta = \frac{2}{5}$, $\csc \theta = \frac{\sqrt{29}}{2}$, $\sec \theta = \frac{\sqrt{29}}{5}$, $\cot \theta = \frac{5}{2}$

3. $\sin \theta = \cos \theta = \frac{\sqrt{2}}{2}$, $\tan \theta = 1$, $\csc \theta = \sqrt{2}$, $\sec \theta = \sqrt{2}$, $\cot \theta = 1$

4. $\sin \theta = \frac{5\sqrt{61}}{61}$, $\cos \theta = \frac{6\sqrt{61}}{61}$, $\tan \theta = \frac{5}{6}$, $\csc \theta = \frac{\sqrt{61}}{5}$, $\sec \theta = \frac{\sqrt{61}}{6}$, $\cot \theta = \frac{6}{5}$

12. Use $\sin^2 \sigma + \cos^2 \sigma = 1$ to find $\cos \sigma$. Then use $\tan \sigma = \frac{\sin \sigma}{\cos \sigma}$. Use $x^2 + y^2 = r^2$ and $\sin \sigma = \frac{y}{r}$ to find x. Then use $\cos \sigma = \frac{x}{r}$ and $\tan \sigma = \frac{y}{x}$.

Supplementary Materials

Study Guide	pp. 189–190
Practice Master	64
Overhead Visual	10

Additional Answers
Written Exercises

1. $\sin\theta = \frac{4}{5}$, $\cos\theta = \frac{3}{5}$, $\tan\theta = \frac{4}{3}$, $\csc\theta = \frac{5}{4}$, $\sec\theta = \frac{5}{3}$, $\cot\theta = \frac{3}{4}$

2. $\sin\theta = \frac{3}{5}$, $\cos\theta = \frac{4}{5}$, $\tan\theta = \frac{3}{4}$, $\csc\theta = \frac{5}{3}$, $\sec\theta = \frac{5}{4}$, $\cot\theta = \frac{4}{3}$

3. $\sin\theta = \frac{\sqrt{2}}{2}$, $\cos\theta = \frac{\sqrt{2}}{2}$, $\tan\theta = 1$, $\csc\theta = \sqrt{2}$, $\sec\theta = \sqrt{2}$, $\cot\theta = 1$

4. $\sin\theta = \frac{\sqrt{5}}{5}$, $\cos\theta = \frac{2\sqrt{5}}{5}$, $\tan\theta = \frac{1}{2}$, $\csc\theta = \sqrt{5}$, $\sec\theta = \frac{\sqrt{5}}{2}$, $\cot\theta = 2$

5. $\sin\theta = \frac{15}{17}$, $\cos\theta = \frac{8}{17}$, $\tan\theta = \frac{15}{8}$, $\csc\theta = \frac{17}{15}$, $\sec\theta = \frac{17}{8}$, $\cot\theta = \frac{8}{15}$

6. $\sin\theta = \frac{4}{5}$, $\cos\theta = \frac{3}{5}$, $\tan\theta = \frac{4}{3}$, $\csc\theta = \frac{5}{4}$, $\sec\theta = \frac{5}{3}$, $\cot\theta = \frac{3}{4}$

(continued on p. 578)

In Exercises 20–25, use the diagram at the right. Find the lengths of the sides and the measures of the angles that are not given. Leave your answers in simplest radical form.

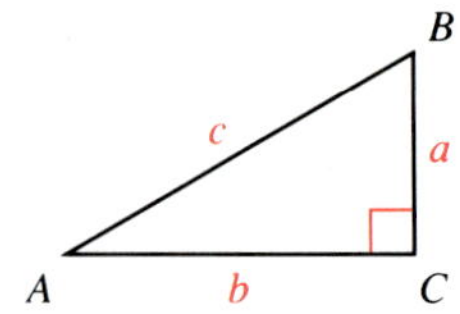

20. $a = 6$, $\angle A = 30°$ — $\angle B = 60°$, $c = 12$, $b = 6\sqrt{3}$

21. $b = 2$, $\angle A = 45°$ — $\angle B = 45°$, $a = 2$, $c = 2\sqrt{2}$

22. $c = 10$, $\angle A = 45°$ — $\angle B = 45°$, $a = 5\sqrt{2}$, $b = 5\sqrt{2}$

23. $c = 20$, $\angle A = 60°$ — $\angle B = 30°$, $b = 10$, $a = 10\sqrt{3}$

24. $a = 3$, $c = 6$ — $\angle A = 30°$, $\angle B = 60°$, $b = 3\sqrt{3}$

25. $a = 4$, $b = 4$ — $\angle A = 45°$, $\angle B = 45°$, $c = 4\sqrt{2}$

In Exercises 26–29, find the length of *x*. Leave your answers in simplest radical form.

B

26. $\frac{15\sqrt{2} + 5\sqrt{6}}{2}$

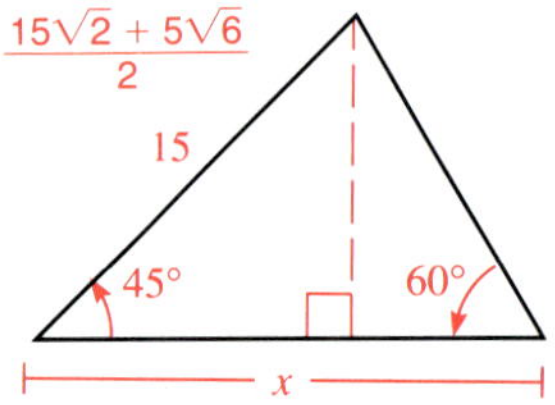

27. $6\sqrt{3} + 6$

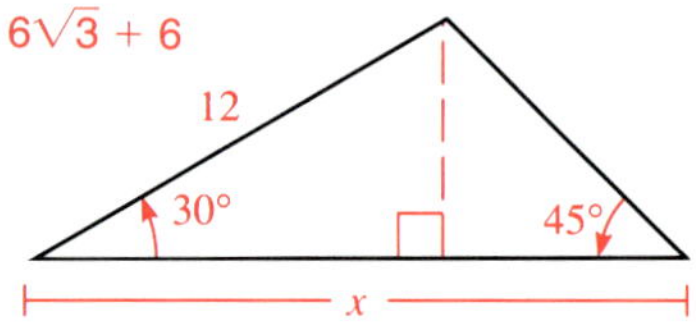

28. $24 - 8\sqrt{3}$

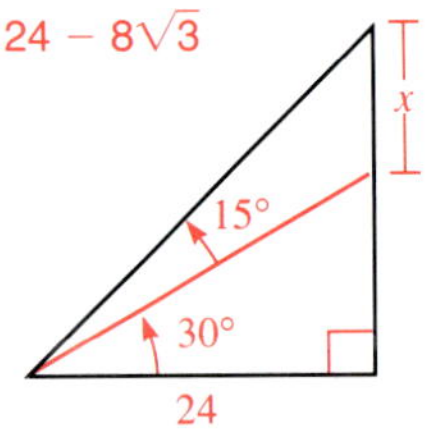

29.

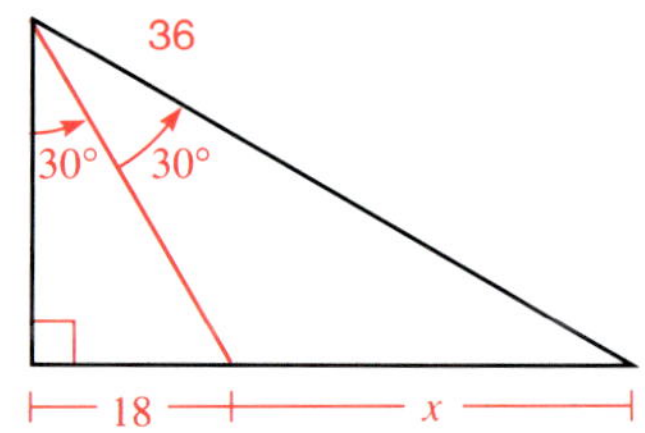

30. Prove that $\cot\theta = \frac{\cos\theta}{\sin\theta}$.

31. The three *Pythagorean identities* in trigonometry are as follows:

$$\sin^2\theta + \cos^2\theta = 1 \qquad \tan^2\theta + 1 = \sec^2\theta \qquad 1 + \cot^2\theta = \csc^2\theta$$

We have shown the first of these to be true for every acute angle θ. Use this first Pythagorean identity to prove that the other two identities are true for every acute angle θ.

32. $P(x, y)$ and $Q(x', y')$ are two points on the terminal side of an acute angle θ in standard position. Let $OP = r$ and $OQ = s$. Use similar triangles to show that $\frac{y}{r} = \frac{y'}{s}$, $\frac{x}{r} = \frac{x'}{s}$, and $\frac{y}{x} = \frac{y'}{x'}$. (This shows that using point Q to define the trigonometric functions θ will yield the same values as using point P.)

C

33. Angle θ is acute and $\sin\theta = u$. Express the other five trigonometric functions of θ in terms of u. (*Hint:* Recall that $\sin^2\theta + \cos^2\theta = 1$.)

34. Angle ϕ is acute and $\cos\phi = v$. Express the other five trigonometric functions of ϕ in terms of v.

12-3 Trigonometric Functions of General Angles

Objective To define trigonometric functions of general angles.

In Lesson 12-2, you learned the definitions for the six trigonometric functions for acute angles. Now you'll extend these definitions to angles of any measure.

As you did with acute angles, place the angle θ in standard position, as shown in each diagram below. Choose a point $P(x, y)$ on the terminal side of θ in each quadrant and let r be the distance OP.

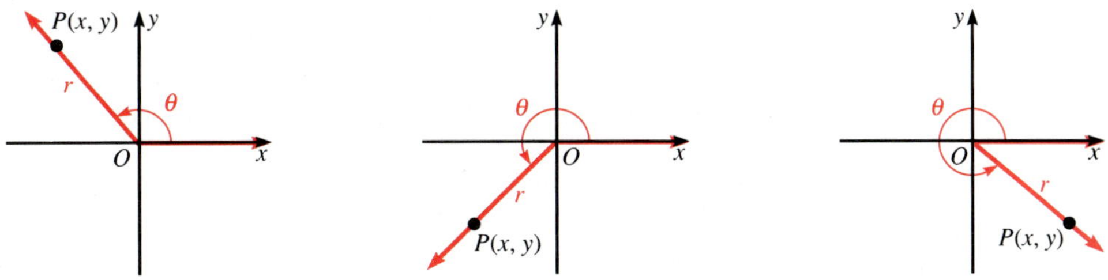

For any angle θ and any point (x, y) on the terminal side:

$\sin\theta = \frac{y}{r}$ $\cos\theta = \frac{x}{r}$ $\tan\theta = \frac{y}{x}$, if $x \neq 0$

$\csc\theta = \frac{r}{y}$, if $y \neq 0$ $\sec\theta = \frac{r}{x}$, if $x \neq 0$ $\cot\theta = \frac{x}{y}$, if $y \neq 0$

Example 1 Find the values of the six trigonometric functions of an angle θ in standard position whose terminal side passes through $(8, -15)$.

Solution First make a sketch similar to the third one above.
Here $x = 8$ and $y = -15$. Since r is the distance between $(0, 0)$ and $(8, -15)$, you can find r by using the distance formula (page 402).

$$r = \sqrt{8^2 + (-15)^2} = \sqrt{64 + 225} = \sqrt{289} = 17$$

$$\sin\theta = \frac{y}{r} = -\frac{15}{17} \qquad \cos\theta = \frac{x}{r} = \frac{8}{17} \qquad \tan\theta = \frac{y}{x} = -\frac{15}{8}$$

$$\csc\theta = \frac{r}{y} = -\frac{17}{15} \qquad \sec\theta = \frac{r}{x} = \frac{17}{8} \qquad \cot\theta = \frac{x}{y} = -\frac{8}{15}$$

In the definitions above, r is always positive. Therefore, the signs of the functions of θ are determined by the signs of x and y, and these signs depend only on the quadrant of the terminal side of θ, as shown on the next page.

Teaching Suggestions p. T123

Suggested Extensions p. T123

Warm-Up Exercises

The measure of an angle in standard position is given. State the coordinates of a point on the terminal side. Answers may vary.

1. 0° (1, 0)
2. 90° (0, 1)
3. 180° (−1, 0)
4. −450° (0, −1)
5. 45° (1, 1)
6. −60° $(1, -\sqrt{3})$

Motivating the Lesson

Just as there is no need to restrict the measures of angles, the definitions of the trigonometric functions need not be restricted to acute angles. In this lesson the definitions will be extended to angles of any measure.

Chalkboard Examples

1. When an angle θ is in standard position its terminal side passes through $(-3, -4)$. Find the values of the six trigonometric functions of θ.
   Since $x = -3$, $y = -4$ and r is the distance between $(0, 0)$ and $(-3, -4)$, $r = \sqrt{(-3)^2 + (-4)^2} = 5$.
   $\sin\theta = -\frac{4}{5}$
   $\cos\theta = -\frac{3}{5}$
   $\tan\theta = \frac{4}{3}$
   $\csc\theta = -\frac{5}{4}$
   $\sec\theta = -\frac{5}{3}$
   $\cot\theta = \frac{3}{4}$

(continued)

Chalkboard Examples
(continued)

2. Determine which functions are defined for a 90° angle and find their values.
Place the angle in standard position and choose a point such as (0, 1) on the terminal side. Then $x = 0$ and $y = 1$, $\sin 90° = \frac{1}{1} = 1$; $\cos 90° = \frac{0}{1} = 0$; $\tan 90°$ is undefined since $x = 0$.
$\csc 90° = \frac{1}{1} = 1$; $\sec 90°$ is undefined since $x = 0$.
$\cot 90° = \frac{0}{1} = 0$

4. Write sin 130° as a function of an acute angle.
The reference angle of 130° is 180° − 130° = 50°. Since 130° is a second-quadrant angle, its sine is positive.
$\therefore \sin 130° = \sin 50°$

5. Find the exact value of tan 135°.
The reference angle of 135° is 45°. Since 135° is a second-quadrant angle, the tangent is negative.
$\tan 135° = -\tan 45° = -1$

6. Find the five other trigonometric functions of θ if $\sin \theta = -\frac{3}{4}$ and $180° < \theta < 270°$. $\left(-\frac{3}{4}\right)^2 + \cos^2 \theta = 1$;
$\cos \theta = -\sqrt{\frac{7}{16}} = -\frac{\sqrt{7}}{4}$;
$\tan \theta = \frac{3\sqrt{7}}{7}$; $\csc \theta = -\frac{4}{3}$;
$\sec \theta = -\frac{4\sqrt{7}}{7}$; $\cot \theta = \frac{\sqrt{7}}{3}$

Function value	Quadrant of θ I	II	III	IV
$\sin \theta$, $\csc \theta$	+	+	−	−
$\cos \theta$, $\sec \theta$	+	−	−	+
$\tan \theta$, $\cot \theta$	+	−	+	−

The sine and cosine functions are defined for all angles, but the other four functions are undefined for certain quadrantal angles.

Example 2 Determine which functions are defined for a 180° angle and find their values.

Solution Place the angle in standard position. Choose a point on the terminal side, such as $P(-1, 0)$. Then $x = -1$, $y = 0$, and $r = 1$.

$\sin 180° = \frac{y}{r} = \frac{0}{1} = 0$

$\cos 180° = \frac{x}{r} = \frac{-1}{1} = -1$

$\tan 180° = \frac{y}{x} = \frac{0}{-1} = 0$

$\csc 180° = \frac{r}{y}$ is undefined since $y = 0$.

$\sec 180° = \frac{r}{x} = \frac{1}{-1} = -1$

$\cot 180° = \frac{x}{y}$ is undefined since $y = 0$.

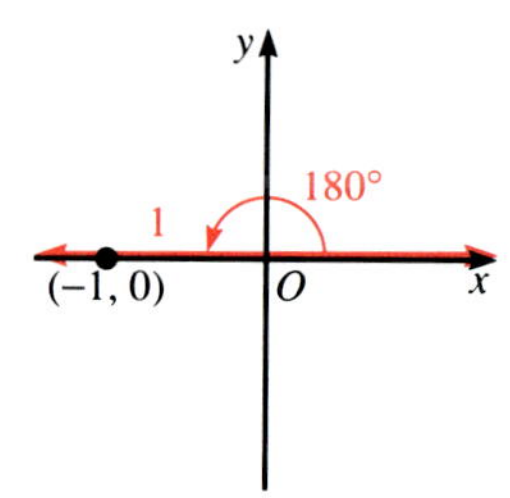

If θ is not a quadrantal angle, there is a unique acute angle α, corresponding to θ, formed by the terminal side of θ and the positive or negative x-axis. When θ is in standard position, we call α the **reference angle** of θ. Diagrams like those below are helpful in finding reference angles.

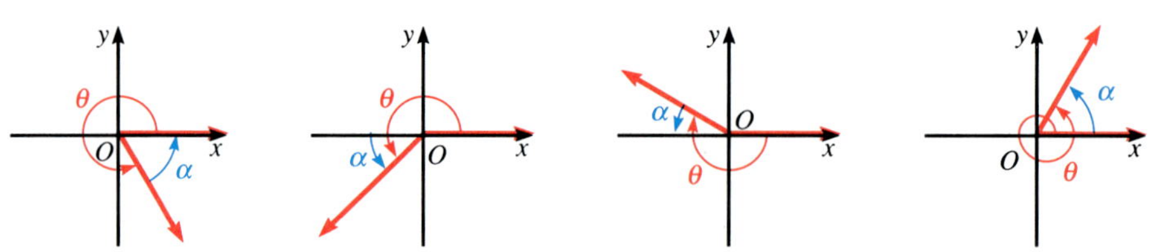

$\theta = 300°$, $\alpha = 60°$ | $\theta = 225°$, $\alpha = 45°$ | $\theta = -210°$, $\alpha = 30°$ | $\theta = 420°$, $\alpha = 60°$

Example 3 Find the measure of the reference angle α for each given angle θ.

a. $\theta = 140°$ **b.** $\theta = 300°10'$ **c.** $\theta = -135°$

Solution

a. Since θ is in quadrant II: $\alpha = 180° - \theta = 180° - 140° = 40°$

b. Since θ is in quadrant IV: $\alpha = 360° - \theta$
$= 360° - 300°10'$
$= 359°60' - 300°10' = 59°50'$

c. Find the positive angle that is coterminal with $-135°$.
$-135° + 360° = 225°$
Since 225° is in quadrant III: $\alpha = \theta - 180° = 225° - 180° = 45°$

You can use the reference angle to draw a **reference triangle** for any angle θ and then use the reference triangle to find values for the trigonometric functions of θ. For example, from right triangle OMP shown at the right, you see that $\cos \alpha = \frac{|x|}{r}$.

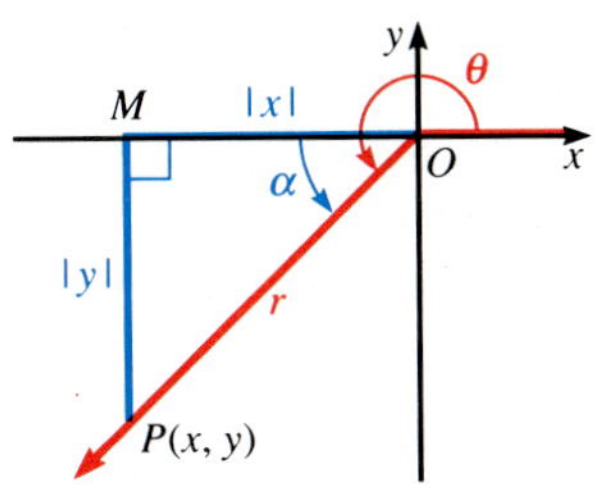

Since $\cos \theta = \frac{x}{r}$, you have

$$\cos \theta = \frac{x}{r} = \pm\frac{|x|}{r} = \pm\cos \alpha.$$

Similar results hold for the other trigonometric functions and for angles in other quadrants. Therefore, each trigonometric function of θ is either equal to the same function of α or equal to its opposite. That is, $\sin \theta = \pm\sin \alpha$, $\cos \theta = \pm\cos \alpha$, and so forth, where α is the reference angle of θ. The correct sign is determined by the quadrant of θ.

Example 4 Write cos 200° as a function of an acute angle.

Solution The reference angle of 200° is $200° - 180° = 20°$.
Since 200° is a third-quadrant angle, its cosine is negative.
$\therefore \cos 200° = -\cos 20°$. ***Answer***

Example 5 Find the exact value of the following.

a. tan 330° **b.** csc (−225°)

Solution

a. The reference angle of 330° is 30°.
Since 330° is a fourth-quadrant angle, its tangent is negative.
$\therefore \tan 330° = -\tan 30° = -\frac{\sqrt{3}}{3}$ ***Answer***

b. The reference angle of −225° is 45°.
Since −225° is a second-quadrant angle, its cosecant is positive.
$\therefore \csc(-225°) = \csc 45° = \sqrt{2}$ ***Answer***

Check for Understanding

Oral Exs. 1–12: use after Example 3.
Oral Exs. 13–20: use after Example 1.
Oral Exs. 21–28: use after the chart on page 562.
Oral Exs. 29–32: use after the box on page 561.

Guided Practice

1. Find the values of the six trigonometric functions of an angle θ in standard position if $P(-3, 4)$ is on the terminal side. $\sin \theta = \frac{4}{5}$, $\cos = -\frac{3}{5}$, $\tan \theta = -\frac{4}{3}$, $\cot \theta = -\frac{3}{4}$, $\sec \theta = -\frac{5}{3}$, $\csc \theta = \frac{5}{4}$
2. Give values of the six trigonometric functions of 360°. If any value is undefined, say so. sin 360° = 0, cos 360° = 1, tan 360° = 0, sec 360° = 1, cot 360° is undefined, csc 360° is undefined
3. Find the measure of the reference angle for the given angle.
   a. 130° 50° **b.** −230° 50°
4. Write as a function of an acute angle.
   a. sin 160° sin 20°
   b. cos (−300°) cos 60°
5. Find the exact value.
   a. sin 315° $-\frac{\sqrt{2}}{2}$
   b. cot (−30°) $-\sqrt{3}$

Summarizing the Lesson

Ask students to state how to find the reference angle for a given angle in each of the four quadrants. Also ask students to state which trigonometric functions have positive values in each of the four quadrants.

Additional Answers
Oral Exercises

13. $\sin\theta = \frac{3}{5}$, $\cos\theta = -\frac{4}{5}$, $\tan\theta = -\frac{3}{4}$, $\csc\theta = \frac{5}{3}$, $\sec\theta = -\frac{5}{4}$, $\cot\theta = -\frac{4}{3}$

14. $\sin\theta = -\frac{4}{5}$, $\cos\theta = \frac{3}{5}$, $\tan\theta = -\frac{4}{3}$, $\csc\theta = -\frac{5}{4}$, $\sec\theta = \frac{5}{3}$, $\cot\theta = -\frac{3}{4}$

15. $\sin\theta = \frac{\sqrt{2}}{2}$, $\cos\theta = \frac{\sqrt{2}}{2}$, $\tan\theta = 1$, $\csc\theta = \sqrt{2}$, $\sec\theta = \sqrt{2}$, $\cot\theta = 1$

16. $\sin\theta = -\frac{\sqrt{2}}{2}$, $\cos\theta = -\frac{\sqrt{2}}{2}$, $\tan\theta = 1$, $\csc\theta = -\sqrt{2}$, $\sec\theta = -\sqrt{2}$, $\cot\theta = 1$

17. $\sin\theta = 1$, $\cos\theta = 0$, $\tan\theta$ is undefined, $\csc\theta = 1$, $\sec\theta$ is undefined, $\cot\theta = 0$.

18. $\sin\theta = 0$, $\cos\theta = 1$, $\tan\theta = 0$, $\csc\theta$ is undefined, $\sec\theta = 1$, $\cot\theta$ is undefined.

19. $\sin\theta = -\frac{2\sqrt{13}}{13}$, $\cos\theta = -\frac{3\sqrt{13}}{13}$, $\tan\theta = \frac{2}{3}$, $\csc\theta = -\frac{\sqrt{13}}{2}$, $\sec\theta = -\frac{\sqrt{13}}{3}$, $\cot\theta = \frac{3}{2}$

20. $\sin\theta = \frac{\sqrt{5}}{5}$, $\cos\theta = -\frac{2\sqrt{5}}{5}$, $\tan\theta = -\frac{1}{2}$, $\csc\theta = \sqrt{5}$, $\sec\theta = -\frac{\sqrt{5}}{2}$, $\cot\theta = -2$

The identities proved for acute angles in Lesson 12-2 are true for general angles θ as well. If $\cos\theta \neq 0$, you have

$$\frac{\sin\theta}{\cos\theta} = \frac{\frac{y}{r}}{\frac{x}{r}} = \frac{y}{x} = \tan\theta \quad \text{and} \quad \frac{\cos\theta}{\sin\theta} = \frac{\frac{x}{r}}{\frac{y}{r}} = \frac{x}{y} = \cot\theta.$$

And, if $P(x, y)$ is at a distance r from the origin on the terminal side of θ, the distance formula gives

$$r = \sqrt{x^2 + y^2} \quad \text{or} \quad r^2 = x^2 + y^2.$$

Therefore, for any angle θ,

$$\sin^2\theta + \cos^2\theta = \frac{y^2 + x^2}{r^2} = 1.$$

Example 6 Find the five other trigonometric functions of θ if $\cos\theta = -\frac{2}{5}$ and $180° < \theta < 360°$.

Solution 1 Use $\sin^2\theta + \cos^2\theta = 1$.

$$\sin^2\theta + \left(-\frac{2}{5}\right)^2 = 1$$

$$\sin^2\theta = 1 - \frac{4}{25} = \frac{21}{25}$$

$$\sin\theta = \pm\sqrt{\frac{21}{25}}$$

$$\sin\theta = \frac{\sqrt{21}}{5} \quad \text{or} \quad \sin\theta = -\frac{\sqrt{21}}{5}$$

Since $\cos\theta$ is negative and $180° < \theta < 360°$, θ is a third-quadrant angle and $\sin\theta$ is negative.

$$\sin\theta = -\frac{\sqrt{21}}{5}$$

$$\tan\theta = \frac{\sin\theta}{\cos\theta} = \frac{-\frac{\sqrt{21}}{5}}{-\frac{2}{5}} = \frac{\sqrt{21}}{2}$$

The three remaining functions are reciprocals of those already known:

$$\sec\theta = -\frac{5}{2} \qquad \csc\theta = -\frac{5}{\sqrt{21}} = -\frac{5\sqrt{21}}{21} \qquad \cot\theta = \frac{2}{\sqrt{21}} = \frac{2\sqrt{21}}{21}$$

Solution 2 If $180° < \theta < 360°$ and $\cos\theta$ is negative, θ must be a third-quadrant angle.

Make a sketch as shown.

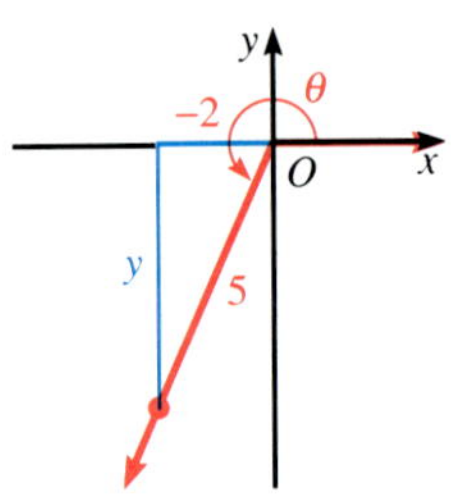

The Pythagorean theorem gives:
$y^2 = 5^2 - (-2)^2 = 21$ or $y = -\sqrt{21}$

Thus, $\sin\theta = -\frac{\sqrt{21}}{5}$ and $\tan\theta = \frac{-\sqrt{21}}{-2} = \frac{\sqrt{21}}{2}$.

The remaining functions are obtained as in Solution 1.

Oral Exercises

State the measure of the reference angle α for each angle θ.

1. 45°

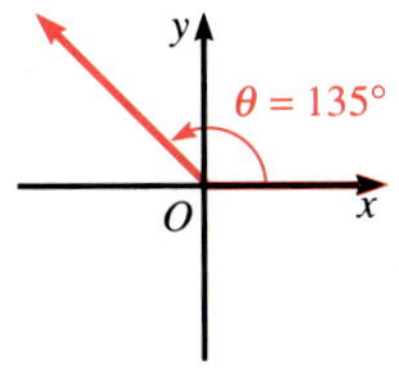

2. 60°

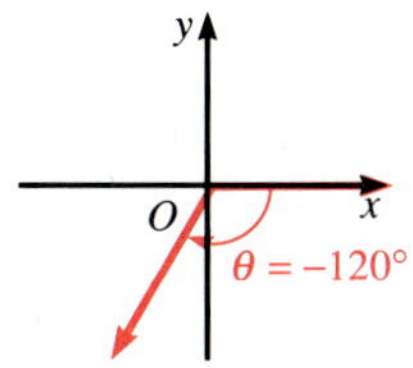

3. 30°

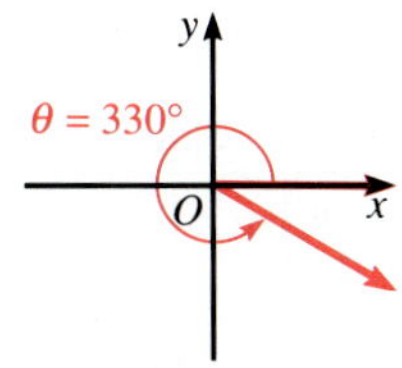

4. 45°

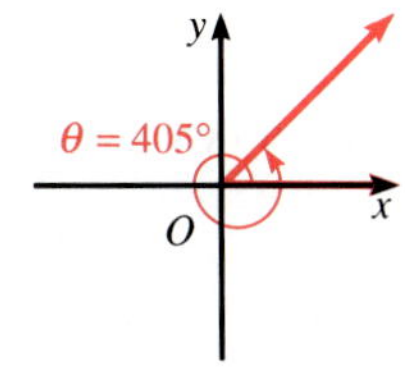

For each angle θ, name its quadrant and its reference angle.

5. $\theta = 170°$ II, 10° **6.** $\theta = 250°$ III, 70° **7.** $\theta = 210°$ III, 30° **8.** $\theta = -70°$ IV, 70°

9. $\theta = -305°$ I, 55° **10.** $\theta = -200°$ II, 20° **11.** $\theta = 460°$ II, 80° **12.** $\theta = -400°$ IV, 40°

Give the values of the six trigonometric functions of θ. If any value is undefined, say so.

13.

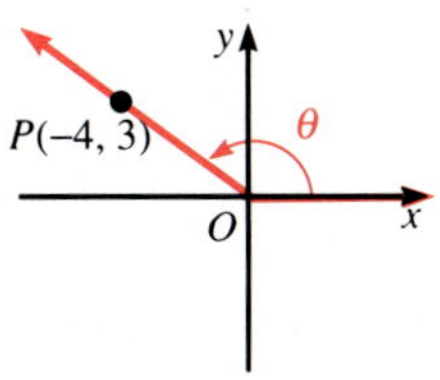

14.

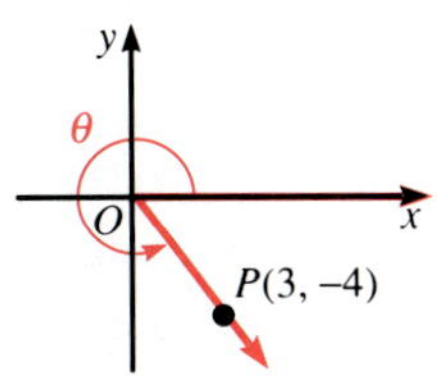

15.

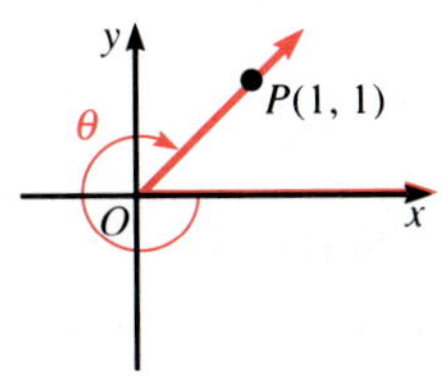

16.

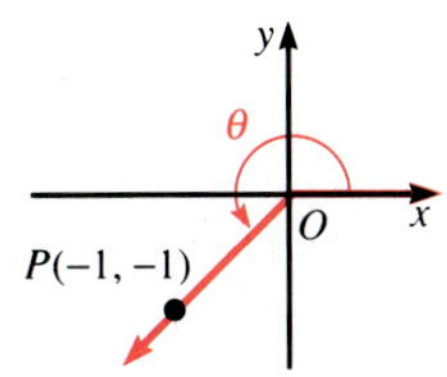

17.

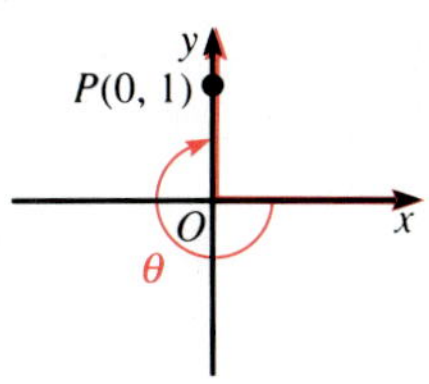

18.

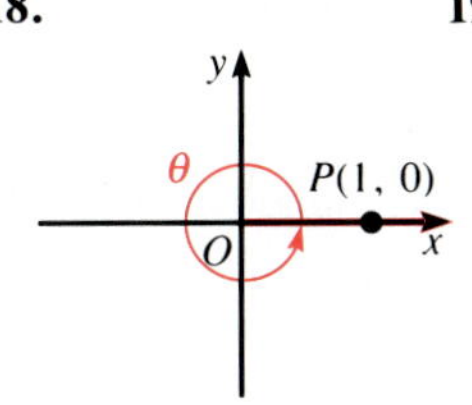

19.

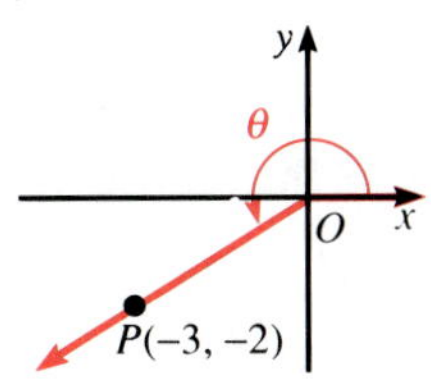

20.

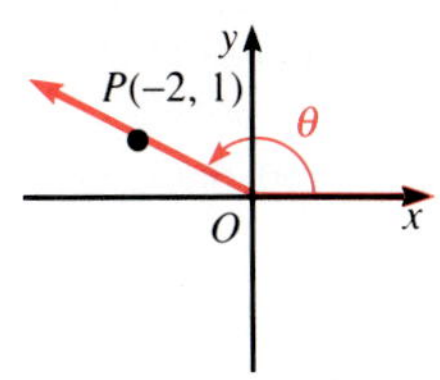

In Exercises 21–28, name the quadrant of θ. (Use the chart on page 562.)

21. $\sin\theta < 0$, $\cos\theta > 0$ IV

22. $\sin\theta < 0$, $\cos\theta < 0$ III

23. $\sin\theta > 0$, $\cos\theta > 0$ I

24. $\sin\theta > 0$, $\cos\theta < 0$ II

25. $\sec\theta > 0$, $\tan\theta < 0$ IV

26. $\csc\theta < 0$, $\cos\theta < 0$ III

27. $\cos\theta < 0$, $180° < \theta < 360°$ III

28. $\tan\theta > 0$, $90° < \theta < 270°$ III

29. If $\cos\theta = \frac{1}{4}$, $\sec\theta =$ __?__ 4

30. If $\sin\theta = -\frac{3}{5}$, $\csc\theta =$ __?__ $-\frac{5}{3}$

31. If $\cot\theta = -3$, $\tan\theta =$ __?__ $-\frac{1}{3}$

32. If $\sec\theta = 1.5$, $\cos\theta =$ __?__ $\frac{2}{3}$

Suggested Assignments

Average Alg. and Trig.
Day 1: 566/1–36
S 559/8–26 even
Day 2: 566/37–67
S 567/Mixed Review

Extended Alg. and Trig.
566/2–76 even
S 567/Mixed Review

Supplementary Materials

Study Guide	pp. 191–192
Practice Master	65
Resource Book	p. 152
Overhead Visual	10

Additional Answers Written Exercises

1. $\sin\theta = -\frac{4}{5}$, $\cos\theta = \frac{3}{5}$, $\tan\theta = -\frac{4}{3}$, $\csc\theta = -\frac{5}{4}$, $\sec\theta = \frac{5}{3}$, $\cot\theta = -\frac{3}{4}$

2. $\sin\theta = \frac{12}{13}$, $\cos\theta = -\frac{5}{13}$, $\tan\theta = -\frac{12}{5}$, $\csc\theta = \frac{13}{12}$, $\sec\theta = -\frac{13}{5}$, $\cot\theta = -\frac{5}{12}$

3. $\sin\theta = \frac{24}{25}$, $\cos\theta = -\frac{7}{25}$, $\tan\theta = -\frac{24}{7}$, $\csc\theta = \frac{25}{24}$, $\sec\theta = -\frac{25}{7}$, $\cot\theta = -\frac{7}{24}$

4. $\sin\theta = -\frac{15}{17}$, $\cos\theta = -\frac{8}{17}$, $\tan\theta = \frac{15}{8}$, $\csc\theta = -\frac{17}{15}$, $\sec\theta = -\frac{17}{8}$, $\cot\theta = \frac{8}{15}$

Additional Answers Written Exercises
(continued)

37. $\sin 330° = -\frac{1}{2}$, $\cos 330° = \frac{\sqrt{3}}{2}$, $\tan 330° = -\frac{\sqrt{3}}{3}$, $\csc 330° = -2$, $\sec 330° = \frac{2\sqrt{3}}{3}$, $\cot 330° = -\sqrt{3}$

38. $\sin 240° = -\frac{\sqrt{3}}{2}$, $\cos 240° = -\frac{1}{2}$, $\tan 240° = \sqrt{3}$, $\csc 240° = -\frac{2\sqrt{3}}{3}$, $\sec 240° = -2$, $\cot 240° = \frac{\sqrt{3}}{3}$

39. $\sin 135° = \frac{\sqrt{2}}{2}$, $\cos 135° = -\frac{\sqrt{2}}{2}$, $\tan 135° = -1$, $\csc 135° = \sqrt{2}$, $\sec 135° = -\sqrt{2}$, $\cot 135° = -1$

40. $\sin(-60°) = -\frac{\sqrt{3}}{2}$, $\cos(-60°) = \frac{1}{2}$, $\tan(-60°) = -\sqrt{3}$, $\csc(-60°) = -\frac{2\sqrt{3}}{3}$, $\sec(-60°) = 2$, $\cot(-60°) = -\frac{\sqrt{3}}{3}$

Written Exercises

Find the values of the six trigonometric functions of an angle θ in standard position whose terminal side passes through point P.

A **1.** $P(3, -4)$ **2.** $P(-5, 12)$ **3.** $P(-7, 24)$ **4.** $P(-8, -15)$

Complete the table. If any value is undefined, say so.

		$\sin\theta$	$\cos\theta$	$\tan\theta$	$\sec\theta$	$\csc\theta$	$\cot\theta$	
5.	0°	? 0	? 1	? 0	? 1	undef ?	?	undef
6.	90°	? 1	? 0	undef ?	undef ?	? 1	?	0
7.	180°	? 0	? −1	? 0	? −1	undef ?	?	undef
8.	270°	? −1	? 0	undef ?	undef ?	? −1	?	0

Find the measure of the reference angle α of the given angle θ.

9. $\theta = 233°$ 53° **10.** $\theta = 126°$ 54° **11.** $\theta = -205°$ 25° **12.** $\theta = -112°$ 68°

13. $\theta = 512°$ 28° **14.** $\theta = 659°$ 61° **15.** $\theta = -725°$ 5° **16.** $\theta = -611°$ 71°

17. $\theta = 96.4°$ 83.6° **18.** $\theta = 134.7°$ 45.3° **19.** $\theta = -184.1°$ 4.1° **20.** $\theta = -344.2°$ 15.8°

21. $\theta = 156°20'$ 23°40′ **22.** $\theta = 213°40'$ 33°40′ **23.** $\theta = 152°30'$ 27°30′ **24.** $\theta = 312°50'$ 47°10′

Write each of the following as a function of an acute angle.

25. cos 216° −cos 36° **26.** tan 334° −tan 26° **27.** sin (−17°) −sin 17°

28. sec (−106°) −sec 74° **29.** cot 287.1° −cot 72.9° **30.** csc 117.8° csc 62.2°

31. cos (−221.9°) −cos 41.9° **32.** sin (−46.6°) −sin 46.6° **33.** tan 265°20′ tan 85°20′

34. cot 137°10′ −cot 42°50′ **35.** sin 212°40′ −sin 32°40′ **36.** sin 324°30′ −sin 35°30′

Find the exact value of the six trigonometric functions of each angle.

37. 330° **38.** 240° **39.** 135° **40.** −60°

41. 315° **42.** 300° **43.** −150° **44.** 480°

First give the quadrant of angle θ. Then find the five other trigonometric functions of θ. Give answers involving radicals in simplest radical form.

B **45.** $\cos\theta = -\frac{8}{17}$, $0° < \theta < 180°$ **46.** $\sin\theta = -\frac{4}{5}$, $90° < \theta < 270°$

47. $\sin\theta = -\frac{5}{13}$, $\cos\theta > 0$ **48.** $\cos\theta = -\frac{3}{5}$, $\sin\theta > 0$

49. $\cos\theta = \frac{2}{3}$, $0° < \theta < 270°$ **50.** $\sin\theta = -\frac{2}{5}$, $-90° < \theta < 180°$

51. $\tan\theta = -\frac{3}{4}$, $\cos\theta < 0$ **52.** $\sec\theta = \frac{13}{5}$, $\sin\theta < 0$

Name all angles θ, $0° \le \theta < 360°$, that make the statement true.

53. $\sin \theta = 0$ 0°, 180°
54. $\cos \theta = 0$ 90°, 270°
55. $\tan \theta = 0$ 0°, 180°
56. $\sin \theta = 1$ 90°
57. $\tan \theta = 1$ 45°, 225°
58. $\cos \theta = -1$ 180°
59. $\sin \theta = -1$ 270°
60. $\sin \theta = \frac{1}{2}$ 30°, 150°
61. $\cos \theta = -\frac{\sqrt{3}}{2}$ 150°, 210°
62. $\sin \theta = \sin 300°$ 300°, 240°
63. $\tan \theta = \tan 15°$ 15°, 195°
64. $\sin \theta = \sin 40°$ 40°, 140°
65. $\cos \theta = \cos 70°$ 70°, 290°
66. $\cos \theta = \cos 100°$ 100°, 260°
67. $\tan \theta = \tan 20°$ 20°, 200°

Write each of the following in terms only of $\sin \theta$ and $\cos \theta$. (*Hint:* A sketch will be useful.)

C 68. $\sin(-\theta)$ $-\sin \theta$
69. $\cos(-\theta)$ $\cos \theta$
70. $\tan(-\theta)$ $-\frac{\sin \theta}{\cos \theta}$
71. $\sin(180° + \theta)$ $-\sin \theta$
72. $\sin(180° - \theta)$ $\sin \theta$
73. $\cos(180° + \theta)$ $-\cos \theta$
74. $\cos(180° - \theta)$ $-\cos \theta$
75. $\tan(180° + \theta)$ $\frac{\sin \theta}{\cos \theta}$
76. $\tan(180° - \theta)$ $-\frac{\sin \theta}{\cos \theta}$

Mixed Review Exercises

Evaluate.

1. $\sum_{k=1}^{8} 5 \cdot 2^{k-1}$ 1275
2. $\sum_{k=1}^{20} (3k + 1)$ 650
3. $\sum_{k=1}^{\infty} 10\left(\frac{3}{5}\right)^{k-1}$ 25

Find the zeros of each function. If the function has no zeros, say so.

4. $f(x) = 3x + 5$ $-\frac{5}{3}$
5. $g(x) = \log(x - 1)$ 2
6. $h(x) = 2x^2 + x - 3$ $-\frac{3}{2}$, 1
7. $F(x) = \frac{x^3 - 4x}{x^2 + 1}$ 0, 2, −2
8. $G(x) = x^2 - 4x + 5$ no real zeros
9. $H(x) = \left(\frac{1}{2}\right)^x$ no real zeros

Career Note / Flight Engineer

Modern aircraft are operated with the aid of many instruments. A flight engineer assists the pilot in checking the plane before takeoff, checks many of the instruments in flight, and records performance levels and fuel usage. Flight engineers report any mechanical problems to the pilot and sometimes make repairs. Flight engineers are licensed by the Federal Aviation Administration.

Most airlines look for candidates with a high school education and at least two years of college. Courses in mathematics and science are important.

41. $\sin 315° = -\frac{\sqrt{2}}{2}$, $\cos 315° = \frac{\sqrt{2}}{2}$, $\tan 315° = -1$, $\csc 315° = -\sqrt{2}$, $\sec 315° = \sqrt{2}$, $\cot 315° = -1$
42. $\sin 300° = -\frac{\sqrt{3}}{2}$, $\cos 300° = \frac{1}{2}$, $\tan 300° = -\sqrt{3}$, $\csc 300° = -\frac{2\sqrt{3}}{3}$, $\sec 300° = 2$, $\cot 300° = -\frac{\sqrt{3}}{3}$
43. $\sin(-150°) = -\frac{1}{2}$, $\cos(-150°) = -\frac{\sqrt{3}}{2}$, $\tan(-150°) = \frac{\sqrt{3}}{3}$, $\csc(-150°) = -2$, $\sec(-150°) = -\frac{2\sqrt{3}}{3}$, $\cot(-150°) = \sqrt{3}$
44. $\sin 480° = \frac{\sqrt{3}}{2}$, $\cos 480° = -\frac{1}{2}$, $\tan 480° = -\sqrt{3}$, $\csc 480° = \frac{2\sqrt{3}}{3}$, $\sec 480° = -2$, $\cot 480° = -\frac{\sqrt{3}}{3}$
45. II; $\sin \theta = \frac{15}{17}$, $\tan \theta = -\frac{15}{8}$, $\csc \theta = \frac{17}{15}$, $\sec \theta = -\frac{17}{8}$, $\cot \theta = -\frac{8}{15}$

(continued on p. 585)

Teaching Suggestions p. T123

Suggested Extensions p. T124

Reading Algebra p. T124

Group Activities p. T124

Warm-Up Exercises

Find the reference angle for the given angle.

1. 218° 38°

2. 481° 59°

3. 340.6° 19.4°

4. 95°20′ 84°40′

Express in decimal degrees.

5. 38°15′ 38.25°

6. 111°24′45″ 111.4125°

Motivating the Lesson

There are many applications of trigonometry in science, engineering, surveying, and so on. To be able to use trigonometry in these fields, students must be able to evaluate trigonometric functions of any angle readily. In this lesson they will learn how to use calculators and trigonometric tables for this purpose.

Chalkboard Examples

1. Find the following.
  a. cos 32.5°
  b. sin 78°10′
  a. Use a calculator or Table 4. cos 32.5° = 0.8434
  b. Use a calculator or Table 5. sin 78°10′ = 0.9787

2. Find sin 28°14′.
  Using a calculator: sin 28°14′ = sin 28.23° = 0.4730
  Using tables: sin 28°10′ < sin 28°14′ < sin 28°20′

12-4 Values of Trigonometric Functions

Objective To use a calculator or trigonometric tables to find values of trigonometric functions.

It is not possible to give *exact* trigonometric function values for most angles. You can, however, get good approximations using a scientific calculator. For example, to find cos 52°, enter 52 and press the cos key to obtain 0.6156615.

If a calculator is not available, you can use Table 4 or Table 5 on pages 814–826 to approximate the functions of an acute angle θ as follows:

1. If $0° \le \theta \le 45°$, find the function name in the *top* row and look *down* to the entry opposite the degree measure of θ at the extreme *left*.
2. If $45° \le \theta \le 90°$, find the function name in the *bottom* row and look *up* to the entry opposite the degree measure of θ at the extreme *right*.

Example 1 Find each function value to four significant digits.

a. sec 34.7° **b.** cos 84°20′

Solution 1 Using a Calculator

a. Many scientific calculators have keys labeled *sin, cos,* and *tan,* but not *sec, csc,* and *cot.* So to find values for secant, cosecant, and cotangent, you may need to use reciprocal functions (page 556).

$$\sec 34.7° = \frac{1}{\cos 34.7°} = \frac{1}{0.822144} = 1.2163319$$

$\therefore$ to four significant digits, sec 34.7° = 1.216. ***Answer***

b. Some calculators operate only with decimal degrees. First you may need to change 84°20′ to decimal degrees by dividing 20′ by 60. (60′ = 1°)

$$\cos 84°20' = \cos 84.333333° = 0.0987408$$

$\therefore$ to four significant digits, cos 84°20′ = 0.0987. ***Answer***

Solution 2 Using Tables

a. In Table 4 look *down* the column under "sec θ" until you see the entry opposite 34.7° at the left. You should find "1.216."

$\therefore$ sec 34.7° = 1.216 ***Answer***

b. In Table 5 look *up* the column over "Cos" until you see the entry opposite 84°20′ at the right. You should find "0.0987."

$\therefore$ cos 84°20′ = 0.0987 ***Answer***

Even though the values given by Table 4 and Table 5 are approximate, it's a common practice to use =, as in Example 1, rather than ≈.

Using a calculator you can easily find function values for angles with measures like 64.72° and 26°27′. However, these measures fall between consecutive entries in the tables. To use the tables, you may either use the table entry nearest the given measure or use *linear interpolation* (Lesson 8-9).

Example 2 Find cos 26°27′

Solution 1 Using a Calculator

You may need to change 26°27′ to decimal degrees by dividing 27′ by 60.

$\therefore$ cos 26°27′ = cos 26.45° = 0.8953234. ***Answer***

Solution 2 Using Tables

Notice that cos 26°20′ > cos 26°27′ > cos 26°30′. Using a vertical arrangement for the linear interpolation, you can write:

		θ	$\cos\theta$		
10′	7′	26°20′	0.8962	d	−0.0013
		26°27′	?		
		26°30′	0.8949		

A negative number because the function value decreases.

$$\frac{7}{10} = \frac{d}{-0.0013};\ d = \frac{7}{10}(-0.0013) = -0.0009$$

$\therefore$ cos 26°27′ = 0.8962 + (−0.0009) = 0.8953 ***Answer***

Example 3 Find the measure of the acute angle θ to the nearest tenth of a degree when $\sin\theta = 0.8700$.

Solution 1 Using a Calculator

To find an acute angle when given one of its function values, you use the inverse function keys ($\sin^{-1}$, $\cos^{-1}$, $\tan^{-1}$, or inv sin, inv cos, inv tan).

If $\sin\theta = 0.8700$, then

$$\theta = \sin^{-1} 0.8700 = 60.458639.$$

$\therefore$ to the nearest tenth of a degree, $\theta = 60.5°$. ***Answer***

Caution: $\sin^{-1} x$ is *a number whose sine is x*. The notation $\sin^{-1} x$ is not exponential notation. It does *not* mean $\frac{1}{\sin x}$.

Solution 2 Using Tables

Reverse the process described in Example 1. Look in the sine columns of Table 4 until you find the entry nearest 0.8700. This entry is 0.8704, and it is opposite 60.5° on the right.

$\therefore$ to the nearest tenth of a degree, $\theta = 60.5°$. ***Answer***

		θ	$\sin\theta$		
10′	4′	28°10′	0.4720	d	0.0026
		28°14′	?		
		28°20′	0.4746		

$\frac{4}{10} = \frac{d}{0.0026}$;

$d = (0.0026)\left(\frac{4}{10}\right)$

$= 0.001$

sin 28°14′ = 0.4720 + 0.001 = 0.4730

3. Find the measure of the acute angle θ to the nearest tenth of a degree given that $\cos\theta = 0.8330$. Using a calculator: $\theta = \cos^{-1} 0.8330 = 33.6°$. Using tables: Find the entry nearest 0.8330. This entry is 0.8329 and is opposite 33.6°. $\theta = 33.6°$

4. Find sin 132.4° and cos 132.4°. Using a calculator: sin 132.4 = 0.7385 cos 132.4° = −0.6743 Using tables: The reference angle is 47.6°. Since 132.4° is a second-quadrant angle, its sine is positive and its cosine is negative. sin 132.4° = sin 47.6° = 0.7385 cos 132.4° = −cos 47.6° = −0.6743

5. Find to the nearest tenth of a degree the two angles satisfying $\cos\theta = 0.7540$ and $0° < \theta < 360°$. Using a calculator: $\theta = \cos^{-1} 0.7540 = 41.1°$. One angle is 41.1°. The other angle is in the fourth quadrant, so its measure is 360° − 41.1°, or 318.9°. Using the tables: First find the reference angle α of θ. Since $\cos\theta = 0.7540$, $\alpha = 41.1°$. Since $\cos\theta > 0$, θ could be in the first or fourth quadrant. Thus, $\theta = 41.1°$ or $\theta = 360° - 41.1° = 318.9°$.

Common Errors

When students are not attentive, they forget that half the trigonometric functions are decreasing between 0° and 90°. Then they may treat a difference in table entries as positive instead of negative. (See Solution 2 of Example 2.) Review the tables to reinforce the fact that the cosine, cotangent, and cosecant are decreasing in the first quadrant.

Check for Understanding

Here is a suggested use of the Oral Exercises to check students' understanding as you teach the lesson.

Oral Exs. 1–12: use after Example 2.
Oral Exs. 13–18: use after Example 3.
Oral Exs. 19–24: use after Example 4.
Oral Ex. 25: use after Example 6.

Guided Practice

Find each function value to four significant digits.

1. cos 42.7° 0.7349
2. tan 50°40′ 1.220
3. sin 16°22′ 0.2818
4. If cos $\theta = 0.5500$, find the measure of the acute angle θ to the nearest tenth of a degree. 56.6°
5. If tan $\theta = 1.200$, find the measure of the acute angle θ to the nearest minute. 50°12′

Sometimes you may be asked to find solutions to the nearest minute instead of to the nearest tenth of a degree. You can use a calculator or tables to do this. To use the tables, you again use linear interpolation.

Example 4 Find the measure of the acute angle θ to the nearest minute when $\cot \theta = 0.0782$.

Solution 1 Using a Calculator

Since $\cot \theta = 0.0782$, $\tan \theta = \dfrac{1}{0.0782} = 12.787724$. Therefore,

$$\theta = \tan^{-1} 12.787724 = 85.52857°.$$

To convert 0.52857° to minutes, multiply by 60 to obtain 31.714191′.
$\therefore$ to the nearest minute, $\theta = 85°32'$. ***Answer***

Solution 2 Using Tables

First locate in Table 5 the nearest values for cot θ that are above and below 0.0782. Then arrange the values as follows.

θ	$\cot \theta$
85°40′	0.0758
?	0.0782
85°30′	0.0787

10′ [85°40′ to 85°30′]; d' [? to 85°30′]; −0.0005 [0.0782 to 0.0787]; −0.0029 [0.0758 to 0.0787]

A negative number because cot 85°40′ < cot 85°30′.

$$\frac{d}{10} = \frac{-0.0005}{-0.0029};\ d = \frac{5}{29}(10) = 2$$

$\therefore$ to the nearest minute, $\theta = 85°30' + 2'$, or $85°32'$. ***Answer***

The trigonometric functions of *any* angle can be found with the help of a calculator or with reference angles.

Example 5 Find sin 147.8° and cos 147.8°.

Solution 1 Using a Calculator

You can obtain the values directly.

$$\sin 147.8° = 0.5329$$
$$\cos 147.8° = -0.8462$$

Solution 2 Using Tables

The reference angle for 147.8° is $180° - 147.8° = 32.2°$. Since 147.8° is a second-quadrant angle, its sine is positive and its cosine is negative. Therefore:

$$\sin 147.8° = \sin 32.2° = 0.5329$$
$$\cos 147.8° = -\cos 32.2° = -0.8462$$

Example 6 Find to the nearest tenth of a degree the measures of two angles satisfying $\cos \theta = -0.7455$ and $0° < \theta < 360°$.

Solution 1 Using a Calculator

You can follow the method of Solution 2 of Example 5 and first find the reference angle α of θ. Or you can work with θ directly:

$$\theta = \cos^{-1}(-0.7455) = 138.20206.$$

Therefore, one angle is 138.2°. (This is a second-quadrant angle.)
Since $\cos \theta < 0$, θ is a second- or third-quadrant angle. To find the other angle, first subtract 138.2° from 180° to obtain a reference angle of 41.8°. Then the third-quadrant angle is $\theta = 180° + 41.8° = 221.8°$.

$\therefore$ the angles are 138.2° and 221.8°. ***Answer***

Solution 2 Using Tables

First find the reference angle α of θ.
Since $\cos \alpha = 0.7455$, $\alpha = 41.8°$.
Because $\cos \theta < 0$, θ is a second- or third-quadrant angle.
Therefore,
$\theta_1 = 180° - \alpha = 180° - 41.8° = 138.2°$ and
$\theta_2 = 180° + \alpha = 180° + 41.8° = 221.8°$. ***Answer***

Oral Exercises

Use a calculator, Table 4, or Table 5 to find each function value to four significant digits.

1. sin 15° 0.2588
2. cos 42° 0.7431
3. tan 22.6° 0.4163
4. cot 34.7° 1.444
5. cos 85.9° 0.0715
6. sin 58.2° 0.8499
7. csc 27°40′ 2.154
8. sec 49°20′ 1.535
9. cot 73°50′ 0.2899
10. tan 56°10′ 1.492
11. sec 12°30′ 1.024
12. csc 47°20′ 1.360

Use a calculator or Table 4 to find the measure of acute angle θ to the nearest tenth of a degree.

13. $\cos \theta = 0.8960$ 26.4°
14. $\sin \theta = 0.6160$ 38.0°
15. $\cot \theta = 7.596$ 7.5°
16. $\tan \theta = 1.632$ 58.5°
17. $\sec \theta = 2.089$ 61.4°
18. $\csc \theta = 1.095$ 66.0°

Use a calculator or Table 5 to find the measure of acute angle θ to the nearest ten minutes.

19. $\sin \theta = 0.5324$ 32°10′
20. $\cos \theta = 0.8511$ 31°40′
21. $\tan \theta = 1.455$ 55°30′
22. $\cot \theta = 0.4950$ 63°40′
23. $\csc \theta = 1.046$ 73°0′
24. $\sec \theta = 5.164$ 78°50′
25. Explain how you would find the measures of two angles θ between 0° and 360° such that $\sin \theta = -0.5000$.

Summarizing the Lesson

In this lesson students have learned to evaluate any trigonometric function of any given angle. Ask the students to explain when to use the inverse function keys on a calculator and when not to.

Additional Answers Oral Exercises

25. First find the reference angle for θ. Then find the third- and fourth-quadrant angles corresponding to the reference angle.

Suggested Assignments

Average Alg. and Trig.
572/2–46 even
S 567/68–76 even

Extended Alg. and Trig.
572/2–46 even
R 573/Self-Test 1

Supplementary Materials

Study Guide	pp. 193–194
Practice Master	66
Test Master	50

Quick Quiz

1. Name two angles, one positive and one negative, that are coterminal with the given angle. Answers may vary.
   a. $-140°$ 220°, −500°
   b. 310° 670°, −50°
2. Express 48°25′ in decimal degrees to the nearest hundredth of a degree. 48.42°
3. Express 56.32° in degrees, minutes, and seconds to the nearest second. 56°19′12″

Written Exercises

8. 2.525 **12.** 0.4473 **16.** 0.9993 **20.** 0.8490

Find each function value to four significant digits.

A

1. tan 15.2° 0.2717 **2.** cos 38.7° 0.7804 **3.** sin 65.2° 0.9078 **4.** cot 81.1° 0.1566

5. cos 31°10′ 0.8557 **6.** sin 46°30′ 0.7254 **7.** sec 54°40′ 1.729 **8.** csc 23°20′

9. csc 7.44° 7.723 **10.** sec 13.22° 1.027 **11.** cos 32.43° 0.8440 **12.** sin 26.57°

13. cos 52°43′ 0.6058 **14.** tan 29°36′ 0.5681 **15.** cot 78°15′ 0.2080 **16.** sin 87°53′

17. sec 111.3° −2.753 **18.** csc 163.4° 3.500 **19.** sin 268.5° −0.9997 **20.** cos 328.1°

21. cot 143°30′ −1.351 **22.** tan 173°20′ −0.1169 **23.** cos 231°30′ −0.6225 **24.** sin 312°40′ −0.7353

Find the measure of the acute angle θ to the nearest tenth of a degree.

25. $\sin \theta = 0.3400$ 19.9° **26.** $\cos \theta = 0.8400$ 32.9° **27.** $\cot \theta = 1.700$ 30.5°

28. $\tan \theta = 1.325$ 53.0° **29.** $\sec \theta = 3.555$ 73.7° **30.** $\csc \theta = 3.000$ 19.5°

Find the measure of the acute angle θ to the nearest minute.

31. $\cos \theta = 0.8621$ 30°27′ **32.** $\sin \theta = 0.2654$ 15°23′ **33.** $\tan \theta = 0.1482$ 8°26′

34. $\cos \theta = 0.2715$ 74°15′ **35.** $\sin \theta = 0.7321$ 47°4′ **36.** $\tan \theta = 2.550$ 68°35′

Find the measures of two angles between 0° and 360° with the given function value. Give answers to the nearest tenth of a degree.

B

37. $\sin \theta = 0.4875$ 29.2°, 150.8° **38.** $\cos \theta = 0.7851$ 38.3°, 321.7° **39.** $\tan \theta = -1.752$ 299.7°, 119.7°

40. $\sin \theta = -0.8300$ 303.9°, 236.1° **41.** $\cos \theta = 0.2524$ 75.4°, 284.6° **42.** $\tan \theta = 0.6182$ 31.7°, 211.7°

Find the measure of an angle θ between 0° and 360° that satisfies the stated conditions. Give answers to the nearest tenth of a degree.

43. $\cos \theta = 0.4275,\ 180° < \theta < 360°$ 295.3° **44.** $\sin \theta = -0.5212,\ 90° < \theta < 270°$ 211.4°

45. $\sin \theta = -0.6118,\ \cos \theta > 0$ 322.3° **46.** $\cos \theta = 0.7815,\ \tan \theta < 0$ 321.4°

Self-Test 1

Vocabulary

degree (p. 549)
initial side (p. 549)
terminal side (p. 549)
positive angle (p. 549)
negative angle (p. 549)
directed angle (p. 549)
standard position (p. 549)
quadrantal angle (p. 549)
coterminal angles (p. 550)
minute, second (p. 550)
sine (p. 555)
cosine (p. 555)
tangent (p. 555)
reciprocal functions (p. 556)
cotangent (p. 556)
secant (p. 556)
cosecant (p. 556)
identity (p. 556)
cofunction (p. 557)
reference angle (p. 562)

1. Name two angles, one positive and one negative, that are coterminal with the given angle. Answers may vary. **Obj. 12-1, p. 549**
a. 250° **b.** −300°
a. 610°, −110°
b. 60°, −660°

2. Express 32°35′ in decimal degrees to the nearest hundredth of a degree. 32.58°

3. Express 74.26′ in degrees, minutes, and seconds to the nearest second. 74°15′36″

4. When angle θ is in standard position, its terminal side passes through $(\sqrt{7}, 3)$. Find the values of the six trigonometric functions of θ. $\sin\theta = \frac{3}{4}$, $\cos\theta = \frac{\sqrt{7}}{4}$, $\tan\theta = \frac{3\sqrt{7}}{7}$, $\csc\theta = \frac{4}{3}$, $\sec\theta = \frac{4\sqrt{7}}{7}$, $\cot\theta = \frac{\sqrt{7}}{3}$ **Obj. 12-2, p. 555**

5. In the right triangle shown at the right, find a, b, and the measure of $\angle B$.

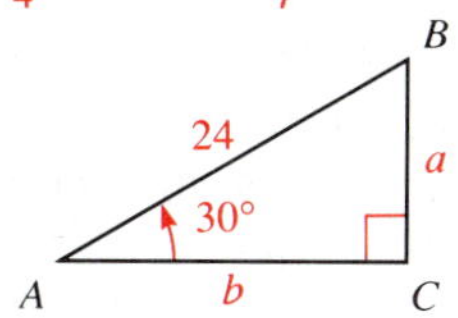

$\angle B = 60°$
$a = 12$
$b = 12\sqrt{3}$

6. Find the reference angle α for an angle with a measure of 300°. 60° **Obj. 12-3, p. 561**
Find the exact values of the six trigonometric functions of 300°, using radicals when necessary.

7. Find the following to four significant digits: **Obj. 12-4, p. 568**
a. cos 36°42′ 0.8018 **b.** csc 64.33° 1.110

8. Find the measure of angle θ to the nearest minute if θ is an acute angle and $\sin\theta = 0.8760$. 61°10′

Check your answers with those at the back of the book.

6. $\sin 300° = -\frac{\sqrt{3}}{2}$, $\cos 300° = \frac{1}{2}$
$\csc 300° = -\frac{2\sqrt{3}}{3}$, $\sec 300° = 2$
$\tan 300° = -\sqrt{3}$, $\cot 300° = -\frac{\sqrt{3}}{3}$

Historical Note / *Tables of Sines*

Among the first historical appearances of trigonometry were "tables of chords," recognizable as sine tables. Values for these early tables, such as the one devised by the Greek astronomer Hipparchus around 140 B.C., were found by using geometric methods to measure lengths of chords of a circle.

Modern trigonometric tables (and also the values found by a calculator or a computer) are computed by using terms of an infinite series. One such series is the *Taylor series* for sin x, where x is given in radian measure (see Lesson 13-1):

$$\sin x = x - \frac{x^3}{3!} + \frac{x^5}{5!} - \frac{x^7}{7!} + \ldots$$

By adding up the terms of this series you can approximate the sine of a given angle to any desired accuracy. For example, using the four terms shown, you find that

$$\sin 1 \approx 1 - \frac{1}{3!} + \frac{1}{5!} - \frac{1}{7!} \approx 0.84147,$$

a value that is correct to five decimal places.

4. When angle θ is in standard position, its terminal side passes through (12, 5). Find the values of the six trigonometric functions of θ.
$\sin\theta = \frac{5}{13}$, $\cos\theta = \frac{12}{13}$,
$\tan\theta = \frac{5}{12}$, $\cot\theta = \frac{12}{5}$,
$\sec\theta = \frac{13}{12}$, $\csc\theta = \frac{13}{5}$

5. In the right triangle shown below, find a, b, and the measure of $\angle B$.

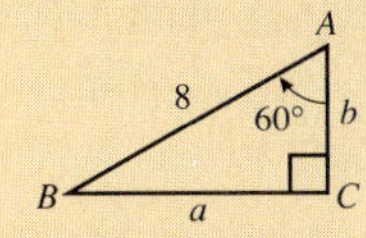

$4\sqrt{3}$, 4, 30°

6. Find the exact values of the six trigonometric functions of 240°. Use radicals when necessary.
$\sin 240° = -\frac{\sqrt{3}}{2}$,
$\cos 240° = -\frac{1}{2}$, $\tan 240° = \sqrt{3}$, $\cot 240° = \frac{\sqrt{3}}{3}$,
$\sec 240° = -2$, $\csc 240° = -\frac{2\sqrt{3}}{3}$

7. Find the following to four significant digits.
a. sin 62°22′ 0.8859
b. sec 14.25° 1.032

8. Find the measure of angle θ to the nearest minute if θ is an acute angle and $\cos\theta = 0.3620$.
68°47′

Teaching Suggestions p. T124

Suggested Extensions p. T125

Warm-Up Exercises

Solve. Give answers to the nearest tenth.

1. $\frac{a}{8} = 0.4500$ {3.6}
2. $\frac{5}{b} = 2.81$ {1.8}
3. $0.036x = 23$ {638.9}
4. $1.122a = 50$ {44.6}
5. $\sin 32° = \frac{b}{6}$ {3.2}

Motivating the Lesson

Ask students how they would answer these questions: How tall is that building? How high is that airplane? How far away is that ship? Tell students that these are a few of the questions that can be answered by solving a right triangle, the topic of today's lesson.

Chalkboard Examples

1. Solve the right triangle given $\angle A = 34.6°$ and $a = 120$. (See the figure above Example 1.)
Since $\angle A + \angle B = 90°$, $\angle B = 55.4°$.
$\tan A = \frac{120}{b}$
$\tan 34.6° = \frac{120}{b}$
$b = \frac{120}{\tan 34.6°} = 174$
$\sin A = \frac{120}{c}$
$\sin 34.6° = \frac{120}{c}$
$c = \frac{120}{\sin 34.6°} = 211$

574

Triangle Trigonometry

12-5 Solving Right Triangles

Objective To find the sides and angles of a right triangle.

Any triangle *ABC* has six measurements associated with it: the lengths of the three sides, denoted by a, b, and c, and the measures of the three angles, denoted by $\angle A$, $\angle B$, and $\angle C$. (Note that the lower-case letters denote the lengths of sides opposite angles labeled by the corresponding capital letters.) If any three of these measurements (other than the three angle measures) are given, then the other three can be found.

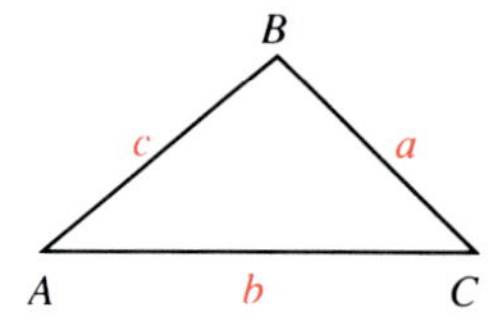

Finding measurements for all the sides and angles of a triangle is called **solving the triangle.**

In the case of a right triangle *ABC*, one angle is 90°. You can solve the triangle if the lengths of two sides, or the length of one side and the measurement of one acute angle, are known.

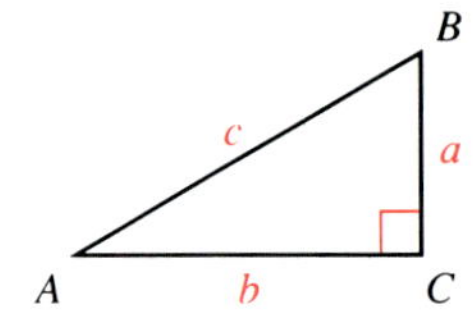

Example 1 Solve the right triangle shown above if $\angle A = 36°$ and $b = 50$.

Solution 1 Since $\angle A + \angle B = 90°$, $\angle B = 90° - \angle A = 90° - 36° = 54°$.

$$\cot A = \frac{b}{a} \qquad \cos A = \frac{b}{c}$$
$$\cot 36° = \frac{50}{a} \qquad \cos 36° = \frac{50}{c}$$
$$a = \frac{50}{\cot 36°} = \frac{50}{1.376} = 36.3 \qquad c = \frac{50}{\cos 36°} = \frac{50}{0.8090} = 61.8$$

$\therefore \angle B = 54°$, $a = 36.3$, and $c = 61.8$ ***Answer***

Solution 2 Again $\angle B = 90° - 36° = 54°$.

$$\tan A = \frac{a}{b} \qquad \sec A = \frac{c}{b}$$
$$\tan 36° = \frac{a}{50} \qquad \sec 36° = \frac{c}{50}$$
$$a = 50(\tan 36°) = 50(0.7265) = 36.3 \qquad b = 50(\sec 36°) = 50(1.236) = 61.8$$

$\therefore \angle B = 54°$, $a = 36.3$, and $c = 61.8$ ***Answer***

In Example 1, Solution 1 or Solution 2 are both well suited for calculator use. However, if tables are used, Solution 2 would be a better choice because long division is avoided.

Measurements of lengths and angles are approximations, as are the values given by tables and calculators. The following is a guide to the corresponding accuracies of length and angle measurements.

An angle measured to		A length measured to
1°	corresponds to	2 significant digits.
0.1° or 10′	corresponds to	3 significant digits.
0.01° or 1′	corresponds to	4 significant digits.

The figure at the right shows a helicopter and a pickup truck. To see the helicopter, the driver's line of sight must be raised, or *elevated,* at an angle θ above the horizontal. This angle θ is called the **angle of elevation** of the helicopter. Similarly, the angle ϕ is called the **angle of depression** of the truck from the helicopter. The angle of elevation and the corresponding angle of depression always have the same measure.

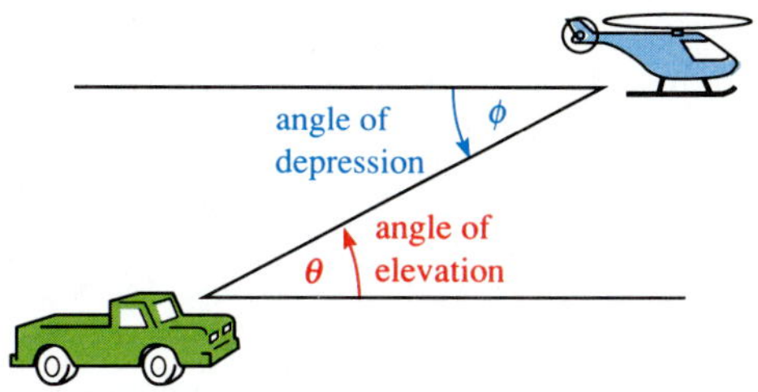

Example 2 A research ship finds that the angle of elevation of a volcanic island peak is 25.6°. After the ship has moved 1050 m closer to the island, the angle of elevation is 31.2°. What is the height of the peak above sea level? (Note that the sketch shown at the right is not drawn to scale.)

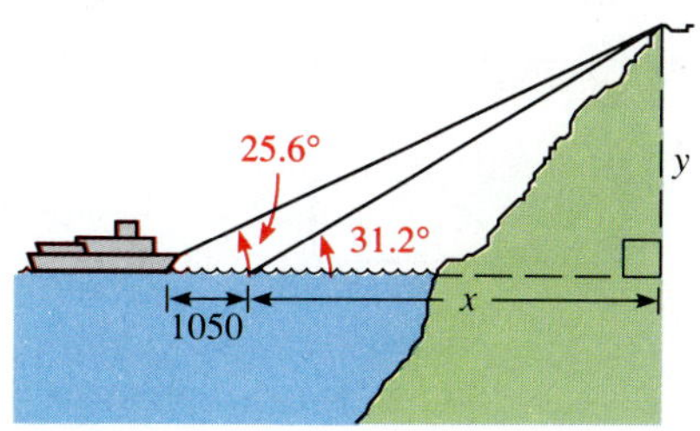

Solution $\cot 31.2° = \dfrac{x}{y}$ and $\cot 25.6° = \dfrac{1050 + x}{y}$

$x = y \cot 31.2°$ $\qquad y \cot 25.6° = 1050 + x$

$x = 1.651y$ $\qquad 2.087y = 1050 + x$

Substitute $1.651y$ for x in the second equation.

$$2.087y = 1050 + 1.651y$$
$$0.436y = 1050$$
$$y = \frac{1050}{0.436} = 2410 \text{ (to three significant digits)}$$

∴ the volcanic peak is 2410 m above sea level. ***Answer***

2. A sailor sights both the bottom and top of a lighthouse that is 20 m high and stands on the edge of a cliff. The two angles of elevation are 35° and 42°. To the nearest meter, how far is the boat from the cliff?

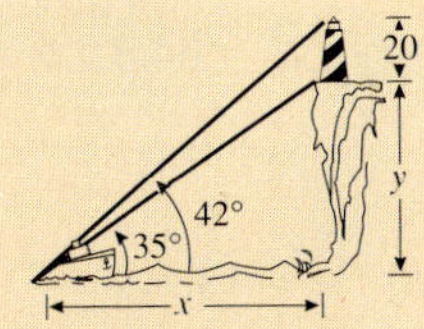

$\tan 35° = \dfrac{y}{x}$

$0.7002 = \dfrac{y}{x}$

$y = 0.7002x$

$\tan 42° = \dfrac{20 + y}{x}$

$0.9004 = \dfrac{20 + y}{x}$

$0.9004x = 20 + y$

$0.9004x = 20 + 0.7002x$

$x = 100$

∴ the boat is 100 m from the cliff.

3. Solve isosceles triangle ABC if $a = 12.4$ and $b = 17.6$. (See the figure in Example 3.)

$\cos B = \dfrac{6.2}{17.6} = 0.3523$

$\angle B = 69.4°$; since $\angle B = \angle C$, $\angle C = 69.4°$.

$\angle A = 180° - (\angle B + \angle C)$
$= 180° - 138.8° = 41.2°$

$\angle B = \angle C = 69.4°$, $\angle A = 41.2°$

Check for Understanding

Oral Exs. 1–6: use after Example 1.

Oral Exs. 7–10: use before Example 2.

Guided Practice

Draw a right triangle ABC with the given parts. Then solve the triangle. Give lengths to three significant digits and angle measures to the nearest tenth of a degree.

1. $\angle C = 90°$, $\angle B = 64.8°$, $b = 38.5$
   $\angle A = 25.2°$, $a = 18.1$, $c = 42.5$
2. $\angle C = 90°$, $a = 19.2$, $c = 47.8$
   $\angle A = 23.7°$, $\angle B = 66.3°$, $b = 43.8$
3. $\angle C = 90°$, $\angle A = 21.1°$, $c = 40.5$
   $\angle B = 68.9°$, $a = 14.6$, $b = 37.8$
4. $\angle C = 90°$, $\angle B = 83.2°$, $c = 127$
   $\angle A = 6.8°$, $a = 15.0$, $b = 126$
5. $\angle C = 90°$, $b = 480$, $a = 550$
   $c = 730$, $\angle A = 48.9°$, $\angle B = 41.1°$

Summarizing the Lesson

In this lesson students have learned to solve right triangles. Ask the students to explain how to solve a right triangle if the length of one leg and the measure of one acute angle are given.

Using a Calculator

Students learned to use their calculators to evaluate trigonometric functions in Lesson 12-4. They will find this helpful in the Written Exercises and the Problems, which require values of trigonometric functions and potentially tedious or difficult computations.

The altitude drawn in the diagram at the right below divides isosceles triangle ABC into two congruent right triangles. Constructing such an altitude will help you to solve any isosceles triangle.

Example 3 Solve isosceles triangle ABC if $a = 31.0$ and $c = 42.5$.

Solution Make a sketch as shown.

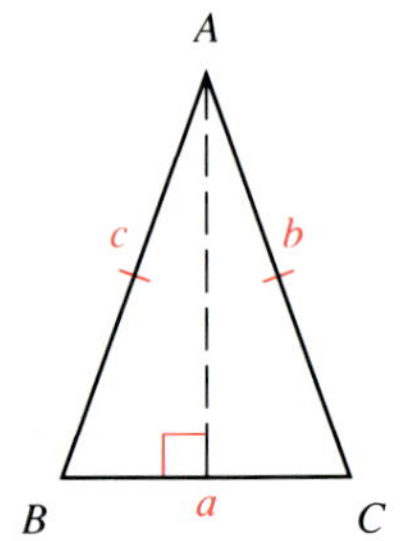

In the right triangle on the left in the diagram, the side adjacent to $\angle B$ has length $\frac{1}{2}a$, or 15.5. So

$$\cos B = \frac{15.5}{42.5} = 0.3647 \text{ and } \angle B = 68.6°.$$

Since ΔABC is isosceles,

$$\angle C = \angle B = 68.6° \text{ and } b = c = 42.5$$

$$\angle A = 180° - (\angle B + \angle C)$$
$$= 180° - 137.2° = 42.8°$$

$\therefore \angle B = 68.6°$, $\angle C = 68.6°$, $\angle A = 42.8°$, and $b = 42.5$ ***Answer***

Oral Exercises

Give an equation that can be used to find the value of x.

1. $\sin 41° = \frac{x}{16}$

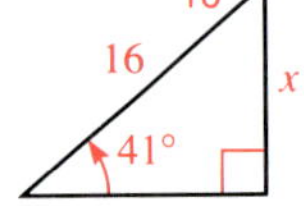

2. $\cot 35° = \frac{x}{16}$

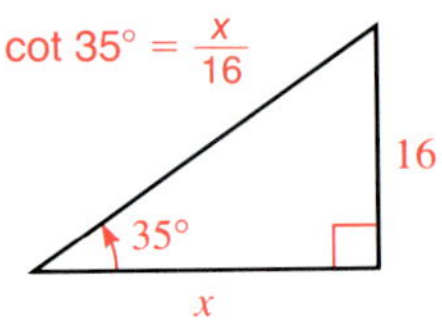

3. $\sec 65° = \frac{x}{10}$

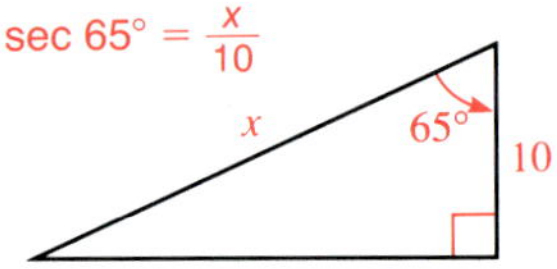

4. $\cos 38° = \frac{x}{12}$

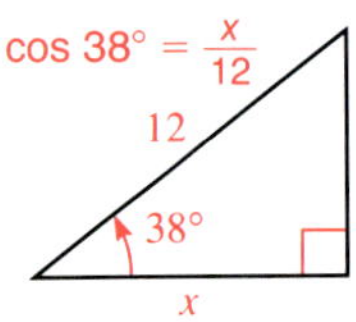

5. $\tan x° = \frac{8}{12}$

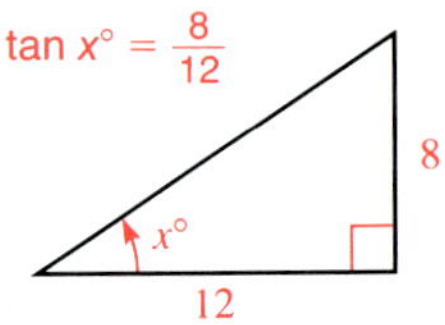

6. $\cos x° = \frac{11}{44}$

7. The angle of depression of Q from P measures 25°. Therefore, the angle of elevation of P from Q measures _?_. 25°

8. What fact from geometry did you use in answering the question in Exercise 7? When parallel lines are cut by a transversal, the alternate interior angles are congruent.

9. Give each angle to the nearest 10′.
a. 42°13′21″ 42°10′ **b.** 140°9′ 140°10′ **c.** 0°11′59″ 0°10′

10. Give each number to three significant digits.
a. 6.758 6.76 **b.** 1284.56 1280 **c.** 0.39975 0.400

Written Exercises

Give lengths to three significant digits and angle measures to the nearest tenth of a degree or nearest ten minutes. You may wish to use a calculator.

Solve each right triangle *ABC*.

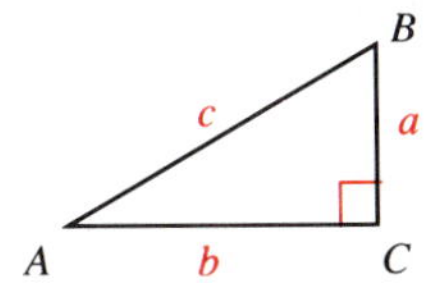

A

1. $\angle A = 36.2°$, $c = 68$
2. $\angle B = 15.8°$, $c = 12.2$
3. $\angle B = 65.4°$, $a = 2.35$
4. $\angle A = 82.1°$, $b = 246$
5. $\angle B = 48.3°$, $b = 74.7$
6. $\angle A = 24.0°$, $a = 5.25$
7. $a = 230$, $c = 320$
8. $a = 52.5$, $b = 28.0$
9. $a = 0.123$, $b = 0.315$
10. $b = 3.90$, $c = 42.5$
11. $\angle B = 58°10'$, $c = 420$
12. $\angle A = 38°40'$, $c = 42.5$
13. $\angle A = 15°30'$, $a = 4.50$
14. $\angle B = 67°20'$, $a = 450$
15. $\angle A = 30°50'$, $b = 53.5$
16. $\angle B = 85°10'$, $b = 0.620$

B

17. The radius of circle O is 13 cm and the length of $\overline{AB}$ is 10 cm. Find the measure of $\angle AOB$. 45.2°

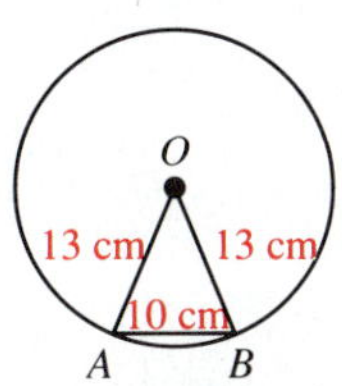

Ex. 17

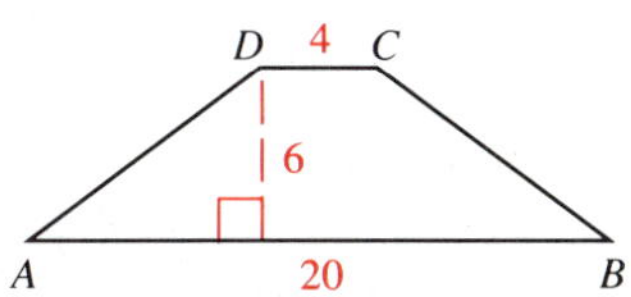

Ex. 18

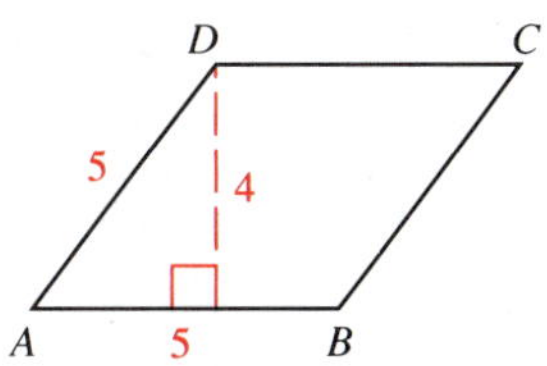

Ex. 19

18. The height of an isosceles trapezoid is 6 units and the bases have lengths 4 units and 20 units. Find the measures of the angles. $\angle A = \angle B = 36.9°$; $\angle C = \angle D = 143.1°$
19. A rhombus has sides 5 units long and its height is 4 units. Find its angles. $\angle A = \angle C = 53.1°$; $\angle B = \angle D = 126.9°$
20. a. Find the perimeter of a regular pentagon inscribed in a unit circle. 5.88
    b. Find the perimeter of a regular pentagon circumscribed about a unit circle. 7.27

(*Hint:* Use the shaded right triangles in the figures below.)

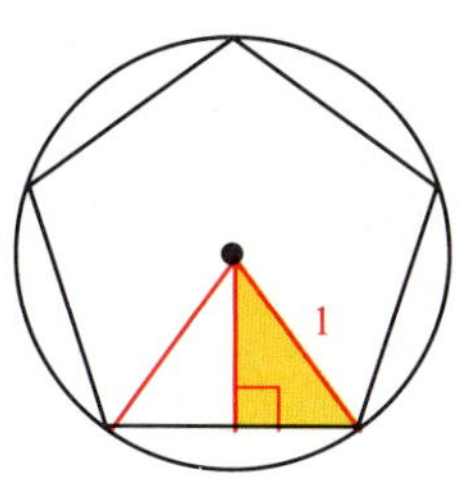

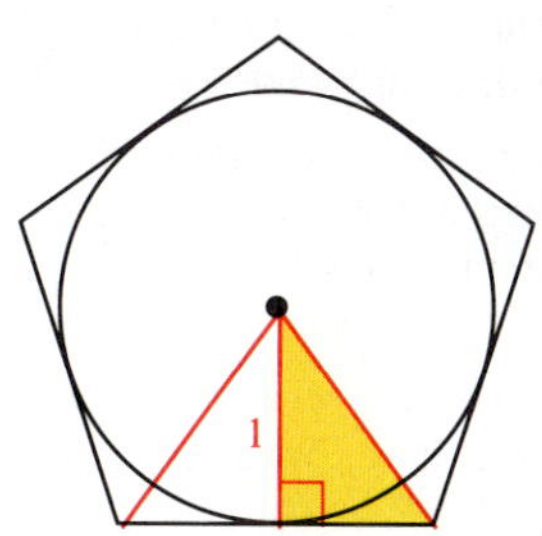

Suggested Assignments

Average Alg. and Trig.
Day 1: 577/1–19 odd
578/*P*: 1, 7
R 573/Self-Test 1
Day 2: 578/*P*: 2–14 even
S 579/Mixed Review

Extended Alg. and Trig.
577/3–21 mult. of 3
578/*P*: 2–16 even
S 579/Mixed Review

Additional Answers
Written Exercises

1. $\angle B = 53.8°$, $a = 40.2$, $b = 54.9$
2. $\angle A = 74.2°$, $a = 11.7$, $b = 3.32$
3. $\angle A = 24.6°$, $b = 5.13$, $c = 5.65$
4. $\angle B = 7.9°$, $a = 1770$, $c = 1790$
5. $\angle A = 41.7°$, $a = 66.6$, $c = 100$
6. $\angle B = 66.0°$, $b = 11.8$, $c = 12.9$
7. $b = 222$, $\angle A = 46.0°$, $\angle B = 44.0°$
8. $c = 59.5$, $\angle A = 61.9°$, $\angle B = 28.1°$
9. $c = 0.338$, $\angle A = 21.3°$, $\angle B = 68.7°$
10. $a = 42.3$, $\angle A = 84.7°$, $\angle B = 5.3°$
11. $\angle A = 31°50'$, $a = 222$, $b = 357$
12. $\angle B = 51°20'$, $b = 33.2$, $a = 26.6$
13. $\angle B = 74°30'$, $b = 16.2$, $c = 16.8$
14. $\angle A = 22°40'$, $b = 1080$, $c = 1170$
15. $\angle B = 59°10'$, $a = 31.9$, $c = 62.3$
16. $\angle A = 4°50'$, $a = 0.052$, $c = 0.622$

21. Repeat Exercise 20 for a regular polygon having 10 sides. **a.** 6.18 **b.** 6.50

22. Repeat Exercise 20 for a regular polygon having 20 sides. **a.** 6.26 **b.** 6.34

23. Explain why the number 2π lies between the answers to the (a) and (b) parts of Exercises 20, 21, and 22. 2π, the circumference of a unit circle, is greater than the perimeter of the inscribed polygon and less than the perimeter of the circumscribed polygon.

C 24. **a.** Repeat Exercise 20 for a polygon having n sides.

b. Use the results of part (a) to tell what number is approached by $n \sin\left(\frac{180}{n}\right)^\circ$ as n gets larger and larger. What number is approached by $n \tan\left(\frac{180}{n}\right)^\circ$ as n gets larger and larger?

a. $2n \sin\left(\frac{180}{n}\right)^\circ$, $2n \tan\left(\frac{180}{n}\right)^\circ$ **b.** Both approach π.

Problems

Give lengths to three significant digits and angle measures to the nearest tenth of a degree. You may wish to use a calculator.

1. 31.7° **2.** 53.1°

A 1. What is the angle of elevation of the sun when a tree 6.25 m tall casts a shadow 10.1 m long?

2. A boy flying a kite is standing 30 ft from a point directly under the kite. If the string to the kite is 50 ft long, find the angle of elevation of the kite.

3. A cable 4 m long is attached to a pole. The cable is staked to the ground 1.75 m from the base of the pole. Find the angle that the cable makes with the ground. 64.1°

4. How far from the base of a building is the bottom of a 30 ft ladder that makes an angle of 75° with the ground? 7.76 ft

5. The angle of elevation of the summit of a mountain from the bottom of a ski lift is 33°. A skier rides 1000 ft on this ski lift to get to the summit. Find the vertical distance between the bottom of the ski lift and the summit. 545 ft

6. The approach pattern to an airport requires pilots to set an 11° angle of descent toward the runway. If a plane is flying at an altitude of 9500 m, at what distance (measured along the ground) from the airport must the pilot start the descent? 48,900 m

7. The distance from the point directly under a hot air balloon to the point where the balloon is staked to the ground with a rope is 285 ft. The angle of elevation up the rope to the balloon is 48°. Find the height of the balloon. 317 ft

8. Opposite corners of a small rectangular park are joined by diagonal paths, each 360 m long. What are the dimensions of the park if the paths intersect at a 65° angle? 193 m by 304 m

Supplementary Materials

Study Guide	pp. 195–196
Practice Master	67
Computer Activity	27

Additional Answers
Written Exercises

(continued from p. 560)

7. $\sin\theta = \frac{\sqrt{10}}{10}$, $\cos\theta = \frac{3\sqrt{10}}{10}$, $\tan\theta = \frac{1}{3}$, $\csc\theta = \sqrt{10}$, $\sec\theta = \frac{\sqrt{10}}{3}$, $\cot\theta = 3$

8. $\sin\theta = \frac{\sqrt{2}}{2}$, $\cos\theta = \frac{\sqrt{2}}{2}$, $\tan\theta = 1$, $\csc\theta = \sqrt{2}$, $\sec\theta = \sqrt{2}$, $\cot\theta = 1$

33. $\cos\theta = \sqrt{1-u^2}$, $\tan\theta = \frac{u\sqrt{1-u^2}}{1-u^2}$, $\csc\theta = \frac{1}{u}$, $\sec\theta = \frac{\sqrt{1-u^2}}{1-u^2}$, $\cot\theta = \frac{\sqrt{1-u^2}}{u}$

34. $\sin\phi = \sqrt{1-v^2}$, $\tan\phi = \frac{\sqrt{1-v^2}}{v}$; $\csc\phi = \frac{\sqrt{1-v^2}}{1-v^2}$, $\sec\phi = \frac{1}{v}$, $\cot\phi = \frac{v\sqrt{1-v^2}}{1-v^2}$

B **9.** A pendulum in a grandfather clock is 160 cm long. The horizontal distance between the farthest points in a complete swing is 65 cm. Through what angle does the pendulum swing? 23.4°

10. A camping tent is supported by a rope stretched between two trees at a height of 210 cm. If the sides of the tent make an angle of 55° with the level ground, how wide is the tent at the bottom? 294 cm

11. From the top of a 135 ft observation tower, a park ranger sights two forest fires on opposite sides of the tower. If their angles of depression are 42.5° and 32.6°, how far apart are the fires? 358 ft

12. From a point 250 m from the base of a vertical cliff, the angles of elevation to the top and bottom of a radio tower on top of the cliff are 62.2° and 59.5°. How tall is the tower? 49.8 m

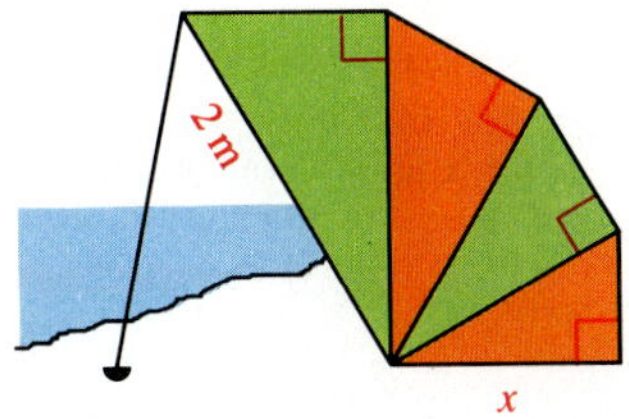

Ex. 13

13. The side of the beach shelter shown at the right is made up of four 30°–60°–90° triangles. Find the dimension marked x. 1.13 m

14. A pendulum 50 cm long is moved 26° from the vertical. How much is the lower end of the pendulum raised? 5.06 cm

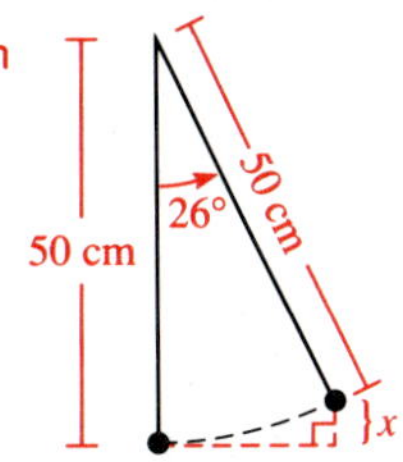

Ex. 14

C **15.** Two observers 1600 m apart on a straight, flat road measure the angles of elevation of a helicopter hovering over the road between them. If these angles are 32.0° and 50.5°, how high is the helicopter? 660 m

16. The pilot of a hot air balloon sees a field straight ahead that he knows is 1000 ft long. The angles of depression to the ends of the field are 25.4° and 34.7°. What is the height of the balloon and its horizontal distance from the nearer end of the field? height: 1510 ft; distance: 2180 ft

17. In a molecule of carbon tetrachloride, the four chlorine atoms are at the vertices of a regular tetrahedron, with the carbon atom in the center. What is the angle between two of the carbon-chlorine bonds (shown in red in the figure)? 109.4°

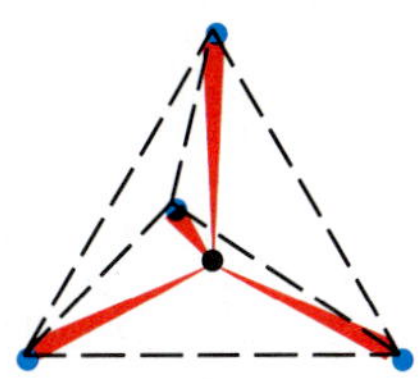

Ex. 17

Mixed Review Exercises

Give the exact values of the six trigonometric functions of each angle. If any function is not defined for the angle, say so.

1. 450° **2.** −135° **3.** 330° **4.** −240°

Write in simplest form without negative exponents.

5. $\frac{x^{-1}-1}{x-x^{-1}}$ $\frac{1}{x+1}$

6. $\sqrt{50m^4}$ $5m^2\sqrt{2}$

7. $\frac{1}{y+2}+\frac{4}{y^2-4}$ $\frac{1}{y-2}$

8. $(-5a^3b)^2(2a^2b)^3$ $200a^{12}b^5$

9. $(x^{2/3})^{-3/4}$ $\frac{1}{x^{1/2}}$

10. $(2u-3)(u^2+u-2)$ $2u^3-u^2-7u+6$

Additional Answers
Mixed Review Exercises

1. $\sin 450° = 1$, $\cos 450° = 0$, $\tan 450° =$ undef., $\csc 450° = 1$, $\sec 450° =$ undef., $\cot 450° = 0$

2. $\sin(-135°) = -\frac{\sqrt{2}}{2}$, $\cos(-135°) = -\frac{\sqrt{2}}{2}$, $\tan(-135°) = 1$, $\csc(-135°) = -\sqrt{2}$, $\sec(-135°) = -\sqrt{2}$, $\cot(-135°) = 1$

3. $\sin 330° = -\frac{1}{2}$, $\cos 330° = \frac{\sqrt{3}}{2}$, $\tan 330° = -\frac{\sqrt{3}}{3}$, $\csc 330° = -2$, $\sec 330° = \frac{2\sqrt{3}}{3}$, $\cot 330° = -\sqrt{3}$

4. $\sin(-240°) = \frac{\sqrt{3}}{2}$, $\cos(-240°) = -\frac{1}{2}$, $\tan(-240°) = -\sqrt{3}$, $\csc(-240°) = \frac{2\sqrt{3}}{3}$, $\sec(-240°) = -2$, $\cot(-240°) = -\frac{\sqrt{3}}{3}$

Teaching Suggestions p. T125

Suggested Extensions p. T125

Warm-Up Exercises

$\triangle ABC$ is a right triangle with right angle at C. Find the missing side to three significant digits.

1. $a = 2, b = 3, c =$ ___?___ 3.61
2. $a = b = 5, c =$ ___?___ 7.07
3. $a = 10, c = 20, b =$ ___?___ 17.3
4. $a = 5, b = 8, c =$ ___?___ 9.43
5. $b = 100, c = 150, a =$ ___?___ 112

Motivating the Lesson

Pose the following question to students: Suppose you and a friend set off hiking in different directions. The two of you walk at your respective paces, and at a given time you both stop. How far apart are you? Tell students that this question can be answered using the law of cosines, the topic of today's lesson.

Thinking Skills

As the law of cosines is stated in the box, it can be used to find a side of a triangle. Once the law is *transformed,* as shown after Example 1, a whole new perspective is gained; the law is now a formula for finding an angle of a triangle. Changing form or expressing a relationship in a different way can give added insight.

12-6 The Law of Cosines

Objective To use the law of cosines to find sides and angles of triangles.

If you solve the equations

$$\sin \theta = \frac{y}{r} \quad \text{and} \quad \cos \theta = \frac{x}{r}$$

for x and y you obtain the following useful fact.

If θ is an angle in standard position and P is a point on its terminal side, then the coordinates of P are $(r \cos \theta, r \sin \theta)$, where $r = OP$.

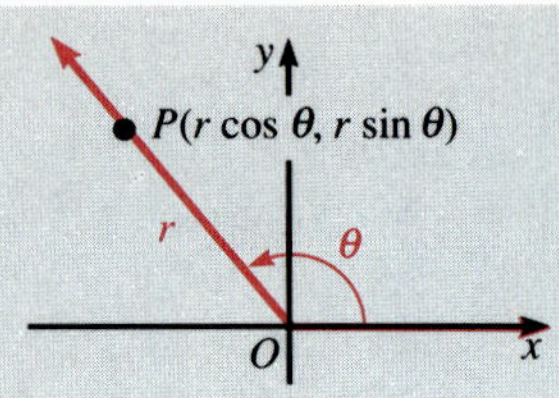

This fact is used to prove the *law of cosines.*

The Law of Cosines

In any triangle ABC,

$$c^2 = a^2 + b^2 - 2ab \cos C$$
$$b^2 = a^2 + c^2 - 2ac \cos B$$
$$a^2 = b^2 + c^2 - 2bc \cos A$$

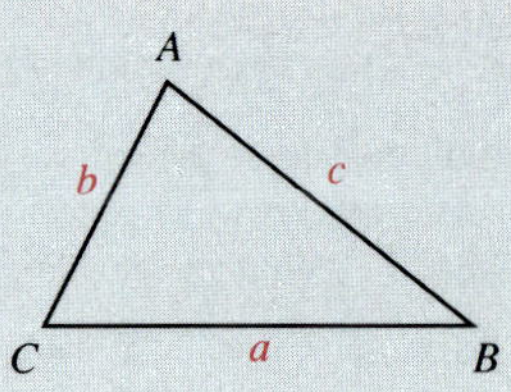

To prove the first form of the law of cosines, draw a coordinate system with $\angle C$ in standard position. (The diagrams below show an acute $\angle C$ and an obtuse $\angle C$ on such a coordinate system.) Then apply the distance formula to AB.

$$\begin{aligned} c^2 &= (b \cos C - a)^2 + (b \sin C - 0)^2 \\ &= b^2 \cos^2 C - 2ab \cos C + a^2 + b^2 \sin^2 C \\ &= a^2 + b^2(\cos^2 C + \sin^2 C) - 2ab \cos C \end{aligned}$$

But, since $\cos^2 C + \sin^2 C = 1$, the equation becomes

$$c^2 = a^2 + b^2 - 2ab \cos C.$$

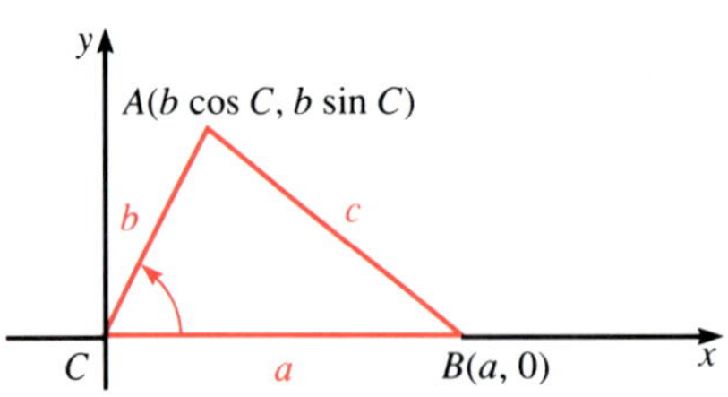

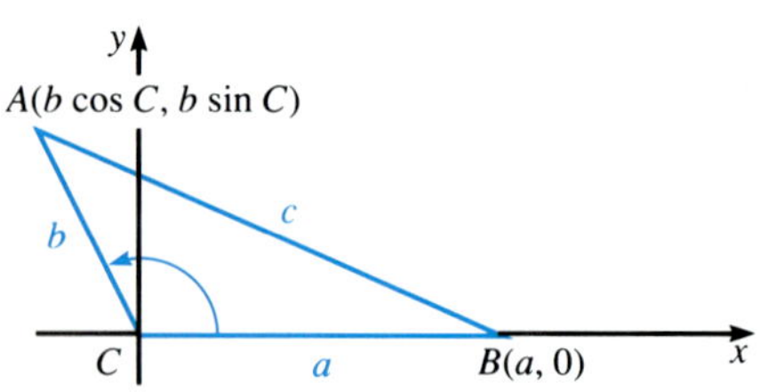

The other two forms of the law of cosines can be proved by repeating the process just shown with $\angle A$ and $\angle B$ in turn put into standard position. Notice that if $\angle C$ is a right angle, $\cos C = 0$, and the law of cosines reduces to the Pythagorean theorem.

Example 1 In $\triangle ABC$, $a = 10$, $b = 13$, and $\angle C = 70°$. Find c to three significant digits.

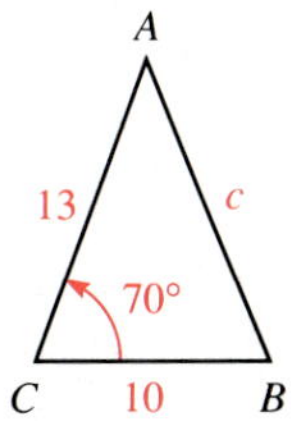

Solution

$$c^2 = a^2 + b^2 - 2ab \cos C$$
$$c^2 = 10^2 + 13^2 - 2(10)(13) \cos 70°$$
$$= 100 + 169 - 260(0.3420)$$
$$= 269 - 88.92$$
$$= 180.08$$
$$c = \sqrt{180.08} = 13.42$$

$\therefore$ to three significant digits, $c = 13.4$. ***Answer***

By rewriting the equation

$$c^2 = a^2 + b^2 - 2ab \cos C$$

in the form

$$\cos C = \frac{a^2 + b^2 - c^2}{2ab},$$

you can use the law of cosines to find the measure of an angle of a triangle when you know the lengths of the sides.

Example 2 A triangular-shaped lot has sides of length 50 m, 120 m, and 150 m. Find the largest angle of the lot to the nearest tenth of a degree.

Solution The largest angle of a triangle is opposite the longest side. Make a sketch. Let $a = 50$, $b = 120$, and $c = 150$. Substitute in this formula:

$$\cos C = \frac{a^2 + b^2 - c^2}{2ab}$$
$$\cos C = \frac{50^2 + 120^2 - 150^2}{2(50)(120)}$$
$$= \frac{2500 + 14{,}400 - 22{,}500}{12{,}000}$$
$$= \frac{-5600}{12{,}000}$$
$$= -0.4667$$

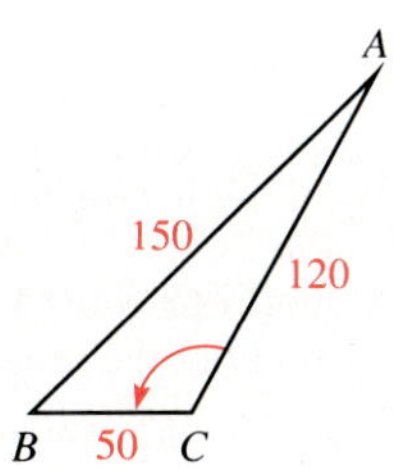

$\therefore \angle C = 117.8°$. ***Answer***

You may find it helpful to use a calculator to solve problems like the ones illustrated in Examples 1 and 2.

Chalkboard Examples

1. In $\triangle ABC$, $b = 10$, $c = 14$, and $\angle A = 54°$. Find a to three significant digits.

$$a^2 = b^2 + c^2 - 2bc \cos A$$
$$= 10^2 + 14^2 - 2(10)(14) \cos 54°$$
$$= 296 - 164.6$$
$$= 131.4$$
$$a = 11.5$$

2. Middleton is 8 km east of Mandra and Deerfield is 10 km northwest of Mandra. How far is Middleton from Deerfield?

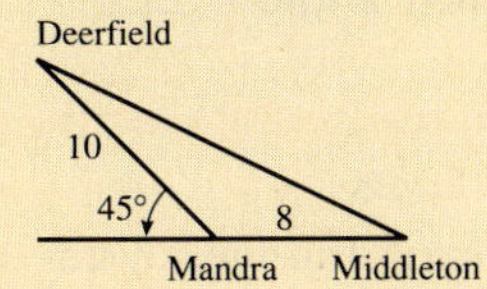

$$c^2 = 10^2 + 8^2 - 2(10)(8) \cos 135°$$
$$c^2 = 164 - 160 \cos 135°$$
$$c^2 = 277.13$$
$$c = 16.6$$

$\therefore$ it is 16.6 km from Middleton to Deerfield.

Check for Understanding

Here is a suggested use of the Oral Exercises to check students' understanding as you teach the lesson.

Oral Exs. 1–6: use after Example 2.

Using a Calculator

As students saw in Examples 1 and 2, solving problems using the law of cosines involves a long series of computations. The keys for the basic operations of addition, multiplication, and so forth, as well as the cosine and inverse cosine keys, are all put into service when using a calculator with the law of cosines.

Guided Practice

Find the indicated part of $\triangle ABC$. Give angles to the nearest 0.1° and lengths to three significant digits.

1. $b = 12$, $c = 10$, $\angle A = 38°$, $a =$ __?__ 7.41
2. $a = 30$, $c = 37$, $\angle B = 102.6°$, $b =$ __?__ 52.5
3. $a = 14$, $b = 15$, $c = 18$, $\angle A =$ __?__ 49.2°
4. $a = 18$, $b = 29$, $c = 16$, largest angle = __?__ 116.9°
   smallest angle = __?__ 29.5°

Common Errors

When using the law of cosines, students sometimes do not take the negative sign into account for an obtuse angle. When reteaching, remind students that the cosine of an acute angle is positive and that the cosine of an obtuse angle is negative.

Summarizing the Lesson

In this lesson students have learned to use the law of cosines to solve a triangle. Ask the students to state the parts of a triangle that must be known in order to use the law of cosines.

Oral Exercises

Use the law of cosines to give an equation involving the side or angle labeled x.

1.

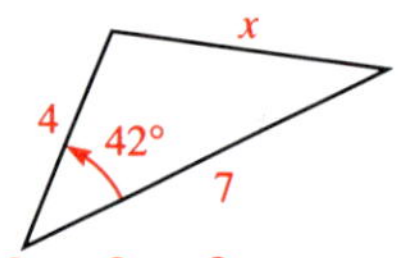

1. $x^2 = 4^2 + 7^2 - 2(4)(7)\cos 42°$

2.

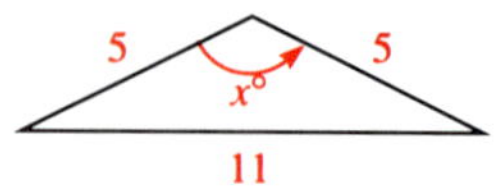

2. $\cos x° = \frac{5^2 + 5^2 - 11^2}{2(5)(5)}$

3.

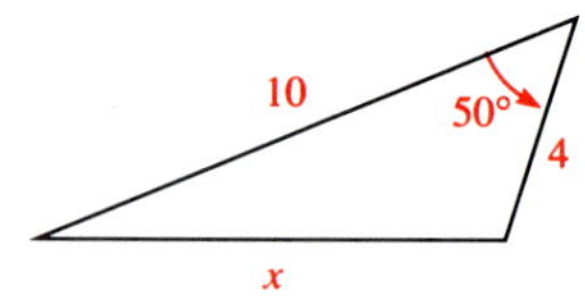

3. $x^2 = 10^2 + 4^2 - 2(10)(4)\cos 50°$

4.

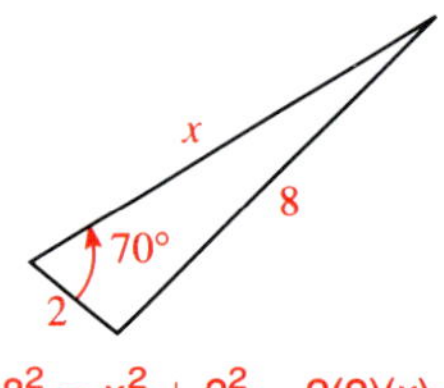

4. $8^2 = x^2 + 2^2 - 2(2)(x)\cos 70°$

5.

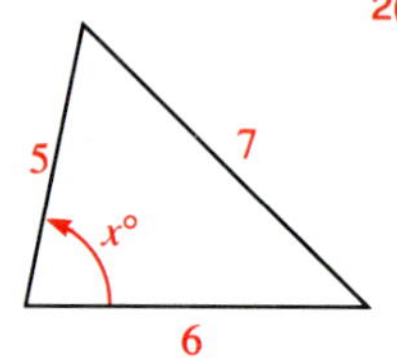

5. $\cos x° = \frac{5^2 + 6^2 - 7^2}{2(5)(6)}$

6.

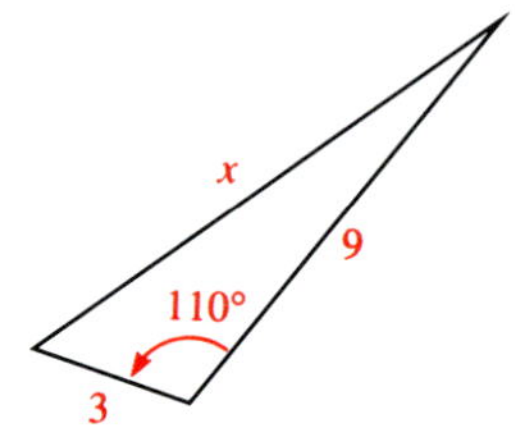

6. $x^2 = 3^2 + 9^2 - 2(3)(9)\cos 110°$

Written Exercises

Find lengths to three significant digits and the measures of the angles to the nearest tenth of a degree. You may wish to use a calculator.

In Exercises 1–12, find the indicated part of $\triangle ABC$.

A

1. $a = 6$, $b = 7$, $\angle C = 20°$, $c =$ __?__ 2.46
2. $b = 12$, $c = 17$, $\angle A = 74°$, $a =$ __?__ 17.9
3. $c = 15$, $a = 13$, $\angle B = 83°$, $b =$ __?__ 18.6
4. $b = 3$, $a = 4$, $\angle C = 40°$, $c =$ __?__ 2.57
5. $c = 15$, $b = 30$, $\angle A = 140°$, $a =$ __?__ 42.6
6. $a = 100$, $c = 200$, $\angle B = 150°$, $b =$ __?__ 291
7. $a = 8$, $b = 10$, $c = 12$, $\angle B =$ __?__ 55.8°
8. $a = 9$, $b = 10$, $c = 15$, $\angle C =$ __?__ 104.1°
9. $a = 13$, $b = 30$, $c = 40$, smallest angle = __?__ 13.8°
10. $a = 30$, $b = 20$, $c = 40$, largest angle = __?__ 104.5°
11. $a = 1.6$, $b = 0.9$, $c = 1.8$, largest angle = __?__ 87.4°
12. $a = 1.2$, $b = 2.4$, $c = 2.0$, smallest angle = __?__ 29.9°

B

13. Given a circle O, chord $AB = 10.1$, chord $BC = 15.5$ and $\angle ABC = 26°10'$. Find the length of chord AC. 7.83
14. A parallelogram has sides 6 cm and 8 cm and a 65° angle. Find the lengths of the diagonals. (Recall that adjacent angles of a parallelogram are supplementary.) 7.71 cm 11.9 cm
15. Three circles with radii 25 in., 36 in., and 30 in. are externally tangent to each other. Find the angles of the triangle formed by joining their centers. 69.2°, 51.1°, 59.7°

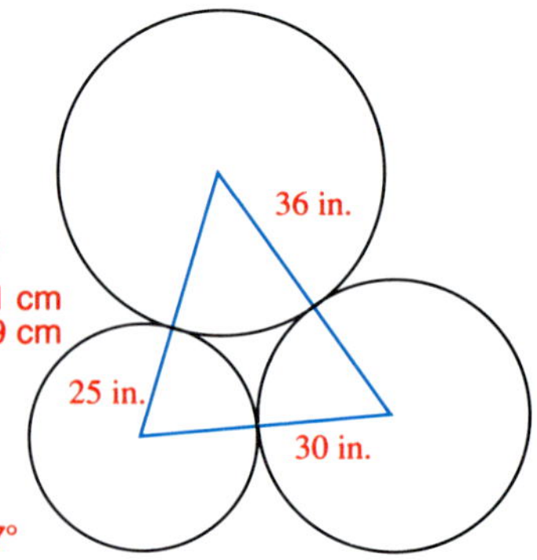

16. Find the lengths of the sides of a parallelogram whose diagonals intersect at a 35° angle and have lengths 6 and 10. (Recall that the diagonals of a parallelogram bisect each other.) 3.07, 7.65

17. In a parallelogram having sides of lengths a and b, let the diagonals have lengths d_1 and d_2. Show that $d_1^2 + d_2^2 = 2(a^2 + b^2)$. (*Hint:* Apply the law of cosines twice and use the fact that $\cos (180° - \theta) = -\cos \theta$.)

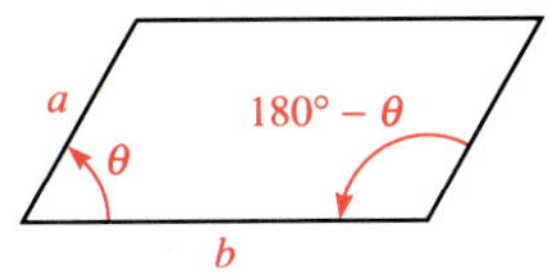

C **18.** In $\triangle ABC$, $a = 2$, $b = 4$, and $c = 3$. Find the length of the median from A to $\overline{BC}$. 3.39

19. Prove that $\cos 2\theta = 1 - 2 \sin^2 \theta$ for $0° < \theta < 90°$ as follows. First, using the triangle shown, express b^2 in two ways: (1) by using the law of cosines and (2) by first finding $\frac{b}{2}$ from the shaded right triangle. Then set the two expressions equal.

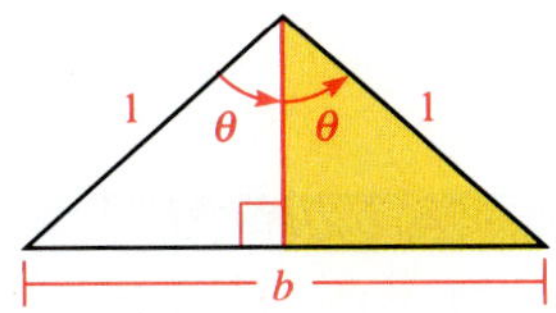

Problems

Find lengths to three significant digits and measures of angles to the nearest tenth of a degree. You may wish to use a calculator.

A **1.** A ranger in an observation tower can sight the north end of a lake 15 km away and the south end of the same lake 19 km away. The angle between these two lines of sight is 104°. How long is the lake? 26.9 km

2. Two planes leave an airport at the same time, one flying due west at 500 km/h and the other flying due southeast at 300 km/h. What is the distance between the planes two hours later? 1490 km

3. A triangular-shaped lot of land has sides of length 130 m, 150 m, and 80 m. What are the measures of the angles? 32.2°, 87.8°, 60.0°

4. Two streets meet at an angle of 52°. If a triangular lot has frontages of 60 m and 65 m on the two streets, what is the perimeter of the lot? 180 m

5. Newtown is 8 mi east of Oldtown and Littleton is 10 mi northwest of Oldtown. How far is Newtown from Littleton? 16.6 mi

6. An oil tanker and a cruise ship leave port at the same time and travel straight-line courses at 10 mi/h and 25 mi/h, respectively. Two hours later they are 40 mi apart. What is the angle between their courses? 49.5°

7. A baseball diamond is a square 90 ft on a side. The pitcher's mound is 60.5 feet from home plate. How far is it from the mound to first base? 63.7 ft

8. A water molecule consists of two hydrogen atoms and one oxygen atom joined as in the diagram. The distance from the nucleus of each hydrogen atom to the nucleus of the oxygen atom is 9.58×10^{-9} cm, and the bond angle θ is 104.8°. How far are the nuclei of the hydrogen atoms from each other? 1.52×10^{-8} cm

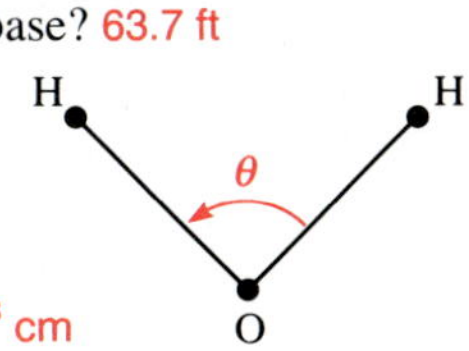

Suggested Assignments

Average Alg. and Trig.
Day 1: 582/1–17 odd
583/*P*: 2, 7
S 578/*P*: 3, 5
Day 2: 583/*P*: 1, 3, 8, 10
Assign with Lesson 12-7.

Extended Alg. and Trig.
582/2–18 even
583/*P*: 2, 7, 8, 10
S 579/*P*: 15

Supplementary Materials

Study Guide	pp. 197–198
Practice Master	68

Additional Answers Written Exercises

17. $d_1^2 = a^2 + b^2 - 2ab \cos \theta$; $d_2^2 = a^2 + b^2 - 2ab \cos (180° - \theta)$; $d_1^2 + d_2^2 = a^2 + b^2 - 2ab \cos \theta + a^2 + b^2 + 2ab \cos \theta = 2a^2 + 2b^2 = 2(a^2 + b^2)$

19. (1) $b^2 = 1^2 + 1^2 - 2(1)(1) \cos 2\theta = 2 - 2 \cos 2\theta$

(2) $\sin \theta = \frac{\frac{b}{2}}{1} = \frac{b}{2}$; $b = 2 \sin \theta$; $b^2 = 4 \sin^2 \theta$; $2 - 2 \cos 2\theta = 4 \sin^2 \theta$; $-2 \cos 2\theta = -2 + 4 \sin^2 \theta$; $\cos 2\theta = 1 - 2 \sin^2 \theta$

Using a Computer

The exercises require students to write programs that solve triangles. Note that Exercises 1–4 deal with right triangles, while Exercises 5–6 deal with general triangles. Also note that the computer's built-in trigonometric and inverse trigonometric functions use radian measures of angles, so that conversions to and from degree measures are necessary.

Additional Answers
Computer Exercises

2. a. $c = 64.4$, $\angle A = 53.8°$, $\angle B = 36.2°$
b. $b = 37$, $\angle A = 38.6°$, $\angle B = 51.4°$
c. $a = 6.86$, $\angle A = 49.2°$, $\angle B = 40.8°$

4. a. $b = 59.2$, $c = 69.8$, $\angle B = 58°$
b. $a = 65.7$, $c = 68$, $\angle A = 75°$
c. $a = 9.31$, $b = 3.35$, $\angle A = 70.2°$

6. a. $b = 20.58$, $\angle C = 40°$
b. $a = 48.33$, $\angle C = 25.9°$
c. $b = 8.45$, $\angle A = 52°$

B **9.** A large park in the shape of a parallelogram has diagonal paths that meet at a 60° angle. If the diagonal paths are 12 km and 20 km long, find the perimeter and the area of the park. perimeter = 45.4 km, area = 103.8 km^2

10. A flagpole 4 m tall stands on a sloping roof. A support wire 5 m long joins the top of the pole to a point on the roof 6 m up from the bottom of the pole. At what angle is the roof inclined to the horizontal? 34.2°

11. A vertical pole 20 m tall standing on a 15° slope is braced by two cables extending from the top of the pole to two points on the ground, 30 m up the slope and 30 m down the slope. How long are the cables? 31.5 m and 40.1 m

C **12.** The measures of two sides of a parallelogram are 50 cm and 80 cm, and one diagonal is 90 cm long. How long is the other diagonal? 98.5 cm

Computer Exercises

For students with some programming experience.

1. Given a right triangle ABC with $\angle C = 90°$ and the lengths of two sides, write a program that will find the length of the other side and the degree measures of the two acute angles. (*Hint:* You can find the measure of the acute angles by using a ratio of the lengths of the legs as the argument for the computer's built-in inverse tangent function ATN. This function gives an angle measure in radians (see Lesson 13-1). To convert radian measure R to degree measure D you can use the formula D = R*180/3.14159.)

2. Run the program in Exercise 1 to find the length of the remaining side and the measures of the remaining angles in each right triangle.

a. $a = 52$, $b = 38$ **b.** $a = 29.6$, $c = 47.4$ **c.** $b = 5.93$, $c = 9.07$

3. Given a right triangle ABC with $\angle C = 90°$ and the length of one side and the measure of one acute angle, write a program that will find the lengths of the other two sides and the measure of the third angle. (*Hint:* Remember that the computer's functions SIN(X), COS(X), and TAN(X) require an argument X in radians. To convert degree measure D to radian measure R you can use the formula R = D*3.14159/180.)

4. Run the program in Exercise 3 to find the lengths of the remaining sides and the measure of the remaining angle in each triangle.

a. $\angle A = 32°$, $a = 37$ **b.** $\angle B = 15°$, $b = 17.6$ **c.** $\angle B = 19.8°$, $c = 9.89$

5. Given the degree measure of two angles and the lengths of two sides of a triangle, write a program that will find the degree measure of the third angle and the length of the other side.

6. Run the program in Exercise 5 to find the measures of the remaining sides and angle in each triangle.

a. $\angle A = 68°$, $\angle B = 72°$, $c = 13.91$, $a = 20.06$
b. $\angle A = 51.3°$, $\angle B = 102.8°$, $c = 27.05$, $b = 60.39$
c. $\angle B = 115.6°$, $\angle C = 12.4°$, $a = 7.385$, $c = 2.012$

Reading Algebra / *Making a Sketch*

Read the following problem and try to visualize how the points, angles, and distances are related.

> Two points, X and Y, are separated by a swamp. To find the distance between them, Kelly walks 150 m from X to another point, Z. She estimates $\angle ZXY$ to be 45° and $\angle XZY$ to be 60°. Find the approximate distance from X to Y.

Can you picture the relative positions of X, Y, and Z? Many problems dealing with distance, angles, and geometric shapes are difficult to think about if you have only words to look at. Making a sketch as you read can help you understand the given information and see what is being asked for. For example, a sketch like the one below shows all the essential information in the problem stated above.

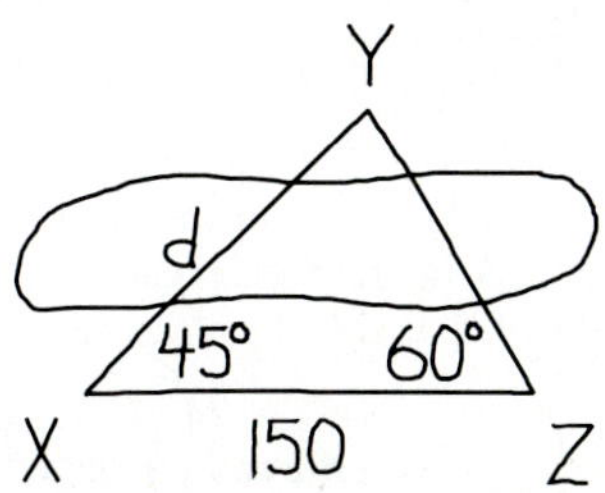

Notice that your sketch for a problem can be useful even if it is not to scale. For example, in the sketch above, there is no need to draw the angles exactly to scale, though we have shown them both as acute, with the larger one at Z.

Most diagrams in your textbook are drawn to scale. When a diagram is drawn for your book, the artist begins with complete information about dimensions. But you usually won't know the exact proportions of the shapes in your sketch until you have solved the problem. So don't wait until your solution is complete to make a drawing. Start right away with a rough sketch; you can improve it later as you discover more information.

Exercises

For each problem listed below, draw a sketch showing the essential information, labeling segments and angles with their given measures, and using variables of your choice to represent unknowns.

1. Problem 5, page 589
2. Problem 10, page 590
3. Problem 1, page 595
4. Problem 12, page 596

Additional Answers
Written Exercises
(continued from p. 567)

46. III; $\cos\theta = -\frac{3}{5}$, $\tan\theta = \frac{4}{3}$, $\csc\theta = -\frac{5}{4}$, $\sec\theta = -\frac{5}{3}$, $\cot\theta = \frac{3}{4}$

47. IV; $\cos\theta = \frac{12}{13}$, $\tan\theta = -\frac{5}{12}$, $\csc\theta = -\frac{13}{5}$, $\sec\theta = \frac{13}{12}$, $\cot\theta = -\frac{12}{5}$

48. II; $\sin\theta = \frac{4}{5}$, $\tan\theta = -\frac{4}{3}$, $\csc\theta = \frac{5}{4}$, $\sec\theta = -\frac{5}{3}$, $\cot\theta = -\frac{3}{4}$

49. I; $\sin\theta = \frac{\sqrt{5}}{3}$, $\tan\theta = \frac{\sqrt{5}}{2}$, $\csc\theta = \frac{3\sqrt{5}}{5}$, $\sec\theta = \frac{3}{2}$, $\cot\theta = \frac{2\sqrt{5}}{5}$

50. IV; $\cos\theta = \frac{\sqrt{21}}{5}$, $\tan\theta = -\frac{2\sqrt{21}}{21}$, $\csc\theta = -\frac{5}{2}$, $\sec\theta = \frac{5\sqrt{21}}{21}$, $\cot\theta = -\frac{\sqrt{21}}{2}$

51. II; $\sin\theta = \frac{3}{5}$, $\cos\theta = -\frac{4}{5}$, $\csc\theta = \frac{5}{3}$, $\sec\theta = -\frac{5}{4}$, $\cot\theta = -\frac{4}{3}$

52. IV; $\sin\theta = -\frac{12}{13}$, $\cos\theta = \frac{5}{13}$, $\tan\theta = -\frac{12}{5}$, $\csc\theta = -\frac{13}{12}$, $\cot\theta = -\frac{5}{12}$

Teaching Suggestions p. T125

Suggested Extensions p. T126

Warm-Up Exercises

Find the altitude h to three significant digits.

1.

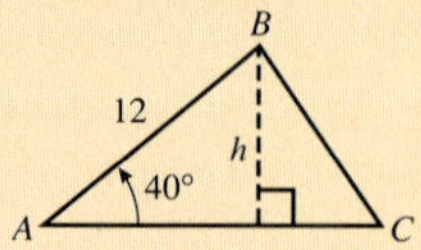

7.71

2.

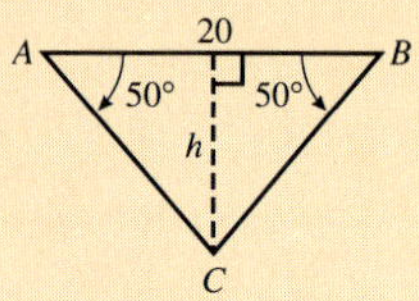

11.9

3.

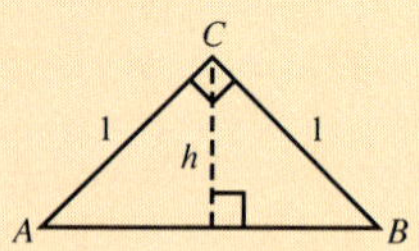

0.707

4.

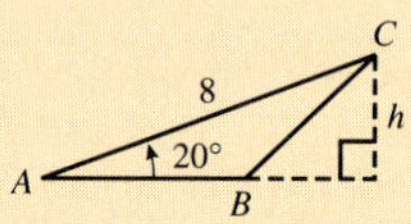

2.74

Motivating the Lesson

Ask students to name the triangle congruence properties from geometry. (SSS, ASA, SAS, and AAS) Then ask to which of these properties does the law of cosines apply? (SSS and SAS)

To complete the goal of solving all triangles, students need a law to handle the ASA and AAS properties. This law, called the law of sines, is the topic of today's lesson.

12-7 The Law of Sines

Objective To use the law of sines to find sides and angles of triangles.

The *law of sines,* like the law of cosines, allows you to use given information about three measurements of a triangle to determine the other measurements. For example, if you know the lengths of two sides and the measure of the included angle, you can find the length of the third side by using the law of cosines. But if the given angle is not the included angle, the law of cosines cannot be used and you need to use the law of sines.

The Law of Sines

In any triangle ABC,

$$\frac{\sin A}{a} = \frac{\sin B}{b} = \frac{\sin C}{c}$$

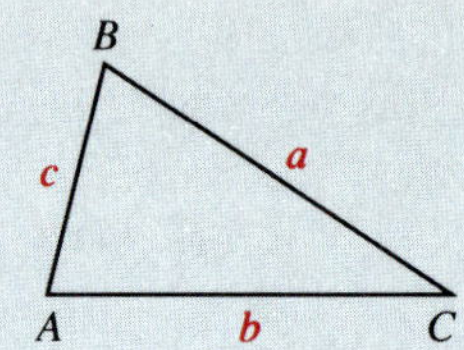

To prove the law of sines, you first find a formula for the area, K, of $\triangle ABC$. Draw a coordinate system with $\angle A$ in standard position. ($\angle A$ may be acute or obtuse.) Let h be the height of the triangle measured from B.

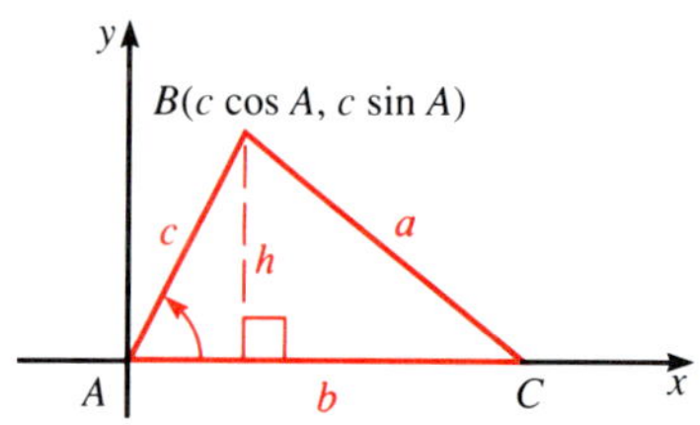

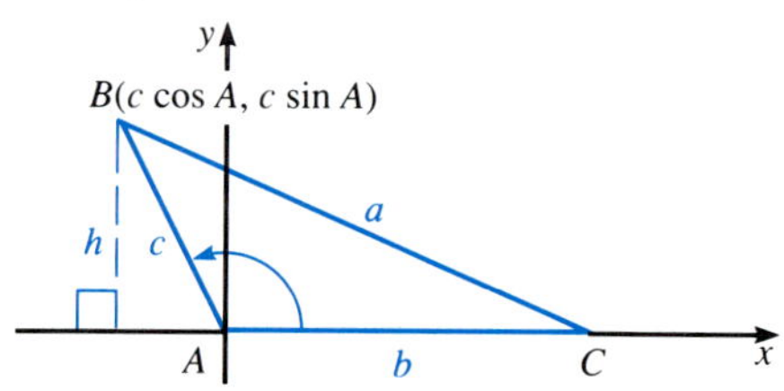

Since $K = \frac{1}{2}$(base)(height), you can write

$$K = \frac{1}{2}bh.$$

Since the y-coordinate of B equals h, you know $h = c \sin A$. Therefore,

$$K = \frac{1}{2}bc \sin A.$$

You can find two other formulas for the area K of $\triangle ABC$ by replacing $\angle A$ in turn with $\angle B$ and $\angle C$:

$$K = \frac{1}{2}ac \sin B \qquad K = \frac{1}{2}ab \sin C$$

Therefore, by substitution,

$$\frac{1}{2}bc \sin A = \frac{1}{2}ac \sin B = \frac{1}{2}ab \sin C.$$

By dividing each expression by $\frac{1}{2}abc$, you obtain the law of sines shown above.

Example 1 In $\triangle ABC$, $\angle A = 40°$ and $a = 15$.

a. Find $\angle B$ if $b = 20$. **b.** Find $\angle B$ if $b = 8$.

Solution Use the formula $\frac{\sin B}{b} = \frac{\sin A}{a}$.

a. $\frac{\sin B}{20} = \frac{\sin 40°}{15}$

$\sin B = \frac{20 \sin 40°}{15}$

$= \frac{20(0.6428)}{15}$

$= 0.8571$

$\angle B = 59°$ or $\angle B = 121°$

Since $40° + 59° = 99°$ and $40° + 121° = 161°$, and both $99°$ and $161°$ are less than $180°$, there are two solutions.

$\therefore \angle B = 59°$ or $\angle B = 121°$. ***Answer***

b. $\frac{\sin B}{8} = \frac{\sin 40°}{15}$

$\sin B = \frac{8 \sin 40°}{15}$

$= \frac{8(0.6428)}{15}$

$= 0.3428$

$\angle B = 20°$ or $\angle B = 160°$

$160°$ is not a solution because $\angle A + \angle B = 40° + 160° = 200°$, which is greater than $180°$.

$\therefore \angle B = 20°$. ***Answer***

If you are given two sides of a triangle and the angle opposite one of the sides, there may be two solutions, one solution, or no solution. You'll learn why in Lesson 12-8.

Example 2 A 123 ft support wire for a transmitting tower makes an angle of 61° with the ground. This wire is to be replaced by a new wire whose angle with the ground is 46°. How long will the new wire be?

Solution Let $\angle A = 46°$, and let c = the length of the new wire. $\angle ACB = 180° - 61° = 119°$, and $a = 123$ (ft).

$\frac{\sin A}{a} = \frac{\sin ACB}{c}$

$\frac{\sin 46°}{123} = \frac{\sin 119°}{c}$

$c = \frac{123 \sin 119°}{\sin 46°}$

$= \frac{123(0.8746)}{0.7193}$

$= 149.6$

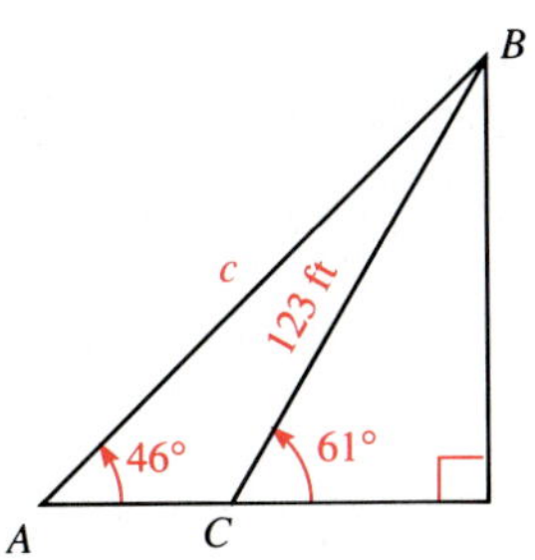

$\therefore$ the new wire will be 149.6 ft long. ***Answer***

Chalkboard Examples

1. In $\triangle PQR$, $\angle Q = 52°$, $p = 3$, and $q = 4$. Find $\angle P$ to the nearest tenth of a degree.

$\frac{\sin Q}{q} = \frac{\sin P}{p}$

$\frac{\sin 52°}{4} = \frac{\sin P}{3}$

$3(0.7880) = 4 \sin P$

$\sin P = 0.5910$

$\angle P = 36.2°$ or $143.8°$

But $143.8° + 52° > 180°$.

$\therefore \angle P = 36.2°$

2. The captain of a ship sighted a reef located so that the angle between the line of sight to the reef and the ship's course was 20°. After sailing 8 km, this angle was observed to be 70°. At that point, how far was the ship from the reef, to the nearest tenth of a kilometer?

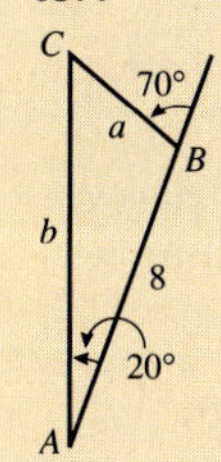

$\angle ABC = 110°$ and $\angle ACB = 50°$

$\frac{\sin A}{a} = \frac{\sin C}{c}$

$\frac{\sin 20°}{a} = \frac{\sin 50°}{8}$

$a = \frac{8 \sin 20°}{\sin 50°}$

$a = \frac{8(0.3420)}{0.7660} = 3.6$

$\therefore$ the ship is 3.6 km from the reef.

Check for Understanding

Here is a suggested use of the Oral Exercises to check students' understanding as you teach the lesson.
Oral Exs. 1–6: use before Example 1.

Using a Calculator

Although the calculations with the law of sines are generally not as complex as with the law of cosines, a calculator is certainly useful. Students should be made aware of the fact that when the law of sines is used to find a missing angle, the inverse sine key on a calculator gives only an acute angle. The angle's supplement, which has the same sine value, may also be a solution.

Guided Practice

Find the indicated part of $\triangle ABC$ to three significant digits or to the nearest tenth of a degree. If there are two solutions, give both.

1. $a = 52$, $\angle A = 74°$, $\angle C = 38°$, $c =$ __?__ 33.3
2. $b = 18$, $c = 32$, $\angle C = 100°$, $\angle B =$ __?__ 33.6°
3. $a = 29$, $b = 74$, $\angle A = 18°$, $\angle B =$ __?__ 52.0° or 128.0°
4. $c = 73$, $\angle A = 27°$, $\angle B = 34°$, $a =$ __?__ 37.9
5. Find the exact value of $\frac{a}{b}$ in $\triangle ABC$ if $\sin A = \frac{1}{2}$ and $\cos B = \frac{3}{4}$. $\frac{2\sqrt{7}}{7}$

Oral Exercises

Given triangle *ABC* with sides *a*, *b*, and *c*, use the law of sines to find an expression equivalent to the given one.

1. $\frac{b}{a}$ $\frac{\sin B}{\sin A}$
2. $\frac{a}{\sin A}$ $\frac{b}{\sin B}$
3. $\frac{\sin A}{\sin B}$ $\frac{a}{b}$
4. $\frac{a \sin B}{b \sin A}$ 1
5. $\frac{a \sin B}{b}$ $\sin A$
6. $\frac{\cos A}{\sin A}$ if $\angle C = 90°$ $\frac{b}{a}$

Written Exercises

Find the indicated part of $\triangle ABC$ to three significant digits or to the nearest tenth of a degree. If there are two solutions, give both. You may wish to use a calculator.

A

1. $a = 14$, $\angle A = 25°$, $\angle B = 75°$, $b =$ __?__ 32.0
2. $c = 12$, $\angle A = 42°$, $\angle C = 69°$, $a =$ __?__ 8.60
3. $b = 3.40$, $\angle A = 110°$, $\angle C = 50°$, $a =$ __?__ 9.34
4. $a = 2.60$, $\angle B = 60°$, $\angle C = 100°$, $c =$ __?__ 7.49
5. $c = 35$, $\angle A = 38°$, $\angle C = 102°$, $b =$ __?__ 23.0
6. $b = 130$, $\angle B = 95°$, $\angle C = 35°$, $a =$ __?__ 100
7. $a = 4$, $b = 3$, $\angle A = 40°$, $\angle B =$ __?__ 28.8°
8. $a = 4.5$, $b = 6.0$, $\angle B = 35°$, $\angle A =$ __?__ 25.5°
9. $a = 4.0$, $c = 6.4$, $\angle C = 125°$, $\angle B =$ __?__ 24.2°
10. $a = 18$, $b = 12$, $\angle A = 110°$, $\angle C =$ __?__ 31.2°
11. $a = 5$, $c = 7$, $\angle A = 42°$, $\angle C =$ __?__ 69.5° or 110.5°
12. $b = 15$, $c = 11$, $\angle C = 40°$, $\angle B =$ __?__ 61.2° or 118.8°

In Exercises 13–18, find the exact value of $\frac{a}{b}$ in $\triangle ABC$ without using tables or calculators. (Recall that $\sin^2 \theta + \cos^2 \theta = 1$.)

Sample $\sin A = \frac{3}{4}$, $\cos B = \frac{2}{3}$

Solution The law of sines can be written in the form $\frac{a}{b} = \frac{\sin A}{\sin B}$.
You need to find $\sin B$. Since $\sin^2 B + \cos^2 B = 1$, you have $\sin^2 B = 1 - \cos^2 B = 1 - \left(\frac{2}{3}\right)^2 = \frac{5}{9}$. Therefore, $\sin B = \frac{\sqrt{5}}{3}$.

By substitution, $\frac{a}{b} = \frac{\sin A}{\sin B} = \frac{\frac{3}{4}}{\frac{\sqrt{5}}{3}} = \frac{9}{4\sqrt{5}} = \frac{9\sqrt{5}}{20}$.

13. $\sin A = \frac{2}{3}$, $\cos B = \frac{4}{5}$ $\frac{10}{9}$

14. $\cos A = \frac{3}{5}$, $\sin B = \frac{4}{5}$ 1

15. $\cos A = \frac{5}{13}$, $\cos B = \frac{3}{5}$ $\frac{15}{13}$

16. $\cos A = \frac{8}{17}$, $\cos B = \frac{15}{17}$ $\frac{15}{8}$

17. $\cos A = \frac{\sqrt{3}}{2}$, $\cos B = \frac{\sqrt{2}}{2}$ $\frac{\sqrt{2}}{2}$

18. $\cos A = \frac{1}{2}$, $\tan B = 1$ $\frac{\sqrt{6}}{2}$

Show that the following formulas are true for any triangle *ABC*.

B 19. $\frac{\sin A + \sin B}{\sin B} = \frac{a + b}{b}$

20. $\frac{\sin A - \sin B}{\sin B} = \frac{a - b}{b}$

21. Prove the formula $\sin 2\theta = 2 \sin \theta \cos \theta$ for $0° < \theta < 90°$. Use the law of sines and the fact that $\sin (90° - \theta) = \cos \theta$.

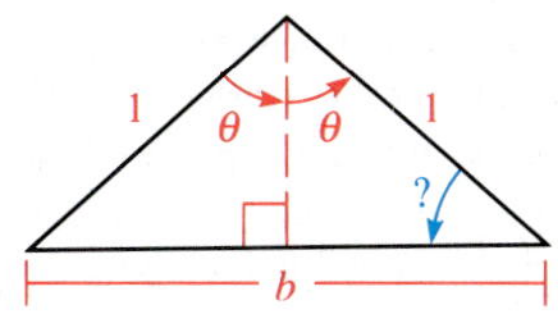

22. Use the law of sines to show that the bisector of an angle of a triangle divides the opposite side in the same ratio as the two adjacent sides; that is,

$$\frac{a}{b} = \frac{x}{y}.$$

(*Hint:* Use the law of sines in $\triangle BNC$ and $\triangle ANC$ and the fact that $\sin (180° - B) = \sin B$.)

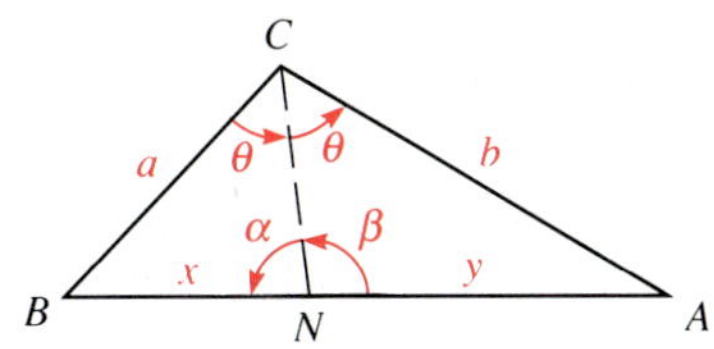

Problems

Give answers to three significant digits. You may wish to use a calculator.

A 1. Two angles of a triangle measure 32° and 53°. The longest side is 55 cm. Find the length of the shortest side. 29.3 cm

2. Two angles of a triangle measure 75° and 51°. The side opposite the 75° angle is 25 in. How long is the shortest side? 20.1 in.

3. How long is the base of an isosceles triangle if each leg is 27 cm and each base angle measures 23°? 49.7 cm

4. A college football pennant is in the shape of an isosceles triangle. The base is 16 in. long. The sides meet at an angle of 35°. How long are the sides? 26.6 in.

5. A fire is sighted from two ranger stations that are 5000 m apart. The angles of observation to the fire measure 52° from one station and 41° from the other station. Find the distance along the line of sight to the fire from the closer of the two stations. 3280 m or 17,200 m

6. Two surveyors are on opposite sides of a swamp. To find the distance between them, one surveyor locates a point T that is 180 m from her location at point P. The angles of sight from T to the other surveyor's position, R, measure 72° for $\angle RPT$ and 63° for $\angle PTR$. How far apart are the surveyors? 227 m

Summarizing the Lesson

In this lesson students have learned to use the law of sines to solve triangles. Ask the students to state the parts of a triangle that must be known to use the law of sines. (Remind them to be careful of the special case of two sides and a nonincluded angle.)

Suggested Assignments

Average Alg. and Trig.
Day 1: 588/2–12 even
Assign with Lesson 12-6.
Day 2: 588/9–19 odd, 20
589/*P*: 2, 3, 5, 6, 8, 9
S 590/Mixed Review

Extended Alg. and Trig.
588/1–19 odd, 20
589/*P*: 2, 6, 8–11
S 590/Mixed Review

Supplementary Material

Study Guide	pp. 199–200
Practice Master	69
Resource Book	p. 153

Additional Answers Written Exercises

19. $\frac{\sin A}{a} = \frac{\sin B}{b}$; $\frac{\sin A}{\sin B} = \frac{a}{b}$;
$\frac{\sin A}{\sin B} + 1 = \frac{a}{b} + 1$;
$\frac{\sin A}{\sin B} + \frac{\sin B}{\sin B} = \frac{a}{b} + \frac{b}{b}$;
$\frac{\sin A + \sin B}{\sin B} = \frac{a + b}{b}$

(continued)

Additional Answers Written Exercises

(continued)

20. $\frac{\sin A}{a} = \frac{\sin B}{b}$; $\frac{\sin A}{\sin B} = \frac{a}{b}$; $\frac{\sin A}{\sin B} - 1 = \frac{a}{b} - 1$; $\frac{\sin A}{\sin B} - \frac{\sin B}{\sin B} = \frac{a}{b} - \frac{b}{b}$; $\frac{\sin A - \sin B}{\sin B} = \frac{a - b}{b}$

21. $\sin \theta = \frac{\frac{b}{2}}{1} = \frac{b}{2}$; $b = 2 \sin \theta$; $\frac{\sin 2\theta}{b} = \frac{\sin (90^\circ - \theta)}{1}$; $\sin 2\theta = b \cos \theta$; $b = \frac{\sin 2\theta}{\cos \theta}$; $\frac{\sin 2\theta}{\cos \theta} = 2 \sin \theta$; $\sin 2\theta = 2 \sin \theta \cos \theta$

22. In $\triangle BNC$, $\frac{a}{\sin \alpha} = \frac{x}{\sin \theta}$, or $\frac{a}{x} = \frac{\sin \alpha}{\sin \theta}$. In $\triangle ANC$, $\frac{b}{\sin \beta} = \frac{y}{\sin \theta}$, or $\frac{b}{y} = \frac{\sin \beta}{\sin \theta}$. But, $\beta = 180^\circ - \alpha$, so $\sin \beta = \sin(180^\circ - \alpha) = \sin \alpha$. Thus, $\frac{\sin \alpha}{\sin \theta} = \frac{\sin \beta}{\sin \theta}$, or $\frac{a}{x} = \frac{b}{y}$. Therefore, $\frac{a}{b} = \frac{x}{y}$.

7. Two markers are located at points A and B on opposite sides of a lake. To find the distance between the markers, a surveyor laid off a base line, $\overline{AC}$, 25 m long and found that $\angle BAC = 85^\circ$ and $\angle BCA = 66^\circ$. Find AB. 47.1 m

8. From a hang glider approaching a 5000 ft clearing the angles of depression of the opposite ends of the field measure 24° and 30°. How far is the hang glider from the nearer end of the field? 19,500 ft

B 9. A loading ramp 5 m long makes a 25° angle with the level ground beneath it. The ramp is replaced by another ramp 15 m long. Find the angle that the new ramp makes with the ground. 8.1°

10. From the top of an office building 72 ft high, the angle of elevation to the top of an apartment building across the street is 31°. From the base of the office building, the angle of elevation is 46°. How tall is the apartment building? 172 ft

11. The captain of a freighter 6 km from the nearer of two unloading docks on the shore finds that the angle between the lines of sight to the two docks is 35°. If the docks are 10 km apart, how far is the tanker from the farther dock? 14.3 km

12. Two wires bracing a transmission tower are attached to the same location on the ground. One wire is attached to the tower 10 m above the other. The longer wire is 30 m long and the angle of inclination of the shorter wire is 53°. To the nearest tenth of a degree, what angle does the longer wire make with the ground? 64.6°

C 13. To find the height of a tree across a river, Carlos laid off a base line 70 m long and measured the angles shown. He found the angle of elevation of the top of the tree from A to be 10°. How tall is the tree? 13.1 m

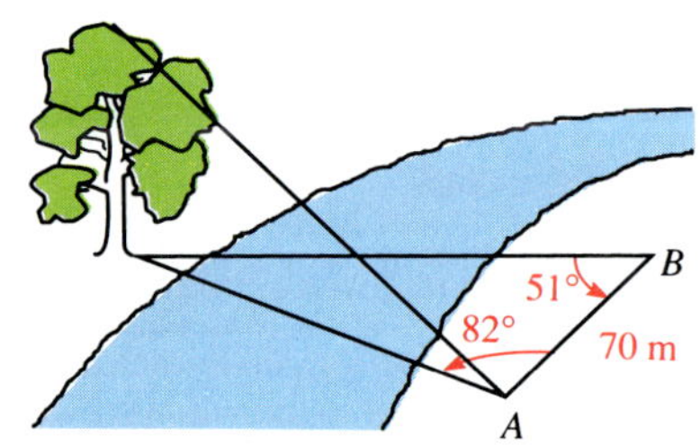

Mixed Review Exercises

Give the following function values to four significant digits.

1. cos 78.2° 0.2045
2. tan (−43°10′) −0.9380
3. sin 213.7° −0.5548
4. csc 305°45′ −1.232
5. cot (−101.9°) 0.2107
6. sec 488°34′ −1.604

Write an equation for each figure described.

7. The circle with center (−1, 4) and radius 3. $(x + 1)^2 + (y - 4)^2 = 9$
8. The line through (−3, 5) and (2, 0). $x + y = 2$
9. The parabola with focus (0, −2) and directrix $y = 2$. $y = -\frac{1}{8}x^2$

12-8 Solving General Triangles

Objective To solve any given triangle.

The problem of solving triangles can be divided into four cases:

SSS: Given three sides.

SAS: Given two sides and the included angle.

SSA: Given two sides and the angle opposite one of them.

ASA and **AAS:** Given two angles and one side.

You may remember that SSS, SAS, ASA, and AAS are used in geometry to prove triangles congruent. However, knowing two sides and a nonincluded angle (the SSA case) is not enough to prove two triangles congruent. You will see later in this lesson that when you try to solve a triangle given these measurements, they may not determine a triangle.

In the examples that follow, lengths will be found to three significant digits and angle measures to the nearest tenth of a degree.

Example 1 (SSS case) Solve $\triangle ABC$ if $a = 4$, $b = 6$, and $c = 5$.

Solution Use the law of cosines to find one of the angles, say $\angle A$.

$$\cos A = \frac{b^2 + c^2 - a^2}{2bc}$$

$$\cos A = \frac{6^2 + 5^2 - 4^2}{2 \cdot 6 \cdot 5}$$

$$= \frac{36 + 25 - 16}{60}$$

$$= \frac{45}{60} = 0.7500$$

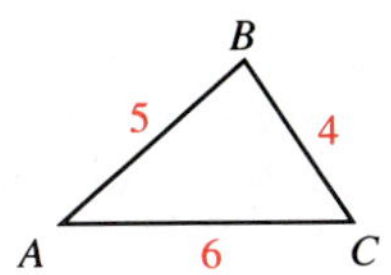

$\therefore \angle A = 41.4°$

To find another angle, it is usually easier to use the law of sines.

$$\frac{\sin C}{c} = \frac{\sin A}{a}$$

$$\sin C = \frac{c \sin A}{a}$$

$$\sin C = \frac{5 \sin 41.4°}{4} = \frac{5(0.6613)}{4} = 0.8266$$

Since $\angle C$ is not the largest angle, it must be acute.
$\therefore \angle C = 55.8°$

Since $\angle A + \angle B + \angle C = 180°$, you know $\angle B = 180° - \angle A - \angle C$.
So by substitution, $\angle B = 180° - 41.4° - 55.8° = 82.8°$.

$\therefore \angle A = 41.4°$, $\angle B = 82.8°$, and $\angle C = 55.8°$. ***Answer***

Teaching Suggestions p. T126

Using Manipulatives p. T126

Suggested Extensions p. T126

Warm-Up Exercises

For $\triangle ABC$, identify the given combination of sides and angles as SSS, SAS, ASA, AAS, or SSA.

1. $\angle A = 40°$, $\angle C = 110°$, $c = 12$ AAS
2. $a = 12$, $c = 13$, $b = 12$ SSS
3. $c = 2.8$, $\angle A = 25°$, $\angle B = 95°$ ASA
4. $b = 10$, $\angle A = 100°$, $c = 6$ SAS
5. $b = 18$, $\angle B = 70°$, $a = 24$ SSA

Motivating the Lesson

Ask students how they would answer the following questions: How far is a plane from its departure point after several course and speed changes? How long must the support wires for an antenna be if there are restrictions on the angles they make with the antenna and the ground? Tell students that these are some of the questions that can be answered by applying the various methods for solving triangles, the topic of today's lesson.

Common Errors

In an SSA situation where the given angle is acute, students often forget to check for the possible second solution. Emphasize that when the law of sines is used to obtain a second angle of the triangle, there are two angles, one acute and one obtuse, that have the same sine value. The obtuse angle may produce a second solution.

Chalkboard Examples

1. Solve $\triangle DEF$ if $d = 5$, $e = 6$, and $f = 7$.

   $d^2 = e^2 + f^2 - 2ef\cos D$

   $5^2 = 6^2 + 7^2 - 2(6)(7)\cos D$

   $\cos D = \dfrac{5^2 - 6^2 - 7^2}{-2(6)(7)}$

   $\angle D = 44.4°$

   $e^2 = d^2 + f^2 - 2df\cos E$

   $6^2 = 5^2 + 7^2 - 2(5)(7)\cos E$

   $\angle E = 57.1°$

   $\angle F = 180° - 44.4° - 57.1° = 78.5°$

2. Solve $\triangle ABC$ if $a = 5$, $b = 6$, and $\angle C = 80°$.

   $c^2 = a^2 + b^2 - 2ab\cos C$

   $c^2 = 5^2 + 6^2 - 2(5)(6)\cos 80°$

   $c^2 = 50.6$

   $c = 7.11$

   $\dfrac{\sin B}{b} = \dfrac{\sin C}{c}$

   $\dfrac{\sin B}{6} = \dfrac{\sin 80°}{7.11}$

   $\sin B = 0.8310$

   $\angle B = 56.2°$

   $\angle A = 180° - 80° - 56.2° = 43.8°$

Example 2 (SAS case) Solve $\triangle ABC$ if $a = 8$, $c = 7$, and $\angle B = 31.8°$.

Solution Use the law of cosines to find the third side, b.

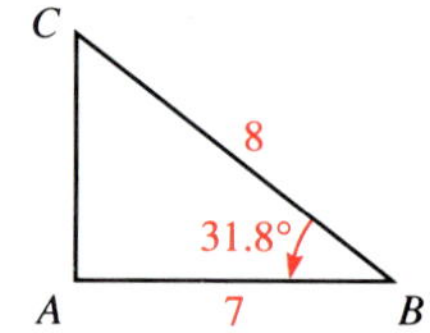

$$b^2 = a^2 + c^2 - 2ac\cos B$$
$$= 8^2 + 7^2 - 2(8)(7)\cos 31.8°$$
$$= 64 + 49 - 112(0.8499)$$
$$= 17.8$$
$$b = \sqrt{17.8} = 4.2$$

Use the law of sines to find the measure of $\angle C$, the smaller of the two remaining angles.

$$\frac{\sin C}{c} = \frac{\sin B}{b}$$
$$\sin C = \frac{c\sin B}{b}$$
$$= \frac{7\sin 31.8°}{4.2}$$
$$= \frac{7(0.5270)}{4.2} = 0.8783$$

Since $\angle C$ is not the largest angle, it must be acute.

$\therefore\ \angle C = 61.4°$

$$\angle A = 180° - 31.8° - 61.4°$$
$$= 86.8°$$

$\therefore\ \angle A = 86.8°$, $b = 4.2$, and $\angle C = 61.4°$. ***Answer***

The SSA case is often called the *ambiguous case* because there are six possible outcomes, four if $\angle A$ is acute and two if $\angle A$ is right or obtuse. These possibilities are illustrated below and on the opposite page.

$\angle A < 90°$

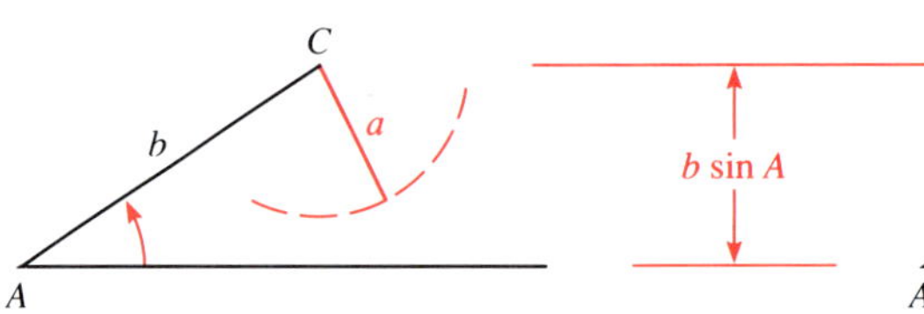

$a < b\sin A$
No solution

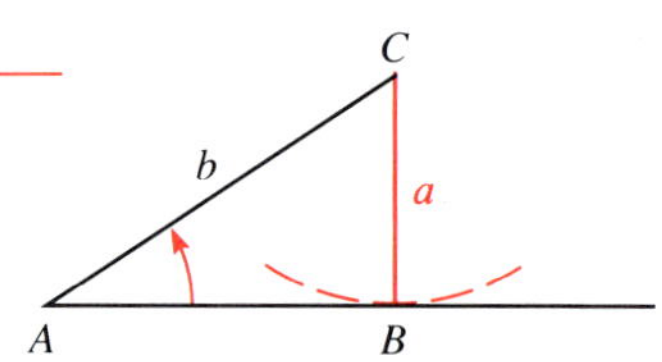

$a = b\sin A$
One solution

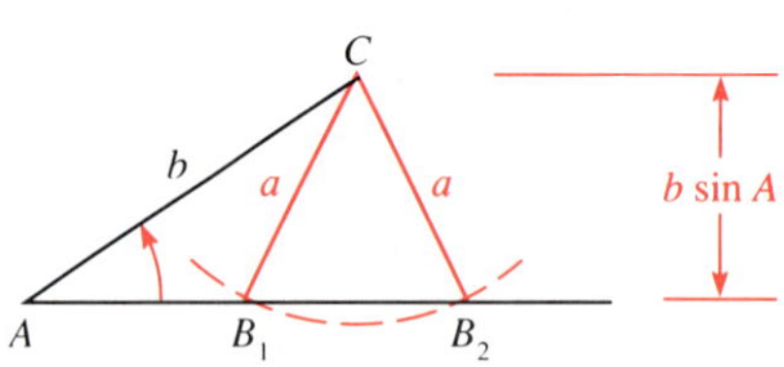

$b\sin A < a < b$
Two solutions

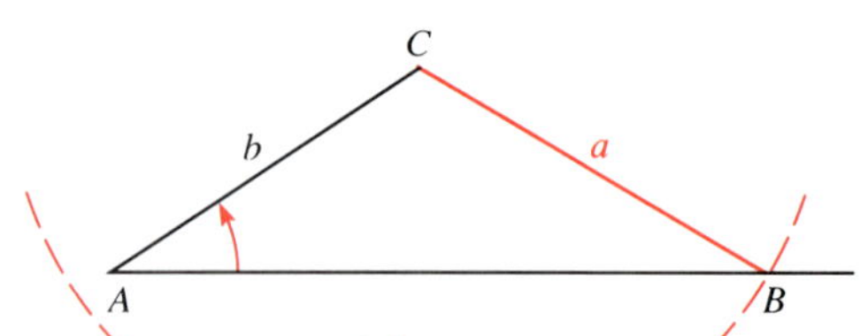

$a \ge b$
One solution

$\angle A \geq 90°$

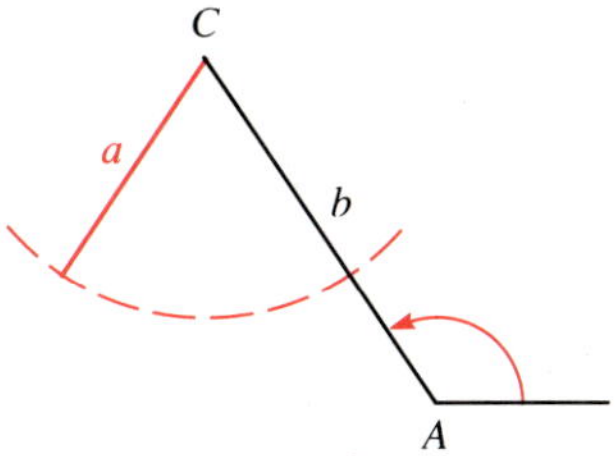

$a \leq b$

No solution

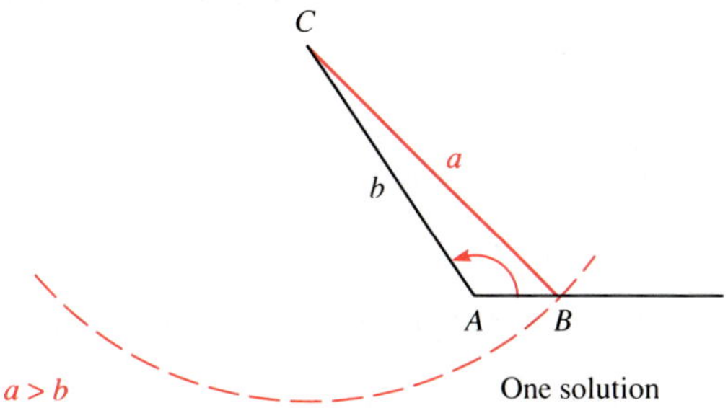

$a > b$

One solution

Example 3 (SSA case) Solve $\triangle ABC$ if $b = 22$, $c = 30$, and $\angle B = 30°$.

Solution

$$\frac{\sin C}{c} = \frac{\sin B}{b}$$

$$\sin C = \frac{c \sin B}{b}$$

$$= \frac{30 \sin 30°}{22}$$

$$= \frac{30(0.5000)}{22} = 0.6818$$

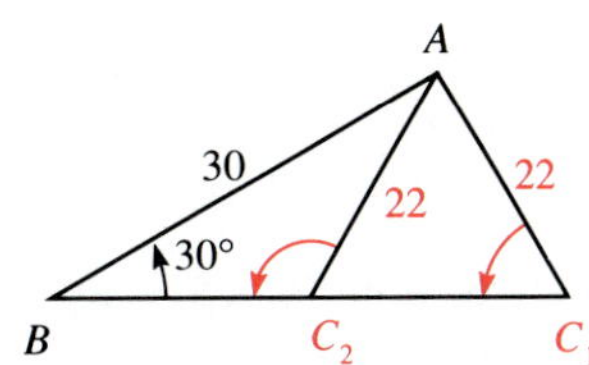

$\angle C = 43.0°$

$\angle A = 180° - 30° - 43.0° = 107.0°$

$$a = \frac{b \sin A}{\sin B}$$

$$= \frac{22 \sin 107.0°}{\sin 30°}$$

$$= \frac{22(0.9563)}{0.5000}$$

$a = 42.1$

or

$\angle C = 180° - 43.0° = 137.0°$

$\angle A = 180° - 30° - 137.0° = 13.0°$

$$a = \frac{b \sin A}{\sin b}$$

$$= \frac{22 \sin 13.0°}{\sin 30°}$$

$$= \frac{22(0.2250)}{0.5000}$$

$a = 9.90$

$\therefore \angle A = 107.0°$, $\angle C = 43.0°$, and $a = 42.1$
or $\angle A = 13.0°$, $\angle C = 137.0°$, and $a = 9.90$ ***Answer***

Example 4 (AAS case) Solve $\triangle ABC$ if $a = 40$, $\angle A = 45°$, and $\angle C = 55°$.

Solution $\angle B = 180° - \angle A - \angle C = 180° - 45° - 55° = 80°$

$$b = \frac{a \sin B}{\sin A} = \frac{40 \sin 80°}{\sin 45°} = \frac{40(0.9848)}{0.7071}$$

$b = 55.7$

$$c = \frac{a \sin C}{\sin A} = \frac{40 \sin 55°}{\sin 45°} = \frac{40(0.8192)}{0.7071}$$

$c = 46.3$

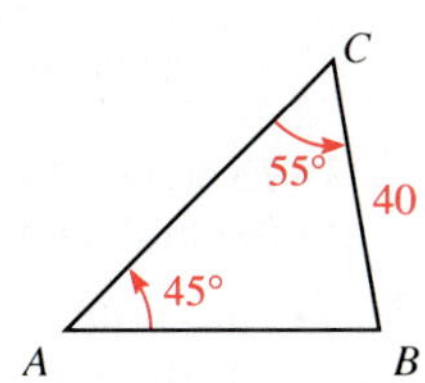

$\therefore b = 55.7$, $c = 46.3$, and $\angle B = 80°$. ***Answer***

3. Solve $\triangle ABC$ if $a = 15$, $c = 18$, and $\angle A = 32°$.

$$\frac{\sin 32°}{15} = \frac{\sin C}{18}$$

$$\sin C = \frac{18 \sin 32°}{15}$$

$= 0.6359$

$\angle C = 39.5°$ or

$\angle C = 140.5°$

$\angle B = 180° - 39.5° - 32° = 108.5°$ or $\angle B = 180° - 140.5° - 32° = 7.5°$

$$\frac{\sin A}{a} = \frac{\sin B}{b}$$

$$b = \frac{a \sin B}{\sin A}$$

$$b = \frac{15 \sin 108.5°}{\sin 32°}$$

$= 26.8$ or

$$b = \frac{15 \sin 7.5°}{\sin 32°}$$

$= 3.7$

$\angle B = 108.5°$

$\angle C = 39.5°$,

$b = 26.8$, or $\angle B = 7.5°$,

$\angle C = 140.5°$, $b = 3.7$

4. Solve $\triangle ABC$ if $\angle A = 20°$, $\angle B = 60°$ and $c = 10$.

$\angle C = 180° - 20° - 60° = 100°$

$$\frac{\sin C}{c} = \frac{\sin A}{a}$$

$$\frac{\sin 100°}{10} = \frac{\sin 20°}{a}$$

$$a = \frac{10 \sin 20°}{\sin 100°}$$

$a = 3.47$

$$\frac{\sin C}{c} = \frac{\sin B}{b}$$

$$\frac{\sin 100°}{10} = \frac{\sin 60°}{b}$$

$$b = \frac{10 \sin 60°}{\sin 100°}$$

$b = 8.79$

Check for Understanding

Here is a suggested use of the Oral Exercises to check students' understanding as you teach the lesson.
Oral Exs. 1–20: use after Example 4.

Guided Practice

Solve each $\triangle ABC$. Give lengths to three significant digits and angle measures to the nearest 0.1°. There may be two solutions or none.

1. $a = 18.5$, $b = 16.3$, $\angle C = 42.2°$ $c = 12.7$, $\angle A = 78.2°$, $\angle B = 59.6°$
2. $\angle A = 21.6°$, $b = 29.7$, $a = 10.7$ no solution
3. $a = 17$, $b = 15$, $c = 13$ $\angle A = 74.4°$, $\angle B = 58.2°$, $\angle C = 47.4°$
4. $b = 23.4$, $c = 14.7$, $\angle C = 37.2°$
   (1) $\angle B = 74.2°$, $\angle A = 68.6°$, $a = 22.6$ or
   (2) $\angle B = 105.8°$, $\angle A = 37.0°$, $a = 14.6$
5. A biologist tracking an animal with a radio transmitting collar detects the animal in a line that is 27.5° off the trail. After moving 155 m down the trail, the line to the animal is 38.8° off the trail. Assuming the animal hasn't moved, find the distance QA between the biologist and the animal.

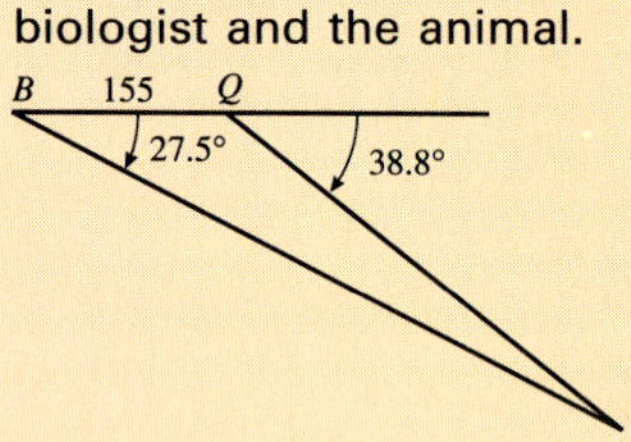

365 m

Oral Exercises

Given the following measurements, tell which law you would use first to solve the triangle.

1. SSS cosine **2.** SAS cosine **3.** SSA sine **4.** AAS sine **5.** ASA sine

Outline a strategy for solving each triangle.

Sample $a = 20$, $c = 30$, $\angle B = 40°$

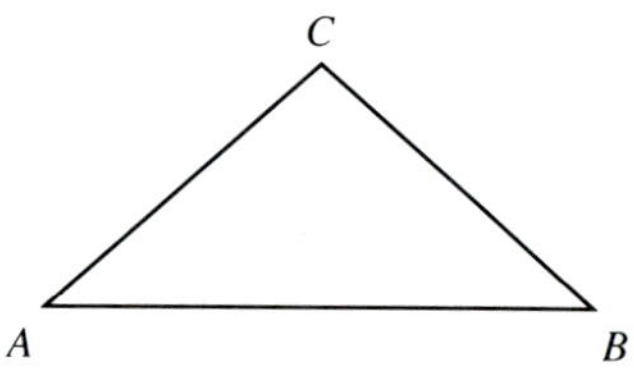

Solution
1. Use the law of cosines to find b.
2. Use the law of sines to find $\angle A$.
3. Use $\angle C = 180° - \angle A - \angle B$ to find $\angle C$.

6. $a = 17$, $\angle B = 20°$, $\angle C = 60°$
7. $a = 8$, $c = 13$, $\angle B = 150°$
8. $a = 5$, $b = 7$, $c = 11$
9. $b = 15$, $\angle B = 100°$, $\angle C = 25°$
10. $b = 10$, $c = 13$, $\angle A = 120°$
11. $a = 10$, $b = 25$, $c = 30$
12. $a = 30$, $b = 20$, $\angle A = 130°$
13. $b = 20$, $c = 18$, $\angle B = 120°$
14. $a = 9$, $b = 10$, $\angle B = 40°$
15. $a = 12$, $b = 11$, $\angle A = 35°$
16. $b = 14$, $c = 18$, $\angle B = 50°$
17. $b = 15$, $c = 14.5$, $\angle C = 70°$
18. In $\triangle ABC$, if $\angle A$ is an obtuse angle, what can you conclude about $\angle C$? It is acute.
19. In $\triangle ABC$, if b is the longest side, what can you conclude about $\angle A$? It is acute.
20. Explain why the SSA case is called the ambiguous case. The given conditions may determine one triangle, two triangles, or no triangle.

Written Exercises

Give lengths to three significant digits and angle measures to the nearest tenth of a degree. You may wish to use a calculator.

A **1–12.** Solve the triangles given in Oral Exercises 6–17. If there are two solutions, find both. If there are no solutions, say so.

In Exercises 13 and 14, recall that a median is the segment from a vertex to the midpoint of the opposite side.

B **13.** In $\triangle ABC$, $a = 4$, $b = 5$, and $\angle C = 110°$. Find the length of the median to the longest side. 2.61

14. In $\triangle ABC$, $c = 10$ and $\angle A = \angle B = 40°$. Find the length of the median to $\overline{AC}$. 7.79

Solve Exercises 15 and 16 by using both the law of sines and the law of cosines. You may wish to use a calculator.

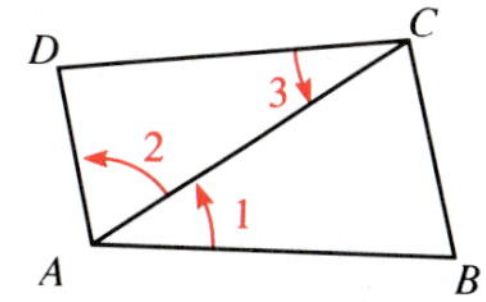

15. In quadrilateral $ABCD$, $\angle 1 = 30°$, $\angle 2 = 80°$, $\angle B = 80°$, $AD = 8$, and $AB = 10$. Find the length of $\overline{CD}$. 12.0

16. In quadrilateral $ABCD$, $\angle B = 110°$, $\angle D = 75°$, $\angle 3 = 35°$, $AB = 5$, and $BC = 6$. Find the length of $\overline{CD}$. 8.78

17. Find the lengths of the diagonals of the trapezoid shown below.

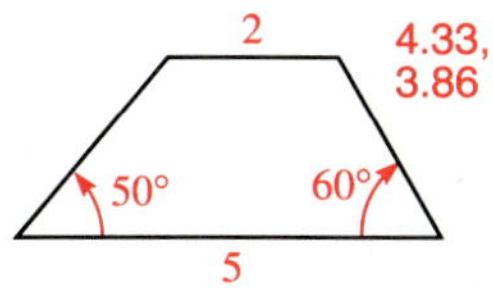

18. Find the lengths of the diagonals of the quadrilateral shown below.

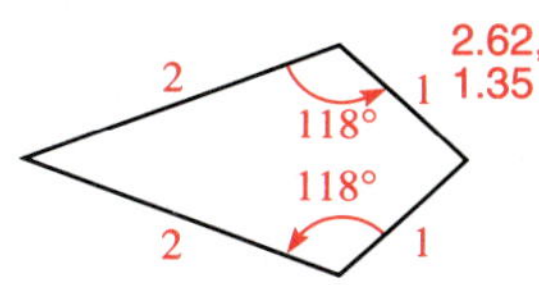

C 19. The angles of a triangle measure 57°, 60°, and 63°. The longest side is 5 units longer than the shortest side. Find the lengths of the three sides. 80.1, 82.7, 85.1

20. One angle of a triangle has measure 65°. The side opposite this angle has length 10. If the area of the triangle is 20, find the perimeter. 25.0

Problems

Give lengths to three significant digits and angle measures to the nearest tenth of a degree. You may wish to use a calculator.

A 1. Two planes leave an airport at the same time, one flying at 300 km/h and the other at 420 km/h. The angle between their flight paths is 75°. After three hours, how far apart are they? 1350 km

2. Jan is flying a plane on a triangular course at 320 mi/h. She flies due east for two hours and then turns right through a 65° angle. How long after turning will she be exactly southeast of where she started? 4.13 h

3. Two cables of length 300 m and 270 m extend from the top of a television antenna to the level ground on opposite sides of the antenna. The longer cable makes an angle of 48° with the ground. Find the acute angle that the shorter cable makes with the ground and the distance between the cables along the ground. 55.7°, 353 m

4. A vertical television mast is mounted on the roof of a building. From a point 750 ft from the base of the building, the angles of elevation to the bottom and top of the mast measure 34° and 50° respectively. How tall is the mast? 388 ft

5. Jim is flying a plane from Midville to Vista, a distance of 500 km. Because of a thunderstorm, he had to fly 17.5° off course for 300 km. How far is he now from Vista, and through what angle should he turn to fly directly there? 232 km, 40.4°

Using a Calculator

The exercises and problems in this lesson contain a mix of applications involving the law of cosines and the law of sines. Students have seen in Lessons 12-6 and 12-7 how useful a calculator can be in such situations.

Summarizing the Lesson

In this lesson students have learned to solve any triangle, including the ambiguous case. Ask the students to describe what can happen in the ambiguous case.

Suggested Assignments

Average Alg. and Trig.
Day 1: 594/1–8
595/*P*: 1–3
S 584/*P*: 9
Day 2: 594/9–13, 15
596/*P*: 6, 8, 10, 14
S 590/*P*: 7, 10, 11

Extended Alg. and Trig.
Day 1: 594/1–19 odd
S 584/*P*: 9
S 590/*P*: 13
Day 2: 595/*P*: 1, 2, 6–10, 14
S 577/20, 22, 23

Supplementary Materials

Study Guide	pp. 201–202
Practice Master	70
Computer Activity	28

Additional Answers Oral Exercises

see p. 602

Additional Answers Written Exercises

see p. 603

6. A monument consists of a flagpole 15 m tall standing on a mound in the shape of a cone with vertex angle 140°. How long a shadow does the pole cast on the cone when the angle of elevation of the sun is 62°? 10.5 m

B 7. From the top of a building 10 m tall, the angle of elevation to the top of a flagpole is 11°. At the base of the building, the angle of elevation to the top of the flagpole is 39°. Find the height of the flagpole. 13.2 m

8. A communication satellite is in orbit 35,800 km above the equator. It completes one orbit every 24 hours, so that from Earth it appears to be stationary above a point on the equator. If this point has the same longitude as Houston, find the measure of θ, the satellite's angle of elevation from Houston. The latitude of Houston is 29.7°N. The radius of Earth is 6400 km. 55.4°

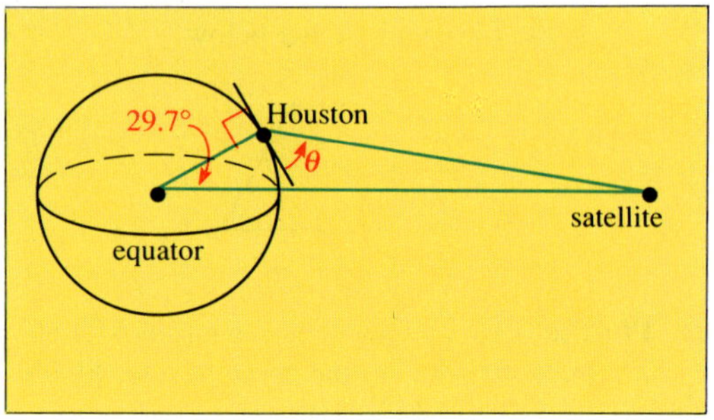

9. In Problem 8, what is the greatest latitude from which a signal can travel to the satellite in a straight line? 81.3°N or 81.3°S

10. From the top of an observation post that is 90 m high, a ranger sights a campsite at an angle of depression of 10°. Turning in a different direction, the ranger sees another campsite at an angle of depression of 13°. The angle between these two lines of sight is 35°. How far apart are the campsites? 298 m

C 11. In quadrilateral $ABCD$, $AB = 3$, $BC = 4$, $CD = 5$, and $DA = 6$. The length of diagonal $\overline{BD}$ is 7. Find the length of the other diagonal. 5.51

12. From Base Camp, located 2000 m above sea level, the angle of elevation of Camp A is 25° and the angle of elevation of the summit of Mount Snow is 41°. From Camp A the angle of elevation of the summit is 53°. If Camp A is 3 km closer to the summit (air distance) than Base Camp is, how high is the summit above sea level? 6.77 km

13. In quadrilateral $ABCD$ shown at the left below, find the length of $\overline{CD}$. 60.1 in.

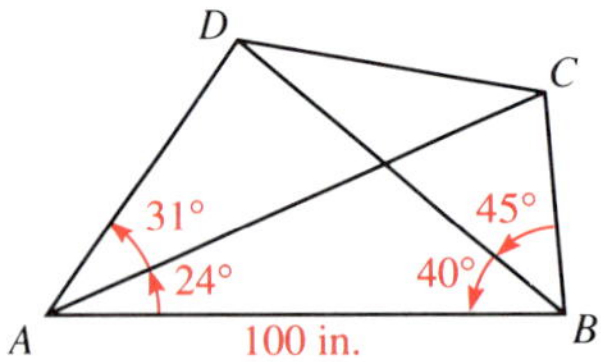

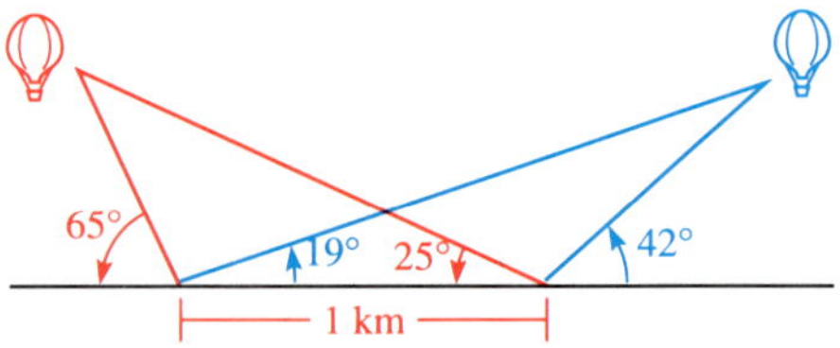

14. Two balloons are moored directly over a straight, level road. The diagram on the right above shows the angle of elevation of the balloons from two observers on the road one kilometer apart. How far apart are the balloons? Which balloon is higher, and by how many meters?
1900 m; the left balloon, by 38 m

12-9 Areas of Triangles

Objective To apply triangle area formulas.

You know that the area of a triangle can be found by using the formula

$$K = \frac{1}{2}(\text{base})(\text{height}).$$

Here are some other area formulas for triangles.

The area K of $\triangle ABC$ is given by each of the formulas listed below.

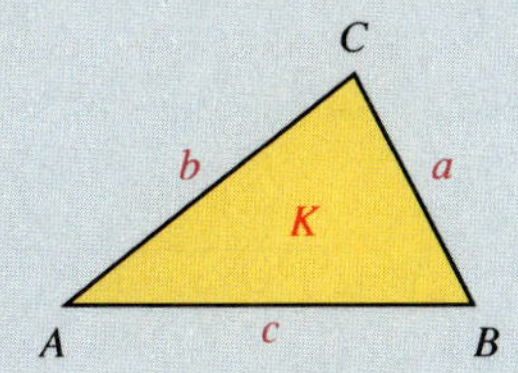

$K = \frac{1}{2}bc \sin A$ $\quad K = \frac{1}{2}ac \sin B$ $\quad K = \frac{1}{2}ab \sin C$

$K = \frac{1}{2}a^2 \frac{\sin B \sin C}{\sin A}$ $\quad K = \frac{1}{2}b^2 \frac{\sin A \sin C}{\sin B}$ $\quad K = \frac{1}{2}c^2 \frac{\sin A \sin B}{\sin C}$

$K = \sqrt{s(s-a)(s-b)(s-c)}$, where $s = \frac{1}{2}(a + b + c)$

The first three formulas are useful for finding areas of triangles in the SAS case. They were derived in Lesson 12-7. The formulas in the second row above are useful in the AAS case. You can prove them using the law of sines in the form $b = \frac{a \sin B}{\sin A}$. By substituting $\frac{a \sin B}{\sin A}$ for b in $K = \frac{1}{2}ab \sin C$, the equation becomes $K = \frac{1}{2}a^2 \frac{\sin B \sin C}{\sin A}$. The other formulas on the second row are proved similarly. The formula on the third row is called **Hero's formula** and is used to find areas in the SSS case. Its proof is outlined in Exercises 17–19 on pages 599–600.

Example 1 (SAS case) Find the area of $\triangle ABC$ if $b = 32$, $c = 27$, and $\angle A = 108°$.

Solution
$$K = \frac{1}{2}bc \sin A = \frac{1}{2}(32)(27) \sin 108°$$
$$= 432\ (0.9511)$$
$$= 411$$

$\therefore$ the area is 411 square units. ***Answer***

Teaching Suggestions p. T126

Suggested Extensions p. T126

Warm-Up Exercises

In $\triangle ABC$, $\angle A = 30°$, $\angle B = 60°$, and $c = 10$. Find the value of each expression.

1. $\angle C$ 90°
2. a 5
3. b $5\sqrt{3}$
4. $ab \sin C$ $25\sqrt{3}$
5. $ac \sin B$ $25\sqrt{3}$
6. $bc \sin A$ $25\sqrt{3}$

Motivating the Lesson

Tell students to suppose they owned a triangular parcel of land. With nothing more than a good tape measure, they could make the measurements to find the area. This is because Hero's formula, one of the triangle area formulas in today's lesson, requires only the lengths of the sides to calculate the area of a triangle.

Chalkboard Examples

1. Find the area of $\triangle ABC$ if $b = 5$, $c = 8$, and $\angle A = 115°$.
   $K = \frac{1}{2}(5)(8) \sin 115°$
   $K = 18.1$
2. Find the area of $\triangle ABC$ if $\angle A = 40°$, $\angle B = 70°$, and $c = 2$.
   $\angle C = 180° - 40° - 70° = 70°$
   $K = \frac{1}{2}(2)^2\frac{\sin 40° \sin 70°}{\sin 70°}$
   $= \frac{1}{2}(2)^2(0.6428)$
   $= 1.29$

(continued)

Chalkboard Examples
(continued)

3. A triangular garden has sides of lengths 20 m, 30 m, and 30 m. Find its area.

$s = \frac{1}{2}(a + b + c)$

$= \frac{1}{2}(20 + 30 + 30)$

$= 40$

$s - a = 20,\ s - b = 10,$
$s - c = 10$

$K = \sqrt{s(s-a)(s-b)(s-c)}$

$= \sqrt{40(20)(10)(10)}$

$= 200\sqrt{2} = 282.8$

$\therefore$ the area is 282.8 m^2.

4. Find the area of $\triangle ABC$ if $\angle A = 40°$, $a = 15$, and $b = 20$.

This triangle was solved partially in Example 1(*a*) on page 587. $\angle C = 81°$ or $\angle C = 19°$.

$K = \frac{1}{2}ab \sin C$

$K = \frac{1}{2} \cdot 15 \cdot 20 \sin 81° = 148,$

or

$K = \frac{1}{2} \cdot 15 \cdot 20 \sin 19° = 48.8$

Check for Understanding

Here is a suggested use of the Oral Exercises to check students' understanding as you teach the lesson.
Oral Exs. 1–9: use after Example 4.

Using a Calculator

A look at the box on page 597 shows the complexity of the computations involved in the area formulas. Hero's formula, for example, requires ten distinct arithmetic operations. The use of a calculator will facilitate the arithmetic and increase the likelihood of correct answers.

Example 2 (ASA case) Find the area of the triangle shown at the right.

Solution Let $\angle A = 100°$ and $\angle B = 65°$.
Then $\angle C = 180° - 100° - 65° = 15°$.

Use the formula $K = \frac{1}{2}c^2 \frac{\sin A \sin B}{\sin C}$.

$K = \frac{1}{2}(2.2)^2 \frac{\sin 100° \sin 65°}{\sin 15°}$

$= 2.42 \frac{(0.9848)(0.9063)}{(0.2588)}$

$= 8.35$

$\therefore$ the area is 8.35 m^2. ***Answer***

Example 3 (SSS case) A triangular city lot has sides of lengths 50 ft, 60 ft, and 80 ft. Find its area.

Solution Use Hero's formula: $K = \sqrt{s(s-a)(s-b)(s-c)}$, where

$$s = \frac{1}{2}(a + b + c)$$
$$= \frac{1}{2}(50 + 60 + 80) = 95.$$

Then: $s - a = 95 - 50 = 45$
$s - b = 95 - 60 = 35$
$s - c = 95 - 80 = 15$

Now substitute: $K = \sqrt{s(s-a)(s-b)(s-c)}$
$= \sqrt{95(45)(35)(15)} = 75\sqrt{399} = 1498$

$\therefore$ to three significant digits, the area is 1500 ft^2. ***Answer***

In the ambiguous case (SSA), the triangle must be partially solved before the area, if any, can be found.

Example 4 (SSA case) Find the area of $\triangle ABC$ if $b = 22$, $c = 30$ and $\angle B = 30°$.

Solution In Example 3, page 593, two triangles fit the given data, one with $\angle A = 107.0°$ and the other with $\angle A = 13.0°$. Therefore:

$\angle A = 107.0°$	or	$\angle A = 13.0°$
$K = \frac{1}{2}bc \sin A$		$K = \frac{1}{2}bc \sin A$
$= \frac{1}{2}(22)(30) \sin 107.0°$		$= \frac{1}{2}(22)(30) \sin 13.0°$
$= 330(0.9563)$		$= 330(0.2250)$
$= 316$		$= 74.3$

$\therefore$ the area is 316 square units or 74.3 square units. ***Answer***

Oral Exercises

Three parts of $\triangle ABC$ are given. Tell which formula you would use to find the area of the triangle.

1. $\angle A = 25°$, $\angle B = 50°$, $b = 30$
2. $a = 18$, $b = 10$, $\angle C = 45°$
3. $a = 6$, $c = 14$, $\angle B = 62°$
4. $\angle A = 70°$, $b = 15$, $c = 36$
5. $a = 12$, $b = 8$, $c = 12$
6. $\angle B = 20°$, $\angle C = 115°$, $c = 16$
7. $c = 15$, $b = 22$, $\angle A = 150°$
8. $a = 7$, $b = 12$, $c = 8$
9. Suppose $a = 12$, $b = 11$, and $\angle B = 60°$. Explain why this is an example of the ambiguous case. $\angle B$ is not the included angle. There may be one triangle, two triangles, or no triangle.

Written Exercises

Give lengths to three significant digits and angle measures to the nearest tenth of a degree. You may wish to use a calculator.

A 1–8. Find the areas of each triangle described in Oral Exercises 1–8.

9. Find the area of a parallelogram that has a 45° angle and sides with lengths 10 and 18. 127
10. Find the area of a rhombus that has perimeter 48 and an angle of 55°. 118

B 11. The area of $\triangle ABC$ is 36 square units. If $\angle B = 30°$ and $c = 8$, find the value of a. 18

12. The area of $\triangle ABC$ is 20 square units. If $\angle A = 130°$ and $b = 6$, find the value of c. 8.70
13. A triangle has area 48 cm^2 and its shorter sides have lengths 9 cm and 12 cm. Find the largest angle of the triangle. 117.3°
14. The angles of a triangle are 25°, 45°, and 110°. What is the length of the longest side if the area of the triangle is 75 square units? 21.7
15. Find the area of a regular octagon whose sides are 10 cm long. 483 cm^2
16. Find the area of a regular octagon inscribed in a unit circle. 2.83

Exercises 17–19 make up a proof of Hero's formula.
Exercises 17 and 18 are used in Exercise 19 on the following page.

17. Show that $K^2 = \frac{1}{4}a^2b^2 \sin^2 C$.
18. Recall that $s = \frac{1}{2}(a + b + c)$, so that $2s = a + b + c$.

    Therefore, $2(s - a) = 2s - 2a = a + b + c - 2a = -a + b + c$.
    Show that $2(s - b) = a - b + c$ and $2(s - c) = a + b - c$.

Guided Practice

Find the area of $\triangle ABC$ to three significant digits.

1. $a = 7$, $b = 8$, $c = 9$ 26.8
2. $a = 12$, $\angle B = 70°$, $\angle C = 50°$ 59.8
3. $a = 10$, $b = 12$, $\angle C = 43.2°$ 41.1
4. Find the area of a rhombus that has perimeter 100 and a 30° angle. 312.5

Summarizing the Lesson

In this lesson students learned to find the area of triangles using various formulas. Ask the students to state which case (SAS, AAS, or SSS) is associated with each of the area formulas in the box on page 597.

Suggested Assignments

Average Alg. and Trig.
599/1–16
S 600/Mixed Review
R 600/Self-Test 2

Extended Alg. and Trig.
599/1–16
S 600/Mixed Review
R 600/Self-Test 2

Additional Answers
Oral Exercises

1. $K = \frac{1}{2}b^2\frac{\sin A \sin C}{\sin B}$
2. $K = \frac{1}{2}ab \sin C$
3. $K = \frac{1}{2}ac \sin B$
4. $K = \frac{1}{2}bc \sin A$
5. $K = \sqrt{s(s-a)(s-b)(s-c)}$
6. $K = \frac{1}{2}c^2\frac{\sin A \sin B}{\sin C}$
7. $K = \frac{1}{2}bc \sin A$
8. $K = \sqrt{s(s-a)(s-b)(s-c)}$

C 19. Justify each of the statements from (a)–(i) below. Use the results of Exercises 17 and 18 where necessary.

a. $a^2 + b^2 - c^2 = 2ab \cos C$ Law of cosines

b. $(a^2 + b^2 - c^2)^2 = 4a^2b^2 \cos^2 C = 4a^2b^2(1 - \sin^2 C)$ $\sin^2 C + \cos^2 C = 1$

c. $(a^2 + b^2 - c^2)^2 = 4a^2b^2 - 4a^2b^2 \sin^2 C = 4a^2b^2 - 16K^2$ Ex. 17

d. $16K^2 = 4a^2b^2 - (a^2 + b^2 - c^2)^2$ Add. prop. of equality

e. $16K^2 = [2ab + (a^2 + b^2 - c^2)][2ab - (a^2 + b^2 - c^2)]$ $x^2 - y^2 = (x + y)(x - y)$

f. $16K^2 = [(a + b)^2 - c^2][c^2 - (a - b)^2]$ See below.

g. $16K^2 = [(a + b) + c][(a + b) - c][c + (a - b)][c - (a - b)]$ $x^2 - y^2 = (x + y)(x - y)$

h. $16K^2 = [2s][2(s - c)][2(s - b)][2(s - a)]$ Ex. 18

i. $K = \sqrt{s(s - a)(s - b)(s - c)}$ Div. prop. of equality; prop. of square roots

f. Assoc. prop. for add; $x^2 + 2xy + y^2 = (x + y)^2$; $x^2 - 2xy + y^2 = (x - y)^2$

Mixed Review Exercises

Find the indicated part of $\triangle ABC$ to three significant digits or to the nearest tenth of a degree.

1. $a = 8$, $b = 5$, $\angle C = 62°$, $c = \underline{\ ?\ }$ 7.17
2. $\angle A = 100°$, $\angle B = 30°$, $b = 15$, $a = \underline{\ ?\ }$ 29.5
3. $a = 6$, $b = 4$, $c = 9$, $\angle C = \underline{\ ?\ }$ 127.2°
4. $a = 20$, $\angle B = 76°$, $\angle C = 48°$, $b = \underline{\ ?\ }$ 23.4

Find the five other trigonometric functions of θ.

5. $\sin \theta = \frac{1}{2}$, $90° < \theta < 180°$
6. $\cos \theta = -\frac{3}{4}$, $180° < \theta < 270°$
7. $\tan \theta = -1$, $270° < \theta < 360°$
8. $\sin \theta = \frac{5}{13}$, $0° < \theta < 90°$

Self-Test 2

Vocabulary solving a triangle (p. 574) · angle of elevation (p. 575) · angle of depression (p. 575) · law of cosines (p. 580) · law of sines (p. 586) · ambiguous case (p. 592)

Give lengths to three significant digits and angle measures to the nearest tenth of a degree.

1. In $\triangle ABC$, $\angle C = 90°$, $\angle B = 27.3°$, and $a = 30$. Find $\angle A$ and sides b and c. $\angle A = 62.7°$, $b = 15.5$, $c = 33.8$ **Obj. 12-5, p. 574**
2. In $\triangle DEF$, $d = 18$, $e = 24$ and $\angle F = 42°$. Find side f. 16.1 **Obj. 12-6, p. 580**
3. In $\triangle XYZ$, $\angle X = 36°$, $x = 14$, and $z = 23.5$. Find $\angle Z$. $\angle Z = 80.6°$ or 99.4° **Obj. 12-7, p. 586**
4. Solve $\triangle ABC$ if $a = 15$, $b = 12$, and $c = 26$. **Obj. 12-8, p. 591**
5. Find the area of $\triangle RST$ if $r = 9$, $s = 12$, and $\angle T = 53.7°$. 43.5 **Obj. 12-9, p. 597**

Check your answers with those at the back of the book.

4. $\angle A = 17.5°$, $\angle B = 13.9°$, $\angle C = 148.6°$

Additional Answers Written Exercises

1. 240 2. 63.6
3. 37.1 4. 254
5. 45.3 6. 34.2
7. 82.5 8. 26.9

17. $K = \frac{1}{2}ab \sin C$; $K^2 = \left(\frac{1}{2}ab \sin C\right)^2 = \frac{1}{4}a^2b^2 \sin^2 C$

18. $2(s - b) = 2s - 2b = a + b + c - 2b = a - b + c$; $2(s - c) = 2s - 2c = a + b + c - 2c = a + b - c$

Additional Answers Mixed Review Exercises

5. $\cos \theta = -\frac{\sqrt{3}}{2}$, $\tan \theta = -\frac{\sqrt{3}}{3}$, $\csc \theta = 2$, $\sec \theta = -\frac{2\sqrt{3}}{3}$, $\cot \theta = -\sqrt{3}$

6. $\sin \theta = -\frac{\sqrt{7}}{4}$, $\tan \theta = \frac{\sqrt{7}}{3}$, $\csc \theta = -\frac{4\sqrt{7}}{7}$, $\sec \theta = -\frac{4}{3}$, $\cot \theta = \frac{3\sqrt{7}}{7}$

7. $\sin \theta = -\frac{\sqrt{2}}{2}$, $\cos \theta = \frac{\sqrt{2}}{2}$, $\csc \theta = -\sqrt{2}$, $\sec \theta = \sqrt{2}$, $\cot \theta = -1$

8. $\cos \theta = \frac{12}{13}$, $\tan \theta = \frac{5}{12}$, $\csc \theta = \frac{13}{5}$, $\sec \theta = \frac{13}{12}$, $\cot \theta = \frac{12}{5}$

Biographical Note / *Edmund Halley*

A comet usually is visible for only a few weeks or months on its path through the solar system. This fact makes its motion difficult to analyze and predict. Early astronomers debated whether comets moved in straight paths or along some regular curve. The Englishman Edmund Halley (1656–1742) solved this problem by describing comets' orbits as ellipses with predictable paths. He claimed that the comets of 1531, 1607, and 1682 were in fact the same comet and that this comet could be seen from Earth approximately every seventy-five years. Later Halley suggested that the comets reported in 1305, 1380, and 1456 were also sightings of this single comet.

Halley predicted the comet's next return for December, 1758. Off schedule by just a few days, the comet appeared in the same part of the sky he had predicted. Although Halley did not live to see this proof of his theory, the scientific world recognized his achievement by naming the comet after him. Halley's comet was last visible from Earth in 1986 and is not due to be seen again until 2061.

Mathematical problems, as well as the positions of the planets, the size of the universe, and stellar motion, also intrigued

Halley. He published papers on topics ranging from higher geometry and the roots of equations to the computation of logarithms and trigonometric functions. One result of his interest in social statistics was the first explanation of how mortality tables could be used to calculate life insurance premiums. Halley was appointed as astronomer royal in 1720. He continued his work well into his old age.

Challenge

One day a visitor to a distant land stood at a fork in the road. The visitor knew that each path led to a different village. The inhabitants of one village always told the truth. The inhabitants of the other village always lied. Luckily, a local citizen happened to be along and agreed to answer one and only one question. The visitor thought for awhile and finally asked a simple question that was guaranteed to indicate the right path to the village of truth tellers, no matter which village the person was from.

What was the question? Explain the visitor's reasoning for asking that particular question.

Supplementary Materials

Study Guide	pp. 203–204
Practice Masters	71–72
Test Master	51
Resource Book	p. 154

Quick Quiz

Give lengths to three significant digits and angle measures to the nearest tenth of a degree.

1. In $\triangle ABC$, $\angle C = 90°$, $a = 13$, and $b = 20$. Solve the triangle. $c = 23.9$, $\angle A = 33.0°$, $\angle B = 57.0°$.
2. In $\triangle DEF$, $d = 9$, $e = 12$, and $\angle F = 100°$. Find side f. 16.2
3. In $\triangle XYZ$, $\angle X = 31°$, $x = 46$, and $z = 50$. Find $\angle Z$. 34.0° or 146.0°
4. Solve $\triangle ABC$ if $a = 10$, $b = 15$, and $c = 20$. $\angle C = 104.5°$, $\angle B = 46.6°$, $\angle A = 28.9°$
5. Find the area of $\triangle RST$ if $r = 8.2$, $s = 12.7$, and $\angle T = 44.6°$. 36.6

Additional Answer
Challenge

The visitor should point to one of the roads and ask, "What would you answer if I asked you whether this is the path to the village of truth-tellers?" By asking this question the visitor must get a truthful answer since a truth-teller would answer correctly, while a liar would have to lie about his answer to the question "Is this the path to the village of truth-tellers?" and thus, in lying about a lie, end up telling the truth.

Additional Answers
Oral Exercises for p. 594

Answers may vary.

6. Find $\angle A$ by $\angle A = 180° - \angle B - \angle C$. Use the law of sines to find b and c.
7. Use the law of cosines to find b. Use the law of sines to find $\angle A$. Find $\angle C$ by $\angle C = 180° - \angle A - \angle B$.
8. Use the law of cosines to find $\angle A$. Use the law of sines to find $\angle B$. Find $\angle C$ by $\angle C = 180° - \angle A - \angle B$.
9. Find $\angle A$ by $\angle A = 180° - \angle B - \angle C$. Use the law of sines to find a and c.
10. Use the law of cosines to find a. Use the law of sines to find $\angle B$. Find $\angle C$ by $\angle C = 180° - \angle A - \angle B$.
11. Use the law of cosines to find $\angle A$. Use the law of sines to find $\angle B$. Find $\angle C$ by $\angle C = 180° - \angle A - \angle B$.
12. Use the law of sines to find $\angle B$. Find $\angle C$ by $\angle C = 180° - \angle A - \angle B$. Use the law of sines to find c.
13. Use the law of sines to find $\angle C$. Find $\angle A$ by $\angle A = 180° - \angle B - \angle C$. Use the law of sines to find a.
14. Use the law of sines to find $\angle A$. Find $\angle C$ by $\angle C = 180° - \angle A - \angle B$. Use the law of sines to find c.
15. Use the law of sines to find $\angle B$. Find $\angle C$ by $\angle C = 180° - \angle A - \angle B$. Use the law of sines to find c.

Chapter Summary

1. Angles may be measured in *degrees,* where one degree is $\frac{1}{360}$ of a complete revolution. The measure is *positive* if the rotation of the angle's *initial side* onto its *terminal side* is counterclockwise. The measure is *negative* if the rotation is clockwise.
2. Let $P(x, y)$ be a point other than the origin on the terminal side of an angle θ in standard position, and let $r = OP$. Then the *trigonometric functions* of θ are:

$$\sin \theta = \frac{y}{r} \qquad \cos \theta = \frac{x}{r} \qquad \tan \theta = \frac{y}{x}$$

$$\csc \theta = \frac{r}{y} \qquad \sec \theta = \frac{r}{x} \qquad \cot \theta = \frac{x}{y}$$

   If θ is acute, the numbers x, y, and r can be thought of as being the lengths of the legs and hypotenuse, respectively, of a right triangle.
3. Scientific calculators give good approximations to the values of the trigonometric functions of any angle. Tables giving values of the functions of acute angles can also be used because each angle θ has an acute *reference angle* α such that $\sin \theta = \pm\sin \alpha$, $\cos \theta = \pm\cos \alpha$, and so forth.
4. To *solve a triangle* is to find the measures of all of its sides and angles when three of the measures (including at least one side) are known. Right triangles can be solved using the definitions of the trigonometric functions. *Any* triangle ABC can be solved with the help of the law of cosines (see page 580) and the law of sines (page 586). Depending on the information given, there may be one solution, two solutions, or no solution.
5. The area K of $\triangle ABC$ can be found by using one of the formulas given on page 597.

Chapter Review

Give the letter of the correct answer.

1. Which of the following is coterminal with $-210°$? 12-1

a. 30° **b.** 150° **c.** $-150°$ **d.** 210°

2. Express 45°14′42″ in decimal degrees.

a. 45.452° **b.** 45.542° **c.** 45.245° **d.** 45.45°

3. Find the product of all six trigonometric functions of an angle θ in standard position whose terminal side passes through $(-\sqrt{3}, 1)$. 12-2

a. -1 **b.** 1 **c.** $\sqrt{3}$ **d.** $\frac{\sqrt{3}}{2}$

4. In $\triangle ABC$, $\angle A = 60°$, $b = 10$, and $\angle B = 90°$. Find c.
   a. $5\sqrt{3}$ (b.) 5 c. 10 d. 20
5. Find x in the diagram at the right.
   a. $16\sqrt{2}$ (b.) $6\sqrt{2}$
   c. $4\sqrt{3}$ d. $8\sqrt{6}$

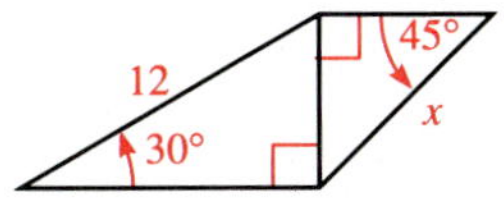

6. Express cos 230° as a function of an acute angle. 12-3
   a. −cos 40° b. −sin 40° (c.) −cos 50° d. cos 50°
7. Find $\tan \theta$ if $\cos \theta = \frac{8}{17}$ and $\sin \theta > 0$.
   (a.) $\frac{15}{8}$ b. $-\frac{15}{17}$ c. $-\frac{8}{15}$ d. $-\frac{15}{8}$
8. Give the exact value of sin 300°.
   a. $-\frac{1}{2}$ b. $\frac{1}{2}$ (c.) $-\frac{\sqrt{3}}{2}$ d. $\frac{\sqrt{3}}{2}$
9. Find sec 116.7° to four decimal places. 12-4
   a. 1.119 b. 2.226 c. −1.119 (d.) −2.226
10. Find θ to the nearest ten minutes if $\cot \theta = -0.7046$ and $0° < \theta < 180°$.
    a. 54°50′ (b.) 125°10′ c. 144°50′ d. −56°50′
11. In $\triangle DEF$, $\angle E = 90°$, $e = 10$, and $d = 8$. Find $\angle F$ to the nearest tenth of a degree. 12-5
    (a.) 36.9° b. 53.1° c. 38.7° d. 44.2°
12. In $\triangle ABC$, $\angle B = 90°$, $\angle C = 55°$, and $c = 12$. Find b to the nearest tenth.
    (a.) 14.6 b. 6.9 c. 20.9° d. 9.8
13. The base angles of an isosceles trapezoid are 45° and the bases have lengths 15 and 25 units. Find the height of the trapezoid.
    (a.) 5 b. 5.2 c. 10.2 d. 5.3
14. In $\triangle ABC$, $a = 6$, $b = 10$, and $c = 7$. Find the measure of the largest angle to the nearest tenth of a degree. 12-6
    a. 43.5° (b.) 100.3° c. 36.2° d. 46.5°
15. Two planes leave the airport at noon, one traveling east at 300 km/h and the other traveling northwest at 450 km/h. How far apart are they at 2 PM?
    a. 637 km b. 821 km (c.) 1390 km d. 1080 km
16. In $\triangle ABC$, find $\angle A$ to the nearest tenth of a degree if $\angle B = 95°$, $b = 15.4$, and $a = 8.0$. 12-7
    a. 32.1° b. 56.8° (c.) 31.2° d. 92.3°
17. Wires of lengths 20 m and 30 m extend from the top of a tower to the ground on the same side of the tower. The shorter wire makes an angle of 42° with the ground. What angle do the wires make with each other?
    (a.) 15.5° b. 63.5° c. 26.5° d. 48.2°

(Chapter Review continues on the next page.)

16. Use the law of sines to find $\angle C$. Find $\angle A$ by $\angle A = 180° - \angle B - \angle C$. Use the law of sines to find a.
17. Use the law of sines to find $\angle B$. Find $\angle A$ by $\angle A = 180° - \angle B - \angle C$. Use the law of sines to find a.

Additional Answers
Written Exercises for p. 594

1. $\angle A = 100°$, $b = 5.90$, $c = 14.9$
2. $b = 20.3$, $\angle A = 11.4°$, $\angle C = 18.6°$
3. $\angle A = 19.7°$, $\angle B = 28.2°$, $\angle C = 132.1°$
4. $\angle A = 55.0°$, $a = 12.5$, $c = 6.44$
5. $a = 20.0$, $\angle B = 25.7°$, $\angle C = 34.3°$
6. $\angle A = 18.2°$, $\angle B = 51.3°$, $\angle C = 110.5°$
7. $\angle B = 30.7°$, $\angle C = 19.3°$, $c = 12.9$
8. $\angle C = 51.2°$, $\angle A = 8.8°$, $a = 3.53$
9. $\angle A = 35.3°$, $\angle C = 104.7°$, $c = 15.1$
10. $\angle B = 31.7°$, $\angle C = 113.3°$, $c = 19.2$
11. $\angle C = 80.0°$, $\angle A = 50.0°$, $a = 14.0$, or $\angle C = 100.0°$, $\angle A = 30.0°$, $a = 9.14$
12. $\angle B = 76.4°$, $\angle A = 33.6°$, $a = 8.54$, or $\angle B = 103.6°$, $\angle A = 6.4°$, $a = 1.72$

18. In $\triangle ABC$, $\angle B = 30°$ and $c = 20$. For what value(s) of b will the triangle have two solutions? 12-8
a. $0 < b < 10$ (b.) $10 < b < 20$ c. $b < 20$ d. $b = 10$

19. Find x in the diagram at the right.
a. 15.8 b. 18.4
c. 19.8 (d.) 20.7

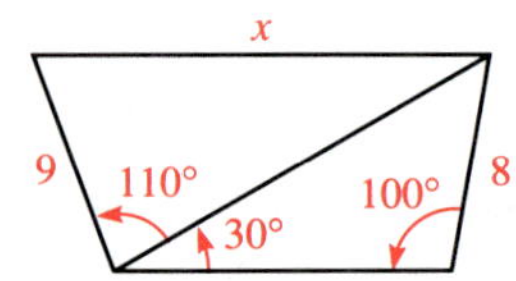

20. Find the area of $\triangle ABC$ if $a = 8$, $b = 15$, and $\angle C = 40°$. 12-9
a. 77.1 b. 60.0 c. 45.9 (d.) 38.6

Alternate Test p. T24

Supplementary Materials

Test Masters	52, 53
Resource Book	p. 155

Chapter Test

1. Express 13.24° in degrees, minutes, and seconds. 13°14′24″ 12-1
2. Find two angles, one positive and one negative, that are coterminal with 285°. 645°, −75°
3. Find x in the diagram at the right. $6\sqrt{6}$ 12-2

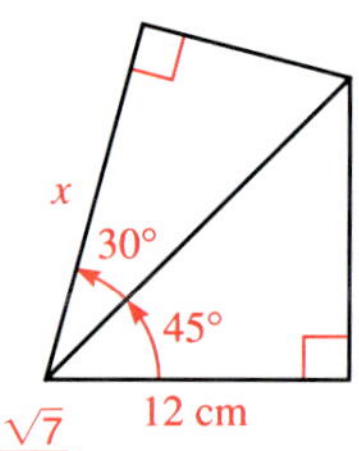

5. $\sin 240° = \frac{-\sqrt{3}}{2}$, $\cos 240° = -\frac{1}{2}$, $\tan 240° = \sqrt{3}$, $\csc 240° = \frac{-2\sqrt{3}}{3}$, $\sec 240° = -2$, $\cot 240° = \frac{\sqrt{3}}{3}$

4. If $\cos \theta = -\frac{3}{4}$ and $\sin \theta < 0$, find $\tan \theta$. $\frac{\sqrt{7}}{3}$ 12-3
5. Give the exact values of the six trigonometric functions of 240°.
6. a. Find sec 145.8°. −1.209 12-4
   b. If $\cos \theta = 0.5606$ and $90° < \theta < 360°$, find θ to the nearest tenth of a degree. 304.1°
7. The angle of elevation from an observer on the street to the top of a building is 55.6°. If the observer is 150 ft from the base of the building, how tall is the building? 219 ft 12-5
8. A compass with legs 3 in. long is opened to measure the diameter of a circle. If the diameter is 5 in., what is the angle between the legs of the compass? 112.9° 12-6

Give lengths to three significant digits and angle measures to the nearest tenth of a degree.

9. Solve $\triangle DEF$ if $\angle D = 32°$, $\angle E = 108°$, and $f = 12$. $\angle F = 40°$, $d = 9.89$, $e = 17.8$ 12-7
10. Solve $\triangle ABC$ if $\angle A = 40°$, $a = 6$, and $b = 8$. See below. 12-8
11. Find the area of a triangular plot of land if the sides have length 200 m, 150 m, and 100 m. 7260 m^2 12-9

10. $\angle B = 59°$, $\angle C = 81°$, $c = 9.22$ or $\angle B = 121.0°$, $\angle C = 19°$, $c = 3.04$

Preparing for College Entrance Exams

Strategy for Success

When solving problems involving trigonometry, area, or distance, you may find it helpful to draw a sketch that shows the given information. Do not make any assumptions in drawing the figure; use only the given information. Use any available space in your test booklet or scrap paper, but avoid making any stray marks on your answer sheet.

Decide which is the best of the choices given and write the corresponding letter on your answer sheet.

1. Find the next term of this sequence: 1, 3, 7, 13, 21, . . . C
(A) 27 **(B)** 37 **(C)** 31 **(D)** 33 **(E)** 29

2. Evaluate $64 \sum_{k=1}^{6} 2^{-k}$. C
(A) 64 **(B)** 128 **(C)** 63 **(D)** 1 **(E)** $\frac{31}{32}$

3. The infinite series $1 - \frac{5}{4} + \frac{25}{16} - \frac{125}{64} + \ . \ . \ .$ C
(A) has the common difference $\frac{5}{4}$ **(B)** has the sum $\frac{4}{9}$ **(C)** has no sum
(D) has the sum -4 **(E)** is neither arithmetic nor geometric

4. Which function(s) is (are) not defined for $\phi = -90°$? C
I. $\cos \phi$ II. $\sec \phi$ III. $\tan \phi$
(A) II only **(B)** I and II only **(C)** II and III only
(D) I and III only **(E)** I, II, and III

5. If $\csc \phi = 3$ and $90° < \phi < 270°$, find $\cot \phi$. A
(A) $-2\sqrt{2}$ **(B)** $\frac{\sqrt{2}}{4}$ **(C)** $-\frac{\sqrt{2}}{4}$ **(D)** -8 **(E)** $-\frac{\sqrt{2}}{3}$

6. In $\triangle XYZ$, $\angle X = 45°$, $\angle Z = 30°$, and $z = 8$. Find x. B
(A) $\frac{8}{3}\sqrt{6}$ **(B)** $8\sqrt{2}$ **(C)** $4\sqrt{2}$ **(D)** $4\sqrt{6}$ **(E)** $32\sqrt{2}$

7. Find the area of a triangle with sides of lengths 5, 6, and 7. D
(A) $2\sqrt{6}$ **(B)** $3\sqrt{15}$ **(C)** $2\sqrt{14}$ **(D)** $6\sqrt{6}$ **(E)** $\frac{21\sqrt{2}}{2}$

8. In $\triangle RST$, $\angle R = 137.4°$, $t = 15$, and $s = 12$. Find r to the nearest integer. E
(A) 10 **(B)** 17 **(C)** 21 **(D)** 23 **(E)** 25

13 Trigonometric Graphs; Identities

Objectives

13-1 To use radians to measure angles.

13-2 To define the circular functions.

13-3 To use periodicity and symmetry in graphing functions.

13-4 To graph the sine, cosine, and related functions.

13-5 To graph the tangent, cotangent, secant, cosecant, and related functions.

13-6 To simplify trigonometric expressions and to prove identities.

13-7 To use formulas for the sine and cosine of a sum or difference.

13-8 To use the double-angle and half-angle formulas for the sine and cosine.

13-9 To use addition, double-angle, and half-angle formulas for the tangent.

Assignment Guide

See p. T58 for Key to the format of the Assignment Guide

Day	Average Algebra	Average Algebra and Trigonometry	Extended Algebra	Extended Algebra and Trigonometry
1		**13-1** 610/3–60 mult. of 3 611/*P*: 1–7 odd **S** 612/*Mixed Review*		**13-1** 610/3–60 mult. of 3 611/*P*: 1–11 odd **S** 612/*Mixed Review*
2		**13-2** 617/1–27 odd **S** 611/*P*: 2, 6		**13-2** 617/1–27 odd, 28–30 **S** 612/*P*: 10, 12
3		**13-3** 621/2–20 even **S** 623/*Mixed Review*		**13-3** 621/2–20 even, 21–26, 29 **S** 623/*Mixed Review*
4		**13-4** 628/1–6, 13–20, 27–32 **S** 622/11–19 odd		**13-4** 628/7–12, 21–26, 33–38 **S** 623/27, 38
5		**13-4** 628/7–12, 21–26, 33–38 **S** 612/*P*: 8		**13-4** 629/39–44 **13-5** 633/1–16
6		**13-5** 633/1–16		**13-5** 633/17–28 **S** 633/*Mixed Review*
7		**13-5** 633/17–22 **S** 633/*Mixed Review*		**13-6** 639/1–47 odd **R** 634/*Self-Test 1*
8		**13-6** 639/1–37 odd **R** 634/*Self-Test 1*		**13-6** 640/24–56 even **S** 629/47, 48
9		**13-6** 640/26–56 even		**13-7** 643/1–33 odd **S** 645/*Mixed Review*
10		**13-7** 643/1–29 odd **S** 645/*Mixed Review*		**13-8** 649/2–40 even **S** 644/30, 34

Assignment Guide *(continued)*

Day	Average Algebra	Average Algebra and Trigonometry	Extended Algebra	Extended Algebra and Trigonometry
11		**13-8** 649/2–36 even **S** 644/16–20 even, 31, 32		**13-9** 652/3–24 mult. of 3, 25–32 **S** 653/*Mixed Review*
12		**13-9** 652/1–27 odd **S** 653/*Mixed Review*		*Prepare for Chapter Test* **R** 654/*Self-Test 2* **R** 655/*Chapter Review*
13		*Prepare for Chapter Test* **R** 654/*Self-Test 2* **R** 655/*Chapter Review*		*Administer Chapter 13 Test*
14		*Administer Chapter 13 Test*		

Supplementary Materials Guide

For Use with Lesson	Practice Masters	Tests	Study Guide (Reteaching)	Resource Book: Tests	Resource Book: Practice Exercises	Resource Book: Applications (A) Enrichment (E) Technology (T)
13-1			pp. 205–206			p. 216 (A)
13-2	Sheet 73		pp. 207–208			
13-3			pp. 209–210		p. 158	
13-4	Sheet 74		pp. 211–212			pp. 246–247 (T)
13-5	Sheet 75	Test 54	pp. 213–214		p. 159	pp. 248–249 (T)
13-6			pp. 215–216		p. 160	
13-7	Sheet 76		pp. 217–218			
13-8			pp. 219–220			
13-9	Sheet 77	Test 55	pp. 221–222		p. 161	
Chapter 13		Tests 56, 57		pp. 69–72	p. 162	p. 232 (E)

Overhead Visuals

For Use with Lessons	Visual	Title
13-4	B	Multi-Use Packet 2
13-2, 13-3, 13-7	10	Angles
13-4	11	Graphing $y = c + a \cos bx$

Software

Software	Algebra Plotter Plus	Using Algebra Plotter Plus	Computer Activities	Test Generator
	Function Plotter, Circular Function Quiz	Enrichment, pp. 40–41 Activity Sheets, pp. 60–63	Activities 29, 30, 37	189 test items
For Use with Lessons	13-4, 13-5, 13-9	13-4, 13-5, 13-9	13-4, Chapter 13	all lessons

Strategies for Teaching

Using Manipulatives and Using Technology

Calculators and computers serve two very important functions in this chapter. Computer simulation makes it possible for students to examine some concepts (trigonometric functions, for example) in depth and to test a variety of hypotheses. Also, the graphic presentation of data is made much easier if a graphing calculator or a computer is available. Students are able to substitute into equations, different values for variables and then able to see the affects that these changes have on the resulting graphs. This enables students to better understand the characteristics of any given trigonometric function.

Manipulatives are helpful in this chapter so that students may go beyond working with algorithms to using concrete approaches to understand the concepts underlying the symbolic procedures.

13-1 Radian Measure

Students can make their own radian protractors.

Trace a regular protractor on heavy paper. Use the fact that 0.2 radians $\approx 11.5°$ to make marks on the paper protractor. Pick up the regular protractor, and cut out the tracing. Now label the marks with the correct decimal radian measure. To help students grasp radian measure meaningfully, they should now use their radian protractors to measure given angles and to draw angles of given radian measures.

13-2 Circular Functions

Some students may have discovered an ingenious way to convert between degrees and radians using their calculators. To convert from degrees to radians, set the calculator in degree mode. Press either the sine or cosine key. Now switch to radian mode, and press the inverse of that function. The result is the decimal radian equivalent of the original degree measure. A similar process converts radians to degrees. To try this method, direct students to Exercises 29–44 on page 611. Ask why the sine is better to use than the tangent (sine is defined for all angles, tangent isn't), and why calculators give incorrect results, using the sine, if the angle is not between $-90°$ and $90°$ (the inverse keys give only the principal value of the inverse function).

The diagram on page 617 can be used as the plan for making a device for direct-reading of cosines of real numbers. On graph paper with large squares, lay out x- and y-axes and a unit circle. Glue this paper to a piece of cardboard and cut out the circle. Put the cardboard circle upright, so that it is perpendicular to the floor. (Pin it to a bulletin board, for example.) Place a strip of graph paper to form a number line with its origin at A, parallel to the y-axis. For a plumb line, cut a piece of string and tie a weight to one end. To find the cosine of 0.65, follow the example below.

Use a piece of light string or thread. Place the end at A. Stretch the string along the vertical number line through A, making the second endpoint of a segment 0.65 units long. Leaving the end at A fixed, wrap the string around the cardboard unit circle. Where the end of the 0.65 segment lies on the unit circle, drop the plumb line. The line will cross the x-axis; simply read the coordinate and you have the value of cos 0.65. Repeat the procedure for any other cosine. Ask students how to use this apparatus to read sines.

13-4 Graphs of the Sine and Cosine

See the Exploration on page 844 for an activity in which students explore finding the sum of two sine curves by using graphs. In this activity they also investigate the connection to light and sound.

References to Strategies

PE: Pupil's Edition **TE:** Teacher's Edition **RB:** Resource Book

Problem Solving Strategies

TE: p. 608

Applications

PE: pp. 606, 634–635 (music); p. 609 (physics); pp. 611–612 (word problems)
TE: pp. 606–607
RB: p. 216

Nonroutine Problems

PE: p. 612 (Probs. 11–13); p. 617 (Exs. 28–32); p. 623 (Exs. 29, 30, Challenge); p. 629 (Exs. 39–48); p. 633 (Ex. 29); p. 635 (Challenge 1, 2); p. 644 (Exs. 34–37); p. 649 (Ex. 43); p. 653 (Exs. 33–35)
TE: pp. T129–T131 (Sugg. Extensions, Lessons 13-5, 13-7, 13-9)

Communication

PE: p. 617 (Ex. 27, convincing argument)
TE: p. T129 (Reading Algebra); p. 624 (translating); p. T130 (Sugg. Extension, Lesson 13-6)

Thinking Skills

PE: p. 640 (Exs. 49–58); p. 644 (Exs. 15–20, 25–30, 35); p. 649 (Exs. 33–40); p. 653 (Exs. 25–28, 32, 34–35, proofs)
TE: pp. 613, 631, 641, 650

Explorations

PE: p. 623 (Challenge); p. 844 (sine curves)

Connections

PE: pp. 606, 634–635 (Music); pp. 608, 618, 647, 651 (Geometry); pp. 609, 611–612, 634 (Physics); p. 611 (Geography); pp. 611–617, 619, 624–633, 636–654 (Trigonometry); p. 618 (History)

Using Technology

PE: pp. 606, 608, 614–615, 624–625, 629, 631, 633; p. 645 (Exs.)
TE: pp. 606, 625, 629–630, 633, 636, 645, T128
RB: pp. 246–249
Computer Activities: pp. 77–82, 100–102
Using Algebra Plotter Plus: pp. 40–41, 60–63

Using Manipulatives/Models

TE: pp. T127–T128 (Lessons 13-1, 13-2)
Overhead Visuals: B, 10, 11

Cooperative Learning

TE: p. T131 (Lesson 13-7)

Teaching Resources

For use in implementing the teaching strategies referenced on the previous page.

Application
***Resource Book,* p. 216**

Application—Graphing Sums of Functions (For use with Chapter 13)

The sum of the functions f and g is defined as $(f + g)(x) = f(x) + g(x)$, where the domain of $f + g$ is the intersection of the domains of f and g.

1. Consider the functions $f(x) = \frac{1}{2}x^2$ and $g(x) = \frac{1}{2}x + 1$, whose graphs are shown below.

   a. State a rule for the sum:

   $(f + g)(x) =$ ____________

   b. Complete the table below.

x	$f(x)$	$g(x)$	$(f + g)(x)$
-3	____	____	____
-2	____	____	____
-1	____	____	____
0	____	____	____
1	____	____	____
2	____	____	____
3	____	____	____

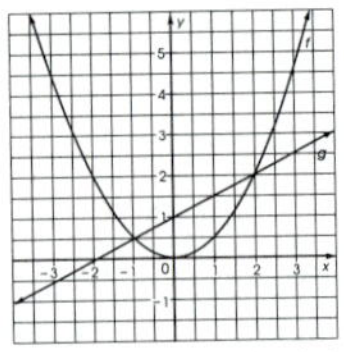

   c. Sketch the graph of $f + g$ using the same axes on which the graphs of f and g are drawn. Notice that for a given x-value you can find the y-value of $f + g$ by adding the corresponding y-values of f and g.

2. The sound wave generated by a tuning fork can be represented by a simple sine wave, but if two different tones are sounded at the same time, the resulting periodic wave is the sum of the two sound waves. For example, the sound wave generated by playing the notes middle C and high C at the same time could be represented by the sum $y = \cos x + \sin 2x$. Sketch the graph of this sum using the graphs of $y = \cos x$ and $y = \sin 2x$ (shown at the right) by adding the corresponding y-values for a given x-value.

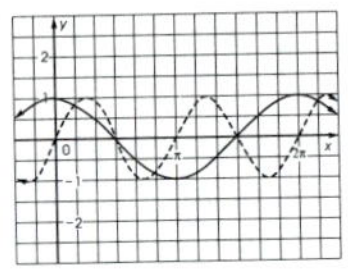

3. The French mathematician Fourier proved that *every* periodic function can be reproduced by a sum of simple sine waves. He also worked with infinite sums of sine waves, including

   the "sawtooth" wave: $y = \sin x + \frac{1}{2}\sin 2x + \frac{1}{3}\sin 3x + \ldots + \frac{1}{N}\sin Nx + \ldots$ (N is an integer),

   and the "square" wave: $y = \sin x + \frac{1}{3}\sin 3x + \frac{1}{5}\sin 5x + \ldots + \frac{1}{N}\sin Nx + \ldots$ (N is an odd integer).

   If you have a graphing calculator or a computer with graphics software, try graphing each of these waves using partial sums. That is, first graph $y = \sin x$, then $y = \sin x + \frac{1}{2}\sin 2x$, then $y = \sin x + \frac{1}{2}\sin 2x + \frac{1}{3}\sin 3x$, and so on. Use these graphs to predict the appearance of the "sawtooth" and "square" waves as N becomes large.

216 APPLICATIONS

Enrichment/Exploration
***Resource Book,* p. 232**

Pairing Points (For use with Chapter 13)

Each of the circles has the equation $x^2 + y^2 = 1$. The line APT is fixed at point A. As it rotates counterclockwise about A, line APT intersects the circle at P and the x-axis at T.

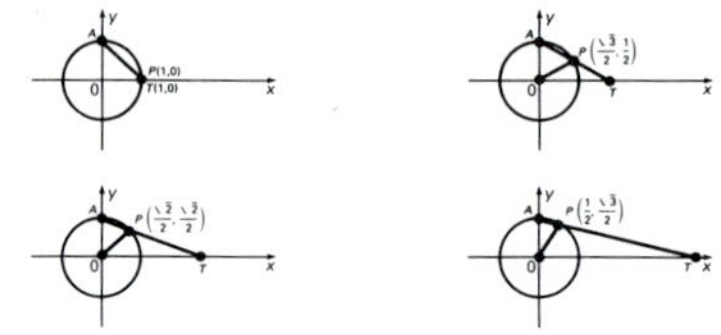

1. Complete the following chart.

Coordinates of P	Coordinates of T
(1, 0)	
$\left(\frac{\sqrt{3}}{2}, \frac{1}{2}\right)$	
$\left(\frac{\sqrt{2}}{2}, \frac{\sqrt{2}}{2}\right)$	
$\left(\frac{1}{2}, \frac{\sqrt{3}}{2}\right)$	

2. If point P has coordinates (x, y) and T has coordinates $(t, 0)$, find a formula for t in terms of x and y. You should be able to check your formula using the data from Problem 1.

____

232 ENRICHMENT

Using Technology
***Resource Book,* p. 246**

Using a Computer or a Graphing Calculator

To complete these activities, you should use a computer with graphics software (such as ALGEBRA PLOTTER PLUS) or a graphing calculator.

Exploring Sine and Cosine Curves (For use with Lesson 13–4)

1. Graph these equations: $y = \sin(x)$, $y = \cos(x)$

   (*Note:* If you are using the ALGEBRA PLOTTER PLUS disk, you should change the x-axis scale from integers to multiples of π.)

   a. The *period* of each curve is the shortest distance along the x-axis over which the curve has one complete up-and-down cycle. In terms of π, what is the period of the sine curve? ________ of the cosine curve? ________

   b. The *maximum* value attained by each curve is the y-coordinate of the highest points on the curve, while the *minimum* value is the y-coordinate of the lowest points. What are the maximum and minimum values of the sine curve? ______, ______ of the cosine curve? ______, ______

   c. The *amplitude* of each curve is half the difference between the maximum and minimum values of the curve. What is the amplitude of the sine curve? ________ of the cosine curve? ________

   d. As you can see from your answers to parts (a)–(c), the sine and cosine curves share the same basic features. In fact, if all points on the cosine curve were shifted to the right some distance d, the translated cosine curve would coincide with the sine curve. In terms of π, what is the smallest possible value of d? ________

2. Graph these equations: $y = 1 + \sin(x)$, $y = -2 + \cos(x)$

   Using π when giving each period, complete the table below.

	Equation	*Period*	*Maximum value*	*Minimum value*	*Amplitude*
a.	$y = 1 + \sin(x)$	____	____	____	____
b.	$y = -2 + \cos(x)$	____	____	____	____

3. What effect does the constant c have on the graphs of $y = c + \sin(x)$ and $y = c + \cos(x)$?

____

4. Try to complete the table below before graphing the given equations.

	Equation	*Period*	*Maximum value*	*Minimum value*	*Amplitude*
a.	$y = -3 + \sin(x)$	____	____	____	____
b.	$y = 4 + \cos(x)$	____	____	____	____

   Now examine the graphs of the given equations to confirm the entries in the table.

(continued)

246 USING TECHNOLOGY

Using Technology
***Resource Book,* p. 247**

Exploring Sine and Cosine Curves (continued)

5. Graph these equations: $y = \frac{1}{2}\sin(x)$, $y = 2\cos(x)$

   Using π when giving each period, complete the table below.

	Equation	*Period*	*Maximum value*	*Minimum value*	*Amplitude*
a.	$y = \frac{1}{2}\sin(x)$	____	____	____	____
b.	$y = 2\cos(x)$	____	____	____	____

6. What effect does the positive constant a have on the graphs of $y = a\sin(x)$ and $y = a\cos(x)$?

____

7. Try to complete the table below before graphing the given equations.

	Equation	*Period*	*Maximum value*	*Minimum value*	*Amplitude*
a.	$y = 3\sin(x)$	____	____	____	____
b.	$y = \frac{3}{2}\cos(x)$	____	____	____	____

   Now examine the graphs of the given equations to confirm the entries in the table.

8. Graph these equations: $y = \sin(2x)$, $y = \cos\left(\frac{1}{2}x\right)$

   Using π when giving each period, complete the table below.

	Equation	*Period*	*Maximum value*	*Minimum value*	*Amplitude*
a.	$y = \sin(2x)$	____	____	____	____
b.	$y = \cos\left(\frac{1}{2}x\right)$	____	____	____	____

9. What effect does the positive constant b have on the graphs of $y = \sin(bx)$ and $y = \cos(bx)$?

____

10. Try to complete the table below before graphing the given equations.

	Equation	*Period*	*Maximum value*	*Minimum value*	*Amplitude*
a.	$y = \sin\left(\frac{2}{3}x\right)$	____	____	____	____
b.	$y = \cos(3x)$	____	____	____	____

   Now examine the graphs of the given equations to confirm the entries in the table.

11. If you have the ALGEBRA PLOTTER PLUS disk, use the "Circular Function Quiz" program to see if you can correctly determine the equations of random sine and cosine curves drawn by the computer.

USING TECHNOLOGY 247

Using Technology
Using Algebra Plotter Plus, p. 62

Tangent, Cotangent, Secant, and Cosecant

Algebra Plotter Plus | Book 2: Lesson 13-5

Use the Function Plotter program of Algebra Plotter Plus. Change the x-axis scale to multiples of π by selecting SCALE, moving the cursor to the "Multiples of pi" option, and selecting "Yes."

1. Graph this equation: $y = \tan(x)$

a. The graph has vertical *asymptotes*, which are lines that the graph approaches but does not cross. One such asymptote is the line $x = 0.5\pi$. Give the equations of all other asymptotes that appear on the screen.

____

b. As the graph indicates, the tangent function is *periodic* (that is, repeating). You can determine the period by finding the distance between consecutive points where the graph crosses the x-axis. In terms of π, what is this period? _____

2. Graph this equation: $y = \cot(x)$

a. The graph of the cotangent function has vertical asymptotes. Give the equations of all asymptotes that appear on the screen.

____

b. In terms of π, what is the period of the cotangent function? _____

3. Graph these equations: $y = \sin(x)$

$y = \csc(x)$

The cosecant function is defined to be the reciprocal of the sine function. Use this fact as well as the graphs of the sine and cosecant functions to answer the questions.

a. The range of $y = \sin(x)$ is $\{y: -1 \le y \le 1\}$. What is the range of $y = \csc(x)$? ____________

b. The period of the sine function is 2π. What is the period of the cosecant function? _____

c. Whenever the graph of $y = \sin(x)$ crosses the x-axis, what happens to the graph of $y = \csc(x)$?

____

4. Graph this equation: $y = \cos(x)$

a. The secant function is defined to be the reciprocal of the cosine function. Use this fact as well as the results of Exercise 3 to sketch the graph of $y = \sec(x)$.

b. Use the Function Plotter to graph $y = \sec(x)$ to confirm your sketch.

c. Why are the graphs of $y = \cos(x)$ and $y = \sec(x)$ tangent to each other whenever x is a multiple of π?

____

____

(continued)

Using Technology
Using Algebra Plotter Plus, p. 63

Tangent, Cotangent, Secant, and Cosecant *(continued)*

5. Graph these equations: $y = \tan(x)$

$y = -1 + \tan(x)$

$y = 2 + \tan(x)$

Describe the effect of the constant c on the graph of $y = c + \tan(x)$.

____

6. Using your answer to Exercise 5, sketch the graph of $y = -2 + \cot(x)$. Then graph $y = -2 + \cot(x)$ to confirm your sketch.

7. Graph these equations: $y = \csc(x)$

$y = 2\csc(x)$

$y = \frac{1}{2}\csc(x)$

Describe the effect of the positive constant a on the graph of $y = a\csc(x)$.

____

8. Using your answer to Exercise 7, sketch the graph of $y = \frac{3}{2}\sec(x)$. Then graph $y = \frac{3}{2}\sec(x)$ to confirm your sketch.

9. Graph these equations: $y = \cot(x)$

$y = \cot(2x)$

$y = \cot(\frac{1}{2}x)$

Describe the effect of the positive constant b on the graph of $y = \cot(bx)$.

____

10. Using your answer to Exercise 9, predict the period of $y = \tan(\frac{2}{3}x)$. _____

Graph $y = \tan(\frac{2}{3}x)$ to confirm your prediction.

Using Models/Exploration
Overhead Visual 11, Sheets 1 and 2

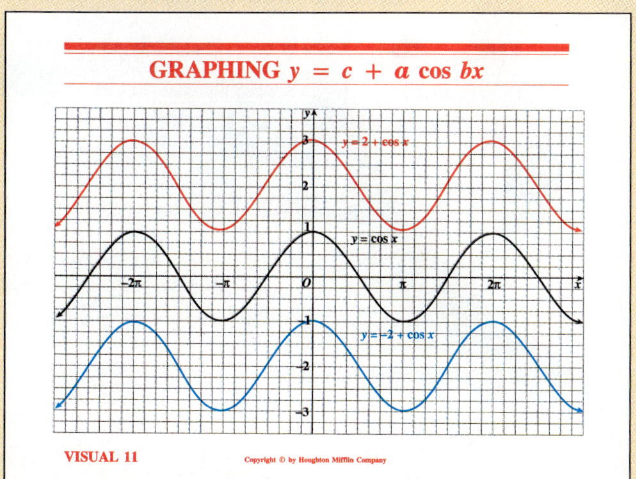

Using Models/Exploration
Overhead Visual 11, Sheets 1 and 3

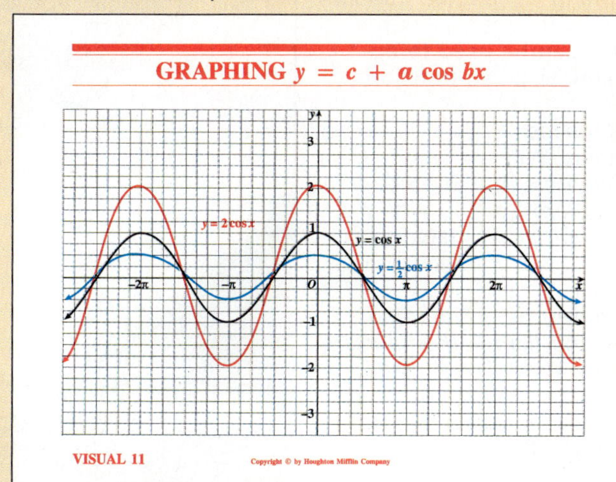

13 Trigonometric Graphs; Identities

The upper graph shows the sine waves for three guitar strings vibrating at different frequencies. The lower graph shows the sum of the two solid sine waves.

606

Application

A tuning fork produces a tone that can be represented by a sine wave, but a single musical note would require a combination of tuning forks to duplicate. For instance, when a piano key is struck, the wire not only vibrates as a whole with a fundamental frequency, but also in halves, in thirds, and so on. A graph of the wire's vibration is a *sum* of sine waves.

The analysis of a vibrating string attracted the attention of 18th-century mathematicians. The work of these mathematicians generated controversy among them. The debate ended with a discovery by Joseph de Fourier: Every function can be represented as an infinite sum of sine and cosine functions. Such sums are known now as Fourier series.

Group Activities

Students might want to investigate how an electronic synthesizer simulates different instruments.

Using a computer or graphing calculator, students can see how finite Fourier series can be used to approximate given functions.

References

Bennett, William Ralph, Jr. *Scientific and Engineering Problem-Solving with the Computer.* Englewood Cliffs, NJ: Prentice-Hall, 1976.

Kline, Morris. *Mathematics for the Nonmathematician.* New York: Dover, 1985.

Support Materials

Resource Book p. 216

Circular Functions and Their Graphs

13-1 Radian Measure

Objective To use radians to measure angles.

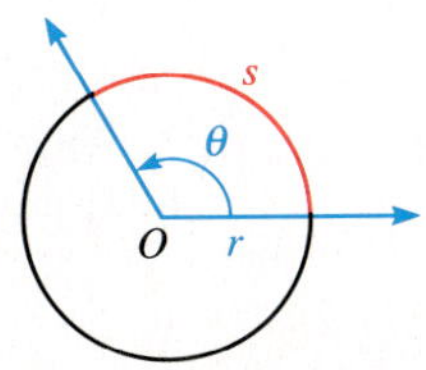

In advanced mathematics angles are usually measured in *radians* rather than degrees. To define the radian measure of an angle θ, let O be a circle of radius r centered at the vertex of θ, as shown at the right. Let s be the length of the arc of O intercepted by θ. The arc length s is considered to be positive if θ is a positive angle and negative if θ is a negative angle.

The **radian measure** of θ is defined to be the ratio of arc length s to radius r:

$$\theta = \frac{s}{r}$$

The symbol θ here stands for the *measure* of angle θ. When no unit of angle measure is specified, you can assume that radian measure is being used. Therefore, "$\theta = 2$" means "the measure of θ is 2 radians," and the length of the arc of O intercepted by θ is twice the length of the radius r of circle O.

If θ is a complete revolution, $s = 2\pi r$ and $\theta = \frac{2\pi r}{r} = 2\pi$. If θ is a complete revolution, the degree measure of θ is 360°. Likewise, an angle that is half a revolution has degree measure 180° and radian measure π. The conversion formula

$$180° = \pi \text{ radians}$$

can be used to convert from one system of measure to the other:

$$1° = \frac{\pi}{180} \text{ radians} \qquad 1 \text{ radian} = \frac{180°}{\pi} \approx 57.3°$$

Example 1 **a.** Express 85° in radians. **b.** Express $\frac{\pi}{3}$ radians in degrees.

Solution **a.** $85° = 85 \cdot \frac{\pi}{180}$ radians

$= \frac{17\pi}{36}$ radians

b. $\frac{\pi}{3}$ radians $= \frac{\pi}{3} \cdot \frac{180°}{\pi} = 60°$

$\left(\text{or } \frac{\pi}{3} \text{ radians} = \frac{1}{3} \cdot \pi \text{ radians} = \frac{1}{3} \cdot 180° = 60°\right)$

Teaching References

Lesson Commentary, pp. T127–T132

Assignment Guide, pp. T69–T71

Supplementary Materials
- Practice Masters 73–77
- Tests 54–57
- Resource Book
  - Practice Exercises, pp. 158–162
  - Tests, pp. 69–72
  - Enrichment Activity, p. 232
  - Application, p. 216
- Study Guide, pp. 205–222
- Computer Activities 29–30
- Test Generator
- Disk for Algebra

Alternate Test, p. T25

Teaching Suggestions p. T127

Suggested Extensions p. T127

Warm-Up Exercises

Draw a circle. Draw and label the center, a central angle, and the arc intercepted by the central angle.

Motivating the Lesson

Tell students that airplanes often fly "great circle" routes. These are arcs of circles with centers at the center of the Earth. The length of such an arc, the central angle that intercepts this arc, and the radius of the Earth are closely related. This relationship is one of the topics in today's lesson.

Problem Solving Strategies

By *experimenting* with their calculators, some students will discover an easy way to convert between degrees and radians. This method involves setting the calculator in degree mode, pressing a trigonometric function key, switching to radian mode, and then pressing the inverse trigonometric function key.

Chalkboard Examples

1. a. Express 135° in radians.

   b. Express $\frac{4\pi}{5}$ in degrees.

   a. $135° = 135 \cdot \frac{\pi}{180} = \frac{3\pi}{4}$

   b. $\frac{4\pi}{5} = \frac{4\pi}{5}\left(\frac{180}{\pi}\right) = 144°$

2. a. Express 34° in radians to the nearest hundredth of a radian.

   $34° = 34 \cdot \frac{\pi}{180} = \frac{34(3.1416)}{180} = 0.59$

   b. Express 3 radians in degrees to the nearest tenth of a degree.

   $3 = \left(3 \cdot \frac{180}{\pi}\right)^{\circ} = \left(\frac{3 \cdot 180}{3.1416}\right)^{\circ} = 171.9°$

Here are the degree and radian measures of some frequently used angles.

Degree measure	0°	30°	45°	60°	90°	120°	135°	150°	180°
Radian measure	0	$\frac{\pi}{6}$	$\frac{\pi}{4}$	$\frac{\pi}{3}$	$\frac{\pi}{2}$	$\frac{2\pi}{3}$	$\frac{3\pi}{4}$	$\frac{5\pi}{6}$	π

Scientific calculators accept both degree and radian measurements in decimal form. You can convert between degrees and radians on a calculator by using the π key or a decimal approximation, such as 3.1416, for π.

Example 2 **a.** Express 35° in radians to the nearest hundredth of a radian.

b. Express 2 radians in degrees to the nearest tenth of a degree.

Solution **a.** $35° = 35 \cdot \frac{\pi}{180}$ radians

$\approx \frac{35(3.1416)}{180}$ radians

≈ 0.61 radians ***Answer***

b. 2 radians $= 2 \cdot \frac{180°}{\pi}$

$\approx \frac{2 \cdot 180°}{3.1416}$

$\approx 114.6°$ ***Answer***

Using radian measure you can write simple formulas for finding the length of an arc and the area of a sector, as shown below. In the formulas, θ denotes radian measure.

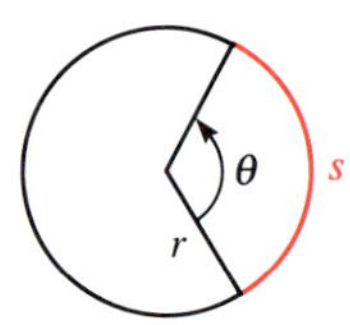

Arc length

$s = r\theta$

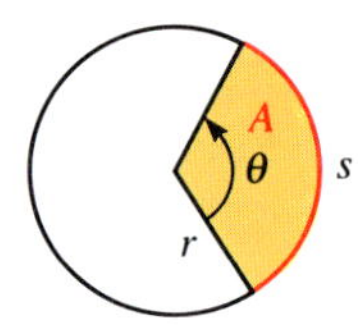

Area of a sector

$A = \frac{1}{2}r^2\theta$ or $A = \frac{1}{2}rs$

The formula $s = r\theta$ is equivalent to $\theta = \frac{s}{r}$, which was used to define radian measure. The area formulas follow from the fact that the areas of sectors are proportional to the measures of their central angles.

$$\frac{\text{area of sector}}{\text{area of circle}} = \frac{\text{measure of central angle of sector}}{\text{measure of central angle of circle}}$$

or $$\frac{A}{\pi r^2} = \frac{\theta}{2\pi}.$$

Solving for A gives $$A = (\pi r^2)\left(\frac{\theta}{2\pi}\right) = \frac{1}{2}r^2\theta.$$

Using $s = r\theta$ gives $$A = \frac{1}{2}r^2\theta = \frac{1}{2}r(r\theta) = \frac{1}{2}rs.$$

Example 3 A central angle of a circle of radius 3 cm measures 1.5 radians. Find **(a)** the length of the intercepted arc and **(b)** the area of the related sector.

Solution

a. $s = r\theta$
$= (3)(1.5) = 4.5$ (cm)

b. $A = \frac{1}{2}r^2\theta$
$= \frac{1}{2}(3^2)(1.5) = 6.75$ (cm^2)

Suppose that a particle moves with a constant speed v around a circle of radius r cm. If the particle travels s centimeters in t seconds, its speed is

$$v = \frac{s}{t} \text{ (cm/s)}.$$

Similarly, if the particle moves through an angle of θ radians in t seconds, its *angular* speed, denoted by ω, is

$$\omega = \frac{\theta}{t} \text{ (radians/s)}.$$

Since arc length s and radian measure θ are related by the formula $s = r\theta$, the speed v and angular speed ω of the particle are related by the formula

$$v = \frac{s}{t} = r\frac{\theta}{t} = r\omega.$$

Therefore, the particle's speed is r times its angular speed.

Example 4 A Ferris wheel 60 ft in diameter makes one revolution in 3 min. Find **(a)** the speed of a seat on the rim and **(b)** the angular speed.

Solution In 1 revolution the seat travels the distance of the circumference of the wheel ($s = 2\pi r$) and through 360°, or 2π radians.

a. $v = \frac{s}{t} = \frac{\pi \cdot 60}{3}$
$= 20\pi \approx 62.8$ (ft/min)

b. $\omega = \frac{\theta}{t} = \frac{2\pi}{3}$
≈ 2.1 (radians/min)

Oral Exercises

Express each degree measure in radians. Leave your answers in terms of π.

1. 180° π **2.** 360° 2π **3.** $-90°$ $-\frac{\pi}{2}$ **4.** $-720°$ -4π

3. A central angle of a circle of radius 3 cm measures 80°. Find **(a)** the length of the intercepted arc, and **(b)** the area of the related sector.
$r = 3$ and $\theta = 80° = 80 \cdot \frac{\pi}{180} = \frac{4\pi}{9} = 1.40$
a. $s = r\,\theta = (3)(1.4) = 4.2$ (cm)
b. $A = \frac{1}{2}r^2\,\theta = \frac{1}{2}(3^2)(1.4) = 6.3$ (cm^2)

Check for Understanding

Here is a suggested use of the Oral Exercises to check students' understanding as you teach the lesson.
Oral Exs. 1–8: use before Example 2.
Oral Exs. 9–13: use after Example 3.

Guided Practice

Express in radians, first in terms of π and then to the nearest hundredth of a radian.

1. $255°$ $\frac{17\pi}{12}$; 4.45
2. $330°$ $\frac{11\pi}{6}$; 5.76
3. $-405°$ $-\frac{9\pi}{4}$; -7.07
4. $189°$ $\frac{21\pi}{20}$; 3.30

Express each radian measure in degrees. If necessary, round your answer to the nearest tenth of a degree.

5. $\frac{2\pi}{9}$ $40°$
6. $\frac{8\pi}{5}$ $288°$
7. 5 $286.5°$
8. -1.4 $-80.2°$
9. An arc of a circle has a central angle of 2 radians. If the radius is 3, find the length of the intercepted arc and the area of the related sector. 6; 9
10. The radius of a record is 15 cm. If the record's angular speed is $\frac{10\pi}{9}$ radians/s, find the speed of a point on the rim. $\frac{50\pi}{3}$ cm/s
11. If the record in Exercise 10 is played for 31 min, how far does a point 6 cm from the center travel? $12{,}400\pi$ cm

Express each radian measure in degrees.

5. $\frac{\pi}{2}$ $90°$ **6.** -2π $-360°$ **7.** $-\pi$ $-180°$ **8.** 3π $540°$

Give the radian measure of ϕ and θ.

9.

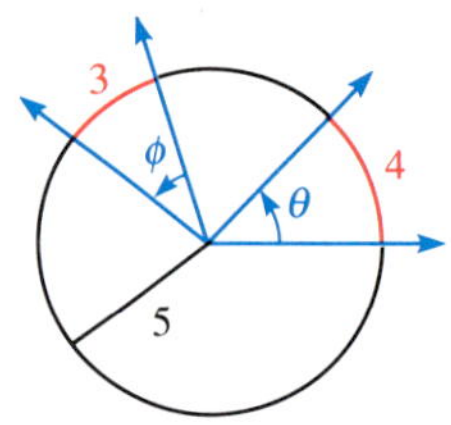

$\phi = \frac{3}{5} = 0.6$
$\theta = \frac{4}{5} = 0.8$

10.

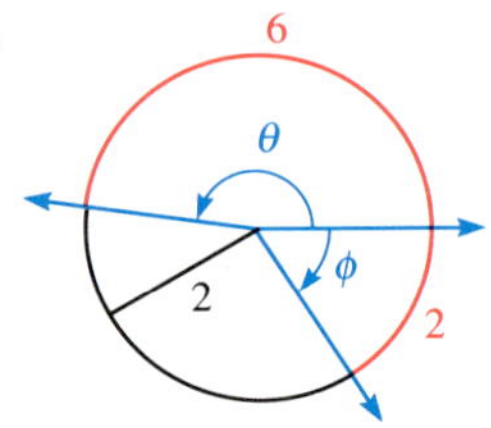

$\phi = 1$
$\theta = 3$

Give the arc length s and the area A of the sector. Angle measures are given in radians.

11.

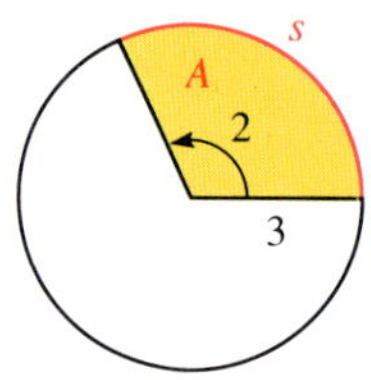

$s = 6$
$A = 9$

12.

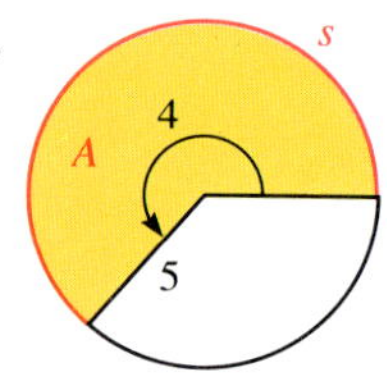

$s = 20$
$A = 50$

13. If a circle has radius r, how long is the arc that is intercepted by an angle with radian measure 1? r

Written Exercises

Express each degree measure in radians. Leave your answer in terms of π.

A

1. $45°$ $\frac{\pi}{4}$ **2.** $30°$ $\frac{\pi}{6}$ **3.** $60°$ $\frac{\pi}{3}$ **4.** $270°$ $\frac{3\pi}{2}$

5. $-120°$ $-\frac{2\pi}{3}$ **6.** $135°$ $\frac{3\pi}{4}$ **7.** $150°$ $\frac{5\pi}{6}$ **8.** $-180°$ $-\pi$

9. $-330°$ $-\frac{11\pi}{6}$ **10.** $240°$ $\frac{4\pi}{3}$ **11.** $-315°$ $-\frac{7\pi}{4}$ **12.** $495°$ $\frac{11\pi}{4}$

Express each radian measure in degrees. In Exercises 25–28, leave your answers in terms of π.

13. $\frac{\pi}{6}$ $30°$ **14.** $-\frac{\pi}{4}$ $-45°$ **15.** $\frac{\pi}{3}$ $60°$ **16.** $-\frac{\pi}{2}$ $-90°$

17. $\frac{4\pi}{3}$ $240°$ **18.** $\frac{5\pi}{6}$ $150°$ **19.** $-\frac{7\pi}{6}$ $-210°$ **20.** $\frac{3\pi}{4}$ $135°$

21. 3π $540°$ **22.** $-\frac{7\pi}{4}$ $-315°$ **23.** $-\frac{7\pi}{2}$ $-630°$ **24.** $\frac{9\pi}{4}$ $405°$

25. 4 $\frac{720°}{\pi}$ **26.** 3 $\frac{540°}{\pi}$ **27.** -2 $-\frac{360°}{\pi}$ **28.** $-\frac{1}{2}$ $-\frac{90°}{\pi}$

Express each degree measure in radians. Give answers to the nearest hundredth of a radian.

29. $10°$ 0.17 **30.** $-50°$ −0.87 **31.** $80°$ 1.40 **32.** $300°$ 5.24

33. $48°$ 0.84 **34.** $265°$ 4.63 **35.** $-174°$ −3.04 **36.** $255°$ 4.45

Express each radian measure in degrees. Give answers to the nearest tenth of a degree.

37. 3 171.9° **38.** −2.5 −143.2° **39.** 0.4 22.9° **40.** −1.6 −91.7°

41. −1.5 −85.9° **42.** 5.5 315.1° **43.** 15 859.4° **44.** 8 458.4°

In Exercises 45–56, r, s, θ, and area A are as shown in the adjacent figure. Angle θ is measured in radians. Find the missing measures.

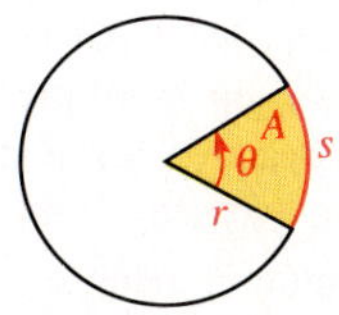

45. $r = 4$, $\theta = 1$, $s = \underline{\ ?\ }$, $A = \underline{\ ?\ }$ 4; 8

46. $r = 5$, $\theta = 2.5$, $s = \underline{\ ?\ }$, $A = \underline{\ ?\ }$ 12.5; 31.25

47. $r = 4$, $s = 12$, $\theta = \underline{\ ?\ }$, $A = \underline{\ ?\ }$ 3; 24

48. $r = 5$, $s = 30$, $\theta = \underline{\ ?\ }$, $A = \underline{\ ?\ }$ 6; 75

50. 3; 6 52. 2.4; 1.44

B **49.** $r = 5$, $A = 15$, $\theta = \underline{\ ?\ }$, $s = \underline{\ ?\ }$ 1.2; 6

50. $r = 2$, $A = 6$, $\theta = \underline{\ ?\ }$, $s = \underline{\ ?\ }$

51. $s = 10$, $\theta = 2.5$, $r = \underline{\ ?\ }$, $A = \underline{\ ?\ }$ 4; 20

52. $s = 1.2$, $\theta = 0.5$, $r = \underline{\ ?\ }$, $A = \underline{\ ?\ }$

53. $r = 2$, $A = 3$, $s = \underline{\ ?\ }$, $\theta = \underline{\ ?\ }$ 3; $\frac{3}{2}$

54. $r = 8$, $A = 6$, $s = \underline{\ ?\ }$, $\theta = \underline{\ ?\ }$

55. $A = 8$, $\theta = 4$, $r = \underline{\ ?\ }$, $s = \underline{\ ?\ }$ 2; 8

56. $A = 6$, $\theta = 3$, $r = \underline{\ ?\ }$, $s = \underline{\ ?\ }$

54. 1.5; 0.1875 56. 2; 6

The angle α has the given radian measure. Find, to the nearest hundredth of a radian, the measure of the angle between 0 and 2π that is coterminal with α.

57. $\alpha = 8.12$ 1.84 **58.** $\alpha = 6.55$ 0.27 **59.** $\alpha = -3$ 3.28 **60.** $\alpha = -20$ 5.13

Problems

Solve. Give answers to two significant digits.

The *latitude* of point P shown in the drawing of Earth at the right is the measure of $\angle EOP$. The radius of Earth is about 4000 miles.

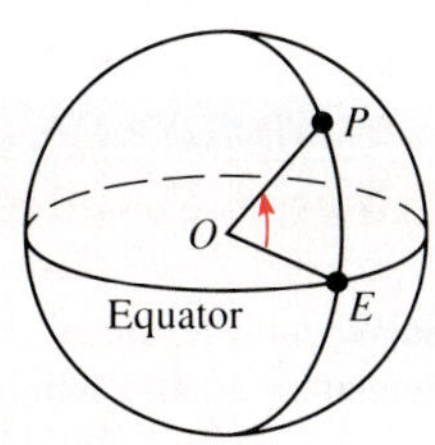

A **1.** The latitude of Houston is 29.8°N. How far is it from the equator? 2100 mi

2. The latitude of New York is 40.7°N. How far is it from the North Pole? 3400 mi

3. A phonograph record with diameter 12 in. rotates at 45 rev./min. Find the distance that a point on the rim travels in one minute. 1700 in.

4. An automobile tire with diameter 2 ft turns 10 times per second. How far does a point on the tread travel in one minute? 3800 ft

Summarizing the Lesson

In this lesson students learned how to measure angles in radians and how to find lengths of arcs and areas of sectors for given central angles of circles. Ask students to explain why radians are considered a more "natural" system of angle measurement than degrees.

Suggested Assignments

Average Alg. and Trig.
610/3–60 mult. of 3
611/*P*: 1–7 odd
S 612/Mixed Review

Extended Alg. and Trig.
610/3–60 mult. of 3
611/*P*: 1–11 odd
S 612/Mixed Review

Supplementary Materials

Study Guide pp. 205–206

5. A Ferris wheel with diameter 50 ft makes one revolution every two minutes. What is the speed of a seat on the rim of the wheel? 79 ft/min

6. A merry-go-round 40 ft in diameter makes four revolutions every minute. What is the speed of a seat on the rim? 500 ft/min

7. From an observation point on Earth, the angle subtended by the sun 150,000,000 km away is about 0.0093 radians. Approximate the diameter of the sun by finding the length of the red arc. 1,400,000 km

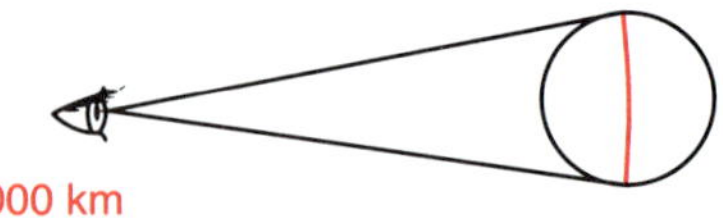

8. The diameter of the moon is approximately 2200 mi. How far is the moon from the Earth if the angle it subtends from an observation point on Earth measures about 0.518°? (Use the diagram in Problem 7.) 240,000 mi

B 9. The angular speed of the larger pulley wheel shown at the right is 50 rev./min. Find the angular speed and rim speed of the smaller wheel. 250π radians/min; 4700 cm/min

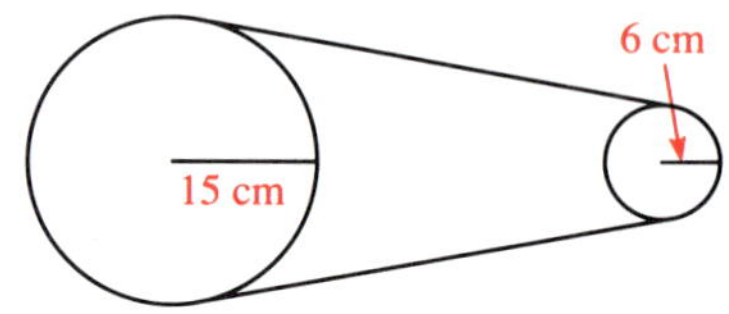

10. A clock pendulum swings back and forth once every two seconds. If its length is one meter and the greatest angle that it makes with the vertical is 12°, how many kilometers does the bottom end of the pendulum travel in one day? 36 km

11. What is the angular speed in radians per second of the hour hand of a clock? If the hand is 5 cm long, how fast is the tip moving in millimeters per second? 0.00015 radians/s; 0.0073 mm/s

12. Each point on Earth's surface (except the poles) moves in a circle as Earth rotates on its axis. What is the angular speed in radians per second of the point where you are located? 7.3×10^{-5} radians/s

C 13. A cylindrical water tank 4 ft in diameter is lying on its side as shown. What percent of its capacity is used when it contains water to a depth of one foot? Approximately 20%

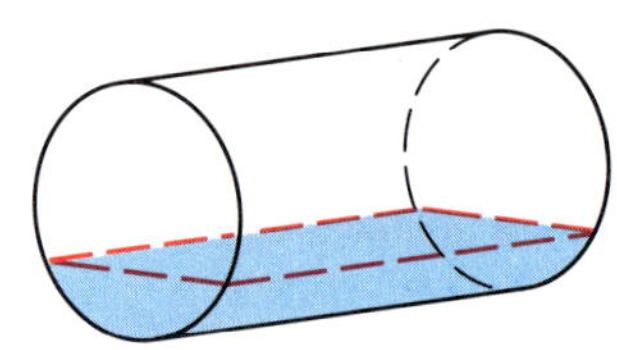

Mixed Review Exercises

Solve each triangle. Give lengths to three significant digits and angle measures to the nearest tenth of a degree.

1. $a = 8$, $b = 5$, $\angle C = 42°$ — $c = 5.44$, $\angle A = 100°$, $\angle B = 38°$
2. $\angle A = 75°$, $\angle B = 30°$, $b = 12$ — $a = 23.2$, $c = 23.2$, $\angle C = 75°$
3. $b = 10$, $c = 7$, $\angle B = 104°$ — $a = 5.64$, $\angle A = 33.2°$, $\angle C = 42.8°$
4. $a = 5$, $b = 6$, $c = 9$ — $\angle A = 31.6°$, $\angle B = 39.0°$, $\angle C = 109.4°$

5–8. Find the area of each triangle in Exercises 1–4. Give answers to three significant digits. **5.** 13.4 **6.** 134 **7.** 19.2 **8.** 14.1

13-2 Circular Functions

Objective To define the circular functions.

There are many nongeometric uses of the trigonometric functions. For example, the first equation shown below describes the shape of an ocean wave (height y and horizontal distance x in meters), and the second equation gives the line voltage V of ordinary alternating current (time t in seconds). In neither case do angles play any part.

$$y = 2.5 \sin (0.03x) \qquad V = 170 \cos (120\pi t)$$

These examples suggest that sin s and cos s can be defined with s representing a *number* rather than an angle.

To develop these definitions, let O be the unit circle $x^2 + y^2 = 1$, and let A be the point (1, 0), as shown at the right. Given any real number s, start at A and measure $|s|$ units around O in a counterclockwise direction if $s \geq 0$, and in a clockwise direction if $s < 0$, arriving at a point $P(x, y)$. The sine of s and the cosine of s are then defined by the coordinates of point P. The tangent, cotangent, secant, and cosecant functions are defined in terms of sine and cosine.

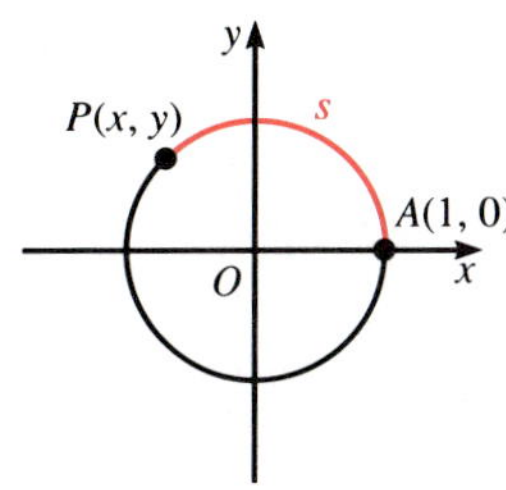

$\sin s = y$	$\cos s = x$
$\tan s = \dfrac{\sin s}{\cos s}$ if $\cos s \neq 0$	$\cot s = \dfrac{\cos s}{\sin s}$ if $\sin s \neq 0$
$\sec s = \dfrac{1}{\cos s}$ if $\cos s \neq 0$	$\csc s = \dfrac{1}{\sin s}$ if $\sin s \neq 0$

These functions are sometimes called *circular functions* because a circle is used in their definition.

Example 1 For some number s, the point $P\left(\frac{2}{3}, -\frac{\sqrt{5}}{3}\right)$ is s units from $A(1, 0)$ along a unit circle. Find the exact values of the six circular functions of s.

Solution

$$\sin s = -\frac{\sqrt{5}}{3} \qquad \cos s = \frac{2}{3}$$

$$\tan s = \frac{-\frac{\sqrt{5}}{3}}{\frac{2}{3}} = -\frac{\sqrt{5}}{2} \qquad \cot s = \frac{\frac{2}{3}}{-\frac{\sqrt{5}}{3}} = -\frac{2}{\sqrt{5}} = -\frac{2\sqrt{5}}{5}$$

$$\sec s = \frac{1}{\frac{2}{3}} = \frac{3}{2} \qquad \csc s = \frac{1}{-\frac{\sqrt{5}}{3}} = -\frac{3}{\sqrt{5}} = -\frac{3\sqrt{5}}{5}$$

Teaching Suggestions p. T127

Suggested Extensions p. T128

Using Manipulatives p. T128

Warm-Up Exercises

Each ordered pair satisfies the equation $x^2 + y^2 = 1$. Find the value of a.

1. $(-1, a)$ 0 **2.** $(a, 0)$ ± 1

3. $(0.6, a)$ ± 0.8

4. (a, a) $\pm\frac{\sqrt{2}}{2}$

Motivating the Lesson

Ask students if they think that sines and cosines are related only to angles. Then tell them that the vibration of a guitar string or the vertical movement of a person on a Ferris wheel can be described by sine functions, and these functions have nothing to do with angles. The circular functions, similar to the trigonometric functions but not determined by angles, are the topic of today's lesson.

Thinking Skills

Stress the relationship between circular and trigonometric functions. (Refer to the discussion at the top of page 614.) Because the two types of functions are completely analogous, what is true for the trigonometric functions is also true for the circular functions. (See the discussion following Example 5.) Students should be able to *transfer* their knowledge of the properties of trigonometric functions directly to the circular functions.

Chalkboard Examples

1. For some number s, the point $P\left(-\frac{3}{5}, \frac{4}{5}\right)$ is s units from $A(1, 0)$ along a unit circle. Find the exact values of the six circular functions of s.
$\sin s = \frac{4}{5}$, $\cos s = -\frac{3}{5}$,
$\tan s = -\frac{4}{3}$, $\cot s = -\frac{3}{4}$,
$\sec s = -\frac{5}{3}$, $\csc s = \frac{5}{4}$

2. Find sin 1.16 to four decimal places. Use a calculator or Table 6. $\sin 1.16 = 0.9168$

3. Find $\sin \frac{\pi}{4}$, $\cos \frac{\pi}{4}$, and $\tan \frac{\pi}{4}$. Since $\frac{\pi}{4}$ radians $= 45°$, we have $\sin \frac{\pi}{4} = \frac{\sqrt{2}}{2}$, $\cos \frac{\pi}{4} = \frac{\sqrt{2}}{2}$, $\tan \frac{\pi}{4} = 1$.

4. Find x if $\cos x = 0.5234$ and $0 \le x \le \frac{\pi}{2}$.
Using a calculator:
$x = \cos^{-1} 0.5234 = 1.02$
Using tables: Use Table 6.
Read across from 0.5234 in the cos θ column.
$\cos 1.02 = 0.5234$
Therefore $x = 1.02$.

Check for Understanding

Here is a suggested use of the Oral Exercises to check students' understanding as you teach the lesson.
Oral Exs. 1–3: use before Example 1.
Oral Exs. 4–7: use after Example 1.
Oral Exs. 8–18: use before Example 2.

There is a close relationship between the circular functions and the corresponding trigonometric functions. The diagram at the right shows a central angle θ, in standard position, that intercepts $\overset{\frown}{AP}$. Notice that $\sin \theta = \frac{y}{1} = y = \sin s$. Also, the radian measure of θ is $\frac{s}{1}$, or s. Therefore, $\sin s$ equals the sine of any angle having radian measure s. Similarly, the other circular functions of the number s are equal to the corresponding trigonometric functions of any angle having radian measure s.

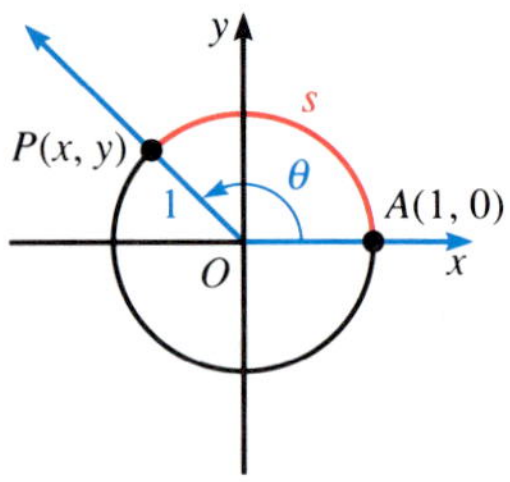

You can use a calculator or a table to find values of circular functions. Scientific calculators can be set to work with radians, and Table 6 gives values of trigonometric functions for angles with radian measure between 0 and $\frac{\pi}{2}$.

Example 2 Find sin 1.44 to four decimal places.

Solution 1 Using a Calculator

Set the calculator in radian mode.

$\sin 1.44 = 0.9915$ to four decimal places. ***Answer***

Solution 2 Using Tables

Use Table 6 to find the sine of an angle having radian measure 1.44.

$\sin 1.44 = 0.9915$ ***Answer***

When multiples of $\frac{\pi}{6}$ and $\frac{\pi}{4}$ are involved, use exact values of the trigonometric functions unless you are asked for a decimal approximation.

Example 3 Find the exact values of the six trigonometric functions of $\frac{\pi}{6}$.

Solution Since $\frac{\pi}{6}$ radians $= 30°$, use values from a 30°-60°-90° triangle (page 558).

$$\sin \frac{\pi}{6} = \frac{1}{2} \qquad \cos \frac{\pi}{6} = \frac{\sqrt{3}}{2} \qquad \tan \frac{\pi}{6} = \frac{\sqrt{3}}{3}$$

$$\csc \frac{\pi}{6} = 2 \qquad \sec \frac{\pi}{6} = \frac{2\sqrt{3}}{3} \qquad \cot \frac{\pi}{6} = \sqrt{3}$$

If you are given the value of a trigonometric function of x, you can use a calculator or Table 6 to find the value of x to two decimal places when x is between 0 and $\frac{\pi}{2}$.

Example 4 Find x to two decimal places if $\sec x = 1.975$ and $0 \le x \le \frac{\pi}{2}$.

Solution 1 Using a Calculator

Since $\sec x = 1.975$, $\cos x = \frac{1}{1.975} = 0.5063291$. Set the calculator in radian mode and use the inverse function key $\cos^{-1}$ (or inv cos):

$$x = \cos^{-1} 0.5063291 = 1.0398738.$$

$\therefore x = 1.04$ to two decimal places. ***Answer***

Solution 2 Using Tables

Use Table 6 and read across from 1.975 in the sec θ column. You should find that $\sec 1.04 = 1.975$.

$\therefore x = 1.04$ ***Answer***

For some given function values, you may be able to express x exactly in terms of π.

Example 5 Find the exact value of x if $\cos x = 0.5$ and $0 \le x \le \frac{\pi}{2}$.

Solution Recall from the geometry of a 30°-60°-90° triangle that $\cos 60° = 0.5$. An angle with degree measure 60° has radian measure $\frac{\pi}{3}$, and $0 \le \frac{\pi}{3} \le \frac{\pi}{2}$.

$\therefore x = \frac{\pi}{3}$ ***Answer***

The equation of the unit circle centered at the origin is $x^2 + y^2 = 1$. By definition, $x = \cos s$ and $y = \sin s$. Therefore, by substitution,

$$\cos^2 s + \sin^2 s = 1.$$

This is the same identity that was proved on page 556 for functions of angles.

For the rest of this book we will follow the usual practice of referring to the circular functions as trigonometric functions. It will be clear from the context whether we are talking about functions of angles or functions of numbers.

Oral Exercises

3. For any point $P(x, y)$ on the unit circle, $|y| \le 1$. Since $\sin s = y$, $-1 \le \sin s \le 1$.

Sketch a circle with equation $x^2 + y^2 = 1$.

1. How many points on the circle have the property that $\cos s = 0$? 2
2. How many points on the circle have the property that $\cos s = 1$? 1
3. Explain why the value of $\sin s$ is always between -1 and 1, inclusive.

Guided Practice

1. For some number s, the point $P\left(\frac{1}{2}, \frac{\sqrt{3}}{2}\right)$ is s units from $A(1, 0)$ along a unit circle. Find the exact values of the six circular functions of s.
$\sin s = \frac{\sqrt{3}}{2}$, $\cos s = \frac{1}{2}$,
$\tan s = \sqrt{3}$, $\cot s = \frac{\sqrt{3}}{3}$,
$\sec s = 2$, $\csc s = \frac{2\sqrt{3}}{3}$

Find the sine, cosine, and tangent of each number to four decimal places.

2. 2.94 0.2002, −0.9797, −0.2044
3. −0.14 −0.1395, 0.9902, −0.1409

Find the exact values of the six trigonometric functions of the given numbers. If a function value is undefined, say so.

4. $-\frac{\pi}{3}$ 5. $\frac{5\pi}{4}$ 6. $\frac{5\pi}{2}$

	4.	5.	6.
sin	$-\frac{\sqrt{3}}{2}$	$-\frac{\sqrt{2}}{2}$	1
cos	$\frac{1}{2}$	$-\frac{\sqrt{2}}{2}$	0
tan	$-\sqrt{3}$	1	*
csc	$-\frac{2\sqrt{3}}{3}$	$-\sqrt{2}$	1
sec	2	$-\sqrt{2}$	*
cot	$-\frac{\sqrt{3}}{3}$	1	0

*Undefined

Summarizing the Lesson

In this lesson students have learned the definitions of the circular functions. Ask students to explain in what way the circular functions are different from the trigonometric functions.

Additional Answers
Oral Exercises

4. $\sin s = \frac{4}{5}$, $\cos s = \frac{3}{5}$, $\tan s = \frac{4}{3}$, $\cot s = \frac{3}{4}$, $\sec s = \frac{5}{3}$, $\csc s = \frac{5}{4}$

5. $\sin s = \frac{3}{5}$, $\cos s = -\frac{4}{5}$, $\tan s = -\frac{3}{4}$, $\cot s = -\frac{4}{3}$, $\sec s = -\frac{5}{4}$, $\csc s = \frac{5}{3}$

6. $\sin s = -\frac{12}{13}$, $\cos s = -\frac{5}{13}$, $\tan s = \frac{12}{5}$, $\cot s = \frac{5}{12}$, $\sec s = -\frac{13}{5}$, $\csc s = -\frac{13}{12}$

7. $\sin s = -\frac{5}{13}$, $\cos s = \frac{12}{13}$, $\tan s = -\frac{5}{12}$, $\cot s = -\frac{12}{5}$, $\sec s = \frac{13}{12}$, $\csc s = -\frac{13}{5}$

Suggested Assignments

Average Alg. and Trig.
617/1–27 odd
S 611/*P*: 2, 6

Extended Alg. and Trig.
617/1–27 odd, 28–30
S 612/*P*: 10, 12

Supplementary Materials

Study Guide	pp. 207–208
Practice Master	73
Overhead Visual	10

Additional Answers
Written Exercises

1. $\sin s = -\frac{4}{5}$, $\cos s = \frac{3}{5}$, $\tan s = -\frac{4}{3}$, $\cot s = -\frac{3}{4}$, $\sec s = \frac{5}{3}$, $\csc s = -\frac{5}{4}$

Give the exact values of the six trigonometric functions of the number s. Each circle is a unit circle centered at the origin.

4.

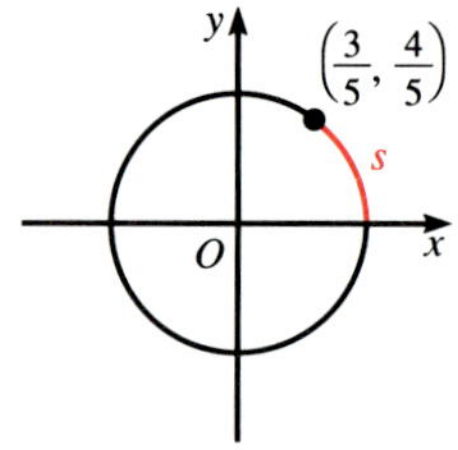

5.

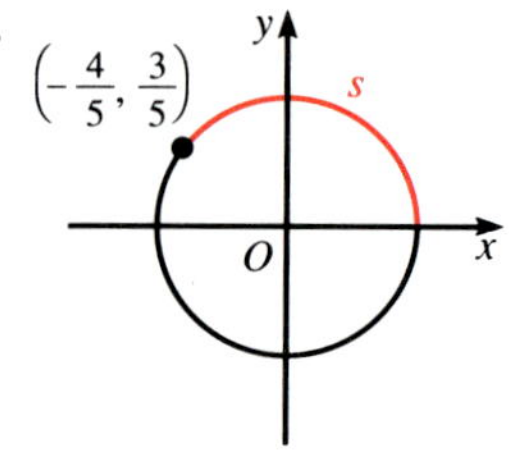

6.

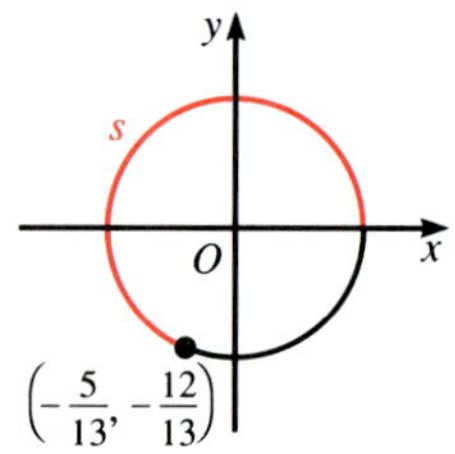

7.

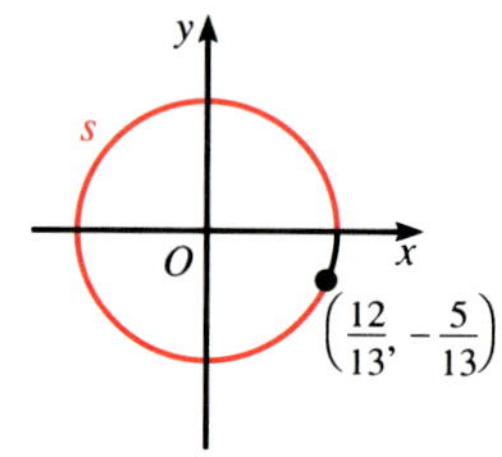

Complete the table by using *positive* or *negative* for each entry.

	8. $\sin s$	**9.** $\cos s$	**10.** $\tan s$	**11.** $\cot s$	**12.** $\sec s$	**13.** $\csc s$
$0 < s < \frac{\pi}{2}$	? +	? +	? +	? +	? +	? +
$\frac{\pi}{2} < s < \pi$	? +	? −	? −	? −	? −	? +
$\pi < s < \frac{3\pi}{2}$	? −	? −	? +	? +	? −	? −
$\frac{3\pi}{2} < s < 2\pi$	? −	? +	? −	? −	? +	? −

14. Complete the following statements showing that $\cos(s + \pi) = -\cos s$ and $\sin(s + \pi) = -\sin s$:
The length of $\widehat{APQ}$ of the unit circle shown at the right is __?__. Therefore, in terms of s, the coordinates of Q are (__?__, __?__). Therefore, __?__ $= -x = -\cos s$, and __?__ $= -y = -\sin s$.
$s + \pi$; $\cos(s + \pi)$; $\sin(s + \pi)$; $\cos(s + \pi)$; $\sin(s + \pi)$

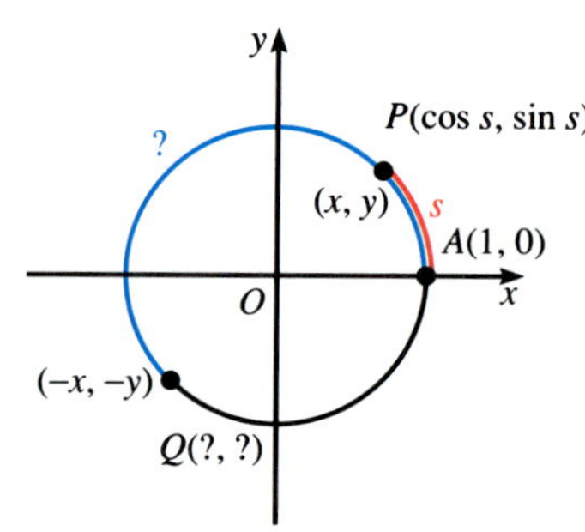

Tell the value of each of the following functions. If the function is undefined for the given number, say so.

15. $\sin \frac{\pi}{2}$ 1

16. $\tan \frac{3\pi}{2}$ undefined

17. $\cos(-\pi)$ −1

18. $\cot\left(-\frac{\pi}{2}\right)$ 0

Written Exercises

For some number s, the point P is s units from $A(1, 0)$ along the unit circle $x^2 + y^2 = 1$. Find the exact values of the six trigonometric functions of s.

A **1.** $P\left(\frac{3}{5}, -\frac{4}{5}\right)$ **2.** $P\left(-\frac{5}{13}, \frac{12}{13}\right)$ **3.** $P\left(\frac{\sqrt{7}}{4}, \frac{3}{4}\right)$ **4.** $P\left(-\frac{2}{5}, -\frac{\sqrt{21}}{5}\right)$

Find the sine, cosine, and tangent of each number to four decimal places.

5. 1.25 **6.** 1.4 **7.** 0.25 **8.** 0.7

Find the exact values of the six trigonometric functions of the given number. If a function is undefined for the number, say so.

9. $\frac{\pi}{3}$ **10.** $\frac{\pi}{4}$ **11.** $\frac{3\pi}{2}$ **12.** $\frac{3\pi}{4}$ **13.** $\frac{11\pi}{6}$

14. π **15.** $\frac{8\pi}{3}$ **16.** $\frac{15\pi}{4}$ **17.** $-\frac{5\pi}{6}$ **18.** $-\frac{3\pi}{4}$

Find the number x if $0 \le x \le \frac{\pi}{2}$ and x has the given function value.

B **19.** $\cos x = 0.7248$ 0.76 **20.** $\csc x = 1.018$ 1.38 **21.** $\sin x = \frac{\sqrt{3}}{2}$ 1.05 or $\frac{\pi}{3}$ **22.** $\tan x = 1$ 0.79 or $\frac{\pi}{4}$

Verify the identity $\cos^2 s + \sin^2 s = 1$ for the given value of s.

23. $\frac{3\pi}{4}$ **24.** $\frac{7\pi}{6}$ **25.** 0.7 **26.** 1

27. Explain how to use Table 6 to find an approximate value for sin 7.

In Exercises 28–31, use the diagram at the right.

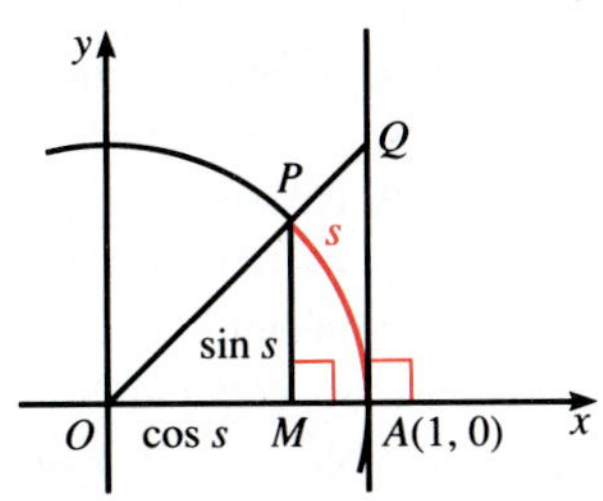

28. Show that if $0 < s < \frac{\pi}{2}$, then $\tan s = AQ$.
(*Hint:* $\triangle OAQ \sim \triangle OMP$.)

29. Show that if $0 < s < \frac{\pi}{2}$, then $\sec s = OQ$.
(See the hint for Exercise 28.)

C **30.** Show that if $0 < s < \frac{\pi}{2}$, then $\sin s < s$.

31. Show that if $0 < s < \frac{\pi}{2}$, then $s < \tan s$.
(*Hint:* Compare the areas of sector OAP and $\triangle OAQ$.)

32. Show that if $0 < s < \frac{\pi}{2}$, then $1 < \frac{s}{\sin s} < \frac{1}{\cos s}$.
(*Hint:* Start by combining the inequalities in Exercises 30 and 31.)

2. $\sin s = \frac{12}{13}$, $\cos s = -\frac{5}{13}$, $\tan s = -\frac{12}{5}$, $\cot s = -\frac{5}{12}$, $\sec s = -\frac{13}{5}$, $\csc s = \frac{13}{12}$

3. $\sin s = \frac{3}{4}$, $\cos s = \frac{\sqrt{7}}{4}$, $\tan s = \frac{3\sqrt{7}}{7}$, $\cot s = \frac{\sqrt{7}}{3}$, $\sec s = \frac{4\sqrt{7}}{7}$, $\csc s = \frac{4}{3}$

4. $\sin s = -\frac{\sqrt{21}}{5}$, $\cos s = -\frac{2}{5}$, $\tan s = \frac{\sqrt{21}}{2}$, $\cot s = \frac{2\sqrt{21}}{21}$, $\sec s = -\frac{5}{2}$, $\csc s = -\frac{5\sqrt{21}}{21}$

5. sin 1.25 = 0.9490, cos 1.25 = 0.3153, tan 1.25 = 3.0096

6. sin 1.4 = 0.9854, cos 1.4 = 0.1700, tan 1.4 = 5.7979

7. sin 0.25 = 0.2474, cos 0.25 = 0.9689, tan 0.25 = 0.2553

8. sin 0.7 = 0.6442, cos 0.7 = 0.7648, tan 0.7 = 0.8423

9. $\sin\frac{\pi}{3} = \frac{\sqrt{3}}{2}$, $\cos\frac{\pi}{3} = \frac{1}{2}$, $\tan\frac{\pi}{3} = \sqrt{3}$, $\cot\frac{\pi}{3} = \frac{\sqrt{3}}{3}$, $\sec\frac{\pi}{3} = 2$, $\csc\frac{\pi}{3} = \frac{2\sqrt{3}}{3}$

10. $\sin\frac{\pi}{4} = \frac{\sqrt{2}}{2}$, $\cos\frac{\pi}{4} = \frac{\sqrt{2}}{2}$, $\tan\frac{\pi}{4} = 1$, $\cot\frac{\pi}{4} = 1$, $\sec\frac{\pi}{4} = \sqrt{2}$, $\csc\frac{\pi}{4} = \sqrt{2}$

11. $\sin\frac{3\pi}{2} = -1$, $\cos\frac{3\pi}{2} = 0$, $\tan\frac{3\pi}{2}$ is undefined, $\cot\frac{3\pi}{2} = 0$, $\sec\frac{3\pi}{2}$ is undefined, $\csc\frac{3\pi}{2} = -1$

Additional Answers Written Exercises (continued)

12. $\sin\frac{3\pi}{4} = \frac{\sqrt{2}}{2}$,

$\cos\frac{3\pi}{4} = -\frac{\sqrt{2}}{2}$,

$\tan\frac{3\pi}{4} = -1$, $\cot\frac{3\pi}{4} = -1$,

$\sec\frac{3\pi}{4} = -\sqrt{2}$, $\csc\frac{3\pi}{4} = \sqrt{2}$

13. $\sin\frac{11\pi}{6} = -\frac{1}{2}$,

$\cos\frac{11\pi}{6} = \frac{\sqrt{3}}{2}$,

$\tan\frac{11\pi}{6} = -\frac{\sqrt{3}}{3}$,

$\cot\frac{11\pi}{6} = -\sqrt{3}$,

$\sec\frac{11\pi}{6} = \frac{2\sqrt{3}}{3}$,

$\csc\frac{11\pi}{6} = -2$

14. $\sin \pi = 0$, $\cos \pi = -1$, $\tan \pi = 0$, $\cot \pi$ is undefined, $\sec \pi = -1$, $\csc \pi$ is undefined

15. $\sin\frac{8\pi}{3} = \frac{\sqrt{3}}{2}$,

$\cos\frac{8\pi}{3} = -\frac{1}{2}$,

$\tan\frac{8\pi}{3} = -\sqrt{3}$,

$\cot\frac{8\pi}{3} = -\frac{\sqrt{3}}{3}$,

$\sec\frac{8\pi}{3} = -2$,

$\csc\frac{8\pi}{3} = \frac{2\sqrt{3}}{3}$

16. $\sin\frac{15\pi}{4} = -\frac{\sqrt{2}}{2}$,

$\cos\frac{15\pi}{4} = \frac{\sqrt{2}}{2}$,

$\tan\frac{15\pi}{4} = -1$,

$\cot\frac{15\pi}{4} = -1$,

$\sec\frac{15\pi}{4} = \sqrt{2}$,

$\csc\frac{15\pi}{4} = -\sqrt{2}$

(continued on p. 655)

Biographical Note / *Maria Goeppert Mayer*

Maria Goeppert Mayer (1906–1972) was the first woman to receive the Nobel Prize for theoretical physics. She began her career at the University of Göttingen as a mathematics student, but she soon started working on quantum mechanics.

In the same year that she married the American chemist Joseph Mayer, she finished her Ph.D. dissertation on the theory of light quanta.

After the Mayers moved to Baltimore, Maria was unable to get a university appointment. Instead, she decided to help other scientists without pay.

In early 1946 the Mayers accepted appointments at the University of Chicago. There Maria was surrounded by some of the greatest names in physics, and she set out to learn nuclear theory and study the origin of elements.

Along with the physicist Hans Jensen she wrote the book *Elementary Theory of Nuclear Shell Structure*.

Historical Note / *A Trigonometric Identity*

One of the most famous questions in the history of mathematics, asked by the Greeks about 300 B.C., was the following: Given an arbitrary angle, can this angle be *trisected* (that is, subdivided into three equal angles) with only a straightedge and a compass? This question remained unanswered until the 19th century, when the following apparently unrelated theorem was proved:

> From a given unit length, a segment whose length is a root of a cubic equation with no rational root cannot be constructed.

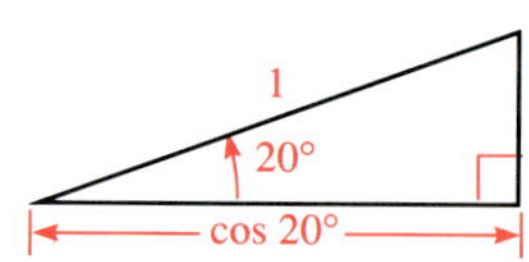

By a trigonometric identity: $\cos 3\theta = 4\cos^3\theta - 3\cos\theta$.

If $\theta = 20°$, then $3\theta = 60°$ and $\cos 3\theta = \frac{1}{2}$. Therefore,

$$\frac{1}{2} = 4\cos^3 20° - 3\cos 20°.$$

If you let $x = \cos 20°$, this equation becomes

$$0 = 8x^3 - 6x - 1.$$

But by applying the rational root theorem (page 382), you find that this equation has no rational root. Therefore, by the theorem above, a segment whose length is a root of this equation, such as cos 20°, cannot be constructed. If you could trisect a 60° angle, you could construct a 20° angle, and by dropping a perpendicular to one side of this angle from a point on the other side a unit distance from the vertex, you could construct a segment of length cos 20°. It follows that a 60° angle cannot be trisected.

13-3 Periodicity and Symmetry

Objective To use periodicity and symmetry in graphing functions.

Because the circumference of the unit circle is 2π, the numbers s and $s + 2\pi$ are associated with the same point $P(x, y)$ on the unit circle, as shown in the diagram at the right. Therefore,

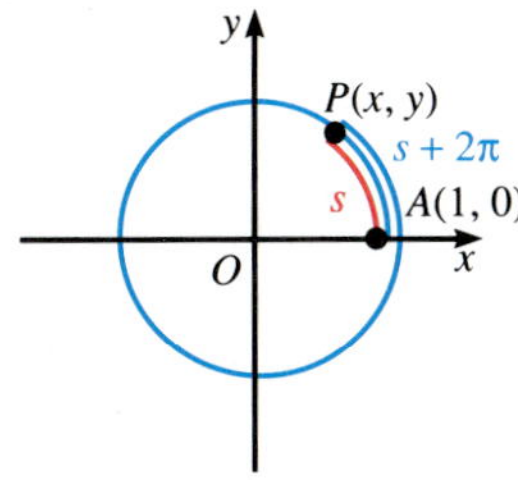

$$\sin(s + 2\pi) = \sin s \quad \text{and} \quad \cos(s + 2\pi) = \cos s.$$

These identities tell you that the sine and cosine functions have *period* 2π.

In general, a function is **periodic** if for some positive constant p,

$$f(x + p) = f(x)$$

for every x in the domain of f. The smallest such p is the **period** of f.

Example 1 Part of the graph of a function f having period 2 is shown. Graph f in the interval $-5 \le x \le 5$.

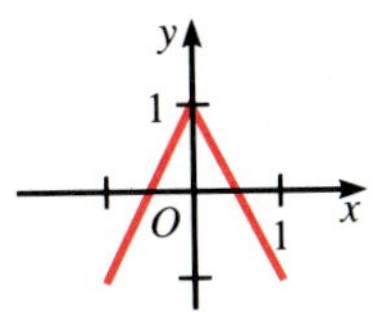

Solution Since the given part of the graph covers one period, repeat it to the right until you reach 5 and to the left until you reach -5.

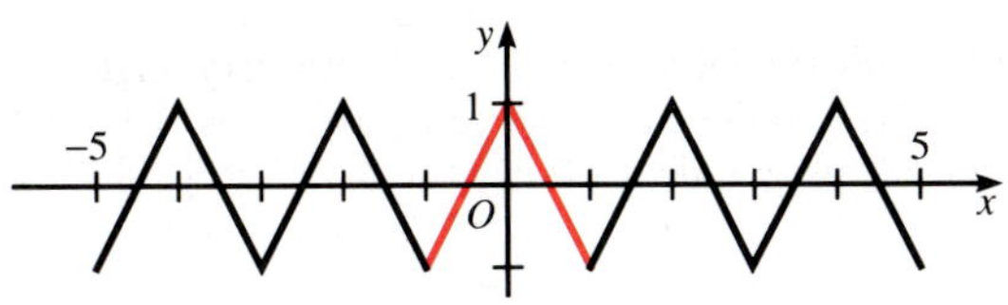

From the diagram at the right you can see that

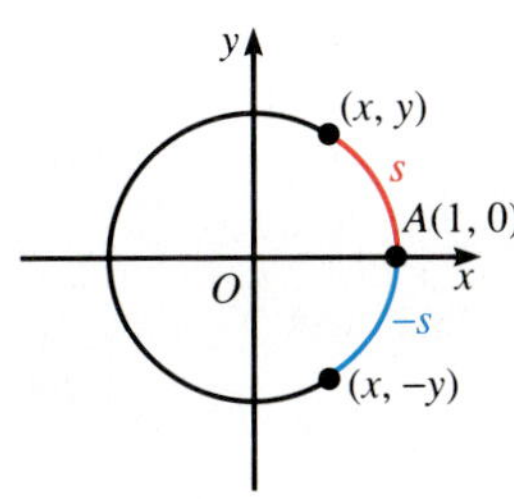

$$\sin(-s) = -y = -\sin s$$

and $$\cos(-s) = x = \cos s.$$

These identities tell you that the sine is an *odd* function and that the cosine is an *even* function.

In general,

f is an **odd function** if $f(-x) = -f(x)$

and f is an **even function** if $f(-x) = f(x)$

for every x in the domain of f. Unlike an integer, a function does not have to be even or odd; it can be neither.

Teaching Suggestions p. T128

Suggested Extensions p. T128

Warm-Up Exercises

Let $f(x) = x^3 - 2x$ and $g(x) = |x| + 3$. Find each function value.

1. $f(-2)$ -4 **2.** $f(2)$ 4

3. $-f(2)$ -4 **4.** $g(-3)$ 6

5. $g(3)$ 6 **6.** $-g(3)$ -6

Motivating the Lesson

Tell students that if they kept a record of their daily activities for a period of time, they would see many repetitive patterns. Today's lesson involves the repetitive, or periodic, nature of the circular functions.

Chalkboard Examples

1. Part of the graph of a function f having period 3 is shown. Graph f in the interval $-3 \le x \le 6$.

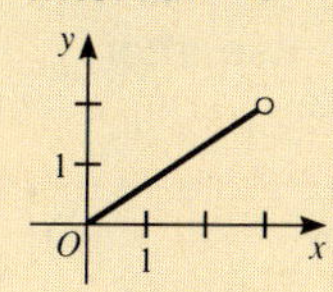

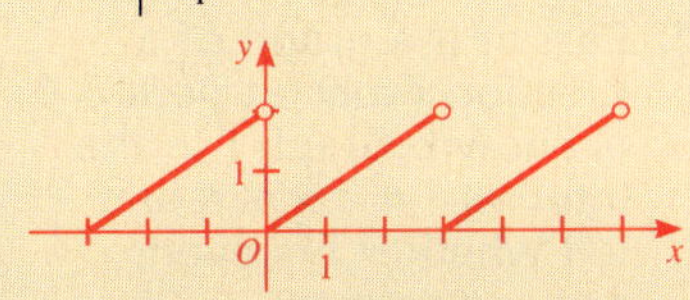

2. Determine whether the given function is odd, even, or neither.
   a. $f(x) = x^4 - 1$
   $f(-x) = (-x)^4 - 1 = f(x)$, even
   b. $g(x) = \dfrac{x}{x^2 - 1}$
   $g(-x) = \dfrac{-x}{(-x)^2 - 1} = \dfrac{-x}{x^2 - 1} = -g(x)$, odd

(continued)

Chalkboard Examples

(continued)

3. Part of the graph of a function f is shown below. Graph f in the interval $-2 \le x \le 2$ given that f is **(a)** even, **(b)** odd.

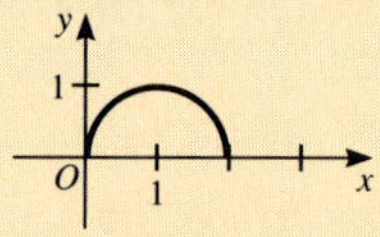

   a. Reflect the given part of the graph in the y-axis.

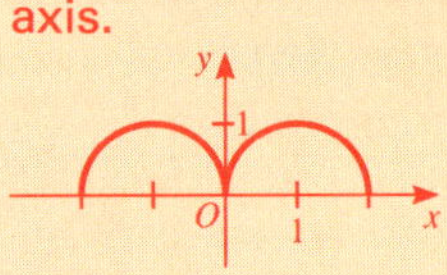

   b. Reflect the given part of the graph in the origin.

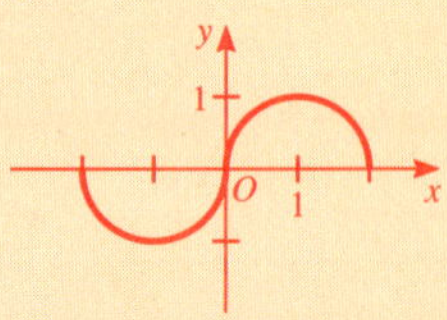

Check for Understanding

Oral Exs. 1–9: use after Example 3.

Guided Practice

1. Part of the graph of a function f having period 4 is shown. Graph f in the interval $-4 \le x \le 4$ and tell whether f is even, odd, or neither.

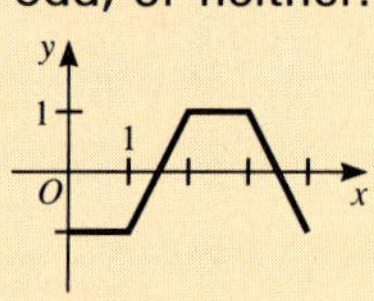

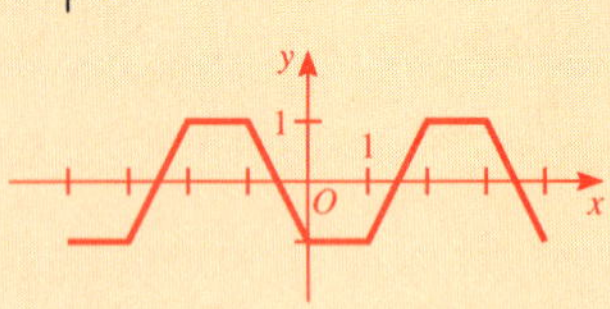

neither

Example 2 Determine whether each function is even, odd, or neither.

a. $f(x) = \dfrac{x^2}{x^4 + 1}$ **b.** $g(x) = x^3 - 4x$ **c.** $h(x) = x^2 + x$

Solution

a. $f(-x) = \dfrac{(-x)^2}{(-x)^4 + 1} = \dfrac{x^2}{x^4 + 1} = f(x)$

$\therefore f$ is even. ***Answer***

b. $g(-x) = (-x)^3 - 4(-x) = -x^3 + 4x = -(x^3 - 4x) = -g(x)$

$\therefore g$ is odd. ***Answer***

c. $h(-x) = (-x)^2 + (-x) = x^2 - x$

Notice that $x^2 - x \ne x^2 + x$ and $x^2 - x \ne -(x^2 + x)$.

$\therefore h$ is neither even nor odd. ***Answer***

The diagrams below illustrate the effect of evenness or oddness on the graph of a function.

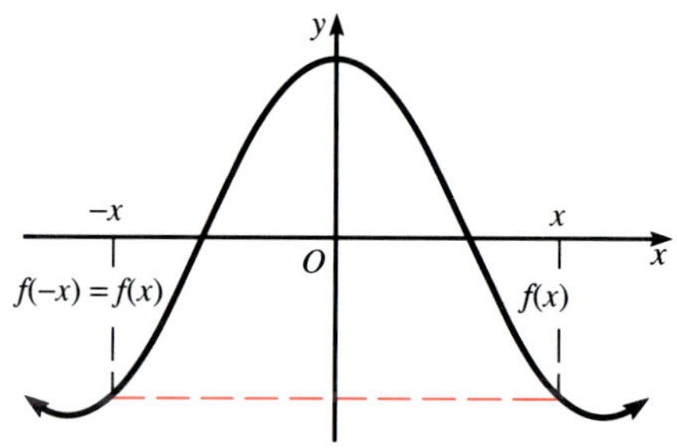

When f is even, the graph is symmetric with respect to the y-axis. That is, if the point (x, y) is on the graph of f, then so is the point $(-x, y)$.

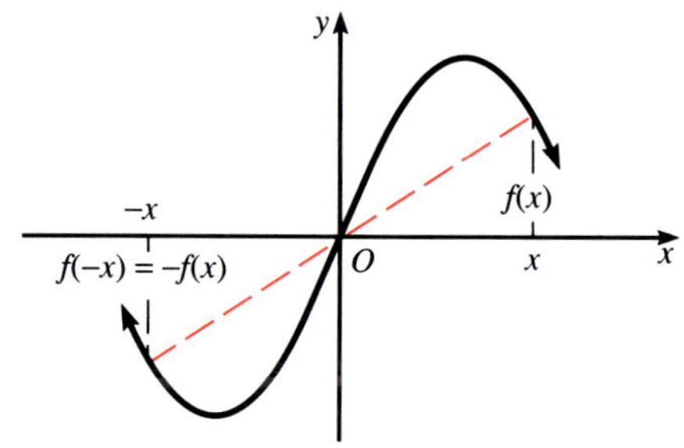

When f is odd, the graph is symmetric with respect to the origin. That is, if the point (x, y) is on the graph of f, then so is the point $(-x, -y)$.

Example 3 Part of the graph of a function f is shown at the right. Graph f in the interval $-3 \le x \le 3$ assuming that f is **(a)** even and **(b)** odd.

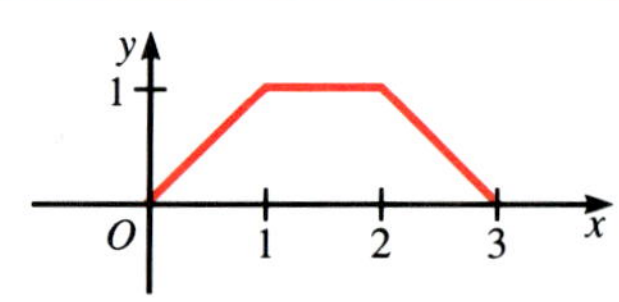

Solution

a. Reflect the given part of the graph in the y-axis.

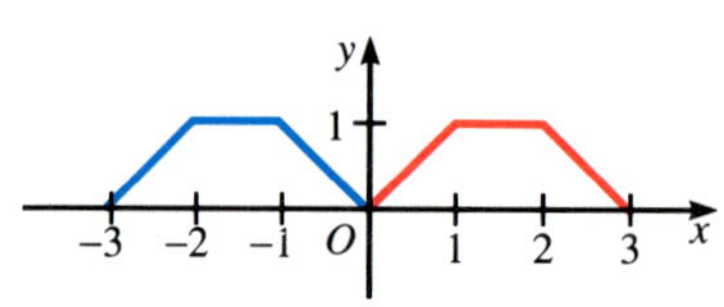

b. Reflect the given part of the graph in the origin.

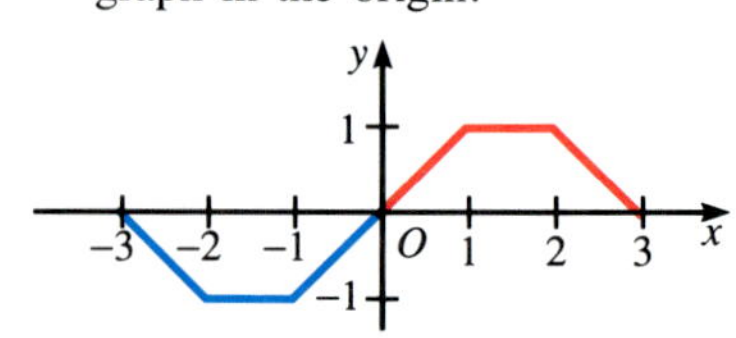

Oral Exercises

Tell the period of the periodic function whose graph is shown. Is the function even, odd, or neither?

1.

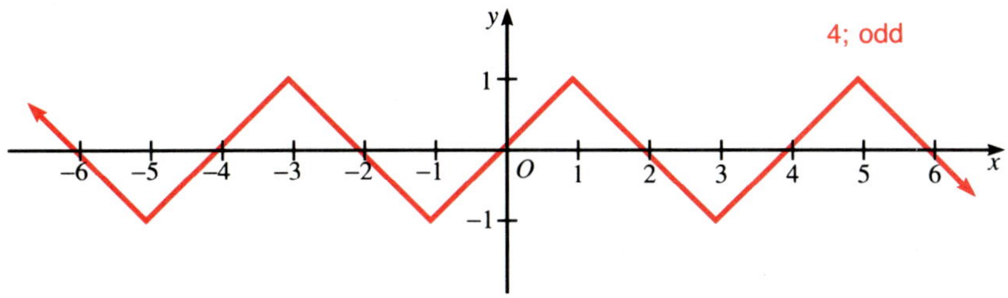

4; odd

2.

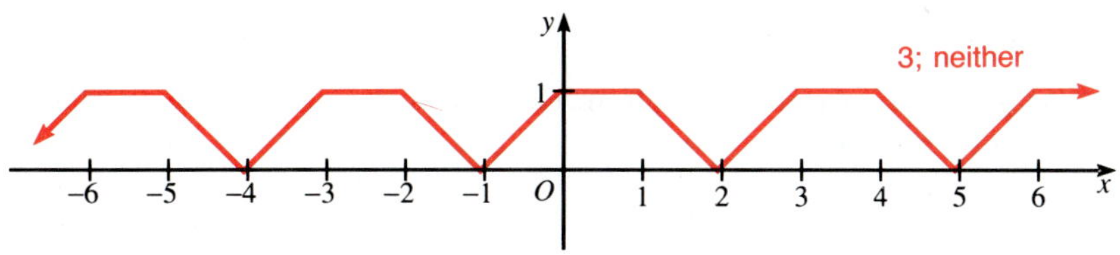

3; neither

3.

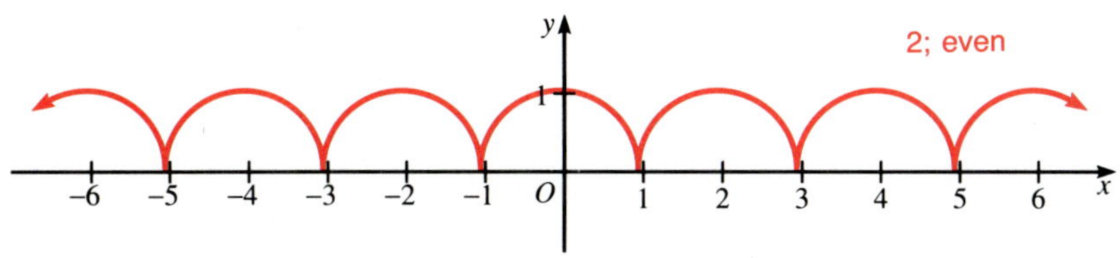

2; even

Tell whether each function is even, odd, or neither.

4. $g(x) = 2x^3 - 3x$ odd

5. $h(x) = 2x^4 - 3x^2 + 5$ even

6. $f(x) = 4x^3 - 2x^2$ neither

7. $t(x) = |x|$ even

8. $s(x) = \dfrac{x}{x^2 + 1}$ odd

9. $r(x) = x^4 - 2x^3 + x^2 - 2$ neither

Written Exercises

Part of the graph of a function *f* having the given period *p* is shown. Graph *f* in the indicated interval and tell whether *f* is even, odd, or neither.

A **1.**

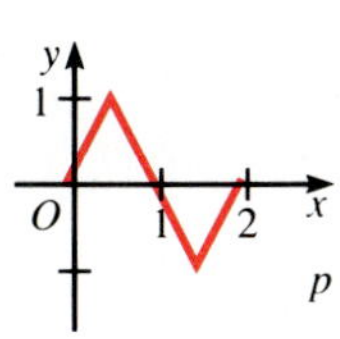

odd

$p = 2; -4 \le x \le 4$

2.

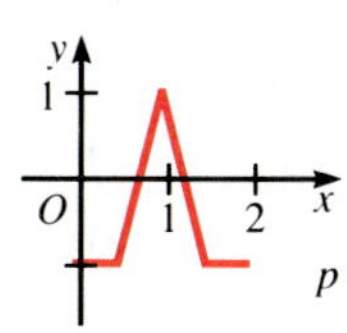

even

$p = 2; -4 \le x \le 4$

2. Part of the graph of a function f is shown. Graph f in the interval $-4 \le x \le 4$ assuming f is (a) even and (b) odd.

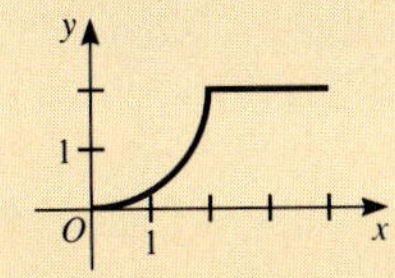

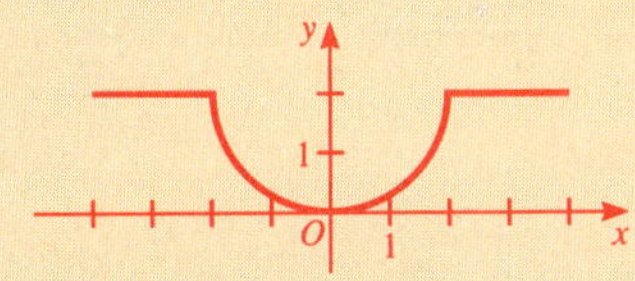

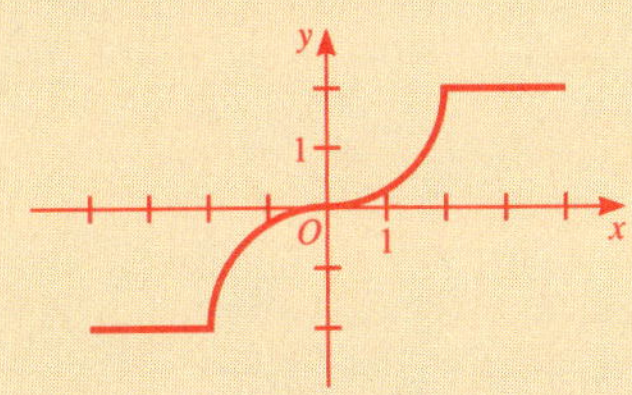

Determine whether each function g is even, odd, or neither.

3. $g(x) = 3 - 2x$ neither

4. $g(x) = x^2 - 2$ even

5. $g(x) = x|x|$ odd

Summarizing the Lesson

In this lesson students learned to use periodicity and symmetry in graphing functions. Ask students to state the definitions of periodic functions and even and odd functions.

Suggested Assignments

Average Alg. and Trig.
621/2–20 even
S 623/Mixed Review

Extended Alg. and Trig.
621/2–20 even, 21–26, 29
S 623/Mixed Review

Supplementary Materials

Study Guide	pp. 209–210
Resource Book	p. 158
Overhead Visual	10

Additional Answers Written Exercises

2.

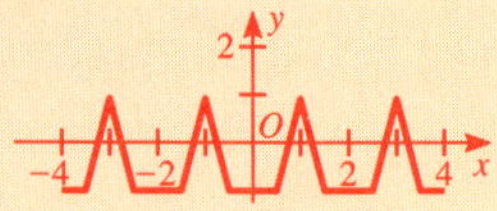

4.

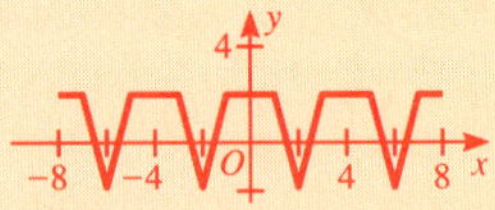

6.

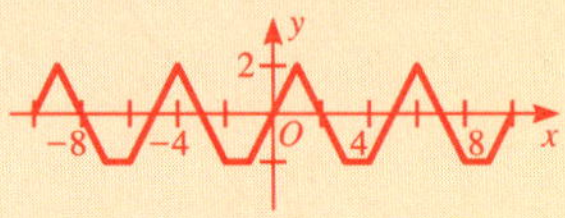

8. a.

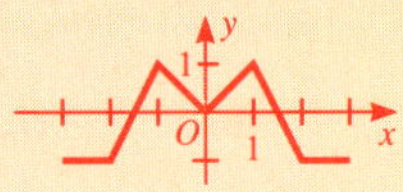

b.

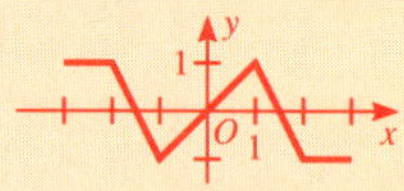

10. a.

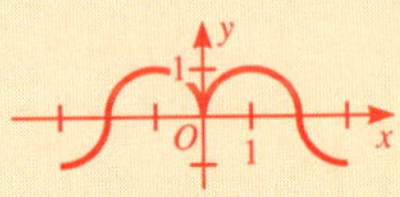

b.

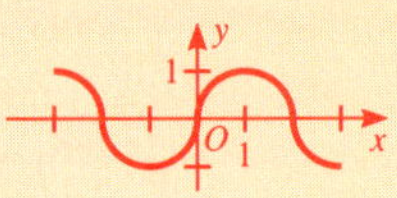

3.

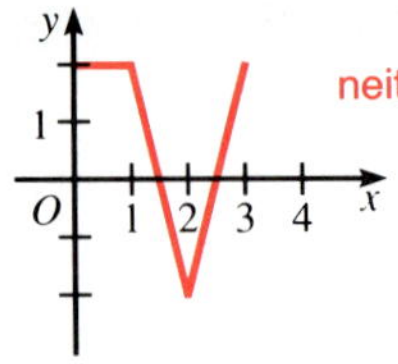

neither

$p = 3; -6 \le x \le 6$

4.

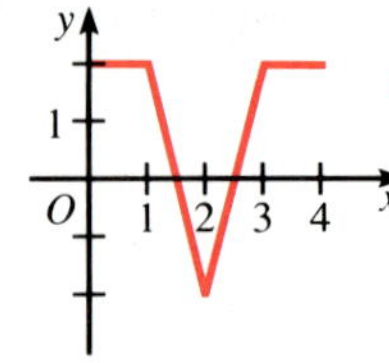

even

$p = 4; -8 \le x \le 8$

5.

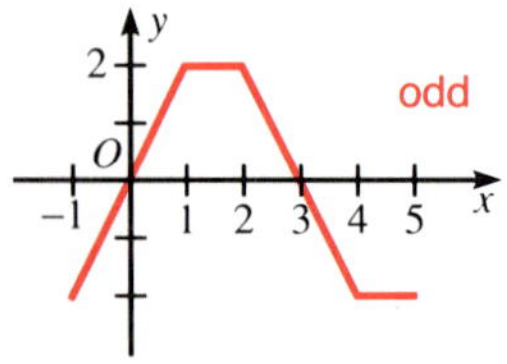

odd

$p = 6; -6 \le x \le 6$

6.

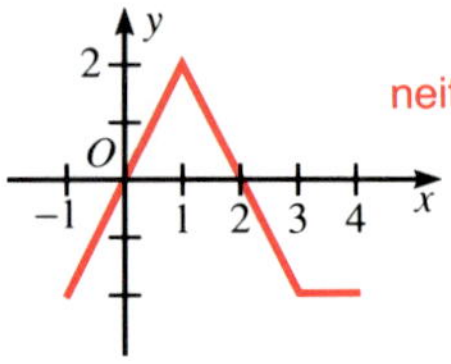

neither

$p = 5; -10 \le x \le 10$

Part of the graph of a function f is given. Graph f in the interval $-3 \le x \le 3$ assuming f is (a) even and (b) odd.

7.

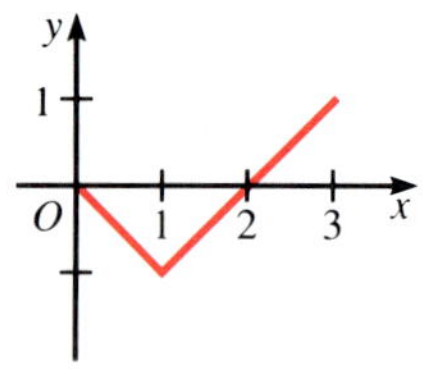

8.

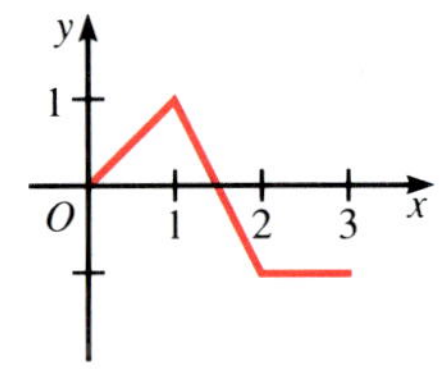

9.

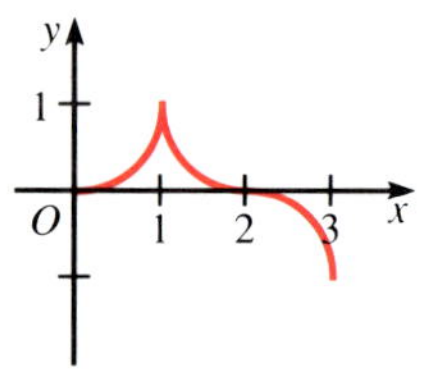

10.

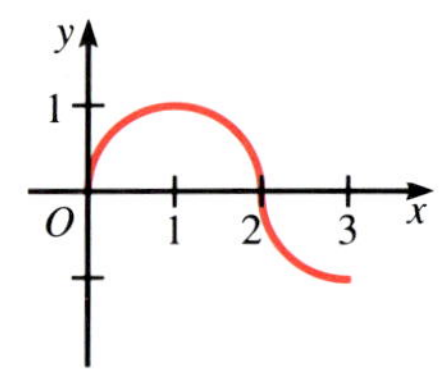

Determine whether each function f is even, odd, or neither.

11. $f(x) = x^3 + x$ odd

12. $f(x) = x^3 + 5x + 3$ neither

13. $f(x) = x^4 + x^3$ neither

14. $f(x) = x^4 + 2x^2 + 1$ even

15. $f(x) = \dfrac{x^2}{x^2 + 4}$ even

16. $f(x) = x\sqrt{x^2 + 1}$ odd

B **17.** $f(x) = x \sin x$ even

18. $f(x) = x \cos x$ odd

19. $f(x) = \sin x \cos x$ odd

20. $f(x) = \sin x + \cos x$ neither

21. Show that the product of two odd functions is even.

22. Show that the product of an even function and an odd function is odd.

23. Show that if two functions have period p then the function that is their sum has a period that divides p. (*Hint:* Let $h(x) = f(x) + g(x)$ where f and g have period p.)

24. Show that the sum of an even function and an odd function is neither even nor odd.

25. Show that $\sin 2x$ has period π. (*Hint:* Find the smallest positive p such that $\sin 2(x + p) = \sin 2x$.)

26. Show that $\cos 2x$ has period π. (See the hint for Exercise 25.)

27. Show that $\cos \frac{1}{2}x$ has period 4π.

28. Show that $\sin \frac{1}{2}x$ has period 4π.

C 29. Show that if $f(x)$ has period p, then $f(kx)$ has period $\frac{p}{k}$ for any $k > 0$.

30. For any function $f(x)$, define two functions $O(x)$ and $E(x)$ as follows:

$$O(x) = \frac{f(x) - f(-x)}{2} \quad \text{and} \quad E(x) = \frac{f(x) + f(-x)}{2}$$

a. Show that $O(x)$ is odd and $E(x)$ is even.

b. Show that $O(x) + E(x) = f(x)$. (Any function can be written as the sum of an odd function and an even function in this way.)

c. Find $O(x)$ and $E(x)$ if $f(x) = x^3 - 3x^2 + 2x + 1$. $O(x) = x^3 + 2x$; $E(x) = -3x^2 + 1$

d. Find $O(x)$ and $E(x)$ if $f(x) = e^x$. (These functions are called the hyperbolic sine and cosine of x and are written $\sinh x$ and $\cosh x$.)
$O(x) = \frac{e^x - e^{-x}}{2}$; $E(x) = \frac{e^x + e^{-x}}{2}$

Mixed Review Exercises

Graph each equation.

1. $y = x^2 + 6x + 11$
2. $y = 2^x - 1$
3. $4x - 3y = 12$
4. $y = \log_{1/2} x$
5. $x^2 + y^2 + 2y = 8$
6. $9x^2 - y^2 = 9$

Solve.

7. $\sqrt{7x + 4} = x + 2$ $\{0, 3\}$
8. $2x + 7 = 1 - x$ $\{-2\}$
9. $x^3 - 2x^2 + 2x - 4 = 0$ $\{2, \pm i\sqrt{2}\}$
10. $\frac{x^2}{12} + 1 = \frac{2x}{3}$ $\{2, 6\}$
11. $2x^2 + x - 10 = 0$ $\left\{-\frac{5}{2}, 2\right\}$
12. $\frac{1}{x - 1} + \frac{1}{x} = \frac{3}{x^2 - x}$ $\{2\}$

Challenge

The function $f(x) = \sin \frac{\pi}{2}x$ has period 4, and the function $g(x) = \cos \frac{\pi}{3}x$ has period 6. Is the function $h(x) = f(x) + g(x)$ periodic? If so, what is its period? yes; 12

Additional Answers
Mixed Review Exercises

1.
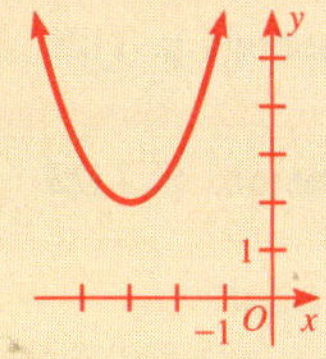

2.
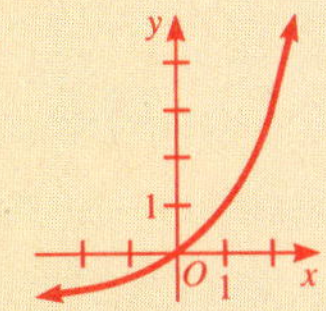

3.
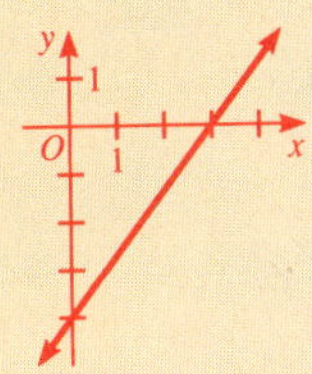

4.
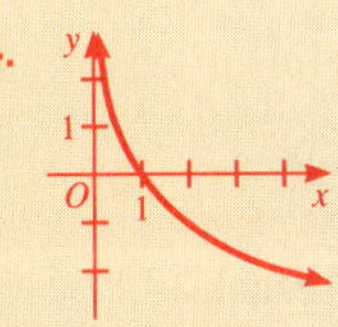

5.
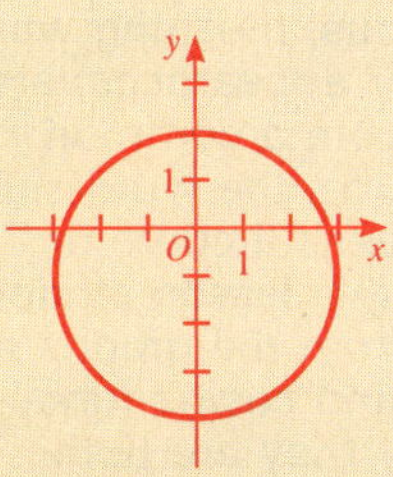

6.
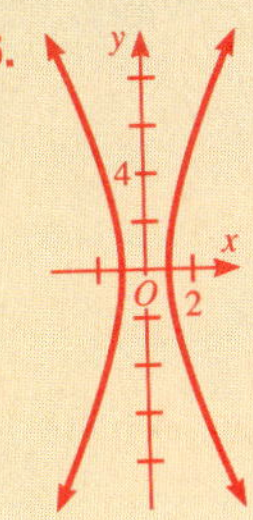

Explorations p. 844

Teaching Suggestions p. T129

Suggested Extensions p. T129

Warm-Up Exercises

Evaluate if $x = \pi$.

1. $\sin 3x$ 0
2. $2 \cos x$ -2
3. $\sin \frac{x}{2} - 2$ -1
4. $3 + \cos 2x$ 4
5. $4 \sin \frac{x}{3}$ $2\sqrt{3}$
6. $\frac{1}{2} \cos x - 1$ $-\frac{3}{2}$
7. $5 + \frac{3}{2} \sin x$ 5
8. $2 \cos \frac{x}{4}$ $\sqrt{2}$
9. $3 \sin \frac{x}{6} - \frac{1}{2}$ 1
10. $3 + 5 \cos 3x$ -2

Motivating the Lesson

Tell students the words sine and sinuous, meaning wavy in form or appearance, are derived from a common Latin source, the word sinus, which means a curve or bend. In this lesson students will see from the graphs of the sine and cosine functions why they are both called sinusoidal—wavy, curved, bent.

13-4 Graphs of the Sine and Cosine

Objective To graph the sine, cosine, and related functions.

To obtain the graphs of $y = \sin x$ and $y = \cos x$, you can use a computer or a graphing calculator. Otherwise a table of values for $0 \le x \le 2\pi$ is needed.

x	0	$\frac{\pi}{6}$	$\frac{\pi}{3}$	$\frac{\pi}{2}$	$\frac{2\pi}{3}$	$\frac{5\pi}{6}$	π	$\frac{7\pi}{6}$	$\frac{4\pi}{3}$	$\frac{3\pi}{2}$	$\frac{5\pi}{3}$	$\frac{11\pi}{6}$	2π
	0	0.52	1.05	1.57	2.09	2.62	3.14	3.67	4.19	4.71	5.24	5.76	6.28
$\sin x$	0	0.50	0.87	1	0.87	0.50	0	-0.50	-0.87	-1	-0.87	-0.50	0
$\cos x$	1	0.87	0.50	0	-0.50	-0.87	-1	-0.87	-0.50	0	0.50	0.87	1

The graphs shown below are obtained by plotting the ordered pairs $(x, \sin x)$ and $(x, \cos x)$ from the table above and by joining the points with smooth curves.

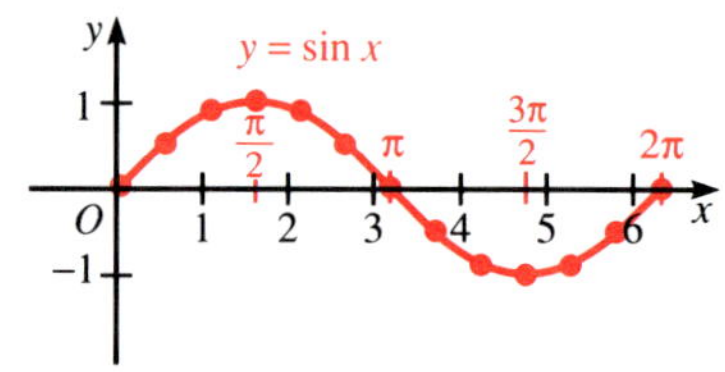

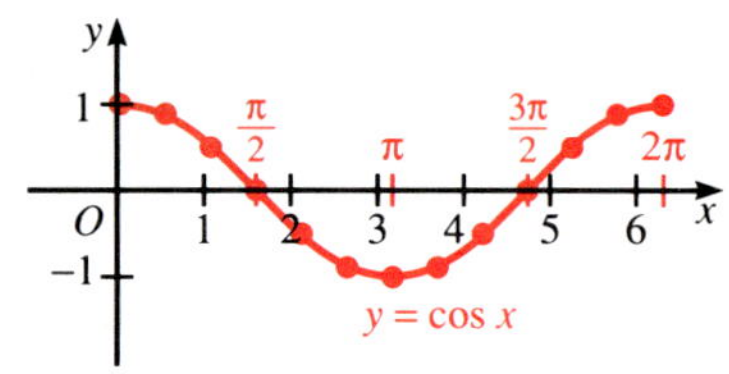

Since both the sine and cosine functions have period 2π, their complete graphs consist of the above graphs repeated over and over, as shown below.

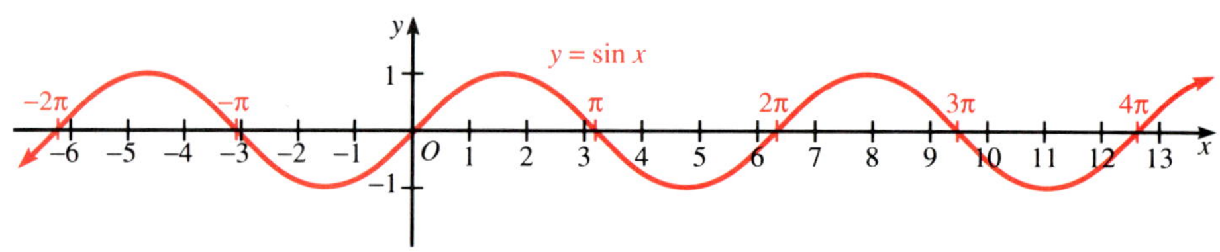

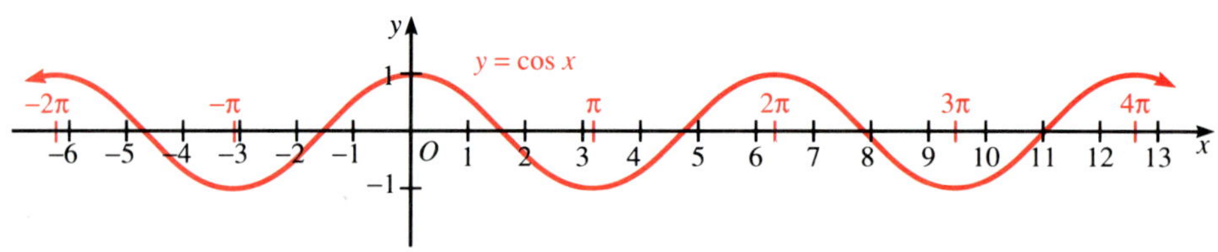

Notice that the sine curve is symmetric with respect to the origin, since sine is an odd function. The cosine curve is symmetric with respect to the y-axis, since cosine is an even function. Notice also that if the cosine curve were shifted $\frac{\pi}{2}$ units to the right, it would coincide with the sine curve.

If you have a computer or a graphing calculator, you can investigate the graphs of functions of the form $y = c + \sin x$, such as those shown at the right. Notice that the graph of the sine function is shifted upward c units if $c > 0$, and it is shifted downward $|c|$ units if $c < 0$. (This also applies to functions of the form $y = c + \cos x$.)

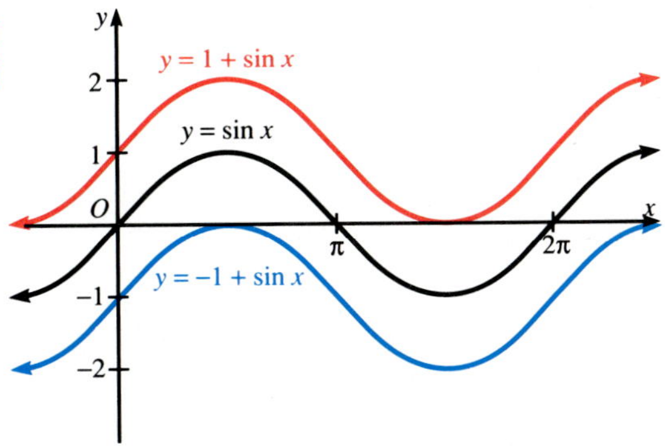

The diagram at the right shows the graphs of three functions of the form $y = a \sin x (a > 0)$. In each case the value of the function varies from a minimum value of $-a$ to a maximum value of a. (This also applies to functions with the form $y = a \cos x$.) The positive number a is called the **amplitude** of $a \sin x$ (or $a \cos x$). Therefore, the amplitude of $2 \sin x$ is 2, of $\sin x$ is 1, and of $\frac{1}{2} \sin x$ is $\frac{1}{2}$.

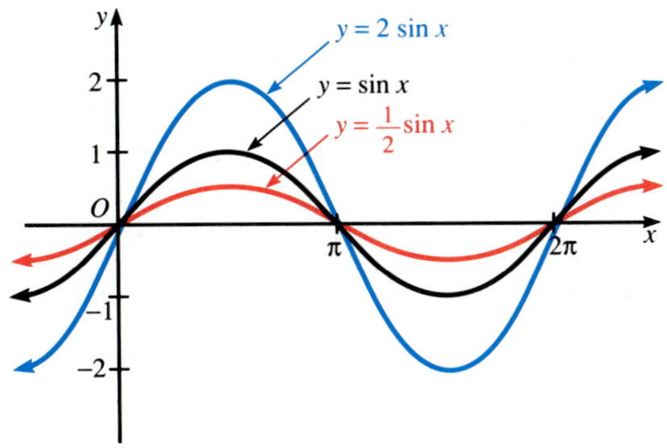

Example 1 Graph $y = 2 + 3 \sin x$.

Solution The sine curve is shifted upward 2 units, and its amplitude is 3.

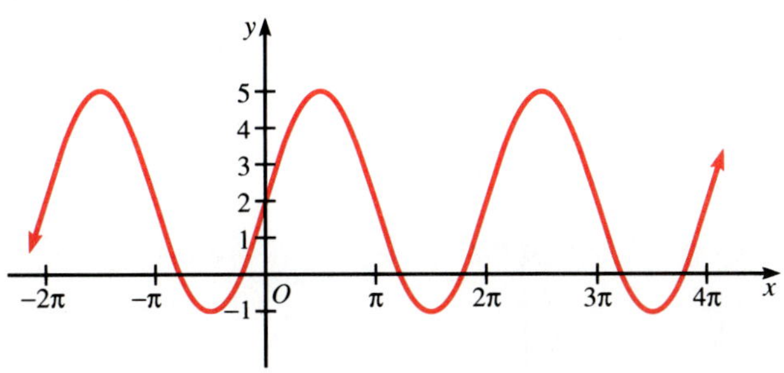

Notice in the solution of Example 1 that the function has a maximum value of 5 and a minimum value of -1. In general, any function of the form $y = c + a \sin x$ (or $y = c + a \cos x$) has maximum value $M = c + a$ when $\sin x = 1$ (or $\cos x = 1$) and minimum value $m = c - a$ when $\sin x = -1$ (or $\cos x = -1$).

Using a Computer or a Graphing Calculator

Students can use a computer or graphing calculator in their initial investigations of the graphs of the circular functions as well as in the exercises where graphs are required. Advise students to choose the limits of both the axes judiciously so that they can get a reasonable view of the graph. For example, to see a full period of the graph of $y = 3 + 2 \sin \pi x$, students should set the limits of the x-axis at 0 and $\frac{2\pi}{\pi}$, or 2. Also, since sine takes on values between -1 and 1 inclusive, students should set the limits of the y-axis at $3 + 2(-1)$, or 1, and $3 + 2(1)$, or 5.

Support Materials
Disk for Algebra
Menu Item: Function Plotter
Resource Book pp. 246–247

Chalkboard Examples

1. Graph $y = 2 \cos x - 1$.
   The cosine curve is shifted down 1 unit, and the amplitude is 2.

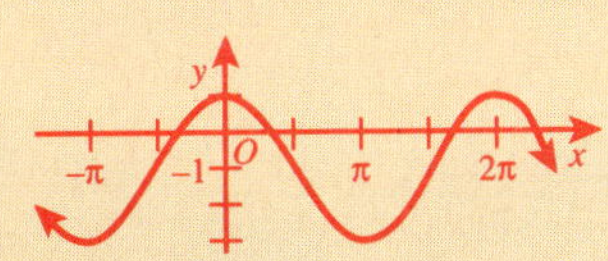

2. Graph $y = 3 \sin 2x$.
   The amplitude is 3 and the period is $\frac{2\pi}{2}$, or π.

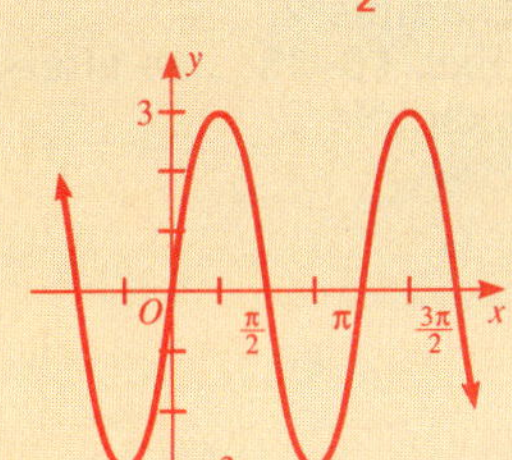

(continued)

Chalkboard Examples

(continued)

3. Each graph that follows has an equation of the form $y = c + a \cos bx$ or $y = c + a \sin bx$. Find an equation of each graph.

a.

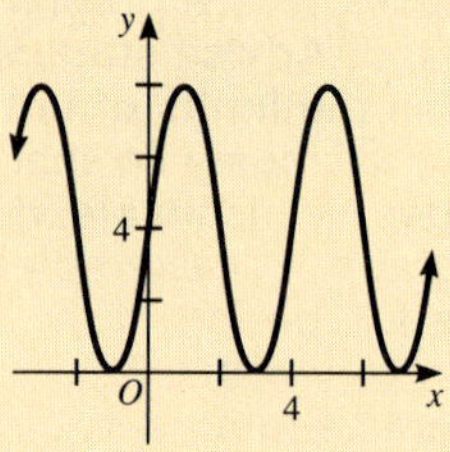

The maximum is 8 and the minimum is 0; $a = 4$. The period is 4, so $b = \frac{\pi}{2}$. The graph is a sine curve.

$y = 4 + 4 \sin \frac{\pi}{2}x$

b.

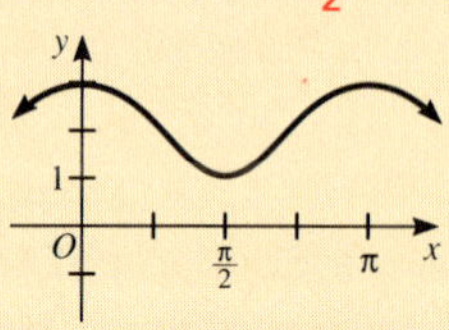

Since the maximum, M, is 3 and the minimum, m, is 1, $a = \frac{M - m}{2} = \frac{3 - 1}{2} = 1$. The period is π so $b = 2$. $c = \frac{M + m}{2} = \frac{3 + 1}{2} = 2$. The graph is a cosine curve. $y = 2 + \cos 2x$

Check for Understanding

Oral Exs. 1–9: use after Example 2.
Oral Exs. 10–13: use after Example 3.

Since

$$\frac{M + m}{2} = \frac{(c + a) + (c - a)}{2} = \frac{2c}{2} = c,$$

the average of the extreme values is the vertical shift in the sine (or cosine) curve. Also, since

$$\frac{M - m}{2} = \frac{(c + a) - (c - a)}{2} = \frac{2a}{2} = a,$$

half the difference between the extreme values is the amplitude of the curve.

While the a in $y = a \sin x$ causes a *vertical* stretch or shrinking in the graph of $y = \sin x$, the b in $y = \sin bx$ causes a *horizontal* stretch or shrinking. In other words, the coefficient b affects the period.

For example, to find the period of $f(x) = \sin 3x$, look at $f(x + p)$: $f(x + p) = \sin 3(x + p) = \sin (3x + 3p)$. Then $f(x + p) = f(x)$ if and only if $\sin (3x + 3p) = \sin 3x$. This equation holds if $3p = 2\pi$. Therefore, $p = \frac{2\pi}{3}$. The graph of $y = \sin 3x$ is shown below with the graph of $y = \sin x$.

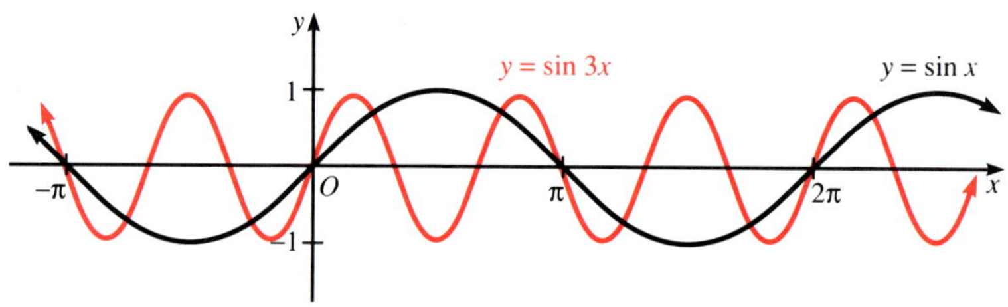

For positive real numbers a and b, the functions $f(x) = a \sin bx$ and $f(x) = a \cos bx$ have amplitude a and period $\frac{2\pi}{b}$.

Example 2 Graph $y = 2 \cos \frac{\pi}{2}x$.

Solution The amplitude is 2, and the period is $\frac{2\pi}{\frac{\pi}{2}}$, or 4.

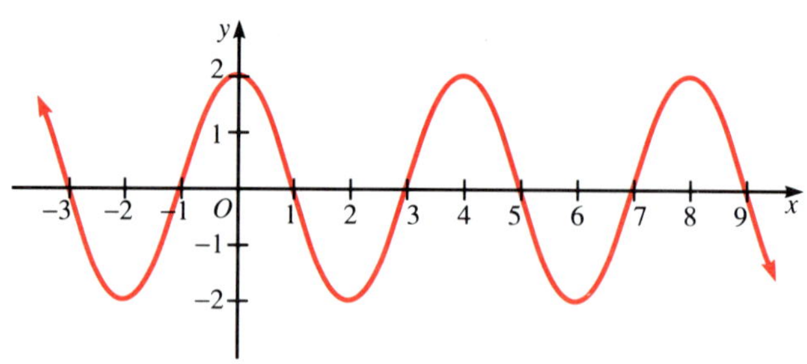

Example 3 Each graph shown below has an equation of the form $y = c + a \sin bx$ or $y = c + a \cos bx$. Find an equation of each graph.

a.

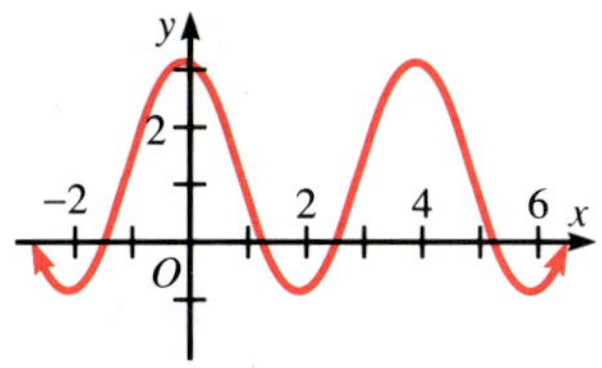

b.

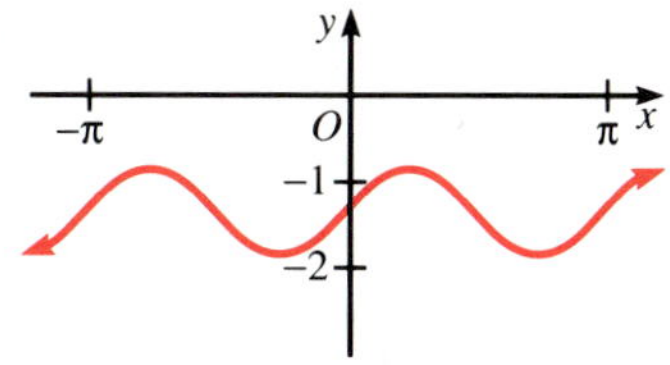

Solution

a. The graph is the same as the one in Example 2 except that it has been shifted upward 1 unit.

$\therefore$ the equation is $y = 1 + 2 \cos \frac{\pi}{2}x$. ***Answer***

b. Since the graph shows a function with period π, $\frac{2\pi}{b} = \pi$ and $b = 2$. Since the maximum value M is -1 and the minimum value m is -2,

$$c = \frac{M + m}{2} = \frac{-1 + (-2)}{2} = -\frac{3}{2},$$

and
$$a = \frac{M - m}{2} = \frac{-1 - (-2)}{2} = \frac{1}{2}.$$

Finally, since the graph is halfway between the maximum and minimum values of the function at $x = 0$, the graph is a sine curve.

$\therefore$ the equation is $y = -\frac{3}{2} + \frac{1}{2} \sin 2x$. ***Answer***

Oral Exercises

Identify each curve as sine or cosine. Then give its amplitude and period.

1.

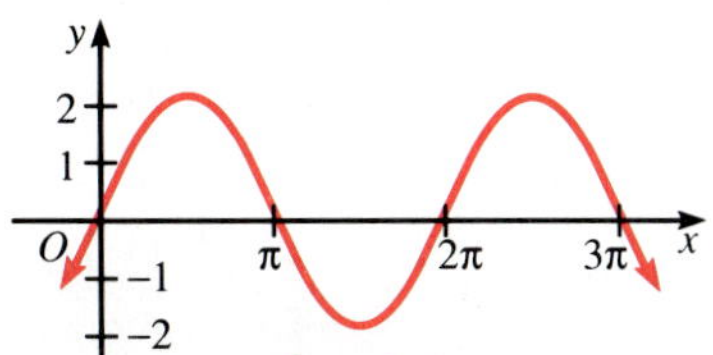

sine; 2; 2π

2.

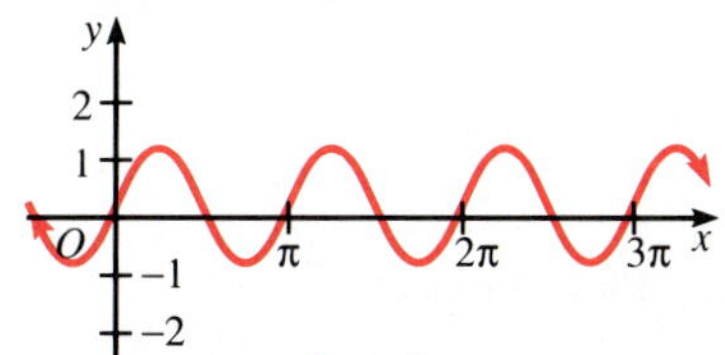

sine; 1; π

3.

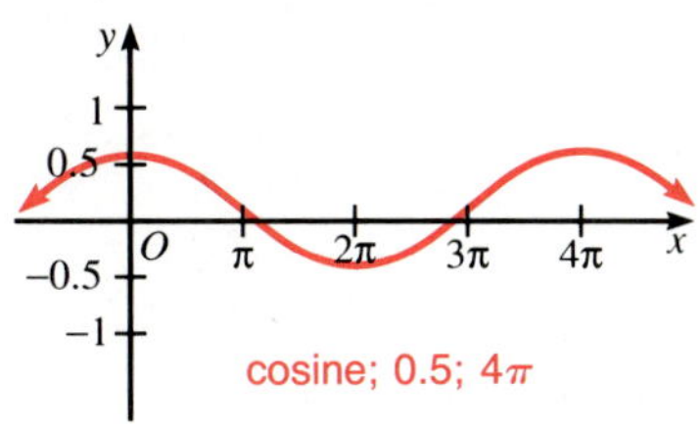

cosine; 0.5; 4π

4.

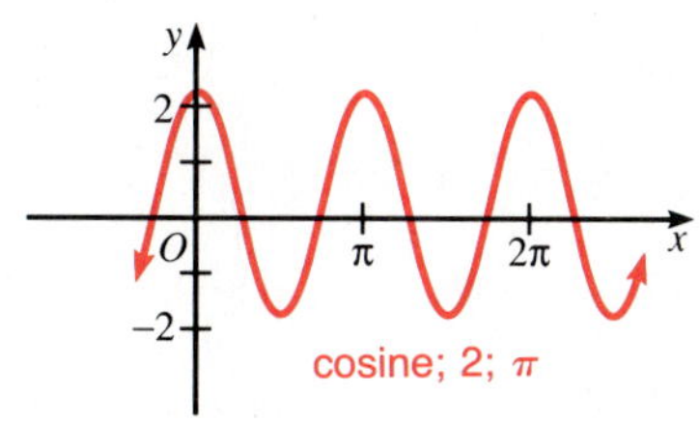

cosine; 2; π

Guided Practice

In each exercise, find: (**a**) the amplitude, (**b**) the maximum and minimum values, and (**c**) the period. Then (**d**) graph the function.

1. $y = 3 \sin 2x$
a. 3 **b.** 3, -3 **c.** π
d.

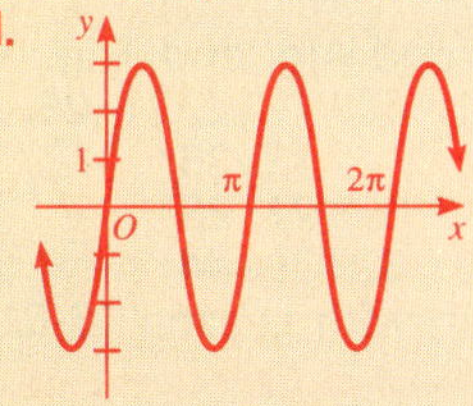

2. $y = 2 \cos \pi x - 1$
a. 2 **b.** 1, -3 **c.** 2
d.

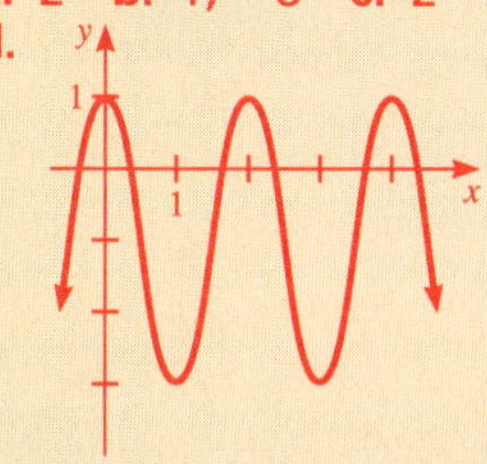

3. Find an equation of the sine curve with maximum value $\frac{1}{2}$, minimum value $-\frac{1}{2}$, and period 6π.
$y = \frac{1}{2} \sin \frac{1}{3}x$

Summarizing the Lesson

In this lesson students learned to graph the sine and cosine functions as well as variations in these functions. Ask students to describe the effect of a, b, and c on the graph of $y = c + a \sin bx$.

Supplementary Materials

Study Guide	pp. 211–212
Practice Master	74
Computer Activity	29
Overhead Visual	11

Suggested Assignments

Average Alg. and Trig.
Day 1: 628/1–6, 13–20, 27–32
S 622/11–19 odd
Day 2: 628/7–12, 21–26, 33–38
S 612/*P*: 8

Extended Alg. and Trig.
Day 1: 628/7–12, 21–26, 33–38
S 623/27, 28
Day 2: 629/39–44
Assign with Lesson 13-5.

Additional Answers Written Exercises

13. $y = 2 \sin x$

14. $y = \sin 2x$

15. $y = \frac{1}{2} \cos \frac{1}{2}x$

16. $y = 2 \cos 2x$

17. $y = \sin \pi x$

18. $y = 3 \cos \frac{\pi}{4}x$

19. $y = 3 + 3 \cos \frac{\pi}{4}x$

20. $y = -1 + \cos \frac{2\pi}{5}x$

28.

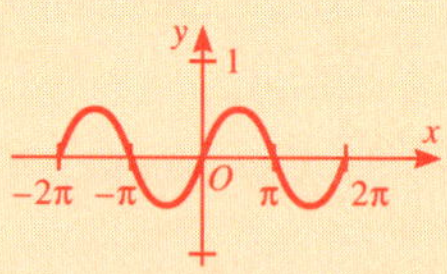

30.

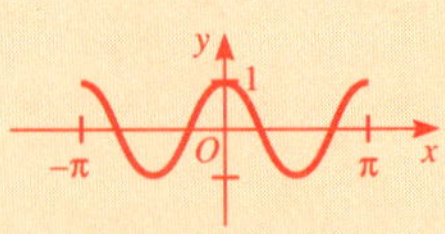

32.

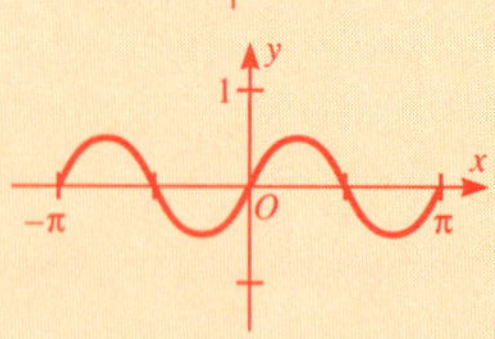

34.

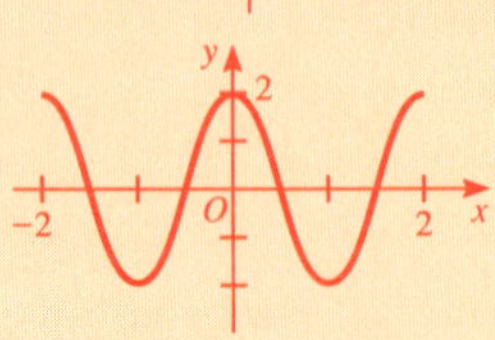

5.

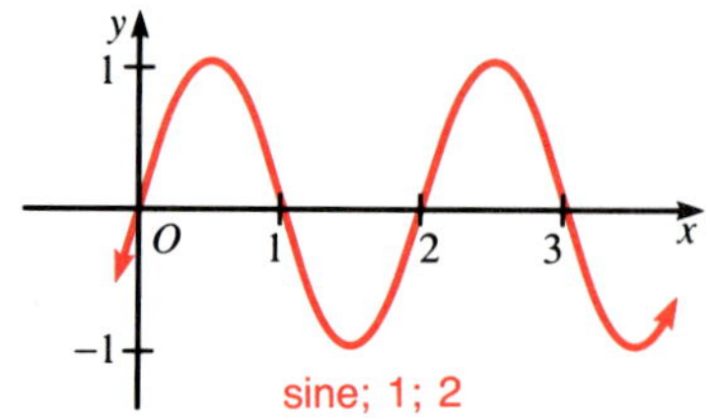

6.

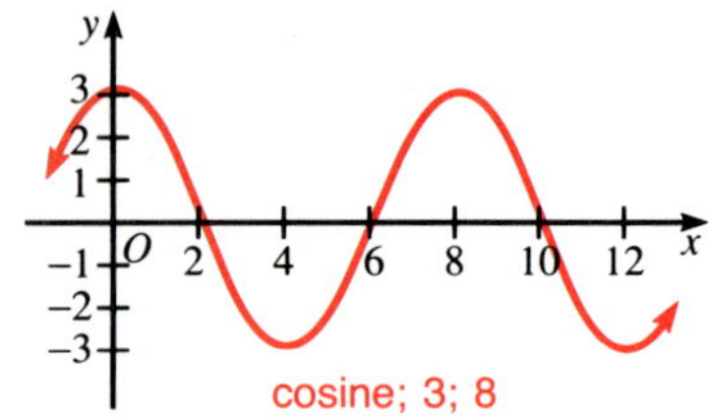

7.

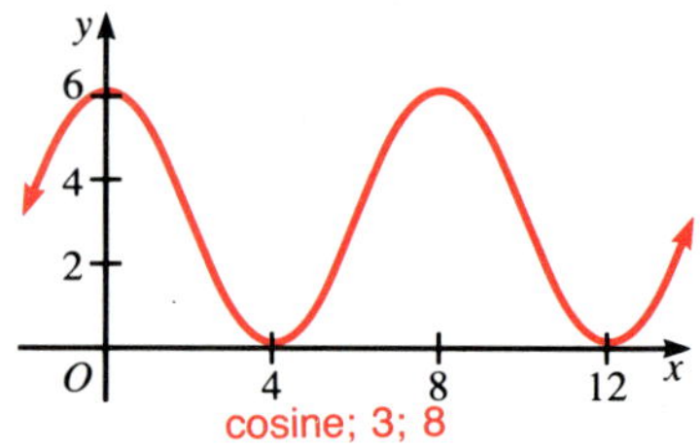

8.

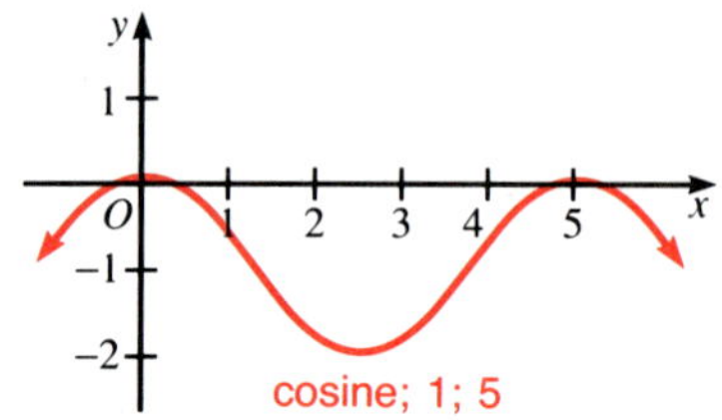

9. Explain why the b in $y = \sin bx$ has no effect on the maximum and minimum values of the sine function. $-1 \le \sin \theta \le 1$ for all θ, so $-1 \le \sin bx \le 1$ for all x.

In Exercises 10–13, match each description with the letter of the function that fits the description.

10. The amplitude is 3. d

11. The period is $\frac{\pi}{2}$. b

12. The minimum value is 1. c

13. The period is 2. e

a. $y = 3 + \sin 2\pi x$

b. $y = 2 \cos 4x$

c. $y = 2 + \sin 3x$

d. $y = 3 \cos 2x$

e. $y = 1 + \sin \pi x$

Written Exercises

In Exercises 1–12:
a. Find the amplitude of each function.
b. Find the maximum and minimum values.
c. Find the period.

3. 1; 1, −1; $\frac{\pi}{2}$
6. $\frac{1}{2}$; $\frac{1}{2}$, $-\frac{1}{2}$; π
9. 2; 2, −2; 4
12. 2; 5, 1; 2

A

1. $y = 2 \cos x$ 2; 2, −2; 2π

2. $y = \frac{1}{2} \sin x$ $\frac{1}{2}$; $\frac{1}{2}$, $-\frac{1}{2}$; 2π

3. $y = \sin 4x$

4. $y = \cos 2x$ 1; 1, −1; π

5. $y = 3 \sin 3x$ 3; 3, −3; $\frac{2\pi}{3}$

6. $y = \frac{1}{2} \sin 2x$

7. $y = \frac{1}{2} \sin 2\pi x$ $\frac{1}{2}$; $\frac{1}{2}$, $-\frac{1}{2}$; 1

8. $y = 2 \cos \pi x$ 2; 2, −2; 2

9. $y = 2 \cos \frac{\pi}{2}x$

10. $y = 3 \sin \frac{\pi}{4}x$ 3; 3, −3; 8

11. $y = \cos 2\pi x - 1$ 1; 0, −2; 1

12. $y = 2 \sin \pi x + 3$

13–20. Find an equation of each curve pictured in Oral Exercises 1–8.

Find an equation of the form $y = c + a \sin bx$ that satisfies the given conditions. (M is the maximum value of the function y, and m is the minimum value.)

Sample $M = 5,\ m = 1$, period π

Solution

1. $a = \frac{M - m}{2} = \frac{5 - 1}{2} = 2$
2. $c = \frac{M + m}{2} = \frac{5 + 1}{2} = 3$
3. $\frac{2\pi}{b} = \pi$, so $b = 2$

$\therefore\ y = 3 + 2 \sin 2x.$ ***Answer***

22. $y = \sin \frac{1}{4}x$
24. $y = 2 \sin \frac{2\pi}{3}x$
26. $y = -3 + 3 \sin \frac{\pi}{2}x$

21. $M = 3,\ m = -3$, period 2π $y = 3 \sin x$

22. $M = 1,\ m = -1$, period 8π

23. $M = 2,\ m = -2$, period 2 $y = 2 \sin \pi x$

24. $M = 2,\ m = -2$, period 3

25. $M = 7,\ m = 1$, period π $y = 4 + 3 \sin 2x$

26. $M = 0,\ m = -6$, period 4

27–32. Graph each function in Exercises 1–6. Show at least two periods. You may wish to verify your graphs on a computer or a graphing calculator.

B **33–38.** Graph each function in Exercises 7–12. Show at least two periods. You may wish to verify your graphs on a computer or a graphing calculator.

36.

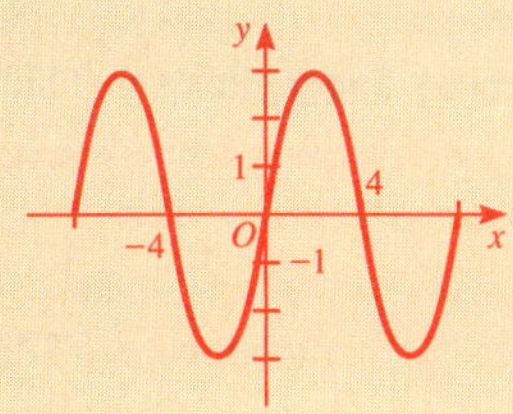

38.

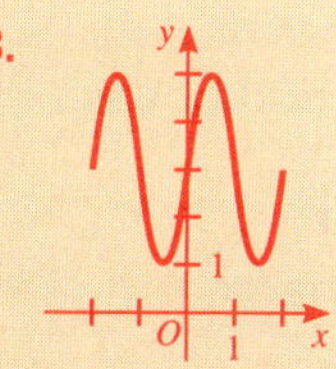

40.

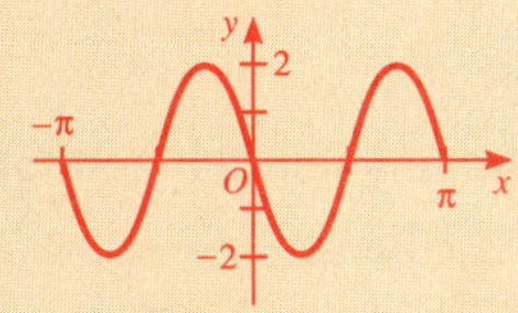

42.

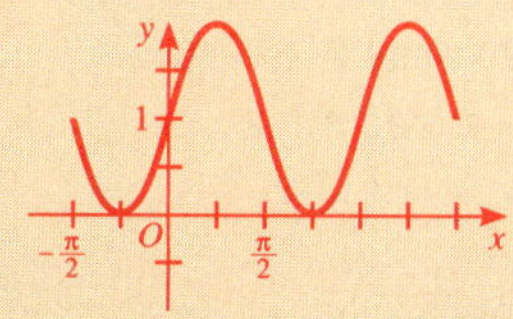

44.

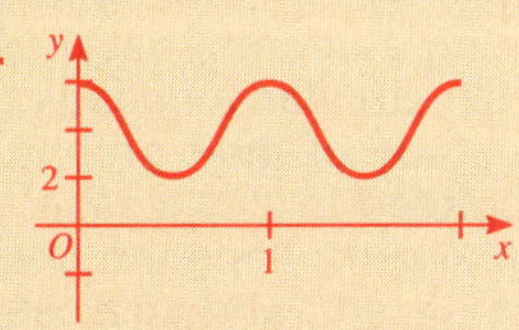

In a function of the form $y = \sin (x - d)$ or $y = \cos (x - d)$, the number d determines the *phase shift* of the graph. In general, for any function f, the graph of $y = f(x - d)$ is obtained by shifting the graph of $y = f(x)$ to the right d units if $d > 0$ or to the left $|d|$ units if $d < 0$.

The graph of $y = -f(x)$ is obtained by reflecting the graph of $y = f(x)$ in the x-axis.

Use the information given in the preceding paragraphs to sketch the graph of each of the following equations. Show at least two periods.

39. $y = \cos \left(x - \frac{\pi}{4}\right) - 1$

40. $y = -2 \sin 2x$

41. $y = -\cos 2\left(x + \frac{\pi}{3}\right)$

C **42.** $y = -\sin (2x - \pi) + 1$

43. $y = -2 \cos (4x + \pi) + \frac{1}{2}$

44. $y = -2 \cos (2\pi x - \pi) + 4$

45. Show that the graph of $y = \cos \left(x - \frac{\pi}{2}\right)$ is a sine curve.

46. Explain how the graphs of $y = \sin x$ and $y = \cos \left(x + \frac{\pi}{2}\right)$ are related.

47. Use the equation $\sin 2x = 2 \sin x \cos x$ to show that the graph of $y = \sin x \cos x$ is a sine curve. Sketch the graph.

48. Use the equation $\cos 2x = 1 - 2 \sin^2 x$ to show that the graph of $y = \sin^2 x$ is a cosine curve. Sketch the graph.

Using a Computer

You may want to have students find the equation for a given sine or cosine curve, as was done in Written Exercises 13–20. The program "Circular Function Quiz" on the Disk for Algebra will draw a random sine or cosine curve and let students visually experiment to find the corresponding equation.

Support Materials
Disk for Algebra
Menu Item: Circular Function Quiz
Resource Book p. 247

Teaching Suggestions p. T129

Reading Algebra p. T129

Suggested Extensions p. T129

Warm-Up Exercises

State the period of each function.

1. $y = 2 \sin x$ 2π
2. $y = 3 + \cos 2x$ π
3. $y = \frac{1}{2} \cos \pi x$ 2
4. $y = \sin \frac{x}{2} - 1$ 4π

Is the function defined for $x = \frac{\pi}{2}$? for $x = \pi$?

5. $y = \tan x$ no; yes
6. $y = \csc x$ yes; no
7. $y = \sec x$ no; yes
8. $y = \cot x$ yes; no

Motivating the Lesson

Remind students that there are six circular functions. The study of their graphs is completed in this lesson by graphing the tangent, cotangent, secant, and cosecant.

Using a Computer or a Graphing Calculator

Here and in the graphing exercises students will need to be alert to values for which a function is undefined (that is, where asymptotes occur) and to the variations due to the constants in the equation.

Support Materials
Disk for Algebra
Menu Item: Function Plotter
Resource Book pp. 248–249

13-5 Graphs of the Other Functions

Objective To graph the tangent, cotangent, secant, cosecant, and related functions.

The tangent function is odd, has period π, and has asymptotes. These properties are investigated below and will be useful in graphing the function.

Example 1 Determine whether the tangent function is even, odd, or neither.

Solution Since sine is an odd function and cosine is even,

$$\tan(-x) = \frac{\sin(-x)}{\cos(-x)} = \frac{-\sin x}{\cos x} = -\tan x.$$

$\therefore$ tangent is an odd function. ***Answer***

In addition to being odd, tangent has period π. To see why, use the fact that $\sin(x + \pi) = -\sin x$ and $\cos(x + \pi) = -\cos x$ (see Oral Exercise 14, page 616):

$$\tan(x + \pi) = \frac{\sin(x + \pi)}{\cos(x + \pi)} = \frac{-\sin x}{-\cos x} = \tan x$$

Another property of tangent comes from the fact that $\tan x$ is undefined when $\cos x = 0$ $\left(\text{that is, when } x = \frac{k\pi}{2} \text{ where } k \text{ is an odd integer}\right)$. To see what happens to $\tan x$ as x *approaches* $\frac{\pi}{2}$, for example, look at the diagram. For $0 < x < \frac{\pi}{2}$, you know $\tan x = AQ$ (see Exercise 28 on page 617). The nearer x is to $\frac{\pi}{2}$, the closer P is to B, and the larger AQ is. Therefore, as x approaches $\frac{\pi}{2}$, $\tan x$ increases without bound.

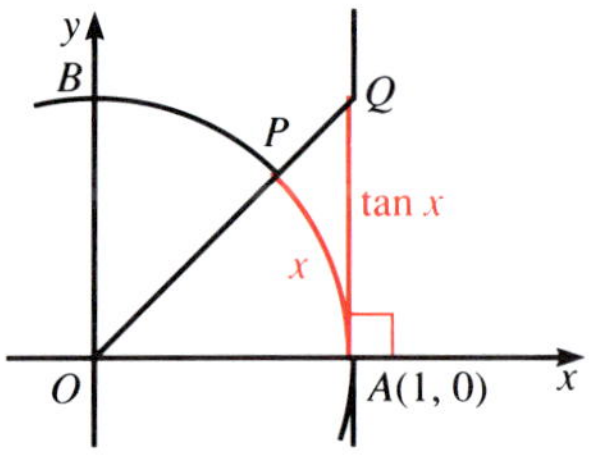

A short table of values is used to start the graph of $y = \tan x$ in the figure below. The dashed line is an *asymptote* of the graph (recall Lesson 9-5).

x	$\tan x$
0	0
$\frac{\pi}{6} \approx 0.52$	0.58
$\frac{\pi}{4} \approx 0.79$	1.00
$\frac{\pi}{3} \approx 1.05$	1.73
$\frac{\pi}{2} \approx 1.57$	—

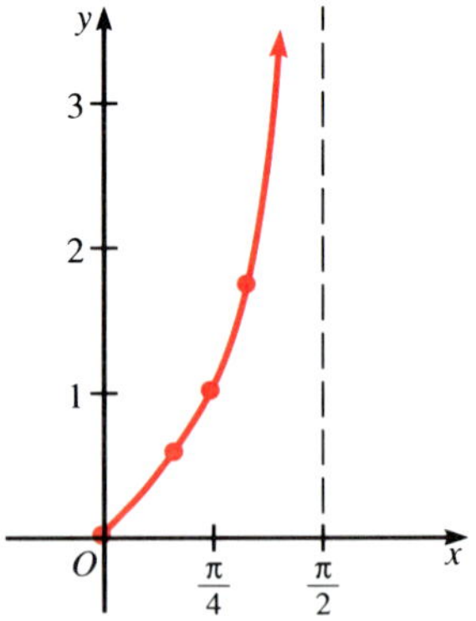

Since tangent is odd, the graph on the previous page can be reflected in the origin to obtain the graph of one period of the function, as shown in Figure 1. The complete graph of the tangent function, shown in Figure 2, is obtained by using the fact that the period of the tangent function is π.

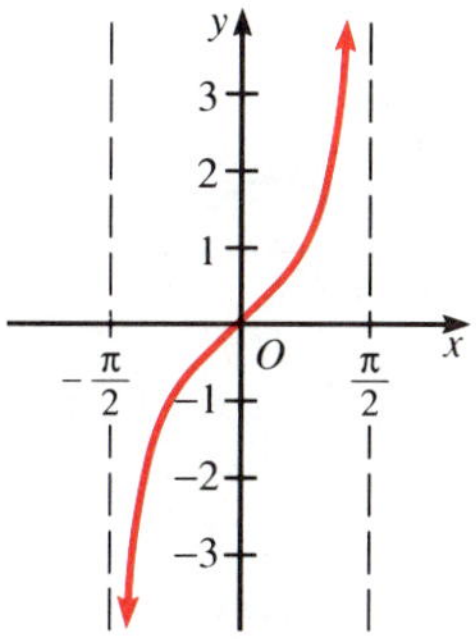

Figure 1

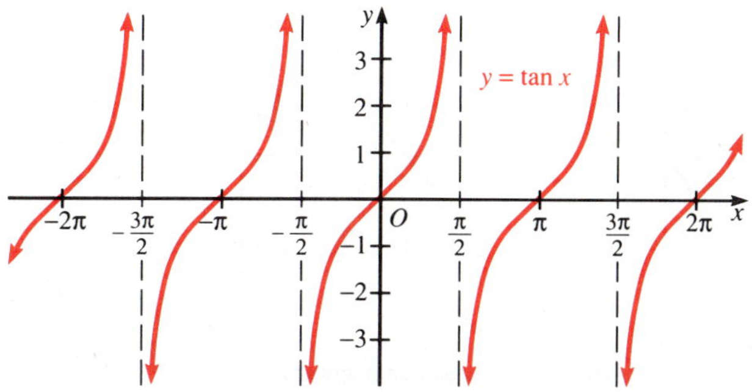

Figure 2

You can use a computer or a graphing calculator to explore the graphs of functions of the form $y = c + a \tan bx$, for various values of a, b, and c. Remember (see Lesson 13-4) that the number c produces a vertical shift in the curve, a produces a vertical stretching or shrinking, and b produces a change in the period.

Example 2 Graph one period of the function $y = 2 \tan \frac{1}{2}x$.

Solution

1. Since $b = \frac{1}{2}$, the period of the function is $\frac{\pi}{\frac{1}{2}} = 2\pi$. Asymptotes occur when $\frac{1}{2}x = \frac{\pi}{2}$ and $\frac{1}{2}x = -\frac{\pi}{2}$. Solving these equations gives asymptotes at $x = \pi$ and $x = -\pi$.

2. Since $a = 2$, there is a vertical stretch of the tangent curve. For example, since $\tan x = 1$ when $x = \frac{\pi}{4}$, $2 \tan \frac{1}{2}x = 2$ when $\frac{1}{2}x = \frac{\pi}{4}$ $\left(\text{that is, when } x = \frac{\pi}{2}\right)$. Similarly, $2 \tan \frac{1}{2}x = -2$ when $x = -\frac{\pi}{2}$. The graph of one period of the function is shown at the right.

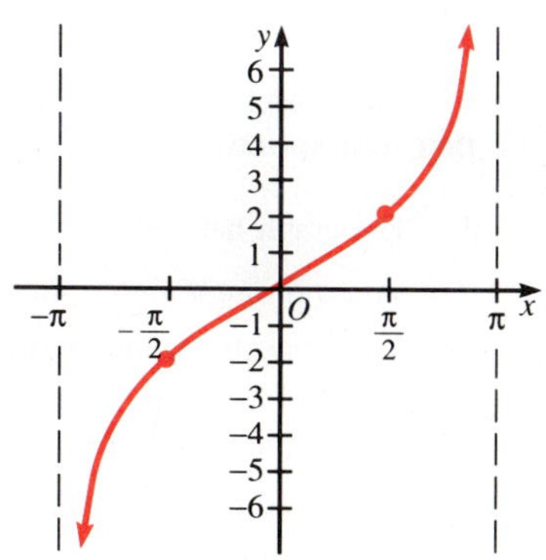

Thinking Skills

Notice that the effect of each constant *a*, *b*, or *c* is the same here, for tangent and the other circular functions, as it was for the sine and cosine functions studied earlier. It is not necessary to start anew here; what was learned earlier now may be *transferred* to a new context.

Chalkboard Examples

1. Determine whether the secant function is odd, even, or neither.
$\sec x = \frac{1}{\cos x}$; $\sec(-x) = \frac{1}{\cos(-x)} = \frac{1}{\cos x}$
$\sec(-x) = \sec x$; even

2. Sketch the graph of $y = \cot 2x$. The period is $\frac{\pi}{2}$. Cot $2x$ is not defined when x is 0 or a multiple of $\frac{\pi}{2}$.

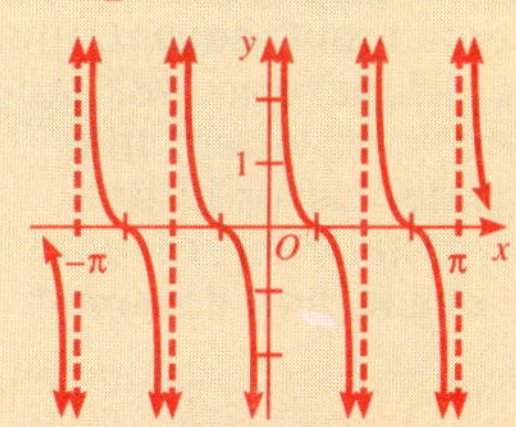

Check for Understanding

Here is a suggested use of the Oral Exercises to check students' understanding as you teach the lesson.
Oral Exs. 1–16: use at the end of the lesson.

Guided Practice

1. Determine if the function $y = \tan x + \cot x$ is even, odd, or neither. odd
2. Sketch the graph of each function.
   a. $y = 2 \sec x$
   b. $y = \tan 2x$

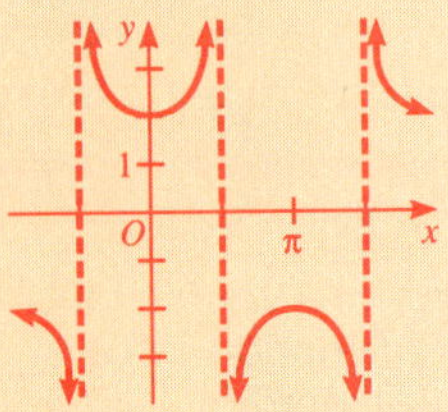

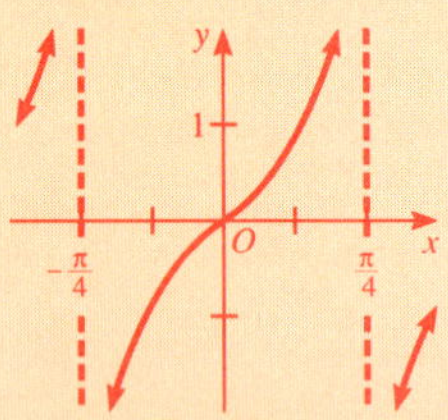

Summarizing the Lesson

In this lesson students learned to graph the tangent, cotangent, secant, and cosecant functions. Ask students to state where the asymptotes occur in the graphs of tangent, cotangent, secant, and cosecant.

Suggested Assignments

Average Alg. and Trig.
Day 1: 633/1–16
Day 2: 633/17–22
S 633/Mixed Review

Extended Alg. and Trig.
Day 1: 633/1–16
Assign with Lesson 13-4.
Day 2: 633/17–28
S 633/Mixed Review

Supplementary Materials

Study Guide	pp. 213–214
Practice Master	75
Test Master	54
Resource Book	p. 159

To graph the cotangent, secant, and cosecant functions or variations of them, you can use methods similar to those discussed in this lesson for tangent. The graphs of $y = \cot x$, $y = \sec x$, and $y = \csc x$ are shown below in red along with the graphs of their reciprocal functions in black.

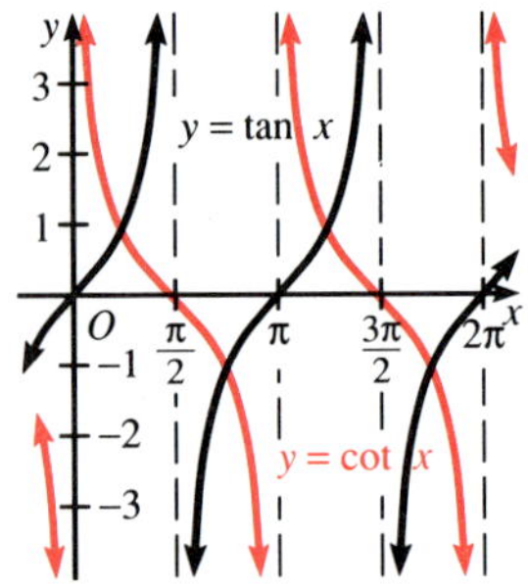

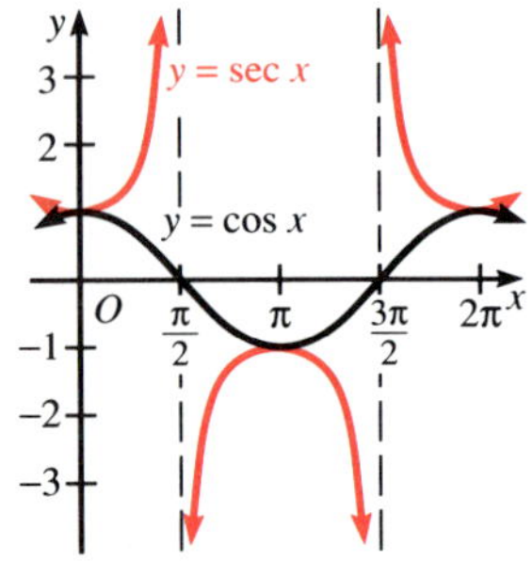

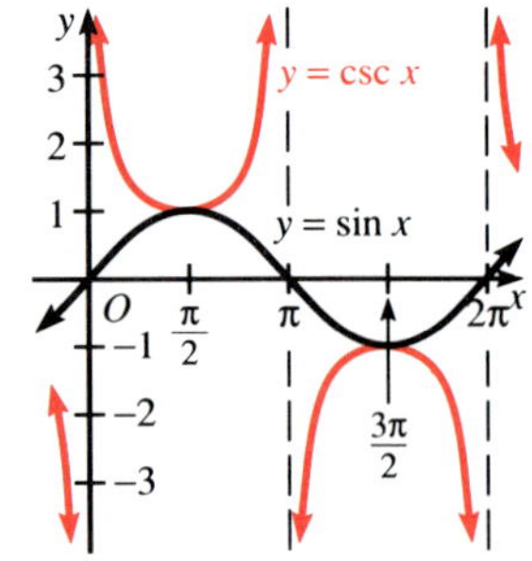

Notice from the graphs that cotangent has period π and that secant and cosecant have period 2π.

Oral Exercises

Give the period of each function.

Sample $y = \csc 4x$ **Solution** Since $b = 4$, the period is $\frac{2\pi}{4} = \frac{\pi}{2}$.

1. $y = \sec 2x$ π
2. $y = 3 \csc x$ 2π
3. $y = \frac{1}{2} \cot x$ π
4. $y = \tan \frac{1}{3}x$ 3π
5. $y = 2 \csc \pi x$ 2
6. $y = \cot 2x + 1$ $\frac{\pi}{2}$
7. $y = \sec \frac{1}{2}x - 1$ 4π
8. $y = 3 \tan \frac{\pi}{4}x$ 4

For which numbers x, $0 \le x < 2\pi$, are the following functions *not* defined?

9. $\tan x$ $\frac{\pi}{2}, \frac{3\pi}{2}$
10. $\sec x$ $\frac{\pi}{2}, \frac{3\pi}{2}$
11. $\cot x$ $0, \pi$
12. $\csc x$ $0, \pi$

For each description, give the letter of the function that has a graph fitting that description.

13. The graph has $x = \frac{\pi}{2}$ as an asymptote and passes through the origin. e
14. The graph is symmetric in the y-axis and passes through the point $(0, 3)$. c
15. The graph does not intersect either axis. d
16. The graph has the y-axis and the line $x = \frac{\pi}{2}$ as asymptotes. a

a. $y = \cot 2x$
b. $y = \tan \frac{x}{2}$
c. $y = 3 \sec x$
d. $y = 3 \csc x$
e. $y = 2 \tan x$
f. $y = 2 \cot \pi x$

Written Exercises

Use the fact that the cosine is even and the sine is odd to determine if the following are even, odd, or neither.

A **1.** cot x odd **2.** sec x even **3.** csc x odd **4.** tan x + sec x neither

5. sec x − csc x neither **6.** x csc x even **7.** x^2 cot x odd **8.** sec x tan x odd

9–16. Sketch the graph of at least one period of the functions given in Oral Exercises 1–8. **17.** No; the tangent function has no maximum or minimum value.

17. Does the tangent function have an amplitude? Justify your answer.

18. Does the secant function have an amplitude? Justify your answer.
No; the secant function has no maximum or minimum value.

Graph each pair of functions in the same coordinate plane for $-\frac{\pi}{2} < x < \frac{\pi}{2}$. You may wish to verify your graphs on a computer or a graphing calculator.

B **19.** $y = \tan x$; $y = -\tan x$ **20.** $y = \sec x$; $y = -\sec x$

21. Graph $x = \tan y$ for $-\frac{\pi}{2} < y < \frac{\pi}{2}$. **22.** Graph $x = \cot y$ for $0 < y < \pi$.

Graph each of the following over the interval $0 \le x \le 2\pi$. (*Hint:* See Exercises 39–44 on page 629.)

23. $y = \tan\left(x - \frac{\pi}{6}\right)$ **24.** $y = \cot\left(x - \frac{\pi}{4}\right)$ **25.** $y = \csc\left(x + \frac{\pi}{2}\right)$

26. $y = \sec\left(x - \frac{\pi}{4}\right)$ **27.** $y = \cot\left(x - \frac{\pi}{4}\right) + 1$ **28.** $y = \tan\left(x + \frac{\pi}{4}\right) - 1$

C **29.** Graph $y = \sin x$, $y = x$, and $y = \tan x\left(0 < x < \frac{\pi}{2}\right)$ in the same coordinate plane. Explain why the resulting figure illustrates the inequality $\sin x < x < \tan x$. (See also Exercises 30 and 31 on page 617.)

Mixed Review Exercises

Give the exact values of the six trigonometric functions of each number. If a function is not defined for the number, say so.

1. $\frac{5\pi}{2}$ **2.** $-\frac{4\pi}{3}$ **3.** $\frac{19\pi}{6}$ **4.** $-\frac{5\pi}{4}$

Evaluate if $a = -2$ and $b = 4$.

5. $\frac{a-b}{a+b}$ -3 **6.** $\log_b(-a)$ $\frac{1}{2}$ **7.** $\sum_{n=1}^{5} b \cdot a^{n-1}$ 44 **8.** $b^{1/a}$ $\frac{1}{2}$

Using a Computer

The program "Circular Function Quiz" on the Disk for Algebra will draw a random tangent, cotangent, secant, or cosecant curve and let students visually experiment to find the corresponding equation. Note that the type of graph to be drawn is set by using the "Other" option on the program's menu.

Support Materials
Disk for Algebra
Menu Item: Circular Function Quiz
Resource Book p. 249

Additional Answers
Written Exercises

10.

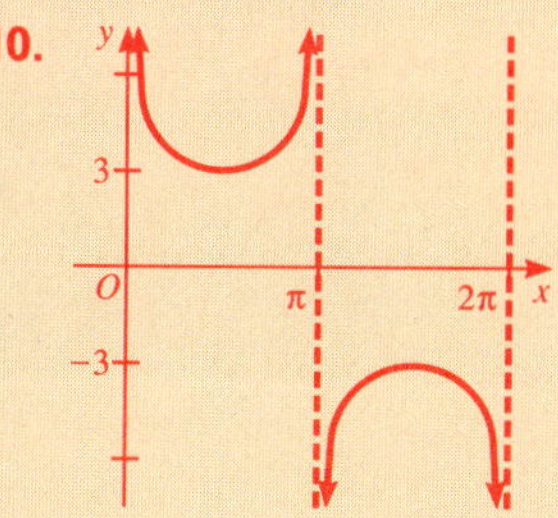

12.

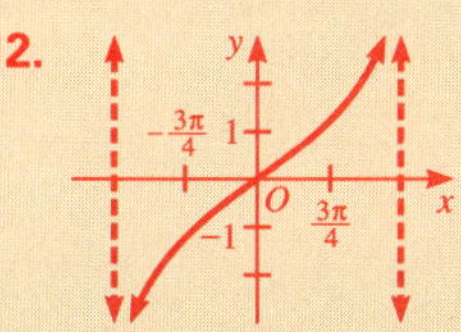

14.

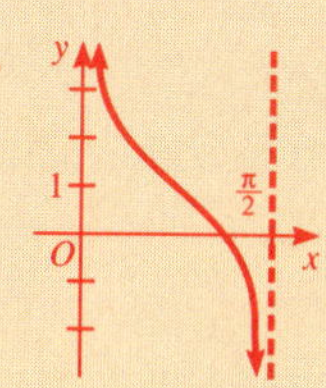

(continued)

Additional Answers Written Exercises
(continued)

16.
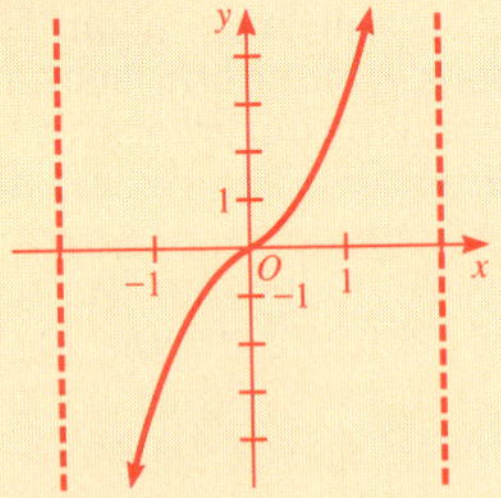

20.
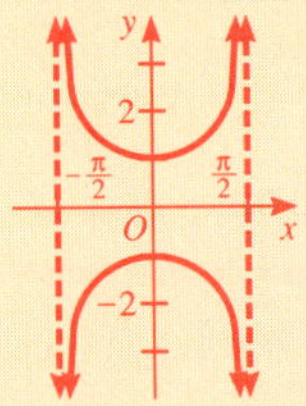

22.
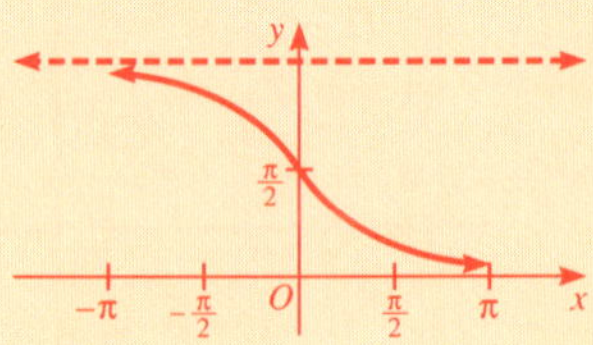

24.
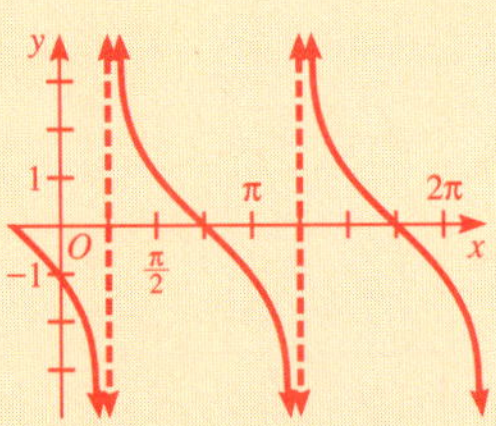

26.
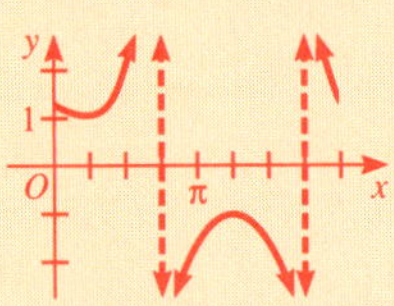

28.
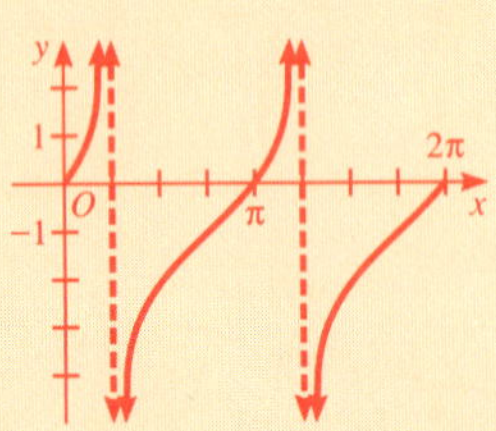

Self-Test 1

Vocabulary

radian measure (p. 607)	periodic, period (p. 619)
arc length (p. 608)	odd function (p. 619)
area of a sector (p. 608)	even function (p. 619)
circular functions (p. 613)	amplitude (p. 625)

1. Express in radians. **Obj. 13-1, p. 607**
  a. 165° (Leave your answer in terms of π.) $\frac{11\pi}{12}$
  b. 256° (Give your answer to the nearest hundredth of a radian.) 4.47

2. Express in degrees. Give your answers to the nearest tenth of a degree.
  a. $-\frac{5\pi}{6}$ $-150°$ **b.** 2.5 143.2°

3. Find the exact values of the six trigonometric functions of $\frac{3\pi}{4}$. **Obj. 13-2, p. 613**

4. Find the sine, cosine, and tangent of 1.44 to four decimal places.

5. Determine whether each function is even, odd, or neither. **Obj. 13-3, p. 619**
  a. $f(x) = x^4 - 3x^2 + 5$ even   **b.** $g(x) = 2x^2 + 6x + 4$ neither

6. Find the amplitude, the maximum and minimum values, and the period of $y = 1 + 3 \sin 2x$. Then sketch the graph, showing at least two periods. 3; 4, -2; π **Obj. 13-4, p. 624**

7. Sketch the graph of $y = \frac{3}{2} \cot \frac{\pi}{2}x$ over the interval $0 < x < 2$. **Obj. 13-5, p. 630**

Check your answers with those at the back of the book.

Application / *Frequencies in Music*

The musical tone produced by a vibrating string, a guitar string for example, can be represented mathematically by a graph. The y-coordinates of the graph represent the displacement of a fixed point in the middle of the string at time t. If you ignore the complicated secondary motions of the string, this graph will be a *sinusoid,* a graph resembling a sine or cosine curve.

The fundamental period of the tone graphed at the top of the next page is $\frac{1}{440}$ s, and its *frequency,* defined as the reciprocal of the period, is therefore 440 Hz (hertz, or cycles per second).

The frequency of a vibrating string is determined by its tension, which the guitarist changes when tuning the instrument, and by its length, which the guitarist controls while playing by pressing the string against one of the metal ridges, or *frets*, on the neck of the guitar.

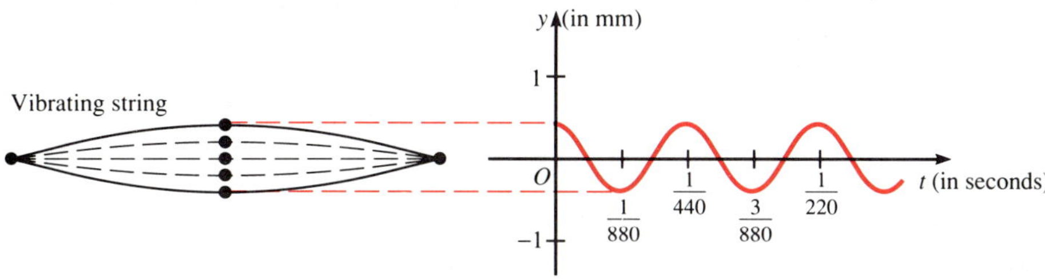

The C major scale beginning at middle C ends on C′, the note one octave above. As the graphs of C and C′ indicate, the fundamental period of C′ is exactly half that of C, and thus its frequency (528 Hz) is exactly twice that of C (264 Hz). The table below shows other relationships among the frequencies of the C major scale. On the left are listed the names of each pair of notes, or *interval*.

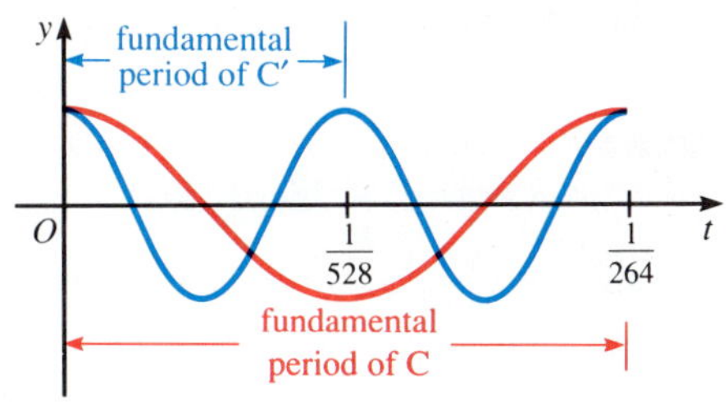

Name of note	C	D	E	F	G	A	B	C′	Ratio of frequencies
Frequency (Hz)	264	297	330	352	396	440	495	528	
Octave									2:1
Fifth									3:2
Fourth									4:3
Major third									5:4
Minor third									6:5

Exercises

1. Suppose a string tuned to F vibrates with an amplitude of 0.5 mm. Write an equation of the form $y = a \cos bt$, where y is in millimeters and t is in seconds, for the motion of a point in the middle of the string. $y = \frac{1}{2} \cos 704\pi t$
2. The frequency of vibration of a string with a given tension is inversely proportional to its length. For example, if you halve the length of the string, its frequency doubles and its pitch goes up by one octave.
   a. A guitar string tuned to G (396 Hz) is shortened to $\frac{3}{4}$ of its original length. What will its frequency be? What note will it play? 528; C′
   b. A guitar string 30 cm long is tuned to E (330 Hz). How long is the vibrating part of the string when the guitarist plays G on this string? 25 cm

Quick Quiz

1. Express 117° in radians (a) in terms of π and (b) to the nearest hundredth of a radian.
   a. $\frac{13\pi}{20}$ **b.** 2.04
2. Express in degrees to the nearest tenth of a degree.
   a. $-\frac{3\pi}{4}$ −135°
   b. 0.5 28.6°
3. Find the exact values of the six trigonometric functions of $-\frac{\pi}{4}$.
   $\sin\left(-\frac{\pi}{4}\right) = -\frac{\sqrt{2}}{2}$;
   $\cos\left(-\frac{\pi}{4}\right) = \frac{\sqrt{2}}{2}$;
   $\tan\left(-\frac{\pi}{4}\right) = -1$;
   $\csc\left(-\frac{\pi}{4}\right) = -\sqrt{2}$;
   $\sec\left(-\frac{\pi}{4}\right) = \sqrt{2}$;
   $\cot\left(-\frac{\pi}{4}\right) = -1$
4. Determine whether the function is odd, even, or neither.
   a. $y = x + \sin x$ odd
   b. $y = x + \cos x$ neither
5. State the amplitude, the maximum and minimum values, and the period of $y = \frac{3}{2} \sin 3x - 1$. Sketch two periods of the graph.
   $\frac{3}{2}; \frac{1}{2}; -\frac{5}{2}; \frac{2\pi}{3}$

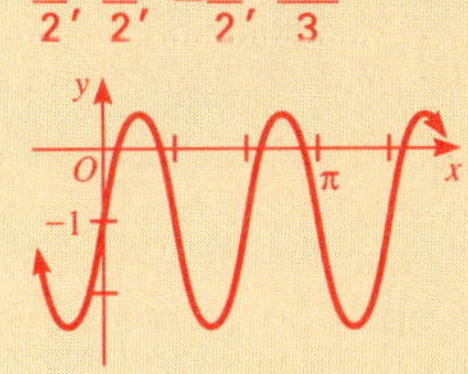

Teaching Suggestions p. T130

Suggested Extensions p. T130

Warm-Up Exercises

Is the equation an identity? (See page 38.) If not, give a counterexample.

1. $(x + 1)^2 = x^2 + 1$
   No; $(2 + 1)^2 = 9$, but $2^2 + 1 = 5$
2. $\ln e^x = x$ Yes
3. $\sqrt{x^2} = x$
   No; $\sqrt{(-1)^2} = 1 \neq -1$
4. $|-2x| = 2|x|$ Yes
5. $1 + 2(1 - x) + x = 3 - x$
   Yes

Motivating the Lesson

Tell each student to look at his or her calculator. How many keys does it have for the trigonometric functions? The answer is probably three, although there are six trigonometric functions. Because of trigonometric identities, the topic of today's lesson, students can still use their calculators to evaluate all six functions.

Chalkboard Examples

1. Simplify $\csc\theta - \cos\theta\cot\theta$.
   $\csc\theta - \cos\theta\cot\theta$
   $= \frac{1}{\sin\theta} - \cos\theta\frac{\cos\theta}{\sin\theta}$
   $= \frac{1 - \cos^2\theta}{\sin\theta}$
   $= \frac{\sin^2\theta}{\sin\theta} = \sin\theta$
2. Write $\sec\theta$ in terms of $\sin\theta$.
   $\sec\theta = \frac{1}{\cos\theta}$
   $\cos\theta = \pm\sqrt{1 - \sin^2\theta}$
   $\sec\theta = \pm\frac{1}{\sqrt{1 - \sin^2\theta}}$

Trigonometric Identities

13-6 The Fundamental Identities

Objective To simplify trigonometric expressions and to prove identities.

You learned earlier that a trigonometric identity is an equation such as

$$\sin^2\alpha + \cos^2\alpha = 1$$

that holds for *all* values of α, except those for which either side of the equation is undefined. This is true of all trigonometric identities. You are already familiar with most of the *fundamental identities* listed below.

The Reciprocal Identities

$$\sin\alpha = \frac{1}{\csc\alpha} \qquad \sin\alpha\csc\alpha = 1 \qquad \csc\alpha = \frac{1}{\sin\alpha}$$

$$\cos\alpha = \frac{1}{\sec\alpha} \qquad \cos\alpha\sec\alpha = 1 \qquad \sec\alpha = \frac{1}{\cos\alpha}$$

$$\tan\alpha = \begin{cases} \dfrac{1}{\cot\alpha} \\ \dfrac{\sin\alpha}{\cos\alpha} \end{cases} \qquad \tan\alpha\cot\alpha = 1 \qquad \cot\alpha = \begin{cases} \dfrac{1}{\tan\alpha} \\ \dfrac{\cos\alpha}{\sin\alpha} \end{cases}$$

The Cofunction Identities

$$\sin\theta = \cos(90° - \theta) \qquad \cos\theta = \sin(90° - \theta)$$

$$\tan\theta = \cot(90° - \theta) \qquad \cot\theta = \tan(90° - \theta)$$

$$\sec\theta = \csc(90° - \theta) \qquad \csc\theta = \sec(90° - \theta)$$

The Pythagorean Identities

$$\sin^2\alpha + \cos^2\alpha = 1 \qquad 1 + \tan^2\alpha = \sec^2\alpha \qquad 1 + \cot^2\alpha = \csc^2\alpha$$

The reciprocal, cofunction, and Pythagorean identities can be used to simplify trigonometric expressions and to prove other identities.

Example 1 Simplify $\cos\alpha + \tan\alpha\sin\alpha$.

Solution Write the functions in terms of sines and cosines. Then simplify.

$$\cos\alpha + \tan\alpha\sin\alpha = \cos\alpha + \frac{\sin\alpha}{\cos\alpha}\sin\alpha = \frac{\cos^2\alpha + \sin^2\alpha}{\cos\alpha} = \frac{1}{\cos\alpha} = \sec\alpha$$

Because $\tan \alpha$ is not defined if $\cos \alpha = 0$, the trigonometric identity $\cos \alpha + \tan \alpha \sin \alpha = \sec \alpha$ is valid only if $\cos \alpha \neq 0$, that is, only if $\alpha \neq 90° + n \cdot 180°$, where n is an integer. In general, such restrictions are understood to apply even though they are not specifically mentioned.

Example 2 Write $\tan x$ in terms of $\sin x$.

Solution You know $\tan x = \frac{\sin x}{\cos x}$. You must now find $\cos x$ in terms of $\sin x$. Use the first Pythagorean identity.

$$\sin^2 x + \cos^2 x = 1$$
$$\cos^2 x = 1 - \sin^2 x$$
$$\cos x = \pm\sqrt{1 - \sin^2 x}$$

$$\therefore \tan x = \frac{\sin x}{\cos x} = \pm\frac{\sin x}{\sqrt{1 - \sin^2 x}}.$$ ***Answer***

The value of x determines which sign is used in the answer to Example 2. If x is the measure of a first- or fourth-quadrant angle, $\sin x$ and $\tan x$ have the same sign and the + is used. If x is the measure of a second- or third-quadrant angle, $\sin x$ and $\tan x$ have opposite signs and the − is used.

You can usually use one of two general strategies in proving identities.

General Strategies for Proving Identities

1. Simplify the more complicated side of the identity until it is identical to the other side.
2. Transform both sides of the identity into the same expression.

Caution: It is incorrect to "work across the = sign." For example, do not cross multiply. To do so assumes that what is being proved is true.

Here are some special strategies that can be helpful in proving identities.

Special Strategies for Proving Identities

1. Express functions in terms of sines and cosines (see Example 1).
2. Look for expressions to which the Pythagorean identities can be applied. For example: $\tan^2 x + 1$ (= $\sec^2 x$), and $1 - \sin^2 x$ (= $\cos^2 x$).
3. Use factoring. For example: $\sin^2 x = 1 - \cos^2 x = (1 + \cos x)(1 - \cos x)$.
4. Combine terms on each side of the identity into a single fraction.
5. Multiply one side of the equation by an expression equal to 1 (see Example 4).

3. Prove that $\tan^2 \theta + \cos^2 \theta + \frac{1}{\csc^2 \theta} = \sec^2 \theta$.

$$\tan^2 \theta + \cos^2 \theta + \frac{1}{\csc^2 \theta} = \tan^2 \theta + \cos^2 \theta + \sin^2 \theta = \tan^2 \theta + 1 = \sec^2 \theta$$

4. Prove that $\frac{1 + \cot^2 \theta}{\sec^2 \theta} = \cot^2 \theta$.

$$\frac{1 + \cot^2 \theta}{\sec^2 \theta} = \frac{\csc^2 \theta}{\sec^2 \theta} = \frac{\frac{1}{\sin^2 \theta}}{\frac{1}{\cos^2 \theta}} = \frac{\cos^2 \theta}{\sin^2 \theta} = \cot^2 \theta$$

5. Prove that $\sec \beta \csc \beta = \tan \beta + \cot \beta$.

$$\sec \beta \csc \beta = \frac{1}{\cos \beta} \cdot \frac{1}{\sin \beta} = \frac{1}{\cos \beta \sin \beta}$$
$$\tan \beta + \cot \beta = \frac{\sin \beta}{\cos \beta} + \frac{\cos \beta}{\sin \beta} = \frac{\sin^2 \beta + \cos^2 \beta}{\cos \beta \sin \beta} = \frac{1}{\cos \beta \sin \beta}$$

Check for Understanding

Here is a suggested use of the Oral Exercises to check students' understanding as you teach the lesson.

Oral Exs. 1–9: use after Example 1.
Oral Exs. 10–13: use after Example 2.
Oral Ex. 14: use after Example 5.

Guided Practice

Simplify.

1. $\cos\theta \csc\theta \tan\theta$ 1
2. $\frac{1}{\cot^2\theta} - \frac{1}{\cos^2\theta}$ -1
3. $\frac{1-\cos^2\alpha}{\sin\alpha\cos\alpha}$ $\tan\alpha$
4. $\csc(t^2) - \cos(t^2)\cot(t^2)$ $\sin(t^2)$
5. Write $\cot x$ in terms of $\sin x$. $\cot x = \pm\frac{\sqrt{1-\sin^2 x}}{\sin x}$
6. Prove: $\cot^2\phi - \cos^2\phi = \cot^2\phi\cos^2\phi$. $\cot^2\phi\cos^2\phi = (\csc^2\phi - 1)\cos^2\phi = \frac{\cos^2\phi}{\sin^2\phi} - \cos^2\phi = \cot^2\phi - \cos^2\phi$

Summarizing the Lesson

In this lesson students learned to simplify trigonometric expressions and to prove identities. Ask students to state some of the strategies that can be helpful in proving identities.

Example 3 Prove that $\tan^2\alpha + \csc^2\alpha - \cot^2\alpha = \sec^2\alpha$.

Solution Use general strategy 1.

$\tan^2\alpha + \csc^2\alpha - \cot^2\alpha \stackrel{?}{=} \sec^2\alpha$

$\tan^2\alpha + \frac{1}{\sin^2\alpha} - \frac{\cos^2\alpha}{\sin^2\alpha}$ Use the reciprocal identities.

$\tan^2\alpha + \frac{1-\cos^2\alpha}{\sin^2\alpha}$ Use special strategy 4.

$\tan^2\alpha + \frac{\sin^2\alpha}{\sin^2\alpha}$ Use special strategy 2.

$\tan^2\alpha + 1$ Simplify.

$\sec^2\alpha = \sec^2\alpha$ Use special strategy 2.

Example 4 Prove that $\frac{1-\cos t}{\sin t} = \frac{\sin t}{1+\cos t}$.

Solution Remember, you cannot cross multiply. Use special strategy 5 and multiply the left side by 1 in the form $\frac{1+\cos t}{1+\cos t}$.

$\frac{1-\cos t}{\sin t} \stackrel{?}{=} \frac{\sin t}{1+\cos t}$

$\frac{(1-\cos t)(1+\cos t)}{\sin t(1+\cos t)}$ Use special strategy 5.

$\frac{1-\cos^2 t}{\sin t(1+\cos t)}$ Use FOIL.

$\frac{\sin^2 t}{\sin t(1+\cos t)}$ Use special strategy 2.

$\frac{\sin t}{1+\cos t} = \frac{\sin t}{1+\cos t}$ Simplify.

Example 5 Prove that $(\csc x - 1)(\sin x + 1) = \cos x \cot x$.

Solution Use general strategy 2.

$(\csc x - 1)(\sin x + 1) \stackrel{?}{=} \cos x \cot x$

$\csc x \sin x + \csc x - \sin x - 1 \quad | \quad \cos x \cdot \frac{\cos x}{\sin x}$

$1 + \csc x - \sin x - 1 \quad | \quad \frac{\cos^2 x}{\sin x}$

$1 + \frac{1}{\sin x} - \sin x - 1 \quad | \quad \frac{1-\sin^2 x}{\sin x}$

$\frac{1}{\sin x} - \sin x = \frac{1}{\sin x} - \sin x$

Since $(\csc x - 1)(\sin x + 1) = \frac{1}{\sin x} - \sin x$ and $\cos x \cot x = \frac{1}{\sin x} - \sin x$, you have proved the identity.

There are often several ways of proving an identity. For instance, look back to the first step in Example 3. There you could use a Pythagorean identity and replace $\csc^2 \alpha - \cot^2 \alpha$ by the number 1, since $1 + \cot^2 \alpha = \csc^2 \alpha$. Then you would immediately obtain $\tan^2 \alpha + 1 = \sec^2 \alpha$.

Suggested Assignments

Average Alg. and Trig.
Day 1: 639/1–37 odd
R 634/Self-Test 1
Day 2: 640/26–56 even

Extended Alg. and Trig.
Day 1: 639/1–47 odd
R 634/Self-Test 1
Day 2: 640/24–56 even
S 629/47, 48

Supplementary Materials

Study Guide	pp. 215–216
Resource Book	p. 160

Oral Exercises

Simplify.

1. $1 - \cos^2 \phi$ $\sin^2 \phi$
2. $\csc^2 \theta - 1$ $\cot^2 \theta$
3. $\sin^2 \alpha - 1$ $-\cos^2 \alpha$
4. $1 - \sec^2 x$ $-\tan^2 x$
5. $\tan \alpha \cos \alpha$ $\sin \alpha$
6. $\cot \theta \sin \theta$ $\cos \theta$
7. $\frac{\cos x}{\sec x}$ $\cos^2 x$
8. $\frac{\csc x}{\sin x}$ $\csc^2 x$
9. $\frac{\cot x}{\tan x}$ $\cot^2 x$

Express the first function in terms of the second.

10. $\sin t$; $\cos t$ $\pm\sqrt{1 - \cos^2 t}$
11. $\sec \theta$; $\tan \theta$ $\pm\sqrt{1 + \tan^2 \theta}$
12. $\tan x$; $\sec x$ $\pm\sqrt{\sec^2 x - 1}$
13. $\cos x$; $\sin x$ $\pm\sqrt{1 - \sin^2 x}$
14. Explain why it is incorrect to prove the following identity by dividing both sides of the equation by $\cos x$. To do so would be "working across the = sign."

$$\frac{\cos x}{\sin^2 x} = \cos x \csc^2 x$$

Written Exercises

Simplify.

A

1. $\sin \alpha \sec \alpha$ $\tan \alpha$
2. $\cos \theta \csc \theta$ $\cot \theta$
3. $\frac{\tan x}{\sin x}$ $\sec x$
4. $\frac{\cos t}{\cot t}$ $\sin t$
5. $(1 - \sin \theta)(1 + \sin \theta)$ $\cos^2 \theta$
6. $(\sec \phi - 1)(\sec \phi + 1)$ $\tan^2 \phi$
7. $\cos \alpha \csc \alpha \tan \alpha$ 1
8. $\cot t \sec t \sin t$ 1
9. $\frac{\sin x}{\csc x} + \frac{\cos x}{\sec x}$ 1
10. $\frac{\csc x}{\sin x} - \frac{\cot x}{\tan x}$ 1
11. $\frac{\sec \theta}{\cos \theta} - 1$ $\tan^2 \theta$
12. $\frac{\sec t}{\cos t} - \sec t \cos t$ $\tan^2 t$
13. $\frac{1 + \tan^2 \alpha}{\tan^2 \alpha}$ $\csc^2 \alpha$
14. $\frac{1 + \tan^2 \theta}{1 + \cot^2 \theta}$ $\tan^2 \theta$
15. $\cos^2 50° + \sin^2 50°$ 1
16. $1 + \tan^2 50°$ $\sec^2 50°$
17. $\csc^2 t - \cot^2 t$ 1
18. $\sec^2 350° - \tan^2 350°$ 1

Additional Answers
Written Exercises

See Solution Key for proofs.

19. $\dfrac{\sin\alpha\cos\alpha}{1-\sin^2\alpha}$ $\tan\alpha$

20. $\dfrac{1-\cos^2\theta}{\sin\theta\cos\theta}$ $\tan\theta$

21. $\sec x - \sin x\tan x$ $\cos x$

22. $\cos^2 t(\cot^2 t + 1)$ $\cot^2 t$

23. $\dfrac{1-\sin^2\phi}{1-\sin\phi} - 1$ $\sin\phi$

24. $\dfrac{\sec^2\alpha - 1}{\sec\alpha + 1} + 1$ $\sec\alpha$

Write the first function in terms of the second.

25. $\sec x$; $\sin x$ $\pm\dfrac{1}{\sqrt{1-\sin^2 x}}$

26. $\csc x$; $\cos x$ $\pm\dfrac{1}{\sqrt{1-\cos^2 x}}$

27. $\tan t$; $\sin t$ $\pm\dfrac{\sin t}{\sqrt{1-\sin^2 t}}$

28. $\tan t$; $\cos t$ $\pm\dfrac{\sqrt{1-\cos^2 t}}{\cos t}$

29. $\tan\alpha\sec\alpha$; $\sin\alpha$ $\dfrac{\sin\alpha}{1-\sin^2\alpha}$

30. $\tan\alpha\sec\alpha$; $\cos\alpha$ $\pm\dfrac{\sqrt{1-\cos^2\alpha}}{\cos^2\alpha}$

Prove each identity.

31. $\sin^2 x(1+\cot^2 x) = 1$

32. $\sin^2\alpha(\csc^2\alpha + \sec^2\alpha) = \sec^2\alpha$

33. $\cos^2\theta - \sin^2\theta = 1 - 2\sin^2\theta$

34. $\cos^2 t - \sin^2 t = 2\cos^2 t - 1$

35. $\cot\phi + \tan\phi = \csc\phi\sec\phi$

36. $\sec^2\alpha + \csc^2\alpha = \sec^2\alpha\csc^2\alpha$

37. $\cos^4 x - \sin^4 x = \cos^2 x - \sin^2 x$

38. $\cos^2\theta\tan^2\theta + \sin^2\theta\tan^2\theta + 1 = \sec^2\theta$

Simplify.

B 39. $\dfrac{\tan t + \cot t}{\sec^2 t}$ $\cot t$

40. $\dfrac{\sec\alpha - \cos\alpha}{\tan^2\alpha}$ $\cos\alpha$

41. $\cot\theta(\cos\theta\tan\theta + \sin\theta)$ $2\cos\theta$

42. $\sin\phi + \cos\phi\cot\phi$ $\csc\phi$

43. $\dfrac{\tan^2 x}{\sec x + 1} + 1$ $\sec x$

44. $\dfrac{\sec t + \csc t}{1+\tan t}$ $\csc t$

45. $\dfrac{\tan\theta}{1+\sec\theta} + \dfrac{1+\sec\theta}{\tan\theta}$ $2\csc\theta$

46. $\dfrac{\sin\alpha}{1+\cos\alpha} + \dfrac{1+\cos\alpha}{\sin\alpha}$ $2\csc\alpha$

47. $(\cos t + \sin t)^2 + (\cos t - \sin t)^2$ 2

48. $(1+\tan x)^2 + (1-\tan x)^2$ $2\sec^2 x$

Prove each identity.

49. $\dfrac{1+\cos x}{1-\cos x} = \dfrac{\sec x + 1}{\sec x - 1}$

50. $\dfrac{\csc\phi}{\csc\phi - 1} + \dfrac{\csc\phi}{\csc\phi + 1} = 2\sec^2\phi$

51. $(\cot\alpha + \tan\alpha)^2 = \csc^2\alpha\sec^2\alpha$

52. $\sec t + \csc t = (\tan t + \cot t)(\cos t + \sin t)$

53. $\dfrac{\sin x}{1+\cos x} = \csc x - \cot x$

54. $\dfrac{\sec y - \tan y}{1-\sin y} = \sec y$

55. $\dfrac{\cos t}{1-\sin t} = \dfrac{1}{\sec t - \tan t}$

56. $\dfrac{\sec\theta - 1}{\tan\theta} = \dfrac{\tan\theta}{\sec\theta + 1}$

C 57. $\dfrac{\sin^2\phi + 2\cos\phi - 1}{\sin^2\phi + 3\cos\phi - 3} = \dfrac{1}{1-\sec\phi}$

58. $\sqrt{\dfrac{1-\cos x}{1+\cos x}} = |\csc x - \cot x|$

13-7 Trigonometric Addition Formulas

Objective To use formulas for the sine and cosine of a sum or difference.

If you were asked to evaluate $\sin 75°$, you could use a calculator or tables to find an approximate answer. To four decimal places, $\sin 75° = 0.9659$. But an exact answer is sometimes needed. Notice that $\sin 75°$ can be expressed as follows:

$$\sin 75° = \sin (45° + 30°).$$

Therefore, if $\sin (45° + 30°)$ can be evaluated in terms of trigonometric functions of 45° and 30°, you can find an exact expression for $\sin 75°$.

The following identities are called *trigonometric addition formulas*.

Addition Formulas for the Sine and Cosine

$$\sin (\alpha + \beta) = \sin \alpha \cos \beta + \cos \alpha \sin \beta$$
$$\sin (\alpha - \beta) = \sin \alpha \cos \beta - \cos \alpha \sin \beta$$
$$\cos (\alpha + \beta) = \cos \alpha \cos \beta - \sin \alpha \sin \beta$$
$$\cos (\alpha - \beta) = \cos \alpha \cos \beta + \sin \alpha \sin \beta$$

Caution: $\sin (\alpha + \beta) \neq \sin \alpha + \sin \beta$. Here is a counterexample:

$$\sin 45° + \sin 30° = \frac{\sqrt{2}}{2} + \frac{1}{2} = 1.207,$$

but

$$\sin (45° + 30°) = \sin 75° = 0.9659.$$

Example 1 Find the exact value of each of the following.

a. $\sin 75°$ **b.** $\cos 195°$

Solution

a.
$$\begin{aligned}\sin 75° &= \sin (45° + 30°)\\ &= \sin 45° \cos 30° + \cos 45° \sin 30°\\ &= \frac{\sqrt{2}}{2} \cdot \frac{\sqrt{3}}{2} + \frac{\sqrt{2}}{2} \cdot \frac{1}{2}\\ &= \frac{\sqrt{6} + \sqrt{2}}{4} \quad \textit{Answer}\end{aligned}$$

b.
$$\begin{aligned}\cos 195° &= \cos (240° - 45°)\\ &= \cos 240° \cos 45° + \sin 240° \sin 45°\\ &= \left(-\frac{1}{2}\right) \cdot \frac{\sqrt{2}}{2} + \left(-\frac{\sqrt{3}}{2}\right) \cdot \frac{\sqrt{2}}{2}\\ &= \frac{-\sqrt{2} - \sqrt{6}}{4}, \text{ or } -\frac{\sqrt{2} + \sqrt{6}}{4} \quad \textit{Answer}\end{aligned}$$

Teaching Suggestions p. T130

Suggested Extensions p. T130

Group Activities p. T131

Warm-Up Exercises

Find each function value.

1. $\cos 90°$ 0
2. $\sin (-45°)$ $-\frac{\sqrt{2}}{2}$
3. $\cos (-180°)$ -1
4. $\sin 510°$ $\frac{1}{2}$
5. $\cos 330°$ $\frac{\sqrt{3}}{2}$
6. $\sin (-270°)$ 1

Motivating the Lesson

Students already know the exact values for sine and cosine for angles that are multiples of 30° or 45°. Tell students that the exact value for, say, sin 75° can also be determined. Such a value is found using the trigonometric addition formulas, the topic of today's lesson.

Thinking Skills

The trigonometric addition formulas should be *memorized.* If students will take a few minutes, however, to examine the boxed formulas on this page, they will probably notice *patterns* that will aid their *recall* of the formulas.

Chalkboard Examples

1. Find the exact value of each of the following.
   a. $\cos 75°$
   $\cos 75° =$
   $\cos (45° + 30°) =$
   $\cos 45° \cos 30° -$
   $\sin 45° \sin 30° =$
   $\frac{\sqrt{2}}{2} \cdot \frac{\sqrt{3}}{2} - \frac{\sqrt{2}}{2} \cdot \frac{1}{2} =$
   $\frac{\sqrt{6} - \sqrt{2}}{4}$
   b. $\sin 105°$
   $\sin 105° =$
   $\sin (60° + 45°) =$
   $\sin 60° \cos 45° +$
   $\cos 60° \sin 45° =$
   $\frac{\sqrt{3}}{2} \cdot \frac{\sqrt{2}}{2} + \frac{1}{2} \cdot \frac{\sqrt{2}}{2} =$
   $\frac{\sqrt{6} + \sqrt{2}}{4}$
2. Prove the identity $\sin\left(\frac{\pi}{2} - x\right) = \cos x$.
   $\sin\left(\frac{\pi}{2} - x\right) = \sin \frac{\pi}{2} \cos x -$
   $\cos \frac{\pi}{2} \sin x = 1 \cdot \cos x -$
   $0 \cdot \sin x = \cos x$

Check for Understanding

Here is a suggested use of the Oral Exercises to check students' understanding as you teach the lesson.
Oral Exs. 1–10: use before Example 1.
Oral Exs. 11–16: use after Example 1.

Guided Practice

Find the exact value.

1. $\sin 165°$ $\frac{\sqrt{6} - \sqrt{2}}{4}$
2. $\cos 255°$ $\frac{\sqrt{2} - \sqrt{6}}{4}$

Example 2 Prove the identity $\cos\left(\frac{3\pi}{2} - \theta\right) = -\sin \theta$.

Solution
$$\cos\left(\frac{3\pi}{2} - \theta\right) = \cos \frac{3\pi}{2} \cos \theta + \sin \frac{3\pi}{2} \sin \theta$$
$$= 0 \cdot \cos \theta + (-1) \sin \theta$$
$$= -\sin \theta \quad \checkmark$$

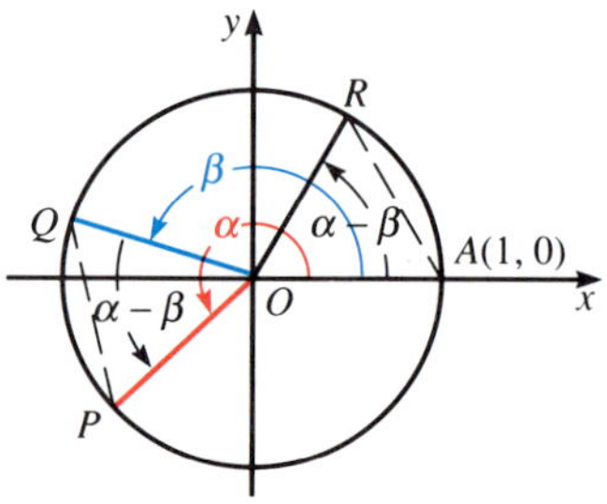

To prove the formula for $\cos (\alpha - \beta)$, place the angles α, β, and $\alpha - \beta$ in standard position and let P, Q, and R be, respectively, the points where their terminal sides intersect the unit circle. The coordinates of the points shown in the diagram are

$P(\cos \alpha, \sin \alpha)$, $Q(\cos \beta, \sin \beta)$,
$R(\cos (\alpha - \beta), \sin (\alpha - \beta))$, and $A(1, 0)$.

Since $\overparen{QP}$ and $\overparen{AR}$ both have central angles of measure $\alpha - \beta$, they are congruent. Therefore chords $\overline{QP}$ and $\overline{AR}$ are congruent and $(QP)^2 = (AR)^2$. Now use the distance formula (page 402) and simplify the result using the Pythagorean identity $\sin^2 \theta + \cos^2 \theta = 1$.

$$(QP)^2 = (\cos \alpha - \cos \beta)^2 + (\sin \alpha - \sin \beta)^2$$
$$= \cos^2 \alpha - 2 \cos \alpha \cos \beta + \cos^2 \beta + \sin^2 \alpha - 2 \sin \alpha \sin \beta + \sin^2 \beta$$
$$= 2 - 2(\cos \alpha \cos \beta + \sin \alpha \sin \beta)$$
$$(AR)^2 = [\cos (\alpha - \beta) - 1]^2 + [\sin (\alpha - \beta) - 0]^2$$
$$= \cos^2 (\alpha - \beta) - 2 \cos (\alpha - \beta) + 1 + \sin^2 (\alpha - \beta)$$
$$= 2 - 2 \cos (\alpha - \beta)$$

Since $(QP)^2 = (AR)^2$,
$$2 - 2(\cos \alpha \cos \beta + \sin \alpha \sin \beta) = 2 - 2 \cos (\alpha - \beta),$$
or
$$\cos (\alpha - \beta) = \cos \alpha \cos \beta + \sin \alpha \sin \beta.$$

To derive the formula for $\cos (\alpha + \beta)$, use the formula for $\cos (\alpha - \beta)$ and the fact that $\cos (-\beta) = \cos \beta$ and $\sin (-\beta) = -\sin \beta$.
$$\cos (\alpha + \beta) = \cos [\alpha - (-\beta)]$$
$$= \cos \alpha \cos (-\beta) + \sin \alpha \sin (-\beta)$$
$$= \cos \alpha \cos \beta - \sin \alpha \sin \beta$$

To prove the formula for $\sin (\alpha + \beta)$, use these cofunction identities:
$$\cos (90° - \theta) = \sin \theta \quad \text{and} \quad \sin (90° - \phi) = \cos \phi.$$
The proof is obtained by writing $\sin (\alpha + \beta)$ as a cosine and simplifying.
$$\sin (\alpha + \beta) = \cos [90° - (\alpha + \beta)] = \cos [(90° - \alpha) - \beta]$$
$$= \cos (90° - \alpha) \cos \beta + \sin (90° - \alpha) \sin \beta$$
$$= \sin \alpha \cos \beta + \cos \alpha \sin \beta.$$

The proof of the formula for $\sin (\alpha - \beta)$ is left as Exercise 33.

Oral Exercises

Use a trigonometric addition formula to expand each expression. Do not evaluate.

Sample 1 $\cos(45° - 30°) = \cos 45° \cos 30° + \sin 45° \sin 30°$

1. $\cos(60° - 45°)$
2. $\sin(30° + 45°)$
3. $\sin(45° - 30°)$
4. $\cos(60° + 45°)$
5. $\cos\left(\frac{\pi}{4} + \frac{\pi}{3}\right)$
6. $\sin\left(\frac{\pi}{2} - \frac{\pi}{6}\right)$

Give each angle as a sum or difference of multiples of 30°, 45°, or 60°. Answers may vary.

7. 105° $60° + 45°$
8. 15° $45° - 30°$
9. 165° $120° + 45°$
10. 285° $240° + 45°$

Simplify to a trigonometric function of a single angle. Then give the exact value of the function.

Sample 2 $\cos 100° \cos 40° + \sin 100° \sin 40° = \cos(100° - 40°) = \cos 60° = \frac{1}{2}$

11. $\cos 20° \cos 70° - \sin 20° \sin 70°$ $\cos 90° = 0$
12. $\sin 100° \cos 10° - \cos 100° \sin 10°$ $\sin 90° = 1$
13. $\sin \frac{3\pi}{4} \cos \frac{\pi}{4} + \cos \frac{3\pi}{4} \sin \frac{\pi}{4}$ $\sin \pi = 0$
14. $\cos \frac{\pi}{2} \cos \frac{\pi}{6} + \sin \frac{\pi}{2} \sin \frac{\pi}{6}$ $\cos \frac{\pi}{3} = \frac{1}{2}$

Simplify to a trigonometric function of a single angle.

15. $\cos 2\theta \cos \theta + \sin 2\theta \sin \theta$ $\cos \theta$
16. $\sin 3\phi \cos 2\phi - \cos 3\phi \sin 2\phi$ $\sin \phi$

Written Exercises

Find the exact value of each of the following.

A

1. $\cos 105°$ $\frac{\sqrt{2} - \sqrt{6}}{4}$
2. $\sin 15°$ $\frac{\sqrt{6} - \sqrt{2}}{4}$
3. $\sin 165°$ $\frac{\sqrt{6} - \sqrt{2}}{4}$
4. $\cos 285°$ $\frac{\sqrt{6} - \sqrt{2}}{4}$
5. $\cos 195°$ $-\frac{\sqrt{6} + \sqrt{2}}{4}$
6. $\sin 255°$ $-\frac{\sqrt{6} + \sqrt{2}}{4}$

Simplify to a trigonometric function of a single angle. Then give the exact value if possible. (See Sample 2 above.) 8. $\sin 45°; \frac{\sqrt{2}}{2}$ 9. $\sin 210°; -\frac{1}{2}$ 10. $\cos(-90°); 0$

7. $\cos 20° \cos 40° - \sin 20° \sin 40°$ $\cos 60°; \frac{1}{2}$
8. $\sin 50° \cos 5° - \cos 50° \sin 5°$
9. $\sin 130° \cos 80° + \cos 130° \sin 80°$
10. $\cos 50° \cos 140° + \sin 50° \sin 140°$
11. $\cos \frac{\pi}{4} \cos \frac{\pi}{12} + \sin \frac{\pi}{4} \sin \frac{\pi}{12}$ $\cos \frac{\pi}{6}; \frac{\sqrt{3}}{2}$
12. $\sin \frac{\pi}{3} \cos \frac{\pi}{12} - \cos \frac{\pi}{3} \sin \frac{\pi}{12}$ $\sin \frac{\pi}{4}; \frac{\sqrt{2}}{2}$
13. $\sin 3\theta \cos \theta - \cos 3\theta \sin \theta$ $\sin 2\theta$
14. $\cos 2\phi \cos \phi - \sin 2\phi \sin \phi$ $\cos 3\phi$

3. $\sin 160° \cos 20° + \cos 160° \sin 20°$ 0
4. $\sin \frac{\pi}{10} \sin \frac{\pi}{10} + \cos \frac{\pi}{10} \cos \frac{\pi}{10}$ 1
5. $\cos \frac{3\pi}{4} \cos \frac{\pi}{12} - \sin \frac{3\pi}{4} \sin \frac{\pi}{12}$ $-\frac{\sqrt{3}}{2}$

Summarizing the Lesson

In this lesson students learned to use the sum and difference formulas for sine and cosine. Ask students to state each formula.

Suggested Assignments

Average Alg. and Trig.
643/1–29 odd
S 645/Mixed Review

Extended Alg. and Trig.
643/1–33 odd
S 645/Mixed Review

Supplementary Materials

Study Guide	pp. 217–218
Practice Master	76
Overhead Visual	10

Additional Answers
Oral Exercises

1. $\cos 60° \cos 45° + \sin 60° \sin 45°$
2. $\sin 30° \cos 45° + \cos 30° \sin 45°$
3. $\sin 45° \cos 30° - \cos 45° \sin 30°$
4. $\cos 60° \cos 45° - \sin 60° \sin 45°$
5. $\cos \frac{\pi}{4} \cos \frac{\pi}{3} - \sin \frac{\pi}{4} \sin \frac{\pi}{3}$
6. $\sin \frac{\pi}{2} \cos \frac{\pi}{6} - \cos \frac{\pi}{2} \sin \frac{\pi}{6}$

Additional Answers
Written Exercises
See Solution Key for proofs.

Prove each identity.

15. $\sin\left(\frac{\pi}{2} + \theta\right) = \cos\theta$

16. $\cos\left(\frac{\pi}{2} - \theta\right) = \sin\theta$

17. $\cos(\pi - x) = -\cos x$

18. $\sin(\pi + x) = -\sin x$

19. $\sin\left(\frac{3\pi}{2} - x\right) = -\cos x$

20. $\cos\left(\frac{3\pi}{2} + x\right) = \sin x$

Simplify to a trigonometric function of a single angle.

B 21. $\sin\left(\frac{\pi}{6} + \theta\right) + \sin\left(\frac{\pi}{6} - \theta\right)$ $\cos\theta$

22. $\cos\left(\frac{\pi}{3} + \theta\right) + \cos\left(\frac{\pi}{3} - \theta\right)$ $\cos\theta$

23. $\cos(x - y)\cos y - \sin(x - y)\sin y$ $\cos x$

24. $\sin(x + y)\cos y - \cos(x + y)\sin y$ $\sin x$

Prove each identity.

25. $\frac{\sin(\alpha + \beta)}{\cos\alpha\cos\beta} = \tan\alpha + \tan\beta$

26. $\frac{\sin(\alpha + \beta)}{\sin\alpha\sin\beta} = \cot\alpha + \cot\beta$

27. $\cos\theta\cos\phi(\tan\theta + \tan\phi) = \sin(\theta + \phi)$

28. $\sin\theta\sin\phi(\cot\theta\cot\phi - 1) = \cos(\theta + \phi)$

29. $2\sin\alpha\cos\beta = \sin(\alpha - \beta) + \sin(\alpha + \beta)$

30. $2\cos\alpha\cos\beta = \cos(\alpha - \beta) + \cos(\alpha + \beta)$

In Exercises 31 and 32, $\sin\alpha = -\frac{3}{5}$, $\cos\beta = \frac{8}{17}$, and α and β are third- and fourth-quadrant angles, respectively. (*Hint:* First find $\cos\alpha$ and $\sin\beta$.)

31. Find **(a)** $\sin(\alpha + \beta)$, **(b)** $\cos(\alpha + \beta)$, and **(c)** the quadrant of $\alpha + \beta$. $\frac{36}{85}$; $-\frac{77}{85}$; II

32. Find **(a)** $\sin(\alpha - \beta)$, **(b)** $\cos(\alpha - \beta)$, and **(c)** the quadrant of $\alpha - \beta$. $-\frac{84}{85}$; $\frac{13}{85}$; IV

33. Show that $\sin(\alpha - \beta) = \sin\alpha\cos\beta - \cos\alpha\sin\beta$.
[*Hint:* $\alpha - \beta = \alpha + (-\beta)$.]

34. Use a trigonometric addition formula to derive a formula for $\sin 2\theta$.

35. Prove that $\tan(\pi - \alpha) = -\tan\alpha$. $\left(\textit{Hint: } \tan(\pi - \alpha) = \frac{\sin(\pi - \alpha)}{\cos(\pi - \alpha)}\right)$.

C 36. Show that in the figure at the left below, $h = \frac{d\sin\alpha\sin\beta}{\sin(\alpha + \beta)}$.

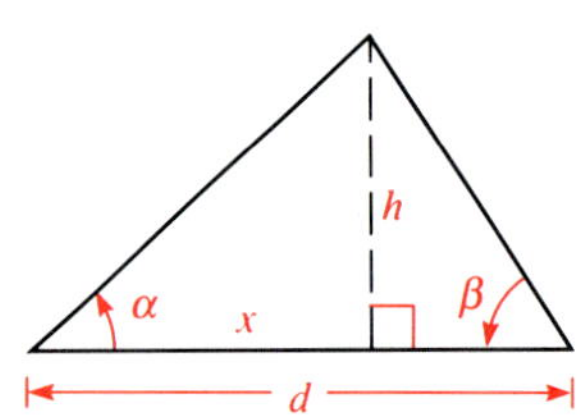

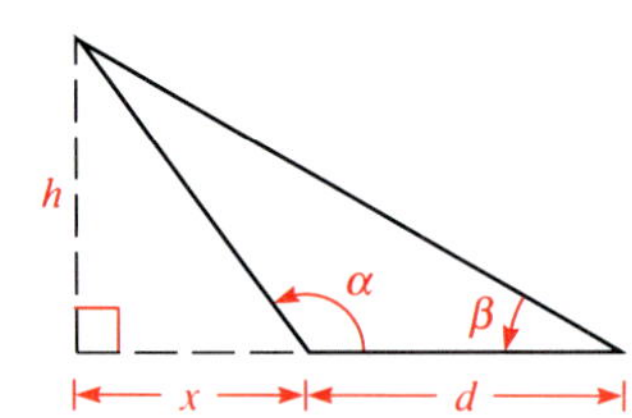

37. Show that in the figure at the right above, $h = \frac{d\sin\alpha\sin\beta}{\sin(\alpha + \beta)}$.
(*Hint:* See Exercise 36.)

Mixed Review Exercises

Graph each function. Show at least two periods.

1. $y = \frac{1}{2} \sin 2x$
2. $y = \tan \frac{\pi}{2}x$
3. $y = 3 \cos \pi x$

Simplify.

4. $\sin^2 x(1 + \cot^2 x)$ 1
5. $x^6(x^{n-2})^3$ x^{3n}
6. $\sqrt[3]{16x^6}$ $2x^2\sqrt[3]{2}$
7. $\dfrac{\log_2 x}{\log_2 \frac{1}{x}}$ -1
8. $\dfrac{1 + \sqrt{3}}{1 - \sqrt{3}}$ $-2 - \sqrt{3}$
9. $\sqrt{\dfrac{1 - \left(-\frac{5}{13}\right)}{2}}$ $\dfrac{3\sqrt{13}}{13}$
10. $\left(\dfrac{x^{2/3}}{x^{1/2}}\right)^6$ x
11. $\dfrac{1}{x} + \dfrac{1}{x^2 - x}$ $\dfrac{1}{x - 1}$
12. $\csc x - \cos x \cot x$ $\sin x$

Computer Exercises

For students with some programming experience.

1. Using only the fact that $\sin 5° = 0.087155743$ and the addition formula for sine on page 641, write a program that will display a table of degree measures and their sines in 5° increments for angles from 5° to 90°. Note that $\sin 10° = \sin (5° + 5°) = \sin 5° \cos 5° + \cos 5° \sin 5°$. Do *not* use the computer's SIN (X) or COS (X) built-in functions. To find cosines use the identity
$$\cos \theta = \sqrt{1 - \sin^2 \theta},\ 0° \le \theta \le 90°.$$
Since the computer will produce sine values having 8 or 9 digits, you can improve the appearance of the table by rounding these values to the nearest ten thousandth using the computer's greatest-integer function INT:
```
100 LET X=INT(10000*X+0.5)/10000
```
Note that, without the 0.5 being added before INT is applied, the computer would merely truncate, or cut off, the value of X after four decimal places.
2. Modify the program in Exercise 1 to include cosine and tangent values. For the tangent, use the fact that
$$\tan \theta = \frac{\sin \theta}{\cos \theta},\ 0° \le \theta < 90°.$$
Run this program to display a table of sines, cosines, and tangents for angles from 5° to 85°. Compare your values with those in Table 6.
3. Given the *sine* of an angle, write a program that will compute the sine of any given multiple of that angle that is less than 90°. For example, if $\sin \theta = 0.224935$ is entered, the program should be able to find, say, $\sin 6\theta$. Again, do *not* use any of the computer's built-in functions.
4. Run the program in Exercise 3 to find the following. **4. a.** 0.890966414 **b.** 1
   a. $\sin 7\theta$ if $\sin \theta = 0.156422$
   b. $\sin 15\theta$ if $\sin \theta = 0.104528$

Additional Answers
Mixed Review Exercises

1.
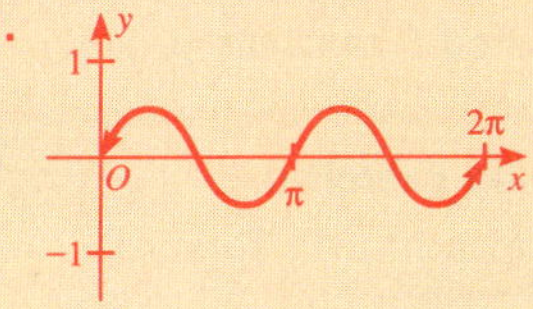

2.
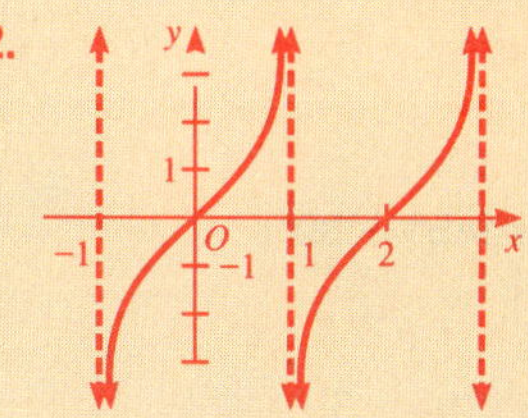

3.
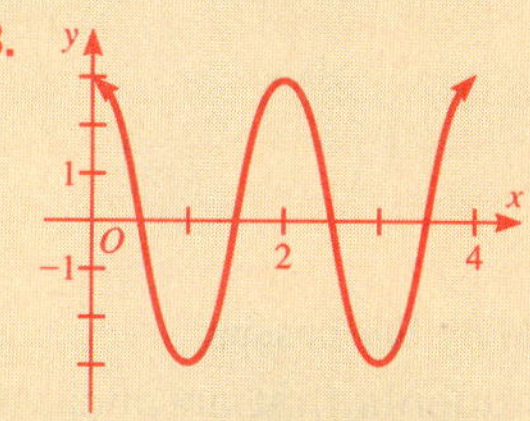

Using a Computer

These exercises use the addition formula for sine as well as some fundamental trigonometric identities. For Exercises 1 and 2, encourage students to keep the appearance of the computer generated tables as neat as possible by using TABs in the PRINT statements.

Teaching Suggestions p. T131

Suggested Extensions p. T131

Warm-Up Exercises

Use an addition formula to write an equivalent expression. Do not evaluate.

1. sin (60° − 45°)
   sin 60° cos 45° − cos 60° sin 45°
2. cos (45° − 30°)
   cos 45° cos 30° + sin 45° sin 30°
3. sin (30° + 30°)
   sin 30° cos 30° + sin 30° cos 30°
4. cos (60° + 60°)
   cos 60° cos 60° − sin 60° sin 60°

Motivating the Lesson

Tell students that an ancient question of geometry concerned whether a given angle could be trisected with compass and straightedge. Hundreds of years later this question was answered using a formula for cos 3α. This formula, in turn, depends on a formula for cos 2α, one of the topics of today's lesson. (An angle *cannot* be trisected with a compass and straightedge.)

13-8 Double-Angle and Half-Angle Formulas

Objective To use the double-angle and half-angle formulas for the sine and cosine.

You can obtain formulas for the sine and cosine of twice an angle by substituting α for β in the formulas for $\sin(\alpha+\beta)$ and $\cos(\alpha+\beta)$:

$$\begin{aligned}\sin 2\alpha &= \sin(\alpha+\alpha)\\ &= \sin\alpha\cos\alpha + \cos\alpha\sin\alpha\\ &= 2\sin\alpha\cos\alpha\end{aligned}\qquad\qquad \begin{aligned}\cos 2\alpha &= \cos(\alpha+\alpha)\\ &= \cos\alpha\cos\alpha - \sin\alpha\sin\alpha\\ &= \cos^2\alpha - \sin^2\alpha\end{aligned}$$

These formulas are called the *double-angle formulas* for sine and cosine.

Two other useful formulas for $\cos 2\alpha$ can be obtained by using the Pythagorean identity $\sin^2\alpha + \cos^2\alpha = 1$.

$$\begin{aligned}\cos 2\alpha &= \cos^2\alpha - \sin^2\alpha\\ &= (1-\sin^2\alpha) - \sin^2\alpha\\ &= 1 - 2\sin^2\alpha\end{aligned}\qquad\qquad \begin{aligned}\cos 2\alpha &= \cos^2\alpha - \sin^2\alpha\\ &= \cos^2\alpha - (1-\cos^2\alpha)\\ &= 2\cos^2\alpha - 1\end{aligned}$$

Double-Angle Formulas for Sine and Cosine

$$\sin 2\alpha = 2\sin\alpha\cos\alpha\qquad\qquad \begin{aligned}\cos 2\alpha &= \cos^2\alpha - \sin^2\alpha\\ \cos 2\alpha &= 1 - 2\sin^2\alpha\\ \cos 2\alpha &= 2\cos^2\alpha - 1\end{aligned}$$

Example 1 If $\sin\alpha = \frac{3}{5}$ and α is in the second quadrant, find $\sin 2\alpha$ and $\cos 2\alpha$.

Solution Make a sketch. You can see that if $\sin\alpha = \frac{3}{5}$, then $\cos\alpha = -\frac{4}{5}$.

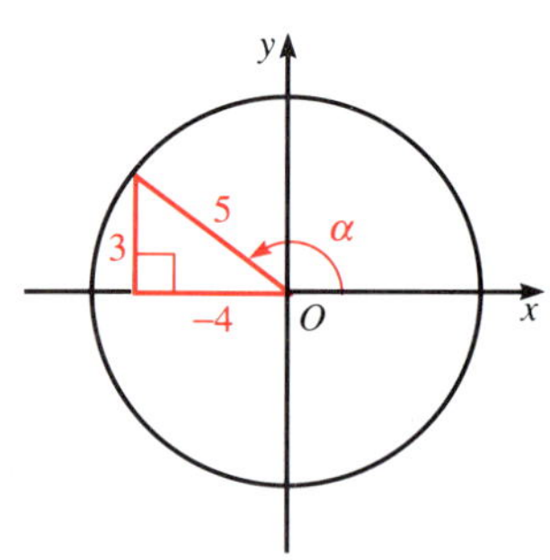

$$\begin{aligned}\sin 2\alpha &= 2\sin\alpha\cos\alpha\\ &= 2\left(\frac{3}{5}\right)\left(-\frac{4}{5}\right) = -\frac{24}{25}\end{aligned}$$

$$\begin{aligned}\cos 2\alpha &= \cos^2\alpha - \sin^2\alpha\\ &= \left(-\frac{4}{5}\right)^2 - \left(\frac{3}{5}\right)^2 = \frac{7}{25}\end{aligned}$$

$\therefore \sin 2\alpha = -\frac{24}{25}$ and $\cos 2\alpha = \frac{7}{25}$. ***Answer***

Formulas for the sine and cosine of half an angle can be derived by replacing α by $\frac{\theta}{2}$ in two of the formulas for $\cos 2\alpha$.

$$\cos 2\alpha = 1 - 2\sin^2 \alpha \qquad \cos 2\alpha = 2\cos^2 \alpha - 1$$

$$\cos 2\left(\frac{\theta}{2}\right) = 1 - 2\sin^2\left(\frac{\theta}{2}\right) \qquad \cos 2\left(\frac{\theta}{2}\right) = 2\cos^2\left(\frac{\theta}{2}\right) - 1$$

$$\cos\theta = 1 - 2\sin^2\frac{\theta}{2} \qquad \cos\theta = 2\cos^2\frac{\theta}{2} - 1$$

$$\sin^2\frac{\theta}{2} = \frac{1-\cos\theta}{2} \qquad \cos^2\frac{\theta}{2} = \frac{1+\cos\theta}{2}$$

Now take the square root of both sides of these equations to obtain the *half-angle formulas*.

Half-Angle Formulas for Sine and Cosine

$$\sin\frac{\theta}{2} = \pm\sqrt{\frac{1-\cos\theta}{2}} \qquad \cos\frac{\theta}{2} = \pm\sqrt{\frac{1+\cos\theta}{2}}$$

Notice that both of the half-angle formulas above contain a plus-or-minus sign. You must decide which sign to use by determining in which quadrant $\frac{\theta}{2}$ is located. Example 2 illustrates.

Example 2 Find the exact value of: **a.** $\sin\frac{7\pi}{8}$ **b.** $\cos 165°$

Solution Since $\frac{7\pi}{8}$ is a second-quadrant angle, $\sin\frac{7\pi}{8}$ is positive. Since 165° is a second-quadrant angle, cos 165° is negative.

a. $\frac{7\pi}{8} = \frac{1}{2}\left(\frac{7\pi}{4}\right)$

$$\sin\frac{7\pi}{8} = \sin\frac{1}{2}\left(\frac{7\pi}{4}\right) = +\sqrt{\frac{1-\cos\frac{7\pi}{4}}{2}} = \sqrt{\frac{1-\frac{\sqrt{2}}{2}}{2}} = \sqrt{\frac{2-\sqrt{2}}{4}} = \frac{1}{2}\sqrt{2-\sqrt{2}} \quad \textbf{\textit{Answer}}$$

b. $165° = \frac{1}{2}(330°)$

$$\cos 165° = \cos\frac{1}{2}(330°) = -\sqrt{\frac{1+\cos 330°}{2}} = -\sqrt{\frac{1+\frac{\sqrt{3}}{2}}{2}} = -\sqrt{\frac{2+\sqrt{3}}{4}} = -\frac{1}{2}\sqrt{2+\sqrt{3}} \quad \textbf{\textit{Answer}}$$

Chalkboard Examples

1. If $\sin\alpha = \frac{4}{5}$ and α is in the second quadrant, find $\sin 2\alpha$ and $\cos 2\alpha$.

   Since $\sin\alpha = \frac{4}{5}$, and α is in the second quadrant, $\cos\alpha = -\frac{3}{5}$.

   $\sin 2\alpha = 2\cdot\left(\frac{4}{5}\right)\left(-\frac{3}{5}\right) = -\frac{24}{25}$

   $\cos 2\alpha = \left(-\frac{3}{5}\right)^2 - \left(\frac{4}{5}\right)^2 = -\frac{7}{25}$

2. Find the exact values of $\sin\frac{5\pi}{8}$ and $\cos\frac{5\pi}{8}$.

   Since $\frac{5\pi}{8} = \frac{1}{2}\left(\frac{5\pi}{4}\right)$ and $\frac{5\pi}{8}$ is a second-quadrant angle, $\sin\frac{5\pi}{8}$ is positive and $\cos\frac{5\pi}{8}$ is negative.

   $\sin\frac{5\pi}{8} = \sqrt{\frac{1-\cos\frac{5\pi}{4}}{2}} = \sqrt{\frac{1+\frac{\sqrt{2}}{2}}{2}} = \frac{\sqrt{2+\sqrt{2}}}{2}$

   $\cos\frac{5\pi}{8} = -\sqrt{\frac{1+\cos\frac{5\pi}{4}}{2}} = -\sqrt{\frac{1-\frac{\sqrt{2}}{2}}{2}} = -\frac{\sqrt{2-\sqrt{2}}}{2}$

3. Prove the identity $\frac{\sin 2\alpha}{1-\cos 2\alpha} = \cot\alpha$.

   $\frac{\sin 2\alpha}{1-\cos 2\alpha} = \frac{2\sin\alpha\cos\alpha}{1-(\cos^2\alpha-\sin^2\alpha)} = \frac{2\sin\alpha\cos\alpha}{\sin^2\alpha+\sin^2\alpha} = \frac{\cos\alpha}{\sin\alpha} = \cot\alpha$

(continued)

Chalkboard Examples (continued)

4. If $180° < \alpha < 360°$ and $\cos \alpha = -\frac{12}{13}$, find the following.
   a. $\sin \frac{\alpha}{2}$ b. $\cos \frac{\alpha}{2}$
   Since $180° < \alpha < 360°$, $90° < \frac{\alpha}{2} < 180°$. Therefore $\sin \frac{\alpha}{2} > 0$ and $\cos \frac{\alpha}{2} < 0$.
   a. $\sin \frac{\alpha}{2} = \sqrt{\frac{1 - \cos \alpha}{2}} = \sqrt{\frac{1 - \left(-\frac{12}{13}\right)}{2}} = \sqrt{\frac{25}{26}} = \frac{5\sqrt{26}}{26}$
   b. $\cos \frac{\alpha}{2} = -\sqrt{\frac{1 + \cos \alpha}{2}} = -\sqrt{\frac{1 + \left(-\frac{12}{13}\right)}{2}} = -\sqrt{\frac{1}{26}} = -\frac{\sqrt{26}}{26}$

Check for Understanding

Oral Exs. 1–12: use after Example 4.

Guided Practice

Simplify to a trigonometric function of a single angle. Do not evaluate.

1. $2 \sin 17° \cos 17°$ $\sin 34°$
2. $\sqrt{\frac{1 + \cos 100°}{2}}$ $\cos 50°$
3. Find the exact value of $\cos^2 15° - \sin^2 15°$. $\frac{\sqrt{3}}{2}$

If $90° < \alpha < 270°$ and $\sin \alpha = -\frac{4\sqrt{5}}{9}$, find the following.

4. $\cos \alpha$ $-\frac{1}{9}$
5. $\cos 2\alpha$ $-\frac{79}{81}$
6. $\sin 2\alpha$ $\frac{8\sqrt{5}}{81}$

Example 3 If $180° < \alpha < 360°$, and $\cos \alpha = -\frac{5}{13}$, find the exact value of each of the following.

a. $\sin \frac{\alpha}{2}$ **b.** $\cos \frac{\alpha}{2}$

Solution Since $180° < \alpha < 360°$, $90° < \frac{\alpha}{2} < 180°$. Therefore, $\frac{\alpha}{2}$ is a second-quadrant angle and $\sin \frac{\alpha}{2}$ is positive and $\cos \frac{\alpha}{2}$ is negative.

$$\textbf{a. } \sin \frac{\alpha}{2} = +\sqrt{\frac{1 - \cos \alpha}{2}} = \sqrt{\frac{1 - \left(-\frac{5}{13}\right)}{2}} = \sqrt{\frac{9}{13}} = \frac{3\sqrt{13}}{13}$$

$$\textbf{b. } \cos \frac{\alpha}{2} = -\sqrt{\frac{1 + \cos \alpha}{2}} = -\sqrt{\frac{1 + \left(-\frac{5}{13}\right)}{2}} = -\sqrt{\frac{4}{13}} = -\frac{2\sqrt{13}}{13}$$

Example 4 Prove the following identity:

$$\cos \theta \csc \theta - \sin \theta \sec \theta = 2 \cot 2\theta.$$

Solution

$$\cos \theta \csc \theta - \sin \theta \sec \theta \stackrel{?}{=} 2 \cot 2\theta$$

$$\frac{\cos \theta}{\sin \theta} - \frac{\sin \theta}{\cos \theta}$$

$$\frac{\cos^2 \theta - \sin^2 \theta}{\sin \theta \cos \theta}$$

$$2\, \frac{\cos^2 \theta - \sin^2 \theta}{2 \sin \theta \cos \theta}$$

$$2\, \frac{\cos 2\theta}{\sin 2\theta}$$

$$2 \cot 2\theta = 2 \cot 2\theta \quad \checkmark$$

Oral Exercises

Simplify to a trigonometric function of a single angle. Do not evaluate.

1. $2 \sin \theta \cos \theta$ $\sin 2\theta$
2. $2 \cos^2 \theta - 1$ $\cos 2\theta$
3. $\cos^2 x - \sin^2 x$ $\cos 2x$
4. $1 - \cos^2 x$ $\sin^2 x$
5. $\sin \phi \cos \phi$ $\frac{1}{2} \sin 2\phi$
6. $2 \sin^2 \phi - 1$ $-\cos 2\phi$
7. $2 \sin 35° \cos 35°$ $\sin 70°$
8. $1 - 2 \sin^2 22.5°$ $\cos 45°$
9. $\sqrt{\frac{1 + \cos 80°}{2}}$ $\cos 40°$
10. $\sqrt{\frac{1 - \cos \pi}{2}}$ $\sin \frac{\pi}{2}$
11. $2 \sin 5x \cos 5x$ $\sin 10x$
12. $\cos^2 4x - \sin^2 4x$ $\cos 8x$

Written Exercises

Simplify to a trigonometric function of a single angle. Do not evaluate.

A

1. $\cos^2 10° - \sin^2 10°$ $\cos 20°$
2. $2 \sin 105° \cos 105°$ $\sin 210°$
3. $1 - 2 \sin^2 4°$ $\cos 8°$
4. $2 \cos^2 \pi - 1$ $\cos 2\pi$
5. $\sqrt{\frac{1 - \cos \alpha}{2}}$, $0° < \alpha < 180°$ $\sin \frac{\alpha}{2}$
6. $\sqrt{\frac{1 + \cos \alpha}{2}}$, $0° < \alpha < 180°$ $\cos \frac{\alpha}{2}$
7. $1 - 2 \sin^2 \frac{\theta}{2}$ $\cos \theta$
8. $2 \cos 2\alpha \sin 2\alpha$ $\sin 4\alpha$
9. $\cos^2 2t - \sin^2 2t$ $\cos 4t$
10. $2 \cos^2 \frac{x}{2} - 1$ $\cos x$
11. $\sqrt{\frac{1 - \cos 2\alpha}{2}}$, $0 < \alpha < \pi$ $\sin \alpha$
12. $\sqrt{\frac{1 + \cos 2\alpha}{2}}$, $-\frac{\pi}{2} < \alpha < \frac{\pi}{2}$ $\cos \alpha$

Find the exact value of each of the following.

13. $2 \sin 15° \cos 15°$ $\frac{1}{2}$
14. $2 \cos^2 22.5° - 1$ $\frac{\sqrt{2}}{2}$
15. $\cos^2 \frac{\pi}{12} - \sin^2 \frac{\pi}{12}$ $\frac{\sqrt{3}}{2}$
16. $\cos^2 \frac{\pi}{8} + \sin^2 \frac{\pi}{8}$ 1
17. $1 - 2 \sin^2 105°$ $-\frac{\sqrt{3}}{2}$
18. $2 \sin 165° \cos 165°$ $-\frac{1}{2}$
19. $\cos 22.5°$ $\frac{1}{2}\sqrt{2 + \sqrt{2}}$
20. $\sin 75°$ $\frac{1}{2}\sqrt{2 + \sqrt{3}}$
21. $\sin \frac{7\pi}{12}$ $\frac{1}{2}\sqrt{2 + \sqrt{3}}$
22. $\cos \frac{5\pi}{8}$ $-\frac{1}{2}\sqrt{2 - \sqrt{2}}$
23. $\sin 112.5°$ $\frac{1}{2}\sqrt{2 + \sqrt{2}}$
24. $\cos 202.5°$ $-\frac{1}{2}\sqrt{2 + \sqrt{2}}$

In Exercises 25–28, $0 < \alpha < 180°$ and $\cos \alpha = -\frac{4}{5}$. Find the following.

25. $\cos 2\alpha$ $\frac{7}{25}$
26. $\sin 2\alpha$ $-\frac{24}{25}$
27. $\sin \frac{\alpha}{2}$ $\frac{3\sqrt{10}}{10}$
28. $\cos \frac{\alpha}{2}$ $\frac{\sqrt{10}}{10}$

In Exercises 29–32, $180° < \alpha < 360°$ and $\cos \alpha = \frac{8}{17}$. Find the following.

29. $\sin 2\alpha$ $-\frac{240}{289}$
30. $\cos 2\alpha$ $-\frac{161}{289}$
31. $\cos \frac{\alpha}{2}$ $-\frac{5\sqrt{34}}{34}$
32. $\sin \frac{\alpha}{2}$ $\frac{3\sqrt{34}}{34}$

Prove each identity.

B

33. $\cos^4 \theta - \sin^4 \theta = \cos 2\theta$
34. $(\sin \theta + \cos \theta)^2 = 1 + \sin 2\theta$
35. $\sin 4x = 4 \sin x \cos x \cos 2x$
36. $\sin x + \sin x \cos 2x = \sin 2x \cos x$
37. $\frac{\sin 2x}{1 + \cos 2x} = \tan x$
38. $\frac{\sin 2x}{1 - \cos 2x} = \cot x$
39. $\cot \alpha + \tan \alpha = 2 \csc 2\alpha$
40. $\csc 2\alpha + \cot 2\alpha = \cot \alpha$
41. Graph $y = \cos^2 x - \sin^2 x$. Give the period and amplitude of the graph. period: π; amplitude: 1
42. Graph $y = \sin x \cos x$. Give the period and amplitude of the graph. period: π; amplitude: $\frac{1}{2}$

C

43. Express $\cos 4\theta$ in terms only of $\cos \theta$. $8 \cos^4 \theta - 8 \cos^2 \theta + 1$

Summarizing the Lesson

In this lesson students learned to use the double-angle and half-angle formulas for sine and cosine. Ask students to state each formula.

Suggested Assignments

Average Alg. and Trig.
649/2–36 even
S 644/16–20 even, 31, 32

Extended Alg. and Trig.
649/2–40 even
S 644/30, 34

Supplementary Materials

Study Guide pp. 219–220

Additional Answers Written Exercises

See Solution Key for proofs.

41.

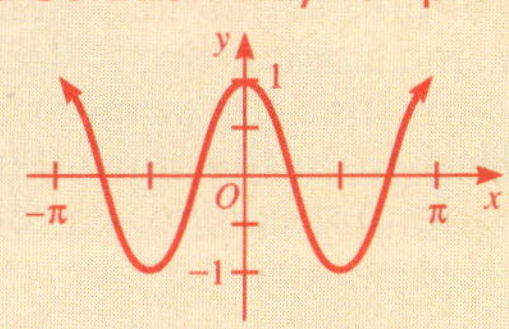

42.

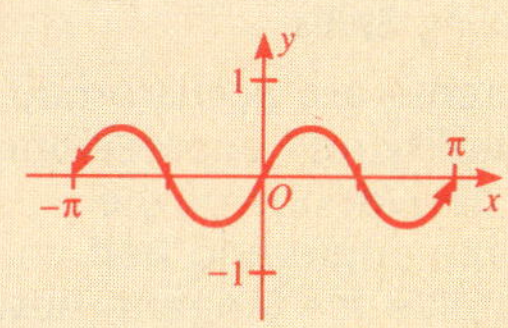

Teaching Suggestions p. T131

Suggested Extensions p. T131

Warm-Up Exercises

Write a formula for the given expression in terms of $\sin\alpha$, $\cos\alpha$, $\sin\beta$, and $\cos\beta$.

1. $\sin(\alpha+\beta)$ $\sin\alpha\cos\beta+\cos\alpha\sin\beta$
2. $\cos(\alpha-\beta)$ $\cos\alpha\cos\beta+\sin\alpha\sin\beta$
3. $\sin 2\alpha$ $2\sin\alpha\cos\alpha$
4. $\cos\frac{\beta}{2}$ $\pm\sqrt{\frac{1+\cos\beta}{2}}$
5. $\tan\alpha$ $\frac{\sin\alpha}{\cos\alpha}$

Motivating the Lesson

Tell students that the tangent has double-angle and half-angle formulas, just as sine and cosine do. In fact, the formulas for tangent are derived from the formulas for sine and cosine, as today's lesson will show.

Thinking Skills

Students are accustomed to "simplifying" expressions. In the derivations of the double-angle and half-angle formulas for tangent, however, the expressions are deliberately made more complicated. Reassure the students at these points that the ends (that is, the formulas to be obtained) justify the means.

13-9 Formulas for the Tangent

Objective To use addition, double-angle, and half-angle formulas for the tangent.

An addition formula for the tangent can be obtained by using the addition formulas for the sine and cosine. If $\tan(\alpha+\beta)$, $\tan\alpha$, and $\tan\beta$ are defined, then:

$$\begin{aligned}\tan(\alpha+\beta) &= \frac{\sin(\alpha+\beta)}{\cos(\alpha+\beta)}\\ &= \frac{\sin\alpha\cos\beta+\cos\alpha\sin\beta}{\cos\alpha\cos\beta-\sin\alpha\sin\beta}\\ &= \frac{\frac{\sin\alpha\cos\beta}{\cos\alpha\cos\beta}+\frac{\cos\alpha\sin\beta}{\cos\alpha\cos\beta}}{\frac{\cos\alpha\cos\beta}{\cos\alpha\cos\beta}-\frac{\sin\alpha\sin\beta}{\cos\alpha\cos\beta}} \quad \text{Divide numerator and denominator by } \cos\alpha\cos\beta.\\ &= \frac{\tan\alpha+\tan\beta}{1-\tan\alpha\tan\beta}\end{aligned}$$

To obtain a formula for $\tan(\alpha-\beta)$, use the formula for $\tan(\alpha+\beta)$ and the fact that $\tan(-\beta)=-\tan\beta$.

$$\tan(\alpha-\beta)=\tan[\alpha+(-\beta)]=\frac{\tan\alpha+\tan(-\beta)}{1-\tan\alpha\tan(-\beta)}=\frac{\tan\alpha-\tan\beta}{1+\tan\alpha\tan\beta}$$

Addition Formulas for the Tangent

$$\tan(\alpha+\beta)=\frac{\tan\alpha+\tan\beta}{1-\tan\alpha\tan\beta}$$

$$\tan(\alpha-\beta)=\frac{\tan\alpha-\tan\beta}{1+\tan\alpha\tan\beta}$$

Example 1 Find the exact value of $\tan 285°$.

Solution

$$\begin{aligned}\tan 285° = \tan(240°+45°) &= \frac{\tan 240°+\tan 45°}{1-\tan 240°\tan 45°}\\ &= \frac{\sqrt{3}+1}{1-(\sqrt{3})(1)}\\ &= \frac{\sqrt{3}+1}{1-\sqrt{3}}\cdot\frac{1+\sqrt{3}}{1+\sqrt{3}}\\ &= \frac{\sqrt{3}+3+1+\sqrt{3}}{1-3}\\ &= \frac{4+2\sqrt{3}}{-2} = -2-\sqrt{3} \quad \textit{Answer}\end{aligned}$$

A double-angle formula for the tangent can be obtained by substituting α for β in the formula $\tan(\alpha + \beta)$.

$$\tan 2\alpha = \tan(\alpha + \alpha) = \frac{\tan\alpha + \tan\alpha}{1 - \tan\alpha\tan\alpha} = \frac{2\tan\alpha}{1 - \tan^2\alpha}$$

Double-Angle Formula for the Tangent

$$\tan 2\alpha = \frac{2\tan\alpha}{1 - \tan^2\alpha}$$

A formula for $\tan\frac{\theta}{2}$ can be obtained by using the half-angle formulas for sine and cosine.

$$\tan\frac{\theta}{2} = \frac{\sin\frac{\theta}{2}}{\cos\frac{\theta}{2}} = \frac{\pm\sqrt{\frac{1-\cos\theta}{2}}}{\pm\sqrt{\frac{1+\cos\theta}{2}}} = \pm\sqrt{\frac{1-\cos\theta}{1+\cos\theta}}$$

A second formula for $\tan\frac{\theta}{2}$ can be derived by using the fact that $2\cos^2\alpha = 1 + \cos 2\alpha$.

$$\tan\alpha = \frac{\sin\alpha}{\cos\alpha} = \frac{2\sin\alpha\cos\alpha}{2\cos\alpha\cos\alpha} = \frac{2\sin\alpha\cos\alpha}{2\cos^2\alpha} = \frac{\sin 2\alpha}{1+\cos 2\alpha}$$

By substituting $\frac{\theta}{2}$ for α you will obtain the second formula given in the chart below. The derivation of the third formula is left as Exercise 32.

Half-Angle Formulas for the Tangent

$$\tan\frac{\theta}{2} = \pm\sqrt{\frac{1-\cos\theta}{1+\cos\theta}} \qquad \tan\frac{\theta}{2} = \frac{\sin\theta}{1+\cos\theta} \qquad \tan\frac{\theta}{2} = \frac{1-\cos\theta}{\sin\theta}$$

Example 2 Find the exact value of $\tan\frac{5\pi}{12}$.

Solution Use the formula $\tan\frac{\theta}{2} = \frac{1-\cos\theta}{\sin\theta}$ with $\theta = \frac{5\pi}{6}$.

$$\tan\frac{5\pi}{12} = \tan\frac{1}{2}\left(\frac{5\pi}{6}\right) = \frac{1-\cos\frac{5\pi}{6}}{\sin\frac{5\pi}{6}} = \frac{1-\left(-\frac{\sqrt{3}}{2}\right)}{\frac{1}{2}} = 2 + \sqrt{3} \quad \textit{Answer}$$

Chalkboard Examples

1. Find the exact value of $\tan 195°$.

$\tan 195° = \tan(45° + 150°)$

$= \frac{\tan 45° + \tan 150°}{1 - \tan 45° \tan 150°}$

$= \frac{1 + \left(-\frac{\sqrt{3}}{3}\right)}{1 - (1)\left(-\frac{\sqrt{3}}{3}\right)}$

$= \frac{3-\sqrt{3}}{3+\sqrt{3}} \cdot \frac{3-\sqrt{3}}{3-\sqrt{3}}$

$= \frac{9 - 6\sqrt{3} + 3}{9-3}$

$= \frac{12 - 6\sqrt{3}}{6} = 2 - \sqrt{3}$

2. Find the exact value of $\tan\frac{7\pi}{8}$.

Use the formula $\tan\frac{\theta}{2} = \frac{1-\cos\theta}{\sin\theta}$ with $\theta = \frac{7\pi}{4}$.

$\tan\frac{7\pi}{8} = \frac{1-\cos\frac{7\pi}{4}}{\sin\frac{7\pi}{4}} =$

$\frac{1-\frac{\sqrt{2}}{2}}{-\frac{\sqrt{2}}{2}} = \frac{\sqrt{2}-2}{\sqrt{2}} =$

$1 - \sqrt{2}$

Check for Understanding

Here is a suggested use of the Oral Exercises to check students' understanding as you teach the lesson.

Oral Exs. 1–11: use after Example 2.

Guided Practice

Simplify to a trigonometric expression of a single angle. Do not evaluate.

1. $\dfrac{\tan 25° + \tan 15°}{1 - \tan 25° \tan 15°}$ $\tan 40°$

2. $\dfrac{2 \tan 18°}{1 - \tan^2 18°}$ $\tan 36°$

Find the exact value.

3. $\tan 255°$ $2 + \sqrt{3}$

4. $\tan \frac{7\pi}{12}$ $-2 - \sqrt{3}$

5. If $\tan \alpha = -\frac{3}{4}$, find $\tan 2\alpha$. $-\frac{24}{7}$

Summarizing the Lesson

In this lesson students learned to use the sum, difference, double-angle, and half-angle formulas for tangent. Ask students to state each of these formulas.

Suggested Assignments

Average Alg. and Trig.
652/1–27 odd
S 653/Mixed Review

Extended Alg. and Trig.
652/3–24 mult. of 3, 25–32
S 653/Mixed Review

Supplementary Materials

Study Guide	pp. 221–222
Practice Master	77
Test Master	55
Computer Activity	30
Resource Book	p. 161

Oral Exercises

10. Since $\tan 22.5° = \tan \frac{45°}{2}$, use a half-angle formula for tangent: $\tan 22.5° = \dfrac{\sin 45°}{1 + \cos 45°}$

Simplify to a trigonometric function of a single angle.

1. $\dfrac{\tan 3\theta + \tan \theta}{1 - \tan 3\theta \tan \theta}$ $\tan 4\theta$

2. $\dfrac{\tan 3\theta - \tan \theta}{1 + \tan 3\theta \tan \theta}$ $\tan 2\theta$

3. $\dfrac{2 \tan x}{1 - \tan^2 x}$ $\tan 2x$

4. $\dfrac{2 \tan 4x}{1 - \tan^2 4x}$ $\tan 8x$

5. $\dfrac{\sin \frac{\alpha}{2}}{1 + \cos \frac{\alpha}{2}}$ $\tan \frac{\alpha}{4}$

6. $\dfrac{1 - \cos 2\alpha}{\sin 2\alpha}$ $\tan \alpha$

7. $\dfrac{\sin x}{1 - \cos x}$ $\cot \frac{x}{2}$

8. $\dfrac{1 + \cos x}{\sin x}$ $\cot \frac{x}{2}$

9. $\dfrac{2 \sin x \cos x}{1 - 2 \sin^2 x}$ $\tan 2x$

10. Explain how you would find the exact value of $\tan 22.5°$.

11. Explain how you would find the exact value of $\tan \frac{\pi}{12}$.

Since $\tan \frac{\pi}{12} = \tan \dfrac{\frac{\pi}{6}}{2}$, use a half-angle formula for tangent: $\tan \frac{\pi}{12} = \dfrac{\sin \frac{\pi}{6}}{1 + \cos \frac{\pi}{6}}$.

Written Exercises

Simplify to a trigonometric function of a single angle. Do not evaluate.

A

1. $\dfrac{\tan 80° - \tan 20°}{1 + \tan 80° \tan 20°}$ $\tan 60°$

2. $\dfrac{\tan 15° + \tan 30°}{1 - \tan 15° \tan 30°}$ $\tan 45°$

3. $\dfrac{\tan 140° + \tan 40°}{1 - \tan 140° \tan 40°}$ $\tan 180°$

4. $\dfrac{\tan 250° - \tan 25°}{1 + \tan 250° \tan 25°}$ $\tan 225°$

5. $\dfrac{\tan \frac{7\pi}{12} + \tan \frac{\pi}{4}}{1 - \tan \frac{7\pi}{12} \tan \frac{\pi}{4}}$ $\tan \frac{5\pi}{6}$

6. $\dfrac{\tan \frac{7\pi}{6} - \tan \frac{\pi}{3}}{1 + \tan \frac{7\pi}{6} \tan \frac{\pi}{3}}$ $\tan \frac{5\pi}{6}$

7. $\dfrac{2 \tan 22.5°}{1 - \tan^2 22.5°}$ $\tan 45°$

8. $\dfrac{2 \tan 67.5°}{1 - \tan^2 67.5°}$ $\tan 135°$

Find the exact value of each of the following.

9. $\tan 75°$ $2 + \sqrt{3}$
10. $\tan 15°$ $2 - \sqrt{3}$
11. $\tan 165°$ $-2 + \sqrt{3}$
12. $\tan 105°$ $-2 - \sqrt{3}$
13. $\tan \frac{11\pi}{12}$ $-2 + \sqrt{3}$
14. $\tan \frac{3\pi}{8}$ $\sqrt{2} + 1$
15. $\tan 22.5°$ $\sqrt{2} - 1$
16. $\tan 112.5°$ $-\sqrt{2} - 1$
17. $\tan \frac{11\pi}{8}$ $\sqrt{2} + 1$
18. $\tan \frac{5\pi}{8}$ $-\sqrt{2} - 1$
19. $\cot 195°$ $2 + \sqrt{3}$
20. $\cot 255°$ $2 - \sqrt{3}$

In Exercises 21–24, $\tan \alpha = -\frac{8}{15}$, $\tan \beta = \frac{3}{4}$, and α is a second-quadrant angle. Find the value of each expression.

21. $\tan (\alpha + \beta)$ $\frac{13}{84}$
22. $\tan (\alpha - \beta)$ $-\frac{77}{36}$
23. $\tan 2\alpha$ $-\frac{240}{161}$
24. $\tan \frac{\alpha}{2}$ 4

Prove each identity.

B **25.** $\tan(\alpha + \pi) = \tan\alpha$ **26.** $\tan(\alpha - \pi) = \tan\alpha$

27. $\tan\left(\frac{\pi}{4} - \alpha\right) = \frac{1 - \tan\alpha}{1 + \tan\alpha}$ **28.** $\tan\left(\frac{\pi}{4} + \alpha\right) = \frac{1 + \tan\alpha}{1 - \tan\alpha}$

29. Express $\cot(\alpha + \beta)$ in terms of $\cot\alpha$ and $\cot\beta$. **29.** $\frac{\cot\alpha\cot\beta - 1}{\cot\alpha + \cot\beta}$ **30.** $\frac{\cot\alpha\cot\beta + 1}{\cot\beta - \cot\alpha}$

30. Express $\cot(\alpha - \beta)$ in terms of $\cot\alpha$ and $\cot\beta$.

31. Use the formula derived in Exercise 29 to find a formula for $\cot 2\alpha$ in terms of $\cot\alpha$. $\frac{\cot^2\alpha - 1}{2\cot\alpha}$

32. Prove the following identity:

$$\tan\frac{\theta}{2} = \frac{1 - \cos\theta}{\sin\theta}$$

C **33.** **a.** Find $\sin\frac{7\pi}{8}$ and $\cos\frac{7\pi}{8}$ using half-angle formulas for sine and cosine. $\frac{1}{2}\sqrt{2 - \sqrt{2}}$; $-\frac{1}{2}\sqrt{2 + \sqrt{2}}$

b. Using the answers found in part (a), show that

$$\tan\frac{7\pi}{8} = -\sqrt{3 - 2\sqrt{2}}.$$

c. Using a half-angle formula for tangent, show that

$$\tan\frac{7\pi}{8} = 1 - \sqrt{2}.$$

d. Show that the answers to parts (b) and (c) are equal.

Prove each identity.

34. $\frac{2\tan\frac{x}{2}}{1 + \tan^2\frac{x}{2}} = \sin x$ **35.** $\tan 3\theta = \frac{3\tan\theta - \tan^3\theta}{1 - 3\tan^2\theta}$

Mixed Review Exercises

Prove each identity.

1. $(\sin x - \cos x)^2 = 1 - \sin 2x$ **2.** $\cos\left(\frac{3\pi}{2} - x\right) = -\sin x$

3. $\sec x - \sin x \tan x = \cos x$ **4.** $1 - \sin 2x \tan x = \cos 2x$

Solve each inequality and graph the solution set.

5. $|x - 3| < 2$ **6.** $x^2 > 2x + 3$ **7.** $2 \le 3x + 5 \le 14$

8. $x^3 < 3x^2$ **9.** $|2x + 1| \ge 3$ **10.** $7 - 4x > -1$

11. $3x - 5 > 13$ **12.** $x^2 - 2x \le 3$ **13.** $|2x| \ge -1$

Additional Answers Written Exercises

See Solution Key for proofs.

Additional Answers Mixed Review Exercises

1. $(\sin x - \cos x)^2 = \sin^2 x - 2\sin x\cos x + \cos^2 x = 1 - 2\sin x\cos x = 1 - \sin 2x$

2. $\cos\left(\frac{3\pi}{2} - x\right) = \cos\frac{3\pi}{2}\cos x + \sin\frac{3\pi}{2}\sin x = 0 \cdot \cos x + (-1)\sin x = -\sin x$

3. $\sec x - \sin x\tan x = \frac{1}{\cos x} - \sin x\left(\frac{\sin x}{\cos x}\right) = \frac{1 - \sin^2 x}{\cos x} = \frac{\cos^2 x}{\cos x} = \cos x$

4. $1 - \sin 2x\tan x = 1 - 2\sin x\cos x\left(\frac{\sin x}{\cos x}\right) = 1 - 2\sin^2 x = \cos 2x$

5. $\{x: 1 < x < 5\}$

6. $\{x: x < -1 \text{ or } x > 3\}$

7. $\{x: -1 \le x \le 3\}$

8. $\{x: x < 3, x \ne 0\}$

9. $\{x: x \le -2 \text{ or } x \ge 1\}$

10. $\{x: x < 2\}$

11. $\{x: x > 6\}$

12. $\{x: -1 \le x \le 3\}$

13. {real numbers}

Additional Answers Self-Test 2

2. $\sec\theta - \tan\theta\sin\theta = \frac{1}{\cos\theta} - \left(\frac{\sin\theta}{\cos\theta}\right)\sin\theta = \frac{1-\sin^2\theta}{\cos\theta} = \frac{\cos^2\theta}{\cos\theta} = \cos\theta$

Quick Quiz

1. Simplify to a trigonometric function of a single angle:
$\cos x + \sin^2 x \sec x$ $\sec x$

2. Prove:
$\frac{1+\tan^2\alpha}{1+\cot^2\alpha} = \tan^2\alpha$

$\frac{1+\tan^2\alpha}{1+\cot^2\alpha} = \frac{\sec^2\alpha}{\csc^2\alpha} = \frac{\sin^2\alpha}{\cos^2\alpha} = \tan^2\alpha$

3. Write $\sin 35° \cos 55° + \cos 35° \sin 55°$ as a function of a single angle. Then give the exact value. $\sin 90°$; 1

4. Simplify to a trigonometric function of a single angle:
$\cos^2 3\alpha - \sin^2 3\alpha$ $\cos 6\alpha$

If $0° < \theta < 180°$ and $\cos\theta = \frac{15}{17}$, find the following.

5. $\sin 2\theta$ $\frac{240}{289}$

6. $\cos\frac{\theta}{2}$ $\frac{4\sqrt{17}}{17}$

7. $\tan\frac{\theta}{2}$ $\frac{1}{4}$

Simplify to a trigonometric function of a single angle. Do not evaluate.

8. $\frac{\tan 50° + \tan 200°}{1 - \tan 50° \tan 200°}$ $\tan 250°$

9. $\frac{2\tan 101°}{1-\tan^2 101°}$ $\tan 202°$

Self-Test 2

1. Simplify to a trigonometric function of a single angle: **Obj. 13-6, p. 636**

$$\cos x + \tan x \sin x \quad \sec x$$

2. Prove the following identity: $\sec\theta - \tan\theta\sin\theta = \cos\theta$.
3. Write $\cos 310° \cos 260° - \sin 310° \sin 260°$ as a trigonometric function of a single angle. Then give the exact value of the function. $\cos 570°$; $-\frac{\sqrt{3}}{2}$ **Obj. 13-7, p. 641**
4. Simplify to a trigonometric function of a single angle: **Obj. 13-8, p. 646**

$$2\sin 5\alpha \cos 5\alpha \quad \sin 10\alpha$$

5. If $180° < \theta < 360°$, and $\cos\theta = -\frac{21}{29}$, find the following:

a. $\cos 2\theta$ $\frac{41}{841}$ **b.** $\sin 2\theta$ $\frac{840}{841}$ **c.** $\cos\frac{\theta}{2}$ $-\frac{2\sqrt{29}}{29}$

Simplify to a trigonometric function of a single angle. Do not evaluate.

6. $\frac{\tan 50° - \tan 40°}{1 + \tan 50° \tan 40°}$ $\tan 10°$ **7.** $\frac{2\tan 35°}{1-\tan^2 35°}$ $\tan 70°$ **Obj. 13-9, p. 650**

8. $\frac{1-\cos 130°}{\sin 130°}$ $\tan 65°$

Check your answers with those at the back of the book.

Chapter Summary

1. The *radian measure* of an angle θ is $\frac{s}{r}$, where r is the radius of any circle centered at the vertex of θ and s is the length of the arc intercepted by θ. To convert from one system of measure to another, you use the conversion formula $180° = \pi$ radians.

   Circular arc length is given by
   $$s = r\theta,$$
   and sector area by
   $$A = \tfrac{1}{2}r^2\theta,$$
   where θ denotes the radian measure of the central angle.
2. *Circular functions* of the number s are equal to the corresponding trigonometric functions of an angle having radian measure s. If angle θ has radian measure s, then $\sin\theta = \sin s$, $\cos\theta = \cos s$, and so on. Therefore, the values of circular functions can be found by using a calculator or tables.

3. A function f has *period* p $(p > 0)$ if $f(x + p) = f(x)$ for every x in the domain of f. The sine, cosine, secant, and cosecant have period 2π; the tangent and cotangent have period π.
4. For every x in the domain of f, a function f is *even* if $f(-x) = f(x)$, and *odd* if $f(-x) = -f(x)$. The sine, cosecant, tangent, and cotangent are odd functions; the cosine and secant are even functions.
5. The graphs of $y = a \sin bx$ and $y = a \cos bx$ each have amplitude a and period $\frac{2\pi}{b}$.
6. The fundamental *trigonometric identities* (page 636), as well as the *addition, double-angle,* and *half-angle formulas* are used to simplify trigonometric expressions and to prove other identities.

Chapter Review

Write the letter of the correct answer.

1. Express $\frac{13\pi}{4}$ radians in degrees. **13-1**

a. 495° **b.** 225° **c.** 585° **d.** 135°

2. Express 150° in radians.

a. $\frac{7\pi}{3}$ **b.** $\frac{5\pi}{6}$ **c.** $\frac{7\pi}{6}$ **d.** $\frac{7\pi}{12}$

3. Find the exact value of $\csc\left(-\frac{7\pi}{6}\right)$. **13-2**

a. $\frac{\sqrt{3}}{2}$ **b.** $-\frac{\sqrt{3}}{2}$ **c.** -2 **d.** 2

4. If $\tan s = \frac{\sqrt{3}}{3}$ and $-\pi < s < 0$, then $s =$ __?__.

a. $-\frac{4\pi}{3}$ **b.** $-\frac{2\pi}{3}$ **c.** $-\frac{5\pi}{6}$ **d.** $-\frac{7\pi}{6}$

5. If $f(x) = \frac{x}{x^2 - 9}$, then f is __?__. **13-3**

a. odd **b.** even **c.** neither odd nor even **d.** both even and odd

6. If $f(x)$ has period 8, then $f(2x)$ has period __?__.

a. 16 **b.** 8 **c.** 4 **d.** 2

7. Which of the following is an equation of the cosine curve shown at the right? **13-4**

a. $y = 2 \cos \frac{2x}{3}$ **b.** $y = 2 \cos \frac{3\pi x}{2}$

c. $y = 2 \cos \frac{2\pi x}{3}$ **d.** $y = 2 \cos 3\pi x$

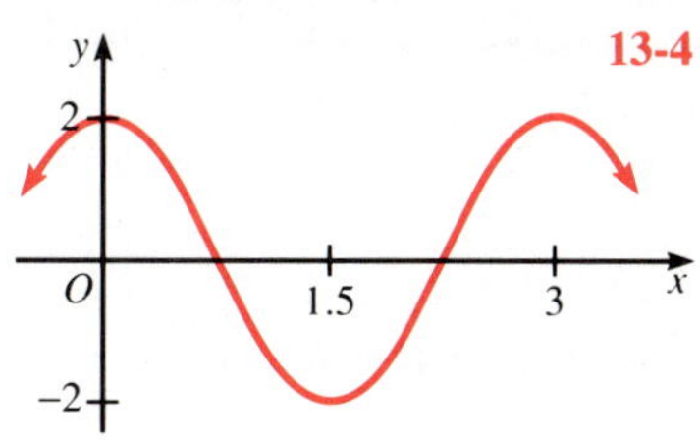

Additional Answers Written Exercises

(continued from p. 618)

17. $\sin\left(-\frac{5\pi}{6}\right) = -\frac{1}{2}$, $\cos\left(-\frac{5\pi}{6}\right) = -\frac{\sqrt{3}}{2}$, $\tan\left(-\frac{5\pi}{6}\right) = \frac{\sqrt{3}}{3}$, $\cot\left(-\frac{5\pi}{6}\right) = \sqrt{3}$, $\sec\left(-\frac{5\pi}{6}\right) = -\frac{2\sqrt{3}}{3}$, $\csc\left(-\frac{5\pi}{6}\right) = -2$

18. $\sin\left(-\frac{3\pi}{4}\right) = -\frac{\sqrt{2}}{2}$, $\cos\left(-\frac{3\pi}{4}\right) = -\frac{\sqrt{2}}{2}$, $\tan\left(-\frac{3\pi}{4}\right) = 1$, $\cot\left(-\frac{3\pi}{4}\right) = 1$, $\sec\left(-\frac{3\pi}{4}\right) = -\sqrt{2}$, $\csc\left(-\frac{3\pi}{4}\right) = -\sqrt{2}$

23. $\cos^2\left(\frac{3\pi}{4}\right) + \sin^2\left(\frac{3\pi}{4}\right) = \left(-\frac{\sqrt{2}}{2}\right)^2 + \left(\frac{\sqrt{2}}{2}\right)^2 = \frac{1}{2} + \frac{1}{2} = 1$

24. $\cos^2\left(\frac{7\pi}{6}\right) + \sin^2\left(\frac{7\pi}{6}\right) = \left(-\frac{\sqrt{3}}{2}\right)^2 + \left(-\frac{1}{2}\right)^2 = \frac{3}{4} + \frac{1}{4} = 1$

25. $\cos^2 0.7 + \sin^2 0.7 \approx (0.7648)^2 + (0.6442)^2 \approx 0.5849 + 0.4150 = 0.9999 \approx 1$

26. $\cos^2 1 + \sin^2 1 \approx (0.5403)^2 + (0.8415)^2 \approx 0.2919 + 0.7081 = 1$

27. Subtract $2\pi(\approx 6.28)$ from 7 to get 0.72; look up sin 0.72.

28–32. See Solution Key

8. Which of the following is an equation of a sine curve with maximum value 3, minimum value -1 and period $\frac{\pi}{3}$?

(a.) $y = 2 \sin 6x + 1$ b. $y = 2 \sin 6\pi x + 1$
c. $y = 2 \sin 3x + 1$ d. $y = 2 \sin 3x$

9. Find the period of the function defined by $y = 2 \sec \pi x$. 13-5

a. π b. 2π c. 1 (d.) 2

10. Which of the following statements is *not* true?

a. $\cot x$ is odd b. $\sec x$ is even (c.) $\tan x$ is even d. $\csc x$ is odd

11. Simplify $\sec \theta - \tan \theta \sin \theta$. 13-6

a. $\sin^2 \theta$ b. $\cos^2 \theta$ c. $\sec \theta$ (d.) $\cos \theta$

12. If $\sin t = u$, then $\tan t =$ __?__.

(a.) $\pm \frac{u}{\sqrt{1 - u^2}}$ b. $\pm \frac{1}{\sqrt{1 - u^2}}$ c. $\pm \frac{1}{1 - u^2}$ d. $\pm \frac{u}{\sqrt{u^2 - 1}}$

13. Write $\sin 2\pi \cos \frac{\pi}{6} - \cos 2\pi \sin \frac{\pi}{6}$ as a trigonometric function of a single angle and evaluate. 13-7

a. $\frac{\sqrt{3}}{2}$ b. $-\frac{\sqrt{3}}{2}$ c. $\frac{1}{2}$ (d.) $-\frac{1}{2}$

14. Find the exact value of $\sin 165°$.

a. $\frac{\sqrt{6} + \sqrt{2}}{4}$ (b.) $\frac{\sqrt{6} - \sqrt{2}}{4}$ c. $\frac{\sqrt{2} - \sqrt{6}}{4}$ d. $-\frac{\sqrt{2} + \sqrt{6}}{4}$

15. If $\sin \theta = \frac{2}{3}$, then $\cos 2\theta =$ __?__. 13-8

a. $\frac{1}{3}$ b. $\frac{5}{9}$ c. $-\frac{1}{9}$ (d.) $\frac{1}{9}$

16. Evaluate $\cos \frac{7\pi}{8}$.

a. $-\frac{\sqrt{2 - \sqrt{2}}}{2}$ b. $\frac{\sqrt{2 + \sqrt{2}}}{2}$ c. $\frac{\sqrt{2 - \sqrt{2}}}{2}$ (d.) $-\frac{\sqrt{2 + \sqrt{2}}}{2}$

17. Simplify $\tan\left(\alpha - \frac{\pi}{4}\right) \tan\left(\alpha + \frac{\pi}{4}\right)$. 13-9

a. 1 (b.) -1 c. $\tan 2\alpha$ d. $\frac{1}{2} \tan 2\alpha$

18. If θ is a fourth-quadrant angle and $\cos \theta = \frac{8}{17}$, find $\tan \frac{\theta}{2}$.

(a.) $-\frac{3}{5}$ b. $\frac{3}{5}$ c. $-\frac{15}{8}$ d. $\frac{8}{15}$

Chapter Test

1. Express $-120°$ in radians. $-\frac{2\pi}{3}$ 13-1

2. Express $\frac{7\pi}{10}$ in degrees. $126°$

3. The central angle of a sector of a circle having diameter 8 in. measures $65°$. Find the following:
   a. the length, s, of the intercepted arc. $s \approx 4.5$ in.
   b. the area, A, of the sector. $A \approx 9$ in.^{2}

4. Find the exact values of the six trigonometric functions of $\frac{4\pi}{3}$. 13-2

5. Determine if $f(x) = x \sin x - \tan x$ is even, odd, or neither. neither 13-3

6. Find the amplitude, the maximum and minimum values, and the period of the function $y = 3 \cos 4x$. $3;\ 3,\ -3;\ \frac{\pi}{2}$ 13-4

7. Find an equation of the sine curve shown at the right. $y = 3 \sin \frac{\pi}{4} x$

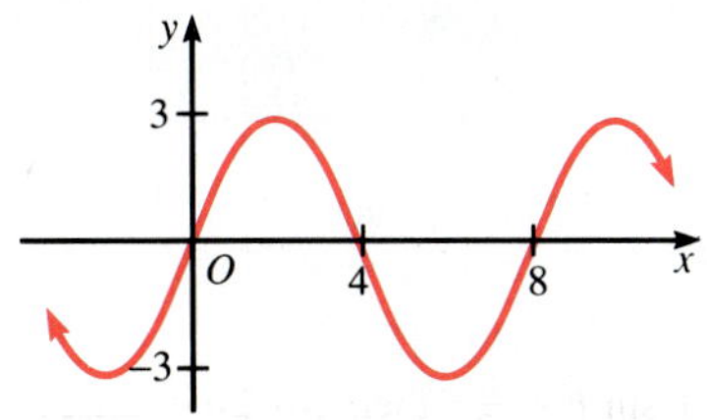

8. Sketch one period of the graph of $y = 2 \cot \pi x$. 13-5

9. Prove the following identity: 13-6

$$\csc \theta - \cot \theta \cos \theta = \sin \theta$$

10. Prove the following identity:

$$\tan x + \cot x = \sec x \csc x$$

11. Simplify the following expression: 13-7

$$\cos(\alpha + 45°) + \sin(\alpha - 45°)\ \ 0$$

12. If $\sin \alpha = \frac{8}{17}$ and $\cos \alpha < 0$, find the following: 13-8
   a. $\sin 2\alpha$ $-\frac{240}{289}$
   b. $\cos 2\alpha$ $\frac{161}{289}$
   c. $\sin \frac{\alpha}{2}$ $\frac{4\sqrt{17}}{17}$

13. If $\sin \theta = -\frac{5}{13}$ and $\cos \theta > 0$, find the following: 13-9
   a. $\tan 2\theta$ $-\frac{120}{119}$
   b. $\tan \frac{\theta}{2}$ $-\frac{1}{5}$

Additional Answers
Chapter Test

4. $\sin \frac{4\pi}{3} = -\frac{\sqrt{3}}{2}$, $\cos \frac{4\pi}{3} = -\frac{1}{2}$, $\tan \frac{4\pi}{3} = \sqrt{3}$, $\cot \frac{4\pi}{3} = \frac{\sqrt{3}}{3}$, $\sec \frac{4\pi}{3} = -2$, $\csc \frac{4\pi}{3} = -\frac{2\sqrt{3}}{3}$

8.

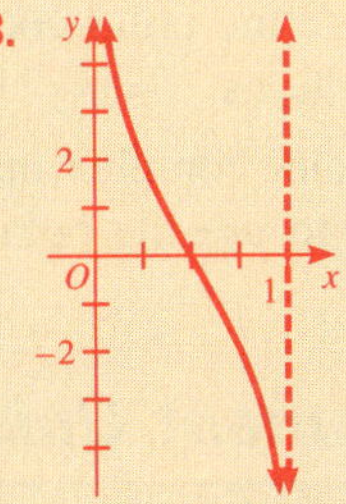

9. $\csc \theta - \cot \theta \cos \theta = \frac{1}{\sin \theta} - \left(\frac{\cos \theta}{\sin \theta}\right) \cos \theta = \frac{1 - \cos^2 \theta}{\sin \theta} = \frac{\sin^2 \theta}{\sin \theta} = \sin \theta$

10. $\tan x + \cot x = \frac{\sin x}{\cos x} + \frac{\cos x}{\sin x} = \frac{\sin^2 x + \cos^2 x}{\cos x \sin x} = \frac{1}{\cos x \sin x} = \frac{1}{\cos x} \cdot \frac{1}{\sin x} = \sec x \csc x$

Alternate Test p. T25

Supplementary Materials

Test Masters	56, 57
Resource Book	p. 162

14 Trigonometric Applications

Objectives

14-1 To define vector operations and apply the resultant of two vectors.

14-2 To find vectors in component form and to apply the dot product.

14-3 To define polar coordinates and graph polar equations.

14-4 To plot complex numbers in the complex plane and to use the polar form of complex numbers.

14-5 To use De Moivre's theorem.

14-6 To evaluate expressions involving the inverse cosine and inverse sine.

14-7 To evaluate expressions involving the inverse trigonometric functions.

14-8 To solve trigonometric equations.

Assignment Guide

See p. T58 for Key to the format of the Assignment Guide

Day	Average Algebra	Average Algebra and Trigonometry	Extended Algebra	Extended Algebra and Trigonometry
1		**14-1** 664/1–15 odd 665/*P*: 2–8 even **S** 665/*Mixed Review*		**14-1** 664/1–21 odd 665/*P*: 1–9 odd **S** 665/*Mixed Review*
2		**14-2** 669/1–14, 15–29 odd **S** 665/*P*: 7, 9		**14-2** 669/1–14, 15–31 odd **S** 665/*P*: 6, 8
3		**14-3** 678/1–26 **S** 679/*Mixed Review* **R** 671/*Self-Test 1*		**14-3** 678/1–28 **R** 671/*Self-Test 1*
4		**14-4** 683/2–30 even **S** 678/27–30		**14-3** 678/29–36, *Extra:* 1–3 **S** 679/*Mixed Review*
5		**14-5** 687/1–8, 9–21 odd **S** 688/*Mixed Review*		**14-4** 683/2–30 even, 31 **S** 670/30, 32
6		**14-6** 692/1–20 **S** 687/16, 18, 20 **R** 688/*Self-Test 2*		**14-5** 687/1–8, 10–22 even **S** 688/*Mixed Review*
7		**14-7** 695/1–22 **S** 696/*Mixed Review*		**14-6** 692/2–16 even, 18–24 **S** 687/17, 19, 21 **R** 688/*Self-Test 2*
8		**14-8** 699/1–27 odd **S** 696/23–26		**14-7** 695/1–17 odd, 19–26 **S** 696/*Mixed Review*
9		**14-8** 700/14–28 even, 29–42 **R** 700/*Self-Test 3*		**14-8** 699/1–27 odd **S** 696/27, 28

Assignment Guide *(continued)*

Day	Average Algebra	Average Algebra and Trigonometry	Extended Algebra	Extended Algebra and Trigonometry
10		*Prepare for Chapter Test* **R** 703/*Chapter Review*		**14-8** 700/14–28 even, 29–45 odd **R** 700/*Self-Test 3*
11		*Administer Chapter 14 Test*		*Prepare for Chapter Test* **R** 703/*Chapter Review*
12				*Administer Chapter 14 Test* *Read 16-1* 769/1–14

Supplementary Materials Guide

For Use with Lesson	Practice Masters	Tests	Study Guide (Reteaching)	Resource Book		
				Tests	Practice Exercises	Applications (A) Enrichment (E)
14-1			pp. 223–224			p. 217 (A)
14-2	Sheet 78		pp. 225–226		p. 163	
14-3			pp. 227–228			
14-4			pp. 229–230			
14-5	Sheet 79	Test 58	pp. 231–232		p. 164	
14-6			pp. 233–234			
14-7	Sheet 80		pp. 235–236			
14-8	Sheet 81	Test 59	pp. 237–238		p. 165	
Chapter 14		Tests 60, 61		pp. 73–76	p. 166	p. 233 (E)
Cum. Rev. 12–14		Test 62				

Overhead Visuals

For Use with Lessons	Visual	Title
14-3	B	Multi-Use Packet 2
14-5	10	Angles
14-1, 14-2	12	Vectors

Software

Software	Algebra Plotter Plus	Using Algebra Plotter Plus	Computer Activities	Test Generator
	Function Plotter	Enrichment, p. 41	Activity 31	168 test items
For Use with Lessons	14-8	14-8	14-8	all lessons

Strategies for Teaching

Cooperative Learning and Using Technology

Groups of algebra students can use manipulatives, solve nonroutine problems, discuss assigned problems, make a model, or explore a concept in depth. The make-up of the group, the size of the group, and the structure of assigned tasks will vary with different classes of students and with their previous experience in group learning settings. Some teachers may wish to leave many decisions to the students, while other teachers may find that detailed directions and organization are necessary. Teachers may want to try different-sized groups and different types of assignments before deciding which of several alternatives will work best in a particular class.

The role of the teacher in cooperative learning situations will no longer be that of the dispenser of knowledge. The teacher will introduce the topic to be explored by the group, give clear directions and check to see that they are understood, specify expectations, and then act as a resource person. The teacher will encourage communication among group members, and assist in the development of the group's cooperation skills. The teacher will provide ways in which students can summarize and communicate the results of their explorations and activities, and will provide extensions of the topic for those groups that finish more quickly than others. While the role of the teacher takes on new directions in a classroom that is involved in cooperative learning activities, planning for this new role must be carried out as carefully as for a lecture and assignment lesson.

Calculators and computers are useful in this chapter to facilitate computations using the trigonometric functions. See the Exploration on page 845 for an activity in which students use graphing software to explore various types of polar equations.

14-3 Polar Coordinates

Students may enjoy playing a polar coordinate version of Tic-Tac-Toe. It goes well with three or four individual players or two teams of two players, each alternating turns.

Play on copies of the board shown below. As a student places *X* or *O*, the coordinates must be called out, giving the angle measure in radians and the distance from the center. Thus a legal move would be $\left(\frac{\pi}{3}, 3\right)$ for the red *X* shown. A player wins with four in a row on a given circle, four in a row on a given radius, or four in a row in a spiral.

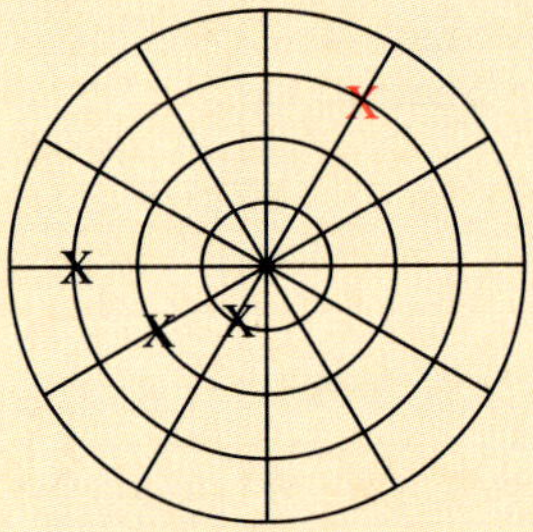

14-7 Other Inverse Functions

Students can benefit from some explanation of how to use their calculators to evaluate expressions that include inverse trigonometric functions.

By this point students should know whether they have inverse sine, cosine, and tangent keys, or whether they must use a separate inverse key with the appropriate function key. Show with examples such as $\text{Sin}^{-1}\ 1.2$ and $\text{Cos}^{-1}\ -2$ the error messages calculators will produce.

Show the sequence of keys required to evaluate an expression such as tan (Sec^{-1} 2):

$$2 \rightarrow \frac{1}{x} \rightarrow \text{INV} \rightarrow \cos \rightarrow \tan$$

Trying to evaluate an expression such as Tan^{-1} (tan 135°) may help students understand the restrictions on the inverse functions better, since the value is not 135°.

14-8 Trigonometric Equations

When the solutions of a trigonometric equation are to be expressed in radians, students can use the Function Plotter program to solve the equation.

For example, to find the primary solutions of the equation $3 \sec x = \csc x$ (see Exercise 15 on page 700), students can graph the equations $y = 3 \sec x$ and $y = \csc x$ together and determine the x-coordinates of all points where the two graphs intersect on the interval $0 \leq x < 2\pi$.

References to Strategies

PE: Pupil's Edition **TE:** Teacher's Edition **RB:** Resource Book

Problem Solving Strategies

TE: p. 662 (draw a diagram)

Applications

PE: pp. 658–705 (word problems); pp. 672–674 (physics)
TE: pp. 658, 660–663, 671; p. T133 (Sugg. Extension, Lesson 14-1)
RB: p. 217

Nonroutine Problems

PE: p. 647 (Challenge 9–15); p. 670 (Exs. 31, 32); p. 684 (Exs. 35–37); p. 688 (Ex. 23); p. 696 (Ex. 29)
TE: pp. T133–T136 (Sugg. Extensions, Lessons 14-1, 14-2, 14-3, 14-5, 14-8)

Communication

PE: pp. 684, 699 (convincing argument)
TE: p. T131 (Reading Algebra)

Thinking Skills

PE: p. 670 (Exs. 31, 32); p. 684 (Ex. 36); p. 692 (Exs. 24, 25); p. 696 (Ex. 29, proofs)
TE: p. 677

Explorations

PE: pp. 679 (spirals); p. 845 (polar coordinate equations)

Connections

PE: pp. 658, 661–663 (Navigation); pp. 659–705 (Trigonometry); pp. 662, 672–674 (Physics); pp. 680–684 (Geometry); p. 684 (History)

Using Technology

PE: p. 670 (Exs.); pp. 683, 698, 845; pp. 701–702 (Computer Key-In)
TE: pp. 671, 682, 697, 701, T136
Computer Activities: pp. 83–85
Using Algebra Plotter Plus: p. 41

Using Manipulatives/Models

Overhead Visuals: B, 10, 12

Teaching Resources

For use in implementing the teaching strategies referenced on the previous page.

Application
Resource Book, p. 217

Application—Radar (For use with Chapter 14)

Radar is a word derived from **radio detecting and ranging.** Radar has many uses. For example, it is used by air traffic controllers to locate planes and by meteorologists to chart weather patterns.

Radar operates by transmitting a radio signal of short duration and then detecting the echoes returning from targets that reflect the signal. Because radio signals travel at a constant speed (in fact, the speed of light), the time between transmitting the signal and receiving the echo determines the distance from the radar to the target. This distance d is given by the formula

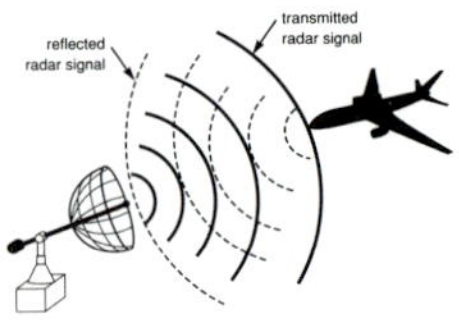

$$d = \frac{1}{2}c\,\Delta T$$

where $c = 186{,}000$ mi/s (the speed of light) and ΔT is the elapsed time (in seconds) between transmitting the signal and receiving the echo.

1. Using the familiar formula $d = rt$, explain the derivation of the above equation for the distance from a radar to a target: ______

2. How far away is the target if the time between transmitting the signal and receiving the echo is
   a. 1 s? ______
   b. 0.01 s? ______
   a. 10^{-6} s (that is, one millionth of a second, or one microsecond)? ______

3. Suppose radio-wave echoes have been recorded from each of the following targets. Find the amount of time required between transmitting the signal and receiving the echo.
   a. A ship 6 mi away: ______
   b. A plane 75 mi away: ______
   c. The moon, which is 249,000 mi from Earth: ______
   d. Venus, when it was 27,000,000 mi from Earth: ______

4. Suppose a radar system sends out a brief signal, waits for the echo, and then transmits the next signal. For this radar system the maximum range of a signal is 140 mi. Also, the radar antenna rotates through 360°, and 7200 signals (20 each degree) are sent during each rotation.
   a. If a second signal cannot be transmitted until the echo from the first signal has been received, what is the maximum time needed between successive transmissions (to the nearest ten thousandth)? ______

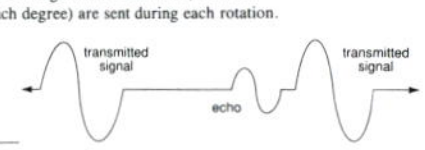

   b. Using the rounded maximum time found in part (a), how many seconds (to the nearest tenth) are required for each rotation? ______

Resource Book, ALGEBRA, Structure and Method, Book 2

APPLICATIONS 217

Enrichment/Nonroutine Problems
Resource Book, p. 233

Graphs (For use with Chapter 14)

For each graph, find the coordinates of the labeled points. Give x-values in radians.

1. $y = \dfrac{1 - \sin x}{x}$

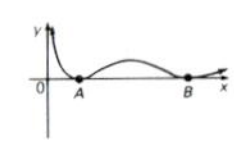

2. $y = \tan x + \sin x$

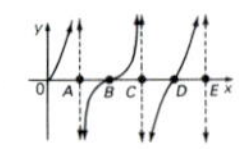

3. $y = \tan x + \cos x$

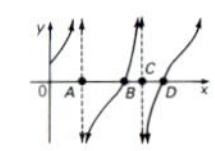

4. $y = \sin x + \cos 2x$

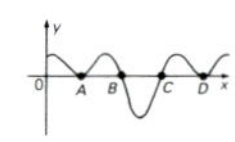

5. $y = \sin x - \cos 2x$

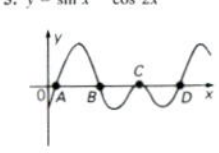

6. $y = \sin(2^x)$

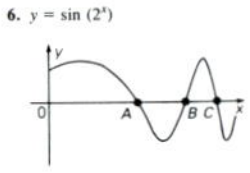

Resource Book, ALGEBRA, Structure and Method, Book 2

ENRICHMENT 233

Using Technology
Computer Activities, p. 83

ACTIVITY 31. *Solving Trigonometric Equations* (*for use with Section 14-8*)

Directions: Write all answers in the spaces provided.

PROBLEM

Use the computer to solve the trigonometric equation $2 \sin x \cos x = 1$.

PROGRAM

```
10 PRINT
20 FOR D = 0 TO 359
30 LET X = D * 3.141593 / 180
40 LET L = 2 * SIN(X) * COS(X)
50 LET R = 1
60 IF ABS(L - R) > .0001 THEN 80
70 PRINT "X = "; D; " DEGREES"
80 NEXT D
90 END
```

PROGRAM CHECK

Type in the program. To test whether you entered it correctly, run the program. (Be patient. The computer may take a minute or so to finish running the program.) The computer should print:

```
X = 45 DEGREES
X = 225 DEGREES
```

ANALYSIS

BASIC requires that radians be used to evaluate trigonometric functions, so line 30 converts degrees to radians. The left and right sides of the equation are evaluated in lines 40 and 50 for radian values corresponding to the degree values from 0° to 359°. If the left and right sides are within 0.0001 of each other, they are assumed to be equal (line 60), and the computer prints the solution in degrees (line 70).

(continued)

Computer Activities, ALGEBRA AND TRIGONOMETRY, Structure and Method, Book 2

83

Using Technology
Computer Activities, p. 84

(*Activity 31 continued*)

USING THE PROGRAM

Modify lines 40 and 50 for the given equation. Then run the program and record your solutions.

	Basic statements	*Solutions*
1. $2\sin^2 x = \cos x + 1$	40 LET L = 2 * SIN(X) ↑ 2	
	50 LET R = COS(X) + 1	______
2. $\sin x = 1 - \cos x$	40 LET L = ______	
	50 LET R = ______	______
3. $\sin^2 x = \cos^2 x$	40 ______	
	50 ______	______
4. $2\sin^2 x - \sin x = 1$	40 ______	
	50 ______	______

5. Solve the equation in Exercise 4 without using the computer. Do your work in the space below.

APPLICATION

Professor Trigge wanted to challenge her algebra students by giving them a trigonometry problem that did not have an integral solution. She asked them to find a first-quadrant angle for which 4 times the cosine of the angle, minus 3 times the sine of the angle, is equal to 2. Professor Trigge asked for the solution to the nearest tenth of a degree.

1. Write an equation that represents the problem. Then modify lines 40 and 50 of the program accordingly.

   Equation: ______

   40 ______

   50 ______

(continued)

84
Computer Activities, ALGEBRA AND TRIGONOMETRY, Structure and Method, Book 2

Connection
Study Guide, p. 229

14–4 The Geometry of Complex Numbers

Objective: To plot complex numbers in the complex plane and to use the polar form of complex numbers.

Vocabulary

Complex plane The rectangular coordinate system when it is used to represent complex numbers. The horizontal axis is called the *real axis*, and the vertical axis is called the *imaginary axis*. The point (x, y) corresponds to the complex number $x + yi$.

Polar (or **trigonometric**) **form** The *polar*, or *trigonometric*, form of a nonzero complex number $z = x + yi$ is $z = r(\cos\theta + i\sin\theta)$ where $r = \sqrt{x^2 + y^2}$ and θ is an angle such that $\cos\theta = \frac{x}{r}$ and $\sin\theta = \frac{y}{r}$. The number r is called the *absolute value* of z (that is, $r = |z|$) and sometimes the *modulus* of z. Angle θ, chosen so that $0° \le \theta < 360°$, is called the *amplitude*, or *argument*, of z.

Multiplication and division of complex numbers in polar form If $w = a(\cos\alpha + i\sin\alpha)$ and $z = b(\cos\beta + i\sin\beta)$, then:

1. $wz = ab[\cos(\alpha + \beta) + i\sin(\alpha + \beta)]$
2. $\frac{w}{z} = \frac{a}{b}[\cos(\alpha - \beta) + i\sin(\alpha - \beta)]$

Example 1 Plot $w = 2 + 3i$, $z = 3 - i$, $w + z$, and $w - z$ in the complex plane.

Solution

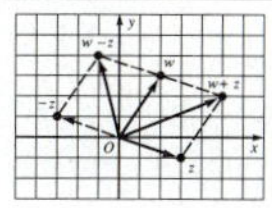

Plot w, z, $w + z$, and $w - z$ in the complex plane.

1. $w = 2 + i, z = 3$
2. $w = 4 + 2i, z = 4 - 2i$
3. $w = -1 + i, z = 3 - i$
4. $w = 3i, z = 3 + 2i$
5. $w = 3 + i, z = -2 - 2i$
6. $w = -3 - 2i, z = -1 + 3i$

Example 2 Let $w = 4(\cos 25° + i\sin 25°)$ and $z = 2(\cos 110° + i\sin 110°)$. Find (a) wz and (b) $\frac{w}{z}$ in polar form with $0° \le \theta < 360°$.

Solution

a. $wz = (4)(2)[\cos(25° + 110°) + i\sin(25° + 110°)]$
$= 8(\cos 135° + i\sin 135°)$

b. $\frac{w}{z} = \frac{4}{2}[\cos(25° - 110°) + i\sin(25° - 110°)]$
$= 2[\cos(-85°) + i\sin(-85°)]$
$= 2(\cos 275° + i\sin 275°)$

Study Guide, ALGEBRA AND TRIGONOMETRY, Structure and Method, Book 2
 229

Using Models
Overhead Visual B

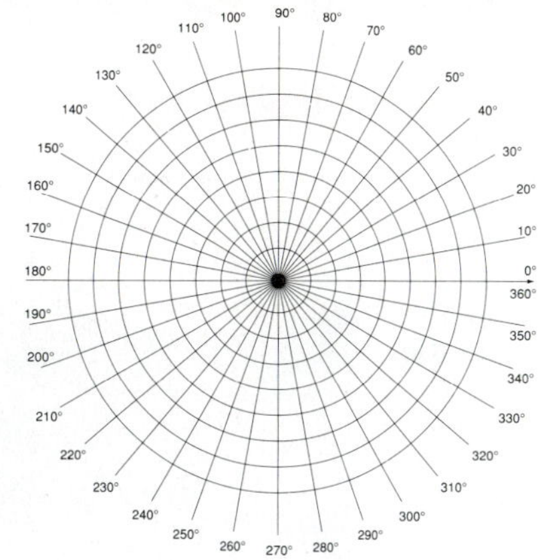

Using Models
Overhead Visual 10, Sheets 1 and 3

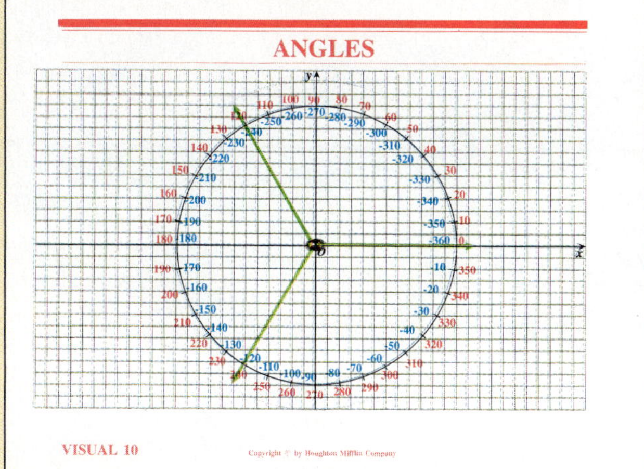

Using Models
Overhead Visual 12, Sheets 1–4

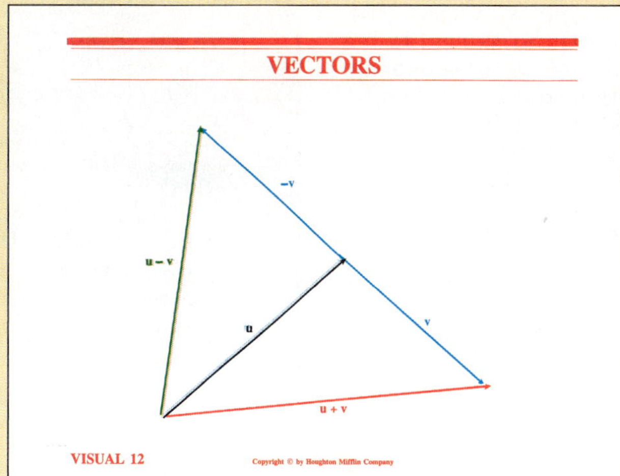

Application

The word *radar* is an acronym for "radio detection and ranging." A radar system operates by emitting a narrow beam of radio waves and detecting any reflection of the beam. The position of an object that reflects the beam is determined both by the direction of the reflected beam and by the elapsed time between the beam's emission and detection (which fixes the distance between the object and the radar system).

Ships use radar for such above-water purposes as detecting other ships and navigating along coastlines. When something below the water must be detected, sonar is used instead. The operation of a sonar system is similar to that of a radar system, except that sound waves take the place of radio waves.

Research Activities
Students can research and report on the many uses of radar, such as forecasting the weather, tracking satellites and spacecraft, and detecting speed-limit violators.

Students who like to fish may want to explain to the class how the small sonar devices sold in sporting goods stores help fishermen determine water depth and location of fish.

Support Materials
Resource Book p. 217

14 Trigonometric Applications

A ship's radar uses a polar coordinate system. The location of a point is determined by its direction and distance from the center (O), called the pole.

658

Vectors

14-1 Vector Operations

Objective To define vector operations and apply the resultant of two vectors.

The *velocity* of an airplane is described by giving both its *speed* and its *direction*. For example, you might say its velocity is 600 km/h northeast. Any quantity that has both magnitude (size) and direction is called a **vector quantity.** You can represent a vector quantity by an arrow, called a **vector,** whose length is proportional to the magnitude of the quantity, and whose direction gives the direction of the vector quantity (see Figure 1).

A vector is often labeled with the magnitude of the vector quantity it represents or the magnitude is indicated by giving a scale. For example, the arrow on the diagram at the left in Figure 1 represents a velocity of 600 km/h northeast. The arrow on the diagram at the right is $4\frac{1}{2}$ times as long as the given 100-m scale and therefore represents a displacement of 450 m due east.

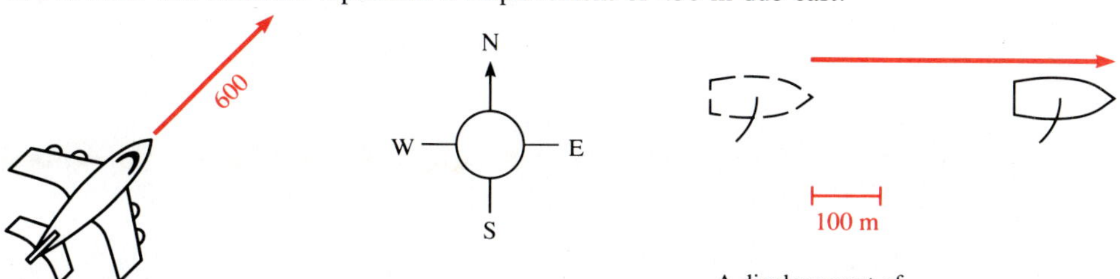

A velocity of 600 km/h northeast

A displacement of 450 m due east

Figure 1

The symbol $\overrightarrow{AB}$ (read "vector AB") denotes the vector extending from point A, the **initial point,** to point B, the **terminal point.** Boldface letters such as **u** and **v** are also used to denote vectors. (When handwriting vectors, use letters with bars or small arrows over them: $\bar{u}$, $\bar{v}$, or $\vec{u}$, $\vec{v}$, in place of boldface.) Two vectors that have the same magnitude and same direction, such as $\overrightarrow{AB}$ and $\overrightarrow{CD}$ in the diagram, are called **equivalent vectors,** and you write $\overrightarrow{AB} = \overrightarrow{CD}$.

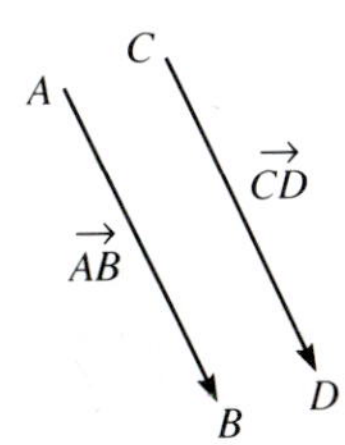

In the vector diagram at the right, a plane flies from A to B and then from B to C. Since the plane could get to the same final position by flying directly from A to C, $\overrightarrow{AC}$ is called the *sum,* or *resultant,* of $\overrightarrow{AB}$ and $\overrightarrow{BC}$. That is,

$$\overrightarrow{AC} = \overrightarrow{AB} + \overrightarrow{BC}.$$

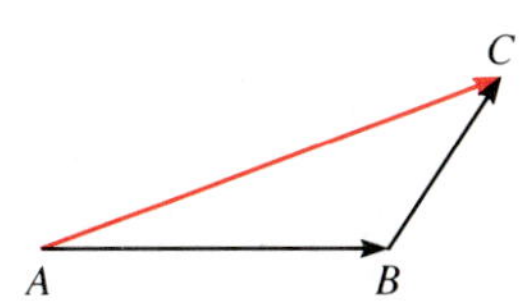

Teaching References

Lesson Commentary, pp. T132–T136

Assignment Guide, pp. T70–T72

Supplementary Materials

- Practice Masters 78–81
- Tests 58–62
- Resource Book
  - Practice Exercises pp. 163–166
  - Tests, pp. 73–76
  - Enrichment Activity, p. 233
  - Application, p. 217
- Study Guide, pp. 223–238
- Computer Activity 31
- Test Generator

Alternate Test, p. T26

Cumulative Review, p. T32

Teaching Suggestions p. T132

Reading Algebra p. T132

Suggested Extensions p. T133

Using Manipulatives p. T133

Warm-Up Exercises

Use the law of cosines or the law of sines to find the indicated side or angle of $\triangle ABC$.

1. $a = 8$, $b = 12$, $\angle C = 72°$, $c =$ __?__ 12.2

2. $a = 10$, $c = 5$, $\angle A = 122°$, $\angle C =$ __?__ 25.1°

3. $a = 6$, $b = 8$, $c = 12$, $\angle C =$ __?__ 117.3°

4. $b = 20$, $\angle A = 40°$, $\angle B = 85°$, $a =$ __?__ 12.9

5. $a = 25$, $b = 40$, $\angle C = 120°$, $c =$ __?__ 56.8

To find the resultant of any two vectors, you use the following fact: For any vector **v** and point P, there is exactly one vector with initial point P that has the same direction and length as **v**. We will often treat vectors as movable arrows and label this vector **v** as well. Therefore, we can *place the initial point of* **v** *at* P by constructing the unique vector with initial point P that is equivalent to **v**. In general, you have the following methods for finding the resultant.

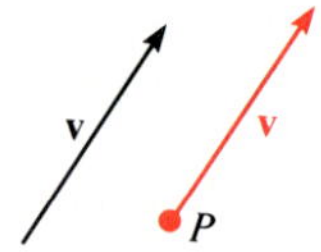

Figure 2

Vector Addition

Given two vectors **u** and **v**, you can find their **sum,** or **resultant,** by using either of the following two methods. These methods show that vector addition is commutative: **u** + **v** = **v** + **u**.

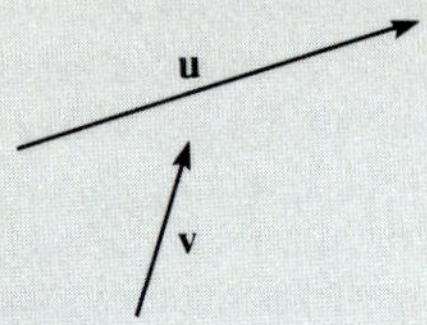

The Triangle Method

Place the initial point of **v** at the terminal point of **u.** Then **u** + **v** is the vector extending from the initial point of **u** to the terminal point of **v.**

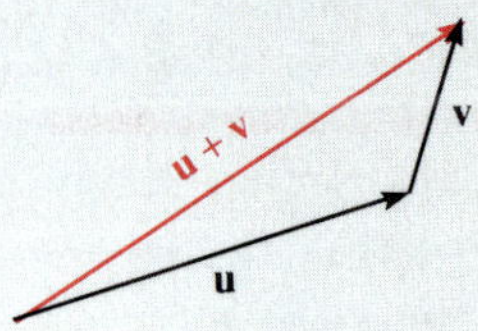

The Parallelogram Method

Form a parallelogram with **u** and **v** as adjacent sides starting from a common point. Then **u** + **v** extends from that point to the opposite vertex of the parallelogram.

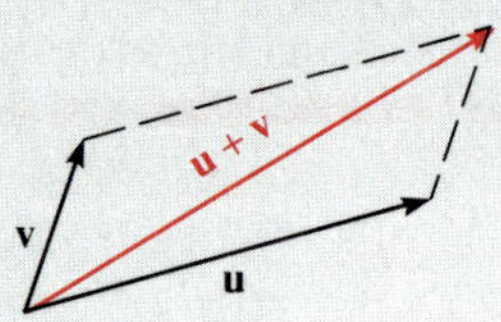

Example 1 For the following two vector quantities, make a scale drawing showing the vectors and their sum:

A 10 km trip due west followed by a 15 km trip southeast.

Solution Let **u** be the vector representing a 10 km trip west and **v** be the vector representing a 15 km trip southeast. Their sum or resultant is shown below.

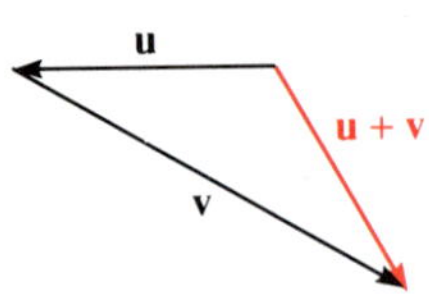

The chart at the top of the next page tells you how to multiply a vector by a real number, or *scalar*.

Motivating the Lesson

Tell the students that when an airplane is flying, many forces are acting on it at once. The engines push the plane one direction, for example, while the wind may push it in another. The navigator uses vectors, the topic of today's lesson, to represent such forces so that the direction of the plane can be controlled.

Chalkboard Examples

1. Make a scale drawing showing the vectors and their sum: A biker travels 12 mi due south and then 20 mi southwest.

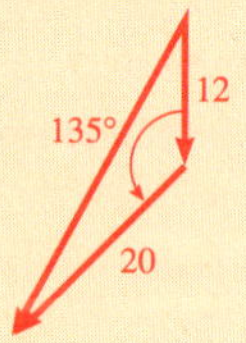

2. Vector **u** has magnitude 8 and bearing 180°. Vector **v** has magnitude 12 and bearing 135°. Sketch **u** + **v** and **u** − **v**.

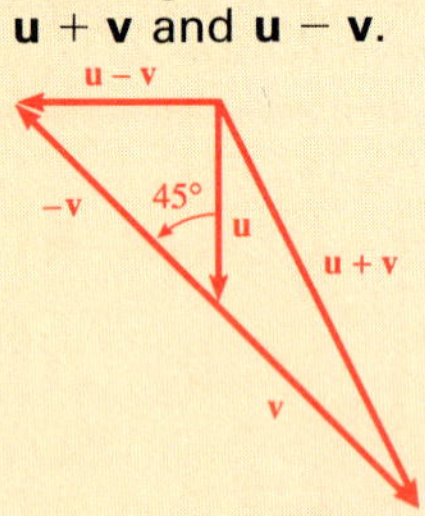

Scalar Multiplication

To multiply the vector **v** by the real number t, multiply the length of **v** by $|t|$ and reverse the direction if $t < 0$.

This rule is illustrated in Figure 3. Notice that $(-1)\mathbf{v}$ is written as $-\mathbf{v}$.

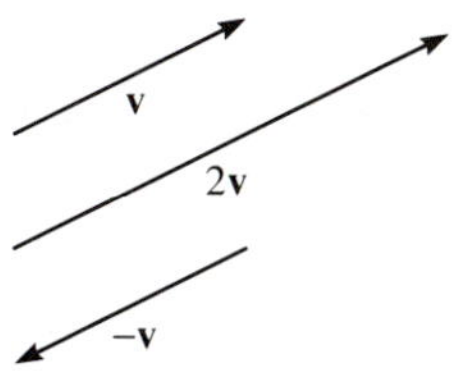

Figure 3

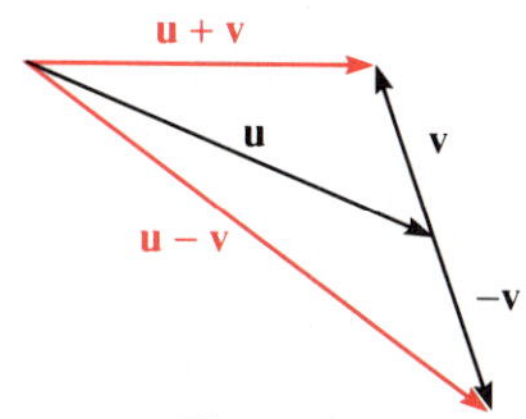

Figure 4

The difference $\mathbf{u} - \mathbf{v}$ is defined to be $\mathbf{u} + (-\mathbf{v})$. (See Figure 4.)

You can think of a point as a vector of length 0. This zero vector is denoted by **0.** For any vector **u,** you have $\mathbf{u} - \mathbf{u} = \mathbf{0}$.

The length or magnitude of a vector **v** is also called its **norm** and is denoted by $\|\mathbf{v}\|$. For example, if **v** is an airplane velocity vector, then $\|\mathbf{v}\|$ is the speed of the airplane. A vector with a norm of 1 is called a **unit vector.**

The direction of a vector **v** is often given by its *bearing*.

The **bearing** of a vector **v** is the angle measured clockwise from due north around to **v.**

Some terms used in navigation are illustrated in Figure 5. You should be aware that a craft may not travel in the direction in which it is headed because of the effect of winds and currents. For example, if you row with a heading of 90° (due east) across a river flowing from north to south, the bearing of your *true course* will be somewhere between 90° and 180°.

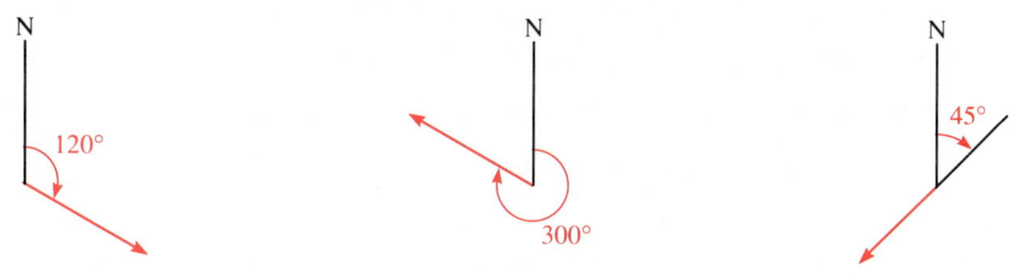

Bearing 120° Heading 300° Wind from 45°

Figure 5

3. Find the magnitude of **w** to three significant digits and its bearing to the nearest tenth of a degree.

u: magnitude 140, bearing 70°
v: magnitude 80, bearing 330°
w: **u** + **v**

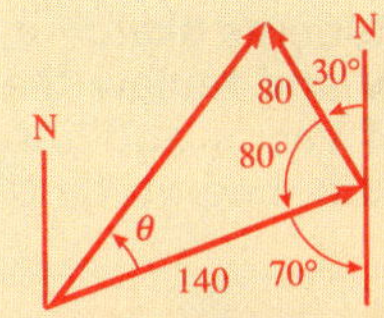

$\|\mathbf{w}\|^2 = 140^2 + 80^2 - 2(80)(140)\cos 80°$

$\|\mathbf{w}\| = 149$

$\frac{\sin\theta}{80} = \frac{\sin 80°}{149}$

$\sin\theta = 0.5288$

$\theta = 31.9°$

The bearing of **w** is $70° - \theta = 38.1°$.

4. The air speed of a light plane is 180 km/h and its heading is 130°. A wind of 40 km/h is blowing from 50°. Find the plane's ground speed and the bearing of its true course.

$\|\mathbf{v}\|^2 = 180^2 + 40^2 - 2(180)(40)\cos 80°$
$= 34{,}000 - 14{,}400(0.1736)$
$= 177.4$ km/h

To find the bearing of the true course use the law of sines.

$\frac{\sin\theta}{40} = \frac{\sin 80°}{177.4}$

$\theta = 12.8°$

The bearing of the true course is $130° + 12.8° = 142.8°$.

Example 2 Use vectors **u** and **v** as shown below. Vector **u** has magnitude 4 and bearing 90° and vector **v** has magnitude 8 and bearing 225°. Sketch **(a)** **u** + **v** and **(b)** **u** − **v**.

Solution **a.** **u** + **v**

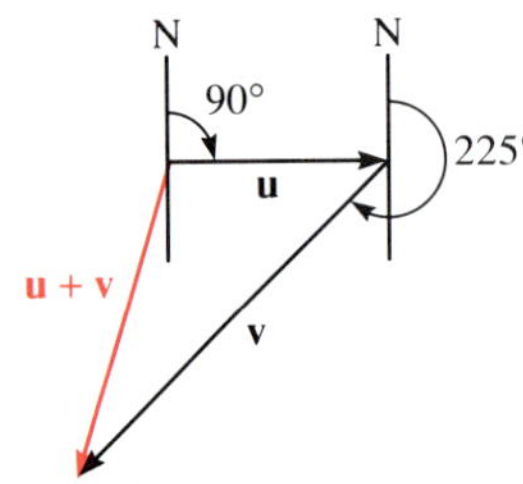

b. Use the fact that **u** − **v** = **u** + (−**v**).

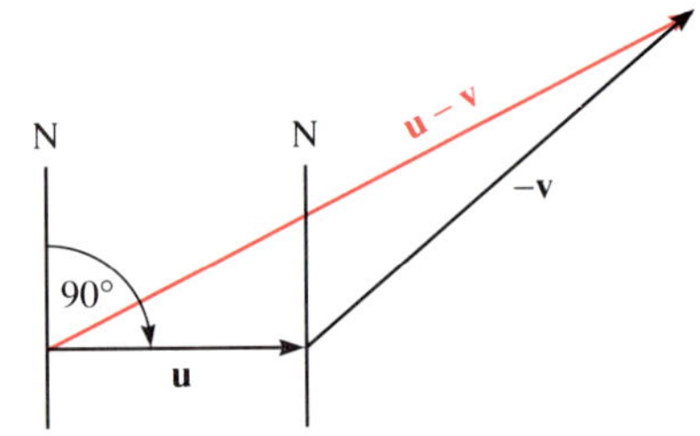

Example 3 The magnitude and direction of vectors **u** and **v** are given below. If **w** = **u** + **v**, find the magnitude of **w** to three significant digits and its bearing to the nearest tenth of a degree.

u: magnitude 7, bearing 135°
v: magnitude 5, bearing 245°

Solution The figure at the right shows the vector quantities involved. The vector **w** is the resultant vector. Since $65° + \alpha + 45° = 180°$, $\alpha = 70°$.
Use the law of cosines to find $\|\mathbf{w}\|$, the magnitude of **w**.

$$\|\mathbf{w}\|^2 = 7^2 + 5^2 - 2(7)(5)\cos 70°$$
$$= 49 + 25 - (70)(0.342)$$
$$= 50.06$$
$$\|\mathbf{w}\| = 7.08 \text{ (to three significant digits)}$$

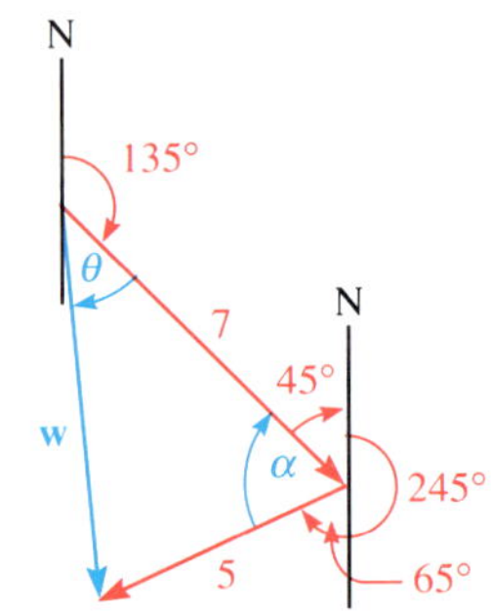

To find the bearing of vector **w**, first use the law of sines to find θ:

$$\frac{\sin\theta}{5} = \frac{\sin 70°}{7.08}$$

$$\sin\theta = \frac{5\sin 70°}{7.08} = \frac{5(0.940)}{7.08} = 0.6640$$

$$\theta = 41.6°$$

Then the bearing of **w** is $135° + \theta = 135° + 41.6° = 176.6°$.

∴ the magnitude of **w** is 7.08 and its bearing is 176.6°. ***Answer***

Example 4 The air speed (speed in still air) of a light plane is 220 km/h, and its heading is 150°. A wind of 45 km/h is blowing from 40°. Find the plane's ground speed (the speed relative to the ground) and the bearing of its true course.

Common Errors

Given a vector **u** with magnitude 4 and bearing 180° and a vector **v** with magnitude 3 and bearing 90°, students may mistakenly think that **u** + **v** = 7, with a bearing of 270°. When reteaching, have students sketch **u**, **v**, and **u** + **v** and review the definition of vector addition. Show that the magnitude of **u** + **v** is 5.

Problem Solving Strategies

Although students have used *diagrams* in problem solving, now you can tell them that diagrams are more than an aid. They are absolutely essential for solving vector problems.

Check for Understanding

Here is a suggested use of the Oral Exercises to check students' understanding as you teach the lesson.
Oral Exs. 1–16: use before the discussion of bearing on page 661.

Guided Practice

1. Make a scale drawing showing the vectors and their sum: A hiker travels 12 km southeast and then 6 km south.

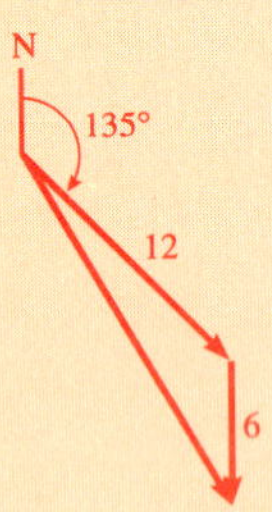

Solution The figure at the right shows the velocity vectors involved. The vector **v** is the true-course velocity. To find the ground speed $\|\mathbf{v}\|$, use the law of cosines. Since $\alpha + 30° + 40° = 180°$, $\alpha = 110°$.

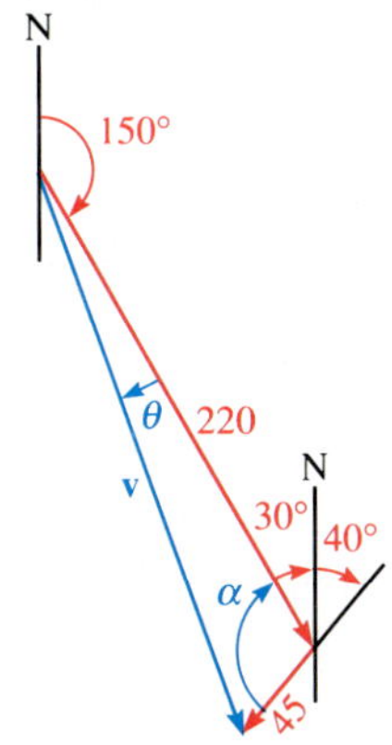

$$\begin{aligned}\|\mathbf{v}\|^2 &= 220^2 + 45^2 - (2)(220)(45)\cos 110° \\ &= 48{,}400 + 2025 + 6772 \\ &= 57{,}197 \\ \|\mathbf{v}\| &= 239 \quad \text{(km/h)}\end{aligned}$$

To find the bearing of **v**, the true course, first use the law of sines to find θ.

$$\frac{\sin\theta}{45} = \frac{\sin 110°}{239}$$

$$\sin\theta = \frac{45(0.9397)}{239} = 0.177$$

$$\theta = 10.2°$$

Then the bearing of **v** is $150° + \theta = 150° + 10.2° = 160.2°$.

$\therefore$ the ground speed of the plane is 239 km/h and the bearing of the true course is 160.2°. ***Answer***

2. Refer to Exercise 1. Find the hiker's distance and bearing from the starting point after traveling 18 km. Give the distance to the nearest 0.1 km and the bearing to the nearest 0.1°.
16.8 km; 149.6°
3. A plane's air speed is 350 km/h and its heading is 190°. If it is flying directly into a 45 km/h wind, find the plane's ground speed.
305 km/h
4. Vectors **u** and **v** have bearings of 30° and 90° and each has magnitude 12. Find the magnitude and bearing of **(a)** $\mathbf{u} + \mathbf{v}$ and **(b)** $\mathbf{u} - \mathbf{v}$.
a. $12\sqrt{3}$; 60°
b. 12; 330°

Summarizing the Lesson

In this lesson students learned to find sums and scalar multiples of vectors and to use vectors to solve problems. Ask students to describe how to add two vectors and how to multiply a vector by a scalar.

Oral Exercises

Exercises 1–16 refer to parallelogram *ABCD*. In Exercises 1–10, determine whether the statement is true or false.

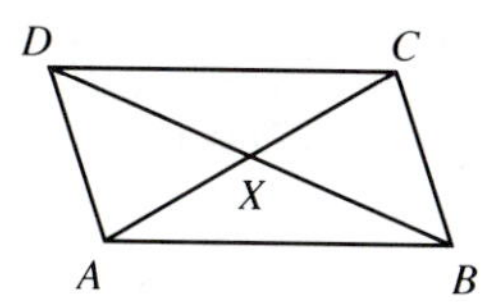

1. $\overrightarrow{DC} + \overrightarrow{CX} = \overrightarrow{DX}$ T
2. $\overrightarrow{XB} = \overrightarrow{XD}$ F
3. $\overrightarrow{XD} + \overrightarrow{XB} = 0$ T
4. $2\overrightarrow{XB} = \overrightarrow{DB}$ T
5. $\overrightarrow{BA} + \overrightarrow{AD} = \overrightarrow{BD}$ T
6. $\overrightarrow{DC} + \overrightarrow{CB} = \overrightarrow{DB}$ T
7. $\overrightarrow{AX} + \overrightarrow{XB} + \overrightarrow{DA} = 0$ F
8. $\overrightarrow{BD} - \overrightarrow{AD} = \overrightarrow{BA}$ T
9. $2\overrightarrow{DX} + \overrightarrow{BC} = \overrightarrow{DC}$ T
10. $\overrightarrow{BA} + \overrightarrow{DA} = \overrightarrow{CA}$ T

Find a vector joining two lettered points in the diagram that is equivalent to the given vector. Answers may vary.

11. $2\overrightarrow{XD}$ $\overrightarrow{BD}$
12. $-\overrightarrow{AC}$ $\overrightarrow{CA}$
13. $\overrightarrow{CD} - \overrightarrow{DA}$ $\overrightarrow{BD}$
14. $\overrightarrow{AD} + \overrightarrow{DC} + \overrightarrow{CX} + \overrightarrow{XB}$ $\overrightarrow{AB}$
15. $\overrightarrow{AD} + \overrightarrow{CX}$ $\overrightarrow{BX}$
16. $\overrightarrow{DC} - \overrightarrow{CB}$ $\overrightarrow{AC}$

Suggested Assignments

Average Alg. and Trig.
664/1–15 odd
665/*P*: 2–8 even
S 665/Mixed Review

Extended Alg. and Trig.
664/1–21 odd
665/*P*: 1–9 odd
S 665/Mixed Review

Supplementary Materials

Study Guide pp. 223–224
Overhead Visual 12

Additional Answers Written Exercises

2.

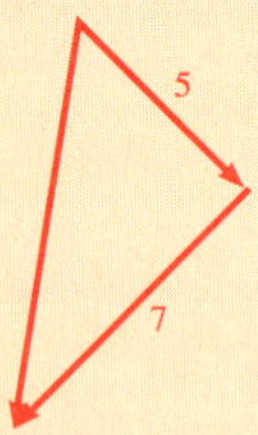

4.

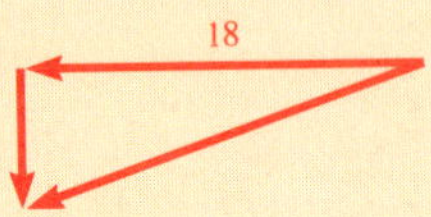

6.

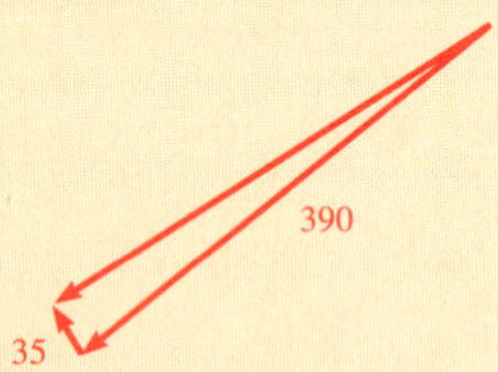

8.

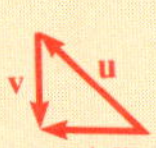

Written Exercises

In Exercises 1–6, two vector quantities are given. Make a scale drawing showing the two vectors and their sum.

A

1. A 9 km trip east followed by a 4 km trip southwest.
2. A 5 km trip southeast followed by a 7 km trip southwest.
3. The heading of a plane is northwest, and its speed is 350 km/h. A wind of 50 km/h is blowing from the west.
4. The heading of a fishing boat is due west, and its speed is 18 km/h. The current is flowing at 6 km/h due south.
5. A small boat sails for 15 km bearing 60° and then sails for 10 km bearing 120°.
6. The air speed of a plane is 390 km/h and its heading is 230°. A wind of 35 km/h is blowing from 150°.

In the diagram at the right, vector u has magnitude 3 and bearing 315°, vector v has magnitude 2 and bearing 180°, and vector w has magnitude 3 and bearing 90°. Use these vectors to sketch the following vectors.

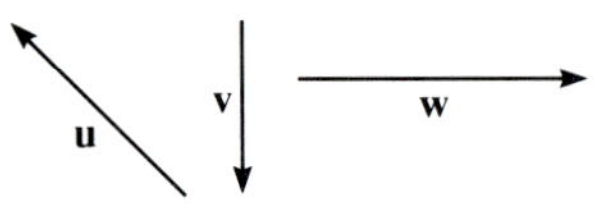

7. **u** + **w**
8. **u** + **v**
9. **u** − **v**
10. **u** − **w**
11. **u** + 2**v**
12. 2**u** + **w**
13. **u** + 2**v** − **w**
14. **w** − **u** − 2**v**

In Exercises 15–22, the magnitude and direction of vectors u and v are given. Find the magnitude of w to three significant digits and its bearing to the nearest tenth of a degree.

15. **u:** magnitude 117, bearing 130°
v: magnitude 102, bearing 220°
w: **u** + **v** 155; 171.2°

16. **u:** magnitude 218, bearing 22°
v: magnitude 170, bearing 112°
w: **u** − **v** 276; 344.1°

B

17. **u:** magnitude 136, bearing 220°
v: magnitude 197, bearing 300°
w: 2**u** − **v** 307; 180.8°

18. **u:** magnitude 1850, bearing 125°
v: magnitude 2960, bearing 25°
w: **u** + 2**v** 5890; 43.0°

19. **u:** magnitude 460, bearing 0°
v: magnitude 712, bearing 130°
w: **u** + **v** 545; 90°

20. **u:** magnitude 23.0, bearing 215°
v: magnitude 14.5, bearing 105°
w: **u** + **v** 22.6; 177.9°

21. **u:** magnitude 3.62, bearing 25°
v: magnitude 14.5, bearing 105°
w: **u** + **v** 15.5; 92.1°

22. **u:** magnitude 621, bearing 305° 2190;
v: magnitude 336, bearing 15° 141.8°
Find **w** if 3(**w** + **u**) = 2(**w** − **v**).

10. 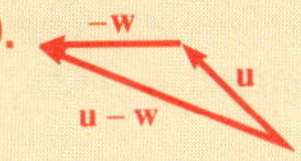

12. 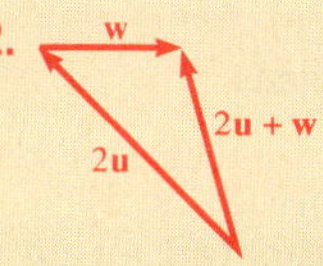

14. 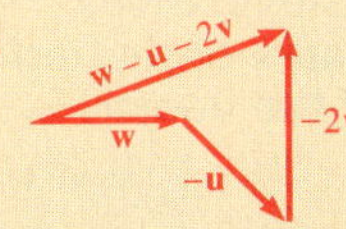

Problems

Solve. Find magnitudes to three significant digits and bearings to the nearest tenth of a degree.

A **1.** A plane flies 285 km due west and then 320 km due north. What are its distance and bearing from the starting point? 429 km; 318.3°

2. A ship leaves port and sails 175 km due southeast and then 168 km due southwest. What are the distance and bearing of the port from the ship? 243 km; 358.8°

3. To fly due north at 580 km/h when there is a 45 km/h wind blowing from due east, with what heading and speed should a plane travel? 4.4°; 582 km/h

4. Repeat Problem 3 assuming that the wind is blowing from 130°. 3.6°; 552 km/h

5. A plane's air speed is 270 km/h, and its heading is 70°. Find its ground speed and true-course bearing if a wind of 50 km/h is blowing from 330°. 283 km/h; 80.0°

6. Repeat Problem 5 assuming that the wind is blowing from due north. 257 km/h; 80.5°

7. Two ships leave port at noon. One ship's speed is 15 km/h, and its heading is 75°. The other ship's speed is 24 km/h, and its heading is 155°. How far apart are the ships at 3 P.M. and what is the bearing of each from the other? 78 km; first ship is on a bearing of 9.6° from the second. Second is on a bearing of 189.6° from the first.

B **8.** A plane leaves Mapleton bound for Logan Beach, which is 600 km away at a bearing of 25° from Mapleton. To avoid a storm, the pilot first flies 140 km on a 325° course. From there the pilot can fly the plane directly to Logan Beach in still air. What heading should be used and how far must the plane fly? 72.2°; 544 km

9. Los Angeles Airport bears 140° from San Francisco Airport and is 540 km away. A pilot is planning a direct flight from San Francisco Airport to Los Angeles Airport to leave at 2 P.M. The plane's air speed will be 640 km/h, and there will be a 60 km/h wind blowing from 290°. What should the compass heading be, and what is the plane's estimated time of arrival (ETA) to the nearest minute? 142.7°; ETA of 2:47 P.M.

Mixed Review Exercises

Determine whether the given function is odd, even, or neither.

1. $f(x) = \frac{x+2}{x-3}$ neither **2.** $y = \cos x - \sin x$ neither **3.** $f(x) = \frac{6x^3 - x}{x^2 + 1}$ odd

Solve.

4. $x^2 + 6x = 8$ $\{-3 \pm \sqrt{17}\}$ **5.** $|3z - 7| = 5$ $\left\{\frac{2}{3}, 4\right\}$ **6.** $(3x - 2)^2 = 36$ $\left\{-\frac{4}{3}, \frac{8}{3}\right\}$

7. $\log_9 x = \frac{3}{2}$ $\{27\}$ **8.** $\sqrt{4t + 9} = t - 3$ $\{10\}$ **9.** $p^2 + \frac{11p}{2} = 3$ $\left\{-6, \frac{1}{2}\right\}$

Teaching Suggestions p. T133

Suggested Extensions p. T133

Warm-Up Exercises

A unit vector has its initial point at the origin and makes the indicated angle with the positive x-axis. Find the coordinates of the terminal point of the vector.

1. $0°$ $(1, 0)$
2. $-90°$ $(0, -1)$
3. $30°$ $\left(\frac{\sqrt{3}}{2}, \frac{1}{2}\right)$
4. $-135°$ $\left(-\frac{\sqrt{2}}{2}, -\frac{\sqrt{2}}{2}\right)$
5. $150°$ $\left(-\frac{\sqrt{3}}{2}, \frac{1}{2}\right)$
6. $-240°$ $\left(-\frac{1}{2}, \frac{\sqrt{3}}{2}\right)$

Motivating the Lesson

Remind students that they have so far dealt with vectors in terms of magnitude and direction. Today's lesson deals with vectors in terms of components.

14-2 Vectors in the Plane

Objective To find vectors in component form and to apply the dot product.

In a coordinate plane we denote by **i** the vector having initial point (0, 0) and terminal point (1, 0). We denote by **j** the vector having initial point (0, 0) and terminal point (0, 1). If point P has coordinates (a, b), then $\overrightarrow{OP}$ is the sum or resultant vector of $a\mathbf{i} + b\mathbf{j}$ and $\overrightarrow{OP} = a\mathbf{i} + b\mathbf{j}$ (see the diagram). In fact, every vector in the plane can be expressed in terms of **i** and **j**.

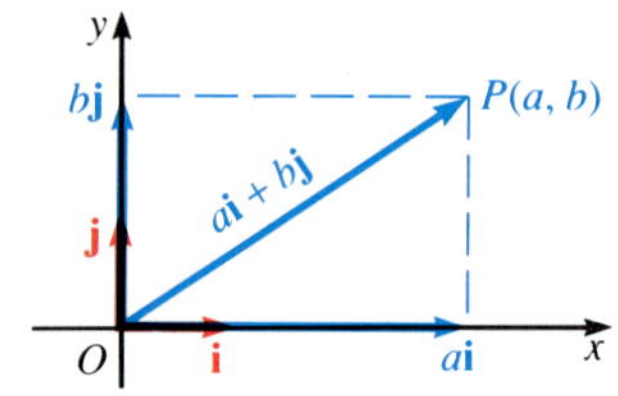

If $\mathbf{u} = a\mathbf{i} + b\mathbf{j}$, the number a is the ***x*-component** of **u** and the number b is the ***y*-component** of **u**. When **u** is written in the form $a\mathbf{i} + b\mathbf{j}$, vector **u** is said to be in **component form.**

Example 1 Given the points $A(-3, 4)$ and $B(2, 1)$, express $\overrightarrow{AB}$ in component form.

Solution 1 Draw an equivalent vector, say $\overrightarrow{OP}$, with its initial point at the origin, as shown below. Then P has coordinates $(5, -3)$, and $\overrightarrow{AB} = \overrightarrow{OP} = 5\mathbf{i} - 3\mathbf{j}$.

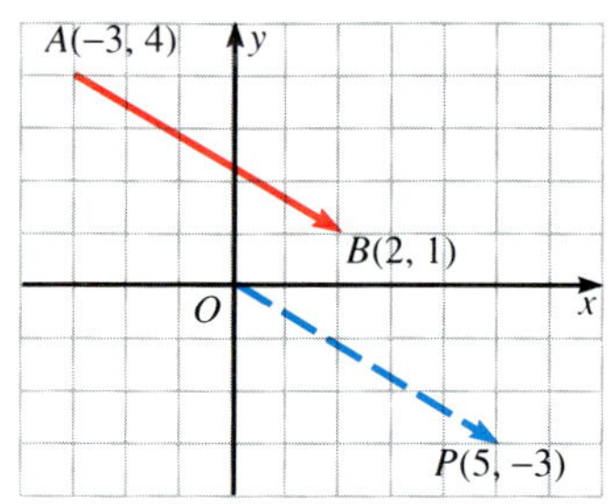

Solution 2 Notice that the x-component of $\overrightarrow{AB}$ is the difference of the x-coordinates of B and A. The y-component is the difference in the y-coordinates of B and A. $\therefore \overrightarrow{AB} = [2 - (-3)]\mathbf{i} + [1 - 4]\mathbf{j} = 5\mathbf{i} - 3\mathbf{j}$.

Example 2 Find the coordinates of B if A has coordinates $(3, -1)$ and $\overrightarrow{AB} = -\mathbf{i} + 2\mathbf{j}$.

Solution Draw the vector $\overrightarrow{OP} = -\mathbf{i} + 2\mathbf{j}$. Point P is 1 unit to the left and 2 units above O. If $\overrightarrow{AB} = \overrightarrow{OP}$, then B must be 1 unit to the left and 2 units above A.

$\therefore B$ has coordinates $(2, 1)$. ***Answer***

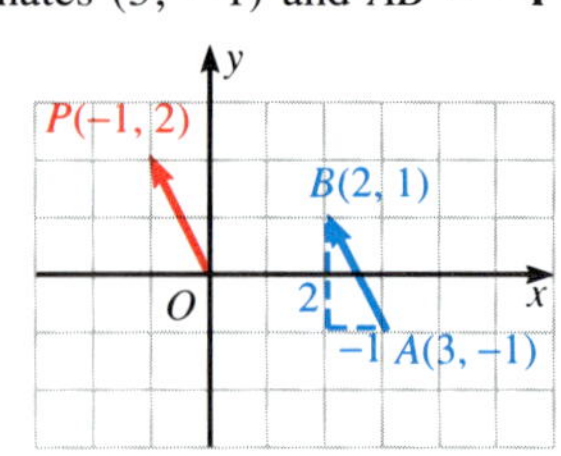

The concepts of Lesson 14-1 can also be expressed in terms of components.

If $\mathbf{u} = a\mathbf{i} + b\mathbf{j}$, $\mathbf{v} = c\mathbf{i} + d\mathbf{j}$ and t is a scalar, then:

$$\mathbf{u} = \mathbf{v} \text{ if and only if } a = c \text{ and } b = d$$
$$\mathbf{u} + \mathbf{v} = (a + c)\mathbf{i} + (b + d)\mathbf{j}$$
$$t\mathbf{u} = ta\mathbf{i} + tb\mathbf{j}$$
$$\|\mathbf{u}\| = \sqrt{a^2 + b^2}$$

The last statement, $\|\mathbf{u}\| = \sqrt{a^2 + b^2}$, results from applying the Pythagorean theorem to the vector diagram showing $\mathbf{u} = a\mathbf{i} + b\mathbf{j}$ (see the diagram below). The vectors $a\mathbf{i}$ and $b\mathbf{j}$ are perpendicular, and since $\mathbf{i}$ and $\mathbf{j}$ are unit vectors, $\|a\mathbf{i}\| = |a|$ and $\|b\mathbf{j}\| = |b|$.

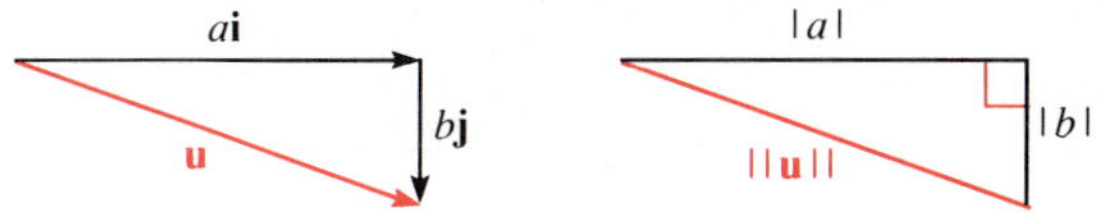

Example 3 Let $\mathbf{u} = \mathbf{i} + 2\mathbf{j}$, $\mathbf{v} = 4\mathbf{i} - 2\mathbf{j}$, and $\mathbf{w} = 2\mathbf{u} - \mathbf{v}$.

a. Find $\mathbf{w}$ in component form.
b. Draw a diagram showing $\mathbf{u}$, $\mathbf{v}$, and $\mathbf{w}$.
c. Find $\|\mathbf{w}\|$.
d. Find the angle γ that $\mathbf{w}$ makes with the positive x-axis.

Solution

a. $\mathbf{w} = 2\mathbf{u} - \mathbf{v} = 2(\mathbf{i} + 2\mathbf{j}) - (4\mathbf{i} - 2\mathbf{j})$
$= 2\mathbf{i} + 4\mathbf{j} - 4\mathbf{i} + 2\mathbf{j}$
$= -2\mathbf{i} + 6\mathbf{j}$

b.

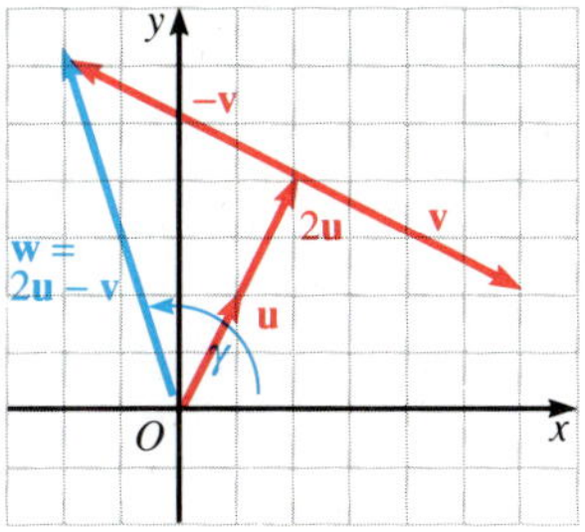

c. $\|\mathbf{w}\| = \sqrt{(-2)^2 + 6^2} = \sqrt{40} = 2\sqrt{10}$

d. From part (a), $\mathbf{w}$ has x-component -2 and y-component 6. So

$$\tan \gamma = \frac{6}{-2} = -3$$

and $\gamma = 108.4°$.

Example 4 The vector $\mathbf{u}$ makes an angle of 143° with the positive x-axis and $\|\mathbf{u}\| = 5$. Express $\mathbf{u}$ in component form.

Solution Let $\mathbf{u} = a\mathbf{i} + b\mathbf{j}$. If the initial point of $\mathbf{u}$ is placed at the origin, the terminal point will have coordinates (a, b). From the definitions of the trigonometric functions, you have

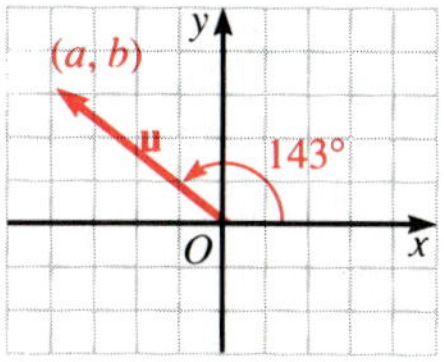

$$\cos 143° = \frac{a}{\|\mathbf{u}\|} \quad \text{and} \quad \sin 143° = \frac{b}{\|\mathbf{u}\|}.$$

(Solution continues on the next page.)

Chalkboard Examples

1. Given the points $A(-2, -3)$ and $B(-1, -1)$, express $\overrightarrow{AB}$ in component form. $\overrightarrow{AB} = [-1 - (-2)]i + [-1 - (-3)]j = i + 2j$

2. Let $\mathbf{u} = 2\mathbf{i} - 3\mathbf{j}$, $\mathbf{v} = 3\mathbf{i} + 2\mathbf{j}$, and $\mathbf{w} = 3\mathbf{u} + \mathbf{v}$.

a. Find $\mathbf{w}$ in component form.
$\mathbf{w} = 3(2\mathbf{i} - 3\mathbf{j}) + 3\mathbf{i} + 2\mathbf{j} = 9\mathbf{i} - 7\mathbf{j}$

b. Find $\|\mathbf{w}\|$.
$\|\mathbf{w}\| = \sqrt{81 + 49} = \sqrt{130}$

c. Find the angle γ that $\mathbf{w}$ makes with the positive x-axis.
$\tan \gamma = -\frac{7}{9}$
$\gamma = 322.1°$

3. The vector $\mathbf{u}$ makes an angle of 115° with the positive x-axis and $\|\mathbf{u}\| = 5$. Express $\mathbf{u}$ in component form.
$\mathbf{u} = a\mathbf{i} + b\mathbf{j}$
$\cos 115° = \frac{a}{\|\mathbf{u}\|}$ and
$\sin 115° = \frac{b}{\|\mathbf{u}\|}$
$a = -2.11$ and
$b = 4.53$
$\mathbf{u} = -2.11\mathbf{i} + 4.53\mathbf{j}$

4. Find the angle between $\mathbf{u} = 2\mathbf{i} + 3\mathbf{j}$ and $\mathbf{v} = 4\mathbf{i} + \mathbf{j}$.
Since $\mathbf{u} \cdot \mathbf{v} = \|\mathbf{u}\|\,\|\mathbf{v}\| \cos \theta$,
$\cos \theta = \frac{2 \cdot 4 + 3 \cdot 1}{\sqrt{2^2 + 3^2}\sqrt{4^2 + 1^2}} = \frac{11}{\sqrt{13 \cdot 17}} = 0.7399.$
$\theta = 42.3°$

(continued)

Chalkboard Examples

(continued)

5. Find a unit vector orthogonal to $\mathbf{u} = 3\mathbf{i} + 2\mathbf{j}$.
Let $\mathbf{v} = x\mathbf{i} + y\mathbf{j}$. Then $\mathbf{u} \cdot \mathbf{v} = 3x + 2y = 0$. One solution is $x = -2$ and $y = 3$. Then $-2\mathbf{i} + 3\mathbf{j}$ is orthogonal to $3\mathbf{i} + 2\mathbf{j}$. Since any nonzero vector divided by its norm has length 1, the vector $-\frac{2\sqrt{13}}{13}\mathbf{i} + \frac{3\sqrt{13}}{13}\mathbf{j}$ is a unit vector orthogonal to $\mathbf{u}$.

Check for Understanding

Oral Exs. 1–4: use before Example 3.
Oral Exs. 5–12: use before Example 5.

Guided Practice

1. For $A(-2, -1)$ and $B(2, -5)$, express AB in component form. $4\mathbf{i} - 4\mathbf{j}$
2. Find the coordinates of A given $B(5, 2)$ and $\overrightarrow{AB} = 3\mathbf{i} - \mathbf{j}$. $(2, 3)$
3. Find s and t if $(2s + 3)\mathbf{i} - 5\mathbf{j} = 8\mathbf{i} + (s + t)\mathbf{j}$.
$s = \frac{5}{2}$; $t = -\frac{15}{2}$

Solve: $\cos 143° = \frac{a}{\|\mathbf{u}\|}$ $\qquad$ $\sin 143° = \frac{b}{\|\mathbf{u}\|}$

$$\begin{aligned} a &= \|\mathbf{u}\| \cdot \cos 143° & b &= \|\mathbf{u}\| \cdot \sin 143° \\ &= 5(-0.7986) & &= 5(0.6018) \\ &= -3.99 & &= 3.01 \end{aligned}$$

Therefore, $\mathbf{u} = -3.99\mathbf{i} + 3.01\mathbf{j}$. ***Answer***

The angle between two nonzero vectors is the angle θ, $0° \le \theta \le 180°$, that the vectors determine when their initial points are placed together. This angle is used to define the *dot product* of two vectors.

The **dot product** of two nonzero vectors **u** and **v** is defined to be

$$\mathbf{u} \cdot \mathbf{v} = \|\mathbf{u}\| \, \|\mathbf{v}\| \cos \theta$$

where θ is the angle between **u** and **v.**

Caution: The dot product is a real number, not a vector.

The following useful theorem is proved in Exercise 31, page 670.

Theorem

If $\mathbf{u} = a\mathbf{i} + b\mathbf{j}$ and $\mathbf{v} = c\mathbf{i} + d\mathbf{j}$, then

$$\mathbf{u} \cdot \mathbf{v} = ac + bd.$$

Example 5 Use the dot product to find the angle between **u** and **v** if $\mathbf{u} = -2\mathbf{i} + \mathbf{j}$ and $\mathbf{v} = 4\mathbf{i} + 3\mathbf{j}$.

Solution By the definition of dot product, $\mathbf{u} \cdot \mathbf{v} = \|\mathbf{u}\| \cdot \|\mathbf{v}\| \cos \theta$. Solving for $\cos \theta$ gives

$$\cos \theta = \frac{\mathbf{u} \cdot \mathbf{v}}{\|\mathbf{u}\| \cdot \|\mathbf{v}\|} = \frac{(-2) \cdot 4 + 1 \cdot 3}{\sqrt{(-2)^2 + 1^2}\sqrt{4^2 + 3^2}}$$

$$= \frac{-5}{5\sqrt{5}} = \frac{-\sqrt{5}}{5} = -0.4472.$$

Therefore, $\theta = 116.6°$. ***Answer***

Two vectors are called **orthogonal** if either vector is **0** or if they are perpendicular. Since $\cos 90° = 0$, the dot product of two vectors will be 0 whenever they are orthogonal.

Vectors **u** and **v** are orthogonal if and only if $\mathbf{u} \cdot \mathbf{v} = 0$.

Example 6 Find a unit vector orthogonal to $\mathbf{u} = 4\mathbf{i} - 3\mathbf{j}$.

Solution Let $\mathbf{v} = x\mathbf{i} + y\mathbf{j}$. Then $\mathbf{u}$ and $\mathbf{v}$ are orthogonal if and only if

$$\mathbf{u} \cdot \mathbf{v} = 4x - 3y = 0.$$

One solution of this equation is (3, 4), since $4(3) - 3(4) = 0$. So the vector $\mathbf{v} = 3\mathbf{i} + 4\mathbf{j}$ is orthogonal to $\mathbf{u}$. To find a unit vector orthogonal to $\mathbf{u}$, use the fact that any nonzero vector divided by its norm has length 1. So,

$$\frac{\mathbf{v}}{\|\mathbf{v}\|} = \frac{3\mathbf{i} + 4\mathbf{j}}{\sqrt{3^2 + 4^2}} = \frac{3\mathbf{i} + 4\mathbf{j}}{5} = \frac{3}{5}\mathbf{i} + \frac{4}{5}\mathbf{j}$$

is a unit vector orthogonal to $\mathbf{u}$. ***Answer***

Oral Exercises

Find *s* and *t*, given the following.

1. $s\mathbf{i} + t\mathbf{j} = 5\mathbf{i} - 2\mathbf{j}$ $s = 5; t = -2$
2. $-7\mathbf{i} + t\mathbf{j} = s\mathbf{i} + 7\mathbf{j}$ $s = -7; t = 7$
3. $(s + 2)\mathbf{i} + \mathbf{j} = 6\mathbf{i} + t\mathbf{j}$ $s = 4; t = 1$
4. $4s\mathbf{i} - t\mathbf{j} = 8\mathbf{j}$ $s = 0; t = -8$

If $\mathbf{u} = 3\mathbf{i} - 2\mathbf{j}$ and $\mathbf{v} = 4\mathbf{i} + 5\mathbf{j}$, find the following.

5. $\mathbf{u} + \mathbf{v}$ $7\mathbf{i} + 3\mathbf{j}$
6. $\mathbf{v} - \mathbf{u}$ $\mathbf{i} + 7\mathbf{j}$
7. $2\mathbf{v}$ $8\mathbf{i} + 10\mathbf{j}$
8. $2\mathbf{v} + \mathbf{u}$ $11\mathbf{i} + 8\mathbf{j}$
9. $\|\mathbf{u}\|$ $\sqrt{13}$
10. $\|\mathbf{v}\|$ $\sqrt{41}$
11. $\mathbf{u} \cdot \mathbf{v}$ 2
12. $\mathbf{v} \cdot \mathbf{u}$ 2

Written Exercises

Refer to the diagram at the right and express each vector in component form.

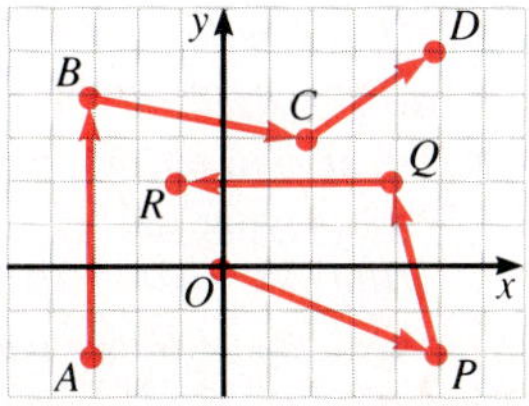

A

1. $\overrightarrow{OP}$ $5\mathbf{i} - 2\mathbf{j}$
2. $\overrightarrow{PQ}$ $-\mathbf{i} + 4\mathbf{j}$
3. $\overrightarrow{QR}$ $-5\mathbf{i}$
4. $\overrightarrow{AB}$ $6\mathbf{j}$
5. $\overrightarrow{BC}$ $5\mathbf{i} - \mathbf{j}$
6. $\overrightarrow{CD}$ $3\mathbf{i} + 2\mathbf{j}$

Find the coordinates of *B*, given the following.

7. $A(4, -3)$, $\overrightarrow{AB} = 2\mathbf{i} - 4\mathbf{j}$ (6, −7)
8. $A(-5, 1)$, $\overrightarrow{AB} = 2\mathbf{i} + \mathbf{j}$ (−3, 2)

Find the coordinates of *A*, given the following.

9. $B(4, 5)$, $\overrightarrow{AB} = 2\mathbf{i} - \mathbf{j}$ (2, 6)
10. $B(0, -4)$, $\overrightarrow{AB} = 6\mathbf{i} - \mathbf{j}$ (−6, −3)

Find *s* and *t*, given that

11. $(s + 1)\mathbf{i} + 5\mathbf{j} = 2\mathbf{i} + (t - 2)\mathbf{j}$ $s = 1; t = 7$
12. $(3s - 2)\mathbf{i} + t\mathbf{j} = s\mathbf{i} + (2t - 3)\mathbf{j}$ $s = 1; t = 3$
13. $(s + t)\mathbf{i} + (s - t)\mathbf{j} = 5\mathbf{i} - \mathbf{j}$ $s = 2; t = 3$
14. $(s + 2t)\mathbf{i} + (s - t)\mathbf{j} = \mathbf{i}$ $s = \frac{1}{3}; t = \frac{1}{3}$

4. Let $\mathbf{u} = 2\mathbf{j}$, $\mathbf{v} = 3\mathbf{i} - \mathbf{j}$, and $\mathbf{w} = \mathbf{v} - 4\mathbf{u}$.
   a. Find $\mathbf{w}$ in component form. $3\mathbf{i} - 9\mathbf{j}$
   b. Find $\|\mathbf{w}\|$. $3\sqrt{10}$
   c. Find the angle to the nearest tenth of a degree that $\mathbf{w}$ makes with the positive x-axis. 288.4°
   d. Find a unit vector orthogonal to $\mathbf{w}$. $\frac{3\sqrt{10}}{10}\mathbf{i} + \frac{\sqrt{10}}{10}\mathbf{j}$
5. Use the dot product to find the angle between $\mathbf{u} = -\mathbf{i} + \mathbf{j}$ and $\mathbf{v} = 9\mathbf{i} + 12\mathbf{j}$. 81.9°

Summarizing the Lesson

In this lesson students learned to express vectors in component form and to use the dot product of two vectors. Ask students to state two formulas for the dot product, one involving magnitudes and directions and the other involving components.

Suggested Assignments

Average Alg. and Trig.
669/1–14, 15–29 odd
S 665/*P*: 7, 9

Extended Alg. and Trig.
669/1–14, 15–31 odd
S 665/*P*: 6, 8

Supplementary Materials

Study Guide	pp. 225–226
Practice Master	78
Resource Book	p. 163
Overhead Visual	12

Additional Answers Written Exercises

16. c.

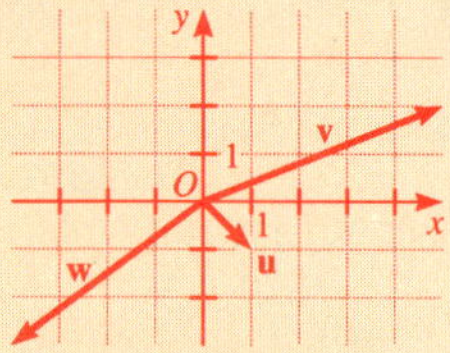

18. c.

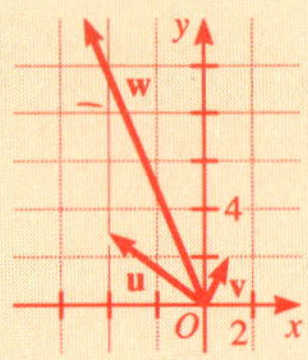

31. By the law of cosines, $\|\mathbf{u}-\mathbf{v}\|^2 = \|\mathbf{u}\|^2 + \|\mathbf{v}\|^2 - 2\|\mathbf{u}\|\,\|\mathbf{v}\|\cos\theta = \|\mathbf{u}\|^2 + \|\mathbf{v}\|^2 - 2\mathbf{u}\cdot\mathbf{v}$; $\mathbf{u}\cdot\mathbf{v} = \frac{1}{2}[\|\mathbf{u}\|^2 + \|\mathbf{v}\|^2 - \|\mathbf{u}-\mathbf{v}\|^2] = \frac{1}{2}[a^2 + b^2 + c^2 + d^2 - (a-c)^2 - (b-d)^2] = \frac{1}{2}[a^2 + b^2 + c^2 + d^2 - a^2 + 2ac - c^2 - b^2 + 2bd - d^2] = \frac{1}{2}(2ac + 2bd) = ac + bd$

32. Let $\mathbf{u} = a\mathbf{i} + b\mathbf{j}$; $\mathbf{u}\cdot\mathbf{i} = a\cdot 1 + b\cdot 0 = a$ and $\mathbf{u}\cdot\mathbf{j} = a\cdot 0 + b\cdot 1 = b$; $(\mathbf{u}\cdot\mathbf{i})\mathbf{i} + (\mathbf{u}\cdot\mathbf{j})\mathbf{j} = a\mathbf{i} + b\mathbf{j} = \mathbf{u}$

In Exercises 15–18:

a. Find w in component form.
b. Draw a diagram showing the vectors u, v, and w.
c. Find $\|\mathbf{w}\|$.
d. Find the angle γ, $0° \le \gamma \le 360°$, that w makes with the positive x-axis.

15. $\mathbf{u} = 4\mathbf{i} + 3\mathbf{j}$, $\mathbf{v} = 2\mathbf{i} - \mathbf{j}$; $\mathbf{w} = \mathbf{u} + \mathbf{v}$ **a.** $6\mathbf{i} + 2\mathbf{j}$ **c.** $2\sqrt{10}$ **d.** 18.4°

16. $\mathbf{u} = \mathbf{i} - \mathbf{j}$, $\mathbf{v} = 5\mathbf{i} + 2\mathbf{j}$; $\mathbf{w} = \mathbf{u} - \mathbf{v}$ **a.** $-4\mathbf{i} - 3\mathbf{j}$ **c.** 5 **d.** 216.9°

17. $\mathbf{u} = 4\mathbf{i} + 3\mathbf{j}$, $\mathbf{v} = 2\mathbf{i} - \mathbf{j}$; $\mathbf{w} = 2\mathbf{u} - \mathbf{v}$ **a.** $6\mathbf{i} + 7\mathbf{j}$ **c.** $\sqrt{85}$ **d.** 49.4°

18. $\mathbf{u} = -4\mathbf{i} + 3\mathbf{j}$, $\mathbf{v} = \mathbf{i} + 2\mathbf{j}$; $\mathbf{w} = 2\mathbf{u} + 3\mathbf{v}$ **a.** $-5\mathbf{i} + 12\mathbf{j}$ **c.** 13 **d.** 112.6°

Use the dot product to find the angle between u and v.

19. $\mathbf{u} = 4\mathbf{i} + 3\mathbf{j}$, $\mathbf{v} = 2\mathbf{i} - \mathbf{j}$ 63.4°

20. $\mathbf{u} = 5\mathbf{i} - 3\mathbf{j}$, $\mathbf{v} = 2\mathbf{i} + 4\mathbf{j}$ 94.4°

21. $\mathbf{u} = 2\mathbf{i} - \mathbf{j}$, $\mathbf{v} = \mathbf{i} + 3\mathbf{j}$ 98.1°

22. $\mathbf{u} = 2\mathbf{i} + 3\mathbf{j}$, $\mathbf{v} = 6\mathbf{i} - 4\mathbf{j}$ 90°

Find a unit vector orthogonal to v. Answers may vary.

23. $\mathbf{v} = \mathbf{i} + \mathbf{j}$ $\frac{\sqrt{2}}{2}\mathbf{i} - \frac{\sqrt{2}}{2}\mathbf{j}$

24. $\mathbf{v} = 5\mathbf{i} - 12\mathbf{j}$ $\frac{12}{13}\mathbf{i} + \frac{5}{13}\mathbf{j}$

25. $\mathbf{v} = 3\mathbf{i} - 4\mathbf{j}$ $\frac{4}{5}\mathbf{i} + \frac{3}{5}\mathbf{j}$

26. $\mathbf{v} = 2\mathbf{i} + 4\mathbf{j}$ $\frac{2\sqrt{5}}{5}\mathbf{i} - \frac{\sqrt{5}}{5}\mathbf{j}$

In Exercises 27–30, α is the angle that vector u makes with the positive x-axis, and β is the angle that vector v makes with the positive x-axis.

a. Express u and v in component form.
b. Find u · v using the definition of dot product.
c. Find u · v using the theorem on page 668.

Give answers to three significant digits.

B

27. $\|\mathbf{u}\| = 8$, $\alpha = 20°$; $\|\mathbf{v}\| = 17$, $\beta = 80°$

28. $\|\mathbf{u}\| = 6$, $\alpha = 15°$; $\|\mathbf{v}\| = 10$, $\beta = 65°$

29. $\|\mathbf{u}\| = 12.5$, $\alpha = 68°$; $\|\mathbf{v}\| = 18.0$, $\beta = 116°$

30. $\|\mathbf{u}\| = 8.70$, $\alpha = 57°$; $\|\mathbf{v}\| = 6.60$, $\beta = -28°$

27. a. $\mathbf{u} = 7.52\mathbf{i} + 2.74\mathbf{j}$; $\mathbf{v} = 2.95\mathbf{i} + 16.7\mathbf{j}$ **b.** 68.0 **c.** 67.9

28. a. $\mathbf{u} = 5.80\mathbf{i} + 1.55\mathbf{j}$; $\mathbf{v} = 4.23\mathbf{i} + 9.06\mathbf{j}$ **b.** 38.6 **c.** 38.6

29. a. $\mathbf{u} = 4.68\mathbf{i} + 11.6\mathbf{j}$; $\mathbf{v} = -7.89\mathbf{i} + 16.2\mathbf{j}$ **b.** 151 **c.** 151

30. a. $\mathbf{u} = 4.74\mathbf{i} + 7.30\mathbf{j}$; $\mathbf{v} = 5.83\mathbf{i} - 3.10\mathbf{j}$ **b.** 5.00 **c.** 5.00

C

31. Let $\mathbf{u} = a\mathbf{i} + b\mathbf{j}$ and $\mathbf{v} = c\mathbf{i} + d\mathbf{j}$. Prove that $\mathbf{u}\cdot\mathbf{v} = ac + bd$ by applying the law of cosines to the triangle shown. [*Hint:* $\|\mathbf{u} - \mathbf{v}\|^2 = (a - c)^2 + (b - d)^2$].

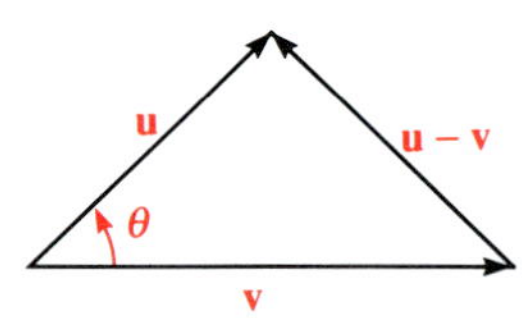

32. Prove that for any vector $\mathbf{u}$, $\mathbf{u} = (\mathbf{u}\cdot\mathbf{i})\mathbf{i} + (\mathbf{u}\cdot\mathbf{j})\mathbf{j}$.

Computer Exercises

For students with some programming experience.

1. Write a program to find the magnitude of a vector given in component form, and the angle in degrees (measured counterclockwise) that it makes with the positive x-axis. Be aware that the function ATN(Z) returns the angle in *radians* in Quadrant I or IV whose tangent is Z. If the vector is vertical, you cannot use the ATN function to find the direction angle.

2. Run the program in Exercise 1 to find the magnitude of each vector and the angle it makes with the positive x-axis.

a. $7\mathbf{i} - 6\mathbf{j}$ **b.** $-15\mathbf{i} + 8\mathbf{j}$ **c.** $-57.3\mathbf{i} - 28.9\mathbf{j}$

3. Write a program to find the resultant of two vectors. Given the magnitudes of two vectors **u** and **v** and the angle in degrees that each makes with the positive x-axis, the program should find the magnitude of $\mathbf{u} + \mathbf{v}$ and the angle that it makes with the positive x-axis. (*Hint:* Convert the given vectors to component form and add them. Then use the program of Exercise 1.)

4. Run the program in Exercise 3 to find the sum of each pair **u** and **v**. α and β are the angles that **u** and **v** make with the positive x-axis.

a. $\|\mathbf{u}\| = 12.5$, $\alpha = 26°$; $\|\mathbf{v}\| = 17.4$, $\beta = 115°$ 21.6009675; 79.6485928°

b. $\|\mathbf{u}\| = 7.03$, $\alpha = -15°$; $\|\mathbf{v}\| = 9.22$, $\beta = 78°$ 11.2980008; 39.5829305°

c. $\|\mathbf{u}\| = 52.8$, $\alpha = 129°$; $\|\mathbf{v}\| = 43.3$, $\beta = 216°$ 70.0145352; 167.140754°

d. $\|\mathbf{u}\| = 431$, $\alpha = 65°$; $\|\mathbf{v}\| = 256$, $\beta = 190°$ 353.163902; 101.425964°

e. $\|\mathbf{u}\| = 30$, $\alpha = 210°$; $\|\mathbf{v}\| = 40$, $\beta = 300°$ 50.0000001; 263.130102°

Self-Test 1

Vocabulary vector quantity (p. 659), vector (p. 659), initial point (p. 659), terminal point (p. 659), equivalent vectors (p. 659), sum of vectors (p. 660), resultant (p. 660), parallelogram rule (p. 660), difference of vectors (p. 661), norm (p. 661), unit vector (p. 661), bearing (p. 661), x-component (p. 666), y-component (p. 666), component form (p. 666), dot product (p. 668), orthogonal (p. 668)

1. If **u** has bearing 37° and magnitude 5, and **v** has bearing 125° and magnitude 6, find the magnitude of $2\mathbf{u} + \mathbf{v}$ and find its bearing to the nearest degree. 12; 67° **Obj. 14-1, p. 659**

2. An ocean current flows due south at 15 km/h. If a ship is going to travel through this current on a course with heading 130° at a speed of 20 km/h, find its true-course bearing. 151.2°

3. Find **w** in component form if $\mathbf{u} = 4\mathbf{i} + 3\mathbf{j}$, $\mathbf{v} = -2\mathbf{i} + 7\mathbf{j}$, and $\mathbf{w} = 2\mathbf{u} + 3\mathbf{v}$. 2i + 27j **Obj. 14-2, p. 666**

4. Use the dot product to find the angle between $\mathbf{u} = 5\mathbf{i} + 2\mathbf{j}$ and $\mathbf{v} = 8\mathbf{i} - 3\mathbf{j}$ to the nearest tenth of a degree. 42.4°

Check your answers with those at the back of the book.

Additional Answers
Computer Exercises

2. a. magnitude = 9.21954447; angle = 319.398705°
   b. magnitude = 17; angle = 151.927513°
   c. magnitude = 64.1755406; angle = 206.764687°

Using a Computer

Students will be using the component form of vectors in these exercises. Review how to convert from degrees to radians in order to use the computer's built-in trigonometric functions.

Quick Quiz

1. If **u** has bearing 165° and magnitude 3 and **v** has bearing 75° and magnitude 2, find the magnitude and bearing of $2\mathbf{v} - \mathbf{u}$. 5; 38.1°
2. A pilot wants to fly a plane on a true-course bearing of 100° and with a ground speed of 350 km/h. If there is a 50 km/h wind from the southwest, find the approximate heading and air speed for the plane. 107.3°; 324 km/h
3. Find **w** in component form and $\|\mathbf{w}\|$ if $\mathbf{w} = 2\mathbf{u} + \mathbf{v}$, $\mathbf{u} = 4\mathbf{i} - \mathbf{j}$ and $\mathbf{v} = -2\mathbf{i} + \mathbf{j}$. $6\mathbf{i} - \mathbf{j}$; $\sqrt{37}$
4. Find the angle between **u** and **v** if $\mathbf{u} = 7\mathbf{i} - \mathbf{j}$ and $\mathbf{v} = -\mathbf{i} - \mathbf{j}$. 126.9°

Application / *Force, Work, and Energy*

In the metric system the basic unit of the vector quantity *force* is the **newton** (N). Near the surface of Earth, the force exerted by gravity on a mass of m kg has a magnitude of approximately $9.8m$ newtons, directed downward. When you work with forces, you often use the following principle:

> When an object is at rest or moving with constant velocity, the sum of the forces acting on it is **0**. (If the sum of the forces is not **0**, the object will accelerate, and its velocity will vary.)

Example 1 A loading ramp makes an angle of 25° with the horizontal. A 120 kg crate slides down the ramp with constant velocity. What is the frictional force acting on the crate?

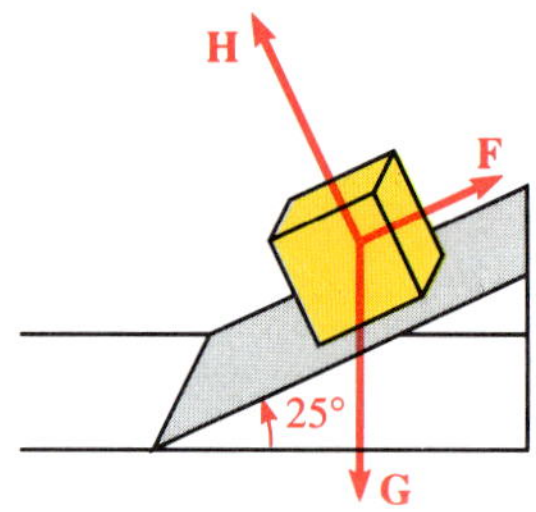

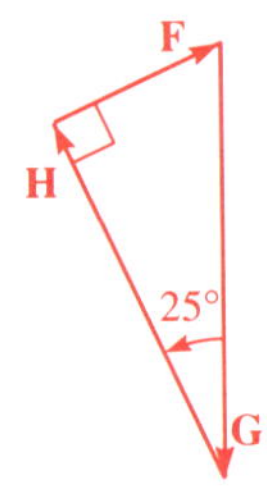

Solution The figure at the left above shows the three forces acting on the crate. The gravitational force $\mathbf{G}$ has magnitude

$$\|\mathbf{G}\| = 9.8 \times 120 = 1176 \text{ N}.$$

The forces $\mathbf{F}$ and $\mathbf{H}$ are respectively parallel and perpendicular to the ramp. $\mathbf{F}$ is the frictional force you need to find. Since the crate is moving with constant velocity, the sum of $\mathbf{G}$, $\mathbf{H}$, and $\mathbf{F}$ is $\mathbf{0}$, as indicated in the vector diagram at the right above. From the diagram,

$$\frac{\|\mathbf{F}\|}{\|\mathbf{G}\|} = \sin 25°.$$

If you solve this equation for $\|\mathbf{F}\|$, you obtain

$$\|\mathbf{F}\| = \|\mathbf{G}\| \sin 25° = (1176)(0.4226) = 497.$$

$\therefore$ the frictional force is 497 N up the ramp. ***Answer***

When you exert a force $\mathbf{F}$ on an object and move it from A to B, you do *work* and expend *energy*. If $\mathbf{F}$ has the same direction as $\mathbf{d} = \overrightarrow{AB}$, the **work** W done is

$$W = \|\mathbf{F}\|\,\|\mathbf{d}\|.$$

This is also the **energy** expended in moving the object from A to B.

The basic unit of work and energy is the *joule* (J). A **joule** is the work done when a force of one newton moves an object one meter in the direction of the force. Another unit of work and energy is the *kilowatt-hour* (kW · h). One **kilowatt-hour** is 3.6×10^6 J.

Example 2 How much energy in kilowatt-hours is needed to lift a 2500 kg elevator 80 m?

Solution The motor lifting the elevator must overcome the force of gravity and therefore must exert an upward force **F** of magnitude

$$2500 \times 9.8 = 2.45 \times 10^4 \text{ N}.$$

The displacement **d** is also upward and has norm 80 m. Therefore, the work done is

$$\begin{aligned} W &= \|\mathbf{F}\|\,\|\mathbf{d}\| \\ &= (2.45 \times 10^4) \times 80 \\ &= 1.96 \times 10^6 \text{ J}. \end{aligned}$$

$\therefore$ the energy needed is

$$\frac{1.96 \times 10^6}{3.6 \times 10^6} = 0.544 \text{ kW} \cdot \text{h}. \quad \textbf{\textit{Answer}}$$

Suppose now that a constant force **F** moves an object from A to B but makes an angle θ with $\mathbf{d} = \overrightarrow{AB}$ (see the diagram). In this case only the part of **F** in the direction of **d** is effective in doing work. The magnitude of this part is $\|\mathbf{F}\| \cos \theta$, and the work done by **F** is therefore

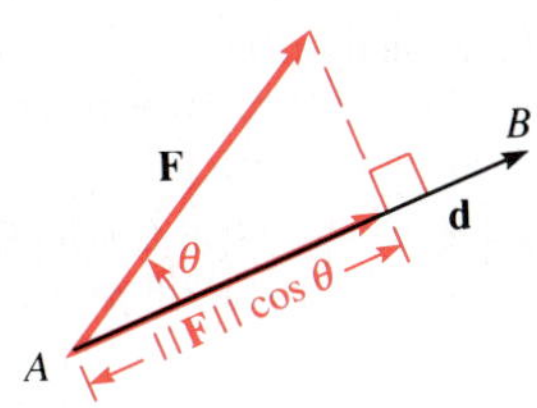

$$\|\mathbf{F}\| \cos \theta \times \|\mathbf{d}\|,$$

or

$$\|\mathbf{F}\|\,\|\mathbf{d}\| \cos \theta.$$

Since $\|\mathbf{F}\|\,\|\mathbf{d}\| \cos \theta = \mathbf{F} \cdot \mathbf{d}$, the work is given by

$$W = \mathbf{F} \cdot \mathbf{d}.$$

Example 3 Find the work done by the force $\mathbf{F} = 6\mathbf{i} - 3\mathbf{j}$ in moving an object from $A(0, 1)$ to $B(5, 2)$. The force is in newtons and the distance is in meters.

Solution $\mathbf{d} = \overrightarrow{AB} = (5 - 0)\mathbf{i} + (2 - 1)\mathbf{j} = 5\mathbf{i} + \mathbf{j}$

$$\begin{aligned} \therefore W = \mathbf{F} \cdot \mathbf{d} &= (6\mathbf{i} - 3\mathbf{j}) \cdot (5\mathbf{i} + \mathbf{j}) \\ &= 6 \cdot 5 - 3 \cdot 1 = 27 \text{ J} \quad \textbf{\textit{Answer}} \end{aligned}$$

Additional Answers
Written Exercises

(continued from p. 679)

6.

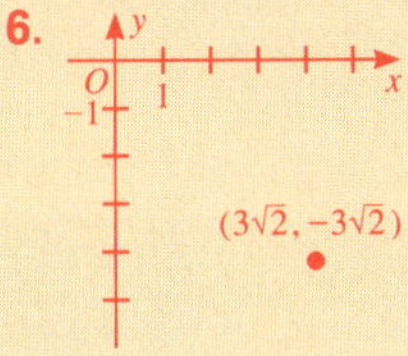

8.

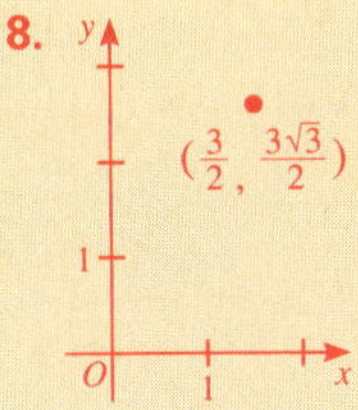

32.

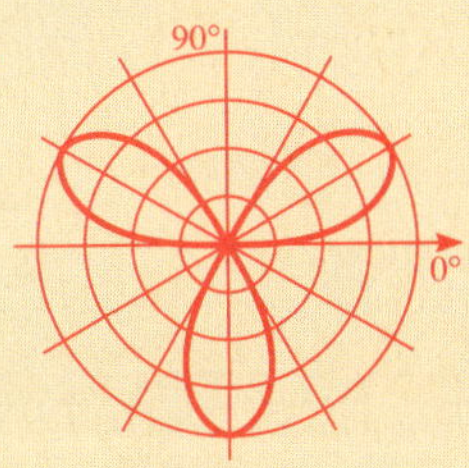

34.

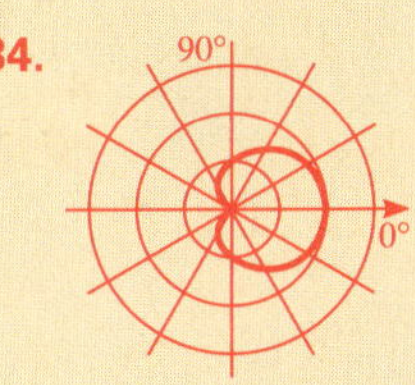

36.

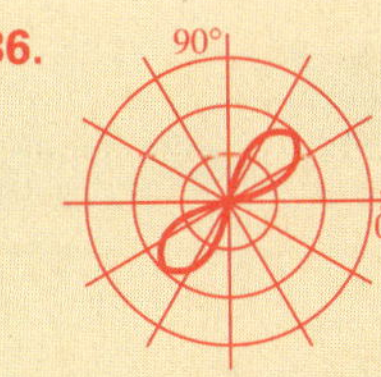

Additional Answers
Extra (from p. 679)

1.

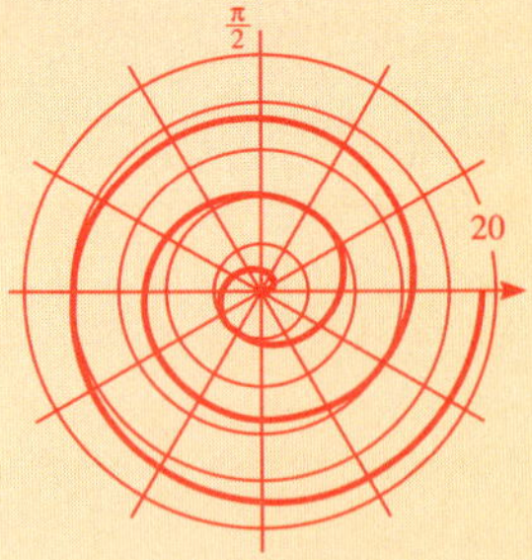

2.

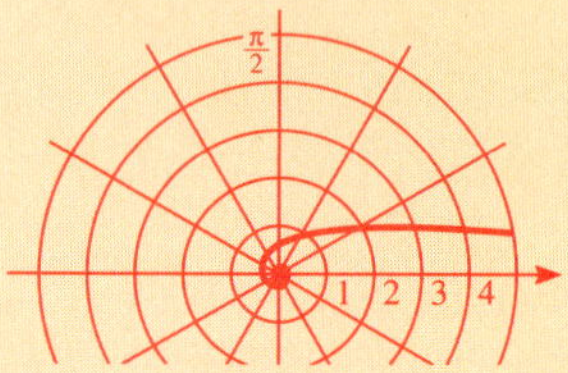

3.

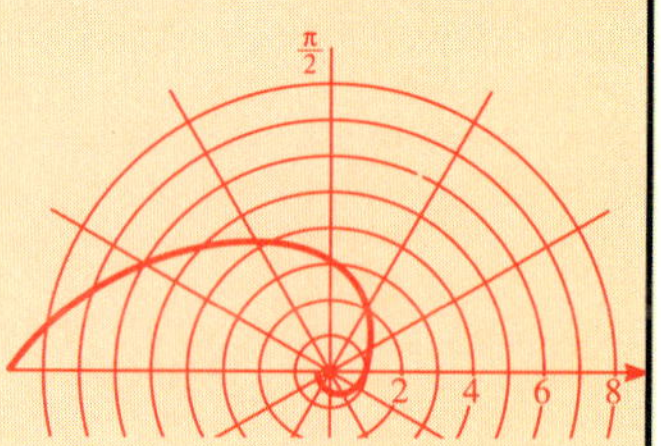

Exercises

Forces are in newtons and distances are in meters. Give work in joules unless kilowatt-hours are called for.

In Exercises 1 and 2, the two given forces act on an object. What additional force is needed to keep the object at rest?

A

1. $\mathbf{F}_1 = 3\mathbf{i} - 2\mathbf{j}$ $-2\mathbf{i} - \mathbf{j}$
$\mathbf{F}_2 = -\mathbf{i} + 3\mathbf{j}$

2. $\mathbf{F}_1 = 5\mathbf{i} - \mathbf{j}$ $-3\mathbf{i}$
$\mathbf{F}_2 = -2\mathbf{i} + \mathbf{j}$

In Exercises 3 and 4, find the work done by the force F in moving an object from *A* to *B*.

3. $\mathbf{F} = 3\mathbf{i} - \mathbf{j}$
$A(-2, 0)$, $B(2, 3)$ 9 J

4. $\mathbf{F} = 5\mathbf{i}$
$A(-3, 2)$, $B(5, -1)$ 40 J

5. How much work is done by the force $\mathbf{F} = 3\mathbf{i} - \mathbf{j}$ in moving an object **(a)** from $A(-2, 1)$ to $B(3, 3)$ to $C(5, 0)$? **(b)** from $A(-2, 1)$ to $C(5, 0)$? **a.** 22 J **b.** 22 J

6. What is the combined work done by the forces $\mathbf{F} = 2\mathbf{i} + 5\mathbf{j}$ and $\mathbf{G} = 4\mathbf{i} - 2\mathbf{j}$ in moving an object from $A(-1, 0)$ to $B(4, 1)$? What is the work done by $\mathbf{F} + \mathbf{G}$ in moving the object from A to B? 33 J; 33 J

In Exercises 7 and 8, the given forces act on an object. Find the magnitude and bearing of the additional force F that will keep the object at rest.

7. $\mathbf{F}_1$: 10 N, bearing 120°
$\mathbf{F}_2$: 5 N, bearing 180° 13.2 N; 319°

8. $\mathbf{F}_1$: 6 N, bearing 300°
$\mathbf{F}_2$: 15 N, bearing 30° 16.2 N; 188.2°

9. The mass of a loaded helicopter is 1500 kg. How much energy does its engine expend in ascending vertically for 100 m? 1.47×10^6 J

10. A ramp makes a 20° angle with the horizontal. A 50 kg crate slides down the ramp with constant velocity. Find the frictional force acting on the crate. 168 N

11. How much energy does a 75 kg man expend in climbing up a 12 m ladder that makes a 78° angle with the horizontal? 8.63×10^3 J

B

12. A conveyer belt carries coal up a 22° slope through a distance of 5 km measured along the slope. How many kilowatt-hours of energy are used in transporting 120 metric tons of coal? (1 metric ton = 1000 kg.) 612 kW · h

13. A sled is pulled for 200 m up a 23° incline by a force of 120 N making a 40° angle with the horizontal. Find the work done. 2.30×10^4 J

In Exercises 14 and 15, use the fact that the tension in a wire, cable, or rope is the magnitude of the force it exerts.

14. A 20 kg mirror is hung from the middle of a wire as shown. Find the tension in the wire if each half of it makes a 30° angle with the horizontal. 196 N

15. A 200 kg load is suspended from two cables that make angles of 40° and 60° with the horizontal. Find the tension in each cable. 995 N; 1525 N

Polar Coordinates and Complex Numbers

14-3 Polar Coordinates

Objective To define polar coordinates and graph polar equations.

If you know how far away an object is and in what direction, you can locate the object. This is the principle of the *polar coordinate system,* which is the coordinate system used to show the position of stars on a map such as the one shown at the right.

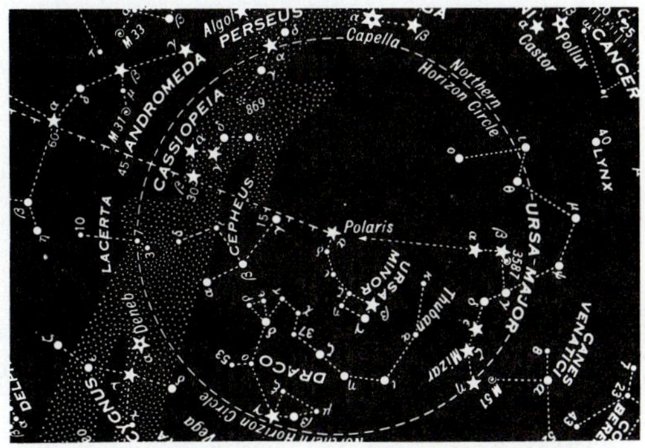

A **polar coordinate system** consists of a point O called the **pole,** and a ray, called the **polar axis,** having O as its endpoint (see Figure 1). The **polar coordinates** of a point P are an ordered pair (r, θ), where $r = OP$ and θ is the measure of an angle from the polar axis to the segment $\overline{OP}$.

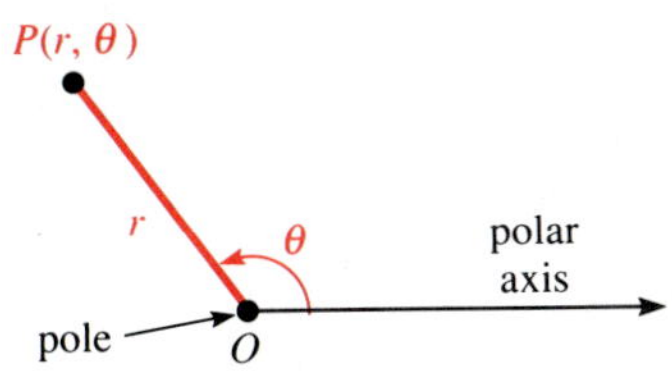

Figure 1

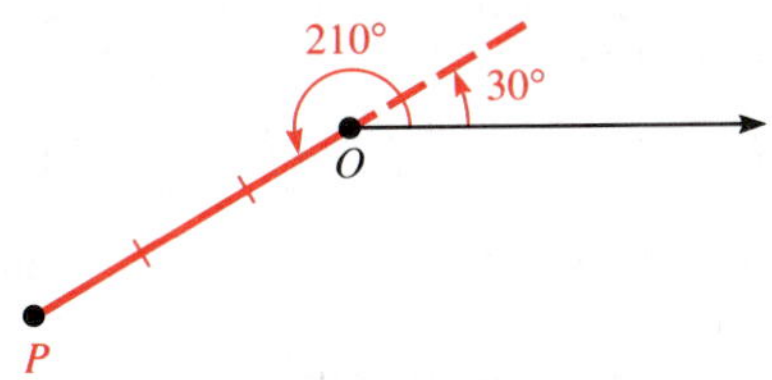

Figure 2

Each point has many pairs of polar coordinates. For example, $(3, 210°)$, $(3, -150°)$, and $(3, 570°)$ are all polar coordinates of the point P in Figure 2. *Negative* values of r are used to indicate that P is on the ray *opposite* to the terminal side of θ. Therefore, $(-3, 30°)$ and $(-3, 390°)$ are also polar coordinates of P in Figure 2.

Example 1 Graph these points in the same polar coordinate system:

$A(3, 120°)$
$B(-3, 45°)$
$C(4, -30°)$

Solution

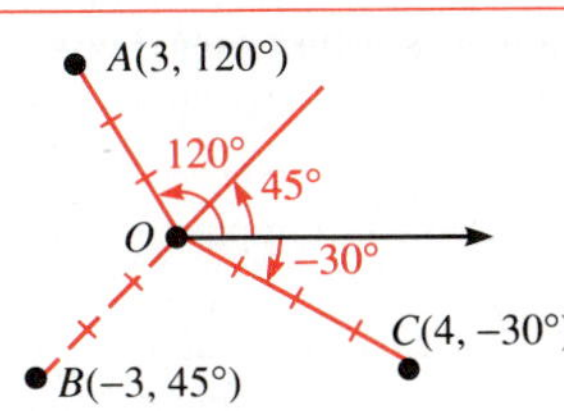

Explorations p. 845

Teaching Suggestions p. T133

Suggested Extensions p. T134

Group Activities p. T134

Warm-Up Exercises

Find the distance of each point from the origin.

1. $(2, 2)$ $2\sqrt{2}$
2. $(-3, 4)$ 5
3. $(-5, -12)$ 13
4. $(4, -6)$ $2\sqrt{13}$
5. $(-1, 0)$ 1

Find the value of r for each value of θ if $r = 2 \sin \theta$.

6. $\theta = 90°$ 2
7. $\theta = 180°$ 0
8. $\theta = 45°$ $\sqrt{2}$
9. $\theta = 210°$ -1
10. $\theta = 240°$ $-\sqrt{3}$

Motivating the Lesson

Tell the students that a weather forecaster refers to a "weather radar" to determine the direction and distance of local storms. In so doing, the forecaster is using polar coordinates, the topic of today's lesson.

Chalkboard Examples

1. Graph the points $A(4, 60°)$ and $B(-3, 45°)$.

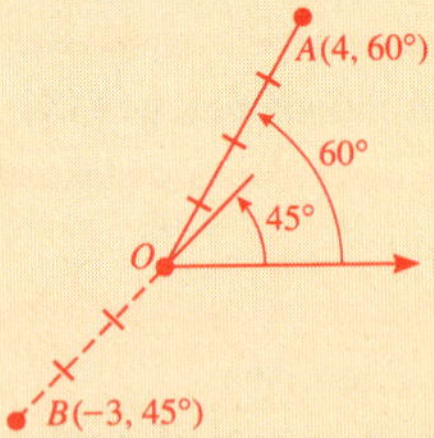

2. Convert:
   a. $A(-3, 3)$ to polar coordinates.
   b. $B(-2, 30°)$ to rectangular coordinates.

   a. $r = \pm\sqrt{(-3)^2 + (3)^2} = \pm 3\sqrt{2}$. Choose $3\sqrt{2}$ so that the terminal side passes through A and θ is a second-quadrant angle. Choose 135° for θ. Thus the polar coordinates are $(3\sqrt{2}, 135°)$.

   b. $\cos 30° = \frac{\sqrt{3}}{2}$ and $\sin 30° = \frac{1}{2}$. So $x = (-2)\left(\frac{\sqrt{3}}{2}\right)$ and $y = (-2)\left(\frac{1}{2}\right)$. The rectangular coordinates of B are $(-\sqrt{3}, -1)$.

3. Graph $r = \sin 3\theta$.

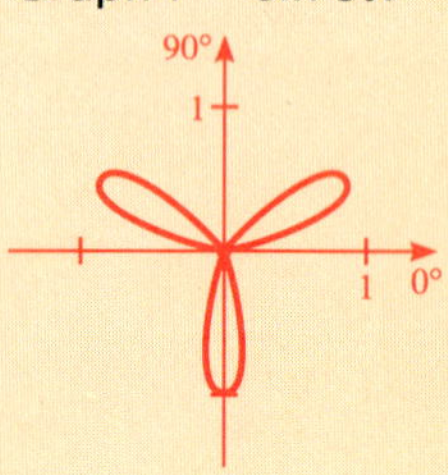

When polar coordinates and rectangular coordinates are used together, the polar axis is taken to coincide with the nonnegative x-axis, as shown in the diagram. The following formulas, which can be derived using the diagram, will help you to convert from one coordinate system to the other.

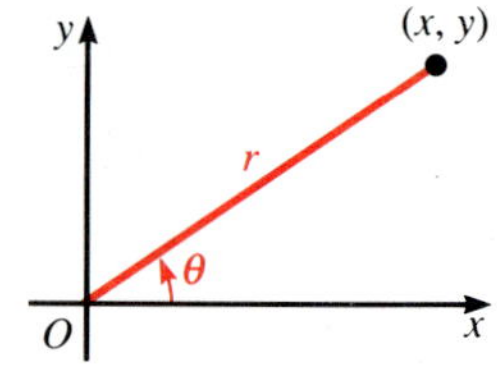

Coordinate-System Conversion Formulas

From polar to rectangular	*From rectangular to polar*
$x = r\cos\theta$	$r = \pm\sqrt{x^2 + y^2}$
$y = r\sin\theta$	$\cos\theta = \frac{x}{r}, \quad \sin\theta = \frac{y}{r}$

Example 2 **a.** Convert $A(-3, 30°)$ to rectangular coordinates.
b. Convert $B(4, -4)$ to polar coordinates.

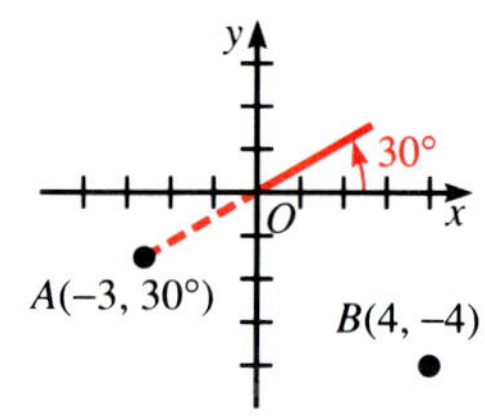

Solution **a.** $x = r\cos\theta = -3\cos 30° = -3\left(\frac{\sqrt{3}}{2}\right) = -\frac{3\sqrt{3}}{2}$

$y = r\sin\theta = -3\sin 30° = -3\left(\frac{1}{2}\right) = -\frac{3}{2}$

rectangular coordinates: $\left(-\frac{3\sqrt{3}}{2}, -\frac{3}{2}\right)$

b. $r = \pm\sqrt{x^2 + y^2} = \pm\sqrt{(4)^2 + (-4)^2} = \pm 4\sqrt{2}$.
If you choose the positive value of r, then the terminal side of θ passes through B and B is a fourth quadrant angle. Since

$$\sin\theta = \frac{-4}{4\sqrt{2}} = -\frac{\sqrt{2}}{2},$$

you can use 315° for θ.

polar coordinates: $(4\sqrt{2}, 315°)$

Choosing the negative value of r would result in answers like $(-4\sqrt{2}, 135°)$ or $(-4\sqrt{2}, -225°)$, which are also correct.

Just like with graphs of equations in x and y, the graph of a polar equation in r and θ is the set of all points (r, θ) that satisfy the equation. To graph polar equations, you use your knowledge of the trigonometric functions.

Example 3 Graph $r = 3\cos 3\theta$.

Solution To graph an equation in polar coordinate form, a polar coordinate graph should be used. Construct a table for θ.

θ	3θ	$\cos 3\theta$	$3 \cos 3\theta$
0°	0°	1.0000	3.00
15°	45°	0.7071	2.12
30°	90°	0.0000	0.00
45°	135°	−0.7071	−2.12
60°	180°	−1.0000	−3.00
75°	225°	−0.7071	−2.12
90°	270°	0.0000	0.00
105°	315°	0.7071	2.12
120°	360°	1.0000	3.00
135°	405°	0.7071	2.12
150°	450°	0.0000	0.00
165°	495°	−0.7071	−2.12
180°	540°	−1.0000	−3.00

Because the equation involves 3θ it is helpful to consider values of θ at intervals of 15°.

As θ increases from 180° to 360° the graph is repeated. The complete graph, called a *three-leafed rose,* is shown at the right.

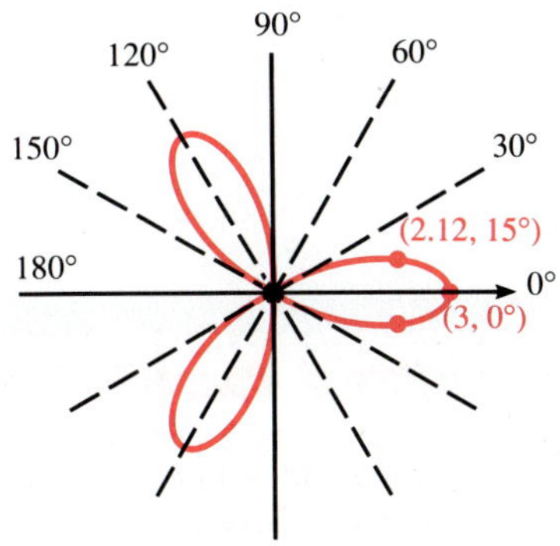

Sometimes a polar equation can be graphed more easily by using its rectangular form. The conversion formulas given on the preceding page can be used to transform equations from one system to another.

Example 4 **a.** Transform $r(1 + \cos \theta) = 3$ to a rectangular-coordinate equation.

b. Transform $x^2 + y^2 = 3y$ to a polar-coordinate equation.

Solution **a.**

$$r(1 + \cos \theta) = 3$$
$$r + r \cos \theta = 3$$
$$r = 3 - r \cos \theta$$
$$\sqrt{x^2 + y^2} = 3 - x$$
$$x^2 + y^2 = 9 - 6x + x^2$$
$$y^2 = 9 - 6x \quad \textbf{\textit{Answer}}$$

b.

$$x^2 + y^2 = 3y$$
$$r^2 = 3r \sin \theta$$
$$r = 3 \sin \theta$$

Answer

4. a. Transform $r^2(\cos \theta + \sin \theta) \cdot (\cos \theta - \sin \theta) = 1$ to rectangular coordinates.

$r^2(\cos \theta + \sin \theta) \cdot (\cos \theta - \sin \theta) = r^2 \cos^2 \theta - r^2 \sin^2 \theta = 1$

$\therefore x^2 - y^2 = 1$

b. Transform $x^2 + y^2 = 2xy$ to polar coordinates.

$x^2 + y^2 = r^2$ and $2xy = 2(r \cos \theta)(r \sin \theta) = r^2 \sin 2\theta.$

$\therefore r^2 = r^2 \sin 2\theta$ and $1 = \sin 2\theta.$

Thinking Skills

After completing Written Exercises 23–30, students should be able to *visualize* (before actually graphing) what the graphs of simple polar equations will look like. For example, for any nonzero constant k, the graph of $r = k$ is a circle centered at the pole, the graph of $\theta = k$ is a line passing through the pole, and so on.

Check for Understanding

Here is a suggested use of the Oral Exercises to check students' understanding as you teach the lesson.

Oral Exs. 1–8: use after Example 2.

Oral Exs. 9–12: use before Example 1.

Guided Practice

Plot the point and find rectangular coordinates in simplest radical form for:

1. $(-8, 300°)$
   $(-4, 4\sqrt{3})$

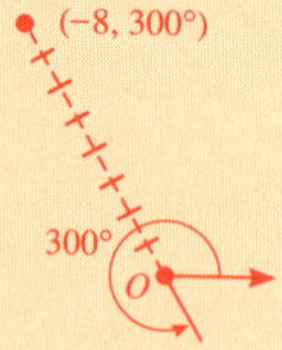

2. $(6, 45°)$
   $(3\sqrt{2}, 3\sqrt{2})$

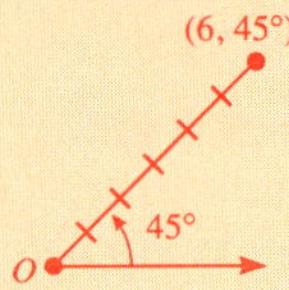

Find a pair of polar coordinates for each point.

3. $(5\sqrt{2}, -5\sqrt{2})$
4. $(-\sqrt{3}, 3)$

3, 4. Answers may vary. Examples are given.
3. $(10, -45°)$, $(-10, 135°)$
4. $(2\sqrt{3}, 120°)$, $(-2\sqrt{3}, -60°)$

Find a polar equation of the curve whose rectangular-coordinate equation is given.

5. $x = -3$
   $r \cos \theta = -3$
6. $x^2 + y^2 = 10y$
   $r = 10 \sin \theta$

Find a rectangular coordinate equation for each polar equation.

7. $\theta = 90°$
   $x = 0$
8. $r = 3$
   $x^2 + y^2 = 9$
9. $r \sin \theta + 4 = 0$
   $y = -4$

Oral Exercises

Give a pair of polar coordinates for each pair of rectangular coordinates.

1. $(-3, 0)$ $(3, 180°)$
2. $(0, -5)$ $(5, 270°)$
3. $(-5, 5)$ $(5\sqrt{2}, 135°)$
4. $(0, 0)$ $(0, 0°)$

Give a pair of rectangular coordinates for each pair of polar coordinates.

5. $(5, 180°)$ $(-5, 0)$
6. $(-3, 90°)$ $(0, -3)$
7. $(\sqrt{2}, 45°)$ $(1, 1)$
8. $(6, -135°)$ $(-3\sqrt{2}, -3\sqrt{2})$

Give three pairs of polar coordinates (r, θ) for each point, including one with negative r and one with negative θ. Answers may vary.

9. A
10. B
11. C
12. D

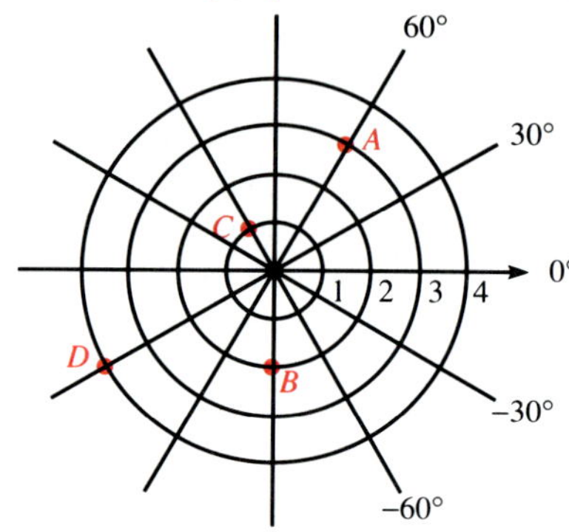

9. $(3, 60°)$, $(-3, 240°)$, $(3, -300°)$
10. $(2, 270°)$, $(-2, 90°)$, $(2, -90°)$
11. $(1, 120°)$, $(-1, 300°)$, $(1, -240°)$
12. $(4, 210°)$, $(-4, 30°)$, $(4, -150°)$

Written Exercises

3. $\left(\frac{3}{2}, -\frac{3\sqrt{3}}{2}\right)$ 4. $\left(\frac{5\sqrt{2}}{2}, -\frac{5\sqrt{2}}{2}\right)$

Plot the point whose polar coordinates are given and find its rectangular coordinates. Give answers in simplest radical form.

A

1. $(4, 30°)$ $(2\sqrt{3}, 2)$
2. $(2, 45°)$ $(\sqrt{2}, \sqrt{2})$
3. $(-3, 120°)$
4. $(-5, 135°)$
5. $(7, -60°)$ $\left(\frac{7}{2}, -\frac{7\sqrt{3}}{2}\right)$
6. $(6, -45°)$ $(3\sqrt{2}, -3\sqrt{2})$
7. $(-6, -150°)$ $(3\sqrt{3}, 3)$
8. $(-3, -120°)$ $\left(\frac{3}{2}, \frac{3\sqrt{3}}{2}\right)$

Find a pair of polar coordinates for each pair of rectangular coordinates.

9. $(4, 0)$ $(4, 0°)$
10. $(0, -3)$ $(3, 270°)$
11. $(-2, 2)$ $(2\sqrt{2}, 135°)$
12. $(\sqrt{3}, -1)$ $(2, 330°)$
13. $(-1, \sqrt{3})$ $(2, 120°)$
14. $(-\sqrt{2}, -\sqrt{2})$ $(2, 225°)$
15. $(-\sqrt{5}, \sqrt{5})$ $(\sqrt{10}, 135°)$
16. $(-\sqrt{2}, \sqrt{6})$ $(2\sqrt{2}, 120°)$

Find a polar equation for each rectangular-coordinate equation.

17. $x = 5$ $r \cos \theta = 5$
18. $y = -3$ $r \sin \theta = -3$
19. $y = x$ $\sin \theta = \cos \theta$
20. $x^2 + y^2 = 9$ $r = 3$
21. $x^2 + y^2 = 6y$ $r = 6 \sin \theta$
22. $x^2 + y^2 + 8x = 0$ $r + 8 \cos \theta = 0$

Find a rectangular-coordinate equation for each polar equation.

23. $r = 2$ $x^2 + y^2 = 4$
24. $\theta = 120°$ $y = -x\sqrt{3}$
25. $r \sin \theta = 2$ $y = 2$
26. $r \cos \theta = -1$ $x = -1$

B

27. $r = 2 \sin \theta$ $x^2 + y^2 = 2y$
28. $r + 2 \cos \theta = 0$ $x^2 + y^2 + 2x = 0$
29. $r(1 - \cos \theta) = 2$ $y^2 = 4 + 4x$
30. $r(1 - \sin \theta) = 1$ $x^2 = 2y + 1$

Graph using the method of Example 3.

31. $r = 2\cos 2\theta$ (Four-leafed rose)

32. $r = 4\sin 3\theta$ (Three-leafed rose)

33. $r = 1 + \sin\theta$

34. $r = 1 + \cos\theta$

(These heart-shaped curves are called *cardioids*.)

C **35.** $r^2 = 2\cos 2\theta$

36. $r^2 = 4\sin 2\theta$

(These curves are called *lemniscates*.)

Mixed Review Exercises

Find the quotient and the remainder.

1. $\dfrac{x^4 - 2x^2 + 3}{x + 2}$ $x^3 - 2x^2 + 2x - 4 + \dfrac{11}{x + 2}$

2. $\dfrac{3x^3 + x^2 - 3x - 1}{3x + 1}$ $x^2 - 1$

3. $\dfrac{6x^2 - 4x + 7}{2x + 2}$ $3x - 5 + \dfrac{17}{2x + 2}$

4. $\dfrac{1 - 2x + 3x^2 - 4x^3}{x - 2}$ $-4x^2 - 5x - 12 + \dfrac{-23}{x - 2}$

Graph each equation.

5. $x^2 + 6x - y + 7 = 0$

6. $x^2 + y^2 - 4x + 2y - 4 = 0$

7. $7x + 5y = 10$

8. $y = \log_2 x$

9. $x = -3$

10. $3x^2 + y^2 = 27$

Extra / Spirals

Some polar equations have graphs that are spirals. In these equations the polar angle θ is usually measured in radians. For example, the *spiral of Archimedes* has an equation of the form $r = c\theta$, c a constant, θ in radians. The *logarithmic spiral* has the equation $r = \frac{c}{\theta}$. And the *equiangular spiral* has an equation of the form $r = cb^{\theta}$, c and b constants.

Sketch the graph of each polar equation below for the given values of θ, using radian measure for θ. First determine what the largest value of r will be, so that you can use a convenient scale for r.

1. $r = \theta$, $0 \le \theta \le 6\pi$

2. $r = \dfrac{1}{\theta}$, $0.2 \le \theta \le 4\pi$

3. $r = 2^{\theta}$, $-\pi \le \theta \le \pi$ (*Hint:* Choose values of θ for which 2^{θ} is easy to compute, such as $\theta = -3, -2, \ldots, 2, 3$. Use the fact that 3 radians is a bit less than half a complete rotation to draw the corresponding angles.)

Summarizing the Lesson

In this lesson students have learned about the polar coordinate system and about graphing polar equations. Ask students to compare the polar and rectangular coordinate systems.

Suggested Assignments

Average Alg. and Trig.
678/1–26
S 679/Mixed Review
R 671/Self-Test 1

Extended Alg. and Trig.
Day 1: 678/1–28
R 671/Self-Test 1
Day 2: 678/29–36, *Extra*: 1–3
S 679/Mixed Review

Supplementary Materials

Study Guide pp. 227–228

Additional Answers
Written Exercises

2.

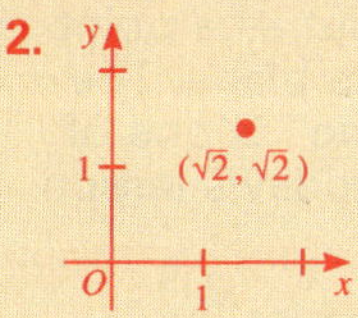

4.

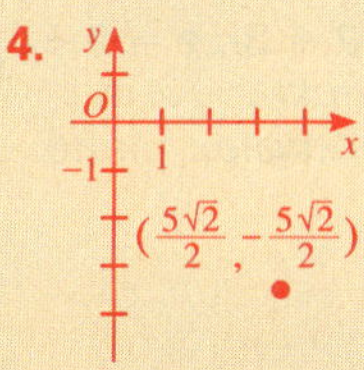

(continued on p. 673)

Additional Answers
Extra

(See p. 674.)

Teaching Suggestions p. T134

Suggested Extensions p. T134

Warm-Up Exercises

Express in $a + bi$ form if $w = -2 + i$ and $z = 3 - 2i$.

1. $w + z$ $1 - i$
2. $z - w$ $5 - 3i$
3. wz $-4 + 7i$
4. $\frac{1}{w}$ $-\frac{2}{5} - \frac{1}{5}i$
5. $\frac{z}{w}$ $-\frac{8}{5} + \frac{1}{5}i$
6. z^2 $5 - 12i$

Motivating the Lesson

Tell the students that a real number is represented as a point on a number line. In contrast, a complex number, having both a real and an imaginary part, is represented not as a point on a number line but as a point in a coordinate plane. The "complex plane" is one of the topics of today's lesson.

Chalkboard Examples

1. Plot $w = -2 + 3i$, $z = 2 + 2i$, $w + z$, and $w - z$ in the same complex plane.

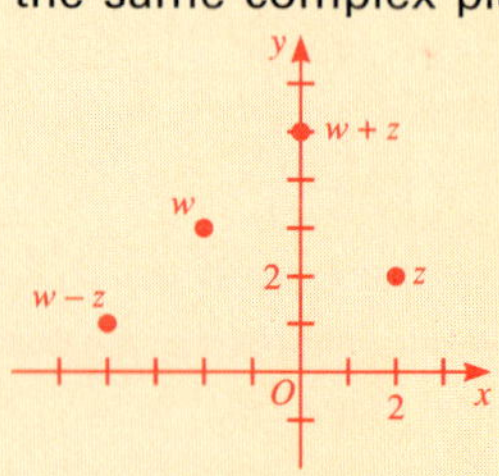

14-4 The Geometry of Complex Numbers

Objective To plot complex numbers in the complex plane and to use the polar form of complex numbers.

You know that every complex number can be written in the form $x + yi$, where x and y are real numbers, and $i^2 = -1$. You can represent complex numbers in a coordinate plane by letting $x + yi$ correspond to the point (x, y), as shown in the diagram.

When the plane is used for this purpose, it is called the **complex plane.** The horizontal axis is called the **real axis,** and the vertical axis is called the **imaginary axis.**

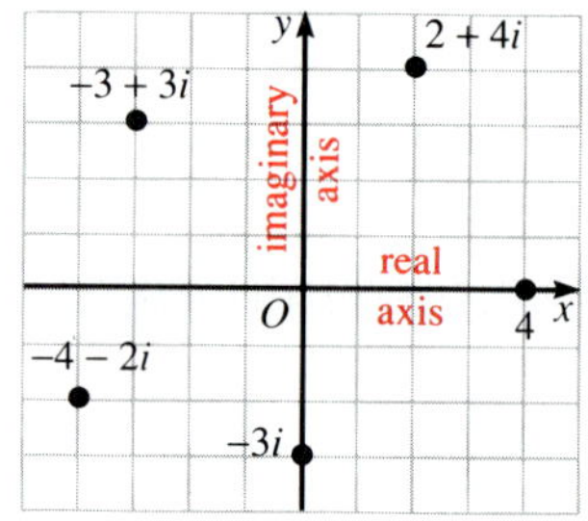

Let $z = x + yi$ be any complex number. Geometrically, the **absolute value** of z, $|z| = \sqrt{x^2 + y^2}$, is the distance from the origin O to the point z; that is, $|z|$ is the magnitude of the vector $\overrightarrow{Oz}$. This is illustrated in Figure 1 below, which also shows the **conjugate** of z, $\bar{z} = x - yi$, and the negative of z, $-z = -x - yi$.

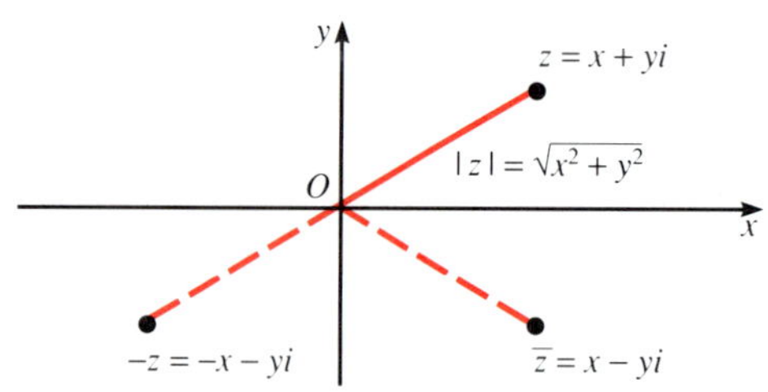

Figure 1

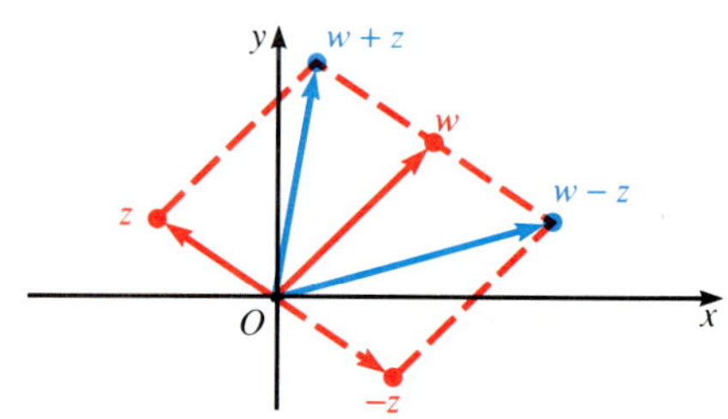

Figure 2

The sum and difference of $w = u + vi$ and $z = x + yi$ can be constructed by adding and subtracting the vectors $\overrightarrow{Ow}$ and $\overrightarrow{Oz}$, as shown in Figure 2.

Example 1 Plot $w = -2 + 4i$, $z = 3 - i$, $w + z$, and $w - z$ in the complex plane.

Solution

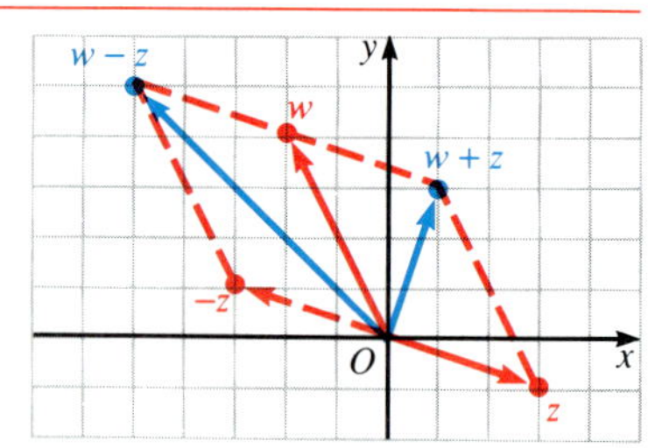

By introducing polar coordinates into the complex plane, you can express the real and imaginary parts of the complex number $z = x + yi$ in terms of its distance and direction from the origin. If the vector $\overrightarrow{Oz}$ has length r and makes an angle θ with the positive real axis, as shown at the right, then

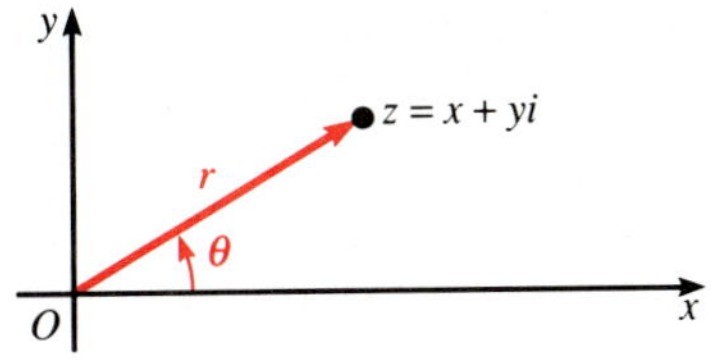

$$x = r\cos\theta \quad \text{and} \quad y = r\sin\theta.$$

Therefore, the complex number $z = x + yi$ can be written as

$$z = (r\cos\theta) + (r\sin\theta)i = r(\cos\theta + i\sin\theta).$$

The **polar,** or **trigonometric, form** of a nonzero complex number $z = x + yi$ is given by

$$z = r(\cos\theta + i\sin\theta),$$

where $r = \sqrt{x^2 + y^2}$, and θ is an angle such that $\cos\theta = \frac{x}{r}$ and $\sin\theta = \frac{y}{r}$.

The angle θ is called the **amplitude,** or **argument,** of z. In this book, angle θ will be chosen such that $0° \le \theta < 360°$. The absolute value, $r = |z|$, is sometimes called the **modulus** of z.

Example 2

a. Write $-4 + 5i$ in polar form with absolute value in simplest radical form and amplitude to the nearest tenth of a degree.

b. Write -6 in polar form.

c. Write $4(\cos 300° + i\sin 300°)$ in $x + yi$ form using radicals.

d. Write $3(\cos 20° + i\sin 20°)$ in $x + yi$ form to four significant digits.

Solution

a. $r = \sqrt{(-4)^2 + 5^2} = \sqrt{41}$

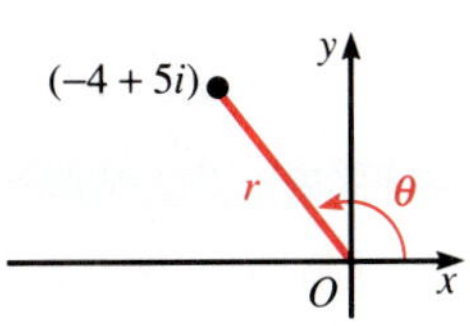

$\sin\theta = \frac{5}{\sqrt{41}} = 0.7809$

So the reference angle of θ is 51.3°.
Since θ is a second-quadrant angle,
$\theta = 180° - 51.3° = 128.7°$.
$\therefore -4 + 5i = \sqrt{41}\,(\cos 128.7° + i\sin 128.7°)$.

b. $r = |-6| = 6$ and $\theta = 180°$.
$\therefore -6 = 6(\cos 180° + i\sin 180°)$.

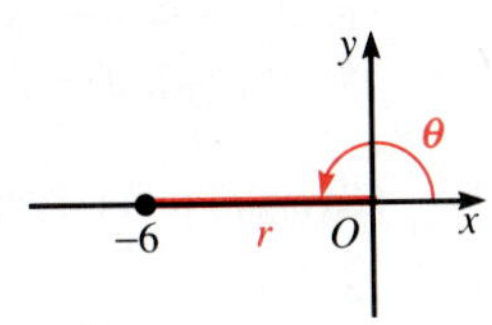

c. $\cos 300° = \frac{1}{2}$ and $\sin 300° = -\frac{\sqrt{3}}{2}$

$\therefore 4(\cos 300° + i\sin 300°) = 4\left(\frac{1}{2} - \frac{\sqrt{3}}{2}i\right) = 2 - 2i\sqrt{3}$.

d. $\cos 20° = 0.9397$ and $\sin 20° = 0.3420$

$\therefore 3(\cos 20° + i\sin 20°) = 3(0.9397 + 0.3420i) = 2.819 + 1.026i$.

2. a. Express $-4 + 3i$ in polar form.
$r = \sqrt{(-4)^2 + 3^2} = 5$
$\sin\theta = \frac{3}{5} = 0.6$
Since θ is a second quadrant angle, $\theta = 143.1°$.
$\therefore -4 + 3i = 5(\cos 143.1° + i\sin 143.1°)$

b. Express $3(\cos 135° + i\sin 135°)$ in $x + yi$ form.
Since $\cos 135° = -\frac{\sqrt{2}}{2}$ and $\sin 135° = \frac{\sqrt{2}}{2}$,
$3(\cos 135° + i\sin 135°) = -\frac{3\sqrt{2}}{2} + \frac{3\sqrt{2}}{2}i$

3. Let $w = 3(\cos 35° + i\sin 35°)$ and $z = 1.6(\cos 80° + i\sin 80°)$.
Find wz and $\frac{w}{z}$ in polar form.
$wz = 3 \times 1.6(\cos(35° + 80°) + i\sin(35° + 80°)) = 4.8(\cos 115° + i\sin 115°)$
$\frac{w}{z} = \frac{3}{1.6}(\cos(35° - 80°) + i\sin(35° - 80°))$
$= 1.875(\cos(-45°) + i\sin(-45°))$
$= 1.875(\cos 315° + i\sin 315°)$

Check for Understanding

Oral Exs. 1–8: use after Example 2.
Oral Exs. 9–12: use before Example 3.

Guided Practice

1. If $w = 3 + i$ and $z = 2 - 2i$, plot w, z, $w + z$, and $w - z$ in the complex plane.

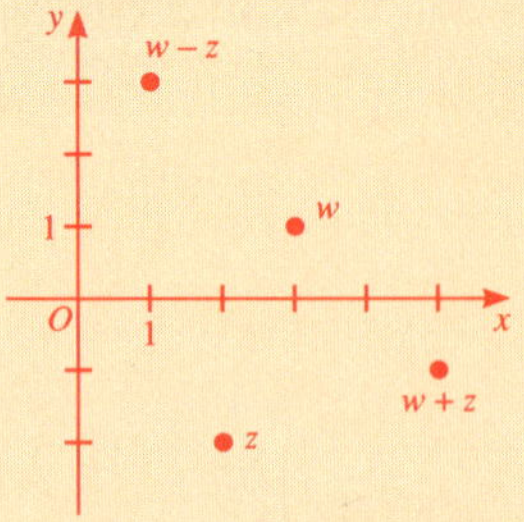

2. Find wz and $\frac{w}{z}$ in polar form if $w = 8(\cos 95° + i \sin 95°)$ and $z = 2(\cos 25° + i \sin 25°)$. $wz = 16(\cos 120° + i \sin 120°)$ $\frac{w}{z} = 4(\cos 70° + i \sin 70°)$
3. Write $2(\cos 25° + i \sin 25°)$ in $x + yi$ form to four significant digits. $1.813 + 0.8452i$
4. Write $-3 + 2i$ in polar form with absolute value in simplest radical form and amplitude to the nearest tenth of a degree. $\sqrt{13}(\cos 146.3° + i \sin 146.3°)$

Using a Calculator

Students need to evaluate sines and cosines, then multiply these by decimal numbers. These are simple but tedious operations, easily done with a calculator.

Summarizing the Lesson

In this lesson students learned about the complex plane and the polar form of complex numbers. Ask the students whether polar form or $x + yi$ form is better for (a) adding and (b) multiplying complex numbers.

The first part of the following theorem tells you that to multiply two complex numbers, you multiply their absolute values and add their amplitudes. The second part tells you that to divide two complex numbers, you divide their absolute values and subtract their amplitudes.

If $w = a(\cos \alpha + i \sin \alpha)$ and $z = b(\cos \beta + i \sin \beta)$, then

$$(1)\ wz = ab[\cos(\alpha + \beta) + i \sin(\alpha + \beta)]$$

$$(2)\ \frac{w}{z} = \frac{a}{b}[\cos(\alpha - \beta) + i \sin(\alpha - \beta)].$$

Proof Here is a proof of part (1). A proof of part (2) is asked for in Exercise 36.

$$\begin{aligned} wz &= a(\cos \alpha + i \sin \alpha) \cdot b(\cos \beta + i \sin \beta) \\ &= ab(\cos \alpha \cos \beta + i \cos \alpha \sin \beta + i \sin \alpha \cos \beta - \sin \alpha \sin \beta) \\ &= ab[\cos \alpha \cos \beta - \sin \alpha \sin \beta + i(\sin \alpha \cos \beta + \cos \alpha \sin \beta)] \\ &= ab[\cos(\alpha + \beta) + i \sin(\alpha + \beta)] \quad \checkmark \end{aligned}$$

Example 3 Let $w = 3.1(\cos 35° + i \sin 35°)$ and $z = 2.4(\cos 140° + i \sin 140°)$. Find wz and $\frac{w}{z}$ in polar form with $0° \le \theta < 360°$.

Solution

$$\begin{aligned} wz &= (3.1)(2.4)[\cos(35° + 140°) + i \sin(35° + 140°)] \\ &= 7.44[\cos 175° + i \sin 175°] \quad \textbf{\textit{Answer}} \end{aligned}$$

$$\begin{aligned} \frac{w}{z} &= \frac{3.1}{2.4}[\cos(35° - 140°) + i \sin(35° - 140°)] \\ &= 1.29[\cos(-105°) + i \sin(-105°)] \\ &= 1.29[\cos 255° + i \sin 255°] \quad \textbf{\textit{Answer}} \end{aligned}$$

Oral Exercises

Give each complex number in $x + yi$ form.

1. $5(\cos 90° + i \sin 90°)$ $5i$
2. $4(\cos 180° + i \sin 180°)$ -4
3. $\sqrt{2}(\cos 45° + i \sin 45°)$ $1 + i$
4. $\sqrt{6}(\cos 135° + i \sin 135°)$ $-\sqrt{3} + i\sqrt{3}$

Give the polar form of each complex number.

5. 3 $3(\cos 0° + i \sin 0°)$
6. $5i$ $5(\cos 90° + i \sin 90°)$
7. $\sqrt{3} - i$ $2(\cos 330° + i \sin 330°)$
8. $-1 - i$ $\sqrt{2}(\cos 225° + i \sin 225°)$

Tell what happens to the absolute value and the amplitude of a complex number when it is multiplied by each of the following numbers.

9. i
10. -1
11. -2
12. $-i$

Written Exercises

Plot w, z, $w + z$, and $w - z$ in the complex plane.

A **1.** $w = 1 - i,\ z = 4$ **2.** $w = -2 - 3i,\ z = -2 + 3i$

3. $w = 3 + i,\ z = 3 - i$ **4.** $w = 2i,\ z = 2 + i$

5. $w = 4 - 3i,\ z = -2 + 5i$ **6.** $w = -1 + 3i,\ z = 4 - i$

7. $wz = 10(\cos 110° + i \sin 110°)$; $\frac{w}{z} = 2.5(\cos 310° + i \sin 310°)$

Find wz and $\frac{w}{z}$ in polar form with $0° \le \theta < 360°$.

7. $w = 5(\cos 30° + i \sin 30°),\ z = 2(\cos 80° + i \sin 80°)$

8. $w = 4(\cos 0° + i \sin 0°),\ z = 3(\cos 130° + i \sin 130°)$

9. $w = 4.5(\cos 150° + i \sin 150°),\ z = 1.2(\cos 315° + i \sin 315°)$

10. $w = 6.3(\cos 160° + i \sin 160°),\ z = 2.0(\cos 210° + i \sin 210°)$

10. $wz = 12.6(\cos 10° + i \sin 10°)$; $\frac{w}{z} = 3.15(\cos 310° + i \sin 310°)$

Write in $x + yi$ form using radicals.

11. $3(\cos 30° + i \sin 30°)$ $\frac{3\sqrt{3}}{2} + \frac{3}{2}i$ **12.** $2(\cos 120° + i \sin 120°)$ $-1 + i\sqrt{3}$

13. $4.5(\cos 120° + i \sin 120°)$ $-2.25 + 2.25i\sqrt{3}$ **14.** $4.0(\cos 240° + i \sin 240°)$ $-2 - 2i\sqrt{3}$

Write in $x + yi$ form to four significant digits. You may wish to use a calculator.

15. $5(\cos 36° + i \sin 36°)$ $4.045 + 2.939i$ **16.** $3(\cos 20° + i \sin 20°)$ $2.819 + 1.026i$

17. $2.2(\cos 150° + i \sin 150°)$ $-1.905 + 1.1i$ **18.** $0.8(\cos 250° + i \sin 250°)$ $-0.2736 - 0.7518i$

Write in polar form with absolute value in simplest radical form and amplitude to the nearest tenth of a degree.

19. $1 - i\sqrt{3}$ **20.** $2\sqrt{2}(1 - i)$ **21.** $2\sqrt{3}(-1 + i)$ **22.** $12 + 5i$

23. $-3 - 4i$ **24.** $5 - i$ **25.** $3 - 2i$ **26.** $-\sqrt{5} + i$

In Exercises 27–30:

a. Use the theorem on page 682 to find wz.

b. Convert the answer in part (a) to $x + yi$ form and simplify.

c. Convert w and z to $x + yi$ form.

d. Multiply w and z in part (c) and simplify.

(Leave your answers to parts (b) and (d) in simplest radical form. They should agree.)

B **27.** $w = 3(\cos 120° + i \sin 120°),\ z = 6(\cos 150° + i \sin 150°)$

28. $w = 2(\cos 60° + i \sin 60°),\ z = 7(\cos 30° + i \sin 30°)$

29. $w = 2(\cos 30° + i \sin 30°),\ z = 4(\cos 120° + i \sin 120°)$

30. $w = 8(\cos 150° + i \sin 150°),\ z = 4(\cos 60° + i \sin 60°)$

Suggested Assignments

Average Alg. and Trig.
683/2–30 even
S 678/27–30

Extended Alg. and Trig.
683/2–30 even, 31
S 670/30, 32

Supplementary Materials

Study Guide pp. 229–230

Additional Answers
Oral Exercises

9. The absolute value is unchanged; the amplitude is changed from θ to $90° + \theta$.

10. The absolute value is unchanged; the amplitude is changed from θ to $180° + \theta$.

11. The absolute value is doubled; the amplitude is changed from θ to $180° + \theta$.

12. The absolute value is unchanged; the amplitude is changed from θ to $270° + \theta$.

Additional Answers
Written Exercises

2.

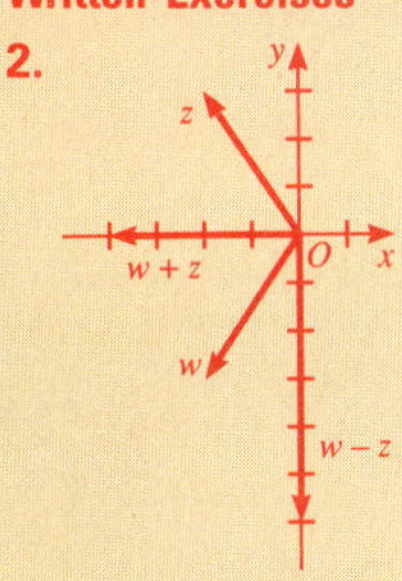

4.

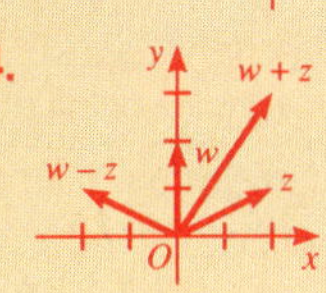

(continued)

Additional Answers Written Exercises (continued)

6.

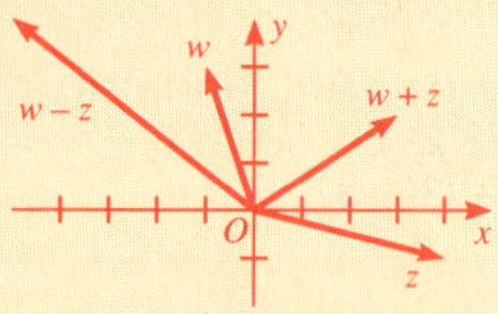

8. $wz = 12(\cos 130° + i \sin 130°)$; $\frac{w}{z} = \frac{4}{3}(\cos 230° + i \sin 230°)$

9. $wz = 5.4(\cos 105° + i \sin 105°)$; $\frac{w}{z} = 3.75(\cos 195° + i \sin 195°)$

19. $2(\cos 300° + i \sin 300°)$

20. $4(\cos 315° + i \sin 315°)$

21. $2\sqrt{6}(\cos 135° + i \sin 135°)$

22. $13(\cos 22.6° + i \sin 22.6°)$

23. $5(\cos 233.1° + i \sin 233.1°)$

24. $\sqrt{26}(\cos 348.7° + i \sin 348.7°)$

25. $\sqrt{13}(\cos 326.3° + i \sin 326.3°)$

26. $\sqrt{6}(\cos 155.9° + i \sin 155.9°)$

27. **a.** $18(\cos 270° + i \sin 270°)$ **b.** $-18i$ **c.** $w = -\frac{3}{2} + \frac{3\sqrt{3}}{2}i$; $z = -3\sqrt{3} + 3i$ **d.** $-18i$

28. **a.** $14(\cos 90° + i \sin 90°)$ **b.** $14i$ **c.** $w = 1 + i\sqrt{3}$, $z = \frac{7\sqrt{3}}{2} + \frac{7}{2}i$ **d.** $14i$

29. **a.** $8(\cos 150° + i \sin 150°)$ **b.** $-4\sqrt{3} + 4i$ **c.** $w = \sqrt{3} + i$, $z = -2 + 2i\sqrt{3}$ **d.** $-4\sqrt{3} + 4i$

30. **a.** $32(\cos 210° + i \sin 210°)$ **b.** $-16\sqrt{3} - 16i$ **c.** $w = -4\sqrt{3} + 4i$, $z = 2 + 2i\sqrt{3}$ **d.** $-16\sqrt{3} - 16i$

In Exercises 31 and 32, let w and z be any two nonzero complex numbers.

31. Show that $|wz| = |w| \cdot |z|$. (*Hint:* Use polar form.)

32. Use geometry to show that $|w + z| \le |w| + |z|$.

In Exercises 33–36, let w and z be any two complex numbers.

33. Show that $z\bar{z} = |z|^2$. (*Hint:* Use $z = x + yi$.)

34. Show that $\frac{1}{z} = \frac{\bar{z}}{|z|^2}$. (*Hint:* Use Exercise 33.)

34. Let z be a complex number; $\frac{1}{z} = \frac{1 \cdot \bar{z}}{z \cdot \bar{z}} = \frac{\bar{z}}{|z|^2}$

C **35.** Show that if $z = b(\cos \beta + i \sin \beta)$, then

$$\frac{1}{z} = \frac{1}{b}(\cos \beta - i \sin \beta).$$

(*Hint:* Use Exercise 34.)

36. Prove the formula for $\frac{w}{z}$ stated on page 682.

(*Hint:* $\frac{w}{z} = w \cdot \frac{1}{z}$. Use Exercise 35 and the product formula.)

37. Explain why the shaded triangles in the diagram at the right are similar.

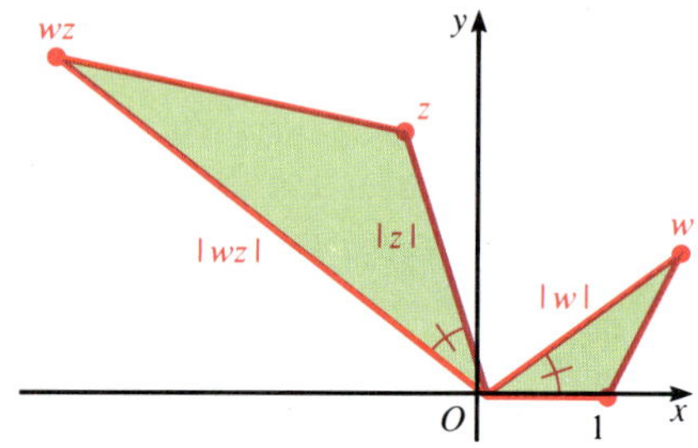

Biographical Note / *Henri Poincaré*

Henri Poincaré (1854–1912), whose broad vision encompassed all branches of mathematics, was interested also in the psychology of mathematical discovery and the role of beauty and harmony in mathematical thought.

Poincaré was born in Nancy, France. His mathematical genius was recognized while he was still in school. His ability to solve complex problems without hesitation was coupled with a physical clumsiness that limited him in sports and geometry. His drawings were in fact so bad that he would have to label them, "this is a horse, this is a tree. . . . "

During his lifetime Poincaré published more than 500 papers on new mathematics as well as more than thirty books. This outpouring revolutionized practically all branches of mathematical physics, theoretical physics, and astronomy as they existed at that time.

14-5 De Moivre's Theorem

Objective To use De Moivre's theorem.

You can use the product formula (formula (1), page 682) to find powers of complex numbers in polar form. For example, if $z = r(\cos \theta + i \sin \theta)$, then

$$z^2 = z \cdot z = r(\cos \theta + i \sin \theta) \cdot r(\cos \theta + i \sin \theta)$$
$$= r \cdot r[\cos (\theta + \theta) + i \sin (\theta + \theta)]$$
$$\therefore z^2 = r^2 (\cos 2\theta + i \sin 2\theta)$$

Also,
$$z^3 = z \cdot z^2 = [r(\cos \theta + i \sin \theta)][r^2 (\cos 2\theta + i \sin 2\theta)]$$
$$= r \cdot r^2[\cos (\theta + 2\theta) + i \sin (\theta + 2\theta)]$$
$$\therefore z^3 = r^3(\cos 3\theta + i \sin 3\theta)$$

These results can be generalized.

De Moivre's Theorem

If $z = r(\cos \theta + i \sin \theta)$, and n is a positive integer, then

$$z^n = r^n(\cos n\theta + i \sin n\theta).$$

Example 1 Use De Moivre's Theorem to find $(-1 + i\sqrt{3})^8$. Give the answer in $x + yi$ form.

Solution Let $z = -1 + i\sqrt{3}$. Then write z in polar form.

$$|z| = 2, \text{ so } z = 2\left(-\frac{1}{2} + i\,\frac{\sqrt{3}}{2}\right).$$

Since $-\frac{1}{2} = \cos 120°$ and $\frac{\sqrt{3}}{2} = \sin 120°$, you have

$$z = 2(\cos 120° + i \sin 120°).$$

Now use De Moivre's Theorem.

$$z^8 = 2^8[\cos (8 \cdot 120°) + i \sin (8 \cdot 120°)]$$
$$= 256(\cos 960° + i \sin 960°)$$
$$= 256[\cos (240° + 2 \cdot 360°) + i \sin (240° + 2 \cdot 360°)]$$
$$= 256(\cos 240° + i \sin 240°) \qquad \text{Simplify.}$$
$$= 256\left(-\frac{1}{2} - i\frac{\sqrt{3}}{2}\right) \qquad \text{Convert to } x + yi \text{ form.}$$
$$= -128 - 128i\sqrt{3}$$

$\therefore (-1 + i\sqrt{3})^8 = -128 - 128i\sqrt{3}$ ***Answer***

Teaching Suggestions p. T134

Suggested Extensions p. T135

Warm-Up Exercises

Express in polar form if $w = 3(\cos 45° + i \sin 45°)$ and $z = 4(\cos 30° + i \sin 30°)$.

1. wz $12(\cos 75° + i \sin 75°)$
2. $\frac{w}{z}$ $\frac{3}{4}(\cos 15° + i \sin 15°)$
3. w^2 $9(\cos 90° + i \sin 90°)$
4. w^3 $27(\cos 135° + i \sin 135°)$
5. z^2 $16(\cos 60° + i \sin 60°)$
6. z^3 $64(\cos 90° + i \sin 90°)$

Motivating the Lesson

Remind students that the number 1 has two square roots, namely 1 and −1. Although 1 has only one *real* cube root, it has two other *complex* cube roots, as today's lesson on De Moivre's theorem will show.

Chalkboard Examples

1. Find $(\sqrt{2} - i\sqrt{2})^5$

$z = \sqrt{2} - \sqrt{2}i$, $|z| = 2$, so $z = 2\left(\frac{\sqrt{2}}{2} - \frac{\sqrt{2}}{2}i\right)$.

$z = 2(\cos 315° + i \sin 315°)$

$z^5 = 2^5[\cos (5 \cdot 315°) + i \sin (5 \cdot 315°)]$

$= 2^5(\cos 135° + i \sin 135°)$

$= 32\left(-\frac{\sqrt{2}}{2} + i\frac{\sqrt{2}}{2}\right)$

$= -16\sqrt{2} + 16i\sqrt{2}$

(continued)

Chalkboard Examples (continued)

2. Find the cube roots of $1 - i$.
Put $1 - i$ into polar form: $\sqrt{2}\left(\frac{\sqrt{2}}{2} - \frac{\sqrt{2}}{2}i\right) = \sqrt{2}(\cos 315° + i \sin 315°)$
Let $z = r(\cos\theta + i\sin\theta)$ be a cube root of $1 - i$.
Then $z^3 = 1 - i$ and $r^3(\cos 3\theta + i \sin 3\theta) = \sqrt{2}(\cos 315° + i \sin 315°)$.
$r = 2^{1/6}$, $\cos 3\theta = \cos 315°$ and $\sin 3\theta = \sin 315°$.
$3\theta = 315° + k \cdot 360°$ or $\theta = 105° + k \cdot 120°$
Let $k = 0$, 1, and 2:
$2^{1/6}(\cos 105° + i \sin 105°)$
$2^{1/6}(\cos 225° + i \sin 225°)$
$2^{1/6}(\cos 345° + i \sin 345°)$.

3. Find the cube roots of unity in polar form and in $x + yi$ form.
$\cos 0° + i \sin 0° = 1$
$\cos 120° + i \sin 120° = -\frac{1}{2} + i\frac{\sqrt{3}}{2}$
$\cos 240° + i \sin 240° = -\frac{1}{2} - i\frac{\sqrt{3}}{2}$

Check for Understanding

Oral Ex. 1: use before Example 1.
Oral Exs. 2–3: use after Example 1.

Guided Practice

Find each power in $x + yi$ form.

1. $(-1 - i)^6$ $-8i$
2. $(\sqrt{2} + i\sqrt{2})^{12}$ -4096
3. $(\sqrt{6} - i\sqrt{2})^9$ $8192i\sqrt{2}$
4. $(1 + i\sqrt{3})^4$ $-8 - 8i\sqrt{3}$

Example 2 Find the cube roots of $2 - 2i$
a. in polar form;
b. in $x + yi$ form, with x and y in simplest radical form.

Solution **a.** Write $2 - 2i$ in polar form:

$$2 - 2i = 2\sqrt{2}\left(\frac{\sqrt{2}}{2} - i\frac{\sqrt{2}}{2}\right) = 2\sqrt{2}(\cos 315° + i \sin 315°)$$

Let $z = r(\cos\theta + i\sin\theta)$ be a cube root of $2 - 2i$.
Then $z^3 = 2 - 2i$ and by De Moivre's Theorem

$$z^3 = r^3(\cos 3\theta + i \sin 3\theta) = 2\sqrt{2}(\cos 315° + i \sin 315°).$$

From this equation,

$$r^3 = 2\sqrt{2} = 2^{3/2}, \text{ so } r = 2^{1/2} = \sqrt{2}$$

Also, $\cos 3\theta = \cos 315°$ and $\sin 3\theta = \sin 315°$.

Therefore, 3θ differs from 315° by an integral multiple of 360°:

$$3\theta = 315° + k \cdot 360° \ (k \text{ is an integer})$$
$$\theta = 105° + k \cdot 120° \ (k \text{ is an integer})$$

Substituting 0, 1, and 2 for k gives

$$\theta = 105°, 225°, 345°$$

(Other values of k give angles coterminal with one of these angles.)

$\therefore$ $2 - 2i$ has these three cube roots:

$\sqrt{2}(\cos 105° + i \sin 105°)$

$\sqrt{2}(\cos 225° + i \sin 225°)$

$\sqrt{2}(\cos 345° + i \sin 345°)$

b. The root $\sqrt{2}(\cos 225° + i \sin 225°) = \sqrt{2}\left(-\frac{\sqrt{2}}{2} - i\frac{\sqrt{2}}{2}\right)$
$= -1 - i.$

To write the other two roots in simplest radical form, you can use the methods of Lesson 13-7 to obtain:

$$\sqrt{2}(\cos 105° + i \sin 105°) = \left(\frac{1 - \sqrt{3}}{2}\right) + i\left(\frac{1 + \sqrt{3}}{2}\right)$$

$$\sqrt{2}(\cos 345° + i \sin 345°) = \left(\frac{1 + \sqrt{3}}{2}\right) - i\left(\frac{1 - \sqrt{3}}{2}\right)$$

The nth roots of the number 1 are usually called the ***n*th roots of unity.** By following the pattern of Example 2, we can show that there are n nth roots of unity, given by this formula:

$$\cos\frac{k \cdot 360°}{n} + i \sin\frac{k \cdot 360°}{n} \quad (k = 0, 1, 2, \ldots, n - 1)$$

Example 3 Find the fifth roots of unity in polar form.

Solution Let $n = 5$ in the formula at the bottom of the preceding page. Then the five fifth roots of unity are as follows:

$$1$$
$$\cos 72° + i \sin 72°$$
$$\cos 144° + i \sin 144°$$
$$\cos 216° + i \sin 216°$$
$$\cos 288° + i \sin 288°$$

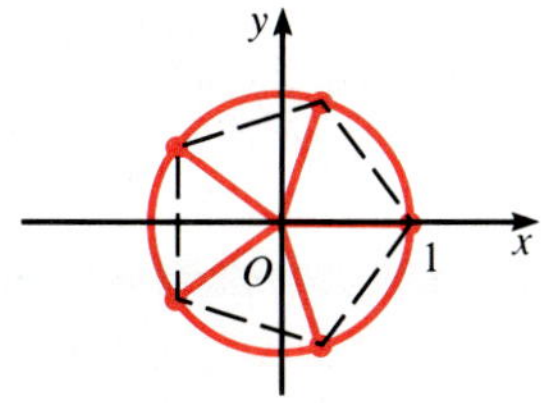

As shown in the diagram, the fifth roots of unity are evenly spaced around the unit circle. Therefore, they are the vertices of a regular pentagon inscribed in the circle.

Oral Exercises

$z^2 = 4(\cos 80° + i \sin 80°)$; $z^5 = 32(\cos 200° + i \sin 200°)$

1. If $z = 2(\cos 40° + i \sin 40°)$, give the polar form of z^2 and z^5.
2. Give the polar form of each complex number. (see below)
   a. i **b.** i^6 **c.** The two square roots of i
3. Let $z = 1 + i$. Give each complex number in $x + yi$ form.
   a. z^2 $2i$ **b.** z^4 -4 **c.** z^8 16 **d.** z^9 $16 + 16i$

2. a. $1(\cos 90° + i \sin 90°)$ b. $1(\cos 180° + i \sin 180°)$
c. $1(\cos 45° + i \sin 45°)$; $1(\cos 225° + i \sin 225°)$

Written Exercises

Use De Moivre's theorem to find each power. Give answers in $x + yi$ form.

A
1. $(\sqrt{3} + i)^4$ $-8 + 8i\sqrt{3}$
2. $(-1 - i)^6$ $-8i$
3. $(-1 + i)^9$ $-16 + 16i$
4. $(1 - i)^7$ $8 + 8i$
5. $(1 - i\sqrt{3})^7$ $64 - 64i\sqrt{3}$
6. $(1 + i\sqrt{3})^{10}$ $-512 - 512i\sqrt{3}$
7. $(-\sqrt{3} - i)^8$ $-128 - 128i\sqrt{3}$
8. $(-\sqrt{3} + i)^9$ $-512i$

Find the required roots in polar form.

9. The ninth roots of unity
10. The 10th roots of unity
11. The cube roots of $4\sqrt{3} - 4i$
12. The cube roots of $-4 + 4i\sqrt{3}$

Find the required roots in $x + yi$ form.

B
13. The cube roots of unity $1, -\frac{1}{2} \pm i\frac{\sqrt{3}}{2}$
14. The fourth roots of unity $1, i, -1, -i$
15. The sixth roots of unity
16. The eighth roots of unity
17. The cube roots of -1 $-1, \frac{1}{2} \pm i\frac{\sqrt{3}}{2}$
18. The cube roots of i $-i, \pm\frac{\sqrt{3}}{2} + \frac{1}{2}i$
19. The fourth roots of $-6 + 6i\sqrt{3}$
20. The fourth roots of $-6 - 6i\sqrt{3}$
21. Plot the complex number $z = 1 + i$. Then plot the following.
    a. z^2 **b.** z^3 **c.** z^4 **d.** z^5 **e.** z^6 **f.** z^7

5. Give two of the fifteenth roots of unity in polar form.
Examples: $\cos 24° + i \sin 24°$, $\cos 48° + i \sin 48°$

6. Find the fourth roots of -16 in $x + yi$ form.
$\sqrt{2} + i\sqrt{2}$, $-\sqrt{2} + i\sqrt{2}$, $-\sqrt{2} - i\sqrt{2}$, $\sqrt{2} - i\sqrt{2}$

Summarizing the Lesson

In this lesson students have learned to use De Moivre's theorem. Ask the students to state this theorem.

Suggested Assignments

Average Alg. and Trig.
687/1–8, 9–21 odd
S 688/Mixed Review

Extended Alg. and Trig.
687/1–8, 10–22 even
S 688/Mixed Review

Supplementary Materials

Study Guide	pp. 231–232
Practice Master	79
Test Master	58
Resource Book	p. 164
Overhead Visual	10

Additional Answers Written Exercises

9. $\cos 0° + i \sin 0°$, $\cos 40° + i \sin 40°$, $\cos 80° + i \sin 80°$, $\cos 120° + i \sin 120°$, $\cos 160° + i \sin 160°$, $\cos 200° + i \sin 200°$, $\cos 240° + i \sin 240°$, $\cos 280° + i \sin 280°$, $\cos 320° + i \sin 320°$

(continued on p. 702)

C **22.** Let $w = \cos 60° + i \sin 60°$. **(a)** Show that w is a sixth root of unity. **(b)** Show that the other sixth roots of unity are: w^2, w^3, w^4, w^5, w^6.

23. An nth root of unity, such as w in Exercise 22, whose powers give all of the other nth roots of unity, is said to be **primitive.** Show that w^5 is another primitive sixth root of unity. (*Hint:* Show that the six powers of w^5, namely $(w^5)^1$, $(w^5)^2$, . . . , $(w^5)^6$, simplify to the sixth roots of unity. For example, $(w^5)^2 = w^{10} = w^6 \cdot w^4 = 1 \cdot w^4 = w^4$.)

Mixed Review Exercises

Let u = 3i + j, v = 2i − 4j, and w = u + v.

1. Find **w** in component form. $5i - 3j$

2. Draw a diagram showing **u, v,** and **w.**

3. Find $\|\mathbf{w}\|$. $\sqrt{34}$

4. Find the angle γ that **w** makes with the positive x-axis. 329.0°

Plot each pair of polar coordinates and find its rectangular coordinates.

5. (4, 30°) $(2\sqrt{3}, 2)$

6. (3, −120°) $\left(-\frac{3}{2}, -\frac{3\sqrt{3}}{2}\right)$

7. (−2, 150°) $(\sqrt{3}, -1)$

Self-Test 2

Vocabulary

pole (p. 675)	conjugate (p. 680)
polar axis (p. 675)	polar form (p. 681)
polar coordinates (p. 675)	trigonometric form (p. 681)
complex plane (p. 680)	amplitude (p. 681)
real axis (p. 680)	argument (p. 681)
imaginary axis (p. 680)	modulus (p. 681)
absolute value (p. 680)	nth roots of unity (p. 686)

1. Find the rectangular coordinates of (4, −60°). Give your answer in simplest radical form. $(2, -2\sqrt{3})$ **Obj. 14-3, p. 675**

2. Find a pair of polar coordinates (r, θ) for the point with rectangular coordinates (−12, 5). Give r to the nearest tenth and θ to the nearest tenth of a degree. (13.0, 157.4°)

3. Plot $w = 4 + 5i$, $z = -6 + 2i$, $w + z$, and $w - z$ in the complex plane. **Obj. 14-4, p. 680**

4. Find wz and $\frac{w}{z}$ in polar form if $w = 6(\cos 35° + i \sin 35°)$ and $z = 3(\cos 70° + i \sin 70°)$. $wz = 18(\cos 105° + i \sin 105°)$; $\frac{w}{z} = 2(\cos 325° + i \sin 325°)$

5. Write $3 - 2i$ in polar form with absolute value to three significant digits and amplitude to the nearest tenth of a degree. $3.61(\cos 326.3° + i \sin 326.3°)$

6. Use De Moivre's Theorem to find $(-\sqrt{3} + i)^6$ in $x + yi$ form. -64 **Obj. 14-5, p. 685**

Quick Quiz

1. Find the rectangular coordinates of (−2, 135°). Give your answer in simplest radical form. $(\sqrt{2}, -\sqrt{2})$

2. Find polar coordinates for the point (4, 5). Give r to the nearest tenth and θ to the nearest tenth of a degree. (6.4, 51.3°)

3. Plot $w = 1 + 3i$, $z = -4 + 3i$, $w + z$, and $w - z$ in the complex plane.

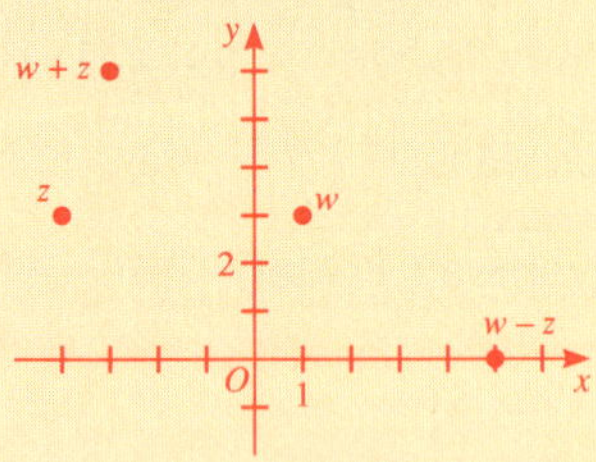

4. Find wz and $\frac{w}{z}$ in polar form if $w = 15(\cos 40° + i \sin 40°)$ and $z = 5(\cos 80° + i \sin 80°)$.
$wz = 75(\cos 120° + i \sin 120°)$
$\frac{w}{z} = 3(\cos 320° + i \sin 320°)$

5. Write $2 - 3i$ in polar form with absolute value to three significant digits and amplitude to the nearest tenth of a degree.
$3.61(\cos 303.7° + i \sin 303.7°)$

6. Use De Moivre's theorem to find $(-\sqrt{3} + 3i)^4$ in $x + yi$ form.
$-72 + 72i\sqrt{3}$

Inverse Functions

14-6 The Inverse Cosine and Inverse Sine

Objective To evaluate expressions involving the inverse cosine and inverse sine.

The equation $\cos x = y$ assigns many values of x to each value of y. For example, $\cos x = \frac{1}{2}$ if x has any of the values $\frac{\pi}{3}$, $-\frac{\pi}{3}$, or any number differing from one of these by an integral multiple of 2π (see below).

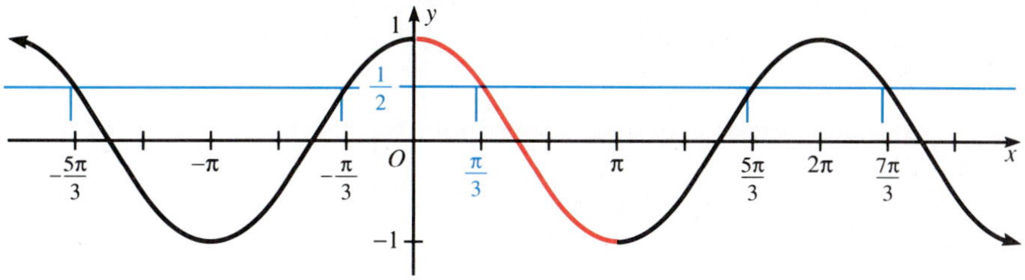

If, however, the domain of the cosine function is restricted to the interval $0 \le x \le \pi$, then only one value of x corresponds to each value of y. This new function, called Cosine, is a *one-to-one function* (recall Lesson 10-2).

By studying the graph of $y = \text{Cos } x$ at the right, you can see that for each number u such that $-1 \le u \le 1$, there is a unique number v $(0 \le v \le \pi)$ such that $\text{Cos } v = u$. For example, if $u = \frac{1}{2}$, then $v = \frac{\pi}{3}$. The number v is denoted by $\mathbf{Cos^{-1}}\ u$ and is called the **inverse cosine** of u. $\left(\text{Cos}^{-1} u \text{ does } \textit{not} \text{ mean } \frac{1}{\text{Cos } u}.\right)$ The inverse cosine is also called the **Arc cosine,** denoted **Arccos.**

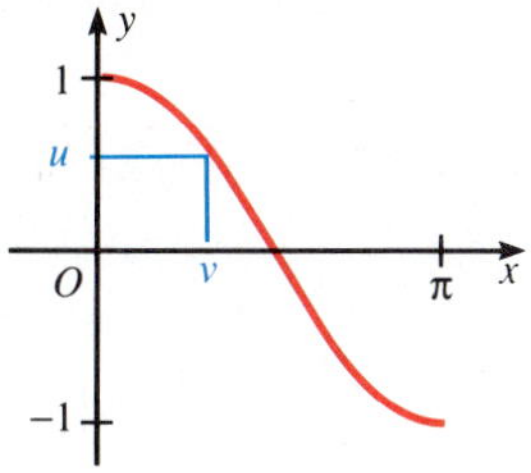

Example 1 Find the value of each of the following.

a. $\text{Cos}^{-1}(-1)$ **b.** $\text{Cos}^{-1} 0$ **c.** $\text{Cos}^{-1}\left(-\frac{\sqrt{2}}{2}\right)$

Solution **a.** Let $v = \text{Cos}^{-1}(-1)$.
Then $\cos v = -1$ and $0 \le v \le \pi$. So $v = \pi$ and therefore

$$\text{Cos}^{-1}(-1) = \pi. \quad \textit{\textbf{Answer}}$$

(Solution continues on the next page.)

Teaching Suggestions p. T135

Suggested Extensions p. T135

Warm-Up Exercises

Find all values of x, $0 \le x < 2\pi$, that satisfy each equation.

1. $\cos x = 0$ $\frac{\pi}{2}, \frac{3\pi}{2}$

2. $\sin x = -\frac{1}{2}$ $\frac{7\pi}{6}, \frac{11\pi}{6}$

3. $\cos x = \frac{\sqrt{2}}{2}$ $\frac{\pi}{4}, \frac{7\pi}{4}$

4. $\sin x = \frac{\sqrt{3}}{2}$ $\frac{\pi}{3}, \frac{2\pi}{3}$

5. $\cos x = -1$ π

Motivating the Lesson

Remind students of the horizontal-line test (page 465) for determining whether a function has an inverse function. Neither the sine nor cosine function "passes" this test, yet inverse functions for sine and cosine can be defined, as today's lesson will show.

Common Errors

If students have a flawed understanding of the inverse sine and cosine functions, they may think that, according to the definition of inverse functions, $\text{Sin}^{-1}(\sin x) = x$ and $\text{Cos}^{-1}(\cos x) = x$ for all values of x. When reteaching, use counterexamples like $\text{Sin}^{-1}(\sin \pi) = \text{Sin}^{-1} 0 = 0$ and $\text{Cos}^{-1}(\cos 2\pi) = \text{Cos}^{-1} 1 = 0$.

Chalkboard Examples

1. Find the value of each of the following.

   a. $\text{Sin}^{-1} 1$

   Let $u = \text{Sin}^{-1} 1$. Then $\sin u = 1$ and $u = \frac{\pi}{2}$.

   $\therefore \text{Sin}^{-1} 1 = \frac{\pi}{2}$.

   b. $\text{Sin}^{-1} 0$

   Let $u = \text{Sin}^{-1} 0$. Then $\sin u = 0$ and $u = 0$.

   $\therefore \text{Sin}^{-1} 0 = 0$.

   c. $\text{Sin}^{-1}\left(-\frac{1}{2}\right)$

   Let $u = \text{Sin}^{-1}\left(-\frac{1}{2}\right)$. Then $\sin u = -\frac{1}{2}$ and $u = -\frac{\pi}{6}$.

   $\therefore \text{Sin}^{-1}\left(-\frac{1}{2}\right) = -\frac{\pi}{6}$.

2. Find the value of each of the following.

   a. $\text{Sin}^{-1}\left(\sin \frac{11\pi}{6}\right)$

   $\text{Sin}^{-1}\left(\sin \frac{11\pi}{6}\right) = \text{Sin}^{-1}\left(-\frac{1}{2}\right) = -\frac{\pi}{6}$

   b. $\cos\left(\text{Sin}^{-1}\left(-\frac{4}{5}\right)\right)$

   Let $v = \text{Sin}^{-1}\left(-\frac{4}{5}\right)$. Then $\sin v = -\frac{4}{5}$ and $-\frac{\pi}{2} \le v \le \frac{\pi}{2}$. $\cos v = \pm\sqrt{1 - \left(-\frac{4}{5}\right)^2} = \pm\frac{3}{5}$. Since $\sin v$ is negative, v is in Quadrant IV and $\cos v$ is positive.

   $\therefore \cos\left(\text{Sin}^{-1}\left(-\frac{4}{5}\right)\right) = \frac{3}{5}$.

b. Let $v = \text{Cos}^{-1} 0$. Then $\cos v = 0$ and $0 \le v \le \pi$.

So $v = \frac{\pi}{2}$ and therefore $\text{Cos}^{-1} 0 = \frac{\pi}{2}$. ***Answer***

c. Let $v = \text{Cos}^{-1}\left(-\frac{\sqrt{2}}{2}\right)$. Then $\cos v = -\frac{\sqrt{2}}{2}$ and $0 \le v \le \pi$.

So $v = \frac{3\pi}{4}$ and therefore $\text{Cos}^{-1}\left(-\frac{\sqrt{2}}{2}\right) = \frac{3\pi}{4}$. ***Answer***

The graph of $y = \text{Cos } x$, which is the red curve in the diagram at the right, has as its inverse the graph of

$$y = \text{Cos}^{-1} x,$$

which is the blue curve in the diagram. Notice that the blue curve is the reflection across the line $y = x$ of the red curve.

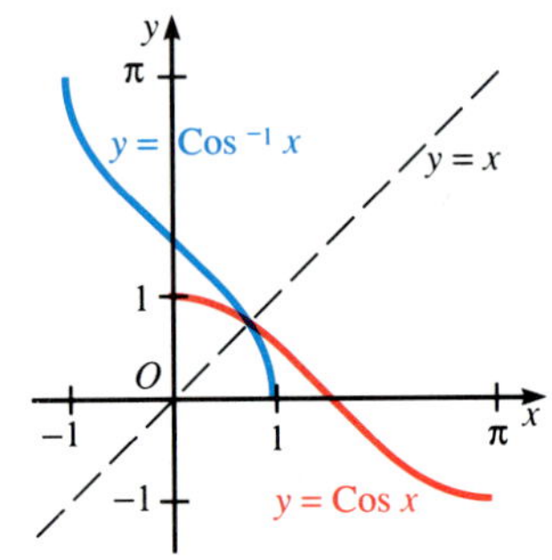

The definition of the inverse cosine can be restated as follows.

$$y = \text{Cos}^{-1} x \text{ if and only if } \cos y = x \text{ and } 0 \le y \le \pi.$$

In defining the **inverse sine** function, denoted $\textbf{Sin}^{-1}\, \boldsymbol{x}$, we reason as above. However, to obtain a *one-to-one function* Sine, we restrict the domain of the sine function to the interval $-\frac{\pi}{2} \le x \le \frac{\pi}{2}$.

$$y = \text{Sin}^{-1} x \text{ if and only if } \sin y = x \text{ and } -\frac{\pi}{2} \le y \le \frac{\pi}{2}.$$

The diagram at the right shows the graph of $y = \text{Sin}^{-1} x$ in blue and of $y = \text{Sin } x$ in red. The inverse sine function is also called the **Arc sine,** denoted **Arcsin.**

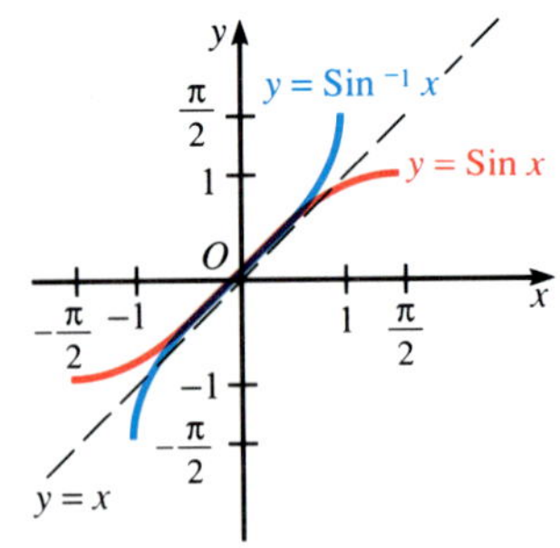

Example 2 Find the value of each of the following.

a. $\text{Sin}^{-1}\left(\cos\frac{5\pi}{3}\right)$ **b.** $\text{Cos}^{-1}\left[\cos\left(\frac{7\pi}{6}\right)\right]$

c. $\sin\left(\text{Sin}^{-1}\frac{1}{9}\right)$ **d.** $\cos\left[\text{Sin}^{-1}\left(-\frac{1}{5}\right)\right]$

Solution **a.** $\text{Sin}^{-1}\left(\cos\frac{5\pi}{3}\right) = \text{Sin}^{-1}\left(\frac{1}{2}\right) = \frac{\pi}{6}$ ***Answer***

b. $\text{Cos}^{-1}\left[\cos\left(\frac{7\pi}{6}\right)\right] = \text{Cos}^{-1}\left(-\frac{\sqrt{3}}{2}\right) = \frac{5\pi}{6}$ ***Answer***

c. Let $v = \text{Sin}^{-1}\frac{1}{9}$.

Then by the definition of the inverse sine, $\sin v = \frac{1}{9}$.

$\therefore \sin\left(\text{Sin}^{-1}\frac{1}{9}\right) = \frac{1}{9}$ ***Answer***

d. Let $v = \text{Sin}^{-1}\left(-\frac{1}{5}\right)$. Then $\sin v = -\frac{1}{5}$, and $-\frac{\pi}{2} \le v \le \frac{\pi}{2}$.

You want to find $\cos v$.

$$\cos v = \pm\sqrt{1 - \sin^2 v} = \pm\sqrt{1 - \left(-\frac{1}{5}\right)^2} = \pm\frac{2\sqrt{6}}{5}$$

Since $\sin v$ is negative and $-\frac{\pi}{2} \le v \le \frac{\pi}{2}$, $\cos v$ is positive.

$\therefore \cos\left[\text{Sin}^{-1}\left(-\frac{1}{5}\right)\right] = \frac{2\sqrt{6}}{5}$ ***Answer***

Although the values of the inverse sine and inverse cosine functions are real numbers, it is often convenient to think of them as measures of angles in radians. For example, to find $\cos\left[\text{Sin}^{-1}\left(-\frac{1}{5}\right)\right]$, you might make a sketch showing $v = \text{Sin}^{-1}\left(-\frac{1}{5}\right)$ as an angle with a sine of $-\frac{1}{5}$. Since $-\frac{\pi}{2} \le v \le \frac{\pi}{2}$, you can represent v as a fourth-quadrant angle. Since $\sin v = -\frac{1}{5}$, the terminal side of v passes through a point 5 units from O with y-coordinate -1. By the Pythagorean theorem, the x-coordinate of this point is $2\sqrt{6}$, and so $\cos v = \frac{2\sqrt{6}}{5}$. Therefore, $\cos\left[\text{Sin}^{-1}\left(-\frac{1}{5}\right)\right] = \frac{2\sqrt{6}}{5}$.

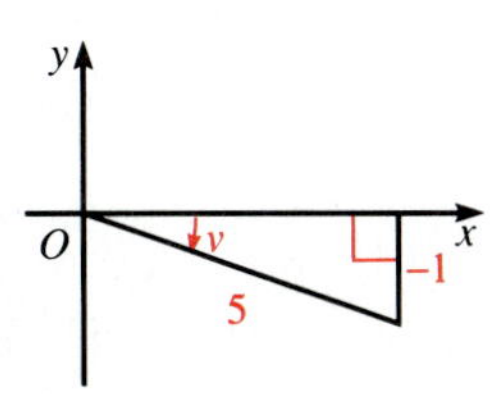

In some applications the values of inverse trigonometric functions may be given in degrees. For example, you might think of $\text{Cos}^{-1}\left(-\frac{1}{2}\right)$ as "the angle between 0° and 180° whose cosine is $-\frac{1}{2}$," and give its value as 120°.

Check for Understanding

Here is a suggested use of the Oral Exercises to check students' understanding as you teach the lesson.
Oral Exs. 1–8: use before Example 2.

Guided Practice

Find the value of each of the following.

1. $\text{Cos}^{-1} 0$ $\frac{\pi}{2}$
2. $\text{Sin}^{-1}\left(-\frac{\sqrt{2}}{2}\right)$ $-\frac{\pi}{4}$
3. $\sin(\text{Sin}^{-1} 0.6)$ 0.6
4. $\cos(\text{Sin}^{-1} 0.6)$ 0.8
5. $\text{Cos}^{-1}\left(\cos\left(-\frac{\pi}{3}\right)\right)$ $\frac{\pi}{3}$
6. $\text{Sin}^{-1}\left(\cos\left(-\frac{5\pi}{6}\right)\right)$ $-\frac{\pi}{3}$

Summarizing the Lesson

In this lesson students have learned about the inverse sine and cosine functions. Ask students to state the domain and range of each inverse function.

Suggested Assignments

Average Alg. and Trig.
692/1–20
S 687/16, 18, 20
R 688/Self-Test 2

Extended Alg. and Trig.
692/2–16 even, 18–24
S 687/17, 19, 21
R 688/Self-Test 2

Supplementary Materials

Study Guide pp. 233–234

Oral Exercises

Give the value of each of the following.

1. $\text{Sin}^{-1}(-1)$ $-\frac{\pi}{2}$
2. $\text{Cos}^{-1} 1$ 0
3. $\text{Sin}^{-1} 0$ 0
4. $\text{Sin}^{-1}\left(-\frac{1}{2}\right)$ $-\frac{\pi}{6}$
5. $\text{Cos}^{-1}\left(\frac{\sqrt{3}}{2}\right)$ $\frac{\pi}{6}$
6. $\text{Cos}^{-1}\left(\frac{\sqrt{2}}{2}\right)$ $\frac{\pi}{4}$
7. $\text{Sin}^{-1} 1$ $\frac{\pi}{2}$
8. $\text{Sin}^{-1} \frac{\sqrt{2}}{2}$ $\frac{\pi}{4}$

Written Exercises

Find the value of each of the following.

A

1. $\text{Sin}^{-1} \frac{\sqrt{3}}{2}$ $\frac{\pi}{3}$
2. $\text{Cos}^{-1}\left(-\frac{\sqrt{2}}{2}\right)$ $\frac{3\pi}{4}$
3. $\text{Cos}^{-1}\left(-\frac{\sqrt{3}}{2}\right)$ $\frac{5\pi}{6}$
4. $\text{Sin}^{-1}\left[\sin\left(\frac{5\pi}{6}\right)\right]$ $\frac{\pi}{6}$
5. $\text{Cos}^{-1}\left(\cos \frac{2\pi}{3}\right)$ $\frac{2\pi}{3}$
6. $\text{Sin}^{-1}\left(\sin \frac{3\pi}{2}\right)$ $-\frac{\pi}{2}$
7. $\text{Cos}^{-1}\left(\cos \frac{5\pi}{3}\right)$ $\frac{\pi}{3}$
8. $\text{Sin}^{-1}\left(\cos \frac{3\pi}{4}\right)$ $-\frac{\pi}{4}$
9. $\text{Cos}^{-1}\left[\sin\left(-\frac{\pi}{3}\right)\right]$ $\frac{5\pi}{6}$
10. $\cos\left[\text{Sin}^{-1}\left(-\frac{\sqrt{3}}{2}\right)\right]$ $\frac{1}{2}$
11. $\sin\left[\text{Cos}^{-1}\left(-\frac{\sqrt{3}}{2}\right)\right]$ $\frac{1}{2}$
12. $\sin\left[\text{Cos}^{-1}\left(-\frac{4}{5}\right)\right]$ $\frac{3}{5}$
13. $\cos\left[\text{Sin}^{-1}\left(-\frac{5}{13}\right)\right]$ $\frac{12}{13}$
14. $\cos\left(\text{Sin}^{-1} \frac{3}{4}\right)$ $\frac{\sqrt{7}}{4}$
15. $\sin\left[\text{Cos}^{-1}\left(-\frac{2}{5}\right)\right]$ $\frac{\sqrt{21}}{5}$

B

16. $\cos\left(\text{Sin}^{-1} \frac{1}{2} + \text{Cos}^{-1} \frac{1}{2}\right)$ 0
17. $\sin\left[\text{Cos}^{-1}\left(-\frac{1}{2}\right) - \text{Sin}^{-1} \frac{1}{2}\right]$ 1

Express without using trigonometric or inverse trigonometric functions.

Sample $\cos(2\ \text{Sin}^{-1} x)$

Solution Let $y = \text{Sin}^{-1} x$. Then $\sin y = x$, where $-\frac{\pi}{2} \le y \le \frac{\pi}{2}$.

So $\cos y = \sqrt{1 - \sin^2 y} = \sqrt{1 - x^2}$.

$\therefore \cos(2\ \text{Sin}^{-1} x) = \cos 2y = \cos^2 y - \sin^2 y = 1 - 2x^2$ ***Answer***

18. $\sin(\text{Cos}^{-1} x)$ $\sqrt{1 - x^2}$
19. $\cos(\text{Sin}^{-1} x)$ $\sqrt{1 - x^2}$
20. $\cos(2\ \text{Cos}^{-1} x)$ $2x^2 - 1$
21. $\sin(2\ \text{Sin}^{-1} x)$ $2x\sqrt{1 - x^2}$
22. $\sin(\text{Sin}^{-1} u + \text{Sin}^{-1} v)$ $u\sqrt{1 - v^2} + v\sqrt{1 - u^2}$
23. $\cos(\text{Sin}^{-1} 1 - \text{Sin}^{-1} v)$ v

Prove the following.

C

24. $\text{Cos}^{-1} \frac{3}{5} - \text{Cos}^{-1} \frac{4}{5} = \text{Sin}^{-1} \frac{7}{25}$
25. $\text{Cos}^{-1} \frac{5}{13} - \text{Sin}^{-1} \frac{4}{5} = \text{Cos}^{-1} \frac{63}{65}$

14-7 Other Inverse Functions

Objective To evaluate expressions involving the inverse trigonometric functions.

Figures 1 and 2 below suggest the following definitions of the **inverse tangent** function (denoted $\mathbf{Tan^{-1}}\,x$) and the **inverse cotangent** function (denoted $\mathbf{Cot^{-1}}\,x$).

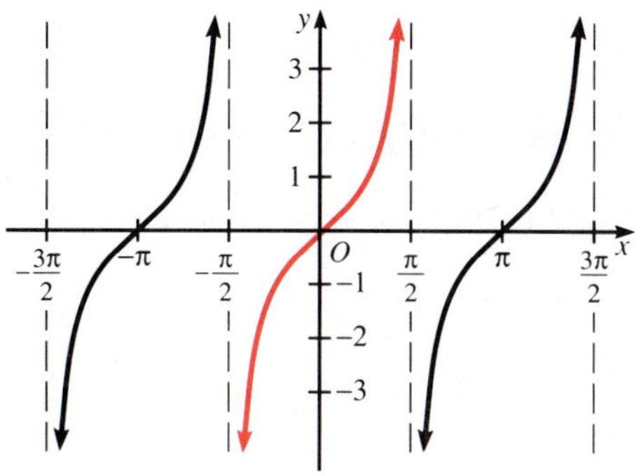

$y = \tan x$

Figure 1

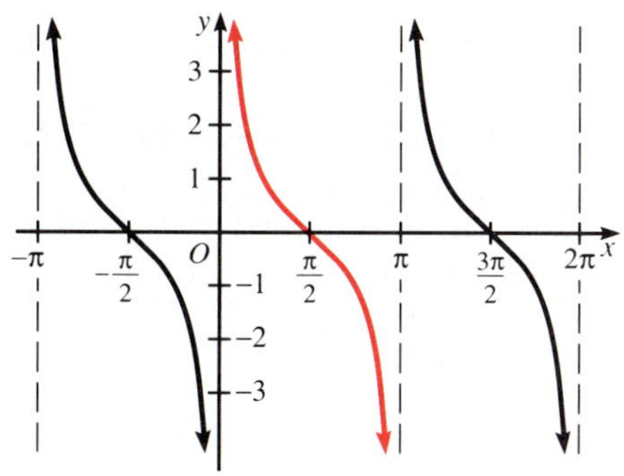

$y = \cot x$

Figure 2

$y = \mathrm{Tan}^{-1}\,x$	$y = \mathrm{Cot}^{-1}\,x$
if and only if	if and only if
$\tan y = x$ and $-\frac{\pi}{2} < y < \frac{\pi}{2}$.	$\cot y = x$ and $0 < y < \pi$.

The graphs of these functions, shown below, are the reflections of the red curves of Figures 1 and 2 across the line $y = x$.

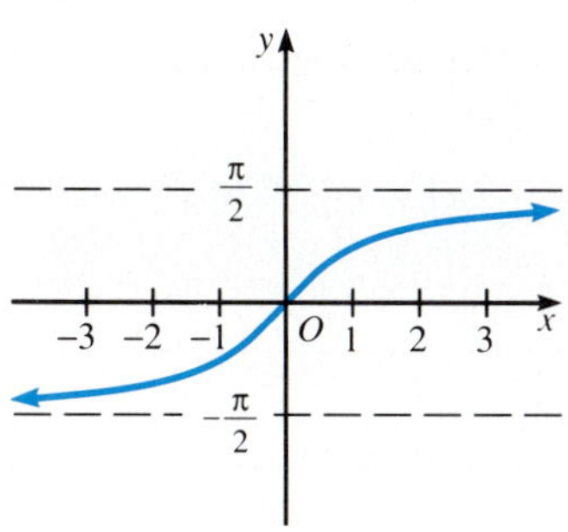

$y = \mathrm{Tan}^{-1}\,x$

Figure 3

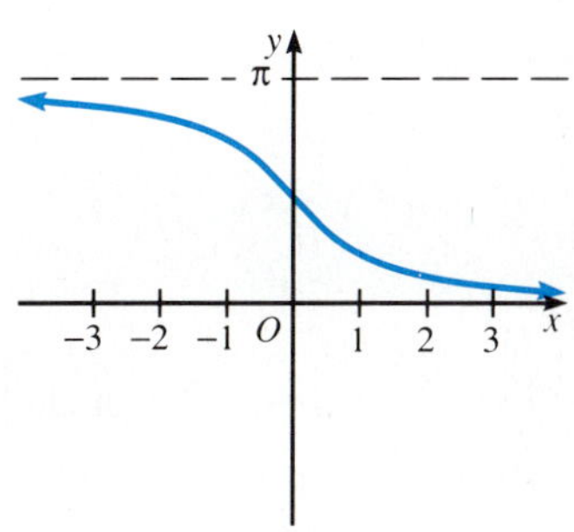

$y = \mathrm{Cot}^{-1}\,x$

Figure 4

Teaching Suggestions p. T135

Suggested Extensions p. T136

Warm-Up Exercises

Tell whether the given functions are inverses.

1. $f(x) = x + 1$, $g(x) = x - 1$ yes
2. $f(x) = x^2$, $g(x) = \sqrt{x}$ no
3. $f(x) = \frac{1}{x}$, $g(x) = \frac{1}{x}$ yes
4. $f(x) = 10^x$, $g(x) = \log x$ yes
5. $f(x) = \sin x$, $g(x) = \mathrm{Sin}^{-1}\,x$ no
6. $f(x) = \mathrm{Cos}\,x$, $g(x) = \mathrm{Cos}^{-1}\,x$ yes

Motivating the Lesson

Remind students that since sine and cosine are not one-to-one functions, they do not have inverse functions unless their domains are restricted. This applies to the other trigonometric functions, whose inverses will be defined today.

Chalkboard Examples

1. Find each value.
   a. $\mathrm{Tan}^{-1}\,0$ Let $u = \mathrm{Tan}^{-1}\,0$. Then $\tan u = 0$ and $u = 0$. $\therefore \mathrm{Tan}^{-1}\,0 = 0$
   b. $\mathrm{Tan}^{-1}\sqrt{3}$ Let $u = \mathrm{Tan}^{-1}\sqrt{3}$. Then $\tan u = \sqrt{3}$ and $u = \frac{\pi}{3}$. $\therefore \mathrm{Tan}^{-1}\sqrt{3} = \frac{\pi}{3}$.
   c. $\mathrm{Sec}^{-1}\sqrt{2}$ Let $u = \mathrm{Sec}^{-1}\sqrt{2}$. Then $\sec u = \sqrt{2}$ and $u = \frac{\pi}{4}$. $\therefore \mathrm{Sec}^{-1}\sqrt{2} = \frac{\pi}{4}$.

(continued)

Chalkboard Examples (continued)

2. Find the value of $\cot\left(\text{Csc}^{-1}\frac{13}{12}\right)$.

$\text{Csc}^{-1}\frac{13}{12} = \text{Sin}^{-1}\frac{12}{13}$

Let $v = \text{Sin}^{-1}\frac{12}{13}$. Then $\sin v = \frac{12}{13}$ and $\cos v = \sqrt{1-\left(\frac{12}{13}\right)^2} = \frac{5}{13}$.

$\cot\left(\text{Csc}^{-1}\frac{13}{12}\right) = \dfrac{\frac{5}{13}}{\frac{12}{13}} = \frac{5}{12}$

3. Find the value of $\tan\left(\text{Tan}^{-1}\left(-\frac{4}{3}\right) + \text{Tan}^{-1}\left(-\frac{1}{7}\right)\right)$.

$\tan\left(\text{Tan}^{-1}\left(-\frac{4}{3}\right) + \text{Tan}^{-1}\left(-\frac{1}{7}\right)\right) = \dfrac{-\frac{4}{3}+\left(-\frac{1}{7}\right)}{1-\left(-\frac{4}{3}\right)\left(-\frac{1}{7}\right)} = \dfrac{-\frac{4}{3}-\frac{1}{7}}{1-\frac{4}{21}} = -\frac{31}{17}$

Check for Understanding

Here is a suggested use of the Oral Exercises to check students' understanding as you teach the lesson.
Oral Exs. 1–8: use before Example 3.

Example 1 Find the value of each of the following.

a. $\text{Tan}^{-1}(-1)$ **b.** $\text{Cot}^{-1} 0$ **c.** $\text{Tan}^{-1}\left(-\frac{\sqrt{3}}{3}\right)$

Solution **a.** Let $y = \text{Tan}^{-1}(-1)$. Then $\tan y = -1$ and $-\frac{\pi}{2} < y < \frac{\pi}{2}$.

So $y = -\frac{\pi}{4}$ and therefore $\text{Tan}^{-1}(-1) = -\frac{\pi}{4}$. ***Answer***

b. Let $y = \text{Cot}^{-1} 0$. Then $\cot y = 0$ and $0 < y < \pi$.

So $y = \frac{\pi}{2}$ and therefore $\text{Cot}^{-1} 0 = \frac{\pi}{2}$. ***Answer***

c. Let $y = \text{Tan}^{-1}\left(-\frac{\sqrt{3}}{3}\right)$. Then $\tan y = -\frac{\sqrt{3}}{3}$ and $-\frac{\pi}{2} < y < \frac{\pi}{2}$.

So $y = -\frac{\pi}{6}$ and therefore $\text{Tan}^{-1}\left(-\frac{\sqrt{3}}{3}\right) = -\frac{\pi}{6}$. ***Answer***

Example 2 Show that if $x > 0$, then $\text{Cot}^{-1} x = \text{Tan}^{-1}\frac{1}{x}$.

Solution Let $y = \text{Cot}^{-1} x$.

Then $\cot y = x$, and since x is positive, $0 < y < \frac{\pi}{2}$.

So $\tan y = \frac{1}{\cot y} = \frac{1}{x}$ and $0 < y < \frac{\pi}{2}$.

Since $\tan y = \frac{1}{x}$, $y = \text{Tan}^{-1}\frac{1}{x}$.

Therefore, by substitution, $\text{Cot}^{-1} x = \text{Tan}^{-1}\frac{1}{x}$.

The result of Example 2 uses the fact that the tangent and cotangent are reciprocal functions. This suggests that we define the **inverse secant** function (denoted $\mathbf{Sec^{-1}}\,x$) in terms of the inverse cosine and the **inverse cosecant** function ($\mathbf{Csc^{-1}}\,x$) in terms of the inverse sine.

If $|x| \geq 1$,

$$\text{Sec}^{-1} x = \text{Cos}^{-1}\frac{1}{x} \quad \text{and} \quad \text{Csc}^{-1} x = \text{Sin}^{-1}\frac{1}{x}.$$

Example 3 Find the value of $\sin\left[\text{Sec}^{-1}\left(\frac{4}{3}\right)\right]$.

Solution By the definition above, $\text{Sec}^{-1}\left(\frac{4}{3}\right) = \text{Cos}^{-1}\left(\frac{3}{4}\right)$.

Let $y = \text{Cos}^{-1}\left(\frac{3}{4}\right)$. Then $\cos y = \frac{3}{4}$ and $0 < y < \pi$.

So $\sin y = \sqrt{1 - \left(\frac{3}{4}\right)^2} = \sqrt{\frac{7}{16}} = \frac{\sqrt{7}}{4}$. Since $\cos y$ is positive and $0 < y < \pi$, $\sin y$ is positive.

$\therefore \sin\left[\text{Sec}^{-1}\left(\frac{4}{3}\right)\right] = \frac{\sqrt{7}}{4}$.

Example 4 Find the value of $\cot\left[\text{Cot}^{-1}\frac{4}{5} + \text{Sin}^{-1}\left(-\frac{4}{5}\right)\right]$.

Solution Let $a = \text{Cot}^{-1}\frac{4}{5}$ and $b = \text{Sin}^{-1}\left(-\frac{4}{5}\right)$. Since

$$\cot(a + b) = \frac{1}{\tan(a + b)} = \frac{1 - \tan a \tan b}{\tan a + \tan b},$$

you need to find $\tan a$ and $\tan b$. Since $a = \text{Cot}^{-1}\frac{4}{5}$, $\cot a = \frac{4}{5}$ and so $\tan a = \frac{5}{4}$. An easy way to find $\tan b$ is to think of b as the measure of an angle between $-\frac{\pi}{2}$ and $\frac{\pi}{2}$ whose sine is $-\frac{4}{5}$. Then make a sketch like the one at the right. Since $z = 3$, you have $\tan b = -\frac{4}{3}$.

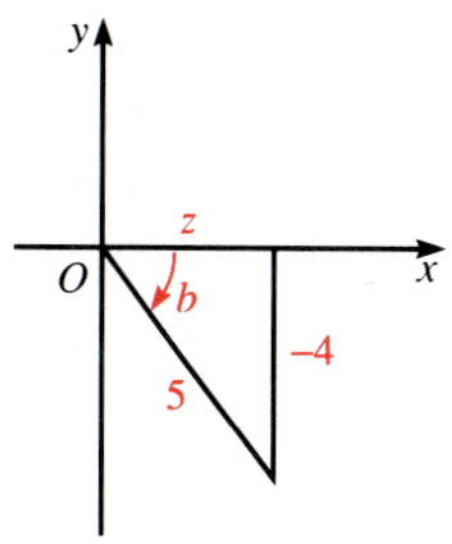

$$\therefore \cot\left[\text{Cot}^{-1}\frac{4}{5} + \text{Sin}^{-1}\left(-\frac{4}{5}\right)\right] = \frac{1 - \left(\frac{5}{4}\right)\left(-\frac{4}{3}\right)}{\frac{5}{4} + \left(-\frac{4}{3}\right)} = -32. \quad \textit{Answer}$$

Oral Exercises

Give the value of each of the following.

1. $\text{Cot}^{-1}(-1)$ $\frac{3\pi}{4}$ **2.** $\text{Tan}^{-1}(\sqrt{3})$ $\frac{\pi}{3}$ **3.** $\text{Cot}^{-1} 0$ $\frac{\pi}{2}$ **4.** $\text{Csc}^{-1}(-1)$ $-\frac{\pi}{2}$

5. $\text{Tan}^{-1}\frac{\sqrt{3}}{3}$ $\frac{\pi}{6}$ **6.** $\text{Cot}^{-1}\frac{\sqrt{3}}{3}$ $\frac{\pi}{3}$ **7.** $\text{Csc}^{-1}(-2)$ $-\frac{\pi}{6}$ **8.** $\text{Sec}^{-1} 2$ $\frac{\pi}{3}$

Written Exercises

Find the value of each of the following.

A **1.** $\text{Csc}^{-1} 2$ $\frac{\pi}{6}$ **2.** $\text{Sec}^{-1}(-1)$ π **3.** $\text{Cot}^{-1}\left(\cot\frac{5\pi}{6}\right)$ $\frac{5\pi}{6}$

4. $\text{Sec}^{-1}\left[\sec\left(-\frac{\pi}{4}\right)\right]$ $\frac{\pi}{4}$ **5.** $\text{Cot}^{-1}\left[\cot\left(-\frac{3\pi}{2}\right)\right]$ $\frac{\pi}{2}$ **6.** $\text{Tan}^{-1}\left[\tan\frac{11\pi}{6}\right]$ $-\frac{\pi}{6}$

Guided Practice

Find the value of each of the following.

1. $\text{Tan}^{-1}(-\sqrt{3})$ $-\frac{\pi}{3}$

2. $\text{Cot}^{-1}\frac{\sqrt{3}}{3}$ $\frac{\pi}{3}$

3. $\text{Sec}^{-1}\left(-\frac{2\sqrt{3}}{3}\right)$ $\frac{5\pi}{6}$

4. $\text{Csc}^{-1}\left(\csc\left(-\frac{3\pi}{4}\right)\right)$ $-\frac{\pi}{4}$

5. $\sin\left(\text{Tan}^{-1}\frac{4}{3}\right)$ $\frac{4}{5}$

6. $\cos(\text{Sec}^{-1}\sqrt{2})$ $\frac{\sqrt{2}}{2}$

7. $\cos(2(\text{Tan}^{-1} 2\sqrt{2}))$ $-\frac{7}{9}$

8. $\cos\left(\text{Csc}^{-1}(-1) + \text{Sec}^{-1}\frac{17}{15}\right)$ $\frac{8}{17}$

Summarizing the Lesson

In this lesson students learned about the inverse tangent, cotangent, secant, and cosecant functions. Ask students to state the domain and range of the inverse tangent and inverse cotangent functions. Also ask students how the inverse secant and inverse cosecant functions are defined.

Suggested Assignments

Average Alg. and Trig.
695/1–22
S 696/Mixed Review

Extended Alg. and Trig.
695/1–17 odd, 19–26
S 696/Mixed Review

Supplementary Materials

Study Guide pp. 235–236
Practice Master 80

Additional Answers
Written Exercises

27.

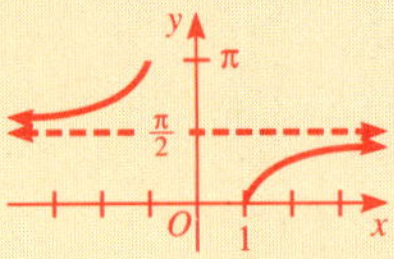

28.

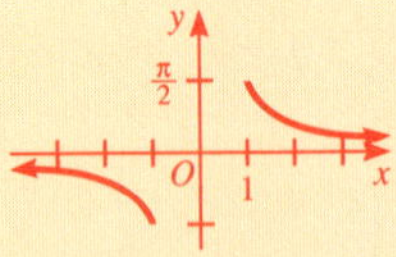

7. $\text{Cot}^{-1}\left(\tan \frac{3\pi}{4}\right)$ $\frac{3\pi}{4}$
8. $\text{Tan}^{-1}\left[\cot\left(-\frac{5\pi}{6}\right)\right]$ $\frac{\pi}{3}$
9. $\tan(\text{Cot}^{-1} 2)$ $\frac{1}{2}$
10. $\sin\left(\text{Csc}^{-1} 3\right)$ $\frac{1}{3}$
11. $\tan\left[\text{Sin}^{-1}\left(-\frac{3}{5}\right)\right]$ $-\frac{3}{4}$
12. $\cot\left[\text{Sin}^{-1} \frac{12}{13}\right]$ $\frac{5}{12}$
13. $\sin\left(\text{Tan}^{-1} 2\right)$ $\frac{2\sqrt{5}}{5}$
14. $\cos\left(\text{Cot}^{-1} \frac{1}{2}\right)$ $\frac{\sqrt{5}}{5}$

B

15. $\tan\left(\text{Sin}^{-1} \frac{5}{13} + \text{Tan}^{-1} \frac{8}{15}\right)$ $\frac{171}{140}$
16. $\cos\left(\text{Tan}^{-1} \frac{4}{3} - \text{Sin}^{-1} \frac{5}{13}\right)$ $\frac{56}{65}$
17. $\sin(2\ \text{Cot}^{-1} 2)$ $\frac{4}{5}$
18. $\tan\left[2\ \text{Cos}^{-1}\left(-\frac{3}{5}\right)\right]$ $\frac{24}{7}$

Express without using trigonometric or inverse trigonometric functions. (*Hint:* See the Sample on page 692.) 21. $\frac{\sqrt{x^2+1}}{x^2+1}$ 22. $\frac{\sqrt{x^2+1}}{x^2+1}$

19. $\csc(\text{Sin}^{-1} x)$ $\frac{1}{x}$
20. $\cot(\text{Tan}^{-1} x)$ $\frac{1}{x}$
21. $\cos(\text{Tan}^{-1} x)$
22. $\sin(\text{Cot}^{-1} x)$
23. $\sin(2\ \text{Tan}^{-1} x)$ $\frac{2x}{x^2+1}$
24. $\cos(2\ \text{Cot}^{-1} x)$ $\frac{x^2-1}{x^2+1}$
25. $\tan(\text{Cot}^{-1} x + \text{Cot}^{-1} y)$ $\frac{x+y}{xy-1}$
26. $\cot(\text{Tan}^{-1} x - \text{Tan}^{-1} y)$ $\frac{xy+1}{x-y}$

C

27. Graph $y = \text{Sec}^{-1} x$.
28. Graph $y = \text{Csc}^{-1} x$.
29. In Example 2 we showed that

$$\text{Cot}^{-1} x = \text{Tan}^{-1} \frac{1}{x} \text{ if } x > 0.$$

If $x < 0$, then

$$\text{Cot}^{-1} x = \pi + \text{Tan}^{-1} \frac{1}{x}.$$

A proof of this identity starts like this: Let $y = \text{Cot}^{-1} x$.
Then $\cot y = x$ and, since $x < 0$, $\frac{\pi}{2} < y < \pi$.
Let $z = y - \pi$.
Then $\cot(z + \pi) = x$ and $-\frac{\pi}{2} < z < 0$.
Complete the proof using the fact that $\cot(z + \pi) = \cot z$.

Since $\cot(z+\pi) = \cot z$, $\cot z = x$; $x < 0$, so $\tan z = \frac{1}{\cot z} = \frac{1}{x}$ and $-\frac{\pi}{2} < z < 0$, so $z = \text{Tan}^{-1} \frac{1}{x}$; $\pi + \text{Tan}^{-1} \frac{1}{x} = \pi + z = y = \text{Cot}^{-1} x$.

Mixed Review Exercises

Determine whether each sequence is arithmetic or geometric. Then find the specified term of the sequence.

1. 7, 12, 17, . . . ; t_{21} arithmetic; 107
2. −3, −6, −12, . . . ; t_8 geometric; −384
3. 9, 18, 36, . . . ; t_{11} geometric; 9216
4. 5, −1, −7, . . . ; t_{15} arithmetic; −79

Write each series using sigma notation. Then find the sum.

5. $3 + 12 + 48 + \cdots + 3072$
6. $5 + 8 + 11 + \cdots + 302$
7. $-5 - 1 + 3 + \cdots + 71$
8. $100 + 20 + 4 + \cdots$

Additional Answers
Mixed Review Exercises

5. $\sum_{k=0}^{5} 3 \cdot 4^k = 4095$
6. $\sum_{n=1}^{100} (2 + 3n) = 15{,}350$
7. $\sum_{n=1}^{20} (4n - 9) = 660$
8. $\sum_{k=0}^{\infty} 100(5^{-k}) = 125$

14-8 Trigonometric Equations

Objective To solve trigonometric equations.

To solve equations involving trigonometric functions you usually use algebraic transformations and trigonometric identities. The strategy used in Examples 1 and 2 is to transform the equation so that it involves only one trigonometric function.

Example 1 Solve $\cos 2x = \sin x$, $0 \le x < 2\pi$.

Solution Of the three formulas for $\cos 2x$, it is best to use $\cos 2x = 1 - 2\sin^2 x$ so that $\sin x$ will be the only function in the transformed equation.

$$\cos 2x = \sin x$$

$1 - 2\sin^2 x = \sin x$ — Make one side equal 0.

$2\sin^2 x + \sin x - 1 = 0$ — Factor.

$(2\sin x - 1)(\sin x + 1) = 0$ — Solve.

$2\sin x - 1 = 0$ or $\sin x + 1 = 0$

$\sin x = \frac{1}{2}$ $\qquad$ $\sin x = -1$

$x = \frac{\pi}{6}, \frac{5\pi}{6}$ $\qquad$ $x = \frac{3\pi}{2}$

$\therefore$ the solution set over the interval $0 \le x < 2\pi$ is:

$$\left\{\frac{\pi}{6}, \frac{5\pi}{6}, \frac{3\pi}{2}\right\}$$ ***Answer***

Example 2 Solve $\sin^2 \theta = \cos^2 \theta$. Find **(a)** the solutions in the interval $0° \le \theta < 360°$ and **(b)** the formulas giving general solutions.

Solution **a.** $\sin^2 \theta = \cos^2 \theta$ — Divide both sides by $\cos^2 \theta$ $(\cos \theta \ne 0)$.

$\frac{\sin^2 \theta}{\cos^2 \theta} = 1$ — Use the identity $\frac{\sin \theta}{\cos \theta} = \tan \theta$.

$\tan^2 \theta = 1$ — Take the square root of both sides.

$\tan \theta = \pm 1$ — Solve.

$\tan \theta = 1$ or $\tan \theta = -1$

$\theta = 45°, 225°$ $\qquad$ $\theta = 135°, 315°$

$\therefore$ the solution set over the interval $0° \le \theta < 360°$ is:

$$\{45°, 135°, 225°, 315°\}$$ ***Answer***

Notice in the first step of part (a) that you may divide by $\cos^2 \theta$ because there is no solution of the equation for which $\cos \theta = 0$.

Teaching Suggestions p. T136

Suggested Extensions p. T136

Warm-Up Exercises

Solve.

1. $4x^2 = 3$ $\left\{\pm\frac{\sqrt{3}}{2}\right\}$
2. $x^2 = x$ $\{0, 1\}$
3. $2x^2 + x - 1 = 0$ $\left\{-1, \frac{1}{2}\right\}$
4. $x = \frac{1}{x}$ $\{\pm 1\}$
5. $x - \sqrt{1 - x^2} = 0$ $\left\{\frac{\sqrt{2}}{2}\right\}$

Motivating the Lesson

Remind students that when solving a polynomial, fractional, or radical equation, they can expect a finite number of solutions. For a trigonometric equation, however, there can be infinitely many solutions. Solving such equations is the topic of today's lesson.

Using a Calculator

Some equations have special angles for their solutions, and students can determine these mentally. For other than the "special" angles, students may want to use the inverse trigonometric function keys on their calculators.

Chalkboard Examples

1. Solve $\sin \theta = 3\cos \theta$ for θ; $0° \le \theta < 360°$.
If $\sin \theta = 3\cos \theta$, then $\tan \theta = 3$. $\theta = 71.6°$ or $251.6°$.

(continued)

b. To find formulas giving general solutions, you should recognize that all other solutions differ from 45° and 135° by integral multiples of 180°, the period of the tangent function. Therefore, all solutions are of this form:

$$\theta = 45° + k \cdot 180°, \ k \text{ is an integer}$$
$$\theta = 135° + k \cdot 180°, \ k \text{ is an integer}$$

$\therefore$ the set of general solutions is:

$$\{\theta: \theta = 45° + k \cdot 180° \quad \text{or} \quad \theta = 135° + k \cdot 180°\}$$

Answer

Solutions in a specified interval, usually $0 \le x < 2\pi$ or $0° \le \theta \le 360°$ as in Examples 1 and 2, are sometimes called **primary solutions.** Formulas that name all possible solutions, such as the ones in the second part of Example 2, give the **general solutions.**

Sometimes a calculator or tables must be used to solve a trigonometric equation.

Example 3 Solve $3 \cos^2 \theta - 8 \cos \theta + 4 = 0$ for θ, $0° \le \theta < 360°$.

Solution

$3 \cos^2 \theta - 8 \cos \theta + 4 = 0$ Factor.

$(\cos \theta - 2)(3 \cos \theta - 2) = 0$ Solve.

$\cos \theta - 2 = 0$ or $3 \cos \theta - 2 = 0$

$\cos \theta = 2$ $\quad$ $\cos \theta = \frac{2}{3} = 0.6667$

No solution, since $-1 \le \cos \theta \le 1$.

To the nearest tenth of a degree, a calculator or table gives

$\theta = 48.2°$.

Another solution in the given interval is $\theta = 311.8°$.

$\therefore$ the solution set over the interval $0° \le \theta < 360°$ is:

$\{48.2°, 311.8°\}$ ***Answer***

Example 4 Solve $\sin (3\theta - 20°) = \frac{1}{2}$ for θ, $0° \le \theta < 360°$.

Solution Let $a = 3\theta - 20°$. Then $\sin a = \frac{1}{2}$, so

$$a = 30° + k \cdot 360° \quad \text{or} \quad a = 150° + k \cdot 360°, \text{ where } k \text{ is an integer.}$$

Therefore, by substitution,

$3\theta - 20° = 30° + k \cdot 360°$	or	$3\theta - 20° = 150° + k \cdot 360°$
$3\theta = 50° + k \cdot 360°$		$3\theta = 170° + k \cdot 360°$
$\theta = 16.7° + k \cdot 120°$		$\theta = 56.7° + k \cdot 120°$

Chalkboard Examples (continued)

2. Solve $\cos 2x = 3 \cos x + 1$. Give (a) the solutions in the interval $0 \le x < 2\pi$ and (b) formulas giving general solutions.
   a. $\cos 2x = 3 \cos x + 1$
   $2 \cos^2 x - 1 = 3 \cos x + 1$
   $2 \cos^2 x - 3 \cos x - 2 = 0$
   $(2 \cos x + 1) \cdot (\cos x - 2) = 0$
   $2 \cos x + 1 = 0$ or $\cos x - 2 = 0$
   $\cos x = -\frac{1}{2}$ or no solution
   $x = \frac{2\pi}{3}, \frac{4\pi}{3}$
   b. The formulas for all the solutions are $\frac{2\pi}{3} + 2\pi n$ and $\frac{4\pi}{3} + 2\pi n$, n an integer.
3. Solve $6 \cos^2 \theta - 1 = \cos \theta$ for θ; $0° \le \theta < 360°$.
   $6 \cos^2 \theta - \cos \theta - 1 = 0$
   $(3 \cos \theta + 1) \cdot (2 \cos \theta - 1) = 0$
   $3 \cos \theta + 1 = 0$ or $2 \cos \theta - 1 = 0$
   $\theta = 109.5°, 250.5°$ or $\theta = 60°, 300°$
   So $\theta = 60°, 109.5°, 250.5°$, or $300°$
4. Solve $\sin (\theta + 40°) = \frac{1}{2}$ for θ, $0° \le \theta < 360°$.
   Since $\sin (\theta + 40°) = \frac{1}{2}$, $\theta + 40° = 150°$ or $390°$. $(40° \le \theta + 40° < 400°)$ Therefore $\theta = 110°$ or $350°$.

Substituting 0, 1, and 2 for k gives

$\theta = 16.7°, 136.7°, 256.7°$ or $\theta = 56.7°, 176.7°, 296.7°$

$\therefore$ the solution set over the interval $0° \le \theta < 360°$ is:

$\{16.7°, 56.7°, 136.7°, 176.7°, 256.7°, 296.7°\}$. *Answer*

Oral Exercises

How many solutions does each equation have in the interval $0 \le x < 2\pi$?

1. $\cos x = \frac{4}{5}$ 2
2. $\sin x = -1$ 1
3. $\sin^2 x = \frac{1}{9}$ 4
4. $\cos x(\tan x - 2) = 0$ 4
5. $\cos x(\sin x - 2) = 0$ 2
6. $\cos 2x = \frac{1}{2}$ 4
7. $\sin \frac{x}{9} = \frac{1}{3}$ 1
8. $\cos \frac{x}{9} = \frac{1}{3}$ 0
9. In solving $2 \sin 2x = 1$, would you use a double-angle formula? If not, what *would* you do? No. Solve for $y = 2x$; then divide answers by 2 to find x.
10. In solving $2 \cos\left(x - \frac{\pi}{4}\right) = 1$, would you use an addition formula? If not, what *would* you do? No. Solve for $y = x - \frac{\pi}{4}$; then add $\frac{\pi}{4}$ to the answers to find x.

Outline a strategy for solving each of the following equations.

11. $\cos \theta = \sin \theta$ $\tan \theta = 1$
12. $\tan x = \cos x$
13. $\sin 2x = \cos x$
14. $2 \cos (\theta - 25°) = 1$ $\cos (\theta - 25°) = \frac{1}{2}$
15. $\sin 3\theta = 1$ $3\theta = \frac{\pi}{2}$
16. $\sin 4x = \cos 2x$

Written Exercises

Find the primary solution of each equation for $0° \le \theta < 360°$ or $0 \le x < 2\pi$. Then find formulas giving the general solution.

Give your answer to the nearest tenth of a degree if the variable is θ. Give your answer to the nearest hundredth of a radian if the variable is x, unless you can express the answer in terms of π.

A

1. $4 \sin \theta = 3$
2. $7 \cos \theta = 2$
3. $2 \cos^2 x = 1$
4. $4 \sin^2 x = 3$
5. $\sin \theta + \cos \theta = 0$
6. $\csc \theta = \sec \theta$
7. $\csc \theta = 16 \sin \theta$
8. $5 \sin x + 2 \cos x = 0$
9. $\cot\left(x - \frac{\pi}{8}\right) = 1$
10. $2 \sin\left(x + \frac{\pi}{4}\right) = 1$
11. $\sin (\theta + 25°) = 0$
12. $\tan (\theta - 20°) = \sqrt{3}$

Check for Understanding

Oral Exs. 1–8: use after Example 2.
Oral Exs. 9–16: use after Example 4.

Guided Practice

Give answers to the nearest tenth of a degree if the variable is θ or to the nearest hundredth of a radian if the variable is x, unless you can express the answer in terms of π.

Solve for $0° \le \theta < 360°$ or $0 \le x < 2\pi$. In Exercises 1 and 2, also give formulas for the general solution.

1. $5 \sin \theta = 4$ $\{53.1°, 126.9°\}$; $\{\theta: \theta = 53.1° + k \cdot 360°$ or $\theta = 126.9° + k \cdot 360°\}$
2. $\sec x = 9 \cos x$ $\{1.23, 1.91, 4.37, 5.05\}$; $\{x: x = 1.23 + k\pi$ or $x = 1.91 + k\pi\}$
3. $\cos^2 \theta = 3 \sin^2 \theta$ $\{30°, 150°, 210°, 330°\}$
4. $\sin 2\theta + \sin \theta = 0$ $\{0°, 120°, 180°, 240°\}$

Summarizing the Lesson

In this lesson students have learned to solve trigonometric equations. Ask students to explain why such equations can have infinitely many solutions.

Additional Answers
Oral Exercises

(See p. 703.)

Suggested Assignments

Average Alg. and Trig.
Day 1: 699/1–27 odd
S 696/23–26
Day 2: 700/14–28 even, 29–42
R 700/Self-Test 3

Extended Alg. and Trig.
Day 1: 699/1–27 odd
S 696/27, 28
Day 2: 700/14–28 even, 29–45 odd
R 700/Self-Test 3

Additional Answers Written Exercises

(See p. 703.)

Supplementary Materials

Study Guide	pp. 237–238
Practice Master	81
Test Master	59
Computer Activity	31
Resource Book	p. 165

Quick Quiz

Find the value of each of the following.

1. $\text{Cos}^{-1}\left(-\frac{\sqrt{3}}{2}\right)$ $\frac{5\pi}{6}$
2. $\text{Sin}^{-1}(\sin \pi)$ 0
3. $\cos\left(\text{Sin}^{-1}\frac{3}{4}\right)$ $\frac{\sqrt{7}}{4}$
4. $\sec\left(\text{Tan}^{-1}\frac{12}{5}\right)$ $\frac{13}{5}$
5. $\tan(2\ \text{Cot}^{-1} 3)$ $\frac{3}{4}$
6. $\text{Cos}^{-1}\left(\sin\left(-\frac{\pi}{4}\right)\right)$ $\frac{3\pi}{4}$
7. Solve $3\cos 2\theta = 7\cos\theta$ for θ if $0° \le \theta < 360°$. Give each solution to the nearest 0.1°. {109.5°, 250.5°}

Solve for θ, $0° \le \theta < 360°$, or for x, $0 \le x < 2\pi$.

13. $4\cos\theta = 3\sec\theta$ {30°, 150°, 210°, 330°}
14. $3\tan\theta = \cot\theta$ {30°, 150°, 210°, 330°}
15. $3\sec x = \csc x$ {0.32, 3.46}
16. $2\cos 2\theta = 1$
17. $\sec^2\theta = 2$
18. $2\cot^2\theta = 1$
19. $\sin 2x - \cos x = 0$
20. $\sin 2\theta = 2\sin\theta$
21. $\cos 2\theta = \cos\theta$
22. $\cos 2x = \sin x$
23. $2\sin 2\theta = \sqrt{3}$
24. $2\cos x = \csc x$
25. $2\tan^2\theta + \tan\theta - 3 = 0$
26. $6\sin^2\theta = \sin\theta + 1$
27. $\cos 2x = 11\sin x + 6$ $\left\{\frac{7\pi}{6}, \frac{11\pi}{6}\right\}$
28. $\cos 2x + 7\sin x + 3 = 0$ $\left\{\frac{7\pi}{6}, \frac{11\pi}{6}\right\}$

B

29. $\tan^2\left(x - \frac{\pi}{3}\right) = 1$ $\left\{\frac{\pi}{12}, \frac{7\pi}{12}, \frac{13\pi}{12}, \frac{19\pi}{12}\right\}$
30. $2\cos^2\left(x + \frac{\pi}{6}\right) = 1$ $\left\{\frac{\pi}{12}, \frac{7\pi}{12}, \frac{13\pi}{12}, \frac{19\pi}{12}\right\}$
31. $\sin 3x + \cos 3x = 0$
32. $\tan^2 3x = 3$
33. $\sin 4\theta = \cos 2\theta$
34. $\cos 3x + \cos x = 0$
35. $\sin 2x = \cot x$
36. $2\sin 2\theta = \tan\theta$ {60°, 120°, 240°, 300°}
37. $\cos 2\theta = 2\sin^2\theta$ {30°, 150°, 210°, 330°}
38. $\cos 2\theta + 2\cos^2\theta = 0$
39. $\cos(2\theta + 10°) = -\frac{1}{2}$ {55°, 115°, 235°, 295°}
40. $\sin(3\theta - 25°) = \frac{\sqrt{3}}{2}$
41. $\sin(3x + \pi) = -\frac{\sqrt{2}}{2}$ {39°, 129°, 219°, 309°}
42. $\tan\left(2x - \frac{\pi}{3}\right) = 1$ $\left\{\frac{7\pi}{24}, \frac{19\pi}{24}, \frac{31\pi}{24}, \frac{43\pi}{24}\right\}$

C

43. $\tan(\theta + 32°) = \cot(\theta - 20°)$
44. $\sin(\theta - 10°) = \cos\theta$ {50°, 230°}
45. $\cos 2\theta = \sin\theta - 1$ {51.3°, 128.7°}
46. $\cot^2 x - \cos^2 x = \cot^2 x \cos^2 x$ $\{x: 0 < x < \pi \text{ or } \pi < x < 2\pi\}$

Self-Test 3

Vocabulary

one-to-one (p. 689)
inverse cosine, $\text{Cos}^{-1} x$ (p. 689)
inverse sine, $\text{Sin}^{-1} x$ (p. 690)
inverse tangent, $\text{Tan}^{-1} x$ (p. 693)
inverse cotangent, $\text{Cot}^{-1} x$ (p. 693)
inverse secant, $\text{Sec}^{-1} x$ (p. 694)
inverse cosecant, $\text{Csc}^{-1} x$ (p. 694)
primary solution (p. 698)
general solution (p. 698)

Find the value of each of the following.

1. $\text{Sin}^{-1}\left(-\frac{\sqrt{3}}{2}\right)$ $-\frac{\pi}{3}$
2. $\text{Sin}^{-1}\left(\sin\frac{7\pi}{6}\right)$ $-\frac{\pi}{6}$ **Obj. 14-6, p. 689**
3. $\text{Cos}^{-1}\left[\sin\left(-\frac{\pi}{3}\right)\right]$ $\frac{5\pi}{6}$
4. $\cos\left[\text{Sin}^{-1}\left(-\frac{4}{5}\right)\right]$ $\frac{3}{5}$
5. $\text{Cot}^{-1}\left(-\frac{\sqrt{3}}{3}\right)$ $\frac{2\pi}{3}$
6. $\sec\left(\text{Tan}^{-1}\frac{8}{15}\right)$ $\frac{17}{15}$ **Obj. 14-7, p. 693**
7. $\text{Csc}^{-1}\frac{2\sqrt{3}}{3}$ $\frac{\pi}{3}$
8. $\cos\left[\text{Sec}^{-1}(-5)\right]$ $-\frac{1}{5}$
9. Solve the equation $3\sin^2\theta - \cos^2\theta = 0$, $0° \le \theta < 360°$. Give θ to the nearest tenth of a degree. {30°, 150°, 210°, 330°} **Obj. 14-8, p. 697**

Computer Key-In

Mathematicians have used the inverse tangent function and the infinite series for $\text{Tan}^{-1}x$ shown below to approximate the irrational number π.

$$\text{Tan}^{-1}x = x - \frac{x^3}{3} + \frac{x^5}{5} - \frac{x^7}{7} + \cdots + (-1)^{n-1}\frac{x^{2n-1}}{2n-1} + \cdots \qquad \textbf{(1)}$$

This series, which was obtained by the Scottish mathematician James Gregory in 1671, gives the value of $\text{Tan}^{-1}x$ for any real number x with $|x| \le 1$. If $x = 1$, equation (1) becomes

$$\frac{\pi}{4} = \text{Tan}^{-1}1 = 1 - \frac{1}{3} + \frac{1}{5} - \frac{1}{7} + \cdots$$

By summing a large number of terms on the right and multiplying the sum by 4, you can approximate π. If you use 1000 terms, however, the approximation is accurate to only three decimal places.

You can use equation (1) to obtain a more efficient method for approximating π by applying the following formula. Let a and b be real numbers with $|a| < 1$ and $|b| < 1$. Then

$$\text{Tan}^{-1}a + \text{Tan}^{-1}b = \text{Tan}^{-1}\left(\frac{a+b}{1-ab}\right).$$

If a and b are chosen so that $\frac{a+b}{1-ab} = 1$, then

$$\text{Tan}^{-1}a + \text{Tan}^{-1}b = \frac{\pi}{4}.$$

Since a and b are less than 1 in absolute value, you can compute $\text{Tan}^{-1}a$ and $\text{Tan}^{-1}b$ using equation (1) and then add these values and multiply by 4 to find π. For example, you can use $a = \frac{1}{2}$ and $b = \frac{1}{3}$ since $\dfrac{\frac{1}{2}+\frac{1}{3}}{1-\frac{1}{2}\cdot\frac{1}{3}} = 1$.

Using 15 terms in equation (1), you obtain for π the value

$$4(0.463647609 + 0.321750551) = 3.141592640,$$

which is correct to seven decimal places. (The value of π to ten places is 3.1415926536.)

The computer program which follows prints an approximation for π when you enter a value of N (the number of terms used in equation (1)) and values of a and b.

```
10 INPUT "ENTER VALUES FOR N, A, B: "; N, A, B
20 LET X=A
30 GOSUB 110
40 LET P=S
50 LET X=B
60 GOSUB 110
70 LET P=P+S
80 PRINT "AN APPROXIMATE VALUE FOR PI IS: ";
```

(Program continues on the next page.)

Computer Key-In Commentary

The statement GOSUB 110 causes the computer to go to the subroutine at line 110 and execute statements until a RETURN statement is encountered. A subroutine is a part of the main program that performs a particular task. It can be used by the main program wherever needed.

Note that the main idea here is a technique for approximating π. Make sure students understand that the values used for A and B when running the program must satisfy $\frac{A+B}{1-AB} = 1$.

Additional Answers Computer Key-In

1. Answers may vary. Sample results are given.
   a. 3.14260537, 3.14158951, 3.14159267
   b. 3.14196328, 3.14159257, 3.14159302
   c. 3.17253103, 3.13905576, 3.14186692
   d. 3.14163420, 3.14159255, 3.14159256
   e. $N = 5$, d; $N = 10$, b; $N = 15$, a
3. Let $x = \text{Tan}^{-1}a$ and $y = \text{Tan}^{-1}b$ so that $\tan x = a$ and $\tan y = b$. Then $\tan(x+y) = \frac{\tan x + \tan y}{1 - \tan x \tan y} = \frac{a+b}{1-ab}$. If $-\frac{\pi}{2} < x + y < \frac{\pi}{2}$, then $\text{Tan}^{-1}(\tan(x+y)) = x + y = \text{Tan}^{-1}a + \text{Tan}^{-1}b = \text{Tan}^{-1}\left(\frac{a+b}{1-ab}\right)$.

Additional Answers
Written Exercises

(continued from p. 687)

10. $\cos 0° + i \sin 0°$, $\cos 36° + i \sin 36°$, $\cos 72° + i \sin 72°$, $\cos 108° + i \sin 108°$, $\cos 144° + i \sin 144°$, $\cos 180° + i \sin 180°$, $\cos 216° + i \sin 216°$, $\cos 252° + i \sin 252°$, $\cos 288° + i \sin 288°$, $\cos 324° + i \sin 324°$

11. $2(\cos 110° + i \sin 110°)$, $2(\cos 230° + i \sin 230°)$, $2(\cos 350° + i \sin 350°)$

12. $2(\cos 40° + i \sin 40°)$, $2(\cos 160° + i \sin 160°)$, $2(\cos 280° + i \sin 280°)$

15. $1, \frac{1}{2} \pm i\frac{\sqrt{3}}{2}, -\frac{1}{2} \pm i\frac{\sqrt{3}}{2}, -1$

16. $1, i, -1, -i, \frac{\sqrt{2}}{2} \pm i\frac{\sqrt{2}}{2}, -\frac{\sqrt{2}}{2} \pm i\frac{\sqrt{2}}{2}$

19. $\frac{\sqrt[4]{108}}{2} + i\frac{\sqrt[4]{12}}{2}, -\frac{\sqrt[4]{108}}{2} - i\frac{\sqrt[4]{12}}{2}, -\frac{\sqrt[4]{12}}{2} + i\frac{\sqrt[4]{108}}{2}, \frac{\sqrt[4]{12}}{2} - i\frac{\sqrt[4]{108}}{2}$

20. $\frac{\sqrt[4]{12}}{2} + i\frac{\sqrt[4]{108}}{2}, -\frac{\sqrt[4]{12}}{2} - i\frac{\sqrt[4]{108}}{2}, -\frac{\sqrt[4]{108}}{2} + i\frac{\sqrt[4]{12}}{2}, \frac{\sqrt[4]{108}}{2} - i\frac{\sqrt[4]{12}}{2}$

```
90 PRINT 4*P
100 END
110 LET S=X
120 FOR J=2 TO N
130 LET T=(-1)∧(J-1)*X∧(2*J-1)/(2*J-1)
140 LET S=S+T
150 NEXT J
160 RETURN
```

Exercises

1. For each pair (a, b) given below, run the program for $N = 5$, $N = 10$, and $N = 15$, and record the approximations for π. (Note that you must enter the fractions as decimals; for the repeating decimals, enter as many places as your computer will accept.)

a. $\left(\frac{1}{4}, \frac{3}{5}\right)$ **b.** $\left(\frac{6}{11}, \frac{5}{17}\right)$ **c.** $\left(\frac{5}{6}, \frac{1}{11}\right)$ **d.** $\left(\frac{2}{5}, \frac{3}{7}\right)$

e. For a given value of N, which pair above gives the best approximation?

2. a. Find another pair (a, b) that meets the requirements. (*Hint:* Choose an a and solve for b.) Answers may vary; for example, $a = \frac{1}{5}$, $b = \frac{2}{3}$.

b. Run the program for your pair, using $N = 15$. Record the approximation. Answers may vary. 3.14159298

3. Prove the formula for $\text{Tan}^{-1}\, a + \text{Tan}^{-1}\, b$. *Hint:* Let $x = \text{Tan}^{-1}\, a$ and $y = \text{Tan}^{-1}\, b$, so that $\tan x = a$ and $\tan y = b$. Use the addition formula for the tangent (page 650) to find $\tan (x + y)$. Then use the fact that $\text{Tan}^{-1} (\tan (x + y)) = x + y$ if $-\frac{\pi}{2} < x + y < \frac{\pi}{2}$.

Chapter Summary

1. *Vector quantities* have both magnitude and direction. A given vector quantity can be represented by an arrow called a *vector*. Vectors can be added, subtracted, and multiplied by real numbers. Two vectors are *equivalent* if they have the same magnitude and direction. If two vectors are perpendicular or if either one is **0**, then they are *orthogonal*.
2. A vector can be expressed in component form, $\mathbf{u} = a\mathbf{i} + b\mathbf{j}$, where the numbers a and b are the x- and *y-components*, respectively, of **u**. The *dot product* of two vectors can be used to find the angle between them.
3. The *polar coordinates* of a point P are (r, θ), where r is the distance from point O, called the pole, to point P, and θ is the measure of the angle that $\overrightarrow{OP}$ makes with the *polar axis*.
4. The complex number $z = x + yi$ can be represented by the point (x, y) in the *complex plane*. The *polar*, or *trigonometric, form* of z is
$$z = r(\cos \theta + i \sin \theta),$$
where (r, θ) are the polar coordinates of the point z.

5. Powers and roots of complex numbers can be found with the help of *De Moivre's theorem:*

$$z^n = r^n (\cos n\theta + i \sin n\theta).$$

6. When the domains of the trigonometric functions are restricted to certain intervals, they become one-to-one, and the *inverse trigonometric functions* can be defined. For example:

$y = \text{Sin}^{-1} x$ if and only if $\sin y = x$ and $-\frac{\pi}{2} \le y \le \frac{\pi}{2}$;

$y = \text{Cos}^{-1} x$ if and only if $\cos y = x$ and $0 \le y \le \pi$;

$y = \text{Tan}^{-1} x$ if and only if $\tan y = x$ and $-\frac{\pi}{2} < y < \frac{\pi}{2}$.

7. A *trigonometric equation* can be solved by using algebraic transformations and trigonometric identities.

Chapter Review

Write the letter of the correct answer.

1. A ship leaves port on a heading of 172° with a speed of 25 km/h. The current is flowing 6 km/h at a heading of 200°. What is the actual heading of the ship to the nearest tenth of a degree? 14-1
a. 5.3° **b.** 166.7° **(c.)** 177.3° **d.** 194.7°

2. Vector **u** has magnitude 6 and bearing 40°. Vector **v** has magnitude 10 and bearing 160°. Find $\|\mathbf{v} - \mathbf{u}\|$ to the nearest hundredth.
a. 4.00 **b.** 8.72 **(c.)** 14.00 **d.** 11.66

3. Find the coordinates of A, given $B(4, 3)$ and $\overrightarrow{AB} = -5 + 4i$. 14-2
(a.) $(9, -1)$ **b.** $(-1, 7)$ **c.** $(9, -7)$ **d.** $(1, -1)$

4. Find a unit vector orthogonal to $9\mathbf{i} - 12\mathbf{j}$.
a. $\frac{\sqrt{2}}{2}\mathbf{i} + \frac{\sqrt{2}}{2}\mathbf{j}$ **b.** $4\mathbf{i} + 3\mathbf{j}$ **c.** $-\frac{3}{5}\mathbf{i} + \frac{4}{5}\mathbf{j}$ **(d.)** $-\frac{4}{5}\mathbf{i} - \frac{3}{5}\mathbf{j}$

5. Find a polar equation of the curve whose rectangular-coordinate equation is $x + y - 2 = 0$. 14-3
a. $r = 2 \sin \theta$ **b.** $r = 2 \cos \theta$ **(c.)** $r = \frac{2}{\cos \theta + \sin \theta}$ **d.** $r = \frac{-2}{\cos \theta + \sin \theta}$

6. Find the rectangular coordinates of the point $(5, -60°)$.
(a.) $\left(\frac{5}{2}, \frac{-5\sqrt{3}}{2}\right)$ **b.** $\left(\frac{5\sqrt{3}}{2}, -\frac{5}{2}\right)$ **c.** $\left(-\frac{5}{2}, -\frac{5\sqrt{3}}{2}\right)$ **d.** $\left(-\frac{5\sqrt{3}}{2}, -\frac{5}{2}\right)$

7. Express $2 - 2i\sqrt{3}$ in polar form. 14-4
a. $2(\cos 300° + i \sin 300°)$ **b.** $2(\cos 120° + i \sin 120°)$
c. $4(\cos 60° + i \sin 60°)$ **(d.)** $4(\cos 300° + i \sin 300°)$

8. Find $\frac{w}{z}$ if $w = 10(\cos 83° + i \sin 83°)$ and $z = 5(\cos 113° + i \sin 113°)$.
a. $2i$ **(b.)** $\sqrt{3} - i$ **c.** $2\sqrt{3} - 2i$ **d.** $1 - i\sqrt{3}$

Additional Answers Oral Exercises
(for p. 699)

12. $\tan x = \frac{\sin x}{\cos x}$,
$\frac{\sin x}{\cos x} = \cos x$;
$\sin x = \cos^2 x$
$\sin x = 1 - \sin^2 x$
$\sin^2 x + \sin x - 1 = 0$, etc.

13.
$2 \sin x \cos x - \cos x = 0$;
$\cos x(2 \sin x - 1) = 0$;
$\cos x = 0$ or
$\sin x = \frac{1}{2}$, etc.

16. $2 \sin 2x \cos 2x - \cos 2x = 0$;
$\cos 2x(2 \sin 2x - 1) = 0$;
$\cos 2x = 0$ or
$\sin 2x = \frac{1}{2}$; $2x = \frac{\pi}{2}$ or $\frac{3\pi}{2}$; $2x = \frac{\pi}{6}$ or $\frac{5\pi}{6}$, etc.

Additional Answers Written Exercises
(for p. 699)

1–12. k is an integer.

1. $\{48.6°, 131.4°\}$
$\{\theta: \theta = 48.6° + k \cdot 360°$ or $\theta = 131.4° + k \cdot 360°\}$

2. $\{73.4°, 286.6°\}$
$\{\theta: \theta = 73.4° + k \cdot 360°$ or $\theta = 286.6° + k \cdot 360°\}$

3. $\left\{\frac{\pi}{4}, \frac{3\pi}{4}, \frac{5\pi}{4}, \frac{7\pi}{4}\right\}$
$\left\{x: x = \frac{(2k+1)\pi}{4}\right\}$

4. $\left\{\frac{\pi}{3}, \frac{2\pi}{3}, \frac{4\pi}{3}, \frac{5\pi}{3}\right\}$
$\left\{x: x = \frac{\pi}{3} + k\pi \text{ or } x = \frac{2\pi}{3} + k\pi\right\}$

5. $\{135°, 315°\}$
$\{\theta: \theta = 135° + k \cdot 180°\}$

(continued)

Additional Answers Written Exercises

(continued)

6. {45°, 225°}
{θ: θ = 45° + k · 180°}
7. {14.5°, 165.5°, 194.5°, 345.5°}
{θ: θ = 14.5° + k · 180° or θ = 165.5° + k · 180°}
8. {2.76, 5.90}
{x: x = 2.76 + kπ}
9. $\left\{\frac{3\pi}{8}, \frac{11\pi}{8}\right\}$
$\left\{x: x = \frac{3\pi}{8} + k\pi\right\}$
10. $\left\{\frac{7\pi}{12}, \frac{23\pi}{12}\right\}$
$\left\{x: x = \frac{7\pi}{12} + 2k\pi \text{ or } x = \frac{23\pi}{12} + 2k\pi\right\}$
11. {155°, 335°}
{θ: θ = 155° + k · 180°}
12. {80°, 260°}
{θ: θ = 80° + k · 180°}
16. {30°, 150°, 210°, 330°}
17. {45°, 135°, 225°, 315°}
18. {54.7°, 125.3°, 234.7°, 305.3°}
19. $\left\{\frac{\pi}{6}, \frac{\pi}{2}, \frac{5\pi}{6}, \frac{3\pi}{2}\right\}$
20. {0°, 180°}
21. {0°, 120°, 240°}
22. $\left\{\frac{\pi}{6}, \frac{5\pi}{6}, \frac{3\pi}{2}\right\}$
23. {30°, 60°, 210°, 240°}
24. $\left\{\frac{\pi}{4}, \frac{5\pi}{4}\right\}$
25. {45°, 123.7°, 225°, 303.7°}
26. {30°, 150°, 199.5°, 340.5°}
31. $\left\{\frac{\pi}{4}, \frac{7\pi}{12}, \frac{11\pi}{12}, \frac{5\pi}{4}, \frac{19\pi}{12}, \frac{23\pi}{12}\right\}$

9. Express $\left(-\frac{\sqrt{2}}{2} + \frac{\sqrt{2}}{2}i\right)^{10}$ in $x + yi$ form. **14-5**

a. 1 **b.** −1 **c.** i **(d.)** $-i$

10. Which of the following is not a cube root of $-i$?

a. i **(b.)** $-i$ **c.** $\frac{\sqrt{3}}{2} - \frac{1}{2}i$ **d.** $-\frac{\sqrt{3}}{2} - \frac{1}{2}i$

11. Find the value of $\text{Sin}^{-1}\left[\sin\left(-\frac{4\pi}{3}\right)\right]$. **14-6**

(a.) $\frac{\pi}{3}$ **b.** $-\frac{\pi}{3}$ **c.** $\frac{2\pi}{3}$ **d.** $\frac{5\pi}{6}$

12. Express $\cos(2\ \text{Sin}^{-1} x)$ without using trigonometric or inverse trigonometric functions.

a. $2x\sqrt{1 - x^2}$ **(b.)** $1 - 2x^2$ **c.** $2x^2 - 1$ **d.** $\sqrt{1 - x^2} - x$

13. Find the value of $\text{Csc}^{-1}\left(-\frac{2\sqrt{3}}{3}\right)$. **14-7**

a. $\frac{2\pi}{3}$ **b.** $\frac{\pi}{3}$ **(c.)** $-\frac{\pi}{3}$ **d.** does not exist

14. Find the value of $\sin[2\ \text{Tan}^{-1}(-2)]$.

a. $\frac{4}{5}$ **(b.)** $-\frac{4}{5}$ **c.** $\frac{\sqrt{5}}{5}$ **d.** $-\frac{4\sqrt{5}}{5}$

15. Which of the following is the solution set of $2 \sin 2\theta = 1$ over the interval $0° \le \theta < 360°$? **14-8**

a. {30°, 150°} **b.** {15°, 30°, 75°, 150°}
c. {15°, 75°} **(d.)** {15°, 75°, 195°, 255°}

16. How many solutions of $4\cos^2(2x + \pi) = 3$ are there in the interval $0 \le x < 2\pi$?

a. 2 **b.** 4 **(c.)** 8 **d.** 16

Chapter Test

1. Two planes leave the airport at 7:30 A.M., each traveling at 250 km/h. Their headings are 165° and 220°. How far apart are they at 9:30 A.M. (to the nearest integer)? What is the bearing of each from the other to the nearest degree? 462 km, 283°, 103° **14-1**

2. Let $\mathbf{u} = 4\mathbf{i} + 3\mathbf{j}$ and $\mathbf{v} = -12\mathbf{i} + 5\mathbf{j}$. **14-2**

a. Find $\|3\mathbf{u} + \mathbf{v}\|$. 14 **b.** Find the angle between $\mathbf{u}$ and $\mathbf{v}$. 120.5°

3. Copy and complete the table. **14-3**

Polar coordinates	(3, 120°)	? (4, 270°)	(−6, 135°)	? $(\sqrt{10}, 135°)$
Rectangular coordinates	? $\left(-\frac{3}{2}, \frac{3\sqrt{3}}{2}\right)$	(0, −4)	? $(3\sqrt{2}, -3\sqrt{2})$	$(-\sqrt{5}, \sqrt{5})$

4. Sketch the polar curve $r = 2 \sin 2\theta$.

5. Let $w = 9(\cos 125° + i \sin 125°)$ and $z = 3(\cos 230° + i \sin 230°)$. Find and simplify wz and $\frac{w}{z}$. 14-4

6. Find the five fifth roots of $-i$ in polar form. 14-5

7. Find the value of $\cos\left[2 \text{ Sin}^{-1}\left(-\frac{24}{25}\right)\right]$. $-\frac{527}{625}$ 14-6

8. Find the value of sec $[\text{Cot}^{-1}(-6)]$. $-\frac{\sqrt{37}}{6}$ 14-7

9. Find the general solution of $\sin 2x = \tan x$. $\left\{x: x = k\pi \text{ or } x = \frac{(2k+1)\pi}{4}\right\}$ 14-8

10. Solve $\sin 2\theta = 2 \sin^2 \theta$, $0° \le \theta < 360°$. {0°, 45°, 180°, 225°}

Mixed Problem Solving

Solve. If a problem has no solution, explain why.

A

1. A jewelry store manager bought some necklaces for $6250 and sold all but 40 for a total of $7130. If the purchase price was $3 less than the selling price, find the number sold. 460 necklaces were sold
2. Rico won a prize of $5 one week, $10 the second week, $15 the third week, and so on for 52 weeks. How much did he win in all? $6890
3. Three oxygen atoms and two nitrogens atoms contain 38 protons. Five oxygen atoms and three nitrogen atoms contain 61 protons. Find the number of protons in one nitrogen atom. 7
4. Find the maximum possible area of a rectangular garden enclosed with 36 m of fencing. 81 m^2
5. A snail moves 2 cm in 1 min. In each succeeding minute it travels half as far as in the previous minute. How far would it travel (a) during the eighth minute? (b) if it could travel forever? **a.** $\frac{1}{64}$ cm **b.** 4 cm
6. A mixture of raisins and peanuts is worth $56. Raisins and peanuts are worth $3/kg and $5/kg, respectively. Find all integral possibilities for the number of kilograms of each used.
7. It takes Rachel 7 h to split a cord of wood. Meg can do the job in 5 h. At 9:00 A.M. Rachel began to work. At 11:00 A.M. Meg joined her. At what time was the job finished? 1:05 P.M.

B

8. A small plane made a round trip of 900 km in 6 h with a 30 km/h head and tail wind. Find the plane's air speed to the nearest whole number. 156 km/h
9. The area of an equilateral triangle varies directly as the square of its perimeter. An equilateral triangle with perimeter 18 has area $9\sqrt{3}$. Find the perimeter of a triangle with area $3\sqrt{3}$. $6\sqrt{3}$
10. The half-life of Carbon-11 is 20 min. How many minutes does it take for a 50 g sample to decay to 5 g? approx. 66 min.
11. The top of a vertical tree broken by the wind hit the ground 10 ft from the base of the tree. The angle formed by the ground and the treetop was 25°. Find the original height of the tree. 15.7 ft.

32. $\left\{\frac{\pi}{9}, \frac{2\pi}{9}, \frac{4\pi}{9}, \frac{5\pi}{9}, \frac{7\pi}{9}, \frac{8\pi}{9}, \frac{10\pi}{9}, \frac{11\pi}{9}, \frac{13\pi}{9}, \frac{14\pi}{9}, \frac{16\pi}{9}, \frac{17\pi}{9}\right\}$

33. {15°, 45°, 75°, 135°, 195°, 225°, 255°, 315°}

34. $\left\{\frac{\pi}{4}, \frac{\pi}{2}, \frac{3\pi}{4}, \frac{5\pi}{4}, \frac{3\pi}{2}, \frac{7\pi}{4}\right\}$

35. $\left\{\frac{\pi}{4}, \frac{\pi}{2}, \frac{3\pi}{4}, \frac{5\pi}{4}, \frac{3\pi}{2}, \frac{7\pi}{4}\right\}$

36. {0°, 60°, 120°, 180°, 240°, 300°}

40. {28.3°, 48.3°, 148.3°, 168.3°, 268.3°, 288.3°}

41. $\left\{\frac{\pi}{12}, \frac{\pi}{4}, \frac{3\pi}{4}, \frac{11\pi}{12}, \frac{17\pi}{12}, \frac{19\pi}{12}\right\}$

Additional Answers Chapter Test

1. 46 km; bearing of first from second is 102.5°; bearing of second from first is 282.5°
4. 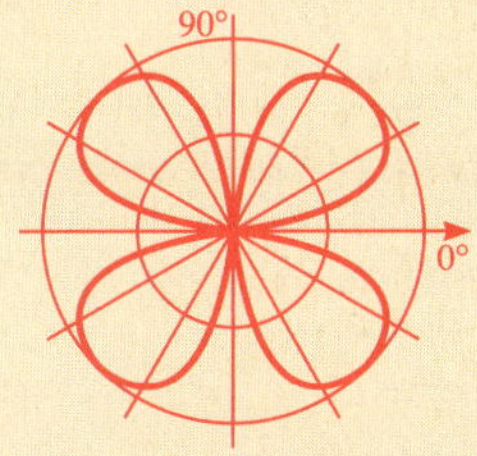

5. $wz = 27(\cos 355° + i \sin 355°)$
   $\frac{w}{z} = 3(\cos 255° + i \sin 255°)$
6. $1(\cos 54° + i \sin 54°)$, $1(\cos 126° + i \sin 126°)$, $1(\cos 198° + i \sin 198°)$, $1(\cos 270° + i \sin 270°)$, $1(\cos 342° + i \sin 342°)$

Additional Answers Mixed Problem Solving

6. 2 kg raisins and 10 kg peanuts, 7 kg raisins and 7 kg peanuts; 12 kg raisins and 4 kg peanuts; 17 kg raisins and 1 kg peanuts.

Supplementary Materials

Test Masters 60, 61, 62
Resource Book p. 166

Alternate Test p. T26

Additional Answers Cumulative Review

2. $\sin\theta = \frac{5\sqrt{29}}{29}$, $\csc\theta = \frac{\sqrt{29}}{5}$, $\cos\theta = -\frac{2\sqrt{29}}{29}$, $\sec\theta = -\frac{\sqrt{29}}{2}$, $\tan\theta = -\frac{5}{2}$, $\cot\theta = -\frac{2}{5}$

7. $\sin\left(-\frac{13\pi}{6}\right) = -\frac{1}{2}$, $\csc\left(-\frac{13\pi}{6}\right) = -2$, $\cos\left(-\frac{13\pi}{6}\right) = \frac{\sqrt{3}}{2}$, $\sec\left(-\frac{13\pi}{6}\right) = \frac{2\sqrt{3}}{3}$, $\tan\left(-\frac{13\pi}{6}\right) = -\frac{\sqrt{3}}{3}$, $\cot\left(-\frac{13\pi}{6}\right) = -\sqrt{3}$

Preparing for College Entrance Exams

Strategy for Success

Approximately one third of the multiple-choice questions in the SAT mathematical sections are quantitative comparison questions. These types of questions, like Questions 1–6 below, emphasize the concepts of equalities, inequalities, and estimation. Read the directions carefully and, if possible, practice answering questions like these before the exams.

Questions 1–6 each consist of two quantities, one in Column A and one in Column B. Write the letter A if the quantity in Column A is greater; B if the quantity in Column B is greater; C if the two quantities are equal; and D if the relationship cannot be determined from the information given.

	Column A	*Column B*
1.	x D	$\|x\|$
2.	$\|a - b\|$ C	$\|b - a\|$
3.	$\csc \frac{10\pi}{3}$ A	$\sec \frac{10\pi}{3}$
4.	$\tan(-x)$ C	$-\tan x$
5.	$\sqrt{1 + \tan^2\theta}$ A	$\sqrt{1 - \cos^2\theta}$
6.	$\sin(180° + \theta)$ C	$\cos(90° + \theta)$

Decide which is the best of the choices given and write the corresponding letter on your answer sheet.

7. Find a unit vector orthogonal to $\mathbf{u} = 6\mathbf{i} - 8\mathbf{j}$. D

(A) $\frac{3}{5}\mathbf{i} + \frac{4}{5}\mathbf{j}$ **(B)** $\frac{3}{5}\mathbf{i} - \frac{4}{5}\mathbf{j}$ **(C)** $4\mathbf{i} - 3\mathbf{j}$ **(D)** $\frac{4}{5}\mathbf{i} + \frac{3}{5}\mathbf{j}$ **(E)** $\frac{4}{5}\mathbf{i} - \frac{3}{5}\mathbf{j}$

8. Express $\sin(2\,\text{Tan}^{-1} x)$ without using trigonometric or inverse trigonometric functions. E

(A) $\frac{x}{\sqrt{x^2+1}}$ **(B)** $\frac{2x}{\sqrt{1-x^2}}$ **(C)** $\frac{x^2-1}{x^2+1}$ **(D)** $\frac{2x\sqrt{1-x^2}}{1-2x^2}$ **(E)** $\frac{2x}{x^2+1}$

9. How many solutions does $\cot x + \tan x = -2$ have in the interval $0 \le x < 2\pi$? C

(A) 0 **(B)** 1 **(C)** 2 **(D)** 3 **(E)** 4

10. Which polar equation is that of a circle with center (0, 1) and radius 1? C

(A) $r\cos\theta = 2$ **(B)** $r = \cos 2\theta$ **(C)** $r = 2\sin\theta$ **(D)** $r = 1\csc\theta$ **(E)** $r^2 = 2\sin 2\theta$

Cumulative Review *(Chapters 12–14)*

1. Name two angles, one positive and one negative, that are coterminal with 105°. 465°, −255°
2. Find the values of the six trigonometric functions of an angle θ in standard position whose terminal side passes through the point $(-2, 5)$.
3. A truck has an 8-foot loading ramp inclined at a 20° angle with respect to the ground. How many feet above the ground is the top of the ramp? Give your answer to three significant digits. 2.74 ft.
4. In ΔABC, $a = 5$, $b = 9$, and $c = 6$. Find $\angle A$, $\angle B$, and $\angle C$ to the nearest tenth of a degree. $\angle A = 31.6°$, $\angle B = 109.5°$, $\angle C = 38.9°$
5. Find the area of ΔABC in Exercise 4. 14.1
6. Write −200° in radians. $-\frac{10\pi}{9}$
7. Find the exact values of the six trigonometric functions of $-\frac{13\pi}{6}$.
8. Find the amplitude, the maximum and minimum values, and the period of $y = 5\cos\frac{\pi}{3}x - 2$. 5; 3 and −7; 6
9. Write $\frac{\cos\theta}{\sin^3\theta}$ in terms of $\cot\theta$. $\cot\theta + \cot^3\theta$ or $\cot\theta(1 + \cot^2\theta)$
10. Prove: **a.** $\sin x = \frac{\sec x}{\cot x + \tan x}$ **b.** $\cot(\alpha + \beta) = \frac{\cot\alpha\cot\beta - 1}{\cot\alpha + \cot\beta}$
11. If $\sin\phi = -\frac{3}{5}$ and $90° < \phi < 270°$, find:

    a. $\cos\frac{\phi}{2}$ $\frac{-\sqrt{10}}{10}$ **b.** $\cos 2\phi$ $\frac{7}{25}$ **c.** $\tan(180° - \phi)$ $-\frac{3}{4}$ **d.** $\sin(90° + \phi)$ $-\frac{4}{5}$
12. The vector **u** has magnitude 10 and bearing 200°. Vector **v** has magnitude 8 and bearing 35°. Find the magnitude of **u** − **v** to three significant digits and its bearing to the nearest tenth of a degree. 17.8; 206.7°
13. Find a unit vector orthogonal to **v** = 2**i** − **j**. $\frac{\sqrt{5}}{5}i + \frac{2\sqrt{5}}{5}j$
14. Use DeMoivre's theorem to express $(2 - 2i\sqrt{3})^5$ in $x + yi$ form. $512 + 512i\sqrt{3}$
15. Find the cube roots of $-8i$ in $x + yi$ form. $2i$, $-\sqrt{3} - i$, $\sqrt{3} - i$

Find the value of each of the following.

16. $\text{Cos}^{-1}\left(\tan\frac{3\pi}{4}\right)$ π
17. $\sin\left(\text{Cos}^{-1}\frac{1}{2} - \text{Cot}^{-1}\frac{4}{3}\right)$ $\frac{4\sqrt{3} - 3}{10}$
18. $\tan\left[2\,\text{Sin}^{-1}\left(-\frac{5}{13}\right)\right]$ $-\frac{120}{119}$

Solve for x, $0 \le x < 2\pi$, or for θ, $0° \le \theta < 360°$.

19. $2\csc^2 x + 3\csc x = 2$ $\left\{\frac{7\pi}{6}, \frac{11\pi}{6}\right\}$
20. $\sin\left(x + \frac{\pi}{3}\right) = -\frac{1}{2}$ $\left\{\frac{5\pi}{6}, \frac{3\pi}{2}\right\}$
21. $\cos 2\theta = 3\cos\theta + 1$ {120°, 240°}
22. $3\tan^2\theta - 1 = 0$ {30°, 150°, 210°, 330°}

10. a. $\frac{\sec x}{\cot x + \tan x} = \frac{\frac{1}{\cos x}}{\frac{\cos x}{\sin x} + \frac{\sin x}{\cos x}} = \frac{\frac{1}{\cos x}}{\frac{\cos^2 x + \sin^2 x}{\sin x\cos x}} = \frac{1}{\cos x}\cdot\frac{\sin x\cos x}{\cos^2 x + \sin^2 x} = \frac{\sin x}{\cos^2 x + \sin^2 x} = \frac{\sin x}{1} = \sin x$

b. $\cot(\alpha + \beta) = \frac{1}{\tan(\alpha + \beta)} = \frac{1 - \tan\alpha\tan\beta}{\tan\alpha + \tan\beta} = \frac{1 - \frac{1}{\cot\alpha}\cdot\frac{1}{\cot\beta}}{\frac{1}{\cot\alpha} + \frac{1}{\cot\beta}} = \frac{\frac{\cot\alpha\cot\beta - 1}{\cot\alpha\cot\beta}}{\frac{\cot\beta + \cot\alpha}{\cot\alpha\cot\beta}} = \frac{\cot\alpha\cot\beta - 1}{\cot\alpha + \cot\beta}$

15 Statistics and Probability

Objectives

15-1 To display data using frequency distributions, histograms, and stem-and-leaf plots, and to compute measures of central tendency.

15-2 To compute measures of dispersion and, together with measures of central tendency, to describe and compare distributions using these statistics.

15-3 To recognize and analyze normal distributions.

15-4 To draw a scatter plot, determine the correlation coefficient, and use the regression line for a set of ordered pairs of data.

15-5 To apply fundamental counting principles.

15-6 To find the number of permutations of the elements of a set.

15-7 To find the combinations of a set of elements.

15-8 To specify sample spaces and events for random experiments.

15-9 To find the probability that an event will occur.

15-10 To identify mutually exclusive and independent events and find the probability of such events.

Assignment Guide

See p. T58 for Key to the format of the Assignment Guide

Day	Average Algebra	Average Algebra and Trigonometry	Extended Algebra	Extended Algebra and Trigonometry
1	**15-1** 711/1–16 **S** 712/*Mixed Review:* 4–6		**15-1** 711/1–17 **S** 712/*Mixed Review,* 4–6	
2	**15-2** 717/1–8, 9–19 odd		**15-2** 717/1–8, 9–19 odd **S** 712/18	
3	**15-3** 722/1–10 **S** 723/*Mixed Review*		**15-3** 722/1–11 **S** 723/*Mixed Review*	
4	**15-4** 727/1–12		**15-4** 727/1–13 **S** 718/20	
5	**15-5** 732/1–10, 11–15 odd **S** 733/*Mixed Review* **R** 729/*Self-Test 1*		**15-5** 732/1–12, 14, 16 **S** 733/*Mixed Review* **R** 729/*Self-Test 1*	
6	**15-6** 737/1–20		**15-6** 737/1–20 **S** 723/12	
7	**15-6** 737/21–28 **15-7** 740/1–15 odd		**15-6** 737/21–30 **15-7** 740/2–10 even, 11–16	
8	**15-7** 740/17–25 odd **S** 741/*Mixed Review*		**15-7** 740/17–22, 23–27 odd **S** 741/*Mixed Review*	
9	**15-8** 744/1–11 odd **R** 742/*Self-Test 2*		**15-8** 744/1–9 odd, 10–14 **R** 742/*Self-Test 2*	

Assignment Guide *(continued)*

Day	Average Algebra	Average Algebra and Trigonometry	Extended Algebra	Extended Algebra and Trigonometry
10	**15-9** 748/1–11 **S** 750/*Mixed Review*		**15-9** 748/1–11 **S** 750/*Mixed Review*	
11	**15-10** 759/1–15 odd **R** 761/*Self-Test 3*		**15-10** 759/1–8 **S** 750/13, 14	
12	*Prepare for Chapter Test* **R** 764/*Chapter Review*		**15-10** 760/9–13, 16, 17 **R** 761/*Self-Test 3*	
13	*Administer Chapter 15 Test*		*Prepare for Chapter Test* **R** 764/*Chapter Review*	
14			*Administer Chapter 15 Test*	

Supplementary Materials Guide

For Use with Lesson	Practice Masters	Tests	Study Guide (Reteaching)	Resource Book: Tests	Resource Book: Practice Exercises	Resource Book: Applications (A) Enrichment (E) Thinking Skl. (TS)
15-1			pp. 239–240			p. 218 (A)
15-2	Sheet 82		pp. 241–242			p. 268 (TS)
15-3			pp. 243–244		p. 167	
15-4	Sheet 83	Test 63	pp. 245–246			
15-5			pp. 247–248		p. 168	
15-6	Sheet 84		pp. 249–250			
15-7	Sheet 85	Test 64	pp. 251–252		p. 169	
15-8			pp. 253–254			
15-9	Sheet 86		pp. 255–256			
15-10	Sheet 87	Test 65	pp. 257–258		p. 170	
Chapter 15		Tests 66, 67		pp. 77–80	p. 171	p. 234 (E)

Overhead Visuals

For Use with Lessons	Visual	Title
15-4	B	Multi-Use Packet 2
15-3, 15-9	13	The Normal Curve

Software

Software	Algebra Plotter Plus	Using Algebra Plotter Plus	Computer Activities	Test Generator
	Statistics Spreadsheet, Sampling Experiment	Activity Sheets, pp. 64–66	Activities 32–34	210 test items
For Use with Lessons	15-2, 15-4, 15-9	15-2, 15-4, 15-9	15-2, 15-4, 15-10	all lessons

Strategies for Teaching

Making Connections

Throughout this book, students meet techniques for dealing with data. In addition to the integration of statistical work throughout the text, in this chapter formal lessons on statistical techniques introduce students to frequency distributions, statistical measures, stem-and-leaf and box-and-whisker plots, and the normal distribution. In Lesson 15-4 (page 724) they draw scatter plots, determine correlation coefficients, and use regression lines to make predictions from data they gather (page 729). In the Application feature (page 752) they test hypotheses and solve problems using random sampling techniques.

Students are prepared for future coursework in statistics and probability with lessons (beginning on page 730) covering fundamental counting principles, permutations and combinations, sample spaces, and the probability of mutually exclusive and independent events.

15-3 The Normal Distribution

Lead a discussion of what types of information would, and would not, be expected to yield distributions that are normal. Examples that would yield such distributions are the height and the weight of sixteen-year-olds. Examples that would not are the number of math credits among Algebra 2 students and the per capita income of Americans.

15-5 Fundamental Counting Principle

The multiplication process of the first counting principle forms the backbone of this lesson. Offer a variety of diagrammatic techniques to help students apply this principle correctly. Use successive boxes to be filled with numbers of possibilities, as in Example 1. Another useful way to analyze multi-step selections is to use tree diagrams, which have an inherent clarity for most students. Make sure students understand that they add when working with mutually exclusive choices only.

Have students apply counting principles to questions about some well-known codes.

1. Letters in Morse Code are formed from combinations of from one to four dots and dashes. How many "letters" can be formed this way? 30
2. In binary code only ones and zeros are used to represent information. Some computers use an "eight bit byte," meaning that they can recognize a given string of eight digits consisting of ones and zeros. How many symbols can be coded using 8-bit bytes? 256
3. Have students read about the ASCII code and its relation to binary code.

15-6 Permutations

In permutations, order is important. Discuss elections, where it makes a difference which office one is elected to; Zip Codes, where the same digits used in a different order identify two different locations; and telephone area codes, where the order of the given digits certainly makes a difference. Take time to introduce factorial notation, since this may be the first exposure to it for many students. Continue to use boxes to be filled with the appropriate number of selections, and tree diagrams, in leading up to a derivation of the permutation formulas.

15-9 Probability

Wildlife biologists sometimes use the "capture-recapture method" to estimate the population of a species in a given area. In simplest terms, consider the example of a bass population in a lake. Suppose 25 bass are caught, tagged, and released back into the lake. We assume that they spread evenly throughout the bass population. Later, bass are again caught. This time 3 bass are tagged out of 40 bass caught.

1. What is the probability that a bass is tagged, in percent? (This is based on the second catch.) 7.5%

2. What is the total population of bass in the lake? (The ratio of the 25 tagged bass to the total bass population should be the same as the probability in Exercise 10. about 333

See the Exploration on page 846 for an activity in which students explore probability with experiments.

References to Strategies

PE: Pupil's Edition **TE:** Teacher's Edition **RB:** Resource Book

Problem Solving Strategies

TE: p. 720 (simplify the problem)

Applications

PE: pp. 708–762 (word problems)
TE: pp. 708–710, 714–715, 721–722, 724–726, 730–732, 734–735, 739–740, 742–744, 746–747, 754–757; pp. T138, T149 (Sugg. Extensions, Lessons 15-5, 15-9)
RB: p. 218

Nonroutine Problems

PE: p. 712 (Exs. 17, 18); p. 718 (Ex. 20); p. 723 (Exs. 11, 12); pp. 728–729 (Exs. 9–13); pp. 732–733 (Exs. 17–20, Challenge); p. 741 (Exs. 27–30); p. 753 (Challenge 1–8)
TE: pp. T138–T141

Communication

PE: pp. 723, 729 (convincing argument); p. 751 (Reading Algebra)
TE: pp. T137, T140 (Reading Algebra); p. T138 (Teaching Sugg., Lesson 15-4)

Thinking Skills

PE: p. 741 (Ex. 28, proof)
TE: pp. 715, 738
RB: p. 268

Explorations

PE: p. 733 (Challenge); p. 753 (random numbers); p. 846 (experimental probability)

Connections

PE: p. 718 (Electricity)

Using Technology

PE: pp. 710, 712; pp. 742, 750 (Exs.); p. 762 (Computer Key-In)
TE: pp. 711–712, 716, 727, 742, 750, 752–753, 762, T139
RB: pp. 250–252
Computer Activities: pp. 86–93
Using Algebra Plotter Plus: pp. 64–66

Using Manipulatives/Models

TE: p. T140 (Lesson 15-8)
Overhead Visuals: B, 13

Cooperative Learning

TE: p. T41 (Lesson 15-10)

Teaching Resources

For use in implementing the teaching strategies referenced on the previous page.

Application
Resource Book, p. 218

Application—Coding and Decoding (For use with Chapter 15)

Cryptography, the science of making and breaking codes, has been important both to the military and to private industry for many centuries. Though mathematicians now use computer technology to develop codes that are very difficult to break, some of the age-old methods of coding still challenge the best code-cracking methods.

1. One of the most common methods of coding is a simple letter substitution, where each letter is replaced by another letter in the alphabet. Dating back to the time of Julius Caesar, letter substitution is now the basis for the modern-day "cryptograms" that often appear in puzzle books. This code is relatively easy to break using a frequency chart for letters in English text.

Letter	*Frequency*	*Letter*	*Frequency*	*Letter*	*Frequency*
E	13%	H, S	6%	B, G, W	1.5%
T	9%	D	4%	V	1%
A, O	8%	L	3.5%	K, J, X	0.5%
N	7%	C, M, U	3%	Q, Z	0.2%
I, R	6.5%	F, P, Y	2%		

Use the frequency table above to decode the following:

H omia ughu my yupturgta dc h its math rhi itktp lb dhrf ub muy bpmlmihx amotiymbiy.

____

2. Codes sometimes require the use of a key word or words to decipher. For example, suppose the first paragraph of the Gettysburg Address is used as a key. Each letter is numbered sequentially: F = 1, O = 2, U = 3, R = 4, S = 5, and so on (see the complete first sentence of the address below). The word "code" could then be represented either by "6 2 81 9" or by "54 7 94 14." Notice that by the end of a paragraph there are many different numbers that could be used for some letters, which makes it difficult to crack the code by frequency analysis.

1 5 10 13 18 23 26 29 36 43 48 50 54 63 67
↓ ↓ ↓ ↓ ↓ ↓ ↓ ↓ ↓ ↓ ↓ ↓ ↓ ↓ ↓
Four score and seven years ago our fathers brought forth on this continent, a new nation,

73 82 84 91 94 103 108 119 126 132 139
↓ ↓ ↓ ↓ ↓ ↓ ↓ ↓ ↓ ↓ ↓
conceived in Liberty, and dedicated to the proposition that all men are created equal.

Use the first sentence of the Gettysburg Address as a key to decode the following:

126 63 31 47 14 126 68 105 52 54 35 58 13 136 120 16

140 39 9 33 11 25 1 69 51 65 113 73 85 101 128 76 127 5

____

3. There are many codes in use for reasons other than sending "secret" messages. Investigate and write a brief report on how each of the following codes are used: **(a)** Braille, **(b)** Morse code, **(c)** ASCII code, and **(d)** UPC (bar codes).

218 APPLICATIONS

Enrichment
Resource Book, p. 234

Comparing Test Scores (For use with Chapter 15)

Mario goes to high school in California, and his cousin Maria is in the same year at a school in Texas. Last year they both took Algebra II. Mario got an 85 on his final exam, and Maria got an 80 on her final exam. Talking on the phone, Mario teased Maria, declaring that his test score was better than hers.

Maria decided to investigate. Her teacher and Mario's teacher gladly provided her with the entire list of exam scores in their respective classes. She arranged the scores according to rank and obtained the following results.

Mario's Class: 92, 91, 90, 90, 90, 88, 87, 85, 85, 84, 83, 80, 75, 70, 65

Maria's Class: 90, 85, 82, 80, 80, 80, 78, 77, 75, 75, 73, 71, 70, 65, 60, 60, 60

Using these figures, Maria computed the mean, the variance, and the standard deviation for each class. Her results are as follows. Later (in Exercise 1) you will perform the same calculations to see if her figures are correct. (Maria always rounded to one decimal place.)

	M	s^2	σ
Mario's Class:	83.7	60.1	7.8
Maria's Class:	74.2	74.7	8.6

Maria next performed the following calculations, using a formula she found in an elementary statistics book.

$$\text{Mario: } X = \frac{85 - 83.7}{7.8} = 0.17 \qquad \text{Maria: } X = \frac{80 - 74.2}{8.6} = 0.67$$

Maria learned in the statistics book that these results are called *standard scores*. The figures meant that her score was 0.67 standard deviations above the average score of her class, whereas Mario's score was only 0.17 standard deviations above the average score of his class.

Maria happily called Mario and teased him, declaring that her test score was really better than his.

1. Perform the necessary calculations to see if Maria's figures are correct.

2. Do you agree that standard scores are preferable to test scores (sometimes called *raw scores*) for comparing scores from different sources? Give the reasons for your opinion.

234 ENRICHMENT

Thinking Skills
Resource Book, p. 268

Thinking Skills (For use after Chapter 15)

Reasoning and drawing inferences

1. Two competing companies, A and B, each have ten employees. The employee salaries are given in the table at the right. Calculate the mean salary for:

a. Company A ____________

b. Company B ____________

Employee Salaries	
Company A	Company B
$16,000	$8,000
$16,000	$10,000
$18,000	$10,000
$18,000	$10,000
$18,000	$10,000
$20,000	$12,000
$20,000	$12,000
$20,000	$12,000
$20,000	$12,000
$24,000	$94,000

2. If you were looking for a job with the two companies of Exercise 1 and were told what the mean salary at each company was (without being given the actual salaries), what conclusion might you draw?

____

3. Would you be likely to draw a different conclusion in Exercise 2 if you were told the actual salaries of the employees at each company? Explain.

____

4. Using the table of Exercise 1, calculate the median salary for:

a. Company A ____________ **b.** Company B ____________

5. For the companies of Exercise 1, why is the median a better indicator of "average" salary than the mean? ______________________________

6. Although the mean is commonly used to indicate the central tendency of a distribution, it can be a misleading statistic unless paired with a measure of the spread of the distribution. Using the table of Exercise 1, calculate the range of salaries for:

a. Company A ____________ **b.** Company B ____________

7. If an "average" spread of a distribution is desired, the range is an inadequate statistic. Why? ______________________________

8. Using the table and results of Exercise 1, calculate the standard deviation of salaries for:

a. Company A ____________ **b.** Company B ____________

Synthesis

9. Although the standard deviation gives an "average" spread of a distribution, it is a difficult statistic to calculate by hand. Try to develop and describe (in detail) some other, less complicated statistic for measuring "average" spread.

____

10. Using the table of Exercise 1 and the new statistic described in Exercise 9, calculate the value of the new statistic for:

a. Company A ____________ **b.** Company B ____________

268 THINKING SKILLS

Using Technology/Exploration
Using Algebra Plotter Plus, p. 66

Sampling

Algebra Plotter Plus | Book 2: Lesson 15-9 Application

Use the Sampling Experiment program of Algebra Plotter Plus.

1. Printed on each bag of Royal Wheat Flour is the claim that the bag contains 2300 g of flour. An independent testing company weighs the contents of 6 bags of flour and finds the mean weight to be only 2250 g, with a standard deviation of 100. To determine the likelihood of obtaining such a result, select the DEFINE option and enter the following numbers:

Population mean: 2300 — Value to test: 2250
Population standard deviation: 100 — Least sample mean shown: 2000
Sample size: 6 — Greatest sample mean shown: 2600

Note that, since the standard deviation of the population is not known, you must assume that the sample standard deviation found by the testing company applies to the population as well.

2. Select SINGLE SAMPLE from the program menu. Have the program produce a random sample three times, and record the results below.

	Sample mean	*Sample standard deviation*
(1)	________	________
(2)	________	________
(3)	________	________

Why do the sample means differ from one another and from the population mean? ______________________________

3. You will learn more about the likelihood of a sample mean being 2250 g or lower by having the program produce a large number of samples. Select MANY SAMPLES, and let the program run until it reaches about 100 samples. Record the following information:

Percent less than 2250: _____

Mean of sample means: _____ Standard deviation of sample means: _____

Suppose that the percent of samples with mean less than 2250 is 10%. This means that even if the Royal Wheat Company's claim is true, about 10% of the samples of the size used by the testing company would have means less than 2250 g. A probability of 0.10 is not considered unlikely enough to prove that Royal Wheat is making a false claim.

4. To make a more definitive test of Royal Wheat's claim, the testing company weighs the contents of 25 bags of flour. The new sample mean is 2265 g, with a standard deviation of 95. Run the "Sampling Experiment" program with the new data for as long as it is practical to do so.

a. What percent of the samples had a mean less than 2265? _____

b. Are you more or less confident than you were in Exercise 3 that Royal Wheat is making a true claim when they say each bag contains an average of 2300 g? Explain why.

____

66 USING ALGEBRA PLOTTER PLUS

Using Technology/Exploration
***Resource Book*, p. 250**

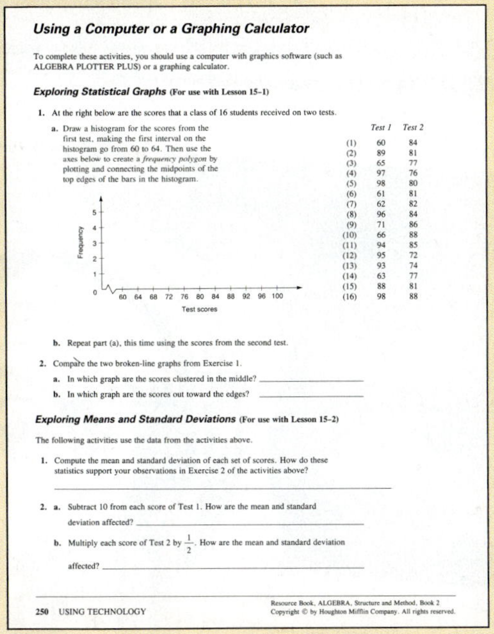

Using a Computer or a Graphing Calculator

To complete these activities, you should use a computer with graphics software (such as ALGEBRA PLOTTER PLUS) or a graphing calculator.

Exploring Statistical Graphs **(For use with Lesson 15–1)**

1. At the right below are the scores that a class of 16 students received on two tests.

a. Draw a histogram for the scores from the first test, making the first interval on the histogram go from 60 to 64. Then use the axes below to create a *frequency polygon* by plotting and connecting the midpoints of the top edges of the bars in the histogram.

	Test 1	*Test 2*
(1)	60	84
(2)	89	81
(3)	65	77
(4)	97	76
(5)	98	80
(6)	61	81
(7)	62	82
(8)	96	84
(9)	71	86
(10)	66	88
(11)	94	85
(12)	95	72
(13)	93	74
(14)	63	77
(15)	88	81
(16)	98	88

Frequency
0 1 2 3 4 5
60 64 68 72 76 80 84 88 92 96 100
Test scores

b. Repeat part (a), this time using the scores from the second test.

2. Compare the two broken-line graphs from Exercise 1.

a. In which graph are the scores clustered in the middle? ____________

b. In which graph are the scores out toward the edges? ____________

Exploring Means and Standard Deviations **(For use with Lesson 15–2)**

The following activities use the data from the activities above.

1. Compute the mean and standard deviation of each set of scores. How do these statistics support your observations in Exercise 2 of the activities above?

____

2. a. Subtract 10 from each score of Test 1. How are the mean and standard deviation affected? ____________

b. Multiply each score of Test 2 by $\frac{1}{2}$. How are the mean and standard deviation affected? ____________

250 USING TECHNOLOGY

Resource Book, ALGEBRA, Structure and Method, Book 2

Using Technology/Exploration
***Resource Book*, p. 251**

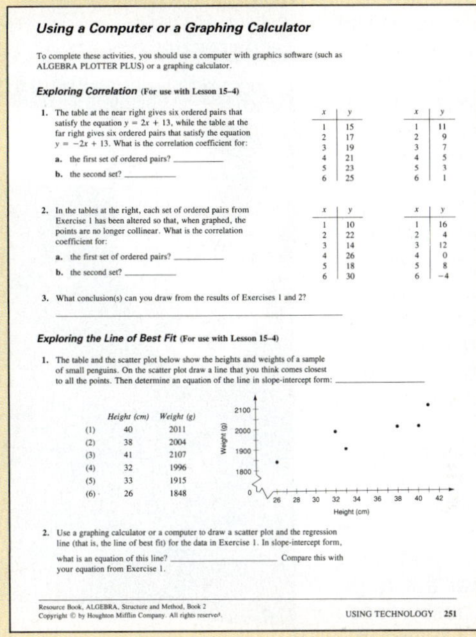

Using a Computer or a Graphing Calculator

To complete these activities, you should use a computer with graphics software (such as ALGEBRA PLOTTER PLUS) or a graphing calculator.

Exploring Correlation **(For use with Lesson 15–4)**

1. The table at the near right gives six ordered pairs that satisfy the equation $y = 2x + 13$, while the table at the far right gives six ordered pairs that satisfy the equation $y = -2x + 13$. What is the correlation coefficient for:

a. the first set of ordered pairs? ________

b. the second set? ________

x	y
1	15
2	17
3	19
4	21
5	23
6	25

x	y
1	11
2	9
3	7
4	5
5	3
6	1

2. In the tables at the right, each set of ordered pairs from Exercise 1 has been altered so that, when graphed, the points are no longer collinear. What is the correlation coefficient for:

a. the first set of ordered pairs? ________

b. the second set? ________

x	y
1	10
2	22
3	14
4	26
5	18
6	30

x	y
1	16
2	4
3	12
4	0
5	8
6	−4

3. What conclusion(s) can you draw from the results of Exercises 1 and 2?

____

Exploring the Line of Best Fit **(For use with Lesson 15–4)**

1. The table and the scatter plot below show the heights and weights of a sample of small penguins. On the scatter plot draw a line that you think comes closest to all the points. Then determine an equation of the line in slope-intercept form: ____________

	Height (cm)	*Weight (g)*
(1)	40	2011
(2)	38	2004
(3)	41	2107
(4)	32	1996
(5)	33	1915
(6)	26	1848

Weight (g)
0 1800 1900 2000 2100
26 28 30 32 34 36 38 40 42
Height (cm)

2. Use a graphing calculator or a computer to draw a scatter plot and the regression line (that is, the line of best fit) for the data in Exercise 1. In slope-intercept form, what is an equation of this line? ____________ Compare this with your equation from Exercise 1.

Resource Book, ALGEBRA, Structure and Method, Book 2

USING TECHNOLOGY 251

Using Models
***Overhead Visual* 13, Sheets 1–4**

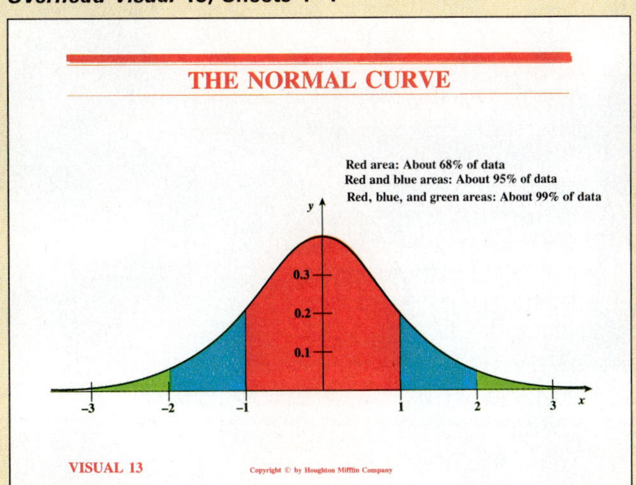

Application

A combination lock may prevent a person from entering a restricted area, but other safeguards are needed to control electronic access to confidential records. For example, automatic teller machines require bank customers to enter personal identification numbers before gaining access to their accounts.

Passwords are used on most computer systems, especially those found in business and government. When compared with the possible combinations for a lock, the number of possible passwords for a computer system is typically much greater. This is in large part due to the fact that one computer might be used to "pick the lock" of another computer, so that the number of possible passwords for the computer being invaded must far exceed the ability of the invading computer to try the possibilities.

Research Activities

Interested students might research some of the problems associated with computer security, including "hackers," industrial espionage, the use and misuse of electronic fund transfers, and the computerized invasion of personal privacy.

References

Baker, Richard H. *The Computer Security Handbook.* Blue Ridge Summit, PA: TAB Professional and Reference Books, 1985.

Support Materials

Resource Book p. 218

15 Statistics and Probability

A more appropriate name for a combination lock might be "permutation lock." To open such a lock, you need to know the order in which to dial the numbers.

Statistics

15-1 Presenting Statistical Data

Objective To display data using frequency distributions, histograms, and stem-and-leaf plots, and to compute measures of central tendency.

The results of an algebra test are shown in the table at the right. This table is called a **frequency distribution** because it shows how many times a given score was attained.

A **histogram,** shown below, can be used to display a frequency distribution. To draw a histogram, you first group the data into convenient intervals. (The intervals used for the test scores, for example, are 50–59, 60–69, 70–79, and so on.) For each interval you then construct a rectangle having a width equal to the interval width and a height equal to the frequency of the data falling within the interval.

Score	Number of Students
99	2
98	1
95	3
94	1
93	1
90	2
88	2
87	3
84	5
83	3
82	1
80	2
78	1
75	2
74	1
73	2
68	1
65	1
56	1

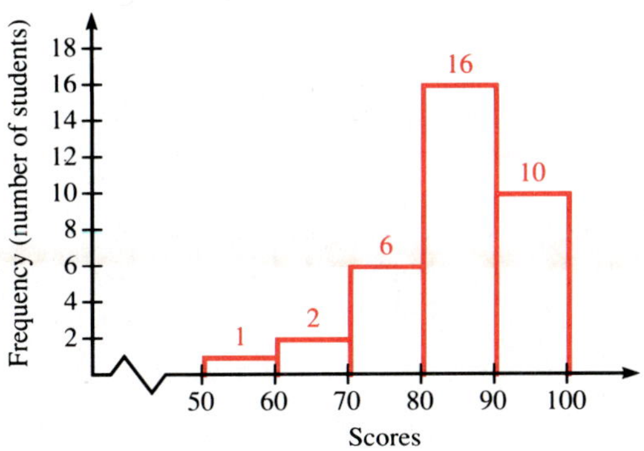

Example 1 Use the histogram above to answer the following questions.

a. Which interval contains the most test scores?
b. Which contains the fewest test scores?
c. How many test scores are 80 or above?
d. How many test scores are below 70?

Solution

a. The interval 80–89, with 16 scores, contains the most.
b. The interval 50–59, with 1 score, contains the least.
c. Add the frequencies for the intervals 80–89 and 90–99: $16 + 10 = 26$
d. Add the frequencies for the intervals 50–59 and 60–69: $1 + 2 = 3$

Teaching References

Lesson Commentary, pp. T136–T141
Assignment Guide, pp. T70–T72
Supplementary Materials
- Practice Masters 82–87
- Tests 63–67
- Resource Book
  - Practice Exercises, pp. 167–171
  - Tests, pp. 77–80
  - Enrichment Activity, p. 234
  - Application, p. 218
- Study Guide, pp. 239–258
- Computer Activities 32–33
- Test Generator
- Disk for Algebra

Alternate Test, p. T27

Teaching Suggestions p. T137

Reading Algebra p. T137

Suggested Extensions p. T137

Warm-Up Exercises

Seven students had the following scores on a quiz: 82, 76, 88, 82, 91, 82, 73.

1. Find the arithmetic average of the scores. 82
2. What number occurs most frequently? 82
3. Arrange the scores in order from least to greatest. What score appears exactly in the middle of the list? 82

Motivating the Lesson

Tell students that statements like the following are often found in the news: "Rainfall is 8 inches below normal." "The median income has increased by $750 in the past year." What do such statements mean? Measures of "normal" or "average" are the topic of today's lesson.

Chalkboard Examples

1. Refer to the histogram on page 709.
   a. How many scores are between 70 and 90? 22
   b. How many students took the test? 35
   c. In what interval does the middle of the distribution occur? 80–90
2. Draw a stem-and-leaf plot for the following exercise times: 26, 32, 40, 36, 28, 32, 44, 36, 28, 32, 45, 47

2	6, 8, 8
3	2, 2, 2, 6, 6
4	0, 4, 5, 7

3. Find the mode, median, and mean of the distribution from Chalkboard Example 2.
   Mode: The most frequent exercise time is 32.
   Median: Since there are 12 exercise times, the median is the average of the sixth and seventh: $\frac{32 + 36}{2}$, or 34.
   Mean: Since the sum of the 12 exercise times is 426, the mean is 426 ÷ 12, or 35.5.

Check for Understanding

Oral Exs. 1–4: use after the discussion of measures of central tendency.
Oral Exs. 5–8: use after Example 4.

A **stem-and-leaf plot,** which is another way of displaying data, actually includes the data in the display.

Example 2 Draw a stem-and-leaf plot for the test scores on the previous page.

Solution You first obtain the *stems* by using only the tens' digit of each score. These are written, in order, to the left of a vertical line. Then, for each score, you record the *leaf,* or units' digit, to the right of its stem.

Stem	Leaf
5	6
6	5, 8
7	3, 3, 4, 5, 5, 8
8	0, 0, 2, 3, 3, 3, 4, 4, 4, 4, 4, 7, 7, 7, 8, 8
9	0, 0, 3, 4, 5, 5, 5, 8, 9, 9

Notice that a stem-and-leaf plot has the shape of a histogram drawn horizontally.

Numbers used to describe a set of data are called **statistics.** Three different statistics are often used to measure the *central tendency* of a distribution: the *mode,* the *median,* and the *mean.*

The **mode** of a distribution is the number that occurs most frequently. From the stem-and-leaf plot of Example 2 you can see that the mode of the test scores is 84.

The **median** of a distribution is the middle number: There are as many numbers greater than the median as there are less than it. Since the stem-and-leaf plot of Example 2 contains 35 scores, the median is the 18th score counting from either the top or the bottom of the distribution. The median of the test scores is therefore 84.

The **mean** of a distribution is the *arithmetic average* of the numbers. Since the sum of the 35 scores in the stem-and-leaf plot of Example 2 is 2933, the mean is $\frac{2933}{35}$, or 83.8.

Computers and calculators, especially graphing calculators, are very useful when working with statistics.

Example 3 Use the distribution of ages given in the stem-and-leaf plot at the right to find **(a)** the mode, **(b)** the median, and **(c)** the mean (M).

1	8, 9, 9
2	2, 2, 2, 5
3	1, 1, 1, 6, 6, 6, 6
4	0, 0, 3, 3, 3

Solution

a. The age that occurs most often is 36. Therefore, the mode is 36.

b. There are 19 ages, so the 10th age from the top (or bottom) is the median. Therefore the median is 31.

c. $$M = \frac{18 + 2(19) + 3(22) + 25 + 3(31) + 4(36) + 2(40) + 3(43)}{19}$$
$$\approx 31.2$$

Finding the mode and median of a distribution is not always as straightforward as finding the mean. When a distribution has two or more numbers that occur most often with equal frequency, each of these numbers is a mode of the distribution. Also, when the count of the numbers in a distribution is even, there are two "middle" numbers, and their average is the median of the distribution.

Example 4 Find **(a)** the mode and **(b)** the median of the following distribution:

5, 6, 6, 6, 8, 10, 13, 13, 13, 16, 18, 20

Solution

a. Since both 6 and 13 occur three times, each is a mode.

b. The two middle scores are 10 and 13.

$\therefore$ the median is $\frac{10 + 13}{2}$, or 11.5.

Oral Exercises

In Exercises 1–4, use the histogram at the right.

1. Find the mode. 27
2. How many scores are 28 or above? 16
3. How many scores are there altogether? 35
4. Find the median. 27

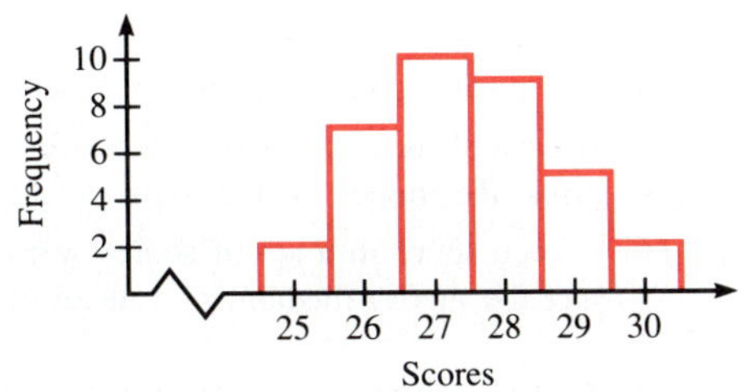

In Exercises 5–8, use the stem-and-leaf plot at the right below.

5. Find the mode. 40
6. Find the median. 32
7. In computing the mean, by what number do you divide? 20
8. Suppose another 32 was in the distribution. Then what would the mode be? 32 and 40 would both be modes.

1	2, 2, 7, 8
2	3, 4, 5, 5, 9
3	2, 2, 2, 6, 8
4	0, 0, 0, 0, 6
5	0

Written Exercises

In Exercises 1–4, draw a stem-and-leaf plot for the given distribution.

A 1. 16, 54, 23, 38, 22, 22, 40, 46, 52, 19, 20

2. 61, 38, 55, 65, 66, 42, 61, 48, 50, 39, 62, 61

3. 2, 5, 13, 28, 61, 9, 18, 10, 52, 34, 28, 42, 19, 28, 7

4. 123, 129, 132, 135, 140, 151, 152, 160, 166, 168, 168, 169

Using a Computer or a Graphing Calculator

Histograms can be drawn using a computer or a graphing calculator. You may wish to have students do this for Exercises 1–4.

Support Materials

Disk for Algebra
Menu Item: Statistics Spreadsheet
Resource Book p. 250

Guided Practice

1. Draw a stem-and-leaf plot for the distribution: 44, 51, 41, 63, 66, 58, 49, 58, 46

4	1, 4, 6, 9
5	1, 8, 8
6	3, 6

2. For the frequency distribution shown, find the mode, median, and mean.

Age	Frequency
14	7
15	6
16	3
17	4

14; 15; 15.2

Summarizing the Lesson

In this lesson students learned to display data and to measure the central tendency of data. Ask students to state the definitions of mode, median, and mean and to give an example of data for which one of these measures of central tendency is significantly different from the other two.

Suggested Assignments

Average Algebra
711/1–16
S 712/Mixed Review 4–6

Extended Algebra
711/1–17
S 712/Mixed Review 4–6

Supplementary Materials

Study Guide pp. 239–240

Using a Calculator

You may wish to have students use a calculator when finding the mean of a distribution.

Additional Answers
Written Exercises

1.

1	6, 9
2	0, 2, 2, 3
3	8
4	0, 6
5	2, 4

2.

3	8, 9
4	2, 8
5	0, 5
6	1, 1, 1, 2, 5, 6

3.

0	2, 5, 7, 9
1	0, 3, 8, 9
2	8, 8, 8
3	4
4	2
5	2
6	1

4.

12	3, 9
13	2, 5
14	0
15	1, 2
16	0, 6, 8, 8, 9

5. a. 22 **b.** 23 **c.** 32

6. a. 61 **b.** 58 **c.** 54

7. a. 28 **b.** 19 **c.** 23.7

8. a. 168 **b.** 151.5 **c.** 149.4

11.

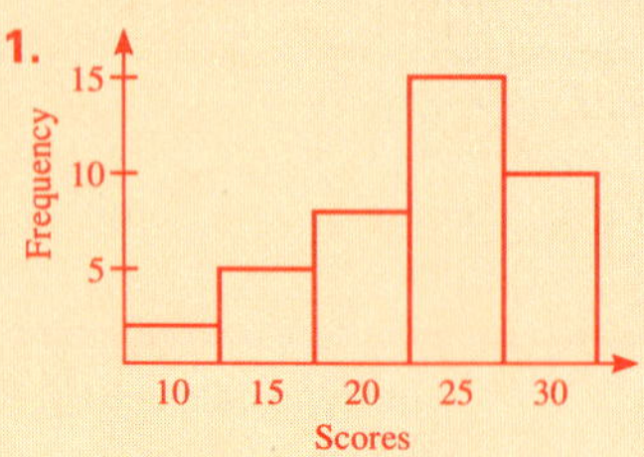

5–8. Find **(a)** the mode, **(b)** the median, and **(c)** the mean for each distribution given in Exercises 1–4. A calculator may be helpful.

9. The frequency distribution of the heights of the members of a high school basketball team is shown at the right. Find **(a)** the mode, **(b)** the median, and **(c)** the mean. **a.** 184 **b.** 184 **c.** 183

Height (cm)	Frequency
175	1
178	1
180	2
181	1
184	3
185	2
188	1
192	1

10. The average monthly temperatures, in degrees Celsius, for Indianapolis during a recent year were −0.05, 2.00, 6.28, 12.72, 18.61, 24.33, 27.44, 26.39, 22.00, 15.89, 8.94, and 2.17. For this distribution find **(a)** the median and **(b)** the mean. **a.** 14.31 **b.** 13.89

11. Draw a histogram for the frequency distribution of quiz scores shown at the right.

12. For the frequency distribution shown at the right, find **(a)** the mode, **(b)** the median, and **(c)** the mean. **a.** 25 **b.** 25 **c.** 23.25

Score	Frequency
10	2
15	5
20	8
25	15
30	10

B **13.** James has test scores of 82, 73, 76, and 92. What must he score on a fifth test if his average test score is to be 82? 87

14. The mean of 12 numbers is 15. What is the sum of the numbers? 180

15. If each score in a set of scores were increased by 5 points, how would this affect the mode, median, and mean of these scores? Each would increase by 5.

16. If each score in a set of scores were reduced by 40%, how would this affect the mode, median, and mean of these scores? Each would be reduced by 40%.

C **17.** At high school A, the mean score of 50 students on a science test is 75. At high school B, the mean score of 40 students on the same test is 80. What is the mean score of the 90 students? 77.2

18. The mean of m scores is M_1. The mean of n scores is M_2.

a. What is the mean of $m + n$ scores? $\frac{mM_1 + nM_2}{m + n}$

b. Under what conditions is the combined mean equal to $\frac{M_1 + M_2}{2}$? Either $M_1 = M_2$ or $m = n$.

Mixed Review Exercises

Convert the given rectangular coordinates to polar coordinates.

1. $(-3, 3\sqrt{3})$ $(6, 120°)$ **2.** $(\sqrt{6}, \sqrt{2})$ $(2\sqrt{2}, 30°)$ **3.** $(0, 3)$ $(3, 90°)$

Write in simplest form without negative exponents.

4. $\left(\frac{m^2n^{-3}}{m^{-4}n}\right)^{-2}\left(\frac{mn^{-1}}{m^3n}\right)$ $\frac{n^6}{m^{14}}$ **5.** $\frac{w^2 - 3w - 4}{3w^2 - 10w - 8}$ $\frac{w + 1}{3w + 2}$ **6.** $8^{\sqrt{2}} \cdot 2^{\sqrt{8}}$ $2^{5\sqrt{2}}$

15-2 Analyzing Statistical Data

Objective To compute measures of dispersion and, together with measures of central tendency, to describe and compare distributions using these statistics.

In Lesson 15-1 you learned that the median of a distribution divides the data into two halves. Finding the median of each half of the data therefore divides the data into quarters. The median of the *lower half* of the data is called the **first quartile** of the distribution. The median of the *upper half* of the data is called the **third quartile.** Thus one fourth of the numbers in the distribution are less than the first quartile, and three fourths are less than the third quartile.

Example 1 For the distribution of test scores shown in the stem-and-leaf plot at the right find **(a)** the median, **(b)** the first quartile, and **(c)** the third quartile.

Stem	Leaves
6	4, 6, 6
7	1, 3, 8, 8
8	0, 5, 7, 9
9	2, 2, 8
10	0, 0, 0

Solution

a. With 17 scores in the distribution, the median is the 9th score counting from either end: 85.

b. With 9 scores (including the median) in the lower half of the distribution, the first quartile is the 5th score counting from the low end: 73.

c. With 9 scores (including the median) in the upper half of the distribution, the third quartile is the 5th score counting from the high end: 92.

In some distributions the numbers are clumped together, while in others the numbers are spread out. One way to measure the *dispersion* of a set of data is to find the **range,** which is the difference between the largest and smallest numbers in the set. For example, the range of the distribution in Example 1 is 100 − 64, or 36.

A diagram known as a **box-and-whisker plot** can be used to show the median, the first and third quartiles, and the range of a distribution.

Example 2 Draw a box-and-whisker plot for the distribution in Example 1.

Solution

1. First show the median, the first and third quartiles, and the lowest and highest scores as dots below a number line.

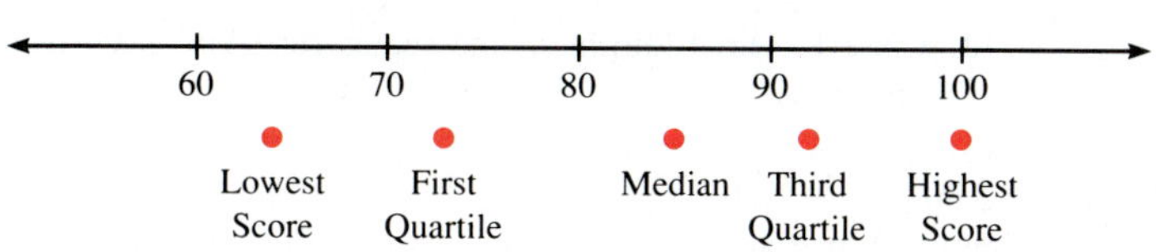

(Solution continues on the next page.)

Teaching Suggestions p. T137

Suggested Extensions p. T137

Warm-Up Exercises

Stem	Leaves
2	8, 8, 9
3	0, 1, 5, 5, 5
4	1, 4

For the distribution shown in the stem-and-leaf plot, find:

1. the mode 35
2. the median 33
3. the mean 33.6
4. the difference between the largest and smallest items of data 16

Motivating the Lesson

Ask the students to consider two algebra classes that had nearly equal mean scores on a certain test. In one class the individual scores were clumped together near the mean, while in the other the scores were spread far apart. Obviously the two classes are not as alike as the mean scores would seem to suggest. Measuring the dispersion of data is one of the topics of today's lesson.

Chalkboard Examples

1. For the distribution of scores shown in the stem-and-leaf plot find the median, the first quartile, and the third quartile.

7	4, 6, 9
8	1, 2, 8, 9
9	1, 3, 7

There are 10 scores. The median is the average of the fifth and sixth scores: $\frac{82 + 88}{2}$, or 85. The first quartile is the third score from the bottom: 79. The third quartile is the third score from the top: 91.

2. Draw a box-and-whisker plot for the distribution in Chalkboard Example 1.

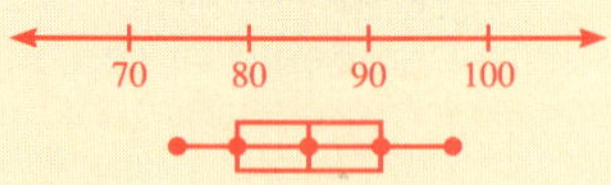

3. Find the mean of the scores in Chalkboard Example 1. Since the sum of the 10 scores is 850, the mean is 850 ÷ 10, or 85.

4. Find the variance and standard deviation of the scores in Chalkboard Example 1. Since the sum of the squares of the deviations from the mean is 532, the variance is 532 ÷ 10, or 53.2. The standard deviation is $\sqrt{53.2}$, or 7.3.

2. Next draw a narrow rectangular box with its shorter sides containing the two quartile dots. Then draw a line segment through the median dot parallel to the shorter sides. Finally, draw line segments, or "whiskers," from the first- and third-quartile dots to the lowest- and highest-score dots, respectively.

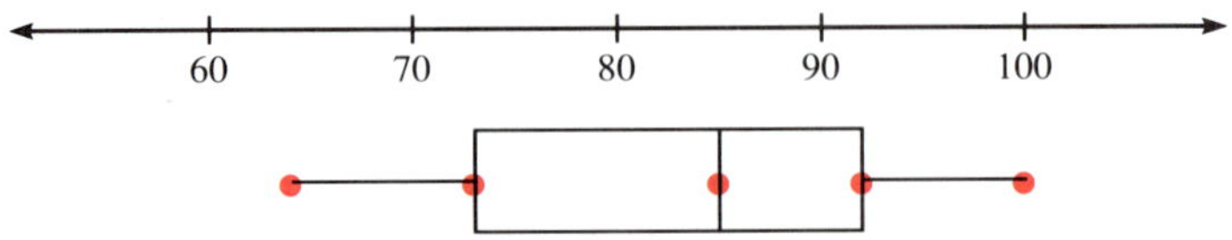

In a box-and-whisker plot, the box encloses the middle half of the distribution and the whiskers indicate the spread of data through the lowest and highest fourths of the distribution. The distance between the two extreme points represents the range of the distribution.

By analyzing the box-and-whisker plots for two distributions you can easily make comparisons between them.

Example 3 Two classes took the same algebra test. The results are shown in the box-and-whisker plots below.

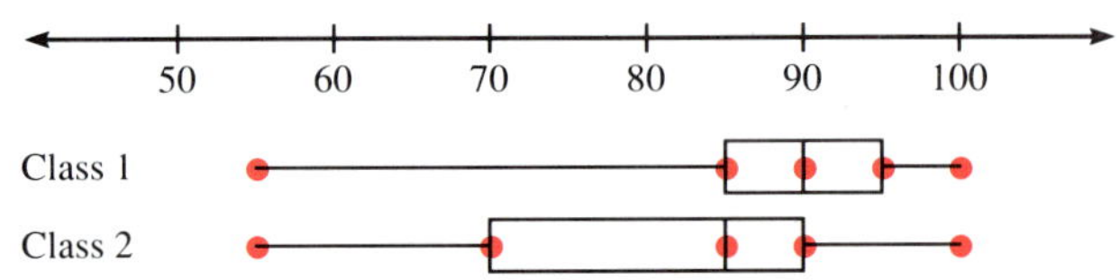

a. Which class has the higher median?

b. Which class has the smaller range?

c. For which class are the scores in the middle half closer together?

d. Which class has the better set of scores?

Solution

a. Class 1: The median of Class 1, which is 90, is greater than the median of Class 2, which is 85.

b. Both classes have the same range: 100 − 55, or 45.

c. Class 1: The box for Class 1 is shorter than the box for Class 2. (Note that by ignoring the extreme scores, you may get a better picture than the range provides of the "typical" spread in the scores.)

d. Class 1: Three fourths of the scores for Class 1 are at or above the median for Class 2.

Besides the range, two other statistics are often used to measure the dispersion of a distribution. They are called the *variance* and the *standard deviation* of the distribution.

The results of a history test are shown in the table at the right. Since the sum of the 8 scores is 632, the mean of the scores is 632 ÷ 8, or 79. The table also shows how much each score differs from the mean. These differences are called *deviations* from the mean.

Score	Deviation from mean
94	+15
89	+10
86	+7
82	+3
75	−4
72	−7
71	−8
63	−16

To get an idea of how the scores are generally scattered about the mean, you can make the deviations positive and find their average. (To make the deviations positive, mathematicians choose to square them rather than take their absolute values.) The result of computing the average of the squares of the deviations is called the *variance* of the distribution.

If $x_1, x_2, \ldots, x_n$ are n numbers and M is their mean, then the **variance** of the distribution is

$$\frac{(x_1 - M)^2 + (x_2 - M)^2 + \cdots + (x_n - M)^2}{n}.$$

Example 4 Find the variance of the distribution of the test scores shown above.

Solution $$\text{Variance} = \frac{15^2 + 10^2 + 7^2 + 3^2 + (-4)^2 + (-7)^2 + (-8)^2 + (-16)^2}{8}$$

$$= \frac{768}{8} = 96 \quad \textit{Answer}$$

The principal square root of the variance is called the **standard deviation.** It is denoted by the Greek letter sigma (σ):

$$\sigma = \sqrt{\frac{\text{sum of the squares of the deviations from the mean}}{\text{number of elements in the distribution}}} = \sqrt{\text{variance}}$$

From Example 4 you can see that the standard deviation of the history test scores is $\sqrt{96}$, or 9.8 to the nearest tenth.

The mean, the median, and the mode each indicate where the center of a distribution is. They are called **measures of central tendency.** The range, the first and third quartiles, the variance, and the standard deviation each indicate how scattered a distribution is. They are called **measures of dispersion.**

A useful way to characterize a distribution is to give a measure of central tendency and a measure of dispersion for it. The mean and standard deviation are the most common pair of measures used.

Example 5 The stem-and-leaf plot at the right shows the distribution of the heights, in centimeters, of the members of a high school basketball team. Find the mean and standard deviation for the distribution.

17	5, 8
18	0, 0, 1, 4, 4, 4, 5, 5, 8
19	2

(Solution is on the next page.)

Check for Understanding

Here is a suggested use of the Oral Exercises to check students' understanding as you teach the lesson.

Oral Exs. 1–4: use after the definition of range.

Oral Ex. 5: use after Example 2.

Oral Exs. 6–13: use after Example 3.

Oral Ex. 14: use after the discussion of standard deviation.

Thinking Skills

Following the procedure outlined in the chart at the top of page 716 gives an accurate and efficient routine to use for finding standard deviations. With complicated arithmetic processes, it is very effective to *analyze* the procedures to be done and to *organize* them in algorithmic form. This helps avoid errors and makes the work less overwhelming.

Guided Practice

1. Find the median, first quartile, third quartile, and range for the stem-and-leaf plot.

1	8
2	2, 5, 7
3	3

25; 22; 27; 15

2. Make a box-and-whisker plot for the distribution in Exercise 1.

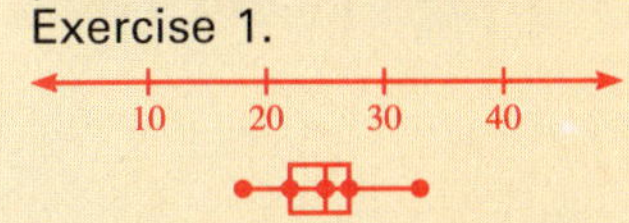

3. Find the mean, variance, and standard deviation for the following distribution: 7, 10, 10, 12, 13, 14
11; 5.3; 2.3

Summarizing the Lesson

In this lesson students learned several ways to measure the dispersion of a distribution and to compare two or more distributions by examining stem-and-leaf plots. Ask students to explain why knowing the dispersion of a distribution is as important as knowing the central tendency.

Suggested Assignments

Average Algebra
717/1–8, 9–19 odd

Extended Algebra
717/1–8, 9–19 odd
S 712/18

Supplementary Materials

Study Guide	pp. 241–242
Practice Master	82
Computer Activity	32

Using a Computer

The program "Statistical Spreadsheet" on the Disk for Algebra will compute the mean and standard deviation for a given distribution.

Support Materials
Disk for Algebra
Menu Item: Statistics Spreadsheet
Resource Book p. 250

Solution Use a table like the one shown below to organize your computations.

Height	Frequency	Height × Frequency	Deviation	$(\text{Deviation})^2$	$(\text{Deviation})^2 \times$ Frequency
175	1	175	−8	64	64
178	1	178	−5	25	25
180	2	360	−3	9	18
181	1	181	−2	4	4
184	3	552	+1	1	3
185	2	370	+2	4	8
188	1	188	+5	25	25
192	1	192	+9	81	81
Sums	12	2196			228

After completing the first three columns, you can compute the mean.

$\therefore M = \frac{2196}{12} = 183$ ***Answer***

After completing the last three columns, you can compute the standard deviation.

$\therefore \sigma = \sqrt{\frac{228}{12}} \approx 4.4$ ***Answer***

Oral Exercises

5. a. Lowest value; 50 b. First quartile; 60 c. Median; 75 d. Third quartile; 85 e. Highest value; 100

In Exercises 1–4, use the stem-and-leaf plot at the right.

1	0, 0, 5, 6
2	0, 4, 5, 5, 5, 5, 7, 9
3	7, 8, 9, 9, 9

1. Find the median. 25
2. Find the first quartile. 20
3. Find the third quartile. 37
4. Find the range. 29
5. Use the box-and-whisker plot at the right to state what each of the following points represents and to give its value. (See above.)
   a. *A* **b.** *B* **c.** *C* **d.** *D* **e.** *E*

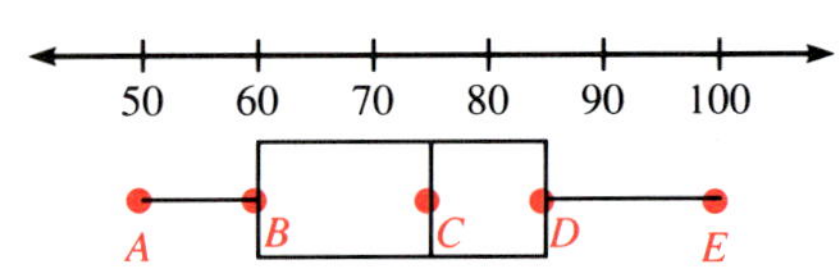

In Exercises 6–13, use the box-and-whisker plots below.

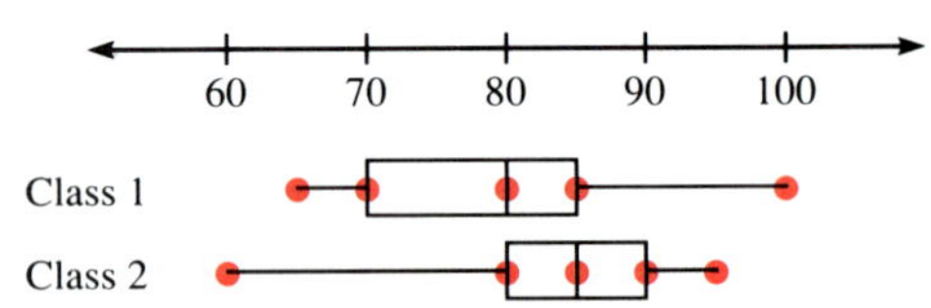

Which class has:

6. the highest score? 1
7. the lowest score? 2
8. the smaller range? Same
9. the higher median? 2
10. the higher first quartile? 2
11. the higher third quartile? 2
12. scores in the middle half closer together? 2
13. the better set of scores? 2
14. What can you say about distributions A and B if they have the same mean but the standard deviation of A is greater than that of B? The data in distribution A is more spread out from the mean.

Written Exercises

For each stem-and-leaf plot find (a) the median, (b) the first quartile, (c) the third quartile, and (d) the range.

A **1.**

0	2, 3, 7
1	0, 4, 4, 8
2	2, 5, 5, 5, 9, 9
3	0, 4, 4, 8, 8
4	0, 1, 7

a. 25 b. 14 c. 34 d. 45

2.

5	9
6	3, 3, 7
7	1, 6, 6, 6
8	0, 0, 4, 4, 8
9	2, 5, 5

a. 78 b. 69 c. 86 d. 36

3–4. Make a box-and-whisker plot for each of the distributions in Exercises 1 and 2.

In Exercises 5–8, use the box-and-whisker plots at the right and justify your answers.

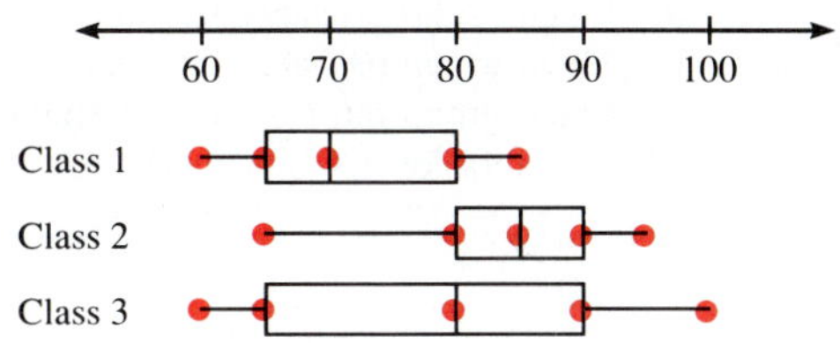

5. Which class has the highest median? 2

6. Which class has the smallest range? 1

7. For which class are the scores in the middle half closest together? 2

8. Which class has the best set of scores? 2

For the distributions in Exercises 9–14 find (a) the mean, (b) the variance, and (c) the standard deviation to the nearest tenth. 11. a. 57 b. 305.3 c. 17.5

9. 1, 4, 6, 6, 7, 8, 8, 8 a. 6 b. 5.3 c. 2.3

10. 3, 4, 5, 5, 5, 5, 6, 9 a. 5.3 b. 2.7 c. 1.6

11. 34, 42, 44, 70, 73, 79

12. 8, 15, 38, 64, 85, 102 a. 52 b. 1205.7 c. 34.7

13. 42, 46, 50, 50, 52, 54, 56 a. 50 b. 19.4 c. 4.4

14. 37, 38, 41, 45, 45, 47, 48 a. 43 b. 16.3 c. 4.0

15. In a golf tournament the 18-hole totals for the top nine golfers were 67, 69, 70, 70, 71, 72, 73, 73, and 74. Find **(a)** the mean, **(b)** the variance, and **(c)** the standard deviation of the scores to the nearest tenth of a stroke. (See below.)

16. The highest temperatures for the first eleven days of June in St. Louis were 70, 68, 63, 66, 70, 70, 73, 87, 70, 68, and 65. Find **(a)** the mean, **(b)** the variance, and **(c)** the standard deviation of the temperatures to the nearest tenth of a degree. (See below.)

15. a. 71 b. 4.4 c. 2.1 16. a. 70 b. 36 c. 6

Additional Answers
Written Exercises

3.

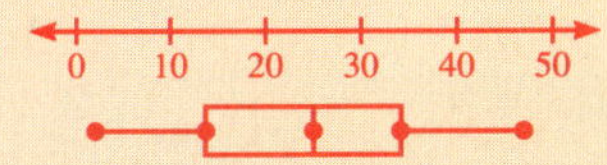

4.

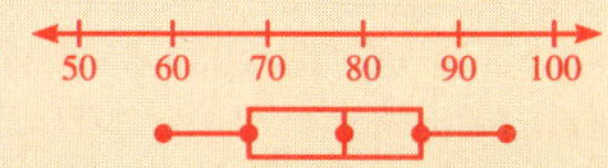

B 17. If each score in a set of scores was increased by 10 points, how would this affect the mean and the standard deviation? *M* is increased by 10; σ is unchanged.

18. If each score in a set of scores was multiplied by 2, how would this affect the mean and the standard deviation? Both are doubled.

19. Make up frequency distributions of four scores between 0 and 100 that have the given mean and standard deviation. Answers may vary.

a. $M = 80$; $\sigma < 3$ 78, 80, 80, 82   **b.** $M = 80$; $\sigma > 20$ 20, 100, 100, 100   **c.** $M = 80$; $\sigma = 10$ 70, 70, 90, 90

C 20. The means and standard deviations of an algebra test given to four different groups of 100 students each are listed in the table at the right. You would need more information to answer the following questions with certainty. However, based on the data given, respond to the questions and tell why you gave your response. 1; *M* is lowest, and σ is largest.

Group	Mean	σ
1	60	16
2	70	12
3	70	14
4	80	6

a. Which group probably has the lowest individual score?

b. Which group probably has the smallest range? 4; σ is smallest.

c. Which group most closely resembles the total population of 400 students? 2 or 3; both have mean equal to the overall mean, and their standard deviations are "average."

Career Note / *Electrician*

Most people tend to take electricity for granted. When you turn on a light or use an electrical appliance, you usually don't think about the power that flows through the wires hidden in the walls of your home. Electricians, however, are very much concerned with getting that power safely to where it is needed.

An electrician's work generally involves either installations or maintenance. When wiring is placed in a building under construction, an electrician must install outlets and switches in each room, run wires through metal or plastic tubing (called conduit) in the walls, and connect the wires to circuit boxes. Once the wiring is installed, it must be maintained through periodic inspection and repair. An electrician deals with problems ranging from blown fuses to short circuits.

A knowledge of mathematics is important to electricians. For example, to increase the current in a circuit, an electrician must increase voltage or

decrease resistance. This fact is expressed in Ohm's Law:

$$I = \frac{E}{R}$$

where I measures current (in amperes), E measures voltage (in volts), and R measures resistance (in ohms).

15-3 The Normal Distribution

Objective To recognize and analyze normal distributions.

The histogram at the right shows the annual rainfall in a Nebraska county over a period of 78 years. Each bar in the histogram shows the number of years in which the annual rainfall was within a given range. For example, the leftmost bar shows that for one year out of the 78 the rainfall was between 11 and 14 inches.

The two histograms below give the weights of 100 pennies and the lifetimes of 433 light bulbs.

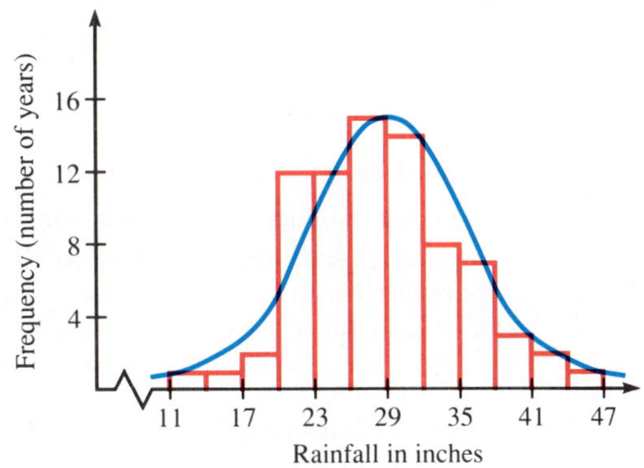

Figure 1

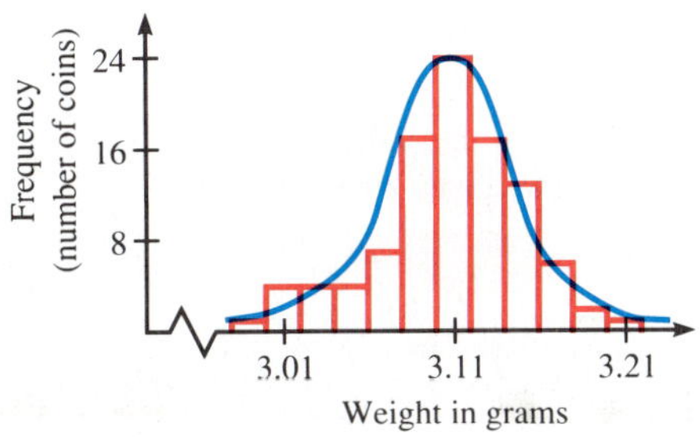

Figure 2

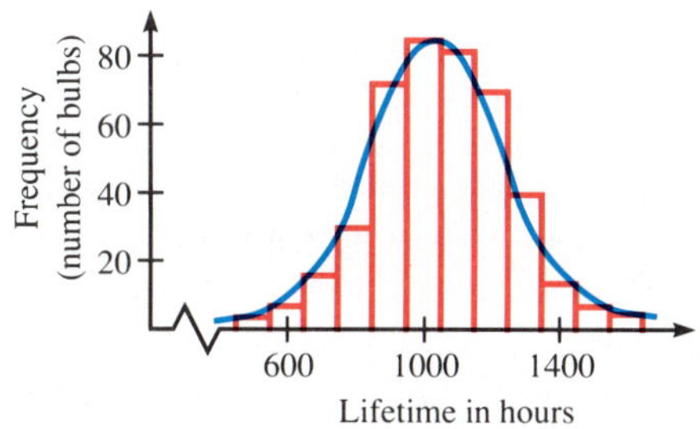

Figure 3

When each of these histograms is approximated by drawing a smooth curve (shown in blue), the resulting curves resemble each other. Such ''bell-shaped'' curves often appear when various sets of data are plotted. For example, all of the following frequency distributions might be represented by bell-shaped curves: the heights of all eleventh-grade girls in a large city, the acidity levels of soil samples from all farms in a rural county, and the achievement test scores for all high school students in a state.

Certain distributions represented by bell-shaped curves are known as *normal distributions*. A normal distribution with mean equal to zero and standard deviation equal to one is called the **standard normal distribution.** The bell-shaped curve that represents a standard normal distribution is called the **standard normal curve.**

The figure at the top of the next page shows the standard normal curve. Note that the x-coordinates represent the number of standard deviations from the mean.

Teaching Suggestions p. T137

Suggested Extensions p. T138

Warm-Up Exercises

Write a brief description of how to determine each statistical measure.

1. mean Divide the sum of data items by the number of data items.
2. variance Divide the sum of the squares of the deviations from the mean by the number of data items.
3. range Find the difference between the largest and smallest data items.
4. median Find the data item that occurs in the middle of an ordered listing of the data items.
5. standard deviation Find the square root of the variance.
6. mode Find the data item that occurs most frequently.

Motivating the Lesson

Tell the students that in many distributions the data tend to be clumped near the middle and taper off at the extremes. The graphs for such distributions look like bell-shaped curves. These distributions, known as normal distributions, are the topic of today's lesson.

Problem-Solving Strategies

A problem that involves finding an area under the standard normal curve is best solved by breaking the area into parts, thus *simplifying the problem.* Once the component areas are known, the solution of the problem usually requires adding or subtracting the parts.

Chalkboard Examples

1. In a standard normal distribution, what fraction of the data is more than 2.0?
$A(x \geq 2.0) = 0.5 - A(2.0)$
$= 0.5 - 0.4772$
$= 0.0228$
2. In a standard normal distribution, what percent of the data is more than two standard deviations from the mean? Since the standard normal curve is symmetric with respect to the y-axis, the combined area under the curve to the left of $x = -2$ and to the right of $x = 2$ is twice the area under the curve to the right of $x = 2$. Use the result of Chalkboard Example 1 to get:
$A(x \leq -2 \text{ or } x \geq 2)$
$= 2 \cdot A(2)$
$= 2(0.0228)$
$= 0.0456$, or 4.56%

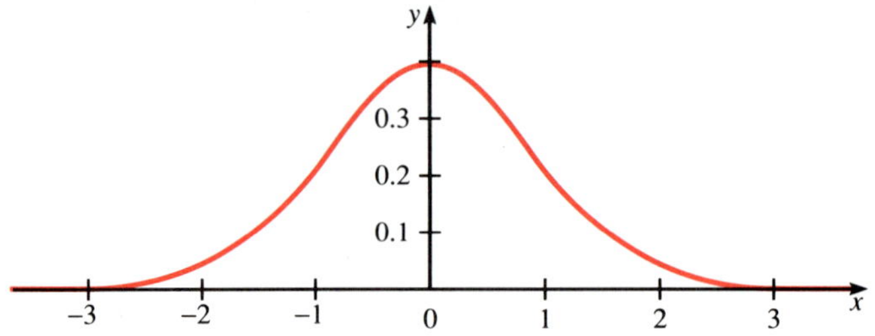

The standard normal curve has the following properties:

1. It is symmetric with respect to the y-axis.
2. It approaches the x-axis asymptotically as $|x|$ increases.
3. The total area under the curve and above the x-axis is equal to 1.

For any set of data that are approximately normally distributed, the standard normal curve can be used to estimate the fraction of the data falling between two given values, a and b. The fraction is simply the shaded area shown in the figure at the right.

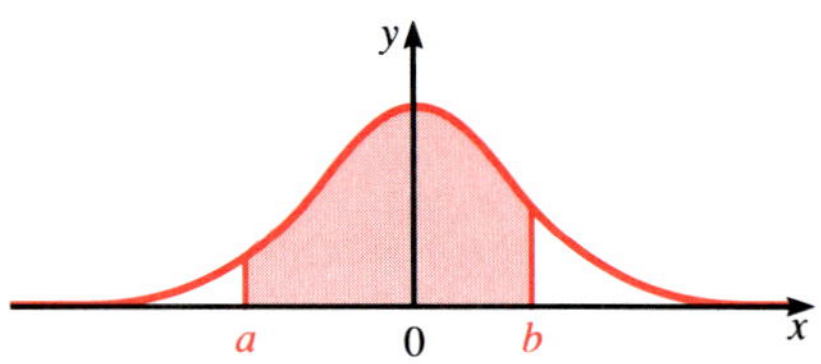

To find this area, extensive tables of area measures for the standard normal curve have been developed. One such table, shown below, gives area measures from 0 to x, at intervals of 0.2, for $0 \leq x \leq 4.0$.

Area under the Standard Normal Curve for $0 \leq x \leq 4.0$

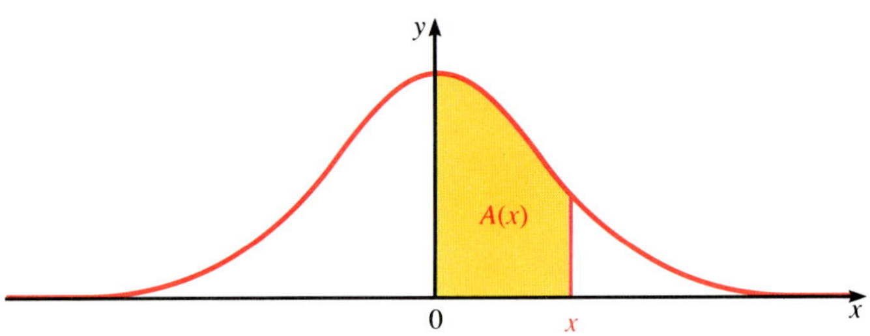

x	Area, $A(x)$	x	Area, $A(x)$	x	Area, $A(x)$
0.0	0.0000	1.4	0.4192	2.8	0.4974
0.2	0.0793	1.6	0.4452	3.0	0.4987
0.4	0.1554	1.8	0.4641	3.2	0.4993
0.6	0.2257	2.0	0.4772	3.4	0.4997
0.8	0.2881	2.2	0.4861	3.6	0.4998
1.0	0.3413	2.4	0.4918	3.8	0.4999
1.2	0.3849	2.6	0.4953	4.0	0.5000

Example 1 In a standard normal distribution, what fraction of the data is between 2.6 and 2.8?

Solution The fraction desired is equal to the area under the standard normal curve bounded by $x = 2.6$ and $x = 2.8$. This is the difference between the area from 0 to 2.8 and the area from 0 to 2.6. Therefore:

$$\begin{aligned} A(2.6 \le x \le 2.8) &= A(2.8) - A(2.6) \\ &= 0.4974 - 0.4953 \\ &= 0.0021 \quad \textbf{\textit{Answer}} \end{aligned}$$

Example 2 In a standard normal distribution, what percent of the data is within two standard deviations from the mean?

Solution The mean is 0 and the standard deviation is 1. Since the standard normal curve is symmetric with respect to the y-axis, the area under the curve between -2 and 2 is twice the area between 0 and 2. Therefore:

$$\begin{aligned} A(-2 \le x \le 2) &= 2 \cdot A(2) \\ &= 2(0.4772) \\ &= 0.9544, \text{ or } 95.44\% \quad \textbf{\textit{Answer}} \end{aligned}$$

Notice in Example 2 that slightly more than 95% of the data in a standard normal distribution are within two standard deviations of the mean. Likewise you can show that slightly more than 68% of the data are within one standard deviation of the mean.

Because all normal distributions are similar to the standard normal distribution, you can use the standard normal curve to answer questions about any normal distribution. Suppose a normal distribution has mean M and standard deviation σ. Then the percent of data falling between $M + a\sigma$ and $M + b\sigma$ is just the area between $x = a$ and $x = b$ under the standard normal curve.

To obtain a from the number $M + a\sigma$, you would first subtract M and then divide by σ. In general, if z is a given value from a normal distribution with mean M and standard deviation σ, then the corresponding value from the standard normal distribution is x, where

$$x = \frac{z - M}{\sigma}.$$

The number x is called the **standardized value** corresponding to z. It gives the number of standard deviations that z is from the mean of the distribution.

Example 3 The heights of a certain group of adults are normally distributed with a mean of 180 cm and a standard deviation of 8 cm.

a. Find the percent of the group having a height greater than 196 cm.
b. Find the percent of the group having heights between 172 cm and 180 cm.

(Solution is on the next page.)

3. A cereal manufacturer claims that Wheat Treat Cereal packages have a mean mass of 340 g and a standard deviation of 1.5 g. Assuming that the masses are normally distributed, find the percent of packages in each category.
a. mass < 336.7 g
b. 339.1 g $<$ mass $<$ 340.3 g

a. $\frac{336.7 - 340}{1.5} = -2.2$

$A(x \le -2.2)$
$= 0.5 - A(2.2)$
$= 0.5 - 0.4861$
$= 0.0139$
about 1.4%

b. $\frac{339.1 - 340}{1.5} = -0.6$

$\frac{340.3 - 340}{1.5} = 0.2$

$A(-0.6 \le x \le 0.2)$
$= A(-0.6 \le x \le 0)$
$+ A(0 \le x \le 0.2)$
$= A(0.6) + A(0.2)$
$= 0.2257 + 0.0793$
$= 0.305$
about 30.5%

Check for Understanding

Here is a suggested use of the Oral Exercises to check students' understanding as you teach the lesson.
Oral Exs. 1–3: Use with the histograms on page 719.
Oral Exs. 4–6: Use before Example 3.

Guided Practice

Exercises 1 and 2 refer to standard normal distributions.

1. Find the percent of the data that lie:
   a. between the mean and −0.4. about 15.5%
   b. between −0.8 and +0.8. about 57.6%
   c. at least 1.6 standard deviations below the mean. about 5.5%
2. The weights of babies born at a certain hospital average 8 lb 1 oz, with a standard deviation of 12 oz. Assuming that the weights are normally distributed, find the percent of babies with weights that are:
   a. more than 8 lb 13 oz 15.87%
   b. less than 6 lb 9 oz 2.28%
   c. between 7 lb 5 oz and 8 lb 13 oz 68.26%

Summarizing the Lesson

In this lesson students learned to answer questions about normal distributions by using areas under the standard normal curve. Ask students to explain the purpose of finding the standardized value x for a given value z from a normal distribution.

Suggested Assignments

Average Algebra
722/1–10
S 723/Mixed Review

Extended Algebra
722/1–11
S 723/Mixed Review

Solution **a.** The standardized value corresponding to 196 is:

$$x = \frac{196 - M}{\sigma} = \frac{196 - 180}{8} = 2$$

The area under the standard normal curve to the right of $x = 2$ is $0.5 - 0.4772$, or 0.0228.

$\therefore$ about 2.3% of the group has height greater than 196 cm.

b. The standardized values corresponding to 172 and 180 are:

$$\frac{172 - 180}{8} = -1 \quad \text{and} \quad \frac{180 - 180}{8} = 0$$

The area under the standard normal curve between −1 and 0 is the same as the area between 0 and 1, which is 0.3413.

$\therefore$ about 34% of the group has height between 172 cm and 180 cm.

Oral Exercises

In Exercises 1–3, use the histograms on page 719.

1. In Figure 1, about what fraction of the 78 years had rainfall between 32 and 38 inches? About $\frac{1}{5}$
2. In Figure 2, what percent of the 100 pennies had a weight of 3.11 g? 24%
3. In Figure 3, about what fraction of the 433 light bulbs had lifetimes of more than 1000 h? About $\frac{1}{2}$
4. What is the area under the standard normal curve to the left of the y-axis? 0.5
5. If the area under the standard normal curve between $x = 0$ and $x = 0.5$ is 0.1915, what percent of the data in a standard normal distribution is within one-half standard deviation from the mean? 38.3%
6. For a normal distribution with mean 12 and standard deviation 4, what is the standardized value corresponding to 20? 2

Written Exercises

In Exercises 1–9, use the table on page 720.

A

1. In a standard normal distribution, what percent of the data is between the mean and 0.4? 15.54%
2. In a standard normal distribution, what percent of the data is between −0.2 and 0.2? 15.86%
3. In a standard normal distribution, what percent of the data is between three and four standard deviations below the mean? 0.13%
4. In a standard normal distribution, what percent of the data is within three standard deviations from the mean? 99.74%

5. The mean weight of a loaf of bread was found by sampling to be 455 g, with a standard deviation of 5 g. Assuming a normal distribution, find the percent of loaves with weights that are:

a. less than 450 g 15.87% b. greater than 445 g 97.72%
c. greater than 470 g 0.13% d. between 450 g and 460 g 68.26%

6. A college aptitude test is scaled so that its scores approximate a normal distribution with a mean of 500 and a standard deviation of 100. Find the percent of the students taking the test who are expected to score:

a. above 800 points 0.13% b. less than 400 points 15.87%
c. between 700 and 900 points 2.28% d. between 800 and 820 points 0.06%

7. The mean life of a certain kind of light bulb is 900 h with a standard deviation of 30 h. Assuming the lives of the light bulbs are normally distributed, find the percent of the light bulbs that will last:

a. less than 900 h 50% b. more than 984 h 0.26%

8. A type of coin in circulation for the past 10 years has a mean weight of 0.22 oz and a standard deviation of 0.01 oz. Assuming a normal distribution, find the percent of these coins with weights that are:

a. greater than 0.24 oz 2.28% b. less than 0.206 oz 8.08%

B 9. Assuming a standard normal distribution, find a number k such that the percent of data less than k is:

a. 50% 0 b. 100% 4 c. 15.87% −1 d. 84.13% 1

10. Assuming a standard normal distribution, explain why the fraction of data greater than a constant k equals 1 minus the fraction of data less than k. The total area under the standard normal curve is 1.

C 11. Assuming a standard normal distribution, explain why the fraction of data having absolute value greater than a nonnegative constant k equals twice the difference between 1 and the fraction of data less than k.

12. The equation of the standard normal curve is

$$y = \frac{1}{\sqrt{2\pi}} e^{(-1/2)x^2}.$$

The base of the natural logarithm, e, is equal to 2.71828 . . . . Use a calculator or a table of logarithms to compute the y-coordinate of the point on the standard normal curve for:

a. $x = 0$ 0.3989 b. $x = 1$ 0.2420 c. $x = 2$ 0.0540

Mixed Review Exercises

For the line containing the given points, find (a) the slope and (b) an equation in standard form.

1. (4, −7), (0, 5) a. −3
2. (3, 2), (7, 5) a. $\frac{3}{4}$
3. (0, 0), (−2, 1) a. $-\frac{1}{2}$
4. (−2, 6), (1, 8) a. $\frac{2}{3}$
5. (−5, −1), (3, −1) a. 0
6. (−3, −2), (−1, −4) a. −1

Additional Answers Written Exercises

11. Since the standard normal curve is symmetric with respect to the y-axis, the fraction of data greater than k equals the fraction of data less than $-k$. Therefore, since the fraction of data greater than k equals the difference between 1 and the fraction of data less than k (see Ex. 10), the combined fractions of data greater than k or less than $-k$ is twice the difference between 1 and the fraction of data less than k.

Supplementary Materials

Study Guide	pp. 243–244
Resource Book	p. 167
Overhead Visual	13

Additional Answers Mixed Review Exercises

1. b. $3x + y = 5$
2. b. $3x - 4y = 1$
3. b. $x + 2y = 0$
4. b. $-2x + 3y = 22$
5. b. $y = -1$
6. b. $x + y = -5$

Teaching Suggestions p. T138

Suggested Extensions p. T138

Warm-Up Exercises

The equation of a line is $11x + 5y = 80$.

1. Is the given point on the line?
   a. (4, 7.2) yes b. (−0.4, 17) no
2. Complete the number pair so that its graph is on the line.
   a. (_?_, 2.8) 6 b. (10, _?_) −6
3. What is the slope of the line? $-\frac{11}{5}$
4. Going from left to right, does the line slant upward or downward? downward

Motivating the Lesson

Ask students to assume that during a rainy period they notice that their grades go up. Would they say that rain causes good grades or is there something else involved, such as fewer recreational distractions? Today's lesson involves correlation, a way of measuring whether two sets of data are related.

Chalkboard Examples

Chalkboard Examples 1–5 refer to the table below.

Rainfall per month	Matt's GPA
1.5	3.0
2.0	3.4
1.8	3.2
2.6	3.5
2.5	3.6

15-4 Correlation

Objective To draw a scatter plot, determine the correlation coefficient, and use the regression line for a set of ordered pairs of data.

The director of sales for a national distributor of prerecorded videocassettes wants to predict future sales. Although such a prediction could be based on past sales alone, the sales director believes that sales are tied to the number of households that have a VCR.

The table at the right gives the nationwide data on VCR households and videocassette sales for five consecutive years. To determine whether a mathematical relationship exists between these two sets of data, you can treat the data as ordered pairs and plot them to obtain a graph called a **scatter plot.**

VCR Households (in millions)	Prerecorded Videocassette Sales (in millions)
8.3	9.5
15.0	22.0
23.5	52.0
32.5	84.0
45.8	110.0

Example 1 Draw a scatter plot of the data in the table above.

Solution Plot the ordered pairs (8.3, 9.5), (15.0, 22.0), (23.5, 52.0), (32.5, 84.0), and (45.8, 110.0) in a coordinate plane, as shown below.

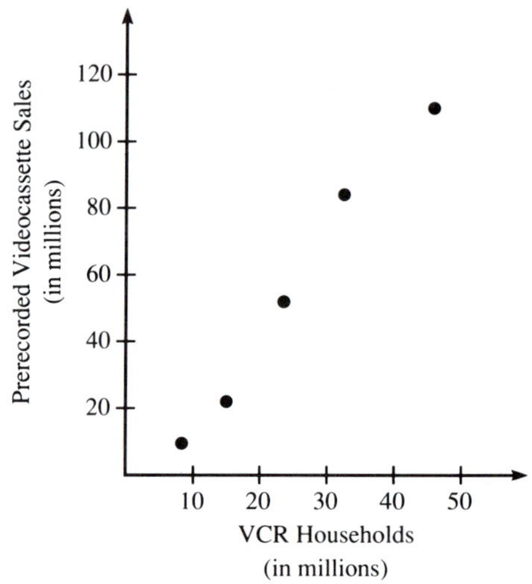

Notice that the points in the scatter plot of Example 1 are very close to being collinear. A statistic called the *correlation coefficient* is used to characterize how closely the points in a scatter plot cluster about a line.

Given a set of ordered pairs (x, y), the **correlation coefficient,** denoted by r_{xy} or merely r, of the ordered pairs is

$$r = \frac{M_{xy} - M_x \cdot M_y}{\sigma_x \cdot \sigma_y},$$

where M_x and σ_x are the mean and standard deviation of the x-values,
M_y and σ_y are the mean and standard deviation of the y-values,
and M_{xy} is the mean of the products of the ordered pairs.

Because of the way in which the correlation coefficient is defined, its value for any set of ordered pairs is always between -1 and 1, inclusive. The more the ordered pairs cluster about a line with a positive slope, the closer r is to 1. Similarly, the more the ordered pairs cluster about a line with a negative slope, the closer r is to -1. If the ordered pairs tend not to be collinear at all, the correlation coefficient is close to 0. These three situations are illustrated in the figures below.

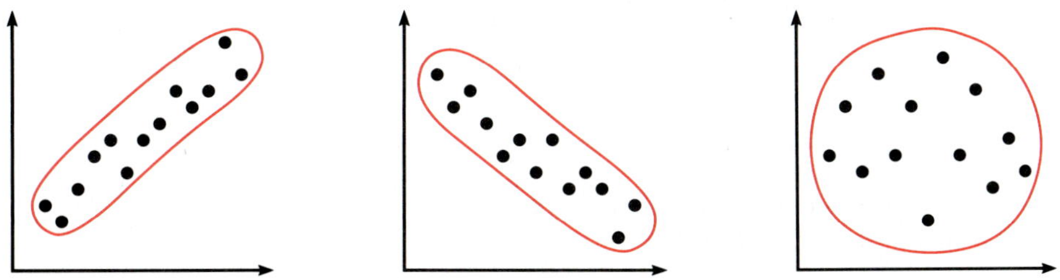

High positive correlation: r is close to 1.

High negative correlation: r is close to -1.

Relatively no correlation: r is close to 0.

Example 2 **a.** Using the scatter plot shown in Example 1, describe the nature of the correlation.

b. Suppose x represents the number of VCR households and y represents the number of videocassette sales. Determine the correlation coefficient of the ordered pairs from the table on the previous page given that $M_{xy} \approx 1879.77$, $M_x \approx 25.02$, $M_y = 55.5$, $\sigma_x \approx 13.19$, and $\sigma_y \approx 37.50$.

Solution **a.** Since the ordered pairs tend to cluster about a line with a positive slope, there is a *high positive correlation* between VCR households and videocassette sales.

b. The correlation coefficient of the ordered pairs is

$$r = \frac{M_{xy} - M_x \cdot M_y}{\sigma_x \cdot \sigma_y} \approx \frac{1879.77 - (25.02)(55.5)}{(13.19)(37.50)} \approx 0.99.$$

1. Draw a scatter plot of the data in the table.

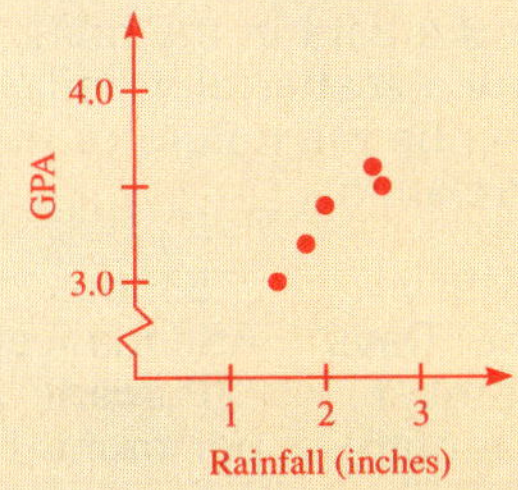

2. Describe the correlation between amount of rainfall and Matt's grade point average. high positive

3. Suppose x represents the amount of rainfall and y represents Matt's GPA. Determine the correlation coefficient of the ordered pairs from the table given that $M_{xy} \approx 7.03$, $M_x = 2.08$, $M_y = 3.34$, $\sigma_x \approx 0.42$, and $\sigma_y \approx 0.22$.
$r \approx \frac{7.03 - (2.08)(3.34)}{(0.42)(0.22)} \approx 0.90$

4. Using the result of Chalkboard Example 3, determine an equation for the regression line relating x and y.
The line contains the point (2.08, 3.34) and has slope $0.90\left(\frac{0.22}{0.42}\right)$, or 0.47.
$y - 3.34 = 0.47(x - 2.08)$
$y = 0.47x + 2.36$

5. Find Matt's expected GPA when the rainfall one month is 1.2 in.
$y = 0.47(1.2) + 2.36$
$= 2.924$
$\therefore$ Matt's expected GPA is about 2.9.

Check for Understanding

Oral Exs. 1–3: use before Example 2.

Guided Practice

1. For the data in the table, draw a scatter plot and describe the nature of the correlation.

Sunny School Days (Number per month)	School Attendance (Percent per month)
15	88
12	92
10	93
14	89
8	96

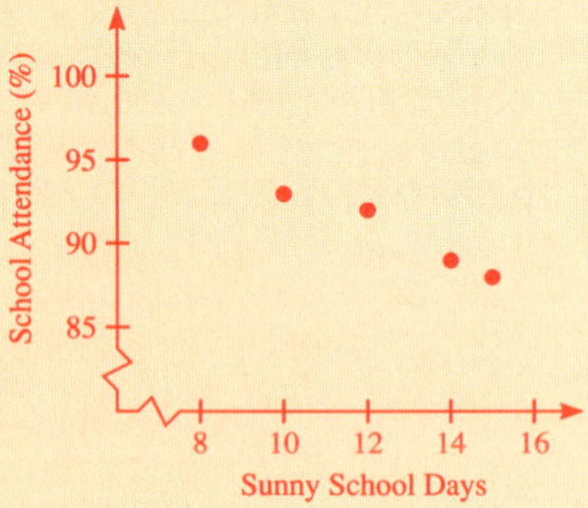

high negative

2. In Exercise 1, suppose x represents the number of sunny school days and y represents the percent school attendance. Given that $M_{xy} = 1073.6$, $M_x = 11.8$, $M_y = 91.6$, $\sigma_x \approx 2.56$, and $\sigma_y \approx 2.87$, find (a) the value of the correlation coefficient for the ordered pairs and (b) an equation of the regression line relating x and y.
   a. -0.99
   b. $y = -1.11x + 104.71$
3. Use the equation of the regression line from Exercise 2 to predict the percent school attendance for a month in which there are 20 sunny school days. 82.5%

A few words of caution are in order here: Although the correlation coefficient measures the strength of the linear relationship between two variables, a high correlation does *not* necessarily imply a cause-and-effect relationship. For instance, research may show a high positive correlation between the body weights and achievement test scores among children. Even if these two variables are linearly related, the correlation between them does not suggest that children should overeat to improve their achievement scores.

A more reasonable conclusion is that body weights and achievement scores are tied to a third variable, such as age. Therefore, as age increases, so do body weights and achievement scores.

When the correlation between two variables is high, you can draw a line, called the **regression line,** that best fits the known values of the variables. The figure at the right, for instance, shows the regression line drawn for the scatter plot of Example 1.

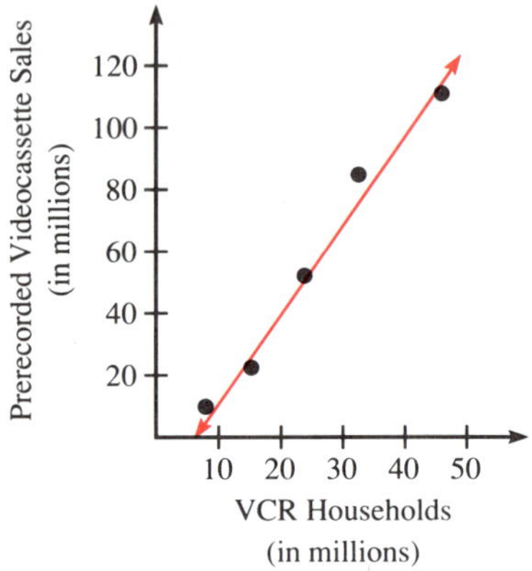

Since the regression line stays reasonably close to the points in a scatter plot, its equation can be used to estimate the value of one variable for a given value of the other. For example, the number of expected videocassette sales could be determined for any given number of VCR households.

To find the equation of a regression line, you use the following fact, proved in a more advanced study of statistics.

For a set of ordered pairs (x, y), the regression line relating x and y contains the point (M_x, M_y) and has as its slope $r\left(\frac{\sigma_y}{\sigma_x}\right)$.

Example 3

a. Using part (b) of Example 2, determine an equation for the regression line relating x and y.

b. Find the expected number of videocassette sales when the number of VCR households is 60 million.

Solution

a. The regression line contains the point $(M_x, M_y) = (25.02, 55.5)$.

The slope of the regression line is: $r\left(\frac{\sigma_y}{\sigma_x}\right) \approx 0.99\left(\frac{37.50}{13.19}\right) \approx 2.81$.

To obtain the equation of the regression line, use the point-slope form of the equation of a line:

$$y - 55.5 = 2.81(x - 25.02)$$
$$y = 2.81x - 14.81 \quad \textbf{\textit{Answer}}$$

b. Substituting 60 for x in the equation from part (a), you get a value of $2.81(60) - 14.81$, or 153.79, for y.

$\therefore$ the number of expected videocassette sales is about 154 million. ***Answer***

Note that the answer to part (b) of Example 3 is only approximate. In a more advanced study of statistics, you will learn how to measure the accuracy of an approximation obtained from a regression line. Generally speaking, the closer $|r|$ is to 1, the greater the accuracy of the approximation.

Oral Exercises

1. **a.** When two variables have a high __?__ correlation, one variable increases as the other variable increases. positive
   b. When two variables have a high __?__ correlation, one variable decreases as the other variable increases. negative
2. For each value of the correlation coefficient, tell whether the linear relationship between x and y is strong or weak.
   a. $r_{xy} = 0.05$ weak **b.** $r_{xy} = -0.95$ strong **c.** $r_{xy} = -0.10$ weak **d.** $r_{xy} = 0.90$ strong
3. For each pair of variables, tell whether you think the correlation is positive, negative, or close to zero.
   a. A high school student's height and weight positive
   b. A car's age and its value negative
   c. A team's standing in its conference and the attendance at its games positive
   d. A state's monthly temperature averages and precipitation totals close to zero
   e. A company's advertising budget and its volume of sales positive

Written Exercises

In Exercises 1–4, draw a scatter plot. Then describe the nature of the correlation.

A 1.

Congressional Elections (for 6 consecutive elections)	
Registered to Vote (percent)	Actually Voted (percent)
70	55
68	55
62	45
63	46
64	49
64	46

2.

Sales of Single-Family Homes (for 6 consecutive months)	
Median Price (in thousands of $)	Number Sold (in thousands)
99	710
95	740
98	720
97	730
105	640
108	650

Summarizing the Lesson

In this lesson students learned to find and interpret the correlation coefficient for a given set of ordered pairs of data and to make predictions about the data using the equation of the regression line. Ask students to explain the difference between causation and correlation.

Suggested Assignments

Average Algebra
727/1–12

Extended Algebra
727/1–13
S 718/20

Using a Computer or a Graphing Calculator

You may wish to have students use a computer or a graphing calculator to draw scatter plots and regression lines.

Support Materials
Disk for Algebra
Menu Item: Statistics Spreadsheet
Resource Book p. 251

Additional Answers Written Exercises

2.

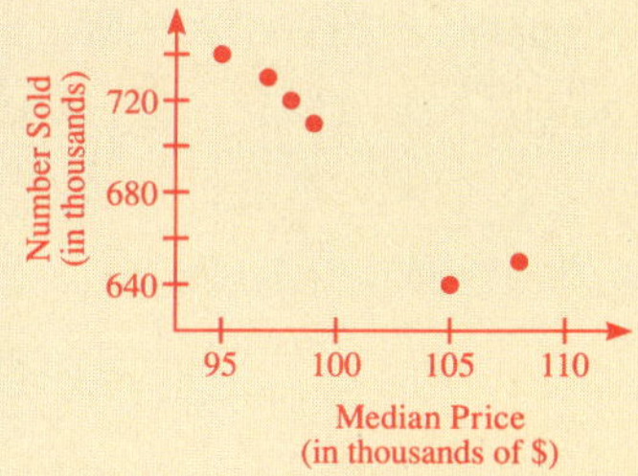

High negative correlation

Additional Answers Written Exercises

4.

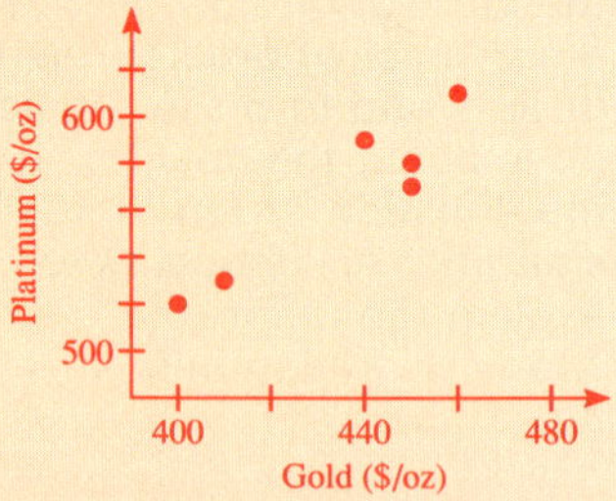

Positive correlation

Supplementary Materials

Study Guide	pp. 245–246
Practice Master	83
Test Master	63
Computer Activity	33

3.

Moviegoing (for 7 consecutive years)	
Average Ticket Price (in dollars)	Admissions (in billions)
2.80	1.06
2.90	1.18
3.20	1.20
3.40	1.20
3.60	1.06
3.70	1.02
3.90	1.09

4.

Price of Precious Metals (for 7 consecutive months)	
Gold (in $ per troy oz)	Platinum (in $ per oz)
400	520
410	530
440	590
460	610
450	570
450	580
460	610

5. In Exercise 1, suppose x represents the percent registered to vote and y represents the percent that actually voted. Given that $M_{xy} \approx 3226.33$, $M_x \approx 65.17$, $M_y \approx 49.33$, $\sigma_x \approx 2.85$, and $\sigma_y \approx 4.19$, find **(a)** the value of the correlation coefficient for the ordered pairs and **(b)** an equation of the regression line relating x and y. **a.** 0.96 **b.** $y = 1.41x - 42.56$

6. In Exercise 2, suppose x represents the median price of single-family homes and y represents the number sold. Given that $M_{xy} \approx 69{,}893.33$, $M_x \approx 100.33$, $M_y \approx 698.33$, $\sigma_x \approx 4.61$, and $\sigma_y \approx 38.91$, find **(a)** the value of the correlation coefficient for the ordered pairs and **(b)** an equation of the regression line relating x and y. **a.** -0.95 **b.** $y = -8.02x + 1502.98$

7. The regression equation for two chemistry tests is $y = 0.7x + 15$, where x represents a student's score on the first test and y represents the student's score on the second test. What is the predicted score on the second test for a student who missed that test but scored 90 on the first test? 78

8. The regression equation for people on a weight-reducing diet is $y = 1.8x + 1.6$, where x is the number of weeks on the diet and y is the number of pounds of weight lost. If a person wants to lose 10 lb, how long should that person stay on the diet? $4\frac{2}{3}$ weeks

B

9. The correlation between two tests given to 100 students in the junior class is 0.70. The mean for the first test was 125 with standard deviation 20. The mean of the second test is 80 with standard deviation 10. A student took the first test and scored 100. What is your best estimate of this student's score on the second test? 71

10. The correlation between the values and the ages of 50 automobiles is -0.85. The average value of these automobiles is \$8000 with standard deviation \$1000. The average age is 4 years with standard deviation 2 years. If you had no specific information about an automobile except that its age was 2 years, what is your best estimate of its value? \$8850

11. State-by-state data on annual per capita income and annual spending on education per student correlate above 0.80 for all the states in the United States. On the basis of this data it is argued that a state should raise educational spending to increase per capita income. Is this a valid argument? Explain your answer. Not valid; no cause-and-effect relationship necessarily exists.

12. The correlation between mental age and height was 0.08 for a large sample of boys of the same age. For a large sample of boys ranging in age from 6 to 15 years old, this correlation was 0.78. How do you explain the difference in these results? How would you answer the question, "Are height and mental age related?" Mental age and height both increase with age; they are *not* causally related.

13. Using two real-world variables of your choosing, form a hypothesis concerning their relationship. Test the hypothesis by gathering data for the two variables and calculating the correlation coefficient. Then write a paragraph indicating whether or not the data supported your hypothesis.

Self-Test 1

Vocabulary

frequency distribution (p. 709)	measures of central tendency (p. 715)
histogram (p. 709)	measures of dispersion (p. 715)
stem-and-leaf plot (p. 710)	standard normal distribution (p. 719)
statistics (p. 710)	standard normal curve (p. 719)
mode, median, mean (p. 710)	standardized value (p. 721)
first, third quartile (p. 713)	scatter plot (p. 724)
range (p. 713)	correlation coefficient (p. 725)
box-and-whisker plot (p. 713)	regression line (p. 726)
variance (p. 715)	
standard deviation (p. 715)	

1. To the nearest whole number, find the mode, median, and mean for the distribution of numbers given in the stem-and-leaf plot below. 28; 27; 26 **Obj. 15-1, p. 709**

1	4, 7, 7, 8
2	1, 2, 2, 5, 6, 8, 8, 8
3	2, 3, 4, 6, 6, 7

2. Find the first and third quartile scores for the distribution in Exercise 1. 21; 33 **Obj. 15-2, p. 713**

3. To the nearest whole number, find the variance and standard deviation for the distribution in Exercise 1. 51; 7

4. In a standard normal distribution, what fraction of the data is between $x = -2.0$ and $x = 1.0$? (Use the table on page 720.) 81.85% **Obj. 15-3, p. 719**

5. The mean weight of the lobsters in a certain bay is 3 kg with a standard deviation of 0.5 kg. What percent of the lobsters weigh more than 4.1 kg? (Use the table on page 720.) 1.39%

6. Draw a scatter plot for the ordered pairs given in the table at the right. Then describe the nature of the correlation between x and y. Negative correlation **Obj. 15-4, p. 724**

x	y
3	12
5	10
8	7
9	5

Quick Quiz

1. To the nearest whole number, find the mode, median, and mean for the distribution in the stem-the-leaf plot below.

2	2, 4, 4
3	0, 5, 6, 8
4	3, 5

24; 35; 33

2. Find the first and third quartiles for the distribution in Exercise 1. 24; 38

3. To the nearest whole number, find the variance and standard deviation for the distribution in Exercise 1. 64; 8

4. In a standard normal distribution, what fraction of the data is between $x = -1$ and $x = 1.6$? 0.7865

5. The mean age in a horse herd is 12 years, with a standard deviation of 2.4 years. What percent of the horses are more than 14.88 years old? 11.51%

6. Draw a scatter plot for the data in the table. Describe the correlation between x and y.

x	y
1	6
4	5
6	4
7	3
10	2

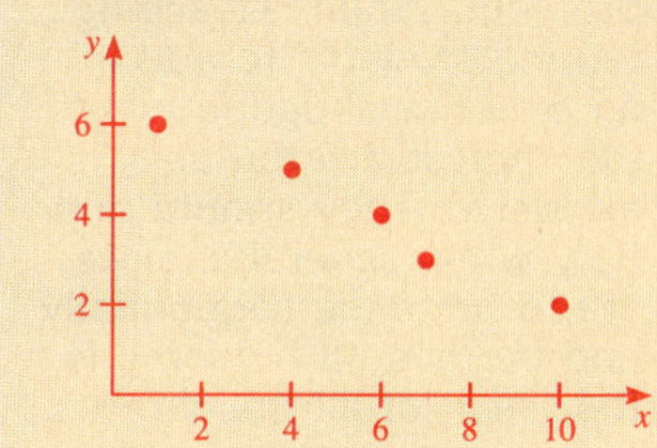

high negative

Teaching Suggestions p. T138

Suggested Extensions p. T138

Warm-Up Exercises

Simplify.

1. $4 \cdot 3 \cdot 2 \cdot 1$ 24
2. $10 \cdot 9 \cdot 8$ 720
3. 3^4 81
4. 2^5 32
5. Evaluate $n(n-1)(n-2)(n-3)$ if $n = 12$. 11,880
6. Find n if $n(n-1)(n-2)\ldots 3 \cdot 2 \cdot 1 = 5040$. 7

Motivating the Lesson

Ask the students to suppose that a state has 20 million vehicles. Would three letters and three digits give enough license plate possibilities so that each vehicle can have a different plate? A question like this can be answered by using the counting principles discussed in today's lesson.

Common Errors

Sometimes students do not notice that there are two fundamental counting principles. One involves multiplication and the other addition. Students may use the wrong operation for a given problem. When reteaching, remind students to add in an either/or situation where the choices are mutually exclusive. They should multiply when several choices are to be made sequentially and the overall outcome is the result of all the choices.

Counting

15-5 Fundamental Counting Principles

Objective To apply fundamental counting principles.

A local moped dealer sells 6 different models of mopeds. Each model is available in 3 colors. Therefore, there are

$$6 \cdot 3, \text{ or } 18,$$

different combinations of model and color that can be ordered. This illustrates a *fundamental counting principle*.

If one selection can be made in m ways, and for each of these a second selection can be made in n ways, then the number of ways the two selections can be made is $m \times n$.

Example 1 How many *odd* 2-digit whole numbers less than 70 are there?

Solution A diagram such as ☐ ☐ is useful to help analyze the problem.

There are *six* possible selections for the tens' digit: 1, 2, 3, 4, 5, and 6. Write 6 in the first box: [6] ☐.

There are *five* possible selections for the units' digit: 1, 3, 5, 7, and 9. Write 5 in the second box: [6] [5].

∴ by the fundamental counting principle stated above, there are $6 \cdot 5$, or 30, odd 2-digit whole numbers less than 70. ***Answer***

To get to school, Rita can either walk or take a bus. These are *mutually exclusive* choices. That is, she can either ride or walk to school, but not both. If she chooses to ride, there are two possible bus routes she can use, and if she walks, there are three routes she can take. Thus there are $2 + 3$, or 5, possible routes for Rita's trip to school. This illustrates another useful counting principle, the additive rule for mutually exclusive possibilities.

If the possibilities being counted can be grouped into *mutually exclusive* cases, then the total number of possibilities is the *sum* of the number of possibilities in each case.

Example 2 How many positive integers less than 100 can be written using the digits 6, 7, 8, and 9?

Solution Consider two mutually exclusive cases: (1) the 1-digit integers and (2) the 2-digit integers.

	T	U	
1-digit integers	—	4	4
2-digit integers	4	4	$4 \cdot 4 = 16$
			Total = 20

$\therefore$ there are 20 positive integers less than 100 that can be written using the digits 6, 7, 8, and 9. ***Answer***

Example 3 How many license plates of 3 symbols (letters and digits) can be made using at least one letter in each?

Solution There are three mutually exclusive cases: license plates with one letter, two letters, or three letters.

1-letter case: There are 26 possibilities for the letter and 10 for each digit. Thus there are $26 \cdot 10 \cdot 10$, or 2600, letter-digit-digit combinations. Since the letter can be in any of three positions on the plate (left, middle, or right), there are $3 \cdot 2600$, or 7800, possible 1-letter plates.

2-letter case: There are $26 \cdot 26 \cdot 10$, or 6760, letter-letter-digit combinations. Since the digit can be in any of three positions, there are $3 \cdot 6760$, or 20,280, possible 2-letter plates.

3-letter case: There are $26 \cdot 26 \cdot 26$, or 17,576, possible 3-letter plates.

$\therefore$ the number of plates is the sum of the three mutually exclusive possibilities: $7800 + 20{,}280 + 17{,}576$, or 45,656. ***Answer***

Oral Exercises

1. Elena can wear one of 2 blouses and one of 5 scarves. How many blouse-scarf combinations are available to her? 10
2. There are 3 trails on the north face of Mount Ezra and 2 trails on the south face of Mount Ezra. How many routes are there going up the north face and down the south face? 6
3. Kelly must buy hamburger rolls for a cookout. She can buy them in one of 4 supermarkets or one of 3 bakery shops. In how many ways can Kelly run her errand? 7
4. George can choose among 15 different flavors of ice cream, 6 different flavors of sherbet, and 5 different flavors of frozen yogurt. In how many ways can he choose a single dessert? 26

Chalkboard Examples

Pat's Pizza Palace will prepare pizza with a thin crust, with a thick crust, or in deep-dish style. There are eight choices of toppings.

1. In how many ways can you choose a one-topping pizza?
   $3 \cdot 8 = 24$ ways
2. How many odd numbers between 10 and 1000 start and end with the same digit?
   2-digit integers: There are 5 choices for the ones' digit and one choice for the tens' digit for each ones' digit.
   $1 \cdot 5 = 5$
   3-digit integers:
   $1 \cdot 10 \cdot 5 = 50$
   $\therefore$ there are $5 + 50 = 55$ such integers.
3. How many license plates of 2 symbols (letters and digits) can be made using at least one letter in each?
   one-letter case: There are $26 \cdot 10$, or 260, letter-digit possibilities. Since each possibility results in two plates (with the letter first or with the digit first), there are $2 \cdot 260$, or 520, possible one-letter plates.
   two-letter case: There are $26 \cdot 26$, or 676, possible two-letter plates.
   $\therefore$ there are $520 + 676$, or 1196, possible plates in all.

Check for Understanding

Here is a suggested use of the Oral Exercises to check students' understanding as you teach the lesson.
Oral Exs. 1–4: use after Example 3.

Guided Practice

1. Identification labels are composed of four letters. How many different labels are possible? 456,976 labels
2. Adele can take one of three buses to work or she can ride one of two trains and then walk along one of four different routes from the train station to her office. In how many ways can Adele go to work? 11 ways
3. Brenda's school offers 5 English courses, 4 math courses, and 4 science courses. How many schedules are possible if Brenda chooses a course in each subject? 80 schedules
4. One hundred cards are numbered from 1 to 100. How many ways are there of choosing two cards if the first card is not returned to the deck? 9900 ways
5. Repeat Exercise 4 if the first card is returned to the deck before the second card is chosen. 10,000 ways

Written Exercises

A

1. How many even 2-digit positive integers less than 50 are there? 20
2. How many odd 2-digit positive integers greater than 20 are there? 40
3. How many odd 3-digit positive integers can be written using the digits 2, 3, 4, 5, and 6? 50
4. How many even 3-digit positive integers can be written using the digits 1, 2, 4, 7, and 8? 75
5. In how many ways can you select one algebra book, one geometry book, and one calculus book from a collection of 8 different algebra books, 5 different geometry books, and 3 different calculus books? 120
6. A student council has 5 seniors, 4 juniors, 3 sophomores, and 2 freshmen as members. In how many ways can a 4-member council committee be formed that includes one member of each class? 120
7. In how many different ways can a 10-question true-false test be answered if every question must be answered? 1024
8. In how many different ways can a 10-question true-false test be answered if it is all right to leave questions unanswered? 59,049
9. How many ways are there to select 3 cards, one after the other, from a deck of 52 cards if the cards are not returned to the deck after being selected? 132,600
10. How many ways are there to write a 3-digit positive integer using the digits 1, 3, 5, 7, and 9 if no digit is used more than once? 60

B

11. How many 7-digit telephone numbers can be created if the first digit must be 8, the second must be 5, and the third must be 2 or 3? 20,000
12. How many positive odd integers less than 10,000 can be written using the digits 3, 4, 6, 8, and 0? 125
13. How many license plates of 3 symbols (letters and digits) can be made using at least 2 letters for each? 37,856
14. How many license plates of 4 symbols can be made using 2 letters and 2 digits? 405,600
15. How many multiples of 3 less than 100 can be formed from the digits 1, 4, 5, 7, and 8? (*Hint:* The sum of the digits of any multiple of 3 is also a multiple of 3.) 12
16. How many multiples of 3 less than 1000 can be formed from the digits 2, 5, and 9? 11

C

17. Suppose you have totally forgotten the combination to your locker. There are three numbers in the combination, and you're sure each number is different. The numbers on the lock's dial range from 0 to 35. If you test one combination every 12 seconds, how long will it take to test all possible combinations? 8568 min $\approx$ 6 days

include the base units adenine, thymine, cytosine, and guanine (A, T, C, and G). The sequence of base units along a strand of DNA encodes genetic information. In how many different sequences can A, T, C, and G be arranged along a short strand of DNA that has only 8 base units? 65,536

19. Protein molecules are made up of many amino acid residues joined end-to-end. Proteins have different properties, depending on the sequence of amino acid residues in the molecules. If there are 20 naturally occurring amino acids, how many different sequences of amino acid residues can occur in a 6-residue-long fragment of a protein molecule? 64,000,000

20. How many 3-letter code words can be formed if at least one of the letters is to be chosen from the vowels *a, e, i, o,* and *u*? 8315

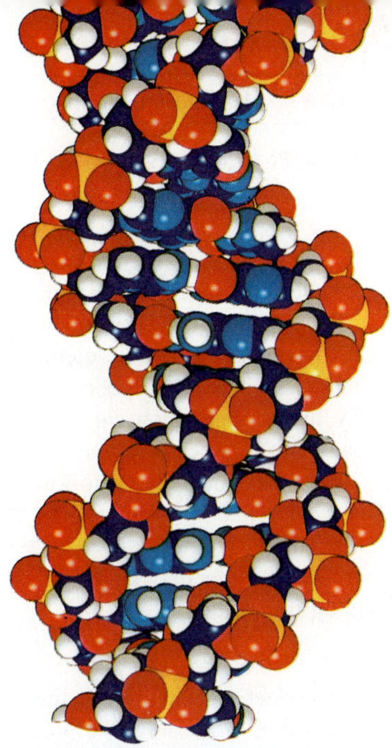

Mixed Review Exercises

Evaluate if $x = -2$ and $y = 8$.

1. $\frac{y^2}{x^5}$ -2
2. $|x - y|$ 10
3. y^x $\frac{1}{64}$
4. $\sqrt{xy}$ $4i$

Use the given root to solve each equation completely.

5. $x^3 + x^2 - 2x - 2 = 0;\ \sqrt{2}$ $\{\pm\sqrt{2}, -1\}$
6. $x^3 - 4x^2 + 3x - 12 = 0;\ -i\sqrt{3}$ $\{\pm i\sqrt{3}, 4\}$
7. $2x^3 - 11x^2 + 4x + 5 = 0;\ 5$ $\left\{5, 1, -\frac{1}{2}\right\}$
8. $x^3 + 6x^2 + 16x + 96 = 0;\ 4i$ $\{\pm 4i, -6\}$

Challenge

If you have two colors of paint available, in how many ways can you paint the faces of a cube so that each face has one of the two colors? (Two colorings are the same if one can be turned so that it matches the other exactly. You don't have to use both colors.) 10

Summarizing the Lesson

In this lesson students learned to use the fundamental counting principles. Ask students to state each counting principle and give an example to illustrate it.

Suggested Assignments

Average Algebra
732/1–10, 11–15 odd
S 733/Mixed Review
R 729/Self-Test 1

Extended Algebra
732/1–12, 14, 16
S 733/Mixed Review
R 729/Self-Test 1

Supplementary Materials

Study Guide pp. 247–248
Resource Book p. 168

Teaching Suggestions p. T139

Suggested Extensions p. T139

Warm-Up Exercises

1. Show all possible arrangements of the letters in the word CAT. How many are there? CAT, CTA, ATC, ACT, TAC, TCA; 6
2. Show all possible arrangements of the letters in the word STOP that begin with a consonant. How many are there? How many form actual words? STOP, STPO, SPOT, SPTO, SOPT, SOTP, TOPS, TOSP, TPSO, TPOS, TSOP, TSPO, POST, POTS, PSTO, PSOT, PTOS, PTSO; 18; 5

Motivating the Lesson

Ask the students to suppose they know the three numbers of their locker combinations but cannot remember the order in which to dial them. Arranging numbers or other objects in a definite order involves permutations, the topic of today's lesson.

Common Errors

Some students may think that an expression like ${}_9P_4$ means $\frac{9!}{4!}$. When reteaching, stress that the denominator in the formula for ${}_nP_r$ is $(n - r)!$, not $r!$.

15-6 Permutations

Objective To find the number of permutations of the elements of a set.

The letters a, b, and c can be arranged in six different ways:

abc	*bac*	*cab*
acb	*bca*	*cba*

Each of these arrangements is called a *permutation* of the letters a, b, and c. When the elements of a set are arranged in a definite order, the arrangement is called a **permutation** of the elements.

To determine the number of permutations of the letters a, b, and c without actually listing them, you can use the first fundamental counting principle stated in Lesson 15-5. Since any one of the three letters may be written first, a 3 appears in the first box of the following diagram:

$$\boxed{3}\,\boxed{}\,\boxed{}$$

After the first letter has been selected, the second must be selected from the remaining two letters:

$$\boxed{3}\,\boxed{2}\,\boxed{}$$

Only one selection remains for the last letter:

$$\boxed{3}\,\boxed{2}\,\boxed{1}$$

Therefore the number of permutations of the letters a, b, and c is

$$3 \cdot 2 \cdot 1, \text{ or } 6.$$

Example 1 Find the number of permutations of the four letters p, q, r, and s.

Solution Complete a diagram: $\boxed{4}\,\boxed{3}\,\boxed{2}\,\boxed{1}$

$\therefore$ the number of permutations is $4 \cdot 3 \cdot 2 \cdot 1$, or 24. ***Answer***

Recall from Lesson 11-8 that for any positive integer n, the product

$$n \cdot (n-1) \cdot (n-2) \cdot \ \ldots \ \cdot 3 \cdot 2 \cdot 1$$

can be written as $n!$ (read "n factorial"). As you have seen, there are 3! permutations of 3 objects and 4! permutations of 4 objects. In general:

The number of permutations of n objects is $n!$.

The next example illustrates what happens when not all the elements of a set are used to form permutations.

Example 2 In how many ways can the letters in the word JUSTICE be arranged using only 5 letters at a time?

Solution The first choice can be any one of the 7 letters, the second can be any one of the 6 remaining letters, and so on:

7	6	5	4	3

$\therefore$ there are $7 \cdot 6 \cdot 5 \cdot 4 \cdot 3$, or 2520, permutations of the 7 letters in JUSTICE when the letters are taken 5 at a time. ***Answer***

The symbol ${}_nP_r$ is used to indicate the number of permutations of n objects taken r at a time. From Example 2 you can see that ${}_7P_5 = 2520$. This result was obtained by filling five boxes with a decreasing sequence of consecutive integers, starting with 7. When you find ${}_nP_r$, the diagram has r boxes to be filled:

n	$n-1$	$n-2$	$\cdots$	$n-(r-1)$

Therefore $${}_nP_r = \underbrace{n(n-1)(n-2)\cdots[n-(r-1)]}_{r \text{ factors}}$$

Notice that when $r = n$, the final factor in the equation above, $n - (r - 1)$, becomes $n - (n - 1)$, or 1. Therefore,

$${}_nP_n = n!$$

Example 3 From a set of 9 different books, 4 are to be selected and arranged on a shelf. How many arrangements are possible?

Solution Find the number of permutations of 9 books taken 4 at a time:

$${}_9P_4 = 9 \cdot 8 \cdot 7 \cdot 6 = 3024 \quad \textit{Answer}$$

In Example 3, ${}_9P_4$ can be rewritten as

$$\frac{9 \cdot 8 \cdot 7 \cdot 6 \cdot 5 \cdot 4 \cdot 3 \cdot 2 \cdot 1}{5 \cdot 4 \cdot 3 \cdot 2 \cdot 1} = \frac{9!}{5!}.$$

In general, when the expression ${}_nP_r$ is multiplied by $\frac{(n-r)\ldots 2 \cdot 1}{(n-r)\ldots 2 \cdot 1}$, it becomes

$$\frac{n(n-1)\ldots[n-(r-1)](n-r)[n-(r+1)]\ldots 2 \cdot 1}{(n-r)[n-(r+1)]\ldots 2 \cdot 1}.$$

Therefore, you have the following formula.

$$\mathbf{{}_nP_r = \frac{n!}{(n-r)!}}$$

Chalkboard Examples

1. In how many ways can the letters of the word SPECIAL be arranged using **(a)** all the letters and **(b)** only four letters at a time?
   a. $7! = 5040$ ways
   b. ${}_7P_4 = 7 \cdot 6 \cdot 5 \cdot 4 = 840$ ways
2. From 1000 contest entries, one entry is chosen for first prize, one for second prize, and one for third prize. In how many ways can the winners be selected?
   ${}_{1000}P_3 = 1000(999)(998) = 997{,}002{,}000$ ways
3. In how many ways can 3 apples, 2 oranges, 4 pears, and 1 banana be given to 10 children if each child receives a piece of fruit?
   $\frac{10!}{3!2!4!} = \frac{10 \cdot 9 \cdot 8 \cdot 7 \cdot 6 \cdot 5}{3 \cdot 2 \cdot 1 \cdot 2 \cdot 1} =$ 12,600 ways

Check for Understanding

Here is a suggested use of the Oral Exercises to check students' understanding as you teach the lesson.
Oral Exs. 1–4: use before Example 2.
Oral Exs. 5–8: use after the formula for ${}_nP_r$.
Oral Exs. 9–12: use after Example 4.

Guided Practice

1. Evaluate $\frac{7!}{4!}$. 210
2. Find ${}_nP_r$ if $n = 6$ and $r = 3$. 120
3. In how many ways can 7 different objects be arranged in a line on a shelf? 5040
4. In how many ways can the letters of the word TUESDAY be arranged using only 4 letters at a time? 840

Summarizing the Lesson

In this lesson students learned to find the number of permutations of the elements of a set. Ask students to explain the difference between finding the number of permutations of the letters in the word HOP versus the word HOOP.

Suggested Assignments

Average Algebra
Day 1: 737/1–20
Day 2: 737/21–28
Assign with Lesson 15-7.

Extended Algebra
Day 1: 737/1–20
S 723/12
Day 2: 737/21–30
Assign with Lesson 15-7.

If you consider the letters *a* and *a* to be distinct, the number of permutations of *a*, *a*, and *b* taken 3 at a time is 3!, or 6. These 6 permutations are listed below:

aab	*aba*	*baa*
aab	*aba*	*baa*

In contrast, the letters *a*, *a*, and *b*, taken 3 at a time, give only 3 distinguishable permutations:

aab *aba* *baa*

Each of these permutations corresponds to two of the original 6 permutations, because there are 2!, or 2, permutations of *a*, *a*. Therefore, if *P* is the number of permutations of *a*, *a*, and *b*,

$$P \cdot 2! = 3!, \quad \text{so} \quad P = \frac{3!}{2!} = 3.$$

In general, when *n* elements of a set are taken *n* at a time and n_1 of the elements are alike, the number of permutations, *P*, is $\frac{n!}{n_1!}$. A similar relationship holds when two or more repeated elements occur.

If a set of *n* elements has n_1 elements of one kind alike, n_2 of another kind alike, and so on, then the number of permutations, *P*, of the *n* elements taken *n* at a time is given by

$$P = \frac{n!}{n_1!n_2!\cdots}.$$

Example 4 Find the number of ways the letters in the word HUBBUB can be arranged.

Solution Since there are 6 letters, 2 of which are U's and 3 are B's,

$$P = \frac{n!}{n_1!n_2!} = \frac{6!}{2!3!} = 60. \quad \textit{Answer}$$

Oral Exercises

Evaluate.

1. $5!$ 120
2. $\frac{10!}{9!}$ 10
3. $\frac{6!}{(6-3)!}$ 120
4. $\frac{4!}{2!2!}$ 6
5. ${}_4P_4$ 24
6. ${}_{10}P_3$ 720
7. ${}_{100}P_2$ 9900
8. ${}_7P_0$ 1

Tell the number of ways the letters in each word can be arranged.

9. FOR 6
10. ALL 3
11. SOME 24
12. NONE 12

Written Exercises

Evaluate each of the following.

A **1.** $3!5!$ 720 **2.** $3(5!)$ 360 **3.** $\frac{8!}{(8-3)!}$ 336 **4.** $\frac{8!}{3!5!}$ 56

Find $_nP_r$ for the given values of n and r.

5. $n = 7$, $r = 7$ 5040 **6.** $n = 5$, $r = 2$ 20 **7.** $n = 6$, $r = 1$ 6 **8.** $n = 8$, $r = 4$ 1680

9. In how many ways can 6 different books be arranged on a shelf? 720

10. In how many ways can 8 people be lined up in a row for a photograph? 40,320

11. In how many ways can 3 cards from a deck of 52 cards be laid in a row face up? 132,600

12. In how many ways can 4 of 7 different kinds of bushes be planted along one side of a house? 840

13. In how many ways can the letters of the word MONDAY be arranged using all 6 letters? 720

14. In how many ways can the letters of the word TODAY be arranged using only 3 of the letters at a time? 60

Find the number of ways the letters of each word can be arranged.

B **15.** ADDEND 120 **16.** BEEKEEPER 3024 **17.** ROTOR 30

18. DEEMED 60 **19.** MISSISSIPPI 34,650 **20.** ALBUQUERQUE 1,663,200

21. How many different signals can be made by displaying five flags all at one time on a flagpole? The flags differ only in color: two are red, two are white, and one is blue. 30

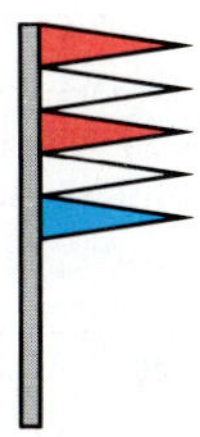

22. How many different signals could be made if the flags of Exercise 21 were of the following colors: three red, one white, and one blue? 20

23. In how many ways can 3 identical emeralds, 2 identical diamonds, and 2 different opals be arranged in a row in a display case? 420

24. In how many ways can 3 red, 4 blue, and 2 green pens be distributed to 9 students seated in a row if each student receives one pen? 1260

25. Show that $_6P_4 = 6(_5P_3)$. **26.** Show that $_5P_r = 5(_4P_{r-1})$.

27. Show that $_nP_5 - {_nP_4} = (n-5)_nP_4$.

28. Show that $_nP_r - {_nP_{r-1}} = (n-r)_nP_{r-1}$.

C **29.** Solve for n: $_nP_5 = 14(_nP_4)$. 18 **30.** Solve for n: $_nP_3 = 17(_nP_2)$. 19

31. Find the number of 6-letter permutations that can be formed from the letters in the word SHOPPER. 2520

Supplementary Materials

Study Guide pp. 249–250
Practice Master 84

Additional Answers Written Exercises

25. $6(_5P_3) = 6\left(\frac{5!}{2!}\right) = \frac{6 \cdot 5!}{2!} = \frac{6!}{2!} = {_6P_4}$

26. $5(_4P_{r-1}) = 5\left(\frac{4!}{[4-(r-1)]!}\right) = \frac{5 \cdot 4!}{(5-r)!} = \frac{5!}{(5-r)!} = {_5P_r}$

27. $_nP_5 - {_nP_4} = \frac{n!}{(n-5)!} - \frac{n!}{(n-4)!} = \frac{n!}{(n-5)!} \cdot \frac{n-4}{n-4} - \frac{n!}{(n-4)!} = \frac{n![(n-4)-1]}{(n-4)!} = \frac{(n-5)n!}{(n-4)!} = (n-5)_nP_4$

28. $_nP_r - {_nP_{r-1}} = \frac{n!}{(n-r)!} - \frac{n!}{(n-(r-1))!} = \frac{n!}{(n-r)!} \cdot \frac{(n-(r-1))}{(n-(r-1))} - \frac{n!}{(n-(r-1))!} = \frac{n!(n-(r-1)-1)}{(n-(r-1))!} = \frac{n!(n-r)}{(n-(r-1))!} = (n-r)_nP_{r-1}$

Suggested Extensions p. T139

Warm-Up Exercises

Evaluate ${}_nP_r$ for the given value of n and r.

1. $n = 6, n = 1$ 6
2. $n = 6, r = 6$ 720
3. $n = 8, r = 4$ 1680
4. $n = 10, r = 2$ 90
5. How many positive odd numbers less than 1000 without repeated digits are there? 360

Motivating the Lesson

Ask the students if they have ever participated in a scavenger hunt. The participants are expected to gather as many items that appear on a given list as possible, but the order in which they gather the items does not matter. Such a situation involves combinations, the topic of today's lesson.

Thinking Skills

Using the formulas for permutations and combinations is not particularly difficult. The challenge sometimes is deciding which formula is appropriate for a given situation. Students need to *analyze* problems to see if order does or does not matter. The analogies of electing officers or choosing a committee can provide a sound basis of comparison.

15-7 Combinations

Objective To find the combinations of a set of elements.

Set B is a **subset** of set A if each member of B is also a member of A. Therefore, every set is a subset of itself. The empty set, $\emptyset$, is considered to be a subset of every set.

Example 1 For the 3-letter set {P, Q, R}, find:

a. all the subsets. **b.** the 2-letter subsets.

Solution **a.** {P, Q, R}, {P, Q}, {P, R}, {Q, R}, {P}, {Q}, {R}, $\emptyset$ ***Answer***
b. {P, Q}, {P, R}, {Q, R} ***Answer***

Consider the 2-letter subsets of the 4-letter set {S, T, U, V}:

{S, T}, {S, U}, {S, V}, {T, U}, {T, V}, and {U, V}.

These subsets are also known as the *combinations* of the letters S, T, U, and V taken two at a time. An r-element subset of a set of n elements is called a **combination** of n elements taken r at a time.

The symbol ${}_nC_r$ denotes the number of combinations of n elements taken r at a time. For example, since there are six 2-letter subsets of the 4-letter set {S, T, U, V}, ${}_4C_2 = 6$.

Recall that the *order* of the elements matters in a permutation: ST and TS, for example, are different permutations. Order is not important in a combination, however: {S, T} and {T, S} are the same combination.

For a given combination there may be a number of different permutations. For example, there are ${}_2P_2$, or 2!, permutations possible for each of the 2-letter combinations of the letters S, T, U, and V. So the total number of permutations of the 4 letters taken 2 at a time is just the number of combinations multiplied by the number of permutations per combination:

$${}_4P_2 = ({}_4C_2)({}_2P_2).$$

In general,
$${}_nP_r = ({}_nC_r)({}_rP_r),$$

and therefore
$${}_nC_r = \frac{{}_nP_r}{{}_rP_r}.$$

Recall from Lesson 15-6 that ${}_nP_r = \dfrac{n!}{(n-r)!}$ and ${}_rP_r = r!$. Therefore:

$${}_nC_r = \frac{{}_nP_r}{{}_rP_r} = \frac{n!}{r!\,(n-r)!}$$

Example 2 Find the number of combinations of the letters in the word SOLVE, taking them **(a)** 5 at a time and **(b)** 2 at a time. List each combination.

Solution

a. ${}_5C_5 = \frac{5!}{5!0!} = 1$ {S, O, L, V, E}

b. ${}_5C_2 = \frac{5!}{2!3!} = 10$ {S, O} {S, L} {S, V} {S, E} {O, L} {O, V} {O, E} {L, V} {L, E} {V, E}

Example 3 In how many ways can a committee of 6 be chosen from 5 teachers and 4 students if:

a. all are equally eligible?

b. the committee must include 3 teachers and 3 students?

Solution

a. There are 9 people eligible for the committee:

$${}_9C_6 = \frac{9!}{6!3!} = 84 \quad \textbf{\textit{Answer}}$$

b. There are ${}_5C_3$ ways to choose 3 teachers and ${}_4C_3$ ways to choose 4 students. Use the first fundamental counting principle stated on page 730 to find the number of ways of selecting the committee:

$${}_5C_3 \cdot {}_4C_3 = \frac{5!}{3!2!} \cdot \frac{4!}{3!1!} = 10 \cdot 4 = 40 \quad \textbf{\textit{Answer}}$$

Example 4 A standard deck of cards consists of 4 suits (clubs, diamonds, hearts, and spades) of 13 cards each. How many 5-card hands can be dealt that include 4 cards from the same suit and one card from a different suit?

Solution There are 4 suits with ${}_{13}C_4$ possible combinations from any one suit:

$$4({}_{13}C_4) = 4\left(\frac{13!}{4!9!}\right) = 4(715) = 2860$$

The remaining card must come from the other 3 suits having $52 - 13$, or 39, cards:

$${}_{39}C_1 = \frac{39!}{1!38!} = 39$$

$\therefore$ the number of 5-card hands that include exactly 4 cards from any one suit is $2860 \cdot 39$, or 111,540. ***Answer***

Oral Exercises

Tell whether each of the following is a combination or a permutation.

1. Nine books placed in a row on a shelf. Permutation
2. Three books selected from a collection of 20 books. Combination
3. An arrangement of the letters in the word BOOK. Permutation

Chalkboard Examples

1. List all two-letter subsets of the set $\{J, K, L, M\}$.
$\{J, K\}, \{J, L\}, \{J, M\}, \{K, L\}, \{K, M\}, \{L, M\}$
2. How many combinations of the letters in the word ABSOLUTE are there, taking the letters three at a time?
${}_8C_3 = \frac{8!}{3!5!} = 56$
3. In how many ways can a student select 6 courses from 7 electives and 5 core courses if 3 electives and 3 core courses must be taken?
${}_7C_3 \cdot {}_5C_3 = \frac{7!}{3!4!} \cdot \frac{5!}{3!2!}$
$= 35 \cdot 10 = 350$
4. From a standard deck of cards, how many hands of 5 cards all from the same suit can be dealt?
There are 4 suits with ${}_{13}C_5$ possible combinations from any one suit:
$4({}_{13}C_5) = 4\left(\frac{13!}{5!8!}\right)$
$= 4(1287) = 5148$

Check for Understanding

Here is a suggested use of the Oral Exercises to check students' understanding as you teach the lesson.

Oral Exs. 1–3: use after the discussion of order.

Oral Exs. 4–7: use after Example 2.

Guided Practice

1. List all three-letter subsets of the set $\{P, Q, R, S\}$. $\{P, Q, R\}$, $\{P, Q, S\}$, $\{P, R, S\}$, $\{Q, R, S\}$

Evaluate.

2. ${}_5C_2$ 10 3. ${}_9C_4$ 126
4. ${}_{12}C_9$ 220 5. ${}_{12}C_{11}$ 12
6. How many different five-player teams can be formed from eight people? 56 teams
7. A deck of cards consists of five yellow cards numbered 1 to 5 and five green cards numbered 1 to 5. How many six-card hands having four yellow cards and two green cards can be dealt? 50 hands

Summarizing the Lesson

In this lesson students learned to find the number of combinations of the elements in a set. Ask students to explain the difference between combinations and permutations.

Suggested Assignments

Average Algebra
Day 1: 740/1–15 odd
Assign with Lesson 15-6.
Day 2: 740/17–25 odd
S 741/Mixed Review
Extended Algebra
Day 1: 740/2–10 even, 11–16
Assign with Lesson 15-6.
Day 2: 740/17–22, 23–27 odd
S 741/Mixed Review

Evaluate.

4. $\frac{5!}{4!1!}$ 5
5. $\frac{7!}{3!4!}$ 35
6. ${}_4C_2$ 6
7. ${}_6C_4$ 15

Written Exercises

A

1. For the 2-letter set {J, K}, find:
   a. all the subsets. {J, K}, {J}, {K}, ∅
   b. the subsets containing fewer than 2 letters. {J}, {K}, ∅
2. For the 4-digit set {1, 3, 5, 7}, find:
   a. the 3-digit subsets. {1, 3, 5}, {1, 3, 7}, {1, 5, 7}, {3, 5, 7}
   b. the subsets in which the sum of the digits is at least 9. {3, 7}, {5, 7}, {1, 3, 5}, {1, 3, 7}, {1, 5, 7}, {3, 5, 7}, {1, 3, 5, 7}

Evaluate.

3. ${}_5C_3$ 10
4. ${}_6C_1$ 6
5. ${}_8C_6$ 28
6. ${}_7C_4$ 35
7. ${}_{10}C_8$ 45
8. ${}_9C_2$ 36
9. ${}_{12}C_5$ 792
10. ${}_{100}C_2$ 4950
11. How many combinations can be formed from the letters in EIGHT, taking them:
    a. 4 at a time? 5 b. 3 at a time? 10 c. 2 at a time? 10
12. How many combinations can be formed from the letters in HEXAGON, taking them:
    a. 6 at a time? 7 b. 4 at a time? 35 c. 2 at a time? 21
13. A volleyball team has 12 members, one coach, and 2 managers. How many different combinations of 7 people can be chosen to kneel in the front row of the team picture? 6435
14. A sample of 4 mousetraps taken from a batch of 100 mousetraps is to be inspected. How many different samples could be selected? 3,921,225
15. In a group of 10 people, each person shakes hands with everyone else once. How many handshakes are there? 45
16. You can order a hamburger with cheese, onion, pickle, relish, mustard, lettuce, tomato, or mayonnaise. How many different combinations of the "extras" can you order, choosing any four of them? 70

B

17. Seven points lie on a circle, as shown in the figure at the right. How many inscribed triangles can be constructed having any three of these points as vertices? 35

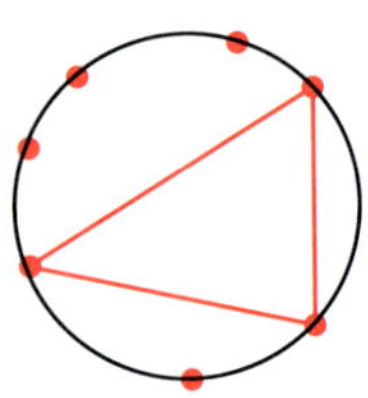

18. Ten points lie on the circumference of a circle. How many inscribed quadrilaterals can be drawn having these points as vertices? 210
19. A school club has 15 boys and 16 girls as members. How many different 6-person committees can be selected from the membership if equal numbers of boys and girls are to be selected? 254,800

20. The junior and senior class councils each have 10 members. In how many ways can a prom committee be formed if it is to consist of 3 seniors and 2 juniors selected from the two class councils? 5400

Exercises 21–24 refer to a standard deck of cards as described in Example 4, page 739.

21. How many 13-card hands having exactly 11 diamonds can be dealt? 57,798

22. How many 13-card hands having exactly 11 cards from any suit can be dealt? 231,192

23. How many 5-card hands having exactly 3 aces and 2 other cards can be dealt? 4512

24. How many 7-card hands having exactly 3 spades, 3 clubs, and 1 heart can be dealt? 1,063,348

25. In how many ways can 4 or more students be selected from 8 students? 163

26. In how many ways can up to 4 students be selected from 6 girls and 5 boys if each selection must have an equal number of girls and boys? 180

C 27. Prove that ${}_nC_r = {}_nC_{n-r}$ by using the formula ${}_nC_r = \frac{n!}{r!(n-r)!}$.

28. Prove that ${}_nC_r = {}_{n-1}C_{r-1} + {}_{n-1}C_r$. (*Hint:* $n! = n \cdot (n-1)!$)

29. a. Refer to the binomial theorem on page 541. What is the relationship between the binomial coefficients in the expansion of $(a+b)^n$ and the values of ${}_nC_r$ for $r = 0, 1, 2, \ldots, n$? ${}_nC_r$ is the coefficient of $a^{n-r}b^r$.

b. Based on the results of part (a), rewrite the binomial theorem using sigma notation (see Lesson 11-4) and ${}_nC_r$. $(a+b)^n = \sum_{r=0}^{n} {}_nC_r a^{n-r}b^r$

30. Show that the total number of subsets of a set having n elements is 2^n. (*Hint:* Each member of the set is or is not selected in forming a subset. Recall that the empty set is defined to be a subset of every set.)

Mixed Review Exercises

Find the value of each function if $x = 4$.

1. $f(x) = \frac{x+6}{2x-3}$ 2

2. $g(x) = |1 - x|$ 3

3. $h(x) = (x+2)^2$ 36

4. $F(x) = -3$ -3

5. $G(x) = -2x + 1$ -7

6. $H(x) = x^4 + 5x^3 - 12x^2 + 6$ 390

Simplify.

7. $\sqrt{72} - \sqrt{18} + \sqrt{8}$ $5\sqrt{2}$

8. $\log_6 2 + \log_6 3$ 1

9. $\frac{1}{m+2} + \frac{6}{m^2 - 2m - 8}$ $\frac{1}{m-4}$

10. $(4 - 3i)(2 + i)$ $11 - 2i$

11. $\frac{2t^2 + 3t - 9}{t^2 - 9}$ $\frac{2t-3}{t-3}$

12. $\left(\frac{\sqrt{13}+2}{3}\right)\left(\frac{\sqrt{13}-2}{3}\right)$ 1

Supplementary Materials

Study Guide	pp. 251–252
Practice Master	85
Test Master	64
Resource Book	p. 169

Additional Answers Written Exercises

27. ${}_nC_r = \frac{n!}{r!(n-r)!}$; ${}_nC_{n-r} = \frac{n!}{(n-r)![n-(n-r)]!} = \frac{n!}{(n-r)!r!} = {}_nC_r$

28. ${}_{n-1}C_{r-1} + {}_{n-1}C_r = \frac{(n-1)!}{(r-1)!(n-r)!} + \frac{(n-1)!}{r!(n-r-1)!} = \frac{r[(n-1)!]}{r!(n-r)!} + \frac{(n-r)[(n-1)!]}{r!(n-r)!} = \frac{(r+n-r)[(n-1)!]}{r!(n-r)!} = \frac{n!}{r!(n-r)!} = {}_nC_r$

30. Each of the n members is or is not selected in forming a subset. There are two choices for each of n members of the set. By the first of the fundamental counting principles on page 730, there are 2^n subsets.

Using a Computer

The exercises require students to write a program to calculate $n!$, then to use this in a program to calculate ${}_nC_r$. Most students with programming experience already know how to keep a "running sum" (that is, a sum that is increased with each pass through the loop). Point out to the students that these exercises require the use of a "running product."

Quick Quiz

1. How many positive integers less than 1000 can be formed using the digits 0, 1, and 2?
   26 integers
2. In how many ways can the letters of the word PARSLEY be arranged using **(a)** all the letters and **(b)** four letters at a time?
   a. 5040 ways
   b. 840 ways
3. Find the number of permutations of all the letters in the word POPPY.
   20
4. Six seniors and five juniors volunteered to serve on a committee.
   a. How many three-person committees are possible?
      165 committees
   b. How many three-person committees are possible if at least one senior must be chosen?
      155 committees

Computer Exercises

For students with some programming experience.

2. b. 6.2270208×10^9 c. $2.09227899 \times 10^{13}$ d. $2.43290201 \times 10^{18}$

1. Write a program that will calculate $n!$ for an input value of n.
2. Run the program in Exercise 1 to find $n!$ for each given value of n.
   a. 10 3,628,800 **b.** 13 **c.** 16 **d.** 20
3. Use the program in Exercise 1 as a subroutine in a program that will compute the value of ${}_nC_r$, for input values of n and r. Have the program check first to make sure n and r are positive integers with $n \geq r$.
4. Run the program in Exercise 3 to find the value of ${}_nC_r$ for each given pair of values.
   a. $n = 12$, $r = 3$ 220 **b.** $n = 18$, $r = 9$ 48,620
5. Modify the program in Exercise 3 so that you can compute ${}_{52}C_r$ for values of r up to 15. (Note that 52! is too large a number for most microcomputers to handle. You can compute $52 \cdot 51 \cdot \ldots \cdot [52 - (r - 1)]$ and then divide by $r!$.) Use your program to compute the following numbers.
   a. ${}_{52}C_5$, the number of different 5-card hands that can be dealt from a 52-card deck. 2,598,960
   b. ${}_{52}C_{13}$, the number of possible bridge hands. 6.3501356×10^{11}
   c. Considering your answer to part (b), do you think it is likely that there is some possible bridge hand that has *never* been dealt? It is likely.

Self-Test 2

Vocabulary

fundamental counting principles (p. 730)	permutation (p. 734)
mutually exclusive (p. 730)	subset (p. 738)
	combination (p. 738)

1. How many even 3-digit positive integers can be written using the digits 2, 4, 5, 6, and 8? 100 — **Obj. 15-5, p. 730**
2. A menu lists 4 appetizers, 7 main courses, and 5 desserts. How many complete meals can be chosen from this menu? 140
3. In how many ways can 5 toy blocks of different color be stacked? 120 — **Obj. 15-6, p. 734**
4. How many different permutations can be formed using all the letters in the word MEMBER? 180
5. In how many ways can a 3-person committee be selected from a group of 8 people? 56 — **Obj. 15-7, p. 738**
6. How many combinations can be formed from the letters in LEAST, taking them 2 at a time? 10

Compare your answers with those at the back of the book.

Probability

15-8 Sample Spaces and Events

Objective To specify sample spaces and events for random experiments.

Suppose you toss a coin and it lands "heads up." You repeat the "experiment" and the coin lands "tails up." An experiment in which you do not necessarily get the same outcome when you repeat it under the same conditions is called a **random experiment.**

The set of all possible outcomes of a random experiment is known as the **sample space** for the experiment. When you toss a coin, there are only two possible outcomes: heads (H) or tails (T). The sample space for the coin tossing experiment is therefore $\{H, T\}$.

Any subset of possible outcomes for an experiment is known as an **event.** When an event is a single element of the sample space, it is often called a **simple event.** $\{H\}$ and $\{T\}$ are simple events for the coin tossing experiment.

Example 1 For the rolling of a single die, specify:

a. the sample space for the experiment.
b. the event that a number greater than 2 results.
c. the event that an odd number results.

Solution **a.** $\{1, 2, 3, 4, 5, 6\}$ **b.** $\{3, 4, 5, 6\}$ **c.** $\{1, 3, 5\}$

Suppose you roll two dice, one red and one blue. An outcome in this experiment can be represented by the ordered pair (r, b), where r is the number showing on the red die and b is the number on the blue die.

The sample space for the two-dice experiment, shown as a graph in the figure at the right, contains $6 \cdot 6$, or 36, elements. Point A, with coordinates $(2, 3)$, represents the outcome "red die shows 2 and blue die shows 3."

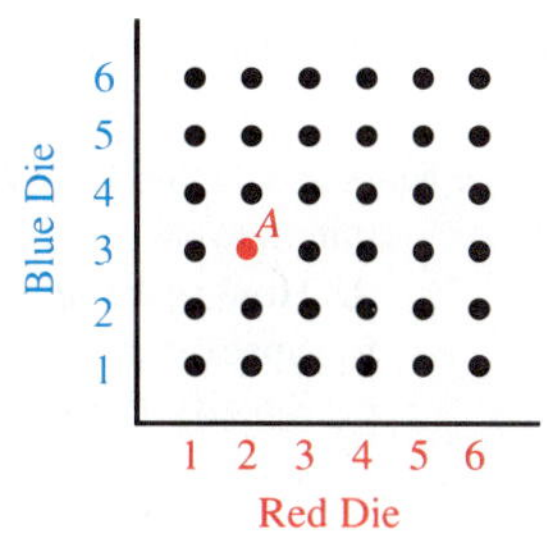

Example 2 For the two-dice experiment, specify each event.

a. The sum of the numbers showing on the two dice is 7.
b. The red die shows 5.

Solution **a.** $\{(1, 6), (2, 5), (3, 4), (4, 3), (5, 2), (6, 1)\}$
b. $\{(5, 1), (5, 2), (5, 3), (5, 4), (5, 5), (5, 6)\}$

Teaching Suggestions p. T139

Using Manipulatives p. T140

Suggested Extensions p. T140

Warm-Up Exercises

List all 2-element and 3-element subsets of $\{A, B, C, D\}$. $\{A, B\}$, $\{A, C\}$, $\{A, D\}$, $\{B, C\}$, $\{B, D\}$, $\{C, D\}$, $\{A, B, C\}$, $\{A, B, D\}$, $\{A, C, D\}$, $\{B, C, D\}$

Motivating the Lesson

Point out that listing the outcomes for tic-tac-toe is easy: X wins, O wins, or the two players draw. Assigning a number to each square and listing all sequences of X's and O's is a more formidable task. Creating lists of possible outcomes is the topic of today's lesson.

Chalkboard Examples

One box contains a red marble and a blue marble. A second box contains a red marble, a blue marble, a yellow marble, and a green marble. One marble is drawn from each box.

1. Specify the sample space. {(R, R), (R, B), (R, Y), (R, G), (B, R), (B, B), (B, Y), (B, G)}
2. Specify the event that exactly one marble is blue. {(R, B), (B, R), (B, Y), (B, G)}
3. Specify the event that at least one marble is red. {(R, R), (R, B), (R, Y), (R, G), (B, R)}

Check for Understanding

Oral Exs. 1–2: use after Example 2.

Guided Practice

A coin is tossed and a die is rolled.

1. Specify the sample space. {(H, 1), (H, 2), (H, 3), (H, 4), (H, 5), (H, 6), (T, 1), (T, 2), (T, 3), (T, 4), (T, 5), (T, 6)}
2. Specify the event that the coin comes up heads. {(H, 1), (H, 2), (H, 3), (H, 4), (H, 5), (H, 6)}
3. Specify the event that the die comes up even. {(H, 2), (H, 4), (H, 6), (T, 2), (T, 4), (T, 6)}
4. Specify the event that the coin is heads and the die is even. {(H, 2), (H, 4), (H, 6)}
5. Specify the event that the coin is heads or the die is even. {(H, 1), (H, 2), (H, 3), (H, 4), (H, 5), (H, 6), (T, 2), (T, 4), (T, 6)}

Summarizing the Lesson

In this lesson students learned to produce lists of possible outcomes for random experiments. Ask students to state the definition of a random experiment.

Suggested Assignments

Average Algebra
744/1–11 odd
R 742/Self-Test 2

Extended Algebra
744/1–9 odd, 10–14
R 742/Self-Test 2

Supplementary Materials

Study Guide	pp. 253–254
Practice Master	86

Additional Answers Written Exercises

(See p. 751.)

Oral Exercises

1. Each of the letters A, B, C, D, E, F, and G is written on a separate card. The cards are then shuffled, and one card is drawn.
   a. Specify the sample space for this experiment. {A, B, C, D, E, F, G}
   b. Specify the event that a vowel is drawn. {A, E}
2. A penny and a quarter are tossed. Let (p, q) represent each outcome.
   a. Specify the sample space for this experiment. {(H, H), (H, T), (T, H), (T, T)}
   b. Specify the event that exactly one head is up. {(H, T), (T, H)}

Written Exercises

4. {(3, 6), (4, 5), (4, 6), (5, 4), (5, 5), (5, 6), (6, 3), (6, 4), (6, 5), (6, 6)}
5. {(1, 1), (1, 2), (1, 3), (1, 4), (1, 5), (1, 6), (2, 1), (2, 2), (2, 3), (2, 4), (3, 1), (3, 2), (3, 3), (4, 1), (4, 2), (5, 1), (6, 1)}

For Exercises 1–6, refer to the two-dice experiment described in this lesson. Specify each event.

A

1. Both dice show the same number. {(1, 1), (2, 2), (3, 3), (4, 4), (5, 5), (6, 6)}
2. The sum of the numbers showing on the two dice is 6. {(1, 5), (2, 4), (3, 3), (4, 2), (5, 1)}
3. The product of the numbers showing on the two dice is 12. {(2, 6), (3, 4), (4, 3), (6, 2)}
4. The sum of the numbers showing on the two dice is greater than 8. (See above.)
5. The product of the numbers showing on the two dice is less than 10. (See above.)
6. The number showing on the red die is greater than the number showing on the blue die. {(2, 1), (3, 1), (4, 1), (5, 1), (6, 1), (3, 2), (4, 2), (5, 2), (6, 2), (4, 3), (5, 3), (6, 3), (5, 4), (6, 4), (6, 5)}

B

7. Two bags contain marbles. The first bag contains a red and a blue marble. The second bag contains a red, a blue, a yellow, and a green marble. One marble is drawn from each bag.
   a. Specify the sample space. {(R, R), (R, B), (R, Y), (R, G), (B, R), (B, B), (B, Y), (B, G)}
   b. Specify the event that at least one red marble is drawn. {(R, R), (R, B), (R, Y), (R, G), (B, R)}
   c. Specify the event that neither marble drawn is blue. {(R, R), (R, Y), (R, G)}
8. A red, a blue, and a green die are tossed. Let (r, b, g) represent each outcome. 8. b. {(1, 1, 1), (2, 2, 2), (3, 3, 3), (4, 4, 4), (5, 5, 5), (6, 6, 6)}
   a. How many elements are in the sample space? 216
   b. Specify the event that all three dice show the same number. (See above.)
   c. Specify the event that the sum of the numbers showing on the red and blue dice is less than the number showing on the green die. {(1, 1, 3), (1, 1, 4), (1, 1, 5), (1, 1, 6), (1, 2, 4), (1, 2, 5), (1, 2, 6), (1, 3, 5), (1, 3, 6), (1, 4, 6), (2, 1, 4), (2, 1, 5), (2, 1, 6), (2, 2, 5), (2, 2, 6), (2, 3, 6), (3, 1, 5), (3, 1, 6), (3, 2, 6), (4, 1, 6)}

For Exercises 9–16, refer again to the two-dice experiment. The ordered pair (r, b) represents an element in the sample space. Specify each event.

9. $r + b = 9$ {(3, 6), (4, 5), (5, 4), (6, 3)}
10. $b > r$ (See below.)
11. $r = 2b$ {(2, 1), (4, 2), (6, 3)}
12. $r - b = 2$ {(3, 1), (4, 2), (5, 3), (6, 4)}

C

13. $b > 4$ or $r < 3$
14. $r + b = 4$ or $r = b$
15. $r + b$ is prime, and b is a perfect square. {(1, 1), (2, 1), (4, 1), (6, 1), (1, 4), (3, 4)}
16. r is a multiple of 2, and $r - b$ is a multiple of 2. {(2, 2), (2, 4), (2, 6), (4, 2), (4, 4), (4, 6), (6, 2), (6, 4), (6, 6)}

10. {(1, 2), (1, 3), (1, 4), (1, 5), (1, 6), (2, 3), (2, 4), (2, 5), (2, 6), (3, 4), (3, 5), (3, 6), (4, 5), (4, 6), (5, 6)}

15-9 Probability

Objective To find the probability that an event will occur.

Consider the toss of a single coin. The sample space contains two elements, H and T. If the coin is fair, these two outcomes are *equally likely* to occur. That is, if you repeated the coin toss a great many times, you would expect heads about half of the time and tails about half of the time. Therefore, the probability of the event $\{H\}$, denoted by $P(H)$, equals the probability of the event $\{T\}$, denoted by $P(T)$:

$$P(H) = P(T) = \frac{1}{2}.$$

Now consider the rolling of a die. There are six possible outcomes, and the sample space is $\{1, 2, 3, 4, 5, 6\}$. If the die is fair, the six simple events $\{1\}$, $\{2\}$, $\{3\}$, $\{4\}$, $\{5\}$, and $\{6\}$ are equally likely. Therefore:

$$P(1) = P(2) = P(3) = P(4) = P(5) = P(6) = \frac{1}{6}.$$

If $\{a_1, a_2, a_3, \ldots, a_n\}$ is a sample space containing n equally likely outcomes, then the probability of each simple event is $\frac{1}{n}$:

$$P(a_1) = P(a_2) = P(a_3) = \cdots = P(a_n) = \frac{1}{n}.$$

Returning to the roll of a die, suppose A is the event that the outcome is an odd number. Event A is therefore the subset $\{1, 3, 5\}$ of the sample space $\{1, 2, 3, 4, 5, 6\}$. The probability of A is defined as the sum of the probabilities of the simple events that make up A. Therefore:

$$P(A) = P(1) + P(3) + P(5) = \frac{1}{6} + \frac{1}{6} + \frac{1}{6} = \frac{3}{6} = \frac{1}{2}.$$

If the sample space for an experiment consists of n equally likely outcomes, and if k of them are in event E, then:

$$P(E) = \underbrace{\frac{1}{n} + \frac{1}{n} + \cdots + \frac{1}{n}}_{k \text{ addends}} = \frac{k}{n}.$$

If the event E contains all elements of the sample space, then $P(E) = \frac{n}{n} = 1$; that is, the event is certain to occur. If event E contains no elements of the sample space, then $P(E) = \frac{0}{n} = 0$; that is, the event is certain *not* to occur.

Explorations p. 846

Teaching Suggestions p. T140

Reading Algebra p. T140

Suggested Extensions p. T140

Warm-Up Exercises

Write the fraction in lowest terms that represents what part of $\{1, 2, 3, 4, \ldots, 100\}$ each given subset is.

1. $\{2, 4, 6, 8, \ldots, 100\}$ $\frac{1}{2}$
2. $\{5, 10, 15, 20, \ldots, 100\}$ $\frac{1}{5}$
3. $\{1, 4, 9, 16, \ldots, 100\}$ $\frac{1}{10}$
4. {positive factors of 100} $\frac{9}{100}$
5. {prime numbers less than 100} $\frac{1}{4}$

Motivating the Lesson

Point out to the students that the world is full of uncertainty. From games of chance, which were the origin of the formal study of probability, to corporate sales strategies and decisions about military tactics, probability is widely applied. Basic concepts of probability are introduced in today's lesson.

Chalkboard Examples

1. A spinner is divided into tenths. The sections are numbered 1 to 10. If the spinner is spun, find the probability that the number is **(a)** 2, **(b)** a multiple of 4, **(c)** between 0 and 30, and **(d)** $\frac{1}{2}$.

   a. $\frac{1}{10}$ **b.** $P(4 \text{ or } 8) = \frac{2}{10} = \frac{1}{5}$ **c.** 1 **d.** 0

2. A bag contains 8 brown socks and 6 black socks. If two socks are randomly drawn, what is the probability that they match?

   $P(2 \text{ brown}) + P(2 \text{ black}) = \frac{{}_8C_2}{{}_{14}C_2} + \frac{{}_6C_2}{{}_{14}C_2} = \frac{28}{91} + \frac{15}{91} = \frac{43}{91}$

3. A machine cuts strips of paper with lengths that are normally distributed. The mean length is 6 inches with a standard deviation of $\frac{1}{16}$ in. What is the probability that a strip of paper is more than $6\frac{1}{8}$ in. long?

   The standardized value corresponding to $6\frac{1}{8}$ is $x = \frac{6\frac{1}{8} - 6}{\frac{1}{16}} = 2$. For a random length, l, $P\left(l > 6\frac{1}{8}\right) = P(x > 2) = 0.5 - A(2) = 0.5 - 0.4772 = 0.0228$.

Example 1 A die is rolled. Find the probability of each event.

a. Event A: The number showing is less than 5.
b. Event B: The number showing is between 2 and 6.

Solution The sample space is $\{1, 2, 3, 4, 5, 6\}$. Since the six outcomes are equally likely, the probability of each simple event is $\frac{1}{6}$.

a. Event $A = \{1, 2, 3, 4\}$

$\therefore P(A) = \frac{4}{6} = \frac{2}{3}$ ***Answer***

b. Event $B = \{3, 4, 5\}$

$\therefore P(B) = \frac{3}{6} = \frac{1}{2}$ ***Answer***

Example 2 Two dice are rolled and the numbers are noted. Find the probability of each event.

a. Event A: The sum of the numbers is less than 5.
b. Event B: The sum of the numbers is 4 or 5.

Solution Look at the figure on page 743. Notice that there are 36 possible outcomes when two dice are rolled.

a. Event $A = \{(1, 1), (1, 2), (1, 3), (2, 1), (2, 2), (3, 1)\}$

$\therefore P(A) = \frac{6}{36} = \frac{1}{6}$ ***Answer***

b. Event $B = \{(1, 3), (1, 4), (2, 2), (2, 3), (3, 1), (3, 2), (4, 1)\}$

$\therefore P(B) = \frac{7}{36}$ ***Answer***

Example 3 There are 12 tulip bulbs in a package. Nine will yield yellow tulips and three will yield red tulips. If two tulip bulbs are selected at random, find the probability of each event.

a. Event Q: Both tulips will be red.
b. Event R: One tulip will be yellow and the other red.

Solution **a.** Since the bulbs are selected at random, all possible pairs are equally likely. Thus:

$$P(Q) = \frac{\text{number of ways to pick a pair of red bulbs}}{\text{number of ways to pick any pair of bulbs}}$$

There are 3 red bulbs. The number of ways of picking 2 of them is:

$${}_3C_2 = \frac{3!}{2!1!} = 3.$$

There are 12 bulbs in all. The number of ways of picking 2 of them is:

$${}_{12}C_2 = \frac{12!}{2!10!} = 66.$$

$\therefore P(Q) = \frac{3}{66} = \frac{1}{22}$ ***Answer***

b. The number of ways of picking one yellow bulb and one red bulb is:

$${}_9C_1 \cdot {}_3C_1 = \frac{9!}{1!8!} \cdot \frac{3!}{1!2!} = 9 \cdot 3 = 27.$$

$\therefore P(R) = \frac{27}{66} = \frac{9}{22}$ ***Answer***

Consider a randomly chosen element of a normal distribution with mean M and standard deviation σ. The probability of this element falling between the values $M + a\sigma$ and $M + b\sigma$ is just the area under the standard normal curve between $x = a$ and $x = b$ (see Lesson 15-3).

Example 4 The lifetimes of 60-watt light bulbs are normally distributed with mean 1000 hours and standard deviation 100 hours. What is the probability that a bulb just put in a lamp will last more than 1100 hours?

Solution The diagram at the right shows the hours h of light bulb life and the standardized values x that correspond to them. The probability that $h > 1100$ is the same as the area under the standard normal curve to the right of $x = 1$. Therefore:

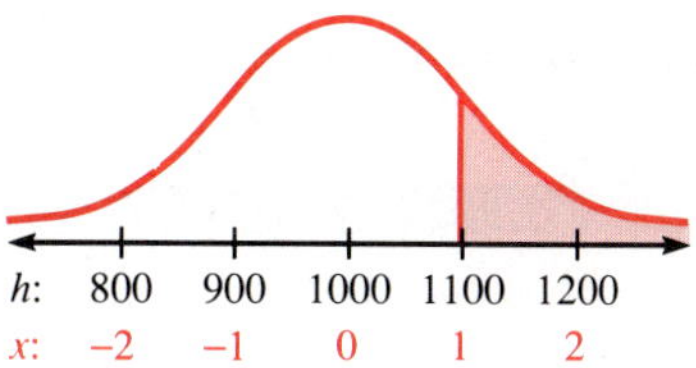

$$\begin{aligned} P(h > 1100) &= P(x > 1) \\ &= 0.5 - A(1) \\ &= 0.5 - 0.3413 \quad \text{Use the table on page 720.} \\ &= 0.1587 \end{aligned}$$

$\therefore$ the probability that the bulb lasts more than 1100 hours is about 0.16.

Example 5 For a certain airline, the trip times for daily flights between two cities are normally distributed. The mean trip time is 144 minutes with a standard deviation of 15 minutes. What is the probability that a flight takes less than 2 hours?

Solution The standardized value corresponding to 2 hours, or 120 minutes, is:

$$x = \frac{120 - M}{\sigma} = \frac{120 - 144}{15} = -\frac{24}{15} = -1.6.$$

Therefore, for a random trip time t:

$$\begin{aligned} P(t < 120) &= P(x < -1.6) \\ &= P(x > 1.6) \quad \text{The standard normal curve is symmetric with respect to the } y\text{-axis.} \\ &= 0.5 - A(1.6) \\ &= 0.5 - 0.4452 \quad \text{Use the table on page 720.} \\ &= 0.0548 \end{aligned}$$

$\therefore$ the probability that a flight takes less than 2 hours is about 0.05.

Check for Understanding

Here is a suggested use of the Oral Exercises to check students' understanding as you teach the lesson.

Oral Exs. 1–8: use after Example 1.

Oral Exs. 9–11: use after Example 2.

Guided Practice

1. A stack of 26 cards consists of the hearts and clubs from a bridge deck. A card is randomly drawn. Find the probability that it is **(a)** a king, **(b)** a diamond, **(c)** a club, **(d)** a 7 or an 8, and **(e)** a red ace.
   a. $\frac{1}{13}$ **b.** 0 **c.** $\frac{1}{2}$ **d.** $\frac{2}{13}$ **e.** $\frac{1}{26}$
2. Two coins are tossed. Find the probability that **(a)** both show heads, **(b)** the coins match, and **(c)** there is at least one tail.
   a. $\frac{1}{4}$ **b.** $\frac{1}{2}$ **c.** $\frac{3}{4}$
3. Two dice are rolled. Find the probability that **(a)** the sum is 10, **(b)** the sum is at least 10, and **(c)** exactly one die shows a 4.
   a. $\frac{1}{12}$ **b.** $\frac{1}{6}$ **c.** $\frac{5}{18}$

Summarizing the Lesson

In this lesson students learned to find the probability of an event. Ask students to explain the importance of the phrase "equally likely" in the boxed statements on page 745.

Suggested Assignments

Average Algebra
748/1–11
S 750/Mixed Review

Extended Algebra
748/1–11
S 750/Mixed Review

Supplementary Materials

Study Guide pp. 255–256
Overhead Visual 13

Oral Exercises

A number wheel is divided into 16 congruent sectors, numbered 1–16. You spin the wheel and record the number. Find the probability of each event.

1. It is a 7. $\frac{1}{16}$
2. It is an odd number. $\frac{1}{2}$
3. It is a 4 or a 9. $\frac{1}{8}$
4. It is a 1, 2, or 3. $\frac{3}{16}$
5. It is between 0 and 17. 1
6. It is less than 13. $\frac{3}{4}$
7. It is less than 3 or greater than 12. $\frac{3}{8}$
8. It is a 19. 0

Two dice are rolled. Find the probability of each event. (Use the figure on page 743 for the sample space of this experiment.)

9. The numbers showing on the two dice are equal. $\frac{1}{6}$
10. The sum of the numbers showing is 7. $\frac{1}{6}$
11. The sum of the numbers showing is 13. 0

Written Exercises

A

1. A box contains 10 slips of paper numbered 1–10. A slip of paper is drawn at random from the box and the number is noted. Find the probability of each event.
   a. It is a 2. $\frac{1}{10}$
   b. It is an odd number. $\frac{1}{2}$
   c. It is less than 4. $\frac{3}{10}$
   d. It is less than 11. 1
   e. It is greater than 8. $\frac{1}{5}$
   f. It is between 3 and 4. 0
2. A letter is selected at random from those in the word TRIANGLE. Find the probability of each event.
   a. It is a vowel. $\frac{3}{8}$
   b. It is a consonant. $\frac{5}{8}$
   c. It is from the first half of the alphabet. $\frac{5}{8}$
   d. It is between F and Q in the alphabet. $\frac{1}{2}$
3. One marble is drawn at random from a bag containing 4 white, 6 red, and 6 green marbles. Find the probability of each event.
   a. It is white. $\frac{1}{4}$
   b. It is white, red, or green. 1
   c. It is red or green. $\frac{3}{4}$
   d. It is not red. $\frac{5}{8}$
4. One card is drawn at random from a 52-card deck. Find the probability of each event.
   a. It is an ace. $\frac{1}{13}$
   b. It is a diamond. $\frac{1}{4}$
   c. It is black. $\frac{1}{2}$
   d. It is the king of clubs. $\frac{1}{52}$
   e. It is a red queen. $\frac{1}{26}$
   f. It is a black heart. 0
5. Two coins are tossed. Find the probability of each event.
   a. Both come up tails. $\frac{1}{4}$
   b. At least one coin comes up heads. $\frac{3}{4}$
   c. The coins match. $\frac{1}{2}$
   d. The coins don't match. $\frac{1}{2}$

6. A 10-speed bicycle is given as a door prize. A total of 220 tickets numbered 1–220 are sold. If the winning number is chosen at random, find the probability of each event.
   a. The winning number is less than 101. $\frac{5}{11}$
   b. The winning number is greater than 50. $\frac{17}{22}$
   c. The winning number is between 10 and 21. $\frac{1}{22}$
7. Of 1,000,000 income tax returns received, all are checked for arithmetic accuracy and 9000 returns, selected at random, are checked thoroughly. For a given return, find the probability of each event.
   a. It is checked for arithmetic accuracy. 1
   b. It is checked thoroughly. $\frac{9}{1000}$
   c. It is not checked thoroughly. $\frac{991}{1000}$
8. Two dice are rolled. Find the probability of each event.
   a. The sum of the numbers showing on the dice is 10. $\frac{1}{12}$
   b. The sum is at least 10. $\frac{1}{6}$
   c. Exactly one die shows a 4. $\frac{5}{18}$
   d. At least one die shows a 4. $\frac{11}{36}$

B 9. Three coins are tossed. Find the probability of each event.
   a. All come up heads. $\frac{1}{8}$
   b. All come up tails. $\frac{1}{8}$
   c. At least one comes up tails. $\frac{7}{8}$
   d. Exactly two come up heads. $\frac{3}{8}$
10. Evangeline makes necklaces from beads of four different colors: red, white, yellow, and turquoise. There are twice as many red beads as white beads, and twice as many white beads as yellow beads or turquoise beads. If Evangeline chooses a bead at random from a box containing all the beads, what is the probability she chooses:
   a. a red bead? $\frac{1}{2}$
   b. a white bead? $\frac{1}{4}$
   c. either a yellow bead or a turquoise bead? $\frac{1}{4}$

11. A bag contains 2 red, 4 yellow, and 6 blue marbles. Two marbles are drawn at random. Find the probability of each event.
   a. Both are red. $\frac{1}{66}$
   b. Both are yellow. $\frac{1}{11}$
   c. Both are blue. $\frac{5}{22}$
   d. One is red and one is yellow. $\frac{4}{33}$
   e. Neither is red. $\frac{15}{22}$
   f. Neither is blue. $\frac{5}{22}$
12. Two cards are drawn at random from a 52-card deck. Find the probability of each event.
   a. Both are hearts. $\frac{1}{17}$
   b. Both are jacks. $\frac{1}{221}$
   c. Neither is red. $\frac{25}{102}$
   d. Neither is a spade. $\frac{19}{34}$

Additional Answers Mixed Review Exercises

1. 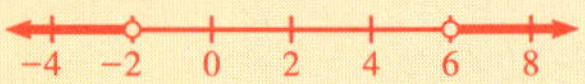

2. 

3. 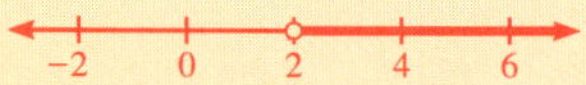

4. 

5. 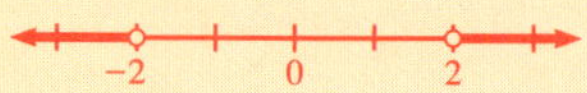

6. 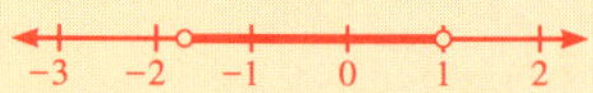

Using a Computer

Simulation is an activity that students enjoy. By using the random number generator and loops, students will write programs to simulate flipping coins and rolling dice. Interested students may want to write programs for other simulations once they see the basic idea involved.

Explain to students the meaning of the phrase "in relative terms" used in Exercises 3 and 6. This means that the absolute value of the difference between the actual number and expected number of heads or doubles should be divided by the expected number. For example, in 20 tosses of a coin, 10 heads are expected; if the computer obtains 8 heads, then the *absolute* difference is 2 but the *relative* difference is $\frac{2}{10}$, or 20%.

In Exercises 13 and 14, use the table on page 720.

13. For a certain math class, the mean time required to complete homework assignments is 36 minutes with a standard deviation of 10 minutes. Assuming that the completion times are normally distributed, what is the probability that the time required to complete a given homework assignment in this class will be:

a. less than 20 minutes? 0.0548 **b.** more than one hour? 0.0082

14. A certain brand of cereal claims that the mean number of raisins in each box is 74 with a standard deviation of 5. Assuming that the raisins are normally distributed, what is the probability that a given box of this cereal will contain:

a. fewer than 65 raisins? 0.0359 **b.** more than 80 raisins? 0.1151

Mixed Review Exercises

Solve each inequality and graph the solution set.

1. $|t - 2| > 4$ $\{t: t < -2 \text{ or } t > 6\}$
2. $x^2 - x \le 2$ $\{x: -1 \le x \le 2\}$
3. $4 - 3n < -2$ $\{n: n > 2\}$
4. $-3 \le 2x - 3 \le 5$ $\{x: 0 \le x \le 4\}$
5. $4y^2 > 16$ $\{y: y < -2 \text{ or } y > 2\}$
6. $|3k + 1| < 4$ $\left\{k: -\frac{5}{3} < k < 1\right\}$

7. How many permutations are there of all the letters in PAPER? 60

8. In how many ways can a study group of four be selected from a class of 24 students? 10,626

Computer Exercises

For students with some programming experience.

Answers will vary.

1. Write a program that will use the computer's random number generator (RND) to simulate the tossing of a coin a given number of times. Have the computer print the number of times that heads come up.

2. Run the program in Exercise 1 for each number of tosses and record the number of heads. **a.** 20 **b.** 80 **c.** 200 **d.** 800

3. When a coin is tossed a number of times, the *expected* number of heads is half the number of tosses. Which of the runs in Exercise 2 comes closest, in relative terms, to the expected number of heads?

4. Write a program that will use the RND function to simulate rolling two dice a given number of times. Have the computer print the number of "doubles" (that is, when the numbers showing on the dice are the same).

5. Run the program in Exercise 4 for each number of rolls and record the number of doubles. **a.** 12 **b.** 72 **c.** 120 **d.** 720

6. When two dice are rolled a number of times, the *expected* number of doubles is one sixth the number of rolls. Which of the runs in Exercise 5 comes closest, in relative terms, to the expected number of doubles?

Reading Algebra / *Probability*

When English words are used as technical terms in algebra they often have a meaning different from their usual meaning in the language. It is particularly important in reading about probability and statistics to distinguish the mathematical usage of a term from its common use, since many of the terms have mathematical meanings that are close to, but not identical to, their meanings in common speech. In the brief lesson that follows, you will be introduced to the concept of *expectation*. As you read the lesson and do the exercises, notice how the meaning of expectation in the study of probability is related to its ordinary meaning.

Often each simple event in a sample space can be measured by a number or score. For example, if you toss two coins you can count the number of heads as the score; it will be 0, 1, or 2. The *expectation* of the score is found by multiplying the score of each simple event by the probability of that event and adding up these products. Here is how the computation looks for the coin tossing experiment:

Event				
Coin 1	Coin 2	Score	Probability	Product
H	H	2	$\frac{1}{4}$	$\frac{1}{2}$
H	T	1	$\frac{1}{4}$	$\frac{1}{4}$
T	H	1	$\frac{1}{4}$	$\frac{1}{4}$
T	T	0	$\frac{1}{4}$	0

The expectation is $\frac{1}{2} + \frac{1}{4} + \frac{1}{4} + 0$, or 1. Therefore, the *expected* number of heads is 1.

Exercises

1. If you toss four coins, what is the expected number of heads? 2
2. What is the probability that you get two heads on a toss of four coins? $\frac{3}{8}$
3. If you repeat the experiment of tossing four coins many times, will you score two heads more than half the time or less than half the time? Less than half
4. When you roll a die with faces numbered 1 to 6, what is your expected score? Will you ever get this score? $3\frac{1}{2}$; no
5. Suppose you roll the die many times and find the average (mean) of all your scores. Can you predict what this average will approximately be? $3\frac{1}{2}$
6. In what sense does the expectation tell you what to expect? Expectation is an average score.

Additional Answers
Written Exercises for p. 744

13. {(1, 1), (1, 2), (1, 3), (1, 4), (1, 5), (1, 6), (2, 1), (2, 2), (2, 3), (2, 4), (2, 5), (2, 6), (3, 5), (3, 6), (4, 5), (4, 6), (5, 5), (5, 6), (6, 5), (6, 6)}

14. {(1, 1), (1, 3), (2, 2), (3, 1), (3, 3), (4, 4), (5, 5), (6, 6)}

Using a Computer

Statistics are commonly used both to make claims and to dispute them. For example, a light bulb manufacturer might claim that for a certain type of bulb the average life is 1000 hours. A consumer group, on the other hand, may question that claim after receiving complaints about such bulbs "burning out" in much shorter periods of time.

Should the manufacturer's claim be accepted or rejected? To answer this question, a researcher for the consumer group must perform an experiment: With the manufacturer's claim serving as the *hypothesis* of the experiment, the researcher gathers a sample of the manufacturer's bulbs and determines their average life.

Suppose the average life of the sample is 850 hours. Although this average is certainly less than the manufacturer's claim, the researcher cannot reject the claim outright: If different samples had been taken, their averages would inevitably vary. Perhaps the researcher's sample simply fell at the low end of the conceivable variations.

The researcher must now determine whether or not the difference between the experimental average and the hypothetical average occurred "by chance." In other words, the researcher needs to know the *probability* of obtaining a sample average of 850 when the population average is 1000.

If the probability turns out to be very low (say, less than 0.05), the result of the experiment is considered to

Application / *Sampling*

A factory produces 200,000 light bulbs during each 8-hour shift. The factory managers want to know how many defective bulbs are being produced. If the number is too high, production methods must be changed to improve the output. One way to determine the number of defective bulbs is to test every one produced. Although this would give the managers the information they need, it would also be too expensive and time-consuming.

Another method is to test only a *sample* of the bulbs to get some idea of what the entire *population* of light bulbs is like. Testing only a relatively small sample is certainly less expensive than testing the entire population, but can the managers of the factory have confidence in the results? That is, do the results in a sample provide a good estimate of the results in the population as a whole? Studies have shown that the answer to this question is: It depends on how samples are drawn from the population.

The best estimates are obtained by drawing random samples. A sample from a population is a *random sample* provided:

1. Each element in the population is equally likely to be in the sample.
2. The selection of each element is independent of the selection of every other element.

It is not always easy to obtain a random sample, however. For example, if each element in a population is to be selected on an "equally likely" basis, then all elements must be known and available. This is usually not the case with large populations, such as all the eligible voters in the United States.

Because random samples are difficult to obtain, nonrandom samples are often used to characterize populations. Here are a few alternative sampling procedures.

Convenience sampling:	Pick the elements in the population that are readily at hand.
Self-selected sampling:	Allow elements in the population to volunteer to be in the sample.
Judgment sampling:	Have an expert choose a sample believed to be representative of the population.

Compared to random sampling, these and other sampling procedures often yield less dependable results. They may, however, be sufficient for whatever purpose the population is being sampled.

Exercises

Which of the following samples is random? If nonrandom, explain why.

1. A random-number generator activates a device that tests light bulbs coming off the assembly line.

a. Testing is done throughout the 8-hour shift each day. Random

b. Testing is done only during the first 2 hours of production each day. Not random

2. A 20-student committee is to be selected from a student body of 500.
   a. Twenty names are drawn from a box containing all the students' names, each written on a separate slip of paper. Random
   b. One name is selected at random from a roster of all the students. The remaining selections are made by picking every 25th name on the roster following the first one selected. Not random
3. A TV-talk-show host lets members of the audience who raise their hands voice their opinions on an issue. Not random
4. A newspaper reporter asks passersby their opinions on an issue. Not random

Describe how samples could be obtained in each of the following situations. Also describe any limitations of the sampling techniques that you propose. Answers will vary.

5. The editorial staff of a magazine needs to know how the subscribers would react to a change in the layout of the magazine.
6. A national news organization wants to know how the current national economy is affecting typical American families.
7. A researcher wants to investigate whether optimists are generally more healthy than pessimists.
8. A school superintendent wants to know whether the voters in the school district will support a bond issue to build a new school.

Extra / *Random Numbers*

Computers use arithmetic to generate sequences of ''random'' numbers that can be used to simulate experiments such as the tossing of a coin. Since these sequences are completely determined by the first number chosen, the numbers are not truly random (in fact, they are often called *pseudo*random numbers), but they can still be used to program realistic simulations.

One common method for generating random numbers is illustrated at the right. First choose a multiplier, 23, and a starting value, 809. Then repeat the following steps: Multiply by 23, save the last three digits, and record the hundreds' digit. The sequence of random numbers is 6, 9, 1, 3, 4, ...

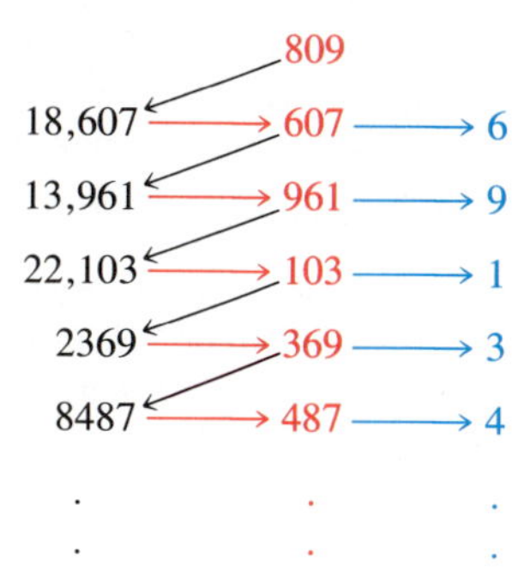

Notice that the rightmost two digits must be discarded, since the units' digit repeats the pattern 7, 1, 3, 9, and the tens' digit is always even. Since the last three digits are saved from each product, the sequence of ''random'' digits must repeat after at most 1000 steps. Actual computer routines save many more digits, and they provide a sequence of 7- or 8-digit pseudorandom numbers that will repeat only after millions of numbers have been drawn.

be *statistically significant:* The researcher can reasonably reject the hypothesis. Otherwise the hypothesis stands.

Many research problems are "resolved" by forming hypotheses and statistically testing them. The program "Sampling Experiment" on the Disk for Algebra can be used to explore this activity further.

Support Materials
Disk for Algebra
Menu Item: Sampling Experiment
Resource Book p. 252

Teaching Suggestions p. T141

Group Activities p. T141

Suggested Extensions p. T141

Warm-Up Exercises

Consider the experiment of rolling a die. Give an example of events A and B that satisfy the given requirements. Answers will vary.

1. $P(A) = 0$ A is the event that the die shows a number greater than 6.
2. $P(A) = 0.5$ A is the event that the die shows an even number.
3. $P(A) = 1$ A is the event that the die shows a number from 1 to 6.
4. $P(A \text{ and } B) = 0$ A is the event that the die shows an even number, and B is the event that the die shows an odd number.
5. $P(A \text{ or } B) \neq P(A) + P(B)$ A is the event that the die shows an even number, and B is the event that the die shows a multiple of 3.

Motivating the Lesson

Present to the students the "birthday problem": What is the minimum number of people needed before the probability of any two of them having the same birthday is at least 0.5? (The answer is 23.) The solution of the "birthday problem" involves mutually exclusive and independent events, topics in today's lesson.

15-10 Mutually Exclusive and Independent Events

Objective To identify mutually exclusive and independent events and find the probability of such events.

Consider the sets A and B where

$$A = \{1, 2, 3, 4\} \text{ and } B = \{3, 4, 5, 6\}.$$

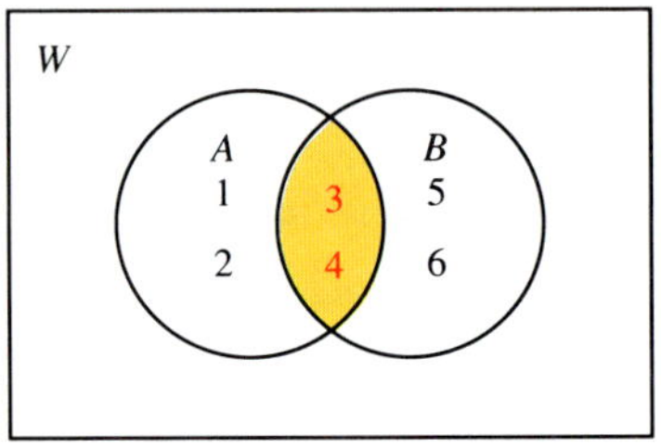

Notice that these sets have two elements in common, namely 3 and 4. This fact can be illustrated through the use of a diagram like the one at the right. (Such diagrams are called *Venn diagrams* in honor of the English mathematician John Venn, who used them extensively to picture sets and set relations.)

In the Venn diagram, sets A and B are shown as subsets of the set W of whole numbers. Notice that the region where sets A and B overlap is shaded. The shaded region represents $\{3, 4\}$, the set of elements that belong to *both* A and B. This set is called the *intersection* of A and B. You can use the symbol $\cap$ to indicate an intersection of sets:

$$\{1, 2, 3, 4\} \cap \{3, 4, 5, 6\} = \{3, 4\}.$$

In general, if A and B are any sets, then the set whose members are the elements belonging to both A and B is called the **intersection** of A and B and is denoted by $A \cap B$.

The sets $\{1, 2, 3, 4\}$ and $\{5, 6, 7, 8\}$ have no elements in common. That is,

$$\{1, 2, 3, 4\} \cap \{5, 6, 7, 8\} = \emptyset.$$

Sets that have no elements in common are called **disjoint** sets.

In the Venn diagram at the right, notice that the shading extends throughout sets A and B. The shaded region represents $\{1, 2, 3, 4, 5, 6\}$, the set containing all the elements of A together with all the elements of B. In other words, the shaded region represents the set consisting of all the elements in *at least one* of the sets A and B. This set is called the *union* of A and B. You can use the symbol $\cup$ to indicate a union of sets:

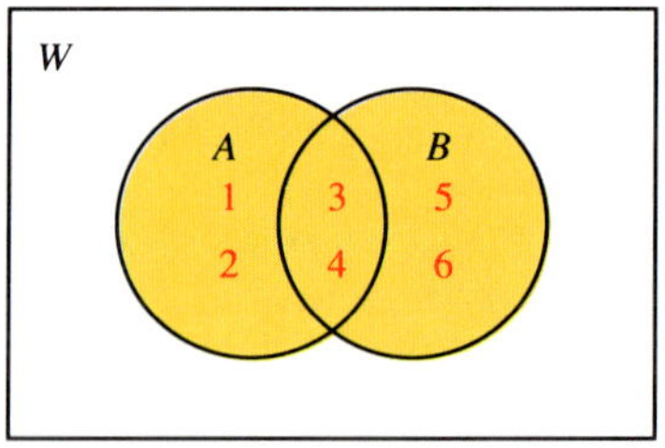

$$\{1, 2, 3, 4\} \cup \{3, 4, 5, 6\} = \{1, 2, 3, 4, 5, 6\}.$$

In general, if A and B are any sets, then the set whose members are the elements belonging to either A or B (or both) is called the **union** of A and B and is denoted by $A \cup B$.

Example 1 Specify each of the following sets by listing its elements. Use the Venn diagram at the right.

a. $A \cap B$ **b.** $A \cup B$
c. $B \cap C$ **d.** $B \cup C$
e. $A \cap D$ **f.** $(A \cup B) \cap C$

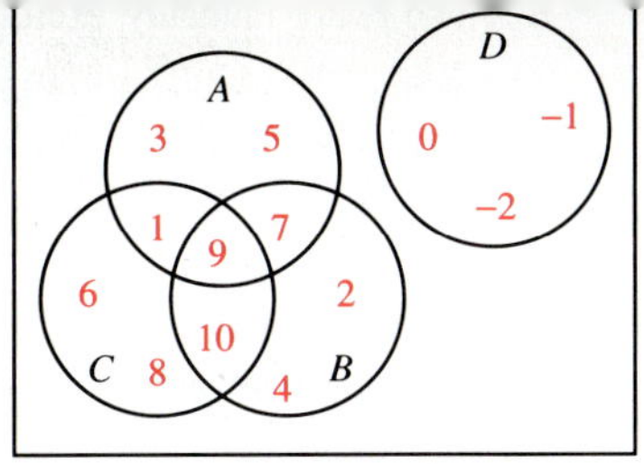

Solution

a. $\{7, 9\}$
b. $\{1, 2, 3, 4, 5, 7, 9, 10\}$
c. $\{9, 10\}$
d. $\{1, 2, 4, 6, 7, 8, 9, 10\}$
e. $\emptyset$
f. $\{1, 9, 10\}$

The Venn diagram below shows the sample space S for the experiment of drawing a number at random from $\{1, 2, 3, 4, 5, 6\}$. Every subset of S is an event, including S itself and the empty set, $\emptyset$. These two events have probabilities 1 and 0, respectively.

For any sample space S,

$$\mathbf{P(S) = 1} \quad \text{and} \quad \mathbf{P(\emptyset) = 0.}$$

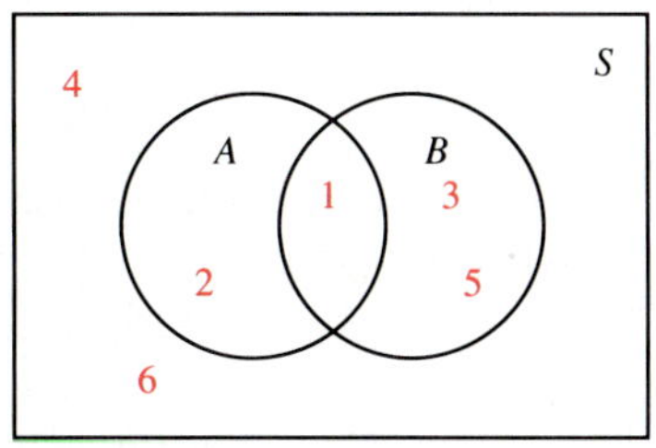

The Venn diagram also shows events A and B in S. Event A, $\{1, 2\}$, is the drawing of a number less than 3. Event B, $\{1, 3, 5\}$, is the drawing of an odd number. Note that

$$P(A) = \tfrac{2}{6} \text{ and } P(B) = \tfrac{3}{6}.$$

The probability that either A or B occurs is $P(A \cup B)$. Since $A \cup B = \{1, 2, 3, 5\}$,

$$P(A \cup B) = \tfrac{4}{6}.$$

To see how $P(A \cup B)$, $P(A)$, and $P(B)$ are related, first note that $P(A \cap B) = \tfrac{1}{6}$, since $A \cap B = \{1\}$. Then:

$$P(A \cup B) = \frac{4}{6} = \frac{2}{6} + \frac{3}{6} - \frac{1}{6} = P(A) + P(B) - P(A \cap B).$$

For any two events A and B in a sample space,

$$\mathbf{P(A \cup B) = P(A) + P(B) - P(A \cap B).}$$

If two events have no elements in common (that is, if their intersection is empty), the events are called **mutually exclusive.** For such events, $P(A \cap B) = P(\emptyset) = 0$. Thus, $P(A \cup B) = P(A) + P(B) - 0 = P(A) + P(B)$.

Chalkboard Examples

1. Specify each of the following sets by roster, referring to the Venn diagram below.

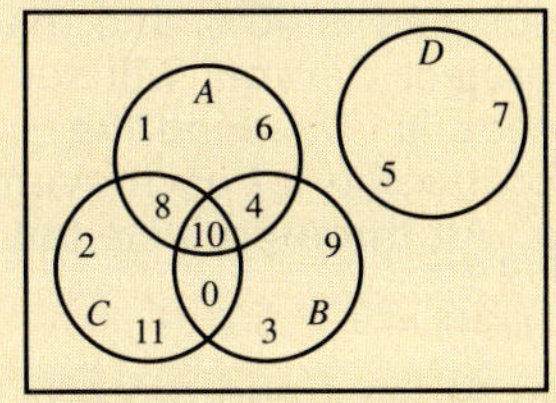

a. $A \cap C$ **b.** $A \cup C$
c. $B \cap C$ **d.** $B \cup C$
e. $B \cap D$ **f.** $(A \cup B) \cap C$
a. $\{8, 10\}$
b. $\{0, 1, 2, 4, 6, 8, 10, 11\}$
c. $\{0, 10\}$
d. $\{0, 2, 3, 4, 8, 9, 10, 11\}$
e. $\emptyset$ **f.** $\{0, 8, 10\}$

2. From a choice of 5 novels and 5 biographies, 3 books will be selected. What is the probability that at least two are novels? $\frac{5}{12}$

3. A bag contains 6 United States pennies and 8 Canadian pennies. A coin is drawn, not replaced, and another coin is drawn. A is the event that the first coin is Canadian; B is the event that the second coin is Canadian. Are the events dependent or independent? dependent

(continued)

(continued)

4. The probability that project Alpha will succeed is 60%, that project Beta will succeed is 45%, and that project Gamma will succeed is 70%. Find the probability of each event.
 a. All three projects fail. $0.4(0.55)(0.3) = 0.066$; 6.6%
 b. At least two projects succeed.
 $P(\alpha \text{ fails}) = 0.4(0.45)(0.7) = 0.126$
 $P(\beta \text{ fails}) = 0.6(0.55)(0.7) = 0.231$
 $P(\gamma \text{ fails}) = 0.6(0.45)(0.3) = 0.081$
 $P(\text{all succeed}) = 0.6(0.45)(0.7) = 0.189$
 $0.126 + 0.231 + 0.081 + 0.189 = 0.627$
 62.7%

Check for Understanding

Here is a suggested use of the Oral Exercises to check students' understanding as you teach the lesson.
Oral Exs. 1–3: use before Example 2.
Oral Exs. 4–5: use after Example 3.

If A and B are mutually exclusive events,

$$P(A \cup B) = P(A) + P(B).$$

Example 2 From a group of 10 seniors and 8 juniors, 3 students are to be selected at random to form a committee. What is the probability that at least 2 seniors are selected?

Solution The selection will include at least 2 seniors if either exactly 2 seniors and 1 junior are selected (event M) *or* 3 seniors are selected (event N). Since the two events are mutually exclusive, $P(M \cup N) = P(M) + P(N)$.

$$P(M) = \frac{{}_{10}C_2 \cdot {}_8C_1}{{}_{18}C_3} = \frac{45 \cdot 8}{816} = \frac{360}{816} = \frac{15}{34}$$

$$P(N) = \frac{{}_{10}C_3}{{}_{18}C_3} = \frac{120}{816} = \frac{5}{34}$$

$$P(M \cup N) = \frac{15}{34} + \frac{5}{34} = \frac{20}{34} = \frac{10}{17}$$

$\therefore$ the probability of at least 2 seniors being selected is $\frac{10}{17}$. ***Answer***

Consider the experiment of tossing a coin twice. The sample space for the experiment is $\{(H, H), (H, T), (T, H), (T, T)\}$. Let A be the event that a head is obtained on the first toss. Let B be the event that a head is obtained on the second toss. Then:

$$A = \{(H, H), (H, T)\} \text{ and } B = \{(H, H), (T, H)\}.$$

Since the experiment has four possible outcomes, the probabilities of the events are:

$$P(A) = \frac{2}{4} = \frac{1}{2} \text{ and } P(B) = \frac{2}{4} = \frac{1}{2}.$$

Notice that A and B are not mutually exclusive, since $A \cap B = \{(H, H)\}$. However, the two events are *independent,* because obtaining a head on the first toss has nothing to do with obtaining a head on the second toss. In general, if the occurrence of one event has no effect on the probability of another event, the two events are said to be **independent.**

Since $P(A \cap B) = \frac{1}{4}$, you can see the following pattern:

$$P(A) \cdot P(B) = P(A \cap B)$$

$$\frac{1}{2} \cdot \frac{1}{2} = \frac{1}{4}$$

Two events A and B are independent if and only if:

$$P(A \cap B) = P(A) \cdot P(B).$$

Example 3 A bag contains 4 green and 4 blue buttons. Two buttons are drawn at random from the bag as follows:

a. The first button drawn is put back into the bag before the second button is drawn.

b. The first button drawn is *not* put back into the bag before the second button is drawn.

Let A be the event that the first button drawn is green.
Let B be the event that the second button drawn is blue.
Are the events in each case dependent or independent?

Solution **a.** $P(A) = \frac{4}{8} = \frac{1}{2}$

$P(B) = \frac{4}{8} = \frac{1}{2}$

$P(A \cap B) = \frac{{}_4C_1 \cdot {}_4C_1}{{}_8C_1 \cdot {}_8C_1} = \frac{4 \cdot 4}{8 \cdot 8} = \frac{16}{64} = \frac{1}{4}$

$P(A) \cdot P(B) = \frac{1}{2} \cdot \frac{1}{2} = \frac{1}{4} = P(A \cap B)$

$\therefore$ events A and B are independent. (Since the first button drawn is put back, the result of the first draw does *not* affect the outcome of the second draw.) ***Answer***

b. $P(A) = \frac{1}{2}$

Event B can come about in two mutually exclusive ways: Either you draw a green button first and then draw one of the four blue buttons among the seven remaining, or you draw a blue button first and then draw one of the other three blue buttons among the seven remaining. Therefore:

$$P(B) = P(\text{first green, then blue}) + P(\text{first blue, then blue})$$
$$= \frac{4}{8} \cdot \frac{4}{7} + \frac{4}{8} \cdot \frac{3}{7} = \frac{28}{56} = \frac{1}{2}$$

The event $A \cap B$ is equivalent to drawing a green button first and then a blue one. That is,

$$P(A \cap B) = P(\text{first green, then blue}) = \frac{4}{8} \cdot \frac{4}{7} = \frac{2}{7}.$$

$$P(A) \cdot P(B) = \frac{1}{2} \cdot \frac{1}{2} = \frac{1}{4} \neq P(A \cap B)$$

$\therefore$ events A and B are dependent. (In this case, the result of the first draw does affect the second draw.) ***Answer***

Suppose an event A is a subset of a sample space S. It is sometimes useful to consider the **complement** of A, which consists of the elements in S that are *not* members of A. The symbol for the complement of an event A is $\overline{A}$. Recall that $P(S) = 1$. Since $A \cup \overline{A} = S$, $P(A \cup \overline{A}) = 1$. Since A and $\overline{A}$ are mutually exclusive, $P(A \cup \overline{A}) = P(A) + P(\overline{A})$. Therefore, $P(A) + P(\overline{A}) = 1$ and $P(\overline{A}) = 1 - P(A)$.

Guided Practice

The grades in a college math course are given below.

Grades	Graduate	Under-Graduate
A	2	0
B	5	4
C	8	8
D	2	2
F	3	2

A student is selected at random from the class.

1. Find each probability.
a. The student is a graduate student. $\frac{20}{36} = \frac{5}{9}$
b. The student received an A or a B. $\frac{11}{36}$

2. Name two mutually exclusive events.
For example, the student chosen is an undergraduate and the student chosen received an A.

3. Tell whether each pair of events is independent.
a. The student is an undergraduate; the student received an F. no
b. The student is a graduate student; the student received a B. yes
c. The student is an undergraduate; the student received an A, a B, or a C. yes

Summarizing the Lesson

In this lesson students learned to find the probability of mutually exclusive and independent events. Ask students to explain what the terms "mutually exclusive" and "independent" mean.

Suggested Assignments

Average Algebra
759/1–15 odd
R 761/Self-Test 3

Extended Algebra
Day 1: 759/1–8
S 750/13, 14
Day 2: 760/9–13, 16, 17
R 761/Self-Test 3

Supplementary Materials

Study Guide	pp. 257–258
Practice Master	87
Test Master	65
Computer Activity	34
Resource Book	p. 170

Example 4 Based on past performances in the school band, the probability that Ryan will be selected to perform in the county band is $\frac{3}{4}$, that Faye will be selected is $\frac{2}{3}$, and that Chung will be selected is $\frac{1}{2}$. Suppose that the selection of one student does not affect another student's chances. Find the probabilities of each of the following events.

a. Ryan and Chung are selected, but Faye is not.
b. At least one of the three is selected.

Solution Let R be the event that Ryan is selected for the county band, F the event that Faye is selected, and C the event that Chung is selected. Then:

$$P(R) = \frac{3}{4},\ P(F) = \frac{2}{3},\ \text{and } P(C) = \frac{1}{2}.$$

a. The probability that Faye is *not* selected is:

$$P(\overline{F}) = 1 - P(F) = 1 - \frac{2}{3} = \frac{1}{3}.$$

$$\therefore P(R) \cdot P(C) \cdot P(\overline{F}) = \frac{3}{4} \cdot \frac{1}{2} \cdot \frac{1}{3} = \frac{1}{8} \quad \textbf{\textit{Answer}}$$

b. Let D be the event that *at least one* of the three students is selected, and let E be the event that *none* of them is selected. Note that $D = \overline{E}$.

$$\begin{aligned} P(D) &= P(\overline{E}) \\ &= 1 - P(E) \\ &= 1 - P(\overline{R}) \cdot P(\overline{F}) \cdot P(\overline{C}) \quad \left\{ E \text{ is equivalent to } \overline{R}, \overline{F}, \text{ and } \overline{C} \text{ occurring together.} \right. \\ &= 1 - \frac{1}{4} \cdot \frac{1}{3} \cdot \frac{1}{2} \\ &= 1 - \frac{1}{24} = \frac{23}{24} \end{aligned}$$

$\therefore$ the probability of at least one being selected is $\frac{23}{24}$. ***Answer***

Oral Exercises

In Exercises 1–3, use the Venn diagram at the right. 1. a. {a, c, d, e, f, g, h}

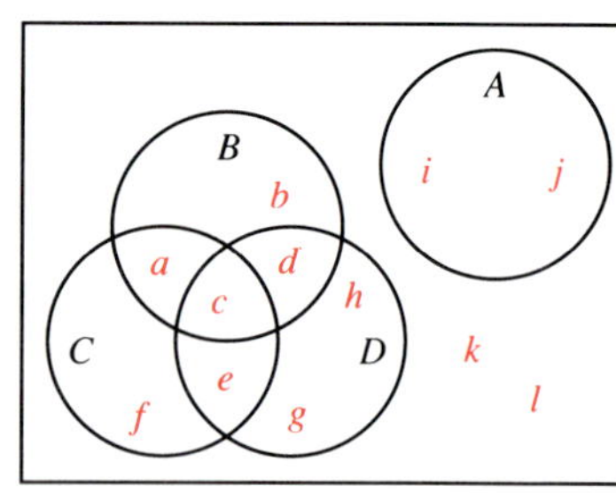

1. Specify each of the following sets by listing its elements.
   a. $C \cup D$ (See above.)
   b. $B \cap D$ {c, d}
   c. $B \cap C$ {a, c}
   d. $(B \cup C) \cap A$ ∅
2. Find the probability of each event.
   a. B $\frac{1}{3}$
   b. $\overline{A}$ $\frac{5}{6}$
   c. $C \cap D$ $\frac{1}{6}$
   d. $A \cup B$ $\frac{1}{2}$
3. Tell whether the events are mutually exclusive.
   a. A and B Yes
   b. B and C No
   c. B and D No
   d. $(B \cup D)$ and A Yes
   e. $(A \cup B)$ and D No
   f. $\overline{A}$ and B No

pendent events.

a. $P(A) = \frac{1}{16}$; $P(B) = \frac{1}{4}$; $P(A \cap B) = \frac{1}{4}$ Dependent

b. $P(A) = \frac{1}{3}$; $P(B) = \frac{3}{4}$; $P(A \cap B) = \frac{1}{4}$ Independent

c. $P(A) = 0.3$; $P(B) = 0.7$; $P(A \cap B) = 0.05$ Dependent

d. $P(A) = 0.4$; $P(B) = 0.5$; $P(A \cap B) = 0.2$ Independent

5. Identify the events in each problem as independent or dependent, and then find the probability requested.

a. In a box there are 3 pink and 3 green erasers. You pick one at random, replace it, and select another at random. What is the probability that both erasers are green? Independent; $\frac{1}{4}$

b. There are two 60-watt light bulbs and two 75-watt light bulbs in a box. You pick one at random and place it in your study lamp. Then you select another at random. What is the probability that you first picked a 75-watt and then a 60-watt light bulb? Dependent; $\frac{1}{3}$

Written Exercises

A

1. Tanya randomly guesses a whole number from 1 to 10. Find the probability of each event.

a. She guesses a number less than 6. $\frac{1}{2}$

b. She guesses an odd number. $\frac{1}{2}$

c. She guesses an odd number less than 6. $\frac{3}{10}$

2. A single marble is drawn from a bag containing 3 red, 5 white, and 2 blue marbles. Find the probability of each event.

a. A red or blue marble is drawn. $\frac{1}{2}$

b. A blue or white marble is drawn. $\frac{7}{10}$

c. A red, white, or blue marble is drawn. 1

3. The names of 3 seniors, 4 juniors, and 5 sophomores are placed in a bowl. One name is drawn at random, set aside, and a second name is then drawn at random. Find the probability of each event.

a. The first name drawn is a junior and the second is a senior. $\frac{1}{11}$

b. Both names drawn are sophomores. $\frac{5}{33}$

4. There are 3 red, 2 blue, and 3 yellow crayons in a box. Jeff randomly selects one, returns it to the box, and then randomly selects another. Find the probability of each event.

a. The first crayon selected is blue and the second is yellow. $\frac{3}{32}$

b. Both crayons selected are red. $\frac{9}{64}$

5. A red and a green die are rolled. Let A be the event that the red die shows 2 and B be the event that the sum of the numbers showing is 8.

a. Find the probability of A, B, $A \cup B$, and $A \cap B$. $\frac{1}{6}$; $\frac{5}{36}$; $\frac{5}{18}$; $\frac{1}{36}$

b. Are A and B independent events? No

Written Exercises

6. a. {(1, *H*), (2, *H*), (3, *H*), (4, *H*), (5, *H*), (6, *H*), (1, *T*), (2, *T*), (3, *T*), (4, *T*), (5, *T*), (6, *T*)}

b. $A = \{(5, H), (6, H), (5, T), (6, T)\}$
$B = \{(1, H), (2, H), (3, H), (4, H), (5, H), (6, H)\}$
$A \cup B = \{(1, H), (2, H), (3, H), (4, H), (5, H), (6, H), (5, T), (6, T)\}$
$A \cap B = \{(5, H), (6, H)\}$

6, and let B be the event that the coin shows heads.

a. Specify the sample space for the experiment.
b. Specify the simple events in A, B, $A \cup B$, and $A \cap B$.
c. Find the probability of A, B, $A \cup B$, and $A \cap B$. $\frac{1}{3}$; $\frac{1}{2}$; $\frac{2}{3}$; $\frac{1}{6}$
d. Are A and B mutually exclusive? Are they independent? No; yes

7. Two dice are rolled. Find the probability of each event.

a. The sum of the numbers showing is 10. $\frac{1}{12}$
b. Either the sum of the numbers showing is 4 or both numbers are 4. $\frac{1}{9}$

8. Two cards are drawn at the same time from a 52-card deck. Find the probability of each event.

a. Both cards are jacks. $\frac{1}{221}$
b. Both cards are sixes. $\frac{1}{221}$
c. Either both cards are jacks or both are sixes. $\frac{2}{221}$

9. A coin is tossed three times. Find the probability of each event.

a. At least two tosses come up tails. $\frac{1}{2}$
b. At least one toss comes up heads. $\frac{7}{8}$

10. From a 52-card deck a card is drawn and then replaced. After the deck is shuffled, a second card is drawn. Find the probability of each event.

a. The first card is a 5 and the second is a 6. $\frac{1}{169}$
b. Both cards are clubs. $\frac{1}{16}$

B **11.** If a six-volume set of books is placed on a shelf at random, what is the probability that the books will be arranged in either correct or reverse order? $\frac{1}{360}$

12. When Carlos shoots a basketball, the probability of a basket is 0.4. When Brad shoots, the probability of a basket is 0.45. What is the probability that at least one basket is made if they each take one shot? 0.67

13. Four managers, five engineers, and one lawyer are in a meeting when a message arrives for one of them. Of these people, three managers, two engineers, and the lawyer are women. Find the probability of each event.

a. The message is for an engineer. $\frac{1}{2}$
b. The message is for a man. $\frac{2}{5}$
c. The message is not for a male engineer. $\frac{7}{10}$

14. A card is drawn from a 52-card deck. If A is the event that the card is red, B the event that it is an ace, and C the event that it is a 9 or 10, which are independent events?

a. A and B Indep. **b.** A and C Indep. **c.** B and C Dep.

15. The probability that Karla will ask Frank to be her tennis partner is $\frac{1}{4}$, that Juan will ask Frank is $\frac{1}{3}$, and that Roger will ask Frank is $\frac{3}{4}$. Find the probability of each event.

a. Karla and Juan ask him. $\frac{1}{12}$

b. Juan and Roger ask him, but Karla doesn't. $\frac{3}{16}$

c. At least two of the three ask him. $\frac{19}{48}$

d. At least one of the three asks him. $\frac{7}{8}$

16. The probability of rain on a certain day is 65% in Yellow Falls and 40% in Copper Creek. Find the probability of each event.

a. It will rain in Yellow Falls but not in Copper Creek. 0.39

b. It will rain in both towns. 0.26

c. It will rain in neither town. 0.21

d. It will rain in at least one of the towns. 0.79

C **17.** When it is not snowing, the probability of the Colorado Drifters winning a football game is $\frac{7}{10}$. When it is snowing, the probability of the Drifters winning is $\frac{2}{5}$. The probability of snow on any day in January is $\frac{1}{2}$.

a. What is the probability of the Drifters winning on January 17? $\frac{11}{20}$

b. If the Drifters won the game on January 17, what is the probability that the team played in snow? $\frac{4}{11}$

Self-Test 3

Vocabulary

random experiment (p. 743)
sample space (p. 743)
event (p. 743)
simple event (p. 743)
equally likely events (p. 745)
intersection (p. 754)
disjoint (p. 754)
union (p. 754)
mutually exclusive events (p. 755)
independent events (p. 756)
complement (p. 757)

Solve.

1. Each of the letters in the word SQUARE is written on a different card, and the six cards are shuffled. One card is drawn at random. **Obj. 15-8, p. 743**

a. Specify the sample space for the drawing. {S, Q, U, A, R, E}

b. Specify the event that a vowel is drawn. {U, A, E}

2. One chip is drawn at random from a box containing 3 green, 5 blue, 2 white, and 8 red chips. What is the probability that the chip drawn is white? $\frac{1}{9}$ **Obj. 15-9, p. 745**

3. Two dice are rolled. **Obj. 15-10, p. 754**

a. Find the probability that either or both numbers showing are 6's or their sum is greater than 8. $\frac{7}{18}$

b. Are the two events independent? No

c. Are the two events mutually exclusive? No

Check your answers with those at the back of the book.

Quick Quiz

1. One box contains red chips and blue chips and another box contains blue chips and green chips. One chip is drawn at random from each box.

a. Specify the sample space. {(R, B), (R, G), (B, B), (B, G)}

b. Specify the event that the selected chips are not the same color. {(R, B), (R, G), (B, G)}

2. Each of the letters of the word EXCELLENT is written on a different card. The cards are shuffled and one card is drawn. Find the probability of each event.

a. An E is drawn. $\frac{1}{3}$

b. An S is drawn. 0

c. A letter from the first half of the alphabet is drawn. $\frac{2}{3}$

3. Two cards are drawn at random from a 52-card bridge deck. Let A be the event that both cards are 2's and B be the event that both are red cards.

a. Find $P(A)$, $P(B)$, $P(A \cup B)$, and $P(A \cap B)$. $\frac{1}{221}; \frac{25}{102}; \frac{55}{221}; \frac{1}{1326}$

b. Are the events mutually exclusive? no

c. Are the events independent? no

Computer Key-In Commentary

The program is a realistic simulation of the processes involved in waiting in a line. You might discuss the advantages and disadvantages of simulations: They save time and money and are safe, but they can't include all the factors involved in the actual situation. In running this simulation, students are able to control a number of factors, thus making it possible to see the effects of changes in a particular variable.

Computer Key-In

Computers are often used to simulate real-world events. A typical simulation involves waiting lines, such as those found in banks and grocery stores.

The program below simulates a single waiting line. The user controls four variables: the duration of the simulation (DUR), the maximum length of the line (MAX), the probability that a customer arrives at any given moment (PROB), and the average service time for a customer (AST).

During the course of the simulation, the computer keeps a running count of the number of customers in line (CIL), the number served (CS), and the number lost because the line was too long (CL). The final status of these counts is reported at the end of the simulation.

```
10 PRINT "THIS PROGRAM SIMULATES A WAITING LINE."
20 INPUT "ENTER THE DURATION (IN MINUTES) OF THE
   SIMULATION: "; DUR
30 INPUT "ENTER THE MAXIMUM LENGTH OF THE WAITING
   LINE: "; MAX
40 INPUT "ENTER THE PROBABILITY THAT A CUSTOMER WILL
   ARRIVE AT ANY GIVEN MINUTE: "; PROB
50 INPUT "ENTER THE AVERAGE SERVICE TIME (IN MINUTES)
   FOR A CUSTOMER: "; AST
60 LET CIL=0
70 LET CS=0
80 LET CL=0
90 LET ST=0
100 FOR T=1 TO DUR
110 IF RND(1)>PROB THEN 140
120 IF CIL=MAX THEN LET CL=CL+1
130 IF CIL<MAX THEN LET CIL=CIL+1
140 IF CIL>0 AND ST=0 THEN LET ST=AST
150 IF ST>0 THEN LET ST=ST-1
160 IF CIL>0 AND ST=0 THEN LET CS=CS+1
170 IF CIL>0 AND ST=0 THEN LET CIL=CIL-1
180 NEXT T
190 PRINT "NUMBER OF CUSTOMERS SERVED: "; CS
200 PRINT "NUMBER OF CUSTOMERS LOST: "; CL
210 PRINT "NUMBER LEFT IN LINE: "; CIL
220 END
```

Exercises

2. As the probability of a customer arriving increases, more customers are lost.

1. Holding DUR, MAX, and AST constant, run the simulation for the following values of PROB. Record the results. Answers will vary.

   a. 0.1 **b.** 0.3 **c.** 0.5 **d.** 0.7 **e.** 0.9

2. Based on the results of Exercise 1, what conclusion can you draw?

Chapter Summary

1. *Frequency distributions, histograms, stem-and-leaf plots,* and *box-and-whisker plots* are four ways of displaying statistical data.
2. The *mode,* the *median,* and the *mean* are used to describe the *central tendency* of a frequency distribution.
3. The *range,* the *variance,* and the *standard deviation* are used to describe the *dispersion* of a frequency distribution.
4. The *variance* of a distribution is defined as follows:
$$\text{variance} = \frac{(x_1 - M)^2 + (x_2 - M)^2 + \cdots + (x_n - M)^2}{n}$$
where $x_1, x_2, \ldots, x_n$ are n numbers and M is their mean.
5. The *standard deviation,* σ, is the principal square root of the variance:
$$\sigma = \sqrt{\text{variance}}.$$
6. The *standard normal curve* is symmetric with respect to the y-axis and approaches the x-axis asymptotically as $|x|$ increases. The total area under the curve and above the x-axis equals 1. For a normal distribution with mean M and standard deviation σ, the fraction of the data falling between $M + a\sigma$ and $M + b\sigma$ is the area under the standard normal curve between $x = a$ and $x = b$.
7. The *correlation coefficient, r,* between two variables x and y is a measure of how closely the ordered pairs (x, y) cluster about a line called the *regression line*. The closer the points (x, y) are to being collinear, the closer $|r|$ is to 1.
8. The number of *permutations* of n things taken r at a time is given by:
$${}_nP_r = n(n - 1)(n - 2) \cdots [n - (r - 1)] = \frac{n!}{(n - r)!}$$
9. The number of *combinations* of n things taken r at a time is given by:
$${}_nC_r = \frac{n!}{r!(n - r)!}$$
10. The set of possible outcomes of a random experiment is called the *sample space* of the experiment. Any subset of the sample space is an *event*. An event consisting of a single outcome is a *simple event*.
11. If the sample space for an experiment consists of n equally likely outcomes, the probability of each simple event is $\frac{1}{n}$. If the event E contains k elements of this event space,
$$P(E) = \frac{1}{n} + \frac{1}{n} + \cdots + \frac{1}{n} = \frac{k}{n}.$$
12. Two events, A and B, are *independent* if and only if
$$P(A \cap B) = P(A) \cdot P(B).$$

13. In general, for any two events A and B in a sample space,

$$P(A \cup B) = P(A) + P(B) - P(A \cap B).$$

If the events have no simple events in common (their intersection is empty), the events are said to be *mutually exclusive*. For mutually exclusive events,

$$P(A \cup B) = P(A) + P(B).$$

Chapter Review

Write the letter of the correct answer.

1. Nine quiz scores were 6, 7, 7, 8, 8, 9, 9, 10, and 10. What is the median of these scores? 15-1

a. 7 **b.** 8 **c.** 9 **d.** 10

2. On four tests, Jean got the following scores: 100, 79, 86, and 91. What is the variance for this distribution? 15-2

a. 89 **b.** 15.3 **c.** 58.5 **d.** 234

3. Under the standard normal curve, the area bounded by the x- and y-axes and the line $x = 3$ is 0.4987. In a standard normal distribution, what percent of the data is between -3 and 3? 15-3

a. 49.87% **b.** 24.94% **c.** 0% **d.** 99.74%

4. The regression line for two variables x and y is $y = 0.45x - 32$. Predict the value of x that is associated with a value of 13 for y. 15-4

a. 10 **b.** 100 **c.** -13 **d.** 553

5. How many positive integers less than 25 can be formed by using the digits 1, 2, and 3? 15-5

a. 3 **b.** 6 **c.** 9 **d.** 12

6. In how many ways can 3 cards be selected from a 12-card deck if the selected cards are not returned to the deck?

a. 220 **b.** 1728 **c.** 1440 **d.** 1320

7. How many different permutations can be made using all the letters in the word ATLANTA? 15-6

a. 420 **b.** 5040 **c.** 840 **d.** 2520

8. In how many ways can 5 players be chosen from 9 players? 15-7

a. 3024 **b.** 36 **c.** 126 **d.** 15,120

9. If two coins are tossed, what does $\{(H, H), (H, T), (T, H), (T, T)\}$ represent? 15-8

a. the event that at least one head turns up

b. the event that exactly one tail turns up

c. the sample space for the experiment

10. Three coins are tossed. What is the probability that at least one tail turns up? 15-9

a. $\frac{1}{8}$ **b.** $\frac{3}{8}$ **c.** $\frac{7}{8}$ **d.** 1

11. If A and B are mutually exclusive events, which of the following relationships applies? 15-10

a. $P(A \cup B) = P(A) \cdot P(B)$ b. $P(A \cap B) \neq 0$
c. $P(A \cap B) = P(A) \cdot P(B)$ (d.) $P(A \cup B) = P(A) + P(B)$

12. A dime and a penny are tossed. Let A be the event that the dime turns up heads, and let B be the event that the penny turns up tails. What is $P(A \cap B)$?

a. 1 b. $\frac{3}{4}$ c. $\frac{1}{2}$ (d.) $\frac{1}{4}$

Chapter Test

1. Draw a stem-and-leaf plot for the set of data below. Then find the mode, median, and mean to the nearest whole number. 25; 25; 24 15-1

13, 25, 17, 16, 31, 28, 25, 17, 25, 32, 33

2. Draw a box-and-whisker plot for the data in Exercise 1. Then compute the variance and standard deviation to the nearest whole number. 45; 7 15-2

3. The mean score on a history test was 74.3 and the standard deviation was 5.7. Assuming the scores were normally distributed, what fraction of the scores were above 80? $\frac{4}{25}$ 15-3

4. Draw a scatter plot for the data shown at the right. Then state whether there is a high positive correlation, a high negative correlation, or relatively no correlation between price and demand. High neg. corr. 15-4

Price	Demand
12	260
15	250
19	210
24	180

5. How many whole numbers less than 500 can be formed using the digits 1, 2, 4, and 5? 68 15-5

6. In how many ways can you arrange 5 pictures in one row on a wall? 120 15-6

7. How many different signals can be made by displaying four pennants, all at one time, on a vertical flagpole? The pennants are identical except for color: three are blue and one is red. 4

8. How many 4-person committees can be formed from a group of 8 people? 70 15-7

9. Two dice are rolled. Specify each of the following events: 15-8
   a. The dice show the same number. {(1, 1), (2, 2), (3, 3), (4, 4), (5, 5), (6, 6)}
   b. The sum of the numbers showing is 4. {(1, 3), (2, 2), (3, 1)}

10. One card is drawn at random from a 52-card deck. Find the probability of each event. 15-9
    a. It is a spade. $\frac{1}{4}$ b. It is an ace. $\frac{1}{13}$ c. It is a 9 or 10. $\frac{2}{13}$

11. There are 3 red, 3 white, and 3 green marbles in a bag. Two are drawn at random. Let A be the event that at least one marble is red. Let B be the event that both marbles are the same color. 15-10
    a. Find $P(A)$, $P(B)$, $P(A \cap B)$, and $P(A \cup B)$. $\frac{7}{12}$; $\frac{1}{4}$; $\frac{1}{12}$; $\frac{3}{4}$
    b. Are A and B mutually exclusive events? Are they independent? No; no

Alternate Test p. T27

Supplementary Materials

Test Masters	66, 67
Resource Book	p. 171

16 Matrices and Determinants

Objectives

16-1 To learn and apply matrix terminology.

16-2 To find sums and differences of matrices and products of a scalar and a matrix.

16-3 To find the product of two matrices.

16-4 To solve problems using matrices.

16-5 To find the determinant of a 2 × 2 or 3 × 3 matrix.

16-6 To solve systems of equations using inverses of matrices.

16-7 To evaluate third-order determinants using expansion by minors.

16-8 To use the properties of determinants to simplify the expansion of determinants by minors.

16-9 To solve systems of equations using determinants.

Assignment Guide

See p. T58 for Key to the format of the Assignment Guide

Day	Average Algebra	Average Algebra and Trigonometry	Extended Algebra	Extended Algebra and Trigonometry
1			**16-1** 769/1–20	*Administer Chapter 14 Test* *Read 16-1* 769/1–14
2			**16-2** 773/1–18 **S** 769/21, 22	**16-1** 769/15–22 **S** 769/*Mixed Review* **16-2** 773/1–23 odd
3			**16-2** 773/19–21, 23 **16-3** 777/1–7	**16-3** 777/1–25 odd **S** 778/*Mixed Review*
4			**16-3** 778/8–28 even **S** 778/*Mixed Review*	**16-4** 782/1–3, 11–17 785/*Extra:* 1, 3, 5
5			**16-4** 782/1–10 785/*Extra:* 1, 3, 5	**16-5** 789/1–15 **S** 789/*Mixed Review* **R** 784/*Self-Test 1*
6			**16-4** 783/11–17 **S** 778/29, 30	**16-6** 792/1–18 **S** 778/27, 29
7			**16-5** 789/1–15 **S** 789/*Mixed Review* **R** 784/*Self-Test 1*	**16-7** 796/1–17 **S** 797/*Mixed Review* **R** 793/*Self-Test 2*
8			**16-6** 792/1–18 **S** 789/17	**16-8** 800/1–14 **S** 797/18
9			**16-7** 796/1–17 **S** 797/*Mixed Review* 1, 3–5 **R** 793/*Self-Test 2*	**16-9** 804/1–13 odd, 15, 16 **S** 805/*Mixed Review* **R** 805/*Self-Test 3*
10			**16-8** 800/1–14 **S** 797/18	*Prepare for Chapter Test* **R** 806/*Chapter Review*

Assignment Guide *(continued)*

Day	Average Algebra	Average Algebra and Trigonometry	Extended Algebra	Extended Algebra and Trigonometry
11			**16-9** 804/1–13 odd, 15, 16 **S** 805/*Mixed Review:* 1–3, 5, 6 **R** 805/*Self-Test 3*	*Administer Chapter 16 Test*
12			*Prepare for Chapter Test* **R** 805/*Chapter Review*	
13			*Administer Chapter 16 Test*	

Supplementary Materials Guide

For Use with Lesson	Practice Masters	Tests	Study Guide (Reteaching)	Resource Book		
				Tests	Practice Exercises	Mixed Review (MR) Applications (A) Enrichment (E)
16-1			pp. 259–260			p. 219 (A)
16-2	Sheet 88		pp. 261–262			
16-3	Sheet 89		pp. 263–264			
16-4	Sheet 90	Test 68	pp. 265–266		p. 172	
16-5			pp. 267–268			
16-6	Sheet 91		pp. 269–270		p. 173	
16-7			pp. 271–272			
16-8	Sheet 92		pp. 273–274			
16-9	Sheet 93	Test 69	pp. 275–276		p. 174	
Chapter 16		Tests 70, 71		pp. 81–84	p. 175	p. 235 (E)
Cum. Rev. 13–16	Sheet 94					
Cum. Rev. 12–16				pp. 85–87		
Cum. Rev. 9–16					pp. 176–177	
Cum. Rev. 8–16				pp. 88–91		
Cum. Rev. 1–16	Sheets 95–97			pp. 92–99		pp. 186–189 (MR)

Software

Software	Algebra Plotter Plus	Using Algebra Plotter Plus	Computer Activities	Test Generator
	Matrix Reducer	Enrichment, p. 42	Activities 35, 36	189 test items
For Use with Lessons	16-4, 16-6	16-6	16-4, 16-6	all lessons

Strategies for Teaching

Making Connections and Cooperative Learning

Connections between mathematics and other curriculum areas occur throughout this chapter, such as the social science connection (e.g., pages 779–784) and the historical connection (e.g., page 793).

Connections between algebra and geometry are made in the Exploration on page 847. In this activity students use matrix multiplication to explore geometric transformations.

Cooperative learning groups may help your students when they are attempting to apply their knowledge to real-world applications. In recent years, many teachers have used cooperative learning groups with students of all ability levels, and have noticed exceptional growth in students' mathematical understanding and attitudes. Students take responsibility for their own learning, actively do mathematics with other students, improve their use of mathematical language as they work, share their ideas with others, become less fearful of making errors in a small group, become more confident in their own abilities through successful group participation, learn that a group can often solve problems that an individual cannot solve, learn how to work cooperatively, learn how to learn, and develop more positive attitudes about mathematics.

16-4 Applications of Matrices

Most students will enjoy this section. Although there is new content here, this is primarily an applications section. Be sure students see how to write the matrix for a communications network. You may want to go slowly as you introduce the square of a communications or dominance matrix. There will be students who do not see the significance of the square without patient explanation, repetition, and good examples. Make sure students see the difference between A^2 and $A + A^2$, the first representing two-stage communication or dominance, the second two-stage or less.

Set up a communications network with its own communications matrix in the classroom. Depending on class size, first write, or have the class write, a 4×4 or 5×5 communications matrix. Distribute copies to the class. Write a secret message to be passed through the communication network. Have students pass the message around in accordance with the communications matrix. After a set time, stop to assess the progress of the message and to interpret what has happened in relation to the matrix.

For a matrix activity with a local flair, have students create their own dominance matrices. They can use the records of local school teams or look in the newspaper to find out which teams won and lost in the past few games.

16-6 Inverses of Matrices

Have the students work in groups to determine what is wrong with the following solution.
Solve for A.

$$\begin{bmatrix} 3 & 2 \\ 2 & 2 \end{bmatrix} A = \begin{bmatrix} 5 & 0 \\ -1 & -1 \end{bmatrix}$$

$$\begin{bmatrix} 1 & -1 \\ -1 & \frac{3}{2} \end{bmatrix}\begin{bmatrix} 3 & 2 \\ 2 & 2 \end{bmatrix} A = \begin{bmatrix} 5 & 0 \\ -1 & -1 \end{bmatrix}\begin{bmatrix} 1 & -1 \\ -1 & \frac{3}{2} \end{bmatrix}$$

$$\begin{bmatrix} 3-2 & 2-2 \\ -3+3 & -2+3 \end{bmatrix} A = \begin{bmatrix} 5-0 & -5+0 \\ -1+1 & 1-\frac{3}{2} \end{bmatrix}$$

$$\begin{bmatrix} 1 & 0 \\ 0 & 1 \end{bmatrix} A = \begin{bmatrix} 5 & -5 \\ 0 & -\frac{1}{2} \end{bmatrix}$$

$$A = \begin{bmatrix} 5 & -5 \\ 0 & -\frac{1}{2} \end{bmatrix}$$

Matrix multiplication is not commutative.

$\begin{bmatrix} 5 & 0 \\ -1 & -1 \end{bmatrix}\begin{bmatrix} 1 & -1 \\ -1 & \frac{3}{2} \end{bmatrix}$ should be $\begin{bmatrix} 1 & -1 \\ -1 & \frac{3}{2} \end{bmatrix}\begin{bmatrix} 5 & 0 \\ -1 & -1 \end{bmatrix}$

References to Strategies

PE: Pupil's Edition **TE:** Teacher's Edition **RB:** Resource Book

Problem Solving Strategies

TE: p. T144 (Sugg. Extension, Lesson 16-6)

Applications

PE: p. 766 (computer graphics); pp. 779–782 (communications)
TE: pp. 766, 770, 780–782, T142
RB: p. 219

Nonroutine Problems

PE: p. 773 (Exs. 21–28); p. 778 (Exs. 27–30); pp. 783–784 (Exs. 17, 18); p. 789 (Exs. 19, 20); p. 797 (Ex. 19); p. 800 (Exs. 13–15); p. 805 (Exs. 17–19)
TE: p. T142 (Sugg. Extensions, Lesson 16-2, 16-3)

Communication

TE: pp. T142–T143 (Sugg. Extensions, Lessons 16-1, 16-5)

Thinking Skills

PE: p. 773 (Exs. 21–28); p. 778 (Exs. 27–30); p. 805 (Exs. 17–19, proofs)

Explorations

PE: pp. 785–786; p. 847 (matrix multiplication, geometric transformations)

Connections

PE: p. 766 (Art); pp. 779–782 (Communication); p. 793 (History); pp. 830–831, 847 (Geometry); pp. 838–839 (Discrete Math)

Using Technology

PE: pp. 766, 786
TE: pp. 766, 786; p. T142 (Sugg. Extension, Lesson 16-1)
Computer Activities: pp. 94–99
Using Algebra Plotter Plus: p. 42

Cooperative Learning

TE: p. T143 (Lesson 16-4)

Teaching Resources

For use in implementing the teaching strategies referenced on the previous page.

Application
Resource Book, p. 219

Application—Drawing in Perspective (For use with Chapter 16)

A typical drawing of a cube is shown in the upper figure at the right. Notice that the edges of the cube connecting the front face to the back face are all inclined at a 45° angle and are therefore parallel. As a result, the back face of the cube is exactly the same size as the front face.

In the perspective drawing of a cube shown in the lower figure at the right, the receding lines all converge at one point, called the "vanishing point," on the horizon. This drawing depicts more realistically how a three-dimensional object would appear to the eye, with the back face smaller than the front face.

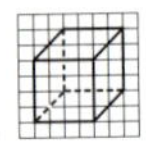

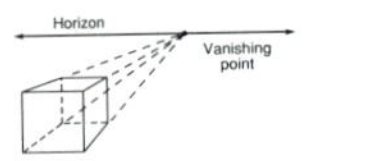

1. An art student made a perspective drawing of the hallway in his high school. Follow the receding lines in the ceiling, floor, and walls to locate the vanishing point. Label this point V.

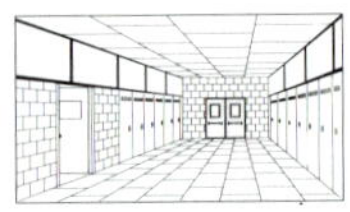

2. Draw a 3-D letter H in two ways: **(a)** by completing the letter using receding edges all inclined at a 45° angle and **(b)** by completing the letter using receding lines converging at the vanishing point V.

a.

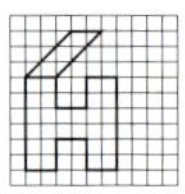

b.

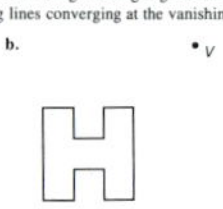

3. In two-point perspective, the receding lines converge at two vanishing points on the horizon, as shown in the figure at the right. Redraw the cube below using two-point perspective. That is, $\overrightarrow{BA}$, $\overrightarrow{GF}$, and $\overrightarrow{DE}$ should meet at one vanishing point, while $\overrightarrow{BC}$, $\overrightarrow{GD}$, and $\overrightarrow{FE}$ should meet at another.

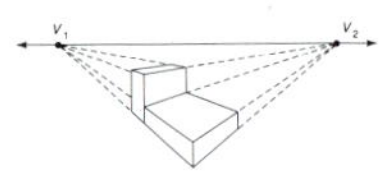

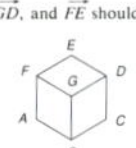

Resource Book, ALGEBRA, Structure and Method, Book 2

APPLICATIONS 219

Enrichment
Resource Book, p. 235

The Square Roots of a Matrix (For use with Chapter 16)

For each matrix A show that $A^2 = I$, where $I = \begin{bmatrix} 1 & 0 \\ 0 & 1 \end{bmatrix}$.

1. $A = \begin{bmatrix} -1 & 0 \\ 0 & -1 \end{bmatrix}$ ____________

2. $A = \begin{bmatrix} 3 & -4 \\ 2 & -3 \end{bmatrix}$ ____________

3. $A = \begin{bmatrix} i & 2 \\ 1 & -i \end{bmatrix}$ ____________

4. $A = \begin{bmatrix} 1 & 0 \\ 3 & -1 \end{bmatrix}$ ____________

A square matrix A is said to be a square root of a matrix B if $A^2 = B$. Each matrix in Exercises 1–4 is a square root of the identity matrix I. Therefore the matrix I has many square roots.

5. If $A^2 = I$, what are two possible values for det A? ____________

6. Show that if $\begin{bmatrix} x & y \\ z & -x \end{bmatrix}^2 = I$, then $x^2 + yz = 1$. Is the converse true?

____

7. If $A^2 = B$, show that $(\det A)^2 = \det B$ for 2×2 matrices A and B.

____

8. Show that the matrix $\begin{bmatrix} \sqrt{1} & 0 & 0 \\ 0 & \sqrt{2} & 0 \\ 0 & 0 & \sqrt{3} \end{bmatrix}$ is a square root of $\begin{bmatrix} 1 & 0 & 0 \\ 0 & 2 & 0 \\ 0 & 0 & 3 \end{bmatrix}$.

____

9. Show that matrix $A = \begin{bmatrix} 2 & 1 \\ 1 & 2 \end{bmatrix}$ is a square root of matrix $B = \begin{bmatrix} 5 & 4 \\ 4 & 5 \end{bmatrix}$ and find another square root of B.

____

10. Find a square root of the matrix $\begin{bmatrix} 5 & 3 \\ 3 & 2 \end{bmatrix}$. ____________

Resource Book, ALGEBRA, Structure and Method, Book 2

ENRICHMENT 235

Application
Study Guide, p. 265

16–4 Applications of Matrices

Objective: To solve problems using matrices.

Vocabulary

Communication matrix A matrix representing routes by which data can be transmitted and received.

Example 1 Write the matrix that illustrates the network of computers shown at the right.

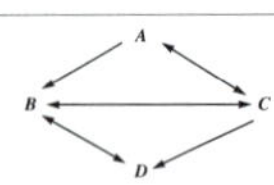

Solution The arrows in the drawing indicate the direction of communication in the network. When writing the matrix for this network, use a "1" to indicate that direct communication from one computer to another is possible and a "0" to indicate that direct communication is not possible. For example, Computer A cannot send data to itself directly, so a "0" goes in the first row, first column of the matrix. Likewise, Computer A can send data to Computer B directly, so a "1" goes in the first row, second column of the matrix.

From computer \ To computer	A	B	C	D
A	0	1	1	0
B	0	0	1	1
C	1	1	0	1
D	0	1	0	0

$$\begin{bmatrix} 0 & 1 & 1 & 0 \\ 0 & 0 & 1 & 1 \\ 1 & 1 & 0 & 1 \\ 0 & 1 & 0 & 0 \end{bmatrix} = X$$

Matrix X is called a *communication matrix*.

Example 2 With regard to the network of computers given in Example 1, find the matrix that represents the number of ways that data can be sent from one computer to another using exactly one relay.

Solution The required matrix is the square of matrix X from the solution of Example 1:

$$X^2 = X \cdot X = \begin{bmatrix} 0 & 1 & 1 & 0 \\ 0 & 0 & 1 & 1 \\ 1 & 1 & 0 & 1 \\ 0 & 1 & 0 & 0 \end{bmatrix} \begin{bmatrix} 0 & 1 & 1 & 0 \\ 0 & 0 & 1 & 1 \\ 1 & 1 & 0 & 1 \\ 0 & 1 & 0 & 0 \end{bmatrix} = \begin{array}{c} \\ A \\ B \\ C \\ D \end{array} \begin{array}{c} \begin{matrix} A & B & C & D \end{matrix} \\ \begin{bmatrix} 1 & 1 & 1 & 2 \\ 1 & 2 & 0 & 1 \\ 0 & 2 & 2 & 1 \\ 0 & 0 & 1 & 1 \end{bmatrix} \end{array}$$

The "1" in the first row, first column of X^2, for example, means that Computer A can send data to itself via Computer C. Likewise, the "2" in the first row, fourth column means that Computer A can send data to Computer D via either Computer B or Computer C.

Study Guide, ALGEBRA AND TRIGONOMETRY, Structure and Method, Book 2

265

Application
Study Guide, p. 266

16–4 Applications of Matrices (continued)

Example 3 Use Examples 1 and 2 to solve these problems.

a. Which computer can transmit directly to the greatest number of computers in the network?

b. Which computer can receive messages from the greatest number of computers in the network?

c. Find the matrix that represents the total number of routes that data can be transmitted from one computer to another using no more than 2 steps (that is, one relay point).

Solution

a. The row entries in matrix X represent routes on which data can be transmitted directly. After finding the sum of the numbers in each row, you can see that Computer C (with three routes) can send messages to the greatest number of computers.

b. The column entries in matrix X represent routes on which data can be received directly. After finding the sum of the numbers in each column, you can see that Computer B (with three routes) can receive messages from the greatest number of computers.

c. Matrix X represents the number of direct message routes, and matrix X^2 represents the number of routes using one relay station. So the sum $X + X^2$ represents the number of routes using *no more than* 2 steps.

$$X + X^2 = \begin{bmatrix} 0 & 1 & 1 & 0 \\ 0 & 0 & 1 & 1 \\ 1 & 1 & 0 & 1 \\ 0 & 1 & 0 & 0 \end{bmatrix} + \begin{bmatrix} 1 & 1 & 1 & 2 \\ 1 & 2 & 0 & 1 \\ 0 & 2 & 2 & 1 \\ 0 & 0 & 1 & 1 \end{bmatrix} = \begin{bmatrix} 1 & 2 & 2 & 2 \\ 1 & 2 & 1 & 2 \\ 1 & 3 & 2 & 2 \\ 0 & 1 & 1 & 1 \end{bmatrix}$$

For Exercises 1–7, use the radio network shown at the right. (*Note*: A "hook" indicates that two line segments do *not* intersect.)

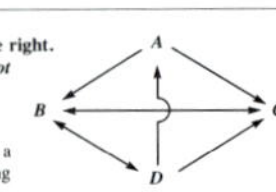

1. Write the matrix that illustrates this network.
2. Find the matrix that represents the number of routes a message can be sent from one station to another using exactly one relay.
3. Name the station(s) that can send messages on the greatest number of routes having exactly one relay.
4. Name the station(s) that can receive messages on the greatest number of routes having exactly one relay.
5. Find the matrix that represents the number of routes a message can be sent from one station to another using at most one relay.
6. Name the station(s) that can send messages on the greatest number of routes having at most one relay.
7. Name the station(s) that can receive messages on the greatest number of routes having at most one relay.

266
Study Guide, ALGEBRA AND TRIGONOMETRY, Structure and Method, Book 2

Using Technology
Computer Activities, p. 94

ACTIVITY 35. *Matrix Multiplication* *(for use with Section 16-4)*

Directions: Write all answers in the spaces provided.

PROBLEM

Use the computer to find the matrix product: $\begin{bmatrix} 1 & 2 \\ 3 & 4 \\ 5 & 6 \end{bmatrix}\begin{bmatrix} 11 & 12 & 13 & 14 \\ 15 & 16 & 17 & 18 \end{bmatrix}$

PROGRAM •

```
10  DIM M(10, 10), N(10, 10), P(10, 10)
20  READ R1, C1, C2
30  FOR R = 1 TO R1
40  FOR C = 1 TO C1
50  READ M(R, C)
60  NEXT C
70  NEXT R
80  FOR R = 1 TO C1
90  FOR C = 1 TO C2
100  READ N(R, C)
110  NEXT C
120  NEXT R
130  PRINT
140  FOR R = 1 TO R1
150  FOR C = 1 TO C2
160  LET P(R, C) = 0
170  FOR T = 1 TO C1
180  LET P(R, C) = P(R, C) + M(R, T) • N(T, C)
190  NEXT T
200  PRINT P(R, C); "  ";
210  NEXT C
220  PRINT
230  PRINT
240  NEXT R
300  DATA 3, 2, 4
400  DATA 1, 2, 3, 4, 5, 6
500  DATA 11, 12, 13, 14, 15, 16, 17, 18
600  END
```

PROGRAM CHECK

Type in the program. To test whether you entered it correctly, run the program. The computer should print:

```
41   44   47   50

93   100   107   114

145   156   167   178
```

(continued)

Using Technology
Computer Activities, p. 95

(Activity 35 continued)

ANALYSIS

Recall that the product of two matrices is defined only if the number of columns of the first matrix equals the number of rows of the second matrix. The data in line 300 indicate that the first matrix has 3 rows and 2 columns and the second matrix has 2 rows and 4 columns. The elements of the first matrix are listed by rows in line 400. The elements of the second matrix are listed by rows in line 500.

USING THE PROGRAM

Use the computer to find the product of the following 2×10 and 10×3 matrices:

$$\begin{bmatrix} 1 & 2 & 3 & 4 & 5 & 6 & 7 & 8 & 9 & 10 \\ 11 & 12 & 13 & 14 & 15 & 16 & 17 & 18 & 19 & 20 \end{bmatrix} \begin{bmatrix} 100 & 90 & 80 \\ 70 & 60 & 50 \\ 40 & 30 & 20 \\ 10 & 89 & 88 \\ 87 & 86 & 85 \\ 84 & 83 & 82 \\ 81 & 79 & 1 \\ 1 & 1 & 1 \\ 1 & 1 & 1 \\ 1 & 2 & 2 \end{bmatrix}$$

First, modify the program.

```
300  DATA  2, 10, 3
400  DATA  1, 2, 3, 4, 5, 6, 7, 8, 9, 10
410  DATA  11, 12, 13, 14, 15, 16, 17, 18, 19, 20
500  DATA  100, 90, 80, 70, 60, 50, 40, 30, 20
510  DATA  10, 89, 88, 87, 86, 85, 84, 83, 82
520  DATA  81, 79, 1, 1, 1, 1, 1, 1, 1, 1, 2, 2
```

Note that you may use any number of DATA statements to list the elements of a given matrix as long as the rows are listed in order.

a. Run the program. The product is: [ ]

b. What are the dimensions of the product matrix? ________

c. The product of an $m \times n$ matrix and an $n \times p$ matrix has dimensions ________.

(continued)

Using Technology
Computer Activities, p. 96

(Activity 35 continued)

APPLICATION

Exciting Lighting Co. makes ceiling fixtures, wall fixtures, and hanging fixtures. Each fixture requires sockets, wires, connectors, and globes. The materials matrix below gives the number of parts required for each type of fixture.

	Sockets	Wires	Connectors	Globes
Ceiling fixtures	2	3	6	1
Wall fixtures	1	3	2	1
Hanging fixtures	3	4	9	3

1. An order is placed for 83 ceiling fixtures, 117 wall fixtures, and 49 hanging fixtures. Run the program to determine how many of each part are needed to fill the order. (You should find the product of a 1×3 matrix and a 3×4 matrix. Remember also to delete any extra DATA lines left over from the previous exercise.)

	Sockets	Wires	Connectors	Globes
Parts needed	________	________	________	________

2. Exciting Lighting anticipates that its summer production levels will be 950 ceiling, 820 wall, and 550 hanging fixtures in June; 1400 ceiling, 975 wall, and 480 hanging fixtures in July; and 460 ceiling, 400 wall, and 250 hanging fixtures in August. Run the program to determine the inventory necessary for each type of part at the beginning of each of the three months. (You should find the product of a 3×3 matrix and a 3×4 matrix.)

	Sockets	Wires	Connectors	Globes
June	________	________	________	________
July	________	________	________	________
August	________	________	________	________

Using Technology
Using Algebra Plotter Plus, p. 42

Book 2, Lesson 16-6 • Finding the Inverse of a Matrix

Have students use the Matrix Reducer program to find inverses of matrices. The steps to be followed are:

1. Start with a square matrix A.
2. Augment A with an identity matrix of the same size.
3. Perform row operations on the augmented matrix until A becomes an identity matrix. What was formerly an identity matrix is now A^{-1}.

For example, to find the inverse of the matrix $\begin{bmatrix} 5 & 3 \\ 2 & 1 \end{bmatrix}$, perform row operations on the augmented matrix $\left[\begin{array}{cc|cc} 5 & 3 & 1 & 0 \\ 2 & 1 & 0 & 1 \end{array}\right]$ until it becomes $\left[\begin{array}{cc|cc} 1 & 0 & -1 & 3 \\ 0 & 1 & 2 & -5 \end{array}\right]$. The matrix $\begin{bmatrix} -1 & 3 \\ 2 & -5 \end{bmatrix}$ is the inverse of the matrix $\begin{bmatrix} 5 & 3 \\ 2 & 1 \end{bmatrix}$. Students can use this procedure to check their answers for Written Exercises 1–8 on page 792.

Students can also use the Matrix Reducer program to solve systems of equations, such as in Exercises 1–6 on page 786. For example, to solve the system:

$$\begin{aligned} x+y+z&=0 \\ x-y+z&=2 \\ x-y-z&=10 \end{aligned}$$

students must rewrite the system in matrix form:

$$\underset{A}{\begin{bmatrix} 1 & 1 & 1 \\ 1 & -1 & 1 \\ 1 & -1 & -1 \end{bmatrix}} \underset{X}{\begin{bmatrix} x \\ y \\ z \end{bmatrix}} = \underset{B}{\begin{bmatrix} 0 \\ 2 \\ 10 \end{bmatrix}}$$

The system is solved by finding A^{-1} and then multiplying both sides of the equation by A^{-1} because of the following justification:

$$\begin{aligned} AX &= B \\ A^{-1}AX &= A^{-1}B \\ IX &= A^{-1}B \\ X &= A^{-1}B \end{aligned}$$

Students can use the Matrix Reducer program to find A^{-1}:

$$A^{-1} = \begin{bmatrix} \frac{1}{2} & 0 & \frac{1}{2} \\ \frac{1}{2} & -\frac{1}{2} & 0 \\ 0 & \frac{1}{2} & -\frac{1}{2} \end{bmatrix}$$

The solution of the system is:

$$X = A^{-1}B = \begin{bmatrix} \frac{1}{2} & 0 & \frac{1}{2} \\ \frac{1}{2} & -\frac{1}{2} & 0 \\ 0 & \frac{1}{2} & -\frac{1}{2} \end{bmatrix}\begin{bmatrix} 0 \\ 2 \\ 10 \end{bmatrix} = \begin{bmatrix} 5 \\ -1 \\ -4 \end{bmatrix}$$

Application

If the image of a three-dimensional object is to appear realistic when displayed on a computer screen, the 3-D coordinates of the points on the object's surface must be converted to 2-D coordinates using a matrix transformation. The computer also uses matrices when rotating the image to create the illusion of movement. This process is complicated, however, by the "hidden line" problem: Only a portion of the object's surface can be seen at any given angle, so the computer must be programmed not to plot any hidden points.

Computer-generated images of three-dimensional objects are increasingly being used by film makers, especially in science fiction movies. The usefulness of the technique is not limited to entertainment, though. Engineers rely upon computer-aided design (CAD) programs to view in three dimensions the design plans for automobiles and airplanes, for example.

Group Activities

Students with programming experience may want to investigate the use of matrix transformations to rotate, translate, and reflect images on a computer screen.

References

Myers, Roy E. *Microcomputer Graphics.* Reading, MA: Addison-Wesley Publishing Company, 1982.

Support Materials

Resource Book p. 219

16 Matrices and Determinants

In today's movies artificial landscapes may be created by computers using fractals. Rotations of graphic images in 3-D use matrix transformations.

Matrices

16-1 Definition of Terms

Objective To learn and apply matrix terminology.

A **matrix** (plural, matrices) is a rectangular array of numbers enclosed by brackets. Here are some examples of matrices:

$$\begin{bmatrix} 2 & 0 \\ 7 & 1\frac{1}{2} \\ -3 & 19 \end{bmatrix} \qquad \begin{bmatrix} 3 & 0 & 9 \\ 0 & -2 & 0 \end{bmatrix} \qquad \begin{bmatrix} \frac{4}{5} \\ 16 \end{bmatrix} \qquad \begin{bmatrix} \sqrt{2} & 5 & -0.34 \end{bmatrix}$$

The numbers in a matrix are called the **elements** of the matrix. The number of **rows** (horizontal) and the number of **columns** (vertical) determine the **dimensions** of the matrix.

When giving the dimensions of a matrix, you always write the number of rows first. The matrices above are, in order, a 3×2 (read "three by two") matrix, a 2×3 matrix, a 2×1 matrix, and a 1×3 matrix.

Some special matrices are illustrated below.

row matrix: exactly one row

$$\begin{bmatrix} 3 & -10 & \frac{1}{2} \end{bmatrix}$$

column matrix: exactly one column

$$\begin{bmatrix} 6 \\ -2 \end{bmatrix}$$

square matrix: an $n \times n$ matrix

$$\begin{bmatrix} 1 & -1 \\ 3 & 0 \end{bmatrix}$$

In a square matrix the number of rows and the number of columns are equal.

Capital letters are used to name matrices. Subscripts are used to indicate dimensions. For example, $A_{2\times5}$ denotes a 2×5 matrix named A. If all the elements of a matrix are zeros, the matrix is called a **zero matrix.** Therefore, the matrix at the right is a 3×3 zero matrix, denoted by $O_{3\times3}$.

$$\begin{bmatrix} 0 & 0 & 0 \\ 0 & 0 & 0 \\ 0 & 0 & 0 \end{bmatrix}$$

Example 1 Write the zero matrix denoted by $O_{2\times4}$.

Solution $O_{2\times4}$ has 2 rows and 4 columns.
Each element is zero.

$$\therefore O_{2\times4} = \begin{bmatrix} 0 & 0 & 0 & 0 \\ 0 & 0 & 0 & 0 \end{bmatrix} \quad \textit{Answer}$$

Teaching References

Lesson Commentary, pp. T141–T144
Assignment Guide, pp. T71–T72
Supplementary Materials
- Practice Masters 88–94
- Tests 68–71
- Resource Book
  - Practice Exercises, pp. 172–177
  - Tests, pp. 81–84
  - Enrichment Activity, p. 235
  - Application, p. 219
  - Mixed Review, pp. 186–189
- Study Guide, pp. 259–276
- Computer Activities 35–36
- Test Generator
- Disk for Algebra

Alternate Test, p. T27
Cumulative Review, p. T33

Teaching Suggestions p. T141

Suggested Extensions p. T142

Warm-Up Exercises

Solve.

1. $4x + 9 = 5$ {−1}
2. $5 - 3x = -7$ {4}
3. $2x + 7 = 1 - x$ {−2}
4. $x + y = 1$
   $x - y = 5$ {(3, −2)}
5. $2x - 3y = 10$
   $3x + y = -7$ {(−1, −4)}

Motivating the Lesson

Point out to the students that they have seen many examples of information that is presented in table or chart form. These include mathematical tables (such as the trigonometric tables in this book), income tax tables, weather charts, and stock market listings. A matrix, the topic of today's lesson, has the arrangement of data in rows and columns in common with these tables and charts.

Chalkboard Examples

1. Let $A = \begin{bmatrix} 1 & 2 & 3 & 4 \\ 5 & 6 & 7 & 8 \\ 9 & 10 & 11 & 12 \end{bmatrix}$
   a. Give the dimensions of A. 3×4
   b. In which row and which column is the element 7? row 2, column 3
2. Write the matrix $O_{3\times1}$. $\begin{bmatrix} 0 \\ 0 \\ 0 \end{bmatrix}$
3. If $\begin{bmatrix} w & x+y \\ x-y & z \end{bmatrix} = \begin{bmatrix} 3 & 4 \\ 5 & 6 \end{bmatrix}$, find the value of each variable. $w = 3, x = 4.5, y = -0.5, z = 6$
4. If a and b are real numbers and $\begin{bmatrix} x+a & y-b \end{bmatrix} = \begin{bmatrix} 0 & 1 \end{bmatrix}$, find the values of x and y. $x = -a, y = b + 1$

Two matrices are equal if and only if they have the same dimensions and the elements in all corresponding positions are equal. For example,

$$\begin{bmatrix} 3 & 2 & \frac{5}{5} \\ -5 & 0 & \frac{4}{6} \end{bmatrix} = \begin{bmatrix} 3 & 2 & 1 \\ -5 & 0 & \frac{2}{3} \end{bmatrix} \quad \text{but} \quad \begin{bmatrix} 6 & 2 \\ \sqrt{2} & -4 \end{bmatrix} \neq \begin{bmatrix} 6 & 2 \\ -4 & \sqrt{2} \end{bmatrix}$$

Example 2 Find the value of each variable.

$$\begin{bmatrix} x+5 & -1 \\ 4 & 6 \end{bmatrix} = \begin{bmatrix} 2 & -1 \\ 4 & 3y \end{bmatrix}$$

Solution Since the matrices are equal, elements in corresponding positions are equal:

$$x + 5 = 2 \quad \text{and} \quad 3y = 6$$
$$\therefore \quad x = -3 \quad \text{and} \quad y = 2 \quad \textit{Answer}$$

Oral Exercises

Let $A = \begin{bmatrix} 3 & -4 & \frac{1}{2} & 0 \\ 9 & 0 & -3 & 5 \\ 2 & 6 & 1 & -6 \end{bmatrix}$. **Give each of the following.**

1. The elements in row 2. 9, 0, −3, 5
2. The elements in column 1. 3, 9, 2
3. The elements in column 3. $\frac{1}{2}$, −3, 1
4. The elements in row 3. 2, 6, 1, −6
5. The number of rows in A. 3
6. The number of columns in A. 4

Give the dimensions of each matrix.

7. $\begin{bmatrix} 9 & 2 & -4 \\ 3 & \frac{1}{2} & 6 \end{bmatrix}$ 2×3
8. $\begin{bmatrix} 4 & 1 \\ 6 & 0 \\ 9 & -4 \end{bmatrix}$ 3×2
9. $\begin{bmatrix} 12 \end{bmatrix}$ 1×1
10. $\begin{bmatrix} 6 \\ 0 \\ -8 \\ 1 \end{bmatrix}$ 4×1
11. $\begin{bmatrix} a & b \\ c & d \end{bmatrix}$ 2×2
12. $\begin{bmatrix} 4 & 9 & -3 & 0 & 2 \end{bmatrix}$ 1×5
13. $\begin{bmatrix} 4 \\ -5 \\ 3 \\ 14 \end{bmatrix}$ 4×1
14. $\begin{bmatrix} 0 & 0 & 0 & 0 \\ 0 & 0 & 0 & 0 \\ 0 & 0 & 0 & 0 \end{bmatrix}$ 3×4

List the matrices in Exercises 7–14 above that can be described as follows.

15. a row matrix 9, 12
16. a column matrix 9, 10, 13
17. a square matrix 9, 11
18. a zero matrix 14

Written Exercises

Write the zero matrix denoted by each of the following.

A **1.** $O_{1\times 2}$ $\begin{bmatrix} 0 & 0 \end{bmatrix}$ **2.** $O_{3\times 1}$ $\begin{bmatrix} 0 \\ 0 \\ 0 \end{bmatrix}$ **3.** $O_{2\times 5}$ $\begin{bmatrix} 0 & 0 & 0 & 0 & 0 \\ 0 & 0 & 0 & 0 & 0 \end{bmatrix}$ **4.** $O_{4\times 3}$ $\begin{bmatrix} 0 & 0 & 0 \\ 0 & 0 & 0 \\ 0 & 0 & 0 \\ 0 & 0 & 0 \end{bmatrix}$

Find the value of each variable.

5. $O_{2\times 2} = \begin{bmatrix} x & y \\ z & 0 \end{bmatrix}$ $x = y = z = 0$

6. $O_{3\times 2} = \begin{bmatrix} 0 & y \\ x & 0 \\ 0 & z \end{bmatrix}$ $x = y = z = 0$

7. $O_{2\times 3} = \begin{bmatrix} x+4 & 5y & 10-z \\ 0 & 0 & 0 \end{bmatrix}$ $x = -4,\ y = 0,\ z = 10$

8. $\begin{bmatrix} 5-x \\ y+6 \\ -3z \end{bmatrix} = O_{3\times 1}$ $x = 5,\ y = -6,\ z = 0$

9. $\begin{bmatrix} 1 \\ 0 \\ x \end{bmatrix} = \begin{bmatrix} y \\ 0 \\ -4 \end{bmatrix}$ $x = -4,\ y = 1$

10. $\begin{bmatrix} 2 & x & 0 \end{bmatrix} = \begin{bmatrix} y & 4 & 0 \end{bmatrix}$ $x = 4,\ y = 2$

11. $\begin{bmatrix} 3 & 2 \\ -1 & 0 \\ x & y \end{bmatrix} = \begin{bmatrix} 3 & 2 \\ -1 & 0 \\ 8 & -5 \end{bmatrix}$ $x = 8,\ y = -5$

12. $\begin{bmatrix} 4 & x & -3 & y \\ 19 & -1 & 5 & 0 \end{bmatrix} = \begin{bmatrix} z & -1 & -3 & 6 \\ 19 & -1 & 5 & 0 \end{bmatrix}$ $x = -1,\ y = 6,\ z = 4$

B **13.** $\begin{bmatrix} x-3 \\ 12 \\ 0 \end{bmatrix} = \begin{bmatrix} 8 \\ y \\ z+4 \end{bmatrix}$ $x = 11,\ y = 12,\ z = -4$

14. $\begin{bmatrix} 2x & 4 \\ -1 & 8 \end{bmatrix} = \begin{bmatrix} -6 & y+2 \\ z & 8 \end{bmatrix}$ $x = -3,\ y = 2,\ z = -1$

15. $\begin{bmatrix} x+y & 1 \\ 0 & x-y \end{bmatrix} = \begin{bmatrix} 2 & 1 \\ 0 & 8 \end{bmatrix}$ $x = 5,\ y = -3$

16. $\begin{bmatrix} 8-y & 0 \\ x+2z & x+z \end{bmatrix} = \begin{bmatrix} -3y & 0 \\ -5 & -1 \end{bmatrix}$ $x = 3,\ y = -4,\ z = -4$

17. $\begin{bmatrix} 2x+3y \\ x-y \end{bmatrix} = \begin{bmatrix} 3 \\ 4 \end{bmatrix}$ $x = 3,\ y = -1$

18. $\begin{bmatrix} 2x-y \\ 3x+5y \end{bmatrix} = \begin{bmatrix} 6 \\ 22 \end{bmatrix}$ $x = 4,\ y = 2$

Let a, b, c, and d be nonzero real numbers. Find the values of x and y in each of the equations below.

19. $\begin{bmatrix} ax \\ by \end{bmatrix} = \begin{bmatrix} 1 \\ 0 \end{bmatrix}$ $x = \frac{1}{a}$, $y = 0$

20. $\begin{bmatrix} ax+y \\ bx+y \end{bmatrix} = \begin{bmatrix} 1 \\ 0 \end{bmatrix}$ $x = \frac{1}{a-b};\ a \neq b$, $y = \frac{-b}{a-b};\ a \neq b$

21. $\begin{bmatrix} ax+b & cy+d \end{bmatrix} = \begin{bmatrix} 1 & 0 \end{bmatrix}$ $x = \frac{1-b}{a},\ y = \frac{-d}{c}$

22. $\begin{bmatrix} ax+by & cx+dy \end{bmatrix} = \begin{bmatrix} 1 & 1 \end{bmatrix}$ $x = \frac{d-b}{ad-bc},\ y = \frac{a-c}{ad-bc};\ ad - bc \neq 0$

Mixed Review Exercises

Solve each triangle. Give lengths to three significant digits and angle measures to the nearest tenth of a degree.

1. $a = 7,\ b = 10,\ \angle C = 49°$ $\angle A = 44.3°,\ \angle B = 86.7°,\ c = 7.56$

2. $c = 8,\ \angle C = 90°,\ \angle B = 30°$ $\angle A = 60°,\ a = 4\sqrt{3},\ b = 4$

3. $a = \frac{3\sqrt{2}}{2},\ b = \frac{3\sqrt{2}}{2},\ \angle C = 90°$ $\angle A = 45°,\ \angle B = 45°,\ c = 3$

4. $a = 24,\ b = 17,\ \angle A = 72°$ $\angle B = 42.4°,\ \angle C = 65.6°,\ c = 23.0$

5–8. Find the area of each triangle in Exercises 1–4 above. Give answers to three significant digits. **5.** 26.4 **6.** 13.9 **7.** 2.25 **8.** 186.0

Check for Understanding

Here is a suggested use of the Oral Exercises to check students' understanding as you teach the lesson.

Oral Exs. 1–18: use after the boxed definitions on page 767.

Guided Practice

1. Write the zero matrix $O_{3\times 3}$.

$\begin{bmatrix} 0 & 0 & 0 \\ 0 & 0 & 0 \\ 0 & 0 & 0 \end{bmatrix}$

Find the value of each variable.

2. $\begin{bmatrix} x & 0 \\ y+3 & 2z+1 \end{bmatrix} = O_{2\times 2}$

$x = 0,\ y = -3,\ z = -\frac{1}{2}$

3. $\begin{bmatrix} x+y & 3 \\ x-y & 5 \end{bmatrix} = \begin{bmatrix} 7 & 3 \\ 1 & 5 \end{bmatrix}$

$x = 4,\ y = 3$

Summarizing the Lesson

In this lesson students learned matrix terminology. Ask the students how many columns, how many rows, and how many elements an $m \times n$ matrix has. Also ask the students what must be true about m and n for an $m \times n$ matrix to be a row matrix, a column matrix, or a square matrix.

Suggested Assignments

Extended Algebra
769/1–20

Extended Alg. and Trig.
Day 1: 769/1–14
Day 2: 769/15–22
S 769/Mixed Review
Assign with Lesson 16-2.

Supplementary Materials

Study Guide pp. 259–260

Teaching Suggestions p. T142

Suggested Extensions p. T142

Warm-Up Exercises

Name the property illustrated by each statement.

1. $a + b = b + a$
   Commutative Prop. of Add.
2. $a(bc) = (ab)c$
   Associative Prop. of Mult.
3. $a(b + c) = ab + ac$
   Distributive Prop.
4. $a + (-a) = 0$
   Inverse Prop. of Add.
5. $1 \cdot a = a$
   Identity Prop. of Mult.

Motivating the Lesson

Ask the students to suppose that the sales director for a large company has just received reports showing sales of the company's various products in different regions of the country. To get an overall picture, the sales director needs to summarize these reports by adding the sales figures for corresponding products together. This is the idea behind adding matrices, one of the topics of today's lesson.

16-2 Addition and Scalar Multiplication

Objective To find sums and differences of matrices and products of a scalar and a matrix.

The **sum of matrices** having the same dimensions is the matrix whose elements are the sums of the corresponding elements of the matrices being added.

Example 1 Simplify: $\begin{bmatrix} 4 & 7 \\ 0 & -2 \\ 1 & -6 \end{bmatrix} + \begin{bmatrix} 2 & -5 \\ -3 & 2 \\ 0 & -4 \end{bmatrix}$

Solution

$$\begin{bmatrix} 4 & 7 \\ 0 & -2 \\ 1 & -6 \end{bmatrix} + \begin{bmatrix} 2 & -5 \\ -3 & 2 \\ 0 & -4 \end{bmatrix} = \begin{bmatrix} 4+2 & 7+(-5) \\ 0+(-3) & (-2)+2 \\ 1+0 & (-6)+(-4) \end{bmatrix}$$

$$= \begin{bmatrix} 6 & 2 \\ -3 & 0 \\ 1 & -10 \end{bmatrix} \quad \textbf{Answer}$$

Since matrices with different dimensions do not have corresponding elements, *addition of matrices of different dimensions is not defined.* In each set of $m \times n$ matrices, addition is both commutative and associative. (Exercises 21 and 22 ask you to prove these facts for the set of 2×2 matrices.)

For each set of $m \times n$ matrices, $O_{m \times n}$ is the *identity* for addition. For example, if $A_{2\times 2} = \begin{bmatrix} a & b \\ c & d \end{bmatrix}$, then

$$A_{2\times 2} + O_{2\times 2} = \begin{bmatrix} a & b \\ c & d \end{bmatrix} + \begin{bmatrix} 0 & 0 \\ 0 & 0 \end{bmatrix} = \begin{bmatrix} a+0 & b+0 \\ c+0 & d+0 \end{bmatrix} = \begin{bmatrix} a & b \\ c & d \end{bmatrix} = A_{2\times 2}.$$

For each set of $m \times n$ matrices, $O_{m \times n}$ is the **identity** for addition.

$$\boldsymbol{A_{m\times n} + O_{m\times n} = O_{m\times n} + A_{m\times n} = A_{m\times n}.}$$

The *additive inverse of matrix A* is the matrix $-A$. Each element of $-A$ is the opposite of its corresponding element in A. For example,

$$\text{if } A = \begin{bmatrix} 2 & -1 \\ 3 & 0 \end{bmatrix}, \text{ then } -A = \begin{bmatrix} -2 & 1 \\ -3 & 0 \end{bmatrix}.$$

For each set of $m \times n$ matrices, the **additive inverse** of A is the matrix $-A$.

$$\boldsymbol{A_{m\times n} + (-A_{m\times n}) = -A_{m\times n} + A_{m\times n} = O_{m\times n}.}$$

As with real numbers, *subtraction of matrices* is defined in terms of addition.

For each set of $m \times n$ matrices, **subtraction** is defined as follows.

$$A_{m\times n} - B_{m\times n} = A_{m\times n} + (-B_{m\times n}).$$

Example 2 Let $A = \begin{bmatrix} 6 & -2 \\ 5 & 4 \end{bmatrix}$ and $B = \begin{bmatrix} 4 & 3 \\ 5 & -2 \end{bmatrix}$. Find $A - B$.

Solution Apply the definition of subtraction: $A - B = A + (-B)$.

$$\begin{bmatrix} 6 & -2 \\ 5 & 4 \end{bmatrix} - \begin{bmatrix} 4 & 3 \\ 5 & -2 \end{bmatrix} = \begin{bmatrix} 6 & -2 \\ 5 & 4 \end{bmatrix} + \begin{bmatrix} -4 & -3 \\ -5 & 2 \end{bmatrix} = \begin{bmatrix} 2 & -5 \\ 0 & 6 \end{bmatrix}$$

Answer

The properties of addition of matrices are summarized below.

Properties of Addition of Matrices

Let A, B, and C be $m \times n$ matrices. Let $O_{m\times n}$ be the $m \times n$ zero matrix.

1. Closure Property — $A + B$ **is an** $m \times n$ **matrix.**
2. Commutative Property — $A + B = B + A$
3. Associative Property — $(A + B) + C = A + (B + C)$
4. Identity Property — $A + O_{m\times n} = O_{m\times n} + A = A$
5. Inverse Property — $A + (-A) = -A + A = O_{m\times n}$

In matrix algebra a real number is called a **scalar.** The **scalar product** of a real number r and a matrix A is the matrix rA. Each element of rA is r times its corresponding element in A. In the expression $2A$, the number 2 is a scalar and $2A$ (which is another matrix) is the scalar product of 2 and A.

Example 3 Let $A = \begin{bmatrix} 2 & 0 \\ -2 & 3 \end{bmatrix}$. Find $3A$ and rA.

Solution

$$3A = 3\begin{bmatrix} 2 & 0 \\ -2 & 3 \end{bmatrix} = \begin{bmatrix} 3 \cdot 2 & 3 \cdot 0 \\ 3(-2) & 3 \cdot 3 \end{bmatrix} = \begin{bmatrix} 6 & 0 \\ -6 & 9 \end{bmatrix} \quad \textit{Answer}$$

$$rA = r\begin{bmatrix} 2 & 0 \\ -2 & 3 \end{bmatrix} = \begin{bmatrix} r \cdot 2 & r \cdot 0 \\ r(-2) & r \cdot 3 \end{bmatrix} = \begin{bmatrix} 2r & 0 \\ -2r & 3r \end{bmatrix} \quad \textit{Answer}$$

The properties of scalar multiplication are summarized in the chart at the top of the next page.

Chalkboard Examples

Let $A = \begin{bmatrix} 1 & 3 & -2 \\ 4 & 0 & -5 \end{bmatrix}$ and $B = \begin{bmatrix} 3 & 1 & -5 \\ 0 & -7 & 2 \end{bmatrix}$.

1. Find $A + B$.

$\begin{bmatrix} 1+3 & 3+1 & -2+(-5) \\ 4+0 & 0+(-7) & -5+2 \end{bmatrix} = \begin{bmatrix} 4 & 4 & -7 \\ 4 & -7 & -3 \end{bmatrix}$

2. Find $A - B$.

$\begin{bmatrix} 1-3 & 3-1 & -2-(-5) \\ 4-0 & 0-(-7) & -5-2 \end{bmatrix} = \begin{bmatrix} -2 & 2 & 3 \\ 4 & 7 & -7 \end{bmatrix}$

3. Find $-2B$.

$\begin{bmatrix} -2(3) & -2(1) & -2(-5) \\ -2(0) & -2(-7) & -2(2) \end{bmatrix} = \begin{bmatrix} -6 & -2 & 10 \\ 0 & 14 & -4 \end{bmatrix}$

4. Solve $3X + \begin{bmatrix} -2 & 1 \\ 3 & 0 \end{bmatrix} = 2\begin{bmatrix} -1 & -4 \\ 0 & 6 \end{bmatrix}$.

$3X + \begin{bmatrix} -2 & 1 \\ 3 & 0 \end{bmatrix} = \begin{bmatrix} -2 & -8 \\ 0 & 12 \end{bmatrix}$

$3X = \begin{bmatrix} -2 & -8 \\ 0 & 12 \end{bmatrix} + \begin{bmatrix} 2 & -1 \\ -3 & 0 \end{bmatrix}$

$3X = \begin{bmatrix} 0 & -9 \\ -3 & 12 \end{bmatrix}$

$X = \frac{1}{3}\begin{bmatrix} 0 & -9 \\ -3 & 12 \end{bmatrix} = \begin{bmatrix} 0 & -3 \\ -1 & 4 \end{bmatrix}$

Check for Understanding

Oral Exs. 1–8: use after Example 2.
Oral Exs. 9–13: use after Example 3.

Common Errors

When subtracting matrices, some students may forget to add the inverse of each element of the second matrix to the corresponding element of the first. When reteaching, have these students rewrite the subtraction problem as an addition problem after finding the complete inverse of the second matrix.

Guided Practice

Simplify.

1. $\begin{bmatrix} 5 & -2 \\ 3 & -7 \end{bmatrix} + \begin{bmatrix} 4 & 2 \\ -1 & -1 \end{bmatrix}$ $\begin{bmatrix} 9 & 0 \\ 2 & -8 \end{bmatrix}$

2. $-\frac{1}{2}\begin{bmatrix} 4 & 0 \\ -2 & 1 \\ 0 & 6 \end{bmatrix}$ $\begin{bmatrix} -2 & 0 \\ 1 & -\frac{1}{2} \\ 0 & -3 \end{bmatrix}$

3. $\begin{bmatrix} 7 \\ 2 \\ -8 \end{bmatrix} - \begin{bmatrix} 9 \\ -5 \\ -1 \end{bmatrix}$ $\begin{bmatrix} -2 \\ 7 \\ -7 \end{bmatrix}$

4. $\begin{bmatrix} 7 & -1 \\ 2 & -3 \end{bmatrix} - 2\begin{bmatrix} 0 & 4 \\ -5 & -9 \end{bmatrix}$ $\begin{bmatrix} 7 & -9 \\ 12 & 15 \end{bmatrix}$

5. $O_{1\times 3} + 2\begin{bmatrix} 1 & 3 & -2 \end{bmatrix}$ $\begin{bmatrix} 2 & 6 & -4 \end{bmatrix}$

Properties of Scalar Multiplication

Let A and B be $m \times n$ matrices. Let $O_{m\times n}$ be the $m \times n$ zero matrix, and let p and q be scalars.

1. Closure Property — pA **is an** $m \times n$ **matrix.**
2. Commutative Property — $pA = Ap$
3. Associative Property — $p(qA) = (pq)A$
4. Distributive Property — $(p + q)A = pA + qA$; $p(A + B) = pA + pB$
5. Identity Property — $1 \cdot A = A$
6. Multiplicative Property of -1 — $(-1)A = -A$
7. Multiplicative Property of 0 — $0 \cdot A = O_{m\times n}$

An equation in which the variable stands for a matrix is called a **matrix equation.** You can solve some simple matrix equations by using matrix addition and scalar multiplication.

Example 4 Solve for the matrix X: $2X + 2\begin{bmatrix} 1 & -2 \\ 0 & 3 \end{bmatrix} = 8\begin{bmatrix} \frac{1}{2} & 0 \\ -1 & \frac{1}{4} \end{bmatrix}$

Solution Find the scalar products:

$$2X + \begin{bmatrix} 2 & -4 \\ 0 & 6 \end{bmatrix} = \begin{bmatrix} 4 & 0 \\ -8 & 2 \end{bmatrix}$$

Add $\begin{bmatrix} -2 & 4 \\ 0 & -6 \end{bmatrix}$, the inverse of $\begin{bmatrix} 2 & -4 \\ 0 & 6 \end{bmatrix}$, to each side.

$$2X + \begin{bmatrix} 2 & -4 \\ 0 & 6 \end{bmatrix} + \begin{bmatrix} -2 & 4 \\ 0 & -6 \end{bmatrix} = \begin{bmatrix} 4 & 0 \\ -8 & 2 \end{bmatrix} + \begin{bmatrix} -2 & 4 \\ 0 & -6 \end{bmatrix}$$

$$2X + O_{2\times 2} = \begin{bmatrix} 2 & 4 \\ -8 & -4 \end{bmatrix}$$

$$X = \frac{1}{2}\begin{bmatrix} 2 & 4 \\ -8 & -4 \end{bmatrix} = \begin{bmatrix} 1 & 2 \\ -4 & -2 \end{bmatrix}$$

The check is left for you. ***Answer***

Oral Exercises

Give the value of each element of the sum or difference.

$$\begin{bmatrix} 4 & 0 \\ 2 & 1 \end{bmatrix} + \begin{bmatrix} 12 & -4 \\ 9 & 3 \end{bmatrix} = \begin{bmatrix} a & b \\ c & d \end{bmatrix} \qquad \begin{bmatrix} 9 & 0 \\ -2 & 7 \end{bmatrix} - \begin{bmatrix} 3 & 7 \\ 8 & -5 \end{bmatrix} = \begin{bmatrix} e & f \\ g & h \end{bmatrix}$$

1. a 16 **2.** b −4 **3.** c 11 **4.** d 4 **5.** e 6 **6.** f −7 **7.** g −10 **8.** h 12

Give the value of each element of the scalar product.

$$-2[15 \quad -2 \quad 0 \quad 8 \quad \sqrt{3}] = [p \quad q \quad r \quad s \quad t]$$

9. p −30 **10.** q 4 **11.** r 0 **12.** s −16 **13.** t $-2\sqrt{3}$

Written Exercises

7. $\begin{bmatrix} -1 & 3 & 4 \\ 0 & 0 & 0 \end{bmatrix}$ **8.** $\begin{bmatrix} -14 & 3 & -20 \\ -3 & 5 & -5 \end{bmatrix}$ **12.** $\begin{bmatrix} -15 & 0 & 5 \\ -30 & -10 & -20 \end{bmatrix}$

Simplify.

A

1. $\begin{bmatrix} 4 & 1 \\ -2 & 0 \end{bmatrix} + \begin{bmatrix} 0 & 3 \\ 2 & 3 \end{bmatrix}$ $\begin{bmatrix} 4 & 4 \\ 0 & 3 \end{bmatrix}$

2. $\begin{bmatrix} 2 & -3 \\ -2 & 10 \end{bmatrix} + \begin{bmatrix} -2 & 3 \\ 2 & -8 \end{bmatrix}$ $\begin{bmatrix} 0 & 0 \\ 0 & 2 \end{bmatrix}$

3. $\begin{bmatrix} 8 & 0 \\ 4 & 10 \end{bmatrix} + \begin{bmatrix} 3 & 4 \\ -1 & 6 \end{bmatrix}$ $\begin{bmatrix} 11 & 4 \\ 3 & 16 \end{bmatrix}$

4. $\begin{bmatrix} -8 & 5 \\ 3 & 0 \end{bmatrix} + \begin{bmatrix} 4 & -2 \\ -2 & 9 \end{bmatrix}$ $\begin{bmatrix} -4 & 3 \\ 1 & 9 \end{bmatrix}$

5. $\begin{bmatrix} 4 & 7 \\ 2 & -1 \\ 0 & 5 \end{bmatrix} - \begin{bmatrix} 0 & 5 \\ 2 & 4 \\ 3 & -2 \end{bmatrix}$ $\begin{bmatrix} 4 & 2 \\ 0 & -5 \\ -3 & 7 \end{bmatrix}$

6. $\begin{bmatrix} 5 & 8 \\ -2 & 14 \\ 0 & -6 \end{bmatrix} - \begin{bmatrix} 5 & 4 \\ 4 & -2 \\ -5 & -5 \end{bmatrix}$ $\begin{bmatrix} 0 & 4 \\ -6 & 16 \\ 5 & -1 \end{bmatrix}$

7. $\begin{bmatrix} 0 & -2 & 4 \\ 3 & -2 & -6 \end{bmatrix} - \begin{bmatrix} 1 & -5 & 0 \\ 3 & -2 & -6 \end{bmatrix}$

8. $\begin{bmatrix} 4 & 6 & -8 \\ -2 & 5 & 0 \end{bmatrix} - \begin{bmatrix} 18 & 3 & 12 \\ 1 & 0 & 5 \end{bmatrix}$

9. $6\begin{bmatrix} 7 & 8 \\ -4 & 0 \end{bmatrix}$ $\begin{bmatrix} 42 & 48 \\ -24 & 0 \end{bmatrix}$

10. $12\begin{bmatrix} 3 & 0 \\ -5 & 2 \end{bmatrix}$ $\begin{bmatrix} 36 & 0 \\ -60 & 24 \end{bmatrix}$

11. $O_{2\times3} + \begin{bmatrix} 3 & 5 & 2 \\ -2 & 8 & 0 \end{bmatrix}$ $\begin{bmatrix} 3 & 5 & 2 \\ -2 & 8 & 0 \end{bmatrix}$

12. $O_{2\times3} - 5\begin{bmatrix} 3 & 0 & -1 \\ 6 & 2 & 4 \end{bmatrix}$

13. $2\begin{bmatrix} 5 & -2 \\ -3 & 4 \\ 0 & 6 \end{bmatrix} + \begin{bmatrix} 1 & 7 \\ 0 & -4 \\ 6 & 5 \end{bmatrix}$ $\begin{bmatrix} 11 & 3 \\ -6 & 4 \\ 6 & 17 \end{bmatrix}$

14. $\begin{bmatrix} 4 & 2 \\ 1 & -3 \end{bmatrix} + \begin{bmatrix} -4 & -2 \\ -1 & 3 \end{bmatrix}$ $\begin{bmatrix} 0 & 0 \\ 0 & 0 \end{bmatrix}$

B

15. $3\begin{bmatrix} 2 \\ 2 \\ 5 \end{bmatrix} + 5\begin{bmatrix} 0 \\ 6 \\ 3 \end{bmatrix}$ $\begin{bmatrix} 6 \\ 36 \\ 30 \end{bmatrix}$

16. $\begin{bmatrix} 3 & 0 \\ 5 & -4 \\ 0 & -3 \end{bmatrix} - 4\begin{bmatrix} 10 & -2 \\ 2 & 4 \\ -7 & 0 \end{bmatrix}$ $\begin{bmatrix} -37 & 8 \\ -3 & -20 \\ 28 & -3 \end{bmatrix}$

Solve each equation for matrix *X*.

17. $X + \begin{bmatrix} 3 & 2 \\ 1 & 0 \end{bmatrix} = \begin{bmatrix} 6 & 3 \\ 7 & -1 \end{bmatrix}$ $\begin{bmatrix} 3 & 1 \\ 6 & -1 \end{bmatrix}$

18. $X + \begin{bmatrix} 0 & 4 \\ 9 & -1 \end{bmatrix} = \begin{bmatrix} 2 & 0 \\ -1 & 2 \end{bmatrix}$ $\begin{bmatrix} 2 & -4 \\ -10 & 3 \end{bmatrix}$

19. $X + 3\begin{bmatrix} -3 & 2 \\ 0 & -1 \end{bmatrix} = \begin{bmatrix} -4 & 10 \\ 12 & 0 \end{bmatrix}$ $\begin{bmatrix} 5 & 4 \\ 12 & 3 \end{bmatrix}$

20. $X - \begin{bmatrix} 0 & 5 \\ 1 & -2 \end{bmatrix} = 3\begin{bmatrix} 2 & -3 \\ 3 & 1 \end{bmatrix}$ $\begin{bmatrix} 6 & -4 \\ 10 & 1 \end{bmatrix}$

Let p and q be scalars and let A, B, and C be 2×2 matrices. Prove these properties for all 2×2 matrices.

C

21. $A + B = B + A$

22. $(A + B) + C = A + (B + C)$

23. $(-1)A = -A$

24. $pA = Ap$

25. $(p + q)A = pA + qA$

26. $(pq)B = p(qB)$

27. $p \cdot O_{2\times2} = O_{2\times2}$

28. $0 \cdot A = O_{2\times2}$

Summarizing the Lesson

In this lesson students learned to add and subtract matrices and to multiply a matrix by a scalar. Ask the students to state in general terms how each of these operations is performed.

Suggested Assignments

Extended Algebra
Day 1: 773/1–18
S 769/21, 22
Day 2: 773/19–21, 23
Assign with Lesson 16-3.

Extended Alg. and Trig.
773/1–23 odd
Assign with Lesson 16-1.

Supplementary Materials

Study Guide pp. 261–262
Practice Master 88

Explorations p. 847

Teaching Suggestions p. T142

Suggested Extensions p. T142

Warm-Up Exercises

Let $A = \begin{bmatrix} 2 & -1 \\ 3 & 2 \end{bmatrix}$ and $B = \begin{bmatrix} 0 & 2 \\ 1 & 1 \end{bmatrix}$.

Write the indicated matrix.

1. $A + B$ $\begin{bmatrix} 2 & 1 \\ 4 & 3 \end{bmatrix}$
2. $A - B$ $\begin{bmatrix} 2 & -3 \\ 2 & 1 \end{bmatrix}$
3. $B - A$ $\begin{bmatrix} -2 & 3 \\ -2 & -1 \end{bmatrix}$
4. $2B - 3A$ $\begin{bmatrix} -6 & 7 \\ -7 & -4 \end{bmatrix}$
5. X, if $A + X = B$ $\begin{bmatrix} -2 & 3 \\ -2 & -1 \end{bmatrix}$
6. X, if $3X = A$ $\begin{bmatrix} \frac{2}{3} & -\frac{1}{3} \\ 1 & \frac{2}{3} \end{bmatrix}$

Motivating the Lesson

Ask the students to state the commutative property of multiplication. Point out that this property is often taken for granted when real numbers are multiplied. In the case of matrix multiplication, however, there is no commutative property, as today's lesson will show.

Chalkboard Examples

Multiply if possible. If not, say so.

1. $\begin{bmatrix} 1 & -1 \\ 2 & 5 \end{bmatrix}\begin{bmatrix} 3 \\ 2 \end{bmatrix}$

$\begin{bmatrix} 1(3) + (-1)2 \\ 2(3) + 5(2) \end{bmatrix} = \begin{bmatrix} 1 \\ 16 \end{bmatrix}$

2. $\begin{bmatrix} -4 \\ 1 \end{bmatrix}\begin{bmatrix} 1 & 2 \\ 0 & -3 \end{bmatrix}$

cannot multiply

16-3 Matrix Multiplication

Objective To find the product of two matrices.

The method of multiplying two matrices is illustrated by this example.

A discount department store sells jeans, sweaters, and shirts. The prices in dollars are shown in the 1×3 matrix, A, below.

$$A = \begin{bmatrix} \text{Jeans} & \text{Sweaters} & \text{Shirts} \\ 22 & 15 & 6 \end{bmatrix}$$

In June, sales of jeans, sweaters, and shirts numbered in order 20, 30, and 48. In July, sales were 15, 20, and 72. Sales are displayed in the 3×2 matrix, B, below.

$$\begin{matrix} & \text{June} & \text{July} \end{matrix}$$
$$B = \begin{bmatrix} 20 & 15 \\ 30 & 20 \\ 48 & 72 \end{bmatrix}$$

In June, income from sales of jeans was \$22 · 20; from sweaters, \$15 · 30; and from shirts, \$6 · 48. Income from sales in June and July is shown below:

June: $22 \cdot 20 + 15 \cdot 30 + 6 \cdot 48$

July: $22 \cdot 15 + 15 \cdot 20 + 6 \cdot 72$

The monthly income from these sales is shown in matrix S.

$$S = \begin{bmatrix} \text{June} & \text{July} \\ (440 + 450 + 288) & (330 + 300 + 432) \end{bmatrix} = \begin{bmatrix} 1178 & 1062 \end{bmatrix}$$

Matrix S is the *product* of matrices A and B:

$$S = A \times B = \begin{bmatrix} 22 & 15 & 6 \end{bmatrix}\begin{bmatrix} 20 & 15 \\ 30 & 20 \\ 48 & 72 \end{bmatrix}$$
$$= \begin{bmatrix} (22 \cdot 20 + 15 \cdot 30 + 6 \cdot 48) & (22 \cdot 15 + 15 \cdot 20 + 6 \cdot 72) \end{bmatrix}$$
$$= \begin{bmatrix} (440 + 450 + 288) & (330 + 300 + 432) \end{bmatrix}$$
$$= \begin{bmatrix} 1178 & 1062 \end{bmatrix}$$

Each element in S is obtained by multiplying elements of the *row* in A with elements of a *column* in B. This example suggests the following definition.

The **product of matrices** $A_{m \times n}$ and $B_{n \times p}$ is the $m \times p$ matrix whose element in the ath row and bth column is the sum of the products of corresponding elements of the ath row of A and the bth column of B.

Two matrices can be multiplied only if the second dimension of the first matrix equals the first dimension of the second matrix. For example,

$$\text{if } A_{1\times2} = \begin{bmatrix} 3 & -1 \end{bmatrix} \text{ and } B_{2\times3} = \begin{bmatrix} 1 & 0 & 2 \\ 0 & 1 & -1 \end{bmatrix},$$

$$\text{then } A_{1\times2} \cdot B_{2\times3} = \begin{bmatrix} 3 & -1 \end{bmatrix}\begin{bmatrix} 1 & 0 & 2 \\ 0 & 1 & -1 \end{bmatrix} = \begin{bmatrix} 3 & -1 & 7 \end{bmatrix};$$

$$\text{but } B_{2\times3} \cdot A_{1\times2} = \begin{bmatrix} 1 & 0 & 2 \\ 0 & 1 & -1 \end{bmatrix}\begin{bmatrix} 3 & -1 \end{bmatrix} \text{ is not defined.}$$

The product $A_{1\times2} \cdot B_{2\times3}$ is a 1×3 matrix. However, matrix multiplication is not defined for $B_{2\times3} \cdot A_{1\times2}$ because there are more elements in each row of B (three elements) than there are in each column of A (one element).

Example 1 Multiply: $\begin{bmatrix} 3 & 1 & 0 \\ -1 & 4 & 2 \end{bmatrix}\begin{bmatrix} -1 & 0 \\ 2 & 3 \\ 5 & 0 \end{bmatrix}$

Solution A 2×3 matrix times a 3×2 matrix gives a 2×2 matrix.

$$\begin{bmatrix} 3 & 1 & 0 \\ -1 & 4 & 2 \end{bmatrix}\begin{bmatrix} -1 & 0 \\ 2 & 3 \\ 5 & 0 \end{bmatrix} = \begin{bmatrix} & \\ & \end{bmatrix}$$

row 1 × column 1:
$3(-1) + 1 \cdot 2 + 0 \cdot 5 =$

$$\begin{bmatrix} 3 & 1 & 0 \\ -1 & 4 & 2 \end{bmatrix}\begin{bmatrix} -1 & 0 \\ 2 & 3 \\ 5 & 0 \end{bmatrix} = \begin{bmatrix} -1 & \\ & \end{bmatrix}$$

row 1 × column 2:
$3 \cdot 0 + 1 \cdot 3 + 0 \cdot 0 =$

$$\begin{bmatrix} 3 & 1 & 0 \\ -1 & 4 & 2 \end{bmatrix}\begin{bmatrix} -1 & 0 \\ 2 & 3 \\ 5 & 0 \end{bmatrix} = \begin{bmatrix} -1 & 3 \\ & \end{bmatrix}$$

row 2 × column 1:
$(-1)(-1) + 4 \cdot 2 + 2 \cdot 5 =$

$$\begin{bmatrix} 3 & 1 & 0 \\ -1 & 4 & 2 \end{bmatrix}\begin{bmatrix} -1 & 0 \\ 2 & 3 \\ 5 & 0 \end{bmatrix} = \begin{bmatrix} -1 & 3 \\ 19 & \end{bmatrix}$$

row 2 × column 2:
$(-1)0 + 4 \cdot 3 + 2 \cdot 0 =$

$\therefore$ the product is $\begin{bmatrix} -1 & 3 \\ 19 & 12 \end{bmatrix}$. ***Answer***

Powers of square matrices are defined just like powers of real numbers.

$$A^2 = A \cdot A, \qquad A^3 = A \cdot A \cdot A, \qquad A^n = \underbrace{A \cdot A \cdot A \cdot \ldots \cdot A}_{n \text{ factors}}$$

An $n \times n$ matrix whose *main diagonal* from upper left to lower right has all elements 1, while all other elements are 0, is called an **identity matrix** and is denoted by $I_{n\times n}$. Here are the 2×2 and 3×3 identity matrices:

$$I_{2\times2} = \begin{bmatrix} 1 & 0 \\ 0 & 1 \end{bmatrix} \text{ and } I_{3\times3} = \begin{bmatrix} 1 & 0 & 0 \\ 0 & 1 & 0 \\ 0 & 0 & 1 \end{bmatrix}$$

3. $\begin{bmatrix} 0 & 2 \\ -2 & 4 \\ 3 & 1 \end{bmatrix} \times \begin{bmatrix} 1 & -5 & 4 \\ -2 & 0 & -6 \end{bmatrix}$

$$\begin{bmatrix} 0(1) + 2(-2) & 0(-5) + 2(0) & 0(4) + 2(-6) \\ -2(1) + 4(-2) & -2(-5) + 4(0) & -2(4) + 4(-6) \\ 3(1) + 1(-2) & 3(-5) + 1(0) & 3(4) + 1(-6) \end{bmatrix} =$$

$$\begin{bmatrix} -4 & 0 & -12 \\ -10 & 10 & -32 \\ 1 & -15 & 6 \end{bmatrix}$$

Common Errors

Students may forget that the second dimension of the first matrix must equal the first dimension of the second matrix in order to multiply two matrices. When reteaching, have students give examples of pairs of matrices that can be multiplied and examples of pairs that cannot.

Students may also need to be reminded that matrix multiplication is not commutative. When reteaching, show the students that $\begin{bmatrix} 2 & 0 \\ 3 & -1 \end{bmatrix}\begin{bmatrix} 2 & 4 \\ 0 & -2 \end{bmatrix} = \begin{bmatrix} 4 & 8 \\ 6 & 14 \end{bmatrix}$ but $\begin{bmatrix} 2 & 4 \\ 0 & -2 \end{bmatrix} \times \begin{bmatrix} 2 & 0 \\ 3 & -1 \end{bmatrix} = \begin{bmatrix} 16 & -4 \\ -6 & 2 \end{bmatrix}$.

Check for Understanding

Oral Exs. 1–10: use before Example 1.
Oral Exs. 11–16: use after Example 1.

Guided Practice

State whether the matrices can be multiplied. If so, find the product.

1. $\begin{bmatrix} 3 & -5 \\ -1 & -2 \end{bmatrix}\begin{bmatrix} 4 & 0 \\ 0 & -1 \\ 1 & -2 \end{bmatrix}$

no

2. $\begin{bmatrix} -2 \\ -3 \\ 1 \end{bmatrix}\begin{bmatrix} 1 & 2 & 3 \end{bmatrix}$

yes; $\begin{bmatrix} -2 & -4 & -6 \\ -3 & -6 & -9 \\ 1 & 2 & 3 \end{bmatrix}$

3. $\begin{bmatrix} 1 & 2 & 3 \end{bmatrix}\begin{bmatrix} -2 \\ -3 \\ 1 \end{bmatrix}$

yes; $\begin{bmatrix} -5 \end{bmatrix}$

4. $\begin{bmatrix} 3 & 2 & 1 \\ 1 & 2 & -4 \\ 0 & 1 & 0 \end{bmatrix}\begin{bmatrix} 4 & 1 \\ 1 & -1 \\ 0 & -2 \end{bmatrix}$

yes; $\begin{bmatrix} 14 & -1 \\ 6 & 7 \\ 1 & -1 \end{bmatrix}$

Summarizing the Lesson

In this lesson students learned to multiply two matrices. Ask the students to state the circumstances under which two matrices *cannot* be multiplied.

You can show (see Exercise 27) that for any $n \times n$ square matrix A,

$$I_{n\times n} \cdot A = A \cdot I_{n\times n} = A.$$

Multiplication of a matrix by a zero matrix yields a zero matrix. In particular, if A is any $n \times n$ square matrix and $O_{n\times n}$ is the zero matrix,

$$O_{n\times n} \cdot A = A \cdot O_{n\times n} = O_{n\times n}.$$

A major difference between real number multiplication and matrix multiplication is the fact that $CD = O_{n\times n}$ does *not* imply that either $C = O_{n\times n}$ or $D = O_{n\times n}$. For example,

$$\text{if } C = \begin{bmatrix} -3 & 1 \\ 6 & -2 \end{bmatrix} \text{ and } D = \begin{bmatrix} 2 & 1 \\ 6 & 3 \end{bmatrix},$$

$$\text{then } CD = \begin{bmatrix} -3 & 1 \\ 6 & -2 \end{bmatrix}\begin{bmatrix} 2 & 1 \\ 6 & 3 \end{bmatrix} = \begin{bmatrix} 0 & 0 \\ 0 & 0 \end{bmatrix} = O_{2\times 2}.$$

Caution: Unlike real number multiplication, matrix multiplication is *not* in general commutative. Here is a counterexample:

$$\text{Let } A = \begin{bmatrix} 0 & 3 \\ 1 & -1 \end{bmatrix} \text{ and } B = \begin{bmatrix} 2 & 0 \\ 1 & 4 \end{bmatrix}.$$

$$\text{Then } AB = \begin{bmatrix} 0 & 3 \\ 1 & -1 \end{bmatrix}\begin{bmatrix} 2 & 0 \\ 1 & 4 \end{bmatrix} = \begin{bmatrix} 3 & 12 \\ 1 & -4 \end{bmatrix}.$$

$$\text{But } BA = \begin{bmatrix} 2 & 0 \\ 1 & 4 \end{bmatrix}\begin{bmatrix} 0 & 3 \\ 1 & -1 \end{bmatrix} = \begin{bmatrix} 0 & 6 \\ 4 & -1 \end{bmatrix}.$$

$\therefore AB \neq BA$

Be sure to compute a product in the order in which it is written.

Although the commutative property does not apply to matrix multiplication, the associative and distributive properties do hold. These and other properties for multiplication of square matrices are summarized below.

Properties of Matrix Multiplication

Let A, B, and C be $n \times n$ matrices. Let $I_{n\times n}$ be the identity matrix and $O_{n\times n}$ be the zero matrix.

1. Associative Property	$(AB)C = A(BC)$
2. Distributive Property	$A(B + C) = AB + AC$ $(B + C)A = BA + CA$
3. Identity Property	$I_{n\times n} \cdot A = A \cdot I_{n\times n} = A$
4. Multiplicative Property of $O_{n\times n}$	$O_{n\times n} \cdot A = A \cdot O_{n\times n} = O_{n\times n}$

Oral Exercises

Give the dimensions of each matrix and tell whether the matrices can be multiplied.

1. $\begin{bmatrix} 3 & 5 \\ 0 & 4 \end{bmatrix}\begin{bmatrix} -1 & 3 \\ 5 & 7 \end{bmatrix}$ yes; 2×2, 2×2

2. $\begin{bmatrix} 6 & 2 & 0 \\ 1 & 0 & 8 \\ 4 & -2 & 5 \end{bmatrix}\begin{bmatrix} 2 & 0 \\ 1 & -3 \\ 0 & 4 \end{bmatrix}$ yes; 3×3, 3×2

3. $\begin{bmatrix} 2 \\ 7 \\ -4 \end{bmatrix}\begin{bmatrix} 1 & 0 & 6 & 4 \end{bmatrix}$ yes; 3×1, 1×4

4. $\begin{bmatrix} 8 & 1 & 0 \end{bmatrix}\begin{bmatrix} 5 & 7 & -2 \\ 0 & 4 & 1 \end{bmatrix}$ no; 1×3, 2×3

5. $\begin{bmatrix} 0 & 0 \\ 0 & 0 \\ 0 & 0 \\ 0 & 0 \end{bmatrix}\begin{bmatrix} 3 & 5 & 9 & 0 \end{bmatrix}$ no; 4×2, 1×4

6. $\begin{bmatrix} -2 & 3 & 0 & 1 \end{bmatrix}\begin{bmatrix} 1 & 0 & 0 \\ 0 & 1 & 0 \\ 0 & 0 & 1 \end{bmatrix}$ no; 1×4, 3×3

Give the dimensions of the product of the matrices.

7. $\begin{bmatrix} 3 & 2 & 1 & 8 \end{bmatrix}\begin{bmatrix} 4 \\ 0 \\ 3 \\ -5 \end{bmatrix}$ 1×1

8. $\begin{bmatrix} 6 & 2 & 8 \\ -1 & 0 & 3 \end{bmatrix}\begin{bmatrix} 9 & 0 \\ 3 & 0 \\ 1 & -4 \end{bmatrix}$ 2×2

9. $\begin{bmatrix} \frac{2}{3} & 0 \\ \frac{1}{5} & \frac{1}{2} \end{bmatrix}\begin{bmatrix} \frac{1}{6} & \frac{1}{10} \\ \frac{1}{3} & 0 \end{bmatrix}$ 2×2

10. $\begin{bmatrix} 0 \\ 3 \\ 5 \end{bmatrix}\begin{bmatrix} 2 & -2 & 0 & 8 & 3 \end{bmatrix}$ 3×5

Give the value of each variable shown in the computation of the product.

$$\begin{bmatrix} -4 & 0 & 2 \\ 5 & 3 & -2 \end{bmatrix}\begin{bmatrix} -1 \\ 5 \\ 1 \end{bmatrix} = \begin{bmatrix} (-4)(-1) + a \cdot 5 + 2c \\ 5(-1) + b \cdot 5 + (-2)d \end{bmatrix} = \begin{bmatrix} e \\ f \end{bmatrix}$$

11. a 0 12. b 3 13. c 1 14. d 1 15. e 6 16. f 8

Written Exercises

Multiply.

A

1. $\begin{bmatrix} 3 & 1 \end{bmatrix}\begin{bmatrix} 4 \\ 6 \end{bmatrix}$ $[18]$

2. $\begin{bmatrix} 0 & -3 & 4 \end{bmatrix}\begin{bmatrix} 1 \\ 0 \\ -4 \end{bmatrix}$ $[-16]$

3. $\begin{bmatrix} 4 \\ -2 \end{bmatrix}\begin{bmatrix} 3 & 0 & -1 & 5 \end{bmatrix}$ $\begin{bmatrix} 12 & 0 & -4 & 20 \\ -6 & 0 & 2 & -10 \end{bmatrix}$

4. $\begin{bmatrix} 0 & 3 \\ 5 & -1 \end{bmatrix}\begin{bmatrix} -1 \\ 3 \end{bmatrix}$ $\begin{bmatrix} 9 \\ -8 \end{bmatrix}$

5. $\begin{bmatrix} 3 & 0 \\ 1 & 2 \end{bmatrix}\begin{bmatrix} -1 & 8 \\ 0 & 3 \end{bmatrix}$ $\begin{bmatrix} -3 & 24 \\ -1 & 14 \end{bmatrix}$

6. $\begin{bmatrix} 0 & 8 \\ 3 & 1 \\ -1 & 5 \end{bmatrix}\begin{bmatrix} 3 & 1 & -2 \\ 0 & 8 & -5 \end{bmatrix}$ 6. $\begin{bmatrix} 0 & 64 & -40 \\ 9 & 11 & -11 \\ -3 & 39 & -23 \end{bmatrix}$

Suggested Assignments

Extended Algebra
Day 1: 777/1–7
Assign with Lesson 16-2.
Day 2: 778/8–28 even
S 778/Mixed Review

Extended Alg. and Trig.
777/1–25 odd
S 778/Mixed Review

Supplementary Materials

Study Guide	pp. 263–264
Practice Master	89
Computer Activity	35

Additional Answers
Written Exercises

7. $\begin{bmatrix} -3 & 0 & 6 \\ 15 & 12 & 2 \\ 3 & 0 & -6 \end{bmatrix}$

8. $\begin{bmatrix} 1 & 0 & 2 \\ 2 & -1 & 4 \\ 0 & -3 & 1 \end{bmatrix}$

9. $\begin{bmatrix} -1 & 2 & 2 & -1 \\ 4 & -1 & -1 & -3 \\ -4 & 4 & 4 & 0 \end{bmatrix}$

10. $\begin{bmatrix} 0 & 2 & 0 \\ -1 & 1 & 1 \\ 1 & 0 & 0 \end{bmatrix}$

11. $\begin{bmatrix} a + 2g & b + 2h & c + 2i \\ -2a + d & -2b + e & -2c + f \\ d & e & f \end{bmatrix}$

12. $\begin{bmatrix} 2r + x & 2s + y & 2t + z \\ 5r - u & 5s - v & 5t - w \\ -2u + 4x & -2v + 4y & -2w + 4z \end{bmatrix}$

(continued)

Additional Answers Written Exercises (continued)

13. $AB = \begin{bmatrix} 2 & 1 \\ -1 & -1 \end{bmatrix}$; $BA = \begin{bmatrix} 0 & 1 \\ 1 & 1 \end{bmatrix}$

14. $BC = \begin{bmatrix} 0 & -1 \\ -2 & 1 \end{bmatrix}$; $CB = \begin{bmatrix} 0 & 2 \\ 1 & 1 \end{bmatrix}$

15. $(AB)C = A(BC) = \begin{bmatrix} -4 & 1 \\ 2 & -1 \end{bmatrix}$

16. $AC + BC =$ $(A + B)C = \begin{bmatrix} -2 & 1 \\ -2 & 0 \end{bmatrix}$

17. $(A + B)(A - B) = \begin{bmatrix} 0 & 1 \\ 1 & 3 \end{bmatrix}$; $A^2 - B^2 = \begin{bmatrix} 2 & 1 \\ -1 & 1 \end{bmatrix}$

18. $(A + B)^2 = \begin{bmatrix} 2 & 1 \\ 1 & 1 \end{bmatrix}$; $A^2 + 2AB + B^2 = \begin{bmatrix} 4 & 1 \\ -1 & -1 \end{bmatrix}$

19. $O_{2\times2} \cdot A =$ $A \cdot O_{2\times2} = \begin{bmatrix} 0 & 0 \\ 0 & 0 \end{bmatrix}$

20. $I_{2\times2} \cdot C =$ $C \cdot I_{2\times2} = \begin{bmatrix} -2 & 0 \\ 0 & 1 \end{bmatrix}$

24. $\begin{bmatrix} -1 & 0 & 0 \\ -3 & 8 & -3 \\ 0 & 0 & -1 \end{bmatrix}$

25. $\begin{bmatrix} 0 & 0 & -1 \\ -1 & 4 & -2 \\ 1 & 0 & -1 \end{bmatrix}$

26. $\begin{bmatrix} 1 & 0 & 0 \\ 3 & -8 & 3 \\ 0 & 0 & 1 \end{bmatrix}$

7. $\begin{bmatrix} 0 & 2 & -1 \\ 4 & 1 & 0 \\ 0 & -1 & 2 \end{bmatrix}\begin{bmatrix} 4 & 3 & 0 \\ -1 & 0 & 2 \\ 1 & 0 & -2 \end{bmatrix}$

8. $\begin{bmatrix} 1 & 0 & 0 \\ 2 & 0 & -1 \\ 0 & 1 & 0 \end{bmatrix}\begin{bmatrix} 1 & 0 & 2 \\ 0 & -3 & 1 \\ 0 & 1 & 0 \end{bmatrix}$

9. $\begin{bmatrix} 1 & 2 \\ 3 & -1 \\ 0 & 4 \end{bmatrix}\begin{bmatrix} 1 & 0 & 0 & -1 \\ -1 & 1 & 1 & 0 \end{bmatrix}$

10. $\begin{bmatrix} 0 & 2 & 0 \\ 0 & 1 & 1 \\ 1 & -1 & 0 \end{bmatrix}\begin{bmatrix} 1 & 1 & 0 \\ 0 & 1 & 0 \\ -1 & 0 & 1 \end{bmatrix}$

11. $\begin{bmatrix} 1 & 0 & 2 \\ -2 & 1 & 0 \\ 0 & 1 & 0 \end{bmatrix}\begin{bmatrix} a & b & c \\ d & e & f \\ g & h & i \end{bmatrix}$

12. $\begin{bmatrix} 2 & 0 & 1 \\ 5 & -1 & 0 \\ 0 & -2 & 4 \end{bmatrix}\begin{bmatrix} r & s & t \\ u & v & w \\ x & y & z \end{bmatrix}$

In Exercises 13–20, let $A = \begin{bmatrix} 1 & 2 \\ 0 & -1 \end{bmatrix}$, $B = \begin{bmatrix} 0 & -1 \\ 1 & 1 \end{bmatrix}$, $C = \begin{bmatrix} -2 & 0 \\ 0 & 1 \end{bmatrix}$.

B

13. Find AB and BA.

14. Find BC and CB.

15. Find $(AB)C$ and $A(BC)$.

16. Find $AC + BC$ and $(A + B)C$.

17. Find $(A + B)(A - B)$ and $A^2 - B^2$.

18. Find $(A + B)^2$ and $A^2 + 2AB + B^2$.

19. Find $O_{2\times2} \cdot A$ and $A \cdot O_{2\times2}$.

20. Find $I_{2\times2} \cdot C$ and $C \cdot I_{2\times2}$.

Let $A = \begin{bmatrix} 2 & -3 \\ 2 & 1 \end{bmatrix}$, $B = \begin{bmatrix} \frac{1}{8} & \frac{3}{8} \\ -\frac{1}{4} & \frac{1}{4} \end{bmatrix}$, and $D = \begin{bmatrix} 1 & 0 & -1 \\ 0 & 2 & -1 \\ 1 & 0 & 0 \end{bmatrix}$. Find each product.

21. AB 21. $\begin{bmatrix} 1 & 0 \\ 0 & 1 \end{bmatrix}$

22. BA 22. $\begin{bmatrix} 1 & 0 \\ 0 & 1 \end{bmatrix}$

23. D^2 23. $\begin{bmatrix} 0 & 0 & -1 \\ -1 & 4 & -2 \\ 1 & 0 & -1 \end{bmatrix}$

24. D^3

25. $(-D)^2$

26. $(-D)^3$

Prove the following properties for all 2 × 2 matrices.

C

27. $A \cdot I_{2\times2} = I_{2\times2} \cdot A = A$

28. $A \cdot O_{2\times2} = O_{2\times2} \cdot A = O_{2\times2}$

29. $(AB)C = A(BC)$

30. $A(B + C) = AB + AC$

Mixed Review Exercises

Find the real solutions of each system.

1. $2x^2 + y^2 = 5$, $x^2 = 1 - y^2$ none

2. $3s + 4t = 2$, $s - 2t = 4$ (2, −1)

3. $y^2 - x^2 = 24$, $2x - 3 = y$ (5, 7); (−1, −5)

4. $4m + 2n = -8$, $3m - n = -11$ (−3, 2)

5. $xy - 15 = 0$, $y^2 - x^2 = -16$ (5, 3); (−5, −3)

6. $a + b - c = -2$, $2a - b + c = 5$, $a + 3b + 2c = 2$ (1, −1, 2)

Solve.

7. $\sqrt{2x + 3} = x$ {3}

8. $\frac{m^2 + 2m}{3} = \frac{1 - m}{2}$ $\left\{\frac{-7 \pm \sqrt{73}}{4}\right\}$

9. $3x + 8 = 3 - 2x$ {−1}

16-4 Applications of Matrices

Objective To solve problems using matrices.

Matrices are used to solve a variety of problems in the social and physical sciences. Two applications of matrices are introduced in this lesson, *communication networks* and *dominance relations*.

Consider a network of four computers as shown in Figure 1. The arrows indicate the directions in which data can be transmitted and received.

A Communication Network

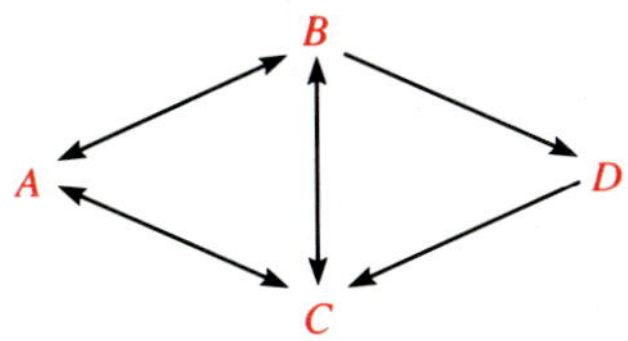

Figure 1

Computer A, for example, can transmit directly to computers B and C, but not to D. Computer A can receive directly from computers B and C, but not from D. Matrix X below illustrates this network. A "1" indicates that direct transmission is possible; a "0" that direct transmission is not possible.

To Computer

From Computer $\begin{array}{c c} & \begin{array}{cccc} A & B & C & D \end{array} \\ \begin{array}{c} A \\ B \\ C \\ D \end{array} & \begin{bmatrix} 0 & 1 & 1 & 0 \\ 1 & 0 & 1 & 1 \\ 1 & 1 & 0 & 0 \\ 0 & 0 & 1 & 0 \end{bmatrix} \end{array} = X$

For example, the "1" in the first row and second column indicates that computer A can transmit data directly to computer B. The "0" in the third row and fourth column indicates that computer C can't transmit data directly to computer D.

In general, rows of matrix X represent the directions data can be transmitted; columns represent the sources from which data may be received. Therefore, computer C can receive data from computers A, B, and D (third column).

Compare each row and column of matrix X with the corresponding point of Figure 1. Matrix X is called a **communication matrix.**

In addition to sending data directly, it is possible to send data from one point to another via a relay point. For example, computer A can send data to computer D via computer B. It can be proved that matrix X^2 represents the number of routes one computer can send data to a second computer in the network through exactly one relay point.

Teaching Suggestions p. T142

Group Activity p. T143

Suggested Extensions p. T143

Warm-Up Exercises

For the matrix $A = \begin{bmatrix} 1 & 1 & 0 \\ 0 & 0 & 1 \\ 1 & 0 & 1 \end{bmatrix}$ find the given matrix.

1. $A \times A$, or A^2 $\begin{bmatrix} 1 & 1 & 1 \\ 1 & 0 & 1 \\ 2 & 1 & 1 \end{bmatrix}$

2. A^3 $\begin{bmatrix} 2 & 1 & 2 \\ 2 & 1 & 1 \\ 3 & 2 & 2 \end{bmatrix}$

3. $A + A^2$ $\begin{bmatrix} 2 & 2 & 1 \\ 1 & 0 & 2 \\ 3 & 1 & 2 \end{bmatrix}$

4. $A + A^2 + A^3$ $\begin{bmatrix} 4 & 3 & 3 \\ 3 & 1 & 3 \\ 6 & 3 & 4 \end{bmatrix}$

Motivating the Lesson

Point out to students that effective communication networks are essential for modern societies. Whether a given network involves computers linked by phone, emergency services linked by radio, or broadcasting stations linked by satellite, it can be analyzed using a communication matrix, one of the topics of today's lesson.

Chalkboard Examples

1. If a communication network is represented by the diagram below, in how many ways can A send a message to B using no more than one relay point?

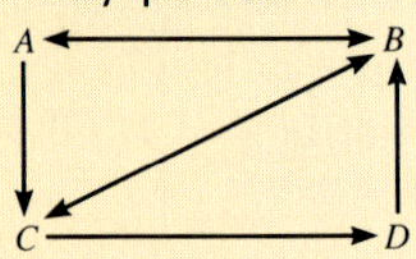

If the communication matrix is A, then $A =$

$$\begin{bmatrix} 0 & 1 & 1 & 0 \\ 1 & 0 & 1 & 0 \\ 0 & 1 & 0 & 1 \\ 0 & 1 & 0 & 0 \end{bmatrix}$$

$A + A^2 =$

$$\begin{bmatrix} 0 & 1 & 1 & 0 \\ 1 & 0 & 1 & 0 \\ 0 & 1 & 0 & 1 \\ 0 & 1 & 0 & 0 \end{bmatrix} + \begin{bmatrix} 1 & 1 & 1 & 1 \\ 0 & 2 & 1 & 1 \\ 1 & 1 & 1 & 0 \\ 1 & 0 & 1 & 0 \end{bmatrix} = \begin{bmatrix} 1 & \textcircled{2} & 2 & 1 \\ 1 & 2 & 2 & 1 \\ 1 & 2 & 1 & 1 \\ 1 & 1 & 1 & 0 \end{bmatrix}$$; 2 ways

Compare matrix X^2 below with Figure 1. Trace the possible routes for sending data from computer B to computer C via one relay point. There are two possible routes:

(1) from B to A, and then to C;

(2) from B to D, and then to C.

The element in the second row and third column of X^2 is "2," the number of routes data can be sent from B to C using exactly one relay point. The "1" written in the fourth row and first column of X^2 indicates there is exactly one route from D to A using exactly one relay point. Verify this by checking Figure 1.

To Computer

$$\text{From Computer} \quad \begin{array}{c} \\ A \\ B \\ C \\ D \end{array} \begin{array}{c} \begin{matrix} A & B & C & D \end{matrix} \\ \begin{bmatrix} 2 & 1 & 1 & 1 \\ 1 & 2 & 2 & 0 \\ 1 & 1 & 2 & 1 \\ 1 & 1 & 0 & 0 \end{bmatrix} \end{array} = X^2$$

Notice that some of the elements on the main diagonal of X^2 are not zero. For example, the "2" in the first row and first column indicates that computer A can send data to itself by two different routes using exactly one relay point.

Matrix X represents direct communication and matrix X^2 represents communication using 2 steps (with exactly one relay). It can be proved that X^3 represents the number of routes data can be sent from one point to another using 3 steps (with exactly two relay points).

The matrix $X + X^2$ represents the total number of routes data can be sent from one point to another in the network using no more than two steps. The matrix $X + X^2 + X^3$ shows the total number of routes data can be sent using no more than three steps.

Example 1 Use Figure 1 and matrices X and X^2 to solve these problems.

a. Which computer can send data directly to the most points?

b. Which computer can receive data directly from the most points?

c. Find the matrix that represents the total number of routes data can be sent from one computer to another using no more than 2 steps (that is, one relay point).

Solution

a. The row entries in matrix X represent routes on which data can be transmitted directly. Therefore computer B (with three routes) can send data to the most points. ***Answer***

b. The column entries in matrix X represent routes on which data can be received directly. Therefore computer C (with three routes) can receive data from the most points. ***Answer***

c. X represents the number of direct data communication routes. X^2 represents the number of routes using 2 steps. So $X + X^2$ represents the number of routes using no more than 2 steps.

$$X + X^2 = \begin{bmatrix} 0 & 1 & 1 & 0 \\ 1 & 0 & 1 & 1 \\ 1 & 1 & 0 & 0 \\ 0 & 0 & 1 & 0 \end{bmatrix} + \begin{bmatrix} 2 & 1 & 1 & 1 \\ 1 & 2 & 2 & 0 \\ 1 & 1 & 2 & 1 \\ 1 & 1 & 0 & 0 \end{bmatrix} = \begin{bmatrix} 2 & 2 & 2 & 1 \\ 2 & 2 & 3 & 1 \\ 2 & 2 & 2 & 1 \\ 1 & 1 & 1 & 0 \end{bmatrix}$$

A **dominance relation** exists between members of a group when, between any two members, one dominates the other.

Consider a four-team volleyball tournament. The Broncos beat the Mavericks and the Knights, but lost to the Raiders. The Mavericks beat only the Knights. The Raiders beat the Broncos and the Mavericks. The Knights beat only the Raiders.

Matrix M characterizes these dominances. A "1" indicates that the team named in the corresponding row beat the team in the corresponding column. The number of games won is indicated in the rows. The number of games lost is indicated in the columns.

	Broncos	Knights	Mavericks	Raiders
Broncos	0	1	1	0
Knights	0	0	0	1
Mavericks	0	1	0	0
Raiders	1	0	1	0

$= M$

For example, the Broncos won two (row one) and lost one (column one). Notice that the teams with the best records are the Broncos and Raiders, both of which won two games and lost one. To pick a tournament winner, you must look beyond these direct dominances because the Broncos and Raiders are tied at that level.

A member x is said to have a second-stage dominance over a member z when x dominates y and y dominates z. Second-stage dominances are shown in matrix M^2.

	Broncos	Knights	Mavericks	Raiders
Broncos	0	1	0	1
Knights	1	0	1	0
Mavericks	0	0	0	1
Raiders	0	2	1	0

$= M^2$

The fourth row shows that the Raiders have a second-stage dominance over the Mavericks and a pair of second-stage dominances over the Knights. In other words, the Raiders beat two other teams that, in turn, beat the Knights.

2. In a tennis tournament Adam beats Bill and Chad, but loses to Derek; Bill beats Chad and Derek; Chad beats only Derek. Write the dominance matrix for the tournament results and find the matrix that shows the second-stage dominances. Then determine the tournament winner by using both direct and second-stage dominances.

The dominance matrix M is

	A	B	C	D
A	0	1	1	0
B	0	0	1	1
C	0	0	0	1
D	1	0	0	0

The second-stage dominances are given by M^2:

	A	B	C	D
A	0	0	1	2
B	1	0	0	1
C	1	0	0	0
D	0	1	1	0

The winner is determined by $M + M^2$:

	A	B	C	D
A	0	1	2	2
B	1	0	1	2
C	1	0	0	1
D	1	1	1	0

The row sums show that Adam, with 5 direct and second-stage dominances, is the winner.

Check for Understanding

Here is a suggested use of the Oral Exercises to check students' understanding as you teach the lesson.

Oral Exs. 1–4: use after the introduction to communication matrices on page 779.

Guided Practice

A communication network for 4 ships is shown.

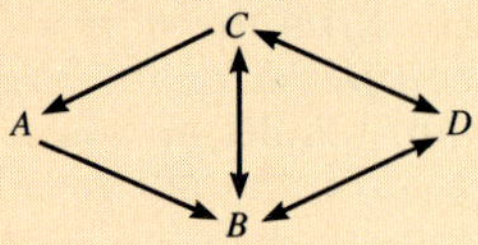

1. Write the communication matrix.

$$\begin{array}{c} \\ A \\ B \\ C \\ D \end{array}\begin{array}{c} \begin{array}{cccc} A & B & C & D \end{array} \\ \begin{bmatrix} 0 & 1 & 0 & 0 \\ 0 & 0 & 1 & 1 \\ 1 & 1 & 0 & 1 \\ 0 & 1 & 1 & 0 \end{bmatrix} \end{array}$$

2. Which ship can send messages to the greatest number of other ships? *C*
3. Write the matrix that represents the number of routes messages can be sent from one ship to another using exactly one relay.

$$\begin{array}{c} \\ A \\ B \\ C \\ D \end{array}\begin{array}{c} \begin{array}{cccc} A & B & C & D \end{array} \\ \begin{bmatrix} 0 & 0 & 1 & 1 \\ 1 & 2 & 1 & 1 \\ 0 & 2 & 2 & 1 \\ 1 & 1 & 1 & 2 \end{bmatrix} \end{array}$$

4. Find the number of ways a message can be sent from Ship *C* to Ship *B* using no more than one relay. 3

Summarizing the Lesson

In this lesson students learned to use matrices when solving problems about communication networks and dominance relations. Ask the students to explain why the elements of the main diagonal of a communication or dominance matrix are zeros.

From the sum of M and M^2, the winner of the tournament can be decided.

$$\begin{array}{l} \\ \text{Broncos} \\ \text{Knights} \\ \text{Mavericks} \\ \text{Raiders} \end{array}\begin{array}{c} \begin{array}{cccc} \text{Broncos} & \text{Knights} & \text{Mavericks} & \text{Raiders} \end{array} \\ \begin{bmatrix} 0 & 2 & 1 & 1 \\ 1 & 0 & 1 & 1 \\ 0 & 1 & 0 & 1 \\ 1 & 2 & 2 & 0 \end{bmatrix} \end{array} = M + M^2$$

The row sums of $M + M^2$ show that the Raiders, with a total of 5 direct and second-stage dominances, are the tournament winners.

Oral Exercises

Matrix Y illustrates a communication network. Use matrix Y to answer the questions below.

$$\text{From Point}\quad \begin{array}{c} \\ A \\ B \\ C \\ D \end{array}\begin{array}{c} \text{To Point} \\ \begin{array}{cccc} A & B & C & D \end{array} \\ \begin{bmatrix} 0 & 0 & 0 & 1 \\ 1 & 0 & 0 & 1 \\ 1 & 1 & 0 & 0 \\ 1 & 0 & 1 & 0 \end{bmatrix} \end{array} = Y$$

1. Name the destinations to which each point can send data.
   a. Point *A* *D* **b.** Point *B* *A, D* **c.** Point *C* *A, B* **d.** Point *D* *A, C*
2. Which point(s) can send data to the most points? *B, C, D*
3. Name the sources from which each point can receive data.
   a. Point *A* *B, C, D* **b.** Point *B* *C* **c.** Point *C* *D* **d.** Point *D* *A, B*
4. Which point(s) can receive data from the most points? *A*

Written Exercises

Use the radio network shown below for Exercises 1–3.

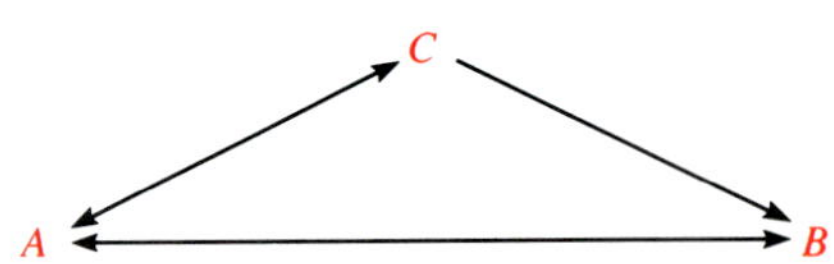

A 1. Write the matrix that illustrates this network.

2. Find the matrix that represents the number of ways a message can be sent from one station to another using exactly one relay.

3. Find the matrix that represents the number of routes a message can be sent from one station to another using at most one relay.

Answers:

1. $$\begin{array}{c} \\ A \\ B \\ C \end{array}\begin{array}{c} \begin{array}{ccc} A & B & C \end{array} \\ \begin{bmatrix} 0 & 1 & 1 \\ 1 & 0 & 0 \\ 1 & 1 & 0 \end{bmatrix} \end{array}$$

2. $$\begin{array}{c} \\ A \\ B \\ C \end{array}\begin{array}{c} \begin{array}{ccc} A & B & C \end{array} \\ \begin{bmatrix} 2 & 1 & 0 \\ 0 & 1 & 1 \\ 1 & 1 & 1 \end{bmatrix} \end{array}$$

3. $$\begin{array}{c} \\ A \\ B \\ C \end{array}\begin{array}{c} \begin{array}{ccc} A & B & C \end{array} \\ \begin{bmatrix} 2 & 2 & 1 \\ 1 & 1 & 1 \\ 2 & 2 & 1 \end{bmatrix} \end{array}$$

Use the computer network described below for Exercises 4–10.

Computer A sends data to computer B. Computer B sends and receives data from computers A, C, and D. Computer E receives data from computers C and D.

4. Draw this communication network.
5. Write the matrix that illustrates this network.
6. Which computer(s), if any, cannot send data to other computers in the network? *E*
7. Find the matrix that represents the number of routes data can be sent from one computer to another using exactly one relay.
8. Name the computer(s) that can receive data from the most points via one relay. *A, B, C, D*
9. Name the computer(s) that can send data to the most points via one relay. *B*
10. Find the matrix that represents the number of routes data can be sent from one computer to another using at most one relay.

Consider the situation described below for Exercises 11–12.

In a tennis tournament Jane beats Cindy, but loses to Yoko and Anna. Cindy beats Yoko and Anna. Yoko beats Jane and Anna. Anna beats only Jane.

B 11. Write the dominance matrix for the above tournament results and find the matrix that shows the second-stage dominances for this tournament.

12. Find the matrix that shows the tournament winner using both direct and second-stage dominances. Name the winner.

Use the communication network at the right for Exercises 13–16.

13. Write the matrix for the communication network shown and find the matrix that represents the number of ways a message can be sent from one point to another via one relay.

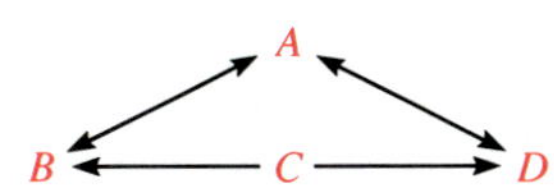

14. Find the matrix that represents the number of ways a message can be sent from one point to another using at most one relay.
15. Find the matrix that represents the number of ways a message can be sent from one point to another using exactly two relays.
16. Find the matrix that represents the number of ways a message can be sent from one point to another using at most two relays.
17. In the communication network at the right, a "hook" indicates that two line segments do not intersect. Find the number of ways each point can send a message back to itself using no more than two relays. *2 ways*

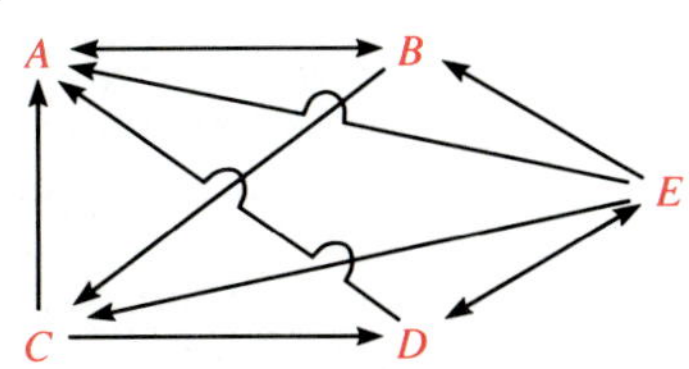

Suggested Assignments

Extended Algebra
Day 1: 782/1–10
785/*Extra*: 1, 3, 5
Day 2: 783/11–17
S 778/29, 30

Extended Alg. and Trig.
782/1–3, 11–17
785/*Extra*: 1, 3, 5

Supplementary Materials

Study Guide	pp. 265–266
Practice Master	90
Test Master	68
Resource Book	p. 172

Additional Answers
Written Exercises

4. $A \leftrightarrow B \leftrightarrow C$, $B \updownarrow D$, $C \downarrow E$, $D \rightarrow E$

5.
$$\begin{array}{c} \\ A \\ B \\ C \\ D \\ E \end{array}\begin{array}{c} \begin{array}{ccccc} A & B & C & D & E \end{array} \\ \begin{bmatrix} 0 & 1 & 0 & 0 & 0 \\ 1 & 0 & 1 & 1 & 0 \\ 0 & 1 & 0 & 0 & 1 \\ 0 & 1 & 0 & 0 & 1 \\ 0 & 0 & 0 & 0 & 0 \end{bmatrix} \end{array}$$

7.
$$\begin{array}{c} \\ A \\ B \\ C \\ D \\ E \end{array}\begin{array}{c} \begin{array}{ccccc} A & B & C & D & E \end{array} \\ \begin{bmatrix} 1 & 0 & 1 & 1 & 0 \\ 0 & 3 & 0 & 0 & 2 \\ 1 & 0 & 1 & 1 & 0 \\ 1 & 0 & 1 & 1 & 0 \\ 0 & 0 & 0 & 0 & 0 \end{bmatrix} \end{array}$$

10.
$$\begin{array}{c} \\ A \\ B \\ C \\ D \\ E \end{array}\begin{array}{c} \begin{array}{ccccc} A & B & C & D & E \end{array} \\ \begin{bmatrix} 1 & 1 & 1 & 1 & 0 \\ 1 & 3 & 1 & 1 & 2 \\ 1 & 1 & 1 & 1 & 1 \\ 1 & 1 & 1 & 1 & 1 \\ 0 & 0 & 0 & 0 & 0 \end{bmatrix} \end{array}$$

(continued)

Additional Answers Written Exercises (continued)

11. $D = \begin{array}{c} \\ J \\ C \\ Y \\ A \end{array}\begin{array}{c} \begin{matrix} J & C & Y & A \end{matrix} \\ \begin{bmatrix} 0 & 1 & 0 & 0 \\ 0 & 0 & 1 & 1 \\ 1 & 0 & 0 & 1 \\ 1 & 0 & 0 & 0 \end{bmatrix} \end{array}$;

$D^2 = \begin{array}{c} \\ J \\ C \\ Y \\ A \end{array}\begin{array}{c} \begin{matrix} J & C & Y & A \end{matrix} \\ \begin{bmatrix} 0 & 0 & 1 & 1 \\ 2 & 0 & 0 & 1 \\ 1 & 1 & 0 & 0 \\ 0 & 1 & 0 & 0 \end{bmatrix} \end{array}$;

12. $D + D^2 = \begin{array}{c} \\ J \\ C \\ Y \\ A \end{array}\begin{array}{c} \begin{matrix} J & C & Y & A \end{matrix} \\ \begin{bmatrix} 0 & 1 & 1 & 1 \\ 2 & 0 & 1 & 2 \\ 2 & 1 & 0 & 1 \\ 1 & 1 & 0 & 0 \end{bmatrix} \end{array}$;

Cindy is the winner.

13. $\begin{bmatrix} 0 & 1 & 0 & 1 \\ 1 & 0 & 0 & 0 \\ 0 & 1 & 0 & 1 \\ 1 & 0 & 0 & 0 \end{bmatrix}$; $\begin{bmatrix} 2 & 0 & 0 & 0 \\ 0 & 1 & 0 & 1 \\ 2 & 0 & 0 & 0 \\ 0 & 1 & 0 & 1 \end{bmatrix}$

14. $\begin{bmatrix} 2 & 1 & 0 & 1 \\ 1 & 1 & 0 & 1 \\ 2 & 1 & 0 & 1 \\ 1 & 1 & 0 & 1 \end{bmatrix}$

15. $\begin{bmatrix} 0 & 2 & 0 & 2 \\ 2 & 0 & 0 & 0 \\ 0 & 2 & 0 & 2 \\ 2 & 0 & 0 & 0 \end{bmatrix}$

16. $\begin{bmatrix} 2 & 3 & 0 & 3 \\ 3 & 1 & 0 & 1 \\ 2 & 3 & 0 & 3 \\ 3 & 1 & 0 & 1 \end{bmatrix}$

18. Matrix operations used to analyze communication networks and dominance relations are similar. For each concept listed below name the comparable concept from the other application area.

	Communication Networks	Dominance Relations
a.	A can send data to B.	? *A* dominates *B* directly.
b.	? *B* can send messages to *A*.	A is dominated by B.
c.	B is a relay point from A to C	? *A* dominates *B* and *B* dominates *C*.
d.	?	B wins more games in a tournament than any other team.
e.	?	A has a second-stage dominance over B.
f.	C can send messages to more stations than any other station in the network using at most one relay.	?

d. *B* can send messages directly to the most stations.
e. *A* can send messages to *B* via a relay point.
f. When the direct and second-stage dominances are added, *C* has the greatest number.

Self-Test 1

Vocabulary

matrix (p. 767)
elements of a matrix (p. 767)
row of a matrix (p. 767)
column of a matrix (p. 767)
dimensions of a matrix (p. 767)
row matrix (p. 767)
column matrix (p. 767)
square matrix (p. 767)
zero matrix (p. 767)
sum of matrices (p. 770)
additive inverse of a matrix (p. 770)
subtraction of matrices (p. 771)
scalar, scalar product (p. 771)
matrix equation (p. 772)
product of matrices (p. 774)
identity matrix (p. 775)
communication matrix (p. 779)
dominance relations (p. 781)

Find the value of each variable in Exercises 1 and 2.

1. $O_{2\times 2} = \begin{bmatrix} 0 & y-5 \\ x+1 & z^2 \end{bmatrix}$ $x = -1, y = 5, z = 0$

2. $\begin{bmatrix} 3x+2 \\ 10 \\ 0 \end{bmatrix} = \begin{bmatrix} 8 \\ -y \\ z-4 \end{bmatrix}$ $x = 2, y = -10, z = 4$

Obj. 16-1, p. 767

Let $A = \begin{bmatrix} 1 & 4 \\ 0 & 3 \end{bmatrix}$ **and** $B = \begin{bmatrix} -1 & 2 \\ 1 & 0 \end{bmatrix}$. **Find the following.**

3. $A + B$ **3.** $\begin{bmatrix} 0 & 6 \\ 1 & 3 \end{bmatrix}$ **4.** $A - B$ **4.** $\begin{bmatrix} 2 & 2 \\ -1 & 3 \end{bmatrix}$ **5.** $3B$ **5.** $\begin{bmatrix} -3 & 6 \\ 3 & 0 \end{bmatrix}$ **Obj. 16-2, p. 770**

6. AB **7.** A^2 **8.** $A - 2B$ **Obj. 16-3, p. 774**

9. a. Write the communication matrix for the network shown. **Obj. 16-4, p. 779**

b. Use matrices to find the number of routes a message can be sent from point A to C using at most one relay point. 3 routes

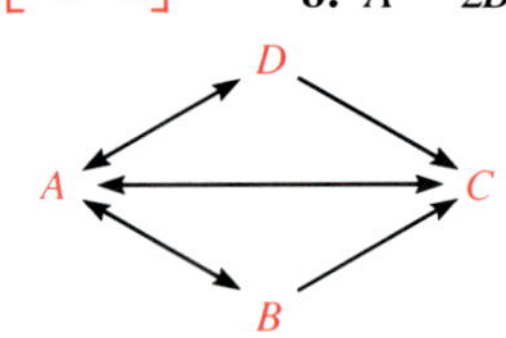

Extra / *Augmented Matrices*

When solving systems of linear equations in three variables, you learned a method called Gaussian elimination (see Lesson 9-9). This method involves transforming a given system into an equivalent triangular system.

Although you learned to do Gaussian elimination by working with the complete equations of a system, you can also solve the system by working only with the numbers in the equations. The coefficients and constants are put in a matrix called the *augmented matrix* for the system. For example, the augmented matrix for the system

$$\begin{aligned} x + y - 2z &= 7 \\ -x + 4y + 3z &= 2 \\ 2x - 3y + 2z &= -2 \end{aligned}$$

is

$$\begin{bmatrix} 1 & 1 & -2 & 7 \\ -1 & 4 & 3 & 2 \\ 2 & -3 & 2 & -2 \end{bmatrix}$$

Recall that as you transform a system into triangular form, you use linear combinations of the system's equations. That is, you multiply one equation by a nonzero constant and add the result to another equation. This technique can also be applied to an augmented matrix.

> To obtain an augmented matrix in triangular form, you replace any row with the sum of that row and any other row multiplied by a nonzero constant.

Example 1 Solve this system (see Example 2, page 445):

$$\begin{aligned} x + y - 2z &= 7 \\ -x + 4y + 3z &= 2 \\ 2x - 3y + 2z &= -2 \end{aligned}$$

Solution

1. Write the augmented matrix. The goal is to obtain a matrix in triangular form (that is, with zeros in the lower left corner).
$$\begin{bmatrix} 1 & 1 & -2 & 7 \\ -1 & 4 & 3 & 2 \\ 2 & -3 & 2 & -2 \end{bmatrix}$$
2. Replace the second row with the sum of the first and the second rows.
$$\begin{bmatrix} 1 & 1 & -2 & 7 \\ 0 & 5 & 1 & 9 \\ 2 & -3 & 2 & -2 \end{bmatrix}$$
3. Replace the third row with the sum of the third row and -2 times the first row.
$$\begin{bmatrix} 1 & 1 & -2 & 7 \\ 0 & 5 & 1 & 9 \\ 0 & -5 & 6 & -16 \end{bmatrix}$$

(Solution continues on the next page.)

Quick Quiz

1. If $\begin{bmatrix} x & 2z - 1 \\ x + y & 0 \\ 3 & 4 \end{bmatrix} = \begin{bmatrix} 7 & z + 4 \\ 12 & 0 \\ 3 & w \end{bmatrix}$, find the value of each variable. $w = 4, x = 7, y = 5, z = 5$

Let $A = \begin{bmatrix} 5 & 0 \\ -1 & -2 \end{bmatrix}$ and $B = \begin{bmatrix} -3 & 1 \\ 0 & 4 \end{bmatrix}$.

Find the following.

2. $A + 2B$ $\begin{bmatrix} -1 & 2 \\ -1 & 6 \end{bmatrix}$
3. $O_{2\times 2} - A$ $\begin{bmatrix} -5 & 0 \\ 1 & 2 \end{bmatrix}$
4. AB $\begin{bmatrix} -15 & 5 \\ 3 & -9 \end{bmatrix}$
5. BA $\begin{bmatrix} -16 & -2 \\ -4 & -8 \end{bmatrix}$
6. Write the communication matrix for the network shown. Find the matrix that shows the number of ways a message can be sent from one point to another via one relay.

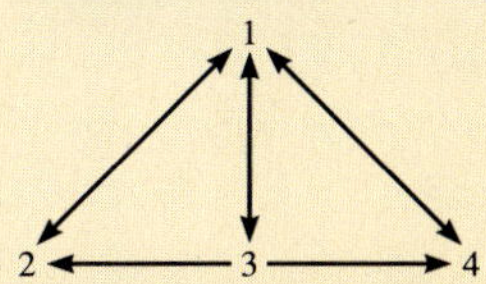

To (columns 1 2 3 4), From (rows 1 2 3 4):

$$\begin{bmatrix} 0 & 1 & 1 & 1 \\ 1 & 0 & 0 & 0 \\ 1 & 1 & 0 & 1 \\ 1 & 0 & 0 & 0 \end{bmatrix} = A;$$

$$A^2 = \begin{bmatrix} 3 & 1 & 0 & 1 \\ 0 & 1 & 1 & 1 \\ 2 & 1 & 1 & 1 \\ 0 & 1 & 1 & 1 \end{bmatrix}$$

Using a Computer

The program "Matrix Reducer" on the Disk for Algebra lets students solve, step by step, a system of linear equations using Gaussian elimination.

Support Materials
Disk for Algebra
Menu Item: Matrix Reducer

4. Replace the third row with the sum of the second and third rows.
$$\begin{bmatrix} 1 & 1 & -2 & 7 \\ 0 & 5 & 1 & 9 \\ 0 & 0 & 7 & -7 \end{bmatrix}$$

The third row tells you that $7z = -7$, so $z = -1$.

The second row tells you that $5y + z = 9$. Since $z = -1$, you have $5y + (-1) = 9$, or $y = 2$.

The first row tells you that $x + y - 2z = 7$. Since $y = 2$ and $z = -1$, you have $x + 2 - 2(-1) = 7$, or $x = 3$.

$\therefore$ the solution of the system is $(3, 2, -1)$. ***Answer***

Example 2 Solve this system:
$$\begin{aligned} 2x - y + 3z &= 5 \\ 4x + y &= 13 \\ 3x + 4z &= 4 \end{aligned}$$

Solution

1. Write the augmented matrix.
$$\begin{bmatrix} 2 & -1 & 3 & 5 \\ 4 & 1 & 0 & 13 \\ 3 & 0 & 4 & 4 \end{bmatrix}$$

2. Replace the second row with the sum of the second row and -2 times the first row.
$$\begin{bmatrix} 2 & -1 & 3 & 5 \\ 0 & 3 & -6 & 3 \\ 3 & 0 & 4 & 4 \end{bmatrix}$$

3. Replace the third row with the sum of the third row and $-\frac{3}{2}$ times the first row.
$$\begin{bmatrix} 2 & -1 & 3 & 5 \\ 0 & 3 & -6 & 3 \\ 0 & \frac{3}{2} & -\frac{1}{2} & -\frac{7}{2} \end{bmatrix}$$

4. Replace the third row with the sum of the third row and $-\frac{1}{2}$ times the second row.
$$\begin{bmatrix} 2 & -1 & 3 & 5 \\ 0 & 3 & -6 & 3 \\ 0 & 0 & \frac{5}{2} & -5 \end{bmatrix}$$

From the third row you get $z = -2$. Then from the second row you get $y = -3$. Finally, from the first row you get $x = 4$.

$\therefore$ the solution of the system is $(4, -3, -2)$. ***Answer***

Computer software can be very useful in reducing augmented matrices to triangular form.

Exercises

Solve each system by using an augmented matrix.

1. $x + y + z = 0$ (5, −1, −4)
$x - y + z = 2$
$x - y - z = 10$

2. $x + 2y - z = 0$ (−2, 3, 4)
$-x + y = 5$
$x + 2z = 6$

3. $x - 2y + z = 9$ (1, −2, 4)
$3x + y = 1$
$-2x - 3y - z = 0$

4. $x + 2y + 3z = 3$
$2x + 3y + 4z = 2$
$-3x - 5y + 2z = 4$
(−4, 2, 1)

5. $2x + y + z = 3$
$4x + 3z = 5$
$3x + 2y = 1$
(−1, 2, 3)

6. $2x - y - 2z = 5$
$3x - y = 1$
$5x + 4z = -2$
(2, 5, −3)

16-5 Determinants

Objective To find the determinant of a 2×2 or 3×3 matrix.

A real number called the *determinant* is associated with each *square* matrix. The determinant of a matrix is usually displayed in the same form as the matrix, but with vertical bars rather than brackets enclosing the elements. The number of elements in any row or column is called the **order** of the determinant.

Determinant of a 2 × 2 matrix

Let $A = \begin{bmatrix} a & b \\ c & d \end{bmatrix}$. Then the determinant of A, denoted by det A, is defined as follows:

$$\det A = \begin{vmatrix} a & b \\ c & d \end{vmatrix} = ad - bc$$

Since the determinant of A has 2 rows and 2 columns, det A is of order 2.

Example 1 Evaluate $\begin{vmatrix} 5 & 2 \\ 4 & -1 \end{vmatrix}$.

Solution $\begin{vmatrix} 5 & 2 \\ 4 & -1 \end{vmatrix} = 5(-1) - (2)(4) = -13$ ***Answer***

The determinant of a 3×3 matrix is defined below.

Determinant of a 3 × 3 matrix

Let $B = \begin{bmatrix} a_1 & b_1 & c_1 \\ a_2 & b_2 & c_2 \\ a_3 & b_3 & c_3 \end{bmatrix}$. Then det B is defined as follows:

$$\det B = \begin{vmatrix} a_1 & b_1 & c_1 \\ a_2 & b_2 & c_2 \\ a_3 & b_3 & c_3 \end{vmatrix}$$
$$= a_1b_2c_3 + a_2b_3c_1 + a_3b_1c_2 - a_3b_2c_1 - a_2b_1c_3 - a_1b_3c_2$$

Teaching Suggestions p. T143

Reading Algebra p. T143

Suggested Extensions p. T143

Warm-Up Exercises

Simplify.

1. $(-4)(0) - (2)(1)$ -2
2. $(-2)(3) - (5)(-1)$ -1
3. $(4)(-6) - (-3)(8)$ 0
4. $36 + 40 + 2 - 12 - 15 - 16$ 35
5. $12 + (-4) + (-60) - (-32) - 10 - (-9)$ -21

Motivating the Lesson

Tell students that the determinant of a square matrix, the topic of today's lesson, is like the discriminant of a quadratic equation. Just as the discriminant indicates the nature of a quadratic equation's roots, the determinant indicates whether a square matrix has an inverse.

Chalkboard Examples

Evaluate each determinant.

1. $\begin{vmatrix} 3 & -4 \\ -1 & -2 \end{vmatrix}$
   $3(-2) - (-1)(-4) = -6 - 4 = -10$
2. $\begin{vmatrix} 1 & 2 \\ -3 & -4 \end{vmatrix}$
   $1(-4) - (-3)(2) = -4 + 6 = 2$
3. $\begin{vmatrix} 1 & 0 & 4 \\ 1 & -1 & 3 \\ 0 & 5 & -2 \end{vmatrix}$
   $1(-1)(-2) + 0 + 4(1)(5) - 0 - 1(3)(5) - 0 = 2 + 20 - 15 = 7$

(continued)

Chalkboard Examples

(continued)

4. $\begin{vmatrix} -3 & 1 & 0 \\ 1 & 0 & -2 \\ -2 & 4 & 1 \end{vmatrix}$

$0 + 1(-2)(-2) + 0 - 0 - (-3)(-2)(4) - 1(1)(1) = 4 - 24 - 1 = -21$

Common Errors

Students often make sign errors when evaluating a 3×3 determinant. When discussing the method shown at the top of this page, stress the fact that you use the *opposites* of the products of the elements on the left-to-right upward diagonals.

Check for Understanding

Oral Exs. 1–8: use after Example 1.

Guided Practice

Evaluate.

1. $\begin{vmatrix} 3 & 5 \\ 1 & 4 \end{vmatrix}$ 7

2. $\begin{vmatrix} 0 & 8 \\ 1 & 10 \end{vmatrix}$ −8

3. $\begin{vmatrix} 2 & -5 & 3 \\ 0 & 8 & 1 \\ -5 & 4 & 0 \end{vmatrix}$ 137

4. $\begin{vmatrix} 2 & 3 & -1 \\ -3 & 0 & -8 \\ 11 & -4 & 6 \end{vmatrix}$ −286

Summarizing the Lesson

In this lesson students learned to evaluate determinants of 2×2 and 3×3 matrices. Ask the students to describe the method for evaluating such determinants.

A convenient method for finding the six terms needed to evaluate a 3×3 determinant is shown below.

1. Copy the first two columns of the matrix in order to the right of the third column.

$$\begin{array}{ccc|cc} a_1 & b_1 & c_1 & a_1 & b_1 \\ a_2 & b_2 & c_2 & a_2 & b_2 \\ a_3 & b_3 & c_3 & a_3 & b_3 \end{array}$$

2. Multiply each element in the first row of the original matrix by the other two elements on the left-to-right downward diagonal. These products are the first three terms of the determinant.

$$\begin{array}{ccc|cc} a_1 & b_1 & c_1 & a_1 & b_1 \\ a_2 & b_2 & c_2 & a_2 & b_2 \\ a_3 & b_3 & c_3 & a_3 & b_3 \end{array}$$

$$a_1b_2c_3 + a_3b_1c_2 + a_2b_3c_1$$

3. Multiply each element in the last row of the original matrix by the other two elements on the left-to-right upward diagonal. The opposites of these products are the last three terms of the determinant.

$$\begin{array}{ccc|cc} a_1 & b_1 & c_1 & a_1 & b_1 \\ a_2 & b_2 & c_2 & a_2 & b_2 \\ a_3 & b_3 & c_3 & a_3 & b_3 \end{array}$$

$$- a_3b_2c_1 - a_1b_3c_2 - a_2b_1c_3$$

Caution: This method works *only* for 3×3 determinants.

Example 2 Evaluate $\det A = \begin{vmatrix} 2 & -1 & 1 \\ 5 & 3 & -1 \\ 0 & 4 & 2 \end{vmatrix}$.

Solution

$$\begin{array}{ccc|cc} 2 & -1 & 1 & 2 & -1 \\ 5 & 3 & -1 & 5 & 3 \\ 0 & 4 & 2 & 0 & 4 \end{array}$$

$\det A = 12 + 0 + 20 - 0 - (-8) - (-10) = 50$ ***Answer***

Recall that the *main diagonal* of a square matrix runs from the upper left to the lower right. A matrix whose only nonzero elements are on the main diagonal is called a **diagonal matrix.** As you can easily verify for 2×2 and 3×3 diagonal matrices, the determinant of a diagonal matrix is the product of the elements on the main diagonal.

Oral Exercises

Evaluate.

1. $\begin{vmatrix} 2 & 0 \\ 0 & 1 \end{vmatrix}$ 2

2. $\begin{vmatrix} 0 & 0 \\ 0 & 0 \end{vmatrix}$ 0

3. $\begin{vmatrix} 1 & 2 \\ 2 & 5 \end{vmatrix}$ 1

4. $\begin{vmatrix} 2 & 1 \\ -2 & 1 \end{vmatrix}$ 4

5. $\begin{vmatrix} 8 & -6 \\ 5 & 3 \end{vmatrix}$ 54

6. $\begin{vmatrix} 2 & 0 \\ 0 & 2 \end{vmatrix} + \begin{vmatrix} 0 & 1 \\ -1 & 0 \end{vmatrix}$ 5

7. $\begin{vmatrix} 0 & 5 \\ -1 & 0 \end{vmatrix} - \begin{vmatrix} 2 & 1 \\ 1 & 2 \end{vmatrix}$ 2

8. $\begin{vmatrix} 3 & 0 \\ 0 & -2 \end{vmatrix} \div \begin{vmatrix} 1 & 0 \\ 0 & 2 \end{vmatrix}$ −3

Written Exercises

Evaluate.

A

1. $\begin{vmatrix} 3 & 1 \\ 4 & 3 \end{vmatrix}$ 5
2. $\begin{vmatrix} 2 & 10 \\ 0 & 3 \end{vmatrix}$ 6
3. $\begin{vmatrix} 5 & -3 \\ 3 & 1 \end{vmatrix}$ 14
4. $\begin{vmatrix} 15 & 0 \\ 0 & 4 \end{vmatrix}$ 60
5. $\begin{vmatrix} 0 & 8 \\ -4 & 0 \end{vmatrix}$ 32
6. $\begin{vmatrix} 3 & 9 \\ 2 & 6 \end{vmatrix}$ 0
7. $\begin{vmatrix} 2 & -5 & 3 \\ 0 & 8 & 1 \\ -5 & 4 & 0 \end{vmatrix}$ 137
8. $\begin{vmatrix} 3 & 0 & 1 \\ 5 & 2 & 2 \\ -2 & 0 & 4 \end{vmatrix}$ 28
9. $\begin{vmatrix} 12 & -9 & 13 \\ 0 & 0 & 8 \\ -9 & 2 & 1 \end{vmatrix}$ 456
10. $\begin{vmatrix} -10 & 0 & 0 \\ 0 & 4 & 0 \\ 0 & 0 & -12 \end{vmatrix}$ 480
11. $\begin{vmatrix} 5 & 0 & 0 \\ 0 & -6 & 0 \\ 0 & 0 & 2 \end{vmatrix}$ −60
12. $\begin{vmatrix} 0 & 2 & -3 \\ 3 & 5 & -3 \\ 1 & 2 & 0 \end{vmatrix}$ −9

B

13. $\begin{vmatrix} 8 & -1 & -5 \\ 0 & 9 & 6 \\ -2 & 0 & 3 \end{vmatrix}$ 138
14. $\begin{vmatrix} 2 & 3 & -1 \\ -3 & 0 & -8 \\ 11 & -4 & 6 \end{vmatrix}$ −286
15. Show that for any 2×2 matrix, A, if all the elements in one row or one column are zeros, then $\det A = 0$.
16. Show that for any 3×3 matrix, B, if all the elements in one row or one column are zeros, then $\det B = 0$.
17. Let A be any 2×2 matrix. Show that if the two rows are the same, then $\det A = 0$.
18. Let A be any 2×2 matrix. Show that if any row or column of A is multiplied by a scalar r, then $\det A$ is multiplied by r.

C

19. Let r be a scalar, and let A be any 2×2 matrix. Show that $\det (rA) = r^2 \det A$.
20. Let A and B be 2×2 matrices. Show that $\det AB = \det A \cdot \det B$.

Mixed Review Exercises

3. $\frac{13s - 5}{3(s+1)(s-1)}$ 6. $\frac{y-2}{y-5}$

Write in simplest form without zero or negative exponents.

1. $\frac{t^2 - 3t + 2}{6 - 5t + t^2}$ $\frac{t-1}{t-3}$
2. $\frac{a^2b^3}{a^3b^7}$ $\frac{1}{ab^4}$
3. $\frac{3}{s+1} + \frac{4}{3s-3}$
4. $\sqrt[3]{a^2} \cdot \sqrt[3]{a^4}$ a^2
5. $\frac{\frac{1}{c} - 1}{c - \frac{1}{c}}$ $-\frac{1}{c+1}$
6. $\frac{y+5}{y-4} \div \frac{y^2 - 25}{y^2 - 6y + 8}$
7. $(-9pr^4q^3)(4p^2r^{-4}q^3)$ $-36p^3q^6$
8. $\frac{x^3 - 1}{x^2 + 3x - 4}$ $\frac{x^2 + x + 1}{x + 4}$
9. $(2^{\pi})^{-1/\pi}$ $\frac{1}{2}$

Suggested Assignments

Extended Algebra
789/1–15
S 789/Mixed Review
R 784/Self-Test 1

Extended Alg. and Trig.
789/1–15
S 789/Mixed Review
R 784/Self-Test 1

Supplementary Materials

Study Guide pp. 267–268

Teaching Suggestions p. T143

Suggested Extensions p. T144

Warm-Up Exercises

Let $A = \begin{bmatrix} 5 & -4 \\ 3 & -2 \end{bmatrix}$,

$B = \begin{bmatrix} -1 & 2 \\ -\frac{3}{2} & \frac{5}{2} \end{bmatrix}$, and $C = \begin{bmatrix} 1 \\ 2 \end{bmatrix}$.

1. Evaluate det A. 2
2. Evaluate det B. $\frac{1}{2}$
3. Find AB. $\begin{bmatrix} 1 & 0 \\ 0 & 1 \end{bmatrix}$
4. Find BC. $\begin{bmatrix} 3 \\ \frac{7}{2} \end{bmatrix}$
5. Solve this system:
   $5x - 4y = 1$
   $3x - 2y = 2$ $(3, \frac{7}{2})$

Motivating the Lesson

Remind students of the inverse property of multiplication: For every nonzero real number a there is a unique inverse $\frac{1}{a}$ such that the product of a and $\frac{1}{a}$ is 1. A similar situation exists for every matrix A with a nonzero determinant, as today's lesson will show.

Chalkboard Examples

1. Find the inverse of $\begin{bmatrix} -9 & 5 \\ 6 & -3 \end{bmatrix}$.

$\begin{vmatrix} -9 & 5 \\ 6 & -3 \end{vmatrix} = 27 - 30 = -3$

$\begin{bmatrix} -9 & 5 \\ 6 & -3 \end{bmatrix}^{-1}$

$= -\frac{1}{3}\begin{bmatrix} -3 & -5 \\ -6 & -9 \end{bmatrix} = \begin{bmatrix} 1 & \frac{5}{3} \\ 2 & 3 \end{bmatrix}$

16-6 Inverses of Matrices

Objective To solve systems of equations using inverses of matrices.

The product of the two matrices below is the identity matrix, $I_{2\times 2}$:

$$\begin{bmatrix} 2 & 1 \\ 3 & -2 \end{bmatrix}\begin{bmatrix} \frac{2}{7} & \frac{1}{7} \\ \frac{3}{7} & -\frac{2}{7} \end{bmatrix} = \begin{bmatrix} 1 & 0 \\ 0 & 1 \end{bmatrix}$$

For any two matrices A and B, if $AB = BA = I$, then A and B are called **inverse matrices.** The symbol A^{-1} denotes the inverse of matrix A. Therefore, $B = A^{-1}$ and $A = B^{-1}$.

Does every nonzero 2×2 matrix, A, have an inverse, A^{-1}?

Let $A = \begin{bmatrix} a & b \\ c & d \end{bmatrix}$. Find $A^{-1} = \begin{bmatrix} x & u \\ y & v \end{bmatrix}$ such that $AA^{-1} = I_{2\times 2}$.

$$\begin{bmatrix} a & b \\ c & d \end{bmatrix}\begin{bmatrix} x & u \\ y & v \end{bmatrix} = \begin{bmatrix} ax + by & au + bv \\ cx + dy & cu + dv \end{bmatrix} = \begin{bmatrix} 1 & 0 \\ 0 & 1 \end{bmatrix}$$

The equation above is true if and only if

$$ax + by = 1 \qquad au + bv = 0$$
$$cx + dy = 0 \qquad cu + dv = 1.$$

If $ad - bc \neq 0$, these systems may be solved to obtain the following:

$$x = \frac{d}{ad - bc} \qquad u = \frac{-b}{ad - bc}$$
$$y = \frac{-c}{ad - bc} \qquad v = \frac{a}{ad - bc}$$

Substitute these values in the matrix A^{-1}.

$$A^{-1} = \begin{bmatrix} \frac{d}{ad - bc} & \frac{-b}{ad - bc} \\ \frac{-c}{ad - bc} & \frac{a}{ad - bc} \end{bmatrix} = \frac{1}{ad - bc}\begin{bmatrix} d & -b \\ -c & a \end{bmatrix},\ ad - bc \neq 0.$$

You can verify that $AA^{-1} = A^{-1}A = I_{2\times 2}$. Since $ad - bc$ is the determinant of A, you have this general result.

If $A = \begin{bmatrix} a & b \\ c & d \end{bmatrix}$ and det $A \neq 0$, then

$$\mathbf{A^{-1} = \frac{1}{\det A}\begin{bmatrix} d & -b \\ -c & a \end{bmatrix}}.$$

If det A = 0, then A has no inverse.

Example 1 Find A^{-1} if $A = \begin{bmatrix} 6 & 2 \\ -3 & 1 \end{bmatrix}$.

Solution Since det $A = 12 \neq 0$, A has an inverse. Here is what the general result says to do to find A^{-1}: Interchange a and d, replace both b and c with their opposites, and then multiply the resulting matrix by $\frac{1}{\det A}$.

$$A^{-1} = \frac{1}{12}\begin{bmatrix} 1 & -2 \\ 3 & 6 \end{bmatrix} = \begin{bmatrix} \frac{1}{12} & -\frac{1}{6} \\ \frac{1}{4} & \frac{1}{2} \end{bmatrix} \quad \textit{Answer}$$

You can use the inverse of a matrix to solve a system of linear equations

$$\begin{aligned} ax + by &= e \\ cx + dy &= f \end{aligned}$$

where a, b, c, d, e, and f are real numbers. You can write this system using the definitions of matrix multiplication and equality of matrices.

$$\begin{bmatrix} a & b \\ c & d \end{bmatrix}\begin{bmatrix} x \\ y \end{bmatrix} = \begin{bmatrix} e \\ f \end{bmatrix}$$

This equation is in the form $AX = B$ where the *matrix of coefficients*

$$A = \begin{bmatrix} a & b \\ c & d \end{bmatrix}, \quad X = \begin{bmatrix} x \\ y \end{bmatrix}, \quad \text{and} \quad B = \begin{bmatrix} e \\ f \end{bmatrix}.$$

Therefore, if A^{-1} exists you have:

$$\begin{aligned} A^{-1}(AX) &= A^{-1}(B) \\ (A^{-1}A)X &= A^{-1}B \\ IX &= A^{-1}B \\ X &= A^{-1}B \end{aligned}$$

Thus if det $A \neq 0$, so that A^{-1} exists, the system has a unique solution, $X = A^{-1}B$.

Example 2 Use matrices to solve this system:

$$\begin{aligned} 2x - 3y &= 1 \\ x + 4y &= 6 \end{aligned}$$

Solution Write the matrix equation $AX = B$: $\begin{bmatrix} 2 & -3 \\ 1 & 4 \end{bmatrix}\begin{bmatrix} x \\ y \end{bmatrix} = \begin{bmatrix} 1 \\ 6 \end{bmatrix}$

Since det $A = 11 \neq 0$, A^{-1} exists, and the solution of the equation $AX = B$ is $X = A^{-1}B$. So find A^{-1}, the inverse of the matrix of coefficients.

$$A^{-1} = \frac{1}{11}\begin{bmatrix} 4 & 3 \\ -1 & 2 \end{bmatrix}$$

Then $\begin{bmatrix} x \\ y \end{bmatrix} = \frac{1}{11}\begin{bmatrix} 4 & 3 \\ -1 & 2 \end{bmatrix}\begin{bmatrix} 1 \\ 6 \end{bmatrix} = \frac{1}{11}\begin{bmatrix} 22 \\ 11 \end{bmatrix} = \begin{bmatrix} 2 \\ 1 \end{bmatrix}$.

$\therefore$ the solution of the system is (2, 1). ***Answer***

Determine whether the system has a unique solution. If it does, find the solution.

2. $3x + 2y = 12$
$12x + 8y = -3$

$\begin{bmatrix} 3 & 2 \\ 12 & 8 \end{bmatrix}\begin{bmatrix} x \\ y \end{bmatrix} = \begin{bmatrix} 12 \\ -3 \end{bmatrix}$

$\begin{bmatrix} 3 & 2 \\ 12 & 8 \end{bmatrix}^{-1}$ does not exist since $\begin{vmatrix} 3 & 2 \\ 12 & 8 \end{vmatrix} = 0$; no unique solution

3. $x - 3y = 1$
$x + 3y = -11$

$\begin{bmatrix} 1 & -3 \\ 1 & 3 \end{bmatrix}\begin{bmatrix} x \\ y \end{bmatrix} = \begin{bmatrix} 1 \\ -11 \end{bmatrix}$

$\begin{bmatrix} 1 & -3 \\ 1 & 3 \end{bmatrix}^{-1} = \frac{1}{6}\begin{bmatrix} 3 & 3 \\ -1 & 1 \end{bmatrix}$

$\begin{bmatrix} x \\ y \end{bmatrix} = \frac{1}{6}\begin{bmatrix} 3 & 3 \\ -1 & 1 \end{bmatrix}\begin{bmatrix} 1 \\ -11 \end{bmatrix} = \frac{1}{6}\begin{bmatrix} -30 \\ -12 \end{bmatrix} = \begin{bmatrix} -5 \\ -2 \end{bmatrix}$

Solution is $(-5, -2)$.

Check for Understanding

Here is a suggested use of the Oral Exercises to check students' understanding as you teach the lesson.

Oral Exs. 1–6: use after Example 1.

Oral Exs. 7–13: use after Example 2.

Guided Practice

Find the inverse of each matrix if it exists. If not, say so.

1. $\begin{bmatrix} -3 & 4 \\ 1 & -2 \end{bmatrix}$

   $\frac{1}{2}\begin{bmatrix} -2 & -4 \\ -1 & -3 \end{bmatrix}$

2. $\begin{bmatrix} 6 & -16 \\ -3 & 8 \end{bmatrix}$

   no inverse

3. $\begin{bmatrix} 0 & -2 \\ 2 & -1 \end{bmatrix}$

   $\frac{1}{4}\begin{bmatrix} -1 & 2 \\ -2 & 0 \end{bmatrix}$

Use matrices to solve each system. If there is no unique solution, say so.

4. $x + 3y = 2$
   $2x + 5y = 7$
   $(11, -3)$

5. $3x + 5y = 18$
   $x - 3y = 13$
   $\left(\frac{17}{2}, -\frac{3}{2}\right)$

6. $-4x + 12y = 7$
   $3x - 9y = 1$
   no unique solution

Summarizing the Lesson

In this lesson students learned to find inverses of matrices and to use them in solving systems of equations. Ask the students to describe how to invert a 2×2 matrix.

Suggested Assignments

Extended Algebra
792/1–18
S 789/17

Extended Alg. and Trig.
792/1–18
S 778/27, 29

Oral Exercises

Tell whether each matrix has an inverse.

1. $\begin{bmatrix} 4 & -10 \\ -1 & 3 \end{bmatrix}$ yes
2. $\begin{bmatrix} 6 & -8 \\ 3 & -4 \end{bmatrix}$ no
3. $\begin{bmatrix} 10 & 5 \\ 9 & 4 \end{bmatrix}$ yes
4. $\begin{bmatrix} 0 & 16 \\ 0 & -12 \end{bmatrix}$ no
5. $\begin{bmatrix} 6 & -3 \\ -10 & 5 \end{bmatrix}$ no
6. $\begin{bmatrix} 8 & -5 \\ 3 & -2 \end{bmatrix}$ yes

Tell whether each system has a unique solution.

7. $5x + y = 1$, $4x + 2y = 3$ yes
8. $4x + 2y = -6$, $6x + 3y = -9$ no
9. $3x - 2y = 2$, $3x + 2y = -9$ yes
10. $6x = 3$, $2x + y = 5$ yes
11. $2x + 4y = 12$, $3x + 6y = 20$ no
12. $8x - y = 10$, $y = 5$ yes
13. If a system of equations does not have a unique solution, can you conclude that the system has no solution? Explain. No; the system could have infinitely many solutions.

Written Exercises

Find the inverse of each matrix. If the matrix has no inverse, say so.

A

1. $\begin{bmatrix} 2 & 1 \\ -1 & 2 \end{bmatrix}$ $\begin{bmatrix} \frac{2}{5} & -\frac{1}{5} \\ \frac{1}{5} & \frac{2}{5} \end{bmatrix}$
2. $\begin{bmatrix} 5 & 3 \\ 2 & 1 \end{bmatrix}$ $\begin{bmatrix} -1 & 3 \\ 2 & -5 \end{bmatrix}$
3. $\begin{bmatrix} 4 & 6 \\ 2 & 3 \end{bmatrix}$ No inverse
4. $\begin{bmatrix} -3 & 1 \\ 4 & -2 \end{bmatrix}$ $\begin{bmatrix} -1 & -\frac{1}{2} \\ -2 & -\frac{3}{2} \end{bmatrix}$
5. $\begin{bmatrix} 1 & 0 \\ 0 & 1 \end{bmatrix}$ $\begin{bmatrix} 1 & 0 \\ 0 & 1 \end{bmatrix}$
6. $\begin{bmatrix} 0 & 9 \\ -1 & 3 \end{bmatrix}$ $\begin{bmatrix} \frac{1}{3} & -1 \\ \frac{1}{9} & 0 \end{bmatrix}$
7. $\begin{bmatrix} 3 & -1 \\ 6 & 2 \end{bmatrix}$ $\begin{bmatrix} \frac{1}{6} & \frac{1}{12} \\ -\frac{1}{2} & \frac{1}{4} \end{bmatrix}$
8. $\begin{bmatrix} 0 & 0 \\ 5 & -7 \end{bmatrix}$ No inverse

Use matrices to find the solution of each system of equations. If a system has no unique solution, say so.

9. $2x - y = 6$, $3x + 2y = -19$ $(-1, -8)$
10. $3x + 2y = 5$, $3x - y = 2$ $(1, 1)$
11. $2x + 2y = 16$, $x - 3y = -4$ $(5, 3)$
12. $4x - 6y = 8$, $-6x + 9y = -12$ No unique solution
13. $x + y = 6$, $2x = 4$ $(2, 4)$
14. $4x + 3y = 1$, $4x + y = -5$ $(-2, 3)$
15. $x + y = 1$, $2x + 2y = 5$ No unique solution
16. $x = 3$, $-x + y = 2$ $(3, 5)$

Solve each equation for matrix X. (*Hint:* These equations are in the form $AX = B$.)

B

17. $\begin{bmatrix} 1 & 3 \\ 2 & 7 \end{bmatrix}X = \begin{bmatrix} 4 & 1 \\ -2 & 3 \end{bmatrix}$ $\begin{bmatrix} 34 & -2 \\ -10 & 1 \end{bmatrix}$
18. $\begin{bmatrix} 4 & 4 \\ 3 & 4 \end{bmatrix}X = \begin{bmatrix} 1 & 0 \\ 0 & 1 \end{bmatrix}$ $\begin{bmatrix} 1 & -1 \\ -\frac{3}{4} & 1 \end{bmatrix}$
19. $\begin{bmatrix} 4 & 2 \\ 1 & 1 \end{bmatrix}X - \begin{bmatrix} 0 & \frac{1}{2} \\ -1 & \frac{3}{2} \end{bmatrix} = \begin{bmatrix} 2 & 3 \\ 0 & 1 \end{bmatrix}$ $\begin{bmatrix} 2 & -\frac{3}{4} \\ -3 & \frac{13}{4} \end{bmatrix}$
20. $\begin{bmatrix} 5 & 3 \\ 3 & 2 \end{bmatrix}X - \begin{bmatrix} 1 & 3 \\ 2 & 1 \end{bmatrix} = \begin{bmatrix} 1 & 2 \\ -1 & 3 \end{bmatrix}$ $\begin{bmatrix} 1 & -2 \\ -1 & 5 \end{bmatrix}$

Self-Test 2

Vocabulary determinant (p. 787) · order of a determinant (p. 787) · diagonal matrix (p. 788) · inverse matrices (p. 790)

In Exercises 1 and 2 evaluate each determinant.

1. $\begin{vmatrix} 3 & 0 \\ 5 & -1 \end{vmatrix}$ −3 **2.** $\begin{vmatrix} 2 & -2 & 0 \\ 1 & 1 & -2 \\ 0 & -3 & 1 \end{vmatrix}$ −8 **Obj. 16-5, p. 787**

3. Find A^{-1} if $A = \begin{bmatrix} 1 & -1 \\ 0 & 3 \end{bmatrix}$. $\begin{bmatrix} 1 & \frac{1}{3} \\ 0 & \frac{1}{3} \end{bmatrix}$ **Obj. 16-6, p. 790**

4. Solve for matrix A.

$$\begin{bmatrix} 2 & 1 \\ -1 & 4 \end{bmatrix} A = \begin{bmatrix} 3 & -1 \\ -6 & 5 \end{bmatrix}. \quad \begin{bmatrix} 2 & -1 \\ -1 & 1 \end{bmatrix}$$

Check your answers with those at the back of the book.

Historical Note / *Matrices*

Matrices are a relatively new concept in mathematics. They were not devised until 1857 when the British mathematician Arthur Cayley (1821–1895) began to use them in the study of linear transformations.

Cayley was working on linear transformations of the type

$$x_1 = ax + by$$
$$y_1 = cx + dy$$

which map an ordered pair (x, y) onto another ordered pair (x_1, y_1). In this transformation a, b, c, and d are real numbers. Cayley noticed that the entire transformation was determined by these four real numbers, which could be arranged in the square array

$$\begin{bmatrix} a & b \\ c & d \end{bmatrix}.$$

When the product of two such arrays is defined as in this chapter, the matrix corresponding to a *sequence* of transformations acting one after the other is the *product* of their matrices. Further analysis led Cayley to formulate most of the rules of the algebra of matrices.

A linear transformation maps any point (x, y) in the plane onto an image point (x_1, y_1), and Cayley saw that the matrix of a transformation gives geometric information about this mapping. For example, the image of a polygon is another polygon, generally with different sides and angles.

However, the area of the image is always equal to the area of the original figure multiplied by the absolute value of the determinant of the matrix of the transformation.

Supplementary Materials

Study Guide	pp. 269–270
Practice Master	91
Computer Activity	36
Resource Book	p. 173

Quick Quiz

Evaluate each determinant.

1. $\begin{vmatrix} -1 & 5 \\ 7 & 3 \end{vmatrix}$ −38

2. $\begin{vmatrix} 0 & 1 & 2 \\ 3 & -1 & 1 \\ 2 & -2 & 3 \end{vmatrix}$ −15

3. Find $\begin{bmatrix} -3 & -2 \\ -4 & -3 \end{bmatrix}^{-1}$.

$\begin{bmatrix} -3 & 2 \\ 4 & -3 \end{bmatrix}$

4. Solve for X:

$\begin{bmatrix} 2 & 1 \\ 5 & 3 \end{bmatrix} X = \begin{bmatrix} 2 & 1 \\ -1 & 0 \end{bmatrix}$

$\begin{bmatrix} 7 & 3 \\ -12 & -5 \end{bmatrix}$

Teaching Suggestions p. T144

Warm-Up Exercises

Evaluate the determinant.

1. $\begin{vmatrix} 2 & -3 \\ 4 & -5 \end{vmatrix}$ 2

2. $\begin{vmatrix} -6 & 2 \\ 3 & -1 \end{vmatrix}$ 0

3. $\begin{vmatrix} 2 & 3 & 1 \\ -1 & -2 & 4 \\ 1 & -3 & 2 \end{vmatrix}$ 39

4. $\begin{vmatrix} 2 & 0 & 2 \\ -1 & 1 & 0 \\ 0 & 3 & -1 \end{vmatrix}$ −8

Motivating the Lesson

Remind students that they have already learned to evaluate a third-order determinant by using a rather cumbersome formula. Although the terms of the formula can be found by multiplying matrix elements along diagonals, this approach *only* works with 3×3 matrices. Today's lesson presents another method of evaluating third-order determinants that also applies to higher-order determinants.

Chalkboard Examples

1. Expand $\begin{vmatrix} 3 & 2 & 4 \\ -1 & 5 & 1 \\ 0 & 2 & 6 \end{vmatrix}$ by minors of the first row.

$3\begin{vmatrix} 5 & 1 \\ 2 & 6 \end{vmatrix} - 2\begin{vmatrix} -1 & 1 \\ 0 & 6 \end{vmatrix} + 4\begin{vmatrix} -1 & 5 \\ 0 & 2 \end{vmatrix} = 3(30 - 2) - 2(-6 - 0) + 4(-2 - 0) = 3(28) - 2(-6) + 4(-2) = 88$

Working with Determinants

16-7 Expansion of Determinants by Minors

Objective To evaluate third-order determinants using expansion by minors.

It is important to distinguish between matrices and determinants. A matrix is a rectangular array of numbers. A determinant, often denoted by a square array of numbers, is in fact a real number associated with a square matrix. The process of evaluating a determinant is called "expanding the determinant."

You have already learned to evaluate a determinant of order 2 or 3. In this lesson you will learn a method of expansion that is applicable to determinants of order n when $n \geq 3$. This method is called "expansion by minors."

The **minor** of an element in a determinant is the determinant resulting from the deletion of the row and column containing the element. Therefore in the determinant

$$\begin{vmatrix} 3 & -1 & 0 \\ 8 & 4 & 2 \\ 0 & -2 & 4 \end{vmatrix},$$

the minor of the element 3 is

$$\begin{vmatrix} \not{3} & \not{-1} & \not{0} \\ \not{8} & 4 & 2 \\ \not{0} & -2 & 4 \end{vmatrix}, \quad \text{or} \quad \begin{vmatrix} 4 & 2 \\ -2 & 4 \end{vmatrix}.$$

The minor of the element 8 is

$$\begin{vmatrix} \not{3} & -1 & 0 \\ \not{8} & \not{4} & \not{2} \\ \not{0} & -2 & 4 \end{vmatrix}, \quad \text{or} \quad \begin{vmatrix} -1 & 0 \\ -2 & 4 \end{vmatrix}.$$

By definition (see page 787), the value of a third-order determinant is given by

$$\begin{vmatrix} a_1 & b_1 & c_1 \\ a_2 & b_2 & c_2 \\ a_3 & b_3 & c_3 \end{vmatrix} = a_1b_2c_3 + a_2b_3c_1 + a_3b_1c_2 - a_3b_2c_1 - a_2b_1c_3 - a_1b_3c_2$$

$$= a_1(b_2c_3 - b_3c_2) - b_1(a_2c_3 - a_3c_2) + c_1(a_2b_3 - a_3b_2) \quad \text{Rearranging terms and factoring}$$

$$= a_1\begin{vmatrix} b_2 & c_2 \\ b_3 & c_3 \end{vmatrix} - b_1\begin{vmatrix} a_2 & c_2 \\ a_3 & c_3 \end{vmatrix} + c_1\begin{vmatrix} a_2 & b_2 \\ a_3 & b_3 \end{vmatrix}.$$

Notice that the three determinants are the minors of a_1, b_1, and c_1, named below by A_1, B_1, and C_1, respectively. Therefore,

$$\begin{vmatrix} a_1 & b_1 & c_1 \\ a_2 & b_2 & c_2 \\ a_3 & b_3 & c_3 \end{vmatrix} = a_1A_1 - b_1B_1 + c_1C_1.$$

The right-hand expression is the expansion of this determinant by minors *about the first row.*

A determinant may be expanded by minors about any row or column.

To Expand a Determinant by Minors

1. Multiply each element of a given row or column by its minor.
2. Add the number of the row and the number of the column for each element. If the sum is odd, multiply the product obtained in Step 1 by -1.
3. Add the products to obtain the value of the determinant.

Example 1 Expand $\begin{vmatrix} 2 & -6 & 3 \\ 4 & 5 & -2 \\ 3 & -1 & 0 \end{vmatrix}$ by the minors of the first row.

Solution Multiply each element of the first row by its minor. Also multiply the product for the element in the first row, second column by -1, since the sum $1 + 2$, or 3, is odd. Then add the products.

$$\begin{vmatrix} 2 & -6 & 3 \\ 4 & 5 & -2 \\ 3 & -1 & 0 \end{vmatrix} = 2\begin{vmatrix} 5 & -2 \\ -1 & 0 \end{vmatrix} + (-1)(-6)\begin{vmatrix} 4 & -2 \\ 3 & 0 \end{vmatrix} + 3\begin{vmatrix} 4 & 5 \\ 3 & -1 \end{vmatrix}$$

$$= 2(0 - 2) + 6(0 + 6) + 3(-4 - 15)$$

$$= -4 + 36 + (-57) = -25 \quad \textbf{\textit{Answer}}$$

You can shorten the work of expanding a determinant by minors by choosing a row or column having the greatest number of zeros as elements.

Example 2 Expand $\begin{vmatrix} 1 & 3 & 0 \\ -1 & 4 & 2 \\ 2 & 1 & 0 \end{vmatrix}$ by the minors of the third column.

Solution

$$\begin{vmatrix} 1 & 3 & 0 \\ -1 & 4 & 2 \\ 2 & 1 & 0 \end{vmatrix} = 0(C_1) + (-1)2\begin{vmatrix} 1 & 3 \\ 2 & 1 \end{vmatrix} + 0(C_3)$$

$$= 0 + (-2)(1 - 6) + 0$$

$$= (-2)(-5) = 10 \quad \textbf{\textit{Answer}}$$

2. Expand $\begin{vmatrix} -2 & 0 & -1 \\ 3 & 5 & 4 \\ 1 & 0 & 2 \end{vmatrix}$ by minors of the second column.

$\begin{vmatrix} -2 & 0 & -1 \\ 3 & 5 & 4 \\ 1 & 0 & 2 \end{vmatrix} =$

$(-1)(0)(B_1) + 5\begin{vmatrix} -2 & -1 \\ 1 & 2 \end{vmatrix}$

$+ (-1)(0)(B_3) = 0 +$
$5(-3) + 0 = -15$

3. Expand $\begin{vmatrix} 1 & 0 & 1 & -1 \\ 1 & 2 & 0 & -4 \\ -5 & 3 & -1 & 0 \\ 1 & 1 & 1 & 0 \end{vmatrix}$ by minors of the fourth column.

$-(-1)\begin{vmatrix} 1 & 2 & 0 \\ -5 & 3 & -1 \\ 1 & 1 & 1 \end{vmatrix} +$

$(-4)\begin{vmatrix} 1 & 0 & 1 \\ -5 & 3 & -1 \\ 1 & 1 & 1 \end{vmatrix} -$

$0 + 0 = \begin{vmatrix} 1 & 2 & 0 \\ -5 & 3 & -1 \\ 1 & 1 & 1 \end{vmatrix} -$

$4\begin{vmatrix} 1 & 0 & 1 \\ -5 & 3 & -1 \\ 1 & 1 & 1 \end{vmatrix} =$

$1\begin{vmatrix} 3 & -1 \\ 1 & 1 \end{vmatrix} - 2\begin{vmatrix} -5 & -1 \\ 1 & 1 \end{vmatrix} +$

$0 - 4\left(1\begin{vmatrix} 3 & -1 \\ 1 & 1 \end{vmatrix} - 0 +\right.$

$\left. 1\begin{vmatrix} -5 & 3 \\ 1 & 1 \end{vmatrix}\right) =$

$(4 - 2(-4)) - 4(4 + (-8)) =$
$12 - 4(-4) = 28$

Check for Understanding

Oral Exs. 1–6: use after Example 1.

Guided Practice

Evaluate by expanding as indicated.

1. $\begin{vmatrix} 1 & 2 & 3 \\ 2 & -3 & 1 \\ 1 & 2 & 4 \end{vmatrix}$; row one −7

2. $\begin{vmatrix} 10 & -1 & -2 \\ 12 & 8 & 1 \\ 0 & 1 & 0 \end{vmatrix}$; column two −34

3. $\begin{vmatrix} 1 & 2 & 3 \\ 3 & 2 & 10 \\ -1 & -1 & 2 \end{vmatrix}$; row three −21

Common Errors

Students may forget Step 2 on page 795 when expanding a determinant by minors. Point out that in the expansion by minors of an *odd* row or column, every other term starting with the *second* is multiplied by −1. For an *even* row or column, every other term starting with the *first* is multiplied by −1.

When asked to expand a given determinant about any row or column, students may have difficulty making the decision as to which row or column to use. Remind the students that their work can be shortened by expanding about a row or column having zeros as elements.

The method of expansion by minors is used to define determinants of order $n \geq 3$. For example, consider the matrix

$$M = \begin{bmatrix} a_1 & b_1 & c_1 & d_1 \\ a_2 & b_2 & c_2 & d_2 \\ a_3 & b_3 & c_3 & d_3 \\ a_4 & b_4 & c_4 & d_4 \end{bmatrix}.$$

Let $A_1 = \begin{vmatrix} b_2 & c_2 & d_2 \\ b_3 & c_3 & d_3 \\ b_4 & c_4 & d_4 \end{vmatrix}$, $B_1 = \begin{vmatrix} a_2 & c_2 & d_2 \\ a_3 & c_3 & d_3 \\ a_4 & c_4 & d_4 \end{vmatrix}$,

$C_1 = \begin{vmatrix} a_2 & b_2 & d_2 \\ a_3 & b_3 & d_3 \\ a_4 & b_4 & d_4 \end{vmatrix}$, and $D_1 = \begin{vmatrix} a_2 & b_2 & c_2 \\ a_3 & b_3 & c_3 \\ a_4 & b_4 & c_4 \end{vmatrix}$.

The determinant of M is defined as

$$\det M = a_1A_1 - b_1B_1 + c_1C_1 - d_1D_1.$$

Note that A_1, B_1, C_1, and D_1 are the minors of a_1, b_1, c_1, and d_1, respectively. Thus they are minors of the first row of this 4×4 determinant. As with third-order determinants, determinants of order four may be expanded by minors about any row or any column. The minors of a fourth-order determinant are each third-order determinants. In general, minors of a determinant of order n are determinants of order $n - 1$. Since the expansion of higher order determinants can be tedious, computers are often used to evaluate them.

Oral Exercises

Give the requested element or determinant using the given expansion by minors.

$$\begin{vmatrix} 0 & 1 & 2 \\ 1 & -1 & -3 \\ -2 & 4 & 5 \end{vmatrix} = a_1 \begin{vmatrix} -1 & -3 \\ b_3 & 5 \end{vmatrix} - b_1 \begin{vmatrix} a_2 & -3 \\ -2 & 5 \end{vmatrix} + 2C_1$$

1. a_1 0
2. b_3 4
3. b_1 1
4. a_2 1
5. C_1 $\begin{vmatrix} 1 & -1 \\ -2 & 4 \end{vmatrix}$
6. Complete the computation to evaluate the determinant given above. 5

Written Exercises

Expand each determinant about the given row or column.

A 1. $\begin{vmatrix} 2 & 1 & 3 \\ -2 & 1 & 4 \\ 1 & 2 & 5 \end{vmatrix}$; row one −7

2. $\begin{vmatrix} 2 & 3 & 1 \\ 0 & -1 & -4 \\ 5 & 0 & 0 \end{vmatrix}$; row three −55

3. $\begin{vmatrix} -2 & -2 & 0 \\ 0 & 2 & 3 \\ 1 & -1 & 1 \end{vmatrix}$; column one −16

4. $\begin{vmatrix} 4 & 0 & -5 \\ 9 & 1 & 11 \\ -4 & 0 & 3 \end{vmatrix}$; column two −8

5. $\begin{vmatrix} -1 & 3 & 4 \\ 0 & 1 & 0 \\ 3 & 6 & -2 \end{vmatrix}$; row two −10

6. $\begin{vmatrix} 2 & 0 & 0 \\ 10 & 1 & -1 \\ 3 & 0 & 4 \end{vmatrix}$; row one 8

7. $\begin{vmatrix} 1 & -1 & 0 \\ 2 & 3 & -1 \\ 10 & -4 & 0 \end{vmatrix}$; column three 6

8. $\begin{vmatrix} 0 & -2 & 4 \\ 1 & 3 & -1 \\ 0 & 0 & 5 \end{vmatrix}$; column one 10

Expand each determinant about any row or column.

9. $\begin{vmatrix} 1 & 0 & -1 \\ 0 & 1 & 2 \\ -1 & 2 & 1 \end{vmatrix}$ −4

10. $\begin{vmatrix} 0 & 2 & 1 \\ 4 & 0 & -2 \\ -1 & 3 & 1 \end{vmatrix}$ 8

11. $\begin{vmatrix} 0 & 3 & 1 \\ 4 & -1 & 2 \\ 0 & 2 & 0 \end{vmatrix}$ 8

12. $\begin{vmatrix} 1 & -1 & 2 \\ 3 & 5 & 1 \\ 4 & 2 & -2 \end{vmatrix}$ −50

Solve each equation for x.

B 13. $\begin{vmatrix} 2 & 1 & 1 \\ 1 & 2 & 1 \\ x & 1 & 2 \end{vmatrix} = 4$ 1

14. $\begin{vmatrix} 1 & 3 & 4 \\ 5 & 15 & 10 \\ -1 & x & 2 \end{vmatrix} = 80$ 5

15. $\begin{vmatrix} 12 & -7 & 19 \\ 0 & x & 5 \\ -9 & 3 & 43 \end{vmatrix} = 135$ 0

16. $\begin{vmatrix} 4 & -3 & 9 \\ 8 & x & -4 \\ 1 & 5 & 3 \end{vmatrix} = 545$ 7

Find the value of each determinant.

17. $\begin{vmatrix} 5 & 0 & 0 & 0 \\ 3 & -3 & 0 & 0 \\ 8 & 2 & 2 & 0 \\ -4 & 10 & 3 & 2 \end{vmatrix}$ −60

18. $\begin{vmatrix} 1 & 2 & 3 & 4 \\ 5 & 6 & 7 & 8 \\ 9 & 10 & 11 & 12 \\ 13 & 14 & 15 & 16 \end{vmatrix}$ 0

C 19. Let A be a 3×3 matrix. Show that if any two rows of A are alike, then $\det A = 0$.

Mixed Review Exercises

Simplify.

1. $125^{-2/3}$ $\frac{1}{25}$
2. $\cos \frac{3\pi}{4}$ $-\frac{\sqrt{2}}{2}$
3. $\frac{8!}{3!5!}$ 56
4. $\sqrt{-2} \cdot \sqrt{-8}$ −4
5. $\log_3 9\sqrt{3}$ $\frac{5}{2}$
6. $\sec^2 20° - \tan^2 20°$ 1

Summarizing the Lesson

In this lesson students learned to evaluate third-order determinants using expansion by minors. Ask the students to describe this method.

Suggested Assignments

Extended Algebra
796/1–17
S 797/Mixed Review, 1, 3–5
R 793/Self-Test 2

Extended Alg. and Trig.
796/1–17
S 797/Mixed Review
R 793/Self-Test 2

Supplementary Materials

Study Guide pp. 271–272

Additional Answers Written Exercises

19. $\begin{vmatrix} a & b & c \\ a & b & c \\ d & e & f \end{vmatrix} = a\begin{vmatrix} b & c \\ e & f \end{vmatrix} - a\begin{vmatrix} b & c \\ e & f \end{vmatrix} + d\begin{vmatrix} b & c \\ b & c \end{vmatrix} = a(bf - ec) - a(bf - ec) + d(bc - bc) = 0;$

$\begin{vmatrix} a & b & c \\ d & e & f \\ a & b & c \end{vmatrix} = a\begin{vmatrix} e & f \\ b & c \end{vmatrix} - d\begin{vmatrix} b & c \\ b & c \end{vmatrix} + a\begin{vmatrix} b & c \\ e & f \end{vmatrix} = a(ec - bf) - d(bc - bc) + a(bf - ec) = aec - abf + abf - aec = 0;$

$\begin{vmatrix} a & b & c \\ d & e & f \\ d & e & f \end{vmatrix} = a\begin{vmatrix} e & f \\ e & f \end{vmatrix} - d\begin{vmatrix} b & c \\ e & f \end{vmatrix} + d\begin{vmatrix} b & c \\ e & f \end{vmatrix} = a(ef - ef) - d(bf - ec) + d(bf - ec) = 0$

Teaching Suggestions p. T144

Warm-Up Exercises

Evaluate.

1. $\begin{vmatrix} 1 & 0 & -2 \\ 2 & 0 & 3 \\ -3 & 0 & 1 \end{vmatrix}$ 0

2. $\begin{vmatrix} -1 & 2 & 1 \\ -1 & 2 & 1 \\ 3 & 1 & -2 \end{vmatrix}$ 0

3. $\begin{vmatrix} -2 & 3 & 1 \\ 0 & -1 & 2 \\ 1 & 0 & -1 \end{vmatrix}$ 5

4. $\begin{vmatrix} 3 & -2 & 1 \\ -1 & 0 & 2 \\ 0 & 1 & -1 \end{vmatrix}$ −5

5. $\begin{vmatrix} -4 & 6 & 2 \\ 0 & -1 & 2 \\ 1 & 0 & -1 \end{vmatrix}$ 10

Motivating the Lesson

Tell the students that to use determinants in applications, such as solving systems of equations, it is important to have simple, efficient techniques for evaluating them. "Shortcut" methods for evaluating higher-order determinants are the topic of today's lesson.

16-8 Properties of Determinants

Objective To use the properties of determinants to simplify the expansion of determinants by minors.

Determinants have properties that you can use to simplify their evaluation. Although third-order determinants are used as examples in this lesson, the properties apply to determinants of any order.

Property 1. If each element in any row (or any column) is 0, then the determinant is equal to 0. For example,

$$\begin{vmatrix} 3 & 5 & -2 \\ 0 & 0 & 0 \\ -1 & 2 & 4 \end{vmatrix} = 0\begin{vmatrix} 5 & -2 \\ 2 & 4 \end{vmatrix} - 0\begin{vmatrix} 3 & -2 \\ -1 & 4 \end{vmatrix} + 0\begin{vmatrix} 3 & 5 \\ -1 & 2 \end{vmatrix}$$
$$= 0 + 0 + 0 = 0.$$

Property 2. If two rows (or two columns) of a determinant have corresponding elements that are equal, then the determinant is equal to 0. For example,

$$\begin{vmatrix} 8 & -4 & 3 \\ 6 & 2 & 1 \\ 6 & 2 & 1 \end{vmatrix} = 8\begin{vmatrix} 2 & 1 \\ 2 & 1 \end{vmatrix} - (-4)\begin{vmatrix} 6 & 1 \\ 6 & 1 \end{vmatrix} + 3\begin{vmatrix} 6 & 2 \\ 6 & 2 \end{vmatrix}$$
$$= 8(0) + 4(0) + 3(0) = 0.$$

Property 3. If two rows (or two columns) of a determinant are interchanged, then the resulting determinant is the opposite of the original determinant. For example, expand these determinants by minors about the second row:

$$\begin{vmatrix} 1 & 2 & 4 \\ 3 & 1 & 0 \\ -2 & -1 & 2 \end{vmatrix} = (-1)3\begin{vmatrix} 2 & 4 \\ -1 & 2 \end{vmatrix} + 1\begin{vmatrix} 1 & 4 \\ -2 & 2 \end{vmatrix} - 0\begin{vmatrix} 1 & 2 \\ -2 & -1 \end{vmatrix}$$
$$= -3(8) + 10 + 0 = -14;$$

$$\begin{vmatrix} 2 & 1 & 4 \\ 1 & 3 & 0 \\ -1 & -2 & 2 \end{vmatrix} = (-1)(1)\begin{vmatrix} 1 & 4 \\ -2 & 2 \end{vmatrix} + 3\begin{vmatrix} 2 & 4 \\ -1 & 2 \end{vmatrix} - 0\begin{vmatrix} 2 & 1 \\ -1 & -2 \end{vmatrix}$$
$$= (-1)(10) + 3(8) + 0 = 14.$$

Property 4. If each element in one row (or one column) of a determinant is multiplied by a real number k, then the determinant is multiplied by k. For example, expand these determinants by minors about the first row:

$$\begin{vmatrix} 1 & 2 & 1 \\ -1 & 4 & -2 \\ 2 & -1 & 1 \end{vmatrix} = 1\begin{vmatrix} 4 & -2 \\ -1 & 1 \end{vmatrix} - 2\begin{vmatrix} -1 & -2 \\ 2 & 1 \end{vmatrix} + 1\begin{vmatrix} -1 & 4 \\ 2 & -1 \end{vmatrix}$$
$$= 2 - 2(3) - 7 = -11;$$

$$\begin{vmatrix} 5\cdot1 & 5\cdot2 & 5\cdot1 \\ -1 & 4 & -2 \\ 2 & -1 & 1 \end{vmatrix} = 5\cdot1\begin{vmatrix} 4 & -2 \\ -1 & 1 \end{vmatrix} - 5\cdot2\begin{vmatrix} -1 & -2 \\ 2 & 1 \end{vmatrix} + 5\cdot1\begin{vmatrix} -1 & 4 \\ 2 & -1 \end{vmatrix}$$
$$= 5\left(1\begin{vmatrix} 4 & -2 \\ -1 & 1 \end{vmatrix} - 2\begin{vmatrix} -1 & -2 \\ 2 & 1 \end{vmatrix} + 1\begin{vmatrix} -1 & 4 \\ 2 & -1 \end{vmatrix}\right)$$
$$= 5(-11) = -55.$$

Property 5. If each element of one row (or one column) is multiplied by a real number k and if the resulting products are then added to the corresponding elements of another row (or another column), then the resulting determinant equals the original one. For example, expand these determinants by minors about the first column:

$$\begin{vmatrix} -4 & -1 & 1 \\ 6 & -3 & 2 \\ -6 & 3 & -1 \end{vmatrix} = -4\begin{vmatrix} -3 & 2 \\ 3 & -1 \end{vmatrix} - 6\begin{vmatrix} -1 & 1 \\ 3 & -1 \end{vmatrix} - 6\begin{vmatrix} -1 & 1 \\ -3 & 2 \end{vmatrix}$$
$$= -4(-3) - 6(-2) - 6(1) = 18;$$

$$\begin{vmatrix} -4 + 2(-1) & -1 & 1 \\ 6 + 2(-3) & -3 & 2 \\ -6 + 2(3) & 3 & -1 \end{vmatrix} = \begin{vmatrix} -6 & -1 & 1 \\ 0 & -3 & 2 \\ 0 & 3 & -1 \end{vmatrix} = -6\begin{vmatrix} -3 & 2 \\ 3 & -1 \end{vmatrix} - 0\begin{vmatrix} -1 & 1 \\ 3 & -1 \end{vmatrix} + 0\begin{vmatrix} -1 & 1 \\ -3 & 2 \end{vmatrix}$$
$$= -6(-3) + 0 + 0 = 18.$$

Properties 1–5 can be used to simplify the expansion of determinants. Notice how transforming the first column in the example for Property 5 simplified the expansion. A similar approach is taken in the example below, where the first row is reduced to 1 0 0 to simplify the expansion.

Example 1 Evaluate $\begin{vmatrix} 15 & 0 & 2 \\ 45 & -1 & 3 \\ 30 & 2 & 8 \end{vmatrix}$.

Solution Factor 15 from the elements in the first column. (Property 4)

$$\begin{vmatrix} 15 & 0 & 2 \\ 45 & -1 & 3 \\ 30 & 2 & 8 \end{vmatrix} = \begin{vmatrix} 15(1) & 0 & 2 \\ 15(3) & -1 & 3 \\ 15(2) & 2 & 8 \end{vmatrix} = 15\begin{vmatrix} 1 & 0 & 2 \\ 3 & -1 & 3 \\ 2 & 2 & 8 \end{vmatrix}$$

Multiply the first column by -2 and add the results to the third column. (Property 5)

$$15\begin{vmatrix} 1 & 0 & 2 \\ 3 & -1 & 3 \\ 2 & 2 & 8 \end{vmatrix} = 15\begin{vmatrix} 1 & 0 & 2 + (-2) \\ 3 & -1 & 3 + (-6) \\ 2 & 2 & 8 + (-4) \end{vmatrix} = 15\begin{vmatrix} 1 & 0 & 0 \\ 3 & -1 & -3 \\ 2 & 2 & 4 \end{vmatrix}$$

(Solution continues on the next page.)

Chalkboard Examples

Evaluate. Use Properties 1–5.

1. $\begin{vmatrix} 15 & 10 & -5 \\ 14 & 7 & 7 \\ 3 & 1 & 1 \end{vmatrix}$

Factor row 1 (Prop. 4).

$5\begin{vmatrix} 3 & 2 & -1 \\ 14 & 7 & 7 \\ 3 & 1 & 1 \end{vmatrix}$

Factor row 2 (Prop. 4).

$35\begin{vmatrix} 3 & 2 & -1 \\ 2 & 1 & 1 \\ 3 & 1 & 1 \end{vmatrix}$

Multiply column 3 by -1 and add the result to column 2 (Prop. 5).

$35\begin{vmatrix} 3 & 3 & -1 \\ 2 & 0 & 1 \\ 3 & 0 & 1 \end{vmatrix}$

Expand by minors about column 2.

$35\left(-3\begin{vmatrix} 2 & 1 \\ 3 & 1 \end{vmatrix} + 0\begin{vmatrix} 3 & -1 \\ 3 & 1 \end{vmatrix} - 0\begin{vmatrix} 3 & -1 \\ 2 & 1 \end{vmatrix}\right) =$

$35(-3(-1)) = 105$

2. $\begin{vmatrix} 3 & 8 & -1 \\ 0 & -1 & 0 \\ -6 & 5 & 2 \end{vmatrix}$

Factor -3 from column 1 (Prop. 4).

$-3\begin{vmatrix} -1 & 8 & -1 \\ 0 & -1 & 0 \\ 2 & 5 & 2 \end{vmatrix}$

Columns 1 and 3 have corresponding elements that are equal; determinant equals 0 (Prop. 2).

Check for Understanding

Oral Exs. 1–4: use before Example 1.

Guided Practice

Evaluate each determinant. Use Properties 1–5.

1. $\begin{vmatrix} 6 & 9 & 3 \\ 4 & 2 & 6 \\ -15 & 20 & 10 \end{vmatrix}$ −1440

2. $\begin{vmatrix} 4 & 12 & 18 \\ 3 & 9 & 13 \\ 2 & 6 & 28 \end{vmatrix}$ 0

3. $\begin{vmatrix} 3 & 0 & 0 \\ 0 & 3 & 0 \\ 0 & 0 & 3 \end{vmatrix}$ 27

Summarizing the Lesson

In this lesson students learned to use the properties of determinants to facilitate evaluating determinants. Ask the students to state the five properties of determinants.

Suggested Assignments

Extended Algebra
800/1–14
S 797/18

Extended Alg. and Trig.
800/1–14
S 797/18

Supplementary Materials

Study Guide pp. 273–274
Practice Master 92

Expand by minors about the first row.

$$15\begin{vmatrix} 1 & 0 & 0 \\ 3 & -1 & -3 \\ 2 & 2 & 4 \end{vmatrix} = 15\left(1\begin{vmatrix} -1 & -3 \\ 2 & 4 \end{vmatrix} - 0 + 0\right)$$

$$= 15(2 - 0 + 0) = 30 \quad \textit{Answer}$$

Oral Exercises

Use the properties of determinants to find the value of each determinant below, given that

$$\begin{vmatrix} \mathbf{0} & \mathbf{3} & \mathbf{1} \\ \mathbf{1} & \mathbf{2} & \mathbf{0} \\ \mathbf{-1} & \mathbf{2} & \mathbf{1} \end{vmatrix} = \mathbf{1}.$$

1. $\begin{vmatrix} 1 & 2 & 0 \\ 0 & 3 & 1 \\ -1 & 2 & 1 \end{vmatrix}$ −1 **2.** $\begin{vmatrix} 0 & 3 & 3 \\ 1 & 2 & 0 \\ -1 & 2 & 3 \end{vmatrix}$ 3 **3.** $\begin{vmatrix} 0 & 3 & 1 \\ 1 & 2 & 0 \\ -1 & 5 & 2 \end{vmatrix}$ 1 **4.** $\begin{vmatrix} 0 & 3 & 1 \\ p & 2 & 0 \\ -p & 5 & 2 \end{vmatrix}$ p

Written Exercises

Evaluate each determinant. Use Properties 1–5 to simplify the work.

A **1.** $\begin{vmatrix} 41 & 13 & -5 \\ 0 & 5 & 9 \\ 41 & 13 & -5 \end{vmatrix}$ 0 **2.** $\begin{vmatrix} -2 & 0 & 12 \\ 4 & 0 & 0 \\ 3 & 0 & 6 \end{vmatrix}$ 0 **3.** $\begin{vmatrix} 2 & -1 & 1 \\ 1 & 3 & -3 \\ 7 & -8 & 8 \end{vmatrix}$ 0

4. $\begin{vmatrix} 5 & 0 & 0 \\ 0 & 5 & 0 \\ 0 & 0 & 5 \end{vmatrix}$ 125 **5.** $\begin{vmatrix} 2 & -4 & 6 \\ 5 & 0 & -8 \\ 3 & -6 & 9 \end{vmatrix}$ 0 **6.** $\begin{vmatrix} -1 & 4 & 12 \\ 1 & -2 & -12 \\ 8 & 16 & 6 \end{vmatrix}$ −204

7. $\begin{vmatrix} 1 & 5 & 10 \\ 0 & -5 & -8 \\ 1 & 5 & -2 \end{vmatrix}$ 60 **8.** $\begin{vmatrix} 26 & 29 & 29 \\ 25 & 28 & 26 \\ 25 & 30 & 27 \end{vmatrix}$ 101 **9.** $\begin{vmatrix} 5 & 1 & 1 \\ 1 & 5 & 1 \\ 1 & 1 & 5 \end{vmatrix}$ 112

B **10.** $\begin{vmatrix} 4 & 0 & 0 & 0 \\ 0 & 4 & 0 & 0 \\ 0 & 0 & 4 & 0 \\ 0 & 0 & 0 & 4 \end{vmatrix}$ 256 **11.** $\begin{vmatrix} 1 & -1 & -1 & 1 \\ 1 & 1 & 1 & -1 \\ -1 & -1 & 1 & 1 \\ -1 & 1 & -1 & -1 \end{vmatrix}$ 0 **12.** $\begin{vmatrix} 1 & 1 & 1 & -1 \\ 2 & 2 & -2 & 2 \\ -3 & -3 & 3 & 3 \\ -4 & 4 & 4 & 4 \end{vmatrix}$ −192

C **13.** Explain how Properties 1 and 5 can be used to justify Property 2.

14. Prove Property 2 for third-order determinants.

15. Prove Property 5 for second-order determinants.

16-9 Cramer's Rule

Objective To solve systems of equations using determinants.

Determinants may be used to solve systems of linear equations. Consider this system of two linear equations in two variables:

$$ax + by = e$$
$$cx + dy = f$$

Using the linear-combination method of Lesson 3-5, you can verify that

$$x = \frac{de - bf}{ad - bc} \quad \text{and} \quad y = \frac{af - ce}{ad - bc}$$

provided that $ad - bc \neq 0$. Note that the denominators are equal to the determinant of coefficients,

$$D = \begin{vmatrix} a & b \\ c & d \end{vmatrix}.$$

The numerators are equal to the determinants D_x and D_y where

$$D_x = \begin{vmatrix} e & b \\ f & d \end{vmatrix} \quad \text{and} \quad D_y = \begin{vmatrix} a & e \\ c & f \end{vmatrix}.$$

Note that D_x is formed by replacing the column of coefficients of x in D with the column of constants. Similarly, D_y is formed by replacing the column of coefficients of y in D with the column of constants.

By substitution,

$$x = \frac{\begin{vmatrix} e & b \\ f & d \end{vmatrix}}{\begin{vmatrix} a & b \\ c & d \end{vmatrix}} = \frac{D_x}{D} \quad \text{and} \quad y = \frac{\begin{vmatrix} a & e \\ c & f \end{vmatrix}}{\begin{vmatrix} a & b \\ c & d \end{vmatrix}} = \frac{D_y}{D}.$$

Example 1 Use determinants to solve this system:
$$2x - y = 6$$
$$3x + 5y = 22$$

Solution

$$x = \frac{D_x}{D} = \frac{\begin{vmatrix} 6 & -1 \\ 22 & 5 \end{vmatrix}}{\begin{vmatrix} 2 & -1 \\ 3 & 5 \end{vmatrix}} = \frac{30 + 22}{10 + 3} = \frac{52}{13} = 4$$

$$y = \frac{D_y}{D} = \frac{\begin{vmatrix} 2 & 6 \\ 3 & 22 \end{vmatrix}}{\begin{vmatrix} 2 & -1 \\ 3 & 5 \end{vmatrix}} = \frac{44 - 18}{10 + 3} = \frac{26}{13} = 2$$

$\therefore$ the solution of the system is (4, 2). ***Answer***

Teaching Suggestions p. T144

Warm-Up Exercises

Solve each system.

1. $2x - y = 6$
   $3x + 5y = 22$ (4, 2)
2. $x - 2y = 5$
   $x + y = -7$ (−3, −4)
3. $2x + 3y = 3$
   $4x + 7y = 1$ (9, −5)
4. $x - 2y - 2z = 2$
   $2x + 4y + 3z = 1$
   $3x - y - 2z = 3$
   (2, −3, 3)

Motivating the Lesson

Tell the students that the need to solve systems of equations is widespread. Problems in science, industry, business, and government generate multitudes of systems of equations, sometimes involving hundreds of variables. Streamlined techniques for solving such systems are the subject of ongoing research. Today's lesson introduces Cramer's rule, an efficient method for solving systems of linear equations.

Chalkboard Examples

Solve using Cramer's rule.

1. $5x - 4y = 1$
   $2x - 3y = 6$

$$x = \frac{\begin{vmatrix} 1 & -4 \\ 6 & -3 \end{vmatrix}}{\begin{vmatrix} 5 & -4 \\ 2 & -3 \end{vmatrix}} = \frac{-3 + 24}{-15 + 8} = -3$$

$$y = \frac{\begin{vmatrix} 5 & 1 \\ 2 & 6 \end{vmatrix}}{\begin{vmatrix} 5 & -4 \\ 2 & -3 \end{vmatrix}} = \frac{30 - 2}{-15 + 8} = -4$$

$\therefore$ the solution is $(-3, -4)$.

2. $x + y + z = 5$
   $3x - z = 2$
   $x - y + 2z = 0$

$$D = \begin{vmatrix} 1 & 1 & 1 \\ 3 & 0 & -1 \\ 1 & -1 & 2 \end{vmatrix} = -11$$

$$D_x = \begin{vmatrix} 5 & 1 & 1 \\ 2 & 0 & -1 \\ 0 & -1 & 2 \end{vmatrix} = -11$$

$$D_y = \begin{vmatrix} 1 & 5 & 1 \\ 3 & 2 & -1 \\ 1 & 0 & 2 \end{vmatrix} = -33$$

$$D_z = \begin{vmatrix} 1 & 1 & 5 \\ 3 & 0 & 2 \\ 1 & -1 & 0 \end{vmatrix} = -11$$

$$x = \frac{D_x}{D} = \frac{-11}{-11} = 1$$

$$y = \frac{D_y}{D} = \frac{-33}{-11} = 3$$

$$z = \frac{D_z}{D} = \frac{-11}{-11} = 1$$

$\therefore$ the solution is $(1, 3, 1)$.

Similarly, determinants can be used to solve a system of three linear equations in three variables:

$$a_1x + b_1y + c_1z = d_1$$
$$a_2x + b_2y + c_2z = d_2$$
$$a_3x + b_3y + c_3z = d_3$$

Let $D = \begin{vmatrix} a_1 & b_1 & c_1 \\ a_2 & b_2 & c_2 \\ a_3 & b_3 & c_3 \end{vmatrix} \neq 0$, $D_x = \begin{vmatrix} d_1 & b_1 & c_1 \\ d_2 & b_2 & c_2 \\ d_3 & b_3 & c_3 \end{vmatrix}$,

$D_y = \begin{vmatrix} a_1 & d_1 & c_1 \\ a_2 & d_2 & c_2 \\ a_3 & d_3 & c_3 \end{vmatrix}$, and $D_z = \begin{vmatrix} a_1 & b_1 & d_1 \\ a_2 & b_2 & d_2 \\ a_3 & b_3 & d_3 \end{vmatrix}$.

Then $x = \frac{D_x}{D}$, $y = \frac{D_y}{D}$, and $z = \frac{D_z}{D}$.

This method can be generalized, as indicated below. Known as Cramer's Rule, it was named for the Swiss mathematician Gabriel Cramer (1704–1752).

Cramer's Rule

The solution of a system of n linear equations in n variables is given by

$$x = \frac{D_x}{D},\ y = \frac{D_y}{D},\ \ldots,$$

where D is the determinant of the matrix of coefficients of the variables ($D \neq 0$) and D_x, D_y, . . . , are derived from D by replacing the coefficients of x, y, . . . , respectively, by the constants.

Example 2 Use Cramer's rule to solve:

$$x - 3y - 2z = 9$$
$$3x + 2y + 6z = 20$$
$$4x - y + 3z = 25$$

Solution

$$D = \begin{vmatrix} 1 & -3 & -2 \\ 3 & 2 & 6 \\ 4 & -1 & 3 \end{vmatrix} = 1(6 + 6) + 3(9 - 24) - 2(-3 - 8) = -11$$

$$D_x = \begin{vmatrix} 9 & -3 & -2 \\ 20 & 2 & 6 \\ 25 & -1 & 3 \end{vmatrix} = 9(6 + 6) + 3(60 - 150) - 2(-20 - 50) = -22$$

$$D_y = \begin{vmatrix} 1 & 9 & -2 \\ 3 & 20 & 6 \\ 4 & 25 & 3 \end{vmatrix} = 1(60 - 150) - 9(9 - 24) - 2(75 - 80) = 55$$

$$D_z = \begin{vmatrix} 1 & -3 & 9 \\ 3 & 2 & 20 \\ 4 & -1 & 25 \end{vmatrix} = 1(50 + 20) + 3(75 - 80) + 9(-3 - 8) = -44$$

$$x = \frac{D_x}{D} = \frac{-22}{-11} = 2 \qquad y = \frac{D_y}{D} = \frac{55}{-11} = -5 \qquad z = \frac{D_z}{D} = \frac{-44}{-11} = 4$$

$\therefore$ the solution of the system is $(2, -5, 4)$. ***Answer***

Cramer's rule provides a method for finding a solution of a system of n linear equations in n variables. Recall that such a system has a unique solution if and only if $D \neq 0$. When $D = 0$ the system may have no solution or it may have infinitely many solutions.

To distinguish between inconsistent and dependent equations, you must consider the value of D_y. If $D = 0$ and $D_y \neq 0$, the equations are inconsistent and their graphs are parallel. If $D = 0$ and $D_y = 0$, the equations are dependent and their graphs coincide. The examples below illustrate systems with consistent equations, inconsistent equations, and dependent equations.

$x + y = 1$
$2x + y = 4$

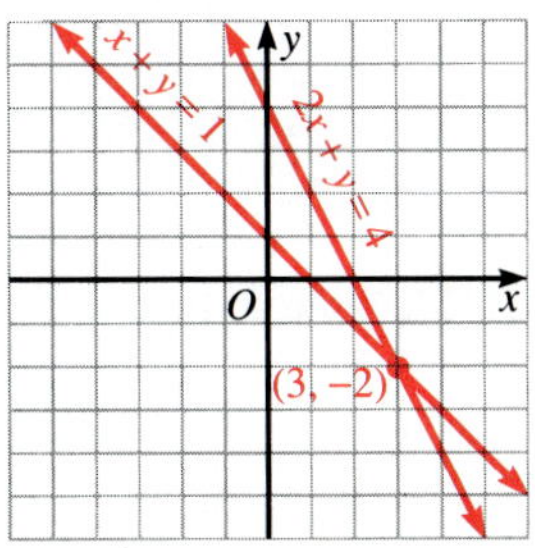

$D \neq 0$

Consistent equations; graphs intersect.

$2x - y = -1$
$2x - y = 4$

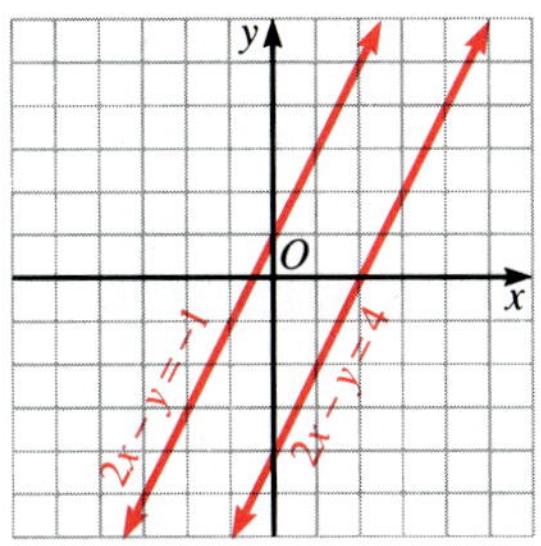

$D = 0$
$D_y \neq 0$

Inconsistent equations; graphs are parallel.

$3x + 2y = 2$
$6x + 4y = 4$

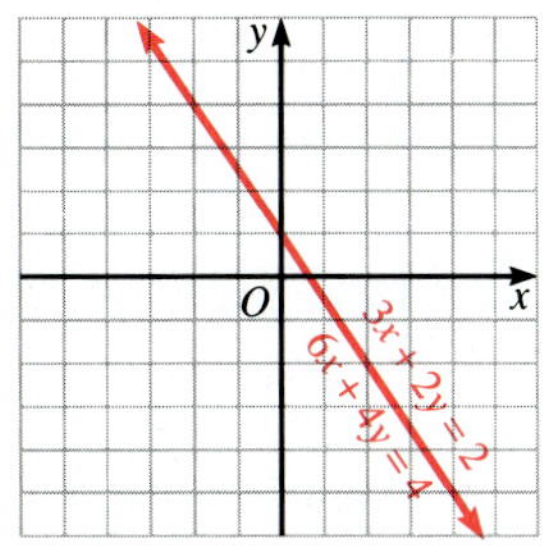

$D = 0$
$D_y = 0$

Dependent equations; graphs coincide.

Example 3 Determine the number of solutions for each system.

a. $2x + y = -6$
$x - 3y = 9$

b. $x - 2y = 3$
$-2x + 4y = 1$

Solution **a.** $D = \begin{vmatrix} 2 & 1 \\ 1 & -3 \end{vmatrix} = -6 - 1 = -7 \neq 0$

Since $D \neq 0$, the equations are consistent,
$\therefore$ the system has one solution. ***Answer***

b. $D = \begin{vmatrix} 1 & -2 \\ -2 & 4 \end{vmatrix} = 4 - 4 = 0 \qquad D_y = \begin{vmatrix} 1 & 3 \\ -2 & 1 \end{vmatrix} = 1 - (-6) = 7$

Since $D = 0$ and $D_y \neq 0$, the equations are inconsistent.
$\therefore$ the system has no solution. ***Answer***

Determine the number of solutions each system has.

3. $4x - 8y = 9$
$-x + 2y = 0$

$D = \begin{vmatrix} 4 & -8 \\ -1 & 2 \end{vmatrix} = 0$

$D_y = \begin{vmatrix} 4 & 9 \\ -1 & 0 \end{vmatrix} = 9$

no solution

4. $3x + 15y = 6$
$2x + 10y = 4$

$D = \begin{vmatrix} 3 & 15 \\ 2 & 10 \end{vmatrix} = 0$

$D_y = \begin{vmatrix} 3 & 6 \\ 2 & 4 \end{vmatrix} = 0$

more than one solution

Check for Understanding

Oral Exs. 1–16: use after Example 2.

Guided Practice

Solve using Cramer's rule.

1. $x - 2y = 21$
$3x + y = 7$
$(5, -8)$

2. $2x - z = 0$
$6x + y + z = 2$
$x + y + 4z = 0$
$(-2, 18, -4)$

3. $x + y + 3z = 0$
$4x + y = -4$
$x - 6z = 2$
$(-2, 4, -\frac{2}{3})$

Determine whether each system has no solution, one solution, or more than one solution.

4. $4x - 6y = 7$
$-6x + 9y = 11$
no solution

5. $x + 2y = -1$
$2x + y = 5$
one solution

Summarizing the Lesson

In this lesson students learned to use Cramer's rule to solve systems of linear equations. Ask the students to state Cramer's rule.

Suggested Assignments

Extended Algebra
804/1–13 odd, 15, 16
S 805/Mixed Review, 1–3, 5, 6
R 805/Self-Test 3

Extended Alg. and Trig.
804/1–13 odd, 15, 16
S 805/Mixed Review
R 805/Self-Test 3

Supplementary Materials

Study Guide	pp. 275–276
Practice Master	93
Test Master	69
Resource Book	p. 174

Additional Answers
Oral Exercises

5. $\begin{vmatrix} 2 & 1 & 1 \\ -4 & 2 & 2 \\ 1 & 2 & -1 \end{vmatrix}$

6. $\begin{vmatrix} 1 & 2 & 1 \\ 1 & -4 & 2 \\ 2 & 1 & -1 \end{vmatrix}$

7. $\begin{vmatrix} 1 & 1 & 2 \\ 1 & 2 & -4 \\ 2 & 2 & 1 \end{vmatrix}$

8. $\begin{vmatrix} 4 & 3 & 2 \\ 2 & -1 & 1 \\ 0 & 2 & -2 \end{vmatrix}$

9. $\begin{vmatrix} -1 & 4 & 2 \\ 2 & 2 & 1 \\ 4 & 0 & -2 \end{vmatrix}$

10. $\begin{vmatrix} -1 & 3 & 4 \\ 2 & -1 & 2 \\ 4 & 2 & 0 \end{vmatrix}$

Oral Exercises

For each system give the entries for columns one, two, and three of the determinant D.

1. $\begin{aligned} x + y + z &= 2 \\ x + 2y + 2z &= -4 \\ 2x + 2y - z &= 1 \end{aligned}$ $\begin{vmatrix} 1 & 1 & 1 \\ 1 & 2 & 2 \\ 2 & 2 & -1 \end{vmatrix}$

2. $\begin{aligned} -x + 3y + 2z &= 4 \\ 2x - y + z &= 2 \\ 4x + 2y - 2z &= 0 \end{aligned}$ $\begin{vmatrix} -1 & 3 & 2 \\ 2 & -1 & 1 \\ 4 & 2 & -2 \end{vmatrix}$

3. $\begin{aligned} 2x - 3z &= 0 \\ 3x + y &= 15 \\ 4y + z &= -8 \end{aligned}$ $\begin{vmatrix} 2 & 0 & -3 \\ 3 & 1 & 0 \\ 0 & 4 & 1 \end{vmatrix}$

4. $\begin{aligned} 5x - y &= -17 \\ -2x + 3y + 3z &= 12 \\ z &= 0 \end{aligned}$ $\begin{vmatrix} 5 & -1 & 0 \\ -2 & 3 & 3 \\ 0 & 0 & 1 \end{vmatrix}$

Give the elements in columns one, two, and three of the indicated determinants.

5. Exercise 1, D_x
6. Exercise 1, D_y
7. Exercise 1, D_z
8. Exercise 2, D_x
9. Exercise 2, D_y
10. Exercise 2, D_z
11. Exercise 3, D_x
12. Exercise 3, D_y
13. Exercise 3, D_z
14. Exercise 4, D_x
15. Exercise 4, D_y
16. Exercise 4, D_z

Written Exercises

Use Cramer's rule to solve each system.

A

1. $\begin{aligned} x - 5y &= 2 \\ 2x + y &= 4 \end{aligned}$ (2, 0)

2. $\begin{aligned} 2x + 3y &= 7 \\ 3x + 4y &= 10 \end{aligned}$ (2, 1)

3. $\begin{aligned} 2x + y - z &= 2 \\ x + y + z &= 7 \\ x + 2y + z &= 4 \end{aligned}$ (5, −3, 5)

4. $\begin{aligned} x - 3y - 2z &= 9 \\ 3x + 2y + 6z &= 20 \\ 4x - y + 3z &= 25 \end{aligned}$ (2, −5, 4)

5. $\begin{aligned} x + y - z &= 0 \\ 2x + y - 2z &= 1 \\ 3x - 4z &= 1 \end{aligned}$ (3, −1, 2)

6. $\begin{aligned} x + y + z &= -2 \\ 2x - y - z &= -1 \\ 3x + 2y + 4z &= -15 \end{aligned}$ (−1, 4, −5)

7. $\begin{aligned} x - y + z &= 3 \\ y + z &= 3 \\ -y + z &= 1 \end{aligned}$ (2, 1, 2)

8. $\begin{aligned} x + 3y + z &= 0 \\ x + 4z &= -2 \\ -6y + z &= 1 \end{aligned}$ $\left(2, -\frac{1}{3}, -1\right)$

9. $\begin{aligned} 2x + y - z &= 3 \\ 4x - y + 4z &= 0 \\ -3y + 2z &= 6 \end{aligned}$ (2, −4, −3)

10. $\begin{aligned} x - 2y &= 8 \\ x + 2z &= 3 \\ -y - 2z &= 2 \end{aligned}$ $\left(2, -3, \frac{1}{2}\right)$

Determine whether each system has no solution, one solution, or more than one solution.

11. $\begin{aligned} 2x - y &= 5 \\ -2x + y &= -1 \end{aligned}$ No solution

12. $\begin{aligned} 3x + 9y &= 14 \\ 6x + 18y &= 28 \end{aligned}$ More than one solution

13. $4x + 3y = 16$ No solution
$8x + 6y = 10$

14. $5x - 9y = 15$ One solution
$-2x + 5y = 7$

Use Cramer's rule to solve each system.

B 15. $\begin{aligned} 2a - 3b + c - 3d &= -1 \\ a + 2b + 3c - d &= 1 \\ 3a + 5b + 6c &= 4 \\ 3a - b - 2d &= 6 \end{aligned}$ (2, 2, −2, −1)

16. $\begin{aligned} p - 2q + r &= -4 \\ 2p - r + s &= -3 \\ q + 2r - 3s &= -14 \\ 3p - 3q + 2r &= -13 \end{aligned}$ (−4, −1, −2, 3)

Prove the following statements about this system: $y = m_1x + b_1$
$y = m_2x + b_2$

C 17. $D \neq 0$ if and only if $m_1 \neq m_2$.

18. $D = 0$ and $D_y \neq 0$ if and only if $m_1 = m_2 \neq 0$ and $b_1 \neq b_2$.

19. $D = 0$ and $D_y = 0$ if and only if $m_1 = m_2$ and $b_1 = b_2$.

Mixed Review Exercises

Perform the indicated matrix operations.

1. $\begin{bmatrix} 5 & 2 \\ 3 & -1 \end{bmatrix} + \begin{bmatrix} 1 & -2 \\ 0 & 5 \end{bmatrix}$ $\begin{bmatrix} 6 & 0 \\ 3 & 4 \end{bmatrix}$

2. $4\begin{bmatrix} 2 & -1 \\ 3 & 0 \end{bmatrix}$ $\begin{bmatrix} 8 & -4 \\ 12 & 0 \end{bmatrix}$

3. $\begin{bmatrix} -1 & 2 \\ -2 & 3 \end{bmatrix}\begin{bmatrix} 0 & 1 \\ 4 & 1 \end{bmatrix}$ $\begin{bmatrix} 8 & 1 \\ 12 & 1 \end{bmatrix}$

Graph each function.

4. $y = \frac{1}{2} \sin 4x$

5. $y = \frac{2}{3}x - 4$

6. $y = -2x^2 + 8x - 7$

Self-Test 3

Vocabulary minor of an element (p. 794) expansion of a determinant (p. 795)

1. Expand by minors: $\begin{vmatrix} 3 & 0 & 0 \\ 0 & 2 & 1 \\ -2 & 4 & -1 \end{vmatrix}$ −18 **Obj. 16-7, p. 794**

2. Evaluate: $\begin{vmatrix} 2 & 1 & 0 & -2 \\ -1 & 0 & 0 & 4 \\ 2 & 3 & 6 & -1 \\ 0 & 1 & 2 & 2 \end{vmatrix}$ −2 **Obj. 16-8, p. 798**

3. Use Cramer's rule to solve: $\begin{aligned} 2x + y - z &= 1 \\ x + 3y + z &= 10 \\ 3x - 2y - 8z &= -3 \end{aligned}$ (−3, 5, −2) **Obj. 16-9, p. 801**

11. $\begin{vmatrix} 0 & 0 & -3 \\ 15 & 1 & 0 \\ -8 & 4 & 1 \end{vmatrix}$

12. $\begin{vmatrix} 2 & 0 & -3 \\ 3 & 15 & 0 \\ 0 & -8 & 1 \end{vmatrix}$

13. $\begin{vmatrix} 2 & 0 & 0 \\ 3 & 1 & 15 \\ 0 & 4 & -8 \end{vmatrix}$

14. $\begin{vmatrix} -17 & -1 & 0 \\ 12 & 3 & 3 \\ 0 & 0 & 1 \end{vmatrix}$

15. $\begin{vmatrix} 5 & -17 & 0 \\ -2 & 12 & 3 \\ 0 & 0 & 1 \end{vmatrix}$

16. $\begin{vmatrix} 5 & -1 & -17 \\ -2 & 3 & 12 \\ 0 & 0 & 0 \end{vmatrix}$

Quick Quiz

1. Expand by minors.
$\begin{vmatrix} 1 & 2 & 3 \\ 5 & 0 & -2 \\ -1 & 0 & 4 \end{vmatrix}$ −36

2. Evaluate.
$\begin{vmatrix} 1 & 2 & 0 & -3 \\ 1 & 0 & -1 & 1 \\ 5 & 0 & 6 & 2 \\ -2 & 0 & 2 & -2 \end{vmatrix}$ 0

3. Solve using Cramer's rule.
$\begin{aligned} 2x + y + z &= -4 \\ 4x - y - 3z &= -5 \\ -3y + 2z &= 3 \end{aligned}$
$(-\frac{3}{2}, -1, 0)$

Chapter Summary

1. Two *matrices* are equal if and only if they have the same *dimensions* and all corresponding *elements* are equal.
2. Matrices having the same dimensions can be added (or subtracted) by adding (or subtracting) corresponding elements. The properties of addition of matrices are listed on page 771.
3. In matrix algebra, a real number is called a *scalar*. The *scalar product* of a real number r and a matrix A is the matrix rA, each of whose elements is r times the corresponding element of A. The properties of scalar multiplication are listed on page 772.
4. The product of matrices $A_{m \times n}$ and $B_{n \times p}$ is the $m \times p$ matrix whose element in the ath row and bth column is the sum of the products of corresponding elements of the ath row of A and the bth column of B. Unlike multiplication of real numbers, matrix multiplication is not commutative. Moreover, $A \times B = O$ does not imply that $A = O$ or $B = O$. The properties of matrix multiplication are listed on page 776.
5. Problems in the physical and social sciences can be solved using matrices.
6. For each $n \times n$ matrix A there is a real number called the *determinant* of A. For 2×2 matrices of the form $\begin{bmatrix} a & b \\ c & d \end{bmatrix}$, the determinant is $ad - bc$. Determinants for higher-order matrices can be defined and expanded by minors about the elements of any row or column. Determinants have properties that can simplify expansion by minors. These properties are listed on pages 798 and 799.
7. If $AB = BA = I$, matrices A and B are said to be inverses of each other. If $A = \begin{bmatrix} a & b \\ c & d \end{bmatrix}$ and $\det A \neq 0$, then $A^{-1} = \frac{1}{\det A}\begin{bmatrix} d & -b \\ -c & a \end{bmatrix}$.
8. The solution of a system of n linear equations in n variables is given by $x = \frac{D_x}{D}$, $y = \frac{D_y}{D}$, . . . , where D is the determinant of the matrix of coefficients of the variables ($D \neq 0$) and D_x, D_y, . . . , are derived from D by replacing the coefficients of x, y, . . . , respectively, by the constants. This method of solution is called *Cramer's rule*.

Chapter Review

Write the letter of the correct answer.

1. Solve for x and y: $\begin{bmatrix} 4 - x & 0 & 3y \end{bmatrix} = O_{1 \times 3}$ **16-1**

a. $x = 4$, $y = -3$ **b.** $x = -4$, $y = 0$ **c.** $x = -4$, $y = 3$ **d.** $x = 4$, $y = 0$

In Exercises 2 and 3, let $A = \begin{bmatrix} -2 & 3 \\ -1 & 6 \end{bmatrix}$ and $B = \begin{bmatrix} 3 & -2 \\ -1 & 4 \end{bmatrix}$.

2. Find $B - 2A$. 16-2

a. $\begin{bmatrix} 5 & -5 \\ 0 & -2 \end{bmatrix}$ **(b.)** $\begin{bmatrix} 7 & -8 \\ 1 & -8 \end{bmatrix}$ **c.** $\begin{bmatrix} 7 & -3 \\ 1 & -8 \end{bmatrix}$ **d.** $\begin{bmatrix} 1 & 3 \\ -3 & 8 \end{bmatrix}$

3. Find BA. 16-3

a. $\begin{bmatrix} -9 & -16 \\ 19 & -22 \end{bmatrix}$ **b.** $\begin{bmatrix} -3 & 8 \\ -9 & -22 \end{bmatrix}$ **(c.)** $\begin{bmatrix} -4 & -3 \\ -2 & 21 \end{bmatrix}$ **d.** $\begin{bmatrix} -9 & -16 \\ -9 & -26 \end{bmatrix}$

4. Find the matrix that shows the number of ways a message can be sent from one point to another using exactly one relay. 16-4

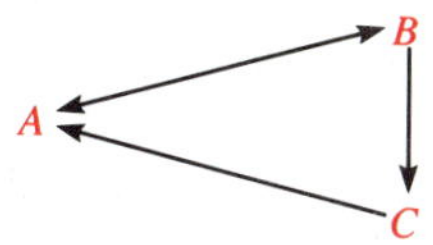

(a.) $\begin{bmatrix} 1 & 0 & 1 \\ 1 & 1 & 0 \\ 0 & 1 & 0 \end{bmatrix}$ **b.** $\begin{bmatrix} 1 & 1 & 0 \\ 0 & 1 & 1 \\ 1 & 0 & 0 \end{bmatrix}$ **c.** $\begin{bmatrix} 0 & 1 & 0 \\ 1 & 0 & 1 \\ 1 & 0 & 0 \end{bmatrix}$ **d.** $\begin{bmatrix} 0 & 0 & 1 \\ 1 & 1 & 0 \\ 0 & 1 & 1 \end{bmatrix}$

5. Evaluate $\begin{vmatrix} 4 & 2 \\ -1 & 3 \end{vmatrix}$. 16-5

a. -14 **b.** 10 **c.** -10 **(d.)** 14

6. Evaluate $\begin{vmatrix} -1 & 1 & 0 \\ 1 & 2 & 2 \\ 1 & 0 & 0 \end{vmatrix}$.

a. -1 **(b.)** 2 **c.** 3 **d.** -2

7. Find the inverse of $\begin{bmatrix} -2 & 1 \\ 0 & 6 \end{bmatrix}$. 16-6

(a.) $\begin{bmatrix} -\frac{1}{2} & \frac{1}{12} \\ 0 & \frac{1}{6} \end{bmatrix}$ **b.** $\begin{bmatrix} \frac{1}{2} & \frac{1}{12} \\ 0 & \frac{1}{6} \end{bmatrix}$ **c.** $\begin{bmatrix} 6 & 0 \\ -1 & -2 \end{bmatrix}$ **d.** $\begin{bmatrix} \frac{1}{2} & -\frac{1}{12} \\ 0 & -\frac{1}{6} \end{bmatrix}$

8. Expand this determinant by minors about the first row: $\begin{vmatrix} -1 & 0 & 1 \\ 2 & 1 & 0 \\ 1 & -1 & 3 \end{vmatrix}$ 16-7

a. 0 **b.** 12 **(c.)** -6 **d.** 6

9. Evaluate the determinant $\begin{vmatrix} 3 & 2 & -2 \\ 1 & 0 & 2 \\ -12 & 3 & 8 \end{vmatrix}$. 16-8

a. 40 **b.** -40 **c.** -44 **(d.)** -88

10. Use Cramer's rule to solve this system: 16-9

$$\begin{aligned} 2x + y + 3z &= -2 \\ 5x + 2y &= 5 \\ 2y + 3z &= -13 \end{aligned}$$

(a.) $\{(3, -5, -1)\}$ **b.** $\{(3, 5, -1)\}$ **c.** $\{(3, 5, 1)\}$ **d.** $\{(-3, -5, 1)\}$

807

Alternate Test p. T27

Supplementary Materials

Test Masters	70, 71
Resource Book	p. 175

Additional Answers
Chapter Test

2. $\begin{bmatrix} 0 & 2 & 8 \\ 5 & -6 & 2 \\ 4 & 2 & -1 \end{bmatrix}$

3. $\begin{bmatrix} -4 & 0 & 2 \\ 5 & -4 & 0 \\ -4 & 4 & -1 \end{bmatrix}$

4. $\begin{bmatrix} 8 & 4 & 12 \\ 0 & -4 & 4 \\ 16 & -4 & 0 \end{bmatrix}$

5. $\begin{bmatrix} 6 & -1 & -7 \\ -10 & 9 & -1 \\ 4 & -7 & 2 \end{bmatrix}$

6. $\begin{bmatrix} 1 & 6 & 8 \\ -5 & 8 & -2 \\ -13 & 9 & 19 \end{bmatrix}$

7. $\begin{bmatrix} 9 & 8 & -14 \\ -35 & 33 & 19 \\ 15 & -18 & 4 \end{bmatrix}$

Chapter Test

1. Find the value x and y: $O_{1\times 3} = \begin{bmatrix} 2x-6 \\ 0 \\ y+4 \end{bmatrix}$ $x = 3;\ y = -4$ **16-1**

In Exercises 2–7, let $A = \begin{bmatrix} 2 & 1 & 3 \\ 0 & -1 & 1 \\ 4 & -1 & 0 \end{bmatrix}$ **and** $B = \begin{bmatrix} -2 & 1 & 5 \\ 5 & -5 & 1 \\ 0 & 3 & -1 \end{bmatrix}$.

Express the following as a single matrix.

2. $A + B$ **3.** $B - A$ **16-2**

4. $4A$ **5.** $A - 2B$

6. AB **7.** B^2 **16-3**

8. Find the number of routes that a message might take from point A to point B using at most one relay point. 2 routes **16-4**

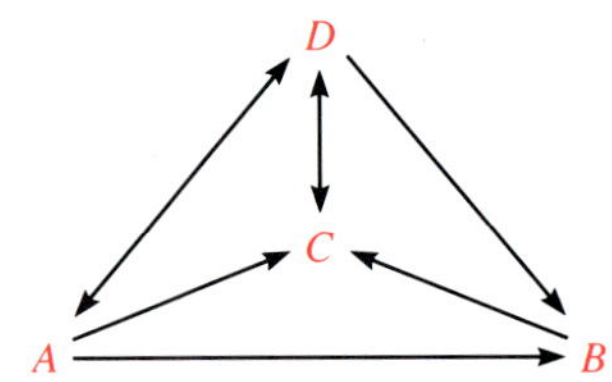

Evaluate these determinants.

9. $\begin{vmatrix} 6 & 3 \\ -2 & -2 \end{vmatrix}$ -6 **10.** $\begin{vmatrix} 1 & 0 & -1 \\ 3 & 1 & 1 \\ 0 & 2 & 2 \end{vmatrix}$ -6 **16-5**

11. Solve this equation for A: $\begin{bmatrix} 2 & 3 \\ -1 & 0 \end{bmatrix} A = \begin{bmatrix} 7 & 3 \\ 4 & 0 \end{bmatrix}$ $\begin{bmatrix} -4 & 0 \\ 5 & 1 \end{bmatrix}$ **16-6**

12. Expand this determinant by minors about the third row: $\begin{vmatrix} 3 & 0 & -1 \\ 2 & -1 & 1 \\ 1 & -1 & 0 \end{vmatrix}$ 4 **16-7**

Use the properties of determinants to evaluate these determinants.

13. $\begin{vmatrix} 10 & 5 & -5 \\ 16 & 8 & -8 \\ 3 & 0 & -4 \end{vmatrix}$ 0 **14.** $\begin{vmatrix} 4 & 3 & -1 \\ -3 & -3 & 5 \\ 5 & 5 & 3 \end{vmatrix}$ -34 **16-8**

15. Use Cramer's rule to solve this system: $\begin{aligned} 4x - 3y &= 1 \\ 6x - 8z &= 1 \\ y - 2z &= 0 \end{aligned}$ $\left(-\frac{1}{2}, -1, -\frac{1}{2}\right)$ **16-9**

Preparing for College Entrance Exams

Strategy for Success

It usually takes less time to solve an equation than to substitute all of the possible solutions. However, if you can eliminate possibilities that have the wrong sign, that are obviously too large or too small, or that do not meet the stated or implied conditions of the problem, then it may be quicker to substitute the remaining choices in the given equation to find the correct solution.

Decide which is the best of the choices given and write the corresponding letter on your answer sheet.

1. If $A = \begin{bmatrix} 1 & -1 \\ 2 & 4 \end{bmatrix}$, for which matrix X is $AX \neq XA$? D

(A) $\begin{bmatrix} 1 & -1 \\ 2 & 4 \end{bmatrix}$ **(B)** $\begin{bmatrix} 1 & 0 \\ 0 & 1 \end{bmatrix}$ **(C)** $\begin{bmatrix} 0 & 0 \\ 0 & 0 \end{bmatrix}$

(D) $\begin{bmatrix} 4 & 2 \\ -1 & 1 \end{bmatrix}$ **(E)** $\begin{bmatrix} \frac{2}{3} & \frac{1}{6} \\ -\frac{1}{3} & \frac{1}{6} \end{bmatrix}$

2. Find the inverse of $\begin{bmatrix} 5 & -6 \\ 2 & 4 \end{bmatrix}$ if it exists. C

(A) $\begin{bmatrix} \frac{5}{32} & \frac{1}{16} \\ -\frac{3}{16} & \frac{1}{8} \end{bmatrix}$ **(B)** $\begin{bmatrix} \frac{1}{8} & \frac{1}{16} \\ -\frac{3}{16} & \frac{5}{32} \end{bmatrix}$ **(C)** $\begin{bmatrix} \frac{1}{8} & \frac{3}{16} \\ -\frac{1}{16} & \frac{5}{32} \end{bmatrix}$

(D) $\begin{bmatrix} \frac{1}{8} & -\frac{3}{16} \\ \frac{1}{16} & \frac{5}{32} \end{bmatrix}$ **(E)** does not exist

3. In how many ways can the letters in the word INITIAL be arranged? B

(A) 120 **(B)** 840 **(C)** 1680 **(D)** 2520 **(E)** 5040

4. Two dice are tossed. Find the probability that both the sum and the product of the numbers are less than six. B

(A) $\frac{5}{18}$ **(B)** $\frac{2}{9}$ **(C)** $\frac{1}{3}$ **(D)** $\frac{1}{6}$ **(E)** $\frac{1}{4}$

5. There are 80 sophomores, 68 juniors, and 72 seniors enrolled at a high school. In how many ways can a 7-member student council be formed if two representatives are chosen from each grade and then the president is elected from the entire student body? E

(A) ${}_{80}C_2 + {}_{68}C_2 + {}_{72}C_2 + {}_{220}C_1$ **(B)** ${}_{80}C_2 \cdot {}_{68}C_2 \cdot {}_{72}C_2 \cdot {}_{220}C_1$

(C) ${}_{80}P_2 + {}_{68}P_2 + {}_{72}P_2 + {}_{214}P_1$ **(D)** ${}_{80}P_2 \cdot {}_{68}P_2 \cdot {}_{72}P_2 \cdot {}_{214}P_1$

(E) ${}_{80}C_2 \cdot {}_{68}C_2 \cdot {}_{72}C_2 \cdot {}_{214}C_1$

Table 1 / *Squares and Square Roots*

N	N^2	$\sqrt{N}$	$\sqrt{10N}$	N	N^2	$\sqrt{N}$	$\sqrt{10N}$
1.0	1.00	1.000	3.162	**5.5**	30.25	2.345	7.416
1.1	1.21	1.049	3.317	**5.6**	31.36	2.366	7.483
1.2	1.44	1.095	3.464	**5.7**	32.49	2.387	7.550
1.3	1.69	1.140	3.606	**5.8**	33.64	2.408	7.616
1.4	1.96	1.183	3.742	**5.9**	34.81	2.429	7.681
1.5	2.25	1.225	3.873	**6.0**	36.00	2.449	7.746
1.6	2.56	1.265	4.000	**6.1**	37.21	2.470	7.810
1.7	2.89	1.304	4.123	**6.2**	38.44	2.490	7.874
1.8	3.24	1.342	4.243	**6.3**	39.69	2.510	7.937
1.9	3.61	1.378	4.359	**6.4**	40.96	2.530	8.000
2.0	4.00	1.414	4.472	**6.5**	42.25	2.550	8.062
2.1	4.41	1.449	4.583	**6.6**	43.56	2.569	8.124
2.2	4.84	1.483	4.690	**6.7**	44.89	2.588	8.185
2.3	5.29	1.517	4.796	**6.8**	46.24	2.608	8.246
2.4	5.76	1.549	4.899	**6.9**	47.61	2.627	8.307
2.5	6.25	1.581	5.000	**7.0**	49.00	2.646	8.367
2.6	6.76	1.612	5.099	**7.1**	50.41	2.665	8.426
2.7	7.29	1.643	5.196	**7.2**	51.84	2.683	8.485
2.8	7.84	1.673	5.292	**7.3**	53.29	2.702	8.544
2.9	8.41	1.703	5.385	**7.4**	54.76	2.720	8.602
3.0	9.00	1.732	5.477	**7.5**	56.25	2.739	8.660
3.1	9.61	1.761	5.568	**7.6**	57.76	2.757	8.718
3.2	10.24	1.789	5.657	**7.7**	59.29	2.775	8.775
3.3	10.89	1.817	5.745	**7.8**	60.84	2.793	8.832
3.4	11.56	1.844	5.831	**7.9**	62.41	2.811	8.888
3.5	12.25	1.871	5.916	**8.0**	64.00	2.828	8.944
3.6	12.96	1.897	6.000	**8.1**	65.61	2.846	9.000
3.7	13.69	1.924	6.083	**8.2**	67.24	2.864	9.055
3.8	14.44	1.949	6.164	**8.3**	68.89	2.881	9.110
3.9	15.21	1.975	6.245	**8.4**	70.56	2.898	9.165
4.0	16.00	2.000	6.325	**8.5**	72.25	2.915	9.220
4.1	16.81	2.025	6.403	**8.6**	73.96	2.933	9.274
4.2	17.64	2.049	6.481	**8.7**	75.69	2.950	9.327
4.3	18.49	2.074	6.557	**8.8**	77.44	2.966	9.381
4.4	19.36	2.098	6.633	**8.9**	79.21	2.983	9.434
4.5	20.25	2.121	6.708	**9.0**	81.00	3.000	9.487
4.6	21.16	2.145	6.782	**9.1**	82.81	3.017	9.539
4.7	22.09	2.168	6.856	**9.2**	84.64	3.033	9.592
4.8	23.04	2.191	6.928	**9.3**	86.49	3.050	9.644
4.9	24.01	2.214	7.000	**9.4**	88.36	3.066	9.695
5.0	25.00	2.236	7.071	**9.5**	90.25	3.082	9.747
5.1	26.01	2.258	7.141	**9.6**	92.16	3.098	9.798
5.2	27.04	2.280	7.211	**9.7**	94.09	3.114	9.849
5.3	28.09	2.302	7.280	**9.8**	96.04	3.130	9.899
5.4	29.16	2.324	7.348	**9.9**	98.01	3.146	9.950
5.5	30.25	2.345	7.416	**10**	100.00	3.162	10.000

Table 2 / Cubes and Cube Roots

N	N^3	$\sqrt[3]{N}$	$\sqrt[3]{10N}$	$\sqrt[3]{100N}$	N	N^3	$\sqrt[3]{N}$	$\sqrt[3]{10N}$	$\sqrt[3]{100N}$
1.0	1.000	1.000	2.154	4.642	**5.5**	166.375	1.765	3.803	8.193
1.1	1.331	1.032	2.224	4.791	**5.6**	175.616	1.776	3.826	8.243
1.2	1.728	1.063	2.289	4.932	**5.7**	185.193	1.786	3.849	8.291
1.3	2.197	1.091	2.351	5.066	**5.8**	195.112	1.797	3.871	8.340
1.4	2.744	1.119	2.410	5.192	**5.9**	205.379	1.807	3.893	8.387
1.5	3.375	1.145	2.466	5.313	**6.0**	216.000	1.817	3.915	8.434
1.6	4.096	1.170	2.520	5.429	**6.1**	226.981	1.827	3.936	8.481
1.7	4.913	1.193	2.571	5.540	**6.2**	238.328	1.837	3.958	8.527
1.8	5.832	1.216	2.621	5.646	**6.3**	250.047	1.847	3.979	8.573
1.9	6.859	1.239	2.668	5.749	**6.4**	262.144	1.857	4.000	8.618
2.0	8.000	1.260	2.714	5.848	**6.5**	274.625	1.866	4.021	8.662
2.1	9.261	1.281	2.759	5.944	**6.6**	287.496	1.876	4.041	8.707
2.2	10.648	1.301	2.802	6.037	**6.7**	300.763	1.885	4.062	8.750
2.3	12.167	1.320	2.844	6.127	**6.8**	314.432	1.895	4.082	8.794
2.4	13.824	1.339	2.884	6.214	**6.9**	328.509	1.904	4.102	8.837
2.5	15.625	1.357	2.924	6.300	**7.0**	343.000	1.913	4.121	8.879
2.6	17.576	1.375	2.962	6.383	**7.1**	357.911	1.922	4.141	8.921
2.7	19.683	1.392	3.000	6.463	**7.2**	373.248	1.931	4.160	8.963
2.8	21.952	1.409	3.037	6.542	**7.3**	389.017	1.940	4.179	9.004
2.9	24.389	1.426	3.072	6.619	**7.4**	405.224	1.949	4.198	9.045
3.0	27.000	1.442	3.107	6.694	**7.5**	421.875	1.957	4.217	9.086
3.1	29.791	1.458	3.141	6.768	**7.6**	438.976	1.966	4.236	9.126
3.2	32.768	1.474	3.175	6.840	**7.7**	456.533	1.975	4.254	9.166
3.3	35.937	1.489	3.208	6.910	**7.8**	474.552	1.983	4.273	9.205
3.4	39.304	1.504	3.240	6.980	**7.9**	493.039	1.992	4.291	9.244
3.5	42.875	1.518	3.271	7.047	**8.0**	512.000	2.000	4.309	9.283
3.6	46.656	1.533	3.302	7.114	**8.1**	531.441	2.008	4.327	9.322
3.7	50.653	1.547	3.332	7.179	**8.2**	551.368	2.017	4.344	9.360
3.8	54.872	1.560	3.362	7.243	**8.3**	571.787	2.025	4.362	9.398
3.9	59.319	1.574	3.391	7.306	**8.4**	592.704	2.033	4.380	9.435
4.0	64.000	1.587	3.420	7.368	**8.5**	614.125	2.041	4.397	9.473
4.1	68.921	1.601	3.448	7.429	**8.6**	636.056	2.049	4.414	9.510
4.2	74.088	1.613	3.476	7.489	**8.7**	658.503	2.057	4.431	9.546
4.3	79.507	1.626	3.503	7.548	**8.8**	681.472	2.065	4.448	9.583
4.4	85.184	1.639	3.530	7.606	**8.9**	704.969	2.072	4.465	9.619
4.5	91.125	1.651	3.557	7.663	**9.0**	729.000	2.080	4.481	9.655
4.6	97.336	1.663	3.583	7.719	**9.1**	753.571	2.088	4.498	9.691
4.7	103.823	1.675	3.609	7.775	**9.2**	778.688	2.095	4.514	9.726
4.8	110.592	1.687	3.634	7.830	**9.3**	804.357	2.103	4.531	9.761
4.9	117.649	1.698	3.659	7.884	**9.4**	830.584	2.110	4.547	9.796
5.0	125.000	1.710	3.684	7.937	**9.5**	857.375	2.118	4.563	9.830
5.1	132.651	1.721	3.708	7.990	**9.6**	884.736	2.125	4.579	9.865
5.2	140.608	1.732	3.733	8.041	**9.7**	912.673	2.133	4.595	9.899
5.3	148.877	1.744	3.756	8.093	**9.8**	941.192	2.140	4.610	9.933
5.4	157.464	1.754	3.780	8.143	**9.9**	970.299	2.147	4.626	9.967
5.5	166.375	1.765	3.803	8.193	**10**	1000.000	2.154	4.642	10.000

Table 3 / Common Logarithms of Numbers

N	0	1	2	3	4	5	6	7	8	9
10	0000	0043	0086	0128	0170	0212	0253	0294	0334	0374
11	0414	0453	0492	0531	0569	0607	0645	0682	0719	0755
12	0792	0828	0864	0899	0934	0969	1004	1038	1072	1106
13	1139	1173	1206	1239	1271	1303	1335	1367	1399	1430
14	1461	1492	1523	1553	1584	1614	1644	1673	1703	1732
15	1761	1790	1818	1847	1875	1903	1931	1959	1987	2014
16	2041	2068	2095	2122	2148	2175	2201	2227	2253	2279
17	2304	2330	2355	2380	2405	2430	2455	2480	2504	2529
18	2553	2577	2601	2625	2648	2672	2695	2718	2742	2765
19	2788	2810	2833	2856	2878	2900	2923	2945	2967	2989
20	3010	3032	3054	3075	3096	3118	3139	3160	3181	3201
21	3222	3243	3263	3284	3304	3324	3345	3365	3385	3404
22	3424	3444	3464	3483	3502	3522	3541	3560	3579	3598
23	3617	3636	3655	3674	3692	3711	3729	3747	3766	3784
24	3802	3820	3838	3856	3874	3892	3909	3927	3945	3962
25	3979	3997	4014	4031	4048	4065	4082	4099	4116	4133
26	4150	4166	4183	4200	4216	4232	4249	4265	4281	4298
27	4314	4330	4346	4362	4378	4393	4409	4425	4440	4456
28	4472	4487	4502	4518	4533	4548	4564	4579	4594	4609
29	4624	4639	4654	4669	4683	4698	4713	4728	4742	4757
30	4771	4786	4800	4814	4829	4843	4857	4871	4886	4900
31	4914	4928	4942	4955	4969	4983	4997	5011	5024	5038
32	5051	5065	5079	5092	5105	5119	5132	5145	5159	5172
33	5185	5198	5211	5224	5237	5250	5263	5276	5289	5302
34	5315	5328	5340	5353	5366	5378	5391	5403	5416	5428
35	5441	5453	5465	5478	5490	5502	5514	5527	5539	5551
36	5563	5575	5587	5599	5611	5623	5635	5647	5658	5670
37	5682	5694	5705	5717	5729	5740	5752	5763	5775	5786
38	5798	5809	5821	5832	5843	5855	5866	5877	5888	5899
39	5911	5922	5933	5944	5955	5966	5977	5988	5999	6010
40	6021	6031	6042	6053	6064	6075	6085	6096	6107	6117
41	6128	6138	6149	6160	6170	6180	6191	6201	6212	6222
42	6232	6243	6253	6263	6274	6284	6294	6304	6314	6325
43	6335	6345	6355	6365	6375	6385	6395	6405	6415	6425
44	6435	6444	6454	6464	6474	6484	6493	6503	6513	6522
45	6532	6542	6551	6561	6571	6580	6590	6599	6609	6618
46	6628	6637	6646	6656	6665	6675	6684	6693	6702	6712
47	6721	6730	6739	6749	6758	6767	6776	6785	6794	6803
48	6812	6821	6830	6839	6848	6857	6866	6875	6884	6893
49	6902	6911	6920	6928	6937	6946	6955	6964	6972	6981
50	6990	6998	7007	7016	7024	7033	7042	7050	7059	7067
51	7076	7084	7093	7101	7110	7118	7126	7135	7143	7152
52	7160	7168	7177	7185	7193	7202	7210	7218	7226	7235
53	7243	7251	7259	7267	7275	7284	7292	7300	7308	7316
54	7324	7332	7340	7348	7356	7364	7372	7380	7388	7396

Table 3 / Common Logarithms of Numbers

N	0	1	2	3	4	5	6	7	8	9
55	7404	7412	7419	7427	7435	7443	7451	7459	7466	7474
56	7482	7490	7497	7505	7513	7520	7528	7536	7543	7551
57	7559	7566	7574	7582	7589	7597	7604	7612	7619	7627
58	7634	7642	7649	7657	7664	7672	7679	7686	7694	7701
59	7709	7716	7723	7731	7738	7745	7752	7760	7767	7774
60	7782	7789	7796	7803	7810	7818	7825	7832	7839	7846
61	7853	7860	7868	7875	7882	7889	7896	7903	7910	7917
62	7924	7931	7938	7945	7952	7959	7966	7973	7980	7987
63	7993	8000	8007	8014	8021	8028	8035	8041	8048	8055
64	8062	8069	8075	8082	8089	8096	8102	8109	8116	8122
65	8129	8136	8142	8149	8156	8162	8169	8176	8182	8189
66	8195	8202	8209	8215	8222	8228	8235	8241	8248	8254
67	8261	8267	8274	8280	8287	8293	8299	8306	8312	8319
68	8325	8331	8338	8344	8351	8357	8363	8370	8376	8382
69	8388	8395	8401	8407	8414	8420	8426	8432	8439	8445
70	8451	8457	8463	8470	8476	8482	8488	8494	8500	8506
71	8513	8519	8525	8531	8537	8543	8549	8555	8561	8567
72	8573	8579	8585	8591	8597	8603	8609	8615	8621	8627
73	8633	8639	8645	8651	8657	8663	8669	8675	8681	8686
74	8692	8698	8704	8710	8716	8722	8727	8733	8739	8745
75	8751	8756	8762	8768	8774	8779	8785	8791	8797	8802
76	8808	8814	8820	8825	8831	8837	8842	8848	8854	8859
77	8865	8871	8876	8882	8887	8893	8899	8904	8910	8915
78	8921	8927	8932	8938	8943	8949	8954	8960	8965	8971
79	8976	8982	8987	8993	8998	9004	9009	9015	9020	9025
80	9031	9036	9042	9047	9053	9058	9063	9069	9074	9079
81	9085	9090	9096	9101	9106	9112	9117	9122	9128	9133
82	9138	9143	9149	9154	9159	9165	9170	9175	9180	9186
83	9191	9196	9201	9206	9212	9217	9222	9227	9232	9238
84	9243	9248	9253	9258	9263	9269	9274	9279	9284	9289
85	9294	9299	9304	9309	9315	9320	9325	9330	9335	9340
86	9345	9350	9355	9360	9365	9370	9375	9380	9385	9390
87	9395	9400	9405	9410	9415	9420	9425	9430	9435	9440
88	9445	9450	9455	9460	9465	9469	9474	9479	9484	9489
89	9494	9499	9504	9509	9513	9518	9523	9528	9533	9538
90	9542	9547	9552	9557	9562	9566	9571	9576	9581	9586
91	9590	9595	9600	9605	9609	9614	9619	9624	9628	9633
92	9638	9643	9647	9652	9657	9661	9666	9671	9675	9680
93	9685	9689	9694	9699	9703	9708	9713	9717	9722	9727
94	9731	9736	9741	9745	9750	9754	9759	9763	9768	9773
95	9777	9782	9786	9791	9795	9800	9805	9809	9814	9818
96	9823	9827	9832	9836	9841	9845	9850	9854	9859	9863
97	9868	9872	9877	9881	9886	9890	9894	9899	9903	9908
98	9912	9917	9921	9926	9930	9934	9939	9943	9948	9952
99	9956	9961	9965	9969	9974	9978	9983	9987	9991	9996

Table 4 / *Trigonometric Functions of θ*

(θ in decimal degrees)

θ Degrees	*θ Radians*	*sin θ*	*cos θ*	*tan θ*	*cot θ*	*sec θ*	*csc θ*		
0.0	.0000	.0000	1.0000	.0000	undefined	1.000	undefined	1.5708	**90.0**
0.1	.0017	.0017	1.0000	.0017	573.0	1.000	573.0	1.5691	89.9
0.2	.0035	.0035	1.0000	.0035	286.5	1.000	286.5	1.5673	89.8
0.3	.0052	.0052	1.0000	.0052	191.0	1.000	191.0	1.5656	89.7
0.4	.0070	.0070	1.0000	.0070	143.2	1.000	143.2	1.5638	89.6
0.5	.0087	.0087	1.0000	.0087	114.6	1.000	114.6	1.5621	89.5
0.6	.0105	.0105	.9999	.0105	95.49	1.000	95.49	1.5603	89.4
0.7	.0122	.0122	.9999	.0122	81.85	1.000	81.85	1.5586	89.3
0.8	.0140	.0140	.9999	.0140	71.62	1.000	71.62	1.5568	89.2
0.9	.0157	.0157	.9999	.0157	63.66	1.000	63.66	1.5551	89.1
1.0	.0175	.0175	.9998	.0175	57.29	1.000	57.30	1.5533	**89.0**
1.1	.0192	.0192	.9998	.0192	52.08	1.000	52.09	1.5516	88.9
1.2	.0209	.0209	.9998	.0209	47.74	1.000	47.75	1.5499	88.8
1.3	.0227	.0227	.9997	.0227	44.07	1.000	44.08	1.5481	88.7
1.4	.0244	.0244	.9997	.0244	40.92	1.000	40.93	1.5464	88.6
1.5	.0262	.0262	.9997	.0262	38.19	1.000	38.20	1.5446	88.5
1.6	.0279	.0279	.9996	.0279	35.80	1.000	35.81	1.5429	88.4
1.7	.0297	.0297	.9996	.0297	33.69	1.000	33.71	1.5411	88.3
1.8	.0314	.0314	.9995	.0314	31.82	1.000	31.84	1.5394	88.2
1.9	.0332	.0332	.9995	.0332	30.14	1.001	30.16	1.5376	88.1
2.0	.0349	.0349	.9994	.0349	28.64	1.001	28.65	1.5359	**88.0**
2.1	.0367	.0366	.9993	.0367	27.27	1.001	27.29	1.5341	87.9
2.2	.0384	.0384	.9993	.0384	26.03	1.001	26.05	1.5324	87.8
2.3	.0401	.0401	.9992	.0402	24.90	1.001	24.92	1.5307	87.7
2.4	.0419	.0419	.9991	.0419	23.86	1.001	23.88	1.5289	87.6
2.5	.0436	.0436	.9990	.0437	22.90	1.001	22.93	1.5272	87.5
2.6	.0454	.0454	.9990	.0454	22.02	1.001	22.04	1.5254	87.4
2.7	.0471	.0471	.9989	.0472	21.20	1.001	21.23	1.5237	87.3
2.8	.0489	.0488	.9988	.0489	20.45	1.001	20.47	1.5219	87.2
2.9	.0506	.0506	.9987	.0507	19.74	1.001	19.77	1.5202	87.1
3.0	.0524	.0523	.9986	.0524	19.08	1.001	19.11	1.5184	**87.0**
3.1	.0541	.0541	.9985	.0542	18.46	1.001	18.49	1.5167	86.9
3.2	.0559	.0558	.9984	.0559	17.89	1.002	17.91	1.5149	86.8
3.3	.0576	.0576	.9983	.0577	17.34	1.002	17.37	1.5132	86.7
3.4	.0593	.0593	.9982	.0594	16.83	1.002	16.86	1.5115	86.6
3.5	.0611	.0610	.9981	.0612	16.35	1.002	16.38	1.5097	86.5
3.6	.0628	.0628	.9980	.0629	15.89	1.002	15.93	1.5080	86.4
3.7	.0646	.0645	.9979	.0647	15.46	1.002	15.50	1.5062	86.3
3.8	.0663	.0663	.9978	.0664	15.06	1.002	15.09	1.5045	86.2
3.9	.0681	.0680	.9977	.0682	14.67	1.002	14.70	1.5027	86.1
4.0	.0698	.0698	.9976	.0699	14.30	1.002	14.34	1.5010	**86.0**
4.1	.0716	.0715	.9974	.0717	13.95	1.003	13.99	1.4992	85.9
4.2	.0733	.0732	.9973	.0734	13.62	1.003	13.65	1.4975	85.8
4.3	.0750	.0750	.9972	.0752	13.30	1.003	13.34	1.4957	85.7
4.4	.0768	.0767	.9971	.0769	13.00	1.003	13.03	1.4940	85.6
4.5	.0785	.0785	.9969	.0787	12.71	1.003	12.75	1.4923	85.5
4.6	.0803	.0802	.9968	.0805	12.43	1.003	12.47	1.4905	85.4
4.7	.0820	.0819	.9966	.0822	12.16	1.003	12.20	1.4888	85.3
4.8	.0838	.0837	.9965	.0840	11.91	1.004	11.95	1.4870	85.2
4.9	.0855	.0854	.9963	.0857	11.66	1.004	11.71	1.4853	85.1
5.0	.0873	.0872	.9962	.0875	11.43	1.004	11.47	1.4835	**85.0**
5.1	.0890	.0889	.9960	.0892	11.20	1.004	11.25	1.4818	84.9
5.2	.0908	.0906	.9959	.0910	10.99	1.004	11.03	1.4800	84.8
5.3	.0925	.0924	.9957	.0928	10.78	1.004	10.83	1.4783	84.7
5.4	.0942	.0941	.9956	.0945	10.58	1.004	10.63	1.4765	84.6
5.5	.0960	.0958	.9954	.0963	10.39	1.005	10.43	1.4748	84.5
5.6	.0977	.0976	.9952	.0981	10.20	1.005	10.25	1.4731	84.4
5.7	.0995	.0993	.9951	.0998	10.02	1.005	10.07	1.4713	84.3
5.8	.1012	.1011	.9949	.1016	9.845	1.005	9.895	1.4696	84.2
5.9	.1030	.1028	.9947	.1033	9.677	1.005	9.728	1.4678	84.1
6.0	.1047	.1045	.9945	.1051	9.514	1.006	9.567	1.4661	**84.0**
		cos θ	*sin θ*	*cot θ*	*tan θ*	*csc θ*	*sec θ*	*θ Radians*	*θ Degrees*

Table 4 / *Trigonometric Functions of θ*

(θ in decimal degrees)

θ *Degrees*	θ *Radians*	*sin* θ	*cos* θ	*tan* θ	*cot* θ	*sec* θ	*csc* θ		
6.0	.1047	.1045	.9945	.1051	9.514	1.006	9.567	1.4661	**84.0**
6.1	.1065	.1063	.9943	.1069	9.357	1.006	9.411	1.4643	83.9
6.2	.1082	.1080	.9942	.1086	9.205	1.006	9.259	1.4626	83.8
6.3	.1100	.1097	.9940	.1104	9.058	1.006	9.113	1.4608	83.7
6.4	.1117	.1115	.9938	.1122	8.915	1.006	8.971	1.4591	83.6
6.5	.1134	.1132	.9936	.1139	8.777	1.006	8.834	1.4574	83.5
6.6	.1152	.1149	.9934	.1157	8.643	1.007	8.700	1.4556	83.4
6.7	.1169	.1167	.9932	.1175	8.513	1.007	8.571	1.4539	83.3
6.8	.1187	.1184	.9930	.1192	8.386	1.007	8.446	1.4521	83.2
6.9	.1204	.1201	.9928	.1210	8.264	1.007	8.324	1.4504	83.1
7.0	.1222	.1219	.9925	.1228	8.144	1.008	8.206	1.4486	**83.0**
7.1	.1239	.1236	.9923	.1246	8.028	1.008	8.091	1.4469	82.9
7.2	.1257	.1253	.9921	.1263	7.916	1.008	7.979	1.4451	82.8
7.3	.1274	.1271	.9919	.1281	7.806	1.008	7.870	1.4434	82.7
7.4	.1292	.1288	.9917	.1299	7.700	1.008	7.764	1.4416	82.6
7.5	.1309	.1305	.9914	.1317	7.596	1.009	7.661	1.4399	82.5
7.6	.1326	.1323	.9912	.1334	7.495	1.009	7.561	1.4382	82.4
7.7	.1344	.1340	.9910	.1352	7.396	1.009	7.463	1.4364	82.3
7.8	.1361	.1357	.9907	.1370	7.300	1.009	7.368	1.4347	82.2
7.9	.1379	.1374	.9905	.1388	7.207	1.010	7.276	1.4329	82.1
8.0	.1396	.1392	.9903	.1405	7.115	1.010	7.185	1.4312	**82.0**
8.1	.1414	.1409	.9900	.1423	7.026	1.010	7.097	1.4294	81.9
8.2	.1431	.1426	.9898	.1441	6.940	1.010	7.011	1.4277	81.8
8.3	.1449	.1444	.9895	.1459	6.855	1.011	6.927	1.4259	81.7
8.4	.1466	.1461	.9893	.1477	6.772	1.011	6.845	1.4242	81.6
8.5	.1484	.1478	.9890	.1495	6.691	1.011	6.765	1.4224	81.5
8.6	.1501	.1495	.9888	.1512	6.612	1.011	6.687	1.4207	81.4
8.7	.1518	.1513	.9885	.1530	6.535	1.012	6.611	1.4190	81.3
8.8	.1536	.1530	.9882	.1548	6.460	1.012	6.537	1.4172	81.2
8.9	.1553	.1547	.9880	.1566	6.386	1.012	6.464	1.4155	81.1
9.0	.1571	.1564	.9877	.1584	6.314	1.012	6.392	1.4137	**81.0**
9.1	.1588	.1582	.9874	.1602	6.243	1.013	6.323	1.4120	80.9
9.2	.1606	.1599	.9871	.1620	6.174	1.013	6.255	1.4102	80.8
9.3	.1623	.1616	.9869	.1638	6.107	1.013	6.188	1.4085	80.7
9.4	.1641	.1633	.9866	.1655	6.041	1.014	6.123	1.4067	80.6
9.5	.1658	.1650	.9863	.1673	5.976	1.014	6.059	1.4050	80.5
9.6	.1676	.1668	.9860	.1691	5.912	1.014	5.996	1.4032	80.4
9.7	.1693	.1685	.9857	.1709	5.850	1.015	5.935	1.4015	80.3
9.8	.1710	.1702	.9854	.1727	5.789	1.015	5.875	1.3998	80.2
9.9	.1728	.1719	.9851	.1745	5.730	1.015	5.816	1.3980	80.1
10.0	.1745	.1736	.9848	.1763	5.671	1.015	5.759	1.3963	**80.0**
10.1	.1763	.1754	.9845	.1781	5.614	1.016	5.702	1.3945	79.9
10.2	.1780	.1771	.9842	.1799	5.558	1.016	5.647	1.3928	79.8
10.3	.1798	.1788	.9839	.1817	5.503	1.016	5.593	1.3910	79.7
10.4	.1815	.1805	.9836	.1835	5.449	1.017	5.540	1.3893	79.6
10.5	.1833	.1822	.9833	.1853	5.396	1.017	5.487	1.3875	79.5
10.6	.1850	.1840	.9829	.1871	5.343	1.017	5.436	1.3858	79.4
10.7	.1868	.1857	.9826	.1890	5.292	1.018	5.386	1.3840	79.3
10.8	.1885	.1874	.9823	.1908	5.242	1.018	5.337	1.3823	79.2
10.9	.1902	.1891	.9820	.1926	5.193	1.018	5.288	1.3806	79.1
11.0	.1920	.1908	.9816	.1944	5.145	1.019	5.241	1.3788	**79.0**
11.1	.1937	.1925	.9813	.1962	5.097	1.019	5.194	1.3771	78.9
11.2	.1955	.1942	.9810	.1980	5.050	1.019	5.148	1.3753	78.8
11.3	.1972	.1959	.9806	.1998	5.005	1.020	5.103	1.3736	78.7
11.4	.1990	.1977	.9803	.2016	4.959	1.020	5.059	1.3718	78.6
11.5	.2007	.1994	.9799	.2035	4.915	1.020	5.016	1.3701	78.5
11.6	.2025	.2011	.9796	.2053	4.872	1.021	4.973	1.3683	78.4
11.7	.2042	.2028	.9792	.2071	4.829	1.021	4.931	1.3666	78.3
11.8	.2059	.2045	.9789	.2089	4.787	1.022	4.890	1.3648	78.2
11.9	.2077	.2062	.9785	.2107	4.745	1.022	4.850	1.3631	78.1
12.0	.2094	.2079	.9781	.2126	4.705	1.022	4.810	1.3614	**78.0**
		cos θ	*sin* θ	*cot* θ	*tan* θ	*csc* θ	*sec* θ	θ *Radians*	θ *Degrees*

Table 4 / *Trigonometric Functions of θ*

(θ in decimal degrees)

θ Degrees	*θ Radians*	*sin θ*	*cos θ*	*tan θ*	*cot θ*	*sec θ*	*csc θ*		
12.0	.2094	.2079	.9781	.2126	4.705	1.022	4.810	1.3614	**78.0**
12.1	.2112	.2096	.9778	.2144	4.665	1.023	4.771	1.3596	77.9
12.2	.2129	.2113	.9774	.2162	4.625	1.023	4.732	1.3579	77.8
12.3	.2147	.2130	.9770	.2180	4.586	1.023	4.694	1.3561	77.7
12.4	.2164	.2147	.9767	.2199	4.548	1.024	4.657	1.3544	77.6
12.5	.2182	.2164	.9763	.2217	4.511	1.024	4.620	1.3526	77.5
12.6	.2199	.2181	.9759	.2235	4.474	1.025	4.584	1.3509	77.4
12.7	.2217	.2198	.9755	.2254	4.437	1.025	4.549	1.3491	77.3
12.8	.2234	.2215	.9751	.2272	4.402	1.025	4.514	1.3474	77.2
12.9	.2251	.2233	.9748	.2290	4.366	1.026	4.479	1.3456	77.1
13.0	.2269	.2250	.9744	.2309	4.331	1.026	4.445	1.3439	**77.0**
13.1	.2286	.2267	.9740	.2327	4.297	1.027	4.412	1.3422	76.9
13.2	.2304	.2284	.9736	.2345	4.264	1.027	4.379	1.3404	76.8
13.3	.2321	.2300	.9732	.2364	4.230	1.028	4.347	1.3387	76.7
13.4	.2339	.2317	.9728	.2382	4.198	1.028	4.315	1.3369	76.6
13.5	.2356	.2334	.9724	.2401	4.165	1.028	4.284	1.3352	76.5
13.6	.2374	.2351	.9720	.2419	4.134	1.029	4.253	1.3334	76.4
13.7	.2391	.2368	.9715	.2438	4.102	1.029	4.222	1.3317	76.3
13.8	.2409	.2385	.9711	.2456	4.071	1.030	4.192	1.3299	76.2
13.9	.2426	.2402	.9707	.2475	4.041	1.030	4.163	1.3282	76.1
14.0	.2443	.2419	.9703	.2493	4.011	1.031	4.134	1.3265	**76.0**
14.1	.2461	.2436	.9699	.2512	3.981	1.031	4.105	1.3247	75.9
14.2	.2478	.2453	.9694	.2530	3.952	1.032	4.077	1.3230	75.8
14.3	.2496	.2470	.9690	.2549	3.923	1.032	4.049	1.3212	75.7
14.4	.2513	.2487	.9686	.2568	3.895	1.032	4.021	1.3195	75.6
14.5	.2531	.2504	.9681	.2586	3.867	1.033	3.994	1.3177	75.5
14.6	.2548	.2521	.9677	.2605	3.839	1.033	3.967	1.3160	75.4
14.7	.2566	.2538	.9673	.2623	3.812	1.034	3.941	1.3142	75.3
14.8	.2583	.2554	.9668	.2642	3.785	1.034	3.915	1.3125	75.2
14.9	.2601	.2571	.9664	.2661	3.758	1.035	3.889	1.3107	75.1
15.0	.2618	.2588	.9659	.2679	3.732	1.035	3.864	1.3090	**75.0**
15.1	.2635	.2605	.9655	.2698	3.706	1.036	3.839	1.3073	74.9
15.2	.2653	.2622	.9650	.2717	3.681	1.036	3.814	1.3055	74.8
15.3	.2670	.2639	.9646	.2736	3.655	1.037	3.790	1.3038	74.7
15.4	.2688	.2656	.9641	.2754	3.630	1.037	3.766	1.3020	74.6
15.5	.2705	.2672	.9636	.2773	3.606	1.038	3.742	1.3003	74.5
15.6	.2723	.2689	.9632	.2792	3.582	1.038	3.719	1.2985	74.4
15.7	.2740	.2706	.9627	.2811	3.558	1.039	3.695	1.2968	74.3
15.8	.2758	.2723	.9622	.2830	3.534	1.039	3.673	1.2950	74.2
15.9	.2775	.2740	.9617	.2849	3.511	1.040	3.650	1.2933	74.1
16.0	.2793	.2756	.9613	.2867	3.487	1.040	3.628	1.2915	**74.0**
16.1	.2810	.2773	.9608	.2886	3.465	1.041	3.606	1.2898	73.9
16.2	.2827	.2790	.9603	.2905	3.442	1.041	3.584	1.2881	73.8
16.3	.2845	.2807	.9598	.2924	3.420	1.042	3.563	1.2863	73.7
16.4	.2862	.2823	.9593	.2943	3.398	1.042	3.542	1.2846	73.6
16.5	.2880	.2840	.9588	.2962	3.376	1.043	3.521	1.2828	73.5
16.6	.2897	.2857	.9583	.2981	3.354	1.043	3.500	1.2811	73.4
16.7	.2915	.2874	.9578	.3000	3.333	1.044	3.480	1.2793	73.3
16.8	.2932	.2890	.9573	.3019	3.312	1.045	3.460	1.2776	73.2
16.9	.2950	.2907	.9568	.3038	3.291	1.045	3.440	1.2758	73.1
17.0	.2967	.2924	.9563	.3057	3.271	1.046	3.420	1.2741	**73.0**
17.1	.2985	.2940	.9558	.3076	3.251	1.046	3.401	1.2723	72.9
17.2	.3002	.2957	.9553	.3096	3.230	1.047	3.382	1.2706	72.8
17.3	.3019	.2974	.9548	.3115	3.211	1.047	3.363	1.2689	72.7
17.4	.3037	.2990	.9542	.3134	3.191	1.048	3.344	1.2671	72.6
17.5	.3054	.3007	.9537	.3153	3.172	1.049	3.326	1.2654	72.5
17.6	.3072	.3024	.9532	.3172	3.152	1.049	3.307	1.2636	72.4
17.7	.3089	.3040	.9527	.3191	3.133	1.050	3.289	1.2619	72.3
17.8	.3107	.3057	.9521	.3211	3.115	1.050	3.271	1.2601	72.2
17.9	.3124	.3074	.9516	.3230	3.096	1.051	3.254	1.2584	72.1
18.0	.3142	.3090	.9511	.3249	3.078	1.051	3.236	1.2566	**72.0**
		cos θ	*sin θ*	*cot θ*	*tan θ*	*csc θ*	*sec θ*	*θ Radians*	*θ Degrees*

Table 4 / *Trigonometric Functions of θ*

(θ in decimal degrees)

θ Degrees	θ Radians	sin θ	cos θ	tan θ	cot θ	sec θ	csc θ		
18.0	.3142	.3090	.9511	.3249	3.078	1.051	3.236	1.2566	**72.0**
18.1	.3159	.3107	.9505	.3269	3.060	1.052	3.219	1.2549	71.9
18.2	.3177	.3123	.9500	.3288	3.042	1.053	3.202	1.2531	71.8
18.3	.3194	.3140	.9494	.3307	3.024	1.053	3.185	1.2514	71.7
18.4	.3211	.3156	.9489	.3327	3.006	1.054	3.168	1.2497	71.6
18.5	.3229	.3173	.9483	.3346	2.989	1.054	3.152	1.2479	71.5
18.6	.3246	.3190	.9478	.3365	2.971	1.055	3.135	1.2462	71.4
18.7	.3264	.3206	.9472	.3385	2.954	1.056	3.119	1.2444	71.3
18.8	.3281	.3223	.9466	.3404	2.937	1.056	3.103	1.2427	71.2
18.9	.3299	.3239	.9461	.3424	2.921	1.057	3.087	1.2409	71.1
19.0	.3316	.3256	.9455	.3443	2.904	1.058	3.072	1.2392	**71.0**
19.1	.3334	.3272	.9449	.3463	2.888	1.058	3.056	1.2374	70.9
19.2	.3351	.3289	.9444	.3482	2.872	1.059	3.041	1.2357	70.8
19.3	.3368	.3305	.9438	.3502	2.856	1.060	3.026	1.2339	70.7
19.4	.3386	.3322	.9432	.3522	2.840	1.060	3.011	1.2322	70.6
19.5	.3403	.3338	.9426	.3541	2.824	1.061	2.996	1.2305	70.5
19.6	.3421	.3355	.9421	.3561	2.808	1.062	2.981	1.2287	70.4
19.7	.3438	.3371	.9415	.3581	2.793	1.062	2.967	1.2270	70.3
19.8	.3456	.3387	.9409	.3600	2.778	1.063	2.952	1.2252	70.2
19.9	.3473	.3404	.9403	.3620	2.762	1.064	2.938	1.2235	70.1
20.0	.3491	.3420	.9397	.3640	2.747	1.064	2.924	1.2217	**70.0**
20.1	.3508	.3437	.9391	.3659	2.733	1.065	2.910	1.2200	69.9
20.2	.3526	.3453	.9385	.3679	2.718	1.066	2.896	1.2182	69.8
20.3	.3543	.3469	.9379	.3699	2.703	1.066	2.882	1.2165	69.7
20.4	.3560	.3486	.9373	.3719	2.689	1.067	2.869	1.2147	69.6
20.5	.3578	.3502	.9367	.3739	2.675	1.068	2.855	1.2130	69.5
20.6	.3595	.3518	.9361	.3759	2.660	1.068	2.842	1.2113	69.4
20.7	.3613	.3535	.9354	.3779	2.646	1.069	2.829	1.2095	69.3
20.8	.3630	.3551	.9348	.3799	2.633	1.070	2.816	1.2078	69.2
20.9	.3648	.3567	.9342	.3819	2.619	1.070	2.803	1.2060	69.1
21.0	.3665	.3584	.9336	.3839	2.605	1.071	2.790	1.2043	**69.0**
21.1	.3683	.3600	.9330	.3859	2.592	1.072	2.778	1.2025	68.9
21.2	.3700	.3616	.9323	.3879	2.578	1.073	2.765	1.2008	68.8
21.3	.3718	.3633	.9317	.3899	2.565	1.073	2.753	1.1991	68.7
21.4	.3735	.3649	.9311	.3919	2.552	1.074	2.741	1.1973	68.6
21.5	.3752	.3665	.9304	.3939	2.539	1.075	2.729	1.1956	68.5
21.6	.3770	.3681	.9298	.3959	2.526	1.076	2.716	1.1938	68.4
21.7	.3787	.3697	.9291	.3979	2.513	1.076	2.705	1.1921	68.3
21.8	.3805	.3714	.9285	.4000	2.500	1.077	2.693	1.1903	68.2
21.9	.3822	.3730	.9278	.4020	2.488	1.078	2.681	1.1886	68.1
22.0	.3840	.3746	.9272	.4040	2.475	1.079	2.669	1.1868	**68.0**
22.1	.3857	.3762	.9265	.4061	2.463	1.079	2.658	1.1851	67.9
22.2	.3875	.3778	.9259	.4081	2.450	1.080	2.647	1.1833	67.8
22.3	.3892	.3795	.9252	.4101	2.438	1.081	2.635	1.1816	67.7
22.4	.3910	.3811	.9245	.4122	2.426	1.082	2.624	1.1798	67.6
22.5	.3927	.3827	.9239	.4142	2.414	1.082	2.613	1.1781	67.5
22.6	.3944	.3843	.9232	.4163	2.402	1.083	2.602	1.1764	67.4
22.7	.3962	.3859	.9225	.4183	2.391	1.084	2.591	1.1746	67.3
22.8	.3979	.3875	.9219	.4204	2.379	1.085	2.581	1.1729	67.2
22.9	.3997	.3891	.9212	.4224	2.367	1.086	2.570	1.1711	67.1
23.0	.4014	.3907	.9205	.4245	2.356	1.086	2.559	1.1694	**67.0**
23.1	.4032	.3923	.9198	.4265	2.344	1.087	2.549	1.1676	66.9
23.2	.4049	.3939	.9191	.4286	2.333	1.088	2.538	1.1659	66.8
23.3	.4067	.3955	.9184	.4307	2.322	1.089	2.528	1.1641	66.7
23.4	.4084	.3971	.9178	.4327	2.311	1.090	2.518	1.1624	66.6
23.5	.4102	.3987	.9171	.4348	2.300	1.090	2.508	1.1606	66.5
23.6	.4119	.4003	.9164	.4369	2.289	1.091	2.498	1.1589	66.4
23.7	.4136	.4019	.9157	.4390	2.278	1.092	2.488	1.1572	66.3
23.8	.4154	.4035	.9150	.4411	2.267	1.093	2.478	1.1554	66.2
23.9	.4171	.4051	.9143	.4431	2.257	1.094	2.468	1.1537	66.1
24.0	.4189	.4067	.9135	.4452	2.246	1.095	2.459	1.1519	**66.0**
		cos θ	sin θ	cot θ	tan θ	csc θ	sec θ	θ Radians	θ Degrees

Table 4 / Trigonometric Functions of θ

(θ in decimal degrees)

θ Degrees	θ Radians	$\sin\theta$	$\cos\theta$	$\tan\theta$	$\cot\theta$	$\sec\theta$	$\csc\theta$		
24.0	.4189	.4067	.9135	.4452	2.246	1.095	2.459	1.1519	**66.0**
24.1	.4206	.4083	.9128	.4473	2.236	1.095	2.449	1.1502	65.9
24.2	.4224	.4099	.9121	.4494	2.225	1.096	2.439	1.1484	65.8
24.3	.4241	.4115	.9114	.4515	2.215	1.097	2.430	1.1467	65.7
24.4	.4259	.4131	.9107	.4536	2.204	1.098	2.421	1.1449	65.6
24.5	.4276	.4147	.9100	.4557	2.194	1.099	2.411	1.1432	65.5
24.6	.4294	.4163	.9092	.4578	2.184	1.100	2.402	1.1414	65.4
24.7	.4311	.4179	.9085	.4599	2.174	1.101	2.393	1.1397	65.3
24.8	.4328	.4195	.9078	.4621	2.164	1.102	2.384	1.1380	65.2
24.9	.4346	.4210	.9070	.4642	2.154	1.102	2.375	1.1362	65.1
25.0	.4363	.4226	.9063	.4663	2.145	1.103	2.366	1.1345	**65.0**
25.1	.4381	.4242	.9056	.4684	2.135	1.104	2.357	1.1327	64.9
25.2	.4398	.4258	.9048	.4706	2.125	1.105	2.349	1.1310	64.8
25.3	.4416	.4274	.9041	.4727	2.116	1.106	2.340	1.1292	64.7
25.4	.4433	.4289	.9033	.4748	2.106	1.107	2.331	1.1275	64.6
25.5	.4451	.4305	.9026	.4770	2.097	1.108	2.323	1.1257	64.5
25.6	.4468	.4321	.9018	.4791	2.087	1.109	2.314	1.1240	64.4
25.7	.4485	.4337	.9011	.4813	2.078	1.110	2.306	1.1222	64.3
25.8	.4503	.4352	.9003	.4834	2.069	1.111	2.298	1.1205	64.2
25.9	.4520	.4368	.8996	.4856	2.059	1.112	2.289	1.1188	64.1
26.0	.4538	.4384	.8988	.4877	2.050	1.113	2.281	1.1170	**64.0**
26.1	.4555	.4399	.8980	.4899	2.041	1.114	2.273	1.1153	63.9
26.2	.4573	.4415	.8973	.4921	2.032	1.115	2.265	1.1135	63.8
26.3	.4590	.4431	.8965	.4942	2.023	1.115	2.257	1.1118	63.7
26.4	.4608	.4446	.8957	.4964	2.014	1.116	2.249	1.1100	63.6
26.5	.4625	.4462	.8949	.4986	2.006	1.117	2.241	1.1083	63.5
26.6	.4643	.4478	.8942	.5008	1.997	1.118	2.233	1.1065	63.4
26.7	.4660	.4493	.8934	.5029	1.988	1.119	2.226	1.1048	63.3
26.8	.4677	.4509	.8926	.5051	1.980	1.120	2.218	1.1030	63.2
26.9	.4695	.4524	.8918	.5073	1.971	1.121	2.210	1.1013	63.1
27.0	.4712	.4540	.8910	.5095	1.963	1.122	2.203	1.0996	**63.0**
27.1	.4730	.4555	.8902	.5117	1.954	1.123	2.195	1.0978	62.9
27.2	.4747	.4571	.8894	.5139	1.946	1.124	2.188	1.0961	62.8
27.3	.4765	.4586	.8886	.5161	1.937	1.125	2.180	1.0943	62.7
27.4	.4782	.4602	.8878	.5184	1.929	1.126	2.173	1.0926	62.6
27.5	.4800	.4617	.8870	.5206	1.921	1.127	2.166	1.0908	62.5
27.6	.4817	.4633	.8862	.5228	1.913	1.128	2.158	1.0891	62.4
27.7	.4835	.4648	.8854	.5250	1.905	1.129	2.151	1.0873	62.3
27.8	.4852	.4664	.8846	.5272	1.897	1.130	2.144	1.0856	62.2
27.9	.4869	.4679	.8838	.5295	1.889	1.132	2.137	1.0838	62.1
28.0	.4887	.4695	.8829	.5317	1.881	1.133	2.130	1.0821	**62.0**
28.1	.4904	.4710	.8821	.5340	1.873	1.134	2.123	1.0804	61.9
28.2	.4922	.4726	.8813	.5362	1.865	1.135	2.116	1.0786	61.8
28.3	.4939	.4741	.8805	.5384	1.857	1.136	2.109	1.0769	61.7
28.4	.4957	.4756	.8796	.5407	1.849	1.137	2.103	1.0751	61.6
28.5	.4974	.4772	.8788	.5430	1.842	1.138	2.096	1.0734	61.5
28.6	.4992	.4787	.8780	.5452	1.834	1.139	2.089	1.0716	61.4
28.7	.5009	.4802	.8771	.5475	1.827	1.140	2.082	1.0699	61.3
28.8	.5027	.4818	.8763	.5498	1.819	1.141	2.076	1.0681	61.2
28.9	.5044	.4833	.8755	.5520	1.811	1.142	2.069	1.0664	61.1
29.0	.5061	.4848	.8746	.5543	1.804	1.143	2.063	1.0647	**61.0**
29.1	.5079	.4863	.8738	.5566	1.797	1.144	2.056	1.0629	60.9
29.2	.5096	.4879	.8729	.5589	1.789	1.146	2.050	1.0612	60.8
29.3	.5114	.4894	.8721	.5612	1.782	1.147	2.043	1.0594	60.7
29.4	.5131	.4909	.8712	.5635	1.775	1.148	2.037	1.0577	60.6
29.5	.5149	.4924	.8704	.5658	1.767	1.149	2.031	1.0559	60.5
29.6	.5166	.4939	.8695	.5681	1.760	1.150	2.025	1.0542	60.4
29.7	.5184	.4955	.8686	.5704	1.753	1.151	2.018	1.0524	60.3
29.8	.5201	.4970	.8678	.5727	1.746	1.152	2.012	1.0507	60.2
29.9	.5219	.4985	.8669	.5750	1.739	1.154	2.006	1.0489	60.1
30.0	.5236	.5000	.8660	.5774	1.732	1.155	2.000	1.0472	**60.0**
		$\cos\theta$	$\sin\theta$	$\cot\theta$	$\tan\theta$	$\csc\theta$	$\sec\theta$	θ Radians	θ Degrees

Table 4 / *Trigonometric Functions of θ*

(θ in decimal degrees)

θ *Degrees*	*θ* *Radians*	*sin θ*	*cos θ*	*tan θ*	*cot θ*	*sec θ*	*csc θ*		
30.0	.5236	.5000	.8660	.5774	1.732	1.155	2.000	1.0472	**60.0**
30.1	.5253	.5015	.8652	.5797	1.725	1.156	1.994	1.0455	59.9
30.2	.5271	.5030	.8643	.5820	1.718	1.157	1.988	1.0437	59.8
30.3	.5288	.5045	.8634	.5844	1.711	1.158	1.982	1.0420	59.7
30.4	.5306	.5060	.8625	.5867	1.704	1.159	1.976	1.0402	59.6
30.5	.5323	.5075	.8616	.5890	1.698	1.161	1.970	1.0385	59.5
30.6	.5341	.5090	.8607	.5914	1.691	1.162	1.964	1.0367	59.4
30.7	.5358	.5105	.8599	.5938	1.684	1.163	1.959	1.0350	59.3
30.8	.5376	.5120	.8590	.5961	1.678	1.164	1.953	1.0332	59.2
30.9	.5393	.5135	.8581	.5985	1.671	1.165	1.947	1.0315	59.1
31.0	.5411	.5150	.8572	.6009	1.664	1.167	1.942	1.0297	**59.0**
31.1	.5428	.5165	.8563	.6032	1.658	1.168	1.936	1.0280	58.9
31.2	.5445	.5180	.8554	.6056	1.651	1.169	1.930	1.0263	58.8
31.3	.5463	.5195	.8545	.6080	1.645	1.170	1.925	1.0245	58.7
31.4	.5480	.5210	.8535	.6104	1.638	1.172	1.919	1.0228	58.6
31.5	.5498	.5225	.8526	.6128	1.632	1.173	1.914	1.0210	58.5
31.6	.5515	.5240	.8517	.6152	1.625	1.174	1.908	1.0193	58.4
31.7	.5533	.5255	.8508	.6176	1.619	1.175	1.903	1.0175	58.3
31.8	.5550	.5270	.8499	.6200	1.613	1.177	1.898	1.0158	58.2
31.9	.5568	.5284	.8490	.6224	1.607	1.178	1.892	1.0140	58.1
32.0	.5585	.5299	.8480	.6249	1.600	1.179	1.887	1.0123	**58.0**
32.1	.5603	.5314	.8471	.6273	1.594	1.180	1.882	1.0105	57.9
32.2	.5620	.5329	.8462	.6297	1.588	1.182	1.877	1.0088	57.8
32.3	.5637	.5344	.8453	.6322	1.582	1.183	1.871	1.0071	57.7
32.4	.5655	.5358	.8443	.6346	1.576	1.184	1.866	1.0053	57.6
32.5	.5672	.5373	.8434	.6371	1.570	1.186	1.861	1.0036	57.5
32.6	.5690	.5388	.8425	.6395	1.564	1.187	1.856	1.0018	57.4
32.7	.5707	.5402	.8415	.6420	1.558	1.188	1.851	1.0001	57.3
32.8	.5725	.5417	.8406	.6445	1.552	1.190	1.846	.9983	57.2
32.9	.5742	.5432	.8396	.6469	1.546	1.191	1.841	.9966	57.1
33.0	.5760	.5446	.8387	.6494	1.540	1.192	1.836	.9948	**57.0**
33.1	.5777	.5461	.8377	.6519	1.534	1.194	1.831	.9931	56.9
33.2	.5794	.5476	.8368	.6544	1.528	1.195	1.826	.9913	56.8
33.3	.5812	.5490	.8358	.6569	1.522	1.196	1.821	.9896	56.7
33.4	.5829	.5505	.8348	.6594	1.517	1.198	1.817	.9879	56.6
33.5	.5847	.5519	.8339	.6619	1.511	1.199	1.812	.9861	56.5
33.6	.5864	.5534	.8329	.6644	1.505	1.201	1.807	.9844	56.4
33.7	.5882	.5548	.8320	.6669	1.499	1.202	1.802	.9826	56.3
33.8	.5899	.5563	.8310	.6694	1.494	1.203	1.798	.9809	56.2
33.9	.5917	.5577	.8300	.6720	1.488	1.205	1.793	.9791	56.1
34.0	.5934	.5592	.8290	.6745	1.483	1.206	1.788	.9774	**56.0**
34.1	.5952	.5606	.8281	.6771	1.477	1.208	1.784	.9756	55.9
34.2	.5969	.5621	.8271	.6796	1.471	1.209	1.779	.9739	55.8
34.3	.5986	.5635	.8261	.6822	1.466	1.211	1.775	.9721	55.7
34.4	.6004	.5650	.8251	.6847	1.460	1.212	1.770	.9704	55.6
34.5	.6021	.5664	.8241	.6873	1.455	1.213	1.766	.9687	55.5
34.6	.6039	.5678	.8231	.6899	1.450	1.215	1.761	.9669	55.4
34.7	.6056	.5693	.8221	.6924	1.444	1.216	1.757	.9652	55.3
34.8	.6074	.5707	.8211	.6950	1.439	1.218	1.752	.9634	55.2
34.9	.6091	.5721	.8202	.6976	1.433	1.219	1.748	.9617	55.1
35.0	.6109	.5736	.8192	.7002	1.428	1.221	1.743	.9599	**55.0**
35.1	.6126	.5750	.8181	.7028	1.423	1.222	1.739	.9582	54.9
35.2	.6144	.5764	.8171	.7054	1.418	1.224	1.735	.9564	54.8
35.3	.6161	.5779	.8161	.7080	1.412	1.225	1.731	.9547	54.7
35.4	.6178	.5793	.8151	.7107	1.407	1.227	1.726	.9529	54.6
35.5	.6196	.5807	.8141	.7133	1.402	1.228	1.722	.9512	54.5
35.6	.6213	.5821	.8131	.7159	1.397	1.230	1.718	.9495	54.4
35.7	.6231	.5835	.8121	.7186	1.392	1.231	1.714	.9477	54.3
35.8	.6248	.5850	.8111	.7212	1.387	1.233	1.710	.9460	54.2
35.9	.6266	.5864	.8100	.7239	1.381	1.235	1.705	.9442	54.1
36.0	.6283	.5878	.8090	.7265	1.376	1.236	1.701	.9425	**54.0**
		cos θ	*sin θ*	*cot θ*	*tan θ*	*csc θ*	*sec θ*	*θ* *Radians*	*θ* *Degrees*

Table 4 / *Trigonometric Functions of θ*

(θ in decimal degrees)

θ *Degrees*	*θ* *Radians*	*sin θ*	*cos θ*	*tan θ*	*cot θ*	*sec θ*	*csc θ*		
36.0	.6283	.5878	.8090	.7265	1.376	1.236	1.701	.9425	**54.0**
36.1	.6301	.5892	.8080	.7292	1.371	1.238	1.697	.9407	53.9
36.2	.6318	.5906	.8070	.7319	1.366	1.239	1.693	.9390	53.8
36.3	.6336	.5920	.8059	.7346	1.361	1.241	1.689	.9372	53.7
36.4	.6353	.5934	.8049	.7373	1.356	1.242	1.685	.9355	53.6
36.5	.6370	.5948	.8039	.7400	1.351	1.244	1.681	.9338	53.5
36.6	.6388	.5962	.8028	.7427	1.347	1.246	1.677	.9320	53.4
36.7	.6405	.5976	.8018	.7454	1.342	1.247	1.673	.9303	53.3
36.8	.6423	.5990	.8007	.7481	1.337	1.249	1.669	.9285	53.2
36.9	.6440	.6004	.7997	.7508	1.332	1.250	1.666	.9268	53.1
37.0	.6458	.6018	.7986	.7536	1.327	1.252	1.662	.9250	**53.0**
37.1	.6475	.6032	.7976	.7563	1.322	1.254	1.658	.9233	52.9
37.2	.6493	.6046	.7965	.7590	1.317	1.255	1.654	.9215	52.8
37.3	.6510	.6060	.7955	.7618	1.313	1.257	1.650	.9198	52.7
37.4	.6528	.6074	.7944	.7646	1.308	1.259	1.646	.9180	52.6
37.5	.6545	.6088	.7934	.7673	1.303	1.260	1.643	.9163	52.5
37.6	.6562	.6101	.7923	.7701	1.299	1.262	1.639	.9146	52.4
37.7	.6580	.6115	.7912	.7729	1.294	1.264	1.635	.9128	52.3
37.8	.6597	.6129	.7902	.7757	1.289	1.266	1.632	.9111	52.2
37.9	.6615	.6143	.7891	.7785	1.285	1.267	1.628	.9093	52.1
38.0	.6632	.6157	.7880	.7813	1.280	1.269	1.624	.9076	**52.0**
38.1	.6650	.6170	.7869	.7841	1.275	1.271	1.621	.9058	51.9
38.2	.6667	.6184	.7859	.7869	1.271	1.272	1.617	.9041	51.8
38.3	.6685	.6198	.7848	.7898	1.266	1.274	1.613	.9023	51.7
38.4	.6702	.6211	.7837	.7926	1.262	1.276	1.610	.9006	51.6
38.5	.6720	.6225	.7826	.7954	1.257	1.278	1.606	.8988	51.5
38.6	.6737	.6239	.7815	.7983	1.253	1.280	1.603	.8971	51.4
38.7	.6754	.6252	.7804	.8012	1.248	1.281	1.599	.8954	51.3
38.8	.6772	.6266	.7793	.8040	1.244	1.283	1.596	.8936	51.2
38.9	.6789	.6280	.7782	.8069	1.239	1.285	1.592	.8919	51.1
39.0	.6807	.6293	.7771	.8098	1.235	1.287	1.589	.8901	**51.0**
39.1	.6824	.6307	.7760	.8127	1.230	1.289	1.586	.8884	50.9
39.2	.6842	.6320	.7749	.8156	1.226	1.290	1.582	.8866	50.8
39.3	.6859	.6334	.7738	.8185	1.222	1.292	1.579	.8849	50.7
39.4	.6877	.6347	.7727	.8214	1.217	1.294	1.575	.8831	50.6
39.5	.6894	.6361	.7716	.8243	1.213	1.296	1.572	.8814	50.5
39.6	.6912	.6374	.7705	.8273	1.209	1.298	1.569	.8796	50.4
39.7	.6929	.6388	.7694	.8302	1.205	1.300	1.566	.8779	50.3
39.8	.6946	.6401	.7683	.8332	1.200	1.302	1.562	.8762	50.2
39.9	.6964	.6414	.7672	.8361	1.196	1.304	1.559	.8744	50.1
40.0	.6981	.6428	.7660	.8391	1.192	1.305	1.556	.8727	**50.0**
40.1	.6999	.6441	.7649	.8421	1.188	1.307	1.552	.8709	49.9
40.2	.7016	.6455	.7638	.8451	1.183	1.309	1.549	.8692	49.8
40.3	.7034	.6468	.7627	.8481	1.179	1.311	1.546	.8674	49.7
40.4	.7051	.6481	.7615	.8511	1.175	1.313	1.543	.8657	49.6
40.5	.7069	.6494	.7604	.8541	1.171	1.315	1.540	.8639	49.5
40.6	.7086	.6508	.7593	.8571	1.167	1.317	1.537	.8622	49.4
40.7	.7103	.6521	.7581	.8601	1.163	1.319	1.534	.8604	49.3
40.8	.7121	.6534	.7570	.8632	1.159	1.321	1.530	.8587	49.2
40.9	.7138	.6547	.7559	.8662	1.154	1.323	1.527	.8570	49.1
41.0	.7156	.6561	.7547	.8693	1.150	1.325	1.524	.8552	**49.0**
41.1	.7173	.6574	.7536	.8724	1.146	1.327	1.521	.8535	48.9
41.2	.7191	.6587	.7524	.8754	1.142	1.329	1.518	.8517	48.8
41.3	.7208	.6600	.7513	.8785	1.138	1.331	1.515	.8500	48.7
41.4	.7226	.6613	.7501	.8816	1.134	1.333	1.512	.8482	48.6
41.5	.7243	.6626	.7490	.8847	1.130	1.335	1.509	.8465	48.5
41.6	.7261	.6639	.7478	.8878	1.126	1.337	1.506	.8447	48.4
41.7	.7278	.6652	.7466	.8910	1.122	1.339	1.503	.8430	48.3
41.8	.7295	.6665	.7455	.8941	1.118	1.341	1.500	.8412	48.2
41.9	.7313	.6678	.7443	.8972	1.115	1.344	1.497	.8395	48.1
42.0	.7330	.6691	.7431	.9004	1.111	1.346	1.494	.8378	**48.0**
		cos θ	*sin θ*	*cot θ*	*tan θ*	*csc θ*	*sec θ*	*θ* *Radians*	*θ* *Degrees*

Table 4 / Trigonometric Functions of θ

(θ in decimal degrees)

θ Degrees	θ Radians	sin θ	cos θ	tan θ	cot θ	sec θ	csc θ		
42.0	.7330	.6691	.7431	.9004	1.111	1.346	1.494	.8378	**48.0**
42.1	.7348	.6704	.7420	.9036	1.107	1.348	1.492	.8360	47.9
42.2	.7365	.6717	.7408	.9067	1.103	1.350	1.489	.8343	47.8
42.3	.7383	.6730	.7396	.9099	1.099	1.352	1.486	.8325	47.7
42.4	.7400	.6743	.7385	.9131	1.095	1.354	1.483	.8308	47.6
42.5	.7418	.6756	.7373	.9163	1.091	1.356	1.480	.8290	47.5
42.6	.7435	.6769	.7361	.9195	1.087	1.359	1.477	.8273	47.4
42.7	.7453	.6782	.7349	.9228	1.084	1.361	1.475	.8255	47.3
42.8	.7470	.6794	.7337	.9260	1.080	1.363	1.472	.8238	47.2
42.9	.7487	.6807	.7325	.9293	1.076	1.365	1.469	.8221	47.1
43.0	.7505	.6820	.7314	.9325	1.072	1.367	1.466	.8203	**47.0**
43.1	.7522	.6833	.7302	.9358	1.069	1.370	1.464	.8186	46.9
43.2	.7540	.6845	.7290	.9391	1.065	1.372	1.461	.8168	46.8
43.3	.7557	.6858	.7278	.9424	1.061	1.374	1.458	.8151	46.7
43.4	.7575	.6871	.7266	.9457	1.057	1.376	1.455	.8133	46.6
43.5	.7592	.6884	.7254	.9490	1.054	1.379	1.453	.8116	46.5
43.6	.7610	.6896	.7242	.9523	1.050	1.381	1.450	.8098	46.4
43.7	.7627	.6909	.7230	.9556	1.046	1.383	1.447	.8081	46.3
43.8	.7645	.6921	.7218	.9590	1.043	1.386	1.445	.8063	46.2
43.9	.7662	.6934	.7206	.9623	1.039	1.388	1.442	.8046	46.1
44.0	.7679	.6947	.7193	.9657	1.036	1.390	1.440	.8029	**46.0**
44.1	.7697	.6959	.7181	.9691	1.032	1.393	1.437	.8011	45.9
44.2	.7714	.6972	.7169	.9725	1.028	1.395	1.434	.7994	45.8
44.3	.7732	.6984	.7157	.9759	1.025	1.397	1.432	.7976	45.7
44.4	.7749	.6997	.7145	.9793	1.021	1.400	1.429	.7959	45.6
44.5	.7767	.7009	.7133	.9827	1.018	1.402	1.427	.7941	45.5
44.6	.7784	.7022	.7120	.9861	1.014	1.404	1.424	.7924	45.4
44.7	.7802	.7034	.7108	.9896	1.011	1.407	1.422	.7906	45.3
44.8	.7819	.7046	.7096	.9930	1.007	1.409	1.419	.7889	45.2
44.9	.7837	.7059	.7083	.9965	1.003	1.412	1.417	.7871	45.1
45.0	.7854	.7071	.7071	1.0000	1.000	1.414	1.414	.7854	**45.0**
		cos θ	sin θ	cot θ	tan θ	csc θ	sec θ	θ Radians	θ Degrees

Table 5 / *Trigonometric Functions of θ*

(θ in degrees and minutes)

Angle	Sin	Cos	Tan	Cot	Sec	Csc	
0° 00′	.0000	1.0000	.0000	- - - -	1.000	- - - -	90° 00′
10′	.0029	1.0000	.0029	343.8	1.000	343.8	50′
20′	.0058	1.0000	.0058	171.9	1.000	171.9	40′
30′	.0087	1.0000	.0087	114.6	1.000	114.6	30′
40′	.0116	.9999	.0116	85.94	1.000	85.95	20′
50′	.0145	.9999	.0145	68.75	1.000	68.76	10′
1° 00′	.0175	.9998	.0175	57.29	1.000	57.30	89° 00′
10′	.0204	.9998	.0204	49.10	1.000	49.11	50′
20′	.0233	.9997	.0233	42.96	1.000	42.98	40′
30′	.0262	.9997	.0262	38.19	1.000	38.20	30′
40′	.0291	.9996	.0291	34.37	1.000	34.38	20′
50′	.0320	.9995	.0320	31.24	1.001	31.26	10′
2° 00′	.0349	.9994	.0349	28.64	1.001	28.65	88° 00′
10′	.0378	.9993	.0378	26.43	1.001	26.45	50′
20′	.0407	.9992	.0407	24.54	1.001	24.56	40′
30′	.0436	.9990	.0437	22.90	1.001	22.93	30′
40′	.0465	.9989	.0466	21.47	1.001	21.49	20′
50′	.0494	.9988	.0495	20.21	1.001	20.23	10′
3° 00′	.0523	.9986	.0524	19.08	1.001	19.11	87° 00′
10′	.0552	.9985	.0553	18.07	1.002	18.10	50′
20′	.0581	.9983	.0582	17.17	1.002	17.20	40′
30′	.0610	.9981	.0612	16.35	1.002	16.38	30′
40′	.0640	.9980	.0641	15.60	1.002	15.64	20′
50′	.0669	.9978	.0670	14.92	1.002	14.96	10′
4° 00′	.0698	.9976	.0699	14.30	1.002	14.34	86° 00′
10′	.0727	.9974	.0729	13.73	1.003	13.76	50′
20′	.0756	.9971	.0758	13.20	1.003	13.23	40′
30′	.0785	.9969	.0787	12.71	1.003	12.75	30′
40′	.0814	.9967	.0816	12.25	1.003	12.29	20′
50′	.0843	.9964	.0846	11.83	1.004	11.87	10′
5° 00′	.0872	.9962	.0875	11.43	1.004	11.47	85° 00′
10′	.0901	.9959	.0904	11.06	1.004	11.10	50′
20′	.0929	.9957	.0934	10.71	1.004	10.76	40′
30′	.0958	.9954	.0963	10.39	1.005	10.43	30′
40′	.0987	.9951	.0992	10.08	1.005	10.13	20′
50′	.1016	.9948	.1022	9.788	1.005	9.839	10′
6° 00′	.1045	.9945	.1051	9.514	1.006	9.567	84° 00′
10′	.1074	.9942	.1080	9.255	1.006	9.309	50′
20′	.1103	.9939	.1110	9.010	1.006	9.065	40′
30′	.1132	.9936	.1139	8.777	1.006	8.834	30′
40′	.1161	.9932	.1169	8.556	1.007	8.614	20′
50′	.1190	.9929	.1198	8.345	1.007	8.405	10′
7° 00′	.1219	.9925	.1228	8.144	1.008	8.206	83° 00′
10′	.1248	.9922	.1257	7.953	1.008	8.016	50′
20′	.1276	.9918	.1287	7.770	1.008	7.834	40′
30′	.1305	.9914	.1317	7.596	1.009	7.661	30′
40′	.1334	.9911	.1346	7.429	1.009	7.496	20′
50′	.1363	.9907	.1376	7.269	1.009	7.337	10′
8° 00′	.1392	.9903	.1405	7.115	1.010	7.185	82° 00′
10′	.1421	.9899	.1435	6.968	1.010	7.040	50′
20′	.1449	.9894	.1465	6.827	1.011	6.900	40′
30′	.1478	.9890	.1495	6.691	1.011	6.765	30′
40′	.1507	.9886	.1524	6.561	1.012	6.636	20′
50′	.1536	.9881	.1554	6.435	1.012	6.512	10′
9° 00′	.1564	.9877	.1584	6.314	1.012	6.392	81° 00′
	Cos	Sin	Cot	Tan	Csc	Sec	Angle

Table 5 / Trigonometric Functions of θ

(θ in degrees and minutes)

Angle	Sin	Cos	Tan	Cot	Sec	Csc	
9° 00′	.1564	.9877	.1584	6.314	1.012	6.392	**81° 00′**
10′	.1593	.9872	.1614	6.197	1.013	6.277	**50′**
20′	.1622	.9868	.1644	6.084	1.013	6.166	**40′**
30′	.1650	.9863	.1673	5.976	1.014	6.059	**30′**
40′	.1679	.9858	.1703	5.871	1.014	5.955	**20′**
50′	.1708	.9853	.1733	5.769	1.015	5.855	**10′**
10° 00′	.1736	.9848	.1763	5.671	1.015	5.759	**80° 00′**
10′	.1765	.9843	.1793	5.576	1.016	5.665	**50′**
20′	.1794	.9838	.1823	5.485	1.016	5.575	**40′**
30′	.1822	.9833	.1853	5.396	1.017	5.487	**30′**
40′	.1851	.9827	.1883	5.309	1.018	5.403	**20′**
50′	.1880	.9822	.1914	5.226	1.018	5.320	**10′**
11° 00′	.1908	.9816	.1944	5.145	1.019	5.241	**79° 00′**
10′	.1937	.9811	.1974	5.066	1.019	5.164	**50′**
20′	.1965	.9805	.2004	4.989	1.020	5.089	**40′**
30′	.1994	.9799	.2035	4.915	1.020	5.016	**30′**
40′	.2022	.9793	.2065	4.843	1.021	4.945	**20′**
50′	.2051	.9787	.2095	4.773	1.022	4.876	**10′**
12° 00′	.2079	.9781	.2126	4.705	1.022	4.810	**78° 00′**
10′	.2108	.9775	.2156	4.638	1.023	4.745	**50′**
20′	.2136	.9769	.2186	4.574	1.024	4.682	**40′**
30′	.2164	.9763	.2217	4.511	1.024	4.620	**30′**
40′	.2193	.9757	.2247	4.449	1.025	4.560	**20′**
50′	.2221	.9750	.2278	4.390	1.026	4.502	**10′**
13° 00′	.2250	.9744	.2309	4.331	1.026	4.445	**77° 00′**
10′	.2278	.9737	.2339	4.275	1.027	4.390	**50′**
20′	.2306	.9730	.2370	4.219	1.028	4.336	**40′**
30′	.2334	.9724	.2401	4.165	1.028	4.284	**30′**
40′	.2363	.9717	.2432	4.113	1.029	4.232	**20′**
50′	.2391	.9710	.2462	4.061	1.030	4.182	**10′**
14° 00′	.2419	.9703	.2493	4.011	1.031	4.134	**76° 00′**
10′	.2447	.9696	.2524	3.962	1.031	4.086	**50′**
20′	.2476	.9689	.2555	3.914	1.032	4.039	**40′**
30′	.2504	.9681	.2586	3.867	1.033	3.994	**30′**
40′	.2532	.9674	.2617	3.821	1.034	3.950	**20′**
50′	.2560	.9667	.2648	3.776	1.034	3.906	**10′**
15° 00′	.2588	.9659	.2679	3.732	1.035	3.864	**75° 00′**
10′	.2616	.9652	.2711	3.689	1.036	3.822	**50′**
20′	.2644	.9644	.2742	3.647	1.037	3.782	**40′**
30′	.2672	.9636	.2773	3.606	1.038	3.742	**30′**
40′	.2700	.9628	.2805	3.566	1.039	3.703	**20′**
50′	.2728	.9621	.2836	3.526	1.039	3.665	**10′**
16° 00′	.2756	.9613	.2867	3.487	1.040	3.628	**74° 00′**
10′	.2784	.9605	.2899	3.450	1.041	3.592	**50′**
20′	.2812	.9596	.2931	3.412	1.042	3.556	**40′**
30′	.2840	.9588	.2962	3.376	1.043	3.521	**30′**
40′	.2868	.9580	.2994	3.340	1.044	3.487	**20′**
50′	.2896	.9572	.3026	3.305	1.045	3.453	**10′**
17° 00′	.2924	.9563	.3057	3.271	1.046	3.420	**73° 00′**
10′	.2952	.9555	.3089	3.237	1.047	3.388	**50′**
20′	.2979	.9546	.3121	3.204	1.048	3.356	**40′**
30′	.3007	.9537	.3153	3.172	1.049	3.326	**30′**
40′	.3035	.9528	.3185	3.140	1.049	3.295	**20′**
50′	.3062	.9520	.3217	3.108	1.050	3.265	**10′**
18° 00′	.3090	.9511	.3249	3.078	1.051	3.236	**72° 00′**
	Cos	**Sin**	**Cot**	**Tan**	**Csc**	**Sec**	**Angle**

Table 5 / Trigonometric Functions of θ

(θ in degrees and minutes)

Angle	Sin	Cos	Tan	Cot	Sec	Csc	
18° 00′	.3090	.9511	.3249	3.078	1.051	3.236	**72° 00′**
10′	.3118	.9502	.3281	3.047	1.052	3.207	**50′**
20′	.3145	.9492	.3314	3.018	1.053	3.179	**40′**
30′	.3173	.9483	.3346	2.989	1.054	3.152	**30′**
40′	.3201	.9474	.3378	2.960	1.056	3.124	**20′**
50′	.3228	.9465	.3411	2.932	1.057	3.098	**10′**
19° 00′	.3256	.9455	.3443	2.904	1.058	3.072	**71° 00′**
10′	.3283	.9446	.3476	2.877	1.059	3.046	**50′**
20′	.3311	.9436	.3508	2.850	1.060	3.021	**40′**
30′	.3338	.9426	.3541	2.824	1.061	2.996	**30′**
40′	.3365	.9417	.3574	2.798	1.062	2.971	**20′**
50′	.3393	.9407	.3607	2.773	1.063	2.947	**10′**
20° 00′	.3420	.9397	.3640	2.747	1.064	2.924	**70° 00′**
10′	.3448	.9387	.3673	2.723	1.065	2.901	**50′**
20′	.3475	.9377	.3706	2.699	1.066	2.878	**40′**
30′	.3502	.9367	.3739	2.675	1.068	2.855	**30′**
40′	.3529	.9356	.3772	2.651	1.069	2.833	**20′**
50′	.3557	.9346	.3805	2.628	1.070	2.812	**10′**
21° 00′	.3584	.9336	.3839	2.605	1.071	2.790	**69° 00′**
10′	.3611	.9325	.3872	2.583	1.072	2.769	**50′**
20′	.3638	.9315	.3906	2.560	1.074	2.749	**40′**
30′	.3665	.9304	.3939	2.539	1.075	2.729	**30′**
40′	.3692	.9293	.3973	2.517	1.076	2.709	**20′**
50′	.3719	.9283	.4006	2.496	1.077	2.689	**10′**
22° 00′	.3746	.9272	.4040	2.475	1.079	2.669	**68° 00′**
10′	.3773	.9261	.4074	2.455	1.080	2.650	**50′**
20′	.3800	.9250	.4108	2.434	1.081	2.632	**40′**
30′	.3827	.9239	.4142	2.414	1.082	2.613	**30′**
40′	.3854	.9228	.4176	2.394	1.084	2.595	**20′**
50′	.3881	.9216	.4210	2.375	1.085	2.577	**10′**
23° 00′	.3907	.9205	.4245	2.356	1.086	2.559	**67° 00′**
10′	.3934	.9194	.4279	2.337	1.088	2.542	**50′**
20′	.3961	.9182	.4314	2.318	1.089	2.525	**40′**
30′	.3987	.9171	.4348	2.300	1.090	2.508	**30′**
40′	.4014	.9159	.4383	2.282	1.092	2.491	**20′**
50′	.4041	.9147	.4417	2.264	1.093	2.475	**10′**
24° 00′	.4067	.9135	.4452	2.246	1.095	2.459	**66° 00′**
10′	.4094	.9124	.4487	2.229	1.096	2.443	**50′**
20′	.4120	.9112	.4522	2.211	1.097	2.427	**40′**
30′	.4147	.9100	.4557	2.194	1.099	2.411	**30′**
40′	.4173	.9088	.4592	2.177	1.100	2.396	**20′**
50′	.4200	.9075	.4628	2.161	1.102	2.381	**10′**
25° 00′	.4226	.9063	.4663	2.145	1.103	2.366	**65° 00′**
10′	.4253	.9051	.4699	2.128	1.105	2.352	**50′**
20′	.4279	.9038	.4734	2.112	1.106	2.337	**40′**
30′	.4305	.9026	.4770	2.097	1.108	2.323	**30′**
40′	.4331	.9013	.4806	2.081	1.109	2.309	**20′**
50′	.4358	.9001	.4841	2.066	1.111	2.295	**10′**
26° 00′	.4384	.8988	.4877	2.050	1.113	2.281	**64° 00′**
10′	.4410	.8975	.4913	2.035	1.114	2.268	**50′**
20′	.4436	.8962	.4950	2.020	1.116	2.254	**40′**
30′	.4462	.8949	.4986	2.006	1.117	2.241	**30′**
40′	.4488	.8936	.5022	1.991	1.119	2.228	**20′**
50′	.4514	.8923	.5059	1.977	1.121	2.215	**10′**
27° 00′	.4540	.8910	.5095	1.963	1.122	2.203	**63° 00′**
	Cos	Sin	Cot	Tan	Csc	Sec	Angle

Table 5 / *Trigonometric Functions of θ*

(θ in degrees and minutes)

Angle	Sin	Cos	Tan	Cot	Sec	Csc	
27° 00′	.4540	.8910	.5095	1.963	1.122	2.203	**63° 00′**
10′	.4566	.8897	.5132	1.949	1.124	2.190	**50′**
20′	.4592	.8884	.5169	1.935	1.126	2.178	**40′**
30′	.4617	.8870	.5206	1.921	1.127	2.166	**30′**
40′	.4643	.8857	.5243	1.907	1.129	2.154	**20′**
50′	.4669	.8843	.5280	1.894	1.131	2.142	**10′**
28° 00′	.4695	.8829	.5317	1.881	1.133	2.130	**62° 00′**
10′	.4720	.8816	.5354	1.868	1.134	2.118	**50′**
20′	.4746	.8802	.5392	1.855	1.136	2.107	**40′**
30′	.4772	.8788	.5430	1.842	1.138	2.096	**30′**
40′	.4797	.8774	.5467	1.829	1.140	2.085	**20′**
50′	.4823	.8760	.5505	1.816	1.142	2.074	**10′**
29° 00′	.4848	.8746	.5543	1.804	1.143	2.063	**61° 00′**
10′	.4874	.8732	.5581	1.792	1.145	2.052	**50′**
20′	.4899	.8718	.5619	1.780	1.147	2.041	**40′**
30′	.4924	.8704	.5658	1.767	1.149	2.031	**30′**
40′	.4950	.8689	.5696	1.756	1.151	2.020	**20′**
50′	.4975	.8675	.5735	1.744	1.153	2.010	**10′**
30° 00′	.5000	.8660	.5774	1.732	1.155	2.000	**60° 00′**
10′	.5025	.8646	.5812	1.720	1.157	1.990	**50′**
20′	.5050	.8631	.5851	1.709	1.159	1.980	**40′**
30′	.5075	.8616	.5890	1.698	1.161	1.970	**30′**
40′	.5100	.8601	.5930	1.686	1.163	1.961	**20′**
50′	.5125	.8587	.5969	1.675	1.165	1.951	**10′**
31° 00′	.5150	.8572	.6009	1.664	1.167	1.942	**59° 00′**
10′	.5175	.8557	.6048	1.653	1.169	1.932	**50′**
20′	.5200	.8542	.6088	1.643	1.171	1.923	**40′**
30′	.5225	.8526	.6128	1.632	1.173	1.914	**30′**
40′	.5250	.8511	.6168	1.621	1.175	1.905	**20′**
50′	.5275	.8496	.6208	1.611	1.177	1.896	**10′**
32° 00′	.5299	.8480	.6249	1.600	1.179	1.887	**58° 00′**
10′	.5324	.8465	.6289	1.590	1.181	1.878	**50′**
20′	.5348	.8450	.6330	1.580	1.184	1.870	**40′**
30′	.5373	.8434	.6371	1.570	1.186	1.861	**30′**
40′	.5398	.8418	.6412	1.560	1.188	1.853	**20′**
50′	.5422	.8403	.6453	1.550	1.190	1.844	**10′**
33° 00′	.5446	.8387	.6494	1.540	1.192	1.836	**57° 00′**
10′	.5471	.8371	.6536	1.530	1.195	1.828	**50′**
20′	.5495	.8355	.6577	1.520	1.197	1.820	**40′**
30′	.5519	.8339	.6619	1.511	1.199	1.812	**30′**
40′	.5544	.8323	.6661	1.501	1.202	1.804	**20′**
50′	.5568	.8307	.6703	1.492	1.204	1.796	**10′**
34° 00′	.5592	.8290	.6745	1.483	1.206	1.788	**56° 00′**
10′	.5616	.8274	.6787	1.473	1.209	1.781	**50′**
20′	.5640	.8258	.6830	1.464	1.211	1.773	**40′**
30′	.5664	.8241	.6873	1.455	1.213	1.766	**30′**
40′	.5688	.8225	.6916	1.446	1.216	1.758	**20′**
50′	.5712	.8208	.6959	1.437	1.218	1.751	**10′**
35° 00′	.5736	.8192	.7002	1.428	1.221	1.743	**55° 00′**
10′	.5760	.8175	.7046	1.419	1.223	1.736	**50′**
20′	.5783	.8158	.7089	1.411	1.226	1.729	**40′**
30′	.5807	.8141	.7133	1.402	1.228	1.722	**30′**
40′	.5831	.8124	.7177	1.393	1.231	1.715	**20′**
50′	.5854	.8107	.7221	1.385	1.233	1.708	**10′**
36° 00′	.5878	.8090	.7265	1.376	1.236	1.701	**54° 00′**
	Cos	Sin	Cot	Tan	Csc	Sec	Angle

Table 5 / Trigonometric Functions of θ

(θ in degrees and minutes)

Angle	Sin	Cos	Tan	Cot	Sec	Csc	
36° 00′	.5878	.8090	.7265	1.376	1.236	1.701	54° 00′
10′	.5901	.8073	.7310	1.368	1.239	1.695	50′
20′	.5925	.8056	.7355	1.360	1.241	1.688	40′
30′	.5948	.8039	.7400	1.351	1.244	1.681	30′
40′	.5972	.8021	.7445	1.343	1.247	1.675	20′
50′	.5995	.8004	.7490	1.335	1.249	1.668	10′
37° 00′	.6018	.7986	.7536	1.327	1.252	1.662	53° 00′
10′	.6041	.7969	.7581	1.319	1.255	1.655	50′
20′	.6065	.7951	.7627	1.311	1.258	1.649	40′
30′	.6088	.7934	.7673	1.303	1.260	1.643	30′
40′	.6111	.7916	.7720	1.295	1.263	1.636	20′
50′	.6134	.7898	.7766	1.288	1.266	1.630	10′
38° 00′	.6157	.7880	.7813	1.280	1.269	1.624	52° 00′
10′	.6180	.7862	.7860	1.272	1.272	1.618	50′
20′	.6202	.7844	.7907	1.265	1.275	1.612	40′
30′	.6225	.7826	.7954	1.257	1.278	1.606	30′
40′	.6248	.7808	.8002	1.250	1.281	1.601	20′
50′	.6271	.7790	.8050	1.242	1.284	1.595	10′
39° 00′	.6293	.7771	.8098	1.235	1.287	1.589	51° 00′
10′	.6316	.7753	.8146	1.228	1.290	1.583	50′
20′	.6338	.7735	.8195	1.220	1.293	1.578	40′
30′	.6361	.7716	.8243	1.213	1.296	1.572	30′
40′	.6383	.7698	.8292	1.206	1.299	1.567	20′
50′	.6406	.7679	.8342	1.199	1.302	1.561	10′
40° 00′	.6428	.7660	.8391	1.192	1.305	1.556	50° 00′
10′	.6450	.7642	.8441	1.185	1.309	1.550	50′
20′	.6472	.7623	.8491	1.178	1.312	1.545	40′
30′	.6494	.7604	.8541	1.171	1.315	1.540	30′
40′	.6517	.7585	.8591	1.164	1.318	1.535	20′
50′	.6539	.7566	.8642	1.157	1.322	1.529	10′
41° 00′	.6561	.7547	.8693	1.150	1.325	1.524	49° 00′
10′	.6583	.7528	.8744	1.144	1.328	1.519	50′
20′	.6604	.7509	.8796	1.137	1.332	1.514	40′
30′	.6626	.7490	.8847	1.130	1.335	1.509	30′
40′	.6648	.7470	.8899	1.124	1.339	1.504	20′
50′	.6670	.7451	.8952	1.117	1.342	1.499	10′
42° 00′	.6691	.7431	.9004	1.111	1.346	1.494	48° 00′
10′	.6713	.7412	.9057	1.104	1.349	1.490	50′
20′	.6734	.7392	.9110	1.098	1.353	1.485	40′
30′	.6756	.7373	.9163	1.091	1.356	1.480	30′
40′	.6777	.7353	.9217	1.085	1.360	1.476	20′
50′	.6799	.7333	.9271	1.079	1.364	1.471	10′
43° 00′	.6820	.7314	.9325	1.072	1.367	1.466	47° 00′
10′	.6841	.7294	.9380	1.066	1.371	1.462	50′
20′	.6862	.7274	.9435	1.060	1.375	1.457	40′
30′	.6884	.7254	.9490	1.054	1.379	1.453	30′
40′	.6905	.7234	.9545	1.048	1.382	1.448	20′
50′	.6926	.7214	.9601	1.042	1.386	1.444	10′
44° 00′	.6947	.7193	.9657	1.036	1.390	1.440	46° 00′
10′	.6967	.7173	.9713	1.030	1.394	1.435	50′
20′	.6988	.7153	.9770	1.024	1.398	1.431	40′
30′	.7009	.7133	.9827	1.018	1.402	1.427	30′
40′	.7030	.7112	.9884	1.012	1.406	1.423	20′
50′	.7050	.7092	.9942	1.006	1.410	1.418	10′
45° 00′	.7071	.7071	1.000	1.000	1.414	1.414	45° 00′
	Cos	Sin	Cot	Tan	Csc	Sec	Angle

Table 6 / Trigonometric Functions of θ

(θ in radians)

θ Radians	θ Degrees	sin θ	cos θ	tan θ	cot θ	sec θ	csc θ
0.00	0° 00′	0.0000	1.000	0.0000	Undefined	1.000	Undefined
.01	0° 34′	.0100	1.000	.0100	100.0	1.000	100.0
.02	1° 09′	.0200	0.9998	.0200	49.99	1.000	50.00
.03	1° 43′	.0300	0.9996	.0300	33.32	1.000	33.34
.04	2° 18′	.0400	0.9992	.0400	24.99	1.001	25.01
0.05	2° 52′	0.0500	0.9988	0.0500	19.98	1.001	20.01
.06	3° 26′	.0600	.9982	.0601	16.65	1.002	16.68
.07	4° 01′	.0699	.9976	.0701	14.26	1.002	14.30
.08	4° 35′	.0799	.9968	.0802	12.47	1.003	12.51
.09	5° 09′	.0899	.9960	.0902	11.08	1.004	11.13
0.10	5° 44′	0.0998	0.9950	0.1003	9.967	1.005	10.02
.11	6° 18′	.1098	.9940	.1104	9.054	1.006	9.109
.12	6° 53′	.1197	.9928	.1206	8.293	1.007	8.353
.13	7° 27′	.1296	.9916	.1307	7.649	1.009	7.714
.14	8° 01′	.1395	.9902	.1409	7.096	1.010	7.166
0.15	8° 36′	0.1494	0.9888	0.1511	6.617	1.011	6.692
.16	9° 10′	.1593	.9872	.1614	6.197	1.013	6.277
.17	9° 44′	.1692	.9856	.1717	5.826	1.015	5.911
.18	10° 19′	.1790	.9838	.1820	5.495	1.016	5.586
.19	10° 53′	.1889	.9820	.1923	5.200	1.018	5.295
0.20	11° 28′	0.1987	0.9801	0.2027	4.933	1.020	5.033
.21	12° 02′	.2085	.9780	.2131	4.692	1.022	4.797
.22	12° 36′	.2182	.9759	.2236	4.472	1.025	4.582
.23	13° 11′	.2280	.9737	.2341	4.271	1.027	4.386
.24	13° 45′	.2377	.9713	.2447	4.086	1.030	4.207
0.25	14° 19′	0.2474	0.9689	0.2553	3.916	1.032	4.042
.26	14° 54′	.2571	.9664	.2660	3.759	1.035	3.890
.27	15° 28′	.2667	.9638	.2768	3.613	1.038	3.749
.28	16° 03′	.2764	.9611	.2876	3.478	1.041	3.619
.29	16° 37′	.2860	.9582	.2984	3.351	1.044	3.497
0.30	17° 11′	0.2955	0.9553	0.3093	3.233	1.047	3.384
.31	17° 46′	.3051	.9523	.3203	3.122	1.050	3.278
.32	18° 20′	.3146	.9492	.3314	3.018	1.053	3.179
.33	18° 55′	.3240	.9460	.3425	2.920	1.057	3.086
.34	19° 29′	.3335	.9428	.3537	2.827	1.061	2.999
0.35	20° 03′	0.3429	0.9394	0.3650	2.740	1.065	2.916
.36	20° 38′	.3523	.9359	.3764	2.657	1.068	2.839
.37	21° 12′	.3616	.9323	.3879	2.578	1.073	2.765
.38	21° 46′	.3709	.9287	.3994	2.504	1.077	2.696
.39	22° 21′	.3802	.9249	.4111	2.433	1.081	2.630
0.40	22° 55′	0.3894	0.9211	0.4228	2.365	1.086	2.568
.41	23° 30′	.3986	.9171	.4346	2.301	1.090	2.509
.42	24° 04′	.4078	.9131	.4466	2.239	1.095	2.452
.43	24° 38′	.4169	.9090	.4586	2.180	1.100	2.399
.44	25° 13′	.4259	.9048	.4708	2.124	1.105	2.348
0.45	25° 47′	0.4350	0.9004	0.4831	2.070	1.111	2.299
.46	26° 21′	.4439	.8961	.4954	2.018	1.116	2.253
.47	26° 56′	.4529	.8916	.5080	1.969	1.122	2.208
.48	27° 30′	.4618	.8870	.5206	1.921	1.127	2.166
.49	28° 05′	.4706	.8823	.5334	1.875	1.133	2.125

Table 6 / Trigonometric Functions of θ

(θ in radians)

θ Radians	θ Degrees	sin θ	cos θ	tan θ	cot θ	sec θ	csc θ
0.50	28° 39′	0.4794	0.8776	0.5463	1.830	1.139	2.086
.51	29° 13′	.4882	.8727	.5594	1.788	1.146	2.048
.52	29° 48′	.4969	.8678	.5726	1.747	1.152	2.013
$\frac{\pi}{6}$ → **.53**	30° 22′	.5055	.8628	.5859	1.707	1.159	1.978
.54	30° 56′	.5141	.8577	.5994	1.668	1.166	1.945
0.55	31° 31′	0.5227	0.8525	0.6131	1.631	1.173	1.913
.56	32° 05′	.5312	.8473	.6269	1.595	1.180	1.883
.57	32° 40′	.5396	.8419	.6410	1.560	1.188	1.853
.58	33° 14′	.5480	.8365	.6552	1.526	1.196	1.825
.59	33° 48′	.5564	.8309	.6696	1.494	1.203	1.797
0.60	34° 23′	0.5646	0.8253	0.6841	1.462	1.212	1.771
.61	34° 57′	.5729	.8196	.6989	1.431	1.220	1.746
.62	35° 31′	.5810	.8139	.7139	1.401	1.229	1.721
.63	36° 06′	.5891	.8080	.7291	1.372	1.238	1.697
.64	36° 40′	.5972	.8021	.7445	1.343	1.247	1.674
0.65	37° 15′	0.6052	0.7961	0.7602	1.315	1.256	1.652
.66	37° 49′	.6131	.7900	.7761	1.288	1.266	1.631
.67	38° 23′	.6210	.7838	.7923	1.262	1.276	1.610
.68	38° 58′	.6288	.7776	.8087	1.237	1.286	1.590
.69	39° 32′	.6365	.7712	.8253	1.212	1.297	1.571
0.70	40° 06′	0.6442	0.7648	0.8423	1.187	1.307	1.552
.71	40° 41′	.6518	.7584	.8595	1.163	1.319	1.534
.72	41° 15′	.6594	.7518	.8771	1.140	1.330	1.517
.73	41° 50′	.6669	.7452	.8949	1.117	1.342	1.500
.74	42° 24′	.6743	.7385	.9131	1.095	1.354	1.483
0.75	42° 58′	0.6816	0.7317	0.9316	1.073	1.367	1.467
.76	43° 33′	.6889	.7248	.9505	1.052	1.380	1.452
.77	44° 07′	.6961	.7179	.9697	1.031	1.393	1.437
.78	44° 41′	.7033	.7109	.9893	1.011	1.407	1.422
$\frac{\pi}{4}$ → **.79**	45° 16′	.7104	.7038	1.009	.9908	1.421	1.408
0.80	45° 50′	0.7174	0.6967	1.030	0.9712	1.435	1.394
.81	46° 25′	.7243	.6895	1.050	.9520	1.450	1.381
.82	46° 59′	.7311	.6822	1.072	.9331	1.466	1.368
.83	47° 33′	.7379	.6749	1.093	.9146	1.482	1.355
.84	48° 08′	.7446	.6675	1.116	.8964	1.498	1.343
0.85	48° 42′	0.7513	0.6600	1.138	0.8785	1.515	1.331
.86	49° 17′	.7578	.6524	1.162	.8609	1.533	1.320
.87	49° 51′	.7643	.6448	1.185	.8437	1.551	1.308
.88	50° 25′	.7707	.6372	1.210	.8267	1.569	1.297
.89	51° 00′	.7771	.6294	1.235	.8100	1.589	1.287
0.90	51° 34′	0.7833	0.6216	1.260	0.7936	1.609	1.277
.91	52° 08′	.7895	.6137	1.286	.7774	1.629	1.267
.92	52° 43′	.7956	.6058	1.313	.7615	1.651	1.257
.93	53° 17′	.8016	.5978	1.341	.7458	1.673	1.247
.94	53° 52′	.8076	.5898	1.369	.7303	1.696	1.238
0.95	54° 26′	0.8134	0.5817	1.398	0.7151	1.719	1.229
.96	55° 00′	.8192	.5735	1.428	.7001	1.744	1.221
.97	55° 35′	.8249	.5653	1.459	.6853	1.769	1.212
.98	56° 09′	.8305	.5570	1.491	.6707	1.795	1.204
.99	56° 43′	.8360	.5487	1.524	.6563	1.823	1.196

Table 6 / *Trigonometric Functions of θ*

(θ in radians)

θ Radians	θ Degrees	sin θ	cos θ	tan θ	cot θ	sec θ	csc θ
1.00	57° 18′	0.8415	0.5403	1.557	0.6421	1.851	1.188
1.01	57° 52′	.8468	.5319	1.592	.6281	1.880	1.181
1.02	58° 27′	.8521	.5234	1.628	.6142	1.911	1.174
1.03	59° 01′	.8573	.5148	1.665	.6005	1.942	1.166
1.04	59° 35′	.8624	.5062	1.704	.5870	1.975	1.160
$\frac{\pi}{3}$ →							
1.05	60° 10′	0.8674	0.4976	1.743	0.5736	2.010	1.153
1.06	60° 44′	.8724	.4889	1.784	.5604	2.046	1.146
1.07	61° 18′	.8772	.4801	1.827	.5473	2.083	1.140
1.08	61° 53′	.8820	.4713	1.871	.5344	2.122	1.134
1.09	62° 27′	.8866	.4625	1.917	.5216	2.162	1.128
1.10	63° 02′	0.8912	0.4536	1.965	0.5090	2.205	1.122
1.11	63° 36′	.8957	.4447	2.014	.4964	2.249	1.116
1.12	64° 10′	.9001	.4357	2.066	.4840	2.295	1.111
1.13	64° 45′	.9044	.4267	2.120	.4718	2.344	1.106
1.14	65° 19′	.9086	.4176	2.176	.4596	2.395	1.101
1.15	65° 53′	0.9128	0.4085	2.234	0.4475	2.448	1.096
1.16	66° 28′	.9168	.3993	2.296	.4356	2.504	1.091
1.17	67° 02′	.9208	.3902	2.360	.4237	2.563	1.086
1.18	67° 37′	.9246	.3809	2.428	.4120	2.625	1.082
1.19	68° 11′	.9284	.3717	2.498	.4003	2.691	1.077
1.20	68° 45′	0.9320	0.3624	2.572	0.3888	2.760	1.073
1.21	69° 20′	.9356	.3530	2.650	.3773	2.833	1.069
1.22	69° 54′	.9391	.3436	2.733	.3659	2.910	1.065
1.23	70° 28′	.9425	.3342	2.820	.3546	2.992	1.061
1.24	71° 03′	.9458	.3248	2.912	.3434	3.079	1.057
1.25	71° 37′	0.9490	0.3153	3.010	0.3323	3.171	1.054
1.26	72° 12′	.9521	.3058	3.113	.3212	3.270	1.050
1.27	72° 46′	.9551	.2963	3.224	.3102	3.375	1.047
1.28	73° 20′	.9580	.2867	3.341	.2993	3.488	1.044
1.29	73° 55′	.9608	.2771	3.467	.2884	3.609	1.041
1.30	74° 29′	0.9636	0.2675	3.602	0.2776	3.738	1.038
1.31	75° 03′	.9662	.2579	3.747	.2669	3.878	1.035
1.32	75° 38′	.9687	.2482	3.903	.2562	4.029	1.032
1.33	76° 12′	.9711	.2385	4.072	.2456	4.193	1.030
1.34	76° 47′	.9735	.2288	4.256	.2350	4.372	1.027
1.35	77° 21′	0.9757	0.2190	4.455	0.2245	4.566	1.025
1.36	77° 55′	.9779	.2092	4.673	.2140	4.779	1.023
1.37	78° 30′	.9799	.1994	4.913	.2035	5.014	1.021
1.38	79° 04′	.9819	.1896	5.177	.1931	5.273	1.018
1.39	79° 39′	.9837	.1798	5.471	.1828	5.561	1.017
1.40	80° 13′	0.9854	0.1700	5.798	0.1725	5.883	1.015
1.41	80° 47′	.9871	.1601	6.165	.1622	6.246	1.013
1.42	81° 22′	.9887	.1502	6.581	.1519	6.657	1.011
1.43	81° 56′	.9901	.1403	7.055	.1417	7.126	1.010
1.44	82° 30′	.9915	.1304	7.602	.1315	7.667	1.009
1.45	83° 05′	0.9927	0.1205	8.238	0.1214	8.299	1.007
1.46	83° 39′	.9939	.1106	8.989	.1113	9.044	1.006
1.47	84° 14′	.9949	.1006	9.887	.1011	9.938	1.005
1.48	84° 48′	.9959	.0907	10.98	.0911	11.03	1.004
1.49	85° 22′	.9967	.0807	12.35	.0810	12.39	1.003

Table 6 / Trigonometric Functions of θ

(θ in radians)

θ Radians	θ Degrees	sin θ	cos θ	tan θ	cot θ	sec θ	csc θ
1.50	85° 57′	0.9975	0.0707	14.10	0.0709	14.14	1.003
1.51	86° 31′	.9982	.0608	16.43	.0609	16.46	1.002
1.52	87° 05′	.9987	.0508	19.67	.0508	19.70	1.001
1.53	87° 40′	.9992	.0408	24.50	.0408	24.52	1.001
1.54	88° 14′	.9995	.0308	32.46	.0308	32.48	1.000
1.55	88° 49′	0.9998	0.0208	48.08	0.0208	48.09	1.000
1.56	89° 23′	.9999	.0108	92.62	.0108	92.63	1.000
1.57	89° 57′	1.000	.0008	1256	.0008	1256	1.000

$\frac{\pi}{2}$ →

Acknowledgments

Book designed by Ligature, Inc.
Technical Art by Precision Graphics
Illustrations by Gary Torrisi
Cover design by Ligature, Inc.
Cover and chapter opener photos by John Payne Photo, Ltd.;
Chin Wai-Lam/Image Bank/Chicago (background photo)

Photos

xiv Peter Chapman
xv Peter Chapman
14 Cleo Freelance Photography
25 NASA/Peter Arnold, Inc.
39 NASA
43 Friend/Denny/Stock Shop
51 Philip Jon Bailey/Picture Cube
64 Michael L. Abramson/Woodfin Camp
65 Nancy Sheehan
116 Russ Kinne/Comstock
133 Norman Owen Tomalin/Bruce Coleman
141 Gabe Palmer/After Image
151 Greg Mancuso/After Image
159 John Marshall
183 Bob McKeever/Tom Stack and Associates
200 Shostal Associates
221 Index Stock
243 Dave Bartuff/FPG
251 David Stoecklin/Stock Market
280 James F. Palka/Nawrocki
292 Manley Features/Shostal Associates
304 Coco McCoy/Rainbow
337 Norman Prince
344 Jeffrey W. Myers/FPG
356 Index Stock
362 John Payne
422 Hubertus Kanus/Shostal Associates
423 Steve Dunwell
425 NASA
436 Index Stock
472 Arjen Nerkaik/Stock Market
487 Jerome Wexler/Uniphoto
493 Fred Whitehead/Nawrocki
514 FPG
535 David Witbeck/Picture Cube
567 Jordan Coonrad/Light Images
578 Philip Jon Bailey/Stock Boston
609 Bob Daemmrich/Stock Boston
634 Eric Carle/Shostal Associates
673 Ron Jantz/Folio, Inc.
675 © Rand McNally, and Co., R.L.
717 Nancy A. Santullo/Stock Market
718 John Scheiber/Stock Market
733 Tom Stack and Associates
749 Shostal Associates
760 Marty Heitner/Taurus Photo
774 Dan McCoy/Rainbow
781 Bob Daemmrich

Introducing Explorations

The following sixteen pages provide you with activities for exploring various concepts of algebra. The activities give you a chance to discover for yourself some of the ideas presented in this textbook. They carefully lead you to interesting conclusions and applications; they will make some of the abstract concepts of algebra easier to understand.

Some of the questions in these activities are open-ended. They ask you to describe, explain, analyze, design, summarize, write, predict, check, generalize, and recognize patterns. Often there is more than one correct way to answer a question. You can work on these activities by yourself or in small groups. You will need to use the materials listed below.

With these explorations we hope you enjoy exploring algebra!

Use With Lessons	Titles	Materials	Pages
1-1	Exploring Irrational Numbers	compass, ruler, graph paper, string, scissors	832
2-1	Exploring Inequalities	calculator	833
3-8	Exploring Functions	graph paper	834
4-5, 4-6	Exploring Polynomial Factors	algebra tiles	835
5-7	Exploring Continued Fractions	calculator	836
6-2	Exploring Radicals	calculator	837
7-5	Exploring Quadratic Equations	computer or graphing calculator	838
8-1	Exploring Direct Variation	springs and weights	839
9-6	Exploring Circles and Ellipses	computer	840
10-2	Exploring Powers and Roots	computer or graphing calculator	841
11-7	Exploring Pascal's Triangle	none needed	842
12-2	Exploring Trigonometric Ratios	protractor, straightedge, metric ruler, scientific calculator	843
13-4	Exploring Sine Curves	ruler, graph paper	844
14-3	Exploring Polar Coordinate Equations	computer or graphing calculator	845
15-9	Exploring Probability with Experiments	coins, computer	846
16-3	Exploring Matrices in Geometry	graph paper	847

Purpose

To have students explore how geometric constructions and models can be used to locate some irrational numbers on the number line.

Materials

compass, ruler, graph paper, string, scissors

Background

From geometry, students should be familiar with constructions as well as with the Pythagorean theorem. However, you may want to briefly review these concepts.

For Discussion

Discuss how these activities suggest that each irrational number corresponds to one and only one point on the number line. Each construction or model determines one and only one length.

Additional Answers

1–4.

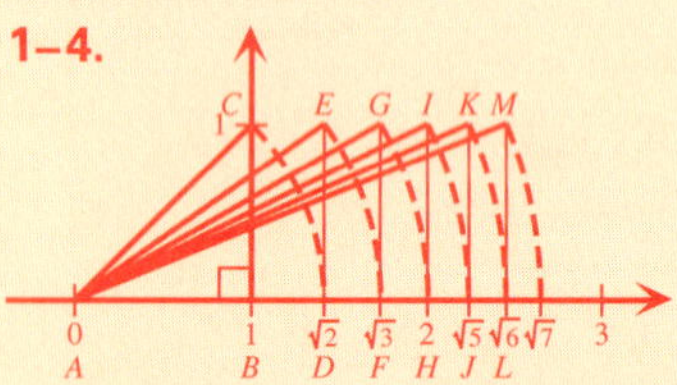

6. Continue the constructions done in Exercise **4** for one more triangle; draw an arc the length of $\sqrt{5}$ that crosses the number line to the left of 0.

7. Start with a circle with diameter 3 units to locate 3π, with diameter $\frac{1}{2}$ unit to locate $\frac{\pi}{2}$.

Explorations

Exploring Irrational Numbers

Use before Lesson 1-1

In this activity, you will explore locating irrational numbers on a number line.

Explore Square Roots

1. a. Draw a large number line on graph paper. Label point A at 0 and point B at 1. Draw $\overline{BC}$ perpendicular to the number line at B and having length 1. Draw $\overline{AC}$. What kind of triangle have you constructed? isosceles right triangle

b. Find the length of the legs and hypotenuse of $\triangle ABC$. $AB = 1$, $BC = 1$, $AC = \sqrt{2}$

2. a. Use a compass to locate point D on the number line with $AD = AC$. Point D should be to the right of point B. Draw $\overline{DE}$ perpendicular to the number line at D and having length 1. Draw $\overline{AE}$. Find the lengths of the legs and hypotenuse of $\triangle ADE$.

b. What is the coordinate of point D? Write it on your number line. $\sqrt{2}$

2a. $AD = \sqrt{2}$, $DE = 1$, $AE = \sqrt{3}$

3. a. Use a compass to locate point F on the number line with $AF = AE$. Point F should be to the right of point D. Draw $\overline{FG}$ perpendicular to the number line at F and having length 1. Draw $\overline{AG}$. Find the lengths of the legs and hypotenuse of $\triangle AFG$.

b. What is the coordinate of point F? Write it on your number line. $\sqrt{3}$

3a. $AF = \sqrt{3}$, $FG = 1$, $AG = \sqrt{4} = 2$

4. Continue this pattern, drawing three new triangles and locating square roots on the number line. Which irrational numbers have you located? $\sqrt{2}$, $\sqrt{3}$, $\sqrt{5}$, $\sqrt{6}$, $\sqrt{7}$

Explore π

5. a. Draw a large number line on graph paper. Cut a string to form a circle with diameter 1 unit, as shown at the right. Use the formula $C = \pi d$ to find the circumference of the circle. π

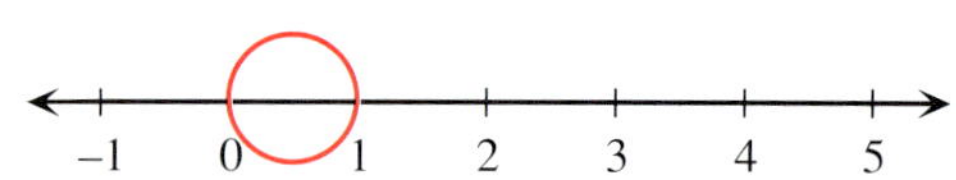

b. Now place the string on your number line, with one end at zero. At what point on the number line does the other end of the string lie? $\pi \approx 3.14$

Use What You Have Observed

6. Describe how you can locate $\sqrt{8}$ and $-\sqrt{5}$ on the number line by using geometric constructions.

7. Using the method of Exercise **5**, how can you locate 3π and $\frac{\pi}{2}$ on the number line?

Exploring Inequalities

Use with Lesson 2-1

In this activity, you will explore statements about inequalities. These statements may be false in general and yet be true for certain replacements of the variables.

Explore Multiplication

1. Investigate this statement: If $a < b$, then $ac < bc$. A calculator will be useful.

a. First choose a value for c such that $c > 0$. Choose various values for a and b so that you consider all possible cases. The table below will help you to organize your work. Make a table and fill in the values you choose.

a	b	c	ac	bc	$ac < bc$ true?
choose $a > 0$	choose $b > 0$				
choose $a < 0$	choose $b > 0$				
choose $a < 0$	choose $b < 0$				
$a = 0$					
	$b = 0$				

b. Next choose a value for c such that $c < 0$, and make a new table.

c. Now let $c = 0$. Is $ac < bc$ true? Why? No; multiplying by 0 results in a product of 0.

d. For which case(s) is the statement true? For which case(s) is it false?

2. Investigate this statement: If $a > b$ and $c > d$, then $ac > bd$.

a. Do you think this statement is always true? Answers will vary.

b. Choose values for a, b, c, and d until you find a case for which the statement is false.

Explore Reciprocals

3. Investigate this statement: If $a > b$, then $\frac{1}{a} < \frac{1}{b}$.

a. Choose $a > 0$ and $b > 0$. Does the statement appear to be always true? Yes

b. Choose $a < 0$ and $b < 0$. Does the statement appear to be always true? Yes

c. Find a case for which the statement is *not* true. $3 > -2$ and $\frac{1}{3} \not< -\frac{1}{2}$.

Use What You Have Observed

4. Decide whether this generalization is true or false: If two numbers are larger than two other numbers, then the product of the larger numbers is greater than the product of the smaller numbers. Explain your answer.
False. See Exercise **2.**

5. What is the error in this solution?

$$\frac{1}{x+2} > \frac{1}{2}$$
$$x + 2 < 2$$
$$x < 0$$

Invalid transformation. See Exercise **3c.**

Purpose

To have students discover the effect of multiplication and finding reciprocals when applied to inequalities. Use with Exercises **25–30** in Lesson 2-1.

Materials

calculators

Background

Students sometimes find difficulty in deciding if a statement is true or false when inequalities are given only in terms of variables. This activity gives students an opportunity to explore different cases by assigning values to the variables.

For Exercise **2,** display students' counterexamples on the board. Then as a class or in groups, have students look for patterns, categorize their results, and write summary statements about the cases for which this statement is false.

Calculators with reciprocal keys may be helpful in Exercise **3.**

Further Exploration

Students can also explore the following statements:

1. If $a > b$ and $b < 0$, then $a^2 > b^2$. True for $|b| < |a|$

2. If $a < b$ and $b > 0$, then $a^2 < b^2$. True for $|b| > |a|$

Additional Answers

1. d. If $a < b$ and $c > 0$, then $ac < bc$ is true. If $a < b$ and $c \le 0$, then $ac < bc$ is false.

2. b. Example: $a = -3$, $b = -4$, $c = 2$, $d = 1$; $-6 \not> -4$

Purpose

To have students explore step functions.

Background

The *rounding-down function* is also called the *floor function* or the *greatest integer function,* while the *rounding-up function* is called the *ceiling function.* Step functions are used in many applications, and they are an important topic of study in discrete mathematics.

Additional Answers

1. Divide the weight by 16 to change ounces to pounds, and determine what whole numbers of pounds the given weight is between.

3.

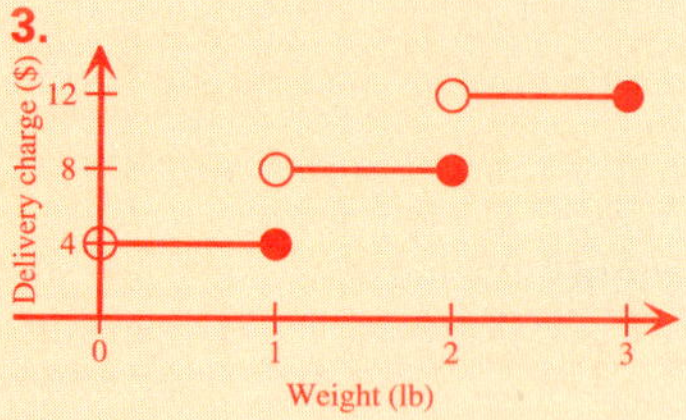

6.

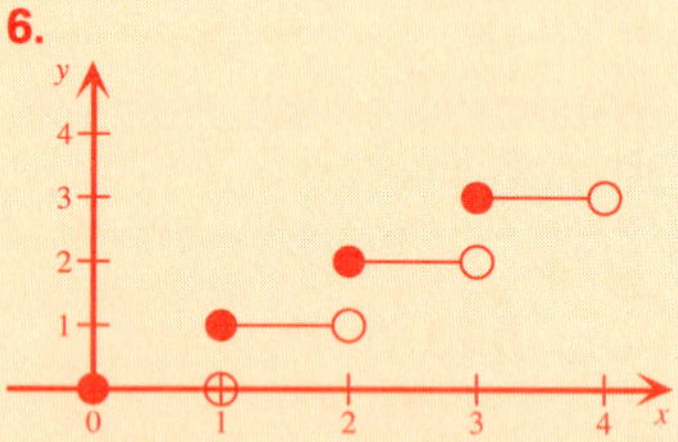

8. Given any number x, $f(x) = x$ if x is an integer; $f(x) =$ the next larger integer if x is not an integer.

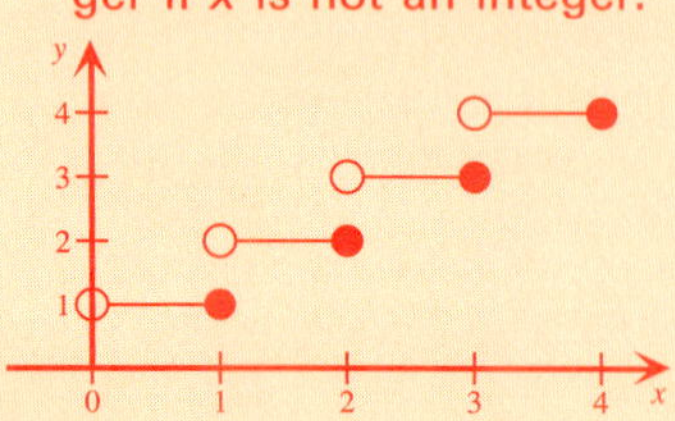

Exploring Functions

Use with Lesson 3-8

In this activity, you will explore properties and graphs of two special functions.

Explore a Real-Life Function

An overnight carrier service will deliver a package weighing 1 lb or less for $4. The delivery charge for a package weighing more than 1 lb but not more than 2 lb is $8. The charge for a package weighing more than 2 lb but not more than 3 lb is $12.

1. If the weight of a package is given in ounces, how can you determine the delivery charge for the package?

2. Give the delivery charge for packages of the following weights.

a. 1 oz; 2 oz; 3 oz; $5\frac{1}{2}$ oz; 8 oz; 14 oz; 16 oz   Each costs $4.

b. 17 oz; 20 oz; $26\frac{1}{2}$ oz; 30 oz; $31\frac{1}{2}$ oz; 32 oz   Each costs $8.

c. 33 oz; 40 oz; 42 oz; 45 oz; 46 oz; 48 oz   Each costs $12.

3. Draw a graph with weights shown on the horizontal axis and delivery charges shown on the vertical axis for all packages weighing from 0 lb through 3 lb.

4. Based on the results of Exercises **2** and **3,** is every weight assigned to exactly one delivery charge? Is the delivery charge a *function* of the weight of the package?   Yes; Yes

Explore a Rounding Function

5. The following rule describes a particular function.

Given any number x,
$f(x) = x$ if x is an integer
$f(x) =$ the next integer less than x if x is not an integer

Find $f(x)$ for each of the following values of x.

a. 2   2   **b.** $3\frac{1}{2}$ 3   **c.** $3.\overline{6}$ 3   **d.** $\sqrt{2}$ 1

6. Graph the function described in Exercise **5** for all x such that $0 \leq x < 4$.

Use What You Have Observed

7. Why might the two functions considered here be called *step* functions?
The graphs of these functions look like steps, rather than points, lines, or curves.

8. Refer to the rule for the rounding-down function given in Exercise **5.** Describe a similar rounding-up function and graph it for $0 < x \leq 4$.

Exploring Polynomial Factors

Use before Lessons 4-5 and 4-6

You can use square and rectangular tiles like those shown below to represent algebraic quantities. In this activity, you will use them to factor polynomials.

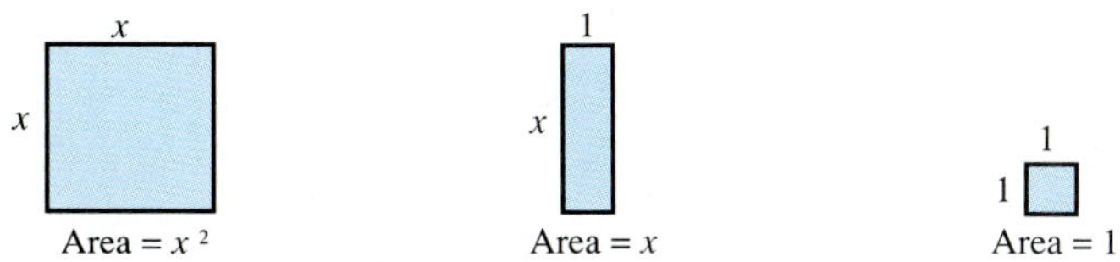

We will let blue tiles represent positive quantities (x^2, x, and 1), and let red tiles represent negative quantities ($-x^2$, $-x$, and -1).

Explore Finding Factors

1. Represent $3x^2 + 6x$ with three blue x-by-x tiles and six blue 1-by-x tiles as shown at the right.

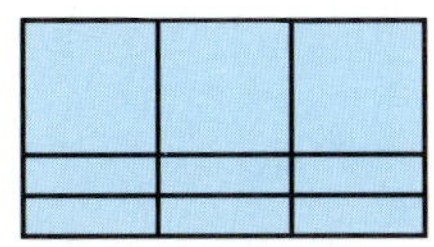

a. What are the lengths of the sides of the rectangle? $3x$, $x + 2$

b. How would you express $3x^2 + 6x$ in factored form? $3x(x + 2)$

2. Use tiles to factor $4x^2 + 8x + 3$.

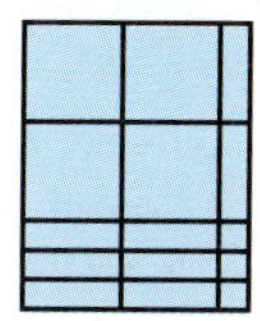

a. What tiles can you use to represent the given polynomial?

b. Arrange the tiles to form a rectangle as shown.

c. What are the lengths of the sides of the rectangle? $2x + 1$, $2x + 3$

d. What are the factors of $4x^2 + 8x + 3$? $2x + 1$, $2x + 3$

2a. 4 blue x-by-x tiles, 8 blue 1-by-x tiles, and 3 blue 1-by-1 tiles

3. Use tiles to factor $x^2 + x - 12$. **3a.** 1 blue x-by-x tile, 1 blue 1-by-x tile, and

a. What tiles can you use to represent the given polynomial? 12 red 1-by-1 tiles

b. Try to form a rectangle using only these tiles. Is it possible? No

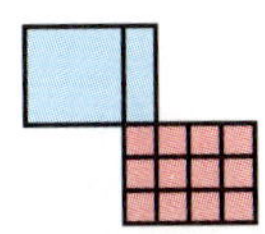

c. Arrange your tiles as shown at the right. To complete the rectangle, add 3 blue 1-by-x tiles along the top and 3 red 1-by-x tiles along the left side. Why can we do this without changing the value of the polynomial?

d. What are the lengths of the sides of the rectangle? $x + 4$, $x - 3$

e. What are the factors of $x^2 + x - 12$? $x + 4$, $x - 3$

3c. because adding $3x$ and $-3x$ to the polynomial is equivalent to adding zero

Use What You Have Observed

4. Use tiles to factor each polynomial.

a. $5x^2 - 10x$ $5x(x - 2)$ **b.** $x^2 + 9x + 14$ $(x + 7)(x + 2)$

c. $x^2 - 5x + 6$ $(x - 3)(x - 2)$ **d.** $3x^2 + 7x + 2$ $(x + 2)(3x + 1)$

e. $x^2 - 16$ $(x + 4)(x - 4)$ **f.** $x^2 - 2x + 1$ $(x - 1)^2$

Purpose

To use a physical geometric model in finding factors of certain polynomials.

Materials

Tiles in two colors as follows: square tiles of two dimensions—x by x and 1 by 1—and rectangular tiles of dimension 1 by x

Background

For Exercise **3b,** help students see that other arrangements of the 12 red 1-by-1 tiles are possible (such as a 2×6 rectangle), but the only arrangement that allows the correct combination of 1-by-x tiles is a 3×4 rectangle.

Cooperative Learning

Students can work in groups of 3 or 4, taking turns and/or discussing which tiles to choose for building the rectangles.

Further Exploration

1. Students can determine the polynomial (and its factors) represented by a given model. Example:

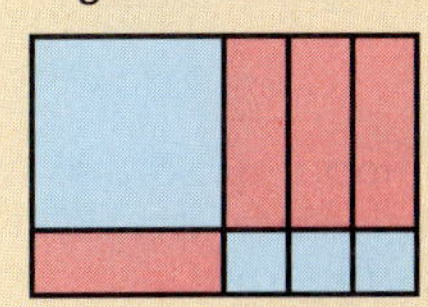

$x^2 - 4x + 3 = (x - 3)(x - 1)$

2. Students may think that all polynomials like those in the activity can be factored. Have them use tiles to try to factor the following.

a. $x^2 + 16$ not factorable

b. $x^2 + 2x + 2$ not factorable

Exploring Continued Fractions

Use with Lesson 5-7

The two expressions below are continued fractions. Example A is finite, or terminating; Example B is infinite, or nonterminating. Both are *simple* continued fractions because each numerator is 1. Nonsimple continued fractions have at least one numerator that is not 1. We will explore simple continued fractions.

Example A $2 + \cfrac{1}{3 + \cfrac{1}{4 + \cfrac{1}{2}}}$ **Example B** $1 + \cfrac{1}{2 + \cfrac{1}{2 + \cfrac{1}{2 + \cdots}}}$

Explore Finite Continued Fractions

1. Evaluate Example A above to find a simple fraction. $\frac{67}{29}$

2. Find a simple fraction for each continued fraction below.

a. $2 + \cfrac{1}{3 + \cfrac{1}{4}}$ $\frac{30}{13}$ **b.** $3 + \cfrac{1}{2 + \cfrac{1}{1 + \cfrac{1}{6 + \cfrac{1}{2}}}}$ $\frac{144}{43}$

3. A simple continued fraction can be represented as $[a_1, a_2, a_3, \ldots, a_n]$, where a_1 through a_{n-1} are the whole numbers preceding the plus signs and a_n is the last denominator. Example A above is [2, 3, 4, 2]. Write the two continued fractions in Exercise **2** in this form. [2, 3, 4]; [3, 2, 1, 6, 2]

4. Write each simple continued fraction below in expanded form and evaluate it.

a. [2, 1, 3, 4] **b.** [4, 2, 3]

a. $2 + \cfrac{1}{1 + \cfrac{1}{3 + \cfrac{1}{4}}} = \frac{47}{17}$ **b.** $4 + \cfrac{1}{2 + \cfrac{1}{3}} = \frac{31}{7}$

Explore Infinite Continued Fractions

5. Example B can be expressed as $[1, \overline{2}]$, which means [1, 2, 2, 2, ...]. Using a calculator, evaluate the following. Each is a number of terms of $[1, \overline{2}]$.

a. [1, 2] 1.5 **b.** [1, 2, 2] 1.4 **c.** [1, 2, 2, 2] $1.41\overline{6}$

d. [1, 2, 2, 2, 2] ≈ 1.4138 **e.** [1, 2, 2, 2, 2, 2] ≈ 1.4143

f. Analyze your answers in parts **a–e.** Each successive value is a closer approximation of what irrational number (square root)? $\sqrt{2}$

6. Evaluate the first several terms of each continued fraction below to find an irrational number the fraction appears to represent.

a. $[1, \overline{1, 2}]$ $1.732 \approx \sqrt{3}$ **b.** $[2, \overline{2, 4}]$ $2.449 \approx \sqrt{6}$

Use What You Have Observed

7. Based on these explorations, make a conjecture about the kind of numbers represented by finite continued fractions and infinite continued fractions. finite: rational numbers; infinite: irrational numbers

Purpose

To have students explore continued fractions.

Background

A continued fraction is an expression of the form

$$a_1 + \cfrac{b_1}{a_2 + \cfrac{b_2}{a_3 + \cfrac{b_3}{a_4 + \cdots}}}$$

The values of a_i and b_i are usually restricted to integers. If $b_i = 1$ and $a_i > 0$ (except a_1 may also equal 0), the fraction is called a simple continued fraction (SCF). Every finite SCF represents a rational number. Every infinite SCF represents an irrational number. If the infinite SCF is periodic, as are those considered here, the irrational it represents is a root of a quadratic equation.

Exercise **5** is related to Exercise **27,** page 240.

Further Exploration

Students familiar with the quadratic formula can explore the following method for finding the irrational number that a periodic SCF represents:

For $x = [2, \overline{2, 4}]$, Let $x = 2 + \frac{1}{y}$. Then $y = 2 + \cfrac{1}{4 + \cfrac{1}{y}}$

Simplify to find a quadratic equation: $4y^2 - 8y - 2 = 0$.

Find the positive root: $y = \frac{2 + \sqrt{6}}{2}$

Substitute for y in the first equation and solve for x.

$$x = 2 + \cfrac{1}{\cfrac{2 + \sqrt{6}}{2}} = \sqrt{6}$$

Exploring Radicals

Use before Lesson 6-2

In this activity, you will use a calculator to explore some properties of radicals.

Explore Addition and Subtraction

1. Use a calculator to evaluate each pair of expressions.

a. $\sqrt{64} + \sqrt{36}$ and $\sqrt{64 + 36}$ 14; 10 **b.** $\sqrt{15} - \sqrt{8}$ and $\sqrt{15 - 8}$ 1.045; 2.646

c. $\sqrt[3]{27} + \sqrt[3]{125}$ and $\sqrt[3]{27 + 125}$ 8; 5.337 **d.** $\sqrt[3]{27} - \sqrt[3]{64}$ and $\sqrt[3]{27 - 64}$ −1; −3.332

2. Based on the results of Exercise **1**, does each of the following generalizations appear to be true if $\sqrt[n]{a}$ and $\sqrt[n]{b}$ are any real numbers?

a. $\sqrt[n]{a} + \sqrt[n]{b} = \sqrt[n]{a + b}$ No **b.** $\sqrt[n]{a} - \sqrt[n]{b} = \sqrt[n]{a - b}$ No

Explore Multiplication and Division

3. Use a calculator to evaluate each pair of expressions.

a. $\sqrt{64} \cdot \sqrt{9}$ and $\sqrt{64 \cdot 9}$ 24; 24 **b.** $\dfrac{\sqrt{24}}{\sqrt{2}}$ and $\sqrt{\dfrac{24}{2}}$ 3.464; 3.464

c. $\sqrt[3]{7} \cdot \sqrt[3]{2}$ and $\sqrt[3]{7 \cdot 2}$ 2.410; 2.410 **d.** $\dfrac{\sqrt[3]{27}}{\sqrt[3]{8}}$ and $\sqrt[3]{\dfrac{27}{8}}$ 1.5; 1.5

4. Based on the results of Exercise **3**, does each of the following generalizations appear to be true if $\sqrt[n]{a}$ and $\sqrt[n]{b}$ are any real numbers?

a. $\sqrt[n]{a} \cdot \sqrt[n]{b} = \sqrt[n]{ab}$ Yes **b.** $\dfrac{\sqrt[n]{a}}{\sqrt[n]{b}}$ and $\sqrt[n]{\dfrac{a}{b}}$ if $b \neq 0$ Yes

Explore Roots and Powers

5. Use a calculator to evaluate each pair of expressions.

a. $\sqrt[3 \cdot 2]{64}$ and $\sqrt[3]{\sqrt{64}}$ 2; 2 **b.** $\sqrt[2 \cdot 3]{18}$ and $\sqrt{\sqrt[3]{18}}$ 1.619; 1.619 **c.** $\sqrt[3]{8^2}$ and $(\sqrt[3]{8})^2$ 4; 4 **d.** $\sqrt{2^3}$ and $(\sqrt{2})^3$ 2.828; 2.828

6. Based on the results of Exercise **5**, does each of the following generalizations appear to be true if each radical represents a real number?

a. $\sqrt[nq]{b} = \sqrt[n]{\sqrt[q]{b}}$ Yes **b.** $\sqrt[n]{b^m} = (\sqrt[n]{b})^m$ Yes

Use What You Have Observed

7. Mentally evaluate each of the following.

a. $\sqrt{9 + 16}$ 5 **b.** $\sqrt{100 - 64}$ 6 **c.** $\sqrt{16 \cdot 36}$ 24 **d.** $\sqrt[3]{\dfrac{125}{64}}$ d. $\dfrac{5}{4} = 1.25$

Purpose

To have students explore properties of radicals.

Materials

calculator

Background

If their calculators do not have a cube root key, students can use y^x where $x = 0.3333333$. Some calculators may give an error message for the second expression in **1d** because the radicand is negative. If so, students can use the fact that $\sqrt[3]{-37} = -\sqrt[3]{37}$ to evaluate the expression. You may want to have students verify several other cases similar to those in Exercises **3** and **5**.

Further Exploration

Students can explore fourth roots by finding square roots of square roots or using the y^x key with $x = 0.25$.

Purpose

To have students explore the relationship between values of a, h, and k in $y - k = a(x - h)^2$ and transformations of the graphs.

Materials

graphing calculator or computer with graphing software

Background

Most computer graphing software requires that an equation be input in the form $y = f(x)$. Therefore, an equation of the form $y - k = a(x - h)^2$ would have to be input as $y = a(x - h)^2 + k$.

In terms of geometric transformations, the graph of $y = a(x - h)^2 + k$ can be obtained from the graph of $y = x^2$ as follows.

1. Translate (slide) the graph $|h|$ units—to the right if $h > 0$, to the left if $h < 0$.
2. Translate the graph $|k|$ units up if $k > 0$, down if $k < 0$.
3. Dilate (stretch/shrink) vertically the graph by a factor of $|a|$. Also if $a < 0$ rotate the graph 180° about its vertex.

Additional Answers

6. a. Graph $y = 2x^2$. Then slide the graph 3 units right and 1 unit up.
   b. Graph $y = 2x^2$. Then slide the graph 7 units left and 3 units down.
7. a. It would be narrower.
   b. It would be wider, and it would open downward instead of upward.

Exploring Quadratic Equations

Use before Lesson 7-5

In this activity, you will use a graphing calculator or a computer with graphing software to explore equations of the form

$$y - k = a(x - h)^2.$$

The form above is equivalent to

$$y = a(x - h)^2 + k.$$

If you let $a = 1$, $h = 0$, and $k = 0$, the equation becomes

$$y = x^2.$$

The graph of this equation is shown at the right.

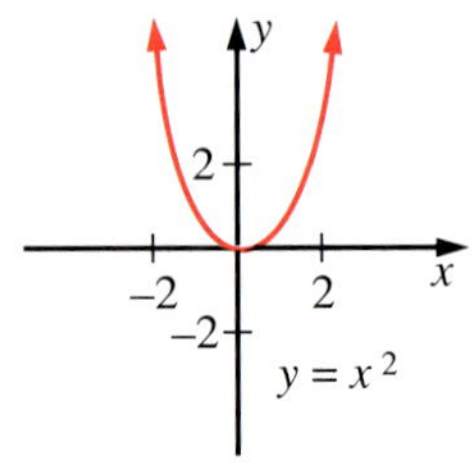

Explore Changes in the Value of a

1. Use the software to graph $y = ax^2$ for several values of a, with $a > 0$. For example, let $a = \frac{1}{4}, \frac{1}{3}, \frac{1}{2}$, 2, 3, and 4. What do you observe?
   For larger values of a, the graph is narrower.
2. Now graph $y = ax^2$ for several values of a, with $a < 0$.
   a. In which direction does the graph of $y = ax^2$ open if $a > 0$? If $a < 0$? upward, downward
   b. What can you conclude about the shape of the graph from the value of $|a|$? The shapes of all the graphs are in general similar, but for larger values of $|a|$, the graph is narrower.

Explore Changes in the Value of h

3. a. Graph $y = 2(x - h)^2$ for $h = 0$.
   b. Increase the value of h by 1. Graph this equation.
   c. Increase the value of h by 1 several more times. Graph each equation. Describe what you observe.
   The new graph is the previous graph slid 1 unit to the right.
4. Repeat Exercise **3,** but decrease the value of h. What do you observe?
   The new graph slides 1 unit to the left.

Explore Changes in the Value of k

5. a. Graph $y = 2(x - h)^2 + k$ for $h = 0$ and $k = 0$.
   b. Increase the value of k by 1 several times. Graph each equation.
   c. Decrease the value of k by 1 several times. Graph each equation.
   d. Describe what you observe. Increasing the value of k slides the graph up, decreasing it slides the graph down.

Use What You Have Observed

6. How can you obtain the graph of each equation from the graph of $y = 2x^2$?
   a. $y = 2(x - 3)^2 + 1$ b. $y = 2(x + 7)^2 - 3$
7. How would you expect the graph of each of the following to differ from the graph of $y = 2(x - 3)^2 + 1$?
   a. $y = 4(x - 3)^2 + 1$ b. $y = -\frac{1}{2}(x - 3)^2 + 1$

Exploring Direct Variation

Use before Lesson 8-1

In this activity, you will explore a special type of linear function, called a direct variation, by using springs and weights.

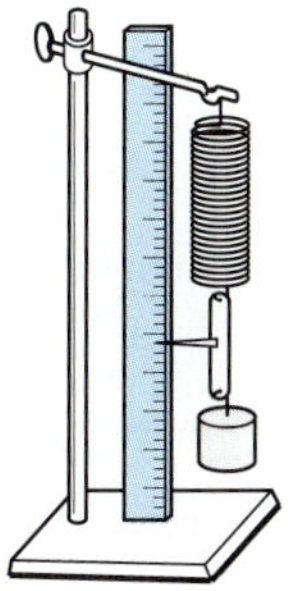

Explore with a Spring

1–4. Answers will vary.

1. Hang the first spring from the stand. Find its length to the nearest 0.1 cm.

2. Attach a weight to the spring. Record the weight to the nearest 0.01 kg. Measure and record the new length of the spring. Compute and record the difference between this length and the length you found in step **1.** (This is called the *stretch* of the spring.)

3. Repeat step **2** for two different weights.

Weight (kg)	Original Length (cm)	New Length (cm)	Stretch (cm)	Stretch ÷ Weight

4. Complete the table by finding stretch ÷ weight.

5. Write a statement about stretch ÷ weight for the spring. It is a constant.

Explore with More Springs

6. Repeat Exercises **1–5** for a second spring. Does your last answer change? No; a different quotient may result but the quotient is still a constant in each case.

7. Repeat Exercises **1–5** for a third spring. Does your last answer change? No; see Exercise **6.**

Use What You Have Observed

8. Use the data you recorded in the first table to predict the stretch for the first spring for a weight $1\frac{1}{2}$ times that of the heaviest weight used. Check by measuring. Answers will vary.

9. Exercise **8** required you to think of stretch (of the spring) as a function of weight. This function can be described by the equation $y = kx$. Refer to the table for the questions below.
   a. In which column are the values of x (the independent variable) located? first column
   b. In which column are the corresponding values of y (the dependent variable) located? fourth column
   c. Where in the table was the value of k determined? fifth column

Purpose

To have students discover the relationship of direct variation by using weights and springs.

Materials

three springs and stand, three weights (can be obtained from science lab), calculator

Background

Explain that the activity introduces a relationship, direct variation, used extensively in both math and science. Point out that they may already know of direct variation relationships intuitively, such as depth of water and water pressure.

Students should observe that the entry in the last column is a constant for each of the tables. Explain that this means that the length of stretch varies directly as the weight attached to the spring and that the constant in each case is the constant of direct variation.

Cooperative Learning

This activity is well-suited for small groups.

For Discussion

1. Will the direct variation relationship continue, as heavier and heavier weights are used? Explain. No; the capability of a spring to stretch is limited.
2. Why does the constant in the third column change from table to table? Different springs have different capability to stretch.

Exploring Circles and Ellipses

Use before Lesson 9-6

You have already seen in Lesson 9-2 that the following equation represents a circle with center (h, k) and radius r.

$$(x - h)^2 + (y - k)^2 = r^2$$

Graphing software makes it easy to explore such graphs. But you might have to graph the equation in the following two parts.

$$y = \sqrt{r^2 - (x - h)^2} + k \qquad y = -\sqrt{r^2 - (x - h)^2} + k$$

Explore Circles

1. Let $(h, k) = (3, 1)$ and $r = 2$. Using graphing software, graph the equation above. What do you observe? The graph is a circle with center (3, 1) and radius 2.

2. Choose several different sets of values for (h, k) and r and graph each equation. What do you observe? In each case, the center is at (h, k) and the radius is r.

Explore Ellipses

You have already seen that the following equation represents an ellipse with center (0, 0), horizontal axis of length $2a$, and vertical axis of length $2b$.

$$\frac{x^2}{a^2} + \frac{y^2}{b^2} = 1$$

Now you will graph equations of this form.

$$\frac{(x - h)^2}{a^2} + \frac{(y - k)^2}{b^2} = 1$$

You might have to graph this equation in the following two parts.

$$y = b\sqrt{1 - \frac{(x - h)^2}{a^2}} + k \qquad y = -b\sqrt{1 - \frac{(x - h)^2}{a^2}} + k$$

3. Let $(h, k) = (5, 2)$, $a = 3$, and $b = 4$. Graph the equation. What do you observe?

4. Choose several different sets of values of (h, k), a, and b and graph each equation. What do you observe?

Use What You Have Observed

5. From the graph of the unit circle $x^2 + y^2 = 1$, describe how you can obtain the graph of each equation below.

a. $x^2 + y^2 = 4$ **b.** $x^2 + y^2 = \frac{1}{16}$ **c.** $(x - (-2))^2 + (y - 4)^2 = 9$

d. $\frac{x^2}{4} + \frac{y^2}{9} = 1$ **e.** $\frac{(x - 1)^2}{4} + \frac{(y - 3)^2}{9} = 1$ **f.** $\frac{(x - 2)^2}{\frac{1}{4}} + \frac{(y - (-3))^2}{16} = 1$

Purpose

To have students explore equations and graphs of circles and ellipses. Use before Lesson 9-6.

Background

The first part of the activity reviews Lesson 9-2 and enables students to become familiar with using the graphing software to graph the kinds of equations being considered here. Exercise **5** requires students to think of geometric transformations in an informal way. They should readily grasp the concept that the unit circle $x^2 + y^2 = 1$ can be transformed to circles of various sizes and to ellipses by stretching or shrinking and by sliding.

Additional Answers

3. The graph is an ellipse with center (5, 2), horizontal minor axis of length 6 and vertical major axis of length 8.

4. Each graph is an ellipse with center (h, k) horizontal axis of length $2a$, and vertical axis of length $2b$.

5.
- **a.** double the radius
- **b.** divide the radius by 4
- **c.** triple the radius and slide the circle 2 units left and 4 units up
- **d.** Stretch the unit circle horizontally by a factor of 2 and vertically by a factor of 3.
- **e.** place center at (1, 3) and stretch as in part **d**
- **f.** place center at (2, −3), shrink horizontally by a factor of $\frac{1}{2}$, and stretch vertically by a factor of 4

Exploring Powers and Roots

Use before Lesson 10-2

In this activity, you will see how graphs can be used to explore and investigate powers and roots. You will need a graphing calculator or graphing software.

Explore Powers

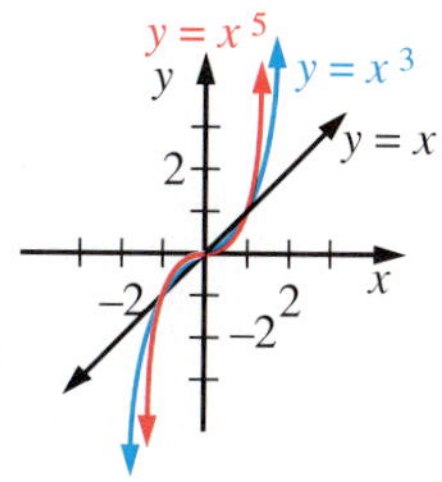

1. Use the software to graph $y = x$, $y = x^3$, and $y = x^5$. If the graphing software you are using can display more than one graph at the same time, your combined graph will look like the figure at the right.

a. In which quadrants do all three graphs lie? first and third

b. Which equation's graph is closest to the y-axis for $0 < x < 1$?

c. Which equation's graph is closest to the y-axis for $x > 1$? $y = x^5$

1. b. $y = x$

2. Think about the graphs $y = x^{13}$ and $y = x^{15}$. (Do not graph them yet.)

a. In which quadrants will these graphs lie? first and third

b. Which equation's graph will be closer to the y-axis for $0 < x < 1$? for $x > 1$? $y = x^{13}$; $y = x^{15}$

c. Graph the equations to see whether your answers to part **b** were correct.

d. Change the scale to examine the curves for $0 < x < 1$ and $0 < y < 0.001$. Then examine your curves for $1 < x < 1.001$ and $1 < y < 1.01$. Are your conclusions in part **b** verified in these magnified views? Yes

3. Graph the three equations $y = x^2$, $y = x^4$, and $y = x^6$.

a. In which quadrants do these graphs lie? first and second

b. Which equation's graph is closest to the y-axis for $0 < x < 1$? for $x > 1$? $y = x^2$; $y = x^6$

c. Which is larger, $(0.9)^2$ or $(0.9)^6$? Which is larger, $(1.01)^2$ or $(1.01)^6$? $(0.9)^2$; $(1.01)^6$

Explore Roots

4. Graph the equation $y = x^2 - 3$.

a. Between what pairs of integers does this graph cross the x-axis? 1 and 2, −1 and −2

b. What are the roots of the equation $x^2 - 3 = 0$? $\sqrt{3}, -\sqrt{3}$

c. What is your best (decimal) estimate of $\sqrt{3}$? (Hint: Estimate where the graph crosses the positive x-axis.)

d. Obtain a more precise estimate of $\sqrt{3}$ by using the zoom feature. approx. 1.732

Use What You Have Observed

5. For $a > 0$ where n and m are positive integers, what conditions are necessary for $a^m > a^n$? $a > 1$ and $m > n$ or $0 < a < 1$ and $m < n$

6. Use graphing software to find the root(s) of $x^2 + 4x + 1 = 0$ to three decimal places. −3.732; −0.268

Purpose

To lead students to discover numerical relations among powers and roots by exploring computer-generated graphs of power functions and quadratic functions. Use as an activity preceding Lesson 10-2.

Materials

Computer with graphing software or graphing calculator

Background

Graphing software that will display multiple graphs is best for Exercises **1–3.** Alternatively, printouts of the individual graphs may be used.

Most students understand that coordinate graphs connect algebra to geometry, but they need experiences like those provided by this activity to begin to fully appreciate how such mathematical connections can be used to gain information about one area of mathematics from another.

Further Exploration

Students can repeat Exercises **1–3** for $x < -1$ and $-1 < x < 0$. Then they should answer the following question: For $a < 0$, where m and n are positive integers, what conditions are necessary for $a^m > a^n$? $a < -1$ and $m < n$ for m and n odd and $m > n$ for m and n even or $-1 < a < 0$ and $m > n$ for m and n odd and $m < n$ for m and n even.

Purpose

To have students explore patterns in Pascal's triangle.

Background

Blaise Pascal (1623–1662), a French mathematician and philosopher, did not discover Pascal's triangle. However, he did discover many of its properties. The triangle first appeared in 1544 in the work of German mathematician Michael Stifel.

Students will use Pascal's triangle in Lesson 11-7 to generate coefficients for binomial expansions. Another surprising pattern is the connection of Pascal's triangle to combinatorics, namely, the $(r + 1)st$ element in the nth row is the number of combinations of n things taken r at a time, which is given by $\frac{n!}{r!(n-r)!}$.

For Discussion

If students are already familiar with probability, Pascal's triangle can be connected to that topic. Divide each element in row n by the sum of the elements in row n.

1	row 0
$\frac{1}{2}$ $\frac{1}{2}$	row 1
$\frac{1}{4}$ $\frac{1}{2}$ $\frac{1}{4}$	row 2
$\frac{1}{8}$ $\frac{3}{8}$ $\frac{3}{8}$ $\frac{1}{8}$	row 3
⋮	

Now each row provides the probabilities for any sample space equivalent to coin tossing experiments (binomial distributions). For example, row 2 above indicates that when 3 coins are tossed the probabilities are $\frac{1}{4}$ for 2 heads, $\frac{1}{2}$ for 1 head and 1 tail, and $\frac{1}{4}$ for 2 tails.

Exploring Pascal's Triangle

Use before Lesson 11-7

The following triangular array of numbers is called Pascal's triangle. Notice that, except for the 1's, each number in the triangle is the sum of the two numbers above it. For example, 15 = 10 + 5.

Triangle	
1	Row 0
1 1	1
1 2 1	2
1 3 3 1	3
1 4 6 4 1	4
1 5 10 10 5 1	5
1 6 15 20 15 6 1	6
⋮	

Explore Sets of Numbers

1. Complete Rows 7 and 8 of Pascal's triangle. Row 7: 1, 7, 21, 35, 35, 21, 7, 1; Row 8: 1, 8, 28, 56, 70, 56, 28, 8, 1
2. Find a diagonal in Pascal's triangle that consists of the set of positive integers. The second diagonal from the left (or right)
3. Find a diagonal that consists of the set of triangular numbers. The third diagonal from the left (or right) (1, 3, 6, 10, 15, ...)
4. Rearrange Pascal's triangle so the rows align on the left. Then find the sum along each diagonal as shown at right. What sequence of numbers is formed? Fibonacci sequence (1, 1, 2, 3, 5, 8, ...)

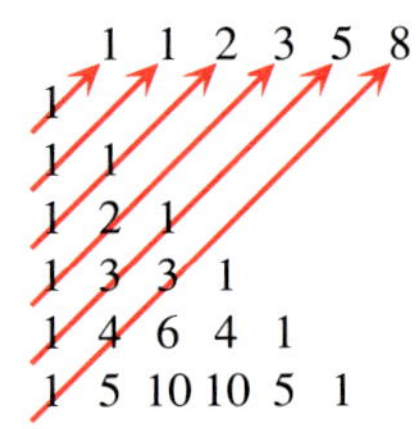

5. Find the sums along each row of Pascal's triangle. What sequence of numbers is formed? Express each number as a power of 2. $2^0, 2^1, 2^2, 2^3, 2^4, ..., 2^n$

Explore Patterns

6. Alternate – and + signs between elements in any row (except row 0). What is the sum? 0
7. Find the sum of the squares of all the elements in any row (except row 0). Can you find the resulting sum elsewhere in Pascal's triangle? Yes; it will always appear as the middle element of a later row.

Use What You Have Observed

8. Many patterns like those above have been discovered about Pascal's triangle, and new ones are still being discovered. Explore the triangle further to discover a pattern on your own. Answers will vary.

Exploring Trigonometric Ratios

Use before Lesson 12-2

In this activity, you will explore the trigonometric ratios tangent, sine, and cosine, abbreviated tan, sin, and cos. Each ratio is defined below in terms of the lengths of the sides of a right triangle.

$$\tan x = \frac{\text{opp.}}{\text{adj.}}$$
$$\sin x = \frac{\text{opp.}}{\text{hyp.}}$$
$$\cos x = \frac{\text{adj.}}{\text{hyp.}}$$

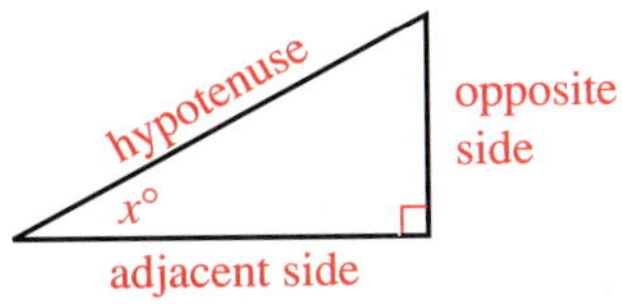

Explore the Ratios in Right Triangles
Answers will vary slightly.

1. Make a chart similar to the one below to record your results.

Right Triangle	opp. 30°	adj. 30°	hyp.	tan 30°	sin 30°	cos 30°
ABC						
ADE						
AFG						

2. Use a protractor and a straightedge to draw a 30° angle, keeping one side horizontal.

3. Create three right triangles so that the right angle is formed with the horizontal side, as shown. Make $\overline{AB}$ at least 15 cm long.

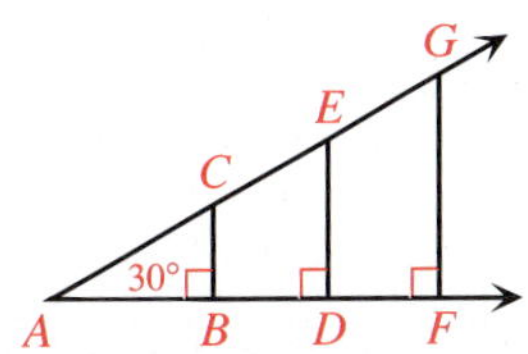

4. For each right triangle, measure to the nearest mm the side opposite the 30° angle, the side adjacent to the 30° angle, and the hypotenuse. Record the measurements in your chart.

5. Use a calculator to compute to the nearest hundredth the ratios for the last three columns of your chart. Record the results.

6. Use the trigonometric function keys on your calculator to find tan 30°, sin 30°, and cos 30°. Compare the values with those in your table.

7. Repeat Exercise **1–6** for angles of 45° and 60°.

Use What You Have Observed

8. For each of the last three columns in each chart, how do the calculated ratios compare? Why does this make sense?

Purpose

To develop the tangent, sine, and cosine ratios through the measurement of sides of various similar right triangles.

Materials

protractor, metric ruler, calculator with trigonometric functions

Background

In Lesson 12-2 students will be asked to show that the ratios $\frac{x}{r}$, $\frac{y}{r}$, and $\frac{x}{y}$ depend only on the angle θ. (See Ex. 32, p. 560.) In this activity students explore this functional relationship using several specific angle measures. They make the connection between the proportionality of the sides of similar triangles and the fact that the angle measure determines the trigonometric ratios.

Further Exploration

Square the values for sin x and cos x in each table, and add the two numbers. What are the results? Can you explain? Result is always 1. In a right triangle, $a^2 + b^2 = c^2$ if c is the length of the hypotenuse, and a and b are the lengths of the legs. Dividing by c^2 gives $\left(\frac{a}{c}\right)^2 + \left(\frac{b}{c}\right)^2 = 1$, which is $\sin^2 x + \cos^2 x = 1$.

Additional Answers

8. For each table, the ratios should be equal in each column. Since the triangles are similar, the sides are proportional.

Purpose

To have students explore graphical addition of sine functions and the connection of sine curves to light and sound.

Materials

compass

Background

Page 606 can be used to introduce this activity.

The graphical addition of y-values can be completed more easily by using a compass instead of a ruler. Make sure students realize that addition of a negative y-value requires a downward displacement.

For Exercise **6,** explain that in most cases, constructive and destructive interference occur simultaneously. The combined effect is the appearance of bands of light or beat frequencies.

Additional Answers

5. a.

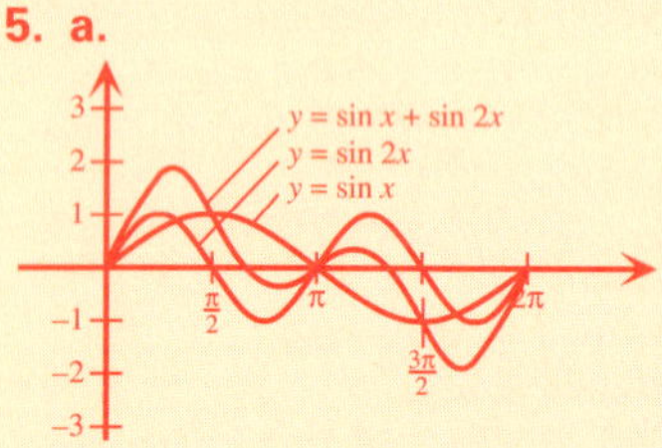

b.

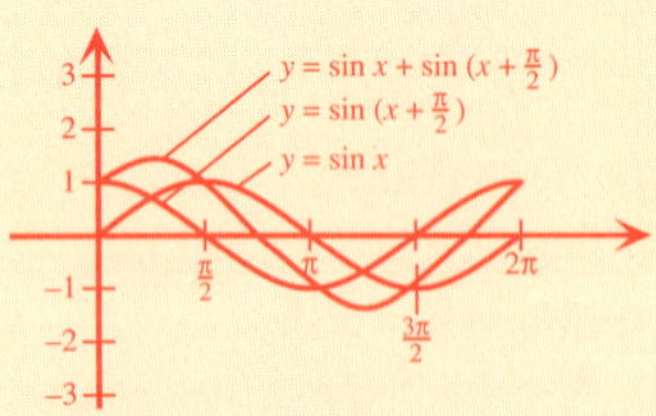

6. Destructive interference, which decreases the volume of sound or the brightness of light.

Exploring Sine Curves

Use with Lesson 13-4

Light and sound travel in waves that can be represented by sine curves. When waves of the same type travel in the same plane at the same time, the resultant light or sound is the "sum" of the waves. In this activity, you will explore finding the sum of two sine curves by using graphs and investigate the connection to light and sound.

Explore Addition

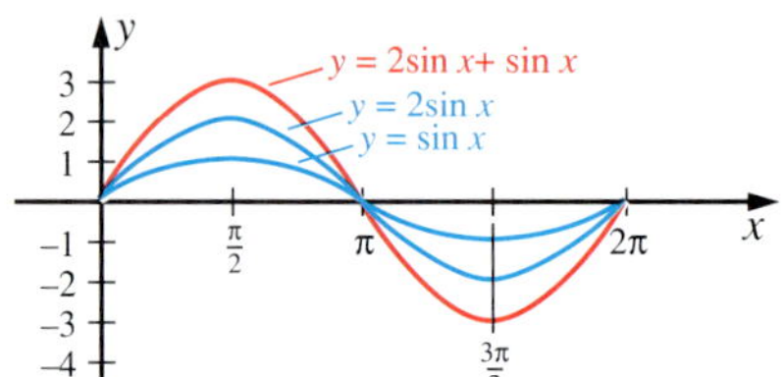

Figure 1

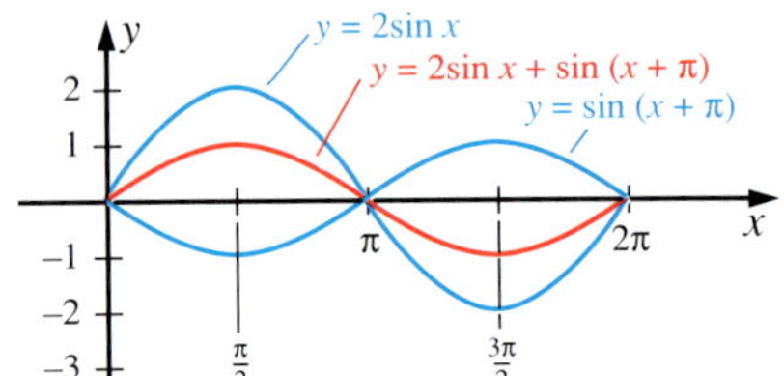

Figure 2

1. Using a ruler, measure the y-value at $x = \frac{\pi}{2}$ for $y = \sin x$ in Figure 1. By sliding the ruler upward add this y-value to the y-value for $y = 2 \sin x$. Notice that the resulting sum is the y-value for $y = 2 \sin x + \sin x$.

2. For several values of x between 0 and π in Figure 1, repeat the addition procedure described in Exercise **1.**

3. For values of x between π and 2π in Figure 1, how will the graphical addition procedure be different from that described in Exercise **1**? Why?
Slide the ruler down instead of up; a negative y-value is being added.

4. For several values of x in Figure 2, verify that the y-value of $y = 2 \sin x + \sin (x + \pi)$ is the sum of the y-values of $y = 2 \sin x$ and $y = \sin (x + \pi)$.

Use What You Have Observed

5. In each of the following, graph both equations (for $0 < x \leq 2\pi$) on the same set of axes. Then graphically find the sine curve that is the sum of the two curves.

a. $y = \sin x$; $y = \sin 2x$ **b.** $y = \sin x$; $y = \sin (x + \frac{\pi}{2})$

6. When light or sound waves are in phase, as are $y = 2 \sin x$ and $y = \sin x$ in Figure 1, the result is called *constructive interference*. Constructive interference increases the volume of sound or the brightness of light. In Figure 2 the waves represented by $y = \sin (x + \pi)$ and $y = 2 \sin x$ are out of phase. What do you think this condition is called and what might its effect be on light or sound?
Destructive interference, which decreases the volume of sound or the brightness of light.

Exploring Polar Coordinate Equations

Use with Lesson 14-3

In this activity, you will use a graphing calculator or a computer with graphing software to explore various types of polar equations.

Explore Equations of Lines

1. Graph $r \cos \theta = a$ for several positive values of a, for example, $a = \frac{1}{4}, \frac{1}{2}$, 1, 2, 4. The graph for $a = 1$ is shown below. What do you observe?
Each graph is a line perpendicular to the polar axis.

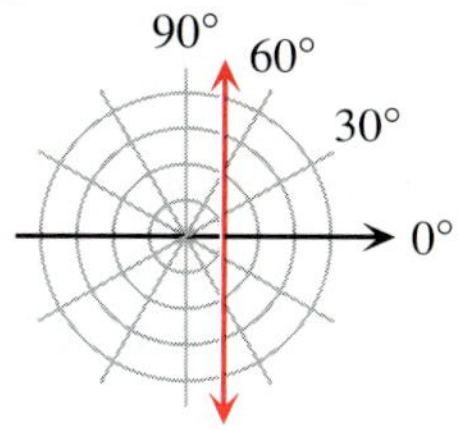

2. Graph $r \cos \theta = a$ for several negative values of a. What do you observe?
Each graph is a line perpendicular to the extension of (the ray opposite to) the polar axis.

3. Graph $r \sin \theta = a$ for several nonzero values of a. What do you observe?
Each graph is a line parallel to the polar axis, above it if $a > 0$ and below it if $a < 0$.

4. Graph $r = k$ for the following values of k.

a. $k = 1, 2, 3$ **b.** $k = -1, -2, -3$

c. Compare the graphs in part **a** with the graphs in part **b.** How are they related? They are the same.

a, b. Each graph is a circle with center at the pole and radius k.

5. Graph $p = r \cos (\theta - k)$ for several values of p and k, for example, $(p, k) =$ (1, 60°), (1, 45°), (1, 30°), (−2, 45°), and (−2, 30°). What do you observe?
Each graph is a line that passes through (p, k) but not through the pole.

Explore Other Polar Equations

6. Graph $r = 2a \cos \theta$ for several values of a. What do you observe? Each graph is a circle that passes through the pole, has radius $|a|$, and has its center on the polar axis or its extension.

7. Graph $r = 2a \sin \theta$ for several values of a. What do you observe? Each graph is a circle that passes through the pole, has radius $|a|$, and has its center on the 90° ray or its extension.

8. Graph $r = 2 \cos b\theta$ for $b = 2$ and $b = 3$. What do you observe? The first graph is a four-leafed rose, the second is a three-leafed rose.

Use What You Have Observed

9. Based on your results in Exercises **6–8,** what would you predict to be the graphs for $r = 2 \sin b\theta$ where $b = 2$ and $b = 3$?
Same as for **8** but rotated 90° counterclockwise.

Purpose

To have students explore graphing polar equations.

Materials

graphing calculator or computer with graphing software

Background

Ensure that students understand how to input data for polar coordinate graphs. Some software packages display polar graphs with the angle measures given in degrees so angle measures must be input in degrees. Yet for calculation of values of trigonometric functions, the angle measures must be converted to radians in the data table of the software.

Note that polar graphs and parametric equations are related as follows: If $r = f(\theta)$, then this curve is defined parametrically to be $x(t) = f(t) \cos t$ and $y(t) = f(t) \sin t$. These equations allow you to use parametric graphing utilities to graph polar equations.

For example, to graph Ex. **1,** first solve for r. $r = \frac{a}{\cos \theta}$. Let $x = r \cos \theta = \frac{a}{\cos \theta} \cdot \cos \theta = a$. Let $y = r \sin \theta = \frac{a}{\cos \theta} \cdot \sin \theta = a \tan \theta$. So, enter these equations as $x = a$ and $y = a \tan t$.

For Discussion

Regarding Ex. **4,** if your calculator or graphing program can display multiple graphs, do not graph parts **a** and **b** together. Ask students why this is important.

Purpose

To have students explore the relationship between theoretical and experimental probabilities.

Background

Students should discover that experimental probabilities only approximate theoretical probabilities. However, in theory, as the number of trials increases the approximations approach the theoretical values.

A computer with the programming language BASIC or a table of random digits can be used.

Different versions of BASIC employ the preprogrammed function RND(X) in different ways. In some versions the greatest integer function INT(X) may have to be used in conjunction with RND(X) to generate random digits. For example, if RND(X) generates a random decimal between 0 and 1, the following program will print 100 random digits 0–9.

```
10  For X = 1 TO 100
20  LET Y = RND(X)
30  PRINT INT(10 * Y)
40  NEXT X
50  END
```

Some versions of BASIC require the random number generator be "reseeded." The following line will do so for the program above.

```
15  RANDOMIZE TIMER
```

Cooperative Learning

This activity progresses from individual work to group work to class work to increase the number of experiments used to find probability.

Exploring Probability With Experiments

Use after Lesson 15-9

In theory, if a coin is tossed, the probability of "heads" is $\frac{1}{2}$. What happens in practice? You can actually toss a coin a number of times and record the results, or you can simulate a coin toss by using random digits.

Explore Tossing a Coin

1. a. Toss a coin 20 times and record your results for the number of heads and the number of tails. Based on your experiment, what is the probability of "heads"? Of "tails"? *Results will vary.*

b. Compare your results with a classmate. Are they exactly the same? Are they approximately the same? Should they be the same? *Results will vary.*

c. Combine your results with your classmate's results. Find the probability of "heads" based on these 40 coin tosses. Is it close to $\frac{1}{2}$? *Yes*

d. Combine the results of all the students in your class, and find the probability of "heads" based on these results. Is it even closer to $\frac{1}{2}$? *Yes*

e. Compare experimental probability with theoretical probability and try to answer questions like these. Which is more accurate? more realistic? more reliable? more valid? more believable? *Discussion will vary.*

Explore Simulating a Coin Toss

2. Use the random number function RND on a computer to generate a list of 100 random digits 0–9. Let each even digit represent "heads" and each odd digit represent "tails." How many heads resulted from this simulation? Is the probability of "heads" approximately $\frac{1}{2}$? *Approx. 50; Yes*

3. Repeat Exercise **2** by generating a new list of 100 random digits. Are the results exactly the same? Is the probability approximately $\frac{1}{2}$? *No; Yes*

Use What You Have Observed

If a baseball player has a batting average of .300, in theory, the player should get 3 hits in every 10 at bats. The player's probability of a hit is $\frac{300}{1000} = \frac{3}{10}$.

4. Design and describe a simulation, using random digits, for the next 100 at bats of a player whose batting average is .300. *Possible answer: Generate 100 random digits 0–9, let 0, 1, and 2 represent a hit and 3–9 represent an out.*

5. Carry out the simulation you designed in Exercise **4.** What is the probability of a hit? *Approx.* $\frac{3}{10}$

Exploring Matrices in Geometry

Use with Lesson 16-3

In this activity, you will use matrix multiplication to explore geometric transformations. You can use matrices to represent geometric figures such as points and triangles.

Point (−1, 2) Matrix $\begin{bmatrix} -1 \\ 2 \end{bmatrix}$

Triangle with vertices at (−1, 2), (0, 0), and (2, 1) Matrix $\begin{bmatrix} -1 & 0 & 2 \\ 2 & 0 & 1 \end{bmatrix}$

Explore Transforming Figures

1. Let $B = \begin{bmatrix} -1 & 0 & 2 \\ 2 & 0 & 1 \end{bmatrix}$ represent a triangle and $A = \begin{bmatrix} 1 & 0 \\ 0 & -1 \end{bmatrix}$ represent a geometric transformation. Draw B on graph paper. Find AB and draw the resulting triangle on the same axes. How are the two triangles related?
AB is the result of flipping (reflecting) B over the x-axis.

2. Find AB for a few more triangles, for example,

$$B = \begin{bmatrix} 2 & 3 & 1 \\ 3 & 5 & 2 \end{bmatrix}, B = \begin{bmatrix} -2 & -3 & -1 \\ -3 & -5 & -2 \end{bmatrix}, \text{ and } B = \begin{bmatrix} 0 & 0 & 3 \\ 2 & -1 & 1 \end{bmatrix}.$$

Does multiplying by A flip (reflect) each triangle over the x-axis? Yes

3. Now let $A = \begin{bmatrix} 0 & -1 \\ 1 & 0 \end{bmatrix}$ and let B represent matrices for triangles as before. Find AB for several triangles. Graph the triangles represented by B and AB. What kind of geometric transformation does A seem to be?
a turn (rotation) of 90° counterclockwise

4. Next we will represent figures and transformations in a slightly different way.

Let $A = \begin{bmatrix} 1 & 0 & 2 \\ 0 & 1 & -1 \\ 0 & 0 & 1 \end{bmatrix}$ represent a transformation and let $B = \begin{bmatrix} -1 & 0 & 2 \\ 2 & 0 & 1 \\ 1 & 1 & 1 \end{bmatrix}$ represent the same triangle as in Exercise **1.** (In both of these matrices, the third row is included only to make matrix multiplication possible.) Find AB. Graph B and AB. What kind of geometric transformation does A seem to be?
a slide (translation) 2 units to the right and 1 unit down

Use What You Have Observed

5. Describe the transformation each matrix below represents.

a. $A = \begin{bmatrix} -1 & 0 \\ 0 & 1 \end{bmatrix}$
a flip (reflection) over the y-axis

b. $A = \begin{bmatrix} -1 & 0 \\ 0 & -1 \end{bmatrix}$
a turn (rotation) of 180°

6. Write a transformation matrix for a slide of 2 units to the left and 3 units up.

Purpose

To have students explore an application of matrices. Use with Lesson 16-3 as an application of matrix multiplication to geometry.

Background

The transformations explored here (reflection, rotation, and translation) are all that is needed to fully characterize the congruence relation since these are the transformations that preserve distance and angle measures. It is also possible to represent dilations (size changes) with matrices. Size changes preserve angle measures but not distance.

The rotation matrices in Exercises **3** and **5b** are special cases. In order to generalize the rotation transformation, the sin and cos functions would have to be used in the elements of rotation matrices.

In Exercises **4** and **6,** make sure students understand that the third row in each matrix functions merely as a placeholder to make the matrices conformable for multiplication.

Further Exploration

Students can represent quadrilaterals with matrices having 4 columns, one for each vertex, and explore transforming these figures by matrix multiplication.

Additional Answers

6. $\begin{bmatrix} 1 & 0 & -2 \\ 0 & 1 & 3 \\ 0 & 0 & 1 \end{bmatrix}$

Portfolio Projects

In the side columns you will find a combination of hints, comments, answers, and ideas for extensions.

Chapter 1

If there were no constraints, the job would take 108 min, and two workers could complete the job in 54 min. With constraints, the job will take longer.

To visualize the precedence rules, students might draw a flowchart. Students might also find a timeline helpful in assigning the tasks according to the specified work rules.

1. Answers will vary. There is more than one way to complete the job in 61 min depending on how long workers are allowed to remain idle. For example, one worker does A, B, C, H, and J in this order with no breaks. The other does D, E, F, G, I, and K with an 8 min break before K.
2. Answers will vary. For example, allow workers to work together on a task, or to work on multiple tasks simultaneously.

Chapter 2

Answers will vary. If you assume that the ball is hit only when it is over home plate, the ball is within hitting range for only (0.41 s/726 in.) • 17 in. ≈ 0.01 s, since home plate is 17 in. long and the ball travels the 726 in. to the plate in 0.41 s. The batter can begin to swing up to almost 0.1 s late, or 0.41 s – 0.29 s = 0.12 s after the pitcher releases the ball.

Portfolio Projects

To make a portfolio, an artist selects a variety of original work to represent the range of his or her skills. Each of the following projects will give you a chance to create a finished product that you will be proud to add to your algebra portfolio.

The projects will help you develop your ability to present and communicate your ideas. They will also help you develop your problem-solving and reasoning abilities as you make connections between what you know and what is new. Your individual insight and creativity will help shape the mathematics you discover.

Let these projects be springboards for further exploration. Feel free to expand them to include new questions or areas of interest that arise. Most of all, have fun!

Efficiency Expert (Chapter 1)

Two warehouse workers do the tasks outlined in the table to assemble and pack a chest of drawers. Certain tasks must be done before others: A before B; B before C; D before E; C and E before I; C, F, G, and H before J; and K, of course, must be done last.

Task	Time
A. Install drawer supports on sides.	10 min
B. Attach sides to top and bottom.	15 min
C. Attach back panel to assembly.	8 min
D. Assemble door.	8 min
E. Attach hinges to door.	9 min
F. Assemble top drawer.	12 min
G. Assemble middle drawer.	12 min
H. Assemble bottom drawer.	12 min
I. Attach door.	6 min
J. Insert and adjust drawers.	10 min
K. Pack the completed chest.	6 min

1. Set up a work schedule that takes the least time. Use the following work rules to assign the tasks to the two workers:

   Rule 1. Work on only one task at a time.

   Rule 2. Take breaks only between tasks.

   Rule 3. Do not work together on any task.

2. Alter the rules so that the job takes less time. Develop a new work schedule.

Swing Time (Chapter 2)

In major league baseball, a fastball travels the distance (60 ft 6 in.) from the pitcher to the catcher in 0.41 s. On average, a batter connects with the ball 0.28 s after the start of the swing. Using these figures, estimate how long the ball is in range to be hit by the batter. In other words, by how much time can the batter be off in the start of the swing and still hit the ball? Draw a diagram and explain your reasoning.

Parametric Graphs (Chapter 3)

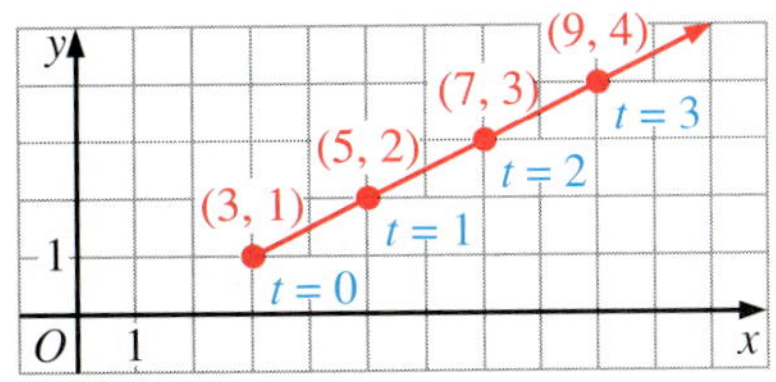

The graph shows the path of an object moving at a constant speed in a straight line. Each point on the graph represents the position of the object at a different time. Notice that the x-coordinate of the starting point is 3 and that the object moves 2 units horizontally in each unit of time. The *parametric equation* $x = 3 + 2t$ gives the object's x-coordinate in terms of time t, which is called a *parameter*.

1. What is the parametric equation for y?
2. How would the parametric equations change if the object moved twice as fast? Write the new equations.

Another object moving at a constant speed in a straight line has the parametric equations $x = 1 + t$ and $y = 10 - t$.

3. What is an equation of the path in terms of x and y?
4. Do the paths of the two objects ever cross? Do the objects ever meet? Explain.
5. To the nearest unit of time, determine when the two objects are closest together.

45° Paths (Chapter 4)

Materials: *Graph paper, straightedge*

Interesting patterns result from the study of *45° paths*. A 45° path originates at a vertex on the boundary of an m by n grid. The path continues in a straight line along diagonals until it hits the grid boundary. Then it "bounces off" at the same angle (45°) and continues until it either reaches a corner of the grid or returns to its starting point, whichever occurs first. A path that begins at a corner and ends at a corner is called a *corner-to-corner path*. A path that begins at a vertex that is not at a corner and then returns to its starting point is called a *closed path*. The length of any 45° path is the total number of grid squares through which it passes.

Corner-to-Corner Path

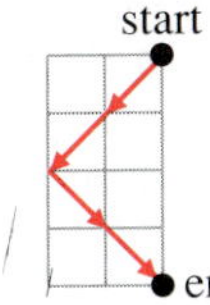

length = 4 units

Closed Path

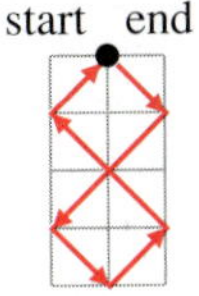

length = 8 units

1. **a.** Draw all corner-to-corner and all closed paths for each of the following grids: 3 by 6, 4 by 6, 4 by 8, 6 by 8, 6 by 9.
   b. For each grid, record the lengths of all corner-to-corner paths and all closed paths.
2. Look for a pattern in your answers to Exercise 1.
   a. Predict the lengths of all the corner-to-corner paths for any m by n grid.
   b. Predict which m by n grids have closed paths and what the lengths of the paths are.

Chapter 3

1. $y = 1 + t$
2. $x = 3 + 6t$; $y = 1 + 2t$
3. $x + y = 10$
4. The paths cross at (7, 3), but the objects do not meet because the first object arrives at the intersection point at time $t = 2$ while the second object arrives at time $t = 6$.
5. $t = 3$. To obtain this value, you might suggest that students use a ruler to measure distances on a scale drawing. Alternatively, you can show students how to express distance in terms of time: $d = \text{sqr}\{[(3 + 2t) - (1 + t)]^2 + [(1 + t) - (10 - t)]^2\} = \sqrt{5t^2 - 32t + 85}$. Using a graphing calculator, students can then find the lowest point on the graph.

Chapter 4

1. **b.** The table below gives the lengths of the paths for each grid.

Grid	Corner-to-corner	Closed
3 by 6	6	12
4 by 6	12	24
4 by 8	8	16
6 by 8	24	24
6 by 9	18	36

2. **a.** The length of a corner-to-corner path is the least common multiple of the grid dimensions.
   b. The length of a closed path is twice the length of a corner-to-corner path on the same grid. If the grid dimensions are relatively prime, no closed paths exist.

Travel Safety (Chapter 5)

Materials: *Almanac*

Which do you think is safer, traveling by car or by plane? Travel safety is calculated in two ways, as deaths per *passenger-mile* and deaths per *passenger-hour*. A **passenger-mile** is equal to one person traveling one mile. A **passenger-hour** is equal to one person traveling for one hour. For example, suppose a family of four travels 120 miles. This represents 4 • 120 = 480 passenger-miles. By car, at an average speed of 40 mi/h, the family's trip takes 3 h. By plane, at an average speed of 400 mi/h, it takes only 0.3 h. By car the trip represents 4 • 3 = 12 passenger-hours, but by plane it represents only 4 • 0.3 = 1.2 passenger-hours.

1. Use an almanac to find out how many deaths resulted from car accidents and how many from plane crashes in the United States in a recent year.
   a. Calculate the number of deaths per billion (10^9) passenger-miles for cars and for planes.
   b. Using the average speeds given above, calculate the number of deaths per billion passenger-hours by car and by air.
   c. Which method of travel seems to be safer? Explain.
   d. The airline industry claims that statistics show that air travel is much safer than driving. Do you agree? Which statistic do you think the industry uses?
2. Find recent statistics for travel by car, plane, bus, and train. Calculate the safety of each using passenger-miles and passenger-hours. Which seems to be the safest way to travel? Explain.

Square Roots of Complex Numbers (Chapter 6)

You can use a scientific calculator to find the square root of a real number. To find the square root of a complex number, you can use algebra. For example, to find the square root of $3 - 4i$, you will have to find $z = a + bi$ such that $z^2 = 3 - 4i$.

$$z^2 = 3 - 4i$$
$$(a + bi)^2 = 3 - 4i$$
$$a^2 - b^2 + 2abi = 3 - 4i$$
$$a^2 - b^2 = 3,\ 2ab = -4$$

1. Solve the pair of simultaneous equations $a^2 - b^2 = 3$ and $2ab = -4$. Use the results to find all values of z.
2. Write four different complex numbers. Use the demonstrated technique to find all their square roots.
3. Using your results in Exercise 2, draw a conclusion about how many square roots a complex number has and how the roots are related. How does this compare to real numbers?

Chapter 5

Note: The statistics used in this project can vary considerably from year to year.

1. The table below gives statistics for 1987.

	Passenger miles (billions)	Deaths
Car	2578.7	24,909
Plane	329.1	253
Bus	116.5	34
Train	12.1	16

   a. cars: 9.66; planes: 0.77
   b. cars: 386; planes: 308
   c. Planes are safer.
   d. Although both pairs of statistics from (a) and (b) support the airline industry's claim, the deaths per billion passenger-hours are almost equal for cars and planes. The airline industry may prefer to use deaths per billion passenger-miles.

2. The table below is derived from the table above and assumes average speeds of 40 mi/h for cars, 400 mi/h for planes, 20 mi/h for buses, and 50 mi/h for trains.

	Deaths per billion pass.-mi	Deaths per billion pass.-h
Car	9.66	386
Plane	0.77	308
Bus	0.29	6
Train	1.32	66

Chapter 6

1. $z = 2 - i$ or $z = -2 + i$
2. Answers will vary.
3. Complex numbers have two squares roots that are opposites. The same is true for real numbers.

Population Growth (Chapter 7)

Materials: *Calculator, graph paper*

Biologists who study how populations grow over time sometimes use mathematical models to help them. One such model is based on this quadratic equation:

$$y = x + rx(1 - x/C)$$

In the model, x represents the current size of the population, y represents the size of the next generation, and r and C are constants. While r is a growth rate (which measures the population's reproductive ability), C is the environment's carrying capacity (which is the maximum number of individuals that the population's environment can sustain). To see how the model works, let $r = 1$ and $C = 5000$. If the size of the population is $x = 1000$, then the size of the first generation is:

$$y = 1000 + 1 \cdot 1000 \cdot (1 - 1000/5000) = 1800$$

For the second generation, let $x = 1800$ and recalculate y:

$$y = 1800 + 1 \cdot 1800 \cdot (1 - 1800/5000) = 2952$$

The table and graph below show what happens to the population over several generations. Notice how the population approaches the carrying capacity.

Using a calculator, explore what happens to a population for other values of r, such as 1.5, 2, and 3. Always let $x = 1000$ initially, and keep $C = 5000$. Write a report (including tables and graphs) of your findings.

Generation	Population
0	1000
1	1800
2	2952
3	4161
4	4859
5	4996

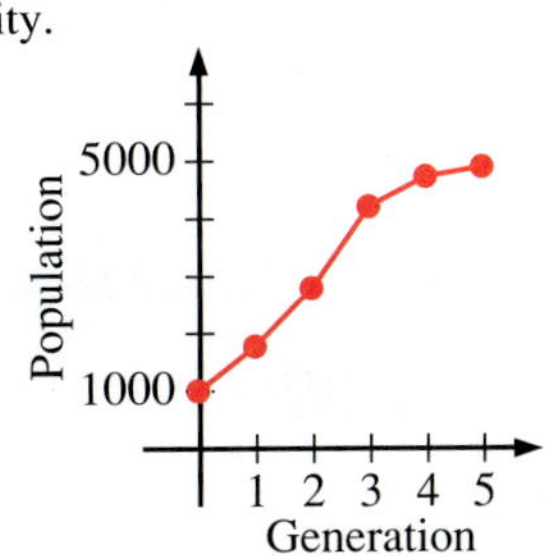

Fitting a Function (Chapter 8)

Using a simple "trick," you can find a polynomial function whose graph goes through any given set of points. For example, the graph of the following function goes through the points (1, 3), (2, 5), and (3, –1):

$$f(x) = 3 \cdot \frac{(x-2)(x-3)}{(1-2)(1-3)} + 5 \cdot \frac{(x-1)(x-3)}{(2-1)(2-3)} + (-1) \cdot \frac{(x-1)(x-2)}{(3-1)(3-2)}$$

1. Simplify $f(x)$. Verify by substitution that $f(1) = 3$, $f(2) = 5$, and $f(3) = -1$.
2. Explain why this function works.
3. Find a polynomial function g that satisfies $g(-2) = 1$, $g(0) = 3$, $g(1) = -2$, and $g(4) = 3$. Show your calculations and simplify the polynomial.
4. What conditions must be met by a set of points to make this "trick" work?

Chapter 7

As the values of r increase, the long-term behavior of the population becomes more chaotic. The table below shows three situations: (1) for $r = 1.5$, the population eventually approaches the carrying capacity; (2) for $r = 2$, the population cycles between two values, one larger and one smaller than the carrying capacity; and (3) for $r = 3$, the population changes erratically.

Successive populations for		
$r = 1.5$	$r = 2$	$r = 3$
1000	1000	1000
2200	2600	3400
4048	5096	6664
5204	4900	11
4885	5096	43
5053	4901	169
4972	5095	660
5014	4901	2378
4993	5095	6119
5003	4901	2011
4998	5095	5617

Chapter 8

1. $f(x) = -4x^2 + 14x - 7$; $f(1) = -4 \cdot 1 + 14 \cdot 1 - 7 = 3$; $f(2) = -4 \cdot 4 + 14 \cdot 2 - 7 = 5$; $f(3) = -4 \cdot 9 + 14 \cdot 3 - 7 = -1$
2. Consider what happens when $x = 1$: The function's last two terms, $5 \cdot \frac{(x-1)(x-3)}{(2-1)(2-3)}$ and $(-1) \cdot \frac{(x-1)(x-2)}{(3-1)(3-2)}$, have a value of 0, while the first term takes on the desired value of 3.
3. $g(x) = \dfrac{11x^3 - 25x^2 - 76x + 54}{18}$
4. No two points can have the same x-coordinate.

Chapter 9

2. Since P' is the reflection of P over the crease, the crease is the perpendicular bisector of the segment PP'. So every point on the crease is equidistant from P and P'. But only one of these points is equidistant from P and l (and therefore is on the parabola). That is the point directly above P'. To find this point, construct a perpendicular to l at P' and mark the point of intersection with the crease.
3. The parabola will widen.

Chapter 10

After step 2, three-quarters of a drop of food coloring is left. After step 3, three-quarters of three-quarters, or 0.5625, of a drop is left. The number of steps will vary. With dark red food coloring, the process may take 15 steps.

1. In step 1 the pollutant enters the pond. In steps 2 and 3 clean stream water replaces some of the polluted pond water. In step 4 so much stream water has replaced pond water that the pollution level is too low to measure.

   Some pollutant is always present, although the amount approaches zero over time.
2. The amount of pollutant will increase until it reaches an equilibrium value.

   To modify the simulation, remove one-quarter cup of the mixture and stir in one-quarter cup of water containing one drop of food coloring at each step.

Paper Folding a Parabola (Chapter 9)

Materials: *Wax paper*

1. a. Draw a horizontal line l near the bottom of a large piece of wax paper and mark a point P several inches above l.

   (a)

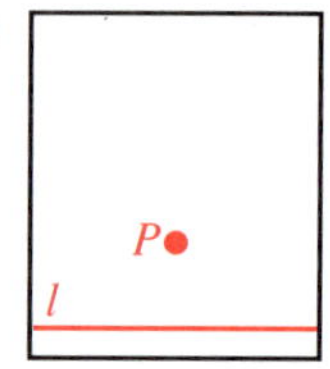

   (b)

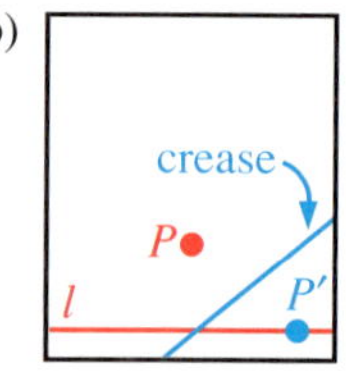

   b. Fold the paper so that P coincides with a point P' near the right end of l. Open the paper and notice the crease that has been formed.

   c. Fold the paper so that P coincides with another point on l, this time a little farther to the left than before. Continue folding the paper in this way until P has traveled all the way across l.

   (c)

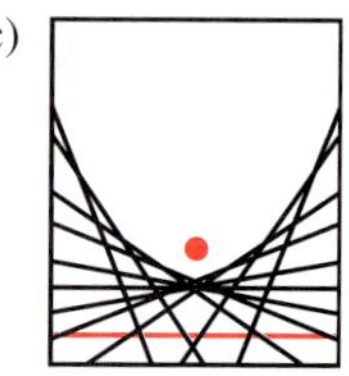

2. The shape outlined by the creases is a parabola, with each crease being tangent to the parabola. Referring to the diagram for step (b) of Exercise 1, describe how to locate the point on the crease that lies on the parabola with focus P and directrix l. Explain your reasoning.
3. Predict what will happen to the shape of the parabola if you increase the distance between P and l. Then, using another piece of wax paper, form a second parabola to confirm your prediction.

A Model of Exponential Decay (Chapter 10)

Materials: *Quarter-cup measure, beaker or clear jar, medicine dropper, food coloring*

A factory releases a pollutant into a pond. Suppose that each day one fourth of the pond water is replaced by fresh water from a stream that enters the pond on one side and leaves the pond on another side. Gradually, the pollutant is diluted and carried away. Use the following steps to simulate this situation.

Step 1 Start with one cup of water (representing the pond) in a beaker. Stir in one drop of food coloring (representing the pollutant).

Step 2 Remove one-quarter cup of the mixture and stir in one-quarter cup of clean water. How much of the food coloring is left?

Step 3 Remove another quarter cup of mixture and stir in another quarter cup of clean water. How much of the food coloring is left?

Step 4 Continue this process until the food coloring is no longer visible. How many steps does this take?

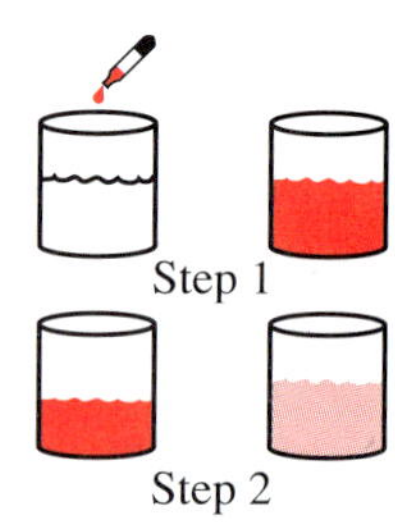

1. Interpret each step of the simulation in terms of the pollutant in the pond. Is the pollutant ever completely removed from the pond?
2. Suppose the factory releases the same amount of pollutant into the pond every day. What effect do you think this will have on the amount of pollutant present in the pond over time? Modify the simulation to test your conjecture.

The Weekly Allowance (Chapter 11)

Katerina has been negotiating with her parents for a raise in her weekly allowance. They offer her the following options:

(1) Beginning now, she can have a one-cent raise in her allowance each week.
(2) After one year, and at the end of each year thereafter, she can have a one-dollar raise in her weekly allowance.

Katerina must stick with the option she chooses for at least five years.

1. Which option should Katerina choose? Explain your reasoning.
2. Katerina bargains for a two-cent raise every week. If her parents agree, should she choose this option? Explain your reasoning.

Indirect Measurement (Chapter 12)

Materials: *Protractor, navigation compass*

Some measurements, such as the width of a pond or the height of a tree, are not easy to find with a ruler or tape measure. Since ancient times people have used trigonometry to find hard-to-obtain measurements *indirectly*.

Choose a local landmark, such as a pond, a tree, a tall statue, or a building, and then devise a method to measure its width or height using trigonometry. If possible, measure your chosen landmark in several different ways or from several different spots and compare the results that you get.

You may find some of the following measuring techniques helpful:

- To measure a distance over flat ground, determine the length of your stride and then pace off the distance you want to know.
- To find an angle of elevation, attach a weighted string to a protractor. Use the edge of the protractor to sight the top of the landmark. Be sure you understand that the angle of elevation is equal to the angle between the 90° mark on the protractor and the point where the weighted string crosses the protractor.

- To find an angle formed by your line of sight to two points on the same level, hold a compass in your hand while looking at one of the points. Note the compass bearing (the degree measure of your line of sight from north). Then turn to sight the other point and again note the compass bearing. Use the two bearings to find the angle through which you turned.

In an illustrated report, describe your method and present your data, your calculations, your results, and any conclusions you draw about their accuracy.

Get the Beat (Chapter 13)

A musical note can be modeled by a sine curve. The reciprocal of the curve's period is the note's *frequency* measured in *hertz* (Hz), the number of cycles per second.

When two notes with very nearly the same frequency are played together, the resulting tone seems to pulse, or *beat*, with the tone's volume repeatedly rising and falling. For example, beats occur if notes of frequency 8 Hz and 10 Hz are played together. The equation $y = \sin 16\pi x + \sin 20\pi x$ (x is in seconds) describes the resulting tone. (Note: Although humans cannot hear tones of such low frequency, these low frequencies will be easier to work with as an example.)

1. How are the numbers 16π and 20π related to the frequencies 8 Hz and 10 Hz?

You can use a graphing calculator or computer with graphing software to see the beats in the graph of the equation given above. Be sure to use radian mode.

2. Set the viewing window for $-0.25 \le x \le 1.25$ and $-2.5 \le y \le 2.5$. Graph the equation $y = \sin 16\pi x + \sin 20\pi x$. What do you think the graph's changing amplitude has to do with the tone's beat?

3. How many beats do you see in the graph of $y = \sin 16\pi x + \sin 20\pi x$ for $-0.25 \le x \le 1.25$? What is the frequency of the beats? (*Hint*: How many seconds are represented by this interval of the x-axis?) Make a conjecture about the relationship of the beat frequency and the frequencies of the two notes.

4. Use graphs to test your conjecture with other pairs of notes having frequencies that are close together, such as 12 Hz and 15 Hz, 20 Hz and 21 Hz, and 35 Hz and 39 Hz. (You may want to use an interval of one second on the x-axis.)

The Stunt (Chapter 14)

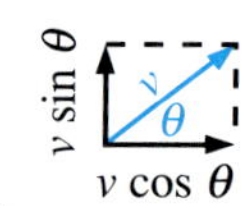

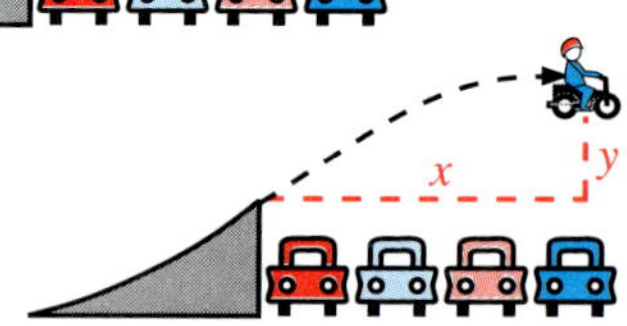

A stuntman on a motorcycle will attempt to launch himself off an angled ramp, pass over 20 cars parked side-by-side, and then land on a ramp on the other side. The x- and y-components of the launch vector (shown in blue) give the stuntman his horizontal and vertical motion. His vertical position, however, is also affected by gravity. If v is the stuntman's launch speed (in ft/s) and θ is the angle the ramp makes with the horizontal, then the stuntman's horizontal and vertical positions (in ft) t seconds after launch are:

$$x = (v \cos \theta)t \qquad \text{and} \qquad y = \underbrace{(v \sin \theta)t}_{\text{vertical motion from launch vector}} - 16t^2 \longleftarrow \text{effect of gravity on vertical motion}$$

The stuntman must decide at what angle to build the ramp and what launch speed to use. Find some combinations of ramp angles and launch speeds that allow the stuntman to complete his jump successfully. Assume each car is 6 ft wide.

Chapter 13

1. A frequency of 8 Hz means that the period is 1/8. Since the period of $y = \sin bx$ is $2\pi/b$, $1/8 = 2\pi/b$ and $b = 16\pi$. Likewise, a frequency of 10 Hz means $1/10 = 2\pi/b$ and $b = 20\pi$.
2. The amplitude of a sound wave determines the sound's volume. Since the tone's volume repeatedly changes from high to low, the tone is heard as a sequence of beats.
3. Three beats are shown. The frequency is 3 beats in 1.5 s, or 2 beats/s. The beat frequency equals the absolute value of the difference between the frequencies of the notes.
4. For 12 Hz and 15 Hz, the beat frequency is 3 beats/s. For 20 Hz and 21 Hz, the beat frequency is 1 beat/s. For 35 Hz and 39 Hz, the beat frequency is 4 beats/s.

As an extension to this project, you might have students explore the envelope that contains the graph of, say, $y = \sin 16\pi x + \sin 20\pi x$. For this graph, the envelope consists of two curves given by $y = \pm 2 \cos 2\pi x$.

Chapter 14

Note: Although students can perform the calculations for this project on a scientific calculator, they may enjoy using a graphing calculator in parametric mode more.

Answers will vary. One possible answer is a ramp angle of 15° and a launch speed of 88 ft/s (or 60 mi/h). In this case, the stuntman will be in the air about 1.4 s and land about 121 ft from his launching point.

Simpson's Paradox (Chapter 15)

A baseball player's batting average is given by the ratio $\frac{\text{number of hits}}{\text{number of times at bat}}$. The result is expressed as a decimal rounded to three places, although the decimal point is often omitted. For example, a ball player who had 64 hits in 400 times at bat has a batting average of $\frac{64}{400} = .160$, or 160.

Here are statistics for two baseball players for the months of May and June:

Player	May	June
Kapinsky	7 hits; 20 times at bat	28 hits; 100 times at bat
Mendoza	32 hits; 100 times at bat	5 hits; 20 times at bat

1. **a.** Find the batting average of each player for May and for June separately. Based on these averages, which player is the better hitter?
   b. Find the batting average of each player for the months of May and June combined. Based on these averages, which player is the better hitter?
2. The results of Exercise 1 are an example of a statistical phenomenon called *Simpson's paradox*. Find another set of statistics that displays the same behavior. Explain what it is about the data that causes Simpson's paradox.

Transition Matrices (Chapter 16)

In a certain metropolitan area, 300,000 people live in the city and 450,000 people live in the suburbs. Census data from the last 20 years show that each year 4% of those who live in the city move to the suburbs, while 2% of those who live in the suburbs move into the city. The two matrices shown below display this information. The right matrix, which shows the population shifts, is a *transition matrix*.

$$\text{Population}\ \begin{matrix} \text{City} & \text{Suburb} \\ [\,300{,}000 & 450{,}000\,] \end{matrix} = P_0 \qquad \text{From}\ \begin{matrix} \text{City} \\ \text{Suburb} \end{matrix} \overset{\begin{matrix}\text{To}\\ \text{City} \quad \text{Suburb}\end{matrix}}{\begin{bmatrix} 0.96 & 0.04 \\ 0.02 & 0.98 \end{bmatrix}} = T$$

2% of suburbanites move to the city. 98% of suburbanites stay in the suburbs.

1. Find the matrix product $P_1 = P_0T$, and verify that P_1 gives the city and suburban populations one year from now.
2. **a.** Find $P_2 = P_1T$, $P_3 = P_2T$, and $P_4 = P_3T$. What does each matrix represent?
   b. Use a calculator or computer with matrix capabilities to find $P_n = P_{n-1}T$ for $n = 5, 6, 7, \ldots$. What happens? Interpret the results.
3. Suppose 3% of the suburban dwellers move to the city each year. What effect does this have on the results of Exercise 2?

Chapter 15

1. The table shows that Kapinsky is the better hitter each month, but Mendoza is the better hitter overall.

Player	May	June	Overall
Kapinsky	.350	.280	.292
Mendoza	.320	.250	.308

2. Note that Mendoza had most of his at-bats when he was batting well (May), while Kapinsky had most of his at-bats when he was batting poorly (June).

Chapter 16

1. $P_1 = [297{,}000\ \ 453{,}000]$; this matrix is correct because the city population will be 96% of 300,000 plus 2% of 450,000, or 297,000, while the suburban population will be 98% of 450,000 plus 4% of 300,000, or 453,000.
2. **a.** $P_2 = [294{,}180\ \ 455{,}820]$; $P_3 = [291{,}529\ \ 458{,}471]$; $P_4 = [289{,}037\ \ 460{,}963]$; these matrices represent the city and suburban populations 2, 3, and 4 years from now.
   b. The city population decreases and the suburban population increases from year to year. The changes get smaller over time; after 200 years or so, the populations reach a steady state of 250,000 in the city and 500,000 in the suburbs.
3. Now the city population increases and the suburban population decreases over time. They eventually reach a steady state of 321,429 in the city and 428,571 in the suburbs.

Appendix

A-1 Common Logarithms: Notation and Interpolation

Objective To find logarithms and antilogarithms using interpolation.

You can use common logarithms to find approximations for complex algebraic expressions such as

$$\frac{\sqrt{93.8}}{0.0592 \times 3.67}$$

You may recall from Lesson 10-6 that every common logarithm has two parts, a *characteristic* and a *mantissa* (see page 479). To use Table 3 as an aid in computation, the mantissa must remain positive. Consider

$$\begin{aligned} \log 0.0217 &= \log (2.17 \times 10^{-2}) \\ &= \log 2.17 + \log 10^{-2} \\ &= 0.3365 - 2. \end{aligned}$$

To retain the positive mantissa, the characteristic -2 is usually written as $8.0000 - 10$, giving

$$\log 0.0217 = 8.3365 - 10.$$

In some cases it may be more convenient to use other differences such as $28.0000 - 30$ or $98.0000 - 100$ to represent -2. Thus, $\log 0.0217$ might be written as $28.3365 - 30$ or as $98.3365 - 100$.

Table 3 gives logarithms of numbers with three significant digits. By interpolating (see Lesson 8-9) you can find better approximations for (1) the logarithms of numbers with four significant digits and (2) the antilogarithms of numbers that fall between entries in the body of the table.

Example Find: **a.** $\log 26.83$ **b.** antilog $(6.7382 - 10)$

Solution **a.** Since $\log 26.83 = \log 2.683 + 1$, interpolate using Table 3 to find $\log 2.683$.

		x	$\log x$		
0.010		2.690	0.4298		0.0017
	0.003	2.683	log 2.683	d	
		2.680	0.4281		

$$\frac{d}{0.0017} = \frac{0.003}{0.010}$$

$$d = \frac{3}{10} \times 0.0017 \approx 0.0005$$

The difference d is rounded to four decimal places because mantissas in Table 3 are accurate to only four places.

Thus, $\log 2.683 \approx 0.4281 + 0.0005$
$= 0.4286$

$\therefore \log 26.83 \approx 1.4286$ ***Answer***

b. To find antilog $(6.7382 - 10)$, locate in the body of Table 3 the consecutive entries 7380 and 7390, between which the mantissa 7382 lies. Make note of the corresponding four-digit sequences 5.470 and 5.480 and interpolate as follows:

		x	$\log x$		
0.010		5.480	0.7388		0.0008
	c	antilog 0.7382	0.7382	0.0002	
		5.470	0.7380		

$$\frac{c}{0.010} = \frac{0.0002}{0.0008}$$
$$c = 0.0025 \approx 0.003$$

antilog $0.7382 \approx 5.470 + 0.003$
$= 5.473$

$\therefore$ antilog $(6.7382 - 10) \approx 5.473 \times 10^{-4}$
$= 0.0005473$ ***Answer***

Note that the value of c was rounded to one significant figure because reverse interpolation in a four-place table yields at most four significant digits for the number whose logarithm is given. With practice, much of the interpolation can be performed mentally.

Written Exercises

6. 8.5863 − 10
9. 9.6957 − 10

Find each logarithm using interpolation.

A

1. log 3.475 0.5410 **2.** log 8.612 0.9351 **3.** log 77.68 1.8903

4. log 239.6 2.3795 **5.** log 0.5624 9.7500 − 10 **6.** log 0.03857

7. log 68244 4.8341 **8.** log 138760 5.1423 **9.** log 0.4962

10. log 0.001634 7.2132 − 10 **11.** log 48560 4.6862 **12.** log 79820 4.9021

Find each antilogarithm to four significant digits using interpolation.

15. 24.23
18. 0.00001356

13. antilog 0.6968 4.975 **14.** antilog 0.7728 5.926 **15.** antilog 1.3843

16. antilog 3.8402 6922 **17.** antilog 8.9531 − 10 0.08976 **18.** antilog 5.1322 − 10

19. antilog 9.9296 − 10 0.8504 **20.** antilog 7.2388 − 10 0.001733 **21.** antilog 6.8041 − 10 0.0006370

A-2 Common Logarithms: Computation

Objective To use common logarithms to calculate products, quotients, powers, and roots.

For common logarithms, the first two laws stated on page 473 become

$$(1)\ \log p \times q = \log p + \log q \quad \text{and}$$
$$(2)\ \log p \div q = \log p - \log q.$$

When using logarithms in computations, you can use these laws to replace the operations of multiplication and division with addition and subtraction.

You can avoid making mistakes when computing with logarithms by arranging your work according to these rules:

1. Estimate the answer.
2. List numbers to be added or subtracted in vertical columns.
3. Align equality symbols and decimal points vertically.
4. Write all characteristics before looking up the mantissas.
5. Indicate the operations (+, −, ×) to be performed.
6. Label each step so that if you must check back, you will know what you are checking.

The purpose of making a preliminary estimate of the answer is to avoid the common error of misplacing the decimal point.

Example 1 below does not require interpolation. In general, examples and exercises *do* require interpolation and will be worked to four-digit accuracy.

Example 1 Simplify 0.0192×4370.

Solution

1. Let $N = 0.0192 \times 4370$.
2. Estimate the value of N.

   $0.0192 \approx 0.02$
   $4370 \approx 4000$

   $\therefore N \approx 0.02 \times 4000 = 80$ (estimate)

3. $$\begin{aligned}\log N &= \log (0.0192 \times 4370)\\ &= \log 0.0192 + \log 4370\end{aligned}$$

$$\begin{aligned}\log 0.0192 &= 8.2833 - 10\\ \log 4370 &= 3.6405 \qquad +\\ \hline \log N &= 11.9238 - 10\\ N &= \text{antilog } 1.9238 = 83.9\end{aligned}$$

$\therefore 0.0192 \times 4370 = 83.9$ ***Answer***

Note that the answer and the estimate agree to one significant digit.

Only common logarithms of numbers that are powers of 10 can be stated exactly. Thus, although 0.0192×4370 is only approximately equal to 83.9, the "=" sign will be used for simplicity.

Example 2 Simplify $\frac{-7.394}{91.47}$.

Solution

1. Since you know the answer will be negative, let $N = \frac{7.394}{91.47}$ to make it easier to compute with logarithms.
2. Estimate the value of N.
$$\frac{7.394}{91.47} \approx \frac{7}{9 \times 10^1} \approx 0.7 \times 10^{-1} = 0.07$$
3. $\log N = \log \frac{7.394}{91.47} = \log 7.394 - \log 91.47$

$$\begin{aligned} \log 7.394 &= 0.8688 && \text{(by interpolation)} \\ \underline{\log 91.47} &\underline{= 1.9613} \quad - && \text{(by interpolation)} \end{aligned}$$

To obtain a difference in which the mantissa is positive, write 0.8688 as $10.8688 - 10$.

$$\begin{aligned} \log 7.394 &= 10.8688 - 10 \\ \log 91.47 &= 1.9613 \qquad - \\ \hline \log N &= 8.9075 - 10 \\ N &= 0.08082 \quad \text{(by interpolation)} \end{aligned}$$

$\therefore \frac{-7.394}{91.47} = -0.08082$ ***Answer***

Check the answer against the estimate.

You can use the third law of logarithms, $\log p^n = n \log p$, to compute powers and roots.

Example 3 Simplify $(0.239)^7$.

Solution

1. Let $N = (0.239)^7$.
2. To estimate, round 3 factors up and 4 factors down.
$N \approx (0.2)^4(0.3)^3 = (0.0016)(0.027) = 0.0000432$ (estimate)
3. $\log N = 7 \log 0.239$

$$\begin{aligned} \log 0.239 &= 9.3784 - 10 \\ & 7 \quad \times \\ \hline \log N &= 65.6488 - 70 \\ &= 5.6488 - 10 \\ N &= \text{antilog}\,(5.6488 - 10) = 0.00004454 \end{aligned}$$

$\therefore (0.239)^7 = 0.00004454$ ***Answer***

Example 4 Simplify $\sqrt[3]{0.482}$.

Solution

1. Let $N = \sqrt[3]{0.482} = (0.482)^{1/3}$.
2. Estimate the value of N.

   $(0.7)^3 = 0.343; \quad (0.8)^3 = 0.512$

   $\therefore 0.7 < N < 0.8$
3. $\log N = \frac{1}{3}(0.482)$

   $\log 0.482 = 9.6830 - 10$

   To ensure that the negative part of the characteristic remains an integer after it is multiplied by $\frac{1}{3}$, write $9.6830 - 10$ as $29.6830 - 30$.

   $$\begin{array}{rl} \log 0.482 = 29.6830 - 30 & \\ \frac{1}{3} & \times \\ \hline \log N = 9.8943 - 10 & \end{array}$$

   $N = \text{antilog}\,(9.8943 - 10) = 0.7841$

$\therefore \sqrt[3]{0.482} = 0.7841$ ***Answer***

Example 5 applies the laws of logarithms to several operations at once.

Example 5 Simplify $\sqrt[5]{\frac{(81.2)^2}{0.59 \times 367}}$. Give the answer to four significant digits.

Solution

1. Estimate: $\sqrt[5]{\frac{(81.2)^2}{0.59 \times 367}} \approx \sqrt[5]{\frac{81^2}{6 \times 36}} = \sqrt[5]{\frac{3^8}{3^3 \cdot 2^3}}$

   $\approx \sqrt[5]{\frac{3^5}{2^3}} \approx \sqrt[5]{\frac{243}{8}} \approx \sqrt[5]{30} \approx \sqrt[5]{32} = 2$
2. Let $H = (81.2)^2$, $K = 0.59 \times 367$, and $N = \left(\frac{H}{K}\right)^{1/5}$.

   Then $\log N = \frac{1}{5}(\log H - \log K)$.

   $\log H = 2 \log 81.2$

   $$\begin{array}{rl} \log 81.2 = 1.9096 & \\ 2 & \times \\ \hline \log H = 3.8192 & \end{array}$$

   $\log K = \log 0.59 + \log 367$

   $$\begin{array}{rl} \log 0.59 = 9.7709 - 10 & \\ \log 367 = 2.5647 & + \\ \hline \log K = 2.3356 & \end{array}$$

   $$\begin{array}{rl} \log H = 3.8192 & \\ \log K = 2.3356 & - \\ \hline \log H - \log K = 1.4836 & \\ \frac{1}{5} & \times \\ \hline \log N = 0.2967 & \end{array}$$

   antilog $0.2967 = 1.980$

$\therefore \sqrt[5]{\frac{(81.2)^2}{0.59 \times 367}} = 1.980$ ***Answer***

Written Exercises

Given that $\log A = 3.6100$, $\log B = 8.1234 - 10$, and $\log C = 1.2000$, find the following.

A 1. $\log AC$ 4.8100 2. $\log BC$ 9.3234 − 10 3. $\log \frac{A}{B}$ 5.4866 4. $\log \frac{A}{C}$ 2.4100

5. $\log 100B$ 0.1234 6. $\log 0.01A$ 1.6100 7. $\log \frac{1000}{C}$ 1.8000 8. $\log \frac{0.001}{B}$ 8.8766 − 10

Given that $\log A = 13.600$, $\log B = 8.1310 - 10$, and $\log C = 3.2148$, find the following.

9. $\log B^2$ 6.2620 − 10 10. $\log C^{10}$ 32.148 11. $\log \sqrt{C}$ 1.6074 12. $\log \sqrt[3]{A}$ 4.5333

13. $\log C^{3/4}$ 2.4111 14. $\log \sqrt[3]{B^2}$ 8.7540 − 10 15. $\log \frac{\sqrt{A}}{\sqrt[3]{C}}$ 5.7284 16. $\log \frac{B}{\sqrt{A}}$ 1.331 − 10

Use Table 3 to evaluate each expression to three significant digits. Do not interpolate.

17. 281×0.94 264 18. 0.749×0.562 0.421 19. 943×804 758,000 20. 0.321×8.35 2.68

21. $\frac{117}{3.26}$ 35.9 22. $\frac{4960}{54,600}$ 0.0908 23. $\frac{0.013}{427}$ 0.0000304 24. $\frac{0.526}{0.049}$ 10.7

25. $(2.81)^{10}$ 30,700 26. $(90.3)^5$ 6,000,000,000 27. $(0.395)^7$ 0.00150 28. $(0.0431)^2$

29. $(21.4)^{1/3}$ 2.78 30. $(8.26)^{2/3}$ 4.09 31. $\sqrt{12.4}$ 3.52 32. $\sqrt[3]{1.36}$ 1.11

28. 0.00186

Use Table 3 to evaluate each expression to four significant digits. Interpolate when necessary. 41. 0.7757 43. 48,270,000,000,000

B 33. 0.572×6370 3643 34. 23.64×1.47 34.75 35. 0.8215×0.051 0.04190

36. $\frac{0.1496}{0.582}$ 0.2570 37. $\frac{137}{5242}$ 0.02614 38. $\frac{0.05274}{0.9412}$ 0.05603

39. $2.16 \times 46.73 \times 134.5$ 13,580 40. $94 \times 0.05 \times 1.728$ 8.120 41. $1.8 \times 32 \times 0.01347$

42. $(12.4)^3(3.86)^2$ 28,410 43. $(37.4)^5(812)^2$ 44. $(42.3)^4(0.0016)^2$ 8.192

45. $\sqrt{\frac{427}{0.592}}$ 26.86 46. $\sqrt[3]{\frac{(2.37)^2}{1.15}}$ 1.697 47. $\sqrt{\frac{(2.03)^3}{97.5 \times 1.98}}$ 0.2082

C 48. $\sqrt[5]{\frac{(527.4)^2}{1542 \times (0.2592)^2}}$ 4.851 49. $\sqrt{\frac{12,560^3}{79.21 \times 8004}}$ 1768

50. $92.37^{0.562}$ 12.73 51. $0.3142^{1.273}$ 0.2291

52. Using $\pi \approx 3.142$, find V when $V = \frac{4}{3}\pi r^3$ and $r = 0.1526$. 0.01488

53. Determine the number of digits in 3^{51} without using a calculator. 25

A-3 Discrete Mathematics

Objective To explore some topics in discrete mathematics.

We frequently associate the study of mathematics with people who intend to pursue careers in mathematics, engineering, science, and teaching. People in other disciplines such as social science and business management are now turning to mathematical models to solve problems that arise in their respective careers. Many of these problems involve counting a finite number of objects, completing a process in a finite number of steps, drawing flow charts or diagrams to represent choices that must be made, or determining by a logical method the most efficient solution of a problem. These and other skills have been grouped under the name *discrete mathematics*.

You have already studied some discrete mathematics topics, for example, functions and relations (Lessons 3-8 and 3-10), counting techniques (Lessons 15-5 through 15-7), and algorithm analysis (page 215).

Mathematical induction, a method of proof by which statements involving the set of natural numbers can be established, is an important topic in discrete mathematics (refer to page 523 for an explanation of mathematical induction). To illustrate, we pose and solve the chessboard problem. Is it possible to cover with triominoes a chessboard having 2^n squares on a side and one square removed? A triomino is a set of three squares in the shape of an "el." We will show that it is possible and that the number of triominoes needed is $\frac{1}{3}(2^{2n} - 1)$. Note that if it is possible to cover the board, then $2^n \cdot 2^n - 1$ squares must be covered by 3-square triominoes. That is, it would take exactly $\frac{1}{3}(2^{2n} - 1)$ triominoes to do it.

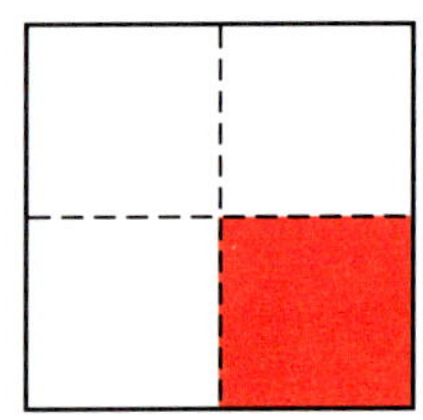

We will prove what is said above by mathematical induction. To begin, let $n = 1$. We have a 2×2 chessboard that can be covered by 1 triomino as shown in the figure at the right. Assume that it is possible to cover a board with dimensions $2^k \times 2^k$. We must prove that the number of triominoes needed for a board with dimensions $2^{k+1} \times 2^{k+1}$ is $\frac{1}{3}(2^{2(k+1)} - 1)$.

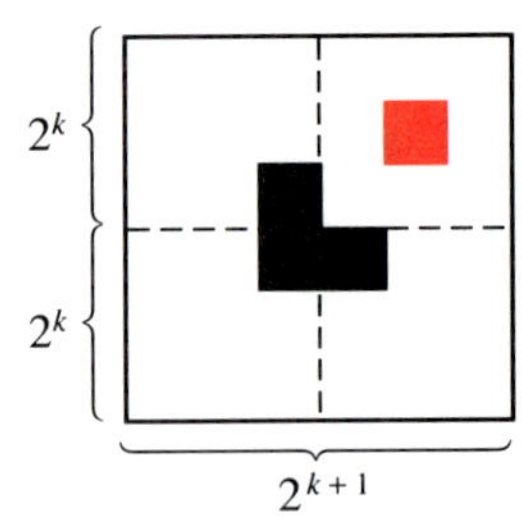

We divide the board into four congruent squares, each with dimensions $2^k \times 2^k$. We place 1 triomino in the center of the board as shown. Each of the four squares is now missing one square. Thus, by the induction hypothesis, the total number of triominoes needed is:

$$\begin{aligned} 4[\tfrac{1}{3}(2^{2k} - 1)] + 1 &= \tfrac{1}{3}[4 \cdot 2^{2k} - 4] + 1 \\ &= \tfrac{1}{3}[(2^2 \cdot 2^{2k} - 1) - 3] + 1 \\ &= \tfrac{1}{3}[2^{2k+2} - 1] - 1 + 1 \\ &= \tfrac{1}{3}[2^{2(k+1)} - 1] \end{aligned}$$

We have shown that the statement is true for $n = 1$ and that if it is true for a natural number k, it is also true for $k + 1$. Thus, by the principle of mathematical induction, the statement is true for all natural numbers.

Another interesting topic in discrete mathematics is *graph theory*. The map in Figure 1 shows seven bridges across the River Pregel in the 18th century city of Konigsberg. Citizens wondered if it was possible to take a walk in such a way as to cross each bridge exactly once.

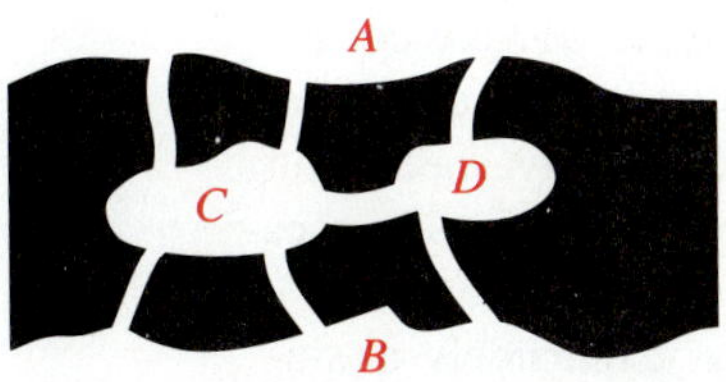

Figure 1

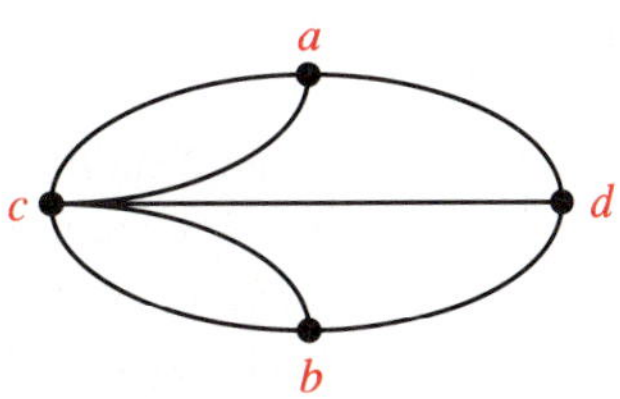

Figure 2

In 1736 the Swiss mathematician Leonard Euler visited the city and showed that such a walk was not possible. He replaced the actual configuration with a graph as in Figure 2. The land masses A, B, C, and D were represented by vertices a, b, c, and d. The seven bridges were represented by the seven edges. The question then became: Is there a path P in the graph that traverses each edge exactly once?

We can show by contradiction that there is no such P. To do so, we call a vertex even (odd) if an even (odd) number of edges meet there. Suppose that such a path P exists and let v be any vertex at which P neither begins nor ends. Whenever a point traversing P comes into v, it must also leave. Thus v must be an even vertex. If P begins at one vertex and ends at a different one, then we see that there are exactly two odd vertices. If P begins and ends at the same vertex, then there are no odd vertices. Figure 2 shows four odd vertices. This contradiction shows that no such P exists. You cannot walk through the city of Konigsberg and cross each bridge exactly once.

Graphs such as the one in Figure 2 are used to illustrate such things as crystals, networks, circuits, and organizational management.

A-4 Preparing for College Entrance Exams

If you plan to attend college, you will most likely be required to take college entrance examinations. Some of these exams attempt to measure the extent to which your verbal and mathematical reasoning skills have been developed. Others test your knowledge of specific subject areas. Usually the best preparation for college entrance examinations is to follow a strong academic program in high school, to study, and to read as extensively as possible.

The following test-taking strategies may prove useful:

- Familiarize yourself with the test you will be taking well in advance of the test date. Sample tests, with accompanying explanatory material, are available for many standardized tests. By working through this sample material, you become comfortable with the types of questions and directions that will appear on the test and you develop a feeling for the pace at which you must work in order to complete the test.
- Find out how the test is scored so that you know whether it is advantageous to guess.
- Skim sections of the test before starting to answer the questions, to get an overview of the questions. You may wish to answer the easiest questions first. In any case, do not waste time on questions you do not understand; go on to those that you do.
- Mark your answer sheet carefully, checking the numbering on the answer sheet about every five questions to avoid errors caused by misplaced answer markings.
- Write in the test booklet if it is helpful; for example, cross out incorrect alternatives and do mathematical calculations.
- Work carefully, but do not take time to double-check your answers unless you finish before the deadline and have extra time.
- Arrive at the test center early and come well prepared with any necessary supplies such as sharpened pencils and a watch.

College entrance examinations that test general reasoning abilities, such as the Scholastic Aptitude Test, usually include questions dealing with basic algebraic concepts and skills. The College Board Achievement Tests in mathematics (Level I and Level II) include many questions on algebra. The following second-year algebra topics often appear on these exams. For each of the topics listed on pages 865 – 867, a page reference to the place in your textbook where this topic is discussed has been provided. As you prepare for college entrance exams, you may wish to review the topics on these pages.

Types of Numbers

Integers (p. 1)
Rational, irrational, and real numbers (pp. 1, 273, 283–285)
Imaginary and complex numbers (pp. 288–290, 292–294)
Polar form of complex numbers (pp. 680–682, 685–687)

Properties of Real Numbers

Properties of equality (p. 14)
Properties of addition and multiplication (pp. 15, 27)
Properties of order (p. 59)
Proving properties of real numbers (pp. 81–84, 88–89)

Sets and Logic

Set notation (pp. 1, 37, 61, 153)
Quantifiers (p. 20)
Logic (pp. 95–97)
Union and intersection, Venn diagrams (pp. 754–755)

Solving Equations

Linear equations (pp. 37–39)
Linear systems in two variables (pp. 124–128, 801, 803)
Linear systems in three variables (pp. 444–447, 802–803)
Equations with fractional coefficients (pp. 242–244)
Fractional equations (pp. 247–249)
Radical equations (pp. 277–279)
Quadratic equations, completing the square (pp. 307–309)
Quadratic equations, the quadratic formula (pp. 311–313, 317–319)
Equations in quadratic form (pp. 322–323)
Polynomial equations (pp. 194–196, 377–379)
Finding rational roots (pp. 382–383)
Approximating irrational roots (pp. 386–388)
Quadratic systems (pp. 439–441)
Exponential equations (pp. 460–461, 480)
Logarithmic equations (pp. 475, 479)
Trigonometric equations (pp. 697–699)

Solving Inequalities

Linear inequalities (pp. 59–61)
Combined inequalities (pp. 65–66)
Inequalities involving absolute value (pp. 73–74, 76–77)
Systems of inequalities (pp. 135–137)
Polynomial inequalities (pp. 202–203)

Factoring

Prime factorization of integers (pp. 179–180)
Perfect square trinomials, difference of squares (pp. 183–184)
Sum and difference of cubes (p. 184)
Quadratic polynomials (pp. 188–189)
Using the factor theorem (pp. 372–374)

Graphing

The number line (pp. 1–2)
Inequalities (pp. 60, 65–66, 73, 76–77)
Using a sign graph (p. 203)
The coordinate plane (pp. 107–108)
Linear equations (pp. 108–110)
Slope (pp. 112–114)
Intercepts, parallel and perpendicular lines (pp. 119–120)
Systems of linear equations (p. 124)
Linear inequalities and systems of linear inequalities (pp. 135–137)
Parabolas (pp. 326–330, 412–415)
Polynomial equations (pp. 386–388)
Circles (pp. 407–409)
Ellipses (pp. 418–421)
Hyperbolas (pp. 426–429)
Conics with center not at the origin (pp. 432–433)
Quadratic systems (pp. 436–437, 439–441)
Exponential functions (pp. 459–460)
Graphing the inverse of a function (pp. 464–465)
Logarithmic functions (p. 468)
Periodicity and symmetry (pp. 619–620)
Sine and cosine, period and amplitude (pp. 624–627)
Tangent, cotangent, secant, cosecant (pp. 630–632)
Inverse trigonometric functions (pp. 690, 693)

Functions

Functions and relations, domain and range (pp. 141–143, 153–155)
Linear functions (pp. 146–148)
Quadratic functions (pp. 333–335)
Polynomial functions (pp. 167–169, 386–388)
Division of polynomials (pp. 340–341, 344–345)
Interpolation (pp. 391–393)
Composition, inverses (pp. 463–465)
Trigonometric functions (pp. 555–558, 561–564)
Circular functions (pp. 613–615)
Trigonometric identities (pp. 598–601)
Inverse trigonometric functions (pp. 689–691, 693–695)

Exponents and Logarithms

Laws of exponents (pp. 171, 212, 455, 459–460)
Laws of logarithms (pp. 473–475)
Change of base (p. 481)

Sequences and Series

Arithmetic and geometric sequences (pp. 501–503, 507–508, 510–512)
Series, sigma notation (pp. 518–520)
Sums of finite arithmetic and geometric series (pp. 525–526)
Sums of infinite geometric series (pp. 531–532)
Expansion of binomials, the binomial theorem (pp. 537–538, 540–542)

Vectors

Operations (pp. 659–661)
The norm (p. 661)
Components (pp. 666–668)
The dot product (p. 668)

Statistics, Counting, Probability

Measures of central tendency, measures of dispersion
(pp. 710–711, 713–716)
Counting principles (pp. 730–731)
Permutations (pp. 734–736)
Combinations (pp. 738–739)
Computing probabilities (pp. 745–747, 754–758)

Solving Problems

Translating words into symbols (pp. 43–45)
Using linear equations (pp. 49–52)
Using inequalities (pp. 69–70)
Using systems (pp. 131–132)
Using linear functions (pp. 147–148)
Using polynomial equations (pp. 198–199)
Using rational expressions (pp. 243–244, 248–249)
Maximum and minimum problems (pp. 340–341)
Using direct variation and proportions (pp. 351–353)
Using inverse and joint variation (pp. 358–360)
Problems on exponential growth and decay (pp. 483–485)
Finding distances and angles using right triangles (pp. 574–576)
Using the law of cosines and the law of sines (pp. 580–581, 586–587, 591–593)
Using vectors (pp. 659–663, 672–673)
Using the normal distribution (pp. 719–722, 747)
Using matrices (pp. 779–782)

Types of Questions

The types of questions you can expect may include five-choice *Multiple–Choice* questions, four-choice *Quantitative Comparison* questions, or *Grid–in* questions. Here is an example of a **Multiple–Choice** question:

If $4(x - 1) = 20$, what is the value of $x + 3$?
(A) 6
(B) 7
(C) 8
(D) 9 *the correct answer*
(E) 10

Quantitative Comparison questions give you two quantities and ask you to compare them. Here is an example of a Quantitative Comparison question:

Column A	Column B
4^{10}	$4^{11} - 4^{10}$

You must choose whether the quantity in Column A is greater, the quantity in column B is greater, the two quantities are equal, or if you cannot determine which is greater from the information given. For this example, the quantity in Column A is greater.

Grid–in questions emphasize active problem solving and critical thinking by asking you to grid the answer directly on the answer sheet rather than recognize it from among the choices. Using the same example as the Multiple–Choice question, the answer is still 9, but you would have to grid 9 on the answer sheet rather than choose (D).

Calculator Use

Some college entrance exams allow students to use calculators. If calculator use is permitted, bring a familiar calculator with you to the test center, but don't plan to use it for every problem. First decide how to solve the problem, and then decide if a calculator will help with the computation. For example, suppose you are given that $P(x) = x^2 - 2x - 3$. If you are asked to find $P(x + 1)$, a calculator will not help. You should substitute $(x + 1)$ for x in $P(x)$ to get $P(x + 1) = (x + 1)^2 - 2(x + 1) - 3 = x^2 - 4$. If you are asked to find $P(-4)$, however, using a calculator may help. Substitute (-4) into the expression $x^2 - 2x - 3$ to get $P(-4) = (-4)^2 - 2(-4) - 3 = 21$. In general, calculators may help solve problems involving computations, data, number patterns, or guess-and-check problem solving.

Glossary

abscissa (p. 107): The first coordinate in an ordered pair of real numbers associated with a point in the coordinate plane; also called *x-coordinate*.

absolute value of a complex number (pp. 298, 680): The absolute value of the complex number $z = x + yi$ is defined as $|z| = \sqrt{x^2 + y^2}$, the distance from z to the origin in the complex plane.

absolute value of a number (p. 3): For each real number a, $|a| = a$ if $a > 0$, $|a| = -a$ if $a < 0$. On a number line, $|a|$ is the distance between the graph of the number a and the origin.

additive inverse of a matrix (p. 770): The additive inverse of matrix A is matrix $-A$. Each element of $-A$ is the opposite of its corresponding element in A.

amplitude of a complex number (p. 681): For a nonzero complex number in the form $z = n(\cos\theta + i\sin\theta)$, the angle θ is the amplitude of z; also called *argument* of z.

amplitude of a periodic function (p. 625): For a function of the form $y = a\sin x$ $(a > 0)$ or $y = a\cos x$, the positive number a is called the amplitude of the function.

angle (p. 549): A figure formed by two rays that have the same endpoint.

angle of elevation [depression] (p. 575): A measure of the deviation of one's line of sight from the horizontal.

antilogarithm (p. 479): If $\log x = a$, then the number x is called the antilogarithm of a.

arithmetic means (pp. 287, 508): The terms between two given terms of an arithmetic sequence. A single arithmetic mean between two numbers is called *the arithmetic mean,* or the *average,* of the two numbers.

arithmetic sequence (p. 501): A sequence in which the difference between any two successive terms (called the *common difference*) is constant. Also called *arithmetic progression*.

arithmetic series (p. 518): *See under* series.

associated equation (p. 135): The equation from which an inequality is derived by replacing the equals sign by an inequality symbol.

axes (p. 107): Two number lines, one horizontal and one vertical, intersecting at a point O.

axiom (p. 81): A statement that is assumed to be true; also called *postulate*.

base of a power (p. 7): The repeated factor of the power. In a^5, a is the base.

bearing of a vector (p. 661): The angle measured clockwise from due north around to the vector.

binomial (p. 174): A polynomial that has two terms.

binomial expansion (p. 537): The sum of terms that results from multiplying out a power of a binomial.

Boolean algebra (p. 95): The algebra of logic.

box-and-whisker plot (p. 713): A diagram that shows the median, quartiles, and range of a distribution.

Cartesian coordinate system (p. 107): A rectangular coordinate system.

central conics (p. 432): Circles, ellipses, and hyperbolas.

characteristic (p. 479): The integer part of a logarithm.

circle (p. 407): The set of all points in a plane that are a fixed distance (the *radius*) from a fixed point (the *center*).

coefficient (p. 167): The constant (or numerical) factor in a variable.

column matrix (p. 767): A matrix consisting of only one column.

combination (p. 738): An r-element subset of a set of n elements is a combination of n elements taken r at a time.

communication matrix (p. 779): A matrix representing routes by which data can be transmitted and received.

complement of an event (p. 757): If an event A is a subset of a sample space S, the complement of A consists of the elements of S that are not members of A.

completing the square (p. 308): Changing the form of a quadratic equation to make the left side a trinomial square.

complex fraction (p. 238): A fraction whose numerator or denominator (or both) has one or more fractions or powers with negative exponents.

complex number (p. 292): A number of the form $a + bi$, where a and b are real. The number a is called the *real part* of $a + bi$; b is called the *imaginary part*.

complex plane (p. 680): The rectangular coordinate system when it is used to represent complex numbers.

component form of a vector (p. 666): When the vector $\mathbf{u}$ is written in the form $a\mathbf{i} + b\mathbf{j}$, it is in component form. The number a is the x-component, and b is the y-component.

composite of two functions (p. 463): The function whose value at x is $f(g(x))$ is called the composite of the functions f and g. The operation that combines f and g to produce their composite is called *composition*.

conics, or **conic sections** (p. 407): Plane curves having second-degree equations.

conjugate of a complex number (p. 680): The conjugate of the number $z = x + yi$ is $\bar{z} = x - yi$.

conjugates (p. 274, 294): Expressions of the form $a\sqrt{b} + c\sqrt{d}$ and $a\sqrt{b} - c\sqrt{d}$. The numbers $a + bi$ and $a - bi$ are *complex conjugates*.

conjunction (p. 65): A sentence formed by joining two sentences with the word *and*.

consistent equations (p. 128): The equations in a system that has at least one solution.

constant (p. 39): A number.

constant of variation (pp. 351, 358): In a direct variation, specified by $y = mx$ ($m \neq 0$), m is the constant of variation. In an inverse variation, specified by $xy = k$ ($k \neq 0$), k is the constant of variation. Also called *constant of proportionality*.

converse (p. 82): The statement that results from interchanging the hypothesis and the conclusion of an if-then statement.

coordinate of a point (p. 1): The real number paired with that point on a number line.

coordinates (p. 107): The unique ordered pair of real numbers associated with each point in the plane.

corollary (p. 81): A theorem that can be proved easily from another theorem.

correlation coefficient (pp. 724–725): A statistic used to determine how closely the points in a scatter plot cluster about a line.

cosecant (p. 556): The cosecant of an angle θ is the reciprocal of sin θ.

cosine (p. 555): For an acute angle θ in standard position, with $P(x, y)$ a point (not the origin) on the terminal side and r = the distance OP,

$$\cos\theta = \frac{x}{r}.$$

cotangent (p. 556): The cotangent of an angle θ is the reciprocal of tan θ.

coterminal angles (p. 550): Two angles whose terminal sides coincide when the angles are in standard position.

counterexample (p. 82): In algebra, a single numerical example that makes a statement false.

cube root (p. 260): A cube root of a number b is a solution of the equation $x^3 = b$.

degree (p. 549): $1° = \frac{1}{360}$ of a complete revolution.

degree of a monomial (p. 167): The sum of the degrees of the variables in the monomial.

degree of a polynomial (p. 167): The greatest of the degrees of its terms after it has been simplified.

degree of a polynomial equation (p. 377): The degree of a polynomial equation of the form $P(x) = 0$ is the degree of $P(x)$.

degree of a variable in a monomial (p. 167): The number of times the variable occurs as a factor in the monomial.

dense (p. 285): A set S of real numbers is dense if between any two numbers in the set there is a member of S.

dependent equations (p. 128): The equations in a consistent system that has infinitely many solutions.

determinant (p. 787): A real number associated with a square matrix and usually displayed in the same form.

diagonal matrix (p. 788): A matrix whose only nonzero elements are on the main diagonal.

difference of vectors (p. 661): The difference $\mathbf{u} - \mathbf{v}$ is defined to be $\mathbf{u} + (-\mathbf{v})$.

dimensions of a matrix (p. 767): The number of rows (horizontal) and the number of columns (vertical).

direct variation (p. 351): A linear function defined by an equation of the form $y = mx$ ($m \neq 0$). The number m is called the *constant of variation*.

directed angle (p. 549): An angle generated by the rotation of a ray (the *initial side*) onto another ray (the *terminal side*).

discriminant (p. 318): The expression $b^2 - 4ac$ is called the discriminant of the quadratic equation $ax^2 + bx + c = 0$.

disjoint sets (p. 754): Sets that have no elements in common.

disjunction (p. 65): A sentence formed by joining two sentences with the word *or*.

division algorithm (p. 364): Dividend = Quotient × Divisor + Remainder

domain of a function (p. 141): *See* function.

domain of a variable (p. 9): The set whose members may serve as replacements for the variable; also called *replacement set*.

dominance relation (p. 781): A dominance relation exists between members of a group when, between any two members, one dominates the other.

dot product (p. 668): The dot product of two nonzero vectors $\mathbf{u}$ and $\mathbf{v}$ is defined to be $\mathbf{u} \cdot \mathbf{v} = \|\mathbf{u}\| \|\mathbf{v}\| \cos \theta$, where θ is the angle between $\mathbf{u}$ and $\mathbf{v}$.

ellipse (pp. 418–419): The set of all points P in the plane such that the sum of the distances from P to two fixed points (the *foci*) is a given constant.

empty set (p. 38): The set with no members; also called the *null set*.

equation (p. 6): A sentence formed by placing an equals sign between two expressions, called *sides* of the equation.

equivalent equations (p. 37): Equations that have the same solution set over a given domain.

equivalent expressions (p. 16): Two expressions that are equal for every value of each variable they contain.

equivalent inequalities (p. 60): Inequalities with the same solution set.

equivalent systems (p. 124): Systems of equations that have the same solution set.

equivalent vectors (p. 659): Two vectors that have the same magnitude and the same direction.

even function (pp. 337, 619): A function f is even if $f(-x) = f(x)$ for every x in the domain of f.

event (p. 743): Any subset of possible outcomes for an experiment.

exponent (p. 7): A positive exponent tells how many times the base occurs as a factor of a power.

exponential equation (p. 460): An equation in which a variable appears in an exponent.

exponential form (p. 456): A radical expression is in exponential form when it is expressed as a power or a product of powers.

exponential function (p. 460): If $b > 0$ and $b \neq 1$, the function defined by $y = b^x$ is called the exponential function with base b.

extraneous root (p. 247): A root of a transformed equation that is not a root of the original equation.

factor a polynomial (p. 183): To express the polynomial as a product of other polynomials taken from a specified factor set.

factor set (pp. 179, 183): The set from which numbers or polynomials are chosen as factors.

factorial notation (p. 540): The symbol $r!$, read "r factorial," is defined as $r! = r(r - 1)(r - 2) \cdot \ldots \cdot 3 \cdot 2 \cdot 1$ if r is a positive integer, and as 1 if $r = 0$.

field (pp. 14–15): A set of numbers having the associative, closure, commutative, distributive, identity, and inverse properties.

finite sequence (p. 501): A sequence that has a limited number of terms.

fractional equation (p. 247): An equation in which a variable occurs in a denominator.

frequency distribution (p. 709): A display of data that shows the number of occurrences of each item.

function (pp. 141, 153): A correspondence between two sets, D and R, that assigns to each member of D exactly one element of R. The *domain* of the function is D. The *range* is the set of all members of R assigned to at least one member of D.

geometric means (pp. 287, 511): The terms between two given terms of a geometric sequence. *The geometric mean* of a and b is generally considered as $\sqrt{ab}$ if a and b are positive and $-\sqrt{ab}$ if a and b are negative.

geometric sequence (p. 502): A sequence in which the ratio of every pair of successive terms (called the *common ratio*) is constant. Also called *geometric progression*.

geometric series (p. 518): *See under* series.

graph of a function (p. 142): The set of all points (x, y) such that x is in the domain of the function and the rule of the function assigns y to x.

graph of an open sentence in two variables (p. 108): The set of all points in the coordinate plane whose coordinates satisfy the sentence.

graph of an ordered pair (p. 107): The point in an xy-coordinate plane associated with an ordered pair of real numbers.

graph of a real number (p. 1): The point on a number line paired with that number.

graph of a system of inequalities (p. 137): Points satisfying all the inequalities in the system.

greatest common factor (GCF) (pp. 179, 180): The GCF of two or more integers is the greatest integer that is a factor of each. The GCF of two or more monomials is the common factor that has the greatest degree and the greatest numerical coefficient.

greatest common factor (GCF) of two or more polynomials (p. 190): The common factor having the greatest degree and the greatest constant factor.

half-plane (p. 135): The set of all points on one side of a given line (the *boundary*).

histogram (p. 709): A bar graph displaying a frequency distribution.

hyperbola (p. 426): The set of all points P in the plane such that the difference between the distances from P to two fixed points (the *foci*) is a given constant.

identity (p. 38): An equation that is satisfied by all values of the variable.

identity function (p. 463): The function $I(x) = x$.

identity matrix (p. 775): An $n \times m$ matrix whose main diagonal from upper left to lower right has all elements 1, while all other elements are 0.

imaginary axis (p. 680): The vertical axis in the complex plane.

imaginary number (p. 288): The number i, defined as $\sqrt{-1}$.

imaginary numbers (p. 288): All numbers of the form $a + bi$, $b \neq 0$. Numbers of the form bi, $b \neq 0$, are called *pure imaginary numbers*.

inconsistent equations (p. 128): The equations in a system that has no solution.

independent events (p. 756): Two events are independent if the occurrence of one event does not affect the probability of the other.

inequality (p. 6): A sentence formed by placing an inequality symbol between two expressions.

inequality symbol (p. 6): One of the symbols $<$, $>$, $\leq$, $\geq$, and $\neq$.

infinite sequence (p. 501): A sequence that has an unlimited number of terms.

initial point (p. 659): The starting point of a vector.

integers (p. 1): Members of the set $\{\ldots, -3, -2, -1, 0, 1, 2, 3, \ldots\}$.

intersection of sets (p. 754): The intersection of sets A and B is the set whose members are elements of both A and B.

inverse functions (p. 464): The functions f and g are inverse functions if $f(g(x)) = x$ for all x in the domain of g and $g(f(x)) = x$ for all x in the domain of f.

inverse matrices (p. 790): For any two matrices A and B, if $AB = BA =$ the identity matrix, then A and B are inverse matrices.

inverse variation (p. 358): A function defined by an equation of the form $xy = k$ or $y = \frac{k}{x}$ $(x \neq 0, k \neq 0)$. The number k is called the *constant of variation*.

irrational numbers (p. 1): Real numbers that are not rational, such as $\sqrt{7}$ and π.

irreducible polynomial (p. 189): A polynomial that has more than one term and cannot be expressed as a product of polynomials of lower degree taken from a given factor set.

joint variation (p. 359): When a quantity varies directly as the product of two or more other quantities, the variation is called joint variation.

least common multiple (LCM) (pp. 179, 180): The LCM of two or more integers is the least positive integer having each as a factor. The LCM of two or more monomials is the common multiple having the least degree and the least positive numerical coefficient.

least common multiple (LCM) of two or more polynomials (p. 190): The common multiple having least degree and least positive constant factor.

like radicals (p. 270): Two radicals with the same index and radicand.

linear combination (p. 125): The result of adding two linear equations.

linear equation in two variables (p. 109): Any equation that can be expressed in the form $Ax + By = C$ (A and B not both 0).

linear function (p. 146): A function f that can be defined by $f(x) = mx + b$, where x, m, and b are any real numbers. If $m = 0$, f is a *constant function*.

linear inequality (p. 135): A sentence obtained by replacing the equals sign in a linear equation by an inequality symbol.

linear system (p. 124): A set of linear equations in the same two variables.

logarithm (p. 469): If b and N are positive numbers ($b \neq 1$), then $\log_b N = k$ if and only if $b^k = N$.

logarithmic function with base 2 (p. 468): The inverse of $f(x) = 2^x$, $b > 0$ and $b \neq 1$.

mantissa (p. 479): The decimal or fractional part of a logarithm.

mapping diagram (p. 141): A diagram that pictures a correspondence between two sets.

mathematical model (p. 198): An equation that represents a real-life problem.

matrix (pp. 447, 767): A rectangular array of numbers, enclosed by brackets. The numbers in the matrix are its *elements*.

mean (p. 710): The arithmetic average of the numbers in a distribution.

measures of central tendency (p. 715): The mean, median, and mode of a distribution.

measures of dispersion (p. 715): The range, first and third quartiles, variance, and standard deviation of a distribution.

median (p. 710): The middle number in a distribution or the average of the two middle numbers.

minor (p. 794): The minor of an element in a determinant is the determinant resulting from the deletion of the row and column containing the element.

mode (p. 710): The number that occurs most frequently in a distribution.

monomial (p. 167): A constant, a variable, or a product of a constant and one or more variables.

multiple roots, multiple zeros (p. 196): Roots of equations or zeros of functions that arise from repeated factors.

mutually exclusive events (p. 755): Events that have no elements in common.

natural logarithm function (p. 489): The logarithm function with base e; $e \approx 2.71828$.

natural numbers (p. 1): Members of the set $\{1, 2, 3, \ldots\}$.

negative angle (p. 549): An angle generated by a clockwise rotation.

negative numbers (p. 2): Numbers less than zero.

norm (p. 661): The length or magnitude of a vector.

***n*th root** (p. 260): An nth root of b is a solution of the equation $x^n = b$, where n is a positive integer.

***n*th roots of unity** (p. 686): The nth roots of the number 1.

null set (p. 38): *See* empty set.

numerical expression (p. 6): A symbol or group of symbols used to represent a number; also called *numeral*.

objective function (p. 159): In linear programming, the expression to be maximized or minimized.

odd function (pp. 337, 619): A function f is odd if $f(-x) = -f(x)$ for every x in the domain of f.

one-to-one function (p. 460): A function f is called one-to-one if for every p and q in the domain of f, $f(p) = f(q)$ implies $p = q$.

open sentence (pp. 37, 101): An equation or inequality that contains one or more variables.

opposites (p. 3): On a number line, numbers that are the same distance from the origin but on opposite sides of it.

ordered pair (p. 101): A pair of numbers having a definite order.

ordinate (p. 107): The second coordinate in an ordered pair of real numbers associated with a point in the coordinate plane; also called *y-coordinate*.

origin (pp. 2, 107): On the number line, the graph of 0. In a plane rectangular coordinate system, the meeting point of the axes; the graph of (0, 0).

orthogonal vectors (p. 668): Vectors **u** and **v** are orthogonal if and only if $\mathbf{u} \cdot \mathbf{v} = 0$.

parabola (p. 412): The set of all points equidistant from a fixed line (the *directrix*) and a fixed point not on the line (the *focus*).

Pascal's triangle (p. 537): A triangular array of numbers representing the coefficients of the expansion of $(a + b)^n$, where n is a nonnegative integer.

periodic function (p. 619): A function is periodic if, for some positive constant p, $f(x + p) = f(x)$ for every x in the domain of f. The smallest such p is the *period* of f.

permutation (p. 734): An arrangement of the elements of a set in a definite order.

plane rectangular coordinate system (p. 107): A system for locating the point associated with any ordered pair by reference to two number lines intersecting at right angles.

polar coordinate system (p. 675): A system that consists of a point O, called the *pole,* and a ray, called the *polar axis,* having O as its endpoint.

polar coordinates (p. 675): The polar coordinates of a point P are an ordered pair (r, θ), where $r = OP$ and θ is the measure of an angle from the polar axis to $\overline{OP}$.

polar form of a complex number (p. 681): The polar form of a nonzero complex number $z = x + yi$ is $z = r(\cos\theta + i\sin\theta)$, where $r = \sqrt{x^2 + y^2}$ and θ is an angle such that $\cos\theta = \frac{x}{r}$ and $\sin\theta = \frac{y}{r}$.

polynomial (p. 167): A monomial or a sum of monomials. The monomials are called the *terms* of the polynomial.

polynomial equation (p. 194): An equation that is equivalent to one with a polynomial as one side and 0 as the other side.

polynomial inequality (p. 202): An inequality equivalent to an inequality with a polynomial as one side and 0 as the other side.

positive angle (p. 549): An angle generated by a counterclockwise rotation.

positive numbers (p. 2): Numbers greater than zero.

postulate (p. 81): *See* axiom.

power (p. 7): A product of equal factors.

prime factorization (p. 179): The process of writing a positive integer as a product of primes.

prime number, or **prime** (p. 179): An integer greater than 1 whose only positive factors are itself and 1.

prime polynomial (p. 189): An irreducible polynomial with integral coefficients is prime if the greatest common factor of its coefficients is 1.

product of matrices (p. 774): The product of matrices $A_{m\times n}$ and $B_{n\times p}$ is the $m \times p$ matrix whose element in the ath row and bth column is the sum of the products of corresponding elements of the ath row of A and the bth column of B.

proof (p. 81): Logical reasoning from hypothesis to conclusion.

proportion (p. 352): An equality of ratios. In the proportion $\frac{y_1}{x_1} = \frac{y_2}{x_2}$, the numbers x_1 and y_2 are the *means*; y_1 and x_2 are the *extremes*.

quadrant (p. 107): One of the four regions into which a plane rectangular coordinate system is divided by its axes.

quadrantal angle (p. 549): An angle whose terminal side lies on a coordinate axis.

quadratic equation in one variable (p. 307): A polynomial equation in the form $ax^2 + bx + c = 0$.

quadratic form (p. 322): An equation is in quadratic form if it can be written as $a[f(x)]^2 + b[f(x)] + c = 0$.

quadratic formula (p. 311): Solutions of the quadratic equation $ax^2 + bx + c = 0$ $(a \neq 0)$ are given by the formula $x = \frac{-b \pm \sqrt{b^2 - 4ac}}{2a}$.

quadratic function (p. 333): A function that can be written in either of the forms $f(x) = ax^2 + bx + c$ $(a \neq 0)$ or $f(x) = a(x - h)^2 + k$ $(a \neq 0)$.

quadratic polynomial, or **second-degree polynomial** (p. 188): A polynomial of the form $a^2 + bx + c$ $(a \neq 0)$. The *quadratic term* is ax^2, the *linear term* is bx, the *constant term* is c.

quadratic system (p. 436): A system containing only quadratic equations or a combination of linear and quadratic equations in the same two variables.

quadratic trinomial (p. 188): A quadratic polynomial in which a, b, and c are all nonzero integers.

quartile (p. 713): The median of the lower half of a set of data is the first quartile; the median of the upper half is the third quartile.

quotient (p. 7): The result of dividing one number by another.

radian (p. 607): An angular measure. In a circle of radius r, centered at the vertex of an angle θ, the radian measure of θ is defined as the ratio of the length of s, the arc intercepted by θ, to the radius r: $\theta = \frac{s}{r}$.

radical (p. 261): The symbol $\sqrt[n]{b}$ is called a radical; n is the *index*, and b is the *radicand*.

radical equation (p. 277): An equation that contains a radical with a variable in the radicand.

range of a function (p. 141): *See* function.

range of a set of data (p. 713): The difference between the largest and smallest numbers in the set.

rational algebraic expression (p. 227): An algebraic expression that can be expressed as a quotient of polynomials. Also called *rational expression*.

rational function (p. 228): A function defined by a simplified rational expression in one variable.

rational numbers (p. 1): Numbers that are the result of dividing an integer by a nonzero integer.

real axis (p. 680): The horizontal axis in the complex plane.

real numbers (p. 1): The set consisting of all the rational and irrational numbers.

reciprocal (p. 15): Each nonzero real number has a reciprocal, or *multiplicative inverse,* such that the product of the number and its reciprocal is 1.

reference angle (p. 562): If θ is not a quadrantal angle, there is a unique angle α corresponding to θ such that either $\theta + \alpha$ or $\theta - \alpha$ is an integral multiple of 180°; α is called the *reference angle* of θ.

regression line (p. 726): A line that relates two variables having a high correlation.

relation (p. 153): Any set of ordered pairs.

repeating decimal (p. 284): When the division process is used to express a rational number as a decimal, the result is a repeating decimal if a digit or block of digits repeats endlessly as the remainder.

resultant (pp. 659–660): The sum of two vectors.

root (solution) of a polynomial equation (p. 194): A value of the variable that satisfies the equation.

row matrix (p. 767): A matrix consisting of only one row.

sample space (p. 743): The set of all possible outcomes of a random experiment.

scalar (p. 771): In matrix algebra, a real number.

scalar product (p. 771): The scalar product of a real number r and a matrix A is the matrix rA, in which each element is r times its corresponding element in A.

scatter plot (p. 724): A graph of ordered pairs used to determine whether there is a mathematical relationship between two sets of data.

scientific notation (p. 221): A method of representing a number in the form $m \times 10^n$, where $1 \le m < 10$ and n is an integer.

secant (p. 556): The secant of an angle θ is the reciprocal of $\cos \theta$.

sequence (p. 501): A function whose domain consists of consecutive positive integers. Each corresponding value is a *term* of the sequence.

series (p. 518): The indicated sum of the terms of a sequence. An *arithmetic series* is a series whose related sequence is arithmetic. A *geometric series* is a series whose related sequence is geometric.

sides of an equation (p. 6): The two expressions joined by the equals sign.

significant digit (p. 221): A significant digit of a number written in decimal form is any nonzero digit or any zero that has a purpose other than placing the decimal point.

similar (like) monomials (p. 167): Monomials that are identical or that differ only in their coefficients.

similar terms (p. 23): Terms that contain the same variable factor; also called *like terms*.

simple event (p. 743): A single element of a sample space.

simplified polynomial (p. 167): A polynomial in which no two terms are similar.

sine (p. 555): For an acute angle θ in standard position, with $P(x, y)$ a point (not the origin) on the terminal side and r = the distance OP, $\sin \theta = \frac{y}{r}$.

slope (p. 112): For any two distinct points (x_1, y_1) and (x_2, y_2) on line L, the slope of L is $\frac{y_2 - y_1}{x_2 - x_1}$.

solution set (pp. 37, 101): The set of all solutions of an open sentence that belong to a given domain of the variable. For an open sentence in two variables, the set of all ordered pairs satisfying the sentence.

square matrix (p. 767): A matrix with the same number of rows and columns.

square root (p. 259): A square root of a number b is a solution of the equation $x^2 = b$. The *principal square root* of b is the positive square root, $\sqrt{b}$.

standard deviation (p. 715): The principal square root of the variance of a distribution, denoted by σ.

standard normal curve (p. 719): The bell-shaped curve that represents a standard normal distribution.

standard normal distribution (p. 719): A normal distribution with mean equal to 0 and standard deviation equal to 1.

standard position (p. 549): An angle is in standard position when its initial side coincides with the positive x-axis.

statistics (p. 710): Numbers used to describe a set of data.

stem-and-leaf plot (p. 710): A way of displaying the data in a frequency distribution.

subset (p. 738): Set B is a subset of set A if each member of B is also a member of A.

subtraction of matrices (p. 771): For each set of $m \times n$ matrices, $A_{m \times n} - B_{m \times n} = A_{m \times n} + (-B_{m \times n})$.

sum of matrices (p. 770): The sum of matrices having the same dimensions is the matrix whose elements are the sums of the corresponding elements of the matrices being added.

sum of vectors (p. 660): For vectors $\mathbf{u}$ and $\mathbf{v}$, the sum $\mathbf{u} + \mathbf{v}$ is the vector extending from the initial point of $\mathbf{u}$ to the terminal point of $\mathbf{v}$.

summand (p. 518): In a series written in sigma notation, for example $\sum_{n=1}^{10} n^2$, the general term n^2 is called the summand. The letter n is called the *index*. The first and last values of n are called the *limits of summation*.

summation sign (p. 518): The Greek letter Σ (sigma), used to write a series in abbreviated form.

synthetic division (p. 368): A method that uses only coefficients to display the process of dividing a polynomial in x by $x - c$. Also called *synthetic substitution*.

tangent (p. 555): For an acute angle θ in standard position, with $P(x, y)$ a point (not the origin) on the terminal side and r = the distance OP, $\tan \theta = \frac{y}{x}$.

tautology (p. 97): A compound statement that is true for all truth values of its component statements.

terminal point (p. 659): The ending point of a vector.

terminating decimal (p. 284): When the division process is used to express a rational number as a decimal, the result is a terminating decimal if the division ends with no remainder. Also called *finite decimal*.

theorem (p. 81): A statement that can be proved.

translation (p. 408): Sliding a graph to a new position in the coordinate plane without changing its shape.

trigonometric form of a complex number (p. 681): *See* polar form.

trinomial (p. 174): A polynomial that has three terms.

uniform motion (p. 44): Motion at a constant speed.

union of sets (p. 754): The union of sets A and B is the set whose members are the elements belonging to either A or B (or both).

unique (p. 14): Identified with one and only one member of the set under consideration.

unit vector (p. 661): A vector with a norm of 1.

value of a numerical expression (p. 6): The number represented by the expression.

values of a function (p. 142): Members of the range of the function.

values of a variable (p. 9): The members of the domain of the variable.

variable (p. 9): A symbol used to represent any member of a given set.

variance of a distribution (p. 715): The average of the squares of the deviations from the mean of the distribution.

vector (p. 659): An arrow representing a vector quantity.

vector quantity (p. 659): Any quantity that has both magnitude and direction.

vertex of a parabola (p. 326): The point where the parabola crosses its axis.

whole numbers (p. 1): Members of the set $\{0, 1, 2, 3, \ldots\}$.

***x*-axis, *y*-axis** (p. 107): A horizontal and a vertical number line meeting at right angles, used in graphing points in the plane.

***x*-coordinate** (p. 107): *See* abscissa.

***x*-intercept** (pp. 119, 329): The x-coordinate of the point where a line or curve intersects the x-axis.

***y*-coordinate** (p. 107): *See* ordinate.

***y*-intercept** (pp. 119, 329): The y-coordinate of the point where a line or curve intersects the y-axis.

zero of a function (p. 195): A number r is a zero of a function f if $f(r) = 0$.

zero matrix (p. 767): A matrix whose elements are all zeros.

Index

Abel, Niels Henrik, 385
Abscissa, 107
Absolute value, 3
  of complex numbers, 294, 298–300, 680
  definition, 9, 89
  in open sentences, 73–79
  theorems involving, 89–91
Accuracy in measurement, 221–222
Addition
  cancellation property, 82
  of fractions, 235–237
  of matrices, 770–771
  of polynomials, 168
  of radicals, 270–273
  of real numbers, 21–25
  of vectors, 660
Addition formulas, trigonometric, 641–644, 650
Addition properties, 15
  of equality, 14
  of matrices, 771
  of order, 59
Additive inverse, 15
  of a matrix, 770
Air speed, 131, 662
Algebraic expressions(s), 9
  equivalent, 16
  rational, 227–229
Algorithm, division, 364, 367
Amplitude
  of a complex number, 681
  of sine and cosine functions, 625
Analytic geometry, 401–453
Angle(s)
  coterminal, 550, 551
  of depression, 575
  directed, 549
  of elevation, 575
  initial side, 549
  naming, 551
  negative, 549
  positive, 549
  quadrantal, 549
  reference, 562–563
  in standard position, 549
  terminal side, 549
Angle measure
  conversion equations for, 607
  decimal degrees, 550, 554
  degrees, minutes, seconds, 550–551, 554
  radians, 607–612
  relation to arc length, 607, 608
Antilogarithm, 479
Applications
  astronomy, 224–225, 350, 424–425, 436, 548
  architecture, 100
  communication networks, 779–781
  compound interest, 483–484
  dominance relations, 781–782
  electrical circuits, 253–254, 258
  exponential growth and decay, 454, 483–488
  force, work, and energy, 672–674
  geometry, 39, 44, 50–51, 198–199, 312–313. *See also* Problems.
  graphics, 766
  light, 166, 210, 258, 306
  light and sound waves, 844
  linear programming, 159–161
  logarithms, 478–482, 489–491
  LORAN, 426
  navigation, 426, 658, 661–663
  music, 606, 634–635
  patterns in nature, 500
  planetary orbits, 424–425
  radiocarbon dating, 493–494
  sampling, 752–753
  trigonometric, 659–705
  See also Modeling, Problems, *and* Formulas.
Approximations
  decimal, of powers having irrational exponents, 459
  irrational roots, 386–389
  length and angle measurements, 575
  by linear interpolation, 391–395
  of measurements, 221–222
  of π, 701–702
  using regression line, 726–727
Arc cosine, 689
Arc length, 607, 608, 609
Arc sine, 690
Arc tangent, 693 (not so named)
Archimedes, 431
Areas
  parabolic section, 431
  sector, 608–609
  triangle, 586, 597–600
  under standard normal curve, 720–722
Argument of a complex number, 681
Associative properties, 15
Astronomical unit, 224–225
Asymptote(s), 230, 231, 630
Average, 508, 710
Axiom, 81
  See also Properties of real numbers.
Axis (Axes), 107
  of ellipse, 419
  imaginary, 680
  polar, 675, 676
  real, 680
  of symmetry of a parabola, 326, 412

Base
  of a power, 7
  of logarithms, 468, 481, 482
BASIC, 5
Bearing, 661
between, meaning of, 65
Binomial(s), 174
  containing radicals, 274–276
  powers of, 537–539
Binomial expansion, 537
  general, 540–543
Binomial theorem, 541
Biographical notes
  Ch'in Chiu-Shao, 220
  Halley, Edmund, 601
  Hamilton, Alice, 544
  Just, Ernest Everett, 177
  Kovalevski, Sonya, 32
  Mayer, Maria Goeppert, 618
  Poincaré, Henri, 684
  Steinmetz, Charles, 316
Bombelli, Rafael, 297
Boole, George, 95
Boolean algebra, 95–97
Boundary, 135
Box-and-whisker plot, 713–714
Briggs, Henry, 472

Calculator Key-In
base e exponentials, 492
conversion: degrees/minutes/ seconds to decimal degrees, 554
factorials, 543
finding roots, 263, 269
order of operations, 12
scientific notation, 225–226
Calculators
graphing, 114, 119, 138, 231, 327, 336, 386, 387, 410, 416, 421, 423, 430, 436, 441, 464, 471, 624, 625, 631, 710, 838, 840, 841, 845
suggestions for when to use, 10, 285, 419, 427, 437, 459, 478–480, 485, 489, 550, 568–571, 608, 614, 615, 698, 710
use in exercises, 24, 30, 35, 224, 240, 268, 314, 316, 324, 481, 482, 486–488, 514, 528, 534, 571, 572, 577, 578, 582, 583, 588, 589, 594, 595, 599, 683, 833, 836, 837, 843
Cancellation property
of addition, 82
of multiplication, 82
Cancellation rule. *See under* Fractions.
Cardano, Girolamo, 297
Cardioid, 679 (Exs. 33, 34)
Cayley, Arthur, 793
Career notes
automotive engineer, 64
computer graphics artist, 423
electrician, 718
flight engineer, 567
marine biologist, 522
statistician, 337
Cartesian coordinate system, 107
Challenge (problems), 48, 140, 158, 325, 363, 411, 492, 516, 601, 623, 733
Chapter reviews, 55–56, 93–94, 162–163, 207–208, 255–256, 302–303, 346–347, 397–398, 451–453, 496–497, 545–546, 602–604, 655–656, 703–704, 764–765, 806–807
Chapter summaries, 55, 92, 161–162, 206, 254–255, 301, 346, 396–397, 450–451, 495, 544–545, 602, 654–655, 702–703, 763–764, 806
Chapter tests, 57, 94, 164, 208, 257, 303, 348, 398, 453, 497, 546, 604, 657, 704, 765, 808
Characteristic of common logarithm, 479
Checking solutions, 38, 247, 278
graphing, 109, 137, 203
polynomial division, 365
radical equations, 278
word problems, 50, 51, 69, 70, 198, 313, 371
Circles, 407–411
center, 407
equations, 407–409
exploring graphs of, 840
radius, 407
translating, 408
unit, 613
Closed
under addition, 19
under division, 35 (Exs. 31–36)
under multiplication, 19
under subtraction, 25 (Exs. 51–56)
Closure properties of real numbers, 15, 19
Coefficient(s), 113, 167
fractional, 242–246
and roots of quadratic equations, 338–341
Cofunctions, 557
College Entrance Exams (preparations), 99, 209, 305, 399, 499, 605, 706, 809, 864–867
Combinations, 738–741
Common difference, in arithmetic sequence, 501
Common ratio, in geometric sequence, 502
Communication networks, 779–781
Commutative properties, 15
Complement of an event, 757
Completeness, property of, 283
Complex fractions, 238–241
Complex numbers, 292–296
absolute value, 294, 298–300, 680
addition and subtraction of, 293, 680
amplitude, or argument, 681
conjugate, 294, 298, 299, 680
equality of, 293
geometry of, 680–684
historical note, 297
operations with, 293–294, 680, 682
polar/trigonometric form, 681
powers and roots of, 685–688
product, definition of, 293
real and imaginary parts, 292, 681
square roots of, 442
sum, definition of, 293
Complex plane, 680
Components, x- and y-, 666
Composite of two functions, 463
Composition of functions, 463–464
Computer Exercises, 5, 32, 64, 75, 122, 187, 321, 395, 406, 530, 584, 645, 670, 742, 750
Computer graphing ideas (techniques), 114, 119, 160, 327, 386, 387, 436, 460, 464, 624, 625, 631, 838, 840, 841, 845
See also Calculators, graphing.
Computer Key-In
approximating π, 701–702
the Euclidean algorithm, 215
factoring trinomials, 193
locating zeros, 389–390
simulating an experiment, 762
solving an inequality, 80
Conclusion, 81
Conditional, 95, 96
Conic sections, 407
circles, 407–411
ellipses, 418–423, 424–425, 432
hyperbolas, 426–431, 432–433
parabolas, 412–416, 431
Conics
central, 432
with center not at the origin, 432–435
See also Conic sections.
Conjugate hyperbolas, 430
Conjugates, 274, 294, 298, 680
Conjunction, 65, 66
and absolute value, 73
symbol, 95
Consecutive integers, 45
Consecutive numbers, 45
Constant, 39, 167
degree of, 167
of proportionality, 352, 358
of variation, 351, 358
Constant term, 188
Constraint, 159
Converse, 82
Coordinate(s)
of point on number line, 1, 2,
polar, 675–676
x-, y-, 107
Coordinate system
Cartesian, 107

plane rectangular, 107
polar, 675–676
Corollary, 81–82
Correlation coefficient, 724–726
Correspondence, one-to-one, 107
Cosecant function, 556, 561, 613
graph, 632
inverse, 694
Cosine function, 555, 561
as function of real numbers, 613
graph, 624–629, 689
inverse, 689–690, 691
Cosines, law of, 580
Cotangent function, 556, 561, 613
graph, 632, 693
inverse, 693
Counterexample, 82
Counting, fundamental principles, 730–733
Cramer, Gabriel, 802
Cramer's rule, 801–805
Critical thinking skills. *See* Thinking skills.
Cube root(s), 260
approximating, 394–395 (linear interpolation)
table, 811
Cubes, sum and difference of, 184
Cumulative reviews, 165, 348–349, 547, 707
Curve
area under, 720, 721
bell-shaped, 719
standard normal, 719–722

Data analysis. *See under* Statistics.
Decimals
repeating, 284
terminating, 284
converting rational numbers to, 283–284
finite, 284
infinite, 284, 285
nonrepeating, 285
Degree
of a monomial, 167
of a polynomial, 167, 168
of a variable in a monomial, 167
Degree in angle measure, 549, 550, 554
De Moivre's theorem, 685–688
Denominator, rationalizing, 265–266, 274–275
Density, 285
Descartes, René, 107
Descartes' rule of signs, 378, 379
Determinant(s), 787–789
element, 787
expansion by minors, 794–800
order, 787
properties, 787, 798–799
of 2 × 2 matrix, 787
of 3 × 3 matrix, 787
use in solving systems, 801–805
Diagonal, main, in matrix, 775, 788
Diagonal of polygon, 506
Difference(s), 7
of cubes, 184
of matrices, 771
of rational expressions, 235–237
of squares, 183
of vectors, 661
Digits, significant, 221–222
Directrix, 412
Dirichlet, 152
Discrete mathematics, 862–863
See also Algorithm, division; Counting, fundamental principles; Functions; Induction, mathematical; Matrix; Probability; Relations; Sets; *and* Statistics.
Discriminant, 317–318
Disjunction, 65, 66
and absolute value, 73, 74
symbol, 95
Dispersion of a set of data, 713
Distance formula, 401–402
Distribution, frequency, 709–729
normal, 719–723
Distributive properties, 15, 23, 34
Division
checking, 365
of complex numbers, 682
definition of, 33
fractions, rule for dividing, 232
long, 364–365
missing terms in, 365
of polynomials, 364–367
quotient in, 7, 33, 232
of rational expressions, 232–234
of real numbers, 33–35
synthetic, 368–370
Domain
of a function, 141
of a relation, 153
of a variable, 9
Dominance relations, 781–782
Dot product, 668
Double-angle formulas
sine and cosine, 646–647
tangent, 651

***e*,** 489
Eccentricity, 422, 424
Electrical circuits, 253–254
Ellipse, 418–423, 424–425
center, 419
equations, 418–421, 432
focal radii, 418
focus (foci), 418
graph, 419–420, 840
extent, 419
intercepts, 419
symmetry, 419
major axis, 419
minor axis, 419
Energy, 672, 673
Enrichment. *See* Applications, Biographical notes, Career notes, Challenge, Explorations, Extras, Historical notes.
Equality of real numbers, 14
Equation(s), 6
associated, 135
of circles, 407–409
cubic, 307
depressed, 374
of ellipses, 418–421, 432
equivalent, 37
exponential, 460–462, 480
fractional, 247–252
with fractional coefficients, 242–246
of hyperbolas, 426–429, 432
logarithmic, 475
matrix, 772–773
in one variable, 37–42, 307
of parabolas, 326–330, 412–415
quadratic, 307–321
in quadratic form, 322–325
radical, 277–281
sides of, 6
transforming, 37–38
trigonometric, 697–700
See also Linear equations, Polynomial equations, Quadratic equations, Systems of equations.
Equivalence, 95, 96–97
Error analysis (cautions), 34, 142, 247, 259, 289, 319, 365, 637, 668, 788
Estimation
of angle measures, 551, 553

in problem solving, 371
by sampling, 752–753
using scientific notation, 222–224
Euclid, 182
Euclidean algorithm, 215
Euler, Leonhard, 297
Evaluating an expression, 9–10
Event(s), 743
complement of, 757
equally likely, 745
independent, 756–757
mutually exclusive, 755–756
simple, 743
Expectation, 751
Explorations
circles and ellipses, 840
continued fractions, 836
direct variation, 839
functions, 834
inequalities, 833
irrational numbers, 832
matrices in geometry, 847
Pascal's triangle, 842
polar coordinate equations, 845
polynomial factors, 835
powers and roots, 841
probability, 846
quadratic equations, 838
radicals, 837
sine curves, 844
trigonometric ratios, 843
Exponential functions, 455–467
applications of, 483–488
Exponent(s), 7
irrational, 459
laws of, 171, 212, 217
negative, 216, 217, 218
rational, 455–458
real number, 459–462
zero, 216
Expression
algebraic, 9
in exponential form, 456
numerical, 6
rational, 227
Extrapolation, linear, 395
Extras
augmented matrices, 785–786
conjugates and absolute value, 298–300
graphing rational functions, 230–231
graphing sequences, 517
growth of functions, 462
induction, 523–524
irrationality of $\sqrt{2}$, 273
logical symbols: quantifiers, 20
random numbers, 753
spirals, 679
symbolic logic: Boolean algebra, 95–97

Factor set, 179, 183
Factor(s), 7
exploring polynomial, 835
greatest common (GCF), 179, 180, 190
greatest monomial, 183
integral, 179
linear, 194
Factor theorem, 373
Factorization
complete, 189–190
prime, 179–181
Factorial notation, 540–541, 734
Fibonacci, 42
Field, 14–15
Focal radii, 418, 426
Focus (foci)
of an ellipse, 418
of a hyperbola, 426
of a parabola, 412
FOIL method of multiplying binomials, 174–175
Force, 672, 673
Formula(s), 39
area of triangle, 586, 597, 599
change of base in logarithms, 481
circular-arc length, 608
circular-sector area, 608
combinations, 738
compound interest, 483–484, 487, 491 (Ex. 51)
coordinate system conversion, 676
distance, 402
double-angle, 646, 651
doubling-time growth, 484
geometric, 39, 40, 41, 47, 830–831
half-angle, 647, 651
half-life decay, 485, 493
Hero's, 597, 599–600
midpoint, 402
multiplying and dividing complex numbers, 682
*n*th term of sequence, 507, 510
Ohm's law, 253, 718
permutations, 735, 736
probability, 745, 755, 756
projectile motion, 39, 199, 201
quadratic, 311
rate of change, 147
from science, 39, 40, 41, 134, 199, 201, 253, 351, 353, 359, 472 (Ex. 43), 484, 485, 493, 609, 672
solving, 39, 41–42
trigonometric addition, 641, 650
uniform motion, 44
Fractals, 535, 536, 766
Fractional coefficients, 242–246
Fractions
adding and subtracting, 235–237
cancellation rule, 211
complex, 238–241
continued, 240, 836
division rule for, 232
in equations, 242–252
least common denominator, 235
multiplying and dividing, 211, 232–234
simplifying, 211–214, 227–229
Frequency, 634
Frequency distribution, 709
Functions, 141–152, 153–154
circular, 613–617
composite, 463
composition of, 463–464
constant, 146
domain of, 141
double zero of, 196
exploring, 834
exponential, 455–467
graphs. *See* Graphs of functions.
greatest-integer, 145
growth of, 462
historical note, 152
identity, 463–464
inverse, 464–465, 689–696
linear, 146–152
logarithmic, 468–477
natural logarithm, 489–492
notation for, 142
objective, 159
odd and even, 619–620
one-to-one, 460, 465, 689
periodic, 619–623
quadratic, 333–337, 340–341
range of, 141
rational, 228, 230–231
signum, 145
trigonometric, 549–574
values of, 142, 334, 335
zero(s) of, 195, 196, 386
Fundamental theorem of algebra, 377

Gauss, Karl Friedrich, 297, 377, 445
Gaussian elimination, 445–446, 785–786
Geometry
analytic, 401–453
of complex numbers, 680–684
exploring transformations, 847
formulas, xviii, 39, 40, 41, 47, 330–331
problems. *See under* Problems.
of quadratic systems, 436–438
Golden rectangle, 315
Graphs
checking, 109
circles, 407–411, 840
conjunction of inequalities, 65–66
disjunction of inequalities, 65–66
ellipses, 419–420, 840
exploring, 834, 838, 840, 841, 844, 845
of functions. *See* Graphs of functions.
hyperbolas, 427–429
linear equations in two variables, 107–111
linear inequalities, 135–139
linear systems, 124, 803
on number line, 1–3, 59–61
open sentence involving absolute values, 73–74, 76–77
open sentence in two variables, 108
of opposites, 2–3
ordered pairs, 107–108
parabolas, 326–337, 412–416
polar equations, 676–679, 845
quadratic systems, 436–438, 439–440
of relations, 153, 154, 155
of sequences, 517
sign, 203
solution set of inequality, 59, 61, 65–66, 202–203
Graphs of functions, 142, 143, 146
amplitude, 625
cotangent, secant, and cosecant, 632
exponential, 460, 468
inverse of a function, 464–465
inverse trigonometric, 689–690, 693
logarithmic, 468
odd and even, 620
periodic, 619–620
phase shift, 629
polynomial, 386–388
quadratic, 333–335, 340
rational, 230–231
sine and cosine, 624–629, 689, 690
symmetric, 620
tangent, 630–631, 693
Greatest common factor (GCF), 179, 180, 190
Greatest monomial factor, 183
Greek letters, xvi, 551
Ground speed, 131, 662
Growth and decay, exponential, 483–488

Half-angle formulas
sine and cosine, 647–648
tangent, 651
Half-life, 485
Half-plane, 135
closed, 136
open, 135
Harmonic mean, 252
Head wind, 131
Hero's formula, 597, 599–600
Hertz (Hz), 634
Hipparchus, 573
Histogram, 709, 710
Historical notes
area of a parabolic section, 431
complex numbers, 297
functions, 152
linkages, 68
logarithms, 472
matrices, 793
prime numbers, 182
roots of polynomial equations, 385
tables of sines, 573
a trigonometric identity, 618
word problems, 42
Horizontal-line test, 465
Hyperbola(s), 426–431, 432–433
asymptotes, 427
center, 428
conjugate, 430
equations, 426–429, 432
focal radii, 426
focus (foci), 426
graphs, 427–429
Hypothesis, 81

***i*,** 288
Identities, trigonometric, 556, 564
addition formulas, 641–644
cofunction, 636
fundamental, 636–640
proving, 637–639
Pythagorean, 556, 560, 636
reciprocal, 636
Identity, 38
Identity properties of real numbers, 15
Imaginary numbers, 288–291
Independent events, 756–757
Index of radical, 261
Index of summation, 518–519
Induction, mathematical, 523–524
Inequalities
combined, 65–68
equivalent, 60
exploring, 833
with fractional coefficients, 242–246
graphs of, 59–61, 65–67, 73–74, 76–77, 135–137, 202–203
linear in two variables, 135–139
in one variable, 59–64, 242, 244, 245
polynomial, 202–205
reversing direction, 60
symbols for, 2, 6
systems of, 137, 139
theorems involving, 88–89
transforming, 60
word problems, 69–72
Inequality, 6
sides of, 6
Integers, 1
congruent, 178
even and odd, 45
perfect, 182
relatively prime, 182
Intercepts
of an ellipse, 49
of a line, 119
of a parabola, 329
Intermediate-value theorem, 387
Interpolation
inverse, 391
linear, 391–395
in log tables, 856–857
Inverse functions, 689–696
cosecant (Csc^{-1}), 694
cosine (Cos^{-1}), 689–690, 691
cotangent (Cot^{-1}), 693
secant (Sec^{-1}), 694
sine (Sin^{-1}), 690–691
tangent (Tan^{-1}), 693
Inverse properties of real numbers, 15

Inverse variation, 358–362, 429
Inverses of matrices, 787–792
Irrational numbers, 1, 273, 283, 285, 832

Joint variation, 359–362
Joule (J), 673

Kepler, Johann, 424
Kilowatt-hour, 673

Latitude, 611
Latus rectum, 416
Law(s)
 of cosines, 580
 of exponents, 171, 212, 217
 of logarithms, 473–477
 of sines, 586
Least common denominator (LCD), 235
Least common multiple (LCM), 179, 180, 190
Leibniz, 42, 152
Lemniscates, 679 (Exs. 35, 36)
Light-year, 224–225 (Exs. 9, 10)
Line(s)
 finding equation, 118–122
 horizontal, 112
 parallel, 119, 124
 perpendicular, 119–120
 regression, 726–727
 slope of, 112–117
 vertical, 112
Linear combination, 125, 785
Linear equation(s)
 consistent, 128, 803
 Cramer's rule, 802
 dependent, 128, 446, 803
 inconsistent, 128, 446–447, 803
 intercept form, 122
 one variable, 49–54, 307
 point-slope form, 118
 simultaneous solution, 124
 slope-intercept form, 119
 standard form, 118
 systems in three variables, 444–449, 785–786, 802–803
 systems in two variables, 124–134, 791–792, 801
 two-point form, 122
 in two variables, 107–111
  graphs, 107–111
  solution set, 101
Linear function, 146
Linear programming, 159–161
Linear term, 188, 189
Linkage, 68
Logarithm(s), 468–470, 856–861
 change of base, 481, 482
 characteristic of, 479
 common, 478, 856–861
 computation with, 478–482, 858–861
 laws of, 473–477
 mantissa of, 479
 Napier's, 472
 natural, 472
 table of, 812–813
Logarithmic functions, 468–477
 applications of, 478–488
Logic, 20, 95–97
Logically equivalent statements, 96–97

Mantissa, 479
Mapping diagram, 141, 153
Matrix (matrices), 447, 767–808
 addition of, 770, 771
 additive inverse, 770
 applications, 779–784
 augmented, 785–786
 column, 767
 communication, 779–781
 determinant of, 787–788
 diagonal, 788
 dimensions, 767
 dominance relations, 779
 elements of, 767
 equal, 768
 exploring, 847
 history, 793
 identity, 770, 775
 inverses of, 787–792
 main diagonal, 775, 788
 powers of, 775–776
 product of, 774–776
 row, 767
 scalar multiplication, 771–772
 square, 767, 775–776
 subtraction of, 771
 sum of, 770
 zero, 767, 776
Matrix equation(s), 772–773
Maximum and minimum values, 335, 340–341
Mean(s)
 arithmetic, 287, 508
 deviations from, 715
 of a distribution, 710
 geometric, 287, 352, 511
 harmonic, 252
Mean proportional, 355
Measure(s)
 of central tendency, 710–711, 715
 degree, 550–551
 of dispersion, 715–716
 radian, 607–608
Median of a distribution, 710, 711, 713
Members of a set, 1
Midpoint formula, 402–403
Minor, 794
Minute, 550
Mixed Problem Solving, 98, 304, 498, 705
Mixed Review Exercises, *at end of each odd-numbered lesson.*
Mode of a distribution, 710, 711
Modeling, mathematical, 147–148, 150–152, 159–161, 183, 198–199, 340–341, 483–485, 779–782
Modulus, 681
Monomial(s), 167
 degree of, 167
 factoring, 180
 quotient of, 212
 similar, or like, 167
Multiple
 integral, 179
 least common, 179, 180, 190
Multiple-choice exercises
 See College Entrance Exams, Chapter reviews.
Multiple roots, 196
Multiplication
 cancellation property, 82
 of complex numbers, 682
 of fractions, 211, 232–234
 matrix, 774–776, 847
 of matrix by scalar, 771–772
 polynomial by monomial, 172, 173
 polynomial by polynomial, 174–176
 of rational expressions, 232–234
 of real numbers, 27–31
 using laws of exponents, 171–173
 of vector by scalar, 660–661
Multiplication property of equality, 14
Multiplicative inverse, 15
 of a matrix, 790
Multiplicative properties of 0 and −1, 27

Napier, John, 472
Natural logarithms, 472
Negation, 95, 96
Negative exponents, 216, 217, 218
Newton (N), 672
Norm, 661
Normal distribution, 719–723
Notation
factorial, 540–541, 734
functional, 142
scientific, 218, 221–225
set, 153
sigma, 518–520
for solution set of inequality, 61
for a vector, 659
***n*th root(s),** 260
principal, 260
of unity, 686–687
Number line, 1
graphing on, 1–5, 202–203
Numbers
absolute value, 3
comparing, 2
complex, 292–296, 680–684
consecutive, 45
integers, 1, 45
irrational, 1, 283, 285, 832
natural, 1
opposite of, 2–3
pentagonal, 505
prime, 179, 182
rational, 1, 283–284
real, 1–2. *See also* Real numbers.
triangular, 505, 530
whole, 1
Numerical expression(s), 6
simplifying, 7–8
value of, 6

Ohm's law, 253, 718
Open sentence(s), 37
involving absolute values, 73–79
solution, or root, 37
solving, 37, 101
in two variables, 101–111
See also Equations, Inequalities.
Opposite of a product, property of, 28
Opposite of a sum, property of, 29
Opposites, properties of, 15
Opposites of real numbers, 2–3
Order
of operations, 8, 233
properties of, 59
in the real numbers, 2
theorems involving, 88–89
Ordered pair, 101, 107, 157
graphing, 107–108
Ordered triple, 444
Ordinate, 107
Origin, 1, 107

Parabola, 326
area of section, 431
axis, 326, 412
directrix, 412
equations, 326–330, 412–415
focus, 412
graphs, 326–337, 412–415
intercepts, 329
latus rectum, 416
vertex, 326, 334, 412
Parallel lines, 119, 124
Parallelogram method, 660
Parsec, 224–225 (Exs. 9, 10)
Pascal's triangle, 537–538, 842
Patterns
in multiplying and factoring polynomials, 175, 183, 184
in sequences, 501, 502, 503
Peaucellier, 68
Percent, 243
Period, 619
frequency, 634, 635
Permutations, 734–737
with repeated elements, 736
Perpendicular lines, 119–120, 403
Pi (π), 607, 608, 701–702, 832
Plane
complex, 680
xy-coordinate, 107
Plotting a point, 107
Point(s)
coordinates of, 107, 675
distance between, 401–402
lattice, 411
plotting, 107
Polar coordinates, 675–676
exploring equations with, 845
Pole, 675
Polynomial equations, 194–201, 372–389
bounds on roots, 389 (Exs. 27, 28)
degree of, 377
double root, 196
irrational roots, 386–389
multiple roots, 196, 374, 377
in problem solving, 198–201
rational roots, 382–385
root, or solution, 194
solution by factoring, 194–195, 374
Polynomial inequality, 202
graph of solution set, 202–203
solution, 202–205
Polynomial(s), 167
adding, 168
degree of, 167, 168
division, 364–367
factoring, 183–192, 835
irreducible, 189
multiplying by monomial, 172
multiplying by polynomial, 174
prime, 189
quadratic, 188–189
simplified, 167
simplifying, 168, 169
special products of, 175
subtracting, 168–169
terms of, 167, 168, 188
Portfolio Projects, 848–855
Postulates, 81
See also under Properties.
Power(s), 7
of a complex number, 685
exploring, 837, 841
with irrational exponent, 459
of a power, 171
of a product, 171
Preparing for College Entrance Exams. *See* College Entrance Exams.
Prime, 179, 182
Prime factorization, 179–181
Prime polynomial, 189
Probability, 743–750
expectation, 751
exploring, 846
Problem solving strategies
checking solution, 50, 51, 69, 70, 199, 313, 371
estimating answer, 371
insufficient information, 51, 53
making a table, 49, 51–52, 132, 244, 248–249
making a sketch, 50, 51–52, 198, 312, 576, 585
plan, 49–50, 371
recognizing a pattern, 175, 183, 184, 501, 502, 503
unneeded information, 51–52, 53
Problems
aircraft, 131–132, 133–134, 248–249, 662–663, 665
areas of triangles, 597–599

checking answers, 50, 51, 69, 70, 199, 313, 371
communication networks, 779–781, 782–784
digit, 103, 106
distance/rate/time, 44, 47, 48, 51–52, 53, 246, 248–249, 250–251
dominance relations, 781–782, 783, 784
electrical circuits, 253–254
force, 672–674
geometric, 44, 46–48, 52–53, 71–72, 198–199, 200–201, 357, 362, 443–444, 449, 506, 534–536
growth and decay, 483–488, 493–494
historical note, 42
latitude, 611
maximum and minimum, 159–161, 340–341, 343–345
mixture, 244, 245, 246
money, 102–103, 105–106, 448, 449
music, 634–635
navigation, 661–663, 665
plan for solving, 49–50
probability, 746–750, 759–761
science, 151, 199–201, 224–225, 344, 351–354, 356–357, 359–362, 392, 394, 443, 472, 487, 579, 733
speed, 609, 611–612
three-dimensional, 39, 47, 444, 579, 611, 612
using fractional equations, 242–252
using inequalities, 69–72
using polynomial equations, 198–202
using systems of equations, 131–134, 442–443, 448–449
trigonometric, 578–579, 583–584, 589–590, 595–596, 599, 611–612
work, 243, 245, 246, 248, 250, 251
See also Applications, Challenge (problems).

Product(s), 7
of complex numbers, 293
dot, 668
estimating, 222
factors of, 7
polynomial and monomial, 172, 173
polynomial and polynomial, 174–176
power of, 171
of powers, 171
of rational expressions, 232–234
of real numbers, 27–31
scalar, 771

Progressions, 501–517
arithmetic, 501
geometric, 502

Proof, 16, 18–19, 81–91
by analytic geometry, 403, 405–406, 411
indirect, 273
informal (convincing argument), 62, 63, 87, 117, 157, 181, 182, 201, 204, 287, 321, 367, 379, 389, 482, 530, 600, 699, 723, 729

Properties of matrices and determinants
addition of matrices, 771
determinants, 798–799
matrix multiplication, 776
scalar multiplication, 772
See also Properties of real numbers.

Properties of real numbers
associative, 15
cancellation, 82, 87
closure, 15, 19
commutative, 15
completeness, 283
density, 285
distributive, 15, 23, 34, 86
of equality
addition, 14
multiplication, 14
reflexive, 14
symmetric, 14
transitive, 14
identity, 15
inverse, 15
multiplicative, of -1, 27, 84
multiplicative of zero, 27, 82
opposite of a product, 28, 83, 85
opposite of a sum, 29, 84
of opposites, 15, 85
of order
addition, 59
comparison, 59
multiplication, 59, 88–89
transitive, 59
of radicals, 261, 264
of reciprocals, 15
zero-product, 83, 86, 194

Proportion(s), 352
extremes, 352
means, 352
See also Variation.

Pythagoras, 273

Pythagorean theorem, 401

Quadrants, 107

Quadratic equation(s), 307
completing the square, 307–310, 319
discriminant, 317–318
double root, 317, 318
exploring graphs of, 838
quadratic formula, 311–315, 319
roots, nature of, 317–318
roots and coefficients, 338–341
solving, 307–321
systems of, 436–438, 439–443
See also Polynomial equations.

Quadratic formula, 311

Quadratic functions, 333–337, 340–341

Quadratic term, 188

Quantifiers, 20

Quartiles, 713–714

Quotient(s), 7, 33
estimating, 222
of monomials, 211–214
of rational expressions, 232–234
simplified, 212

Radian, 607
related to degree, 607–608

Radical(s), 261
binomials containing, 274–276
exploring, 837
properties of, 261, 264, 837
simplifying, 264–269
sums of, 270–273
equations containing, 277–281

Radical equations, 277–281

Radical sign, 259, 261

Radicand, 261

Radiocarbon dating, 493–494

Radius (radii)
of a circle, 407
focal, of ellipse, 418

Random experiment, 743

Random numbers, 753

Random sample, 752

Range
of a frequency distribution, 713, 714
of a function, 141
of a relation, 153

Rate of change, 147
Rational expressions, 227–241
products and quotients, 232–234
in simplest form, 227
sums and differences, 235–237
Rational functions, 228
graphing, 230–231
Rational numbers. *See under* Numbers.
Rational root theorem, 382, 385
Reading algebra, xiv–xv
independent study, 26
making a sketch, 585
probability, 751
problem solving, 371
symbols, xvi, 178
Real numbers, 1–2
decimal representations of, 283–285
equality of, 14
graphs of, 1–3
irrational, 285, 832
rational, 283–284
roots of, 259–263
set of, 1, 283
See also Properties of real numbers.
Reciprocal(s), 15, 33
property of, 15
Recorde, Robert, 42
Reflexive property, 14
Region, feasible, 160
Regression line, 726–727
Relations, 153–157
domain, 153
dominance, 781–782
graphs, 153, 154, 155
range, 153
Remainder theorem, 372
Replacement set, 9
Resultant, 659–660
Reviews. *See* Chapter reviews, Chapter tests, College Entrance Exams, Mixed Problem Solving, Mixed Review Exercises, Self-Tests.
Rhind Papyrus, 42
Root(s)
approximating irrational, 386–389, 836, 841
bounds on, 389 (Exs. 27, 28)
and coefficients, 338–341
conjugate imaginary, 318
cube, 260, 394–395
double, 196, 317, 318
extraneous, 247, 278
finding rational, 318
multiple, 196
*n*th, 260
of open sentence, 37
of polynomial equation, 194, 377–379, 382–383, 385, 386–388
and radicals, 259–282
square, 259, 288, 394–395
theorems regarding, 266
Roots of unity, 687–688
primitive, 688 (Ex. 22)
Rose
three-leafed, 677, 679 (Ex. 32)
four-leafed, 679 (Ex. 31)

Sample space, 743
Sampling, 752–753
Scalar, 660, 661, 771
Scalar product, 660–661, 771–772
Scatter plot, 724
Science. *See* Applications, Formulas, Problems.
Scientific notation, 218, 221–225, 478
Secant function, 556, 561, 613
graph, 632
inverse, 694
Second, 550
Self-Tests
Chap. 1: 13, 36, 54
Chap. 2: 72, 79, 92
Chap. 3: 123, 140, 158
Chap. 4: 177, 192, 205
Chap. 5: 226, 241, 252
Chap. 6: 282, 297
Chap. 7: 316, 325, 345
Chap. 8: 363, 381, 396
Chap. 9: 417, 435, 450
Chap. 10: 467, 477, 493
Chap. 11: 516, 536, 543
Chap. 12: 572, 600
Chap. 13: 634, 654
Chap. 14: 671, 688, 700
Chap. 15: 729, 742, 761
Chap. 16: 784, 793, 805
Sequence(s), 501–517
arithmetic, 501, 507–509
Fibonacci, 505
finite, 501
formula for *n*th term, 503, 507–508, 510
geometric, 502, 510–514
graphing, 517
infinite, 501
term of, 501
Series, 518–536
arithmetic, 518
geometric, 518
infinite geometric, 531–536
sums of, 525–536
Taylor, 573
Set(s)
dense, 285
disjoint, 754
empty, 38, 738
factor, 179, 183
intersection of, 754
null, 38
of primes, 179
replacement, 9
solution, 37, 61
union of, 754
Set notation, 153
Sigma notation, 518–520
Signs, Descartes' rule of, 378, 379
Sign graph, 203
Significant digit(s), 221–222
Simplest radical form, 266
Simplifying an expression, 7–8
Simultaneous solution, 124
Sine function(s), 555, 561
exploring addition of, 844
as function of real numbers, 613
graph, 624–629, 690
inverse, 690–691
Sines, law of, 586
Sinusoid, 634
Slope, 112
Solution(s)
of conjunction, 65, 66
of disjunction, 65, 66
of equations and inequalities having fractional coefficients, 242–244
of equations containing radicals, 277–279
of exponential equations, 460–461, 480
of linear equations in two variables, 108–110
of linear systems, 124–130, 444–449, 785–786
of open sentences, 37, 101–103
of open sentences involving absolute values, 73–74, 76–77
of polynomial inequalities, 202–203
primary and general, 698
of quadratic equations, 307–321
of quadratic systems, 436–437, 439–441

of trigonometric equations, 697–700
Speed, 609
angular, 609
Spirals, 679
Square(s)
completing the, 307–310, 319
difference of, 183
trinomial, 183
Square root(s), 259
approximating, 394–395
exploring, 832
of negative numbers, 259
principal, 259
table, 810
Standard deviation, 714, 715
Standard normal curve, 719–722
Standardized value in normal distribution, 721
Statistics
analyzing data, 391–395, 713–718
correlation, 724–729
gathering data, 752–753
normal distribution, 719–723
presenting data, 709–712
Stem-and-leaf plot, 710
Subset, 1, 738
Substitution, 8
synthetic, 373
transformation by, 125
Substitution principle, 8
Subtraction
definition, 22
of matrices, 771
of polynomials, 168–169
of rational expressions, 235–237
of real numbers, 22–25
Sum(s), 7, 19
of complex numbers, 293
of cubes, 184
of first *n* terms
arithmetic series, 525–526
geometric series, 526
of infinite geometric series, 531–536
opposite of, 29
of radicals, 270–273
of rational expressions, 235–237
of roots, 338–340
of two matrices, 770
of vectors, 659, 660
Summand, 518
Summation
limits of, 519
sign for, 518
Symbolic logic, 95–97
Symbols
absolute value, 3
division, 7
equality, 6
greater than, less than, 2
grouping, 7
history of, 42
inequality, 2, 6
logical, 20
multiplication, 7, 8
table of, xvi
words into, 43–48
Symmetric property, 14
Symmetry, axis of, 326, 412
Synthetic division, 368–370
Synthetic substitution, 373
Systems of equations
consistent, 128, 803
dependent, 128, 446, 803
equivalent, 124–125
homogeneous, 448
inconsistent, 128, 446–447, 803
infinitely many solutions, 127–128
linear in three variables, 444–449, 785–786, 802–803
linear in two variables, 124–134, 791–792, 801
quadratic, 436–438, 439–443
triangular form, 445–446, 785
used to solve word problems, 131–134, 442–443, 448–449
Systems of inequalities, 137, 139

Tables
common logarithms, 812–813
cubes and cube roots, 811
of measurement, xvii
squares and square roots, 810
in statistics, 709, 715, 716, 720, 724
symbols, xvi
trigonometric functions, 814–830
truth, 95–97
Tail wind, 131
Tangent function, 555, 561
as function of real numbers, 613
graph, 630–631, 693
inverse, 693–694
Tautology, 97
Technology. *See* Calculator Key-In, Calculators, Computer Exercises, Computer graphing ideas, Computer Key-In, *and* Science.
Term(s), 7
constant, in quadratic polynomial, 188
like, 23
linear, in quadratic polynomial, 188, 189
of polynomial, 167, 168
quadratic, in quadratic polynomial, 188
of sequence, 501
similar, 23
Tests. *See* Chapter tests, Self-Tests, College Entrance Exams.
Theorem(s), 81
about absolute value, 89
binomial, 541
about complex numbers, 682
conjugate root, 377, 381
De Moivre's, 685
Descartes' rule of signs, 378
dot product of vectors, 668
equation of line, 114
factor, 373
fundamental, of algebra, 377
about graphs and related equations, 108
intermediate-value, 387
about order, 89
about parallel and perpendicular lines, 119
proving, 81–90
Pythagorean, 401
converse of, 401
about radicals, 266
rational root, 382, 385
remainder, 372
roots of polynomial equation, 377–379
roots of quadratic equation, 338, 339
slope of line, 113
sum of infinite geometric series, 532
Thinking skills
Analysis. *See* Data analysis, Error analysis, Patterns, *and* Problem solving strategies.
Applying concepts. *See* Applications, Enrichment, Formulas, *and* Modeling.
Interpreting. *See* Patterns, Problem solving strategies, Reading Algebra, *and* Statistics.
Reasoning and inferencing. *See* Hypothesis, Induction, Logic, Probability, Proof, *and* Set(s).

Recall and transfer. *See* Applications, Enrichment, Formulas, Modeling, Mixed Problem Solving, *and* Problems.
Spatial perception. *See* Geometry, Modeling, *and* Problems, three-dimensional.
Synthesis. *See* Algorithm, division; Conjunction; Disjunction; Enrichment; Euclidean Algorithm; Mixed Problem Solving; Problems; *and* Proof.
See also Explorations.
Transformations
exploring geometric, 847
that produce equivalent equations, 37
that produce equivalent inequalities, 60
that produce equivalent systems, 125
Transitive property, 14
Translating
a graph, 327–328, 408, 432
words into symbols, 43–48, 70
Triangle(s)
ambiguous case, 592–593
areas of, 586, 597–600
reference, 563
solving general, 591–596
solving right, 574–579
Trigonometry, triangle, 574–600
Trigonometric functions
of acute angles, 555–560, 843
cofunction relationships, 557–558
of general angles, 561–567
identities, 636–654
inverses, 689–696
of real numbers, 613–617
tables, 568–571 (use of)
30°, 60°, 45° angles, 557–558
values of, 568–572
Trinomial(s), 174
factoring, 188–189
perfect square, 183
quadratic, 188
Truth table, 95–97

Union of sets, 754
unique, meaning of, 14
Unit circle, 613
Unit vector, 661

Value(s)
of function, 142, 334, 335
of numerical expression, 6
of variable, 9
Variable, 9
domain of, 9
replacement set of, 9
values of, 9
Variance, 714, 715
Variation
constant of, 351, 358
direct, 351–357, 839
inverse, 358–362, 429
joint, 359–362
Vertical line test, 154, 433
Vector(s), 659
addition, 660
in component form, 666
difference of, 661
dot product of, 668
equivalent, 659
initial and terminal points, 659, 666
norm of, 661
notation for, 659
orthogonal, 668–669
in the plane, 666–671
resultant of, 659–660
scalar multiplication, 660–661
sum, 659, 660
unit, 661
zero, 661
Vector quantity, 659
Venn, John, 754
Venn diagrams, 754–755
Vertex, 326, 412
Vertical-line test for function, 154

Watt, James, 68
Wind speed, 131
Word problems. *See* Problems.
Work, 672–673

***x*-axis,** 107
***x*-coordinate,** 107
***x*-intercept(s)**
of a graph of a polynomial function, 386
of a line, 119
of a parabola, 329

***y*-axis,** 107
***y*-coordinate,** 107
***y*-intercept**
of a line, 119
of a parabola, 329

Zero(s)
absolute value of, 3
degree of, 167
division by, 33
double, 196
as exponent, 216
of a function, 195, 386, 392
graph as origin, 1
multiplicative property, 27, 82
multiple, 196
Zero-product property, 83, 194

Answers to Selected Exercises

Chapter 1 Basic Concepts of Algebra

Written Exercises, pages 4–5 **1.** -6 **3.** 0 **5.** -1 **7.** -4 **9.** $-\frac{1}{2}$ **11.** $0 > -6$ **13.** $-3 < -1$ **15.** $6 > -5$ **17.** $0 > -4$

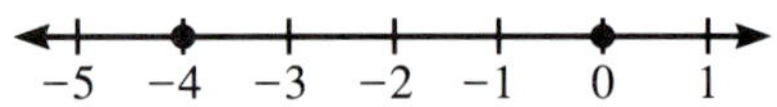

19. $\frac{1}{2} > -\frac{3}{2}$

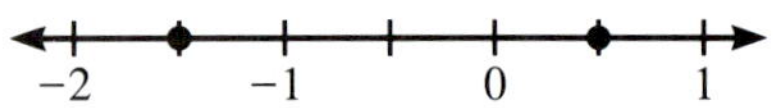

21. $0.5 > -1.5$

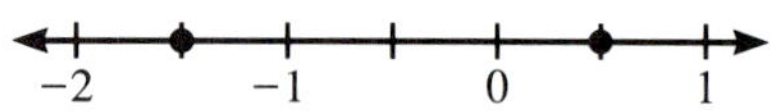

23. -5 **25.** 3 **27.** $-5, -1, 0, 2, 4$ **29.** $-2, -\frac{3}{2}, -\frac{1}{2}, 2, \frac{5}{2}$ **31.** $-2.6, -1.8, -1.6, -0.6$ **33.** -1 **35.** -3.6 **37.** $\frac{1}{3}$ **39.** $-1, 7$

Mixed Review Exercises, page 5 **1.** False **2.** False **3.** False **4.** True **5.** False **6.** False **7.** True **8.** True **9.** True **10.** True **11.** True **12.** False

Computer Exercises, page 5 **2. a.** $-8 < -7$ **b.** $0.1 > 0.01$ **c.** $0 = 0$ **d.** $3.65 > 3.56$ **e.** $4 > -4$ **f.** $800 < 8000$ **4. a.** 5.5 **b.** 0 **c.** 1.565 **d.** -5.16 **e.** 9.055 **f.** 1

Written Exercises, pages 10–12 **1.** $=$ **3.** $>$ **5.** $=$ **7.** $<$ **9.** $=$ **11. a.** 11 **b.** 1 **c.** 5 **13. a.** 44 **b.** 180 **c.** 84 **15.** 2 **17.** 3 **19.** $\frac{2}{3}$ **21.** 28 **23.** 3 **25.** 19 **27.** 343 **29.** 100 **31.** $\frac{7}{2}$ **33.** 625 **35.** $\frac{1}{3}$ **37.** 14 **39.** 32 **41.** 8 **43.** 90 **45.** $18 \div 2 - 3 \cdot (2 + 1) = 0$ **47.** $[6 - (5 - 3)] \cdot 2 = 8$ **49.** $(3^2 - 2)^2 - 4 \cdot (3 + 3) = 25$ **51. a.** 9, 36, 100, 225 **b.** 9, 36, 100, 225 **53.** Answers may vary; $3 \cdot 3 + (3 + 3) \div 3$

Note that for all Calculator Key-Ins your calculator may give answers to more places or fewer places than shown in these Answers to Selected Exercises. Also, some calculators round the last displayed digit, while others simply discard digits that cannot be displayed. Answers given here are rounded to the number of places shown.

Calculator Key-In, page 12 **1.** 45 **3.** 15 **5.** 4.25 **7.** 3.965517 **9.** 0.7777778 **11.** 22.7375

Self-Test 1, page 13

1.

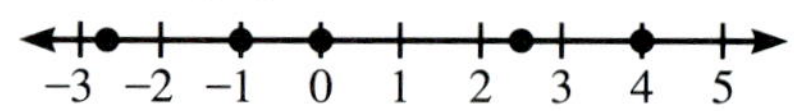

2. $-2 < -\frac{1}{2}$ **3.** False **4.** True **5.** True **6.** True **7.** 12 **8.** 4 **9.** 16 **10.** 3

Written Exercises, pages 17–19 **1.** 150 **3.** 0 **5.** $4z + 1$ **7.** 0 **9.** $2a$ **11.** True **13.** False **15.** True **17. a.** Dist. prop. **b.** Assoc. prop. of mult. **c.** Prop. of reciprocals **d.** Ident. prop. of mult. **19. a.** Comm. prop. of add. **b.** Assoc. prop. of add. **c.** Prop. of opposites **d.** Ident. prop. of add. **e.** Prop. of reciprocals **21. a.** Assoc. prop. of add. **b.** Ident. prop. of mult. **c.** Dist. prop. **d.** Ident. prop. of mult. **e.** Dist. prop. **23. a.** Add. prop. of = **b.** Assoc. prop. of add. **c.** Prop. of opposites **d.** Ident. prop. of add. **e.** Mult. prop. of = **f.** Assoc. prop. of mult. **g.** Prop. of reciprocals **h.** Ident. prop. of mult. **25. a.** Closed under addition **b.** Closed under multiplication **27. a.** Closed under addition **b.** Closed under multiplication **29. a.** Closed under addition **b.** Closed under multiplication **31. a.** Closed under addition **b.** Closed under multiplication **33. a.** Not closed under addition $\left(\frac{2}{3} + \frac{2}{3} = \frac{4}{3}\right)$ **b.** Closed under multiplication **35.** No; Prop. of reciprocals **37.** No; Ident. prop. of add.

Mixed Review Exercises, page 20 **1.** $>$ **2.** $<$ **3.** $<$ **4.** $=$ **5.** $<$ **6.** $<$

Extra, page 20 **1.** $\forall_x\ 2x = x + x$ **3.** $\forall_x \forall_y\ x + y = y + x$ **5.** False **7.** True

Written Exercises, pages 24–25 **1.** -21 **3.** 34 **5.** 25 **7.** 7.4 **9.** -23 **11.** 30.7 **13.** -2 **15.** 8 **17.** 19 **19.** -9 **21.** -21 **23.** $4p - 8q$ **25.** $a - 2b + 3$ **27.** $10c + 6$ **29.** $11x - 2$ **31.** $-3p - q$ **33.** $-5s + 9$ **35.** 9 **37.** -3 **39.** 4 **41.** 10.1 **43.** 152.7°C **45.** 3822 ft below sea level **47.** $74\frac{5}{8}$ **49.** Answers may vary; $7 - 2 \neq 2 - 7$ **51.** No; $0 - 1 = -1$ **53.** No; $1 - 2 = -1$ **55.** Yes

Reading Algebra, page 26 **1.** Solve certain equations in one variable. **3.** An equation or inequality containing a variable; equations having the same solution set over a given domain **5.** {3} **7.** Table 7, p. 830

Written Exercises, pages 30–31 **1.** -210 **3.** 12 **5.** -2.4 **7.** $60xy$ **9.** $-6abc$ **11.** 10 **13.** 96 **15.** 96 **17.** $-2 + 4x + 6x^2$ **19.** $-3z^2 + 2z - 1$ **21.** $-9xy$ **23.** $-5k - 24$ **25.** $2x - 2y$ **27.** $-18a^3 - 16$ **29.** $2tw - 13t - 14$ **31. a.** 3 **b.** -3 **c.** 0 **33. a.** 0 **b.** 0 **c.** 4 **35.** 1. Ident. prop. of mult. 2. Prop. of opp. of a prod. 3. Comm. prop. of mult. 4. Trans. prop. of = 5. Symm. prop. of =

Mixed Review Exercises, page 31 **1.** 0 **2.** 1 **3.** 7 **4.** -1 **5.** -4 **6.** 14 **7.** Comm. prop. of add. **8.** Ident. prop. of add. **9.** Symm. prop. of = **10.** Prop. of reciprocals **11.** Assoc. prop. of mult. **12.** Dist. prop.

Computer Exercises, page 32 **2. a.** 3 **b.** 380.585 **c.** -127.375 **d.** -212.28352 **4. a.** 5.897 **b.** -186.542125 **c.** 29.921848 **d.** 392.66634

Written Exercises, page 35 **1.** 3 **3.** 8 **5.** -18 **7.** 4 **9.** 48 **11.** $-\frac{3}{2}$ **13.** 8 **15.** $-\frac{2}{5}$ **17.** -54 **19.** $-3x^2 - 9$ **21.** $n^2 - 1$ **23.** $6c^2 - 4c - 1$ **25. a.** 2 **b.** 2 **c.** 0 **27. a.** -1 **b.** $\frac{3}{2}$ **c.** -15 **29.** Answers may vary; $4 \div 2 \neq 2 \div 4$ **31.** Yes **33.** Yes **35.** Yes

Self-Test 2, page 36 **1. a.** Comm. prop. of add. **b.** Assoc. prop. of add. **c.** Prop. of opposites **d.** Indent. prop. of add. **e.** Prop. of reciprocals **2.** 2 **3.** -10 **4.** 8 **5.** -2.8 **6.** -9 **7.** $3cd$ **8.** 25 **9.** 2 **10.** -2 **11.** $3x^2 - 5x + 1$ **12.** -1

Written Exercises, pages 40–42 **1.** {3} **3.** {12} **5.** {6} **7.** {2} **9.** {2} **11.** ∅ **13.** {3} **15.** {−7.5} **17.** {−4} **19.** {3} **21.** {real numbers} **23.** {4} **25.** Yes; No **27.** Yes; Yes **29.** No; Yes **31.** $x = \frac{5}{2}y + 5$ **33.** $p = \frac{I}{rt}$ **35.** $x = \frac{y - b}{m}$ **37.** $w = \frac{1}{2}p - l$ **39.** $x = 2b + \frac{c}{a}$ **41.** $v = \frac{s}{t} + \frac{1}{2}gt$ **43.** $h = \frac{2}{\pi}$ **45.** $t = 1.5$ **47.** $b_1 = 28$ **49.** $s = 15$ **51.** $h = \frac{A}{2\pi r} - r$ **53.** $a = \frac{S(1 - r)}{1 - r^n}$

Mixed Review Exercises, page 42 **1.** 8 **2.** -6 **3.** 25 **4.** 4 **5.** -12 **6.** -2 **7.** $x - 8$ **8.** $4ab$ **9.** $-x^2y^3$ **10.** $-3c + 4d$ **11.** 0 **12.** $m - 1$

Written Exercises, pages 46–48 **1.** $(60 - w)$ ft **3.** $\frac{1}{2}x + 2$ **5. a.** $l(l - 2)$ cm^2 **b.** $(4l - 4)$ cm **7.** $9r$ mi **9.** $\left(120 - \frac{a}{3}\right)^\circ$ **11.** $\frac{1}{3}x(2x - 3)^2$ **13.** $(4t - 250)$ dollars **15.** $(0.15q + 1.80)$ dollars **17–29.** Answers may vary; examples are given. **17.** Let x = the length of shortest side; $x + (x + 2) + (x + 4) + (x + 6) = 60$ **19.** Let x = the regular price; $x + 2(x - 2) = 41$ **21.** Let g = the amount of money that Greg has; $g + (g + 12) + 2(g + 12) = 124$ **23.** Let c = the car's speed; $2c + 2\left(\frac{2}{3}c\right) = 140$ **25.** Let b = the measure of $\angle B$; $(b + 20) + b + 2b + 4b = 360$ **27.** Let c = the number of kilograms of cashews; $7c + 9(20 - c) = 7.80(20)$ **29.** Let r = Kevin's speed going; $\frac{320}{r} + \frac{1}{3} = \frac{320}{r - 4}$

Challenge, page 48 Answers may vary; $23 = 2^3 + 2^3 + 1^3 + 1^3 + 1^3 + 1^3 + 1^3 + 1^3 + 1^3$

Problems, pages 52–54 **1.** Amy: \$11; Maria: \$19 **3.** 180 **5.** 84 ft × 49 ft **7.** 51° **9.** 12:15 P.M. **11.** 34 dimes, 6 quarters **13.** not enough information **15.** no solution **17.** $5\frac{1}{2}$% and $8\frac{1}{2}$%; extra information **19.** 2:30 P.M. **21.** 80 ppm/h **23.** winner: 728 votes; loser: 673 votes

Mixed Review Exercises, page 54 **1.** {−2} **2.** {3} **3.** {−1} **4.** -20 **5.** 100 **6.** 26 **7.** -1 **8.** -3 **9.** 2

Self-Test 3, page 54 **1.** {5} **2.** {2} **3.** $b = 2m - a$ **4.** $2(n + n^2)$ **5.** $4x + 10$ **6.** $2\frac{1}{2}$ h

Chapter Review, pages 55–56 **1.** c **3.** d **5.** a **7.** c **9.** d **11.** c **13.** b **15.** c **17.** a

Chapter 2 Working with Inequalities

Written Exercises, pages 62–63

1. $\{x: x > 2\}$

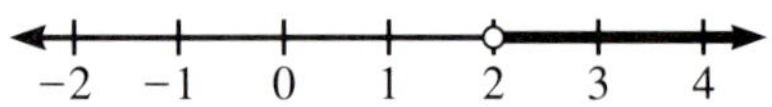

3. $\{t: t < 3\}$

5. $\{x: x > -2\}$

7. $\{t: t < -3\}$

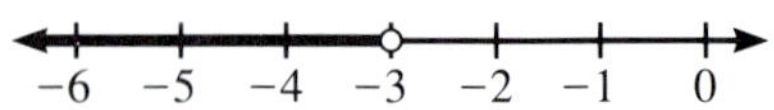

9. $\{s: s > -1\}$

11. $\{y: y > 4\}$

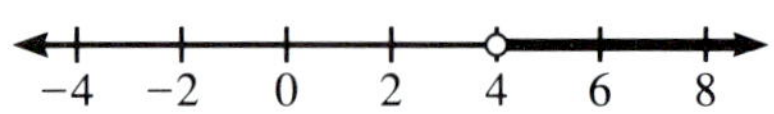

13. $\{h: h > -1\}$

15. $\{u: u > -3\}$

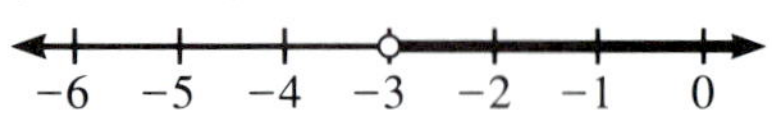

17. ∅ **19.** $\{y: y > -1\}$

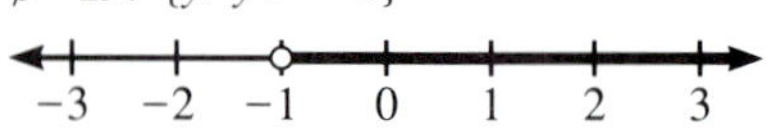

21. $\{k: k < 1\}$ **23.** {real numbers} **25.** True **27.** False; $-1 < 0$, but $(-1)^2 > 0^2$. **29.** True **31.** True **33.** If you multiply any number between 0 and 1 by itself, the result is smaller than the original number.

Mixed Review Exercises, page 63 **1.** $16t - 10$ **2.** -60 **3.** $-16p^3$ **4.** 12 **5.** $-4cd + 3$ **6.** 2 **7.** 14 **8.** 54 **9.** $a^2 - 5a + 7$ **10.** $8x - 5y$ **11.** -5 **12.** $11 - 6m$

Computer Exercises, page 64 **2. a.** {1, 2, 3, 4, 5, 6} **b.** {5, 6, 7} **c.** ∅ **4. a.** No solutions **b.** {−10, −9, −8, −7, −6, −5, −4, −3, −2, −1, 0, 1} **c.** {1}

Written Exercises, pages 67–68

1. $\{x: 3 \le x < 5\}$

3. $\{t: t < 1 \text{ or } t \ge 3\}$

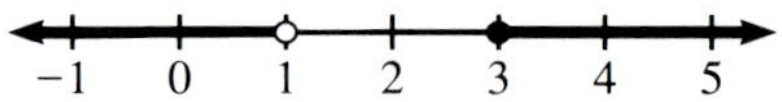

5. $\{y: y \ge 3\}$

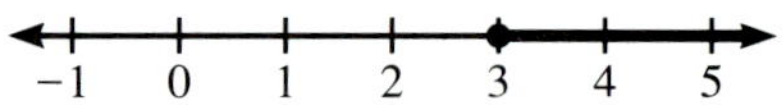

7. {real numbers}

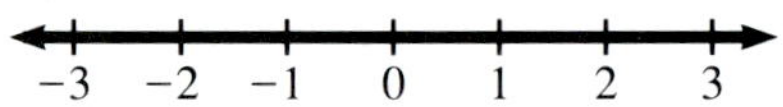

9. $\{x: 2 \le x < 5\}$

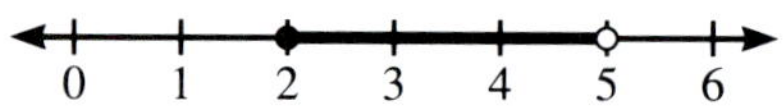

11. $\{r: -2 < r < 2\}$

13. $\{z: z \le 3 \text{ or } z > 5\}$

15. $\{t: t \ge 3 \text{ or } t < 2\}$

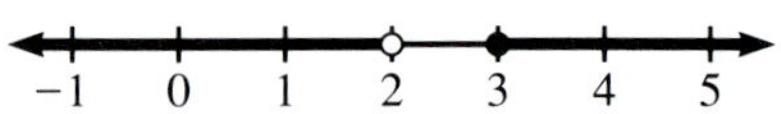

17. ∅

19. $\{k: -1 < k < 3\}$

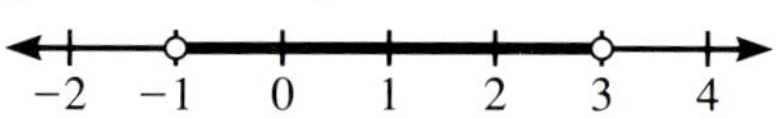

21. $\{d: 9 \le d < 15\}$

23. {real numbers}

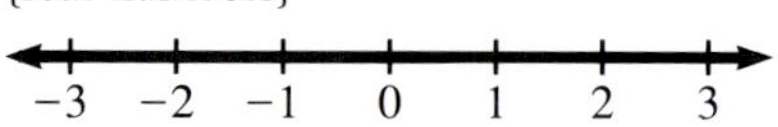

25. $\{x: -1 < x < 8\}$

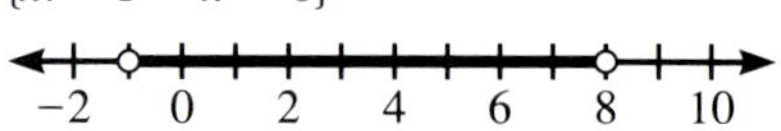

27. {real numbers}

29. $\left\{t: 0 < t \le \frac{9}{2}\right\}$ **31.** $\left\{t: t > -\frac{4}{3}\right\}$
33. $\{x: -2 \le x \le 4\}$
35. $\{y: -2 < y < -1 \text{ or } 3 < y < 4\}$

Problems, pages 71–72 **1.** At most 64 **3.** 9 cm **5.** {5, 7, 9}, {7, 9, 11} **7.** 93 or more **9.** At most 8 cm **11.** At least 160 min **13.** At most $2\frac{1}{2}$ min **15.** Width: at least 13 m; length: at least 18 m

Mixed Review Exercises, page 72
1. $\{x: x \ge -2\}$

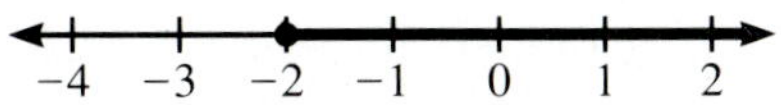

2. $\{t: t \le -4 \text{ or } t \ge 1\}$

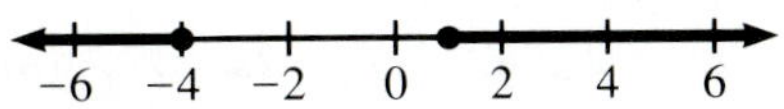

3. $\{m: m < -5\}$

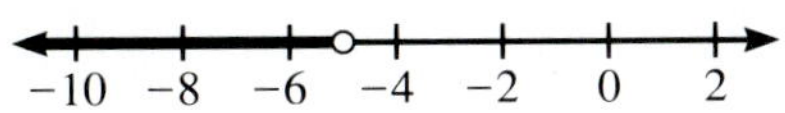

4. $\emptyset$
5. $\{d: d > 1\}$

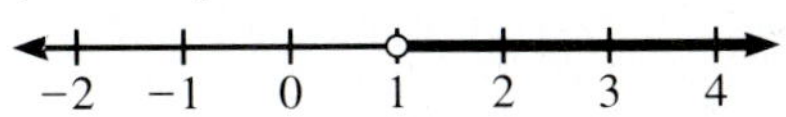

6. $\{p: 0 < p < 3\}$

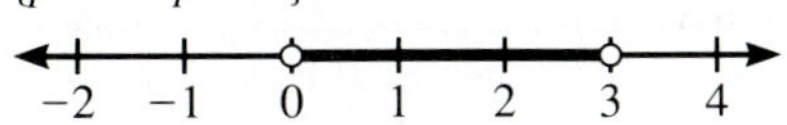

7. 4 **8.** 10 **9.** 21 **10.** 21

Self-Test 1, page 72
1. $\{m: m \le 3\}$

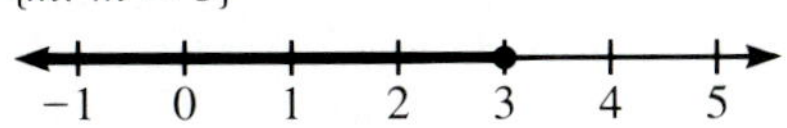

2. $\{y: y > -2\}$

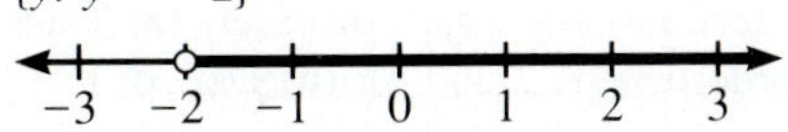

3. $\{x: x < 5\}$

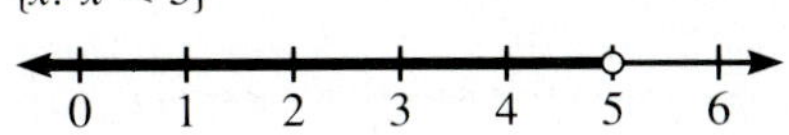

4. $\{n: n \le -1\}$

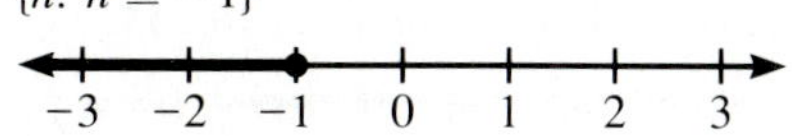

5. $\{a: -1 < a < 1\}$

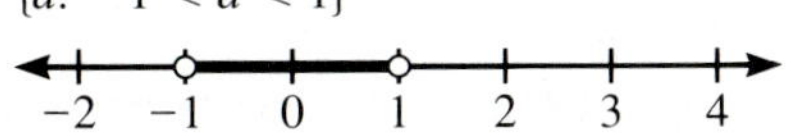

6. $\{c: c < -3 \text{ or } c \ge 3\}$

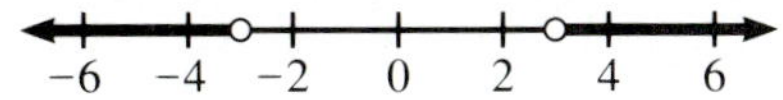

7. 83 or more

Written Exercises, page 75
1.

3.

5.

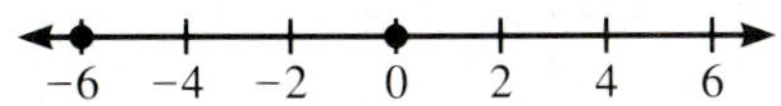

7.

9. $\{t: -4 < t < -1\}$

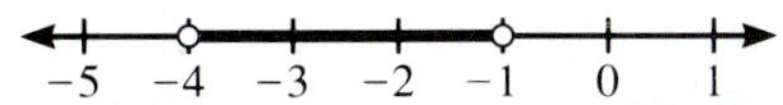

11. $\left\{\frac{5}{2}\right\}$

13. $\{x: x \le 1 \text{ or } x \ge 5\}$

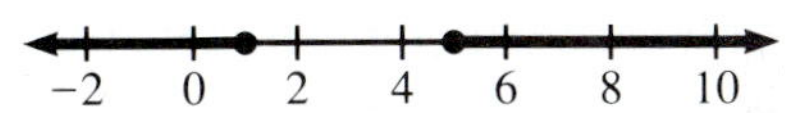

15. {real numbers}

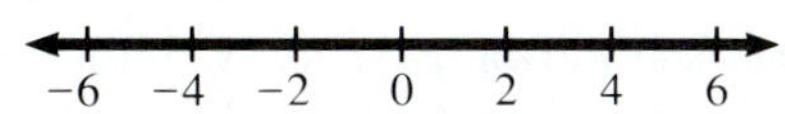

17. $\{t: 0 \le t \le 4\}$

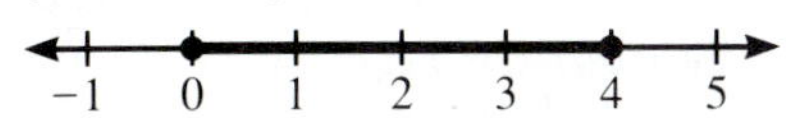

19. $\{-1, -9\}$ **21.** $\{u: -1 \le u \le 2\}$
23. $\left\{d: \frac{3}{2} \le d \le 2\right\}$ **25.** $\left\{t: t < \frac{4}{3} \text{ or } t > 2\right\}$
27. $\emptyset$
29.

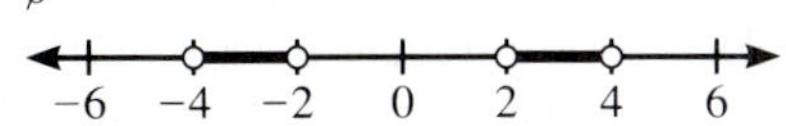

31.

33. $\{x: -3 \le x \le 1\}$

Computer Exercises, page 75 **2. a.** $x = -3, -2, -1, 0, 1, 2, 15, 16, 17, 18, 19$
b. $x = -12, -11, -10, -9, -8, -7, -6, -5, -4, -3, -2$

Written Exercises, pages 78–79 **1.** $\{-4, 4\}$
3. $\{t: -3 < t < 3\}$ **5.** $\{u: u \le -2 \text{ or } u \ge 2\}$
7. $\{y: -1 < y < 5\}$ **9.** $\{t: -1 \le t \le 4\}$
11. $\{r: r < -7 \text{ or } r > 3\}$ **13.** $\{-1, 2\}$
15. $\{p: p \le -4 \text{ or } p \ge -1\}$ **17.** $\{t: 1 \le t \le 5\}$
19. $\left\{x: \frac{1}{2} < x < \frac{5}{2}\right\}$ **21.** $\left\{f: -\frac{13}{3} < f < -\frac{5}{3}\right\}$

23. $\{t: -8 < t < 2\}$
25. $\{x: x \le a - c \text{ or } x \ge a + c\}$
27. $\{x: 0 < x < 2c\}$
29. $\left\{x: x \le \frac{-a-c}{b} \text{ or } x \ge \frac{-a+c}{b}\right\}$
31. $\left\{x: x < -\frac{a-c}{b} \text{ or } x > \frac{a-c}{b}\right\}$
33. $\left\{x: -a - \frac{c}{b} < x < -a + \frac{c}{b}\right\}$

Mixed Review Exercises, page 79
1. $\{x: -1 \le x \le 3\}$ **2.** $\{1, 4\}$ **3.** $\{c: c > 3\}$
4. $\{y: y \neq -1\}$ **5.** $\{w: w \le -2 \text{ or } w \ge 2\}$
6. $\{p: p < -3 \text{ or } p \ge 2\}$ **7.** $\{z: z \le -2\}$
8. $\{-4\}$ **9.** $\{m: -2 < m < 4\}$

Self-Test 2, page 79 **1.** $|3x + 2| \le 4$
2. $\{x: x < -2 \text{ or } x > 1\}$

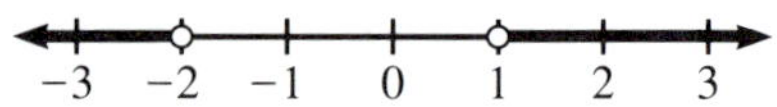

3. $\{y: 2 \le y \le 4\}$

4. $|x + 2| \ge 5$ **5.** $\{m: 2 \le m \le 6\}$
6. $\{n: n < -2 \text{ or } n > -1\}$

Computer Key-In, page 80 **1.** $-1.5 < x < 4$
3. $x < -0.2$ or $x > 0.6$ **5.** $-8 < x < 5$
7. No solution

Written Exercises, pages 85–87 $(-1)^2 = 1^2$, but $-1 \neq 1$ **3.** $|0 - 1| \neq |0| - |1|$
5. (1) Comm. prop. of mult.; (2) Proved in Example 4; (3) Comm. prop. of mult. **7.** (1) Def. of subtr.; (2) Assoc. prop. of add.; (3) Prop. of opp.; (4) Ident. prop. of add. **9.** (1) Given; (2) Def. of u^2; (3) Prop. of recip.; (4) Mult. prop. of eq.; (5) Assoc. prop. of mult.; (6) Prop. of recip.; (7) Ident. prop. of mult.
11. (1) Given; (2) Prop. of recip.; (3) Mult. prop. of eq.; (4) Mult. prop. of 0; (5) Assoc. prop. of mult.; (6) Prop. of recip.; (7) Ident. prop. of mult.; (8) Steps 2–7 **13.** (1) Given, prop. of recip.; (2) Def. of div.; (3) Dist. prop. of mult. with respect to subtr.; (4) Def. of div.
15. (1) $a \neq 0$ (Given); (2) $a \cdot \frac{1}{a} = 1$ (Prop. of recip.); (3) $\frac{1}{a} \cdot \frac{1}{\frac{1}{a}} = 1$ (Prop. of recip.); (4) $\therefore a = \frac{1}{\frac{1}{a}}$ (Steps 2 and 3 and prop. of recip. uniqueness) **17.** (1) $b \neq 0$ (Given); (2) $(-b)\left(-\frac{1}{b}\right) = \frac{1}{b} \cdot b$ (Prod. of opp.) (3) $(-b)\left(-\frac{1}{b}\right) = 1$ (Prop. of recip.); (4) $\therefore -\frac{1}{b} = \frac{1}{-b}$ (Prop. of recip. uniqueness)
19. (1) $b \neq 0$ (Given); (2) $\frac{-a}{-b} = (-a)\left(\frac{1}{-b}\right)$ (Def. of div.); (3) $\frac{-a}{-b} = (-a)\left(-\frac{1}{b}\right)$ (Proved in Exercise 17); (4) $\frac{-a}{-b} = a \cdot \frac{1}{b}$ (Product of opp. (Ex. 6)); (5) $\therefore \frac{-a}{-b} = \frac{a}{b}$ (Def. of div.)
21. (1) $b \neq 0$, $d \neq 0$ (Given);
(2) $\frac{a}{b} \cdot \frac{c}{d} = \left(a \cdot \frac{1}{b}\right) \cdot \left(c \cdot \frac{1}{d}\right)$ (Def. of div.);
(3) $\frac{a}{b} \cdot \frac{c}{d} = (ac)\left(\frac{1}{b} \cdot \frac{1}{d}\right)$ (Assoc. and Comm. prop. of mult.); (4) $\frac{a}{b} \cdot \frac{c}{d} = (a \cdot c)\left(\frac{1}{bd}\right)$ (Proved in Ex. 20); (5) $\therefore \frac{a}{b} \cdot \frac{c}{d} = \frac{ac}{bd}$ (Def. of div.)
23. (1) $c \neq 0$, $d \neq 0$ (Given);
(2) $\frac{a}{b} \div \frac{c}{d} = \frac{a}{b} \cdot \frac{1}{\frac{c}{d}}$ (Def. of div.);
(3) $\frac{a}{b} \div \frac{c}{d} = \frac{a}{b} \cdot \frac{d}{c}$ (Proved in Ex. 22);
(4) $\therefore \frac{a}{b} \div \frac{c}{d} = \frac{ad}{bc}$ (Proved in Ex. 21) **25.** Let the numbers be a and b. If $ab = 0$, then at least one of a and b must be 0 by the zero-product property. But a and b are given to be nonzero.

Written Exercises, pages 90–91 **1.** (1) Given; (2) Second mult. prop. of order; (3) Mult. prop. of -1 **3.** (1) Given; (2) Add. prop. of order; (3) Given; (4) Add. prop. of order; (5) Trans. prop. of order (Steps 2 and 4) **5.** (1) Given; (2) Second mult. prop. of order; (3) Given; (4) Second mult. prop. of order; (5) Trans. prop. of order (Steps 2 and 4); (6) Def. of a^2 and b^2
7. (1) $a > 0$ (Given); (2) $a + (-a) > 0 + (-a)$ (Add. prop. of order); (3) $0 > 0 + (-a)$ (Prop. of opp.); (4) $\therefore 0 > -a$, or $-a < 0$ (Ident. prop. of add.) **9.** (1) $a > 0$, $b > 0$ (Given); (2) $ab > 0 \cdot b$ (First mult. prop. of order); (3) $\therefore ab > 0$ (Mult. prop. of 0) **11.** (1) $a > 0$, $b > 0$ (Given); (2) $a \cdot \frac{1}{b} > 0 \cdot \frac{1}{b}$ (First mult. prop. of order (assuming $\frac{1}{b} > 0$)); (3) $a \cdot \frac{1}{b} > 0$ (Mult. prop. of 0); (4) $\therefore \frac{a}{b} > 0$ (Def. of div.)

13. (1) $a > 0$, $b < 0$ (Given); (2) $a \cdot \frac{1}{b} < 0 \cdot \frac{1}{b}$ (Second mult. prop. of order (assuming $\frac{1}{b} < 0$)); (3) $a \cdot \frac{1}{b} < 0$ (Mult. prop. of 0); (4) $\therefore \frac{a}{b} < 0$ (Def. of div.) **15.** (1) $a > 0$, $b > 0$, $\frac{1}{a} > \frac{1}{b}$ (Given); (2) $ab > 0$ (Exercise 9); (3) $ab \cdot \frac{1}{a} > ab \cdot \frac{1}{b}$ (First mult. prop. of order); (4) $\left(a \cdot \frac{1}{a}\right)b > a\left(b \cdot \frac{1}{b}\right)$ (Assoc. and Comm. prop. of mult.); (5) $1 \cdot b > a \cdot 1$ (Prop. of recip.); (6) $\therefore b > a$, or $a < b$ (Ident. prop. of mult.). **17.** (1) $a < b$ (Given); (2) $a + a < a + b$ and $a + b < b + b$ (Add. prop. of order); (3) $2a < a + b$ and $a + b < 2b$ (Simplification); (4) $\therefore a < \frac{a+b}{2}$ and $\frac{a+b}{2} < b$, or $a < \frac{a+b}{2} < b$ (First mult. prop. of order (using the fact $\frac{1}{2} > 0$))

Mixed Review Exercises, page 91 **1.** True **2.** True **3.** True
4. {2, 4}

5. $\{d: d \geq -1\}$

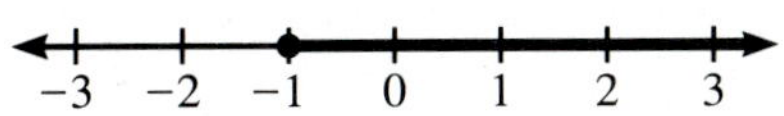

6. $\{y: -1 < y < 3\}$

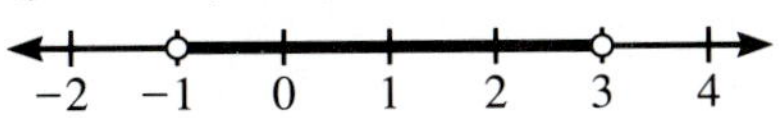

7. {2}

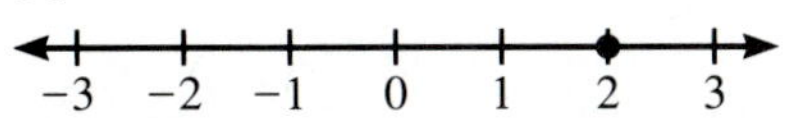

8. $\emptyset$ **9.** $\{k: k < -4 \text{ or } k \geq 4\}$

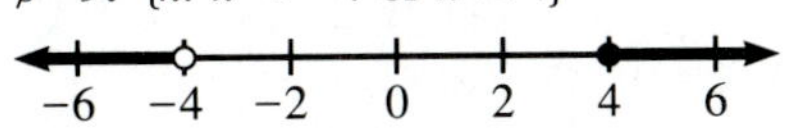

Self-Test 3, page 92 **1.** (1) Given; (2) Prop. of opp.; (3) Add. prop. of eq.; (4) Assoc. prop. of add.; (5) Prop. of opp.; (6) Ident. prop. of add.; (7) Def. of subtr. **2.** (1) $ab = c$, $b \neq 0$ (Given); (2) $\frac{1}{b}$ is a real number (Prop. of recip.); (3) $ab \cdot \frac{1}{b} = c \cdot \frac{1}{b}$ (Mult. prop. of eq.); (4) $a\left(b \cdot \frac{1}{b}\right) = c \cdot \frac{1}{b}$ (Assoc. prop. of mult.); (5) $a \cdot 1 = c \cdot \frac{1}{b}$ (Prop. of recip.); (6) $a = c \cdot \frac{1}{b}$ (Ident. prop. of mult.); (7) $\therefore a = \frac{c}{b}$ (Def. of div.). **3.** (1) $a > 1$ (Given); (2) $a > 0$ (Trans. prop. of order (using the fact $1 > 0$)); (3) $a \cdot a > a \cdot 1$ (First mult. prop. of order); (4) $a \cdot a > a$ (Ident. prop. of mult.); (5) $\therefore a^2 > a$ (Def. of a^2)

Chapter Review, pages 93–94 **1.** b **3.** a **5.** b **7.** c **9.** b **11.** d **13.** c

Extra, page 97 **1.** true **3.** true **5.** true **7.** true **9.** true **11.** p is true and $\sim q$ is true; $p \wedge \sim q$ is true. **13.** Logically equivalent **15.** Not a tautology **17.** Tautology

Mixed Problem Solving, page 98 **1.** At least 14 quarters **3.** 32 h **5.** \$1600 at 5%; \$2400 at 8% **7.** $\frac{32}{d}$ **9.** 60° **11.** 3 lb **13.** 7 cm × 12 cm

Preparing for College Entrance Exams, page 99 **1.** C **3.** C **5.** D **7.** A

Chapter 3 Linear Equations and Functions

Written Exercises, pages 104–105

1. $\left\{(-1, 3), \left(0, \frac{7}{3}\right), (2, 1)\right\}$
3. $\left\{\left(-1, \frac{1}{2}\right), (0, 0), (2, -1)\right\}$ **5.** $\left\{(-1, -1), \left(0, -\frac{5}{9}\right), \left(2, \frac{1}{3}\right)\right\}$
7. $\left\{\left(-2, \frac{11}{3}\right), \left(1, \frac{5}{3}\right), \left(3, \frac{1}{3}\right)\right\}$
9. $\left\{(-2, 1), \left(1, -\frac{1}{2}\right), \left(3, -\frac{3}{2}\right)\right\}$
11. $\left\{\left(-2, -\frac{13}{9}\right), \left(1, -\frac{1}{9}\right), \left(3, \frac{7}{9}\right)\right\}$
13. 6, 4, 3 **15.** $-\frac{7}{2}, \frac{7}{5}, -11$
17. $-\frac{1}{2}, -1, 2$ **19.** $-\frac{1}{4}, 1, 2$ **21.** 5
23. -2 **25.** -3
27. {(0, 4), (1, 3), (2, 2), (3, 1), (4, 0)}
29. {(0, 15), (1, 11), (2, 7), (3, 3)}
31. {(0, 6), (3, 4), (6, 2), (9, 0)}
33. {(1, 1), (1, 2), (1, 3), (2, 1), (2, 2), (3, 1)}
35. {(1, 1), (1, 2), (1, 3), (2, 1), (2, 2), (3, 1), (3, 2), (4, 1)} **37.** {(1, 3), (6, 2), (9, 1)}

39. Let $N = 10t + u$. Then $K = 10u + t$. $N - K = (10t + u) - (10u + t) =$ $10t - t + u - 10u = 9t - 9u = 9(t - u)$. So $N - K$ is an integral multiple of 9.

Problems, pages 105–106 **1.** 3 \$5 and 3 \$20; or 7 \$5 and 2 \$20; or 11 \$5 and 1 \$20; or 15 \$5 and 0 \$20 **3.** 7 dimes and 1 quarter; or 2 dimes and 3 quarters **5.** 4 cm, 4 cm, 4 cm and 7 cm; or 5 cm, 5 cm, 5 cm and 4 cm; or 6 cm, 6 cm, 6 cm and 1 cm **7.** 34 nickels, 3 dimes, and 0 quarters; 27 nickels, 4 dimes, and 1 quarter; or 20 nickels, 5 dimes, and 2 quarters; or 13 nickels, 6 dimes, and 3 quarters; or 6 nickels, 7 dimes and 4 quarters **9.** 41, 51, 61, 63, 71, 73, 81, 83, 85, 91, 93, 95 **11.** 10, 11, 12, 13, 20, 21 **13.** 4, 15 and 8; 5, 12 and 10; 6, 9 and 12; 7, 6, and 14

Mixed Review Exercises, page 106 **1.** −26 **2.** −36 **3.** 7 **4.** −25 **5.** −2 **6.** 12 **7.** 9 **8.** −7

9. $\{y: -1 < y \leq 4\}$

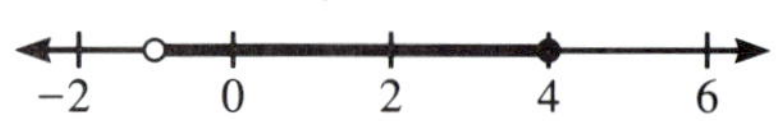

10. $\{m: m < 2 \text{ or } m > 4\}$

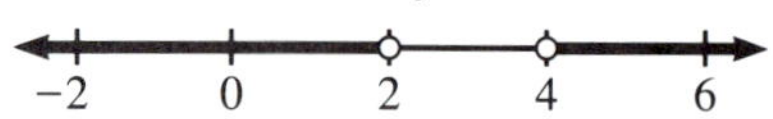

11. $\{n: n \geq 4\}$

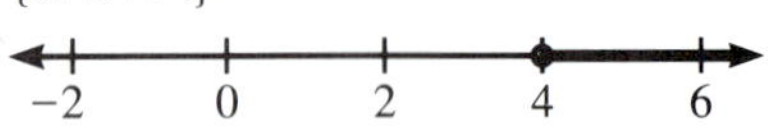

Written Exercises, page 111

1.

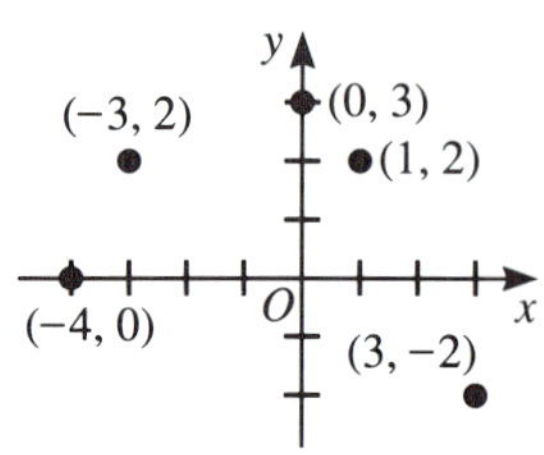

3.

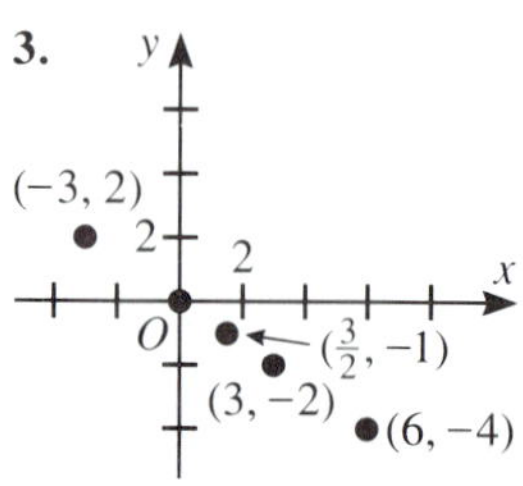

5.

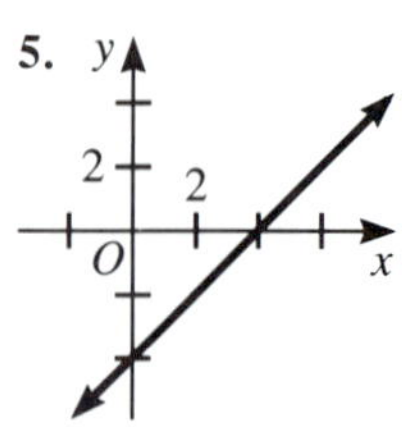

7.

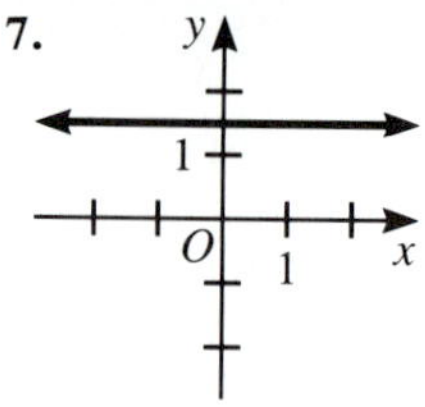

9.

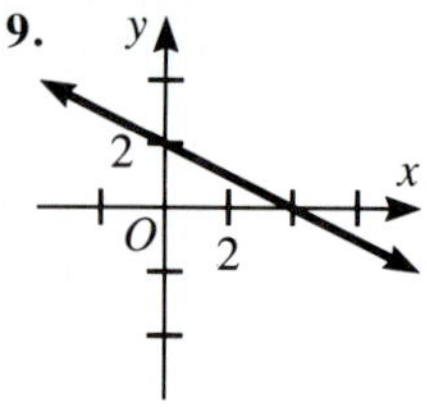

11.

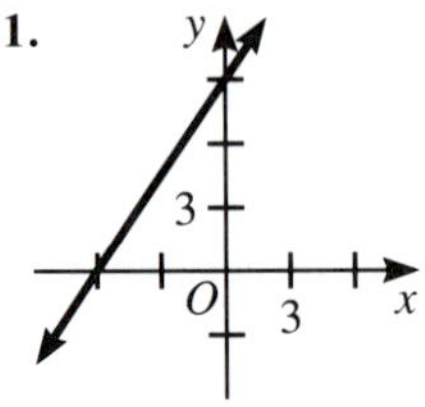

13.

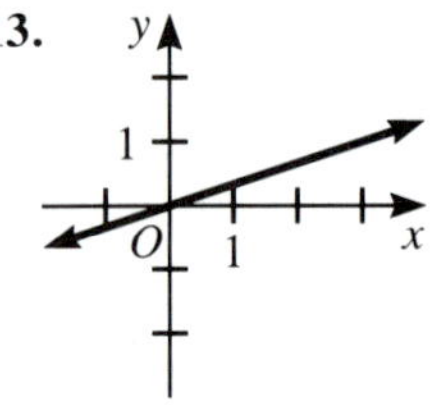

15.

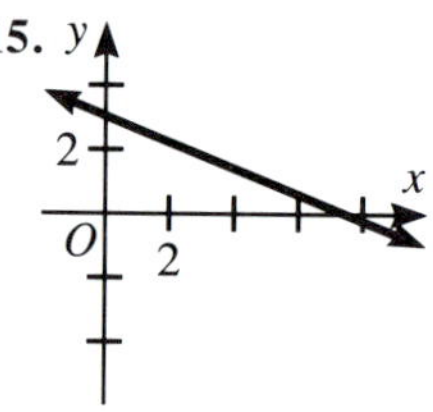

17.

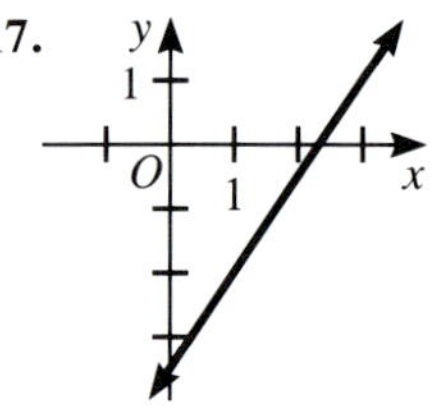

19.

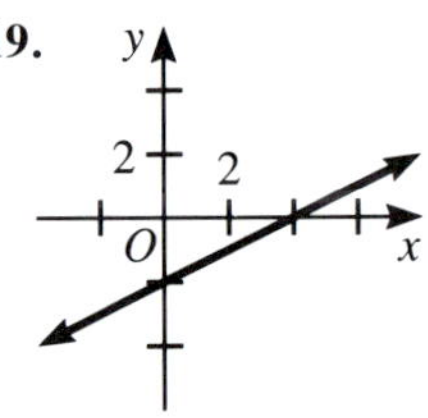

21.

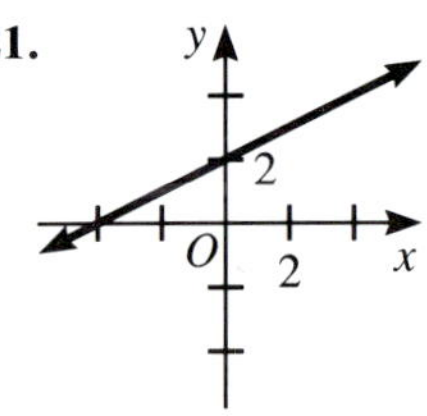

23. 2 **25.** 0

27. (5, −2)

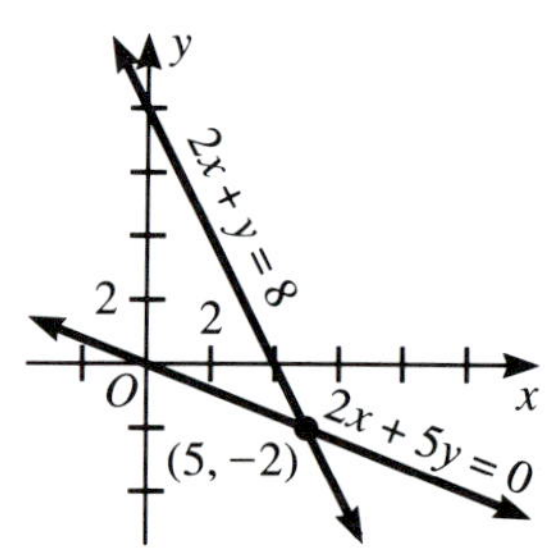

29. (−3, 5)

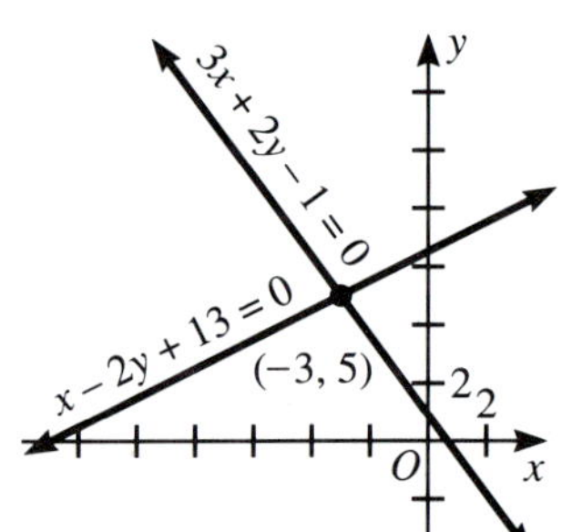

31.
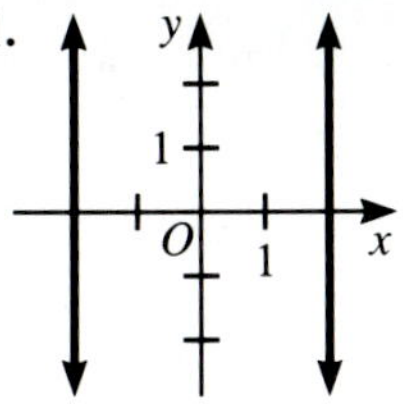

33.
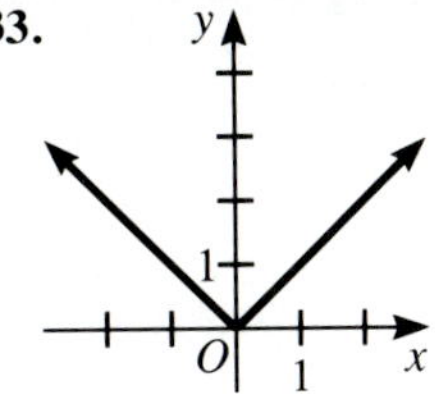

35.
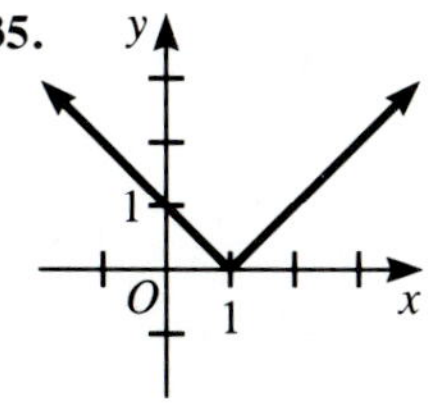

37.
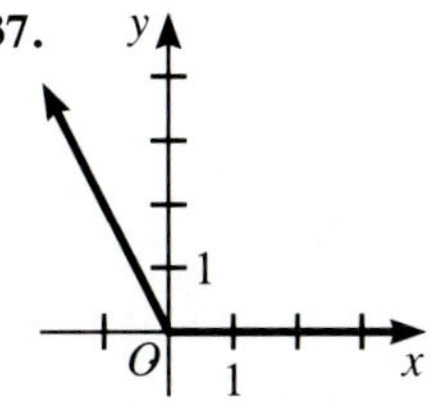

39.
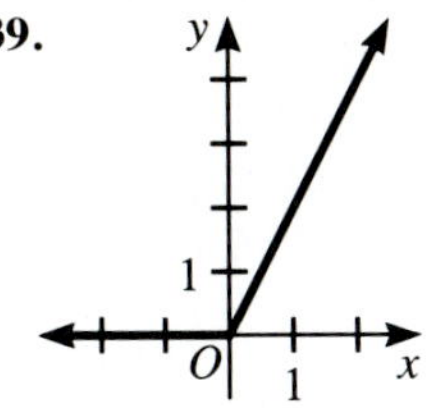

Written Exercises, pages 116–117 **1.** 2 **3.** $-\frac{2}{3}$ **5.** vertical **7.** 4 **9.** $-\frac{4}{5}$ **11.** -1 **13.** -1 **15.** $-\frac{1}{2}$ **17.** 1 **19.** $\frac{1}{3}$ **21.** $-\frac{3}{2}$ **23.** 6 **25–35.** Points chosen may vary. Examples are given.

25.
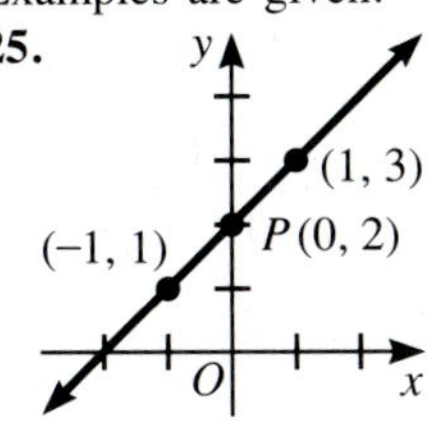

27.
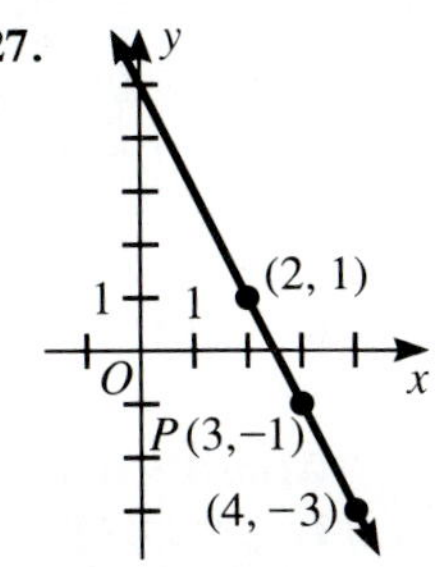

29.
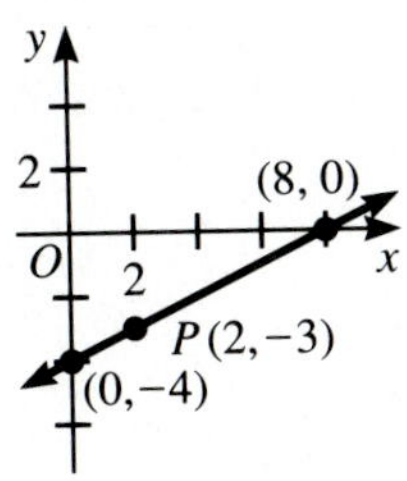

31.
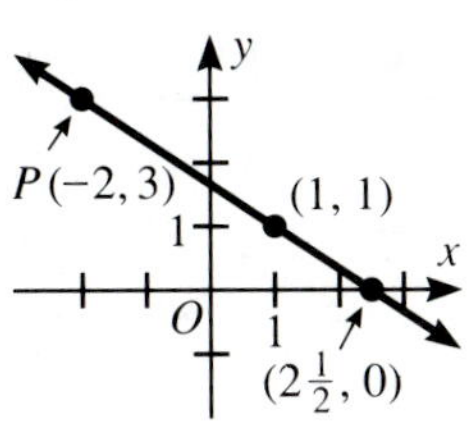

33.
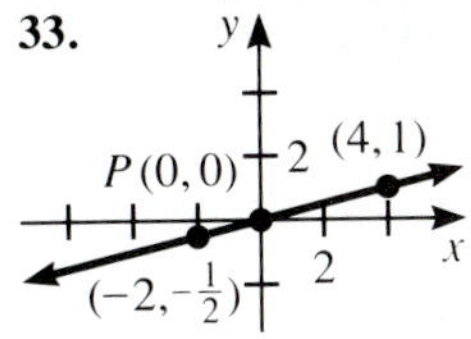

35.
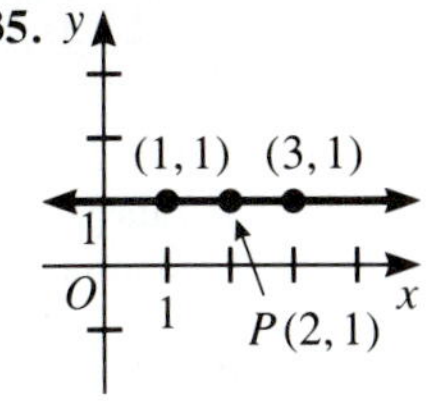

37. 60 ft **39.** 6 **41.** $\frac{3}{2}$ **43.** 1 **45.** 2 **47.** -4 **49.** $y - y_1 = m(x - x_1)$; $y - y_1 = mx - mx_1$; $mx - y = mx_1 - y_1$; $A = m$, $B = -1$, $C = mx_1 - y_1$ **51.** slope $= -\frac{A}{B} = -\frac{m}{-1} = m$ **53.** The coordinates of $Q(x', y')$ satisfy the equation of L.

Mixed Review Exercises, page 117 **1.** -3 **2.** 1 **3.** 2 **4.** -2

5.
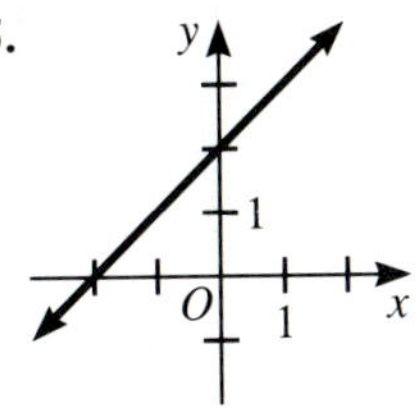

6.
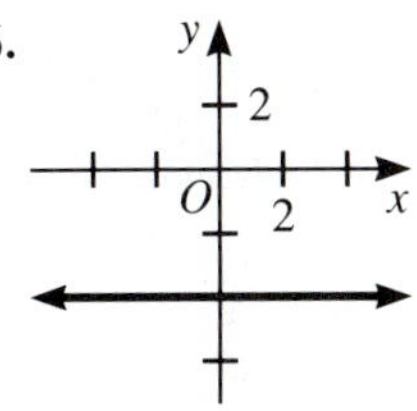

7.
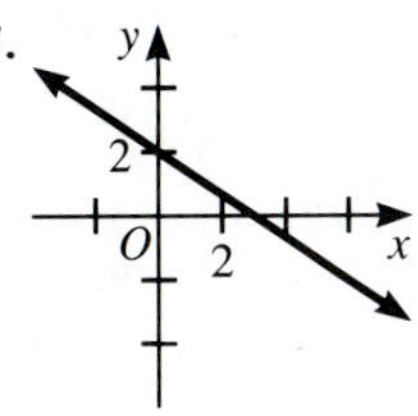

8.
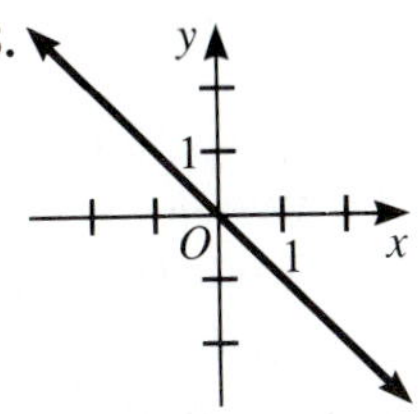

9.
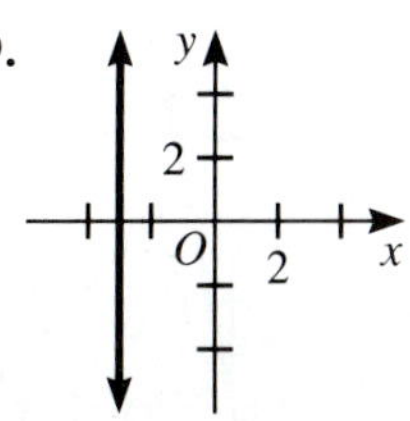

10.
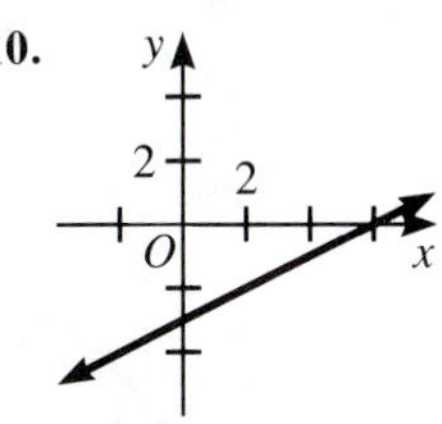

Written Exercises, pages 121–122
1. $x - y = -1$ **3.** $2x + y = 10$ **5.** $x - 2y = 1$ **7.** $x - 5y = 19$ **9.** $y = -1$ **11.** $2x - 5y = -24$ **13.** $x + y = 2$ **15.** $x - 2y = -3$ **17.** $6x - 5y = 3$ **19.** $2x + 5y = 0$ **21.** $x + y = 1$ **23.** $2x + 3y = 0$ **25.** $x = -2$ **27.** $y = \frac{1}{2}$ **29.** $6x - 6y = 7$ **31. a.** $x + y = 3$ **b.** $x - y = -3$ **33. a.** $x - 2y = 8$ **b.** $2x + y = -4$ **35. a.** $x + 2y = 2$ **b.** $2x - y = 4$ **37. a.** $y = 1$ **b.** $x = -4$

39. $y = 4$ **41.** $x = 0$ **43.** $y = 6$
45. $3x + 4y = 12$ **47.** $x + y = -1$
49. Parallelogram, not a rectangle
51. Parallelogram, rectangle **53.** Not a parallelogram

Computer Exercises, pages 122–123
2. a. 4.5 **b.** −2.28571429 **c.** vertical
4. $y = 1.72x + 17.2$ **6. a.** $y = 4.5x - 14.5$
b. $y = -2.28571429x - 4.42857143$ **c.** $x = 7$

Self-Test 1, page 123
1. $\left\{\left(-3, -\frac{21}{2}\right), \left(-1, -\frac{13}{2}\right), \left(2, -\frac{1}{2}\right)\right\}$
2. 10 jazz, 1 classical; 7 jazz, 3 classical; 4 jazz, 5 classical; 1 jazz, 7 classical
3.
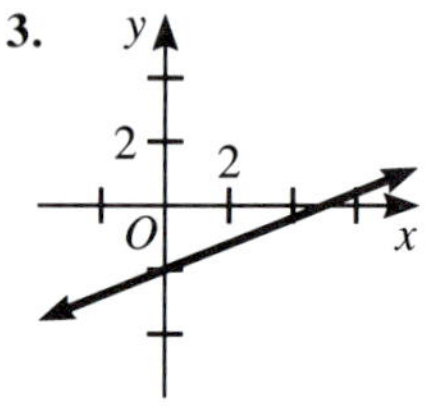

4.
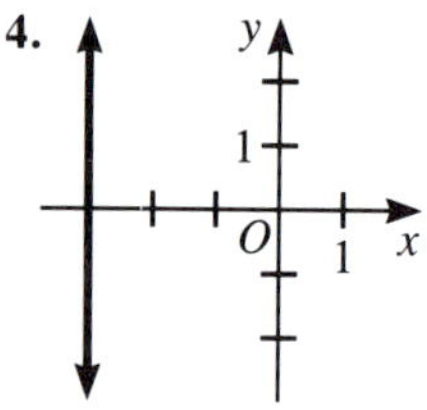

5. $\frac{3}{2}$ **6.** $-\frac{5}{4}$ **7.** 0 **8.** $y - 4 = \frac{1}{6}(x - 2)$ or $y - 3 = \frac{1}{6}(x + 4)$ **9.** $2x - y = -8$ **10.** $y = 1$

Written Exercises, pages 129–130 **1.** (−1, 3)
3. (3, 1) **5.** (3, 1) **7.** (−2, −2) **9.** (2, 4)
11. $\left(-\frac{9}{2}, 5\right)$
13. (3.5, −0.5)
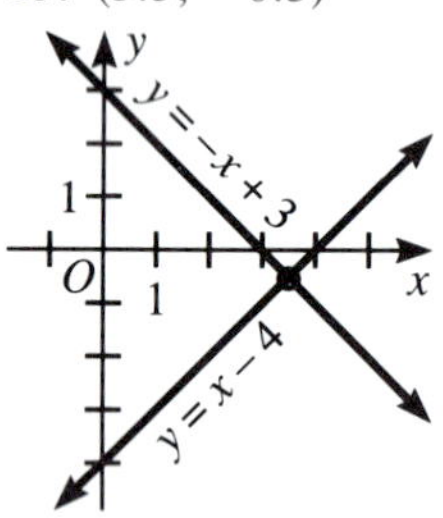

15. (2, 8)
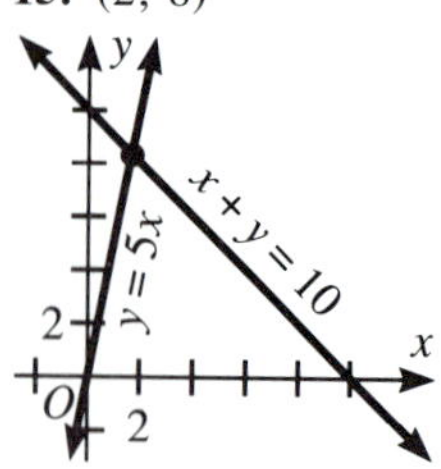

17. $\left(\frac{27}{11}, -\frac{5}{11}\right)$ **19.** (0, 2) **21.** $\left(\frac{23}{4}, -\frac{1}{2}\right)$
23. $\{(x, y): 2x - y = 1\}$; examples: (0, −1), $\left(\frac{1}{2}, 0\right)$, (2, 3) **25.** $\left(-\frac{2}{3}, -\frac{1}{3}\right)$
27. No solution **29.** inconsistent
31. consistent **33.** (2, −3) **35.** (−2, 3)
37. $\left(\frac{1}{2}, 2\right)$ **39.** $(-1, a + b)$; $a \neq b$
41. $\left(\frac{de - bf}{ad - bc}, \frac{af - ce}{ad - bc}\right)$; $ad \neq bc$
43. $\left(\frac{b_2 - b_1}{m_1 - m_2}, \frac{m_1b_2 - m_2b_1}{m_1 - m_2}\right)$ is a common solution, and it is made up of real numbers since $m_1 \neq m_2$ and hence $m_1 - m_2 \neq 0$.

Mixed Review Exercises, page 130 **1. a.** $\frac{1}{2}$
b. $x - 2y = -2$ **2. a.** 3 **b.** $3x - y = 5$
3. a. $\frac{2}{3}$ **b.** $2x - 3y = 0$ **4. a.** 0 **b.** $y = -1$
5. a. −1 **b.** $x + y = -1$ **6. a.** $-\frac{4}{3}$
b. $4x + 3y = 0$ **7.** $m = 1$, $b = -4$
8. $m = -\frac{5}{3}$, $b = 2$ **9.** $m = 2$, $b = 1$
10. $m = 0$, $b = 5$ **11.** $m = 3$, $b = 2$
12. $m = -\frac{1}{6}$, $b = \frac{2}{3}$

Problems, pages 132–134 **1.** 12 \$20 bills, 3 \$50 bills **3.** 55°, 55°, 70° **5.** 9 pairs of jeans, 12 shirts **7.** air speed, 165 mi/h; wind speed, 15 mi/h **9.** first minute, \$2.80; add. minute, \$1.20 **11.** \$5000 at 15%; \$3000 at 6%
13. air speed, 60 km/h; distance 105 km
15. $v_0 = 8$ m/s; $a = 5$ m/s^2 **17.** weekly charge, \$120; charge per mile, \$.25
19. a.
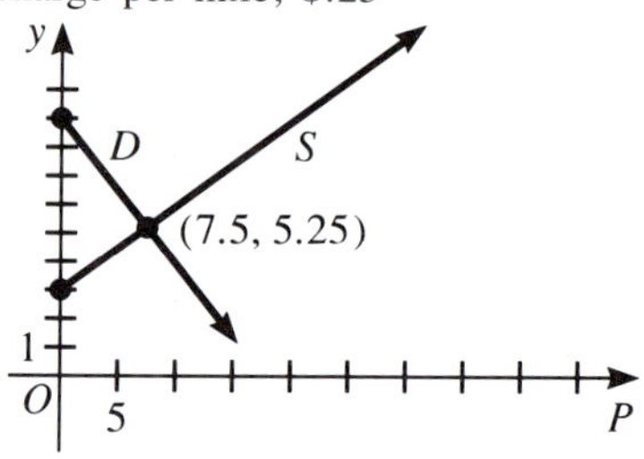

b. \$7.50 **c.** Supply is greater than demand.

Written Exercises, pages 138–139
1.
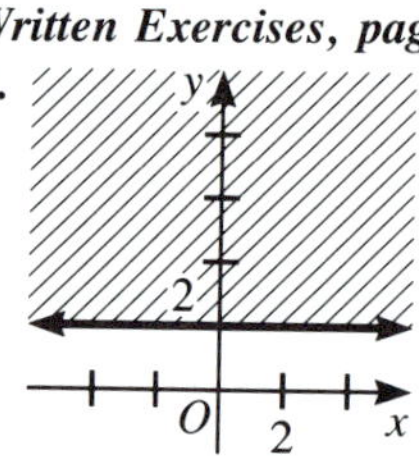

3.
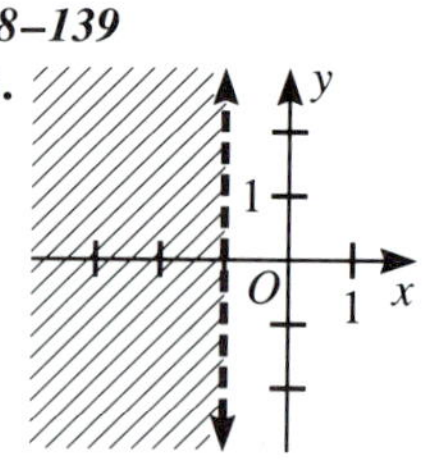

5.
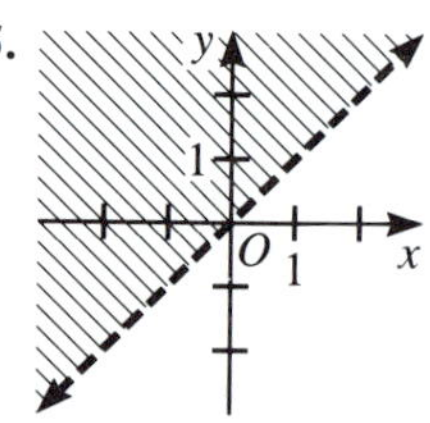

7.
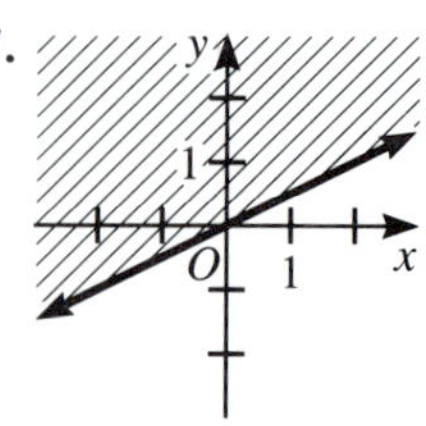

9. 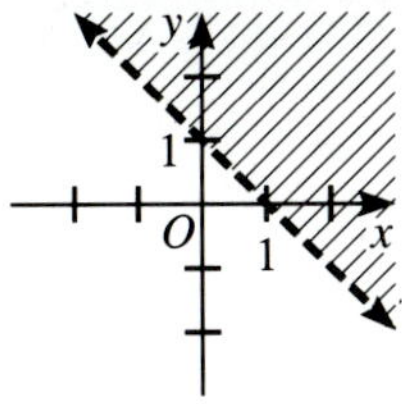

11. 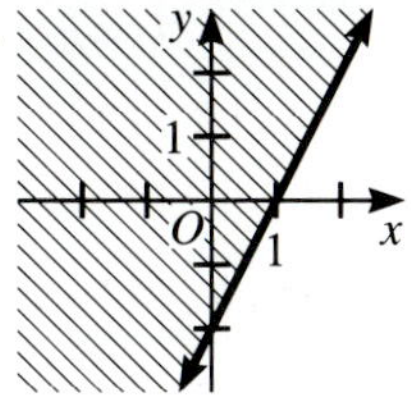

13. 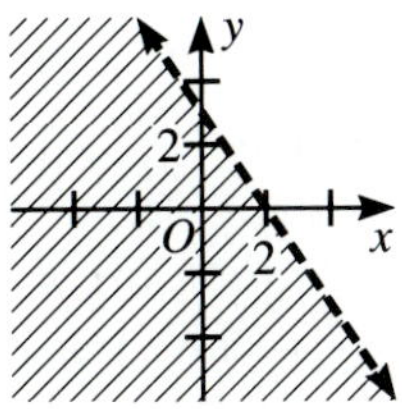

15. 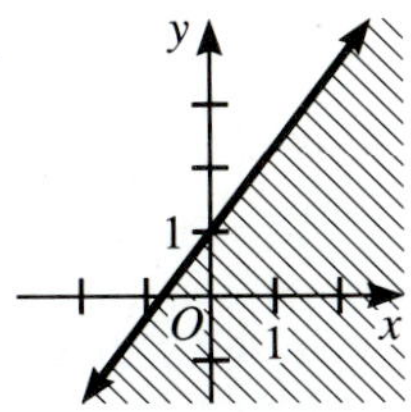

17. 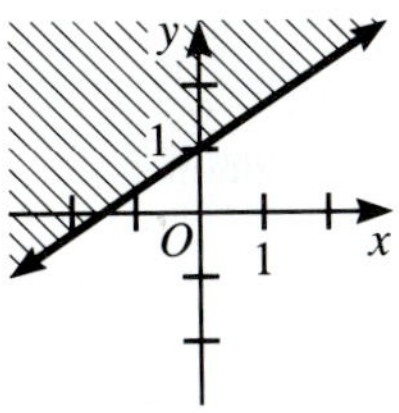

19. 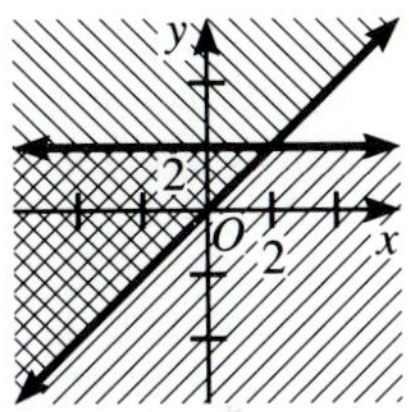

21. 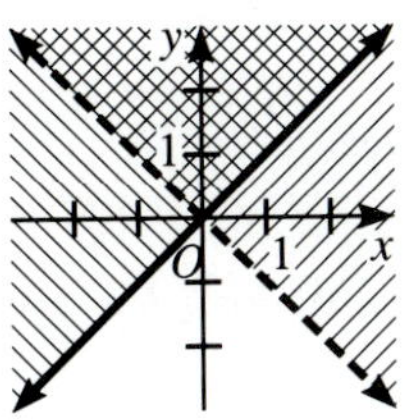

23. 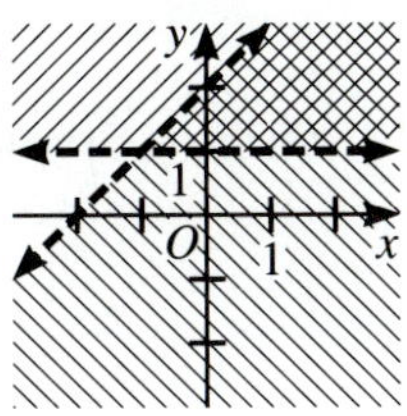

25. 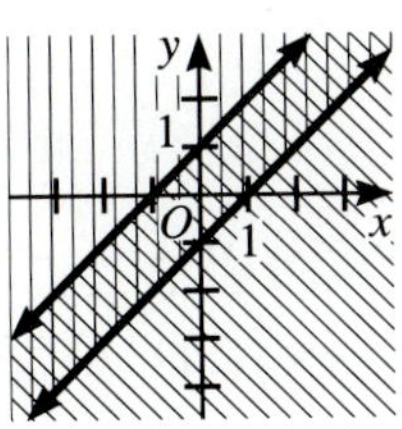

27. 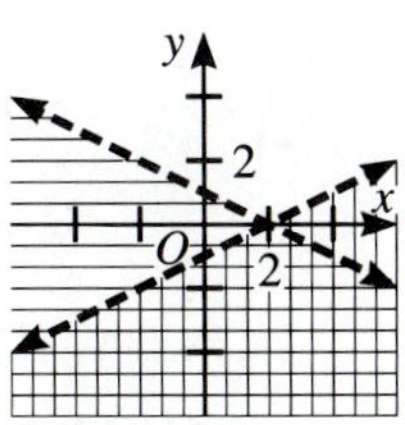

29. 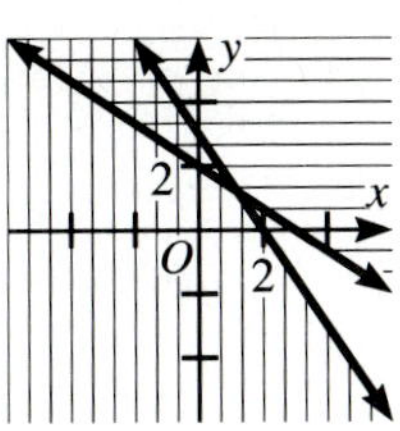

31. 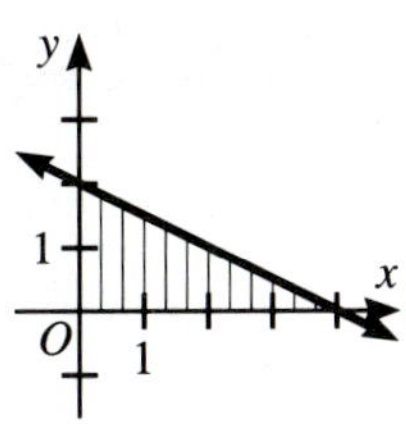

33. 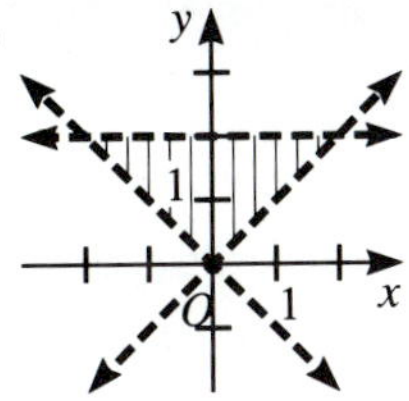

35. 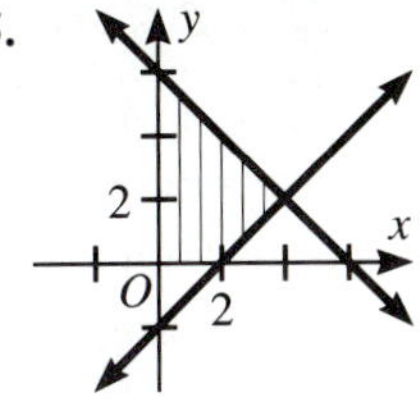

37. 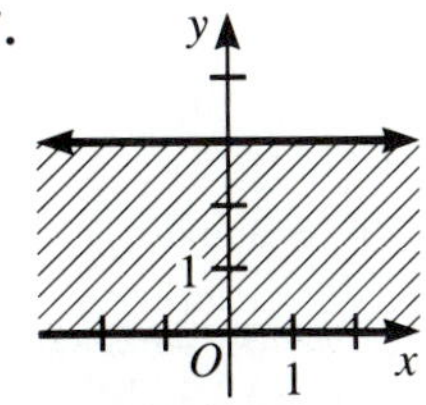

39. 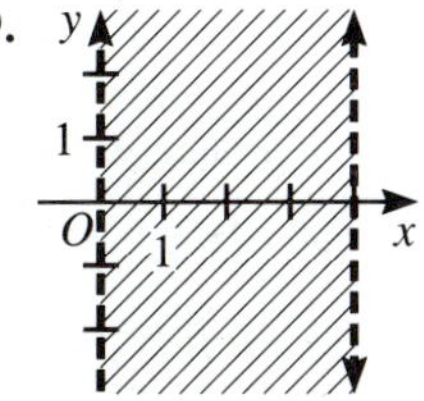

41. 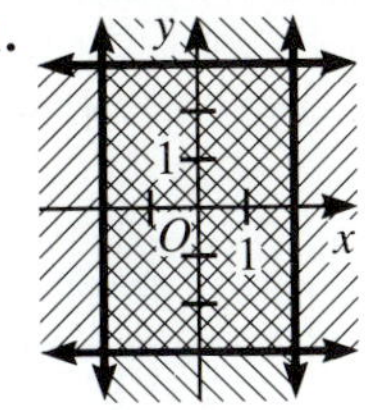

43. 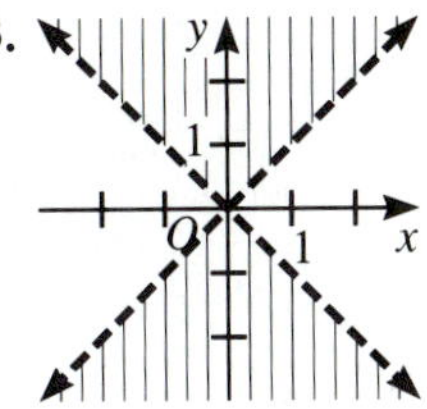

45. 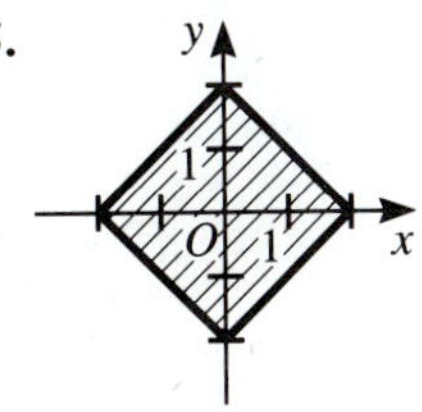

Mixed Review Exercises, page 139 **1.** $(-2, 1)$ **2.** $(0, -4)$ **3.** $(3, -4)$

4. $x - y = -4$

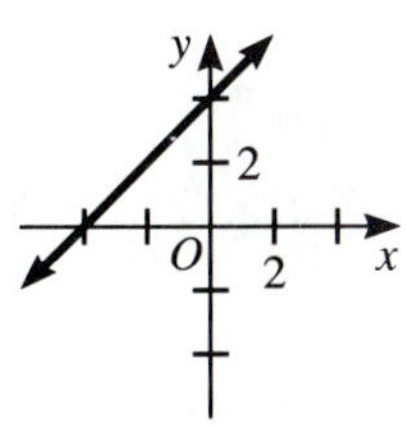

5. $x = 2$

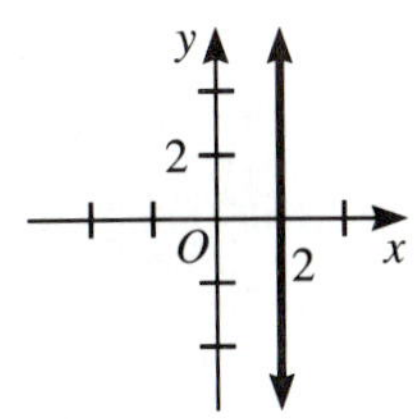

6. $4x + y = 0$

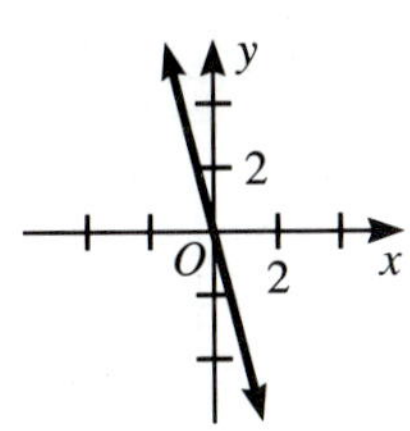

7. $2x - 3y = -6$

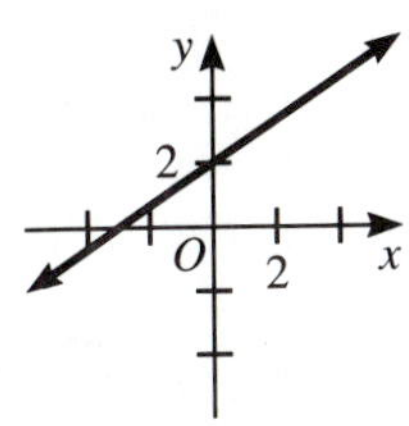

8. $y = 7$

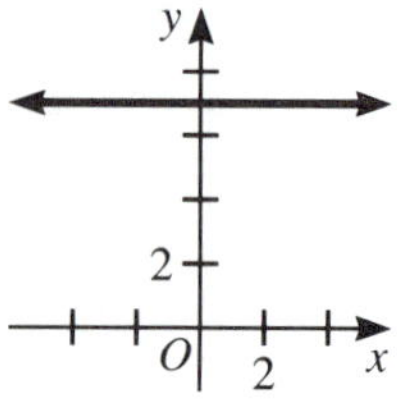

9. $x + 2y = -6$

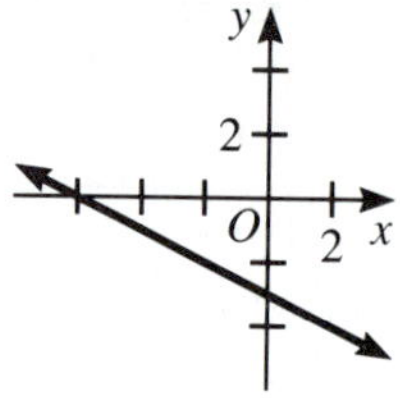

Self-Test 2, p. 140 **1.** (6, 12) **2.** No solution **3.** $\{(x, y): 4x + 3y = 10\}$; examples: $\left(0, \frac{10}{3}\right)$, $\left(\frac{5}{2}, 0\right)$, (1, 2) **4.** (−2, 7) **5.** wind speed, 15 mi/h; air speed, 105 mi/h **6.** Bret, \$1760; Sandra, \$1800

7.

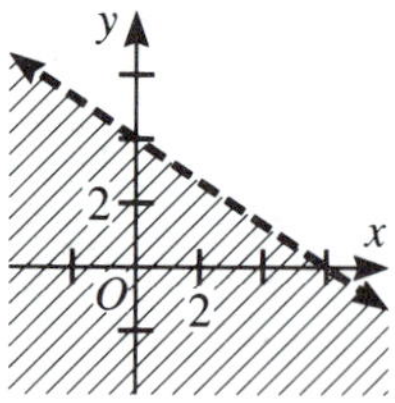

8.

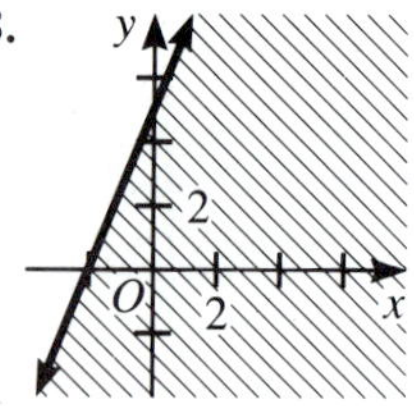

9.

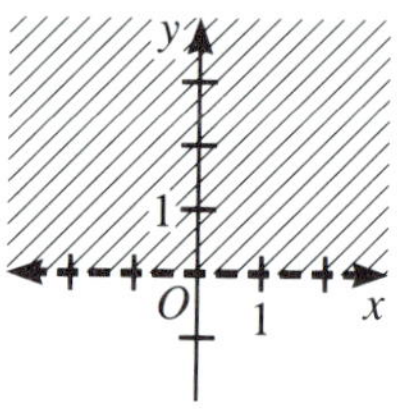

10.

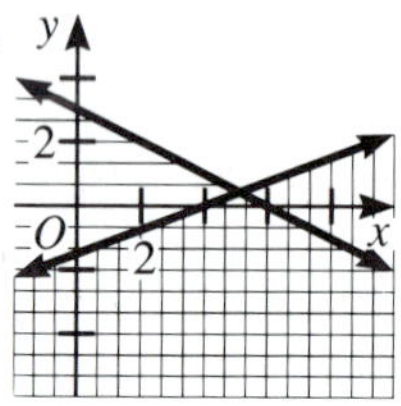

Written Exercises, pages 144–145
1. {−3, −1, 1, 3} **3.** {−2, 2} **5.** {−2, 0, 4} **7.** {0, 1, 4} **9.** {0, 12} **11.** {−1, 0, 1}

13.

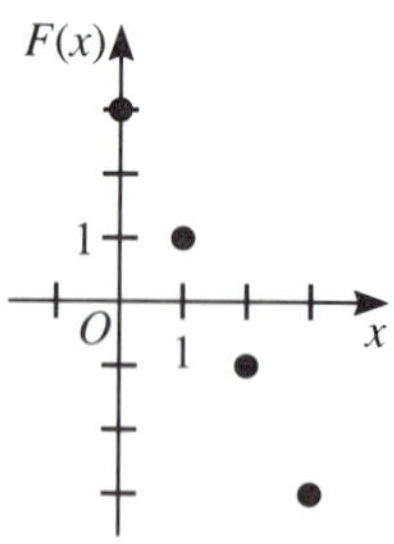

15.

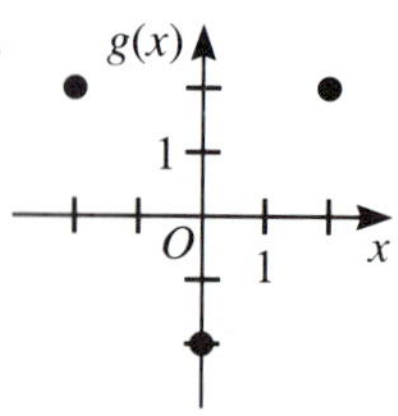

17.

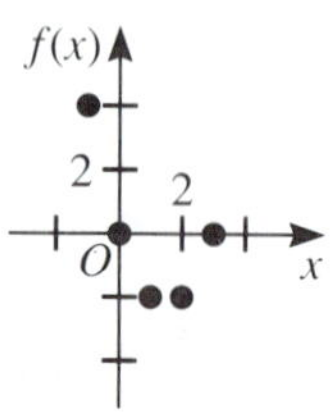

19.

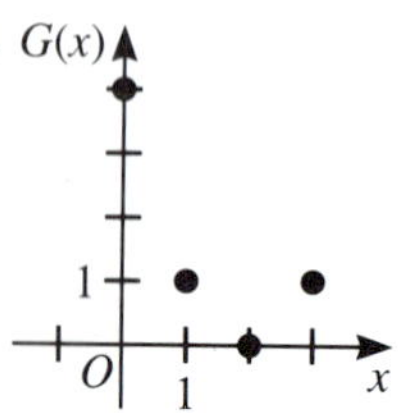

21.

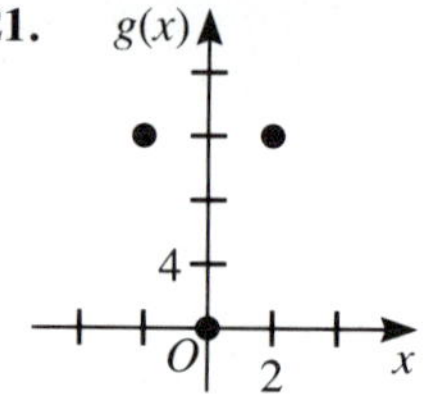

23.

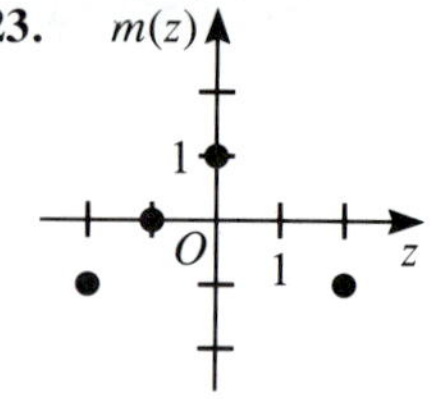

25. 2 **27.** 2 **29.** {real numbers}
31. {real numbers}
33. $\{x: x \neq -2 \text{ and } x \neq -3\}$ **35.** 0, 1
37. 8, −5 **39.** 8, −1 **41.** $-\frac{24}{25}$ **43.** −2
45. $2x + h$
47. a. $f(0) = f(0 + 0) = f(0) + f(0)$. Since $f(0) = f(0) + f(0)$, $f(0)$ is additive identity, i.e. $f(0) = 0$
b. $f(2a) = f(a + a) = f(a) + f(a) = 2f(a)$
c. $f(a + (-a)) = f(a) + f(-a)$; $f(0) = f(a) + f(-a)$; from part **(a)**, $f(0) = 0$; $0 = f(a) + f(-a)$; $f(-a) = -f(a)$
49. $s(0) = 0$, but $f(0)$ is undefined

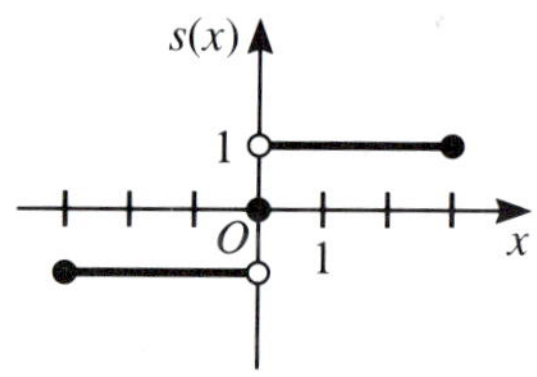

Written Exercises, pages 149–150
1. $f(x) = 2x + 3$ **3.** $f(x) = 3x + 1$ **5.** $f(x) = \frac{1}{2}x - 2$ **7.** $f(x) = 2x + 1$ **9.** $f(x) = 2x + 3$ **11.** $f(x) = -\frac{3}{2}x + 5$ **13.** $f(x) = 2x + 1$ **15.** $f(x) = -3$ **17.** $f(x) = 3x - 1$ **19.** $f(x) = -\frac{3}{2}x$ **21.** $f(x) = -\frac{3}{2}x - \frac{1}{2}$ **23.** $g(0) = -5$; $g(-1) = -8$ **25.** $g(0) = 2$; $g(3) = \frac{1}{2}$ **27.** $f(-3) = -8$; $f(10) = \frac{41}{3}$ **29.** $f(10) = -10$; $f(20) = -\frac{70}{3}$ **31.** m

Problems, pages 150–152 **1. a.** \$3150 **b.** 3 years, 8 months **3.** \$235 **5.** \$275
7. a. 179 lb **b.** 70 days
9. a. $c(f) = \frac{5}{9}f - \frac{160}{9}$ **b.** 37°C **c.** −40°
11. a. $C(n) = 4.5 + 0.062n$
b. $C(n) = 10.1 + 0.055n$

11. c.

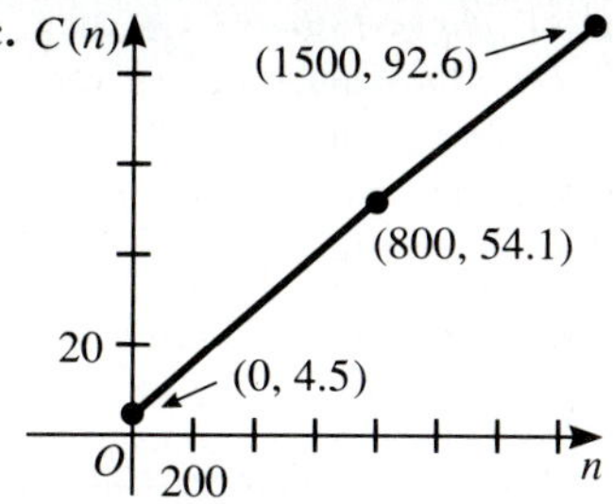

Mixed Review Exercises, page 152

1.

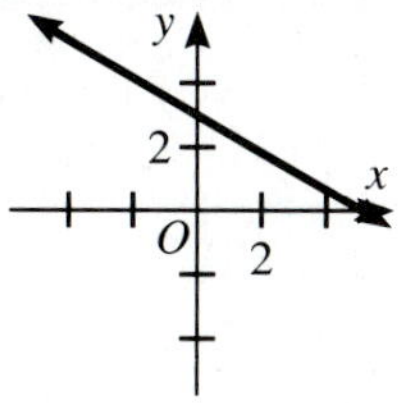

2.

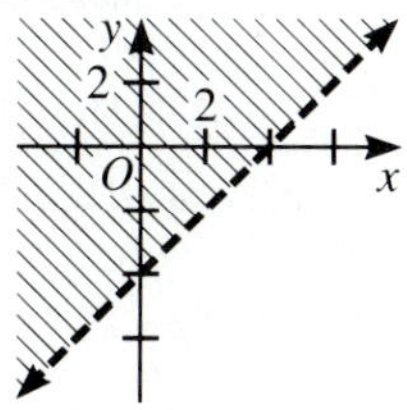

3.

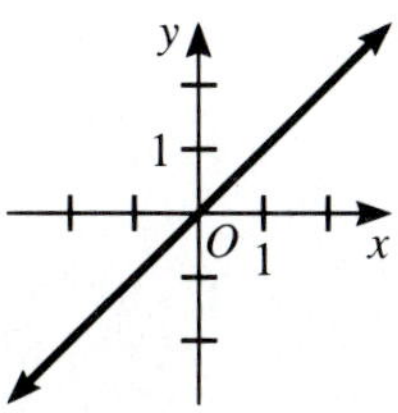

4.

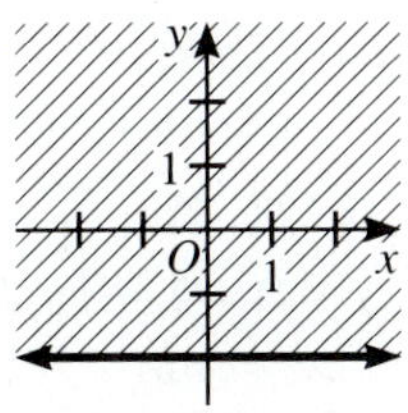

5.

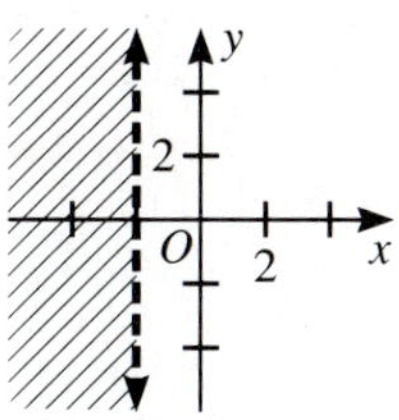

6.

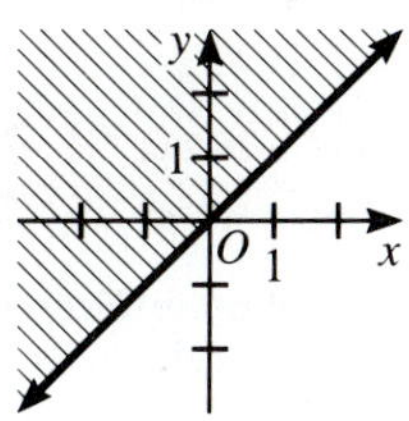

7.

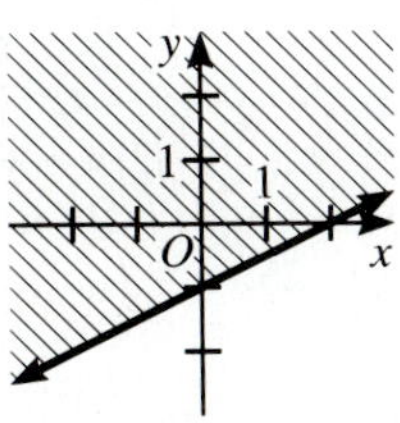

8.

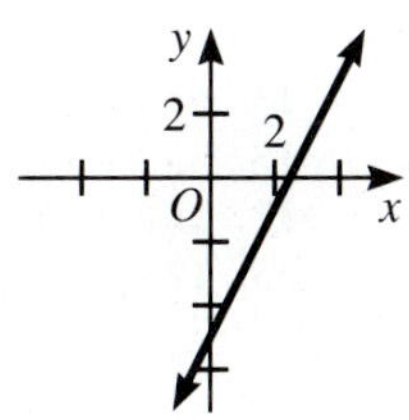

9.

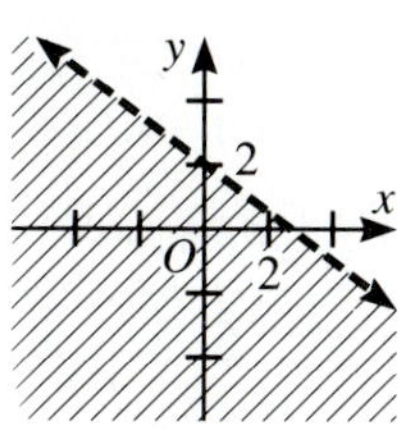

10. $\{-4, -1, 2\}$ **11.** $\{0, 1\}$ **12.** $\{5, 6, 7\}$
13. $\{-7, -2, 3\}$

Written Exercises, pages 156–157

1. not a function

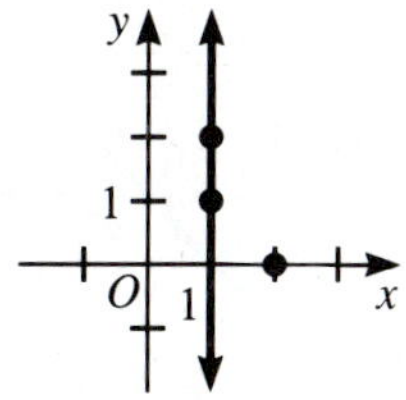

3. function

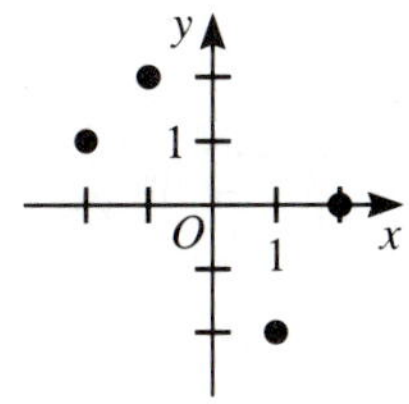

5. not a function

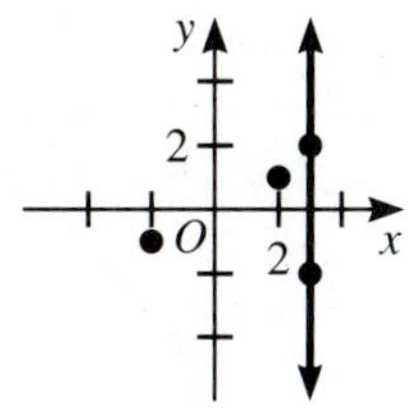

7. $D = \{-1, 0, 1\}$; no

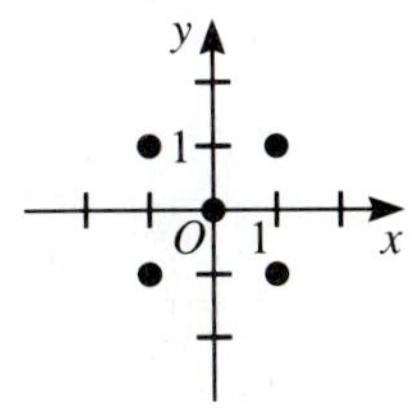

9. $D = \{0, 1, 2, 3\}$; no
11. $D = \{-2, -1, 0, 1, 2\}$; no

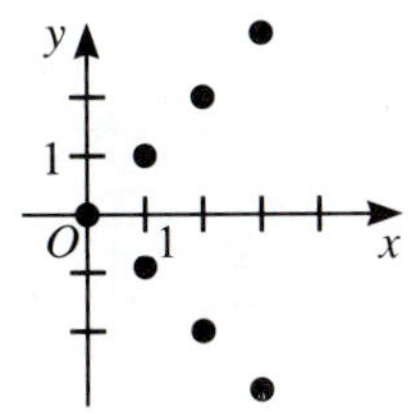

Ex. 9

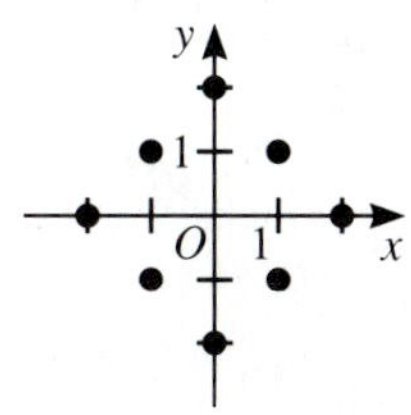

Ex. 11

13. $D = \{-2, -1, 0, 1, 2\}$; yes
15. $D = \{-2, -1, 0, 1, 2\}$; no

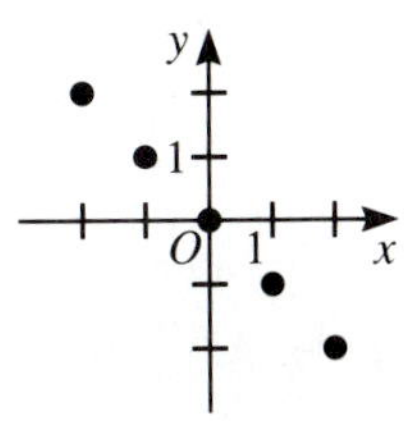

Ex. 13

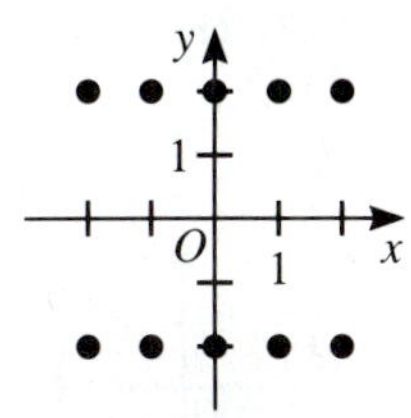

Ex. 15

17. $D = \{-2, -1, 1, 2\}$; no
19. $D = \{-2, -1, 1, 2\}$; yes

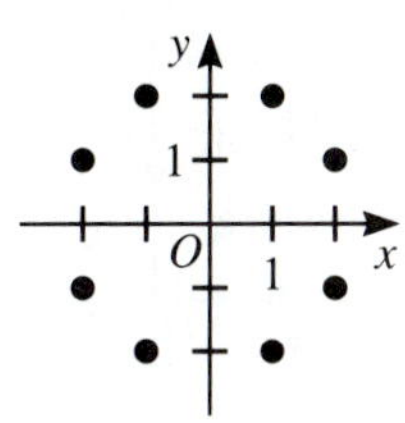

Ex. 17

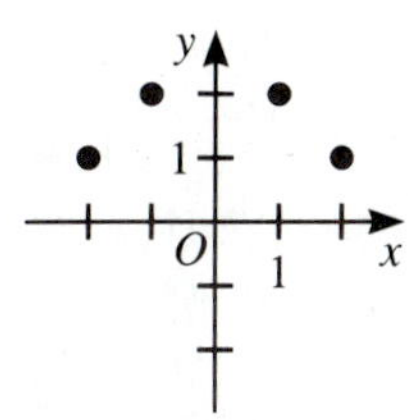

Ex. 19

21. $f \neq g$ **23.** $f = g$

25. yes

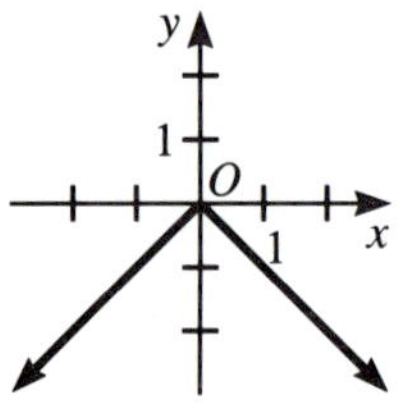

27. yes

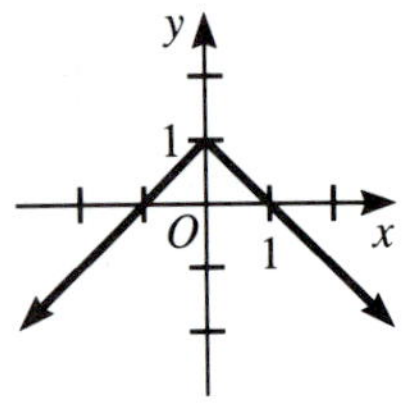

29. yes

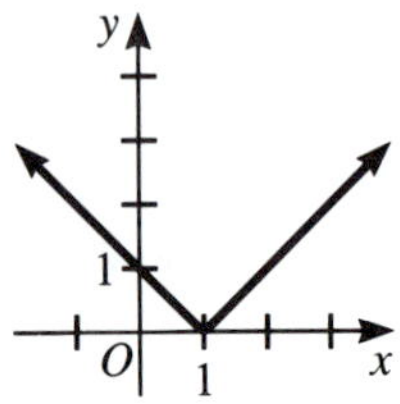

31. no

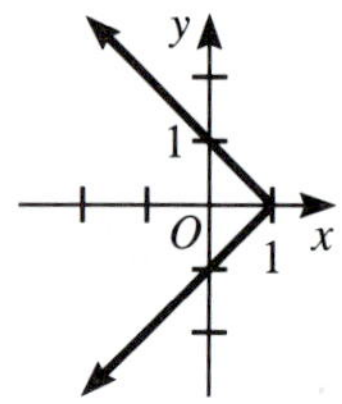

33. no

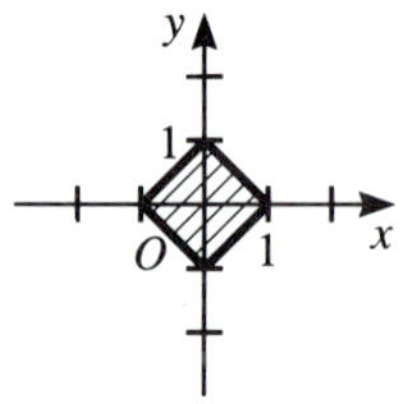

35. no

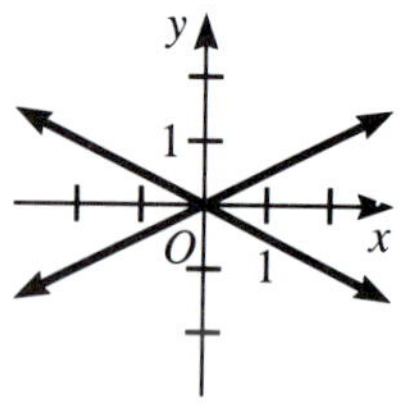

37. a. {{5, 3}, 5}, {5, {5, 3}}, {5, {3, 5}}
b. (4, 1) **c.** {a, {{a, b}, {a, b, c}}}

Self-Test 3, page 158 **1.** $R = \{1, 3, 7\}$
2. $D = \left\{x: x \neq \frac{5}{2}\right\}$

3.

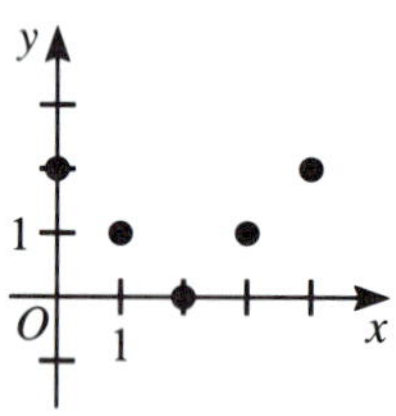

4. $f(x) = -\frac{2}{3}x + 1$

5. $f(x) = -2x + 5$

6. $f(x) = 0.2x + 8.5$

7. \$710

8. no
9. $D = \{-2, -1, 0, 1, 2\}$; yes

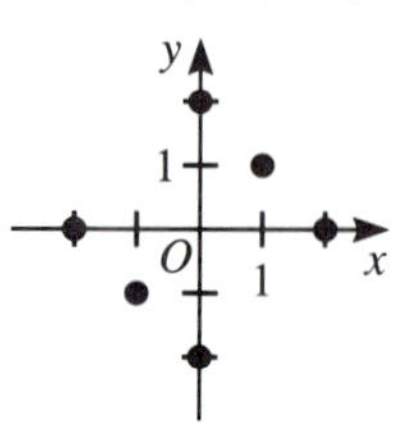

Ex. 8

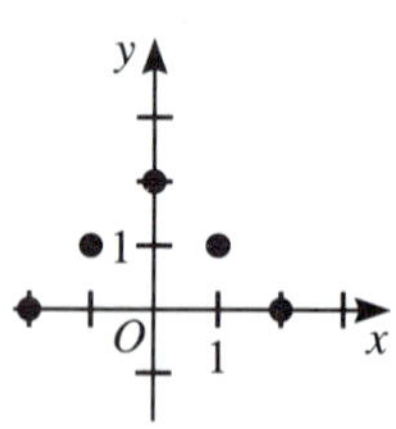

Ex. 9

Application, pages 159–161 **1.** max. value = 470; min. value = 0 **3.** max. value = 121; min. value = 15 **5.** 12 packages of Trailblazer mix and 24 packages of Frontier mix

Chapter Review, pages 162–163 **1.** c **3.** d **5.** b **7.** a **9.** c **11.** d **13.** b **15.** d **17.** c

Cumulative Review (Chapters 1–3), page 165
1. -3 **3.** $5x - y$ **5.** 4 h

7. $\{y: y \leq -3\}$

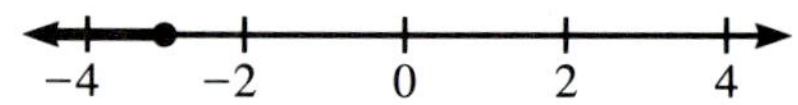

9. $\{-4, 1\}$

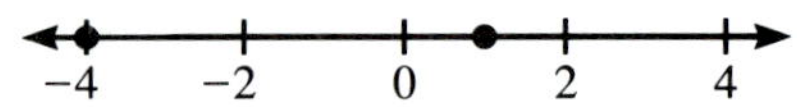

11. $\{n: -2 \leq n \leq 2\}$

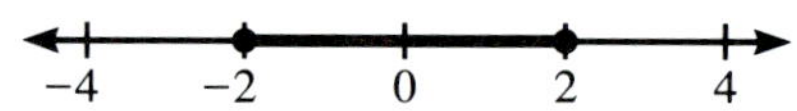

13. $\{(-1, -2), \left(0, -\frac{1}{2}\right), (1, 1)\}$ **15.** 0
17. $7x - 2y = -22$ **19.** $5x - 4y = 3$
21. $(2, -1)$ **23.** $\left(-\frac{1}{2}, \frac{1}{2}\right)$ **25.** \$8 **27.** -13

Chapter 4 Products and Factors of Polynomials

Written Exercises, page 170 **1.** $x^2 - 2x + 2$; 2
3. $3x^3 - x$; 3 **5.** $9x^2y^2 - 3x^2 + 5xy^2$; 4
7. $5x^3y^2z^2 + 6x^2yz^3 - xyz$; 7 **9. a.** $7m - 1$
b. $3m - 7$ **11. a.** $2t^2 - 3t - 13$ **b.** $-13t - 1$
13. a. $5v^3 + v^2 - 1$ **b.** $5v^3 - v^2 - 4v + 3$
15. a. $5x^2 - 2xy + 7y^2$ **b.** $x^2 - 2xy + y^2$
17. $13x^2 - 6x - 2$ **19.** $m^2 + 21$ **21.** $7ax$
23. $12p^2 - 21pq - 20q^2$ **25.** $a = -2, b = 4, c = 3, d = 2$ **27.** $a = -1, b = 3, c = 2$

Mixed Review Exercises, page 170
1. $f(x) = -2x + 2$ **2.** $f(x) = x + 5$
3. $f(x) = -\frac{1}{2}x + \frac{5}{2}$ **4. a.** -1 **b.** $x + y = 4$
5. a. $\frac{1}{4}$ **b.** $x - 4y = -7$ **6. a.** $-\frac{2}{3}$
b. $2x + 3y = 0$ **7. a.** $-\frac{3}{4}$ **b.** $3x + 4y = -2$
8. a. $\frac{2}{5}$ **b.** $2x - 5y = -43$ **9. a.** 0
b. $y = 5$

Written Exercises, page 173 **1.** $6z^5$ **3.** $-t^{12}$ **5.** $3x^3y^3$ **7.** $2u^5v^5$ **9.** $16a^6b^4$ **11.** $-27p^3q^{12}r^6$ **13.** z^6 **15.** s^8t^9 **17.** $3y^4 - 6y^3 + 9y$ **19.** $r^3s^2 - 2r^2s^3 - rs^4$ **21.** z^{2n} **23.** x^{2m} **25.** r^{3h} **27.** $t^n + t^{n+1} + t^{n+2}$ **29.** $p^{m+1} + p^m$ **31.** $z^{2m} + z^m - 1$ **33.** t^{n^2} **35.** $\{5\}$ **37.** $\{2\}$

Written Exercises, pages 175–176
1. $6v^2 - 13v - 5$ **3.** $12z^2 - 7z - 12$ **5.** $9x^2 + 60x + 100$ **7.** $25y^2 - 4$ **9.** $14t^2 - 3t - 2$ **11.** $1 - 81t^2$ **13.** $3x^2 - 2xy - 8y^2$ **15.** $6p^2 + 5pq - 6q^2$ **17.** $x^4 - 9$ **19.** $s^6 + 2s^3t^3 + t^6$ **21.** $t^3 - t^2 - 2t$ **23.** $x^3y - 2x^2y^2 + xy^3$ **25.** $2c^3 - 5c^2 + c + 2$ **27.** $x^3 + 5x^2 + x - 10$ **29.** $y^6 - 5y^4 + 7y^2 - 2$ **31.** $x^4 + 3x - 2$ **33.** $a^4 - 4a^2b^2 - ab^3 - 2b^4$ **35.** $p^{2n} - 2p^n + 1$ **37.** $r^{2n} + r^ns^n - 2s^{2n}$ **39.** $a^3 - 3a^2b + 3ab^2 - b^3$ **41.** $a^3 + b^3$ **43.** $a^4 - b^4$ **45.** $x^4 - y^4$ **47.** $x^4 + 4$ **49.** 4 **51.** 4 **53.** -2 **55.** 4, -4

Mixed Review Exercises, page 176
1. $2x^2 - 11x + 10$ **2.** a^6b^9 **3.** $-12m^3n^4$ **4.** $7y - 11$ **5.** $5cd - 6c + 12d$ **6.** $8p^3 - 4p^2 + 20p$ **7.** $-u^{11}$ **8.** $9z^3 + 5z^2 - 4z - 8$ **9.** $-10y$ **10.** $2x^6$

Self-Test 1, page 177 **1.** $x^2 + x - 1$ **2.** $t - 6$ **3.** $12p^4q^4$ **4.** t^{2n} **5.** $8a^2 + 2ab - 15b^2$ **6.** $2y^3 + 3y^2 + 4y - 3$ **7.** $9x^2 - 12x + 4$ **8.** $16c^2 - 9$

Reading Algebra, page 178 **1. a.** True **b.** True **c.** False **d.** True **e.** False **3. a.** True **b.** True **c.** False **5.** No, they might be opposites. **7.** $5|(19 - 4)$ **9. a.** True **b.** False **c.** True **d.** True **11.** odd integers

Written Exercises, pages 181–182 **1.** $2^2 \cdot 5 \cdot 7$ **3.** prime **5.** $3^2 \cdot 7^2$ **7.** $2^2 \cdot 7^2 \cdot 13$ **9.** 5; 140 **11.** 12; 432 **13.** 28; 840 **15.** 1; 315 **17.** $3p^2$; $45p^3q$ **19.** $17y^2z$; $204xy^2z^2$ **21.** $22h^2k^2r$; $440h^3k^2r^2$ **23.** 7; $42abc$ **25.** $13p^2q^2r^2$; $78p^3q^3r^3$ **27.** 126
29. $1 + 2 + 4 + 8 + 16 + 31 + 62 + 124 + 248 = 496$ **31. a.** $1, p, q, pq$ **b.** $1, p, q, p^2, q^2, pq, p^2q, pq^2, p^2q^2$ **c.** $1, p, q, p^2, q^2, pq, p^3, p^2q, pq^2, q^3, p^3q, p^2q^2, pq^3, p^3q^2, p^2q^3, p^3q^3$

Written Exercises, pages 185–186
1. $16x^2(x - 4)$ **3.** $(t + 9)^2$ **5.** $(4k - 1)(4k + 1)$ **7.** $(2y + 5)^2$ **9.** $(4x - 5)(4x + 5)$ **11.** $(11s - 3t)^2$ **13.** $(6p - 7q)(6p + 7q)$ **15.** $s(t - 1)(t + 1)$ **17.** $(t - 3)(t^2 + 3t + 9)$ **19.** $2rs(2r + s)(4r^2 - 2rs + s^2)$ **21.** $(x + 2)(y - 3)$ **23.** $(x - 2)(y - 3)$ **25.** $(q + 2)(p - 2)$ **27.** $(a - 2)(b + 1)$ **29.** $(x - 3 + 2y)(x - 3 - 2y)$ **31.** $(u - v + 1)(u + v - 1)$ **33.** $(x^2 - y)^2$ **35.** $(a^2 + b)(a^4 - a^2b + b^2)$ **37.** $(2s - 3)(2s + 3)(4s^2 + 9)$ **39.** $(x - y)(x^2 + xy + y^2)(x + y)(x^2 - xy + y^2)$ **41.** $(u + v)(u - v - 2)$ **43.** $2q(3p^2 + q^2)$ **45.** $(s + t)(s^2 + t^2)$ **47.** $4ab(3a^2 + b^2)(a^2 + 3b^2)$ **49.** $(x^n - 1)(x^n + 1)$ **51.** $(x^n + y^n)^2$ **53.** $x^{2n}(x^n - y^n)(x^n + y^n)$ **55.** $(x^2 - x + 1)(x^2 + x + 1)$ **57.** $(x^2 + y^2 - xy)(x^2 + y^2 + xy)$

Mixed Review Exercises, page 187
1. $x^2 - 6x + 9$ **2.** $2y^2 - 5y - 12$ **3.** $2m^3n - 3m^2n^2$ **4.** $4c - 9$ **5.** $a^3 - 6a^2 + 2a - 12$ **6.** $8w^5$ **7.** $6x$; $72x^3$ **8.** $15a^2b^3$; $150a^3b^4$ **9.** $21mp^2$; $126mp^3$

Computer Exercises, page 187
2. a. $(x - 28y)(x - 28y)$
b. $(3x + 24y)(3x + 24y)$
c. $64x^2 + 56xy + 49y^2$; not a perfect square
d. $(25x - 21y)(25x - 21y)$
4. a. $(12x - 2y)(12x - 2y)$
b. $(12x + 2y)(12x - 2y)$
c. $(12x + 2y)(12x + 2y)$ **d.** $144x^2 + 0xy + 4y^2$; not a perfect square or diff. of squares

Written Exercises, pages 191–192
1. $(x - 8)(x - 1)$ **3.** $(z - 9)(z - 2)$ **5.** $(r + 2)(r + 10)$ **7.** prime **9.** $(s - 2)(s - 18)$ **11.** $(x + 4)(x - 3)$ **13.** $(t - 7)(t + 5)$ **15.** $(3z + 1)(z + 1)$ **17.** $-(s - 4)(s + 2)$ **19.** $(x - 6y)(x + 5y)$ **21.** prime **23.** $(2t - 1)(t + 3)$ **25.** $(p - 3)(3p + 2)$ **27.** $(2x - 3y)(3x + y)$ **29.** $(2h - 3k)(h + 5k)$ **31.** $(6x - 5)(x + 2)$ **33.** prime **35.** prime **37.** $x(2x - y)(2x + 5y)$ **39.** $4pq(q - 2)(q^2 + 2q + 4)$ **41.** $(r - 2s)(r + 2s)(r^2 + 4s^2)$ **43.** $(x - 2)(x + 2)(x^2 + 1)$ **45.** $x - 2$; $(x - 1)(x - 2)^2$ **47.** $t(t + 1)$; $t(t + 1)(t + 4)(t - 3)$ **49.** $p - q$; $(p - q)(p^2 + pq + q^2)$ **53.** $(a^n - b^n)(a^n + b^n)(a^{2n} + b^{2n})$ **55.** $(x^n - 2)(x^n + 2)(x^n - 1)(x^n + 1)$ **57.** $4(x + 2)(x^2 - 3x + 1)$ **59.** $x^2 - x + 1$

Self-Test 2, page 192 **1.** $2 \cdot 3^2 \cdot 5 \cdot 11$ **2.** $2^3 \cdot 3 \cdot 5^3$ **3.** $18ab$; $108a^2b^3c^2$ **4.** $2p^2q^2$; $24p^3q^3r^2$ **5.** $3x(3x - 2)(3x + 2)$ **6.** $(3t + 1)^2$

7. $(5s - 2t)(25s^2 + 10st + 4t^2)$
8. $(a + 1)(b - 1)$ **9.** $2(3t - 1)(t + 1)$
10. $(4z - 7)(z + 2)$

Computer Key-In, page 193
1. $(3x + 8)(4x - 15)$ **3.** $(x - 8)(6x - 7)$
5. $(x - 1)(2x + 10)$

Written Exercises, page 196 **1.** $\{1, 4\}$
3. $\{0, -1, 2\}$ **5.** $\{1$ (d.r.), 3 (d.r.)$\}$ **7.** $\{1, 3\}$
9. $\{-1, 0, 1\}$ **11.** $\{0, 2$ (d.r.)$\}$ **13.** $\left\{\frac{1}{3}, 1\right\}$
15. $\left\{-2, \frac{3}{2}\right\}$ **17.** $\left\{-3, \frac{2}{3}\right\}$ **19.** $\{-6, 12\}$
21. $\{2, 8\}$ **23.** $\left\{-1, \frac{4}{3}\right\}$ **25.** $\{-2, 1$ (d.r.)$\}$
27. $\{-3, 0$ (d.r.), 3 (d.r.)$\}$
29. $\{-1$ (d.r.), 1 (d.r.)$\}$
31. $\{-3, -1, 0$ (d.r.), 1, 3$\}$ **33.** $\{0, 3$ (d.r.)$\}$
35. $\left\{-1, \frac{1}{3}\right\}$ **37.** $\{1, -4\}$ **39.** $\{1$ (d.r.), 2$\}$
41. $\{-2$ (t.r.), 2 (t.r.)$\}$ **43.** $(-1$ (d.r.), 1 (d.r.)$\}$
45. 2, 4, 6 **47.** 0 (d.z.), $\frac{2}{3}$ (d.z.)
49. 1 (t.z.), 5 **51.** $\left\{-\frac{b}{a}, \frac{b}{a}\right\}$ **53.** $\{-a, 1, -1\}$
55. $x = 0$ or $y = 0$

Mixed Review Exercises, page 197 **1.** $(x - 3)^2$
2. prime **3.** $(2n + 3)(4m - 5)$
4. $(4a - 5b)(4a + 5b)$ **5.** $2x(4x - 3)(3x - 2)$
6. $(2u + 1)(4u^2 - 2u + 1)$ **7.** $z^3 - z^2$
8. $16a^2 + 40a + 25$ **9.** $5m + 1$
10. $p^4 - p^2 - 6$ **11.** -1 **12.** $-27c^6d^3$

Problems, pages 199–201 **1.** -9 or 8
3. -13 and -11 or 11 and 13 **5.** 9 cm by 13 cm **7.** 8 m and 11 m **9.** 9 ft **11.** 15 m and 20 m **13.** 11 cm by 14 cm **15.** 490 m; 20 s **17.** 40 s **19.** 3062.5 m **21.** No
23. 343 cm^3 **25.** 6, 8, 10 **27.** 18 m by 38 m
29. 2.5 m

Written Exercises, pages 204–205
1. $\{x: 2 < x < 5\}$ **3.** $\{x: -2 \le x \le 2\}$
5. $\{z: z < 0$ or $z > 4\}$ **7.** $\{y: y \le -3$ or $y \ge 3\}$
9. $\{x: 1 < x < 4\}$ **11.** $\{t: t < 3$ or $t > 6\}$
13. $\left\{z: -\frac{3}{2} \le z \le \frac{5}{2}\right\}$ **15.** {real numbers}
17. $\{s: -3 \le s \le 4\}$ **19.** $\{x: -4 < x < 0$ or $x > 4\}$ **21.** $\{y: y < -3$ or $0 < y < 2\}$
23. $\{x: -2 < x < 0$ or $2 < x < 4\}$
25. $\{x: x < -2$ or $x > 2\}$ **27.** $\{x: -3 \le x \le -1$ or $1 \le x \le 3\}$ **29.** $\{x: x < -2$ or $x > 3\}$
31. $\{0, 2\}$ **33.** $p \ne q$

Mixed Review Exercises, page 205
1. $\left\{-\frac{3}{2}, 1\right\}$ **2.** $\{5\}$ **3.** $\{-4, 4\}$ **4.** $\{0\}$
5. $\{1, 4\}$ **6.** $\left\{-\frac{1}{2}, 4\right\}$ **7.** $\{8\}$ **8.** $\{0, 2\}$
9. $\left\{-\frac{1}{3}\right\}$ **10.** $\{0, 1\}$ **11.** $\{0, 4\}$ **12.** $\{11\}$

Self-Test 3, page 205 **1.** $\{-3, 14\}$
2. $\left\{-1, \frac{2}{3}\right\}$ **3.** 3 m **4.** $\left\{x: x < -5 \text{ or } x > \frac{3}{2}\right\}$
5. $\{x: 1 \le x \le 4\}$

Chapter Review, pages 207–208 **1.** c **3.** d
5. a **7.** c **9.** d **11.** b **13.** b **15.** c **17.** a
19. a

Preparing for College Entrance Exams, page 209 **1.** B **3.** C **5.** A **7.** D

Chapter 5 Rational Expressions

Written Exercises, pages 213–214 **1.** $3x^2$
3. $-\frac{3p}{q}$ **5.** $\frac{3uv}{5}$ **7.** $\frac{27r^3}{s^6}$ **9.** $\frac{3s^3}{t^3}$ **11.** $\frac{3x}{2y}$
13. $\frac{t^2}{r^2}$ **15.** $\frac{9v}{u^2}$ **17.** s^2 **19.** $\frac{z^2}{x^2}$ **21.** $\frac{4x}{3}$
23. $-\frac{c^9}{d^6h^3}$ **25.** b **27.** p^{n-1} **29.** t^n
31. $(p - q) + (r - s) = (p + r) - (q + s)$; yes
33. 1. Def. of division; 2. Comm. prop. and Assoc. prop. of mult.; 3. Prop. of the reciprocal of a product; 4. Def. of division

Mixed Review Exercises, page 215
1. $\{x: x < -4$ or $x > 2\}$ **2.** $\{y: y \le -2\}$
3. $\{w: -2 < w < 2\}$ **4.** $\{p: p < 4\}$
5. $\{t: -4 \le t \le -1\}$ **6.** $\left\{a: -1 < a < \frac{3}{2}\right\}$
7. $\{c: -1 < c < 3\}$ **8.** $\{m: m \le -1$ or $m \ge 2\}$
9. $\{d: d < -3$ or $d > -1\}$
10. $\{g: g \le 0$ or $g \ge 1\}$ **11.** $\{n: 1 \le n \le 2\}$
12. $\{u: u \le 2$ or $u > 3\}$

Computer Key-In, page 215 **1. a.** 48 **b.** 2
c. 37 **d.** 6

Written Exercises, pages 218–220 **1.** $\frac{3}{5}$ **3.** 9
5. 12 **7.** $\frac{25}{2}$ **9.** 7×10^{-4} **11.** $6x^2y^{-3}$
13. 5.96 **15.** 0.0072 **17.** -0.216
19. 0.5625 **21.** $\frac{3y}{x^2}$ **23.** $\frac{1}{st^3}$ **25.** u^2v **27.** $\frac{x^4}{4y^4}$
29. $\frac{y^2}{3}$ **31.** $\frac{x^2}{8y^2}$ **33.** $\frac{1}{x^5}$ **35.** $\frac{p}{q}$ **37.** $\frac{y^2}{x^2}$

39. $\frac{1}{u^3v^6}$ **41.** $\frac{5x^3}{y^6}$ **43** and **45.** Answers may vary. Examples are given. **43.** $(1 + 2)^{-1} = 3^{-1} = \frac{1}{3}$; $1^{-1} + 2^{-1} = 1 + \frac{1}{2} = \frac{3}{2}$; $\frac{1}{3} \neq \frac{3}{2}$
45. $2 \cdot 3^{-1} = \frac{2}{3}$; $\frac{1}{2 \cdot 3} = \frac{1}{6}$; $\frac{2}{3} \neq \frac{1}{6}$
47. $x^2 - 4x + 2$ **49.** $4x^2 - 5x + 1$
51. $x^2 - 4x + 4$

Written Exercises, page 223 **1.** 7.5×10^3
3. 6.08×10^{-1} **5.** 1.005×10^1
7. 3.20×10^{-2} **9.** 6.55×10^5
11. 5.60×10^{-4} **13.** 5000 **15.** 0.0043
17. 67,500 **19.** 0.00750 **21.** $>$ **23.** $>$
25. 100 **27.** 0.03 **29.** 4×10^2 **31.** 2.6×10^3

Problems, pages 224–225 **1.** U.S.A.: 244,000,000; 9,360,000; China: 1,060,000,000; 9,600,000; Italy: 57,400,000; 301,000; World: 5,030,000,000; 149,000,000 **3. a.** 4.85% **b.** 21.1% **c.** 1.14% **5.** \$10,200 **7.** no greater than 96 cm **9.** 1 AU $\approx 4.85 \times 10^{-6}$ parsecs; 1 parsec $\approx 2.06 \times 10^5$ AU or 3.26 light years; 1 light year $\approx 6.33 \times 10^4$ AU
11. 0.389 AU; 30.3 AU **13.** 1.9×10^{22}

Mixed Review Exercises, page 225 **1.** $\frac{2x}{3y^3}$
2. $\frac{9a^2}{4b^2}$ **3.** $\frac{4m^3}{5n^2}$ **4.** $\frac{c^3}{d^3}$ **5.** $\frac{q^2}{p}$ **6.** -1 **7.** $\frac{7}{5}$
8. $\frac{v^3}{u^2}$ **9.** $\frac{x^3}{y^2}$ **10.** c^2d^2 **11.** $\frac{3}{2}$ **12.** $\frac{b^4}{a^6}$

Calculator Key-In, page 226 **1.** 1×10^9
3. 1×10^{-10} **5.** 1.081×10^{-4} **7.** 9.517×10^0

Self-Test 1, page 226 **1.** $\frac{3s^2}{2t}$ **2.** $\frac{4y^8}{wx^2}$ **3.** $\frac{q^3r^2}{p^2}$
4. $\frac{y^4z^5}{x^2}$ **5.** $x^2 - 4$ **6.** 4.8260×10^2
7. 2.10×10^{-4} **8.** 3609 **9.** 0.05400
10. 90,000

Written Exercises, pages 228–229 **1.** $\frac{x-3}{2x}$
3. $\frac{u-2}{u}$ **5.** -1 **7.** $\frac{s+t}{s-t}$ **9.** $\frac{x-2}{x-4}$
11. $\frac{1-2y}{1+2y}$ **13.** $\frac{r-1}{r-4}$ **15.** -1 **17.** $\frac{z^2+1}{z^2}$
19. $(x + a)^2$ **21.** reals except 0 and 9; ± 3
23. reals except 1; ± 2 **25.** reals except -2, 0, and 2; -3, $\frac{3}{2}$ **27.** reals except -1 and 1; 2
29. $\frac{x+1}{x-1}$ **31.** $\frac{1}{x+y}$ **33.** $\frac{s^2+t^2}{(s+t)(s-t)}$
35. $\frac{x^2+xy+y^2}{x^2+y^2}$ **37.** $\frac{x+y+2}{x+y-2}$ **39.** $\frac{x+y+z}{x+y-z}$
41. $\frac{x^2+xy+y^2}{x+y}$ **43.** $\frac{x^n-y^n}{x^n+4y^n}$

Extra, page 231

1.
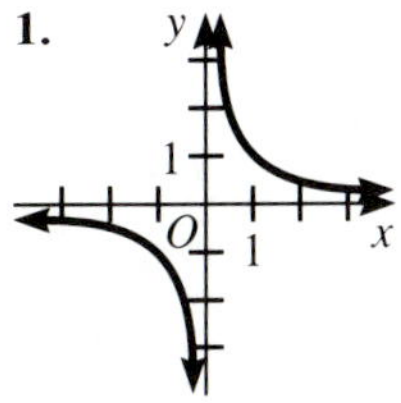

3.
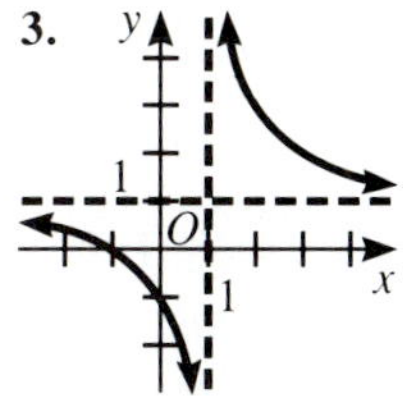

5.
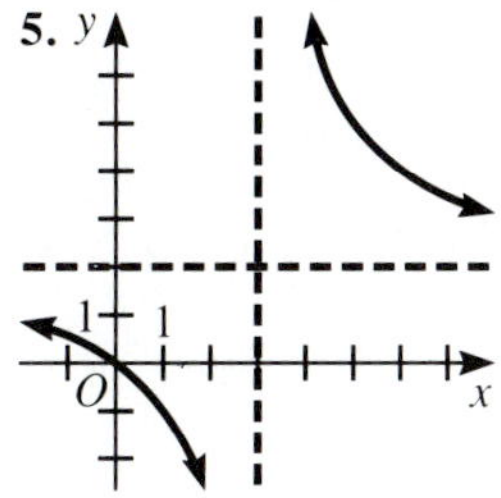

7.
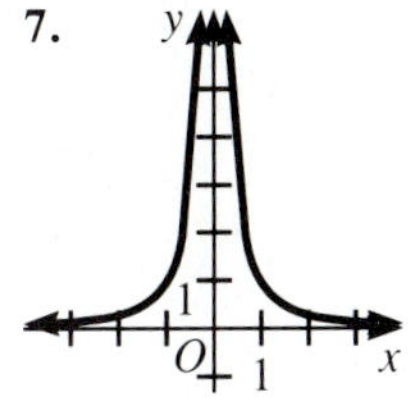

9.
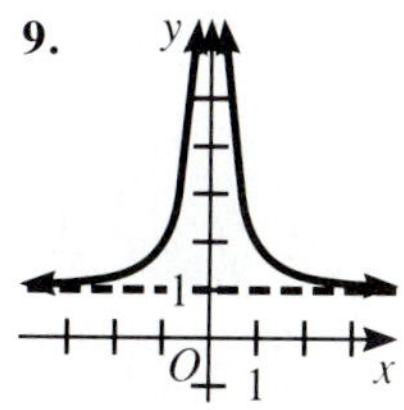

11.
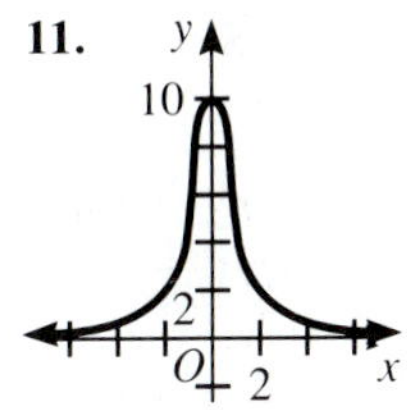

Written Exercises, page 234 **1.** $-\frac{8}{9}$ **3.** x
5. $12t$ **7.** $3y$ **9.** $\frac{r}{3}$ **11.** $\frac{x}{(x-2)(x-1)}$
13. $\frac{2u+1}{u+2}$ **15.** $\frac{x+2}{x-2}$ **17.** $\frac{p-q}{p+q}$ **19.** $\frac{u}{v-u}$
21. $\frac{x+y}{x+3y}$ **23.** $\frac{a^2+b^2}{a+b}$

Mixed Review Exercises, page 234
1. 5.4×10^{-4} **2.** 6.34×10^7 **3.** 1×10^{-1}
4. 3.281×10^3 **5.** $\frac{x+3}{x+1}$ **6.** $\frac{3}{3y-2}$ **7.** $\frac{2c^2}{3d^3}$
8. $\frac{a-1}{a+1}$ **9.** $\frac{3}{2}$ **10.** $\frac{9}{16z^5}$ **11.** $\frac{t^2+1}{2(t+1)(t-1)}$
12. $u^2 - u + 1$

Written Exercises, page 237 **1.** $-\frac{1}{4}$ **3.** $\frac{3}{7}$
5. $\frac{5}{9}$ **7.** $\frac{t}{2}$ **9.** $\frac{z^2+1}{z^2}$ **11.** $\frac{1}{6}$ **13.** $\frac{p^2+4q^2}{2p^3q^4}$
15. $\frac{(a+b)(a-b)}{ab}$ **17.** $\frac{a^2+b^2+c^2}{abc}$

19. $\frac{2(y - z)}{yz}$ **21.** $\frac{8}{(z + 4)(z - 4)}$
23. $\frac{2}{(t + 1)(t - 1)}$ **25.** $-\frac{2}{(x + 1)(x - 1)^2}$
27. $-\frac{2}{(s + 1)^2(s - 1)}$ **29.** $\frac{2b}{a - b}$
31. $\frac{2y}{(x + y)(x - y)}$ **33.** $\frac{5x}{(x - 2)(x + 2)(x - 3)}$
35. $\frac{3u}{(2u - v)(2u + v)(u - v)}$ **37.** $\frac{2a}{x + a}$
41. $A = 2$; $B = -1$

Written Exercises, pages 239–241 **1.** 2 **3.** 4 **5.** x **7.** $-ab$ **9.** $\frac{v + u}{uv}$ **11.** $\frac{y - x}{y + x}$
13. $\frac{h^2 - h + 1}{h}$ **15.** $\frac{st + 1}{t}$ **17.** $-\frac{y}{y + 3}$
19. $-a$ **21.** $\frac{x + 1}{x}$ **23.** $\frac{a - b}{a + b}$ **25.** $\frac{x - 1}{x}$
27. a. 1.500 **b.** 1.400 **c.** 1.417 **d.** 1.414
29. $-\frac{1}{(x + 1)(x + h + 1)}$ **31.** $-\frac{2x + h}{x^2(x + h)^2}$
33. $\frac{x}{2x + 1}$ **35.** $\frac{x - 1}{x}$

Mixed Review Exercises, page 241 **1.** $\frac{y + x}{x^2y^2}$
2. $\frac{4(a + 3)}{3a^2}$ **3.** $\frac{8v^2}{15u^2}$ **4.** $\frac{t^2 + t + 2}{(t + 2)(t - 2)}$ **5.** $-\frac{b}{2a}$
6. $\frac{x^2 + y^2}{2xy}$ **7.** $\frac{3x - 6}{8}$ **8.** -1 **9.** $\frac{2(a - 2)}{3b}$
10. $(3, -2)$ **11.** $(1, 3)$ **12.** $\left(-2, \frac{1}{2}\right)$

Self-Test 2, page 241 **1.** $\frac{t + 1}{t}$ **2.** $\frac{x - 1}{x}$
3. $\frac{p^7r^5}{9s^5}$ **4.** $\frac{u + 2v}{u - 2v}$ **5.** $\frac{(b - a)^2}{a^2b^2}$
6. $\frac{2}{(2x + 1)(2x - 1)^2}$ **7.** $\frac{1}{2}$ **8.** h

Written Exercises, page 245 **1.** $\left\{\frac{9}{2}\right\}$ **3.** $\{12\}$
5. $\{z: z \le 3\}$ **7.** $\{r: r > -2\}$ **9.** $\{2, 4\}$
11. $\left\{-\frac{3}{2}, 2\right\}$ **13.** $\left\{-\frac{3}{2}, \frac{3}{5}\right\}$ **15.** $\left\{\frac{2}{3}, -\frac{1}{2}\right\}$
17. $\left\{\frac{3}{2}, -\frac{5}{3}\right\}$ **19.** $\left\{\frac{2}{3}, \frac{3}{2}\right\}$
21. $\{y: -3 < y < 1\}$ **23.** $\left\{t: t \le -2 \text{ or } t \ge \frac{5}{2}\right\}$

Problems, pages 245–246 **1.** 60 **3.** 500
5. $13\frac{1}{3}$ h **7.** 6 L **9.** 36 km **11.** \$400 at 4.5%, \$1800 at 7% **13.** 0.8 L of 7.5%; 1 L of 12% **15.** 135 h **17.** 75 km at 45 km/h; 75 km at 75 km/h **19.** $2\frac{5}{8}$ h **21.** 160 km

Written Exercises, pages 249–250 **1.** $\{4\}$ **3.** $\{-3\}$ **5.** $\{9\}$ **7.** $\{2\}$ **9.** $\{-7\}$ **11.** $\{6\}$ **13.** no solution **15.** $\{4\}$ **17.** $\{2, 3\}$ **19.** $\{1, 3\}$ **21.** $\{0, 3\}$ **23.** $\{-4\}$ **25.** no solution
27. $\{1, -3\}$ **29.** $\left\{-2, \frac{1}{2}\right\}$

Problems, pages 250–252 **1.** 4 and 12 **3.** 5 **5.** 20 h **7.** 35 mi/gal **9.** 2.5 km/h **11.** 40 **13.** 250 km/h **15.** 48 mi/h **17.** 75 km/h **19.** 5 h **21.** 20

Mixed Review Exercises, page 252 **1.** $-x - 1$
2. $\frac{2t - 1}{2}$ **3.** $\frac{4m^3}{3n}$ **4.** $\frac{1}{8z^2}$ **5.** $\frac{a(b^2 + a)}{b(b + a^2)}$
6. $\frac{c + 3}{c + 2}$ **7.** $\{4\}$ **8.** $\{x: x \ge -2\}$ **9.** $\{-3, 5\}$
10. $\{x: x > -6\}$ **11.** $\{4\}$ **12.** $\left\{x: -6 < x < \frac{4}{3}\right\}$

Self-Test 3, page 252 **1.** $\{4\}$ **2.** $\left\{-\frac{6}{5}, 1\right\}$
3. $3\frac{1}{3}$ h **4.** no solution **5.** $\{-4\}$ **6.** 120 km/h

Application, page 254 **1.** 1.5 A **3.** 10 Ω

Chapter Review, pages 255–256 **1.** b **3.** a **5.** b **7.** d **9.** a **11.** b **13.** d **15.** c **17.** b

Chapter 6 Irrational and Complex Numbers

Written Exercises, pages 262–263 **1. a.** 4 **b.** -4 **c.** Not real **d.** 2 **3. a.** 9 **b.** -9 **c.** Not real **d.** 3 **5. a.** 0.1 **b.** Not real **c.** 0.1 **d.** -0.1 **7. a.** 7 **b.** 7 **c.** 7 **d.** -7
9. a. $\frac{1}{8}$ **b.** $\frac{1}{8}$ **c.** $-\frac{1}{4}$ **d.** $-\frac{1}{4}$ **11. a.** 10 **b.** 10^2 **c.** 10^3 **d.** 10^{10} **13. a.** $|a|$ **b.** a^2 **c.** a^2 **d.** $|a|$ **15.** ± 12 **17.** None **19.** $\pm\frac{2}{3}$
21. $\pm\frac{5}{4}$ **23.** $\pm\frac{1}{2}$ **25.** ± 3 **27.** $x \ge -5$
29. All reals **31.** All reals **33. a.** $x \ge -1$ **b.** $x \ge 1$ **c.** All reals **d.** $x \le -1$ or $x \ge 1$ **35.** $-3 \le x \le 0$ or $x \ge 3$ **37.** $0 \le x \le 1$

Mixed Review Exercises, page 263 **1.** $\{0, 4\}$ **2.** $\{-3, 0\}$ **3.** $\{3\}$ **4.** $\{2, 3\}$ **5.** $\{-3, 2\}$
6. $\left\{-\frac{1}{2}, 1\right\}$ **7.** $\{5\}$ **8.** $\left\{-4, \frac{3}{2}\right\}$ **9.** No solution **10.** No solution

Calculator Key-In, page 263 **1.** 105 **3.** 0.99 **5.** 2×10^5 **7.** Both expressions equal 0.51763809 on a calculator.

Written Exercises, pages 267–269 **1.** $2\sqrt{13}$ **3.** $9\sqrt{2}$ **5.** 14 **7.** $\frac{2\sqrt{2}}{3}$ **9.** $\frac{2\sqrt{3}}{3}$ **11.** $2\sqrt{2}$ **13.** $3\sqrt{5}$ **15.** $6\sqrt{35}$ **17.** 2 **19.** $5\sqrt[3]{2}$ **21.** $\frac{\sqrt[3]{10}}{2}$ **23.** 3 **25.** 28 **27.** $3\sqrt[3]{20}$ **29.** $\frac{\sqrt[3]{45}}{3}$ **31. a.** $4\sqrt{2}$ **b.** $2\sqrt[3]{4}$ **c.** $2\sqrt[4]{2}$ **d.** 2 **33.** 6.24 **35.** 7.66 **37.** 3.46 **39.** $3|x|\sqrt{2}$ **41.** $5a\sqrt[3]{3a^2}$ **43.** $\frac{|x|\sqrt{y}}{y^2}$ **45.** $\frac{3\sqrt[3]{2ab^2}}{2b^2}$ **47.** $4\sqrt{a+b}$ **49.** $|a+1|\sqrt{2}$ **51.** $\frac{1}{6}$ **53.** $\frac{\sqrt[3]{3}}{2}$ **55.** -1 **57.** $12\sqrt[3]{12}$ **59.** $\frac{3x\sqrt{3x}}{|y|}$ **61.** $\frac{3\sqrt{5a}}{a}$ **63.** $\frac{3\sqrt[3]{2y}}{y}$ **65.** $\frac{1}{2x}$

Calculator Key-In, page 269 **1.** 5.47 **3.** 0.46 **5.** No, 10 is not a power of 2. **7.** 7.83 **9.** 107.81

Written Exercises, pages 272–273 **1.** $8\sqrt{2}$ **3.** $2\sqrt{3}$ **5.** Not possible **7.** $6 + 7\sqrt{6}$ **9.** $\sqrt{2} + 3\sqrt{7}$ **11.** $5\sqrt[3]{2} + 2\sqrt[3]{5}$ **13.** $\frac{2\sqrt{15}}{5}$ **15.** $\frac{5\sqrt{6}}{6}$ **17.** $\frac{3\sqrt[3]{4}}{2}$ **19.** $4 + 2\sqrt{5}$ **21.** $3\sqrt{5} + 10\sqrt{3}$ **23.** -36 **25.** $-\sqrt{3}$ **27.** 17 **29.** $3 - \sqrt{3}$ **31.** $10 - 2\sqrt[3]{10}$ **33.** $\sqrt[3]{6} + 3\sqrt[3]{18}$ **35.** $-x\sqrt{2x}$ **37.** $(p + r)\sqrt{pr}$ **39.** $\frac{7\sqrt{10a}}{10}$ **41.** $3w\sqrt{2} + 2w^2\sqrt{3}$ **43.** $\sqrt{102}$

Mixed Review Exercises, page 273 **1.** $2|x|\sqrt{7x}$ **2.** $\frac{9}{7}$ **3.** $\frac{a}{xy}$ **4.** $\frac{3\sqrt{10x}}{8y^2}$ **5.** $3 - x$ **6.** $-3\sqrt[3]{5}$ **7.** $9x^2 - 1$ **8.** $\frac{\sqrt{x}}{|x|}$ **9.** $2xy^2\sqrt[3]{x}$

Written Exercises, pages 275–276 **1.** 2 **3.** $8 + 2\sqrt{7}$ **5.** $5 + 4\sqrt{2}$ **7.** $\frac{4+\sqrt{3}}{13}$ **9.** $9 - 2\sqrt{14}$ **11.** $-6 + 5\sqrt{3}$ **13.** $\sqrt{5} - \sqrt{2}$ **15.** 4 **17.** $34 - 2\sqrt{3}$ **19.** $\frac{5\sqrt{3} - 3\sqrt{5}}{2}$ **21.** $27 + 4\sqrt{35}$ **23.** 7 **25.** $21 - 6\sqrt{10}$ **27.** $\frac{1}{2}$ **29.** $60 - 60\sqrt{2} + 12\sqrt{3} - 12\sqrt{6}$ **31.** $-2 - \sqrt{5}$ **33.** $-\frac{\sqrt{2}}{2}$ **37. a.** $2\sqrt{5} + 3\sqrt{2}$ **b.** $\frac{2\sqrt{5} - 3\sqrt{2}}{2}$ **c.** $\frac{2\sqrt{5} - 3\sqrt{2}}{2}$ **39.** 6 **41.** 1 **43.** $\frac{2w}{w-1}$ **45.** $\sqrt{a}$ **47.** It equals $2a^3 + 6ab$, which is rational.

Written Exercises, pages 280–281 **1.** {7} **3.** {36} **5.** {−4, 4} **7.** {21} **9.** {−4} **11.** {0, 8} **13.** {2} **15.** {3} **17.** {2} **19. a.** {4} **b.** $\{2\sqrt{5}\}$ **21. a.** {1} **b.** $\{\sqrt{2}\}$ **23. a.** {9} **b.** $\{-3 - 3\sqrt{2}\}$ **25.** {4} **27.** No solution **29.** {22} **31.** {0} **33. a.** 30 mi **b.** $h = \frac{2}{3}d^2$ **35. a.** $10 - 5\sqrt{2}$ **b.** $2\sqrt{6}$ **37. b.** 12 cm **39. b.** $3\sqrt[6]{2}$

Mixed Review Exercises, page 282 **1.** $\sqrt{5} + \sqrt{2}$ **2.** $6\sqrt{2} - 6\sqrt{3}$ **3.** $a^2 + ab + b^2$ **4.** $\frac{4+\sqrt{6}}{5}$ **5.** m^8n^9 **6.** $15 + 10\sqrt{2}$ **7.** $\frac{1}{x-1}$ **8.** $6\sqrt{3}$ **9.** $18 - 36\sqrt{2} + 2\sqrt{3} - 4\sqrt{6}$ **10.** $\frac{2\sqrt{6}}{3}$ **11.** $5y^3 - 13y^2 + 14y + 8$ **12.** $4\sqrt[3]{5}$

Self-Test 1, page 282 **1.** $\frac{7}{11}$ **2.** -0.5 **3.** 13 **4.** $x \le 4$ **5.** $5\sqrt{6}$ **6.** $\frac{2a^2\sqrt{b}}{3b^2}$ **7.** $|x-3|\sqrt{2}$ **8.** $3\sqrt[3]{2}$ **9.** $\frac{3}{2}$ **10.** $9 + 14\sqrt{3}$ **11.** $6\sqrt{2} + 8$ **12.** {9} **13.** {6}

Written Exercises, pages 286–287 **1. a.** R **b.** Ir **3. a.** Ir **b.** R **5. a.** R **b.** R **c.** Ir **7.** 0.625 **9.** $1.\overline{857142}$ **11.** $\frac{253}{50}$ **13.** $\frac{118}{25}$ **15.** $\frac{4}{9}$ **17.** $\frac{5}{6}$ **19.** $\frac{26}{11}$ **21.** $\frac{1010}{333}$
23–33. Answers may vary. Examples are given. **23. a.** 0.15 **b.** $\frac{\sqrt{2}}{10}$ **25. a.** 2.5 **b.** $\sqrt{6.5}$ **27. a.** 3.9 **b.** $\sqrt{15.5}$ **29. a.** $\frac{127}{144}$ **b.** 0.87515115111 . . . **31. a.** 0.45 **b.** $\frac{\sqrt{5}}{5}$ **33. a.** $x = \sqrt{2}$, $y = -\sqrt{2}$ **b.** $x = \sqrt{2}$, $y = \sqrt{3}$ **35.** (1) and (4) **37.** Answers may vary. For example: $\frac{1}{10}, \frac{1}{16}, \frac{1}{20}, \frac{1}{50}$

Written Exercises, pages 290–291 **1.** $9i$ **3.** $-24i$ **5.** $2i\sqrt{5}$ **7.** $6i\sqrt{2}$ **9.** -6 **11.** $7i$ **13.** $-5\sqrt{2}$ **15.** -49 **17.** -1 **19.** -2 **21.** -3 **23.** $2i$ **25.** $-\frac{i\sqrt{5}}{5}$ **27.** $-\frac{i\sqrt{3}}{2}$ **29.** $-2i$ **31.** $\{\pm 12i\}$ **33.** $\{\pm 7i\}$ **35.** $\{\pm 2i\sqrt{3}\}$ **37. a.** $11i$ **b.** -30 **39. a.** $-2i\sqrt{2}$ **b.** 30 **41. a.** $5i\sqrt{2}$ **b.** -12 **43.** $-6a$ **45.** -2 **47.** $4ic\sqrt{3} - 3ic\sqrt{5}$ **49.** $-6ri\sqrt{r}$ **51.** $5x^2i - 4x^2i\sqrt{x}$ **53.** $-1, -i, 1, i$: these 4 values repeat in order.

Mixed Review Exercises, page 291 **1.** $\{14\}$ **2.** $\{-5, 3\}$ **3.** $\{2\}$ **4.** $\{13\}$ **5.** No solution **6.** $\{-8\}$ **7.** $\{2, 3\}$ **8.** $\{-2, 2\}$ **9.** $\{5\}$ **10.** R **11.** R **12.** Ir **13.** Ir

Written Exercises, pages 295–296 **1.** $10 - 5i$ **3.** $-3 - 9i$ **5.** $-18 + 11i$ **7.** $-4 + 3i$ **9.** $4 + 8i$ **11.** 10 **13.** 58 **15.** $-37 - 12i$ **17.** $-17 - i$ **19.** 21 **21.** $-12 - 16i$ **23.** $-2 - 2i\sqrt{3}$ **25.** 169 **27.** 7 **29.** $\frac{3}{5} - \frac{4}{5}i$ **31.** $\frac{3}{5} + \frac{1}{5}i$ **33.** $-\frac{3}{5} + \frac{4}{5}i$ **35.** $\frac{17}{19} - \frac{6\sqrt{2}}{19}i$ **37.** $\frac{2}{13} - \frac{3}{13}i$ **39.** $-\frac{\sqrt{3}}{9} - \frac{\sqrt{6}}{9}i$ **41.** $\frac{11}{10} + \frac{27}{10}i$ **43.** $\frac{3}{7} + \frac{2\sqrt{3}}{7}i$ **45.** $(2 + i)^2 - 4(2 + i) + 5 =$ $3 + 4i - 8 - 4i + 5 = 0$ **47.** ***b.*** $-\frac{\sqrt{2}}{2} - \frac{\sqrt{2}}{2}i$

Self-Test 2, page 297 **1.** $0.\overline{384615}$ **2.** $\frac{19}{110}$ **3.** Answers may vary. Examples: 10.02; 10.010110111 . . . **4.** -9 **5.** $5i\sqrt{7}$ **6.** $-2\sqrt{15}$ **7.** $\frac{i}{2}$ **8.** $\{\pm 3i\sqrt{5}\}$ **9.** $3 + 4i$ **10.** $-3 - 4i$ **11.** $-14 + 13i\sqrt{3}$ **12.** $-\frac{10}{17} + \frac{11}{17}i$

Extra, page 300 **1.** 5 **3.** 2 **5.** 1 **7.** $-\frac{4}{25} - \frac{3}{25}i$ **9.** $-\frac{i}{2}$ **11.** $\frac{\sqrt{2}}{2} - \frac{\sqrt{2}}{2}i$ **15. a.** $\frac{(\bar{z} - z)i}{2}$

Chapter Review, pages 302–303 **1.** c **3.** b **5.** a **7.** c **9.** d **11.** b **13.** a **15.** b

Mixed Problem Solving, page 304 **1.** 3 lemons, 6 limes; 8 lemons, 2 limes **3.** 27 **5.** \$2000 at 6% and \$5000 at 10%. **7.** $\frac{27}{12}$ **9.** length 6 cm; width 3 cm **11.** 11:36 A.M.

Preparing for College Entrance Exams, page 305 **1.** B **3.** D **5.** B **7.** C

Chapter 7 Quadratic Equations and Functions

Written Exercises, pages 309–310 **1. a.** $\{\pm\sqrt{3}\}$ **b.** $\{1 \pm \sqrt{3}\}$ **c.** $\left\{\frac{1 \pm \sqrt{3}}{2}\right\}$ **3. a.** $\{\pm 4\}$ **b.** $\{-3, -11\}$ **c.** $\left\{-1, -\frac{11}{3}\right\}$ **5. a.** $\{\pm 2i\}$ **b.** $\{-7 \pm 2i\}$ **c.** $\left\{-\frac{7}{2} \pm i\right\}$ **7.** $\{7 \pm 2\sqrt{3}\}$ **9.** $\{-6 \pm 9\sqrt{2}\}$ **11.** $\{7 \pm 2i\}$ **13.** $\{1 \pm \sqrt{6}\}$ **15.** $\{-3 \pm \sqrt{11}\}$ **17.** $\{-10 \pm 10i\}$ **19.** $\{2 \pm \sqrt{5}\}$ **21.** $\left\{-1 \pm \frac{\sqrt{2}}{2}\right\}$ **23.** $\{3 \pm i\sqrt{11}\}$ **25.** $\left\{\frac{1}{2} \pm \frac{\sqrt{5}}{2}\right\}$ **27.** $\left\{-\frac{2}{3}, -1\right\}$ **29.** $\{1 \pm i\sqrt{3}\}$ **31.** $\left\{2 \pm \frac{\sqrt{6}}{3}\right\}$ **33.** $\left\{-\frac{1}{2}, 3\right\}$ **35.** $\{-3 \pm \sqrt{5}\}$ **37.** $\{3\}$ **39.** $\{8\}$ **41.** $\left\{-\frac{b}{2} \pm \frac{\sqrt{b^2 - 4c}}{2}\right\}$

Mixed Review Exercises, page 310 **1.** $5i\sqrt{2}$ **2.** $14 - 5i$ **3.** $54\sqrt{2}$ **4.** $5\sqrt{2}$ **5.** -24 **6.** $12 - i$ **7.** $\frac{7\sqrt{10}}{10}$ **8.** $-\frac{x}{3y}$ **9.** $\frac{x - 2}{x + 3}$

Written Exercises, pages 313–314 **1.** $\{-3 \pm \sqrt{5}\}$ **3.** $\{2 \pm 3i\}$ **5.** $\left\{\frac{2}{5}, -1\right\}$ **7.** $\left\{-\frac{6}{5} \pm \frac{2}{5}i\right\}$ **9.** $\left\{\frac{-1 \pm \sqrt{13}}{6}\right\}$ **11.** $\left\{\frac{-1 \pm \sqrt{15}}{2}\right\}$ **13.** $\left\{\frac{4 \pm \sqrt{10}}{3}\right\}$ **15.** $\left\{\frac{2 \pm \sqrt{10}}{2}\right\}$ **17.** $\left\{\frac{5}{2} \pm \frac{\sqrt{7}}{2}i\right\}$ **19.** $\{-1.24, 3.24\}$ **21.** $\{-0.83, 2.83\}$ **23.** $\{-1.41, -0.59\}$ **25.** $\{\pm 3\}$ **27.** $\{1\}$ **29.** $\left\{\frac{\sqrt{2} \pm \sqrt{6}}{2}\right\}$ **31.** $\{\sqrt{2} \pm 1\}$ **33.** $\left\{-2\sqrt{2}, -\frac{\sqrt{2}}{2}\right\}$ **35.** $\{-2i, i\}$ **37.** $\{2 + i, 1 + i\}$ **39.** $\{-i, 7i\}$

Problems, pages 314–315 **1.** 1.66 **3.** 1.62 **5.** 50 m by 100 m **7.** 1.68 m **9.** 18.94 mm **11.** 1.39 **13.** 5.37 **15.** Going out, 45 km/h; returning, 30 km/h

Self-Test 1, page 316 **1.** $\{3 \pm \sqrt{7}\}$ **2.** $\left\{\frac{-9 \pm \sqrt{57}}{6}\right\}$ **3.** $\left\{\frac{5 \pm \sqrt{37}}{2}\right\}$ **4.** $\left\{-\frac{1}{4} \pm \frac{\sqrt{3}}{4}i\right\}$ **5.** 0.70 cm by 4.30 cm **6.** 3 ft

Written Exercises, pages 320–321 **1.** two different irrational real roots **3.** two conjugate imaginary roots **5.** two different irrational real roots **7.** two conjugate imaginary roots **9.** two different rational real roots **11.** two different irrational real roots **13.** $\{1, 5\}$ **15.** $\left\{\frac{2}{3}\right\}$ **17.** $\{-7\}$ **19.** $\{-9, 11\}$ **21.** $\{2 \pm \sqrt{6}\}$

23. $\left\{-\frac{5}{2}, 3\right\}$ **25.** $\{2 \pm \sqrt{2}\}$ **27.** $\left\{\frac{3}{2} \pm \frac{3\sqrt{3}}{2}i\right\}$ **29.** $\{3\}$ **31. a.** $k = 2$ **b.** $k < 2$ **c.** $k > 2$ **33. a.** $k = \pm 2$ **b.** $-2 < k < 2$ **c.** $k < -2$ or $k > 2$ **35. a.** $k = \pm 2$ **b.** $-2 < k < 2$ **c.** $k < -2$ or $k > 2$ **37.** $\frac{1}{2}$

Mixed Review Exercises, page 321

1. $\{2 \pm 3\sqrt{2}\}$ **2.** $\{-1, 4\}$ **3.** $\left\{\frac{3 \pm \sqrt{7}}{2}\right\}$ **4.** $\left\{\frac{1}{2} \pm \frac{\sqrt{19}}{2}i\right\}$ **5.** $\left\{-\frac{4}{3}, \frac{7}{2}\right\}$ **6.** $\left\{\frac{-2 \pm \sqrt{7}}{3}\right\}$ **7.** x^6 **8.** $6x\sqrt{2x}$ **9.** $\frac{1}{x^2}$ **10.** x^3y^4 **11.** $\frac{\sqrt{x}}{x^2}$ **12.** $\frac{y^2}{x}$ **13.** $5x$ **14.** $\frac{y^6}{x^4}$

Computer Exercises, page 321 **2. a.** real; irrational **b.** real, rational **c.** imaginary **4. a.** 5, −2.33333334 **b.** 0.192582404, −5.1925824 **c.** 5.10977223, −4.10977223 **6. a.** $-1 \pm 2.64575131i$ **b.** 2.39718086, −1.14718086 **c.** $1.5 \pm 1.11803399i$

Written Exercises, pages 324–325

1. a. $\{1, -2\}$ **b.** $\left\{1, \frac{5}{2}\right\}$ **c.** $\{\pm 1, \pm 2\}$ **3. a.** $\left\{-\frac{1}{6}, 1\right\}$ **b.** $\left\{\pm\frac{3\sqrt{2}}{2}, \pm 1\right\}$ **c.** $\left\{-\frac{1}{3}, 2\right\}$ **5. a.** $\{\pm i, \pm 2\}$ **b.** $\{16\}$ **7. a.** $\{\pm 3i, \pm 2\}$ **b.** $\left\{\pm\frac{1}{3}i, \pm\frac{1}{2}\right\}$ **9.** $\{25, 36\}$ **11.** $\{\pm 2, \pm 3\}$ **13.** $\{-7, 11\}$ **15.** $\{4\}$ **17.** $\{25\}$ **19.** $\{4\}$ **21.** $\{\pm 2.91\}$ **23.** $\{0.08\}$ **25.** $\{-1.19, 1, 2, 4.19\}$ **27.** $144x^4 - 73x^2 + 4 = 0$ **29.** $\{10, 250\}$ **31.** $\{\pm 2, \pm 3, \pm i\}$

Self-Test 2, page 325 **1.** Imaginary, conjugate **2.** Real, unequal, rational **3.** Real, unequal, irrational **4.** Real, unequal, rational **5.** $\{2 \pm \sqrt{5}\}$ **6.** $\{-1, 4\}$ **7.** $\{4 \pm \sqrt{10}\}$ **8.** $k > \frac{4}{3}$ **9.** $\{\pm 2i, \pm\sqrt{3}\}$ **10.** $\left\{\frac{4}{3}\right\}$

Written Exercises, pages 331–332

1.

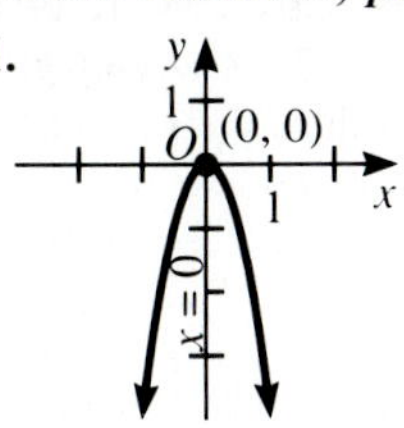

3.

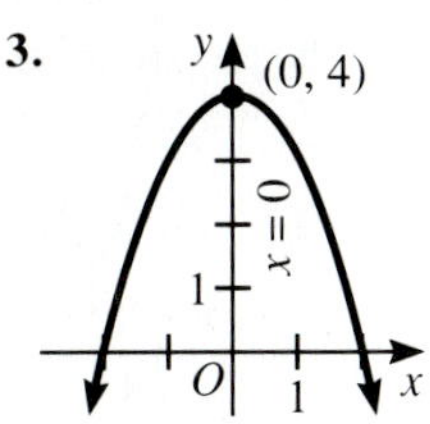

5.

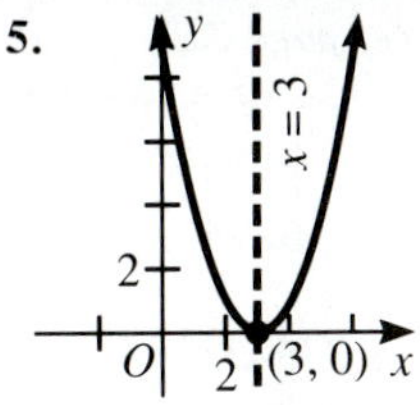

7.

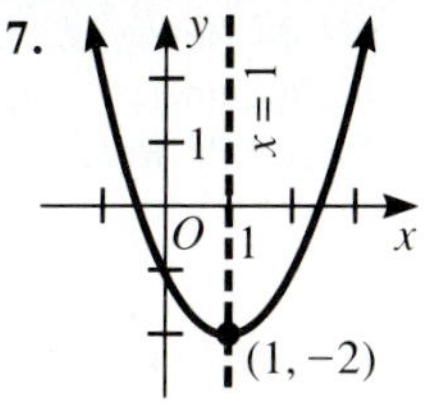

9.

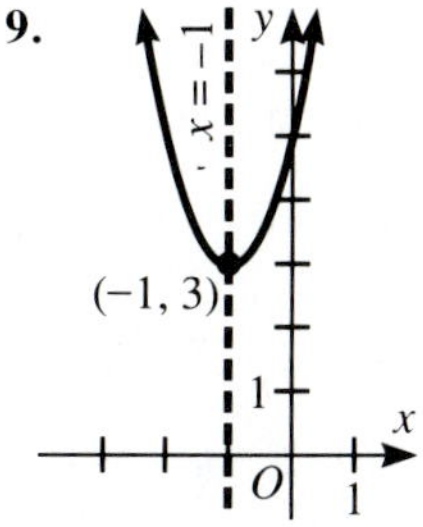

11.

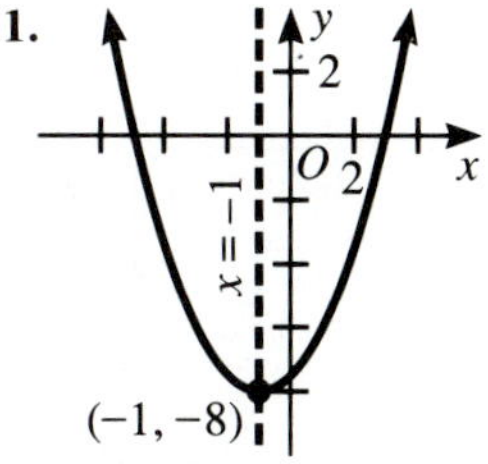

13.

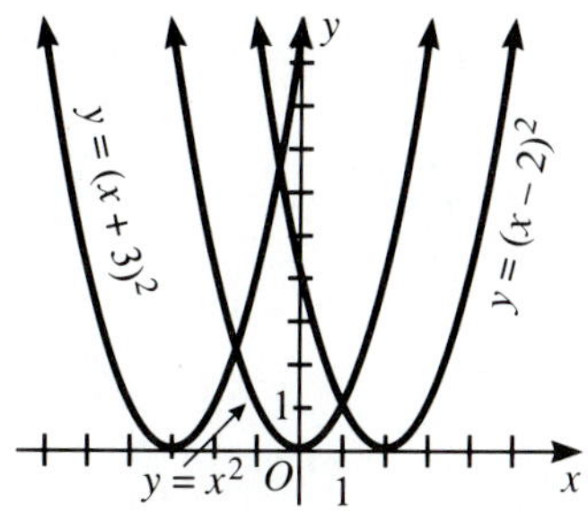

15.

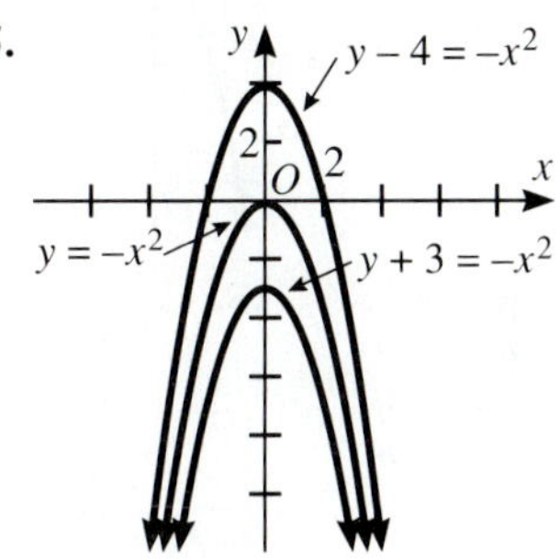

17.

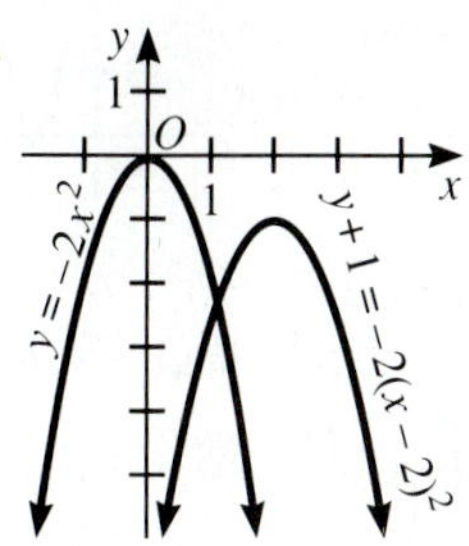

19. $y + 3 = \frac{1}{2}(x - 4)^2$ **21.** $y = \frac{1}{3}x^2$ **23.** $y - 5 = -\frac{1}{3}(x - 3)^2$ **25.** $y - 2 = -2(x - 4)^2$ **27.** 5 **29.** 2 **31.** $a = 2$, $k = 3$

Mixed Review Exercises, page 332

1. Imaginary, conjugate **2.** Real, double, rational **3.** Real, unequal, irrational **4.** $\{\pm 1, \pm 2\}$ **5.** $\left\{\frac{1}{2} \pm i\right\}$ **6.** $\{4\}$

Written Exercises, pages 336–337

1.
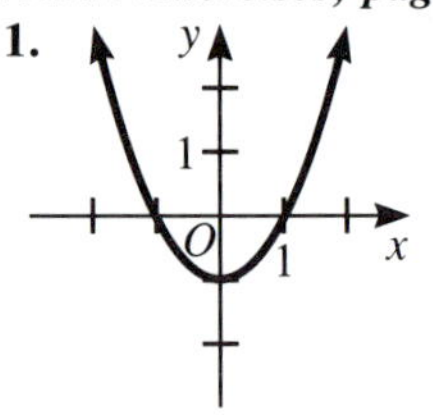

3.
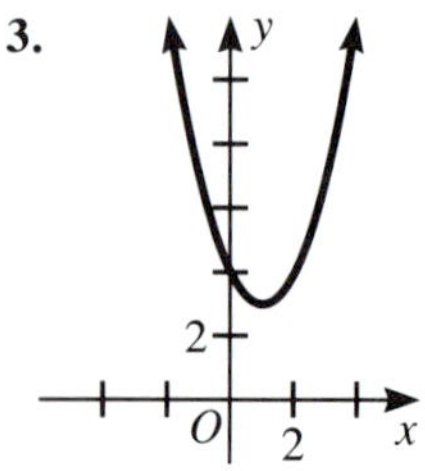

5.
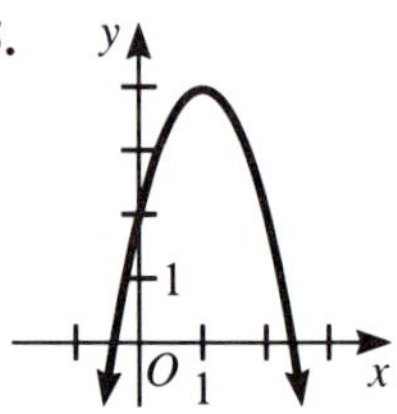

7.
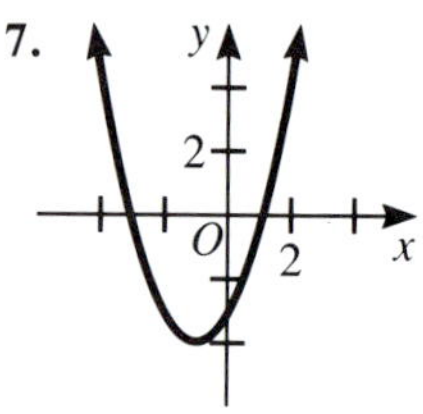

9.
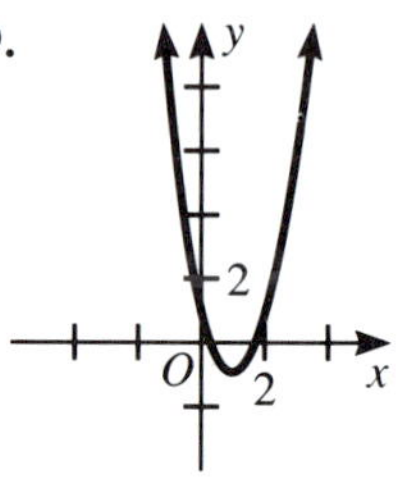

11.
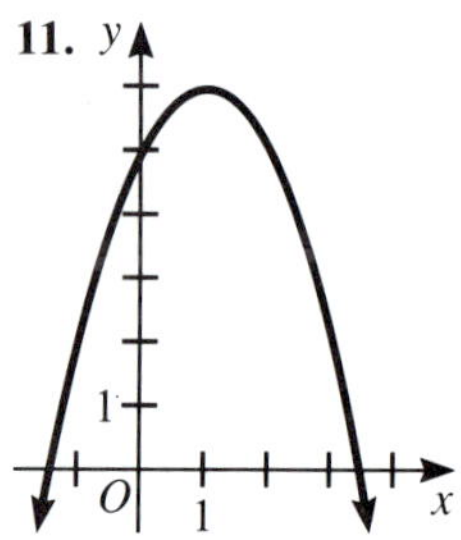

13.
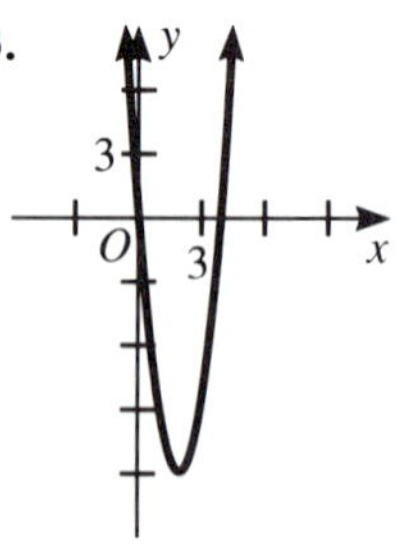

15.
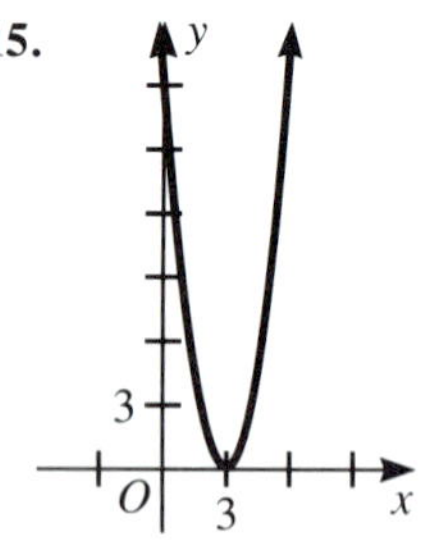

17.
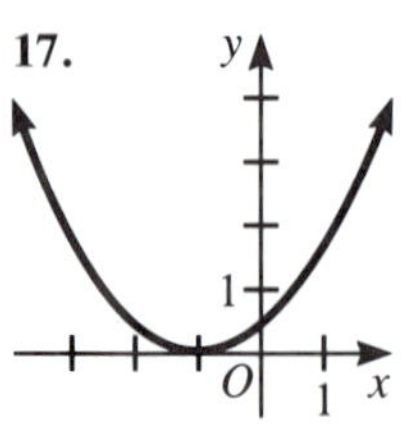

19. min., -8; $(-2, -8)$ **21.** min., $-\frac{5}{2}$; $\left(\frac{3}{2}, -\frac{5}{2}\right)$ **23.** max., $\frac{49}{4}$; $\left(-\frac{3}{2}, \frac{49}{4}\right)$

25. a. $(2, -7)$ **b.** {real numbers} **c.** $\{y: y \geq -7\}$ **d.** $2 \pm \sqrt{7}$ **27. a.** $(-1, 9)$ **b.** {real numbers} **c.** $\{y: y \leq 9\}$ **d.** $-4, 2$ **29. a.** $(1, -6)$ **b.** {real numbers} **c.** $\{y: y \geq -6\}$ **d.** $1 \pm \sqrt{6}$ **31. a.** $(1, -72)$ **b.** {real numbers} **c.** $\{y: y \geq -72\}$ **d.** $-5, 7$ **33. a.** $(-6, 9)$ **b.** {real numbers} **c.** $\{y: y \leq 9\}$ **d.** $-9, -3$ **35. a.** $(-1, 2)$ **b.** {real numbers} **c.** $\{y: y \leq 2\}$ **d.** $-3, 1$ **37. a.** neg. **b.** pos. **c.** pos. **d.** zero **39. a.** zero **b.** pos. **c.** pos. **d.** pos. **41.** $y - \frac{4c - b^2}{4} = \left(x - \frac{-b}{2}\right)^2$

Written Exercises, pages 342–343

1. $\left\{\frac{3}{2} \pm \frac{\sqrt{15}}{2}i\right\}$ **3.** $\left\{-\frac{3}{2}, 2\right\}$ **5.** $\left\{\frac{3}{4} \pm \frac{\sqrt{7}}{4}i\right\}$ **7.** $x^2 - 7x + 10 = 0$ **9.** $2x^2 - x - 10 = 0$ **11.** $x^2 - 3 = 0$ **13.** $x^2 - 2x - 2 = 0$ **15.** $9x^2 - 6x - 1 = 0$ **17.** $x^2 + 5 = 0$ **19.** $x^2 - 6x + 10 = 0$ **21.** $x^2 - 10x + 27 = 0$ **23.** $8x^2 - 4x + 3 = 0$ **25.** $f(x) = -10x^2 + 40x - 30$ **27.** $f(x) = \frac{1}{2}x^2 - 4x$ **29.** $f(x) = -\frac{1}{3}x^2 + \frac{4}{3}x + \frac{32}{3}$ **31.** $f(x) = \frac{2}{3}x^2 - \frac{8}{3}x - \frac{10}{3}$ **33.** $f(x) = -\frac{2}{3}x^2 - \frac{8}{3}x + \frac{10}{3}$ **35.** $f(x) = \frac{2}{5}x^2 - \frac{12}{5}x + 2$ **37. a.** $(3 + i\sqrt{2})^2 - 6(3 + i\sqrt{2}) + 11 = 7 + 6i\sqrt{2} - 18 - 6i\sqrt{2} + 11 = 0$ **b.** $3 - i\sqrt{2}$

Problems, pages 343–345 **1.** $20 - x$; $-x^2 + 20x$; 100 **3.** 400 **5.** 625 cm^2 **7. a.** \$210; \$227.50; $(20 + n)(8 - 0.10n)$ **b.** 30 **9.** 25 passengers **11.** 3.75 amps; 225 watts **13.** $26{,}666\frac{2}{3}$ m^2 **15. a.** 40 ft **b.** Yes

Mixed Review Exercises, page 345

1.
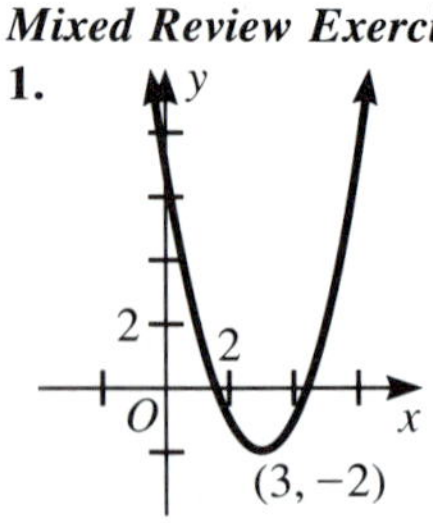

3.
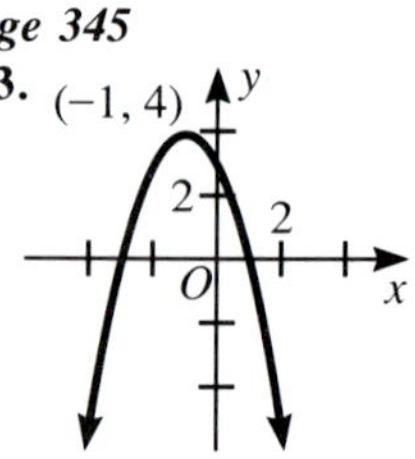

5.
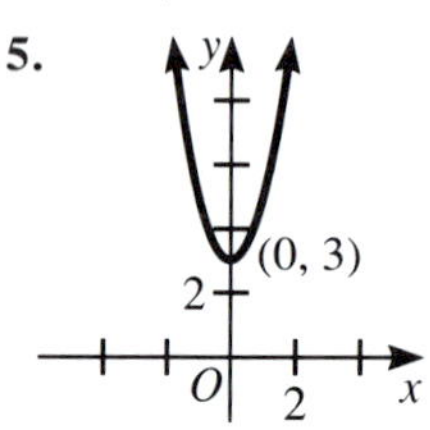

7. {real numbers}; ± 3 **8.** {real numbers}; $\frac{8}{3}$ **9.** {real numbers}; $-3 \pm \sqrt{11}$

10. {real numbers}; no zeros **11.** {real numbers}; no real zeros **12.** $\left\{x: x \neq \frac{5}{2}\right\}$; $-\frac{1}{4}$

Self-Test 3, page 345 **1.** $(-3, 4)$; $x = -3$ **2.** $(0, -6)$, $(6, 0)$, $(2, 0)$ **3.** $y - 1 = 3(x + 2)^2$ **4.** vertex $(-3, -1)$; axis $x = -3$ **5.** D = {real numbers}; $R = \{y: y \geq -6\}$; zeros: $3 \pm \sqrt{6}$ **6.** Answers may vary. $2x^2 - 2x - 1 = 0$ **7.** 36 ft

Chapter Review, pages 346–347 **1.** c **3.** a **5.** a **7.** a **9.** c **11.** b **13.** a **15.** d

Cumulative Review (Chapters 4–7), pages 348–349 **1.** $x^2 - 9x + 8$ **3.** $x^8y^4z^{12}$ **5.** $(2x - 3)(x - 4)$ **7.** $(4a - 3b)(4a + 3b)$ **9.** $\{-1, 4\}$ **11.** $\{-3, 0, 3\}$ **13.** 9 and 11 or -9 and -11 **15.** $\frac{3x}{4y^3}$ **17.** $\frac{a}{a + 2}$ **19.** $\frac{1}{2y}$ **21.** $\{2\}$ **23.** \$1200 at 5%; \$2800 at 8% **25.** $\frac{3\sqrt{6}}{2}$ **27.** $\frac{\sqrt{7} + \sqrt{3}}{2}$ **29.** $\frac{5}{13} - \frac{12}{13}i$ **31.** $\left\{\frac{3 \pm \sqrt{2}}{2}\right\}$ **33.** $k > 3$ **35.** vertex $(3, 1)$; axis $x = 3$ **37.** -9

Chapter 8 Variation and Polynomial Equations

Written Exercises, pages 354–355 **1.** 10 **3.** 20 **5.** 75 **7.** 50 **9.** 6 **11.** $6\sqrt{2}$

Problems, pages 356–357 **1.** 15 m **3.** \$6525 **5.** 175 ft **7.** 12,500 **9.** 25 kg **11.** 115 L **13.** 13 cm **15.** 96 ft **17.** 2 amperes **19.** $\frac{\pi}{6}$; $V = \frac{\pi}{6}d^3$

Mixed Review Exercises, page 357 **1.** $\left\{\frac{5 \pm \sqrt{13}}{6}\right\}$ **2.** $\{4\}$ **3.** $\left\{0, \frac{5}{2}\right\}$ **4.** $\{-8, 4\}$ **5.** $\left\{\frac{1}{2}\right\}$ **6.** $\left\{-4, \frac{1}{2}\right\}$ **7.** $\{-1 \pm \sqrt{2}\}$ **8.** $\{6\}$ **9.** $\left\{\frac{1 \pm 2\sqrt{3}}{3}\right\}$

Written Exercises, pages 360–361 **1.** 1 **3.** 12 **5.** 36 **7.** 8 **9.** 0.5

Problems, pages 361–362 **1.** 750 kHz **3.** 864 BTU **5.** 6 lux **7.** 576 r/m; about 61.7 r/m **9.** 1.0125 mm **11.** 378 kg **13.** About 35,800 km

Self-Test 1, page 363 **1. a.** $\frac{1}{7}$ **b.** 3 **2.** 216 **3.** 68.6 kPa **4. a.** 15 **b.** $\frac{1}{2}$ **5.** 9 **6.** 90 **7.** 800 r/min

Written Exercises, pages 366–367 **1.** $x + 1 + \frac{-6}{x + 2}$ **3.** $x + 4 + \frac{-2}{2 - x}$ **5.** $2t - 3 + \frac{4}{2t + 1}$ **7.** $x^2 + 2x - 4 + \frac{-2}{x - 3}$ **9.** $2s^2 - 8s + 3 + \frac{1}{s + 4}$ **11.** $2t^2 - t + 3 + \frac{4}{3t + 2}$ **13.** $5z + 3$ **15.** $2x^2 - x + 3$ **17.** $3u^2 - 2 + \frac{8u + 8}{3u^2 + 2u + 2}$ **19.** $x^2 + 2ax + \frac{a^3}{x + 2a}$ **21.** $3t^2 - ct - c^2 + \frac{c^3t + 2c^4}{2t^2 + ct + c^2}$ **23.** $x^2 + x + 1 + \frac{2x^3}{x^4 - x^2 + 1}$ **25.** $x^2 - a^2 + \frac{2a^4}{x^2 + a^2}$ **27.** $x^4 - ax^3 + a^3x - a^4$ **29.** $x^3 - x^2 - 10$ **31.** $-\frac{9}{2}$ **33.** $c = 1$; $k = -2$

Mixed Review Exercises, page 367 **1.** 12 **2.** 16 **3.** $3\sqrt{3}$ **4.** 25.6 **5.** $\frac{16}{3}$ **6.** 1 **7.** $\sqrt{3}$ **8.** 0.4

Written Exercises, page 370 **1.** $3x^2 + x + 3 + \frac{4}{x - 2}$ **3.** $x^2 - 2$ **5.** $t^3 - 2 + \frac{3}{t + 5}$ **7.** $2s^3 - s^2 - 3s - 2$ **9.** $x^4 + x^3 + x^2 + x + 1$ **11.** $2x^3 - x^2 + x - 2$ **13.** $x^2 - 2x + 3 + \frac{-5}{2x + 1}$ **15.** $2t^3 + 3t^2 + 2t - 2$ **17.** $z^2 + (-2 + 2i)z - 4i + \frac{3}{z - 2i}$ **19.** $Q(x) = 2x^2 - x + 3$; $R = -5$ **21.** $Q(z) = 2z^2 - (3 + 4i)z + 6i$; $R = 2$ **23.** 4

Reading Algebra, page 371 **3.** \$126.50; 7% account **5. a.** 6 h **b.** less than; Tom's time is greater than Sue's **c.** 7.5 h

Written Exercises, pages 375–376 **1.** 5 **3.** -10 **5.** 13 **7.** 10 **9.** Yes **11.** No **13.** Yes **15.** Yes **17.** $\{-3, \pm\sqrt{3}\}$ **19.** $\{-4, 2 \pm i\}$ **21.** $x^3 - 7x + 6 = 0$ **23.** $x^4 + 2x^3 - 5x^2 - 6x = 0$ **25.** $2x^3 - 3x^2 - 11x + 6 = 0$ **27.** $x^4 - x^3 - 12x^2 - 4x + 16 = 0$

29. $\{-1, 4, -2, 2\}$ 31. $\left\{1, -\frac{2}{3}, -1 \pm \sqrt{2}\right\}$
35. 3 37. 2

Mixed Review Exercises, page 376
1. $x + 1 + \frac{-3}{x+2}$ 2. $y - 1$ 3. $2u - 3$
4. $3w^2 + 2 + \frac{-1}{w^2 - 2}$ 5. $a^4 - a^3 - 3a^2 + 3a$
6. $c + 2$ 7. Imag. 8. Rat. 9. Irrat.

Written Exercises, pages 380–381
1. $x^3 + x^2 + 25x + 25 = 0$
3. $x^3 + 4x^2 + 6x + 4 = 0$ 5. $-2i$ 7. $1 + 2i$
9. $\{-1 \pm 2i, 2\}$ 11. $\{3 \pm i, \pm\sqrt{10}\}$
13–19. Answers are given in this order: pos. real, neg. real, imaginary. 13. 1, 1, 2
15. 0, 2, 2; or 0, 0, 4 17. 1, 2, 2; or 1, 0, 4
19. 3, 2, 0; 1, 2, 2; 3, 0, 2; or 1, 0, 4
21. $x^4 - 2x^3 + 6x^2 - 8x + 8 = 0$

Self-Test 2, page 381 1. $x^2 - 2x + 3$
2. $2x^3 + x^2 + 2x - 1 + \frac{2}{x-2}$ 3. 16
4. Yes 5. $2x^3 - 5x^2 - 4x + 3 = 0$
6. a. 3 or 1 pos.; 1 neg. b. $\left\{2 \pm i, 1, -\frac{1}{2}\right\}$
7. $x^3 + 6x + 20 = 0$

Written Exercises, pages 384–385
1. $\{-3, 1, 2\}$ 3. none 5. $\{-1, 3, -1 \pm 2i\}$
7. $\left\{\frac{1}{2}, -2 \pm i\right\}$ 9. $\left\{\frac{3}{2}, 2 \pm \sqrt{2}\right\}$
11. $\left\{-2, \frac{2}{3}, \pm i\right\}$ 23. $\left\{-\frac{3}{2}, \pm i\sqrt{2}\right\}$
25. $\left\{\frac{5}{2}, \pm 2i\right\}$

Mixed Review Exercises, page 385 1. $\{3\}$
2. $\left\{\frac{2}{3} \pm \frac{\sqrt{2}}{3}i\right\}$ 3. $\{4\}$ 4. $\{-1\}$ 5. $\left\{-\frac{2}{3}, \frac{1}{4}\right\}$
6. $\{\pm 2, \pm i\sqrt{2}\}$ 7. $\{\pm i, 2\}$ 8. $\{-3, 1 \pm \sqrt{2}\}$
9. $\{2 \pm i, -1\}$ 10. $\{1 \pm i, -2, 3\}$

Written Exercises, pages 388–389
1. 2.5 3. −1.5 5. 2 7. −1.5, −0.5, 2
9. 1.5, 2.5 11. 2.5 13. −1.5 15. 2.1
17. −1.5, −0.3, 1.9 19. 1.5, 2.8 21. $c = 3$
23. $c = \frac{1}{2}$ 25. 1.53 29. 4; −2

Computer Key-In, page 390 1. a. $x \approx -0.7$, 2.0, 2.7 b. $x \approx 2.73$ 3. 1.149

Written Exercises, pages 394–395
1. 99 million 3. 198 million 5. 1916
7. 1957 9. 1.090 kg/m^3 11. 1.179 kg/m^3
13. 216 m 15. 2070 m

Computer Exercises, page 395 3. 60, 68, 251, 259 5. a. $f(2.8) = 23.2$; exact value = 21.952
b. $f(7) = 14$; exact value = 14
c. $f(13.71) = 0.08145$; exact value = 0.0729394603
d. $f(14.3) = 606292.576$; exact value = 599967.756

Self-Test 3, page 396 1. The only possible rational roots, ±1 and ±3, do not satisfy the equation. 2. $\left\{-\frac{1}{2}, 1, \pm 2i\right\}$ 3. −1.7
4. 2.315 5. a. 33.2°C b. 0.15 min

Chapter Review, pages 397–398 1. c 3. d
5. b 7. c 9. c 11. c 13. c

Preparing for College Entrance Exams, page 399 1. B 3. E 5. B 7. C 9. E

Chapter 9 Analytic Geometry

Written Exercises, pages 404–405 1. 13; $\left(\frac{13}{2}, 6\right)$ 3. $\sqrt{74}$; $\left(-\frac{5}{2}, \frac{5}{2}\right)$ 5. $2\sqrt{5}$; (4, 4)
7. $\frac{13}{3}$; $\left(\frac{7}{6}, 0\right)$ 9. $11\sqrt{2}$; $\left(\frac{11}{2}, \frac{11}{2}\right)$ 11. 3;
$\left(0, \frac{1}{2}\right)$ 13. 6; $(1, \sqrt{3})$ 15. $|a|$; $\left(\frac{a}{2}, b\right)$
17. $2\sqrt{a^2 + b^2}$; (b, a) 19. (6, 10) 21. (10, 6)
23. $(-h, -k)$ 25. yes; yes; $\frac{17}{2}$ 27. yes; no
29. collinear 31. not collinear

Problems, pages 405–406 1. $3x + 5y - 8 = 0$
3. $5x - 4y - 11 = 0$ 5. $\left(\frac{8}{3}, 0\right)$, $\left(0, \frac{8}{5}\right)$

Mixed Review Exercises, page 406
1. Graph is a line passing through (0, −2) and (4, 0). 2. Graph is a parabola with vertex (0, 3) and axis $x = 0$. 3. Graph is a parabola with vertex (−2, 1) and axis $x = -2$. 4. 8
5. $\frac{1}{2}$ 6. −4

Computer Exercises, page 406
2. a. 7.07106782 b. 18.4390889
c. 294.49618 4. a. 22.471793 b. 150.512173

Written Exercises, pages 410–411
1. $(x - 3)^2 + y^2 = 9$
3. $(x - 2)^2 + (y + 5)^2 = 64$ 5. $x^2 + y^2 = 144$

7. $(x - 6)^2 + (y - 1)^2 = 2$

9.

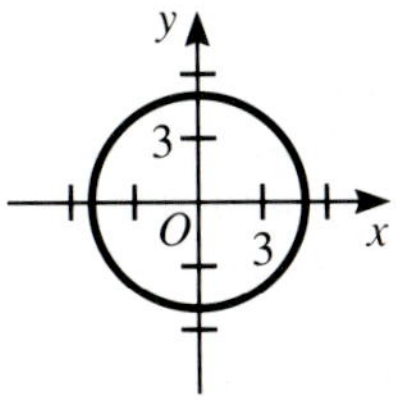

11.

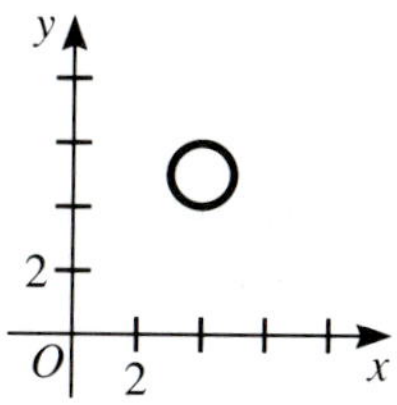

13.

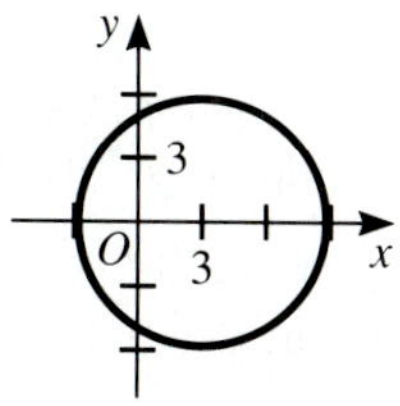

15. (0, 0); 4
17. (0, −4); 4
19. (2, −1); 3
21. no graph
23. $\left(-\frac{3}{2}, 2\right)$; $\frac{5}{2}$

25.

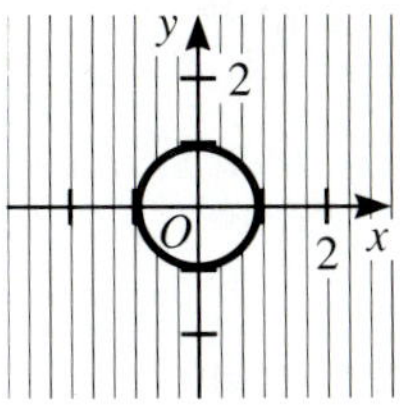

27.

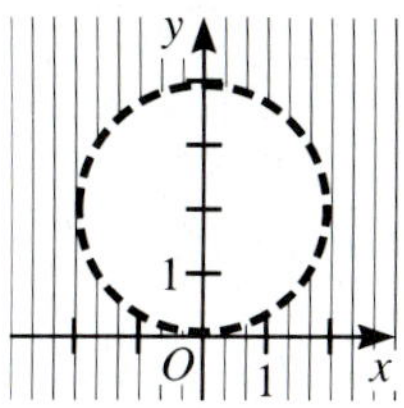

29.

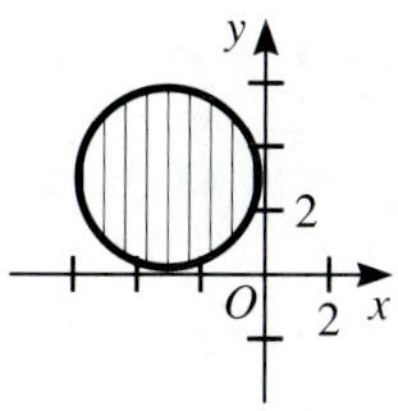

31. (2, 3); 2 **33.** $\left(1, -\frac{1}{4}\right)$; $\frac{\sqrt{17}}{4}$
35. $x^2 + (y - 5)^2 = 25$
37. $(x - 1)^2 + (y - 4)^2 = 2$
39. $(x + 2)^2 + (y - 4)^2 = 16$
41. $(x - 5)^2 + (y + 4)^2 = 16$

Written Exercises, pages 415–416 **1.** $F(4, 7)$ **3.** D: $y = 4$ **5.** D: $y = -8$ **7.** $y - 2 = -\frac{1}{8}x^2$ **9.** $y = -\frac{1}{16}x^2$ **11.** $x = \frac{1}{4}y^2$ **13.** $x - 1 = -\frac{1}{4}(y - 2)^2$ **15.** $y - 2 = \frac{1}{8}(x - 3)^2$ **17.** (0, 0); $\left(0, \frac{3}{2}\right)$; $y = -\frac{3}{2}$; $x = 0$ **19.** (−1, 2); (0, 2); $x = -2$; $y = 2$ **21.** (−2, 1); (−2, −1); $y = 3$; $x = -2$ **23.** $\left(-\frac{3}{2}, 3\right)$; $\left(\frac{1}{2}, 3\right)$; $x = -\frac{7}{2}$; $y = 3$ **25.** (−5, −2); $\left(-5, -\frac{3}{2}\right)$; $y = -\frac{5}{2}$; $x = -5$

27.

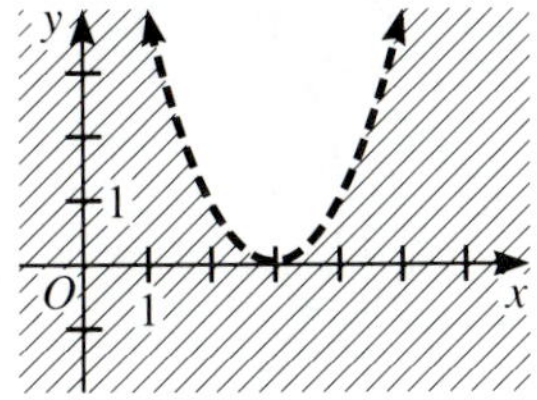

29.

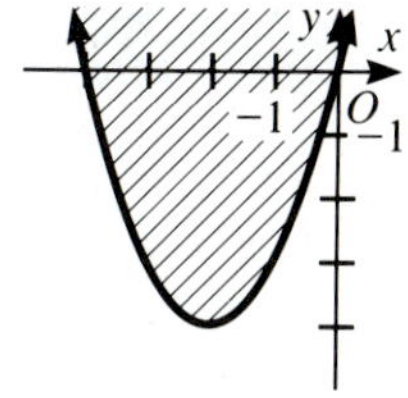

31.

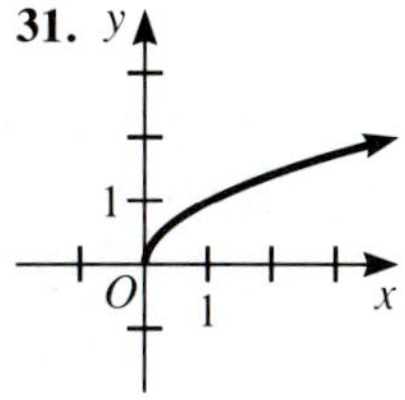

33.

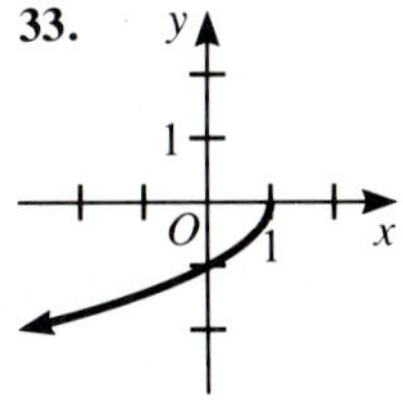

Mixed Review Exercises, page 417 **1.** 0; 8; (1, 2) **2.** 4; $2\sqrt{17}$; (2, −2) **3.** $-\frac{1}{7}$; $5\sqrt{2}$; $\left(\frac{1}{2}, \frac{7}{2}\right)$ **4.** $2x - 3y = 5$ **5.** $x + y = -1$
6. $(x + 1)^2 + (y - 3)^2 = 16$

Self-Test 1, page 417 **1.** $\sqrt{65}$ **2.** $\left(\frac{5}{2}, 8\right)$
3. not collinear **4.** $(x - 5)^2 + (y + 4)^2 = 49$
5. (−6, 2); $2\sqrt{2}$ **6.** $x + 4 = \frac{1}{16}(y - 3)^2$
7. (−5, 2); (−5, −2); $y = 6$; $x = -5$
8. vertex (−1, −3); focus (1, −3); directrix $x = -3$; axis $y = -3$

Written Exercises, pages 421–423
1. $(-\sqrt{5}, 0)$, $(\sqrt{5}, 0)$ **3.** $(-4\sqrt{2}, 0)$, $(4\sqrt{2}, 0)$ **5.** $(0, -\sqrt{6})$, $(0, \sqrt{6})$
7. $(0, -2\sqrt{5})$, $(0, 2\sqrt{5})$ **9.** $(0, -2\sqrt{2})$, $(0, 2\sqrt{2})$ **11.** (0, −2), (0, 2)
13. $\left(-\frac{2\sqrt{2}}{3}, 0\right)$, $\left(\frac{2\sqrt{2}}{3}, 0\right)$
15. $\frac{x^2}{25} + \frac{y^2}{4} = 1$ **17.** $\frac{x^2}{4} + \frac{y^2}{2} = 1$
19. $\frac{x^2}{81} + \frac{y^2}{45} = 1$ **21.** $\frac{x^2}{128} + \frac{y^2}{144} = 1$
25. a. 0 **b.** 0 **c.** At the center **d.** A circle
27. $\frac{4\sqrt{5}}{3}$ m (≈ 3 m)

Application, page 425 **1.** max: 1.017 AU ≈ 153 million km; min: 0.983 AU ≈ 147 million km **3.** ≈1.0001:1
5. ≈1.78:1; since the ratio is much larger, the orbit is much more elongated than that of Earth.

Written Exercises, pages 430–431

1. a.

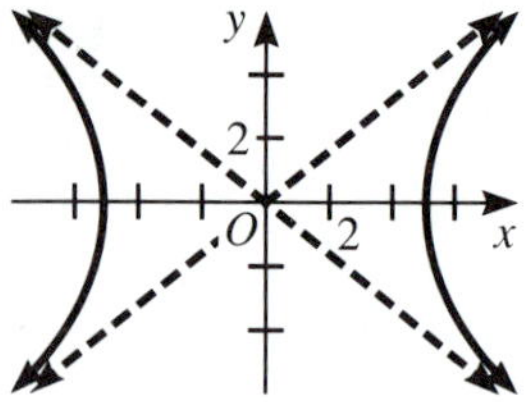

b. $(-\sqrt{41}, 0)$, $(\sqrt{41}, 0)$

3. a.

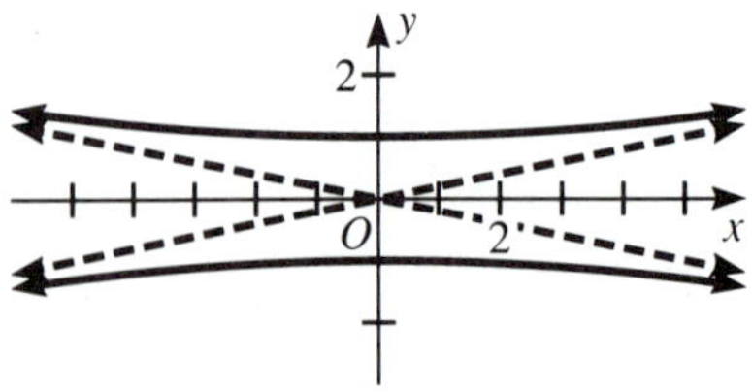

b. $(0, -\sqrt{26})$, $(0, \sqrt{26})$

5. a.

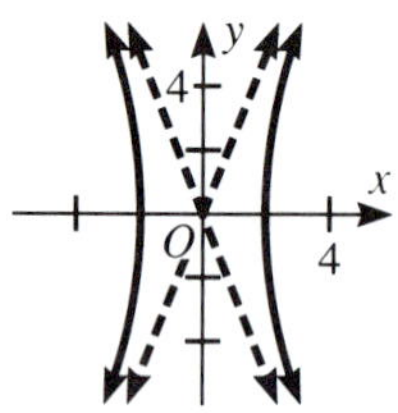

b. $(-\sqrt{29}, 0)$, $(\sqrt{29}, 0)$

7. a.

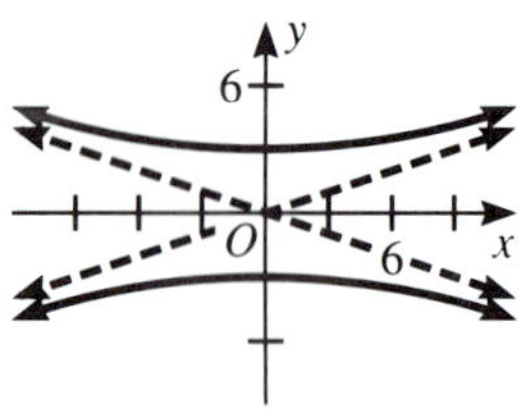

b. $(0, -3\sqrt{10})$, $(0, 3\sqrt{10})$

9. a.

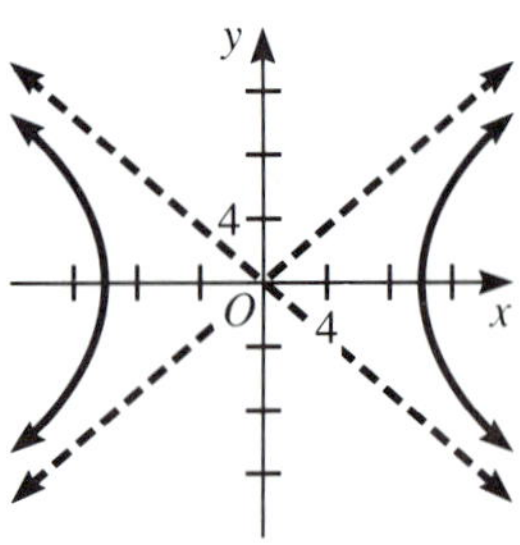

b. $(-5\sqrt{7}, 0)$, $(5\sqrt{7}, 0)$

11. a.

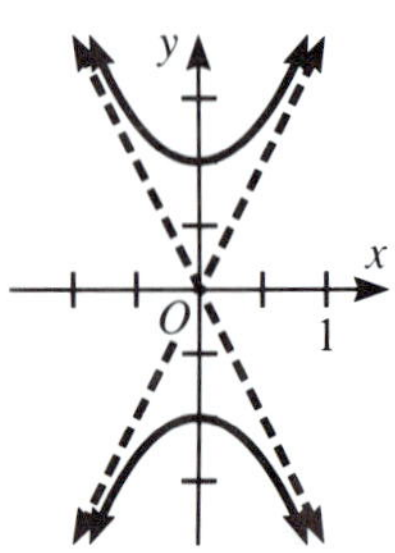

b. $\left(0, -\frac{\sqrt{5}}{2}\right)$, $\left(0, \frac{\sqrt{5}}{2}\right)$

13. $\frac{y^2}{25} - \frac{x^2}{39} = 1$ **15.** $\frac{y^2}{9} - \frac{x^2}{4} = 1$

17. $\frac{y^2}{9} - \frac{x^2}{1} = 1$

Mixed Review Exercises, page 431

1.

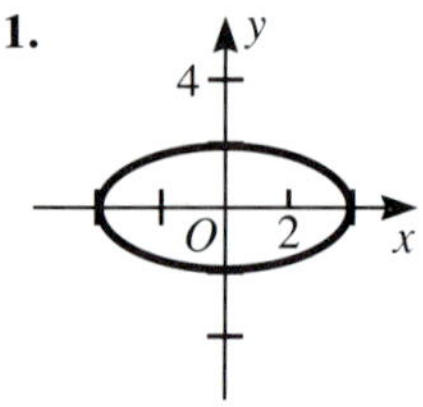

3.

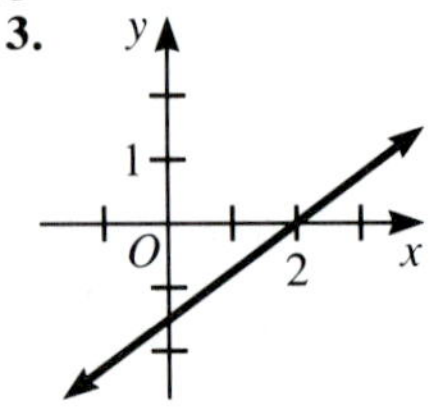

5. yes **6.** yes **7.** no **8.** yes **9.** yes **10.** no

Written Exercises, pages 434–435

1. $\frac{(x-6)^2}{16} + \frac{(y-3)^2}{25} = 1$

3. $\frac{(x+3)^2}{7} + \frac{y^2}{16} = 1$

5. $\frac{(x-2)^2}{25} + \frac{(y+3)^2}{9} = 1$

7. $\frac{(x-4)^2}{1} - \frac{(y+2)^2}{15} = 1$

9. $\frac{(y+5)^2}{4} - \frac{(x-3)^2}{5} = 1$

11. $\frac{(y+5)^2}{9} - \frac{(x-5)^2}{7} = 1$ **13.** hyperbola; $(1, -3)$; $(1 + \sqrt{5}, -3)$, $(1 - \sqrt{5}, -3)$
15. circle; $(3, 8)$; radius 4 **17.** ellipse; $(-2, 3)$; $(-6, 3)$, $(2, 3)$
19. $20x^2 + 20y^2 - 32xy - 9 = 0$
21. a. parabola **b.** circle **c.** ellipse **d.** hyperbola **e.** circle: $A = B$; parabola: $AB = 0$; ellipse: $AB > 0$ and $A \neq B$; hyperbola: $AB < 0$

Self-Test 2, page 435

1.

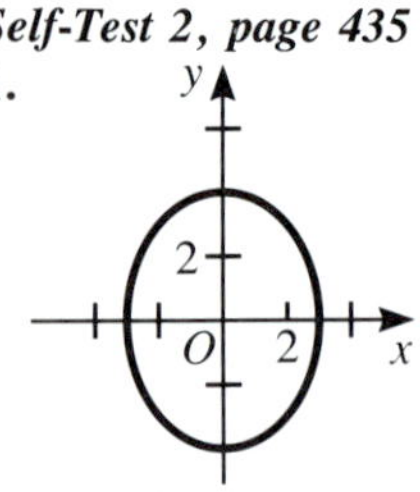

2. $\frac{x^2}{16} + \frac{y^2}{7} = 1$

3.

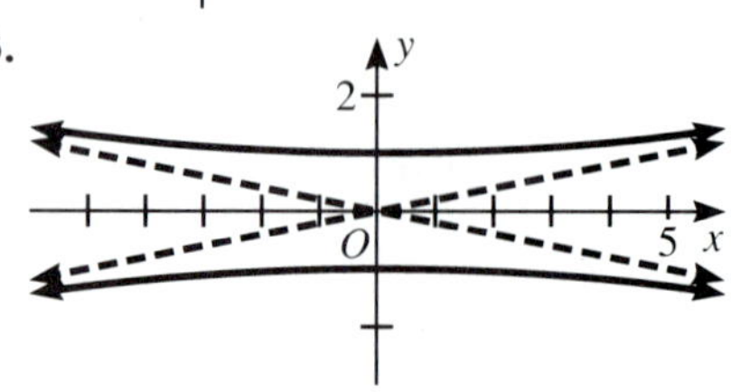

4. $\frac{y^2}{4} - \frac{x^2}{12} = 1$

5. hyperbola; see figure at right.

6. $\frac{x^2}{36} + \frac{(y-2)^2}{11} = 1$

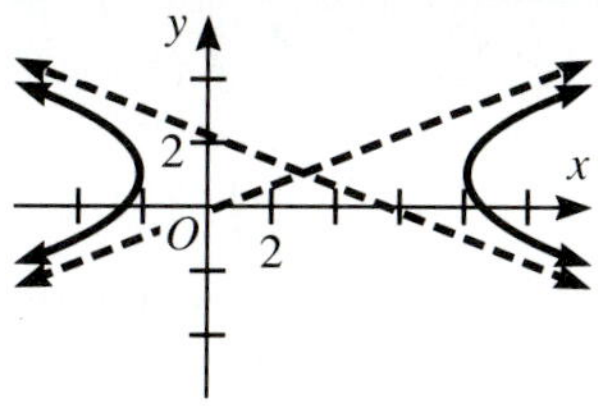

Written Exercises, page 438
1. 2 **3.** 0 **5.** 1 **7.** 4 **9.** 3
11. (2.5, 1.5) **13.** (4.5, 2), (−4.5, −2), (2, 4.5), (−2, −4.5) **15.** (−2, 0), (2.5, 1), (2.5, −1)

Mixed Review Exercises, page 438
1. $y - 1 = \frac{1}{8}(x-2)^2$ **2.** $\frac{x^2}{6} + \frac{y^2}{4} = 1$
3. $\frac{y^2}{4} - \frac{x^2}{5} = 1$ **4.** $(x+3)^2 + (y-4)^2 = 25$
5. {(1, 5)} **6.** {(2, −1)} **7.** {(3, −3)}

Written Exercises, pages 441–442
1. {(−4, 11), (2, −1)} **3.** $\{(\sqrt{3}, 3), (-\sqrt{3}, 3)\}$ **5.** {(8, −7)} **7.** {(2, 4), (4, 2)}
9. {(−43, 9), (5, −3)} **11.** {(4, 3), (4, −3), (−4, 3), (−4, −3)} **13.** $\{(\sqrt{21}, 2), (\sqrt{21}, -2), (-\sqrt{21}, 2), (-\sqrt{21}, -2)\}$ **15.** No real solution **17.** {(3, −2), (−3, 2), (2, −3), (−2, 3)} **19.** {(−6, −1), (2, 3), (−3, −2)}
21. $2 + i, -2 - i$ **23.** $3 + 2i, -3 - 2i$
25. $3 + 4i, -3 - 4i$

Problems, pages 442–443 **1.** 5 and 11
3. 35 m by 43 m **5.** 5 ft by 12 ft **7.** 57
9. 30 m by 45 m **11.** 3.7 km/h and 4.7 km/h
13. $\left(\frac{4}{5}, \frac{3}{5}\right)$

Written Exercises, pages 447–448
1. {(−10, 14, −2)} **3.** {(3, −2, −4)}
5. {(−4, −2, 1)} **7.** {(−1, 0, −4)}
9. {(−1, −3, 5)} **11.** {(1, 2, 2)}
13. {(1, 3, 2)} **15.** {(−1, 2, 3)}
17. $\left\{\left(1, -1, -\frac{1}{2}\right)\right\}$ **19. a.** $\{(5 + z, 2 + 2z, z)\}$
b. Examples: (5, 2, 0), (8, 8, 3), (3, −2, −2)
23. $\{(2z, 5z, z)\}$ **25.** $\{(0, 2z, z)\}$

Problems, pages 448–449 **1.** 6 nickels, 8 dimes, 5 quarters **3.** 2 carats, 4 carats, 8 carats
5. 30°, 70°, 80° **7.** 48 patrons, 103 sponsors, 175 donors **9.** $y = 3x^2 - 2x - 1$
11. $x^2 + y^2 - 4x + 6y + 9 = 0$

Mixed Review Exercises, page 449
1. $\{(\sqrt{2}, 1), (\sqrt{2}, -1), (-\sqrt{2}, 1), (-\sqrt{2}, -1)\}$ **2.** {(2, 5)} **3.** $\left\{\left(\frac{5}{4}, \frac{3}{2}\right), (3, -2)\right\}$
4. $\frac{x^2}{9y^4}$ **5.** $12x\sqrt{x}$ **6.** $-12x^5y^5$ **7.** $12x^2$
8. $\frac{5x^2}{3y^5}$ **9.** $\frac{3x}{5y^3}$ **10.** $\frac{3x^2\sqrt{3}}{2}$ **11.** $\frac{y^4}{x^5}$

Self-Test 3, page 450
1. b. 4 **c.** {(4.5, 2), (4.5, −2), (−4.5, 2), (−4.5, −2)} **2. b.** 1 **c.** {(3, 2.5)} **3. b.** 2
c. {(3, 0), (−3, 0)}
4. {(0, 5), (0, −5)} **5.** {(−9, −11), (1, −1)}
6. {(2, −1, 3)} **7.** {(2, −3, 4)}

Chapter Review, pages 451–453 **1.** d **3.** c
5. a **7.** c **9.** c **11.** a **13.** c **15.** d **17.** d
19. c

Chapter 10 Exponential and Logarithmic Functions

Written Exercises, page 458 **1.** 9 **3.** $\frac{1}{7}$ **5.** 8
7. 8 **9.** $-\frac{1}{5}$ **11.** $\frac{1}{2}$ **13.** −4 **15.** $\frac{1}{5}$ **17.** $\frac{1}{2}$
19. 49 **21.** $x^{3/2}y^{5/2}$ **23.** $a^{-1}b^{3/2}$ **25.** $a^{-5}b^{5/2}$
27. $8a^{-1/2}b^{-3/2}$ **29.** $2\sqrt[3]{2}$ **31.** $\sqrt{2}$ **33.** $\sqrt[4]{2}$
35. 3 **37.** x **39.** $x^{1/12}$ **41.** $a^2 - 2a$
43. a. {16} **b.** {5} **45. a.** $\left\{\frac{1}{25}\right\}$ **b.** $\left\{\frac{1}{200}\right\}$
47. {−56} **49.** {±11}

Mixed Review Exercises, page 458 **1.** $\frac{x+2}{x+1}$
2. $\frac{2\sqrt{2xy}}{y^2}$ **3.** $2a^2 - a + 1$ **4.** $-8c^4$ **5.** ab^3
6. $2n^2 - 5n - 12$

Written Exercises, pages 461–462 **1. a.** $9^{\sqrt{2}}$
b. $9^{\sqrt{2}}$ **c.** 9 **d.** 81 **3.** 100^{π} **5.** 6^{π}
7. $\frac{1}{10{,}000}$ **9.** 1 **11.** $\frac{1}{2}$ **13.** 1 **15.** $\frac{2-\sqrt{3}}{2}$
17. $\frac{1}{3^{\pi}}$ **19.** {−3} **21.** $\left\{-\frac{5}{3}\right\}$ **23.** $\left\{\frac{2}{3}\right\}$
25. No solution **27.** {2} **29.** {3} **31.** {3}

33.

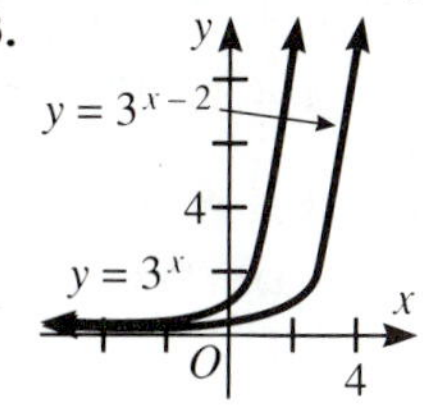

35.

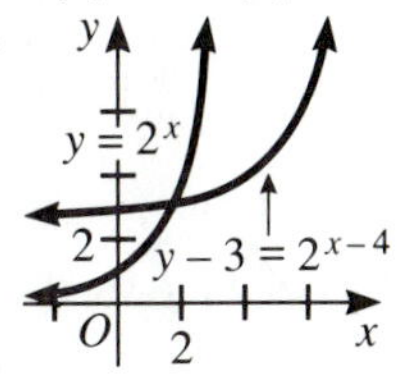

37. {6} **39.** {0}

Written Exercises, pages 466–467 **1. a.** $\frac{5}{2}$ **b.** -4 **c.** $-\frac{3}{2}$ **d.** $\frac{x-3}{2}$ **3. a.** $\frac{3}{2}$ **b.** 1 **c.** not real **d.** $\frac{\sqrt{x}}{2}$ **5. a.** 3 **b.** not real **c.** $\sqrt{x-3}$ **d.** $\sqrt[4]{x}$ **7.** no **9.** no

11. $f^{-1}(x) = \frac{x+3}{2}$ **13.** $f^{-1}(x) = \sqrt[3]{x}$

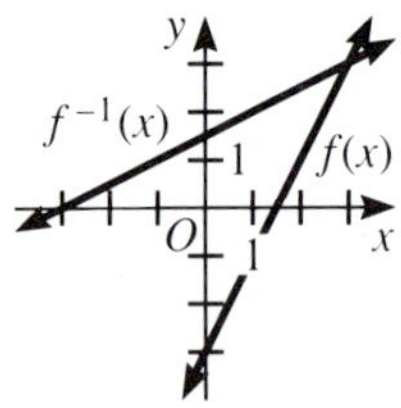

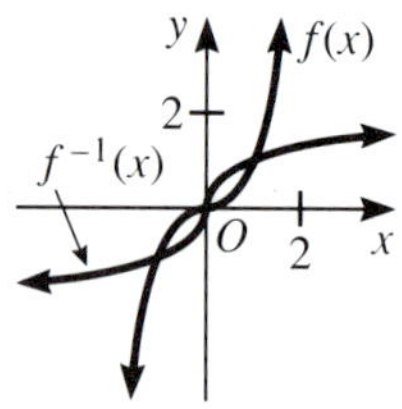

15. $g^{-1}(x) = \frac{8}{\sqrt[3]{x}}$ **17.** no inverse **19.** no inverse **21.** no inverse **23. c.** 1; 2; 3; -1 **d.** f: $D = \{\text{reals}\}$, $R = \{y\colon y > 0\}$; f^{-1}: $D = \{x\colon x > 0\}$, $R = \{\text{reals}\}$ **25.** $m = 1$, $b = 0$; or $m = -1$

Mixed Review Exercises, page 467 **1.** $\frac{1}{125}$ **2.** $7\sqrt{2}$ **3.** $-5 - 12i$ **4.** 81 **5.** -24 **6.** $\frac{1}{16}$ **7.** 9 **8.** $5 - 4\sqrt{2}$ **9.** -3 **10.** 128 **11.** $1 + 2i$ **12.** 2

Self-Test 1, page 467 **1. a.** $2x^{2/3}y^{-1/3}$ **b.** $6^{-1/3}$ **2. a.** $\frac{25\sqrt{5}}{2}$ **b.** $x^2y\sqrt[6]{xy^5}$ **3.** {34} **4. a.** $2^{-5\sqrt{2}}$ **b.** $2^{-8\sqrt{5}}$ **5.** {3} **6. a.** 13 **b.** 5 **c.** $6\sqrt{x} + 1$ **d.** $\sqrt{6x+1}$

7. $f(g(x)) = 3\left(\frac{x+7}{3}\right) - 7 = x + 7 - 7 = x$;

$g(f(x)) = \frac{(3x-7)+7}{3} = \frac{3x}{3} = x$

Written Exercises, pages 470–472 **1.** 3 **3.** 4 **5.** 0 **7.** -2 **9.** $\frac{3}{2}$ **11.** $\frac{1}{4}$ **13.** $\frac{2}{3}$ **15.** -3 **17.** $-\frac{2}{3}$ **19.** {49} **21.** $\left\{\frac{1}{3}\right\}$ **23.** $\left\{\frac{1}{8}\right\}$ **25.** {9} **27.** $\left\{\frac{1}{49}\right\}$ **29.** $\{x\colon x > 0 \text{ and } x \neq 1\}$

31. a. $3 + 2 = 5$ **b.** $\frac{1}{2} + \frac{3}{2} = 2$ **c.** $\log_b M + \log_b N = \log_b MN$ **33. a.** $\log_6 x$ **b.** 2; $-\frac{1}{2}$ **c.** f: $D = \{\text{reals}\}$, $R = \{y\colon y > 0\}$; f^{-1}: $D = \{x\colon x > 0\}$, $R = \{\text{reals}\}$

35.

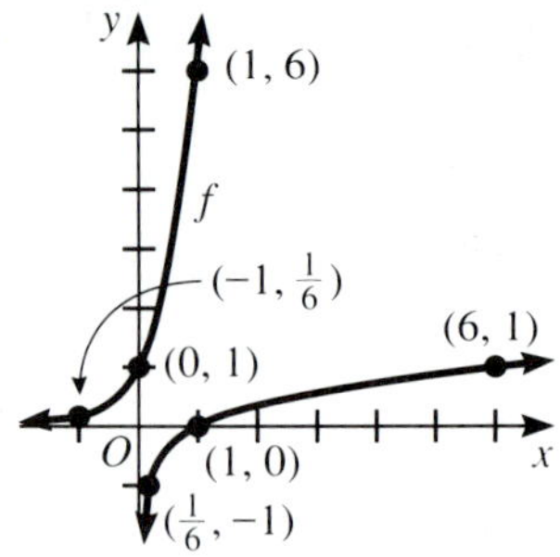

37.

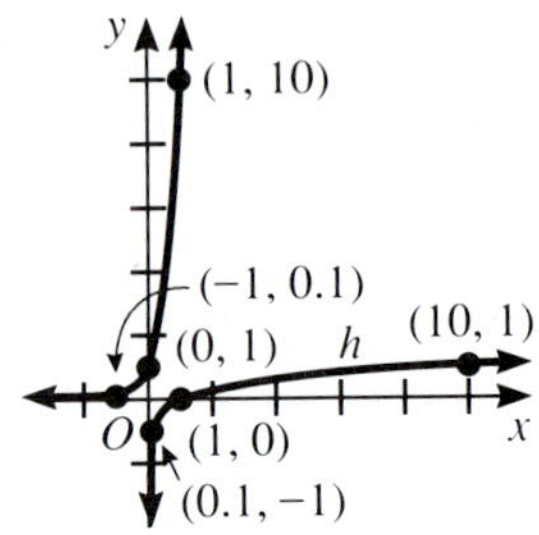

39. {3} **41.** positive **43. a.** 120 dB **b.** 10^4

Written Exercises, pages 476–477 **1.** $6 \log_2 M + 3 \log_2 N$ **3.** $\log_2 M + \frac{1}{2}\log_2 N$ **5.** $4 \log_2 M - 3 \log_2 N$ **7.** $\frac{1}{2}\log_2 M - \frac{3}{2}\log_2 N$ **9.** 1.90 **11.** 0.15 **13.** 0.90 **15.** 0.35 **17.** -0.95 **19.** -0.22 **21.** $\log_4 p^5q$ **23.** $\log_3 \frac{A^4}{\sqrt{B}}$ **25.** $\log_2 8MN$ **27.** $\log_5 \frac{5}{x^3}$ **29.** 2 **31.** $\frac{3}{2}$ **33.** {45} **35.** {1} **37.** {6} **39.** {±5} **41. a.** 6 **b.** $\frac{1}{4}$ **c.** 1 **45.** {3} **47.** {2} **49.** $\{\sqrt{85}\}$ **51.** {2}

Mixed Review Exercises, page 477 **1.** $\left\{\frac{\sqrt{2}}{2}\right\}$ **2.** $\{-3, 1, 2\}$ **3.** {5} **4.** {4} **5.** $\left\{-\frac{3}{2}\right\}$ **6.** {7} **7.** $\left\{-\frac{1}{2}\right\}$ **8.** {4} **9.** $\{2 \pm \sqrt{5}\}$ **10.** 3 **11.** 2 **12.** 1 **13.** 1

Self-Test 2, page 477 **1. a.** $3^4 = 81$ **b.** $6^3 = 216$ **2. a.** $\log_5 625 = 4$ **b.** $\log_{25} 125 = \frac{3}{2}$ **3. a.** 3 **b.** 12 **4.** {3} **5.** $\frac{5}{3}\log_2 M + 2\log_2 N$ **6.** -1.40 **7.** {3}

Written Exercises, pages 481–482 **1.** 1.79 **3.** 0.00792 **5.** 575 **7.** 33.7 **9.** 7.13 **11.** 692 **13.** 0.0158 **15. a.** $\frac{\log 30}{\log 3}$ **b.** 3.10

17. a. $\frac{\log 56}{\log 5.6}$ **b.** 2.34 **19. a.** $-\frac{\log 5}{\log 30}$ **b.** −0.473 **21. a.** $\frac{\log 60}{2 \log 3.5}$ **b.** 1.63 **23.** $\left\{\frac{7}{4}\right\}$ **25.** $\left\{\frac{5}{6}\right\}$ **27.** {6740} **29.** {21.6} **31.** {2.19} **33.** {1.03} **35.** 3.17 **37.** 3.36 **39.** {0.631, 1.46} **41. a.** 2; $\frac{1}{2}$ **b.** 3; $\frac{1}{3}$ **c.** They are reciprocals. $\log_b a = \frac{\log_a a}{\log_a b} = \frac{1}{\log_a b}$

Problems, pages 486–488 **1. a.** \$1120 **b.** \$1254.40 **c.** \$1404.93 **d.** \$3105.85 **3. a.** \$1125.51 **b.** \$1266.77 **c.** \$1425.76 **d.** \$3262.04 **5. a.** \$10,000 **b.** \$8000 **c.** \$6400 **d.** \$1342.18 **7. a.** $4N_0$ **b.** $32N_0$ **c.** $(2^{w/3})N_0$ **9. a.** 25 kg **b.** 6.25 kg **c.** $100\left(\frac{1}{2}\right)^{y/6000}$ kg **11.** about 9 years **13.** 12.6% **15.** 0.997; 9.09×10^{-13} **17.** 11.4% **19.** 19.7% **21. a.** \$1,065,552.45 **b.** \$1,066,086.39 **c.** \$1,066,091.81

Mixed Review Exercises, page 488

1. $\left\{0, \frac{1}{2}, 1\right\}$ **2.** {1, 2} **3.** $\{\sqrt{2}, 2, 4\}$ **4.** {0, 6} **5.** $\left\{\frac{1}{3}, \frac{1}{2}, \frac{2}{3}\right\}$ **6.** $\{1, \sqrt{2}, 2\}$ **7.** 1 **8.** 2 **9.** $\frac{1}{6}$ **10.** 1 **11.** $\frac{5}{12}$ **12.** 3

Written Exercises, pages 490–491 **1.** $e^{2.08} = 8$ **3.** $e^{-1.10} = \frac{1}{3}$ **5.** ln 20.1 = 3 **7.** $\ln 1.65 = \frac{1}{2}$ **9.** 2 **11.** −3 **13.** 0 **15.** 5 **17.** ln 12 **19.** $\ln \frac{9}{5}$ **21.** $\ln 10e^3$ **23.** $\{e^3\}$ **25.** $\{4 + e^{-1}\}$ **27.** $\{\pm e^{9/2}\}$ **29.** {ln 2} **31.** {ln 5} **33.** {2 + ln 2} **35.** {ln 9} **37.** $\left\{\frac{3}{5} \ln 10\right\}$ **39.** $\{e\}$ **41.** {2} **43.** {ln 3, ln 4} **45.** $D = \{x: x \neq 0\}$; R = {reals} **47.** $D = \{x: x > 5\}$; R = {reals}

Calculator Key-In, page 492 **1. a.** \$1061.84 **b.** \$1127.50 **c.** \$1822.12 **3. a.** 1; 0.99; 0.96; 0.91; 0.85; 0.78; 0.70; 0.61; 0.53; 0.44; 0.37 **5.** 2.718056; e

Self-Test 3, page 493 **1.** {650} **2.** {5.81} **3.** {1.07} **4.** \$3277.23 **5.** $\ln\left(\frac{5}{e^2}\right)^{1/3}$ $\left(\text{or } \frac{\ln 5 - 2}{3}\right)$

Application, pages 493–494 **1.** 6600 **3. a.** 95.3% **b.** 93.0%

Chapter Review, pages 496–497 **1.** c **3.** d **5.** b **7.** a **9.** b **11.** a **13.** d **15.** b

Mixed Problem Solving, page 498 **1.** 10 mL **3.** CA:45; NY:34; NC:11 **5.** −9 **7.** \$1500 **9.** $A = \frac{C^2}{4\pi}$ **11.** Larger plant: 8.4 h; smaller: 12.4 h **13.** 4 h

Preparing for College Entrance Exams, page 499 **1.** C **3.** E **5.** A **7.** E

Chapter 11 Sequences and Series

Written Exercises, pages 504–506 **1.** A; 8, 5 **3.** G; 625, 3125 **5.** A; 30, 38 **7.** N; $\frac{1}{25}, \frac{1}{36}$ **9.** G; $4^{9/2}, 4^{11/2}$ **11.** 7, 11, 15, 19; A **13.** 1, 3, 9, 27; G **15.** $-\frac{1}{4}, \frac{1}{2}$, −1, 2; G **17.** log 2, log 3, log 4, log 5; N **19. a.** A **b.** G **21.** 32, 44 **23.** 20, 18 **25.** 63; 127 **27.** 21, 34 **29.** 48, 71 **31. a.** 15, 21 **b.** 55 **33. a.** 9 **b.** 35 **35. a.** 16, 31; No **b.** 57

Mixed Review Exercises, page 506

1.

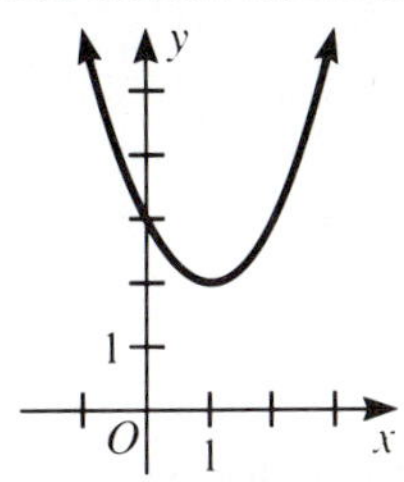

2.

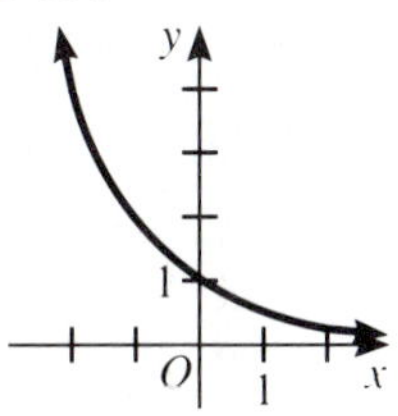

7. (2, −3) **8.** $\left(-\frac{3}{2}, \frac{7}{2}\right)$, (1, 1) **9.** (2, 1), (−2, 1), (2, −1), (−2, −1)

Written Exercises, page 509 **1.** $t_n = 8n + 16$ **3.** $t_n = 4 - 7n$ **5.** $t_n = 4n + 3$ **7.** 104 **9.** 52 **11.** −902 **13.** 87.5 **15.** 8 **17.** −61 **19.** 2 **21.** $\frac{3}{2}$ **23. a.** −7, 13 **b.** −12, 3, 18 **c.** −15, −3, 9, 21 **25. a.** 19, 27 **b.** 17, 23, 29 **c.** 15.8, 20.6, 25.4, 30.2 **27.** 101 **29.** 300 **31.** 16 **33.** $\frac{a+b}{2} - a = b - \frac{a+b}{2} = \frac{b-a}{2}$

Written Exercises, pages 513–514
1. $t_n = 2 \cdot 3^{n-1}$ **3.** $t_n = (\sqrt{2})^{n-1}$
5. $t_n = 64\left(-\frac{3}{4}\right)^{n-1}$ **7.** 39,366 **9.** $\frac{5}{256}$
11. $\frac{5}{128}$ **13.** $\frac{16}{3}$ **15.** 5120 **17.** y^{39} **19.** 4
21. 3 **23.** 10; 20; 40 **25.** $\sqrt[5]{2}, \sqrt[5]{4}, \sqrt[5]{8}, \sqrt[5]{16}$ **27.** A; $t_n = 2n$ **29.** A;
$t_n = 25 + 8(n-1)$ **31.** G; $t_n = 200\left(-\frac{1}{2}\right)^{n-1}$
33. A; $t_n = (n+1)a + (2n-1)$ **35.** A;
$t_n = -10n$ **37.** $t_n = \frac{n+1}{n^2}$
39. $\frac{\sqrt{ab}}{a} = \frac{b}{\sqrt{ab}} = \sqrt{\frac{b}{a}}$

Problems, pages 514–515 **1.** \$21,000
3. \$35.43 **5.** 14 **7.** 192 million **9. a.** 19
b. 15 **11.** \$1601.87 **13.** B; B

Mixed Review Exercises, page 515
1. $\{x: x < -2 \text{ or } x > 7\}$ **2.** $\{x: -3 \le x \le 2\}$
3. $\{x: -4 \le x < 2\}$
4. $\{x: -2 < x < 0 \text{ or } x > 2\}$ **5.** $\{x: x \le 2\}$
6. $\{x: -5 < x < -1\}$ **7.** $x + y = 1$
8. $y = -\frac{1}{4}x^2$ **9.** $\frac{x^2}{4} + \frac{y^2}{3} = 1$

Self-Test 1, page 516 **1. a.** G; $\frac{27}{8}, \frac{81}{16}$ **b.** A;
$-11, -15$ **c.** N; $\frac{5}{11}, \frac{6}{13}$ **d.** A; 2.2, 2.5
2. a. $t_n = 43 - 7n$ **b.** -90 **3. a.** 4 **b.** 2, 6
4. a. $t_n = 48\left(-\frac{1}{2}\right)^{n-1}$ **b.** $-\frac{3}{32}$ **5. a.** $4\sqrt{2}$
b. 4, 8 **6.** \$58.32

Extra, page 517 **1.** 6 **3.** -2
5. {real numbers}; {positive integers}
7. {real numbers}; {positive integers}

Written Exercises, pages 521–522
1. $11 + 12 + 13 + 14 + 15 + 16$
3. $2 + 4 + 8 + 16 + 32 + 64$
5. $1 - \frac{1}{2} + \frac{1}{3} - \frac{1}{4} + \frac{1}{5} - \frac{1}{6}$
7. $2 + 1 + 0 + 1 + 2 + 3$ **9.** $\sum_{n=1}^{500} 2n$
11. $\sum_{n=1}^{20} n^3$ **13.** $\sum_{n=1}^{99} \frac{n}{n+1}$ **15.** $\sum_{n=1}^{100} (2n-1)$
17. $\sum_{n=0}^{6} 2^n$ **19.** $\sum_{n=1}^{\infty} \frac{1}{n}$ **21.** $\sum_{n=1}^{\infty} 27\left(-\frac{1}{3}\right)^n$
23. $\sum_{n=0}^{3} 6(-2)^n$ **25.** $\sum_{n=1}^{\infty} \frac{1}{n^2}$ **27.** $\sum_{n=20}^{199} 5n$
29. $\sum_{n=0}^{\infty} \left(\frac{1}{4}\right)^n$ **31.** $\frac{j+4}{j+8}$ **33.** $\sum_{k=1}^{4} k \log 5 =$
$\log 5 + 2\log 5 + 3\log 5 + 4\log 5 =$
$10 \log 5 = \log 5^{10}$ **35.** 21

Written Exercises, pages 527–528 **1.** 670
3. 3420 **5.** 25,250 **7.** 3925 **9.** 15,250
11. 893 **13.** 255 **15.** $-29{,}524$ **17.** $\frac{4095}{4096}$
19. a. -300 **b.** $\frac{3069}{64}$ **21.** 1960 **23.** 247,500
25. a. 420 **b.** 2,097,150 **27.** 3240
29. a. $\sum_{k=1}^{n} 2^{k-1} = \frac{2^0(1-2^n)}{1-2} = \frac{1-2^n}{-1} = 2^n - 1$
b. 20 **31.** 16, 48, 144; geometric
33. Answers will vary. **a.** 0.00000095
b. 0.00002656 **c.** 0.00001427 **35.** -50
37. 2,097,360

Problems, pages 528–530 **1.** 880 **3.** 156
5. 2046 **7.** Plan B; \$58.83 more **9.** B; A
11. \$13,971.64 **13. a.** $T_n = \frac{n(n+1)}{2}$
b. $T_n = \sum_{k=1}^{n} k$ **15.** the square of an odd number

Mixed Review Exercises, page 530 **1.** $\frac{\sqrt{2}}{4}$
2. 0 **3.** -2 **4.** 4 **5.** -1 **6.** $2\sqrt{17}$ **7.** 16
8. -1 **9.** -2 **10.** $\frac{3}{2}$ **11.** 4 **12.** (3, 0) and
$(-3, 0)$

Computer Exercises, page 530 **2. a.** 506
b. 1.99998957 **c.** 0.740049503

Written Exercises, pages 533–534 **1.** 48
3. 16.2 **5.** no sum **7.** no sum
9. $\frac{9}{2}(\sqrt{3} - 1)$ **11.** 4 **13.** 8, 10, $\frac{21}{2}, \frac{85}{8}, \frac{341}{32}$;
$S \approx 11$; $S = \frac{32}{3}$ **15.** $\frac{3}{4}, \frac{3}{8}, \frac{9}{16}, \frac{15}{32}, \frac{33}{64}$; $S \approx \frac{1}{2}$;
$S = \frac{1}{2}$ **17.** $\frac{1}{3}$ **19.** $\frac{1040}{333}$ **21.** $\frac{1}{2}$ **23.** 8, $\frac{8}{3}, \frac{8}{9}$
25. 40, $-\frac{40}{3}, \frac{40}{9}$ **29. a.** 3 **b.** 32

Problems, pages 534–536 **1.** 72 ft
3. $48(2 + \sqrt{2})$ cm **5.** 288 cm² **7.** 480 m
9. a. 0.18889 in. **b.** 249.24 in. **c.** 250 in.

11. a. 20, 100, $4 \cdot 5^{n-2}$; $\frac{1}{81}$, $\frac{1}{729}$, $\left(\frac{1}{9}\right)^{n-1}$; $\frac{20}{81}$, $\frac{100}{729}$, $\frac{4 \cdot 5^{n-2}}{9^{n-1}} = \frac{4}{9}\left(\frac{5}{9}\right)^{n-2}$ **b.** 2 **13.** $\frac{2\sqrt{3}}{5}$

Self-Test 2, page 536 **1. a.** G; $\sum_{n=0}^{3} 27\left(-\frac{2}{3}\right)^n$ **b.** N; $\sum_{n=1}^{15} \frac{n}{n+1}$ **c.** A; $\sum_{n=1}^{7} (23 - 3n)$ **d.** N; $\sum_{n=1}^{\infty} n^2$ **2. a.** 1080 **b.** −45 **3. a.** 315 **b.** $\frac{130}{81}$ **4. a.** $\frac{125}{8}$ **b.** no sum

Written Exercises, page 539
1. $x^3 + 3x^2y + 3xy^2 + y^3$
3. $c^5 - 5c^4d + 10c^3d^2 - 10c^2d^3 + 5cd^4 - d^5$
5. $a^8 + 8a^7 + 28a^6 + 56a^5 + 70a^4 + 56a^3 + 28a^2 + 8a + 1$ **7.** $243t^5 + 1620t^4 + 4320t^3 + 5760t^2 + 3840t + 1024$
9. $x^{12} - 6x^{10} + 15x^8 - 20x^6 + 15x^4 - 6x^2 + 1$
11. $p^6 + 3p^4q^3 + 3p^2q^6 + q^9$
13. $136x^2y^{15} + 17xy^{16} + y^{17}$ **15.** $167{,}960x^9$
17. $2a^7 + 42a^5b^2 + 70a^3b^4 + 14ab^6$
19. $a^6 + 6a^5b + 15a^4b^2 + 20a^3b^3 + 15a^2b^4 + 6ab^5 + b^6$ **21.** 179.52 **23. a.** 1, 2, 4, 8, 16, 32, 64 **b.** 2^n

Mixed Review Exercises, page 539 **1.** N; $\frac{2n}{2n+1}$ **2.** G; $\frac{9}{16}\left(\frac{4}{3}\right)^{n-1}$ **3.** A; $11 - 4n$ **4.** G; $-24\left(-\frac{1}{2}\right)^{n-1}$ **5.** 275 **6.** 24 **7.** 378

Written Exercises, pages 542–543 **1.** 720 **3.** 20 **5.** 100 **7.** 120 **9.** $n + 1$ **11.** $\frac{n^2 + n}{2}$
13. a. $a^{14} + 14a^{13}b + 91a^{12}b^2 + 364a^{11}b^3$
b. $a^{14} - 14a^{13}b + 91a^{12}b^2 - 364a^{11}b^3$
15. a. $a^{19} + 19a^{18}b + 171a^{17}b^2 + 969a^{16}b^3$
b. $x^{19} - 38x^{18}y + 684x^{17}y^2 - 7752x^{16}y^3$
17. $15{,}504a^{15}b^5$ **19.** $1001s^4t^{10}$ **21.** $4032a^4b^5$
23. $1120c^8d^4$ **25.** False; $a = 2$, $b = 3$ for example **27.** False; $n = 3$ for example
29. a. $1 + \frac{1}{2}x - \frac{1}{8}x^2 + \frac{1}{16}x^3$ **b.** 1.4375

Calculator Key-In, page 543 **1.** 3,628,800
3. a. 2.433×10^{18} **b.** 9.075×10^9

Self-Test 3, page 543
1. $x^5 - 10x^4 + 40x^3 - 80x^2 + 80x - 32$
2. $256a^4 + 768a^3 + 864a^2 + 432a + 81$
3. $1024x^{10} + 15{,}360x^9y + 103{,}680x^8y^2$
4. $-35a^8b^3$ **5.** $-1320x^8y^3$

Chapter Review, pages 545–546 **1.** b **3.** b **5.** c **7.** a **9.** c **11.** c **13.** a

Cumulative Review, page 547 **1.** 3
3. $\{2 \pm 3i, -1 \pm \sqrt{2}\}$ **5.** $x = \frac{1}{16}y^2 + 4$
7. (1, −2) **9.** (2, −1, −2) **11.** 128
13. $\left\{-\frac{1}{2}\right\}$ **15.** {27} **17.** 2.29 **19.** G; $t_n = 108\left(-\frac{2}{3}\right)^{n-1}$
21. $x^4 - 4x^3y^2 + 6x^2y^4 - 4xy^6 + y^8$

Chapter 12 Triangle Trigonometry

Written Exercises, pages 552–554
1. a. second quadrant **b.** third quadrant

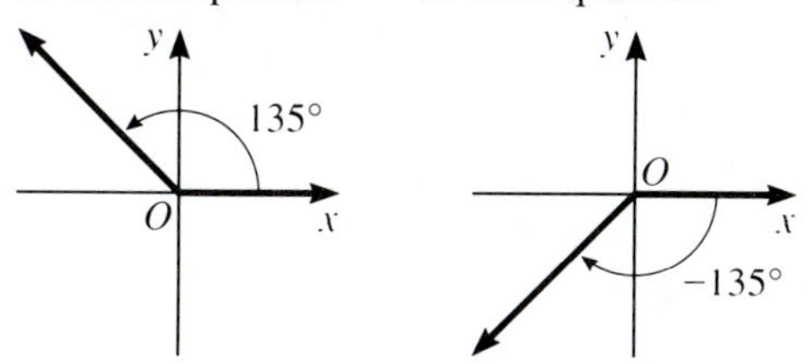

3. a. fourth quadrant **b.** first quadrant
5. quadrantal **7.** fourth quadrant **9.** second quadrant **11.** quadrantal **13.** 240° **15.** −270°
17. 576° **19.** 45°, 405°, 765°, −315°
21. 60°, 240°, 420°, −120°
23–29. Answers may vary. Examples are given. **23. a.** $35° + n \cdot 360°$ **b.** 395°, −325°
25. a. $-100° + n \cdot 360°$ **b.** 260°, −460°
27. a. $160° + n \cdot 360°$ **b.** 160°, −200°
29. a. $280° + n \cdot 360°$ **b.** 280°, −80°
31. 15.5° **33.** 72.8° **35.** 25.75° **37.** 45.31°
39. 25°24′ **41.** 44°54′ **43.** 34°24′36″
45. 23°40′12″ **47.** 315° **49.** 37° **51.** 30°
53. 240° **55.** Any angle between 0° and 22.5°.
57. Any angle between 67.5° and 90°. **59.** Any angle between 30° and 45°. **61.** Any angle between 0° and 180°. **63.** Any angle between 360° and 540°. **65.** Any angle between 450° and 900°.

Mixed Review Exercises, page 554 **1.** $40x^3y^2$
2. $252a^6b^2$ **3.** $78m^{22}n^6$ **4.** −6 **5.** 1
6. $3\sqrt{2} - 4$ **7.** $\frac{2}{3}$ **8.** $14 - 5i$ **9.** 0 **10.** 3
11. $\frac{7\sqrt{6}}{9}$ **12.** $\frac{3}{2}$

Calculator Key-In, page 554 **1.** 58°36′
3. 86°25′48″ **5.** 36.4266667° **7.** 73.8736111°

Written Exercises, pages 559–560

1. $\sin\theta = \frac{4}{5}$; $\cos\theta = \frac{3}{5}$; $\tan\theta = \frac{4}{3}$; $\cot\theta = \frac{3}{4}$; $\sec\theta = \frac{5}{3}$; $\csc\theta = \frac{5}{4}$ **3.** $\sin\theta = \frac{\sqrt{2}}{2}$; $\cos\theta = \frac{\sqrt{2}}{2}$; $\tan\theta = 1$; $\cot\theta = 1$; $\sec\theta = \sqrt{2}$; $\csc\theta = \sqrt{2}$ **5.** $\sin\theta = \frac{15}{17}$; $\cos\theta = \frac{8}{17}$; $\tan\theta = \frac{15}{8}$; $\cot\theta = \frac{8}{15}$; $\sec\theta = \frac{17}{8}$; $\csc\theta = \frac{17}{15}$ **7.** $\sin\theta = \frac{\sqrt{10}}{10}$; $\cos\theta = \frac{3\sqrt{10}}{10}$; $\tan\theta = \frac{1}{3}$; $\cot\theta = 3$; $\sec\theta = \frac{\sqrt{10}}{3}$; $\csc\theta = \sqrt{10}$ **9.** $\frac{4}{5}$; $\frac{3}{4}$ **11.** $\frac{1}{2}$; $\frac{\sqrt{3}}{3}$ **13.** $\frac{8}{15}$; $\frac{\sqrt{161}}{8}$ **15.** $\frac{\sqrt{15}}{4}$; $\sqrt{15}$ **17.** 50° **19.** 10° **21.** $\angle B = 45°$, $a = 2$, $c = 2\sqrt{2}$ **23.** $\angle B = 30°$, $b = 10$, $a = 10\sqrt{3}$ **25.** $\angle A = 45°$, $\angle B = 45°$, $c = 4\sqrt{2}$ **27.** $6\sqrt{3} + 6$ **29.** 36 **33.** $\cos\theta = \sqrt{1 - u^2}$, $\tan\theta = \frac{u\sqrt{1-u^2}}{1-u^2}$, $\cot\theta = \frac{\sqrt{1-u^2}}{u}$, $\sec\theta = \frac{\sqrt{1-u^2}}{1-u^2}$, $\csc\theta = \frac{1}{u}$

Written Exercises, pages 566–567

1. $\sin\theta = -\frac{4}{5}$, $\cos\theta = \frac{3}{5}$, $\tan\theta = -\frac{4}{3}$, $\cot\theta = -\frac{3}{4}$, $\sec\theta = \frac{5}{3}$, $\csc\theta = -\frac{5}{4}$ **3.** $\sin\theta = \frac{24}{25}$, $\cos\theta = -\frac{7}{25}$, $\tan\theta = -\frac{24}{7}$, $\cot\theta = -\frac{7}{24}$, $\sec\theta = -\frac{25}{7}$, $\csc\theta = \frac{25}{24}$ **5.** 0; 1; 0; 1; undef.; undef. **7.** 0; −1; 0; −1; undef.; undef. **9.** 53° **11.** 25° **13.** 28° **15.** 5° **17.** 83.6° **19.** 4.1° **21.** 23°40′ **23.** 27°30′ **25.** −cos 36° **27.** −sin 17° **29.** −cot 72.9° **31.** −cos 41.9° **33.** tan 85°20′ **35.** −sin 32°40′ **37.** $\sin 330° = -\frac{1}{2}$, $\cos 330° = \frac{\sqrt{3}}{2}$, $\tan 330° = -\frac{\sqrt{3}}{3}$, $\cot 330° = -\sqrt{3}$, $\sec 330° = \frac{2\sqrt{3}}{3}$, $\csc 330° = -2$ **39.** $\sin 135° = \frac{\sqrt{2}}{2}$, $\cos 135° = -\frac{\sqrt{2}}{2}$, $\tan 135° = -1$, $\cot 135° = -1$, $\sec 135° = -\sqrt{2}$, $\csc 135° = \sqrt{2}$ **41.** $\sin 315° = -\frac{\sqrt{2}}{2}$, $\cos 315° = \frac{\sqrt{2}}{2}$, $\tan 315° = -1$, $\cot 315° = -1$, $\sec 315° = \sqrt{2}$, $\csc 315° = -\sqrt{2}$ **43.** $\sin(-150°) = -\frac{1}{2}$, $\cos(-150°) = -\frac{\sqrt{3}}{2}$, $\tan(-150°) = \frac{\sqrt{3}}{3}$, $\cot(-150°) = \sqrt{3}$, $\sec(-150°) = -\frac{2\sqrt{3}}{3}$, $\csc(-150°) = -2$ **45.** second; $\sin\theta = \frac{15}{17}$, $\tan\theta = -\frac{15}{8}$, $\cot\theta = -\frac{8}{15}$, $\sec\theta = -\frac{17}{8}$, $\csc\theta = \frac{17}{15}$ **47.** fourth; $\cos\theta = \frac{12}{13}$, $\tan\theta = -\frac{5}{12}$, $\cot\theta = -\frac{12}{5}$, $\sec\theta = \frac{13}{12}$, $\csc\theta = -\frac{13}{5}$ **49.** first; $\sin\theta = \frac{\sqrt{5}}{3}$, $\tan\theta = \frac{\sqrt{5}}{2}$, $\cot\theta = \frac{2\sqrt{5}}{5}$, $\sec\theta = \frac{3}{2}$, $\csc\theta = \frac{3\sqrt{5}}{5}$ **51.** second; $\sin\theta = \frac{3}{5}$, $\cos\theta = -\frac{4}{5}$, $\cot\theta = -\frac{4}{3}$, $\sec\theta = -\frac{5}{4}$, $\csc\theta = \frac{5}{3}$ **53.** 0°, 180° **55.** 0°, 180° **57.** 45°, 225° **59.** 270° **61.** 150°, 210° **63.** 15°, 195° **65.** 70°, 290° **67.** 20°, 200° **69.** $\cos\theta$ **71.** $-\sin\theta$ **73.** $-\cos\theta$ **75.** $\frac{\sin\theta}{\cos\theta}$

Mixed Review Exercises, page 567 **1.** 1275 **2.** 650 **3.** 25 **4.** $-\frac{5}{3}$ **5.** 2 **6.** $-\frac{3}{2}$, 1 **7.** 0, 2, −2 **8.** no real zeros **9.** no real zeros **10.** no real zeros

Written Exercises, page 572 **1.** 0.2717 **3.** 0.9078 **5.** 0.8557 **7.** 1.729 **9.** 7.723 **11.** 0.8440 **13.** 0.6058 **15.** 0.2080 **17.** −2.753 **19.** −0.9997 **21.** −1.351 **23.** −0.6225 **25.** 19.9° **27.** 30.5° **29.** 73.7° **31.** 30°27′ **33.** 8°26′ **35.** 47°4′ **37.** 29.2°, 150.8° **39.** 299.7°, 119.7° **41.** 75.4°, 284.6° **43.** 295.3° **45.** 322.3°

Self-Test 1, pages 572–573 **1.** Answers may vary. **a.** 610°, −110° **b.** 60°, −660° **2.** 32.58° **3.** 74°15′36″ **4.** $\sin\theta = \frac{3}{4}$, $\cos\theta = \frac{\sqrt{7}}{4}$, $\tan\theta = \frac{3\sqrt{7}}{7}$, $\cot\theta = \frac{\sqrt{7}}{3}$, $\sec\theta = \frac{4\sqrt{7}}{7}$, $\csc\theta = \frac{4}{3}$ **5.** $a = 12$, $b = 12\sqrt{3}$, $\angle B = 60°$ **6.** 60°; $\sin 300° = -\frac{\sqrt{3}}{2}$, $\cos 300° = \frac{1}{2}$, $\tan 300° = -\sqrt{3}$, $\cot 300° = -\frac{\sqrt{3}}{3}$, $\sec 300° = 2$, $\csc 300° = -\frac{2\sqrt{3}}{3}$ **7. a.** 0.8018 **b.** 1.110 **8.** 61°10′

Lessons 12-5 to 12-9: Note that because of roundoff error, answers obtained by using the trigonometric tables occasionally differ slightly from the answers given here, which are correct to the number of significant digits shown.

Written Exercises, pages 577–578
1. $\angle B = 53.8°$; $a = 40.2$; $b = 54.9$
3. $\angle A = 24.6°$; $b = 5.13$; $c = 5.65$
5. $\angle A = 41.7°$; $a = 66.6$; $c = 100$ **7.** $b = 222$; $\angle A = 46.0°$; $\angle B = 44.0°$ **9.** $c = 0.338$; $\angle A = 21.3°$; $\angle B = 68.7°$ **11.** $\angle A = 31°50'$; $a = 222$; $b = 357$ **13.** $\angle B = 74°30'$; $b = 16.2$; $c = 16.8$ **15.** $\angle B = 59°10'$; $a = 31.9$; $c = 62.3$ **17.** 45.2° **19.** 53.1°, 126.9° **21. a.** 6.18 **b.** 6.50

Problems, pages 578–579 **1.** 31.7° **3.** 64.1° **5.** 545 ft **7.** 317 ft **9.** 23.4° **11.** 358 ft **13.** 1.13 m **15.** 660 m **17.** 109.4°

Mixed Review Exercises, page 579
1. $\sin 450° = 1$, $\cos 450° = 0$, $\tan 450°$ and $\sec 450°$ undef., $\csc 450° = 1$, $\cot 450° = 0$
2. $\sin(-135°) = \cos(-135°) = -\frac{\sqrt{2}}{2}$, $\tan(-135°) = \cot(-135°) = 1$, $\sec(-135°) = \csc(-135°) = -\sqrt{2}$
3. $\sin 330° = -\frac{1}{2}$, $\cos 330° = \frac{\sqrt{3}}{2}$, $\tan 330° = -\frac{\sqrt{3}}{3}$, $\cot 330° = -\sqrt{3}$, $\sec 330° = \frac{2\sqrt{3}}{3}$, $\csc 330° = -2$
4. $\sin(-240°) = \frac{\sqrt{3}}{2}$, $\cos(-240°) = -\frac{1}{2}$, $\tan(-240°) = -\sqrt{3}$, $\cot(-240°) = -\frac{\sqrt{3}}{3}$, $\sec(-240°) = -2$, $\csc(-240°) = \frac{2\sqrt{3}}{3}$
5. $-\frac{1}{x+1}$ **6.** $5m^2\sqrt{2}$ **7.** $\frac{1}{y-2}$ **8.** $200a^{12}b^5$ **9.** $\frac{1}{x^{1/2}}$ **10.** $2u^3 - u^2 - 7u + 6$

Written Exercises, pages 582–583 **1.** 2.46 **3.** 18.6 **5.** 42.6 **7.** 55.8° **9.** 13.8° **11.** 87.4° **13.** 7.83 **15.** 69.2°, 51.1°, 59.7°

Problems, pages 583–584 **1.** 26.9 km **3.** 32.2°, 87.8°, 60.0° **5.** 16.6 mi **7.** 63.7 ft **9.** perimeter = 45.4 km, area = 103.8 km^2 **11.** 31.5 m, 40.1 m

Computer Exercises, page 584
2. a. $\angle A = 53.8°$; $\angle B = 36.2°$; $c = 64.4$
b. $\angle A = 38.6°$; $\angle B = 51.4$; $b = 37.0$
c. $\angle A = 49.2°$; $\angle B = 40.8°$; $a = 6.86$
4. a. $\angle B = 58°$; $c = 69.8$; $b = 59.2$
b. $\angle A = 75°$; $c = 68.0$; $a = 65.7$
c. $\angle A = 70.2°$; $b = 3.35$; $a = 9.31$
6. a. $\angle C = 40°$; $b = 20.58$ **b.** $\angle C = 25.9°$; $a = 48.33$ **c.** $\angle A = 52.0°$; $b = 8.451$

Written Exercises, pages 588–589 **1.** 32.0 **3.** 9.34 **5.** 23.0 **7.** 28.8° **9.** 24.2° **11.** 69.5° or 110.5° **13.** $\frac{10}{9}$ **15.** $\frac{15}{13}$ **17.** $\frac{\sqrt{2}}{2}$

Problems, pages 589–590 **1.** 29.3 cm **3.** 49.7 cm **5.** 3280 m **7.** 47.1 m **9.** 8.1° **11.** 14.3 km **13.** 13.1 m

Mixed Review Exercises, page 590 **1.** 0.2045 **2.** −0.9380 **3.** −0.5548 **4.** −1.232 **5.** 0.2107 **6.** −1.604 **7.** $(x+1)^2 + (y-4)^2 = 9$ **8.** $x + y = 2$ **9.** $y = -\frac{1}{8}x^2$

Written Exercises, pages 594–595
1. $\angle A = 100°$, $b = 5.9$, $c = 14.9$
3. $\angle A = 19.7°$, $\angle B = 28.2°$, $\angle C = 132.1°$
5. $a = 20.0$, $\angle B = 25.7°$, $\angle C = 34.3°$
7. $\angle B = 30.7°$, $\angle C = 19.3°$, $c = 12.9$
9. $\angle A = 35.3°$, $\angle C = 104.7°$, $c = 15.1$
11. $\angle C = 80.0°$, $\angle A = 50.0°$, $a = 14.0$; or $\angle C = 100.0°$, $\angle A = 30.0°$, $a = 9.14$ **13.** 2.61 **15.** 12.0 **17.** 4.33, 3.86 **19.** 80.1, 82.7, 85.1

Problems, pages 595–596 **1.** 1350 km **3.** 55.7°, 353 m **5.** 232 km, 40.4° **7.** 13.2 m **9.** 81.3° N or 81.3° S **11.** 5.51 **13.** 60.1 in.

Written Exercises, pages 599–600 **1.** 240 **3.** 37.1 **5.** 45.3 **7.** 82.5 **9.** 127 **11.** 18 **13.** 117.3° **15.** 483 cm^2
17. $K^2 = \left(\frac{1}{2}ab \sin C\right)^2 = \frac{1}{4}a^2b^2 \sin^2 C$
19. a. law of cosines **b.** $\sin^2 C + \cos^2 C = 1$ **c.** Ex. 17 **d.** Add. prop. of eq. **e.** $x^2 - y^2 = (x+y)(x-y)$ **f.** Assoc. prop. of add., factoring **g.** see (e) **h.** Ex. 18 **i.** Div. prop. of eq., prop. of square roots

Mixed Review Exercises, page 600 **1.** 7.17 **2.** 29.5 **3.** 127.2° **4.** 23.4

5. $\cos\theta = -\frac{\sqrt{3}}{2}$, $\tan\theta = -\frac{\sqrt{3}}{3}$, $\cot\theta = -\sqrt{3}$, $\sec\theta = -\frac{2\sqrt{3}}{3}$, $\csc\theta = 2$ **6.** $\sin\theta = -\frac{\sqrt{7}}{4}$, $\tan\theta = \frac{\sqrt{7}}{3}$, $\cot\theta = \frac{3\sqrt{7}}{7}$, $\sec\theta = -\frac{4}{3}$, $\csc\theta = -\frac{4\sqrt{7}}{7}$ **7.** $\sin\theta = -\frac{\sqrt{2}}{2}$, $\cos\theta = \frac{\sqrt{2}}{2}$, $\cot\theta = -1$, $\sec\theta = \sqrt{2}$, $\csc\theta = -\sqrt{2}$ **8.** $\cos\theta = \frac{12}{13}$, $\tan\theta = \frac{5}{12}$, $\cot\theta = \frac{12}{5}$, $\sec\theta = \frac{13}{12}$, $\csc\theta = \frac{13}{5}$

Self-Test 2, page 600 **1.** $\angle A = 62.7°$, $b = 15.5$, $c = 33.8$ **2.** 16.1 **3.** $\angle Z = 80.6°$ or 99.4° **4.** $\angle A = 17.5°$, $\angle B = 13.9°$, $\angle C = 148.6°$ **5.** 43.5

Chapter Review, pages 602–604 **1.** b **3.** b **5.** b **7.** a **9.** d **11.** a **13.** a **15.** c **17.** a **19.** d

Preparing for College Entrance Exams, page 605 **1.** C **3.** C **5.** A **7.** D

Chapter 13 Trigonometric Graphs; Identities

Written Exercises, pages 610–611 **1.** $\frac{\pi}{4}$ **3.** $\frac{\pi}{3}$ **5.** $-\frac{2\pi}{3}$ **7.** $\frac{5\pi}{6}$ **9.** $-\frac{11\pi}{6}$ **11.** $-\frac{7\pi}{4}$ **13.** 30° **15.** 60° **17.** 240° **19.** −210° **21.** 540° **23.** −630° **25.** $\frac{720°}{\pi}$ **27.** $-\frac{360°}{\pi}$ **29.** 0.17 **31.** 1.40 **33.** 0.84 **35.** −3.04 **37.** 171.9° **39.** 22.9° **41.** −85.9° **43.** 859.4° **45.** 4; 8 **47.** 3; 24 **49.** 1.2; 6 **51.** 4; 20 **53.** 3; $\frac{3}{2}$ **55.** 2; 8 **57.** 1.84 **59.** 3.28

Problems, pages 611–612 **1.** 2100 mi **3.** 1700 in. **5.** 79 ft/min **7.** 1,400,000 km **9.** 250π radians/min; 4700 cm/min **11.** 0.00015 radians/s; 0.0073 mm/s **13.** Approximately 20%

Mixed Review Exercises, page 612 **1.** $c = 5.44$, $\angle A = 100.0°$, $\angle B = 38.0°$ **2.** $a = 23.2$, $c = 23.2$, $\angle C = 75°$ **3.** $a = 5.64$, $\angle A = 33.2°$, $\angle C = 42.8°$ **4.** $\angle A = 31.6°$, $\angle B = 39.0°$, $\angle C = 109.4°$ **5.** 13.4 **6.** 134 **7.** 19.2 **8.** 14.1

Written Exercises, page 617 **1.** $\sin s = -\frac{4}{5}$, $\cos s = \frac{3}{5}$, $\tan s = -\frac{4}{3}$, $\cot s = -\frac{3}{4}$, $\sec s = \frac{5}{3}$, $\csc s = -\frac{5}{4}$ **3.** $\sin s = \frac{3}{4}$, $\cos s = \frac{\sqrt{7}}{4}$, $\tan s = \frac{3\sqrt{7}}{7}$, $\cot s = \frac{\sqrt{7}}{3}$, $\sec s = \frac{4\sqrt{7}}{7}$, $\csc s = \frac{4}{3}$ **5.** 0.9490; 0.3153; 3.010 **7.** 0.2474; 0.9690; 0.2553 **9.** $\sin\frac{\pi}{3} = \frac{\sqrt{3}}{2}$, $\cos\frac{\pi}{3} = \frac{1}{2}$, $\tan\frac{\pi}{3} = \sqrt{3}$, $\cot\frac{\pi}{3} = \frac{\sqrt{3}}{3}$, $\sec\frac{\pi}{3} = 2$, $\csc\frac{\pi}{3} = \frac{2\sqrt{3}}{3}$ **11.** $\sin\frac{3\pi}{2} = -1$, $\cos\frac{3\pi}{2} = 0$, $\tan\frac{3\pi}{2}$ and $\sec\frac{3\pi}{2}$ undef., $\cot\frac{3\pi}{2} = 0$, $\csc\frac{3\pi}{2} = -1$ **13.** $\sin\frac{11\pi}{6} = -\frac{1}{2}$, $\cos\frac{11\pi}{6} = \frac{\sqrt{3}}{2}$, $\tan\frac{11\pi}{6} = -\frac{\sqrt{3}}{3}$, $\cot\frac{11\pi}{6} = -\sqrt{3}$, $\sec\frac{11\pi}{6} = \frac{2\sqrt{3}}{3}$, $\csc\frac{11\pi}{6} = -2$ **15.** $\sin\frac{8\pi}{3} = \frac{\sqrt{3}}{2}$, $\cos\frac{8\pi}{3} = -\frac{1}{2}$, $\tan\frac{8\pi}{3} = -\sqrt{3}$, $\cot\frac{8\pi}{3} = -\frac{\sqrt{3}}{3}$, $\sec\frac{8\pi}{3} = -2$, $\csc\frac{8\pi}{3} = \frac{2\sqrt{3}}{3}$ **17.** $\sin\left(-\frac{5\pi}{6}\right) = -\frac{1}{2}$, $\cos\left(-\frac{5\pi}{6}\right) = -\frac{\sqrt{3}}{2}$, $\tan\left(-\frac{5\pi}{6}\right) = \frac{\sqrt{3}}{3}$, $\cot\left(-\frac{5\pi}{6}\right) = \sqrt{3}$, $\sec\left(-\frac{5\pi}{6}\right) = -\frac{2\sqrt{3}}{3}$, $\csc\left(-\frac{5\pi}{6}\right) = -2$ **19.** 0.76 **21.** $\frac{\pi}{3}$ **23.** $\left(\frac{\sqrt{2}}{2}\right)^2 + \left(-\frac{\sqrt{2}}{2}\right)^2 = \frac{1}{2} + \frac{1}{2} = 1$ **25.** $(0.6442)^2 + (0.7648)^2 \approx 1$

Written Exercises, pages 621–623

1. odd

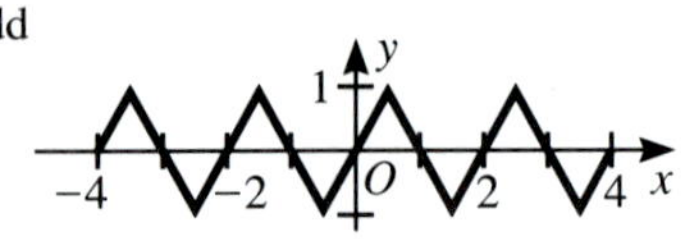

3. neither

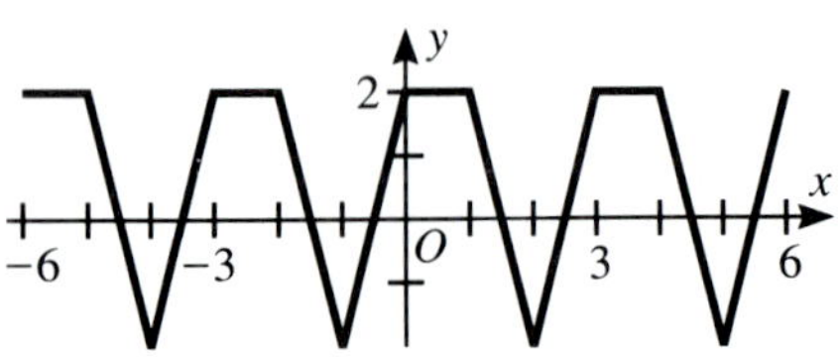

5. odd

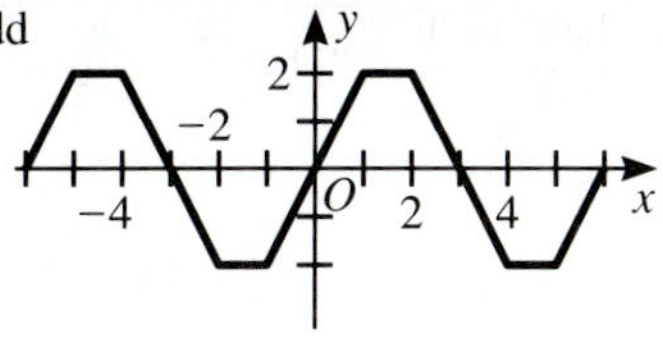

7. a.

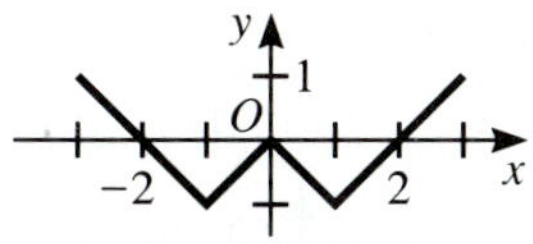

b.

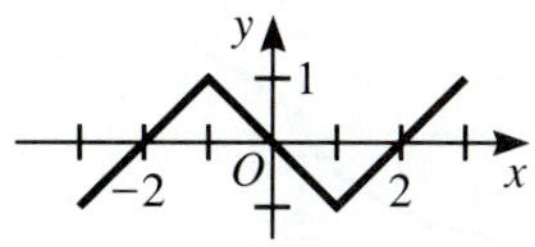

9. a.

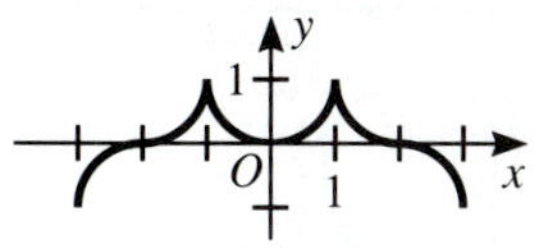

b.

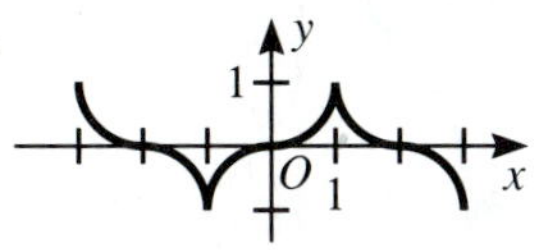

11. odd **13.** neither **15.** even **17.** even **19.** odd

Mixed Review Exercises, page 623

1.

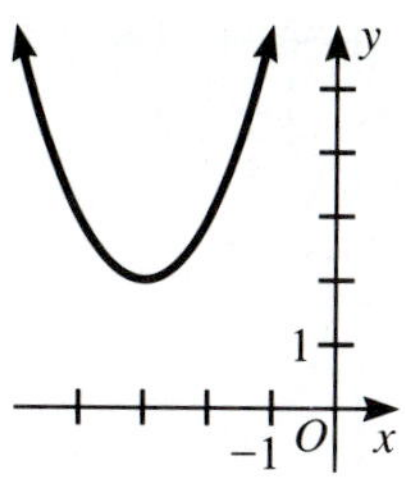

3.

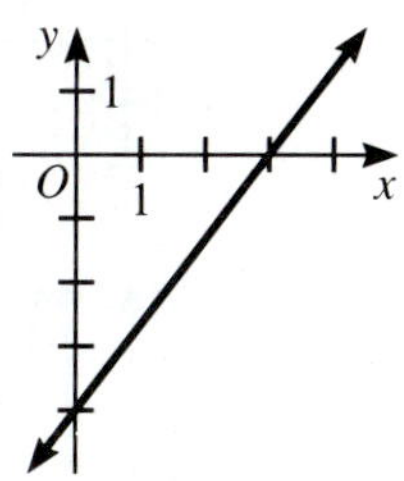

5.

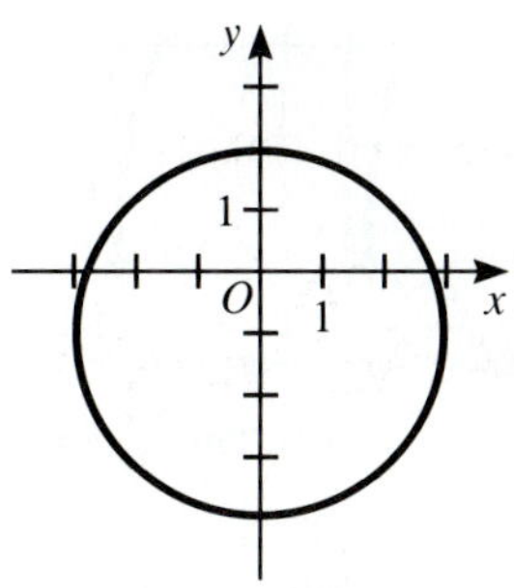

7. $\{0, 3\}$ **8.** $\{-2\}$ **9.** $\{2, \pm i\sqrt{2}\}$ **10.** $\{2, 6\}$ **11.** $\left\{-\frac{5}{2}, 2\right\}$ **12.** $\{2\}$

Written Exercises, pages 628–629 **1. a.** 2 **b.** 2, −2 **c.** 2π **3. a.** 1 **b.** 1, −1 **c.** $\frac{\pi}{2}$ **5. a.** 3 **b.** 3, −3 **c.** $\frac{2\pi}{3}$ **7. a.** $\frac{1}{2}$ **b.** $\frac{1}{2}$, $-\frac{1}{2}$ **c.** 1 **9. a.** 2 **b.** 2, −2 **c.** 4 **11. a.** 1 **b.** 0, −2 **c.** 1 **13.** $y = 2 \sin x$ **15.** $y = \frac{1}{2} \cos \frac{1}{2}x$ **17.** $y = \sin \pi x$ **19.** $y = 3 + 3 \cos \frac{\pi}{4}x$ **21.** $y = 3 \sin x$ **23.** $y = 2 \sin \pi x$ **25.** $y = 4 + 3 \sin 2x$

27.

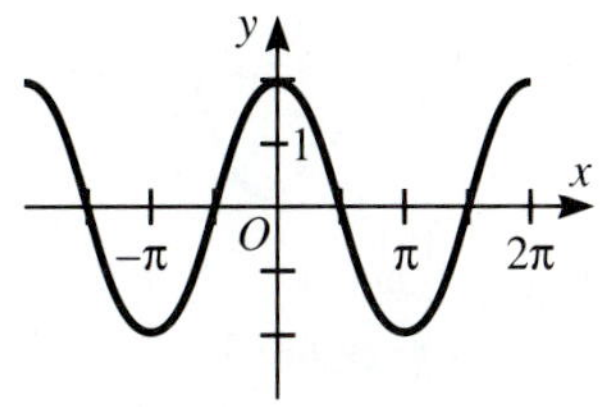

29.

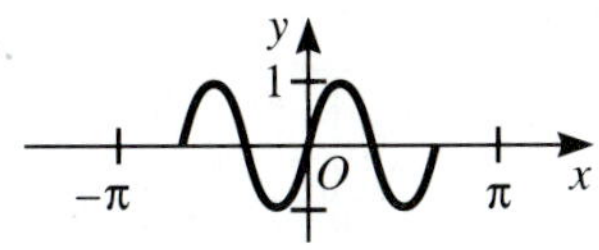

31.

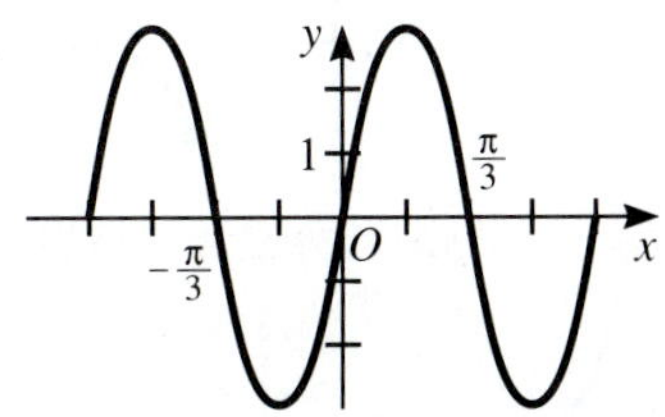

33.

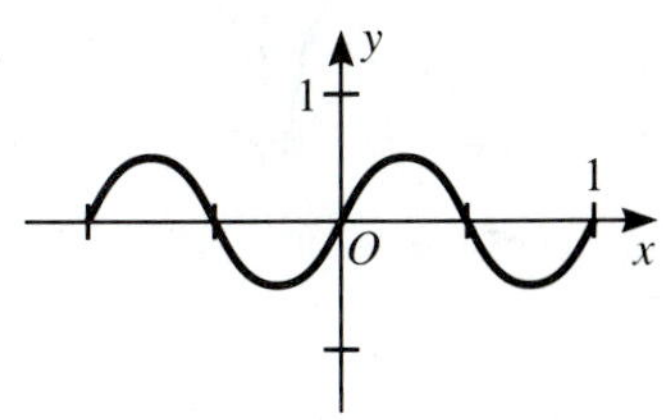

35.

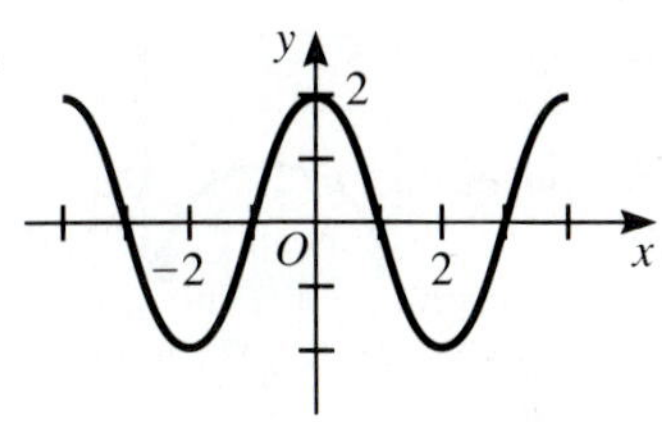

37.

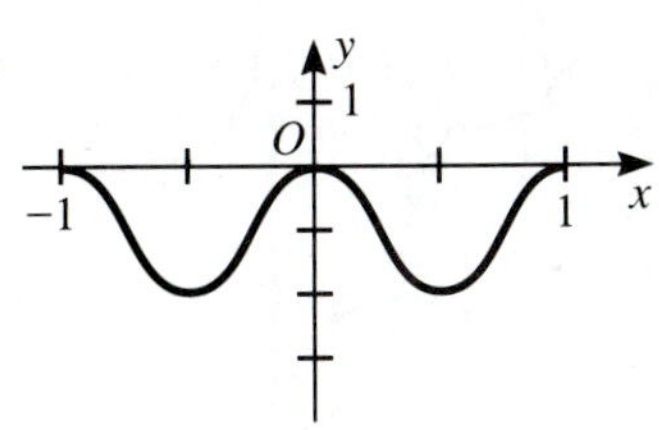

39.

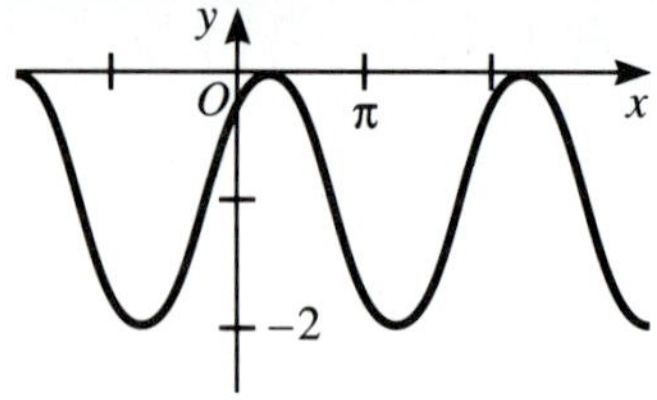

41.

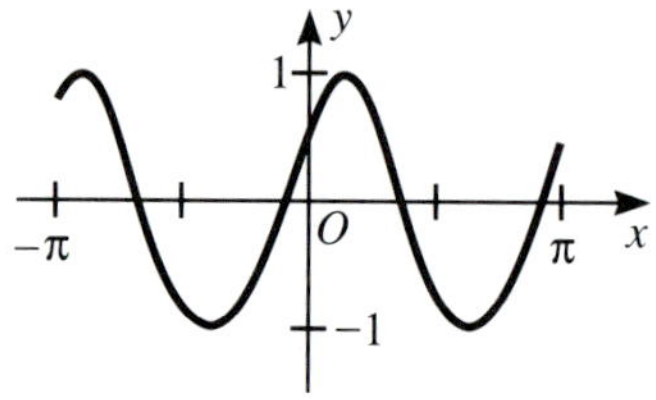

Written Exercises, page 633

1. odd
3. odd
5. neither
7. odd

9.

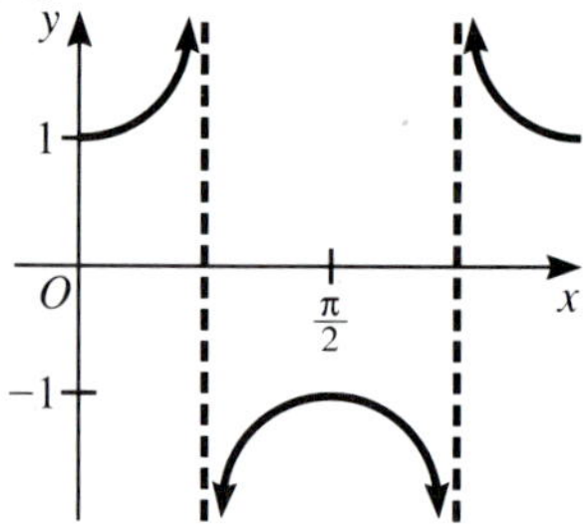

11.

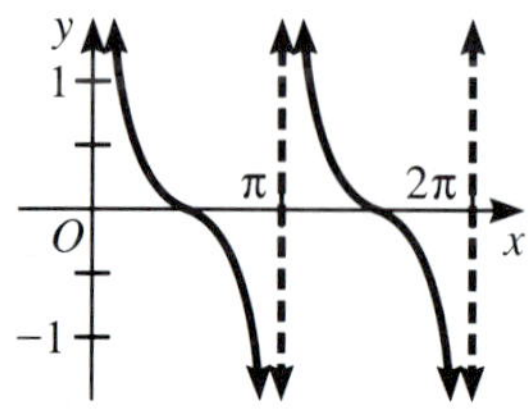

13.

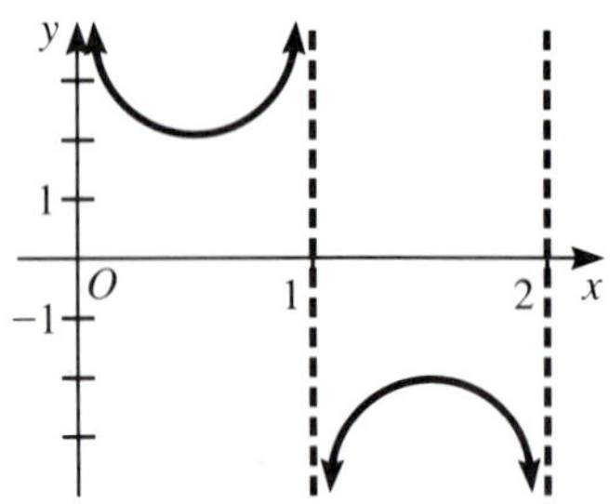

15.

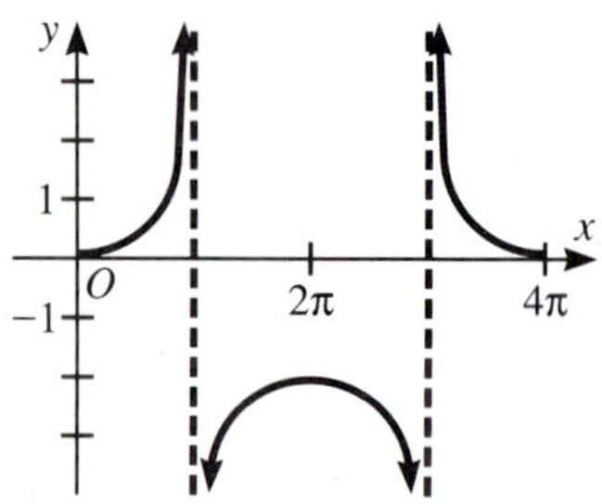

17. No; tangent has no maximum or minimum value.

19.

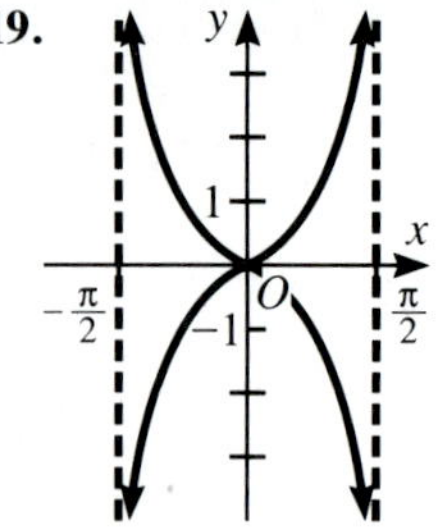

21.

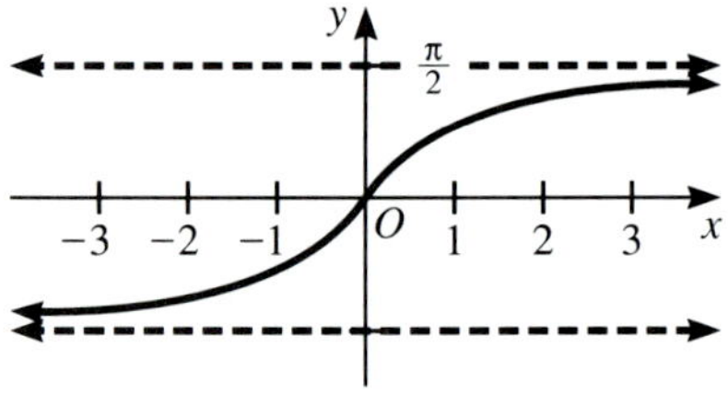

23.

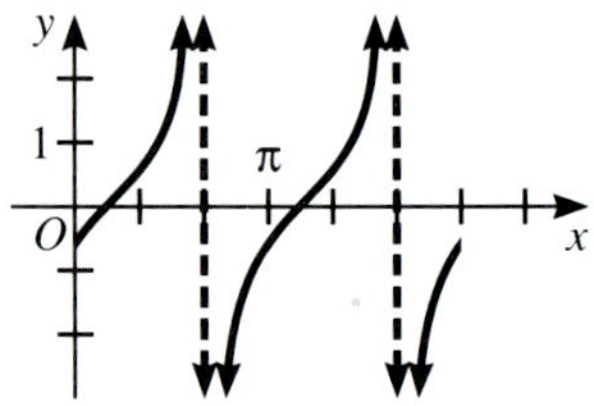

25.

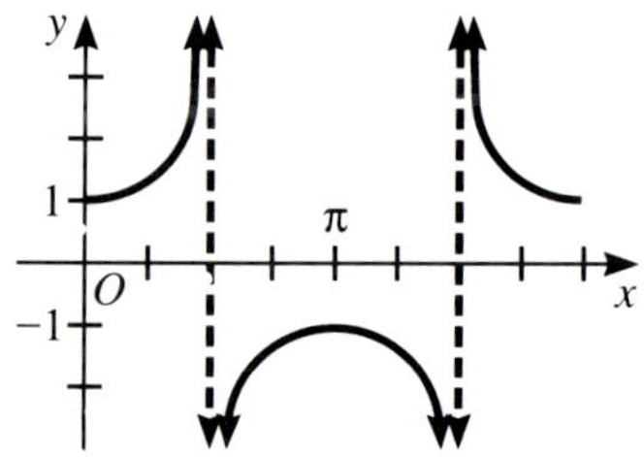

27.

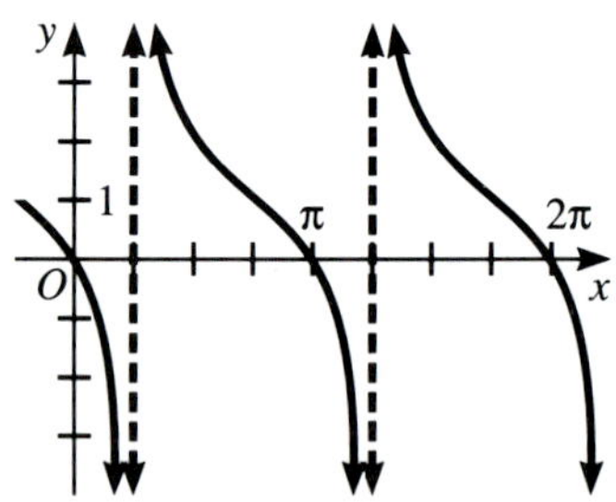

Mixed Review Exercises, page 633

1. $\sin \frac{5\pi}{2} = 1$; $\cos \frac{5\pi}{2} = 0$; $\tan \frac{5\pi}{2}$ and $\sec \frac{5\pi}{2}$ undef.; $\cot \frac{5\pi}{2} = 0$; $\csc \frac{5\pi}{2} = 1$

2. $\sin\left(-\frac{4\pi}{3}\right) = \frac{\sqrt{3}}{2}$; $\cos\left(-\frac{4\pi}{3}\right) = -\frac{1}{2}$; $\tan\left(-\frac{4\pi}{3}\right) = -\sqrt{3}$; $\cot\left(-\frac{4\pi}{3}\right) = -\frac{\sqrt{3}}{3}$; $\sec\left(-\frac{4\pi}{3}\right) = -2$; $\csc\left(-\frac{4\pi}{3}\right) = \frac{2\sqrt{3}}{3}$

3. $\sin \frac{19\pi}{6} = -\frac{1}{2}$; $\cos \frac{19\pi}{6} = -\frac{\sqrt{3}}{2}$; $\tan \frac{19\pi}{6} = \frac{\sqrt{3}}{3}$; $\cot \frac{19\pi}{6} = \sqrt{3}$; $\sec \frac{19\pi}{6} = -\frac{2\sqrt{3}}{3}$; $\csc \frac{19\pi}{6} = -2$

4. $\sin\left(-\frac{5\pi}{4}\right) = \frac{\sqrt{2}}{2}$; $\cos\left(-\frac{5\pi}{4}\right) = -\frac{\sqrt{2}}{2}$; $\tan\left(-\frac{5\pi}{4}\right) = -1$; $\cot\left(-\frac{5\pi}{4}\right) = -1$; $\sec\left(-\frac{5\pi}{4}\right) = -\sqrt{2}$; $\csc\left(-\frac{5\pi}{4}\right) = \sqrt{2}$ **5.** -3

6. $\frac{1}{2}$ **7.** 44 **8.** $\frac{1}{2}$

Self-Test 1, page 634 **1. a.** $\frac{11\pi}{12}$ **b.** 4.47 **2. a.** $-150°$ **b.** $143.2°$ **3.** $\sin \frac{3\pi}{4} = \frac{\sqrt{2}}{2}$, $\cos \frac{3\pi}{4} = -\frac{\sqrt{2}}{2}$, $\tan \frac{3\pi}{4} = -1$, $\cot \frac{3\pi}{4} = -1$, $\sec \frac{3\pi}{4} = -\sqrt{2}$, $\csc \frac{3\pi}{4} = \sqrt{2}$

4. 0.9915, 0.1304, 7.6018 **5. a.** even **b.** neither **6.** 3; 4, -2; π

7.

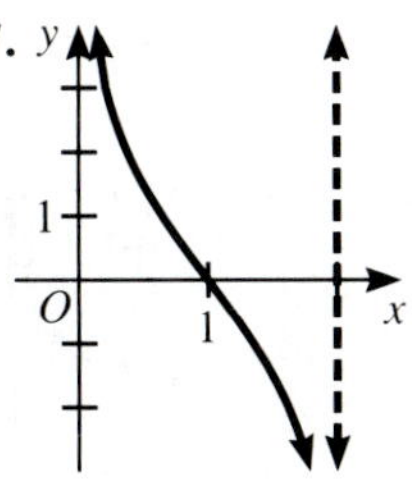

Application, page 635 **1.** $y = \frac{1}{2} \cos 704\pi t$ **2. a.** 528; C′ **b.** 25 cm

Written Exercises, pages 639–640 **1.** $\tan \alpha$ **3.** $\sec x$ **5.** $\cos^2 \theta$ **7.** 1 **9.** 1 **11.** $\tan^2 \theta$ **13.** $\csc^2 \alpha$ **15.** 1 **17.** 1 **19.** $\tan \alpha$ **21.** $\cos x$ **23.** $\sin \phi$ **25.** $\pm\frac{1}{\sqrt{1 - \sin^2 x}}$ **27.** $\pm\frac{\sin t}{\sqrt{1 - \sin^2 t}}$ **29.** $\frac{\sin \alpha}{1 - \sin^2 \alpha}$ **39.** $\cot t$ **41.** $2 \cos \theta$ **43.** $\sec x$ **45.** $2 \csc \theta$ **47.** 2

Written Exercises, pages 643–644

1. $\frac{\sqrt{2} - \sqrt{6}}{4}$ **3.** $\frac{\sqrt{6} - \sqrt{2}}{4}$ **5.** $-\frac{\sqrt{6} + \sqrt{2}}{4}$ **7.** $\cos 60°$; $\frac{1}{2}$ **9.** $\sin 210°$; $-\frac{1}{2}$ **11.** $\cos \frac{\pi}{6}$; $\frac{\sqrt{3}}{2}$ **13.** $\sin 2\theta$ **21.** $\cos \theta$ **23.** $\cos x$ **31. a.** $\frac{36}{85}$ **b.** $-\frac{77}{85}$ **c.** 2nd

Mixed Review Exercises, page 645

4. 1 **5.** x^{3n} **6.** $2x^2\sqrt[3]{2}$ **7.** -1 **8.** $-2 - \sqrt{3}$ **9.** $\frac{3\sqrt{13}}{13}$ **10.** x **11.** $\frac{1}{x - 1}$ **12.** $\sin x$

Written Exercises, page 649 **1.** $\cos 20°$ **3.** $\cos 8°$ **5.** $\sin \frac{\alpha}{2}$ **7.** $\cos \theta$ **9.** $\cos 4t$ **11.** $\sin \alpha$ **13.** $\frac{1}{2}$ **15.** $\frac{\sqrt{3}}{2}$ **17.** $-\frac{\sqrt{3}}{2}$ **19.** $\frac{1}{2}\sqrt{2 + \sqrt{2}}$ **21.** $\frac{1}{2}\sqrt{2 + \sqrt{3}}$ **23.** $\frac{1}{2}\sqrt{2 + \sqrt{2}}$ **25.** $\frac{7}{25}$ **27.** $\frac{3\sqrt{10}}{10}$ **29.** $-\frac{240}{289}$ **31.** $-\frac{5\sqrt{34}}{34}$ **41.** period π, amp. 1 **43.** $8 \cos^4 \theta - 8 \cos^2 \theta + 1$

Written Exercises, pages 652–653 **1.** $\tan 60°$ **3.** $\tan 180°$ **5.** $\tan \frac{5\pi}{6}$ **7.** $\tan 45°$ **9.** $2 + \sqrt{3}$ **11.** $-2 + \sqrt{3}$ **13.** $-2 + \sqrt{3}$ **15.** $\sqrt{2} - 1$ **17.** $\sqrt{2} + 1$ **19.** $2 + \sqrt{3}$ **21.** $\frac{13}{84}$ **23.** $-\frac{240}{161}$ **29.** $\frac{\cot \alpha \cot \beta - 1}{\cot \alpha + \cot \beta}$ **31.** $\frac{\cot^2 \alpha - 1}{2 \cot \alpha}$ **33. a.** $\frac{1}{2}\sqrt{2 - \sqrt{2}}$; $-\frac{1}{2}\sqrt{2 + \sqrt{2}}$

Mixed Review Exercises, page 653

5. $\{x: 1 < x < 5\}$ **6.** $\{x: x < -1 \text{ or } x > 3\}$ **7.** $\{x: -1 \le x \le 3\}$ **8.** $\{x: x < 3, x \ne 0\}$ **9.** $\{x: x \le -2 \text{ or } x \ge 1\}$ **10.** $\{x: x < 2\}$ **11.** $\{x: x > 6\}$ **12.** $\{x: -1 \le x \le 3\}$ **13.** {real numbers}

Self-Test 2, page 654 **1.** $\sec x$ **3.** $\cos 570°$; $-\frac{\sqrt{3}}{2}$ **4.** $\sin 10\alpha$ **5. a.** $\frac{41}{841}$ **b.** $\frac{840}{841}$ **c.** $-\frac{2\sqrt{29}}{29}$ **6.** $\tan 10°$ **7.** $\tan 70°$ **8.** $\tan 65°$

Chapter Review, pages 655–656 **1.** c **3.** d **5.** a **7.** c **9.** d **11.** d **13.** d **15.** d **17.** b

Chapter 14 Trigonometric Applications

Written Exercises, page 664

1.

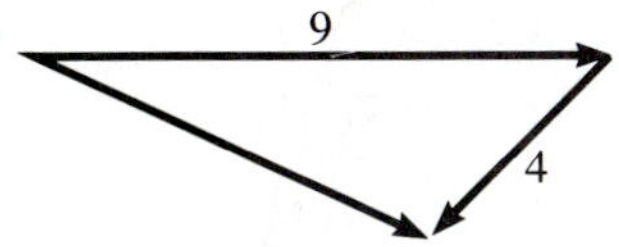

3.

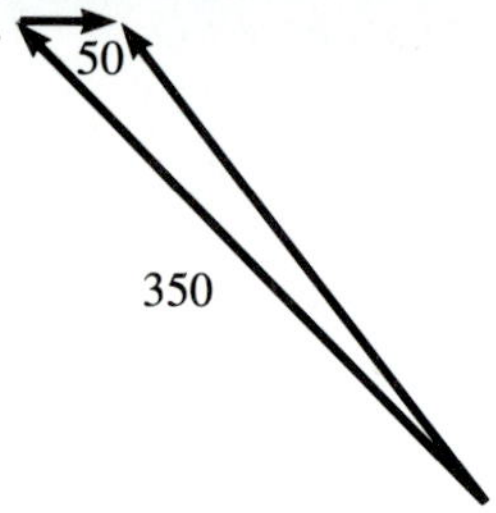

5.

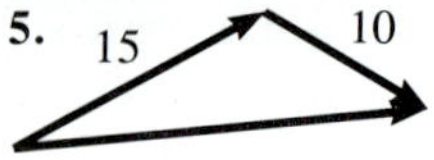

7.

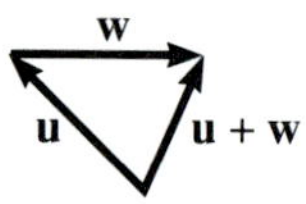

9.

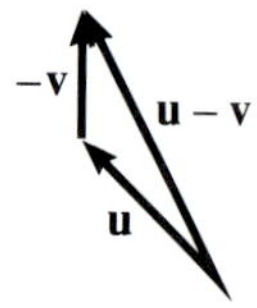

11.

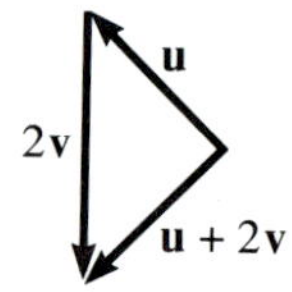

13.

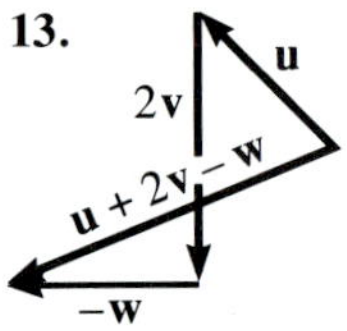

15. 155; 171.2°
17. 307; 180.8°
19. 545; 90°
21. 15.5; 92.1°

Problems, page 665 **1.** 429 km; 318.3° **3.** 4.4°; 582 km/h **5.** 283 km/h; 80.0° **7.** 78 km; 2nd to 1st, 9.6°; 1st to 2nd, 189.6° **9.** 142.7°; ETA 2:47 P.M.

Mixed Review Exercises, page 665 **1.** neither **2.** neither **3.** odd **4.** $\{-3 \pm \sqrt{17}\}$ **5.** $\left\{\frac{2}{3}, 4\right\}$ **6.** $\left\{-\frac{4}{3}, \frac{8}{3}\right\}$ **7.** {27} **8.** {10} **9.** $\left\{-6, \frac{1}{2}\right\}$

Written Exercises, pages 669–670 **1.** $5\mathbf{i} - 2\mathbf{j}$ **3.** $-5\mathbf{i}$ **5.** $5\mathbf{i} - \mathbf{j}$ **7.** $(6, -7)$ **9.** $(2, 6)$ **11.** $s = 1, t = 7$ **13.** $s = 2, t = 3$ **15. a.** $6\mathbf{i} + 2\mathbf{j}$ **c.** $2\sqrt{10}$ **d.** 18.4° **17. a.** $6\mathbf{i} + 7\mathbf{j}$ **c.** $\sqrt{85}$ **d.** 49.4° **19.** 63.4° **21.** 98.1° **23.** $\frac{\sqrt{2}}{2}\mathbf{i} - \frac{\sqrt{2}}{2}\mathbf{j}$ **25.** $\frac{4}{5}\mathbf{i} + \frac{3}{5}\mathbf{j}$ **27. a.** $\mathbf{u} = 7.52\mathbf{i} + 2.74\mathbf{j}$; $\mathbf{v} = 2.95\mathbf{i} + 16.7\mathbf{j}$ **b.** 68.0 **c.** 67.9 **29. a.** $\mathbf{u} = 4.68\mathbf{i} + 11.6\mathbf{j}$; $\mathbf{v} = -7.89\mathbf{i} + 16.2\mathbf{j}$ **b.** 151 **c.** 151

Self-Test 1, page 671 **1.** 12; 67° **2.** 151.2° **3.** $2\mathbf{i} + 27\mathbf{j}$ **4.** 42.4°

Application, page 674 **1.** $-2\mathbf{i} - \mathbf{j}$ **3.** 9 J **5. a.** 22 J **b.** 22 J **7.** 13.2 N; 319° **9.** 1.47×10^6 J **11.** 8.63×10^3 J **13.** 2.30×10^4 J **15.** 995 N; 1525 N

Written Exercises, pages 678–679

1. $(2\sqrt{3}, 2)$ **3.** $\left(\frac{3}{2}, -\frac{3\sqrt{3}}{2}\right)$ **5.** $\left(\frac{7}{2}, -\frac{7\sqrt{3}}{2}\right)$ **7.** $(3\sqrt{3}, 3)$ **9.** (4, 0°) **11.** $(2\sqrt{2}, 135°)$ **13.** (2, 120°) **15.** $(\sqrt{10}, 135°)$ **17.** $r \cos \theta = 5$ **19.** $\sin \theta = \cos \theta$ **21.** $r = 6 \sin \theta$ **23.** $x^2 + y^2 = 4$ **25.** $y = 2$ **27.** $x^2 + y^2 = 2y$ **29.** $y^2 = 4 + 4x$

31.

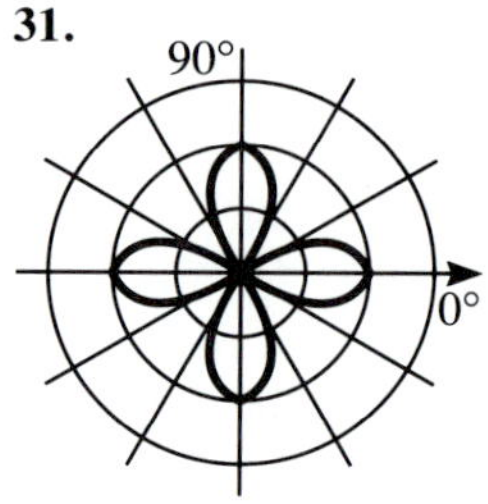

33.

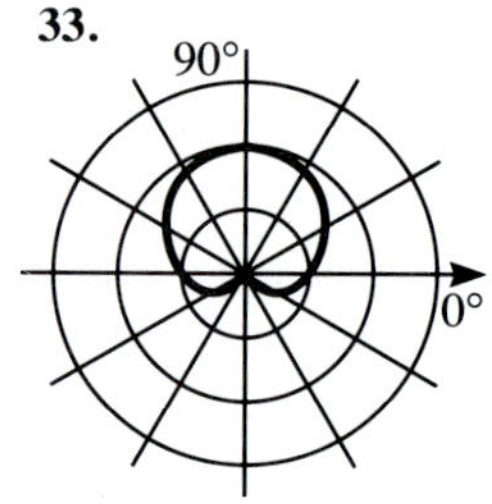

Mixed Review Exercises, page 679

1. $x^3 - 2x^2 + 2x - 4 + \frac{11}{x + 2}$ **2.** $x^2 - 1$ **3.** $3x - 5 + \frac{17}{2x + 2}$ **4.** $-4x^2 - 5x - 12 + \frac{-23}{x - 12}$

5.

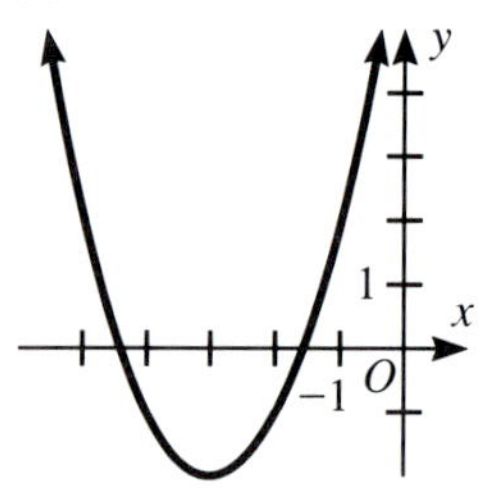

7.

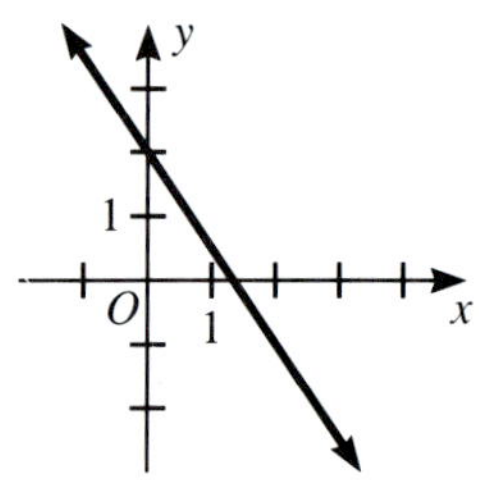

Written Exercises, pages 683–684

1.

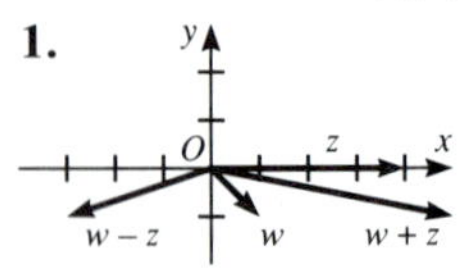

3.

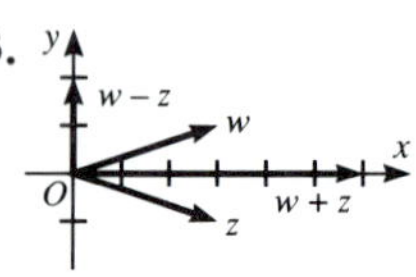

5.

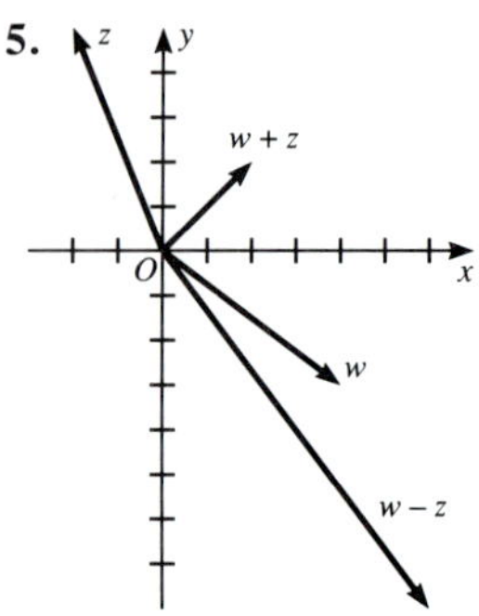

7. $10(\cos 110° + i \sin 110°)$; $2.5(\cos 310° + i \sin 310°)$ **9.** $5.4(\cos 465° + i \sin 465°)$; $3.75(\cos 195° + i \sin 195°)$ **11.** $\frac{3\sqrt{3}}{2} + \frac{3}{2}i$ **13.** $-2.25 + 2.25i\sqrt{3}$ **15.** $4.045 + 2.939i$ **17.** $-1.905 + 1.1i$ **19.** $2(\cos 300° + i \sin 300°)$ **21.** $2\sqrt{6}(\cos 135° + i \sin 135°)$ **23.** $5(\cos 233.1° + i \sin 233.1°)$ **25.** $\sqrt{13}(\cos 326.3° + i \sin 326.3°)$ **27. a.** $18(\cos 270° + i \sin 270°)$ **b.** $-18i$ **c.** $w = -\frac{3}{2} + \frac{3\sqrt{3}}{2}i$, $z = -3\sqrt{3} + 3i$ **d.** $-18i$ **29. a.** $8(\cos 150° + i \sin 150°)$ **b.** $-4\sqrt{3} + 4i$ **c.** $w = \sqrt{3} + i$, $z = -2 + 2i\sqrt{3}$ **d.** $-4\sqrt{3} + 4i$

Written Exercises, pages 687–688 **1.** $-8 + 8i\sqrt{3}$ **3.** $-16 + 16i$ **5.** $64 - 64i\sqrt{3}$ **7.** $-128 - 128i\sqrt{3}$ **9.** $\cos 0° + i \sin 0°$; $\cos 40° + i \sin 40°$; $\cos 80° + i \sin 80°$; $\cos 120° + i \sin 120°$; $\cos 160° + i \sin 160°$; $\cos 200° + i \sin 200°$; $\cos 240° + i \sin 240°$; $\cos 280° + i \sin 280°$; $\cos 320° + i \sin 320°$ **11.** $r_1 = 2(\cos 110° + i \sin 110°)$; $r_2 = 2(\cos 230° + i \sin 230°)$; $r_3 = 2(\cos 350° + i \sin 350°)$ **13.** 1, $-\frac{1}{2} \pm \frac{i\sqrt{3}}{2}$ **15.** ± 1, $\frac{1}{2} \pm \frac{i\sqrt{3}}{2}$, $-\frac{1}{2} \pm \frac{i\sqrt{3}}{2}$ **17.** -1, $\frac{1}{2} \pm \frac{i\sqrt{3}}{2}$ **19.** $\frac{\sqrt[4]{108}}{2} + \frac{i\sqrt[4]{12}}{2}$, $\frac{-\sqrt[4]{108}}{2} - \frac{i\sqrt[4]{12}}{2}$, $\frac{-\sqrt[4]{12}}{2} + \frac{i\sqrt[4]{108}}{2}$, $\frac{\sqrt[4]{12}}{2} - \frac{i\sqrt[4]{108}}{2}$

Mixed Review Exercises, page 688 **1.** $5\mathbf{i} - 3\mathbf{j}$ **3.** $\sqrt{34}$ **4.** $329.0°$ **5.** $(2\sqrt{3}, 2)$ **6.** $\left(-\frac{3}{2}, -\frac{3\sqrt{3}}{2}\right)$ **7.** $(\sqrt{3}, -1)$

Self-Test 2, page 688 **1.** $(2, -2\sqrt{3})$ **2.** $(13.0, 157.4°)$

3.

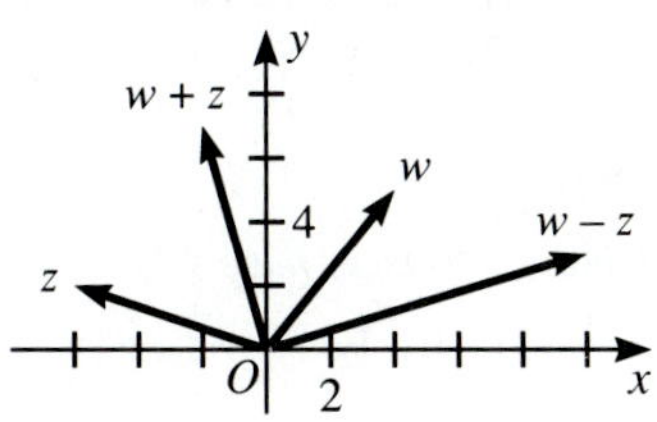

4. $wz = 18(\cos 105° + i \sin 105°)$, $\frac{w}{z} = 2(\cos 325° + i \sin 325°)$ **5.** $3.61(\cos 326.3° + i \sin 326.3°)$ **6.** -64

Written Exercises, page 692 **1.** $\frac{\pi}{3}$ **3.** $\frac{5\pi}{6}$ **5.** $\frac{2\pi}{3}$ **7.** $\frac{\pi}{3}$ **9.** $\frac{5\pi}{6}$ **11.** $\frac{1}{2}$ **13.** $\frac{12}{13}$ **15.** $\frac{\sqrt{21}}{5}$ **17.** 1 **19.** $\sqrt{1 - x^2}$ **21.** $2x\sqrt{1 - x^2}$ **23.** v

Written Exercises, pages 695–696 **1.** $\frac{\pi}{6}$ **3.** $\frac{5\pi}{6}$ **5.** $\frac{\pi}{2}$ **7.** $\frac{3\pi}{4}$ **9.** $\frac{1}{2}$ **11.** $-\frac{3}{4}$ **13.** $\frac{2\sqrt{5}}{5}$ **15.** $\frac{171}{140}$ **17.** $\frac{4}{5}$ **19.** $\frac{1}{x}$ **21.** $\frac{\sqrt{x^2 + 1}}{x^2 + 1}$ **23.** $\frac{2x}{x^2 + 1}$ **25.** $\frac{x + y}{xy - 1}$

Mixed Review Exercises, page 696
1. Arithmetic; 107 **2.** Geometric; -384 **3.** Geometric; 9216 **4.** Arithmetic; -79 **5.** $\sum_{n=0}^{5} 3(4^n)$; 4095 **6.** $\sum_{n=1}^{100} (2 + 3n) = 15{,}350$ **7.** $\sum_{n=0}^{19} (-5 + 4n)$; 660 **8.** $\sum_{k=0}^{\infty} 100(5^{-k}) = 125$

Written Exercises, pages 699–700 **1.** $\{48.6°, 131.4°\}$ **3.** $\left\{\frac{\pi}{4}, \frac{3\pi}{4}, \frac{5\pi}{4}, \frac{7\pi}{4}\right\}$ **5.** $\{135°, 315°\}$ **7.** $\{14.5°, 165.5°, 194.5°, 345.5°\}$ **9.** $\left\{\frac{3\pi}{8}, \frac{11\pi}{8}\right\}$ **11.** $\{155°, 335°\}$ **13.** $\{30°, 150°, 210°, 330°\}$ **15.** $\{0.32, 3.46\}$ **17.** $\{45°, 135°, 225°, 315°\}$ **19.** $\left\{\frac{\pi}{6}, \frac{\pi}{2}, \frac{5\pi}{6}, \frac{3\pi}{2}\right\}$ **21.** $\{0°, 120°, 240°\}$ **23.** $\{30°, 60°, 210°, 240°\}$ **25.** $\{45°, 123.7°, 225°, 303.7°\}$ **27.** $\left\{\frac{7\pi}{6}, \frac{11\pi}{6}\right\}$ **29.** $\left\{\frac{\pi}{12}, \frac{7\pi}{12}, \frac{13\pi}{12}, \frac{19\pi}{12}\right\}$ **31.** $\left\{\frac{\pi}{4}, \frac{7\pi}{12}, \frac{11\pi}{12}, \frac{5\pi}{4}, \frac{19\pi}{12}, \frac{23\pi}{12}\right\}$ **33.** $\{15°, 45°, 75°, 135°, 195°, 225°, 255°, 315°\}$ **35.** $\left\{\frac{\pi}{4}, \frac{\pi}{2}, \frac{3\pi}{4}, \frac{5\pi}{4}, \frac{3\pi}{2}, \frac{7\pi}{4}\right\}$ **37.** $\{30°, 150°, 210°, 330°\}$ **39.** $\{55°, 115°, 235°, 295°\}$ **41.** $\left\{\frac{\pi}{12}, \frac{\pi}{4}, \frac{3\pi}{4}, \frac{11\pi}{12}, \frac{17\pi}{12}, \frac{19\pi}{12}\right\}$ **43.** $\{39°, 129°, 219°, 309°\}$ **45.** $\{51.3°, 128.7°\}$

Self-Test 3, page 700 **1.** $-\frac{\pi}{3}$ **2.** $-\frac{\pi}{6}$ **3.** $\frac{5\pi}{6}$ **4.** $\frac{3}{5}$ **5.** $\frac{2\pi}{3}$ **6.** $\frac{17}{15}$ **7.** $\frac{\pi}{3}$ **8.** $-\frac{1}{5}$ **9.** $\{30°, 150°, 210°, 330°\}$

Chapter Review, pages 703–704 **1.** c **3.** a **5.** c **7.** d **9.** d **11.** a **13.** c **15.** d

Mixed Problem Solving, page 705 **1.** 460 **3.** 7 **5. a.** $\frac{1}{64}$ cm **b.** 4 cm **7.** 1:05 P.M. **9.** $6\sqrt{3}$ **11.** 15.7 ft

Preparing for College Entrance Exams, page 706 **1.** D **3.** A **5.** A **7.** D **9.** C

Cumulative Review, page 707 **1.** 465°, −255° for example **3.** 2.74 ft **5.** 14.1 **7.** $\sin\left(-\frac{13\pi}{6}\right) = -\frac{1}{2}$, $\cos\left(-\frac{13\pi}{6}\right) = \frac{\sqrt{3}}{2}$, $\tan\left(-\frac{13\pi}{6}\right) = -\frac{\sqrt{3}}{3}$, $\cot\left(-\frac{13\pi}{6}\right) = -\sqrt{3}$, $\sec\left(-\frac{13\pi}{6}\right) = \frac{2\sqrt{3}}{3}$, $\csc\left(-\frac{13\pi}{6}\right) = -2$ **9.** $\cot\theta + \cot^3\theta$ **11. a.** $-\frac{\sqrt{10}}{10}$ **b.** $\frac{7}{25}$ **c.** $-\frac{3}{4}$ **d.** $-\frac{4}{5}$ **13.** $\frac{\sqrt{5}}{5}\mathbf{i} + \frac{2\sqrt{5}}{5}\mathbf{j}$ **15.** $2i$, $-\sqrt{3} - i$, $\sqrt{3} - i$ **17.** $\frac{4\sqrt{3} - 3}{10}$ **19.** $\left\{\frac{7\pi}{6}, \frac{11\pi}{6}\right\}$ **21.** {120°, 240°}

Chapter 15 Statistics and Probability

Written Exercises, pages 711–712

1.

Stem	Leaves
1	6, 9
2	0, 2, 2, 3
3	8
4	0, 6
5	2, 4

3.

Stem	Leaves
0	2, 5, 7, 9
1	0, 3, 8, 9
2	8, 8, 8
3	4
4	2
5	2
6	1

5. a. 22 **b.** 23 **c.** 32 **7. a.** 28 **b.** 19 **c.** 23.7 **9. a.** 184 **b.** 184 **c.** 183

11.

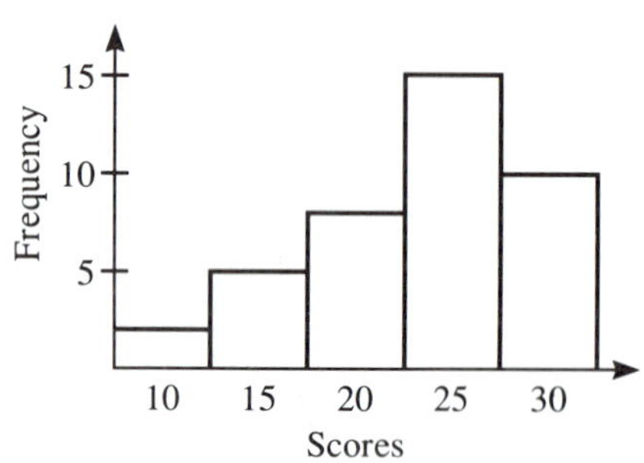

13. 87 **15.** Each would be increased by 5. **17.** 77.2

Mixed Review Exercises, page 712 **1.** (6, 120°) **2.** $(2\sqrt{2}, 30°)$ **3.** (3, 90°) **4.** $\frac{n^6}{m^{14}}$ **5.** $\frac{w + 1}{3w + 2}$ **6.** $2^{5\sqrt{2}}$

Written Exercises, pages 717–718 **1. a.** 25 **b.** 14 **c.** 34 **d.** 45

3.

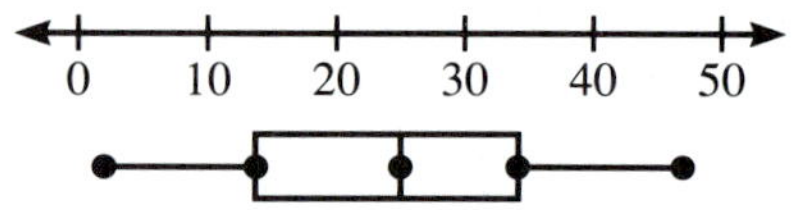

5. 2 **7.** 2 **9. a.** 6 **b.** 5.3 **c.** 2.3 **11. a.** 57 **b.** 305.3 **c.** 17.5 **13. a.** 50 **b.** 19.4 **c.** 4.4 **15. a.** 71 **b.** 4.4 **c.** 2.1 **17.** M is increased by 10; σ is unchanged. **19.** Answers may vary. **a.** 78; 80; 80; 82 **b.** 20; 100; 100; 100 **c.** 70; 70; 90; 90

Written Exercises, pages 722–723 **1.** 15.54% **3.** 0.13% **5. a.** 15.87% **b.** 97.72% **c.** 0.13% **d.** 68.26% **7. a.** 50% **b.** 0.26% **9. a.** 0 **b.** 4 **c.** −1 **d.** 1

Mixed Review Exercises, page 723 **1. a.** −3 **b.** $3x + y = 5$ **2. a.** $\frac{3}{4}$ **b.** $3x - 4y = 1$ **3. a.** $-\frac{1}{2}$ **b.** $x + 2y = 0$ **4. a.** $\frac{2}{3}$ **b.** $2x - 3y = -22$ **5. a.** 0 **b.** $y = -1$ **6. a.** −1 **b.** $x + y = -5$

Written Exercises, pages 727–728 **1.** High positive **3.** No correlation **5. a.** 0.96 **b.** $y = 1.41x - 42.56$ **7.** 78 **9.** 71 **11.** No; there may be no cause-and-effect relationship.

Self-Test 1, page 729 **1.** 28; 27; 26 **2.** 21; 33 **3.** 51; 7 **4.** 81.85% **5.** 1.39% **6.** High negative

Written Exercises, pages 732–733 **1.** 20 **3.** 50 **5.** 120 **7.** 1024 **9.** 132,600 **11.** 20,000 **13.** 37,856 **15.** 12 **17.** 8568 min, or about 6 days **19.** 64,000,000

Mixed Review Exercises, page 733 **1.** −2 **2.** 10 **3.** $\frac{1}{64}$ **4.** $4i$ **5.** $\{\pm\sqrt{2}, -1\}$ **6.** $\{\pm i\sqrt{3}, 4\}$ **7.** $\left\{5, 1, -\frac{1}{2}\right\}$ **8.** $\{\pm 4i, -6\}$

Written Exercises, page 737 **1.** 720 **3.** 336 **5.** 5040 **7.** 6 **9.** 720 **11.** 132,600 **13.** 720

15. 120 **17.** 30 **19.** 34,650 **21.** 30 **23.** 420 **29.** 18 **31.** 2520

Written Exercises, pages 740–741
1. a. {J, K}, {J}, {K}, ∅ **b.** {J}, {K}, ∅ **3.** 10 **5.** 28 **7.** 45 **9.** 792 **11. a.** 5 **b.** 10 **c.** 10 **13.** 6435 **15.** 45 **17.** 35 **19.** 254,800 **21.** 57,798 **23.** 4512 **25.** 163 **29. a.** ${}_nC_r$ is the coefficient of $a^{n-r}b^r$.
b. $(a + b)^n = \sum_{r=0}^{n} {}_nC_r a^{n-r}b^r$

Mixed Review Exercises, page 741 **1.** 2 **2.** 3 **3.** 36 **4.** −3 **5.** −7 **6.** 390 **7.** $5\sqrt{2}$ **8.** 1 **9.** $\frac{1}{m-4}$ **10.** $11 - 2i$ **11.** $\frac{2t-3}{t-3}$ **12.** 1

Self-Test 2, page 742 **1.** 100 **2.** 140 **3.** 120 **4.** 180 **5.** 56 **6.** 10

Written Exercises, page 744 **1.** {(1, 1), (2, 2), (3, 3), (4, 4), (5, 5), (6, 6)} **3.** {(2, 6), (3, 4), (4, 3), (6, 2)} **5.** {(1, 1), (1, 2), (1, 3), (1, 4), (1, 5), (1, 6), (2, 1), (2, 2), (2, 3), (2, 4), (3, 1), (3, 2), (3, 3), (4, 1), (4, 2), (5, 1), (6, 1)} **7. a.** {(R, R), (R, B), (R, Y), (R, G), (B, R), (B, B), (B, Y), (B, G)} **b.** {(R, R), (R, B), (R, Y), (R, G), (B, R)} **c.** {(R, R), (R, Y), (R, G)} **9.** {(3, 6), (4, 5), (5, 4), (6, 3)} **11.** {(2, 1), (4, 2), (6, 3)} **13.** {(1, 1), (1, 2), (1, 3), (1, 4), (1, 5), (1, 6), (2, 1), (2, 2), (2, 3), (2, 4), (2, 5), (2, 6), (3, 5), (3, 6), (4, 5), (4, 6), (5, 5), (5, 6), (6, 5), (6, 6)} **15.** {1, 1), (2, 1), (4, 1), (6, 1), (1, 4), (3, 4)}

Written Exercises, pages 748–750 **1. a.** $\frac{1}{10}$ **b.** $\frac{1}{2}$ **c.** $\frac{3}{10}$ **d.** 1 **e.** $\frac{1}{5}$ **f.** 0 **3. a.** $\frac{1}{4}$ **b.** 1 **c.** $\frac{3}{4}$ **d.** $\frac{5}{8}$ **5. a.** $\frac{1}{4}$ **b.** $\frac{3}{4}$ **c.** $\frac{1}{2}$ **d.** $\frac{1}{2}$ **7. a.** 1 **b.** $\frac{9}{1000}$ **c.** $\frac{991}{1000}$ **9. a.** $\frac{1}{8}$ **b.** $\frac{1}{8}$ **c.** $\frac{7}{8}$ **d.** $\frac{3}{8}$ **11. a.** $\frac{1}{66}$ **b.** $\frac{1}{11}$ **c.** $\frac{5}{22}$ **d.** $\frac{4}{33}$ **e.** $\frac{15}{22}$ **f.** $\frac{5}{22}$ **13. a.** 0.0548 **b.** 0.0082

Mixed Review Exercises, page 750
1. $\{t: t < -2 \text{ or } t > 6\}$ **2.** $\{x: -1 \le x \le 2\}$ **3.** $\{n: n > 2\}$ **4.** $\{x: 0 \le x \le 4\}$ **5.** $\{y: y < -2 \text{ or } y > 2\}$ **6.** $\left\{k: -\frac{5}{3} < k < 1\right\}$ **7.** 60 **8.** 10,626

Application, pages 752–753 **1. a.** Random **b.** Not random **3.** Not random **5–8.** Answers will vary.

Written Exercises, pages 759–761 **1. a.** $\frac{1}{2}$ **b.** $\frac{1}{2}$ **c.** $\frac{3}{10}$ **3. a.** $\frac{1}{11}$ **b.** $\frac{5}{33}$ **5. a.** $\frac{1}{6}, \frac{5}{36}, \frac{5}{18}, \frac{1}{36}$ **b.** No **7. a.** $\frac{1}{12}$ **b.** $\frac{1}{9}$ **9. a.** $\frac{1}{2}$ **b.** $\frac{7}{8}$ **11.** $\frac{1}{360}$ **13. a.** $\frac{1}{2}$ **b.** $\frac{2}{5}$ **c.** $\frac{7}{10}$ **15. a.** $\frac{1}{12}$ **b.** $\frac{3}{16}$ **c.** $\frac{19}{48}$ **d.** $\frac{7}{8}$ **17. a.** $\frac{11}{20}$ **b.** $\frac{4}{11}$

Self-Test 3, page 761 **1. a.** {S, Q, U, A, R, E} **b.** {U, A, E} **2.** $\frac{1}{9}$ **3. a.** $\frac{7}{18}$ **b.** No **c.** No

Chapter Review, pages 764–765 **1.** b **3.** d **5.** c **7.** a **9.** c **11.** d

Chapter 16 Matrices and Determinants

Written Exercises, page 769
1. $\begin{bmatrix} 0 & 0 \end{bmatrix}$ **3.** $\begin{bmatrix} 0 & 0 & 0 & 0 & 0 \\ 0 & 0 & 0 & 0 & 0 \end{bmatrix}$
5. $x = y = z = 0$ **7.** $x = -4, y = 0, z = 10$ **9.** $x = -4, y = 1$ **11.** $x = 8, y = -5$ **13.** $x = 11, y = 12, z = -4$ **15.** $x = 5, y = -3$ **17.** $x = 3, y = -1$ **19.** $x = \frac{1}{a}, y = 0$ **21.** $x = \frac{1-b}{a}, y = -\frac{d}{c}$

Mixed Review Exercises, page 769
1. $\angle A = 44.3°, \angle B = 86.7°, c = 7.56$ **2.** $\angle A = 60°, a = 4\sqrt{3}, b = 4$ **3.** $\angle A = 45°, \angle B = 45°, c = 3$ **4.** $\angle B = 42.4°, \angle C = 65.6°, c = 23.0$ **5.** 26.4 **6.** 13.8 **7.** 2.25 **8.** 186.0

Written Exercises, page 773
1. $\begin{bmatrix} 4 & 4 \\ 0 & 3 \end{bmatrix}$ **3.** $\begin{bmatrix} 11 & 4 \\ 3 & 16 \end{bmatrix}$ **5.** $\begin{bmatrix} 4 & 2 \\ 0 & -5 \\ -3 & 7 \end{bmatrix}$
7. $\begin{bmatrix} -1 & 3 & 4 \\ 0 & 0 & 0 \end{bmatrix}$ **9.** $\begin{bmatrix} 42 & 48 \\ -24 & 0 \end{bmatrix}$
11. $\begin{bmatrix} 3 & 5 & 2 \\ -2 & 8 & 0 \end{bmatrix}$ **13.** $\begin{bmatrix} 11 & 3 \\ -6 & 4 \\ 6 & 17 \end{bmatrix}$ **15.** $\begin{bmatrix} 6 \\ 36 \\ 30 \end{bmatrix}$

17. $\begin{bmatrix} 3 & 1 \\ 6 & -1 \end{bmatrix}$ **19.** $\begin{bmatrix} 5 & 4 \\ 12 & 3 \end{bmatrix}$

Written Exercises, pages 777–778

1. $\begin{bmatrix} 18 \end{bmatrix}$ **3.** $\begin{bmatrix} 12 & 0 & -4 & 20 \\ -6 & 0 & 2 & -10 \end{bmatrix}$

5. $\begin{bmatrix} -3 & 24 \\ -1 & 14 \end{bmatrix}$ **7.** $\begin{bmatrix} -3 & 0 & 6 \\ 15 & 12 & 2 \\ 3 & 0 & -6 \end{bmatrix}$

9. $\begin{bmatrix} -1 & 2 & 2 & -1 \\ 4 & -1 & -1 & -3 \\ -4 & 4 & 4 & 0 \end{bmatrix}$

11. $\begin{bmatrix} a+2g & b+2h & c+2i \\ -2a+d & -2b+e & -2c+f \\ d & e & f \end{bmatrix}$

13. $AB = \begin{bmatrix} 2 & 1 \\ -1 & -1 \end{bmatrix}, BA = \begin{bmatrix} 0 & 1 \\ 1 & 1 \end{bmatrix}$

15. $(AB)C = \begin{bmatrix} -4 & 1 \\ 2 & -1 \end{bmatrix}, A(BC) = \begin{bmatrix} -4 & 1 \\ 2 & -1 \end{bmatrix}$

17. $(A + B)(A - B) = \begin{bmatrix} 0 & 1 \\ 1 & 3 \end{bmatrix}, A^2 - B^2 = \begin{bmatrix} 2 & 1 \\ -1 & 1 \end{bmatrix}$

19. $O_{2\times 2} \cdot A = A \cdot O_{2\times 2} = \begin{bmatrix} 0 & 0 \\ 0 & 0 \end{bmatrix}$ **21.** $\begin{bmatrix} 1 & 0 \\ 0 & 1 \end{bmatrix}$

23. $\begin{bmatrix} 0 & 0 & -1 \\ -1 & 4 & -2 \\ 1 & 0 & -1 \end{bmatrix}$ **25.** $\begin{bmatrix} 0 & 0 & -1 \\ -1 & 4 & -2 \\ 1 & 0 & -1 \end{bmatrix}$

Mixed Review Exercises, page 778 **1.** No solution **2.** (2, −1) **3.** (5, 7), (−1, −5) **4.** (−3, 2) **5.** (5, 3), (−5, −3) **6.** (1, −1, 2) **7.** {3} **8.** $\left\{\frac{-7 \pm \sqrt{73}}{4}\right\}$ **9.** {−1}

Written Exercises, pages 782–784

1. $\begin{array}{c} \\ A \\ B \\ C \end{array}\begin{array}{c} \begin{matrix} A & B & C \end{matrix} \\ \begin{bmatrix} 0 & 1 & 1 \\ 1 & 0 & 0 \\ 1 & 1 & 0 \end{bmatrix} \end{array}$

3. $\begin{array}{c} \\ A \\ B \\ C \end{array}\begin{array}{c} \begin{matrix} A & B & C \end{matrix} \\ \begin{bmatrix} 2 & 2 & 1 \\ 1 & 1 & 1 \\ 2 & 2 & 1 \end{bmatrix} \end{array}$

5. $\begin{array}{c} \\ A \\ B \\ C \\ D \\ E \end{array}\begin{array}{c} \begin{matrix} A & B & C & D & E \end{matrix} \\ \begin{bmatrix} 0 & 1 & 0 & 0 & 0 \\ 1 & 0 & 1 & 1 & 0 \\ 0 & 1 & 0 & 0 & 1 \\ 0 & 1 & 0 & 0 & 1 \\ 0 & 0 & 0 & 0 & 0 \end{bmatrix} \end{array}$

7. $\begin{bmatrix} 1 & 0 & 1 & 1 & 0 \\ 0 & 3 & 0 & 0 & 2 \\ 1 & 0 & 1 & 1 & 0 \\ 1 & 0 & 1 & 1 & 0 \\ 0 & 0 & 0 & 0 & 0 \end{bmatrix}$

9. B

11. $\begin{array}{c} \\ J \\ C \\ Y \\ A \end{array}\begin{array}{c} \begin{matrix} J & C & Y & A \end{matrix} \\ \begin{bmatrix} 0 & 1 & 0 & 0 \\ 0 & 0 & 1 & 1 \\ 1 & 0 & 0 & 1 \\ 1 & 0 & 0 & 0 \end{bmatrix} \end{array}; \begin{bmatrix} 0 & 0 & 1 & 1 \\ 2 & 0 & 0 & 1 \\ 1 & 1 & 0 & 0 \\ 0 & 1 & 0 & 0 \end{bmatrix}$

13. $\begin{array}{c} \\ A \\ B \\ C \\ D \end{array}\begin{array}{c} \begin{matrix} A & B & C & D \end{matrix} \\ \begin{bmatrix} 0 & 1 & 0 & 1 \\ 1 & 0 & 0 & 0 \\ 0 & 1 & 0 & 1 \\ 1 & 0 & 0 & 0 \end{bmatrix} \end{array}; \begin{bmatrix} 2 & 0 & 0 & 0 \\ 0 & 1 & 0 & 1 \\ 2 & 0 & 0 & 0 \\ 0 & 1 & 0 & 1 \end{bmatrix}$

15. $\begin{bmatrix} 0 & 2 & 0 & 2 \\ 2 & 0 & 0 & 0 \\ 0 & 2 & 0 & 2 \\ 2 & 0 & 0 & 0 \end{bmatrix}$

17. 2

Self-Test 1, page 784 **1.** $x = -1, y = 5, z = 0$ **2.** $x = 2, y = -10, z = 4$

3. $\begin{bmatrix} 0 & 6 \\ 1 & 3 \end{bmatrix}$ **4.** $\begin{bmatrix} 2 & 2 \\ -1 & 3 \end{bmatrix}$ **5.** $\begin{bmatrix} -3 & 6 \\ 3 & 0 \end{bmatrix}$

6. $\begin{bmatrix} 3 & 2 \\ 3 & 0 \end{bmatrix}$ **7.** $\begin{bmatrix} 1 & 16 \\ 0 & 9 \end{bmatrix}$ **8.** $\begin{bmatrix} 3 & 0 \\ -2 & 3 \end{bmatrix}$

9. a. $\begin{array}{c} \\ A \\ B \\ C \\ D \end{array}\begin{array}{c} \begin{matrix} A & B & C & D \end{matrix} \\ \begin{bmatrix} 0 & 1 & 1 & 1 \\ 1 & 0 & 1 & 0 \\ 1 & 0 & 0 & 0 \\ 1 & 0 & 1 & 0 \end{bmatrix} \end{array}$ **b.** 3

Extra, page 786 **1.** (5, −1, −4) **3.** (1, −2, 4) **5.** (−1, 2, 3)

Written Exercises, page 789 **1.** 5 **3.** 14 **5.** 32 **7.** 137 **9.** 456 **11.** –60 **13.** 138

Mixed Review Exercises, page 789 **1.** $\frac{t-1}{t-3}$ **2.** $\frac{1}{ab^4}$ **3.** $\frac{13s-5}{3(s+1)(s-1)}$ **4.** a^2 **5.** $-\frac{1}{c+1}$ **6.** $\frac{y-2}{y-5}$ **7.** $-36p^3q^6$ **8.** $\frac{x^2+x+1}{x+4}$ **9.** $\frac{1}{2}$

Written Exercises, page 792

1. $\begin{bmatrix} \frac{2}{5} & -\frac{1}{5} \\ \frac{1}{5} & \frac{2}{5} \end{bmatrix}$ **3.** No inverse

5. $\begin{bmatrix} 1 & 0 \\ 0 & 1 \end{bmatrix}$ **7.** $\begin{bmatrix} \frac{1}{6} & \frac{1}{12} \\ -\frac{1}{2} & \frac{1}{4} \end{bmatrix}$

9. (–1, –8) **11.** (5, 3) **13.** (2, 4) **15.** No unique solution

17. $\begin{bmatrix} 34 & -2 \\ -10 & 1 \end{bmatrix}$ **19.** $\begin{bmatrix} 2 & -\frac{3}{4} \\ -3 & \frac{13}{4} \end{bmatrix}$

Self-Test 2, page 793 **1.** –3 **2.** –8

3. $\begin{bmatrix} 1 & \frac{1}{3} \\ 0 & \frac{1}{3} \end{bmatrix}$ **4.** $\begin{bmatrix} 2 & -1 \\ -1 & 1 \end{bmatrix}$

Written Exercises, pages 796–797 **1.** –7 **3.** –16 **5.** –10 **7.** 6 **9.** –4 **11.** 8 **13.** 1 **15.** 0 **17.** –60

Mixed Review Exercises, page 797 **1.** $\frac{1}{25}$ **2.** $-\frac{\sqrt{2}}{2}$ **3.** 56 **4.** –4 **5.** $\frac{5}{2}$ **6.** 1

Written Exercises, page 800 **1.** 0 **3.** 0 **5.** 0 **7.** 60 **9.** 112 **11.** 0

Written Exercises, pages 804–805 **1.** (2, 0) **3.** (5, –3, 5) **5.** (3, –1, 2) **7.** (2, 1, 2) **9.** (2, –4, –3) **11.** No solution **13.** No solution **15.** (2, 2, –2, –1)

Mixed Review Exercises, page 805

1. $\begin{bmatrix} 6 & 0 \\ 3 & 4 \end{bmatrix}$ **2.** $\begin{bmatrix} 8 & -4 \\ 12 & 0 \end{bmatrix}$ **3.** $\begin{bmatrix} 8 & 1 \\ 12 & 1 \end{bmatrix}$

4.

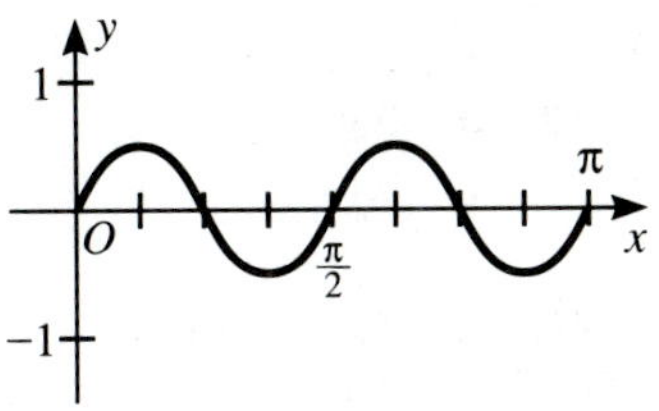

5.

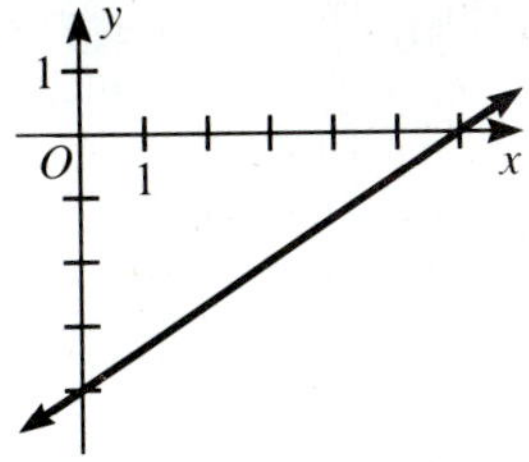

6.

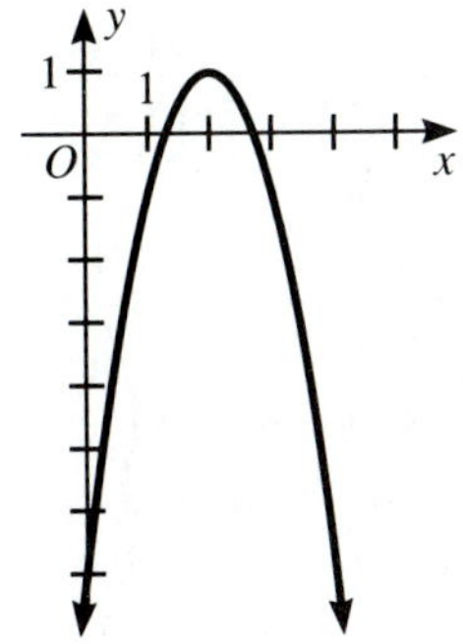

Self-Test 3 **1.** –18 **2.** –2 **3.** (–3, 5, –2)

Chapter Review, pages 806–807 **1.** d **3.** c **5.** d **7.** a **9.** d

Preparing for College Entrance Exams, page 809 **1.** D **3.** B **5.** E

Explorations

Page 832 ***(1-1)*** **1. a.** isosceles right triangle **b.** $AB = 1$, $BC = 1$, $AC = \sqrt{2}$ **3. a.** $AF = \sqrt{3}$, $FG = 1$, $AG = 2$ **b.** $\sqrt{3}$ **5. a.** π **b.** $\pi \approx 3.14$ **7.** Start with a circle with diameter 3 units to locate 3π, with diameter $\frac{1}{2}$ unit to find $\frac{\pi}{2}$.

Page 833 ***(2-1)*** **1. c.** No; multiplying by 0 results in a product of 0. **d.** If $a < b$ and $c > 0$, then $ac < bc$ is true. If $a < b$ and $c \le 0$, then $ac < bc$ is false. **3. a.** yes **b.** yes **c.** $3 > -2$ and $\frac{1}{3} \not< -\frac{1}{2}$ **5.** Invalid transformation. See Exercise 3c.

Page 834 ***(3-8)*** **1.** Divide the weight by 16, and find which whole numbers of pounds the weight is between. **5. a.** 2 **b.** 3 **c.** 3 **d.** 1 **7.** The graphs of these functions look like steps rather than points, lines, or curves.

Page 835 ***(4-5, 4-6)*** **1. a.** $3x$, $x + 2$ **b.** $3x(x + 2)$ **3. a.** 1 blue x-by-x tile, 1 blue 1-by-x tile, and 12 red 1-by-1 tiles **b.** no **c.** because adding $3x$ and $-3x$ to the polynomial is equivalent to adding zero **d.** $x + 4$, $x - 3$ **e.** $x + 4$, $x - 3$

Page 836 ***(5-7)*** **1.** $\frac{67}{29}$ **3.** [2, 3, 4]; [3, 2, 1, 6, 2] **5. a.** 1.5 **b.** 1.4 **c.** $1.41\overline{6}$ **d.** ≈ 1.4138 **e.** ≈ 1.4143 **f.** $\sqrt{2}$ **7.** finite: rational numbers; infinite: irrational numbers

Page 837 ***(6-2)*** **1. a.** 14; 10 **b.** 1.045; 2.646 **c.** 8; 5.337 **d.** -1; -3.332 **3. a.** 24; 24 **b.** 3.464; 3.464 **c.** 2.410; 2.410 **d.** 1.5; 1.5 **5. a.** 2; 2 **b.** 1.619; 1.619 **c.** 4; 4 **d.** 2.828; 2.828 **7. a.** 5 **b.** 6 **c.** 24 **d.** $\frac{5}{4} = 1.25$

Page 838 ***(7-5)*** **1.** For larger values of a, the graph is narrower. **3. c.** The new graph is the previous graph slid 1 unit to the right. **5. d.** Increasing the value of k slides the graph up, decreasing it slides the graph down. **7. a.** It would be narrower. **b.** It would be wider, and it would open downward instead of upward.

Page 839 ***(8-1)*** **5.** It is a constant. **7.** No; a different quotient may result but the quotient is still a constant in each case. **9. a.** first column **b.** fourth column **c.** fifth column

Page 840 ***(9-6)*** **1.** The graph is a circle with center (3, 1) and radius 2. **3.** The graph is an ellipse with center (5, 2), horizontal minor axis of length 6, and vertical major axis of length 8. **5. a.** double the radius **b.** divide the radius by 4 **c.** triple the radius and slide the circle 2 units left and 4 units up **d.** stretch the circle horizontally by a factor of 2 and vertically by a factor of 3 **e.** place center at (1, 3) and stretch as in part (d) **f.** place center at (2, –3), shrink horizontally by a factor of $\frac{1}{2}$, and stretch vertically by a factor of 4

Page 841 ***(10-2)*** **1. a.** first and third **b.** $y = x$ **c.** $y = x^5$ **3. a.** first and second **b.** $y = x^2$; $y = x^6$ **c.** $(0.9)^2$; $(1.01)^6$ **5.** $a > 1$ and $m > n$, or $0 < a < 1$ and $m < n$

Page 842 ***(11-7)*** **1.** Row 7: 1, 7, 21, 35, 35, 21, 7, 1; Row 8: 1, 8, 28, 56, 70, 56, 28, 8, 1 **3.** third diagonal from the left or right (1, 3, 6, 10, 15, ...) **5.** $2^0, 2^1, 2^2, 2^3, 2^4, \ldots, 2^n$ **7.** Yes; it will always appear as the middle element of a later row.

Page 844 ***(13-4)*** **3.** Slide the ruler down instead of up; a negative y-value is being added.

Page 845 ***(14-3)*** **1.** Each graph is a line perpendicular to the polar axis. **3.** Each graph is a line parallel to the polar axis, above it if $a > 0$ and below it if $a < 0$. **5.** Each graph is a line that passes through (p, k) but not through the pole. **7.** Each graph is a circle that passes through the pole, has radius $|a|$, and has its center on the 90° ray or its extension. **9.** Same as for Ex. 8, but rotated 90° counterclockwise.

Page 846 ***(15-9)*** **1. c.** yes **d.** yes **3.** no; yes **5.** Approx. $\frac{3}{10}$

Page 847 ***(16-3)*** **1.** AB is the result of flipping (reflecting) B over the x-axis. **3.** a turn (rotation) of 90° counterclockwise **5. a.** a flip (reflection) over the y-axis **b.** a turn (rotation) of 180°